以珍贵树种大径材培育为主导的多功能近自然森林经营技术研究

蔡道雄　贾宏炎 ◆ 主编

Study on close-to-nature forest management technology oriented by large diameter timber cultivation of valuable tree species

中国林业科学研究院热带林业实验中心成立40周年论文集

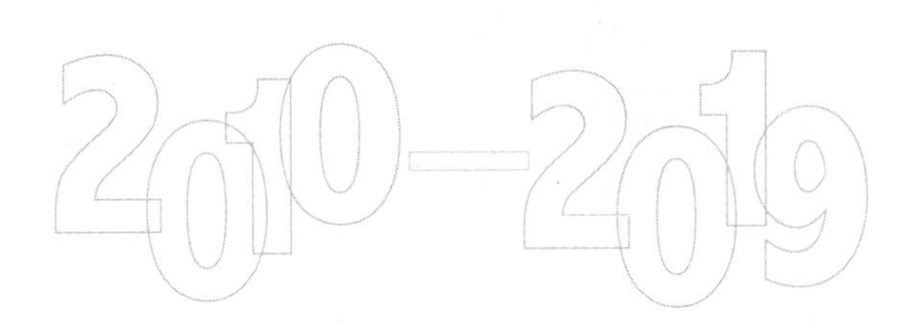

中国林業出版社
CFPH China Forestry Publishing House

本书编委会

主　编　蔡道雄　贾宏炎

副主编　许基煌　孙文胜　明安刚

编　委　(以姓氏笔画为序)

邓硕坤　韦菊玲　卢立华　许基煌　孙文胜
孙冬婧　刘志龙　苏建苗　吴思达　劳庆祥
陈建全　陈　琳　明安刚　郝　建　郑　路
贾宏炎　郭文福　唐继新　谌红辉　曾　冀
蔡道雄

图书在版编目(CIP)数据

以珍贵树种大径材培育为主导的多功能近自然森林经营技术研究 / 蔡道雄，贾宏炎主编. — 北京：中国林业出版社，2019.11

ISBN 978-7-5219-0312-6

Ⅰ. ①以… Ⅱ. ①蔡… ②贾… Ⅲ. ①珍贵树种-栽培技术-文集②珍贵树种-森林经营-文集 Ⅳ. ①S79-53②S75-53

中国版本图书馆 CIP 数据核字(2019)第 248012 号

以珍贵树种大径材培育为主导的多功能近自然森林经营技术研究

责任编辑：张　华

出版发行　中国林业出版社(100009　北京市西城区德内大街刘海胡同 7 号)
邮　　编　100009
电　　话　(010)83143566
印　　刷　北京中科印刷有限公司
版　　次　2019 年 11 月第 1 版
印　　次　2019 年 11 月第 1 次印刷
开　　本　889mm×1194mm　1/16
印　　张　59
字　　数　2050 千字
定　　价　398.00 元

FOREWORD 序

中国林业科学研究院热带林业实验中心(以下简称热林中心)是从事我国热带南亚热带林业科学研究与实验、技术示范与推广的国家级林业实验基地。热林中心正式批复建设于1979年，是在原广西壮族自治区国营大青山林场的基础上成立的，经营面积1.9万hm^2，森林蓄积量145万m^3。40年来，在上级部门的正确领导和地方各级政府的大力支持下，经过热林中心全体职工的艰苦奋斗，在林木种质资源收集保存与开发利用、珍优树种良种繁育、珍优树种大径材培育、多功能近自然森林经营、人工林生态系统服务功能研究等方面取得了丰硕的成果，并在我国广西、广东、海南、云南、福建等地广泛应用推广，为我国地带性林业发展和生态文明建设提供了重要的技术支撑。

热林中心以建设世界一流林业科技创新实验基地为目标，坚持以科学研究、试验示范和技术推广为首要任务，自成立以来，先后承担了各类科研项目201项，获得各级科技成果奖励37项，其中，获得国家科技进步一等奖1项、二等奖4项，省部级科技进步奖32项，发表论文601篇，出版专著23部，获得专利授权19项，制订国家标准及林业行业标准13项，审定或认定林木良种6个。建有红椎、西南桦、柚木、格木、降香黄檀等热带南亚热带珍贵优良阔叶树种试验示范林约6000hm^2。

40年来，热林中心在科研条件和平台建设方面取得了显著成效，建立了广西大青山森林综合型国家长期科研基地、广西友谊关森林生态系统国家定位观测研究站、国家林业和草原局热带珍贵树种繁育利用研究中心等一大批科研平台，现已成为林业科学研究与实验的最理想场所之一。

值此热林中心成立40周年之际，热林中心编辑出版了论文集。该论文集共选录了热林中心近10年来(2010—2019年)200篇论文或摘要，全面展示了热林中心在珍优树种良种选育技术、珍优树种高效培育、多功能近自然森林经营、人工林生态系统服务功能以及森林资源现代化管理技术等方面的研究成果。我欣然为之写序，衷心祝愿热林中心抓住新时代、新形势下的新机遇，继往开来，努力奋斗，勇于创新，再创辉煌，努力建设世界一流的林业科学实验基地，并为我国热带南亚热带地区林业高质量发展和生态文明建设作出更加卓越的贡献!

中国林业科学研究院院长

国际林业研究组织联盟副主席，研究员，博士生导师

PREFACE 前言

中国林业科学研究院热带林业实验中心(简称热林中心)成立于1979年，是中国林业科学研究院下属的国家级林业科研示范基地，主要承担我国热带南亚热带林业科学研究与实验，组装配套推广林业科技成果，为林业发展和生态建设提供科技支撑和示范等任务。40年来，热林中心立足于我国热带南亚热带地区，以服务国家林业发展大局和搭建林业科学试验研究平台为己任，经过广大科技人员的艰苦奋斗和不懈努力，中心的科技创新能力和科学研究综合实力显著增强，建成了以珍贵优良树种培育为主的多学科研究、产学研结合的试验示范基地，为地域性林业发展和生态文明建设提供科技支撑和示范样板，为建设世界一流林业科技创新基地的目标奠定坚实基础。

值此热林中心成立40周年之际，回顾过去，系统地总结科学研究和科技推广的成果及经验，展望未来，为热带南亚热带乃至世界林业现代化建设作出更大贡献。

1 历史沿革与发展定位

改革开放之初，我国科学研究事业百废待兴，1978年全国科学大会的召开让林业科技工作者无比振奋。当时中国林业科学研究院根据1978—1985年全国科学技术发展规划纲要关于“建立若干农、林、牧、渔现代化综合科学试验基地”的要求，计划建立广西大青山南亚热带珍贵用材树种现代化综合科学试验基地，得到了国家林业局以及广西区政府的大力支持，1979年9月中国林业科学研究院决定在广西大青山林场的基础上成立广西大青山实验局，1979年3月获得国家科学技术委员会和农村工作委员会的批准，1990年更名为中国林业科学研究院热带林业实验中心。

成立之初，热林中心定位为一个以热带南亚热带珍贵特用树种营林科学技术中间实验为主的实验基地，通过试验林和示范林的营建，成为从小试、中试到成果推广应用的桥梁纽带。热林中心的主要任务是研究热带南亚热带地区林业现代化的具体途径、内容、方法及规律，将热林中心建成林业现代化生产的样板和热带南亚热带珍贵特用树种综合科学实验基地。1987年7月中国林业科学研究院进一步明确热林中心的性质是林业科学技术成果转化为生产力的中间试验场所，是林业科技成果的示范基地，为林业现代化建设提供科学经营管理的样板，特别强调将林业科技成果组装配套试验进行再创新。

20世纪90年代，热林中心强调科研工作面向林业主战场，为建立林业两大体系服务，积极推广科技成果，组装配套了“热带南亚热带珍贵阔叶树种人工林培育”等八大示范样板。进入21世纪，根据国家林业和生态建设的要求，热林中心及时调整科研重点，在搞好热带南亚热带珍优阔叶树引种栽培技术研究的同时，主动向林木种质资源保存、石漠化综合治理、科学实验林生态定位观测等领域拓展，明确了中心的六大研究领域：①珍贵优良树种引种驯化、良种选育、栽培技术研究；②林木种质资源收集保存与利用技术研究；③人工林生态服务功能研究；④石漠化地区植被恢复技术及生态功能研究；⑤森林资源现代化经营管理技术研究与示范；⑥植物新品种测试服务。“十二五”以后，中心进一步明确了今后科研工作的指导思想和工作定位，即立足于我国热带南亚热带地区，围绕“以珍贵树种大径材培育为主导的多功能近自然森林经营技术体系”和“森林质量精准提升关键技术”，着重开展珍贵树种大径材培育技术、森林生态系统服务功能提升、珍贵树种良种选育和基地建设、森林多功能经营、森林质量精准提升关键技术等五个方面的研究工作。以实用技术研发为主，科研和营林紧密结合，加强与国内外相关研究机构合作，积极应用国内外先进成果和装备，强化集成创新，建设一流的林业先进技术示范样板和科学研究平台。

2 科研能力建设

近十年来，热林中心以珍贵树种大径材培育为主导的多功能近自然森林经营技术研究为主线，在科学研究、学科建设、科研人才队伍建设、科研平台建设、科研成果及其组装配套和示范推广等方面取得了可喜的成绩。

2.1 人才队伍建设

热林中心高度重视科技人才队伍建设，人才结构得到逐步优化，专业技术人员的比例从1979年的4.5%上升至2019年的56.2%。经过多年培养和科技人员自身的学习、实践、锻炼，一大批科技人才迅速成长起来，在林业科学研究、成果推广等工作中担当重任。截至2019年7月，在职职工334人，其中专业技术人员197人，具有高级技术职称人员42人，具有博士学位11人，硕士学位45人；享受国务院政府特殊津贴1人；特聘客座研究员4人。

2.2 科研平台建设

热林中心经营面积1.9万hm^2，其中有林面积1.5万hm^2，森林活立木蓄积145万m^3，森林覆盖率84.4%。近十年来，中心引进和吸收多功能近自然森林经营理论与技术，全面实施森林可持续经营，推进珍贵树种培育示范基地建设，林种、树种结构明显优化，尤其是珍贵优良树种的数量和面积大幅度增加，珍优阔叶树比例从中心成立之初的10%提高到现在的40%。营造了树种多样、模式丰富、林龄齐全、档案完备的珍优阔叶树种科学试验示范林6000hm^2，建成了我国南方树种最多、规模最大的珍优树种人工林试验示范区，为林业科学研究搭建了理想的平台。同时建有广西大青山森林综合型国家长期科研基地、国家林业和草原局热带珍贵树种繁育利用研究中心、国家林业和草原局东盟林业合作研究中心、广西友谊关森林生态系统国家定位观测研究站、广西凭祥竹林生态系统国家定位观测研究站、国家林业和草原局新品种测试中心华南分中心等科研平台。

2.3 主要科研成果

热林中心成立40年来，先后主持和参加的科研项目201项，其中国际合作6项，国家级项目52项，省部级项目95项。“十二五”以来，广西自然科学基金项目和国家自然科学基金项目均实现零突破，标志着中心承担科研项目水平逐步提高。先后发表科技论文601篇，其中以热林中心为第一单位发表论文388篇，特别是“十一五”以来，发表科技论文的数量和质量逐年提高。依托中心发表学术论文213篇，发表学位论文43篇，其中，博士论文26篇，硕士论文17篇。主编或参编专著23部，获得专利授权19项，制订国家标准及林业行业标准13项，审定或认定林木良种6个。主持或参与获得各类科技奖励37项，其中，“棕榈藤的研究”获国家科技进步一等奖，“我国南方人工用材林林业局(场)森林资源现代化经营管理技术”“林木种质资源收集保存与利用研究”“天然林保护与生态恢复技术”获国家科技进步二等奖。

2.4 成果示范推广

中心成立以来，充分利用实验基地优势，集成创新配套推广应用科技成果。经中间试验筛选出红椎、格木、西南桦、柚木、降香黄檀等50多个热带南亚热带珍贵树种和石山造林树种大规模应用于我国南方林业发展和生态建设，取得了良好的生态、社会和经济效益。

近十年，中心先后组织科技人员100多人次深入广西崇左市的天等县，百色市的乐业、田阳、凌云县，河池市的都安县等岩溶山地生态脆弱地区开展石灰岩山地的植被恢复与生态重建，共营建试验示范林5万多亩；并总结出了“林果”“林药”“林竹”“林草”“林藤”5个生态与经济结合较好的石漠治理高效栽培模式。此外，通过实施“红椎西南桦等珍贵优良树种大径材培育技术创新集成与示范”“华南岩溶地区生态经济型高效经营模式集成推广”等国家级林业推广项目，将成熟的技术在区内外进行进一步的推广应用，扩大了珍优树种的种植规模，起到了良好的示范带动作用，为区域性林业经济、环境建设和社会发展作出了应有的贡献。

3　主要学术成就

近十年来，中心围绕建设成为世界一流林业科技创新实验基地的目标，立足资源和区位两大优势，抓住出人才、出成果、出效益三个重点，强化科技创新、森林资源培育与管护，着重开展热带南亚热带主要珍优树种良种选育技术、珍优树种高效培育、多功能近自然森林经营、人工林生态系统服务功能以及森林资源现代化管理技术等研究工作，科技创新能力稳步提升。

3.1　热带南亚热带主要珍优树种良种选育技术研究

热林中心成立以来，十分重视珍优树种的良种选育工作，先后开展与珍优树种相关的育种科研项目有22项，有4项科研成果获得奖励，其中“林木种质资源收集保存与利用研究”获得2005年国家科技进步二等奖，“广西马尾松育种群体建立与应用”获2016年广西科技进步一等奖，西南桦“青山1号”等4个良种通过了省级良种审定。营建了热带南亚热带主要珍优树种的种质资源库与良种基地，解决了一批主要珍优树种良种壮苗的繁育技术问题，实现了林木良种壮苗生产的规模化。

(1)林木引种驯化及种质资源保存库的建设

依托大青山石山树木园和夏石引种树木园，对热带南亚热带地区的珍稀频危树种进行了引种驯化，对引进的86个珍稀频危树种进行了幼苗培育及造林观察，总结评价了引种材料的保存及生长情况，重点对降香黄檀等50个树种进行了物候、生长和适应性的观测评价，为这些树种的推广种植提供了重要的技术支撑。

中心充分利用我国南方珍贵树种的种质资源，建成林木种质资源收集保存库53.8hm^2，收集林木种质材料2096份，树种包括西南桦、光皮桦、柚木、红椎、江南油杉、格木、土沉香及竹藤类，部分树种是木材兼食品、药用与生物质能源等多用途的树种。“中国林业科学研究院热带林业实验中心热带与南亚热带珍贵树种林木种质资源库”2016年被国家林业和草原局认定为国家林木种质资源库，并规划扩建珍贵树种的种质资源收集库100hm^2，收集32种热带与南亚热带的主要珍贵树种优良种质资源，保存种质资源5000份以上。

(2)良种基地建设

热林中心现拥有良种基地面积323hm^2，其中种子园、母树林面积150hm^2，西南桦、柚木、红椎、格木、米老排、火力楠、马尾松等树种年产种量可达10t以上。中心种质资源丰富，先后获得了国家级良种基地“中国林业科学研究院凭祥国家西南桦、柚木良种基地”与广西壮族自治区级良种基地“中国林业科学研究院热带林业实验中心珍贵乡土树种良种基地”，为珍贵树种的推广提供了良种保障。

(3)良种壮苗培育技术研究

良种壮苗是优质高效人工林培育的基本前提，热林中心选用优树材料开展了西南桦、柚木、红椎等珍贵树种的嫁接、扦插和组培快繁技术研究，其中西南桦、光皮桦、柚木的优良无性系组培快繁技术已达工厂化生产水平。以红椎、山白兰、西南桦、灰木莲和格木等颇具发展潜力的树种为研究对象，研发了成熟的工厂化容器苗育苗技术，尤其是轻基质网袋容器苗培育技术，有效提高了珍贵树种的造林成活率，为其推广种植发挥了重要作用。热林中心现建有苗圃200亩，年产100万组培苗的生产线，可培育苗木500万株以上。

3.2　珍优树种近自然高效培育与资源管理技术研究

珍贵优良树种培育技术研究与推广示范是实验基地的主要任务之一。40年来，在前期开展了引种驯化及小规模造林试验研究，撑握主要珍贵树种的生物生态学及林学特性的基础上，进一步研究探索以近自然森林经营理论为指导的培育珍贵树种优质大径材多功能人工林的关键技术和森林资源现代化管理技术，为构建热带南亚热带珍贵优良树种人工林高效培育和森林可持续经营管理提供科学依据。

(1)珍贵优良阔叶树种培育技术

围绕珍贵树种多功能近自然林经营技术体系建设，以主要珍优造林树种为对象，深入研究格木、柚木、灰木莲等不同树种对光照、气温、水分等环境因子生态适应性及人工林生长过程规律，为造林立地精准选择、混交林树种科学配置、大径材培育目标制定等技术提供了理论依据；森林抚育技术研究方面，研究了格木、米老排、柚木、红椎等主要树种林木树冠生长规律、节子与分枝特性及其与林分密度关系，为优质大径材培育技术规程研制提供了科学依据；同时着重探讨了米老排、红椎、降香黄檀等树种天然更新特性及人工促进天然更新技术，其中米老排人工林迹地更新及幼林抚育技术有较高的生产实用价值，已推广使用。

(2)马尾松人工林近自然高效培育技术

马尾松是热林中心重要的乡土造林树种，是脂材兼用，具有较高经济价值的商品林树种，在培育技术研究方面，40年来除继续开展林分密度对林分生产力及产脂量影响的研究外，重点开展了马尾松近自然化改造技术研究，主要研究在马尾松人工纯林引入红椎、格木、香梓楠、米老排和火力楠等乡土珍贵优良树种以改善林分树种结构组成，加快先锋树种人工群落向结构优化、生态稳定的珍贵树种近自然林转化的技术改造，进行林分结构改造对马尾松毛虫危害控制的技术探索，为马尾松人工林的可持续经营及低效人工针叶林向生态高效、经济高值化经营提供技术基础。

(3)多功能森林经营作业法技术研究与示范

热林中心是国家林业和草原局下属20家全国森林经营样板基地之一，打造出7个森林经营类型：①珍贵树种大径材经营类型；②针叶人工纯林近自然化改造经营类型；③马尾松脂材兼用林近自然化经营类型；④速生桉—珍贵树种混交经营类型；⑤松杉人工针叶林经营类型；⑥速生阔叶树纯林经营类型；⑦退化天然次生林改造经营类型。

近年来，中心开展了多功能森林经营作业法技术研究与示范，总结提出了热林中心7种经营类型15种典型经营模式作业法：包括红椎大径材目标树单株择伐作业法、马尾松(杉木)—硬阔类针阔混交林目标树单木择伐作业法、马尾松—速生阔叶树相对同龄林皆伐作业法、降香黄檀—速生桉混交林择伐作业法和米老排人工纯林抚育性渐伐作业法等。

近年来，中心继续加大树种和林种结构调整，大力发展珍贵树种及其混交林，探索珍贵树种丛植混交造林模式，引进目标树经营和近自然林经营技术，培育多功能森林，探索适合热带南亚热带可持续多功能的森林经营技术与模式，建立健全长效推进森林经营工作的制度和政策体系，带动推进区域森林经营的科学化、集约化、专业化，显著提高森林经营水平和森林经营多种效益，努力打造成科技支撑能力强、森林经营水平高、辐射带动效果好的样板基地。

(4)林业机械化研究

为加强林业机械化、专业化建设，提升森林经营水平和多功能近自然森林经营技术，保证相关技术要求及经营措施得到有效落实，提高森林经营项目实施速度、效率和质量，中心高度重视林业机械化建设，引进应用林业机械设备，并组建营林专业队，在营林生产中开展定向采伐抚育、珍贵树种修枝整形及索道反坡集材等机械化作业，大大提高了作业质量、生产效率和安全作业水平，并大幅减少了对人工的依赖。中心相继组建了割灌除草专业队、目标树选择专业队、油锯定向采伐专业队和索道集材专业队4个营林专业施工队伍。同时加强了对专业队的技术培训，先后聘请了国内外教授、专家对专业队进行理论指导和现场培训，有效提高了专业队的理论水平和业务技能。通过不断生产实践和总结，大幅提高了专业队的理论水平、营林施工技能、施工质量和生产效率。

3.3 南亚热带主要人工林生态系统服务功能研究

(1)南亚热带主要树种人工林生态系统服务功能长期定位观测研究

依托广西友谊关森林生态系统国家定位观测研究站和广西凭祥竹林生态系统国家定位观测研究站，开展了马尾松、红椎等南亚热带主要树种和粉单竹、中华大节竹等森林或竹林的水文、土壤、气象和生物等要素的长期定位观测研究，每年为国家林业和草原局陆地生态系统定位研究网络提供大量区域人工林和竹林可靠而规范化的科学观测数据，为国家区域生态发展战略的制定和参与国际谈判及履约等提供科学数据与支撑。此外，通过长期的定位观测研究，了解和掌握了主要树种人工林的碳、氮、水循环及其耦合过程、森林植物群落种群空间分布格局、植物多样性与稳定性、土壤水热特征及其对气象要素的响应、森林降水再分配及冠层淋溶等的规律，为南亚热带人工林可持续经营提供了科学依据和技术支撑。

(2)南亚热带主要人工林多功能协同提升关键技术研究

通过开展不同树种、模式、经营技术对人工林多功能的影响研究，分析了人工林不同器官及生物组分的碳素密度，建立了铁力木、格木、米老排、红椎、西南桦等珍优树种的生物量方程，探明了主要树种与经营模式及经营技术对人工林生产力、生物量和碳储量等的响应，以及生态系统生物量和碳储量的分配格局，阐明了不同树种人工林生态系统的碳库特征，为区域人工林多功能协同提升提供了优良树种、经营模式及经营的关键技术。

(3)森林立地类型划分与质量评价研究

以林班为尺度，以主成分分析法确定立地划分主导因子，依据主导因子及其分级标准将中心的林地进行了精准立地类型划分，同时，采用综合评价法对立地质量进行了评价，并根据评价结果，提出了林地科学利用与树种布局规划，可为热林中心适地适树安排造林树种及精准提升林分质量提供坚实的理论基础。

4 未来科研设想

热林中心作为国家级林业科技创新实验示范基地，是一个具有多学科并举、产学研结合的研究平台，尤其在珍贵树种良种选育、培育技术以及人工林生态服务功能长期定位研究等方面具有得天独厚的优势。今后一段时期，热林中心将紧紧围绕以珍贵树种大径材培育为主导的多功能近自然森林经营技术体系建设，在热带南亚热带主要造林树种种质资源收集保存、评价利用、良种选育、珍贵树种和马尾松大径材培育、人工林生态功能研究和森林资源现代化管理、林业机械化、珍贵优良树种木材开发利用等领域集中开展科学研究和技术创新。重点开展以下5个方面的研究工作，为我国热带南亚热带林业发展作出新贡献。

4.1 主要树种种质资源收集保存、评价利用与良种选育

重点开展热带南亚热带主要树种林木种质资源的收集保存与评价利用，主要造林树种良种选育和良种壮苗繁育技术研究等，重点解决主要树种造林良种化和苗木规模化繁育问题，着实推进我国热带南亚热带林木种质资源库、国家和广西良种基地建设。

4.2 主要珍贵优良树种大径材培育关键技术研究

以建设珍贵优良树种木材资源战略储备为目标，以发展材质优良、市场价值高、培育前途大的珍贵优良树种资源为重点，因地制宜，合理布局，科学发展，积极稳妥地培育珍贵优良树种资源。探索主要用材树种尤其是珍贵树种人工林森林质量精准提升技术，结合我国的国情和林情，借鉴“近自然森林经营”理论和技术，结合中心实际，研发高效低成本的混交林营建、单株目标树经营管理等森林质量精准提升综合技术。

4.3 人工林生态功能评估与多功能协同提升技术研究

开展不同树种和不同经营模式下林分生态过程和生态系统服务功能提升、南亚热带典型人工林多目标经营技术与效益评价、松杉人工林近自然化改造技术与效益分析和珍贵树种生理和生态学特性等研究，重点开展人工林大径材培育同物种多样性、固碳增汇和地力维持等生态功能协同提升技术研究，为人工林森林质量精准提升提供技术支撑。

4.4 人工林多功能近自然森林经营全周期作业技术体系研建

重点推进森林全生命周期多功能经营林分作业法研建，完善以珍贵树种大径材培育为主导的多

功能近自然森林经营技术体系；编制科学可靠的热林中心多功能可持续森林经营方案。加快推进森林资源信息化、集约化、现代化和智能化管理，提高林业机械化水平，不断提升林业生产效率和森林经营水平。

4.5　珍贵树种木材材性改良与高效增值加工与利用技术研究

围绕热带南亚热带珍贵树种木材高值化利用问题，重点开展主要珍贵树种木材材性及影响因素研究，为珍贵树种木材材性改良技术研发和定向培育提供科学依据。开展珍贵树种木材高效增值加工与高效利用技术攻关，实现珍贵用材价值最大化。

5　关于本书

在热林中心成立40周年之际，我们系统地整理了近10年来(2010—2019年)热林中心科技人员发表的论文并出版。由于论文数量较多，论文集篇幅有限，我们对论文进行了适当筛选。入选的论文主要以热林中心为第一单位发表的研究论文，其中，科学引文索引(SCI)、中国科学引文数据库(CSCD)和中文核心期刊的科技论文收录全文，其余科技论文收录文章摘要，以热林中心为平台开展研究发表的论文以附录的形式收录文章标题。全书共收录200篇论文或摘要，所有入选论文和摘要均征得相关编辑部和作者的同意，并标明了全部作者、全部署名单位和原发表出处，在此对本论文集贡献论文的全体作者致以诚挚的谢意。论文集分为热带南亚热带主要珍优树种良种选育技术研究，珍优树种近自然高效培育与资源管理技术研究，热带南亚热带主要人工林生态系统服务功能研究三大部分，分别由卢立华、郭文福、谌红辉、苏建苗、曾冀、郝建、陈琳、郑路、唐继新、劳庆祥负责整理、编排。本书的策划、前言的写作以及统稿由明安刚、孙冬婧、韦菊玲、吴思达、刘志龙、邓硕坤、陈建全负责，审定工作由蔡道雄、贾宏炎、许基煌和孙文胜负责。

由于时间匆忙，编写过程中难免出现错漏之处，欢迎广大科技工作者批评指正！

编者

2019年10月

CONTENTS 目录

第二部分 珍优树种近自然高效培育与资源管理技术研究

第三部分　热带南亚热带主要人工林生态系统服务功能研究

第一部分

热带南亚热带主要珍优树种良种选育技术研究

Growth and Nutrient Efficiency of *Betula alnoides* Clones in Response to Phosphorus Supply

CHEN Lin[1], JIA Hongyan[1], ZENG Jie[2], BERNARD Dell[3]

([1] *The Experimental Center of Tropical Forestry*, *CAF*, *Pingxiang* 532600, *Guangxi*, *China*;

[2] *Research Institute of Tropical Forestry*, *CAF*, *Guangzhou* 510520, *Guangdong*, *China*;

[3] *Division of Research & Development*, *Murdoch University*, *Western Australia* 6150, *Australia*)

Abstract: As phosphorus deficiency limits the productivity of many plantation forests in Asia, there is considerable interest in developing phosphorus-efficient clones for the region through targeted breeding programs. Therefore, we determined growth, nutrient concentrations and nutrient absorption and utility efficiencies of four *Betula alnoides* clones (C5, C6, 1-202 and BY1) in response to six phosphorus levels of 0, 17, 52, 70, 140 and 209mg P/plant coded as P1 to P6, respectively. Maximum growth occurred in the P4, P5 and P6 plants since they had the largest height, biomass, leaf area and branch number. Phosphorus application increased the phosphorus concentrations of all clones. Nutrient loading was achieved with the P6 treatment because growth and biomass were not significantly higher, but root, stem and leaf phosphorus concentrations were approximately twice those of P4 plants. Clone BY1 had the highest phosphorus-efficiency, and is recommended for field application due to its maximum root collar diameter, biomass, root/shoot ratio, leaf area, nutrient absorption and utility efficiency among the four clones. The findings will help to improve the nutrient efficiency of this species in plantation forestry in Asia.

Key words: growth; nutrient concentrations; nutrient absorption and utility efficiencies; phosphorus

1 INTRODUCTION

Phosphorusis a major element constraining the growth of fast-growing species, particularly in phosphorus-poor soils[1,2]. Applying phosphorus fertilizer is a common practice to overcome the phosphorus limitation and to promote the successful establishment and productivity of tree species[3-5]. However, only 10% ~ 25% of inorganic phosphorus applied in fertilizer is taken up by plants, which can lead to serious soil and water pollution[6] if the bulk of the phosphorus is not strongly adsorbed by soil surfaces. Thus, efforts to develop high nutrient-efficient plants through breeding programs have been undertaken exploring inter-and intra-specific variations for plant growth, physiology and nutrient efficiency[7-10].

Phosphorus efficiency has been shown to be closely linked to the ability to absorb phosphorus from the soil and the efficiency of allocation of phosphorus within the plant[11,12]. For example, superior clones of white spruce (*Picea glauca*) had faster height growth, longer needles, different tannin structure and distribution, greater N, P and K use efficiencies, higher root growth potential and net photosynthesis compared to the parent family[7]. The nutrient-efficient hybrid aspen clone (*Populus tremula* × *P. tremuloides*) with high growth saved around 5% of nutrients in the short term, but different nutrient storage strategies between clones may be regarded as a possible way to save nutrients in the long run[13]. For two Chinese fir clones (*Cunninghamia lanceolata*) with high phosphorus efficiency, the adaption of these clones to low phosphorus condition was attributed to increased phosphorus acquisition and utilization efficiencies[14].

Betula alnoides Buch.-Ham. ex D. Don is a broad-leaved and fast-growing species in southern China which is favored for reforestation in recent years because of its wood quality and ecological functions[15]. The nitrogen loading requirement of *B. alnoides* has been studied[16], however the phosphorus requirement of this species being considered for further reforestation is unknown. Generally, the soils of southern China limit the growth of plantation tree species[17]. Therefore, we investigated the growth and nutrition responses of four *B. alnoides* clones to phosphorus. The objectives of this study are to: ① determine the optimal phosphorus requirement for maximum growth and nutrient uptake of *B. alnoides*. ② identify any superior clones with high phosphorus efficiency and favorable biomass production in order to improve the establishment success of *B. alnoides* in the field.

2 MATERIALS AND METHODS

Four clones of *Betula alnoides* Buch. -Ham. ex D. Don, coded as C5, C6, 1-202 and BY1, were selected with good performance in the nursery and recent employment in reforestation. Clones C5, C6 and BY1 were from Longzhou County and clone 1-202 was from Baise City, in Guangxi, China. Clonal plants were initially grown in a mixture of 60% composted bark, 30% composted sawdust and 10% charred bark (bark charcoal) in the nursery of the Experimental Center of Tropical Forestry, CAF at Pingxiang City, Guangxi, China. After two months, healthy plants with equal height of about 4 cm were selected on April 25, 2011, the roots rinsed clean by tap water followed by deionized water, then transplanted into plastic pots (17.5cm × 11cm × 12cm, height, upper and bottom diameters) filled with a high-pressure steam pasteurized mixture of peat, vermiculite, perlite (v : v : v, 3 : 2 : 2). The pots were lined with two plastic bags to prevent water and nutrient leaching. During the experimental period, the average daily light intensity, temperature and relative humidity of the greenhouse ranged from 72 to 376 μmol photon/m^2 s, 26 to 34℃ and 55 to 80%, respectively. To avoid pests and disease, the media surface was sprayed weekly with 0.2% carbendazim or chlorpyrifos solution from 2 weeks after transplanting.

A split-plot experiment design was conducted with four replications, the main plot consisted of six phosphorus levels: 0, 17, 52, 70, 140 and 209mg P/plant, coded as P1, P2, P3, P4, P5 and P6, respectively, and the subplot included four *B. alnoides* clones: C5, C6, 1-202 and BY1. Each subplot had 12 plants, thus a total of 288 plants (12 plants × 6 P treatments × 4 replications) for each clone. The plants were watered with equal amounts of deionized water by the method of Chen et al.[18] Fertilization started at eighteen days after transplanting. Plants were supplied with 50ml of nutrient solutions once a week for 12 weeks. The basal nutrient solution contained 6 mM KNO_3, 4 mM $Ca(NO_3)_2 \cdot 4H_2O$, 2 mM $MgSO_4$, 0.1 mM Fe-EDTA, 0.05 mM H_3BO_3, 0.01 mM $MnCl_2 \cdot 4H_2O$, 0.0008 mM $ZnCl_2$, 0.0003 mM $CuCl_2 \cdot 2H_2O$, 0.0001 and mM MoO_3. Phosphorus was supplied as $NaH_2PO_4 \cdot 2H_2O$ as a series of concentrations which doubled every three weeks from initial 0, 0.25, 0.75, 1, 2 and 3 mM to final 0, 2, 6, 8, 12 and 24 mM for the six treatments, respectively. The nutrient solutions were poured onto the surface of the potting mix carefully avoiding any contact with the shoot.

Just prior to fertilization, 10 representative plants were chosen from each clone and combined to determine the initial dry weight (0.07g/plant on average), and then for analyzing the initial plant concentrations of total nitrogen (12.83g/kg), phosphorus (7.83g/kg) and potassium (15.79g/kg). One week after the last fertilization, the height, root collar diameter and number of branches were measured. Furthermore, five plants were randomly selected from each subplot and then divided into roots, stems and leaves separately. The whole plants were emerged in tap water and washed carefully by hand to remove the fine roots from the peat mixture. Leaf area was measured for each plant according to Chen et al[18]. After that, these components were separately composited for each subplot, and dried at 80℃ for 48 h to determine root, stem and leaf dry weight, and ground by a portable crusher for subsequent chemical analysis. Plant material was wet-digested in a block digester using H_2SO_4-$HClO_4$ mixture solutions. The digests were analyzed for total N by the titration method, total P by the molybdenum blue method and total K by atomic absorption spectroscopy[18].

Normality of data was checked with one-sample K-S test, and homogeneity of variance test was done. These tests confirmed that the data were of normal distribution andequal variance. Two-way ANOVAs were then conducted with general linear model to examine the main and interaction effects of phosphorus and clone at aspects of growth, nutrient concentrations, nutrient absorption efficiencies (NAE, PAE and KAE) and nutrient utility efficiencies (NUE, PUE and KUE) by SPSS 16.5 (SPSS Institute Inc. 2003). Significant treatment means were further compared by Duncan's multiple range tests at the 5% level. The NAE, PAE and KAE were defined as nitrogen, phosphorus and potassium contents in every plant, and NUE, PUE and KUE as dry weight divided by nitrogen, phosphorus and potassium contents of each plant, respectively.

3 RESULTS

3.1 Growth performance

Plant height, root collar diameter, biomass, leaf area and number of branches differed significantly with phosphorus supply and clone($P<0.01$, Table 1). Additionally, there was an interaction between phosphorus supply and clone for height and branch number($P<0.05$, Table 1) but not for the other growth parameters($P>0.05$, Table 1).

Table 1 Effects of phosphorus supply and clone on growth of *Betula alnoides*

	Growth indices					
	Height (cm)	Root collar diameter(mm)	Biomass (g/plant)	Root and shoot ratio	Leaf area (cm^2/plant)	Branch number
Phosphorus supply						
P1	18.2(0.4)d	2.31(0.04)e	0.87(0.23)d	0.35(0.01)a	19.2(3.9)d	6.5(0.2)d
P2	34.8(0.4)c	3.40(0.04)d	3.66(0.23)c	0.22(0.01)b	139.3(4.0)c	16.1(0.2)c
P3	37.5(0.4)b	3.73(0.04)c	4.34(0.23)b	0.21(0.01)b	170.7(4.0)b	17.0(0.2)b
P4	38.5(0.4)ab	3.84(0.04)c	4.59(0.23)ab	0.19(0.01)b	184.9(3.4)a	18.0(0.2)a
P5	39.1(0.4)a	3.98(0.04)b	4.89(0.23)ab	0.20(0.01)b	192.4(4.0)a	18.1(0.2)a
P6	39.3(0.4)a	4.18(0.04)a	5.07(0.23)a	0.20(0.01)b	190.1(4.0)a	18.2(0.2)a
Clone						
C5	32.1(0.3)d	3.35(0.04)c	3.08(0.34)c	0.21(0.01)b	128.6(3.5)c	14.9(0.1)c
C6	36.5(0.3)a	3.70(0.04)b	3.95(0.34)b	0.23(0.01)b	159.0(3.1)a	16.4(0.1)a
1-202	34.2(0.3)c	3.43(0.04)c	3.57(0.34)bc	0.21(0.01)b	149.0(3.2)b	15.6(0.1)b
BY1	35.4(0.3)b	3.82(0.04)a	5.01(0.34)a	0.28(0.01)a	161.2(3.1)a	15.5(0.1)b
ANOVA						
Phosphorus supply	0.000**	0.000**	0.000**	0.000**	0.000**	0.000**
Clone	0.000**	0.000**	0.000**	0.000**	0.000**	0.000**
Phosphorus supply × Clone	0.002**	0.153	0.922	0.077	0.259	0.025*

Notes: P1, P2, P3, P4, P5 and P6 treatments represent supply of 0, 17, 52, 70, 140 and 209mg P/plant, respectively. Figures in parentheses are standard errors. Values followed by the same letters in a column are not significantly different among treatments at 0.05 level according to Duncan's multiple range tests. "*" and "**" represent significant difference among treatments at 0.05 and 0.01 levels, respectively.

With increasein phosphorus supply, the root and shoot ratio decreased significantly from 0.35 in P1 to 0.22 in P2 but was not affected at higher phosphorus supply(Table 1). However, other growth parameters increased gradually and then remained stable, with plateau values reached in P4 for height, biomass, leaf area and number of branches. The P6 plants had the highest root collar diameter (Table 1).

Compared with the control (P1), phosphorus additions increased the height, root collar diameter, biomass, leaf area and number of branches by 91%~116%, 47%~81%, 321%~483%, 626%~902% and 148%~180%, respectively(Table 1). Of the four clones, clone C5 had the poorest growth performance. Clone BY1 had superior root collar diameter, biomass, root and shoot ratio, and leaf area, being approximately 14%, 63%, 33% and 25% higher than those of clone C5, respectively. However, clone C6 had maximal height and number of branches, about 14% and 10% higher, respectively, than those of clone C5 (Table 1).

3.2 Nutrient concentrations

Therewere significant effects of phosphorus supply on root, stem and leaf nutrient concentrations($P<0.05$, Table 2). However, the clones did not differ with respect to phosphorus concentrations($P>0.05$, Table 2) but differed strongly for nitrogen concentrations in root and leaf as well as potassium concentrations in leaf and stem ($P<0.01$, Table 2). Overall, there were significant interactions between phosphorus supply and clone for root, stem and leaf phosphorus concentrations as well as for stem nitrogen concentration($P<0.05$, Table 2).

The root, stem and leaf nutrient concentrations showed a similar trend among clones as the amount of phosphorus addition increased. For example, phosphorus concentrations increased gradually with the increase in phosphorus supply, reaching the maximum at P6(Table 2). In contrast, stem and leaf nitrogen concentrations decreased rapidly and then tended to be stable above P2. The potassium concentrations gradually increased from P1 to P3, and then declined in the roots or remained stable in the stem and leaf potassium. Of four clones, clone BY1 had the lowest leaf nitrogen and stem potassium concentrations but highest leaf potassium concentration(Table 2).

Table 2 Effects of phosphorus supply and clone on nutrient concentrations of *Betula alnoides*

	Nutrient concentration(g/kg)								
	Root			Stem			Leaf		
	N	P	K	N	P	K	N	P	K
Phosphorus									
P1	14.0(0.3)a	1.2(0.1)e	13.7(0.4)c	12.8(0.3)a	0.8(0.1)f	13.8(0.3)c	19.2(0.3)a	1.1(0.1)e	16.2(0.2)c
P2	12.6(0.3)b	1.5(0.1)e	15.0(0.4)b	8.1(0.3)b	1.2(0.1)e	15.3(0.3)b	17.6(0.3)b	1.6(0.1)d	17.5(0.2)b
P3	12.4(0.3)b	2.5(0.1)d	16.2(0.4)a	7.8(0.3)b	2.5(0.1)d	16.5(0.3)a	18.2(0.3)b	2.2(0.1)c	17.6(0.2)b
P4	11.7(0.3)bc	3.5(0.1)c	15.7(0.4)ab	7.9(0.3)b	3.3(0.1)c	16.7(0.3)a	17.9(0.3)b	2.4(0.1)c	17.7(0.2)b
P5	11.4(0.3)c	5.7(0.1)b	14.6(0.4)bc	7.6(0.3)b	5.2(0.1)b	16.9(0.3)a	17.6(0.3)b	3.6(0.1)b	18.1(0.2)b
P6	11.0(0.3)c	7.1(0.1)a	13.7(0.4)c	7.3(0.3)b	6.4(0.1)a	16.9(0.3)a	17.3(0.3)b	4.9(0.1)a	18.7(0.2)a
Clone									
C5	12.8(0.3)a	3.7(0.1)	14.7(0.3)	8.7(0.3)	3.3(0.1)	16.7(0.2)a	18.5(0.2)a	2.7(0.1)	17.2(0.2)b
C6	11.2(0.3)b	3.4(0.1)	14.4(0.3)	8.3(0.3)	3.2(0.1)	16.2(0.2)ab	18.6(0.2)a	2.7(0.1)	17.4(0.2)b
1-202	12.3(0.3)a	3.6(0.1)	14.9(0.3)	9.1(0.3)	3.1(0.1)	15.9(0.2)bc	18.4(0.2)a	2.6(0.1)	17.3(0.2)b
BY1	12.3(0.3)a	3.6(0.1)	15.2(0.3)	8.3(0.3)	3.3(0.1)	15.3(0.2)c	16.5(0.2)b	2.6(0.1)	18.6(0.2)a
ANOVA									
Phosphorus	0.000**	0.000**	0.000**	0.000**	0.000**	0.000**	0.000**	0.000**	0.000**
Clone	0.001**	0.277	0.262	0.122	0.256	0.000**	0.000**	0.900	0.000**
Phosphorus×Clone	0.656	0.002**	0.670	0.005**	0.020*	0.092	0.998	0.002**	0.191

Notes: P1, P2, P3, P4, P5 and P6 treatments represent supply of 0, 17, 52, 70, 140 and 209mg P/plant, respectively. Figures in parentheses are standard errors. Values followed by the same small letters in a column are not significantly different among treatments at $P<0.05$ according to Duncan's multiple range tests. "*" and "**" represent significant difference among treatments at 0.05 and 0.01 levels, respectively.

3.3 Nutrient absorption and utility efficiencies

Nitrogen, phosphorus and potassium absorption efficiencies (NAE, PAE and KAE) and utility efficiencies (NUE, PUE and KUE) were all significantly influenced by phosphorus supply ($P<0.01$, Table 3). Moreover, there were marked differences in the NAE, PAE, KAE and NUE ($P<0.01$, Table 3), unlike for PUE and KUE, among clones($P>0.05$, Table 3). However, the only significant interaction between phosphorus supply and clone was in KUE($P<0.05$, Table 3).

Withthe increase in phosphorus supply, NAE, PAE and KAE of the four clones increased first, and then NAE and KAE tended to be constant after P3 or P4, while the PAE increased continuously(Table 3). As for nutrient utility efficiencies, PUE and KUE decreased gradually as phosphorus addition increased, and then maintained stable after P5 and P2, respectively. However, NUE increased with the increase in phosphorus supply up to P2 and then was unchanged. Notably, clone BY1 showed the highest NAE, NUE, PAE and KAE among the four clones (Table 3), which was in agreement with its superior growth performance(Table 1).

Table 3 Effects of phosphorus supply and clone on nutrient absorption efficiency(nitrogen, NAE; phosphorus, PAE; potassium, KAE) and nutrient utility efficiency(nitrogen, NUE; phosphorus, PUE; potassium, KUE) of *Betula alnoides*

	NAE (mg/plant)	PAE (mg/plant)	KAE (mg/plant)	NUE (g/mg)	PUE (g/mg)	KUE (g/mg)
Phosphorus supply						
P1	13.68(2.64)c	0.97(1.47)e	12.83(3.64)d	0.06(0.00)c	1.08(0.03)a	0.07(0.00)a
P2	49.34(2.53)b	5.14(1.25)d	59.67(3.64)c	0.07(0.00)ab	0.70(0.03)b	0.06(0.00)b
P3	59.15(2.53)a	9.96(1.25)c	73.53(3.64)b	0.07(0.00)ab	0.43(0.03)c	0.06(0.00)b
P4	62.04(2.53)a	12.91(1.20)c	77.99(3.64)ab	0.07(0.00)b	0.35(0.03)d	0.06(0.00)b
P5	64.74(2.53)a	22.13(1.20)b	83.89(3.64)ab	0.07(0.00)ab	0.22(0.03)e	0.06(0.00)b
P6	64.65(2.53)a	30.15(1.25)a	87.22(3.64)a	0.08(0.00)a	0.18(0.03)e	0.06(0.00)b
Clone						
C5	43.47(2.07)c	11.13(1.13)b	51.59(2.97)c	0.07(0.00)c	0.48(0.03)	0.06(0.00)
C6	53.61(2.07)b	13.55(0.98)b	66.50(2.97)b	0.07(0.00)b	0.48(0.02)	0.06(0.00)
1-202	49.37(2.13)b	12.36(0.98)b	59.25(2.97)bc	0.07(0.00)bc	0.48(0.02)	0.06(0.00)
BY1	62.61(2.07)a	17.13(1.06)a	86.07(2.97)a	0.08(0.00)a	0.54(0.03)	0.06(0.00)
ANOVA						
Phosphorus supply	0.000**	0.000**	0.000**	0.000**	0.000**	0.000**
Clone	0.000**	0.001**	0.000**	0.000**	0.203	0.587
Phosphorus supply×Clone	0.950	0.555	0.825	0.636	0.526	0.014*

Notes: P1, P2, P3, P4, P5 and P6 treatments represent supply of 0, 17, 52, 70, 140 and 209mg P/plant, respectively. Figures in parentheses are standard errors. Values followed by the same small letters in a column are not significantly different among treatments at $P<0.05$ according to Duncan's multiple range tests. "*" and "**" represent significant difference among treatments at 0.05 and 0.01 levels, respectively.

4 DISCUSSION

Phosphorus additionstrongly promoted growth of the four *B. alnoides* clones, indicating that phosphorus deficiency was a major limiting factor for plant development in the nursery environment[19] similar to studies on aspen and its hybrids[20], and three *Leucadendron* cultivars [21]. However, the root and shoot ratio of *B. alnoides* plants showed little response to phosphorus addition in the present study. This might be attributed to an insensitive response of root and shoot ratio to variable amounts of phosphorus where the nitrogen supply remains constant[22].

The survival and early performance of outplanted nursery stock at the establishment stage are likely to be more influenced by internal nutrient retranslocation than by the current nutrient supply, hence the benefit of nutrient loading in nursery production[23-25]. The P4, P5 and P6 treatments all resulted in maximum growth measured as height, biomass, leaf area and number of branches. This implies that the P4 treatment(70mg P/plant) was sufficient for the growth of *B. alnoides* clones. Furthermore, as the P6 treatment (209mg P/plant) did not cause phosphorus toxicity symptoms even though phosphorus concentrations were twice those in P4 plants, this suggests that *B. alnoides* clones had high phosphorus uptake ability, and nutrient loading can be achieved at the highest phosphorus treatment[26,27]. One probable mechanism to prevent the accumulation of inorganic phosphorus(Pi) reaching toxic concentrations in *B. alnoides* was the conversion of Pi into organic storage compounds, e.g. phytic acid[28]. As the phosphorus supply in soils in south China is commonly limiting for tree growth[29,30], it is recommended that luxury phosphorus additions be applied to nursery stock prior to outplanting in nutrient-poor or weed-prone sites. Apart from the amount of fertilizer, the nutrient addition ratio and rate should also be considered to improve nutrient efficiency in the future[31,32].

Variations often occur in growth and nutrient use efficiency among clones, which provides a potential advantage for developing nutrient-efficient plants through breeding programs[33]. For example, some clones of poplar(*Populus deltoides* × *P. cathayana* and *P. deltoides* × *P. simonii*) were selected for reforestation under low phosphorus stress because they absorbed more phosphorus and thus grew better than other clones through increasing rhizosphere acidification and affinity of roots to capture phosphate[34,35]. Natural hybrid clones (e. g. *Eucalyptus* PF1) were progressively replaced by artificial hybrid clones(*E. urophylla* × *E. grandis*) around Pointe-Noire in Congo since they had lower biomass and nutrient use efficiencies but greater nutrient contents loss when harvesting compared to artificial hybrid clones[36]. In *Betula*, clone BY1 had the largest root collar diameter, dry weight, leaf area, root and shoot ratio among the four clones tested. The allocation of more resources to the root, so as to absorb more water and nutrients to support plant growth, helps to explain the greater productivity in this clone[37]. This can also be confirmed from its higher nutrient absorption efficiencies and NUE[23,24]. Clone BY1 had lower leaf nitrogen concentration but higher leaf potassium concentration compared with other clones, suggesting that nitrogen dilution and synergism between phosphorus and potassium ions were occurred in the fast-growing clones of *B. alnoides*.

5 CONCLUSIONS

Thegrowth, nutrient concentrations and nutrient efficiencies were more influenced by phosphorus supply than clones or the interaction effect. Although phosphorus supply with 70mg P/plant(P4) was sufficient for *B. alnoides* plants, supply with 209mg P/plant(P6) is recommended for higher quality clones of this species, so they have greater fitness to endure phosphorus-deficient soils commonly distributed in south China after out-planting. Clone BY1 performed the best with the maximum biomass, root and shoot ratio and internal nutrient supply. Further research on physiological plasticity, such as leaf traits, organic acid and mycorrhizal responses of *B. alnoides* clones to phosphorus supply should be undertaken to better understand the adaptive mechanisms of plants to low phosphorus contents of growing media or soils[30,38-41].

6 ACKNOWLEDGMENTS

Thisstudy was supported by The Ministry of Science and Technology of China(2012BAD21B0102). We thank Cai-Lan Meng at the Experimental Center of Tropical Forestry, CAF, for her assistance in the nursery. We are also grateful to Le-Su Yang and Bin Yu at the Research Institute of Tropical Forestry, CAF, for nutrient determinations.

REFERENCES

[1]ELSER J J, BRACKEN M E, CLELAND E E, et al. Global analysis of nitrogen and phosphorus limitation of primary producers in freshwater, marine and terrestrial ecosystems[J]. Ecology Letters, 2007, 10(12): 1135-1142.

[2]VITOUSEK P M, PORDER S, HOULTON B Z, et al. Terrestrial phosphorus limitation: mechanisms, implications, and nitrogen-phosphorus interactions[J]. Ecological Applications, 2010, 20(1): 5-15.

[3]BROWN K R, VAN DEN DRIESSCHE R. Effects of nitrogen and phosphorus fertilization on the growth and nutrition of hybrid poplars on Vancouver Island[J]. New Forests, 2005, 29(1): 89-104.

[4]DESROCHERS A, VAN DEN DRIESSCHE R, THOMAS B R. NPK fertilization at planting of three hybrid poplar clones in the boreal region of Alberta[J]. Forest Ecology and Management, 2006, 232(1): 216-225.

[5]SINGH B, SINGH G. Phosphorusenhanced establishment, growth, nutrient uptake, and productivity of *dalbergia sissoo* seedlings maintained at varying soil water stress levels in an indian arid zone[J]. Journal of Sustainable Forestry, 2011, 30(6): 480-495.

[6]SYERS J K, JOHNSTON A E. CURTIN D. Efficiency of soil and fertilizer phosphorus use: Reconciling changing concepts of soil phosphorus behaviour with agronomic information[R]. FAO Fertilizer and Plant Nutrition Bulletin 2008, 18.

[7]LAMHAMEDI M S, CHAMBERLAND H, BERNIER P Y, et al. Clonal variation in morphology, growth, physiology, anatomy and ultrastructure of container-grown white spruce somatic plants [J]. Tree physiology, 2000, 20 (13): 869-880.

[8]MARI S, JANSSON G, JONSSON A. Genetic variation in nutrient utilization and growth traits in *Picea abies* seedlings [J]. Scandinavian Journal of Forest Research, 2003, 18 (1): 19-28.

[9] PLASSARD C, DELL B. Phosphorus nutrition of mycorrhizal trees[J]. Tree Physiology, 2010, 30(9): 1129-1139.

[10] URICH R, CORONEL I, SILVA D, et al. Intraspecific variability in Commelina erecta: response to phosphorus addition[J]. Canadian Journal of Botany, 2003, 81(9): 945-955.

[11] HIDAKA A, KITAYAMA K. Allocation of foliar phosphorus fractions and leaf traits of tropical tree species in response to decreased soil phosphorus availability on Mount Kinabalu, Borneo[J]. Journal of Ecology, 2011, 99(3): 849-857.

[12] VAN DE WIEL C C, VAN DER LINDEN C G, SCHOLTEN O E. Improving phosphorus use efficiency in agriculture: opportunities for breeding[J]. Euphytica, 2016, 207(1): 1-22.

[13] RYTTER L, STENER L G. Clonal variation in nutrient content in woody biomass of hybrid aspen[J]. Silva Fennica, 2003, 37(3): 313-324.

[14] WU P, MA X, TIGABU M, et al. Root morphological plasticity and biomass production of two Chinese fir clones with high phosphorus efficiency under low phosphorus stress[J]. Canadian Journal of Forest Research, 2011, 41(2): 228-234.

[15] ZENG J, GUO W F, ZHAO Z G, et al. Domestication of *Betula alnoides* in China: current status and perspectives[J]. Forestry Research, 2006, 19(3): 379-384. [in Chinese]

[16] CHEN L, ZENG J, XU D P, et al. Effect of exponential nitrogen loading on the growth and foliar nutrient status of *Betula alnoides* seedlings[J]. Silvae Sinicae, 2010, 46(5): 35-40. [in Chinese]

[17] XU D, DELL B, MALAJCZUK N, et al. Effects of P fertilisation on productivity and nutrient accumulation in a *Eucalyptus grandis* × *E. urophylla* plantation in southern China[J]. Forest Ecology and Management, 2002, 161(1): 89-100.

[18] CHEN L, ZENG J, JIA H Y, et al. Growth and nutrient uptake dynamics of *Mytilaria laosensis* seedlings under exponential and conventional fertilizations[J]. Soil Science and Plant Nutrition, 2012, 58(5): 618-626.

[19] WARREN C R, ADAMS M A. Phosphorus affects growth and partitioning of nitrogen to Rubisco in *Pinus pinaster*[J]. Tree Physiology, 2002, 22(1): 11-19.

[20] LIANG H, CHANG S X. Response of trembling and hybrid aspens to phosphorus and sulfur fertilization in a Gray Luvisol: growth and nutrient uptake[J]. Canadian Journal of Forest Research, 2004, 34(7): 1391-1399.

[21] SILBER A, GANMORE-NEUMANN R, BEN-JAACOV J. The response of three *Leucadendron* cultivars (Proteaceae) to phosphorus levels[J]. Scientia Horticulturae, 2000, 84(1): 141-149.

[22] GARRISH V, CERNUSAK L A, WINTER K, et al. Nitrogen to phosphorus ratio of plant biomass versus soil solution in a tropical pioneer tree, *Ficus insipida*[J]. Journal of Experimental Botany, 2010, 61(13): 3735-3748.

[23] BOWN H E, WATT M S, CLINTON P W, et al. The influence of N and P supply and genotype on N remobilization in containerized *Pinus radiata* plants[J]. Ciencia e Investigación Agraria, 2012, 39(3): 505-520.

[24] FOLK R S, GROSSNICKLE S C. Stock-type patterns of phosphorus uptake, retranslocation, net photosynthesis and morphological development in interior spruce seedlings[J]. New Forests, 2000, 19(1): 27-49.

[25] SALIFU K F, TIMMER V R. Optimizing nitrogen loading of *Picea mariana* seedlings during nursery culture[J]. Canadian Journal of Forest Research, 2003, 33(7): 1287-1294.

[26] SALIFU K F, JACOBS D F, BIRGE Z K. Nursery nitrogen loading improves field performance of bareroot oak seedlings planted on abandoned mine lands[J]. Restoration Ecology, 2009, 17(3): 339-349.

[27] TIMMER V R. Exponential nutrient loading: a new fertilization technique to improve seedling performance on competitive sites[J]. New Forests, 1997, 13(1-3): 279-299.

[28] SCHACHTMAN D P, REID R J, AYLING S M. Phosphorus uptake by plants: from soil to cell[J]. Plant Physiology, 1998, 116(2): 447-453.

[29] Tan L, He Y J, Qin L, et al. Characteristics of nutrient contents and storages in *Castanopsis hystrix* and *Betula alnoides*[J]. Guihaia, 2015, 35(1): 69-74. [in Chinese]

[30] XIE X, WENG B, CAI B, et al. Effects of arbuscular mycorrhizal inoculation and phosphorus supply on the growth and nutrient uptake of *Kandelia obovata* (Sheue, Liu & Yong) seedlings in autoclaved soil[J]. Applied Soil Ecology, 2014, 75: 162-171.

[31] ISAAC M E, HARMAND J M, DREVON J J. Growth and nitrogen acquisition strategies of *Acacia senegal* seedlings under exponential phosphorus additions[J]. Journal of Plant Physiology, 2011, 168(8): 776-781.

[32] KELLY J M, ERICSSON T. Assessing the nutrition of juvenile hybrid poplar using a steady state technique and a mechanistic model[J]. Forest Ecology and Management 2003, 180(1): 249-260.

[33] BALIGAR V C, FAGERIA N K, HE Z L. Nutrient use efficiency in plants[J]. Communications in Soil Science

and Plant Analysis, 2001, 32(7-8): 921-950.

[34] ZHANG H C, WANG G P, XU T Z, et al. Effects of rhizosphere acidification on phosphorus efficiency in clones of poplar [J]. Chinese Journal of Applied Ecology, 2003a, 14(10): 1607-1611. [in Chinese]

[35] ZHANG H C, WANG G P, Xu T Z, et al. Phosphate uptake characteristics of kinetics and phosphorus efficiency in clones of poplar [J]. Scientia Silvae Sinicae, 2003b, 39(6): 40-46. [in Chinese]

[36] SAFOU-MATONDO R, DELEPORTE P, LACLAU J P, et al. Hybrid and clonal variability of nutrient content and nutrient use efficiency in Eucalyptus stands in Congo [J]. Forest Ecology and Management, 2005, 210(1): 193-204.

[37] ZHANG Y, ZHOU Z, YANG Q. Genetic variations in root morphology and phosphorus efficiency of *Pinus massoniana* under heterogeneous and homogeneous low phosphorus conditions [J]. Plant and Soil, 2013, 364(1-2): 93-104.

[38] PANG J, TIBBETT M, DENTON M D, et al. Variation in seedling growth of 11 perennial legumes in response to phosphorus supply [J]. Plant and Soil, 2010, 328(1-2): 133-143.

[39] RICHARDSON A E, LYNCH J P, RYAN P R, et al. Plant and microbial strategies to improve the phosphorus efficiency of agriculture [J]. Plant and Soil, 2011, 349(1-2): 121-156.

[40] ZHAO T T, ZHAO N X, GAO Y B, Ecophysiological response in leaves of *Caragana microphylla* to different soil phosphorus levels [J]. Photosynthetica, 2013, 51(2): 245-251.

[41] NIU Y F, CHAI R S, JIN G L, et al, Responses of root architecture development to low phosphorus availability: a review [J]. Annals of Botany, 2013, 112: 391-408.

[原载: Journal of Forestry Research, 2018, 29(01)]

Growth and Nutrient Uptake Dynamics of *Mytilaria laosensis* Seedlings Under Exponential and Conventional Fertilizations

CHEN Lin[1], ZENG Jie[2], JIA HongYan[1], ZENG Ji[1], GUO WenFu[1], CAI DaoXiong[1]

([1] *The Experimental Center of Tropical Forestry, Chinese Academy of Forestry, Pingxiang* 532600, *Guangxi, China*;
[2] *Research Institute of Tropical Forestry, Chinese Academy of Forestry, Guangzhou* 510520, *Guangdong, China*)

Abstract: The growth characteristics and nutrient uptake dynamics of *Mytilaria laosensis* Lec. seedlings treated weekly with conventional and exponential fertilizations were investigated at intervals of 3 weeks for 12 weeks in the greenhouse. Leaf area and pigment compositions were also examined at the final harvest. The fertility treatments (mg nitrogen/seedling) included two conventional (50C and 100C) and four exponential (50E, 100E, 200E and 400E) fertilizations, and no fertilization (0) as control. The biomass and nutrient contents of *M. laosensis* seedlings increased exponentially with time. Steady-state nutrition of nitrogen (N) and phosphorus (P) were achieved under exponential fertilization treatment of 50mg N/seedling (50E) and conventional fertilization treatment of 100mg N/seedling (100C), resulting from simultaneous increase of their biomass and nutrient contents. The nutrient uptake efficiency continuously increased over time in conventionally fertilized seedlings, but it increased initially and declined or remained stable from 11 weeks after the transplanting in the exponential fertilized seedlings. At the end of the experiment, the conventionally fertilized seedlings performed remarkably better than all exponentially fertilized seedlings except for seedlings in the exponential treatment of 200mg N/seedling (200E) in height, root collar diameter and biomass. The optimum N and P uptake occurred in 200E seedlings because their N and P contents were 71%/60% and 14%/9% higher than both conventionally fertilized seedlings (50C/100C) without significant differences of growth performance between them. The leaf areas and chlorophyll contents of seedlings increased significantly with the increase of fertilizer levels and nearly peaked at the range from 100 to 200mg N/seedling, whereas the delivery schedule (conventional and exponential) had little effect on leaf areas and chlorophyll contents of seedlings at the same nutrient level (50 or 100mg N/seedling). The findings will provide evidence to make guidelines on fertilization for nursery production of *M. laosensis*, and help understand the nutrient demands for this species and further benefit the development of its plantations.

Key words: exponential fertilization; growth performance; *Mytilaria laosensis*; nutrient uptake efficiency; steady-state nutrient

1 INTRODUCTION

Mytilaria laosensis Lec. is an important broad-leaved tree species indigenous to the tropical and subtropical areas of south-east Asia and south China. It has been widely used in the establishment of ecological benefit oriented forests as well as fast-growing and high-yield plantations due to its high productivity and good quality of timber[1]. However, poor seedling quality usually due to unreasonable fertilization has been observed in the practice. As shown in Figure 1, chlorosis and irregular green patches were obvious on leaves of five-month-old *M. laosensis* seedlings grown in light media containing 80% pine bark and 20% sawdust in the nursery and the entire seedlings grew weak with fewer leaves, lower height and fewer branches, which is perhaps related to imbalanced nutrients such as nitrogen, it is thus important for improvement of seedling quality to identify nitrogen requirement of this species. In addition, better matching fertilizer application to plant requirements also has environmental benefits in reducing nutrient losses to the atmosphere or water[2].

Traditionally, nutrients were applied in a single or constant rate of fertilization, which might result in early toxicity due to over-fertilization and later nutrient deficiency in the growing season[3], and nutrient uptake efficiency could be reduced since applied nutrients were not taken up immediately by seedlings and might be lost by leaching[4]. Therefore, supplying full-element solutions to plant in lower nutrient concentration but higher nutrient addition rate during the exponential growth phase may generally provide a better rationale to build steady-state nutrition in seedlings, which subsequently induced maximum plant growth[5-8].

Figure 1 The symptoms of nutrient imbalance in five-month-old *Mytilaria laosensis* seedlings grown in light media containing 80% pine bark and 20% sawdust in the nursery.

Exponential rather than constant fertilizer addition can promote early root development and better match relative growth and nutrient consumption during the exponential growth phase of seedlings[3,9,10]. Exponential fertilization has been successfully applied to several conifer and deciduous species. However, results of these studies varied in different species or even within the same species. For example, it was concluded in some species such as *Pinus resinosa* Ait.[7], *Picea mariana* (Mill.) B. S. P.[10] and *Cunninghamia lanceolata* (lamb) Hook[3] that exponential fertilization was superior to conventional fertilization in promoting seedling growth with equal amount of nitrogen. However, in other studies about species such as *C. lanceolata*[11] and *P. mariana*[12], exponentially fertilized seedlings exhibited remarkably lower biomass than conventionally fertilized seedlings with the same dose of fertilizer, or there was no significant difference in biomass between them for *Prosopis chilensis* Mol.[9], *Larix kampferi* Sarg. and its hybrid larch[13], *Tsuga heterophlla* (Raf.) Sarg.[14], *Querbus rubra* L.[15] and *Quercus ilex* L.[16]. This is probably because species have their distinct biological characters and nutrient requirements on the one hand. On the other hand, different doses of fertilizer were involved for comparisons of seedling growth and nutrient status between exponential and conventional fertilizations in this kind of the studies.

In previous studies comparing exponential and conventional fertilizations, usually one dosage of fertilizer was involved for comparison of seedling growth and nutrient status between them[12,13,15], while things might be different when more levels of fertilizer were used. Additionally, nutrient uptake efficiencies and physiological responses to both kinds of fertilization were hardly compared in these studies. The objective of the present study is to compare the growth and nutrient dynamics, leaf area and pigment compositions of *M. laosensis* seedlings exposed to conventional and exponential fertilizations along a broad range of N supply, and to identify fertility targets for this species.

2 MATERIALS AND METHODS

2.1 Plant materials and seedling management

Mytilaria laosensis seeds were sown in sands in the nursery of Experimental Center of Tropical Forestry (CAF) in June 2009. Two months later, when 4cm high, seedlings were washed with tap water and transplanted to 370ml plastic pots filled with an autoclaved mixture of peat, vermilite and perlite (3 : 2 : 2 by volume). To prevent water and nutrient leaching, two plastic bags were placed in the bottom of pots. Each pot was irrigated to 90% of field capacity determined gravimetrically when transplanting[17]. The desired amount of fertilizer was injected to individual pots once a week using a mechanical pipettor. Supplemental irrigation was supplied by periodic weighing of pots so as to bring medium moistures back to 90% of field cavity. 24 pots were fitted into one tray and trays were arranged onto a greenhouse bench and rotated every two weeks to reduce the edge effects. The daily mean air temperature in the greenhouse ranged from 24℃ to 28℃ and a relative humility from 54% to 80% during the experiment.

2.2 Experimental design

A randomized complete block design was arranged with four replications and 36 seedlings (one and a half of trays) per treatment replication. The seedlings were supplied with a commercial water-soluble fertilizer mixture (Plant Products as 20N : 20 phosphorus pentoxide (P_2O_5) : 20 potassium oxide (K_2O) plus micronutrient, Co.

Ltd., Brampton, Ontario) to create seven treatments including two conventional (constant, C) fertilizations (50 and 100mg N/seedling), four exponential (E) fertilizations (50, 100, 200 and 400mg N/seedling) and no fertilizer (0) as control. Fertilization started two weeks after transplanting and continued for 12 weeks. The conventional treatment was calculated and supplied at a constant weekly rate (4.17 or 8.33mg N/seedling). Weekly applications of exponential fertilization were based on the following equations described by Timmer and Miller[18].

$$N_T = N_s(e^{rt} - 1)$$

where N_T is the desired total amount of N (50, 100, 200 or 400mg) for a seedling, N_s is the initial content of N in a seedling (4.38mg N/seedling) determined at the beginning of the fertilization, t is the number of fertilizer applications through the experimental period ($t = 12$), and r is the relative addition rate needed to increase N_s to a final level $N_T + N_s$. Once r was determined for the fertilization period (0.210, 0.264, 0.320 and 0.377 for 50E, 100E, 200E and 400E, respectively), the quantity of N to be added on a specific day (N_t) was calculated from the following equation:

$$N_t = N_s(e^{rt} - 1) - N_{t-1}$$

where N_{t-1} is the cumulative amount of N added up to and including the previous application. The weekly nitrogen applications were shown in Table 1.

Table 1 Schedule of nitrogen amount (mg N/seedling) added weekly

Treatments	3rd week	4th week	5th week	6th week	7th week	8th week	9th week	10th week	11th week	12th week	13th week	14th week	Total
0	0	0	0	0	0	0	0	0	0	0	0	0	0
50C	4.17	4.17	4.17	4.17	4.17	4.17	4.17	4.17	4.17	4.17	4.17	4.17	50
100C	8.33	8.33	8.33	8.33	8.33	8.33	8.33	8.33	8.33	8.33	8.33	8.33	100
50E	1.02	1.26	1.56	1.92	2.37	2.92	3.60	4.45	5.49	6.77	8.35	10.30	50
100E	1.33	1.73	2.25	2.93	3.81	4.97	6.47	8.42	10.97	14.29	18.61	24.24	100
200E	1.65	2.28	3.14	4.32	5.95	8.20	11.29	15.56	21.43	29.52	40.66	56.00	200
400E	2.01	2.93	4.27	6.22	9.07	13.22	19.28	28.11	40.99	59.76	87.13	127.03	400

2.3 Harvesting measurements

In August 2009, initial dry mass (0.0975g/seedling) and nutrient concentrations (44.94mg/g for N; 7.13mg/g for P and 13.65mg/g for K) were determined for the extra-transplanted 70 seedlings. Root collar diameter and height were measured for each seedling on 5, 8, 11 and 14 weeks after transplanting. Additionally seven seedlings per experimental unit were destructively sampled and divided into roots and shoots. All tissues were dried at 70℃ for 48h and weighted. Samples were then composited by treatment replication and milled for nutrient analysis. Plant samples were wet-digested in a block digester using a sulfuric acid-perchloric acid ($H_2SO_4 - HClO_4$) mixture solution. The digest was analyzed for total N by diffusion distillation, P by molybdenum blue method and K by atomic absorption. Nutrient uptake efficiency (N, P and K) was estimated by the following equation[19].

$$\text{Nutrient uptake efficiency}(\%) = \frac{\text{Nutrient uptake(fertilized plot)} - \text{Nutrient uptake(control plot)}}{\text{Nutrient rate applied}} \times 100$$

At the end of the experiment, four randomly selected seedlings from each experimental unit were measured for leaf area. All leaves of each unit were separately collected and scanned to obtain leaf photographs with Canon Image Class MF 4100 scanner. Leaf pixels were examined using Photoshop Microsoft. Hence, leaf area could be calculated according to the closely positive relationships between the leaf area and pixel[20]. Two pieces of fully expanded leaves were also collected from each seedling and composited by treatment replication to determine pigment compositions. The chlorophyll and carotenoid contents were estimated spectrophotometrically in 80% acetone extract and calculated according to the method of Lichtenthaler and Wellburn[21].

2.4 Statistical analysis

A one-way analysis of variance was conducted on growth and nutritional data using SPSS 16.0 software. Significant treatment means were further separated by Duncan multiple range tests at the 5% level. Regression analysis was used to examine changes in the growth performance and nutrient contents of *M. laosensis*seedlings over time with linear, expotential models etc., and optimium model was determined according to square correlation coefficients for each treatment.

3 RESULTS AND DISCUSSION

3.1 Seeding growth dynamics

As shown in Figure 2, the root collar diameter, height and biomass of *M. laosensis* seedlings increased significantly over time. In particular the biomass obviously matched exponential functions under all fertilization treatments ($R^2 \geqslant 0.9$, equations not shown). The root/shoot ratio mainly increased as experiment proceeded for both conventional fertilization treatments, while it only increased within 11 weeks after transplanting and then mostly decreased obviously for all exponential fertilization treatments. This was probably because nutrient limitation occurred gradually at later stage for the former, while only at earlier stage for the latter since nutrient limitation generally resulted in allocations of more

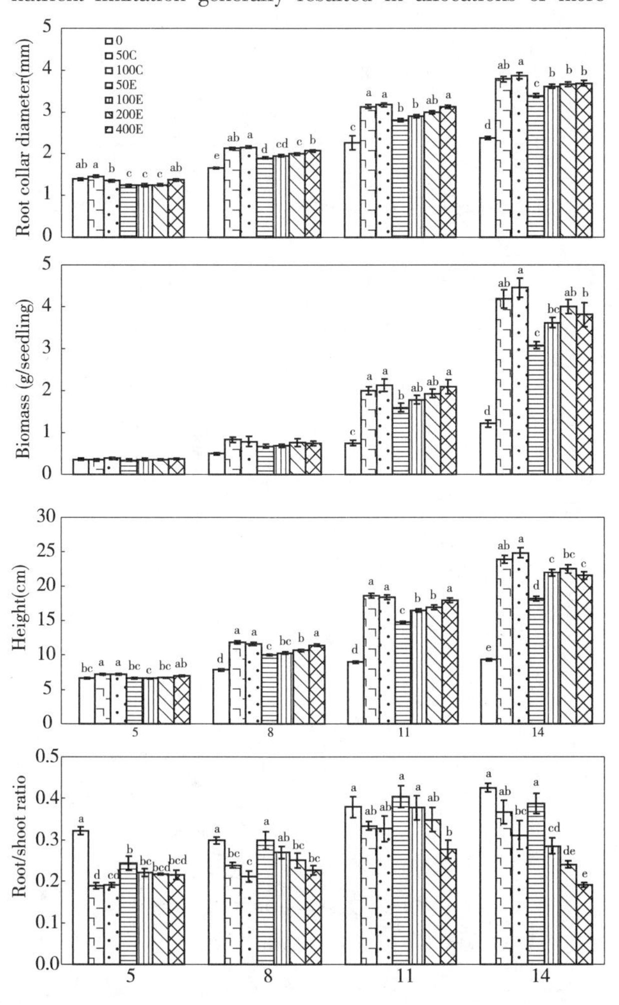

Figure 2 Growth dynamics of *Mytilaria laosensis* seedlings in different nitrogen treatments

Notes: Paired treatments without the same letter are significantly different (P<0.05) according to Duncan's Multiple Range Test at 5, 8, 11 and 14 weeks after transplanting, respectively. Error bars represent standard errors.

biomass to roots, which was also inferred from the root/shoot ratio dynamics of the control.

The root collar diameter, height and root/shoot ratio were affected significantly by treatments in all four times of measurements(P<0.05), but no obvious effect on biomass was tested until 11 weeks after transplanting (Figure 2). There were no significant differences between 50C and 100C for all growth traits throughout the experiment. The conventionally fertilized seedlings performed remarkably better than all exponentially fertilized seedlings except for the 200E in height, root collar diameter and biomass at the end of the experiment (Figure 2), probably due to relatively higher nutrient uptake efficiency under conventional fertilizations (Figure 3). This was in accordance to Everett et al.[11] on Douglas fir[*Pseudotsuga mnziesii* var. *glauca*(Beissn.)Franco] in which exponentially fertilized seedlings were smaller than conventionally fertilized seedlings although 25% more nitrogen were applied to the former. There was also concern that salt stress to seedlings might happen in the later stages at high rates of exponential fertilization. As shown in Figure 2, the highest exponential dose treatment (400mg N/seedling) decreased biomass by 9 and 14% respectively, relative to both conventional fertilization treatments. It appeared that the exponential fertilization improved nutrient efficiency only at a suitable fertilizer level. The root/shoot ratio tended to decline with the increase of nitrogen amount applied, this was consistent with those depicted in *Cunninghamia lanceolata*[3]. As for comparisons of seedling growth under conventional and exponential fertilization regimes with equivalent dose of fertilizers, the 50C and 100C fertilized seedlings had higher root collar diameter, height and biomass than that of 50E and 100E treated seedlings at any time. This may be attributed to that the exponential treatment allocated more nutrients and energies into root early in the experiment because of the relatively low nutrient inputs[9,11]; therefore the exponential fertilization could not meet the nutrient demand for early growth of the seedlings. With the increase in applied fertilizer, exponential fertilization accumulated more nutrients to the aboveground, so no significant difference in the root/shoot ratio appeared between two groups of equal fertility treatments since 11 weeks after transplanting. However, the root collar diameter, height and biomass was increased by 7%/12%, 13%/31% and 23%/36% in both kinds of conventional fertilized seedlings (50C/100C), respectively, in comparison with their relevant exponential fertilized seedlings at the end of the experiment. That resulted probably from the fact that almost 21 and 24% of the total N applied through two lower exponential fertilizer rates(50E and 100E) in 7d before the final measurement; therefore it seems that insufficient time was provided for the seedlings to absorb and respond to that dosage.

3.2 Nutrient uptake dynamics

The nutrient contents of *M. laosensis* seedlings were exponentially increasing for the duration of the experiment under all treatments except for the control($R^2>0.9$, exponential functions not shown, Table 2), which was in agreement with the findings of Xu and Timmer[3] for *Cunninghamia lanceolata* seedlings. The N concentration in 50C declined significantly as the experiment proceeded ($P<0.05$), inferring that the constant low rate addition schedule generally resulted in growth dilution. Similar results have been noted for *Picea mariana*[12] and two larch species[13]. However, the exponential fertilization treatments induced an increase or stable in the N and P concentrations, confirming that exponential fertilization can be effectively free of growth dilution. Surprisingly, 100C fertilized seedlings exhibited steady-state nutrition for both N and P, inferred from the fact that concentration of these nutrient elements remained relatively constant over time(Table 2). This normally resulted from synchronous increases of growth and nutrient uptake[6]. The 200E treatment might induce luxury nutrient consumption because it raised N and P contents by 71%/60% and 14%/10% without significant differences of biomass from the best fertilization treatments (50C and 100C)[18,22-23]. The K concentration exhibited stable or a reduction over time in all treatments, which presumably resulted from the interactions between K^+ and other ions[22].

The fertilization regimes induced different patterns of the nutrient uptake efficiency (Figure 3). Conventionally fertilized seedlings exhibited an increasing trend of nutrient uptake efficiency over time, but for the exponential fertilization treatments, the nutrient uptake efficiency increased initially, and declined or maintained stable since 11 weeks after transplanting(Figure 3). When seedlings applied with the same dose of nitrogen were compared, higher nutrient uptake efficiency was seen in seedlings conventionally fertilized at the last harvest, although exponential fertilized seedlings had greater nutrient uptake efficiency than the conventionally fertilized seedlings at former three times. This was inconsistent with the study of Dumroese *et al.*[4] on *Pinus monticola* Dougl. ex D. Don in which exponential fertilization increased seedling N uptake efficiency by 75% although exponentially fertilized seedlings received 45% less fertilizer than conventionally fertilized seedlings during the experiment. This was closely related to the difference of the fertilizer addition dose between exponential and conventional fertilizations in both studies. These results exemplified the need to modify the exponential fertilization through reducing the last addition amount and compensating it to the initial times, to avoid lower late-season nutrient up-

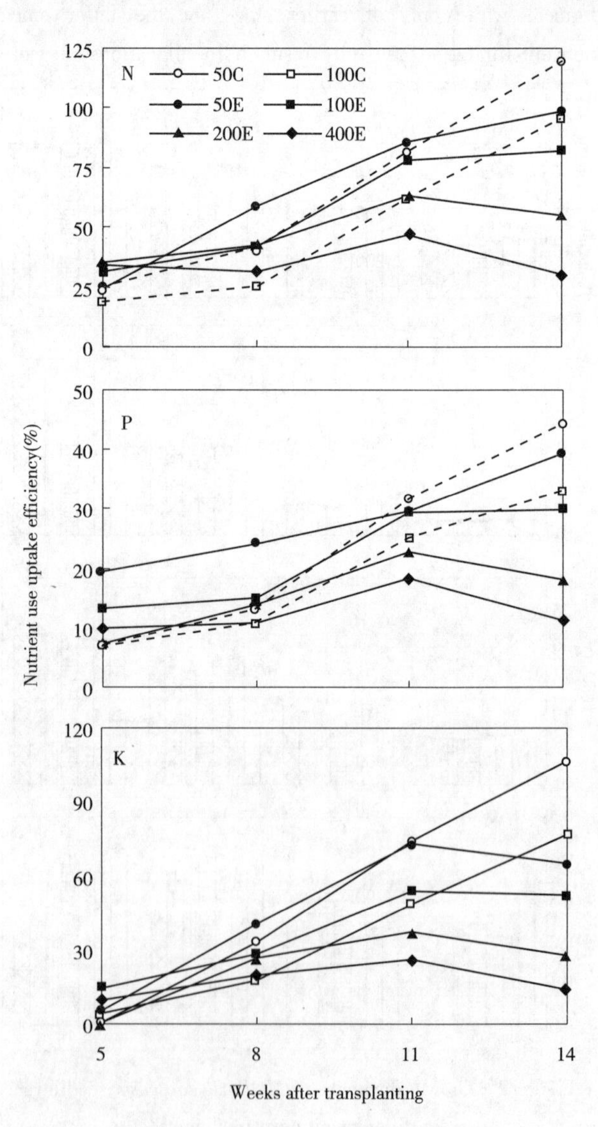

Figure 3 The dynamics of nutrient uptake efficiency(N, P and K) of *Mytilaria laosensis* seedlings in different nitrogen treatments

take efficiency of this species[22]. For example, Imo and Timmer[9] showed that the modified exponential fertilization was more effective in promoting nutrient uptake of *Prosopis chilensis* seedlings.

Table 2 Nutrient dynamics of *Mytilaria laosensis* seedlings in different nitrogen treatments

Weeks after transplanting(wk)	Treatments	Nutrient concentration(mg/g)			Nutrient content(mg/seedling)		
		N	P	K	N	P	K
5	0	16.51(0.43)dA	1.71(0.09)cA	17.05(0.47)	5.92(0.38)dB	0.61(0.04)cAB	6.11(0.35)D
	50C	26.40(0.51)abA	2.90(0.13)b	18.05(0.21)A	9.02(0.78)abD	0.99(0.07)bC	6.18(0.57)D
	100C	27.66(0.27)a	3.63(0.19)a	18.26(0.29)B	10.54(0.81)aC	1.39(0.16)aD	6.98(0.65)D
	50E	20.42(1.39)c	2.82(0.28)b	18.67(0.46)A	6.86(0.70)cdD	0.94(0.09)bD	6.28(0.56)D
	100E	20.95(1.12)cB	2.57(0.18)bC	18.78(0.28)	7.58(0.78)bcdD	0.92(0.04)bD	6.75(0.39)C
	200E	24.21(0.51)bB	2.40(0.05)bC	17.60(0.64)AB	8.42(0.48)bcC	0.84(0.05)bcC	6.14(0.47)D
	400E	24.31(0.30)bC	2.72(0.08)bC	18.32(0.34)	9.08(0.40)abC	1.01(0.04)bC	6.83(0.24)C
8	0	14.33(0.40)cB	1.67(0.25)cA	17.34(0.70)	7.06(0.37)cB	0.80(0.08)cA	8.53(0.39)cC
	50C	20.92(1.39)bB	2.78(0.33)b	18.69(0.83)A	17.27(1.68)abC	2.26(0.25)abC	15.41(1.21)aC
	100C	25.03(1.76)a	3.97(0.38)a	19.91(0.76)A	19.79(3.43)aC	3.18(0.63)aC	15.84(2.90)aC
	50E	20.39(1.72)b	2.97(0.37)b	18.18(0.29)A	13.59(1.01)bC	1.97(0.22)bC	12.22(0.82)abC
	100E	20.64(0.69)bB	2.83(0.20)bBC	18.24(0.48)	14.15(0.76)abC	1.93(0.09)bC	12.53(0.75)abC
	200E	23.65(0.97)abB	3.18(0.25)abB	18.62(0.31)A	17.83(1.76)abC	2.36(0.18)abC	14.20(1.71)aC
	400E	25.47(0.35)aC	3.48(0.21)abB	19.69(0.68)	18.90(1.44)abC	2.60(0.32)abC	14.68(1.50)aC
11	0	10.01(0.65)fC	0.66(0.05)fB	16.36(0.32)d	7.58(1.02)eB	0.49(0.03)eB	12.22(0.94)eB
	50C	19.16(0.43)eBC	2.85(0.07)d	17.66(0.40)bcdA	38.13(1.36)cB	5.66(0.13)cB	35.24(1.87)bcB
	100C	25.35(0.61)c	4.11(0.15)b	20.02(0.27)aA	53.74(2.74)bB	8.70(0.53)bB	42.53(2.48)aB
	50E	18.02(0.52)e	2.29(0.05)e	17.02(0.37)cdA	28.72(1.12)dB	3.67(0.21)dB	27.15(1.16)dB
	100E	22.91(0.52)dB	3.38(0.16)cAB	17.62(0.36)bcd	41.02(2.90)cB	6.00(0.14)cB	31.44(1.70)cdB
	200E	28.27(0.77)bA	4.08(0.10)bA	18.27(0.51)bcA	54.49(1.35)bB	7.86(0.24)bB	35.21(0.98)bcB
	400E	31.98(1.13)aB	5.09(0.17)aA	18.95(0.65)ab	66.32(3.45)aB	10.59(0.77)aB	39.32(2.18)abB
14	0	9.52(0.10)gC	0.67(0.02)eB	16.60(0.74)b	11.54(0.79)dA	0.81(0.07)dA	19.99(0.91)dA
	50C	17.11(0.76)fC	2.47(0.17)d	15.42(1.01)bB	71.17(1.95)cA	10.41(1.24)cA	64.28(4.61)bA
	100C	23.94(0.48)d	3.42(0.05)c	18.77(0.28)aB	106.70(6.17)bA	15.22(0.60)bA	83.52(3.67)aA
	50E	19.73(0.56)e	3.06(0.07)c	15.28(0.92)bB	60.90(3.01)cA	9.42(0.14)cA	46.88(1.74)cA
	100E	25.89(0.82)cA	3.83(0.18)bA	17.49(0.71)ab	93.45(2.02)bA	13.84(0.57)bA	63.35(3.90)bA
	200E	30.48(0.76)bA	4.15(0.17)bA	16.52(0.31)bB	121.94(6.05)aA	16.68(1.36)bA	66.23(4.10)bA
	400E	34.71(0.34)aA	5.43(0.05)aA	17.47(0.35)ab	132.21(9.45)aA	20.65(1.38)aA	66.76(5.69)bA

Notes: Paired treatments without the same letter in the same column are significantly different($P<0.05$) according to Duncan's Multiple Range Test, capital and lower letters refer to comparisons among and within 5, 8, 11 and 14 weeks after transplanting, respectively. The figures in parentheses are standard errors.

3.3 Leaf area and pigment compositions

The leaf area of *M. laosensis* seedlings increased with the increased amount of N applied when the N application was lower than 100mg N/seedling, and remained stable thereafter, then declined when the applied N amount was more than 200mg N/seedling (Figure 4). There were no significant difference in the leaf area between conventionally and exponentially fertilized seedlings which were applied with equal amounts of nitrogen($P>0.05$), but significant differences were observed between seedlings treated with two different N amounts, indicating that leaf area was influenced by N levels rather than fertilization regimes for this species.

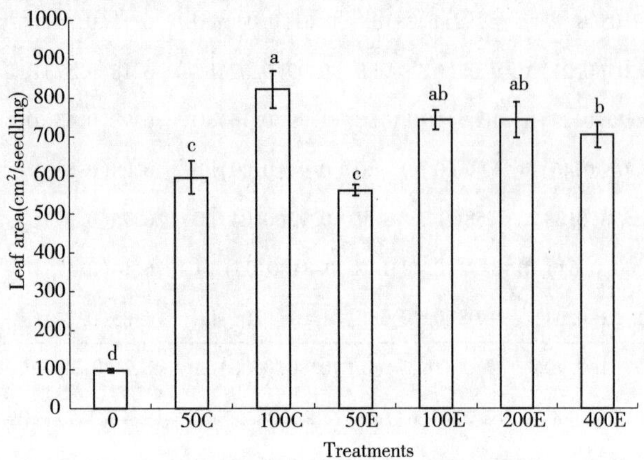

Figure 4 Comparisons of leave areas of *Mytilaria laosensis* seedlings in different nitrogen treatments

Paired treatments without the same letter are significantly different($P<0.05$) according to Duncan's Multiple Range Test at the end of the experiment. Error bars represent standard errors.

The pigment contents of *M. laosensis* seedlings were significantly affected by fertilization treatments except for carotenoid content and carotenoid/chlorophyll (Car/Chl) ratio($P<0.05$, Table 3). As the amount of N increased, chlorophyll contents showed an increasing trend when N amounts were lower than 100mg N/seedling, and were then stable, indicating that suitable fertilization could increase Chl contents. This was due to close relationship between concentrations of N and total Chl[24]. In contrast, the chlorophyll a/b (Chla/b) ratio declined with the increased amounts of N applied. Similar to the leaf area, the Chl contents were affected by the N amounts rather than by fertilization regimes because significant differences were observed in Chl contents between seedlings treated with different N levels, while no significant differences between conventionally and exponentially fertilized seedlings which were applied with equal amounts of N fertilizer.

Table 3 Effects of different nitrogen levels on pigment compositions of *Mytilaria laosensis* seedlings

Treatments	Chla(mg/g)	Chlb(mg/g)	Chla/b	Total chlorophyll (mg/g)	Carotenoid (mg/g)	Car(Chl)
0	0.252b (0.013)	0.393d (0.028)	0.645a (0.018)	0.645c (0.041)	0.135 (0.005)	0.211 (0.007)
50C	0.254b (0.010)	0.423d (0.022)	0.604ab (0.027)	0.677c (0.028)	0.138 (0.004)	0.204 (0.005)
100C	0.280ab (0.010)	0.546bc (0.037)	0.516cd (0.021)	0.825ab (0.046)	0.136 (0.018)	0.169 (0.026)
50E	0.254b (0.008)	0.475cd (0.035)	0.543bc (0.040)	0.729bc (0.035)	0.126 (0.013)	0.177 (0.024)
100E	0.282ab (0.009)	0.594ab (0.025)	0.477cd (0.020)	0.877a (0.030)	0.143 (0.020)	0.166 (0.026)
200E	0.301a (0.020)	0.609ab (0.034)	0.494cd (0.011)	0.911a (0.053)	0.143 (0.021)	0.162 (0.028)
400E	0.289ab (0.007)	0.640a (0.028)	0.451d (0.003)	0.929a (0.021)	0.176 (0.001)	0.190 (0.006)

Notes: Paired treatments without the same letter in the same column are significantly different($P<0.05$) according to Duncan's Multiple Range Test at the end of the experiment. The figures in parentheses are standard errors.

4 CONCLUSIONS

In the present study, biomass of *M. laosensis* seedlings was maximized at 50C and 100C, and steady-state nutrition were achieved under 50E and 100C, resulting from simultaneous exponential increase of their biomass and nutrient contents over time. Exponential fertilization with 200mg N/seedling induced luxury nutrient consumption because it raised seedling N and P contents by 71%/60% and 14%/9% without significant differences of biomass from the best fertilization treatment (50C/100C). Since higher internal nutrient reserves benefit root development and exploitation during the critical establishment period, and thus improved field performance of holm oak seedlings[16], it is worth examining whether this effect was obvious in the case of *M. laosensis* seedlings.

When comparing leaf area and pigment compositions of conventionally and exponentially fertilized seedlings treated with the equivalent N doses, no significant differences were seen between them, indicating that these indexes were influenced by different fertilizer amounts rather than the methods of fertilization. These findings will not only provide evidence to make guideline on fertilization for nursery production of *M. laosensis*—for instance, the total N amount recommended for this species is 50~100mg N/seedling under conventional fertilization or 200mg N/seedling under exponential fertilization in the practices—but also help us understand the nutrient demands for this species and further benefit the development of its plantations.

ACKNOWLEDGMENTS

This research was financially supported by the Experimental Center of Tropical Forestry, CAF(RL2011-05). We thank Cai-Lan Meng at the Experimental Center of Tropical Forestry, CAF for their assistance in the nursery. We are also grateful to Le-Su Yang and Bin Yu at the Research Institute of Tropical Forestry, CAF for nutrient determinations.

REFERENCES

[1] HUANG Z T, WANG S F, JIANG Y M, et al. Exploitation and utilization prospects of eximious native tree species *Mytilaria laosensis* [J]. J. Guangxi Agric. Sci., 2009, 40(9): 1220-1223. in Chinese with English abstract.

[2] DONG W Y, QIN J, LI J Y, et al. Interactions between soil water content and fertilizer on growth characteristics and biomass yield of Chinese white poplar (*Populus tomentosa* Carr.) seedlings [J]. Soil Sci. Plant Nutr., 2011, 57: 303-312.

[3] XU X J, TIMMER V R. Biomass and nutrient dynamics of Chinese fir seedlings under conventional and exponential fertilization regimes [J]. Plant Soil, 1998, 203: 313-322.

[4] DUMROESE R K, PAGE-DUMROESE D S, SALIFU K F, et al. Exponential fertilization of *Pinus monticola* seedlings: nutrient uptake efficiency, leaching fractions, and early outplanting performance [J]. Can. J. For. Res., 2005, 35: 2961-2967.

[5] INGESTAD T. New concepts on soil fertility and plant nutrition as illustrated by research on forest trees and stands [J]. Geoderma, 1987, 40: 237-252.

[6] INGESTAD T, LUND A B. Theory and techniques for steady state mineral nutrition and growth of plants [J]. Scand. J. For. Res., 1986, 1: 439-453.

[7] TIMMER V R, ARMSTRONG G. Growth and nutrition of containerized *Pinus resinosa* at exponentially increasing nutrient additions [J]. Can. J. For. Res., 1987, 17: 644-647.

[8] ZABEK L M, PRESCOTT C E. Steady-state nutrition of hybrid poplar grown from un-rooted cuttings [J]. New For., 2007, 34: 13-23.

[9] IMO M, TIMMER V R. Nitrogen uptake of mesquite seedlings at conventional and exponential fertilization schedules [J]. Soil Sci. Soc. Am. J., 1992, 56: 927-934.

[10] TIMMER V R, ARMSTRONG G, MILLER B D. Steady-state nutrient preconditioning and early outplanting performance of containerized black spruce seedlings [J]. Can. J. For. Res., 1991, 21: 585-594.

[11] EVERETT K T, HAWKINS B J, KIISKILA S. Growth and nutrient dynamics of Douglas-fir seedlings raised with exponential or conventional fertilization and planted with or without fertilizer [J]. Can. J. For. Res., 2007, 37: 2552-2562.

[12] QUORESHI M, TIMMER V R. Growth, nutrient dynamics, and ectomycorrhizal development of container-grown *Picea mariana* seedlings in response to exponential nutrient loading [J]. Can. J. For. Res., 2000, 30: 191-201.

[13] QU L, QUORESHI A M, KOIKE T. Root growth characteristics, biomass and nutrient dynamics of seedlings of two larch species raised under different fertilization regimes [J]. Plant Soil, 2003, 255: 293-302.

[14] HAWKINS B J, BURGESS D, MITCHELL A K. Growth and nutrient dynamics of western hemlock with conventional or exponential greenhouse fertilization and planting in different fertility conditions [J]. Can. J. For. Res., 2005, 35: 1002-1016.

[15] SALIFU K F, JACOBS D F. Characterizing fertility targets and multi-element interactions in nursery culture of *Quercus rubra* seedlings [J]. Ann. For. Sci., 2006, 63: 231-237.

[16] OLIET J A, TEJADA M, SALIFU K F, et al. Performance and nutrient dynamics of holm oak (*Quercus ilex* L.) seedlings in relation to nursery nutrient loading and post-transplant fertility [J]. Eur. J. Forest Res., 2009, 128: 253-263.

[17] TIMMER V R, ARMSTRONG G. Growth and nutrition of containerized *Pinus resinosa* seedlings at varying moisture

regimes[J]. New For., 1989, 3: 171-180.

[18]TIMMER V R, MILLER B D. Effects of contrasting fertilization and moisture regimes on biomass, nutrients, and water relations of container grown red pine seedlings. New For., 1991, 5: 335-348.

[19]SILVEIRA M L, HABY V A, LEONARD A T. Response of coastal bermudagrass yield and nutrient uptake efficiency to nitrogen sources[J]. Agron. J., 2007, 99: 707-714.

[20]CHEN L, ZENG J, XU D P, et al. Effects of exponential nitrogen loading on growth and foliar nutrient status of *Betula alnoides* seedlings[J]. Sci. Silv. Sin., 2010, 46(5): 35-40. in Chinese with English abstract.

[21]LICHTENTHALER H K, WELLBURN A R. Determination of total carotenoids and chlorophyll a and bof leaf extracts in different solvents[J]. Biochem. Soc. Trans., 1983, 603: 591- 592.

[22]BIRGE Z K D, SALIFU K F, JACOBS D F. Modified exponential nitrogen loading to promote morphological quality and nutrient storage of bareroot-cultured *Quercus rubra* and *Quercus alba* seedlings[J]. Scand. J. For. Res., 2006, 21: 306-316.

[23]SALIFU K F, TIMMER V R. Optimizing nitrogen loading of *Picea mariana* seedlings during nursery culture[J]. Can. J. For. Res., 2003, 33: 1287-1294.

[24]MICHELSEN A, JONASSON S, SLEEP D, et al. Shoot biomass, $\delta^{13}C$, nitrogen and chlorophyll responses of two arctic dwarf shrubs to in situ shading, nutrient application and warming simulating climatic change[J]. Oecologia, 1996, 105: 1-12.

[原载：Soil Science and Plant Nutrition, 2012, 58(5)]

Macronutrient Deficiencies in *Betula alnoides* Seedlings

CHEN Lin[1,2], ZENG Jie[1], XU Daping[1], ZHAO Zhigang[1], GUO Junjie[1]

([1]*Research Institute of Tropical Forestry, Chinese Academy of Forestry, Longdong, 510520, Guangzhou, China;*
[2]*Experimental Center of Tropical Forestry, Chinese Academy of Forestry, PingXiang, 536000, Guangxi, China*)

Abstract: The decline in seedling quality and production of birch tree(*Betula alnoides*) is often associated with nutrient stress. visual foliar symptoms, growth performance, pigment compositions and nutrient interaction of birch seedlings in response to six macronutrient deficiencies were studied. visual foliar symptoms were most obvious for no nitrogen(-N), no potassium(-K), and no magnesium(-Mg) seedlings, but were not apparent for no calcium(-Ca) seedlings. Apart from no-Mg and no sulfur(-S) treatments, seedlings lacking other nutrients showed decreases in the majority of growth measurements but an increase in root/shoot ratio. Phosphorous deficiency had no effect on all fractions of pigments, while N and K deficiencies resulted in significant reductions of chlorophyll-a(chla), chlorophyll-b(chlb), total chlorophyll(chl), and carotenoid(car) but increases in chla/chlb and car/chl ratios. Vector analyses showed that N deficiency not only decreased leaf N concentration but also increased leaf P concentration, possibly because of the antagonism between both ions. Similarly, K, Ca and S deficiencies induced a slight decrease in leaf N concentration that may be explained by a synergism between N and these ions. This kind of nutrient interaction also occurred between P and S in-P seedlings or between Mg and Ca in -Mg seedlings. The findings provide a guideline for diagnosing major macronutrient deficiencies of seedlings. The theoretical foundation for silviculture of *B. alnoides* will help improve seedling quality and accelerate the sustainable plantation development of this species.

Key words: foliar symptom; growth performance; nutrient interaction; pigment composition

1 INTRODUCTION

Soil and water loss has been increasing in most areas of south China over the last two decades due to the excessive deforestation and heavy vegetation disturbances. This result in serious nutrient poverty of soils, especially for N, P, K and Mg[1]. A consequence is a substantial decrease in forest quality and production, particularly for plantation species. While the nutrient availability of sites influences growth of trees in plantations, the nutrient status of seedlings is also closely related to seedling quality[2-4] as it influences the survival rate of young seedlings, initial growth performance and resistance to water stress, low temperature and disease[3-7]. Nutrient deficiency experiments using tree seedlings are often used as a means of estimating the optimization of nutritional management, evaluating the effects of nutrient deficiencies on seedling morphology and physiology[8-11].

Betula alnoides, also known as birch tree, is a fast-growing indigenous hardwood species in the northern India, Myanmar, Indochina and south China. It appears to be the most southerly species of *Betula* in the northern hemisphere. Some studies had shown that *B. alnoides* can grow well on all sorts of soils with pH ranging from 4.2 to 6.5, probably because it is a strong deep-rooted species and able to form ectomycorrhiza or vesicular-arbuscular mycorrhiza[12-14]. Therefore, this species often plays an important role in water conservation, long-term maintenance of land fertility and biodiversity of forest ecosystems. Due to its well-formed stem, moderate density and beautiful texture, the wood of *B. alnoides* is commonly used for high-quality floorboards, furniture, and room decorations[14]. Plantation of *B. alnoides* have been rapidly expanding in south China, from 100 ha in the mid-1990s to more than 50000 ha in 2005[14]. Despite its rise as an important economical and environmental species, little is known about the nutritional requirements of *B. alnoides*, which may limit the optimal silviculture of this species.

The objective of this studywas to identify the visual symptoms of *B. alnoides* seedlings in response to deficiencies of six major macronutrient elements. The responses of *B. alnoides* seedlings to nutrient deficiencies with regard to growth attributes, pigment composition concentration and mineral element contents were studied as these parameters relate to improving the nutritional managements of *B. alnoides* plantations.

2 MATERIALS AND METHODS

Healthy tissue-cultured seedlings of *B. alnoides* with almost uniform height of 14.5cm were used in sand culture trials and grown in a ventilated greenhouse at the Research Institute of Tropical Forestry on the March 2009. Quartz sand of 1 ~ 2mm diameter was washed with water and soaked in 1% HCl for 24h to remove impurities, organic matter and nutrient residues. The sand was then leached with running tap water, rinsed with deionised water and finally wind dried. The seedlings were transplanted into 18cm × 13cm × 16cm plastic pots filled with equal weight of quartz sand and watered on alternate days. To prevent the runoff of water or leaching of nutrients, white plastic bags were placed in the plastic pots. The irrigation regimes were measured by an initial watering test in which water was poured slowly into a pot filled with quartz sand using a cylinder. When water was 2cm from the bottom of the pot, the volume of water added was the irrigation regime. This experiment was replicated four times. The real irrigation regime was influenced by seedling growth demand and weather condition. Fungi diseases were prevented by injecting a 0.2% solution of carbendazin or thiophanate into quartz sand every one or two weeks. The average daytime temperature of the greenhouse ranged from 24℃ to 29℃, natural light intensity from 110 to 140 μmol photon/m^2 · s and a relative humility from 48% to 71%. The positions of the pots were rotated every two weeks to minimize edge effects.

The randomizedcomplete block design was arranged with four blocks of seven treatments: i. e.: complete nutrient mix (Control), minus nitrogen(-N), minus phosphorus(-P), minus potassium(-K), minus calcium(-Ca), minus magnesium(-Mg), and minus sulfur(-S) nutrient solutions (n = 28). Each experiment unit (plot) consisted of 10 seedlings, totalling 280 seedlings in all treatments. The treatments lasted for 12 weeks and started two weeks after transplanting when all the seedlings were well established. Seedlings were supplied with 80 ml of half-strength nutrient solutions (Table 1) once a week on the first six weeks and full-strength thereafter. The test concluded when all the treatments exhibited foliar symptoms.

Table 1 Chemical composition of complete and various macronutrient-lacking nutrient solutions

No.	Sources	Ml/L						
		Control	-N	-P	-K	-Ca	-Mg	-S
1	1mol/L KNO_3	5	—	6	—	5	6	6
2	1mol/L $Ca(NO_3)_2 \cdot 4H_2O$	5	—	4	5	—	4	4
3	1mol/L $MgSO_4$	2	2	2	2	2	—	—
4	1mol/L KH_2PO_4	1	—	—	—	1	1	1
5	0.5mol/L K_2SO_4	—	5	—	—	—	3	—
6	0.5mol/L $CaCl_2$	—	10	—	10	—	—	—
7	1mol/L $NaH_2PO_4 \cdot 2H_2O$	—	10	—	10	—	—	—
8	0.01mol/L $CaSO_4 \cdot 2H_2O$	—	20	—	—	—	—	—
9	1mol/L $Mg(NO_3)_2 \cdot 6H_2O$	—	—	—	—	—	—	2
10	0.05mol/L Fe-EDTA	2	2	2	2	2	2	2
11	Arnon micronutrient solutions	1	1	1	1	1	1	1
	0.05mol/L H_3BO_3							
	0.01mol/L $MnCl_2 \cdot 4H_2O$							
	0.77mmol/L $ZnCl_2$							
	0.32mmol/L $CuCl_2 \cdot 2H_2O$							
	0.09mmol/L H_2MoO_4(85%~90% MoO_3)							

Notes: The pH of all nutrient solutions was adjusted to 6.0 with 1mol/L NaOH or HCl.

The seedlings were observed daily for symptoms of deficiency and various visual symptoms were recorded. The root collar diameter, height and number of branches of all the seedlings were measured every three weeks after treatments. At the end of the experiment, three of 10 seedlings per plot were randomly selected and separated from the quartz sand by gently washing with deionized water. The seedlings were divided into leaves, stems and roots. The leaf samples were scanned to obtain leaf photographs and leaf pixel were examined using Photoshop Microsoft. Hence, leaf area could be calculated according to the positive relationships between the leaf area and pixel. Different parts of plants were oven dried at 65℃ for 48h for dry mass determination. Dried leaf samples were then ground for subsequent chemical analysis. Plant materials were digested in a block digester using a H_2SO_4 and K_2SO_4-$CuSO_4$ mixture catalyst for total N by diffusion method while H_2NO_3-$HClO_4$ mixture solution for P, K, Ca, Mg and S by inductively coupled plasma optical emission spectrometry. Another three plants in each plot were randomly selected, from which the third or fourth leaf from the apex in each plant were randomly collected at the end of the experiment for the pigment composition measurements. The chlorophyll and carotenoid contents of fully-expanded leaves were estimated spectrophotometrically in 80% acetone extract. The absorbance was measured at 470, 646 and 663nm for the estimation of chlorophyll-a(chla), chlorophyll-b(chlb), total chlorophyll and carotenoid and their contents were calculated according to the method of Lichtenthaler and Wellburn[15].

Analysis of variance(ANOVA) was conducted tostudy the effects on growth, pigment compositions and leaf nutrient status of the seedlings using SPSS 11.5(2003). Significant means were separated by Duncan's multiple range test at $P < 0.05$. Leaf nutrient conditions between different treatments were characterized by vector analysis following a technique described by Salifu and Timmer[16], where the nutrient content (x), nutrient concentration (y) and dry mass(z) were plotted in a single monogram(Figure 1) satisfying the function $x=f(y, z)$. The technique facilitates interpretation of plant nutrient status by simultaneously comparing growth and associated nutrient responses by normalizing status of the reference treatment to 100. Diagnosis is based on both vector direction and magnitude in terms of increasing(+), decreasing(−) or unchanging(0) parameter status(Figure 1), identifying cases of possible depletion, dilution, sufficiency, deficiency, luxury consumption, and excess of nutrients, respectively, induced by treatments. The orientation of the major vector commonly signifies an enrichment response, while the magnitude of the major vector reflects the largest relative increase or decrease in nutrient conditions of the reference from the compared point status[17].

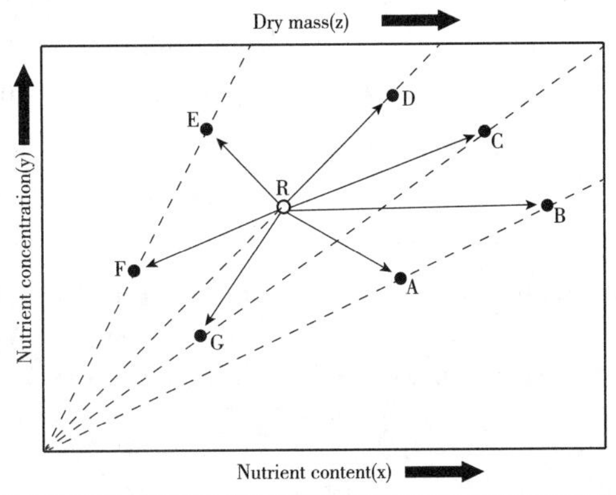

Vector shift	Relative change in			Interpretation	Possible diagnosis
	z	y	x		
A	+	-	+	Dilution	Growth dilution
B	+	0	+	Sufficiency	Steady-state
C	+	+	+	Deficiency	Limiting
D	0	+	+	Luxury consumption	Accumulation
E	-	+ +	±	Excess	Toxic accumulation
F	-	-	-	Excess	Antagonistic
G	0, +	-	-	Depletion	Retranslocation

Figure 1 Vector interpretation of directional changes in dry mass(z), nutrient concentration(y) and nutrient content(x) of plants. Reference point(R) represents status of seedlings normalized to 100. Diagnosis is based on vector shifts(A to G) which characterize an increase(+), decrease(−) or no change(0) in dry mass and nutrient status relative to the reference status as described in the box below[16-17].

3 RESULTS

3.1 visual deficiency symptoms

−N seedlings showed deficiency symptoms 21 days after treatment (DAT). At first, middle areas of the relatively old basal leaves turned yellow and spread gradually across the entire leaves. This symptom then extended towards the younger apical leaves. The entire seedling grew slowly with fewer leaves, smaller leaf size, lower height and fewer branches compared to the control. Interveinal chlorosis occurred at 45 DAT. As seedlings aged, terminal buds gradually turned brown and died. Irregular brown patches were seen on all leaves at the end of the experiment.

The height growth of −P seedlings was stunted at 28 DAT. As deficiency progressed, leaves and stems gradually exhibited fairly dark green andpurplish red coloration, respectively. Interveinal chlorosis was observed on a few old leaves by the end of the experiment.

The symptoms of −K seedlings werefirst apparent on old basal leaves, which exhibited chlorosis. Seedling height growth was significantly suppressed at 25 DAT. As the seedlings grew, a discoloration similar to iron rust developed, especially along edges of the fourth leaf and gradually extended to the lower leaves. When K deficiency became more severe, this iron-rust discoloration transitioned to reddish-brown necrosis with entire leaves twisting upward. At 45 DAT, interveinal chlorosis was observed and reddish-brown necrotic patches appeared on the upper leaves.

visible symptoms of Ca deficiency in Ca-deprived seedlings were not apparent. Although seedling height was less than the control at 21 DAT, no other visual symptom was apparent.

The seedlings were highly responsive to Mg deficiency. At 21 DAT, an opaque green discoloration appeared between the veins of the fourth or fifth leaf and spread toward older basal leaves. The opaque green areas of leaves developed all sizes of bronze patches and entire leaves curled downward at severe stage of deficiency. At the end of the experiment, interveinal chlorosis appeared on the leaves of mid-lower part of seedlings and most of the damaged leaves were dropping off.

Seedlings growth of the −S treatment was relatively slow and new leaves exhibiting a light green discoloration at 36 DAT. As the severity of deficiency continued, seedlings height was increasingly suppressed.

3.2 Seedling growth

There were no significant differences in root collar diameter, height and number of branches of *B. alnoides* seedlings between different treatments at 14 DAT (Table 2). However, profound differences ($P<0.05$) were evident at each of the three subsequent dates of examination; inferring that nutrient deficiency treatments substantially hindered growth performances of *B. alnoides* seedlings.

Table 2 shows that the growth of *B. alnoides* seedlings is affected by N, P, K and Ca. The At the end of the experiment, the root collar diameter, height and number of branches of all the nutrient deficiency treatments were significantly lower than the control except for the number of branches of −Mg treatment and the root collar diameter of −S treatment.

Table 2 Effects of different nutrient deficiencies on growth of *B. alnoides* seedlings

Growth indexes	DAT (days)	Control	−N	−P	−K	−Ca	−Mg	−S
Root collar diameter (mm)	14	1.73 (0.04)	1.78 (0.04)	1.76 (0.03)	1.73 (0.03)	1.80 (0.03)	1.83 (0.04)	1.73 (0.03)
	35	2.15ab (0.05)	2.14b (0.04)	2.14b (0.04)	2.23ab (0.04)	2.14b (0.04)	2.28a (0.05)	2.28a (0.04)
	56	3.23a (0.05)	2.36e (0.05)	2.76c (0.04)	2.59d (0.05)	2.83c (0.06)	3.01b (0.06)	3.22a (0.05)
	77	3.72a (0.06)	2.14f (0.05)	3.03d (0.06)	2.63e (0.05)	3.34c (0.06)	3.54b (0.07)	3.76a (0.07)
Height (cm)	14	14.2 (0.26)	14.2 (0.29)	14.7 (0.22)	14.4 (0.27)	14.9 (0.23)	14.6 (0.27)	14.5 (0.26)
	35	22.1a (0.45)	16.3d (0.34)	22.2a (0.36)	18.8c (0.36)	20.2b (0.31)	22.4a (0.41)	22.5a (0.36)

(续)

Growth indexes	DAT (days)	Control	-N	-P	-K	-Ca	-Mg	-S
Height(cm)	56	37. 4a (0. 45)	19. 2f (0. 42)	29. 9d (0. 55)	25. 2e (0. 33)	32. 5c (0. 51)	33. 2bc (0. 60)	34. 1b (0. 49)
	77	48. 8a (0. 76)	20. 7e (0. 42)	32. 8c (0. 57)	27. 4d (0. 31)	41. 9b (0. 69)	41. 8b (0. 69)	41. 3b (0. 79)
No. of branches	14	0	0	0	0	0	0	0
	35	10a (0. 23)	5d (0. 30)	9b (0. 22)	9bc (0. 30)	8c (0. 30)	10a (0. 25)	11a (0. 26)
	56	14a (0. 21)	6d (0. 31)	11c (0. 31)	10c (0. 28)	12b (0. 28)	14a (0. 29)	14a (0. 20)
	77	18a (0. 25)	6d (0. 36)	11c (0. 33)	10c (0. 31)	16b (0. 37)	18a (0. 31)	17b (0. 28)

Values followed by the same letter within rows are not significantly different at $P < 0.05$ according to Duncan's Multiple Range Test. The figures within parentheses are standard errors.

Analysis of variance for leaf area indicated that nutrient deficiency caused great reduction of leaf area (Figure 2). Except for −Mg and −S treatments, the seedling biomasses of the other nutrient deficiency treatments were significantly lower than that of the control. In the case of root/shoot ratio, the nutrient deficiency treatments were all significantly higher than the control except for the −Mg treatment.

3.3 Pigment compositions

No difference occurred in the pigment compositions between −P and control treatments (Table 3). Both −N and −K treatments led to significant declines in chlorophyll and carotenoid contents but increases in chla/b and car/chl ratios compared with the control. As for −Ca, −Mg and −S treatments, the contents of chlb, total chlorophyll, and carotenoid were lower than the control. There were no profound differences in the chla content, chla/b ratio and car/chl ratio between them and the control (Table 3).

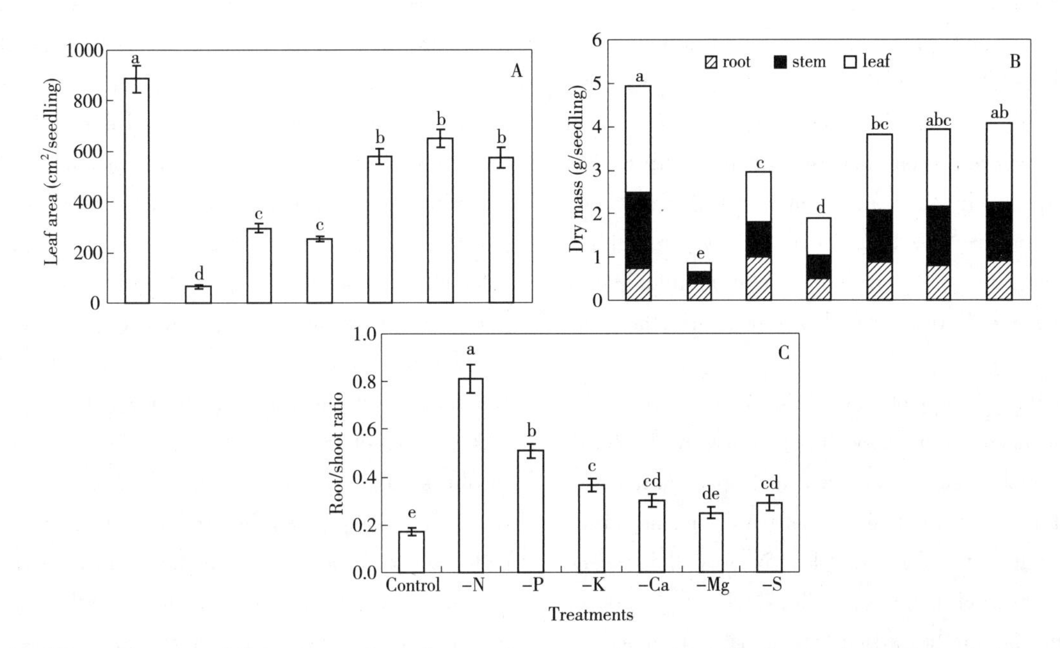

Figure 2 Effects of nutrient deficiencies on the leaf area, dry mass and root/shoot (leaf + stem) ratio of *B. alnoides* seedlings. Paired treatments without the same letter aresignificantly different ($P<0.05$) according to Duncan's Multiple Range Test. Error bars represent standard errors.

Table 3 Effects of different nutrient deficiencies on pigment compositions of *B. alnoides* seedlings

Treatments	Chla (mg/g)	Chlb (mg/g)	Chla/b	Total chlorophyll (mg/g)	Carotenoid (mg/g)	Car/Chl
Control	0.269a (0.014)	0.537a (0.013)	0.504c (0.034)	0.806ab (0.014)	0.154ab (0.002)	0.191c (0.003)
-N	0.147c (0.003)	0.173d (0.007)	0.854a (0.051)	0.320e (0.005)	0.077e (0.002)	0.242a (0.003)
-P	0.270a (0.013)	0.572a (0.012)	0.473c (0.024)	0.842a (0.019)	0.161a (0.001)	0.192c (0.003)
-K	0.214b (0.009)	0.313c (0.015)	0.689b (0.038)	0.527d (0.019)	0.112d (0.003)	0.213b (0.003)
-Ca	0.261a (0.012)	0.459b (0.011)	0.569c (0.028)	0.720c (0.017)	0.143c (0.003)	0.199c (0.004)
-Mg	0.262a (0.011)	0.483b (0.015)	0.543c (0.025)	0.745c (0.020)	0.145bc (0.002)	0.196c (0.004)
-S	0.258a (0.016)	0.461b (0.024)	0.561c (0.015)	0.719c (0.039)	0.139c (0.005)	0.193c (0.004)

Values followed by the same letter within columns are not significantly different at $P<0.05$ according to Duncan's Multiple Range Test. The figures within parentheses are standard errors.

3.4 Nutrient interaction

Nis the major vector and the leaf N content, leaf N concentration, and leaf mass of -N seedlings were declined by 94.9%, 40.9% and 91.4%, respectively, compared with the control (Figure 3A). Leaf P content and leaf P concentration increased 25.6% and 13.7% times, respectively, indicating that P was also important for the growth of -N seedlings. P was, thus, recognized as the secondary vector since its vector magnitude was smaller than the N vector and larger than the other vectors.

Whilethe primary nutrients of the various nutrient deficiency treatments to limit seedling growth were the corresponding deficient elements omitted from their nutrient solutions, the secondary limiting nutrient elements varied between treatments. For example, N was the second limiting nutrient of the -K, -Ca, and -S seedlings, while S and Ca were the second limiting nutrient influencing the growth of -P and -Mg seedlings respectively (Figure 3, B-F).

4 DISCUSSION

Betula alnoides seedlings were most affected by the -N treatment. This was consistent with the fact that N is one of the most important and commonly limiting nutrients for tree growth. Lack of N decreased height, root collar diameter and number of branches of seedlings compared with the control. This is possibly because new cells elongation and division are severely limited under N deficiency[18]. The root/shoot ratio is often increased due to N deficiency[9,10,19]. This is observed in the present study and was in line with the nitrogen-carbon balance concept according to Thornley[20]. N is an essential element for the formation of chlorophyll, which is a photosensitive catalyst in the process of photosynthesis[21], which caused leaf chlorosis and induced declines of 92.7% and 82.8% of the leaf area and biomass of the seedlings respectively in comparison with the control (Figure 2). The car/chl ratio of the -N seedlings was significantly higher than the control. Thus, N deficiency apparently induced an increase in the light protection but a decrease in the absorbance ability[22]. We also found that N deficiency induced not only a high reduction of leaf N concentration but also an increase in leaf P concentration (Figure 3A). This interaction indicates a possible antagonism between N and P.

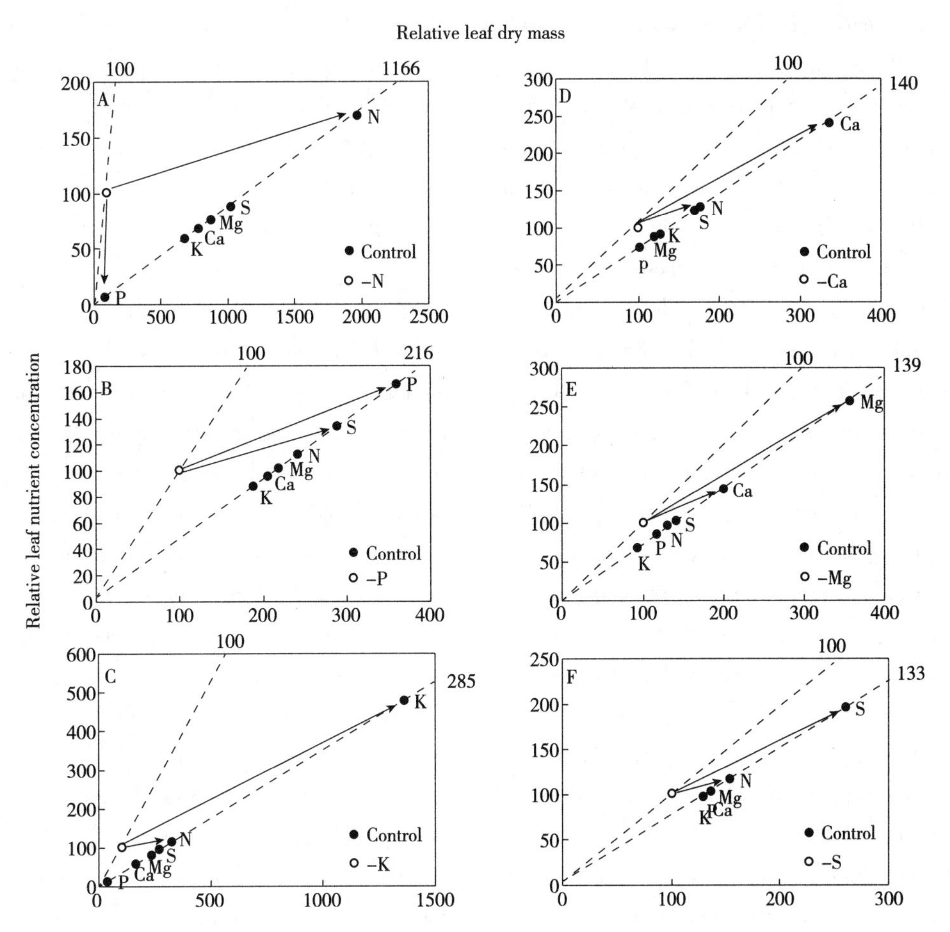

Figure 3 Vector nomogram of relative changes in the leaf nutrient concentration, leaf nutrient content and leaf dry mass between *B. alnoides* seedlings supplied with different nutrient deficiency solutions.

The leaf N, P, K, Ca, Mg, and S concents of the control are 61. 11, 5. 04, 58. 40, 33. 84, 11. 38, and 7. 59mg, respectively. The leaf N, P, K, Ca, Mg, and S concentrations of the control are 24. 92, 2. 06, 23. 86, 13. 73, 4. 65, and 3. 09g/kg, respectively. Status of reference treatments(open symbols) at 77 DAT were equalized to 100 to allow nutrient comparison on a common base.

Phosphorus deficiency decreased seeding height, root collar diameter, number of branches, leaf area and biomass. This is related to its central roles in energy metabolism, synthesis of organic compounds such as nucleic acids, phospholipids and nucleotides[18]. The root/shoot ratio of −P seedlings increased compared to the control, as occurs in certain plants, i. e. , *Pinus resinosa*[8] and *B. pendula*[19]. However, for *Santalum album* and *Pistacia lentiscus*, P deficiency did not affect root/shoot ratio but affected the root morphology[9-10]. It seems that root morphology is more sensitive than biomass allocation to P deficiency. In this study, we found dark green pigment in the leaves of P-deficient seedlings although no significant difference was detected in chla and chlb concentrations between −P treatment and the control (Table 3). Probably cell elongation was more affected than chlorophyll content and this caused an increase in the chlorophyll content per leaf area[23]. This could be further explained by the great differences of special leaf area (the ratio of leaf area to leaf dry mass) between −P seedlings and control, the special leaf area of the former was only one third that of the latter. This observation was also reported for *Betula papyrifera*[24] and *Khava ivorensis*[11]. Of all the nutrient deficiency treatments, only the −P treatment failed to affect leaf pigment compositions. Seedling leaf color was not obviously distinguished from the control. Their stems were purplish red, apparently due to formation and accumulation of anthocyanin pigments. P deficiency in birch differed from the report by Gopikumar and Varghese [25] for *Tectona grandis* in which the −P seedlings exhibited purple-bronze patches on older leaves. −P treatment not only led to a decrease in leaf P concentration but also reduced leaf S concentration due to the synergism of both ions.

Foliar symptoms of K deficient seedlings included chlorotic patches appearing on leaves during the initial stage and the entire leaves distorting upward, as a result of faster growth in the middle of leaves[18]. Potassium, as an enzyme activator in many important metabolic processes in plants, is important for phloem transport, osmotic balance and photosynthesis[11]. Hence, K deficiency slowed growth and resulted in dramatic reductions in the root collar diameter, height, number of branches, leaf area and biomass of seedlings. Although Thornley[20] and Dewar[26] predicted that K shortage would l a limitation limit the uptake of CO_2 and then reduction of allocation of biomass to the belowground parts, there were contrasting findings from different studies suggesting that differences in the root/shoot ratio probably existed between species[19]. In this work, the root/shoot ratio increased significantly for the −K seedlings. In the case of pigment compositions, K deficiency led to significant decreases in the chla, chlb, total chlorophyll and carotenoid contents but increases in the chla/chlb and car/chl ratios, in the same way as N deficiency. K deficiency resulted in decreases in leaf K concentration, as well as leaf N concentration, indicating the synergism between both ions.

Discolouration, a symptom of −Ca deficiency, first appeared on young stems and young leaves of *K. ivorensis* because Ca is a relatively phloem immobile element in plants[11], while discolouration occurred on all foliage of *B. papyrifera* with basal leaves often paler[24]. However, the foliar symptoms of the −Ca seedlings were not apparent in the present study. This was probably because a small amount of Ca(1.6×10^{-5} g/kg) was rinsed from the quartz sand when watering or fertilizing, which could satisfy the basic Ca requirements for *B. alnoides* seedlings. Alternatively, the −Ca seedlings might need more time to exhibit the foliar symptoms. It is known that Ca plays an important role in cell wall and plasma membrane stabilization[19]. Therefore, Ca deficiency can induce failure in cell extension. In the present study, severe reductions in the root collar diameter, height, number of branches, leaf area and biomass as well as an increase in the root/shoot ratio were observed in response to Ca-shortage, which were in accordance with *S. album*[9]. As to pigment compositions, Ca had more effects on the chlb, total chlorophyll and carotenoid contents than chla content, chla/chlb ratio and car/chl ratio. Moreover, leaf Ca and leaf N concentrations decreased in the −Ca seedlings as a result of the synergism between Ca and N ions. This was also found in other plants such as *K. invorensis*[11].

The characteristic symptom for Mg deficiency was interveinal chlorosis as Erdmann *et al.*[24] and Jeyanny *et al.*[11] had demonstrated. Mg deficiency also resulted in severe suppression of root collar diameter, height and leaf area. This might be explained by the role of this element in photosynthesis as an activator of a large number of enzymes, including ribulose bisphosphate carboxylase and ribulose-5-bisphosphate kinase[19]. With regard to the number of branches, biomass and root/shoot ratio, they were relatively less impaired by the-Mg treatment. Mg is a component of the chlorophyll molecule. Mg deficiency led to remarkable reductions of the chlb and total chlorophyll contents rather than the chla and carotenoid contents. The indicates that Mg-limitation has more effects on the chlb content than on the other pigment compositions. Mg deficiency caused a significant decrease in leaf Mg concentration as well as leaf Ca concentration, inferring the synergism between Mg and Ca ions.

Discolouration was also observed on young leaves of the −S seedlings since S is not readily mobile, as with Ca, in plants. Sulphur is a constitute of the amino acids cysteine and methionine[23]. Thus, S deficiency is likely to inhibit protein synthesis. This led to reductions in the height, number of branches and leaf area of the-S seedlings. However, the root collar diameter and biomass were less influenced by the-S treatment. The S-limitation resulted in an increase in the root/shoot ratio of seedlings. This is in accordance with the findings by Ericsson[19] for *B. pendula*. However, effects of S deficiency on the root morphology rather than on root/shoot ratio were found similar to those of P deficiency in *S. album*[9]. Leaf S concentration decreased as well as leaf N concentration in the −S seedlings. This seems to be related to the synergism of these two ions.

5 CONCLUSIONS

The present study indicated that the foliar symptoms of −N, −K, and −Mg seedlings were the most apparent. Symptoms included leaf discoloration, bronze patches, leaftwisting, reddish-brown necrosis, and interveinal chlorosis. The height, root collar diameter, number of branches, leaf area as well as biomass of *B. alnoides* seedlings significantly decreased and the root/shoot ratio considerably increased with N, P, K or Ca deficiencies. Reductions in the chlb and total chl contents were evident under all nutrient deficiencies except for P deficiency. Vector analysis revealed that the synergism and antagonism between foliar nutrient elements commonly existed in nutrient-deficient seedlings. This work provides a guideline for diagnosing major macronutrient deficiencies in *B. alnoides* seedlings. It offers theoretical foundation for improving silviculture and sustainability of *B. alnoides* plantation.

ACKNOWLEGEMENTS

This research was financially supported by The Ministry of Science and Technology(2006BAD24B09-02C). We are grateful to KQ Lin and E Sha for assistance in the nursery and LS Yang and B Yu for nutrient determinations.

REFERENCES

[1]MACLEOD D A. 2004, Managing fertility of the red soils for forage production[C]. Pp. 50-60 in Scott JM et al. (Eds.) Conservation, Forages for the Red Soils Area of China Proceedings of an International Workshop, Jianyang, Fujian Province, P. R. China, 6-9 October 1997. Australian Centre for International Agricultural Research, Canberra.

[2]ELLIOTT K J, WHITE A S. Effects of light, nitrogen, and phosphorus on red pine seedling growth and nutrient useefficiency [J]. Forest Science, 1994, 40(1): 47-58.

[3]CLOSE D C, BAIL I, HUNTER S, et al. Effects of exponential nutrient-loading on morphological and nitrogen characteristics and on after-planting performance of *Eucalyptus globulus* seedlings [J]. Forest Ecology and Management, 2005, 205: 397-403.

[4]BIRGE Z K D, SALIFU K F, JACOBS D F. Modified exponential nitrogen loading to promote morphological quality and nutrient storage of bareroot-cultured *Quercus rubra* and *Quercus alba* seedlings [J]. Scandinavian Journal of Forest Research, 2006, 21: 306-316.

[5]VAN DEN DRIESSCHE R. Effects of nutrients on stock performace in the forest [M]. In: Van den Driessche R (Ed.), Florida: Mineral Nutritional in Conifer Seedlings. CRC Press, 1991: 229-260.

[6]QUORESHI A M, TIMMER V R. Early outpalnting performance of nutrient-loaded containerized black spruce seedlings inoculated with *Laccaria bicolor*: a bioassay study [J]. Canadian Journal of Forest Research, 2000, 30: 744-752.

[7]FLOISTAD IS, Kohmamn K. Influence of nutrient supply on spring frost hardiness and time of bud break in Norway spruce [*Picea abies* (L.) Karst.] seedlings [J]. New Forests, 2004, 27: 1-11.

[8]HAWKINS B J, BURGESS D, MITCHELL A K. Growth and nutrient dynamics of western hemlock with conventional or exponential greenhouse fertilization and planting in different fertility conditions [J]. Canadian Journal of Forest Research, 2005, 35: 1002-1016.

[9]TIMMER V R, ARMSTRONG G. Diagnosing nutritional status of containerized tree seedlings: comparative plant analyses [J]. Soil Science Society of America Journal, 1987, 51: 1082-1086.

[10]BARRETT D R, FOX J E D. *Santalum album*: Kernel composition, morphological and nutrient characteristics of pre-parasitic seedlings under various nutrient regimes [J]. Annals of Botany, 1997, 79: 59-66.

[11]TRUBAT R, CORTINA J, VILAGROSA A. Plant morphology and root hydraulics are altered by nutrient deficiency in *Pistacia lentiscus*(L.)[J]. Trees, 2006, 20: 334-339.

[12]JEYANNY V, A B RASIP A G, WAN RASIDAH K, et al. Effects of macronutrient deficiencies on the growth and vigor of *Khaya ivorensis* seedlings [J]. Journal of Tropical Forest Science, 2009, 21(2): 73-80.

[13]ZENG J, ZHENG H S, WENG Q J. Geographic distributions and ecological conditions of *Betula alnoides* in China [J]. Forest Research, 1999, 12(5): 479-484. (in Chinese with English abstract).

[14]GONG M Q, WANG F Z, Chen Y, et al. Mycorrhizal dependency and inoculant effects on the growth of *Betula alnoides* Seedlings [J]. Forest Research, 2000, 13(1): 8-14. (in Chinese with English abstract).

[15]ZENG J, GUO W F, ZHAO Z G, et al. Domestication of *Betula alnoides* in China: current status and perspectives [J]. Forest Research, 2006, 19(3): 379-384. (in Chinese with English abstract).

[16]LICHTENTHALER H K, WELLBURN A R. Determination of total carotenoids and chlorophyll a and b of leaf extracts in different solvents [J]. Biochemical Society Transactions (London), 1983, 603: 591-592.

[17]SALIFU K F, TIMMER V R. Optimizing nitrogen loading of *Picea mariana* seedlings during nursery culture [J]. Canadian Journal of Forest Research, 2003, 33: 1287-1294.

[18]HAASE D L, ROSE R. Vector analysis and its use for interpreting plant nutrient shifts in response to silvicultural treatments [J]. Forest Science, 1995, 41(1): 54-66.

[19]WANG Z. Plant Physiology [M]. Beijing: Chinese Agricultural University Press, 2000: 84-87.

[20]ERICSSON T. Growth and shoot: root ratio of seedlings in relation to nutrient availability [J]. Plant and Soil,

1995, 168-169: 205-214.

[21] THORNLEY J H M. A balanced quantitative model for root: shoot ratios in vegetative plants[J]. Annals of Botany, 1972, 36: 431-441.

[22] LIU Y Q, SUN X Y, WANG Y, et al. Effects of shades on the photosynthetic characteristics and chlorophyll fluorescence parameters of *Urtica dioica* [J]. Acta Ecologica Sinica, 2007, 27: 3457-3464. (in Chinese with English abstract).

[23] GUO S L, YAN X F, BAI B, et al. Effects of nitrogen supply on photosynthesis in larch seedlings [J]. Acta Ecologica Sinica, 2005, 25 (6): 1291 - 1298. (in Chinese with English abstract).

[24] LU J L. Plant Nutrition [M]. Second edition. Beijing: Chinese Agricultural University Press, 2003: 23-76.

[25] ERDMANN G G, METZGER F T, OBERG R R. Macronutrient deficiency symptoms in seedlings of four northern hardwoods [R]. General Technical Report NC-53. 40p USDA Forest Service, North Central Forest Experimental Station. St. Paul. MN, 1979.

[26] GOPIKUMAR K, VARGHESE V. Sand culture studies of teak (*Tectona grandis*) in relation to nutritional deficiency symptoms, growth and vigour [J]. Journal of Tropical Forest Science, 2004, 16(1): 46-61.

[27] DEWAR RC. A root-shoot partitioning model based on carbon-nitrogen-water interactions and Munch phloem flow [J]. Functional Ecology, 1993, 7: 356-368.

[原载: Journal of Tropical Forest Science, 2010, 22(4)]

西南桦轻基质网袋容器苗基质选择试验

郭文福[1] 曾 杰[2] 黎 明[1] 蒙彩兰[1] 曾 冀[1]
([1] 中国林业科学研究院热带林业实验中心，广西凭祥 532600；
[2]中国林业科学研究院热带林业研究所，广东广州 510520)

摘 要 采用沤制松树皮、沤制锯末、炭化锯末 3 种材料设置 9 个轻基质配方开展西南桦网袋容器苗培育试验，依据苗高、地径、根系生长表现及生物量等指标进行苗木综合评价，旨在筛选出适合西南桦苗木培育的轻基质配方。结果表明：不同配方之间的苗高、地径、根系和生物量均差异极显著($P<0.01$)；以配方Ⅴ(50%沤制松树皮+50%炭化锯末)和Ⅵ(25%沤制松树皮+75%炭化锯末)的育苗效果最好，Ⅲ、Ⅷ和Ⅺ的效果最差，与炭化锯末比沤制锯末更利于西南桦苗木生长有关。建议生产上采用沤制松树皮和炭化锯末的混合基质培育西南桦苗木，若添加沤制锯末，其比例不宜超过 20%。

关键词 西南桦；轻基质；容器苗

Light Media Selection for Growing Container Seedlings of *Betula alnoides*

GUO Wenfu[1], ZENG Jie[2], LI Ming[1], MENG Cailan[1], ZENG Ji[1]
([1] *The Experimental Center of Tropical Forestry*, *CAF*, *Pingxiang* 532600, *Guangxi*, *China*
[2] *Research Institute of Tropical Forestry*, *CAF*, *Guangzhou* 510520, *Guangdong*, *China*)

Abstract: Light media selection trials were conducted for growing container seedlings of *Betula alnoides* with 9 media containing different ratio of composted pine bark and sawdust or charring sawdust, and comprehensively assessed based on traits of growth performance such as height, diameter, root and biomass. It was shown that there existed greatly significant differences of all these traits between light media ($P<0.01$). Among these light media, Media Ⅴ and Ⅵ were the best for growing container seedlings of this species, while Media Ⅲ, Ⅷ and Ⅸ were the worst. This is due to the fact that the charring sawdust is more suitable for *B. alnoides* than the composted sawdust. We suggested that mixed light media containing composted pine bark and charring sawdust be used in practice for growing container seedlings of this species, and if composted sawdust added, its ratio be less than 20 percent.

Key words: *Betula alnoides*; light media; container seedling

西南桦(*Betula alnoides* Buch. -Ham. ex D. Don)是我国热带南亚热带地区的一个珍贵优良乡土阔叶树种，具有重要的生态和经济价值。近 10 年来，西南桦在云南、广西等地得到迅速推广，造林面积已逾 6.67 万 hm^2。优质壮苗是人工林培育的基础，对于西南桦之类大规模发展的树种而言，壮苗培育尤为重要。容器育苗是获得优质壮苗的主要方法，是营造速生丰产林的重要环节之一。容器育苗的发展趋势是采用轻基质、小容器，实现苗木的工厂化和自动化大规模生产[1,2]。轻基质网袋容器苗不仅适于工厂化管理，而且其运输和造林施工效率明显提高，在我国劳动力日益缺乏、劳力成本逐步增加的情况下，开展轻基质网袋容器苗培育研究显得尤为迫切。黎明等(2007)曾用黄心土与松树皮配置的半轻基质培育西南桦苗木，取得了良好的效果，由于其不适合工厂化生产[3]，蒙彩兰等(2007)进行了西南桦轻基质网袋育苗的初步尝试[4]，其育苗效果有待进一步提高。因此笔者以沤制松树皮、沤制锯末以及炭化锯末等常用材料配置 9 个轻基质配方，旨在通过试验筛选出适合培育西南桦网袋容器苗的轻基质配方，为西南桦苗木工厂化生产提供技术支撑。

1 苗圃地概况

试验地点位于中国林业科学研究院热带林业实验中心林木种苗示范基地，距离广西凭祥市约1.5km。地理位置为106°47′E，22°19′N，海拔250m，属南亚热带季风气候，年均气温20.5~21.7℃，≥10℃年积温6000~7600℃；年降水量1200~1500mm，相对湿度约80%，干湿季节明显，降水主要集中在4~9月份。

2 材料与方法

2.1 材料

种子采自广西百色市凌云县伶站乡。轻基质材料包括经沤制的松树皮、锯末以及炭化锯末，材料来自热带林业实验中心林区，应用蒙彩兰等[4](2007)的方法依照表1中的配方生产轻基质并装入网袋容器。采用翁启杰等(2004)的方法进行播种育苗[5]，于7月初播种，9月中旬选用生长高度基本一致的幼苗移植入轻基质网袋。

表1 参试轻基质配方的组成及其配比

配方号	沤制树皮	沤制锯末	炭化锯末
Ⅰ	75	25	/
Ⅱ	50	50	/
Ⅲ	25	75	/
Ⅳ	75	/	25
Ⅴ	50	/	50
Ⅵ	25	/	75
Ⅶ	70	20	10
Ⅷ	45	35	20
Ⅸ	20	50	30

注：体积比，单位为%。

2.2 试验设计

按随机区组设计铺设试验，9个处理(表1)，4次重复，每个小区50株。

2.3 幼苗管理

移苗30d后开始施尿素或复合肥，每10d施肥1次，浓度从0.3%起逐渐加大，但不能超过1%，施肥后适当喷清水，清洗叶面肥液，以免产生肥害。每半月用波尔多液或多菌灵药液交替喷洒1次以预防病害，其水分管理与一般苗木生产相似，应注意对苗木进行空气修根。

2.4 苗木生长观测

出圃前每个小区随机抽取30株苗木测定苗高、地径；依据平均苗高和地径选取5株苗木测定根系生长和生物量。

2.5 数据处理与分析

采用SPSS 11.0进行方差分析。由于各处理间方差不齐性，采用Tamhane's T2进行成对t检验[6]。

3 结果与分析

3.1 轻基质配方对西南桦苗高、地径生长的影响

参试的9个轻基质配方处理间西南桦苗高、地径和径高比均存在极显著差异($P<0.01$，下同)。苗高和地径生长均以配方Ⅴ和Ⅵ最佳(图1A，1B)，配方Ⅲ、Ⅷ和Ⅸ最差；而配方Ⅲ、Ⅷ和Ⅸ的径高比最大(图1C)，显著高于配方Ⅱ、Ⅵ($P<0.05$，下同)。用沤制松树皮与沤制锯末配置的3个配方中，配方Ⅰ的苗高、地径以及配方Ⅱ的苗高、径高比显著高于配方Ⅲ；用沤制松树皮与炭化锯末配置的3个配方间苗高、地径和径高比均差异不显著($P>0.05$，下同)；而用沤制松树皮、沤制锯末及炭化锯末配置的3个配方中，仅配方Ⅶ的苗高与配方Ⅷ和Ⅸ差异极显著。

3.2 轻基质配方对西南桦幼苗根系生长的影响

各轻基质配方间苗木根长差异极显著，以配方Ⅴ的苗木根系最长，其次为配方Ⅵ，配方Ⅰ、Ⅲ、Ⅷ和Ⅸ的苗木根系最短(图1D)，极显著地低于前两者，其次为配方Ⅱ，显著低于配方Ⅴ。而无论用沤制松树皮与沤制锯末配置的3个配方(Ⅰ、Ⅱ和Ⅲ)之间，还是用沤制松树皮与炭化锯末配置的3个配方(Ⅳ、Ⅴ和Ⅵ)之间，抑或用3种材料配置的配方(Ⅶ、Ⅷ和Ⅸ)之间，其苗木根长均差异不显著。

3.3 轻基质配方对西南桦幼苗生物量的影响

9个轻基质配方间苗木总生物量及根冠比差异极显著。苗木总生物量以配方Ⅴ为最大，其次为配方Ⅵ，而配方Ⅲ、Ⅷ、Ⅸ为最小(图1E)；根冠比以配方Ⅲ为最大，显著高于Ⅰ、Ⅱ、Ⅷ和Ⅸ。用沤制松树皮和沤制锯末配置的3个配方间总生物量差异不显著。用沤制松树皮和炭化锯末配置的3个配方间苗木总生物量和根冠比均差异不显著。用沤制松树皮、锯末以及炭化锯末配置的3个配方中，仅配方Ⅶ的苗木总生物量与Ⅷ差异极显著。

3.4 轻基质配方的综合评价

应用坐标综合评定法依据苗木生长表现和生物量指标，对各基质配方进行综合评价。由于各处理间

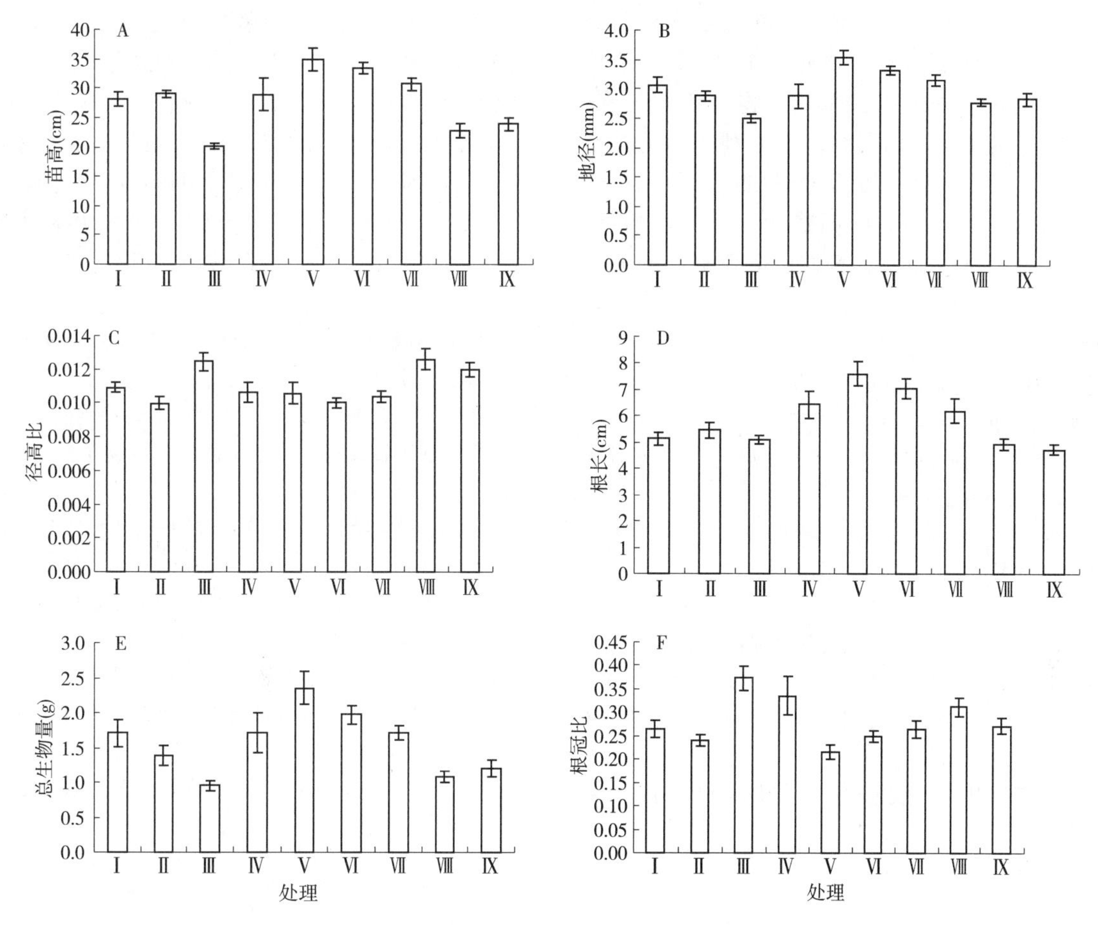

图1 9种配方轻基质西南桦网袋容器苗的生长与生物量比较

地下部分生物量差异不显著，因此剔除地上部分生物量和根冠比，仅用总生物量作为反应苗木干物质积累方面的指标。首先将数据进行标准化处理，具体方法是：求出各处理(i)各指标(j)的均值，找出每个指标的最大值，将该指标所有处理的值除以最大值即得标准化值(a_{ij})；然后按照下述公式计算各处理(i)到标准点的距离平方(P_{i2})：

$$P_{i2}=\sum_{j}(1-a_{ij})^2$$

最后依照P_{i2}大小排序(表2)。

由表2可知，配方Ⅴ的育苗效果最佳，其次是配方Ⅵ和Ⅶ，配方Ⅲ、Ⅷ和Ⅸ效果最差。

表2 应用9种配方轻基质培育西南桦网袋容器苗的综合评定

处理	苗高	地径	径高比	根长	总生物量	P_{i2}	排序
Ⅰ	0.0379	0.0180	0.0178	0.1053	0.0760	0.2550	5
Ⅱ	0.0288	0.0349	0.0430	0.0788	0.1685	0.3541	6
Ⅲ	0.1787	0.0886	0.0001	0.1082	0.3533	0.7290	9
Ⅳ	0.0292	0.0354	0.0252	0.0236	0.0746	0.1879	4
Ⅴ	0.0000	0.0000	0.0260	0.0000	0.0000	0.0260	1
Ⅵ	0.0019	0.0039	0.0426	0.0056	0.0264	0.0804	2
Ⅶ	0.0145	0.0120	0.0312	0.0339	0.0734	0.1650	3
Ⅷ	0.1220	0.0470	0.0000	0.1231	0.2946	0.5867	8
Ⅸ	0.0992	0.0400	0.0022	0.1423	0.2384	0.5220	7

4 结论与讨论

确定基质材料及其配比是轻基质容器育苗的关键环节之一。大量研究表明，轻基质种类和配比往往导致容器苗生长的显著差异[7,8]。本研究利用林区丰富的松树皮和锯末资源适当加工作为基质材料开展试验，9个配方中Ⅲ、Ⅷ和Ⅸ的育苗效果最差，与基质的通气性差、持水量过高有关；从沤制松树皮和沤制锯末的混合基质以及3种材料的混合基质配方的比较分析可以看出，沤制锯末比例越大，其苗木生长越差；这是由于西南桦不耐水浸，淋水过后，由于沤制锯末吸水能力强，往往形成基质中水分含量过高，影响西南桦的生长；郑海水等(1998)亦发现基质含水量过高往往导致西南桦苗木生长不良且易感病[9]，这亦与香樟、乳源木莲和马褂木的研究相符[10]；配方Ⅴ和Ⅵ的育苗效果最佳，其苗高、地径、根系等生长表现最好，而且干物质积累最多，说明松树皮与炭化锯末的混合基质有利于西南桦苗木生长，锯末经炭化后一方面可增加基质的无机养分含量，而且对养分具有吸附作用，保肥性能好，另一方面改善了基质的贮水特性，提高基质的通气性，有利于根系生长和养分吸收。总体而言，松树皮的有机质以及有效氮、磷、钾含量较高，而且具有较强黏结能力，使基质形成较大孔隙度，保持基质疏松、通气、透水，适当加入炭化锯末，有利于苗木生长。因此建议生产上采用树皮、炭化锯末的混合基质(如松树皮50%+炭化锯末50%)培育西南桦苗木，若添加沤制锯末，其比例不宜超过20%。

参考文献

[1]侯元兆. 现代林业育苗的概念与技术[J]. 世界林业研究，2007，20(4)：24-29.

[2]刘建明. 热带树种育苗技术现状与发展[J]. 广东林业科技，2003，19(2)：30-34.

[3]黎明，郭文福，蔡道雄，等. 以松皮粉为基质的西南桦容器苗培育技术[J]. 福建林业科技，2007，34(1)：43-45，49.

[4]蒙彩兰，黎明，郭文福. 西南桦轻基质网袋容器育苗技术[J]. 林业科技开发，2007，21(6)：104-105.

[5]翁启杰，曾杰，郑海水. 西南桦育苗技术研究[J]. 林业实用技术，2004(5)：20-22.

[6]苏金明，傅荣华，周建斌，等. 统计软件SPSS for Windows实用指南[M]. 北京：电子工业出版社，2000：401-417.

[7]君晓阳，李德芬，金天喜，等. 云南樟刺槐不同基质容器育苗比较试验[J]. 山地农业生物学报，2003，22(2)：122-126.

[8]邓煜，刘志峰. 温室容器育苗基质及苗木生长规律的研究[J]. 林业科学，2005，36(5)：33-39.

[9]郑海水，曾杰，翁启杰. 西南桦育苗基质选择试验初报[J]. 林业科技通讯，1998(10)：23-25.

[10]金国庆，周志春，胡红宝. 3种乡土阔叶树种容器育苗技术研究[J]. 林业科学研究，2005，18(4)：387-392.

[原载：种子，2010，29(10)]

穗条生根剂育苗基质和季节对西南桦扦插生根的影响

郭文福　蒙彩兰

（中国林业科学研究院热带林业实验中心，广西凭祥　532600）

摘　要　以无性系采穗圃的枝条为材料，开展了西南桦扦插生根促进剂与穗条组合、扦插基质以及扦插季节3个试验，研究不同生根促进剂及其浓度、穗条幼嫩程度、扦插基质以及季节对西南桦扦插生根的影响。结果表明，各种试验因素均对生根观测指标有显著影响；扦插育苗最优方案为以带顶芽的嫩枝为扦插穗条，用800～1000mg/kg生根粉ABT6和IBA浸泡处理，扦插基质采用含炭化树皮比例达80%的轻基质，选择夏、秋两季进行扦插育苗，其扦插生根率可达96.67%以上。

关键词　西南桦；生根促进剂(ABT6和IBA)；轻基质；扦插育苗

Effects of Root-inducing Regulator, Propagation Medium and Season on Rooting of *Betula alnoides* Cuttings

GUO Wenfu, MENG Cailan

(*Experimental Centre of Tropical Forestry*, *Chinese Academy of Forestry*, *Pingxiang* 532600, *Guangxi*, *China*)

Abstract: The cuttings from scion plucking orchard were used to test the combinations of cutting shoots with root-inducing regulators, media and cutting seasons in order to study the effects of different root-inducing regulators (ABT6 and IBA) and their concentrations, the lignifying degree of cutting shoots, the media types and propagation seasons on rooting ability, rooting number per shoot and rooting length of *Betula alnoides*. The result showed that tremendous significant differences were observed among different treatments, and the rooting percentage could be higher than 96.67% under the optimal cutting measure when conducted in the season of summer or autumn by using young shoots with apical bud as cutting shoots, treated with a higher concentration of ABT6 and IBA (800～1000mg/kg), and planting in the organic light medium which consisted of 80% carbonized pine bark.

Key words: *Betula alnoides*; root-inducing regulators; ABT6; IBA; light medium; rooting by cutting propagation

西南桦(*Betula alnoides* D. Don)是我国西南各省(自治区、直辖市)造林规模最大的乡土珍贵用材树种。西南桦具有材性优良、容易加工、色泽淡红、不易变形等木材特性，是高级家具、室内装修、木地板的优质材料，经济价值甚高。由于目前种子主要来源于天然林，随着天然林资源日益贫乏，加剧了造林规模不断扩大与供种量逐渐减少的矛盾；而且，实生苗造林的林木个体间遗传变异较大，林分木材单产较低。随着良种选育工作的深入，西南桦人工林发展也向“有性选育，无性利用”途径发展。近年来，国内已开展了西南桦无性繁殖技术研究，组织培养和扦插育苗快繁技术研究取得较快的进展[1-5]。作者单位西南桦组培育苗技术已基本上实现了规模化生产。无性系造林及其区域化试验正在陆续展开，但组培育苗技术要求较高，加之成本比扦插育苗高，因此，研究成本较低的西南桦扦插育苗技术仍有必要。由于轻基质育苗可提高苗木质量，提高造林成活，减轻种植的劳动强度，故为近年来应用较多的先进育苗技术[6]，而用树皮锯末为主要原料的育苗轻基质进行扦插育苗的研究少有报道。本试验针对西南桦扦插育苗的主要影响因素，于2009年开展了生根促进剂、穗条幼嫩程度、扦插季节和基质配方等因素对插穗生根影响的试验，旨在解决轻基质扦插育苗的关键技术，为扦插苗的规模化生产提供技术指导。

1 试验地概况

试验地设在广西凭祥市中国林业科学研究院热带林业实验中心种苗基地(106°44′E，22°07′N)，海拔250m，具南亚热带季风气候特点，终年温暖湿润，年平均气温21.5℃，年降水量1200~1400mm，干湿季节明显，4~9月为雨季，年均湿度82.3%，年均日照时数1512h。苗圃地势较开阔，排水良好，育苗用水为所在地职工生活用水，属硬质水，pH为7.2。

2 材料与方法

2.1 材料

扦插穗条采自广西凭祥市中国林业科学研究院热带林业实验中心自建的采穗圃。采穗母株用混系组培苗培育，年龄为1~3年。选择生长健壮、无病虫害、粗细匀称、生长期2~3个月龄的枝条作为试验穗条。扦插育苗基质的主要原料包括堆沤期为8~12个月的马尾松树皮和马尾松锯末(以下分别简称为“松树皮”和“松锯末”)或炭化过的松树皮和松锯末(以下简称“炭化树皮”和“炭化锯末”)、黄心土和河沙。生根促进剂为生根粉ABT6(中国林业科学研究院研制生产)和吲哚丁酸(IBA)。

2.2 穗条采集、处理及扦插后的管理

将采集的枝条从顶部向下依次按每10~15cm一段剪取，至多取3段，分别为3种不同木质化程度的穗条，即未木质化的嫩枝、半木质化的绿枝和木质化的硬枝。每段插穗基部剪成马蹄形，切口平滑，留24个半张叶。修剪好的插条用600倍多菌灵溶液消毒3~5min，然后在插穗基部蘸生根促进剂，并立即插入育苗基质中。采用直插法扦插，插穗扦插深度为2~3cm，扦插后用1200倍70%甲基托布津可湿性粉剂配制的药液均匀喷施。试验在全光自动间隙喷雾扦插育苗床上进行。此后每隔1周喷1次杀菌剂(多菌灵或百菌清，交替使用)；扦插2周后，喷1次叶面肥。叶面肥分别为磷酸二氢钾、尿素、硫酸复合肥，3种肥料交替使用，肥液浓度为0.10%~0.15%，1个月后提高到0.20%。约50d插条生根后移至大田培育。大田管理方法与实生苗培育措施相同[7]。

2.3 试验设计及数据处理

2.3.1 生根促进剂与插穗组合选择试验

考虑ABT6生根粉、IBA以及插穗木质化程度3个因素，各设3个水平(表1)，按正交表L9(34)设计试验，每处理小区扦插30根枝条。试验于2009年3月进行。扦插基质配比为炭化树皮80%+松锯末10%+炭化锯末10%(按体积比，下同)。扦插50d后调查生根率和平均每穗条生根条数(以下简称“生根数”)。试验数据处理按正交试验分析方法，找出影响最大的试验因素及最佳的试验处理，并进行方差分析。

2.3.2 扦插基质选择试验

采用随机区组试验设计，试验处理为4种不同的扦插基质，即M_1(炭化树皮80%+松锯末10%+炭化锯末10%)、M_2(松树皮30%+松锯末50%+炭化锯末20%)、M_3(黄心土)和M_4(河沙)，3次重复，每小区扦插30根嫩枝穗条。用ABT6生根粉液(800mg/kg)处理插穗。试验开始时间同“2.3.1”，扦插50d后调查插穗的生根率、生根数和平均每穗条生根长度(下简称“根长”)。试验数据采用SPSS11.0软件进行方差分析和Duncan多重比较。

表1 试验因素及水平

因素组成	代号	试验水平
ABT6 生根粉(mg/kg)	A_1	600
	A_2	800
	A_3	1000
IBA(mg/kg)	B_1	600
	B_2	800
	B_3	1000
穗条木质化程度	C_1	嫩枝(未木质化，带顶芽)
	C_2	绿枝(半木质化，中段)
	C_3	硬枝(木质化，下段)

2.3.3 扦插季节选择试验

采用考虑因素交互作用的双因素随机区组试验设计，试验处理为不同扦插季节以及不同IBA浓度。于2009年的4个季节用嫩枝进行扦插试验，共9次，其中，春季2次(2、3月份)，夏季3次(4、5、6月份)，秋季3次(7、8、9月份)和冬季1次(11月份)。每批次的插穗分别用IBA的4种不同浓度(400、600、800、1000mg/kg)处理。扦插基质同“2.3.1”。扦插50d后，调查生根率。采用多因素方差分析法进行数据分析，数据处理软件同“2.3.2”。

3 结果及分析

3.1 生根促进剂和不同木质化插穗对生根的影响

正交试验数据分析结果(表2)表明：3个试验因素中，穗条幼嫩程度对西南桦扦插生根的影响最大，

嫩枝的平均生根率达93.33%，硬枝仅为28.89%。方差分析结果(表3)表明：穗条的木质化程度对生根率的影响极显著，2种生根剂不同处理水平对生根率的影响未达显著水平；穗条的不同木质化程度对插穗生根数影响最大，但3个因素对生根数的影响均未达显著水平。从表2看出：促进穗条生根率的最佳处理组合是A3B3C1；提高生根数的最佳处理组合是A2B3C1。正交试验的9个处理中，生根率最高达96.67%(A2B3C1和A3B2C1)，可见，扦插穗条使用带顶芽的嫩枝是最适宜的。因为幼嫩程度高的枝条含有较高的内源激素，对分化再生根有利，因此，推测最优处理组合的生根率应达96.7%以上。

表2 生根促进剂和穗条对西南桦扦插生根的影响

试验号	A (ABT6)	B (IBA)	C (穗条类型)	X (生根率/%)	Y (生根数/条)
1	1(A_1)	1(B_1)	1(C_1)	86.67	5.73
2	1(A_1)	2(B_2)	2(C_2)	56.67	4.66
3	1(A_1)	3(B_3)	3(C_3)	26.67	4.20
4	2(A_2)	2(B_2)	3(C_3)	30.00	4.21
5	2(A_2)	3(B_3)	1(C_1)	96.67	6.58
6	2(A_2)	1(B_1)	2(C_2)	53.33	5.17
7	3(A_3)	3(B_3)	2(C_2)	70.00	4.34
8	3(A_3)	1(B_1)	3(C_3)	30.00	4.12
9	3(A_3)	2(B_2)	1(C_1)	96.67	5.67
X_1(%)	56.67	56.67	93.33		
X_2(%)	60.00	61.11	60.00		
X_3(%)	65.56	64.45	28.89		
X极差(%)	8.89	7.78	64.44		
Y_1(条)	4.86	5.01	5.99		
Y_2(条)	5.32	4.85	4.72		
Y_3(条)	4.71	5.04	4.18		
Y极差(条)	0.61	0.19	1.81		

3.2 基质对扦插生根的影响

方差分析表明，4种不同扦插基质对插穗生根率、生根条数和根长的影响均达极显著水平($P<0.01$)，说明扦插基质配方因素对扦插生根有极重要的影响。进一步通过Duncan检验(表4)表明，4种基质中以M_1处理的扦插生根效果最好，其生根率、每穗生根数及根长均为最大值，且与其他处理的差异达到极显著水平；生根效果较好的是M_2；其他两种处理的生根效果较差，且M_3和M_4处理间的生根效果差异不显著。

3.3 扦插季节及激素浓度对扦插生根的影响

不同扦插季节和IBA浓度对穗条生根影响的双因素方差分析结果表明，不同扦插季节和激素浓度，以及两因素的互作对插穗生根的影响均达极显著程度($p<0.01$，表5)。Duncan检验结果(表5)表明，夏季扦插成活率最高，成活率达90.2%，其次是秋季，成活率87.9%，春季的成活率达85.1%，最差是冬季，成活率只有65.8%；IBA浓度较大的处理(800mg/kg和1000mg/kg)的生根率较高，且与其他2个浓度的处理有极显著差异。

表3 生根促进剂和穗条对西南桦扦插生根影响的方差分析结果

变异来源	自由度	均方差		F		Fa
		生根率(%)	根数	生根率(%)	根数	
ABT6	2	61.33	0.24	7.32	1.44	F0.05=19.0
IBA	2	46.51	0.11	5.55	0.69	F0.01=99.0
穗条	2	3114.60	2.52	371.67	15.30	
误差	2	8.38	0.17			
总计	8	3230.82	3.04			

表 4　基质对西南桦扦插生根的影响

处理号	生根率(%)	1 级根数(条)	1 级根长(cm)
M1	95.93a	10.44a	4.07a
M3	75.89b	7.44b	1.94b
M2	36.67c	4.51c	1.32b
M4	31.12c	3.15c	1.17b
平均值	59.90	6.39	2.13

注：表中数据为各处理的平均值；表中同列不同字母表示处理间差异极显著($P<0.01$)，下同。

表 5　季节和 IBA 浓度对生根率影响的 Duncan 检验结果

因素	水平	代号	生根率(%)	显著性($P<0.01$)
季节	春季	1	85.10	b
	夏季	2	90.17	c
	秋季	3	87.91	b c
	冬季	4	65.79	a
IBA(mg/kg)	400	1	69.69	a
	600	2	82.47	b
	800	3	86.98	c
	1000	4	89.82	c

表 6　季节和 IBA 浓度对生根率影响

(续)

因素	水平	代号	生根率(%)
季节	春季	1	85.10b
	夏季	2	90.17c
	秋季	3	87.91bc
	冬季	4	65.79a
IBA(mg/kg)	400	1	69.69a
	600	2	82.47b
	800	3	86.98c
	1000	4	89.82c

4　结论及讨论

(1)穗条的木质化程度对扦插生根有极显著影响，木质化程度越低，穗条越幼嫩，其内源激素水平越高，扦插生根效果越好；生根促进剂对生根也有较大的影响。2 种生根促进剂和穗条幼嫩程度对扦插生根率影响的最佳组合是 $A_3B_3C_1$，即采用幼嫩穗条并通过浓度为 1000mg/kg 的 ABT6 和 IBA 处理后，可取得较好的生根效果，其生根率可达 96.67%。

(2)西南桦扦插育苗季节以夏季较为理想，夏季嫩枝扦插生根率可达 90%以上；秋季扦插效果次之，也是扦插育苗的合适季节，在以春季造林为主的地区，值得推荐。因为秋季育苗其苗木至次年春季出圃造林时苗木规格比较适中，如在夏季育苗，次年春季出圃时苗木过高，对造林成活率有不利影响。

(3)西南桦扦插育苗的基质对扦插效果的影响显著。扦插基质中炭化树皮比例达 80%的 M_1 基质生根率达 95%以上，而西南桦普通有根苗育苗基质(M_2)和河沙(M_4)则生根效果较差；因为 M_2 中含较多颗粒细小的锯末，通气性稍差；M_4为河沙，保水性略有不足[5]，说明选择保水性与通气性能均衡的扦插基质非常必要。

参考文献

[1]刘英，曾炳山. 西南桦以芽繁芽组织快繁技术研究[J]. 林业科学研究，2003，16(6)：715-719.

[2]陈伟，施季森，方镇坤，等. 西南桦不同种源扦插生根能力比较[J]. 南京林业大学学报(自然科学版)，2004(284)：29-33.

[3]陈存及，刘春霞，陈登雄，等. 光皮桦扦插繁殖技术研究[J]. 福建林学院学报，2002，22(2)：101-104.

[4]刘德朝. 西南桦良种采穗圃营建技术试验研究[J]. 林业勘察设计，2006(1)：101-103.

[5]许洋，许传森. 主要造林树种网袋容器育苗轻基质技术[J]. 林业实用技术，2006(10)：37-40.

[6]翁启杰，曾杰，郑海水. 西南桦育苗技术研究[J]. 林业实用技术，2004(05)：20-21.

[7]王凌晖，韦原莲，允辉，等. 植物生长调节剂对西南桦苗木生长的影响[J]. 广西植物，2009，22(5)：458-462.

[原载：林业科学研究，2011，24(06)]

不同种源麻栎种子和苗木地理变异趋势面分析

刘志龙[1,3] 虞木奎[2] 马 跃[1] 唐罗忠[3] 方升佐[3]
([1]中国林业科学研究院热带林业实验中心，广西凭祥 532600；
[2]中国林业科学研究院亚热带林业研究所，浙江富阳 311400；
[3]南京林业大学森林资源与环境学院，江苏南京 210037)

摘 要 对36个种源的麻栎种子和苗木性状进行了测定，应用趋势面分析这些性状在经纬2维方向的地理变异。结果表明：①种子百粒重、长度和宽度总体表现双向渐变趋势，随经度增高而增大，随纬度增高而减小，主要受到经度的控制；以西南到东北为中间地带，可溶性糖含量向东南表现先下降后上升的趋势，向西北则相反；淀粉含量从西北到东南呈逐渐减小的趋势。②种源间苗高、地径和生物量均呈双向渐变，经正向变异且变化幅度较大，纬负向变异且变化幅度较小，经度影响大于纬度；木质素含量北部大于南部，在经度的影响下，北部以西北部最高，南部以东南部最高。

关键词 麻栎；地理变异；趋势面分析

Atrendsurface Analysis of Geographic Variation of Seed and Seedling of Different *Quercus acutissima* Provenances

LIU Zhilong[1,3], YU Mukui[2], MA Yue[1], TANG Luozhong[3], FANG Shengzuo[3]
([1]*The Experimental Center of Tropical Forestry, CAF, Pingxiang* 532600, *Guangxi, China*;
[2]*Research Institute of Subtropical Forestry, CAF, Fuyang* 311400, *Zhejiang, China*;
[3]*College of Forest Resources and Environment, Nanjing Forestry University*, 210037, *Jiangsu, China*)

Abstract: The characteristic index of seed and seedling from thirty-six *Quercus acutissima* Carr. provenances were measured, which the geographical variations were analyzed by trend surface analysis in two dimension of latitude and longitude. The results showed that: ①Geographic variation pattern of seed length, seed width and 100-seed weight showed a bidirectional trend, e. g. increasing with the increase of longitude, decreasing gradually with the increase of latitude; Southwest to northeast be considered as the intermediate zone, content of soluble sugar showed decreases and then increases trend toward southeast, while toward northwest showed a contrary changing tendency. Content of starch showed a decreases gradually trend form northwest to southeast. ②Geographic pattern of variation of seedling height, ground diameter and biomass showed a bidirectional trend, positive variation in longitude, while negative variation in longitude latitude, the effect of latitude greater than longitude. The lignin content in the woods of northern higher than southern, under the effect of longitude, northwestern was highest in northern, while southeastern was highest in southern.

Key words: *Quercus acutissima*; qeographic variation; trend surface analysis

麻栎(*Quercus acutissima*)隶属于壳斗科(Fagaceae)栎属(*Quercus*)，是优良的硬阔叶能源和用材树种，具有耐瘠薄土壤和良好的水土保持功能，适合优质木炭开发等优点，极具开发潜力，是一种理想的可再生的“绿色能源”[1]。生物质能源作为可再生环境友好型能源受到世界各地广泛重视，麻栎作为理想的能源树种成为研究热点[2-4]。林木种内地理变异是普遍的客观现象，种子和苗木性状的变异是对这种复杂环境的一种适应[5-7]。麻栎在我国分布广泛，分布区内不同的气候及土壤条件，致使形成大量不同的遗传类型。目前，国内对麻栎的研究已有了一定基础[8-12]，但关于麻栎种源地理变异趋势面分析还未见报道。作为一种早期的测定手段，对麻栎种子和苗期地理变异研究可以在很短时间内取得关于该树种地理变异的格局、大小与趋势的一些重要资料，为麻栎的遗传改良及种子生产和调拨提供参考。

1 麻栎种源试验概况

1.1 试验材料收集

采集麻栎自然分布区13省(自治区、直辖市)36个种源，各种源产地名称、地理位置和气候状况见表1。每个采种地15~20株母树种子等量混合，作为该种源种子。置于0~5℃的冷库中沙藏，第二年春播种育苗。

1.2 试验地地理气候概况

试验地设在安徽省滁州市南谯区红琊山林场，位于皖东江淮之间，1118°04′E，32°10′N，处于华中湿润带向华北半干旱温带的过渡地区，四季分明；年平均气温15.2℃，年降水量1041.6mm，地带性植被类型为落叶阔叶林；境内岗峦起伏、海拔100~300m；土壤多为泥质岩、石灰岩发育的普通黄棕壤，土层浅薄。

1.3 试验设计

育苗试验点采用完全随机区组设计，3次重复。培筑苗床，床宽1.2m，床长20m，每个小区10行，每行播种15粒种子，行距10cm。各试验小区除草、施肥等抚育措施一致。

2 研究方法

2.1 种子性状测定

每个种源抽取100粒种子，重复3次，共计300粒，用电子天平测定种子百粒重，然后用游标卡尺分别测量种子的长度(纵向)和宽度(横向)。将种子在105±2℃烘干至恒重并粉碎备用，蛋白质含量用考马斯亮蓝法测定[13]，可溶性糖含量和淀粉含量用蒽酮显色法测定[14]。

2.2 苗木性状测定

2.2.1 生长量和生物量

苗木生长停止后，每小区选择30株测量苗高和地径，3次重复。根据调查结果，每个种源选择10株平均苗木进行生物量测定，起苗时要保证根系完整，将苗木的根、茎和叶分开后，在105±2℃下烘干至恒重，用电子天平分别称量。

2.2.2 热值和木材化学组分

样品在105±2℃下烘干至恒重，热值用美国Parr6300氧弹式量热仪测定，纤维含量用意大利VELP FIWE3/6粗纤维测定仪测定，木质素含量按照GB/T2677.8-94方法测定，所有样品均重复3次。

2.3 数据分析方法

采用Excel2003和SPSS16.0统计软件将所测各性状进行方差分析、差异显著性检验、趋势面分析等。种源试验的趋势面分析就是用多元回归的方法来拟合出一个种源生长量与地理因子(经度、纬度及其二者的各种组合)的曲面方程[15]。设试验有 n 个种源，测定各种源平均值为 $yi(i=1, 2, \cdots, n_{\circ})$，当 x_1 和 x_2 的最高次数为2时，二次趋势面方程的模型[16]为：$y=\beta_0+\beta_1x_1+\beta_2+x_2+\beta_3+x_1^2+\beta_4x_2^2+\beta_5x_1x_2+\varepsilon_{ij}$ 其中：β 为回归系数；x_1 为经度；x_2 为纬度；ε_{ij} 为随机变量。拟合精度为回归平方和U占总平方和S的比值百分数，即C=U/S×100，最后用Surfer8.0做出等值线-趋势面图[17]。

3 结果与分析

3.1 种子性状趋势面分析

不同麻栎种源间种子形态、百粒重和营养成分含量方差分析结果表明(表2)，种源间种子形态、百粒重和营养成分含量都达到极显著水平，说明各种源的种子存在一定的地理变异。

表1 参试麻栎种源地理坐标和气候状况

采种地点	地理坐标				年平均气温(℃)	年降水量(mm)	无霜期(d)	年日照时(h)	≥10℃年积温(℃)
	N(°	′)	E(°	′)					
广西融水	25	06	109	08	19.3	1824.8	320	1379.7	6258.4
广东乐昌	25	08	113	21	19.6	1522.6	300	1499.7	6386.5
贵州榕江	25	56	108	30	18.1	1200.6	310	1312.5	5717.1
湖南新宁	26	26	110	50	17.5	1327.1	311	1495.2	5510.5
贵州三穗	26	53	108	52	14.9	1147.1	276	1263.3	5437.4
贵州黄平	27	52	107	38	16.3	1050.6	258	1317.9	5500.5
浙江龙泉	28	01	119	07	17.6	1669.7	263	1850.5	5572.6

（续）

采种地点	地理坐标				年平均气温(℃)	年降水量(mm)	无霜期(d)	年日照时(h)	≥10℃年积温(℃)
	N(°	′)	E(°	′)					
湖南长沙	28	12	113	04	17.2	1361.6	263	1677.1	5457.8
四川泸州	28	28	105	37	17.8	1188.5	265	1290.4	5770.1
湖南常德	29	03	111	41	16.8	1274.2	265	1246.4	5189.1
浙江开化	29	09	118	23	16.4	1901.4	250	1785.1	5125.4
湖南岳阳	29	13	113	03	15.6	1331.6	279	1651.9	4958.6
湖南桑植	29	28	110	03	14.5	1530.6	235	1996.6	5200.3
浙江建德	29	29	119	16	17.4	1600.7	254	1760.8	5270.1
安徽休宁	29	33	118	02	16.1	1773.4	231	1931.5	5100.2
浙江富阳	29	57	119	46	16.1	1463.8	232	1858.7	5064.5
安徽池州	30	11	117	28	16.5	1556.9	220	1931.5	5129.1
安徽黄山	30	16	118	07	15.5	1670.3	237	1930.4	4856.4
安徽泾县	30	26	118	13	15.6	1556.4	240	2114.3	4950.8
湖北浠水	30	27	115	13	16.9	1350.7	257	1895.6	5405.8
安徽太湖	30	37	116	24	16.4	1363.5	249	1936.2	5214.2
湖北远安	30	48	111	42	15.5	1100.5	241	1878.5	4895.4
安徽潜山	30	55	116	41	16.3	1336.2	242	1234.7	5177.1
安徽六安	31	29	116	05	15.2	1085.2	242	2220.4	5004.5
江苏句容	32	04	118	51	15.1	1105.3	229	2018.3	4859.2
四川万源	32	05	108	03	14.7	1169.3	240	1474.4	4242.6
安徽滁州	32	10	118	04	16.5	1000.8	218	2238.8	4800.4
湖北襄樊	32	13	112	04	15.6	1012.8	248	2013.6	4272.4
四川广元	32	16	105	27	16.2	1063.8	264	1398.8	4515.6
陕西汉中	32	49	106	30	14.5	858.60	235	1710.4	4028.3
河南南召	33	35	112	24	14.8	839.50	216	1978.8	4685.4
山东平邑	35	13	117	22	13.2	784.80	212	2589.2	5100.4
山东费县	35	15	117	58	13.6	856.40	197	2533.4	4279.3
山东蒙阴	35	19	117	34	12.8	820.30	200	2257.6	4380.3
山东沂水	35	48	118	44	12.3	782.10	205	2414.7	4183.7
山西方山	37	34	112	02	11.9	712.8	125	2508.1	4056.2

表 2　不同麻栎种源间种子性状的差异和变异系数

种子性状	平均值	变异幅度	标准差	*F* 值	变异系数
种子长度(cm)	2.00	1.69 ~ 2.26	0.27	31.27**	13.75
种子宽度(cm)	1.78	1.06 ~ 2.50	0.32	33.39**	17.78
种子百粒重(g)	341.73	83.55 ~ 599.01	131.65	201.97**	38.52
可溶性糖含量(mg/g)	0.08	0.05 ~ 0.11	0.02	4.24**	19.25
淀粉含量(%)	3.72	27.82 ~ 47.16	0.59	8.48**	15.74
蛋白质含量(mg/g)	26.63	21.50 ~ 34.70	3.52	6.56**	13.21

分别以表 2 中种子性状为因变量，以经度和纬度为自变量进行二元二次趋势面分析，趋势面回归方程和拟合度见表 3。由表 3 可以看出，种子百粒重、种子长度和宽度拟合度较高，并到达极显著水平，可以揭示其空间变化趋势。种子营养成分含量拟合度较低，其中蛋白质含量回归方程不显著($P=0.58$)，此方程无意义，不能揭示其空间变化趋势。

表 3　麻栎种子性状二元二次趋势面分析回归方程

性状	趋势面回归方程	拟合度(%)	P 值
种子百粒重	$Y=17425.746-374.381\ X_1+197.831\ X_2+2.043\ X_{12}-2.097\ X_{12}+0.380\ X_{22}$	84.55	0.00
种子长度	$Y=18.938-0.445\ X_1+0.426\ X_2+0.00263\ X_{12}-0.00382\ X_{12}-0.00031\ X_{22}$	76.71	0.00
种子宽度	$Y=21.504-0.441\ X_1+0.213\ X_2+0.00267\ X_{12}-0.00396\ X_{12}+0.00320\ X_{22}$	62.72	0.00
可溶性糖含量	$Y=1.544-0.0215\ X_1-0.0236\ X_2+0.000058X_{12}+0.000315\ X_{12}-0.000173\ X_{22}$	41.33	0.03
淀粉含量	$Y=162.272-3.4275\ X_1+2.0648\ X_2+0.0183\ X_{12}-0.02115\ X_{12}+0.00481\ X_{22}$	45.41	0.01
蛋白质含量	$Y=434.834-5.6017\ X_1-5.0749\ X_2+0.0514\ X_{12}+0.0687\ X_{12}-0.02923\ X_{22}$	15.50	0.58

注：X_1：经度；X_2：纬度，下同。

一次多项式趋势面为平面，二次为抛物面、椭圆面或双曲面，三次以上的趋势面为更复杂的图形。当样本数足够大时，拟合程度应随着最高次数的增加而增大。从图 1 可以看出，种子百粒重的变异呈双向渐变，随经度的增高，种子百粒重逐渐增大，且变化较快；随纬度的增高，种子百粒重逐渐减小，且变化较慢；经度的影响大于纬度，在 E112°～118°表现尤为明显。三次趋势面分析较二次分析结果更能表现局部变化规律，随纬度的增加，在东北和西北部具有而下降的趋势，在西南则由先下降后上升的趋势。从图 2 和图 3 可以看出，种子长度和宽度二次趋势面总体表现相似的变化趋势，主要受到经度的控制，东部变化幅度比西部大，东南部大于东北，且变化较大，西南部的种子长度小于西北部，但种子宽度大于西北，且变化较小。从三次趋势面还可以看出，西北部表现有随纬度的增加而下降的趋势。

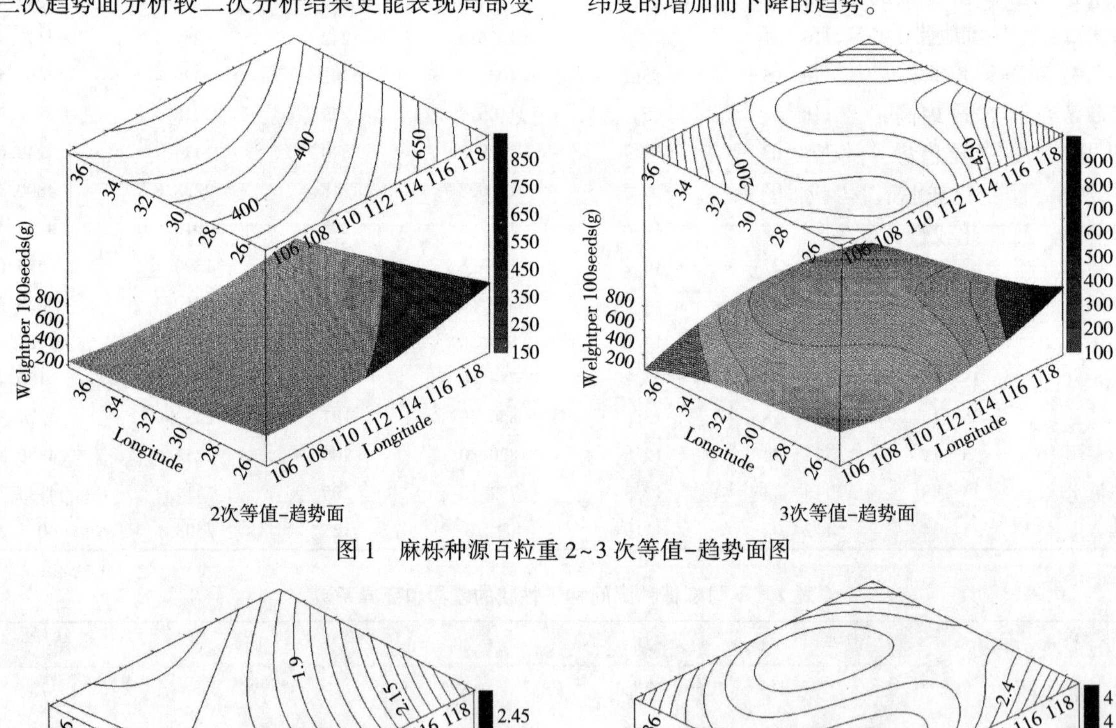

图 1　麻栎种源百粒重 2～3 次等值-趋势面图

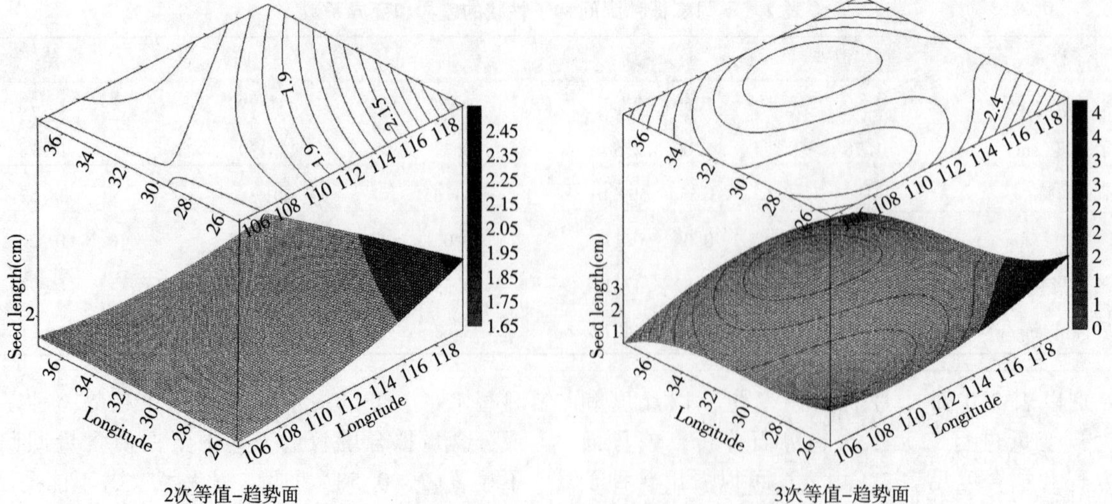

图 2　麻栎种源种子长度 2～3 次等值-趋势面图

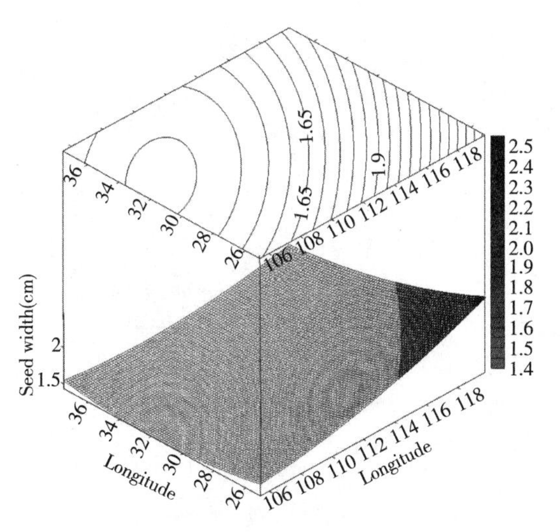

2次等值-趋势面　　3次等值-趋势面

图 3　麻栎种源种子宽度 2~3 次等值-趋势面图

分别对可溶性糖和淀粉含量进行二元三次趋势面分析，拟合度分别达到 68.45%和 73.48%。从图 4 可以看出，可溶性糖含量的变异呈双向渐变，由西南到东北(定为中间地带)，向东南部表现先下降和上升的趋势，但西北部表现相反的趋势，总体而言，东南部可溶性糖含量最高，西北部最低。淀粉含量也呈双向渐变的变异趋势，随经度的增高，淀粉含量逐渐减小，且变化较快；随纬度的增高，淀粉含量逐渐增加，且变化较慢；总体来说，淀粉含量从西北到东南呈逐渐减小的趋势，经度的影响大于纬度。

3.2　苗木性状趋势面分析

3.2.1　生长量和生物量

分别以苗高、地径和生物量为因变量，以经度和纬度为自变量进行二元二次趋势面分析，由表 4 可以看出，趋势面回归方程拟合精度和显著性均符合统计要求。

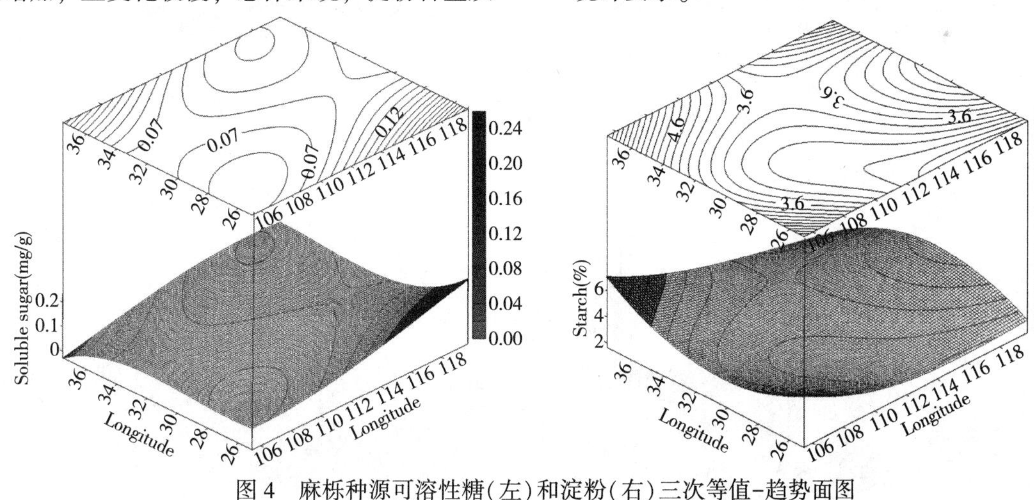

图 4　麻栎种源可溶性糖(左)和淀粉(右)三次等值-趋势面图

表 4　麻栎苗木性状二元二次趋势面分析回归方程

性状	趋势面回归方程	拟合度(%)	*P* 值
苗高	$Y=1939.256-38.575X_1+15.753X_2+0.216X_{12}-0.289X_{12}+0.226X_{22}$	70.72	0.00
地径	$Y=14.024-0.315X_1+0.242X_2+0.00188X_{12}-0.00303X_{12}+0.0011X_{22}$	68.52	0.00
生物量	$Y=983.875-19.079X_1+6.128X_2+0.109X_{12}-0.154X_{12}+0.155X_{22}$	70.72	0.00

苗高(图 5)、地径(图 6)和生物量(图 7)的变异呈双向渐变，经正向变异，东部种源要优于西部种源，且变化幅度较大；纬负向变异，南部种源好于北部种源，且变化幅度较小；总体来说，表现从东南到西北呈逐渐减小的趋势，经度的影响较纬度更大些。综上研究表明，苗高、地径和生物量主要受经度的控制，东部变化幅度比西部大，东南部大于东北部，且变化较大，西南部大于西北部，且变化较小。

变量进行二元二次趋势面分析。

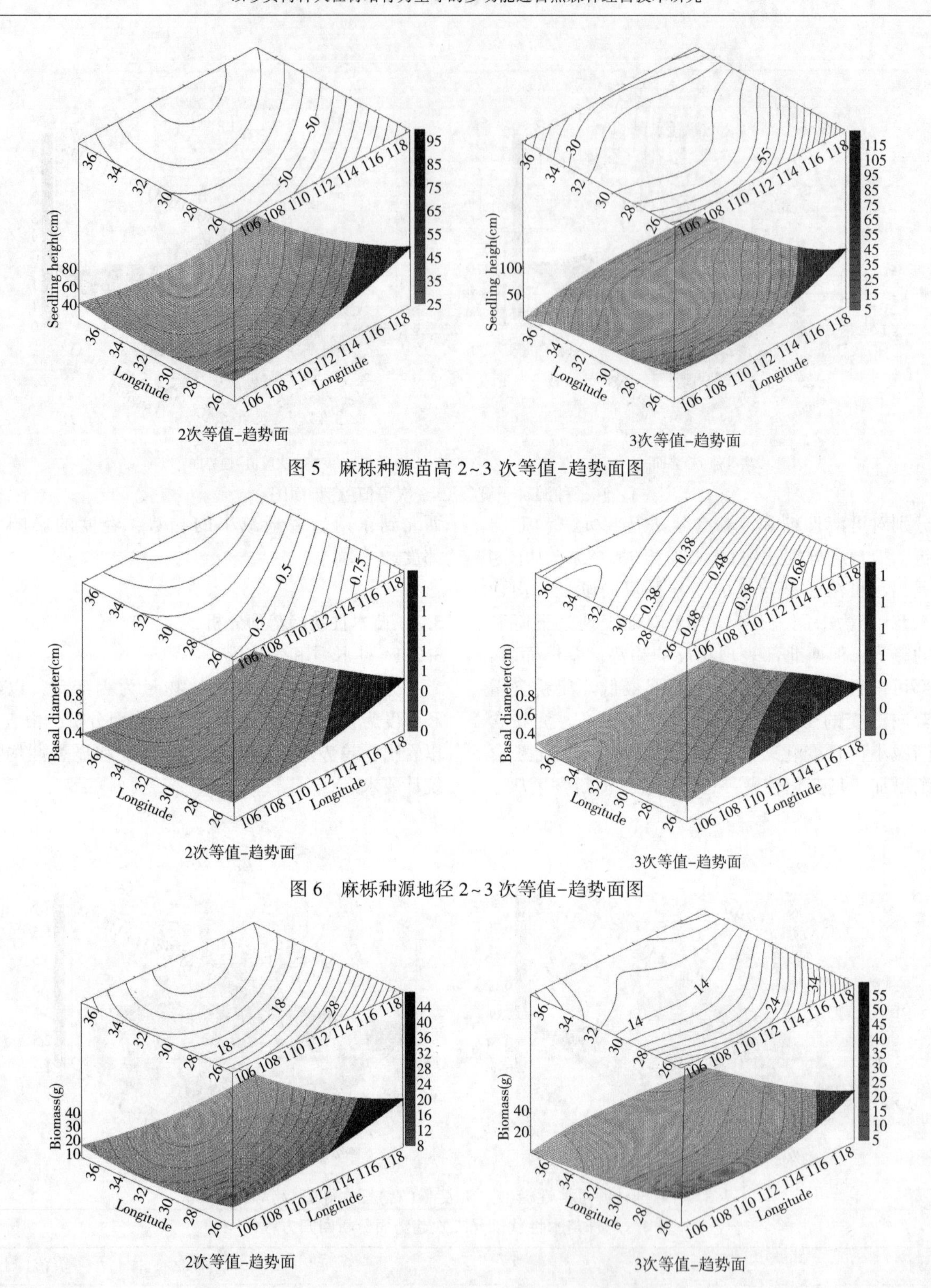

图5 麻栎种源苗高2~3次等值-趋势面图

图6 麻栎种源地径2~3次等值-趋势面图

图7 麻栎种源生物量2~3次等值-趋势面图

3.2.2 热值和茎木材化学组分

分别以热值(叶、茎和根)和茎部木材化学组分(木质素和综纤维素)为因变量，以经度和纬度为自变量进行二元二次趋势面分析。由表5可知，木质素含量回归方程达极显著差异水平，但拟合精度不高。各组分热值和纤维素含量回归方程不显著，此方程无意义，对木质素含量进行四次多项式趋势面分析，拟合度>60%，可以模拟其地理变异趋势。从图8可以看出，木质素主要受到纬度的控制，北部种源明显大于南部种源，在经度的影响下，北部以西北部最高，南部以东南部最高。

表5 麻栎种源热值和木材化学组分二元二次趋势面分析回归方程

性状	趋势面回归方程	拟合度 C(%)	*P* 值
茎热值	$Y=114.062-1.586X_1-0.426X_2+0.008X_{12}-0.005X_{12}+0.016X_{22}$	21.79	0.551
根热值	$Y=51.525-0.049X_1-0.633X_2+0.0009X_{12}+0.008X_{12}-0.0039X_{22}$	27.26	0.349
叶热值	$Y=-20.316+0.910X_1-0.638X_2-0.007X_{12}+0.021X_{12}-0.029X_{22}$	29.67	0.412
木质素	$Y=556.434-7.541X_1-6.691X_2+0.0247X_{12}+0.059X_{12}+0.0051X_{22}$	55.70	0.000
纤维素	$Y=397.38-3.779X_1-8.237X_2+0.006X_{12}+0.079X_{12}-0.008X_{22}$	45.66	0.120

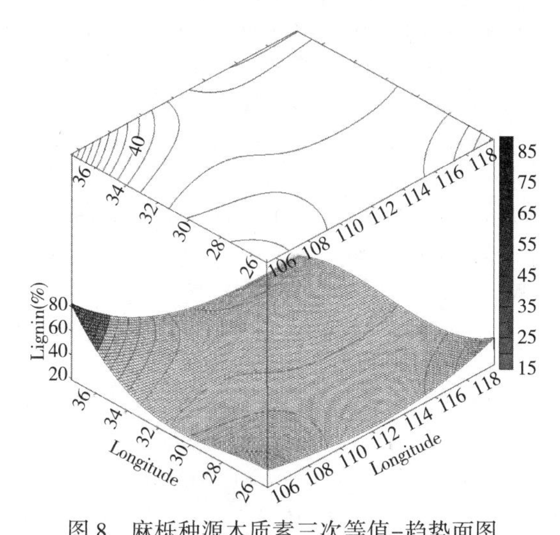

图8 麻栎种源木质素三次等值-趋势面图

4 结论与讨论

对我国36个麻栎地理种源种子性状的研究表明：麻栎种源间种子形态特征、营养成分含量在种源间差异极显著，产地效应较为明显，东南部分布区(例如浙江富阳、广东乐昌等)的种子往往大于西北部(例如山西方山、陕西汉中等)的种子，这与我国麻栎产区的地理位置及气候条件有关。除蛋白质含量外，麻栎种源存在明显的地理变异趋势，种子百粒重呈双向渐变趋势，从东南到西北呈逐渐减小。

种子形态和贮藏营养物质多少间接反映了出苗率及幼苗生长的好坏，麻栎苗期生长性状的变异具有一定的地理变异规律，研究表明：麻栎种源间苗木生长量、生物量和木材化学组分差异极显著。苗高、地径和生物量的变异呈双向渐变，经正向变异，东部的种源要优于西部的种源，随经度的增高增大，且变化较快；纬负向变异，南部的种源要好于北部的种源，随纬度的增高逐渐减小，且变化较慢；木质素含量主要受到纬度的控制，北部种源明显大于南部种源，在经度的影响下，北部以西北部最高，南部以东南部最高。

物种在长期的进化过程中为适应复杂多变的环境而产生与之相适应的遗传变异，同一树种不同产地的母树为了适应环境变化也会产生遗传变异，并将稳定的遗传变异性状反映在种子的各种品质中[18]。本研究所采集的麻栎种子产地跨度大，产地间的地理位置及生态因子存在很大梯度性变化，伴随着这些因子梯度变化，麻栎群体中受环境选择影响的形态性状在表达上将呈现梯度性变异。与传统方法相比，趋势面分析在地理变异的线性相关不显著的情况下，往往能收集到可能存在的非线性相关信息，从而能够更精细地刻画出种源的地理变异趋势[19]。此外，高次趋势方程拟合优度较高，呈现不同的带状或中心变异模式，能发现一些与拟合趋势值相比，反映出异常高值或低值的变异小群体，因此有必要对这些地区加密取样点，进一步寻找林分小群体的差异，从而提高改良效益。

本研究只是以种子和苗期性状为指标进行分析，有可能此时各种源遗传性状尚未得到充分表达与稳定，这些早期表现优良的性状是否与晚期相关尚需进一步作观测，但是这些结果对后续研究奠定了重要基础。此外，由于麻栎结实过程还存在“大小年”现象，且种子采集过程中也存在诸多限制因素，因而难以对不同产地间种子地理变异进行准确的描述。若能结合分子水平方面进一步揭示麻栎种源间遗传变异规律，进一步揭示麻栎种源间地理变异规律，这些工作还有待于进一步研究。

参考文献

[1] LIU Z L, YU M K, TANG L Z, et al. Progress and Utilization Countermeasure of *Quercus acutissima*. Forest By Product and Speciality in China, 2009, 12: 93-96.

[2] SHAO H, CHU L. Resource evaluation of typical energy plants and possible functional zone planning in China [J]. Biomass and Bioenergy, 2008, 32: 283-288.

[3] CHOW J, KOPP R J, PORTNEY P R. Energy resources and global development. Science, 2003, 302: 1528-1531.

[4] FANG S Z, XUE J H, TANG L Z. Biomass production and carbon sequestration potential in poplar plantations with different management patterns. Journal of Environment

Management, 2007, 85: 672-679.

[5]SHU X, YANG Z L, YANG X, et al. Variation in seed characters of *Magnolia officinalis* from different locations. Forest Research, 2010, 23(3): 457-461.

[6]WANG Q Y, JIA H B, SHANG J. Geographic variation and genetic performance of *Picea koraiensis* in growth and wood characteristics. Journal of Forestry Research, 2005, 16(2): 93-96.

[7]TONG Z K, ZENG Y R, SI J P. Variation, heredity and selection of effective ingredients in*Magnolis officinalis* of different provenances. Journal of Forestry Research, 2002, 13(1): 7-11.

[8] TANG L Z, IU Z L, YU M K, et al. Nutrient accumulation and allocation of aboveground parts in *Quercus acutissima* plantations under two site conditions in Anhui, China. Chinese Journal of Plant Ecology, 2010, 34(6): 661-670.

[9]WANG B, YU M K, SHAN Q H, et al. Photosynthetic characters of *Quercus acutissima* from different provenances under effects of salt stress. Chinese Journal of Applied E-cology, 2009, 20(8): 1817-1824.

[10]XU F, GUO W H, XU W H, et al. Effects of light intensity on growth and photosynthesis of seedlings of *Quercus acutissima* and *Robinia pseudoacacia*. Acta Ecologica Sinica, 2010, 30(12): 3098-3107.

[11]LIU Z L, YU M K, TANG L Z, et al. Variation and cluster analyses of morphological characters and nutrient content of *Quercus acutissima* seed from different provenances. Journal of Plant Resources and Environment, 2009, 18(1): 36-41.

[12]ZHAO D, ZHU G Q, TANG L Z, et al. Tissue culture and rapid propagation of *Quercus acutissima*. Chinese Agricultural Science Bulletin, 2010, 26(15): 168-171.

[13]LI H S. Technology and principle of plants physiological and biochemical. Beijing: Hher Eucation Press, 2000.

[14]WANG J Y, AO H, ZHANG J. Tchnology and principle of plants physiological and biochemical experiment. Harbin: Northeast Frestry Uiversity Press, 2003.

[15]YANG C Q. Research of the variation and selection of geographic provenances of platycladus orientalis and the variation of their progeny. Master degree thesis of Shandong Agricultural University, 2005.

[16]TANG Q Y, FENG M G. Data processing system of DPS. Science press, 2006.

[17]WANG J, BAI S B, CHEN Y. Geography information cartography of Surfer 8. Beijing: Chinese Mps Press, 2004.

[18]YUE H F, SHAO W H, JING Z H, et al. Geographic variation of seed characters of *Castanopsis sclerophylla*. Forest Research, 2010, 23(3): 453-456.

[19]LI J Y, RAO L B, YANG W Z. Trend surface analysis of geographic variations for the seedling height of Chinese Wing-Nut tree(*Pterocarya stenoptera*) provenance. Journal of central south forestry university, 2003, 23 (1): 15-21.

[原载：生态学报，2011，31(22)]

储藏方法对灰木莲种子储藏时间和发芽率的影响

卢立华[1] 蒙彩兰[1] 何日明[1] 苏 勇[2]
([1]中国林业科学研究院热带林业实验中心，广西凭祥 532600；
[2]南宁良凤江国家森林公园，广西南宁 530031)

摘 要 采用6种储藏方法对灰木莲种子进行储藏试验，结果表明，A、B处理的储藏期最长，都达到了16个月，而C、D、E、F处理的储藏期都很短，分别经52、52、32、64d储藏，种子或发了芽，或丧失发芽力。储藏效果最好的处理为A，经2个月储藏，发芽率为74.8%，6个月仍达63.5%；其次为处理B，储藏2个月的发芽率为74.5%，6个月为51%；所有处理的种子发芽率都随储藏期增加而下降，方差分析表明，储藏期对种子发芽率的影响达到极显著差异。

关键词 灰木莲；储藏方法；储藏期；发芽率

Effect of Different Storage Treatments on Storage Period and Germination Rate of *Manglietia glauca* Seeds

LU Lihua[1], MENG Cailan[1], HE Riming[1], SU Yong[2]
([1] *The Experimental Center of Tropical Forestry, Chinese Academy of Forestry, Pingxiang* 532600, *Guangxi, China*;
[2] *Nanning Liangfengjiang National Forest Park, Nanning* 530031, *Guangxi, China*)

Abstract: Atrail was carried out on the seeds of *Manglietia glauca* using six storage methods. The results showed that the storage periods of treatments A and B were the longest, reaching up to 16 months, while the storage periods of treatments C, D, E and F were very short since their seeds bud or dead after 52, 52, 32, and 64d, respectively. The best treatment was treatment A, with 74.8% and 63.5% of germination rate after 2 and 6 months, respectively. The second was treatment B, with74.5% and 51% of germination rate after 2 and 6 months, separately. The germination rate of all seeds decreased with the increase of storage period. The variance analysis showed that the effect of storage period on seed germination rate was extremely significant.

Key words: *Manglietia glauca*; storage method; storage period; germination rate

灰木莲(*Manglietia glauca*)为热带常绿阔叶树种，原产越南、印度尼西亚、爪哇等地，在我国无自然分布。灰木莲树干通直，材质优良，生长快，抗性强，20世纪60年代初在广东、广西、海南、福建等地引种，生长甚佳，幼林年平均树高、胸径生长量超过1.0m和1.0cm，最早于1964年在广西国营高峰林场从越南引进种植11hm^2，目前林分最大胸径达60cm，树高35m。灰木莲树形挺拔美观，花大洁白，有芳香，且其大树移植容易成活[1]，故也是园林绿化优良树种。

在越南，灰木莲5~6年生即可结果，果熟很快开裂，须及时采收，种子无明显休眠期，宜随采随播[2]。灰木莲在我国引种已有较长历史，并有一定规模，但个体结实不理想，目前，仅见杨耀海等[3]和林捷等[4]关于灰木莲能正常开花结实的报道。2009年在广西、广东所进行的调查表明，各种年龄的灰木莲都只花不实。中国林业科学研究院热带林业实验中心树木园于1980年引种的灰木莲，生长表现良好，胸径超过30cm，但至今仍是花而不实。据了解，一直以来，我国灰木莲的生产用种主要来自越南。

按照Roberts[5]对种子的分类，灰木莲种子属顽拗性种子。Dickie等[6]对205种顽拗性种子统计发现它们多无休眠性，而且萌发迅速。通常情况下，顽拗性种子忌干燥与低温，贮藏方法多采用适温保湿贮藏延长其寿命[5]。目前，灰木莲广泛采用的储藏方法为常温保湿沙藏，储藏期仅30d左右，这与热带地区植物受高温多湿环境的影响，常形成种子寿命短的特性的结论一致[5]。

中国林业科学研究院热带林业实验中心与越南毗邻，近10年来每年都从越南引进数量不等的种子，因无有效的储藏方法，常出现种子引进多浪费、引进少影响生产的情况。此外，引入(9月底10月初)即播，并非灰木莲育苗的最佳时间，因此，时育苗芽苗上袋不久就进入冬季，气温较低，作为热带树种的灰木莲，在低温下生长量很低，甚至不长，导致其当年育苗，于翌年当地造林最适宜季节(3~4月份)出圃时，因苗木过小(仅20cm左右)、过嫩而致造林效果不佳。若多培育一年出圃，又因苗期过长(1.5年)、苗木过于高大(达到1.0m左右)，而致育苗和造林成本增加，并因苗木穿根严重，影响造林成活率和幼林生长。要解决上述问题，找到能有效延长灰木莲种子储藏期的方法为唯一途径。而目前，我国对灰木莲的研究仅局限于其生物学特性、育苗基质、育苗技术、光合生理特性、引种驯化及生长表现等方面[7-13]，对灰木莲种子贮藏技术的研究仍为空白，而灰木莲种子主要靠进口，不仅引进困难，价格也较高，而且引进即播不是灰木莲理想的播种时间。可见，无论从减少种子进口次数、科学安排播种时间、调节种子丰缺、减少种子浪费，或是从培育优质苗木、降低育苗和造林成本、提高造林质量等考虑，开展这一研究都十分必要和迫切。

1 材料与方法

1.1 材料

供试种子为2009年9月22日从越南引进。23日对种子清洗、除杂后用0.2%~0.3%高锰酸钾溶液浸泡2h，用清水冲洗干净，在通风的房内摊开、阴干备用。

1.2 试验设计

储藏试验于2009年9月24日开始，采用6种储藏处理方法，A处理：将种子直接装入塑料袋内密封置于5℃的冰箱中储藏；B处理：将种子加入30%的湿沙混合均匀后装入塑料袋内密封放置在5℃的冰箱中储藏；C处理：将种子直接装入塑料袋内密封置于室内储藏；D处理：将种子直接装入布袋内封口置于室内储藏；E处理：将种子加入30%的湿沙混合均匀后装入塑料袋内密封置于室内储藏；F处理：将种子加入30%的湿沙并混合均匀后装入布袋内封口置于室内储藏。

1.3 取样与观测

A、B两处理每次取样间隔为2个月；C、D处理为4d；E、F处理为8d。每次取样前先将种子搅拌拌均匀后各取种子400粒，均分为4份(即4次重复)，分别摊放于底铺双层滤纸的培养皿中，并用蒸馏水浸湿，加盖后置于恒温培养箱中。以往研究表明，80%以上热带树种种子萌发的适宜温度为24~30℃[14]，本研究采用的催芽温度定为25℃。在种子的整个催芽过程中，每天定时观察1次，种子开始发芽后，记录发芽时间及发芽粒数，取出发芽种子后继续催芽、观察与记录；直至连续3d没有观察到发芽种子为本次催芽试验结束。

1.4 数据处理

发芽率统计描述和方差分析采用Excel和SPSS11.5软件进行。发芽率计算公式：

发芽率(%)=催芽期内发芽种子总数/供试种子数×100%

2 结果与分析

2.1 不同处理灰木莲种子储藏期比较

试验开始前对参试批次种子的发芽率进行测定，结果表明，该批种子储藏前的发芽率为84.5%。从图1可见，储藏方法不同灰木莲种子的储藏期存在明显差异，在所设置的6种处理中以A、B处理的储藏期最长，经储藏16个月后种子才完全丧失发芽力，经6个月的储藏时它的发芽率还能达到50%以上，储藏效果良好；而常温条件下，辅助其他储藏手段(C、D、E、F处理)的方法储藏期都很短，分别储藏52、52、32、64d后种子或者已经发芽或者丧失活力。可见，无论是塑料袋密封或布袋封口、拌沙或不拌沙、采用常温储藏都无法很有效地延长灰木莲种子的储藏期，而只有采用在5℃的冰箱中贮藏这一适当低温的储藏方法，才能有效地延长灰木莲种子的储藏期，实现灰木莲种子的跨年度储藏，从而有效解决灰木莲种子随采随播所产生的一系列问题。

2.2 不同处理灰木莲种子萌发的影响

从图1可见，各处理的灰木莲种子发芽率都随储藏时间增加而下降，但不同处理的幅度不一样。其中以用塑料袋密封的发芽率降幅比用布袋封口的低；

不拌沙的降幅比拌沙的低。

方差分析结果表明，各处理经不同时间的储藏对灰木莲种子发芽率的影响都达到了极显著差异水平。从图1可见，储藏效果最好是A处理，经2个月储藏，种子发芽率达74.8%；4个月为70.5%；6个月仍达到63.5%；储藏到8个月降为30%；10、12、14、16个月分别为23.5%、10.5%、7.5%、2.5%；其次是B处理，储藏2个月发芽率为74.5%；4个月为69.8%；6个月的发芽率为51%；储藏到8个降为26.3%；10、12、14、16个月分别为19.0%、7.0%、4.5%、1.3%；上述结果表明，用5℃低温储藏灰木莲种子以储藏6个月左右为好，超过6个月后发芽率较低，如A处理储藏8个月的发芽率不及储藏6个月时的一半，B处理亦基本一致，也降低了接近一半的发芽率。储藏到16个月2种处理的发芽率都已经很低，已不宜使用。而C、D、E、F处理的储藏效果都很差，储藏8d，发芽率分别为60.2%、45.8%、60.5%、40.7%，与储藏6个月的A处理比较，分别下降了3.3%、17.7%、3.0%、22.8%，储藏期不到2个月或完成了发芽，或发芽率很低(5%以下)，基本失去了利用价值，可见，C、D、E、F四种处理都不能解决灰木莲播种、保种等所存在的问题，只有A、B两种处理基本能解决这些问题。

多重比较结果(图1)表明，灰木莲种子储藏前的发芽率与各种处理不同储藏期的发芽率比较都达到了极显著差异。

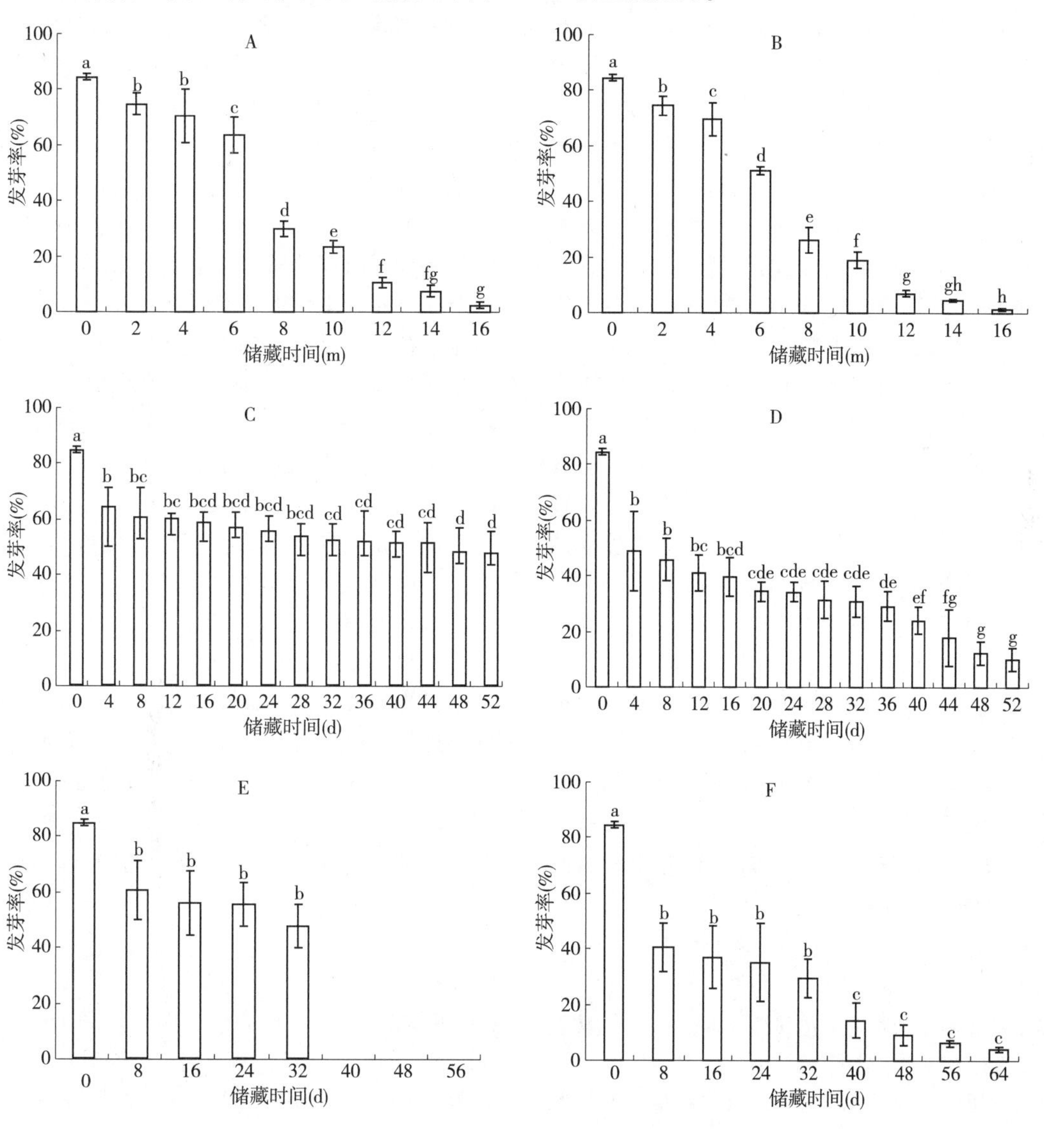

图1

注：储藏时间m为月，d为天

A处理(图1A)储藏2个月与储藏4个月的发芽率比较没有达到显著差异，说明灰木莲种子在5℃的温度下储藏，其在2~4个月的生理过程比较存在明显差异，故发芽率较为稳定，但与随后不同储藏期的种子发芽率比较都达到了显著或极显著差异；储藏4个月与储藏6个月比较达显著差异，与随后不同储藏期的发芽率比较都达到了极显著差异；储藏6个月与随后不同储藏时间的种子发芽率比较都达到了极显著差异，储藏8个月与储藏10个月比较达到了显著差异，与随后不同储藏时间的发芽率比较达到了极显著差异；储藏10个月与随后不同储藏时间的发芽率比较都达到了极显著差异；储藏12个月与储藏16个月比较达显著差异，与储藏14个月比较差异不显著，储藏14个月与储藏16个月比较差异不显著，这主要是储藏12个月后灰木莲种子的发芽率已经很低之故。

B处理(图1B)灰木莲种子储藏前的发芽率与不同储藏期的发芽率比较都达到了极显著的差异；储藏2个月与4个月的发芽率比较达显著的差异，而与随后不同储藏期的种子发芽率比较都达到了极显著差异；储藏4、6、8、10个月的种子发芽率与它们随后不同储藏时间的种子发芽率比较都达到了极显著差异；12个月与16个月比较达显著差异，而与14个月比较差异不显著；14个月与16个月比较差异不显著。而C、D、E、F四种处理的储藏期都较短，对解决灰木莲种子的储藏问题没有现实意义，多重比较结果可参见(图1C、D、E、F)。

灰木莲种子储藏前与经各种处理储藏后的首次发芽率比较都达到了极显著的差异；说明灰木莲种子一经储藏其发芽率就极显著的下降，这主要与灰木莲种子的特点有关，因为它没有后熟期，成熟种子，发芽条件具备，种子萌发的一系列生理过程即开始，在从越南采种处理完成到进入我国的半个多月时间里，灰木莲种子采用常温保湿沙藏，这种储藏方法与在沙床中催芽无异，致使部分种子已处于即将萌芽的状态，而这一生理过程是不可逆的，突然采用所设置的方法储藏，使即将萌芽的种子失去了发芽所需要的某些条件，萌发过程受到抑制，而处于萌芽状态的种子可溶性养分较丰富，易致种子变质而丧失发芽力，导致了种子储藏前与储藏后的发芽率降幅达到极显著差异。

3 结论与讨论

温度、湿度和氧气是影响种子储藏期和发芽率的主要因素。灰木莲种子是顽拗性种子，对脱水敏感，含水量过低会造成严重的脱水伤害，种子活力下降[15]，故忌干燥；温度过低也会对种子造成伤害；氧气量也是影响种子储藏期的重要因素，氧气充足，种子的呼吸作用强，种内物质转化快，可溶性养分丰富，得到的能量多，对种子储藏不利。因此，种子正常含水量、适当低温和缺氧环境是延长灰木莲种子储藏期的主要条件。

在所设置的6种处理中以A、B处理的储藏期最长，经储藏16个月后种子才完全丧失发芽力，经6个月的储藏时它的发芽率还能达到50%以上，储藏效果良好；而常温条件下，辅助其他储藏手段(C、D、E、F处理)的方法储藏期都很短，分别储藏52、52、32、64d后种子或者已经发芽或者丧失活力。温和缺氧环境是延长灰木莲种子储藏期的主要条件。在设置的6种储藏处理中，A处理基本满足灰木莲储藏的要件，故其储藏期最长，同期内的发芽率最高。B处理与A处理相近，但因在种子中混入了湿沙，储藏湿度提高，使氧更易透过种皮进入种内及二氧化碳更易排出，此外，加入湿沙增加了空间和空隙，相当于增加氧气量，它们都使种子呼吸作用加强，而致部分种子寿命缩短，故其储藏期虽与A处理一致，但同期发芽率比A处理差。而C、D、E、F四种处理的储藏要件都不全，所以效果都很差。如C、E处理为限氧，且C处理限氧程度比E稍强，所以其储藏期比E处理长些，而E基本与常规的沙藏方法相近，故其结果与常规沙藏差不多，储藏1个月左右种子基本发芽完成；而D、F处理为限湿，F处理比D处理限湿程度高，由于F处理容易失水，尤其是直接与布袋接触的种子，失水受损的概率较高，故在同等储藏期内，其种子发芽率比D处理差，但因其在储藏期内种子萌发的条件很难满足，故少量种子的储藏期比D处理长。

参考文献

[1]贾宏炎，农瑞红．灰木莲大苗移植技术[J]．广西林业科学，2006，35(1)：34-35.

[2]杨耀海，刘明义，常森有．灰木莲引种栽培试验研究[J]．西南林学院学报，2007，27(3)：29-32.

[3]杨耀海，刘明义，常森有．灰木莲引种栽培试验研究[J]．西南林学院学报，2007，27(3)：29-32.

[4]林捷，叶功富，沈德炎，等．灰木莲和子京在闽南山地的引种表现[J]．林业科技开发，2004，18(1)：18-20.

[5]ROBERTS E H. Predicting the storage life of seeds[J]. Seed Sci& Technol，1973(1)：499-514.

[6]DICKIE J B，PRITCHARD H W. Systematic and evolutionary aspects of desiccation tolerance in seeds[C]//Black M，eds. Desiccation and Survival in Plants：Drying With our Dying，Wallingford：CAB International，2002：239-259.

[7]傅家瑞．顽拗性种子的贮藏及其种质保存问题[J]．种

子，1988，7(3)：51-53.
[8]韦秀芳．探索种子发芽与贮藏的关系[J]．种子，1984，3(1)：4-5.
[9]曾冀，卢立华，贾宏炎．灰木莲生物学特性及引种栽培[J]．林业实用技术，2010(10)：20-21.
[10]蔡道雄，贾宏炎，郭文福，等．灰木莲容器苗培育基质筛选试验[J]．浙江林业科技，2006，26(5)：36-38.
[11]陈伟光，张卫强，李召青，等．木荷等6种阔叶树种光合生理特性比较[J]．广东林业科技，2010，26(4)：12-17.
[12]林沐恩．灰木莲的引种栽培效果与育苗技术[J]．引进与咨询，2005(11)：62-63.
[13]张运宏，戴瑞坤，欧世坤．灰木莲等6种阔叶树种的早期生长表现[J]．广东林业科技，2004，20(1)：43-46.
[14]杨期和，尹小娟，叶万辉，等．顽拗型种子的生物学特性及种子顽拗性的进化[J]．生态学杂志，2006，25(1)：79-86.
[15]杨丽洲，冯志坚，周兵，等．不同处理方法对短序润楠种子发芽的影响[J]．广东林业科技，2010，26(3)：55-58.

[原载：种子，2011，30(10)]

林木苗期营养诊断与施肥研究进展

陈 琳[1] 曾 杰[2] 贾宏炎[1] 徐大平[2] 蔡道雄[1]

([1]中国林业科学研究院热带林业实验中心，广西凭祥 532600；
[2]中国林业科学研究院热带林业研究所，广东广州 510520)

摘 要 林木苗期营养诊断与施肥研究对于指导培育壮苗、提高苗木质量和造林成活率至关重要。文中详细介绍了形态分析、植物组织分析、生理生化等几种主要的林木营养诊断方法在国内外林木苗期诊断试验中的应用情况，从施肥方式、施肥量以及元素配比等方面对国内外林木苗期施肥研究进行总结，并展望未来林木苗期营养诊断与施肥研究的发展方向。

关键词 林木；苗期；营养诊断；施肥

Advances in Nutrient Diagnosis and Fertilization of Tree Seedlings

CHEN Lin[1], ZENG Jie[2], JIA Hongyan[1], XU Daping[2], CAI Daoxiong[1]

(*[1]Experimental Center of Tropical Forestry, Chinese Academy of Forestry, Pingxiang 532600, Guangxi, China;*
[2]Research Institute of Tropical Forestry, Chinese Academy of Forestry, Guangzhou 510520, Guangdong, China)

Abstract: The recent research on nutrient diagnosis and fertilization of tree seedlings has shown beneficial to robust seedling cultivation, seedling quality improvement and plantation survival rate increase. In this paper, we provided an overview of the application of tree nutrient diagnosis including morphological analysis, plant tissue analysis and phisiological and biochemical analysis, summarized the current state of research knowledge regarding seedling fertilization in terms of fertilization method and amount and nutrient element ratio. Finally, the problems were pointed out which need to be further study.

Key words: tree; seedling; nutrient diagnosis; fertilization

苗木是植树造林的基础和保障，苗木质量直接影响着造林成活率、早期生长表现以及对于水分胁迫、低温和病虫害等的抗性[1-3]。然而在苗木生产过程中，人们大多凭经验施肥，导致或用量过大，或不注意肥料之间的配合，浪费现象很普遍，甚至造成“肥害”或不足而常使苗木出现生理病症。开展幼苗营养诊断和施肥研究，掌握其养分缺乏或过量时的形态、生长和生理生化反应，探明幼苗的养分需求，是制定合理施肥方案的前提，将有利于苗木的优质、增产，进而促进我国人工林的健康、快速发展。

1 林木苗期营养诊断

林木营养诊断是指通过叶片症状观察、林木生长调查、养分及生理指标测定或土壤养分分析等，对树体内的养分状况进行评价。目前，林木营养诊断方法各式各样[4,5]，而基于盆栽试验的形态分析法、植物组织分析法和生理生化分析法在国内外林木苗期诊断研究中最为常用。

1.1 形态分析法

形态分析法是根据植物的叶色、长势等外观形态变化对其营养丰缺状况进行诊断，其优点是直观、简单和快速，但具有一定的滞后性，一旦苗木表现出明显的缺素症状时，损失很难弥补，况且不同元素可能引发相似的症状，更增大了诊断的难度，降低了诊断结果的准确性，故通常与其他营养诊断方法配合使用。仲崇禄等[6]以6种森林土壤为试验基质，运用形态分析法和土壤分析法对巨桉(*Eucalyptus grandis*)苗进行了13个缺素处理的盆栽试验，揭示了巨桉苗在6种土壤上的缺素表现，并确定了这些土壤的营养限制因子。Sujatha等[7]和Jeyanny等[8]综合运用形态分析法和植物叶片分析法分别探究了柚木(*Tectona grandis*)和科特迪瓦桃花芯木(*Khaya ivorensis*)因大量或微量元素缺乏引起的各种形态、生长及养分浓度变化。Chen等[9]采用形态分析法和生理生化法探讨了大量元素缺乏对西南桦(*Betula alnoides*)幼苗生长的影响，指出缺氮、缺钾和缺镁处理下幼苗的叶片症状最明显，其叶绿素和类胡萝卜素含量均显著低于全素处理，并且运用向量分析法分析了元素间的交互作用。

1.2 植物组织分析法

植物组织分析法是根据植物养分浓度与生长之间的相互关系，将植物养分浓度作为判断植物养分状况的重要指标，主要包括临界浓度值法、综合营养诊断法和向量分析法。Shedley[10]运用临界浓度值法得出了巨桉苗在3种氮源供应下的适宜叶片氮浓度，分别为2.9%~4.7%(硫酸铵)、2.6%~4.3%(硝酸铵)和2.4%~3.3%(硝酸钙)。张旭东等[11]以珍稀乡土树种福建柏(*Fokienia hodginsii*)的1年生播种苗为研究对象，对苗期的综合营养诊断技术进行了系统研究，成功确定了综合营养诊断图解法及指数法的营养诊断标准，并验证了综合营养诊断方法不受采样时期和部位的限制。虽然综合营养诊断法强调了元素间适宜配比对苗木生长的作用，克服了临界浓度法只考虑养分浓度的缺点，但其诊断指标的确定需要耗费大量的人力和物力以获取养分和产量数据，而且无法消除稀释效应和养分平衡的影响。因此，为了克服上述2种诊断方法的不足，国外普遍使用向量分析法揭示植物体内的养分供应状况(亏缺、充足、奢侈消耗、毒害、稀释、拮抗和耗竭等)。Timmer和Armstrong[12]研究了3种基质湿度(92%，73%和57%)下美国红松(*Pinus resinosa*)容器苗的生长及养分状况变化，通过向量分析发现，虽然不同处理间幼苗养分浓度差异不显著，但低湿度水平抑制了苗木对养分的吸收，进而导致幼苗的养分含量和生物量显著下降。Miller和Hawkins[13]对内陆云杉(*Picea glauca*×*P. engelmannii*)中3个速生家系和3个慢生家系进行了氮素施肥研究，通过向量分析得出了慢生家系对氮素吸收不足、速生家系对氮素吸收过量的结论。Salifu和Timmer[14]通过设置7个氮素梯度确定了黑云杉苗(*Picea mariana*)的适宜施氮量为30~64mg/株，并通过向量分析法验证了在此施氮量范围内苗木体内养分充足。此外，向量分析法的应用有助于探讨营养元素间的交互作用。例如，Chen等[9]运用向量分析法揭示了缺氮不仅导致西南桦幼苗叶片氮浓度降低，而且引起叶片磷浓度升高，缺磷和缺镁分别引起叶片硫和钙的缺乏，而缺钾、缺钙和缺硫均导致幼苗叶片氮浓度降低，这是因为不同元素之间发生了拮抗或协同作用。

1.3 生理生化分析法

营养失调不仅会引起苗木形态、生长及养分含量发生变化，一些生理生化指标(光合色素含量、气体交换参数、氨基酸含量及形态解剖结构等)也会随之改变。因此，通过运用现代生物技术手段检测生理生化指标，即生理生化分析法正日益成为研究苗木养分平衡的重要手段。Kopriva等[15]通过测定氨基酸含量探讨了氮和硫对杂交杨(*Populus tremula*×*P. alba*，INRA Clone7171B4)硫酸盐合成的调节作用，发现氮或硫的缺乏导致与硫酸盐合成相关的一系列酶含量发生明显变化。Trubat等[16]和Gloser等[17]又通过测定根系水流导度或光合等参数，分别报道了氮或磷缺乏对黄连木(*Pistacia lentiscus*)、挪威云杉(*Picea abies*)和北美香柏(*Thuja occidentalis*)产生的一系列生理反应，从不同角度揭示了植物对养分浓度变化的响应。陈屏昭和王磊[18]对缺硫导致的脐橙(*Citrus sinensis*)叶片光合特性进行了研究，发现缺硫脐橙的光合能力降低，可能是由叶绿体发育不全或特性功能蛋白含量不足所致。理永霞等[19]研究了氮、磷、钾缺乏对桉树U6(*Eucalyptus* U6)无性系幼苗形态、生长及生理指标(叶绿素含量、光合速率及根系活力)的影响，揭示出桉树幼苗均在缺氮情况下表现最差。虽然关于营养失衡引起植物各生理生化指标发生变化的研究日益增多，但是在生产上可供进行林木营养诊断的生理生化指标尚未确立，不同树种及元素配比下的诊断标准还需要进一步探究。

2 林木苗期施肥研究

由于苗期施肥周期短、占地面积小、成本低且试验条件容易控制，因此林木苗期施肥试验不仅有助于改善苗木质量，提高育苗效率，而且对于探究养分与林木生长及生理的关系和改进营林措施具有指导作用。国内外关于林木苗期施肥的研究多见于氮、磷和钾3种元素，着重探讨施肥方式、施肥用量、元素配比等。

2.1 施肥方式

传统施肥方式通常采用一次性或衡量施肥法。

自 Ingestad 和 Lund[20] 提出植物稳态矿质营养平衡理论以来，采用何种施肥方式能有效地提高苗木的养分利用率逐渐成为人们关注的焦点，并在此基础上不断衍生出指数施肥法、修正指数施肥法以及指数营养载荷法[21]，大量有关苗木施肥方式的对比研究由此展开。Imo 和 Timmer[22] 对不同施肥方式下牧豆树(*Prosopis chilensis*)的养分浓度动态变化进行了研究，发现一次性和衡量施肥均导致苗木体内的养分浓度不断下降，指数施肥处理的苗木在施肥后期才表现出养分浓度稳定，而修正指数施肥处理的苗木养分浓度则一直保持稳定。Qu 等[23] 比较了不同施肥处理对 2 种落叶松(*Larix*)根系生长、生物量及养分动态的影响后发现，不同氮素水平对苗木的根系生长及生物量影响显著，指数施肥处理苗木的养分浓度不断升高，而常规施肥处理的则明显下降，相同施肥量情况下指数施肥比常规施肥更有利于苗木对氮素的吸收。苗木接种菌根将有助于提高苗木质量，传统施肥和除草剂的应用虽然也促进苗木的生长，但会抑制菌根的形成。Quoreshi 和 Timmer[24] 运用指数营养载荷法进行施肥试验，结果不仅增加了黑云杉苗的氮和磷含量，而且促进了菌根的形成。

我国通常采用一次性或衡量施肥法，自 20 世纪 80 年代初郑槐明和贾慧君等将植物稳态矿质营养理论和相关技术引入我国以来，相继对小叶杨(*Populus simonii*)、泡桐(*Paulownia tomentosa*)、湿地松(*Pinus elliottii*)以及西南桦等树种的幼苗进行了较为系统的指数施肥研究[25-27]。大量研究表明，遵循苗木或林木生长规律(生长模型)，按照指数模型法之类探讨施肥方案，将有助于增强施肥的科学性、合理性。

2.2 施肥用量

苗木生长量一般随施肥量的增加而增加，但当施肥量达到一定水平后，继续增加施肥量可能导致生长量下降，因此施肥临界值的确定成为合理施肥的关键。国外关于施肥用量的研究主要是围绕着氮、磷两种元素展开的。例如，Larimer 和 Struve[28] 对 9 月生的红栎(*Quercus rubra*)和红糖槭(*Acer rubrum*)进行了为期 1 年的氮素施肥试验，依据幼苗的生长表现及氮在苗木体内的分布情况，认为红栎和红糖槭的适宜施氮浓度分别为 400 和 200～400mg/L。Manter 等[29] 对 1 年生的道格拉斯杉木(*Pseudot sugamenziesii* var. *menziesii*)设置了 5 个氮素处理，施肥 4 个月后发现当施氮浓度为 560mg/L 时，叶片氮浓度可达 12.72mg/g，此时幼苗的光合速率和生物量最大。Graciano 等[30] 探讨了 3 月生的巨桉苗在 3 种不同土壤上对氮、磷的需求，施肥周期为 3 个月，研究结果表明，施氮浓度为 77mg/kg(土)时，幼苗的生物量在 3 种基质上均差异不显著；施磷浓度为 200～400mg/kg(土)时，苗木的生长表现最佳，但此施磷浓度远低于 Thomas 等[31] 得出的 1000mg/kg(土)的适宜施磷浓度，其原因在于虽然两者研究的树种相同，但其供试苗木大小、所选基质、施肥周期以及养分梯度设置均存在着差异。

我国关于施肥量的研究亦主要集中在氮和磷两种元素上。20 世纪 90 年代以来，国内林业工作者先后开展了尾叶桉(*Eucalyptus urophylla*)[32]、水曲柳(*Fraxinus mandshurica*)[33] 和落叶松(*Larix olgensis*)[34-35] 等树种的苗期施肥研究，从生长和养分吸收、光合特性以及生物量积累与分配等方面探讨了不同树种的适宜施氮量或施磷量，认为水曲柳和落叶松的适宜硝酸铵浓度均为 8mmol/L，而尾叶桉的适宜施磷浓度为 20mg/kg(土)，这些研究结果为制定合理的施肥方案奠定了基础。

2.3 元素配比

适宜的元素配比有利于促进苗木的正常生长，提高苗木产量和改善品质，同时还能明显提高肥料吸收和利用率，降低成本，减少对环境的污染。近年来，我国开展了大量关于林木苗期配方施肥的研究。例如，刘水娥等[36] 研究了不同 N，P，K 配比对马占相思(*Acacia mangium*)苗期生长的影响，得到 N，P，K 最佳配比为 190：25：160(mg/L)。曹帮华等[37] 对三倍体毛白杨(*Populus tomentosa*)苗期 N，P，K 不同配方施肥效应进行了研究，发现地径、苗高、光合和蒸腾对应的最佳配方不完全一致，基于苗木培育中地茎是最重要指标，认为较理想的施肥配方为 N(9g/株)+P_2O_5(5g/株)+K_2O(5g/株)。梁坤南等[38] 以 N，P，K，Mg，Z(沸石)作为施肥试验因子，研究了不同施肥处理对柚木苗(*Tectona grandis*)生长和干物质质量的影响，其中 N(4g/株)+P(7.5g/株)+K(0g/株)+Mg(2g/株)+Z(20g/株)是 8 个施肥处理中最好的，苗高和地径生长分别是不施肥处理的 3.58 和 2.74 倍。薛丹等[39] 通过对杨树(*Populus deltoids*'Lux'×*P. deltoides* 'Havard')扦插苗的配方施肥研究表明，施用 N(17mg/L)、P(10mg/L)及 K(8.5mg/L)对根、皮、枝、叶及总生物量的促进作用最显著，分别比对照提高了 294%、66%、96%、126%和 118%。针对不同树种确定适宜的元素配比对于进一步研制各树种的高效专用肥具有重要意义，我国南方各省(自治区、直辖市)已经在桉树[40]、杨树[41] 等树种高效专用肥的研制方面取得了卓有成效的进展。不同的是，国外单纯对于营养元素配比的研究相对较少，仅见海岸松(*Pinus pinaster*)[42]、铁杉(*Tsuga heterophylla*)[43]、垂枝桦(*Betula pendula*)和欧洲桦(*B. pubescens*)[44]等树种苗期施肥配比研究，可能因为国外对于配方肥的研制已经发展到了商品阶段。

3　问题与讨论

营养诊断方法的选择是诊断结果准确与否的关键，而苗期适当施肥不仅能够增加苗木体内的养分积累，提高造林后苗木对杂草的竞争能力，而且不会对环境造成破坏。回顾国内外关于林木苗期营养诊断与施肥研究可以看出，我国目前尚有大量主要造林树种，尤其是珍贵乡土阔叶树种尚未系统地开展过营养诊断与施肥研究，而且此领域的研究较发达国家尚有较大差距，主要体现在以下几个方面：一是多采用单一诊断方法，而且向量分析法、生理生化分析法等诊断方法的应用尚不多见；二是对不同苗木施肥方式的对比研究相对较少，指数施肥法的应用仅集中在泡桐、湿地松和西南桦等少数几个树种上；三是重视苗期施肥研究，而大多忽略了对苗木造林效果的评价。针对我国林木苗期营养诊断与施肥研究中存在的这些问题，建议加强以下几个方面的研究。

(1)基于我国目前大力发展各地乡土珍贵阔叶树种的具体情况，综合运用形态分析法、植物组织分析法及生理生化分析法广泛开展不同树种的营养诊断研究，建立健全相关诊断指标体系将是林木经营管理过程之中的重要一环。

(2)我国通常采用一次性或衡量施肥法进行苗木施肥，大大降低了苗木对养分的吸收效率，为此开展不同苗木施肥方式的对比研究十分必要。许多研究表明，指数营养载荷法有利于提高苗木造林早期对干旱、养分贫瘠以及杂草竞争等恶劣环境的适应性，进而提高造林成活率，但由于不同树种的生态生物学特性存在差异，应用指数营养载荷法得出的适宜施肥范围[45-46]、施肥效果以及施肥持续时间[47-48]也会有不同，因此加强不同树种在这些方面的研究将有利于提高施肥的有效性，在增加产量的同时节约成本，符合广大林农的生产愿景。

(3)除了地径、苗高、根体积、生物量、一级根的数量、适宜的叶片养分比值以及健康状况之外，造林后苗木的早期生长表现也是一项重要的苗木质量考核指标[49,50]。国外一直重视用造林效果来验证苗木质量[1,51]，而我国林木苗期施肥研究主要围绕如何提高苗高、地径、生物量、根体积、水分及养分状况、酶活性等形态和生理指标来提高苗木质量，忽略了苗木造林效果的研究。因此，今后在探讨各种形态及生理指标与苗木生长关系的同时，强化不同形态和生理品质的苗木与造林效果的关系研究是将来林木苗期施肥研究的一个发展方向。

参考文献

[1]CLOSE D C, BAIL I, HUNTER S, et al. Effects of exponential nutrient-loading on morphological and nitrogen characteristics and on after-planting performance of *Eucalyptus globulus* seedlings[J]. Forest Ecology and Management, 2005, 205(1-3): 397-403.

[2] FLOISTAD I S, KOHMAMN K. Influence of nutrient supply on spring frost hardiness and time of bud break in Norway spruce[*Picea abies*(L.)Karst.]seedlings[J]. New Forests, 2004, 27(1): 1-11.

[3]VAN DEN DRIESSCHE R. Effects of Nutrients on Stock Performance in the Forest[M]// Van den Driessche R (ed.). Mineral Nutritional in Conifer Seedlings. Florida: CRC Press, 1991, 229-260.

[4]陈竣，李贻铨，杨承栋．中国林木施肥与营养诊断研究现状[J]．世界林业研究，1998(3)：58-65.

[5]唐菁，杨承栋，康红梅．植物营养诊断方法研究进展[J]．世界林业研究，2005，18(6)：45-48.

[6]仲崇禄，REDDELL P. 六种土壤类型上巨桉苗缺素试验[J]．林业科学研究，1994，7(6)：704-708.

[7]SUJATHA M P. Micronutrient deficiencies in teak(*Tectona grandis*) seedlings: foliar symptoms, growth performance and remedial measures[J]. Journal of Tropical Forest Science, 2008, 20(1): 29-37.

[8]JEYANNY V, AB RASIP A G, WAN RASIDAH K, et al. Effects of macronutrient deficiencies on the growth and vigor of *Khaya ivorensis* seedlings[J]. Journal of Tropical Forest Science, 2009, 21(2): 73-80.

[9]CHEN L, ZENG J, XU D P, et al. Macronutrient dificiency symptoms in*Betula alnoides* seedlings[J]. Journal of Tropical Forest Science, 2010, 22(4): 403-413.

[10]SHEDLEY E, DELL B, GROVE T. Diagnosis of nitrogen deficiency and toxicity of *Eucalyptus globulus* seedlings by foliar analysis [J]. Plant and Soil, 1995, 177(2): 183-189.

[11]张旭东，董林水，周金星，等．珍稀乡土树种福建柏苗期 DRIS 营养诊断[J]．生态学报，2005，25(5)：1165-1170.

[12]TIMMER V R, ARMSTRONG G. Growth and nutrition of containerized *Pinus resinosa* seedlings at varying moisture

regimes[J]. New Forests, 1989, 3(2): 171-180.

[13]MILLER B D, HAWKINS B J. Nitrogen uptake and utilization by slow- and fast-growing families of interior spruce under contrasting fertility regimes[J]. Canadian Journal of Forest Research, 2003, 33(6): 959-966.

[14]SALIFU K F, TIMMER V R. Optimizing nitrogen loading of *Picea mariana* seedlings during nursery culture[J]. Canadian Journal of Forest Research, 2003, 33(7): 1287-1294.

[15]KOPRIVA S, HARTMANN T, MASSARO G, et al. Regulation of sulfate assimilation by nitrogen and sulfur nutrition in poplar trees[J]. Trees, 2004, 18(3): 320-326.

[16]TRUBAT R, CORTINA J, VILAGROSA A. Plant morphology and root hydraulics are altered by nutrient deficiency in *Pistacia lentiscus*(L.)[J]. Trees, 2006, 20(3): 334-339.

[17]GLOSER V, SEDLACEK P, GLOSER J. Consequences of nitrogen deficiency induced by low external N concentration and by patchy N supply in*Picea abies* and *Thuja occidentalis*[J]. Trees, 2009, 23(1): 1-9.

[18]陈屏昭，王磊. 缺硫对脐橙叶片光合特性和叶绿素荧光参数的影响[J]. 生态学杂志，2006，25(5)：503-506.

[19]理永霞，茶正早，罗微，等. 桉树幼苗缺素症状的研究[J]. 土壤通报，2009，40(2)：290-293..

[20]INGESTAD T, LUND A B. Theory and techniques for steady-state mineral nutrition and growth of plants[J]. Scandinavian Journal of Forest Research, 1986, 1(1-4): 439-453.

[21]TIMMER V R. Exponential nutrient loading: a new fertilization technique to improve seedling performance on competitive sites[J]. New Forests, 1996, 13(1-3): 275-295.

[22]IMO M, TIMMER V R. Vector diagnosis of nutrient dynamics in mesquite seedlings[J]. Forest Science, 1997, 43(2): 268-273.

[23]QU L, QUORESHI A M, KOIKE T. Root growth characteristics, biomass and nutrient dynamics of seedlings of two larch species raised under different fertilization regimes[J]. Plant and Soil, 2003, 255(1): 293-302.

[24]QUORESHI M, TIMMER V R. Growth, nutrient dynamics, and ectomycorrhizal development of container-grown*Picea mariana* seedlings in response to exponential nutrient loading[J]. Canadian Journal of Forest Research, 2000, 30(2): 191-201.

[25]郑槐明，贾慧君. 植物稳态矿质营养理论与技术研究及展望[J]. 林业科学，1999，35(1)：94-103.

[26]贾慧君，郑槐明，李江南，等. 稳态营养与Pt菌根化在湿地松育苗中的应用[J]. 林业科学，2004，40(1)：41-46.

[27]陈琳，曾杰，徐大平，等. 氮素营养对西南桦幼苗生长及叶片养分状况的影响[J]. 林业科学. 2010，46(5)：35-40.

[28]LARIMER J, STRUVE D. Growth, dry weight and nitrogen distribution of red oak and 'autumn flame' red maple under different fertility levels[J]. Journal of Environmental Horticulture, 2002, 20(1): 28-35.

[29]MANTER D K, KAVANAGH K L, ROSE C L. Growth response of Douglas-fir seedlings to nitrogen fertilization: importance of Rubisco activation state and respiration rates[J]. Tree physiology, 2005, 25(8): 1015-1021.

[30]GRACIANO C, GOYA J F, FRANGI J L, et al. Fertilization with phosphorus increases soil nitrogen absorption in young plants of Eucalyptus grandis[J]. Forest Ecology and Management, 2006, 236(2-3): 202-210.

[31]THOMAS D S, MONTAGU K D, CONROY J P. Leaf inorganic phosphorus as a potential indicator of phosphorus status, photosynthesis and growth of *Eucalyptus grandis* seedlings[J]. Forest Ecology and Management, 2006, 223(1-3): 267-274.

[32]徐大平. 不同磷水平对不同种源尾叶桉的生长和养分吸收的影响[J]. 热带亚热带土壤科学，1997，6(2)：76-81.

[33]吴楚，王政权，范志强，等. 不同氮浓度和形态比例对水曲柳幼苗叶绿素合成、光合作用以及生物量分配的影响[J]. 植物生态学报，2003，27(6)：771-779.

[34]郭盛磊，阎秀峰，白冰，等. 供氮水平对落叶松幼苗光合作用的影响[J]. 生态学报，2005，25(6)：1291-1297.

[35]吴楚，王政权，孙海龙，等. 氮磷供给对长白落叶松叶绿素合成、叶绿素荧光和光合速率的影响[J]. 林业科学，2005，41(4)：31-36.

[36]刘水娥，张方秋，陈祖旭，等. N、P、K营养元素不同配比对马占相思苗期生长的影响[J]. 林业科学研究，2002，15(2)：163-168.

[37]曹帮华，巩其亮，齐清. 三倍体毛白杨苗期不同配方施肥效应的研究[J]. 山东农业大学学报(自然科学版)，2004，35(4)：512-516.

[38]梁坤南，潘一峰，刘文明. 柚木苗期多因素施肥试验[J]. 林业科学研究，2005，18(5)：535-540.

[39]薛丹，陈金林，于彬，等. 杨树苗木配方施肥试验[J]. 南京林业大学学报(自然科学版)，2009，33(5)：37-40.

[40]曾令海，连辉明，何波祥，等. 广东省林科院桉树专用肥规模化推广效果评价[J]. 广东林业科技，2006，2(4)：59-62.

[41]袁巍，皮兵. 林业有机高效系列专用肥的推广与应用

[J]. 湖南林业科技, 2005, 32(2): 62-63.

[42] WARREN C R, ADAMS M A. Phosphorus affects growth and partitioning of nitrogen to rubisco in*Pinus pinaster*[J]. Tree physiology, 2002, 22(1): 11-19.

[43] HAWKINS B J, HENRY G, KING J. Response of western hemlock crosses to nitrogen and phosphorus supply [J]. New Forests, 2000, 20(2): 135-143.

[44] PORTSMUTH A, NIINEMETS U. Interacting controls by light availability and nutrient supply on biomass allocation and growth of *Betula pendula* and *B. pubescens* seedlings [J]. Forest Ecology and Management, 2006, 227(1-2): 122-134.

[45] BIRGE Z K D, SALIFU K F, JACOBS D F. Modified exponential nitrogen loading to promote morphological quality and nutrient storage of bareroot-cultured*Quercus rubra* and *Quercus alba* seedlings [J]. Scandinavian Journal of Forest Research, 2006, 21(4): 306-316.

[46] SALIFU K F, JACOBS D F, BIRGE Z K. Nursery nitrogen loading improves field performance of bareroot oak seedlings planted on abandoned mine lands [J]. Restoration Ecology, 2009, 17(3): 339-349.

[47] WAY D A, SEEGOBIN S D, SAGE R F. The effects of carbon and nutrient loading during nursery culture on the growth of black spruce seedlings: a six-year field study [J]. New Forests, 2007, 34(3): 307-312.

[48] HEISKANEN J, LAHTI M, LUORANEN J, et al. Nutrient loading has a transitory effect on the nitrogen status and growth of outplanted norway spruce seedlings [J]. Silva Fennica, 2009. 43(2): 249-260.

[49] 李国雷, 刘勇, 祝燕, 等. 国外苗木质量研究进展 [J]. 世界林业研究, 2011, 24(2): 27-35.

[50] WILSON B C, JACOBS D F. Quality assessment of temperate zone deciduous hardwood seedlings [J]. New Forests, 2006, 31(3): 417-433.

[51] IMO M, TIMMER V R. Growth and nitrogen retranslocation of nutrient loaded*Picea mariana* seedlings planted on boreal mixedwood sites [J]. Canadian Journal of Forest Research, 2001, 31(8): 1357-1366.

[原载: 世界林业研究, 2012, 25(03)]

西南桦组培苗培育的轻基质筛选

贾宏炎[1] 曾 杰[2] 黎 明[1] 蒙彩兰[1] 郭文福[1]
([1]中国林业科学研究院热带林业实验中心，广西凭祥 532600；
[2]中国林业科学研究院热带林业研究所，广东广州 510520)

摘 要 以黄心土为对照，应用3种轻基质开展了5个西南桦无性系的组培苗培育试验。结果表明：基质、无性系及其交互作用对西南桦组培苗的存活率以及绝大多数的生长指标影响显著($P<0.05$)或极显著($P<0.01$)，仅无性系间地上部分生物量差异不显著以及交互作用对地径影响不显著($P>0.05$)；各无性系对4种基质的适应性存在差异，其中B5无性系对所有基质均表现出较为一致的适应性；基质对组培苗存活的影响比对苗木生长的影响小得多，应用黄心土和轻基质培育5个无性系组培苗，其平均存活率均在85%以上，而基质间西南桦组培苗的生长表现大多差异显著或极显著；随着轻基质中沤制松树皮比例减少或炭化锯末比例增多，西南桦组培苗各生长指标大致呈现递增趋势。整体而言，应用基质A3(沤制松树皮25%+炭化锯末75%)培育西南桦组培苗效果最好，适应的无性系最多，建议生产上采用基质A3培育西南桦组培苗。

关键词 西南桦；轻基质；无性系；组培苗

Light Media Selection for Growing Micro-propagated Seedlings of *Betula alnoides*

JIA Hongyan[1], ZENG Jie[2], LI Ming[1], MENG Cailan[1], GUO Wenfu[1]
([1]*Experimental Center of Tropical Forestry, CAF, Pingxiang 532600, Guangxi, China*;
[2]*Research Institute of Tropical Forestry, CAF, Guangzhou 510520, Guangdong, China*)

Abstract: Light media selection trial was conducted for cultivating micro-propagated seedlings of *Betula alnoides*. It was indicated that the media, clones and their interactions had significant ($P<0.05$) or extremely significant ($P<0.01$) influences on survival rate and majority of growth traits of the seedlings, but the effects of differences of above-ground biomass and their interactions on ground diameter were not significant($P>0.05$). The three light media were remarkably better than soil for the growth of *B. alnoides* seedlings. The flexibilities of five clones to these light media were also quite different, for example, The clone B5 adapted to all The media. As a whole, The medium with 25 percent of pine skin and 75 percent of carbonized saw dust was the best for cultivating micro-propagated seedlings of *B. alnoides*, and was recommended to be used in the practice.

Key words: *Betula alnoides*; light growing media; clone; micro-propagated seedling

西南桦(*Betula alnoides* Buch. -Ham. ex D. Don)是我国热带、南亚热带地区的一个乡土珍贵用材树种，其适应性强、生长迅速、材质优良，深受林农、私营投资商、林业企事业单位的喜爱，已在该地区广为栽培[1]，近10余年来推广面积达8万hm^2。由于采用实生苗造林出现林分参差不齐，林木分化严重的现象，中国林业科学研究院西南桦课题组通过多年的良种选育、繁育研究，初步选育出一批优良无性系，并开发出其组培技术，2005年以来在生产上得到了初步推广，取得了良好的示范效果。

轻基质育苗以其适合工厂化生产、苗木便于运输和造林施工等特点[2]，已在南方许多树种上得以应用[3-7]。当前，国内外趋向于采用农林废弃物作为轻基质，树皮和锯末作为育苗基质在发达国家普遍应用[8-9]。笔者曾采用沤制松树皮、沤制锯末及炭化锯末等配置9个轻基质配方，开展了西南桦实生苗培育的基质选择[10]，亦应用蒙彩兰等[6]研发的轻基质培育西南桦组培苗，但是在生产中发现不同批次的西南桦组培苗质量不稳定，可能与无性系不同有关。因此，本试验[10]的研究基础上，以黄心土基质为对照，选取5个西南桦无性系开展轻基质配比研究，依据存活率、苗高、地径、生物量等指标，揭示不同轻基质配方对西南桦无性系组培苗生长的影响以及不同无性系对基质的适应性，筛选出适宜轻基质配方，为培育高产优质的无性系人工林奠定基础。

1 材料与方法

1.1 苗圃地概况

试验苗圃位于广西凭祥市中国林科院热带林业实验中心，属南亚热带季风气候，终年温暖湿润，年平均气温21.5℃，极端最高温39.8℃，最低温-1.5℃，≥10℃的积温6000~7600℃，全年日照时数1218~1620h，年降水量1200~1400mm，干湿季节明显，4~9月份为雨季，相对湿度80%。

1.2 试验设计

试验设基质和无性系2个因子，参试基质为黄心土100%(A0)、沤制松树皮75%+炭化锯末25%(A1)、沤制松树皮50%+炭化锯末50%(A2)和沤制松树皮25%+炭化锯末75%(A3)，5个无性系为Ly4、A12、白云、Q_2和102，分别用B1、B2、B3、B4和B5表示。采用随机区组设计，首先是无性系随机排列，在每个无性系内基质亦随机排列，重复4次，每个小区50株。所用轻基质材料的处理及网袋容器制作的方法参照文献[6]进行，西南桦无性系试管苗由本单位组培室提供。

1.3 苗期管理及生长指标观测

2010年4月，将试管苗炼苗18d后，用清水将其培养基洗净，选用高度基本一致的幼苗，移栽到轻基质网袋内，及时浇定根水，保持基质湿润。每周施尿素或复合肥1次，浓度为2000~4000mg/L，视苗木生长情况逐渐加大施肥浓度，其他管护措施同一般生产性育苗。8月初调查苗木移植的存活率。每个小区随机抽取30株苗木调查苗高、地径，根据苗木的平均高和地径，选取5株苗木测定及生物量。

1.4 数据处理与分析

采用SPSS13.0进行数据处理和分析，包括双因素方差分析、单因素方差分析以及Duncan多重比较。依据上述生长指标，应用布雷金多性状综合评定法对各基质进行综合评价[11]。

2 结果与分析

2.1 基质对西南桦组培苗移植存活率的影响

由表1可以看出，各基质处理以及B1、B2、B3等3个无性系的西南桦幼苗平均存活率均在90%以上，B4、B5无性系的平均存活率亦超过85%。双因素方差分析结果表明，基质、无性系及其交互作用对西南桦苗木存活率影响均极显著($P<0.01$)。进一步对每个无性系各基质处理的苗木存活率进行单因素方差分析得出，对于同一无性系，不同基质对苗木存活率的影响各异。对于无性系B1和B5而言，参试基质处理间存活率差异不显著($P>0.05$)；而对于B2、B3和B4来说，基质处理间差异显著($P<0.05$)。多重比较分析发现，对于无性系B2，A0、A2和A3基质处理间苗木保存率差异不显著，而均显著高于A1；对于无性系B3，A0、A1和A2基质处理间差异不显著，而均显著高于A3；而对于无性系B4，A2显著高于A0和A1，而A3则与这些处理间差异不显著。

表1 不同基质处理对西南桦不同无性系组培苗存活率的影响 (%)

基质	无性系				
	B1	B2	B3	B4	B5
A0	91.0 (4.397)	94.0 (2.582)a	95.5 (2.082)a	85.0 (5.944)b	90.0 (5.164)
A1	95.0 (1.826)	90.0 (3.266)b	96.0 (1.633)a	85.5 (5.323)b	90.5 (5.323)
A2	95.5 (2.082)	96.0 (1.826)a	97.0 (0.816)a	96.0 (0.816)a	87.5 (4.203)
A3	92.5 (2.887)	97.0 (1.826)a	90.5 (3.697)b	90.5 (3.697)ab	84.0 (1.633)

注：括号内的数值为标准差。同列有小写字母相同或未标注字母表示差异不显著，不相同字母表示差异显著($P<0.05$)。

2.2 基质对西南桦组培苗生长的影响

通过双因素方差分析得出，基质处理间西南桦苗高、地径和生物量均差异极显著($P<0.01$)；无性系间地径、地下部分生物量差异极显著，苗高差异显著($P<0.05$)，地上部分生物量差异不显著($P>0.05$)；基质和无性系的交互作用则地径差异不显著，苗高和地上部分生物量差异显著，地下部分生物量差异极显著。单因素方差分析结果表明，对于同一无性系，基质处理间苗高、地径和地上部分生物量均差异显著或极显著，而基质对地下部分生物量的影响则因无性系而异。

从图1可以看出，对于所有无性系而言，3个轻基质处理的地径、苗高、地上部分生物量均显著高于黄心土(A0)。如，应用A3基质培育B1、B2和B4无性系幼苗，其苗高分别比黄心土提高2.02~2.18倍，地上部分生物量亦提高2.55~3.11倍。3个轻基质处理间比较，地径差异不显著，苗高和地上部分生物量的差异则因无性系而异。对于B1和B2，A3基质处理的苗高和地上部分生物量显著高于A1和A2；对于B3，A2和A3显著高于A1；对于B4，A3显著高于A1，两者与A2差异不显著；对于B5，3个轻基质处理间差异不显著。

而基质对地下部分生物量的影响小得多。对于无性系B4和B5，所有基质间地下部分生物量差异不显著；无性系B1的地下部分生物量在黄心土与3个轻基质处理间差异不显著，A2显著低于A1和A3；对于无性系B2，黄心土处理的地下部分生物量显著低于A3，与A1和A2差异不显著；无性系B3的地下部分生物量在黄心土、A2和A3间差异不显著，而三者显著高于A1。

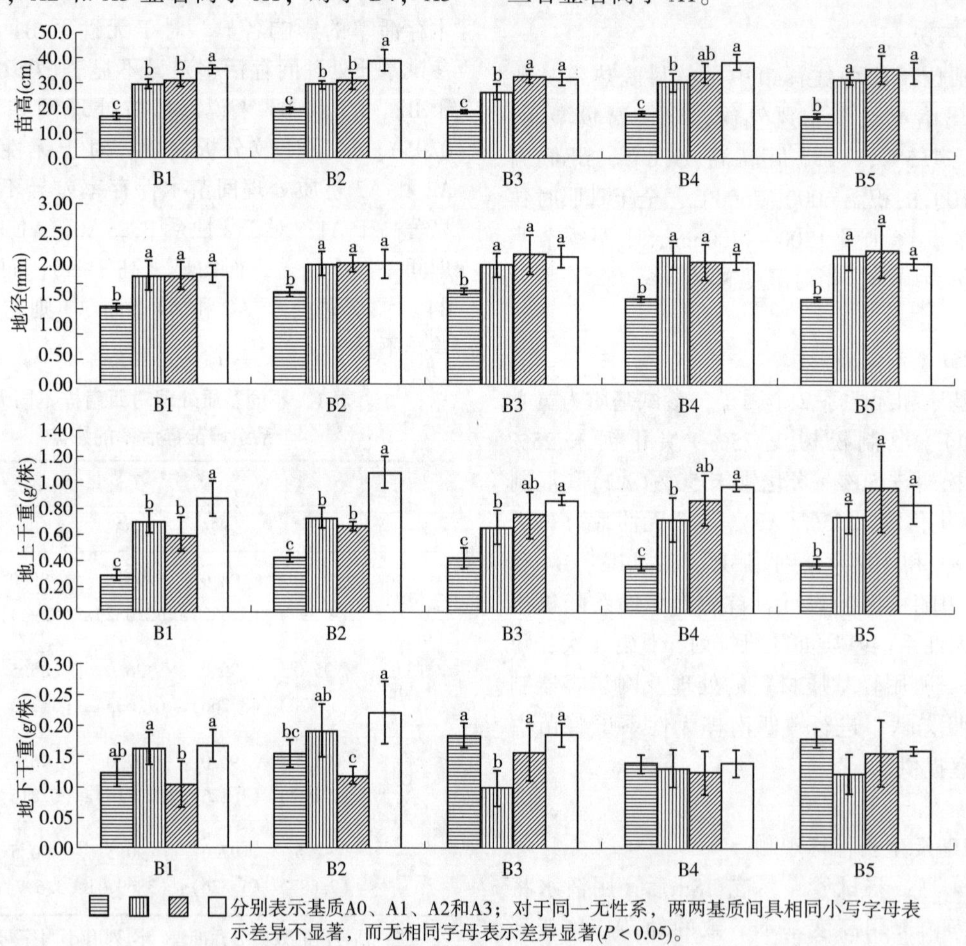

分别表示基质A0、A1、A2和A3；对于同一无性系，两两基质间具相同小写字母表示差异不显著，而无相同字母表示差异显著($P<0.05$)。

图1　西南桦无性系在各基质中生长的比较

2.3　基质间组培苗培育效果的综合比较

依据西南桦组培苗存活率和生长性状指标，应用布雷金综合评定法对每个无性系的各基质处理进行综合评价。具体方法是：首先在每个无性系4种基质处理中找出每个性状的最大值X_{ijmax}(i和j分别表示性状和无性系，均为1，2，3，4，5)，然后将该性状在所有基质处理中相应的数值X_{ijk}(k表示基质，为1，2，3，4)除以X_{ijmax}得到Y_{ijk}，求算各性状此值之和的平方根即为综合评价值$A_{jk}=\sqrt{\sum_{i}^{5} Y_{ijk}}$。$A_{jk}$愈大说明应用基质$k$培育无性系$j$组培苗的效果愈好，愈小则育苗效果愈差。由表2可知，对于所有无性系而言，黄心土基质的育苗效果最差，而轻基质的育苗效果因无性系而异。其中，无性系B1、B2、B3和B4的育苗效果均以A3基质为最好，其次为A2和A1；B5则以A2为最好，其次为A3和A1。由此可见，基质A3适应于较多西南桦无性系组培苗生长，且育苗综合效果最好，A2次之，A0效果最差。

表 2 西南桦无性系在不同基质中生长的综合评价(布雷金综合评定法)

无性系	基质	存活率(%)	苗高(cm)	地径(mm)	地上干重(g/株)	地下干重(g/株)	求和	求和的平方根	排名
B1	A0	0.953	0.458	0.709	0.325	0.731	3.175	1.782	4
	A1	0.995	0.833	0.989	0.798	0.976	4.591	2.143	2
	A2	1.000	0.870	0.984	0.682	0.617	4.152	2.038	3
	A3	0.969	1.000	1.000	1.000	1.000	4.969	2.229	1
B2	A0	0.969	0.495	0.761	0.396	0.701	3.322	1.823	4
	A1	0.928	0.749	0.995	0.676	0.869	4.217	2.054	2
	A2	0.990	0.791	1.000	0.616	0.543	3.940	1.985	3
	A3	1.000	1.000	0.995	1.000	1.000	4.995	2.235	1
B3	A0	0.985	0.566	0.721	0.483	0.989	3.744	1.935	4
	A1	0.990	0.814	0.921	0.759	0.532	4.016	2.004	3
	A2	1.000	1.000	1.000	0.870	0.839	4.709	2.170	2
	A3	0.933	0.968	0.986	1.000	1.000	4.887	2.211	1
B4	A0	0.885	0.461	0.667	0.373	0.993	3.379	1.838	4
	A1	0.891	0.790	1.000	0.734	0.936	4.351	2.086	3
	A2	1.000	0.891	0.953	0.887	0.886	4.617	2.149	2
	A3	0.943	1.000	0.958	1.000	1.000	4.900	2.214	1
B5	A0	0.994	0.472	0.644	0.398	1.000	3.509	1.873	4
	A1	1.000	0.886	0.964	0.764	0.683	4.297	2.073	3
	A2	0.967	1.000	1.000	1.000	0.867	4.834	2.199	1
	A3	0.928	0.959	0.905	0.864	0.894	4.551	2.133	2

3 结论与讨论

本研究表明，基质间西南桦组培苗的存活率和生长指标均存在极显著或显著差异。相对而言，基质对组培苗存活的影响比对幼苗生长的影响小得多。无论是黄心土还是轻基质，5 个参试无性系的组培苗存活率基本上超过 85%，而且 B1 和 B5 两个无性系组培苗的存活率在基质间差异不显著；而基质间各无性系组培苗的生长表现大多差异显著或极显著。采用 3 种轻基质培育 5 个西南桦组培苗的整体效果均显著优于黄心土，与西南桦喜透水性和透气性强的基质或土壤有关[12]。良好的通气性有利于根部的气体交换，对于壮苗培育至关重要[8]。黄心土相对容易板结，尽管在本试验中对幼苗存活率影响较小，但不利于西南桦幼苗生长，而 3 种轻基质由树皮和炭化锯末配置而成，排水和透气性良好，有利于幼苗生长[9]。

5 个西南桦无性系间除了地上部分生物量差异不显著，其他指标存在极显著或显著差异。基质和无性系的交互作用除了地径差异不显著外，其余指标均差异极显著或显著，说明参试的 5 个无性系对各基质的适应性不同，其中无性系 B5 对黄心土基质和 3 种轻基质的适应性较为一致，换言之，其组培苗培育对基质的适应性较广。

从整体上看，应用基质 A3(沤制松树皮 25%+炭化锯末 75%)培育西南桦组培苗效果最好，适应的无性系最多，其次为基质 A2(沤制松树皮 50%+炭化锯末 50%)，二者均明显优于 A1(沤制松树皮 75%+炭化锯末 25%)，与笔者以往培育西南桦实苗的基质筛选结果略有不同[10]，后者以 A2 为最佳基质，其次为 A3，可能与组培苗和实生苗的差异有关。从轻基质材料来源和加工难易程度等方面考虑，锯末较松树皮来源广泛，加工简便且成本较低；因此，建议生产上采用基质 A3 培育西南桦组培苗。

参考文献

[1]曾杰，郭文福，赵志刚，等．我国西南桦研究的回顾与展望[J]．林业科学研究，2006，19(3)：379-384.

[2]张建国，王军辉，许洋，等．网袋容器苗育苗新技术[M]．北京：科学出版社，2007.

[3]蔡道雄，贾宏炎，黎明，等．非洲桃花心木轻基质容器培育试验[J]．林业实用技术，2007，3：18-20.

[4]何贵平，麻建强，冯建民，等．珍贵用材树种柏木轻基质容器育苗试验研究[J]．林业科学研究，2010，23(1)：134-137.

[5]贾宏炎，黎明，郭文福．马尾松和湿加松轻基质网袋容器育苗试验[J]．林业科技，2009，34(2)：16-18.

[6]蒙彩兰，黎明，郭文福．西南桦轻基质网袋容器育苗技术[J]．林业技术开发，2007，21(6)：104-105.

[7]尹晓阳，李德芬，金天喜，等．云南樟、刺槐不同基质容器育苗比较试验[J]．山地农业生物学报，2003，22(2)：122-126.

[8]LANDIS T D. Containers and growing media [M]. The Container Tree Nursery Manual (Vol.2), Agriculture Handbook, 674, Washington DC: USDepartment of Agriculture, Forest Service, 1990, 41-85.

[9]RILEY L E, STEINFELD D. Effects of bare root nursery practices on tree seedling root development: an evolution of cultural practices at J. Herbert Stone nursery [J]. New Forests, 2005, 30: 107-126.

[10]郭文福，曾杰，黎明，等．西南桦轻基质网袋容器苗基质选择试验[J]．种子，2010，29(10)：62-64.

[11]周永学，苏晓华，樊军锋．引种欧洲黑杨无性系苗期生长测定与选择[J]．西北农林科技大学学报(自然科学版)，2004，32(10)：102-106.

[12]郑海水，曾杰，翁启杰．西南桦育苗基质选择试验初报[J]．林业科技通讯，1998，10：23-25.

[原载：林业科学研究，2012，25(02)]

追肥量、时间与次数对柚木幼林生长的影响

李运兴[1] 蔡道雄[1] 欧耀成[1] 劳庆祥[1] 陆 毅[1] 窦福元[2]
([1]中国林业科学研究院热带林业实验中心，广西凭祥 532600；
[2]玉林市林业种苗管理站，广西玉林 537000)

摘 要 为掌握追肥量、时间与次数对柚木生长的影响，在2年生柚木纯林和柚木格木混交林中，布设了肥种为复混肥(N_{10}-P_{10}-K_8)、3因素(追肥量、时间与次数)、8处理的施肥试验。试验结果显示：3年生和4年生时追肥量分别以1000g/株·年和2000g/株·年为最好，低于此量时树高、胸径生长量随追肥量的增加而增加，高于此量，树高、胸径生长量都有所减少；施肥时间5月树高、胸径生长量好于6或7月；施肥次数2次树高、胸径生长量大于1次，不同地块施肥效果不同；纯林与混交林施肥效果没差异。

关键词 柚木；追肥量；时间；次数；生长；立地

The Effects of Fertilizing, Time and Times on Growth of Young Tree of *Tectona grandis*

LI Yunxing[1], CAI Daoxiong[1], LAO qingxiang[1], OU Yaocheng[1], LU Yi[1], DOU Fuyuan[2]
([1]*The Experimental Centre of Tropical Forestry*, *Chinese Academy of Forestry*, *Pingxing* 532600, *Guangxi*, *China*;
[2]*Yulin Forestry Seedling Management Station*, *Yulin* 53700, *Guangxi*, *China*)

Abstract: In order to master the effects offertilizing, time and times on growth of *Tectona grandis*, we set the experiments of compounded fertilizer(N_{10}-P_{10}-K_8), 3 factors(fertilizing , time, times) and 8 treatments during pure forest of 2a of *Tectona grandis* and mingled forest of *Tectona grandis* and *Erythrophloeum fordii* . The results showed that: the best amounts of fertilizing were 1000g/plant per year of 3a and 2000g/plant per year of 4a. When it's lower than this level, the increments of height and DBH will be increased by the addition of fertilizing, the increments of height and DBH will be reduced when it's higher than this level, the increments of height and DBH when fertilized in May were better than that in June or July, the increments of them when fertilized for two times were larger than one time, different sites had different effects of fertilization; There are no differences of effects of fertilization between pure forest and mingled forest.

Key words: *Tectona grandis*; fertilizing; time; times; growth; site

柚木(*Tectona grandis*)属马鞭草科落叶或半落叶大乔木，树高40m，胸径1~2m，干形通直。木材为环孔材，心材大，淡褐色，容重0.6~0.72g/cm³，纹理直，结构细致而美观，坚韧而有弹性，不翘不裂；含油质，强度大，耐浸，是航海、军需、建筑、车厢、家具、雕刻、铸造木模、贴面板等珍贵优良用材树种之一[1-2]。原产于印度、缅甸、泰国和印度尼西亚，我国引种于广东、广西、海南、云南、福建、台湾等地，已有170多年历史；当前可用木材资源锐减，社会需求增加，交易价格攀升，造林积极性不断提高，发展势头大。由于柚木天然和资源分布等方面原因，以往柚木方面的研究侧重于种质资源引进和种源家系选育方面[3-8]，在栽培施肥方面的研究较少，梁坤南[9]柚木幼苗期施肥、潘一峰[10]柚木幼林施肥试验结果认为：施用复合肥或含N、P、K的配方肥都可以大幅度提高树高和胸径的生长量，但除肥料种类外，施肥时间和次数也是影响肥效的重要因子，由于土壤系肥力的一种易变体，不同区域土壤气候条件相差甚大，为了解本地区柚木人工幼林对施肥量的需求和施肥时间、次数对柚木生长的影响，开展了本试验，其结论对柚木的集约经营有较大的参考和启示作用。

1 材料和方法

1.1 试验地概况

施肥试验布设于青山实验场场部附近，平而河畔、阶地上，近旁有水泥道路，可及度极高，施工十分便利，系较典型的四旁地；地形开阔，阳光充足，海拔150m，年平均气温为21.5℃，≥10℃积温7500℃，月平均气温≥22℃有7个月，最热月(7月)平均气温28.3℃，极端最高温40.3℃，最冷气温13.5℃，极端最低温-1.5℃，年降水1220~1380mm，土壤为砂泥岩发育的红赤红壤，质地为轻黏土，0~40cm混合土样，有机质23.63g/kg，pH(H_2O)4.97，全N值1.45g/kg、全P值0.81g/kg，全K值20.5g/kg，碱解N值98.0mg/kg、速效P值0.95mg/kg、速效K值51.8mg/kg，盐基饱和度32.28%，前茬为丢荒地，主要植被有：山芝麻(*Helicteres angustifolia* L.)、白茅[*Imperatam cylindrica*(L.)Beauv.]，五节芒[*Miscanthus floridulus*(Labill.)Warb.]，马缨丹(*Lantana camara* L.)等，主要布设有柚木纯林和柚木+格木混交林两种林分，株行距：3m×3m，造林密度：75株/亩，混交林按1∶1列间混交方式进行。2007年12月林地清理挖穴整地，2008年1月施基肥，2008年4月2日用1年生柚木和格木实生苗造林，柚木、格木造林成活率分别为93.8%和87.3%，造林后头2年每年在4月和7月松土1次，全铲草2次，第1年7月追肥1次，第2年追肥5月、7月追肥2次，柚木、格木保存率分别为：91.9%、85.4%；2年生柚木、格木平均树高、胸径分别为：4.2m、1.64m、4.0cm、1.46cm。

1.2 对象及设计

2年生柚木幼林，造林前施(含P_2O_5为12.0%)磷肥作基肥：1000.0g/株，1年生林追施配方肥($N_{10}-P_{10}-K_8$)：50.0g/株；2年生林追施配方肥($N_{10}-P_{10}-K_8$)：500.0g/株；3年和4年生时按表1处理进行追肥，因素1，肥种：配方肥($N_{10}-P_{10}-K_8$)；因素2，肥量：3水平；因素3，时间：4或5月和6或7月2水平；因素4，施肥次数，2水平，共8处理(小区)，具体见下表1所列，每小区株数：5(横)×7(纵)=35株，6次重复，按随机区组试验设计方法实施。

表1 不同施肥处理

处理代号	A	B	C	D	E	F	G	H
3年生施肥量(g)	500.0	1000.0	500.0	1000.0	250.0+250.0	500.0+500.0	750.0+750.0	0
4年生施肥量(g)	1000.0	2000.0	1000.0	2000.0	500.0+500.0	1000.0+1000.0	1500.0+1500.0	0
月份	4或5(1次)		6或7(1次)		4或5(1次)+6或7(1次)			不施肥

1.3 施肥方法和调查统计

按照试验处理，施肥前先进行铲草松土，松土范围：内径0.60m外径1.00m圆环内面积，松土深0.30m，在松土地内开半月形环型沟(深0.15m、宽0.20m)进行覆土式施肥；在布设试验时和施肥后每月底对树高(H)、胸径(D)、地径指标进行测定；用SPSS 17.0软件进行相关统计分析。

2 结果与分析

2.1 施肥对树高、径生长的影响

2.1.1 不同处理

经测定和统计各处理树高结果如表2所列，树高均值(H均值)3年生时大小依次为：F>G>E>D>B>H(对照)>A>C，4年生时，大小依次为：B>F>D>A>E>C>G>H(对照)，树高生长量，3年生时大小排列顺序为：F>E > D > G > B > A > C > H(对照)，4年生时，大小依次为：B>C>A>D>F>E > H > G，方差分析显示，树高均值、生长量处理间差异显著($P<0.05$)；胸径均值3年生时大小依次为：F>D>G>E>B>A>H>C，4年生时，大小依次为：F>B>D>G>C>E>A>H(对照)，胸径生长量3年生时大小依次为：F>E>G>B>A>H=D>C，4年生时，大小依次为：F>B>C>D>G>A>E>H，方差分析显示，胸径均值、生长量处理间差异显著($P<0.05$)；由于均值指标没减除施肥前本底，其反映的结果未能真实体现处理间的实际差异，生长量指标扣减施肥前本底，较好地表达处理间差异，对树高、胸径生长量指标统计结果显示，施肥能较大程度地增加树高胸径生长，在3年生与4年生时增长幅度分别可达：18.6%、17.8%和21.1%、90.7%。不同施肥措施的施肥效果不一：恰当的措施(F处理)效果较好，措施不当(G处理)肥效降低甚至有害。

表 2 不同施肥处理树高、胸径统计

林龄(年)	指标	处理							
		A	B	C	D	E	F	G	H
3/4	H 均值(m)	5.52/7.04	5.6/7.46	5.37/6.94	5.74/7.16	5.84/6.96	6.17/7.36	5.9/6.90	5.56/6.57
	H 生长量(m)	1.74/1.52	1.75/1.86	1.68/1.57	1.85/1.42	1.87/1.12	1.98/1.19	1.77/1.00	1.67/1.01
	D 均值(cm)	5.79/7.33	5.87/7.90	5.62/7.52	6.12/7.88	5.98/7.40	6.33/8.37	6/7.58	5.76/6.82
	D 生长量(cm)	1.78/1.54	1.81/2.03	1.45/1.90	1.66/1.76	1.92/1.42	2.01/2.04	1.82/1.58	1.66/1.06

表 3 不同施肥时间、次数树高、胸径

林龄(年)	指标	施肥次数		
		0	1	2
3/4	H 均值(m)	5.56/7.17	5.56/7.16	5.97/7.67
	H 生长量(m)	1.67/1.61	1.75/1.60	1.87/1.70
	D 均值(m)	5.76/7.57	5.85/7.66	6.09/8.08
	D 生长量(m)	1.66/1.81	1.68/1.81	1.92/1.99

表 4 2 年不同月份气象因子统计

指标	月份											
	1	2	3	4	5	6	7	8	9	10	11	12
月均温(℃)	15.5	14.3	18.3	21.7	25.7	27.4	28.1	27.3	25.3	23.8	19.6	11.9
月降水(mm)	29.8	73.3	50.9	187.0	224.2	118.8	194.2	241.8	243.8	145.5	153.8	3.0
日照(h)	78.8	64.8	61.1	95.0	170.7	160.0	201.8	168.9	180.4	172.5	143.3	113.9

2.1.2 不同施肥时间、次数

由于不同时期的气候情况差别明显，柚木生长又需要有一个适宜的气候条件，不同时节其生长组织活跃程度不一，树高、胸径生长差异大，选择一个合适的时间施肥尤为必要，表 2 中 3 年生时 A 处理的树高生长量、胸径生长量都较 C 处理大，B 处理的胸径生长量都较 D 处理大，4 年生时 B 处理的树高生长量、胸径生长量都较 D 处理大，比较可见：在 4 或 5 月份施肥效果好于 6 或 7 月，这是由下述原因引起的，5 月均温(表 4)已大于 24.0℃一般植物组织活跃所需温度，在降水、光照保证的条件下，各器官生长趋于旺盛，此时施肥，可以及时提供充足养分，4、5 月比 6、7 月份施肥，养分充足生长期长，高、径生长量自然就大。

施肥次数的多少，不仅关系到营林成本的高低，而且对林木的生长可能存在直接影响。表 3 为按施肥次数统计的试验结果，结果显示在肥量相同的情况下，2 次施肥的树高均值、生长量、胸径均值、生长量都好于 1 次施肥，这是因为 1 次施入大量肥后，林木不能充分吸收利用，部份肥料被流失所致。

2.1.3 不同施肥量

为比较不同施肥量对树高生长的影响，按照施肥量的大小对试验数据整理统计得表 5，从树高均值指标看，3 年生时，施肥量越大树高就越大，即：1500g 处理 > 1000g 处理 > 500g 处理 > 不施肥处理，4 年生时，施肥量以 2000g/株处理为最大，在小于此量时，树高均值随着施肥量的增加而增加，超过此量时，树高均值则减少。树高生长量指标，并不是随着施肥量的增加而增大，而是在 3 年生时以 1000g 处理、4 年生时以 2000g 为最好，在小于此量时，树高生长量随着肥量的增加而增加，超过此量时，树高生长量则减少。胸径均值、生长量指标：在 3 年生时 1000g 处理 > 1500g 处理 > 500g 处理 > 不施肥处理，在 4 年生时 2000g 处理 > 3000g 处理 > 1000g 处理 > 不施肥处理，变化趋势与树高指标相一致：在 3 年生时、4 年生时分别以 1000g 和 2000g 处理为最好，在小于此量时，胸径均值和生长量随着肥量的增加而增加，超过此量则减少。

表5 不同施肥量树高、胸径统计

林龄(年)	指标	施肥量(g)			
		0+0	500+1000	1000+2000	1500+3000
3/4	树高均值(m)	5.56/6.98	5.58/7.03	5.84/7.33	5.9/6.90
	树高生长量(m)	1.67/1.42	1.76/1.45	1.86/1.49	1.77/1.00
	胸径均值(m)	5.76/7.42	5.8/7.57	6.11/8.04	6.02/7.58
	胸径生长量(m)	1.66/1.66	1.72/1.77	1.83/1.93	1.82/1.56

2.2 不同地段(重复)和林分类型施肥效果

不同地段立地因子差别较大，影响着林木的生长，施用配方肥后增加的仅是土壤的N、P、K含量；在不同地段施入同一种配方肥，施肥效果可能不同，为此很有必要进行不同地段的施肥效果分析，表6为6个施肥地段施肥后的树高、胸径均值和生长量，表7为最优施肥处理(F)6个施肥地段施肥后的树高、胸径均值和生长量；表7数据反映出，同一施肥措施在不同的施肥地段其施肥效果不同，1地段的施肥效果好于其他地段，地段间树高生长量、胸径生长量差异显著($P<0.05$)；表6数据显示出纯林与混交林柚木的树高生长量、胸径生长量差异不显著($P>0.05$)。

表6 不同地段树高、胸径情况

林龄(年)	指标	重复					
		纯林			混交林		
		1	2	3	4	5	6
3/4	树高均值(m)	7.45/8.82	4.47/5.87	5.51/6.78	5.38/6.54	5.08/5.73	6.38/7.67
	树高生长量(m)	1.79/1.37	1.53/1.40	1.69/1.27	1.72/1.16	1.67/0.65	2.16/1.29
	胸径均值(m)	7.42/10.14	5.34/5.51	5.61/7.15	5.63/7.18	5.17/6.08	6.40/7.98
	胸径生长量(m)	2.11/2.72	1.30/0.17	1.50/1.54	1.68/1.55	1.67/0.91	2.31/1.58

表7 最优施肥处理不同地段树高、胸径

林龄(年)	指标	重复					
		1	2	3	4	5	6
3/4	树高均值(m)	7.70/9.83	6.65/7.65	5.86/7.16	5.34/6.38	5.33/6.26	7.71/8.38
	树高生长量(m)	2.13/2.13	2.00/1.00	1.68/1.30	1.42/1.04	2.13/0.93	2.53/0.67
	胸径均值(m)	7.65/11.50	5.40/6.95	6.16/8.68	5.28/7.48	5.57/7.44	6.97/8.38
	胸径生长量(m)	2.38/3.85	1.70/1.55	1.76/2.52	1.52/2.20	2.27/1.87	2.43/1.41

3 结论与思考

恰当的施肥能增加柚木幼树的树高和胸径生长，施肥效果的大小受影响于施用肥料的种类、施肥量、施肥时间、施肥次数的多少，此外，立地因素也对肥效作用明显。

肥料的种类对施肥的效果起着决定性的作用，大多数植物对大量元素N、P、K等的需求[11]较大，在试验中柚木也表现出相一致的情况，在造林前后1~4年，通过施用磷肥和配方肥，树高胸径生长量都较传统粗放经营(不施肥)有较大幅度的增加；据相关研究认为，柚木除对N、P、K大量元素有需求外，对钙、镁等中量元素和微量元素也有较高需要[12,13]，因此柚木施肥除需进行N、P、K大量元素方面的试验研究外，尚有必要在不同立地进行添加钙、镁等元素方面的施肥研究，其研究结论将对柚木的种植培育有较大意义。

施肥量、施肥时间、施肥次数系决定施肥效果的重要因素[10]，适量施肥效果最好，肥量不足，效果差，过量施肥则会产生肥害；施肥时间宜选择在柚木生长开始进入活跃期时及时进行；一次性大量施入，土壤养分没能被及时吸收利用，会造成流失浪费，降低肥效。

立地包括了气候因素和土壤因素，雨后土壤湿润时施肥效果好于干燥时[10]；地段不一样，土壤养分差别较大，养分充足和过低施肥肥效果都差，此外土壤物理性质也系影响肥效的重要原因，因此应做到因地施肥。造林后柚木树高、胸径生长量大于格木，柚木纯林与柚木格木混交林不同施肥处理的柚木树高、胸径生长量差异不显著，说明，光照强度不是影响柚木幼林(郁闭前)生长的主要原因，混交树种格木与柚木的种间关系比较协调，不会出现偏害竞争。

参考文献

[1]广西林业局，广西林学会．阔叶树种造林技术[M]．南宁：广西人民出版社，1980.

[2]中国树木志编委会．中国主要造林树种造林技术[M]．北京：中国林业出版社，1981.

[3]邝炳朝，郑淑珍，罗明雄，等．柚木种源主要性状聚合遗传值的评价[J]．林业科学研究，1996，9(1)：7-14.

[4]李运兴．柚木家系试验[J]．广西林业科学，2001，30(1)：50-52.

[5]梁坤南，白嘉雨，周再知，等．珍贵树种柚木良种繁育发展概况[J]．广东林业科技，2006，22(3)：85-89.

[6]马华明，梁坤南，周再知．我国柚木的研究与发展[J]．林业科学研究，2003，16(6)：768-773.

[7]张荣贵，蓝猛，乔光明，等．红河州柚木种源试验五年评价[J]．林业科学研究，1999，12(2)：190-196.

[8]梁坤南，赖猛，黄桂华，等．10个柚木种源27年生长与适应性[J]．中南林业科技大学学报，2011，31(4)：8-12.

[9]梁坤南，潘一峰，刘文明．柚木苗期多因素施肥试验[J]．林业科学研究．2005，6(5).

[10]潘一峰，刘文明．酸性土壤改良对不同种源的柚木生长的影响[J]．热带亚热带土壤科学，1997，6(1)：9-14.

[11]沈其荣．土壤肥料学通论[M]．北京：高等教育出版社，2001.

[12]周再知．酸性土壤柚木钙素营养研究[D]．中国林业科学研究院，2009.

[13]周再知，梁坤南，徐大平，等．钙与硼、氮配施对酸性土壤上柚木无性系苗期生长的影响[J]．林业科学，2010，46(5)：102-108.

[14]李娜，曹继钊，唐黎明，等．不同施肥方式和施肥量对桉树生长量影响初探[J]．广西林业科学，2009，38(2)：102-106.

[原载：中南林业科技大学学报，2012，32(09)]

西南桦优树选择技术研究

刘光金 谌红辉 郭文福 贾宏炎 马 跃

（中国林业科学研究院热带林业实验中心，广西凭祥 532600）

摘 要 以胸径、树高、材积为数量评价指标，结合干形、冠高比和分枝粗细形质指标，采用5株优势木对比法开展了西南桦人工林的优树选择技术研究。通过对西南桦人工林10~21年生不同林龄阶段的优树选择标准的总结，提出了优树选择标准，即优树胸径、树高、单株材积分别大于5株对比优势木平均值的9%~11%、5%~7%、26%~31%以上，形质指标综合得分大于7.5，共选择出西南桦优树37株，入选率为33.9%。西南桦优树的选择标准在实际应用中可根据林分状况适当调整，选优既要考虑其生长性状，同时又要注意材性、抗性的选择以保证西南桦育种群体的遗传多样性。

关键词 西南桦；人工林；优树；优势木；选择标准

Study on Criterion for Superior Trees of *Betula alnoides*

LIU Guangjin，CHEN Honghui，GUO Wenfu，JIA Hongyan，MA Yue

（*Experimental Center of Tropical Forestry*，*Chinese Academy of Forestry*，*Pingxiang* 532600，*Guangxi*，*China*）

Abstract：Based on quantity indexes（DBH，tree height and individual volume）in combination with quality indexes（stem form，crown/height ratio and branch size），the selective criteria of superior trees for *Betula alnoides* Buch. -Ham. Ex D. Don was established using a method of five dominant trees. By analysing the superior tree selective criteria for *B. alnoides* plantation during different age stages ranged from 10 to 21 years，we proposed that the strategy of superior tree selection for this species involved as follows：the DBH，tree height，and individual volume of superior tree were 9-11%，5-7% and 26-31%，respectively greater than that of five dominant trees，and the multiple shape score was more than 7.5. Thirty seven in 109 candidate trees were selected according to this criteria and the selected ratio was 33.9%. The superior tree selective criteria should be modified in the practice according to different stand conditions. Since the *B. alnoides* plantation were managed for large-size timber production，the growth index should be regarded as the principal factor for selection of superior trees so as to obtain high genetic gain. On the other hand，the wood property and resistance also should be considered to insure the genetic diversity of *B. alnoides* breeding population.

Key words：*Betula alnoides*；plantation；superior trees；dominant trees；superior trees selected critetia

西南桦（*Betula alnoides* Buch. -Ham. ex D. Don）为桦木科（Betulaceae）珍贵乡土速生用材树种，适应性强、生态效益好，广泛分布于我国热带、南亚热带地区，广泛用于高档建筑和高档家具制作[1]。随着西南桦人工林的规模化发展，良种壮苗成为亟待解决的问题。目前，西南桦种苗繁育的嫁接、扦插技术与无性系组培工厂化育苗技术均已成熟[2-6]，如果能选择出优良的种质资源，短期内即可应用于营林生产。为了获取优良的种质资源，必须开展表型优异的优树选择技术研究。对于入选的优树，除开展种子园营建与家系评价外，如利用光照与激素刺激树干基部，还能获取复壮能力较好的叶芽，易于进行无性系的组培快繁与利用。

优树选择最理想的林分是树龄一致，立地条件相同，性状已经充分表现出来的林分，且没有非目的树种干扰。为了通过优树选择获取优良的种质资源，许多林业科技工作者对不同树种的优树选择技术开展过研究[7-10]。通过对西南桦种源家系试验的总结，证明西南桦个体间存在显著的遗传变异[11-12]。如果经人工高强度选择，提高选择差，可获得较好的遗传增益。西南桦为珍贵用材树种，市场需求旺盛，因而天然林遭到过度采伐，现存林木多数为散生孤立木，很难找到面积较大、保存完好的林分，不利于西南桦的选优工作。为此，本研究针对不同林龄的西南桦人工林开展优树选择技术研究，对于开发利用西南桦优异种质资源，促进西南桦无性系林业发展，具有重要的理论意义与实用价值。

1 材料与方法

1.1 试验林概况

西南桦选优林分位于广西凭祥市中国林业科学研究院热带林业实验中心下属伏波实验场，地理位置106°43′E，22°06′N，海拔500m，低山；年均气温19.9℃，年降水1400mm，属南亚热带季风气候区；土壤为花岗岩发育成的红壤，土层厚1m以上。为了使西南桦选优标准有较好的适用性，按林龄序列选择了3片人工林进行选优，各林分基本状况见表1。

1.2 优树选择与评价方法

1.2.1 选优目标与评价指标

优树指的是同一林分相同立地条件下，生长量、材性、干形、适应性、抗逆性等方面远远超过同种、同龄的树木[13]。本次以生长优异、形质优良、无病虫害的西南桦优树为选择目标，以胸径、树高、材积为数量评价指标，以干形、冠高比和分枝因素为形质评价指标，综合评价进行优树选择，探讨西南桦人工林优树选择方法与标准。

根据干形、冠高比和分枝因素对树木形质影响的权重，以10分为满分，各指标评分标准见表2。

表1 西南桦选优林分林生长概况

林龄(年)	林分组成	郁闭度	平均胸径(cm)	平均树高(m)	蓄积(m^3/hm^2)	林分密度(株/hm^2)
10	纯林	0.7	14.5	13.7	90.9	750
15	纯林	0.6	22.4	18.3	165.2	450
21	纯林	0.6	27.1	19.6	169.0	300

表2 西南桦形质指标评分标准

干形			冠高比			侧枝		
通直、圆满	微弯	较弯	1/4	2/4	3/4	细小	中等	粗大
5.0	4.0	3.0	3.0	2.5	2.0	2.0	1.5	1.0

1.2.2 优树选择方法

常用的优树选择方法有优势木对比法、小标准地法、绝对生长量法、标准差法等[14]。本次西南桦候选林分生长良好，同时群体优良性状具有较高的遗传力水平，为优树表型选择提供了条件。根据西南桦的生长规律，同时考虑选择的外业成本与准确性，本次采用5株优势木生长指标对比法，结合形质指标综合评分进行优树选择。

在优良林分中，进行实地调查，发现生长性状和形质性状特别优良的单株，即编号标定为候选优树，然后以此候选优树为中心，15m半径范围内选择仅次于候选优树的5株优势木。测量候选优树和5株优势木的胸径、树高和冠高比，并对干形、冠高比和分枝情况进行评分。

1.2.3 数据统计分析方法

(1)单株材积(V)计算公式[15]：

$$V=0.52764\times10^{-4}D^{1.88216}H^{1.00931}$$

式中：V为单株材积；D为胸径；H为树高。

(2)配对t检验方法公式[16]：

$$t=\frac{\sum d/n}{\sqrt{[\sum d^2-(\sum d)^2/n]/n(n-1)}}$$

式中：t表示显著差异性检验计算值；d表示候选优树评价指标值与5株优势木相应评价指标平均值的差值；n表示t检验的配对样本数。

1.2.4 数据收集与统计

本次共选择了109株候选优树和545株优势木，109株候选优树根据其胸径、树高计算单株材积，545株优势木按其所在的候选优树小区计算平均胸径、平均树高、平均单株材积。候选优树与优势木统计结果见表3。

表3 候选优树与优势木生长指标

林龄(年)	候选优树株数(株)	候选优树平均值			5株优势木平均值		
		胸径(cm)	树高(m)	材积(m^3)	胸径(cm)	树高(m)	材积(m^3)
10	50	19.3	16.6	0.240	17.8	15.6	0.192
15	25	28.9	20.9	0.652	24.9	19.0	0.456
21	34	33.6	23.6	0.977	28.8	20.6	0.646

2 结果与分析

2.1 候选优树与优势木平均值差异性比较

为了初步评价候选优树的优异性，利用 t 检验方法将候选优树的生长指标与5株优势木的平均值进行了差异性比较。由表4可见：候选优树胸径、树高、单株材积 t 检验结果均大于临界值，表明候选优树生长性状明显优于对比优势木，大部分候选优树都有入选的可能。

表4 候选优树与优势木平均值显著差异性 t 检验

林龄(年)	优树样本数	t 检验值		
		胸径(cm)	树高(m)	材积(m^3)
10	50	10.234	6.357	7.967
15	25	4.319	4.411	4.238
21	34	4.355	3.843	4.360

注：$t_{0.05}(50)=2.009$；$t_{0.05}(25)=2.06$；$t_{0.05}(34)=2.032$。

2.2 优树入选标准的确定

2.2.1 优树入选生长指标标准的确定

优树生长指标的入选标准应该在优势木平均值 x 的基础上加上一定的附加值△，当候选树的生长指标超过 x+△时入选，否则，不能入选。

△取值方法分两步确定：第一步，以优势木平均值的5%为起点，对候选优树与增加5%后的优势木平均值进行 t 检验，每增加5%检验1次，直到 t 检验结果不显著为止，以此初步确定△取值范围。经统计比较得出西南桦各林龄的候选优树性状△值初步取值范围：10年生林分胸径、树高、材积的△取值范围分别为(11%~15%)x、(0%~5%)x、(25%~30%)x；15年生林分胸径、树高、材积的△取值范围分别为(6%~10%)x、(6%~10%)x、(26%~30%)x；21年生林分胸径、树高、材积的△取值范围分别为(11%~15%)x、(6%~10%)x、(31%~35%)x。

第二步，在△取值范围内，以1%的步长逐步增加优势木平均值后再与候选优树相应性状值进行 t 检验，直至不显著为止，则倒数第2次的增加值即为△临界值。通过再次 t 检验计算，得出西南桦优树各性状的附加值△：10年生林分胸径、树高、材积的△取值分别为11%x、5%x、26%x；15年生林分胸径、树高、材积的△取值分别为9%x、7%x、28%x；21年生林分胸径、树高、材积的△取值分别为11%x、6%x、31%x(表5)。

表5 西南桦人工林优树选择标准与入选株数

林龄(年)	优树生长指标大于优势木比率(%)			形质指标综合评分	入选株数(株)
	胸径(cm)	树高(m)	材积(m^3)		
10	11	5	26	9.0	17
15	9	7	28	8.5	10
21	11	6	31	7.5	10

2.2.2 综合评分法优树入选标准的确定

在野外初选优树时，对候选优树干形、冠高比、分枝粗细3个形质表型分级评分。干形的评分范围是3.0~5.0，冠高比的评分范围是2.0~3.0，分枝的评分范围是1.0~2.0，三者相加最高分值为10分，最低分值为3分。干形、冠高比和分枝综合得分分别与6~9给定值单样本 t 检验，直到 t 检验不显著或给定值大于候选优树得分平均值为止，t 检验结果见表6。

经以上统计分析后，西南桦优树形质指标评价标准为：10年生优树干形、分枝与冠高比相加的分值应达到9分以上，15年生优树干形、分枝与冠高比相加的分值应达到8.5分以上，21年生优树干形、分枝与冠高比相加的分值应达到7.5分以上(表6)。

表6 不同林龄候选优树形质指标综合得分与给定值的 t 检验结果

林龄(年)	形质指标给定值						
	6	6.5	7	7.5	8	8.5	9
10	43.0619	36.3124	29.5629	22.8134	16.0639	9.3143	2.5648
15	20.2469	16.7318	13.2167	9.7016	6.1866	2.6715	
21	14.5315	10.8604	7.1893	3.5182			

注：$t_{0.05}(50)=2.009$；$t_{0.05}(25)=2.060$；$t_{0.05}(34)=2.032$。

2.3 入选优树的确定

通过上述计算，确定了西南桦优树的生长量指标和形质指标入选标准，生长量指标包含胸径、树高和单株材积3个性状，形质指标综合干形、分枝粗细和冠高比3个性状，综合考虑生长量指标和形质指标的双重标准，各林龄的优树选择标准如下：

10年生林分优树胸径、树高、单株材积分别大于5株对比优势木平均值的11%、5%、26%，形质指标综合得分大于9.0。50株候选树17株入选，入选率为34.0%。

15年生林分优树胸径、树高、单株材积分别大于5株对比优势木平均值的9%、7%、28%，形质指标综合得分大于8.5。24株候选树10株入选，入选率为41.7%。

21年生林分优树胸径、树高、单株材积分别大于5株对比优势木平均值的11%、6%、31%，形质指标综合得分大于7.5。35株候选树10株入选，入选率为28.6%。

根据以上选择标准，109株候选优树中，有37株候选树符合上述标准入选，入选率为33.9%。

3 结论与讨论

本文以优势木对比法开展了西南桦人工林的优树选择研究，以胸径、树高、材积为数量评价指标，综合干形、冠高比和分枝粗细形质评价指标，采用相应的数据统计分析方法确定了不同林龄的优树选择标准，共选出优树37株，入选率为33.9%。

经过分析比较，各林龄的优树选择标准较为接近，为保证选择标准的通用性与实用性，对林龄达10年生以上的林分，可采用较为统一的选择标准，即优树胸径、树高、单株材积分别大于5株对比优势木平均值的9%～11%、5%～7%、26%～31%以上，形质指标综合得分大于7.5。因优势木的生长状况与林分年龄、密度、遗传基础及立地条件相关，在实际使用时可在一定范围内调整选择标准。

西南桦人工林培育目标以高价值大中径材为主[17,18]，在选优时首先应考虑其生长指标，以便缩短其工艺成熟期而提高经济效益，同时注意材性与抗性等质量性状的选择，对各表型性状进行科学的评价分析，以保证西南桦育种群体的遗传多样性。

参考文献

[1]朱积余，廖培来．广西名优经济树种[M]．北京：中国林业出版社，2006：66-69.

[2]赵志刚，曾杰，郭丽云，等．西南桦嫁接试验[J]．林业科技，2006，31(1)：18-19.

[3]郭俊杰，赵志刚，曾杰，等．西南桦花枝嫁接人工制种技术[J]．浙江林业科技，2010，30(5)：72-75.

[4]郭文福，蒙彩兰．穗条生根剂育苗基质和季节对西南桦扦插生根的影响[J]．林业科学研究，2011，24(6)：788-791.

[5]谌红辉，蒙彩兰，农淑霞，等．西南桦嫩枝扦插育苗技术研究[J]．林业实用技术，2009(12)：20-21.

[6]谌红辉，曾杰，贾宏炎．西南桦叶芽离体培养再生植株技术[J]．林业实用技术，2007(10)：21-22.

[7]陈强，周跃华，常恩福，等．西南桦优树选择的研究[J]．浙江林业院学报，2005，22(3)：91-295.

[8]陈健波，张照远，项东云，等．邓恩桉优树的选择标准[J]．应用研究，2008，22(1)：17-19.

[9]周建云，杨祖山，郭军战．栓皮栎优树选择标准和方法的初步研究[J]．西北农林科技大学学报(自然科学版)，2003，31(3)：151-154.

[10]杨培华，郭俊荣，谢斌，等．油松优树选择方法的研究[J]．西北植物学报，2000，20(5)：720-726.

[11]曾杰，郭文福，赵志钢，等．我国西南桦研究的回顾与展望[J]．林业科学研究，2006，19(3)：379-384.

[12]陈伟．西南桦速生单株选择[J]．植物资源与环境学报，2005，14(4)：30-35.

[13]沈熙环．林木育种学[M]．北京：中国林业出版社，2007：41.

[14]GB10018-1988．主要针叶造林树种优树选择技术[S].

[15]广西林业勘察设计院、广西林学分院．森林调查手册[R]，1986.

[16]宴姝，胡德活，韦如萍，等．南洋楹优树选择标准研究[J]．林业科学研究，2011，24(2)：272-276.

[17]蔡道雄，贾宏炎，卢立华，等．论我国南亚热带珍优乡土阔叶树种大径材人工林的培育[J]．林业科学研究，2007，20(2)：165-169.

[18]郭文福，蔡道雄，贾宏炎，等．马尾松与红椎等3种阔叶树种营造混交林的生长效果[J]．林业科学研究，2010，23(6)：839-844.

[原载：林业科学研究，2012，25(04)]

麻栎种源间苗期生长性状及遗传稳定性差异分析

刘志龙[1] 方升佐[2] 虞木奎[3] 唐罗忠[2] 陈厚荣[1] 李武志[1]

([1]中国林业科学研究院热带林业实验中心，广西凭祥 532600；[2]南京林业大学森林资源与环境学院，江苏南京 210037；
[3]中国林业科学研究院亚热带林业研究所，浙江富阳 311400)

摘 要 麻栎(*Quercus acutissima* Carr.)在我国分布广泛，麻栎苗木生长和遗传稳定性与育苗地点的地理气候和土壤条件密切相关，研究其苗木生长及遗传稳定性差异对选择优良种源和适生区十分必要。本研究用36个麻栎种源在3个试验点进行育苗试验，对苗高和地径生长量及其遗传稳定性进行了调查与分析。结果表明：苗高和地径生长量在种源间和试验点间均存在极显著差异，种源×试验点互作上也表现为互作效应极显著，说明开展种源区域试验是必要的；根据苗木生长量的平均数、回归系数和回归离差的统计结果，可以将36种源分成4种类型：高产稳定型、高产不稳定型、低产稳定型、低产不稳定型。

关键词 麻栎；种源；区域化；苗期生长；遗传稳定性

The Difference Analysis of Growth Traits and Genetic Stability Among Different *Quercus acutissima* Provenances

LIU Zhilong[1], FANG Shengzuo[2], YU Mukui[3], TANG Luozhong[2], CHEN Hourong[1], LI Wuzhi[1]

([1]*The Experimental Center of Tropical Forestry, CAF, Pingxiang* 532600, *Guangxi, China*;
[2]*College of Forest Resources and Environment, Nanjing Forestry University, Nanjing* 210037, *Jiangsu, China*;
[3]*Research Institute of Subtropical Forestry, CAF, Fuyang* 311400, *Zhejiang, China*)

Abstract: *Quercus acutissima* is widely distributed in China, the growth of *Q. acutissima* seedlings is closely related to the geographical and climatic condition of seedlings experimental site. Therefore, in order to select the superior provenances and suitable distribution areas, it is very necessary to study the difference of growth characters and genetic stability of different *Q. acutissima* provenances. Cultivation experiment of thirty-six *Q. acutissima* provenances was studied in three experimental sites, and growth increment of seedlings and growth stability were also investigated and analysised. The results showed that: seedling heigh and ground diameter were very significant difference in provenances and experimental sites. There was a significant interaction between provenance and experimental site, which indicated that the necessary of provenance regionalization test. Based on mean value, regression coefficient and deviation from regression of seedling growth increment, thirty-six *Q. acutissima* provenances were divided into high-yield stable type, high-yield unstable type, slow-yield stable type, slow-yield unstable type.

Key words: *Quercus acutissima*; provenance; regionalization; seedling growth; genetic stability

生物质能源作为可再生、环境友好型能源受到世界各地的广泛重视，成为研究热点之一[1-3]。生物质能源的发展，能够推进能源替代，缓解资源与环境的压力，被认为是解决全球能源危机的最理想途径之一[2-6]。麻栎是重要的生物质能源树种之一，栎炭具有色泽光亮、易燃、燃烧彻底而持久、热值高、无烟、碳素含量高、比重大、火力强的优点，是理想的生活和工业原料，极具开发潜力[7,8]。

麻栎在我国分布广泛，已有悠久栽培利用的历史，因其生态效益、经济效益和社会效益显著而被群众所喜欢。近年来，有关麻栎育苗[9-11]、抚育管理[12-14]、抗逆生理[15,16]、木材性质[17,18]的研究报道较多，但是关于麻栎种源区域化育苗试验的研究还未见报道。为了分析不同种源麻栎区域化育苗生长性状及遗传稳定性差异，本研究在3个试验点进行36个麻栎地理种源育苗试验，通过调查比较分析，探讨不同种源麻栎在不同试验点的生长差异和遗传稳定性，以期为麻栎优良种源的选择和推广奠定基础。

1 材料与方法

1.1 采种点和采种林分的确定

参考麻栎分布区的特点，在麻栎天然分布区内以经度、纬度网格式为主，适当考虑各大山系、等温线、等雨线等综合因素，采集麻栎自然分布区13省(自治区、直辖市)36个种源，各种源产地名称、地理位置和气候状况见表1。采种林分主要为当地起源的林分，林龄30年以上，采种母树要求为林分的平均木以上，生长正常、干形通直、无病虫害，选采的母树之间相距在50m以上。每个采种地15~20株母树种子等量混合，作为该种源种子。种子收集后置于0~5℃的冷库中沙藏，第二年春天发放到各试验点播种育苗，开展苗期试验。

表1 参试麻栎种源采种地点地理坐标

采种地点	地理坐标				年平均气温(℃)	年降水量(mm)	无霜期(d)
	N(°	′)	E(°	′)			
广西融水	25	06	109	08	19.3	1824.8	320
广东乐昌	25	08	113	21	19.6	1522.6	300
湖南新宁	26	26	110	50	17.5	1327.1	311
贵州三穗	26	53	108	52	14.9	1147.1	276
贵州黄平	27	52	107	38	16.3	1050.6	258
浙江龙泉	28	01	119	07	17.6	1669.7	263
湖南长沙	28	12	113	04	17.2	1361.6	263
四川泸州1	28	28	105	37	17.8	1188.5	265
四川泸州2	28	30	105	48	17.8	1188.5	265
湖南常德	29	03	111	41	16.8	1274.2	265
浙江开化	29	09	118	23	16.4	1901.4	250
湖南岳阳	29	13	113	03	15.6	1331.6	279
湖南桑植	29	28	110	03	14.5	1530.6	235
浙江建德	29	29	119	16	17.4	1600.7	254
安徽休宁	29	33	118	02	16.1	1773.4	231
浙江富阳	29	57	119	46	16.1	1463.8	232
安徽池州	30	11	117	28	16.5	1556.9	220
安徽黄山	30	16	118	07	15.5	1670.3	237
安徽泾县	30	26	118	13	15.6	1556.4	240
湖北浠水	30	27	115	13	16.9	1350.7	257
安徽太湖	30	37	116	24	16.4	1363.5	249
湖北远安	30	48	111	42	15.5	1100.5	241
安徽潜山	30	55	116	41	16.3	1336.2	242
安徽六安	31	29	116	05	15.2	1085.2	242
江苏句容	32	04	118	51	15.1	1105.3	229
四川万源	32	05	108	03	14.7	1169.3	240
安徽滁州	32	10	118	04	16.5	1000.8	218
湖北襄樊	32	13	112	04	15.6	1012.8	248
四川广元	32	16	105	27	16.2	1063.8	264
陕西汉中	32	49	106	30	14.5	858.60	235
河南南召	33	35	112	24	14.8	839.50	216
山东平邑	35	13	117	22	13.2	784.80	212
山东费县	35	15	117	58	13.6	856.40	197
山东蒙阴	35	19	117	34	12.8	820.30	200
山东沂水	35	48	118	44	12.3	782.10	205
山西方山	37	34	112	02	11.9	712.8	125

(续)

注：调查时间为2007年。

1.2 育苗试验点的确定

依据麻栎天然分布区和适生区同时设点的原则，选择了3个试验点，各苗圃地理气候状况见表2。

1.3 育苗试验设计

各育苗试验点均采用完全随机区组设计，3次重复。试验地选择光照条件好、土壤疏松、排水良好的圃地，培筑苗床，床宽1.2m，床长20m，每个小区10行，每行播种15粒种子，行距10cm。四周设有同树种1m宽保护行，以防止边缘效应。

1.4 数据收集及分析方法

1.4.1 苗期生长量测定

苗木生长停止后，每个小区随机选择30株生长正常的苗木作为观测株，分别用钢卷尺和游标卡尺测量苗高和地径，分别精确到0.1cm和0.01cm。

1.4.2 多点联合方差分析

Excel2003和SPSS16.0软件对3个试验地点用苗高和地径的种源均值进行方差分析，分析种源、地点、种源×地点是否存在着显著差异。

1.4.3 遗传稳定性分析

采用Eberhart &Russell模型中回归系数表征种源对环境的反应，用作稳定性相对测度。分析不同种源在不同环境下的稳定性。一般将 $bi=1$ 的种源定义为平均稳定型，适应于广泛的生境。$bi<1$ 的种源其生长量超过平均稳定型，特别适应在不利的生境条件下生长。$bi>1$ 的种源低于平均稳定型，在立地好的地区表现好，而在立地差的地方表现差[19]。

表 2　三个育苗个试验点地理坐标及气候状况

试验点	地理坐标		年平均气温	年降水量	无霜期	土壤类型	气候带类型
	E(°　′)	N(°　′)	(℃)	(mm)	(d)		
安徽滁州	118°04′	32°10	14.5	1000.8	218	普通黄棕壤	北亚热带向暖温带
浙江长乐	119°50′	30°20′	16.1	1463.3	234	疏松红黄壤	亚热带南缘季风
江苏镇江	119°32′	32°16′	15.1	1105.3	229	冲积砂壤土	北亚热带季风

2　结果与分析

2.1　苗木生长量差异分析

由于各种源对地理气候和土壤条件适应性不同，不同种源在各试验点苗高(图 1)和地径(图 2)生长量差异很大，从图 1 和图 2 可以看出，总体上，长乐试验点表现最好，滁州试验点次之，镇江试验点最差。长乐试验点苗高平均值是滁州试验点的 1.95 倍，是镇江试验点的 4.82 倍；地径也是长乐试验点最大，分别是滁州试验点的 2.02 倍、镇江试验点的 3.58 倍。

安徽滁州试验点调查显示，建德、龙泉、乐昌、榕江和开化种源表现最好，苗高和地径分别是当地种源(滁州)平均苗高的 1.31～1.56 倍和 1.31～1.57 倍，是最差种源的 2.48～2.94 倍和 2.00～2.40 倍；江苏镇江育苗试验点适宜采用乐昌、开化和太湖种源，其苗高和地径生长量远远高于其他种源，苗高分别是该点平均苗高的 2.08、1.79 和 1.73 倍，地径分别是该点平均地径的 1.26、1.25 和 1.24 倍。在浙江长乐试育苗验点，整体生长表现都较高。龙泉、乐昌、富阳、建德和开化种源的苗高和地径生长量较大。苗高分别是该点平均苗高的 1.29、1.27、1.22、1.21 和 1.19 倍，地径分别比该点平均地径高出的 26.54%、24.69%、24.07%、23.15% 和 17.90%。总体而言，3 个试验点优良种源的生长量排名顺序相似，生长表现优良的主要是浙江、广东、江苏和安徽种源，生长表现较差的主要是山东、四川、山西和湖北种源。

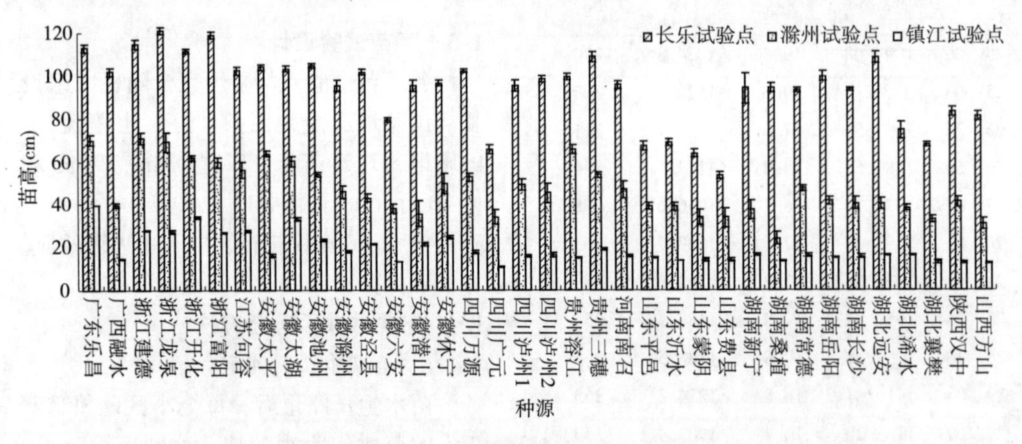

图 1　不同种源麻栎 3 个试验点 1 年生苗苗高生长量比较

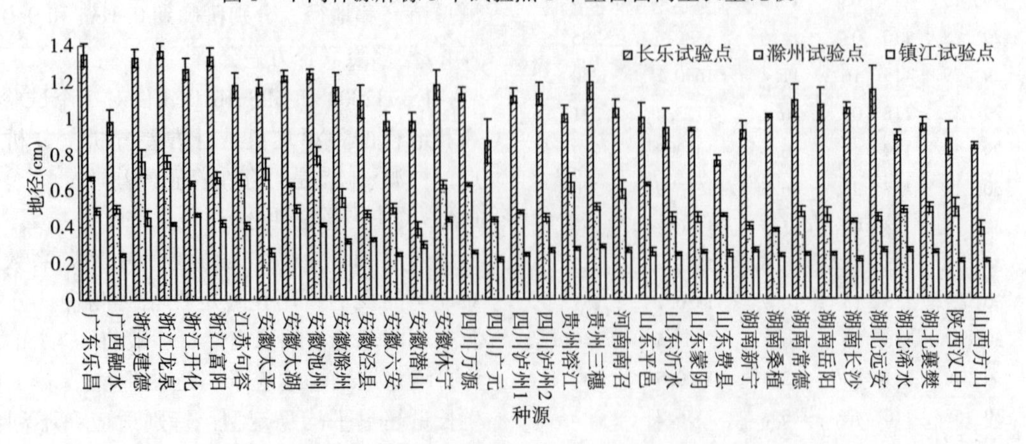

图 2　不同种源麻栎 3 个试验点 1 年生苗地径生长量比较

2.2 苗木生长量方差分析

对 3 个试验点苗高和地径生长量进行方差分析(表 3)，F 检验说明：种源间、试验点间、种源与试验点间的交互作用均达到了差异极显著水平，说明种源、环境、种源×环境互作等效应对麻栎苗木生长性状影响是显著的。由表 3 还可以看出，林木自身的基因型差异很可能是导致种源间生长量上的差异，表明麻栎的苗期选择具有较大潜力，开展种源试验对麻栎良种选育和改良是有意义的。同时，同一种源麻栎对不同的育苗环境的适应性也有所不同，所以开展区域化育苗试验是必要的。

表 3 不同种源麻栎苗高和地径多点方差分析

变异来源	性状	自由度	均方	*F* 值
种源	苗高	35	1005.798	93.14652**
	地径	35	0.105265	18.57910**
地点	苗高	2	144964.6	13425.11**
	地径	2	18.27151	3224.888**
种源×地点	苗高	70	176.3774	16.33423**
	地径	70	0.015385	2.71544**

2.3 苗木生长量遗传稳定性分析

以各种源生长量性状主效应值为横坐标，以非参数稳定性统计量 *bi* 为纵坐标，作描述各种源遗传稳定性能的二维散点图(图 1 和图 2)。

从散点分布图可以看出，*bi*＝1.0 附近有：广东乐昌、浙江开化、安徽滁州、太湖和池州、江苏句容、四川万源和泸州、河南南召、湖南长沙和常德。说明这些种源稳定性中等，有较广泛的适应区，在有利的条件下，这些种源能有一定的生长量，其中广东乐昌、安徽池州、浙江开化和江苏句容种源苗高均值较大，表现稳定，属于稳定高产型种源。稳定性系数 *bi*>1.1 的种源有：浙江龙泉、建德和富阳、安徽太平、湖北远安和汉中、贵州三穗等。这些种源在有利的生长环境下生长潜力很大，具有特殊的适应性，在特定的条件下会有较大的增益，因此，这几个种源可在适宜的地理气候条件地区推广。稳定性系数 *bi*<0.9 的种源主要有：山西方山、四川广元、湖北襄樊和浠水、山东费县、平邑、蒙阴和沂水等种源。这些种源表现较高的稳定性，他们生长较少受生长环境的影响。根据以上参数，可以将参试种源分成 4 种类型：①高产稳定型；②高产不稳定型；③低产稳定型；④低产不稳定型。

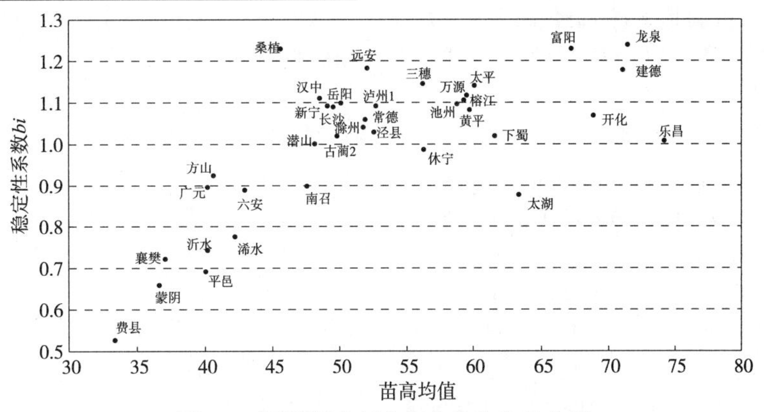

图 3 不同种源麻栎苗高生长稳定性分析

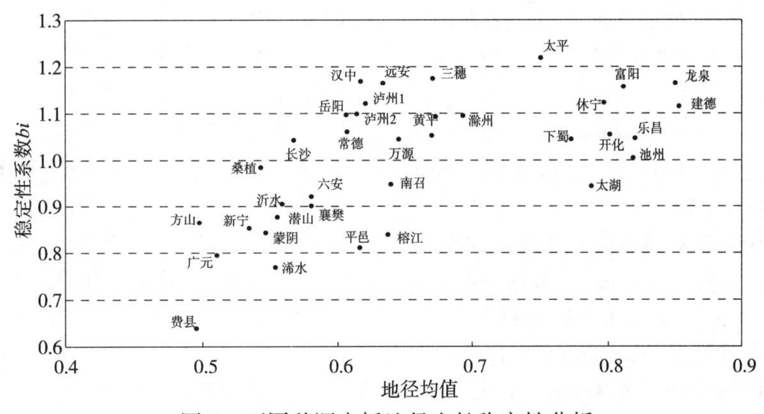

图 4 不同种源麻栎地径生长稳定性分析

3 结论与讨论

麻栎苗木生长量在种源间差异极显著，产地效应明显。麻栎分布区南部(浙江、江苏、广东和安徽等)的苗木生长量普遍大于北部(山东、山西、陕西和湖北等)，这与种源的地理位置、土壤和气候条件有关，已有相关论文阐述[9,10,20]，本文重点分析地点效应和种源×地点互作效应。实际育种过程中，基因型×环境的互作效应不同程度地存在，通过对种源稳定性的研究，有助于了解某一特定种源对环境的适应特性，为种源的选择和应用提供理论依据[21]。

在本研究中，苗高和地径生长量在试验点、种源×试验点互作上均存在极显著差异，说明苗木生长量性状还受地点效应的控制。育苗环境对苗木生长和种源测定的影响较大，长乐试验点水热条件较好，圃地肥力较高，促进了苗木生长量的提高。镇江试验点育苗圃地土壤肥力稍差而影响苗木的生长。此外，同一种源对不同育苗环境的适应性也不尽相同，这也说明了仅通过某一点的试验观测结果进行优良种源的选择是不科学的。

当将一个树种引种至新环境，其生长会发生较大的变化，有时难以预测，研究树种地理种源间的生长和适应性变异，为试验地区筛选经济性与适应性表现较佳种源，是全分布区种源试验的重要内容[22-23]。在一个区试点筛选出的优良种源不能简单地推广应用于造林区立地生境迥异的其他地区，而应根据区域试验分别不同地区推广，在不同试验点，选择适宜的种源很有意义[24]。本研究表明，不同育苗地点麻栎生长量因子的遗传稳定性大小不一，生长量上表现好的种源，其生长性状的遗传稳定性大小不一定表现好，这说明环境对生长和遗传稳定性的重要影响，与前人的研究结果一致[25-27]。

综合种源稳定性参数和生长量，可将麻栎分为4个种类型：①高产稳定型；②高产不稳定型；③低产稳定型；④低产不稳定型。高产稳定型可全面大力推广，速生不稳定型可在较好立地条件上表现良好的种源以充分利用潜在的种源与地点互作，提高苗木和幼林的生长量，慢生稳定型宜进一步观测、谨慎应用，慢生变化型建议暂不推广[28-30]。

本研究虽然从我国麻栎主要分布区选择了有代表性的种源，但由于客观条件的限制，没有全分布区收集种源，有待于进一步扩大研究范围，使研究结果更科学，代表性更强。植物生长是一个动态过程，这个过程中的每一点都可能与环境发生交互作用，而且交互作用可能决定最后结果[31]。本研究只做了麻栎种源间苗期试验，笔者已经开展了不同种源麻栎区域化造林试验研究，这些还有待于继续长期的观测研究。

参考文献

[1]SHIRO S. Recent progress in supercritical fluid science for biofuel production from woody biomass [J]. For Stud China, 2006, 8(3): 9-15.

[2] EDWARD M W, ASDRE P C. Bioerergy potential from forestry in 2050[J]. Climatic Change, 2007, 81(3-4): 353-390.

[3]SHAO H B, CHU L Y. Resource evaluation of typical energy plants and possible functional zone planning in China [J]. Biomass and Bioenergy, 2008, 32(4): 283-288.

[4]李顺龙，王耀华，宋维明. 发展林木生物质能源对二氧化碳减排的作用[J]. 东北林业大学学报，2009，37(4)：83-85.

[5] JOHN W B, MARK A T. Eco-environmental impact of bioenergy production [J]. Journal of Resources and Ecology, 2010, 1(2)110-116.

[6] ERICSSON K, NILSSON L. Assessment of the potential biomass supply in Europe using a resource-focused approach [J]. Biomass and Bioenergy, 2006, 30(1): 1-15.

[7]刘志龙，虞木奎，唐罗忠，等. 麻栎资源研究进展及开发利用对策[J]. 中国林副特产，2009，6：93-96.

[8]于一苏，于光明，李丛，等. 皖东丘陵地区麻栎栽培模式及栽培技术[J]. 林业科技开发，2009，23(4)：81-84.

[9]刘志龙，虞木奎，唐罗忠，等. 不同种源麻栎种子形态特征和营养成分含量的差异及聚类分析[J]. 植物资源与环境学报，2009，18(1)：36-41.

[10]刘志龙，虞木奎，唐罗忠，等. 不同地理种源麻栎苗期变异和初步选择[J]. 林业科学研究，2009，22(4)：486-492.

[11]胡永栋，吴江，姜金波，等. 麻栎种子不同贮藏方法的对比试验[J]. 防护林科技，2009，3：40-41.

[12]刘志龙，虞木奎，方升佐，等. 江淮丘陵地区麻栎人工林地上部分能量现存量研究[J]. 南京林业大学学报，2009，33(4)：62-66.

[13]唐罗忠，虞木奎，严春风，等. 立地条件及抚育措施对麻栎人工林生长的影响[J]. 福建林学院学报，2008，28(2)：130-135.

[14]唐罗忠，刘志龙，虞木奎，等. 两种立地条件下麻栎人工林地上部分养分的积累和分配[J]. 植物生态学报，2010，34(6)：661-670.

[15]王标，虞木奎，孙海菁，等. 盐胁迫对不同种源麻栎叶片光合特征的影响[J]. 应用生态学报，2009，20(8)：1817-1824.

[16]张晓磊，马凤云，陈益泰，等. 水涝胁迫下不同种源麻栎生长与生理特性变化[J]. 西南林学院学报，2010，30(3)：16-19.
[17]刘志龙，虞木奎，方升佐，等. 壳斗科4个树种木材基本密度及燃烧值的变异分析[J]. 江西农业大学学报，2009，31(4)：674-678.
[18]LIU Z L，FANG S Z，TANG L Z，et al. Spatial and vertical variation in calorific value of two Quercus species and its correlation to wood chemical components [C]. Conference on China Technological Development of Renewable Energy Source，2010，171-178.
[19]EBERHART S A，RUSELL W A. Stability parameters for comparing varieties. Crop Science，1966，6(1)：36-40.
[20]王标，虞木奎，王臣，等. 不同种源麻栎苗期生长性状差异及聚类分析[J]. 植物资源与环境学报，2008，17(4)：1-8.
[21]汤玉喜，吴敏，唐洁，等. 湘北平原美洲黑杨无性系的丰产性与遗传稳定性[J]. 林业科技开发，2010，24(6)：18-22.
[22]刘青华，金国庆，张蕊，等. 24年生马尾松生长、形质和木材基本密度的种源变异与种源区划[J]. 林业科学，2009，45(10)：55-61.
[23]毛爱华，李建祥，张超英，等. 19年生侧柏种源变异及选择研究[J]. 北京林业大学学报，2010，32(1)：63-68.
[24]周志春，范辉华，金国庆，等. 木荷地理遗传变异和优良种源初选[J]. 林业科学研究，2006，19(6)：718-724.
[25]徐有明，林汉，班龙海，等. 不同环境下火炬松种源造纸材材性遗传差异与遗传稳定性分析[J]. 林业科学，2008，44(6)：157-163.
[26]王军辉，顾万春，李斌，等. 桤木优良种源/家系的选择研究—生长的适应性和遗传稳定性分析[J]. 林业科学，2000，36(3)：59-66.
[27]李火根，黄敏仁，潘惠新，等. 美洲黑杨新无性系生长遗传稳定性分析[J]. 南京林业大学学报，1997，15(6)：1-5.
[28]郝玉宝，胡德活，梁机，等. 杉木优良家系区域化测定研究[J]. 广东林业科技，2010，26(1)：1-6.
[29]焦月玲，周志春，余能健，等. 南方红豆杉苗木性状种源分化和育苗环境对苗木生长的影响[J]. 林业科学研究，2007，20(3)：363-369.
[30]刘永红，樊军锋，杨培华，等. 油松种子园子代变异和遗传稳定性的分析[J]. 东北林业大学学报，2009，37(1)：6-7，17.
[31]苏兵强. 杉木优良家系木材密度及生长遗传稳定性分析[J]. 福建林业科技，2001，(S1)：1-4.

[原载：南京林业大学学报(自然科学版)，2012，36(05)]

坡位对灰木莲生长的影响

卢立华[1] 何日明[1] 农瑞红[1] 赵志刚[2]
([1]中国林业科学研究院热带林业实验中心，广西凭祥 532600；
[2]中国林业科学研究院热带林业研究所，广东广州 510520)

摘 要 对原产地越南尚河和引种地(广西凭祥市的白云和哨平)10年生灰木莲人工林不同坡位的生长及土壤进行了调查。结果表明：所有地点不同坡位的生长表现为：下坡>中坡>上坡，中坡、下坡的胸径、树高、蓄积生长量与上坡比，白云分别提高了30.35%、49.00%；34.78%、56.03%；122.97%、236.15%。哨平分别提高了30.07%、53.44%；39.00%、54.54%；125.09%、245.76%；尚河分别提高19.75%、35.08%；11.08%、27.49%；56.58%、119.21%。南亚热带白云和哨平坡位对灰木莲生长的影响大于原产地北热带尚河的影响。土壤显著影响灰木莲林分生长，表明灰木莲是一个土壤敏感型树种。方差分析结果表明，不同坡位的灰木莲生长差异极显著($P<0.01$)。多重比较结果显示，不同坡位之间的胸径、树高、蓄积生长量差异显著。相关分析结果表明，胸径、树高、蓄积生长量与坡位呈极显著的线性相关，但不同地点的直线斜率存在差异，南亚热带大于北热带，与3个地点生长量受坡位影响的顺序一致，可见坡位对灰木莲生长的影响程度因气候条件而异。

关键词 灰木莲；坡位；生长量

The Effect of Slope Position on the Growth of *Manglietia glauc*

LU Lihua[1], HE Riming[1], NONG Ruihong[1], ZHAO Zhigang[2]
([1]*Experimental Center of Tropical forestry, Chinese Academy of Forestry, Pingxiang 532600, Guangxi, China*;
[2]*Research Institute of Tropical Forestry, Chinese Academy of Forestry, Guangzhou 510520, Guangdong, China*)

Abstract: The survey was conducted on the growth of 10a *Manglietia glauc* and their soil condition in different slope position separately from three sites: the original place Shanghe, Laojie province, vietnam and the introduction places Baiyun and Shaoping in Pingxiang, Guangxi, China. Results showed that: the growth of *Manglietia glaucin* different sites showed down slope>middle slope>upper slope. Compared the down/upper slope with upper slope in different sites, the DBH, tree height and volume growth were risen by 30.35%, 49.00%; 34.78%, 56.03%; 122.97%, 236.15% in Baiyun, by 30.07%, 53.44%; 39.00%, 54.54%; 125.09%, 245.76% in Shanghe, and by 19.75% 35.08%; 11.08% 27.49%; 56.58% 119.21% in Shaoping, respectively. Baiyun and Shaoping located in south subtropical zone, had higher effect on the growth of *Manglietia glauc* than that in its original place Shanghe located in north tropical zone. The effect of soil condition in *Manglietia glauc* stands showed the same regularity. This implied that *Manglietia glauc* is a soil sensitive species. Result of variance analysis showed that slope position on the growth of *Manglietia glauc* reached a highly significant difference($P<0.01$). Multiple comparisons showed that the DBH, tree height and volume growth had a highly significant difference in different slopes. The correlation analysis result confirmed that the DBH, tree height and volume growth were highly significant linear correlative with slope position. Straight line figures in three locations showed, the growth linear gradient of *Manglietia glauc* was alike in the same climatic zones, but differed in the different climatic zones, the growth in the south subtropical zone was higher than that in the north tropical zone, the order of the growth linear gradient was the same as the trend of the effect on slope in different climatic zones. It is explained that the degree of effect on slope in *Manglietia glauc* growth varied in different habitat condition.

Key words: *Manglietia glauc*; slope position; the growth of *Manglietia glauc*

灰木莲(*Manglietia glauca* Blume)属木兰科(Magnoliaceae)常绿阔叶大乔木，树干通直，高大挺拔，材质优良，木材为散孔材，干材密度为0.611g/m^{3}纹理细致，易加工，切面光滑美丽，干燥速度较快，干缩均匀，为易于干燥类木材[1]，是建筑、家具和胶合板等的优良用材[2]。灰木莲树形优美，花大洁白，有芳香，移植容易成活[3]，它既是优良用材树种，也是形、景、味俱佳的园林绿化树种。因此，在我国不仅作为优良速生用材树种进行发展，而且已被广泛应用于园林绿化中。灰木莲原产于越南、印度尼西亚爪哇等热带地区，具有早期较速生的特点。研究表明，灰木莲树高的速生期在14年生以前，而胸径的速生期在8年生以前[4]。灰木莲在我国没有天然分布，它于20世纪60年代从越南引入我国进行试种[5]，引种在广西南宁市山谷湿润肥沃处生长表现良好，14年生人工林平均树高18m，胸径14cm[6,7]，是一个具有发展前景的好树种。

灰木莲引种到我国已有半个多世纪，其形态特征、栽培技术等早有报道[8]，但对它重视仅始于近十年，因此，对它的研究还不多，在现有文献中以引种表现为主[9-12]，其他方面的报道甚少，尤其是坡位对灰木莲生长影响研究至今未见报道。众所周知，坡位对环境资源起着再分配的作用，因此，生长受坡位影响的树种较多[13-17]，只是不同树种受影响的程度与规律不同而已。我国南方丘陵山地，虽然山体较小，坡面也不长，但有研究表明，坡位对林木生长的影响大于坡度和坡向[18]。传统的造林树种，由于已掌握了其树种的适应性和培育技术，可以通过技术措施来消减一些坡位对生长所造成的影响，而对于灰木莲这一研究相对滞后的树种，则仍缺少相应的技术措施为支撑。此外，以往其他树种开展坡位对生长影响的研究仅仅针对同一地点[13-17]，而本研究包括原产地与引种地共三个地点，更具科学性。在气候带上，涉及北热带及南亚热带，这在以往的研究中也没有类似报道。通过本研究，了解和掌握灰木莲在不同气候带、不同地点种植生长受坡位影响的程度，同时，了解生长量与坡位的相关关系，为灰木莲在我国的发展与科学经营提供参考。

1 材料与方法

1.1 试验地概况

本研究包括两个国家的三个地点。①中国：中国林业科学研究院热带林业实验中心哨平实验场(哨平)和白云实验场(白云)，位于广西壮族自治区西南部的凭祥市(21°57′47″~22°19′27″N；106°39′50″~106°59′30″E)；属南亚热带气候，年均气温21.5℃，≥10℃积温7500℃，年降水量1220~1400mm；哨平土壤为紫色砂岩发育的紫色土，白云土壤为泥质砂岩发育的砖红壤。②越南：保安县尚河乡(尚河)位于越南西北部的老街省(22°05′~22°30′N；104°11′~104°38′E)；属北热带气候，年均气温23.5℃，≥10℃积温8500℃，降水量1500~1990mm，土壤为花岗岩发育的砖红壤。

1.2 研究方法

采用样地调查法，调查林分为10年生的灰木莲人工纯林。造林苗木均为1.5年生的营养苗，种源均来自越南尚河，且为同一批次种子，造林密度2m×2m。于2011年9~10月分别在中国、越南的3个不同试验地中选择坡面较一致、坡向均为东南坡、海拔300~400m、长120m左右的坡面，分别在其上、中、下坡位铺设20m×30m样地，3次重复。然后对样地进行调查，测定灰木莲的树高、胸径，并根据树高和胸径计算其单株材积，同时，对土壤进行调查与采集。为减少土壤分析量，将3个重复土壤的同层次土样按等量混合均匀后用四分法取其混合样作为分析样。3个地点不同坡位土壤养分含量见表1。

表1 试验地土壤养分含量

地点	坡位	土层厚度(cm)	pH	有机质(g/kg)	全N(g/kg)	全P(g/kg)	全K(g/kg)	速效N(mg/kg)	速效P(mg/kg)	速效K(mg/kg)
白云	上坡	0~25	4.09	25.887	1.210	0.305	21.786	111.73	1.58	85.91
		25~60	4.33	11.542	0.951	0.267	23.017	52.33	0.26	61.80
	中坡	0~25	4.17	26.993	1.468	0.316	19.072	115.97	1.16	65.25
		25~60	4.22	11.167	0.996	0.269	19.492	57.99	0.31	51.47
	下坡	0~25	4.18	31.177	2.010	0.377	19.739	154.16	1.62	85.91
		25~60	4.22	17.535	1.017	0.343	22.017	79.20	0.45	68.69

（续）

地点	坡位	土层厚度 (cm)	pH	有机质 (g/kg)	全N (g/kg)	全P (g/kg)	全K (g/kg)	速效N (mg/kg)	速效P (mg/kg)	速效K (mg/kg)
哨平	上坡	0~25	4.33	16.682	0.863	0.204	7.276	60.81	0.99	85.91
		25~60	4.55	6.998	0.693	0.178	7.494	33.94	0.27	57.21
	中坡	0~25	4.42	17.271	0.865	0.191	7.239	70.71	0.68	103.13
		25~60	4.54	10.915	0.866	0.201	8.874	42.43	0.15	98.54
	下坡	0~25	4.42	26.101	0.950	0.208	4.056	103.24	1.15	97.39
		25~60	4.27	20.571	0.735	0.225	4.045	63.64	0.83	64.10
尚河	上坡	0~25	4.33	31.830	1.557	0.310	14.152	161.23	1.72	87.06
		25~60	4.47	20.414	1.037	0.269	13.053	101.83	0.33	66.39
	中坡	0~25	4.35	34.773	1.663	0.319	14.311	176.79	1.03	87.06
		25~60	4.5	16.641	0.950	0.309	14.047	86.27	0.26	56.06
	下坡	0~25	4.44	34.986	1.900	0.418	14.108	185.27	1.15	72.13
		25~60	4.49	19.072	1.123	0.392	14.535	100.41	0.36	60.66

1.3　数据处理与分析

采用PASW18.0进行方差分析，S-N-K法进行多重比较，采用person法进行相关分析。土壤养分采用土壤常规分析。

2　结果与分析

2.1　不同地点及坡位灰木莲的生长量

不同地点与坡位灰木莲生长量观测结果见表2。

表2　不同地点与坡位灰木莲的生长量

地点	项目区组	胸径(cm)			树高(m)			材积(m^3/株)		
		上坡	中坡	下坡	上坡	中坡	下坡	上坡	中坡	下坡
白云	Ⅰ	10.40	13.65	15.7	10.15	13.51	15.78	0.029 5	0.065 9	0.099 2
	Ⅱ	10.53	13.89	15.75	10.21	13.68	15.95	0.029 7	0.066 4	0.100 1
	Ⅲ	10.61	13.56	15.52	9.99	13.74	15.64	0.029 7	0.065 7	0.099 1
	总和	31.54	41.10	46.97	30.35	40.93	47.37	0.088 9	0.198 0	0.298 4
	平均	10.51	13.70	15.66	10.12	13.64	15.79	0.029 6	0.066 0	0.099 5
哨平	Ⅰ	10.46	13.41	15.81	9.73	13.35	14.78	0.027 5	0.060 6	0.093 8
	Ⅱ	10.18	13.49	15.73	9.47	13.51	14.73	0.026 8	0.060 8	0.093 1
	Ⅲ	10.29	13.32	15.92	9.58	13.14	14.96	0.027 1	0.059 6	0.094 1
	总和	30.93	40.22	47.46	28.78	40.00	44.47	0.081 4	0.181 0	0.281 0
	平均	10.31	13.41	15.82	9.59	13.33	14.82	0.027 1	0.060 3	0.093 7
尚河	Ⅰ	12.15	14.36	16.26	12.93	14.17	16.22	0.048 8	0.075 1	0.105 3
	Ⅱ	11.87	14.25	16.03	12.56	13.93	16.1	0.046 8	0.074 1	0.103 6
	Ⅲ	11.99	14.51	16.33	12.71	14.32	16.38	0.048 1	0.075 7	0.106 1
	总和	36.01	43.12	48.62	38.2	42.42	48.7	0.143 7	0.224 9	0.315 0
	平均	12.00	14.37	16.21	12.73	14.14	16.23	0.047 9	0.075 0	0.105 0

从表 2 可见：不同地点与坡位灰木莲的生长量差异较大，在 3 个地点中，尚河的生长量最高，白云的次之，哨平的最差；3 个地点不同坡位灰木莲的生长都表现为下坡>中坡>上坡。不同地点灰木莲生长差异的原因除与其所处地点的水热条件不同有关外，主要由土壤肥力差异较大所致。这可从表 1 得到证实，从表 1 可见：3 个地点的土壤有机质、全 N 及速效 N 含量都以尚河>白云>哨平，同一地点不同坡位土壤有机质、全 N 及速效 N 含量也基本表现为下坡>中坡>上坡。土壤养分高低的总体趋势与林木生长优劣趋势一致，表明土壤肥力状况对灰木莲生长具有最直接的影响，从而也说明灰木莲是一个对土壤肥力较为敏感的树种。为了解不同地点与坡位对灰木莲生长的影响情况，将不同地点中、下坡的生长量与上坡进行比较，计算它们所占的百分率，并计算中、下坡的生长量与上坡比所增加的百分数(表 3)。

表 3　不同地点坡位对灰木莲生长的影响

地点	坡位	胸径(cm)	中、下坡与上坡比(%)	中、下坡比上坡增(%)	树高(m)	中、下坡与上坡比(%)	中、下坡比上坡增(%)	材积(m^3/株)	中、下坡与上坡比(%)	中、下坡比上坡增(%)
白云	上坡	10.51	100.00		10.12	100.00		0.0296	100.00	
	中坡	13.70	130.35	30.35	13.64	134.78	34.78	0.0660	222.97	122.97
	下坡	15.66	149.00	49.00	15.79	156.03	56.03	0.0995	336.15	236.15
哨平	上坡	10.31	100.00		9.59	100.00		0.0271	100.00	
	中坡	13.41	130.07	30.07	13.33	139.00	39.00	0.0603	225.09	125.09
	下坡	15.82	153.44	53.44	14.82	154.54	54.54	0.0937	345.76	245.76
尚河	上坡	12.00	100.00		12.73	100.00		0.0479	100.00	
	中坡	14.37	119.75	19.75	14.14	111.08	11.08	0.0750	156.58	56.58
	下坡	16.21	135.08	35.08	16.23	127.49	27.49	0.1050	219.21	119.21

从表 3 可见：3 个地点中，坡位对灰木莲生长的影响都比较明显，中坡、下坡的胸径、树高、材积生长量与上坡相比：白云分别提高了 30.35%、49.00%，34.78%、56.03%，122.97%、236.15%；哨平分别提高了 30.07%、53.44%，39.00%、54.54%，125.09%、245.76%；尚河分别提高了 19.75%、35.08%，11.08%、27.49%，56.58%、119.21%。引种地白云、哨平坡位对灰木莲生长的影响程度相近，且比原产地尚河所受的影响大，这可能与尚河为北热带，白云与哨平为南亚热带，热带地区的水、热条件优于南亚热带地区有关。此外，可能还与不同地点的坡度、经纬度等的差异使环境因子分异加大有关，这需今后做更进一步的研究。

2.2 灰木莲生长量方差分析

为了解坡位对灰木莲生长影响的程度，对生长量进行方差分析，方差分析结果(表 4)表明：3 个地点灰木莲的树高、胸径、材积生长量在不同区组间的显著性概率(*Sig.*)均大于 0.05，说明各区组间的生长量差异不显著；而灰木莲的树高、胸径、材积生长量在不同坡位间的显著性概率(*Sig.*)均小于 0.01，说明不同地点与坡位对灰木莲生长有极显影响。对不同坡位灰木莲的生长量进行多重比较，结果(表 5)表明：3 个地点灰木莲的胸径、树高和材积生长量不同坡位间差异极显著，说明坡位对灰木莲生长有极显著影响，这结论与钱永平等[18]的研究结论一致。因坡位变化，林木生长的条件(包括：水、肥、气、热条件)会随之发生变化，但本研究中的坡面不长，高差不大，因此，不同坡位的热量条件差异不大，故造成灰木莲在不同坡位生长达极显著差异的主要原因应是土壤的水、肥、气条件，而在同一个坡面的同一种土壤中，土壤的通透性与保水性差别不会太大，所以，土壤的肥力应是制约灰木莲生长的关键因子。

表 4 不同地点不同坡位灰木莲生长方差分析

地 点	项目	误差来源	自由度	平方和	均方差	*F* 值	Sig.
白 云	胸径	区组	2	0.046	0.023	1.418	0.342
		坡位	2	40.437	20.219	1257.12	0.000**
		误差	4	0.064	0.016		
		总和	8	40.547			
	树高	区组	2	0.043	0.022	1.44	0.339
		坡位	2	49.232	24.616	1650.25	0.000**
		误差	4	0.060	0.015		
		总和	8	49.335			
	材积	区组	2	3.412	1.706	3.20	0.148
		坡位	2	45747.596	22873.798	42859.32	0.000**
		误差	4	2.135	0.534		
		区和	8	45753.143			
哨平	胸径	区组	2	0.013	0.007	0.441	0.671
		坡位	2	45.774	22.887	1541.78	0.000**
		误差	4	0.059	0.015		
		总和	8	45.846			
	树高	区组	2	0.006	0.003	0.098	0.908
		坡位	2	43.561	21.780	691.44	0.000**
		误差	4	0.126	0.032		
		总和	8	43.693			
	材积	区组	2	1.866	0.933	0.507	0.637
		坡位	2	41466.974	20733.487	11264.87	0.000**
		误差	4	7.362	1.841		
		总和	8	41476.202			
尚 河	胸径	区组	2	0.094	0.047	6.61	0.053
		坡位	2	26.646	13.323	1903.29	0.000**
		误差	4	0.028	0.007		
		总和	8	26.768			
	树高	区组	2	0.135	0.067	5.25	0.076
		坡位	2	18.611	9.305	725.41	0.000**
		误差	4	0.051	0.013		
		总和	8	18.797			
	材积	区组	2	36.766	18.388	12.11	0.051
		坡位	2	30580.321	15290.160	10074.38	0.000**
		误差	4	6.071	1.518		
		总和	8	30623.158			

注：** 表示极显著差异。

表 5 不同坡位灰木莲胸径、树高和材积多重比较结果

坡位	白云			哨平			尚河		
	胸径(cm)	树高(m)	材积量(m^3/株)	胸径(cm)	树高(m)	材积量(m^3/株)	胸径(cm)	树高(m)	材积量(m^3/株)
上坡	10.51A	10.12A	0.0296A	10.31A	9.59A	0.0271A	12.00A	12.73A	0.0479A
中坡	13.70B	13.64B	0.0660B	13.41B	13.33B	0.0603B	14.37B	14.14B	0.0750B
下坡	15.66B	15.79C	0.0995C	15.82C	14.82C	0.0937C	16.21C	16.23C	0.1050C

注：同列中不同大写字母表示差异极显著($p<0.01$)。

2.3 灰木莲生长量与坡位的相关分析

灰木莲的胸径、树高、材积生长量均随坡位上升而下降，且这种变化具有连续趋势，为了解灰木莲生长与坡位的相关关系，对其进行相关分析。图1表明：灰木莲胸径、树高、材积生长量与坡位具有直线相关关系，相关系数 $R^2>0.94$。可见，灰木莲的生长量与坡位间具有极显著的线性相关关系，进一步证明了坡位与灰木莲生长的密切关系。从图1还发现：同一气候带的白云与哨平，其直线斜率较接近，而不同气候带的直线斜率则相差较大，南亚热带(白云、哨平)的直线斜率大于北热带(尚河)，不同气候带直线斜率大小顺序与其生长量受坡位影响的大小顺序一致，说明气候条件差异对不同坡位灰木莲的生长也会产生一定影响。

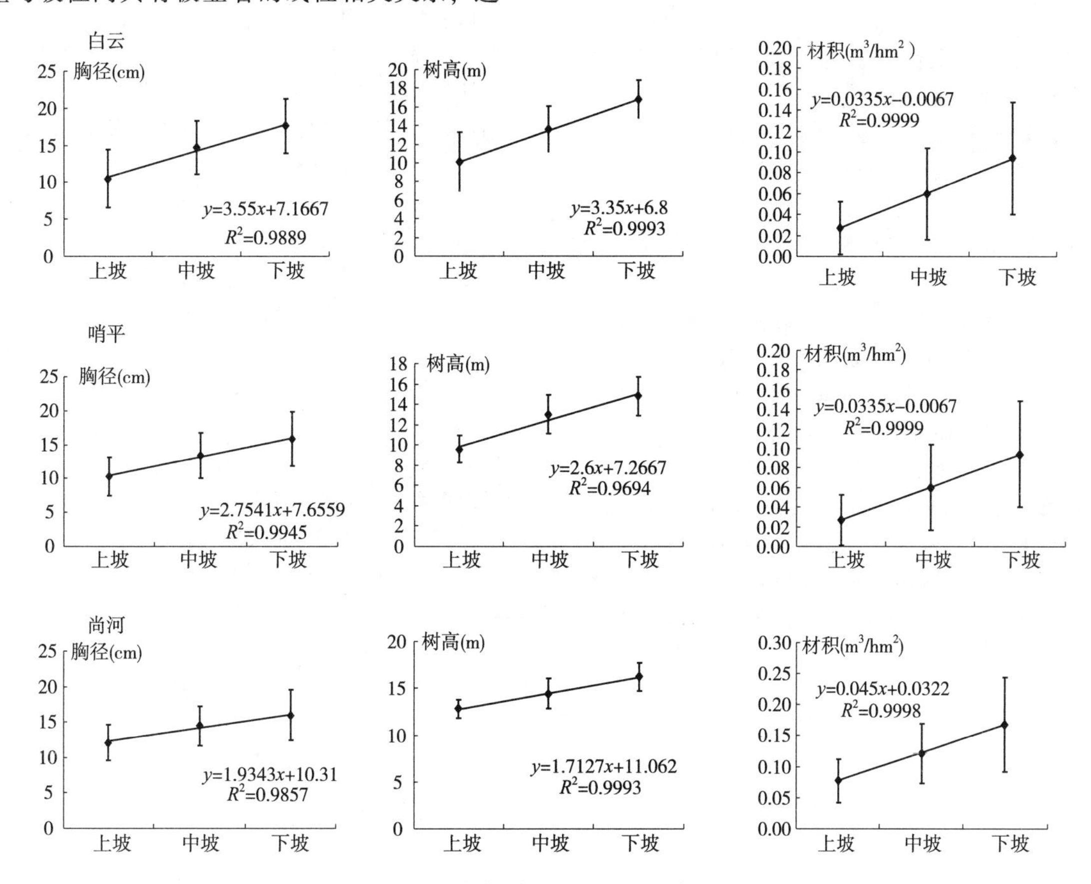

图1 不同地点灰木莲生长量与坡位相关关系

3 结论与讨论

灰木莲为天然分布于越南等热带地区的常绿阔叶用材树种，自20世纪60年代以来，在广西、广东、福建、云南等地均进行了引种，生长表现良好[9-12]，已成为我国热带南亚热带地区引种成功的外来珍贵优良阔叶树种之一[4]，它的引种成功对丰富我国南方人工林树种资源具有重要意义。

研究表明：灰木莲对土壤肥力较敏感，因此，生长受坡位的影响也特别明显。方差分析表明：坡位对灰木莲胸径、树高、材积生长量有极显著影响($P<0.01$)。多重比较表明：不同坡位灰木莲的胸径、树高、材积生长量均达极显著差异；相关分析证实，灰木莲的生长量与坡位具有极显著的直线相关，表明坡位对灰木莲的生长具有极重要影响。灰木莲生长随坡位变化的规律与陈淑容等[18]对楠木和谢天时[15]对水杉的研究结论一致。导致坡位对灰木莲生长产生显著影响的主要原因是坡位影响了土壤养分的再分配，从而导致了灰木莲在不同坡位生长的差异。

灰木莲生长的气候条件不同，其生长量受坡位影响的程度也有差异。研究表明：灰木莲在南亚热带地区引种，其生长量受坡位影响的程度比热带地区大。主要是由于灰木莲为热带树种，更适于在高温、高热、高湿的热带气候条件下生长，而南亚热带地区的年均气温、年积温、降水量都比热带地区低，因此，南亚热带地区的水热条件不是灰木莲生长最适宜的水、热条件，这可从表2生长量的比较得到证实。这一结果说明，在南亚热带地区引种热带树种灰木莲，对立地条件的要求比在原产地更严。

综上可见，立地条件与气候都会对灰木莲的生长产生影响，要确保灰木莲引种成功，并能达到速生丰产，应选择深厚、肥沃的土壤，坡的中下部位及高温多雨的环境种植为佳。

参考文献

[1]李俊贞，黎小波，唐天，等. 灰木莲木材干燥特性研究[J]. 木材工业，2011，25(3)：44-46.
[2]韦善华，唐天，符韵林，等. 灰木莲树皮率、心材率及木材密度研究[J]. 西北林学院学报，2011，26(3)：152-155.
[3]贾宏炎，农瑞红. 灰木莲大树移栽技术[J]. 广西林业科学，2006，35(1)：34-35.
[4]韦善华，覃静，朱贤良，等. 南宁地区灰木莲人工林生长规律研究[J]. 西北林学院学报，2011，26(5)：174-178.
[5]王克建，蔡子良. 热带树种栽培技术[M]. 广西：广西科学技术出版社，2008.
[6]郑万钧. 中国树木志[M]. 北京：中国林业出版社，1983.
[7]刘玉壶. 中国木兰[M]. 北京：科学技术出版社，2003.
[8]潘志刚，游应天. 中国主要外来树种引种栽培[M]. 北京：北京科学技术出版社，1994：333-336.
[9]林捷，叶功富，沈德炎，等. 灰木莲和子京在闽南山地的引种表现[J]. 林业科技开发，2004，18(1)：18-20.
[10]张运宏，戴瑞坤，欧世坤. 灰木莲等6种阔叶树种的早期生长表现[J]. 广东林业科技，2004，20(1)：43-46.
[11]杨耀海，刘明义，常森有. 灰木莲引种栽培试验研究[J]. 西南林学院学报，2007，27(3)：29-32.
[12]曾冀，卢立华，贾宏炎. 灰木莲生物学特性及引种栽培[J]. 林业实用技术，2010(10)：20-21.
[13]汪炳根，卢立华. 同一立地营造不同树种林木生长与土壤理化性质变化研究[J]. 林业科学研究，1995，8(3)：334-339.
[14]方志伟. 不同坡位对香樟生长发育的影响[J]. 福建林业科技，1997，24(4)：20-22.
[15]谢天时. 不同林龄和坡位水杉人工林生长状况的比较[J]. 亚热带农业研究，2007，3(3)：184-189.
[16]刘发茂. 不同坡位木荷人工林生物量及营养结构研究[J]. 福建林业科技，1995，22(增刊)：59-65.
[17]梁淑娟，潘攀，孙志虎，等. 坡位对水曲柳及胡桃楸生长的影响[J]. 东北林业大学学报，2005，33(3)：18-19.
[18]钱永平，李宝银，吴承祯，等. 生态环境因子对阔叶林质量的影响—坡位与林分蓄积关系的研究[J]. 林业资源管理，2009(1)：80-83.
[19]陈淑容. 不同立地因子对楠木生长的影响[J]. 福建林学院学报，2010，30(2)：157-160.

[原载：林业科学研究，2012，25(06)]

望天树苗木分级技术研究

马 跃 谌红辉 李武志 刘志龙 王小宁 蒙彩兰

（中国林业科学研究院热带林业实验中心，广西凭祥 532600）

摘 要 通过对望天树容器苗质量评价研究，提出了以苗高和地径作为该树种苗木分级的质量指标，并总结出了望天树半年生苗木的3级分级标准。Ⅰ级苗：苗高>27.5cm，地径>0.4cm；Ⅱ级苗：20.5cm<苗高≤27.5cm，0.3<cm地径≤0.4cm；Ⅲ级苗：苗高≤20.5cm，地径≤0.3cm。

关键词 望天树；苗木质量；苗木分级

Study on Seedling Grading Techniques of *Parashorea chinensis* Wang Hsie

MA Yue, CHEN Honghui, LI Wuzhi, LIU Zhilong, WANG Xiaoning, MENG Cailan

(*The Experimental Centre of Tropical Forestry, Chinese Academy of Forestry, Pingxiang* 532600, *Guangxi, China*)

Abstract: The seedling quality of *Parashorea chinensis* Wang Hsie container seedling was studied by using the method of gradual cluster analysis. Seedling height and basal diameter were proposed as the main indexes of quality for seedling grading of this species. The following standards were produced for seedlings of the first to the third grades. The seedling with the height of 27.5cm or above, basal diameter of 0.4cm or above were classified as grade one, those with height lower than 27.5cm but higher than or equal to 20.5cm, basal diameter lower than 0.4cm but higher than 0.3cm were classified as grade two, those with height lower than 20.5cm and basal diameter lower than 0.3cm were classified as grade three.

Key words: *Parashorea chinensis*; seedling quality; seedling grading

望天树(*Parashorea chinensis* Wang Hsie)是龙脑香科柳安属常绿大乔木，现为我国一级保护珍稀树种。1974年在西双版纳州勐腊县境内被首次发现，主要分布于我国云南南部、东南部(勐腊、马关、河口)及广西西南部局部地区。望天树生长快，成年后树体高大，干形圆满通直，不分杈，而且材质坚硬、耐腐性强、纹理美观，是制造各种家具及用于造船、桥梁、建筑等的优质木材[1]。

近年来，望天树作为一种优良的造林树种得到重视，中国林科院西双版纳热带植物园对望天树育苗技术及造林方式方面有一些探讨，但是对其苗木质量及苗木分级方面的研究尚未见报道[2]。优良的苗木质量是营造望天树人工林的物质基础，随着望天树作为速生丰产林树种在其适生地区栽培范围的扩大，用规范化、统一的望天树苗木质量标准来指导生产壮苗，可以提高造林成活率和林木生长量[3-6]。本文以望天树实生容器苗群体作为研究对象，对望天树苗木质量评价指标进行筛选，提出望天树造林苗木分级标准，以期为林业生产提供技术参考。

1 材料与方法

1.1 材料

试验地点设在广西凭祥市中国林业科学研究院热带林业实验中心苗圃，地理位置106°47′E，21°57′N，属北热带季风气候，终年温暖湿润，年均温21.5℃，极端最高温39.8℃，最低温-1.5℃，≥10℃的积温6000~7600℃，全年日照时数1218~1620h，年降水量1200~1400mm，干湿季节明显，4~9月份为雨季，相对湿度80%，年蒸发量1200~1600mm，有霜期3~5d，气候适合望天树苗木的生长。供试苗木为半年生望天树容器苗。

1.2 研究方法

在苗圃中随机设置5个观测样方，样方面积1m×1m，每样方随机选定10株，共计50株作为研究对象，测定出圃苗高、地径、根系长、大于5cm长侧根数、大于1cm长侧根数、全株鲜重、地上鲜重、地下鲜重、全株干重、地上干重、地下干重等11个指标，经相关分析后，确定苗木质量指标。采用聚类分析法进行分级，得出望天树苗木分级标准。

运用SPSS16.0统计软件对苗木各生长因子进行相关分析，对选定样苗进行聚类分析。

首先，SPSS根据样苗的实际情况确定3个有代表性的数据作为初始类中心，计算所有样本数据点到3个类中心的欧式距离，SPSS按距3个类中心点距离最短原则，把所有样本分派到各中心点所在的类中，形成一个新的3类，完成一个迭代过程。

欧式距离的计算公式为：

$$EDCLID = \sqrt{\sum_{i=1}^{k}(x_i - y_i)^2}$$

式中：k表示每个样本有k个变量；x_i表示第一个样本在第i个变量上的取值；y_i表示第二个样本在第i个变量上的取值。

接着，SPSS重新确定3个类的中心点，计算每个样本的变量值均值，并以均值点作为类的中心点；最后重复上面两步计算过程，直到达到终止迭代的判断要求为止[7]。

2 结果与分析

2.1 望天树苗木分级指标的确定

望天树苗木苗高、地径、主根长、大于5cm长Ⅰ级侧根数、大于1厘米长侧根数、全株鲜重、地上鲜重、地下鲜重、全株干重、地上干重、地下干重等11个生长指标间的相关分析结果见表1。表1可以看出，苗高、地径与苗木各器官生物量相关性极显著，最能体现苗木的质量，应该为评价苗木质量的最优指标。在实际生产中，苗高、地径是两个反映苗木质量的形态指标，其测定方法简便易行，便于应用，可作为苗木分级指标。

根系指标中，大于1cm长侧根数与其他各指标相关性不显著，大于5cm长Ⅰ级侧根数和主根长与其他指标显著相关，这可能是因为在营养袋中生长的苗木，在主根伸出营养袋之前，须根生长并不旺盛，并且在测定过程中，根系离开营养袋时须根容易受损。本研究中望天树苗木采用容器袋育苗，所以根系指标只作为参考，无需作为分级指标。除根系指标外，苗木其他各形态指标间都达到极显著相关(表1)。

表1 望天树容器苗各形态指标间的相关矩阵

	苗高	地径	主根长	≥5cm	≥1cm	全株鲜重	地上鲜重	地下鲜重	全株干重	地上干重	地下干重
苗高	1.000										
地径	0.952**	1.000									
主根长	0.515*	0.630	1.000								
≥5cm	0.453*	0.519*	0.710**	1.000							
≥1cm	0.408	0.342	0.151	0.065	1.000						
全株鲜重	0.960**	0.986**	0.594**	0.511*	0.310	1.000					
地上鲜重	0.975**	0.979**	0.525*	0.476*	0.301	0.991**	1.000				
地下鲜重	0.831**	0.915**	0.725**	0.557*	0.304	0.932**	0.874**	1.000			
全株干重	0.924**	0.977**	0.649**	0.537*	0.302	0.988**	0.964**	0.962**	1.000		
地上干重	0.954**	0.986**	0.596**	0.511*	0.310	0.994**	0.985**	0.924**	0.993**	1.000	
地下干重	0.782**	0.886**	0.742**	0.565**	0.259	0.904**	0.840**	0.991**	0.948**	0.902**	1.000

2.2 望天树苗木分级标准的计算

由表2及表3的聚类分析结果可以得出，Ⅰ级苗的分级界值为苗高>27.5cm，地径>0.4cm；Ⅱ级苗的分界值为20.5cm<H≤27.5cm，0.3cm<D≤0.4cm；Ⅲ级苗的分界值为H≤20.5cm，D≤0.3cm。其中Ⅰ级苗占总数的8%，Ⅱ级苗占总数的54%，Ⅲ级苗占总数的38%。

通过对分级标准做进一步的方差分析，结果(表4)表明：苗高和地径两者皆达到极显著水平，并且无论苗高或地径，Ⅰ、Ⅱ、Ⅲ级苗之间均存在极显著差异，因而对望天树苗木采用三级数量划分标准是科学合理、切实可行的，其中Ⅰ、Ⅱ级苗为合格苗木，Ⅲ级苗为不合格苗木，不能出圃造林，待留圃继续培育[8-10]。

表 2 望天树苗木聚类分析结果

样苗号	苗高(cm)	地径(cm)	QCL1	QCL2	样苗号	苗高(cm)	地径(cm)	QCL1	QCL2
1	21.4	0.41	1	1.37	26	21.5	0.37	1	1.24
2	11.6	0.28	2	3.44	27	16	0.29	2	0.97
3	24	0.42	1	1.33	28	12	0.23	2	3.10
4	18.5	0.36	2	3.53	29	22	0.40	1	0.75
5	12.5	0.29	2	2.53	30	18	0.26	2	2.99
6	24	0.41	1	1.29	31	15	0.35	2	0.52
7	22.3	0.42	1	0.61	32	24	0.36	1	1.27
8	15.8	0.32	2	0.81	33	24.5	0.34	1	1.80
9	11.5	0.27	2	3.54	34	17.5	0.30	2	2.47
10	23	0.39	1	0.26	35	22	0.34	1	0.86
11	16	0.32	2	1.00	36	22	0.31	1	1.00
12	22	0.39	1	0.75	37	27.5	0.40	3	2.63
13	9.3	0.23	2	5.76	38	30.5	0.32	3	0.93
14	22.8	0.42	1	0.41	39	17.5	0.28	2	2.48
15	24	0.47	1	1.55	40	18.5	0.30	2	3.47
16	11.4	0.28	2	3.63	41	23.5	0.33	1	0.92
17	21	0.39	1	1.75	42	23	0.37	1	0.28
18	15	0.34	2	0.49	43	24.5	0.38	1	1.76
19	24	0.39	1	1.26	44	20.5	0.34	1	2.29
20	22.5	0.45	1	0.74	45	25	0.42	1	2.29
21	22	0.36	1	0.78	46	31.5	0.44	3	1.42
22	19.5	0.34	1	3.27	47	15.5	0.28	2	0.50
23	16	0.29	2	0.97	48	31	0.46	3	1.03
24	20.5	0.40	1	2.25	49	25.5	0.34	1	2.79
25	18	0.34	2	3.00	50	23	0.33	1	0.57

注：表中 QCL1 表示样品所属类，QCL2 表示与中心点的距离。

表 3 聚类分析结果

Ⅰ级样苗号	Ⅱ级样苗号	Ⅲ级样苗号
37、38、46、48	1、3、6、7、10、12、14、15、17、19、20、21、22、24、26、29、32、33、35、36、41、42、43、44、45、49、50	2、4、5、8、9、11、13、16、18、23、25、27、28、30、31、34、39、40、47

表 4 分级苗木苗高 (*H*) 和地径(*D*) 方差分析

等级	苗木数(株)	变异来源	苗高					地径				
			平方和	自由度	均方	*F* 值	*Sig.*	平方和	自由度	均方	*F* 值	*Sig.*
Ⅰ级	4	组间	1073.816	2	536.908	119.826	0.000	0.095	2	0.048	29.176	0.000
Ⅱ级	27	组内	210.594	47	4.481			0.077	47	0.002		
Ⅲ级	19	总变异	1284.410	49				0.172	49			

3 结论与讨论

(1)苗木质量是指苗木在其类型、年龄、生理及活力等方面满足特定立地条件下实现造林目标的程度[11,12]。苗木形态指标是评价苗木质量最主要的依据。其中苗高反映出叶量多少，体现光合能力和蒸腾面积大小；地径与苗木根系大小和抗逆性关系紧密，粗壮的地径具有更强的支撑、抗弯曲能力，在虫害、动物破坏以及高温损害等方面的耐力大于细弱的苗木；苗木重量包括苗木鲜重和干重，它能反映物质积累状况，有效体现苗木生长量的大小；根系是植物的重要器官，造林后苗木能否迅速生根是决定其能否成活的关键，目前常用的指标有根系长度、侧根数等[13-16]。

相关系数是衡量变量之间相关程度的一个量值。苗木是一个复杂的生物体，苗木各部分的生长发育是相互平衡、相互制约的。从统计角度看，苗木形态指标之间存在着很强的依赖关系(相关性)。苗高和地径是反映苗木质量最直观的指标。

(2)对于苗木分级来说，就是利用苗木的分级指标，来划分苗木个体的相似程度。按照生产实际，苗木一般分为Ⅲ级，其中Ⅰ、Ⅱ级苗为合格苗，可出圃上山造林，Ⅲ级苗为不合格苗，应留圃继续培育。在SPSS中，快速聚类分析可以对随机抽样的50棵样苗进行逐步聚类分析，它先对数据进行简单分类，然后逐步调整，得到最终分类。本文利用苗高和地径对望天树苗木数据进行快速聚类分析，建立了3级苗木分级标准，即Ⅰ级苗苗高>27.5cm，地径>0.4cm；Ⅱ级苗，20.5cm<苗高≤27.5cm，0.3cm<地径≤0.4cm；Ⅲ级苗，苗高≤20.5cm，地径≤0.3cm。除苗高和地径等形态指标外，苗木的综合控制条件，即无检疫对象病虫害，无机械损伤，苗木通直，色泽正常，也是影响苗木质量的重要因素。在实际生产中，苗木综合控制条件由肉眼判断，对综合控制条件不合格的苗木直接判定为不合格苗[17]。

(3)由于各地育苗技术条件的差异，造成苗木质量的参差不齐，苗木出圃时间也不一定是6个月，所以，该分级标准仅供各地望天树苗木等级时作参考。

参考文献

[1]郭建华. 国家一级保护植物望天树及其家族[J]. 生物学教学，2000(25)：37-38.
[2]闫兴富，曹敏. 濒危树种望天树大量结实后幼苗的生长和存活[J]. 植物生态学报，2008，32(1)55-64.
[3]郑益兴，冯永刚，彭兴民，等. 印楝1年生苗木生长节律与数量分级标准[J]. 南京林业大学学报自然科学版：2008，32(3)：25-30.
[4]杨斌，周凤林，史富强，等. 铁力木苗木分级研究[J]. 西北林学院学报：2006，21(1)：85-89.
[5]高丽霞，孔旭晖. 红皮云杉苗木质量的研究[J]. 西北林学院学报：1992(02)：21-25.
[6]张树芬，张荣贵. 柚木苗木聚类分级[J]. 林业调查规划，2004，Dec，29(4)：4-7.
[7]倪雪梅. 精通SPSS统计分析[M]. 北京：清华大学出版社，2010.
[8]徐玉梅，王卫斌，景跃波，等. 南方红豆杉容器苗苗木分级研究[J]. 林业调查规划，2008，Feb，33(1)：126-129.
[9]徐金光，解孝满，刘和风. 聚类分析法在苗木质量分级的应用[J]. 山东林业科技，1994(4)：20-21.
[10]杨斌，赵文书，姜远标，等. 思茅松容器苗苗木分究[J]. 西部林业科学，2004，33(1)：32-37.
[11]刘勇，等. 苗木质量调控理论与技术[M]. 北京：中国林业出版社，1999，4.
[12]刘勇. 我国苗木培育理论与技术进展[J]. 世界林业研究，2000，(5)：43-49.
[13]郑天汉. 红豆树苗木质量评价指标的研究[J]. 福建林业科技：2007，34(4)71-73.
[14]李银华，李福双，曲银鹏，等. 河北省国槐苗木分级标准研究[J]. 林业实用技术，2005(12)：6-7.
[15]陈志生，白忠义，吴立群. 系统抽样在苗木调查中的应用[J]. 防护林科技，2000(12)：58-60.
[16]周凤林，李玉媛，史富强，等. 印度紫檀苗木分级研究[J]. 西部林业科学，2004，33(2)：29-33.
[17]国家林业局. 全国森林培育技术标准汇编-种子苗木卷[M]. 北京：中国标准出版社，2003.

[原载：西北林学院学报，2012，27(04)]

基质类型、容器规格和施肥量对红椎容器苗质量的影响

温恒辉 贾宏炎 黎 明 吴光枝 赵 樟 刘 云 蔡道雄

（中国林业科学研究院热带林业实验中心，广西凭祥 532600）

摘 要 采用析因试验设计的方法，研究了基质类型、容器规格和施肥量对2年生红椎容器苗质量的影响。结果表明：传统基质培育的容器苗生长表现最好，半轻基质次之，轻基质较差；容器规格对红椎容器苗的生长影响极显著($P<0.01$)，随着容器规格增大，地径、主根长和总生物量明显提高；复合肥施用量为7.5g/株和12.5g/株时，红椎容器苗的高、地径及总生物量均显著高于2.5g/株施用量，但这3个处理间苗木根系生长和根冠比差异不显著；综合考虑苗木质量和育苗成本等方面，培育2年生红椎容器苗的最佳方案为半轻基质，容器规格为20cm×20cm，复合肥施用量7.5g/株。

关键词 红椎；基质类型；容器规格；复合肥量；容器苗

Effects of Media Type, Container Size and Fertilizer Quantity on Growth and Quality of Container Seedlings of *Castanopsis hystrix*

WEN Henghui, JIA Hongyan, LI Ming, WU Guangzhi, ZHAO Zhang, LIU Yun, CAI Daoxiong

(*The Experimental Center of Tropical Forestry*, *CAF*, *Pingxiang* 532600, *Guangxi*, *China*)

Abstract: An experiment with factorial design was conducted to investigate the effects of media type, container size and compound fertilizer quantity on seedling growth and quality of *Castanopsis hystrix*. The results showed that the container seedlings raised by traditional media(v/v: 100% sandy loam) grew the best, followed by those with the half light media(v/v: 50% of sandy loam +25% of pine bark powder+ 25% of carbonized pine bark), and the seedlings grew the poorest in the light media(v/v: 60% of pine bark powder + 20% of sawdust + 20% of carbonized pine bark); container size had significant effects on the growth performance of container seedlings of *C. hystrix*, and ground diameter, taproot length and total biomass increased obviously with increasement of container size; under application of 7.5 and 12.5g compound fertilizer per seedling, seedling height, ground diameter and total biomass were significantly higher than those with 2.5g fertilizer per seedlings, while there existed no significant differences of seedling root growth and root and shoot ratio between those three treatment; taking seedling quality and cost into consideration, the combination for half light media, container size of 20cm×20cm, 7.5g compound fertilizer per seedling were the best for growing two-year-old seedling of this species.

Key words: *Castanopsis hystrix*; media type; container size; compound fertilizer quantity; container seedling

红椎(*Castanopsis hystrix*)是我国热带南亚热带地区的一个乡土珍贵阔叶树种，其材质坚硬、色泽美观、用途广泛；而且红椎适应性强、生长较为迅速，其林分具有良好的涵养水源、保持水土等生态功能。因此，红椎的推广种植日益受到重视，并被作为该地区重点发展的珍贵阔叶树种之一。

容器苗具有侧须根较多、根系舒展完整，而且又增加造林季节的灵活性，从而显著提高了造林成活率和幼林生长量[1,2]，培育容器苗是人工林高效培育的关键措施之一。在生产实际中发现，培育2年生红椎容器苗与1年生容器苗相比，其苗木质量显著提高，造林后郁闭成林快，能节约幼林抚育成本等，尤其对于营造高效人工林以及林下套种是至关重要的环节之一。由于基质、容器和施肥量直接影响容器苗质量和育苗成本，因此成为种苗学研究的一个热点。近10年来学者对育苗基质等单一方面的研究

较多[3-11]，而有关基质、容器乃至施肥方面的综合研究较少，仅见对木荷、浙江楠等树种的研究[12-13]，但尚未见有关红椎等热带南亚热带树种在这方面的研究报道。为此，笔者采用析因试验设计，系统研究基质类型、容器规格以及施肥量对2年生红椎容器苗质量的影响，为完善红椎容器育苗技术以及营造高效红椎人工林奠定基础。

1　材料与方法

1.1　试验地点及试验材料

育苗试验点位于广西凭祥市中国林业科学研究院热带林业实验中心伏波实验场苗圃(22°02′N，106°39′E)，海拔500m。属于南亚热带季风气候，年均温19.5℃，≥10℃积温6500~7000℃，最热月(7月)平均气温25.8℃，最冷月(1月)平均气温11.7℃，极端最低气温-2℃，年降水量达1400mm，相对湿度83%，4~9月为雨季。

选用1年生红椎的容器苗，苗木规格基本一致，平均苗高、地径分别为20.2cm和3.4mm。松树皮、砂质壤土采集于热带林业实验中心林区内，锯末收集于附近加工厂。松树皮、锯末和炭化松皮的处理与蒙彩兰等的方法类似[14]。采用无纺布容器育苗袋以及山东生产的史丹利复合肥(N：P：K=16：10：7)。

1.2　试验设计与育苗措施

采用基质类型(A)、容器规格(B)和施肥量(C)3个因素的析因试验设计。基质类型设3个水平：A1为砂质壤土100%，以下简称传统基质；A2为砂质壤土50%+沤制松皮粉25%+炭化松树皮25%，以下简称半轻基质；A3为沤制松皮粉60%+沤制锯末20%+炭化松树皮20%，以下简称轻基质。育苗容器为无纺布容器袋，其规格设3个水平(口径×高度)：B1为15cm×20cm；B2为20cm×20cm；B3为25cm×30cm。施用复合肥，其施肥量设置3个水平：C1为2.5g/株；C2为7.5g/株；C3为12.5g/株。按析因试验设计，设置27个试验处理，3次重复，每个试验处理20株。

按照试验设计要求将基质填充至无纺布育苗袋，再整齐地放置在垫有黑色地膜的苗床上。于2011年5月中旬，将1年生红椎苗移入育苗袋内并浇透水。移苗后不盖遮阴网，视天气状况每天浇水1~2次。具体单株施肥方法，将每个水平的复合肥总量分5次施入，每次每株各施1/5，施肥时间为2011年6~10月的每月中旬。其他管理措施与常规育苗方法类似。

1.3　苗木生长观测和统计分析

2011年11月底进行苗木生长观测。对每个处理20株苗木的苗高和地径进行测定，然后依据平均苗高和平均地径选取3株平均苗木，洗除基质，测量其主根长、统计侧根数，于根颈部剪断，按地上、地下部分分别装入纸袋内，经105℃杀青30min，80℃烘干至恒重，取出用电子天平称重，计算总生物量和根冠比。

应用SPSS 16.0统计软件进行多因素方差分析及Duncan多重比较，检验基质类型(A)、容器规格(B)和复合肥施用量(C)的主效应及其交互效应的显著性。

2　结果与分析

2.1　基质类型对红椎容器苗质量的影响

从表1可以看出，应用3种类型基质培育出的2年生红椎容器苗，其高、地径、主根长和侧根数以及总生物量均存在极显著差异($P<0.01$)，且呈现出相同的变化趋势，即传统基质>半轻基质>轻基质，而根冠比则相反，但基质处理间差异不显著($P>0.05$)。传统基质与半轻基质处理间苗高、侧根数以及总生物量均差异显著($P<0.05$)；而其地径和主根长差异不显著，但显著高于轻基质处理；由此可见，采用传统基质培育2年生红椎容器苗，其生长表现最好，其次为半轻基质，轻基质育苗效果则较差。

表1　基质类型对2年生红椎容器苗质量的影响

基质类型	苗高(cm)	地径(mm)	主根长(cm)	侧根数(根)	总生物量(g)	根冠比
传统基质(A1)	78.03　Aa	9.10　Aa	17.39　Aa	23.00　Aa	52.89　Aa	0.36　ns
半轻基质(A2)	71.56　Bb	8.89　Aa	16.81　ABa	19.04　Bb	45.07　Bb	0.38　ns
轻基质(A3)	50.50　Cc	7.30　Bb	14.16　Bb	15.99　Bc	23.43　Cc	0.42　ns

注：同列不同小字母表示在0.05水平上差异显著，不同大写字母表示在0.01水平上差异极显著，ns表示差异不显著(下同)。

2.2　容器规格对红椎容器苗质量的影响

容器规格对2年生红椎容器苗的高、地径、侧根数、总生物量均存在极显著影响，而对主根长、根冠比影响不显著(表2)。随着容器规格增大，地径、主根长和总生物量呈递增趋势，这与容器大、基质养分多有关。B1 处理的红椎容器苗大多数生长指标明显低于 B2 和 B3 处理，其地径、主根长、侧根数和总生物量以及根冠比最小。B2 和 B3 处理的地径、侧根数和总生物量以及根冠比差异不显著，但 B1 与 B2 处理的苗高生长显著高于 B3 处理。从以上比较分析得出，试验中容器规格为 20cm×20cm(B2)或 25cm×30cm(B3)均适合培育 2 年生红椎苗。

2.3 施肥量对红椎容器苗质量的影响

复合肥施用量对 2 年生红椎容器苗的生长和总生物量影响均极显著，对其根冠比影响显著，而对于苗木主根长、侧根数则影响不显著(表 3)。C2 和 C3 处理间的容器苗高、地径和总生物量均差异不显著，而显著高于 C1 处理，其中 C2 处理的苗高、地径和总生物量较 C1 处理分别提高 12.2%、6.4%和 31.7%，C3 处理较 C1 处理分别提高 8.6%、7.3%和 28.7%。C2 和 C3 处理间根冠比差异不显著，而显著低于 C1 处理。由此说明，选择 7.5g/株(C2)和 12.5g/株(C3)复合肥施用量，均可显著促进红椎苗的生长。

2.4 不同因素及其交互作用对 2 年生红椎容器苗质量的影响

从表 4 可以看出，3 个因素对地径、侧根数和总生物量的影响大小排序为基质类型>容器规格>复合肥施用量，对苗高和主根长的影响大小排序为基质类型>复合肥施用量>容器规格，而对于根冠比影响大小排序则为复合肥施用量>容器规格>基质类型，说明基质类型主要影响苗高、地径、主根度、侧根数以及总生物量，而复合肥施用量对红椎容器苗的根冠比影响较大。基质类型与容器规格的交互作用除了对侧根数影响不显著之外，对其余生长指标均产生极显著或显著的影响；基质类型与复合肥施用量的交互作用仅对总生物量具有显著影响。说明在育苗生产上可根据不同基质类型选用适宜的容器规格或相应调整复合肥施用量，从而促进红椎容器苗的生长发育，提高苗木出圃质量。

表 2 容器规格对 2 年生红椎容器苗质量的影响

容器规格	苗高(cm)	地径(mm)	主根长(cm)	侧根数(根)	总生物量(g)	根冠比
B1	68.52 Aa	7.94 Bb	15.73 ns	17.25 Bb	32.69 Ba	0.35 ns
B2	68.52 Aa	8.63 Aa	16.07 ns	21.17 Aa	42.53 Ab	0.39 ns
B3	63.27 Bb	8.74 Aa	16.55 ns	19.61 ABab	46.18 Ac	0.42 ns

表 3 复合肥量对 2 年生红椎容器苗质量的影响

复合肥量	苗高(cm)	地径(mm)	主根长(cm)	侧根数(根)	总生物量(g)	根冠比
C1	62.44 Bb	8.08 Bb	15.41 ns	18.59 ns	33.59 Bb	0.43 Aa
C2	70.03 Aa	8.60 Aa	16.46 ns	19.57 ns	44.23 Aa	0.36 Ab
C3	67.84 Aa	8.67 Aa	16.50 ns	19.86 ns	43.23 Aa	0.36 Ab

表 4 2 年生红椎容器苗生长和质量性状的多因素方差分析(*F* 值)

变异来源	自由度	苗高(cm)	地径(mm)	主根长(cm)	侧根数(根)	总生物量(g)	根冠比
A	2	252.92**	149.99**	4.80**	16.06**	172.64**	1.93
B	2	11.42**	28.91**	0.28	5.07**	36.07**	2.59
C	2	18.97**	14.35**	0.61	0.57	28.12**	3.83*
A×B	4	8.72**	11.61**	3.66*	1.37	11.22**	3.15*
A×C	4	1.25	1.99	1.17	0.53	3.07*	0.27
B×C	4	2.13	0.25	1.26	1.28	0.44	0.79
A×B×C	8	1.06	0.64	0.28	0.25	1.05	1.17

注：A 为基质类型，B 为容器规格，C 为 复合肥施用量。** 和 * 分别表示 0.01 和 0.05 显著水平。

3 结论与讨论

基质类型、容器规格和施肥量是影响容器苗的生长和出圃质量以及育苗成本的重要因子[13]。2 年生红椎容器育苗表明，采用传统基质培育出的红椎容器苗的生长表现优于半轻基质和轻基质，可能与基质通气、吸水及保水性能有关。虽然采用半轻基质培育出的容器苗的高生长显著低于传统基质，但其地径和主根长与传统基质间差异不显著，且容器苗亦符合出圃要求；而且试验中还发现，采用半轻基质培育红椎容器苗，其根系具有较多的外生菌根菌，而传统基质则外生菌根较少，轻基质极少，这也许与本次利用的砂质壤土带有外生菌根菌以及半轻基质的通气性良好有关，而外生菌根菌有利于红椎幼苗生长[15,16]，由此可见，半轻基质对提高红椎苗质量及其后续生长较有利。从生产成本方面考虑，沤制松皮粉、沤制锯末分别为130元/m^3和90元/m^3，炭化松皮为220元/m^3，而砂质壤土仅为40元/m^3。半轻基质育苗成本比传统基质高，但其重量与传统基质相比轻23%，应用半轻基质培育的苗木在运输及其造林等环节可显著降低成本[1]。因此以半轻基质(砂质壤土50%+沤制松皮粉25%+炭化松树皮25%)培育红椎容器苗较适宜。然而整体来看，半轻基质育苗成本仍较高，有待进一步调整基质种类及其配比开展试验，以筛选出既能保证红椎苗出圃质量，又能生产出更低成本的基质组合。

供试的3种容器规格中，以20cm×20cm和25cm×30cm规格无纺布容器袋对2年生红椎容器苗生长较为有利。从节约育苗成本和生产实际考虑，20cm×20cm与25cm×30cm容器规格相比，其容器苗不仅生长表现好、育苗成本也较低，而且达到优质苗木的出圃质量，因此，以20cm×20cm容器规格较适合。

施复合肥是提高苗木生长的一项重要技术措施。本试验结果表明，复合肥施用量对2年生红椎容器苗的生长量和总生物量影响极显著，对根冠比影响显著，而对主根长和侧根数影响较小。当复合肥施用量为7.5g/株和12.5g/株时，其容器苗生长和总生物量均显著增大，而低于这一施肥量时(2.5g/株)，其生长量显著下降。说明适量施用复合肥，能促进红椎容器苗的生长发育，但从节约施肥量角度考虑，本试验2年生红椎容器苗复合肥施用量以7.5g/株更为经济有效。

综合各因素对生长的主效应及其互作效应、育苗成本、造林及菌根形成等方面考虑，培育2年生红椎容器苗的最佳方案为采用半轻基质(砂质壤土50%+沤制松皮粉25%+炭化松树皮25%)，容器规格为20cm×20cm，复合肥施用量为7.5g/株。由于容器苗生长受气候条件、育苗期以及苗圃经营水平等诸多因素的综合影响，对苗木出圃质量要求也不相同，可根据基质类型的变化、选择适宜的容器规格或相应地调整复合肥用量，确定优化育苗方案，为提高红椎容器苗质量提供科技支撑。

参考文献

[1]蔡道雄，黎明，郭文福，等. 灰木莲容器苗培育基质筛选试验[J]. 浙江林业科技，2006，26(5)：36-38.

[2]王月海，房用，史少军，等. 平衡根系无纺布容器苗造林试验[J]. 东北林业大学学报，2008，36(1)：14-15.

[3]贾斌英，徐惠德，刘桂丰，等. 白桦容器育苗的适宜基质筛选[J]. 东北林业大学学报，2009，37(11)：64-67.

[4]邓华平，杨桂娟. 不同基质配方对金叶榆容器苗质量的影响[J]. 林业科学研究，2010，23(1)：138-142.

[5]金国庆，周志春，胡红宝，等. 3种乡土阔叶树种容器育苗技术研究[J]. 林业科学研究，2005，18(4)：387-392.

[6]韦小丽，朱忠荣，尹小阳，等. 湿地松轻基质容器苗育苗技术[J]. 南京林业大学学报，2003，27(5)：55-58.

[7]贾宏炎，黎明，郭文福，等. 马尾松和湿加松轻基质网袋容器育苗试验[J]. 林业科技，2009，34(2)：16-18.

[8]程庆荣. 蔗渣和木屑作尾叶桉容器育苗基质的研究[J]. 华南农业大学学报，2002，23(2)：11-14.

[9]尚秀华，杨小红，彭彦. 不同基质对桉树育苗效果的影响[J]. 热带作物学报，2010，31(7)：1073-1077.

[10]尚秀华，谢耀坚，彭彦，等. 腐熟桉树皮基质对桉树育苗效果的影响[J]. 中南林业科技大学学报，2011，31(6)：33-38.

[11]黎明，郭文福. 红椎容器苗基质试验简报[J]. 广西林业科学，2006，35(1)：31-33.

[12]马雪红，胡根长，冯建国，等. 基质配比、缓释肥量和容器规格对木荷容器苗质量的影响[J]. 林业科学研究，2010，23(4)：505-509.

[13]周志春，刘青华，胡根长，等. 3种珍贵用材树种轻基质网袋容器育苗方案优选[J]. 林业科学，2011，47(10)：172-178.

[14]蒙彩兰，黎明，郭文福. 西南桦轻基质网袋容器育苗技术[J]. 林业科技开发. 2007，21(6)：104-105.

[15]陈应龙，弓明钦，陈羽，等. 外生菌根菌接种对红椎生长及光合作用的影响[J]. 林业科学研究，2001，14(5)：515-522.

[16]陈羽，梁俊峰，周再知，等. 红菇和正红菇菌种接种三个乡土树种的苗期效果[J]. 广东林业科技，2010，26(1)：22-28.

[原载：种子，2012，31(07)]

叶面施肥对灰木莲幼苗生长的影响

陈　琳　卢立华　蒙彩兰　贾宏炎

（中国林业科学研究院热带林业实验中心，广西凭祥　532600）

摘　要　叶面施肥可以有效地提高苗木的产量和质量，是幼苗施肥的重要措施之一。本文开展了灰木莲叶面配方施肥研究，根据幼苗生长表现探讨其适宜的叶面施肥方案。结果表明：除了处理1(对照)和处理2(N90-50-0、P10-130-80、K100-150-100)之外，灰木莲的苗高、地径和生物量均随着时间的推移呈线性递增，而根冠比亦呈上升趋势；叶面施肥显著提高了灰木莲幼苗的高、地径、生物量和叶片数，最佳配方为处理6(N290-170-0、P90-290-240、K300-350-300)，分别比对照提高了31%、20%、34%和17%。

关键词　灰木莲；生长动态；叶面施肥

Effect of Foliar Fertilization on Growth of *Manglietia glauca* Seedlings

CHEN Lin, LU Lihua, MENG Cailan, JIA Hongyan

(*Experimental Center of Tropical Forestry*, *CAF*, *Pingxiang* 532600, *Guangxi*, *China*)

Abstract: Foliar fertilization is an important measure of seedling fertilization and it can effectively improve the yield and quality of seedlings. Different foliar fertilizations were applied in this study to determine the optimum fertilization amount of *Manglietia glauca* seedlings according to the seedling performance. Except for the treatments one(control) and two(N90-50-0、P10-130-80、K100-150-100), the height, root collar diameter and biomass of seedlings increased linearly with time, while there was a rising trend in the root and shoot ratio. The foliar fertilization significantly increased the height, root collar diameter, biomass and leaf number of seedlings and the optimum fertilization treatment was treatment six(N290-170-0、P90-290-240、K300-350-300), which was 31%, 20%, 34% and 17% higher in the height, root collar diameter, biomass and leaf number, respectively than the control.

Key words: *Manglietia glauca*; growth dynamics; foliar fertilization

灰木莲为常绿乔木，树干通直圆满、结构细致、易加工、树形整齐美观，是一个优良用材和城镇绿化树种。灰木莲原产于越南和印度尼西亚，于1960年从越南引种至我国，在我国北回归线以南地区具有广泛的发展前景[1]。在灰木莲轻基质容器育苗过程中，通常采用基肥和根部追肥相结合的方法对苗木进行施肥，然而由于灰木莲育苗往往容器小、幼苗后期叶片大，加之淋洗作用等，不仅降低了根部追肥的利用效率，而且造成了施肥对环境的污染。大量研究表明，叶面施肥可以有效地提高苗木的产量和质量[2,3]。因此，本文根据苗木的需肥特点、育苗进程，针对主要元素N、P、K设置8个施肥方案开展灰木莲幼苗的叶面施肥研究，旨在确定其适宜的施肥方案，为其壮苗培育提供指导。

1　材料与方法

1.1　试验材料

试验所用的灰木莲种子采自越南。2011年11月7日播种，播种前用0.5%高锰酸钾溶液对沙床进行消毒。2011年12月3日，芽苗长出2cm左右开始移苗，将大小一致、生长健康的芽苗移入装有黄心土、草皮泥和树皮(三者比例为7∶1∶2，基质中拌入0.5%钙镁磷肥)的营养袋中，营养袋规格为7.5cm×12cm(口径×高)。育苗基质的pH为6.84，全氮含量为1.11g/kg，有机质26.29g/kg，速效磷46.38mg/kg，速效钾146.73mg/kg。为了防止苗木穿根，营养袋下面垫一层防草布。

1.2　试验设计

采用完全随机区组设计，设置8个处理(表1)，重复3次，每小区32株(置于15孔育苗盘中，每盘间隔排列8株，共4盘)。2012年1月3日(即移苗1个月后)，开始根际追肥，施肥浓度为100mg/L尿素，每周1次，共27次。2012年6月15日至9月15日，开始进行叶面施肥处理，每周施1次，共14次(表1)，以叶面湿透为宜，施肥后约1h再淋水。N为尿素(N46%，四川美丰化工股份有限公司)，P为二水磷酸二氢钠(P19.87%)，K为氯化钾(K52.41%)。试验期间，按照生产上的常规方法进行苗期病虫害防治以及水分管理。

表1 施肥时间和施肥量

处理 (mg/L)	6月15日至7月15日			7月15日至8月15日			8月15日至9月15日		
	N	P	K	N	P	K	N	P	K
1	0	0	0	0	0	0	0	0	0
2	90	10	100	50	130	150	0	80	100
3	140	30	150	80	170	200	0	120	150
4	190	50	200	110	210	250	0	160	200
5	240	70	250	140	250	300	0	200	250
6	290	90	300	170	290	350	0	240	300
7	340	110	350	200	330	400	0	280	350
8	390	130	400	230	370	450	0	320	400

1.3 测定方法和统计分析

2012年6月15日，调查初始苗高和地径分别为15.89cm和3.78mm，根据测量结果选取6株平均苗，测定其平均生物量为1.9068g/株。此后每隔1个月调查1次，共调查4次，调查指标为苗高、地径以及根、茎、叶干重(每个小区取4株，105℃杀青15min，80℃烘干至恒重)。在最后1次测量时，增测其叶片数。

应用SPSS 16.0软件对灰木莲苗木的生长指标进行方差分析和多重比较，检验叶面施肥处理间幼苗生长的差异显著性(α=0.05)，并对各叶面施肥处理下灰木莲幼苗生长指标的动态变化进行回归分析。

2 结果与分析

2.1 叶面施肥对灰木莲地径、苗高生长的影响

8种处理灰木莲的苗高和地径均随时间的推移而呈线性递增($R^2>0.85$)。对于每次调查，不同处理间灰木莲幼苗的高和地径均差异显著($P<0.05$，表2)，从表2可以看出，叶面施肥能够显著提高灰木莲幼苗的高和地径生长，而且随着时间的推移施肥效果愈加明显。在最后1次测量时，灰木莲的苗高和地径随着施肥量的增加而增加，施肥量增至处理6之后保持稳定，即处理6、处理7和处理8的苗高和地径均差异不显著($P>0.05$)(表2)。

表2 不同叶面施肥处理下灰木莲幼苗各生长指标的动态变化

生长指标	处理天数(d)	处理1	2	3	4	5	6	7	8
苗高(cm)	30	21.9 (0.58)c	23.1 (0.63)bc	23.7 (0.61)abc	23.8 (0.73)abc	23.7 (0.61)abc	25.6 (0.62)a	24.7 (0.66)ab	25.2 (0.73)a
	60	23.7 (0.62)c	25.6 (0.66)bc	26.2 (0.69)b	27.0 (0.69)b	27.4 (0.59)b	30.8 (0.65)a	29.7 (0.72)a	31.0 (0.80)a
	90	25.3 (0.62)d	27.5 (0.66)c	28.7 (0.69)bc	29.0 (0.69)bc	30.2 (0.59)b	33.8 (0.65)a	33.6 (0.72)a	34.5 (0.80)a
	120	26.4 (0.87)d	27.8 (0.86)cd	29.6 (0.92)bc	30.7 (0.89)b	30.7 (0.80)b	34.7 (0.95)a	33.8 (0.92)a	35.6 (1.05)a
地径(mm)	30	4.94 (0.07)bc	4.87 (0.06)c	4.98 (0.07)abc	5.07 (0.06)ab	4.98 (0.06)abc	5.14 (0.06)a	5.10 (0.06)ab	5.16 (0.06)a
	60	5.63 (0.07)c	5.69 (0.07)c	5.92 (0.07)b	5.99 (0.07)b	5.99 (0.07)b	6.32 (0.06)a	6.25 (0.07)a	6.22 (0.07)a
	90	5.81 (0.09)c	5.97 (0.09)c	6.44 (0.09)b	6.55 (0.08)b	6.51 (0.09)b	6.98 (0.09)a	7.01 (0.10)a	6.95 (0.10)a
	120	6.16 (0.09)e	6.47 (0.09)d	6.73 (0.09)cd	7.08 (0.09)b	6.95 (0.10)bc	7.37 (0.10)a	7.60 (0.11)a	7.58 (0.10)a

（续）

生长指标	处理天数(d)	处理							
		1	2	3	4	5	6	7	8
生物量(g/株)	30	3.20 (0.54)	3.69 (0.69)	3.87 (0.62)	3.53 (0.44)	3.76 (0.30)	3.78 (0.41)	3.98 (0.41)	3.91 (0.30)
	60	4.95 (0.63)	5.73 (0.61)	5.24 (0.67)	5.10 (0.81)	5.74 (0.48)	6.00 (0.43)	6.10 (0.54)	5.79 (0.62)
	90	5.62 (0.30)c	6.24 (0.45)bc	7.04 (0.82)abc	6.50 (0.39)bc	7.34 (0.05)abc	8.65 (0.69)a	7.81 (0.18)ab	7.71 (0.79)ab
	120	5.45 (0.57)c	6.03 (0.84)bc	7.28 (0.58)abc	8.16 (1.02)ab	7.15 (0.74)abc	9.66 (0.62)a	7.41 (0.31)abc	8.61 (1.29)ab
根冠比	30	0.39 (0.02)	0.40 (0.03)	0.42 (0.03)	0.41 (0.06)	0.38 (0.01)	0.35 (0.01)	0.40 (0.02)	0.32 (0.01)
	60	0.46 (0.01)	0.43 (0.03)	0.47 (0.02)	0.44 (0.04)	0.41 (0.00)	0.45 (0.02)	0.41 (0.03)	0.42 (0.01)
	90	0.49 (0.02)	0.49 (0.05)	0.51 (0.04)	0.50 (0.01)	0.40 (0.03)	0.42 (0.02)	0.42 (0.02)	0.49 (0.01)
	120	0.61 (0.04)	0.64 (0.02)	0.63 (0.03)	0.68 (0.08)	0.66 (0.02)	0.72 (0.04)	0.64 (0.02)	0.64 (0.05)

注：表中字母为多重比较结果，每行中处理间含相同字母表示差异不显著，反之表示差异显著(P<0.05)；括号中数值为标准误。

2.2 叶面施肥对灰木莲幼苗生物量和根冠比的影响

除了处理1(对照)和处理2因施肥量不足导致幼苗的生物量在最后1次测量时(即停止施肥1个月后)未见明显增长之外，其他各处理的幼苗生物量随时间的推移而呈线性增加(R^2>0.8)。根冠比方面，各处理的幼苗根冠比均随时间而不断增加，但开始增加的幅度比较小，最后1次测量时，增加的幅度变大(表2)，这可能是后期养分尤其是N供应不足导致生物量更多地向地下部分分配之故。

除了前2次测量外，8种处理间灰木莲幼苗的生物量均差异显著(P<0.05，表2)，叶面施肥能够显著提高灰木莲幼苗的生物量，而各处理间灰木莲幼苗的根冠比无显著差异(P>0.05，表2)，说明叶面施肥对幼苗的根冠比无显著影响。在最后1次测量时，除了处理2的生物量与对照无显著差异外，其他处理的生物量均显著高于对照，相对于对照提高了31~77%(表2)，这主要是因为叶面施肥显著提高了灰木莲幼苗的根和叶生物量，但对幼苗的茎生物量无显著影响(图1)。生物量分配方面，8种处理灰木莲幼苗的根、茎、叶生物量比值基本相同，均为1.4∶1∶1.2，可见生物量分配次序为根—叶—茎(图1)。

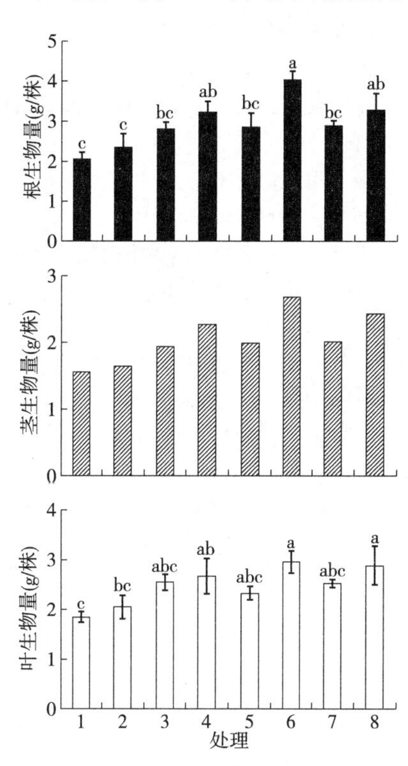

图1 叶面施肥对灰木莲生物量分配的影响

注：图中字母为多重比较结果，处理间含相同字母表示差异不显著，反之表示差异显著(P<0.05)；误差线根据标准误绘制，下同。

2.3 叶面施肥对灰木莲幼苗叶片数的影响

8种叶面施肥处理间灰木莲幼苗的叶片数差异显著(P<0.05)，从图2可知，叶面施肥能够显著增加灰木莲幼苗的叶片数。与灰木莲幼苗的高、地径和生物量随施肥量的变化趋势相似，叶片数亦随着施肥量的增加而不断增加，至处理6之后，其变化较为稳定，处理6、7和处理8三者间差异不显著(图2)。

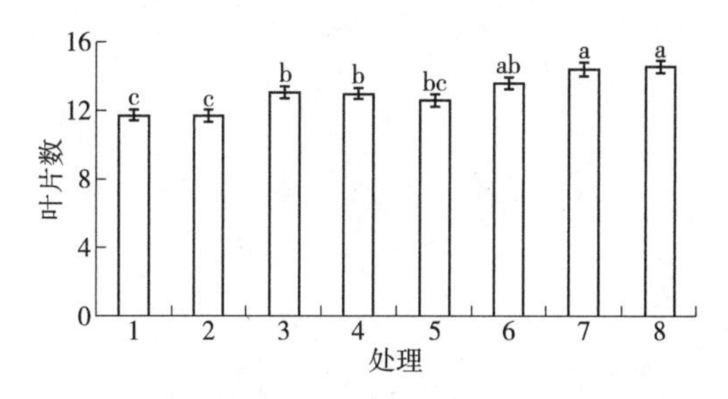

图2 叶面施肥对灰木莲幼苗叶片数的影响

3　结论与讨论

本研究中，各叶面施肥处理灰木莲幼苗的苗高和地径均随着时间的变化(7 月 15 日至 10 月 15 日)呈线性递增，近似“S”型曲线的速生期，与周洪英等[4]对贵州槭的研究结果一致。对 2 年生贵州槭苗定期追施磷酸二氢钾、复合肥、尿素 3 种肥料，发现在幼苗的整个生长期(4 月 27 日至 11 月 26 日)，其地径生长呈上升趋势，苗高生长曲线近似“S”形，然而在幼苗的速生期(6 月 27 日至 9 月 26 日)，地径和苗高均近似线性递增。

叶面施肥可以显著提高灰木莲幼苗的高、地径和生物量，这与马占相思[2]、红豆杉[5,6]和秃杉[3]等树种的研究结果相一致。叶面施肥还可以增加灰木莲幼苗的叶片数，江永清等[7]关于银杏叶面施肥的研究亦发现不同肥料和施肥浓度均对银杏叶的产量有显著影响。综合考虑叶面施肥对灰木莲幼苗的高、地径、生物量和叶面积的影响，确定最佳施肥处理为处理 6、7 和处理 8，然而从经济角度出发，认为处理 6 为最佳叶面施肥配方，可作为轻基质容器育苗施肥的辅助手段。处理 6 的施肥浓度和施肥量均高于白桦无性系[8]，但低于马尾松[9]、落叶松[10]和银杏[11]等树种，一方面由于不同树种或同一树种在不同的生长阶段对营养元素的吸收速率和需要量存在明显差异，另一方面与育苗基质和施肥种类有关。

参考文献

[1]王克建. 热带树种栽培技术[M]. 南宁：广西科学技术出版社，2008：71-72.

[2]潘月芳. 叶面施肥促进马占相思苗木根瘤结瘤量试验[J]. 林业科技开发，2005，19(5)：59-60.

[3]张先动. 根外追肥对秃杉一年生苗木生长的影响[J]. 福建林业科技，2009，36(3)：137-140.

[4]周洪英，邹天才，刘海燕，等. 不同施肥和光照对贵州槭两年生实生苗生长的影响[J]. 种子，2006，25(12)：63-67.

[5]李延群. 根外追肥对南方红豆杉一年生苗木生长的影响[J]. 福建林业科技，2005，32(4)：95-96，101.

[6]张宗勤，董丽芬，蒋明兵，等. 根外追肥对红豆杉生长的影响[J]. 西北林学院学报，2006，21(1)：90-92.

[7]江永清，苏宏斌. 银杏叶面追肥试验初报[J]. 林业科技通讯，1999：30-32.

[8]李天芳，姜静，王雷，等. 配方施肥对白桦不同家系苗期生长的影响[J]. 林业科学，2009，45(2)：60-64.

[9]江兴龙，杨屹梅，王德智. 马尾松百日容器苗培育技术[J]. 林业科技开发，2004，18(1)：32-33.

[10]庞秀谦，崔显军，孙振芳，等. 磷酸二氢钾叶面肥在落叶松育苗中的应用[J]. 林业实用技术，2010，3：22.

[11]康志雄，陈顺伟，金民赞，等. 叶用银杏不同施肥处理效应[J]. 浙江林学院学报，1999，16(3)：265-269.

[原载：种子，2013，32(06)]

柚木组培瓶苗移植育苗基质筛选试验与分析

白灵海[1] 莫慧华[1] 梁坤南[2] 李运兴[1] 农良书[1] 蒙彩兰[1]

([1]中国林业科学研究院热带林业实验中心，广西凭祥 532600；
[2]中国林业科学研究院热带林业研究所，广东广州 510520)

摘 要 以炭化树皮、锯末、炭化锯末、杉表土为材料，进行了12个处理的柚木组培瓶苗移植育苗基质试验，结果显示：对成活率有促进作用的有锯末和杉表土；对保存率有促进作用的的基质只有锯末；炭化树皮、杉表土对苗高、径有促进作用；50%炭化树皮+50%杉表土和80%炭化树皮+10%炭化锯末+10%锯末处理基质移植成活率、保存率大于95.1%，95.0%，1年生苗平均苗高、苗地径在33.4cm和0.68cm以上。

关键词 柚木；组培瓶苗；基质

Selection Experiment and Analysis of Nursery Substrate in Transplanting the Tissue Culture Seedlings of *Tectona grandis*

BAI Linghai[1], MO Huihua[1], LIANG Kunnan[2], LI Yunxing[1], NONG Liangshu[1], MENG Cailan[1]

([1]*The Experimental Center of Tropical Forestry*, *Chinese Academy of Forestry*, *Pingxiang* 532600, *Guangxi*, *China*;
[2]*Research Institute of Tropical Forestry*, *Chinese Academy of Forestry*, *Guangzhou* 510520, *Guangdong*, *China*)

Abstract: We carried out some experiments on different nursery substrates, e. g. carbonized bark, sawdust, carbonized sawdust, topsoil of *Cunninghamia lanceolata*, with 12 different treatments during transplanted the tissue culture seedlings of *Tectona grandis*. Results showed that ①sawdust and topsoil of *Cunninghamia lanceolata* can promote survival rate of the seedlings; ②sawdust is the only substrate that promotes the preserving rate; ③carbonized bark and topsoil of *Cunninghamia lanceolata* can accelerate the growth of seedling height and diameter; ④ the survival rate and the preserving rate of seedling in the treatment of 50% carbonized bark+50% topsoil of *Cunninghamia lanceolata* and the treatment of 80% carbonized bark+10% carbonized sawdust+10% sawdust are greater than 95. 1% and 95. 0%, and the average height and average diameter of one-year- seedlings in the above treatments reach 33. 4cm, 0. 68cm respectively.

Key words: *Tectona grandis*; tissue culture seeding; nurserg substrate

柚木(*Tectona grandis*)属马鞭草科落叶或半落叶大乔木，干形通直。木材为环孔材，心材大，淡褐色，比重0.6~0.72g/cm^3，纹理直，结构细致而美观，坚韧而有弹性，不翘不裂；含油质，强度大，耐浸，是航海、军需、建筑、车厢、家具、雕刻、铸造木模、贴面板等珍贵优良用材树种之一[1,2]。原产于印度、缅甸、泰国和印度尼西亚，我国引种于广东、广西、海南、云南、福建、台湾等地，已有170多年历史；当前可用木材资源锐减，社会需求增加，交易价格攀升，造林积极性不断提高，发展势头大，传统的撒播、圃地移植培育裸根苗法，已不能满足柚木良种壮苗和规模发展的需要；容器苗较裸根苗具有苗木根系损伤少，培育周期短，可造林季节气候灵活等优点，已在部分优良阔叶树种经营中广泛应用[3-7]，并且已取得较好的效益；基质在容器育苗中系种苗的载体，常用的有：表土、心土、泥炭土、珍珠岩、火烧土、炭化树皮、木糠、谷壳等，育苗基质的选择是否恰当关系到育苗的成败，目前仅有柚木实生容器苗表土黄心土培育基质方面的研究报道[8]，而组培瓶苗移植育苗基质方面的研究尚少涉及，因此开展这方面的探索和研究，作用意义大。

1 材料与方法

1.1 参试材料和育苗地点时间

无性系组培瓶苗系热带林业研究所繁育的7544号无性系，基质有炭化树皮、锯末、炭化锯末、杉表土4

种，营养袋规格为：8cm×12cm(径×高)试验地点设在中国林业科学研究院热带林业实验中心苗圃，海拔250m，地理位置106°47′E，21°57′N，属于北热带季风气候，终年温暖湿润，年均温21.5℃，年降水量1200~1400mm，干湿季节明显，4~9月份为雨季(雨热同季)；移植育苗时间为2010年4月7日至2011年5月5日。

1.2　试验设计

按各成分含量梯度不同布设12个处理(表1)，共有12个处理，每处理5×5=25株，设3重复，按随机区组试验设计方法实施。

表1　12处理基质成分

处理	炭化树皮	锯末	炭化锯末	杉表土
1	100			
2				100
3	50			50
4	25	25	25	25
5		50		50
6			100	
7	10		10	80
8	75			25
9	80	10	10	
10	20		80	
11		75		25
12	10	80		10

1.3　日常管理

在整个育苗期间进行病虫害调查与观察，每10天追施0.2%~0.5%浓度复合肥(15.0%N+15.0%P_2O_5+15.0%K_2O)液1次。

1.4　试验测定与统计

移植满30d时进行成活率和苗高调查；在12月底进行保存率、H(苗高)、D(地径)调查；用SPSS 17.0软件进行相关统计分析。

2　结果与分析

2.1　成活率与保存率

移植成活率(表2)是由移植苗本身的再生能力和基质的特性决定的，所有处理成活率都在64.0%以上，成活率大于95.0%的基质有：9、3处理；大于或等于85.0%而小于95.0%的有：4、5处理；85.0%以下的基质有：2、1、7、11、6、8、10、12处理，方差分析显示，处理间差异显著，多重比较(LSD法)见表2所列；按基质分类平均得表3，比较显示，对成活率有促进作用的有锯末和杉表土，其他基质作用不大；保存率各处理间分化较大，最高为处理9，保存率达95.1%，最低的为处理12，仅有61.8%，方差分析显示，处理间差异显著，多重比较见表2所列，按基质分类平均得表3，比较显示，对保存率有促进作用的基质只有锯末1种基质。

表2　各基质处理移植成活率、保存率、苗高、苗径均值及多重比较

成活率(%)			保存率(%)			苗高(cm)			苗径(cm)		
处理	指标	多重比较	处理	指标	多重比较	处理	指标	多重比较	处理	指标	多重比较
9	95.4	a	9	95.1	a	3	33.7	a	3	0.70	a
3	95.1	a	3	95.0	a	9	33.4	a	8	0.68	a
5	86.7	ab	5	86.4	a	8	32.6	a	9	0.67	a
4	85.3	b	4	85.0	ab	4	31.7	a	4	0.66	a
2	84	b	2	78.6	b	5	31.7	a	10	0.62	b
1	77.3	b	1	75.7	b	10	31.5	ab	7	0.60	b
7	72.0	c	11	71.1	bc	2	30.9	b	5	0.59	b
11	71.2	c	7	71.0	c	7	30.7	b	2	0.58	bc
6	69.3	c	6	69.3	c	1	29.9	c	1	0.57	bc
8	68.3	c	8	67.7	c	6	29.4	cd	12	0.55	c
10	65.3	c	10	65.0	c	11	29.0	d	6	0.54	c
12	64	c	12	61.8	c	12	28.3	d	11	0.54	c

表 3　成活率、保存率、苗高、苗径按基质分类统计表

处理	炭化树皮		锯末		炭化锯末		杉表土	
	0	1	0	1	0	1	0	1
成活率(%)	88.89	84.96	84.44	88.31	88.00	87.57	85.17	86.33
保存率(%)	70.67	58.85	61.00	65.30	71.33	61.74	65.59	62.00
苗高(cm)	30.7	31.0	32.6	29.7	32.4	30.4	30.0	31.8
苗径(cm)	0.62	0.63	0.69	0.59	0.69	0.61	0.60	0.67

2.2　苗高

各处理苗高数据见表 2 所列，所有处理苗高都在 28.3cm 以上，大于 30.0cm 的有：3、9、8、4、5、10、2、7 处理；30.00cm 以下的有：1、6、11、12 处理；方差分析显示，处理间差异显著，多重比较见表 2 所示；按基质分类平均得表 3，比较显示，对苗高有促进作用的基质有炭化树皮、杉表土，其他基质作用不大，据分析炭化树皮、杉表土所含有机质较其他基质多，此结果与高是由土壤肥力决定的理论相一致。

2.3　苗径

各处理苗径数据见表 2 所列，所有处理苗径都在 0.54cm 以上，大于或等于 0.60cm 的有：3、8、9、4、10、7 处理；0.60cm 以下的处理有：5、2、1、12、6、11 处理；方差分析显示，处理间差异显著，多重比较见表 2 所列；按基质分类平均得表 3，比较显示，炭化树皮、杉表土对苗径也有促进作用，其他基质作用不大。

2.4　育苗基质选择

移植成活率、苗木保存率、苗高、苗径生长量都直接影响着育苗的产量与质量，为综合考虑，用成活率、保存率、苗高、苗径作为变量对所有处理进行聚类分析。属 1 类的有 2 个处理，属 2 类的有 3 个处理，属 3 类的有 4 个处理，属 4 类的有 3 个处理，具体见表 4 所列。

表 4　各基质聚类分析结果

类别	处理
1	3、9
2	4、5、1
3	2、7、11、6
4	8、10、12

3　结论与建议

(1)组培瓶苗轻基质移植试验结果显示：好的基质处理：50%炭化树皮+50%杉表土和 80%炭化树皮+10%炭化锯末+10%锯末，移植成活率可以达到 95.1%，保存率也在 95.0%以上，1 年生容器苗，苗高 33.4cm，地径 0.68cm 以上，可以满足柚木造林对苗木的质量要求，因此可见，只要管理科学、到位，柚木组培瓶苗轻基质移植育苗是可行的。

(2)适合柚木移植育苗的基质较多，有森林表土、炭化树皮、锯末、炭化锯末等，因此可以根据各地实际情况选择配比使用，但要注意做好基质的腐熟和移植前后的杀菌消毒工作；柚木为好热性树种，当气温稳定在 25℃以上时，可着手进行移植育苗，确保经过较长时间的培育，绝大部分苗木规格达到Ⅱ级苗以上。

(3)轻基质较传统纯泥基质育苗具有营养土质量轻、透性好方面的特点，但部分树种造林后生长不良，白蚁蛀食严重，多次造林结果显示，柚木轻基质苗，只要气候适宜，造林成活率都在 95.6%以上，没有白蚁蛀食现象出现，柚木轻基质育苗比传统育苗有无法比拟的优点，值得推广应用。

参考文献

[1]广西林业局，广西林学会. 阔叶树种造林技术[M]. 南宁：广西人民出版社，1980.

[2]中国树木志编委会. 中国主要造林树种造林技术[M]. 北京：中国林业出版社，1981.

[3]卢立华，黎明. 育苗基质对任豆苗期生长的影响[J]. 林业实用技术，2005(11).

[4]黎明，郭文福. 红椎容器苗基质试验简报[J]. 广西林业科学，2006，35(1)：31-33.

[5]蔡道雄，黎明，郭文福，等. 灰木莲容器苗培育基质筛选试验[J]. 浙江林业科技，2006，26(5)：36-38.

[6]蒙彩兰，黎明，郭文福. 西南桦轻基质网袋容器育苗技术[J]. 林业科技开发，2007，21(6)：104-105.

[7]黎明，郭文福，蔡道雄，等. 以松皮粉为基质的西南桦容器苗培育技术[J]. 福建林业科技，2007，34(1)：43-49.

[8]钟瑜，梁国校，郝海坤，等. 柚木不同基质育苗试验[J]. 广西林业科学，2010，(7)：54-61.

[原载：种子，2013，32(09)]

不同林龄红椎人工林优树选择技术

刘光金　贾宏炎　卢立华　蔡道雄　温恒辉

（中国林业科学研究院热带林业实验中心，广西凭祥　532600）

摘　要　红椎是我国热带南亚热带珍优用材树种，为了挖掘和开发优异种质资源，本文以10、15、22、29、34年生优良林分调查数据为据，选用胸径、树高、单株材积3个生长量指标，结合冠高比、冠幅、干形、分枝角度、侧枝粗细5个形质指标，采用5株优势木对比法和形质指标综合评分法进行优树选择技术研究。结果表明：优树胸径、树高、单株材积应分别超过优势木平均值的15%~26%、5%~7%、41%~64%，形质指标综合得分不低于6.5；选择出优树47株，入选率为40.52%。优树选择标准在实际应用中可根据林分状况在一定范围内调整。红椎人工林主要培养大径材，在选优时应首先考虑其生长量指标，同时也应注意材性、抗性的选择以保证育种群体的遗传多样性。

关键词　红椎；人工林；优树选择技术

Selective Criterion for Superior Tree of *Castanopsis hystrix*

LIU Guangjin, JIA Hongyan, LU Lihua, CAI Daoxiong, WEN Henghui

(*The Experimental Centre of Tropical Forestry*, *CAF*, *Pingxiang* 532600, *Guangxi*, *China*)

Abstract: The five selective criteria of superior trees for *Castanopsis hystrix* A. DC which from 10 to 34 years had been studied by the growth index of DBH(the diameter at breast height), tree height and individual volume and form quality index of stem form, height and branch. It has been shown that as a superior tree, the DBH, tree height and individual volume of superior tree need exceed the average value 15%~26%, 5%~7%, 41%~64% of domain tree respectively. And at the same time, the form quality score of superior tree is not less than 6.5 of domain tree. As this criterion, 47 superior trees were selected from 116 candidate trees, yielding a selected ratio of 40.52%. The chosen specification of superior tree need have a certain range adjustment which based on the site index. Due to need big diameter of the *C. hystrix* A. DC, when selecting superior tree, the growth index is the first prior factor, and also the wood properties and resistance need to be considerable to insure the genetic diversity of breeding population.

Key words: *Castanopsis hystrix*; planted forest; superior tree selection technique

红椎(*Castanopsis hystrix*)为速生丰产乡土树种，是我国南方造林规模最大的珍优阔叶用材树种。其材质优良、加工性能好，广泛适用于高档家具制作和室内装饰，经济价值显著[1]。由于红椎人工林种源多采自天然次生林分，个体分化严重，木材生产力低，亟待良种壮苗技术研究。优质种源、家系、单株等有性选育种质筛选和高产、稳定无性系审定已成为良种壮苗的技术核心，亟须开展表型优异的优树选择技术研究。优树选择有针对性地对目标性状进行高强度的选择，是短时期内提高林木遗传品质的有效手段。林业科技工作者非常重视优质表型变异选择，开展了多个树种的优树选择技术研究[2-4]，获得大量的优良种质。红椎个体间差异显著[5,6]，经人工选择，可获得较好的遗传增益，而针对优质表型选择的优树技术研究少有报道。为此，针对10~34年生5个林龄段的红椎人工林开展优树选择技术研究，目的是运用优质表型选择技术获取优良种质，开展种子园营建与家系评价，并把嫁接、扦插等成熟的无性繁育技术[7-10]应用于营林生产，对于优异红椎种质评价与无性系林业发展具重要的理论意义与实用价值。

1　选优林分与选择方法

1.1　选优林分概况

红椎选优林分位于广西凭祥市中国林业科学研

究院热林中心下属伏波实验场(106°51′E, 22°02′N), 属南亚热带季风气候区, 温暖湿润, 年均温 19.9℃; 土壤为红壤、赤红壤, 土层深度 1m 以上, 林龄分别为 10、15、22、29、34 年。各林分生长状况见表 1。

表 1 红椎选优林分概况

林龄(年)	林分组成	郁闭度	胸径(cm)	树高(m)	蓄积(m^3/hm^2)	林分密度(株/hm^2)
10	纯林	0.7	16.36±0.932	14.24±0.604	100	675
15	纯林	0.6	22.5±1.402	18.41±1.176	165	450
22	纯林	0.6	29.17±1.673	20.41±1.132	215	375
29	纯林	0.6	31.46±2.690	21.6±2.249	230	300
34	纯林	0.6	34.32±1.061	21.85±0.686	260	300

注: 表中胸径、树高的数值为"平均值±标准差"。

1.2 优树选择与评价

1.2.1 选择目标与评价指标

优树是指在相近或相似立地条件下生长量和品质特别优异的单株木[11]。以速生丰产、形质优良的单株木为选择目标, 平价指标选择胸径、树高、单株材积 3 个生长量指标和冠高比、冠幅、干形、分枝角度、侧枝粗细 5 个形质指标。根据形质因子影响权重, 各指标评分标准见表 2。

表 2 红椎形质指标评分标准

冠高比	冠幅(m)	干形	分枝角度(°)	侧枝粗细	赋分
0~0.25	0~5	通直	0~30	细小	2.0
>0.25~0.50	>5~8	微弯	>30~60	中等	1.5
>0.50~0.75	>8~12	较弯	>60~90	粗大	0.5

1.2.2 优树评选方法

采用 5 株优势木对比法, 生长量评选结合形质指标综合评定。实地踏查, 标记候选优势木, 在 15m 半径范围内选择 5 株对比优势木。实测候选优树和对比优势木的生长量指标和形质指标, 并对形质指标进行评分。

1.2.3 数据统计分析

单株材积计算公式[12]:

$$V=0.52764\times10^{-4}D^{1.88216}H^{1.00931}$$

式中: V 为单株材积; D 为胸径; H 为树高。

配对 t 检验计算公式[13]:

$$t=\frac{\sum d/n}{\sqrt{[\sum d^2-(\sum d)^2/n]/n(n-1)}}$$

式中: t 表示差异显著性检验计算值; d 表示候选优树与优势木平均值的差值; n 表示 t 检验的配对数。

1.2.4 数据收集与处理

116 株候选优树和 580 株对比优势木的胸径、树高、单株材积生长指标测算结果见表 3。

2 结果与分析

2.1 遗传表型

胸径、树高、材积、冠高比、冠幅、干形、分枝角度、侧枝粗细 9 个性状进行相关性分析, 结果见表 4。

由表 4 可知, 干形与其他性状间相关性不密切, 说明干形是一个相对独立的性状, 与其他性状不存在基因连锁作用, 但干形与病虫害侵害或风力、冰冻以及栽植密度、疏伐强度等培育技术关系密切。分枝角度仅与胸径、冠幅显著正相关, 而与其他性状相关性不显著, 说明主干生长旺盛、茂密, 分枝角度越大。侧枝粗细与胸径、树高、材积负相关性显著, 与冠幅、树皮厚度正相关显著, 说明相对于主干来说, 侧枝较细的生长量较大, 而侧枝较粗壮的生长量较小。冠幅与胸径、树高、材积、分枝角度、侧枝粗细显著正相关, 而与其他性状相关性不显著。胸径、树高、材积 3 个生长量性状与冠幅显著正相关, 与侧枝粗细显著负相关, 胸径性状相关系数由大到小的排序为: 材积、树高、冠幅、分枝角度、侧枝粗细; 树高性状相关系数由大到小的排序为: 材积、冠幅、侧枝粗细; 材积性状相关系数由大到小的排序为: 胸径、树高、冠幅、侧枝粗细。

表 3　候选优树与优势木生长指标

林龄(年)	候选优树株数	候选优树			5 株优势木		
		胸径(cm)	树高(m)	材积(m^3)	胸径(cm)	树高(m)	材积(m^3)
10	27	20.90±2.154	16.22±0.794	0.27±0.060	17.73±1.339	14.83±0.987	0.18±0.029
15	25	28.94±5.560	20.92±1.465	0.65±0.331	23.79±3.544	19.53±1.498	0.45±0.183
22	22	35.94±5.078	22.42±2.035	1.22±0.320	29.33±4.876	20.56±2.198	0.67±0.235
29	26	38.92±5.563	23.50±3.529	1.30±0.445	32.05±4.314	21.15±2.996	0.82±0.279
34	16	45.03±7.437	23.01±3.091	1.65±0.446	31.42±4.580	21.29±2.318	0.77±0.260

注：表中数值为“平均值±标准差”。

表 4　红椎表型性状的遗传相关性

表型	胸径	树高	材积	冠高比	冠幅	干形	分枝角度	侧枝粗细
胸径	1							
树高	0.753**	1						
材积	0.972**	0.830**	1					
冠高比	-0.043	0.054	-0.069	1				
冠幅	0.458**	0.378**	0.431**	-0.103	1			
干形	0.104	-0.138	0.073	-0.182	0.089	1		
分枝角度	0.288**	0.109	0.183	0.04	0.341**	-0.107	1	
侧枝粗细	-0.155**	-0.088*	-0.129*	0.126*	0.115*	0.06	-0.079	1

注：* 表示在 0.05 水平上差异显著，** 表示在 0.01 水平上差异显著。

2.2　候选优树与优势木平均值差异性比较

候选优树与 5 株对比优势木配对 t 检验结果(表 5)表明：5 个林龄段的候选优树胸径、树高、单株材积 3 个生长指标与对比优势木达到差异极显著水平，候选优树明显优于对比优势木，说明大部分候选优树都有入选的可能。

表 5　候选优树与优势木平均值显著差异性 t 检验

林龄(年)	优树样本数	t 检验值		
		胸径(cm)	树高(m)	单株材积(m^3)
10	27	13.566	8.084	11.608
15	25	9.317	6.121	9.938
22	22	9.465	5.453	9.267
29	26	10.201	5.744	10.071
34	16	9.709	3.834	8.848

注：$t_{0.05}(27)=2.052$；$t_{0.05}(25)=2.060$；$t_{0.05}(22)=2.074$，$t_{0.05}(26)=2.056$；$t_{0.05}(16)=2.12$。

2.3　优树入选标准

2.3.1　生长指标入选标准

候选优树的胸径、树高、单株材积 3 个生长量指标都超过对比优势木平均值与附加值之和，即可入选，否则，不能入选[2-4]。经 t 检验计算，得出各林龄红椎优树生长指标胸径、树高、材积[14,15]。结果见表 6。

表 6　生长指标入选标准与入选数量

林龄(年)	优树生长指标大于优势木比率(%)			入选株数(株)	候选优树(株)	入选率(%)
	胸径(cm)	树高(m)	材积(m^3)			
10	15	6	41	15	27	55.56
15	16	7	46	16	25	64.00
22	25	5	62	15	22	68.18
29	17	7	47	13	26	50.00
34	26	7	64	8	16	50.00

2.3.2　形质指标入选标准

经给定值单样本 t 检验(表 7)，得出各林龄红椎优树形质指标评价入选标准为：10 年生优树形质综合得分应达到 7.5 分以上，15 年生优树形质综合得分应达到 7.5 分以上，22 年生优树形质综合得分应达到 6.5 分以上，29 年生优树形质综合得分应达到 6.5 分以上，34 年生优树形质综合得分应达到 7.5 分以上。

表 7　形质指标综合得分与给定值的 t 检验结果

林龄(年)	形质指标给定值						
	5	5.5	6	6.5	7	7.5	8
10	18.986	15.86	12.734	9.609	6.483	3.357	0.232
15	20.531	16.312	13.562	9.314	6.216	2.671	0.216
22	13.819	11.087	7.354	4.621	1.413		
29	14.115	11.031	7.947	4.863	1.779		
34	12.258	10.425	8.592	6.759	4.926	3.093	1.260

注：$t_{0.05}(27)=2.052$；$t_{0.05}(25)=2.060$；$t_{0.05}(22)=2.074$，$t_{0.05}(26)=2.056$；$t_{0.05}(16)=2.12$。

表 8　红椎人工林优树入选标准及入选率

林龄(年)	优树生长指标大于优势木比率(%)			形质指标	入选株数(株)	候选优树(株)	入选率(%)
	胸径(cm)	树高(m)	材积(m^3)	综合评分			
10	15	6	41	7.5	11	27	40.74
15	16	7	46	7.5	12	25	48.00
22	25	5	62	6.5	7	22	31.81
29	17	7	47	6.5	12	26	46.15
34	26	7	64	7.5	5	16	31.25

2.4　入选优树的确定

红椎优树评选综合考虑胸径、树高、单株材积 3 个生长量指标和冠高比、冠幅、干形、分枝角度、侧枝粗细 5 个形质指标，确定了各林龄段优树入选标准，结果见表 8。

根据评选标准，10 年生林分，入选优树 11 株，入选率为 40.74%；15 年生林分，入选优树 12 株，入选率为 48.00%；22 年生林分，入选优树 7 株，入选率为 31.81%；29 年生林分，入选优树 12 株，入选率为 46.15%；34 年生林分，入选优树 5 株，入选率为 31.25%。

3　结论与讨论

林木早期选择最适宜林龄应达到半个轮伐期[16]，此时林木表型已趋稳定，选择结果可靠。中国林业科学研究院热带林业实验中心规模化发展红椎等珍优用材树种 30 余年，在生产实践中探索了红椎适宜砍伐林龄为 30 年，本研究以此为据对具较高的遗传力水平、表型变异已充分表现、性状趋近稳定的 10~34 年生红椎人工林开展优树选择研究，综合胸径、树高、单株材积 3 个生长指标和冠高比、冠幅、干形、冠高比、分枝粗细 5 个形质指标，确定了各林龄阶段的优树选择标准，共选择出红椎优树 47 株，入选率为 40.52%。由于 5 个林龄段入选标准接近，可采用统一的选择标准，即优树胸径、树高、单株材积超过优势木平均值的 15%~26%、5%~7%、41%~64%，形质指标综合得分不低于 6.5。

优树选择是挖掘优异种质有效的育种技术，在改良林木性状和缩短育种进程方面具重要的意义，尤其在立地条件相似、林木性状充分表现出来的林地开展的优树选择。朱积余等[17]对红椎天然次生林和人工栽培区进行过优树选择，得到了胸径超过优势木平均胸径 16%，树高超过优势木平均树高 9%，材积超过优势木平均材积的 52%的入选标准，与本文优树入选标准相似。因此，此次红椎人工林选优标准亦可在天然林次生林沿用，实用价值较高。如对评选出来的速生丰产、形质优良的优树开展无性系诱导，使其优良性状固化，形成优良无性系，即可用于营林生产，从而促进林业科技成果转化、提高林分生产力，推动珍优用材树种无性系林业的发展。

红椎是一种珍贵用材树种，材质优良、色泽红润、经济效益显著，人工林培育目标以高价值大径材为主，以便取得更好的经济效益[18,19]。由于红椎与栲属其他椎类植物存在天然杂交，致木材伪红，木材品质与木材“真伪”是材性选择与改良难点。随着红椎遗传多样性的研究的开展[20-22]，分子标记辅助选择育种成为红椎甄别和遗传改良的关键技术。本次选优仅以速生丰产单株为目标，结合形质表型开展红椎优树选择研究，以便缩短工艺成熟期而提高经济效益，优质材性与抗性选择将按计划进一步开展，以确保红椎育种群体的真实性与多样性。

参考文献

[1]树木学(南方本)编写委员会. 树木学[M]. 北京：中国林业出版社，2003：316-317.

[2]陈健波，张照远，项东云，等. 邓恩桉优树的选择标准[J]. 应用研究，2008，22(1)：17-19.

[3]周建云，杨祖山，郭军战. 栓皮栎优树选择标准和方法的初步研究[J]. 西北农林科技大学学报：自然科学版，2003，31(3)：151-154.

[4]杨培华，郭俊荣，谢斌，等. 油松优树选择方法的研究[J]. 西北植物学报，2000，20(5)：720-726.

[5]张方秋，朱积余，黄永权，等. 红椎中心分布区种源早期生长研究[J]. 广东林业科技，2005，21(4)：9-12.

[6]朱积余，蒋燚，梁瑞龙. 广西红椎种源/家系造林试验研究初报[J]. 西部林业科学，2005，34(4)：5-9.

[7]邓燕忠，连辉明. 红椎扦插繁殖试验[J]. 广东林业科技，2005，21(3)：45-50.

[8]黄永权，杨胜强，赖旭恩. 红椎嫁接技术试验[J]. 广东林业科技，2005，21(3)：51-53.

[9]梁东成，梁志勇，黄永权. 红椎嫁接“假活”现象原因探讨[J]. 广东林业科技，2005，21(1)：63-65.

[10]蒋燚，唐卫辰，姚广彬. 红椎扦插育苗试验[J]. 西部林业科学，2006，35(1)：40-43.

[11]沈熙环. 林木育种学[M]. 北京：中国林业出版社，2007：41.

[12]郭文福，蔡道雄，贾宏炎，等. 马尾松与红椎等3种阔叶树种营造混交林的生长效果[J]. 林业科学研究，2010，23(2)：839-844.

[13]宴姝，胡德活，韦如萍，等. 南洋楹优树选择标准研究[J]. 林业科学研究，2011，24(2)：272-276.

[14]刘光金，谌红辉，郭文福，等. 西南桦优树选择技术研究[J]. 林业科学研究，2012，25(4)：438-441.

[15]刘光金，黄弼昌，谌红辉，等. 光皮桦优树选择技术[J]. 林业实用技术，2013，(1)：13-14.

[16]KANG H. Juvenile selection in tree breeding：some mathematical models[J]. Silvae genetica，1985，34(2)：75-84.

[17]朱积余，蒋燚，潘文. 广西红椎优树选择标准研究[J]. 广西林业科学，2002，31(3)：109-113.

[18]蔡道雄，贾宏炎，卢立华，等. 论我国南亚热带珍优乡土阔叶树种大径材人工林的培育[J]. 林业科学研究，2007，20(2)：165-169.

[19]蔡道雄，郭文福，贾宏炎. 南亚热带优良珍贵阔叶人工林的经营模式[J]. 林业资源管理，2007，(2)：11-14.

[20]杨峰，李志辉，蒋燚，等. 红椎优良家系ISSR遗传多样性分析[J]. 中南林业科技大学学报，2012，32(6)：123-127.

[21]王蕾，叶志云，蒋燚，等. 利用ISSR技术对优质红椎种质资源遗传多样性的分析[J]. 厦门大学学报：自然科学版，2006，45：91-94.

[22]徐斌，张方秋，潘文，等. 我国红椎天然群体的遗传多样性和遗传结构[J]. 林业科学，2013，49(10)：162-166.

[原载：东北林业大学学报，2014，42(05)]

顶果木天然林优树的选择标准

刘志龙　马　跃　谌红辉　刘光金　李洪果　全昭孔　莫慧华　蒙明君

（中国林业科学研究院热带林业实验中心，广西凭祥　532600）

摘　要　通过对广西顶果木种质资源的调查，实测了3个市(县)89株顶果木的数量性状、质量性状及环境因子，选出候选优树20株。对候选优树6个形质性状进行主成分分析后选择出干形、分枝数和冠径比作为顶果木选优的主要性状。采用基准线法初选优树15株，再以形质指标评分法复选优树9株，入选率45%，可供近期无性繁殖利用。

关键词　顶果木；优树选择；干形；分枝数；冠径比

A Study on Selection Standard for Superior Tree of *Acrocarpus fraxinifolius*

LIU Zhilong, MA Yue, CHENG Honghui, LIU Guangjin, LI Hongguo, QUAN Zhaokong, MO Huihua, MENG Mingjun

(*The Experimental Centre of Tropical Forestry*, *Chinese Academy of Forestry*, *Pingxiang* 532600, *Guangxi*, *China*)

Abstract: On the base of the investigation on the status of *Acrocarpus fraxinifolius* genetic resources in Guangxi Autonomous Region, the quantitative properties, qualitative properties and environmental factors of 89 *Acrocarpus fraxinifolius* in 3 counties were measured and 20 candidate superior trees have been selected. The Principal component analysis of 6 properties of these candidate trees showed that stem form, branch numbers and ratio of crown to diameter could be used as the main indexes for selection of superior trees of *Acrocarpus fraxinifolius*。We have selected 15 preliminary superior trees by baseline method and 9 elite superior trees for the recent clonal propagation utility by form quality index evaluation method, the selection rate was 45%. This paper have proposed standards and methods for superior tree selection of *Acrocarpus fraxinifolius* natural forest in Guangxi Autonomous Region, aim to establish the science foundation of genetic improvement and innovation utilization.

Key words: *Acrocarpus fraxinifolius*; superior tree selection; stem form; ramification number; ratio of crown to diameter

顶果木(*Acrocarpus fraxinifolius* Wight ex Arn.)为苏木科落叶大乔木，高达40m，胸径达150cm，多生在山谷、下坡疏林中，为乔木层上层的常见成分[1]。顶果木干形圆满，材质坚韧，少开裂，锯材加工容易，木纤维细长而壁薄，是纤维工业的好原料。顶果木主要分布在广西西南部、西部，贵州西部和云南南部、西南部到西部，印度、斯里兰卡、印度尼西亚等国也有分布。顶果木早期特别速生，寿命长、衰退慢，是培育大径材的优良树种，具有广阔的应用推广前景[2-5]。

顶果木虽然分布较广，但零星分散，因树干通直，材质好，遭受砍伐比较严重，甚至濒临灭绝，在《中国植物红皮书》中被列为稀有种，国家三级保护。20世纪70~80年代开始，一些学者对顶果木的生物学特性、繁殖栽培和生长等方面进行了研究[6-8]，直到2010年以后，在无性繁殖和栽培技术等研究有些相关报道[9-14]，关于顶果木良种选育方面未见报道。为抢救顶果木珍贵种质资源，我们开展了顶果木的优树选择研究，初步制定出顶果木选优的标准及方法，为该树种遗传改良和开发利用提供理论依据和技术支持。

1　材料与方法

1.1　选优林分概况

根据资料查阅和实地调查，选择在年龄21~40年、长势良好、没有经过负向选择的天然林和天然次生林作为选优林分，确立在广西田林县、龙州县和凭祥市3个地方进行优树选择。选优样区的自然条件及林分状况见表1。

表1 顶果木选优林分基本状况

地点	纬度	经度	海拔(m)	年均气温(℃)	年降水量(mm)	林龄(年)	小环境
龙州弄岗自然保护区	22°28′08″	106°57′08″	202	20.8~22.4	1088~1799	40	沟谷地
田林县浪平乡	24°32′32″	106°15′02″	435	26.8~21.5	1180~1254	40	河口、村旁
凭祥市夏石镇	22°53′28″	106°53′41″	212	19.5~21.4	1062~1772	30	山脚坡地

1.2 选优方法

在调查顶果木天然林分基础上选择出优质林分，再选择出候选优良单株，采用数量指标与形质指标相结合的方法确定优树。数量指标采用基准线法，相同立地条件的优良林分中，用生长锥测出优势木的年龄，以5年为1个龄级，实测其树高、胸径，根据胸径与材积回归曲线方程计算材积[15]，最后计算出该林分优树的树高、胸径及材积的基准线。形质指标采用评分法，通过主成分分析，选出主要因子和各因子的权重，每个因子采用3分制评分，最后算出总分。

1.2.1 优树生长量基准线计算

$$D\text{线(或}H\text{线)}=\overline{D}\text{(或}\overline{H}\text{)}+S$$

式中：S为标准差；$\overline{D}$、$\overline{H}$分别为优势木的胸径、树高年生长量均值，即：

$$\overline{D}=\frac{\sum_{i=1}^{n}(Di/Ai)}{n} \qquad \overline{H}=\frac{\sum_{i=1}^{n}(Hi/Ai)}{n}$$

式中：Di、Hi、Ai分布表示各优势木的胸径、树高和年龄，n为实测的株数。

1.2.2 形质指标的标准化

采用多目标决策的一维比较法，将候选优树形质指标进行标准化处理，冠幅和枝下高采用式(1)换算，干形、分枝、高径比和冠径比采用式(2)换算，指标标准化换算公式为：

$$Y=1-0.9\times(X_{max}-V)/(X_{max}-X_{min}) \quad (1)$$

$$Y=1-0.9\times(X-X_{min})/(X_{max}-X_{min}) \quad (2)$$

式中：X代表候选优树的形质指标测定值；X_{max}和X_{min}代表每个指标的最大值和最小值。

1.3 数据分析

采用Excel统计软件计算候选优树及优势木的树高、胸径和材积的总生长量和年均生长量，形质指标的主成分分析采用SPSS16.0统计软件。

2 结果与分析

2.1 数量指标的确定

本研究中选优林分的各龄级的树高、胸径、材积的生长量见表2。从表2中可以看出：优树的树高、胸径、材积总生长量呈随树龄的上升而增大的趋势，胸径和材积的年均生长量也随树龄的上升而增大，但树高年均生长量则相反。按照1.2.1中基准线计算方法，各龄级优树生长量基准线见表3。凡初选优树的胸径和树高年生长量达到标准线的要求则中选，共选出优树15株。

2.2 形质标准的确定

在顶果木天然林选优过程中，我们调查了优树的干形、冠幅、枝下高、高径比、冠径比、分枝数等6项指标，为了提高选优效率，简化形质指标数量，根据1.2.2中方法，在对各个形质指标进行标准化处理的基础上进行主成分分析，结果见表4。由表4可以看出，由于前3个主成分累计贡献率已达86.573%，其中，第1主成分中干形的特征向量最大，第2主成分中分枝数的特征向量最大，第3主成分中冠径比的特征向量最大。综上表明，干形、分枝数和冠径比是形质指标的决定因子。

表2 顶果木各龄级优势木的总生长量和年均生长量

龄级(年)	株数	范围	总生长量			年均生长量标		
			胸径(cm)	树高(m)	材积(m^3)	胸径(cm)	树高(m)	材积(m^3)
15~20	20	最大	29.20	29.80	0.7976	1.43	1.59	0.0419
		最小	20.80	14.80	0.2505	1.38	0.96	0.0164
		平均	24.53	21.24	0.4929	1.40	1.21	0.0276

（续）

龄级(年)	株数	范围	总生长量			年均生长量标		
			胸径(cm)	树高(m)	材积(m^3)	胸径(cm)	树高(m)	材积(m^3)
21~25	19	最大	36.30	31.50	1.4582	1.46	1.51	0.0591
		最小	29.50	10.90	0.3702	1.43	0.51	0.0173
		平均	32.91	23.69	0.9354	1.45	1.04	0.0408
26~30	21	最大	44.60	33.80	2.2452	1.48	1.19	0.0759
		最小	37.40	18.20	0.9006	1.46	0.71	0.0352
		平均	40.12	26.51	1.5195	1.47	0.97	0.0553
31~35	13	最大	52.70	35.10	2.9730	1.50	1.06	0.0894
		最小	45.70	22.00	1.5903	1.48	0.66	0.0517
		平均	49.50	26.66	2.2463	1.49	0.80	0.0676
36~40	6	最大	61.00	37.10	3.8322	1.51	1.01	0.1044
		最小	54.00	25.20	2.6071	1.50	0.63	0.0724
		平均	56.97	30.87	3.3744	1.50	0.82	0.0892
41~45	10	最大	69.50	47.90	7.2671	1.52	1.08	0.1633
		最小	61.20	24.10	3.1562	1.51	0.53	0.0763
		平均	65.33	30.90	4.4178	1.51	0.72	0.1020

表 3　顶果木优树的年平均生长量标准线

指标	林级(年)					
	15~20	21~25	26~30	31~35	36~40	41~45
胸径(cm/年)	1.42	1.46	1.48	1.50	1.51	1.52
树高(m/年)	1.43	1.25	1.09	0.93	0.97	0.88
材积(m^3/年)	0.0357	0.0498	0.0654	0.0782	0.1017	0.1289

表 4　形质指标的因子负荷量和累计贡献率

因子负荷量	枝下高	冠幅	干形	分枝	高径比	冠径比	累计贡献率
主成分 1	−0.622	0.115	0.852	0.509	−0.768	0.596	34.537
主成分 2	0.166	0.029	−0.897	0.639	0.143	−0.057	55.835
主成分 3	0.116	0.195	−0.188	−0.338	0.464	0.764	86.573

基于上述分析，以干形、分枝数和冠径比作为选择衡量优树的形质指标，将这三个指标联合作为考察标准评定候选优树的形质指标。干形、分枝数指标按 3 分制评分，主要观测树干 0~8m 内弯曲和分枝粗度情况，具体评判方法见表 5。根据表 5 的结果和林分实际情况，赋予干形、冠幅和分枝的权重分别为：0.5、0.3、0.2。三个指标的分值范围均为 1~3 分，分值越小，形质越优良。各优树的形质评分为：0.5×干形得分+0.3×分枝数得分+0.2×冠径比得分，以评分<2 为形质标准。表 5 中优树评判得分标准，在基准线选择的 15 株优树中确定优树 9 株。

表 5　优树干形和分枝数评判得分标准

指标	评判标准	得分
干形	通直、圆满	1
	1 个弯	2
	2 个或 2 个以上的弯	3
分枝	无分枝	1
	分枝细小，最粗分枝与同一位置主干的比小于 1/3	2
	分枝过多，最粗分枝与同一位置主干的比大于 1/2	3

3 结论与讨论

(1)本研究在3个市(县)的顶果木天然次生林中共选出候选优树20株，依据基准线法初选优树15株。根据优树干形和分枝评判得分标准，复选优树9株，入选率45%。顶果木天然林目前在广西分布范围较小，本选择标准限于广西西南部，不能代表其他分布区域的情况，所确定的优树生长量选择标准为最低标准，实际评选时可根据实际情况适当提高入选标准，本选优标准亦可为其他地方顶果木选优标准提供参考。

(2)本研究的林分为天然次生林，林分结构复杂、年龄不同，采用通常的优势木对比木法[16,17]及多性状联合选择[18,19]无法进行选优，因此，采用生长指标结合形质指标的标准线法是目前天然林优树选择最常用的方法。于树成等[20]在水曲柳天然林优树选择中，结合形质指标，采用标准地内样木胸径平均数加一倍标准差即为所选林分的候选优树。白卉等[21]山杨优树初选时采用分龄级制定出优树生长量的绝对标准，即预选优树达到或超过该龄阶平均树高乘以1.1，胸径乘以1.2，所得值为Ⅰ级优树最低标准，形质指标也符合要求时即可入选。以上研究证明，此方法选优方法易于掌握，可靠性高，遗传增益显著提高，对于其他阔叶树种优树选择具有较高的参照价值。此外，在实际选优过程中，有些树种生长年轮不明显，通过生长锥获取树龄困难，郑天汉等[22]通过对红豆树立木生长性状间的关系进行分析，找出材积性状与其他重要性状间的相关性及其相关程度，计算各因子的相对权重系数和综合指数，进行红豆树天然林表型优树选择。

(3)基准线法选出的优树只说明被选树本身生长量的快慢，3个生长指标同时达到入选标准的候选优树数量偏少。因此，还要兼顾形质方面的选择，分析候选优树与周围环境的关系，选择立地条件、龄级范围、生长势等要基本一致的周围对比木，消除环境误差的影响，尽量减少部分速生的优良资源的丢失。此外，在实际调查中，形质指标过多，影响选优效率，需要通过主成分分析法筛选出主要因子和因子权重，此方法计算形质综合得分结果客观、准确、高效。

参考文献

[1]郑万钧. 中国树木志：第二卷[M]. 北京：中国林业出版社，1998.

[2]周全连，李文付. 顶果木栽培技术[J]. 林业科技开发，2007，21(1)：91-92.

[3]朱积余，侯远瑞，刘秀. 广西岩溶地区优良造林树种选择研究[J]. 中南林业科技大学学报，2011，31(3)：81-84.

[4]杨成华. 速生珍稀树种顶果木[J]. 贵州林业科技，1989，17(2)：59-61.

[5]吕曼芳，江德候，秦武明，等. 珍贵树种顶果木的研究现状及趋势[J]. 广西林业科学，2014，1(4)：345-348.

[6]吕福基，袁杰，朱德金. 顶果木的生物学特性及其繁殖栽培[J]. 云南林业科技通讯，1987(1)：27-29.

[7]谢福惠，莫新礼. 速生优良树种—广西顶果木初步研究[J]. 广西植物，1981(1)：31-33.

[8]左辞秋. 顶果木种子解剖与催芽处理[J]. 热带林业科技，1986(1)：28-30.

[9]马跃，刘志龙，谌红辉，等. 顶果木嫩枝扦插育苗试验[J]. 林业科技开发，2012，26(6)：111-112.

[10]谌红辉，王小宁，马跃，等. 顶果木叶芽组织培养繁殖技术[J]. 林业实用技术，2012(20)：26-27.

[11]马跃，谌红辉，刘光金，等. 顶果木嫁接育苗技术[J]. 林业实用技术，2013(5)：34-35.

[12]郝建，全昭孔，农志，等. 顶果木抗旱生理特性研究[J]. 中国农学通报. 2014，30(1)：16-19.

[13]郝建，马小峰，谌红辉，等. 顶果木的耐旱性评价[J]. 西北林学院学报，2013，28(3)：63-66.

[14]吕曼芳. 珍贵树种顶果木人工林生长规律及价值核算研究[D]. 南宁：广西大学，2013.

[15]吕曼芳，梁乃鹏，秦武明，等. 顶果木人工林生长规律的研究[J]. 中南林业科技大学学报，2008，33(8)：43-49.

[16]国家标准局. LY/T 1344-1999 主要针叶造林树种优树选择技术[S]. 北京：中国标准出版社，1999.

[17]翁海龙，陈宏伟，段安安. 国内主要针叶树种优树选择技术研究进展[J]. 福建林业科技，2007，34(3)：250-254.

[18]洪永辉，胡集瑞，林文奖，等. 马尾松种子园无性系亲本多性状联合选择[J]. 南京林业大学学报(自然科学版)，2011，35(6)：23-28.

[19]林能庆. 闽西马尾松优树子代测定及优良单株选择[J]. 南京林业大学学报(自然科学版)，2013，37(5)：31-34.

[20]于树成，张桂芹，王宏，等. 水曲柳优树选择技术[J]. 林业勘查设计，2008(1)：49-50.

[21]白卉，邢亚娟，李春明. 山杨优树选择标准[J]. 中国林副特产，2008(3)：57-58.

[22]郑天汉，兰思仁. 红豆树天然林优树选择[J]. 福建农林大学学报(自然科学版)，2013，42(2)：365-370.

[原载：南京林业大学学报(自然科学版)，2014，38(05)]

顶果木优良家系和优良单株的选择研究

马 跃 刘志龙 谌红辉 刘福妹 麻 静

（中国林业科学研究院热带林业实验中心，广西凭祥 532600）

摘 要 本文以广西壮族自治区范围内顶果木17个家系的实生苗为材料，探讨了顶果木优良家系和优良单株苗期选择的一些关键问题。结果表明：①不同家系间的苗高、地径生长量存在极显著差异。②通过苗高、地径两指标综合选择，选出苗期生长较快的优良家系5个，入选率为29.41%。③采用标准差选择法，选出155株超级苗，入选率为17.2%。

关键词 顶果木；优良家系；优良单株；苗期选择

Seedling Selection of Superior Families and Superior Individual Plant Selection of *Acrocarpus fraxinifolius* Wight

MA Yue, LIU Zhilong, CHEN Honghui, LIU Fumei, MA Jing

(*The Experimental Centre of Tropical Forestry*, *Chinese Academy of Forestry*, *Pingxiang* 532600, *Guangxi*, *China*)

Abstract: Some key problems existed in seedling selection of superior families and superior individual plant of Acrocarpus fraxinifolius Wight were studied with the seedlings of 17 families in the Guangxi Zhuang Autonomous Region. The result s showed that: ①heights and basal diameter increments among different families were significantly different. ②5 superior families were selected with the characteristics of fast growing and the selection ratio is 29.41%. ③155 superior seedlings were selected by standard variance selection way with a selection ratio of 17.2% .

Key words: *Acrocarpus fraxinifolius*; superior families; superior individual plant; seedling selection

顶果木(*Acrocarpus fraxinifolius* Wight ex Arn.)，为苏木科(Caesalpiniaceae)顶果木属植物，被列入国家重点3级保护野生植物名录。它分布于马来西亚、印度、缅甸等热带和南亚热带地区，在我国主要分布于云南南部与广西西部地区。近年来，顶果木作为珍贵乡土用材树种及优良的石山造林树种，受到越来越多的重视，优质种苗的需求也日益增大。通过对现有的遗传材料进行收集整理，从混杂的群体中选择优良家系和优良单株，进行无性系测验，然后再选择其中的优良无性系造林，可以在短期内获得较大的增益。20世纪80年代学者对顶果木生长速度、材质、生态分布、播种育苗等有过报道[1]，近年来，关于顶果木的育苗技术、栽培技术以及无性繁殖技术也有一定的介绍[2-4]。但关于顶果木优良遗传材料选择的研究几乎还是一片空白。本研究在对广西壮族自治区范围内的顶果木天然林资源进行调查的基础上，选出优良母树17株，并对不同家系顶果木进行苗圃育苗及苗期选择，以期选出表现优良的家系及单株，为进一步进行顶果木无性系选育提供材料。

1 材料与方法

1.1 试验地概况

试验地点设在广西凭祥市中国林业科学研究院热带林业实验中心苗圃，106°47′E，21°57′N，属北热带季风气候，终年温暖湿润，年均温21.5℃，极端最高温39.8℃，最低温-1.5℃，≥10℃的积温6000~7600℃，全年日照时数1218~1620h，年降水量1200~1400mm，干湿季节明显，4~9月份为雨季，相对湿度80%，年蒸发量1200~1600mm，有霜期3~5d，气候适合顶果木苗木的生长。

1.2 试验材料

试验所用种子采自广西壮族自治区田林县、凤山县、巴马县等地经过初步选优的17株母树，分株进行采种并编号。播种时，每株的种子作为一个家系，分家系播于轻基质营养袋中育苗。

1.3 试验方法

本试验于2012年12月进行顶果木播种，约20d后经处理的种子在沙床中发芽并长出两片真叶，此时将芽苗移栽入轻基质营养袋中，采用随机区组设计，每个家系为一个小区，每小区育苗60株，重复3次。育苗期间按苗圃常规方法进行施肥、除草、消毒等田间管理。2013年7月对半年生顶果木进行苗木生长调查。分小区测定所有苗木高度(H)、随机抽取10株测量地径(D)。数据统计以小区为单位，利用SPSS 16.0统计软件对苗高、地径进行方差分析和多重比较。

2 结果与分析

2.1 不同家系间的生长量差异

对不同家系顶果木半年生苗的苗高、地径进行单因素方差分析可知，不同家系之间的苗高、地径生长量存在极显著差异(表1)，即不同家系生长性状具有遗传差异，这是顶果木进行优良家系选择的基础。

表1　不同家系间生长量差异显著性检验

项目	变异来源	离差平方和	自由度	均方	*F*值	*Sig.*
苗高	组间	3595.089	16	224.693	4.441	.000
	组内	1720.042	34	50.589		
	总和	5315.131	50			
地径	组间	8.316	16	0.520	3.604	.001
	组内	4.903	34	0.144		
	总和	13.218	50			

在各家系顶果木苗高、地径差异极显著的基础上，对不同家系间生长量进行S-N-K多重比较，如表2所示，苗高可以分为7个组，地径可分为5个组。各组间的家系差异显著，而各组内的家系差异不显著。从苗高来看，组1的5个家系苗高生长最差，可作为早期淘汰对象。组5、组6、组7的苗高生长最好，可作为早期入选家系。组2、3、4的苗高生长处于中等水平，应保留对其后期生长进一步实施观测，视后期生长优劣再作选择或淘汰。从地径来看，组1中的11个家系地径生长量无显著差别，不宜直接淘汰。组4、组5的2个家系地径生长量最好，可作为早期入选家系。

2.2 优良家系的选择

本研究选择苗高、地径两因素作为选优的指标。如果某一家系苗高、地径的平均值大于所有参试家系苗高的平均值，则该家系可以入选优良家系。参试各家系的平均苗高42.49cm，平均地径为2.96mm。据此标准，符合苗高入选条件的家系有14、16、5、15、3、13、11、17号，入选率为47.06%。符合地径入选条件的家系有15、5、2、3、14、17、11号，入选率为41.18%。同时满足苗高和地径入选条件的家系有5、15、3、11、17号，入选率为29.41%。

由表2还可看出，不同家系间苗高生长量较地径生长量差异大，因此按苗高因素将入选的5个优良家系进行排列，17号为最优家系，11号次之，其次分别是3号、15号、5号。

表2　不同家系间苗高、地径多重比较结果

家系		苗高(cm)							家系	地径(mm)				
		1	2	3	4	5	6	7		1	2	3	4	5
9	3	27.03							1	2.36				
1	3	29.87							10	2.45				
4	3	32.50							4	2.61				
8	3	34.32							7	2.61				
6	3	37.75							16	2.61				
12	3		40.61						9	2.66				
7	3		40.82						12	2.75				
10	3		41.05						6	2.84				
2	3		41.95						8	2.93				
14	3			44.36					13	2.96				
16	3			44.44					15	2.97				
5	3			46.25					5		3.21			
15	3				46.47				2		3.23			
3	3					48.98			3			3.39		

（续）

家系		苗高(cm)							家系	地径(mm)				
		1	2	3	4	5	6	7		1	2	3	4	5
13	3						52.58		14			3.47		
11	3						53.82		17				3.59	
17	3							59.60	11					3.76
Sig.		0.107	0.081	0.051	0.084	0.109	0.062	0.103	Sig.	0.103	0.098	0.054	0.073	0.128

2.3 优良单株的选择

为选出基因优良的遗传材料，本研究将从入选的5个优良家系苗中选择表现突出的优良单株，即超级苗。根据常规经验，一般将苗高、地径平均值加上1~3倍标准差作为选择超级苗的标准。但是本试验中，各家系的苗高、地径差异极显著，标准差较大(表3)，因此选择平均值加1倍标准差作为选择标准。

表3　各家系苗高、地径选择标准

家系号	苗高			地径			入选株数	入选率
	平均值	标准差	H+δ	平均值	标准差	D+δ		
15	46.50	14.60	61.10	2.97	0.73	3.70	24	13.3%
5	45.95	13.60	59.55	3.20	0.78	3.98	42	23.3%
3	49.00	14.10	63.10	3.40	0.83	4.23	37	20.6%
17	58.75	19.50	78.25	3.57	0.79	4.36	22	12.2%
11	52.95	17.50	70.45	3.70	1.1	4.80	30	16.7%

优良单株的选择标准即苗高和地径均在平均值加上1倍标准差以上的超级苗木，同时还要考虑形质指标，如干形较差、受病虫害危害或机械损伤者不能入选。经淘汰保留下来的苗木即为定选优良单株。由表3可知，5个优良家系中共有155个单株入选，入选率为17.2%。其中，家系5入选株数最多，为42株，入选率23.3%，家系17入选株数最少，为22株，入选率12.2%。

3　结论与讨论

3.1 结论

(1)顶果木不同家系间苗高、地径生长量存在极显著差异。各家系顶果木苗圃育苗条件相同，由此可认为不同家系间的生长状况差异是由亲本的遗传基础所致。这是顶果木优良家系及优良个体苗期选择的理论依据。

(2)苗高、地径是苗木质量中最直观和最准确的生长量指标，在苗木生产和科研中，常作为苗木分级指标使用。本研究通过苗高指标选择，共选出生长表现较好的优良家系8个，入选率为47.06%。通过地径指标选择，共选出苗期生长较好的家系7个，入选率为41.18%。通过苗高、地径两指标综合选择，选出生长良好的家系5个，入选率为29.41%。其中17号与11号表现尤为突出。

(3)以优良家系内的个体为选择对象，以优良家系平均苗高生长量加上1倍标准差为标准，共选出超级苗155株，入选率为17.2%。

3.2 讨论

由于林木生长发育周期很长，因而通过树木生长早期测定来正确评价林木育种成果为许多林木育种学者所关注[5-10]。遗传学和农林业生产实践的大量事实证明，从混杂的群体中选择优良的单株，通过无性系测验，然后再选择其中的优良无性系造林，可以获得目前科学水平下最大的增益[11]。顶果木为我国三级保护的珍稀树种，现存的大树很少，从现有的母树资源中选择优良的家系，并从苗圃中按一定的比例选择生长优异的优良单株，可较快获得遗传增益明显的无性系材料。目前国内对黑杨、马尾松、池杉、杉木、油松、樟子松、云南松等树种早期选择研究表明[12-16]，根据苗高和胸径生长优势进行苗期选择可能获得可靠的效果。

刘代亿[17]等认为云南松在苗期高生长较地径生长变异大，因此选择时以苗高为主、地径为辅。郑仁华[18]等对福建柏苗期选择的研究也得到相同结论，即福建柏苗期选择以苗高为主、地径为辅。但本研究认为苗高和地径在苗木生产中是两个极为重要的指标，应综合考虑进行苗木分级和选优。

参考文献

[1]左辞秋. 顶果木种子解剖与催芽处理[J]. 热带林业科技, 1986(1): 28-30.

[2]何关顺, 文宝, 何广琼. 乡土速生树种顶果木育苗技术[J]. 广西林业, 2008(5): 32-33.

[3]周全连, 李文付. 顶果木栽培技术[J]. 林业科技开发, 2007, 21(1): 91-92.

[4]马跃, 刘志龙, 谌红辉, 等. 顶果木嫩枝扦插育苗试验[J]. 林业科技开发, 2012, 26(6): 111-112.

[5]朱之悌. 树木的无性繁殖与无性系育种[J]. 林业科学, 1986, 22(3): 280-289.

[6]黄菊生, 王豁然. 世界林木遗传、育种和改良的研究进展和动向[J]. 世界林业研究, 1991, 4(1): 7-11.

[7]李明鹤. 遗传参数在林木早期选择中的应用[J]. 湖北林业科技, 1990(8): 13-18.

[8]叶培忠, 陈岳武. 杉木早期选择的研究[J]. 南京林业大学学报, 1981(1): 100-116.

[9]马常耕. 池杉速生无性系早期选择的研究[J]. 林业科学, 1979, 15(1): 194-198.

[10]陈益泰. 林木早期选择研究新进展[J]. 林业科学研究, 1994, 7(7): 13-22.

[11]武汉市园林科研所, 等. 选择池杉超级苗速生无性系研究小结[J]. 湖北林业科技, 1978(2): 28-33.

[12]郑畹, 舒筱武. 云南松优良种源生长量早期选择的研究[J]. 云南林业科技, 1998, 84(3): 12-17.

[13]王章荣, 陈天华, 周志春, 等. 福建华安马尾松生长早晚期相关及早期选择[J]. 南京林业大学学报, 1987, 11(3): 41-47.

[14]卢国美, 李国锋, 侯振中, 等. 油松生长力早期选择[J]. 河南林业科技, 1994(2): 12-14.

[15]丁振芳, 王景章, 方海峰, 等. 日本落叶松家系早期选择技术[J]. 东北林业大学学报, 1997, 25(3): 65-67.

[16]茹广欣, 张国栓, 冯胜, 等. 黑杨无性系的苗期选择分析[J]. 河南农业大学报, 2002, 36(2): 143-146. 1-9.

[17]刘代亿, 李根前, 李莲芳, 等. 云南松优良家系及优良个体苗期选择研究[J]. 西北林学院学报, 2009, 24(4): 67-72.

[18]郑仁华, 杨宗武, 施季森, 等. 福建柏优树子代苗期性状遗传变异和生长节律研究[J]. 林业科学, 2003, 39(1): 179-183.

[原载: 中国野生植物资源, 2014, 33(06)]

石灰岩山地芸香竹分株育苗基质和时间的选择

刘光金[1] 李洪果[1] 卢立华[1] 蔡道雄[1] 郭起荣[2]

([1]中国林业科学研究院热带林业实验中心，广西凭祥 532600；
[2]国际竹藤中心，国家林业局竹藤科学与技术重点开放实验室，北京 100102)

摘 要 为了探索芸香竹育苗的最适方法及其关键技术，从而提高繁殖系数，采用无纺布袋荫棚繁育的方法，开展了芸香竹分株育苗试验，筛选适宜基质和育苗时间。结果表明：育苗基质对育苗成活率影响显著，100%轻基质，或者50%轻基质混50%岩溶山地土为合适的基质，分株育苗成活率可达70%以上；夏季分株育苗成活率最高，平均为71%，其中7月育苗最高达75%；春季分株育苗平均成活率51%，5月最高可达70%，秋冬季不适合分株育苗。

关键词 芸香竹；育苗基质；育苗季节；石灰岩山地；分株育苗

Effects of Propagation Medium and Season on Offshoot Cultivating Seedlings of *Bonia amplexicaulis*

LIU Guangjin[1], LI Hongguo[1], LU Lihua[1], CAI Daoxiong[1], GUO Qirong[2]

([1]*The Experimental Centre of Tropical Forestry*, *CAF*, *Pingxiang* 532600, *Guangxi*, *China*;
[2]*International Centre for Bamboo and Rattan*, *Beijing*, 100102, *China*)

Abstract: In order to explore the optimum method and some key propagation techniques on *Bonia amplexicaulis* seedling, and to enhance propagation coefficient, offshoot breeding was conducted in the non-woven bag shade canopy, and the optimum breeding medium and season were selected. The results showed that breeding media had significant differences on survival rate of seedlings. The optimum breeding medium was 100% light medium or the mixture of 50% light medium and 50% limestone mountain soil, and the survival rate could be higher than 70%. The survival rate was the highest in summer, the mean value was 71%, and it in July was up to 75%. The survival rate was 51% in spring, and it in May was up to 70%. Autumn and winter was unfit to do offshoot breeding.

Key words: *Bonia amplexicaulis*; propagation medium; seedling season; karst topography bamboo; offshoot cultivating

芸香竹[*Bonia amplexicaulis* (L. C. Chia *et al.*) N. H. Xia]属禾本科竹亚科单枝竹属竹种[1]，为石灰岩山区少见的长势优良，广泛分布于广西、广东、贵州、云南等地的石漠化的重要经济竹种[1-3]。芸香竹根系发达，紧附在石岩山地土壤，少些能扎入岩石缝隙，对石漠化地区生态防护与治理有重要意义；民间还常用以编制围篱、农具、造纸、牛羊饲料，又因竹叶清香用于蒸煮粽子等，表现出优良的食材特性。芸香竹繁殖技术还未见报道，为了提供繁殖系数，丰富石漠化治理生物材料，开展了分株育苗试验。

1 材料与方法

1.1 试验地概况

试验点设在广西凭祥中国林业科学研究院热带林业实验中心苗圃，地理位置106°47′E，21°57′N，属南亚热带季风气候，年平均温21.5℃，极端最高温39.8℃，最低温-1.5℃，≥10℃的积温6000~7600℃，全年日照时数1218~1620h，年降水量1200~1400mm，干湿季节明显，4~9月为雨季，相对湿度80%，年蒸发量1200~1600mm，有霜期3~5d。

1.2　试验材料

供试材料采自广西凭祥中国林业科学研究院热带林业实验中心石山树木园攀爬在石灰岩山体上的健壮的芸香竹。于阴天早晨9：00挑选生长健壮蔸丛，选择1~2年生竹秆，秆径0.5~1.0cm，修剪20~30cm高，小心从秆柄处截断，分为单根带蔸竹秆，竹秆上部要求带有秆节1~2个，采用杯径为15cm的无纺布营养袋，配置5种不同育苗基质，开展分株育苗试验。育苗前1个月采用透光率为50%遮阴网控光；注意适时浇水，手攥基质不滴水时，灌一次透水。培育1个月开始抽枝展叶，进行一般水肥管理；培育2个月后，去除遮阴网，进行全日照管理。

1.3　试验方法

1.3.1　育苗基质

适宜育苗基质试验于2012年5月开展。试验基质为黄心土、岩溶山地土、河沙、100%轻基质，50%轻基质混50%岩溶山地土(体积比)等5种。轻基质为中国林业科学研究院热带林业实验中心轻基质厂生产，主要原料为炭化树皮80%+松锯末10%+炭化锯末10%，堆沤8个月[4]；岩溶山地土采自石山树木园。每个处理定植30个育苗段，重复3次。6个月后，调查育苗成活率。以定植在基质袋中芸香竹竹秆节部侧芽或蔸部鞭芽萌发，抽枝展叶，视为成活株，进行统计分析，不计竹秆干枯、枯死的植株。

1.3.2　育苗季节

分别于2012年夏季(6月、7月、8月)、秋季(9月、10月、11月)、冬季(12月、1月、2月)和2013年的春季(3月、4月、5月)共4个季节，每月进行试验，每次定植30个育苗段。育苗基质选用100%轻基质。6个月后，调查孕笋发竹情况。

2　结果与分析

2.1　育苗基质对成活率的影响

调查结果表明，成活株新发叶9~57片，平均18.4片；叶长6.0~24.9cm，平均16.9cm；叶宽1.1~5.1cm，平均2.6cm。从表1的统计数据可知：5种基质处理的育苗成活率差异显著，100%轻基质和50%轻基质混50%岩溶山地土两个处理，育苗成活率最高，达70%；黄心土和岩溶山地土次之，为40%；河沙处理最低，仅为20%。基质对育苗成活率影响的方差分析结果(表2)说明，不同基质对分株育苗成活率的影响显著($P=0.023<0.05$)，表明育苗基质对育苗成活有重要的影响。100%轻基质和50%轻基质混50%岩溶山地土通透性良好、保水能力强、肥力好，适宜生根、发芽、成活，是合适的分株育苗基质。黄心土、岩溶山地土黏性强，易板结，通透性差，竹苗难以生根。而河沙保水能力差，竹苗易干枯，成活率最低，育苗效果最差。

表1　不同育苗基质中芸香竹的成活情况

编号	基质	成活率(%)	标准差	标准误差
1	黄心土	45	8.66	5
2	岩溶山地	40	20	11.54
3	河沙	20.33	26.27	15.16
4	100%轻基质	73.33	11.54	6.67
5	50%轻基质+50%岩溶山地土	70	17.32	10

表2　不同育苗基质中芸香竹成活率的方差分析

变异来源	平方和	自由度	均方	F值	显著性
组间	5847.6	4	1461.9	4.572	0.023
组内	3197.33	10	319.73		
总变异	9044.93	14			

2.2　育苗季节对成活率的影响

芸香竹分株育苗成活率的育苗时间效应(图1)表明：10月份分株育苗，成活率仅为8%，不适宜分株育苗；11月份、12月份和1月份的均无成活，不能实施分株育苗。2月份、3月份、4月份、8月份和9月份的成活率为50%左右，可以分株育苗；5月份、6月份、7月份的成活率最好，达70%以上，适合分株育苗。对育苗时间的统计分析可知(表3)，夏季分株育苗成活率最高，平均为66%；春季的成活率次之，均值为58.3%；秋、冬两季的成活率最低，为18%、16%左右。育苗月份对成活率影响的方差分析(表4)表明，育苗季节对其成活率的影响显著($P=0.027<0.05$)，说明育苗季节对育苗成活有重要的影响。

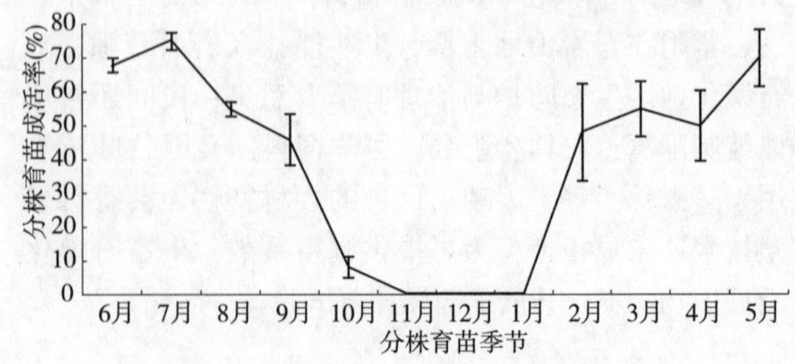

图1　芸香竹分株育苗时间响应

表 3 不同育苗季节芸香竹的成活情况

编号	育苗季节	成活率(%)	标准差	标准误差
1	春季	58.3	10.408	6.009
2	夏季	66.0	10.149	5.859
3	秋季	18.0	24.576	14.189
4	冬季	16.0	27.712	16.000

表 4 不同育苗季节芸香竹成活率的方差分析

变异来源	平方和	自由度	均方	*F* 值	显著性
组间	6214.25	3	2071.417	5.233	0.027
组内	3166.667	8	395.833		
总变异	9380.917	11			

3 结论与讨论

如其他竹类植物一样，芸香竹不常开花[5]，天然次生林下林龄超过 50 年的芸香竹仍未见开花结实，因此，其适花龄应>50 年，有性繁殖困难[5,6]。野外调查时发现，3~9 月，芸香竹一直有笋芽萌发，以笋芽为特征的无性繁殖是其野生种群繁衍、更新的主要方式，而人工育苗技术研究尚未见报道，笔者首次开展的分株育苗试验，最高成活率可达 70%以上，可为其繁苗、栽培与利用提供技术积累。

育苗基质是苗木培育的基础，对苗木成活率和生长影响显著[7-9]。100%轻基质、50%轻基质混 50%石山土两种育苗基质类型分株育苗成活率达 70%以上，为适宜的育苗基质；而黄心土、石山土、河沙等常用基质育苗成活差，说明选择保水性与通气性均衡的育苗基质有利于苗木成活。

育苗季节对成活率影响显著[4,10]。夏季是分株育苗最适季节，成活率达 70%以上，其中 7 月育苗成活率最高；春季次之，成活率为 58%，可以实施分株育苗；秋、冬季不适于分株育苗。蒋能[11]等对单枝竹属单枝竹变种箭秆竹(*Bonia saxatilis* var. *solida*)分株育苗的试验也表明，育苗季节对成活率影响显著，以 3 月份育苗成活率最高，1 月份次之，2 月份最差。因此，筛选最适育苗时间是提高芸香竹分株育苗时提高成活率的关键。

参考文献

[1]XIA Nianhe，CHRIS Stapleton. *Bonia Balansa*，J. Bot. (Morot) 4：29. 1890[J]. Flora of China，2006，22：49-50.

[2]XIA N H. A study of the genus*Bonia*(Gramineae：Bambusoideae)[J]. Kew Bull，1996，51(3)：565-569.

[3]李德铢，郭振华．云南竹亚科一些属种的增订[J]. 云南植物研究，2000，22(2)：43-46.

[4]郭文福，蒙彩兰．穗条生根剂、育苗基质和季节对西南桦扦插生根的影响[J]. 林业科学研究，2011，24(6)：788-791.

[5]杜凡，薛嘉榕，杨宇明，等．15 年来云南竹子的开花现象及其类型研究[J]. 林业科学，2000，36(6)：57-68.

[6]杨南，李福秀，普晓兰，等. 竹类植物育苗技术的研究进展[J]. 竹子研究汇刊，2008，27(3)：37-41.

[7]李明伟，叶维雁，刘鹏，等．不同基质对葡萄柚嫁接苗生长的影响[J]. 经济林研究，2014，32(3)：139-143.

[8]黄桂华，梁坤南，周再知，等．不同基质配方对柚木组培苗移植效果的影响[J]. 中南林业科技大学学报，2014，34(1)：32-36.

[9]宋祥兰，王兰英，邝先松，等．“赣州油”系列油茶容器育苗基质配方试验初探[J]. 中南林业科技大学学报，2014，34(1)：23-26.

[10]胡冬南，万晓敏，谢风，等．光皮树嫩枝扦插繁殖技术[J]. 经济林研究，2013，31(2)：146-150.

[11]蒋能，黄仕训，周太久．三种竹子繁殖特性初探[J]. 广西科学院学报，2002，18(3)：141-144.

[原载：经济林研究，2014，32(04)]

降香黄檀工厂化育苗轻基质筛选试验

贾宏炎 黎 明 曾 冀 蒙彩兰

（中国林业科学研究院热带林业实验中心，广西凭祥 532600）

摘 要 将沤制松树皮、沤制锯末和炭化松树皮及黄心土4种材料按照不同组合与配比制成12种轻基质配方，用于降香黄檀网袋容器育苗试验，依据苗木出圃率、苗高、地径、生物量以及基质成本等指标对其育苗成效进行综合评价，旨在筛选出适合降香黄檀苗木培育的轻基质配方。结果表明，不同配方间苗木出圃率、苗高、地径、地上和地下生物量均存在极显著差异($P<0.01$)；依据苗木出圃率、生长指标以及育苗成本进行综合评价可知，含有松树皮和黄心土的配方12、11和10育苗效果最好，其次是含有沤制松树皮和炭化松树皮的配方5、4，这些基质育苗效果好与其持水量适中有关；含有较多锯末的配方3、8和9育苗效果最差。由于配方12和11因黄心土比例过大不利于轻基质网袋的灌装加工以及配方5因炭化松树皮成本较高，建议采用配方10(沤制松树皮90%+黄心土10%)、配方4(沤制松树皮75%+炭化松树皮25%)2种混合基质进行降香黄檀苗木工厂化生产。若选用有锯末的配方基质，其添加比例不宜超过20%。

关键词 降香黄檀；工厂化育苗；轻基质；筛选试验

Screening Tests on Light Growing Media for Industrial Container Seedling of *Dalbergia odorifera*

JIA Hongyan, LI Ming, ZENG Ji, MENG Cailan

(*Experimental Center of Tropical Forestry, Chinese Academy of Forestry, Pingxiang* 532600, *Guangxi, China*)

Abstract: Twelve kinds of light growing media were prepared in accordance with different combinations and ratios, with four sorts of materials (composted pine bark, sawdust, charred pine bark and forest soil). In order to screen suitable light matrix formula for *Dalbergia odorifera* T. seedling cultivation, the mesh bag and container seedling tests of *D. odorifera* were carried out with the twelve prepared light growing media. The seedling raising efficiency was comprehensively evaluated by taking qualified seedling rate, seedling height, ground diameter, biomass and substrate cost into account. The results show that among different formulation treatments, there existed very significant differences in percentage of out planted seedlings, seedling height, root collar diameter and above-ground biomass and under-ground biomass ($P< 0.01$); Comprehensively taking above factors into consideration, the nursery effects of medium No. 12 (*v/v*, 70% retting pine bark + 30% soil), No. 11 (*v/v*, 80% retting pine bark + 20% soil) and No. 10 (*v/v*, 90% retting pine bark and 10% soil) were the best, followed by medium No. 5 and No. 4, this resulted from suitable water conservation of these media; Medium No. 3, No. 8 and No. 9 containing much sawdust had the poorest production of seedlings. The medium No. 12 and No. 11 were unfavorable for filling and processing light matrix mesh bag due to large proportion yellow soil, and the medium No. 5 had higher cost due to higher price of charred pine bark, so the medium No. 10 and No. 4 were recommended for industrial production of *D. odorifera* container seedlings. And when sawdust was used as light medium for growing seedlings of this tree species, its ratio should not be more than 20%.

Key words: *Dalbergia odorifera*; industrial container seedling; light growing media; screening test

降香黄檀(*Dalbergia odorifera* T. Chen)又称花梨木、黄花梨，为蝶形花科黄檀属半落叶乔木，自然分布于我国海南，广东、广西、福建、云南等地相继引种成功[1,2]。降香黄檀不仅木材坚硬，是制作名贵家具、雕刻、美工装饰品的上等用材，而且从其木材提取的降香油具有降血压、血脂、止血、止痛的功效，亦可作香料中的定香剂[3]。由于降香黄檀木材极为珍贵，市场上供不应求，而其野生资源已近枯竭[4]，近几年来热带、南亚热带地区掀起了其人工林的发展高潮，年苗木需求量逾千万株，降香黄檀已成为该地区重点发展的珍贵树种之一。

尽管降香黄檀萌芽能力强，生产上有人采用培育裸根苗截干造林方式，然而因其造林效果受季节影响大，绝大多数采用容器苗造林。基质选择及其配比是容器育苗成功与否的关键，一直成为容器育苗技术研究的重点[5]。轻基质网袋容器育苗与传统基质容器育苗相比，其具有基质重量轻、容器小，可进行空气修根、适合工厂化生产[6,7]及易运输、造林效果好等优点，已在多个树种上得到应用[8-11]。近年来，陈海军等[12]将泥炭以及锯末、谷壳等农林废弃物应用于降香黄檀轻基质网袋容器苗生产，何琴飞[13]等采用黄心土、椰糠、泥炭、塘泥等配置多种类型基质，筛选降香黄檀容器苗的适宜基质配方。这些研究或因采用不可再生资源泥炭而成本偏高，或因黄心土比例过大难以应用于工厂化生产。本试验采用沤制松树皮、沤制锯末和炭化松树皮及黄心土4种基质材料，配置12种轻基质配方开展降香黄檀育苗试验，依据幼苗生长表现评价其育苗效果，筛选出适合的轻基质配方应用于降香黄檀苗木的工厂化生产，为其人工林大规模发展提供技术支撑。

1　材料与方法

1.1　苗圃地概况

试验圃地位于中国林业科学研究院热带林业实验中心(以下简称“热林中心”)林木种苗示范基地，属南亚热带季风气候，终年温暖湿润，年平均气温21.5℃，极端最高温39.8℃，极端最低温-1.5℃，≥10℃的积温6000~7600℃，全年日照时数1218~1620h，年降水量1200~1400mm，干湿季节明显，4~9月份为雨季，相对湿度约80%。

1.2　试验材料与育苗管理

降香黄檀种子采自热林中心石山树木园。松树皮、黄心土和锯末采集于本单位林区或附近木材加工厂，黄心土需经过筛处理，其他轻基质材料采用参考文献[14]的方法处理，按表1中的配方制成口径4.5cm、长10cm的轻基质网袋容器，装入塑料育苗盘待用。每个育苗盘内有54个圆柱形框架，可对所育苗木进行空气修根，亦可防止苗木间出现串根现象。

播种前将沙床和种子进行消毒。播种前1d，用清水浸种24h。于2013年3月上旬将种子均匀撒播于平整的沙床上，再覆沙1.5~2cm。保持苗床湿润，约10d开始发芽。3月下旬待子叶完全展开时移栽至轻基质网袋容器，采用人工雾状喷淋，具体每次喷淋时间及次数，视天气状况和基质湿度而定。每隔10d施氮肥或复合肥1次，浓度为3000~8000mg/L，施肥浓度逐渐加大。其他措施与常规育苗类似。

1.3　试验设计与铺设

试验采用随机区组设计，按不同基质组合及其配比共设置4组12个轻基质配方(表1)，4次重复，每小区54株，即每个小区1盘。

表1　参试轻基质配方的组分及其配比(体积比)

分组	配方号	沤制松树皮(%)	炭化松树皮(%)	沤制锯末(%)	黄心土(%)
	1	75	15	10	
I	2	50	30	20	
	3	25	45	30	
	4	75	25		
II	5	50	50		
	6	25	75		
	7	75		25	
III	8	50		50	
	9	25		75	
	10	90			10
IV	11	80			20
	12	70			30

1.4 苗木生长指标观测

2013 年 12 月中旬测定苗高和地径，每个小区测 50 株幼苗，根据其平均苗高和地径，选取 5 株平均大小的苗木调查生物量。出圃率依据总体苗高、地径的均值及其标准差，参照平均值减 1 个标准差法[15]，确定出圃苗木生长指标的分界值，统计并计算出圃率。在此研究中，苗高和地径均分别大于或等于 27.5cm 和 4.5mm 的幼苗即达到出圃要求。

1.5 基质理化性质的测定

对于每种配方的基质，每次从制作的基质肠中随机截取 5 小段，每小段长为 40cm，重复 5 次，即每种配方基质取 5 个样，待晾干后测定其理化性质。基质物理性质的测定方法参照文献[16]，采用烘干法测定容重，采用吸水法和浸水法测算出相应的毛管孔隙度、总孔隙度、非毛管孔隙度、最大持水量等。基质化学性质的测定采用常规方法[17]，首先按风干基质与蒸馏水之比 1∶5(W∶V)混合，振荡 40min，过滤后获得基质浸提液，然后分别应用 PHS-3C 型 pH 计和 DDS-11A 数显电导仪测定浸提液的 pH 和电导率；有机质用重铬酸钾氧化—外加热法测定，水解氮用扩散法测定，速效磷用盐酸—硫酸浸提法测定，速效钾用乙酸铵浸提—原子吸收法测定。具体结果见表 2。

表 2 参试轻基质的理化性质

配方号	容重(g/cm^3)	总孔隙度(%)	毛管孔隙度(%)	非毛管孔隙度(%)	最大持水量(%)	pH
1	0.23±0.01ef	69.2±3.1b	58.5±3.0bc	10.6±2.0b	300.2±0.2d	5.0±0.2
2	0.23±0.01ef	66.9±4.8bc	52.3±2.7ef	14.6±6.9b	290.6±0.3de	4.9±0.3
3	0.24±0.01def	66.9±1.9bc	53.8±1.7de	13.2±3.0b	278.0±0.2de	4.8±0.4
4	0.24±0.02de	66.0±4.8bcd	55.3±3.3cde	10.7±4.6b	270.6±0.2e	4.9±0.4
5	0.26±0.02de	60.0±3.9de	48.4±4.5f	12.3±4.6b	234.2±0.1f	4.9±0.3
6	0.27±0.01d	57.5±6.7e	35.3±3.4g	22.2±5.5a	213.4±0.2fg	4.3±0.3
7	0.21±0.02fg	71.7±3.0b	59.0±3.0bc	2.7±0.9b	343.0±0.2c	4.7±0.4
8	0.19±0.01h	77.4±2.5a	61.6±1.1b	15.8±3.2b	406.2±0.2b	5.0±0.3
9	0.17±0.01h	80.8±1.8a	70.4±3.4a	10.4±3.1b	490.2±0.1a	5.1±0.3
10	0.32±0.05c	67.3±6.5bc	52.2±4.9ef	15.1±3.3b	204.4±0.2g	5.0±0.5
11	0.50±0.05b	62.6±4.1cde	57.5±2.4bcd	5.1±3.0c	124.8±0.1h	4.7±0.3
12	0.56±0.03a	60.5±3.5de	55.5±2.1cde	5.0±2.2c	108.2±0.1h	4.7±0.4

配方号	电导率(ms/cm)	有机质(mg/kg)	水解氮(mg/kg)	有效磷(mg/kg)	速效钾(mg/kg)
1	1.2±0.5abc	611.9±13.7d	427.1±28.3c	321.3±10.5e	501.3±6.2c
2	0.9±0.2c	652.4±16.9bc	415.7±20.5c	342.7±6.8d	545.1±7.5a
3	1.0±0.3bc	677.5±28.0a	367.1±15.7d	361.2±18.2c	527.8±8.8b
4	1.1±0.2bc	594.0±17.7d	357.3±21.3d	276.0±13.3f	78.4±1.9fg
5	1.1±0.3bc	650.3±19.9bc	299.3±8.2e	222.7±4.4g	73.6±3.4g
6	1.2±0.3abc	638.7±31.3c	228.3±29.6f	189.9±2.1h	75.3±0.6g
7	1.2±0.3abc	600.4±11.0d	569.3±5.7b	368.8±12.8c	82.8±1.7f
8	1.2±0.7abc	612.0±13.3d	558.0±32.8b	431.9±18.5b	89.5±0.5e
9	0.7±0.2c	670.1±5.7ab	599.0±25.0a	510.3±18.7a	94.3±1.4de
10	1.7±0.4a	192.9±7.4e	301.4±12.1e	195.1±10.5h	97.1±4.1d
11	1.5±0.5ab	122.0±7.8f	196.9±7.5g	132.0±3.9i	49.9±1.4h
12	1.3±0.3abc	94.0±3.4g	187.7±4.6g	93.0±2.9j	35.1±0.2i

1.6 数据处理与分析

应用 SPSS16.0 软件进行数据处理与分析，对各项指标进行方差分析及 Duncan 多重比较，幼苗出圃率进行反正弦转换后进行方差分析。根据上述生长指标，应用布雷金多性状综合评定法对各基质配方进行综合评价[18]。

2 结果与分析

2.1 轻基质配方对降香黄檀幼苗出圃率的影响

不同轻基质配方之间降香黄檀幼苗出圃率差异极显著($P<0.01$，下同)。从表 3 看出，配方 1、4、

表 3　12 种轻基质配方间降香黄檀幼苗出圃率和生长差异

分组	配方号	出圃率(%)	苗高(cm)	地径(mm)	地上部分干重(g/株)	地下部分干重(g/株)
	1	81.1±6.62ab	38.0±1.98cd	5.4±0.38c	3.72±0.64bc	1.25±0.16bcd
Ⅰ	2	75.6±6.47b	34.4±1.14e	5.1±0.20d	3.12±0.50cde	1.03±0.35cd
	3	72.1±10.71b	32.8±1.36ef	5.1±0.27d	2.70±0.21def	0.91±0.17de
	4	83.0±2.16ab	39.9±2.91bc	5.6±0.20bc	3.83±0.38bc	1.20±0.23bcd
Ⅱ	5	80.8±8.24ab	40.2±2.17bc	5.6±0.15bc	3.88±0.62bc	1.40±0.30abc
	6	80.6±9.85ab	37.8±1.78cd	5.6±0.11bc	3.48±0.49bc	1.36±0.18abc
	7	75.5±4.63b	35.7±0.92de	5.1±0.11d	3.35±0.49cd	1.01±0.36cd
Ⅲ	8	55.6±7.68c	30.3±2.16f	4.9±0.13d	2.57±0.36ef	0.85±0.19de
	9	31.6±15.79d	24.8±2.48g	4.5±0.16e	2.22±0.26f	0.59±0.13e
	10	83.1±5.66ab	42.5±1.49ab	5.7±0.12bc	4.19±0.64ab	1.55±0.24ab
Ⅳ	11	83.0±9.45ab	44.0±3.41a	5.8±0.19b	4.61±0.64a	1.74±0.31a
	12	89.7±6.03a	43.6±2.17a	6.1±0.14a	4.84±0.29a	1.71±0.21a

注：平均值±标准差。同列有小写字母相同表示差异不显著，不相同字母表示差异显著($P<0.05$)。

5、6、10、11、12 的苗木出圃率较高，均超过 80%，且这些配方间苗木出圃率差异不显著；其次为配方 2、3 和 7，其出圃率显著低于 12 号配方，但与上述其他配方差异不显著；而配方 8 和 9 的苗木出圃率均在 60%以下，显著低于其他基质配方。比较每个组内各配方间出圃率差异可以看出，对于组 I、II 和 IV，其 3 个配方间苗木出圃率均差异不显著，而组 III 的 3 个配方间差异显著，即随着沤制锯末比例的增大，其苗木出圃率显著减小。

2.2　轻基质配方对降香黄檀幼苗生长的影响

不同轻基质配方对降香黄檀苗高和地径生长均具有极显著的影响。以配方 10、11 和 12 的苗高最大，其次为配方 4、5，而配方 3、8 和 9 的苗高最小，显著低于其他基质配方；地径则以配方 12 为最大，显著高于其他基质配方，其次为配方 4、5、6、10、11，配方 3、8 和 9 地径最小(表 3)。组内各配方比较，对于组 I 和 III，其 3 个配方间苗高和地径均差异显著，苗高和地径随着沤制锯末比例的增加而递减；组 II 的 3 个配方间苗高和地径均差异不显著；而组 IV 的 3 个配方间苗高差异不显著，地径差异显著，地径随着黄心土比例的增加而递增。

各轻基质配方间苗木地上生物量和地下生物量均存在极显著差异。地上生物量和地下生物量以配方 11 和 12 最大，配方 1、4、5、6、10 次之，配方 3、8、9 的地上生物量和地下生物量均最小(表 3)。分析各组内配方间生物量的差异可知，组 II 和 IV 的 3 个配方间地上和地下生物量均差异不显著；组 III 的 3 个配方间差异显著，而组 I 的 3 个配方间地上生物量差异显著，地下生物量差异不显著；组 I 和 III 的生物量均随着沤制锯末比例的增加而递减。

2.3　轻基质生产的成本对比分析

依据目前市场价，沤制松树皮、炭化松树皮、沤制锯末、黄心土的价格分别为 130、260、90 和 50 元/m^3，统一按照 1m^3基质生产 4000 个网带容器，相应的网袋、水电、消毒液和人工等其他费用为 130 元/m^3，计算得出各基质配方的生产成本(表 4)。从表 4 可以看出，轻基质网袋容器的生产成本以配方 6 为最高，其次为配方 5 和配方 3，配方 9 的成本为最低，仅为配方 6 的 64.4%。在轻基质网袋容器生产过程中注意到，配方 11 和 12(黄心土比例分别为 20%和 30%)的基质生产效率均低于其他基质配方，其主要原因是这两个配方中添加黄心土比例过大，增加了基质的紧实度，导致容器灌装机的下料螺杆旋转速率减慢或受阻，从而影响了基质生产的效率。

表4　12种配方的轻基质生产成本核算

配方号	基质成本(元/m^3)	其他费用(元/m^3)	合计(元/m^3)
1	145.5	130	275.5
2	161.0	130	291.0
3	176.5	130	306.5
4	162.5	130	292.5
5	195.0	130	325.0
6	227.5	130	357.5
7	120.0	130	250.0
8	110.0	130	240.0
9	100.0	130	230.0
10	122.0	130	252.0
11	114.0	130	244.0
12	106.0	130	236.0

注：1m^3基质生产4000个轻基质网带容器；其他费用为生产4000个轻基质网袋容器所需的网袋、水电、消毒液和人工等费用之和。

2.4　轻基质配方间降香黄檀育苗效果的综合比较

表5　12种轻基质配方降香黄檀网袋容器苗的综合评价

配方号	出圃率	苗高	地径	地上生物量	地下生物量	成本	求和	求和的平方根	排序
1	0.904	0.864	0.885	0.769	0.718	0.835	4.975	2.230	6
2	0.843	0.782	0.836	0.645	0.592	0.790	4.488	2.118	9
3	0.804	0.745	0.836	0.558	0.523	0.750	4.217	2.053	10
4	0.925	0.907	0.918	0.791	0.690	0.786	5.017	2.240	5
5	0.901	0.914	0.918	0.800	0.809	0.708	5.049	2.247	4
6	0.899	0.859	0.918	0.719	0.782	0.643	4.820	2.195	7
7	0.842	0.811	0.836	0.692	0.580	0.920	4.682	2.164	8
8	0.620	0.689	0.803	0.531	0.489	0.958	4.090	2.022	11
9	0.352	0.564	0.738	0.459	0.339	1.000	3.451	1.858	12
10	0.926	0.966	0.934	0.866	0.895	0.913	5.500	2.345	3
11	0.925	1.000	0.951	0.952	1.000	0.943	5.771	2.402	2
12	1.000	0.991	1.000	1.000	0.983	0.975	5.948	2.439	1

根据苗木出圃率、生长指标与成本，应用布雷金综合评定法对各基质配方的育苗效果进行综合评价。具体方法是：首先在12个基质配方中找出每个性状的最大值，然后将该性状在所有基质配方中相应的数值除以最大值得到标准化值，求算各性状此值之和的平方根，即为综合评价值，其值越大，说明该基质配方的育苗效果越好，反之则越差。由于出圃率和生长指标是越高越好，而实际成本是越低越好，为了便于比较，将实际成本求倒数后再进行上述计算。由表5可知，配方12、11和10的育苗效果最好，其次为配方5和4，配方3、8和9的育苗效果最差。

3　结论与讨论

以往有关降香黄檀轻基质育苗研究中，陈海军等[12]采用锯末35%+炭化锯末15%+泥炭30%+炭化谷壳20%的基质培育降香黄檀1年生容器苗，85%以上的苗木地径和高分别在0.5cm和30cm以上，取得了较好的育苗效果，然而泥炭为不可再生资源，而且成本高。本试验利用南方资源丰富的松树皮、锯末和黄心土，经适当加工作为基质材料培育降香黄檀苗木，培育约9个月的苗木，大部分轻基质配方处理的出圃率即达80%以上，尤其是配方10、11、12和4的出圃率均在83%以上，而且综合评价排名靠前，这些基质配方中松树皮所占比例均超过 70%，说明松树皮可代替泥炭作为降香黄檀育苗基质。

根据苗木出圃率、生长指标以及育苗成本的综合评价结果，组 IV 和 II 轻基质的育苗效果明显优于其他两组。组 IV 中的配方 12、11 和 10 的综合育苗效果位列前 3，其苗高、地径、生物量等生长表现最佳，而且苗木出圃率最高，说明松树皮添加适量黄心土对降香黄檀苗木生长有利。从其高电导率可以看出(表 2)，添加黄心土增加了阳离子交换量，可能补充了矿物质，特别是微量元素，从而有效促进降香黄檀苗木生长。组 II 中的 3 个配方处理均为沤制松树皮和炭化松树皮的混合基质，其基质持水量亦适中，取得了较好的育苗效果。可能由于炭化松树皮的成本高，而配方 6 炭化松树皮的比例大，以致配方 5 和 4 的综合育苗效果分列第 4 和 5 位，而配方 6 降至第 7 位。组 I 和 III 中的配方处理均掺有沤制锯末，育苗成效随着锯末比例的增大而下降，以配方 3(沤制松树皮 25%＋炭化松树皮 45%＋沤制锯末 30%)、8(沤制松树皮 50%＋沤制锯末 50%)和配方 9(沤制松树皮 25%＋沤制锯末 75%)的育苗效果最差。这些基质尽管容重均较轻，非毛管孔隙度大，但淋水后基质持水量过高，不利于根部的气体交换，进而影响根系生长及养分吸收，导致降香黄檀苗木生长不良，这与金国庆等对香樟等 3 种乡土阔叶树种[19-20]的研究结果一致。因此，培育降香黄檀网袋容器苗，若用松树皮和锯末，其掺入锯末的比例不宜超过 20%。

综上所述，降香黄檀育苗效果最好的轻基质配方为配方 12、11 和 10，其次为配方 5 和 4。尽管如此，考虑到基质灌装时配方 12 和 11 的基质由于黄心土比例过大，容易导致机器出现故障，影响生产效率，而配方 5 的基质成本偏高，且出圃率低于配方 4，因此从苗木质量和基质成本及机械化操作等方面综合考虑，建议在降香黄檀工厂化育苗时采用配方 10(沤制松树皮 90%＋黄心土 10%)和配方 4(沤制松树皮 75%＋炭化松树皮 25%)2 种基质。

参考文献

[1]郭文福，贾宏炎．降香黄檀在广西南亚热带地区的引种[J]. 福建林业科技，2006，33(4)：152-155.

[2]孟慧，杨云，冯锦东．降香黄檀引种栽培现状与发展[J]. 广东农业科学，2010(7)：79-80.

[3]王克建，蔡子良．热带树种栽培技术[M]. 南宁：广西科学技术出版社，2008：45-46.

[4]罗文扬，罗萍，武丽琼，等．降香黄檀及其可持续发展对策探讨[J]. 热带农业科学，2009，29(1)：44-46.

[5]韦小丽，朱忠荣，君小阳，等．湿地松轻基质容器苗育苗技术[J]. 南京林业大学学报，2003(5)：55-58.

[6]侯元兆．现代林业育苗的概念与技术[J]. 世界林业研究，2007，20(4)：24-29.

[7]张建国，王军辉，许洋，等．网袋容器苗育苗新技术[M]. 北京：科学出版社，2007.

[8]尚秀华，杨小红，彭彦，等．不同基质对桉树育苗效果的影响[J]. 热带作物学报，2010(7)：1072-1077.

[9]周志春，刘青华，胡根长，等.3 种珍贵用材树种轻基质网袋容器育苗方案优选[J]. 林业科学，2011，47(10)：172-178.

[10]贾宏炎，曾杰，黎明，等．西南桦组培苗培育的轻基质筛选[J]. 林业科学研究，2012，36(5)：33-39.

[11]彭玉华，郝海坤，何琴飞，等．不同基质对印度紫檀幼苗生长的影响[J]. 林业科技开发，2012，26(4)：105-109.

[12]陈海军，官莉莉，赖建明，等．降香黄檀轻基质网袋容器育苗技术[J]. 湖南林业科技，2010，37(2)：59-61.

[13]何琴飞，彭玉华，曹艳云，等．降香黄檀容器育苗基质试验[J]. 林业科技开发，2012，26(6)：92-95.

[14]蒙彩兰，黎明，郭文福．西南桦轻基质网袋容器育苗技术[J]. 林业技术开发，2007，21(6)：104-105.

[15]陈晓波，王继志，叶燕萍，等．蒙古栎苗木分级标准的研究[J]. 北华大学学报(自然科学版)，2002，3(3)：251-254.

[16]刘士哲．现代实用无土栽培技术[M]. 北京：中国农业出版社，2001：151-153.

[17]中国土壤学会农业化学专业委员会．土壤农业化学常规分析方法[M]. 北京：科学出版社，1983.

[18]周永学，苏晓华，樊军锋．引种欧洲黑杨无性系苗期生长测定与选择[J]. 西北农林科技大学学报(自然科学版)，2004，32(10)：102-106.

[19]金国庆，周志春，胡红宝.3 种乡土阔叶树种容器育苗技术研究[J]. 林业科学研究，2005，18(4)：387-392.

[20]孙洁，刘俊，郁培义，等．不同基质配方对降香黄檀幼苗生长生理的影响[J]. 中南林业科技大学学报，2015，35(7)：45-49.

[原载：中南林业科技大学学报，2015，35(11)]

不同基质和育苗容器规格对格木幼苗生长的影响

黎 明 韦叶桥 蒙 愈 冯 海 陈金新

（中国林业科学研究院热带林业实验中心，广西凭祥 532600）

摘 要 筛选适合格木幼苗培育的基质和容器，为完善格木容器育苗技术和营造格木高效人工林提供参考依据。以腐熟松树皮、沤制锯末、炭化松树皮、黄心土及格木林下表土配置5种基质(分别设置为A1、A2、A3、A4和CK)，使用12cm×15cm(B1)和15cm×20cm(B2)两种规格的无纺布容器，开展格木幼苗培育试验，调查不同基质和不同规格容器格木幼苗的地径、苗高和生物量及造林后的成活率和生长表现，分析其造林效果。不同基质、不同规格容器的格木幼苗生长表现及造林效果差异均达显著($P<0.05$，下同)或极显著($P<0.01$)水平，而不同基质与不同规格容器交互作用仅幼苗生物量差异达显著水平。幼苗生长表现和造林效果均以基质A2(松树皮50%+黄心土30%+格木林下表土20%)最好，基质A4(松树皮60%+沤制锯末30%+炭化松树皮10%)最差；容器B2和B1的幼苗造林成活率整体上差异不明显，但前者的生长表现显著优于后者。基质A2的两种规格容器幼苗的造林效果差异不显著($P>0.05$)。生产上宜选用12cm×15cm规格的无纺布容器和松树皮50%+黄心土30%+格木林下表土20%基质培育格木幼苗。

关键词 格木；育苗；基质；容器规格；造林效果

Effectof Different Mediums and Seedling-breeding Container Sizes on Growth Performance of *Erythrophleum fordii* seedlings

LI Ming, WEI Ye-qiao, MENG Yu, FENG Hai, CHEN Jinxin

(*Experimental Centre of Tropical Forestry, CAF, Pingxiang 532600, Guangxi, China*)

Abstract: In order to provide reference for improving seedling-breeding technique of *Erythrophleumfordii* in container and constructing its half-benefit plantations, the optimal growing medium and container for *E. fordii* seedlings were screen out. Five kinds of growing medium (coded as A1, A2, A3, A4, and CK) was made up of composted pine bark, retted sawdust, charred pine bark, subsoil and topsoil of *E. fordii* plantations. And two sizes of non-woven containers (coded as B1 and B2) were used to breed *E. fordii* seedlings; with were 12cm×15cm and15cm×20cm in size respectively. Then the ground diameter, seedling height, biomass, survival rate and growth performance of seedlings breeded with different mediums and containers were investigated, so as to analyze afforestation effects. The results showed that the medium and container size had significant or extremely significant effects on seedlings growth performance and afforestation effects($P<0.05$ or $P<0.01$, the same below). While the interactions between mediums and containers only had significant effect on seedling biomass. In addition, in terms of growth performance and afforestation effect, the growing medium A2 (v/v; 50% composted pine bark+30% subsoil+20% topsoil) was the best, while the medium A4 (v/v; 60% composted pine bark+30% retted sawdust+10% charred pine bark) was the worst. There existed no significant difference in survival rate between containers B1 and B2, while as for seedling growth performance, container B2 was significantly better than that container B1 as a whole. However, There existed no significant difference in afforestation effect between two sizes container ($P>0.05$). The growing medium 50% composted pine bark+30% subsoil+20% topsoil and non-woven container (12cm×15cm in size) were recommended to breed seedlings of *E. fordii*.

Key words: *Erythrophleum fordii*; seedling-breeding; growing medium; container size; afforestation effect

引言

【研究意义】格木(*Erythrophleum fordii* Oliv.)俗称铁木，为苏木科常绿乔木，是我国Ⅱ级重点保护植物[1]，也是我国热带南亚热带地区重点发展的珍贵树种之一。格木自然分布于广西、广东、福建和台湾等地[2]，其木材质坚硬、耐腐性强，广泛应用于制作高档家具、工艺用品及室内装修，还具有较高的药用价值和生态价值[3,4]。近年来中国林业科学研究院热带林业实验中心积极发展格木人工林，其面积已达3000hm^2，在生产实践中，大多采用传统基质和常用规格容器培育10~12个月生苗造林，幼苗较小，造林成活率低且苗木生长缓慢。因此，筛选适合格木幼苗培育的基质和容器，对提高格木幼苗质量和格木人工林营建具有重要意义。【前人研究进展】基质类型和容器大小直接影响苗木质量和育苗成本，国内已对油杉[5]、木荷[6]、台湾相思[7]、印度紫檀[8]及红椎[9]等珍贵树种开展过此方面的研究，对推动珍贵用材树种的发展发挥了重要作用。前人对格木育苗基质、容器规格等相关研究较少，仅见余正国等[10]开展了格木育苗基质试验，从4种基质中筛选出沙质壤土80%+火烧土20%的基质最适合格木幼苗生长。【本研究切入点】目前，未见有关基质和容器规格对格木幼苗生长影响的研究报道。【拟解决的关键问题】选取林区中丰富的松树皮、锯末、黄心土及格木林下表土等材料配置若干类型基质，使用两种规格容器，探讨不同基质和容器规格对1.5年生格木幼苗生长及其造林效果的影响，以期筛选出适合格木育苗的基质和容器，为完善格木容器育苗技术和营造高效人工林提供参考。

1 材料与方法

1.1 育苗和造林地概况

育苗地点广西凭祥中国林业科学研究院热带林业实验中心哨平实验场苗圃位于22°02′52″N，106°53′18″E，海拔200m，属于南亚热带气候，年均气温21.7℃，最热月(7月)平均气温28℃，极端最高气温40.3℃，最冷月(1月)平均气温13.5℃，极端最低气温不低于0℃，年降水量1200~1300mm，干湿季节明显，4~9月为雨季(雨热同季)。造林地选在凭祥市中国林业科学研究院热带林业实验中心夏石引种树木园东南部，属马尾松采伐迹地，坡度约15°，位于坡下部，土壤为老洪积母质发育而成的砖红壤性红壤，pH5.2，土层较厚。

1.2 试验材料

格木种子采自凭祥市中国林业科学研究院热带林业实验中心人工林。基质材料为陈年腐熟松树皮、沤制锯末、炭化松树皮、黄心土和格木林下表土5种，均取自当地木材加工厂和林区。两种规格(口径×高度分别为12cm×15cm、15cm×20cm)的无纺布育苗容器袋购自安庆市怀宁县俊林环保育苗袋厂。

1.3 试验方法

1.3.1 试验设置

采用完全随机区组试验设计，设置基质(A)和容器规格(B)两个因子，其中A因子包括5个水平：对照(CK)为黄心土100%，A1为松树皮30%+黄心土50%+格木林下表土20%，A2为松树皮50%+黄心土30%+格木林下表土20%，A3为松树皮70%+黄心土10%+格木林下表土20%，A4为松树皮60%+沤制锯末30%+炭化松树皮10%，参试基质均添加2%过磷酸钙作基肥；B因子包括两个水平：B1为无纺布容器袋(口径×高度)12cm×15cm，B2为无纺布容器袋15cm×20cm。试验设10个处理，3次重复，共30个小区，每小区80株。育苗试验结束后，各处理选取75株平均苗高、平均地径相当的苗木用于造林试验，采用随机区组试验设计，3次重复，共30个小区，每小区25株。

1.3.2 育苗及造林

按照试验设计要求将基质充分混合，装入无纺布育苗袋，整齐地摆放在苗床上。于2012年5月中旬将经浓硫酸处理[11]的格木种子均匀撒播在沙床上，用木板将种子压平于沙面，覆沙1.5cm，保持苗床湿润，约7d开始发芽。5月下旬子叶完全展开时，移苗至育苗袋内并及时淋水，加盖遮阳网。苗木恢复生长后，施2000~8000mg/L复合肥或氮肥，每10d施1次，施肥浓度逐渐加大。2012年10月下旬揭开遮阳网，其他管护措施与常规育苗相似。2014年2月下旬开展造林试验，采用常规方法进行整地和试验林管护。

1.3.3 指标测定

2013年11月中旬，格木苗龄为1.5年生时，每小区随机抽取50株苗木测定其苗高和地径，选取4株平均苗，测定其地上部分和地下部分生物量及苗木重量。2014年11月中旬调查造林成活率、苗高和地径。

1.4 统计分析

应用SPSS 16.0软件进行双因素方差分析和Duncan's多重比较。数据处理时，造林成活率经反正弦转换。

2　结果与分析

2.1　不同基质、不同规格容器对格木幼苗生长的影响

由表1可知，各基质和容器规格处理间1.5年生格木的苗高、地径、生物量及根冠比均差异极显著($P<0.01$，下同)，基质和容器规格仅对生物量存在显著或极显著的交互作用。多重比较结果表明(表2)，基质A1、A2和A3间的苗高和地上部分生物量差异不显著($P>0.05$，下同)，而均显著高于或大于基质A4和CK($P<0.05$，下同)；基质A2、A3的地径、根冠比、地下部分生物量及总生物量均显著高于或大于基质CK、A1和A4。与容器规格B1相比，规格B2的苗高、地径、地上和地下部分生物量及总生物量分别高或大35.36%、17.09%、72.78%、50.45%和65.86%，且差异均达极显著，而根冠比则下降11.36%。说明容器规格大、基质养分多有利于培养格木壮苗。

表1　1.5年生格木容器苗生长的双因素方差分析

变异来源	自由度	苗高(cm)	地径(mm)	地上部分生物量	地下部分生物量	总生物量	根冠比
A	4	32.98**	64.44**	62.90**	51.86**	73.05**	8.70**
B	1	139.73**	221.34**	259.28**	84.79**	236.71**	13.21**
A×B	4	1.95	1.74	8.16**	3.38*	7.37**	2.36

注：数据后**和*分别表示在0.01、0.05水平下差异显著。下同。

表2　1.5年生格木容器苗的生长表现

因素	水平	苗高(cm)	地径(mm)	地上部分生物量(g/株)	地下部分生物量(g/株)	总生物量(g/株)	根冠比
基质	CK	48.08±6.92b	10.25±0.76b	26.44±6.46b	9.97±2.36b	36.41±8.64c	0.38±0.04b
	A1	52.68±9.28a	10.44±0.73b	30.29±9.01a	11.20±2.12b	41.49±10.90b	0.38±0.07b
	A2	54.98±8.61a	10.99±1.06a	32.22±9.99a	15.01±3.84a	47.23±13.60a	0.48±0.06a
	A3	54.66±11.65a	10.87±0.98a	31.97±11.55a	14.05±3.80a	46.01±15.28a	0.45±0.06a
	A4	35.72±7.07c	8.58±1.07c	13.36±3.78c	5.13±1.56c	18.48±5.33d	0.38±0.02b
容器规格	B1	41.83±6.77	9.42±0.98	19.69±5.17	8.84±3.00	28.53±8.01	0.44±0.07
	B2	56.62±9.39*	11.03±0.90*	34.02±10.11*	13.30±4.62*	47.32±14.45*	0.39±0.05*

注：表中的数据为平均值±标准差；基质处理中同列数据后不同小写字母表示差异显著($P<0.05$)；*表示容器规格间差异显著。下同。

2.2　不同基质和容器规格对格木幼苗造林效果的影响

由表3可知，造林9个月时，不同基质处理的格木幼苗成活率、苗高、地径差异均极显著，容器规格间苗高、地径差异极显著，成活率差异显著，而基质和容器规格的交互作用对苗高、地径、成活率均无显著影响。多重比较结果表明(表4)，基质A2的苗木造林成活率最高，较基质A3、CK、A1和A4分别提高3.5%、4.6%、10.4%和15.7%，其差异显著；基质A2的苗高显著高于基质CK、A1，但三者与基质A3间差异不显著，而均显著高于基质A4；基质A1、A2和A3间的地径差异不显著，而均显著大于基质A4和CK。容器规格B2较规格B1的苗高、地径分别高13.6%和18.3%，且差异显著，而造林成活率仅提高1.8%，差异不显著。

进一步应用T检验比较两种容器规格幼苗的造林效果(表5)发现，5种基质、两种容器规格间的幼苗造林成活率差异均不显著；仅基质A4在两种容器中的幼苗苗高差异显著。换言之，对于基质CK、A1、A2、A3，苗高在两种容器规格幼苗间差异不显著；基质A1、A3和A4在两种容器中的幼苗地径差异显著。

表3　不同基质和容器规格格木幼苗造林9个月后成活率及生长的双因素方差分析

变异来源	自由度	成活率	苗高(cm)	地径(mm)
A	4	64.77**	16.70**	18.99**
B	1	7.46*	17.21**	41.64**
A×B	4	0.21	0.85	0.52

表4　不同基质和容器规格格木幼苗造林9个月后的成活率及生长表现

因素	水平	成活率(%)	苗高(cm)	地径(mm)
基质	CK	93.7±1.7b	84.04±6.46b	17.05±1.75b
	A1	88.7±2.2c	87.59±8.93b	18.74±2.18a
	A2	98.0±1.2a	99.86±10.04a	20.32±2.22a
	A3	94.7±1.9b	92.68±9.03ab	18.94±1.52a
	A4	84.7±1.4d	67.51±11.05c	14.41±2.39c
容器规格	B1	91.1±5.0	80.85±13.79	16.39±2.56
	B2	92.7±5.2	91.82±12.16*	19.39±2.20*

表5　同一基质两种容器规格的幼苗造林效果比较(造林后9个月)

基质	成活率(%)		苗高(cm)		地径(mm)	
	B1	B2	B1	B2	B1	B2
CK	93.0±2.0	94.3±1.4	82.16±4.04	85.93±8.79	15.85±0.82	18.24±1.62
A1	88.0±2.0	89.3±2.5	81.05±2.23	94.13±8.14	17.10±1.40	20.37±1.36*
A2	97.3±1.4	98.7±0.5	93.33±10.73	106.38±3.01	18.85±2.11	21.79±1.17
A3	93.3±1.5	96.0±1.3	89.02±8.44	96.34±9.61	17.80±0.94	20.08±1.01*
A4	84.0±1.0	85.3±1.6	58.69±7.15	76.33±4.57*	12.34±0.21	16.47±1.17*

2.3　育苗成本及其基质重量对比分析

依据陈年腐熟松树皮、炭化松树皮、沤制锯末、格木林下表土和黄心土的价格分别为80、260、90、70和50元/m^3，按照每立方米基质分别可装600、300个12cm×15cm(B1)和15cm×20cm(B2)无纺布容器袋，加上相应的容器、水、肥、药剂及人工等费用，计算得出两种容器规格各基质的育苗成本(表6)。由表6可知，基质A4的育苗成本最高，基质A3和A2次之，对照基质(CK)的成本最低；容器规格B1的育苗成本和基质重量分别为0.71元/株、1.56kg/袋，与容器规格B2相比，其成本每株少0.26元，基质重量每袋轻1.83kg。

表6　不同基质和容器规格1.5年生格木幼苗的培育成本及基质重量

基质	容器规格(B1)		容器规格(B2)	
	成本(元/株)	基质重量(kg/袋)	成本(元/株)	基质重量(kg/袋)
CK	0.67	1.91	0.89	4.05
A1	0.70	1.70	0.94	3.80
A2	0.71	1.63	0.96	3.60
A3	0.72	1.45	0.98	3.10
A4	0.76	1.09	1.07	2.39

3　讨论

本研究参试的5种基质中，以基质A2(松皮50%+黄心土30%+格木林下表土20%)和A3(松树皮70%+黄心土10%+格木林下表土20%)培育格木容器苗效果最佳，其各项生长指标整体上优于其他基质，与其通气、保水、保肥性能良好有关；而基质A4的育苗和造林效果最差，与其掺入较大比例的锯末有关，这是因为沤制锯末吸水能力强，浇水后基质持水量过高，不利于格木苗根部气体交换，进而影响其根系生长和养分吸收，导致苗木生长不良。这与金国庆等[12]对香樟等3个乡土阔叶树种及郭文福等[13]对西南桦的研究结果相似。基质A2的成本较A3还低，因此，采用基质A2培育格木苗木效果最佳，其次为基质A3。造林试验结果表明，造林9个月后，基质A2与A3的苗高和地径均无显著差异，但前者的造林成活率显著高于后者。

容器规格B2的苗高、地径及生物量均显著高于或大于容器规格B1，而根冠比则明显下降，与马雪红等[6]对木荷的研究结果一致，而与蒋云东等[14]对南酸枣等5个阔叶树种的研究结果不一致，后者研究表明，3种规格容器培育的4~5个月龄南酸枣等苗木，其苗高、地径生长无明显差异。这可能与苗木年龄及树种差异有关。在以往的研究中，鲜有开展不同容器规格苗木的造林效果试验，而本研究进一步针对每种育苗基质比较两种容器规格幼苗间的造林效果发现，对于基质A2和CK，容器规格对幼苗造林效果并无显著影响；对于基质A3，容器规格仅显著影响造林后9个月苗木的地径，对苗高和成活率无显著影响。两种容器规格苗木的造林效果变化规律并未与苗期相吻合；与容器规格B2相比，规格B1的育苗成本每株少0.26元，基质重量每袋轻1.83kg。说明容器规格并不是越大越好，也说明造林试验对于此类研究的重要性。由此可见，采用规格B1培育的幼苗，不仅能明显降低育苗成本，而且显著提高造林效率。

4　结论

基质A2的格木幼苗生长表现和造林效果最好，容器B1基质用量少、育苗成本低、造林效率较高。建议生产上使用12cm×15cm规格的无纺布容器和松树皮50%+黄心土30%+格木林下表土20%基质培育格木幼苗。

参考文献

[1]汪松，解焱．中国物种红色名录：第一卷[M]．北京：高等教育出版社，2002.

[2]赵志刚，郭俊杰，沙二，等．我国格木的地理分布与种实表型变异[J]．植物学报，2009，44(3)：388-344.

[3]刘一樵，李士汤．森林植物学(南方本)[M]．北京：中国林业出版社，1993：309-310.

[4]LI N，YU F，YU S S. 2004. Triterpenoids from *Erythrophleum fordii*[J]. Acta Botanica Sinica，46(3)：371-374.

[5]张纪卯．不同基质和容器规格对油杉容器苗生长的影响[J]．福建林学院学报，2001，21(2)：176-180.

[6]马雪红，胡根长，冯建国，等．基质配比、缓释肥量和容器规格对木荷容器苗质量的影响[J]．林业科学研究，2010，23(4)：505-509.

[7]刘永安，陈小勇，王友芳，等．攀西地区台湾相思适宜育苗容器和基质[J]．东北林业大学学报，2012，40(10)：98-102.

[8]杨斌，史富强，付玉嫔，等．印度紫檀容器育苗适宜容器规格、基质和肥料配方[J]．东北林业大学学报，2012，40(2)：26-29.

[9]温恒辉，贾宏炎，黎明，等．基质类型、容器规格和施肥量对红椎容器苗质量的影响[J]．种子，2012，31(7)：75-77.

[10]余正国，罗建华．格木育苗技术试验研究[J]．热带林业，2007，35(1)：22-23.

[11]易观路，罗建华，林国荣，等．不同处理对格木种子发芽的影响[J]．福建林业科技，2004，31(3)：68-70.

[12]金国庆，周志春，胡红宝，等．3种乡土阔叶树种容器育苗技术研究[J]．林业科学研究，2005，18(4)：387-392.

[13]郭文福，曾杰，黎明，等．西南桦轻基质网袋容器苗基质选择试验[J]．种子，2010，29(10)：62-64

[14]蒋云东，王达明，杨德军，等．热区几种阔叶树种的育苗基质和容器规格研究[J]．云南林业科技，2003，28(4)：19-23

[原载：南方农业学报，2015，46(09)]

施肥对白桦树生长及开花结实的影响

刘福妹[1,2] 姜 静[1] 刘桂丰[1]

([1]东北林业大学林木遗传育种国家重点实验室，黑龙江哈尔滨 150040；
[2]中国林业科学研究院热带林业实验中心，广西凭祥 532600)

摘 要 合理施肥是促进林木生长发育的有效措施。本研究是在前期获得显著促进白桦开花结实的配方肥基础上，进一步研究了配方肥和其氮、磷、钾肥元素对白桦生长、开花结实及荧光参数的影响。结果表明：施氮肥与施配方肥的效果一致，均能显著促进白桦的苗高和地径生长，使白桦苗高和地径分别较对照提高1.32倍和1.36倍；施配方肥和氮肥能使70%以上的白桦开花，而施用磷、钾肥的白桦开花率仅为3%和1%，对照没有开花；与对照相比，虽然施肥均明显地提高白桦叶片 F_v/F_m、$\Phi_{PS\,II}$ 和 qP 等荧光参数，降低 NPQ，但施氮肥效果和配方肥的效果更加明显，并且二者还能显著提高白桦叶绿素含量。因此，说明配方肥中的氮是促进白桦生长和开花结实的关键元素，在以生产木材为经营目的的丰产林中可以施加氮肥为主的肥料，在以生产良种为目的的白桦种子园中应该选用配方肥。

关键词 白桦；施肥；生长；叶绿素荧光参数

Effects of Fertilization on the Growth and Flowering of *Betula platyphylla*

LIU Fumei[1,2], JIANG Jing[1], LUI Guifeng[1]

([1]*State Key Laboratory of Forest Genetics and Tree Breeding*, *Northeast Forestry University*, *Harbin* 150040, *Heilong Jiang*, *China*;
[2]*The Experimental Centre of Tropical Forestry*, *Chinese Academy of Forestry Official*, *Pingxiang* 532600, *Guangxi*, *China*)

Abstract: Rational fertilization is an effective measure to promote forest of fast-growing fertility. Based on the selected optimum formula fertilizer (T3) which had the best effect of accelerating the white birch (*Betula platyphylla*) flowering, we studied the effects of T3 and its elements (N, P, K) on the growth, flowering and chlorophyll fluorescence characteristics of white birch. The results showed that: the nitrogen that had the same effect of T3, could significantly promote the vegetative growth of white birch by making the 4-year-old trees reach 1.32 times in height and 1.36 times in ground diameter compared to the no fertilizer (control). Applying the T3 and nitrogen could make more than 70% white birch flower, Compared with only 3%, 1% and 0 by applying only phosphorus, potassium or without fertilizer; Compared with the control, fertilizations could significantly increase the level of F_v/F_m, $\Phi_{PS\,II}$ and qP, decrease the level of NPQ, however, applying the nitrogen and T3 had the best effective, and only the two of them could significantly increase the level of SPAD. Therefore, the nitrogen is a key factor of T3 in promoting the switch of white birch to reproductive growth. Also, oriented applying nitrogen in the high-yield plantation of wood production and oriented applying T3 in white birch seed orchard of improved variety production are good choices.

Key words: *Betula platyphylla*; fertilization; growth; chlorophyll fluorescence

林木施肥是集约经营人工丰产林提高森林生产力的重要技术措施，在林业生产中的应用愈来愈普遍[1]。大量研究表明，根据林地土壤肥力状况合理施肥是促进林木速生丰产的有效途径[2-4]。白桦(*Betula platyphylla*)是落叶乔木，属于桦木科桦木属。它在我国14个省(自治区、直辖市)有分布[5]，其用途多，并且是培育单板类人造板材速生丰产林的首选树种之一[6,7]，因此，研究白桦合理施肥有非常现实的意义。笔者所在实验室从2007年开始综合前人的研究，并考虑到林业上的施肥与农业的不同之处[8-10]，将N、P、K按照不同比例设置了3套配方施肥方案，开展了白桦不同家系的最佳配方筛选研究[11]，选出了适于各种基因型的最佳施肥配方。本研究是在此基础上，将最佳配方中的各个元素按照原来设置的浓度单独处理白桦，通过测定白桦树高、地径、开花结实及叶绿素荧光参数(F_v/F_m、Φ_{PSII}、qP 和 NPQ)，研究施肥对白桦生长发育的影响，为今后根据白桦不同的培育目施用不同营养元素及配方提供参考。

1 材料和方法

1.1 试验材料及设计

选择500株来自同一家系的1年生白桦盆栽实生苗被分成5组(每组100株)，1组施用最佳配方肥(T3)，1组施用等量的水(C)，其他3组分别施用氮肥(N)、磷肥(P)、钾肥(K)；其中T3是通过前期实验获得的能促进白桦提早开花结实的最佳施肥配方肥[12]，而单独施用的氮磷钾肥的浓度则是按照T3中含有相应元素的质量设定的(具体用量见表1)。

1.2 处理方法

试验中的所有白桦苗木均定植在上口直径为40cm、下底直径为30cm、高度为40cm的圆锥体型容器中，并且在容器底部垫一个直径为40cm托盘，防止营养的渗漏和苗木根系生长到外界土壤里。苗木培育基质均为草炭土：河沙：黑土(体积比)= 4：2：2，其pH为6.25，速效氮、磷、钾的质量比分别为193.60mg/kg、37.92mg/kg、164.97mg/kg。

施肥方法[12]：施肥时用水稀释配制成1/1000质量浓度的混合营养液，每年自4月下旬苗木发芽后开始施用至9月中旬结束，2年生的白桦每次每株施肥量均为0.5L，3、4年生的白桦每次每株施肥量均为1L，施肥间隔期为半个月。

表1 不同生长发育时期白桦施肥配方 (g/株)

处理	树龄(年份)	5月1日至6月30日			7月1日至8月14日			8月15日至9月14日	
		氮肥	磷肥	钾肥	氮肥	磷肥	钾肥	磷肥	钾肥
T3	2年(2009)	0.70	0.55	0.43	0.47	0.88	0.72	0.35	0.29
	3年(2010)	2.35	1.42	1.15	1.40	1.40	1.15	1.05	0.86
	4年(2011)	2.63	1.42	1.15	1.58	1.75	1.43	1.40	1.15
N	2年(2009)	0.70			0.47				
	3年(2010)	2.35			1.40				
	4年(2011)	2.63			1.58				
P	2年(2009)		0.55			0.88		0.35	
	3年(2010)		1.42			1.40		1.05	
	4年(2011)		1.42			1.75		1.40	
K	2年(2009)			0.43			0.72		0.29
	3年(2010)			1.15			1.15		0.86
	4年(2011)			1.15			1.43		1.15

注：氮肥是硝铵，磷肥是过磷酸钙，钾肥是硫酸钾，复合肥是磷酸二氢钾。

1.3 试验数据的调查

1.3.1 生长及开花结实的测定

2011年9月末，当白桦停止生长后，分别测定各施肥处理条件下白桦的树高和地径。统计每株白桦雄花序数量，2012年6月初统计结实数。

1.3.2 叶绿素整体含量的测定

2011年8月末，利用SPAD502型叶绿素仪测定白桦在各施肥处理条件下的叶绿素含量。每个处理随机抽选60株白桦，在每株树冠相同位置选择3片功能叶，每个叶片测定3个点的叶绿素含量，然后求平均值。

1.3.3 叶绿素荧光参数的测定

2011年8月末，在每个处理中选择3株白桦苗木，每株选取3片成熟叶，利用PAM-2500(Walz, German')测定其叶绿素荧光参数：PSII最大光化学效率(F_v/F_m)、PSII实际光化学效率(Φ_{PSII})、光化学猝灭系数(qP)和非光化学猝灭系数(NPQ)。此外，测定前叶片均暗适应30min。

1.4 统计分析方法

利用Excel2010整理数据，其中求算开花率、结实率和每株平均果序数量时用到的公式如下：

平均果序数(个/株)= 形成果序个数/参试株数

开花率(%)=(形成雌或雄花序的株数/参试株数)×100%

结实率(%)=(形成果序的株数/参试株数)×100%

利用SPSS16.0软件对获得的数据进行方差分析和多重比较(Duncan)。多重比较均是在$P<0.01$水平上进行的。

2 结果与分析

2.1 配方肥及氮、磷、钾元素对白桦生长的影响

2011 年，在苗木 4 年生时，分别对测定的不同处理白桦苗木的树高和地径的数据进行方差分析(表 2)，发现各性状的差异均达到显著水平($P<0.01$)，说明连续 3 年不同施肥处理对白桦的生长产生显著影响，树高和地径的变异系数分别为 14.0%和 16.1%。

表 2 不同处理条件下 4 年生白桦树高、地径的方差分析和主要变异参数

性状	均值及变异参数			*F* 值
	均值	变幅	变异系数(%)	
树高(m)	3.65	3.24~4.27	14.0	91.81**
地径(mm)	31.74	27.23~37.68	16.1	145.81**

注：表中用“**”标记的为差异达到显著水平。

为进一步说明最佳配方施肥及该配方中各营养元素的处理效果，对各种处理后的白桦各个性状进行了多重比较分析(表 3)。结果表明：在高生长和地径生长上，施氮肥与配方肥处理的白桦差异不显著，但与其他三个处理的差异均显著，二者苗高和地径分别达到了 4.15m、4.27m 和 37.68mm、36.91mm，分别是对照的 1.28、1.32 和 1.38、1.36 倍，说明施氮肥与施配方肥的作用一样，均显著促进白桦高生长和地径生长，并且与磷肥、钾肥及非施肥对照处理的差异达到了极显著水平。

表 3 不同处理条件下 4 年生白桦各性状的多重比较分析

处理	树高(m)			地径(mm)		
T3	4.15	±0.37	B	37.68	±3.22	B
N	4.27	±0.52	B	36.91	±4.49	B
P	3.28	±0.43	A	28.23	±2.43	A
K	3.32	±0.38	A	28.67	±2.35	A
C	3.24	±0.56	A	27.23	±5.17	A

注：不同字母的处理表示在 $P<0.01$ 时差异性显著，相同字母的处理之间差异不显著，下同。

2.2 配方肥及氮、磷、钾元素对白桦开花结实的影响

调查参试白桦通过连续 3 年的单独施加氮、磷、钾肥及最佳配方肥处理后的结实情况发现，除未进行施肥处理的白桦尚未开花外，其他处理均已经进入开花结实期。其中，施用配方肥和氮肥的处理开花率分别达到了 100%和 70%，并且二者的结实率也均在 60%以上；而施磷肥和钾肥的开花率和结实率却处在一个较低的水平，仅为 1%~3%(表 4)，说明单独施用磷肥和钾肥对白桦的提早开花结实的效果很小。

表 4 不同处理条件下 3 年生白桦开花结实情况

处理	开花率(%)	结实率(%)	平均果序数(个/株)
T3	100	92	88.31
N	70	62	41.3
P	3	2	3.13
K	1	0	0
C	0	0	0

注：统计的开花率和结实率均只计算正常生长的苗木，不算已经死亡和遭受虫害的苗木。

2.3 配方肥及氮、磷、钾元素对白桦叶片叶绿素含量的影响

植物的生长发育与植物的光合作用有着必然的联系，而叶绿素作为植物进行光合作用的主要色素，其含量的变化会对植物叶片光合能力产生影响[13]。因此，用叶绿素仪测定了白桦叶片中能够有效反映叶绿素相对含量的 SPAD 值。通过方差分析可知，连续 3 年的施肥处理对白桦叶片中叶绿素含量均产生了极其显著的影响($P<0.01$，$F=1.055E3$)。

进一步进行多重比较分析(表 5)发现，只有 T3 处理和 N 处理的白桦叶绿素含量与对照的差异达到了极显著水平，其相对含量分别为 47.09SPAD 和 47.68SPAD，是对照的 1.88 和 1.90 倍，说明施配方肥和氮肥能显著提高白桦叶绿素的相对含量。

表 5 不同施肥处理下白桦叶绿素相对含量(SPAD)多重比较

处理	叶绿素相对含量(SPAD)		
T3	47.09	±3.76	B
N	47.68	±5.05	B
P	26.41	±3.42	A
K	26.04	±3.47	A
C	25.04	±4.00	A

注：不同字母的处理表示在 $P<0.01$ 时差异性显著，相同字母的处理之间差异不显著。

2.4 配方肥及氮、磷、钾元素对白桦叶绿素荧光参数的影响

叶绿素荧光是光合作用的有效探针[14]，可反映光合机制内一系列重要的调节过程。通过对各种荧光参数的分析，可以得到有关光能利用途径的信息[15]。因此，本研究测定了不同处理白桦的叶绿素荧光参数，通过分析施肥对白桦光合作用的影响来说明光能利用与生长的关系。对各处理白桦 F_v/F_m、Φ_{PSII}、qP 和 NPQ 测定结果进行方差分析，发现各处理白桦叶绿素荧光参数差异均达极显著水平(表6)。

表6 不同施肥处理下白桦叶绿素荧光参数方差分析

性状	自由度	平方和	均方	F
F_v/F_m	4	0.292	0.073	45.458**
Φ_{PSII}	4	1.511	0.378	76.445**
qP	4	29.192	7.298	46.711**
NPQ	4	0.566	0.142	22.402**

F_v/F_m 是 PSII 的最大量子产量，它直观反映了植物潜在的最大光合能力。经过多重比较分析发现(表7)，4个施肥处理的 F_v/F_m 值均显著高于非施肥对照，说明在同等环境下，合理的施肥能使白桦叶片保持良好的生长潜能，这就为白桦的快速生长提供了条件。其中，T3处理和N处理的 F_v/F_m 值是最高的，分别是对照的1.15和1.18倍，说明施用配方肥和氮肥的效果较优。

Φ_{PSII} 是光适应下 PSII 的实际光化学效率[16,17]，直接反映了植物叶片在光下用于电子传递的能量占吸收光能的比例。从表7中可以看出各施肥处理的 Φ_{PSII} 的结果与 F_v/F_m 的类似，非施肥对照的值仍然是最小的，仅0.32；T3处理和N处理白桦的值最大，均为0.59，较对照的提高了84%，说明施配方肥和氮肥能够明显提高白桦叶片对光能的实际利用率。

光化学碎灭系数 qP 反映 PSII 天线色素吸收的光能用于光化学电子传递的比例，不仅在一定程度上反映了 PSII 反应中心的开放程度[18]，而且也能够反映植物光合效率和对光能的利用[19]。如多重比较分析表7所示，各施肥处理白桦的 qP 值仍均显著高于对照的，说明施肥可以显著提高 PSII 反应中心的开放程度，提高白桦苗木光能利用的效率，加快 PSII 的电子传递，其中，施用钾肥的效果是最显著的，其 qP 值达到了0.86。非光化学淬灭 NPQ 反映了 PSII 反应中心对天线色素吸收过量光能后的热耗散能力[20]，即植物无害化热耗散过剩激发能的能力[21]。从荧光参数的多重比较表中可以看出，非施肥对照的 NPQ 值最大，达到了1.42；与对照相比，各施肥处理白桦叶片的 NPQ 值均有显著的降低，即施肥后的热耗散仅为对照的29%~33.1%。这也说明施肥增加了叶片对沟通的利用。

3 结论与讨论

研究表明，将氮、磷、钾等按照适宜比例科学配方，合理施用，能够使林木在短期内有效而迅速地生长发育[3]，本研究是在已获得能显著促进白桦开花结实配方肥的基础上，进一步研究了最佳配方和其氮、磷、钾肥元素的作用。结果表明，最佳配方肥中的氮元素是促进白桦营养生长和生殖生长的主要因素。它不但显著促进白桦的高生长和地径生长，施氮肥的白桦的树高和地径分别达到了4.27m和36.91mm，分别是对照的1.32倍和1.36倍；而且显著提高了白桦的开花率和结实率，分别达到了70%和62%。虽然氮能促进白桦的提早开花结实，但是其效果仍然比不上最佳配方(其开花率和结实率分别为100%和92%)，这说明氮肥是促进白桦营养生长和生殖生长的关键因素，但必须与磷、钾肥互作，才会有更好的效果。

表7 不同施肥处理下白桦叶绿素荧光参数多重比较

处理	F_v/F_m			Φ_{PSII}			qP			NPQ		
T3	0.76	±0.03	C	0.59	±0.02	C	0.83	±0.05	B	0.32	±0.04	A
N	0.78	±0.03	C	0.59	±0.03	C	0.80	±0.05	B	0.27	±0.05	A
P	0.73	±0.03	B	0.52	±0.07	B	0.80	±0.05	B	0.47	±0.09	A
K	0.70	±0.04	B	0.54	±0.03	B	0.86	±0.05	C	0.43	±0.09	A
C	0.66	±0.06	A	0.32	±0.06	A	0.68	±0.09	A	1.42	±0.24	B

在测定各处理白桦的叶绿素荧光参数之前，哈尔滨出现了持续5d的低温天气，自然低温能使杨树和桉树的 F_v/F_m 不同程度的下降[22,23]，而本研究中测定的 F_v/F_m(范围0.66~0.78)均明显低于过去研究中提到的0.80~0.83[24]，说明低温天气也对白桦的 F_v/F_m 产生了一定的影响；施肥处理的白桦 F_v/F_m 均显著高于非施肥处理的，而 F_v/F_m 能作为光抑制和PSII复合体受伤的指标[25]，说明合理施肥能够让叶片保持良好的生活状态，在应对突发的低温时，也表现出更好的抗性。大量研究表明合理的施肥可以改善叶片光合特性[26-29]，这也与本研究的结果一致——施肥后白桦的叶绿素荧光参数均优于非施肥处理的白桦的：Φ_{PSII} 较对照提高了62.5%~84.4%；qP 提高了17.6%~26.5%；NPQ 较对照降低了66.9%~81.0%；并且综合发现，施配方肥和氮肥的白桦叶绿素荧光参数的作用最佳。此外，施配方肥和氮肥均能显著提高白桦叶片叶绿素含量。综合白桦叶绿素含量和叶绿素荧光结果可以发现二者与白桦地径和苗高的生长情况是一致的，说明施配方肥和氮肥可以让白桦叶片保持良好的状态，并激发白桦叶片的最大光合潜能，显著提高叶片的实际光合能力，最终促进白桦的生长和发育。因此，在以生产木材为经营目的的丰产林中可以施加氮肥为主的肥料，在以生产良种为目的的白桦种子园中应该施用筛选出的配方肥。

参考文献

[1]李花兰．施肥对林木生长效应的理论研究[J]．山西林业科技，2012(1)：50-52.

[2]金国庆，余启国，焦月玲，等．配比施肥对南方红豆杉幼林生长的影响[J]．林业科学研究，2007(2)：251-256.

[3]于彬，郭彦青，陈金林．杨树配方施肥技术研究进展[J]．西南林学院学报，2007(2)：85-90.

[4]JEYANNY V，AB RASIP A G，WAN RASIDAH K，et al. Effects of macronutrient deficiencies on the growth and vigour of Khaya lvorensis seedlings [J]. Journal of Tropical Forest Science，2009. 21(2)：73-80.

[5]郑万钧．中国树木志[M]．北京：中国林业出版社，1985，2143.

[6]郁书君，汪天，金宗郁，等．白桦容器栽培试验(I)[J]．北京林业大学学报，2001，23(1)：24-28

[7]俞天珍．白桦在青海省育苗成功的技术措施[J]．农林科技，2005，34(4)：59.

[8]JONSSON A，ERICSSON T，ERIKSSON G，et al. Interfamily variation in nitrogen productivity of Pinus sylvestris seedlings [J]. Scandinavian Journal of Forest Research，1997. 12(1)：1-10.

[9]LI B，MCKEAND S E，ALLEN H L. Genetic variation in nitrogen use efficiency of loblolly pine seedlings [J]. Forest Science，1991，37(2)：613-626.

[10]MCKEAND S E，GRISSOM J E，RUBILAR R，et al. Responsiveness of diverse families of loblolly pine to fertilization：eight- year results from SETRES- 2// Proc 27th South For Tree Impr Conf [J]. Stillwater，2003：30-331.

[11]李天芳，姜静，王雷，等．配方施肥对白桦不同家系苗期生长的影响[J]．林业科学，2009(2)：60-64.

[12]刘福妹，李天芳，姜静，等．白桦最佳施肥配方的筛选及其各元素的作用分析[J]．北京林业大学学报，2012(2)：57-60.

[13]赵鸿彬，张正斌，徐萍，等．$HgCl_2$ 胁迫对小麦幼苗水分利用效率和叶绿素含量的影响．西北植物学报[J]，2007，27(12)：2478-2483.

[14]BAKER N R. A possible role for photosystem Ⅱ in environmental perturbations of photosynthesis [J]. Physiol Plant，1991，81：563- 570.

[15]罗青红，李志军，伍维模，等．胡杨、灰叶胡杨光合及叶绿素荧光特性的比较研究[J]．西北植物学报，2006(5)：983-988.

[16]GENTY B，BRIANTAIS J M，BAKER N R. The relationship between the quantum yield of photosynthetic electron transport and quenching of chlorophyll fluorescence [J]. Biochim Biophys Acta，1989，990：87-92.

[17]张杰，邹学史，杨传平，等．不同蒙古栎种源的叶绿素荧光特性[J]．东北林业大学学报，2005，33(3)：20-21.

[18]胡筑兵，陈亚华，王桂萍，等．铜胁迫对玉米幼苗生长、叶绿素荧光参数和抗氧化酶活性的影响[J]．植物学通报，2006(2)：129-137.

[19]汪月霞，孙国荣，王建波，等．NaCl胁迫下星星草幼苗MDA含量与膜透性及叶绿素荧光参数之间的关系[J]．生态学报，2006(1)：122-129.

[20]BADGER M R，von CAEMMERER S，RUUSKA S，et al. Electron flow to oxygen in higher plants and algae：rates and control of direct photoreduction(Mehler reaction) and rubisco oxygenase [J]. Biological Science，2000，355(1402)：1433-1455

[21]MULLER P，LI X P，NIYOGI K K. Non-Photochemical quenching. A response to excess light energy [J]. Plant Physiol，2001，125(4)：1558-1566.

[22]李春明，于文喜．秋季低温胁迫对杨树叶绿素荧光的影响[J]．林业科技，2011(2)：5-6+22.

[23]靖长柏，张利阳，童再康，等．自然低温胁迫下3种桉树的叶绿素荧光特性研究[J]．浙江林业科技，2011(1)：7-10.

[24] BUTLER W, KITAJIMA M. Fluorescence quenching in photosystem Ⅱ of chloroplasts[J]. Biochim Biophys Acta, 1975, 376:

[25]梁文斌，薛生国，沈吉红，等．锰胁迫对垂序商陆光合特性及叶绿素荧光参数的影响[J]. 生态学报，2010(3)：619-625.

[26]代向阳，徐程扬，马履一．氮磷配比对水曲柳光合作用的影响[J]. 山东林业科技，2006(2)：1-6.

[27]曲道春，江洪，由美娜．氮沉降对香樟叶片光合及叶绿素荧光特性的影响研究[J]. 环境污染与防治，2011(11)：15-19+23.

[28]霍常富，孙海龙，王政权，等．光照和氮营养对水曲柳苗木光合特性的影响[J]. 生态学杂志，2008(8)：1255-1261

[29]王奇峰，徐程扬．氮、磷对107杨叶片光合作用的影响[J]. 西北林学院学报，2007(4)：9-12.

[原载：西北林学院学报，2015，30(02)]

不同嫁接时间和处理对灰木莲嫁接成活的影响

蒙彩兰[1] 刘福妹[1] 卢立华[1] 刘子嘉[2]
([1]中国林业科学研究院热带林业实验中心，广西凭祥 532600
[2]三亚林业科学研究院，海南三亚 572000)

摘 要 研究通过比较在不同月份采用经不同浓度 ABT1 号生根粉处理的硬枝和嫩枝接穗嫁接的灰木莲成活率，探讨不同嫁接时间、不同处理对灰木莲嫁接成活率的影响。结果表明：1~2 月份是灰木莲嫁接的最佳时间，此时嫁接能使灰木莲嫁接成活率达到 48%；灰木莲硬枝嫁接的成活率是嫩枝嫁接的 9~11 倍，因此灰木莲宜采用硬枝接穗进行嫁接；虽然，从全年来看经 ABT1 号生根粉处理的接穗会对灰木莲月均嫁接成活产生不利影响，但在 1~2 月份用 100mg/L 的 ABT1 号生根粉处理硬枝接穗嫁接，可使灰木莲嫁接成活率达到 80%和 75%，是未经 ABT1 号生根粉处理的 1.45 和 1.41 倍；因此，在最适合灰木莲嫁接的 1 月份和 2 月份期间使用适宜浓度的 ABT1 号生根粉能提高灰木莲的嫁接成活率。
关键词 木莲；嫁接；接穗生长；ABT1 生根粉

Effects of Grafting Survival of *Manglietia glanca* Blume under Different Treatments in Different Months

MENG Cailang[1], LIU Fumei[1], LU Lihua[1], LIU Zijia[2]
([1]*The Experimental Centre of Tropical Forestry*, *Chinese Academy of Forestry Official*, *Pingxiang* 532600, *Guangxi*, *China*;
[2]*Sanya Academy of Forestry*, *Sanya* 572000, *Hainan*, *China*)

Abstract: In this study, to explore the effects of grafting survival rate of *Manglietia glanca Blume* under different treatments in different grafting time, we compared the survival rate grafted with hard- and tender- branches under different concentration of ABT 1 rooting powder in different months. The results showed that: January and February was the best time for grafting by making the survival rate up to 48%;, the hard- branch was the appropriate choice for the graft, with which the survival rate grafted was 9~11 times to the rate with tender branches; in Januaryand February, grafted with the hard branches treated with the ABT1 rooting powder of 100 mg/L ould make the survival rate achieve 80% and 75%, which was 1.45 and 1.41 times of the rate grafted without ABT1 rooting powder, although grafting with the scion treated with the ABT1 rooting powder had an adverse impact in the perspective of average rate monthly; therefore using the ABT1 of optimum concentration could significantly improve the survival rate during January and February.
Key words: *Manglietia glanca*; grafting; scion growth; ABT1 rooting powder

灰木莲(*Manglietia glanca* Blume)属于木兰科木莲属，是原产于越南及印度尼西亚等热带区域的优良速生用材树种[1-2]。我国从 1960 年开始相继在广东、广西、云南、福建等地开展了灰木莲的引种栽培，14 年生木莲人工林平均树高可达到 18m，胸径达到 14cm，表明灰木莲在营建速生丰产林方面具有良好的发展前景[2-4]。灰木莲虽然能开花结实，但是其自然结实率极低，无法满足人工林生产建设对种子的需要[5]，因此开展灰木莲无性繁殖技术相关方面的研究具有非常现实的意义。树木嫁接是林业生产上常用的无性繁殖方法之一，它能将优良单株的遗传信息较完整地保留到无性系群体中；同时，若采用已经达到性成熟的枝条作为接穗，嫁接植株就能提前开花结实，缩短结实年限。林木嫁接技术的关键是怎样保证嫁接有较高的成活率，研究表明，嫁接成活与嫁接的时期，嫁接使用的接穗和方法密切相关[6-7]，本研究通过比较在不同月份采用不同浓度的 ABT1

号生根粉对硬枝和嫩枝两种接穗处理后进行嫁接的灰木莲的成活率，探讨不同嫁接时间、不同处理对灰木莲嫁接成活率的影响，并对嫁接成活后的灰木莲生长情况进行分析，以期筛选出最适合灰木莲嫁接的时期和方法，为今后开展灰木莲优良种质资源收集、保存及今后营造灰木莲嫁接种子园等奠定一定的基础。

1　材料和方法

1.1　试验地概况

试验点设在广西壮族自治区凭祥市中国林业科学研究院热带林业实验中心苗圃，地理位置为21°57′47″N，106°48′E，属南亚热带季风气候区，年均气温21.5℃，极端最高温和最低温分别是39.8℃和-1.5℃，无霜期340d，≥10℃的积温6000~7600℃，全年日照时数1218~1620h，年降水量1220~1400mm，降水集中期为4月中旬至8月末[8]。

1.2　试验材料和方法

1.2.1　砧木和接穗的选择

试验中用到的砧木材料均是同一时期培育的、生长旺盛、长势良好、无病虫，且在离地30cm处直径约为8mm的2年生灰木莲幼苗，定植在下底直径为20cm、高度为30cm的圆柱体型容器中，培育基质是黄土：火烧土：树皮=7：2：1，同时添加适量磷肥。

试验中采用的接穗是从5年生灰木莲苗木采集的1年生枝条，要求枝条营养状况良好、芽眼饱满、无病虫害。采集后的枝条按照生长的不同部位，将每段枝条截顶去基后分成中上部和中下部两段，分别作为嫩枝和硬枝接穗备用。

1.2.2　嫁接方法

试验中均用插皮接的方法进行嫁接，即：将砧木在25cm高度截顶，然后在树皮光滑处纵划一刀，将树皮两边适当挑开，再插入接穗，使双方形成层接触；此外，插入接穗应留0.5cm的伤口面在接口之上，利于愈合。嫁接后，用一条长约50cm、宽为砧木直径1.5倍的塑料条捆绑，要求将接口绑紧，保证不露出伤口。

1.2.3　试验设计

试验采用两种不同的接穗，分别蘸取0mg/L、100mg/L和500mg/L 3种不同浓度的ABT1号生根粉后进行嫁接(表1)，整个试验分为6个处理，每个处理有60株苗，分成3个重复，嫁接时间从2012年9月开始到2013年8月结束，每月嫁接一次。

表1　对接穗不同处理方法

处理	接穗	ABT1号生根粉(mg/L)
YS	硬枝接穗	0
NS	嫩枝接穗	0
YY	硬枝接穗	100
NY	嫩枝接穗	100
YW	硬枝接穗	500
NW	嫩枝接穗	500

1.2.4　嫁接后管理

嫁接完成后，立即将苗移入荫棚，避免巨大的天气波动对嫁接苗产生不利影响。同时要适时适量浇水，保持土壤含水量，满足嫁接苗对水分的需求，浇水最好采用滴灌，避免淋湿嫁接点，还要及时清除萌蘖，以免影响接穗的生长。待30d后，将苗从荫棚中移出，进行常规的苗期管护。

1.3　试验数据的调查及处理

2012年9月到2013年8月，每次嫁接完成后30天，统计每个处理的嫁接成活株数，计算成活率。

成活率(%)=(嫁接成活株数/20)×100%

2014年6月份，统计保存率，然后测定每个处理接穗的生长量，同时计算接穗的月均生长量；接穗月均生长量=接穗总生长量/嫁接后成活月份数。

利用SPSS16.0软件对获得的数据进行方差分析和多重比较(Duncan，$P<0.01$)，其中在进行方差分析前对成活率进行标准化处理。

2　结果与分析

2.1　嫁接时间对灰木莲嫁接成活的影响

为了解嫁接时间对灰木莲嫁接成活的影响，对不同月份嫁接的灰木莲成活率进行方差分析，发现各处理间灰木莲的成活率差异达到极显著水平($P<0.01$，$F=61.474$)，说明不同嫁接时间会对灰木莲嫁接的成活率产生极其显著的影响。因此，为筛选出最适合灰木莲嫁接的时间，进一步对各月份嫁接的灰木莲成活率进行多重比较分析(图1)，发现嫁接成活率最高的月份为1月、2月，分别达到了48.1%、47.8%；其次是4月和5月，分别为31.1%和35.2%；而嫁接成活率最低的是6月份和8月份，仅为6%，说明每年中最适合灰木莲嫁接的时间为1月份和2月份，其次是4月份和5月份。

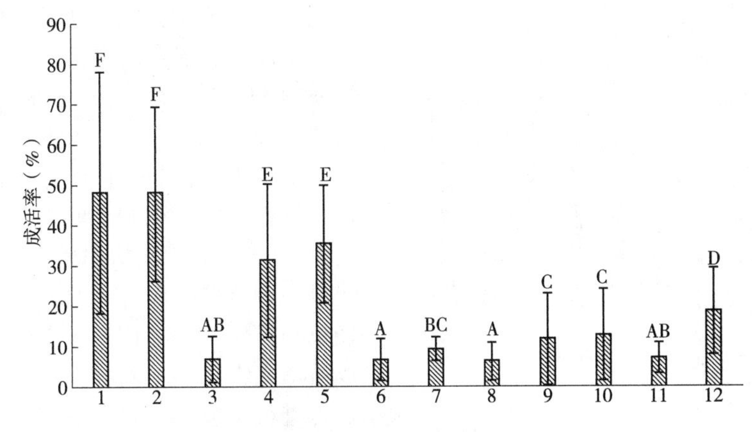

图 1　不同月份灰木莲嫁接成活率

注：不同字母表示不同处理之间差异达到极显著($P<0.01$)，反之则表示差异不显著，下同。

2.2　不同处理对灰木莲嫁接成活的影响

为了解不同接穗在不同浓度 ABT 1 号生根粉影响下的嫁接成活情况，对全年不同处理下灰木莲的嫁接成活率进行方差分析，发现各处理间成活率差异达到极显著水平($P<0.01$，$F=402.733$)，表明不同浓度 ABT1 号生根粉对灰木莲嫁接成活率的影响达到极显著水平。

为进一步比较不同浓度 ABT1 号生根粉对灰木莲嫁接成活的影响，对各处理的成活率进行多重比较分析(图 2)。结果表明不同的接穗和不同的处理方法均对灰木莲嫁接的成活率产生显著影响。采用硬枝接穗蘸取 0mg/L、100mg/L 和 500mg/L 3 种不同浓度的 ABT1 号生根粉后进行嫁接的灰木莲月均成活率分别为 28.2%(YS)、20.8%(YY)和 11.2%(YW)，而采用嫩枝接穗蘸取 3 种不同浓度的 ABT1 号生根粉后的嫁接成活率分别为 0.97%(NS)、1.7%(NY)和 0(NW)，即：采用硬枝接穗进行嫁接的灰木莲成活率均显著高于采用嫩枝嫁接的灰木莲，说明进行灰木莲的嫁接宜选择木质化枝条作为接穗。采用硬枝接穗嫁接的灰木莲在不同浓度 ABT1 号生根粉处理下的成活率差异也达到显著水平，并且呈现随着浓度的升高嫁接成活率会出现下降的趋势，说明在全年整体嫁接时间内 ABT1 号生根粉会对硬枝嫁接月均成活率产生一定的不利影响。此外，笔者发现，在 1 月份和 2 月份，使用 100mg/L 的 ABT1 号生根粉处理硬枝接穗，可以使灰木莲的嫁接成活率分别达到 80%和 75%，而使用清水的嫁接成活率仅为 55%和 53%，说明在合理的时间内使用适宜浓度的 ABT1 号生根粉能够提高灰木莲嫁接成活率。

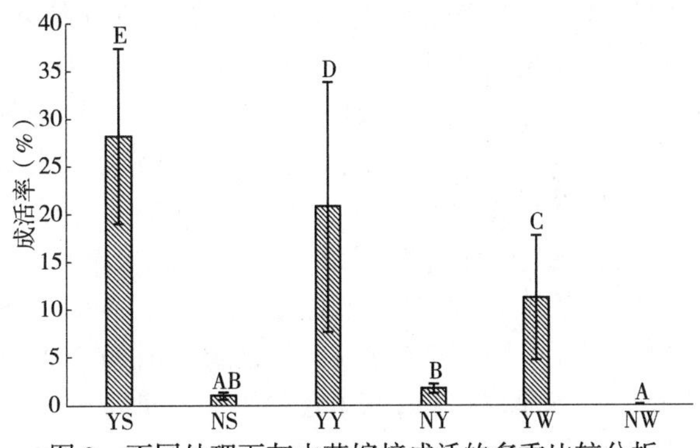

图 2　不同处理下灰木莲嫁接成活的多重比较分析

2.3　不同月份不同处理嫁接成活后接穗的生长情况

为进一步了解不同月份不同处理嫁接成活后灰木莲的生长情况，对所有灰木莲嫁接苗接穗的月均生长量进行方差分析(表 2)，结果发现不同处理间接穗的月均生长量差异不显著，而不同月份间接穗的月均生长量差异达到极显著水平。因此，进一步对不同月份嫁接的灰木莲接穗月均生长量进行方差分析(图 3)。结果发现，生长情况最差的是 6 月份嫁接的灰木莲，接穗月均生长量仅有 1.23cm，其次是 11 月份嫁接的，接穗月均生长量为 1.34cm；生长情况最优的是 10 月份和 2 月份嫁接的灰木莲，其接穗的月均生长量分别达到了 2.62cm 和 2.56cm；即 10 月份和 2 月份嫁接灰木莲的接穗生长要显著优于 6 月份、7 月份和 11 月份嫁接的。

表 2　接穗月均生长量方差分析

变异来源	平方和	自由度	均方	F 值	P 值
月份	59.91	11	5.447	5.562	0.000**
处理	6.875	4	1.719	1.744	0.14
月份*处理	18.725	22	0.851	0.864	0.643

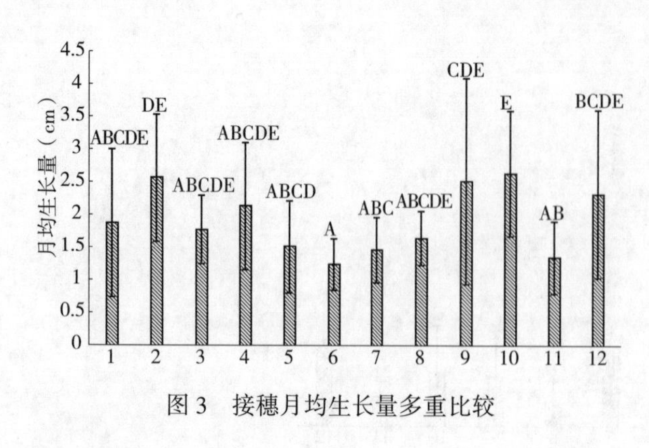

图3 接穗月均生长量多重比较

3 结论与讨论

林木嫁接能够保持和发展母树的优良特性，缩短优良种质的繁殖周期，保存优良的种质资源，因此林木嫁接技术已经成为林木生长发展中的重要环节，被广泛应用在林业生产上[9-11]。大量研究表明嫁接时间对嫁接成活率的影响极大，且不同树种的最佳嫁接时期不同：美国黄松的异砧嫁接试验表明嫁接最佳时间以3月中旬效果最好[6]，垂榆嫁接的最佳嫁接时间在张北为5月15日至5月30日[12]，华山松的硬枝嫁接的最佳时间是2~3月份[13]，本研究中发现灰木莲的嫁接时间以1~2月份为佳，此时嫁接能使灰木莲嫁接成活率达到48%。接穗的选择也是影响嫁接成活的关键因素之一，对美国山核桃嫁接技术的研究表明同一接穗枝条中部的穗条木质化程度好，枝条内细胞活性最高，因此与用顶端的穗条相比，其嫁接成活率最高[11]，本研究也证实了这一结论，同时发现使用硬枝接穗嫁接的灰木莲成活率是同等条件下使用嫩枝接穗嫁接成活率的11~29倍，因此灰木莲的嫁接宜采用硬枝接穗。

ABT生根粉是一种绿色植物生长调节剂，能强化、调控植物内源激素的含量，强化酶的活性，促进大分子的合成，诱导植物不定根或不定芽的形态建成，调节植物新陈代谢[2-4]，促进植物生长，增强植物抗性[14-16]。因此，本研究为提高灰木莲嫁接成活率，尝试用不同浓度的ABT1号生根粉处理接穗再进行插皮嫁接的方法，结果发现从全年来看，使用ABT1号生根粉处理接穗会对灰木莲月均嫁接成活产生不利影响，且随着浓度的升高嫁接成活率的下降趋势越明显；若只关注最适合灰木莲嫁接的1~2月份，则用100mg/L的ABT1号生根粉处理硬枝接穗，可以使灰木莲的嫁接成活率分别达到80%和75%，是采用500mg/L和0mg/L ABT1生根粉处理的嫁接成活率的6.4倍、3.5倍和1.45倍、1.41倍，说明在合理的时间内使用适宜浓度的ABT1号生根粉能显著提高灰木莲的嫁接成活率。不同浓度的ABT1号生根粉能影响灰木莲嫁接的成活率，但对灰木莲嫁接苗接穗的生长没有影响，而嫁接时间会对嫁接苗接穗的生长有一定的影响，且2月份的嫁接苗月均生长量达到了2.56cm。因此，本研究筛选出最适合灰木莲嫁接的条件为：在1~2月份期间采用硬枝接穗醮取100mg/L的ABT1号生根粉进行嫁接。

参考文献

[1]贾宏炎，农瑞红．灰木莲大树移植技术[J]．广西林业科学，2006，35(1)：34-35.

[2]杨耀海，刘明义，常森有．灰木莲引种栽培试验研究[J]．西南林学院学报，2007，27(3)：29-32.

[3]郑万钧．中国树木志[M]．北京：中国林业出版社，1983：438-439.

[4]刘玉壶．中国木兰[M]．北京：科学技术出版社，2003：142-143

[5]卢立华，蒙彩兰，何日明，等．储藏方法对灰木莲种子储藏时间和发芽率的影响[J]．种子，2011(10)：82-85.

[6]张成高，唐德瑞．美国黄松异砧嫁接繁殖研究[J]．西北林学院学报，2005(2)：100-103.

[7]刘志峰，汪加魏，邓世鑫，等．油橄榄夏季芽接技术研究[J]．西北林学院学报，2014(4)：119-122.

[8]QIN W J，WANG Y Q，QIN Z N，et al. Study on the variation characteristics of precipitation concentration degree in Guangxi under the background of global climate becoming warm [J]. Meteorological and Environmental Research, 2010，1(5)：17-21.

[9]刁学良，范志远，邹伟烈，等．东京山核桃砧对美国山核桃嫁接成活率及树体生长结果的影响[J]．西北林学院学报，2006(2)：76-79+96.

[10]王东雪，曾雯珺，江泽鹏，等．遮荫强度对油茶春季芽苗砧嫁接育苗的影响[J]．西北林学院学报，2013(2)：101-104.

[11]马婷，陈宏伟，熊新武，等．砧木、接穗的选择对美国山核桃嫁接成活率及生长的影响[J]．西北林学院学报，2012(4)：141-143+148.

[12]戴福．嫁接时间、接穗粗度与垂榆嫁接成活率关系的研究[J]．河北林业科技，2007(S1)：53-54.

[13]李桐森．华山松嫩枝与硬枝嫁接技术研究[J]．西南林学院学报，2002(2)：68-71+75.

[14]李胜奇．ABT生根粉浸种对青桐种子发芽的影响[J]．西北林学院学报，2008(1)：92-95.

[15]杨荣慧，王延平，段旭昌，等．大果沙棘引种扦插育苗试验研究[J]．西北林学院学报，2004(3)：28-30.
[16]何文广，苏印泉，徐咏梅，等．外源激素影响杜仲叶中次生代谢物含量的研究[J]．西北林学院学报，2009(6)：121-123.

［原载：中国野生植物资源，2015，34(01)］

降香黄檀轻基质容器苗分级标准研究

黎 明 黄柏华 韦叶桥 梁福江 曾 冀

（中国林业科学研究院热带林业实验中心，广西凭祥 532600）

摘 要 探讨适合降香黄檀轻基质容器苗的分级方法和标准，为制定降香黄檀苗木培育技术规程提供参考依据。利用轻基质培育降香黄檀实生苗，于苗龄9个月时测定苗木的株高、地径、地上和地下部分生物量，计算每株苗木的总生物量、高径比及茎根比。应用相关分析法确定降香黄檀苗木的质量分级指标，采用逐步聚类、快速聚类和平均值±标准差划分降香黄檀苗木的分级标准。根据分级标准确定苗木的级别并开展造林试验，比较分析各级别苗木的造林效果，对成活苗木进行重新分级。应用平均值±标准差法分级的降香黄檀合格苗(Ⅰ、Ⅱ级苗)所占比例及造林9个月后各级别苗木的存活情况优于逐步聚类和快速聚类法；将造林9个月后成活的苗木进行重新分级发现，应用平均值±标准差法确定的合格苗有10.5%转变为Ⅲ级苗，有1.8%和37.9%的Ⅲ级苗(不合格苗)分别转变为Ⅰ和Ⅱ级苗，采用逐步聚类和快速聚类两种方法分级的合格苗转变为Ⅲ级苗的比率分别为23.2%和63.7%，Ⅲ级苗转变为合格苗的比率分别为53.0%和80.2%。降香黄檀轻基质容器苗适宜采用平均值±标准差法进行分级，其分级标准为：Ⅰ级苗 $H \geqslant 55.3$cm，$D \geqslant 7.9$mm；Ⅱ级苗 24.4cm $\leqslant H < 55.3$cm，4.7mm $\leqslant D < 7.9$mm；Ⅲ级苗 $H < 24.4$cm，$D < 4.7$mm。

关键词 降香黄檀；轻基质容器苗；苗木分级标准；相关分析；平均值±标准差法；造林效果

Classification of Container Seedlings Growing with Light Media for *Dalbergia odorifera*

LI Ming, HUANG Bohua, WEI Yeqiao, LIANG FuJiang, ZENG Ji

(*Experimental Centre of Tropical Forestry, Chinese Academy Forestry, Pingxiang* 532600, *Guangxi, China*)

Abstract: Grading method and standard for *Dalbergia odorifera* T. Chen seedlings in containers with light media were investigated in order to provide reference for formulating cultivation technique regulations. Taking *D. odorifera* T. Chen seedlings in containers with light media as materials, seedling height(H), ground diameter(D), aboveground biomass and underground biomass were measured when the seedlings were nine-month-old. Then total biomass, seedling height to ground diameter ratio and shoot to root ratio were calculated. Using correlation analysis to determine *D. odorifera* T. Chen seedling quality grading indexes, and using gradual clustering, k-means clustering and mean ± standard deviation to divide seedling grading standard. The seedlings were graded according to the classification standard and afforestation experiment was conducted. Afforestation effects of seedlings at different grades were compared and analyzed. The survival seedlings were re-graded. The proportion of qualified seedlings(grade Ⅰ and Ⅱ) and survival seedlings after nine months of planting graded by mean ± standard deviation were better than those graded by gradual clustering and k-means clustering. Reclassification of survival seedlings nine months after out-planted showed that in mean ± standard deviation method, 10.5% of the qualified seedlings transferred into grade Ⅲ, 1.8% of grade Ⅲ seedlings(unqualified seedlings) transferred into grade Ⅰ, and 37.9% of grade Ⅲ seedlings(unqualified seedlings) transferred into grade Ⅱ. In terms of gradual clustering, 23.2% of the qualified seedlings transferred into grade Ⅲ, 63.7% of grade Ⅲ seedlings transferred into qualified seedlings. In k-mean clustering, 53.0% of the qualified seedlings transferred into grade Ⅲ, and 80.2% of grade Ⅲ transferred into qualified seedlings. Mean ± standard deviation is suitable for classification of *D. odorifera* T. Chen seedlings in containers with light media, and the grading standard is as follows: grade Ⅰ $H \geqslant 55.3$cm, $D \geqslant 7.9$mm; grade Ⅱ 24.4cm $\leqslant H < 55.3$cm,

4.7mm≤D<7.9mm; grade Ⅲ H<24.4cm, D<4.7mm.

Key words: *Dalbergia odorifera*; light media container seedling; seedling grading standard; correlation analysis; mean ± standard deviation method; afforestation effect

引言

【研究意义】降香黄檀(*Dalbergia odorifera* T. Chen)别名黄花梨、花梨木，为国家二级保护植物，天然分布于我国海南，广东、广西、福建及云南等地亦有大规模引种栽培[1,2]。降香黄檀不仅木材坚硬，是制作名贵家具、雕刻工艺品的上等用材，且具有较高的生态价值[3]和药用价值[4]，市场上供不应求，其野生资源已濒临枯竭[5]。发展降香黄檀人工林是解决市场需求与资源枯竭矛盾的最佳方法，但市场上供应的降香黄檀苗木没有统一的分级标准，质量优劣并存，制约了降香黄檀产业的健康发展。与常规容器育苗相比，轻基质网袋容器育苗具有重量轻、容器小、可进行空气修根、适合工厂化生产及运输容易、造林效果好等优点[6]。本课题组结合“珍贵树种轻基质工厂化育苗技术研究与示范”项目，已筛选出适合降香黄檀苗木生长的轻基质配方[7]，且正在大力推广降香黄檀轻基质网袋容器苗造林。由于苗木质量的优劣直接关系造林的成败和林分的生产力[8,9]，因此，苗木分级对于降香黄檀规模化、规范化种植至关重要。【前人研究进展】陈晓波等[10]采用平均值±标准差法对蒙古栎、杨斌等[11]及马跃等[12]应用逐步聚类法分别对铁力木和望天树开展了苗木分级研究。在降香黄檀苗木分级研究方面，方碧江[13]在闽南地区、周双清等[14]在海南应用逐步聚类分析法研究了6个月生降香黄檀常规容器苗的分级标准，其合格苗的指标分别为苗高≥18.9cm、地径≥2.7mm和苗高≥41.7cm、地径≥1.9mm；郭俊杰等[15]研究认为，开展造林效果试验有助于将苗木分级与造林生产紧密结合起来，使分级标准更可靠。【本研究切入点】目前，针对降香黄檀轻基质容器苗进行分级并进行造林效果试验的研究鲜见报道。【拟解决的关键问题】以9个月生降香黄檀轻基质容器苗为研究对象，应用逐步聚类、快速聚类和平均值±标准差法对降香黄檀轻基质容器苗进行分级，比较分析各级别苗木的造林效果，旨在提出降香黄檀轻基质容器苗的分级标准，为制定降香黄檀苗木培育技术规程提供参考依据。

1 材料与方法

1.1 育苗地和造林地概况

育苗地点在广西凭祥市中国林业科学研究院热带林业实验中心白云实验场苗圃，属于南亚热带季风气候，年均气温21.4℃，最热月(7月)平均气温27.6℃，极端最高气温39.8℃，最冷月(1月)平均气温13.1℃，极端最低气温-1.0℃，年降水量1200~1400mm，干湿季节明显，4~9月为雨季(雨热同季)。造林地点在中国林业科学研究院热带林业实验中心哨平实验场(106°53′18″E，22°02′52″N)，海拔230m，属桉树采伐迹地，阳坡，坡度约13°，位于坡中部，土壤为紫色土。

1.2 试验材料

降香黄檀种子采自中国林业科学研究院热带林业实验中心石山树木园。两种轻基质材料松树皮和锯末取自当地木材加工厂或林区。可降解的无纺布网袋卷(制作轻基质网袋容器用)购自安徽安庆市亿科育苗材料有限公司。育苗所用轻基质配方为沤制的松树皮、锯末和碳化松皮按6∶3∶1的体积比例配置而成。

1.3 试验方法

1.3.1 网袋容器制作及苗木管理

松树皮、锯末和碳化松树皮的堆沤及轻基质网袋容器的制作参照蒙彩兰等[16]和郭文福等[17]的方法进行处理。于2014年3月初播种，3月中旬将降香黄檀芽苗移栽到轻基质网袋容器后，保持基质湿润，每周施复合肥或尿素1次，浓度为2000~5000mg/L，根据苗木生长情况逐渐加大施肥浓度，其他措施与常规育苗相似。

1.3.2 样苗选取及指标测定

2014年12月中旬，从苗圃随机抽取835株降香黄檀苗，按照抽取顺序编号及挂牌，测定其苗高和地径，进一步从835株降香黄檀苗中随机抽取415株作为样苗，逐株洗除基质后，将地上和地下部分剪下分别装入纸袋，置于60℃烘箱干燥48h至恒重，用电子天平(精度为0.001g)称量各样苗的地上和地下部分生物量。计算高径比(苗高/地径)，参考陈秋夏等[18]的方法计算其茎根比(地上部分生物量/地下部分生物量)和质量指数(苗木总生物量/高径比+茎根比)。

1.3.3 造林效果试验

应用逐步聚类、快速聚类和标准差±平均值法确定余下420株降香黄檀苗的级别，2015年3月10日在哨平实验场开展造林试验。造林前经炼山、清理林地后，按照株行距 2m×2m、穴规格 50cm×50cm×30cm、施基肥(桐麸与钙镁磷肥按 1∶1 的质量比混合)1.0kg/穴进行种植，造林后其他管理措施与常规生产林相似。试验采用完全随机排列方式，面积约 0.168hm^2。造林 1 周时测定苗高和地径，2015 年 12 月上旬调查存活率。

采用上述 3 种方法对造林 9 个月后保存的苗木进行重新分级，比较三者之优劣，最终确定苗木分级标准。

1.4 统计分析

试验数据应用 SSPS16.0 进行方差分析和相关分析。

2 结果与分析

2.1 苗木分级质量指标的确定

由表 1 可知，降香黄檀苗地下部分生物量与其他质量指标均呈极显著相关($P<0.01$，下同)；苗高和地上生物量与茎根比相关不显著($P>0.05$，下同)，与其他指标呈极显著相关；地径与高径比、茎根比相关不显著，与其他指标呈极显著相关。鉴于地径、苗高指标与其他指标的相关性较密切，且在生产管理中较直观且容易测取，为了便于生产上应用，将其选为降香黄檀轻基质容器苗分级的质量指标。

表 1 降香黄檀轻基质容器苗各指标间的相关性

指标	苗高(cm)	地径(mm)	地上生物量	地下生物量	总生物量	高径比	茎根比
地径	0.704**						
地上生物量	0.813**	0.851**					
地下生物量	0.656**	0.839**	0.894**				
总生物量	0.795**	0.863**	0.996**	0.931**			
高径比	0.720**	-0.046	0.310**	0.133**	0.279**		
茎根比	0.062	-0.024	-0.015	-0.137**	-0.040	0.109*	
质量指数	0.580**	0.874**	0.893**	0.973**	0.925**	0.006	0.128**

注：n=415；** 和 * 分别表示性状间相关极显著($P<0.01$)和显著($P<0.05$)。表 3 同。

2.2 苗木的分级结果

2.2.1 逐步聚类法分级结果

从 415 株样苗中分别找出苗高、地径观测值的最大值和最小值，计算各样苗苗高和地径的标准化值并求和后排序，于变化明显处将样苗初步分为三级，计算各级样苗苗高和地径标准化值的平均值，并以此平均值作为各级样苗的凝聚中心，计算每株样苗与各凝聚中心的距离，最后依据最短距离原则判定样苗所属的级别。经 12 次聚类修订后样苗分级结果保持不变，最终得到Ⅰ、Ⅱ和Ⅲ级样苗的凝聚中心分别为 XⅠ(0.64, 0.74)、XⅡ(0.34, 0.50)和 XⅢ(0.15, 0.28)，并计算每个级别样苗苗高和地径标准方差之和的平方根作为该级别的半径。将凝聚中心和半径按照相同比例尺转绘至方格纸上，即可读出Ⅰ级苗的苗高和地径标准化值下限为 0.52 和 0.60，Ⅱ级苗的苗高和地径标准化值下限为 0.26 和 0.38。将这些值转化为初始值，即获得降香黄檀苗的分级标准(表 2)。

2.2.2 快速聚类法分级结果

参考倪雪梅(2010)的方法，根据实际情况先将聚类数目分为 3 类，再依据 415 株样苗高和地径观测值确定 3 个有代表性的数据作为初始类中心，计算每株样苗与 3 个聚类中心点的欧氏距离，并依照最短距离的原则，将样苗分派到各类中心点所在的类中，形成新的 3 类，即完成第 1 次迭代过程；此后重新计算每个类中各变量的均值，以均值点作为新类的中心点，重复上述两步计算过程，直至达到终止迭代的要求为止；经 6 次迭代过程聚类分析结束，即得到降香黄檀苗木的最终聚类中心、分类结果及其分级标准(表 2)。

2.2.3 平均数±标准差法分级结果

计算 415 株样苗苗高和地径的平均值和标准差，将平均值+标准差和平均值-标准差分别作为Ⅰ和Ⅱ级苗的下限值，即得到降香黄檀苗木的分级标准(表 2)。

由表 2 可知，根据 3 种方法得出的苗木分级结果，各级别间苗高、地径均呈极显著差异($P<0.01$)。相对于逐步聚类和快速聚类法，采用平均值±标准差法进行降香黄檀轻基质容器苗分级，其Ⅲ级苗的比例明显降低，Ⅱ级苗的比例明显升高，Ⅰ级苗比例变化不明显；Ⅰ、Ⅱ级苗所占比例为 78.1%，均明显高于逐步聚类法的 63.8%和快速聚类法的 56.2%。在生产实践中往往将Ⅰ、Ⅱ级苗视为合格苗，Ⅲ级苗为不合格苗。从其比率看，开展降香黄檀轻基质苗木分级以平均值±标准差法优于逐步聚类和快速聚类法。

表 2 9 个月生降香黄檀轻基质容器苗的分级标准及各级别苗木生长指标的差异

分级方法	苗木级别	分级标准	比例(%)	苗高(cm)	地径(mm)
快速聚类	Ⅰ	$H \geq 57.0$cm，$D \geq 8.0$mm	14.0	68.5±8.4aA	8.3±1.1aA
	Ⅱ	57.0cm>$H \geq$36.0cm，8.0mm>$D \geq$6.0mm	42.2	44.3±5.7bB	6.7±1.3bB
	Ⅲ	H<36.0cm，D<6.0mm	43.8	26.5±5.5cC	5.3±1.0cC
逐步聚类	Ⅰ	$H \geq 54.0$cm，$D \geq 7.3$mm	20.7	62.7±11.0aA	8.4±1.0aA
	Ⅱ	54.0cm>$H \geq$34.5cm，7.3mm>$D \geq$5.6mm	43.1	40.3±7.1bB	6.5±0.8bB
	Ⅲ	H<34.5cm，D<5.6mm	36.2	26.3±6.7cC	4.8±0.7cC
平均值±标准差	Ⅰ	$H \geq 55.3$cm，$D \geq 7.9$mm	18.1	61.7±12.3aA	8.7±0.7aA
	Ⅱ	55.3cm>$H \geq$24.4cm，7.9mm>$D \geq$4.7mm	60.0	38.8±10.0bB	6.2±0.9bB
	Ⅲ	H<24.4cm，D<4.7mm	21.9	24.9±8.1cC	4.5±0.8cC

注：$n=415$；H 表示苗高，D 表示地径；同一方法同列数据后不同大、小写字母分别表示差异极显著($P<0.01$)和显著($P<0.05$)。表 4 同。

2.3 造林效果比较

造林是评价苗木质量优劣最有效的方法，质量优的苗木能明显提高造林成活率和幼林生长量。相关分析结果(表 3)表明，造林后 9 个月降香黄檀的苗高、地径与造林前的苗高、地径、高径比呈极显著或显著正相关($P<0.05$，下同)，苗高净增长与苗高、地径呈极显著或显著正相关，与高径比相关不显著，地径净增长与苗高、高径比呈显著正相关，与地径生长相关不显著。说明采用苗高和地径作为降香黄檀轻基质容器苗的分级质量指标合理、可行。

表 3 9 个月降香黄檀苗造林前与造林后生长性状间的相关性

项目(造林前)	造林后 9 个月			
	苗高(cm)	地径(mm)	苗高净增长(cm)	地径净增长(mm)
苗高	0.438**	0.292**	0.185*	0.117*
地径	0.415**	0.327**	0.146**	0.072
高径比	0.261**	0.136*	−0.062	0.111*

注：$n=325$。

比较 3 种方法分级苗木的造林效果(表 4)可知，Ⅰ、Ⅱ和Ⅲ级苗的造林存活率均随苗木级别的升高而呈递减趋势，其中根据逐步聚类和快速聚类法分级标准分级的Ⅰ、Ⅱ级苗造林存活率均在 95.0%以上，明显高于Ⅲ级苗的存活率，而采用平均值±标准差法分级标准分级的降香黄檀苗Ⅰ级苗造林存活率在 95.0%以上，Ⅱ和Ⅲ级苗的造林存活率明显降低，仅 88.7%和 77.0%。3 种方法分级的苗木造林 9 个月后，Ⅰ、Ⅱ和Ⅲ级苗的地径和苗高均呈显著或极显著递减趋势，苗高和地径比造林前有较大幅度增长。

进一步应用 3 种分级方法对造林 9 个月后存活的苗木进行重新分级，结果(表 4)表明，采用平均值±标准差法分级，有 5.3%的Ⅰ级苗和 5.2%的Ⅱ级苗(合计为 10.5%)转变为Ⅲ级苗，1.8%和 37.9%的Ⅲ级苗分别转变为Ⅰ和Ⅱ级苗；采用逐步聚类法分级，有 9.4%的Ⅰ级苗和 13.8%的Ⅱ级苗(合计为 23.2%)转变为Ⅲ级苗，8.6%和 44.4%的Ⅲ级苗(合计为 53.0%)分别转变为Ⅰ和Ⅱ级苗；采用快速聚类法分级，有 14.6%的Ⅰ级苗和 49.1%的Ⅱ级苗(合计为 63.7%)转变为Ⅲ级苗，19.8%和 60.4%的Ⅲ级苗(合计为 80.2%)分别转变为Ⅰ和Ⅱ级苗。可见，应用平均值±标准差法分级苗木造林 9 个月后苗木级别的转变比率均明显低于逐步聚类法和快速聚类法，从造林效果看，平均值±标准差法更适用于对降香黄檀轻基质苗木分级。

表4 造林后9个月各级别降香黄檀轻基质容器苗的生长表现及重新分级结果

分级方法	苗木级别	造林时生长情况			造林9个月后生长情况			重新分级后各级苗比例(%)		
		比例(%)	苗高(cm)	地径(mm)	存活率(%)	苗高(cm)	地径(mm)	Ⅰ	Ⅱ	Ⅲ
快速聚类	Ⅰ	19.5	68.0±7.6aA	8.8±0.7aA	98.3	118.1±31.0aA	18.8±4.6aA	37.8	47.6	14.6
	Ⅱ	46.3	47.9±8.8bB	7.0±0.8bB	97.2	99.9±33.0bB	17.0±4.9bA	6.3	44.6	49.1
	Ⅲ	38.6	30.2±4.5cC	5.7±1.1cC	81.5	70.8±36.8cC	14.6±6.2cB	19.8	60.4	19.8
逐步聚类	Ⅰ	20.1	63.4±9.8aA	8.5±0.8aA	96.9	113.7±30.2aA	18.5±4.7aA	45.3	45.3	9.4
	Ⅱ	46.3	44.1±8.7bB	6.8±0.8bB	95.6	97.7±34.9bB	17.0±5.1bA	31.9	54.3	13.8
	Ⅲ	33.6	28.6±4.2cC	5.3±1.0cC	80.6	67.5±33.8cC	13.8±5.8cB	8.6	44.4	47.0
平均值±标准差	Ⅰ	15.2	66.6±8.4Aa	8.7±0.7 Aa	97.3	117.6±32.1aA	18.9±4.9aA	43.9	50.8	5.3
	Ⅱ	67.6	43.4±10.3bB	6.8±0.9bB	88.7	94.2±34.4bB	16.7±5.1bB	21.4	73.4	5.2
	Ⅲ	18.1	26.4±4.7cC	4.7±0.5cC	77.0	64.5±37.2cC	12.8±6.1cC	1.8	37.9	60.3

注：$n=325$。

3 讨论

本研究相关分析结果表明，降香黄檀轻基质容器苗的苗高、地径与多数苗木质量指标呈显著或极显著相关，而且造林后9个月的苗高、地径及苗高净增长与初始苗高、地径亦呈显著或极显著相关，说明采用苗高和地径作为降香黄檀轻基质容器苗的分级质量指标是合理的。

利用平均值±标准差法确定降香黄檀轻基质苗分级标准，其效果优于逐步聚类和快速聚类法。具体而言，其一，应用平均值±标准差法分级的合格苗(Ⅰ和Ⅱ级苗)所占比例明显大于逐步聚类和快速聚类法，前者合格苗所占比例为78.1%，后两种方法所占比例分别为63.8%和56.2%；其二，逐步聚类和快速聚类法分级的各级别苗木造林后，其Ⅰ和Ⅱ级苗存活率均在95.0%以上且差异不明显，而平均值±标准差法分级的各级别苗木造林后Ⅰ级苗成活率在95.0%以上，与Ⅱ和Ⅲ级苗的造林成活率差异明显；其三，对造林9个月后的存活苗木进行重新分级发现，采用平均值±标准差法分级，其合格苗中仅10.5%转变为Ⅲ级苗，不合格苗(Ⅲ级苗)仅1.8%转变为Ⅰ级苗，而采用逐步聚类和快速聚类两种方法分级，其合格苗转变为Ⅲ级苗的比率分别高达23.2%和63.7%，不合格苗转变为Ⅰ级苗的比率亦分别达8.6%和19.8%。此外，从统计分析的工作量等方面考虑，相对于逐步聚类和快速聚类法，平均值±标准差法的数据统计简单、工作量较小，且也便于生产上应用。

依据本研究确定的降香黄檀轻基质容器苗分级标准，其9个月生合格苗的地径、株高分别在4.7mm和24.4cm以上，地径明显高于方碧江[18]、周双清等[19]的6个月生合格苗(2.7mm和18.9cm；1.9mm和41.7cm)，而苗高则介于后两者之间，可能与种源、基质、气候及苗龄不同等多种因素有关。但考虑到苗高适中的粗壮苗造林效果表现更佳，培育9个月的苗木可能更适合用于造林。

4 结论

降香黄檀轻基质容器苗分级适宜采用平均值±标准差法，即以平均值加标准差作为Ⅰ级苗的地径和苗高下限，平均值减标准差作为Ⅱ级苗的地径和苗高下限，具体分级标准为：Ⅰ级苗 $H\geqslant55.3$cm，$D\geqslant7.9$mm；Ⅱ级苗 $24.4\text{cm}\leqslant H<55.3\text{cm}$，$4.7\text{mm}\leqslant D<7.9\text{mm}$；Ⅲ级苗 $H<24.4$cm，$D<4.7$mm。

参考文献：

[1]倪臻，王凌晖，吴国欣，等．降香黄檀引种栽培技术研究概述[J]．福建林业科技，2008，35(2)：265-268.
[2]孟慧，杨云，冯锦东．降香黄檀引种栽培现状与发展[J]．广东农业科学，2010(7)：79-80.
[3]梁建平．广西珍稀濒危树种[M]．南宁：广西科学技术出版社，2003.
[4]国家药典委员会．中华人民共和国药典2005年版(一部)[M]．北京：化学工业出版社，2005.
[5]罗文扬，罗萍，武丽琼，等．降香黄檀及其可持续发展对策探讨[J]．热带农业科学，2009，29(1)：44-46.
[6]张建国，王军辉，许洋，等．网袋容器苗育苗新技术[M]．北京：科学出版社，2007.
[7]贾宏炎，黎明，曾冀，等．降香黄檀工厂化育苗轻基质筛选试验[J]．中南林业科技大学学报，2015，35(11)：74-79.

[8]刘勇．我国苗木培育理论与技术进展[J]．世界林业研究，2000，13(5)：43-49.
[9]侯元兆．现代林业育苗的概念与技术[J]．世界林业研究，2007，20(4)：24-29.
[10]陈晓波，王继志，叶燕萍，等．蒙古栎苗木分级标准的研究[J]．北华大学学报(自然科学版)，2002，3(3)：251-254.
[11]杨斌，周凤林，史富强，等．铁力木苗木分级研究[J]．西北林学院学报，2006，21(1)：85-89.
[12]马跃，谌红辉，李武志，等．望天树苗木分级技术研究[J]．西北林学院学报，2012，27(4)：153-156.
[13]方碧江．降香黄檀苗木分级标准的探讨[J]．福建热作科技，2009，34(4)：30-32.
[14]周双清，周亚东，吴群富，等．海南降香黄檀种苗分级标准[J]．西部林业科学，2015，44(3)：26-30.
[15]郭俊杰，尚帅斌，汪奕衡，等．热带珍贵树种青梅苗木分级研究[J]．西北林学院学报，2016，31(3)：74-78.
[16]蒙彩兰，黎明，郭文福．西南桦轻基质网袋容器育苗技术[J]．林业科技开发，2007，21(6)：104-105.
[17]郭文福，曾杰，黎明，等．西南桦轻基质网袋容器苗基质选择试验[J]．种子，2010，29(10)：62-64.
[18]陈秋夏，廖亮，郑坚，等．光照强度对青冈栎容器苗生长和生理特征的影响[J]．林业科学，2011，47(12)：53-59.

[原载：南方农业学报，2016，47(12)]

西南桦24个无性系的幼苗叶绿素荧光特性

张　培[1]　郭俊杰[2]　谌红辉[1]　曾　杰[2]

([1]中国林业科学研究院热带林业实验中心，广西凭祥　532600；
[2]中国林业科学研究院热带林业研究所，广东广州　51052)

摘　要　以24个西南桦无性系的组培苗为试验材料，对其叶绿素荧光特性进行了研究。结果表明：无性系间叶绿素荧光参数 F_o、F_v、F_m、F_v/F_o 以及 F_v/F_m 均差异显著($P<0.05$)；FB4、BY-1无性系的 F_v/F_o 和 F_v/F_m 与A3及1-202无性系差异不显著，而显著高于绝大多数其他无性系，说明这些无性系光合效率高；无性系的苗高和地径均与 F_o 呈显著负相关，与 F_v、F_m、F_v/F_o 和 F_v/F_m 大多呈显著正相关；A3、FB4、1-202和BY-1等4个无性系生长表现好且光合效率高，可作为优良无性系在生产上推广应用。

关键词　西南桦；无性系评价；叶绿素荧光

Chlorophyll Fluorescent Characteristics of Twenty-four *Betula alnoides* Clones

ZHANG Pei[1], GUO Junjie[2], CHEN Honghui[1], ZENG Jie[2]

([1]*Experimental Center of Tropical Forestry, CAF, Pingxiang* 532600, *Guangxi, China*;
[2]*Research Institute of Tropical Forestry, CAF, Guangzhou* 510520, *Guangdong, China*)

Abstract: Chlorophyll fluorescence characteristics of seedlings of 24 *Betula alnoides* clones were investigated. The results showed that chlorophyll fluorescence parameters including F_o, F_v, F_m, F_v/F_o and F_v/F_m were significantly different among these clones ($P<0.05$). F_v/F_o, F_v/F_m of FB4 and BY-1clones did not remarkably differ from those of A3, B3 and1-202 ($P>0.05$), while were quite higher than those of majority of other clones, indicating that these clones were of high photosynthetic efficiency. The seedling height and root collar diameter were negatively correlated with F_o, while positively correlated with F_v, F_m, F_v/F_m and F_v/F_o. The above four clones were of good growth performance and high photosynthetic efficiency, and are therefore recommended as prior clones for application in practice.

Key words: *Betula alnoides*; clone selection; chlorophyll fluorescence

叶绿素荧光测定技术是利用植物体内的叶绿素荧光对其光合生理状况进行测定与诊断，能灵敏地反映植物荧光的动态变化与环境的互作关系，还可直接或间接地了解植物光合作用过程，是研究植物光合能力的一个重要手段[1]。叶绿素荧光测定技术具有操作简便、速度快、对植株无损伤以及结果准确等优点，已广泛应用于农业、园艺作物以及林木的培育[2,3,14]。学者们利用叶绿素荧光技术，开展了林木抗逆性评价[4]以及林木种源和无性系选择等研究[5-7,13,15]。

西南桦(*Betula alnoides* Buch. -Ham. ex D. Don)为桦木科(Betulaceae)桦木属(*Betula*)的一个珍贵用材树种，在我国天然分布于云南、广西、贵州和西藏等地。西南桦干形通直，材质优良，且耐干旱贫瘠，是我国热带南亚热带地区的一个主要造林树种[8]。近几年来，随着无性系选育工作不断取得进展，西南桦种植业发展迎来新契机。在以往的研究中，谌红辉等(2013)依据生长指标对4年生西南桦无性系测定林进行了评价，初步筛选出6个优良无性系[9]，然而有关西南桦无性系叶绿素荧光特性研究尚未见报道。开展无性系叶绿素荧光特性研究，并结合其生长表现进行分析，有助于揭示无性系速生性的光合生理基础，提高无性系评价的可靠性[15]。因此，本研究利用叶绿素荧光技术测定不同西南桦无性系的光合差异，分析叶绿素荧光参数与生长指标的相关性，为西南桦优良无性系的选择提供科学依据。

1 材料与方法

研究地位于广东省广州市中国林业科学研究院热带林业研究所苗圃内(23°8′N，113°17′E)。该地属于南亚热带季风气候，年均气温21.9℃，平均相对湿度77%，年均降水量约1720mm。

1.1 材料

试验材料为以黄心土为基质、采用常规方法培育的1年生西南桦无性系组培苗，24个无性系的编号及生长表现详见表1。

表1 24个西南桦无性系1年生幼苗的生长表现

无性系	苗高(cm)	地径(mm)	无性系	苗高(cm)	地径(mm)
A2	85.07±2.93	4.82±0.38	B3	66.53±8.43	4.42±0.22
A3	90.93±2.51	5.59±0.89	C5	69.30±6.92	4.85±0.30
A4	85.40±8.12	5.08±0.72	C6	74.93±1.89	4.96±0.59
A5	48.90±3.44	3.55±0.19	Q1	77.27±5.06	4.60±0.51
A12	74.00±6.61	4.31±0.50	Q2	68.73±3.98	4.14±0.07
A13	70.30±7.85	4.04±0.20	FB02	58.83±8.90	3.81±0.13
A14	87.10±10.29	4.72±0.33	FB4	91.47±10.45	6.13±0.83
A15	54.67±9.09	4.09±0.74	FB4+	61.67±4.83	3.99±0.36
A16	49.07±10.28	3.76±0.18	1-202	102.77±6.86	5.24±0.86
A17	72.33±7.21	4.79±0.43	0104	77.90±6.58	4.11±0.55
B1	58.77±10.84	3.96±0.79	VY-4	80.07±9.92	4.41±0.72
B2	55.27±0.85	4.01±0.33	BY-1	92.97±6.25	5.39±0.28

注：表中数据为平均值±标准差。

1.2 试验方法

选取无病虫害、生长良好的西南桦无性系幼苗，每个无性系选择3株，从每株苗木上选择1片成熟叶片(从顶芽开始第4或5片叶)。2014年10月，选择晴朗无风天气，于上午8：00～11：00，采用OS-30p+便携式叶绿素荧光测定仪(美国Opti-Sciences公司生产)测定所选叶片的各种叶绿素荧光动力学参数。

测定时，先将待测叶片暗适应20min，设置调制光光强，直接测量获取初始荧光F_o，最大荧光F_m，可变荧光F_v，PSII最大光化学效率或原初光能转换效率F_v/F_m和PSII的潜在活性F_v/F_o的数值。

采用SPSS13.0对各无性系的叶绿素荧光参数进行方差分析、Duncan多重比较，并将幼苗生长表现与叶绿素荧光参数进行相关性分析，相关系数采用Pearson系数。

2 结果与分析

2.1 叶绿素荧光参数比较

各无性系的叶绿素荧光参数如表2所示。方差分析结果表明，24个西南桦无性系间各叶绿素荧光参数的差异均达到了显著水平($P<0.05$)，说明西南桦不同无性系对光能利用存在显著差异。初始荧光F_o反映光系统PSII反应中心全部开放即QA(质体醌)全部氧化时的荧光水平，以B1、B2、0104、A16等4个无性系为最大，显著高于大多数无性系，以Q2、A17、B3、A4等4个无性系为最小值。F_m和F_v值分别反映了PSII的电子传递状况和潜在活性，FB4、0104、A3、FB02、BY-1、1-202无性系具有较高的F_m与F_v值。F_v/F_o和F_v/F_m均以FB4、BY-1无性系为最大，与A3、B3以及1-202无性系差异不显著，而显著高于其他无性系。

表2 24个西南桦无性系的叶绿素荧光参数

无性系 Clones	F_o	F_m	F_v	F_v/F_o	F_v/F_m
A2	212.33±4.73defg	828.67±6.43cd	616.33±6.03bc	2.90±0.08bc	0.74±0.01bc
A3	209.00±8.00cdef	1008.00±15.13l	799.00±10.82k	3.83±0.14kl	0.79±0.01ijk

(续)

无性系 Clones	F_o	F_m	F_v	F_v/F_o	F_v/F_m
A4	196.67±6.51abc	811.33±12.50bc	614.67±8.02bc	3.13±0.09cdef	0.76±0.01cdef
A5	214.67±7.64efg	743.00±12.53a	528.33±8.96a	2.46±0.09a	0.71±0.01a
A12	197.67±5.51abc	833.67±11.24d	636.00±15.62cd	3.22±0.16defg	0.76±0.01def
A13	201.33±7.51bcd	924.33±8.02i	723.00±1.00g	3.59±0.13ijk	0.78±0.01hij
A14	206.00±5.29cdef	875.67±9.29fgh	669.67±12.50ef	3.25±0.13defg	0.76±0.01def
A15	216.00±5.57fg	882.00±11.53fgh	666.00±12.49ef	3.09±0.12bcdef	0.76±0.01bcde
A16	228.67±9.02hi	874.00±11.53fg	645.33±13.50de	2.83±0.15b	0.74±0.01b
A17	188.00±6.56a	844.00±12.77de	656.00±19.31def	3.49±0.23ghij	0.78±0.01ghi
B1	231.67±6.66i	984.00±15.72k	752.33±16.50hi	3.25±0.13defg	0.76±0.01efg
B2	230.00±7.94i	885.00±8.89gh	655.00±16.46def	2.85±0.17bc	0.74±0.01bc
B3	192.00±6.08ab	926.00±14.93i	734.00±17.58gh	3.83±0.19kl	0.79±0.01ijk
C5	198.00±7.21abc	862.67±10.02ef	664.67±10.97ef	3.36±0.15fghi	0.77±0.01fgh
C6	197.33±6.81abc	795.00±13.11b	597.67±10.21b	3.03±0.11bcde	0.75±0.01bcde
Q1	221.00±5.57ghi	886.67±12.66gh	665.67±16.17ef	3.01±0.14bcd	0.75±0.01bcde
Q2	187.00±6.00a	805.33±10.02b	618.33±15.82bc	3.31±0.19efgh	0.77±0.01fgh
FB02	217.33±7.51fgh	1007.33±14.74l	790.00±21.93k	3.64±0.23jk	0.78±0.01ghi
FB4	197.67±7.09abc	989.67±15.31kl	792.00±21.66k	4.01±0.25l	0.80±0.01k
FB4+	208.33±4.93cdef	952.00±11.53j	743.67±7.09ghi	3.57±0.06hijk	0.78±0.00ghi
1-202	207.33±2.52cdef	995.67±12.06kl	788.33±10.97jk	3.80±0.06kl	0.79±0.00ijk
0104	229.67±1.53i	996.00±10.54kl	766.33±9.02ij	3.34±0.02fghi	0.77±0.00fgh
VY-4	222.33±8.33ghi	896.67±10.79h	674.33±12.66f	3.04±0.15bcde	0.75±0.01bcd
BY-1	203.67±4.16bcde	1007.67±11.24l	804.00±10.15k	3.95±0.09l	0.80±0.00k

注：表中数据为平均值±标准差；小写字母为多重比较结果，无性系间具相同字母表示差异不显著(P>0.05)，字母不同表示差异显著(P<0.05)。

2.2 生长指标与叶绿素荧光参数的相关性分析

相关性分析结果显示(表3)：西南桦无性系的苗高、地径与绝大部分叶绿素荧光参数相关显著。其中苗高和地径与 F_o 分别呈显著(P<0.05)、极显著(P<0.01)负相关，与 F_v、F_m、F_v/F_m、F_v/F_o 呈正相关，且大多达到了极显著水平；F_v/F_m、F_v/F_o 与 F_o 呈负相关，与 F_v、F_m 呈正相关，且均达到了极显著水平。

表3 西南桦无性系苗高、地径与叶绿素荧光参数的相关性

	苗高	地径	F_o	F_v	F_m	F_v/F_o
地径	0.729**					
F_o	-0.283*	-0.381**				
F_v	0.348**	0.296*	0.078			
F_m	0.284*	0.216	0.261*	0.983**		
F_v/F_o	0.450**	0.468**	-0.478**	0.836**	0.721**	
F_v/F_m	0.454**	0.460**	-0.475**	0.825**	0.711**	0.985**

注：** 表示极显著相关(P<0.01)，* 表示显著相关(P<0.05)。**，P<0.01；*，P<0.05。

3 讨论与结论

叶绿素荧光参数 F_v/F_m 值反映光系统 PSII 的光化学效率，在正常光照条件下，其值波动范围为0.75~0.85，若低于0.75，说明植物受到了光抑制[6]。本文对24个西南桦无性系的幼苗叶绿素荧光参数研究结果显示，所有无性系基本上处于正常值范围内，说明广州10月份的天气状况适合参试的所有西南桦无性系的幼苗生长，未发生明显的光抑制。24个西南桦无性系间叶绿素荧光参数 F_o、F_v、F_m、F_v/F_o 以及 F_v/F_m 均差异显著(P<0.05)。

林定达等(2011)对芳樟无性系[7]以及吕芳德等(2006)对美国山核桃无性系[10]的叶绿素荧光特性研究显示[7,10]：F_v/F_o 和 F_v/F_m 二者的数值越高，所捕获的光能可更有效地转化为植物所需的化学能，这两个参数已被公认为是反映叶片光合效率的重要依据。在本研究中，A3、FB4、1-202和BY-1等4个西南桦无性系的幼苗 F_v/F_o 与 F_v/F_m 明显高于绝大多数其他无性系（$B3$ 无性系例外），说明其 PSII 反应中心的光能捕获效率高，能更有效地将光能转化为其生长所需的化学能。

相关性分析表明：西南桦无性系幼苗的苗高、地径与 F_v/F_o 与 F_v/F_m 呈极显著正相关，说明西南桦无性系的 PSII 光化学功能与其生长表现密切相关，光合活性越强，生长表现则越好，这与张杰等[12]对不同蒙古栎种源叶绿素荧光特性的研究结果类似。在前期研究中，谌红辉等（2013）评价了 4 年生西南桦无性系测定林，从 20 个无性系中选择出 6 个优良无性系[9]，参试的无性系绝大部分与本试验相同，只是其无性系编号依据生长表现优劣排名进行重新编号，经过比对发现，本研究中光合效率高、生长表现好的 BY-1、FB4、1-202 和 A3 在谌红辉等（2013）的研究中分列第 1、2、6 和 9 名。由此可见，生长调查结合叶绿素荧光特性研究有助于提高优良无性系选择的可靠性。

综上所述，A3、FB4、1-202 和 BY-1 等 4 个西南桦无性系的生长表现好且光合效率高。因此，建议生产上采用此 4 个无性系推广造林。

参考文献

[1]林世青，许春辉，张其德，等．叶绿素荧光动力学在植物抗性生理学、生态学和农业现代化中的作用[J]．植物学通报，1992，9(1)：1-16.

[2]孙志勇，季孔庶．干旱胁迫对 4 个杂交鹅掌楸无性系叶绿素荧光特性的影响[J]．西北林学院学报，2010，25(4)：35-39.

[3]刘立云，李艳，杨伟波，等．不同品种油茶叶绿素荧光参数的比较研究[J]．热带作物学报，2012，33(5)：886-889.

[4]蔡晓明，卢宇蓝，施季森．一球悬铃木无性系耐旱性研究[J]．西北林学院学报，2010，25(6)：19-24.

[5]林晗，陈辉，吴承祯，等．千年桐种源间叶绿素荧光特性的比较[J]．福建农林大学学报（自然科学版），2012，41(1)：34-39.

[6]孙红英，曹光球，辛全伟，等．香樟 8 个无性系叶绿素荧光特征比较[J]．福建林学院学报，2010，30(4)：309-313.

[7]林达定，张国防，于静波，等．芳樟不同无性系叶片光合色素含量及叶绿素荧光参数分析[J]，植物资源与环境学报，2011，20(3)：56-61.

[8]曾杰，郭文福，赵志刚，等．我国西南桦研究的回顾与展望[J]．林业科学研究，2006，19(3)：379-384.

[9]谌红辉，贾宏炎，郭文福，等．西南桦无性系测定与评价[J]．林业实用技术，2013(6)：11-13.

[10]吕芳德，徐德聪，蒋瑶．美国山核桃无性系叶绿素的荧光特性[J]．中南林学院学报．2006，26(2)：18-21.

[11]吕芳德，徐德聪，侯红波，等．5 种红山茶叶绿素荧光特性的比较研究[J]．经济林研究，2003，21(4)：4-7.

[12]张杰，邹学忠，杨传平，等．不同蒙古栎种源的叶绿素荧光特性[J]．东北林业大学学报，2005，33(3)：21-22.

[13]杜鹏珍，廖绍波，孙冰，等．班克木幼苗的光合色素及叶绿素荧光特性[J]．中南林业科技大学学报，2014，34(9)：49-54.

[14]温国胜，田海涛，张明如，等．叶绿素荧光分析技术在林木培育中的应用[J]．应用生态学报，2006，17(10)：1973-1977.

[15]赵曦阳，王军辉，张金凤，等．楸树无性系叶绿素荧光及生长特性变异研究[J]．北京林业大学学报，2012，34(3)：41-47.

［原载：西北林学院学报，2016，31(02)］

容器规格和基质配方对红椎幼苗生长及造林效果影响

陈 琳[1] 曾 杰[2] 贾宏炎[1] 蒙彩兰[1] 黎 明[1]

([1]中国林业科学研究院热带林业实验中心，广西凭祥 532600；
[2]中国林业科学研究院热带林业研究所，广东广州 510520)

摘 要 通过对不同容器规格和基质配方条件下1~2年生红椎幼苗生长及造林效果对比分析，为2种年龄红椎苗木培育筛选适宜的容器规格和基质配方，亦为其造林苗龄的选择提供科学依据。采用2种容器规格和11种基质配方分别培育1~2年生红椎苗，测定红椎苗期的生长动态，并开展造林效果试验，调查其早期生长表现，运用方差分析、多重比较等方法，探明不同容器规格和基质配方条件下红椎幼苗生长差异，应用相关分析揭示红椎苗期生长与造林效果的相关性。容器规格和基质配方显著影响1~2年生红椎苗的苗高、地径、根生物量和总生物量($P<0.05$)，且两者的交互作用对1~2年生幼苗的苗高、地径、根生物量以及2年生幼苗的总生物量具有显著影响。1年生红椎苗造林当年，各基质配方间保存率差异不显著($P<0.05$)，而苗高差异显著；造林第2年，各基质配方的造林保存率和幼苗高差异不显著，而地径差异显著。2年生红椎苗造林当年，各容器规格的造林保存率、幼苗高和地径均差异显著，各基质配方的造林保存率差异不显著，而幼树高和地径差异显著。容器规格和基质配方的交互作用对2年生红椎苗造林当年苗高具有显著影响。1年生红椎苗造林，当年幼树高、第2年幼树高和地径与造林前苗高、地径、根生物量和总生物量呈显著正相关，且当年和第2年和造林前幼苗根冠比呈显著负相关。2年生红椎苗造林，容器规格1(8cm×12cm，直径×高)当年幼树高、地径与造林前幼苗高和总生物量呈显著正相关，容器规格2(12cm×15cm，直径×高)当年幼树高与造林前幼苗高、地径、根生物量和总生物量呈显著正相关，而当年地径与造林前总生物量呈显著正相关，表明红椎苗木质量显著影响其前两年造林效果。容器规格与基质配方对1~2年生红椎苗木生长存在交互作用，因此红椎苗期适宜基质配方选择应依据苗龄和容器规格而定。培育1年生红椎苗，建议采用容器规格1和基质配方10(50%沤制树皮+25%黄心土+25%锥林表土)；培育2年生红椎苗则采用容器规格2和基质配方7(50%沤制树皮+50%锥林表土)。在杂灌控制及时的良好立地，采用规格1容器培育1年生红椎苗，其造林效果优于规格2容器培育的2年生红椎苗。

关键词 红椎；苗龄；苗木质量；生长动态；造林效果

Effects of Container Size and Medium Formula on Seedlings Growth and Early Field Performance of *Castanopsis hystrix*

CHEN Lin[1], ZENG Jie[2], JIA Hongyan[1], MENG Cailan[1], LI Ming[1]

([1]*Experimental Centre of Tropical Forestry*, *CAF*, *Pingxiang* 532600, *Guangxi*, *China*;
[2]*Research Institute of Tropical Forestry*, *CAF Guangzhou* 510520, *Guangdong*, *China*)

Abstract: The growth of one- and two-year-old *Castanopsis hystrix* seedlings raised with different media and different sizes of containers were compared, and their out-planted performances were also evaluated, aiming to determine the optimum container size and medium, as well as to provide guidance about selection of suitable seedling age for planting. *C. hystrix* seedlings were raised in containers of two sizes with eleven media. Growth dynamics and early field performance were investigated, respectively. Variance analysis with Duncan's multiple comparison was conducted to test the differences in seedling growth and the early field performance among treatments, and corrrelation analysis was used to determine the relationship between seedling growth and early filed performance. Container size and growing medium significantly influenced height, root collar diameter, root and total biomass of one- and two-year-old *C. hystrix* seedlings. Moreover, their interaction significantly affected height, root collar diameter and root biomass of one- and two-year-old seedlings as well as total biomass of two-year-old seedlings. For one-year-old seedlings, there was significant difference in height but not in survival rate among media in the first year after out-planting. A significant difference was found in root collar diameter but not in height and survival rate in the second year after out-planting. For two-year old seedlings, there were profound effects of container size and medium on height and root collar diameter in the first year after out-planting. Moreover, the interaction of container size and medium influenced obviously the height of two-year-old seedlings in the first year after planting. The height in the first year, height and root collar diameter in the second year after planting showed positive relationships with the height, root collar diameter and total biomass of one-year-old seedlings in the nursery, respectively. However, the height in the first and second year after planting were negatively correlated with the root and shoot ratio in the nursery, respectively. As for two-year-old seedlings, with container size one, their height and root collar diameter at the first year after planting were positively correlated with the height and total biomass in the nursery separately, while with container size two, their height in the first year after planting had a positive relationship with their height, root collar diameter, root and total biomass in the nursery, while the root collar diameter in the first year after planting was positively correlated with total biomass in the nursery, suggesting that seedling quality significantly influenced the field growth performance of *C. hystrix* seedlings within first two years after planting. The interaction of container size and medium had a profound effect on the seedling growth of one-year and two-year old *C. hystrix* seedlings, therefore the optimum medium should be selected according to different seedling ages and container sizes. Container size one (8cm×12cm, diameter and height) and medium ten (50% composted bark, 25% yellow soil and 25% surface soil) were recommended to raise one-year-old *C. hystrix* seedlings, while container size two (12cm×15cm, diameter and height) and medium seven (50% composted bark and 50% surface soil) were suggested to raise two-year old ones according to the seedling growth performance in nursery and after planting. It could be preliminarily concluded that under weed-controlled site, the field performance of one-year-old *C. hystrix* seedlings with container size one at the second year after out-planting was better than those of two-year-old seedlings with container size two at the first year after planting.

Key words: *Castanopsis hystrix*; seedling age; seedling quality; growth dynamics; field performance

随着我国生态公益林、珍贵用材林基地建设、针叶林阔叶化改造等一系列林业工程的实施，我国对容器苗数量和质量的要求不断增加，而容器规格和基质配方是提高容器苗质量的2个重要手段[1]。容器规格影响苗木根系对养分和水分的吸收面积，进而影响苗木形态、生理特性以及早期造林生长表现[2-4]。物理化学性质稳定、持水性和透气性良好的基质配方有利于促进苗木生长[5-8]。虽然国内外已经开展了许多树种的容器规格或基质配方研究，然而适宜的容器规格和基质配方往往因树种而异，且容器规格和基质配方之间存在交互作用[9-11]，因此有必要针对某一树种开展容器苗的容器规格和基质配方研究。此外，国内关于容器规格和基质配方的研究中，往往缺少对苗木造林效果的评价，容易造成苗木培育与造林生产脱节，不利于容器苗的质量评价[12,13]。

红椎(*Castano psishystrix*)是我国热带和亚热带地区重要的乡土阔叶树种，为菌根营养型树种，其幼树耐阴，材质坚重、纹理直、耐腐、易加工，在用材林、水源林和薪炭林建设中具有重要地位[14]。随着红椎种植规模的不断扩大，尤其是红椎被广泛应用于生态公益林营建和改造，其壮苗培育研究日益受到重视。目前，已经开展了1年生红椎容器苗基质初步筛选以及2年生红椎容器苗容器规格、基质配方和施肥研究[15,16]。但是对于2种苗龄红椎苗木质量差异尚不清楚，而苗龄是影响苗木造林效果的重要因素[17,18]。因此，本研究设置了2种容器规格和11种基质配方分别培育1年生和2年生红椎苗，比较不同容器规格和基质配方条件下红椎容器苗生长以及造林早期生长表现，旨在筛选适宜的容器规格和基质配方，并为红椎造林适宜苗龄选择提供科学依据。

1 材料与方法

1.1 苗圃阶段

红椎芽苗购自广西壮族自治区林业科学研究院，为广西博白种源，高约4cm。2012年4月8日，在中国林业科学研究院热带林业实验中心苗圃，开展不同容器规格和基质配方红椎苗生长对比试验。采用裂区试验设计(表1)，主区为11个基质配方，副区为容器规格，每个主小区于规格1(8cm×12cm，直径×高)和规格2(12cm×15cm，直径×高)无纺布袋内分别移栽75株和48株芽苗，重复3次，合计分别为2475株和1584株。移苗前，将黄心土、沤制树皮、炭化树皮、沤制锯末、锥林表土按照试验设计的比例充分混合，装入2种规格的无纺布袋中，置于15孔的塑料穴盘和50cm×50cm(长×宽)的塑料盘中，育苗盘下方铺一层防草布以防止苗木根系穿入土壤。移苗后3个月内，根据天气情况对红椎苗进行遮阴处理。移苗1个月后开始施肥，前2个月淋施0.3%尿素(氮≥46.4%)，每周1次；此后每周施1次0.5%复合肥(硝态氮≥18%、络合型钾≥25%、螯合型中微量元素≥8%)，至11月中旬为止；翌年3月开始，淋施0.5%复合肥(氮≥15%、五氧化二磷≥15%、氧化钾≥15%)，每10天施1次，至10月底结束，2种容器规格红椎苗的施肥量保持一致。试验期间，每10~15d交替喷800~1000倍的多菌灵或甲基托布津。

于2012年5月30日、7月4日、10月6日、2013年1月5日、4月17日、7月6日、8月12日、11月15日、2014年1月6日(即移苗后第53、88、182、273、375、455、492、587、639d)调查红椎苗高和地径(前2次未调查地径)。2013年1月5日和2014年1月6日，每个小区分别选取有代表性的5株1年生和4株2年生红椎苗，将其分为根、茎、叶，于105℃杀青15min，80℃烘48h至恒质量，称量各部分干质量。

表1 基质配方

基质编号	组成
1	100% 黄心土(生产常用)
2	65%沤制树皮+ 15%炭化树皮+ 20%锯末(生产常用)
3	30%沤制树皮+ 70%黄心土
4	50%沤制树皮+ 50%黄心土
5	70%沤制树皮+ 30%黄心土
6	30%沤制树皮+ 70%锥林表土
7	50%沤制树皮+ 50%锥林表土
8	70%沤制树皮+ 30%锥林表土
9	30%沤制树皮+ 35%黄心土+ 35%锥林表土
10	50%沤制树皮+ 25%黄心土+ 25%锥林表土
11	70%沤制树皮+ 15%黄心土+ 15%锥林表土

1.2 造林阶段

2013年1月和2014年2月，分别将1年生红椎苗(容器规格1)和2年生红椎苗(容器规格1和2)在中国林业科学研究院热带林业实验中心伏波实验场的2块采伐迹地(21°57′N，106°47′E)上营建对比试验林。试验地年均气温19.9℃，年均降水量1400mm，海拔500m，2块林地的立地条件基本一致，其土壤为花岗岩发育的红壤，土层厚度1m以上。11个基质配方处理，每个小区种植20株，3次重复。穴大小为50cm×50cm×30cm(长×宽×深)，株行距分别为4m×4m和3m×3m，虽然造林密度不同，但是在造林前2年苗木尚未郁闭，密度效应尚未发生作用，因此不会对本研究结果造成影响。造林后每年进行2次灌草抚育。2013年8月30日和2014年11月25日，分别调查1年生和2年生红椎苗造林当年或第2年幼树保存率、高度和地径。

1.3 统计分析

应用SPSS16.0软件和One-wayANOVA程序分别对红椎苗期生长和造林效果进行多因素和单因素方差分析，检验容器规格和基质配方的主效应和交互效应，并进行Duncan多重比较。分别不同容器规格和苗龄，对不同基质配方处理的红椎苗期生长与其造林效果做Pearson相关分析，评价苗木质量对造林效果的影响。在统计分析时，保存率(百分比数据)经平方根反正弦转换。

2 结果与分析

2.1 苗期生长

容器规格和基质配方的交互作用对1~2年生红椎苗高、地径、根生物量以及2年生红椎苗总生物量具有显著影响($P<0.05$)，而对其他指标影响不显著($P<0.05$)。容器规格对1~2年生红椎苗高、地径、根生物量和总生物量均影响极显著($P<0.01$)，而对根冠比则无显著影响。红椎苗1年生时，容器规格2苗高、地径、根生物量和总生物量比容器规格1分别提高7%、5%、32%和28%(图1)；2年生时，容器规格2苗高、地径、根生物量和总生物量比容器规格1分别提高50%、27%、147%和151%(图2)。

不含相同的小写字母或大写字母表示在$P<0.05$水平上差异显著，误差线代表标准误，下同。

无论苗龄和容器规格，各基质配方间红椎苗高和地径均差异极显著；除了应用规格1容器培育的1年生红椎苗中各基质配方间根生物量和总生物量差异不显著，其他红椎苗在各基质配方间生物量大多差异极显著；对于根冠比而言，各基质配方间大多无显著差异，仅在规格1容器培育的1年生红椎苗中基质2显著大于其他基质。在11种基质配方中，苗高、地径、根生物量和总生物量均以基质1或2为最小，而其最大值出现的基质配方则因容器规格和苗龄而异。如 1 年生红椎苗，容器规格 1 以 10 号基质的苗高和地径为最大，比 2 号基质分别提高 74%和 42%；容器规格 2 以 7、9、10 号基质的苗高、地径、根生物量和总生物量为最大，比 2 号基质分别提高 88% ~ 92%、48% ~ 53%、218% ~ 275% 和 153% ~ 204%(图 1)。对于 2 年生红椎苗而言，容器规格 1 以 6 号基质的苗高、地径、根生物量和总生物量为最大，比 1 号基质分别提高 44%、25%、92%和 105%；容器规格 2 以 7 号基质为最大，比 2 号基质分别提高 67%、39%、239%和 151 %(图 2)。

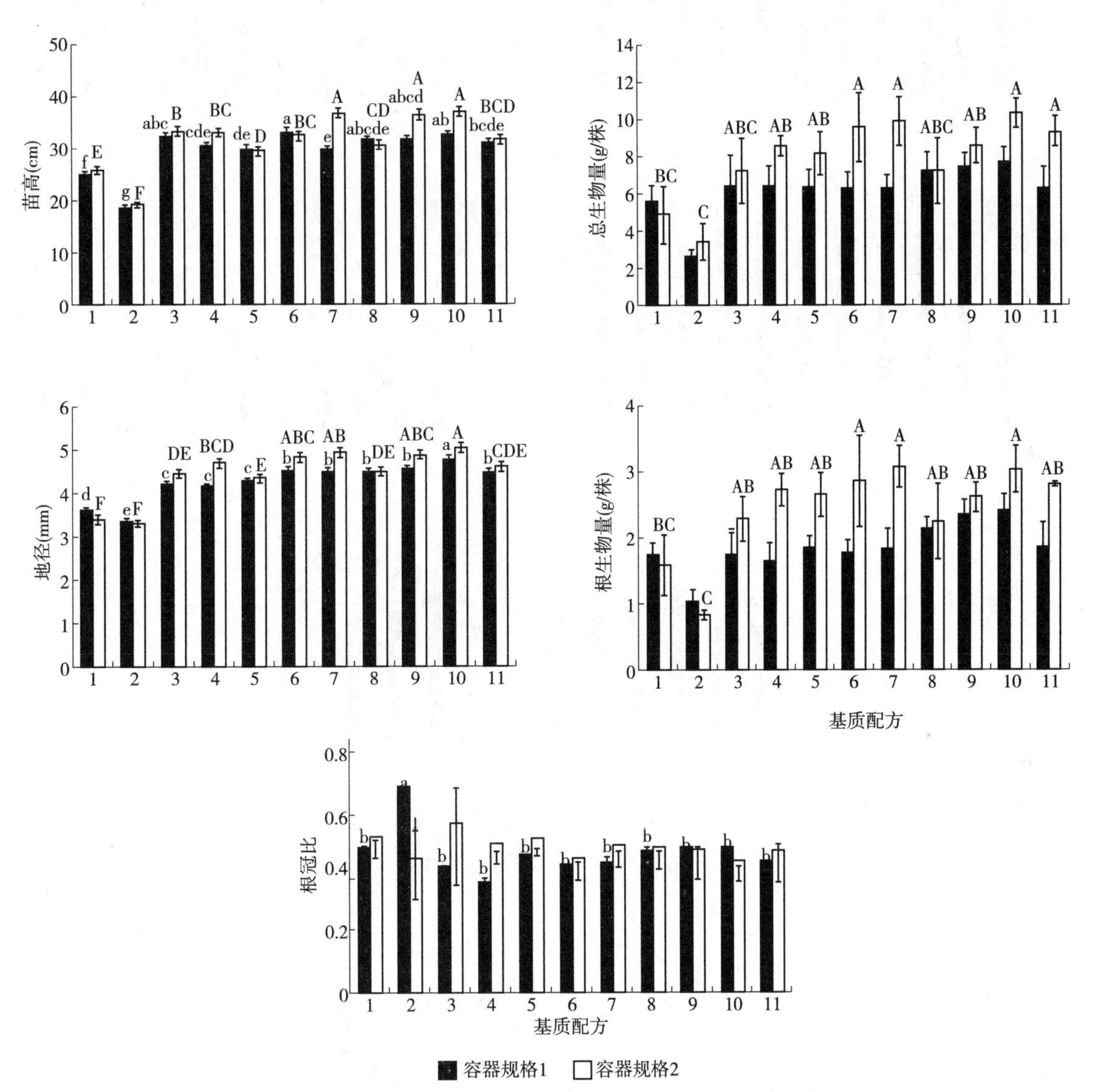

图 1　容器规格和基质配方对 1 年生红椎苗生长的影响

苗高和地径分别在移苗后 53 ~ 273d 和 182 ~ 273d 生长缓慢，2 种容器规格之间苗高和地径差异不显著；移苗后 275d，即约 1 年生时，2 种容器规格之间红椎苗高和地径开始出现极显著差异；移苗后 587d，2 种容器规格之间苗高和地径差异达到最大值，此后保持恒定(图 3)。

2.2　造林效果

1 年生红椎苗造林当年，11 种基质配方间造林保存率差异不显著($P<0.05$)，而苗高差异极显著($P<0.01$)，以 2 号基质苗高最小，比其他基质低 21% ~ 31%；造林第 2 年，各基质配方间保存率和苗高差异均不显著，而地径差异极显著，以 2 号基质地径最小，比其他基质低 16% ~ 29%(图 4)。

容器规格和基质配方的交互作用对 2 年生红椎苗造林当年幼树高具有显著影响，而对保存率和地径无显著影响(图 5)。容器规格对造林保存率、幼树高和地径影响极显著，这 3 次指标容器规格 2 比容器规格 1 分别提高 23%、16% 和 32%。无论哪种容器规格，不同基质配方间保存率差异不显著，但其幼树高

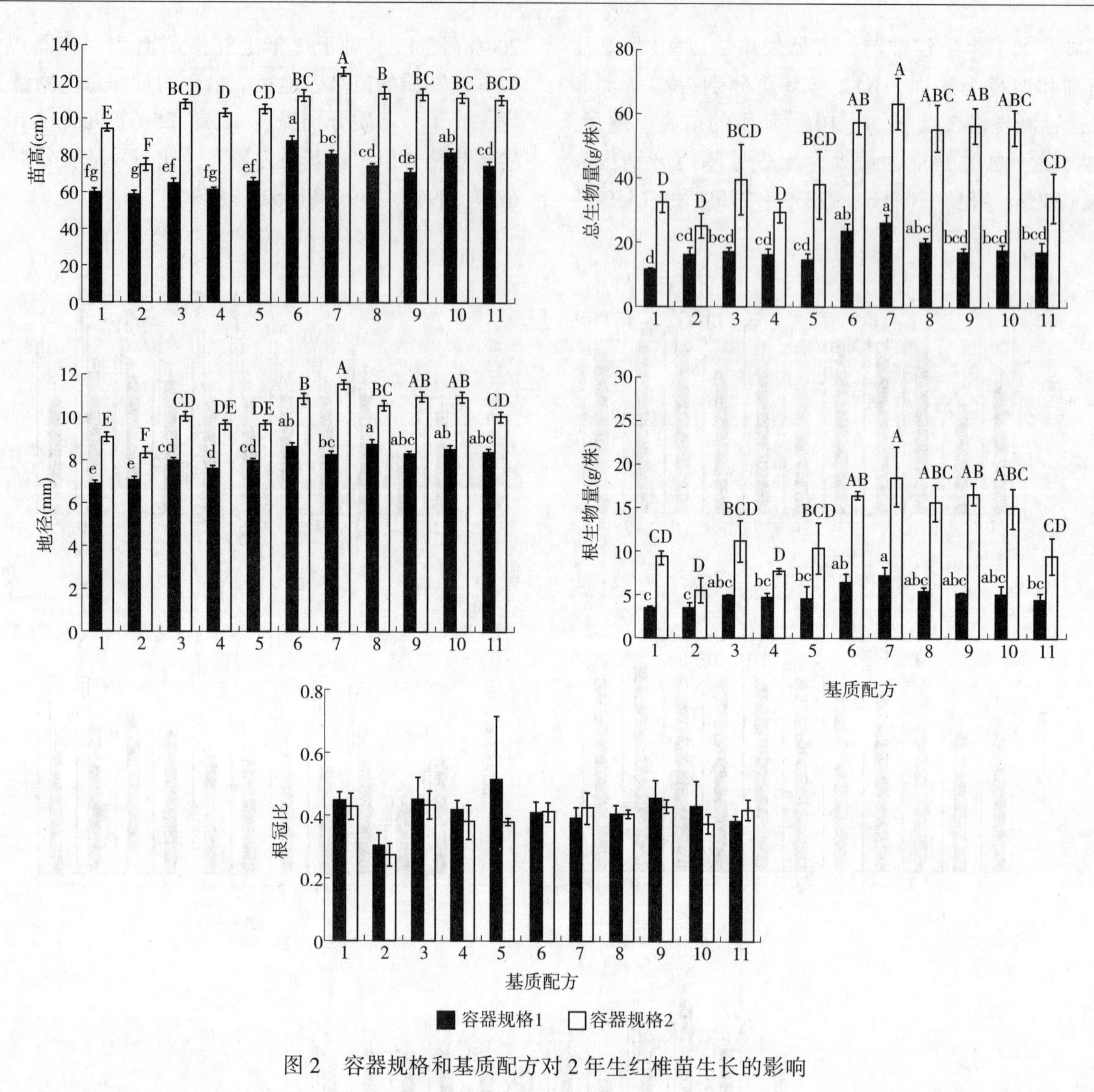

图 2　容器规格和基质配方对 2 年生红椎苗生长的影响

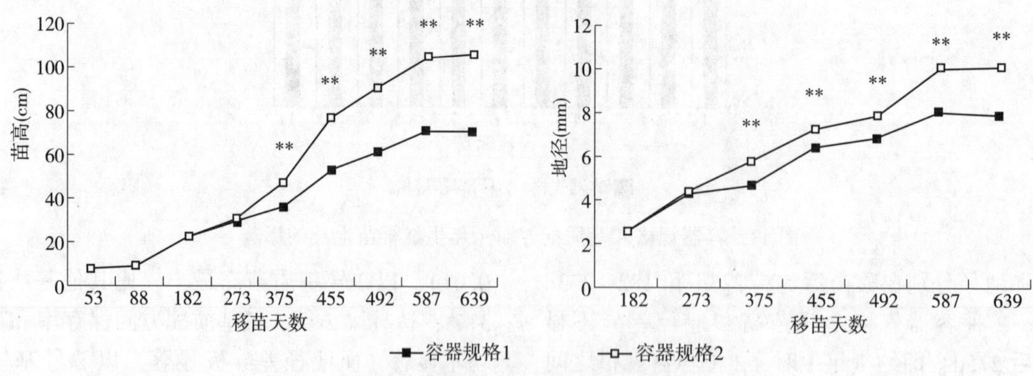

图 3　两种容器规格红椎苗高和地径的动态变化

** 表示在 $P<0.01$ 水平上差异显著。

和地径差异极显著，均以 1 号基质为最小，而其最大值出现的基质配方则因容器规格而异。容器规格 1，以 6、7 和 10 号基质配方的幼树高和地径最大，比 1 号基质分别高 20%~32%和 21%~28%；容器规格 2，以 6、9 和 10 号基质配方的幼树高和地径最大，比 1 号基质分别高 37%~48%和 18%~23 %(图 5)。采用规格 1 容器培育的 1 年生红椎苗在造林第 2 年的生长表现与采用规格 2 容器培育的 2 年生红椎苗造林当年的生长表现相比较，其保存率无显著差异，而前者的平均苗高和地径比后者分别提高 60%和 80%(图 4，图 5)。

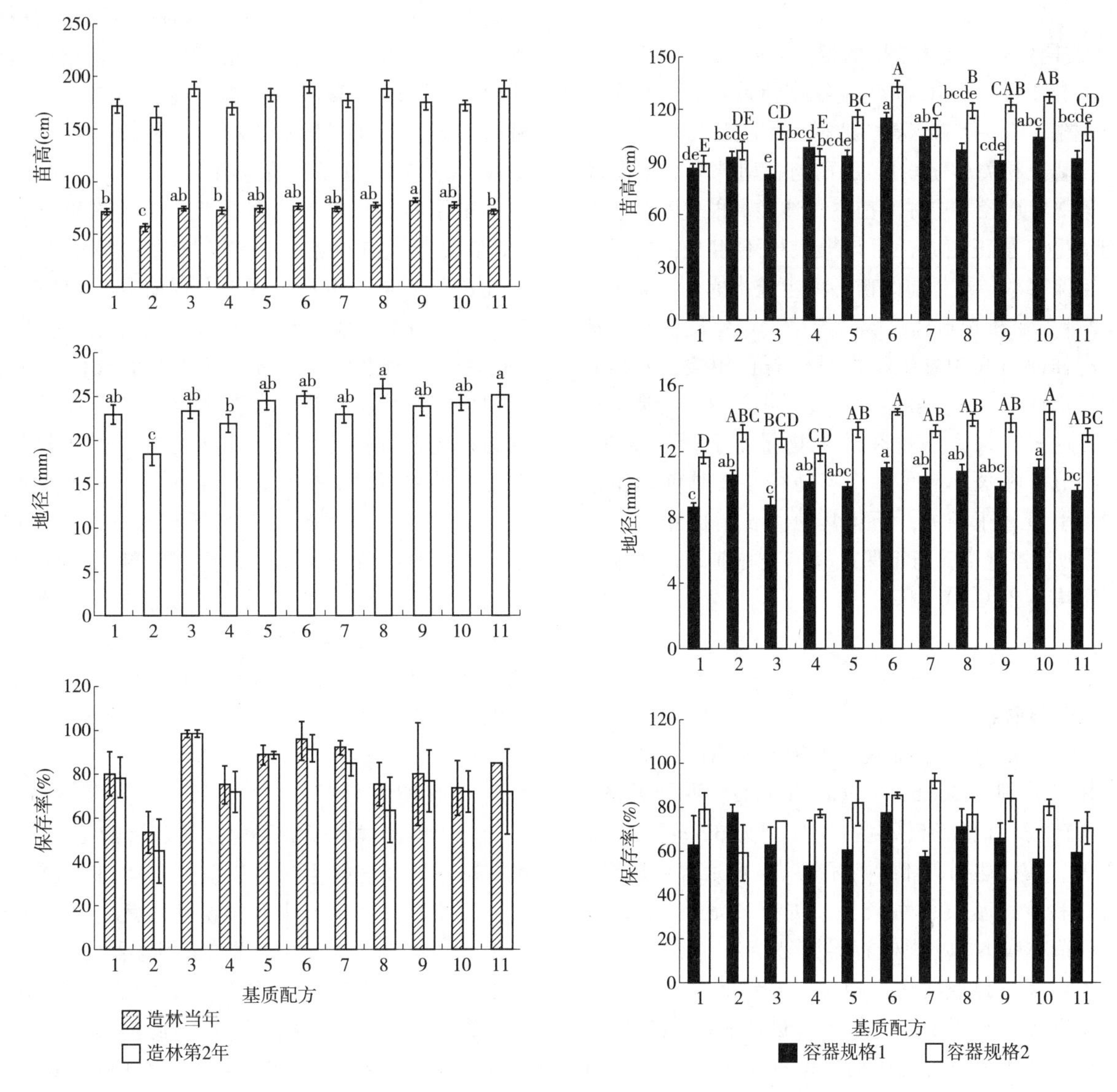

图4 1年生红椎苗(容器规格1)当年和第2年造林效果

图5 2年生红椎苗当年造林效果

表2 红椎苗期生长与造林效果的相关性

造林苗龄	容器规格	造林效果	苗高	地径	总生物量	根生物量	根冠比
1年生	8cm×12cm	当年保存率	0.217	0.248	0.181	0.122	-0.287
		当年苗高	0.786**	0.718**	0.742**	0.717**	-0.434*
		第2年保存率	0.284	0.260	0.221	0.207	-0.258
		第2年苗高	0.722**	0.561**	0.533**	0.425*	-0.411*
		第2年地径	0.644**	0.585**	0.595**	0.601**	-0.322
2年生	8cm×12cm	当年保存率	0.152	-0.070	0.137	0.129	0.072
		当年苗高	0.542**	0.203	0.373*	0.304	-0.117
		当年地径	0.417*	0.174	0.378*	0.237	-0.204
2年生	12cm×15cm	当年保存率	0.300	0.328	0.220	0.268	0.272
		当年苗高	0.427*	0.431*	0.473**	0.445**	0.131
		当年地径	0.284	0.306	0.383*	0.307	-0.136

注：样本数为33，* 和 ** 分别表示在 $P<0.05$ 和 $P<0.01$ 水平上显著相关。

2.3　苗期生长与造林效果的相关分析

由表2可以看出，1年生红椎苗造林，当年和第2年幼树高与造林前苗高、地径、根生物量和总生物量呈显著正相关($P<0.05$)，而与根冠比呈显著负相关；第2年地径与造林前苗高、地径、根生物量和总生物量呈显著正相关，而与根冠比不相关；当年和第2年保存率与造林前幼苗各项生长指标均不相关。

2年生红椎苗造林，容器规格1当年幼树高和地径与造林前苗高和总生物量呈显著正相关，而与其他生长指标不相关。容器规格2当年幼树高与造林前苗高、地径、根生物量和总生物量呈显著正相关，而与根冠比相关不显著；当年地径与造林前幼苗总生物量呈显著正相关，而与其他生长指标相关不显著。2种容器规格，当年保存率与造林前幼苗各项生长指标均不相关(表2)。

3　讨论

3.1　容器规格

1年生红椎苗，容器规格1的平均苗高、地径和总生物量分别为30cm、4.3cm和6.2g，容器规格2的平均苗高、地径和总生物量分别为32cm、4.6cm和7.9g，均已达到造林出圃标准[6]，但是容器规格1所需材料少，基质相对较轻，可有效地降低育苗、运输和造林成本。小规格容器更易使苗木形成根团，有利于提高造林成活率；相反，大规格容器的苗木根密度较小，不利于形成根团，进而影响造林成活率[19,20]，因此在满足生产需要的前提下宜选择容器规格1培育1年生红椎苗进行造林。红椎苗2年生时，容器规格2的苗高、地径、根生物量和总生物量比容器规格1分别提高50%、27%、147%和151%，表明大规格容器更有利于苗木根系发育，从而促进了苗木生长，特别是随着苗龄的增长，当根系生长量超过了小规格容器承受范围时，大容器的促进效应越明显[21]。2年生红椎苗造林当年，容器规格2的造林保存率、幼树高和地径均显著高于容器规格1，可能与大规格容器更有利于苗木根系发育有关，因为容器规格2的根生物量比容器规格1高147%，亦说明采用规格2容器培育2年生红椎苗提高了苗木质量，进而促进了其造林当年的生长，因此宜选择体积较大的规格2容器培育2年生红椎苗。Aphalo等[2]亦发现容器规格不仅影响欧洲白桦(*Betula pendula*)苗木生长和形态，而且影响其造林第5年的生长。然而，Close等[3]研究表明虽然容器规格显著影响蓝桉(*Eucalyptus globulus*)造林当年的生长表现，但对其造林第4年的生长无显著影响，这种差异可能与树种特性有关。本研究中，容器规格对2年生红椎苗造林当年保存率、幼树高和地径均有显著影响，但由于造林时间短，容器规格对红椎苗造林效果影响的时间效应将有待继续观测方能得出结论。

采用容器规格1培育1年生红椎苗和容器规格2培育2年生红椎苗，比较前者在造林第2年和后者在造林当年的生长表现，发现前者平均幼树高和地径分别比后者高60%和80%，说明采用容器规格1培育1年生红椎苗造林第2年的效果优于采用容器规格2培育2年生红椎苗造林当年的效果，可能因为小规格容器培育的苗木在立地适宜且杂灌控制良好的情况下，其生长速度高于大规格容器的苗木生长速度[3,22]。但是在严寒环境或杂灌丛生的立地，宜选择抗性和竞争能力更强、规格较大苗木进行造林[18]。因此，今后将开展红椎2种苗龄在不同立地条件下的造林对比试验，探讨苗龄、容器规格、立地条件等因素交互作用对红椎苗造林效果的影响[23,24]。

3.2　基质配方

由于容器规格和基质配方显著影响1~2年生红椎苗高、地径、根生物量以及2年生红椎苗总生物量，因此建议在生产中红椎苗期适宜基质配方应依据苗龄和容器规格而定。比如采用容器规格1培育1年生红椎苗，适宜基质配方为10号(50%沤制树皮+25%黄心土+25%锥林表土)，与此前筛选的适宜基质配方(75%黄心土+25%锥林表土+磷肥0.5%)相比，其基质重量更轻，且红椎苗生长表现更佳[15]。采用容器规格2培育2年生红椎苗，适宜基质配方为7号(50%沤制树皮+50%锥林表土)，虽然与此前筛选的适宜基质配方(50%砂质壤土+25%沤制松皮粉+25%炭化松树皮)相比，其基质重量相近，但是红椎苗生长表现更佳[16]，因此本研究获得了更适宜红椎育苗的基质配方。无论苗龄和容器规格，本研究的适宜基质配方中均含有锥林表土，这是因为基质配方中添加适当比例的锥林表土有利于菌根的形成，扩大了根系吸收水分和养分的面积，从而促进了红椎苗生长[14]。Kazantseva等[25]亦发现在基质中添加适当比例的森林表土有助于促进道格拉斯冷杉(*Pseudotsuga menziesii* var. *glauca*)幼苗根系形成菌根，降低了幼苗在水分胁迫下的死亡率。但是由于锥林表土资源有限，今后将加强红椎苗菌根菌接种研究[26,27]。进一步比较各基质配方的造林效果，发现10号和7号基质的造林生长表现亦良好，说明苗期筛选出的适宜基质配方经造林检验是可靠的。由红椎苗期生长与造林效果之间相关分析结果可知，红椎苗木质量显著影响其前2年的造林效果，无论苗龄和容器规格，苗高和总生物量可作为评价红椎苗早期造林生长表现的可靠和简易指标[21,22,28]。

4 结论

容器规格与基质配方对1，2年生红椎苗木生长存在交互作用，因此红椎苗期适宜基质配方选择应依据苗龄和容器规格而定。综合红椎苗期生长和早期造林效果，培育1年生红椎苗，适宜采用容器规格1(8cm×12cm，直径×高)和基质配方10(50%沤制树皮+25%黄心土+25%锥林表土)；培育2年生红椎苗，适宜采用容器规格2(12cm×15cm，直径×高)和基质配方7(50%沤制树皮+50%锥林表土)；在杂灌控制及时且立地好的情况下，采用容器规格1培育的1年生红椎苗，其造林效果优于采用容器规格2培育的2年生红椎苗。

参考文献

[1]邓华平，杨桂娟，王正超，等．容器大苗培育技术研究现状[J]．世界林业研究，2011，24(2)：36-41.

[2]APHALO P，RIKALA R. Field performance of silver-birch planting-stock grown at different spacing and in containers of different volume [J]. New Forests，2003，25(2)：93-108.

[3]CLOSE D C，PATERSON S，CORKREY R，et al. Influences of seedling size，container type and mammal browsing on the establishment of *Eucalyptus globulus* in plantation forestry. New forests，2010，39(1)：105-115.

[4]DOMINGUEZ-Lerena S，SIERRA N H，MANZANO I C，et al. Container characteristics influence*Pinus pinea* seedling development in the nursery and field[J]. Forest Ecology and Management，2006，221(1)：63-71.

[5]邓煜，刘志峰．温室容器育苗基质及苗木生长规律的研究[J]．林业科学，2000，36(5)：33-39.

[6]林霞，郑坚，刘洪见，等．不同基质对无柄小叶榕容器苗生长和叶片生理特性的影响[J]．林业科学，2010，46(8)：62-70.

[7]毛世忠，唐文秀，骆文华，等．不同栽培基质对广西火桐幼苗生长及净光合速率的影响[J]．西北林学院学报，2011，26(5)：96-99.

[8]da SILVA R B G，da SILVA M R，SIMÕES D. Substrates and controlled-release fertilizations on the quality of eucalyptus cuttings [J]. Revista Brasileira de Engenharia Agrícola e Ambiental，2014，18(11)：1124-1129.

[9]金国庆，周志春，胡红宝，等.3种乡土阔叶树种容器育苗技术研究[J]．林业科学研究，2005，18(4)：387-392.

[10]周志春，刘青华，胡根长，等.3种珍贵用材树种轻基质网袋容器育苗方案优选[J]．林业科学，2011，47(10)：172-178.

[11] GEPLY O A，BAIYEWU R A，ADEGOKE I A，et al. Effect of different pot sizes and growth media on the agronomic performance of*Jatropha curcas* [J]. Pakistan Journal of Nutrition，2011，10(10)：952-954.

[12]乌丽雅斯，刘勇，李瑞生，等．容器育苗质量调控技术研究评述[J]．世界林业研究，2004，17(2)：9-13.

[13]许飞，刘勇，李国雷，等．我国容器苗造林技术研究进展[J]．世界林业研究，2013，26(1)：64-68.

[14]陈应龙，弓明钦，陈羽，等．外生菌根菌接种对红椎生长及光合作用的影响[J]．林业科学研究，2001，14(5)：515-522.

[15]黎明，郭文福．红椎容器苗基质试验简报[J]．广西林业科学，2006，35(1)：31-33.

[16]温恒辉，贾宏炎，黎明，等．基质类型、容器规格和施肥量对红椎容器苗质量的影响[J]．种子，2012，31(7)：75-77，82.

[17]李国雷，祝燕，李庆梅，等．红松苗龄型对苗木质量和造林效果的影响[J]．林业科学，2012，48(1)：35-41.

[18]JOHANSSON K，NILSSON U，ALLEN H L. Interactions between soil scarification and Norway spruce seedling types [J]. New Forests，2007，33(1)：13-27.

[19] 李永胜，朱锦茹，江波，等．乳源木莲管形容器育苗技术研究[J]．浙江林业科技，2007，27(2)：11-15.

[20] 郑坚，陈秋夏，李效文，等．无柄小叶榕容器苗形态和生理质量评价指标筛选[J]．中国农学通报，2010，26(15)：141-148.

[21]SOUTH D B，HARRIS S W，BARNETT J P，et al. Effect of container type and seedling size on survival and early height growth of *Pinus palustris* seedlings in Alabama，U. S. A[J]. Forest Ecology and Management，2005，204(2)：385-398.

[22]RENOU-WILSON F，KEANE M，FARRELL E P. Effect of planting stocktype and cultivation treatment on the establishment of Norway spruce on cutaway peatlands [J]. New forests，2008，36(3)：307-330.

[23]LANDIS T D，DUMROESE R K. Applying the target plant concept to nursery stock quality. In：Plant quality：a key to success in forest establishment[C]. Proceedings of the National Council for Forest Research and Development (COFORD) conference，Dublin，Ireland(pp. 1-10). 2006.

[24]PINTO J R，MARSHALL J D，DUMROESE R K，et al. Establishment and growth of container seedlings for refores-

tation: a function of stocktype and edaphic conditions[J]. Forest Ecology and Management, 2011, 261 (11): 1876-1884.

[25] KAZANTSEVA O, BINGHAM M, SIMARD S W, et al. Effects of growth medium, nutrients, water, and aeration on mycorrhization and biomass allocation of greenhouse-grown interior Douglas-fir seedlings[J]. Mycorrhiza, 2009, 20(1): 51-66.

[26] ÓSKARSSON Ú. Potting substrate and nursery fertilization regime influence mycorrhization and field performance of *Betula pubescens* seedlings [J]. Scandinavian Journal of Forest Research, 2010, 25(2): 111-117.

[27] REPÁC I, BALANDA M, VENCURIK J, et al. Effects of substrate and ectomycorrhizal inoculation on the development of two-years-old container-grown Norway spruce (*Picea abies* Karst.) seedlings[J]. iForest-Biogeosciences and Forestry, 2015, 8: 487-496.

[28] LI G L, LIU Y, ZHU Y, et al. Influence of initial age and size on the field performance of *Larix olgensis* seedlings[J]. New Forests, 2011, 42(2): 215-226.

[原载: 林业科学, 2017, 53(03)]

不同氮素水平对米老排苗期生长和叶绿素荧光特性的影响

刘福妹[1,2] 劳庆祥[1] 庞圣江[1] 马 跃[1] 陈建全[1] 韦菊玲[1]
([1]中国林业科学研究院热带林业实验中心，广西凭祥 532600
[2]广西友谊关国家森林生态系统定位观测研究站，广西凭祥 532600)

摘 要 为探讨不同氮素水平对米老排苗期生长的影响，采用根部施肥的方法，研究了5个氮素条件下(总施氮量分别是100、200、400、600和800m/株)米老排幼苗生长和荧光特性的差异。结果表明：①米老排的苗高、地径和总生物量均会随着施用氮素水平的增加，呈现先升高后降低的规律，均在N2(200mg/株)时达到最大值(40.52cm、4.98mm和4.07g)，依次较对照处理提高了145.7%、152.8%和158.4%。②氮肥能够促进米老排幼苗叶片叶绿素含量的增加。③随着氮水平的增加，实际光化学效率(Φ_{PSII})、PSII最大光化学效率(F_v/F_m)、光化学猝灭系数(qP)均是先升高后降低，非光化学猝灭系数(NPQ)则先降后升，且均是在N2时达到最大值和最小值，说明适当增加供氮水平可以显著提高米老排幼苗的光合效率。④200mg/株的施氮量最适合米老排幼苗的生长。

关键词 米老排；氮素营养；生长；叶绿素荧光参数

Effects of Nitrogen Fertilization on Growth Perfomance and Chlorophyll Fluorescence Parameters of *Mytilaria laosensis* Seedlings

LIU Fumei[1,2], LAO Qingxian, PANG Shenjiang[1],
MA yue[1], CHEN Jiangquan[1], WEI Juling[1]
(*[1]The Experimental Centre of Tropical Forestry, Chinese Academy of Forestry, Pingxiang* 532600, *Guangxi, China*;
[2]Guangxi Youyiguan Forest Ecosystem Research Station, Pingxiang 532600, *Guangxi, China*)

Abstract: To study growth response of *Mytilaria laosensis* seedlings to nitrogen supply, growth performance and chlorophyll fluorescence of its seedlings were investigated under five levels including N1(100mg/seedling), N2(200mg/seedling), N4(400mg/seedling), N6(600mg/seedling) and N8(800mg/seedling), The results showed that: ①height, ground diameter and total biomass of *Mytilaria laosensis* seedlings increased initially, and then decreased with the increase of nitrogen supply, compared with the control, they all reached the maximum(40.52cm、4.98mm and 4.07g) under N2(200mg/seedling) and increased by 145.7%, 152.8% and 158.4 %, respectively. ②The nitrogen fertilization could significantly enhance the chlorophyll contents of seedlings. ③as the increasing nitrogen supply, the maximum photochemical efficiency of PSII(F_v/F_m), the actual photosynthetic efficiency(Φ_{PSII}) and photochemical quenching(qP) all showed a trend of initial increase followed by decrease, while the non-photochemical quenching(NPQ) did the opposite trend. their maximum and or minimum were all reached under N2, suggested the appropriate nitrogen supply could significantly raise the photosynthetic efficiency of *M. laosensis* seedlings, As a whole, 200mg N/seedling was the most appropriate choice to facilitate seedling growth of this species.

Key words: *Mytilaria laosensis*; nitrogen fertilization; growth; chlorophyll fluorescence

氮被称为“生命元素”，是植物体内许多重要有机化合物的构成成分，也是植物进行光合作用起决定作用的叶绿素的组分，是世界农林业生产中消耗量和浪费量最大的元素之一，被称为生态系统中最为限制植物生长发育的营养元素[1-3]。大量研究表明，外源环境中的氮元素能够影响植物光合生理特性[4-7]，而合理的施用氮肥能够促进植物的生长和发育[8-11]。但是不同植物对氮素水平的适应不同，同一植物不同生长阶段对氮素的需求也不同[3]，只有找到适宜的氮素水平才能保证植物快速、优质地生长，实现高效、低成本的营林目的。因此，研究氮素水平对不同植物生长和发育的影响已经受到越来越多的关注。

米老排(*Mytilaria laosensis* Lec.)别名壳菜果、米显灵、三角枫，属于金缕梅科壳菜果属的常绿乔木，具有生长迅速、干形通直、材质优良、用途广泛和涵养水源等优点，是建筑、家具、造纸、人造板和生态林的优质选择，是我国南亚热带地区重点发展的珍优速生乡土树种之一，具有重要的经济效益和生态效益。我国从20世纪60年代就开展了与米老排相关的研究工作，研究内容集中在育苗技术、造林方案和成果、引种栽培以及生态效益功能这几个方面[12-16]，而有关供氮水平与米老排苗期生长的研究少见报道，更少有与米老排叶绿素荧光特性相关的报道。因此，本研究通过分析5个氮素水平对米老排幼苗生长和叶绿素荧光特性的影响，明确适合米老排苗期生长的最佳施氮量，为米老排的合理施肥提供理论依据；同时也为米老排幼苗对环境响应与适应机理方面的研究提供参考，为进一步完善米老排的高效培育技术，充分发挥米老排的经济和社会效益。

1 材料和方法

1.1 试验材料和设计

本次试验地设置在广西壮族自治区崇左市下属凭祥市(22.1°N，106.7°E)中国—东盟珍贵树种种苗繁育基地的温室大棚中，该地海拔约为240m，年平均温度为21.5℃。

参试的米老排苗木均为实生苗，于2013年11月份在米老排母树林采种，于次年4月份在中国—东盟珍贵树种种苗繁育基地催芽并散播在沙床，出苗后选取生长状态良好、10cm高的幼苗，移植到规格为15cm×15cm(直径×高)的圆柱体花盆中，花盆外用一层透明塑料袋，防止淋溶氮肥流失。培养基质为适合米老排幼苗生长且经消毒处理的苗圃熟土，其速效氮、速效磷和速效钾含量分别是169.00、0.43和37.06mg/kg，全氮、全磷和全钾含量分别是1.18、0.31和6.10g/kg，土壤pH为5.5。

试验设6个处理：包括1个空白对照(非施肥处理)和5个氮素营养处理(总施氮量分别是100、200、400、600和800mg/株)，每个处理有米老排幼苗30株。试验于2014年5月15日开始至2014年10月14日结束，每15d进行一次根部施肥，共进行10次，具体施用量和施用时间如表1所示。氮肥为尿素[$CO(NH_2)_2$]，氮素有效成分比例为46%。

在移苗前，根据Timmer[17]的方法确定培育基质的最大持水量，然后以最大持水量的75%作为移苗时初始水量；在整个实验期间，苗木的维护管理工作由基地工作人员根据幼苗的生长状况、天气情况等统一进行。

1.2 测定方法

1.2.1 苗高、地径和总生物量的调查

2014年10月29日，测定各处理米老排苗木(共计30株)的苗高和地径。2014年10月30日，每个处理选取3株平均苗(即：地径和苗高约为整个处理苗木的平均地径和平均苗高的苗木，下同)，用去离子水洗净，整株置于烘箱中105℃杀青20min，然后70℃下烘48h后称其干质量，记为总生物量。

1.2.2 叶绿素总含量的测定

2014年8月29号，每个处理选6株苗，每株取一片成熟叶，叠加在一起，用圆孔取样法在除去叶主脉的部分打孔取0.2g鲜叶，共3次，然后采用95%乙醇提取法[18]用紫外可见分光光度计(UV-2550, SHIMADZU公司)测定并计算叶绿体色素含量。

表1 米老排苗期氮素水平用量表

处理	施用氮肥总量(mg/株)	2014年5月15日至6月14日		2014年6月15日至10月14日	
		施用次数(次)	每次施用量(mg/株)	施用次数(次)	每次施用量(mg/株)
CK	0	2	0	8	0
N1	100	2	6	8	11
N2	200	2	12	8	22
N4	400	2	24	8	44
N6	600	2	36	8	66
N8	800	2	44	8	89

注：表中施肥日期为每月15日和30日。采用等量施肥方式，但为避免过高浓度的氮素液体胁迫米老排幼苗生长，因此在实验的第一个月内，施氮量减半。

1.2.3 叶绿素荧光参数指标的测定

2014 年 10 月 20 号，每个处理取 3 株平均苗，自顶端数第 5 和第 6 片完全展开叶，采用 PAM-2500 (Walz, Germanv) 测定叶片的 PSII 最大光化学效率 (F_v/F_m) 和实际光化学效率 ($\Phi PSII$)、光化学猝灭系数 (qP) 和非光化学猝灭系数 (NPQ) 四个叶绿素荧光参数。在测定前，所有米老排苗木的叶片均暗适应 30min。

1.3 数据处理

利用 Excel 2010 整理数据，利用 SPSS 16.0 软件在 $P<0.05$ 条件下进行方差分析和多重比较(Dancun 方法)。

2 结果与分析

2.1 不同氮素水平对米老排苗高和地径生长的影响

对米老排苗高和地径数据进行方差分析发现，两个指标的差异均达到显著水平(表 2)，说明不同浓度的氮素营养水平对米老排的生长产生了显著的影响。

表 2 不同处理米老排苗木平均苗高和地径方差分析($P<0.05$)

性状	自由度	平方和	均方	F 值
苗高(cm)	5	7097.140	1419.928	135.715*
地径(mm)	5	48.526	9.705	50.647*

表 3 不同处理米老排苗木平均苗高和地径多重比较

处理	平均苗高 +Sd(cm)	平均地径+Sd(mm)
CK	16.49±3.02a	3.26±0.64b
N1	29.45±4.36b	3.96±0.32c
N2	40.52±2.56d	4.98±0.2d
N4	33.60±3.47c	3.64±0.41c
N6	28.24±2.11b	3.10±0.28b
N8	15.63±3.40a	2.68±0.58a

注：表中不同字母的处理表示在 $P<0.05$ 时差异性显著，相同字母的处理之间差异不显著，下同。

经过多重比较分析发现(表 3)：在平均苗高方面，与非施肥对照相比，只有 N8 处理差异不显著，其他 4 个处理的差异均达到显著水平，且是平均苗高均显著优于对照处理，说明适当施用氮素营养，能够促进苗木苗高的增长；在各施肥处理间，仅 N1 和 N6 处理间差异不显著，说明不同氮素水平对苗高生长的影响不同；其中，生长最好的是 N2 处理，达到了 40.5cm，是对照的 2.46 倍，说明 200mg/株的施氮量对苗木高生长的促进作用最显著；其次是 N4、N1 和 N6 处理，依次是对照处理的 2.04 倍、1.79 倍和 1.71 倍；苗高生长最差的是 N8 处理，仅为 15.63cm，甚至比对照还矮了 0.87cm。

在平均地径方面，与对照相比，各施肥处理(除 N6 外)的差异达均到显著水平，且各处理间的差异也达到了显著水平。所有处理中，地径生长最好的仍是 N2 处理，其平均地径较对照提高了 52.8%，达到了 4.98mm，说明 200mg/株的施氮量也是促进苗木地径生长的最佳选择；其次是 N1 和 N4 处理(二者间差异不显著)，分别较对照提高了 21.4%和 11.7%；生长最差的是 N8 处理，其平均地径较对照降低了 17.8%，仅为 2.68mm，说明 800mg/株的施氮量已经开始在一定程度上阻碍苗木地径的生长。

综合分析不同处理苗木平均苗高和地径多重比较，发现随着施用氮素水平的增加，米老排苗木的生长出现先增加后下降的规律；尤其是在地径生长方面，随着施用量增加到 800mg/株时，其地径的生长反而显著弱于非施肥对照，说明过量施用氮肥会对米老排的生长产生一定的阻碍作用。

2.2 不同氮素水平对米老排苗木总生物量的影响

苗木的总生物量是指苗木在一定时期内通过光合作用净合成的有机物的量。对 6 个处理米老排苗木的全株总生物量进行方差分析，发现在 $P<0.05$ ($F=537.909$) 水平下各处理的差异均达到显著水平。

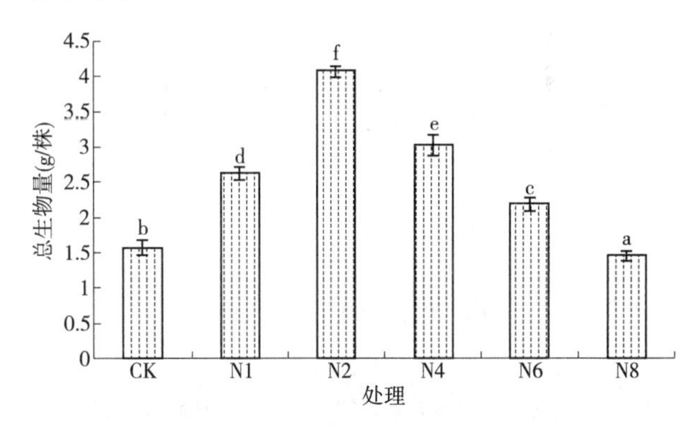

图 1 不同处理米老排苗木全株生物量

进一步对各处理苗木总生物量数据进行多重比较(图 1)。结果表明：各处理的平均总生物量差异均达到显著水平；且与对照相比，除了 N8 外，其他施肥处理较对照均有显著提高。其中，平均总生物量最大的是 N2 处理，达到了 4.07g，较对照提高了 158.4%；最小的是 N8 处理，仅为 1.45g，比对照还降低了 7.8%，说明 200mg/株的施氮量对米老排苗木干物质积累的促进作用最显著，而 800mg/株开始阻碍苗木营养物质的积累。从图 1 也可以发现，随着氮素水平的增加，苗木的平均总生物量也呈现先增加后下降的趋势。

2.3 不同氮素水平对米老排叶绿素总含量的影响

植物的生长发育与植物的光合作用有着必然的联系，而叶绿素作为植物进行光合作用的主要色素，其含量的变化会对植物叶片光合能力产生影响[19]。测定了不同处理苗木叶片的总叶绿素含量，通过方差分析发现，不同处理米老排苗木总叶绿素含量的差异均达到了显著水平（$F=20730$，$P<0.05$）。

进一步进行多重比较分析发现(图2)：施用氮肥处理的米老排苗木均显著高于对照的，说明施氮肥均能够显著提高米老排叶片叶绿素总含量。其中，含量最高的是N4处理，总量达到了1.240mg/株，较对照处理(0.808mg/株)提高了53.5%；其次是N6处理(1.226mg/株)，较对照提高了51.7%；而N8处理虽然是氮素营养施用量最多的处理，达到了800mg/株，但是其叶绿素总含量却是所有施肥处理中增加最少的，仅提高了21.1%；此外，从图2可以得出与地径、苗高和总生物量一样的趋势，即：随着氮素水平的增加，叶绿素总量会先增加后下降。

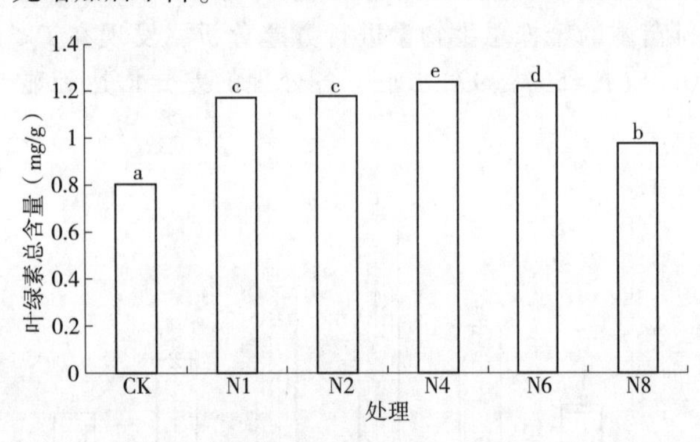

图2　不同处理米老排苗木叶绿素总含量

2.4 不同氮素水平对米老排叶绿素荧光参数的影响

叶绿素荧光是评价植物光合作用能力的有效探针，因此叶绿素荧光参数可以在一定程度上反映光合效率及植物对逆境胁迫的耐受性[20]，本研究测定了不同处理米老排苗木的PSII最大光化学效率(F_v/F_m)、实际光化学效率(Φ_{PSII})、光化学猝灭系数(qP)和非光化学猝灭系数(NPQ)等参数。因N8处理(800mg/株)的米老排成熟叶出现了比较严重的枯萎，所以该处理的苗木不参与测定。因此对除N8外的各处理米老排苗木F_v/F_m、Φ_{PSII}、NPQ和qP的测定结果进行方差分析，结果发现米老排苗期4个叶绿素荧光参数指标的差异显著(表4)。

表4　不同施肥处理下米老排叶绿素荧光参数方差分析

性状	自由度	平方和	均方	F值
F_v/F_m	4	0.483	0.121	195.691*
Φ_{PSII}	4	1.228	0.307	66.796*
qP	4	0.469	0.102	18.669*
NPQ	4	22.68	5.670	51.345*

表5　不同处理下米老排叶绿素荧光参数多重比较*

处理	F_v/F_m	Φ_{PSII}	qP	NPQ
CK	0.75±0.01b	0.33±0.09b	0.70±0.13a	2.38±0.60c
N1	0.74±0.02b	0.52±0.07c	0.84±0.07b	0.82±0.12b
N2	0.79±0.01c	0.69±0.03d	0.92±0.02c	0.22±0.07a
N4	0.76±0.02bc	0.55±0.09c	0.85±0.04bc	0.95±0.40b
N6	0.50±0.04a	0.22±0.04a	0.68±0.04a	1.09±0.08b

F_v/F_m是暗适应下PSII的最大光化学效率，能直观反映了植物最大潜在光能；高等植物的Fv/Fm一般在0.75~0.84，当植物受到胁迫时该参数就变小。通过多重比较(表5)可知：在各施肥处理中，仅N2和N6处理与对照有显著差异；其中，N2显著高于对照处理，可知200mg/株的施氮量能够激发植物的光合潜能，提高光化学效率；N6显著低于对照处理，仅为0.50，说明600mg/株的施氮量对米老排苗木产生了胁迫作用。

Φ_{PSII}是光适应下PSII的实际光化学效率，反映植物叶片将光能转化为化学能的能力[21]。由表5可知：各施肥处理与对照间差异均达到显著水平，且除N6外，其他施肥处理均显著高于对照，表明适宜的氮肥能够对米老排苗木叶片实际光合效率的提高起到促进作用；其中，N2处理的Φ_{PSII}值仍是最大，达到了0.69，比对照提高了109.1%，说明200mg/株的施氮量对苗木光合效率的提高效果最佳；最小的是N6处理(0.22)，较对照降低了33.3%，说明600mg/株的施氮量已经对米老排的光合效率产生了抑制作用，从而影响米老排的生长。

qP反映了PSII天线色素吸收光能并用在光化学电子传递的比例，即qP越大，PSII反应中心的电子活性越大，因此能在一定程度上反映PSII反应中心的开放程度，也能够反映植物光合效率和对光能的利用。不同处理下米老排苗木叶片的qP多重比较分析结果(表5)显示，与对照处理相比，除N6处理外，其他施肥处理的差异均达到了显著水平，且均显著高于对照处理，说明施用适宜浓度的氮素营养能够在一定程度上提高苗木PSII反应中心的电子活性，其中，qP 值最大的仍是 N2 处理，为 0.92，是对照的 1.31 倍，说明 200mg/株施氮量是提高米老排苗木对光能利用效率的最佳选择。

NPQ(非光化学淬灭)能够反映 PSII 反应中心对天线色素吸收过量光能后的热耗散能力[22]。由多重比较结果可知，对照处理的 NPQ 值最大，达到了 2.38；与对照相比，各施肥处理白桦叶片的 NPQ 值均有显著的降低，N2 处理最小，仅为 0.22，较对照降低了 90.8%，表明施用的氮素营养不过量时能够在一定程度上增加叶片对光能的利用，且 200mg/株施氮量的效果最佳。

3 结论与讨论

氮元素作为植物能量代谢和物质代谢的基础，是植物生长和发育的主要限制因子[23]；且增加移栽苗木体内氮素的含量能够促进苗木造林后的早期生长[24-26]；适宜浓度的氮肥能够促进白桦提早开花结实[7,9]；氮肥能够促进长白落叶松、闽楠、香樟和刨花润楠幼苗的生长[27-30]；本研究也发现一定量的氮肥能够显著促进米老排幼苗的生长。

氮素是世界农林业生产中消耗量和浪费量最大的元素之一[23]，因此根据不同树种的不同生长时期确定氮素需求量，提高氮肥的有效性降低成本也是研究的热点。Chen *et al.* [31]比较了传统施肥方式下施用浓度为 50C 和 100C 与指数施肥方式下 50E、100E、200E、400E 四个施肥浓度对米老排生长和养分吸收动态的影响，最终确定采用传统施肥方式，施用浓度为 100C 对增加米老排植株生物量和养分质量分数效果更佳。闫彩霞[32]探索了在 5 个氮素水平下(50、100、200、400、600mg/株)等量施肥、阶段递增施肥和修正指数施肥对米老排苗期生长动态的影响，确定米老排苗期最佳供氮方式为等量施肥法，最佳供氮水平为 200mg/株。本研究也发现 200mg/株是促进米老排幼苗快速增长的最佳施氮水平，与闫彩霞[32]的研究结果一致。。

氮素是叶绿素的重要组成成分之一[20]，外源环境中氮含量会直接影响植物叶片的叶绿素含量，本研究发现施用氮肥后，米老排幼苗叶片的叶绿素含量均较对照有显著提高。叶绿素荧光参数是植物光合机理和光合生理状况的一组常用变量，可反应植物的光合效率及其对逆境胁迫的耐受性。本研究中，N8 处理(800mg/株)的米老排苗木因成熟叶均已经出现不同程度枯萎而未参与测定，已经说明高水平的氮素会损害苗木的正常生长；N6 处理的 F_v/F_m 只达到了 0.50，而 Φ_{PSII} 较对照降低了 33.3%(0.22)，进一步证明 600mg/株的氮素水平已经对米老排苗木的叶片产生了胁迫，抑制植物的光合作用，进而阻碍了植物的生长。

随着氮素水平的增加，不同处理米老排苗木的 F_v/F_m、Φ_{PSII} 和 qP 三个叶绿素荧光参数指标呈现先增加后降低的趋势(NPQ 则呈现相反趋势)，这与米老排苗木的苗高、地径和生物量的变化趋势一致；是因为苗高、地径和生物量这三个常用地衡量植物形态变化的主要指标是幼苗的遗传特性、生理状况及生存环境条件三者之间相互作用的外在表现，当遗传特性和生理状况基本一致时，形态指标就由生存环境决定；这也说明适量增加外源环境中的氮素水平能够提高米老排幼苗的光合效率，促进苗木生长；当超过最佳施用量(200mg/株)但不过量时，多余的氮肥则削弱促进效果，降低氮素的利用效率；当氮素过量时，氮肥就会对苗木叶片形成胁迫，阻碍苗木的生长。

本研究通过对比 5 个氮素水平对米老排苗木生长的影响，证明了 200mg/株的氮素水平是最佳选择，为今后米老排苗木苗期施肥提供了参考依据；但也存在一定的局限性，因为氮素水平为 200~400mg/株时，米老排苗木的具体生长情况还未确定，应该进一步通过研究论证。

参考文献

[1] KIM T H, ITO H, HAYASHI K, et al. New antitumor sesquiter penoids from *Santalum album* of Indian origin [J]. Tetrahedron, 2006, 62: 6981-6989.

[2] PAUL L R, CHAPMAN B K, CHANWAY C P. Nitrogen fixation associated with Suillus tomentosus tuberculate ectomy corrhizac on *Pinus contora* var. *latifolia* [J]. Ann Bot, 2007, 99: 1101-1109.

[3] 周志强，彭英丽，孙铭隆，等. 不同氮素水平对濒危植物黄檗幼苗光合荧光特性的影响[J]. 北京林业大学学报，2015，12：17-23.

[4] 何明，翟明普，曹帮华. 水分胁迫下增施氮、磷对刺槐无性系苗木光合特性的影响[J]. 北京林业大学学报，2009，31(6)：116-120.

[5] 杨自立，马履一，贾忠奎，等. 不同供氮水平对栓皮栎播种苗光响应曲线的影响[J]. 北京林业大学学报，2011，33(5)：56-60.

[6] 郝龙飞，刘婷岩，张连飞，等. 氮素指数施肥对白桦播种苗养分承载和光合作用的影响[J]. 北京林业大学学报，2014，(6)：17-23.

[7] 刘福妹，姜静，刘桂丰. 施肥对白桦树生长及开花结实的影响[J]. 西北林学院学报，2015，30(2)：116-

120，195.
[8]王晓英，王冬梅，黄益宗．不同施氮水平下 AMF 群落对白三叶草生长及养分吸收的影响[J]. 北京林业大学学报，2011，33(2)：143-148.
[9]刘福妹，李天芳，姜静，等．白桦最佳施肥配方的筛选及其各元素的作用分析[J]. 北京林业大学学报，2012，34(2)：57-60.
[10]刘迪，杨秀珍，戴思兰，等．轻型基质栽培条件下氮浓度对独本菊‘紫如意’生长开花的影响[J]. 北京林业大学学报，2014，(4)：102-106.
[11]肖迪，王晓洁，张凯，等．模拟氮沉降对五角枫幼苗生长的影响[J]. 北京林业大学学报，2015，10：50-57.
[12]郭文福，蔡道雄，贾宏炎，等．米老排人工林生长规律的研究[J]. 林业科学研究，2006，19(5)：585-589.
[13]郭文福．米老排人工林生长与立地的关系[J]. 林业科学研究，2009，22(6)：835-839.
[14]郑路，蔡道雄，卢立华，等．南亚热带不同树种人工林生物量及其分配格局[J]. 林业科学研究，2014，27(4)：454-458.
[15]王小燕，薛杨，郭海燕，等．米老排引种栽培技术研究初报[J]. 热带林业，2015，43(2)：25-28.
[16]孙冬婧，温远光，罗应华，等．近自然化改造对杉木人工林物种多样性的影响[J]. 林业科学研究，2015，28(2)：202-208.
[17]TIMMER V R，ARMSTRONG G. Growth and nutrition of containerized*Pinus Resinosa* seedlings at varying moisture regimes [J]. New Forests，1989，3(2)：171-180.
[18]高俊凤．植物生理学实验指导[M]. 北京：高等教育出版社，2006：74-77.
[19]EVANS J R. The photosynthesis and nitrogen relationship in leaves of C_3 plants [J]. Oecologia，1989，78(1)：9-19.
[20]MAXWELL K，JOHSON G N. Chlorophyll fluorescence a practical guide. Journal of Experimental Botany，2000，51(345)，659-668.
[21]GCENTY B，BRIANTAIS J M，BAKER N R. The relationship between the quantum yield of photosynthetic electron transport and quenching of chlorophyll fluorescence [J]. Biochim Biophys Acta，1989，990：87-92.
[22]BADGER M R，von CAEMMERER S，RUUSKA S，et al. Electron flow to oxygen in higher plants and algae：rates and control of direct photoreduction(Mehler reaction) and rubisco oxygenase [J]. Philosophical transactions of the Royal Society of London. Series B，Biological Science，2000，355(1402)：1433-1446.
[23]麻文俊，张守攻，王军辉，等．楸树无性系苗期 N 素利用差异和高产无性系选择[J]. 林业科学，2012，48(10)：157-162.
[24]RYTTER L，ERICSSON T，RYTTER R M. Effects of demand-driven fertilization on nutrient use，root：plant ratio and field performance of *Betula pendula* and *Picea abies*. Scandinavian Journal of Forest Research，2003，18：401-415.
[25]RIKALA R，HEISKANEN J，LAHTI M. Autumn fertilization in the nursery affects growth of Picea abies container seedlings after transplanting. Scandinavian Journal of Forest Research，2004，19：409-414.
[26]HEISKANEN J，LAHTI M，LUORANEN J. Nutrient loading has a transitory effect on the nitrogen status and growth of outplanted Norway spruce seedlings [J]. Silva Fennica，2009，43(2)：249-260.
[27]祝燕，刘勇，李国雷，等．氮素营养对长白落叶松移植苗生长及养分状况的影响[J]. 林业科学，2011，09：168-172.
[28]王东光，尹光天，邹文涛，等．氮素营养对闽楠幼苗生长及光合特性的影响[J]. 林业科学研究，2013，26(1)：70-75.
[29]黄复兴，范川，李晓清，等．施肥对盆栽香樟幼苗细根生长的影响[J]. 西北林学院学报，2013，28(5)：103-108+114.
[30]胡厚臻，侯文娟，潘启龙，等．配方施肥对刨花润楠幼苗生长和光合生理的影响[J]. 西北林学院学报，2015，30(6)：39-45.
[31]CHEN L，ZENG J，JIA H Y，et al. Growth and nutrient up-take dynamics of *Mytilaria laosensis* seedlings under exponential and conventional fertilizations [J]. Soil Science and Plant Nutrition，2012，58(5)：618-626.
[32]闫彩霞，杨锦昌，尹光天，等．供氮方式及水平对米老排苗期生长动态的影响[J]. 东北林业大学学报，2015，43(5)：11-16.

[原载：西北林学院学报，2018，33(01)]

基质配比和缓释肥用量对望天树容器苗的生长效应

庞圣江[1,2] 马 跃[1] 张 培[1] 劳庆祥[1] 杨保国[1] 刘士玲[1]
([1]中国林业科学研究院热带林业实验中心，广西凭祥 532600
[2]广西友谊关森林生态系统国家定位观测研究站，广西凭祥 532600)

摘 要 为了研究望天树育苗的优化方案，给育苗生产中提供理论与技术参考，采用析因实验设计方法，研究基质配比和缓释肥用量对望天树1年生容器苗生长量及其根系发育的影响。结果表明：①随着基质中红壤土比例的降低，望天树容器苗高生长呈降低趋势，地径、生物量及根系发育等指标均呈明显先增高再降低趋势，红壤土所占比例为60%时，地径、全株生物量及根体积指标值为最大。②随着缓释肥用量的增加，望天树容器苗苗高、地径生长表现为波动升高而后减缓的趋势，缓释肥用量为2.5kg/m³时，容器苗高、地径和全株生物量等生长指标值均为最大。③望天树容器苗生长对基质配比和施加缓释肥均较为敏感，但根系生长发育的缓释肥效应不够明显。④本研究综合各因素对望天树1年生容器苗生长的主效应和互作效应，选出望天树容器育苗最佳的优化方案，即红壤土：松树皮：碳化树皮配比基质的体积比为6：2：2时，缓释肥用量2.5kg/m³。

关键词 望天树；容器苗；基质配比；缓释肥；苗高；地径

Effect of Substrate Ratio and Slow-release Fertilizer Dose on the Growth of Containerized *Parashorea chinensis* Seedlings

PANG Shengjiang[1,2], MA Yue[1], ZHANG Pei[1],
LAO Qingxiang[1], YANG Baoguo[1], LIU Shiling[1]
([1]*Experimental Center of Tropical Forestry, CAF, Pingxiang* 536000, *Guangxi, China*;
[2]*Guangxi Youyiguan Forest Ecosystem Research Station, Pingxiang* 532600, *Guangxi, China*)

Abstract: In factorial experiment design, one-year-old non woven fabric containerized *Parashorea chinensis* seedlings cultivated with light substrate were used to study the effect of different substrate ratios and varied doses of slow-release fertilizer on the growth and root trait development, It is aimed to select the optimum scheme about substrate ratio and fertilizer doses for the seedlings. The results showed that: ① The container seedlings height of *Parashorea chinensis* then decreased, the ground diameter, biomass and root index showed a significantly increase at first and then decreased along with the decrease of laterite soil proportion in the matrix. When the proportion of laterite soil reached 60%, the ground diameter, biomass and root characteristics reached the maximum. ② The seedling's height, the ground diameter increased at first, and then decreased with the increasing slow-release fertilizer doses. As soon as applying slow-release fertilizer at doses of 2.5kg/m³, the seedling's height, the ground diameter and biomass corresponding parameters value reached to the largest. ③ The growth traits of the containerized seedlings took an obvious effect on the substrate ratio and slow-release fertilizer doses, while the effect of fertilizer on root growth was not enough to be noticed. ④Optimum scheme was selected according to the main effects and interaction effects of all factors to the 1 year container seedling growth comprehensively, the most suitable substrate is treated as laterite soil : pine bark : Carbonized bark = 6 : 2 : 2), and the application of 2.5kg/m³ slow release fertilizer (APEX).

Key words: *Parashorea chinensis*; containerized seedling; substrate ratio; slow-release fertilizer; seedling height; ground diameter

望天树(*Parashorea chinensis*)为龙脑香科柳安属常绿大乔木，是热带雨林的标志性树种，天然分布于我国广西西南部以及云南南部等局部地区，为我国一级保护珍稀濒危树种[1]。望天树高可达60m，胸径150cm以上，因其具有速生、干形圆满通直、材质坚硬、耐腐性强和纹理美观等优良特性，是制造各种高级家具的优质用材[2]。然而，望天树作为极有发展前途的乡土珍贵阔叶树种，在我国热带地区的造林面积甚小[3]。一方面，因其种子为典型顽拗性种，容易发芽，难贮藏，采种育苗的时间节点要求比较严格；另一方面，苗木主根发达，侧根生长不旺盛，根须稀疏致使移栽的成活率较低。此外，望天树对立地条件要求甚高，造成其用材林的规模化发展存在较大难度。

容器苗因其具有较为完整的根系土团，起苗与运输过程中不伤根系，能够适应不同的造林季节和延长造林时间、有效地提高了人工造林的成活率[4,5]。尤其是立地条件较差的情况下，与裸根苗相比较，容器苗恢复迅速、用于造林有着明显的优势[6,7]。同时，容器育苗过程中，基质配比和增施缓释肥，对苗木生长环境的稳定以及获取充足的养分资源起着决定作用，一直是苗木培育的研究重点[8,9]。

当前，容器育苗主要以质量轻、便于运输的轻型育苗基质，且缓释肥能显著地提高苗木生长质量，有效降低育苗成本和环境污染，成为育苗的首选方案[10,11]。然而，以往轻基质育苗大多数用于松树(*Pinus* spp.)、杉木(*Cunninghamia lanceolata*)和桉树(*Eucalyptus* spp.)等速生用材树种[12-15]；对花榈木(*Ormosia hosiei*)、南方红豆杉(*Taxus wallichiana* var. *mairei*)和木荷(*Schima superba*)等乡土珍贵树种的基质育苗、容器的规格大小、基质配比以及控根技术等相关方面的研究也陆续见报道[16-18]，关于基质配比对闽楠(*Phoebe bournei*)、赤皮青冈(*Cyclobalanopsis gilva*)、狭叶冬青(*Ilex fargesii*)和塔姆岛金花茶(*Camellia tamdaoensis*)容器育苗，及其土壤养分库的缓释肥生长效应也进行了研究探讨[19-22]。

由于植物生物学特性的差异，轻基质容器育苗过程中，基质配比与施肥措施因树种而异[22]。为提高望天树育苗质量，本研究采用1种容器规格、4种基质配比和5种缓释肥用量，开展望天树1年生基质容器育苗试验，探讨不同影响因素及其互作效应对望天树容器苗苗高、地径和生物量以及根系发育等生长测量指标的影响，利用隶属函数方法，对不同处理的轻基质育苗效果进行客观地评价，筛选出适合望天树容器苗培育的优化方案，为望天树优质容器苗培育提供理论与技术参考。

1　材料与方法

1.1　研究材料

研究地位于广西友谊关森林生态系统国家定位观测研究站热林中心苗木繁育基地，坐标为21°57′N，106°47′E，属于南亚热带季风气候区；年均温21℃，极端最低温-1.5℃，最高温39℃，≥10℃年积温6000~7500℃；年均降水量约1400mm，主要集中在4~10月；全年日照时数1200~1600h，有霜期3~5d；气候适宜望天树的生长发育。

供试验苗木为望天树播种实生苗，由中国林业科学研究院热带林业实验中心提供，所选取幼苗高度基本一致(10cm左右)。育苗容器为无纺布育苗袋，口径与高度规格为20cm×15cm；以红壤土、松树皮和碳化树皮作为育苗基质，采用美国Simplot公司生产的爱贝斯(apex)长效缓释肥，其全K、速效P含量均为80g/kg，全N含量为180g/kg，肥料养分释放时间达8个月以上。在育苗基质配比时，缓释肥直接与其搅拌均匀使用。本次容器育苗试验过程中，需要做好各项管理措施及时除草和喷水，保证苗木的生长发育，每相隔1周调换幼苗盆位置，达到减少边缘效应的影响。

1.2　试验方法

本试验于2016年2月初开始进行，试验设计育苗基质配比(P)、缓释肥量(Q)2个因子。设置4个水平的基质配比(P_1：红壤土∶松树皮∶碳化树皮=8∶1∶1；P_2：红壤土∶松树皮∶碳化树皮=6∶2∶2；P_3：红壤土∶松树皮∶碳化树皮=4∶3∶3；P_4：红壤土∶松树皮∶碳化树皮=2∶4∶4)，缓释肥设置5个水平分别为Q_1(1kg/m^3)、Q_2(1.5kg/m^3)、Q_3(2kg/m^3)、Q_4(2.5kg/m^3)和Q_5(3kg/m^3)。采用析因设计方法，共设计20个试验处理(T)，详见表1。每个试验处理10株，重复3次，

表1　望天树育苗基质配比与缓释肥量2因子试验处理

试验因子	因子水平及处理			
基质配比	P_1	P_2	P_3	P_4

（续）

试验因子	因子水平及处理				
缓施肥量	Q_1	Q_2	Q_3	Q_4	Q_5
基质配比×缓施肥量	$P_1 \times Q_1$	$P_1 \times Q_2$	$P_1 \times Q_3$	$P_1 \times Q_4$	$P_1 \times Q_5$
	$P_2 \times Q_1$	$P_2 \times Q_2$	$P_2 \times Q_3$	$P_2 \times Q_4$	$P_2 \times Q_5$
	$P_3 \times Q_1$	$P_3 \times Q_2$	$P_3 \times Q_3$	$P_3 \times Q_4$	$P_3 \times Q_5$
	$P_4 \times Q_1$	$P_4 \times Q_2$	$P_4 \times Q_3$	$P_4 \times Q_4$	$P_4 \times Q_5$

1.3 数据调查与统计分析

2016年11月下旬，每个重复试验抽取8株望天树容器苗，测量其苗高、地径、生物量、根长和根体积等生长指标，各试验调查株数总共为24株；同时，测定叶片、茎、根和全株生物量，使用Excel 2007求取均值，SPSS 16.0软件对望天树容器苗生长性状进行方差分析和多重比较，以检验基质配比(P)、缓释肥用量(Q)主效应及其互作效应的显著性。在方差分析时高径比数据经反正弦转换。

根据模糊数学中隶属函数的求值方法[20]，对望天树容器育苗方案进行优选。各项测定指标的隶属值计算公式：

$$隶属值=(N-N_{min})/(N_{max}-N_{min})$$

式中：N表示某处理指标值；N_{min}最小指标值；N_{max}最大指标值。

如果某个测定指标与幼苗生长等呈反向关系，可利用反隶属函数公式进行计算：

$$隶属值=1-(N-N_{min})/(N_{max}-N_{min})$$ [24]

2 结果与分析

2.1 基质配比和缓释肥用量对望天树容器苗生长的影响

2.1.1 基质配比

从表1可以看出，不同基质配比培育望天树容器苗的生长效应差异显著。随着基质配比中红壤土的比例由80%降低到20%时，其苗高生长呈逐渐降低的趋势；容器苗地径、生物量积累及根体积表现为先增大后减小，红壤土的比例为60%时，幼苗地径的平均生长量、全株生物量和根体积均为最大，分别为5.62mm、4.26g和2.67cm^3。由此可见，红壤土的比例为60%时，容器苗高生长不受影响的情况下，地径生长和全株生物量积累得到显著提高。上述分析结果表明，不同基质配比过程中红壤土、松树皮与碳化树皮为6∶2∶2时，可较好地促进望天树容器苗的生长。

2.1.2 缓释肥量

由表2可知，缓释肥用量对望天树容器苗高、地径生长和生物量积累等生长指标的影响差异显著。随着缓释肥用量的增加，望天树容器苗高、地径生长和全株生物量积累表现为波动升高而后减缓的趋势，其中，缓释肥施用量为2.5kg/m^{3}时，容器苗高、地径和全株生物量等生长指标值均为最大，分别为56.6cm、6.36mm和4.84g。如若继续增加缓释肥用量，其苗高、地径生长和生物量积累则显著地降低。综上所述，培育望天树容器苗较为适宜的缓释肥用量为2.5kg/m^3。

2.1.3 基质配比×缓释肥用量组合

不同基质配给×缓释肥用量组合培育的望天树容器苗生长性状，如表2所示。方差分析结果显示，20个试验处理的望天树容器苗高、地径、高径比以及生物量积累等生长测量指标的差异达到了显著水平($P<0.05$)，说明不同基质配比×缓释肥用量组合培育的望天树容器苗生长性状存在显著差异。从各试验组合培育的望天树容器苗生长性状可知，其苗高、地径生长以及生物量积累等生长指标值均为最大，平均值分别为81.5cm、6.05mm和8.39g，因此，P_2Q_4(红壤土∶松树皮∶碳化树皮=6∶2∶2，缓释肥用量为2.5kg/m^3)组合可用于培育优质容器苗。

表2 不同基质配比对望天树生长和根系发育的影响

因素	水平		苗高(cm)	地径(mm)	高/径比	总根长(cm)	根体积(cm^3)	叶生物量(g)	茎生物量(g)	根生物(g)	全株生物(g)
基质配比	P_1	平均值	45.3a	4.83ab	93.83a	292.58bc	1.45a	1.21c	2.41bc	0.18b	3.80bc
	P_2	平均值	44.6abcd	5.62def	79.45a	306.62bcde	1.73b	1.12abc	2.97cd	0.17b	4.26e
	P_3	平均值	43.9e	5.06abc	84.88bc	280.03a	2.67ef	0.78a	1.55a	0.12a	2.45a
	P_4	平均值	42.3ef	4.75a	89.19cde	312.46def	2.48bcde	1.49cd	2.23b	0.22bc	3.94bcd

（续）

因素	水平	苗高（cm）	地径（mm）	高/径比	总根长（cm）	根体积（cm^3）	叶生物量（g）	茎生物量（g）	根生物（g）	全株生物（g）
缓释肥	Q_1 平均值	48. 7ab	5. 30a	90. 60cde	307. 63a	1. 62a	0. 88ab	1. 75b	0. 13ab	2. 76ab
	Q_2 平均值	47. 4a	5. 52abc	86. 34ab	316. 44ab	1. 89abc	1. 00bc	1. 99bc	0. 15bcd	3. 14cde
	Q_3 平均值	51. 2abcd	6. 22def	84. 36a	338. 15b	2. 24cd	1. 37cde	2. 74bcde	0. 21cde	4. 32def
	Q_4 平均值	56. 6f	6. 36f	95. 68def	309. 39a	2. 72f	1. 54f	3. 07ef	0. 23ef	4. 84f
	Q5 平均值	54. 7cde	6. 30ef	85. 61ab	305. 89a	2. 45cde	0. 83a	1. 66a	0. 12a	2. 61a
基质配比×缓释肥	P_1Q_1 平均值	51. 5ef	5. 58bed	92. 35ab	287. 39a	1. 47ab	1. 61bcd	2. 66bc	0. 21bc	4. 48hij
	P_1Q_2 平均值	58. 6bed	5. 71cde	102. 64cde	300. 08abcd	2. 52ef	1. 33ef	2. 20hi	0. 18b	3. 71f
	P_1Q_3 平均值	53. 5ab	5. 7h	93. 89cdef	292. 07ef	1. 47ab	1. 10k	2. 52b	0. 18b	3. 80fg
	P_1Q_4 平均值	52. 4ab	5. 6gh	93. 54ef	328. 52bcd	2. 15bcdef	1. 26gh	2. 78ef	0. 20bc	4. 24gh
	P_1Q_5 平均值	56. 5cde	5. 73efg	98. 59fg	343. 58bc	2. 18bcdef	1. 39gh	2. 37f	0. 19b	3. 95fg
	P_2Q_1 平均值	43. 6fg	4. 65ab	93. 70efg	279. 39abc	1. 44ab	1. 19k	3. 66k	0. 24bcde	5. 09lijk
	P_2Q_2 平均值	46. 5fg	4. 79cd	97. 04def	295. 84bc	2. 06fg	1. 83ghij	2. 53b	0. 22bc	4. 58hij
	P_2Q_3 平均值	69. 5c	5. 87def	118. 45efg	359. 66def	2. 47ef	1. 33ef	4. 60def	0. 30defg	6. 23jk
	P_2Q_4 平均值	81. 5a	6. 05g	134. 63bed	404. 97cdef	3. 18f	1. 63bcd	6. 36l	0. 40l	8. 39l
	P_2Q_5 平均值	61. 5f	5. 73fg	107. 35fg	329. 73cde	2. 30efg	1. 18kl	3. 35jk	0. 23bcd	4. 76hijk
	P_3Q_1 平均值	47. 4fg	3. 5bcd	135. 43bcde	281. 4def	1. 68bc	1. 45ghi	1. 75cdef	0. 16b	3. 36bcdef
	P_3Q_2 平均值	51. 4dg	5. 64ab	91. 13abc	296. 44defg	2. 89fgh	1. 07ij	3. 07j	0. 21bc	4. 35hi
	P_3Q_3 平均值	53. 6ef	5. 92bcd	90. 51ab	275. 48hi	2. 06fg	0. 69a	1. 99h	0. 13a	2. 81ab
	P_3Q_4 平均值	47. 4cd	5. 15a	92. 08def	309. 58cdef	2. 25defg	1. 66bcd	1. 66cde	0. 17b	3. 49def
	P_3Q_5 平均值	39. 6efg	3. 56bdce	111. 24cd	316. 31abcd	1. 88g	1. 40gh	3. 00ijk	0. 22a	4. 62hij
	P_4Q_1 平均值	38. 6bcd	3. 78cde	104. 63def	301. 893bcd	1. 31cd	1. 02ij	2. 02h	0. 15b	3. 19cde
	P_4Q_2 平均值	64. 6f	5. 84cd	110. 67abcd	322. 83def	1. 16b	1. 93fgh	3. 93kl	0. 29defg	6. 15jk
	P_4Q_3 平均值	50. 6ef	5. 61def	90. 23de	319. 51bcd	2. 25efg	1. 26gh	3. 26jk	0. 23bcd	4. 75hijk
	P_4Q_4 平均值	68. 6c	5. 74efg	119. 48cdef	347. 88cd	1. 97g	0. 41fgh	2. 41f	0. 14a	2. 96bcd
	P_4Q_5 平均值	45. 5cd	3. 66def	124. 32fg	309. 43cde	1. 33cd	0. 78ab	1. 81hi	0. 13a	2. 72a

2.3 不同因素及其交互作用对望天树容器苗质量的影响

从表3可以看出，望天树容器苗各生长指标的双因素效应达到显著水平（P<0. 05），其苗高、地径和全株生物量的双因素交互效应也均达到显著水平（P<0. 05）。根系测量指标中总根长受基质配比的影响比缓释肥用量大，但双因素的交互效应不显著（P>0. 05）。根体积的基质配比、缓释肥用量以及其双因素交互效应均达到显著水平（P<0. 05），与上述相关方差分析结果相一致。由此表明，望天树容器育苗过程中，不同基质配比和缓释肥用量的试验处理，不能利用方差分析结果进行简单地判定。

表3 望天树容器苗生长指标与质量性状的双因素方差分析

变异来源	苗高	地径	全株生物量	总根长	根体积
基质配比	24. 965**	36. 385**	16. 594**	4. 697*	10. 254**
缓施肥量	20. 747**	8. 302**	2. 732**	3. 483	4. 892**
基质配比×缓施肥量	6. 349*	1. 936*	1. 624*	4. 578	3. 281*

注：表中数值为方差分析 F 值，** 表示效应极显著，* 表示效应显著。

2.4 望天树1年生容器育苗方案优选

根据隶属函数的计算方法，综合各因素对1年生望天树容器苗生长性状和互作效应，选取差异达到显著水平的苗高、地径，叶、茎根和全株生物量以及根体积隶属度平均值作为优选标准，为望天树容器育苗选出20种优化方案，详见表4。在综合考虑基质原料和缓释肥成本，培育1年生望天树优质容器苗，最优方案为排名第1的育苗方案(红壤土：松树皮：碳化树皮=6：2：2，缓释肥量为2.5kg/m^3)。

表4 基于指标隶属度值的望天树容器育苗方案优选

基质配比	缓施肥量(kg/m^3)	苗高	地径	叶生物量	茎生物量	根生物量	全株生物量	根体积	隶属度均值	排名
P_1	Q_1	0.47	0.49	0.45	0.45	0.46	0.45	0.66	0.49	20
P_1	Q_2	0.47	0.57	0.43	0.66	0.59	0.54	0.57	0.55	19
P_1	Q_3	0.63	0.60	0.80	0.81	0.76	0.80	0.81	0.74	6
P_1	Q_4	0.60	0.57	0.80	0.69	0.69	0.76	0.69	0.68	10
P_1	Q5	0.55	0.44	0.80	0.78	0.75	0.78	0.73	0.70	9
P_2	Q_1	0.69	0.59	0.83	0.81	0.76	0.80	0.81	0.75	3
P_2	Q_2	0.49	0.63	0.72	0.70	0.62	0.61	0.56	0.62	18
P_2	Q_3	0.83	0.74	0.83	0.81	0.82	0.83	0.77	0.81	2
P_2	Q_4	0.90	0.89	0.86	0.87	0.86	0.88	0.66	0.84	1
P_2	Q5	0.48	0.50	0.70	0.67	0.67	0.69	0.69	0.63	16
P_3	Q_1	0.53	0.34	0.71	0.70	0.71	0.71	0.70	0.63	17
P_3	Q_2	0.58	0.50	0.57	0.72	0.71	0.72	0.70	0.64	14
P_3	Q_3	0.57	0.77	0.50	0.63	0.67	0.70	0.71	0.65	12
P_3	Q_4	0.53	0.49	0.85	0.81	0.80	0.87	0.83	0.74	5
P_3	Q5	0.44	0.41	0.73	0.75	0.71	0.74	0.73	0.64	15
P_4	Q_1	0.45	0.64	0.74	0.75	0.74	0.74	0.70	0.68	11
P_4	Q_2	0.70	0.71	0.76	0.71	0.69	0.72	0.83	0.73	7
P_4	Q_3	0.76	0.61	0.79	0.73	0.74	0.76	0.82	0.75	4
P_4	Q_4	0.51	0.58	0.74	0.79	0.71	0.79	0.77	0.70	8
P_4	Q5	0.65	0.51	0.71	0.69	0.71	0.67	0.76	0.50	13

3 结论与讨论

本研究发现，1年生望天树容器苗生长测量指标和根系发育特征，对不同基质配比的反应情况有所差异。随着基质中红壤土所占比例的降低，容器苗高生长呈逐渐降低的趋势。容器苗地径、生物量积累及根体积等生长指标均呈明显先增高再降低趋势。当红壤土所占比例为60%时，地径、全株生物量及根系发育等生长指标值为最大，即基质配比为红壤土：松树皮：碳化树皮=6：2：2时，最适宜望天树容器苗生长。这与杉木适宜的基质配比为7泥炭：1.5稻壳：1.5木屑[15]，浙江楠(*Phoebe chekiangensis*)为3黄心土：6泥炭：1草木灰[25]，而基质配比为2黄心土：2泥炭：1蛭石时，能够显著促进南方红豆杉容器苗的生长[26]试验结果各异。由于考虑就地取材的便利性以及不同树种生物学特性的不同，对基质配比的要求亦有所差异。本研究中，望天树容器苗对土壤水肥条件的要求比较高，且需要透气性能较好的基质。育苗基质中红壤土的比例(80%)较高时，其饱和持水量高、透气性差，不利于望天树容器苗根系生长发育，若红壤土中配比适量的松树皮和碳化树皮(红壤土：松树皮：碳化树皮=6：2：2)，有利于改善基质的透气性，不影响苗木地上器官生长情况下，根系与基质形成比较紧密的根土团，从而促进望天树容器苗的生长发育。

施用缓释肥是容器育苗重要的技术措施，其既可根据幼苗生长的需要缓慢地释放养分，又减少传统育苗的多次施肥，从而有效地降低育苗成本[7,27]。本研究中，随着缓释肥用量的增加，望天树容器苗高、地径和生物量积累表现为波动升高而后减缓的趋势。缓释肥用量为 2.5kg/m^{3}时，苗高、地径和全株生物量积累等生长指标值均为最大。这与赤皮青冈[20]、塔姆岛金花茶[22]和欧洲云杉(*Picea abies*)[28]容器育苗基质中缓释肥用量的研究结果相似，他们同时指出，缓释肥施用过量时，容器苗生长量与生物量的积累反而有所降低，这可能与苗木地下部分的根系生长受到抑制有关。

本研究中，望天树容器苗生长对基质配比和缓释肥用量均较为敏感，但根系生长发育的缓释肥效应影响相对较小。基质中施用一定量缓释肥可促进望天树容器苗的生长，对容器苗根系特别是根长的生长发育效果的影响较小。原因可能与望天树苗期的根系特征有着密切关系，其主根发达，直接影响到根系生物量与根体积，而根长直接受侧根数量的影响，侧根生长不旺盛，根须稀疏致使施用不同缓释肥量的苗木根长差异不明显，但生产实践发现，望天树容器苗根须数量远优于裸根苗。

望天树容器育苗过程中，不同基质配比、缓释肥用量对苗木生长、生物量积累和根系发育影响甚大；双因素方差分析结果亦表明，容器苗高、地径和全株生物量以及根体积存在着显著交互效应。由此可见，不同基质配比与缓释肥用量处理时，彼此存在着相互关联，故望天树容器育苗的优选方案，不能简单地以单个因素处理水平的分析结果进行组配。本研究综合考虑各因素对 1 年生望天树容器苗生长的主效应和互作效应，利用隶属函数计算方法，筛选出适合培育望天树容器苗最佳的优化方案，即基质配比中红壤土：松树皮：碳化树皮所占比例为 6：2：2，缓释肥用量 2.5kg/m^3。该育苗试验方案的确定，为望天树优质容器苗培育提供理论与技术参考。

参考文献

[1]郭建华. 国家一级保护植物望天树及其家族[J]. 生物学教学，2000，12(25)：37-38.

[2]中国植物志编辑委员会. 中国植物志[M]. 北京：科学出版社，1990，50(2)：126.

[3]唐建维，邹寿青. 望天树人工林林分生长与林分密度的关系[J]. 中南林业科技大学学报，2008，28(4)：83-86.

[4]EDWARD R，KRISTJAN C，ANDREW Park. Root characteristics and growth potential of container and bare-root seedlings of red oak(*Quercus rubra* L.) in Ontario，Canada[J]. New forests，2007，34(2)：163-176.

[5]王月海，房用，史少军，等. 平衡根系无纺布容器苗造林试验[J]. 东北林业大学学报，2008，36(1)：14-15.

[6]马常耕. 世界容器苗研究、生产现状和我国发展对策[J]. 世界林业研究，1994，(5)：33-41.

[7]楚秀丽，孙晓梅，张守攻，等. 日本落叶松容器苗不同控释肥生长效应[J]. 林业科学研究，2012，25(6)：697-702.

[8]贾斌英，徐惠德，刘桂丰，等. 白桦容器育苗的适宜基质筛选[J]. 东北林业大学学报，2009，37(11)：64-67.

[9]邓华平，杨桂娟. 不同基质配方对金叶榆容器苗质量的影响[J]. 林业科学研究，2010，23(1)：138-142.

[10]刘军，姜景民，陈益泰，等. 闽楠种子轻基质容器育苗及优良家系选择[J]. 西北林学院学报，2011，26(6)：70-73.

[11]冯瑜，张国防，李左荣. 模拟不同施肥处理对芳樟树高和地径的影响[J]. 中南林业科技大学学报，2015，35(6)：34-39.

[12]程庆荣. 蔗渣和木屑作尾叶桉容器育苗基质的研究[J]. 华南农业大学学报，2002，23(2)：11-14.

[13]韦小丽，朱忠荣，尹小阳，等. 湿地松轻基质容器苗育苗技术[J]. 南京林业大学学报(自然科学版)，2003，27(5)：55-58.

[14]徐玉梅，唐红燕，张建珠，等. 不同轻基质配方对思茅松容器育苗的影响[J]. 西北林学院学报，2015，30(6)：147-150.

[15]周新华，厉月桥，肖智勇，等. 基质配比、容器规格和缓释肥量对杉木容器育苗的影响[J]. 江西农业大学学报，2017，39(1)：72-81.

[16]周志春，刘青华，胡根长，等.3 种珍贵用材树种轻基质网袋容器育苗方案优选[J]. 林业科学，2011，47(10)：172-178.

[17]王月生，周志春，金国庆，等. 基质配比对南方红豆杉容器苗及其移栽生长的影响[J]. 浙江林学院学报，2007，24(5)：643-646.

[18]马雪红，胡根长，冯建国，等. 基质配比、缓释肥量和容器规格对木荷容器苗质量的影响[J]. 林业科学研究，2010，23(4)：505-509.

[19]王艺，王秀花，吴小林，等. 缓释肥加载对浙江楠和闽楠容器苗生长和养分库构建的影响[J]. 林业科学，2013，49(12)：57-63.

[20]吴小林，张东北，楚秀丽，等. 赤皮青冈容器苗不同基质配比和缓释肥施用量的生长效应[J]. 林业科学研究，2014，27(6)：794-800.

[21]姚德生，何彦峰．狭叶冬青不同基质配比容器育苗试验研究[J]．西北林学院学报，2015，30(6)：156-160.
[22]韦晓娟，梁晓静，李开祥，等．基质配比和缓释肥量对塔姆岛金花茶容器苗质量的影响[J]．中南林业科技大学学报，2017，37(2)：19-23.
[23]NICESE F P. Effect of container type nursery techniques on growth and chlorophyll content of *Acer platanoides* L. and *Liquidambar styraciflua* L. plants [J]. Journal of Food Agriculture & Environment，2006，4(4)：209-213.
[24]王丁，张丽琴，薛建辉．苗木抗旱性综合评价研究——以6种喀斯特造林树种苗木为例[J]．中国农学通报，2011，27(25)：5-12.
[25]邱勇斌，乔卫阳，刘军，等．容器、基质和施肥对浙江楠容器大苗的影响[J]．东北林业大学学报，2016，44(9)：20-23.
[26]王金凤，陈卓梅，刘济祥，等．不同基质及缓释肥对南方红豆杉容器大苗生长的影响[J]．浙江林业科技，2016，36(2)：74-78.
[27]CORTINA J，VILAGROSA A，TRUBAT R. The role of nutrients for improving seedling quality in drylands [J]. New Forests，2013，44(5)：719-732.
[28]HAWKINS B J，BURGESS D，MITCHELL A K. Growth and nutrient dynamics of western hemlock with conventional or exponential greenhouse fertilization and planting in different fertility conditions [J]. Canadian Journal Forest Research，2005，35(4)：1002-1016.

[原载：西北林学院学报，2018，33(06)]

白木香容器苗基质配比与缓释肥施用量的生长效应

庞圣江[1,2] 张 培[1] 马 跃[1] 杨保国[1] 刘士玲[1] 冯昌林[1]

([1]中国林业科学研究院热带林业实验中心，广西凭祥 532600；
[2]广西友谊关森林生态系统国家定位观测研究站，广西凭祥 532600)

摘 要 选取白木香为研究对象，采用析因试验设计方法，系统地研究基质配比与缓释肥施用量对白木香1年生容器苗生长量及其根系发育的影响，并筛选出育苗的优化方案。结果表明：随着基质配比中酸性砖红壤比例的降低，白木香苗高、地径生长和各器官生物量等指标均呈明显先增高再降低趋势；当基质配比中*V*(酸性砖红壤土)：*V*(松树皮)：*V*(碳化树皮)=6：2：2时，苗高、地径和植株生物量指标值为最大；随着缓释肥量的增加，白木香苗高、地径生长和各器官生物量均表现为先升高后减缓的趋势，当缓释肥施用量为2.5kg/m^{3}时，苗高、地径和各器官生物量指标值最大；双因素方差分析发现，1年生白木香容器苗高、地径和植株生物量的双因素交互效应达到极显著水平(P<0.01)，与根体积的双因素效应也显著(P<0.05)；本试验研究综合各因素对白木香1年生容器苗生长的主效应和互作效应，选出白木香容器育苗最佳的优化方案，即*V*(酸性砖红壤土)：*V*(松树皮)：*V*(碳化树皮)=6：2：2时，缓释肥施用量为2.5kg/m^3。

关键词 白木香；基质配比；缓释肥；容器苗；生长

Effect of SubstrateRatio and Slow-release Fertilizer Dose on the Growth of Containerized *Aquilaria sinensis* Seedlings

PANG Shengjiang[1,2], ZHANG Pei[1], MA Yue[1], YANG Baoguo[1], LIU Shiling[1], FENG Changlin[1]

([1]*Experimental Center of Tropical Forestry, CAF, Pingxiang 536000, Guangxi, China*;
[2]*Guangxi Youyiguan Forest Ecosystem Research Station, Pingxiang 532600, Guangxi, China*)

Abstract: In factorial experiment design, one-year-old non woven fabric containerized *Aquilaria sinensis* seedlings cultivated with light substrate were used to study the effect of different substrate ratios and varied doses of slow-release fertilizer on the growth and root trait development, It is aimed to select the optimum scheme about substrate ratio and fertilizer doses for the seedlings. The results showed that: The container seedlings height of *Aquilaria sinensis* then decreased, the ground diameter and biomass index showed a significantly increase at first and then decreased along with the decrease of lateritic red soil proportion in the matrix. When the proportion of lateritic red soil：pine bark：Carbonized bark=6：2：2, the ground diameter and biomass characteristics reached the maximum. The seedling's height, the ground diameter and biomass index increased at first, and then decreased with the increasing slow-release fertilizer doses. As soon as applying slow-release fertilizer at doses of 2.5kg/m^3, the corresponding parameters value reached to the larges. Two-way ANOVA showed that the interactive effects of height, ground diameter and plant biomass of 1-year-old container seedlings were extremely significant(P<0.01), and root volume were also significant(P<0.05). Optimum scheme was selected according to the main effects and interaction effects of all factors to the 1 year container seedling growth comprehensively, the most suitable substrate is treated as lateritic red soil： pine bark：Carbonized bark=6：2：2, and the application of 2.5kg/m^{3}slow release fertilizer.

Key words: *Aquilaria sinensis*; substrate ratio; slow-release fertilizer; containerized seedling; growth

白木香(*Aquilaria sinensis*)又称白木香树、土沉香和莞香，为瑞香科沉香属常绿乔木，天然分布于我国广西、广东和海南等地[1]，是热带南亚热带地区特有的药材和木材兼用珍贵树种。近年来，白木香种植面积不断增长，其苗木培育研究也引起重视[2,3]，如一些林业工作者对白木香种子发育过程[4]、幼苗生长与生物量积累的限制元素[5]以及缺素对幼苗生长和叶片解剖结构的影响[6]等进行了研究。然而，有关利用轻基质配比与施加缓释肥培育白木香苗木试验研究尚未见报道。

由于轻基质容器苗较好地保持根系土团、便于运输且造林不受季节限制[7,8]，与裸根苗相比造林后缓苗快、能够有效地提高造林成活率[9,10]。施加适量缓释肥有利于保持培育苗木生长环境的稳定和获取充足养分元素，一直是苗木繁育试验研究的重点[11,12]。近年来，轻基质育苗已被广泛应用于松树(*Pinus* spp.)、杉木(*Cunninghamia lanceolata*)和桉树(*Eucalyptus* spp.)等用材树种苗木繁育[13-16]；有关基质配比与施加缓释肥培育闽楠(*Phoebe bournei*)、塔姆岛金花茶(*Camellia tamdaoensis*)等珍贵树种苗木也陆续得到推广[17,18]。由于生物学特性的差异，苗木培育的轻基质取材、配比和施肥措施亦因树种而异[20]。本试验选用取材低廉的酸性砖红壤土、松树皮和碳化树皮等作为原材料，应用析因试验设计方法，研究基质配比与缓释肥对白木香1年生容器苗生长效应的影响，优选出育苗方案，旨在为白木香苗木繁育规模化生产提供成本低、环保可行的育苗技术。

1 试验地概况

试验地位于广西友谊关森林生态系统国家定位观测研究站热林中心站点苗圃(21°57′N，106°47′E)，地处广西凭祥市，属于南亚热带季风气候区；干湿季节明显，雨季主要集中在4~10月份，年均降水量约1400mm；年均温21℃，极端最低温-1.5℃，最高温39℃；≥10℃年积温6000~7500℃；全年日照时间1200~1600h，有霜期3~5d；土壤为酸性砖红壤，pH为3.8~6.0，气候与土壤均适宜白木香的生长发育。

2 材料与方法

2.1 试验材料

采用高度相似(约10cm)、生长健壮的白木香实生苗。采用无纺布育苗袋作为育苗容器，袋口直径与高度规格为20cm×15cm。采用试验区域内取材便利的酸性砖红壤土、松树皮、碳化树皮作为试验的育苗基质。采用吉林云天化农业发展有限公司生产的长效缓释肥，其质量分数 ɯ(氮)：ɯ(磷)：ɯ(钾)=30：14：10，总养分质量分数≥54%，肥料养分释放时间长达8个月左右。

2.2 试验方法

本次试验于2017年2月中旬开始进行，育苗试验涉及基质配比(X)、缓释肥(Y)及其2个因素。其中，育苗基质配比设4个水平，即X_1，V(酸性砖红壤土)：V(松树皮)：V(碳化树皮)=8：1：1；X_2，V(酸性砖红壤土)：V(松树皮)：V(碳化树皮)=6：2：2；X_3，V(酸性砖红壤土)：V(松树皮)：V(碳化树皮)=4：3：3；X_4，V(酸性砖红壤土)：V(松树皮)：V(碳化树皮)=2：4：4。

施用缓释肥量设5个水平，分别为Y_1(1kg/m^3)、Y_2(1.5kg/m^3)、Y_3(2kg/m^3)、Y_4(2.5kg/m^3)和Y_5(3kg/m^3)。按照析因试验设计方法，共设置20个试验处理(T)：X_1Y_1，X_1Y_2，X_1Y_3，X_1Y_4，X_1Y_5；X_2Y_1，X_2Y_2，X_2Y_3，X_2Y_4，X_2Y_5；X_3Y_1，X_3Y_2，X_3Y_3，X_3Y_4，X_3Y_5；X_4Y_1，X_4Y_2，X_4Y_3，X_4Y_4，X_4Y_5。每个试验处理12株，重复3次，本试验施用的缓释肥在基质配比过程中，直接与之搅拌均匀。

2.3 数据调查与统计分析

2017年11月底，从每个试验处理中，随机抽取30株生长状况良好的白木香容器苗，测量苗高、地径、根长和根体积等生长指标，测定叶片、茎、根和全株生物量，使用Excel 2007对各试验白木香容器苗生长性状等调查数据进行处理，用SPSS 16.0软件进行方差分析和Duncan多重比较。

根据模糊数学中隶属函数的求值方法[19]，对白木香容器育苗方案进行优选。各项测定指标的隶属值计算公式：

$$隶属值=(N-N_{min})/(N_{max}-N_{min})$$

式中：N表示某处理指标值；N_{min}最小指标值；N_{max}最大指标值。

如果某个测定指标与幼苗生长等呈反向关系，可利用反隶属函数公式进行计算：

$$隶属值=1-(N-N_{min})/(N_{max}-N_{min})$$ [20]

3 结果与分析

3.1 基质配比与缓释肥施用量对白木香容器苗生长的影响

从表1可以看出，4种基质配比的白木香容器苗高、地径等生长表现各不相同，差异达到显著水平。随着基质中V(酸性砖红壤土)：V(松树皮)：V(碳化树皮)的体积比从8：1：1逐渐变至2：4：4时，白木香苗高和地径生长均呈先增加后又降低的趋势；其中，基质中V(酸性砖红壤土)：V(松树皮)：V(碳化树皮)的体积比6∶2∶2时，白木香苗高和地径生长量均为最高。由此说明，基质配比 V(酸性砖红壤土)∶V(松树皮)∶V(碳化树皮)=6∶2∶2时，白木香容器苗的生长表现最优。

由表1可知，轻基质配比过程中，施用缓释肥对白木香容器苗高、地径等生长指标的影响是有差异的。随着缓释肥量的增加，白木香苗高、地径生长测量指标均呈先升高后减缓的趋势，且差异达到显著水平。其中，施用缓释肥量为2.5kg/m^3，白木香容器苗高、地径生长和高径比均达到最高水平，分别为58.80cm、6.36mm和95.68。因此，从各生长测量指标的综合考虑，培育白木香容器苗较为适宜的缓释肥量为2.5kg/m^3。

3.2　基质配比与缓释肥施用量对白木香容器苗生物量的影响

4种轻基质配比对1年生白木香容器苗生物量的影响，有着显著的差异(表1)。白木香容器苗叶、茎、根和植株生物量，随着基质配比中酸性砖红壤比例的降低均呈增加后减少的趋势；其中，基质中 V(酸性砖红壤土)∶V(松树皮)∶V(碳化树皮)=6∶2∶2时，各器官生物量均为最大，与其他基质比例有着显著差异。由此说明，基质配比中 V(酸性砖红壤土)∶V(松树皮)∶V(碳化树皮)=6∶2∶2时，较为适宜白木香容器苗生长。

表1结果显示，白木香容器苗配比的轻基质中，施用不同缓释肥量对植株各器官生物量的影响有着明显的差异。白木香容器苗叶、茎、根和植株生物量，随着配比基质中缓释肥量呈现先增加后减小的趋势。其中，施用缓释肥量为2.5kg/m^3时，白木香容器苗各器官生物量积累均为最高。上述研究结果表明，施用缓释肥量为2.5kg/m^3时，白木香容器苗获得最好的生长效果。

基质配比和缓释肥用量的双因素方差分析发现，1年生白木香容器苗高、地径和植株生物量的双因素交互效应达到极显著水平(P<0.01)，与根体积的双因素效应也显著(P<0.05)。因此，白木香容器育苗过程中，用隶属函数方法综合评判各种组合的效果。

表1　不同基质配比和缓释肥用量的白木香容器苗生长表现

处理	苗高(cm)	地径(mm)	高/径	根体积(cm^3)	叶生物量(g)	茎生物(g)	根生物(g)	植株生物(g)
V(酸性砖红壤土)∶V(松树皮)∶V(碳化树皮)=8∶1∶1	(42.40±4.99)a	(4.85±0.26)a	(87.42±8.36)a	(0.46±0.13)a	(3.79±0.48)a	(2.39±0.71)a	(1.43±0.32)a	(7.62±0.57)a
V(酸性砖红壤土)∶V(松树皮)∶V(碳化树皮)=6∶2∶2	(50.43±5.74)b	(5.82±0.57)b	(86.65±6.40)a	(0.73±0.38)b	(4.78±0.71)c	(3.25±0.49)b	(1.93±0.38)b	(9.97±0.83)c
V(酸性砖红壤土)∶V(松树皮)∶V(碳化树皮)=4∶3∶3	(46.33±4.97)a	(5.65±0.51)b	(82.00±4.29)b	(0.68±0.27)b	(4.06±0.66)ab	(3.01±0.52)b	(1.47±0.22)b	(8.54±1.06)b
V(酸性砖红壤土)∶V(松树皮)∶V(碳化树皮)=2∶4∶4	(43.07±3.28)a	(5.28±0.41)ab	(81.57±5.28)b	(0.59±0.25)bc	(3.91±0.86)a	(2.72±0.46)a	(1.51±0.41)a	(8.14±0.81)a
缓释肥1.0kg/m^3	(47.33±3.55)a	(4.40±0.39)a	(90.60±4.93)a	(0.53±0.22a	(2.83±0.41)a	(2.58±0.46)a	(1.75±0.31)a	(7.16±1.31)a
缓释肥1.5kg/m^3	(42.67±3.53)a	(4.82±0.30)a	(86.34±6.01)a	(0.67±0.20)ab	(3.58±0.38)b	(2.91±0.92)b	(1.62±0.36)b	(8.12±0.64)b
缓释肥2.0kg/m^3	(48.27±3.74)b	(5.97±0.66)b	(84.36±7.86)a	(0.74±0.36)b	(3.67±0.64)bc	(3.06±0.89)bc	(2.12±0.67)b	(8.85±0.96)bc
缓释肥2.5kg/m^3	(58.80±4.92)c	(6.36±0.43)b	(95.68±9.14)b	(0.89±0.33)c	(4.65±0.43)bc	(3.41±1.13)bc	(2.38±0.42)b	(10.45±1.17)bc
缓释肥3.0kg/m^3	(51.70±4.73)b	(6.02±0.54)b	(85.61±8.81)a	(0.72±0.28)b	(3.86±2.65)c	(3.32±0.76)c	(1.86±0.39)b	(9.05±0.78)c

注：表中数据为平均值±标准差；同列不同小写字母表示不同处理下白木香容器苗的形态指标差异显著性(α=0.05)。

3.3 白木香1年生容器育苗方案优选

根据隶属函数的研究方法，充分考虑白木香容器苗各生长测量指标与互作效应，选取苗高、地径，叶片、茎、根和植株生物量以及根体积的隶属度平均值作为优选标准，为白木香容器苗计算出20种育苗方案（表2）。综合考虑轻基质材料和施用缓释肥的成本，培育健壮的白木香容器苗，最优方案应选取排名第1的育苗方案，即V(酸性砖红壤土)：V(松树皮)：V(碳化树皮)=6：2：2，缓释肥量为2.5kg/m³。

表2 基于指标隶属度值的白木香容器育苗方案优选

基质配比	缓施肥(kg/m³)	隶属度值							均值	排名
		苗高	地径	根体积	叶生物量	茎生物量	根生物量	植株生物量		
8：1：1	1	0.66	0.45	0.47	0.49	0.45	0.46	0.45	0.49a	20
8：1：1	1.5	0.47	0.57	0.57	0.43	0.66	0.59	0.54	0.55ab	19
8：1：1	2	0.56	0.53	0.61	0.61	0.63	0.51	0.60	0.58abc	18
8：1：1	2.5	0.60	0.57	0.69	0.80	0.69	0.69	0.76	0.68fgh	8
8：1：1	3	0.63	0.61	0.53	0.69	0.60	0.53	0.64	0.61bcd	16
6：2：2	1	0.53	0.64	0.61	0.62	0.52	0.60	0.60	0.59abc	17
6：2：2	1.5	0.49	0.63	0.56	0.72	0.70	0.62	0.61	0.62bcde	15
6：2：2	2	0.74	0.74	0.83	0.83	0.81	0.82	0.83	0.80kl	2
6：2：2	2.5	0.86	0.90	0.86	0.66	0.87	0.89	0.88	0.85l	1
6：2：2	3	0.81	0.85	0.65	0.66	0.65	0.67	0.66	0.71hi	6
4：3：3	1	0.53	0.34	0.70	0.71	0.70	0.71	0.71	0.63cde	14
4：3：3	1.5	0.58	0.50	0.70	0.57	0.72	0.71	0.72	0.64cdef	12
4：3：3	2	0.57	0.77	0.71	0.50	0.63	0.67	0.70	0.65defg	10
4：3：3	2.5	0.53	0.49	0.83	0.85	0.81	0.80	0.87	0.74ij	4
4：3：3	3	0.44	0.41	0.73	0.73	0.75	0.71	0.74	0.64cdef	13
2：4：4	1	0.45	0.64	0.70	0.74	0.75	0.74	0.74	0.68fgh	9
2：4：4	1.5	0.51	0.58	0.77	0.74	0.79	0.71	0.79	0.70h	7
2：4：4	2	0.70	0.71	0.83	0.76	0.71	0.69	0.72	0.73ij	5
2：4：4	2.5	0.76	0.61	0.82	0.79	0.73	0.74	0.76	0.75jk	3
2：4：4	3	0.50	0.51	0.76	0.71	0.71	0.69	0.67	0.65defg	11

注：表中隶属度均值中小写字母为多重比较结果，同列相同字母表示差异不显著($P>0.05$)，不同字母表示差异显著($P<0.05$)。

4 结论与讨论

本试验研究发现，随着基质配比中酸性砖红壤比例的降低，白木香苗高、地径生长和各器官生物量等指标均呈明显先增高再降低趋势；当基质中V(酸性砖红壤土)：V(松树皮)：V(碳化树皮)=6：2：2时，苗高、地径和植株生物量指标值为最大。在酸性砖红壤、松树皮和碳化树皮的基质配比中，其透气性随着酸性砖红壤比例的降低而升高；说明白木香幼苗期对土壤水分条件要求较高，需要透气性较适宜的育苗基质。较高的酸性砖红壤体积比V(酸性砖红壤土)：V(松树皮)：V(碳化树皮)=8：1：1，轻基质中饱和持水率大，透气性能较差，不利于白木香容器苗根系生长发育。若酸性砖红壤配比适量的松树皮和碳化树皮V(酸性砖红壤土)：V(松树皮)：V(碳化树皮)=6：2：2，能够有效地改善基质透气性，使得根系与基质形成比较紧密的根系土团，促进白木香容器苗根系生长。从考虑植物生物学特性方面的差异以及就地取材的便利性，不同类型树种对基质配比的要求亦有所不同。以往其他基质配比育苗试验表明，适宜杉木轻基质容器苗的基质配比为V(泥炭)：V(稻壳)：V(木屑)=7：1.5：1.5[21]，浙江楠(*Phoebe chekiangensis*)为V(泥炭)：V(黄心土)：V(草木灰)=6：3：1[22]，而基质配比为V(黄心土)：

V(泥炭)：V(蛭石)= 2：2：1时，能够显著地促进南方红豆杉(*Taxus wallichiana* var. *mairei*)容器苗生长[23]。本试验结果显示，白木香1年生容器苗生长受基质中酸性砖红壤、松树皮和碳化树皮体积比例改变的影响较大，尤其是苗高、地径和各器官生物量及其根系发育受酸性砖红壤体积比改变的变化较为敏感，即基质配比为V(酸性砖红壤土)：V(松树皮)：V(碳化树皮)= 6：2：2时，白木香1年生容器苗获得最优的生长效果。

施用缓释肥是基质容器育苗十分重要的技术措施，其既可根据苗木生长过程中养分需求而缓慢释放，又减少传统育苗多次施肥的繁琐环节，从而有效地降低育苗成本[10,24]。本试验研究中，随着缓释肥量的增加，白木香苗高、地径生长和各器官生物量均表现为先升高后减缓的趋势，当缓释肥施用量为2.5kg/m^{3}时，苗高、地径和各器官生物量指标值最大。这与塔姆岛金花茶[18]、赤皮青冈(*Cyclobalanopsis gilva*)[25]和欧洲云杉(*Picea abies*)[26]容器育苗过程中，施用缓释肥量的试验结果相似，他们同时指出，缓释肥施用过量时，容器苗生长量与生物量的积累反而有所降低，这与苗木地下部分的根系生长受到抑制有着密切关联。

本试验结果显示，白木香容器苗生长对基质配比和施加缓释肥均较为敏感，但根系生长发育的缓释肥效应不够明显。其他育苗试验研究中，赤皮青冈[25]、木荷(*Schima superba*)[27]和挪威云杉(*Picea excelsa*)[28]也得出相类似结论。这原因可能与白木香根系生长特征有着关联，其幼苗期主根发达，直接影响到根系生物量与根体积，而根长直接受侧根数量的影响，侧根生长不旺盛，根须稀疏致使施用缓释肥时，幼苗根长差异不明显。

白木香轻基质容器育苗过程中，基质配比、缓释肥量对幼苗生长、生物量和根系发育有着明显的交互效应。在充分考虑各因素，对白木香1年生容器苗生长的交互效应，利用隶属函数研究方法，选出白木香容器育苗的最优方案，即V(酸性砖红壤土)：V(松树皮)：V(碳化树皮)= 6：2：2时，缓释肥施用量2.5kg/m^3。

参考文献

[1]田耀华，原慧芳，倪书邦，等. 沉香属植物研究进展[J]. 热带亚热带植物学报，2009，17(1)：98-104.
[2]杨晓清. 土沉香幼苗氮、磷营养及与水分胁迫关系的研究[D]. 北京：中国林业科学研究院，2013.
[3]贾贤. 土沉香组织培养再生体系的建立[D]. 海口：海南大学，2014.
[4]张丽霞，兰芹英，李海涛，等. 白木香种子脱水耐性的发育变化及贮藏特性[J]. 植物分类与资源学报，2011，33(4)：458-464.
[5]贾晓红，周再知，梁坤南，等. 缺素对土沉香幼苗生长和叶片解剖结构的影响[J]. 福建农林大学学报(自然版)，2015，44(1)：40-45.
[6]王冉，李吉跃，张方秋，等. 不同施肥方法对马来沉香和土沉香苗期根系生长的影响[J]. 生态学报，2011，31(1)：98-106.
[7] EDWARDR，KRISTJANC，ANDREWP. Root characteristics and growth potential of container and bare-root seedlings of red oak(*Quercus rubra* L.) in Ontario，Canada [J]. New forests，2007，34(2)：163-176.
[8]王月海，房用，史少军，等. 平衡根系无纺布容器苗造林试验[J]. 东北林业大学学报，2008，36(1)：14-15.
[9]马常耕. 世界容器苗研究、生产现状和我国发展对策[J]. 世界林业研究，1994(5)：33-41.
[10]楚秀丽，孙晓梅，张守攻，等. 日本落叶松容器苗不同控释肥生长效应[J]. 林业科学研究，2012，25(6)：697-702.
[11]贾斌英，徐惠德，刘桂丰，等. 白桦容器育苗的适宜基质筛选[J]. 东北林业大学学报，2009，37(11)：64-67.
[12]邓华平，杨桂娟. 不同基质配方对金叶榆容器苗质量的影响[J]. 林业科学研究，2010，23(1)：138-142.
[13]韦小丽，朱忠荣，尹小阳，等. 湿地松轻基质容器苗育苗技术[J]. 南京林业大学学报(自然科学版)，2003，27(5)：55-58.
[14]徐玉梅，唐红燕，张建珠，等. 不同轻基质配方对思茅松容器育苗的影响[J]. 西北林学院学报，2015，30(6)：147-150.
[15]周新华，厉月桥，肖智勇，等. 基质配比、容器规格和缓释肥量对杉木容器育苗的影响[J]. 江西农业大学学报，2017，39(1)：72-81.
[16]程庆荣. 蔗渣和木屑作尾叶桉容器育苗基质的研究[J]. 华南农业大学学报，2002，23(2)：11-14.
[17]王艺，王秀花，吴小林，等. 缓释肥加载对浙江楠和闽楠容器苗生长和养分库构建的影响[J]. 林业科学，2013，49(12)：57-63.
[18]韦晓娟，梁晓静，李开祥，等. 基质配比和缓释肥量对塔姆岛金花茶容器苗质量的影响[J]. 中南林业科技大学学报，2017，37(2)：19-23.
[19]姚德生，何彦峰. 狭叶冬青不同基质配比容器育苗试验研究[J]. 西北林学院学报，2015，30(6)：156-160.
[20]王丁，张丽琴，薛建辉. 苗木抗旱性综合评价研究：以6种喀斯特造林树种苗木为例[J]. 中国农学通报，

2011，27(25)：5-12.
[21]周新华，厉月桥，肖智勇，等．基质配比、容器规格和缓释肥量对杉木容器育苗的影响[J]．江西农业大学学报，2017，39(1)：72-81.
[22]邱勇斌，乔卫阳，刘军，等．容器、基质和施肥对浙江楠容器大苗的影响[J]．东北林业大学学报，2016，44(9)：20-23.
[23]王金凤，陈卓梅，刘济祥，等．不同基质及缓释肥对南方红豆杉容器大苗生长的影响[J]．浙江林业科技，2016，36(2)：74-78.
[24]CORTINA J，VILAGROSA A，TRUBAT R. The role of nutrients for improving seedling quality in drylands[J]. New Forests，2013，44(5)：719-732.
[25]吴小林，张东北，楚秀丽，等．酸性砖皮青冈容器苗不同基质配比和缓释肥施用量的生长效应[J]．林业科学研究，2014，27(6)：794-800.
[26]HAWKINS B J，BURGESS D，MITCHELL A K. Growth and nutrient dynamics of western hemlock with conventional or exponential greenhouse fertilization and planting in different fertility conditions[J]. Canadianjournal forest research，2005，35(4)：1002-1016.
[27]马雪红，胡根长，冯建国，等．基质配比、缓释肥量和容器规格对木荷容器苗质量的影响[J]．林业科学研究，2010，23(4)：505-509.
[28]RUFAT J，DEJONG T M. Changes in fine root production and longevity in relation to water and nutrient availability in a Norway spruce stand in northern Sweden [J]. Tree Physiology，2001，21(14)：1057.

[原载：东北林业大学学报，2018，46(11)]

不同氮素水平对米老排苗期生长节律的影响

刘福妹[1,2]　韦菊玲[1]　庞圣江[1]　劳庆祥[1]　陈建全[1]　谌红辉[1]

([1]中国林业科学研究院热带林业实验中心，广西凭祥　532600；
[2]广西友谊关国家森林生态系统定位观测研究站，广西凭祥　532600)

摘　要　为明确米老排苗期生长的最佳施肥时期和施氮量，观测了1年生米老排苗期的生长节律，并拟合获得Logistic生长方程；分析了100mg/株、200mg/株、400mg/株、600mg/株和800mg/株5个氮素营养水平对米老排苗木5~10月份生长节律的影响。结果表明：①米老排苗高和地径的Logistic方程拟合效果显著(R_H^2=0.990，R_D^2=0.978，P<0.005)，生长节律表现出"慢—快—慢"的阶段性节律。②氮素营养对米老排苗高生长的影响要早于地径。③米老排的速生期始于6月中旬，最佳施氮量为200mg/株。

关键词　米老排；氮素营养；Logistic生长方程；生长节律

Effects of Nitrogen Fertilization on Growth Rhythm of *Mytilaria laosensis* Seedlings

LIU Fumei[1,2], WEI Juling[1], PANG Shengjiang[1],
LAO Qingxiang[1], CHEN Jianquan[1], CHEN Honghui[1]

([1]*The Experimental Centre of Tropical Forestry*, *Chinese Academy of Forestry*, *Pingxiang* 532600, *Guangxi*, *China*;
[2]*Guangxi Youyiguan Forest Ecosystem Research Station*, *Pingxiang*, 532600, *Guangxi*, *China*)

Abstract: To explicit the optimal fertilization period and nitrogen doses of *Mytilaria laosensis* seedlings, By using the non-linear models of Logistic, the growth rhythm of one-year-old seedlings were studied, as well as the growth performances (from May to October) under five nitrogen levels [100mg/seedling (N1), 200mg/seedling (N2), 400mg/seedling (N4), 600mg/seedling (N6) and 800mg/seedling (N8)]. The results showed that: ① The fitness of height and ground diameter Logistic equations were up to 0.990(R_H^2) and 0.978(R_D^2), and the growth rhythm showed significant stages of "slow-fast-slow". ②The nitrogen fertilization showed an earlier effect on height than ground diameter. ③the fast-growing stage of seedlings started in mid-june, and 200 mg N/seedling was the most appropriate dose for *M. laosensis*.

Key words: *Mytilaria laosensis*; nitrogen fertilization; logistic equation; growth rhythm

米老排(*Mytilaria laosensis* Lec.)为金缕梅科壳菜果属常绿乔木，具有生长迅速、干形通直、材质优良、木材结构细致、经久耐用等特性，同时是营建水源涵养生态公益林的首选树种，是我国南亚热带地区重点发展的珍优速生乡土树种之一，具有广阔的发展空间[1]。氮元素是植物体内许多重要有机化合物的构成成分，被称为生态系统中最为限制植物生长发育的营养元素[2-4]，也是世界农林业生产中消耗量和浪费量最大的元素之一，提高氮肥的有效性、降低成本一直是研究工作的热点。因此，本研究观测了1年生米老排苗期的生长节律，并拟合获得Logistic生长方程；同时，通过分析5个氮素营养水平对米老排苗木5~10月份生长节律的影响，明确适合米老排苗期生长的最佳施氮量，为米老排苗期施肥和施肥选择合理时期提供理论依据；进一步完善米老排的高效培育技术，充分发挥米老排的经济和社会效益。

1　材料和方法

1.1　试验材料及设计

试验地位于中国林业科学研究院热带林业实验中心的中国—东盟东南亚珍贵树种种苗繁育基地(广西凭祥，22.125°N，106.742°E)。该地属于南亚热带地区(为南亚热带季风气候)，海拔约为246m，年平均温度为22.12℃，7月最高气温37.04℃，1月最低气温-0.36℃，月均相对湿度81.15%，年平均降水量为1267mm，年平均日照时数为1657.37h，年内无霜期为335d(由广西友谊关国家森林生态系统定位观测研究站提供)。

参试的米老排苗木均为实生苗，种实于2013年11月份采于中国林业科学研究院热带林业实验中心白云实验场白云小试区36年生米老排母树林中，米老排种子净度是97%，平均千粒重是146.78g，种子平均长、宽、高依次为11.04cm、5.23cm和5.14cm。于2014年5月份育苗，基质为苗圃熟土(消毒)，速效氮、速效磷和速效钾含量分别是169.90mg/kg、0.45mg/kg和37.26mg/kg，全氮、全磷和全钾含量分别是1.08g/kg、0.29g/kg和6.04g/kg，土壤pH为5.3。

生长规律试验用到的实生苗是直接点播到规格为15cm高×15cm底径的圆柱形花盆容器中；不同氮素处理的苗木则是先撒播在沙床，在出苗后选取生长状态良好的幼苗，移植到相同规格花盆中(外裹一层透明塑料袋)。所有苗木在非处理期间均由工作人员按基地管理方法统一管护。

苗期生长规律试验设计：在播种后，等到米老排幼苗出土开始，选定30株作为1个大样本，每隔10d定期测定1次苗高和地径，持续到12月中旬(苗木基本停止生长)。

氮素营养对苗期生长节律影响试验：设有施氮总量依次为100mg/株、200mg/株、400mg/株、600mg/株和800mg/株5个营养水平处理和1个非施肥对照处理，每个处理30株幼苗，共180株苗；于2014年5月15日开始，每15d进行1次根部施肥，共10次，至2014年10月14日结束(不影响米老排的正常越冬)。具体施氮情况见表1。

表1　氮素水平用量

处理	施肥总量(mg/株)	5月15日至6月14日		6月15日至10月14日	
		施氮次数(次)	每次施氮量(mg/株)	施氮次数(次)	每次施氮量(mg/株)
CK	0	2	0	8	0
N1	100	2	6	8	11
N2	200	2	12	8	22
N4	400	2	24	8	44
N6	600	2	36	8	66
N8	800	2	44	8	89

注：表中施肥日期为每月15日和30日。采用等量施肥方式，但为避免过高浓度的氮素液体胁迫米老排幼苗生长，因此在试验的第一个月内，施氮量减半。氮肥为尿素[$CO(NH_2)_2$]，氮素有效成分比例为46%。

1.2　测定方法

生长规律试验设计：从2014年5月20日到11月30日，大概每10d测1次苗高，共测定20次；因幼苗茎较细嫩，易受伤，因此地径每30d测定1次，共测定7次。

氮素营养对苗期生长节律影响试验：从2014年5月到10月，每月30号测定各处理苗木的苗高和地径，共6次。

1.3　数据处理

利用Excel 2010和SPSS 16.0软件进行相应数据的整理和作图；方差分析和多重比较(Dancun方法)均在$P<0.05$水平下进行。

利用SPSS 16.0软件中非线性回归(NonlinearRegression)模型来对测定的米老排苗期数据进行Logistic方程拟合、方差分析和相应预测值的求解[5]。方程表达示为：

$$Y_{H/D}=k_{H/D}/(1+\alpha e\text{-}\beta t) \quad (1)$$

$$D=\ln\alpha/\beta \quad (2)$$

$$t_1=\ln\frac{\alpha}{3.73205}/\beta \quad (3)$$

$$t_2=\ln\frac{\alpha}{0.26795}/\beta \quad (4)$$

式中：t表示时间(d)，$Y_{H/D}$表示苗高/地径生长量(cm)；其中k、α、β分别为待定系数。$k_{H/D}$表示苗木苗高和地径生长极限，α和β通过三点法求出。D为生长拐点；t_1为由萌动到速生期的时间拐点，t_2为由速生期转为缓慢生长的分界点；二者为连日生长量变化速率最快的两个点；试验中根据t_1和t_2来划分米老排的苗高生长时期，两点之间的时间段称为速生期。

2　结果与分析

2.1　米老排苗高、地径生长节律

利用Logistic方程对1年苗木生长节律进行拟合，获得的拟合方程为：$Y_H=16.296/(1+15.533e^{-0.03t})$和

$Y_D=4.149/(1+8.162e^{-0.023t})$；在 $P<0.005$ 水平下 R_H^2 和 R_D^2 分别达到 0.990 和 0.978，说明利用 Logistic 方程对米老排苗木苗高和地径生长情况的拟合效果非常好。

从米老排苗期生长节律图可发现米老排生长的节律与"S"形曲线非常相符(图 1)，即：苗期高和地径的生长均呈现出明显阶段性的"慢—快—慢"节律。因此，根据拟合及计算结果可将米老排苗木的整个生长时期大致划分生长前期、速生期和生长后期 3 个阶段。由表 2 可知：苗木苗高和地径速生期的起止时间不一样，但持续的时间都非常长，分别为 87d 和 113d，增长了 9.34cm(苗高)和 2.37mm(地径)，达到整个苗期苗高和地径生长量的 58.11%和 59.25%；米老排苗高和地径的生长拐点均出现在 91d，说明尽管速生期开始的时间不同，但是 91d 后苗木的生长速率就会逐渐下降。

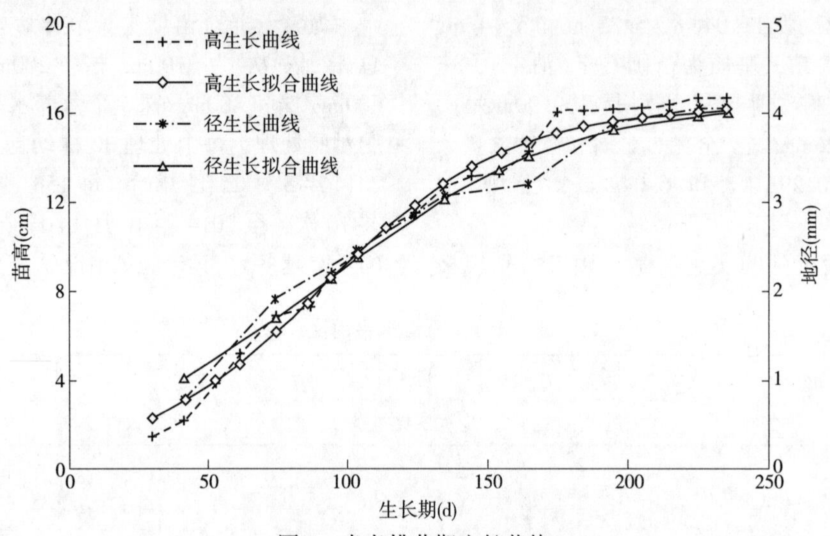

图 1 米老排苗期生长节律

表 2 米老排苗期生长拟合特征值

指标	t_1(d)	t_2(d)	D(d)	生长前期			速生期			生长后期		
				持续时间(d)	累积生长量(cm)	累积百分率(%)	持续时间(d)	累积生长量(cm)	累积百分率(%)	持续时间(d)	累积生长量(cm)	累积百分率(%)
苗高(cm)	48	135	91	48	3.48	21.65	87	9.34	58.11	100	3.25	20.24
地径(mm)	35	148	91	35	0.89	22.30	113	2.37	59.25	87	0.74	18.46

注：t_1、t_2 分别为速生期的起止时间；D 为生长拐点。

2.2 不同氮素水平对米老排苗高生长节律的影响

对不同月份获得的不同处理米老排的苗高数据进行方差分析，发现除 5 月份外，各氮素水平下苗木苗高生长的差异均达到显著水平(表 3)。说明不同浓度的氮素营养从 6 月开始对米老排的生长产生显著影响。

表 3 不同氮素水平下米老排苗木苗高生长节律方差分析

月份	自由度	平方和	均方	*F* 值
5	5	0.598	120	1.453
6	5	41.415	8.283	7.366*
7	5	161.373	32.275	4.864*
8	5	2630.636	526.127	49.08*
9	5	4678.696	935.739	193.621*
10	5	7097.14	1419.428	135.715*

注：表中"*"表示在 $P<0.05$ 水平下差异达到显著水平，下同。

从不同氮素水平下米老排苗期苗高生长节律图可以看出，不同处理的米老排苗木苗高生长均呈现出"S"形。进一步通过多重比较分析(图 2)发现：6 月，除 N1 处理外，其他施肥处理的苗高显著高于非施肥对照，即：苗高生长分化差异出现在施肥早期，说明氮素营养对米老排苗木苗高生长的促进作用见效非常快；到 7 月，只有 N8 处理与对照处理苗木苗高差异不显著，说明高浓度的氮素营养对促进米老排的生长无效；从 8~10 月 3 个月的生长期内，N8 处理始终与对照差异不显著，而其他处理的苗高均显著优于对照处理，尤其是 N2 处理的苗木，从 7 月开始，苗高生长就一直显著高于其他 4 个处理，说明 200mg/株的氮素营养对米老排苗高生长的促进作用最显著。

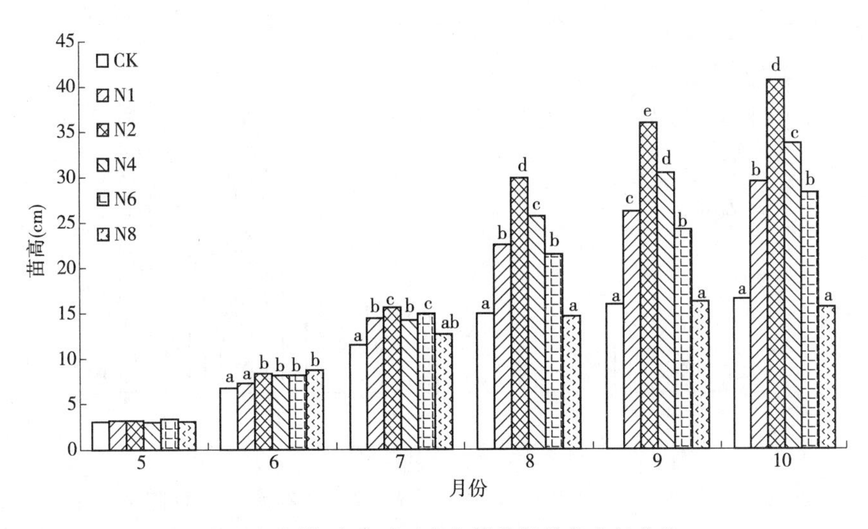

图 2 不同氮素水平下米老排苗期苗高生长节律

注：图中不同字母的处理表示在 $P<0.05$ 时差异性显著(Duncan's)，相同字母的处理之间差异不显著，下同。

纵观5个氮素水平下米老排苗期苗高生长节律图发现：100mg/株的氮素营养(N1)对苗高生长有促进作用，但是效果缓慢，直到7月与对照处理的差异才达到显著水平；200mg/株、400mg/株和600mg/株共3个水平氮素营养对米老排均有促进作用；其中，200v的氮素营养效果最佳，使10月N2处理的苗高达到40.5cm，较对照提高了145.72%；随后，随着氮素营养水平的升高促进效果逐渐减弱；当氮素营养提升到800v时，虽施用初期会对苗高生长有促进作用，使6月份米老排苗高较对照处理的苗木提高了30.3%，但10月份时N8处理的苗高生长最差，仅为15.63cm，甚至比对照还矮了0.87cm，说明800mg/株的氮素营养对米老排生长产生胁迫作用。

2.3 不同氮素水平对米老排地径生长节律的影响

从不同氮素水平下米老排苗木地径方差分析(表4)可知，米老排苗木地径生长差异在8~10月达到显著水平，说明不同氮素营养水平对米老排苗期初期地径生长的影响并不显著，肥效需要慢慢地积累，直到8月才开始产生显著的影响。

表4 不同氮素水平下米老排苗木地径方差分析($P<0.05$)

月份	自由度	平方和	均方	F值
5	5	0.058	0.012	0.231
6	5	1.081	0.216	1.422
7	5	2.023	0.405	1.614
8	5	26.863	5.373	18.098*
9	5	32.144	6.428	19.433*
10	5	48.526	9.705	50.647*

通过多重比较分析(图3)发现：N1处理、N2处理及N4处理的苗木地径生长均显著优于对照处理，说明100mg/株、200mg/株和400mg/株的氮素营养对地径生长有明显的促进作用，其中生长最佳的仍是N2处理，从8~10月，其地径分别是对照处理的1.53~1.65倍，说明200mg/株的氮素营养仍然是促进米老排苗期地径生长的最佳施用量；在所有施肥处理中，只有N8处理的苗木地径与对照处理在8月和9月差异不显著，甚至10月时，地径生长较对照处理(3.26mm)显著降低了17.79%，仅为2.68mm，说明800mg/株的氮素营养已经严重影响了米老排苗木的正常生长。

从不同氮素水平下米老排苗期地径生长节律(图3)可以看出，不同处理的米老排苗木地径生长也呈现"S"形；在8月、9月和10月中，N1、N2、N4、N6和N8共5个处理苗木地径生长会随着氮素水平的增加呈现先增加，在N2处理达到峰值然后再降低的现象，说明适宜的氮素营养能够促进米老排的生长，当营养过量时(介于200mg/株到800mg/株)，促进效用会逐渐被削弱，严重过量时(≥800mg/株)，则会抑制苗木的生长。

3 结论与讨论

(1)大量研究表明Logistic方程能够较好地描述植物的生长过程[6-9]，本研究也采用该方法对1年生米老排实生苗苗高和地径生长规律进行了拟合，发现拟合的效果达到显著水平($R_H^2=0.990$，$R_D^2=0.978$，$P<0.005$)。米老排苗期生长节律具有明显"慢—快—慢"的阶段性，这与覃敏等[9]的研究结果一致。

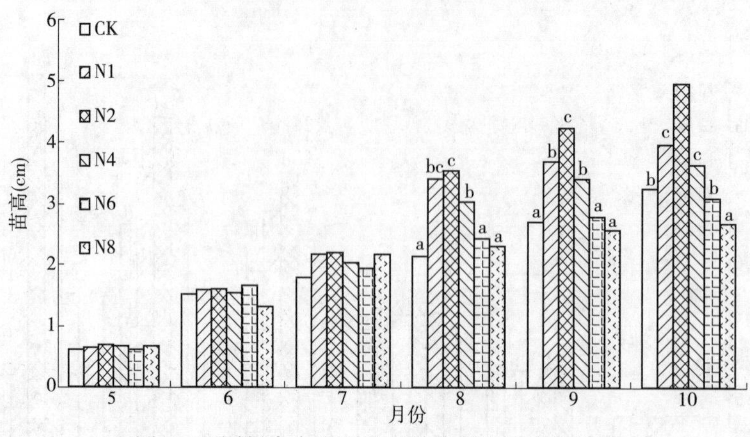

图3 不同氮素水平下米老排苗期地径生长节律

根据拟合获得的苗高和地径 logistic 生长方程将整个生长期大致划分为前期、速生期和长后期 3 个阶段。在生长前期米老排苗木积累了全年生长量的 22%，与钟荣[10]等研究结果相似；米老排的速生期始于 6 月中旬，在速生期米老排积累 60%全年生长量，但苗高生长与地径生长持续的时间不同，分别为 87d 和 113d，明显短于覃敏等拟合的不同种源米老排速生期 128~160d 的持续时间，这与 logistic 方程的精度相关，用来拟合的数据频率越高则精度越高，本研究中用来拟合苗高 logistic 生长方程的数据记录有 20 次，要显著多于后者(6 次)。

(2) 大量研究表明氮肥可以显著促进植物的生长[1,11-13]，本研究也有同样的结论；并且发现200mg/株是促进米老排幼苗快速增长的最佳施氮水平，也与前人[1,14]的研究结果一致。5 个氮素营养水平对米老排苗期生长的促进作用随着施用浓度的增加呈现先增加再下降的趋势，是由于氮素营养的过度施用会影响苗木的生长：当施用量轻度过量时(介于 200mg/株到 800mg/株)，会先逐渐削弱对生长的促进效用，严重过量时(≥800mg/株)，则会抑制苗木的生长，甚至在试验结束后 N8 处理的苗木出现大量的死亡。氮素营养对米老排苗高产生生长分化的影响始于 6 月份，与 logistic 生长方程拟合结果一致，因此在实际生产中用在此阶段宜增加适量外源氮肥的供给，促进米老排的生长壮苗。

参考文献

[1]刘福妹，劳庆祥，庞圣江，等．不同氮素水平对米老排苗期生长和叶绿素荧光特性的影响[J]．西北林学院学报，2018，33(1)：62-67.

[2]KIM T H，ITO H，HAYASHI K，et al. New antitumor sesquiter penoids from *Santalum album* of Indian origin [J]. Tetrahedron，2006，62：6981-6989.

[3]PAUL L R，CHAPMAN B K，CHANWAY C P. Nitrogen fixation associated with Suillus tomentosus tuberculate ectomy corrhizac on *Pinus contora* var. *latifolia* [J]. Ann Bot，2007，99：1101-1109.

[4]周志强，彭英丽，孙铭隆，等．不同氮素水平对濒危植物黄檗幼苗光合荧光特性的影响[J]．北京林业大学学报，2015，37(12)：17-23.

[5]贾乃光．数理统计[M]．北京：中国林业出版社，1999：176.

[6]赖文胜．长序榆一年生播种苗的年生长规律[J]．南京林业大学学报(自然科学版)，2001，25(4)：57-60.

[7]杨耀仙，卞尧荣，姚小华．林木苗期生长灰色模型的选择[J]．林业科学研究，1991，4(2)：211-216.

[8]程龙霞，施曼，祝遵凌．欧洲鹅耳枥一年生播种苗年生长动态探究[J]．中国野生植物资源，2015，34(5)：46-50.

[9]覃敏，尹光天，杨锦昌，等．米老排不同种源苗期生长规律研究[J]．中南林业科技大学学报，2017，37(1)：53-57.

[10]钟荣，郭赋英，徐志文，等．毛桃木莲 1 年生播种苗的年生长规律及育苗技术研究[J]．江西林业科技，2008(4)：18-20，23.

[11]刘福妹，李天芳，姜静，等．白桦最佳施肥配方的筛选及其各元素的作用分析[J]．北京林业大学学报，2012，34(2)：57-60.

[12]刘福妹，姜静，刘桂丰．施肥对白桦树生长及开花结实的影响[J]．西北林学院学报，2015，30(2)：116-120，195.

[13]张烨，覃子海，肖玉菲，等．配方施肥对澳洲茶树幼林生长性状的影响研究[J]．西部林业科学，2018，47(01)：29-33.

[14]闫彩霞，杨锦昌，尹光天，等．供氮方式及水平对米老排苗期生长动态的影响[J]．东北林业大学学报，2015，43(5)：11-16.

[原载：西部林业科学，2018，47(06)]

基于表型性状初步构建格木核心种质

李洪果[1] 陈达镇[2] 劳庆祥[1] 马 跃[1] 陈建全[1] 韦菊玲[1]

([1]中国林业科学研究院热带林业实验中心，广西凭祥 532600；
[2]梧州市林政稽查支队，广西梧州 543002)

摘 要 为构建格木核心种质，去除收集材料中的冗余样品，为格木种质资源的保存、研究和利用提供依据。本研究将114份格木种质按地理区域分组，通过测量果荚和种子的13个表型性状，在系统聚类和优先取样法的基础上，利用遗传多样性法、比例法和对数法3种方法，设定10%、20%和30%3种取样比例，共产生7种取样策略。采用极差符合率(CR)、最小值变化率(CR_{MIN})、最大值变化率(CR_{MAX})、变异系数变化率(VR)、均值差异百分率(MD)、方差差异百分率(VD)、平均值变化率(CR_{MEA})和平均表型多样性指数(M_I)8个参数评价7种策略构建的核心种质，以选出参数最优的核心种质。通过比较核心种质与原始种质的表型多样性指数、符合率、主成分和样品分布图，验证所构建核心种质的代表性。结果表明：①采用比例法在20%的取样比例时形成的核心种质参数最优。②在8个参数分别为100.00%、100.00%、100.00%、138.85%、0.00%、61.54%、99.29%和2.68时，核心种质的最终取样比例为21.05%。③核心种质与原始种质在13个表型性状上的多样性指数经t检验，差异均不显著；均值符合率在97.39%~99.98%，极大值、极小值符合率为100%，多样性指数符合率在90.34%~99.56%。原始种质和核心种质4个主成分的累积贡献率分别为87.382%和88.206%；两者的样品分布图具有相似的分布结构。以上表明获得的24份核心材料较好地代表了原始种质的表型性状变异特征。

关键词 格木；核心种质；表型性状；种质资源

Preliminary Construction of Core Collection of *Erythrophleum fordii* Based on Phenotypic Traits

LI Honguo[1], CHEN Dazhen[2], LAO Qingxiang[1], MA yue[1], CHEN Jianquan[1], WEI Juling[1]

([1]*Experimental Center of Tropical Forestry*, *Chinese Academy of Forestry*, *Pingxiang* 532600, *Guangxi*, *China*;
[2]*Forestry Administration Inspection Detachment of Wuzhou*, *Wuzhou* 543002, *Guangxi*, *China*)

Abstract: Corecollectionof *Erythrophleum fordii* was constructed to retrench germplasms had collected, which will provide a theoretical basis for preservation, utilizing and studying germplasm of *Erythrophleum fordii*. 114 germplasms of *E. fordii* grouping by geographical area were used as test materials. Through the measurement of 13 related phenotypic traits of pods and seeds of *E. fordii*, based on system cluster and preferred method, applied genetic diversity, proportional and logarithmic method, 10%, 20% and 30% sample ratio, total seven sampling strategies were formed. The sampling strategies were tested by the value of CR, CR_{MIN}, CR_{MAX}, VR, MD, VD, CR_{MEA}, M_I. By comparing the phenotypic diversity index, coincidence rate and principal component and distribution map of the core collection and total collection, the core collection was tested by representativeness. The results wereas follows: ①The optimal strategy is: proportional method and 20% sampling ratio. ②The values above are 100.00%, 100.00%, 100.00%, 138.85%, 0.00%, 61.54%, 99.29% and 2.68, respectively. The final sampling ratio of core collection is 21.05%. ③The diversity index of 13 phenotypic traits of core collection and total collection is not significant by t test. The average coincidence rate between the core collection and total collection on the 13 indexes is between 97.39%~99.98%, the maximum and minimum coincidence rate is 100%, and the diversity index coincidence rate is between 90.34%~99.56%. There are 4 principal components of core collection and total collection, the accumulative contribution rates are 87.382% and 88.206%, respectively. The sampling distribution map of core collection and total collection is similar. The obtained core collection of 24 *E. fordii* could satisfactorily represent the phenotypic variation characteristics of 114 *E. fordii*germplasms.

Key words: *Erythrophleum fordii*; core collection; phenotypic traits; germplasm

格木(*Erythrophleum fordii*)为国家二级保护植物，是中国热带亚热带地区珍贵的用材和绿化树种，同时也具有较高的药用价值，木材结构致密坚实，极耐腐，有"铁木"之称，是优良的工业、家具和工艺品用材。长期以来，由于筑路修桥、城市发展等因素造成的过度砍伐、大面积毁林等，导致格木天然林急剧减少，部分地区已经消失[1]。遗传多样性广泛存在于物种及其基因组之中[2]。作为物种应对环境适应性的潜力，其在制定育种策略和濒危物种保存等各个方面有着积极意义[3,4]。表型性状变异的丰富性和均匀程度是遗传多样性的直接体现，利用表型性状的差异来检测种群的遗传多样性是长期以来最简便易行的方法[5,6]。

截至2011年，全世界已异地保存740万份植物资源[7]。不同于作物、草本和花卉等短周期植物，林木种质资源的收集和保存需要更广的地域和更长的观测周期，这就需要有针对性地从该树种众多种质资源中抽检关键样品收集保存，以节约成本和工作量。核心种质即是用最少的样本数最大程度地代表原始种质遗传变异的种质材料集，其作为种质资源管理的有效途径被广泛应用。利用植物的表型性状构建核心种质已在杜仲(*Eucommiaulmoides* Oliver)、新疆野杏(*Armeniaca vulgaris* Lam.)、狗牙根(*Cynodondactylon* Linn.)、水稻(*Oryza sativa* Linn.)、小麦(*Triticum aestivum* Linn.)等林木与作物上得到应用[6,8-11]。持续的观测发现，格木的叶、果荚和种子等器官均存在丰富的表型变异，但对于同一单株而言，格木果荚及种子的表型性状在达到结实稳定期后(20年)更为稳定[11]，较叶片更适合用于核心种质构建。

本研究通过测量114个进入结实稳定期的格木单株的13个果荚及种子表型性状，初步构建格木的核心种质，为濒危植物格木的种质资源收集保存和育种群体建立提供科学依据。

1　结果与分析

1.1　格木核心种质的初步建立

计算8个格木种群的遗传多样性，并根据各个种群的遗传多样性占比和样本数占比，在种质分组、系统聚类和优先取样法的基础上，按预设的取样比例形成7个不同的核心种质子集。

1.1.1　各分组样本数的确定

系统聚类取样策略遵循资源总体的分布结构，不易遗漏关键的表型；优先取样法有利于保存有特异表型的种质材料。在系统聚类取样策略和优先取样法的条件下，设定了7种取样策略的样本数及最终取样比例(表1)。结果显示，7种取样策略下8个种群的Shannon-Weaver指数为0.9209~1.8856，多样性占比为7.81%~15.98%；样本数为13~35，取样比例为11.4%~30.7%。8个种群中，容县、龙圩等多样性指数较高的群体是格木种质资源收集保存的关键区域。此外，种群遗传多样性的高低通常与种群的样本数呈正相关关系，但对比种群样本数占比和遗传多样性占比发现，苍梧、北流、陆川种群的多样性占比远大于其样本数占比，说明这些种群中样本包含的遗传变异可能更为丰富和均匀，在种质资源收集、遴选过程中也应加以重视。

表1　7种取样策略下8个种群的抽取样本数及其比例

种群及代码	多样性指数	样本数	遗传多样性法				比例法				对数法
			多样性占比(%)	10%	20%	30%	样本数占比(%)	10%	20%	30%	
北流(P1)	0.9230	4	7.82	1	2	3	3.51	1	1	1	1
博白(P2)	1.6450	22	13.95	2	3	5	19.30	2	4	7	3
苍梧(P3)	1.4477	8	12.27	1	3	4	7.02	1	2	2	2
合浦P4)	1.6059	21	13.61	2	3	5	18.42	2	4	6	3
龙吁(P5)	1.7473	15	14.81	2	3	5	13.16	2	3	5	3
陆川(P6)	0.9209	3	7.81	1	2	3	2.63	1	1	1	1
浦北(P7)	1.6210	13	13.74	2	3	5	11.40	1	3	4	3
容县(P8)	1.8856	28	15.98	2	4	5	24.56	3	6	8	3
最终取样			—	13	23	35	—	13	24	34	19
最终取样比例			—	11.40	20.18	30.70	—	11.40	21.05	29.82	16.67

1.1.2 取样策略的选定

根据王建成等(2007)对核心种质评价参数有效性的排序，即 $CR > M_I > CR_{MAX}$、$CR_{MIN} > VR > MD > VD > CR_{MEA}$。首先，观察极差符合率(CR)和多样性指数($M_I$)，其中G3、P2和P3三种策略的CR和$M_I$参数明显高于其他策略；其次，在G3、P2和P3策略的最大值变化率(CR_{MAX})、最小值变化率(CR_{MIN})、均值差异百分率(MD)和平均值变化率(CR_{MEA})几乎相等的情况下，P2策略的变异系数变化率(VR)和方差差异百分率(VD)参数最高，分别为：138.85%和61.54%(表2)。因此，选择P2作为构建格木核心种质的最佳策略。该策略下的样本数为24，取样比例为：21.05%。

表2 7种取样策略下核心种质子集的遗传参数比较

核心种质	评价参数							
	CR(%)	CR_{MAX}(%)	CR_{MIN}(%)	VR(%)	MD(%)	VD(%)	CR_{MEA}(%)	M_I
G1	91.77	97.97	101.11	156.40	7.69	61.54	97.48	2.50
G2	95.09	98.70	100.30	132.82	0.00	38.46	99.54	2.66
G3	99.72	100.00	100.27	124.84	0.00	0.00	99.43	2.71
P1	92.20	97.97	100.94	153.67	7.69	46.15	96.56	2.55
P2	100.00	100.00	100.00	138.85	0.00	61.54	99.29	2.68
P3	100.00	100.00	100.00	127.45	0.00	7.69	99.49	2.72
L	94.08	98.70	100.94	141.70	0.00	61.54	99.09	2.61

注：CR：极差符合率；CR_{MAX}：最大值变化率；CR_{MIN}：最小值变化率；VR：变异系数变化率；MD：均值差异百分率；VD：方差差异百分率；CR_{MEA}：平均值变化率；M_I：平均表型多样性指数；G1：遗传多样性法+10%取样比例；G2：遗传多样性法+20%取样比例；G3：遗传多样性法+30%取样比例；P1：比例法+10%取样比例；P2：比例法+20%取样比例；P3：比例法+30%取样比例；L：对数法。

1.2 核心种质的验证

1.2.1 核心种质与原始种质的表型多样性指数 *t* 检验

核心种质与原始种质13个表型性状的多样性指数经 *t* 检验，差异均不显著(表3)。另外，从两个群体均值差的95%置信区间看，所有表型性状的区间跨0(下限为负值，上限为正值)，这也说明两个群体的表型性状均值无显著差异。以上说明核心种质对原始种质表型多样性的保持情况较好，代表性较高。

表3 核心种质与原始种质多样性指数的 *t* 检验

表型性状		代码	*t* 值	差分95%的置信区间	
				下限	上限
果荚	果荚长(mm)	PL	0.623	−0.093	0.058
	果荚指数	PSI	0.948	−0.095	0.089
	果荚宽(mm)	PW	0.684	−0.088	0.059
	果荚面积(mm^2)	PA	0.580	−0.093	0.055
	果荚周长(mm)	PP	0.783	−0.082	0.063
	果荚厚(mm)	PT	0.624	−0.088	0.144
	每荚种子数	PNS	0.978	−0.103	0.101
种子	种子长(mm)	SL	0.943	−0.101	0.095
	种形指数	SSI	0.833	−0.110	0.135
	种子宽(mm)	SW	0.909	−0.099	0.088
	种子面积(mm^2)	SA	0.820	−0.098	0.079
	种子周长(mm)	SP	0.784	−0.094	0.123
	种子厚(mm)	ST	0.850	−0.083	0.100

1.2.2　核心种质与原始种质的符合率检验

核心种质与原始种质的均值符合率在97.31%~99.98%，极大值和极小值的符合率均为100%，多样性指数的符合率在90.34%~99.56%（表4），以上说明构建的格木核心种质具有较强的代表性。核心种质与原始种质的极大值和极小值符合率完全相同，也说明核心种质对特异种质的保存效果较好。

1.2.3　核心种质与原始种质的主成分分析

原始种质和核心种质均有4个特征根大于1的主成分，累积贡献率分别为87.382%和88.206%（表5），说明两者均能解释原始种质80%以上的遗传信息。另外，核心种质的方差贡献率略高于原始种质，也说明构建的核心种质能够排除部分冗余样本，使累积贡献率有所提高。

表4　核心种质与原始种质表型性状的符合率检验

性状	均值			极大值			极小值			多样性指数		
	核心种质	原始种质	符合率（%）	核心种质	原始种质	符合率（%）	核心种质	原始种质	符合率（%）	核心种质	原始种质	符合率（%）
PL（mm）	147.80	150.29	98.35	185.66	185.66	100.00	115.26	115.26	100.00	2.11	2.05	97.08
PSI	3.90	3.93	99.29	4.93	4.93	100.00	3.01	3.01	100.00	2.03	2.05	99.09
PW（mm）	38.44	38.54	99.74	47.76	47.76	100.00	29.05	29.05	100.00	2.25	2.08	92.00
PA（mm^2）	4453.08	4534.14	98.21	6479.23	6479.23	100.00	2735.88	2735.88	100.00	2.22	2.07	92.46
PP（mm）	367.95	372.31	98.83	453.11	453.11	100.00	287.19	287.19	100.00	2.09	2.10	99.56
PT（mm）	8.80	9.04	97.39	13.54	13.54	100.00	6.92	6.92	100.00	1.83	1.98	92.49
PNS	7.27	7.45	97.58	10.10	10.10	100.00	3.60	3.60	100.00	2.07	1.99	95.89
SL（mm）	16.54	16.43	99.34	19.74	19.74	100.00	13.64	13.64	100.00	1.97	2.01	98.00
SSI	1.24	1.23	99.22	1.56	1.56	100.00	1.12	1.12	100.00	2.00	1.91	95.17
SW（mm）	13.39	13.39	99.98	17.20	17.20	100.00	10.90	10.90	100.00	2.22	2.03	90.34
SA（mm^2）	172.73	169.94	98.36	263.59	263.59	100.00	119.33	119.33	100.00	2.16	2.01	92.59
SP（mm）	51.58	51.25	99.34	64.94	64.94	100.00	41.54	41.54	100.00	2.06	2.01	97.71
ST（mm）	5.67	5.81	97.61	7.07	7.07	100.00	4.72	4.72	100.00	1.81	2.08	86.92

注：性状代码见表3。

表5　核心种质与原始种质的主成分分析

成分	原始种质的初始特征值			提取平方和载入			核心种质的初始特征值			提取平方和载入		
	合计	方差（%）	累积（%）	合计	方差（%）	累积（%）	合计1	方差（%）	累积（%）	合计	方差（%）	累积（%）
1	5.834	44.880	44.880	5.834	44.880	44.880	5.994	46.110	46.110	5.994	46.110	46.110
2	2.496	19.198	64.078	2.496	19.198	64.078	2.385	18.344	64.454	2.385	18.344	64.454
3	1.875	14.424	78.502	1.875	14.424	78.502	1.942	14.941	79.396	1.942	14.941	79.396
4	1.154	8.880	87.382	1.154	8.880	87.382	1.145	8.810	88.206	1.145	8.810	88.206
5	0.582	4.481	91.863	—	—	—	0.687	5.283	93.489	—	—	—
6	0.536	4.125	95.988	—	—	—	0.473	3.635	97.124	—	—	—
7	0.401	3.083	99.071	—	—	—	0.316	2.433	99.557	—	—	—
8	0.061	0.468	99.539	—	—	—	0.026	0.200	99.757	—	—	—
9	0.027	0.209	99.749	—	—	—	0.016	0.124	99.881	—	—	—
10	0.016	0.126	99.875	—	—	—	0.008	0.062	99.943	—	—	—
11	0.011	0.086	99.961	—	—	—	0.005	0.038	99.981	—	—	—
12	0.004	0.030	99.991	—	—	—	0.002	0.015	99.996	—	—	—
13	0.001	0.009	100.000	—	—	—	0.001	0.004	100.000	—	—	—

1.2.4 核心种质与原始种质的样品分布

基于主成分的样品分布图可以近似地反映群体的遗传结构，从核心种质和原始种质基于第1、2主成分的样品分布图可知(图1)，原始种质的样品分布相对较为集中，存在一定的遗传相似性和遗传冗余，而核心种质中样品分布的重叠程度得到了显著降低，表明建立的核心种质去除了原始种质中的部分遗传冗余[6]。另外，核心种质和原始种质的样品分布特征也较为类似，进一步验证了所构建核心种质的有效性。

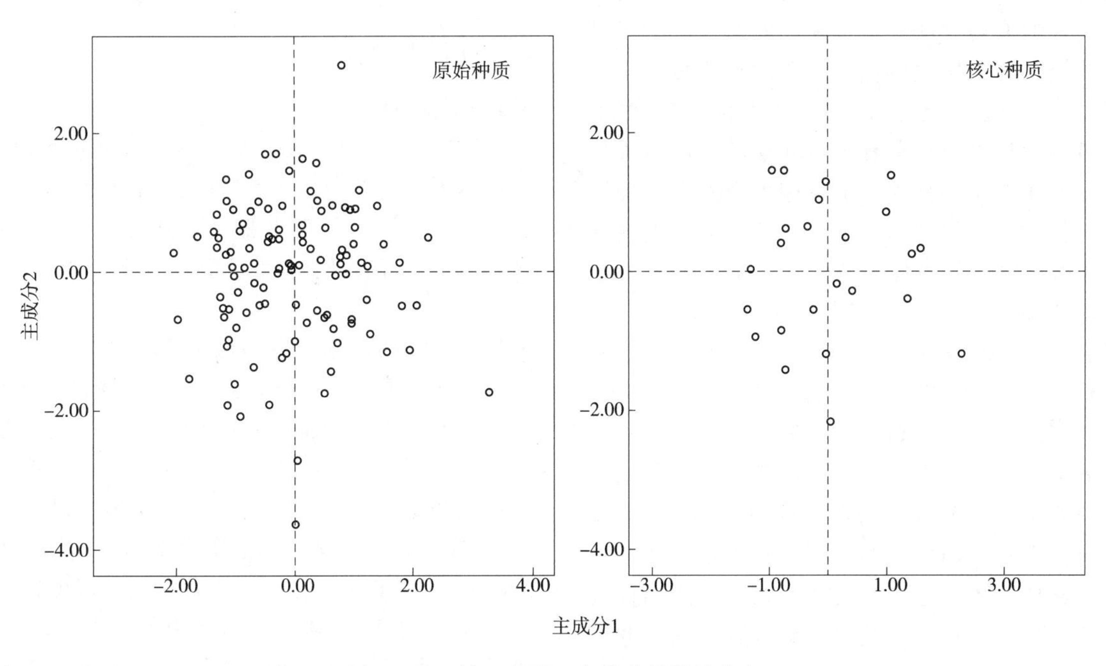

图1 基于第1和第2主成分的样品分布

2 讨论

2.1 种质分组

分组的目的是为了保证核心种质取样的代表性和能反映不同条件下的遗传多样性差异，通常对收集的原始材料按不同的地理来源分组(狗牙根、灰楸)、按光温生态区分组(野生大豆)、按农家品种、育成品种和国外品种分组(谷子)、按早熟、中熟、晚熟分组(大白菜)等[10,12-15]。分组的另一个目的是尽可能消除环境因素的影响。一般而言，基于表型性状构建核心种质的植物材料，往往需要其外在的环境因子尽可能一致。然而，不同于农作物、草本和花卉等，大多数林木，尤其是一些广布种、珍贵用材树种等，通常需要很长的生长周期才能观测到结实等性状，加之因树体高大和引种适应性问题，难以被广泛收集保存于相同的立地条件下。在此背景下，为了加快特异种质的筛选和保存利用，将种质按地理区域分组，把近似一致区域小环境(年均温、降水量、伴生植物、土壤理化性质等)看作相同的立地条件，不失为一种简捷有效的方法。该方法的本质是基于地理分布构建分组核心种质，再将各分组的核心种质进行叠加，最后形成一个汇总的核心样本群体，这在新疆野杏、小麦、黍稷、大白菜等的核心种质构建中均有不同程度的应用[8-11]。

长期观察发现，格木的叶、果荚和种子等器官均存在丰富的表型变异，但对于同一单株而言，格木果荚及种子的表型性状在达到结实稳定期后更为稳定，较叶片更适合用于核心种质构建。格木实生苗通常在15年左右才开始结实，结实前几年种子质量不高，发芽率等指标相对较低，可能伴随有表型指标不稳定的情况，20年后进入结实稳定期，果荚及种子的各项指标相对稳定，未见有发生变化的现象[10,12]。本研究在2017年广西壮族自治区名木古树调查的基础上选取实验材料，树龄最大为1300年以上，最小为37年，果荚及种子表型性状均已进入稳定期，适用于核心种质构建研究。综合前人的方法和格木表型变异的特点，本研究选择按地理区域分组(县级行政区划)，基于格木果荚和种子的表型性状构建格木的核心种质，取得了较好的构建效果。

2.2 格木核心种质构建方法和评价参数的选择

核心种质构建中，系统聚类取样通常优于随机取样，因为它遵循资源总体的分布结构，即使很小的取样比例，也不会发生主要性状类型的遗漏[17]，同时，聚类取样的抽样方法可用均值的 t 检验和符合率检验等方法评价[18]。Yonezawa 和 Morishima(1995)在分组的基础上选取核心材料时，通过比较比例法、对数法和遗传多样性法，认为遗传多样性法最能代表原始种质的多样性[19]。Brown(1995)认为如果各分组的遗传多样性未知[120]，对数法考虑了各分组的大小，是较好的取样方法。本研究在前人方法的基础上加以改进，即采用比例法、对数法和遗传多样性法 3 种方法，在系统聚类和优先取样法的基础上抽取样本，最终达到设定的取样比例，在遵从原始种质资源总体分布结构的同时，最大限度地避免了关键种质特殊性状的遗漏。基于前述方法，在比例法和 20%的取样比例时的取样策略，其 8 个评价参数最优，为最佳取样策略，经验证其代表性较好。

代表性是核心种质最重要的性质，而评价核心种质的代表性，须借助一系列的评价参数，只有确定了核心种质的评价参数，对其取样方法、取样比例等的研究才有判定标准。一般而言，描述数量性状的特征数为均值和变异系数，对于物种这类数量较大的有限总体，均值作为数据的代表值，处于核心种质观察值的中心位置，可作为核心种质的代表值与原始种质群体比较，从而明确两者之间的差异情况。因此，性状均值是一个极其重要的参数[17]。多样性指数常用于离散型指标多样性的符合率检验，将数量性状离散化，考察核心种质与原始种质的表型频率分布特点，从而判断两者的分布结构是否一致，是核心种质构建中最常用的方法[17]。另外，参考水稻核心种质构建中对各个评价参数有效性、稳定性、敏感性研究结果，最终选择了文中的 8 个评价参数，结合胡晋等的核心种质代表性评价标准[9-11]，用于判定核心种质的代表性，得到了较为可靠的研究结论。

2.3 格木核心种质的验证体系

核心种质的代表性经过检验后，常采用一些方法进一步验证所构建核心种质的代表性。表型多样性指数的 t 检验、符合率检验、主成分分析和基于主成分的样品分布图都是应用较广的方法[9-11]。也有学者用性状间的相关系数来衡量推断原始种质中控制性状间相关性的内在共适应基因系统是否被核心种质很好的保留下来[10-20-22]。然而多数情况下，占原始种质样本数 5%~30%的核心种质，往往因样本数显著减少造成统计学意义上的计算偏差、性状间的假相关、观测性状较多冲淡了原本存在强相关关系的性状等因素，导致性状间相关性的丢失。不过在狗牙根、新疆野苹果和秘鲁藜等植物中有性状间相关性基本保持的情况[10-21-22]。因此，本研究最终选择多样性指数的 t 检验、符合率检验、主成分分析和基于主成分的样品分布图 4 种方法验证所构建核心种质的有效性。结果表明，所构建的格木核心种质具备代表性、有效性和实用性等特征。

3 材料与方法

3.1 实验材料

2017 年 10 月、11 月从中国中心分布区(广西)8 个格木天然种群 114 个格木单株中，用高枝剪、登高板等工具从树上采集成熟的果荚及种子，择发育正常、无病虫害的果荚及种子带回实验室测量。

3.2 表型指标、多样性参数及其算法

测量果荚的 7 个表型性状，每个性状测量 10 次，取平均值作为实测值，其中果荚长、果荚指数、果荚宽、果荚面积、果荚周长 5 个性状用万深叶面积分析系统测量，测量结果精确至 0.0001；每荚种子数人工数取；果荚厚度用游标卡尺测量，精确到 0.01mm。测量种子的 6 个表型性状，每个性状测量 25 次取平均值作为实测值，其中种子长、种形指数、种子宽、种子面积、种子周长 5 个性状用万深叶面积分析系统测量，测量结果精确至 0.0001，种子厚度用游标卡尺测量，精确到 0.01mm。

表型性状的均值及其 t 检验在 SPSS19.0 软件中完成。变异系数(CV)计算公式：

$$CV=\frac{\sigma}{\bar{x}}\times 100\%$$

式中：σ 和 $\bar{x}$ 分别为性状的标准差和均值；

多样性指数(H)的计算公式：

$$H=-\sum P_i ln(P_i),\ P_i=n_i/N$$

式中 P_i 表示某表型性状第 i 个代码出现的频率；n_i 为某表型性状出现的次数；N 为样本数；在计算多样性指数前先将测量数据分成 10 级，转换为质量性状，$<\bar{x}-2\sigma$ 为第 1 级，$\geq\bar{x}+2\sigma$ 为第 10 级，每级相差 0.5σ(郑轶琦等，2014)。符合率(C_{OR})的计算公式：

$$C_{OR}=(1-|X_C-X_T|)/X_T\times 100\%$$

式中：X_C为核心种质某性状均值；X_T为相应原始种质某性状的均值。

3.3 核心种质的构建、评价及验证方法

3.3.1 核心种质的构建方法

(1)取样方法。在分组的基础上，采用优先取样策略，从聚类图最低分支水平的各组遗传材料中优

先选取有极值表型性状的株系进入下一轮聚类，如果处于最低分支水平的各遗传材料均无极值性状则随机选择一个进入下一轮聚类，只有一个遗传材料的组则直接进入下一轮聚类[11]，保证每个分组群体在每种取样策略下至少有1份材料入选。

(2)构建方法。在分组和优先取样法的基础上，比较3种取样策略下构建的格木核心种质：遗传多样性法(G法)、比例法(P法)和对数法(L法)。其中遗传多样性法和比例法分别设定10%、20%和30%的取样比例，对数法为单一样本集，总计7种取样策略。遗传多样性法：以各分组群体的遗传多样性占原始种质群体遗传多样性的比例，从各分组中选取相应的样本数，遗传多样性高的群体抽取相应多的样本。反之，抽取相应少的样本，即计算群体水平上的表型多样性指数，群体多样性指数之和即为总的遗传多样性，按各群体所占多样性的比例确定各群体的取样数，每个群体至少有一个样本入选核心种质库，之后按比例增加取样数。比例法：按各分组样本数占总体样本数的多少抽取样本数，每个分组至少有一个样本入选，之后按比例增加取样数。对数法：按各分组样本数的自然对数取样，每个分组至少有一个样本入选。以上方法均在分组和聚类的基础上，采用优先取样策略抽取样本数。

3.3.2 核心种质的评价

采用核心种质研究中11个评价参数中的8个数量性状参数评价筛选构建的7个格木核心种质，主要为变异系数变化率(VR)、极差符合率(CR)、均值差异百分率(MD)、方差差异百分率(VD)、平均值变化率(CR_{MEA})、最小值变化率(CR_{MIN})和最大值变化率(CR_{MAX})、平均表型多样性指数(M_I)8个参数评价筛选核心种质，各参数的计算公式见文献(王建成等，2007)。评价标准：均值差异百分率(MD)<20%，且极差符合率(CR)>80%时，可认为核心种质具有较好的代表性，且MD参数值越小，其他参数值越大，则核心种质的代表性越强[10,19-22]。

3.3.3 核心种质的验证

通过比较核心种质与原始种质多样性指数的t检验、符合率检验、主成分分析和基于主成分的样品分布图验证所构建核心种质的有效性。

参考文献

[1]黄忠良，郭贵仲，张祝平．渐危植物格木的濒危机制及其繁殖特性的研究[J]，生态学报，1997，6(6)：671-676.

[2]SOARES M P，WEISS G. The iron age of host-microbe interactions[J]. EMBO rep.，2015，16(11)：1482-1500.

[3]FORCADA J，Hoffman J I. Climate change selects for heterozygosity in a declining fur seal population[J]. Nature，2014，511(7510)：462-475.

[4]HAKE S，ROSS-IBARRA J. Genetic，evolutionary and plant breeding insights from the domestication of maize[J]，eLife，2015，4：e05861

[5]杨旭，刘飞，张宇，等．利用SSR标记研究茄子种质资源遗传多样性[J]．基因组学与应用生物学，2016，35(12)：3450-3457.

[6]李洪果，杜红岩，贾宏炎，等．利用表型性状构建杜仲雄性资源核心种质[J]．分子植物育种，2018，16(2)：591-601.

[7]王述民，张宗文．世界粮食和农业植物遗传资源保护与利用现状[J]．植物遗传资源学报，2011，12(3)：325-338.

[8]董玉琛，曹永生，张学勇，等．中国普通小麦初选核心种质的产生[J]，植物遗传资源学报，2003，4(1)：1-8.

[9]王建成，胡晋，张彩芳，等．建立在基因型值和分子标记信息上的水稻核心种质评价参数[J]．中国水稻科学，2007，21(1)：51-58.

[10]郑轶琦，郭玢，房淑娟，等．利用表型数据构建狗牙根初级核心种质[J]．草业学报，2014，23(4)：49-60.

[11]刘娟，廖康，曹倩，等．利用表型性状构建新疆野杏种质资源核心种质[J]．果树学报，2015，32(5)：787-796.

[12]徐宁，程须珍，王素华，等．以地理来源分组和利用表型数据构建中国小豆核心种质[J]．作物学报，2008，34(8)：1366-1373.

[13]李丽，何伟明，马连平，等．用EST-SSR分子标记技术构建大白菜核心种质及其指纹图谱库[J]．基因组学与应用生物学，2009，28(1)：76-88.

[14]李秀兰，贾继文，王军辉，等．灰楸形态多样性分析及核心种质初步构建[J]．植物遗传资源学报，2013，14(2)：243-248.

[15]王海岗，贾冠清，智慧，等．谷子核心种质表型遗传多样性分析及综合评价[J]．作物学报，2016，42(1)：19-30)

[16]胡兴雨，王纶，张宗文，等．中国黍稷核心种质的构建[J]．中国农业科学，2008，41(11)：3489-3502)

[17]李自超，张洪亮，孙传清，等．植物遗传资源核心种质研究现状与展望[J]，中国农业大学学报，1999，4(5)：51-62.

[18]YONEZAWA K，MORISHIMA H. Sampling strategies for use in stratified germplasm collections，In：HODGKIN T，BROWN A H D，van HINTUM T J L，et al. Core collec-

tions of plant genetic resources, John Wiley and Sons, Chichester, UK, 1995: 95-108.
[19]HU J, ZHU J, XU H M. Methods of constructing core collections by stepwise clustering with three sampling strategies based on the genotypic values of crops[J]. Theoretical and Applied Genetics, 2000, 101(1-2): 264-268
[20]BROWN A H D. The core collection at the crossroads, In: HODGKIN T, BROWN AHD, van HINTUM TJL, MORALES EAV(eds.), Core collections of plant genetic resources[M]. John Wiley and Sons, Chichester, UK, 1995: 3-19.
[21] ORTIZ R, RUIZTAPIA E N, MUJICASANCHEZ A. Sampling strategy for a core collection of *Peruvian quinoa* germplasm[J]. Theoretical and Applied Genetics, 1998, 96(3-4): 475-483.
[22]刘遵春，张春雨，张艳敏，等. 利用数量性状构建新疆野苹果核心种质的方法[J]. 中国农业科学，2010, 43(2): 358-370.

[原载：分子植物育种，1-18]

8 个格木天然林种源苗期生长差异与早期选择

李洪果[1,2] 潘启龙[1] 刘光金[1] 劳庆祥[1] 刘福妹[1] 张 培[1] 谌红辉[1]
([1]中国林业科学研究院热带林业实验中心，广西凭祥 532600
[2]广西友谊关森林生态系统国家定位观测研究站，广西凭祥 532600)

摘 要 为研究格木不同天然种源家系苗的生长差异，选出相对速生的优良种源，本研究对 8 个格木天然林种源、110 个家系 1 年生的苗高和地径等生长量指标进行测量，结合各家系的种子表型性状进行方差分析、多重比较、相关性分析和聚类分析。结果表明：不同种源的格木苗期生长量差异显著，具备种源选择的基础；不同种源的苗期生长量差异主要来源于种源内，具备从优良种源内选择优良家系的基础；苗期生长量与种子大小、种子厚和千粒重等性状呈极显著正相关关系，大粒、饱满种子可作为优良种源和家系选择的初选参考指标。初步选出浦北、北流种源作为格木优良种源。本研究为格木速生种源及家系的早期选择提供依据。

关键词 格木；种源；苗期生长；苗期选择

Difference Analysis of Growth Traits and Seedling Selection of Eight *Erythrophleum fordii* Provenances

LI Hongguo[1,2], PAN Qilong[1], LIU Guangjin[1],
LAO Qingxiang[1], LIU Fumei[1], ZHANG Pei[1], CHEN Honghui[1]
([1]*Experimental Center of Tropical Forestry*, *Chinese Academy of Forestry*, *Pingxiang*, 532600, *Guangxi*, *China*
[2]*Guangxi Youyiguan Forest Ecosystem Research Station*, *Pingxiang*, 532600, *Guangxi*, *China*)

Abstract: Annual growth and seed traits of eightprovenances, 110 families of *Erythrophleum fordii*were measured to research the difference of growth traits and seedling selection. Based on variance analysis, multi-comparison, correlation analysis and cluster analysis, the results showed that there were significant differences in seedling growth among different provenances, which provided the basis for provenance selection. The difference of seedling growth of different provenances is mainly from within provenances and has the basis of selecting fine families from superior provenances. Correlation analysis revealed that there was a significant positive correlation between seedling growth and seed size, seed thickness and 1000-grain weight. Large and full seeds can be used as a primary reference index for the selection of good provenances and families. Pubei and Beiliu provenance were selected as the optimal provenances. This study provide evidence for optimal provenances and families selection of *Erythrophleum fordii*.

Key words: *Erythrophleum fordii*; provenances; seedling growth; seedling selection

格木(*Erythrophleum fordii*)是中国热带亚热带地区珍贵的用材和绿化树种，也具有较高的药用价值，其木材结构致密坚实，极耐腐，有“铁木”之称，同时属于广西三大硬木和明清七大硬木之一，是优良的工业、家具和工艺品用材。然而，材性优良的同时往往也伴随有慢生和长周期经营的特点，格木心材形成的起始年限在 12 年以上[1]，营林生产上认为格木规格材的经营周期应在 50~100 年以上。

林木在自然选择的长期影响下会产生不同的表型和遗传变异类型，形成不同的种源，不同种源在相同立地条件下的生长差异、生理特性等往往不同，这一现象在云杉、闽楠和翅荚木等树种上均有体现[2,3,4]。一般而言，在水肥条件、培育措施和管理方法等因素相同的条件下，苗木的地径、苗高等指标表现优异，则很可能为优良基因型。樟树 2 年生苗期时选出的20个优良单株，在4年生时仍有11个与2年生时的表现一致[5]。杉木无性系大多数苗期生长性状能稳定遗传，性状早晚的相关性明显[6]。也有研究认为不同种源子代苗期的发育性状与种实的表型性状间存在相关性[7]。因此，苗期选择对于长周期经营的林木来说尤为重要。本研究通过收集格木在中国中心分布区(广西)8个天然林种群的种子，按种源家系育苗，测定了地径、苗高和高径比等生长量指标，结合种子的表型性状，探讨了不同种源格木苗的生长差异，以及与种子表型性状间的相关性，并根据各个种源、家系的苗期生长量指标进行聚类，选出苗期相对速生的种源，以期为格木的种源选择提供相应依据。

1　结果与分析

1.1　不同种源格木苗期生长量和种子表型差异及多重比较

由不同种源格木苗期生长量和种子表型性状差异及多重比较可知(表1)，地径、苗高和高径比3个苗期生长指标中：浦北种源(7)地径生长量均值最大，为4.57mm；陆川种源(6)的最小，为3.69mm；地径最大值和最小值分别出现在北流和容县种源，为5.17mm和3.39mm；8个种群地径生长量均值为4.30mm，从大到小依次为浦北>北流>合浦>龙圩>博白>苍梧>容县>陆川。北流种源(1)的苗高生长量均值最大，为24.62；陆川种源的最小，为18.38；苗高最大值和最小值分别出现在合浦(4)和容县(8)种源，为31.61cm和14.00cm；8个种群苗高生长量从大到小，依次为北流>合浦>龙圩>浦北>博白>苍梧>容县>陆川。高径比反映苗木高度和粗度间的平衡关系，是反映苗木抗性及造林成活率的较好指标，8个种群的高径比从大到小依次为北流>合浦>龙圩>博白>陆川>浦北>苍梧>容县。3个苗期生长量指标的变异系数中，仅苗高的变异系数均值大于15%(16.11%)，说明苗高的变异程度较地径和高径比更为丰富，更容易发生表型分化，地径和高径比不易发生表型分化。以上说明：北流(1)、浦北(7)和合浦(4)种源是苗期生长较快的种源，苍梧(3)和容县(8)种源因高径比相对较小，在造林时可能具有较好的抗性，如抗风能力等。

种子的5个表型指标中：千粒重指标常用于反映种子质量，其中，北流(1)种源的千粒重最大，为1054.03g；陆川(6)种源的千粒重最小，为885.37g，8个种源千粒重指标均值为965.83g，从大到小依次为北流>浦北>容县>龙圩>苍梧>博白>合浦>陆川。种子厚度指标反映种子的饱满程度，8个种源的种子厚度从大到小依次为北流>浦北>博白>龙圩>合浦>容县>苍梧>陆川。种子面积主要反映种子的大小，8个种源的单种子面积从大到小依次为北流>浦北>博白>龙圩>合浦>容县>苍梧>陆川，以上3个指标的排序大致相当。种子长和种子宽指标同样反映种子大小，两者的排序与种子面积的排序类似。5个种子表型性状的变异系数均值都低于15%，说明种子的表型性状变异都比较集中和稳定，不易发生表型分化。以上说明，北流(1)和浦北(7)种源的种子呈现大粒、饱满种子的特征，质量较好。苍梧(3)和陆川(6)种源呈现出小粒、空瘪种子的特征，质量相对较差。

表1　18个种源格木苗期生长量和种子表型性状的差异及多重比较

种群及其代码	参数	D(mm)	H(cm)	H/D	SL(mm)	SW(mm)	SA(mm^2)	ST(mm)	WS(g)
北流(1)	Min	4.05	19.75	48.80	15.61	13.49	156.7	6.22	936.99
	Max	5.17	28.70	63.86	17.02	13.72	174.97	6.57	1113.37
	CV(%)	9.04	14.22	10.40	3.08	0.67	3.91	2.22	6.52
	Mean(SD)	4.52 ±0.41AB	24.62 ±3.50A	54.42 ±5.66	16.27 ±0.5AB	13.62 ±0.09AB	165.42 ±6.48AB	6.40 ±0.14A	1054.03 ±68.77A
博白(2)	Min	3.70	17.11	39.67	14.59	11.92	134.37	5.55	769.22
	Max	4.65	28.20	65.33	17.21	13.94	192.07	6.54	988.52
	CV(%)	5.21	14.68	13.87	4.57	3.84	8.75	4.10	7.08
	Mean(SD)	4.24 ±0.22AB	22.04 ±3.24AB	51.97 ±7.21	15.73 ±0.72B	12.86 ±0.49B	156.24 ±13.66B	5.93 ±0.24AB	899.00 ±63.66BC

（续）

种群及其代码	性状及其代码								
	参数	D(mm)	H(cm)	H/D	SL(mm)	SW(mm)	SA(mm²)	ST(mm)	WS(g)
苍梧(3)	Min	3.77	17.53	42.55	13.64	10.90	119.33	4.78	588.28
	Max	4.44	25.40	57.44	18.62	14.88	209.74	6.29	1263.20
	CV(%)	4.80	12.43	10.00	8.10	8.08	14.73	8.03	18.19
	Mean(SD)	4.20 ±0.20B	20.91 ±2.60AB	49.75 ±4.97	16.41 ±1.33AB	13.39 ±1.08AB	168.04 ±24.75AB	5.61 ±0.45BC	971.18 ±176.66ABC
合浦(4)	Min	3.95	16.13	38.74	14.70	11.83	142.32	4.97	687.32
	Max	4.87	31.61	68.07	17.76	14.44	182.57	6.41	1023.93
	CV(%)	6.30	17.63	14.14	4.92	4.39	7.65	5.76	10.36
	Mean(SD)	4.38 ±0.28AB	23.72 ±4.18A	54.01 ±7.64	15.87 ±0.78AB	12.92 ±0.57B	157.30 ±12.03B	5.79 ±0.33B	893.80 ±92.60C
龙圩(5)	Min	3.72	17.00	43.70	14.66	11.38	132.42	4.83	824.16
	Max	5.09	28.30	60.47	18.12	15.36	213.10	7.07	1228.86
	CV(%)	8.02	14.01	11.54	5.61	8.07	13.00	10.64	12.70
	Mean(SD)	4.35 ±0.35AB	23.06 ±3.23A	53.06 ±6.12	16.57 ±0.93AB	13.59 ±1.10AB	172.34 ±22.40AB	5.89 ±0.63AB	1012.63 ±128.55ABC
陆川(6)	Min	3.67	15.33	41.78	16.19	13.43	174.56	5.01	817.43
	Max	3.71	21.43	57.77	17.52	14	189.88	5.34	953.31
	CV(%)	0.53	16.58	16.06	3.96	2.09	4.20	3.13	7.67
	Mean(SD)	3.69 ±0.02C	18.38 ±3.05B	49.78 ±7.99	16.86 ±0.67AB	13.71 ±0.29AB	182.22 ±7.66A	5.17 ±0.16C	885.37 ±67.94C
浦北(7)	Min	3.94	18.28	38.69	15.93	12.79	154.17	4.86	883.01
	Max	5.08	28.40	59.51	19.11	15.15	228.12	6.92	1174.78
	CV(%)	6.22	13.92	13.21	6.05	4.53	10.59	9.85	9.21
	Mean(SD)	4.57 ±0.28A	22.73 ±3.16A	49.78 ±6.58	17.00 ±1.03A	14.01 ±0.63A	187.07 ±19.81A	6.03 ±0.59AB	1032.03 ±95.04AB
容县(8)	Min	3.39	14.00	36.38	15.22	12.01	142.60	5.07	793.02
	Max	4.82	26.05	58.00	19.74	17.20	263.59	6.64	1341.64
	CV(%)	7.95	14.04	10.82	6.48	7.96	13.81	7.49	12.00
	Mean(SD)	4.17 ±0.33B	20.67 ±2.90AB	49.54 ±5.36	17.00 ±1.10A	13.65 ±1.09AB	179.89 ±24.85AB	5.70 ±0.43B	1013.96 ±121.67ABC
总体	Min	3.39	14.00	36.38	13.64	10.90	119.33	4.78	588.28
	Max	5.17	31.61	68.07	19.74	17.20	263.59	7.07	1341.64
	CV(%)	7.65	16.11	13.10	6.57	6.90	13.17	8.02	12.82
	Mean(SD)	4.30 ±0.33	22.20 ±3.58	51.58 ±6.76	16.40 ±1.08	13.36 ±0.92	169.25 ±22.29	5.83 ±0.47	965.83 ±123.78

注：D：地径；H：苗高；H/D：高径比；SL：种子长；SW：种子宽；SA：种子面积；ST：种子厚；WS：种子千粒重；Min：极小值；Max：极大值；CV：变异系数；Mean：平均值；SD：标准差；两两种群间不同字母表示差异显著($P<0.05$)。

1.2 不同种源格木苗期生长方差分析

对8个格木种源的苗高、地径和高径比方差分析，齐次性检验的结果表明(表2)，苗高、地径和高径比的方差齐次性检验显著性值均大于0.05，说明以上3个指标均适合进行方差分析。方差分析的结果表明(表3)，不同种源格木苗期的地径和苗高生长量差异显著($P<0.05$)；高径比指标差异不显著($P>0.05$)，说明不同种源的苗高和地径生长量存在显著差异，这为苗期选择速生的种源和家系提供了依据。另外，地径、苗高和高径比的方差来源均以组内为主，表明不同格木种源内的家系苗在地径、苗高和高径比上均存在丰富的遗传差异，说明从速生种源内优中选优，进而挑选速生的单株和家系存在可能性，这为后期营建格木速生丰产林提供了方向。

表2 苗期生长性状的方差齐性检验

性状	统计量	种源自由度1	样本自由度2	显著性
D	1.429	7	102	0.202
H	0.575	7	102	0.774
H/D	0.971	7	102	0.457

注：α=0.05；D：地径；H：苗高；H/D：高径比。

表3 8个种源格木苗期生长量的方差分析

性状	方差来源	平方和	自由度	均方	F值	显著性
D(mm)	组间	2.601	7	0.372	4.082	0.001
	组内	9.284	102	0.091		
	总数	11.884	109			
H(cm)	组间	191.075	7	27.296	2.291	0.033
	组内	1215.296	102	11.915		
	总数	1406.371	109			
H/D	组间	373.453	7	53.350	1.170	0.326
	组内	4650.099	102	45.589		
	总数	5023.552	109			

注：α=0.05；D：地径；H：苗高；H/D：高径比。

1.3 格木苗期生长量与种子表型性状间的相关性分析

由苗期生长量与种子表型性状间的相关关系可知(表4)，地径与苗高、种子厚度、千粒重指标呈极显著的正相关关系，苗高与高径比呈极显著正相关、与种子厚度呈显著正相关的关系，高径比与其他指标的相关性不强。另外，种子长、种子宽、种子厚度、种子面积和千粒重等指标，相互间均存在极显著或显著的相关关系。综合来看，千粒重和种子饱满程度显著影响格木苗期的地径和苗高生长量，说明大粒种子、饱满种子很可能伴随有苗期速生的情况。

表4 格木苗期生长量与种子表型性状间的相关性分析

性状	D	H	H/D	SL	SW	SA	ST	WS
D	1							
H	0.578**	1						
H/D	0.137	0.885**	1					
SL	0.094	-0.077	-0.142	1				
SW	0.129	-0.011	-0.079	0.741**	1			
SA	0.113	-0.054	-0.123	0.920**	0.907**	1		
ST	0.290**	0.224*	0.097	-0.237*	-0.285**	-0.274**	1	
WS	0.290**	0.077	-0.075	0.732**	0.672**	0.698**	0.218*	1

注：**：在0.01水平(双侧)上显著相关；*：在0.05水平(双侧)上显著相关；D：地径；H：苗高；H/D：高径比；SL：种子长；SW：种子宽；SA：种子面积；ST：种子厚；WS：种子千粒重。

1.4 不同种源格木苗期生长量聚类分析

由8个格木种源的苗期生长量聚类结果可知(图1)，在欧氏距离约为12处，陆川(6)、苍梧(3)和容县(8)种源聚为一类，其余北流(1)、博白(2)、合浦(4)、龙圩(5)和浦北(7)种源聚为另一类。结合苗高和地径生长量均值的多重比较可知(表1)，前者3个种源在苗期表现为慢生的特征，后者5个种源表现为相对速生的特征，其中，北流(1)和浦北(7)种源苗期的地径和苗高生长量最大，可初步作为格木优良的候选种源。

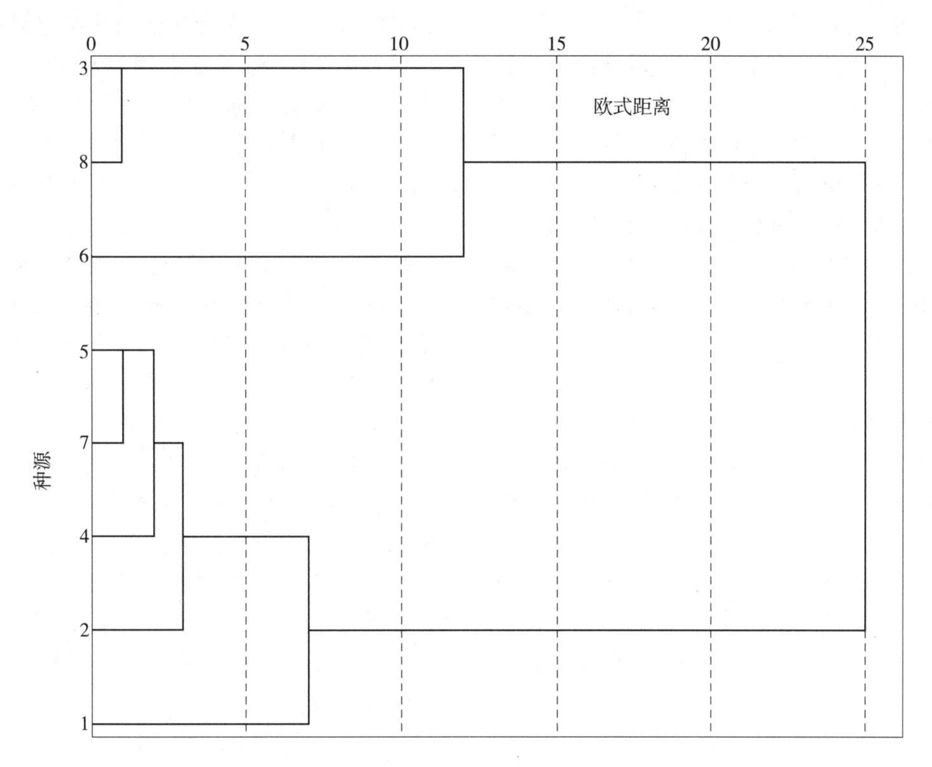

图1 18个格木种源苗期的生长量聚类

2 讨论

苗期的生长量差异是早期选择的重要内容，种源和家系的苗期选择，对于长周期经营的林木育种和生产十分重要。收集不同的资源群体，在人工可控、环境条件一致的情况下播种育苗，减少外部因素差异对苗木生长表现的影响，更好地体现种源间的遗传差异是较为常见的做法。如通过测量不同种源白刺植物在苗期的生长量和光合生理指标，筛选抗旱性较强的种源[8]；通过拟合不同柳树无性系苗期的生长规律，选择优良无性系[9]；通过分析不同种源和家系麻栎苗期生长量的差异，筛选优良种源[10]。

本研究利用8个格木天然林种源的110个家系种子育苗，测量不同家系苗的地径和苗高，计算高径比，并结合种子的表型性状探讨两者之间的关联。结果发现：不同格木种源的苗期生长量指标存在显著差异，且种源内的差异远大于种源间，这为格木种源和家系的选择提供了基础，因为种源间差异显著，才具备种源选择的前提，种源内变异丰富，才有利于从中选择优良家系。变异系数用以衡量性状的变异特征，变异系数越大，性状的离散程度越大，变异越丰富。本研究8个种源的苗高、地径和高径比的变异系数分别为7.65%、16.11%和13.10%，远小于前人研究的6个种群的16.31%、24.43%和22.59%[11]，这可能与种源选择的地域性跨度有关。赵志刚[11]等的研究对象为1个福建种群、2个广东种群和3个广西种群，而本研究的8个种群全部为广西种群，种源间的地理气候因子更为相似，进而影响了生长量变异的丰富程度。另外，5个种子表型性状的变异系数与赵志刚[12]等的研究相似，均小于15%，说明种子表型性状的变异程度和离散程度较低。本研究发现格木的地径、苗高等生长量与种子的大小、千粒重和饱满程度等性状密切相关，这与郝海坤[13]等的研究结果相同，即大粒种子在苗期的生长更快，说明种子大小、饱满程度和千粒重可以作为苗期速生的参考指标。

虽然众多研究表明，苗期选择有其积极的科学意义，可以为种源和家系选择提供一定的理论指导，但不同植物的生长发育特征存在一定差异，格木苗期选择的可靠性和稳定性有待进一步验证，后期可通过布设种源家系试验林，以定期调查的方式，直至格木的径生长和高生长速生期结束(30~50年)，利用获得数据，建立准确的数学模型，修正和指导格木的苗期选择方法。

3 材料与方法

3.1 试验地概况

试验地位于广西友谊关森林生态系统国家定位观测研究站热林中心站点，属南亚热带季风气候。该区属南亚热带季风型的半湿润—湿润气候，主要受东南与西南两季风控制，干湿季节明显，光、水、热等资源丰富，全年日照时数 1218～1620h，年均气温 20.5～21.7℃，≥10℃积温 6000～7600℃；年均降水量 1200～1500mm，年蒸发量 1261～1388mm，相对湿度 80%～84%。

3.2 试验材料和测量方法

2018 年 12 月对生长在广西凭祥市热林中心苗圃（106°44′32″E，22°7′32″N）的 1 年生格木苗进行测量，共调查 8 个种源、110 个家系、2018 株格木苗。原则上每家系共处理 25 粒种子，育苗 25 株，但个别家系因种子质量差、出苗率低或苗期折损等原因，调查时成苗不足 25 株。因此，最终每家系的测量株数在 10～25 株。

种子的 5 个测量指标中，种子长、宽、面积 3 个指标用万深叶面积分析系统测量，每家系种子测量 25 次，取平均值作为实测值，精确到 0.0001，最终保留到 0.01；种子厚度用游标卡尺测量，测量 25 次，取平均值作为实测值，精确到 0.01mm；千粒重指标用万分之一分析天平测量 25 粒种子的质量，精确到 0.0001g，扩大 40 倍记为千粒重，最终保留到 0.01g。苗期生长量 3 个指标中，每家系测量 10～25 株，取平均值作为实测值，其中，地径指标用游标卡尺测量，精确到 0.01mm；苗高指标用米尺测量，精确到 0.01cm；地径和苗高统一单位为毫米（mm）后，计算高径比。

3.3 数据处理和分析

在 Excel 软件中计算测量性状的均值、标准差等，计算极值、高径比、变异系数等指标。在 SPSS 19.0 软件中进行方差齐次性检验、方差分析、多重比较和聚类分析等，其中，聚类分析时先用 Z 得分法将原始测量数据标准化，再采用组间联接法和欧氏距离平方聚类。

参考文献

[1]唐继新，麻静，贾宏炎，等．南亚热带珍稀濒危树种格木生长规律研究[J]．中南林业科技大学学报，2015，35(7)：37-44.

[2]何小勇，柳新红，袁德义，等．不同种源翅荚木的抗寒性[J]．林业科学，2007，43(4)：24-30.

[3]吴际友，黄明军，陈明皋，吴哲，李艳，刘球，程勇，王旭军，等．闽楠种源苗期生长差异与早期选择研究[J]．中南林业科技大学学报，2015，35(11)：1-4.

[4]安三平，许娜，杜彦昌，等．云杉种和种源生长性状早期评价[J]．林业科学研究，2018，31(5)：20-26.

[5]钟永达，袁凡，孟伟伟，等．材用樟树遗传变异与苗期选择[J]．南昌大学学报(理科版)，2016，40(2)：197-204.

[6]饶显生，程书建，刘化桐，等．杉木无性系苗期选择可靠性分析[J]．森林与环境学报，2002，22(1)：82-85.

[7]DANGASUK O G，SEUREI P，GUDU S. Genetic variation in seed and seedling traits in 12 African provenances of *Faidherbia albida*(Del.)[J]. A. Chev. at Lodwar，Kenya，Agrofor. Syst.，1997，37(2)：133-141

[8]李清河，张景波，李慧卿，等．不同种源白刺幼苗生理生长对水分梯度的响应差异[J]．林业科学，2008，44(1)：52-56.

[9]万志兵，冯刚，朱成磊，等．不同柳树无性系一年生生长差异分析[J]．分子植物育种，2018，16(7)：2358-2363.

[10]王标，虞木奎，王臣，等．不同种源麻栎苗期生长性状差异及聚类分析[J]．植物资源与环境学报，2008，17(4)：1-8.

[11]赵志刚，郭俊杰，曾杰，等．濒危树种格木天然群体自由授粉子代苗期生长变异[J]．植物研究，2011，31(1)：100-104.

[12]赵志刚，郭俊杰，沙二，等．中国格木的地理分布与种实表型变异[J]．植物学报，2009，44(3)：338-344.

[13]郝海坤，黄志玲，彭玉华，等．格木种子大小变异及萌发特性[J]．广西林业科学，2016，45(1)：30-34.

[原载：分子植物育种，1-9]

不同种源家系交趾黄檀在凭祥引种及其早期适应性

刘福妹[1,2] 韦菊玲[1] 庞圣江[1] 贾宏炎[1] 洪 舟[3] 谌红辉[1]

([1]中国林业科学研究院热带林业实验中心，广西凭祥 532600；
[2]广西友谊关国家森林生态系统定位观测研究站，广西凭祥 532600；
[3]中国林业科学研究院热带林业研究所，广东广州 510000)

摘 要 探究交趾黄檀的种子活力和苗木生长表现，开展交趾黄檀引种试验，为其在我国南亚热带地区引种栽培提供参考依据。以从泰国和柬埔寨引入的10个种源25个家系交趾黄檀种子为材料，观测不同种源家系交趾黄檀种子在广西凭祥的萌发情况；利用培育的实生苗开展引种试验，调查其造林保存率和苗木生长情况，分析不同种源家系交趾黄檀在南亚热带地区的速生性和适应性。在10个交趾黄檀种源中，柬埔寨暹粒种源的种子活力最弱，发芽率和发芽势仅39.33%和19.33%，显著小于其他9个种源种子的发芽率和发芽势(分别为60.67%~76.95%和49.05%~71.33%)，9个种源间差异不显著($P>0.05$，下同)。在25个家系中，泰国北柳T-KF-4的发芽率达88.00%，泰国北柳T-KF-6的发芽势最高，为82.00%。2016年造林后，种源的平均保存率为88.17%，家系的平均保存率为89.59%；在10个种源和25个家系中，长势较佳的是泰国彭世洛种源(T-PP)和泰国北柳T-KF-1家系，其苗高和地径分别为1.62m、2.89cm和1.70m、3.09cm。2017年，2年生交趾黄檀苗木的平均苗高和胸径分别为2.69m和2.17cm；苗木长势在种源间差异不显著，但在家系间差异显著，其中，泰国巴真T-KI-5的平均苗高达3.15m，泰国巴真T-KI-7的平均胸径达2.60cm。交趾黄檀在广西凭祥引种栽培获得初步成效，且以泰国种源和家系对广西凭祥的适应性更佳。

关键词 交趾黄檀；引种；发芽率；发芽势；生长适应性

A Study on Introduction and Adoption of *Dalbergia cochinchinensis* from Different Provenances and Families in Pingxiang

LIU Fumei[1,2], WEI Juling[1], PANG Shenjiang[1],
JIA Hongyan[1], HONG Zhou[3], CHEN Honghui[1]

([1]*The Experimental Centre of Tropical Forestry, Chinese Academy of Forestry, Pingxiang* 532600, *Guangxi, China*;
[2]*Guangxi Youyiguan Forest Ecosystem Research Station, Pingxiang* 532600, *Guangxi, China*;
[3]*Research Institute of Tropical Forestry, Chinese Academy of Forestry, Guangzhou* 510000, *Guangdong, China*)

Abstract: To provide a useful basis for cultivation of *Dalbergia cochinchinensis* in the south subtropical region of China, we carried out an experiment on introduction to Pingxiang city of Guangxi Provence. Seeds from 10 provenances and 25 families of Thailand and Cambodia were used to study the performances on germination, and seedlings from these seeds were used for forestation. Their survive rate and growth performances were investigated to analyze their adoption in the introduction site. Among the ten provenances, seeds from Siam Reap of Cambodia(C-SR) were the poorest, with the smallest germination(39.33%) and germination potentiality rate(19.33%). While there was no significant difference among the other nine provenances, and their germination and germination potentiality rate were in the range of 60.67%~76.95% and 52.67%~71.33%($P<0.05$), respectively. Among the 25 families, the highest germination and germination potentiality rate was 88.00% and 82.00%, represented in T-KF-4(the family from Chachoengsao of Thailand) and T-KF-6(the family from Chachoengsao of Thailand), respectively. In 2016, the average survival rate of these provenances and families reached up to 88.17% and 89.59%, respectively. No significant difference showed in the survival rate among 10 provenances or 25 families, whereas significant differences showed in their growth performances. Of which, T-PP(Phitsanulok of Thailand) was the best provenance with mean height and mean diameter of 1.62 m and 2.89 cm, followed the best family T-KF-1 (Chachoengsao of Thailand, 1.70 m and 3.09 cm). In 2017, the average height and breast height (DBH) of all these seedlings was 2.69 m and 2.17 cm. There is no significant difference in growth performance among 10 provenances, whereas significant differences showed among 25 families. T-KI-5 (3.15 m) and T-KI-7 (2.60 cm), families both from Prachin Buri of Thailand, performed best on height and DBH, respectively. We achieved a preliminary success of *D. cochinchinensis* cultivation in subtropical China, and trees from Thailand performed better in Pingxiang, Guangxi Provence.

Key words: *Dalbergia cochinchinensis*; introduction; germination rate; germination potentiality rate; growth performances

引言

【研究意义】交趾黄檀(*Dalbergia cochinchinensis* Pierre ex Laness)属蝶形花科(Papilionaceae)黄檀属(*Dalbergia*)树种，原产于泰国、老挝、越南和柬埔寨等东南亚地区，主要分布于海拔400~500m、年均降水量1200~1650mm的半落叶常绿森林中，现存林分多为小面积的稀疏块状或点状分布。在我国，交趾黄檀是名贵的红木用材树种之一，其木材又称大红酸枝、老红木，市场上流通的交趾黄檀木材均源自天然林且全部依赖进口[1]。交趾黄檀天然分布区域狭窄且多处于散生状态，自然更新能力差，长期的过度利用和乱采滥伐已导致资源严重枯竭[2,3]，致使交趾黄檀木材供需矛盾日益突出，制约着相关产业经济的健康发展，已被《濒危野生动植物种国际贸易公约》列入严格限制采伐与出口的树种[4,5]。因此，通过在国内适宜地区引种培育交趾黄檀人工林是缓解木材供需矛盾的重要途径之一。广西凭祥市地处中国南部，与越南北部谅山接壤，为南亚热带季风气候，年均温度22.12℃，年均降水量1267mm，年均日照时数1657.37h，海拔200~500m。在此地开展交趾黄檀的引种栽培，探究交趾黄檀的种子活力和苗木生长表现，对推动我国交趾黄檀引种工作，促进交趾黄檀相关产业开发与应用具有重要意义。【前人研究进展】交趾黄檀具有极高的药用价值和经济效益，茎、叶[6,7]和心材富含一系列具有抗氧化、清除自由基、抑制脂质过氧化、消炎、化血瘀等功效的苯并呋喃[8]、黄酮类[9]、萜类[10]、查尔酮类及其衍生物[11-13]等化学物，在食品工业和医药行业广泛应用[14]；其木材材质优良，带状木纹细腻美观，伴有怡人清香，且木材解剖后会以红色为基调，在大气中氧化、包浆变成褐红、黑红或紫红色，作为高档红木家具的主要用材[15]深受消费者的青睐。尽管我国交趾黄檀的利用和加工历史久远，但直至2006年才开展对该树种的相关研究工作，近十年来的研究重点主要为木材构造及物理性能分析[16,17]、心材化学成分测定[9,10]和木材鉴别[18]。【本研究切入点】目前，我国在交趾黄檀引种方面的研究工作相对滞后，仅见李瑞聪(2017)对12个交趾黄檀种源(2年生)在闽南山地引种情况进行初步研究的报道，鲜见在广西引种研究的报道信息。【拟解决的关键问题】以来源于泰国和柬埔寨两个国家10个种源、25个家系交趾黄檀的种子为材料，观察其在凭祥的萌发情况，并以培育出的实生苗首次在广西凭祥中国林业科学研究院热带林业实验中心(南亚热带季风气候)开展引种试验，调查造林保存率和苗木生长情况，分析了解不同种源和家系交趾黄檀在南亚热带地区的速生性和适应性，为交趾黄檀的引种及开发利用提供参考依据。

1 材料与方法

1.1 交趾黄檀种子原产地与引种地概况

交趾黄檀种子由中国林业科学研究院热带林业研究所提供，采自热带季风气候的泰国和柬埔寨，分布在东经100°16′~105°28′、北纬11°59′~16°49′、海拔7~280m的天然林分中，种源信息如表1所示。种子于2014年12月采摘，用密封袋保存后邮寄回国内。

表1 交趾黄檀种源信息

来源	家系数	编号	海拔(m)	年均气温(℃)	年均降水量(mm)	经度	纬度
柬埔寨马德望	1	C-BP	17	23.8~33.8	1180	103°12′	13°06′
柬埔寨磅湛	1	C-KS	21	25.0~33.9	2150	105°28′	11°59′
柬埔寨菩萨	1	C-PP	39	25.0~30.8	1350	103°53′	12°28′

（续）

来源	家系数	编号	海拔(m)	年均气温(℃)	年均降水量(mm)	经度	纬度
柬埔寨暹粒	1	C-SR	18	23.8~33.8	1180	103°50′	13°22′
泰国北柳	10	T-KF-1-T-KF-10	7	22.0~31.8	1420	101°05′	13°41′
泰国巴真	7	T-KI-1-T-KI-7	19	24.1~34.3	1840	101°40′	14°02′
泰国孔敬	1	T-KK	169	22.5~32.7	1304	102°50′	16°26′
泰国呵叻	1	T-NR	184	23.2~33.0	1161	102°06′	14°59′
泰国彭世洛	1	T-PP	50	22.9~33.2	1412	100°16′	16°49′
泰国依华佬	1	T-NH	280	23.7~32.8	1034	102°28′	16°17′

试验于广西凭祥中国林业科学研究院热带林业实验中心进行。该地属南亚热带地区，为南亚热带季风气候，年均温度22.12℃，月最高气温37.04℃，月最低气温-0.36℃，月均相对湿度81.15%，年均降水量1267mm，年均日照时数1657.37h，年内无霜期335d。育苗地为热带林业实验中心苗圃(106°45′E、22°8′N)，海拔约246m，定植林地为热带林业实验中心青山实验场珍贵树种示范林地(106°43′E、22°13′N)，海拔500m。

1.2 试验方法

1.2.1 发芽试验

2015年10月在广西凭祥热带林业实验中心苗圃进行发芽试验。试验前进行种子催芽，即用30℃温水将种子浸泡24h，再用0.3%高锰酸钾溶液浸种15min进行消毒灭菌，最后用蒸馏水冲洗干净。预处理好的种子置于编号的沙盘中，采用随机区组设计，每个家系随机选取100粒种子，重复3次，共300粒，观测并记录发芽情况。试验期间保持沙盘湿润，注意预防种子霉变。此外，用百粒法测定种子千粒重，若种源有多个家系，则以家系间均值计算(下同)。

发芽率(%)=(已发芽种子数/100)×100

变异系数(%)=(性状标准偏差/性状平均值)×100

1.2.2 播种育苗

造林用交趾黄檀家系种子均用沙床育苗。育苗沙床和用来覆盖种子的细沙在使用前均先用5%多菌灵溶液将沙床浇透进行消毒处理；种子也进行预处理，处理方法同1.2.1。将处理好的种子均匀撒播在沙床上，用厚度约2cm的细沙覆盖后用水将沙床浇透。播种后每天定时定量浇水，保持苗床湿润；每15d喷洒1%多菌灵溶液1次以消毒苗床；即时去除苗床上的杂草和病株。当沙床上交趾黄檀小苗长出3~4片叶或苗高达5cm时，可移植于高15cm、底径10cm的无纺布容器中，容器中基质为草碳土、砂壤土和森林腐殖土按3∶3∶4混合而成。移苗后需对小苗进行遮阴处理(移苗后1~10d用两层遮阴网，此后用一层遮阴网)。

1.2.3 苗期管理

待苗高10~15cm时去除遮阴网。同时，注意防治病虫害，即时除草，适时浇水，保持基质湿润。适时追肥，施肥周期为15~30d，追肥浓度不宜太高，推荐叶面喷施1%复合肥。

1.2.4 造林

于2016年1月进行造林，林地为马尾松间伐林地(保留马尾松64株/hm^2)，开挖规格为40cm×40cm×40cm的明坑，每穴施500g有机肥，然后回填表土。2016年5月末至6月初按设计方案进行造林，采用3株随机区组，共设9个区组，株行距3m×2m，2个月后用竹竿固定幼树主杆防止苗木倒伏。造林后对交趾黄檀林分进行常规抚育，但不再追施肥料。

1.2.5 观测项目

2016年8月，统计造林成活情况；2016年12月，测定造林后交趾黄檀的苗高和地径；2017年12月，测定交趾黄檀的苗高和胸径，若苗高低于1.5m，则仅测定苗高而不测定胸径。

1.3 统计分析

观测数据采用Excel 2010和SPSS 16.0进行统计分析。所有方差分析均在0.05水平上进行，多重比较采用Duncan's进行，发芽率、发芽势和保存率经标准化处理(SPSS 16.0自有标准化程序)后再进行相关性分析。

2 结果与分析

2.1 不同种源和家系交趾黄檀种子在凭祥的萌发情况

由表2可知，10个种源25个家系交趾黄檀种子间的千粒重、发芽率和发芽势3个性状均存在显著差异($P<0.05$，下同)；各性状变异系数变幅为14.28%~26.92%；其中发芽势的变异系数在种源间和家系间均最大，分别为25.52%和26.92%。说明交趾黄檀的种子活力在种源和家系内变异丰富，在广西凭祥引种交趾黄檀具有较大的选择潜力。

表 2 不同种源和家系交趾黄檀种子千粒重和萌发特性的方差分析结果

来源	性状	自由度	平方和	均方	F 值	均值	变幅	变异系数 CV(%)
种源	千粒重	9	827.05	91.894	11.736*	25.78 g	20.91~33.77 g	18.28
	发芽率	9	29.078	3.231	4.675*	66.72%	39.33~76.95%	16.90
	发芽势	9	21.916	2.435	3.039*	54.68%	19.33~71.33%	25.52
家系	千粒重	24	1288.345	53.681	56.32*	25.29 g	15.63~33.77 g	16.72
	发芽率	24	45.492	1.896	3.325*	71.89%	39.33~88.00%	14.28
	发芽势	24	56.967	2.374	6.968*	52.99%	19.33~82.00%	26.92

注 * 表示性状间差异显著($P<0.05$)。表 4 同。

由表 3 可知，在 10 个交趾黄檀种源种子中，柬埔寨暹粒(C-SR)和菩萨(C-PP)种源种子的千粒重分别为 33.77 和 33.75g，显著高于其他 8 个种源种子；其次是泰国巴真(T-KI)种源种子，千粒重为 27.65g。千粒重最小的是柬埔寨马德望(C-BP)种源种子，仅 20.91g。柬埔寨暹粒(C-SR)种子的发芽率和发芽势均最小(分别为 39.33%和 19.33%)，显著小于其他 9 个种源种子的发芽率和发芽势(分别为 60.67%~76.95%和 49.05%~71.33%)。说明柬埔寨暹粒种源种子的活力显著低于其他种源种子。

对比 25 个家系种子发现，千粒重最大的是柬埔寨暹粒(C-SR)和柬埔寨菩萨(C-PP)家系的种子，分别为 33.77 和 33.75g，二者间差异不显著($P>0.05$，下同)，但均显著大于其他家系种子；千粒重最小的是泰国北柳 T-KF-2 家系种子，仅 15.63g，显著小于其他 24 个家系种子。在发芽率方面，泰国北柳 T-KF-4 家系种子最高，达 88.00%；发芽率最低的是柬埔寨暹粒 C-SR 家系种子，仅 39.33%。25 个交趾黄檀家系种子的发芽势变幅为 19.33%~82.00%，其中发芽势最大的是泰国北柳 T-KF-6 家系种子。

表 3 不同种源和家系交趾黄檀种子千粒重和萌发特性的多重比较

来源	编号	千粒重(g)		发芽率(%)		发芽势(%)	
		种源	家系	种源	家系	种源	家系
柬埔寨马德望	C-BP	20.91c	20.91i	60.67a	60.67bc	52.67a	52.67bcdefg
柬埔寨磅	C-KS	22.04bc	22.04hi	76.67a	76.67ab	71.33a	71.33ab
柬埔寨菩萨	C-PP	33.75a	33.75a	65.33a	65.33ab	59.33a	59.33abcde
柬埔寨暹粒	C-SR	33.77a	33.77a	39.33b	39.33c	19.33b	19.33h
泰国北柳	T-KF(均值)	23.18bc		74.27a		53.73a	
	T-KF-1		24.51efg		74.00ab		39.33efgh
	T-KF-2		15.63k		80.00ab		67.33abc
	T-KF-3		22.82ghi		68.67ab		50.00bcdefg
	T-KF-4		18.47j		88.00a		44.00cdefg
	T-KF-5		25.11efg		66.67ab		65.33abcd
	T-KF-6		23.57fgh		85.33ab		82.00a
	T-KF-7		23.40 fgh		72.67ab		29.33gh
	T-KF-8		23.60 fgh		60.67bc		58.00abcde
	T-KF-9		26.63de		78.67ab		41.33defgh
	T-KF-10		28.07cd		68.00ab		60.67abcde
泰国巴真	T-KI(均值)	27.65b		76.95a		49.05a	
	T-KI-1		28.98bc		78.67ab		49.33bcdefg
	T-KI-2		29.83bc		77.33ab		43.33cdefg
	T-KI-3		28.23cd		78.67ab		58.00abcde
	T-KI-4		24.69efg		75.33ab		38.00efgh
	T-KI-5		26.07de		64.67ab		33.33fgh
	T-KI-6		30.58b		85.33ab		62.00abcde
	T-KI-7		25.19efg		78.67ab		59.33abcde

（续）

来源	编号	千粒重(g)		发芽率(%)		发芽势(%)	
		种源	家系	种源	家系	种源	家系
泰国孔敬	T-KK	25.34bc	25.34ef	72.67a	72.67ab	65.33a	65.33abcd
泰国呵叻	T-NR	23.41bc	23.41fgh	73.33a	73.33ab	60.00a	60.00abcde
泰国彭世洛	T-PP	26.27bc	26.27de	63.33a	63.33ab	59.33a	59.33abcde
泰国依华伥	T-NH	21.49bc	21.49hi	64.67a	64.67ab	56.67a	56.67bcdef

注：同列数据后不同小写字母表示差异显著($P<0.05$)。表5同。

2.2 不同种源和家系交趾黄檀的造林情况

由表4可知，交趾黄檀的造林保存率在10个种源间和25个家系间均无显著差异。10个种源的造林平均保存率变幅为77.78%~100.00%，其中，泰国呵叻(T-NR)种源的造林保存率(100.00%)最高(表5)。25个家系的造林平均保存率为89.59%(表4)，其中泰国北柳T-KF-3家系、泰国北柳T-KF-10家系、泰国巴真T-KI-3家系、泰国巴真T-KI-7家系及泰国呵叻T-NR家系的造林保存率均达100.00%(表5)。说明泰国的家系比其他家系更能适应凭祥的种植环境。此外，在造林中还发现，交趾黄檀蘖生能力特别强，且定植早期易出现断头现象，当主枝受到损伤后马上会有新的侧枝长出替代，特别影响干形。

于2016年12月和2017年12月观测交趾黄檀苗木的苗高(1年生和2年生)、地径(1年生)和胸径(2年生)。对获得的数据进行方差分析(表4)发现，在不同种源间，2016年苗木(1年生)的平均苗高和地径(分别为1.41m和2.54cm)差异均达显著水平，2017年苗木(2年生)的平均苗高和胸径(分别为2.69m和2.17cm)差异不显著；在不同家系间，平均苗高和地径(或胸径)连续两年均存在显著差异。从2016—2017年，苗高的变异系数在种源内由9.97%(1年生)降为6.61%(2年生)，在家系内由10.38%(1年生)降为8.43%(2年生)。说明随苗木的生长，交趾黄檀的苗高生长差异在种源间和家系间逐渐减小。

由表5可知，在10个种源中，造林后第1年(2016年)生长状况最好的是泰国彭世洛(T-PP)种源，其苗高和地径分别为1.62m和2.89cm，其次为泰国依华伥(T-NH)种源，其苗高和地径分别为1.58m和2.87cm，二者的苗高均显著高于柬埔寨菩萨(C-PP)、柬埔寨磅湛(C-KS)和柬埔寨马德望(C-BP)3个种源，地径显著优于其他8个种源。说明泰国彭世洛(T-PP)和依华伥(T-NH)种源的苗木在造林初期比其他种源交趾黄檀具有更佳的速生性，且能更快适应凭祥的林地条件。

从表5还可看出，在25个家系中，2016年速生性表现最好的是泰国北柳T-KF-1家系，苗高和胸径分别达1.70m和3.09cm，较表现最差的泰国巴真T-KI-2家系(苗高和胸径分别为1.10m和1.94cm)分别提高54.17%和59.06%；在2017年，各家系的苗高和地径分别为2.23~3.15m和1.58~2.60cm，其中生长表现较佳的是泰国巴真T-KI-5家系(苗高达3.15m)和泰国巴真T-KI-7家系(胸径达2.60cm)。

表4 不同种源和家系交趾黄檀造林保存率的方差分析结果

年份(年)	来源	性状	自由度	平方和	均方	F值	均值	变幅	变异系数(%) CV
2016	种源	造林保存率	9	23.335	1.556	1.653	88.17%	77.78~100.00%	7.95
	家系	造林保存率	24	12.161	1.351	1.372	89.59%	70.37~100.00%	8.78
2016	种源	苗高	9	5.153	0.573	3.580^*	1.41m	1.19~1.62m	9.97
		地径	9	45.286	5.032	8.918^*	2.54cm	2.23~2.89cm	10.39
	家系	苗高	24	13.99	0.583	3.886^*	1.41m	1.10~1.70m	10.38
		地径	24	63.110	6.798	4.782^*	2.54cm	1.94~3.09cm	12.23
2017	种源	苗高	9	7.598	0.844	1.451	2.69m	2.33~3.03m	6.61
		胸径	9	7.39	0.821	1.003	2.17cm	1.87~2.48cm	7.89
	家系	苗高	24	29.916	1.246	2.233^*	2.69m	2.23~3.15m	8.43
		胸径	24	48.022	2.001	2.601^*	2.17cm	1.58~2.60cm	13.58

表 5 不同种源和家系交趾黄檀造林保存率和生长情况的多重比较

来源	编号	保存率(%)		生长指标(2016)		生长指标(2017)	
		种源	家系	苗高(m)	地径(cm)	苗高(m)	胸径(cm)
柬埔寨马德望	C-BP	77.78	77.78	1.29(bcd)bcde	2.23(b)def	2.33cd	1.87abc
柬埔寨磅湛	C-KS	85.19	85.19	1.23(cd)cde	2.25(b)def	2.81abcd	2.20abc
柬埔寨菩萨	C-PP	92.59	92.59	1.19(d)de	2.39(b)cdef	2.56abcd	2.16abc
柬埔寨暹粒	C-SR	92.59	92.59	1.49(ab)abcd	2.60(b)abcde	2.71abcd	2.03abc
柬埔寨均值		87.04	87.04	1.32	2.38	2.61	2.08
泰国北柳	T-KF(均值)	91.67		1.44(abcd)	2.96(b)	2.66	2.16
	T-KF-1		88.89	1.70a	3.09a	2.70abcd	2.33abc
	T-KF-2		92.59	1.55abc	2.95abc	2.51abcd	1.99abc
	T-KF-3		100.00	1.62a	3.03ab	2.79abcd	2.16abc
	T-KF-4		85.19	1.42abcde	2.92abc	2.74abcd	2.33abc
	T-KF-5		92.59	1.23cde	2.33cdef	2.86abcd	2.51ab
	T-KF-6		92.59	1.41abcde	2.76abcde	2.39bcd	1.92abc
	T-KF-7		85.19	1.41abcde	2.79abcd	2.79abcd	2.33abc
	T-KF-8		96.30	1.26bcde	2.52abcdef	2.68abcd	2.13abc
	T-KF-9		83.33	1.43abcd	2.67abcde	2.75abcd	2.50ab
	T-KF-10		100.00	1.36abcde	2.66abcde	2.39bcd	1.58c
泰国巴真	T-KI(均值)	88.69		1.38(abcd)	2.23(b)	2.75	2.08
	T-KI-1		81.48	1.41abcde	2.23def	2.48abcd	1.78bc
	T-KI-2		87.50	1.10e	1.94f	2.76abcd	1.96abc
	T-KI-3		100.00	1.33abcde	2.41bcdef	2.96abc	2.45ab
	T-KI-4		70.37	1.58ab	2.35cdef	2.58abcd	1.71bc
	T-KI-5		88.89	1.47abcd	2.34cdef	3.15a	2.25abc
	T-KI-6		92.59	1.41abcde	2.14ef	2.23d	1.59c
	T-KI-7		100.00	1.38abcde	2.21def	3.01abc	2.60a
泰国孔敬	T-KK	92.59	92.59	1.41(abcd)abcde	2.69(b)abcde	2.64abcd	2.36abc
泰国呵叻	T-NR	100.00	100.00	1.44(abc)abcd	2.44(b)bcdef	2.81abcd	2.37abc
泰国彭世洛	T-PP	77.78	77.78	1.62(a)a	2.89(a)abc	3.03ab	2.48ab
泰国侬华佧	T-NH	85.19	85.19	1.58(a)ab	2.87(a)abc	2.58abcd	2.07abc
泰国均值		89.32	91.15	1.50	2.69	2.76	2.26

同列数据后不同小写字母表示家系间差异显著($P<0.05$)，括号内不同小写字母表示种源间差异差异($P<0.05$)

2.3 不同种源和家系交趾黄檀种子性状、地理气候因子与萌发和生长指标相关性

对凭祥引种不同种源和家系交趾黄檀种子的千粒重、发芽率、发芽势、造林后苗木的生长情况及地理气候因子进行相关性分析，结果(表6)发现，种子发芽率与发芽势、海拔与纬度、2016年(1年生)的生长指标(苗高和地径)与纬度、2017年(2年生)的苗高与2017年(2年生)的胸径呈显著正相关。说明在凭祥引种的交趾黄檀早期生长主要受纬度因素影响。

表 6　不同种源和家系交趾黄檀种子性状、地理气候因子与在凭祥萌发和生长指标的相关性分析

指标	海拔	年均气温	年均降水量	经度	纬度	千粒重	发芽率	发芽势	保存率	2016 年苗高	2016 年地径	2017 年苗高	2017 年胸径
海拔	1.000												
年均气温	−0.139	1.000											
年均降水量	−0.511	0.330	1.000										
经度	−0.119	0.089	0.288	1.000									
纬度	0.668*	0.006	−0.444	−0.666*	1.000								
千粒重	−0.341	−0.260	−0.064	0.144	−0.222	1.000							
发芽率	0.133	−0.116	0.521	−0.168	0.055	−0.521	1.000						
发芽势	0.277	−0.254	0.349	0.022	0.150	−0.557	0.815*	1.000					
保存率	0.237	−0.349	−0.145	0.106	−0.089	0.342	0.117	−0.130	1.000				
2016 年苗高	0.438	0.166	−0.450	−0.687*	0.810*	−0.116	−0.273	−0.283	−0.121	1.000			
2016 年地径	0.349	−0.393	−0.442	−0.580	0.645*	−0.071	−0.140	−0.065	−0.010	0.762*	1.000		
2017 年苗高	−0.037	0.191	0.380	−0.346	0.344	0.126	0.157	0.110	0.070	0.449	0.271	1.000	
2017 年胸径	0.261	−0.219	0.096	−0.382	0.562	0.004	0.355	0.473	0.220	0.340	0.390	0.773*	1.000

注：* 表示显著相关（$P<0.05$）。

3　讨论

交趾黄檀具有极高的经济价值和开发应用前景。但该树种资源主要分布在东南亚等国的天然林分及保护区，我国目前尚无大量人工繁殖和栽培的成功经验，严重影响交趾黄檀在我国的开发利用和产业化进程。林木引种是引入外来树种发展本国林业的重要手段，能在增加森林生态系统生物多样性、实现树木基因资源易地保存的同时，获取乡土树种所不能提供的产品[19]。本研究从泰国和柬埔寨引入 10 个种源 25 个家系交趾黄檀种子，并在广西凭祥开展引种试验，发现不同种源和家系种子间的千粒重、发芽率和发芽势 3 个性状存在显著差异，与李瑞聪[20]、莫世琴等[21]、李科等[22]在福建龙海、海南乐东、广东广州和阳江等地进行交趾黄檀发芽试验的研究结果一致。在柬埔寨，交趾黄檀种子一般即采即播，否则发芽率会显著下降；浸种 24h 后，发芽率可达 70%~80%[3]，而本研究中，除柬埔寨暹粒种源的发芽率为 39.33% 外，其他种源的发芽率为 60.67%~76.95%，说明除柬埔寨暹粒种源外，其他种源交趾黄檀种子对凭祥气候环境的适应性均较佳。

本研究引种的交趾黄檀种源平均造林保存率为 88.17%，其中泰国呵叻种源的保存率达 100.00%；1 年生(2016 年)苗木的平均苗高和地径分别为 1.41m 和 2.54cm，说明交趾黄檀在凭祥具有较强的生长适应性。6 个泰国种源交趾黄檀 1 年生(2016 年)苗木的平均苗高和地径分别为 1.50m 和 2.69cm，分别较柬埔寨种源提高 13.64% 和 13.00%，说明泰国种源在引种早期对凭祥的生长适应性更强。在种植过程中发现，交趾黄檀早期生长迅速，苗木木质化程度不够，抗风性能差，易出现倒伏、断头现象，因此定植早期要进行扶杆，且引种造林地不宜选在风口，应优先选择中、下坡位。

交趾黄檀在柬埔寨西哈努克(Preah Sihanouk)5 年生人工纯林的平均苗高和胸径分别为 5.7m 和 6.7cm[3]，在泰国萨开拉(Sakaerat)12 年生人工纯林的平均苗高和胸径分别为 15.4m 和 10cm[23]，在越南同奈 38 年生人工纯林的平均苗高和胸径分别为 21.8m 和 29.0cm[24]，3 个地点的交趾黄檀苗高和胸径年均生长量均值变幅为 0.57~1.28m 和 0.76~1.34cm。本研究中，2 年生交趾黄檀苗高和胸径的年均生长量分别达 1.35m 和 1.09cm，说明在引种地凭祥交趾黄檀的速生性与原产地基本一致，进一步说明交趾黄檀在广西凭祥的引种获得了初步成效。

本研究结果表明，交趾黄檀引种主要受纬度因素影响。广西凭祥地处 106°43′E、22°13′N，为南亚热带气候区，而引种交趾黄檀的原产地经、纬度分别为 100°16′~105°28′、11°59′~16°49′，说明交趾黄檀的种植范围可向北扩展到南亚热带气候区。本研究中，2 年生(2017 年)交趾黄檀的平均苗高达 2.69m，是福建龙海市(E113°30′、N24°29′)山地引种 2 年生交趾黄檀苗高(1.70m)的 1.58 倍[25]，说明引种地的纬度会在一定程度上影响交趾黄檀的速生性，进一步说明广西凭祥比闽南更适宜交趾黄檀生长。

4 结论

从泰国和柬埔寨引入的10个种源25个家系交趾黄檀种子在广西凭祥引种栽培获得初步成效，且以泰国的种源和家系对广西凭祥的适应性更佳。

参考文献

[1]吴培衍，张荣标，张金文．红木树种新贵——交趾黄檀[J]．福建热作科技，2016，41(4)：51-54.

[2]IDA T P，TAPIO L，ANTO R，et al. An overview of the conservation status of potential plantation and restoration species in Southeast Asia[J]. Aho，2007，26(3)：213-219.

[3]SO T，DELL B. Conservation and utilization of threatened hardwood species through reforestation-an example of *Afzelia xylocarpa*(Kruz.) Craib and *Dalbergia cochinchinensis* Pierre in Cambodia[J]. Pacific Conservation Biology，2010，16(2)：101-116.

[4]BARRETT M，BROWN J，YODER A. Protection for trade of precious rosewood[J]. Nature，2013(499)：29-29.

[5]CITES. Consideration of proposals for amendment of appendices I and II[C]. Bangkok(Thailand)：Convention on International Trade in Endangered Species(CITES)，Co P16 Prop. 2013，60：1-16.

[6]DELANG Oc. The role of medicinal plants in the provision of health care in Lao PDR[J]. Journal of Medicinal Plants Research，2007，1(3)：50-59.

[7]PORNPUTTAPITAK W. Chemical constituents of the branches of Anomianthus dulcis and the branches of *Dalbergia cochinchinensis* Pierre[D]. Bangkok：Silpakorn University，2008：8-29.

[8]SHIROTA O，PATHAK V，SEKITA S，et al. Phenolic constituents from *Dalbergia cochinchinensis*[J]. Journal of Natural Products，2003，66(8)：1128-1131.

[9]LIU R H，WEN X C，SHAO F，et al. Flavonoids from heartwood of *Dalbergia cochinchinensis*[J]. Chinese Herbal Medicines，2016，8(1)，89-93.

[10]LIU R H，LI Y Y，SHAO F，et al. A new chalcone from the heartwood of *Dalbergia cochinchinensis*[J]. Chemistry of Natural Compounds，2016，52(3)，405-408.

[11]AREE T，TIP-PYANG S，SEESUKPHRONRARAK S，et al. 2-(5，7-Dihydroxy-4-oxo-4-H-chromen-3-yl)-5-methoxy-1，4-benzoquinone(isoflavonequinone)[J]. Acta Crystallographic，2003，59(3)：363-365.

[12]AREE T，TIP-PYANG S，PARAMAPOJN S，et al. 3，9-Dimethoxy-6 a，11 a-dihydro-6 H-benzo[4，5]furo[3，2-c]chromene-4，10-diol monohydrate[J]. Acta Crystallographic，2003，59(3)：380-383.

[13]ZHONG Y X，HUANG R M，ZHOU X J，et al. Chemical constituents from the heartwood of *Dalbergia cochinchinensis*[J]. Natural Product Research & Development，2013，25(11)：1515-1518.

[14]RASHEEHA N，IFTIKHAR H，ABDUL T，et al. Antimicrobial activity of the bioactive components of essential oils from Pakistani spices against *Salmonella* and other multidrug resistant bacteria[J]. BMC Complementary and Alternative Medicine，2013，13(1)：265.

[15]霄迪．大红酸枝——交趾黄檀[J]．家具，2009，S1(012)：90-93.

[16]LUO S，WU YQ，HUANG J. Thermal and chemical properties of benzene-alcohol extractives from two species of redwood[C]. International Conference on Biobase Material Science and Engineering(BMSE)，2013：156-160.

[17]JIANG T，LI K，LIU H，et al. The effects of drying methods on extract of *Dalbergia cochinchinensis* Pierre[J]. European Journal of Wood & Wood Products，2016，74(5)，663-669.

[18]ZHANG F D，XU C H，LI M Y，et al. Identification of *Dalbergia cochinchinensis*(CITES Appendix II) from other three *Dalbergia* species using FT-IR and 2D correlation IR spectroscopy[J]. Wood Science & Technology，2016，50(4)：693-704.

[19]王豁然，江泽平．论中国林木引种驯化策略[J]．林业科学，1995，31(4)：367-371.

[20]李瑞聪．交趾黄檀不同家系种子形态特性和发芽率差异[J]．福建热作科技，2016，41(2)：29-33.

[21]莫世琴，林明平，王淑娥．不同产地交趾黄檀种子形态与发芽率研究[J]．热带林业，2016，44(2)：8-10.

[22]李科，洪舟，杨曾奖，等．3不同种源交趾黄檀种子形态及多点发芽率的差异[J]．浙江农林大学学报，2018，35(1)：121-127.

[23]KAMO K，VACHARANGKURA T，TIYANON S，et al. Plant species diversity in tropical planted forests and implication for restoration of forest ecosystems in Sakaerat，northeastern Thailand[J]. Japan Agricultural Research Quarterly Jarq，2002，36(2)：111-118.

[24]NGHIA N H，LUOMAAHO T，HONG L T，et al. Status of forest genetic resources conservation and management in Vietnam[M]. Serdang，Malaysia：Proceeding of the Asia Pacific Forest Genetic Resources Programme(APFORGEN)，2004：290-301.

[25]李瑞聪．闽南山地引种交趾黄檀种源初步表现[J]．绿色科技，2017，(15)：127-129.

[原载：南方农业学报，2019，50(01)]

濒危植物格木天然种群的表型多样性及变异

李洪果[1] 陈达镇[2] 许靖诗[3] 刘光金[1] 庞晓东[4] 叶金辉[5] 莫小文[6] 谌红辉[1]

([1]中国林业科学研究院热带林业实验中心，广西凭祥 532600；[2]梧州市林政稽查支队，广西梧州 543002；
[3]北海市林业技术推广站，广西北海 536000；[4]博白县林业局，广西博白 537600；
[5]北流市民乐镇林业站，广西北流 537403；[6]陆川县沙坡镇林业站，广西陆川 537714)

摘 要 通过对濒危植物格木在我国中心分布区种群的果荚和种子表型性状测量分析，研究其表型多样性、进化和适应性潜力、表型变异规律、表型分化水平及种群聚类特征，结合地理气候信息探讨影响表型变异的关键因子，为格木天然种群的遗传多样性保护恢复和引种栽培提供科学依据。对8个格木天然种群114个单株的13个果荚及种子表型性状进行测量，利用巢式方差分析、多重比较、相关性分析、主成分分析、聚类分析等分析方法，探讨格木天然种群的表型多样性水平、进化和适应性潜力、表型变异规律、表型分化水平、种群聚类特征及其与地理气候因子的相关性。格木表型多样性水平中等，13个性状的Shannon-Weaver指数为1.9111(种形指数)~2.1039(果荚周长)，平均为2.0278；8个种群的Shannon-Weaver指数在0.9209(P6)~1.8856(P8)，均值为1.4747。格木表型性状离散程度较轻，13个性状的变异系数在7.5446%(种子长)~18.8685%(果荚面积)，均值为12.4109%；8个种群的变异系数在8.8529%(P1)~13.9848%(P5)，均值为12.4109%。13个表型性状的极大值为极小值的1.3952(种形指数)~2.8056(每荚种子数)倍，均值为1.7917。巢式方差分析表明，格木13个表型性状在种群间和种群内均存在极显著差异($P<0.01$)，说明这些性状在种群间和种群内均存在丰富的变异；种群内的变异(40.3871%)远大于种群间(11.9489%)。种群平均表型分化系数为21.8579%，各性状的分化系数变化幅度在7.4765%(种形指数)~38.6740%(果荚宽)。相关性分析表明，13个表型性状之间大多存在显著或极显著相关，果荚宽度、种子周长等性状和温度指标呈负相关，与纬度呈正相关，格木在气温高、降水量大的地区往往具有小果荚、小种子的特征，随着温度指标降低和纬度增加，格木果荚及种子有变大的趋势。格木13个表型性状可提炼为4个主成分，累积贡献率为87.382%。8个天然群体基于欧氏距离可分为3类，第1类2个种群呈现果荚小、种子小的特征，该类表型主要出现在格木分布区的南端；第2类5个种群呈现果荚大、种子大的特征，该类表型主要出现在格木分布区的北部，仅1个种群位于分布区的南部；第3类1个种群呈现果荚小、种子少、种子大而薄的特征，该类表型出现在格木南北分布区的中间衔接部。格木的表型多样性处于中等水平，在个体上存在较为丰富的遗传变异。在群体及物种水平上，格木表型性状离散程度较轻，各个性状分布相对集中，较为稳定，不易发生群体分化，拥有特殊表型的格木单株具有较高的保存价值。果荚及种子大小进化及适应性潜力相对较大，种形指数最不易产生变异。种群内的变异是格木表型变异的主要来源，格木的种质资源收集保存可减少种群数量，增加种群内个体数量，并优先收集有特殊表型的单株。容县和龙圩的格木种群是格木遗传多样性保护及种质资源收集保存的重点区域。年均温等温度因子是影响格木表型及分布的最主要因子。果荚及其中种子的大小是格木种群间聚类的主要依据。

关键词 格木；濒危植物；遗传多样性；天然种群；表型性状；种质资源

Phenotypic Genetic Diversity and Variations Research in Natural Populations of *Erythrophleum fordii*, an Endangered Plant Species

LI Hongguo[1], CHEN Dazhen[2], XU Jingshi[3], LIU Guangjin[1],
PANG Xiaodong[4], YE Jinhui[5], MO Xiaowen[6], CHEN Honghui[1]

([1]*Experimental Center of Tropical Forestry*, *Chinese Academy of Forestry*, *Pingxiang* 532600, *Guangxi*, *China*;
[2]*Forestry Administration Inspection Detachment of Wuzhou*, *Wuzhou* 543002, *Guangxi*, *China*;
[3]*Forestry Technology Extension Station of Beihai*, *Beihai* 536000, *Guangxi*, *China*;[4]*Bobai Forestry Bureau*, *Bobai* 537600, *Guangxi*, *China*;
[5]*Minle forestry station of Beiliu*, *Beiliu* 537403, *Guangxi*, *China*;[6]*Shapo forestry station of Luchuan*, *Luchuan* 537714, *Guangxi*, *China*)

Abstract: The phenotypic traits of pod and seed from an endangered plant *Erythrophleum fordii* were measured to investigate the genetic diversity, evolutionary potential, phenotypic variation, phenotypic differentiation and cluster feature of populations. And combing with geographical and climatic informations, the factors, which may affect the phenotypic variation, were discussed. This study provide basis for genetic diversity protection, reestablishment, and introduction of *E. fordii*. The phenotypic characteristics of thirteen pod and seed from 114 individuals of eight populations were measured. Nested analysis of variance, multi-comparison, correlation analysis, principal components analysis, and cluster analysis were used to study the pod and seed genetic diversity, evolutionary potential, phenotypic variation, phenotypic differentiation, clusterfeature of populations, and the relationship between phenotypic traits and climate factors, respectively. *E. fordii* showed a modest level of genetic diversity, the range of Shannon-Weaver index is from 1.9111 (Shape index of seed) to 2.1039 (Perimeter of pod), and the average was 2.0278. The range of Shannon-Weaver index of eight populations is from 0.9209 (P6) ~ 1.8856 (P8), and the average was1.4747. The range of variation of variance among the 13 traits was from 7.5446% (Length of seed) to 18.8685 % (Area of pod), and the average was 12.4109%. The range of variation of variance among the eight populations was from 8.8529% (P1) ~ 13.9848% (P5), and the average was 12.4109%. The range of times of maximum and minimum was from 1.3952 (Shape index of seed) ~ 2.8056 (No. of seeds in pod), and the average was 1.7917. Analysis of nested variance showed that there was a significant differences (P<0.01) among and within populations, suggesting that abundant variation among and within populations. The within-population variation (40.3871%) was much larger than the among-population variation (11.9489%); and the average phenotypic population differentiation coefficient was 21.8579% with a range from 7.4765% (Shape index of seed) to 38.6740% (Width of pod). Correlation analysis revealed that there were significant correlations between most phenotypic traits. The significantly negative correlation between the width of pod, length of seed and perimeter of seed was detected, whereas, there were significantly positive correlations between width of pod and latitude. The small pod and seed phenotypic traits were found in the areas with high temperature and rainfall in general, and the size of pod and seed became bigger and bigger along with the increase of the latitude and the decrease of the temperature. Principal component analysis showed that the four principal components added up to 87.382% of the variation. The eight natural populations were divided into three groups by UPGMA cluster analysis based on euclidean distance calculated from phenotypic traits. The first group was comprised of the small pod and small seed, which located in the south areas. The second group has characteristics of big pod and big seed, which located in the nouth areas, and only one population form the south area. The small pod, big and thin seed, which located in the middle areas, belonged to the third group. *E. fordii* has the modest level of genetic diversity, there were abundant variations among and within populations. The degree of phenotypic traits was slight and stable, and there were low level of population differentiation in *E. fordii*. Within-population variation was the main source of variation. Germplasm collection and preservation will decrease the populations, and increase the individual trees, especially the individuals had the special phenotypic traits. Rongxian and Longxu are the most important region for genetic protection and germplasm resources collection. Temperature factor was the most important factor affect the phenotypic traits and distribution of *E. fordii*. Pod and seed size is the clustering basis of populations.

Key words: *Erythrophleum fordii*; endangered plant species; genetic diversity; natural population; phenotypic traits; germplasm resources

遗传多样性广泛存在于物种及其基因组之中[1]，作为物种应对环境适应性的潜力，其在制定育种策略和濒危物种保存等各个方面有着积极意义[2,3]。植物在进化过程中，不断与周围环境进行选择与被选择的作用，最终形成许多外在形态和内在生理上的适应策略，这些生态适应特征即为表型性状[4]。表型多样性是遗传多样性的直接体现，是基因与环境在长期进化过程中共同作用的结果。利用表型性状的差异来检测种群的遗传多样性是长期以来最简便易行的方法。研究表明，濒危植物和动物较非濒危种往往表现出更少的遗传多样性[5,6]，因此，种群数量对物种的稳定遗传有一定影响，维持种群有效样本的保有量是濒危动植物收集和保护的关键。

格木(*Erythrophleum fordii*)是我国热带亚热带地区珍贵的用材和绿化树种，同时也具有较高的药用价值，木材结构致密坚实，极耐腐，有"铁木"之称，是优良的工业、家具和工艺品用材。格木在我国主要分布在广西、广东中南部、福建南部和台湾东海岸[7]，其中心分布区在广西博白、合浦、陆川一带[8]，也有学者认为越南、老挝是格木的中心分布区，并基于广西的表型变异最丰富，推测我国福建、广东的格木种群是广西种群的自然延伸[9]。但受限于国土面积，越南的格木分布并不多(有记录的仅为

2.5hm^2 人工林），大致位于和广西接壤的越南中北部地区；老挝的资源数量未知，只是根据气候环境适生推测其有分布。所以，从资源数量来看，格木天然林资源中我国应占大多数，广西的天然林资源在我国的格木资源分布中又占多数，加之广西可能是格木在我国的中心分布区，而物种中心分布区的种群往往具有更高的遗传多样性，可作为种质资源收集保存和遗传多样性研究的重点[10]。因此，广西的格木天然种群似可作为的表型多样性及变异规律研究的关键种群，有必要进行更广泛的取材和更深入的研究。

长期以来，由于筑路修桥、城市发展等因素造成的过度砍伐、大面积毁林等，导致格木天然林急剧减少，部分地区已经消失[11]。虽然格木当前已被列为国家二级保护植物，各县市林业局宣传和保护力度较大，同时市场上也杜绝了格木木材交易，但前期破坏性开发造成的格木生境碎片化现状未见有恢复措施，大多数格木仍处于孤立木、散生木和小群落的生存状态，时刻面临雷击、野火和盗伐等的威胁。格木种子自然萌发难和苗期存活率低等特性致使格木天然更新困难，难以仅依靠自身实现种群延续[12]，而经营周期长、木材生长缓慢、自然干形差等原因也造成社会层面上栽植积极性不高，难以通过人们的自发行为实现种群数量的恢复，偶有国有林场等部门用格木造林，也与格木遗传多样性的保护无关。整体而言，格木种群尤其是天然林种群，仍处于种群数量持续减少的状态，现有就地保存的方式不足以保证格木天然种群脱离濒危状态，也不利于格木天然种群的多样性保护和稳定遗传。因此，开展格木在我国中心分布区种群的表型多样性研究，了解不同种群的表型变异特征和多样性水平，参照其地理变异规律在种群样本数量稀少的分布区引入合适的种源造林，对格木天然林种群的多度恢复和遗传多样性保护具有重要意义。

1 材料与方法

1.1 试验材料

在广西壮族自治区 2017 年名木古树调查的基础上，于 2017 年 10、11 月从格木在我国的中心分布区及周边共 8 个县市(苍梧、龙圩、容县、北流、陆川、博白、浦北、合浦)收集 114 个格木单株的果荚及种子。各种群的样本数及地理气候因子信息见表 1。原则上每种群的样本数应在 10 株以上，但部分种群由于样本偏少或采样困难等原因，样本数不足 10 株。用高枝剪、攀登板等工具从树上采集成熟的果荚及种子，择发育正常、无病虫害的果荚及种子带回实验室进行测量。

1.2 表型性状的测量

测量果荚的 7 个表型性状，每个性状测量 10 次，取平均值作为实测值，其中果荚长、果荚指数、果荚宽、果荚面积、果荚周长 5 个性状用万深叶面积分析系统测量，测量结果精确至 0.0001；每荚种子数人工数取；果荚厚度用游标卡尺测量，精确到 0.01mm。测量种子的 6 个表型性状，每个性状测量

表 1　格木 8 个种群的地理位置及主要气候因子

种群及代码	样本数	年均温(℃)	年均相对湿度(%)	年均降水量(mm)	1 月均温(℃)	7 月均温(℃)	经度(E)	纬度(N)
北流(P1)	4	22.1	78	1555.8	13.4	28.4	110.37	22.85
博白(P2)	22	22.4	79	1796.7	14.0	28.3	109.97	22.28
苍梧(P3)	8	21.5	79	1446.6	12.2	28.7	111.23	23.42
合浦(P4)	21	23.0	80	1760.5	14.6	28.9	109.20	21.67
龙圩(P5)	15	21.2	79	1452.9	12.2	28.2	111.27	23.48
陆川(P6)	3	21.8	81	1914.1	13.4	27.9	110.27	22.33
浦北(P7)	13	21.7	82	1705.3	13.1	27.9	109.55	22.27
容县(P8)	28	21.5	79	1635.3	12.8	28.1	110.55	22.87

25次取平均值作为实测值，其中种子长、种形指数、种子宽、种子面积、种子周长5个性状用万深叶面积分析系统测量，测量结果精确至0.0001，种子厚度用游标卡尺测量，精确到0.01mm。经纬度信息来自当地政府门户网站，气象数据来自中国气象数据网(1981-2010)。

1.3　数据处理与分析

1.3.1　巢式方差分析和多重比较

在R3.5.0软件中对13个表型性状进行巢式方差分析和多重比较。其线性模型为：

$$Y_{ijk}=\mu+\alpha_i+\beta_{j(i)}+\varepsilon_{ijk}$$

式中：Y_{ijk}为第i个种群的第j个个体的第k个观测值；μ为总体平均值；α_i为第i个种群的效应值；$\beta_{j(i)}$为第i个种群内第j个个体的效应值，ε_{ijk}为第ijk个观测值的试验误差[13,14]。

1.3.2　相关性分析、主成分分析和聚类分析

在SPSS19.0软件中进行相关性分析、主成分分析和聚类分析等。相关性分析，基于样本表型性状的测量均值，采用Pearson相关系数和双尾检验进行分析。主成分分析，先将样本表型性状的原始测量数据利用Z得分法标准化，以消除量纲和数量级，经KMO检测判断适合主成分分析后(KMO>0.7)，再进行主成分分析。聚类分析，根据主成分分析的结果，基于因子得分系数矩阵，提取贡献率最大的第1和第2主成分，基于组间连接法和欧式距离平方做群体间的聚类分析。

1.3.3　多样性参数的计算

在Excel软件中计算各性状的变异系数(CV)、表型分化系数(Vst)[15]和Shannon-Weaver指数(H)[16]。CV表明性状变异的离散程度，公式为：

$$CV=\frac{\sigma}{x}\times100\%$$

式中：σ和$\bar{x}$分别为性状的标准差和均值。Vst描述表型变异在种群间贡献的大小，公式为：

$$V\text{st}=\sigma^2_{t/s}(\sigma^2_{t/s}+\sigma^2_s)$$

式中：$\sigma^2_{t/s}$和σ^2_s分别为种群间和种群内方差值。Shannon-Weaver指数(H)说明该性状表型变异的丰富及均匀程度，公式为：

$$H=-\sum P_i ln(P_i)$$

式中：$P_i=n_i/N$，P_i表示某表型性状第i个代码出现的频率；n_i为某表型性状出现的次数；N为样本数；在计算*Shannon-Weaver*指数前先将测量数据分成10级，转换为质量性状，$<\bar{x}-2\sigma$为第1级，$\geq\bar{x}+2\sigma$为第10级，每级相差0.5σ。

2　结果与分析

2.1　格木表型性状的多样性

格木种质资源的果荚及种子表型性状特征见表2。13个表型性状变异的变异系数在7.5446%~18.8685%，均值为：12.4109%。果荚的表型变异更加丰富，其中果荚面积的变异系数最大(18.8685%)，变异幅度为：2735.8798~6479.2329mm^2，果荚宽度的变异系数最小(10.3597%)，变异幅度为：29.0547~47.7572mm。种子的6个表型性状中，种子面积的变异系数最大(13.7713%)，变异幅度为：119.3331~263.5889mm^2，种子长度的变异系数最小(7.5446%)，变异幅度为：13.6435~19.7397mm。13个表型性状的变异系数，由大到小分别为：PA>PNS>PT>PP>SA>PSI>PL>ST>PW>SP>SW>SSI>SL。变异系数的分析结果表明：格木的表型性状变异相对集中稳定，物种层面上表型性状的离散程度较轻，尚未形成地理上的种群分化，短期内不易形成亚种或地理小种。

13个表型性状的遗传多样性指数(H)在1.9111~2.1039，均值为：2.0278。其中果荚周长(PP)的遗传多样性指数最高(2.1039)，种形指数(SSI)的遗传多样性指数最低(1.9111)。13个表型性状的遗传多样性从高到低分别为：PP>PW>ST>PA>PSI>PL>SW>SA>SP>SL>PNS>PT>SSI。Shannon-weaver多样性指数的分析结果表明：格木在个体水平上存在较为丰富的表型变异，这为格木制定保护策略提供了方向，即应优先保存各个种群中有特殊表型的单株，因为在表型性状分布较为集中的背景下，特殊表型往往也是稀有表型，具有较高的保护价值；相反，性状出现频率较高的单株可以适当减少保存株数。

植物不同表型性状的极值往往反应该性状在不同环境条件下的进化及适应性潜力。13个表型性状的极大值为极小值的1.3952(种形指数)~2.8056(每荚种子数)倍。果荚的7个表型性状中，每荚种子数(PNS)的极大值与极小值的比值最大(2.8056)，果荚周长(PP)的极大值与极小值的比值最小(1.5777)；种子的6个表型性状中，种子面积(SA)的极大值与极小值的比值最大(2.2089)，种形指数(SSI)的极大值与极小值的比值最小(1.3952)。13个表型性状的进化及适应性潜力从高到低分别为：PNS>PA>SA>PT>PW>PSI>PL>SW>PP>SP>ST>SL>SSI。13个表型性状的极值分析表明，果荚和种子大小等测量性状具有较高的进化或适应性潜力，种子形状的进化及适应性潜力最小。

表2 格木表型性状的多样性

	表型性状	代码	均值	标准差	极大值	极小值	极差	变异系数(%)	多样性指数 *H*
果荚	果荚长(mm)	PL	150.2879	14.7855	185.6618	115.2563	70.4055	12.5586	2.0463
	果荚指数	PSI	3.9294	0.3672	4.9254	3.0073	1.9181	12.9542	2.0480
	果荚宽(mm)	PW	38.5434	3.8500	47.7572	29.0547	18.7025	10.3597	2.0799
	果荚面积(mm^2)	PA	4534.1389	815.7941	6479.2329	2735.8798	3743.3531	18.8685	2.0685
	果荚周长(mm)	PP	372.3136	38.0067	453.1073	287.1886	165.9187	14.6390	2.1039
	果荚厚(mm)	PT	9.0350	1.0521	13.5380	6.9200	6.6180	15.0764	1.9762
	每荚种子数	PNS	7.4542	0.9923	10.1000	3.6000	6.5000	18.4660	1.9884
种子	种子长(mm)	SL	16.4297	1.0759	19.7397	13.6435	6.0961	7.5446	2.0059
	种形指数	SSI	1.2331	0.0612	1.5615	1.1192	0.4423	7.7911	1.9111
	种子宽(mm)	SW	13.3869	0.9237	17.2011	10.8981	6.3030	8.8739	2.0284
	种子面积(mm^2)	SA	169.9411	22.3956	263.5889	119.3331	144.2558	13.7713	2.0144
	种子周长(mm)	SP	51.2452	3.6838	64.9372	41.5409	23.3964	9.9577	2.0134
	种子厚(mm)	ST	5.8118	0.4808	7.0729	4.7160	2.3569	10.4812	2.0776
	均值(标准差)		—	—	—	—	—	12.4109 (3.5527)	2.0278 (0.0497)

表型性状的变异系数分析、多样性分析和极值分析都表明，果荚性状中的果荚大小、种子性状中的种子大小等性状较其他性状变异更为丰富，而种形指数(SSI)表现最为稳定，无论果荚和种子的长、宽及其他性状如何变化，是小粒种子还是大粒种子，种子形状都始终保持稳定。

2.2 格木果荚与种子的表型变异特征(方差分析、均值、标准差和多重比较)

格木13个表型性状在种群间和种群内的方差分析结果见表3，表3表明，13个表型性状在种群内和种群间均存在极显著差异，说明格木果荚及种子的形态性状在种群间和种群内都存在一定程度的变异。

13个表型性状的均值、标准差和多重比较见表4。表4表明，P1种群的果荚长、果荚面积、果荚周长、果荚厚度、种子厚度均为最大，表现为大果荚，饱满种子的特征；P2、P4种群的果荚长、果荚宽、果荚面积、果荚周长、种子长、种子宽、种子面积、种子周长分别为最小和次小，表现为果荚小、种子多，种子小的特征；P3和P5种群的13个表型测量指标均居中，没有特殊的表型特征；P6种群的果荚指数、果荚厚度、每荚种子数、种子厚度均为最小，果荚长、果荚面积和果荚周长等指标较小，而种子长、种子面积、种子周长为最大，表现为果荚小、种子数少、种子大而薄、种子不饱满等败育种子的特征，这与该分布区资源数量稀少，样本长期处于孤立木和小种群状态，结实率低的事实相符；P7种群的果荚指数最大而种形指数最小，表现为果荚长而窄，种子短而宽，这与格木种子在果荚内沿果荚一侧横向排列的方式相符；P8种群的果荚最宽、种形指数最大而果荚指数次小，表现为果荚长而宽，种子短而宽的特征。

2.3 格木种群内果荚与种子的表型变异特征

变异系数CV表示种群内表型性状的变异特征，变异系数越大，表型性状的离散程度越大。由表5可知，8个种群的变异系数在8.8529%（P1）~

13.9848%(P5)，均值为：12.410 9%。在种群水平上：8个种群的13个表型性状变异系数均值均未达到15%，说明格木表型性状的变异水平较低；8个种群的果荚表型性状变异系数均大于种子表型性状的变异系数，说明格木果荚的表型变异较种子更为丰富；8个种群中，P5、P7种群的表型变异相对丰富，P1种群的表型变异水平最低。P1、P2、P3和P4种群的果荚及种子表型的变异系数均未达到15%，说明这些种群的果荚及种子表型性状都处于较低的变异水平。P5、P6、P7和P8种群果荚的7个表型性状变异系数达到了15%以上，说明这4个种群的果荚表型变异相对丰富，其中P6种群果荚的表型变异最为丰富。在物种水平上，除了果荚面积(PA)、果荚厚度(PT)和每荚种子数(PNS)外，其他10个表型性状的变异系数均未达到15%，说明除前述3个性状外，其余10个性状较为稳定；13个表型性状中，果荚面积表型变异最为丰富，而种子长度的表型变异水平最低。格木种子的表型性状无论在种群水平上还是在物种水平上，表型变异系数均未达到15%，说明格木种子的表型变异水平极低。

2.4 格木种群内果荚与种子的表型多样性

Shannon-Weaver指数反映不同表型性状的丰富及均匀程度，8个种群13个表型性状的多样性见表6。由表6可知，8个种群的Shannon-Weaver指数(H)在0.9209(P6)~1.8856(P8)，均值为：1.4747。果荚和种子的表型多样性指数均为容县(P8)种群最大，陆川(P6)种群最小，8个种群的表型多样性指数从高到低分别为：容县(P8)>龙圩(P5)>博白(P2)>浦北(P7)>合浦(P4)>苍梧(P3)>北流(P1)>陆川(P6)。因此，容县和龙圩种群是遗传多样性保护和种质资源收集的重点区域。一般而言，种群表型多样性的高低通常与样本数多少有很大关系，样本数多的种群往往具有更高的表型多样性，对比种群样本数占比和多样性占比发现，苍梧(P3)、北流(P1)、陆川(P6)等样本数少的种群其多样性占比远高于样本数量占比，似可认为这些种群的样本具有更高的表型多样性及变异，在种质资源收集过程中也应予以重视。

表3 格木种群间及种群内各表型性状的方差分析①

性状	均方			F	
	种群间	种群内	随机误差	种群间	种群内
PL(mm)	5170.7122	1964.9422	208.5314	24.7958**	9.4228**
PSI	3.0099	1.1927	0.1649	18.2563**	7.2339**
PW(mm)	913.5415	95.6635	9.1574	99.7596**	10.4465**
PA(mm^2)	31312224.0000	4978819.0000	373185.0000	83.9054**	13.3414**
PP(mm)	51947.6930	11611.0500	2116.3510	24.5459**	5.4864**
PT(mm)	18.1479	10.6669	1.3078	13.8762**	8.1561**
PNS	1.1942	0.1494	0.0225	53.1379**	6.6478**
SL(mm)	106.1374	20.1743	0.7974	133.1012**	25.2995**
SSI	0.1002	0.0819	0.0070	14.2349**	11.6335**
SW(mm)	61.2813	16.4292	0.9404	65.1635**	17.4699**
SA(mm^2)	48009.5522	8637.4056	287.1656	167.1842**	30.0781**
SP(mm)	1240.2003	240.1476	17.1675	72.2410**	13.9885**
ST(mm)	11.7562	4.6597	0.2068	56.8380**	22.5282**

注：①性状代码见表2；**：$P<0.01$。下同。

表 4　8 个格木天然种群 13 个表型性状的均值、标准差及多重比较

种群	性状												
	PL(mm)	PSI	PW(mm)	PA(mm^2)	PP(mm)	PT(mm)	PNS	SL(mm)	SSI	SW(mm)	SA(mm^2)	SP(mm)	ST(mm)
P1	162.6249±10.1831A	4.0234±0.3935ABC	40.6160±2.6676AB	5247.7955±426.0899A	416.2664±44.7787A	9.7553±1.2156A	7.2647±1.1364B	16.4305±1.0778C	1.2106±0.0908BC	13.6063±0.8439BC	167.4708±16.9725C	51.5352±4.0204C	6.4410±0.4701A
P2	144.4897±15.7899C	4.0437±0.5252ABC	35.9639±3.3737E	4059.8494±689.6582D	349.2166±39.0073C	8.8289±1.2175CD	8.2243±1.5401A	15.6605±1.1004D	1.2240±0.0870BC	12.8297±0.9410D	154.9332±19.4795D	48.6160±4.2904D	5.9356±0.4952BC
P3	152.0950±17.9072B	3.8915±0.4922BCD	39.4495±5.0905BCD	4677.6204±977.7450BC	380.0562±60.1821B	8.8184±1.4333CD	7.2500±1.0614B	16.4092±1.5965C	1.2319±0.0949ABC	13.3752±1.4469C	167.9750±29.9671C	51.5340±6.3808C	5.6278±0.6707E
P4	143.0568±16.2821C	4.0522±0.4701AB	35.4698±3.3880E	4016.2991±711.1578D	352.9805±48.7914C	8.6624±0.9686DE	7.2905±1.0565B	15.8766±1.2135D	1.2327±0.1007ABC	12.9280±1.0656D	157.4356±20.2132D	49.1628±4.3682D	5.7936±0.5326CD
P5	154.0226±22.4575B	3.8722±0.4649CD	39.8963±4.7792BC	4791.6477±998.7544B	387.9009±67.6653B	9.5076±1.7492AB	7.3733±1.3881B	16.5675±1.2477C	1.2279±0.0985BC	13.5809±1.5118C	172.2102±28.1711C	52.1227±6.0368BC	5.8802±0.7703BC
P6	143.5110±24.7440C	3.7142±0.5671D	38.6663±3.3608CD	4063.4127±899.6260D	355.4976±59.0345C	8.2093±1.1895E	4.8667±1.3830C	17.2303±1.0360A	1.2386±0.0788AB	13.9509±1.0414AB	187.9067±17.1799A	53.7779±5.2453A	5.0274±0.5006F
P7	154.6753±22.9713B	4.1049±0.6931A	38.1734±5.1085D	4511.0765±1080.4592C	382.0603±64.2912B	9.1277±1.1865BCD	7.4462±1.5354B	16.9098±1.3444B	1.2103±0.0975C	14.0280±1.2443A	186.5124±26.8691A	53.3159±5.3763AB	6.0168±0.7728B
P8	155.1500±20.8677B	3.7514±0.4673D	41.5230±4.3268A	5065.7189±1028.1827A	384.7501±56.3632B	9.2448±1.9450ABC	7.4786±1.2238B	17.0195±1.3452AB	1.2617±0.1187A	13.5830±1.4842C	179.5405±30.2997B	52.7431±5.4582ABC	5.7145±0.6398DE

注：群体及性状代码分别见表 1、2；两两种群间不同字母表示差异显著($P<0.01$)。下同。

表 5 格木 8 个种群的表型性状变异系数

类型	性状	种群								均值	标准差
		P1	P2	P3	P4	P5	P6	P7	P8		
果荚	PL	6.2617	10.9280	11.7737	11.3816	14.5807	17.2419	14.8513	13.4500	12.5586	3.0900
	PSI	9.7803	12.9881	12.6481	11.6011	12.0061	15.2684	16.8847	12.4567	12.9542	2.0600
	PW	6.5679	9.3808	12.9038	9.5518	11.9791	8.6918	13.3824	10.4202	10.3597	2.1508
	PA	8.1194	16.9873	20.9026	17.7068	20.8437	22.1397	23.9512	20.2969	18.8685	4.5695
	PP	10.7572	11.1699	15.8351	13.8227	17.4440	16.6062	16.8275	14.6493	14.6390	2.3908
	PT	12.4609	13.7899	16.2535	11.1817	18.3979	14.4897	12.9989	21.0389	15.0764	3.0882
	PNS	15.6428	18.7262	14.6400	14.4915	18.8260	28.4176	20.6199	16.3640	18.4660	4.2822
	均值(标准差)SD	9.9415 (3.1119)	13.4243 (3.1349)	14.9938 (2.8679)	12.8196 (2.5242)	16.2968 (3.2249)	17.5508 (5.7718)	17.0737 (3.6867)	15.5251 (3.6728)	14.7032 (2.8829)	—
种子	SL	6.5598	7.0266	9.7293	7.6433	7.5310	6.0127	7.9504	7.9039	7.5446	1.0405
	SSI	7.5004	7.1078	7.7035	8.1691	8.0218	6.3620	8.0559	9.4079	7.7911	0.8278
	SW	6.2023	7.3345	10.8178	8.2426	11.1318	7.4648	8.8701	10.9269	8.8739	1.7672
	SA	10.1346	12.5728	17.8402	12.8390	16.3586	9.1428	14.4061	16.8762	13.7713	2.9633
	SP	7.8013	8.8251	12.3817	8.8852	11.5819	9.7536	10.0839	10.3487	9.9577	1.4057
	ST	7.2986	8.3429	11.9176	9.1929	13.0999	9.9574	12.8440	11.1961	10.4812	1.9890
	均值(标准差)SD	7.5828 (1.2649)	8.5350 (1.9226)	11.7317 (3.1316)	9.1620 (1.7193)	11.2875 (2.9969)	8.1156 (1.5838)	10.3684 (2.4479)	11.1100 (2.8002)	9.7366 (2.0903)	—
总平均(标准差)SD		8.8529 (2.7084)	11.1677 (3.5971)	13.4882 (3.4058)	11.1315 (2.8495)	13.9848 (3.9977)	13.1960 (6.4203)	13.9789 (4.6106)	13.4874 (3.9657)	12.4109 (3.5527)	—

表 6 格木 8 个种群表型性状的 Shannon-Weaver 指数

类型	性状	种群							
		P1	P2	P3	P4	P5	P6	P7	P8
果荚	PL	1.0397	1.7550	1.3209	1.6246	1.7670	1.0986	1.4181	2.0440
	PSI	1.3863	1.8977	1.4942	1.8146	1.5868	1.0986	1.8185	1.7915
	PW	1.0397	1.7356	1.4942	1.7002	1.5292	0.6365	1.6052	1.8253
	PA	0.6931	1.5981	1.5596	1.6462	1.7141	0.6365	1.5247	1.9667
	PP	1.0397	1.6580	1.5596	1.5714	1.8989	1.0986	1.5858	1.9057
	PT	0.5623	1.4658	1.3209	1.3282	1.7321	1.0986	1.7380	1.9950
	PNS	1.0397	1.8110	1.0822	1.5917	1.6566	0.6365	1.6977	1.8313
	均值(标准差)SD	0.9715 (0.2494)	1.7030 (0.1324)	1.4045 (0.1613)	1.6110 (0.1377)	1.6978 (0.1128)	0.9006 (0.2287)	1.6269 (0.1255)	1.9085 (0.0893)
种子	SL	1.0397	1.7087	1.4942	1.7317	1.6792	1.0986	1.4986	1.7959
	SSI	1.3863	1.6174	0.9743	1.9015	1.6566	0.6365	1.7380	1.8833
	SW	0.0000	1.5403	1.5596	1.4610	1.9338	1.0986	1.5247	1.8375
	SA	1.0397	1.5506	1.7329	1.4347	1.8065	1.0986	1.6716	1.9116
	SP	0.6931	1.5775	1.6675	1.4265	1.8943	0.6365	1.6260	1.8379
	ST	1.0397	1.4804	1.5596	1.6448	1.8594	1.0986	1.6260	1.8874
	均值(标准差)SD	0.8664 (0.4361)	1.5792 (0.0711)	1.4980 (0.2469)	1.6001 (0.1765)	1.8050 (0.1044)	0.9446 (0.2178)	1.6141 (0.0819)	1.8589 (0.0388)
总平均(标准差)SD		0.9230 (0.3522)	1.6459 (0.1248)	1.4477 (0.2105)	1.6059 (0.1569)	1.7473 (0.1214)	0.9209 (0.2248)	1.6210 (0.1078)	1.8856 (0.0748)
样本数占比(%)		3.5088	19.2982	7.0175	18.4211	13.1579	2.6316	11.4035	24.5614
多样性占(%)		6.9545	12.4013	10.9079	12.0999	13.1653	6.9387	12.2137	14.2074

表7 格木表型性状的方差分量及种群间与种群内表型分化系数

性状	方差分量			方差分量比			表型分化系数(%)
	种群间	种群内	随机误差	种群间	种群内	随机误差	
PL(mm)	36194.9900	208283.8700	211450.8300	7.9387	45.6833	46.3779	14.8049
PSI	21.0695	126.4222	167.1787	6.6957	40.1761	53.1282	14.2851
PW(mm)	6394.7910	10140.3270	9285.6380	24.7661	39.2720	35.9619	38.6740
PA(mm^2)	219185569.0000	527754766.0000	378409610.0000	19.4771	46.8970	33.6260	29.3444
PP(mm)	363633.9000	1230771.3000	2145979.5000	9.7218	32.9049	57.3732	22.8068
PT(mm)	127.0351	1130.6942	1326.1551	4.9164	43.7595	51.3241	10.1003
PNS	8.3594	15.8364	22.7881	17.7920	33.7060	48.5020	34.5489
SL(mm)	742.9620	2138.4770	1936.1330	15.4219	44.3891	40.1890	25.7844
SSI	0.7016	8.6825	17.0953	2.6496	32.7896	64.5608	7.4765
SW(mm)	428.9692	1741.4904	2283.3510	9.6315	39.1011	51.2674	19.7640
SA(mm^2)	336066.9000	915565.0000	697238.1000	17.2442	46.9793	35.7765	26.8503
SP(mm)	8681.4020	25455.6440	41682.7720	11.4500	33.5739	54.9761	25.4309
ST(mm)	82.2936	493.9260	502.2015	7.6309	45.8008	46.5682	14.2816
均值(标准差) Mean(*SD*)	—	—	—	11.9489 (6.2267)	40.3871 (5.3943)	47.6639 (8.8762)	21.8579 (9.0632)

2.5 格木果荚与种子表型性状变异的来源及种群间表型分化

表型分化系数 *V*st 是种群间遗传变异占总变异的百分比，用以估算种群间的表型分化情况。由表7可知，13个格木表型性状在种群间的表型分化系数在38.6740%～7.4765%，平均为：21.8579%。种群间方差分量占总变异的11.9489%；种群内占40.3871%，说明格木表型在种群间和种群内均存在一定程度的变异，但种群内的变异是其表型变异的主要来源，种群内的多样性大于种群间的多样性。

2.6 格木表型性状间的相关性及其与地理气候因子间的相关性

由表8可知，13个表型性状间大多存在显著或极显著相关性。格木果荚长度与果荚指数、果荚宽度、果荚面积、每荚种子数、种子长度种子宽度、种子面积和种子周长呈极显著相关，说明较长的果荚往往也是较大的果荚，其包含的种子数相应较多，且种子较大。果荚指数与果荚宽、种子长、种形指数呈极显著负相关，与果荚周长和每荚种子数呈极显著正相关，这与格木种子在果荚一侧横向排列，种子的长宽分别对应果荚的宽和长相符；果荚宽度与果荚面积、果荚周长、种子长、种形指数、种子宽、种子面积、种子的周长呈极显著正相关，与果荚的厚度显著正相关，说明宽果荚往往也是较大的果荚，并伴随有大粒种子的特征。果荚面积与果荚厚、每荚种子数、种子长、种形指数、种子宽、种子面积、种子周长和种子厚度呈极显著正相关，也说明大果荚伴随大粒种子和饱满种子的特征。果荚周长与每荚种子数、种子长、种子宽、种子面积和种子周长呈极显著正相关。果荚厚度与每荚种子数显著正相关，与种子厚度极显著正相关，说明较厚的果荚往往包含更饱满的种子，可能由于营养较为充足，果荚中的种子数也有相应增加的情况。每荚种子数与种子宽、种子面积和种子周长呈显著负相关，与种子厚度呈极显著正相关，说明果荚中种子越小，每荚种子数越多，种子越饱满果荚也越厚。种子长与种形指数、种子宽、种子面积和种子周长呈极显著正相关，与种子厚度呈极显著负相关；种形指数与种子宽度成极显著负相关；种子宽度与种子面积和种子周长呈极显著正相关，与种子厚度呈极显著负相关；种子面积与种子周长呈极显著正相关，与种子厚度呈极显著负相关；种子周长与种子厚度呈极显著负相关。种子表型性状间的相关性说明，长粒种子、宽粒种子往往也是大粒种子，但是种子过大也伴随有种子变薄而不饱满的情况，这可能意味着畸形或败育种子产生。

表 8 性状间的相关性分析

性状	性状												
	PL	PSI	PW	PA	PP	PT	PNS	SL	SSI	SW	SA	SP	ST
PL	1												
PSI	0.438**	1											
PW	0.572**	-0.481**	1										
PA	0.866**	-0.040	0.880**	1									
PP	0.929**	0.224*	0.705**	0.903**	1								
PT	0.123	-0.087	0.194*	0.179	0.151	1							
PNS	0.436**	0.448**	0.014	0.286**	0.310**	0.237*	1						
SL	0.535**	-0.186*	0.701**	0.665**	0.589**	-0.004	-0.139	1					
SSI	0.054	-0.321**	0.342**	0.252**	0.125	0.027	0.072	0.314**	1				
SW	0.470**	0.040	0.433**	0.457**	0.475**	-0.022	-0.188*	0.741**	-0.404**	1			
SA	0.492**	-0.103	0.587**	0.560**	0.528**	-0.038	-0.195*	0.919**	-0.032	0.909**	1		
SP	0.531**	-0.111	0.628**	0.614**	0.573**	0.007	-0.183*	0.930**	-0.025	0.914**	0.974**	1	
ST	0.041	0.066	-0.021	0.039	0.070	0.409**	0.333**	-0.260**	0.051	-0.296**	-0.294**	-0.283**	1

注：* 在 0.05 水平(双侧)上显著相关；** 在 0.01 水平(双侧)上显著相关。下同。

表 9 格木表型性状与地理生态因子间的相关性

性状	年均温(℃)	年均相对湿度(%)	年均降水量(mm)	1 月均温(℃)	7 月均温(℃)	经度(E)	纬(N)
PL	-0.507	-0.370	-0.693	-0.551	-0.167	0.400	0.592
PSI	0.532	0.184	-0.001	0.443	0.340	-0.597	-0.436
PW	-0.743*	-0.393	-0.564	-0.726*	-0.346	0.680	0.743*
PA	-0.537	-0.546	-0.727*	-0.590	-0.095	0.534	0.680
PP	-0.428	-0.437	-0.681	-0.487	-0.078	0.410	0.578
PT	-0.322	-0.525	-0.702	-0.373	-0.031	0.313	0.506
PNS	0.123	-0.360	-0.419	0.009	0.338	-0.055	0.095
SL	-0.715*	0.490	0.000	-0.559	-0.745*	0.234	0.262
SSI	-0.157	0.008	0.033	-0.140	0.047	0.122	0.041
SW	-0.663	0.467	-0.040	-0.518	-0.752*	0.202	0.264
SA	-0.635	0.635	0.148	-0.453	-0.809*	0.108	0.128
SP	-0.751*	0.446	-0.073	-0.610	-0.738*	0.304	0.339
ST	0.177	-0.421	-0.453	0.086	0.231	-0.121	0.087

格木表型性状与地理生态因子的相关性分析结果见表 9，由表 9 可知，格木果荚及种子性状与地理生态因子间的相关性不强。表型性状中，果荚长度、果荚指数、果荚周长、果荚厚度、每荚种子数、种形指数和种子厚度与所有地理及生态因子间无相关性。地理生态因子中，相对湿度、经度与所有表型性状间无相关性。

果荚宽度与年均温和 1 月均温呈显著负相关，与纬度呈显著正相关；果荚面积与年降水量呈显著负相关；种子长度、种子宽度和种子面积与 7 月均温呈显著负相关；种子周长与年均温和 7 月均温呈显著负相关。以上说明，气温高、降水量大的分布区，格木往往具有果荚较小、种子较小的特征。随着纬度增加，年均温等温度指标降低，果荚宽度有增加的趋势，结合果荚及性状间的相关性分析结果，果荚宽度增加往往伴随着果荚长度、果荚面积、种子长、种子宽和种子面积等测量指标的增加，意味着随着纬度的增加、气温的降低，格木果荚及种子有变大的趋势。

2.7 格木果荚与种子表型性状的主成分分析和聚类分析

样本表型性状的原始测量数据标准化后，KMO 检测值为 0.723(>0.7)，适合做主成分分析。通过对格木 13 个表型性状进行主成分分析(表 10)，得到 4 个特征根大于 1 的主成分，累积贡献率为 87.382%。第 1 主成分主要包含果荚长、果荚宽、果荚面积和果荚周长等与果荚大小有关的性状，以及种子长、种子宽、种子面积和种子周长等与种子大小有关的性状，主要表征果荚及种子大小；第 2 主成分主要包含每荚种子数和种子厚度等，主要表征每个果荚包含种子的粒数和饱满程度；第 3 主成分主要是种形指数，主要表征种子的长宽比；第 4 主成分主要是果荚的厚度，主要表征果荚厚度及其中种子的饱满程度。

表 10 格木表型性状的主成分分析

性状	主成分			
	1	2	3	4
PL	0.780	0.519	−0.243	−0.165
PSI	−0.056	0.468	−0.795	−0.205
PW	0.818	0.072	0.482	0.029
PA	0.872	0.367	0.171	−0.093
PP	0.828	0.444	−0.064	−0.106
PT	0.090	0.428	0.269	0.708
PNS	0.040	0.819	−0.155	−0.117
SL	0.908	−0.246	0.129	−0.075
SSI	0.119	0.159	0.754	−0.485
SW	0.788	−0.356	−0.406	0.271
SA	0.883	−0.356	−0.131	0.100
SP	0.911	−0.319	−0.109	0.117
ST	−0.171	0.613	0.208	0.460
特征值	5.834	2.496	1.875	1.154
贡献率(%)	44.880	19.198	14.424	8.880
累积贡献(%)	44.880	64.078	78.502	87.382

8 个种群间用组间连接法基于欧氏距离聚类，得到种群聚类图 1A。由图 1A 可知，8 个种群根据表型性状可以分为 3 类，第 1 类为 P2 和 P4 种群，主要位于格木分布区的最南端，年均温、1 月和 7 月均温等温度指标均较高，主要特征为果荚狭长且小，其中的种子也较小。第 2 类为 P1、P3、P5、P7 和 P8 种群，该类种群主要位于格木分布区的中部和北部，各项温度指标相对较低，主要特征为果荚及其中的种子均较大。第 3 类为 P6 种群，该种群位于 8 个天然分布区的中间衔接部，该种群在果荚形态上接近第 1 类，属于狭长的小果荚，但其种子形态上又接近第 2 类，属于大粒种子，这可能与其所处的地理位置有关，兼具了第 1、2 类群格木果荚与种子的表型特征；考虑到该种群又有果荚厚度和种子厚度小，每荚种子数少且普遍不饱满等特征，也可能是该种群由于资源数量稀少，发生种群衰退和种子败育的原因。

第 1 和 2 主成分的贡献率分别为 44.880% 和 19.198%，基于第 1、2 主成分绘制散点图。图 1B 表明，格木表型性状与种群的地理分布有一定的相关性，其中 P3、P5、P7 种群受地理因子影响较小，其余 5 个种群受地理因子影响均较大。结合相关性分析和聚类分析的结果，认为格木表型性状主要受年均温等温度因子的影响，格木的果荚及种子呈现出由南至北逐渐增大的趋势。

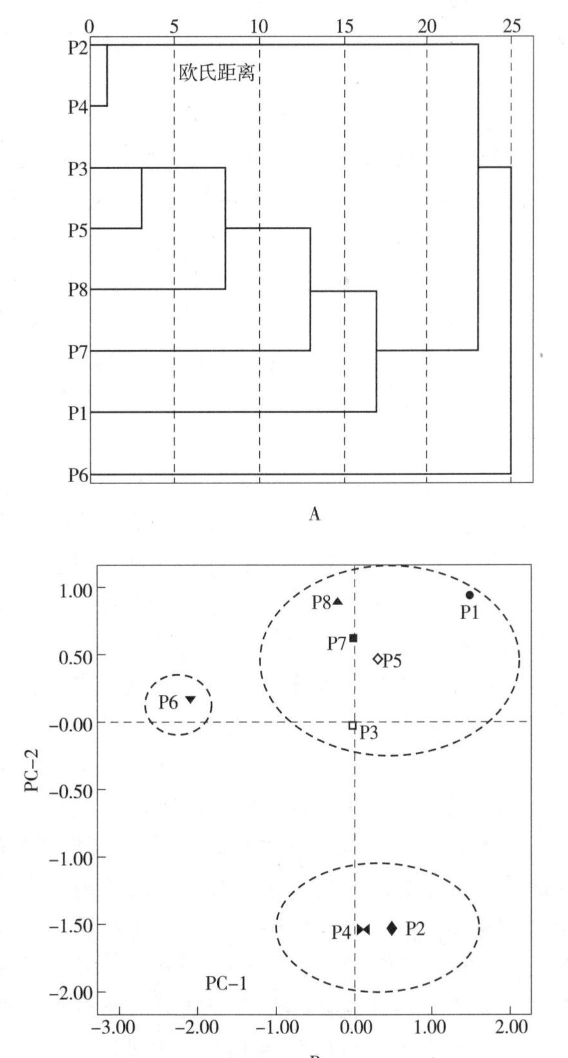

图 1 基于表型性状的格木种群聚类关系和种群位置关系

3 讨论

3.1 格木种质资源表型性状的多样性及变异规律

格木表型多样性水平中等(H=2.0278)，这与同为二级保护植物杜仲(*Eucommia ulmoides*)的雌性资源表型遗传多样性指数接近(2.0160)[17]，大于杜仲雄性资源(1.9181)和新疆野杏(0.9935)[18,19]，小于非濒危种灰楸(2.0437)和西伯利亚杏(2.6480)[19,24]。8个种群的表型多样性指数从高到低分别为：容县>龙圩>博白>浦北>合浦>苍梧>北流>陆川。因此，容县、龙圩等表型多样性高的种群是种质资源收集保存和遗传多样性保护的重点区域，但研究同时也发现北流、陆川等样本数少的种群其多样性占比远高于样本数量占比，这些种群的样本可能具有更高的表型多样性及变异，在种质资源收集过程中也应加以重视。

格木表型性状的离散程度较轻(CV=12.4109%)，各个性状分布相对集中，无论在种群水平上还是物种水平上，格木的表型性状都相对稳定，不易发生种群分化。格木13个表型性状中果荚面积的变异系数最大，种子长度的变异系数最小，此结果与赵志刚等人的研究结果相同，即果荚长宽乘积变异系数最大(16.84%)，种子长度变异系数最小(6.58%)[9]。这与青海云杉(*Picea crassifolia*)(11.17%)和濒危树种水青树(*Tetracentron sinense*)的变异系数接近(12.56%)[21,14]，大于山东稀有种小果白刺(*Nitraria sibirica*)(9.34%)，小于非濒危种砂生槐(*Sophora moorcroftiana*)(19.39%)、西伯利亚杏(20.15%)[23,24]。

变异系数和Shannon-Weaver多样性指数反应种质资源多样性的不同内涵。本研究表明格木种质资源的表型性状在变异程度上具有多样性，但在数量分布上不均衡，各个表型性状分布相对集中，变异性状所占比例较少，说明具有特殊表型的格木种质资源更具保护和保存价值，在种质资源收集保存时应优先收集保存，反之，可适当减少保存株数。

3.2 格木表型性状的变异来源及遗传分化

格木13个表型性状在种群间和种群内均存在极显著差异，说明这些性状在种群间和种群内均存在丰富的变异，格木种群内的变异远大于种群间。这与大多数林木种群的表型变异来源相同，即变异来源主要来源于种群内，种群内的变异大于种群间的变异。虽然格木种群内变异远大于种群间的变异，但种群间的变异反映了地理、生殖隔离上的差异，是种内多样性的重要体现，其意义要大于种群内的变异[25]。格木种群的平均表型分化系数为21.8579%，与海南岛青梅(*Vatica mangachapoi*)(18.31%)[13]近似，较青海云杉(31.20%)、长柄扁桃(*Amygdalus pedunculata*)(45.90%)[26]、水青树(46.69%)、小果白刺(40.71%)等树种的表型分化系数要小，这也说明格木的表型性状不易发生种群分化。格木各性状的分化系数变化幅度为7.4765%(种形指数)~38.6740%(果荚宽度)，说明种子性状最不易发生分化，果荚宽度较其他性状发生分化的可能性要高。格木在物种及种群水平上表型性状分化程度较低可能与所选种群均来自广西，种群间距离较小且从南到北呈连续分布有关，一方面，自然分布范围小，分布区内土壤气候条件等相对一致，通过自然选择实现种群间分化的可能性不高，另一方面在格木自然分布范围内，种群近乎连续分布，种群间基因交流频繁，限制了种群间的遗传分化，即使目前格木生境呈现碎片化的状态，但是这种状态持续的时间相对较短，尚未有足够的时间形成与生殖隔离有关的遗传分化。

植物表型性状是个体基于内在基因对环境变化所做出的形态调整，是植物对环境异质性的一种适应[27]。本研究通过分析格木在物种水平、种群水平上的表型多样性及变异特征，获得了一些科学规律，但尚不能明确这些表型差异有多少是由个体遗传因素造成的，有多少是环境因素所致，后期可通过布设种源家系试验林，待其进入结实性状稳定期后(20年)，采用重新测量的方式进一步明确表型性状的变异来源及其大小，或通过布设长期样地，收集种群的坡向、坡位、土壤理化性质、生境微生物等信息后，选择主要因素采用建立模型的方式确定。

3.3 格木表型变异与地理生态因子间的相关性

不同植物对环境因子的适应机制和敏感程度的差异导致其呈现不同的地理变异规律，如苦楝(*Melia azedarach*)的果核长度、种子宽度等表型性状由南向北有增大的趋势[28]；栀子(*Gardenia jasminoides*)野生居群的叶片表型与经度呈负相关，而果实表型与经度成正相关[29]。格木13个表型性状之间大多存在显著或极显著相关，总体而言，果荚狭小，往往对应种子较小、饱满且每荚种子数较多；果荚宽大，往往具有大粒种子。格木表型性状与地理气候因子的相关性分析表明，大多数表型性状与地理气候因子间的相关性不强。仅果荚宽度、种子周长等性状和温度指标呈负相关，与纬度呈正相关，即格木在气温高、降水量大的地区往往具有小果荚、小种子的特征，随着温度指标降低和纬度增加，格木果荚及种子有变大的趋势。这与赵志刚的研究结果相似，随着海拔升高、气温下降和降水量增加，种子变大。另外，本研究未发现前人研究中格木果荚形态与经度的负相关关系，可能是之前选择的种群从西到东大致沿经度带状连续分布，经度跨度大(9.1°)，本研究所选种群从南至北大致沿纬度带状连续分布，经度跨度较小(2.07°)的原因[9]。

对于大多数生境不连续的濒危物种而言，种群数量大小可能会影响植物的表型。一个种群的有效样本数长期较少且处于孤立状态，可能会因近交衰退造成果实及种子等表型性状的改变，进而导致种实败育、种群灭绝[10]。但是致使发生表型改变的时间跨度一般较长，也因树种和其繁殖特性而异。从民间传承的格木家具、器物、建筑、桥梁等的数量来看，历史上格木在广西的天然分布较多，而格木种群受到大规模干扰基本上是近100年以内的事情。因此，似可认为当前格木因种群数量不同导致表型性状差异的可能性相对较低。然而，目前格木生境碎片化严重，大多处于孤立木、散生木和小群落的生存状态，其主要以虫媒传粉、大粒硬实种子的特性也使得格木难以远距离传播更新。因此，如不采取人为干预，格木的自交率将长期处于较高水平，不利于其种群延续和稳定遗传。

对于林木天然种群而言，其个体年龄往往难以判断，通常根据径级和生长量大致判断其树龄，在含有公孙树的种群中，幼龄林木在到达形态稳定期前可能与成熟林木在表型上存在差异。格木木材密度大、硬度高，属于慢生的长周期树种，实生苗通常在定植15年后开始结实(胸径约12cm)，结实前几年种子质量不高，可能存在种实各项表型指标不稳定的情况，20年以后种子质量相对稳定(胸径约16cm)，种实等各项表型指标也进入稳定期，一般不再发生改变[30,31]，暂未发现格木单株进入结实稳定期后，其种实表型性状有随时间发生变化的证据。本研究是在2017年广西壮族自治区名木古树调查的基础上采样，取样单株大多属于格木名木古树(100年以上)，胸径最大为134cm(龙圩区广平镇调村)，树龄约1300年；胸径最小约为23cm(容县十里镇杨外村)，树龄为37年，其余单株的胸径介于两者之间，均为成熟林木。格木单株进入结实稳定期后，果荚及种子表型性状是否会随径级的增加而改变有待持续观测做进一步的验证。

3.4 格木种质资源的保护策略

格木作为热带亚热带地区的珍优乡土阔叶树种，在优质木材、城市绿化和中药材开发等方面颇具发展前景和应用价值。虽然当前被列为国家二级保护植物，但前期破坏性开发造成的格木生境碎片化现状未见有恢复措施，加之格木天然更新难的特性，导致格木天然林种群的数量仍处于持续减少状态。为更好地保护格木现有的天然种群，实现格木资源的可持续利用，提出如下建议：①鉴于格木表型变异以种群内为主、表型分化水平低和表型性状分布集中、稀有表型占比少等特征，开展格木种质资源收集保存时，可适当减少种群数，增加种群内的个体数，加强对具有特殊表型格木单株的收集。②应持续加强现有格木天然林种群和单株的保护力度，通过收集种质资源，建立格木基因库，同时对数量少的种群开展多度恢复工作，防止种群发生自交衰退。本研究调查时发现，个别种群由于数量太少，似已丧失繁衍能力，如北海市铁山港区种群仅发现6株天然格木，3株为孤立木，果荚和种子极少且虫害严重、发育畸形，未收集到正常的果荚及种子；3株为散生小老头树，均已多年未结实，该种群正濒临灭绝；另一数量少的陆川群体，仅发现2株孤立木和1个盗伐后萌芽形成的次生小群落，该种群结实量少，果荚和种子多数畸形，种子薄而不饱满，这些表型可能是该分布区特殊的地理气候因子所致，但参照铁山港种群，因种群数量长期较少而发生自交衰退和种实败育的可能性也很大。③应加强格木南、北原生分布区衔接地带种群的数量恢复工作，如北流、陆川等种群。调查发现格木南部的分布区，如合浦、博白、浦北分布区的资源数量相对较多，北部分布区，如容县、龙吁、苍梧分布区资源数量也较多，而南北分布区的中间衔接部北流和陆川种群资源数量稀少，格木南北连续的原生分布带有断裂的可能。④铁山港、陆川和北流种群的格木亟须保护和恢复种群数量。基于就近引种的原则，北海市铁山港区和陆川县的格木天然林种群引入具有狭小果荚、小粒饱满种子特征的合浦、博白种源应有更好的适应性。北流市格木天然林数量也很少，但该分布区种子大而饱满，质量较高，仅需利用本地种源增加种群数量，如引种，宜引入大粒的北部种源区种子。其他县市格木种群也宜通过政府干预，利用本地种源，以种植行道树等方式适当增加本地区格木种群的数量。

4 结论

本研究通过对8个格木天然种群13个表型性状的分析，主要得出以下结论：

(1)格木在个体水平上存在较为丰富的遗传变异，与此同时，格木的表型性状相对集中稳定(稀有表型占比少)，不易发生群体分化，具有特殊表型的格木种质资源具有更高的保存价值。

(2)种群内的变异是格木表型变异的主要来源，格木的种质资源收集保存可适当减少种群数量，增加种群内个体数量，优先收集有特殊表型的单株。

(3)容县和龙圩等多样性指数较高的种群是格木种质资源多样性保护和收集保存的重点区域；因样本数较少导致种群多样性指数偏低，但种群多样性占比较高的陆川、北流种群也应广泛收集保存。

(4)格木果荚及种子的表型性状与年均温等温度因子相关性明显，北部分布区普遍为宽大果荚和大粒种子，南部分布区则多为狭长果荚和小粒种子，因此，温度因子应是影响格木表型及分布最主要的因子，这一特征为格木的引种栽培提供了依据。

(5)通过探讨格木的传粉特性、生长特性和种子特征，结合本研究调查的格木天然种群生存现状，认为格木天然林资源的保育，需要人为干预，应及早对资源数量较少的种群开展资源多度恢复工作。

参考文献

[1]SOARES M P，WEISS G. The Iron age of host-microbe interactions[J]. EMBO reports，2015，16(11)：1482-1500.

[2]HAKE S，ROSS-IBARRA J. Genetic，evolutionary and plant breeding insights from the domestication of maize[J]. Elife，2015，4(4)：61-71.

[3]FORCADA J，HOFFMAN J I. Climate change selects for heterozygosity in a declining fur seal population[J]. Nature，2014，511(7510)：462-475.

[4]胡启鹏，郭志华，李春燕，等．植物表型可塑性对非生物环境因子的响应研究进展[J]．林业科学，2008，44(5)：135-142.

[5]SPIELMAN D，BROOK B W，FRANKHAM R. Most species are not driven to extinction before genetic factors impact them. Proceedings of the National Academy of Sciences of the United States of America，2004，101(42)：15261-15264.

[6]PINSKY M L，PALUMBI S R. Meta-analysis reveals lower genetic diversity in overfished populations[J]. Molecular Ecology，2014，23(1)：29-39.

[7]中国科学院植物研究所．中国珍稀濒危植物[M]．上海：上海教育出版社，1989：188-190.

[8]郑万钧．中国树木志[M]．北京：中国林业出版社，1985：1207-1209.

[9]赵志刚，郭俊杰，沙二，等．我国格木的地理分布与种实表型变异[J]．植物学报，2009，44(3)：338-344.

[10]T. L. 怀特，W. T. 亚当斯，D. B. 尼尔．森林遗传学[M]．崔建国，李火根，译．北京：科学出版社，2013：230-250.

[11]黄忠良，郭贵仲，张祝平．渐危植物格木的濒危机制及其繁殖特性的研究[J]．生态学报，1997，6(6)：671-676.

[12]赵志刚．珍稀濒危树种格木保护生物学研究[D]．北京：中国林业科学研究院，2011.

[13]尚帅斌，郭俊杰，王春胜，等．海南岛青梅天然居群表型变异．林业科学，2015，51(2)：154-162.

[14]李珊，甘小洪，憨宏艳，等．濒危植物水青树叶的表型性状变异[J]．林业科学研究，2016，29(5)：687-697.

[15]葛颂，王明庥，陈岳武．用同工酶研究马尾松群体的遗传结构[J]．林业科学，1988，24(4)：399-409.

[16]BUSS G R. Microsatellite and amplified sequence length polymorphisms in cultivated and wild soybean[J]. Genome，1995，38(4)：715-723.

[17]李洪果，杜庆鑫，王淋，等．利用表型数据构建杜仲雌株核心种质[J]．分子植物育种，2017，15(12)：5197-5209.

[18]李洪果，杜红岩，贾宏炎，等．利用表型性状构建杜仲雄性资源核心种质[J]．分子植物育种，2018，16(2)：591-601.

[19]刘娟，廖康，刘欢，等．新疆野杏种质资源表型性状多样性研究[J]．西北植物学报，2015，35(5)：1021-1030.

[20]李秀兰，贾继文，王军辉，等．灰楸形态多样性分析及核心种质初步构建[J]．植物遗传资源学报，2013，14(2)：243-248.

[21]尹明宇，高福玲，乌云塔娜．内蒙古西伯利亚杏种质资源表型多样性研究[J]．植物遗传资源学报，2017，18(2)：242-252.

[22]王娅丽，李毅，陈晓阳．祁连山青海云杉天然群体表型性状遗传多样性分析[J]．林业科学，2008，44(2)：70-77.

[23]董昕，王磊，鲁仪增，等．山东稀有植物小果白刺天然群体表型变异研究[J]．林业科学研究，2017，30(2)：293-299.

[24]林玲，王军辉，罗建，等．砂生槐天然群体种实性状的表型多样性[J]．林业科学，2014，50(4)：137-143.

[25]李斌，顾万春，卢宝明．白皮松天然群体种实性状表型多样性研究[J]．生物多样性，2002，10(2)：181-188.

[26]柳江群，尹明宇，左丝雨，等．长柄扁桃天然种群表型变异[J]．植物生态学报，2017，41(10)：1091-1102.

[27]胡启鹏，郭志华，李春燕，等．植物表型可塑性对非生物环境因子的响应研究进展[J]．林业科学，2008，44(5)：135-142.

[28]陈丽君，邓小梅，丁美美，等．苦楝种源果核及种子性状地理变异的研究[J]．北京林业大学学报，2014，36(1)：15-20.

[29]邓绍勇，曹泉，余林，等．栀子野生居群叶片和果实性状的表型多样性[J]．林业科学研究，2015，28(2)：289-296.

[30]朱积余，廖培来．广西名优经济树种[M]．北京：中国林业出版社，2006：61-64.

[31]唐继新，贾宏炎，麻静，等．南亚热带珍稀濒危树种格木生长规律研究[J]．中南林业科技大学学报，2015，35(7)：37-44.

[原载：林业科学，2019，55(04)]

基于等位基因最大化法初步构建杜仲核心种质

李洪果[1,2] 许基煌[1] 杜红岩[2] 乌云塔娜[2] 刘攀峰[2] 杜庆鑫[2]

([1]中国林业科学研究院热带林业实验中心，广西凭祥 532600；
[2]国家林业局泡桐研究开发中心，河南郑州 450000)

摘　要　构建杜仲核心种质，去除基因库中的遗传冗余，为杜仲种质资源的保存、研究和利用提供依据。以国内外54个地区的887份杜仲种质资源为试验材料，基于9对基因组SSR引物和等位基因数目最大化策略构建杜仲核心种质。利用一些分子生物学软件和统计分析软件，通过等位基因数(n)、平均等位基因数(Na)、平均有效等位基因数(Ne)、平均Shannon-Weaver指数(I)、平均Nei's遗传多样性指数(H)、平均基因型数(Ng)、平均多态信息含量(PIC)等7个遗传多样性参数及其保留比例对所构建核心种质进行评价，结合遗传多样性指数的t检验法和主坐标分析法(PCoA)验证和确认核心种质对原始种质的代表性。9对SSR引物共检测到107个等位基因，平均有效等位基因数(Ne)为5.096，Shannon-Weaver指数(I)为1.812，Nei's遗传多样性指数(H)为0.925，表明杜仲种质资源具有丰富的遗传多样性。887份杜仲种质基于等位基因数目最大化原则得到189份核心种质和698份保留种质。189份核心种质占原始种质样品数的21.3%，保存了原始种质100%的等位基因，9个SSR位点的平均等位基因数(Na)、平均有效等位基因数(Ne)、平均Shannon-Weaver指数(I)、平均Nei's遗传多样性指数(H)、平均基因型数(Ng)、平均多态信息含量(PIC)等遗传多样性参数的保留比例分别为：100%、116.5%、108.7%、101.5%、100%、103.3%，以上参数经t检验，与原始种质在0.01水平上差异不显著，主坐标分析也表明，核心种质与原始种质的样品在分布图上有着相似的分布结构，说明构建的核心种质具有代表性。698份保留种质占原始种质样品数的78.7%，保存了原始种质86.9%的等位基因，9个SSR位点的平均等位基因数(Na)、平均有效等位基因数(Ne)、平均Shannon-Weaver指数(I)、平均Nei's遗传多样性指数(H)、平均基因型数(Ng)、平均多态信息含量(PIC)等遗传多样性参数的保留比例分别为：86.9%、95.7%、96.7%、99.4%、77%、99%，以上参数经t检验，与原始种质在0.01水平上差异不显著。189份核心种质很好的代表了887份原始种质的遗传多样性。核心种质的7个遗传多样性参数均高于保留种质，在杜仲种质资源保存和建立育种群体时应优先使用核心种质。本研究为杜仲优异基因发掘和新品种选育奠定基础。

关键词　杜仲；核心种质；SSR

Preliminary Construction of Core Collection of *Eucommia ulmoides* Oliv. Basedon Allele Number Maximize Strategy

LI Hongguo[1,2], XU Jihuang[1], DU Hongyan[2], WUyun Tana[2], LIU Panfeng[2], DU qingxin[2]

([1]*Experimental Center of Tropical Forestry, Chinese Academy of Forestry, Pingxiang 532600, Guangxi, China;*
[2]*China paulownia Research Center, Zhengzhou 450000, Henan, China*)

Abstract: Core collection of *Eucommia ulmoides* was constructed to retrench germplasm resources in the gene pool, which will provide a theoretical basis for protecting, utilizing and studying germplasm resources of *E. ulmoides*. Based on allele number maximization strategy, core collection of *E. ulmoides* was constructed from 887 total collections, which located in 54 distribution area, by using nine SSR primers and bioinformatics software. Core collection of *E. ulmoides* was assessed by the number of allele (n), average number of allele(N_e), average effective number of allele (N_a), average Shannon's information index (I), Nei's diversity index (H), average genotype number(N), average polymorphism information content (PIC), and their retaining ratio. The representative of the core collection to the total collection was confirmed with t-test and PCoA analysis. 107 alleles(n) were detected in 9 SSR primer pairs, the average of N. was 5.096. High genetic diversity was revealed in the germplasm resources of *E. ulmoides*(I = 1.812, H = 0.925). 189 of 887 core collections were obtained based on allele number maximization strategy, suggesting that the 21.3% of the collection samples contained 100% number of alleles. The retaining ratio of NN, I, H, N, and PIC of nine SSR locus was 100%, 116.5%, 108.7%, 101.5%, 100%, 103.3%, respectively. t-test analysis suggested that there was no significant correlation between the six evaluation parameters of core collection and total collection. This result was further confirmed by the PCoA analysis. 698 of 887 reserve collections were obtained based on allele number maximization strategy, suggesting that the 78.7% of the collection samples contained 86.9% number of allele. The retaining ratio of N_a N_e I, H_g N and PIC of the nine SSR locus was 86.9%, 95.7%, 96.7%, 99.4%, 77%, 99%, respectively. t-test analysis suggested that there was no significant correlation between the six evaluation parameters of reserve collection and total collection. The core collection of *E. ulmoides* was well constructed, which contained all the alleles and genotypes. There was no significant difference in six evaluation para meters of the core collection and total collection. Moreover, the core collection and total collection samples had the similar structure in the distribution map. All the seven evaluation parameters of the core collection were higher than the reserve collection. Therefore, the core collection should be priority option in protecting germplasm resources and constructing breeding population of *E. ulmoides*. This study laid a foundation for the identification of favorable genes and breeding of new varieties of *E. ulmoides*.

Key words: *Eucommia ulmoides*; core collection; SSR

杜仲(*Eucommia ulmoides* Oliv.)是我国特有的珍贵孑遗树种，为二倍体植物，雌雄异株，适生于年均温9~20℃，极端低温不低于-33℃，pH在5.0~8.4的广大区域[1]。杜仲的果、叶皮均含有丰富的天然橡胶，可被开发出橡胶弹性材料、热塑性和热弹性材料等用途的材料；此外，杜仲还是名贵的药用经济树种，具有强筋骨、补肝肾和轻身耐劳等功效[1]。目前，从全国范围内收集的千余份杜仲种质资源由于引种资料匮乏等原因存在种质不清和重复收集等问题，含有一定程度的遗传冗余。另外，在杜仲栽培生产逐步向良种化发展的趋势下，一些栽培表现普通的种质面临淘汰和分布区进一步缩减的压力，存在种质灭失的风险。因此，通过构建核心种质，加强现有杜仲种质资源遗传多样性的利用和保护尤为必要。

植物种质资源是遗传改良的物质基础，随着物种多样性消失日益加剧，各国对遗传资源重要性的认知逐渐加深，相继建立了大量的种质资源库，截至2011年，全世界非原生境保存的植物收集品已达740万份[2]。但是种质资源的不断积累，也提高了种质资源的管理费用，增加了优异种质材料筛选和挖掘的难度。Frankel和Brown提出核心种质的概念，用最少的资源数量和最小的遗传冗余最大限度地代表该物种最丰富的遗传多样性、结构及整个群体的地理分布[3]。核心种质的构建为种质资源的深入评价和有效保护利用开辟了新的途径。早期构建核心种质大多采用形态及农艺性状数据，分子标记技术发明后，因其不受环境影响、精度高、稳定性好等优势，成为构建核心种质的主要手段。

目前，已利用分子标记的方法在新疆野杏(*Armeniaca vulgaris* Lamarck)[4]、桂花(*Osmanthus fragrans* Loureiro)[5]、葡萄(*Vitis vinifera* Linnaeus)[25]、柿子(*Diospyros kaki* Thunberg)[6]、水稻(*Oryza sativa* Linnaeus)[29]等林木、花卉及农作物上建立了核心种质。本研究以887份杜仲资源为试验材料，利用9对基因组SSR引物基于等位基因数最大化原则构建杜仲的核心种质，为杜仲种质资源的保护、利用、种质创新和建立育种群体提供依据。

1　材料与方法

1.1　试验材料

试验材料为以嫁接方式保存在杜仲基因库内54个收集区的887份杜仲种质，不同材料收集区的样本数见表1。将‘华仲1号’‘华仲5~12号’‘华仲16~18号’‘大果1号’及‘密叶杜仲’‘龙拐杜仲’‘大叶杜仲’‘小果杜仲’和‘红木杜仲’等18个有特异表型或在生产中应用较广的种质作为必选材料，见表2。试验材料于春季采集枝条顶端的幼叶装入自封塑料袋中，置于放有干冰的便携式冰箱，带回试验室超低温保存(-80℃)。

表 1 试验材料

收集区编号	样本数	种质来源	收集区编号	样本数	种质来源
1	15	安徽亳州	28	10	日本东京
2	216	北京杜仲公园	29	8	日本岩手
3	12	北京清华大学	30	11	山东青岛
4	64	北京万泉河路	31	11	山西运城
5	7	福建建阳	32	10	陕西安康
6	12	甘肃康县	33	14	陕西略阳
7	10	广东乐昌	34	12	上海
8	10	广西兴安	35	16	四川广元
9	22	贵州遵义	36	16	天津蓟县
10	36	河北安国	37	11	新疆阿克苏
11	9	河南鹤壁	38	6	新疆乌鲁木齐
12	10	河南开封	39	6	新疆伊宁
13	10	河南洛阳	40	4	云南盐津
14	9	河南南阳	41	13	浙江杭州
15	13	河南汝阳	42	7	重庆沙坪坝
16	5	河南三门峡	43	1	河南许昌
17	35	河南商丘	44	1	湖北郧西
18	13	河南新乡	45	1	江西龙南
19	8	河南郑州	46	1	江西南昌
20	18	湖北神农架	47	1	辽宁营口
21	8	湖北武汉	48	3	山东济南
22	28	湖南慈利	49	3	山东临淄
23	12	湖南株洲	50	2	山东青州
24	14	吉林集安	51	3	陕西杨陵
25	17	江苏响水	52	1	新疆喀什
26	10	江西九连山	53	1	江苏南京
27	90	洛阳林业科学研究所	54	1	河南原阳

表 2 必选种质名单

种质编号	收集区编号	种质特征
‘华仲 1 号’	43	良种。耐旱、耐寒、少病虫害。速生、丰产，雄花及杜仲皮产量高
‘华仲 5 号’	22	良种。耐旱、耐寒、少病虫害。速生、丰产，果实及杜仲皮产量高
‘华仲 6 号’	13	良种。少病虫害。果皮含胶量、种仁粗脂肪及 α 亚麻酸含量高，早实、高产、稳产
‘华仲 7 号’	22	良种。果皮含胶量、种仁粗脂肪及亚麻酸含量高，早实、高产、稳产
‘华仲 8 号’	9	良种。果皮含胶量、种仁粗脂肪及亚麻酸含量高，早实、高产、稳产
‘华仲 9 号’	44	良种。果皮含胶量、种仁粗脂肪及亚麻酸含量高，早实、高产、稳产
‘华仲 10 号’	19	良种。果皮含胶量、种仁粗脂肪及亚麻酸含量高，是目前 α-亚麻酸含量最高的良种。早实、高产、稳产
‘华仲 11 号’	27	良种。雄花产量高，速生、丰产，少病虫害
‘华仲 12 号’	13	良种。特殊变异类型(叶片红色或紫红色)。少病虫害
‘大果 1 号’	27	良种及特殊变异类型(果实大)。杜仲胶含量、产果量、产胶量高。早实、高产、稳产、少病虫害
‘密叶杜仲’	13	良种及特殊变异类型(叶片密生)，少病虫害
‘华仲 16 号’	10	良种。高产、稳产、耐旱、耐寒

（续）

种质编号	收集区编号	种质特征
‘华仲 17 号’	4	良种。耐旱、耐寒。果皮含胶量、种仁粗脂肪及亚麻酸含量高，早实、高产、稳产
‘华仲 18 号’	17	良种。耐旱、耐寒。果皮含胶量、种仁粗脂肪及亚麻酸含量高，早实、高产、稳产
‘龙拐杜仲’	13	特殊变异类型(枝条规律性左右曲折)
‘大叶杜仲’	11	特殊变异类型(叶片大)
‘小果杜仲’	27	特殊变异类型(果实小)
‘红木杜仲’	54	特殊变异类型(木材心材呈红色)

注：收集区编号对应的收集区见表 1，下同。

1.2 试验方法

1.2.1 DNA 提取、SSR 引物及 PCR 产物检测

DNA 提取按天根生化科技有限公司生产的 DP320 离心柱型植物基因组 DNA 提取试剂盒说明书进行。提取结束后，取 1μLDNA 产物，用超微量紫外可见光光度计 P330 测量记录 DNA 浓度及纯度，再取 2μLDNA 产物与 6×Loadingbuffer 混匀后，用 1.2% 的琼脂糖凝胶电泳检测。根据已报道的杜仲基因组 SSR 引物，筛选出扩增产物多态性高、重复性好的 9 对引物用于 887 份杜仲种质的分子生物学分析(引物名称及序列见表 3)。

引物合成及 SSR 检测由北京金唯智生物科技有限公司完成。SSR 反应体系参照吴敏的反应体系。荧光 PCR 产物用 ABI3730 测序仪进行毛细管电泳检测，经 Gene-Mapper 软件分析。湖南株洲群体(23)中 10003C 种质基于 9 对引物的毛细管电泳检测结果见图 1，表明扩增的目的片段清晰、准确、可靠。其中 GEU045 引物的结果呈现双峰现象，可能是由于 PCR 循环结束后，延伸时间不够长或延伸温度不够高，导致大部分扩增片段未能添加上额外的腺苷酸(+A)，从而产生的分裂峰，而非四倍体材料或非特异扩增。该现象可以通过优化扩增条件体系(如提高终延伸温度、增加终延伸时间或选择高质量的 Taq 酶)加以改善，但由于毛细管电泳的信号正常，峰形也在可判读范围内，从检测结果的各项指标来看可以用于后续分析，所以未进行重新检测消除该分裂峰，而是统一选择荧光值高的片段作为毛细管电泳检测的最终结果。

1.2.2 核心种质的构建流程

(1)提出必选材料名单。根据现有资料和长期观测的结果，将曾在生产上大面积使用的、在育种中起过重要作用的、有独特利用价值的种质作为必选材料[8]。

(2)核心种质的获得。采用不分组的方法，把所有种质作为整体构建核心种质。将 887 份资源的毛细管电泳检测结果输入 Power Core 软件中，基于等位基因数最大化原则运算获得核心种质[9]。原始种质扣除核心种质后剩余的种质作为保留种质。

(3)核心种质的代表性检验。核心种质构建完成后，通过比较核心种质、保留种质、原始种质之间的遗传多样性参数及其保留比例，结合各群体相关遗传参数 t 检验法检验核心种质的代表性。

表 3 9 对 SSR 引物信息

引物	重复单元	SSR 类型	正向引物(5’-3’)	反向引物(5’-3’)	T℃	产物范围
GEU026	(ATTCCTA)6	p7	CAGCCACTCAATACCGAATC	CACCAGAGGCACAAGCATAA	60	152~194
GEU045	(GATA)11	p4	CATCGTCATTCTTCTCCTTGC	CACGACACACTGAAGATCCA	58	233~301
GEU053	(TTTAAAA)6	p7	GGTATGAGCATGAACGATGG	AACTTCTTCGGTCAACCTTGT	57	195~251
GEU150	(ATATAC)8	p6	CGCACCGAAGCATAGAACTT	CATGGCAATGCAAGCTAGAC	60	250~304
GEU155	(TAC)15	p3	CTGCCGATTCTCAGCCTATG	TTCGTGCACTTGCGAGTATT	60	214~250
GEU158	(GCCATC)8	p6	TCTCCAAGCTACACGCAATG	CTCCACAGTCCATTCAACGA	60	70~154
GEU191	(TCT)12	p3	CCAATCTACGAAGTTGCCAAG	TTGACCGAAGCACTAGGTGA	60	125~164
GEU219	(AAC)10	p3	ATCGATCTGAAGGCGTGATT	ATGGTCCTTGATCGCTCATT	60	208~247
GEU271	(AGA)11	p3	TGCTCTCCTGCTAACTGCAA	GGTGCGAACAGAGAATCGTC	60	142~172

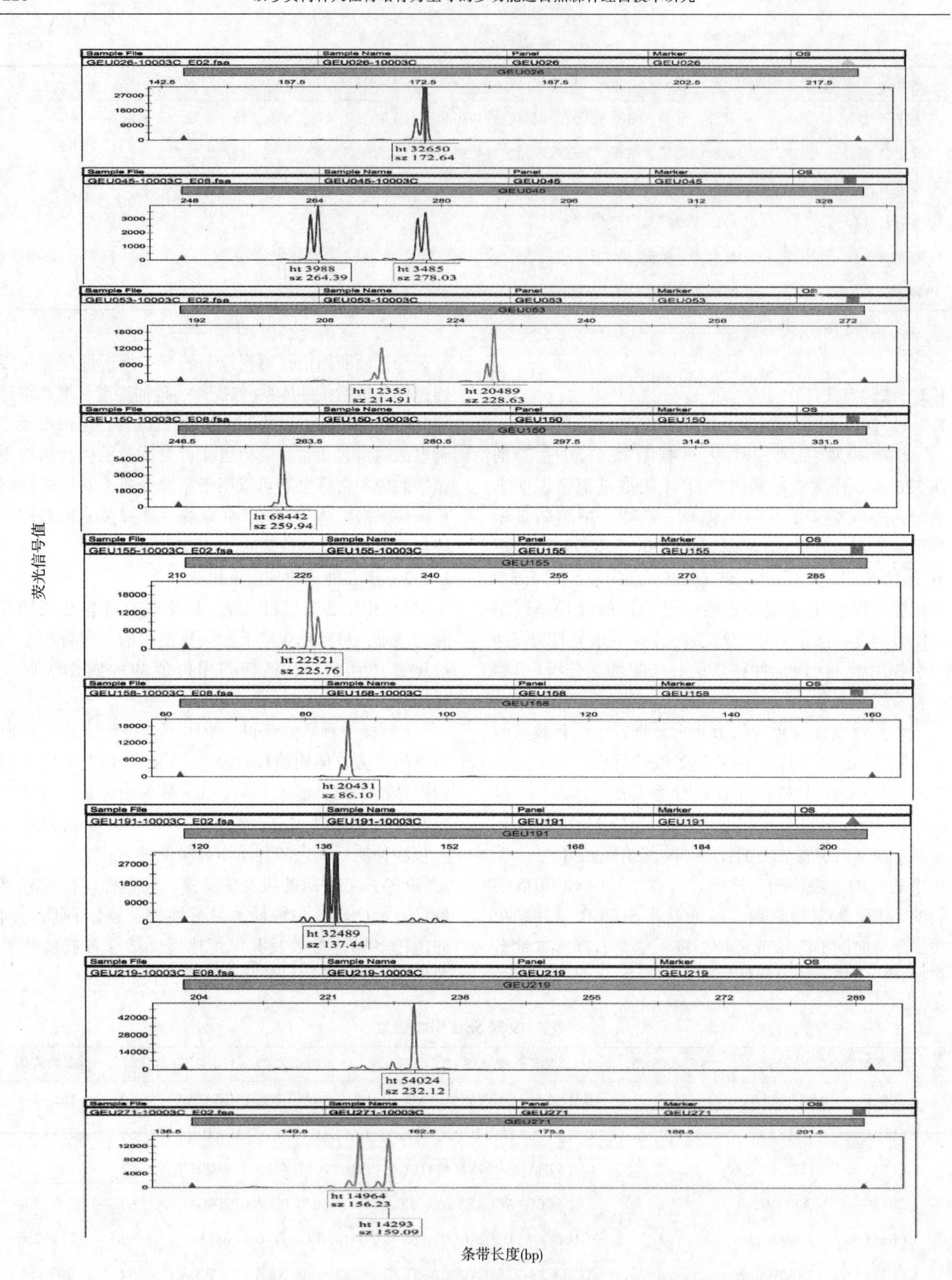

图 1 湖南株洲群体 10003C 样品基于 9 对 PCR 引物扩增的毛细管电泳检测结果

(4)核心种质的确认。核心种质经代表性检验后，采用主坐标分析法(PCoA)生成核心种质、原始种质样品在主坐标中的分布图，对核心种质作进一步确认。

1.3 核心种质代表性检验的评价参数及标准

选择 9 个 SSR 位点检测到的全部等位基因数(n)、平均等位基因数(Na)、平均有效等位基因数(Ne)、平

均 *Shannon*-Weaver 指数(I)、平均 Nei's 遗传多样性指数(H)、平均基因型数(*Ng*)、平均多态信息含量(*PIC*)和以上7个参数的保留比例作为验证核心种质对原始种质代表性的评价参数。*PIC* 通过公式：$PIC = 1 - \sum f_{ij^2}$ 计算，f_{ij} 为第 i 个位点第 j 个等位基因出现的频率[10]。其余6个遗传参数通过 GenAlex6.5 软件和 PowerCore 软件计算。评价标准：等位基因的保留比例需大于 70%，其他遗传参数越大越好[6,11,12]。用 SPSS16.0 软件对核心种质、保留种质和原始种质的群体遗传参数进行 t 检验，两者差异不显著，说明构建的核心种质具有代表性。利用 NTSYS-2.10e 软件计算 SM 相似系数和 PCoA 分析，生成核心种质和原始种质在主坐标上的样品分布图，两者的分布结构越相似，构建的核心种质代表性越好。

2 结果与分析

2.1 核心种质、保留种质和原始种质的遗传多样性比较

887 份原始材料在 Power Core 软件运算得到包含 18 份必选材料在内的 189 份核心种质，取样比例为 21.3%。原始种质、核心种质、保留种质的遗传多样性结果见表 4，由表 4 可知，887 份种质在 9 个 SSR 位点上共扩增出 107 个等位基因，每个位点的平均等位基因数(*Na*)、平均有效等位基因数(*Ne*)、平均 *Shannon*-Weaver 指数(I)、平均 Nei's 遗传多样性指数(H)、平均基因型数(*Ng*)和平均多态信息含量(PIC)分别为：11.889、5.096、1.812、0.925、46.889 和 0.795。核心种质占原始种质 21.3% 的样品，但保存了原始种质 100% 的等位基因和基因型，对其他 5 个遗传多样性参数的保留比例分别为：100%、116.5%、108.7%、101.5%、103.3%。保留种质占原始种质 78.7% 的样品，保存了原始种质 86.9% 的等位基因，对其他 6 个遗传多样性参数的保留比例分别为：86.9%、95.7%、96.7%、99.4%、77.0%、99.0%。根据等位基因保留比例需大于 70%，其他遗传参数越大越好的评价标准[13,14]，本研究构建的核心种质各个遗传参数较优，符合核心种质的要求。

2.2 核心种质、保留种质与原始种质多样性指数的 *t* 检验

为进一步验证核心种质对原始种质的代表性，将核心种质和保留种质对原始种质在 9 个 SSR 位点上的平均等位基因数(Na)、平均有效等位基因数(Ne)、平均 *Shannon*-Weaver 指数(I)、平均 Nei's 遗传多样性指数(H)、平均基因型数(Ng)、平均多态信息含量(PIC)进行 *t* 检验(表 5)，结果表明，核心种质、保留种质的 6 个遗传多样性参数与原始种质相比在 0.01 水平上差异不显著，说明构建的核心种质具有很好的代表性。

表 4 原始种质、核心种质和保留种质的遗传多样性比较

	N	n	Na	Ne	I	H	Ng	PIC
原始种质	887	107	11.889	5.096	1.812	0.925	46.889	0.795
核心种质	189	107	11.889	5.937	1.970	0.939	46.889	0.821
保留比例(%)	21.3	100	100	116.5	108.7	101.5	100	103.3
保留种质	698	93	10.333	4.877	1.753	0.919	36.111	0.787
保留比例(%)	78.7	86.9	86.9	95.7	96.7	99.4	77.0	99.0

注：N：样本数；n：等位基因数；Na：平均等位基因数；Ne：平均有效等位基因数；I：平均 *Shannon*-Weaver 指数；H：平均 Nei's 遗传多样性指数；Ng：平均基因型数；PIC：平均多态信息含量。下同。

表 5 核心种质、保留种质与原始种质的 *t* 检验

评价参数	均值	标准差	标准误	平均差	均差标准差	*F* 值	*t* 值
原始种质	11.890	2.759	0.920				
核心种质	11.890	2.759	0.920	0.000	1.301	1.000	1.000
保留种质	10.330	2.784	0.928	1.556	1.306	0.935	0.251
原始种质	5.096	1.072	0.357				
核心种质	5.937	1.550	0.517	-0.841	0.628	0.071	0.199

（续）

评价参数	均值	标准差	标准误	平均差	均差标准差	F 值	t 值
保留种质	4.877	0.964	0.321	0.218	0.480	0.649	0.656
原始种质	1.812	0.219	0.073				
核心种质	1.970	0.262	0.087	-0.159	0.114	0.418	0.182
保留种质	1.753	0.204	0.068	0.059	0.100	0.757	0.562
原始种质	0.925	0.283	0.009				
核心种质	0.939	0.258	0.009	-0.014	0.013	0.922	0.285
保留种质	0.919	0.029	0.010	0.006	0.014	0.926	0.682
原始种质	46.89	17.047	5.682				
核心种质	46.89	17.047	5.682	0.000	8.036	1.000	1.000
保留种质	36.11	11.667	3.889	10.778	6.886	0.131	0.137
原始种质	0.795	0.046	0.015				
核心种质	0.821	0.048	0.016	-0.026	0.022	0.641	0.263
保留种质	0.787	0.046	0.015	0.008	0.022	0.890	0.712

注：$t_{0.05}=1.9600$，$t_{0.01}=2.5758$。

表 6　核心种质与保留种质的 t 检验

评价参数	均值	标准差	标准误	平均差	均差标准差	F 值	t 值
核心种质	11.890	2.759	0.920				
保留种质	10.330	2.784	0.928	1.556	1.306	0.935	0.251
核心种质	5.937	1.550	0.517				
保留种质	4.877	0.964	0.321	1.060	0.609	0.029	0.105
核心种质	1.970	0.262	0.087				
保留种质	1.753	0.204	0.068	0.218	0.111	0.269	0.067
核心种质	0.939	0.258	0.009				
保留种质	0.919	0.029	0.010	0.020	0.013	0.843	0.148
核心种质	46.89	17.047	5.682				
保留种质	36.11	11.667	3.889	10.778	6.886	0.131	0.137
核心种质	0.821	0.048	0.016				
保留种质	0.787	0.046	0.015	0.034	0.022	0.551	0.145

注：$t_{0.05}=1.9600$，$t_{0.01}=2.5758$。

核心种质与保留种质基于6个遗传多样性参数的t检验结果表明两者在0.01水平上差异不显著(表6)，说明所构建的核心种质不仅有效反映了原始种质的遗传多样性，也没有显著影响保留种质的遗传多样性。从表6还可以看出，核心种质的6个遗传多样性评价参数均高于保留种质，在建立杜仲育种群体时应该优先使用核心种质。

2.3　核心种质的确认

利用主坐标分析法(PCoA)对构建的核心种质进行确认。主坐标分析法是研究数据相似性或差异性的可视化方法，通过一系列的特征值和特征向量进行排序后，找到距离矩阵中最主要的坐标，其结果是数据矩阵的一个旋转，并不改变样品点之间的相互位置关系，通过PCoA可以观察个体或群体间的差异[15]。采用主坐标分析法(PCoA)查看核心种质样品在原始种质样品中的分布情况(图2)，结果表明：核心种质均匀分布在整个主坐标图中，较好地反映了原始样品在主坐标中的分布情况，说明所构建的核心种质具有很好的代表性。

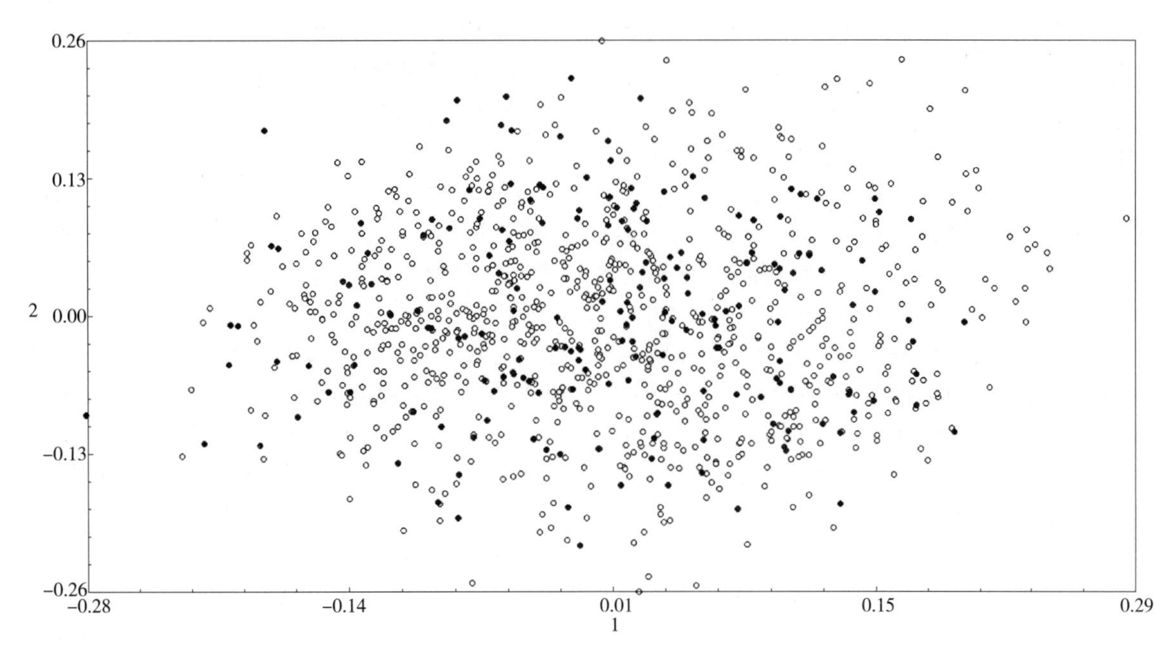

图 2　核心种质和保留种质的主坐标分布图

3　讨论

(1)数据选择。常用的构建核心种质的数据主要是形态标记数据和分子标记数据[16,17]。对多年生的杜仲而言，要准确测量其相关的表型数据往往需要多年的持续观测，尤其是叶、花、果和产量等经济性状数据，还需要种质保存地的立地条件、栽植年份、抚育措施等尽可能一致，以减小因环境因素引起的表型差异。因此，表型数据较难收集且易受环境等因素的影响，构建核心种质的准确性相对分子标记数据要差。分子标记以 DNA 多态性为基础，通常不具生物功能活性，不会或极少受植物生长期和外界环境的影响，较形态标记更适合构建核心种质和遗传多样性评价[17,18]。SSR 标记是多态性最高的遗传标记之一，在基因组中分布着大量的简单重复序列，且同一个位点上重复序列的数量变化很大。因此，本研究选用 9 对多态性较高的 SSR 引物，基于等位基因数最大化策略获得 189 份杜仲核心材料，对杜仲资源深入研究和利用具有重要意义。

(2)种质分组。分组的目的是为了保证取样的代表性和反映不同条件下的遗传多样性差异。在核心种质构建过程中，常将种质材料按照地理起源、植物学分类、生态区等进行分组，在分组的基础上构建核心种质[19,20]。也有采用不分组的方法，如 Van Hintum 等[21]为避免具有共同亲本的种质同时选入核心种质，提出了系谱法；Hu 等[22]为避免在聚类图上选择划分切点的困难，提出了逐步聚类法；Wang 等[23]提出了最小距离逐步聚类法。杜仲是地质史上残留下来的孑遗植物，现存杜仲仅一属一种。我国历史上对杜仲的栽培利用较早，2000 多年来不同地区之间相互引种的情况一直存在。长期相互引种使杜仲地理多样性和遗传多样性的关系逐渐变得模糊，材料间地理起源的差异已不能反映其在遗传及形态上的差异。因此，常见的按植物学分类、地理起源、生态区、品种分类体系等分组的方法并不适用于杜仲分组，本研究采用了不分组的方法，将所有参试群体作为整体构建核心种质，以等位基因数最大化为标准构建杜仲的核心种质，最终用 189 份种质保存了 887 份原始材料中检测到的全部的等位基因，达到了精简杜仲基因库的目的。

(3)构建方法。自核心种质的概念被提出以来，很多学者对其构建方法给出了不同的建议，如等位基因数目最大化法[6]、基于遗传距离的聚类法[24]、稀有等位基因优先法[4]等。迄今为止，等位基因数最大化法是其中最具优势的一种方法，已在一些模式植物和农作物的核心种质构建过程中被采用，如拟南芥[*Arabidopsis thaliana*(Linnaeus) Heynhold][11]、葡萄[25]等。此策略的最大优点在于能够保存遗传多样性最为丰富的种质资源，满足分类学家和遗传学家的需要[26]。Power Core 软件基于等位基因数目最大化策略构建核心种质的有效性已在葡萄、柿子等作物和林木上得到验证和应用[25,6]。因此，本研究采用等位基因数最大化法在 Power Core 软件中获得杜仲核心种质，核心种质对原始种质中 7 个遗传参数的保留比例均在 100%及以上，保留效果较好。

(4)评价参数及代表性检验。代表性是核心种质最重要的性质，常用的核心种质遗传多样性评价参数有：等位基因数、基因型数、Shannon 多样性指数和 Nei's 基因多样性指数。等位基因数被认为是最相关的指标[27,28]，等位基因数最大化意味着保存了遗传多样性最为丰富的种质资源。王建成在研究水稻核心种质的评价参数时认为平均 Shannon-Weaver 多样性指数、平均多态信息含量和平均 Simpson 指数是评价核心种质代表性的重要参数[7]。也有学者将多态位点百分率、观测等位基因数、有效等位基因数等参数用于核心种质的遗传多样性评价[30]。综合前人的研究结果，本研究选择等位基因数、平均等位基因数、平均有效等位基因数、平均 Shannon-Weaver 多样性指数、平均 Nei's 基因多样性指数、平均基因型数和平均多态信息含量等 7 个参数及其保留比例作为核心种质代表性的评价参数，结合不同群体相关遗传参数的 t 检验法和主坐标分析法(PCoA)对构建的核心种质代表性进行验证和确认，效果较好。保留比例是核心种质各个遗传参数占原始种质各个遗传参数的百分比[31,6]。本研究中，核心种质和原始种质基于 9 个 SSR 位点的等位基因数、平均等位基因数、平均基因型数 3 个参数的保留比例为 100%。平均有效等位基因数、平均 Shannon-Weaver 指数、平均 Nei's 遗传多样性指数、平均多态信息含量等 4 个参数的保留比例大于 100%，这主要是由于群体内样本数和等位基因频率改变引起的。这些遗传参数均基于等位基因频率按不同的方法估算获得，用于度量不同群体(原始种质、核心种质、保留种质)在多样性上的遗传冗余情况，而核心种质的构建过程是一个减少出现频率高的等位基因、增加稀有等位基因比例的过程，在去除遗传冗余的过程中，各等位基因频率的不规律增减易导致相应遗传多样性参数保留比例大于 100%，这一现象在新疆野杏[4]、蜡梅[31]、柿子[6]和桂花[5]等的核心种质构建中均存在。根据等位基因保留比例需大于 70%，其他遗传参数越大越好的评价标准[13,14]，在等位基因没有丢失的情况下，本研究构建的核心种质各个遗传参数较优，符合核心种质的要求，对原始种质的遗传多样性保留效果较好。

4 结论

887 份杜仲种质资源基于等位基因数目最大化原则得到 189 份核心种质，占原始种质 21.3%的样品，保存了原始种质 100%的等位基因和基因型。核心种质与原始种质、保留种质的 6 个遗传多样性参数经 t 检验，在 0.01 水平上差异不显著；主坐标分析也表明，核心种质与原始种质在分布图上有着相似的分布结构。以上说明构建的核心种质具有很好的代表性，可为杜仲优异基因资源发掘、育种群体建立和新品种选育提供参考，获得的 189 份核心种质可用于营建杜仲核心种质保存库。

参考文献

[1]杜红岩．中国杜仲橡胶资源与产业发展报告[M]．北京：社会科学文献出版社，2013：1-25.

[2]王述民，张宗文．世界粮食和农业植物遗传资源保护与利用现状[J]．植物遗传资源学报，2008，12(3)：325-338.

[3]FRANKEL O，BROWN A. Plant genetic resources today：a critical appraisal//Holden J H W，Williams J T. [C]. Crop Genetic Resources：Conservation and Evaluation，George Allenand Unwin London，198，249-257.

[4]刘娟，廖康，赵世荣，等．利用 ISSR 分子标记构建新疆野杏核心种质资源[J]．中国农业科学，2015，48(10)：2017-2028.

[5]张维瑞，袁王俊，尚富德．基于 AFLP 分子标记的桂花品种核心种质的构建[J]．西北植物学报，2012，32(7)：1349-1354.

[6]张艳芳．利用形态学和 SSR 标记建立柿核心种质[D]．武汉：华中农业大学博士学位论文，2012.

[7]王建成，胡晋，张彩芳，等．建立在基因型值和分子标记信息上的水稻核心种质评价参数[J]．中国水稻科学，2007，21(1)：51-58.

[8]董玉琛，曹永生，张学勇，等．中国普通小麦初选核心种质的产生[J]．植物遗传资源学报，2003，4(1)：1-8.

[9]KIM K W，CHUNG H K，CHO G T，et al. Power Core：a program applying the advanced M strategy with a heuristic search for establishing core sets[J]. Bioinformatics，2007，23(16)：2155-2162.

[10]BOTSTEIN D，WHITE RL，SKOLNICK M，et al. Construction of genetic linkage map in man using restriction fragment length polymorphism[J]. American Journal of Human Genetics，1980，32(3)：314-331.

[11]BROWN A H D. 2011. Core collections：a practical approach to genetic resources management[J]. Genome，31(2)：818-824.

[12]FRANCO J，CROSSA J，WARBURTON M L，et al. Sampling strategies for conserving maize diversity when forming core subsets using genetic markers[J]. Crop Science，2006，46(2)：854-864.

[13]MCKHANN H I，CAMILLERI C，BÉRARD A，et al. Nested core collections maximizing genetic diversity in Arabidopsis thaliana [J]. Plant Journal for Cell &

Molecular Biology, 2004, 38(1): 193-202.
[14] RONFORT J, BATAILLON T, SANTONI S, et al. Microsatellite diversity and broad scale geographic structure in a model legume: building a set of nested core collection for studying naturally occurring variation in Medicago truncatula[J]. BMC Plant Biology, 2006, 6(1): 28-41.
[15] JIANG X T, PENG X, DENG G H, et al. Illumina sequencing of 16S rRNA tag revealed spatial variations of bacterial communities in a mangrove wetland[J]. Microbial Ecology, 2013, 66(1): 96-104.
[16] 王建成，胡晋，黄歆贤，等. 植物遗传资源核心种质新概念与应用进展[J]. 种子，2008，27(5)：47-50.
[17] 杨汉波，张蕊，王帮顺，等. 基于SSR标记的木盒核心种质构建[J]. 林业科学，2015，53(6)：37-46.
[18] 刘新龙，刘洪博，马丽，等. 利用分子标记数据逐步聚类取样构建甘蔗杂交品种核心种质库[J]. 作物学报，2014，40(11)：1885-1894.
[19] 郑轶琦，郭琰，房淑娟，等. 利用表型数据构建狗牙根初级核心种质[J]. 草业学报，2014，23(4)：49-60.
[20] 李秀兰，贾继文，王军辉，等. 灰楸形态多样性分析及核心种质初步构建[J]. 植物遗传资源学报，2013，14(2)：243-248.
[21] van HINTUM TJ, HAALMAN D. Pedigree analysis for composing a core collection of modern cultivars, with examples from barley (*Hordeum vulgare* s. lat.) [J]. Theor Appl Genet, 1994, 88(1): 70-74.
[22] HU J, ZHU J, XU H M. Methods of constructing core collections by stepwise clustering with three sampling strategies based on the genotypic values of crops[J]. Theoretical and Applied Genetics, 2000, 101: 264-268.
[23] WANG J C, HU J, XU H M, et al. A strategy on constructing core collections by least distance stepwise sampling[J]. Theoretical and Applied Genetics, 2007, 115(1): 1-8.
[24] 何建文，韩世玉. 基于SSR标记不同距离聚类与抽样方法构建辣椒核心种质库[J]. 西南农业学报，2015，28(5)：2199-2204.
[25] 郭大龙，刘崇怀，张君玉，等. 葡萄核心种质的构建[J]. 中国农业科学，2012，45(6)：1135-1143.
[26] MARITA J M, RODRIGUEZ J M, NIENHUIS J. Development of an algorithm identifying maximally diverse core collections[J]. Genetic Resources and Crop Evolution, 2000, 47(5): 515-526.
[27] MOUSADIK A E, PETIT R J. High level of genetic differentiation for allelic richness among populations of the argan tree [Argania spinosa (L.) Skeels] endemic to Morocco [J]. Theoretical and Applied Genetics, 1996, 92(7): 832-839.
[28] PETIt R J, MOUSADIK A E, PONS O. Identifying populations for conservation on the basis of genetic markers [J]. Conservation Biology, 1998, 12(4): 844-855.
[29] 吴敏. 杜仲全基因组SSR标记开发及遗传多样性评价[D]. 北京：中国林业科学研究院硕士学位论文，2014.
[30] 李谋强，师桂英，叶树辉，等. 基于ISSR分子标记数据的兰州百合核心种质构建方法研究[J]. 中国沙漠，2015，35(6)，1573-1578.
[31] 赵冰. 蜡梅种质资源遗传多样性与核心种质构建的研究[D]. 北京：北京林业大学博士学位论文，2008.

[原载：林业科学，2018，54(02)]

几种珍贵热带用材树种的育苗基质筛选

白灵海

（中国林业科学研究院热带林业实验中心，广西凭祥　532600）

摘　要　通过红椎、火力楠、西南桦与米老排4种苗木对4种不同育苗基质的生长适应性分析可知，除红椎与米老排对含塘泥的基质表现出不适应外，各树种对其他3种基质均适应良好，在生产中可推广应用。

关键词　珍贵树种；热带用材树种；育苗基质；筛选

［原载：林业实用技术，2010(05)］

美国火炬松第3代种子园不同家系育苗与造林试验初报

白灵海 刘光金 孙文胜 蒙彩兰

（中国林业科学研究院热带林业实验中心，广西凭祥 532600）

摘 要 通过对10个美国火炬松(*Pinus taeda*)第3代种子园优良家系引种苗期试验研究，结果表明各家系苗高、地径生长存在显著差异。对1年生火炬松苗高、地径生长性状与混合马尾松家系进行 t 检验，表明：火炬松表型好，苗高、地径生长性状显著优于马尾松，经过选育后可显著提高现实增益。

关键词 火炬松；家系；引种

［原载：广西林业科学，2010，39(03)］

美国湿地松第3代种子园不同家系育苗与造林试验研究

刘光金 白灵海 孙文胜 蒙彩兰

（中国林业科学研究院热带林业实验中心，广西凭祥 532600）

摘 要 研究引种美国湿地松第三代种子园10个优良家系苗期遗传变异，为广西速生丰产高产脂湿地松选育提供参考。以马尾松为对照，采用家系小区随机区组的试验方法，研究引种火炬松苗高、地径主要生长性状遗传变异。引种美国湿地松第三代种子园10个优良家系造林定植成活率高，平均保存率96%，表明美国湿地松适应性强，适宜于广西凭祥立地环境；苗期湿地松高生长不显著，表型最好的家系MWV-447，高于对照Masson 1和Masson 2，MWV-443、MWV-444、MWV-446、MWV-448、MWV-451、MWV-452高于对照Masson 2，但低于对照Masson 1，MWV-445、MWV-449、MWV-450最差，低于对照Masson 1和Masson 2；参试10个优良家系地径生长显著优于对照Masson 1和Masson 2，MWV-449最大，为28.08mm；超出对照Masson 142.3%，超出对照Masson 265.9%，MWV-444最小，为24.96mm，超出对照Masson 126.5%，超出对照Masson 247.4%。经方差分析与t检验，美国湿地松优良家系苗期性状差异显著，经过林木改良与选育可筛选出适宜家系推广利用。

关键词 湿地松；第三代种子园家系；引种

［原载：安徽农业科学，2010，38(19)］

兰花槛轻基质网袋容器苗基质选择试验

蒙彩兰　黎　明

（中国林业科学研究院热带林业实验中心，广西凭祥　532600）

摘　要　探讨适合兰花槛培育苗木的轻基质配方，为生产提供技术参考。以松皮、锯末、炭化锯末3种基质设置9种配方，采用随机区组设计，开展兰花槛网袋容器育苗试验，对不同轻基质配方的苗高、地径和侧根及生物量等指标进行苗木综合评定。不同配方之间的苗高、地径、侧根和生物量等均达到显著差异；配方A(沤制松皮75%+沤制锯末25%)、D(沤制松皮75%+炭化锯末25%)、E(沤制松皮50%+炭化锯末50%)、G(沤制松皮70%+沤制锯末20%+炭化锯末10%)的育苗效果排列前4位，其中配方E(沤制松皮50%+炭化锯末50%)的最好，育苗效果最差的为配方C(沤制松皮25%+沤制锯末75%)和配方Ⅰ(沤制松皮20%+沤制锯末50%+炭化锯末30%)。配方E(沤制松皮50%+炭化锯末50%)最适宜培育兰花槛苗木。

关键词　兰花槛；轻基质；网袋容器苗

［原载：安徽农业科学，2010，38(17)］

西南桦轻基质网袋容器苗的质量评价指标

黎 明[1] 曾 杰[2] 郭文福[1] 吴光枝[1]

([1]中国林业科学研究院热带林业实验中心，广西凭祥 532600；
[2]中国林业科学研究院热带林业研究所，广东广州 510520)

摘 要 为制定西南桦轻基质网袋容器苗的分级标准，为营建高效西南桦人工林奠定基础，采用主成分分析确定西南桦轻基质容器苗质量评价的主要指标，应用逐步聚类分析研究其分级标准，并通过造林效果试验对分级结果予以验证。结果表明：西南桦苗木可分为3级，Ⅰ级苗：$H \geqslant 28.5$cm、$D \geqslant 2.0$mm，Ⅱ级苗：28.5cm$>H \geqslant$23.9cm、2.0mm$>D \geqslant$1.5mm，Ⅲ级苗：$H<23.9$cm、$D<1.5$mm；造林后半年时不同级别苗木的造林效果差异明显，Ⅰ、Ⅱ级苗的造林成活率均在85%以上，而两者苗高、地径生长量差异显著（$P<0.05$），Ⅲ级苗的成活率低于75%，其生长量大多极显著低于Ⅰ、Ⅱ级苗($P<0.01$)。

关键词 西南桦；轻基质容器苗；苗木分级；主成分分析；逐步聚类分析

[原载：贵州农业科学，2011，39(06)]

光皮桦优良家系育苗试验初报

刘光金　王小宁　马　跃　李武志

（中国林业科学研究院热带林业实验中心，广西凭祥　532600）

摘　要　以前期收集的广西天峨林朵林场光皮桦优树成熟种子为试材，进行光皮桦优良家系育苗试验。通过统计分析42个优良家系4个月苗龄苗高、地径两个生长性状指标，结果表明：各家系苗期分化明显，苗高、地径差异显著，经苗期评价，初选30、39、17、35、19、32、20、2、7、27、42等11个优势家系，其苗高超过群体49%、46%、29%、28%、24%、22%、23%、21%、20%、20%、17%；地径超过群体22%、4%、4%、18%、4%、19%、17%、12%、19%、4%、10%。研究成果可为广西地区发展光皮桦人工林提供科学依据。

关键词　光皮桦；家系；苗期

［原载：湖北农业科学，2011，50(16)］

顶果木叶芽组织培养繁殖技术

谌红辉　王小宁　马　跃　刘光金　刘志龙

（中国林业科学研究院热带林业实验中心，广西凭祥　532600）

摘　要　本研究采用顶果木叶芽做外植体，研究总结了顶果木的组织培养繁殖技术。结果表明：叶芽外植体用70%酒精消毒10s，再用10%的 H_2O_2 消毒5min效果最好，叶芽外植体转接到诱导分化培养基后，暗培养7d后再全光培养，诱导成功率达55.2%。顶果木适宜的继代增殖培养基为MS+6-BA0.5mg/L+KT0.5mg/L+NAA0.2mg/L，增殖系数可达3.6。组培芽苗采用生根培养基1/2MS+NAA0.5mg/L+IBA0.5mg/L+6-BA0.3mg/L培养，生根率可达82%。

关键词　顶果木；组织培养；叶芽外植体；培养基

［原载：林业实用技术，2012(10)］

香梓楠轻基质育苗试验

麻　静　黄积寿　张万幸

（中国林业科学研究院热带林业实验中心，广西凭祥　532600）

摘　要　香梓楠为热带南亚热带珍贵优良用材树种，使用轻基质网袋育苗，具有重量轻、通气透水性好和方便种植等特点，可以提高造林成活率和降低造林成本。选择树皮和锯末为基质原料，经晒干、粉碎、过筛、堆沤后，按一定比例把拌匀的基质灌装至网袋内制成育苗容器。香梓楠种子经催芽后移植至容器内，注意淋水，保持湿润；每半个月施肥一次，以氮肥和复合肥为主，宜勤施薄施：当苗高约 15cm 时，移疏苗木，并适时进行空气修根，促进侧根的生长；约 11 月中旬以后可实施炼苗，当苗龄达 12 个月、苗高达 45cm 以上时，苗木可出圃造林。

关键词　香梓楠；轻基质；育苗

［原载：森林工程，2012，28(06)］

顶果木嫩枝扦插育苗试验

马 跃[1] 谌红辉[1] 刘志龙[1] 林天龙[2]
([1]中国林业科学研究院热带林业实验中心，广西凭祥 532600；
[2]广西岑王老山国家级自然保护区管理局，广西百色 533000)

摘 要 本文以顶果木半年生优势苗作为采穗母株，对嫩枝扦插育苗技术进行了系统的研究。试验表明：截干高度对母株产穗量影响显著，最适宜截干高度为20~30cm；生根促进剂浓度对插穗生根影响显著，采用GGR的适宜浓度为800~1000mg/kg；扦插时间同样影响插穗生根，一年中适宜的扦插时间为3~7月。

关键词 顶果木；扦插；萌条；生根

[原载：林业科技开发，2012，26(06)]

青钱柳扦插育苗技术研究

王小宁

(中国林业科学研究院热带林业实验中心，广西凭祥 532600)

摘 要 本研究以青钱柳幼树作为采穗母树，对青钱柳扦插育苗的技术作了系统的总结。经过试验可知：截干高度与穗条总产量呈显著的正相关关系，根据母树生长状况与产穗量统计，采穗母树应筛选萌芽力强的优势苗培育，母树截干高度20~30cm为宜。不同植物生长调节剂与浓度处理对扦插穗条的成活率与生根数量影响较大，青钱柳扦插宜采用吲哚丁酸或6号ABT生根粉，浓度配制采用0.1%~0.125%药液为宜，成活率分别达81.1%、83.1%。青钱柳扦插采用轻基质生根效果最佳，青钱柳扦插育苗的时间在4~8月较好，插穗易生根成活。综合分析可知，青钱柳扦插育苗对于快速繁殖生产用苗是一个有效的途径。

关键词 青钱柳；扦插；植物生长调节剂；轻基质；濒危植物

[原载：林业实用技术，2012(04)]

西南桦无性系测定与评价

谌红辉[1] 贯宏炎[1] 郭文福[1] 黄弼昌[2] 吴龙墩[3] 刘光金[1] 秦旭东[4]
([1]中国林业科学研究院热带林业实验中心，广西凭祥 532600；[2]广西林朵林场，广西河池 547300；
[3]广西岑王老山林场，广西百色 533300；[4]广西林业勘测设计院，广西南宁 530000)

摘　要　利用西南桦优树无性系，在广西3个主要栽培区域设置无性系试验林进行测定评价。通过4年生的生长材料分析可知，不同西南桦无性系生长差异显著，筛选出的6个优良无性系，对比实生苗材积增益分别达48.4%~99.3 %，无性系在推广利用前必须开展测定评价。无性系与立地条件的互作效应显著，造林工作中应做到适地适无性系。

关键词　西南桦；无性系；无性系评价；材积增益

[原载：林业实用技术，2013(06)]

红椎容器育苗基质的选择

李忠国[1] 郭文福[1] 蒙彩兰[1] 苏 勇[2]
([1]中国林业科学研究院热带林业实验中心，广西凭祥 532600；
[2]广西南宁市树木园，广西南宁 530031)

摘 要 开展红椎容器育苗基质筛选研究，为培育优质红椎实生苗提供技术支持，为红椎大径级用材林培育奠定良好基础。采用沤制松树皮、锯末、炭化锯末等林业废弃物，作为基质组分，设计10个基质配方，根据苗高、地径、地上生物量和地下生物量等指标对苗木质量进行综合评价。各育苗基质苗木成活率差异不显著，各基质质量综合评价得分值大小顺序为：Ⅵ>Ⅱ>Ⅲ>Ⅳ>Ⅴ>Ⅶ>Ⅷ>Ⅸ>Ⅹ>Ⅰ。基质Ⅵ(泥炭60%、炭化锯末20%、炭化树皮20%)育苗效果最好，基质Ⅱ(黄心土50%、沤制树皮25%、炭化树皮25%)育苗效果次之，基质Ⅰ(黄心土85%、火烧土15%)育苗效果最差。宜选择基质Ⅵ培育优质红椎实生苗。

关键词 红椎；容器育苗；育苗基质

[原载：林业实用技术，2013(10)]

光皮桦优树选择技术

刘光金[1] 黄弼昌[2] 谌红辉[1] 孙浩忠[2] 马 跃[1]
([1]中国林业科学研究院热带林业实验中心，广西凭祥 532600；[2]广西天峨林朵林场，广西天峨 547300)

摘 要 本文采用5株优势木对比法开展了光皮桦木人工林的优树选择技术研究。以胸径、树高、材积为数量评价指标，结合干形、冠高比和分枝粗细形质指标，对13年生光皮桦木人工林优树观测数据进行了总结分析，提出了光皮桦人工林优树选择标准，即优树胸径、树高、单株材积分别大于5株对比优势木平均值的14%、4%、35%，形质指标综合得分大于8.5，共选择出光皮桦优树10株，入选率为30.3%。优树选择即要考虑其生长性状，同时注意材性、抗性的选择以保证光皮桦育种群体的遗传多样性，其选择标准在实际应用中可根据林分状况适当调整。

关键词 光皮桦；人工林；优树；优势木；选择标准

[原载：林业实用技术，2013(01)]

西南桦大树法选优及其增益估测的研究

刘光金

（中国林业科学研究院热带林业实验中心，广西凭祥 532600）

摘　要　以广西凭祥热带林业实验中心11年生西南桦同龄人工纯林为研究材料，采用优树缩差与优良度回归分析，构建西南桦优树缩差估测方程，并探讨了西南桦优树优良度与遗传增益估算相关关系。结果表明：①西南桦优树材积缩差估测方程 $Uv=1.977+0.043Kd$，$r=0.729$，达到显著水平；经21株候选优树和105株优势木实测数据验证，平均相对偏差小于10%，可在实际工作中应用。②通过构建缩差、入选率、选择强度三维换算表，确定了西南桦选优时优树优良度下限为5%，上限为20%，其对应入选率、遗传增益分别为4%、0.5%、10.136%、12.485%。

关键词　西南桦；优树选择；优良单株

［原载：湖北农业科学，2013，52(02)］

顶果木嫁接育苗技术

马 跃 谌红辉 刘光金 刘志龙

（中国林业科学研究院热带林业实验中心，广西凭祥 532600）

摘 要 介绍了广西珍贵乡土树种顶果木优良苗木嫁接繁育的过程。包括嫁接前的砧木苗培育、接穗的选择、嫁接方法以及嫁接后的嫁接苗管理等工作。

关键词 顶果木；嫁接；育苗

［原载：林业实用技术，2013(05)］

望天树嫩枝扦插技术初步研究

马　跃[1]　刘志龙[1]　谌红辉[1]　岳志强[2]

([1]中国林业科学研究院热带林业实验中心，广西凭祥　532600；
[2]中国林业科学研究院科技处，北京　100091)

摘　要　望天树为国家Ⅰ级重点保护野生植物，是我国珍贵的用材树种之一。本文以望天树半年生优势苗作为采穗母株，采用 GGR-6 作为生根促进剂，对望天树嫩枝扦插育苗技术进行了初步研究。结果表明：①截干高度对母株产穗量影响显著，最适宜截干高度为 20～30cm。②生根促进剂浓度对插穗生根影响显著，适宜浓度为 800～1000mg/kg。③扦插时间对插穗生根影响显著，扦插时间为 3～7 月。

关键词　望天树；扦插；截干高度；萌条；生根

［原载：林业实用技术，2013(02)］

檀香紫檀嫁接育苗技术

李洪果 马 跃 谌红辉 贾宏炎 刘光金 刘福姝

（中国林业科学研究院热带林业实验中心，广西凭祥 532600）

摘 要 檀香紫檀是20世纪60年代引入我国的外来树种，材质优良，可用材、观赏兼用，在我国与原产地类似气候条件下的栽培表现为适生。由于繁殖材料极为有限，种苗问题成为制约檀香紫檀在我国进一步发展的瓶颈。本研究以结实量大、种质材料广泛的降香黄檀为砧木嫁接檀香紫檀获得初步成功，嫁接20株，成活6株，为檀香紫檀在我国的繁殖利用开辟了新的途径。

关键词 非本砧嫁接；檀香紫檀；嫁接；红木

［原载：湖南林业科技，2014，41(04)］

广西望天树优树选择标准和方法研究

刘志龙　马　跃　谌红辉　刘光金　李洪果　全昭孔　莫慧华　蒙明君

（中国林业科学研究院热带林业实验中心，广西凭祥　532600）

摘　要　在调查广西境内望天树遗传资源状况基础上，分析望天树天然次生林和人工林的生长特性，提出了望天树天然次生林和人工林选优的标准与方法，共选出优树 18 株，其中，天然次生林 10 株，人工林 8 株。旨在为望天树的遗传改良和创新利用奠定科学基础和技术支持。

关键词　望天树；优树；选择标准

［原载：林业实用技术，2014(09)］

柚木轻基质实生容器苗培育技术

麻　静　李运兴　黄积寿

（中国林业科学研究院热带林业实验中心，广西凭祥　532600）

摘　要　柚木为热带珍贵阔叶树种，近来深受林农、私营投资商的青睐。基于多年的柚木育苗经验，从苗木生产各个环节总结了柚木轻基质容器苗的培育技术，以促进柚木壮苗生产，进而为柚木种植业发展提供苗木保障。

关键词　柚木；种子处理；播种；轻基质；容器苗

［原载：林业实用技术，2014(08)］

土沉香轻基质育苗试验

蒙彩兰　黎　明　贾宏炎

（中国林业科学研究院热带林业实验中心，广西凭祥　532600）

摘　要　用松树皮、锯末和炭化树皮等配制 6 种轻基质，以黄心土和森林表土混合基质为对照，开展土沉香育苗试验，揭示系列基质的理化性质及其对土沉香幼苗生长的影响，从而筛选出适合土沉香幼苗生长的轻基质配方。结果表明：各轻基质处理间的理化性质差异显著（$P<0.05$）或极显著（$P<0.01$），而且其土沉香苗高、地径、地上和地下部分干重及总干重均显著高于对照。其中以基质 T2（沤制松树皮 50%+沤制锯末 25%+炭化树皮 25%）和 T5（沤制松树皮 50%+沤制锯末 40%+鸡粪 10%）育苗效果最好，其苗高分别较对照提高 1.54 倍和 1.40 倍，地径增粗 1.15 倍和 1.07 倍，总干重增加 6.39 倍和 5.27 倍。苗木生长与基质的容重、孔隙度、气水比、有机质和全 N 含量及阳离子交换量显著或极显著相关。建议生产上优先采用基质 T2 培育土沉香苗木，其次为基质 T5。

关键词　土沉香；轻基质；理化性质；网袋容器苗

［原载：种子，2015，34(07)］

芽苗切根对降香黄檀容器苗生长的影响

马　跃　刘志龙　贾宏炎　蒙彩兰　李丽利

（中国林业科学研究院热带林业实验中心，广西凭祥　532600）

摘　要　对降香黄檀(*Dalbergia odorifera* T. Chen)种子胚根不同程度切根处理进行育苗对比试验，分析不同处理切根处理的苗高、地径、根瘤数、生物量等苗木形状差异。试验结果表明：降香黄檀芽苗切根 1/2 时，根生物量、根瘤数、根冠比最大，有利于造林后的成活率及生长量的提高。

关键词　苗木培育；降香黄檀；芽苗切根；容器苗

［原载：林业科技通讯，2016(03)］

不同供氮水平对顶果木苗期生长及氮素分配的影响

陈厚荣[1,2]　马　跃　刘志龙[1,2]　蒙明君[1,2]　明安刚[1,2]
([1]中国林业科学研究院热带林业实验中心，广西凭祥　532600；
[2]广西友谊关森林生态系统国家定位观测研究站，广西凭祥　532600)

摘　要　顶果木(*Acrocarpus fraxinifolius*)作为南亚热带珍贵优良速生用材树种之一，苗期施用氮素对其生长有重要的意义。本研究采用单因素随机设计的盆栽试验，对顶果木苗期采用不同的氮素添加处理的效应进行了研究。结果表明：适量施用氮肥能够有效提高顶果木苗高、地径的生长，增加苗木生物量，且有利于植株对氮素的吸收。不同浓度氮肥对顶果木苗木各生长指标有显著影响，施氮量为6g/盆时，顶果木苗高、地径、总生物量、全株含氮量表现最佳，分别比对照增加155%、61.7%、438.3%、97.6%，结果同时表明，氮肥施用过多时会对苗木生长产生抑制作用。

关键词　顶果木；氮素；苗期生长；氮含量

[原载：广西林业科学，2017，46(03)]

广西米老排地理种源苗期性状变异分析

刘光金　农　志　贾宏炎

（中国林业科学研究院热带林业实验中心，广西凭祥　532600）

摘　要　利用8个米老排(*Mytilaria laosensis* Lecomte)种源82株优势木自由授粉种子及其1年生苗木，开展了种子表型变异和苗期生长性状变异的研究。结果表明：米老排种子表型性状与苗木生长性状存在明显的种源差异。种子表型性状与苗木生长性状显著相关，种重与苗高显著正相关；种长与叶重、茎重、全株湿重显著正相关；种宽与苗高、地径、茎重、全株鲜重显著负相关；依据苗木生长性状聚类分析结果，将8个种源分为3类：第一类靖西、容县，为生长优势种源；第二类龙州、那坡、凭祥、德保，为生长中速种源；第三类东兴、上思，为生长较差种源。

关键词　米老排；性状变异；种源试验

［原载：湖北农业科学，2017，56(09)］

越南白花泡桐引种试验初报

马 跃[1] 叶金山[2] 潘启龙[1] 劳庆祥[1] 谌红辉[1]

([1]中国林业科学研究院热带林业实验中心，广西凭祥 532600；
[2]国家林业局泡桐研究开发中心，河南郑州 450003)

摘 要 对广西凭祥市中国林业科学研究院热带林业实验中心引种的2年生越南白花泡桐幼林进行调查研究，结果表明：越南白花泡桐最适宜生长的是水肥条件较好的下坡，施用桐麸作为底肥能有效促进越南白花泡桐树高和胸径的生长。越南白花泡桐在广西凭祥市生长良好，种植潜力较大。

关键词 白花泡桐；引种试验；树高；胸径

[原载：林业科技通讯，2017(10)]

基于 311-A 最优回归设计研究配施氮磷钾肥对山白兰苗期生长的影响

刘福妹[1] 李武志[1] 庞圣江[1] 郝 建[2] 潘启龙[1] 韦菊玲[1] 黄振声[3]

([1]中国林业科学研究院热带林业实验中心，广西凭祥 532600；
[2]广西友谊关森林生态系统国家定位观测研究站，广西凭祥 532600；
[3]广西国有派阳山林场，广西宁明 532500)

摘 要 本研究采用 311-A 最优回归设计，选当年播种的实生苗为试验材料，通过监测苗木的地径和苗高指标，探讨氮、磷、钾肥三个因素对山白兰生长的影响，以期为山白兰育苗提供科学依据。结果表明：合理的配施氮、磷、钾肥可以显著促进山白兰苗木的生长；建立了山白兰苗期生长量(地径和苗高)与氮、磷、钾肥施用量关系的回归模型，得出氮素、过磷酸钙、硫酸钾分别为 0.83g/株、3.21g/株、1.2g/株和 2.61g/株、0.64g/株、0.24g/株依次为促进地径和苗高增长的最佳组合，分别使二者增长量达到最大值 1.10cm 和 67.2cm。综合考虑，推荐使用苗高组合，可使地径和苗高分别达到 1.04cm 和 67.2cm。

关键词 山白兰；氮磷钾肥；生长；311-A

[原载：林业科技通讯，2018(12)]

第二部分

珍优树种近自然高效培育与资源管理技术研究

林分密度及施肥对马尾松林产脂量的影响

安　宁[1,2]　丁贵杰[1]　谌红辉[2]

([1]贵州大学造林生态研究所，贵州贵阳　550025；
[2]中国林业科学研究院热带林业实验中心，广西凭祥　532600)

摘　要　采用长期定位观测，定时、定株收获的方法，研究了马尾松采脂林5种保留密度及16种施肥组合对产脂量影响。分析结果表明：不同密度松脂产量差异显著，随着密度的增大，单位面积松脂产量也随之增加，呈正相关关系，单株平均年松脂产量则随密度增大而减少，初步认定培育马尾松高产脂林保留密度C~D(750~900株/hm^2)较适合；在中等立地，施肥能有效增加松脂产量，不同施肥组合的松脂产量存在明显差异。较理想的3个施肥组合是：N250g/株+P500g/株、P1000g/株+K300g/株和N250g/株+K150g/株。同时分析了主要气象因子对松脂年产量变化的影响。

关键词　马尾松；松脂产量；林分密度；施肥

The Effect of Stand Density and Fertilization on Resin Yield of *Pinus massoniana*

AN Ning[1,2], DING Guijie[1], CHEN Honghui[2]

([1] *The Institute of Silviculture and Ecology, Guizhou University, Guiyang 550025, Guizhou, China;*
[2] *The Experimental Centre of Tropical Forestry, CAF, Pingxiang 532600, Guangxi, China*)

Abstract: The resin yield of *Pinus massoniana* stand was studied in four kinds of reserve density and combinations of sixteen kinds of fertilizers, by the method of yield at fixed time, sample tree and position. The results showed that resin yield was significantly different among densities, as the density increases, the unit area of resin production also increased, there was a positive correlation, and resin production per plant reduced with increasing density. The results indicated that the stand density C~D(750~900 plant/hm^2) of high yield resin was more suitable. In the middle of site, fertilization could increase the resin yield, the resin yields were significantly different among different fertilizer combinations. Ideal fertilizer combinations were N 250g/plant + P 500g/plant, P 1000g/plant + K 300g/plant and N 250g/plant + K 150g/plant. The effect of meteorological factors on the resin yield was studied too.

Key words: *Pinus massoniana*; resin yield; stand density; fertilization

马尾松(*Pinus massoniana* Lamb.)是我国松树中分布面积最广和采割脂的主要树种之一。广泛分布于秦岭，淮河以南，云贵高原以东17个省(自治区、直辖市)，面积居全国针叶林首位[1]。松香是我国林化产品中的主要出口产品，产量和出口量均居世界首位。我国虽然有20多种松树可以采割松脂，但目前松脂主要采自马尾松。由于天然马尾松林面积逐年下降，可采脂树迅速减少，已不能适应林产化工业的需要，因此，营建人工高产脂原料林是发展的必然趋势。以往对马尾松栽培技术、采伐年龄、地理种源的遗传变异、染色体分析、合理采伐年龄、人工林的营养循环及其遗传改良等方面的研究报道较多[2-8]，但对马尾松采脂林林分保留密度及施肥对马尾松松脂产量影响的系统研究报道很少[9-11]。因此，开展此项研究可为制定马尾松高产脂林的定向培育管理技术体系提供理论依据和技术支撑。

1　试验地概况及研究方法

1.1　试验地概况

试验林设在位于广西凭祥市中国林业科学研究

院热带林业实验中心伏波实验场，地处106°43′E，22°06′N，海拔500m，低山，年平均气温19.9℃，年降水量1400mm属于南亚热带季风气候区，土壤为花岗岩发育成的红壤。土层厚度80cm以上，腐殖质厚度10cm以上。

1.2 试验材料及设计

试验林分来源于伏波实验林场8号和9号林班，分别为1989年和1991年人工营造的马尾松纯林，前茬均为杉木，主伐后明火炼山。

密度试验在1991年人工营造的马尾松林分内进行，开展不同保留密度试验。试验采用随即区组试验设计，试验设5个密度：即A 450株/hm^2，B 600株/hm^2，C 750株/hm^2，D 900株/hm^2，E 1050株/hm^2。重复3次，共15个小区，每个小区面积600m^2，收脂时以小区为单位称重并记录。

施肥试验在1989年所造的马尾松人工林林分内进行。施肥试验采用$L_{16}(4^5)$正交表设计安排正交试验，N，46%的尿素，分别N_1(0g/株)，N_2(250g/株)，N_3(500g/株)，N_4(750g/株)。P，18%的钙镁磷肥，分别为P_1(0g/株)，P_2(500g/株)，P_3(1000g/株)，P_4(1500g/株)。K，60%的氯化钾，分别为K_1(0g/株)，K_2(150g/株)，K_3(300g/株)，K_4(450g/株)。试验设16个施肥组合，重复2次，共32个小区，每个小区面积为400m^2，收脂时以小区为单位称重并记录。

1.3 采脂方法

采脂均采用“下降式”采脂法，在树木的向阳面、节疤少的地方选定剖面；先刮去粗皮，刮至无裂隙、淡红色较致密的树皮层出现为止，然后割中沟及侧沟，割面离地面高2.2m，两侧沟夹角为60°~70°，割面负荷率为50%~60%；每天加割一次新的侧沟，每对侧沟均应与第一对侧沟等长、等深、平行[12]。以上数据都取平均数进行分析，方差分析及多重比较运用SPSS软件[13]。气象资料来自凭祥市气象局。上述试验均于2008年11月开始布置，2009年5月中旬开始调查松脂产量，每隔15d收获一次，试验结束于2009年10月下旬。收脂时按试验要求称重并记录。

2 结果与分析

2.1 不同密度松脂产量年变化

以日期为横坐标，以公顷产量为纵坐标(把每次松脂收集量换算成公顷产量)，绘制不同密度松脂产量年变化曲线，如图1所示。可以看出，每个时间点上都是随着密度的增大，松脂产量也随之增加。

不同密度马尾松松脂产量的年度变化规律为：一年中松脂产量呈抛物线形状，在个别点上略有波动。5种密度的松脂产量年变化曲线走势基本一致。5月份产量较低，随着时间的推移产量逐步增加，在8月份产量达到最高，之后又逐渐下降，直到10月底基本停止。形成这样规律是与马尾松自身生长特性以及气象因子影响有关。当昼夜平均温度在7~10℃以上时马尾松树液开始流动，可以开始采割松脂。但此时温度较低，马尾松生命活动较弱，松脂产量不高。随着温度的升高，马尾松生命活动逐渐加强，新陈代谢变得旺盛，松脂产量也随之增加。

不同密度松脂产量的年度变化曲线在6月30日、7月15日和8月15日这3个点松脂产量略有波动。这是因为除了与自身生长特性相关以外，马尾松产脂量更受气象因子综合作用影响。即使达到适宜温度，但是松脂产量同时也受其他气象因子影响，如日照时数长短、降水量大小、相对湿度大小等。

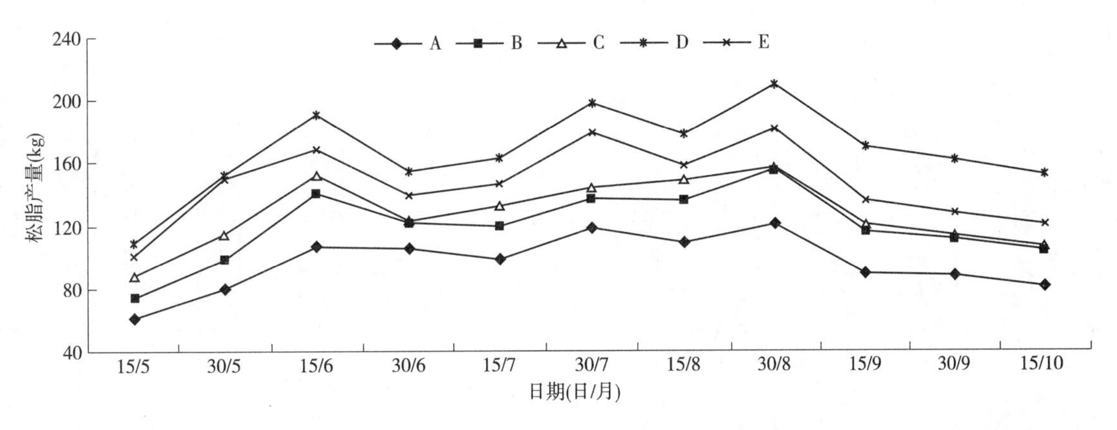

图1 不同密度马尾松林松脂产量年变化曲线

表 1 主要气象因子数据

日期（日/月）	指标			
	气温(℃)	降水量(mm)	日照时数(h)	相对湿度(%)
15/5	25.2	49	54.4	75
31/5	25.9	87	71.5	75
15/6	27.9	194	101.6	72
30/6	28.2	91	84.9	77
15/7	27.9	45	85.7	76
31/7	28.1	76	90.0	77
15/8	28.6	68	83.1	77
31/8	28.6	44	140.0	72
15/9	28.4	19	118.5	72
30/9	26.6	58	77.3	74
15/10	24.4	29	73.2	72

从表 1 分析得出，在 6 月 16 日至 6 月 30 日期间，平均气温相对较高，为 28.2℃，降水量也较充沛，为 91mm，但是降水分布不均匀，多是集中在 6 月 18 号和 6 月 23 号，而其他时间降水很少。水分吸收较少，松节油挥发，导致松脂干固，堵塞树脂道口，松脂便停止流出即造成松脂产量下降，同时水分吸收少又造成马尾松光合作用的强度的下降，同时气温较高，又会造成水分代谢的减弱，导致新陈代谢强度下降，最终松脂产量下降；在 7 月 1 日至 7 月 15 日期间，降水较少，只有 45mm，水分代谢减弱本身就可以造成松脂产量下降，马尾松体内光合作用的酶活性减弱，新陈代谢强度下降，松脂产量也随之下降；在 8 月 1 日至 8 月 15 日期间，降水量为 68mm，但是分布不均匀，造成了松脂产量的下降，进而影响了不同密度松脂产量年变化曲线。最适宜于提高马尾松的松脂产量的气象条件是气温适宜、降雨充沛且均匀、空气湿度较大。

2.2 密度对松脂产量的影响

图 2 可以清晰看出随着密度的增大，单位面积的松脂产量增加。密度 B、C、D、E 松脂的产量分别比密度 A 提高了 24.52%、32.40%、54.33%、74.34%。平均单株松脂年产量随密度增大而下降。因为密度对马尾松林分平均胸径有极显著影响，随密度的增大，林分平均胸径减小，同时大径木比例减小，而小径木比例增大。林分密度从 A 到 E，林分平均胸径由 24cm 下降到 20cm。

马尾松采脂量与采脂树木的胸径呈一元线性回归关系，径级越大，松脂产量越高。因为在同样的采脂负荷率下，径级大的割沟长度明显比径级小的割沟长度要长，割破的树脂道的个数要多，产脂量自然就增加。高密度林分单位面积年产量虽然高于低密度的年产量，但是在单株年产量上却是高密度要低于低密度，如图 3。

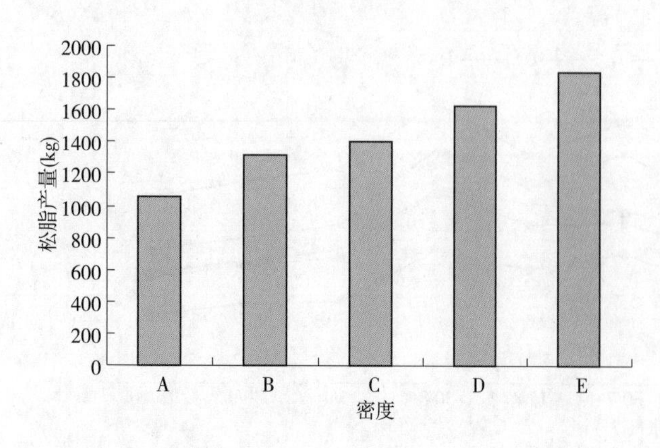

图 2 不同密度马尾松林单位面积松脂年产量

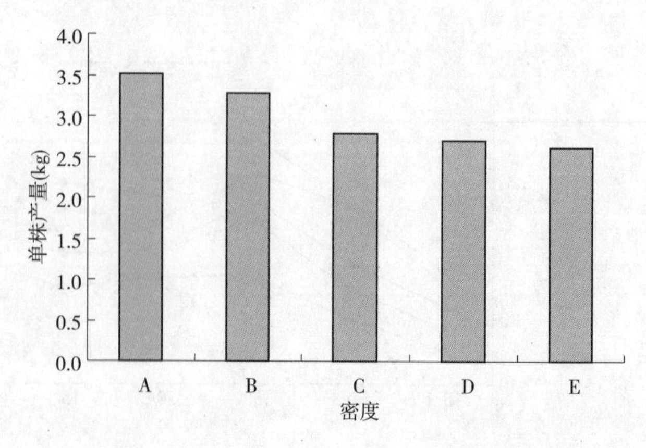

图 3 不同密度马尾松林单株松脂年产量

表 2　不同密度方差分析表

差异源	*SS*	*Df*	*MS*	*F*	*Sig.*
处理间	1098870	4	274717.523	42.773	0.000
处理内	64226	10	6422.648		
总计	1163096	14			

表 2 *Sig.* =0.000<0.05，表明 5 种密度的松脂年产量具有显著差异，而且达到极显著水平。多重比较检验结果表明：在 5 种密度之间只有密度 B 和密度 C 之间差异不显著，其他密度之间松脂年产量均存在显著差异，如表 3。

表 3　不同密度多重比较检验表

密度	A-B	A-C	A-D	A-E	B-C	B-D	B-E	C-D	C-E	D-E
MD	280.28*	282.95*	593.40*	776.49*	2.67	313.12*	496.21*	310.45*	493.54*	183.08*
Sig.	0.002	0.002	.000	0.000	0.968	0.001	0.000	0.001	0.000	0.019

在实际生产中，马尾松培育目标有的是以木材为主，松脂为辅；有的是脂材兼用。密度对马尾松林分平均胸径、单株材积有极显著影响，随密度的减小，林分平均胸径与单株材积增大，同时大径木比例加大，而小径木比例减少，枝下高降低而单位蓄积减小。在获得较大松脂产量同时，也要考虑出材量、生产成本。根据以上分析，初步认定培育马尾松高产脂林林分保留密度控制在 750~900 株/hm^2 较适合。

2.3　施肥对松脂产量的影响

从图 4 可知，施肥对马尾松松脂产量有一定影响。其中 1 号施肥组合为空白对照，在 16 个组合中有 2 号、3 号、5 号、6 号、7 号、9 号、10 号、12 号、15 号、16 号施肥组合的松脂产量均高于 1 号对照，分别比对照提高 5.44%、16.77%、13.23%、22.57%、8.07%、8.33%、4.14%、5.65%、8.6%、2.31%。而其他施肥组合则出现负效应，松脂产量低于空白对照。在中等立地条件下施肥具有明显增效，本试验样地土壤较肥沃，造成少量施肥组合出现微弱负效应。

表 4　施肥组合方差分析表

差异源	*SS*	*Df*	*MS*	*F*	*Sig.*
处理间	11159	15	743.963	1.238	0.038
处理内	9616	16	601.010		
总计	20775	31			

由表 4 *Sig.* =0.038<0.05，表明各组施肥组合的松脂产量差异显著；最好的 3 个组合为 6 号(N：250g/株，P：500g/株)、3 号(P：1000g/株，K：300g/株)和 5 号(N：250g/株，K150g/株)。分别比对照产量提高了 22.57%、16.77%、13.23%。在 0.05 水平，6 号与 1 号、4 号、8 号、11 号、13 号、14 号和 16 号存在显著差异；3 号与 4 号、8 号、13 号和 14 号存在显著差异；5 号与 4 号、8 号和 14 号存在显著差异；其他施肥组合之间差异不显著。南方土壤普遍缺 P，少 K，N 中等。其中尤其以 P 的作用最为明显。

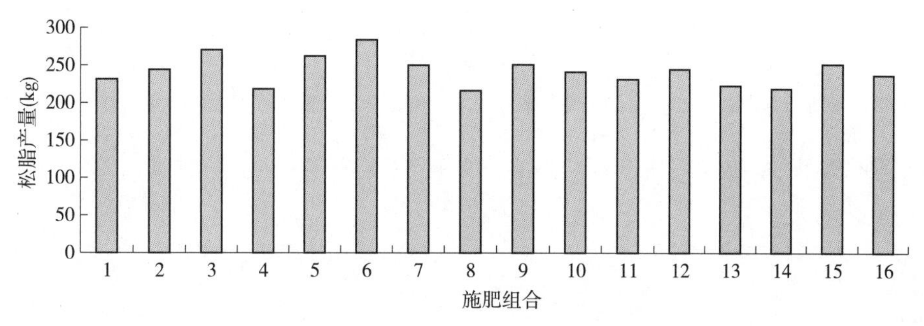

图 4　施肥组合单位面积松脂产量

注：施肥组合 1-16 分别为 $N_1P_1K_1$、$N_1P_2K_2$、$N_1P_3K_3$、$N_1P_4K_4$、$N_2P_1K_2$、$N_2P_2K_1$、$N_2P_3K_4$、$N_2P_4K_3$、$N_3P_1K_3$、$N_3P_2K_4$、$N_3P_3K_1$、$N_3P_4K_2$、$N_4P_1K_4$、$N_4P_2K_3$、$N_4P_3K_2$、$N_4P_4K_1$。

表5 施肥组合多重比较检验表

施肥组合	1-6	3-8	4-6	6-8	6-11	6-13	6-14
MD	-52.29*	54.07*	-65.42*	67.53*	52.59*	60.91*	65.03*
Sig.	0.049	0.042	0.017	0.014	0.048	0.024	0.017

注：表中仅列出差异显著的施肥组合。

3 结论

(1)不同密度间的松脂年产量存在显著差异。密度1050株/hm^2的单位面积产量比密度450株/hm^2的提高了70.14%，但平均单株年产量却下降了25.29%。高密度虽然松脂产量高，但是树木个体生长指标相对低密度要差，在考虑松脂产量增加的同时也需考虑最后的出材量，马尾松采脂林的保留密度应控制在750~900株/hm^2为宜。在本试验所设5个密度，并未出现单位面积松脂产量随密度增大而下降的趋势，所以有必要增加高密度试验做进一步研究。

(2)各施肥组合之间松脂年产量差异显著。有10个施肥组合的单位面积松脂年产量高出空白对照。最好的3个施肥组合为6号(N：250g/株，P：500g/株)、3号(P：1000g/株，K：300g/株)和5号(N：250g/株，K：150g/株)。分别比对照产量提高了22.57%、16.77%、13.23%。本试验仅为1年试验数据，可以确定施用复合肥比单一肥种效果要好，具体哪种施肥组合效果最为明显，还需进一步试验确定。

参考文献

[1]周政贤．中国马尾松[M]．北京：中国林业出版社，2001：14.

[2]丁贵杰，周志春，王章荣．马尾松纸浆材林培育与利用[M]．北京：中国林业出版社，2006.

[3]谌红辉，丁贵杰．马尾松造林密度效应研究[J]．林业科学，2004，40(1)：92-98.

[4]丁贵杰．马尾松人工纸浆材林采伐年龄初步研究[J]．林业科学，2000，36(1)：15-20.

[5]周志春，黄光霖，金国庆．马尾松不同种源对环境的反映函数和优良种源的合理布局[J]．林业科学研究，1999，12(3)：229-236.

[6]姚瑞玲，丁贵杰．不同密度马尾松人工林凋落物及养分归还量的年变化特征[J]．南京林业大学，2006，30(5)：83-86.

[7]周志春，李光荣，黄光霖等．马尾松木材化学组分的遗传控制及其对木材育种的意义[J]．林业科学，2000，36(2)：110-115.

[8]吴启彬．马尾松优良种源内优良林分选择的研究[J]．福建林业科技，2008，35(1)：21-25.

[9]汪佑宏，牛敏，刘杏娥，等．马尾松树脂含量与其解剖特征的关系[J]．林业实用技术，2008，11：5-7.

[10]王长新．马尾松采脂量的相关因子分析[J]．河南科技大学学报(农学版)，2004，24(3)：22-25.

[11]蔡树威，龙伟，杨章旗．马尾松不同种源采脂量与树体因子关系的研究[J]．广西林业科学，2006，12(35)增刊：18-19.

[12]王光仁，朱国发．湿地松产脂量相关因子分析及高产脂单株的选择[J]．林产化工通迅，1999(1)：27-29.

[13]卢纹岱．SPSS for Windows 统计分析[M]．北京：电子工业出版社，2002.

[原载：中南林业科技大学学报，2010，30(09)]

马尾松间伐密度效应研究

谌红辉[1] 方升佐[2] 丁贵杰[3] 许基煌[1] 温恒辉[1]
([1]中国林业科学研究院热带林业实验中心，广西凭祥 532600；
[2]南京林业大学森林资源与环境学院，江苏南京 210037；
[3]贵州大学林学院，贵州贵阳 550025)

摘　要　通过马尾松20年生间伐密度试验林的10年观测资料，分析了不同间伐保存密度对林分生长和经济效益的影响。结果表明：不同间伐保存密度(1200株/hm²，2000株/hm²，2800株/hm²，3400株/hm²)对林分生长、林分结构及材种规格均有显著影响，其中林分胸径、单株材积、冠幅、冠高比随密度增大而减小，高径比、自然稀疏强度随密度增大而增大，密度对树高生长无显著影响。不同密度的林分蓄积量与出材量随林龄增长而差异变小。随着密度增大，小径阶株数率及小径材出材量所占的比例增大，而大径阶株数率与大径阶材种出材量减少。综合效益核算、材种出材量及马尾松人工林生长规律，马尾松人工林进入中龄林期后培育纤维材与中小径材保存密度控制在B~D(2000~3400株/hm²)并在15~17年生时采伐效益较好，培育大、中径材林分保存在A~B(1200~2000株/hm²)间的密度效益较好。

关键词　马尾松；间伐；密度效应；经济分析；出材量

Study of Thinning Density Effects on Masson Pine Plantation

CHEN Honghui[1,2], FANG Shenzuo[1], DING Guijie[3],
XU Jihuang[2], WEN Henghui [2]
(*[1]The Experimental Centre of Tropical Forestry, CAF, Pingxiang* 532600, *Gangxi, China*;
[2]Nanjing Forestry University, Nanjing 210037, *Jiangsu, China*;
[3]College of Forestry, Guizhou University, Guiyang 550025, *Guizhou, China*)

Abstract: The effects of thinning density on growth and economic benefit of masson pine plantation were analyzed by 10 years data of 20-year-old thinning density experiment stand(including treatments of A, B, C, D, their density remained is 1200, 2000, 2800, 3400 trees/hm² respectively), the results showed that it had significant effects on stand growth, stand structure, log type and mill run. With the increase of stand density, diameter at breast height(DBH), single tree volume, crown diameter, ratio of crown length to tree height decreased, but tree height/DBH, self-thinning intensity increased. The thinning density had no significant effect on tree height. With stand age increasing, the difference of stand volume and total mill run of different density stand decreased. Trees percentage of small logs and mill run of small timber increased with the increase of stand density, but which of large timber decreased. So it is difficult to raise stand final productivity by thinning, but it can improve the wood quality. By a synthetic analysis of mill run, economic evaluation and growth law of masson pine plantation, stand density of cultivating short rotation industry timber could be determined as 2000~3400 trees/hm² when the stand age entered the middle-aged stage, to cultivate large or middle diameter timber, which could be determined as 1200~2000 trees/hm².

Key words: massom pine; thinning density; density effects; economic analysis; mill run

关于林分密度效应与控制问题，国外林业科技工作者进行了大量的研究工作[1]。西德 H. Kramer

于1930-1974年对挪威云杉(*Picea abies*)的间伐试验结果分析指出：以采用强度下层抚育间伐法效果较好，其次是中度下层抚育间伐法和强度上层抚育间伐法，以弱度下层抚育间伐法最差，另外根据对比试验同时明确一个重要问题，即抚育间伐同样可以影响优势木的生长。英国G. J. Hamilton在1967-1974年对美国的西加云杉(*Picea sitchensis*)、欧洲赤松(*Pinus sylvestris*)、南欧黑松(*Pinus nlgra* var. *maritima*)等进行了各种形式的行状抚育间伐试验，其目的是查明不同形式的行状抚育间伐对保留木生长的影响。日本川那、斋藤等人对10年生柳杉(*Cryptomeria fortunei*)幼林采用下层抚育间伐的方法进行了不同间伐强度的试验，结果表明，间伐强度越大越能提高单株木的生长量，但有可能影响单位面积的出材量。国内许多学者也开展了这方面的研究[2-11]，树种包括马尾松(*Pinus massoniana*)、杉木(*Cunninghamia lanceolata*)、桉树(*Eucalyptus* spp.)、樟子松(*Pinus sylvestris* var. *mongolica*)等，并获得了可喜的成果。

马尾松是我国南方主要工业用材树种之一，广泛分布于16个省(自治区、直辖市)。造林密度和保留密度是否合理，直接关系到培育目标能否实现，直接影响经营者的经济效益。为了探明南亚热带栽培区科学的马尾松密度调控技术，1991年我们在广西中国林业科学研究院热带林业实验中心伏波实验场设置了间伐试验，根据20年生林分10年(次)的观测材料对不同间伐保存密度的林分生长、材种出材量、经济效益作了定量的分析，为生产部门根据培育目标，选择相应的造林密度和不同时期的保留密度提供了科学依据。

1 研究材料与方法

1.1 试验林概况

试验林设在广西凭祥市中国林业科学研究院热带林业实验中心大青山林区，大青山属十万大山西端余脉，106°43′E，22°06′N，海拔150~1200m，以低山地形为主，年均温21.7℃，降水量1856mm，属南亚热带季风气候区，土壤主要为花岗岩发育成的红壤，间有部分石灰岩土、酸性紫色土和冲积土，土层厚100cm，马尾松主要分布在海拔300~800m地段，立地指数16~22为主。

试验林为1983年春造，西南坡向，坡位中，海拔600m左右，花岗岩红壤，土层厚100cm，立地指数16，初植密度3600株/hm^2，5年生时进行一次卫生清理抚育，1991年春(8年生)时因林分的自然稀疏，少量林木枯死，林分保存密度平均为3400株/hm^2，平均树高为8.11m，平均胸径为9.7cm，平均蓄积为115.12m^3/hm^2。该年春设置不同间伐强度的对比试验，分强度、中度、弱度3种间伐强度与对照4种处理，保存密度分别为A1200株/hm^2、B2000株/hm^2、C2800株/hm^2、D3400株/hm^2，重复6次，共24个试验小区，每小区面积为600m^2。为保证试验条件的一致性，同一重复的试验小区尽量保持在同一水平位置。一直连续观测到20年生，共10次观测资料。

1.3 主要研究方法

实行定时(每年年底)、定株、定位观测记录，测定内容包括胸径、树高、冠幅、枝下高、林木生长状况等。用断面积平均求林分平均胸径，采用Richards曲线拟合胸径与树高关系，然后用林分平均胸径求算林分平均高，其他测树指标均采用实测法计算平均值。按广西马尾松二元材积公式求算单株材积，乘上径阶株数得径阶材积，累计各径阶材积得蓄积量。利用削度方程[3]和原木材积公式[12]计算材种出材量。

2 结果分析

各试验处理、各年的逐年观测资料见表1。

2.1 不同间伐强度对马尾松生长效应的影响

密度是影响林分生产力的三大主要因子(良种、立地与密度)之一，也是最易为人工控制的因素[5]。为了比较不同保存密度的生长差异情况，对试验林逐年调查资料进行了方差分析(表2)。

2.1.1 不同间伐强度对树高生长的影响

密度对林分平均高的影响比较复杂，结论也不一。有些研究表明密度对树高生长有影响，但影响较弱，在相当宽的一个中等密度范围内无显著影响[13]。根据表3统计，同一密度处理的平均树高随林龄的增长而增长，不同密度处理的平均树高生长基本相近，20年生时A、B、C、D各处理的平均高分别为14.80m、14.26m、14.05m、14.07m，树高生长差别低于5.3%。优势高的生长也表现出同样的规律，20年生时A、B、C、D各处理的优势高分别为16.84m、16.65m、16.71m、16.58m，树高生长差别低于1.6%。从表4的方差分析可知，密度对马尾松的平均高与优势高生长均无显著影响。

表 1　不同间伐保存密度林分生长过程

项目	处理	林龄(年)									
		9	10	11	12	13	14	15	16	17	20
平均树高(m)	A	9.12	9.65	10.31	10.62	11.39	11.91	12.29	12.55	13.10	14.80
	B	8.45	9.09	9.80	10.29	10.78	11.33	11.87	12.31	12.77	14.26
	C	8.37	8.87	9.68	10.16	10.71	11.37	11.74	12.06	12.57	14.05
	D	8.63	9.19	9.78	10.24	10.66	11.39	11.65	11.97	12.57	14.07
优势高(m)	A	10.46	11.05	11.89	12.37	12.83	13.43	14.06	14.30	14.86	16.84
	B	10.23	10.98	11.71	12.38	12.67	13.16	13.84	14.05	14.79	16.65
	C	10.25	10.83	11.73	12.16	12.65	13.13	13.98	14.17	14.60	16.71
	D	10.51	11.08	11.73	12.27	12.63	13.15	13.88	14.06	14.64	16.58
平均胸径(cm)	A	13.04	14.16	15.31	16.19	16.97	17.62	18.25	18.81	19.24	21.01
	B	11.27	12.06	12.90	13.44	13.94	14.51	15.04	15.59	16.05	17.55
	C	10.57	11.23	11.90	12.35	12.87	13.34	13.79	14.25	14.77	16.35
	D	10.17	10.75	11.34	11.84	12.34	12.87	13.26	13.78	14.36	16.09
胸径变动系数	A	0.16	0.16	0.17	0.17	0.18	0.19	0.20	0.20	0.20	0.23
	B	0.26	0.26	0.27	0.28	0.29	0.29	0.29	0.29	0.29	0.31
	C	0.30	0.30	0.31	0.32	0.32	0.32	0.33	0.32	0.31	0.32
	D	0.30	0.31	0.32	0.33	0.33	0.33	0.33	0.32	0.31	0.31
单株材积(m^3)	A	0.0634	0.0779	0.0956	0.1089	0.1268	0.1417	0.1550	0.1681	0.1819	0.2393
	B	0.0452	0.0548	0.0665	0.0750	0.0838	0.0943	0.1051	0.1162	0.1265	0.1655
	C	0.0397	0.0468	0.0563	0.0631	0.0716	0.0807	0.0883	0.0965	0.1072	0.1427
	D	0.0380	0.0445	0.0521	0.0588	0.0658	0.0756	0.0816	0.0895	0.1011	0.1386
蓄积(m^3/hm^2)	A	72.96	89.90	110.29	125.62	145.87	163.69	177.69	189.36	202.52	264.83
	B	91.38	110.12	132.86	149.57	166.22	185.71	202.72	215.04	225.80	290.25
	C	111.95	131.72	157.32	175.35	196.76	217.56	236.14	251.32	260.83	330.93
	D	132.98	155.34	179.85	202.68	222.81	247.54	263.51	270.86	281.90	347.96
冠幅(m^3)	A	3.20	3.58	3.35	3.24	3.36	3.20	3.29	3.47	3.56	3.20
	B	2.94	3.09	2.83	2.83	3.00	2.70	2.76	2.85	2.81	2.28
	C	2.82	2.83	2.79	2.64	2.82	2.35	2.52	2.51	2.37	2.14
	D	2.63	2.82	2.26	2.59	2.58	2.22	2.48	2.59	2.53	2.28
冠高比	A	0.65	0.64	0.64	0.61	0.59	0.54	0.50	0.51	0.47	0.44
	B	0.60	0.60	0.61	0.58	0.54	0.50	0.46	0.47	0.45	0.38
	C	0.59	0.58	0.58	0.57	0.53	0.48	0.43	0.47	0.43	0.39
	D	0.55	0.54	0.54	0.52	0.50	0.47	0.41	0.48	0.39	0.36
高径比	A	70.0	68.2	67.3	65.6	67.2	67.6	67.5	65.4	66.7	70.5
	B	75.0	75.4	76.0	76.6	77.4	78.1	78.9	76.3	76.6	81.2
	C	79.2	78.9	81.3	82.2	83.2	85.2	85.1	81.5	81.8	86.0
	D	84.9	85.5	86.3	86.6	86.4	88.6	88.0	85.4	85.9	87.5

2.1.2　不同间伐强度对直径生长的影响

直径是密度对产量效应的基础，同时直径又是材种规格的重要指标，密度对直径的影响有显著的相关性，这一点林学界普遍认同[13]。本试验研究表明，胸径生长量随密度增大而减小。由表 2 可知，不同间伐保存密度的胸径生长量除 C、D 处理间差异不显著外，A 与 B、C、D 及 B 与 C、D 间的差异均呈现出极显著性。20 年生时，A、B、C、D 各处理胸径值分别为 21.01cm、17.55cm、16.35cm、16.09cm，A、B 处理比 D 处理分别大 31.1%、9.4%。

胸径变动系数是反映林分分化与离散程度的重要指标。本试验研究表明，同一密度处理的胸径变动系数随林龄的增大而减小，同一林龄的林分胸径变动系数随密度的增大而增大，然后趋于稳定。由表2中对不同间伐保存密度的胸径变异系数分析可知，A处理与B、C、D处理间差异显著，A、B处理随林龄增大变异系数值有所增大，但各处理间有随林龄增大趋向一致的现象。在20年生时A、B、C、D各处理的胸径变异系数值分别为0.23、0.31、0.32、0.31。这是由于A、B处理密度较C、D小，竞争激烈程度比C、D高密度处理来得迟一些，随着林木个体对生存空间竞争与利用的调和，胸径分布结构趋势于稳定，基本稳定在0.3左右。

表2　不同间伐保存密度试验方差分析结果

项目	方差分析	林龄(年)									
		9	10	11	12	13	14	15	16	17	
平均树高(m)	F值	3.05	2.71	2.06	0.97	2.54	1.93	1.69	1.45	1.29	1.76
	Q检										
优势高(m)	F值	0.77	0.49	0.17	0.29	0.23	0.41	0.21	0.37	0.25	0.15
	Q检										
平均胸径DBH(cm)	F值	53.76**	63.42**	85.07**	86.93**	85.15**	97.35**	92.89**	61.07**	56.51**	84.95**
	Q检	ab ac ad bc bd	ab ac ad bc bd	ab ac ad bc bd	ab ac ad bc bd	ab ac ad bc bd	ab ac ad bc bd	ab ac ad bc bd	ab ac ad bc bd	ab ac ad bc bd	ab ac ad bc bd
胸径变动系数	F值	16.29**	14.89**	16.82**	16.25**	16.66**	16.90**	16.91**	18.44**	15.97**	10.21**
	Q检	ab ac ad	ab ac ad	ab ac ad	ab ac ad	ab ac ad	ab ac ad	ab ac ad	ab ac ad	ab ac ad	ab ac ad
单株材积(m^3)	F值	34.47**	38.56**	51.96**	52.83**	49.84**	56.26**	55.91**	37.37**	35.71**	42.40**
	Q检	ab ac ad	ab ac ad bd	ab ac ad bd	ab ac ad bd	ab ac ad bd	ab ac ad bd	ab ac ad bd	ab ac ad bd	ab ac ad bd	ab ac ad bd
林分蓄积(m^3/hm^2)	F值	25.86**	23.46**	25.15**	22.73**	19.08**	21.99**	15.53**	11.72**	10.94**	7.73**
	Q检	ac ad bc bd cd	ac ad bd cd	ac ad bc bd	ac ad bd	ac ad bd	ac ad bc bd	ac ad bd	ac ad bd	ac ad bd	ac ad bd
冠幅(m^3)	F值	8.94**	18.40**	13.06**	12.99**	2.84	27.51**	2.43	2.75	2.57	11.31**
	Q检	ac ad	ab ac ad	ab ac ad bd cd	ab ac ad		ab ac ad bc bd				ab ac ad
冠高比	F值	14.81**	14.29**	12.44**	5.75**	3.31*	12.32**	4.09*	1.24	2.94	5.49**
	Q检	ab ac ad bd	ac ad bd	ac ad bd	ad bd	ad	ab ac bd	ad			ab ad
高径比	F值	24.42**	33.77**	40.23**	48.89**	41.96**	68.43**	48.87**	68.94**	54.37**	52.08**
	Q检	ac ad bd cd	ab ac ad bd cd	ab ac ad bc bd	ab ac ad bc bd	ab ac ad bc bd	ab ac ad bc bd	ab ac ad bc bd	ab ac ad bc bd	ab ac ad bc bd	ab ac ad bc bd

注：* 显著，** 极显著；d 表示两两间差异显著。

2.1.3　不同间伐强度对材积生长的影响

立木的材积取决于胸径、树高、形数3个因子，密度对3因子均有一定的影响。通过方差分析可知，试验林各密度处理的单株材积生长差异显著，蓄积有随林龄增长差异缩小的趋势。

对不同间伐强度的试验林进行方差分析可知(表2)，除C、D处理单株材积差异不显著外，A与B、C、D间及B与D处理间均表现出差异显著性。对间伐试验林的蓄积生长分析可知，在间伐初期生长差异显著，10年生时D处理比A处理蓄积大61.8%。但在20年生时蓄积差异缩小，蓄积D处理比A处理仅大31.2%。从方差分析的F值随林龄增长逐渐变小的趋势可知，各密度蓄积量随林龄增长差异呈缩小的趋势。

由于林分的蓄积取决于单株材积与株数密度，而这两因子互为消长，达到平衡时遵守产量恒定法则[13]。综合试验林的观测结果，说明单位面积的林地生产力是有一定限度的，不同密度的林分蓄积随时间推移趋向一致，通过间伐提高林地最终生产力可能性很小，但能提高材种规格，在生产中应尽力在不减产的情况下提高木材质量，以获取最佳经济效益。

2.1.4　不同间伐强度对树冠生长的影响

许多研究表明树冠的大小和密度是紧密相关的[13]。经方差分析表明，各处理间冠幅生长差异显著，主要表现在A与B、C、D及B与C、D之间。对试验材料的分析，发现在间伐调整初期，冠幅差异显著，冠幅生长有一定的波动性，随后出现较稳定的差异性。这种情况可能是由于调控初期个体生长空间大，林分尚未完全郁闭，个体生长有一定的差异性，但随时间推移，争夺营养空间逐渐激烈，出现波动，然后逐渐调和，出现一定的稳定性，充分利用营养空间。

对试验林的冠高比分析可知，同一密度级随时间推移，比值逐渐减小，不同密度处理间差异显著，密度越大比值越小，说明自然整枝强烈。

2.1.5 林分密度对干形的影响

立木的高径比是林木的重要形质指标之一，与木材的质量与经济价值密切相关。营造用材林时选择的密度应有利于自然整枝、干形通直饱满和较大的高径比。对试验林的材料分析可知，林分郁闭后同一密度级的高径比随林龄增大而增大，林龄相同时，高径比随密度增大而增大。

对试验林材料分析可知，高径比在间伐后，除A处理外，B、C、D处理的高径比随时间推移有所增大，但各处理间一直呈显著性差异。因此，对于工业用材林我们应适当密植，降低树干尖削度。

2.1.6 不同保存密度对自然稀疏的影响

林分密度调节的核心是自然稀疏，即不断减少林木株数调节生长与繁殖[13]。通过表3对不同密度级的株数自然稀疏率统计分析可知，不同密度级所表现出来的稀疏时间与强度有所不同。对表3不同间伐密度处理的自然稀疏情况统计分析可知，同一密度级随林龄增长，连年自然稀疏率逐步上升，出现一个峰值后开始下降，这种现象应该与间伐后林分恢复郁闭有关。这一特点与其他研究结果相似[14-15]。不同处理出现的稀疏时间有所不同，出现的时间依密度减小而推迟，但出现稀疏峰值的时间基本接近。B、C、D三种处理的自然稀疏分别出现在13、12、12年生，A处理16年生才出现稀疏现象，B、C、D三种处理稀疏峰值时间分别出现在16、16、17年生，自然稀疏率分别为4.50%、6.37%、9.47%(虽然18、19年没统计，可从20年生统计的定期稀疏强度可知)。从统计的总稀疏强度来看，总稀疏强度基本与密度正相关，A、B、C、D 4种处理20年生时总稀疏强度分别为7.14%、20.7%、32.5%、37.3%，依据总稀疏强度可为我们确定间伐强度提供可靠依据。林分自然稀疏后保存密度分别为A1116株/hm^2、B1600株/hm^2、C1890株/hm^2、D2194株/hm^2。

综合试验林的自然稀疏情况分析可知，连年稀疏强度高峰期出现在林分郁闭后的一段时间内，总稀疏强度与密度呈正相关，在间伐施工中应选伐被压木及部分小径级中等木。间伐施工原则应以留优去劣为主，而适当照顾均匀。

表3 不同间伐保存密度的自然稀状况统计

林龄(年)	总稀疏强度(%)				连年稀疏强度(%)			
	A	B	C	D	A	B	C	D
12	0	0	0.6	1.45	0	0	0.6	1.45
13	0	0.84	3.55	4.78	0	0.84	1.81	2.96
14	0	7.44	7.69	7.18	0	2.52	1.84	2.53
15	0	9.92	10.06	11.00	0	2.59	2.48	4.12
16	1.47	14.05	15.98	17.22	1.47	4.50	6.37	7.14
17	1.47	15.70	20.12	24.88	0.00	1.87	4.76	9.47
20	7.14	20.70	32.50	37.30	5.88	10.38	17.99	17.22

注：表中稀疏强度为株数百分率，20年生连年稀疏强度为18~20年3年累计值。

2.2 林分密度对林分结构与材种出材量的影响

密度能影响林分不同时期的自我调控过程与林分生产力[13]，对现实林分密度进行人工调控，使林分结构优化，是实现培育目标的关键技术。

2.2.1 密度对径级株数分布的影响

因密度影响林分直径生长，自然会影响林分的直径结构规律，探讨密度对林分株数按直径分布的影响，对营林工作十分有益。各处理株数按径阶分布情况见表4。

综合表4分析可知，同一密度级的株数最大分布率所处的径阶值随林龄增大而增大。间伐试验林14年生时A、B、C、D各处理所处的径阶值分别为16、14、12、12，20年生时所处的径阶值分别为18、16、14、14。

表 4　不同间伐强度的径级株数分布率

林龄(年)	处理	径阶分布率(%)														
		4	6	8	10	12	14	16	18	20	22	24	26	28	30	32
11	A				5.3	16.1	26.6	27.6	16.8	6.0	1.7					
	B		3.8	10.4	20.8	21.0	20.9	12.3	7.1	2.5						
	C	2.4	9.2	14.0	19.8	21.4	16.9	9.4	5.3	1.5						
	D	3.9	11.6	15.3	20.9	21.2	14.9	8.0	2.7	1.2						
14	A				2.6	5.8	15.1	22.5	22.1	18.7	8.2	3.6	1.2			
	B		2.8	7.7	13.6	16.7	21.1	15.5	9.4	6.8	4.4	1.0				
	C	1.6	6.6	11.7	13.8	19.8	16.7	14.5	8.4	3.7	2.7					
	D	1.0	8.6	11.9	17.1	19.6	16.5	12.6	7.6	2.8	1.5					
17	A				2.0	4.2	10.9	14.4	23.1	14.4	16.4	7.4	4.0	2.0	1.0	
	B			5.6	10.2	16.0	14.9	18.2	13.2	7.0	7.8	4.2	1.0	1.0		
	C		4.7	7.8	11.8	18.7	16.6	13.8	11.5	6.5	4.4	2.3	1.1			
	D		3.3	7.4	16.1	19.4	16.9	14.5	10.1	6.8	2.7	1.0				
20	A					4.2	7.9	10.6	16.3	16.0	13.3	13.6	7.7	4.9	2.2	1.7
	B			4.4	8.3	13.1	13.7	16.0	14.0	9.4	6.8	5.8	4.6	2.7		
	C		1.7	6.2	9.6	15.8	15.9	14.7	12.2	8.9	6.2	3.5	2.4	1.9		
	D		1.1	5.0	13.0	15.3	18.6	12.6	12.0	9.1	6.4	3.5	1.3	1.1		

不同密度级的株数最大分布率径阶值随密度加大而变小。由图 1 可知，观察间伐试验林 A、B、C、D 4 种处理绘成的 20 年生株数分布率曲线图的规律，A、B、C、D 各处理株数最大分布率所处的径阶值分别为 18、16、14、14，C、D 分布状态基本相似。A、B、C、D 4 种处理大于 20 径阶的株数率分别为：43.4%、19.9%、14.8%、12.3%，14～20 径阶株数分布率分别为 50.8%、53.1%、51.7%、52.3%，小于 14 径阶分布率分别为 4.2%、25.8%、33.5%、34.4%。可见不同密度对材种出材量影响很大，随着密度增加，小径阶材种出材量增加，大径阶材种出材量减小。在营林工作中应根据不同的培育目标选择相应的林分保存密度。

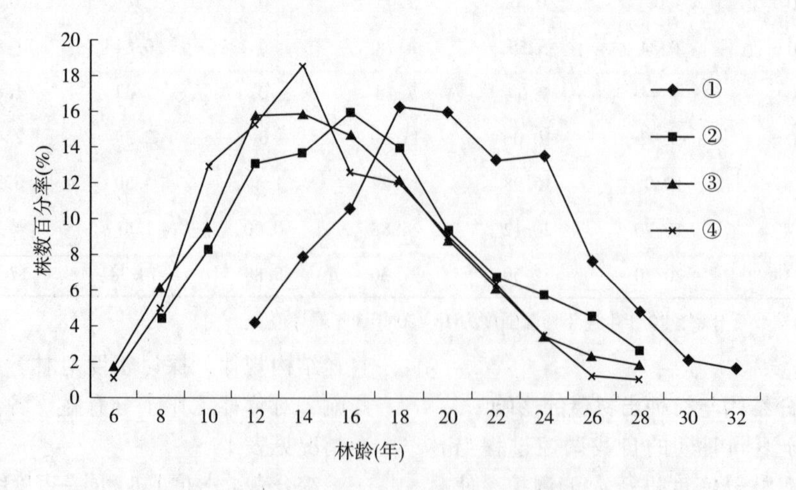

图 1　不同间伐保存密度径阶株数率分布率曲线图(20 年生)

①处理 A；②处理 B；③处理 C；④处理 D

2.2.2 林分密度对材种出材量的影响

综合间伐试验材料分析可知，随着林龄增长不同保存密度处理的出材量差异逐渐变小，但随着密度增大，大径级规格材出材量变小。

计算出材量时我们采取密切联系生产的方法，以南方普遍采用的2m原木检尺长为造材标准，利用马尾松削度方程[3]，求出从地面开始每上升2m处的去皮直径，即为该2m段的小头检尺径，先分径阶求出林分各径级规格材种出材量，再把各径阶径级规格相同的材积相加，得林分各径级规格材的材积，再把所有材种材积相加，得林分总出材量(见表5)。

表5 20年生不同间伐保存密度林分的材种出材量

处理	林分密度 (trees/hm^2)	径阶出材量(m^3/hm^2)				合计	出材率(%)
		4~12	14~18	20~28	≥30		
A	1066	47.60	100.90	78.61	0.85	227.96	86.08
B	1583	85.51	98.45	56.78	0.84	241.58	83.23
C	1900	111.96	113.48	45.35	0.00	270.79	81.83
D	2083	133.24	107.37	42.66	0.00	283.27	81.41

由表5分析可知，总出材量随间伐保存密度增大而增大，但增幅较小，随密度加大出材率有所减小，尤其是大、中径级规格材出材量变小。A、B、C处理的总出材量分别为D处理的81.2%、85.5%、95.1%。20cm以上规格材公顷出材量A、B、C、D各处理分别为79.46m^3、57.62m^3、45.35m^3、42.66m^3。

由表2方差分析可看出，随时间推移，林分蓄积的F值均有变小的趋势，说明不同密度林分蓄积与出材量有趋于相近的趋势，所以想提高林分蓄积不能盲目加大密度。

2.3 不同间伐保存密度林分的经济评价

2.3.1 生产成本与产值计算

根据广西具体生产实践进行成本核算(表6)。生产成本主要包括基本建设投资(苗木、林地清理、整地、栽植、林道、抚育费及10%的间接费)、经营成本(管护、管理)、木材生产(采伐、运输、归堆等)3大类。产值计算以现行市场价格为准，将20年生不同密度不同规格材种乘以相应的价格即得出产值(材种规格计算至检尺径4cm，主要以经济用材进行效益核算)。

表6 20年生不同间伐保存密度林分生产成本与产值统计 (元/hm^2)

处理	基本建设投资	经营成本	木材生产	总计投资	产值	利润
A	2158.87	1000.00	15637.05	18795.92	138493.80	119697.88
B	2158.87	1000.00	16745.95	19904.82	141528.40	121623.58
C	2158.87	1000.00	18014.10	21172.97	148703.40	127530.43
D	2158.87	1000.00	19091.15	22250.02	154534.80	132284.78

2.3.2 不同间伐处理的经济评价

从经济学的角度评价效益必须采用动态分析的方法，才能确定方案的好坏。为了分析各处理间的经济效益差别，按折现率10%的标准，分析净现值(NPV)与内部收益率(IRR)的动态变化规律(图2)。

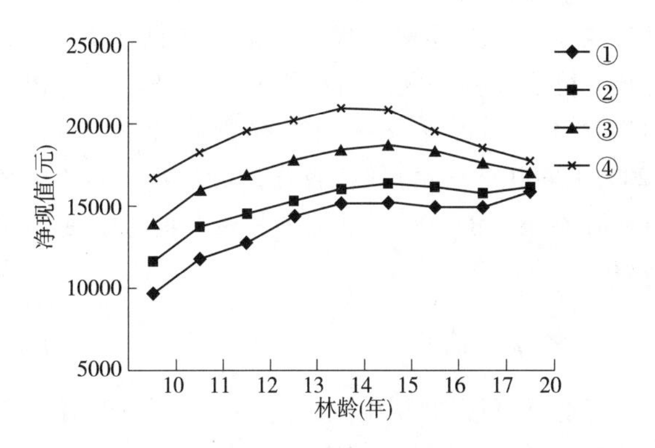

图2 20年生不同保存密度林分的净现值曲线图

①处理A；②处理B；③处理C；④处理D

表 7 不同间伐保存密度林分的经济评价

项目	处理	林龄(年)								
		10	11	12	13	14	15	16	17	20
净现值	A	9654.15	11796.83	12774.46	14383.76	15162.02	15253.28	14943.43	14919.66	15873.59
	B	11626.16	13751.15	14539.87	15321.93	16042.10	16368.90	16160.04	15781.42	16159.83
	C	13893.24	15979.92	16920.61	17799.91	18437.83	18711.92	18361.99	17621.26	17037.84
	D	16681.64	18268.88	19571.70	20209.61	20956.69	20862.26	19559.62	18563.70	17744.55
内部收益率(%)	A	22.0	20.2	18.5	17.1	15.9	14.8	13.9	13.1	11.3
	B	21.5	19.5	17.9	16.5	15.3	14.3	13.6	12.7	11.0
	C	21.2	19.2	17.6	16.2	15.0	14.0	13.3	12.4	10.9
	D	21.0	19.0	17.5	16.2	14.9	14.0	13.2	12.4	10.7

根据表 7 与图 2 净现值曲线变化图可看出，林龄相同时，净现值与密度呈正相关性。同密度级的林分随林龄增长，净现值上升至一定峰值后下降，上升与下降的速率与密度成正相关，随着林龄的增长各处理间净现值的差异呈逐步减小的趋势。A、B、C 处理峰值出现在第 15 年，D 处理出现在 14 年。17 年生后低密度 A、B 处理略有上升的趋势。根据净现值曲线发展趋势，20 年后 A 处理会有大于 B、C、D 处理的趋势。进入成熟林后的变化有待进一步观测。根据 20 年生前不同密度的净现值比较，C、D 高密度处理比 A、B 低密度处理效益好。

再由内部收益率变化可知，不同密度处理的内部收益率相差不大，同一密度级内部收益率随林龄增长有下降的趋势。在 20 年生开始，C、D 处理内部收益率已接近 10%。根据内部收益率变化规律说明林分采伐期愈提前愈能取得较好的经济效益。当然，还要兼顾工艺成熟。

根据净现值曲线变化规律及内部收益率变化特点，说明不同密度林分的利用时间点非常重要。在兼顾工艺成熟的前提下，培育纤维原料林 B、C、D 处理在 15~17 年生时采伐可取得较好的效益，A 处理还有待进一步观测。马尾松人工林进入中龄林期后培育纤维材与中小径材保存密度控制在 B~D (2000~3400 株/hm^2) 效益较好，培育大、中径材林分保存在 A~B(1200~2000 株/hm^2)间的密度效益较好。

3 结论与讨论

经对马尾松 20 年生间伐林 10 年(次)的观测资料分析表明，不同保存密度的林分生长差异非常显著。随着密度增大，胸径、单株材积、冠幅生长量与冠高比减小，表现出与密度间的负相关；高径比与自然稀疏强度随密度增大而增大，但密度对树高生长无显著影响。不同密度的林分蓄积量与出材量随林龄增长差异变小。

不同保存密度对林分结构与材种规格有着显著影响。随着密度增大，小径阶株数分布率、小径级材种出材量及所占比例增加，而大、中径级株数分布率与大、中径级材种出材量减少。因此，通过间伐提高林地最终生产力的可能性很小，但能提高材种规格，从而提高经济效益。

综合效益核算、材种出材量与马尾松人工林生长规律，进入中林期(10 年)后培育短周期工业用材林宜采用的保存密度为 2000~3400 株/hm^2，培育大、中径材宜采用 1200~2000 株/hm^{2}间的保存密度。

为了获得较好的经济效益，人工林间伐的时间、强度与次数是主要决定因素。国外研究者为了减少间伐次数，降低成本，提高采伐量，同时有利于主伐木的生长，多向高强度、长间隔期的方向发展。本次马尾松人工林在 8 年生间伐后，各处理的净现值均呈上升趋势，高密度的 B、C、D 处理净现值的峰值出现在 15 年生左右，也是自然稀疏的高峰期，然后净现值下降，下降速率与密度呈正相关。因此，在净现值开始下降时对各处理再进行一次间伐，研究多次间伐对主伐目标树的径级规格与林分净现值动态变化的影响，有着十分重要的意义。这将在本人开展的马尾松 15 年生的近熟林间伐试验中阐述。在本次间伐试验中，不同密度处理的净现值与蓄积随时间推移有趋向相近的趋势，这说明保持高密度提高林地生产力与效益的可能性很小。因此，在营林工作中应根据培育目标确定相应的林分密度来提高商品林的经济效益。

总之，密度控制技术主要包括造林密度的确定、间伐的强度与次数、间伐间隔期与起止期的确定、疏伐前后林分结构与各测树因子的变化，间伐后的生长预测及间伐作业的效益评估、保留密度和主伐年龄的确定、林分生长预测与材种出材量统计等内容，求得它们的最优组合，实现营林工作的最佳经济效益。这些研究内容要求从不同林龄阶段，对不同现实密度的林分进行科学的密度调控设计并连续观测，总结出科学的人工林密度调控技术规程为营林工作服务。

参考文献

[1]中国林业科学研究院科技情报研究所．森林抚育间伐[M]．北京：中国林业出版社，1981：1-7．

[2]丁贵杰，周政贤，严仁发，等．造林密度对杉木生长进程及经济效果影响的研究[J]．林业科学，1997，33(Sp.1)：67-75.

[3]丁贵杰，周政贤．马尾松不同造林密度和不同利用方式经济效果分析[J]．南京林业大学学报，1996，20(2)：24-29.

[4]童书振，盛炜彤，张建国．杉木林分密度效应研究[J]．林业科学研究，2002，15(1)：66-75.

[5]苏培正．木荷纯林不同抚育间伐强度对比试验[J]．湖北林业科技，2005，35(2)：22-24.

[6]唐守正．同龄纯林自然稀疏规律的研究[J]．林业科学，1993，29(3)：234-241.

[7]曾德慧，姜凤岐等．沙地樟子松人工林自然稀疏规律[J]．生态学报，2000，20(2)：235-242.

[8]张春锋，殷鸣放，孔祥文等．不同间伐强度对人工阔叶红松林生长的影响[J]．辽宁林业科技，2007，33(1)：12-15.

[9]张金文．巨尾按大径材间伐试验研究[J]．林业科学研究，2008，21(4)：464-468.

[10]张水松，陈长发，等．杉木林间伐强度试验20年生长效应的研究[J]．林业科学，2005，41(5)：56-65.

[11]张彩琴，郝敦元，李海平．人工林林分密度最优控制策略的数学模型[J]．东北林业大学学报，2006，2(34)：24-27.

[12]木材编写组．中华人民共和国国家标准—木材．北京：国家标准局，1984，64-67.

[13]孙时轩，沈国舫，王九龄，等．造林学[M].2版．北京：中国林业出版社，1992：133-134.

[14]张大勇，赵松龄．森林自疏过程中密度变化规律的研究[J]．林业科学，1985，2(4)：369-373.

[15]谌红辉，丁贵杰．马尾松造林密度效应研究[J]．林业科学，2004，40(1)：92-99.

[原载：林业科学 2010，46(05)]

马尾松人工同龄纯林自然稀疏规律研究

谌红辉[1,2] 方升佐[1] 丁贵杰[3] 许基煌[2] 温恒辉[2]

([1]南京林业大学森林资源与环境学院，江苏南京 210037；
[2]中国林业科学研究院热带林业实验中心，广西凭祥 532600；
[3]贵州大学林学院，贵州贵阳 550025)

摘 要 通过对马尾松4~20年生不同密度的人工同龄纯林生长资料的分析表明：第1次树高生长分化高峰期在4~7年生，第2次在11~14年生，幼林郁闭后林木直径分化状态比较稳定。中幼林期抚育间伐主要采伐劣等木，进入近熟林期后除采伐劣等木外，还可采伐一部分中等木。间伐施工应以留优去劣为主，适当照顾均匀。总稀疏强度与密度呈正相关性，出现稀疏的时间随密度增大而提前；连年稀疏强度高峰期出现在林分郁闭后的一段时间内。根据总稀疏强度与稀疏时间同密度的关系可确定不同密度林分的间伐强度与时间。利用马尾松人工林观测资料拟合出与立地条件、现存株数密度、林龄三因子相关的自然稀疏模型，从而可推算出不同立地条件的自然稀疏表，为马尾松人工林密度调控提供科学依据。

关键词 马尾松；人工同龄纯林；自然稀疏；间伐

A Study on the Nature Thinning Law of Even-aged Pure Masson Pine Plantation

CHEN Honghui[1,2], FANG Shenzuo[1], DING Guijie[3], XU Jihuang[2], WEN Henghui[2]

([1]*Nanjing Forestry University*, *Nanjing* 210037, *Jiangsu*, *China*;
[2]*The Experimental Centre of Tropical Forestry*, *CAF*, *Pingxiang* 532600, *Guangxi*, *China*;
[3]*College of Forestry*, *Guizhou University*, *Guiyang* 550025, *Guizhou*, *China*)

Abstract: According to analysis of different density *Pinus massoniana* Lamb. plantation growing data of 4~20 years old, it showed that tree height changing peak period during 4~7 years old, the second during 11~14 years old, and the diameter structure stable in closed stand. Thinning is to mainly cut inferior tree during young stands, but a part of intermediate tree could be cut for near mature plantation. Thinning mainly to retain dominant tree, moreover, properly keep homogeneous density. Total nature thinning intensity increased with stand density raised, and the starting period of nature thinning advanced. The peak period of current annual intensity of nature thinning emerged during the period that the stand just closed. By the total self-thinning intensity and thinning time of different density stand, the thinning intensity and time were to be decided. The nature thinning mathematic model relate to site factors, current number of tree and stand age was established. By this, nature thinning condition could be predicted under different site conditions, and which could instruct work of stand density management.

Key words: *Pinus massoniana*; even-aged pure *Pinus massoniana* plantation; nature thinning; thinning

密度是影响人工林生产力的三大主要因子(良种、立地与密度)之一，也是最易人工控制的因素。密度调控的目的是加速林木生长，改善森林卫生状况，提高林木质量，从而提高林分的经济效益与生态效益，其理论基础是基于林分的不同生长发育时期的不同特点、林分的分化与自然稀疏规律、密度与林分生长的关系等方面的内容。密度控制是否合理，关系到林分结构与生产力，从而直接影响培育目标及经济效益。

因此，许多学者对不同树种人工林的密度效应与自然稀疏规律进行过研究[1-14]。但许多研究材料均来自临时样地，缺乏时间上的连续性，取之以空间代替时间的研究方法，使许多结论与相关数学模型出现与实际林分生长不符的现象。为了探明马尾松（*Pinus massoniana* Lamb.）人工同龄纯林的自然稀疏规律，为马尾松人工林的经营提供科学的密度控制技术资料，作者采用对固定标准地连续定位观测的方法，先后对不同初植密度与不同间伐保存密度的试验林进行了长达12年的观测，对马尾松人工同龄纯林的自然稀疏规律进行了总结分析。根据林分密度与时间的相关性拟合出了精度较高的自然稀疏模型，并采用聚类分析的方法对马尾松人工同龄纯林生长过程中林木个体分化规律进行了定量的分析。利用该研究成果，可为马尾松人工林的密度调控提供科学的技术指导。

1 试验材料与方法

1.1 试验地概况

试验林设在广西凭祥市中国林业科学研究院热带林业实验中心大青山林区，大青山属十万大山西端余脉（106°43′E，22°06′N），海拔150~1200m，以低山地形为主，年均气温21.7℃，降水量1856mm，属南亚热带季风气候区，土壤主要为花岗岩发育成的红壤，间有部分石灰岩土、酸性紫色土和冲积土，土层厚100cm，马尾松主要分布在海拔300~800m地段，立地指数16~22为主。

1.2 试验材料及试验设置

1.2.1 13年生马尾松造林密度试验林

1989年在广西凭祥市热带林业实验中心林区设置初植密度分别为：A1667株/hm²、B3333株/hm²、C5000株/hm²、D6667株/hm² 4种处理，4次重复，共16个试验小区，每小区面积为600m²，共11年观测资料。

1.2.2 20年生间伐试验林

设置于广西凭祥市热带林业实验中心林区，该林区1983年春造林，初植密度为4500株/hm²。1991年春（8年生）设置不同间伐强度的试验，分强度、中度、弱度、对照4种间伐处理，保存密度分别为：A1200株/hm²、B2000株/hm²、C2800株/hm²、D3500株/hm²，6次重复，共24个试验小区，每小区面积为600m²，共12年的观测资料。

1.2.3 其他试验数据材料

马尾松人工林其他固定样地多年连续观测资料及专题成果的其他综合材料。

1.3 研究方法

1.3.1 马尾松人工同龄纯林中林木个体生长分化的动态变化

对造林密度试验中效果表现较好的B处理（4~10年生材料）与间伐试验中对照处理D（11~17年生材料）的林木个体的每木逐年调查材料进行统计分析，将林木个体按树高为第一聚类参考指标，胸径为第二聚类参考指标，用类间平方和爬山法进行逐步聚类[15]，将林木个体分成劣等木、中等木、优势木3类，分别用数字1、2、3表示，按树高与胸径生长变化分别定期统计分析一次林木个体等级的动态变化状况，统计结果见表1。

表1 林木个体生长动态变化统计

项目	林龄（年）	株数变化率（%）						定期生长量		
	（A1~A2）	N12	N13	N23	N21	N31	N32	X1	X2	X3
树高（m）	4~7	30.1	0.0	28.1	16.1	2.3	31.3	2.7	3.1	3.2
	7~10	29.2	0.0	17.2	12.3	0.0	32.1	2.1	2.8	3.0
	11~14	44.3	8.2	49.1	13.2	0.0	3.4	0.9	1.2	1.2
	15~17	29.1	3.2	15.3	20.5	0.0	20.3	0.6	0.9	1.4
胸径（cm）	4~7	20.1	1.5	16.1	10.1	1.5	26.3	3.2	5.3	6.8
	7~10	7.2	1.7	9.7	8.3	0.0	15.4	0.9	2.0	3.6
	11~14	4.1	0.0	4.5	6.2	0.0	10.7	0.8	1.4	2.2
	15~17	10.4	0.0	6.3	5.1	0.0	12.8	0.3	0.9	2.0

注：A1~A2表示生长期，N12表示林分下层劣等木向中等木层移动的株数率，N13表示劣等木层向优势木层移动的株数率，N21表示中等木层向劣等木层移动的株数率，N23表示中等木层向优势木层移动的株数率，N31表示优势木层向劣等木层移动的株数率，N32表示优势木层向中等木层移动的株数率；X1表示劣等木定期生长量，X2表示中等木定期生长量，X3表示优势木定期生长量。

1.3.2 自然稀疏规律及自然稀疏模型的研究

依据13年生造林密度试验林的观测材料、20年生不同间伐保存密度试验林的观测材料及其他试验样地的多年观测材料，对不同初植密度的幼龄林及进入中龄期后不同间伐保存密度的林分自然稀疏规律进行研究，总结不同密度的人工林自然稀疏规律。选择林分现存密度、立地条件、林龄作为主要因子，借鉴 Clutter 等[16] 在研究林分株数随时间变化时采用的数学模型：

$$N_2=[N_1^{-\gamma}-\alpha\cdot\gamma/(\beta+1)\cdot(T_2^{(\beta+1)}-T_1^{(\beta+1)})]^{(1/\gamma)}$$

式中：N_2、N_1 为林木株数；T_2、T_1 为林龄；α、γ、β 为参数，建立自然稀疏模型。

2 结果与分析

2.1 自然稀疏过程中林木个体生长动态变化过程分析

林分密度调节的核心是自然稀疏，即不断减少林木株数，调节生长与繁殖[3]。林分郁闭后由于个体间产生空间与资源的竞争，在生长发育过程中会产生林木个体分化，从而出现自然稀疏。

2.1.1 林木个体生长过程中树高生长分化的动态变化

由表1资料分析可知：树高分化主要表现在幼林期的4~7年生阶段与中龄林期的11~14年生阶段。不同林龄阶段，林冠下层的劣等木有一定的比率进入中等木冠层，在11~14年生表现最大，达44.3%，但进入上冠层成为优势木的概率较小，最高仅为8.2%。随着时间的推移，中等木中有少部分下降为被压木，一部分上升为优势木，上升比率在11~14年生时最大，达49.1%。说明马尾松人工林进入速生阶段的中龄期后，林分群体中、下冠层的林木个体为争夺生存资源，竞争十分激烈。优势木在生长过程中退化为被压木的概率几乎为0，退居中层木的概率幼林期(4~7、7~10年生)为30%左右，进入中林期后的11~14年生较稳定，15~17年生时只达20.3%。

综合树高生长动态变化与定期生长量分析可知：树高生长竞争高峰第1次在幼林期的4~7年生，第2次在中龄林期11~14年生阶段。可见，在营林工作中，幼林期清理被压木是可行的，即可淘汰劣等基因，又节省生存空间资源；进入中龄林期后，第1次间伐强度不宜太大，以免损失一部分优良基因；进入近熟林期后，林木个体已充分分化，林分生长稳定，可按间伐强度采伐劣等木和一部分中等木。

2.1.2 林木个体生长过程中胸径生长分化的动态变化

对表1中胸径生长变化概率分析可知：胸径分化后变动的概率非常小，比较稳定。除幼林期4~7年生变化稍大一点外，7年生以后都比较稳定，林冠下层的劣等木进入中等木冠层的比率均小于10.4%，因此在间伐时采伐小径级木是可行的。

2.2 林分密度对自然稀疏的影响

通过对不同密度试验林自然稀疏状况统计分析可知：不同密度级林分所表现出来的稀疏时间与强度有所不同，高密度的林分自然稀疏时间早，自然稀疏强度大。

由表2造林密度试验林自然稀疏材料可知：同一密度级的总自然稀疏强度随林龄增大而增大，林龄相同时，总自然稀疏强度随密度增大而增大；11年生时，A、B、C、D 4种处理总稀疏强度分别为5.2%、13.2%、25.5%、26.1%，林分保存密度分别为A 1580株/hm^2、B 2893株/hm^2、C 3725株/hm^2、D 4927株/hm^2。

对连年稀疏强度分析可知：林分郁闭后一定时期内连年稀疏强度呈上升趋势，达到一定峰值后开始下降，大规模稀疏阶段通常是自然稀疏刚刚开始的一段时间内，这一规律与其他学者的结论类似[14]。低密度的A、B处理11年生时尚未出现明显的稀疏高峰，高密度的C、D处理稀疏高峰出现在8年生左右。这主要是由于此时马尾松个体正进入生长速生时期，个体对营养空间的需要急剧增加，加之此时高密度处理的林分种群密度较大，因此，加剧了种群的自然稀疏，使高密度处理林分种群个体间的竞争在此时表现最强烈，而达到最大的淘汰率。此后，由于各处理的林木树冠生长竞争减缓，趋于稳定，可基本充分利用营养空间，因此，以后的自然稀疏率降低。

表2 不同造林密度对自然稀疏的影响

林龄(年)	总稀疏强度(%)				连年稀疏强度(%)			
	A	B	C	D	A	B	C	D
6	0.0	1.0	2.2	3.2	0.0	1.2	2.2	3.2
7	0.0	3.3	5.3	6.1	0.0	2.3	3.1	3.3
8	0.0	6.2	12.4	12.3	0.0	3.2	7.2	7.5

（续）

林龄(年)	总稀疏强度(%)				连年稀疏强度(%)			
	A	B	C	D	A	B	C	D
9	2.1	7.1	17.1	18.4	2.2	2.1	6.0	7.3
10	4.3	11.0	21.2	23.2	2.3	4.0	5.3	6.2
11	5.2	13.2	25.5	26.1	1.5	3.2	4.2	3.1

对表3不同间伐密度处理试验林的自然稀疏情况统计分析可知，同一密度级随林龄增长，连年自然稀疏率逐步上升，出现一个峰值后开始下降，这种现象应该与间伐后林分恢复郁闭有关。

不同间伐密度处理出现的稀疏时间有所不同，出现的时间依密度减小而推迟，但B、C、D 3种处理出现稀疏峰值的时间基本接近。B、C、D 3种处理的自然稀疏分别出现在13、12、12年生，A处理16年生才出现稀疏现象。B、C、D 3种处理第一个稀疏峰值时间分别出现在16、16、17年生，自然稀疏率分别为4.50%、6.37%、9.47%，越过峰值时间后稀疏率有所下降。但林龄达20年生后，B、C、D 3种处理稀疏强度又开始上升，25年生时自然稀疏率分别为6.03%、6.83%、7.23%，是否有第二个峰值时间有待进一步观测。

从统计的总稀疏强度来看，总稀疏强度基本与密度正相关，A、B、C、D 4种处理20年生时总稀疏强度分别为7.14%、20.7%、32.5%、37.3%，25年生时总稀疏强度分别为21.40%、43.30%、54.40%、60.30%，依据总稀疏强度可为我们确定间伐强度提供可靠依据。

表3　不同间伐保存密度对自然稀疏的影响

林龄(年)	总稀疏强度				连年稀疏强度			
	A	B	C	D	A	B	C	D
12	0	0	0.6	1.45	0	0	0.6	1.45
13	0	0.84	3.55	4.78	0	0.84	1.81	2.96
14	0	7.44	7.69	7.18	0	2.52	1.84	2.53
15	0	9.92	10.06	11.00	0	2.59	2.48	4.12
16	1.47	14.05	15.98	17.22	1.47	4.50	6.37	7.14
17	1.47	15.70	20.12	24.88	0.00	1.87	4.76	9.47
20	8.61	20.70	32.50	40.20	1.96	1.46	5.99	5.74
22	11.40	29.20	38.50	48.30	1.55	5.25	2.30	3.85
25	21.40	43.30	54.40	60.30	3.77	6.03	6.83	7.23

综合以上林木生长分化与自然稀疏规律可知：马尾松人工林的林木个体分化与自然稀疏高峰期主要出现在幼林郁闭后的一段时期内及进入中龄林速生期后的一段时期内，总稀疏强度与密度呈正相关。因此，在营林工作中，当幼林郁闭后及进入中林速生期后应及时进行抚育采伐，间伐小径级被压木及部分中等木。间伐原则以留优去劣为主，适当照顾均匀。

2.3　自然稀疏模型的研究

Clutter等[16]在研究林分株数随时间变化时采用的微分方程与其积分方程如下：

$$dN/N^{dT} = \alpha \cdot T^{\beta} \cdot N^{\gamma} \quad (1)$$

$$N_2 = [N_1^{-\gamma} - \alpha \cdot \gamma/(\beta+1) \cdot (T_2^{(\beta+1)} - T_1^{(\beta+1)})]^{(-1/\gamma)} \quad (2)$$

式(2)中：T_1、T_2为林龄；N_1、N_2分别为林龄T_1、T_2时每公顷的林木株数；α、γ、β为参数。

国内学者引用后认为该数学模型拟合精度较高，且具有较好的生物学意义[1,4]。因此，本文借鉴Clutter等对湿地松建立枯损函数的方法，并加以改善。研究表明：枯损不仅与现存株数(N)及林龄(A)有关，而且与立地条件也有关，因此应将立地因素引入微分方程。因为林分优势高(H_0)是立地指数(SI)、林龄(A)的函数，所以可通过引入变量$T=A \cdot H_0^k$来替代原微分方程中的变量T，得如下方程：

$$N_2=[N_1^{C0}+C_1(A_1^{C2}\cdot H_{01}^{C3}-A_2^{C2}\cdot H_{02}^{C3})]^{1/C^0} \quad (3)$$

式(3)中：$C_0=-\gamma$，$C_1=\alpha\cdot\gamma/(\beta+1)$，$C_2=(\beta+1)$，$C_3=k\cdot(\beta+1)$，$H_{01}$、$H_{02}$为林龄 A_1、A_2 时的优势高。

用马尾松有关材料拟合方程后，求得如下参数；

$C_0=-0.8067$，$C_1=-8.0565E-07$，$C_2=1.5632$，$C_3=0.8588$

其中：相关系数 $r=0.9956$，样本数 $n=319$。

自然稀疏模型经过 F 检验与适用性检验，均符合统计要求。

按建立的自然稀疏模型(3)拟合出 20 指数级马尾松自然稀疏表(表 4)。

表 4　马尾松人工林自然稀疏表(20 指数级)

造林密度（株/hm²）	林龄(年)												
	6	8	10	12	14	16	18	20	22	24	26	28	30
1500	1474	1433	1377	1310	1234	1154	1072	991	913	839	771	708	650
2000	1956	1888	1797	1689	1571	1448	1326	1209	1098	997	904	821	746
2500	2435	2334	2202	2048	1882	1714	1550	1396	1254	1126	1012	911	822
3000	2910	2771	2592	2387	2171	1955	1749	1559	1388	1236	1101	984	882
3500	3381	3200	2970	2710	2440	2176	1929	1704	1504	1329	1177	1045	932
4000	3849	3622	3336	3018	2693	2380	2091	1833	1606	1410	1242	1098	974
4500	4314	4036	3691	3312	2930	2568	2239	1949	1697	1482	1298	1143	1011
5000	4776	4443	4035	3593	3154	2743	2375	2054	1779	1545	1348	1182	1042
5500	5234	4844	4370	3863	3365	2906	2500	2150	1852	1602	1392	1217	1070
6000	5690	5238	4695	4121	3566	3059	2616	2237	1919	1653	1432	1248	1094

3　结论

(1)通过对马尾松人工同龄纯林 4~17 年生的林木个体生长分化的动态变化可知：树高生长分化高峰期第 1 次在 4~7 年生，第 2 次在 11~14 年生，在幼林郁闭后径级结构比较稳定。中幼林期间抚育采伐时，选伐被压木与小径级木，进入近熟林期后可按间伐强度采伐劣等木和一部分中等木；间伐应以留优去劣为主，适当照顾均匀。

(2)经对 4~25 年生不同密度的马尾松试验林的自然稀疏观测分析表明：连年稀疏强度高峰期出现在林分郁闭后的一段时间内；总稀疏强度与密度呈正相关性，出现稀疏的时间随密度增加而提前；根据总稀疏强度与稀疏时间同密度的关系可确定不同密度林分的间伐强度与间伐时间。

(3)借鉴 Clutter 与 Jones 对湿地松建立枯损函数的方法，并加以改善，建立了自然稀疏数学模型$N_2=[N_1^{C0}+C_1(A_1^{C2}\cdot H_{01}^{C3}-A_{2C}^{2}\cdot H_{02}^{C3})]^{1/C0}$。因枯损不仅与现存株数($N$)及林龄($A$)有关，而且与立地条件也有关，该模型将 3 种影响因子引入，能较好地模拟林分的自然稀疏过程，建模所需数据株数(N)、林龄(A)及优势高(H_0)均易从现实林分获得，便于实际运用预测林分生长趋势。国内学者引用后反映拟合精度较高，并且符合生物学规律。本文利用马尾松人工林统计资料拟合出与立地条件、现存株数密度、林龄三因子相关的自然稀疏模型，从而可推算出不同立地条件的自然稀疏表，为马尾松人工林密度调控提供科学依据。

参考文献

[1]丁贵杰．马尾松人工林生长收获模型系统的研究[J]．林业科学，1997，33(Sp.1)：57-66.
[2]黄家荣．人工用材林最优密度控制模型[J]．浙江林学院学报，2001，18(1)：36-40.
[3]李景文，王义文，赵惠勋，等．森林生态学[M]．2 版．北京：中国林业出版社，1994：121-125.
[4]惠刚盈，盛炜彤．我国杉木人工林生长与收获模型系统的研究[J]．世界林业研究，1996，9(专集)：32-53.
[5]苏培正．木荷纯林不同抚育间伐强度对比试验[J]．湖北林业科技，2005(2)：22-24.
[6]童书振，盛炜彤，张建国．杉木林分密度效应研究[J]．林业科学研究，2002，15(1)：66-75.
[7]唐守正．同龄纯林自然稀疏规律的研究[J]．林业科学，1993，29(3)：234-241.
[8]薛立，荻原秋男．纯林自然稀疏研究综述[J]．生态学报，2001，5(21)：834-838.
[9]曾德慧，姜凤岐，等．沙地樟子松人工林自然稀疏规

律[J]. 生态学报，2000，20(2)：235-242.
[10]张春锋，殷鸣放，孔祥文，等. 不同间伐强度对人工阔叶红松林生长的影响[J]. 辽宁林业科技，2007(1)：12-15.
[11]张金文. 巨尾按大径材间伐试验研究[J]. 林业科学研究，2008，21(4)：464-468.
[12]张水松，陈长发，等. 杉木林间伐强度试验20年生长效应的研究[J]. 林业科学，2005，41(5)：56-65.
[13]张彩琴，郝敦元，李海平. 人工林林分密度最优控制策略的数学模型[J]. 东北林业大学学报，2006，2(34)：24-27.
[14]张大勇，赵松龄. 森林自疏过程中密度变化规律的研究[J]. 林业科学，1985，2(4)：369-373.
[15]唐守正. 多元统计分析[M]. 北京：中国林业出版社，1986.
[16] CLUTTER J L，JONES E P. Prediction of growth after thinning in old field slash pine plantation[R]. USDA For Serv Res Paper 1980：SE-217.

[原载：林业科学研究，2010，23(01)]

马尾松与红椎等3种阔叶树种营造混交林的生长效果

郭文福 蔡道雄 贯宏炎 温恒辉

（中国林业科学研究院热带林业实验中心，广西凭祥 532600）

摘 要 对马尾松分别与红椎等3种阔叶树种及与杉木混交试验林的6年生和11年生林分生长数据进行了分析，结果表明：①6年生林分，混交小区及纯林小区中马尾松的平均树高为4.6～4.9m，平均胸径为6.0～7.4cm，不同混交组合间差异不显著，但在马尾松×米老排混交组合中马尾松生长不良；11年生林分，不同处理对马尾松平均胸径和蓄积量均有显著影响，其中，对照(马尾松纯林)的平均胸径和蓄积量最大，分别为12.6cm和121.05m^3/hm^2；马尾松×米老排混交组合中马尾松的平均胸径和蓄积量最小，分别为8.7cm和43.18m^3/hm^2。②11年生林分各树种平均树高顺次为：米老排(14.3m)>红椎(10.2m)>火力楠(9.9m)>杉木(9.1m)>马尾松(8.1m)；马尾松×米老排组合中的马尾松和马尾松×红椎组合中的红椎，因林木生长竞争激烈，需要及时间伐；③适合培育阔叶树种大径材的松阔混交组合有2种，分别为马尾松×红椎和马尾松×火力楠，而米老排以人工纯林方式造林效果好，初植密度建议为1000～1600株/hm^2。

关键词 马尾松；红椎；火力楠；米老排；混交林；生长；径级分布

An Analysis of the Growth and Structure of Mixed Plantations Consisted of *Pinus massoninia* and Broadleaf Species

GUO Wenfu, CAI Daoxiong, JIA Hongyan, WEN Henghui

(*Experimental Centre of Tropical Forestry*, *Chinese Academy of Forestry*, *Pingxiang* 532600, *Guangxi*, *China*)

Abstract: ANOVA was used to analyze the 6 years and 11 years growth data of the *Pinus massoniana*(pine) plantation mixed with *Castanopsis hystrix*, *Michelia macclurei*, *Mytilaria laosensis* or *Cuninghamia lancelata* respectively. The result showed that: ①The differences of tree height(H) and diameter at breast height(DBH) of mixed pine plantation were not significant among treatments at the age of 6, and the mean H and DBH were respectively 4.6～ 4.9m and 6.0～7.4cm, of which the growth was very poor in mixed stand with *Mytilaria laosensis* and the differences of DBH and accumulated stock of mixed pine plantation were significant at 11 year old stands, among which the pure pine stand(the controlled treatment) with 12.6m^3/hm^2 cm in DBH and 121.05m^3/hm^2 in accumulated stock is the best for tree growth, however, the pine stand mixed with *Mytilaria laosensis* is the worst with 8.7 cm in DBH and 43.18 m^3/hm^2 in accumulated stock. ② At the age of 11, the mean H for 4 different species in relevant mixed stands respectively were 14.3m for *Mytilaria laosensis*, 10.2m for *Castanop sis hystrix*, 9.9m for *Michelia macclurei*, 9.1m for *Cuninghamia lancelata* and 8.1m for Pine. The pine trees in stand mixed with *Mytilaria laosensis* and the *Castanopsis hystrix* trees in pine-*Castanopsis hystrix* mixed stand should be thinned to a suitable density. ③Two mixed stands with composition of *P. massoniana*×*Castanopsis hystrix*, and *P. massoniana* ×*Michelia macclurei* are suitable to grow valuable big diameter timber for broadleaf species, and *Mytilaria laosensis* is recommended to plant as pure plantation with the best initiative density of 1000～1600 trees/hm^2.

Key words: *Pinus massoninia*; *Castanopsis hystrix*; *Michelia macclurei*; *Mytilaria laosensis*; mixed stand model; growth; distribution of diameter level

马尾松(*Pinus massoniana* Lamb.)是我国松属(*Pinus* Linn.)树种中分布最广的一种，也是我国亚热带东部湿润地区典型的乡土针叶树种，广泛分布于15个省(自治区、直辖市)，是南方重要的商品用材树种，无论经营马尾松大、中或小径材均有较好的经济效益，但经营马尾松纯林和连茬种植马尾松，对林地生产力及可持续经营都产生一些负面作用，如林地有机养分含量，N、P、K、Ca 等大量元素都有不同程度的降低[1]。营造混交林，特别是马尾松、杉木[*Cunninghamia lanceolata*（Lamk.）Hook]与乡土阔叶树种营造针阔叶混交林，是解决这一问题的可行途径[2-4]。我国南亚热带地区树种丰富，红椎（*Castanopsis hystrix* A. DC.）和火力楠（*Michelia macclurei* Dandy）等优良乡土阔叶树种均是营造马尾松阔叶混交林较为成功的树种[2-6]。营造松阔混交林对马尾松生长及林地生产力的保持和提高有较好的促进作用。优良珍贵阔叶用材林一般以生产大径级用材为经营目的，其经济效益和森林的多种效益较高[7]，而通过营造松阔混交林，既可高效培育优质大径级阔叶用材，又可在短时间内收获经济效益较好的中小径级松材，但相关的试验研究报道较少。本文利用马尾松与米老排(*Mytilaria laosensis* Lecte.)、红椎、火力楠、杉木的混交试验林 6 年生和 11 年生时的观测数据，分析了马尾松在几种混交林分中的生长效果，并评价其对培育优质阔叶用材林的作用，为南亚热带地区高效、可持续培育马尾松用材林和珍贵优良乡土阔叶用材林提供科学依据。

1 试验地概况

试验地位于广西凭祥中国林业科学研究院热带林业实验中心伏波试验场(22°03′49″N，106°50′37″E)，属南亚热带季风气候，年均气温 21.5℃，年均降水量 1400mm。土壤为花岗岩发育的红壤，风化壳厚度 6m 以上，土壤层深 1~2m；轻黏土，20~40cm 土层的土壤密度为 1.20~1.40g/cm^3，土壤呈酸性(pH 4.3~5.1)；土壤肥力状况中上等，有机质含量 21~45g/kg，土壤全 N、全 P、全 K 的含量分别为 0.9~1.1、0.4~0.5、1.8~2.4 g/kg。前茬为杉木林。1989 年 10~12 月在采伐迹地炼山后块状整地，植穴规格为 50cm×50cm×27cm。1990 年 3 月按试验设计完成各树种的种植，初植密度 3333 株/hm^2。2001 年开始间伐，主要伐除混交林小区中所有的马尾松及小部分阔叶树，保留阔叶树，以培育大径级阔叶用材。

2 试验设计和数据收集

采用随机区组试验设计，不同的树种组合，5 个处理，即马尾松×米老排(M_1)、马尾松×红椎(M_2)、马尾松×火力楠(M_3)、马尾松×杉木(M_4)和马尾松纯林(对照，M_5)。马尾松与混交树种比例为 4∶1，每植 4 行马尾松间种 1 行混交树种，垂直行状混交，即行的走向为坡面的流水方向，每小区面积约 0.15hm^2。每区组 3 次重复，同一区组的坡位和坡向均相同，立地条件比较一致。

试验观测因子为树高、枝下高和胸径等。从造林第 1 年至第 11 年期间，间隙性地进行过 9 次固定样地调查，为突出结果的代表性，本文分别取其中 6 年生和 11 年生的生长观测数据进行生长效果分析。因为 6 年生是林分郁闭后林木间开始个体竞争分化的阶段，此时也是生产上的卫生伐起始林龄；11 年生为中龄林的起始阶段，林木间分化较大，一般是第 1 次抚育间伐的开始林龄。本文用数理统计软件 SPSS 中的方差分析及多重比较(LSD)方法[8]，分析不同混交组合林分的生长及其差异性；用 11 年生林分调查数据，分析林木的径级分布规律。

林木材积分别按以下公式计算[9-11]：

马尾松：$V_{松}=0.7143\times10^{-4}D^{1.867008}H^{0.9014632}$；

杉木：$V_{杉}=0.6567\times10^{-4}D^{1.769412}H^{1.069769}$；

米老排：$V_{米}=0.6833\times10^{-4}D^{1.926256}H^{0.8840614}$；

红椎和火力楠：$V_{阔}=0.5276\times10^{-4}D^{1.882161}H^{1.00931}$。

3 结果与分析

3.1 幼龄混交林的生长

对 6 年生林分平均树高和平均胸径进行方差分析，结果表明：马尾松在各处理林分间的平均树高和平均胸径差异不显著。由表 1 看出：马尾松在各混交小区及纯林小区中的平均树高为 4.6~4.9m，变化不大；平均胸径为 6.0~7.4cm，变化较大，M_5的平均胸径最大(7.4cm)，M_1的最小(6.0cm)；总蓄积量中，M_1最大，M_5次之，其他处理间马尾松蓄积量和总蓄积差别不大，说明在此阶段，混交树种对马尾松生长的影响未达显著程度。

在 5~6 年生时，试验林林分开始郁闭。马尾松在郁闭之初的幼林生长阶段，由于林木个体间的竞争才刚开始(除 M_1外)，不同混交组合对马尾松尚未出现促进生长的正面影响，反而因树种选择不当，使马尾松生长受到较大的负面影响，如在 M_1林分中，由于幼林生长速度较快，6 年生米老排平均树高达 8.2m，比马尾松高 3.5m；此外，米老排枝叶非常浓密，叶面指数高，对强阳性树种马尾松来说，生长环境的上方和侧方光照均严重不足，势必影响其正常生长。据观察，M_1林分从第 4 年开始，马尾松枝叶稀疏，并部分开始枯落，甚至出现少量树木枯死的现象，而其他处理林分并无此现象。故马尾松×米老排混交组合不能生产有商品价值的马尾松小径材，但对米老排的生长极为有利，因为它占据了较大的营养空间。

表1 不同处理6年生林分的生长指标

处理号	树种	保存率(%)	保存密(株/hm²)	平均树高(m)	平均胸径(cm)	单株材积(m³)	蓄积量(m³/hm²)	总蓄积量(m³/hm²)
M_1	马尾松	91.7	2 445	4.7	6.0	0.0081	19.8025	42.4119
	米老排	92.0	613	8.2	10.0	0.0369	22.6094	
M_2	马尾松	90.5	2414	4.9	7.3	0.0121	29.2050	31.3936
	红椎	93.5	623	4.8	4.1	0.0035	2.1886	
M_3	马尾松	90.5	2414	4.6	7.0	0.0105	25.3432	29.1322
	火力楠	95.0	633	5.0	5.0	0.0060	3.7890	
M_4	马尾松	93.2	2486	4.7	6.5	0.0094	23.3651	33.0672
	杉木	99.1	660	5.6	7.5	0.0147	9.7021	
M_5	马尾松纯林	92.1	3070	4.8	7.4	0.0124	38.0642	38.0642

3.2 中龄混交林的生长

3.2.1 胸径的生长

由表2可见：M_1中马尾松的平均胸径与其他处理间均差异显著($P=0.05$)，其他各处理间林分中马尾松的平均胸径差异不显著。表3表明：11年生马尾松在不同混交处理间的平均胸径差异极显著。林分中混交树种平均胸径的大小顺序为：米老排(18.1cm)>杉木(12.9 cm)>火力楠(12.2cm)>红椎(10.7cm)。马尾松在M_5(纯林)林分中的平均胸径最大，说明马尾松林在中龄期初，除米老排外，其他混交树种无论是杉木还是红椎和火力楠，对马尾松胸径的生长均未产生显著影响。换言之，此时马尾松对阔叶树的生长也没有产生明显影响。另外，混交林中的米老排与一般生产上密度较大的米老排纯林相比，胸径生长量提高30%以上[10,12]，但此时混交林中的马尾松径级太小，尚无大的经济利用价值。

表2 11年生各混交林分中马尾松的平均胸径和平均树高

处理号	胸径		树高	
	平均值(cm)	差异性	平均值(m)	差异性
M_5	12.6	a	8.6	a
M_2	12.0	a	8.6	a
M_3	11.7	a	8.1	a
M_4	11.2	a	7.8	a
M_1	8.7	b	7.5	b

注：同列相同字母表示差异不显著，不同字母表示差异显著。

表3 11年生各混交林分中马尾松平均胸径、平均树高生长方差分析结果

观测指标	方差源	均方和	自由度	均方	*F*值
平均胸径	组间	27.271	4	6.8177	8.9392**
	组内	7.627	10	0.7627	
	合计	34.898	14		
平均树高	组间	2.88	4	0.720	2.3529
	组内	3.06	10	0.306	
	合计	5.94	14		

注：$F_{0.05}(4, 10)=3.48$，$F_{0.01}(4, 10)=5.99$。

3.2.2 树高的生长

由表2可见：不同混交林分中，马尾松树高间有差别，M_1和M_4林分中马尾松的树高均低于8.0m。与胸径结果分析相似，在米老排混交小区(M_1)中，马尾松生长受到压抑，生长不良。方差分析结果(表3)表明：各处理间马尾松的树高差异不显著，但混交树种间的平均树高相差较大，其大小依次为：米老排(14.3m)>红椎(10.2m)>火力楠(9.9m)>杉木(9.1m)。各混交树种的平均树高均大于相应处理中马尾松的树高，尤其是M_1处理，米老排的树高超过马尾松6.8m，其他树种也高出马尾松近2m。由此可见：11年生混交林分中，不同树种间，尤其是马尾松与阔叶树种间的

竞争开始进入较强烈的阶段，此时要根据林分的经营目标，通过间伐对林木密度进行调控。

3.2.3 材积及蓄积量的生长

本试验各处理的林分立木总蓄积量分别由马尾松蓄积量和混交树种蓄积量2部分组成。表4、5表明：不同的混交组合对马尾松蓄积量的影响达极显著水平；马尾松纯林(M_5)的蓄积量最大，是M_1林分中马尾松蓄积量的2.8倍；M_5与M_3、M_4的马尾松蓄积量差异显著，与M_2的差异不显著。从11年生中龄林马尾松蓄积量来看，混交林造林方式暂无提高马尾松材积生长量的作用。这说明，若以经营马尾松短周期原料林为目的，经营纯林可得到较多的木材产量；从经济收益方面来看，是最佳选择。不同混交组合对林分总蓄积量的影响不显著，M_1处理由于米老排生长量较大，故总蓄积量最大，其他各处理间的总蓄积量差异不大。

表4 11年生各混交林分中马尾松平均蓄积量和林分总蓄积量 (m^3/hm^2)

处理号	马尾松蓄积量		总蓄积量	
	平均值	差异性	平均值	差异性
M_5	121.05	a	121.05	a
M_2	102.40	a，b	127.76	a
M_3	92.42	b	125.08	a
M_4	83.06	b	118.37	a
M_1	43.18	c	159.38	a

表5 11年生各混交林分中马尾松蓄积量和林分总蓄积量方差分析结果

观测指标	方差源	均方和	自由度	均方	*F*值
马尾松蓄积量	组间	10055.9	4	2513.961	13.72**
	组内	1832.9	10	183.286	
	合计	11888.7	14		
林分总蓄积量	组间	3322.2	4	830.5548	1.7
	组内	4889.2	10	488.9196	
	合计	8211.4	14		

3.3 林分结构及林分质量状况

林木径阶分布能反映林分的生长状况和林木间的竞争关系，是林分结构稳定性的重要指标[6,11]。林分的径高比不但反映林木的尖削度等形质指标，同时也反映林木间的生长竞争程度，进而反映林分的稳定性。径高比值越小，则林木尖削度越小，林木间的竞争强度越大，林分稳定性弱；反之尖削度越大，林木间竞争越小，林分稳定性强。

3.3.1 纯林小区和混交小区的马尾松林分结构及质量

由表6、图1看出：马尾松纯林(M_5)的平均胸径最大，径级频率分布图的偏度最小，径级分布最接近正态分布(正态分布的偏度为0，峰度为0)，说明此处理的林木生长正常，林木间的竞争分化不大；与火力楠混交的小区(M_3)，偏度和峰度都最大的，说明径级分布偏离正态分布较大，主要特点是比平均值稍小径级林木出现的频率较高，大于平均直径林木则有较大的分化，因此，认为此类混交组合林木间开始出现较强竞争，环境空间有利于较大径级林木的发展。

表6 11年生各混交林分中马尾松的胸径结构

处理	株数(株)	平均胸径(cm)	标准差	变动系数(%)	偏度	峰度	最小胸径(cm)	最大胸径(cm)
马尾松纯林(M_5)	166	12.1	3.625	30.04	0.286	−0.338	3.8	21.3
马尾松×米老排(M_1)	53	8.1	2.35	29.03	0.319	0.195	3.1	14.2
马尾松×红椎(M_2)	159	11.6	3.272	28.33	0.384	0.010	4	20.7
马尾松×火力楠(M_3)	148	11.1	3.553	32.02	0.533	0.570	4.7	24.7
马尾松×杉木(M_4)	145	10.7	2.921	27.25	0.317	−0.487	5.3	18.5

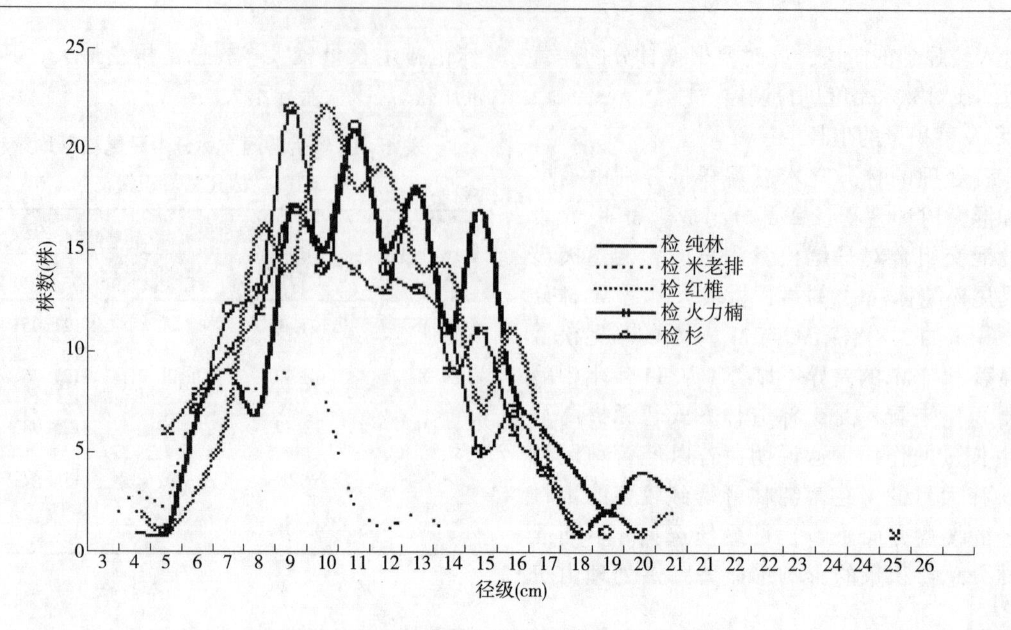

图1 在不同混交林分中马尾松的径级分布

根据11年生林分调查数据统计，不同混交处理(包括对照)马尾松胸径大于11.0cm(即年均生长量1.0cm以上)的株数占总株数的比例，按大小顺次为：M_5(57.8%)>M_2(52.6%)>M_3(47.3%)>M_4(43.4%)>M_1(9.4%)，与马尾松蓄积量的大小顺次相同。可见，马尾松纯林及其与红椎混交林分的大径级木比例较大，与米老排混交(M_1)的最小。各混交组合小区中马尾松林木的径高比，按大小顺次为：M_5(1∶68.3)>M_3(1∶69.0)>M_4(1∶69.4)>M_2(1∶71.4)>M_1(1∶85.2)，说明与米老排混交的马尾松径高比最小，林木的尖削度最小，林木间竞争最强，林分稳定性最差；与红椎混交小区的径高比次之(1∶71.4)；马尾松纯林的林木尖削度最大，林木间竞争不如混交林的剧烈。

据观察，11年生不同混交林分中，M_1组合的米老排生长最快，而马尾松因光照等条件不足，生长不良，枝叶稀疏，叶色发黄，半数以上林木枯死；M_2、M_3和M_4等组合的马尾松生长仍属正常阶段，林分健康状况较好。说明如经营目的树种为马尾松，应选择经营纯林或与红椎、火力楠营造混交林，米老排不合适与其混交。

3.3.2 混交小区的阔叶树种林分结构及其质量

表7表明：混交林中米老排生长速度最快，胸径年均生长达1.6cm，比普通米老排纯林的生长速度快30%左右，胸径变异系数最小。米老排生长速度快主要与其生物学特性有关，即幼龄阶段一般比马尾松生长快，在混交林中，米老排占据了较有利的生长空间。红椎和火力楠的生长速度比相似条件下的纯林高10%~15%[5,13]，但与马尾松的生长竞争相对较小。如图2所示，在林木径级分布规律上，米老排的径级分布较接近正态分布，有较合理的结构；火力楠次之；而红椎的径级分布与正态分布相差较大，胸径变异系数也比较大，即比平均胸径小的林木集中出现在靠近平均胸径的径级处，而大于平均胸径的林木其径级变幅较大，进而表明，红椎林木径级朝大径级方向分化，林分稳定性不如米老排和火力楠的高。

在混交林中，阔叶树种的径高比从小到大为：红椎(1∶94.7)<火力楠(1∶80.9)<米老排(1∶78.9)，由此表明，红椎林木的尖削度小，干形较圆满；火力楠与米老排林木的干形差不多，尖削度较大。也可进一步说明，红椎林分中林木的竞争程度较后两树种的大，故此年龄段的马尾松×红椎混交林要及时进行间伐，调节林木密度使其生长处于最佳状态。

表7 在混交林中米老排等阔叶树种的径级分布

混交树种	株数(株)	平均胸径(cm)	变动系数(%)	偏度	峰度	最小胸径(cm)	最大胸径(cm)
米老排	159	18.1±2.928	16.19	-0.046	-0.637	11.7	25.9
红椎	176	10.1±3.880	38.30	0.386	0.871	1.4	26.3
火力楠	178	11.9±3.176	26.76	0.104	0.987	2.5	22.3

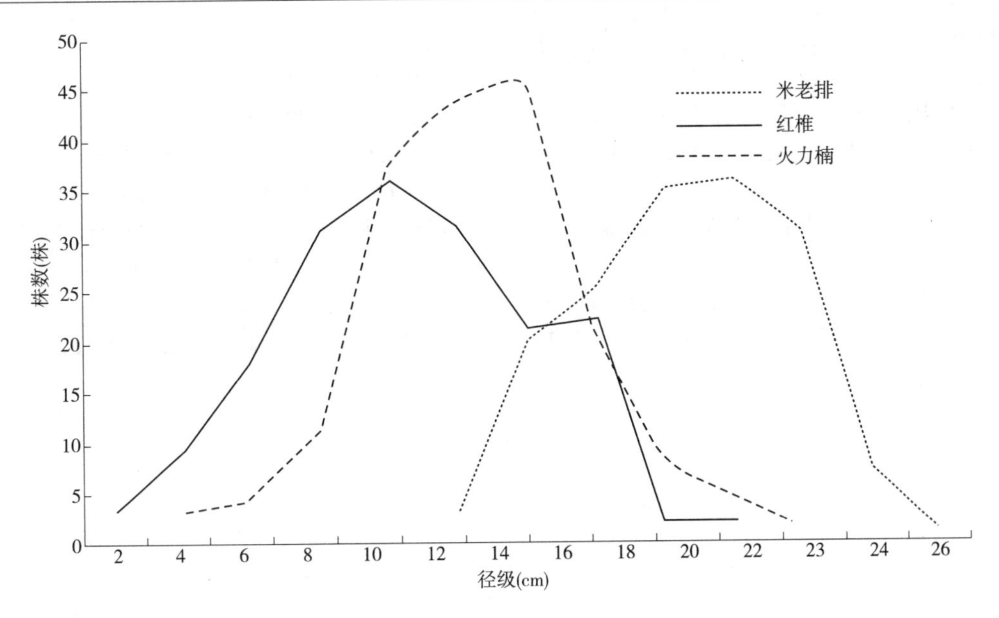

图2　在不同混交林分中阔叶树种林木径级

4　结论及讨论

(1)本试验中，若以经营大径级阔叶材为经营目的的混交林，红椎和火力楠与马尾松营造的针阔混交林，均为较好的混交组合。因为无论是6年生还是11年生，混交林中阔叶树种的生长表现均比一般阔叶纯林好，而此时马尾松的生长尚未受到阔叶树种的激烈竞争，林木生长正常。11年生马尾松×红椎和马尾松×火力楠混交林中，马尾松的平均胸径均达11.1~11.6cm，接近纯林，且年生长量在1.0cm以上，属较快的生长速度，若让其生长至12~15年，然后将马尾松全部或大部分间伐，同时也间伐少量生长不良的阔叶树，生产马尾松中小径级材作为短周期造纸原料或人造板原料，可取得较好的经济效益，达到“以短养长”的经营效果；保留的阔叶树在间伐马尾松后，可取得更大的营养空间，有利于培养阔叶树种优质大径材[7]。

(2)马尾松×米老排混交林分中的米老排生长比一般生产上密度较大的米老排纯林快，6年生林分的平均树高远高于马尾松，11年生林分平均胸径达18.1cm。因马尾松在其树冠下生长不良，大部分枯死，这相当于密度较小的米老排纯林，据此可推知，若以经营米老排大径材为目的，在较好的立地条件下，经营纯林的初植密度一般为1000~1600株/m^2，不宜过密。这样可缩短培育大径材米老排的周期，提高经济效益。

(3)混交树种的选择、混交比例和混交方式的设计是营造混交林的关键技术。从马尾松林分生长效果看，无论是林分郁闭初期的幼龄阶段，还是中龄林开始阶段，所选的几种树种与马尾松营造松阔或松杉混交林，对马尾松的生长均未起到促进林木生长的作用。主要原因，一方面是混交树种选择不当，如阔叶树种米老排由于其早期生长迅速，枝叶浓密，严重压抑马尾松的生长，在林分6年生之前马尾松就开始大量枯死，说明选择米老排营造马尾松阔叶混交林不适当。其他树种与马尾松的混交林，许多研究结果表明是成功的[2-6,11]，至于本试验的11年生林分尚未显示积极的混交效果，可能与混交树种的株数比例过低有关(只占25%)，因为一般马尾松在针阔混交林中，阔叶树种比例大多在30%~40%[2-3,5-6,11]；另一方面，从混交方式上看，行状混交是采取单行还是多行间混交？混交行的走向以坡面流水方向的行状混交方式还是与等高线平行的行状混交？可能其混交效果大小及出现的早晚是不同的，采取何种方式最好，有待进一步研究。

参考文献

[1]黄勤坚．培育马尾松大径材适宜松楠混交模式研究[J]．广西林业科学研究，2004，33(3)：119-123.
[2]曹汉洋，陈金林．杉木马尾松木荷混交林生产力研究[J]．福建林学院学报，2000，20(2)：158-161.
[3]梁建平，龙定健，曹艳云．松阔人工异龄混交林营造技术[J]．广西林业科学，1997，26(13)：106-111.
[4]王道兴．杉木、火力楠混交林效益与营造技术[J]．福建林业科技，1998，25(4)：78-80.
[5]蒋家淡．红椎杉木混交造林效果研究[J]．福建林学院学报，2002，22(4)：329-333.
[6]蔡道雄，贾宏炎，卢立华，等．论我国南亚热带珍优

乡土阔叶树种大径材人工林的培育[J]. 林业科学研究, 2007, 20(2): 165-169.
[7]苏宗明, 傅荣华, 周建斌, 等. 统计软件SPSS for Windows实用指南[M]. 北京: 电子工业出版社, 2000: 418-452.
[8]广西林业勘测设计院, 广西林学分院. 森林调查手册[R]. 1986.
[9]陈永富, 郭文福. 米老排立木材积表及地位指数表的编制[J]. 林业科学研究, 1991, 4(增刊): 117-120.
[10]林俊平. 红椎不同模式造林效果分析[J]. 福建林业科技, 2002, 29(3): 59-61.
[11]郭文福, 李运兴, 卢志芳. 米老排人工林生长规律的研究[J]. 林业科学研究, 2006, 19(5): 585-589.
[12]卢立华, 汪炳根, 何日明. 立地与栽培模式对红椎生长的影响[J]. 林业科学研究, 1999, 12(5): 519-523.

[原载: 林业科学研究, 2010, 23(06)]

马尾松中幼龄林不同施肥处理经济收益分析

唐继新 谌红辉 卢立华 刘 云

(中国林业科学研究院热带林业实验中心，广西凭祥 532600)

摘 要 以重置成本法，按项目投资的财务效益评估手段，对马尾松中、幼龄林不同施肥处理的经济收益进行对比分析，结果表明：①幼龄林不同配比施肥处理的经济收益差异较大，经济收益以P_2处理的最优，NP_1处理的次之，K_2处理的最差。②将磷肥均分成2次施用到幼龄林的经济收益最佳(明显优于单次施入)。③中龄林前17年生(施肥8年后)仍未达经济成熟。④中、幼龄林均以施磷肥处理经济收益最佳。

关键词 马尾松；施肥；经济收益

The Analysis on Economic Profits of the Middle-aged and Young *Pinus passoniana* Plantations in Different Fertilization Treatments

TANG Jixin, CHEN Honghui, LU Lihua, Liu Yun

(*Experimental Center of Tropical Forestry*, *Chinese Academy of Forestry*, *pingxiang* 532600, *Guangxi*, *China*)

Abstract: Based on the replacement cost and the evaluation method of the project investment, the economic profits were compared among the middle-aged and young *Pinus massoniana* plantations with different fertilization treatments, the results showed that: ①the young *Pinus massoniana* plantations with different ratio fertilizer treatments have distinctly different economic profits. In all the economic profits of the treatments, the economic profit from P_2treatment is the first, NP_1 treatment the second, K_2 treatment the last. ②in the young *Pinus massoniana* plantation, the treatment with P fertilizer divided equally into two halves gets the highest economic profits. ③all the middle-age *pinus massoniana* plantations before the age of 17 years(after fertilizing 8 years) haven't reached the economic maturity. ④by the treatment of P fertilizer, the economic profits of the middle-aged and young*Pinus massoniana* plantations were the highest in all the fertilization treatments.

Key words: *Pinus massoniana*; fertilization; economic profits

在特定立地条件下，根据林木生长的养分需求及林地养分供给的实际水平施肥，能取得较好的增产效果[1,2]。中国南方马尾松林区土壤大都缺磷少钾[3]，中国林业科学研究院热带林业实验中心(下文简称热林中心)的马尾松林区土壤亦如此，为实现马尾松林分经营的木材增产、经济增收，自20世纪80年代“七五”时期开始，热林中心进行了马尾松中、幼龄林不同施肥措施的课题研究[3-8]。基于此，采用项目投资的财务效益评估手段，对热林中心马尾松林分不同施肥配比(不同施肥时间)处理的经济收益进行了分析，旨在为中国热带、亚热带马尾松人工林经营的投资决策提供参考。

1 材料与方法

1.1 试验林概况

试验地位于广西凭祥市热林中心伏波实验场金生站(22°02′N，106°41′E)，属北热带季风气候区，年均温19.5℃，年降水量1400mm，相对湿度83%，立地指数18[3,4]。试验林造林技术相似，造林密度为3600株/hm^2中、幼龄林的造林时间分别为1983年和1991年春，间伐时间分别为1996年和2001年底；试验布设时间均为1991年，布设试验时，将中龄林被压木、双叉木等砍除400株/hm^2(蓄积5.22m^3/hm^2)[5]；中龄林于1991年9年生时经修枝及卫生伐。

1.2　试验设计与方法

试验均采用随机区组设计，13个处理(幼林不同配比施肥、不同时间施肥试验分别为11、10个处理)，4次重复[3,5]，各处理水平施肥量详见表1。重复内处理在同一坡面沿等高线排列，且重复内各小区立地(林相)、密度基本一致。试验肥料均为尿素(含N 46%)、钙镁磷肥(含P_2O_5 18%)、氯化钾(含K_2O 60%)，追肥时间为年初3~4月份。幼龄林，①不同配比施肥试验，各处理施肥量均分2次施用，其中：N、K_1、K_2处理于造林后第2、3年各施肥1次，其余处理均为第1、3年均分施肥(且第1次做基肥)；②不同时间施肥试验，CK处理不施肥，N_a处理第2年1次施肥，N_b处理第3年1次施肥，P_a处理作基肥施肥，P_b处理第2年1次施肥，P_c处理第3年1次施肥，P_d处理1/2作基肥(另1/2第3年施肥)，P_e处理于第2、3年均分2次施肥，K_a处理第2年1次施肥，K_b处理第3年1次施肥。中龄林，1992年将肥料1次施入。

1.3　不同施肥配比处理的马尾松林生长情况

马尾松中龄林不同施肥配比处理的平均胸径、平均树高生长情况如表2所示，马尾松幼龄林不同施肥处理的生长情况如表3所示。

表1　马尾松中、幼龄林施肥试验各处理施肥量　　(kg/hm^2)

中龄林施肥试验各处理施肥量					幼龄林不同配比、不同时间施肥试验各处理施肥量				
编号	处理	尿素	钙镁磷	氯化钾	编号	处理	尿素	钙镁磷	氯化钾
1	N_1	225	0	0	1	$N(N_a、N_b)$	217.4	0	0
2	N_2	450	0	0	2	P_1	0	277.8	0
3	P_1	0	720	0	3	$P_2(P_a、P_b、P_c、P_d、P_e)$	0	555.5	0
4	P_2	0	1440	0	4	P_3	0	1 111.1	0
5	P_3	0	2880	0	5	$K_1(K_a、K_b)$	0	0	166.7
6	P_4	0	5760	0	6	K_2	0	0	333.4
7	K	0	0	360	7	NP_1	217.4	277.8	0
8	N_1P_1	225	720	0	8	P_1K_1	0	277.8	166.7
9	N_1K	225	0	360	9	P_2K_2	0	555.5	333.4
10	P_1K	0	720	30	10	NP_2K_1	217.4	555.5	166.7
11	N_1P_3K	225	2880	360	11	CK(CK)	0	0	0
12	N_2P_2K	450	1440	360	—	—	—	—	—
13	CK	0	0	0	—	—	—	—	—

表2　马尾松中龄林不同施肥配比处理的平均胸径、平均树高生长情况

处理	9~17年生时的平均胸径(cm)								9~17年生时的平均树高(m)							
	9	10	11	12	13	14	15	17	9	10	11	12	13	14	15	17
N_1	10.52	11.50	12.32	13.09	13.68	14.22	16.70	18.31	7.57	8.21	8.93	9.39	9.85	10.42	11.10	12.02
N_2	10.45	11.48	12.37	13.16	13.83	14.31	17.37	18.95	7.66	8.33	9.03	9.61	10.13	10.73	11.68	12.32
P_1	10.71	11.64	12.54	13.30	13.89	14.43	16.96	18.50	7.53	8.17	9.13	9.64	10.20	10.93	11.58	12.54
P_2	10.24	11.16	12.00	12.71	13.39	13.94	17.08	18.76	7.35	7.98	8.78	9.38	9.91	10.60	11.47	12.42
P_3	10.31	11.24	12.13	12.81	13.49	14.03	16.79	18.29	7.68	8.28	9.19	9.74	10.28	10.86	11.73	12.49
P_4	10.34	11.25	12.06	12.87	13.54	14.08	17.17	18.76	7.53	8.20	8.88	9.51	10.11	10.75	11.74	12.57
K	9.93	10.78	11.74	12.44	13.07	13.58	16.55	18.17	7.38	8.03	8.76	9.21	9.81	10.42	11.21	12.08
N_1P_1	10.36	11.37	12.25	13.01	13.75	14.33	17.39	19.08	7.87	8.61	9.47	10.04	10.52	11.20	12.06	12.59
N_1K	10.66	11.57	12.36	13.07	13.74	14.23	17.56	19.18	7.78	8.37	9.25	9.80	10.36	11.06	11.82	12.43
P_1K	10.14	11.10	12.14	12.87	13.53	14.11	16.88	18.54	7.45	8.11	9.04	9.62	10.15	10.70	11.55	12.49
N_1P_3K	9.75	10.78	11.77	12.55	13.28	13.90	17.33	19.08	7.29	7.93	8.75	9.36	9.98	10.60	11.58	12.28
N_2P_2K	10.47	11.47	12.37	13.14	13.91	14.45	17.46	19.15	7.77	8.42	9.34	10.06	10.58	11.36	12.13	12.95
CK	10.34	11.19	11.98	12.73	13.29	13.70	16.45	17.83	7.99	8.65	9.43	9.94	10.50	11.10	11.81	12.57

表 3 不同配比施肥处理马尾松幼龄林 11 和 12 年生时的生长情况

处理	平均胸径(cm)		平均树高(m)		处理	平均胸径(cm)		平均树高(m)	
	11	12	11	12		11	12	11	12
N	11.36	13.28	7.98	8.47	N_a	11.24	12.81	7.33	7.57
P_1	12.13	14.07	7.78	8.06	N_b	11.37	12.39	7.43	7.63
P_2	13.62	14.98	8.30	8.79	P_a	12.42	13.42	7.52	7.76
P_3	12.45	13.59	8.08	8.30	P_b	12.40	13.10	7.79	8.10
K_1	12.27	14.68	7.93	8.36	K_b	11.13	12.14	7.69	7.95
K_2	11.07	11.82	7.73	7.99	P_c	11.40	12.98	7.25	7.64
NP_1	12.52	14.66	8.18	8.44	P_d	12.60	13.89	8.14	8.59
P_1K_1	12.31	13.30	7.90	8.33	P_e	13.38	13.97	7.84	8.25
P_2K_2	11.68	14.30	7.94	8.72	K_a	11.55	12.64	7.56	8.00
NP_2K_1	12.13	14.05	7.86	8.61	CK	10.88	11.13	7.54	7.69
CK	11.93	13.36	8.00	8.50	—	—	—	—	—

2 评价指标的确定及营林投资成本的构成

2.1 评价指标的确定

对马尾松人工林施肥效益的评价，选用的评价指标为产量(出材量)、产值、成本、税费及财务经济评价指标[9](内部收益率、净现值、动态投资回收期)。马尾松各材种产量(出材量)，以马尾松削度方程和国家原木材积公式[10]进行计算；产值为马尾松各材种售价与出材量乘积的累计值；热林中心马尾松各材种 4、6、8~12、14~18、20~30cm 径阶(长 2m)；2009 年的均价分别为 330、390、520、620、720 元/m^3；税费“两金一费”按马尾松木材售价的 10%计，装车费、检疫费均为价外费用。

2.2 材种出材量的计算

根据试验固定样地每木检尺所得各径阶的平均胸径、平均树高及径阶株数，利用马尾松削度方程及国家原木材积公式计算，按长 2m、尾径大于 4cm 进行造材(长度不足 2m 的作为废材)，可得马尾松试验林各处理材种的理论出材量(表 4、表 5)，而考虑到林木材质缺陷(如弯曲、折断、节子、扭曲、分杈、心腐、劈裂等)对造材存在影响，及实际造材中存在木材损失，在各处理经济收益的分析时，按各处理理论出材量的 85%进行分析。

表 4 不同施肥配比处理马尾松各材种的理论出材量 (m^3/hm^2)

处理	17 年生中龄林各径级(cm)的出材量						12 年生幼龄林各径级(cm)的出材量					
	4	6	8~12	14~18	20~30	合计	4	6	8~12	14~18	20~30	合计
N_1	1.8	4.1	48.0	72.6	19.4	145.9	N	18.3	21.0	54.8	23.0	1.2
N_2	1.2	4.7	51.8	80.3	18.3	156.3	P_1	13.9	23.7	59.4	26.3	4.8
P_1	2.1	4.0	49.8	65.4	25.9	147.1	P_2	17.4	27.3	68.2	35.4	1.3
P_2	1.7	4.1	48.4	69.2	20.2	143.7	P_3	18.1	20.7	54.1	22.7	1.2
P_3	1.5	5.4	48.6	63.0	18.6	137.1	K_1	15.9	24.8	62.0	32.2	1.2
P_4	2.0	3.8	49.2	68.2	30.4	153.6	K_2	14.1	21.9	39.6	10.7	0
K	1.6	4.6	45.6	64.4	14.1	130.3	NP_1	16.0	25.1	62.7	32.6	1.2
N_1P_1	1.6	4.0	42.6	63.7	25.4	137.3	P_1K_1	17.5	20.1	52.4	22.0	1.2
N_1K	1.8	3.0	44.2	55.6	32.9	137.6	P_2K_2	15.8	24.7	61.7	32.0	1.2
P_1K	1.6	4.9	49.0	71.0	22.4	149.0	NP_2K_1	14.0	23.9	59.9	26.6	4.9
N_1P_3K	1.2	4.0	39.7	62.5	25.3	132.8	CK	18.1	20.7	54.1	22.7	1.2
N_2P_2K	2.0	3.5	47.6	74.5	29.0	156.7	—	—	—	—	—	—
CK	1.8	5.8	51.2	65.8	13.6	138.2	—	—	—	—	—	—

表5 不同时间施肥处理马尾松幼林的理论出材量 (m^3/hm^2)

处理	4	6	8~12	14~18	20~30	合计
N_a	17.9	19.7	47.5	10.0	1.9	97.0
N_b	16.1	19.4	42.7	9.0	0.0	87.2
P_a	11.9	19.6	44.3	15.4	5.8	97.0
P_b	14.2	18.2	47.1	19.7	3.1	102.3
P_c	13.3	21.0	44.7	15.0	1.5	95.5
P_d	11.2	20.4	55.4	23.6	9.6	120.2
P_e	10.5	14.4	56.3	24.6	11.0	116.8
K_a	13.2	20.9	44.6	16.6	0.0	95.3
K_b	16.2	21.6	41.5	9.2	0.0	88.5
CK	15.2	19.2	37.7	4.5	0.0	76.6

2.3 投资成本的构成

因马尾松施肥试验投入成本均发生在20世纪80~90年代，相对现时用工、物价成本变化巨大，本着更为真实、可靠、可比的会计原则，消除物价变动、通胀因素的影响[11]，进而客观地实际反映马尾松施肥营林的投资收益，采用更新重置成本法[12]计算投资成本，投资成本按热林中心及其周边地区近年经营的平均成本计。2009年尿素、钙镁磷肥、氯化钾的均价分别为1900、650、2200元/t；施肥工资由施肥量确定，施肥量<0.5、0.5~1.0、1.0~1.5、1.5~2.0、2.0~3.0、>3.0t/hm^2的施肥工资分别为0、600、800、1000、1200、1400、1600元/hm^2；基肥施肥工资为0元。

12年生马尾松幼龄林不同配比施肥CK处理的总投资成本为2.9788万元/hm^2。其中①营林投资成本1.7535万元/hm^2，其详细构成如表6所示。②林道成本(包括道路修建与维护)525元/hm^2，投入在第12年。③采运成本(包括伐区设计、采伐、造材、剥皮、运输、木材检尺、采伐和运输的证件办理、储木场木材销售及管理等费用，按照100元/hm^3计算)1.1428万元/hm^2，第11和12年分别投入1500和9928元/hm^2。④其他成本525元/hm^2，也是第12年投入。因试验林营林技术相似，仅有肥料、追肥用工、追肥时间不同，其他地租、采运成本等计算标准均一致，所以其他不同配比施肥处理的仅列出相应的结果，马尾松各施肥处理的投资成本如表7所示，马尾松不同时间施肥处理投资成本构成如表8所示。

表6 12年生马尾松幼龄林不同配比施肥对照处理的营林投资成本构成 (元/hm^2)

项目	第1年	第2年	第3年	第4年	第5年	第6年	第7年	第8年	第9年	第10年	第11年	第12年	总计
炼山、清山	600	—	—	—	—	—	—	—	—	—	—	—	600
挖坑、整地	2880	—	—	—	—	—	—	—	—	—	—	—	2880
定植及补植	540	—	—	—	—	—	—	—	—	—	—	—	540
基肥	—	—	—	—	—	—	—	—	—	—	—	—	0
苗木	900	—	—	—	—	—	—	—	—	—	—	—	900
抚育除草	750	750	—	—	—	—	—	—	225	—	—	—	1725
施肥肥料	—	—	—	—	—	—	—	—	—	—	—	—	0
追肥用工	—	—	—	—	—	—	—	—	—	—	—	—	0
病虫害防治	30	30	30	—	—	—	—	—	—	—	—	—	90
护林防火	150	150	150	150	150	150	150	150	150	150	150	150	1800
地租	750	750	750	750	750	750	750	750	750	750	750	750	9000
小计	6600	1680	930	900	900	900	900	900	1125	900	900	900	17535

表7 马尾松不同配比施肥处理投资成本构成 (元/hm^2)

处理	17年生马尾松中龄林的投资成本构成					处理	12年生幼龄林的生产投资成本构成				
	营林	林道	其他	采运	合计		营林	林道	其他	采运	合计
N_1	23062.5	525.0	300.0	17139.2	41026.7	N	19148.2	525.0	300.0	11555.5	31528.7
N_2	23490.0	525.0	300.0	17605.0	41920.0	P_1	18315.6	525.0	300.0	12388.5	31529.1

（续）

处理	17年生马尾松中龄林的投资成本构成					处理	12年生幼龄林的生产投资成本构成				
	营林	林道	其他	采运	合计		营林	林道	其他	采运	合计
P_1	23303.0	525.0	300.0	18097.7	42225.7	P_2	18496.2	525.0	300.0	14216.0	33537.2
P_2	23971.0	525.0	300.0	16868.2	41664.2	P_3	18857.2	525.0	300.0	11428.0	31110.2
P_3	25307.0	525.0	300.0	16620.8	42752.8	K_1	19101.8	525.0	300.0	13068.5	32995.3
P_4	27379.0	525.0	300.0	17643.0	45847.0	K_2	19468.6	525.0	300.0	8835.5	29129.1
K	23427.0	525.0	300.0	15471.0	39723.0	NP_1	18728.8	525.0	300.0	13196.0	32749.8
N_1P_1	23730.5	525.0	300.0	17358.8	41914.3	P_1K_1	18682.4	525.0	300.0	11122.0	30629.4
N_1K	24054.5	525.0	300.0	17059.1	41938.6	P_2K_2	19229.6	525.0	300.0	13009.0	33063.6
P_1K	24295.0	525.0	300.0	17728.7	42848.7	NP_2K_1	19276.2	525.0	300.0	12490.5	32591.7
N_1P_3K	25976.0	525.0	300.0	16231.6	43032.6	CK	17535.0	525.0	300.0	11428.0	29788.0
N_2P_2K	26018.0	525.0	300.0	18737.1	45580.1	—	—	—	—	—	—
CK	22035.0	525.0	300.0	16688.7	39548.7	—	—	—	—	—	—

表8　不同时间施肥处理马尾松幼林的投资成本构成　（元/hm^2）

处理	营林	林道	其他	采运	合计
CK	17535.0	525.0	300.0	8011.0	26371.0
N_a	18548.0	525.0	300.0	9745.0	29118.0
N_b	18548.0	525.0	300.0	8912.0	28285.0
P_a	17896.2	525.0	300.0	9745.0	28466.2
P_b	18496.2	525.0	300.0	10195.5	29516.7
P_c	18496.2	525.0	300.0	9617.5	28938.7
P_d	19096.2	525.0	300.0	11717.0	31038.2
P_e	18496.2	525.0	300.0	11428.0	31349.2
K_a	18501.8	525.0	300.0	9600.5	28927.3
K_b	18501.8	525.0	300.0	9022.5	28349.3

3　结果与分析

在特定立地条件下，根据林地养分的供给实际水平，对林木施肥能取得较好的增产效果。然而，在诸多施肥处理中，何种配比施肥经济收益最佳？何时施肥投资回报最大？同一施肥措施何时收获盈利最强、风险最低、回收最快？这些都是林业经营、投资者最为关注的问题之一。因此，下面以马尾松营林近年的历史投入成本为基础，从项目投资财务效益评估的角度，对马尾松中、幼龄林各施肥措施的经济效益进行对比分析。

3.1　马尾松幼林施肥效益分析

根据马尾松幼龄林配比施肥CK处理投资成本、出材量、各材种均价、税费，以及各年的现金收支情况，可计算出12年生马尾松幼林不施肥CK处理全部投资(贴现率取8%)的现金流量(表9)。由此可见，12年生马尾松幼龄林不施肥CK处理的净收益(各年净现金之和)、年均净收益、净现值、年均净现值分别为18495.7、1541.3、1664.4、138.7元/hm^2，内部收益率为9.46%，动态回收期[9]为11.9年，各指标均表明不施肥的投资处理，不仅能及时回收资本，而且具有一定的获利能力。

表9 12年生马尾松幼林不同配比施肥对照处理的投资现金流量表 （元/hm²）

时间	资金流入	资金流出					净现金流量	净现值流量
	木材收入	营林成本	林道成本	其他成本	采运成本	税费		
第1年	—	6600.0	—	—	—	—	-6600.0	-6600.0
第2年	—	1680.0	—	—	—	—	-1680.0	-1555.6
第3年	—	930.0	—	—	—	—	-930.0	-797.3
第4年	—	900.0	—	—	—	—	-900.0	-714.4
第5年	—	900.0	—	—	—	—	-900.0	-661.5
第6年	—	900.0	—	—	—	—	-900.0	-612.5
第7年	—	900.0	—	—	—	—	-900.0	-567.2
第8年	—	900.0	—	—	—	—	-900.0	-525.1
第9年	—	1125.0	—	—	—	—	-1125.0	-607.8
第10年	—	900.0	—	—	—	—	-900.0	-450.2
第11年	5100.0	900.0	—	—	1500.0	510.0	2190.0	1014.4
第12年	48548.6	900.0	525.0	300.0	9928.0	4854.9	32040.7	13741.7

根据马尾松各施肥处理对应年度投资资金的收支情况，用全部投资的现金流量分析法[13]，可计算、整理出幼龄林、中龄林各施肥处理对应的全部投资经济收益(表10、表11)。不同配比施肥处理经济分析结果(表10)表明：在幼林不同配比施肥处理中，与CK相比，经济收益高于CK的处理有P_1、P_2、K_1、NP_1、P_2K_2、NP_2K_1；不同配比施肥处理之间的经济收益差异较大，以P_2最优，NP_1次之，K_2最差，P_2处理的净收益、年均净收益、年均净现值、内部收益率分别比CK的提高了6820.9元/hm²、568.4元/hm²、199.0元/hm²和1.62%；收益最好的P_2处理的净收益、年均净收益、净现值、年均净现值、内部收益率分别比收益最差的K_2处理增加了1.89344万元/hm²、0.15779万元/hm²、0.84780万元/hm²、706.5元/hm²和7.37%。从资本投资的时间价值考虑时，所有施肥处理(除K_2投资亏本)均实现了盈利。

不同施肥时间处理经济分析结果(表11)表明：在12年生马尾松幼林不同时间施肥处理中，与CK处理相比，各处理均实现了显著的增产、增收(内部收益率至少提高1.13%)，并以磷肥1、3年均分施肥的P_d处理经济最优，但考虑资金投资的时间价值时，仅有P_d、P_e施肥处理实现开始投资获利。

表10 12年生马尾松幼林不同施肥处理投资收益情况

不同配比施肥的投资收益							不同时间施肥的投资收益						
处理	净收益(元/hm²)	年均净收益(元/hm²)	净现(元/hm²)	年均净现值(元/hm²)	内部收益率(%)	动态回收期(年)	处理	净收益(元/hm²)	年均净收益(元/hm²)	净现值(元/hm²)	年均净现值(元/hm²)	内部收益率(%)	动态回收期(年)
N	17 315.8	1443.0	411.8	34.3	8.3	11.0	N_a	10553.4	879.4	-2245.4	-187.1	5.89	+∞
P_1	22388.1	1865.7	2986.4	248.9	10.4	11.0	N_b	7412.2	617.7	-3523.2	-293.6	4.46	+∞
P_2	25316.6	2109.7	4052.3	337.7	11.1	11.8	P_a	13097	1091.4	-857.3	-71.4	7.2	+∞
P_3	17173.5	1431.1	479.3	39.9	8.4	12.0	P_b	13875.63	1156.3	-794.8	-66.2	7.29	+∞
K_1	23604.8	1967.1	3130.5	260.9	10.4	11.8	P_c	10996.6	916.4	-1963.7	-163.6	6.15	+∞
K_2	6382.2	531.9	-4425.6	-368.8	3.7	+∞	P_d	20984.9	1748.7	2294.4	191.2	9.85	11.9
NP_1	24433.2	2036.1	3657.0	304.8	10.8	11.8	P_e	20310.5	1692.5	1720.3	143.4	9.36	11.9
P_1K_1	16315.6	1359.6	198.7	16.6	8.2	12.0	K_a	10845.8	903.8	-2097	-174.8	6.04	+∞
P_2K_2	23267.2	1938.9	2906.7	242.2	10.2	11.8	K_b	7647	637.3	-3402.6	-283.6	4.59	+∞
NP_2K_1	21806.7	1817.2	2257.0	188.1	9.7	11.9	CK	4916	409.7	-4159.7	-346.6	3.33	+∞
CK	18495.7	1541.3	1664.4	138.7	9.5	11.9	—	—	—	—	—	—	—

3.2 马尾松中龄林施肥效益分析

3.2.1 各施肥各处理经济分析

据17年生马尾松中龄林不同配比施肥处理全部投资收益情况(表11)，从静态的经济分析角度来看，各处理经济净收益由高到低的排序为N_2P_2K、P_1、N_2、P_1K、N_1P_1、N_1、N_1K、P_4、P_2、CK、P_3、N_1P_3K、K；不同施肥处理均有净收益，且最低年均净收益为2275.6元/hm^2；各施肥处理与CK相比，经济收益差异明显，如最优处理N_2P_2K、最差处理K的净收益与对照处理CK的差值分别为0.67853万元/hm^2和-0.57572万元/hm^2。从资金时间价值的动态投资角度(净现值)分析，各施肥处理都实现了盈利，其经济净收益高于CK由高至低的排序依次为P_1、N_2P_2K、N_2、N_1P_1、P_1K、N_1、N_1K，其中P_1处理较CK处理净现值、内部收益率分别增加1645.5元/hm^2、0.66%。由动、静态的经济收益对比分析可知，P_4、P_2处理的静态经济收益高于CK处理，但考虑资金的时间使用效益，CK处理的投资净现值高于P_4、P_2处理，营林投资实践中应选择后者，即不施肥。

表11　17年生马尾松中龄林不同配比施肥处理投资收益情况

处理	净收益(元/hm^2)	年均净收益(元/hm^2)	净现值(元/hm^2)	年均净现值(元/hm^2)	内部收益率(%)	动态回收期(年)
N_1	46653.6	2744.3	5776.8	339.8	10.95	16.6
N_2	48374.4	2845.6	6073.1	357.2	11.04	16.6
P_1	50365.2	2962.7	7049.2	414.7	11.51	16.5
P_2	44612.4	2624.3	4969.2	292.3	10.56	16.6
P_3	41552.1	2444.2	3885.3	228.5	10.04	16.7
P_4	45532.5	2678.4	4508.6	265.2	10.23	16.7
K	38684.8	2275.6	3279.6	192.9	9.79	16.7
N_1P_1	47060.9	2768.3	6021.4	354.2	11.07	16.6
N_1K	46237.1	2719.8	5624.0	330.8	10.87	16.6
P_1K	47874.3	2816.1	5968.9	351.1	10.99	16.6
N_1P_3K	40627.0	2389.8	3843.8	226.1	10.09	16.7
N_2P_2K	51227.3	3013.4	6687.3	393.4	11.23	16.6
CK	44442.0	2614.2	5403.7	317.9	10.85	16.6

3.2.2 各施肥处理内时间序列分析

各种施肥处理在不同年度收获的经济效益如何?施肥处理的木材增益与其经济收益的关系如何?这些都是经营、投资者最关注的问题之一。为探索研究这些问题，基于马尾松各施肥处理的蓄积量及各材种出材量的变化，选用净现值、年均净现值、内部收益率等财务指标，对马尾松不同施肥处理的时间序列收益、蓄积增产与经济增值进行了分析，结果如表12、13所示(蓄积量是根据各处理对应年份各径阶平均胸径、平均树高及径阶株数，采用广西马尾松二元材积计算公式[7]计算后再累加的)。

从17年生马尾松中龄林各施肥处理的资金时间价值的动态角度分析可知，到11年生时马尾松各施肥处理的净现值均小于0；施肥3年内各处理的净现值仍低于对照(CK)，各处理因肥效所导致的蓄积增产并未能由此带来经济收益的增加；施肥4年后，N_1P_1处理经济收益开始明显高于对照；施肥5年时，仅有N_1P_3K处理净现值小于0，其余各处理则均能实现大于基准收益率8%的盈余；到17年生各施肥处理(除N_1外)的年均净现值、内部收益均随林龄增加而增加，表明到17年生各处理仍未达经济成熟。从林龄9~17年生期间蓄积、净现值增量分析表明，各施肥处理的蓄积增量均大于对照的，但就净现值增值而言，大于对照的有N_1、N_2、P_1、P_2、N_1P_1、P_1K、N_2P_2K、N_1P_3K处理(即实现了蓄积增产、效益增收)，且P_1K处理的净现值增量比对照的多增加1885.8元/hm^2，其余处理则表现增产歉收；施肥的木材增产效益仍未超过营林投入增量资本的时间价值。数据综合分析结果表明，若以培育中小径材用途为目的的马尾松工业用材林，选择合适的肥种、肥量进行施肥，能较快的实现盈利及较好的投资收益；在各施肥处理中以P_1处理经济效益最佳。

表 12　各施肥处理净现值时间序列分析

处理	9~17 年的增值量		9~17 年的净现值(元/hm²)							
	蓄积量(m^3/hm^2)	净现值(元/hm²)	9 年生	10 年生	11 年生	12 年生	13 年生	14 年生	15 年生	17 年生
N_1	157.6	9317.7	-3541.0	-2730.4	-3042.4	-316.6	-81.5	1370.5	2541.6	5776.8
N_2	165.2	9966.1	-3893.0	-2804.0	-1206.1	24.4	681.1	2532.9	4370.9	6073.1
P_1	166.4	10203.2	-3154.0	-2604.6	-384.3	490.5	1054.3	3437.2	4808.1	7049.2
P_2	164.0	9444.6	-4475.4	-3957.6	-2122.4	-1181.6	-251.3	1465.4	2954.2	4969.2
P_3	162.8	7918.9	-4033.6	-4335.2	-2387.5	-1264.5	-511.2	318.0	2622.5	3885.3
P_4	167.1	8139.3	-3630.8	-4569.6	-3717.6	-2191.5	-858.3	732.3	2523.5	4508.6
K	152.7	8040.1	-4760.5	-4625.0	-3753.3	-2003.9	-1009.7	22.2	1607.2	3279.6
N_1P_1	168.4	9584.9	-3563.5	-2615.6	-537.7	433.6	1752.2	4094.0	4503.6	6021.4
N_1K	154.5	8601.7	-2977.8	-2909.5	-1053.9	662.1	1669.2	2384.5	4442.1	5624
P_1K	173.2	10558.4	-4589.5	-4312.9	-1905.9	-1053.3	997.7	1268.7	3870.2	5968.9
N_1P_3K	161.3	9204.4	-5360.6	-6230.0	-4570.5	-2975.5	-2055.9	-1138.4	1017.1	3843.8
N_2P_2K	166.6	9889.0	-3201.7	-3547.5	-1931.7	-562.5	524.0	2673.8	3676.8	6687.3
CK	146.8	8672.6	-3268.9	-2236.7	-158.3	798.5	1199.1	3061.5	3095	5403.7

表 13　各施肥处理年均净现值、内部收益率时间序列分析

处理	年均净现值(元/hm²)			内部收益率(%)		
	14 年	15 年	17 年	14 年	15 年	17 年
N_1	97.9	169.4	339.8	8.97	9.62	10.95
N_2	180.9	291.4	357.2	9.71	10.61	11.04
P_1	245.5	320.5	414.7	10.26	10.87	11.51
P_2	104.7	196.9	292.3	9.02	9.83	10.56
P_3	22.7	174.8	228.5	8.23	9.62	10.04
P_4	52.3	168.2	265.2	8.50	9.52	10.23
K	1.6	107.1	192.9	8.02	9.05	9.79
N_1P_1	292.4	300.2	354.2	10.63	10.70	11.07
N_1K	170.3	296.1	330.8	9.61	10.65	10.87
P_1K	90.6	258.0	351.1	8.88	10.33	10.99
N_1P_3K	-81.3	67.8	226.1	7.17	8.65	10.09
N_2P_2K	191.0	245.1	393.4	9.75	10.19	11.23
CK	218.7	206.3	317.9	10.07	9.97	10.85

4　结论与讨论

马尾松中、幼林不同配比(不同时间)施肥处理的经济分析结果均表明：不同的施肥处理方式，经济收益差异较大；就单肥种而言，以施磷肥的经济收益最优；混合肥种以氮、磷肥混合效益最佳。幼林磷肥以均分成 2 次施用最佳。马尾松人工林前 17 年生(施肥 8 年后)各处理的投资年均净现值、内部收益率均大致随林龄的增加而增加，各处理林分的经济成熟林龄应该在 17 年之后。在营林施肥投资中增产歉收较为常见，故选择合适的肥种、肥量、施肥时间，才能获得较高的经济收益。

与不施肥对比，马尾松中(幼)龄林部分施肥处理的经济收益在施肥8年(12年)后较大，若以培育中、小径级马尾松工业用材为经营目的，与不施肥的相比，这些施肥处理的经济收益是显著的，但施肥肥效产生的经济效益在后续的年份是否还在延续，这还有待下一步的深入研究。

参考文献

[1]吴立潮，胡曰利，吴晓芙，等．杉木中龄林施肥效应与效益研究[J]．中南林学院学报，1997，17(3)：1-7.

[2]吴晓芙，胡曰利，吴立潮，等．林木施肥的有效立地指数区间与目标肥效[J]．中南林学院学报，1997，17(1)：1-6.

[3]卢立华，蔡道雄，何日明．马尾松幼林林施肥效益综合分析[J]．林业科学，2004，40(4)：99-105.

[4]梁瑞龙，温恒辉．广西大青山马尾松人工林施肥研究[J]．林业科学研究，1992，5(1)：111-115.

[5]谌红辉，温恒辉．马尾松中龄林平衡施肥研究[J]．广西林业科学，2005，34(4)：170-174.

[6]谌红辉，温恒辉．马尾松人工幼林施肥肥效与增益持续性研究[J]．林业科学研究，2000，13(6)：652-658.

[7]谌红辉，温恒辉．马尾松人工中龄林施肥肥效与增益持续性研究[J]．林业科学研究，2001，14(5)：533-539.

[8]梁瑞龙，蒙福祥．广西马尾松人工林施肥效应研究[J]．广西林业科学，1996，25(1)：32-37.

[9]唐继新，韦中绵，韦善华，等．马占相思人工林经济效益评价：以广西南宁地区为例[J]．林业科学研究，2009，22(6)：855-859.

[10]杨锦昌，尹光天，李荣生，等．采收方式对马尾松与黄藤间种林分经济效益的影响[J]．林业科学，2007，43(11)：50-56.

[11]齐新民，丁贵杰，王德炉，等．马尾松纸浆用材林不同培育技术措施经济效益分析[J]．浙江林业科技，2001，21(3)：69-73.

[12]汪海粟．资产评估[M]．北京：高等教育出版社，2003：4.

[13]周慧珍．投资项目评估[M]．第2版．大连：东北财经大学出版社，2000：3.

[原载：林业经济问题，2010，30(05)]

广西大青山米老排人工林经济效益分析

白灵海 唐继新 明安刚 蔡道雄

（中国林业科学研究院热带林业实验中心，广西凭祥 532600）

摘 要 本文基于3块标准样地的调查和7株解析木数据，通过收集当地有关技术经济指标，采用重置成本法和选用年均利润、净现值及内部收益率3个经济分析指标，对广西凭祥市热林中心伏波实验场28年生米老排人工林的经济效益进行评价，结果表明：随着林龄的增加，年均利润逐年增大，第28年投资的利润和年均利润分别高达155366元/hm^{2}和5549元/hm^2；第15年时投资开始盈利，净现值总体上呈先增加后递减趋势，并在第23年达到峰值；内部收益率先不断上升，在第18年达到峰值后开始递减；以最大净现值为判断依据，确定米老排人工林经济成熟龄为第23年，此时年均利润、净现值分别达4953、11392元/hm^2，内部收益率为11.4%，表明经营米老排人工林可获得较高的经济收益。

关键词 米老排；经济效益；树干解析；净现值；内部收益率

Economic Benifit Analysis of 28-year-old *Mytilaria laosensis* Plantations in Daqingshan, Guangxi of China

BAI Linghai, TANG Jixin, MING Angang, CAI Daoxiong

(*Experimental Centerof Tropical Forestry*, *Chinese Academy of Forestry*, *Pingxiang* 532600, *Guangxi*, *China*)

Abstract: In this paper, on the basis of the survey of 3 standard sample plots and the stem analysis of 7 trees, and collecting the local technical economic indicators, the replacement cost method was used and the economic indicators of average annual profit, net present value and internal rate of return were selected to evaluate the economic benefits of 28-year-old Mytilaria plantation in Fubo Experimental Field, Experimental Center of Tropical Forestry of Chinese Academy of Forestry, which locates at Pingxiang city of Guangxi Zhuang Autonomous Region. The results showed that: With the increase of forest age, the annual profits could increase every year, the average annual profit and revenue of the 28th year were respectively as high as 155366 yuan/hm^2 and 5549 yuan/hm^2; the investment became profitable in the 15th year, as a whole net present value increased at first then declined, the peak appeared in the 23rd year; the movement of internal revenue was similar to that of the net present, rising firstly and then declining, the peak appeared in the 18th year; based on the maximum net present basis for determining, the economic maturity age of Mytilaria plantation was in the 23rd year, when the average annual profit, net present value and internal rate of return were respectively reached 4593 yuan/hm^2, 11392 yuan/hm^2, and 11.4%. This means the management of Mytilaria plantation can obtain higher economic return.

Key words: *Mytilaria laosensis*; economic benefit; stem analysis; net present value; internal rate of return

米老排(*Mytilaria laosensis* Lecomte)又名壳菜果、三角枫，为金缕梅科常绿乔木树种，天然分布于我国广东、广西和云南以及越南和老挝等地，具有速生、干形通直圆满、材质优良，兼具改良土壤、保持水土等优点，是建筑、家具、造纸和人造板的优质原料。鉴于该树种的优良特性及广泛用途，已有学者对其进行了育种、栽培、木材材性及利用开发等基础应用研究[1-6]；但有关该树种的投资收益状况未见报道，这在一定程度上影响了米老排树种的发展与推广。本文应用中国林业科学研究院热带林业实验中心(下文简称热林中心)28年生米老排人工林调查资料及解析木数据，对其人工林经营投资的经济效益进行了分析，旨在为该树种的经营与推广提供决策参考。

1 试验地概况

米老排试验地位于广西凭祥市热林中心伏波实验场(21°57′47″～22°19′27″N，106°39′50″～106°59′30″E)，林分郁闭度0.9，保存密度575株/hm²(初植密度1660株/hm²)；海拔600m，属低山丘陵；土壤为由岩浆岩发育而成的红壤，土层较厚，腐殖质5cm；气候属南亚热带季风区，年均气温20.5℃左右，≥10℃积温6500～7000℃，月平均气温≥22.0℃的有6个月，最热月平均气温27.5℃，最冷月平均气温12.0℃，极端最低气温-0.5℃；年降水量1400mm，年蒸发量1260mm；全年的日照时数1200～1300h。

2 研究方法

2.1 样地设置

在长势中等的米老排林分中，设置3块大小为20m×20m的临时标准地；在标准地内，以2cm为一径阶进行每木检尺，并在各径阶伐取1株标准木进行树干解析(两端径阶除外)。标准地内林木的调查汇总结果及解析木资料见表1。

2.2 相关技术经济指标

米老排人工林经营投入主要发生在20世纪80年代，相对现时的物价和经营投资成本变化巨大，为消除物价变动和通胀的影响，本文采用更新重置成本法[7,8]，以2010年热林中心的营林成本、木材采运成本、木材价格及目前广西各种税费的征收标准，作为米老排人工林经营投资分析的依据。

表1 标准地株数径级分布及解析木情况

径阶(cm)	株数(株)	解析木胸径(cm)	解析木树高(m)
16	1	/	/
18	4	18.9	22.3
20	6	20.7	22.0
22	10	22.6	22.1
24	14	23.6	26.5
26	17	26.6	24.3
28	11	28.7	27.9
30	4	30.4	28.3
32	2	/	/
合计	69	/	/

2.2.1 经营投资成本

主要包含以下5个方面：①炼山清山、挖穴、整地、苗木、基肥、定植等造林费用；②营林前期的幼林抚育费用；③营林前期病虫害防治费用；④护林防火、道路维护、地租和管理等年固定费用；⑤采运成本，主要包括伐区设计、采伐、造材、集材、运输、木材检尺、采伐(运输)证件办理和储木场销售管理等费用。米老排人工林经营成本的详细构成见表2。

表2 米老排人工林的经营成本

造林费用(元/hm²)	抚育费用(元/hm²)			病虫害防治(元/hm²)			年固定费用(元/hm²)	采运成本(元/hm²)
	第1年	第2年	第3年	第1年	第2年	第3年		
5435	750	750	750	150	150	150	950	120

2.2.2 木材价格及相关税费

米老排各规格材种价格按热林中心储木场2010年的售价计，木材的税费“两金一费”按各规格材售价的10%计，装车费、检疫费均为价外费用。米老排木材不同径级4～6、8～12、14～18、20～30cm的售价分别为：440、560、710、830元/m³。

2.2.3 经济分析指标的选择

选用税后年均利润、净现值、内部收益率[9]等经济分析指标对米老排人工林营林经济效益进行评价。

2.2.4 投资贴现率的确定

考虑到林业为集生态、社会与经济效益于一体的特殊性行业，各国政府一般均出台相关的产业政策予以扶持，同时参考我国近年的宏观经济形势，物价的变动巨大，CPI均接近或高于6%，故选取8%为投资贴现率。

2.3 林分各龄段出材量计算

根据标准地各解析木的数据，按国家原木材积公式进行造材[10]，并以4～6、8～12、14～18、20～30cm将各材种进行归类，按各径阶解析木(两端径阶出材以相邻径阶出材替代)在各龄段某一材种的出材乘上各径阶的每公顷株数，可计算出每公顷各龄段和各材种的理论出材量(表3)。考虑到林木材质缺陷(如弯曲、节子、扭曲、分杈、心腐、劈裂等)及实际造材中存在木材损失，故在进行各龄段经济收益的分析时，按米老排各龄段理论出材量的95%进行分析。

表 3 米老排各龄段材种理论出材量

林龄(年)	出材量(m^3)				
	规格材	≥20cm	14~18cm	8~12cm	4~6cm
28	319.98	188.85	91.78	34.16	5.19
27	302.61	162.26	103.76	31.78	4.81
26	281.40	136.43	115.03	25.08	4.86
25	265.80	128.78	102.33	29.24	5.45
24	250.76	111.11	102.39	32.34	4.92
23	243.86	101.21	106.46	30.18	6.01
22	224.32	75.56	110.75	32.49	5.52
21	210.33	64.76	104.69	35.33	5.55
20	194.72	60.56	93.89	34.61	5.66
19	180.50	43.65	91.13	39.60	6.12
18	165.50	40.05	80.48	39.69	5.28
17	147.95	18.45	84.81	39.83	4.86
16	132.70	18.45	77.68	30.83	5.74
15	123.56	0.00	80.38	38.35	4.83
14	106.16	0.00	68.29	30.09	7.78

注：林分第 7、15 年的间伐出材分别为 6.7、37.6m^3/hm^2，第 15 年理论出材量包含当年间伐材。

3 结果与分析

为科学分析米老排人工林的投资与收益状况，本文分别从财务分析、投资序列收益和敏感性变化的角度评价其经济效益。

3.1 财务分析

根据表 1~3 的基础数据，采用静态和动态结合的财务分析法，选用年均利润、内部收益率、净现值等经济指标，按 8%的贴现率，从造林年度开始对 28 年生的米老排人工林的经济效果进行了分析。经计算可知：在第 28 年采伐，米老排人工林的年均投资利润、净现值分别为 5549、9112 元/hm^2，内部收益率为 10.3%，其人工林的经营投资具有较强的盈利及偿还债务能力。

3.2 投资序列收益分析

在特定的木材价格市场下，人工林何时收获可使投资利润率最高、利润最大、风险最低、回收最快，均为林业经营、投资者最为关注的结果。本文依据米老排人工林的调查及解析木资料和相关技术经济指标，按 8%的贴现率对米老排人工林的投资序列收益进行分析，其计算结果见表 4。

表 4 米老排人工林投资序列收益

林龄(年)	利润(元/hm^2)	年均利润(元/hm^2)	净现值	内部收益率(%)
28	155366	5549	9112	10.3
27	145420	5386	9476	10.5
26	134100	5158	9551	10.6
25	125907	5036	9981	10.8
24	117359	4890	10270	11.0
23	113917	4953	11392	11.4
22	102419	4655	10938	11.5
21	94989	4523	11124	11.7
20	87882	4394	11255	11.9
19	79336	4176	10887	12.0
18	72700	4039	10816	12.2
17	63567	3739	9846	12.2

（续）

林龄（年）	利润（元/hm²）	年均利润（元/hm²）	净现值	内部收益率（%）
16	44151	2759	5322	10.7
15	34575	2305	3272	9.9
14	9992	714	-4298	4.5

由表4分析可知：若不考虑资本的时间价值，米老排人工林随着林龄的增加，其营林的静态利润（收益）逐年增大，28年生米老排人工林的最大投资利润和年均利润分别达155366元/hm^2和5549元/hm^2；而考虑资本的时间价值，从第15年起米老排人工林的投资开始盈利，净现值总体上呈先增加后递减趋势，在20~23年间产生一定波动；内部收益率先不断上升，达到峰值后开始递减。在现有的木材价格及生产经营成本下，米老排人工林经营投资的净现值和内部收益率的峰值分别形成于第23年和第18年，林分采伐的超前或滞后均导致投资收益的减少；与第23年相比，其他年份的净现值减少137元/hm^2以上，且采伐年份越晚投资收益损失就越多，最高损失可达2280元/hm^2。综合分析表明，在现有人工林经营投资成本与木材价格的条件下，米老排人工林的经济成熟龄为第23年。

3.3 投资敏感性分析

长周期的林业生产经营存在诸多不确定性因素，不同的经营措施（如栽培密度、间伐强度及施肥措施等）、利率水平及木材市场价格皆可引起人工林经营的成本（或产量）变化，进而影响经营投资收益的变化。本文分别选择了经营成本（含造林、营林、采运及税费等成本）、出材量、木材价格和贴现率（即利率）作为敏感性因素，分析各因素的变化对23年生米老排人工林净现值及内部收益率的影响，结果见表5。

敏感性分析结果表明：即使各敏感性因素降低或升高30%，净现值仍≥0；贴现率对净现值的影响最大，当贴现率下降30%时，净现值上升129.9%；价格因素对净现值的影响次之，当价格因素上升（或下降）30%时，净现值变化率达97.5%；净现值受经营成本和出材量的影响程度相当；相对于净现值而言，内部收益率受上述因素的影响变动则较小。综合分析表明：米老排人工林收益（净现值）虽受价格、经营成本、出材量和贴现率的影响较敏感，但其仍有较强的盈利与抗风险能力。

表5　米老排人工林投资敏感性分析

项目	因素变化率（%）	净现值（元/hm²）	净现值变化率（%）	内部收益率（%）	内部收益率变化率（%）
原经营措施	—	11392	—	11.4	—
价格变动	-30	290	97.5	8.1	28.9
	-15	5841	48.7	9.9	12.8
	15	16943	48.7	12.6	10.8
	30	22494	97.5	13.7	20.1
经营成本变动	-30	20310	78.3	14.8	29.6
	-15	15851	39.1	13.0	13.8
	15	6933	39.1	10.0	12.4
	30	2473	78.3	8.7	23.8
出材量变动	-30	2400	78.9	8.9	22.3
	-15	6896	39.5	10.2	10.2
	15	15888	39.5	12.4	8.8
	30	20384	78.9	13.3	16.6
贴现率变动	-30	26188	129.9	11.4	0
	-15	17837	56.6	11.4	0
	15	6408	43.7	11.4	0
	30	2548	77.6	11.4	0

4　结论与讨论

4.1　结论

米老排人工林的投资收益率高于行业的基准收益率 8%，具有较强的盈利与抗风险性能力；贴现率对其经济效益影响最大，价格因素次之，经营成本和产量影响最小。

在现有经营成本、木材价格及利率的条件下，米老排人工林的经济成熟龄为第 23 年，其人工林采伐的提前或延迟均会导致净现值的减少，特别是在现行限额采伐制度下，损失更多。

4.2　讨论

(1)本研究中，虽然 23 年生米老排人工林的净现值和内部收益率不太高，分别为 11392 元/hm^2 和 11.40%，但其收益是基于近年的经营成本(即以更新重置成本)计算而得，所得结果已完全剔除了造林初期至今的通胀及物价变动，故认为长周期的米老排人工林投资收益仍是较为可观的。

(2)区域的立地与气候因素均为影响林木生长的重要因素，本研究以广西热林中心的米老排人工林为研究对象，故研究所得仅能代表本区域米老排人工林的投资收益状况。由于缺乏国内木材市场的参考价，本研究米老排木材按一般杂木的价格计价；此外，从未来的经济社会发展和大径级优质阔叶材的市场供需分析，高品质的米老排木材[5]价格有较大上涨的趋势，因此，米老排人工林的实际投资收益会比本研究的结论好。

(3)以往敏感性分析研究中，相对而言，最为敏感的因素一般为价格，其次是成本和产量；而本研究结果与此不同，最为敏感的因素是贴现率，其次是价格，最不敏感的是成本。引起这一现象的原因，可能与本研究选择的成本计算方法(更新重置成本)和设置的投资贴现率有关。

(4)为了满足国内优质阔叶用材日益增长的消费需求，逐步减少进口国外原木的依赖，建议政府重新调整现有以米老排为代表的阔叶用材林的经济主伐年龄，根据不同的立地及利率制定不同的主伐年龄标准，最终使国内林业企业(或林场)投资经营优质阔叶用材林的利润实现最大化，并提高其投资经营优质阔叶用材林的积极性。

参考文献

[1]郭文福，蔡道雄，贾宏炎，等．米老排人工林生长规律的研究[J]．林业科学研究，2006，19(5)：585-589.

[2]李炎香，谭天泳，黄镜光，等．米老排造林密度初报[J]．林业科学研究，1988，2(1)：206-212.

[3]郭文福，黄镜光．米老排抚育间伐研究[J]．林业科学研究，1991，4(增刊)：76-81.

[4]梁善庆，罗建举．人工林米老排木材化学成分及其在树干高度上的变异[J]．中南林学院学报，2004，24(5)：28-31.

[5]梁善庆，罗建举．人工林米老排木材的物理力学性质[J]．中南林业科技大学学报，2007，27(5)：97-100.

[6]黄正暾，王顺峰，姜仪民，等．米老排的研究进展及其开发利用前景[J]．广西农业科学，2009，40(9)：1220-1223.

[7]汪海粟．资产评估[M]．北京：高等教育出版社，2003：4.

[8]唐继新，谌红辉，卢立华，等．马尾松中幼龄林不同施肥处理经济收益分析[J]．林业经济问题，2010，30(5)：390-396，401.

[9]黄和亮，吴景贤，许少洪，等．桉树工业原料林的投资经济效益与最佳经济轮伐期[J]．林业科学，2007，43(6)：128-133.

[10]杨锦昌，尹光天，李荣生，等．采收方式对马尾松与黄藤间种林分经济效益的影响[J]．林业科学，2007，43(11)：50-56.

[原载：林业科学研究，2011，24(06)]

造林密度对马尾松林分生长与效益的影响研究

谌红辉[1] 丁贵杰[2] 温恒辉[1] 陆 毅[1]
([1]中国林业科学研究院热带林业实验中心，广西凭祥 532600；
[2]贵州大学林学院，贵州贵阳 550025)

摘 要 用造林密度试验林21年的逐年观测资料，分析了造林密度对生长和经济效益的影响。结果表明：① A、B、C、D(1667、3333、5000、6667株/hm²)4种不同造林密度对林分生长、林分结构均有显著影响，其中林分胸径、单株材积，冠幅、冠高比随密度增大而减小，高径比随密度增大而增大，初始间伐期随密度增大而提前，造林密度对树高生长无显著影响。②造林密度对林分蓄积与出材量在前期呈正相关，但随林龄增长趋于相近。21年生林分A、B、C、D各密度的出材量分别为300.13m³/hm²、309.94m³/hm²、303.19m³/hm²、313.30m³/hm²。随着密度增大，小径阶株数百分比及小径材出材量所占的比例增大。③经出材量与效益核算，培育短周期工业用材林宜采用接近B处理的密度(2200~3300株/hm²)，培育大、中径材宜采用接近A处理的密度(1667~2200株/hm²)。

关键词 马尾松；造林密度；密度效应；经济分析；出材量

Study of Thinning Density Effects on Masson Pine Plantation

CHEN Honghui[1], DING Guijie[2], WEN Henghui[1], LU Yi[1]
([1]The Experimental Centre of Tropical Forestry, CAF, Pingxiang 532600, Gangxi, China;
[2]College of Forestry, Guizhou University, Guiyang 550025, Guizhou, China)

Abstract: The effects of planting density on growth and economic benefit of Masson Pine plantation were analyzed by using the data obtained year by year during 21 years from the experiment stand (including treatments of A, B, C, D, their planting density is 1667, 3333, 5000, 6667 trees/hm² respectively), the results showed: ① Planting density has significant effects on stand growth and stand structure. With the increase of planting density, diameter at breast height (DBH), single tree volume, crown diameter, ratio of crown length to tree height decreased, and first thinning time was earlier, but tree height /DBH increased . The planting density had no significant effect on tree height. ②With the increase of planting density, trees percentage of small logs and mill run of small timber increased, and stand total mill run increased before 11-year-old, but then it approximate with stand age growing. Mill run of A, B, C, D is 300.13 m³/hm², 309.94 m³/hm², 303.19 m³/hm², 313.30 m³/hm² respectively at 21-year-old . ③By a synthetic analysis of mill run and economic analysis, planting density of cultivating short rotation industry timber could be determined as 2200~3300 trees/hm², to cultivate large or middle diameter timber, it could be determined as 1667~2 200 trees/hm² .

Key words: *Pinus massoniana* ; planting density; density effects; economic analysis; mill run

马尾松(*Pinus massoniana*)是我国南方主要工业用材树种之一，广泛分布于16个省(自治区、直辖市)。造林密度和保留密度是否合理，直接关系到培育目标能否实现，直接影响经营者的经济效益。密度是影响速生丰产的关键技术之一，因此，密度调控一直是林业研究的难点和热点。许多学者开展了这方面的研究[1-14]。为了探明南亚热带栽培区适宜的马尾松造林密度，1989年我们在中国林业科学研究院热带林业实验中心伏波实验场设置了造林密度试验。根据21年的观测材料对不同造林密度的林分生长、材种出材量、经济效益做了定量的分析，为生产部门根据培育目标，选择相应的造林密度和不同时期的保留密度提供科学依据。

1 试验地概况

试验林设在广西凭祥市中国林业科学研究院热带林业实验中心伏波实验场，106°43′E，22°06′N，海拔500m，低山，年均温19.9℃，降水量1400mm，属南亚热带季风气候区，土壤为花岗岩发育成的红壤，土层厚100cm，马尾松立地指数18，前茬为杉木。

2 试验方法

试验地于1988年主伐杉木清理后，明火炼山，块状整地，1989年1月用1年生马尾松裸根苗定植。造林当年成活率99%以上。试验采用随机区组设计，设如下4个密度处理，即A、B、C、D(造林密度分别为1667株/hm^2、3333株/hm^2、5000株/hm^2、6667株/hm^2)，4次重复，小区面积为20m×30m，实行定时(每年年底)、定株、定位观测记录。测定内容包括胸径、树高、冠幅、枝下高、林木生长状况等，其中胸径全测，其他因子每小区样本数不少于50株。用断面积平均求林分平均胸径，采用Richards曲线拟合胸径与树高关系，然后用林分平均胸径求算林分平均高。其他测树指标均采用实测法计算平均值。按广西马尾松二元材积公式求算单株材积，乘上径阶株数得径阶材积，累计各径阶材积得蓄积量。利用削度方程和原木材积公式计算材种出材量。

3 结果分析

各试验处理的观测资料见表1。

表1 不同造林密度林分生长过程

项目	处理	林龄(年)					
		5	8	10	16	19	21
树高(cm)	A	4.65	7.42	9.15	13.66	14.98	15.56
	B	4.40	7.43	9.21	13.57	14.74	15.47
	C	4.67	7.61	9.52	13.70	15.26	16.07
	D	4.71	7.72	9.43	13.78	15.11	16.12
优势高(m)	A	6.28	9.59	11.35	15.49	17.23	18.26
	B	6.00	9.21	11.10	15.19	16.40	17.94
	C	6.45	9.42	11.31	15.30	17.15	18.23
	D	6.21	9.43	11.08	15.38	16.99	18.50
胸径(cm)	A	7.41	12.62	14.78	18.99	21.27	22.17
	B	6.06	10.18	11.97	16.01	18.33	19.00
	C	6.00	9.30	10.92	14.20	16.70	17.52
	D	5.63	8.49	9.88	13.35	16.08	17.16
胸径变动系数	A	0.30	0.27	0.28	0.31	0.28	0.26
	B	0.36	0.36	0.34	0.38	0.31	0.29
	C	0.34	0.35	0.31	0.36	0.32	0.29
	D	0.33	0.35	0.34	0.39	0.35	0.31
单株材积(m^3)	A	0.0123	0.0501	0.0809	0.1842	0.2477	0.2764
	B	0.0079	0.0334	0.0549	0.1329	0.1844	0.2063
	C	0.0083	0.0290	0.0477	0.1076	0.1607	0.1841
	D	0.0073	0.0247	0.0391	0.0966	0.1484	0.1775
蓄积(m^3/hm^2)	A	20.67	82.50	131.17	252.50	297.00	318.67
	B	26.17	101.83	156.33	280.17	318.17	331.17
	C	41.50	125.83	182.50	290.67	314.83	319.17
	D	47.83	138.33	191.67	293.00	315.67	330.00
冠幅(m)	A	3.17	3.09	3.08	2.86	2.71	2.86
	B	2.69	2.66	2.68	2.30	2.47	2.59
	C	2.66	2.04	2.49	2.39	2.08	2.31
	D	2.48	1.93	2.26	2.11	2.19	2.44

(续)

项目	处理	林龄(年)					
		5	8	10	16	19	21
重叠度	A	1.26	1.19	1.19	0.88	0.69	0.74
	B	1.75	1.72	1.66	0.88	0.84	0.85
	C	2.60	1.54	1.99	1.21	0.69	0.75
	D	2.97	1.77	2.25	1.10	0.80	0.88
冠高比	A	0.88	0.73	0.56	0.39	0.29	0.23
	B	0.87	0.65	0.49	0.32	0.27	0.27
	C	0.82	0.61	0.47	0.31	0.23	0.25
	D	0.79	0.58	0.46	0.29	0.22	0.21
高径比	A	62.8	58.8	62.4	74.23	72.09	71.09
	B	72.6	74.1	78.3	90.98	83.71	83.15
	C	77.8	84.5	88.7	103.32	97.26	94.17
	D	83.7	93.4	97.6	114.65	101.31	96.78

3.1 不同造林密度对林木生长的影响

造林密度的选择是人工林培育的重要措施，对林分在不同时期的林木种群数量有决定性作用，从而影响林分结构与生产力，直接影响培育目标能否实现及经营者的经济效益。为了比较不同造林密度的生长差异情况，我们根据逐年调查资料，进行了方差分析(表2)。

3.1.1 不同造林密度对树高生长的影响

密度对林分平均高的影响比较复杂，结论也不一。有些研究表明密度对树高生长有影响，但影响较弱，在相当宽的一个中等密度范围内无显著影响[12-14]。从表1的树高生长资料与表2的方差分析可知，密度对马尾松林分的平均高与优势高生长无显著影响。21年生时A、B、C、D密度处理的平均树高分别为15.56m、15.47m、16.07m、16.12m。

3.1.2 不同造林密度对胸径生长的影响

直径是密度对产量效应的基础，同时直径又是材种规格的重要指标，密度对直径的影响显著相关，这一点林学界普遍认同[1-2,12-13]。本试验研究表明，林分郁闭后胸径生长量随密度增大而减小，而且呈显著差异。由表2可知，不同造林密度的胸径生长差异随林龄增大而增大，差异主要表现在A与B、C、D及B与D之间。由于5年前林分尚未充分郁闭，处于个体生长阶段，密度对胸径生长影响不显著，5年后开始，林分充分郁闭，密度大的林分个体营养空间小，不利于林木种群的所有个体生长发育。因此，5年生开始，不同密度间胸径生长量一直表现出极显著差异。统计量F值8年生开始迅速上升，在16年生左右达峰值，然后逐渐下降，说明8~19年处于胸径生长差异高峰期。

胸径变动系数是反映林分分化与离散程度的重要指标。同林龄的林分胸径变动数随密度的增大而增大，同一密度处理的胸径变动系数随林龄的增大而呈规律性变化。林分5~10年生胸径变动系数由大逐渐变小，11~16年生变动系数逐渐增大，在16年生左右达峰值，17~21年生变动系数又逐渐变小。说明林木个体竞争经历从平缓到剧烈，又从剧烈到平缓的发展过程。由表2的方差分析可知，7年生时开始呈现出显著差异，在16年生左右到达峰值。16年生前差异主要表现在A与B、D之间，B、C、D之间差异不显著，16年生后主要表现在A与C、D之间。

表2 不同造林密度试验方差分析结果

项目	方差分析	林龄(年)												
		2	3	4	5	6	7	8	9	10	11	16	19	21
树高	F值	1.50	1.57	1.38	0.71	0.61	0.78	0.34	0.35	0.32	0.29	0.07	0.19	0.98
优势高	F值	0.13	1.26	1.33	1.06	0.41	1.89	0.82	0.04	0.36	0.34	0.12	1.09	0.42

（续）

项目	方差分析	林龄(年)												
		2	3	4	5	6	7	8	9	10	11	16	19	21
胸径	*F* 值	1	1.38	2.70	5.07*② ad	12.79** ab ac ad	22.45** ab ac ad bd	41.95** ab ac ad bd	56.88** ab ac ad bd	55.42** ab ac ad bd	68.13** ab ac ad bc bd	98.37** ab cd	46.84** ab ac ad bc bd	23.96** ab ac ad bd
胸径变动系数	*F* 值	1	0.52	0.85	2.10	2.27	3.80* ad	4.06* ab ad	4.07* ab ad	3.65* ab	6.02** ab ad	11.10** ab ac ad	8.49** ac ad	1.97
单株材积	*F* 值	1	1	2.31	3.23	6.19** ab ac ad	8.99** ab ac ad	12.11** ab ac ad	14.37** ab ac ad	17.08** ab ac ad	20.28** ab ac ad	22.49** ab ac ad bd	13.26** ab ac ad	12.38** ab ac ad
蓄积	*F* 值	1	1	8.55** ac ad bd	9.12** ac ad bd	8.87** ac ad bd	9.39** ac ad bd	10.60** ac ad bd	7.74** ac ad	7.61** ac ad	6.79** ac ad	3.24	3.23	0.534
冠幅	*F* 值	1	1	21.74** ab ac ad	5.79* ad	10.86** ac ad	16.16** ab ac ad	13.67** ac ad bd	4.40* ad	5.20* ad	34.12** ab ac ad bd	5.98** ab ad	6.91** ac ad	3.96* ac
重叠度	*F* 值	1	1	29.23** ac ad bc bd	13.38** ac ad bd	10.08** ad bd	8.23** ad bd	1.25	2.43	1.40	1.15	2.32	1.16	1.13
冠高比	*F* 值	1	1	13.90** ac ad bd	38.97** ac ad bc bd	13.64** ac ad	14.10** ab ac ad	11.91** ab ac ad	9.34** ab ac ad	4.99* ab ac ad	10.11** ab ac ad	3.56* ad	6.64** ac ad	1.15
高径比	*F* 值	1	1	7.46** ac ad	34.04** ab ac ad bd	131.92** ab ac ad bc bd cd	146.55** ab ac ad bc bd cd	122.39** ab ac ad bc bd cd	142.55** ab ac ad bc bd cd	126.61** ab ac ad bc bd cd	126.37** ab ac ad bc bd cd	86.14** ab cd	43.05** ab ac ad bc bd	32.36** ab ac ad bc bd

注：* 显著，** 极显著，ab，ac，ad，bc，bd，cd 表示两两间差异显著。

3.1.3　不同造林密度对材积生长的影响

立木的材积取决于胸径、树高、形数 3 个因子，密度对 3 因子均有一定的影响。分析各处理逐年生长资料(表 1)得知，同一密度级的单株材积随林龄的增大而增大，同林龄的单株材积随密度的增大而减小。方差分析表明，不同造林密度单株材积从第 6 年开始表现出极显著差异，统计量 *F* 值在 16 年生左右达峰值，然后逐渐变小，其差异主要表现在 A 与 B、C、D 之间。这种现象是由于 3~5 年生时林分刚郁闭，种群个体间竞争较小，密度对胸径和树高生长影响都很小，所以对单株材积生长影响尚不显著。6 年生时林分充分郁闭，密度间的胸径生长差异已十分明显，因而导致单株材积生长差异显著，但在 16 年生后，由于高密度林分已自然稀疏一部分林木个体，密度级差距变小，单株材积差异有所减小。

林分蓄积生长前期与密度呈正相关，与单株材积生长正好相反，但后期各密度间蓄积趋于相近。由表 2 方差分析可知，4 年生时蓄积生长开始表现出显著差异，4~8 年生时蓄积生长差异随林龄增长而加大，并在 8 年生时达到最大差异，D 密度蓄积生长量比 A 密度大 70.7%，该生长期内差异主要表现在 A 与 B、C、D 及 B 与 C、D 之间，9 年生时蓄积生长差异开始有所减小，但 11 年生时 D 密度蓄积生长量仍比 A 密度大 41.0%，差异主要表现在 A 与 B、C、D 之间，B、C、D 之间差异不显著，16 年生后各密度处理间的蓄积已无显著差异。这说明在营林生产中盲目加大造林密度不能提高蓄积量。

3.1.4　造林密度对冠幅生长的影响

许多研究表明树冠的大小和直径是紧密相关的[13]。从本研究逐年生长资料可知，林龄相同时冠幅随密度增大而减小，但随着林龄增长，16 年生后各密度的冠幅差异开始变小。同一密度级的冠幅有随林龄增大而增大的趋势，但在 6 年生开始，因林分充分郁闭，出现自然整枝，使冠幅大小出现一定的波动，但在 11 年生后各密度冠幅均有逐步变小并趋于稳定的趋势。经方差分析表明，各处理间冠幅生长差异显著，11 年生前主要表现在 A 与 B、C、D 及 B 与 C、D 之间，11 年生后主要表现在 A 与 C、D 之间。

由逐年生长资料可知，同一密度级林分在郁闭后(5年生后)，重叠度变化与冠幅一样有一定的波动，7年生前重叠度随密度增大而增大，7年后规律不明显。经方差分析发现，重叠度4~7年生时，不同造林密度间差异明显，7年后差异不显著，这可能是在一定的营养空间内，林分郁闭后，树冠面积总和趋向一定的饱和值。

3.1.5 造林密度对干形的影响

营造用材林时选择的密度应有利于自然整枝、干形通直饱满和较小的冠高比，林龄相同时冠高比愈小愈好。经资料分析发现，同一密度级林分随林龄增长冠高比减小，林龄相同时冠高比随密度增大而减小，但随林龄增长冠高比差异变小。经方差分析表明，冠高比在4年以后一直表现出显著差异，主要表现在A与B、C、D之间，11年生后冠高比差异开始变小，21年生时已无显著差异。

立木的高径比是林木的重要形质指标之一，与木材的质量与经济价值密切相关。林分郁闭后同一密度级的高径比随林龄增大而增大，林龄相同时，高径比随密度增大而增大，经方差分析表明，各处理间的高径比差异，两两间均显著。因此，对于工业用材我们应适当密植，降低树干尖削度。

3.1.6 不同造林密度的起始间伐期

确定林分直径和断面积连年生长量的变化能明显反映出林分的密度状况，因此，直径和断面积连年生长量的变化可以作为是否需要进行第一次间伐的指标[13]。本研究材料表明，初始间伐期随着密度增大而提前。从表4可知，C、D两种密度从第6年开始胸径与断面积的连年生长量已明显下降，所以在该立地条件下，若培育建筑材，可将第7~8年生定为C、D两种密度的初始间伐期，同样，B密度第7年时断面积连年生长量开始明显下降，可将第8~10年作为B密度的初始间伐期。A密度第11年开始胸径与断面积的连年生长量已明显下降，可将12年作为A密度的初始间伐期。

表4 不同造林密度的胸径与断面积连年生长量 (cm)

项目	处理	林龄(年)											
		3	4	5	6	7	8	9	10	11	12~16①	17~19	20~21
胸径	A	1.77	2.58	2.43	2.27	1.56	1.38	1.34	0.70	0.92	0.67	0.76	0.45
	B	1.60	2.22	2.02	1.69	1.34	0.94	1.09	0.64	0.82	0.67	0.77	0.33
	C	1.72	2.08	1.77	1.27	1.01	0.73	0.96	0.76	0.57	0.55	0.83	0.41
	D	1.70	1.99	1.54	1.13	0.89	0.61	0.77	0.63	0.69	0.58	0.91	0.54
断面积	A	4.21	14.95	23.63	30.45	25.62	25.85	27.96	15.73	21.84	18.21	24.02	15.35
	B	2.56	10.21	16.02	18.32	17.71	14.11	18.10	11.49	15.67	15.00	20.85	9.81
	C	3.48	10.42	14.21	13.23	12.33	9.91	14.30	12.35	9.86	11.03	20.21	11.01
	D	3.34	9.67	11.75	10.99	10.07	7.62	10.45	9.24	10.84	10.89	21.02	14.09

注：①平均生长量。

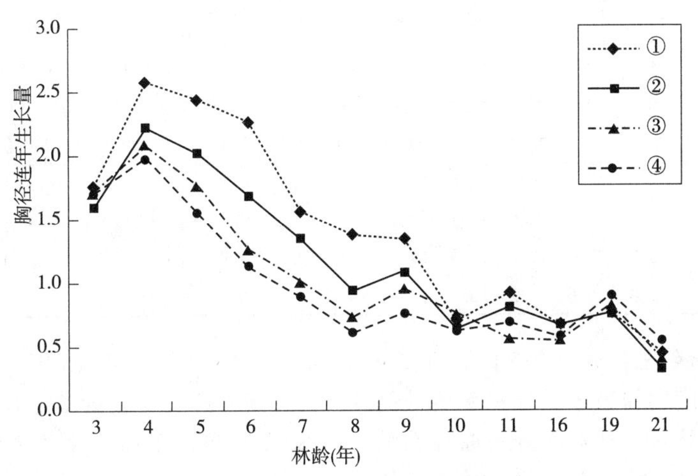

图1 不同造林密度的胸径连年生长量

①处理A；②处理B；③处理C；④处理D

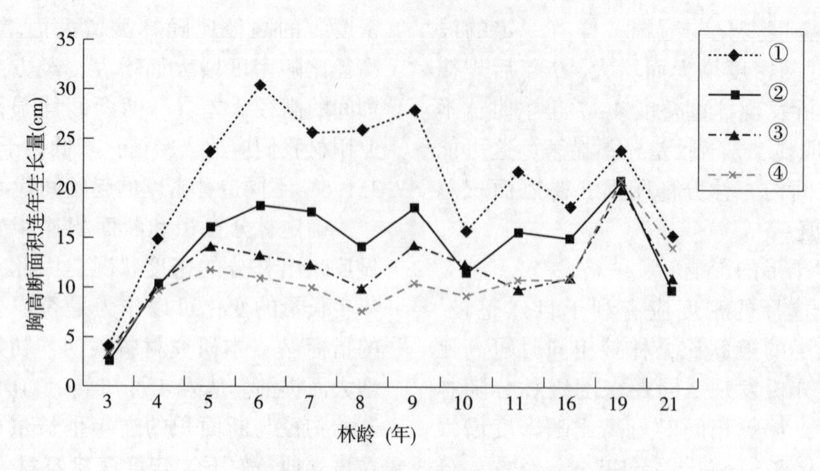

图 2 不同造林密度的胸高断面积连年生长量

①处理 A；②处理 B；③处理 C；④处理 D

3.2 造林密度对林分结构与出材量的影响

造林密度对林分生长各时期的保存密度有着决定性的影响，从而影响林分不同时期的自我调控过程与林分生产力，我们应选择适宜的造林密度以利于林分结构的优化与林分生产力的提高。

3.2.1 密度对径级株数分布的影响

因密度影响林分直径生长，自然会影响林分的直径结构规律，探讨密度对林分株数按直径分布的影响，对营林工作十分有益。各处理株数按径阶分布情况见表5。

表 5 不同密度的径级株数分布率

(%)

林龄(年)	处理	径阶(cm)																		
		2	4	6	8	10	12	14	16	18	20	22	24	26	28	30	32	34	36	38
5	A	5.4	8.7	30.4	33.7	17.5	4.3													
	B	10.2	24.6	37.9	19.9	6.6	0.6													
	C	9.8	25.4	35.6	24.7	4.1	0.4													
	D	10.8	29.1	40.5	17.4	2.2														
11	A		1.3	1.3	4.5	10.8	12.1	15.2	19.7	17.3	10.7	5.5	1.3	0.3						
	B	0.1	2.4	9.8	12.5	17.4	19.0	16.8	9.9	7.1	4.3	0.4	0.3							
	C		3.1	10.3	20.0	20.8	19.4	14.0	7.9	3.4	0.8	0.2	0.1							
	D	0.1	4.2	15.8	24.1	23.1	15.4	9.4	5.6	1.7	0.3	0.3								
16	A			0.3	1.2	4.3	7.6	12.8	14.0	13.4	15.2	12.8	10.4	5.8	1.2	0.3	0.6			
	B			1.4	5.2	10.9	14.3	17.0	15.6	14.5	9.5	6.1	3.4	1.2	0.6	0	0.4			
	C			1.5	8.9	17.4	21.5	17.3	15.0	7.6	6.3	2.6	1.2	0.3	0.3	0.2				
	D			3.7	16.2	19.9	20.4	14.4	10.5	7.4	3.3	3.1	0.8	0.1	0.1	0.1				
21	A					2.2	2.9	5.8	10.5	12.3	11.2	14.9	10.5	12.0	10.1	4.0	2.2	0.7	0.4	0.4
	B				1.0	5.9	6.2	14.1	13.1	18.8	11.3	11.8	8.0	5.9	0.8	2.1	0.3	0.5	0	0.3
	C				2.3	7.3	12.7	16.4	19.4	12.4	9.8	8.9	5.6	2.3	1.6	0.2	0.7	0	0.2	
	D				2.4	10.0	15.2	16.3	16.3	12.8	10.2	6.1	5.2	3.7	1.1	0.2	0.4	0.2		

由表5与同一密度径级株数分布率随林龄动态变化曲线图可知，同一密度的株数最大分布率所处的径阶值随林龄的增大而增大，株数率分布曲线的峭度值减小。说明幼林期林木个体分化小，林木个体主要集中分布在较小的径级范围内，随着林龄增大，林木个体分化加剧，林木个体分布的径级范围加大。A密度5年、11年、16年、21年时的林木主要分布径级范围扩大变化为6~10、10~20、14~24、16~28，B密度5年、11年、16年、21年时的林木主要分布径级范围的扩大变化为4~8、6~16、10~20、10~26。因此，在营林工作中应在中龄期个体分化加剧后，间伐部分小径级木，让林木集中分布在较少

的大径级范围内以提高木材径级规格。

由表5与不同造林密度径级株数分布率随林龄动态变化曲线图可知，不同造林密度林分的径级结构差异表现出幼林期小，中林期大，成熟林期小的变化规律。不同林龄阶段曲线峰值均随密度加大而左移，由近似常态的分布变为顶峰左偏的偏态分布，而且同龄曲线峭度随密度增大而增大。4种密度在5年、11年、16年、21年时的株数最大分布率所处的径阶范围依次为6~8、6~14、12~20、16~22。差异主要表现在A与B、C、D之间，其中B、C、D峰值之间仅相差1~2个径阶。21年生时，A、B、C、D 4种密度处理大于18径阶的株数率分别为66.4%、41.0%、29.3%、27.0%，14~18径阶的株数率分别为28.6%、46.0%、48.2%、45.4%，小于12径阶的株数率分别为5.0%、13.0%、22.5%、23.6%。可见密度对材种出材量影响很大，而且密度增大时林木直径集中分布在较少的径级范围内。因此，营林工作中应根据培育目标选择相应的造林密度。

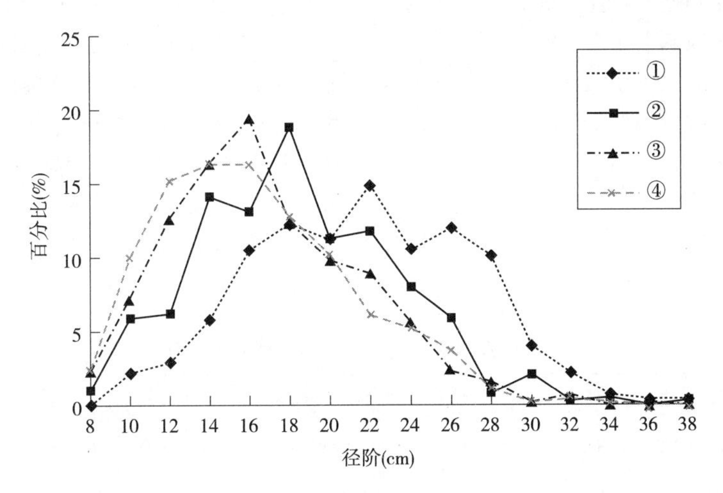

图3 不同造林密度径级株数分布曲线图(21年生)

①处理A；②处理B；③处理C；④处理D

3.2.2 造林密度对材种出材量的影响

计算出材量时采取密切联系生产的方法，以南方普遍采用的2m原木检尺长为造材标准，利用马尾松削度方程[1-3]求出从地面开始每上升2m处的去皮直径，即为该2m段的小头检尺径，先分径阶求出林分各径级规格材种出材量，再把各径阶径级规格相同的材积相加，得林分各径级规格材的材积，再把所有材种材积相加，得林分总出材量，见表6。

表6 不同造林密度林分的材种出材量

林龄(年)	处理	材种出材量(m³/hm²)				合计	出材率(%)
		薪炭材	小径材	中径材	大径材		
11	A	3.31	126.89	2.25	0.00	132.45	84.9
	B	5.30	146.57	0.65	0.00	152.52	85.7
	C	13.66	163.35	0.32	0.00	177.33	87.4
	D	18.85	171.76	0.00	0.00	190.61	87.4
16	A	3.29	172.76	52.52	2.59	231.16	91.7
	B	5.32	222.85	27.36	2.06	257.59	92.0
	C	9.33	244.86	14.77	0.51	269.47	92.9
	D	8.24	250.48	11.31	0.95	270.98	92.5
21	A	1.61	166.06	112.87	19.59	300.13	94.4
	B	3.07	227.45	68.56	10.86	309.94	93.6
	C	5.57	241.83	49.65	6.14	303.19	95.0
	D	6.94	258.60	42.71	5.07	313.32	95.0

注：材种规格：薪炭材检尺径<6cm；小径材6~18cm；中径材20~24cm；大径材≥26cm。

由表6可知，马尾松人工林中、幼林期总出材量随密度增大而增大，但随着林龄增长出材量逐渐接近。4种密度处理11年生时C与D处理相差不大，比A处理出材量分别高34.0%、43.0%，从造林投入考虑可以否定D密度。16年生时4种密度出材量已无显著差异。21年生时除A密度外，B、C、D密度材种仍以小径材为主，A、B、C、D各密度的出材量分别为 300.13m^3/hm^2、309.94m^3/hm^2、303.19m^3/hm^2、313.30m^3/hm^2。

综上所述，从发展的观点来看，培养短周期工业用材(如纸浆材、纤维原料林)造林密度可选取接近B处理的密度，即2200~3300株/hm^2，有利于缩短轮伐期，培育大、中径材造林密度宜选择A密度为参照，即1667~2200株/hm^2。

3.3 不同造林密度的效益评价

生产成本主要包括基本建设投资(苗木、林地清理、整地、栽植、林道、抚育费及10%的间接费)、经营成本(管护、管理)、木材生产(采伐、运输、归堆等)3大类，结合广西具体生产实践进行成本核算，见表7。产值计算以现行市场价格为准，折现率定为10%，将不同密度不同规格材种乘以相应的价格即得出产值(材种规格我们已计算至检尺径4cm，主要以经济用材进行效益核算)。

根据表7与不同造林密度产值随林龄变化曲线图可知，各处理的产值随林龄增长而增长，但随着林龄增长密度与产值的相关性由正相关转化为负相关性。11年生前C、D高密度处理的产值高于A、B低密度处理，11年生后各处理的产值开始接近，16年生时产值差异小于3.6%，且A、B低密度处理的产值开始高于C、D高密度处理，21年生时产值差异扩大至8.4%。21年生时A、B、C、D处理的产值分别为177412.10元/hm^2、173316.30元/hm^2、163627.60元/hm^2、167151.70元/hm^2。

表7 不同造林密度的效益评价(折现率：10%) (元/hm^2，m^3/hm^2,%)

林龄(年)	处理	基本建设投资	经营成本	木材产量	木材生产投资	总计投资	产值	净现值	内部收
11	A	5250.80	825.00	132.45	8789.50	14865.30	68788.70	15768.77	20.5
	B	6754.50	825.00	152.52	10102.30	17681.80	73627.30	15637.54	19.6
	C	8251.80	825.00	177.33	11751.35	20828.15	82327.80	16747.85	19.1
	D	9750.60	825.00	190.61	12566.45	23142.05	84867.80	15989.87	18.6
16	A	5250.80	1200.00	231.16	15025.40	21476.20	127909.00	19206.53	14.1
	B	6754.50	1200.00	257.59	16744.00	24698.50	132510.50	18466.93	13.5
	C	8251.80	1200.00	269.47	17515.55	26967.35	131967.60	16819.69	13.1
	D	9750.60	1200.00	270.98	17614.35	28564.95	131364.80	15304.45	13.0
21	A	5250.80	1575.00	300.13	19508.45	26334.25	177412.10	15915.50	10.8
	B	6754.50	1575.00	309.94	20146.10	28475.60	173316.30	13908.87	10.5
	C	8251.80	1575.00	303.19	19707.35	29534.15	163627.60	11297.74	10.2
	D	9750.60	1575.00	313.32	20364.50	31690.10	167151.70	10322.60	10.2

根据表7与不同造林密度净现值随林龄变化曲线图可知，随着林龄增长密度与净现值的相关性由正相关转化为负相关性。净现值峰值出现的时间随密度减小而推迟，越过峰值后均随林龄增长而下降。11年生前C、D高密度处理的净现值高于A、B低密度处理，16年生后造林密度与净现值呈规律性的负相关性，A、B低密度处理的净现值高于C、D高密度处理，16年生时A、B、C、D处理的净现值分别为19206.53元/hm^2、18466.93元/hm^2、16819.69元/hm^2、15304.45元/hm^2。高密度C、D处理的净现值峰值时间出现在16年生前，而低密度A、B处理的净现值峰值时间出现在16年生后。

根据表7与不同造林密度内部收益率随林龄变化曲线图可知，随着林龄增长各密度处理的内部收益率均呈下降趋势，并且各密度处理的内部收益率趋向接近。11～16 年生前下降较快，平均每年下降 1.1～1.3 个百分点，16 年生后内部收益率下降减缓，各密度处理的内部收益率趋向接近，19 年生内部收益率为 11.3%～12.0%，21 年生时为 10.2%～10.8%。所以，20 年生后采伐经济效益不太理想。

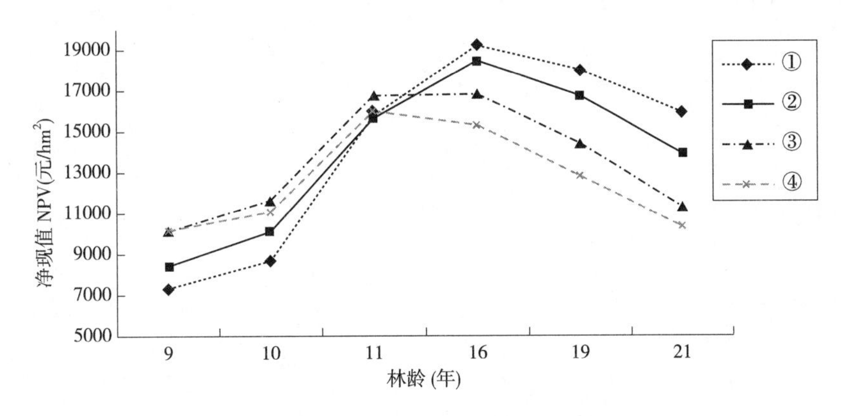

图 4　不同造林密度净现值随林龄变化图

①处理 A；②处理 B；③处理 C；④处理 D

综合林分生长状况与效益评价，高密度 C、D 处理的经济收获期宜在 16 年生左右，低密度的 A、B 处理的经济收获期宜在 19 年生左右。高密度造林很难提高出材量与经济效益。因此，在营林生产中应根据培育目标选择科学的造林密度。培育短周期工业用材林造林密度可选取接近 B 处理的密度，即 2200～3300 株/hm²，培育大、中径材宜选择 A 密度为参照，即 1667～2200 株/hm²，这样既可减少造林与间伐投资，又可尽快成材。

4　结论

经 21 年的观测资料表明，马尾松造林密度对林分生长的影响显著，随着密度增大，胸径、单株材积、冠幅生长量与冠高比减小，表现出与密度间的负相关；高径比随密度增大而增大，但造林密度对树高生长无显著影响。间伐初始期随密度增大而提前。

不同造林密度对林分蓄积与出材量在前期呈正相关，但随林龄增长趋于相近。密度对林分结构与材种出材量影响显著。随着密度增大，小径阶株数分布率、小径级材种出材量及所占比例增加，但大径阶株数分布率与大径阶材种出材量减少

综合林分生长状况与效益评价，高密度 C、D 处理的经济收获期宜在 16 年生左右，低密度的 A、B 处理的经济收获期宜在 19 年生左右。高密度造林很难提高出材量与经济效益。因此，在营林生产中应根据培育目标选择科学的造林密度。从发展的观点来看，培养短周期工业用材(如纸浆材、纤维原料林)造林密度可选取 2200～3300 株/hm²，有利于缩短轮伐期，培育大、中径材造林密度宜选择 1667～2200 株/hm²。

参考文献

[1]丁贵杰，周政贤．马尾松不同造林密度和不同利用方式经济效果分析[J]．南京林业大学学报，1996，20(2)：24-29.

[2]丁贵杰，周政贤，严仁发，等．造林密度对杉木生长进程及经济效果影响的研究[J]．林业科学，1997，33(Sp.1)：67-75.

[3]丁贵杰．马尾松人工林生长收获模型系统的研究[J]．林业科学，1997，33(Sp.1)：57-66.

[4]李景文，王义文，赵惠勋，等．森林生态学[M]．第 2 版．北京：中国林业出版社，1994：122-124.

[5]木材编写组．中华人民共和国国家标准—木材．北京：国家标准局，1984：64-67.

[6]童书振，盛炜彤，张建国．杉木林分密度效应研究[J]．林业科学研究，2002，15(1)：66-75.

[7]唐守正．同龄纯林自然稀疏规律的研究[J]．林业科学，1993，29(3)：234-241.

[8]曾德慧，姜凤岐，等．沙地樟子松人工林自然稀疏规律[J]．生态学报，2000，20(2)：235-242.

[9]张春锋，殷鸣放，孔祥文，等．不同间伐强度对人工阔叶红松林生长的影响[J]．辽宁林业科技，2007，33(1)：12-15.

[10]张金文．巨尾按大径材间伐试验研究[J]．林业科学研究，2008，21(4)：464-468.

[11]张水松，陈长发，等．杉木林间伐强度试验 20 年生长效应的研究[J]．林业科学，2005，41(5)：56-65.

[12]杉木造林密度试验协作组．杉木造林密度试验阶段报告[J]．林业科学，1994，30(5)：419-429.

[13]孙时轩，沈国舫，王九龄，等．造林学[M]．第2版．北京：中国林业出版社，1992：133-134.
[14]张彩琴，郝敦元，李海平．人工林林分密度最优控制策略的数学模型[J]．东北林业大学学报，2006，2(34)：24~27

[原载：林业科学研究，2011，24(04)]

柚木幼树侧芽萌发特性调查与分析

李运兴[1] 蔡道雄[1] 梁坤南[2] 周再知[2] 贾宏炎[1] 劳庆祥[1] 陆 毅[1] 窦福元[3]

([1]中国林业科学研究院热带林业实验中心，广西凭祥 532600；
[2]中国林业科学研究院热带林业研究所，广东广州 510520；
[3]玉林市林业种苗管理站，广西玉林 537000)

摘 要 春季柚木1~3年生幼树在主梢萌动抽生之前，在基部以上3m范围内侧芽大量抽生，为掌握柚木幼树侧芽萌发特性与了解相关影响因子，对柚木无性系测定林、施肥试验林、实生起源同龄混交林幼树开展了侧芽萌发情况调查，对不同重复不同处理不同地块柚木侧芽萌发特性和相关因子进行统计分析，结果显示：柚木抽生侧芽特性受遗传、施肥措施和立地因素的影响，其结论对柚木立地质量评价、无性系品种选择和柚木的集约经营有较大的启示作用。

关键词 柚木；幼树；侧芽萌发；无性系；施肥；混交林；立地

Survey and Analysis on Lateral Bud Germination Characteristics of *Tectona grandis* L. f. Saplings

LI Yunxing[1], CAI Daoxiong[2], LIANG Kunnan[2], ZHOU Zaizhi[2], JIA Hongyan[1]
LAO Qingxiang[1], LU Yi[1], DOU Fuyuan[3]

([1]*The Experimental Centre of Tropical Forestry*, *Chinese Academy of Forestry*, *Pingxing* 532600, *Guangxi*, *China*;
[2]*Research Institute of Tropical Forestry* , *Chinese Academy of Forestry*, *Guangzhou* 510520, *Guangdong*, *China*;
[3]*Yulin Forestry Seedling Management Station*, *Yulin*, 53700 *Guangxi*, *China*)

Abstract: In spring, lateral buds of *Tectona grandis* saplings in 1~3 years germinate quickly at base more than 3m high before main treetop germinating. Survey on bud germination was made in clone test stands, fertilization test stands and same-aged seedling mixed saplings, statistic analysis was made on lateral bud germination characteristics and relative impact factors of *Tectona grandis* in different sites, in different repeated clone tests and in different treatments. The results showed that: Lateral bud germination characteristics were affected by heredity, fertilization and site. This conclusion may play an enlightening role on site quality evaluation, clone species selection and intensive management.

Key words: *Tectona grandis* ; sapling; lateral bud germination; clone; fertilization; mixed stands; site

柚木(*Tectona grandis* L. f.)属马鞭草科落叶或半落叶大乔木，树高40m，胸径1~2m，干形通直。木材为环孔材，心材比例大，淡褐色，比重0.6~0.72g/cm^3，纹理直，结构细致而美观，坚韧而有弹性，不翘不裂；含油质，强度大，耐浸，是航海、军需、建筑、车厢、家具、雕刻、铸造木模、贴面板等珍贵优良用材树种之一[1,2]。原产于印度、缅甸、泰国和印度尼西亚，我国引种于广东、广西、海南、云南、福建、台湾等地，已有170余年历史；当前可用木材资源锐减，社会需求增加，交易价格攀升，造林积极性不断提高，发展势头大。由于柚木天然资源分布和对气候土壤适应性差等方面原因，已往柚木方面的研究侧重于种质资源引进和种源家系选育方面[4-9]，栽培方面的研究[10-13]尚不能满足生产实际需要，培育技术的缺乏已制约着柚木的规模发展；春季柚木1~3年生幼树在主梢萌动抽生之前，在基部以上3m范围内大量抽生，为掌握柚木幼树侧芽萌发特性与了解相关影响因子，对部分幼林开展了调查统计分析，其结论对柚木立地质量评价、无性系品种选择和柚木的集约经营有较大的启示作用。

1 调查对象和方法

1.1 调查对象

调查林分有无性系测定林、多因素施肥试验林、一般柚木格木混交林3种，种植于距凭祥市20km的青山实验场茶陋林区，属平而河河畔低山，海拔150~230m，可及度较高，地形较开阔，阳光充足，年平均气温为21.5℃，≥10℃积温7500℃，月平均气温≥22℃有7个月，最热月(7月)平均气温28.3℃，极端最高温40.3℃，最冷月均气温13.5℃，极端最低温-1.5℃，年降水量1220~1380mm，土壤为砂泥页岩发育的黄赤红壤，质地为黏壤土或粉质黏土，0~40cm土壤混合样本，有机质12.4~20.3g/kg，pH(H_2O)4.6~5.2，全N值1.03~1.51g/kg、全P_2O_5值0.62~0.81g/kg、全K_2O值17.88~23.33g/kg，盐基饱和度26.5%~43.6%，为杉木迹地，主要植被有山芝麻(*Helicteres angustifolia* L.)、茅草(*Imperata cylindrica* Beauv.)、五节芒(*Miscanthus floridulus* Warb.)、大青(*Clerodendron oyrtophyllum* Turcz.)等，株行距3m×3m，初植密度为1125株/hm^2，穴规格60cm×60cm×40cm(长×宽×深)，造林时间：无性系测定林和一般柚木格木混交林为2008年4月、多因素施肥试验林为2009年4月；除施肥试验外林分采取高标准集约方式经营，柚木施1.0kg磷肥作基肥和在第1~2年各追施含N、P、K 0.5kg复混肥，其他树种不施肥，造林后头2年4月和8月除杂灌2次。

(1)无性系测定林。试验林布设于中上山坡，参试无性系15个、实生子代1个，共16个试验处理，单列小区，每列6株，随机区组设计，6重复，重复间上下坡品字形排布，初植密度为1125株/hm^2(其他林分相同)，林木保存率：96.2%，2年生平均树高3.71m，胸径3.36cm。

(2)多因素施肥试验林。施肥处理(小区)有8个(表1)，每处理5(横)×7(纵)=35株，3次重复，排布成品字形，按随机区组试验设计实施，林木保存率98.6%，2年生平均树高3.07m，胸径2.44cm。

表1 不同处理肥料配比情况

处理	基肥(g)			第2年追肥(g)			
1	钙镁磷/1846	石灰/692.3	沸石/461.5	尿素/62.5	钙镁磷/250.0	绿化钾/94.0	沸石/94.0
2	钙镁磷/1846	石灰/692.3	麦饭石/461.5	尿素/62.6	钙镁磷/250.0	绿化钾/94.0	麦饭石 94.0
3	复合肥/3000				复合肥/500.0		
4	钙镁磷/3000				复合肥/500.0		
5	鸡粪/3000				复合肥/500.0		
6	钙镁磷/1846	石灰/692.3	沸石/461.5		复合肥/500.0		
7	钙镁磷/1846	石灰/692.3	麦饭石/461.5		复合肥/500.0		
8	钙镁磷/1846	鸡粪/692.3	沸石/461.5		复合肥/500.0		

(3)一般柚木格木混交林。采取列间混交方式进行(柚木1列+格木1列)，初植密度为1125株/hm^2，2年生林木保存率97.7%，柚木平均高3.02m，平均胸径2.74cm，格木平均高1.80m，平均胸径2.16cm。

1.2 调查统计方法

2010年3月与2011年4月，无性系测定林和多因素施肥试验林按试验设计株数调查；混交林分别在不同地点的中坡设置3块样地进行调查，每块样地对50株柚木的抽生侧芽数量进行调查；长3cm以下芽不计数，抽生侧芽率=抽生侧芽株数/调查总株数，侧芽抽生程度分级：1级(少量)抽生侧芽5个以下；2级(中等)抽生侧芽多于5个，但在10个以下；3级(较多)抽生侧芽多于10个；抽芽指数=Σ(某级侧芽抽生株数×相应侧芽抽生程度分级数)/调查总株数×最高侧芽抽生程度分级数(3级)；用SPSS 17.0统计软件进行均值统计、方差分析和多重比较。

2 结果与分析

2.1 无性系试验

2.1.1 重复间侧芽抽生情况差异

无性系试验重复间侧芽抽生情况见表2，6重复少量抽芽株数比例都在13.64%以上，极差为

25.94%，一般抽芽株数比例为33.84%~46.88%，极差为13.04%，大量抽芽株数比例在13.54%~48.31%间变化，极差为34.77%；各重复各级抽芽比例分布趋势不一：重复1和重复5大量抽芽株数比例偏大；抽生侧芽率第1重复最大，第3重复最小，由下坡向上坡有一定递减趋势，重复间差异显著（$P<0.05$）；平均侧芽数也以第1重复最大，第3重复最小，由下坡向上坡同样有一定递减趋势，重复间差异显著（$P<0.05$）；抽芽指数第1重复最大，第3重复最小，由下坡向上坡同样有一定递减趋势，重复间差异显著（$P<0.05$）。

表2　无性系试验重复间侧芽抽生情况

重复	各级抽芽比例（%）			均值			抽芽指数	抽生侧芽率（%）
	1级	2级	3级	芽数（个）	树高（m）	胸径（cm）		
1	13.64	40.91	45.45	9.63	3.82	3.48	0.773	100.00
2	30.21	45.83	23.96	7.22	3.74	3.23	0.646	96.88
3	39.58	46.88	13.54	6.23	3.49	3.14	0.580	84.38
4	36.46	45.83	17.71	7.06	3.78	3.32	0.604	89.58
5	17.85	33.84	48.31	8.43	3.79	3.40	0.7682	98.17
6	36.50	46.34	17.17	7.60	3.62	3.27	0.6023	91.55

表3　各无性系侧芽抽生情况统计

无性系/有性子代	各级抽芽比例（%）			均值			抽芽指数	侧芽抽生率（%）
	1级	2级	3级	芽数（个）	树高（m）	胸径（cm）		
1	16.67	45.83	37.50	7.39	4.20	3.84	0.736	95.83
2	34.78	39.13	26.09	8.64	3.26	3.01	0.638	95.65
3	17.39	52.17	30.43	10.14	2.77	2.58	0.710	91.38
4	25.00	41.67	33.33	11.96	3.80	3.63	0.694	100.00
5	29.17	54.17	16.67	7.88	2.47	2.23	0.625	100.00
6	21.74	34.78	43.48	8.22	3.71	3.08	0.739	95.65
7	21.43	57.14	21.43	5.43	3.09	2.66	0.667	56.52
8	25.00	54.17	20.83	6.71	3.07	2.70	0.653	100.00
9	20.83	58.33	20.83	7.46	3.12	2.76	0.667	100.00
10	39.13	47.83	13.04	5.26	3.62	3.29	0.580	95.83
11	17.39	56.52	26.09	7.04	3.36	3.05	0.696	95.83
12	25.00	58.33	16.67	5.25	3.82	3.34	0.639	100.00
13	18.75	56.25	25.00	7.30	3.03	2.69	0.688	91.67
14	19.05	19.05	61.90	5.20	2.84	2.50	0.810	90.91
15	40.91	40.91	18.18	7.94	2.95	2.67	0.591	77.27
16	40.91	50.00	9.09	8.18	3.42	3.06	0.561	91.67

2.1.2　无性系间侧芽抽生情况差异

经统计，各无性系各级抽芽比例、侧芽抽生率、单株平均芽数及抽芽指数如表3所列，各无性系少量抽芽株数比例平均值为25.82%，一般抽芽株数比例平均值为47.89%，大量抽芽株数比例平均值为26.29%，可见多数无性系大部分植株属于中度抽芽，大量抽芽和少量抽芽植株都较少，而6、14号无性系重度抽芽株数最多，与其他无性系间区别明显；侧芽抽生率达100.00%的有4、5、8、9、12无性系，抽生率最低的为7号无性系仅为 56.52%，无性系间差异极显著($P<0.01$)；抽芽数最多的无性系为 4 号达 11.96 个，最少的无性系 14 仅为 5.20 个，无性间差异极显著($P<0.01$)；抽芽指数最大的为 14 号无性系 0.810，最小的为 16 号 0.561(实生子代)，无性系间差异极显著($P<0.01$)。从侧芽抽生率、抽芽数和抽芽指数的概念和定义可知，抽芽指数是一个最能反映抽芽严重程度的指标，显然 6 号、14 号是属于抽生侧芽最严重无性系。

2.1.3 高径对抽生侧芽的影响

表 4 为无性系测定试验按不同树高、胸径范围归类统计结果，抽生侧芽率：树高≤2.00m 或胸径≤2.0cm 最少，分别为 50.0% 和 60.12%，树高>3.50m 或胸径>3.5cm 最大，分别为 100.00% 和 99.47%，随树高、胸径的增加而增加趋势明显；芽数均值：树高≤2.00m 或胸径≤2.0cm 最少，分别为 4.5 个和 5.2 个，树高>3.50m 或胸径>3.5cm 最大，分别为 10.9 个和 11.7 个，随树高、胸径的增加而增加趋势也明显；抽芽指数：树高≤2.00m 或胸径≤2.0cm 最少，分别为 0.500 和 0.465，树高>3.50m 或胸径>3.5cm 最大，分别为 0.844 和 0.858，随树高、胸径的增加而增加趋势同样显著。

2.2 施肥试验

2.2.1 重复间侧芽抽生情况差异

施肥试验重复间侧芽抽生情况见表 5，统计显现，施肥试验林抽生侧芽率平均值为：90.54%，且以第 1 重复为最大达 94.82%，第 3 重复最小为 86.74%，重复间差异不显著($P>0.05$)；单株平均抽生侧芽 6.38 个，也以第 1 重复为最多达 7.09 个，第 3 重复最少为 5.00 个，重复间差异不显著($P>0.05$)；抽芽指数：3 重复平均抽芽指数为：0.562，以第 2 重复为最多达 0.624，第 3 重复最少 0.474，重复间差异不显著($P>0.05$)。

2.2.2 施肥处理间侧芽抽生情况差异

经统计，各施肥处理各级抽芽比例及抽芽指数如表 6 所列，统计发现，施肥各处理抽生侧芽率平均值为 91.15%，且以第 1 处理最高都为 97.78%，第 3 处理最低为 75.27%，但处理间差异不显著($P>0.05$)；抽生侧芽数，各处理平均 6.53 个，以 2 处理最高为 8.33 个，5 处理最低为 4.62 个，处理间差异不显著($P>0.05$)；抽芽指数平均值为 0.562，且以第 4 处理最高为 0.660，第 5 处理最低为 0.474，处理间差异不显著($P>0.05$)。

表 4 不同树树高/胸径范围的抽芽株数比例、抽生侧芽率、平均芽数、平均指数

变量范围	各级抽芽比例(%)			芽数(个)	抽芽指数	抽生侧芽率(%)
	1 级	2 级	3 级			
树高≤2.00(m)	50.0	50.0	0.0	4.5	0.500	50.00
2.00<胸径≤3.50(m)	17.54	43.86	38.60	9.0	0.737	98.25
树高>3.50(m)	6.67	33.33	60.00	10.9	0.844	100.00
胸径≤2.0(cm)	60.55	39.45	0.00	5.2	0.465	60.12
2.0<胸径≤3.5(cm)	15.77	50.03	34.20	8.1	0.728	97.33
胸径>3.5(cm)	1.31	40.06	58.68	11.7	0.858	99.47

表 5 施肥试验重复间侧芽抽生情况统计

重复	各级抽芽比例(%)			均值			抽芽指数	抽生侧芽率(%)
	1 级	2 级	3 级	芽数(个)	树高(m)	胸径(cm)		
1	45.90	32.21	21.89	7.09	2.93	2.29	0.587	94.82
2	43.70	25.47	30.83	7.04	3.55	2.76	0.624	90.07
3	66.99	23.85	9.17	5.00	2.72	2.26	0.474	86.74

表 6 各施肥处理侧芽抽生情况

处理	各级抽芽比例(%)			均值			抽芽指数	抽生侧芽率 (%)
	1 级	2 级	3 级	芽数(个)	树高(m)	胸径(cm)		
1	53.33	31.11	15.56	6.40	3.48	2.75	0.541	97.78

（续）

处理	各级抽芽比例(%)			均值			抽芽指数	抽生侧芽率（%）
	1级	2级	3级	芽数(个)	树高(m)	胸径(cm)		
2	27.80	47.45	24.75	8.33	2.67	2.11	0.656	95.40
3	73.62	7.33	19.05	4.79	1.90	1.56	0.485	75.27
4	37.14	27.78	35.08	8.08	2.91	2.25	0.660	90.63
5	62.22	33.33	4.44	4.62	3.60	2.84	0.474	86.67
6	48.34	20.82	30.84	7.53	3.33	2.58	0.608	95.05
7	67.31	19.91	12.78	4.91	3.65	2.98	0.485	92.99
8	47.78	29.68	22.54	7.61	2.99	2.41	0.583	95.40

2.2.3　高、径对抽生侧芽的影响

表7为施肥试验按不同树高、胸径范围归类统计结果，抽生侧芽率：树高≤2.50m或胸径≤2.0cm最少，分别为90.0%和95.7%，树高>3.50m或胸径>2.5cm最大，分别为95.0%和100.0%，随树高的增加而增加趋势明显，其变化趋势与胸径变化规律不一致；芽数均值：树高≤2.50m或胸径≤2.0cm最少，分别为4.19个和5.96个，树高>3.50m或胸径>2.5cm最大，分别为7.37个和7.96个，随树高胸径的增加而增加趋势明显；抽芽指数：树高≤2.50m或2.0<胸径≤2.5cm最少，分别为0.414和0.435，树高>3.50m或胸径>2.5cm最大，分别为0.534和0.547，随树高的增加而增加趋势明显，而与胸径走势不明显。

2.2.4　不同肥种对抽生侧芽抽生的影响

植物对N、P、K等元素都有需求，但N、P、K等对植物不同组织的促进作用是不同的，N主要是促进叶和主梢的萌发和生长，P主要是增进根系的生长，K主要对径生长有极大的促进作用；表8为不同施肥处理肥料组成与抽生侧芽情况统计表，将表中抽芽率、芽数、指数与N、P、K的施用量进行相关分析统计，得到抽芽率与N、P、K的相关系数分别为：-0.959＊＊（“＊＊”表示极显著相关，“＊”表示显著相关，下同）、-0.256、-0.930＊＊；抽芽数与N、P、K的相关系数分别为-0.581、0.330、-0.523；抽芽指数与N、P、K的相关系数分别为-0.533、0.390、-0.476；结果表明抽芽率与N、K间有极显著的负相关关系，抽芽数、抽芽指数与N、P、K的相关关系不显著；为分析基肥和追肥施用化学肥料和土壤改良剂对侧芽抽生的影响，把8处理按是否施用钙镁磷、复合肥、鸡粪、沸石、麦饭石、生石灰的不同，进行归类方差分析统计，结果为：在基肥中施用不同化学肥料的抽芽率差异极显著（$P<0.01$）：施钙镁磷94.37%或鸡粪86.67%显著高于施用复合肥75.23%；在基肥和追肥中是否施用沸石、麦饭石、生石灰的抽芽率、抽芽数、抽芽指数差异不显著（$P>0.05$）。

2.3　实生起源柚木格木混交林

2.3.1　不同地块林木侧芽抽生情况分析

3个地点柚木侧芽抽生情况见表10，抽生侧芽率，地点1（2.6%）<地点2（65.6%）<地点3（68.4%）；侧芽数均值：地点2（0.33个）<地点1（0.34个）<地点3（2.55个），侧芽数地点间差异显著；抽芽指数：地点1（0.333）<地点3（0.360）<地点2（0.482）。

2.3.2　侧芽抽生与树高、胸径关系

3地块中，在第2年生时，以第1地块树高、胸径值最大，第2地块次之，第3地块最小，3地块间树高、胸径值差异显著；相关分析显示：1~3地块侧芽数与树高的相关系数分别为0.088、0.399＊、0.419＊＊；1~3地块侧芽数与胸径的相关系数分别为：0.095、0.270、0.267，显然，侧芽抽生与树高、胸径相关关系不显著。

2.3.3　侧芽抽生与立地关系

从表9中数据可见，第1地块的pH（H_2O）=5.16，比第2、3地块大，第2、3地块属强酸性土，已不适合柚木的生长[6]，第1地块的有机质、全N、全P、全K、速效P、速效K、盐基饱和度明显大于第2、3地块，另一方面，第2、3地块的土壤质地差于第1地块，其透气性极差，不能满足柚木生长需要，可见，立地条件好的地段，侧芽抽生较弱，反之则旺。

表7 不同树树高/胸径范围的抽芽株数比例、抽生侧芽率、平均芽数、平均指数

变量范围	各级抽芽比例(%)			芽数(个)	抽芽指数	抽生侧芽率(%)
	1级	2级	3级			
树高≤2.50(m)	87.13	1.74	11.16	4.19	0.414	90.0
2.50<树高≤3.50(m)	74.75	15.08	10.17	5.58	0.451	91.7
树高>3.50(m)	45.63	48.44	5.93	7.37	0.534	95.0
胸径≤2.0(cm)	59.13	31.4	9.47	5.96	0.501	95.7
2.0<胸径≤2.5(cm)	71.3	26.8	1.91	6.21	0.435	90.2
胸径>2.5(cm)	43.22	49.32	7.46	7.96	0.547	100.0

表8 不同处理侧芽抽生情况与肥料①组成

处理号	芽数(个)	指数	抽芽率(%)	N(g/株)	P(g/株)	K(g/株)	基肥(g)		追肥(g)
							改良剂(f/m)	改良剂(ca)	改良剂类(f/m)
1	6.40	0.541	97.78	28.75	356.32	56.40	461.5/f②	692.3/ca④	94/f
2	8.33	0.656	95.40	28.75	356.32	56.40	461.5/m③	692.3/ca	94/m
3	4.79	0.485	75.27	525.00	525.00	525.00			
4	8.08	0.660	90.63	75.00	585.00	75.00			
5	4.62	0.474	86.67	186.00	180.00	132.90			
6	7.53	0.608	95.05	75.00	388.82	75.00	461.5/f	692.3/ca	
7	4.91	0.485	92.99	75.00	388.82	75.00	461.5/m	692.3/ca	
8	7.61	0.583	95.40	100.62	413.05	88.36	461.5/f		

注：①肥料N、P、K含量按以下计算：复合肥含N15.0%、含 P_2O_5 15.0%、含 K_2O 15.0%，含钙镁磷：P_2O_5 17.0%，鸡粪[3]含N 3.70 %、含 P_2O_5 3.50 %、含 K_2O 1.93 %；② 为施用沸石；③为施用麦饭石 ④为施用石灰。

表9 不同地块土壤化学性质物理性质因子

地点	pH(H_2O)	有机质(g/kg)	全N(g/kg)	全P(g/kg)	全K(g/kg)	速效P(mg/kg)	速效K(mg/kg)	盐基饱和度(%)	质地
1	5.16	23.63	1.45	0.81	23.50	1.07	43.63	88.46	黏壤土
2	4.48	21.37	1.10	0.64	17.88	0.66	27.14	72.62	黏土
3	4.52	18.14	1.02	0.81	16.33	0.59	18.32	52.18	黏土

表10 各地块侧芽抽生情况

地点	各级抽芽比例(%)			均值			抽芽指数	侧芽抽生率(%)
	1级	2级	3级	芽数(个)	树高(m)	胸径(cm)		
1	100.00	0.00	0.00	0.34	3.43	3.11	0.333	2.6
2	63.16	28.95	7.89	0.33	3.16	2.81	0.482	65.6
3	92.11	7.89	0.00	2.55	2.49	2.29	0.360	68.4

3 结论与建议

从试验调查和以上统计结果看，柚木1~3年生幼林侧芽的抽生系的普遍现象，1年生幼林，抽生率为30.70%~60.57%，2~3年生幼林，抽生率为30.70%~60.57%，抽生侧芽率、平均芽数、平均指数的大小受多方面原因的影响。

柚木无性系测定试验侧芽调查统计结果显示：无性系间、无性系与实生苗间、区组间，侧芽抽生率、平均芽数、平均指数差易显著，反映出侧芽的抽生主要受遗传因素的影响，因而除生长指标外，侧芽抽生状况应作为无性系选择的依据指标之一。

多因素施肥试验侧芽调查统计结果：虽然侧芽抽生率、平均芽数、平均指数处理间、重复间差异不显著，但侧芽抽生率与氮、钾肥施用量呈极显著负相关关系，而与沸石等土壤改良剂关系不密切，这与氮素能促进林木主梢（树高）生长[14,15]一般规律相一致，也说明侧芽的抽生同时受土壤养分的作用，不同施肥措施影响到幼树侧芽的抽生，N、P、K等营养元素不足或失衡时，植株生长即出现障碍。

从柚木格木混交林不同地段、无性系测定林的调查分析结果可以看出，地段间、区组间侧芽抽生率、平均芽数、平均指数差易显著，说明，造林立地与侧芽的抽生关系密切，正确选择造林地，可以抑制或减少侧芽的抽生率和降低抽生程度，促进林木的生长[5]。

柚木幼树侧芽的抽生可能对林木树高、径和杆材形成产生负面影响，其影响程度如何有待进一步试验调查研究。

参考文献

[1]广西林业局，广西林学会. 阔叶树种造林技术[M]. 南宁：广西人民出版社，1980.

[2]中国树木志编委会. 中国主要造林树种造林技术[M]. 北京：中国林业出版社，1981.

[3]中国科学院南京土壤研究所. 土壤理化分析[M]. 上海：上海科学技术出版社，1978.

[4]邝炳朝，郑淑珍，罗明雄，等. 柚木种源主要性状聚合遗传值的评价[J]. 林业科学研究，1996，9(1)：7-14.

[5]李运兴. 柚木家系试验[J]. 广西林业科学，2001，30(1)：50-52.

[6]潘一峰，刘文明. 酸性土壤改良对不同种源的柚木生长的影响[J]. 热带亚热带土壤科学，1997，6(1)：9-14.

[7]梁坤南，白嘉雨，周再知，等. 珍贵树种柚木良种繁育发展概况[J]. 广东林业科技，2006，22(3)：85-89.

[8]马华明，梁坤南，周再知. 我国柚木的研究与发展[J]. 林业科学研究，2003，16(6)：768-773.

[9]张荣贵，蓝猛，乔光明，等. 红河州柚木种源试验五年评价[J]. 林业科学研究，1999，12(2)：190-196.

[10]邝炳朝，郑淑珍，罗明雄，等. 柚木苗规格(标准)与经济效益的研究[J]. 林业科学研究，1991，4(6)：589-595.

[11]邝炳朝，郑淑珍，罗明雄. 柚木小棒槌苗贮藏技术的研究[J]. 林业科学研究，1988，1(16)：579-587.

[12]陶国祥. 柚木栽培气候区划的研究[J]. 云南林业科技，1992，(1)：22-26.

[13]卢俊培. 海南柚木立地类型及评价[J]. 林业科学研究，1994，7(6)：677-683.

[14]刘塑，何朝均，何绍彬，等. 不同施肥对麻疯树幼林生长的影响[J]. 四川林业科技，2009，30(4)：53-56.

[15]马跃，马履一，刘勇，等. 追施氮肥对长白落叶松移植苗生长的影响[J]. 林业科技开发，2010，24(3)，60-63.

［原载：中南林业科技大学学报，2011，31(11)］

广西马尾松松脂的化学组成研究

安 宁[1] 丁贵杰[2]

([1]中国林业科学研究院热带林业实验中心，广西凭祥 532600；
[2]贵州大学造林生态研究所，贵州贵阳 550025)

摘 要 用气相色谱法测定不同年龄、不同径级及16种施肥组合的马尾松松脂的化学成分，分析比较各处理之间成分差异。实验结果表明：不同年龄、不同径级及16种施肥组合的马尾松松脂的主要成分基本一致，只是各成分含量略有不同，存在一定差异，含油量有随着年龄以及径级的增大而增加的趋势，2号、3号、4号、14号和16号施肥组合的含油量高于空白对照。各处理马尾松松节油的α-蒎烯含量均很高。

关键词 马尾松；松脂；树脂酸；松节油

Study on Chemical Constituents of Oleoresin from *Pinus msssoniana* in Guangxi

AN Ning[1], DING Guijie[2]

([1]*The Experimental Centre of Tropical Forestry, CAF, Pingxiang 532600, Guangxi, China;*
[2]*The Institute of Silviculture and Ecology in Guizhou University, Guiyang 550025, Guizhou, China*)

Abstract: Abstract: Determination of different ages, different diameter and fertilization of the chemical composition of pine resin, analysis and comparison of the treatment difference between the components. The results show that: the main component of turpentine consistent treatment, only slightly different content, there are some differences. oil content increasing with the age and diameter. 0il content of NO. 2, NO. 3, NO. 4 and NO. 16 fertilization are higher than the control's. α-pinene content of turpentine pine in all treatments of is high.

Key words: *Pinus massoniana*; resin; resin acids; turpentine oil

马尾松(*Pinus massoniana*)是我国松树中分布面积最广和采割松脂的主要树种之一，也是我国亚热带东部湿润地区典型的针叶乡土树种。马尾松经济价值高，是可进行综合利用的典型树种[1]。松脂经过加热蒸馏除去杂质，就可制成松香(rosin)和松节油(turpentine oil)，二者均是重要的化学工业原料[2]。松香是我国林业唯一在世界上有重大影响的大宗出口商品，年创汇达1亿多美元。产量占世界总产量的1/3，出口量占一半以上。本文通过对不同年龄、径级及施肥组合松脂的化学组成进行研究，找出它们之间的差异，以期为营建高产脂人工林提供理论依据。

1 试验地概况及研究材料

1.1 试验地概况

试验林设在位于广西凭祥市的中国林业科学研究院热带林业实验中心伏波实验场，地处106°43′E，22°06′N，海拔500m，低山，年平均气温21.5℃，年降水量1400mm，属于南亚热带季风气候区，土壤为花岗岩发育成的红壤。土层厚度1~2m，腐殖质厚度10cm以上，土壤呈酸性(pH 4.3~5.1)。前茬为杉木(*Cunninghamia lanceolata*)，主伐后明火炼山。

1.2 研究材料

在现有采脂林中采集1995年、1993年、1991年、1989年所造马尾松人工林林分松脂样品进行不同年龄松脂成分研究；采集1991年所造马尾松人工林林分松脂样品进行不同径级松脂成分研究；采集1989年所造马尾松人工林林分松脂样品进行不同施肥组合松脂成分研究。施肥设3因素，4水平具体如下：N，46%的尿素，分别N_1(0g/株)、N_2(250g/株)、N_3(500g/株)、N_4(750g/株)；P，18%的钙镁磷肥，分别为P_1(0g/株)、P_2(500g/株)、P_3(1000g/株)、P_4(1500g/株)；

K，60%的氯化钾，分别为 K_1(0g/株)、K_2(150g/株)、K_3(300g/株)、K_4(450g/株)。

2　实验方法

2.1　松脂含油量及松节油成分的测定

按林业局松脂标准 LY223-81 测定松脂中的含油量。将上述所得松节油直接进气相色谱仪分析。

仪器：日本岛津 GC298AM 气相色谱仪；数据处理系统：C2R3A 积分器；色谱柱：DB21 弹性石英玻璃毛细管柱，ϕ0.28mm×50m；色谱条件：程序升温，80℃（20min）5$\xrightarrow{℃/min}$200℃（30min）；汽化室温度 250℃；检测器温度 230℃；载气 N2；流速 50ml/min；分流比 100∶1；进样量 0.8μL。

2.2　松脂成分测定

将样品从冰箱取出放至室温，搅拌均匀，取少量溶于无水乙醇中，以酚酞做指示剂，用 25% 四甲基氢氧化铵滴至红色，以此溶液进行气相色谱分析。

仪器：日本鸟津 GC29AM 气相色谱仪；数据处理系统：C2R3A 积分器；色谱柱：QF21 弹性石英玻璃毛细管柱，ϕ0.28mm×45m；色谱条件：程序升温，60℃（2min）$\xrightarrow{4℃/min}$230℃（30min）；汽化室温度 250℃；检测器温度 250℃；载气 N2；流速 50ml/min；分流比 100∶1；进样量 0.5μL。

从气相色谱图得到的各组分含量为面积归一法。松脂在同一色谱条件下，单萜和双萜的分子质量相差较大，必须对色谱数据进行校正。宋湛谦等人分析树脂酸用十八烷酸作内标物进行计算[3]。本实验以实测松脂中含油量为校正系数，对色谱数据进行校正，得到松脂中各树脂酸含量[4]。

3　结果与讨论

3.1　不同年龄松脂成分差异研究

从表 1 及表 2 可以看出，不同年龄的松脂所含主要成分较一致，但各种成分的含量有一定差异，且规律性不明显。

表 1　不同年龄松香主要成分及含量

松香主要成分含量(%)	年龄(年)			
	14	16	18	20
海松酸	0.41	0.32	0.47	0.40
湿地松酸	10.39	8.51	7.01	9.13
异海松酸	1.81	1.86	1.82	1.50
左旋海松酸	52.43	40.77	47.70	43.20
去氢枞酸	0.20	0.27	0.21	0.31
枞酸	7.25	17.71	11.57	8.85
新枞酸	9.46	14.95	12.73	17.92

马尾松松脂树脂酸主要化学成分有海松酸、湿地松酸、异海松酸、左旋海松酸、去氢枞酸、枞酸和新枞酸。其中海松酸、异海松酸属于海松酸型树脂酸，其分子结构的 C-13 位上有一个甲基和一个乙烯基，不存在共轭双键，相对含量稳定；而左旋海松酸、枞酸、新枞酸属于枞酸型树脂酸，分子中具有共轭双键，不稳定，各植株变化较大，可能是受到气温变化或遗传差异的影响，这说明在株间进行选优是可行的，且能收到良好的效果。去氢枞酸分子中三环菲结构有一苯环，属稳定结构。4 个年龄中，20 年含油量最高为 15.62%，14 年含油量最低为 13.14%，16 年和 18 年的含油量分别为 14.40% 和 13.57%，有随着年龄增大而增加的趋势。在所含树脂酸中左旋海松酸含量最高，均超过 40%，其中 14 年含量最高达到 50.43%，而 16 年含量最低为 40.77%。新枞酸和去氢枞酸与产脂力呈正相关[5]，20 年的新枞酸含量最高为 17.92%，其次为 16 年的 14.95%，最低为 14 年的 9.46%，而去氢枞酸在 4 个年龄中的含量随着年龄处理的增大依次为 0.20%、0.27%、0.21% 和 0.31%，有随年龄增大而增加的趋势。

表 2　不同年龄松节油主要成分及含量

松节油含量及主要成分含量(%)	年龄(年)			
	14	16	18	20
含量	13.14	14.40	13.57	15.62
α-蒎烯	78.03	53.02	70.35	58.48
莰烯	1.50	0.96	1.22	1.03
β-蒎烯	3.81	10.92	4.35	2.34
月桂烯	—	0.12	0.05	0.23
Δ3-蒈烯	1.47	0.33	0.86	0.51
长叶烯	5.82	22.39	12.80	12.85
石竹烯	4.06	2.78	4.37	3.05

马尾松松节油主要化学成分有 α-蒎烯、莰烯、β-蒎烯、月桂烯、Δ3-蒈烯、长叶烯和石竹烯。α-蒎烯含量高是马尾松松节油的特征，4 个年龄含量均超过

50%，而β-蒎烯含量为4%~10%，β-蒎烯含量变化主要由种子变异引起的[6]。松节油中长叶烯含量高及其与石竹烯相差大，这是高抗马尾松种源的内在特征[7]，这也是优良种源在生长速度和产量上优于一般马尾松的原因之一[8]。4个年龄中，16年的长叶烯含量最高为22.39%，14年的含量最低仅为5.82%，长叶烯与石竹烯含量相差最大的为16年，相差最小为14年。4个年龄的松脂成分并没有较大差异。

3.2 不同径级松脂成分差异研究

从表3及表4可以看出，各径级松香及松节油的主要成分一致，但各成分的含量存在一定差异。

表3 不同径级松香主要成分及含量

松香主要成份含量(%)	径级					
	14cm	16cm	18cm	20cm	22cm	26cm
海松酸	0.73	0.50	0.47	0.45	0.49	0.44
湿地松酸	9.96	10.47	7.01	8.48	9.98	7.76
异海松酸	1.68	1.82	1.82	1.43	1.47	1.56
左旋海松酸	47.20	53.58	47.70	49.64	50.02	45.13
去氢枞酸	0.13	0.21	0.11	0.36	0.29	0.39
枞酸	8.13	8.43	11.57	9.65	8.65	9.99
新枞酸	11.66	9.55	12.73	11.17	10.08	10.25

在6个径级中，26cm含油量最高为18.30%，14cm含油量最低为9.39%，16cm、18cm、20cm、22cm的含油量分别为13.70%、11.21%、17.20%和16.59%，有随径级增大而增加的趋势。在所含树脂酸中左旋海松酸含量最高，均超过40%，其中16cm含量最高达到53.58%，而26cm含量最低为45.13%。新枞酸含量最高是18cm为12.73%，其次为20cm的11.17%，最低是16cm的9.55%，而去氢枞酸在4个年龄中的含量最高是26cm的0.39%，最低为18cm的0.11%。据报道马尾松产脂力与新枞酸和去氢枞酸呈正相关。

表4 不同径级松节油主要成分及含量

松节油含量及主要成分含量(%)	径级					
	14cm	16cm	18cm	20cm	22cm	26cm
含量	9.39	13.70	11.21	17.20	16.59	18.30
α-蒎烯	64.19	63.28	70.35	68.50	47.07	54.36
莰烯	1.23	1.12	1.22	1.06	1.07	0.87
β-蒎烯	2.30	3.41	4.35	4.13	5.98	3.80
月桂烯	0.54	0.31	0.05	0.07	0.20	0.29
Δ3-蒈烯	0.12	0.43	0.86	0.36	1.68	0.21
长叶烯	14.23	16.98	12.80	14.11	26.01	27.25
石竹烯	5.15	3.34	4.37	4.55	3.97	3.23

α-蒎烯在6个径级中只有22cm径级含量为47.07%，其他径级均高于50%。而β-蒎烯含量为4%左右，在各径级中，26cm径级的长叶烯含量最高为27.25%，18cm的含量最低仅为12.80%，Δ3-蒈烯在22cm含量为1.68%，而其他径级Δ3-蒈烯含量都低于1.00%。长叶烯与石竹烯含量相差最大的为26cm径级，相差最小为18cm径级。各径级的松脂成分没有较大差异。

3.3 施肥处理对松脂成分的影响

从表5及表6可以看出，各施肥组合的松脂主要成分一致，但各成分含量不同，存在一定差异。

表5 不同施肥组合松香主要成分及含量

施肥组合	松香主要成分含量(%)						
	海松酸	湿地松酸	异海松酸	左旋海松酸	去氢枞酸	枞酸	新枞酸
1	0.40	9.13	1.50	43.20	0.31	8.85	17.92
2	0.43	9.25	1.77	47.62	0.32	10.93	7.67
3	0.48	9.45	1.46	45.07	0.46	7.67	9.98
4	0.38	10.16	1.62	44.40	0.27	8.07	12.72
5	1.00	8.59	1.72	51.00	0.19	10.47	13.56
6	0.61	9.50	1.83	44.64	0.29	14.56	13.53
7	0.64	9.97	1.80	47.09	0.30	9.06	9.64
8	0.38	8.54	1.54	46.09	0.29	7.20	11.20
9	0.99	9.31	1.81	50.40	0.13	7.94	13.52
10	0.76	10.35	1.53	46.23	0.31	8.30	7.94
11	0.62	7.88	1.43	40.48	0.13	8.99	10.70
12	0.76	8.03	1.70	51.88	0.18	11.73	10.30
13	0.63	10.70	1.61	46.15	0.40	8.19	9.47
14	0.48	14.01	1.84	47.33	0.24	7.87	12.92
15	0.46	9.20	1.59	50.64	0.26	8.08	13.41
16	0.55	9.23	1.57	44.46	0.29	10.29	9.25

注：施肥组合1-16分别为$N_1P_1K_1$、$N_1P_2K_2$、$N_1P_3K_3$、$N_1P_4K_4$、$N_2P_1K_2$、$N_2P_2K_1$、$N_2P_3K_4$、$N_2P_4K_3$、$N_3P_1K_3$、$N_3P_2K_4$、$N_3P_3K_1$、$N_3P_4K_2$、$N_4P_1K_4$、$N_4P_2K_3$、$N_4P_3K_2$、$N_4P_4K_1$，具体各水平见前文。

在16个施肥组合中，2号含油量最高为21.36%，15号含油量最低为11.01%。1号为施肥空白对照其含油量为15.62%，在16个处理中，2号、3号、4号、14号和16号含油量高于1号。树脂酸中左旋海松酸含量最高，均超过40%，其中12号含量最高达到51.88%，而11号含量最低为40.48%。新

表6 不同施肥组合松节油主要成分及含量

施肥组合	松节油含量及主要成分含量(%)							
	含油量	α-蒎烯	莰烯	β-蒎烯	月桂烯	Δ3-蒈烯	长叶烯	石竹烯
1	15.62%	58.48	1.03	2.34	0.23	0.51	12.85	3.05
2	21.36%	51.92	0.97	3.00	0.72	0.05	19.30	5.34
3	19.44%	55.64	0.93	8.94	0.44	0.20	17.80	3.79
4	17.58%	55.18	0.91	7.39	0.46	0.16	15.08	4.41
5	13.77%	55.49	1.15	2.90	0.41	1.55	13.73	5.26
6	14.27%	65.95	1.08	2.12	0.16	0.54	9.81	4.71
7	15.25%	56.12	0.97	9.18	0.35	0.24	20.59	4.58
8	13.90%	54.04	1.00	5.54	0.16	1.03	16.57	3.39
9	15.22%	52.67	1.07	7.05	0.07	0.54	21.06	3.88
10	15.92%	52.08	1.16	2.27	0.44	0.21	17.61	4.51
11	11.98%	58.81	1.12	2.99	0.61	0.08	13.78	4.85
12	17.46%	69.29	1.29	2.57	0.39	0.19	14.79	3.79
13	14.98%	56.43	1.15	2.54	0.37	0.56	19.43	4.51
14	17.49%	52.08	0.83	7.28	0.55	1.04	17.34	4.55
15	11.01%	49.10	0.99	5.38	0.25	0.68	17.99	4.36
16	17.39%	62.19	1.09	2.55	0.28	0.64	11.86	4.28

注：施肥组合1-16同表5。

枞酸含量最高是1号为17.92%，最低是2号的7.67%，而去氢枞酸在施肥处理中的含量最高是13号的0.40%，最低为9号和11号的0.13%，差异不显著。

α-蒎烯含量在16个施肥组合中只有15号含量低于50%，其他施肥处理均高于50%。Δ3-蒈烯含量超过1.00%的有5号、8号和14号，分别为1.55%、1.03%和1.04%。其他处理均低于0.70%，含量不到0.10%的有2号和11号。9号的长叶烯含量最高为21.06%，6号含量最低仅为9.81%，长叶烯与石竹烯含量相差最大的为9号，相差最小为6号。据报道长叶烯含量高而其与石竹烯含量相差大是优良种源生长速度和产量高于一般马尾松的原因之一，而在同种源不同单株上存在的差异是否也存在此规律有待进一步研究。

4 结论

(1)各年龄松脂主要成分一致，只是含量不同。20年含油量最高为15.62%，14年含油量最低为13.14%，含油量有随着年龄增大而增加的趋势。α-蒎烯含量高是马尾松松节油的特征，4个年龄处理α-蒎烯含量均超过50%。

(2)各径级的松脂主要成分一致，各成分含量存在一定差异。26cm含油量最高为18.30%，14cm含油量最低为9.39%，含油量有随着径级增大而增加的趋势。在6个径级中，22cm径级α-蒎烯含量为47.07%，其他径级含量均高于50%。

(3)各施肥组合松脂主要成分一致，但含量不同。在16个处理中，2号含油量最高为21.36%，15号含油量最低为11.01%。1号为施肥空白对照，其含油量为15.62%，在16个处理中，2号、3号、4号、14号和16号含油量高于1号。α-蒎烯的含量除了15号以外，在其他施肥组合中含量均高于50%。Δ3-蒈烯含量超过1%的有5号、8号和14号，其他处理均低于0.70%。

参考文献

[1]周政贤．中国马尾松[M]．北京：中国林业出版社，2001：14.

[2]刘玉春．1995-2000年中国松香的生产、消费和发展趋势[J]．林产化工通讯，2001，35(5)：31-33.

[3]宋湛谦，刘星，梁志勤，等．国外引种松树松脂化学组成的特征[J]．林产化学与工业，1993，13(4)：277-287.

[4]钟国华，梁忠云，沈美英，等．广西湿地松松脂化学组成的研究[J]．林产化学与工业，2001，21(3)：29-33.

[5]何波祥，连辉明，曾令海，等．高脂马尾松优树松脂化学组分及其地理变异的研究广东林业科技，1999，15(4)：1-7.
[6]钟国华，梁忠云，沈美英，等．广西湿地松松脂化学组成的研究[J]．林产化学与工业，2001，21(3)：29-33.
[7]赵振东，李冬梅，胡樨萼，等．抗松材线虫病马尾松种源化学成分与抗性机理研究(第Ⅱ报)[J]．林产化学与工业，2001，21(1)：56-60.
[8]梁忠云，陈海燕，沈美英，等．广西优良品种马尾松松脂的化学组成[J]，林产化工通讯，2002，36(5)：8-10.

[原载：中南林业科技大学学报，2012，32(03)]

红椎人工林生长规律的初步研究

唐继新　白灵海　郭文福　曾　冀　蔡道雄

（中国林业科学研究院热带林业实验中心，广西凭祥　532600）

摘　要　基于林分树干解析资料，对广西大青山 27 年生红椎人工林的生长规律进行分析研究，结果表明：①7～12 年为胸径连年生长量的速生期，第 8 年胸径连年生长量达峰值，第 11 年后连年生长量略呈下降趋势，但连年生长量与平均生长量曲线直到 22-23 年才相交。②4～11 年为树高连年生长量快速增长期，树高连年生长量于第 7、8 年达峰值，随后开始逐渐下降，并在第 12 年与树高平均生长量相交。③11～27 年材积生长一直处于不断增长的阶段，直到 27 年生时，材积生长量仍未达数量成熟，红椎人工林材积的生长潜力大。④利用 Richards 方程拟合红椎人工林胸径、树高和材积的生长规律，拟合优度均在 0.99 以上，拟合效果显著。

关键词　红椎；树干解析；生长规律

Preliminary Study on the Growth Regularity of *Castanpsis hystrix* Plantation

TANG Jixin, BAI Linghai , GUO Wenfu, ZENG Ji , CAI Daoxiong

(*Experimental Centerof Tropical Forestry*, *CAF*, *Pingxiang* 532600, *Guangxi*, *China*)

Abstract: Based on the material of analysis stem, the growth regularity of *Castanpsis Hystrix A. DC* plantation in Daqingshan was analyzed, the results showed that: ① the fast-growing period of annual increment of DBH was from the 7th year to the 12nd year, the growth of annual increment of DBH reached the peak in the 8th year, it declined slightly after 11th years, but the curve of annual increment of DBH and the curve of average growth didn't intersect until the 22nd ～ 23rd year. ② the fast growing period of annual increment of tree height was during the 4th- 11st year, and the annual increment of tree height reached the peak in the 7th and 8th year, after that it gradually decreased, and it intersected with the curve of average growth of tree height in the 12nd years. ③ the fast growing period of volume was during the 11st ～ 27th year, and until the 27th year, the growth of volume hadn't reached the quantitative maturity, the growth of *Castanpsis Hystrix A. DC* plantation volume had a great potential. ④ the Richards equation was good to mode the growth regularity of DBH, tree height and volume of *Castanpsis hystrix* plantation, the goodness of fit were more than 0.99, fitting effect was obvious.

Key words: *Castanpsis hystrix*; analysis stem; growth regularity

红椎（*Castanpsis hystrix*）又名红锥、红栲、红柯，壳斗科常绿阔叶高大乔木，具有速生、材质优、适应性强等特性，是优良的珍贵阔叶用材树种，广西、广东、云南和福建等地为其天然分布及主要人工栽培区[1-3]。近年来随着我国经济的快速发展，中高档大径级阔叶材的需求急剧上升，尤其是大径级珍优阔叶材的供需矛盾日益凸显[4]，为缓解国内珍优阔叶用材的短缺，培育优质的大径级阔叶用材林显得尤为重要。本文依据中国林业科学研究院热带林业实验中心（下文简称热林中心）红椎试验人工林树干解析的资料，对其生长规律进行了分析研究，旨在为红椎人工林大径材的培育及经营提供科学依据。

1　样地概况

样地位于广西凭祥市热林中心伏波实验场（21°57′47″～22°19′27″N，106°39′50″～106°59′30″E）1983 年春营造的红椎试验人工林；林分初植密度 1667 株/hm²，分别于造林后第 7 年、第 15 年和第 19 年进行了 3 次抚育间伐，每次间伐强度为 35%～40%，现有立木密度、郁闭度分别为380 株/hm²和 0.85；林下植被有红椎幼树、杉木（*Cunninghamia lanceolata*）、杜茎山(*Maesa japonica*)、路边青(*Geum aleppicum*)、异叶榕(*Ficus heteromorpha* Hemsl.)等；低山丘陵地貌，海拔460~650m；土壤为由花岗岩发育而成的赤红壤，土层厚度100~150cm，腐殖质层厚度5~10cm；南亚热带季风气候，年均气温20.5℃，≥10℃积温6500℃，最热月平均气温27.5℃，最冷月平均气温12.0℃，极端最低气温-0.5℃；年均降水量1400mm，年蒸发量1260mm；全年日照时数1200~1300h。

2 材料与方法

在27年生长势中等的红椎试验林中，布设面积600m²的临时标准地3块，在每块标准地进行每木检尺，并在其中1块标准地内按不同径阶伐取解析木10株，以2m为一区分段进行树干解析，3块标准地内径级分布及各解析木资料见表1。各龄阶去皮直径取解析木东西、南北向的平均值，各龄阶的树高由树高生长过程曲线图查出，各龄阶的去皮材积利用区分段法求算[5]；各龄阶红椎林分胸径、树高和材积等因子的平均值，依据标准地红椎各径级的株数分布及解析木的树高推算得出。

3 结果与分析

3.1 生长过程分析

经对10株红椎解析木数据的测定，取林分胸径、树高和材积等因子的平均值，进行生长过程分析，结果如表2所示。

表1 标准地内林木径级分布及解析木情况

径阶(cm)	14	16	18	20~22	24~26	28	30	32	34	36	≥38	合计
标准地径级株数分布	4	5	7	17	16	5	3	2	4	3	2	68
解析木株数	1	1	1	1	1	1	1	1	1	1	/	10
解析木胸径(cm)	13.2	15.6	18.1	19.6	24.1	28.2	29.7	31.9	34.0	36.0	/	/
解析木树高(m)	20.2	22.4	20.2	21.2	17.5	21.1	25.6	23.9	24.9	25.8	/	/

表2 红椎树干总生长过程

林龄(年)	去皮胸径(cm)			树高(m)			去皮蓄积(m³)			形数f
	总生长量	平均生长量	连年生长量	总生长量	平均生长量	连年生长量	总生长量	平均生长量	连年生长量	
1	0.00	0.00	0.00	0.52	0.52	0.52	0.00002	0.00002	0.00001	—
2	0.14	0.07	0.14	1.34	0.67	0.82	0.00015	0.00008	0.00014	72.75
3	0.43	0.14	0.29	2.10	0.70	0.87	0.00055	0.00018	0.00039	18.04
4	1.10	0.27	0.66	3.07	0.77	0.97	0.00136	0.00034	0.00082	4.66
5	1.90	0.38	0.80	4.10	0.82	1.03	0.00271	0.00054	0.00134	2.33
6	2.77	0.46	0.88	5.17	0.86	1.07	0.00471	0.00078	0.00200	1.51
7	3.86	0.55	1.09	6.26	0.89	1.08	0.00815	0.00116	0.00344	1.11
8	5.02	0.63	1.15	7.34	0.92	1.08	0.01277	0.00160	0.00462	0.88
9	5.91	0.66	0.89	8.40	0.93	1.06	0.01867	0.00207	0.00591	0.81
10	6.93	0.69	1.02	9.43	0.94	1.03	0.02707	0.00271	0.00840	0.76
11	8.04	0.73	1.11	10.42	0.95	0.99	0.03882	0.00353	0.01175	0.73
12	9.16	0.76	1.13	11.37	0.95	0.95	0.05204	0.00434	0.01321	0.69
13	10.16	0.78	1.00	12.28	0.94	0.91	0.06567	0.00505	0.01363	0.66
14	11.13	0.80	0.97	13.14	0.94	0.86	0.08143	0.00582	0.01576	0.64
15	12.07	0.80	0.94	13.96	0.93	0.82	0.10010	0.00667	0.01867	0.63
16	13.08	0.82	1.01	14.73	0.92	0.77	0.11925	0.00745	0.01916	0.60

（续）

林龄（年）	去皮胸径（cm）			树高（m）			去皮蓄积（m^3）			形数 f
	总生长量	平均生长量	连年生长量	总生长量	平均生长量	连年生长量	总生长量	平均生长量	连年生长量	
17	14.13	0.83	1.04	15.46	0.91	0.73	0.14174	0.00834	0.02249	0.58
18	15.18	0.84	1.06	16.14	0.90	0.68	0.16464	0.00915	0.02290	0.56
19	16.18	0.85	0.99	16.78	0.88	0.64	0.19140	0.01007	0.02677	0.56
20	17.12	0.86	0.95	17.39	0.87	0.60	0.21832	0.01092	0.02691	0.55
21	18.07	0.86	0.95	17.95	0.85	0.57	0.24651	0.01174	0.02820	0.54
22	18.97	0.86	0.91	18.48	0.84	0.53	0.27190	0.01236	0.02539	0.52
23	19.69	0.86	0.72	18.98	0.83	0.50	0.29807	0.01296	0.02618	0.52
24	20.31	0.85	0.62	19.45	0.81	0.47	0.32898	0.01371	0.03090	0.52
25	21.13	0.85	0.82	19.89	0.80	0.44	0.36019	0.01441	0.03122	0.52
26	21.79	0.84	0.66	20.30	0.78	0.41	0.39120	0.01505	0.03101	0.52
27	22.43	0.83	0.64	20.68	0.77	0.38	0.41924	0.01553	0.02803	0.51
带皮	23.22	—	—	20.68	—	—	0.45271	—	—	—

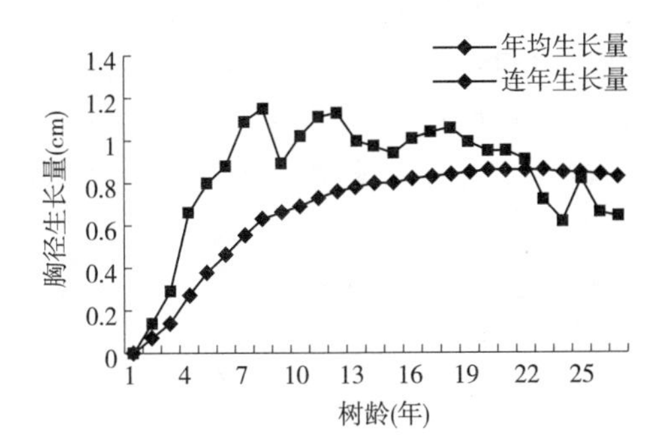

图 1　胸径连年生长量与平均生长量曲线

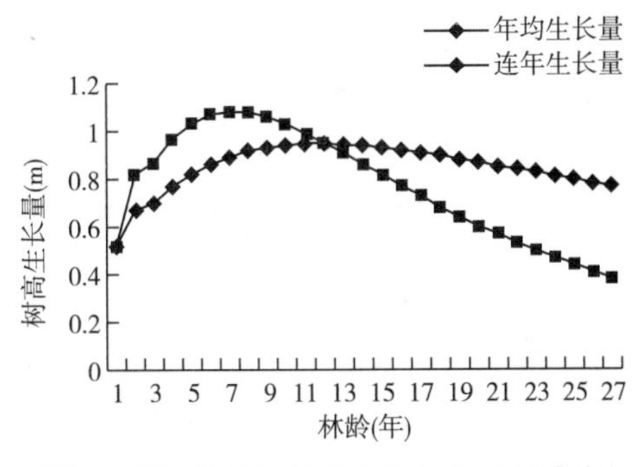

图 2　树高连年生长量与平均生长量曲线

3.1.1　胸径生长过程

由表 2、图 1 分析可知：7～12 年为胸径连年生长量的速生期，但第 9 年胸径连年生长量呈急剧减缓，结合当地的气象资料，分析认为这与红椎生长第 9 年时（1991 年）的气候异常干旱有关；9～13 年间连年生长量呈波浪形曲线，14～22 年间增长相对较平稳，22 年后连年生长开始减弱。其中，7～22 年间连年生长量的值均大于 0.91cm，连年生长量高峰值（1.15cm）在第 8 年；连年生长量与平均生长量曲线相交于 22～23 年生间，此时平均生长量达最大值；14～27 年间，平均生长量基本生长比较平稳，基本维持在 0.80～0.86cm。

3.1.2　树高生长过程

由表 2、图 2 分析可知：前 4～11 年是红椎树高生长的旺盛期，其连年生长量均在 0.97m 以上；在树高整个生长过程中，树高连年生长量峰值在第 7、8 年，并与平均生长量曲线相交于第 12 年，此时平均生长量达最大值 0.95m，此后连年生长量逐渐下降，27 年生的连年生长量为最小，仅为 0.38m。

3.1.3　材积生长过程

由表 2 和图 3 分析可知，11～25 年间红椎材积的生长几乎处于不断快速增长的趋势，其连年生长量均高于 0.01m^3，并在 25 年生时达到最大值（0.03122m^3）。平均生长量一直表现为平稳的增长趋势，直到 27 年生时，材积连年生长量与平均生长量曲线仍未相交，即材积平均生长量尚未到最大值，此时红椎材积还处在旺盛的生长阶段，其仍未达到数量成熟年龄。可见，红椎人工林材积的生长潜力巨大。

3.1.4　形数变化分析

形数为反映林木生长过程的重要因子，由表 2 和图 4 分析可知：红椎形数随林龄的增加而降低，呈典型的反 J 型曲线变化，形数在第 7 年后均小于 1，并在第 22 年后趋于稳定状态，保持在 0.51～0.52；各龄阶形数情况表明红椎的形数较大、树干尖削度较小。

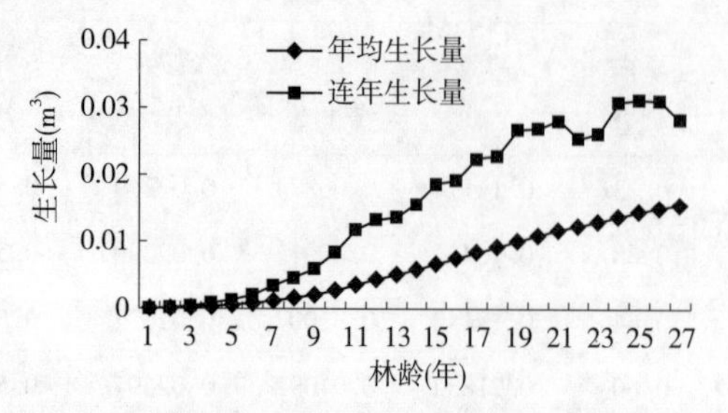

图3　单株材积连年生长量与平均生长量曲线

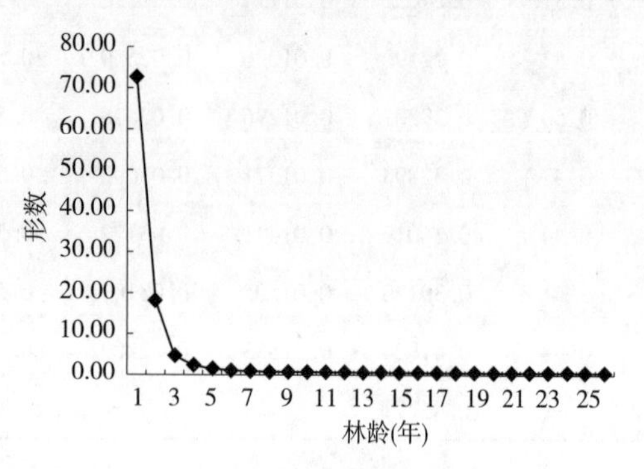

图4　胸高形数变化曲线

3.2　生长特性模拟及预测

为了准确反映及预测红椎人工林胸径、树高和材积的生长规律，并由此制定红椎人工林科学的经营策略，因此本文对红椎人工林胸径、树高和材积的生长过程进行了模拟。

3.2.1　带皮与去皮直径的模型拟合

根据10株解析木各圆盘带皮与去皮数据，运用Evicws软件及普通最小二乘法[6]模拟出了带皮胸径与去皮胸径的线性回归方程：

$$D_1=a+bD_2 \tag{1}$$

式中：D_1、D_2 分别表示带皮和去皮直径；a、b 为参数。经迭代选优后，得到的模型参数为：$a=1.026649$，$b=0.215093$，拟合优度 $R^2=0.99933$，残差平方和 $RSS=11.13382$，样本数119。模型的参数结果表明：带皮直径与去皮直径的模拟效果很好。

3.2.2　胸径、树高和材积的生长模拟及预测

在模拟林分平均胸径、平均树高和平均单株材积的年龄生长模型中，做了多种尝试，最终选定了拟合效果最优的Richards生长模型：

$$y=a(1-e^{-ct})^b \tag{2}$$

式中：a 为林木生长的极值参数；c 为生长速率参数；b 表示与同化作用幂指数有关的参数[8]。根据林分去皮平均胸径、平均树高和平均单株材积的资料，其模型拟合的参数详细情况如表3所示。

由表3分析可知：在红椎林分平均胸径、树高及单株材积的年龄生长模型中，模型的拟合优度均高于0.999，且残差平方和都很小，结果表明胸径、树高、单株材积的拟合效果很好。

表3　林分平均胸径树高材积总生长量与年龄关系的曲线模拟结果

项目	a	b	c	R^2	RSS
胸径	32.46906	2.280034	0.070360	0.999639	0.487816
树高	25.86925	1.576601	0.075018	0.999941	0.059211
材积	1.288804	4.552245	0.056351	0.999942	2.76E-05

注：R^2、RSS 分别表示拟合优度及残差平方和。

4　结论与讨论

4.1　结论

(1)利用Richards方程拟合红椎人工林的平均胸径、平均树高和平均单株材积的年生长规律，拟合优度均在0.99以上，拟合效果显著。

(2)径向生长过程表明：7~12年为胸径连年生长量的速生期，第8年胸径连年生长量达峰值，但连年生长量与平均生长量曲线直到22~23年才相交。为培育红椎大径材，必须及时对林分进行抚育及间伐，以保证林木生长的足够营养空间。

(3)树高生长过程表明：4~11年为树高连年生长量的速生期，树高连年生长量于第7、8年达峰值，随后连年生长量开始逐渐下降，并在第12年与平均生长量相交。

(4)11~27年生，是材积生长的速生期，连年生长量均高于0.01m³，材积连年生长量在25年生达到最大值，但直到27年生时，材积连年生长量与平均生长量曲线仍未相交(即材积数量仍未成熟)，红椎人工林材积的速生期持续时间长，生长潜力大。

4.2　讨论

(1)本研究27年生红椎人工林的平均胸径、平均树高分别为23.22cm、20.68m，与36年生红椎天然林的平均生长水平[8](平均胸径22.5cm、平均树高23.1)相当，表明适当的经营措施(如抚育、间伐)促进红椎树种的生长效果较好。

(2)立地与栽培措施(如初植密度、间伐时间、间伐强度)均为影响林木生长的重要因素，本研究仅选择了热林中心以往典型经营措施的红椎人工林进行分析，故研究所得具有一定的局限性。

(3)27年生红椎试验人工林的解析资料表明，红

椎林分材积生长仍未达数量成熟，故红椎林分成熟时的生长规律如何，还有待下一步的深入研究。

参考文献

[1]赵登科．红椎优质用材林间伐整枝与中期施肥试验[J]．现代农业科技，2009(17)：186-188.

[2]卢立华，汪炳根，何日明．立地与栽培模式对红椎生长的影响[J]．林业科学研究，1999，12(5)：519-523.

[3]朱积余．红椎速生丰产栽培的试验研究[J]．林业科技通讯，1993(2)：8-12.

[4]蔡道雄，贾宏炎，卢立华，等．我国南亚热带珍优乡土阔叶树种大径材人工林的培育[J]．林业科学研究，2007，20(2)：165-169

[5]秦光华，姜岳忠，王卫东，等.1-69 杨人工林生长进程动态分析[J]．甘肃农业大学学报，2004，39(6)：666-670.

[6]王中昭，李丽明．计量经济学实验及例题分析[M]．广西科学技术出版社，2005. 11.

[7]孟宪宇．测树学[M].3 版．北京：中国林业出版社，2006：184.

[8]黄全能．红椎天然林生长规律与生物量的调查研究[J]．福建林业科技，1998，25(2)：20-23.

[原载：中南林业科技大学学报，2012，32(04)]

马尾松近熟林施肥技术研究

谌红辉[1] 丁贵杰[2] 温恒辉[1] 安 宁[1] 蒙明君[1]
([1]中国林业科学研究院热带林业实验中心，广西凭祥 532600；
[2]贵州大学林学院，贵州贵阳 550025)

摘 要 通过对17年生马尾松近熟林施肥试验4年的生长效应分析表明：在南方红壤地区施适量的P、K肥对马尾松近熟林生长有显著的促进作用，但施用过量却有副作用；施N肥短期内对林木生长不利；施肥效果显著时可掩盖林地本身质量对林木生长的影响。根据对各肥种水平效应值的分析可知，最佳施肥处理为每株混施P_2O_5 90g，K_2O 90g，其4年蓄积定期生长量比对照高13.4%。施肥措施对促进小径阶木生长有利。在肥效期内施肥措施与马尾松近熟林的生长效应呈显著相关性。

关键词 马尾松近熟林；施肥；生长效应；方差分析

Fertilization Techniques for Masson Pine Near-mature Plantation

CHEN Honghui[1], DING Guijie[2], WEN Henghui[1], AN Ning[1], MENG Mingjun[1]
([1]*The Experimental Centre of Tropical Forestry*, *CAF*, *Pingxiang* 532600, *Guangxi*, *China*;
[2]*College of Forestry*, *Guizhou University*, *Guiyang* 550025, *Guizhou*, *China*)

Abstract: By 4 years' test data studied on 17-year-old masson pine near-mature fertilization, Which showed that applying N fertilizer was disadvantageous to the growth of plantation in short time in south red soil area, While fertilizing Properly P And K Improved The Growth of Near-Mature Plantation Significantly and Applying Much Were Disadvantageous. It Would Cover Up Site Affections on Growth of Tree When Affecting Is Significantly After Fertilizing. Accord To Analysis of Effects of Different Treatments, It Showed That Optimal Application Amounts of P、K Were P_2O_5 90g、K_2O 90g Per Tree, Compared With The Control, Which Increased The Periodic Increments of Volume 13.4% After 4 Years'. Fertilization Could Promote Smaller Diameter Grade Tree Growing. Regression Analysis Showed That Fertilization Appeared Significant Correlation To Growth Of Near-Mature Forest.

Key words: masson pine near-mature plantation; fertilization; growth effect; analysis of variance

通过林地施肥能改善立地条件，提高林地生产力。因此，许多林业工作者对不同用材树种及不同林龄阶段的人工林开展过施肥技术研究，以此提高用材林的木材产量与材种规格，从而达到提高经济效益的目的[1-3]。马尾松(*Pinus massoniana*)是我国南方主要工业用材树种之一，广泛分布于16个省(自治区、直辖市)。过去，马尾松人工林培育目标主要为短周期工业用材林，因此，科研人员先后对马尾松人工幼龄林与中龄林施肥效应进行了研究总结[4-7]。但随着日益增长的马尾松大径材市场，林木的采伐年龄有所延长，很有必要开展马尾松近熟林的施肥技术研究，以便为马尾松人工大径材林培育关键技术提供科学依据。在马尾松南带产区10~20年生为马尾松人工林的材积速生期，为此，作者选择了17年生马尾松近熟林开展施肥试验，根据林分施肥后连续4年(次)的观测材料，对不同施肥处理的林分生长效应进行定量的分析。

1 试验地概况

试验地位于广西凭祥市大青山林区，地理坐标为106°43′E，22°06′N，属南亚热带季风气候区，年降水量1400mm，年平均气温19℃。林地为花岗岩发育成的红壤，海拔500m，土层厚度1m以上，腐殖质层厚10cm以上，pH4.5，呈弱酸性。因长期的淋溶风化成土作用，缺P少K是该区域性土壤养分状况的主要特点[8]。

试验林为17年生马尾松人工近熟林。前茬为杉木人工林，主伐炼山后进行块状整地，1年生裸根苗定植造林，初植密度为3600株/hm^2，先后在10年生与15年生进行2次间伐，施肥时调整为525株/hm^2。林分平均胸径为25.6cm，平均树高为15.7m，平均蓄积198.1m^3/hm^2。

2 试验设计与施工

试验方案采用正交设计[9]，3因素(肥种)4水平，共16个处理，3次重复，随机区组。试验小区面积为20m×20m，在林木上坡方向开半环状小沟(深10cm左右)，然后将肥料施入，覆土。施肥前及施肥后的每年年终对林木进行测定分析。试验林初始状况如表1所示，各试验因素与水平如表2所示。

表1 试验林初始林分因子统计

林分因子	处理															
	1	2	3	4	5	6	7	8	9	10	11	12	13	14	15	16
胸径(cm)	25.45	25.78	25.63	25.68	26.26	27.45	27.50	25.43	26.85	25.12	25.18	24.97	24.72	25.00	27.50	25.41
树高(m)	15.62	15.67	15.21	16.14	15.95	15.77	16.38	15.39	15.62	15.41	14.92	15.84	16.101	5.17	16.46	15.59
蓄积(m^3/hm^2)	189.9	193.3	190.4	198.6	208.5	220.4	234.3	185.3	212.6	175.0	177.6	190.8	187.6	175.9	236.5	191.7

表2 肥种及各水平施肥量 (g/株)

肥种	水平			
	1	2	3	4
N	0	120	240	360
P_2O_5	90	180	270	
K_2O	0	90	180	270

3 结果与分析

3.1 不同施肥处理与不同肥种水平对马尾松近熟林定期生长量的影响

林木生长状况与林地质量密切相关，施肥对林地质量的影响效果可以根据林木的生长变化来分析。根据表3的林分定期生长量资料与正交试验设计的原理，对各试验因子水平的平均定期生长量进行了统计(表4)。通过表3统计数据分析表明，施肥4年后，以林分蓄积定期生长量为主要参考指标，增产效应最好的为2号处理(N1P2K2)，蓄积定期生长量为75.22m^3/hm^2，比对照处理1号(N1P1K1)高13.40%，最差施肥处理组合为7号处理(N2P3K4)，蓄积定期生长量为55.22m^3/hm^2，比对照处理1号(N1P1K1)低16.75%。通过表4的统计分析可知，各肥种效应最佳水平分别为N1、P2、K2；4年蓄积定期生长量分别为70.23、70.91、71.23m^3/hm^2。综合分析可知，施N肥对马尾松近熟林生长不利，而施用适量的P肥与K肥对促进马尾松近熟林生长有利，但施用过量则有副作用。

表3 不同施肥处理的定期生长量统计

生长期	项目	施肥处理															
		1	2	3	4	5	6	7	8	9	10	11	12	13	14	15	16
1年	胸径(cm)	1.30	1.39	1.64	1.43	1.51	1.90	1.49	1.33	1.47	1.70	1.41	1.62	1.21	1.64	1.49	1.46
2年		1.75	1.88	2.06	1.84	2.06	2.15	1.63	1.79	1.72	2.01	1.88	2.22	1.69	2.09	1.60	1.64
3年		2.15	2.32	2.35	2.24	2.58	2.45	1.80	2.24	1.98	2.30	2.28	2.80	2.19	2.51	1.72	1.82
4年		2.52	2.81	2.83	2.65	3.10	2.60	1.98	2.65	2.21	2.58	2.66	3.22	2.48	2.88	1.81	1.99
1年	树高(m)	0.64	0.83	0.70	0.67	0.86	0.60	0.82	0.68	0.78	0.87	0.69	0.85	0.83	1.06	0.74	0.91
2年		1.58	1.43	1.39	1.42	1.68	1.30	1.45	1.39	1.46	1.45	1.42	1.51	1.81	1.84	1.49	1.64
3年		1.91	1.90	1.64	1.78	1.96	1.51	1.71	1.70	2.01	1.86	1.71	1.84	2.10	2.13	1.70	2.00
4年		2.41	2.10	1.98	1.90	2.25	1.89	2.08	1.99	2.00	2.19	1.95	2.24	2.34	2.41	2.11	2.21

（续）

生长期	项目	施肥处理															
		1	2	3	4	5	6	7	8	9	10	11	12	13	14	15	16
1年	蓄积（m^3/hm^2）	26.77	34.00	32.14	30.46	31.89	35.25	31.98	23.67	31.06	33.10	27.83	30.01	28.82	33.57	29.30	28.84
2年		40.36	48.19	46.04	43.86	43.61	46.54	36.46	38.25	41.79	44.47	43.33	48.48	44.73	47.89	41.62	39.34
3年		54.42	62.13	59.03	56.66	55.43	57.65	48.83	51.86	52.09	55.44	59.53	63.67	60.64	61.46	53.66	51.21
4年		66.33	75.22	72.12	68.32	66.22	68.53	55.22	64.34	63.12	65.32	69.67	75.08	69.78	74.56	66.78	61.88

表4 不同肥种水平的马尾松近熟林定期生长量统计

肥种	水平	胸径(cm)				树高(m)				蓄积(m^3/hm^2)			
		1年	2年	3年	4年	1年	2年	3年	4年	1年	2年	3年	4年
N	1	1.44	1.88	2.27	2.70	0.71	1.46	1.81	2.10	30.84	44.61	58.22	70.23
	2	1.56	1.87	2.27	2.58	0.74	1.46	1.72	2.05	30.70	40.72	53.94	64.38
	3	1.55	1.96	2.34	2.67	0.80	1.46	1.86	2.10	30.504	4.52	57.68	68.50
	4	1.45	1.76	2.06	2.29	0.89	1.70	1.98	2.27	30.13	43.40	56.74	68.25
P	1	1.37	1.81	2.23	2.58	0.78	1.63	1.70	2.05	29.64	42.62	55.65	66.34
	2	1.66	2.03	2.40	2.82	0.84	1.51	1.85	2.15	33.98	46.77	59.17	70.91
	3	1.51	1.75	2.04	2.32	0.74	1.44	1.69	2.03	30.31	41.36	55.76	66.45
	4	1.46	1.87	2.28	2.63	0.78	1.49	1.83	2.09	28.25	42.48	55.81	67.61
K	1	1.52	1.86	2.18	2.44	0.71	1.49	1.78	2.12	29.67	42.39	55.70	66.61
	2	1.50	1.94	2.36	2.74	0.82	1.53	1.85	2.18	31.30	45.48	58.71	71.23
	3	1.52	1.92	2.27	2.64	0.81	1.52	1.87	2.10	30.11	43.49	56.12	68.44
	4	1.46	1.75	2.13	2.42	0.80	1.53	1.86	2.13	31.09	41.88	55.18	65.16

3.2 不同肥种水平对马尾松近熟林生长效应的影响及其肥效持续性分析

林分初始胸径与树高为非试验因素，采用协方差分析方法可排除其对试验结果的影响，从而可正确分析施肥试验因素及各水平的贡献。以参数C表示试验的基础值，即当各施肥试验因素取1水平(对照)，非试验因素初始树高、胸径取0时试验的平均值；效应值即为各试验因子水平比基础值C增大多少[10]。表5表明：①N肥对胸径生长起正效应，但效应微弱，对树高的生长第3年开始表现出负效应，N2，N4水平对蓄积生长量产生的效应值为负，这表明施N肥短期内对马尾松近熟林生长不利。②P肥各水平的效应值表明，P2水平对林木生长有利，P3及P4水平短期内对树高与蓄积生长呈负效应，这表明适当的P肥用量能促进马尾松近熟林生长，过量却有副作用。③K肥各水平与对照K_1相比，K2及K3水平的肥量对胸径、树高、蓄积生长表现出正效应，但K4水平短期内对胸径、蓄积生长均产生负效应，这表明施适量K肥能促进马尾松近熟林生长，过量则有副作用。综合各肥种水平的效应值与营林成本分析，最佳处理组合为N1P2K2，即每株混施N0g、$P_2O_5$90g、K_2O90g。

初始胸径值D。对胸径、树高、蓄积生长均表现出负效应，这表明施肥对小径级木的生长有利(效应值$\Delta=D_0\times\beta$，β为单位效应值，$\beta<0$时，D_0小者效应值大)。树高初始值H_0对胸径、树高、蓄积生长的效应值均为正效应，因树高初始值H_0对林地质量有良好的指示作用，表明林木生长与林地本身质量密切相关。

表 5　不同肥种水平对马尾松近熟林生长的效应值统计

因子	水平	胸径(cm)				树高(m)				蓄积(m^3/hm^2)			
		1年	2年	3年	4年	1年	2年	3年	4年	1年	2年	3年	4年
N	1	0	0	0	0	0	0	0	0	0	0	0	0
	2	0.05	0.01	0.01	0.01	0.07	0.06	-0.08	-0.19	-1.63	-4.07	-3.15	-2.26
	3	0.11	0.07	0.12	0.15	0.09	0.01	0.01	-0.24	0.41	0.81	0.60	0.49
	4	0.01	-0.13	0.10	0.11	0.17	0.24	-0.13	-0.72	-1.26	-1.98	-1.11	0.20
P	1	0	0	0	0	0	0	0	0	0	0	0	0
	2	0.28	0.24	0.28	0.39	0.07	0.12	0.11	0.06	5.31	5.61	4.88	4.31
	3	0.09	-0.04	0.10	0.13	-0.01	-0.15	-0.13	0.01	0.41	0.40	1.86	2.35
	4	0.11	0.05	0.10	0.12	-0.02	-0.17	-0.12	0.03	-0.78	-0.11	1.23	1.88
K	1	0	0	0	0	0	0	0	0	0	0	0	0
	2	-0.03	0.07	0.14	0.18	0.11	0.05	0.24	0.65	-0.15	0.94	1.82	1.99
	3	0.01	0.05	0.11	0.14	0.09	0.03	0.22	0.38	0.95	1.27	1.36	1.69
	4	-0.06	-0.12	0.13	0.13	0.06	0.02	0.11	0.17	-0.14	-3.06	1.22	1.14
D_0		0.05	-0.03	-0.03	-0.04	-0.05	-0.07	-0.08	-0.08	0.80	-0.78	-0.58	-0.53
H_0		0.01	0.03	0.04	0.05	0.03	0.03	0.02	0.02	3.10	4.61	4.84	5.10
基础值		-0.39	2.02	2.41	2.85	1.31	2.81	3.66	4.45	-39.77	-8.85	-6.65	-4.13

表 6　各年度不同肥种水平效应值的方差分析(F 值)

因子	胸径(cm)				树高(m)				蓄积(m^3/hm^2)			
	1年	2年	3年	4年	1年	2年	3年	4年	1年	2年	3年	4年
N	0.35	0.54	1.51	1.86	0.73	1.45	1.39	1.65	0.67	0.75	0.09	0.34
P	1.69	1.10	4.11*	4.47*	0.24	0.70	1.89	1.03	5.52**	3.44*	3.69*	2.40
K	0.10	0.58	0.75	1.52	0.33	0.04	0.94	1.65	0.18	0.66	0.45	0.22
D_0	1.11	0.15	3.01	2.32	0.82	1.38	1.77	2.55	1.27	0.27	0.88	1.22
H_0	0.02	0.08	2.67	0.16	0.14	0.08	1.97	2.83	6.81*	3.48	2.08	2.09

注：* 表示差异显著；* * 表示差异极显著。

从表 6 的方差分析可知(临界值 F_α 取 0.05 水平)，施肥后 4 年内 N 肥与 K 肥对马尾松近熟林生长基本无显著影响。P 肥对树高生长影响不显著，但对胸径与蓄积有显著影响，因 P 肥为缓效肥，肥效后续效果有待进一步观测。对初始值 D_0、H_0 进行协方差分析表明，在施肥后 4 年内，初始值 D_0、H_0 对林木生长各指标无显著影响，说明试验林立地质量基本相近。

3.3　试验因素对生长影响的偏相关分析

通过偏相关分析可区别不同试验因素对林木生长影响的密切程度[10,11]。根据偏相关分析表明，施肥措施与马尾松近熟林生长效应呈显著相关性。

从表 7 偏相关分析结果可知：①短期内 N 肥对胸径的生长无显著的相关性，但对树高、蓄积的生长却表现出显著相关性，表明 N 肥对近熟林树高生长产生的负效应是明显的，短期内这一现象与中龄林施肥效果相似[5,6]；②P 肥对各项生长指标均表现出显著相关性，表明 P 肥的施放及用量水平对马尾松近熟林的生长影响差别明显；③K 肥在施肥 2 年后对林分树高、蓄积的生长有显著的相关性，但对胸径的生长无显著相关性，表明适量的 K 肥能改善立地质量，促进马尾松近熟林的生长。

初始胸径值 D_0 短期内对各项生长指标表现出相关系数为负值，表明施肥能促进小径级木的生长。初始树高值 H_0 在施肥后 4 年内对胸径、树高生长无显著相关性，对蓄积生长相关性 2 年后迅速减小，因

表7　各试验因素对生长影响的偏相关分析(偏相关系数)

因子	胸径(cm)				树高(m)				蓄积(m^3/hm^2)			
	1年	2年	3年	4年	1年	2年	3年	4年	1年	2年	3年	4年
N	0.22	0.27	0.30	0.34	0.31*	0.42*	0.52**	0.49**	0.30	0.32	0.36*	0.43*
P	0.45*	0.37*	0.45*	0.46*	0.19	0.30	0.52**	0.16	0.47**	0.42*	0.35*	0.52**
K	0.12	0.28	0.27	0.23	0.21	0.08	0.26	0.46**	0.16	0.30	0.43*	0.39*
D_0	0.27	-0.10	0.24	0.31	-0.24	-0.32*	0.28	0.48**	0.29	-0.13	-0.45**	-0.48**
H_0	0.03	0.07	0.06	-0.20	0.10	0.07	0.25	0.30	0.55**	0.43**	-0.04	0.17
r复	0.56**	0.50**	0.68**	0.62**	0.46*	0.57**	0.67**	0.86**	0.80**	0.58**	0.74**	0.75**

注：*表示显著相关；**表示极显著相关。

初始树高值H_0对林地质量有一定的指示作用，表明前期林地本身质量是林分生长密切相关的因子，在施肥2年后肥效显著时可掩盖林地质量的作用。从表7中复相关系数变化可知，5种相关因子对马尾松近熟林生长的综合效应是明显的。

4　结论

(1)综合分析表明，在南方红壤地区，施N肥对马尾松近熟林生长无促进作用，施适量的P，K肥对林木生长有显著的促进作用，过量却有副作用。施肥对小径阶木生长有利。

(2)以林分蓄积定期生长量为主要参考指标，根据各肥种水平效应值综合分析，最佳处理组合为N1P2K2，即每株混施N 0g，P_2O_5 90g，K_2O 90g，其4年蓄积定期生长量比对照高13.40%。

(3)根据相关分析表明，施肥措施能改善林地质量，短期内显著促进马尾松近熟林生长。林地质量是林木生长的决定性因子，但肥效高峰期能掩盖林地本身对林木生长的影响。因林木生长周期长，其中P肥为缓效肥，肥效后续反应状况有待进一步观测。

参考文献

[1]孙晓梅，张守攻，祁万宜，等．北亚热带高山区日本落叶松幼龄林施肥技术的研究[J]．林业科学研究，2007，20(1)：68-73.

[2]吴立潮，胡曰利，吴晓芙，等．杉木中龄林施肥效应与效益研究[J]．中南林学院学报，1997，17(3)：1-7.

[3]董健，尤文忠，范俊岗，等．日本落叶松近熟林施肥效应[J]．东北林业大学学报，2002，30(3)：8-12.

[4]谌红辉，温恒辉．马尾松人工幼林施肥肥效与增益持续性研究[J]．林业科学研究，2000，13(6)：652-658.

[5]谌红辉，温恒辉．马尾松人工中龄林施肥肥效与增益持续性研究[J]．林业科学研究，2001，14(5)：533-539.

[6]谌红辉，温恒辉．马尾松中龄林平衡施肥研究[J]．林业科学研究，2005，34(4)：171-174.

[7]梁瑞龙，温恒辉．广西大青山马尾松人工林施肥研究[J]．林业科学研究，1992，5(1)：111-114.

[8]殷细宽．地质学基础[M]．北京：农业出版社，1998.

[9]北京林学院．数理统计[M]．北京：中国林业出版社，1980.

[10]唐守正．多元统计分析方法[M]．北京：中国林业出版社，1986.

[11]李贻铨，张建国，纪建书，等．杉木施肥肥效与增益持续性研究[J]．林业科学研究，1996，19(增刊)：18-26.

[原载：林业资源管理，2012(02)]

顶果木的耐旱性评价

郝　建[1]　马小峰[2]　谌红辉[1]　刘志龙[1]　陈建全[1]，郭文福[1]
([1]中国林业科学研究院热带林业实验中心，广西凭祥　532600；
[2]凭祥出入境检验检疫局，广西凭祥　532600)

摘　要　为了评价顶果木(*Acrocarpus fraxinifolius*)的耐旱性，采取模拟自然极度干旱条件方法，利用盆栽方式，研究顶果木在干旱胁迫期间的生长量、植株体的水分分布。结果表明，停止供水后第11d到第16d的土壤含水量降幅较大，顶果木地径和苗高生长增幅出现明显的下降趋势，地茎生长的降幅尤为明显。因此，干旱胁迫出现在第13d之后，干旱胁迫的临界土壤含水量在20%左右；结合顶果木苗木生长指标分析，第21d的土壤含水量(12.99%)可能已经达到顶果木的萎蔫系数；顶果木的地上部分对外界的水分环境状况响应速度比地下部分快，顶果木叶片和根系的含水量与土壤含水量之间具有显著的相关性，相关系数分别达到0.88和0.91，茎的含水量与土壤含水量之间的相关系数达到0.97的极显著水平。

关键词　顶果木；干旱胁迫；耐旱性；水分分布

Evaluation of Drought Tolerance of *Acrocarpus fraxinifolius*

HAO Jian[1], MA Xiaofeng[2], CHEN Honghui[1], LIU Zhilong[1], CHEN Jianquan[1], GUO Wenfu[1]
([1]*Experimental Center of Tropical Forestry*, *Pingxiang* 532600, *Guangxi*, *china*;
[2]*Pingxiang Entry-exit Inspection and Quarantine Bureau*, *Pingxiang* 532600, *Guangxi*, *china*)

Abstract: In order to evaluate the drought tolerance of *Acrocarpus fraxinifolius*, through mimicing natural extreme drought condition, researched the growth, the plant body moisture distribution of *A. fraxinifolius* during the period of drought stress on using potting way. The results showed that, from 11th day to the 16th day after stopping watering, the ground diameter increment and the nursery height increment of *A. fraxinifolius* began falling, obviously expressed the ground diameter increment. Therefore, drought stress environment appeared 13th day later, and the critical soil water capacity was 20%. Known from the growth and soil water capacity, the soil water capacity of 21th day(12.99%) may have reached the wilting coefficient. The response rate to the soil water capacity of the overground parts was faster than the underground parts. The correlation of leaves and soil water content, root and soil water content were significant, correlation coefficients were 0.88 and 0.91. And the correlation coefficient between the stem moisture content and soil moisture content reached 0.97, extremely significant level.

Key words: *Acrocarpus fraxinifolius*; drought stress; drought tolerance; moisture-distribution

顶果木(*Acrocarpus fraxinifolius*)，属国家三级重点保护树种，是苏木科顶果木属落叶大乔木，树高40m以上，枝下高20m以上，胸径40～80cm，最粗达120cm以上[1]。树干通直圆满，出材率高，特别适合培养大径材；材质轻韧，花纹美丽，心材淡红褐色，少开裂，较耐腐，可作上等家具用材，也是优质纤维用材；树干挺拔、树冠开阔，还可用作行道树、风景树或生态防护树种。分布于广西西部、西南部，贵州西部和云南南部、西南部至西部；适应性较强，对土壤适应范围较广，在石灰岩山地或土山都能生长。在广西龙州县石灰岩山地，22年生顶果木树高达25.6m，胸径35.8cm，单株材积1.1895m³[2]。

从1989年开始，针对顶果木的生物学特性[3]、虫害[4]、育苗造林[5-7]、无性繁殖[8,9]、立地适应性[10-12]等多个方面做了研究。顶果木作为广西岩溶地区的优良造林树种之一，水分对其具有重要影响作用。近年来，由于全球气候变化影响，局部地区易出现久旱等极端气候，对正常造林工作的开展带来一定困难。同时，劳动力紧俏，造林季节劳动力不足，时常导致造林工作滞后，错过雨季造林。在此背景下开展顶果木耐旱性研究，评价其耐旱能力，为干旱地区或非湿润季节造林提供理论依据。

1 材料方法

1.1 材料

长势一致的半年生顶果木轻基质容器苗，平均地径 0.276cm，平均苗高 23.60cm。

1.2 材料处理方法

试验方法参考蔡晓明等研究一球悬铃木(*Platanus occidentalis*)无性系耐旱性[13]和韩艳研究 5 种常绿阔叶树幼苗的耐旱性[14]的研究方法。选取 100 棵长势一致的半年生顶果木轻基质容器苗，定植在规格为 40cm×30cm 育杯中，每杯一株，土壤为质地均匀的林地壤土。前两周，统一正常供水，待苗木恢复正常生长后，停止供水。此时随机抽取 10 株苗木，测量苗高、地径，同时测定育杯中土壤和植株根、茎、叶的含水量。之后每隔 5d 测定一次，直至苗木出现枯死为止。本试验过程均在透光通风挡雨的开放温室内进行。

1.3 指标测量与测定

1.3.1 生长指标测量

停止水分供应时和之后每隔 5d，用钢卷尺测量苗木苗高生长量，单位为 cm，精确至 0.1cm。用游标卡尺测量苗木地径生长量，单位为 cm，精确至 0.01cm。

1.3.2 土壤和植株含水量测定

土壤含水量测定，取洗净的编有号码的有盖铝盒，放在 105±2℃的烘箱中烘干，取出放入干燥器中冷却，天平称得其恒重(A)。将约 10.0g 土样，均匀地平铺在铝盒中，准确称重(B)。将敞开盖子的铝盒放入 105±2℃的恒温烘箱中，盖放在铝盒旁侧烘 6h 左右。将铝盒取出，盖上盖子，置铝盒于干燥器中 20~30min，冷却到室温称重。启开铝盒盖，再烘 2h，冷却，称至恒重(C)。

$$土壤含水量=\frac{B-C}{C-A}\times100\%$$

式中：A—铝盒的重量(g)；B—土样与铝盒的重量(g)；C—烘干土与铝盒的重量(g)。

每次进行植株含水量测定时，随机抽取 3 株顶果木苗木，将单株顶果木苗木叶、茎、根(去除土壤等杂质)分别分开，天平称鲜重。然后放在 60℃烘箱烘 10h，至恒重，天平称其干重。

$$叶的含水量=\frac{叶鲜重-叶干重}{叶鲜重}\times100\%;$$

$$茎的含水量=\frac{茎鲜重-茎干重}{茎鲜重}\times100\%$$

$$根的含水量=\frac{根鲜重-根干重}{根鲜重}\times100\%$$

2 结果与分析

2.1 水分胁迫后土壤含水量的变化动态

图 1 显示的是从试验处理开始，育杯内土壤含水量的变化动态。在停止供水时，育杯内土壤含水量为 39.47%，然后依次递减到第 21d 的 12.99%。第 11d 到第 21d 这 5d 内的土壤含水量降幅最大。从第 1d 开始到第 21d，共 5 次测定结果之间存在极显著差异(F = 244.422，P = 0.0001)。从数据来看，第 11d 到第 16d 土壤含水量降幅为 10.72%，降幅最大。

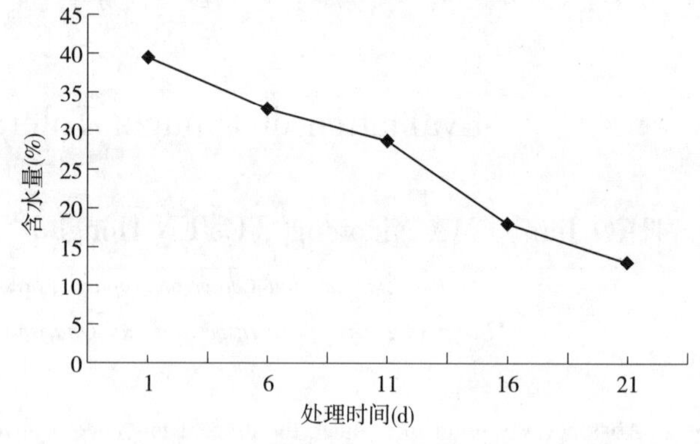

图 1 干旱胁迫后的土壤含水量

2.2 水分胁迫下顶果木生长量的动态变化

如图 2、3 所示，在水分胁迫处理期间顶果木苗高、地径增量分别出现先增加后降低的趋势。在整个水分胁迫处理的共 4 个 5d 时间内的顶果木地径增量之间存在极显著差异(F=67.211，P=0.0001)。除第 1 个和第 2 个 5d 地径增量之间的差异不显著外，其余之间差异均为是极显著。4 个不同时间段内顶果木苗高增量之间存在极显著差异(F = 83.477，P = 0.0001)。除第 1 个和第 3 个 5d 的苗高增量之间的差异不显著外，与其他时间段以及其他时间段之间的差异均是极显著。在停止供水后的第 2 个 5d 内苗高生长出现高峰增量为 8.3cm，分别高于其他 3 个 5d 生长增量。第 3 个 5d 增量开始下降，说明该段时间内生长受到抑制，其增量是 5.8cm，略高于第 1 个 5d 的苗高增量，说明生长开始受抑制的时间可能出现在第 14d 之后。到了第 16d 之后顶果木的苗高生长受抑制更加明显，增量仅为 1.2cm。

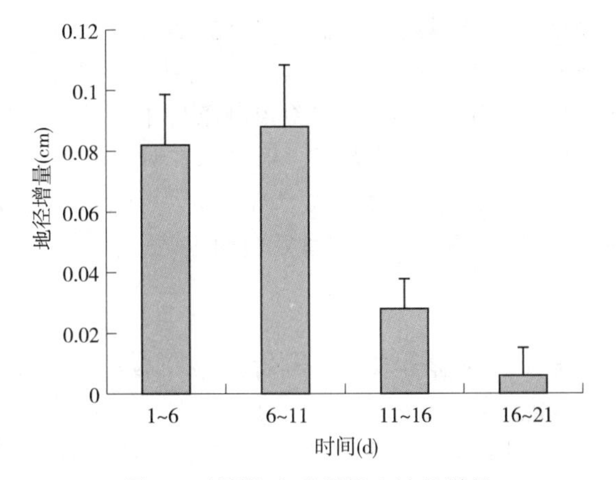

图 2 干旱胁迫后顶果木地径增量

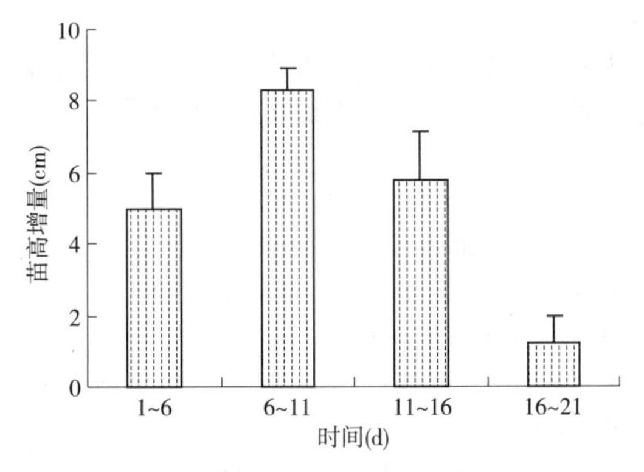

图 3 干旱胁迫后顶果木苗高增量

2.3 水分胁迫下顶果木含水量的动态变化

图 4 显示的是顶果木的叶、茎、根的含水量随着水分胁迫时间的延长而不同程度的下降的结果。叶片含水量的降幅较大，茎的含水量降幅次之，根的含水量降幅较小而且缓慢。顶果木叶、茎、根的含水量在水分胁迫不同时间段均存在极显著差异(叶：$F=15.768$，$P=0.0003$，茎：$F=28.05$，$P=0.0001$，根：$F=27.734$，$P=0.0001$)。测定结果显示，叶片和茎的含水量在第 16d 开始维持在一个相对恒定水平，而同一时期根系的含水量相对较高。所以，顶果木的地上部分(叶片与茎)相对于地下部分(根系)的含水量对水分胁迫响应速度较快。可以直接测定地上部分含水量来考察顶果木的耐旱性。

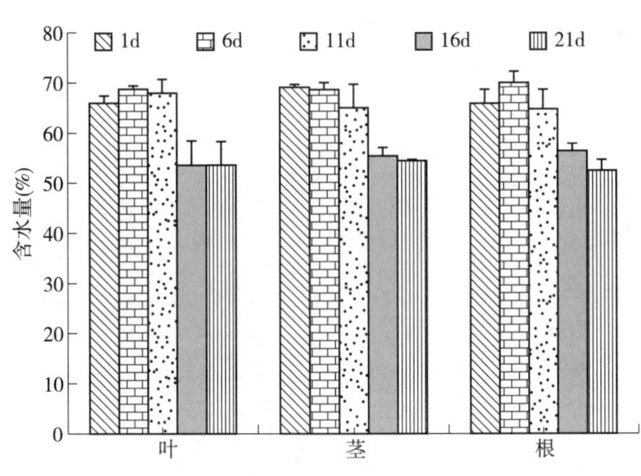

图 4 干旱胁迫下顶果木苗含水量的变化动态

2.4 干旱胁迫下顶果木不同器官含水量与土壤含水量的相关性分析

表 1 顶果木苗木含水量和土壤含水量的相关性分析

指标	土壤含水量	叶含水量	茎含水量	根含水量
土壤含水量	1	0.88*	0.97**	0.91*
叶片含水量	0.88*	1	0.96**	0.96**
茎含水量	0.97**	0.96**	1	0.96**
根含水量	0.91*	0.96**	0.96**	1

注：* $P<0.05$，** $P<0.01$。

表 1 反映的是顶果木不同器官和土壤含水量共 4 个指标之间的相关性分析。顶果木植株含水量与土壤含水量之间呈正相关关系，其叶片和根含水量与土壤含水量之间具有显著的相关性，相关系数分别达到 0.88 和 0.91；茎含水量与土壤含水量之间的相关系数更是达到 0.97 的极显著水平。

3 结论与讨论

植物对环境的适应能力与环境的水分供应水平状况有关。在其生长及发育过程中，经常会遇到水分胁迫情况[15]。同种植物种类在一般情况下，植物体内含水量高低主要取决于土壤水分含量，植物不同生长期对水分的要求和适应能力也不同[16]。在停止供水后的第 11d 到第 16d 内的土壤含水量降幅较大，顶果木地径和苗高生长增幅出现明显的下降趋势，地茎生长的降幅尤为明显。干旱胁迫环境可能出现在第 13d 之后，此时的土壤含水量可能为 20.0%。结合顶果木苗木生长指标分析，第 21d 的土壤含水量降低到 12.99%，可能已经达到顶果木的萎蔫系数。

在干旱胁迫期间，顶果木叶片含水量的降幅较大，茎的含水量降幅次之，根的含水量降幅最小而且缓慢。顶果木叶片以及茎的含水量在第 16d 开始维持在一个相对恒定水平，而同一时期顶果木根系的含水量相对较高。所以，顶果木的地上部分对外界的水分环境状况响应速度比地下部分快。一般耐旱性较强的植物的叶片含水量要比耐旱性弱的植物的高且稳定[17]。所以，可以直接测定顶果木地上部分含水量考察顶果木的体内的水分变化情况。顶果木植株含水量与外界环境的土壤含水量之间呈显著或极显著生物正相关关系。顶果木叶片以及其根系的含水量与外界环境的土壤含水量之间具有显著的相关性，相关系数分别达到0.88和0.91；顶果木茎的含水量与外界环境的土壤含水量之间的相关系数更是达到0.97的极显著水平。

在此次人为控制的试验条件下，在停止供水后的第21d胁迫严重顶果木各种生理机能开始丧失，生长停止开始出现死亡。然而在自然条件下，由于光照强度大、空气湿度小、空气流动速度快等因素，土壤水分的蒸发速度比试验控制条件下要快，一旦发生异常的长期干旱天气，顶果木在野外干旱条件下的正常生长的时间肯定要少于13d，可能缩短至7d左右，其抗旱生长时间也肯定会相应缩短。

参考文献

[1]朱积余，廖培来．广西名优经济树种[M]．中国林业出版社，2006.
[2]梁瑞龙，黄开勇．广西热带岩溶区林业可持续发展技术[M]．中国林业出版社，2010.
[3]杨成华．速生珍稀树种顶果木[J]．贵州林业科技，1989，17(2)：59-61.
[4]顾茂彬，陈佩珍，王春玲．小黄粉蝶的初步研究[J]．广东林业科技，1991，2：33-38.
[5]何关顺，文宝，何广琼．乡土速生树种顶果木育苗技术[J]．广西林业，2008：32-34.
[6]周全连，李文付．顶果木栽培技术[J]．林业科技开发，2007，21(1)：91-93.
[7]黎素平，郭耆，朱昌叁，等．顶果木高产栽培技术[J]．现代农业科技，2011，14：229-230.
[8]黎素平，朱昌叁，廖英汉，等．顶果木扦插繁殖技术研究[J]．林业实用技术，2011，10：25-27.
[9]周传明，秦武明，吕曼芳，等．顶果木离体培养研究[J]．安徽农业科学，2012，40(3)：1457-1458.
[10]吕仕洪，李先琨，陆树华，等．广西岩溶乡土树种育苗及造林研究[J]．广西科学，2006，13(3)：236-240.
[11]吕仕洪，李先琨，陆树华，等．桂西南岩溶地区珍稀濒危树种育苗与造林初报[J]．广西植物，2009，29(2)：222-226.
[12]朱积余，侯远瑞，刘秀．广西岩溶地区优良造林树种选择研究[J]．中南林业科技大学学报，2011，31(3)：81-85.
[13]蔡晓明，卢宇蓝，施季森．一球悬铃木无性系耐旱性研究[J]．西北林学院学报，2010，25(6)：19-24.
[14]韩艳．5种常绿阔叶树幼苗的耐旱性和耐寒性研究[D]．浙江林学院，2009.
[15]东北林学院．森林生态学[M]．北京：中国林业出版社，1981：25-27.
[16]王沙生，高荣孚，吴贯明．植物生理学[M]．北京：中国林业出版社，1981：102-135.
[17]钱瑭璜，雷江丽，庄雪影．3种草本蕨类植物耐旱性研究[J]．西北林学院学报，2012，27(1)：22-27.

[原载：西北林学院学报，2013，28(03)]

马尾松不同径级产脂量及松脂成分差异研究

安　宁[1,2]　丁贵杰[1]　贾宏炎[2]　谌红辉[2]
([1]贵州大学造林生态研究所，贵州贵阳　550025；
[2]中国林业科学研究院热带林业实验中心，广西凭祥　532600)

摘　要　采用长期定位观测，定时、定株收获的方法，研究了马尾松9种径级与产脂量以及6种径级与松脂成分之间的关系。分析结果表明：不同径级产脂量差异显著，随着径级的增大，产脂量也随之增加，呈正相关关系，初步认定马尾松产脂林采割起始径级不小于18cm；不同径级的松脂主要成分一致，各成分含量存在一定差异，含油量有随着径级增大而增加的趋势，α-蒎烯含量高是马尾松松节油的特征。
关键词　马尾松；径级；产脂量；松脂

Study on Resin Yield and Turpentine Differences from Different Diameters of *Pinus massoniana*

AN Ning[1,2], DING Guijie[1], JIA Hongyan[2], CHEN Honghui[2]
([1]*The Institute of Silviculture and Ecology in Guizhou University*, *Guiyang* 550025, *Guizhou*, *China*;
[2] *The Experimental Centre of Tropical Forestry*, *CAF*, *Pingxiang* 532600, *Guangxi*, *china*)

Abstract: The long-term positioning observation, timing, method of harvesting plants, Study on the relationship between nine kinds of DBH and masson pine resin yield, And the relationship between six kinds of DBH and turpentine. The results show that: the resin yield in different DBH are significantly different, Resin yield increase with the DBH increasing, There was a positive correlation between the DBH and the resin yield. Initially identified: massion pine resin tapping forest trails starting DBH level is not less than 18cm; The main component of turpentine is same in different DBH. There are some differences in the content of each component, There are the trend of oil content increases with increasing DBH. High levels of α-pinene is characteristic of masson pine turpentine.
Key words: *Pinus massoniana*; diameter class; resin yield; resin

马尾松(*Pinus massoniana*)是我国松树中分布最广、栽培面积最大，同时是采割松脂的主要树种[1]。松脂主要成分有多种树脂酸和萜类[2]，是重要的化工原料。近年随着珍贵树种的发展，频发的松毛虫危害，造成马尾松林面积下降及可采脂树减少，现有马尾松林已不能满足林产化工业的需要，营建马尾松高产脂人工林是发展的必然趋势。以往对马尾松的研究都集中在栽培技术、生理生化以及基金项目：国家"十二五"科技支撑课题"马尾松速生丰产林定向培育关键技术研究与示范"、贵州省特助人才计划(TZJF-2007年20号)、贵州省重大专项(黔科合重大专项字[2012]6001号、中国林业科学研究院热带林业实验中心主任基金项目(RL2011-06)。遗传育种等方面[3-8]，在马尾松采脂林产脂量的影响因素、松脂组成等方面的研究很少[9-15]。因此，开展此项研究对完善马尾松高产脂林的定向培育技术体系具有重要意义。

1　试验地概况及研究方法

1.1　试验地概况

试验林设在位于广西凭祥市热带林业实验中心伏波实验场，为1991年营造的马尾松纯林，前茬杉木主伐后明火炼山。位于106°43′E，22°06N′，海拔500~520m，低山地貌，年平均气温19.5℃，年降水量1500mm，属于南亚热带季风气候区。土壤为花岗岩发育成的红壤，土层厚度>1m，腐殖质厚度>10cm。

1.2 研究方法

试验设9个处理：14cm、16cm、18cm、20cm、22cm、24cm、26cm、28cm、30cm径级。每个处理30株，收脂时单株称重并记录。采用“下降式”采脂法，具体方法见参考文献[16]。运用SPSS软件进行方差分析及多重比较[17]。试验于2008年布置，2009年在采脂期内进行产脂量数据采集，收脂时按单株称重并记录。

2 实验方法

2.1 松脂含油量及松节油成分的测定

按林业局松脂标准LY223-81测定松脂中的含油量。将所得松节油直接进气相色谱仪分析。方法见参考文献[18]。

2.2 松脂成分测定

样品在广西林业科学研究院林化中心测定，方法见参考文献[18]。

3 结果与分析

3.1 不同径级与产脂量的关系

从图1可以看出马尾松单株年产脂量随着径级的增大表现出上升的趋势，采脂量与采脂树木的胸径呈线性正相关关系，随着树木胸径的增大，单株产脂量随之增加。

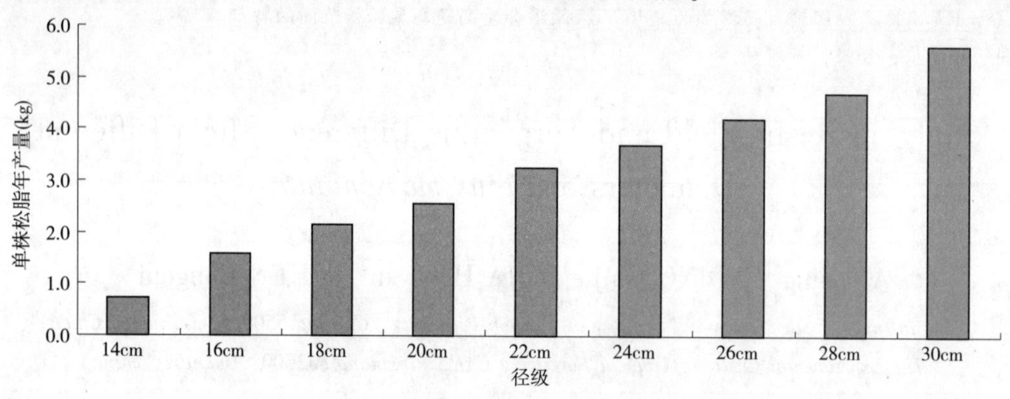

图1 不同径级单株松脂年产量

方差分析结果(表1)表明：不同径级间产脂量差异显著。14cm径级的单株年产脂量为0.73kg；径级20cm的年产脂量为2.55kg，相比产量增加149.38%；26cm径级的单株年产脂量为4.19kg，比20cm径级的产量增加64.29%，而30cm径级的单株年产脂量为5.62kg，比26cm径级的年产脂量增加34.13%(图1)。产量呈现此种趋势是由于径级越大，在相同的剖面负荷率下，剖面的面积越大，割破的树脂道数目也越多，因而产脂量也越大。

表1 不同径级产量方差分析表

处理源	*SS*	*Df*	*MS*	*F*	*Sig.*
处理间	16384.295	8	2048.037	111.052	0.000
处理内	331.958	18	18.442		
总计	16716.253	26			

多重比较(LSD)检验结果(表2)表明：不同径级间产脂量差异显著，具体检验结果及两两关系见表2。所有处理中，只有18cm径级和20cm径级的产脂量差异不显著，而其他径级之间产脂量均存在显著差异。

表2 不同径级多重比较检验表

		14cm	16cm	18cm	20cm	22cm	24cm	26cm	28cm	30cm
14cm	*MD*		14.14*	23.88*	30.51*	42.08*	49.65*	57.78*	66.41*	81.66*
	Sig.		0.001	0.000	0.000	0.000	0.000	0.000	0.000	0.000
16cm	*MD*	−14.14*		9.74*	16.37*	27.94*	35.52*	43.64*	52.27*	67.52*
	Sig.	0.001		0.012	0.000	0.000	0.000	0.000	0.000	0.000
18cm	*MD*	−23.88*	−9.74*		6.63	18.20*	25.77*	33.90*	42.53*	57.78*
	Sig.	0.000	0.012		0.075	0.000	0.000	0.000	0.000	0.000
20cm	*MD*	−30.51*	−16.37*	−6.63		11.57*	19.15*	27.27*	35.90*	51.15*
	Sig.	0.000	0.000	0.075		0.004	0.000	0.000	0.000	0.000

（续）

		14cm	16cm	18cm	20cm	22cm	24cm	26cm	28cm	30cm
22cm	*MD*	-42.08 *	-27.94 *	-18.20 *	-11.57 *		7.57 *	15.70 *	24.33 *	39.58 *
	Sig.	0.000	0.000	0.000	0.004		0.000	0.000	0.000	0.000
24cm	*MD*	-49.65 *	-35.52 *	-25.77 *	-19.15 *	-7.57 *		8.13 *	16.76 *	32.01 *
	Sig.	0.000	0.000	0.000	0.000	0.045		0.032	0.000	0.000
26cm	*MD*	-57.78 *	-43.64 *	-33.90 *	-27.27 *	-15.70 *	-8.13 *		8.63 *	23.88 *
	Sig.	0.000	0.000	0.000	0.000	0.000	0.032		0.024	0.000
28cm	*MD*	-66.41 *	-52.27 *	-42.53 *	-35.90 *	-24.33 *	-16.76 *	-8.63 *		15.25 *
	Sig.	0.000	0.000	0.000	0.000	0.000	0.000	0.024		0.000
30cm	*MD*	-81.66 *	-67.52 *	-57.78 *	-51.15 *	-39.58 *	-32.01 *	-23.88 *	-15.25 *	
	Sig.	0.000	0.000	0.000	0.000	0.000	0.000	0.000	0.000	

在实际生产与采脂操作过程中，对采脂树木的起始径级有相应的要求。采割径级过小，采脂割面过窄，产脂量低，松脂效益低，同时也增加了采脂成本；如果采割松脂起始径级偏大，减少了采脂年限，也会造成总体经济效益的下降。综合考虑马尾松采脂林起始径级应在18~20cm。

3.2 不同径级松脂成分差异研究

从表3及表4可以看出，在6个径级中：松香及松节油的组成一致，但各成分的含量存在一定差异。

马尾松松脂树脂酸主要化学成分有海松酸、异海松酸、湿地松酸、去氢枞酸、左旋海松酸、枞酸和新枞酸。在马尾松各树脂酸中左旋海松酸的含量最高，均超过45%，其中16cm含量达到53.58%，为6个径级中含量最高，而26cm含量最低为45.13%，二者相差8.5%；海松酸除14cm的0.73%外，其他径级含量均在0.44%~0.50%，变化幅度很小；枞酸除18cm的含量为11.57%，其他径级含量均在8.13%~9.99%；新枞酸含量最高是18cm为12.73%，最低是16cm的9.55%，其他径级含量在10.08%~11.66%；而去氢枞酸含量最高是26cm的0.39%，最低为18cm的0.11%。据报道马尾松产脂力与新枞酸和去氢枞酸呈正相关[17]，在本实验中并未表现出规律性。

表3 不同径级松香主要成分及含量

径级	松香主要成分含量(%)						
	海松酸	湿地松酸	异海松酸	左旋海松酸	去氢枞酸	枞酸	新枞酸
14cm	0.73	9.96	1.68	47.20	0.13	8.13	11.66
16cm	0.50	10.47	1.82	53.58	0.21	8.43	9.55
18cm	0.47	7.01	1.82	47.70	0.11	11.57	12.73
20cm	0.45	8.48	1.43	49.64	0.36	9.65	11.17
22cm	0.49	9.98	1.47	50.02	0.29	8.65	10.08
26cm	0.44	7.76	1.56	45.13	0.39	9.99	10.25

表4 不同径级松节油主要成分及含量

径级	松节油含量及主要成分含量(%)							
	含油量	α-蒎烯	莰烯	β-蒎烯	月桂烯	Δ3-蒈烯	长叶烯	石竹烯
14cm	9.39	64.19	1.23	2.30	0.54	0.12	14.23	5.15
16cm	13.70	63.28	1.12	3.41	0.31	0.43	16.98	3.34
18cm	11.21	70.35	1.22	4.35	0.05	0.86	12.80	4.37

（续）

径级	松节油含量及主要成分含量(%)							
	含油量	α-蒎烯	莰烯	β-蒎烯	月桂烯	Δ3-蒈烯	长叶烯	石竹烯
20cm	17.20	68.50	1.06	4.13	0.07	0.36	14.11	4.55
22cm	16.59	47.07	1.07	5.98	0.20	1.68	26.01	3.97
26cm	18.30	54.36	0.87	3.80	0.29	0.21	27.25	3.23

马尾松松节油主要化学成分有α-蒎烯、β-蒎烯、莰烯、Δ3-蒈烯、月桂烯、长叶烯和石竹烯。α-蒎烯含量高是马尾松松节油的特征。松节油在6个径级中的含量分别为9.39%、13.70%、11.21%、17.20%、16.59%和18.30%。14cm含量最低，26cm含量最高，松节油含量有随径级增大而增加的趋势。22cm径级α-蒎烯含量47.07%，其他径级的含量均高于54%；β-蒎烯在22cm径级中含量最高，但各径级中的含量均不超过6%，含量在2.30%～5.98%；莰烯的含量均在1%左右；月桂烯含量最高为14cm的0.54%，最低为18cm的0.05%，长叶烯含量依次为14.23%、16.98%、12.80、14.11%、26.01%和27.25%，22cm和26cm径级的含量明显高于其他径级。Δ3-蒈烯除在22cm含量略高外，其他径级含量均低于1.00%。长叶烯与石竹烯含量相差最大为26cm径级的24.02%，相差最小为18cm径级的7.43%。可以明显看出各径级的松脂成分基本一致，只是含量存在一定差异。

4　结论与讨论

(1)随着径级的增大，马尾松产脂量表现出上升的趋势，呈正相关关系，这与王长新[13]、蔡树威等[14]、舒文波等[19]研究相一致。径级为14cm的马尾松，单株年松脂产量为0.73kg，而30cm径级的松脂产量为5.62kg，增加了4.98kg，各径级产脂量除18cm与20cm差异不显著外，其他径级之间产脂量均差异显著。在实际生产与采脂操作中，考虑到产脂量与树木生长情况，马尾松采脂径级应在18cm以上，这与国家林业局2007年发布的林业行业标准松脂采集技术规程相一致。

(2)马尾松各径级的松脂有相同的组成，各成分含量存在一定差异。马尾松所含树脂酸中左旋海松酸含量最高，均超过45%；含油量有随着径级增大而增加的趋势，14cm含油量最低为9.39%，26cm含油量最高为18.30%。α-蒎烯含量高是马尾松松节油的特征，在6个径级中α-蒎烯含量均高于47%；β-蒎烯在22cm径级中含量最高，为5.98%，β-蒎烯市场价格比α-蒎烯要高，但本试验其含量在各径级中并未表现出规律性。在松香和松节油的组成上一致，只是各组分的含量存在一定差异。

参考文献

[1]周政贤．中国马尾松[M]．北京：中国林业出版社，2001：14.

[2]刘玉春．1995-2000年中国松香的生产、消费和发展趋势[J]．林产化工通讯，2001，35(5)：31-33.

[3]丁贵杰，周志春，王章荣．马尾松纸浆材林培育与利用[M]．北京：中国林业出版社，2006.

[4]谌红辉，丁贵杰．马尾松造林密度效应研究[J]．林业科学．2004，40(1)；92-98.

[5]丁贵杰．马尾松人工纸浆材林采伐年龄初步研究[J]．林业科学，2000，36(1)：15-20.

[6]姚瑞玲，丁贵杰．不同密度马尾松人工林凋落物及养分归还量的年变化特征[J]．南京林业大学，2006，30(5)：83-86.

[7]周志春，李光荣，黄光霖等．马尾松木材化学组分的遗传控制及其对木材育种的意义[J]．林业科学，2000，36(2)：110-115.

[8]吴启彬．马尾松优良种源内优良林分选择的研究[J]．福建林业科技，2008，35(1)：21-25.

[9]陈祖洪，常新民，陈海燕．采脂间隔期对马尾松产脂量影响的试验初探[J]．广西林业科学，2001，30(3)：138-139.

[10]安宁，丁贵杰，谌红辉．林分密度及施肥对马尾松产脂量的影响[J]．中南林业科技大学学报，2010，30(9)：46-50.

[11]汪佑宏，牛敏，刘杏娥，等．马尾松树脂含量与其解剖特征的关系[J]．林业实用技术，2008，11：5-7.

[12]安宁，丁贵杰．广西马尾松松脂的化学组成研究[J]．中南林业科技大学学报，2012，32(3)：59-62.

[13]王长新．马尾松采脂量的相关因子分析[J]．河南科技大学学报(农学版)，2004，24(3)：22-25.

[14]蔡树威，龙伟，杨章旗．马尾松不同种源采脂量与树体因子关系的研究[J]．广西林业科学．2006，12(35)增刊：18-19.

[15]李爱民，全妙华，牛友芽，等．马尾松松脂产量影响因素研究进展[J]．湖南林业科技，2008，35(1)：58-59.
[16]王光仁，朱国发．湿地松产脂量相关因子分析及高产脂单株的选择[J]．林产化工通迅，1999，(1)：27-29.
[17]卢纹岱．SPSS for Windows 统计分析[M]．北京：电子工业出版社，2002.
[18]梁忠云，陈海燕，沈美英，等．广西优良品种马尾松松脂的化学组成[J]．林产化工通迅，2002，36(5)：8-10.
[19]舒文波，杨章旗．马尾松不同采脂年龄和径级产脂量变化特点研究[J]．中南林业科技大学学报，2011，31(11)：39-43.

[原载：中南林业科技大学学报，2015，35(08)]

南亚热带格木珍稀濒危树种生长规律研究

唐继新[1,2]　麻　静[1]　贾宏炎[1]　曾　冀[1]　雷渊才[2]　蔡道雄[1]　郝　建[1]

（[1]中国林业科学研究院热带林业实验中心，广西凭祥　532600；
[2]中国林业科学研究院资源信息研究所，北京　100091）

摘　要　应用30年生格木人工林树干解析资料，对珍稀濒危树种格木人工林的生长规律进行研究，结果表明：①林分中等木胸径生长量的速生期在7~25年，连年生长量最高峰值在第15年，连年生长量与平均生长量曲线在20~26年相交；林分优势木径向生长的速生期在5~28年，连年生长量的最高峰值在第11年，平均生长量的峰值在18~19年。②林分中等木树高生长的速生期在第4~21年，连年生长量的最高峰值在第7年，平均生长量的峰值在第8年，平均生长量与连年生长量曲线在8~21年多次相交；林分优势木树高生长的速生期在4~20年，连年生长量的最高峰值在第4年，平均生长量的峰值在第9年，平均生长量与连年生长量在9~16年多次相交。③中等木与优势木的材积速生期持续时间长，生长潜力大，直至第30年时仍未达数量成熟。④心材形成的起始树龄和起始树干去皮直径分别为12.2年、7.73cm。⑤林分中等木胸径、树高和材积的最优拟合方程分别为Gompertz、Richards、Richards方程，其调整后的拟合优度均≥0.978691，拟合效果显著；林分优势木胸径、树高、材积的最优拟合模型分别为Schumacher、Richards、Richards方程。⑥利用Adobe Acrobat软件距离工具进行树干解析测定，具有测量结果精准、效率高、成本低及复测易于实现等优点，相对于传统手工测量方式是一种较好的新途径。

关键词　格木；树干解析；心材；生长规律；Adobe Acrobat软件；测量工具

Study on the Growth Law of a Rare and Endangered Tree Species of *Erythrophleum fordii* in South Subtropical Area of China

TANG Jixin[1,2], Ma Jing[1], JIA Hongyan[1], ZENG Ji[1,2], LEI Yuancai[2], CAI Daoxiong[1], HAO Jian[1]

([1]*Experimental Center of Tropical Forestry*, *CAF*, *Pingxiang* 532600, *Guangxi*, *China*;
[2]*Research Institute of Forest Resources Information Techniques* , *CAF* , *Beijing* 100091, *China*)

Abstract: Based on the material of stem analysis of standard plots of 30-year-old Erythrophleum fordii Oliv plantation, growth regularity of the plantation was studied. The results showed that ① About the intermediate tree, the fast-growing period of annual increment of DBH was during 7~25th year, the growth of annual increment of DBH reached the peak in the 15th year, and it has intersected with the curve of average growth several times during the 20~26th year. To the dominant tree, the fast-growing period of annual increment of DBH was during 5~28th year, the growth of annual increment of DBH reached the peak in the 11th year, the highest value of average growth of DBH was during the 18~19th year. ② To the intermediate tree, the fast growing period of annual increment of tree height was during the 4~21th year, the average increment of tree high was in the 8th year, the annual increment of tree height reached the peak in the 7th year, then it gradually decreased, while it has intersected with the curve of average growth of tree height several times during the 8~21th year. In the dominant tree, the fast growing period of annual increment of tree height was during the 4~20th year, the average increment of tree high was in the 9th year, the highest annual increment of tree height was in the 4th years, after that it gradually decreased, while it has intersected with the curve of average growth of tree height several times during the 9~16th year. ③Both of the intermediate tree and the dominant tree, their fast growth period of volume have been sustaining a long-term, until the 30th year, both of the growth of volume have not reached the quantitative maturity, the growth of individual volume has a great potential. ④The heartwood initiation age and xylem diameter of the stem were respectively at 12. 2year and 7. 73cm. ⑤To the intermediate tree, the best equations to simulate the growth process of DBH, tree height and volume were the equations of Gompertz, Richards and Richards in respectively, each of the goodness of Adjusted R-squared was more than 0. 978691, and the fitting effects were obvious. While to the dominant tree, the best equations to were Schumacher, Richards and Richards in respectively. ⑥It is a good andnew way to measure analytic trees with the distance tool of Adobe Acrobat software, which has the advantages of accurate results, high efficiency, lowcost and easy retest with the measurement, comparing with the traditional manual measurement method.

Key words: *Erythrophleum fordii*; stem analysis; heartwood; growth regularity; Adobe Acrobat software; measuring tool

长期以来，我国珍优阔叶材基本取自天然林，长期过量的采伐天然林不仅对环境的负面影响日益突出，也使天然林存量资源严重不足[1]，尤其自天然林保护工程实施后，大径级珍优阔叶材供需失衡的结构性矛盾更为凸显，消耗巨额外汇从东南亚、美国、南非等地进口原木，已成近十年大径级原木供给的主要途径。据统计，我国2010年原木进口量已达1.84亿m^3(其中大径级珍贵用材0.34亿m^3)，原木进口已占全球原木贸易总量的1/3[2]；一些国家考虑到自身的利益和国际环保组织的压力，已逐渐限制或禁止原木出口[3]，我国木材安全形势日益严峻[2]。因此，要从根源上解决我国大径级原木供需失衡的结构性矛盾、缓解天然林保护的压力、增加珍优阔叶材的战略储备及提高我国的国际声誉，调整目前用材林的树种结构、加速珍优阔叶材的培育及深化珍优阔叶材的经营技术研究具有重要战略意义。格木(*Erythrophleum fordii* Oliv.)又名孤坟柴、斗登风、赤叶木，属苏木科格木属，是我国二级重点保护的珍稀濒危植物之一[4]，天然分布于我国广西、广东、福建和台湾等地，及越南和老挝等地，为热带典型的珍贵阔叶用材树种，其木材具有材质细腻、防虫蛀、耐腐蚀、比重高、坚硬、心边材区分明显等特点，是高档家具和木质工艺品的优质原料[5-7]。目前已有学者对其进行了种子处理、育苗、造林技术、群落学特征、濒危机制及保护对策、光合作用响应等生物学特性研究[8,9]，但有关该树种胸径、树高、材积、形数的生长规律，尤其是心材的变化规律，目前尚未见报道。本文依据中国林业科学研究院热带林业实验中心(下文简称热林中心)格木试验人工林树干解析的资料，对其生长规律进行研究分析，旨在为格木大径材的培育、经营、心材的利用以及目标树最佳培育径级的确定等提供科学的依据。

1 样地概况

试验样地位于广西凭祥市热林中心白云实验场(22°6′35″N，106°48′01″E)，属南亚热带季风气候区，干湿季明显，太阳总辐射强，全年日照时数1218~1620h，年均气温21.5℃，≥10℃积温7098℃，最热月平均气温27.2℃，最冷月平均气温12.7℃，极端最低气温-0.5℃，年均降水量1379.4mm，年蒸发量1370.2mm。海拔250m，属低山丘陵，坡向为西北；土壤为花岗岩发育而成的赤红壤，土层较厚，腐殖质层厚度3~5cm，(马尾松树种的)立地指数为20。调查对象为格木试验人工纯林，造林时间1982年春，造林苗木为裸根实生苗，造林密度2500株/hm^2，第7年透光伐后保留密度1800株/hm^2；林分分别于第11、16、21、28年历经4次强度为30%~35%的抚育间伐，现林分立木株数密度、郁闭度分别为390株/hm^2和0.85；林下植被主要有格木幼树、杜茎山(*Maesa japonica*)、玉叶金花(*Mussaenda pubuscens*)、三叉苦(*Evodia lepta*)、大青(*Clerodendrum cyrtophyllum*)、五节芒(*Miscanthus floridulus*)、金毛狗(*Cibotium barometz*)等。

2 材料与方法

2.1 试验材料

在30年生的格木试验人工纯林(面积5.0hm^2)中，按坡位、坡向等基本相近处均匀布设5个20m×20m临时标准地，对标准地乔木层进行每木检尺；由于格木为我国重点保护的珍稀濒危树种，现存资源极少，尤其是面积连片(≥1hm^2)30年生之上的非幼龄人工林，国内仅存热林中心1处[6]。为保护试验林及降低采样破坏，故对格木树种的生长过程研究，未按树干解析的常规要求采伐大量解析木，而仅在样地周围按林木生长分级法划分的优势木、中等木和被压木的三个林木等级[10]中各选取2~3株标准木，以2m区分段进行树干解析；5个标准地林分平均胸径分别为23.4、23.3、22.1、21.6、21.9cm，总平均胸径为22.4cm，5个标准地林木总的径级(径阶)分布及解析木如表1所示。

表 1　5 块标准地内的径级分布及解析木情况

林木分级	径阶	径阶株数分布	解析木株数	解析木平均胸径(cm)	解析木平均树高(m)
被压木	16	4	1	16.7	18.8
	18	12	1	17.6	19.1
	20	16	1	20.3	16.7
中等木	22	16	2	21.8	19.0
	24	13	1	23.4	19.6
优势木	26	12	1	26.0	21.4
	28	2	1	27.4	19.6
	30	3	/	/	/
	32	0	/	/	/
	34	1	/	/	/
	合计	78	8	/	/

2.2　研究方法

2.2.1　数据的采集与处理

对经刨光和标记南北直径线的圆盘样品，先用高清扫描仪扫描图像，然后用 Adobe Acrobat 7.0 Professional 软件打开扫描的图像，再用其软件菜单中的“工具(T)/图画标记(U)/线条工具(L)”标记圆盘直径线上各年龄直径对应的端点，最后用其软件菜单中的“工具(T)/测量工具(M)/距离工具(D)”进行测量。距离测量具体的过程：打开 Adobe Acrobat 7.0 Professional 软件“距离工具(T)”的操作界面，在“缩放率”的选项框中选择“1cm＝1cm”的缩放率；勾选“测量标记”，并填写圆盘各年龄测量直径对应的“批注”信息，然后用鼠标点击圆盘各年龄直径线所需测量的直径端点，即可完成各年龄对应直径的测量，在完成各年龄对应直径的测量后，最后利用“距离工具”中的“数据导出功能”将数据导出至 Excel。测量数据为圆盘东西和南北向的带皮直径(diameter outside bark，DO)、去皮直径(xylem diameter，XD)、心材直径(heartwood diameter，HD)、心材年轮数(heartwood ring number，HRN)和圆盘形成层年轮(cambial age，CA)[11]；各龄阶去皮直径(或心材直径)取圆盘截面东西和南北向的均值，各龄阶总边材宽度(total sapwood width，TSW)为树干圆盘截面去皮直径与心材直径的差，各龄阶的树高基于圆盘断面高及生长到该断面高所需年龄数(即形成层年轮数)按内插法求算，各龄阶的去皮材积及心材材积以中央区分段法求算[12]；林分各径级林木平均的胸径、树高和材积(或心材材积)，取标准地对应径级林木的加权平均值。

2.2.2　生长方程的拟合

对林分生长因子回归拟合采用的软件为 Eviews 6.0，其中心材变化规律、圆盘带皮与去皮直径的回归模拟选用一元线性回归模型，对林分中等木与优势木的胸径、树高和材积的回归拟合采用 Logistic、Gompertz、Richards、Korf、Mitscheerlich、Schumacher 和 Weibull 方程；非线性拟合方程的评价指标为调整后的拟合优度(adjusted R^2 statistic，R_a^2)、残差平方和(sum of squared residuals，SSR)、残差标准误差(standard error of the regression，SE)、参数估计值标准误差(Standard error of the estimate)，线性拟合方程的评价指标为 R_a^2、F 统计值的显著性水平[Prob (F-statistic)，Prob.][13]。拟合回归方程公式分别如下：

Logistic：$$Y=\frac{A}{1+Be^{-CT}} \tag{1}$$

Compertz：$$Y=Ae^{(-Be^{-CT})} \tag{2}$$

Richards：$$Y=A\left(1-e^{-BT}\right)^{C} \tag{3}$$

K orf：$$Y=Ae^{-BT^{-C}} \tag{4}$$

Mitscheerlich：$$Y=A(1-Be^{-CT}) \tag{5}$$

Schumacher：$$Y=Ae^{\frac{-B}{T}} \tag{6}$$

Weibull：$$Y=A\left[1-e^{-\left(\frac{T-B}{C}\right)^{D}}\right] \tag{7}$$

一元线性回归模型：

$$XD=aDO+b \tag{8}$$

$$XD=aDO \tag{9}$$

方程(1)～(5)为理论方程，其中：A 表示林木的生长极值参数，B 表示与初始值有关的参数(或生长速率参数)，C 表示生长速率参数(或与同化作用幂指数相关参数)，T 为林龄；方程(6)～(7)为经验方程，其中：A 为林木生长的上渐近值，B 为随机参数，C 为尺度参数，D 为形状参数，T 为林龄[12,14-15]；方程(8)~(9)为线性方程，其 a、b 为随机参数。

3 结果与分析

格木林分中等木胸径、树高、材积和实验形数的生长过程，其结果如表 2 所示。

3.1 生长过程分析

3.1.1 胸径生长过程分析

由表 2 和图 1 分析可知：在造林初期格木幼树的径向生长极缓慢，直至第 4 年其中等木(去皮的)的胸径总生长量仅有 0.42cm，且前 7 年平均胸径小于 3cm。林木个体生长的差异，在第 4~5 年显著，但在第 6~10 年开始显现，且在第 11~30 年逐步增大，此结果与范少辉等人对小黑杨的研究结果[16]类似。此表明：随着林龄的增长，林木生长对空间及养分的竞争进一步增强，由占有资源的不同而导致生长的差异[16]。

由图 2 和表 2 分析可知：①格木林分中等木径向生长的速生期在 7~25 年(此阶段内其连年生长量均≥0.70cm，明显高于其他阶段连年生长量)，连年生长量在第 15 年达最高峰值(1.22cm)，并在第 25 年后趋于减缓(其连年生长量均≤0.69cm)；连年生长量与平均生长量在 20~26 年间两次相交，径向平均生长量的峰值(0.76cm)在 22~25 年。②优势木径向生长的速生期在 5~28 年(此阶段的连年生长量均≥0.74cm)，连年生长量的最高峰值在第 11 年(>1.60cm)，连年生长量在第 19 年后趋缓；连年生长量与平均生长量在 18~22 年间两次相交，径向平均生长量的峰值在 18~19 年。③优势木、中等木的径向生长均呈多峰状，二者的生长趋势基本一致，但前者平均生长量的峰值较后者早 4~7 年。④气象因子为影响优势木生长的外在关键因子，但抚育间伐的经营活动、林分生态系统的健康稳定和林地微生物的多样性等因子亦对其生长有影响。如：在优势木的速生期内，其第 7、8、10、14 年(即 1988、1989、1991、1995 年)的生长状况与对应年份的气候“异常干旱”记录吻合，但其第 12、15 年(即 1993、1996 年)的生长表现则较为复杂，难以用气象或间伐经营活动等单一因素进行解释。⑤在整个生长过程中(除部分年份)，优势木的连年生长量明显高于中等木的连年生长量，二者连年生长量的最大差值(1.0cm)在第 11 年。

表 2 格木人工林中等木生长过程

林龄(年)	去皮胸径(cm)			树高(m)			去皮材积(m^3)			实验形数
	总生长量	平均生长量	连年生长量	总生长量	平均生长量	连年生长量	总生长量	平均生长量	连年生长量	
1	—	—	—	0.43	0.43	0.43	0	0	0	—
2	—	—	—	0.87	0.44	0.44	0.00002	0.000010	0.00002	—
3	—	—	—	1.30	0.43	0.43	0.00009	0.000030	0.00007	—
4	0.42	0.11	0.42	2.24	0.56	0.94	0.00024	0.000060	0.00015	3.09
5	1.01	0.20	0.59	3.18	0.64	0.94	0.00054	0.000108	0.00030	1.02
6	1.64	0.27	0.63	4.05	0.68	0.87	0.00113	0.000188	0.00059	0.71
7	2.40	0.34	0.76	5.60	0.80	1.55	0.00182	0.000260	0.00069	0.44
8	3.47	0.43	1.07	6.73	0.84	1.13	0.00257	0.000321	0.00075	0.26
9	4.65	0.52	1.18	7.49	0.83	0.76	0.00494	0.000549	0.00237	0.26
10	5.63	0.56	0.98	8.15	0.82	0.66	0.00802	0.000802	0.00308	0.27
11	6.33	0.58	0.70	8.97	0.82	0.82	0.01160	0.001055	0.00358	0.29
12	7.25	0.60	0.92	10.15	0.85	1.18	0.01737	0.001448	0.00577	0.3
13	8.26	0.64	1.01	10.99	0.85	0.84	0.02501	0.001924	0.00764	0.31
14	9.40	0.67	1.14	11.82	0.84	0.83	0.03389	0.002421	0.00888	0.31
15	10.62	0.71	1.22	12.41	0.83	0.59	0.04505	0.003003	0.01116	0.31
16	11.60	0.73	0.98	13.01	0.81	0.60	0.05675	0.003547	0.01170	0.31
17	12.44	0.73	0.84	13.60	0.80	0.59	0.06843	0.004025	0.01168	0.32
18	13.27	0.74	0.83	14.63	0.81	1.03	0.08173	0.004541	0.01330	0.31

（续）

林龄（年）	去皮胸径(cm)			树高(m)			去皮材积(m^3)			实验形数
	总生长量	平均生长量	连年生长量	总生长量	平均生长量	连年生长量	总生长量	平均生长量	连年生长量	
19	14.19	0.75	0.92	15.19	0.80	0.56	0.09586	0.005045	0.01413	0.31
20	15.05	0.75	0.86	15.74	0.79	0.55	0.11060	0.005530	0.01474	0.31
21	15.80	0.75	0.75	16.81	0.80	1.07	0.12607	0.006003	0.01547	0.30
22	16.65	0.76	0.85	17.08	0.78	0.27	0.14321	0.00651	0.01714	0.31
23	17.44	0.76	0.79	17.34	0.75	0.26	0.16034	0.006971	0.01713	0.31
24	18.30	0.76	0.86	17.61	0.73	0.27	0.17955	0.007481	0.01921	0.31
25	19.07	0.76	0.77	17.87	0.71	0.26	0.19862	0.007945	0.01907	0.31
26	19.57	0.75	0.50	18.16	0.70	0.29	0.21634	0.008321	0.01772	0.32
27	20.26	0.75	0.69	18.47	0.68	0.31	0.23632	0.008753	0.01998	0.32
28	20.85	0.74	0.59	18.73	0.67	0.26	0.25699	0.009178	0.02067	0.32
29	21.39	0.74	0.54	18.98	0.65	0.25	0.27751	0.009569	0.02052	0.33
30	21.91	0.73	0.52	19.24	0.64	0.26	0.29769	0.009923	0.02018	0.33
带皮	22.49	—	—	19.24	—	—	0.31503	—	—	0.33

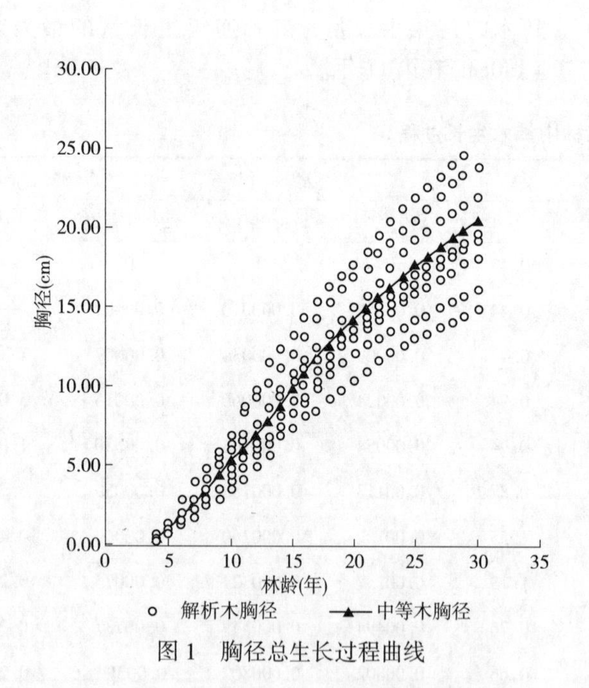

图1　胸径总生长过程曲线

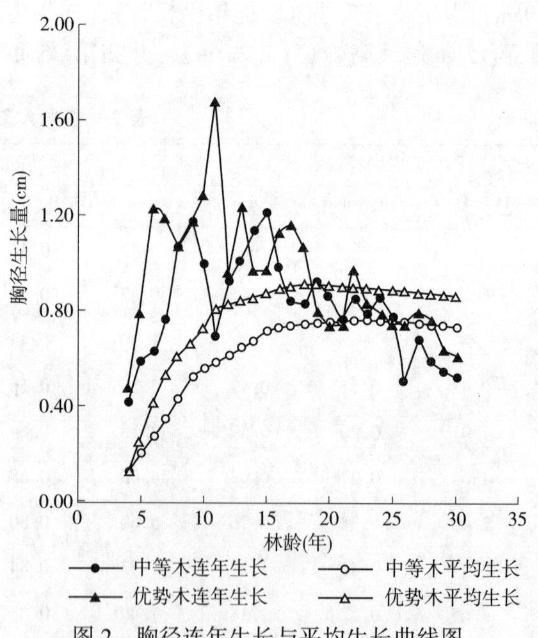

图2　胸径连年生长与平均生长曲线图

3.1.2　树高生长过程分析

由图3分析可知：前7年林木个体间树高生长的差异极小，第8年后生长差异略有显现，但相对胸径生长的差异而言，树高生长的差异相对较小。

由表2和图4综合分析可知：①中等木树高生长的速生期主要在4~21年(该阶段内的连年生长量均≥0.55m，显著高于其他阶段连年生长量)，连年生长量的最高峰值(1.55m)在第7年，在第21年后生长趋缓(连年生长量均≤0.31m)；树高平均生长量的峰值(0.85m)在12~13年。②优势木高生长的速生期在4~20年，连年生长量的最高峰值(>1.6m)在第4年，平均生长量的峰值在第9年。③前3年优势木与中等木的高生长均较缓慢(其年均生长量与连年生长量仅略高于0.4m)，第3年后二者的连年生长量与平均生长量均多次相交，呈多峰状。④在整个生长过程中(除部分年份)，优势木的连年生长量明显高于中等木的连年生长量，二者连年生长量最大的差值(>0.8m)在第4年。

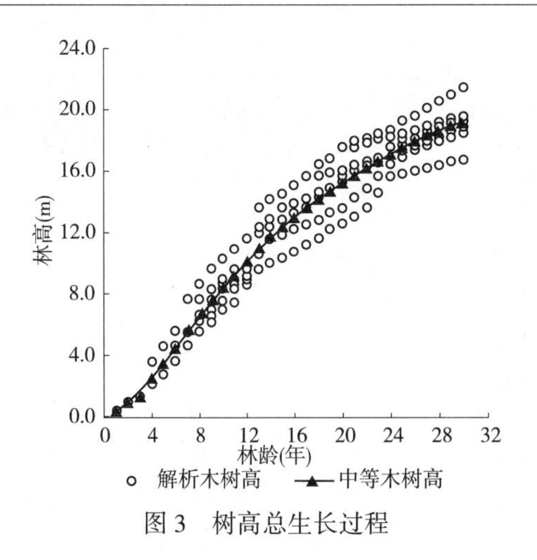

图3　树高总生长过程

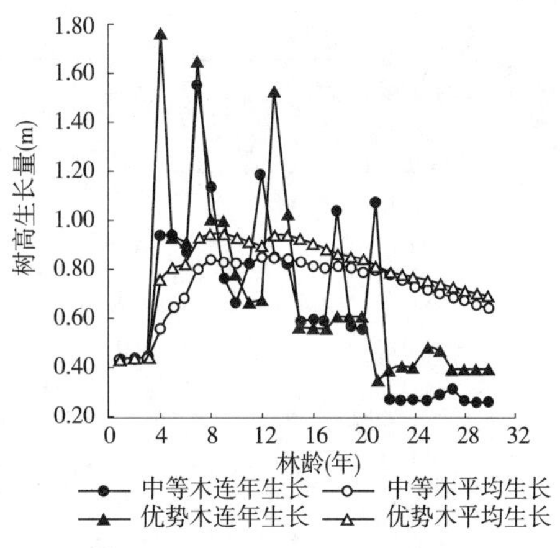

图4　树高连年生长与平均生长曲线图

3.1.3　材积生长过程分析

由图5分析可知：前10年林分平均单株材积的生长总量极小；在前8年林木个体单株材积生长无差异，而直至第9~10年其生长的差异才显现；此后随着林龄的增长林木个体材积生长的差异逐步增大，此与胸径生长的变化趋势规律类似。

由表2和图6分析可知：在前8年，林分中等木与优势木的材积生长均极缓慢，但第9年后开始快速递增；在第15~30年，中等木与优势木材积的连年生长量均高于0.01m^3，且在第28年均达最高峰值；在第8年后，中等木与优势木的平均生长量一直呈平稳的增长；直至第30年，材积的生长仍未达数量未成熟，表明该树种的生长潜力大、经营周期长；第8年后，优势木的连年生长量明显高于中等木的连年生长量，二者连年生长量的最大差值在第28年。

3.1.4　实验形数变化分析

由表2分析可知：在前11年，格木中等木的实验形数基本随林龄的增加而降低，但在第11年后趋于稳定状态(基本维持在0.32~0.33)；中龄阶段的实验形数值远低于主要阔叶乔木树种的平均实验形数值(0.40)[12]。

3.1.5　心材的变化分析

心边材区分明显类珍贵用材树种，其材性优良，通常指其心材，掌握此类树种心材的变化规律，是有效开发利用的基础。故本研究对格木树种心材的起始树龄(heartwoodinitiation age，HIA)、心材起始树干去皮直径(heartwood initiation xylem diameter，HIXD)和心材沿树干垂直变化的规律进行分析研究。

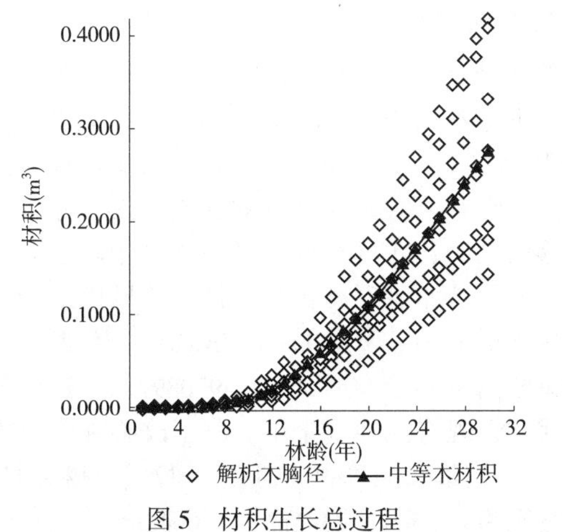

图5　材积生长总过程

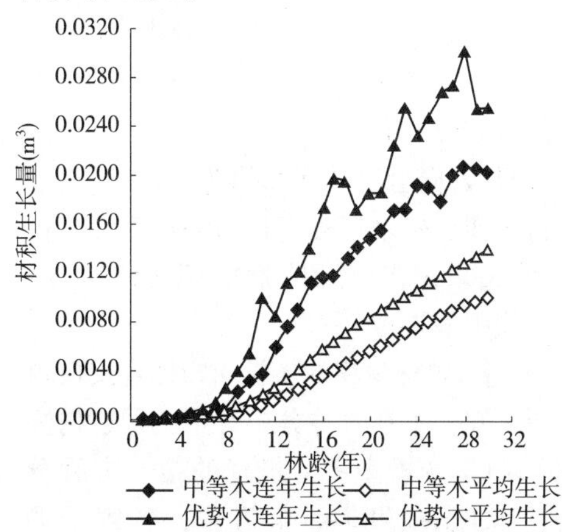

图6　材积连年生长与平均生长曲线图

表3　中等木心材沿树干高度生长过程表

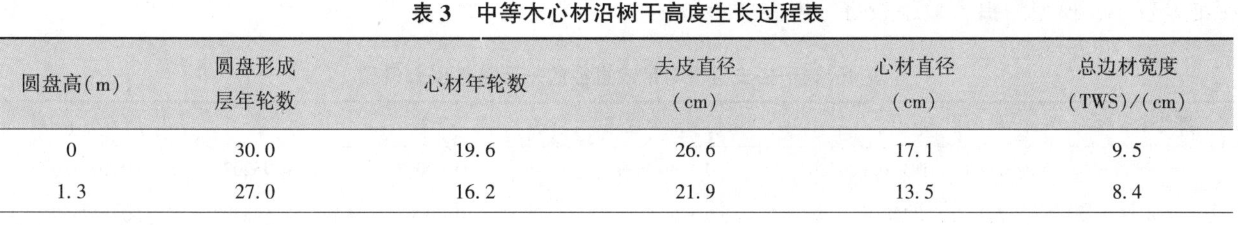

圆盘高(m)	圆盘形成层年轮数	心材年轮数	去皮直径(cm)	心材直径(cm)	总边材宽度(TWS)/(cm)
0	30.0	19.6	26.6	17.1	9.5
1.3	27.0	16.2	21.9	13.5	8.4

（续）

圆盘高(m)	圆盘形成层年轮数	心材年轮数	去皮直径(cm)	心材直径(cm)	总边材宽度(TWS)/(cm)
3.6	24.8	12.6	19.9	11.5	8.4
5.6	23.0	10.7	18.0	9.2	8.8
7.6	21.3	8.5	16.5	7.2	9.3
9.6	18.4	6.0	14.4	4.7	9.7
11.6	16.0	4.1	11.8	2.6	9.2
13.6	12.7	2.2	9.6	1.4	8.2
15.6	9.5	1.0	6.5	0.4	6.1
17.6	5.4	0.4	4.1	0.1	4.0

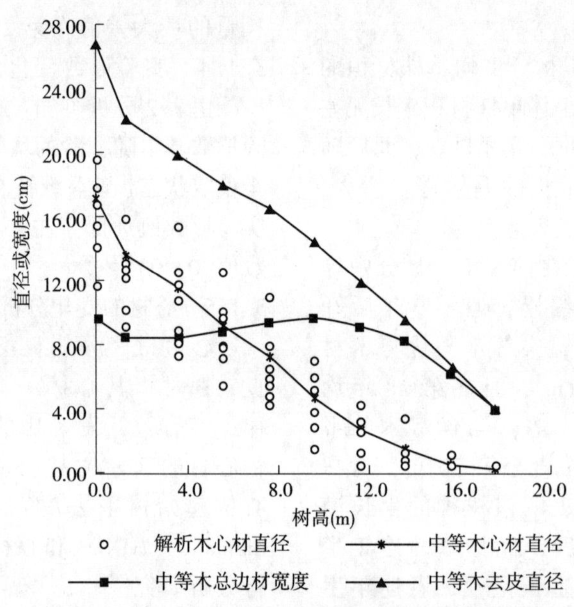

图7　中等木心材直径、总边材宽度随树高的变化

3.2　生长过程模拟及预测

3.2.1　带皮直径与去皮直径的回归拟合

由表4分析可知，格木带皮直径与去皮直径呈极显著的一元线性关系；在一元线性的拟合模型中，带常数项的模型模拟效果较好。

3.2.2　胸径、树高、单株材积的生长回归拟合及预测

由表5分析可知：①在对林分中等木胸径、树高和材积的众多回归拟合模型中，最优的拟合方程分别为Gompertz、Richards、Richards方程，各因子调整后的拟合优度均≥0.978691，且残差平方和与剩余标准差较小，模型的拟合效果较好，模型的参数估计值A表明：在现有的林分立地水平及经营密度下，林分中等木(去皮)胸径、树高、材积的生长极值分别为25.2cm、21.6m、0.76595m^3，在林分的后续经营中，唯有适时调整林分的立木密度，并改善林分土壤的肥力，才可获取更高的林分平均生长量。②对林分优势木胸径、树高、材积的最优拟合模型分别为Schumacher、Richards、Richards方程，各因子的调整后的拟合优度均≥0.970897，参数估计值A表明：在现有立地水平下，若不改善林分土壤的状况，林分优势木(去皮)胸径、树高、材积的生长极值分别为47.5cm、23.5m、1.46648m^3。

表4　树干去皮直径与带皮直径的一元线性回归模型

圆盘样本数	回归模型	调整的拟合优度	概率	残差平方和	残差标准差
81	XD=0.978914DO-0.200586	0.999744	0.0000	0.952079	0.109780
81	XD=0.967268DO	0.999569	0.0000	1.623792	0.142469

表 5 格木生长量与年龄关系的最优回归模拟模型

林木等级	因子	解析木样本数	回归模型	参数估值			模型统计指标		
				A	B	C	调整的拟合优度	残差平方和	残差标准差
中等木	胸径	3	Gompertz	25.17565 (0.930723)	5.002607 (0.341635)	0.114202 (0.007313)	0.978691	78.64137	1.004103
	树高	3	Richards	21.56222 (0.431523)	0.094697 (0.005487)	1.935375 (0.101858)	0.992813	23.60714	0.520909
	材积	3	Richards	0.765945 (0.191323)	0.061843 (0.012929)	5.718613 (1.034693)	0.983209	0.013071	0.012257
优势木	胸径	2	Schumacher	47.51652 (0.604267)	18.93550 (0.260187)		0.996382	11.73713	0.475094
	树高	2	Richards	23.51493 (1.404218)	0.082409 (0.013417)	1.689063 (0.213876)	0.970897	68.99829	1.100226
	材积	2	Richards	1.466477 (0.485904)	0.046317 (0.011667)	4.426531 (0.684680)	0.990829	0.009264	0.012749

注：括号内数值为模型参数估计值在 0.05 显著水平下的标准误。

3.2.3 心材的生长回归拟合

当树木达到一定年龄或直径后，其边材就以相对稳定的速率形成心材[11]。由表 6 分析可知：格木树种的心材年轮数与形成层年龄、心材直径与去皮直径都呈极显著的正相关关系（$P=0.0000$），心材直径与圆盘高度成极显著的负相关关系（$P=0.0000$）；形成层年龄能够解释 92%之上的心材年轮数变化，去皮直径能解释 92%以上的心材直径变化；心材形成速率每年达 0.9417a~1，与以往有关研究结果相吻合（0.5~1.0/年）[11]，心材形成的起始树龄和起始树干去皮直径分别为 12.2 年、7.73cm。

表 6 格木心材线性回归模型

圆盘样本数	回归模型	调整后的拟合优度	概率	心材起始树龄（年）	心材起始去皮直径（cm）
65	$HRN=0.941707CA-11.47407$	0.921899	0.0000	12.2	—
65	$HD=0.925984XD-7.155084$	0.926254	0.0000	—	7.73
65	$HD=-0.980344H_1+13.98280$	0.807675	0.0000	—	—

注：H_1 为树干圆盘截面高度。

4 讨论与结论

4.1 讨论

（1）本研究格木林分第 23 年的生长状况（平均的去皮胸径、树高分别为 17.4cm、17.3m）与越南的同龄格木人工林（平均的胸径、树高分别为 18.8cm、14.8m）[17]相比，两地格木的径向生长水平基本相同，但与格木天然林的生长相比，本研究 30 年生林分的生长状况（胸径值为 22.5cm）与魏识广等人[18]对广东鼎湖山格木天然林 40 年生时的研究结果（胸径值为 23cm）相近。表明适当的经营措施可较好地促进格木的生长，并缩短林木的成熟期。

（2）在全光下，以常规苗木（高度在 0.3~0.5m）新造的格木纯林，其前 3 年的生长极缓慢，并极易受杂草和灌木的干扰，为促进格木幼树更好的生长，造林前 3 年的抚育除草极为重要[16]，因此为使格木幼树更好的生长，并减少其抚育铲草投入的成本，对比开展常规苗木与大苗（高度在 0.5~1.0m）的不同造林技术研究具有重要实践意义；此外，结合格木树种的生长特性，在其径向生长的速生期适时对林分进行抚育间伐，是保证林木快速生长所需的养分和空间关键措施。

（3）受虫害影响，格木纯林顶芽交换频繁，树干分叉现象普遍[6]，这不仅干扰了林木的正常生长，亦影响了林木的干形。故今后，加强格木纯林的近自然化改造，探索格木树种与其他树种的混交栽培、目标树经营及其修枝整形的理论研究，对格木人工林的健康、稳定、可持续经营、高价值无节大径材的培育和林分立地生产力的提高，具有重要的理论和实践意义。

(4)格木林分经营周期远比速生类人工林的经营周期长，直至第30年材积生长仍远未达数量成熟。根据格木树种的生物学特性，开展其与中短轮伐期类树种(如：松、桉类树种)的混交试验，是探索格木人工林长期效益与短期效益有机结合及可持续经营的有效途径。

(5)相对解析木的传统手工测量方式，利用Adobe Acrobat软件的测量工具，不仅可获取的信息多、精准、效率高(个人可完成测量和数据记录的所有工作)、复测易于实现，而且成本低廉。因此，利用Adobe Acrobat软件进行解析木测定是一种较好的新途径。

4.2 结论

(1)格木幼林郁闭前的径向生长极缓慢，郁闭后开始增速。林分中等木径向的速生期在7~25年，连年生长量的最高峰值在第15年，平均生长量的峰值在22~25年，连年生长量曲线与平均生长量曲线在20~26年间相交。林分优势木径向生长的速生期在5~28年，连年生长量的最高峰值在第11年，平均生长量的峰值在18~19年。

(2)格木中等木高生长的速生期在4~21年，连年生长量的最大峰值在第7年，平均生长量的峰值在第8年。优势木高生长的速生期在4~20年，连年生长量的最高峰值在第4年，平均生长量的峰值在第9年。

(3)格木心材形成的起始林龄和起始树干去皮直径分别为12.2年、7.73cm，30年生格木可用的心材量较小。

(4)格木中等木胸径、树高和材积的最优拟合方程分别为Compertz、Richards、Richards方程，其调整后的拟合优度均≥0.978691，拟合效果较好。林分优势木胸径、树高、材积的最优拟合模型分别为Schumacher、Richards、Richards方程。

(5)在格木生长的过程中(除部分年份外)，优势木径向生长与树高生长的连年生长量显著高于中等木的连年生长量；直至第30年，中等木与优势木的材积生长均仍未达数量成熟。

参考文献

[1]蔡道雄，贾宏炎，卢立华，等．我国南亚热带珍优乡土阔叶树种大径材人工林的培育[J]．林业科学研究，2007，20(2)：165-169.

[2]许传德．略论闽、粤、桂木材战略储备基地建设的先导性[J]．林业经济，2012(6)：28-30.

[3]黎云昆．论我国珍贵用材树种资源的培育[J]．绿色中国，2005(16)：24-28.

[4]黄忠良，郭贵仲，张祝平．渐危植物格木的濒危机制及其繁殖特性的研究[J]．生态学报，1997，17(6)：671-676.

[5]赵志刚，郭俊杰，沙二，等．我国格木的地理分布与种实表型变异[J]．植物学报，2009，44(3)：338-344.

[6]赵志刚．珍稀濒危树种格木生物学保护研究[D]．北京：中国林业科学研究院，2011.

[7]赵志刚，郭俊杰，曾杰，等．濒危树种格木天然群体自由授粉子代苗期生长变异[J]．植物研究，2011，31(1)：100-104.

[8]李胜强，许建新，陈波，等．珍稀植物格木的研究进展[J]．广东林业科技，2008，24(6)：61-64.

[9]申文辉，李志辉，彭玉华，等．格木不同种源光合作用光响应分析研究[J]．中南林业科技大学学报，2014，34(6)：13-18.

[10]谭俊，刘志．林木生长分级理论在航空像片林分测定中的应用[J]．辽宁林业科技，1992，(5)：19-22.

[11]王兴昌，王传宽，张全智，等．东北主要树种心材与边材的生长特征[J]．林业科学，2008，44(5)：102-108.

[12]孟宪宇．测树学[M].3版．北京：国林业出版社，2006.

[13]王中昭，李丽明．计量经济学实验及例题分析[M]．广西科学技术出版社，2005.

[14]段爱国，张建国，童书振.6种生长方程在杉木人工林林分直径结构上的应用[J]．林业科学研究，2003，16(4)：423-429.

[15]彭舜磊，王得祥．火地塘林区铁杉生长规律研究[J]．西北农林科技大学学报：自然科学版，2008，36(4)：83-88.

[16]范少辉，冯慧想，张群，等．华北沙地小黑杨人工林生长特性[J]．林业科学，2008，44(3)：29-33.

[17]SEIN C C，MITLOHNER R. *Erythrophloeum fordii Oliver*：ecology and silviculture in Vietnam[M]. Center for International Forestry Research (CIFOR)，2011.

[18]魏识广，李林，刘海岗，等．鼎湖山格木种群动态分析[J]．生态环境，2008，17(1)：285-289.

[原载：中南林业科技大学学报，2015，35(07)]

水与肥对柚木生长的影响

冯　海[1]　李运兴[1]　梁坤南[2]　明安刚[1]　欧耀成[1]　王亚南[1]

([1]中国林业科学研究院热带林业实验中心，广西凭祥　532600；
[2]中国林业科学研究院热带林业研究所，广东广州　510520)

摘　要　为探索水因子、施肥、立地对柚木生长的影响，在造林后第1年起，按裂区试验方法布设了一个包含有2个水处理(主区)和4个N、P、K处理(副区)，3个重复的柚木施肥与水处理裂区试验，在1~3年和13年时对试验进行高径等生长指标调查和分析，结果表明：①在1~3年和13年时施肥处理对柚木生长影响不显著。②水处理在1~3年和13年对柚木生长有极显著影响。③试验重复(不同地段)2年时起，对柚木生长影响显著。④立地和水处理对柚木生长有持续稳定影响，施肥处理与立地、施肥处理与水处理存在显著的互作关系，做到适地适树适肥，才能取得好的经营效果。

关键词　柚木；施肥；水处理；生长效果

The Effects of Water and Fertilizer on the Growth of Teak Plantation

FENG Hai[1], LI Yunxing[1], LIANG Kunnan[2], MING Angang[1],
OU Yaocheng[1], WANG Yanan[1]

([1]*The Experimental Center of Tropical Forestry*, *Chinese Academy of Forestry*, *Pingxiang* 532600, *Guangxi*, *China*;
[2]*Research Institute of Tropical Forestry*, *Chinese Academy of Forestry*, *Guangzhou* 510520, *Guangdong*, *China*)

Abstract: In order to study the effects of water, fertilizer, site on the growth of teak, according to split plot test method, we carried out an experiment about teak fertilization and water treatment (n=3), including two water treatments (main district) and four N, P, K treatments(deputy district) in the first year after planting. A survey was conducted to the height and diameter of the teak in 1~3a and 13a respectively. The results show that: ① Fertilization treatment had no significant effect on the growth of the teak in 1~3a and 13a. ② Water treatment showed very significant effect on the growth of the teak in 1~3a and 13a. ③ The replicated test in different sites showed significant effect on the growth of the teak after 2 years. ④ Site and water treatment have a steady influence on the growth of teak, and there is significant interaction relationship between fertilizer treatment and site and as well as fertilizer treatment and water treatment . Therefore, select adapted tree species planted on the suitable site with appropriate fertilizer is a effective way for forest management.

Key words: *Tectona grandis*; fertilization; water treatment; growth effect

柚木(*Tectona grandis*)属马鞭草科落叶或半落叶大乔木，树高40m，胸径1~2m，干形通直。木材为环孔材，心材大，淡褐色，密度0.60~0.72g/cm^3，纹树直，结构细致而美观，坚韧而有弹性，不翘不裂；含油质，强度大，耐浸，是航海、军需、建筑、车厢、家具、雕刻、铸造木模、贴面板等珍贵优良用材树种之一[1-3]。原产于印度、缅甸、泰国和老挝[4]，我国引种于广东、广西、海南、云南、福建、台湾等地，已有多年历史，当前可用木材资源锐减，社会需求增加，交易价格攀升，造林积极性不断提高，发展势头大，我国的柚木研究也从已往单方面侧重于种质资源引进方面[5]转向全方面的良种选育[6-10]、生理需求、干旱胁迫[11,12]和栽培技术研究方面，丁美华认为土壤水分对柚木生长作用明显[13]，梁坤南[14]柚木幼苗期施肥、潘一峰[15]、李运兴[16]柚木幼林施肥试验结果认为：施用含N、P、K的配方肥都可以在一定程度上提高树高和胸径的生长量，但施肥效果受多因子影响，其中立地状况、肥料种类数量、降水量因素影响最大，为探索水因子、施肥、立地对柚木生长的影响大小及互作关系，深入开展柚木立地质量、营养、水分需求方面研究，对完善培育措施、提高集约经营水平有较大的积极意义。

1 材料与方法

1.1 试验地概况

幼林施肥水处理试验布设于平而河畔低山、青山实验场平架站，地形开阔，阳光充足，海拔150~190m，年平均气温为21.5°C，N10°C积温7500℃，月平均气温N22℃有7个月，最冷月(2月)平均气温13.5℃，极端最低温-1.5℃，最热月(7月)平均气温28.3℃，极端最高温40.3℃，年降水量1220~1380mm，土壤为砂页岩发育的红赤红壤，质地为黏壤土，0~40cm混合土样，有机质15.89g/kg，pH(H_2O)值5.01，全N值2.41g/kg，全P值0.19g/kg，全K值5.22g/kg，碱解N值278.91mg/kg，速效P值2.18mg/kg，速效K值33.7mg/kg，盐基饱和度31.53%。前茬为丢荒地，主要植被有山芝麻、五节芒、五色梅、茅草等，株行距3m×3m，造林密度1125株/hm^2，1997年12月林地清理挖穴整地，1998年1月施基肥，同年3月用1年生柚木实生苗造林，柚木造林成活率为97.9%，造林后3年每年在4月和7月全铲草各1次，第2年年底松土1次，第1年至第3年5月，按试验方案挖环形沟各追施肥1次，1年、2年、3年、13年生柚木平均树高分别为0.62、1.78、3.23、12.43m；2年、3年、13年平均胸径分别为2.07、3.29、13.98cm，13年时，林木保存率达97.5%，试验保存完好。

1.2 试验设计与调查统计

试验按裂区试验设计实施，主区为水因子，分富水区与贫水区2处理，水处理方法：考虑地形对聚集降水的作用，把富水区处理安排在稍凹的地形上布设，造林后第1年4~9月，雨天雨水湿透40cm土层时不灌淋水，阴晴天每7d灌淋水1次，每次20min，至种植穴周围1.0m半径土壤全湿透为止，副区为施肥因子，按施肥种类和量的不同分4个处理，具体见表1和表2所列，每小区5(横)×7(纵)=35株，3次重复。

在清理整地前进行土壤和植被调查，在造林后1月进行成活率调查并及时补植，在1年、2年、3年和13年年终对树高、径(地径或胸径)进行测量；用SPSS 19.0软件对数据进行相关统计分析。

表1 不同施肥处理

年份(年)	基肥/追肥	施肥处理			
		1	2	3	4
1	基肥	不施肥	500g钙镁磷+50g复合肥	500g钙镁磷	500g钙镁磷+100g复合肥+200g复合肥
1	追肥	不施肥	50g复合肥	100g复合肥	200g复合肥
2	追肥	不施肥	100g复合肥	200g复合肥	300g复合肥
3	追肥	不施肥	1000g钙镁磷	300g尿素	300g复合肥

注：钙镁磷含P_2O_5：17%，尿素含N：46%，复合肥含N：15%、P_2O_5：15%、K2O：15%。

表2 不同施肥处理养分差异

年份-基肥或追肥	处理											
	1			2			3			4		
	N	P	K	N	P	K	N	P	K	N	P	K
1年-基肥(g)	0	0	0	7.5	93.5	7.5	15	100	15	30	115	30
1年-追肥(g)	0	0	0	7.5	7.5	7.5	15	15	15	30	30	30
2年-追肥(g)	0	0	0	15	15	15	30	30	30	45	45	45
3年-追肥(g)	0	0	0	0	170	0	138	0	0	45	45	45
分N、P、K小计(g)	0	0	0	30	286	30	198	145	60	150	235	150
养分合计(g)		0			346			403			535	

2 结果与分析

2.1 施肥对柚木生长的影响

2.1.1 对树高影响

试验林 1~3 年及 13 年各副区树高平均值见表 3，从表 3 中数据可以看出，1 年时，各施肥处理树高大小排列为 3>4>2>1；2 年时，4>3>2>1；3 年时，4>3>2>1；13 年时，4>3>1>2。方差分析显示不同年度、各处理间差异不显著(P=0.05，下同)，由此可见，虽有树高随施肥量增加而增加的趋势，但差异不明显。

2.1.2 对径生长影响

试验林 1~3 年及 13 年各副区径平均值见表 3，从表 3 中数据可以看出，2 年时，各施肥处理径大小排列为 4>2>3>1；3 年时，4>3>2>1；13 年时，4>1>3>2。方差分析显示各年度、各处理间差异不显著；由此可见，同样有径随施肥量增加而增加的趋势，但不明显。

2.1.3 对径高比影响

试验林 2~3 年及 13 年各副区径高比平均值见表 3。从表 3 中数据可以看出，2 年时，各施肥处理径高比大小排列为 1>2>4>3；3 年时，4>1=2>3；13 年时，2>1>3>4。方差分析显示各年度、各处理间差异不显著，由此可见，有径高比随施肥量增加而减少的趋势，但不明显。

2.2 水处理对柚木生长的影响

2.2.1 对树高影响

试验林 1~3 年及 13 年各主区树高平均值见表 3。从表 3 中数据可以看出，1~3 年、13 年时，水处理树高>非水处理树高，方差分析显示处理间差异极显著(P<0.01，下同)。由此可见．水处理大于非水处理趋势明显。

2.2.2 对径生长影响

试验林 1~3 年及 13 年各主区径平均值见表 3。从表 3 中数据可以看出，1~3 年、13 年时，水处理径>非水处理径，方差分析显示处理间差异极显著，因此，水处理大于非水处理趋势也明显。

2.2.3 对径高比影响

试验林 2~3 年及 13 年各主区径高比平均值见表 3。从表 3 中数据可以看出，2 年时，水处理径高比〈非水处理径高比；3 年时，水处理径高比>非水处理径高比；13 年时，水处理径高比<非水处理径高比。方差分析显示处理间差异不显著，由此可见，水处理径高比小于非水处理趋势不明显。

表 3 不同年度重复各处理树高、径、径高比

生长年份指标	年份(年)	重复			施肥处理(副区)				水处理(主区)	
		1	2	3	1	2	3	4	0	1
树高(m)	1	0.49	0.74	0.63	0.53	0.59	0.71	0.66	0.54	0.71
	2	2.03	1.85	1.46	1.43	1.73	1.83	2.14	1.41	2.15
	3	3.43	3.2	3.05	2.78	3.05	3.21	3.85	2.51	3.94
	13	15.46	11.81	10.02	11.14	10.10	12.05	12.63	8.79	13.23
胸径(cm)	2	2.18	2.04	2	1.94	2.00	1.98	2.37	1.71	2.43
	3	3.6	3.17	3.11	2.85	3.14	3.17	4.01	2.52	4.06
	13	16.31	13.92	11.7	13.64	12.72	13.27	13.66	10.45	14.89
径高比	2	1.07	1.10	1.37	1.36	1.16	1.08	1.11	1.21	1.13
	3	1.05	0.99	1.02	1.03	1.03	0.99	1.04	1.00	1.03
	13	1.05	1.18	1.17	1.22	1.26	1.10	1.08	1.19	1.13

2.3 重复对柚木生长的影响

2.3.1 对树高影响

试验林 1~3 年及 13 年各重复树高平均值见表 3。从表 3 中数据可以看出，1 年时，各重复树高大小排列为 2>3>1；2~13 年时，1>2>3。方差分析显示各重复 2、3、13 年时树高差异显著，其他年度、指标差异不显著。

2.3.2 对径生长影响

试验林 2~3 年及 13 年各重复胸径平均值见表 3。从表 3 中数据可以看出，2~13 年时，1>2>3，方差分析显示各重复 13 年时胸径差异显著，其他年度差异不显著。

2.3.3 对径高比影响

试验林 2~3 年及 13 年各重复径高比平均值见表 3。从表 3 中数据可以看出，2 年时，各重复径大小排列为 3>2>1；3 年时，1>3>2；13 年时，2>3>1。

方差分析显示各重复2年和13年时径高比差异显著，其他年度、指标差异不显著。

2.4　因子间的交互作用

方差分析结果显示：1~3年时因子间的交互作用不显著，13年时，重复与副区、主区与副区的交互作用显著，为更清晰地显现因子间的互作效应，将13年试验数据按不同施肥处理整理得表4与表5。从表4和5中可以看出2、3、4处理的施肥效果在1水处理和1重复表现最好。

表4　不同处理树高、径、径高比

施肥处理	水处理						重复								
	0	1	0	1	0	1	1			2			3		
	树高（m）	树高（m）	径高（cm）	径高（cm）	径高比	径高比	树高（m）	径高（cm）	径高比	树高（m）	径高（cm）	径高比	树高（m）	径高（cm）	径高比
1	10.26	11.64	12.31	14.4	1.20	1.24	14.62	17.87	1.22	11.43	13.95	1.22	10.89	13.38	1.23
2	8.71	11.43	10.67	14.66	1.23	1.28	14.36	16.86	1.17	11.04	13.38	1.21	9.30	12.15	1.31
3	8.09	13.81	8.78	15.27	1.08	1.11	15.13	16.61	1.10	14.26	15.92	1.12	8.05	8.76	1.09
4	8.90	14.28	11.07	14.80	1.24	1.04	15.86	15.94	1.00	9.85	11.86	1.20	12.03	13.15	1.09

表5　不同处理施肥效果

施肥处理	水处理						重复								
	树高		径高		径高比		1			2			3		
	0	1	0	1	0	1	树高	径高	径高比	树高	径高	径高比	树高	径高	径高比
	效果值	效果值	效果值	效果值	效果值	效果值	效果值	效果值	效果值	效果值	效果值	效果值	效果值	效果值	效果值
1	0	0	0	0	0	0	0	0	0	0	0	0	0	0	0
2	−1.55	−0.21	−1.64	0.26	0.03	0.04	−0.26	−1.01	−0.05	−0.39	−0.57	−0.01	−1.59	−1.23	0.08
3	−2.17	2.17	−3.53	0.87	−0.12	−0.13	0.51	−1.26	−0.12	2.83	1.97	−0.10	−2.84	−4.62	−0.14
4	−1.36	2.64	−1.24	0.40	0.04	−0.20	1.24	−1.93	−0.22	−1.58	−2.09	−0.02	1.14	−0.23	−0.14
均值	−1.69	1.53	−2.14	0.51	−0.02	−0.10	0.50	−1.40	−0.13	0.29	−0.23	−0.04	−1.10	−2.03	−0.07

注：效果值=施肥处理−对照。

2.5　试验因素对生长影响稳定性分析

为了解因子作用的持续性大小，分别对各重复、各水处理、各施肥处理树高、径生长指标进行年度变化分析，结果见表6。数据显示，各试验重复第2~13年各生长指标相关关系极显著（$P<0.01$，下同）；各水处理第1年起各年度生长指标相关关系极显著，各施肥处理年度间生长指标相关关系不显著，说明不同重复（地段）和水处理的生长差异是持续稳定的，而不同施肥处理生长差异年度间变动较大。

表6　各重主区副区生长指标年度变化

因素	处理	1年		2年				3年				13年					
		树高		树高		胸径		树高		胸径		树高		胸径		单株材积	
		$H(m)$	与均值差(%)	$H(m)$	与均值差(%)	$D(cm)$	与均值差(%)	$H(m)$	与均值差(%)	$D(cm)$	与均值差(%)	$H(m)$	与均值差(%)	八/	与均值D/Cm差(%)	$F*$(m³)	与均值差(%)
重复	1	0.49	−21.0	2.03	14.0	2.18	5.1	3.43	6.3	3.6	9.3	15.46	24.4	16.31	16.7	0.1969	58.7
	2	0.74	19.4	1.85	3.9	2.04	−1.6	3.2	−0.8	3.17	−3.7	11.81	−5.0	13.92	−0.4	0.1095	−11.7
	3	0.63	1.6	1.46	−18.0	2.00	−3.5	3.05	−5.5	3.11	−5.6	10.02	−19.4	11.7	−16.3	0.0657	−47.1
主区（水处理）	0	0.54	−13.6	1.41	−20.8	1.71	−17.4	2.51	−22.2	2.52	−23.4	8.79	−20.2	10.45	−17.5	0.0459	−50.7
	1	0.71	13.6	2.15	20.8	2.43	17.4	3.94	22.2	4.06	23.4	13.23	20.2	14.89	17.5	0.1404	50.7

（续）

因素	处理	1年		2年				3年				13年					
		树高		树高		胸径		树高		胸径		树高		胸径		单株材积	
		H(m)	与均值差(%)	H(m)	与均值差(%)	D(cm)	与均值差(%)	H(m)	与均值差(%)	D(cm)	与均值差(%)	H(m)	与均值差(%)	八/与均值D/Cm差(%)		F*(m³)	与均值差(%)
副区（施肥处理）	1	0.53	-14.9	1.43	-19.8	1.94	-6.4	2.78	-13.7	2.85	-13.4	11.14	-3.0	13.64	2.4	0.0992	1.3
	2	0.59	-5.2	1.73	-2.9	2.00	-3.5	3.05	-5.4	3.14	-4.6	10.1	-12.0	12.72	-4.5	0.0782	-20.1
	3	0.71	14.1	1.83	2.7	1.98	-4.5	3.21	-0.4	3.17	-3.7	12.05	5.0	13.27	-0.4	0.1016	3.7
	4	0.66	6.0	2.14	20.1	2.37	14.4	3.85	19.5	4.01	21.8	12.63	10.0	13.66	2.5	0.1128	15.2

3　结论与讨论

试验结果显示，在幼林期，适宜水处理对柚木的高径生长具有极显著的促进作用，这反映了柚木对水分有较高的要求，在方便管理的地段柚木生长期土壤缺水时，加以水处理措施，可以促进生长，不同地区间降水有差异，坡位坡形对降水有重新分配作用，因此加强造林地规划、采取人造小地形、改善土壤涵水措施（如营造混交林[17-20]）都会起到积极的作用。

从施肥对柚木树高、胸径生长作用分析中可见，各处理差异不显著，由于各处理N、P、K含量各不相同，且级差明显，因此可以认为在营造在酸性土上的柚木幼林期1~3年施用仅含N、P、K素的无机肥，施肥效果不明显，这可能原因有两方面，其一，可能系水处理与重复对施肥的互作效应大，降低了处理间的差异程度；其二，施肥处理实际效果小，这结论与印度Gawande，S. R.[21]相一致，梁坤南[14]、周再知[22,23]认为钙对柚木生长有显著作用，可考虑在施肥中加入钙肥，以增加土壤钙的不足，达到增大施肥效果之目的。各重复、各水处理、各施肥处理生长指标年度变化分析结果显示，试验重复（地段）、水处理对柚木生长的影响差异显著且是持续稳定的，在一般情况下，水也是影响立地差异的重要因子之一，因此，反映出立地对柚木生长影响极大：在适宜立地上生长良好，立地不适则生长不良，充分体现出加强立地研究和选择的必要性。

在柚木幼林期1~3年重复与副区、主区与副区的交互作用不显著，13年时，交互作用显著，说明随着林龄的增加，因子间的交互作用表现得越来越明显，在施肥后期，有些处理其施肥效果在某一适合地段，适宜的水分和其他养分条件下，会表现出来，也体现出施肥效果受多因素影响[24]，适地适肥[25]很重要。

土壤N、P、K、有机质等营养元素，肥料成分与施肥量对林木影响大[26]，本研究水处理、施肥处理处理数少（级差和养分配比少），仅考虑了N、P、K和水因素对生长的影响，不能全面反映出林木的实际需求，也有可能因此掩盖了施肥本应存在的理想效果；我国南方山地土壤类型多样，不同类型养分差异大，水分又系影响林木生长的又一重要因子，但在山地灌淋水措施难于实施，人造小地形（如反坡平台、积水沟穴等）更具实际价值[27-33]，在研究中很有必要在不同土壤类型中继续开展包含有Ca、Mg、N、P、K、有机质等营养元素和多种人造小地形因素的综合试验，多方面了解柚木的各种营养需要和对水分要求，完善各种营林措施，促进柚木人工林的健康发展。

参考文献

[1]广西林业局，广西林学会．阔叶树种造林技术[M]．南宁：广西人民出版社，1980.

[2]中国树木志编委会．中国主要造林树种造林技术[M]．北京：中国林业出版社，1981.

[3]DOTANIYA M L, MEENA V D, Manju Lata, et al. Teak plantation-A potential source of Income generation [J]. Popular Kheti, 2013, 1(3): 61-63.

[4]JAYARAMAN K, BHAT K V. Innovations in the management of planted teak forests[C]. Peechi, Kerala, India: KEAKNET KFRI FAO KSCSTE, 2011, p7.

[5]马华明，梁坤南，周再知．我国柚木的研究与发展[J]．林业科学研究，2003，16(6)：768-773.

[6]梁坤南，白嘉雨，周再知，等．珍贵树种柚木良种繁育发展概况[J]．广东林业科技，2006，22(3)：85-89.

[7]邝炳朝，郑淑珍，罗明雄，等．柚木种源主要性状聚合遗传值的评价[J]．林业科学研究，1996，9(1)：7-14.

[8]李运兴．柚木家系试验[J]．广西林业科学，2001，30

(10)：50-52.

[9]张荣贵，蓝猛，乔光明，等．红河州柚木种源试验五年评价[J]．林业科学研究，1999，12(2)：190-196.

[10]梁坤南，赖猛，黄桂华，等.10个柚木种源27年生长与适应性[J]．中南林业科技大学学报，2011，31(4)：8-12.

[11]武勇．干旱胁迫下柚木叶片生理指标的变化[J]．福建林学院学报，2006，26(2)：103-106.

[12]黄桂华，梁坤南，周再知，等．柚木无性系苗期抗寒生理评价与选择[J]．东北林业大学学报，2015，43(9)：12-17.

[13]T美华．海南岛尖峰岭幼龄柚木年生长规律与温度-水分关系的初步研究[J]．林业科学，1982，18(1)：85-92.

[14]梁坤南，潘一峰，刘文明．柚木苗期多因素施肥试验[J]．林业科学研究，2005，18(5)：535-540.

[15]潘一峰，刘文明．酸性土壤改良对不同种源的柚木生长的影响[J]．热带亚热带土壤科学，1997，6(1)：9-14.

[16]李运兴，蔡道雄，欧耀成，等．追肥量、追肥时间与次数对柚木幼林生长的影响[J]．中南林业科技大学学报，2012(9)：16-19.

[17]赖景全．桉树混交林生长和土壤理化性质研究[J]．绿色科技，2014(7)：55-56.

[18]肖文光，王尚明，陈孝，等．核树与厚荚相思混交林生物量及对土壤影响研究[J]．广东林业科技，1999(1)：20-30.

[19]徐小牛，李宏开．马尾松枫香混交林生长及其效应研究[J]．林业科学，1997，33(5)：385-393.

[20]蒋家淡．红椎杉木混交造林效果研究[J]．福建林学院学报，2002，22(4)：329-333.

[21] GAWANDE S R. Stand Density manipulation and fertilization studies on teak[D]. Kerala Agricultural University, Thrissur, 1991：81.

[22]周再知．酸性土壤柚木钙素营养研究[D]．北京：中国林业科学研究院，2009.

[23]周再知，梁坤南，徐大平，等．钙与硼、氮配施对酸性土壤上柚木无性系苗期生长的影响[J]．林业科学，2010，46(5)：102-108.

[24]沈其荣．土壤肥料学通论[M]．北京：高等教育出版社，2001.

[25]王少元，何应同，曾祥福，等．杨树不同土壤立地条件施肥效应的研究[J]．林业科学，1999，35(1)：106-112.

[26]杨承栋．中国主要是造林树种土壤质量演化与调控机理[M]．北京：科学出版社，2009.

[27]杨绍兵，王克勤，陈志中，等．坡耕地反坡水平阶对土壤水N_ P垂直再分配的影响[J]．中国水土保持科学，2011，9(1)：56-60.

[28]崔永忠，李昆，孙永玉，等．元谋干热河谷不同整地措施造林成效研究[J]．西南农业学报，2009，22(5)：1300-1304.

[29]韩冈U，韩恩贤，薄颖生，等．黄土高原不同整地方法造林试验[J]．陕西林业科技，2003(4)：34-37.

[30]张维国，曹丽萍．反坡梯田整地效果的探讨[J]．防护林科技，2008，86(5)：129-130.

[31]虎久强，安永平，李英武．不同整地方法对造林成效影响的比较研究[J]．宁夏师范学院学报，2007，28(3)：110-113.

[32]谷瑞民，雷振民，马翠霞．不同土壤管理措施对山地核桃生长结果的影响[J]．陕西林业科技，2009(2)：58-59，124.

[33]李昆，崔永忠．不同保水措施的赤桉造林成效分析[J]．云南林业科技，1998，82(1)：36-39.

[原载：中南林业科技大学学报，2016，36(08)]

西南桦与红椎混交造林模式评价

郝 建 莫慧华 麻 静 农 友 谌红辉 蔡道雄

（中国林业科学研究院热带林业实验中心，广西凭祥 532600）

摘 要 为了评价西南桦（*Betula alnoides*）与红椎（*Castanopsis hystrix*）混交造林模式的优劣，以单株形质指标、胸径分布、生长量、蓄积量、虫害和林下植被为指标，对25年生西南桦与红椎异龄混交林及13年生西南桦与红椎同龄混交林和相应的西南桦纯林的林分结构、生长量及健康状况进行研究。结果表明：西南桦与红椎混交造林，有利于西南桦干形塑造，可以提高西南桦生长量及单位面积总蓄积量，可以减少西南桦的感虫率及单株的虫孔密度，可以避免恶性杂草对林木生长的影响。

关键词 西南桦；红椎；混交林；综合评价

Evaluation of Mixed Plantation Regimes of *Betula alnoides* and *Castanopsis hystrix*

HAO Jian, MO Huihua, MA Jing, NONG You, CHEN Honghui, CAI Daoxiong

(*Experimental Center of Tropical Forestry*, *Pingxiang* 532600, *Guangxi*, China)

Abstract: In order toevaluate the mixed plantation regimes of *Betula alnoides* and *Castanopsis hystrix*, stand structure stand growth and forest healthy condition of 25-year-olduneven-agedmixed plantation regimes of *B. alnoides* and *C. hystrix*13-year-old even-aged mixed plantation of *B. alnoides* and *C. hystrix*and corresponding pure stands of *B. alnoides* were studied. Indexes of the morphological index, diameter distribution, growth, growing stock, insect attack and undergrowth were measured. Results indicated that *B. alnoides* mixed with *C. hystrix* was conducive to improve stem form of *B. alnoides*, increase growth and total volume per unit area, reduce infestation rate and borer density, avoid the influence of malignant weedson tree growth.

Key words: *Betula alnoides*; *Castanopsis hystrix*; mixed forest; comprehensive evaluation

西南桦（*Betula alnoides*）是我国南亚热带速生乡土用材树种，其树干通直圆满，木材具有密度适中、纹理优美、易于加工等优良特性，被广泛用于木地板制作、高档家具、室内装饰材料，具有较高的经济价值[1]；又因其良好的涵养水源[2]、保持地力[3]、固定碳素[4,5]及生物多样性维持[6]能力，被广泛应用于我国热带、南亚热带优良速生用材林基地建设、低产林改造及生态公益林营建。其造林面积已逾5万 hm^2[7]，发展前景良好。

大规模发展人工纯林导致病虫害频发、生物多样性下降、地力衰退和生产力下降等一系列的生态问题。中国林业科学研究院热带林业实验中心在营建西南桦试验林工作中，一直探索科学合理的西南桦混交林模式。因此，本文通对位于广西凭祥市的中国林业科学研究院热带林业实验中心伏波实验场1989年和2001年营造的西南桦×红椎混交林和西南桦纯林的林分结构、生长量及健康状况进行评价研究，为科学制定西南桦混交林营建方案提供理论依据。

1 材料与方法

1.1 研究区概况

研究区域位于广西凭祥市的中国林业科学研究院热带林业实验中心伏波试验场（106°51′~106°53′E，22°02′~22°04′N）。该地区属南亚热带季风型半湿润—湿润气候，可明显分为干季（10月至翌年3月）和湿季（4~9月）。年均气温20.5~21.7℃，最冷月（1月）平均气温13.5℃，最热月（7月）平均气温27.6℃；年均降水量 1200~1500mm，相对湿度 80%~84%。地貌类型以低山丘陵为主，海拔 430~680m，地带性土壤为花岗岩发育的山地红壤，土层厚度>80cm[5]。

试验地位于 2001 年 2 月营造的西南桦×红椎同龄混交林(初植密度 1667 株/hm²，西南桦与红椎混交比例为 1∶1)，2001 年 4 月营造的西南桦纯林(初植密度 1667 株/hm²)。西南桦×红椎同龄混交林经历了 2005 年和 2009 年两次间伐(间伐强度分别为 20%和 30%，间伐强度均为株数强度)，郁闭度 0.7，保留密度 900 株/hm²。西南桦纯林于 2005 年进行了间伐(间伐强度 25%)，郁闭度 0.6，保留密度约 1000 株/hm²。1989 年 2 月营造的西南桦纯林，初植密度 1667 株/hm²；1989 年 2 月营造的西南桦×马尾松同龄混交林(初植密度 3000 株/hm²，西南桦与马尾松混交比例为 1∶6)，之后由于马尾松严重被压，生长衰退，于 2002 年将马尾松砍除，在西南桦林下补植红椎，此后便形成西南桦×红椎异龄混交林，初植密度 1100 株/hm²，西南桦与红椎混交比例为 1∶2。

1.2　研究方法

2014 年 10 月，在西南桦纯林、西南桦×红椎同龄、异龄混交林中，每种林分中按坡位布置 9 块 20m×30m 的标准地(上、中和下坡各设 3 块)。

生长量调查：对标准地内的西南桦和红椎进行每木检尺，利用胸径尺测定胸径、超声波测高器(Hagl of VERTEX-IV，瑞典)测定树高。

利用干形、冠高比和分枝因素为形质评价指标，对每个标准地内的西南桦形质指标进行综合评价。根据干形、冠高比和分枝因素对树木形质影响的权重，以 10 分为满分，各指标评分标准见表 1[8]，优良木大于等于 7.5 分；中等木 4.5~7.5 分；劣势木小于等于 4.5 分。

表 1　西南桦形质指标评分标准

干形			冠高比			侧枝		
通直、圆满	微弯	较弯	1/4	1/2	3/4	细小	中等	粗大
5.0	4.0	3.0	3.0	2.5	2.0	2.0	1.5	1.0

虫害状况统计：拟木蠹蛾是西南桦人工林的主要害虫，影响西南桦生长量和木材品质。当年拟木蠹蛾幼虫钻蛀树干时，在坑道外面有由粉末状棕色虫粪或与树皮碎屑等组成的隧道，坑道周围树皮也会因此受损而产生颜色差异，而老虫口则由于雨水冲刷等外力作用外坑道及虫粪消失[9]。因此，2014 年 10 月调查样地内西南桦的当年虫口数，及虫口在树干上的高度位置。虫口高度采用超声波树木测高器进行测定。

林下植被统计：在每个标准地内沿对角线设置 3 个 2m×2m 的样方，详细调查样方内灌木、幼树、草本的种名、数量、平均高度、盖度等指标。

1.3　数据分析

单株材积(V)计算公式[10]：

$$V=0.52764\times10^{-4}D^{1.88216}H^{1.00931}$$

式中：V 为单株材积；D 为胸径；H 为树高。

采用 Excel 2013 对数据进行处理。

2　结果与分析

2.1　林木质量

按形质指标综合评分，统计了每个林分中西南桦在优良木、中等木、劣势木三个等级的综合评分和数量分布情况。由表 2 可知，西南桦×红椎同龄及异龄混交林中优良木所占比例均高于同期营造的西南桦纯林，且所占比例均大于 50%，混交林中西南桦的形质指数的综合评分也均高于纯林。因此，西南桦与红椎同龄或者异龄混交造林，均有利于西南桦干形塑造。

表 2　各林分中西南桦形质指标评价

林分	优良木		中等木		劣势木	
	综合评分	比例(%)	综合评分	比例(%)	综合评分	比例(%)
西南桦×红椎同龄混交	8.36±2.62	59.34	6.15±3.68	31.68	3.36±1.89	8.98
西南桦纯林	7.88±3.21	35.51	6.75±3.99	45.62	4.21±2.56	18.87
西南桦×红椎异龄混交	9.28±3.61	62.18	7.03±3.76	30.04	3.86±2.32	7.78
西南桦纯林	8.87±2.51	40.21	6.78±3.38	46.35	3.96±1.68	13.44

径阶分布是林分结构的基本规律之一，可以反映林分生长稳定性及林分株间竞争的主要指标[11]。西南桦在西南桦×红椎混交林与同期营造西南桦纯林中的径阶分布存在较大的差异(图1)。同龄混交林中，西南桦胸径主要分布在20.1~25cm(占50.95%)、25.1~30cm(28.79%)，而同期营造的纯林胸径分布在15.1~20cm(占46.61%)、小于等于15cm(27.94%)；异龄混交林中，西南桦胸径主要分布在20.1~25cm(占28.03%)、25.1~30cm(占31.17%)、大于30cm(25.08%)，而同期营造的纯林胸径分布在20.1~25cm(占39.63%)、25.1~30cm(26.41%)。西南桦与红椎混交效果较好，发挥了混交林改善林分土壤理化性状，提高土壤肥力的特点[12]。因此，西南桦与红椎的同龄及异龄混交林中西南桦的胸径生长优于相应的西南桦纯林。

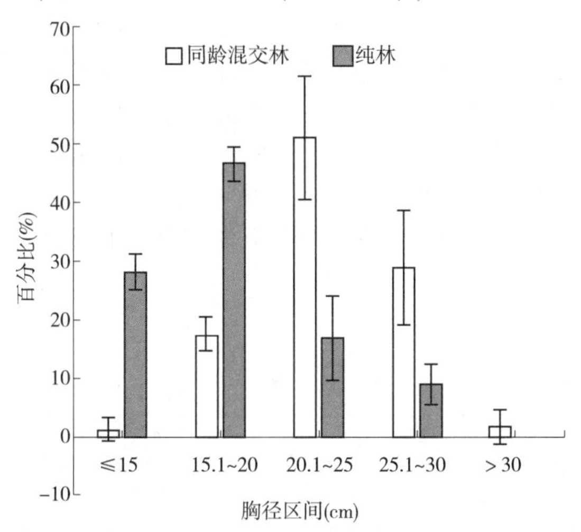

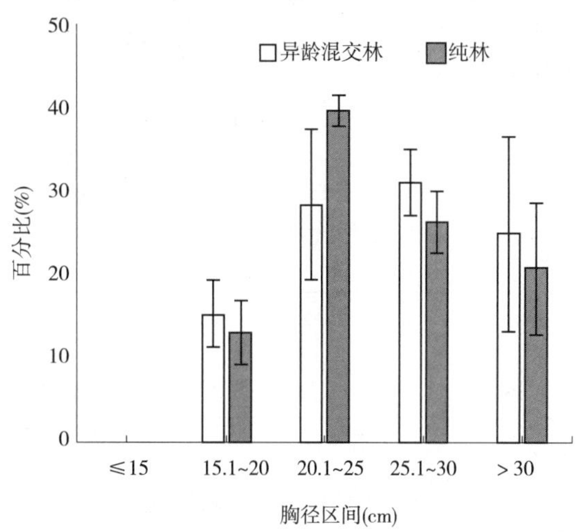

图1 西南桦在西南桦×红椎混交林及西南桦纯林中的径阶分布

2.2 林分生长量

由表3可知，西南桦与红椎的同龄混交、异龄混交林的胸径、树高和单位面积总蓄积量均高于同期营造的西南桦纯林。同龄混交林中西南桦平均胸径及树高比其纯林分别提高32.29%、11.98%，公顷总蓄积提高29.68%；异龄混交林中西南桦平均胸径及树高比其纯林分别提高28.27%、23.00%，公顷总蓄积提高24.43%。混交林有利于调节地表径流，增加土壤有效水，减少降水的无效损失和表土的流失[13]，可以改善林地土壤养分状况和土壤生物学特性[14]，提高林分生长量。

表3 西南桦×红椎混交林及其纯林生长状况比较

林分	树种	年龄(年)	胸径(cm)	树高(m)	公顷蓄积(m^3)
西南桦×红椎	西南桦	13	23.23±3.27	20.56±4.35	207.82
同龄混交	红椎	13	15.76±4.48	15.67±3.68	76.12
西南桦纯林	西南桦	13	17.56±5.16	18.36±4.89	218.96
西南桦×红椎	西南桦	25	25.95±4.18	23.32±5.12	226.72
异龄混交	红椎	12	16.31±3.28	14.86±4.21	103.90
西南桦纯林	西南桦	25	20.23±4.67	18.96±5.23	265.71

2.3 林分健康

对西南桦混交林及纯林样地内西南桦单株的拟木蠹蛾危害状况的调查数据进行统计分析(表4)得出，异龄混交与同龄混交林内西南桦的感虫率均低于同期营造的纯林，同龄混交感虫率降低了25.50%，异龄混交感虫率降低了32.57%；混交林内西南桦的虫害主要发生高度在15m以上，与混交的红椎单株高度相近，均高于纯林(8m以上)；混交林内西南桦单株的虫孔密度均小于纯林的，同龄混交单株虫孔密度降低了30.86%，异龄混交单株虫孔密度降低了31.97%。因此，西南桦与红椎混交对减少西南桦虫害具有一定作用。由于纯林结构简单和环境异质性低的特点，易为昆虫提供充足的食物来源和繁殖环境，对林分造成持续危害[15,16]。而科学营建的混交林，可以改善林分环境和提高林分异质性，降低易感树种密度、调节害虫的种类和丰富度[17,18]，限制其扩散和危害，从而降低虫害发生和危害程度[19]，是预防病虫害和提高林分生产力重要的营林措施[20]。

表4 西南桦×红椎混交林及其纯林拟木蠹蛾为害状况比较

林分	感虫株率(%)	虫害发生高度	虫孔密度(个/株)
西南桦×红椎同龄混交	38.67	15m以上	2.98±0.23
西南桦纯林	51.22	8m以上	4.31±0.45
西南桦×红椎异龄混交	29.34	15m以上	3.15±0.31
西南桦纯林	43.51	8m以上	4.63±0.43

由表5可知，纯林林下植被状况分别是：2001年营造的林下主要灌木种类为4种，盖度约14%，主要杂草种类为3种，盖度约81%；1989年营造的林下：主要灌木种类为4种，盖度约11%，主要杂草种类为3种，盖度约87%。混交林林下植被状况分别是，同龄混交的林下主要灌木种类为3种，盖度约3%，主要杂草种类为2种，盖度约12%；异龄混交林下：主要灌木种类为4种，盖度约6%，主要杂草种类为1种，盖度约5%。西南桦纯林林下植被种类数量及盖度均大于西南桦与红椎同、异龄混交林，与混交林相比，纯林林下杂草的种类及盖度极高，盖度较大的均为深根恶性杂草，对土壤水分及肥力消耗较大。因此，营造西南桦与红椎混交林可以避免恶性杂草对林木生长的影响。

表5 西南桦×红椎混交林及其纯林林下主要植被种类及盖度

西南桦×红椎同龄混交林		西南桦纯林		西南桦×红椎异龄混交林		西南桦纯林	
植被种类	盖度(%)	植被种类	盖度(%)	植被种类	盖度(%)	植被种类	盖度(%)
九节(*Psychotriarubra*)	1	裂叶榕(*Ficuslaceratifolie Levlet van*)	2	九节	1	裂叶榕	2
弓果黍(*Cyrtococcum patens*)	1	三叉苦(*Evodia lepta*)	4	杜茎山(*Maesa japonica*)	1	三叉苦	2
大青(*Clerodendrum cyrtophyllum*)	1	九节	3	弓果黍	1	九节	1
扇叶铁线蕨(*Adiantum flabellulatum*)	5	盐肤木(*Rhus chinensis*)	5	毛果算盘子(*Glochidion eriocarpum*)	3	野牡丹(*Melastoma candidum*)	6
金毛狗(*Cibotium barometz*)	7	五节芒(*Miscanthus floridulu*)	45	扇叶铁线蕨	5	五节芒	39
		粽叶芦(*Thysanolaena maxima*)	30			粽叶芦	38
		扇叶铁线蕨	6			铁芒萁(*Dicranopteris linearis*)	10

3 结论

西南桦与红椎混交造林，具有以下优点：

(1)有利于西南桦干形塑造。西南桦与红椎混交林中优良木所占比例均高于同期营造的西南桦纯林，混交林中西南桦的形质指数的综合评分也均高于纯林。

(2)有利于西南桦生长，单位面积总蓄积量明显高于西南桦纯林。

(3)可以减少西南桦虫害。西南桦混交林内西南桦的感虫率均低于纯林，且西南桦的虫害发生高度在红椎单株高度以上，混交林内西南桦单株的虫孔密度均小于纯林的。

(4)可以避免恶性杂草对林木生长的影响。西南桦纯林林下植被种类数量及盖度均大于西南桦与红椎混交林，纯林林下杂草的种类及盖度极高，盖度较大的均为深根恶性杂草，对土壤水分及肥力消耗较大。

因此，红椎是西南桦理想的混交树种，无论营造西南桦与红椎同龄混交林，还是利用红椎对现有西南桦纯林进行混交改造，都可达到较好的造林效果。

参考文献

[1] ZENG Jie, ZHENG Haishui, WENG Qijie. Betula alnoides-a valuable tree species for tropical and warm-subtropical areas[J]. Forest, Farm, and Community Tree Research Reports, 1999, 4: 60-63.

[2]孟梦，陈宏伟，刘永刚，等．西双版纳西南桦、山桂花人工林水源涵养效能研究[J]．云南林业科技，2002，3：46-49.

[3]蒋云东，周凤林，周云，等．西南桦人工林土壤养分含量变化规律研究[J]．云南林业科技，1999，2：27-31.

[4]李江，陈宏伟，冯弦．云南热区几种阔叶人工林C储量的研究[J]．广西植物，2003，23(4)：294-298.

[5]何友均，覃林，李智勇，等．西南桦纯林与西南桦×红椎混交林碳贮量比较[J]．生态学报，2012，32(23)：7586-7594.

[6]陈宏伟，刘永刚，冯弦，等．西南桦人工林群落物种多样性特征研究[J]．广西林业科学，2002，31(1)：5-11.

[7]曾杰，郭文福，赵志刚，等．我国西南桦研究的回顾与展望[J]．林业科学研究，2006，19(3)：379-384.

[8]刘光金，谌红辉，郭文福，等．西南桦优树选择技术研究[J]．林业科学研究，2012，25(4)：438-441.

[9]王春胜，赵志刚，吴龙敦，等．修枝高度对西南桦拟木蠹蛾为害的影响[J]．西北林学院学报，2013，27(6)：120-123.

[10]广西林业勘察设计院、广西林学分院．森林调查手册[R]. 1986.

[11]黄志森．戴云山红楠种群直径分布规律[J]．福建林学院学报，2010，30(2)：133-136.

[12]王青天．福建柏与马尾松混交造林模式的环境效应与生长分析[J]．西北林学院学报，2013，28(3)：126-130.

[13]姚庆端．不同杉木混交模式土壤肥力及土壤蓄水量研究[J]．福建林学院学报，1996，16(3)：282-286.

[14]叶存旺，翟巧绒，郭梓娟，等．沙棘-侧柏混交林土壤养分、微生物与酶活性的研究[J]．西北林学院学报，2007，22(5)：1-6.

[15]盛炜彤．人工林的生物学稳定性与可持续经营[J]．世界林业研究，2001，14(6)：14-20.

[16]张星耀，吕全，梁军，等．中国森林保护亟待解决的若干科学问题[J]．中国森林病虫，2012，31(5)：1-6.

[17]JACTEL H, BROCKERHOFF E G. Tree diversity reduces herbivory by forest insects[J]. Ecology letters, 2007, 10(9): 835-848.

[18]PLATH M, MOBY K, POTVIN C, et al. Establishment of native tropical timber trees in monoculture and mixed-species plantations: small-scale effects on tree performance and insect herbivory[J]. Forest Ecology and Management, 2011, 261(3): 741-750.

[19] KLINGENBERG M D, SATFFAN LINBGREN B, GILLINGHAM M P, et al. Management response to one insect pest may increase vulnerability to another[J]. Journal of applied ecology, 2010, 47(3): 566-574.

[20]STAMPS W T, MCGRAW R L, GOBSEY L, et al. The ecology and economics of insect pest management in nut tree alley cropping systems in the Midwestern United States[J]. Agriculture, ecosystems & environment, 2009, 131(1): 4-8.

[原载：西北林学院学报，2016，31(02)]

广西大青山西南桦人工林拟木蠹蛾为害的影响因子

庞圣江[1] 唐 诚[2] 张 培[1] 贾宏炎[1] 曾 杰[2]
([1]中国林业科学研究院热带林业实验中心，广西凭祥 532600；
[2]中国林业科学研究院热带林业研究所，广东广州 510520)

摘 要 以广西大青山西南桦(*Betula alnoides*)人工林为研究对象，依据林分类型、立地状况设置53块20m×30m的典型样地，调查了其立地和林分特征以及拟木蠹蛾(*Arbela* spp.)为害状况，揭示了西南桦人工林拟木蠹蛾为害的关键影响因子。偏相关分析表明，林分类型(r=-0.361，P=0.008)、林下植被盖度(r=-0.330，P=0.016)和高度(r=-0.471，P=0.000)以及坡位(r=-0.404，P=0.003)与西南桦林分感虫株率相关显著，是影响西南桦林分拟木蠹蛾为害的关键因子。方差分析显示，西南桦纯林和西南桦×红椎(*Castanopsis hystrix*)混交林拟木蠹蛾为害最为严重，其感虫株率分别为49.35%和43.55%，显著高于西南桦×杉木(*Cunninghamia lanceolata*)(17.63%)和西南桦×灰木莲(*Manglietia glauca*)混交林(14.04%)；林下植被茂盛的林分内拟木蠹蛾为害程度较轻；下坡林分感虫株率高达68%，显著高于上坡和中坡(20%以下)，可能与下坡林地更靠近虫源有关。

关键词 西南桦；拟木蠹蛾；蛀干害虫；林分结构；立地条件

Factors Influencing Attack of *Arbela* spp. in *Betula alnoides* Plantations at Mountain Daqingshan, Guangxi

PANG Shengjiang[1], TANG Cheng[2], ZHANG Pei[1], JIA Hongyan[1], ZENG Jie[2]
(*[1]Experimental Center of Tropical Forestry, CAF, Pingxiang, 536000, Guangxi, China*;
[2]Research Institute of Tropical Forestry, CAF, Guangzhou, 510520, Guangdong, China)

Abstract: Fifty three plots of 20m×30m were set up according to stand type and site condition of *B. alnoides* in Mountain Daqingshan, Guangxi, and stand structure and site properties were investigated as well as attack of *Arbela* spp. so as to reveal status of attack by the pest and the key influencing factors. Partial correlation analysis showed that stand type (r=-0.361, P=0.008), cover (r=-0.330, P=0.016) and height (r=-0.471, P=0.000) of understory vegetation, and slope position (r=-0.404, P=0.003) were significantly correlated with ratio of pest-damaged individuals, these were key factors influencing attack of *Arbela* spp. in *B. alnoides* plantations. Variance analysis revealed that the pure plantations and mixed ones with *Castanopsis hystrix* were attacked the most heavily, and their ratio of damaged individuals were 49.35% and 43.55%, respectively, obviously higher than those in mixed plantations with *Cunninghamia lanceolata* (17.63%) or *Manglietia glauca* (14.04%). The plantations with thicker understory vegetation were damaged more slightly. The pest attacks on lower slopes were high up to 68%, remarkably higher than that on upper and middle slopes, which was perhaps due to the fact that plantations on lower slope were closer to source of the pest.

Key words: *Betula alnoides*; *Arbela* spp.; trunk borer; forest structure; site condition

钻蛀类昆虫为害不仅抑制林木生长、降低木材质量，而且严重时亦影响森林健康和稳定性，是当前世界上人工林可持续经营的主要制约因素和关键问题之一[1,2]。国内外学者对林木钻蛀类害虫的发生、为害及其影响因素进行了大量研究。Jactel *et al.* 研究海岸松(*Pinus pinaster*)人工林松梢螟(*Dioryctria sylvestrella*)为害时发现林缘比林内受害严重，与阔叶树种混交可有效减轻为害[3]；李孟楼等认为，多树种混交能有效控制蛀干害虫为害，但混交林的抗虫效果与伴生树种的抗虫性及其混交比例有关[4]；姚松等研究影响松墨天牛(*Monochamus alternatus*)种群数量的林分因素发现，林下植被盖度、坡向显著影响松墨天牛数量，而且林缘和林内亦差异显著[5]。总体而言，目前有关天牛类等害虫研究较多，而对于诸如拟木蠹蛾类（Metarbelidae）等其他钻蛀类害虫，这方面的研究鲜见报道。由于钻蛀类昆虫发生、为害因其生物学特性和生活习性而异，广泛开展其发生、为害以及影响因子研究，有助于通过营林技术措施实现其长期有效防控，是林业发展急需解决的一个重大难题[6]。

拟木蠹蛾(*Arbela* spp.)为鳞翅目(Lepidoptera)拟木蠹蛾科的一种钻蛀类害虫，1年1个世代，3月中旬至下旬为化蛹期，4月上旬至中旬为羽化期，5月幼龄虫开始为害，取食至12月中旬后逐渐进入越冬期[7]，由于其啃食树皮和木质部而留下伤疤和虫孔，2~3年内均易辨别。目前拟木蠹蛾防治是珍贵树种西南桦大规模发展正面临的关键难题。拟木蠹蛾对西南桦的为害主要表现为，幼虫从西南桦树干伤口、树皮裂缝和树杈处钻蛀导致受害植株生长势削弱，危害严重时容易发生风折断梢，严重影响木材产量和质量[8]。随着西南桦种植业在云南、广西和广东等地大规模发展，拟木蠹蛾对西南桦的为害已为林业科技工作者所关注[9]。王春胜等通过修枝试验发现，修枝可显著降低西南桦修枝段的虫孔数，但对林分感虫株率无显著影响[8]。刘有莲等调查发现，西南桦受拟木蠹蛾为害程度随着海拔的升高而下降，在低海拔地段最为严重[10]。然而有关拟木蠹蛾为害的影响因子综合研究尚未见报道。从林分和立地特征等方面探明拟木蠹蛾发生、为害的关键影响因子，对于其有效防控具有重要意义。因此，本研究以广西大青山西南桦人工林为研究对象，开展拟木蠹蛾为害以及林分结构、立地条件等因素的系统调查，揭示拟木蠹蛾为害及其影响的关键因子，为有效控制拟木蠹蛾为害提供科学依据，将促进西南桦种植业的健康、高效发展。

1 研究地概况

研究地位于广西凭祥市中国林业科学研究院热带林业实验中心(21°58′~22°19′N，106°40′~106°59′E)。该地属于南亚热带季风气候，年均气温21.0℃，年均降水量约1400mm，主要集中在4~10月。该地位于广西大青山林区，是我国最先开展西南桦人工种植的地区。

2 研究方法

2.1 样地设置

2015年4~5月，根据西南桦林分类型、立地状况以及拟木蠹蛾为害情况，典型设置53个20m×30m样地，其中西南桦纯林9个，西南桦×红椎17个，西南桦×马尾松6个，西南桦×杉木15个，西南桦×灰木莲6个。对样地内乔木层树种进行每木检尺，记录树种种类、胸径和树高；依据拟木蠹蛾为害后留下的树皮伤疤和虫孔调查统计每个样地内感虫株率；于每个样地内沿对角线的两个角和中心位置设置3个5m×5m的样方，调查样方内下层植被盖度和高度等。详细调查、记录每个样地的林分类型、林龄、主林层郁闭度、海拔、坡位、坡向和坡度等。

2.2 数据分析

运用SPSS13.0软件对数据进行偏相关分析、方差分析和LSD多重比较。由于林分类型、坡位和坡向为定性因子，为了分析方便，对这些因子进行等级数量化处理，如林分类型，依据其为害程度从高到低依次将西南桦纯林、西南桦与红椎、马尾松、杉木和灰木莲的混交林赋值1~5；坡向，阴坡、半阴坡半阳坡、阳坡分别记为1、2和3；坡位，下坡、中坡和上坡分别记为1、2和3。其中，林分类型3.1(1~5)，林龄15.5年(8~20年)，林分密度652株/hm²(217~1383株/hm²)，平均胸径14.5cm(8.8~24.8cm)，平均树高10.3m(6.9~19.6m)，主林层郁闭度0.8(0.6~0.9)，林下植被盖度56.6%(12.5%~100.0%)，林下植被平均高度76.1cm(20.3~152.5cm)，坡位2.0(1~3)，坡向1.9(1~3)，坡度29.3°(9°~42°)，海拔456.4m(190~670m)，感虫株率30.2%(0~97.3%)。各影响因子后的数据为均值(变化范围)。

为了更清晰地揭示关键因子对西南桦拟木蠹蛾为害的影响，将其进行等级划分，并做比较分析。林下植被盖度划分为4个等级：A≤25%；B>25%和≤50%；C>50%和≤75%；D>75%。林下植被平均高度亦划分为4个等级：A≤40cm；B>40cm和≤80cm；C>80cm和≤120cm；D>120cm。

3 结果与分析

自20世纪70年代末以来，广西大青山林区在各种立地大规模营建西南桦人工林，且种植模式多样，主要包括西南桦纯林以及西南桦×马尾松(*Pinus massoniana*)、西南桦×杉木(*Cunninghamia lanceolata*)、西南桦×灰木莲(*Manglietia glauca*)和西南桦×红椎(*Castanopsis hystrix*)等混交林。在经营实践中发现，该地西南桦林分几乎全部遭受拟木蠹蛾为害，但为害程度差异明显。

3.1　拟木蠹蛾为害的关键影响因子筛选

偏相关分析表明(表1)，林分感虫株率与林分类型、林下植被盖度和平均高度以及坡位呈极显著相关($P<0.01$)或显著相关($P<0.05$)，其相关系数分别为-0.361、-0.330、-0.471和-0.404，说明这4个因子是影响西南桦拟木蠹蛾为害的关键因子。

表1　拟木蠹蛾为害与调查因子的偏相关性分析

项目	偏相关系数	P值
林分类型	-0.361	0.008
林龄	-0.025	0.859
林分密度	0.188	0.178
平均胸径	-0.098	0.486
平均树高	-0.148	0.291
主林层郁闭度	0.054	0.703
林下植被盖度	-0.33	0.016
林下植被平均高度	-0.471	0
坡位	-0.404	0.003
坡向	0.14	0.316
坡度	0.047	0.74
海拔	0.014	0.919

3.2　拟木蠹蛾为害关键影响因子分析

3.2.1　林分类型

5种西南桦林分中，西南桦纯林和西南桦×红椎混交林拟木蠹蛾为害最为严重，感虫株率分别49.35±9.27%和43.55±7.98%；其次为西南桦×马尾松混交林，其感虫株率为25.24±6.13%；而西南桦×杉木和西南桦×灰木莲混交林的感虫率最低，分别仅为17.63±6.65%和14.04±7.51%。方差分析结果显示，5种林分类型间林分感虫株率差异显著($F=3.35$，$P=0.0164$)，西南桦纯林、西南桦×红椎混交林的感虫株率显著高于西南桦×杉木、西南桦×灰木莲混交林，而西南桦×马尾松则与这4种林分差异均不显著($P>0.05$)。

3.2.2　林下植被盖度

随着林下植被盖度的增加，西南桦林分感虫株率呈现先升高后降低的趋势。林下植被盖度介于25%~50%时，西南桦林分拟木蠹蛾为害最为严重，其感虫株率为48.27±3.63%；林下植被盖度在75%以上时，拟木蠹蛾为害程度最轻，感虫株率仅为15.78±12.99%；林下植被盖度在25%以下或为50%~75%时，感虫株率居中，分别为29.86±4.40%和18.41±5.79%方差分析结果显示，4个盖度水平间西南桦林分感虫株率差异极显著($F=5.45$，$P=0.0026$)，林下植被盖度介于25%~50%的林分感虫株率显著高于其他3个盖度水平林分，而后三者间差异不显著。

3.2.3　林下植被平均高度

随着林下植被平均高度的增加，西南桦林分感虫株率大致呈先下降后稳定的趋势。4个高度水平间林分感虫株率差异极显著($F=3.75$，$P=0.0167$)，林下植被高度小于40cm的林分感虫株率为55.39±9.97%，显著高于40~80cm、80~129cm以及大于120cm3个高度水平林分，后三者的感虫株率分别为26.06±5.16%、28.01±5.44%和19.74±3.15%，其差异不显著，换言之，林下植被高度大于40cm时，林分感虫株率没有显著变化。

3.2.4　坡位

随着坡位的升高，西南桦林分拟木蠹蛾为害程度呈递减趋势。3个坡位间林分感虫株率差异极显著($F=4.09$，$P=0.0223$)，下坡林分感虫株率为68.04±21.01%，显著高于中坡(31.59±3.95%)和上坡(17.99±5.73%)，而中坡和上坡间差异不显著。下坡的林分感虫株率高达68%，而上坡则低于20%。

4　结论与讨论

本研究结果显示，西南桦林分感虫株率与林分类型、林下植被盖度和平均高度以及坡位等因子呈显著相关($P<0.05$)。西南桦纯林的拟木蠹蛾为害最为严重，而西南桦与杉木或灰木莲混交能有效减轻拟木蠹蛾对西南桦的为害，这与马尾松×木荷(*Schima superba*)混交显著减少马尾松松梢螟(*Dioryctria splendidella*)为害[11]以及毛白杨(*Populus tomentosa*)×刺槐(*Robinia pseudoacacia*)混交能有效控制光肩星天牛(*Anoplophora glabripennis*)为害[4]是一致的，这可能与混交林内天敌昆虫数量多于纯林有关[12]；然而本研究中亦发现，西南桦与红椎混交并未显著减轻拟木蠹蛾对西南桦林分的为害，换言之，西南桦×红椎混交林的抗虫效果不明显。究其原因，①调查发现，杉木和灰木莲未见拟木蠹蛾为害，而红椎遭受拟木蠹蛾为害，因此，红椎未能像杉木、灰木莲一样对拟木蠹蛾起到阻隔作用；②红椎×西南桦混交林内，红椎受害程度远较西南桦轻，即与红椎相比，拟木蠹蛾更喜欢取食西南桦。这些因素的

综合作用导致西南桦纯林与西南桦×红椎混交林拟木蠹蛾为害程度差异不显著。在木麻黄(*Casuarina* spp.)与相思(*Acacia* spp.)混交林内，由于木麻黄和相思同为拟木蠹蛾的寄主植物，混交林的受害程度甚至较木麻黄纯林更严重[7]。本研究还发现，尽管马尾松亦未见拟木蠹蛾为害，但是西南桦与马尾松混交林内西南桦拟木蠹蛾为害处于中等水平，与其他类型林分差异不显著，其原因颇为复杂，尚待进一步研究。

林下植被亦显著影响西南桦林分拟木蠹蛾为害状况。林下植被盖度为25%~50%的西南桦林分遭受拟木蠹蛾为害最为严重；盖度小于25%以及大于50%的林分，其感虫株率差异不显著，而显著低于盖度为25%~50%的林分。林下植被越高，西南桦林分感虫株率则越低。本研究结果与姚松等(2008)对于马尾松松斑天牛(*Monochamus alternatus*)为害的研究基本一致，他们认为林下植被越茂盛，马尾松纯林或者混交林遭受松斑天牛为害程度越轻[5]。也有学者认为，林下植被盖度和高度越大，湿地松萧氏茎象(*Hylobitelus xiaoi*)为害则越严重[13,14]，这与本研究的结果不一致，可能与两种钻蛀类害虫的生物学特性和生活习性差异有关，松萧氏茎象是一种不耐高温、喜阴湿的昆虫，而拟木蠹蛾则是喜好植被稀少、林内阳光充足的林分。有关西南桦林下植被与拟木蠹蛾为害的作用机制尚待从害虫的生物学特性、生活习性以及林内小气候等方面开展深入研究。

本研究还发现，拟木蠹蛾为害较为严重的林地大多分布在下坡，而上坡为害较轻。以往其他树种的研究发现，下坡或者沟谷沙棘(*Hippophae rhamnoides*)植株的沙棘木蠹蛾(*Holcocerus hippophaecolus*)虫口密度明显高于上坡[15]，与本研究结论相似。究其原因，可能与虫源有关。广西大青山林区下坡的大多数西南桦人工林靠近台湾相思(*Acacia confusa*)防火林带以及附近种植柑橘(*Citrus reticulata*)、荔枝(*Litchi chinensis*)和油茶(*Camellia oleifera*)等经济作物，这些都是拟木蠹蛾的寄主植物[7]，因此，下坡比中、上坡的西南桦遭受拟木蠹蛾为害的概率更大。

综上所述，营建西南桦人工林时应慎重考虑立地、林分结构等因素，通过立地选择、合理混交、适当抚育等营林技术措施，可以避免拟木蠹蛾的严重危害。其一，林地选择和规划上，宜开展造林地周边拟木蠹蛾寄主植物调查，尽量选择虫源较少的林地造林；或者在更大尺度上规划造林地，避免西南桦造林区域内种植拟木蠹蛾寄主植物。其二，营建西南桦人工林宜采用混交模式，选择非拟木蠹蛾寄主树种进行混交，减少西南桦的比例，使伴生树种发挥对拟木蠹蛾的阻隔作用，限制害虫扩散和为害，从而降低拟木蠹蛾的种群密度。其三，幼林抚育方面，西南桦植株高过灌草层后，减少良性杂灌的抚育，保持林下植被茂盛生长，改善林分环境和提高林分异质性，从而减轻西南桦拟木蠹蛾为害。

参考文献

[1]JI L Z, WANG Z, WANG X W, et al. Forest insect pest management and forest management in China: An overview[J]. Environmental Management, 2011, 48: 1107-1121.

[2]张星耀，吕全，梁军，等．中国森林保护亟待解决的若干科学问题[J]．中国森林病虫，2012，31(5)：1-6.

[3]JACTEL H, GOULARD M, MENASSIEU P, et al. Habitat diversity in forest plantations reduces infestations of the pine stem borer *Dioryctria sylvestrella*[J]. Journal of Applied Ecology, 2002, 39: 618-628.

[4]李孟楼，郭新荣，庄世宏，等．混交林的多样性及其光肩星天牛的抗性研究[J]．林业科学，2005，41(1)：157-164.

[5]姚松，汪来发，朴春根，等．林分因素对松墨天牛种群数量的影响[J]．安徽农业大学学报，2008，35(3)：411-415.

[6]张执中．森林昆虫学[M]．北京：中国林业出版社，1997.

[7]康文通．相思拟木蠹蛾生物学特性及防治研究[J]．华东昆虫学报，1998，7(2)：41-44.

[8]王春胜，赵志刚，吴龙敦，等．修枝高度对西南桦拟木蠹蛾为害的影响[J]．西北林学院学报，2012，27(6)：120-123.

[9]曾杰，郭文福，赵志刚，等．我国西南桦研究的回顾与展望[J]．林业科学研究，2006，19(3)：379-384.

[10]刘有莲，庞正轰，苏付保，等．西南桦林相思拟木蠹蛾危害调查[J]．中国森林病虫，2012，31(3)：26-28.

[11]黄文超，黄丽莉．马尾松—木荷混交造林效果的调查研究[J]．林业科学研究，2004，17(3)：316-320.

[12]罗长维，李昆．人工林物种多样性与害虫的控制[J]．林业科学，2006，42(8)：109-115.

[13]罗永松，肖活生，孙江华．萧氏松茎象种群发生与植被盖度的关系[J]．昆虫知识，2004，41(4)：367-370.

[14]温小遂，施明清，匡元玉．萧氏松茎象发生成因及生态控制对策[J]．江西农业大学学报，2004，26(4)：495-498.

[15]路常宽，骆有庆，李镇宇，等．沙棘灌丛林受害程度的预测模型[J]．生态学报，2006，26(2)：503-507.

[原载：东北林业大学学报，2016，44(11)]

大青山林区米老排人工林伐桩萌芽更新研究

庞圣江 张 培 刘福妹 赵 总 杨保国 刘士玲

(中国林业科学研究院热带林业实验中心，广西凭祥 5326002)

摘 要 以33年生米老排伐桩为研究对象，分析了不同伐桩基径(≤30cm、30cm<X≤40cm、40cm<X≤50cm、>50cm)及伐桩高度(≤5cm、5cm<Y≤10cm、10cm<Y≤20cm、>20cm)对其萌芽植株数量、胸径及高度等生长状况的影响。结果表明，4个不同的伐桩基径级之间，30<X≤40cm其萌芽植株数量、胸径及高生长较大；伐桩基径对其萌芽植株数量影响差异显著，但对胸径及高生长影响差异不显著。伐桩高度在≤5cm和5cm<Y≤10cm时，其萌芽植株数量、胸径及高生长较大，且随着伐桩高度的增加而降低。

关键词 米老排；伐桩；伐桩直径；伐桩高度；萌芽更新；人工林

Effeet of Varied Diameter and Height of Stump on Sprout Regeneration of *Mytilaria laosensis* Plantations in Daqingshan

PANG Shengjiang, ZHANG Pei, LIU Fumei, ZHAO Zong, YANG Baoguo, LIU Shiling

(*Experimental Center of Tropical Forestry*, *CAF*, *Pingxiang* 5326002, *Guangxi*, *China*)

Abstract: By taking the stump of 33a *Mytilaria laosensis* as the research object. Analysis of effects of the growth status of quantity、DBH and height for the sprout plants of different stump basal diameter (≤30cm、30cm<X≤40cm、40cm<X≤50cm、>50cm) and the height(≤5cm、5cm<Y≤10cm、10cm<Y≤20cm、>20cm). The result showed that, during the four stump basal diameter, the quantity、DBH and height are the biggest of its sprout plant; The stump basal diameter shows the prominent difference for the effect to the sprout plants, but has no prominent difference to DBH and height. When the stump height is between ≤5cm and 5cm<Y≤10cm , the quantity、DBH and height are the biggest of its sprout plant, it will reduce as long as the stump height increasing.

Key words: *Mytilaria laosensis*; stump; stump diameter; stump height; sprout regeneration; plantation

伐桩萌芽更新是指伐桩上的休眠芽萌发进而生长形成植株[1]。其具有迅速覆盖采伐迹地、防止水土流失和缩短更新周期等特点，已成为林木伐后更新的重要方式之一[2]。国内外林业科技工作者对伐桩萌芽更新进行了大量研究，如P. S. Johnson[3]研究发现，在一定的伐桩基径级范围，北美红栎(*Quercus rubra*)伐桩的萌芽率，随着伐桩基径的增大呈增加后降低的趋势；对糖槭(*Acer saccharum*)的伐桩萌芽更新研究认为，小基径级伐桩萌芽率和平均萌芽植株数量明显高于大基径级伐桩[4]；而李荣[5]等研究辽东栎(*Quercus wutaishanica*)伐桩萌苗的发育规律，表明伐桩萌芽植株数量随着伐桩基径的增加呈现先增加后减少，而随着伐桩高度的增加而增加的趋势；陈梦俅[6]等以伐桩基径以及高度对杉木(*Cunninghamia lanceolata*)萌芽更新的研究发现，伐桩萌芽6个月和1年时，杉木萌芽植株数量、萌芽基径及高度的受伐桩高度的影响差异极显著，而与伐桩基径的影响不显著。总体而言，目前有关林木伐后伐桩萌芽更新的研究报道较多，由于大多数的树种都能由休眠芽生长形成植株的特点，但其持续萌发能力因树种的生物学特性不同而差异较大[7]，因此，在林业领域开展不同树种采伐后伐桩萌芽更新研究，对于人工林近自然化经营具有重要的意义[8]。

米老排(*Mytilaria laosensis*)别称壳菜果、三角枫等，属于金缕梅科(Hamamelidaceae)壳菜果属(*Mytilaria*)的一种常绿阔叶乔木，天然分布于我国广西、广东及云南等地。其具有速生、伐桩萌芽更新能力强、干形通直和木材结构细密等优良特性，已成为南亚热带地区人工造林的主要优良乡土阔叶树种之一。然而，对米老排的研究大多数集中在种质资源收集、生物学特性、人工林生长规律及其生态和经济效益等方面[9-12]，对于如何利用米老排伐桩萌芽更新的相关研究尚未见报道，而这些正是米老排人工林伐后进一步抚育经营的核心问题。为此，本研究以广西大青山林区米老排人工林的伐桩为研究对象，通过分析皆伐后米老排伐桩萌芽规律和萌生植株生长状况，探讨其采伐迹地不同伐桩基径及伐桩高度对其萌芽植株数量、胸径和高度的影响，以期为米老排萌芽更新培育提供依据。

1 材料与方法

1.1 研究地概况

研究地位于广西凭祥市大青山林区，即中国林业科学研究院热林中心林区(21°57′~22°19′N，106°39′~106°59′E)。该区属于南亚热带季风气候，全年日照时数1218~1620h。年均气温21.6℃，年均降水量约1400mm，季节分布差异较大，主要集中在4~10月，相对湿度80%~84%。土层深厚、肥沃，土壤类型为砖红性红壤，原生植被有季雨林和常绿阔叶林。1975年营造的米老排人工林，郁闭度0.7，林分密度1400株/hm^2，平均胸径21.9cm，平均树高24.7m，林木蓄积量447.1m^3/hm^2。

1.2 试验设计

2008年10月，在米老排人工林的中部阳坡方向，设置4块20m×30m样地，调查林木胸径、树高等生长指标。12月进行皆伐后，对样地内伐桩进行统一编号和标记，并调查每木伐桩基径，用于分析不同伐桩基径对米老排伐桩萌芽植株更新的影响；同时，调查伐桩高度(距上坡位的高度)，分析不同伐桩高度对米老排伐桩萌芽植株更新的影响。2011年3月，对样地内的伐桩萌芽进行调查，记录伐桩萌芽生长形成的植株数量、胸径以及高度等指标。将不同伐桩高度划分≤30cm、30cm<X≤40cm、40cm<X≤50cm、>50cm等4个伐桩高度级，伐桩高度划分≤5cm、5cm<Y≤10cm、10cm<Y≤20cm、>20cm等4个伐桩高度级。

1.3 数据处理

采用Excel2003和SPSS13.0统计软件对调查数据进行统计分析，采用方差分析对不同伐桩基径以及高度对米老排伐桩萌芽植株数量(平方根转换)、胸径和高度之间的差异检验，并用Duncan多重比较检验分析伐桩基径及高度对伐桩萌芽植株数量、胸径和高度的差异性。

2 结果与分析

2.1 伐桩基径对米老排伐桩萌芽植株生长状况的影响

米老排伐桩基径分布范围为22.0~58.7cm，平均基径38.3cm，其中，≤30cm伐桩径级占17.68%，30cm<X≤40cm伐桩径级占40.59%，40cm<X≤50cm伐桩径级占25.42%，>50cm伐桩径级占16.32%。米老排的伐桩基径30cm<X≤40cm，平均萌芽株数量最多；伐桩基径40cm<X≤50cm，平均萌芽植株数量次之；伐桩基径>50cm，其平均萌芽植株数量为9株；伐桩基径≤30cm，平均萌芽植株数量最小。方差分析表明，伐桩基径对米老排伐桩萌芽植株数量的影响达到显著差异，采用Duncan多重比较分析表明，4个伐桩基径级的萌芽植株数量的差异达到显著水平(P<0.05)，而40cm<X≤50cm和>50cm伐桩基径级的萌芽植株数量的差异不显著(P>0.05，表1)。

伐桩基径对米老排伐桩萌芽植株直径生长的影响各不相同，其中，伐桩基径>50cm，伐桩萌芽植株的平均胸径最大；伐桩基径40cm<X≤50cm，伐桩萌芽植株平均胸径次之；伐桩基径30cm<X≤40cm，伐桩萌芽植株平均胸径位列第3；而伐桩基径≤30cm，伐桩萌芽植株平均胸径最小，但无显著性差异。

伐桩基径对米老排伐桩萌芽植株高度生长影响的分析表明，伐桩基径30cm<X≤40cm的伐桩萌芽植株平均高度值最大，其次是伐桩基径40cm<X≤50cm的萌芽植株平均高，伐桩基径>50cm，伐桩萌芽植株平均高度次之，而伐桩基径≤30cm，伐桩萌芽植株平均高为最小，但无显著性差异。

表1 伐桩基径对米老排伐桩萌芽更新的影响

伐桩基径(cm)	伐桩数量(株)	平均萌芽株数(株)	萌芽株数(株)	$\overline{D}$(cm)	$\overline{H}$(m)	比例(%)
≤30	28	6a	169	2.27±0.83	4.21±1.93	17.68
30<X≤40	32	12de	388	2.80±0.58	4.92±1.15	40.59

（续）

伐桩基径(cm)	伐桩数量(株)	平均萌芽株数(株)	萌芽株数(株)	$\overline{D}$(cm)	$\overline{H}$(m)	比例(%)
40<X≤50	23	10bc	243	2.92±0.72	4.88±1.28	25.42
>50	17	9bc	156	2.93±0.89	4.43±0.84	16.32

注：X表示伐桩基径，$\overline{D}$平均胸径，$\overline{H}$平均树高，表中$\overline{D}$和$\overline{H}$数据为平均值±标准差；小写字母为多重比较结果，不同字母不同表示差异显著(P<0.05)，下同。

2.2　伐桩高度对米老排伐桩萌芽植株生长状况的影响

米老排伐桩高度分布在3.19～36.45cm，平均17.45cm，其中，≤5cm伐桩径级占24.58%，5cm<Y≤10cm伐桩径级占28.14%，10cm<Y≤20cm伐桩径级占33.26%，>20cm伐桩径级占14.02%。随着米老排伐桩高度的增加，米老排伐桩萌芽植株的数量呈现减少的趋势(表2)。米老排伐桩高度≤5cm与5cm<Y≤10cm的萌芽植株的数量最多，分别为10株和11株，其次是伐桩高度10cm<Y≤20cm的萌芽植株的数量为9株，而伐桩高度>20cm萌芽植株的数量最少(7株)。伐桩的保留高度对米老排伐桩萌芽植株的数量具显著影响($F=5.6618>F_{0.05}$)。Duncan多重比较表明，伐桩高度≤5cm和5cm<Y≤10cm对米老排伐桩萌芽植株的数量差异不显著，而与后两者伐桩保留高度存在显著差异。

随着米老排伐桩高度的增加，米老排伐桩萌芽植株直径生长呈减小的趋势(表2)。米老排伐桩高度≤5cm萌芽植株平均胸径最大(3.04cm)，伐桩高度5cm<Y≤10cm的萌芽植株平均胸径次之，为2.79cm，而10cm<Y≤20cm和>20cm萌芽植株平均胸径最小，分别为1.57cm和1.52cm。伐桩的保留高度对米老排伐桩萌芽植株平均胸径有显著影响，伐桩保留高度≤5cm与5cm<Y≤10cm之间以及10cm<Y≤20cm与>20cm之间萌芽植株平均胸径差异不显著，而前两者与后两者伐桩保留高度萌芽植株的平均胸径则存在显著差异。

随着米老排伐桩高度的增加，米老排萌芽植株高生长也逐渐减低，与直径生长的表现一致(表2)。米老排伐桩高度≤5cm萌芽植株平均高度值最大，高达4.78m，伐桩高度5cm<Y≤10cm次之(4.38m)，其次是伐桩高度10cm<Y≤20cm(2.54m)，而伐桩高度>20cm萌芽植株平均高度值最小，为2.49m。伐桩的保留高度对米老排伐桩萌芽植株平均高度值有显著影响，伐桩保留高度≤5cm与5cm<Y≤10cm之间以及10cm<Y≤20cm与>20cm之间萌芽植株平均高度值差异不显著，而前两者与后两者伐桩保留高度萌芽植株的平均高度值则存在显著差异。

表2　伐桩高度对米老排伐桩萌芽更新的影响

伐桩高度(cm)	伐桩数量(株)	平均萌芽株数(株)	萌芽株数(株)	$\overline{D}$(cm)	$\overline{H}$(m)	比例(%)
≤5	22	10bc	235	3.04±0.63ab	4.78±0.81ab	24.58
5<Y≤10	25	11bc	269	2.79±0.79ab	4.38±1.28ab	28.14
10<Y≤20	34	9de	318	1.57±0.49cd	2.54±0.77cd	33.26
>20	19	7a	134	1.52±0.52cd	2.49±0.94cd	14.02

3　结论与讨论

伐桩基径对米老排伐桩萌芽植株数量差异显著，与伐桩萌芽植株胸径及高度差异不显著。伐桩基径30～40cm的伐桩萌芽植株数量最多，与辽东栎(*Quercus liaotungensis*)[5]、刺槐(*Robinia pseudoacacia*)[13]以及杉木[14]等林分中，中等基径级伐桩的萌芽植株数量最多的研究结果一致，这可能与较小的基径级伐桩一般维持大量萌发需要，且其大多数为生长衰落或储存养分不足的被压木，伐桩萌芽植株数量较少；而较大的基径级伐桩多为生长旺盛且休眠芽较少，因此，伐桩萌芽植株数量也相对较少[14]。伐桩萌芽生长形成的植株位于树干表皮层内，说明伐桩休眠芽一直处于树皮表层内。相关学者研究表明，随着伐桩基径的增大，树皮加厚对不定芽的萌发和生长产生机械阻碍的作用[15]。

伐桩高度对米老排萌芽植株数量、胸径以及高度具显著影响，在4个不同的伐桩高度之间，5～10cm伐桩萌芽植株数量最多，而5cm以下伐桩萌芽植株数量次之，二者的生长状况明显优于其他2个高度级的伐桩，由此可知，≤10cm 为米老排较为适合的采伐高度，具有较高的萌芽率和较好的生长势。与洪长福[15]等对巨尾桉（*Eucalyptus grandis* × *urophylla*）伐桩萌芽更新高度的研究基本一致。伐桩休眠芽与根系距离越短，休眠芽的发育阶段越年轻，生活力越强，伐桩萌芽植株生长旺盛，高速生长期持续时间长，不易出现早熟化的现象[17]。因此，米老排轮伐期经营过程中，为得到生长更加健壮的伐桩萌芽植株，尽量避免早熟化现象的出现，采伐时应尽量降低伐桩高度，使得伐桩萌芽更新的效果更好[14,16]。

当前，我国关于林木伐桩萌芽更新的研究已在刺槐[13]、巨尾桉[16]、杉木[17]、水曲柳（*Fraxinus mandshurica*）[18]、栓皮栎（*Quercus variabilis*）[19]等树种上取得了一定的研究成果。林木伐桩萌芽更新出伐桩基径和伐桩高度 2 个主要影响因素外，立地条件、林龄、采伐季节、轮伐期长短及伐桩保留萌芽植株数量等因子也对林木伐桩萌芽的更新产生影响。同时，许多研究采用了伐桩的存活率、萌芽率、植株生物量、萌芽植株总断面等[15,20,21]指标来衡量林木伐桩的萌芽更新能力，并取得了不同程度的进展。本研究仅就基径和高度对伐桩萌芽植株更新的影响进行了探讨，至于其他相关的影响因子，如立地条件、采伐季节、伐桩植株生物量等对伐桩萌芽更新的影响有待进一步研究。

参考文献

[1]林武星，叶功富，黄金瑞，等．杉木萌芽更新原理及技术述评[J]．福建林业科技，1996，23(2)：19-23.

[2]伊力塔，韩海荣．山西灵空山林区辽东栎萌芽更新规律研究[J]．林业资源管理，2007(4)：57-61.

[3]JOHNSON P S. Growth and structural development of Red Oak Sprout Clumps [J]. Forest Science，1975，21(4)：413-418.

[4]MACDONALD J E，POWELL G R. Relationships between stump sprouting and parent-tree diameter in sugar maple in the 1st year following clear-cutting [J]. Canadian Journal of Forest Research，1983，13(3)：390-394.

[5]李荣，张文辉，何景峰，等．辽东栎伐桩萌苗的发育规律[J]．林业科学，2012，48(3)：82-87.

[6]陈梦俅，田晓萍，曹光球，等．伐桩基径及高度对杉木萌芽更新的影响[J]．亚热带农业研究，2015，11(1)：11-14.

[7]叶镜中，孙多．森林经营学[M]．北京：中国林业出版社，1989：19-27.

[8]李景文，刘世英，王清海，等．三江平原低山丘陵区水曲柳无性更新研究[J]．植物研究，2000，20(2)：215-220.

[9]汪炳根．热带、南亚热带林木种质资源保存与评价[J]．广西林业科学，2001(30)：49-51.

[10]郭文福，蔡道雄，贾宏炎，等．米老排人工林生长规律的研究[J]，林业科学研究，2006，19 (5)：585-589.

[11]明安刚，贾宏炎，陶怡，等．米老排人工林碳素积累特征及其分配格局[J]．生态学杂志，2012，31(11)：2730-2735.

[12]白灵海，唐继新，明安刚，等．广西大青山米老排人工林经济效益分析[J]．林业科学研究，2011，24(6)：784-787.

[13]孙长忠，王开运，任兴俄，等．渭北黄土高原刺槐萌生林生长状况的调查研究[J]．西北林学院学报，1993，8(2)：36-40.

[14]高健，刘峰，叶镜中．伐桩粗度和高度对杉木萌芽更新的影响[J]．安徽农业大学学报，1995，22(2)：145-149.

[15]洪长福，薛瑞山，韩金发．巨尾桉萌芽更新最佳伐根高度的确定[J]．森林工程，2003，19(1)：11-13.

[16]叶镜中．杉木萌芽更新[J]．南京林业大学学报：自然科学版，2007，31(2)：1-4.

[17]荆涛，马万里，KUJIANSUU J，等．水曲柳萌芽更新的研究[J]．北京林业大学学报，2002，24(4)：12-15.

[18]易青春，张文辉，唐德瑞，等．采伐次数对栓皮栎伐桩萌苗生长的影响[J]．西北农林科技大学学报：自然科学版，2013，41(4)：147-154.

[19]黄世能．不同伐桩直径及高度对马占相思萌芽更新影响研究[J]．林业科学研究，1990，3(3)：242-248.

[20]黄世能，郑海水．采伐季节、伐桩直径及采伐工具对大叶相思萌芽更新影响的研究[J]．林业科学研究，1993，6(1)：76-82.

[21]孙黎黎，张文辉，何景峰，等．黄土高原丘陵沟壑区不同生境条件下柠条人工种群无性繁殖与更新研究[J]．西北林学院学报，2010，25(1)：1-6.

[原载：西北林学院学报，2016，31(06)]

经营单位级城市森林可持续经营评价指标体系研究

韦菊玲[1] 陈世清[2] 徐正春[2]

([1]中国林业科学研究院热带林业实验中心，广西凭祥 532600；
[2]华南农业大学林学与风景园林学院，广东广州 510642)

摘 要 明确界定以经营单位级城市森林为研究对象，通过对城市森林研究进展和国内外森林可持续经营指标体系进行研究分析和总结归纳，结合广州城市森林经营单位的实际情况，通过Delphi法和目标法构建了一套3个层次30个指标的城市森林可持续经营指标体系。利用研建的指标体系对广州市流溪河国家森林公园、大岭山林场和大夫山森林公园可持续经营状况进行评价，得出综合评价指数分别为0.8602、0.8762和0.8747，均属于可持续经营状态。在此基础上对影响广州城市森林可持续经营的制约因素进行分析，并提出城市森林可持续经营对策。

关键词 广州市；城市森林；森林经营单位；森林可持续经营；评价指标体系

Evaluation Index System of Sustainable Urban Forest Management Based on Forest Management Unit Level

WEI Juling[1], CHEN Shiqing[2], XU Zhengchun[2]

([1]*Experimental Center of Tropical Forestry, CAF, Pingxiang 532600, Guangxi, China;*
[2]*South China Agricultural University, College of Forestry and landscape Architecture, Guangzhou 510642, Guangdong, China*)

Abstract: Clearly defining the forest management unit level city as the research object, by summarizing and analyzing the situation of urban forest development at home and abroad and index system of sustainable urban forest management, combining the urban forest management conditions in Guangzhou, we used the target method and Delphi method to set up an index system of sustainable urban forest management based on forest unit level, which contains thirty indicators at three levels. Then the current condition of Liuxihe National Forest Park, Dalingshan Forest Farm and Dafushan Forest Park have been evaluated with the index system. The total scores were 0.8602, 0.8762 and 0.8747, belonging to a state of sustainable management. On the basis of analyzing the factors restricting the sustainable forest management in Guangzhou, we put forward some countermeasures for sustainable urban forest management.

Key words: Guangzhou city; urban forest; forest management unit level; sustainable forest management; evaluation index system

近年来，由于不合理的人类活动导致了森林生态系统遭到破坏，森林问题逐渐成为世界各国广泛关注和研究的热点问题，这使得森林传统经营的观念发生了很大的改变，人们都在积极寻找新的林业发展模式。可持续发展战略受到广泛认可[1]。城市森林可持续经营标准与指标体系的研究不仅可以为构建城市森林可持续经营评价系统奠定基础，更是实现城市可持续发展的根本保障。目前，城市森林可持续经营指标体系的研究主要集中在国际、国家、区域水平和森林经营单位4个层次上[2-6]。国际、国家和区域水平这3个层次的指标体系仅仅可以给出在森林经营管理活动中以及森林经营方案制定中需要考虑的方面，只能够提供森林可持续经营评价系统的逻辑框架[7]。也就是说，经营单位水平的标准指标体系才真正具有可行性[8]。目前有关城市森林可持续经营的研究取得的突破在于指标体系的构建不仅仅停留在宏观或定性分析，已经涉及指标的具体量化处理方法和可持续性评价方法，同时对具体的案例进行了研究[9-20]。但指标选择与定量化分析方面还存在较大争议，如何全面科学地反映城市森林多方面的功能结构特征以及如何定量化反映不同方面的功能价值仍是亟待解决的重要问题。另外，一些指标如社会、经济和生态效益方面所需的定量数据均缺乏贮备，资料来源或统计口径不一。因此，以城市森林为对象的经营单位级的可持续经营评价研究具有十分重要的意义。

1 经营单位级城市森林可持续经营指标构建的依据和基本原则

1.1 理论依据

城市森林主要作用是改善城市环境，将满足市民的身心健康和整个城市的发展列为重点考核指标[14]。截至2015年，广州市GDP为18100.41亿元，人口为1350.11万，作为华南地区的中心城市，在当前国家大力推进环境友好型和资源节约型社会建设的号召下，城市森林在广州经济社会发展中表现出的功能和作用，表现为生态效益为主导地位，并兼顾社会效益和经济效益的特殊性[14]。本研究结合广州市城市森林经营的实际，建立了一套可行的经营单位级城市森林可持续经营评价指标体系，为描述、监测和评价经营单位水平森林经营的可持续性提供评价依据和标准，从而引导和推进广州城市森林走可持续经营之路。

1.2 基本原则

构建经营单位级城市森林可持续经营指标体系有五大基本原则：科学可行性原则、系统性原则、综合性原则、实用性原则和定量化原则。

2 经营单位级城市森林可持续经营指标体系的建立

2.1 指标体系的构建方法

目前，指标体系的构建方法主要有Delphi法、目标法、归类法和系统法[21,22]。研究采用目标法和Delphi法对影响城市森林可持续经营的因子进行重要程度的分析选择，同时根据变异系数做进一步的筛选，最终构建了一套3个层次30个指标的森林经营单位级城市森林可持续经营指标体系（表1）。

表1 森林经营单位级城市森林可持续经营指标体系

目标层	标准层	指标层
广州城市森林可持续经营评价	生物多样性保护(B_1)	森林类型多样性指数(C_1)
		生态公益林一、二类面积比例(C_2)
		天然林面积比例(C_3)
		针阔混交林面积比例(C_4)
	森林生产力维持能力(B_2)	林地利用率(C_5)
		乔木林单位面积蓄积量(C_6)
		乔木林成过熟林面积蓄积比例(C_7)
		造林保存率(C_8)
	森林健康与活力(B_3)	病虫害危害的森林面积占有林地面积比例(C_9)
		森林火灾面积占有林地面积比例(C_{10})
		气候和其他自然灾害破坏的森林面积占有林地面积比例(C_{11})
		人为破坏的森林面积占有林地面积比例(C_{12})
		人为干扰的森林面积占有林地面积比例(C_{13})
	水土保持(B_4)	主要用于生态保护目的的林地面积比例(C_{14})
		土壤蓄水能力(C_{15})
		土壤侵蚀严重的林地面积和比例(C_{16})
	森林应对气候变化能力(B_5)	森林覆盖率(C_{17})
		平均生物量(C_{18})
		碳密度(C_{19})

（续）

目标层	标准层	指标层
广州城市森林可持续经营评价	社会经济效益(B_6)	供宣传科普知识和科研教育的林地面积占森林面积比例(C_{20})
		以游憩和旅游为主要目的林地面积占森林总面积比例(C_{21})
		职工素质结构(C_{22})
		职工收入水平(C_{23})
	保障条件和机制(B_7)	职工工资保障(C_{24})
		社会保险参保率(C_{25})
		经济产出效率(C_{26})
		资产负债率(C_{27})
		林权证发证率(C_{28})
		林权纠纷调处率(C_{29})
		森林生态效益补偿金比例(C_{30})

2.2 指标值计算方法

2.2.1 指标评价值(Ci)的无量纲化

无量纲化也叫数据的规格化、标准化，是通过数学变换来消除描述指标(原始变量)量纲影响的方法[9]。评价指标评价值(Ci)反映各指标实际值(Pi)和各指标参照值(Ri)的接近程度。

正指标，即为越大越好的指标的无量纲化：

$$Ci=Pi/Ri$$

逆指标，即为越小越好的指标的无量纲化：

$$Ci=Ri/Pi$$

对于参照值为0的指标：$Ci=1-Pi$

当正指标 $Pi>Ri$，或逆指标 $Pi<Ri$，此时 $Ci=1$，表明该指标已达到理想程度。

2.2.2 指标实际值(Pi)的确定

表2列出了各评价指标的实际值计算方法。

2.2.3 指标参照值(Ri)的确定

指标参照值(Ri)是森林可持续经营的期望值，目的就是为了反映森林经营单位目前的森林资源可持续性状态与理想状态的差距。参照值按以下五种情况进行确定：①理论上容易计算，并与实际基本情况相符的，参照值按理论值确定。②国家、广东省政府、广东省林业厅或广州市林业局已明确制定规划目标的，参照值按规划目标确定。③理论上存在最优期望值，然而短期之内难以实现的，参照值论按最优值结合森林经营单位实际情况而定。④可通过典型试点调查和数据分析进行推算难以确定的理论值。⑤一些经济指标和社会指标采用我国实现小康社会和事业单位要达到的标准值[21]。各评价指标的参照值如表2所示[20,23-27]。

2.2.4 指标值权重确定方法

采用层次分析法，通过运用软件YAAHP0.5.3，计算出各层次权重(表3)。

2.2.5 综合评价指数的计算和等级划分

计算综合评价指数采用加权综合评分法[28]，其计算方法为：

$$A=\sum_{j=1}^{m}\left[\sum_{i=1}^{n}W_i\cdot C_i\right]\cdot W_j$$

式中：A 为加权处理后的综合指数值；C_i 指标层中指标评价值，W_i 为指标层中的指标权重；W_j 为标准层中的指标权重，n 为指标层中指标数量，m 为标准层中的指标数量。

森林经营单位级广州城市森林可持续经营评价等级可以划分为4个等级：综合评价指数值 $A\geqslant 0.8000$ 为可持续；$0.8000>A\geqslant 0.6000$ 为基本可持续；$0.6000>A\geqslant 0.4000$ 弱可持续；$A<0.4000$ 为不可持续。

表2 各评价指标计算方法和参照值汇总表

编号	评价指标	计算公式	参照值
1	森林类型多样性指数(P_1)	$P_1=-\sum_{i=1}^{n}(p_iLnp_i)/Ln(n)$	1
2	生态公益林一、二类面积(P_2)	P_2=生态公益林一二类面积/生态公益林面积×100%	80%
3	天然林面积比例(P_3)	P_3=天然林面积/森林总面积×100%	50%
4	针阔混交林面积比例(P_4)	P_4=针阔混交林面积/有林地面积比例×100%	90%

（续）

编号	评价指标	计算公式	参照值
5	林地利用率(P_5)	P_5=有林地面积/林地总面积×100%	90%
6	乔木林单位面积蓄积量(P_6)	P_6=活立木蓄积量/乔木林面积×100%	112.95m^3/hm^2
7	乔木林成过熟林面积、蓄积比例(P_7)	P_7=成过熟林面积(蓄积)/乔木林面积(蓄积)×100%	面积比(S) 50%，蓄积比(V)71.4%
8	造林保存率(P_8)	P_8=人工造林3年后的保存株数/当年实际造林株数×100%	80%
9	病虫害危害的森林面积占有林地面积比例(P_9)	P_9=病虫害危害强烈和剧烈的森林面积/有林地面积	5‰
10	森林火灾面积占有林地面积比例(P_{10})	P_{10}=森林火灾中度和重度的森林面积/有林地面积	1‰
11	气候和其他自然灾害破坏的森林面积占有林地面积比(P_{11})	P_{11}=气候和其他自然灾害强度和剧烈破坏的森林面积/有林地面积×100%	5%
12	人为破坏的森林面积占有林地面积比例(P_{12})	P_{12}=人为破坏的森林面积/有林地面积×100%	5%
13	人为干扰的森林面积占有林地面积比例(P_{13})	P_{13}=人为干扰的森林面积/有林地面积×100%	5%
14	主要用于生态保护目的的林地面积比例(P_{14})	P_{14}=用于水土资源保护的森林面积/有林地面积×100%	40%
15	土壤蓄水能力(P_{15})	P_{15}=人工林面积比例/1.864+天然林面积比例	1
16	土壤侵蚀严重的林地面积和比例(P_{16})	P_{16}=土壤侵蚀强烈和剧烈的森林面积/有林地面积×100%	5%
17	森林覆盖率(P_{17})	P_{17}=森林总面积/土地总面积×100%	95%
18	平均生物量(P_{18})	$P_{18}=aV+b$	90.64 t/hm^2
19	碳密度(P_{19})	$P_{19} = \sum_{i=1}^{n} B_i CCR_i / S$	44.63 t/hm^2
20	供宣传科普知识和科研教育的林地面积占森林面积比(P_{20})	P_{20}=供宣传科普知识和科研教育的林地面积/森林面积×100%	15%
21	以游憩和旅游为主要目的林地面积占森林总面积比例(P_{21})	P_{21}=以游憩和旅游为主要目的的林地面积/森林总面积×100%	40%
22	职工素质结构(P_{22})	P_{22}=(大专以上文化程度的职工数+中级职称以上职工数)/(职工总人数×2)×100%	85%
23	职工年收入水平(P_{23})	一年总收入	50577元
24	职工工资保障率(P_{24})	P_{24}=[1-(拖欠工资总额/工资总额)]×100%	100%
25	社会保险参保率(P_{25})	P_{25}=参加社会保险的职工数/职工总人数×100%	100%
26	经济产出效率(P_{26})	P_{26}=本年度总经营收入额/本年度总经营投入额×100%	100%
27	资产负债率(P_{27})	P_{27}=负债总额/总资产×100%	50%
28	林权证发证率(P_{28})	P_{28}=发证林地面积/纳入林改范围的集体林地面积×100%	95%
29	林权纠纷调处率(P_{29})	P_{29}=调处山林纠纷宗数/发生纠纷宗数×100%	50%
30	森林生态效益补偿金比例(P_{30})	P_{30}=生态效益补偿金/总经营投入额×100%	50%

表 3 评价指标体系权重表

目标层(A)	标准层(B)	标准层权重	指标层(C)	指标层权重
广州城市森林可持续经营评价	B_1	0.1580	C_1	0.2618
			C_2	0.3390
			C_3	0.1408
			C_4	0.2584
	B_2	0.1947	C_5	0.1669
			C_6	0.4072
			C_7	0.2099
			C_8	0.2160
	B_3	0.1693	C_9	0.3177
			C_{10}	0.4388
			C_{11}	0.0715
			C_{12}	0.0938
			C_{13}	0.0782
	B_4	0.1217	C_{14}	0.4779
			C_{15}	0.2364
			C_{16}	0.2857
	B_5	0.1993	C_{17}	0.3652
			C_{18}	0.3269
			C_{19}	0.3079
	B_6	0.0904	C_{20}	0.1374
			C_{21}	0.2259
			C_{22}	0.1124
			C_{23}	0.1792
			C_{24}	0.1454
			C_{25}	0.0648
			C_{26}	0.0658
			C_{27}	0.0691
	B_7	0.0665	C_{28}	0.3439
			C_{29}	0.2616
			C_{30}	0.3945

3 评价结果与分析

3.1 可持续经营综合评价

由表 4 可知流溪河国家森林公园、大岭山林场和大夫山森林公园可持续经营综合评价指数分别为 0.8602、0.8762 和 0.8747，均属于可持续经营状态。

3.2 经营单位级城市森林可持续经营的制约因素

根据广州城市森林经营单位实际情况，结合可持续经营综合评价结果得出影响广州城市森林可持续经营的主要制约因素有 3 点：

3.2.1 城市森林资源结构不合理

根据表 4 可知针阔混交林面积比例远远低于参照值 90%；广州市桉树和马尾松所占比例也较大，分别为 12.57% 和 9.92%；优势树种组成以其他经济树种为主，占了总面积的 18.03%。以上都说明城市森林资源结构不合理，树种组成单一。

3.2.2 城市森林经营管理整体技术水平不高

一方面，指标“职工素质结构”明显偏低。林场和森林公园职工数量较多，老龄化现象严重，教育水平偏低，林农职工人员文化水平较高者不多，拥有专业优秀的技术人员较少，对新技术新知识的学习兴趣和能力都不强。另一方面，流溪河国家森林公园和大夫山森林公园职工工资水平只达到广东省在岗职工年平均收入的 69% 和 62%，林农职工人员收入明显偏低，直接影响了职工在森林可持续经营中发挥的的主观能动性。解决人为因素的影响是提高广州城市森林可持续经营水平的关键因素之一。

3.2.3 城市森林发展后劲不足

流溪河国家森林公园指标“经济产出效率”偏低。流溪河国家森林公园作为广州市安置水库移民的社会公益性生态林场，还要解决水库移民就业和生活问题。林场停止经营性采伐后，职工、移民大量下岗待业，生活困难，就业问题矛盾突出，迫切寻找新的产业增长点。

3.3 经营单位级城市森林可持续经营对策

3.3.1 进一步优化林分结构，促进生态效益的发挥

广州城市森林树种结构经过几年的调整，总体上还是马尾松和杉木一统天下，阔叶林比重过小，针叶林为主的格局依然。加上城市森林资源总量不足、质量不高、树种比较单一、林相景观较差，这些问题都直接影响到森林生态系统的稳定和生态功能的发挥[14]。因此，广州城市森林的改造应遵循近自然林业理论和技术，套种乡土阔叶树种，将单一同龄针叶纯林逐步改造成为复层异龄针阔混交乡土森林。

3.3.2 进一步重视人才投资，提高人才队伍整体素质

自然保护区，国有林场和森林公园等经营单位都处于远离城市中心的郊区，较难吸引人才。因此，森林经营单位要树立“人才是第一生产力”的观点，实行人才优先的战略布局，对人才资本优先投资，优先积累。同时运用不同的方式对现有人才开展继续教育的投资，例如高等教育、远程教育和职业教育等，还可以利用林场丰富的天然林资源以及丰富的珍贵树种，如穗花杉等，积极与高校或科研单位进行合作，建立科学实验基地，一方面为它们的科研提供了场所，另一方面也能通过与高校或科研单位的合作，建立起产学研相结合具有林场自身特色的经营体系，创新人才引进、培育、使用机制，满足经营单位林业生产和发展对人才的需求。

3.3.3 进一步开展多种经营，丰富森林多功能经营

目前广州城市森林的发展后劲不足主要体现在：发展资金严重不足，主要依靠政府投资，不能依靠其自身的产业循环来发展；林产品结构单一，缺乏优势产业和特色品牌产品，缺乏竞争力[29]。广州森林不仅具有优美的自然景观资源，还具有十分丰富、集中的人文资源，让人置身其中心旷神怡，因此，其丰富的旅游资源具有发展森林旅游业的绝对优势。生态旅游业发展的好坏直接影响到林业职工的收入水平。在偏远郊区还可以拓展特色经济林的经营，如无核黄皮、砂糖橘、荔枝、龙眼、茶叶、竹笋等的经营。依据林业分类经营理论，确定林业产业发展模式，鼓励职工开展养殖业、种植业、森林旅游服务业，一方面可以为林农增加收入，一方面还可以为居民的观赏游憩提供良好的场所。广州市城市森林经营单位应建立和完善生态旅游设施和相应的配套服务设施，积极推介森林旅游产品和旅游线路，创新旅游宣传促销机制，促进行业间的合作，加大宣传力度，加强科技投入，提高旅游和服务质量，促进林业第三产业的发展，做到生态效益最大化的同时还可以兼顾多重效益。

4 结论与讨论

明确界定经营单位级城市森林为研究对象，通过层次分析法、专家咨询法等方法构建了一套以充分发挥森林多种功能为目标，分别从生物多样性保护、森林生产力的维持、森林健康与活力、水土保持、森林应对气候变化能力、社会经济效益的保持，保障条件和机制7个方面提出了更为具体的30个指标在内的评价指标体系。利用研建的指标体系对广州市流溪河国家森林公园，大岭山林场和大夫山森林公园可持续经营综合评价指数分别为0.8602、0.8762和0.8747，均属于可持续经营状态，验证了本指标体系的科学性和可操作性。最后针对经营单位级城市森林可持续经营的制约因素提出了经营对策。

(1)指标参照值的确定仍需进一步探讨。有些指标的参照值是在借鉴前人研究的基础上进行适当的调整，在实际中是很难实现的，只能够为指标量化提供参照数据。另外有些指标以国家、省、市“已明确制定的规划目标”作为参照值，“规划目标”是规划期要实现的阶段性目标，并不一定是“可持续经营”的目标或水平，因此，今后怎样选取更有价值的指标参照值是一个值得深究的问题。

(2)评价指标体系仍需进一步完善。森林经营是一个广泛动态、周期很长过程，指标体系研究任重而道远，难点在于评价指标的筛选和定性指标的定量化分析。另外，研究用同一个指标体系去评价不同的经营类型的适用性，还需在以后的工作中加以探讨。

(3)评价指标的应用具有局限性。由于我国地域辽阔，不同城市具有较大的环境差异使得各个城市森林建设的发展规划目标不同，相应地也将会有不同的属性要求[13]，因此针对不同城市森林特征的具体评价标准也会有很大差别。研究提出的评价指标体系具有地域的局限性，对其他地区不一定具有适用性，如需推广还需要结合各地具体情况进行调整。

表 4　广州城市森林可持续经营评价指标值汇总表

标准层(B)	指标层(C)	流溪河		大岭山		大夫山	
		评价值	得分	评价值	得分	评价值	得分
B_1	C_1	0.7912	0.0327	0.4278	0.0177	0.8301	0.0343
	C_2	1.0000	0.0536	1.0000	0.0536	1.0000	0.0536
	C_3	1.0000	0.0222	1.0000	0.0222	0.8400	0.0187
	C_4	0.1001	0.0041	0.0926	0.0038	0.1903	0.0078
B_2	C_5	1.0000	0.0325	1.0000	0.0325	1.0000	0.0325
	C_6	0.9319	0.0739	0.9318	0.0739	1.0000	0.0793
	C_7	0.7036	0.0288	0.8875	0.0363	0.9943	0.0406
	C_8	1.0000	0.0421	1.0000	0.0421	1.0000	0.0421
B_3	C_9	0.7532	0.0405	0.7143	0.0384	1.0000	0.0538
	C_{10}	1.0000	0.0743	1.0000	0.0743	1.0000	0.0743
	C_{11}	1.0000	0.0121	1.0000	0.0121	1.0000	0.0121
	C_{12}	0.9982	0.0159	1.0000	0.0159	1.0000	0.0159
	C_{13}	1.0000	0.0132	1.0000	0.0132	1.0000	0.0132
B_4	C_{14}	1.0000	0.0582	1.0000	0.0582	1.0000	0.0582
	C_{15}	0.8654	0.0249	0.9080	0.0261	0.7308	0.0210
	C_{16}	1.0000	0.0348	1.0000	0.0348	1.0000	0.0348
B_5	C_{17}	0.8728	0.0635	1.0000	0.0728	0.6923	0.0504
	C_{18}	1.0000	0.0652	1.0000	0.0652	1.0000	0.0652
	C_{19}	1.0000	0.0614	1.0000	0.0614	1.0000	0.0614
B_6	C_{20}	0.0000	0.0000	0.2044	0.0025	0.0000	0.0000
	C_{21}	1.0000	0.0204	1.0000	0.0204	1.0000	0.0204
	C_{22}	0.1249	0.0013	0.4459	0.0045	0.1249	0.0013
	C_{23}	0.6917	0.0112	0.9443	0.0153	0.6185	0.0100
	C_{24}	1.0000	0.0132	1.0000	0.0132	1.0000	0.0132
	C_{25}	1.0000	0.0059	1.0000	0.0059	1.0000	0.0059
	C_{26}	0.4440	0.0026	1.0000	0.0060	1.0000	0.0060
	C_{27}	1.0000	0.0063	1.0000	0.0063	1.0000	0.0063
B_7	C_{28}	1.0000	0.0229	1.0000	0.0229	1.0000	0.0229
	C_{29}	1.0000	0.0174	1.0000	0.0174	1.0000	0.0174
	C_{30}	0.2044	0.0054	0.2888	0.0076	0.0962	0.0025
总计			0.8602		0.8762		0.8747

参考文献

[1]陈世清．广东省国有林场经营理论与实践研究[D]．北京：北京林业大学，2007.

[2]叶绍明，郑小贤，谢伟东．桉树工业人工林林分水平可持续经营指标体系研究[J]．林业经济，2007(7)：31-34.

[3]VAN DEN BOSCH C K. A decade of urban forestry in Europe [J]. Forest Policy and Economics, 2003, 5(2): 173-186.

[4]HUNTER I R. What do people want from urban forestry?: the European experience [J]. Urban Ecosystems, 2003, 5(4): 277-284.

[5]TYRVAINEN L. Economic valuation of urban forest benefits in Finland [J]. Journal of Environmental Management, 2001, 62(1): 75-92.
[6]GERMANN-CHIARI C, SEELAND K. Are urban green spaces optimally distributed to act as social integration? Results of a geographical information system (GIS) approach for urban forestry research [J]. Forest Policy and Economics, 2004, 6: 3-13.
[7]张守攻，朱春全，肖文发．森林可持续经营导论[M]．北京：中国林业出版社，2001.
[8]国家林业局．LY/T1594—2002 中国森林可持续经营标准与指标[S]．北京：中国标准出版社，2002.
[9]郭建宏．福建中亚热带经营单位水平森林可持续经营评价研究[D]．福州：福建农林大学，2003.
[10]郭峰．北沟林场森林资源可持续性评价研究[D]．北京：北京林业大学，2013.
[11]JIM C Y. The urban forestry programme in the heavily built-up milieu of Hong Kong [J]. Cities, 2000, 17(4): 271-283.
[12]沈洪霞．鄂尔多斯市造林总场森林可持续经营评价研究[D]．内蒙古：内蒙古农业大学，2009.
[13]刘昕，孙铭，朱俊，等．上海城市森林评价指标体系[J]．复旦学报(自然科学版)，2004，43(6)：988-994.
[14]吴茂林．广州城市森林分类及经营对策研究[D]．湖南：中南林业科技大学，2007.
[15]王蓉丽．城市森林可持续发展指标体系的建立和应用[J]．华东森林经理，2006，20(2)：44-49.
[16]丁俊．成都城市森林指标体系研究[J]．山西建筑，2007，33(10)：360-361.
[17]门可佩，周萍蒋，梁瑜．构建和谐城市评价体系探讨：以南京市为例[J]．统计教育，2008(3)：34-36.
[18]刘婷．城市森林综合评价指标体系的研究：以武汉市为例[D]．武汉：华中农业大学，2011.
[19]韩明臣．城市森林保健功能指数评价研究：以北宫国家森林公园为例[D]．北京：中国林业科学研究院，2011.
[20]潘丽．生态公益型国有林场可持续发展评价指标体系研究[D]．长沙：中南林业科技大学，2013.
[21]刘代汉，郑小贤．森林经营单位级可持续经营指标体系研究[J]．北京林业大学学报，2004，26(6)：44-48.
[22]李金良，郑小贤，王昕．东北过伐林区林业局级森林生物多样性指标体系研究[J]．北京林业大学学报，2003，25(1)：48-52.
[23]广东省林业厅．广东省林业发展“十二五”规划[R]．2011.
[24]广东省森林资源与生态环境监测中心，广东省森林资源管理总站，广东省林业调查规划院．广东省森林资源二类调查与森林生态状况调查工作操作细则[内部资料][R]．2003.
[25]广州市林业和园林局，广州市财政局．关于印发2012—2016年期间提高生态公益林补偿标准和完善管理机制工作方案的通知(穗林业园林通〔2012〕242号)[Z]．2012.
[26]亢新刚．森林经理学[M]．4版．北京：中国林业出版社，2011.
[27]李海奎，雷渊才．中国森林植被生物量和碳储量评估[M]．北京：中国林业出版社，2010.
[28]崔国发，邢韶华，姬文元，等．森林资源可持续状况评价方法[J]．生态学报，2011，31(19)：5524-5530.
[29]许飞，涂慧萍．广州城市森林可持续经营的策略研究[J]．中南林业调查规划，2007，26(1)：1-3.

[原载：北京林业大学学报，2016，38(09)]

桂西南杉木林分生长对间伐的动态响应

曾 冀[1,2] 雷渊才[2] 蔡道雄[1] 唐继新[1] 明安刚[1]
([1]中国林业科学研究院热带林业实验中心，广西凭祥 532600；
[2]中国林业科学研究院资源信息研究所，北京 100091)

摘 要 以广西凭祥14年生杉木(*Cunninghamia lanceolata*)人工林(保存密度为1219株/hm^2)为对象，设置69%(处理Ⅰ)、60%(处理Ⅱ)、51%(处理Ⅲ)、40%(处理Ⅳ)4个间伐强度处理和不间伐对照，定期观测间伐后9年内胸径、树高、枝下高和冠幅等林分生长指标，并计算单株材积和林分蓄积，从而揭示林分生长对间伐的动态响应，筛选适宜的间伐强度，为杉木人工林近自然化改造提供技术支撑。结果表明：间伐强度显著影响杉木林分胸径、枝下高、单株材积和蓄积生长($P<0.05$)，而对树高和冠幅的影响不显著($P\geq0.05$)；胸径和单株材积以处理Ⅲ为最高，而枝下高和蓄积则以对照为最高。胸径、冠幅和枝下高年均增量在间伐后第1~3年最大，处理间的差异随着时间的推移逐渐缩小。树高、单株材积和林分蓄积年均增量高峰则出现在间伐后第3~5年。根据上述试验结果，为了快速培育杉木大径材，建议在杉木近自然化改造中选择间伐强度约50%(处理Ⅲ)为宜。

关键词 森林经营学；杉木；间伐强度；密度效应；生长动态

Dynamic Growth Response of *Cunninghamia lanceolata* Plantation on Thinning in Southwestern Guangxi, China

ZENG Ji[1,2], LEI Yuancai[2], CAI Daoxiong[1], TANG Jixin[1], MING Angang[1]
([1]*Experimental Centre of Tropical Forestry, Chinese Academy of Forestry, Pingxiang* 532600, *Guangxi, China*;
[2]*Research Institute of Forest Resources Information Techniques, Chinese Academy of Forestry, Beijing* 100091, *China*)

Abstract: A thinning trial was conducted in a 14-year-old plantation of *Cunninghamia lanceolata* at Pingxiang City, Guangxi. The stand density was about 1219 trees per hectare, and four thinning treatments were arranged as 69% (Ⅰ), 60% (Ⅱ), 51% (Ⅲ) and 40% (Ⅳ) as well as a control without thinning. Diameter at breast height (DBH), height, crown base height (CBH), crown width (CW), individual tree volume (ITV) and stand volume (SV) were investigated termly in 9 years after thinning so as to reveal dynamic response of stand growth to these thinning treatments, and screen out suitable thinning intensity for close-to-nature reforestation of Chinese fir plantations. The results showed that thinning intensity significantly ($P<0.05$) affected DBH, CBH, ITV and SV rather than height and CW of *C. lanceolata* plantation, and the highest DBH and ITV were observed in treatment Ⅲ, while the highest CBH and SV in the control; The mean annual increments of DBH, CW and CBH were the highest in 1 to 3 years after thinning, then the differences between them were gradually declined as increase of forest age, while the peak of height, ITV and SV growth occurred in 3 to 5 years after thinning. Based on above findings, it can be recommended that thinning intensity of about 50% (treatment Ⅲ) should be applied for close-to-nature reforestation and management so as to produce large-sized timber of *C. lanceolata* quickly.

Key words: forest management; *Cunninghamia lanceolata*(Chinese fir); thinning intensity; density effect; growth dynamics

杉木(*Cunninghamia lanceolata*)是中国南方特有的速生用材树种，生长快，材质好，用途广，深受群众喜爱，是一个重要的商品材树种[1]。据第7次全国森林资源清查结果显示，中国有杉木人工林面积853.86万hm^2，占全国人工林面积的21.35%，蓄积为6.2亿m^3，木材产量约占商品材的31.64%[2]。但中国木材多以中小径材为主，大径材资源很少，木材供需的结构性矛盾十分突出[3]。近年来，中国各地的实践表明，人工林近自然经营能够实现长期稳定的林木生长和林分发育，是人工培育大径材的有效途径之一[4,5]。培育杉木大径材已为社会所急需，而培育大径材的关键措施在于密度控制[6]。间伐作为一种有效改善林分密度的经营手段，对于提升森林产量有着重要意义。合理间伐能够有效促进林木生长，最佳间伐强度的选择一直是森林经营措施中的关键环节。当前国内有关间伐对杉木林分和单株的影响已开展了一系列研究，如：张水松等[7]研究得出林分间伐可以有效促进杉木个体生长，但不能增加其林分蓄积；孙洪刚等[8]亦得出相似结论，认为间伐不能增加杉木林分断面积。然而上述研究多侧重于杉木间伐后的效果，而有关间伐后杉木生长的动态研究尚少见报道，仅李婷婷等[9]比较了杉木2种强度(47%和61%)间伐后4年内林分蓄积生长量、单木生长量等指标的动态变化。目前，广西杉木的种植面积已逾100万hm^2，总蓄积量约1.2亿m^3，占广西林木蓄积量的20%[10]。本研究以桂西南14年生杉木人工林为对象开展近自然化改造的间伐试验，定期进行生长观测，揭示间伐后杉木林分的生长动态，以期选择合理的间伐强度，为完善杉木人工林近自然经营和大径材培育技术提供科学依据。

1 试验地概况

研究地位于广西凭祥市热林中心伏波实验场(22°03′N，106°51′E)，海拔400m，属于南亚热带季风气候，干湿季节明显，年均气温20.5℃，年均降水量1500mm，年蒸发量1388mm；土壤为花岗岩发育而成的砖红壤。

1993年春季于热林中心伏波实验场采用广西融水种源的1年生裸根苗营造杉木人工纯林，初植密度为2500株/hm^2，分别于1999年和2003年进行了强度约为20%的透光伐和强度约为30%的抚育性间伐。

2 研究方法

2.1 试验设计

2007年10月，选择立地条件和杉木生长情况基本一致的地段开展间伐试验。试验采用随机区组设计，设置4个间伐强度(表1)和1个对照(ck)，4次重复，共20个小区，小区面积为1500m^2。间伐后在杉木林下采用1年生苗均匀套种大叶栎(*Castanopsis fissa*)、红椎(*C. hystrix*)、格木(*Erythrophleum fordii*)、灰木莲(*Manglietia glauca*)、铁力木(*Mesua ferrea*)、枫香(*Liquidambar formosana*)和香梓楠(*Michelia hedyosperma*)等阔叶树种，套种密度为450~525株/hm^2。对照为不间伐、不套种。间伐前，根据单木生长竞争特征将林木分为目标树、特殊目标树、干扰树和一般木，分别进行标记、编号。选择目标树做永久性标记，按照试验设计要求，伐除全部干扰树及部分一般木[4]。此工作在2007年底之前全部完成。

表1 杉木人工林间伐强度的设置

处理	保留株数(株/hm^2)	间伐强度(%)	处理	保留株数(株/hm^2)	间伐强度(%)
Ⅰ	375	69	Ⅳ	732	40
Ⅱ	488	60	CK	1219	0
Ⅲ	594	51			

2.2 生长观测

间伐作业前，于杉木纯林内设置4块400m^2的圆形样地进行本底调查，调查样地内所有杉木的胸径、树高、枝下高和冠幅，包括应用围尺进行每木检尺，用皮尺分东、西、南和北4个方向测定冠幅，采用VERTEX超声波测高器测量树高和枝下高。林分平均保留密度为1219株/hm^2，平均胸径13.71cm，平均树高11.43m，平均蓄积量98.68m^3/hm^2。间伐后于每个小区内布设面积为400m^2的圆形固定样地，分别于2008年、2010年、2012年、2014年底和2016年8月进行定株生长观测。

2.3 数据处理

冠幅为东西、南北2个方向的平均值。单株材积的计算采用：

$$V=g(h+3)f$$

式中：h为全树高(m)；g为胸高断面积(m^2)；f为平均实验形数；杉木取0.429[11]。

采用单因素方差分析和Duncan多重比较检验各间伐强度间杉木林分生长表现及其年均增量的差异。应用SPSS16.0软件进行数据统计分析。由于套种的植株尚未进入主林层，对杉木生长的影响不大，故本研究在分析生长动态时未予考虑。

3 结果与分析

3.1 胸径生长动态

由图1A可以看出，各处理的胸径生长总体上随着时间的推移而呈直线升高的趋势。间伐后第1~7年，处理Ⅰ和处理Ⅲ胸径差异不显著，而在第7~9年，处理Ⅲ显著高于处理Ⅰ($P<0.05$)；间伐后第1~9年，处理Ⅰ和处理Ⅲ的胸径显著高于处理Ⅱ和处理Ⅳ($P<0.05$)，后两者在各时间段的胸径几乎一致。4个间伐处理的胸径在各时间段均显著高于对照($P<0.05$)，以处理Ⅲ为最高，间伐后第1年，较对照高出13.5%，第9年较对照高26.3%。间伐后第9年，中、大径级比例随着间伐强度的增大呈现出先升高后下降的趋势(表2)。4种间伐处理未见径级10cm(含10)以下的林木，而对照有8.72%；比较12~20cm径级林木的分布频率，对照和处理Ⅳ较为接近，分别为70.26%和74.36%，其余3个处理为48.42%~65.38%；径阶22cm以上林木的分布频率，对照和处理Ⅳ亦较为接近，为21.03%和25.64%，而其余处理为34.62%~51.58%。中、大径级林木的比例均以处理Ⅲ为最高。

各处理的胸径年均增量随着时间的推移均呈逐渐降低的趋势(图1B)。间伐后第1~3年，4个间伐处理间胸径年均增量无明显差异，但均显著高于对照($P<0.01$)；间伐后第3~9年，处理Ⅲ的胸径年均增量显著高于处理Ⅳ($P<0.05$)，与处理Ⅰ和处理Ⅱ的处理间差异不显著，4个间伐处理均显著高于对照($P<0.05$)。在间伐后的9年间，处理Ⅲ的胸径年均增量比对照增加了约1倍。

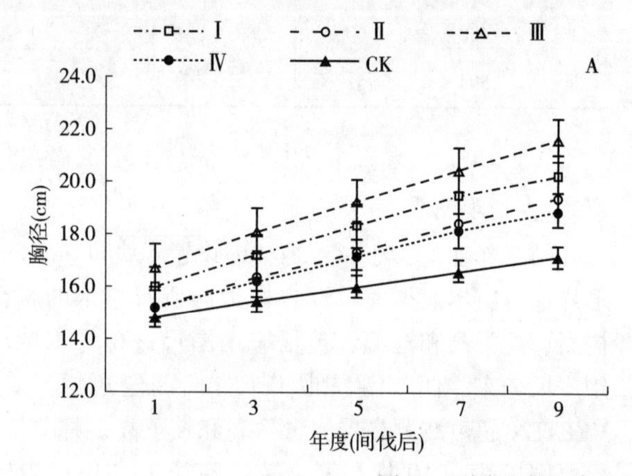

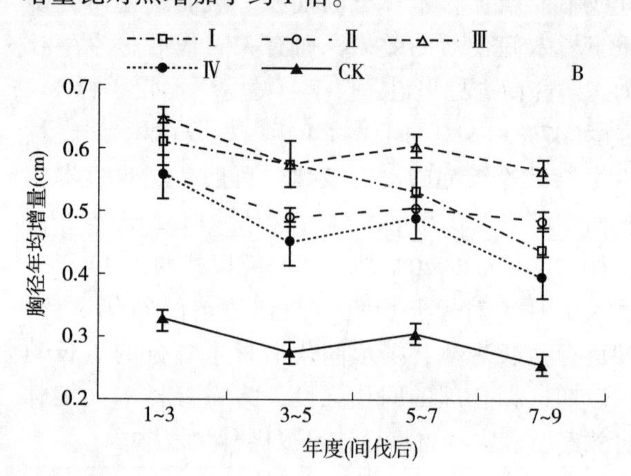

图1　不同强度间伐后杉木人工林胸径生长动态

表2　间伐后第9年杉木人工林的径级分布

径级(cm)	不同处理的林木径级分布频率(%)				
	Ⅰ	Ⅱ	Ⅲ	Ⅳ	ck
8	0.00	0.00	0.00	0.00	1.54
10	0.00	0.00	0.00	0.00	7.18
12	1.67	2.56	1.05	3.42	10.26
14	6.67	5.13	6.32	9.40	15.90
16	10.00	21.79	13.68	18.80	23.08
18	16.67	12.82	12.63	26.50	10.26
20	25.00	23.08	14.74	16.24	10.77
22	20.00	19.23	16.84	9.40	7.69
24	5.00	11.54	13.68	6.84	5.64
26	6.67	1.28	10.53	5.13	3.59
28	5.00	2.56	6.32	2.56	2.05
30	3.33	0.00	3.16	0.85	2.05
32	0.00	0.00	1.05	0.85	0.00

3.2 树高生长动态

从整体上看，各处理间树高生长较为平稳，而间伐后第3~5年是树高的快速增长期(图2A和2B)。间伐后第1~9年间，处理Ⅰ和处理Ⅲ和对照的树高显著高于处理Ⅱ和处理Ⅳ($P<0.05$)。间伐后第1年，树高以处理Ⅲ为最大，处理Ⅱ为最小，前者较后者高10%；第9年则以处理Ⅰ为最大，较处理Ⅱ高14%。间伐后第1~3年，处理Ⅰ的树高年均增量显著高于处理Ⅱ，处理Ⅲ，处理Ⅳ和对照($P<0.05$)；第3~9年，5个处理间的树高年均增量无明显差异。

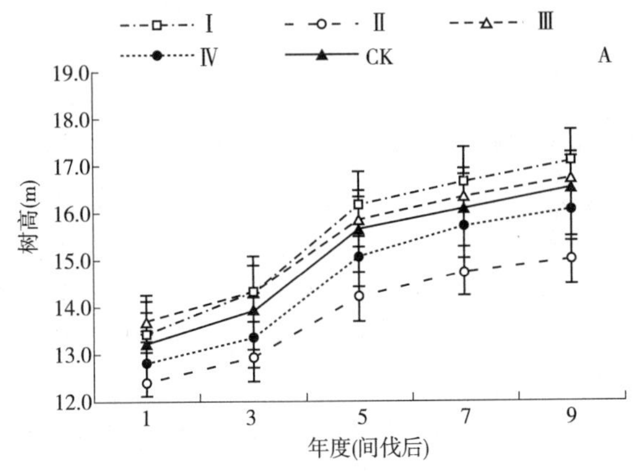

图2 不同强度间伐后杉木人工林树高生长动态

3.3 枝下高动态变化

各间伐处理的枝下高变化亦较为平稳，均显著低于对照(图3A)。处理Ⅰ、处理Ⅲ和处理Ⅳ的枝下高在间伐后各时间段内均非常接近，而处理Ⅱ略微低于此3个处理。对照的枝下高在间伐后第1年高出4个间伐处理0.3~1.0m；随着时间的推移，差距在不断地拉大，到间伐后第9年，对照的枝下高比处理Ⅱ高1.5m。

由图3B可知，各间伐处理和对照的枝下高年均增量随着时间的推移先下降而后趋于平稳。对照的枝下高年均增量在间伐后各时间段均显著高于4个间伐处理($P<0.01$)。间伐后第1~3年，各处理的枝下高增量出现峰值，以处理Ⅲ为最小，仅为对照的1/3。

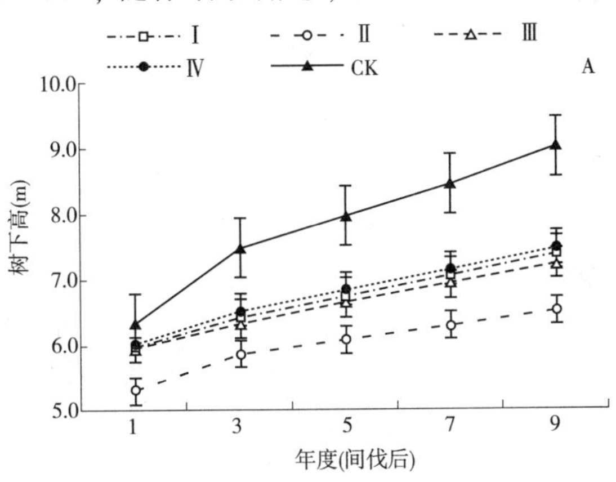

图3 不同强度间伐后杉木人工林枝下高动态变化

3.4 冠幅生长动态

间伐后第1~9年，随着杉木林分保留密度增大，冠幅生长减缓(图4A)。间伐后第1年、第3年、第5年、第7年，处理Ⅰ、处理Ⅱ和处理Ⅲ的冠幅显著大于处理Ⅳ，而4个间伐处理均显著大于对照($P<0.05$)。而间伐后第9年，4个间伐处理的冠幅差异不显著，但仍显著大于对照($P<0.05$)，以处理Ⅱ为最大，是对照的1.4倍。

从图4B中可以看出，处理Ⅰ在间伐后第1~3年出现冠幅生长高峰，随后其增量逐年减小。处理Ⅱ、处理Ⅲ、处理Ⅳ和对照的冠幅年均增量呈平稳或下降趋势，无显著差异。除了处理Ⅰ的冠幅年均增量在间伐后第1~3年显著高于其他各处理外($P<0.01$)，其余各时间段内，各间伐处理的冠幅年均增量与对照差异不显著。在间伐后第1~3年，处理Ⅰ的冠幅年均增量为对照2.4倍。

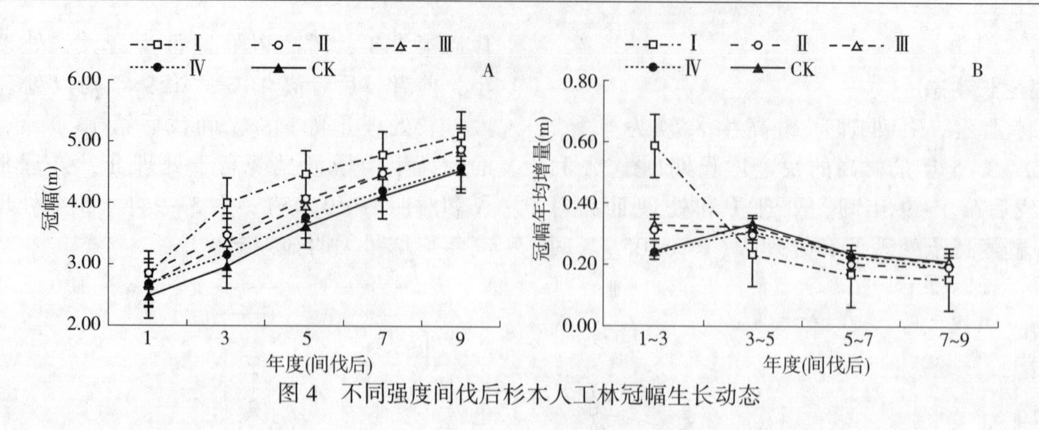

图 4 不同强度间伐后杉木人工林冠幅生长动态

3.5 单株材积年均增长量

结果显示(图 5A)：间伐处理Ⅰ和处理Ⅲ的单株材积在间伐后第 1~9 年均显著大于处理Ⅱ和处理Ⅳ和对照($P<0.05$)。间伐后，处理Ⅲ的单株材积为最高，对照处理的单株材积为最低，随着时间的推移，两者间的差距在逐渐加大。到间伐后第 9 年，处理Ⅲ的单株材积为对照的 1.5 倍。

各处理的单株材积年均增量随时间的推移呈现先增大而后减小或稳定的趋势(图 5B)。间伐后第 1-7 年，处理Ⅰ和处理Ⅲ的单株材积年均增量显著高于处理Ⅱ，处理Ⅳ以及对照($P<0.05$)；而在间伐后第 7-9 年，处理Ⅲ的单株材积年均增量显著高于其他 3 个间伐处理($P<0.01$)。

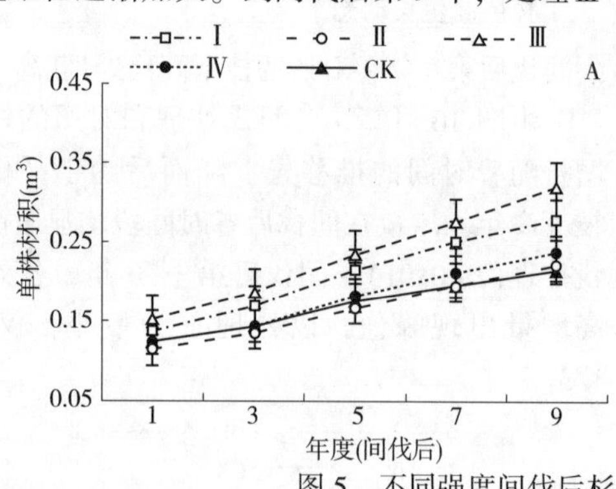

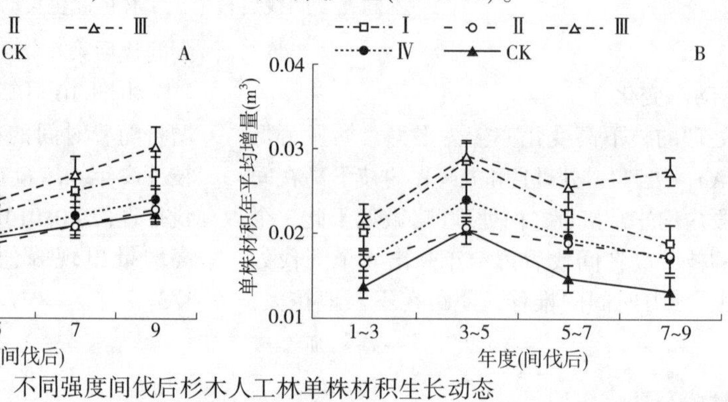

图 5 不同强度间伐后杉木人工林单株材积生长动态

3.6 林分蓄积增长动态

由图 6A 可以看出，处理Ⅲ和处理Ⅳ，处理Ⅰ和处理Ⅱ在间伐后各时间段的蓄积量均几乎相等；4 个间伐处理的蓄积显著低于对照($P<0.01$)，处理Ⅲ和处理Ⅳ的蓄积显著高于处理Ⅰ和处理Ⅱ($P<0.05$)。在间伐后第 5 年、第 7 年、第 9 年，对照处理的蓄积量与处理Ⅲ和处理Ⅳ的差距在逐渐缩小，间伐后第 1 年对照的蓄积是处理Ⅲ的 1.7 倍，第 9 年为处理Ⅲ的 1.5 倍。

各间伐处理的林分蓄积年均增量变化平稳，间伐后第 3~5 年略微增大；而对照的林分蓄积年均增量在间伐后第 1~3 年缓慢增长，第 3~5 年陡然增大，在间伐后第 5~7 年，第 7~9 年缓慢降低(图 6B)。间伐后第 1~9 年，处理Ⅰ和处理Ⅱ的林分蓄积年均增量显著小于对照($P<0.01$)，而处理Ⅲ和处理Ⅳ与对照无显著差异。间伐后第 7~9 年，蓄积年均增量最大的间伐处理Ⅲ为对照的 83.1%。

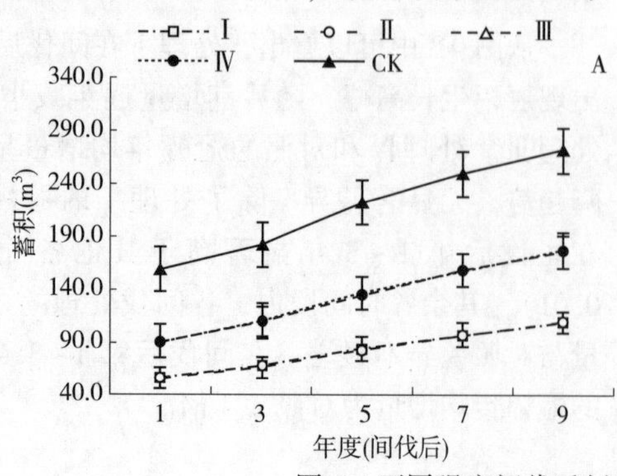

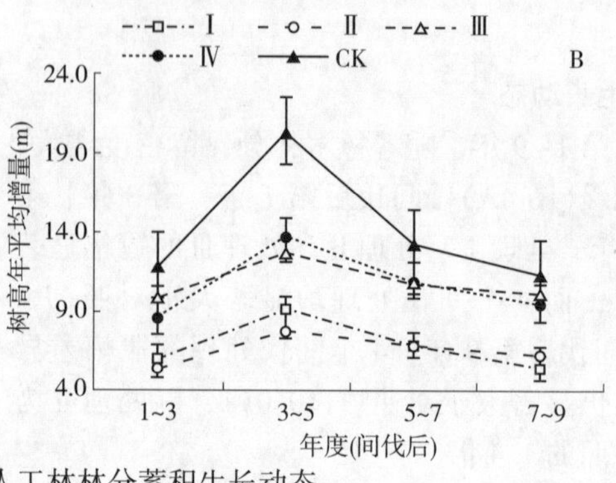

图 6 不同强度间伐后杉木人工林林分蓄积生长动态

4 讨论

选择最佳间伐强度调整林分结构，对于林木生长极为重要[12,13]。本研究中14年生杉木的胸径和冠幅在间伐后呈现快速生长，不论是从林分的胸径均值还是年均增长量来看，4个间伐处理均显著高于对照，且间伐强度越大，其胸径、冠幅生长越快。间伐对杉木生长的促进作用无明显滞后现象，间伐后第1~3年即是其胸径和冠幅的快速增长期，随后6年(至最后1次观测)其胸径增加量仍显著高于对照。CRECENTE-CAMPO等[14]研究得出：强度间伐(32%~46%)和超强度间伐(51%~57%)能使50年生欧洲赤松(*Pinus sylvestris* 的胸径、冠幅增量在短期(4年)内显著增加。MISSON等[15]对挪威云杉 *Piceaabies* 间伐(强度为40%~80%)后25年连续观测研究发现：强度间伐对林木径向生长的促进作用能够持续4~10年。而17年生湿地松(*Pinus elliottii*)间伐后对胸径和冠幅生长的促进作用存在2~4年的滞后现象[16]。由此可见，间伐对林木生长促进作用的起始时间、持续时间因树种、林龄而异；杉木生长对间伐响应快速，间伐对其保留木生长的促进作用持续时间长，其可能的原因在于：其一，杉木为中性偏阴树种，对间伐后光环境变化的适应能力强，因而间伐对其生长促进作用无滞后现象；其二，杉木属窄冠型树种，其冠幅小、冠长大，而且其冠幅生长相对较慢，因而间伐对杉木生长的促进作用持续时间长。

本研究中杉木各间伐处理与对照间的树高年均增量相差不大。此与张水松等[7]对10年生杉木人工林强、中、弱度间伐后20年的生长观测结果一致。然而，周成军等[17]对33年生杉木人工林进行了强度为12.9%~61.5%的择伐试验，发现中度(23.5%)择伐更有利于树高生长。这些研究得出不一致的结果，可能与其杉木年龄、立地条件、生长势等差异有关。不过，大多数学者认为间伐对树高生长影响较小，如，MEDHURST等[18]对不同年龄亮果桉(*Eucalyptus nitens*) 人工林进行强度为14%~72%的间伐研究，亦得出间伐不影响树高生长的结论。4个间伐处理的杉木枝下高年均增量显著低于对照，而4个间伐处理间差异不显著。究其原因，可从自然整枝与林分竞争方面得以解释：枝下高的变化直接取决于自然整枝强度，而自然整枝强度与林分竞争有关；对照林分密度大，生长空间竞争激烈，会加剧其自然整枝[19,20]，而4个处理的间伐强度为40%~69%，其空间竞争差异可能尚不明显。

间伐是增加林木收获量和大径材比例最常见的干预措施[21]。林分单株材积及其年均增长量方面，处理Ⅲ均显著高于对照林分。从林分蓄积来看，处理Ⅰ和处理Ⅱ的蓄积及其年均增量均显著低于对照，而处理Ⅲ和处理Ⅳ的蓄积年均增量与对照无显著差异。随着时间的推移，4个间伐处理的蓄积年均增量与对照的差距在缩小。由此可见，间伐后林木单株材积的增量可弥补因保留木数量减少而造成的林分蓄积增量的降低。国内一些学者研究亦表明，间伐有利于杉木大径级林木的生长，虽然间伐使林木株数减少，但到主伐年龄时其蓄积增量与不间伐相差不大[22,23]。从间伐后第9年林木径级分布还可以看出，间伐处理的中、大径级林木的比例远高于对照，尤其以处理Ⅲ的大径级林木比例最高，有利于大径材生产，其经济效益亦会得到显著提升。

杉木林分的胸径、冠幅和单株材积的增量随间伐强度的增大而增大，而枝下高和林分蓄积则呈相反的变化趋势。杉木人工林近自然化改造不同于传统的人工林经营，其间伐强度高，旨在快速培育大、中径材，且在其间伐后套种珍贵乡土阔叶树种，营造针阔异龄混交林。根据不同间伐处理下杉木单株和林分生长表现，建议选择50%的间伐强度为宜。本研究是在未考虑套种阔叶树影响的情况下进行的，在今后的长期试验中，还应考虑到林下阔叶树逐步进入到主林层后林分的空间结构与环境变化、林木枯损、种间关系等因素对杉木生长的影响，此亦为今后研究的重点。

参考文献

[1]盛炜彤，范少辉．杉木人工林长期生产力保持机制研究[M]．北京：科学出版社，2005.

[2]国家林业局森林资源管理司．第7次全国森林资源清查及森林资源状况[J]．林业资源管理，2010(1)：1-8.

[3]石春娜，王立群．我国森林资源质量变化及现状分析[J]．林业科学，2009，45(11)：90-97.

[4]陆元昌．近自然森林经营理论与实践[M]．北京：科学出版社，2006.

[5]高云昌，张文辉，何景峰，等．黄龙山油松人工林间伐效果的综合评价[J]．应用生态学报，2013，24(5)：1313-1319.

[6]何齐发，杨馥宁，郑小贤．杉木大径材作业级水平经营模式研究[J]．林业资源管理，2008(6)：55-58.

[7]张水松，陈长发，吴克选，等．杉木林间伐强度试验20年生长效应的研究[J]．林业科学，2005，41(5)：56-65.

[8]孙洪刚，张建国，段爱国，等．杉木密度间伐试验林

林分断面积生长效应[J]. 林业科学研究, 2010, 23(1): 6-12.

[9]李婷婷, 陆元昌, 庞丽峰, 等. 杉木人工林近自然经营的初步效果[J]. 林业科学, 2014, 50(5): 90-100.

[10]陈代喜, 陈琴, 蒙跃环, 等. 杉木大径材高效培育技术探讨[J]. 南方农业学报, 2015, 46(2): 293-298.

[11]孟宪宇. 测树学[M]. 2版. 北京: 中国林业出版社, 1996.

[12]李春明, 杜纪山, 张会儒. 抚育间伐对森林生长的影响及其模型研究[J]. 林业科学研究, 2003, 16(5): 636-641.

[13]WARD J S. Intensity of precommercial crop tree release increases diameter growth and survival of upland oaks. [J]. Can J For Res, 2009, 39 (1): 118-130.

[14] CRECENTE - CAMPO F, POMMERENING A, RODRÍGUEZ - SOALLEIRO R. Impacts of thinning on structure, growth and risk of crown fire in a *Pinus sylvestris*, L. plantation in northern Spain[J]. For Ecol Manage, 2009, 257(9): 1945-1954.

[15] MISSON L, VINCKE C, DEVILLEZ F. Frequency responses of radial growth series after different thinning intensities in Norway spruce [*Picea abies* (L.) Karst.] stands [J]. Forest Ecol Manage, 2003, 177 (1/3): 51-63.

[16]SCHEXNAYDER B, CAMILLE J. Growth of a Slash Pine Spacing Study Five Years after Thinning [D]. Baton Rouge: Louisiana State University, 2005.

[17]周成军, 巫志龙, 周新年, 等. 山地杉木人工林不同强度择伐后生长动态仿真[J]. 山地学报, 2012, 30(6): 669-674.

[18]MEDHURST J L, BEADLE C L, NEILSEN W A. Early-age and later-age thinning affects growth, dominance, and intras[J]. Can J For Res, 2001, 31(2): 187-197.

[19]张宁, 张怀清, 林辉, 等. 基于竞争指数的杉木林分生长可视化模拟研究[J]. 林业科学研究, 2013, 26(6): 692-697.

[20]FU Liyong, SUN H, SHARMA R P, et al. *Nonlinear mixed-effects crown width models for individual trees of Chinese fir* (Cunninghamia lanceolata) *in south-central China*[*J*]. *For Ecol Manage*, 2013, 302(6): 210-220.

[21]*HUONG V D, MENDHAM D S, CLOSE D C. Growth and physiological responses to intensity and timing of thinning in short rotation tropical Acacia hybrid plantations in South Vietnam*[*J*]. *For Ecol Manage*, 2016, 380: 232-241.

[22]吴建强, 王懿祥, 杨一, 等. 干扰树间伐对杉木人工林林分生长和林分结构的影响[*J*]. 应用生态学报, 2015, 26(2): 340-348.

[23]叶功富, 涂育合, 林瑞荣, 等. 杉木人工林不同密度管理定向培育大径材[*J*]. 北华大学学报(自然科学版), 2005, 6(6): 544-549.

[原载: 浙江农林大学学报, 2017, 34(05)]

桂西南马尾松人工林生长对不同强度采伐的动态响应

曾　冀[1,2]　雷渊才[2]　贾宏炎[1]　蔡道雄[1]　唐继新[1]
([1]中国林业科学研究院热带林业实验中心，广西凭祥　532600；
[2]中国林业科学研究院资源信息研究所，北京　100091)

摘　要　研究不同强度采伐下马尾松的生长动态，筛选适宜的采伐强度，为马尾松人工林近自然经营提供技术支撑。2007 年 10 月在 14 年生马尾松人工林(保存密度 1100 株/hm^2)内进行采伐试验，设置 4 个采伐强度，即保留密度分别为 225、300、375、450 株/hm^2，以不采伐为对照；其后，自 2008 年开始连续 8 年，每 2 年测定 1 次马尾松的胸径、树高、枝下高和冠面积等生长指标，并计算单株材积和林分蓄积量，应用方差分析和 Duncan 多重比较分析生长指标对不同采伐强度的动态响应。结果表明：采伐强度显著影响林分生长，其中，林分平均胸径、单株材积、冠面积的年均增长量随保留密度增大而减小，但均显著高于对照($P<0.05$)。采伐后第 1~3 年，马尾松冠面积增长量显著高于采伐后期，胸径则在采伐后第 3~5 年最高，而采伐强度对林分树高生长影响不明显。保留密度显著影响林分枝下高和蓄积量的动态变化，其年均增长量随密度增大而递增。5 个处理间林分蓄积年均增长量的差异随林龄的增大而逐渐缩小。马尾松人工林生长对强度采伐的动态响应以树冠最敏感，冠面积首先陡然增大，进而引起胸径的快速生长。树高和枝下高在采伐后年均增量变化相对平稳。4 个采伐强度均显著促进单株材积生长，而仅保留密度为 225 株/hm^{2}的采伐对林分蓄积增长量影响显著。综合比较林分的单株材积和林分蓄积连年增长量，建议在桂西南 15 年生的马尾松人工林近自然经营中宜选择 300 株/hm^{2}的保留密度进行采伐。

关键词　马尾松；强度采伐；密度效应；生长动态

Dynamic Growth Response of *Pinus massoniana* Plantation on Intensive Thinning in Southwestern Guangxi, China

ZENG Ji[1,2], LEI Yuancai[2], JIA Hongyan[1], CAI Daoxiong[1], TANG Jixin[1]
([1]*Tropical Forestry Research Center* , *Chinese Academy of Forestry*, *Pingxiang* 532600, *Guangxi*, *China*;
[2]*Research Institute of Forest Resources Information Techniques*, *Chinese Academy of Forestry*, *Beijing* 100091, *China*)

Abstract: Growth dynamics of *Pinus massoniana* were studied under intensive thinning so as to screen out suitable thinning intensity and provide technical support for close-to-nature management of Masson pine plantations. A thinning trial was conducted in a 14-year-old Masson pine plantation with stand density of 1100 trees per hectare in October, 2007. Four thinning treatments including 225, 300, 375 and 450 trees left per hectare and the control without thinning were arranged. Since 2008, growth performance such as diameter at breast height (DBH), height, height of crown base and crown area were measured every two years, and wood volumes were calculated at the single tree and stand level. These data were analyzed through one-way variance analysis and Duncan multiple range tests to reveal dynamical responses of these index to all sorts of thinning intensities. Stand growth performance were significantly influenced by thinning intensity, mean annual increments of DBH, single tree volume, crown area decreases with increase of stand density of thinning treatments, but they were all significantly higher than those of the control ($P<0.05$). The crown area increments of Masson pine were the highest during year 1 to 3 after thinning, while increments of DBH were the highest during year 3 to 5. No obvious relationship was observed between tree height growth rate and thinning density. Increases of crown base height and stand volume were significantly affected by stand density, and they increased with increase of stand density, while the differences of stand volume decreased with increase of forest age. In the present study, Masson pine's response of crown growth to intensive thinning was the most sensitive and rapid. The crown area increased sharply at first, then resulted in rapid growth of DBH, while tree height and crown base height increased a bit stably. Thinning with four intensities all remarkably influenced the growth of single tree volume, but only thinning with 225 trees left per hectare had significant effect on increment of stand volume. Taking mean annual increments of the single tree volume and stand volume, thinning intensity with 300 trees left per hectare was recommended for close-to-nature management of 15-year-old Masson pine plantations in southwestern Guangxi.

Key words: *Pinus massoniana*; intensive thinning; density effect; growth dynamics

马尾松(*Pinus massoniana* Lamb.)是我国南方亚热带地区分布最广、资源最多的松类树种，不仅具有很高的经济价值，而且在森林生态系统恢复和重建中发挥着重要作用[1]。桂西南地区热量丰富、雨量充沛且雨热同季，是马尾松高产区之一。然而，由于大面积成片马尾松纯林经营，出现虫害频发、生态功能下降等问题[2]。为解决马尾松虫害问题，中国林业科学研究院热带林业实验中心（以下简称热林中心）自2005年以来，陆续开展了马尾松人工林近自然化改造的探索，方法是选择适宜的采伐强度，在林下套种乡土阔叶树进行马尾松人工纯林近自然化改造，促进其向马尾松—阔叶异龄林的转变，并加快马尾松大径材培育速度。本研究以热林中心14年生马尾松人工林为对象，于2007年开始进行近自然化改造的不同强度采伐试验，定期开展生长观测，以期揭示不同强度采伐下马尾松林分的生长动态，为完善马尾松人工林近自然经营提供参考。

早在19世纪，瑞士和斯洛文尼亚即开展近自然森林经营研究，SchÜtz[3]研究了近自然化改造100年后的欧洲橡树林，发现近自然森林经营能很好地解决采伐和天然更新等经营问题。Ward[4]对北美红栎(*Quercus rubra* L.)及Miller[5]对北美红栎、山栎(*Quercus prinus* L.)、美国黑樱桃(*Prunus serotina* Ehrh.)和黄杨木(*Liriodendron tulipifera* L.)的研究均表明，采伐能促进幼龄林的胸径生长。Smith等[6]对北美红栎及Stringer等[7]对白栎(*Quercus fabri* Hance)的研究发现，强度采伐亦能促进成熟林的胸径生长。Trimble[8]对黄杨木和美国黑樱桃及Lamson等[9]对加拿大糖槭(*Acer negundo* Linn.)的一些研究认为，采伐会减缓树高生长；Smith等[10]的研究则认为，采伐对美国黑樱桃、加拿大糖槭的树高生长基本无影响，但对黄杨木的树高生长具促进作用。相对于常规间伐，近自然经营中的采伐强度更大，主要针对林分中的目标树进行单株管理，使目标树在较短的时间内达到目标胸径[11]。从国内有关近自然经营采伐的研究报道可看出，大部分是关于采伐后的林下物种多样性[12]、森林更新[13]和目标树生长[14]等研究，而关于林木生长对采伐的动态响应研究较少，仅李婷婷等[15]研究了杉木人工林近自然化改造后的林分生长，比较了47%和61%两个强度采伐后4年内林分蓄积生长量和单木生长量等动态变化。尽管目前已有一些关于马尾松、杉木间伐方面的研究[16-18]，而对不同强度采伐后的马尾松生长动态鲜有报道。为此，本文利用马尾松近自然改造试验林，以系统研究马尾松生长指标与采伐强度间的相关性，为马尾松人工林近自然化改造工作提供依据和指导。

1 试验地概况

研究地位于广西凭祥市热林中心伏波实验场(22°03″N，106°51″E)，海拔400m，属南亚热带季风气候，干湿季明显，年均气温20.5~21.7℃，年均降水量1200~1500mm，年蒸发量1261~1388mm。土壤为花岗岩发育而成的砖红壤。

1993年，在热林中心伏波实验场用种源为宁明桐棉的1年生裸根苗营造马尾松人工纯林，株行距为2m×2.5m，初植密度为2000株/hm^2，于1999年和2003年分别进行了强度约20%的透光伐和强度约30%的抚育性间伐。

2 研究方法

2.1 试验设计

2007年10月，选择立地条件和生长情况基本一致的马尾松人工林地段，开展不同强度采伐试验，样地概况见表1。试验采用完全随机区组设计，设置80%(Ⅰ)、73%(Ⅱ)、66%(Ⅲ)和59%(Ⅳ)4个采伐强度以及1个对照(未采伐，CK)，4次重复，共20个小区，小区面积为1500m^2。采伐后在马尾松林下挖规格为30cm×30cm×40cm的穴，按照4m×5m株行距均匀套种1年生大叶栎(*Castanopsis fissa* Rehd. et Wils.)、红椎(*Castanopsis hystrix* Miq.)、格木(*Erythrophleum fordii* Oliv.)、灰木莲(*Manglietia glauca* Blume)、铁力木(*Mesua ferrea* L.)、枫香(*Liquidambar formosana* Hance)和香梓楠(*Michelia hedyosperma* Law)等阔叶树种1年生苗450~525株/hm^2，对照为不采伐、不套种。采伐前，根据单木生长竞争因子将林木分为目标树、特殊目标树、干扰树和一般木，分别进行标记、编号。选择目标树做永久性标记，按照试验设计要求，伐除全部干扰树及部分一般木[11]，此工作在2007年底之前全部完成。

表1　样地概况

样地	坡度(°)	海拔(m)	平均胸径(cm)	平均树高(m)	立地指数	郁闭度
Ⅰ$_1$、Ⅰ$_2$、Ⅰ$_3$、Ⅰ$_4$	20~40	350~430	17.73	12.78	22	0.8
Ⅱ$_1$、Ⅱ$_2$、Ⅱ$_3$、Ⅱ$_4$	20~40	350~430	17.65	12.04	20	0.85
Ⅲ$_1$、Ⅲ$_2$、Ⅲ$_3$、Ⅲ$_4$	20~40	350~430	17.45	12.89	22	0.8
Ⅳ$_1$、Ⅳ$_2$、Ⅳ$_3$、Ⅳ$_4$	20~40	350~430	17.33	11.93	20	0.9
CK$_1$、CK$_2$、CK$_3$、CK$_4$	20~40	350~430	17.96	12.64	22	0.9

注：Ⅰ、Ⅱ、Ⅲ和Ⅳ表示处理，其采伐强度分别为80%、73%、66%和59%；下标为区组编号。

2.2　生长观测

采伐作业前，在马尾松纯林内设置3块400m^{2}的圆形样地进行本底调查，调查样地内所有马尾松植株的胸径、树高、枝下高和冠幅。应用围尺进行每木检尺，采用皮尺分东、西、南、北4个方向测定冠幅，VERTEX超声波测高器测量树高和枝下高。林分平均保留密度为1100株/hm^2，平均胸径18.26cm，平均树高11.64m，平均蓄积量150.25m^3/hm^2。采伐后，在每个小区中心位置附近选取一个圆心点布设半径为11.3m、面积400m^{2}的圆形样地，分别于2008年、2010年、2012年、2014年底和2016年8月对所有保留木进行生长观测。

2.3　数据处理

冠面积为东西、南北两个方向冠幅的乘积。

单株材积(V)：　$V=g\times f\times h$

式中：h为全树高(m)；g为胸高断面积(m^2)；f为平均实验形数，马尾松取0.39[19]。

采用单因素方差分析和Duncan多重比较检验不同采伐强度下马尾松林分的胸径、树高、单株材积和蓄积年均增长量及枝下高、冠面积年均变化量，应用SPSS 16.0软件进行数据统计分析。由于套种的植株尚未进入主林层，对马尾松生长的影响较小，故本研究在分析生长动态时未予考虑。

3　结果与分析

3.1　胸径生长动态

采伐试验第9年与第1年比较，各处理的胸径增幅为22.6%(CK)~53.1%(Ⅱ)，胸径总生长增量为4.19(CK)~11.22cm(Ⅱ)，均随着采伐强度的增大呈先递增后逐渐稳定的趋势(表2)。由图1可看出：各采伐处理的胸径生长随着时间的推移均呈现先加快后逐渐减缓的趋势。胸径生长高峰出现在采伐后第3~5年，其胸径年均增长量显著高于采伐后第1~3年、第5~7年和第7~9年，而这3个时段间差异大多不显著。对照的胸径年均增长量随着时间推移变化较小。

表2　采伐后第1年和第9年马尾松林分的生长表现

处理	保留株数	胸径(cm)		树高(m)		枝下高(m)	
	(trees/hm^2)	第1年	第9年	第1年	第9年	第1年	第9年
Ⅰ	225	22.10±0.42a	33.14±0.64a	13.56±0.24a	17.65±0.22a	6.96±0.18a	9.35±0.21c
Ⅱ	300	21.13±0.61a	32.35±0.81a	13.44±0.23a	17.63±0.21a	6.65±0.12a	9.13±0.15c
Ⅲ	375	20.71±0.44a	30.18±0.74b	14.17±0.23a	18.30±0.18a	7.04±0.13a	10.37±0.24b
Ⅳ	450	20.96±0.56a	29.67±0.71b	13.11±0.17a	17.64±0.24a	7.00±0.15a	10.74±0.26ab
CK	1100	18.55±0.25b	22.74±0.37c	13.27±0.08a	17.03±0.09a	6.94±0.08a	11.10±0.15a

处理	保留株数	冠面积(m^2)		单株材积(m^3)		蓄积量(m^3/hm^2)	
	(trees/hm^2)	第1年	第9年	第1年	第9年	第1年	第9年
Ⅰ	225	10.67±0.61b	26.75±2.17a	0.25±0.01a	0.73±0.03a	55.86±5.07c	163.82±10.83c
Ⅱ	300	12.38±0.69a	26.84±1.66a	0.23±0.01a	0.71±0.04ab	69.79±2.88c	213.57±8.59bc

（续）

处理	保留株数 (trees/hm²)	胸径(cm)		树高(m)		枝下高(m)	
		第1年	第9年	第1年	第9年	第1年	第9年
Ⅲ	375	9.14±0.49c	19.00±1.19b	0.23±0.01a	0.64±0.03bc	83.46±7.42bc	235.12±16.38bc
Ⅳ	450	7.74±0.42d	16.35±0.86b	0.22±0.01a	0.60±0.03c	105.24±17.03b	283.13±36.73b
CK	1100	6.19±0.25e	9.66±0.37c	0.17±0.01b	0.34±0.01d	195.38±12.47a	388.21±35.08a

注：表中字母为 Duncan 多重比较结果，处理间具相同字母表示差异不显著（$P \geq 0.05$），不同字母表示差异显著（$P<0.05$）。

同一时间段不同处理间比较，采伐后第3~5年和第7~9年，处理Ⅰ和Ⅱ间的胸径年均增长量差异不显著，但二者显著高于处理Ⅲ和Ⅳ，且采伐后第3~5年处理Ⅲ显著高于处理Ⅳ；采伐后第1~3年和第5~7年，处理Ⅰ和Ⅱ均显著高于处理Ⅳ。4个采伐处理的胸径年均增长量在各时间段均显著高于对照，在胸径生长高峰期（第305年），处理Ⅰ和Ⅱ比对照增加了2.2倍。

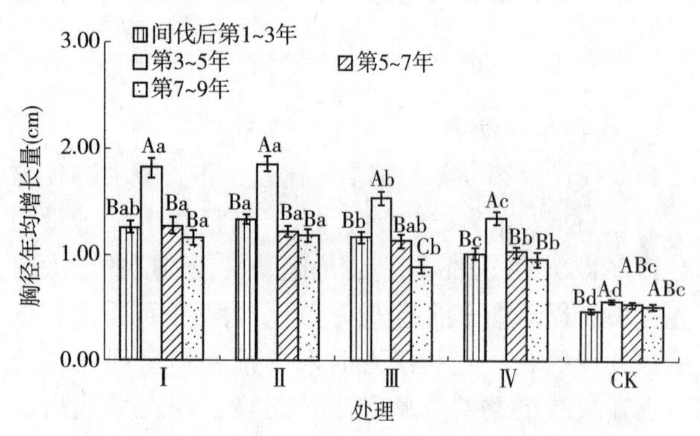

图1 不同采伐强度马尾松人工林胸径年均增长量
大写字母表示同一采伐强度处理年际间差异；
小写字母表示同一年份采伐强度处理间的差异。
具相同字母表示差异不显著（$P \geq 0.05$）；
不同字母表示差异显著（$P<0.05$），下同。

3.2 树高生长动态

各采伐处理树高第9年较第1年增幅为28.3%(CK)~34.6%(Ⅳ)，总生长增量为3.76(CK)~4.53m(Ⅳ)，4个采伐处理的树高增量与CK差异不显著(表2)。由图2可知：各处理的树高年均增长量随时间的推移呈先增大后逐渐稳定或略微减小的趋势，其中，处理Ⅰ、Ⅱ、Ⅲ和对照的生长高峰出现在采伐后第5~7年，而处理Ⅳ出现在采伐后第3~5年。不同处理间比较，在采伐后第1~3年和第7~9年4个采伐处理的树高年均增长量与对照均差异不显著；第3~5年处理Ⅰ、Ⅱ、Ⅲ与对照差异不显著，均显著低于处理Ⅳ；第5~7年处理Ⅲ显著高于对照，而其他采伐处理与对照差异不显著。

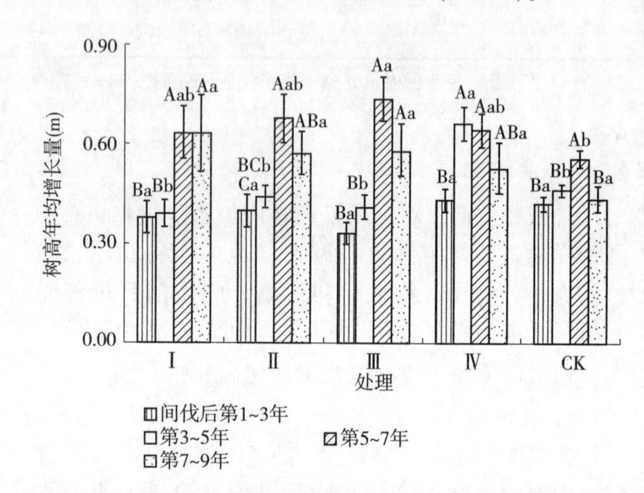

图2 不同采伐强度马尾松人工林树高年均增长量

3.3 枝下高动态变化

由表2可知：从采伐试验开始后的第1~9年枝下高的增幅为34.3%(Ⅰ)~59.9%(CK)，其总增量为2.39(Ⅰ)~4.16m(CK)，均随着采伐强度的增大而减小。各采伐处理的枝下高年均增量在采伐后第1~7年表现得比较平稳，而后显著升高；而对照在第1~5年变化不明显，第5~9年显著提高(图3)。处理间比较，在采伐后第7~9年，处理Ⅲ、Ⅳ的枝下高年均增量显著高于处理Ⅰ、Ⅱ，增量最大的处理Ⅳ比最小的处理Ⅱ高1.1倍；在试验的第5~7年，对照显著高于各采伐处理。

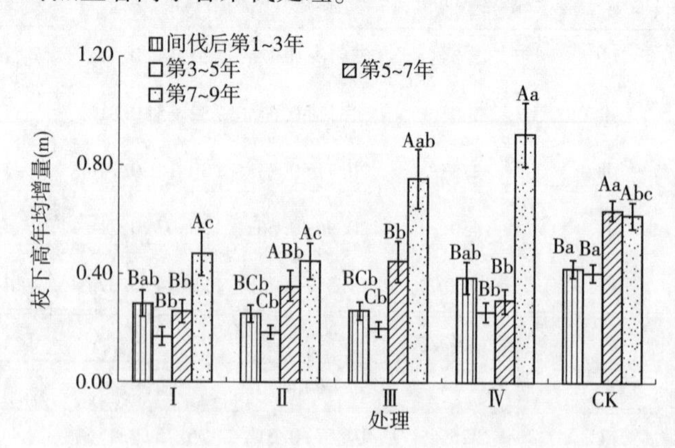

图3 不同采伐强度马尾松人工林枝下高年均增量

3.4 冠面积动态变化

采伐试验后第9年，各处理的冠面积增量为3.47(CK)~16.08m^2(Ⅰ)，第1~9年的增幅为56.1%(CK)~150.7%(Ⅰ)，均随采伐强度的增大而增大。处理Ⅰ的冠面积增量是对照的4.6倍，其采伐后第9年的增幅约是对照的2.7倍(表2)。

处理Ⅰ、Ⅱ的冠面积年均增长量随时间的推移呈先增大后减小的趋势，处理Ⅲ、Ⅳ在采伐后第1~3年冠面积急剧增大，此后增长速度逐渐减缓，对照则一直呈稳定状态(图4)。同一时间段内，冠面积年均增量随采伐强度的增大逐渐增大，均显著大于对照。在采伐后第1~7年，处理Ⅰ、Ⅱ的冠面积增量均高于处理Ⅲ、Ⅳ，而前、后二者间均差异不显著，处理Ⅰ、Ⅱ的冠面积增量在冠幅生长高峰期(第3~5年)是对照的6倍；在采伐后第7~9年，处理Ⅰ的冠面积增量显著高于其他3个采伐处理。

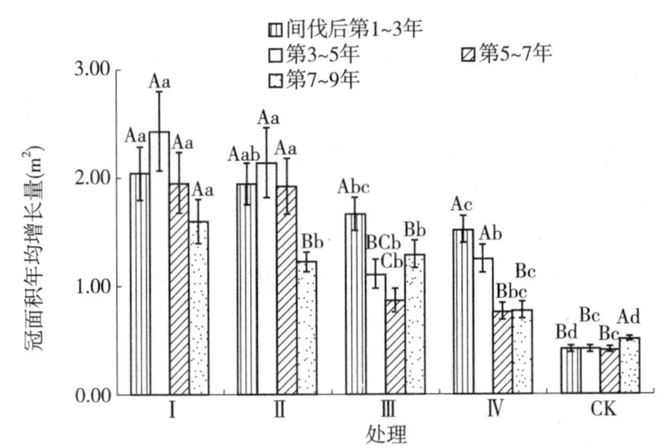

图4　不同采伐强度马尾松人工林冠面积年均增长量

3.5 单株材积生长动态

采伐试验后第9年较第1年的单株材积增幅为100%(CK)~209%(Ⅱ)，单株材积总生长增量为0.17(CK)~0.48m^3(Ⅰ、Ⅱ)，均随采伐强度的加大先增大后趋于稳定(表2)。

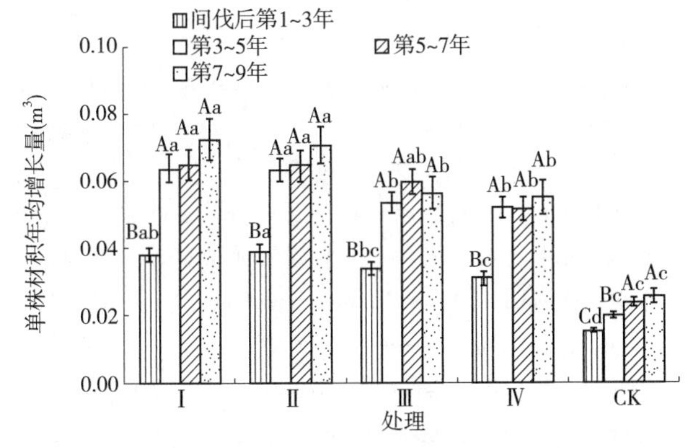

图5　不同采伐强度马尾松人工林单株材积年均增长量

4个采伐处理的单株材积年均增长量在采伐后第3~5年陡然升高，而后趋于稳定或略微增大。对照的单株材积年均增长量在各时间段内尽管差异显著，但其增长量明显低于各采伐处理(图5)。相同时间段内单株材积年均增长量比较，处理Ⅰ和Ⅱ大多显著大于处理Ⅲ和Ⅳ，且4个采伐处理均显著大于对照；各时段均是处理Ⅰ、Ⅱ最高，是对照的2~3倍。

3.6 马尾松林分蓄积动态变化

由表2可知：采伐试验后第9年较第1年的林分单位面积蓄积增幅为98.7%(CK)~206.0%(Ⅱ)，随采伐强度的增大呈先递增后稳定的趋势。第1~9年林分单位面积蓄积增长总量随采伐强度的增大呈递减趋势，对照的最大(192.82m^3/hm^2)，显著高于处理Ⅰ(107.96m^3/hm^2)，与其余采伐处理(143.78~177.90m^3/hm^2)差异不显著。

各采伐处理的林分蓄积年均增长量在采伐后第1~3年缓慢增长，第3~5年均急剧增大，在采伐后第5~9年趋于稳定或略微增大；而对照随时间的推移变化较平稳，年际间差异不显著(图6)。比较同一时段不同处理间林分蓄积年均增长量，采伐后第1~3年和第5~7年，对照显著高于处理Ⅰ、Ⅱ、Ⅲ，与处理Ⅳ差异不显著；4个采伐处理中，仅处理Ⅰ显著低于处理Ⅲ和Ⅳ。采伐后第3~5年和第7~9年，对照的蓄积年均增长量与处理Ⅱ、Ⅲ、Ⅳ差异不显著；在蓄积增长高峰期(采伐后第7~9年)，对照的蓄积年均增长量是处理Ⅰ的1.9倍。

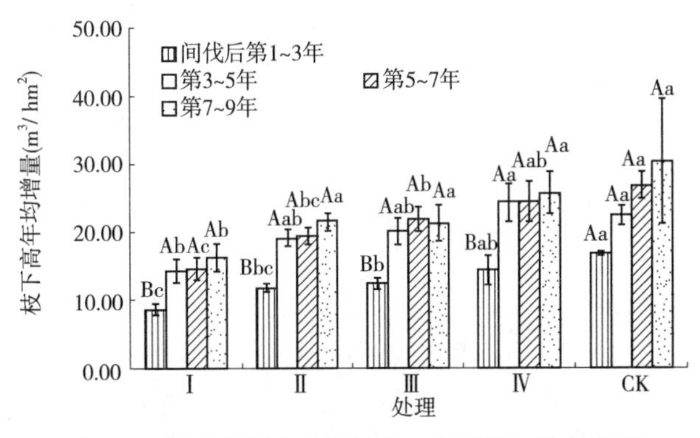

图6　不同采伐强度马尾松人工林蓄积年均增长量

4 讨论

本研究结果表明，14年生马尾松人工林的树冠在采伐后1~3年即呈现快速生长，其冠面积年均增量随时间的推移呈先增大后减缓的趋势。随着树冠的生长，马尾松单株光合作用面积逐渐增加，进而加速了胸径的生长[20,21]。胸径生长的高峰期在采伐后第3~5年，正好在冠面积年均增长量出现高峰期之后。Miller[5]对12~16年生的几种阔叶树选择的目标树进行采伐释放冠幅空间，对胸径和冠幅生长有促进作用，此亦在许多其他树种的研究中得到验证[22-25]。本研究中，各采伐处理对马尾松树高生长的影响不显著，与国外Medhurst等[26]对亮果桉(*Eucalyptus nitens* Maiden)以及国内张水松等[27]对杉木和谌红辉等[28]对马尾松人工林的间伐试验结果一致，这些研究均得出，间伐能显著促进胸径、单株材积、冠幅的生长，而对树高生长无显著影响。

本研究中，除处理CK(保留株数为1100株/hm^2)外，其余4个采伐处理的枝下高年均增量高峰均在采伐后第7~9年。枝下高的增量跟树冠结构关系紧密，树冠越密自然整枝越强，枝下高越大[29]。而处理Ⅳ采伐后，经过7年的时间，林分树冠又开始出现重叠，郁闭度升高，枝下高增量陡然变大，其余3个采伐处理可能树冠尚未相接，枝下高增量平稳。

本研究中，马尾松单株材积的年均增长量呈先逐步增大而后稳定和略微增大的趋势，在采伐后第3~9年，处理Ⅰ(保留株数225株/hm^2)和处理Ⅱ(保留株数300株/hm^2)的单株材积显著高于其他处理。林分的蓄积年均增长量受单株材积和保留株数的共同影响，在采伐后第7~9年，采伐处理Ⅱ(保留株数300株/hm^2)、Ⅲ(保留株数375株/hm^2)、Ⅳ的林分蓄积年均增长量与对照差异不显著，而处理Ⅰ则显著低于对照。大径材比例与林分平均单株材积密切相关，林分平均单株材积越大，说明大径材比例越高。由于大径材的价格是中小径材的数倍，大径材培育是我国未来林业发展的重点方向之一[30]。因此，在森林经营中，培育大径材的同时还要保证林分蓄积量的持续高速增长，在二者之间如何寻求平衡尚有待研究。

林下套种阔叶树对保留木生长存在一定的影响，但本研究未考虑其对上层林木的影响。然而，随着时间的推移可能逐渐出现套种林木与马尾松的水肥竞争。另一方面，阔叶树的凋落物分解快，可以加速林地的养分循环，从而对林下物种多样性、土壤等亦会产生积极影响。强度采伐后，林分的恢复是一个长期的过程，采伐强度越大，则影响的持续时间越长[31]。总而言之，本研究只是采伐后马尾松生长动态研究的阶段性总结，未来阔叶树逐渐进入主林层，马尾松的生长动态必然会受影响；林下植被的变化亦会影响地力，进而影响林分生产力[32-34]，有待长期观测研究。

5 结论

马尾松在强度采伐下，林分空间结构发生变化，树冠生长空间得到释放，冠幅加速生长，林分的胸径和单株材积在采伐后第3~5年陡然增大，年均增长量显著高于对照林分。采伐强度对马尾松树高生长影响不显著，而对枝下高影响显著，呈现出采伐强度越大，枝下高平均增量越小。单株材积增长量随采伐强度的增大显著提高，而林分蓄积量则呈递减趋势；然而，随着林龄的增大，采伐处理间林分蓄积年均增长量的差异逐渐缩小。4个采伐强度的马尾松生长响应大体相近，即采伐后冠幅首先加速生长，随后胸径亦加速生长，只是冠幅和胸径的增幅随着采伐强度的增大而增大，而树高和枝下高变化不大；对照林分的冠幅、胸径、树高生长相对平稳，枝下高增量明显变大。马尾松在强度采伐下，根据林分单株材积和林分蓄积年均增量的比较，得出采伐强度66%的林分蓄积量增长量优于其他处理；采伐保留密度300株/hm^2适于马尾松中龄林近自然化改造，既能快速提高林分蓄积量，又能保证单株材积的快速增长。

参考文献

[1]雷蕾，肖文发，曾立雄，等．不同营林措施对马尾松林土壤呼吸影响[J]．林业科学研究，2015，28(5)：713-719.

[2]马归燕．马尾松天然林病虫害调查[J]．中国森林病虫，2001，20(2)：40-41.

[3]SCHÜTZ J P. Close-to-nature silviculture：is this concept compatible with species diversity? [J]. Forestry，1999，72(4)：359-366.

[4]WARD J S. Crop-tree release increase growth of red oak saw timber：12 year results[C]. Proceedings 16th Central Hardwood Forest Conference，2008. 457-465.

[5]MILLER G W. Effect of crown growing space on the development of young hardwood crop trees[J]. Northern Journal of Applied Forestry，2000，17(1)：25-35.

[6]SMITH H C，Miller G W. Releasing 75-to 80-year-old Appalachian hardwood saw timber trees：5-year DBH response[C]. Proceedings of the 8th Central Hardwood Forest Conference，1991：403-413.

[7]STRINGER J W，Miller G W，Wittwer R F. Applying a crop-tree release in small-saw timber white oak stands [R]. Research Paper-Northeastern Forest Experiment Station，USDA Forest Service NE-620，1988.

[8]TRIMBLE G R. Response to crop tree release by 7-year-old stems of yellow-poplar and black cherry[R]. Research Paper - Northeastern Forest Experiment Station，USDA

Forest Service NE-303, 1973.

[9]LAMSON N I, SMITH H C. Response to crop-tree release: sugar maple, red oak, black cherry, and yellow-poplar saplings in a 9-year-old stand[R]. Research Paper-Northeastern Forest Experiment Station, USDA Forest Service NE-394, 1978.

[10]SMITH H C, LAMSON N I. Precommercial crop-tree release increases diameter growth of Appalachian hardwood sapling[R]. Research Paper - Northeastern Forest Experiment Station, USDA Forest Service NE-534, 1983.

[11]陆元昌. 近自然森林经营理论与实践[M]. 北京: 科学出版社. 2006.

[12]马履一, 李春义, 王希群, 等. 不同强度间伐对北京山区油松生长及其林下植物多样性的影响[J]. 林业科学, 2007, 43(5): 1-9.

[13]李荣, 何景峰, 张文辉, 等. 近自然经营间伐对辽东栎林植物组成及林木更新的影响[J]. 西北农林科技大学学报: 自然科学版, 2011, 39(7): 83-91.

[14]王懿祥, 张守攻, 陆元昌, 等. 干扰树间伐对马尾松人工林目标树生长的初期效应[J]. 林业科学, 2014, 50(10): 67-73.

[15]李婷婷, 陆元昌, 庞丽峰, 等. 杉木人工林近自然经营的初步效果[J]. 林业科学, 2014, 50(5): 90-100.

[16]盛炜彤. 不同密度杉木人工林林下植被发育与演替的定位研究[J]. 林业科学研究, 2001, 14(5): 463-471.

[17]MILLER G W, STRINGER J K. Effect of crown release on tree grade and DBH growth of white oak sawtimber in eastern Kentucky[C]. Proceedings of the 14th Central Hardwood Forest Conference, 2004: 27-44.

[18]徐金良, 毛玉明, 郑成忠, 等. 抚育间伐对杉木人工林生长及出材量的影响[J]. 林业科学研究, 2014, 27(1): 99-107.

[19]孟宪宇. 测树学[M]. 2版. 北京: 中国林业出版社. 1996.

[20]ZHANG J, OLIVER W W, POWERS R F. Long-term effects of thinning and fertilization on growth of red fir in northeastern California[J]. Canadian Journal of Forest Research, 2005, 35(6): 1285-1293.

[21]O'HARA K L, NESMITH J C B, LEONARD L, et al. Restoration of old forest features in coast redwood forests using early-stage variable-density thinning[J]. Restoration Ecology, 2010, 18(s1): 125-135.

[22]CONDÉS S, STERBA H. Derivation of compatible crown width equations for some important tree species of Spain[J]. Forest Ecology & Management, 2005, 217(2): 203-218.

[23]RUSSELL M B, WEISKITTEL A R. Maximum and largest crown width equations for 15 tree species in Maine[J]. Northern Journal of Applied Forestry, 2011, 28(2): 84-91.

[24]公宁宁, 马履一, 贾黎明, 等. 不同密度和立地条件对北京山区油松人工林树冠的影响[J]. 东北林业大学学报, 2010, 38(5): 9-12.

[25]符利勇, 孙华. 基于混合效应模型的杉木单木冠幅预测模型[J]. 林业科学, 2013, 49(8): 65-74.

[26]MEDHURST J L, BEADLE C L, NEILSEN W A. Early-age and later-age thinning affects growth, dominance, and intraspecific competition in *Eucalyptus nitens* plantations[J]. Canadian Journal of Forest Research, 2001, 31(2): 187-197.

[27]张水松, 陈长发, 吴克选, 等. 杉木林间伐强度试验20年生长效应的研究[J]. 林业科学, 2005, 41(5): 56-65.

[28]谌红辉, 方升佐, 丁贵杰, 等. 马尾松间伐的密度效应[J]. 林业科学, 2010, 46(5): 84-91.

[29]孟京辉, 陆元昌, 王懿祥, 等. 海南岛热带天然次生林生长动态研究[J]. 林业科学研究, 2010, 23(1): 77-82.

[30]蔡道雄, 贾宏炎, 卢立华, 等. 我国南亚热带珍优乡土阔叶树种大径材人工林的培育[J]. 林业科学研究, 2007, 20(2): 165-169.

[31]MISSON L, VINCKE C, DEVILLEZ F. Frequency responses of radial growth series after different thinning intensities in Norway spruce [*Picea abies* (L.) Karst.] stands[J]. Forest Ecology and Management, 2003, 177(1-3): 51-63.

[32]盛炜彤. 杉木林的密度管理与长期生产力研究[J]. 林业科学, 2001, 37(5): 2-9.

[33]雷相东, 陆元昌, 张会儒, 等. 抚育间伐对落叶松云冷杉混交林的影响[J]. 林业科学, 2005, 41(4): 78-85.

[34]熊有强, 盛炜彤. 不同间伐强度杉木林下植被发育及生物量研究[J]. 林业科学研究, 1995(4): 408-412.

[原载: 林业科学研究, 2017, 30(02)]

格木人工林节子的分布特征及预测模型

郝　建[1,2]　蒙明君[1]　黄德卫[1]　韦菊玲[1]　李忠国[1]　唐继新[1]　徐大平[2]

([1]中国林业科学研究院热带林业实验中心，广西凭祥　532600；
[2]中国林业科学研究院热带林业研究所，广东广州　510520)

摘　要　分析节子在格木生长过程中的发生、形成及分布特征，同时通过逐步回归分析，筛选出关键因子建立评判节子影响的多元回归模型。以30年生格木(*Erythrophleum fordii*)作为研究对象，利用树干解析方法对其节子的形成及分布特征进行研究。与地理方位相比，坡向是影响格木分枝分布的重要因素；2.0~6.0m的格木树干分布的节子最多，此段是木材利用率最高部分，严重影响格木的利用价值；分枝角度小于60°的分枝形成节子的直径均大于2.5cm，直径越大死节长度越大，节子在木质部的跨度越大；第1~15年是格木形成分枝的高峰期，分枝脱落及伤口愈合集中在第16~25年；在第11~20年间格木形成死节最多，是控制死节形成的关键时期。通过逐步回归分析，筛选出分枝直径(BD)、分枝角度(IA)和分枝年龄(YB)3个关键因子建立了多元回归模型：$y=1.6344x_1+0.06776x_2+0.1648x_3-1.61139$($F$值=106.8697，$P$值=0.0001)。可以利用该模型来预测分枝形成节子后对木材的影响状况。

关键词　格木；节子；分布特征；预测模型

Distribution Characteristics and Predicting Models of Knot in *Erythrophleum fordii* Plantation

HAO Jian[1,2], MENG Mingjun[1], HUANG Dewei[1], WEI Juling[1]
LI Zhongguo[1], TANG Jixin[1], XU Daping[2]

(*[1]Experimental Center of Tropical Forestry, Chinese Academy of Forestry, Pingxiang 532600, Guangxi, China;*
[2]Research Institute of Tropical Forestry, Chinese Academy of Forestry, Guangzhou 510520, Guangdong, China)

Abstract: In order to investigate the formation and distribution characteristics of knots during the process of *Erythrophleum fordii* growth, and screen out the key factors to establish a multiple regression model to predict the effects on branch wood caused by knot formation by step wise regression analysis. 30-year-old *Erythrophleum fordii* plantation was researched by using a stem analysis method. The results indicated that compared with geographical location, slope direction was the important factor influencing the distribution of *E. fordii* branches. Most knots were distributed on the highest utilization region of E. fordii trunks (2.0~6.0m), which seriously reduced the wood utilization value. When branch angle was less than 60°, the knot diameter was greater than 2.5cm. As the branch diameter increased, the length between dead knots and the span of knots in xylem was larger. Peak forming on branches occurred at 1~15 years. Branch wound healing was concentrated at 15~16 years, and most dead knots formed at 11~20 years, which was the critical period for the control of dead knot formation. Diameter of the branch (BD), insertion angle of the branch (IA), and year of birth of the branch (YB) were selected as key factors to establish a multiple regression model by step wise regression analysis, $y = 1.6344x_1 + 0.0677\,6x_2 + 0.164\,8x_3 - 1.61139$ (F = 106.869 7, P = 0.0001). The model was suitable to predict the effects on branch wood caused by knot formation.

Key words: *Erythrophleum fordii*; knot; distribution characteristics; predicting models

树枝是树木不可缺少的组成部分，而节子是树木分枝在木质部的最终存在形式。由于节子固有的特征，破坏了树木木材的连续性和有序性，使得节子成为影响木材加工利用的一种缺陷。木材内纤维的严重不连续性导致极大的应力集中，制约了木材的结构性用途[1]和板材加工性能[2]。传统上利用树干解析方法来研究节子属性[3]。Fujimori[4]对日本扁柏(*Chamaecyparis obtusa*)的分枝和节子发育进行了研究，Makinen[5,6]先后研究了欧洲赤松(*Pinus sylvestris* L.)和垂枝桦(*Betula pendula*)的分枝和节子生长动态并建立相应生长预测模型，Kershaw 等[7]对花旗松(*Pseudotsuga menziesii*)节子半径生长动态和枝条寿命进行研究。

格木(*Erythrophleum fordii*)为苏木科格木属常绿乔木，列入中国植物红皮书第一册的我国二级重点保护的植物之一[8]，天然分布于我国广西、广东、福建和台湾等地区[9,10]。其木材纹理直，结构细密坚实，坚硬，强度大，抗虫蛀，耐水耐腐，故经历千年不朽，气干密度 0.857g/cm³[11]，是珍贵优质硬材，有铁木之称。是家具、造船、车辆、桥梁建筑、机械工业、地板、雕刻的特好用材。格木作为我国南方一种高价值优良用材树种，大径材是其最终经营目标，这就要求对格木人工林的经营管理区别于一般用材树种，需要精细的经营管理措施。人工修枝作为大径材培育过程中一项不可或缺的技术措施，未见格木该方面的研究报道。因此，利用树干解析方法对 30 年生格木节子的分布特征进行研究，分析节子在格木生长过程中的发生、形成及分布特征，同时通过逐步回归分析，筛选出关键因子建立评判节子影响的多元回归模型，为开展格木人工林修枝措施提供依据，奠定格木人工修枝技术研究基础。

1 材料与方法

1.1 试验地概况

试验样地位于广西凭祥市中国林业科学研究院热带林业实验中心白云实验场(106°44′18″~106°50′38″E，22°5′40″~22°7′30″N)，海拔 250m，属低山丘陵，坡向为北。所选试验地为 1982 年春造的格木人工林，坡度 23°，林龄 30 年，造林密度为 2500 株/hm²，经 3 次间伐后，现有立木密度 1110 株/hm²，林分平均胸径(21.5±4.2)cm，林分平均树高(16.8±2.7)m。

1.2 样木选择及指标测定

在 30 年生的格木试验林(人工纯林)中，布设 4 个 20m×20m 临时标准地，进行每木检尺；由于格木为我国重点保护的珍稀濒危树种，现存资源极少，故每个样地选 2 棵平均木作为研究材料，共 8 株，生长数据见表 1。对所选中等木树冠活枝以下树干进行树干剖析研究。在采伐取样前，对每棵采伐木进行编号，并用手持罗盘仪标定正北向和上坡向，带回实验室备用。

表 1 样地平均木生长量概况

所选平均木生长量		
树号	胸径(cm)	树高(m)
1	23.6	17.3
2	21.1	16.9
3	20.4	17.0
4	19.8	16.5
5	22.3	17.6
6	18.6	16.3
7	19.3	16.0
8	20.5	18.2

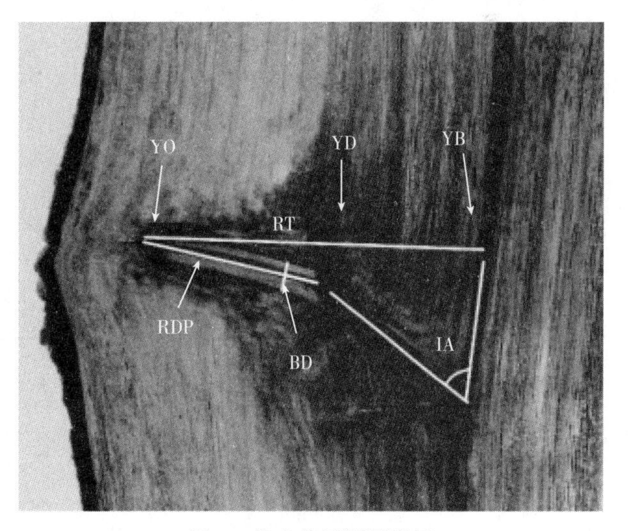

图 1 格木节子测量指标

参考贾炜玮[12]、陈东升[13]和 Heinrich[14]的方法对格木节子进行手工解剖。格木分枝脱落伤口愈合后，在树干表面会形成一个明显的突起，根据其分枝愈合痕的特点定位每个节子的位置，向下 30~40cm锯下含节子木段，在每个含节子木段基部标记北向和上坡向。根据节子位置用锯子过髓心剖开，得到节子的纵剖面，选较好的一面打磨光滑后进行标记测量，最后得到 312 个节子纵剖面材料。

表 2 各测量指标及其定义

指标	意义	单位
分枝高度(BH)	形成节子的分枝发生点到树干基部距离	m
分枝直径(BD)	该分枝形成的节子直径	cm
分枝角度(IA)	该分枝形成的节子与树干之间的夹角	°

（续）

指标	意义	单位
分枝坡向位置(SP)	该分枝形成的节子指向的坡向方位	—
分枝年龄(YB)	形成该节子的分枝发生的年龄	年
分枝死亡年龄(YD)	形成该节子的分枝枯死的年龄	年
伤口愈合年龄(YO)	形成该节子的分枝脱落后伤口愈合时的树龄	年
节子形成所需时间(YN)	形成该节子所需的年限	年
死节长度(RDP)	节子中松散部分的长度	cm
节子发生点到愈合点距离(RT)	形成该节子的分枝发生点到脱落伤口愈合点的水平距离	cm

其中分枝坡向位置采用数量化处理：上=1，右上=2，右=3，右下=4，下=5，左下=6，左=7，左上=8。采用 Excel 2013 对收集数据进行统计分析并作图，DPS 14.50 数据处理软件进行逐步回归和方差分析。

2　结果与分析

2.1　格木节子的空间分布特征

不同高度格木树干的节子分布趋势是随着高度增加而增加，树干高度达到 6m 后向上逐渐减少(图 2A)。节子的分布主要集中在树干的 2.0~8.0m 处，其中 4.1~6.0m 的节子数量最多(29.29%)，此段又是木材利用的核心部分。因此，节子严重降低了格木木材的质量和价值。

图 2B 表示的是不同地理方位格木节子的数量分布，北向分布的分枝最多，占总数的 24.24%，其次是东北与西北方向，分别为 18.18%、15.19%。相对于其他坡向，格木在下坡向分枝数量最多，所占比例为 25.25%，其次为右下的 20.20% 和左下的 19.19%(图 2C)。所选取样地的坡向为北向，即下坡向的地理朝向为北向，格木在该朝向的分枝较多，因为该朝向为格木分枝生长提供更多的空间。因此，立木生长空间分配是影响格木分枝分布的重要因素。

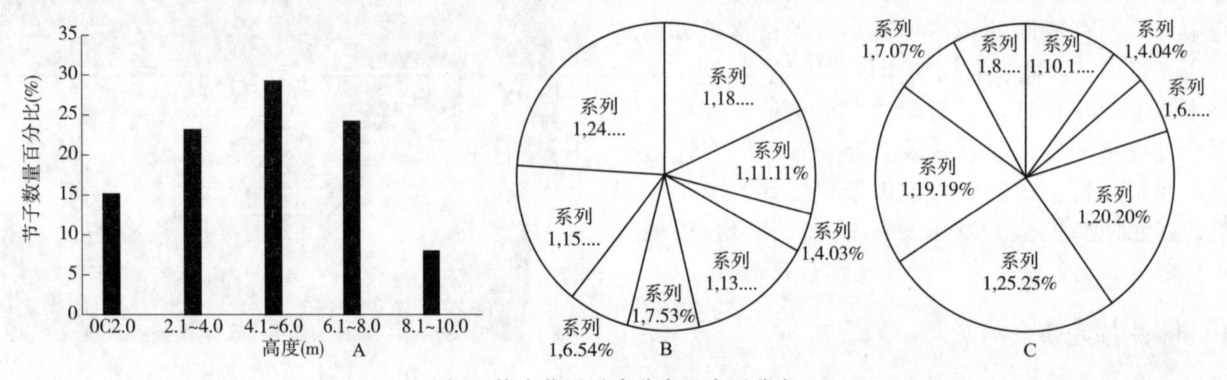

图 2　格木节子垂直分布及水平分布

2.2　格木节子的时间分布特征

由图 3 可知，格木在不同年龄段生成的分枝数量统计结果显示：前 15 年是格木形成分枝高峰时期，在此之后形成的分枝数量明显下降。在第 1~5 年形成分枝数量占总数的 25.24%，第 6~10 年为 29.29%，第 11~15 年为 35.35%，而在第 16~20 年则降低到 7.06%。第 1~15 年各年龄段分枝脱落后形成死节的数量呈先增后降低的趋势(图 3)。第 1~5 年是格木分枝所形成死节的数量占总数的 1.01%，第 6~10 年为 12.11%，第 1~15 年为 25.23%，第 16~20 年为 42.43%，是死节形成的高峰期，之后逐年降低。可以看出第 1~10 年形成死节较多，也是控制死节形成的关键时期。从图 3 可知格木分枝枯死脱落后伤口愈合发生在第 11 年之后，第 11~15 年为 7.07%，第 16~20 年为 21.22%，第 21~25 年是分枝脱落伤口愈合的高峰期，占总数的 51.52%；第 26~29 年降低到 20.20%。因此，格木人工林应该在造林后 10~15 年之前进行一次人工修枝。

从树木分枝的形成、枯死脱落到伤口的愈合，是节子形成的过程。图 3 表明，格木需要 11~15 年形成的节子占的比例最高(33.34%)，其他需要时间段所占比例依次为：1~5 年的为 12.12%，6~10 年的为 26.27%，16~20 年的为 17.17%，21~25 年的为 7.07%，26~29 年的为 4.04%，形成节子需要的时间越久，该节子对木材的影响就越大。

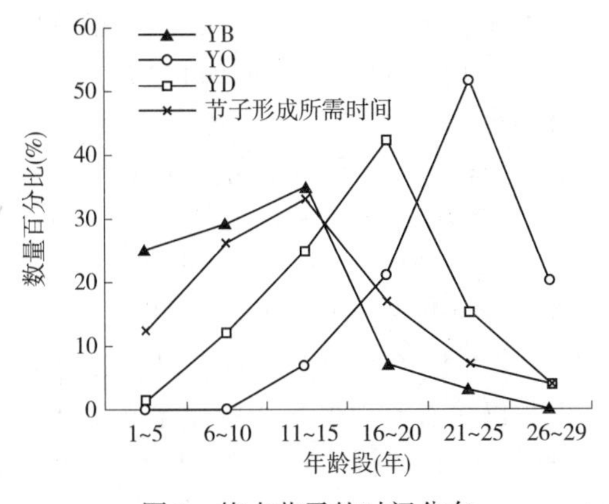

图 3　格木节子的时间分布

2.3　节子发生点到愈合点距离多元线性回归模型的构建

2.3.1　确定关键因子

死节长度(RDP)和节子发生点到愈合点距离(RT)是评判节子对木材影响度的直观指标，而 RDP 又是 RT 的组成部分，因此选择分枝高度(BH)、分枝直径(BD)、分枝角度(IA)、分枝年龄(YB)、愈合年龄(YO)、坡向(SP)作为变量与节子发生点到愈合点距离(RT)进行逐步回归分析。各变量的偏相关系数及检验结果如表 3 所示。

表 3　各自变量偏相关系数及 t 检验结果

项目	变量					
	分枝高度	分枝直径	分枝角度	分枝年龄	愈合年龄	坡向
偏相关系数	-0.0549	0.8244	0.2592	0.3067	-0.1333	0.1141
P	0.5990	0.0001	0.0117	0.0026	0.2003	0.2736

偏相关系数最大的变量是分枝直径(BD)，为 0.8244，其次是分枝年龄(YB)和分枝角度(IA)，分别为 0.3067 和 0.2592。3 个变量与节子发生点到愈合点距离(RT)均呈正相关关系。t 检验结果显示分枝直径和分枝年龄与节子发生点到愈合点距离的关系达到 0.01 的显著水平，分枝角度达到 0.05 的显著水平。因此选择这 3 个因子作为预测节子影响的关键因子。

2.3.2　节子发生点到愈合点距离与关键因子的分析

节子发生点到愈合点距离与关键因子的关系见图 4。计算得出节子发生点到愈合点距离(RT)与分枝直径(BD)的回归方程为：$y=1.6608x+5.3769$。F 检验的结果显示相关性达到了极显著水平($F=270.1650$，$P=0.0001$)。分枝直径越大，节子发生点到愈合点距离越长，节子在木质部的跨度越大，木材质量越低。

分枝角度(IA)与节子发生点到愈合点距离(RT)的回归方程为：$y=-0.09541+14.77612$。F 检验的结果显示相关性达到了显著水平($F=4.3400$，$P=0.0399$)。分枝角度越小，节子发生点到愈合点距离越长。

计算得出节子发生点到愈合点距离(RT)与分枝年龄(YB)的回归方程为：$y=0.5318x-1.25585$。F 检验的结果显示相关性达到了极显著水平($F=32.4572$，$P=0.0001$)。分枝年龄越大，节子在木质部的跨度越大。

2.3.3　建立多元线性回归模型

应用分枝直径(BD)、分枝角度(IA)和分枝年龄(YB)3 个关键因子建立与节子发生点到愈合点距离(RT)的回归模型为：$y=1.6344x_1+0.06776x_2+0.1648x_3-1.61139$。经 F 检验的结果显示相关性达到了极显著水平($F=106.8697$，$P=0.0001$)，利用该模

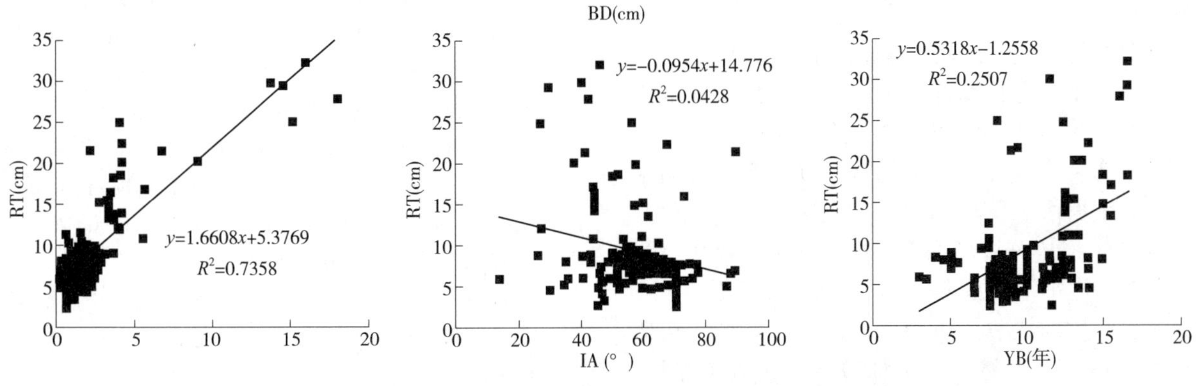

图 4　节子发生点到愈合点距离与分枝直径、分枝角度及分枝年龄的关系

型对测量数据进行计算，拟合效果如图5所示。将计算结果与实际测量结果进行方差分析，显示两者之间差异性极小($F=0$，$P=1$)。因此，在生产实践中，根据调查格木分枝的表观特征(分枝角度、直径)及生长时间等数据，利用该模型可以预测分枝形成节子后对木材的影响状况。

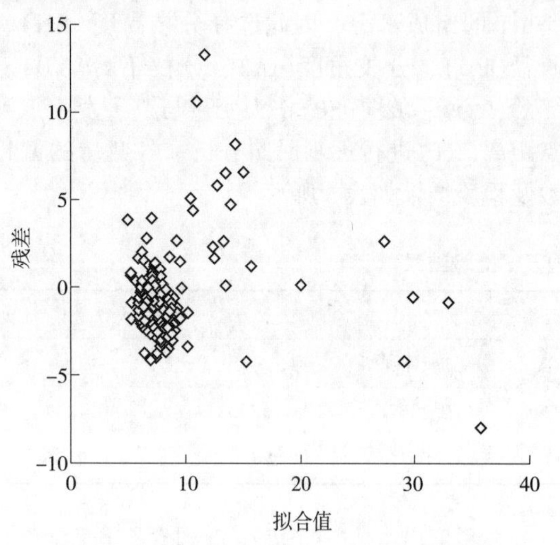

图5　多元线性回归模型残差分布

3　讨论

不同高度格木树干的节子分布呈先增加后减小的趋势，节子的分布主要集中在树干的2.0~8.0m之处，其中4.1~6.0m的节子数量最多(29.29%)，此段又是木材利用的核心部分，严重降低木材价值。依所选样地格木的节子分布与地理方位、坡向的关系统计结果，因坡度和坡向因子影响，下坡向为分枝生长提供更大的空间，格木在此方向分布的分枝数量明显多于其他几个坡向。因此，与地理方位相比，坡向是影响格木分枝分布的重要因子。而樟子松分枝的水平分布呈均匀分布[12]，差异的原因在于其研究的试验地属于缓坡丘陵，坡向对分枝生长影响作用明显弱于南方山地，樟子松分枝的着枝点具有明显的均匀轮生特点，而格木分枝的着枝点没有明确规律性。

从分枝的形成、枯死、脱落及伤口的愈合需要一个过程，由于分枝基部树干的支撑作用，常常出现分枝枯而不落现象，严重延迟了伤口愈合速度，加大病虫害侵害概率，增加死节长度，严重影响木材质量。根据格木节子的时间分布特征分析，第1~15年是格木形成分枝高峰期，格木分枝脱落伤口愈合主要集中在第16~25年，第11~20年开始形成死节较多，是控制死节形成的关键时期。因此，在造林后10~15年，对格木人工林进行一次人工修枝，可以减小格木分枝对木材质量的影响。另外，修枝的枝条粗度应有一个界限[15]，采用合理的修枝方式及修枝伤口处理方法，尽快使切口愈合，防止感染导致木材心腐。

一般认为林木分枝角度是由其遗传特性决定的，分枝角度多变的现象是由于枝条自身重力不同引起，分枝自身重力越大，越倾向于水平[16]。而Cluzeau等[17]认为着枝角度与枝长间并没有明确的关系。Lemieux等[18]认为枝条基径是预测内部节子大小的最好参数，但是仍然有其他的重要的变量来描述节子形态。通过逐步回归分析，筛选出分枝直径(BD)、分枝角度(IA)和分枝年龄(YB)3个关键因子建立了多元回归模型：$y=1.6344x_1+0.06776x_2+0.1648x_3-1.61139$，相关性达到了极显著水平，将计算结果与实际测量结果之间差异性极小。在生产实践中，根据调查格木分枝的表观特征(分枝角度、直径)及生长时间等数据，利用该模型可以预测分枝形成节子后对木材的影响状况。

参考文献

[1] 黄素涌，王建和，吕建雄，等．世界节子研究进展[J]．林产工业，2011，38(5)：3-7.

[2] BODIG J，JAYNE B A. Mechanics of wood and wood composites[M]. Van Nostrand Reinhold，1981：712.

[3] LEMIEUX H，SAMSON M，USENIUS A. Shape and distribution of knots in a sample of *Picea abies* logs[J]. Scandinavian Journal of Forest Research，1997，12(1)：50-56. DOI：10.1080/02827589709355383.

[4] FUJIMORI T. Dynamics of crown structure and stem growth based on knot analysis of a hinokicypress[J]. Forest ecology and management，1993，56(1)：57-68. DOI：10.1016/0378-1127(93)90103-T.

[5] Mäkinen H. Effect of stand density on radial growth of branches of Scots pine in southern and central Finland[J]. Canadian Journal of Forest Research，1999，29(8)：1216-1224. DOI：10.1139/x99-060.

[6] Mäkinen H. Effect of stand density on the branch development of silver birch (*Betula pendula* Roth) in central Finland[J]. Trees，2002，16(4-5)：346-353. DOI：10.1007/s00468-002-0162-x.

[7] KERSHAW Jr J A，MAGUIRE D A，HANN D W. Longevity and duration of radial growth in Douglas-fir branches[J]. Canadian Journal of Forest Research，1990，20(11)：1690-1695. DOI：10.1139/x90-225.

[8] 史军辉，黄忠良，蚁伟民，等．渐危植物格木群落动态及其保护对策[J]．西北林学院学报，2005，20(3)：65-69.

[9] 黄忠良，郭贵仲，张祝平．渐危植物格木的濒危机制

及其繁殖特性的研究[J]. 生态学报, 1997, 17: 671- 676.

[10] 赵志刚, 郭俊杰, 沙二, 等. 我国格木的地理分布与种实表型变异[J]. 植物学报, 2009, 44(3): 338- 344.

[11] 方夏峰, 方柏州. 闽南格木木材物理力学性质的研究[J]. 福建林业科技, 2007, 34(2): 146-147. DOI: 10.13428/j.cnki.fjlk.2007.02.037.

[12] 贾炜玮. 樟子松人工林枝条生长及节子大小预测模型的研究[D]. 哈尔滨: 东北林业大学, 2006.

[13] 陈东升, 金钟跃, 李凤日, 等. 樟子松节子的大小及分布[J]. 东北林业大学学报, 2007, 35(5): 19-21. DOI: 10.13759/j.cnki.dlxb.2007.05.006.

[14] SPIECKER H, HEIN S. Comparative analysis of occluded branch characteristics for *Fraxinus excelsior* and *Acer pseudoplatanus* with natural and artificial pruning [J]. Canadian Journal of Forest Research, 2007, 37(8): 1414-1426. DOI: 10.1139/X06-308.

[15] 方升佐, 徐锡增, 严相进, 等. 修枝强度和季节对杨树人工林生长的影响[J]. 南京林业大学学报(自然科学版), 2000, 24(6): 6-10.

[16] DELEUZE C, Hervé J C, COLIN F, et al. Modelling crown shape of *Picea abies*: spacing effects[J]. Canadian Journal of Forest Research, 1996, 26(11): 1957-1966. DOI: 10.1139/x26-221.

[17] CLUZEAU C, GOFF N L, OTTORINI J M. Development of primary branches and crown profile of *Fraxinus excelsior* [J]. Canadian Journal of Forest Research, 1994, 24(12): 2315-2323. DOI: 10.1139/ x94 -299.

[18] LEMIEUX H, BEAUDOINn M, ZHANG S Y. Characterization and modeling of knots in black spruce (*Picea mariana*) logs[J]. Wood and Fiber Science, 2001, 33(3): 465-4.

[原载: 南京林业大学学报(自然科学版), 2017, 41(03)]

格木人工林不同冠层光合特征

郝　建[1,2,4]　潘丽琴[2,3]　潘启龙[1,4]　李忠国[1,4]　杨桂芳[1,4]

([1]中国林业科学研究院热带林业实验中心，广西凭祥　532600；
[2]中国林业科学研究院热带林业研究所，广东广州　510520；
[3]西南林业大学，云南昆明　650204；
[4]广西友谊关森林生态系统定位观测研究站，广西凭祥　532600)

摘　要　评价格木人工林不同冠层分枝的光合能力，为格木人工林修枝提供光合生理基础，以9年生格木-马尾松混交人工林为研究对象，观测格木人工林不同冠层叶片的光响应曲线、二氧化碳响应曲线及光合因子的日变化。结果表明：格木人工林不同冠层叶片的光合的初始斜率(α)和最大净光合速率(P_{max})的大小顺序相同，依次为：冠层C>冠层B>冠层A，而暗呼吸速率(R_d)大小顺序则相反，高冠层叶片的饱和光强(I_{sat})较大，而光补偿点(I_c)却较低，且3个冠层各参数差异均达到显著或极显著水平；3个冠层叶片的初始羟化效率(α)、光合能力(A_{max})的顺序相同，均为冠层C>冠层B>冠层A，而二氧化碳补偿点(Γ)顺序相反，光呼吸(R_p)变化与光合作用保持着平行关系；3个冠层的光合有效辐射(*PAR*)，蒸腾速率(*Tr*)与叶片温度(*T*)的日变化为单峰型，净光合速率(*Pn*)与气孔导度(Gs)日变化曲线呈双峰型，但峰值出现的时间均不同，3个冠层胞间二氧化碳浓度(*Ci*)的日变化一致，为"U"型曲线，均在13：00出现最低值；通径分析结果表明，光合有效辐射与气孔导度对3个冠层净光合速率的直接通径系数绝对值均较大，表明光合有效辐射及气孔导度是影响格木人工林3个冠层叶片的净光合速率的主要因子。格木高冠层叶片具有更高的光合速率和更强的潜在光合能力，对光合物质的积累能力较强，消耗光合产物较少。

关键词　格木人工林；冠层；光响应曲线；二氧化碳响应曲线

Photosynthetic Characteristics in Different Canopy Positions of an *Erythrophleum fordii* plantation

HAO Jian[1,2,4], PAN Liqin[2,3], PAN Qilong[1,4], LI Zhongguo[1,4], YANG Guifang[1,4]

([1]*Experimental Center of Tropical Forestry, Chinese Academy of Forestry, Pingxiang* 532600, *Guangxi, China*;
[2]*Research Institute of Tropical Forestry, Chinese Academy of Forestry, Guangzhou* 510520, *Guangdong, China*;
[3]*Southwest Forestry University, Kunming*, 650204, *Yunnan, China*;
[4]*Guangxi Youyiguan Forest Ecosystem Research Station, Pingxiang* 532600, *Guangxi, China*)

Abstract: To evaluate the photosynthetic characteristics of different canopy positions and to provide a photosynthetic physiological basis for pruning in an *Erythrophleum fordii* plantation, light intensity, light response curve, CO_2 response curve, and the dynamic changes of photosynthetic factors were measured in different canopy positions of an *E. fordii* × *Pinus massoniana* plantation. Also, photosynthetic capacity of branches in different canopy positions was analyzed and a path analysis was conducted. Results showed that the different canopy positions significantly affected on initial slope photo inhibition ($P<0.05$) and maximum net photosynthetic rate ($P<0.01$), and ranked as canopy C > canopy B > canopy A; however, dark respiration rate was opposite. There was significant difference in the saturation light and light compensation point in different canopy positions ($P<0.01$), the saturation light of higher canopy was greater, but the light compensation point was lower. Initial carboxylation efficiency and photosynthetic capacity higher in the canopy were greater than lower in the canopy with highly significant difference ($P<0.01$), and the carbon dioxide compensation point of different positions in the canopy was in the reverse order. The relationship between photorespiration and photosynthesis was parallel. Daytime change of photosynthetic active radiation, transpiration rate, and leaf temperature in the three canopy positions appeared as a single peak and the diurnal variation net photosynthetic rate and stomatal conductance showed a typical curve with a double peak, but the time of the peaks was different. The diurnal change of intercellular CO_2 concentration in the three canopies was consistent, as a "U" type curve. The path analysis indicated that photosynthetic available radiation and stomatal conductance were the key factors for the net photosynthetic rate in leaves of the three canopy positions of the *E. fordii* plantation. The high canopy position's photosynthetic rate, potential photosynthetic capacity, and photosynthetic material accumulating ability were stronger, and less of the photosynthetic product was consumed higher in the canopy than lower in the canopy, which could provide scientific guidance to determine the pruning intensity in *E. fordii* plantation.

Key words: *Erythrophleum fordii*; canopy; light-response curve of photosynthesis; CO_2 response curve of photosynthesis

树冠是林木进行光合作用的物质基础，是制造干物质的载体。树木对太阳辐射和降水的利用受限于树冠的结构和空间分布特征，不同特征的树冠对能量的接受、利用、传输和分配呈现出空间异质性，导致不同冠层叶片的光合作用存在空间差异[1,2]。国内外针对不同冠层光合作用的研究涉及很多树种，如糖枫(*Acer saccharum*)[3]，欧洲山杨(*Populus tremula*)[4]、小叶椴(*Tilia cordata*)[4]、欧榛(*Corylus avellana*)[4]、杉木(*Cunninghamia lanceolata*)[5]、三倍体毛白杨(*Populus tomentosa*)[6]、樟树(*Cinnamomum camphora*)[7]、杂种落叶松(*Larix gmelinii*)[8]、杨树(*Populus×euramericana*)[9]等树种。目前，针对用材林人工修枝技术的研究，主要集中在人为设置不同修枝强度进行一些对比试验，很少从不同冠层分枝光合能力差异方面进行研究，难以真正科学指导用材林的人工修枝。因此，本研究以9年生格木—马尾松混交人工林为研究对象，在指定方向分3个冠层测定其叶片的光合作用，评判不同冠层分枝叶片的光合能力，为科学制定格木人工林修枝技术提供理论依据。

1 材料与方法

1.1 试验地概况

研究地点位于广西凭祥市中国林业科学研究院热带林业实验中心青山实验场。该地区处于南亚热带季风气候区域内的西南部，与北热带北缘毗邻，属湿润半湿润气候。境内日照充足，雨量充沛，干湿季节明显，光、水、热资源丰富。年均气温为20.5~21.7℃，极端高温为40.3℃，极端低温为-1.5℃；≥10℃活动积温为6000~7600℃。年均降水量为1200~1500mm，年蒸发量为1261~1388mm，相对湿度为80%~84%。主要地貌类型以低山丘陵为主，坡度以25°~30°为多。地带性土壤为砖红壤，土层厚度大于1m。

1.2 供试材料

所选试验地为2006年春营造的格木-马尾松混交人工林，造林密度为2500株/hm²，混交比例为1∶3。设置3个试验样地，在每木检尺的基础上，每个样地选出3株冠形完好的优势木，根据格木枝条生长特性，将树冠分3层(从下至上依次为冠层A、冠层B和冠层C)，通过人工搭竹架来观测不同冠层叶片的光合指标。

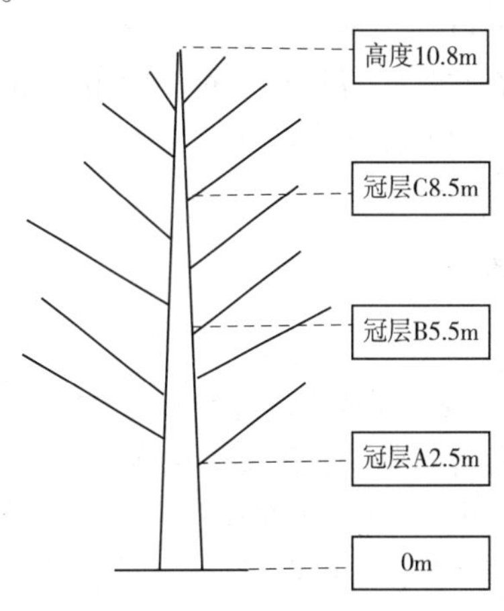

图1 格木不同冠层枝条的分布

1.3 测定方法

于2016年6~7月，在所选3个冠层的指定方位各选取最大枝条上外围的成熟功能叶并标记，用Li-6400XT光合仪(美国Li-cor公司)观测已标记功能叶的光合参数。

光响应曲线测定：测量时间选择天气晴朗的上午9：30~11：30，为保持其他环境因子稳定且适宜，将叶室温度设为28℃，二氧化碳浓度设为400±10μmol/m²·s，用Li-6400红蓝光源控制光强，光强依次设为2000，1800，1600，1400，1200，1000，800，600，400，200，100，80，60，40，30，20，10，5，0μmol/m²·s，并根据该曲线计算出光补偿点、光饱和点及最大净光合速率等重要参数。

二氧化碳响应曲线测定：测量时间仍选择在晴朗天气的上午9：30~11：30，使用LED红蓝光源控制光和有效福射值为光饱和点，设置环境二氧化碳摩尔分数的变化梯度为1800，1500，1200，1000，800，600，400，200，150，100，50，30，20，0μmol/mol，依次测定不同二氧化碳摩尔分数下的光合速率，绘制二氧化碳响应曲线。并根据曲线计算二氧化碳补偿点、饱和点、羧化效率及最大光合能力等指标。

不同冠层光合参数的日变化：选择晴朗天气，在8：00~17：00，将Li-6400XT光合仪叶室条件与自然环境保持一致，每1h测定1次3个冠层已标记功能叶的的净光合速率(P_n，mmol/m²·s，蒸腾速率(T_r，mmol/m²·s)、胞间二氧化碳摩尔分数(C_i，μmol/mol)、气孔导度(G_s，mmol/m²·s)、光合有效辐射(P_{AR}，μmol/m²·s¹)、叶片温度(T,℃)等光合生理指标。

1.4　数据分析

用DPS14.5数据处理软件对格木人工林不同冠层叶片光合特性进行统计分析；利用直角双曲线修正模型分别对其光响应曲线[10,11]和二氧化碳响应曲线[12]进行拟合及参数计算，并通过Excel进行分析作图。

2　结果与分析

2.1　格木人工林不同冠层叶片对光的响应

利用直角双曲线修正模型对格木人工林下、中、上3个冠层叶片的光响应曲线进行拟合，拟合效果较好，确定系数R^2值均在0.99以上，其模型参数如表1所示。由图2可看出，格木人工林不同冠层叶片净光合速率(P_n)均随着光照强度(P_{AR})的增强而升高，但其光响应曲线变化趋势的不同阶段各光合参数存在明显差异。

表1　不同冠层光响应曲线拟合模型参数

冠层	a	b	g	R_d	R^2
冠层A	0.033 971	0.000 112	0.005 958	1.145 9	0.995 5
冠层B	0.049 884	0.000 058	0.004 753	0.484 0	0.996 0
冠层C	0.066 298	0.000 066	0.003 333	0.288 9	0.999 5

注：α、β、γ是3个系数，α是光响应曲线的初始斜率，表示植物在光合作用对光的利用效率，β为修正因子，系数$\gamma=\alpha/P_{max}$，R_d为暗呼吸速率。

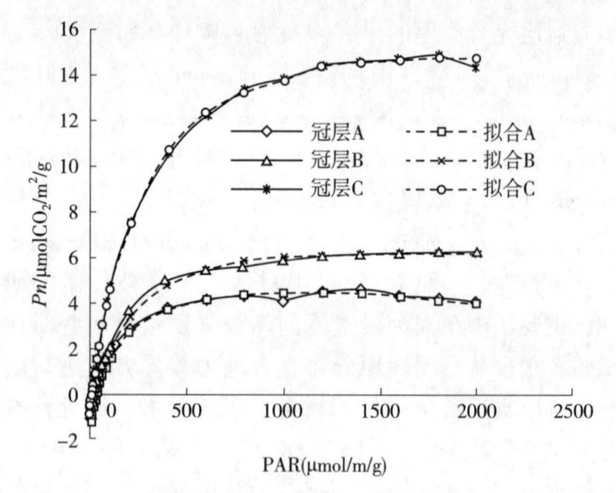

图2　格木不同冠层光响应曲线

表2　不同冠层光响应曲线各参数统计表

冠层	初始斜率 a	最大净光合速率 P_{max} (μmol/m²·s)	饱和光强 I_{sat} (μmol/m²·s)	光补偿点 I_c (μmol/m²·s)	暗呼吸速率 R_d (μmol/m²·s)
冠层A	0.033971±0.0132bc	4.4686±1.12BCc	1067.17±9.47Cc	30.98±6.78Aa	1.15±0.33Aa
冠层B	0.049884±0.0196ab	6.2429±1.86Bb	1700.54±7.39Bb	12.89±5.39Bb	0.48±0.21Bb
冠层C	0.066298±0.0218a	14.7335±3.61Aa	1852.14±11.82Aa	4.42±2.43Cc	0.29±0.15BCc

从表2可知，格木人工林中冠层A、冠层B、冠层C叶片的初始斜率(a)、最大净光合速率(P_{max})和饱和光强值(I_{sat})的顺序相同，均为冠层A<冠层B<冠层C，光补偿点(I_c)和暗呼吸速率(R_d)顺序相反，为冠层A>冠层B>冠层C，且3个冠层各参数差异均达到显著或极显著水平。随着格木人工林冠层高度增加叶片光合的初始斜率增大，最大净光合速率增大，而暗呼吸速率却减小，这表明格木叶片对光合物质的积累随着冠层高度增加而增强，对光合产物的消耗在减少。随着格木人工林冠层高度增加，饱和光强增大，而光补偿点却减小，表明随着冠层越高，格木叶片的光合能力越强，光合作用要求的有效辐射强度范围越宽，对较强及弱光的利用效率越高。

2.2　格木人工林不同冠层叶片对二氧化碳的响应

通过直角双曲线修正模型对格木人工林下、中、上3个冠层叶片的二氧化碳响应曲线进行拟合，确定系数R^2值均在0.99以上，模型参数如表3所示。格木人工林不同冠层叶片二氧化碳响应曲线与光响应曲线有着相同的变化趋势，随着二氧化碳浓度升高，冠层A、冠层B、冠层C叶片的P_n值不断增高后呈平缓趋势，3个冠层P_n差距增加，趋势线逐渐分离(图3)。

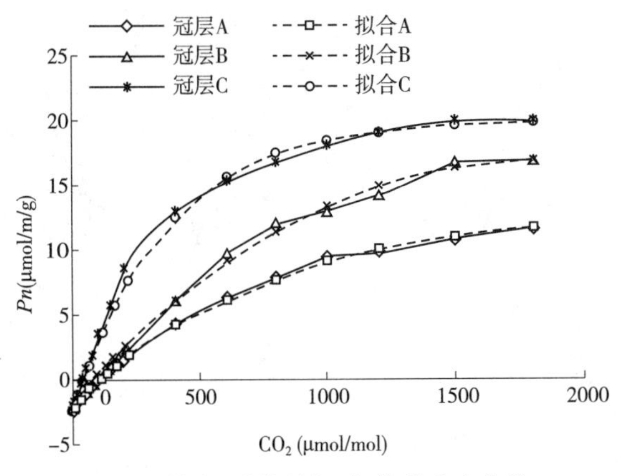

图3 格木不同冠层二氧化碳响应曲线

从表3可知，格木人工林3个冠层叶片的二氧化碳响应拟合曲线各参数的均达到显著或极显著水平。冠层A、冠层B和冠层C叶片的初始羟化效率(a)、光合能力(A_{max})的顺序相同，均为冠层A<冠层B<冠层C，说明高冠层格木叶片在低二氧化碳浓度条件下，具有更高的光合速率，且潜在的光合能力更强；二氧化碳补偿点(Γ)顺序相反，二氧化碳补偿点低，表明叶片具有净光合速率高、产量高的特点，二氧化碳补偿点经常作为评价植物光合性能的一个可靠指标；光呼吸速率(R_p)大小顺序为冠层A<冠层B<冠层C，光呼吸变化与光合作用保持着一定的平行关系[13]，因此，高冠层叶片光合能力较强，光呼吸速率也相应地增高。同时，光呼吸速率和羧化效率共同影响二氧化碳补偿点的变化结果[14]。

表3 不同冠层二氧化碳响应曲线拟合模型参数

冠层	a	b	g	R_p	R^2
冠层A	0.014245	0.00023438	0.00012	0.6476	0.9962
冠层B	0.023058	0.00020346	0.00024	1.5034	0.9970
冠层C	0.071167	0.00009942	0.00211	2.1416	0.9923

注：α、β、γ是3个系数，α为CO_2响应曲线上$C_{isat}=0$处的斜率，即为植物的初始羧化效率，β为修正因子，系数$\gamma=\alpha/A_{max}$，R_p为光呼吸速率。

表4 不同冠层二氧化碳响应曲线各参数统计表

冠层	初始羧化效率 a (mol/m^2·s)	光合能力 A_{max} (mol/m^2·s)	饱和胞间二氧化碳浓度 C_{isa} (mol/m^2·s)	二氧化碳补偿点 Γ (mol/m^2·s)	光呼吸速率 R_p (mol/m^2·s)
冠层A	0.0142±0.0085BCc	11.57±2.67Cc	1913.21±11.65ab	66.21±8.56Aa	0.65±0.32Cc
冠层B	0.0231±0.0103Bb	17.04±4.93ABb	1988.08±10.85a	57.16±7.39ABb	1.50±0.61ABb
冠层C	0.0712±0.0152Aa	19.74±4.03Aa	1758.68±15.67bc	32.25±9.88Cc	2.14±1.53Aa

2.3 格木不同冠层光合参数的日变化

在格木人工林生长旺盛季节，选择晴朗无云天气测定3个冠层同一天内不同时间的光合参数，并绘制分布图(图4)。格木人工林3个冠层的P_{AR}，T_r与T的日变化表现出大致相同的规律，为单峰型，分别在12：00、13：00与14：00达到峰值。P_n与G_s日变化曲线呈双峰型，3个冠层P_n峰值出现的时间均不同，冠层A为10：00与15：00，冠层B为11：00与15：00，冠层C为12：00与15：00；3个冠层A的G_s峰值的时间与冠层B和冠层C不同，冠层A为10：00与15：00，冠层B和冠层C为11：00与16：00。3个冠层C_i的日变化一致，为"U"型曲线，均在13：00出现最低值。3个冠层的各光合参数均随冠层高度增加而增大，C_i与T变化幅度较小。

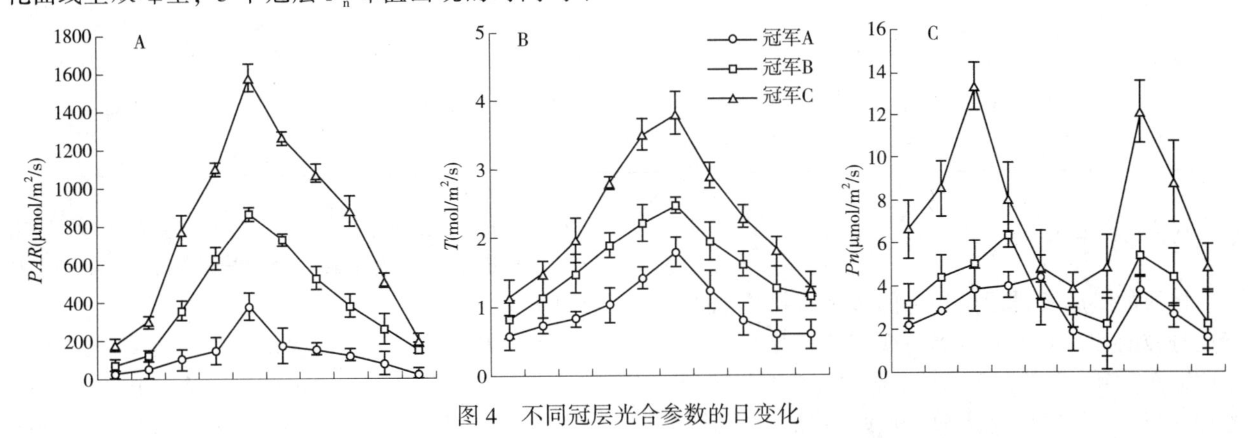

图4 不同冠层光合参数的日变化

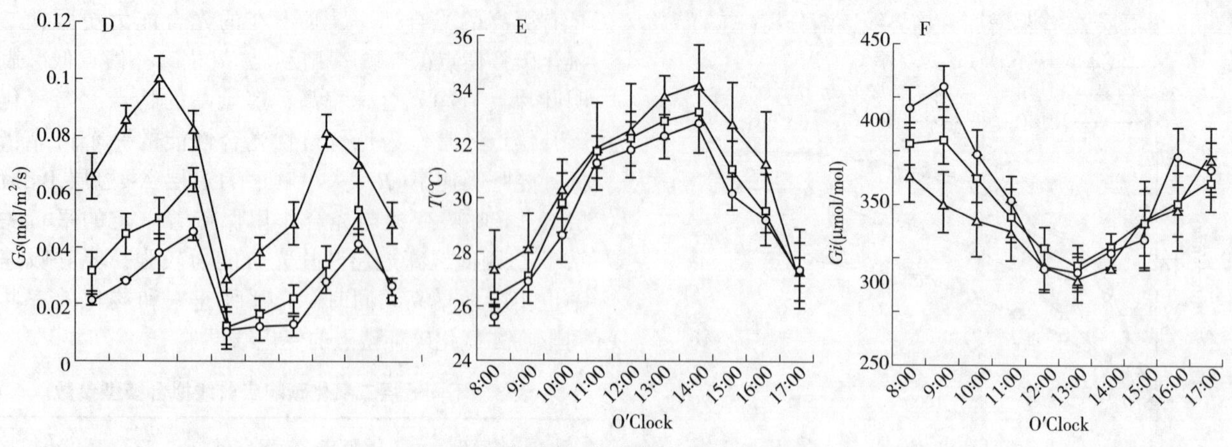

图4　不同冠层光合参数的日变化(续图)

2.4　格木不同冠层净光合速率与光合因子的关系

格木不同冠层叶片净光合速率与各光合因子的通径分析结果如表5所示。冠层A中P_{AR}、T、与G_s对P_n的直接通径系数绝对值较大，其中P_{AR}的直接作用最大。冠层B中G_s与P_{AR}对Pn的直接通径系数绝对值较大，其中G_s的直接作用最大。冠层C中T_r、P_{AR}与G_s对P_n的直接通径系数绝对值较大，其中T_r的直接作用最大。

表5　不同冠层净光合速率与光合因子的通径分析

冠层	因子	直接通径系数	间接通径系数				
			P_{AR}	T_r	G_s	C_i	T
冠层A	P_{AR}	1.1717		0.0804	-0.3763	0.2570	-0.6359
	T_r	0.1113	0.8465		-0.5077	0.2785	-0.7230
	G_s	0.8647	-0.5099	-0.0653		-0.1703	0.2576
	C_i	-0.3468	-0.8683	-0.0894	0.4246		0.8006
	T	-0.8833	0.8434	0.0911	-0.2522	0.3143	
冠层B	P_{AR}	0.4719		-0.0183	-0.2701	-0.1877	0.0214
	T_r	-0.0191	0.4525		-0.3143	-0.1976	0.0224
	G_s	0.8790	-0.1450	0.0068		0.0926	-0.0055
	C_i	0.2153	-0.4113	0.0176	0.3780		-0.0225
	T	0.0239	0.4222	-0.0179	-0.2015	-0.2028	
冠层C	P_{AR}	0.8905		-0.8600	-0.3455	-0.2094	0.3234
	T_r	-0.8961	0.8546		-0.4247	-0.2172	0.3298
	G_s	0.8309	-0.3703	0.4580		0.0880	-0.1115
	C_i	0.2289	-0.8147	0.8503	0.3195		-0.3445
	T	0.3760	0.7660	-0.7861	-0.2463	-0.2097	

3　结论与讨论

树冠的组成及树冠生物量的多少直接影响不同高度分枝的叶片对光辐射能的截获[15]。格木人工林不同冠层受到的遮光强度不同，导致3个冠层接受的光辐射存在差异，高冠层各时间段测定的P_{AR}值要高于低冠层的。通径分析表明，光合有效辐射及气孔导度均是影响格木人工林3个冠层的净光合速率的主要因子。

植物光合作用对不同强度的光照具有一定的适应能力，但是光强度是影响植物光合作用的重要外界因子[16]。冠层上层到下层光辐射强度的衰减会导致不同冠层叶片的光合能力呈现显著差异[17,18]。所处环境光照强度不同，格木不同冠层叶片光合特性表现出了明显的差异。高冠层叶片的光合的初始斜率、最大净光合速率均大于低冠层的，而暗呼吸速率相反，这说明格木高冠层叶片光合物质的积累能力较强，对光合产物的消耗也较少；高冠层叶片要求饱和光强较大，而光补偿点却较小，说明高冠层格木叶片对光辐射强度适应范围较宽，对光的利用效率较高。3 个冠层格木叶片的二氧化碳响应拟合曲线研究结果分析结果显示，高冠层叶片的初始羟化效率、光合能力均高于低冠层的，高冠层格木叶片具有更高的光合速率和更强潜在光合能力。

综合研究结果，高冠层分枝叶片的初始羟化效率、光合能力及对光合物质的积累能力均显著高于中冠层与低冠层。这说明下部冠层枝条长期处在较弱光辐射强度环境中，处在该冠层分枝叶片的光合能力下降。因此，依据不同冠层分枝叶片的光合特性的差异，可以通过人工修枝措施，修除下冠层光合能力较差的分枝，以优化树冠结构，调节树冠内微环境，提高树木的光合作用能力[8]。此种人工修枝措施一方面能促进格木生长，另一方面又可以优化干形，对格木人工林无节良材的培育具有重要指导意义。

参考文献

[1] VIERLING L A, WESSMAN C A. Photosynthetically active radiation heterogeneity within a monodominant Congolese rain forest canopy[J]. Agric For Meteorol, 2000, 103(3): 265-278.

[2] ANTEN N P R. Optimal photosynthetic characteristics of individual plants in vegetation stands and implications for species coexistence[J]. Ann Bot, 2005, 95(3): 495-506.

[3] ELLSWORTH D S, REICH P B. Canopy structure and vertical patterns of photosynthesis and related leaf traits in a deciduous forest[J]. Oecologia, 1993, 96(2): 169-178.

[4] KULLO, NIINEMETSÜ. Distribution of leaf photosynthetic properties in tree canopies: comparison of species with different shade tolerance[J]. Funct Ecol, 1998, 12(3): 472-479.

[5] 张小全，徐德应，赵茂盛，等．CO_2增长对杉木中龄林针叶光合生理生态的影响[J]．生态学报，2012，20(3)：390-396.

[6] 富丰珍，徐程扬，李广德，等．冠层部位对三倍体毛白杨光合生理特性的影响[J]．中南林业科技大学学报，2010，30(3)：95-99.

[7] 黄志宏，田大伦，闫文德，等．城市樟树人工林冠层光合作用的时空特征[J]．中南林业科技大学学报，2011，31(1)：38-46.

[8] 张元元．杂种落叶松人工林冠层光合特征及生长过程机理的研究[D]．哈尔滨：东北林业大学，2012.

[9] 马永春，方升佐．欧美杨 107 不同冠层光合特性的研究[J]．南京林业大学学报：自然科学版，2011，35(4)：39-42.

[10] YE Z P, YU Q. A coupled model of stomatal conductance and photosynthesis for winter wheat[J]. Photosynthetica, 2008, 46(4): 637-640.

[11] 叶子飘．光合作用对光和 CO_2 响应模型的研究进展[J]．植物生态学报，2010，34(6)：727-740.

[12] 叶子飘，于强．光合作用对胞间和大气 CO_2响应曲线的比较[J]．生态学杂志，2009，28(11)：2233-2238.

[13] 董志新，韩清芳，贾志宽，等．不同苜蓿(*Medicago sativa* L.)品种光合速率对光和 CO_2浓度的响应特征[J]．生态学报，2007，27(6)：2272-2278.

[14] 蔡时青，许大全．大豆叶片 CO_2补偿点和光呼吸的关系[J]．植物生理学报，2000，26(6)：545-550.

[15] MONTEITH J L. 2-DOES LIGHT LIMIT CROP PRODUCTION? [J]. Physioll Processes Limiting Plant Prod, 1981: 23-38.

[16] 张其德，唐崇钦，林世青，等．光强度对小麦幼苗光合特性的影响[J]．植物学报，1988，30(5)：508-514.

[17] AMTHOR J S. Scaling CO_2-photosynthesis relationships from the leaf to the canopy[J]. Photosynth Res, 1994, 39(3): 321-350.

[18] HAN Q M, CHIBA Y. Leaf photosynthetic responses and related nitrogen changes associated with crown reclosure after thinning in a young *Chamaecyparis obtusa* stand[J]. J For Res, 2009, 14(6): 349-357.

[原载：浙江农林大学学报，2017，34(05)]

岩溶石山降香黄檀人工林的天然更新

农　友[1,2]　卢立华[1,2]　孙冬婧[1,2]　黄德卫[1,2]　李　华[1,2]　雷丽群[1,2]　明安刚[1,2]

([1]中国林业科学研究院热带林业实验中心，广西凭祥　532600；
[2]广西友谊关森林生态系统国家定位观测研究站，广西凭祥　532600)

摘　要　为深入了解降香黄檀(*Dalbergia odorifera*)人工林天然更新的特点及其与环境因子的关系，应用典型样地法，研究广西西南部岩溶石山降香黄檀人工林的天然更新，分析降香黄檀实生幼苗幼树的结构特征、空间分布格局以及影响其种群更新的环境因子。结果表明：在20个5m×5m的样方中共调查降香黄檀幼苗幼树94株，其平均密度1880株/hm^2，空间分布以随机分布为主。幼苗幼树的个体主要集中在$DBH \leqslant 3$cm和$H \leqslant 4.0$m以内，且胸径随树高的增加而增长；其多度随径级的增加而减少，随高度级的增加先增加后减少，均呈现偏锋型曲线。用Canoco对可能影响降香黄檀幼苗幼树更新的9个环境因子进行主成分分析(PCA)表明，选取的环境因子共解释幼苗幼树分布信息的81.91%，影响降香黄檀幼苗幼树密度的主要因子依次为草本个体数、草本覆盖度、林窗面积、土壤含水率、土壤表层pH、灌木覆盖度、岩石裸露率、土壤有机质含量、灌木个体数。其中，降香黄檀幼苗幼树密度与草本个体数及其覆盖度之间存在极显著正相关($P<0.01$)；土壤表层pH、灌木覆盖度、岩石裸露率与降香黄檀幼苗幼树密度呈负相关。

关键词　岩溶石山；降香黄檀；人工林；天然更新；环境因子

The Natural Regeneration of Karst Plantation of *Dalbergia odorifera*

NONG You[1,2], LU Lihua[1,2], SUN Dongjing[1,2], HUANG Dewei[1,2]
LI Hua[1,2], LEI Liqun[1,2], MING Angang[1,2]

([1]*Experimental Center of Tropical Forestry*, *Chinese Academy of Forestry*, *Pingxiang* 532600, *Guangxi China*;
[2]*Guangxi Youyiguan Forest Ecosystem Research Station*, *Pingxiang* 532600, *Guangxi China*)

Abstract: The natural regeneration of karst artificial forest of Dalbergia odorifera of Southwest Guangxi was studied by using the method of typical plots. We analyzed the structural characteristics, spatial distribution pattern and the environmental factors influencing the population regeneration of the seedlings and saplings. The purpose of the study is to find out the characteristics of natural regeneration of Dalbergia odorifera artificial forest and its relationship with environmental factors. The results showed that: A total number of 94 seedlings and saplings were found in twenty plots, the density was 1880 plants/ha. The abundance of the seedlings and saplings was decreased with diameter while increased first and then decreased with the height class. Seedling and sapling individuals were concentrated within the range of $DBH \leqslant 3$cm and $H \leqslant 4.0$m, and the larger diameter the higher tree height. The spatial distribution of seedlings and saplings was mostly randomly distributed. The principal component analysis of 9 environmental factors showed that, the environmental factors were explained the information of 81.91%. There was a significantly positive correlation between density and the number of individuals of herbs and its coverage ($P<0.01$); soil pH, shrub coverage, the rock exposed rate were negatively correlated with the density of seedlings and saplings.

Key words: karst; *Dalbergia odorifera*; artificial forest; natural regeneration; environmental factors

植物天然更新是一个非常复杂的生态学过程，是指在没有人为因素的参与下，利用林木自身繁殖能力形成新林的过程，与物种自身的生长特性及环境密切相关[1-4]。种子植物的天然更新主要依靠两种方式，即有性繁殖和无性繁殖，其中有性繁殖是其主要方式。从种子产生到幼苗定居是植物天然更新中关键的阶段之一，多种因素会对种子和幼苗的命运产生影响，例如动物取食或病原体侵袭、群落微环境、干扰等[5]。其中，群落微环境对植物的更新至关重要，有研究指出，林下草本层和灌木层在资源竞争中会对幼苗产生不利影响[6]，土壤水分、林窗等因子也会制约幼苗的生长[7]。种子植物的天然更新对种群和群落动态以及物种多样性维持等方面均具有重要影响[8,9]。

降香黄檀(*Dalbergia odorifera*)，俗称海南黄花梨，属蝶形花科(Papilionaceae)黄檀属(*Dalbergia*)植物[10]。降香黄檀在自然条件下生长缓慢，30~40 年方可成材，并随着树龄增长木材价值越来越高。目前降香黄檀在我国海南、广东、广西和福建等地有较大面积的种植，以往对其研究主要涉及其木材[11-12]、营林培育[13-16]等方面，但对其人工林林下幼苗幼树的天然更新状况及其影响机制的研究非常有限。开展森林天然更新机制的研究不仅有助于了解森林生态系统的动态规律，而且对采取合理的经营措施也具有指导意义[17]。为促进这一珍贵树种的可持续发展，有必要对降香黄檀人工林的天然更新展开系统的研究。

本文以中国林业科学研究院热带林业实验中心石山树木园 1979 年引种种植的降香黄檀人工林为对象，对其林下幼苗幼树更新特征及其可能影响因子进行研究，试图深入了解环境因子对降香黄檀幼苗幼树更新的影响，旨在为降香黄檀人工林更新机制的进一步研究奠定基础，为降香黄檀人工林的可持续发展积累材料。

1 研究地区与研究方法

1.1 研究区概况

中国林业科学研究院热带林业实验中心地处南亚热带南缘，与北热带毗邻，属于南亚热带季风气候(22°07′32″N，106°44′34″E)，其“石山树木园”面积 540 亩，以收集保存热带南亚热带岩溶石山树种为主，共引种保存树种 680 种。年平均气温 21.5℃，最冷月(1 月)平均气温为 13.5℃，最热月(7 月)平均气温为 27.6℃，≥10℃年积温 7500℃，年降水量 1400mm。

1.2 样地设置与调查

以中国林业科学研究院热带林业实验中心“石山树木园”为对象，于 2014 年 12 月在降香黄檀人工林下选择典型地段设置 20 个 5m×5m 的调查样方，记录样方内降香黄檀的胸径、树高、坐标，调查样方内灌木和草本的种名、株数、盖度、高度等，记录样方内林窗的长短轴长、岩石裸露度，并在样方内随机选取 5 个点的混合表层土壤(0~10cm)，带回实验室进行分析。

1.3 结构划分

年龄结构是种群动态的重要特征，幼苗研究中常采用高度级来代替年龄结构[2,18]。因研究对象为岩溶区生长缓慢的降香黄檀人工林，本文将 $DBH \leqslant 5$cm 的植株作为幼苗幼树进行调查。将胸径划分为四个等级，Ⅰ 级 $DBH<2$cm，2cm $\leqslant DBH<3$cm 为Ⅱ 级，3cm $\leqslant DBH<4$cm 为Ⅲ 级，4cm $\leqslant DBH \leqslant 5$cm 为Ⅳ 级；将高度划分为六个等级，Ⅰ 级 H<2m，2m $\leqslant H<3$m 为Ⅱ 级，3m $\leqslant H<4$m 为Ⅲ 级，4m $\leqslant H<5$m 为Ⅳ 级，5m $\leqslant H<6$m 为Ⅴ级，Ⅵ 级 $H \geqslant 6$cm。

1.4 点格局分析

点格局分析是把植物个体看成空间上的点，分析其数量特征，基于点对点之间距离的统计，克服传统方法只分析单一尺度上空间分布格局的缺点，最大限度地利用空间点的信息，能较好的描述不同尺度空间格局信息[19]。

本文采用 Wiegand-Molloney's O-*ring* 统计方法分析一定尺度范围内降香黄檀幼苗幼树的空间分布格局。研究尺度 0~5m，设定栅格大小为 1m×1m，圆环宽度为 3m，选择完全空间随机模型(CSR，complete spatial randomness)[20]作为零假设，为了高精密度分析植物种群的空间分布，本研究应用 Monte Carlo 循环 99 次，产生置信度为 99%的包迹线以检验点格局分析的显著性。根据 Monte Carlo 模拟结果，对于空间分布，若 O(r)值位于上包迹线之上，则为聚集分布；若位于上下包迹线之间，则为随机分布；若位于下包迹线之下，则为均匀分布。其原理和计算过程详见参考文献[21,22]。

1.5 数据处理

用 Canoco 4.5 进行主成分分析，用 SPSS 18.0 进行相关性分析，用 Programita(2014 版)进行空间分布格局分析，林窗面积采用椭圆面积公式计算：

$$S = \pi ab/4$$

式中：S 为林窗面积；a 为长轴长；b 为短轴长[23]。

2 结果与分析

2.1 幼苗幼树的结构特征

2.1.1 幼苗幼树的密度结构

本次调查共记录到降香黄檀幼苗幼树 94 株，20 个调查样方中，幼苗幼树植株分布极不均匀，有 6 个样方的幼苗幼树植株密度为 0 株/hm²，2 个样方为 800 株/hm²，5 个样方为 1000~2000 株/hm²，4 个样方为 2000~4000 株/hm²，3 个样方为>4000 株/hm²。幼苗幼树平均密度为 1880 株/hm²。

2.1.2 幼苗幼树的高度结构

以高度级为横坐标，存活个体数为纵坐标，绘制出降香黄檀幼苗幼树的高度级结构图(图 1)。结果显示，降香黄檀幼苗幼树的高度结构呈偏态分布。Ⅲ级幼苗幼树的个体数量最多，Ⅰ级幼苗幼树其次，H≤4.0m 的幼苗幼树占总个体数的 70.21%。

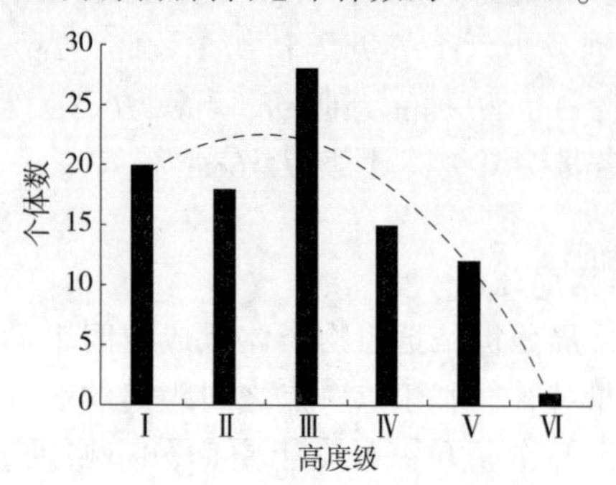

图 1 降香黄檀幼苗幼树的高度级结构

2.1.3 幼苗幼树的径级结构

以胸径级为横坐标，存活个体数为纵坐标，绘制出降香黄檀幼苗幼树的径级结构图(图 2)。结果显示，降香黄檀幼苗幼树的径级结构呈偏态分布。随着径级的增加，幼苗幼树多度逐渐减少，DBH≤3cm 的幼苗幼树占总个体数的 67.02%。

2.2 幼苗幼树的生长

降香黄檀幼苗幼树的胸径和树高具有相同的变化趋势，二者呈直线回归函数相关(图 3)。胸径随高度的增加而增长，反映了幼苗幼树正处于生长阶段。

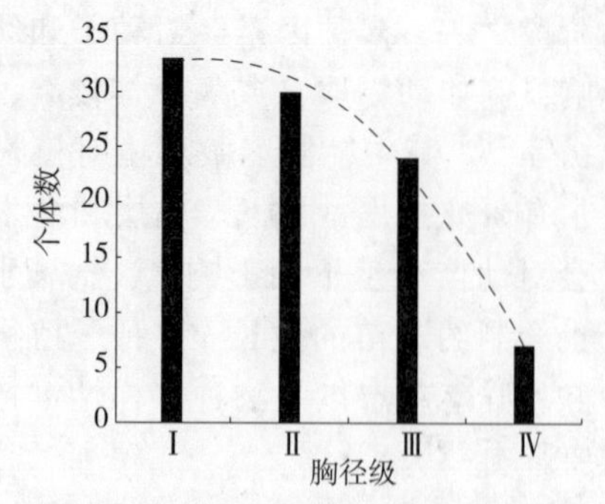

图 2 降香黄檀幼苗幼树的径级结构

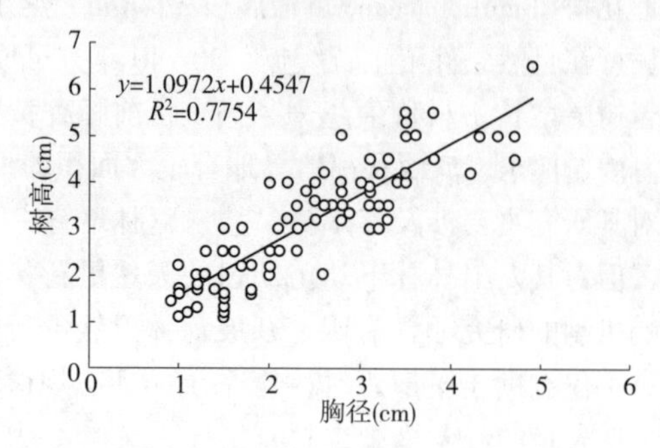

图 3 降香黄檀幼苗幼树胸径与树高的关系

2.3 幼苗幼树的空间分布

选取降香黄檀幼苗幼树个体数大于或等于 5 株的 8 个样方，在 0~5m 的尺度下，对幼苗幼树的空间分布格局进行分析(图 4)。各样方幼苗幼树的空间分布格局各异，呈现出聚集分布和随机分布的特点，并以随机分布为主。3 号样方在 0~1.5m、2.5~5m 的尺度下，呈随机分布；7 号、10 号、13 号样方在 0.5~5m 尺度下，呈随机分布；8 号样方在 1.5~5m 尺度下，呈随机分布；9 号、12 号样方在 2.5~5m 尺度下，呈随机分布；11 号样方在 0~0.5m、1~2.5m 和 3~5m 尺度下，呈随机分布。

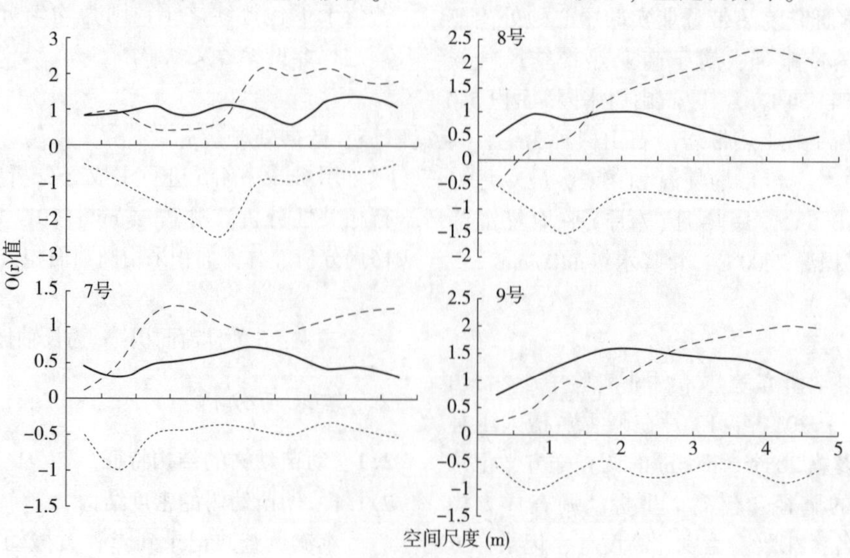

图 4 降香黄檀幼苗幼树空间分布

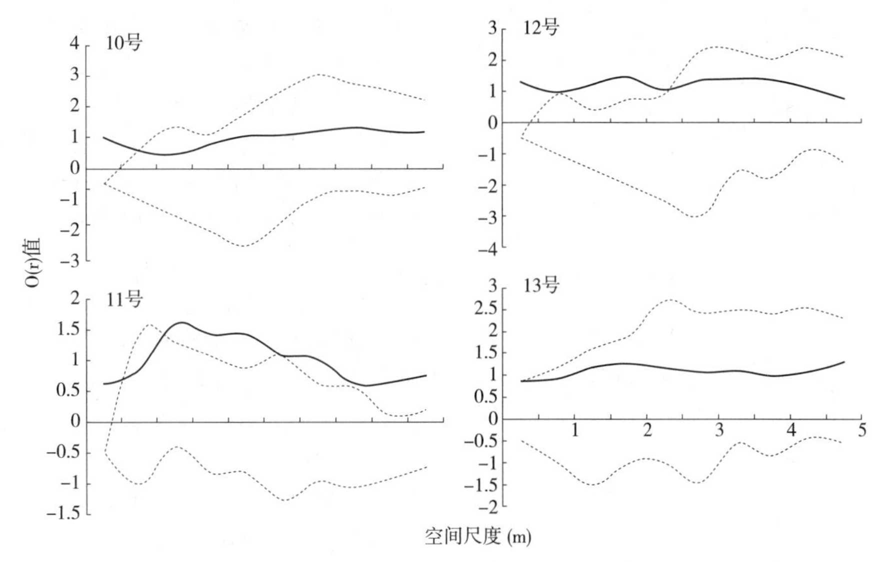

图 4 降香黄檀幼苗幼树空间分布(续图)

2.4 幼苗与环境因子的关系

主成分分析中，9 个环境因子指标共解释降香黄檀幼苗幼树密度信息的 81.91%(表 1)。根据物种与环境因子的关系(图 5)以及相关性的显著性检验结果(表 2)，草本个体数、草本覆盖度、林窗面积、土壤含水率、土壤表层 pH、灌木覆盖度、岩石裸露率、土壤有机质含量、灌木个体数依次是影响降香黄檀幼苗密度的主要因子。其中，降香黄檀幼苗幼树密度与草本个体数及其覆盖度之间存在极显著正相关(P<0.01)；土壤表层 pH、灌木覆盖度、岩石裸露率与降香黄檀幼苗幼树密度呈负相关。对环境因子间的相互关系，岩石裸露率与灌木个体数之间呈显著负相关；林窗面积与草本盖度之间呈显著正相关，与土壤有机质含量之间呈极显著负相关；灌木覆盖度与草本个体数及盖度之间呈极显著负相关；草本覆盖度、个体数与灌木覆盖度及土壤 pH 之间呈显著负相关；灌木个体数与岩石裸露率呈显著负相关。

表 1 PCA 分析的 4 个轴的特征值及累计解释量

排序轴	特征值	累计百分比变化率(%)
1	0.6286	62.86
2	0.1619	79.05
3	0.0286	81.91
4	0.0000	81.91

表 2 物种与环境因子的相关性检验

	Den	R	Gap	SC	HC	SN	HN	W	pH	Soc
Den	1	-0.096	0.272	-0.182	0.593**	0.059	0.619**	0.270	-0.267	0.087
R	-0.096	1	-0.321	0.013	-0.159	-.537*	-0.200	0.015	-0.063	0.028
Gap	0.272	-0.321	1	-0.227	.494*	0.138	0.364	0.063	-0.371	-.570**
SC	-0.182	0.013	-0.227	1	-.458*	0.224	-0.490*	0.117	-0.169	0.142
HC	0.593**	-0.159	0.494*	-0.458*	1	-0.019	0.929**	0.101	-.485*	-0.181
SN	0.059	-0.537*	0.138	0.224	-0.019	1	0.020	-0.372	-0.286	-0.191
HN	0.619**	-0.200	0.364	-0.490*	0.929**	0.02	1	0.044	-.496*	-0.137
W	0.270	0.015	0.063	0.117	0.101	-0.372	0.044	1	-0.188	0.338
pH	-0.267	-0.063	-0.371	-0.169	-.485*	-0.286	-0.496*	-0.188	1	-0.004
Soc	0.087	0.028	-.570**	0.142	-0.181	-0.191	-0.137	0.338	-0.004	1

注：SN：灌木个体数；Gap：林窗面积；Den：幼苗幼树密度；HC：草本盖度；HN：草本个体数；W：土壤含水率；R：岩石裸露率；Soc：土壤有机质含量；SC：灌木盖度；pH：土壤表层 pH。

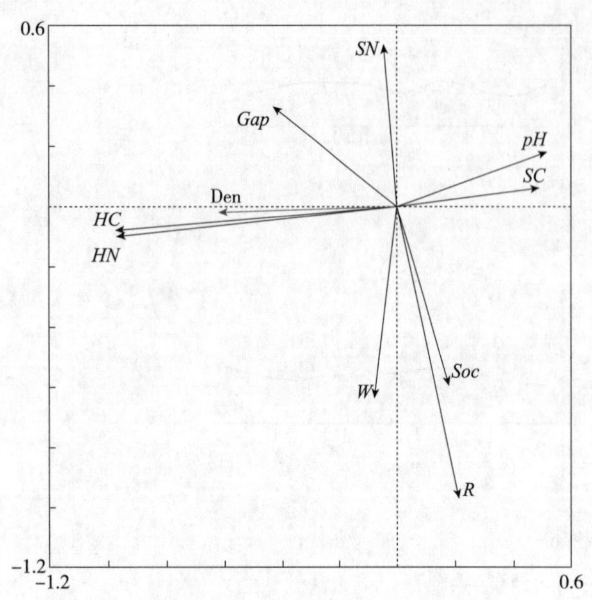

图5　物种因子与环境因子的主成分分析

注：SN：灌木个体数；Gap：林窗面积；Den：幼苗幼树密度；HC：草本盖度；HN：草本个体数；W：土壤含水率；R：岩石裸露率；Soc：土壤有机质含量；SC：灌木盖度；pH：土壤表层pH。

3. 讨论

3.1　幼苗幼树的特征及其空间分布

幼苗幼树的个体数量能够直接反映森林天然更新质量的好坏。本次调查中，降香黄檀幼苗幼树个体数量较少，反映降香黄檀人工林林下更新的状况不是太好。幼苗幼树个体主要集中在 $DBH \leqslant 3$cm 和 $H \leqslant 4.0$m 以内，就种群结构而言，属于稳定性种群，且其胸径和树高二者呈直线回归函数相关，反映了幼苗幼树正处于生长阶段。物种的空间格局和空间关联性在较小尺度上可能受种内竞争、种子扩散方式影响；在较大尺度上可能由物种分布的异质性或斑块性、环境异质性所决定[24,25]。研究指出，在喀斯特森林中，生境的高度异质性是影响物种空间分布格局形成的重要因素之一[26]。一般而言，物种在自然界中的分布主要为聚集分布，然而此次研究发现，降香黄檀幼苗幼树的空间分布格局以随机分布为主。究其原因，岩溶石山林地岩石裸露率大、微环境复杂、土壤分布不均匀，降香黄檀种子成熟扩散过程中，部分种子会散落在裸露的岩石表面而失去萌发机会，降香黄檀的荚果扁平，果荚边缘呈膜状，风媒作用可使其种子趋向随机分布。自身的生物学特性、种子散布、生境异质性可能是导致降香黄檀幼苗幼树呈现随机分布的主要原因，但具体结论仍需进一步论证。

3.2　幼苗幼树更新的影响因素

森林天然更新受到环境梯度、干扰、物种自身的生物学特性及其与周围树种的关系等因素的综合影响[27,28]。从种子产生到幼苗能否成功定居，很大程度上取决于其所处的生境条件[29]。研究表明，林下灌草层对乔木幼苗的生长有强烈的抑制作用[30]，而草本和林冠郁闭度对幼苗的存活和生长也有显著影响[31]。林下灌草层降低了地表光照，改变凋落物的分布，对更新的幼苗产生竞争压力，但也间接影响动物对种子的镊取行为[32]。林窗中的光照、气温都比林下高，有利于更新幼苗的生长和定居[33]。本研究中，草本个体数及其覆盖度与降香黄檀幼苗幼树密度之间存在极显著正相关（$P<0.01$）；土壤表层pH、灌木覆盖度、岩石裸露率与降香黄檀幼苗幼树密度呈负相关。究其原因，岩溶石山区的微环境复杂，岩石裸露的地方，幼苗难以萌发，反之，草本植物大量存在的地方，岩石裸露率低，灌木植物相对较少，土壤的pH相对较低，林窗面积较大，光照较充足，这样的环境为阳性树种降香黄檀林下更新创造了有利条件。然而，降香黄檀的林下更新是一个复杂的过程，各个环境因子之间并不是相互独立而是存在相互促进或抑制的作用，本次研究选取的环境因子只是其中的一部分，因此，为深入了解降香黄檀的天然更新过程，还需要通过对幼苗幼树及环境因子长期全面的定位观测，进行定性和定量分析。

本研究发现，岩溶石山区降香黄檀人工林林下幼苗幼树的个体主要集中在 $DBH \leqslant 3$cm 和 $H \leqslant 4.0$m 以内，数量较少，且以随机分布为主，林下更新效果并不显著，这可能是由于物种自身的生物学特性及环境因子的影响导致，但具体原因需要进一步论证。为了促进其林下天然更新进程，一方面要加强森林植被的保护，另一方面要进行人为的适度干扰，在植物恢复中期阶段适当剔除一些灌木植物，以改善种子发芽和幼苗幼树生长发育的条件；利用天然更新与人工促进天然更新相结合的方式，如人工补播补植，以弥补天然种苗的数量不足，促进其种群的演替进程。

参考文献

[1]朱教君，刘足根，王贺新．辽东山区长白落叶松人工林天然更新障碍分析[J]．应用生态学报，2008，19(04)：695-703.

[2]欧芷阳，苏志尧，彭玉华，等．桂西南喀斯特山地蚬木幼龄植株的天然更新[J]．应用生态学报，2013，24(09)：2440-2446.

[3]马姜明，刘世荣，史作民，等．川西亚高山暗针叶林恢复过程中岷江冷杉天然更新状况及其影响因子[J].

植物生态学报，2009，33(04)：646-657.

[4]李艳丽，杨华，亢新刚，等．长白山云冷杉针阔混交林天然更新空间分布格局及其异质性[J]．应用生态学报，2014，25(02)：311-317.

[5]彭闪江，黄忠良，彭少麟，等．植物天然更新过程中种子和幼苗死亡的影响因素[J]．广西植物，2004，24(2)：113-121.

[6]BEIER C M，HORTON J L，WALKER J F，et al. Carbon limitation leads to suppression of first year oak seedlings beneath evergreen understory shrubs in Southern Appalachian hardwood forests. Plant Ecology，2005，176(1)：131-142.

[7]李宁，白冰，鲁长虎．植物种群更新限制——从种子生产到幼树建成[J]．生态学报，2011，31(21)：6624-6632.

[8]TILMAN D. Competition and biodiversity in spatially structured habitats[J]．Ecology，1994，75(1)：2-16.

[9]HURTT G C，PACALA S W. The consequences of recruitment limitation：reconciling chance，history and competitive differences between plants[J]．Journal of Theoretical Biology，1995，176(1)：1-12.

[10]贾瑞丰．降香黄檀人工促进心材形成的研究[D]．北京：中国林业科学研究院，2014.

[11]吕金阳，罗建举．扫描电镜下降香黄檀木材构造的研究[J]．中南林业科技大学学报，2014，(04)：108-113+117.

[12]余敏，张浩，刘盛全，等．降香黄檀木材 DNA 提取及 rDNA-ITS 序列条形码分子鉴定[J]．林产化学与工业，2014，34(05)：103-108.

[13]吴国欣，王凌晖，梁惠萍，等．氮磷钾配比施肥对降香黄檀苗木生长及生理的影响[J]．浙江农林大学学报，2012，29(02)：296-300.

[14]周双清，周亚东，盛小彬，等．降香黄檀繁殖技术研究进展[J]．林业实用技术，2013，(12)：28-30.

[15]莫慧华．大青山珍贵树种格木、降香黄檀与巨尾桉混交效果初步评价[D]．南宁：广西大学，2013.

[16]贾宏炎，黎明，曾冀，等．降香黄檀工厂化育苗轻基质筛选试验[J]．中南林业科技大学学报，2015，(11)：74-79.

[17]徐振邦，代力民，陈吉泉．长白山红松阔叶混交林森林天然更新条件的研究[J]．生态学报，2001，21(9)：1413-1420.

[18]王彬，王辉，杨君珑．子午岭油松林更新特征研究[J]．西北林学院学报，2009，24(5)：58-60.

[19]DIGGLE P J. Statistical analysis of spatial point patterns. 1983，Academic press.

[20]GETZIN S，WIEGAND T，WIEGAND K，et al. Heterogeneity influences spatial patterns and demographics in forest stands[J]．Journal of Ecology，2008，96：807-820.

[21]WIEGAND T MOLONEY K A．Rings circles and null-models for point pattern analysis in ecology[J]．Oikos，2004，104：209-229.

[22]WIEGAND T MOLONEY K A．A handbook of spatial point pattern analysis in ecology. Chapman and Hall/CRC press，Boca Raton，FL. 2014.

[23]符婵娟，刘艳红，赵本元．神农架巴山冷杉群落更新特点及影响因素[J]．生态学报，2009，29(08)：4179-4186.

[24]ZHANG Z H，HU G，ZHU J D，et al. Spatial patterns and interspecific associations of dominant tree species in two old-growth karst forests，SW China. Ecological Research，2010，25(6)：1151-1160.

[25]袁春明，孟广涛，方向京，等．珍稀濒危植物长蕊木兰种群的年龄结构与空间分布[J]．生态学报，2012，32(12)：3866-3872.

[26]韩文衡，向悟生，叶铎，等．广西木论保护区喀斯特常绿落叶阔叶混交林优势种空间格局及其相关性[J]．应用生态学报，2010，21(11)：2769-2776.

[27]ZHU J，MATSUZAKI T，LEE F，et al. Effect of gap size created by thinning on seedling emergency，survival and establishment in a coastal pine forest. Forest Ecology and Management，2003，182(1)：339-354.

[28]MUSCOLO A，SIDARI M，MERCURIO R. Influence of gap size on organic matter decomposition，microbial biomass and nutrient cycle in Calabrian pine (Pinus laricio，Poiret) stands[J]．Forest Ecology and Management，2007，242(2)：412-418.

[29]THOMAS T Veblen. Tree regeneration response to gaps along a transandean gradient[J]．Ecology，1989，70(3)：541-543.

[30]TABARELLI M，MANTOVANI W，PERES C A. Effects of habitat fragmentation on plant guild structure in the montane Atlantic forest of southeastern Brazil[J]．Biological conservation，1999，91(2)：119-127.

[31]HOWE H F，SMALLWOOD J. Ecology of seed dispersal[J]．Annual review of ecology and systematics，1982，13(1) 201-228.

[32]DENSLOW J S，NEWELL E，ELLISON M. The effect of under-story palms and cyclanths on the growth and survival of Inga seedlings[J]．Biotropica，1991，23：225-234.

[33]昝启杰，李鸣光，王伯荪，等．黑石顶针阔叶混交林演替过程中群落结构动态[J]．应用生态学报，2000，11(1)：1-4.

[原载：中南林业科技大学学报，2017，37(03)]

光皮桦人工林生长规律与种质评价年龄的研究

潘启龙[1] 贾宏炎[1] 杨桂芳[1] 孙浩忠[2] 谌红辉[1] 劳庆祥[1]

([1] 中国林业科学研究院热带林业实验中心，广西凭祥 532600；
[2]广西天峨县国营林朵林场，广西天峨 547300)

摘 要 以桂西北地区光皮桦人工林为研究对象，以更全面反映林分整体生长规律的径阶标准木为材料，通过树干解析及采用 Richards 曲线方程进行建模，对 14 年生光皮桦人工林的胸径、树高和单株材积的生长过程进行了精度较高的拟合，并据此分析了光皮桦人工林胸径、树高及单株材积的生长节律。结果表明：光皮桦胸径、树高及材积与生长年龄的回归关系均达到极显著水平($P<0.001$)，拟合度分别为 86.1%、95.1%、84.1%，模拟方程可用于光皮桦人工林的生长分析和预测。光皮桦人工林的胸径、树高与材积的速生期分别出现在 1~6 年、1~7 年和 6~12 年，根据林分个体分化的初显与稳定两个特征，确定 7~8 年生、12 年生可分别作为光皮桦种质评价的初选年龄与决选年龄。

关键词 光皮桦；树干解析；生长规律；种质评价年龄

Study on Growth Rhythm and Germplasm Resources Evaluating Age of *Betula luminifera* Plantation

PAN Qilong[1], JIA Hongyan[1], YANG Guifang[1], SUN Haozhong[2], CHEN Honghui[1], LAO Qingxiang[1]

([1]*The Experimental Centre of Tropical Foresty*, *CAF*, *Pingxiang* 532600, *Guangxi*, *China*;
[2]*Guangxi Linduo Forest Farm*, *Tiane* 547300, *Guangxi*, *China*)

Abstract: The standard trees of different diameter classes of 14-year-old Betula luminifera plantation in the northwest of Guangxi were selected to stimulate the growth dynamics of *DBH*, tree height and individual volume by stem analysis and Richard's regression analysis. The results showed that there were significant differences in the relationships between *DBH*, tree height, individual volume and tree age ($P<0.001$) and the regression coefficients were 0.86, 0.95 and 0.84 respectively, indicating that these regression models could be used to predict the growth of Betula luminifera plantation. The fast-growing periods of *DBH*, tree height and individual volume were from first to sixth, first to seventh and sixth to twelfth year, respectively. However, the preliminary and final age of evaluating the growth of this species were from seventh to eighth year, and twelfth year according to the characteristics of individual differentiation.

Key words: *Betulal luminifera*; stem analysis; growth rhythm; germplasm resources evaluating age

随着我国对珍贵木材需求的日益增长，加之南方用材树种松、杉、桉比重过大带来的潜在生态风险与市场风险，推广种植珍贵乡土树种具有重要意义。光皮桦(*Betula luminifera* H. Wink)为桦木科(Betulaceae)桦木属(*Betula* Linn.)落叶乔木，是我国亚热带地区广泛分布的珍贵乡土速生用材树种，具有适应性强、生长快、材质优良、用途广泛、经济价值高的特点[1,2]，也是广西主要造林树种之一。已有学者在光皮桦的生物与生态学特性、育种与培育技术等方面开展过相关研究[3-8]。因在良种选育过程中首先要确定目标树种的种质评价年龄标准，而不同树种的种质评价年龄应根据其生长过程与节律来确定。目前尚没有相关文献对光皮桦合理的种质评价年龄进行研究报道。因光皮桦为速生用材树种，根据中华人民共和国国家标准《林木良种审定规范 GB/T14071—93》要求，其种质评价年龄需达到二分之一个轮伐期，按照《云南省森林资源规划设计调查操作细则》(试行)中的标准，光皮桦天然林成熟林组为21~30年，另根据董建文等人的研究，光皮桦人工林速生期比天然林早[9]。因此，本文以达到评价年龄的14年生光皮桦人工纯林为研究对象，通过对各径阶标准木的树干解析，探究光皮桦人工林生长过程与节律，为光皮桦的种质评价年龄与培育技术提供参考。

1 材料与方法

1.1 试验地概况

材料取自广西河池市天峨县林朵林场14年生人工林，107°14′E，24°58′N，海拔750m，低山，属亚热带季风气候区。土壤为砂页岩发育而成的黄壤，土层厚度100cm以上。前茬为杉木林，初植密度为1740株/hm²，经2次抚育间伐，现保留密度为667株/hm²。

1.2 材料收集与处理方法

1.2.1 径阶标准木材料的采集

在光皮桦人工林中选择生长状况具有代表性的林分，采用标准地作业法进行林分调查，标准地面积为600m²，进行每木检尺后按径阶统计株数。根据每木调查的数据，在标准地中选取径阶标准木，伐倒后按树干解析作业要求，以2m为区分段分别截取中间圆盘，另外截取根颈盘与胸径圆盘，采用中央断面区分求积法求算材积，标准木生长因子统计数据见表1，根据径阶标准木材料统计标准地径阶蓄积列于表2。

表1 14年生光皮桦人工林径阶标准木生长因子统计

项目	标准木树号						
	1	2	3	4	5	6	7
胸径(m)	11.3	14.5	16.3	17.5	19.6	21.6	25.5
树高(m)	17.3	21.4	20.6	23.8	21.5	22.3	22.9
材积(m³)	0.0757	0.1501	0.1939	0.2778	0.2780	0.3724	0.4975

表2 14年生光皮桦人工林标准地林分结构情况

项目	径阶(cm)						
	12	14	16	18	20	22	26
平均胸径(cm)	11.9	14.2	15.9	18.0	19.3	21.2	26.2
平均树高(m)	15.2	17.3	18.8	21.2	20.0	20.5	24.0
活枝下高(m)	10.3	11.9	11.6	11.8	11.7	10.0	10.9
株数(n)	7	8	10	3	6	4	2
径阶蓄积(m³)	0.609	1.106	1.860	0.796	1.711	1.396	1.219

1.2.2 材料统计分析方法

(1)生长模型选择与参数求算　采用相关专家认为通用性较好的Richards方程$y=M(1-e^{-kA})^B$为生长模型[10-13]，拟合胸径、树高、材积的生长过程并求算模型参数。式中y为因变量(胸径、树高、材积)，A为自变量(林龄)，M、k、B为模型参数。

(2)利用树干解析方法求算出光皮桦各径阶标准木生长过程，利用DPS软件[14]求出胸径、树高、材积的Richards生长模型的相关参数(见表3)，并绘制曲线图。根据生长模型曲线的变化规律分析光皮桦生长节律，从而确定光皮桦的种质评价年龄；根据径阶标准木主干材积比例与树干高度的关系确定合理的修枝高度。

2 结果与分析

2.1 生长模型的适用性分析

光皮桦胸径、树高、材积Richards生长模型拟合曲线分别见图1、图2、图3。从各图可知，拟合的Richards方程曲线能较好地反映光皮桦林分平均胸径、树高及材积的生长变化规律，从各生长指标的观察值与模型值的对比关系得到证实。用方差分析的方法对模拟效果进行检验表明，光皮桦胸径、树高及材积与生长年龄的回归关系均达到极显著水平($P<0.001$)，拟合度分别为86.1%、95.1%、84.1%(表4)，表明方程具有较高的可靠性。因此，利用Richards方程曲线方程对光皮桦林分平均胸径、树高和材积的生长规律进行拟合是可行的。

表 3　胸径、树高、材积的 Richards 方程及回归假设检验

回归方程	方差来源	平方和	自由度	均方	F 值	P 值
$D=20.2374\ (1-e^{-0.129078t})^{1.146}$	回归 SSR	1677.345	2	838.672	226.086	0.000
	剩余 SSE	270.796	73	3.710		
	总的 SST	1948.141	75	25.975		
	$R=0.9279$	$R^2=0.8610$				
$H=26.0627(1-e^{-0.153527t})^{1.2463}$	回归 SSR	3583.893	2	1791.946	851.745	0.000
	剩余 SSE	183.035	87	2.104		
	总的 SST	3766.928	89	42.325		
	$R=0.9754$	$R^2=0.9514$				
$V=0.399187(1-e^{-0.133252t})^{3.1452}$	回归 SSR	0.499	2	0.250	203.202	0.000
	剩余 SSE	0.095	77	0.001		
	总的 SST	0.594	79	0.008		
	$R=0.9169$	$R^2=0.8407$				

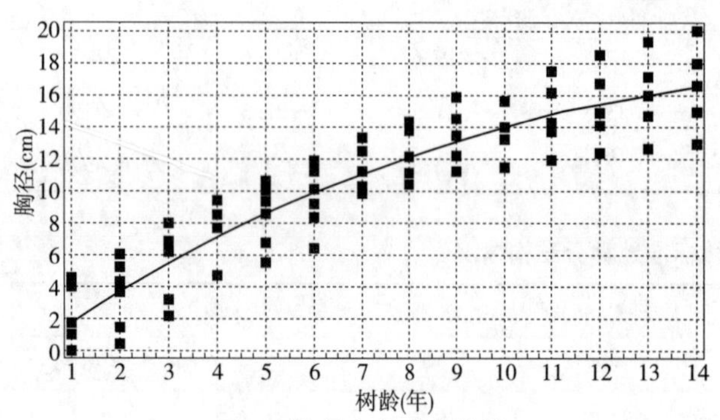

图 1　胸径生长动态曲线

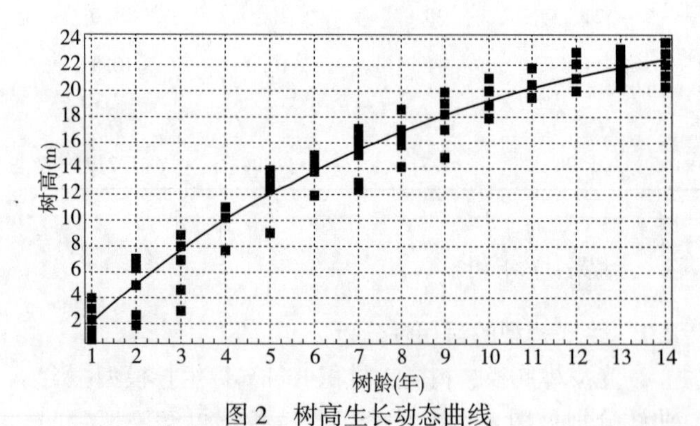

图 2　树高生长动态曲线

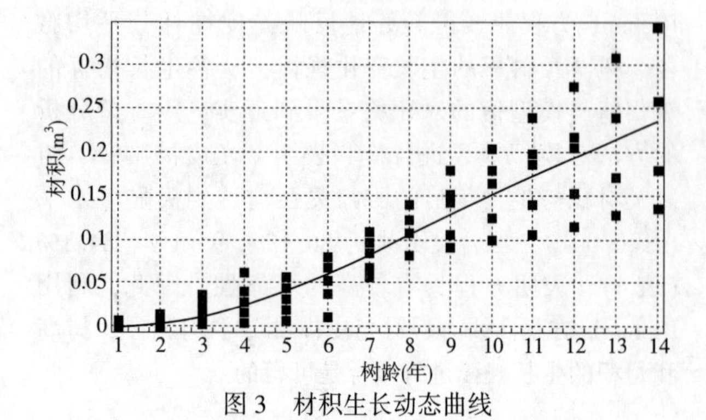

图 3　材积生长动态曲线

2.2　光皮桦人工林生长节律分析

以林龄为变量，对林分平均胸径、树高及材积生长的回归方程进行一阶和二阶求导，分别得出各因子的连年生长量曲线方程与连年生长量变化速度曲线方程，并作几何曲线图(图 4 至图 9)，所得曲线方程如下。

胸径：

$$D'=2.9936(1-e^{-0.129078t})^{0.146}e^{-0.129078t}$$

$$D''=0.05642(1-e^{-0.129078t})^{-0.854}e^{-0.258156t}-0.3864e^{-0.129078t}(1-e^{-0.129078t})^{0.146}$$

树高：

$$H'=4.98686(1-e^{-0.153527t})^{0.2463}e^{-0.153527t}$$

$$H''=0.18857*(1-e^{-0.153527t})^{-0.7537}e^{-0.307054t}-0.765616e^{-0.153527t}(1-e^{-0.153527t})^{0.2463}$$

材积：

$$V'=0.01673(1-e^{-0.133252t})^{2.1452}e^{-0.133252t}$$

$$V''=0.047823(1-e^{-0.133252t})^{1.1452}e^{-0.266504}-0.022293(1-e^{-0.133252t})^{2.1452}e^{-0.133252t}$$

由图4、图5分析可知，光皮桦胸径生长呈现前期生长较快、后期较慢的规律。从图 4 可见，光皮桦人工林的速生期出现较早，根据模型求算，当 $t=1.1$ 年时，胸径连年生长量达到极大值 1.9cm，此后逐年递减；从曲线变化趋势预测，14 年以后连年生长量保持在 0.5cm 以下，且降幅较小，进入平稳生长期。另外，由图 5 分析可知，胸径连年生长量变化速率表现出前期下降迅速，后期保持较小的变化速率，10 年生后下降幅度小于 0.1cm，从而与图 4 体现的变化规律一致。

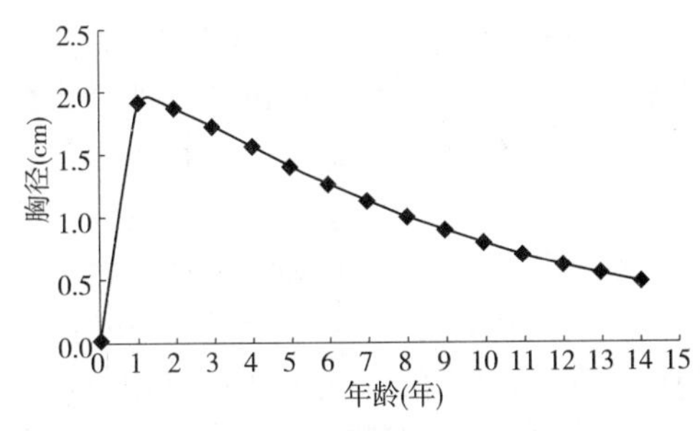

图 4　胸径连年生长量曲线

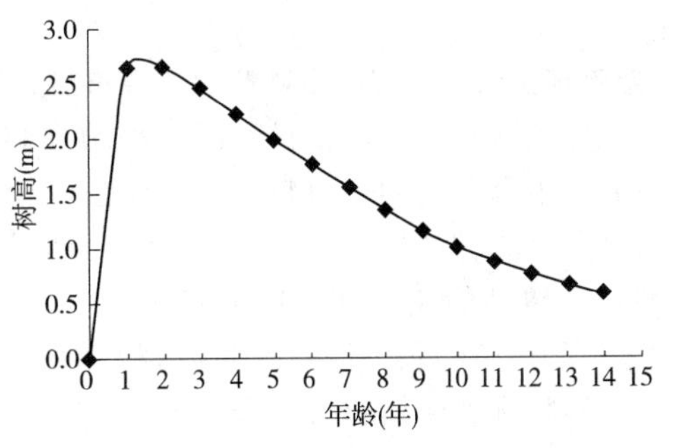

图 6　树高连年生长量曲线

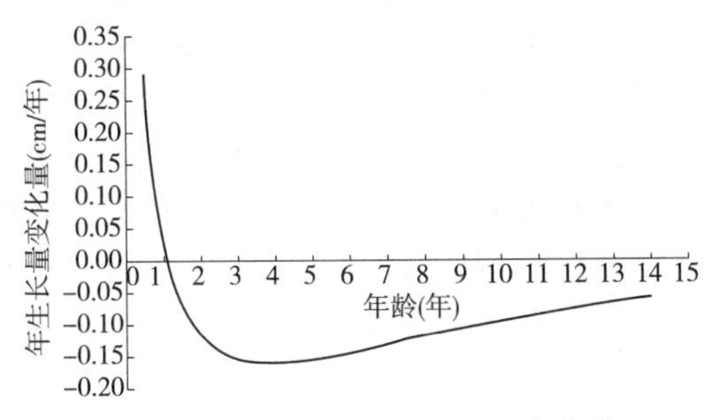

图 5　胸径连年生长量变化速率曲线

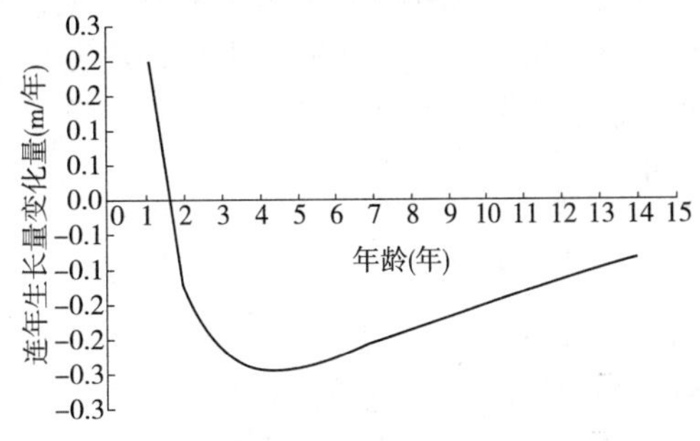

图 7　树高连年生长量变化速率曲线

依次以每连续 3 年的胸径连年生长量为样本组(即样本组 1 林龄为 1、2、3，样本组 2 为林龄为 2、3、4，…，下同)，依次对相邻前后 2 个样本组的均值差异性检验可见，从第 7 年起，胸径连年生长量与前 3 年差异显著($P=0.0062$)，由此说明第 7 年后的胸径生长量明显小于前 6 年，1~6 年为光皮桦胸径生长的速生期，从第 7 年生开始，光皮桦胸径生长进入缓增期。

根据图 6、图 7 分析可知，光皮桦树高连年生长量曲线及连年生长量变化曲线表现出的变化规律与胸径变化状况基本一致，反映了树高与胸径生长节律的同步性，可划分为快–慢两个生长时期。当 $t=1.4$ 年时，树高连年生长量达到极大值 2.7m，此后逐年减小。根据图 6 生长曲线变化趋预测，14 年后树高每年生长在 0.6m 以下，且变化幅度较小，从图 7 可知，12 年时树高连年生长量下降幅度已小于 0.1m。依次以每连续 3 年的树高连年生长量为样本组(与胸径分组方法相同)，对前后相邻样本组的均值差异性检验可知，从第 8 年生开始，树高连年生长量均值与前面 3 年差异显著($P=0.0044$)，由此说明 1~7 年为光皮桦树高生长的速生期，此后光皮桦树高生长进入缓增期。与天然光皮桦林不同[9,15]，光皮桦人工林高生长高峰期出现得早且持续时间较短，其生长节律特点类似速生桉人工林[16,17]，通过人工培育可促使光皮桦胸径、树高的速生期更早，连年生长量的峰值更大。

从图 8、图 9 分析可知，材积连年生长量曲线、连年生长量变化速率曲线的变化规律与胸径、树高有所不同，呈现为慢—快—慢的规律。8 年生前材积连年生长量随林龄增长逐年上升，当 $t=8.6$ 年时，材积连年生长量达到极大值 $0.0234m^3$，然后材积连年生长量开始缓慢下降。依次以每连续 3 年的材积连年生长量分组，材积样本组经均值差异性检验可知，从第 6 年起材积连年生长量均值与前面 3 年差异显著($P=0.0209$)，从第 13 年起，材积连年生长量均值与最大组(林龄为 8、9、10)均值存在显著差异($P=0.0312$)，说明第 6 年生开始材积生长进入速生期，延续生长到第 13 年生后，材积生长量开始明显下降。由此表明，1~5 年光皮桦人工林材积生长较慢，6~12 年为光皮桦材积速生期，第 13 年之后材积进入缓增期。

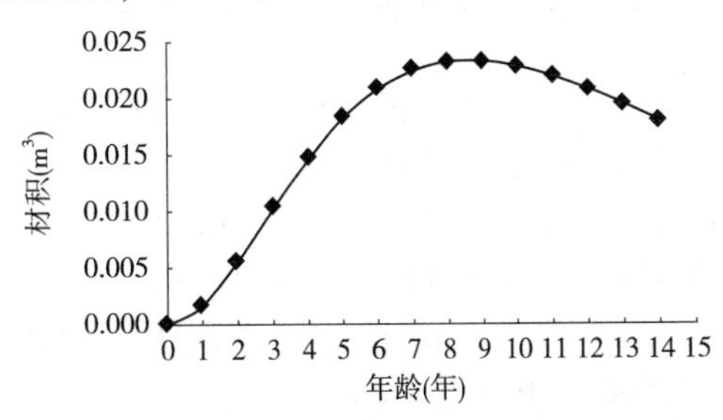

图 8　胸径连年生长量曲线

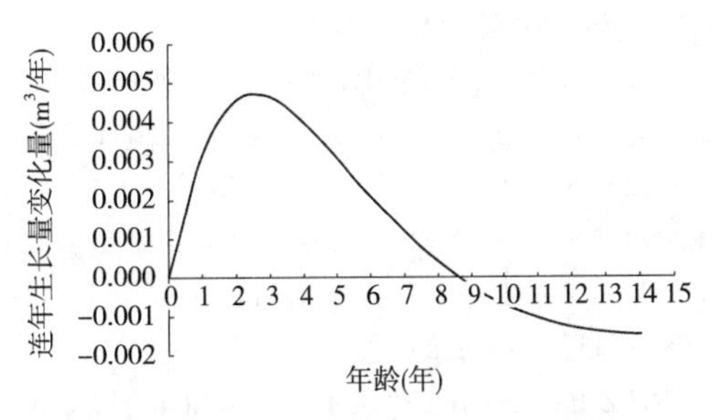

图 9　胸径连年生长量变化速率曲线

2.3 光皮桦生长节律与种质材料的评价年龄分析

从光皮桦胸径、树高及材积的生长节律特点可知，7~8年生可作为种质材料的初选年龄，12年生可作为决选年龄。由于7年生与8年生前为光皮桦人工林胸径和树高的速生期，7~8年生后林木胸径、树高生长高峰期已过，林木已表现出相对稳定的生长势和干形特征，个体差异初步显现，因此，在造林后的第7~8年，即可对光皮桦优树或其他种质评价林进行初选。6~12年为光皮桦材积速生期，第12年生后，材积生长高峰期已过，林木干形基本定型，分化程度已经很明显，因此12年生可作为种质材料的决选年龄。

3 结论与讨论

利用Richards曲线方程对光皮桦人工林的平均胸径、树高和材积的生长规律进行拟合，经检验，回归关系均达到极显著水平，可用于光皮桦人工林的生长分析和预测。

根据14年生光皮桦人工林胸径、树高及材积的生长节律分析可以确定其种质评价年龄，为良种审(认定)提供参考依据。7~8年生后林木胸径、树高生长高峰期已过，林木个体差异初步显现。已有相关研究表明，5~10年生胸径和树高性状的早期选择效率较高[18-21]。因此，7~8年生可作为光皮桦种质评价的初选年龄。6~12年为光皮桦材积速生期，12年生后材积生长高峰期已过，林木个体的分化已基本定型，因此，12年生可作为种质材料的决选年龄。

因为光皮桦人工林培育时间较短，对于光皮桦的生长规律以前相关学者主要利用天然林的生长数据进行分析，其研究结论与人工林有差异。主要表现为天然林状态下胸径、树高与材积生长高峰期出现比人工林要晚，持续时间长[9,22,23]。本研究利用14年生光皮桦人工林材料进行研究，掌握人工林生长规律，能对光皮桦的良种选育与培育技术提供理论指导。

参考文献

[1]陈存及，陈伙法，梁一池，等. 阔叶树种栽培[M]. 北京：中国林业出版社，2000.

[2]邹琴. 光皮桦研究进展综述[J]. 安徽农学通报，2011，17(12)：43-44.

[3]赵文君，丁访军，周凤娇，等. 贵州西部光皮桦天然次生林生物循环特征[J]. 中南林业科技大学学报，2016，36(9)：114-118.

[4]郑仁华，邹绍荣. 光皮桦优树子代性状遗传变异及选择[J]. 植物资源与环境学报，2004，13(2)：44-48.

[5]冯元璋. 光皮桦组培苗移植技术试验[J]. 防护林科技，2010，96(3)：9-11.

[6]周凤娇，丁访军，潘明亮，等. 贵州西部地区光皮桦的生长规律[J]. 贵州农业科学，2011，39(9)：170-173.

[7]刘光金，黄弼昌，谌红辉，等. 光皮桦优树选择技术[J]. 林业实用技术，2013，(1)：13-15.

[8]邵增明. 光皮桦良种选育报告[J]. 安徽林业科技，2014，40(1)：33-34.

[9]董建文，陈慈禄，陈东阳，等. 光皮桦栽培生物学特性研究[J]. 江西农业大学学报，2001，23(2)：220-223.

[10]KNOWE S A, FOSTER G S. Application of growth models for simulating genetic gain of loblolly pine[J]. For Sci, 1989, 35: 211- 228.

[11]SPRINZ P T , TABERT C B, Strub M R. Height _ age trends from Arkansas seed source study[J]. For Sci, 1989, 35: 677- 691.

[12]段爱国，张建国，童书振. 6种生长方程在杉木人工林林分直径结构上的应用[J]. 林业科学研究，2003，16(4)：423-429.

[13]周元满，谢正生，刘新田. Richards函数在桉树无性系林分生长预测上的应用研究[J]. 西南农业大学学报，2005，27(2)：240-243.

[14]唐启义. DPS数据处理系统[M]. 北京：科学出版社，2010.

[15]陈奕良，谢正成，张俊红，等. 天然光皮桦树干生长特性初步研究[J]. 浙江林业科技，2009，29(4)：73-77.

[16]周义罡，王瑞辉，谢裕宏，等. 架空线下速生超高树木树高生长规律研究[J]. 中南林业科技大学学报，2016，23(2)：75-78.

[17]陆玉云，宋永全. 滇南地区尾巨桉生长量预测[J]. 林业调查规划，2010，35(1)：18-22.

[18]GWAZE D P, BRIDGWATER P E. Determining the optimum selection age for diameter and height in loblolly pine[J]. For Forest Genetics, 2002, 9: 159-165.

[19]WENG Y H, TOSH K J, PARK Y S, et al. Age-related trends in genetic parameters for jack pine and theirimplications for early Selection[J]. For Silvae Genetica, 2007, 56: 242-251.

[20]赖猛. 落叶松无性系遗传评价与早期选择研究[D]. 北京：中国林业科学研究院，2014.

[21]丁振芳，王景章，方海峰，等. 日本落叶松家系早期选择技术[J]. 东北林业大学学报，1997，25(3)：65-67.

[22]梁跃龙，吴钦树，杨清培等．江西九连山自然保护区光皮桦生长规律研究[J]．江西林业科技，2010，(4)：4-6.
[23]周凤娇，丁访军，潘明亮，等．贵州西部地区光皮桦的生长规律[J]．贵州农业科学，2011 ，39(9) ：170-17.

[原载：中南林业科技大学学报，2017，37(04)]

留萌措施对米老排伐桩萌芽条生长与生物量的影响

庞圣江[1,2] 张 培[1] 刘福妹[1] 杨保国[1] 刘士玲[1] 赵 总[1]
([1]中国林业科学研究院热带林业实验中心，广西凭祥 532600；
[2]广西友谊关森林生态系统国家定位观测研究站，广西凭祥 532600)

摘 要 以广西大青山林区米老排人工林主伐后的伐桩萌芽条为研究对象，分析不同萌芽条保留数量对米老排伐桩萌芽条生长与生物量积累的影响。结果表明，不同留萌处理措施，随着米老排伐桩萌芽条保留数量的增加，萌芽条的高度与胸径生长呈减小趋势，保留1株萌芽条更具有生长优势；萌芽条器官生物量大小均表现为主干>枝条>叶；萌芽条的总生物量呈现先增加后减小的趋势，保留2株萌芽条的总生物量积累最大，但保留1株萌芽条个体生物量积累明显高于其他处理的单株生物量。建议在米老排伐桩萌芽更新过程中，萌芽条生长1~2年再进行植株的密度调控处理，如培养薪炭林为主，每个伐桩保留1~2株生长健壮萌芽条；若需培育木材为目的，保留1株生长健壮萌芽条为宜。

关键词 米老排；伐桩；萌芽；生长量；生物量；人工林

Effects of Reserved Sprout Measurement on Sprout Growth and Biomass Accumulation of *Mytilaria laosensis*

PANG Shengjiang[1,2], ZHANG Pei[1], LIU Fumei[1]
YANG Baoguo[1], LIU Shiling[1], ZHAO Zong[1]
([1]*Experimental Center of Tropical Forestry*, *CAF*, *Pingxiang*, 536000, *Guangxi*, *China*;
[2]*Guangxi Youyiguan Forest Ecosystem Research Station*, *Pingxiang* 532600, *Guangxi*, *China*)

Abstract: In this paper, effects of sprouting number of *Mytilaria laosensis* on the seedlings growth and biomass accumulation were investigated with 120 stumps in Daqingshan forest region. Different treatments of removing sprouts were applied on the stumps after clear cutting: no removal of sprouts serving as control, reserving 1 sprout, 2 sprouts or 3 sprouts per stump with all other sprouts removed. The results showed that four treatments measures profoundly influenced sprout growth; the mean height and diameter of sprouts were decreased with the increasing of sprout number per stump, the 1 sprout per stump reserved facilitated the seedling growth; the biomass of sprout various components under four treatments was ranked as showed: trunk>branch>leaf; With the reserved sprout numbers per stump increased, the total sprout biomass increased at first and then reduced, and the 2 sprouting seedlings reserved produced the most biomass, however, The accumulation of biomass of the 1 sprout per stump reserved was significantly higher than that of other treated individual biomass. Reservation of no more than 2 sprouts per stump would achieve the different purposes to the plantation cultivation.

Key words: *Mytilaria laosensis*; stump; sprout; growth; biomass; plantation

林木的伐桩萌芽更新，具有能够迅速恢复林地的覆盖、减少水土流失、缩短更新周期等生态功能[1,2]；同时，由于伐桩萌芽条早期的速生性，能够有效地降低人工造林的成本[3]，从而实现潜力巨大的生物质产量[4]。长期以来，大多数林木采伐后任其自然更新，致使伐桩萌芽条数量过多，萌芽植株生长不良，从而导致伐后林地生产力难以恢复。国内外林业研究者开展不同树种伐桩萌芽条数量对生长量与生物量的影响探讨，如Bullarda [5]等对2个柳树(*Salix* spp.)品种短期轮作矮林生物质产量潜力研究发现，人工除萌蘖处理有利于萌芽条生物量的积累。Lockhart[6]等研究间伐5年后樱皮栎(*Quercus pagoda*)伐桩萌芽更新发现，随着萌芽条的生长，其对光照、水肥等生存资源的竞争加剧，生长较弱的萌芽条出现枯萎等自稀疏现象。Johansson[7]研究发现欧洲垂枝桦(*Betula pendula*)位于阳坡的伐桩萌芽条数量较多，且生长状况明显优于阴坡。李荣[8]等研究指出辽东栎(*Quercus liaotungensis*)伐桩上1年生的萌芽条生长速度较快，而后高度生长呈逐渐降低的趋势；易青春[9]等研究认为栓皮栎(*Quercus variabilis*)伐桩萌芽条数量对生长量和生物量有着显著的影响；伐桩保留1个粗壮的萌芽条更有利植株生长，保留3个萌芽条生物量积累最多。截至目前，关于栎属(*Quercus* spp.)伐桩保留萌芽措施对生长与生物量影响的研究较多，但其他树种伐桩萌芽条抚育经营的研究报道较少。由于林木的伐桩萌发能力、萌芽条生长与生物量的积累因树种生物学特性而异，广泛开展不同林木的伐桩萌芽条保留数量对生长和生物量的影响研究，利用营林技术措施实现林地生产力的恢复，是林业发展亟待解决的一个关键问题[10,11]。

米老排(*Mytilaria laosensis*)属于金缕梅科(Hamamelidaceae)壳菜果属(*Mytilaria*)常绿阔叶乔木，在我国天然分布于广西、广东及云南等地。米老排生长迅速、干形通直、出材率高和伐桩萌芽更新能力强，是我国热带南亚热带地区人工造林中发展前景较好的优良乡土阔叶树种。以往相关米老排的研究中，汪炳根[12]等简要介绍了米老排的生物学特性；郭文福[13]等对米老排人工中幼龄林的生长表现进行初步探讨，并确定其材积生长的数量成熟龄；白灵海[14]等对米老排人工林产生的经济效益进行了分析；庞圣江[15]等对米老排伐桩保留高度与基径大小对萌芽条数量、胸径和高生长的影响进行了研究；然而，有关米老排伐桩萌芽条保留数量对其生长与生物量的影响尚未见报道，开展伐桩萌芽条的保留数量研究，并结合其生长表现与生物量积累进行分析，有助于提高米老排伐后的抚育经营水平，尽早实现米老排人工林伐后林地生产力恢复。

本研究以大青山林区米老排人工林主伐后伐桩萌芽条为研究对象，分析不同萌芽条保留数量对其生长与生物量积累的影响，确定合理的伐桩萌芽条保留数量，为不同经营管理目标的米老排人工林提供参考。

1 材料与方法

1.1 试验区概况

本研究试验地位于广西凭祥市大青山林区白云小试区，21°56′~22°18′N，106°38′~106°57′E。该试验地属于南亚热带季风气候，年均温21.5℃，年均降水量1420mm，雨季分布差别明显，主要集中在4~10月，相对湿度82%，表土腐殖质丰富，土层深厚肥沃，土壤类型为红壤土。1975年造林，密度为1400株/hm^2，郁闭度为0.8，平均树高25.2m，平均胸径24.8cm，蓄积量约445m^3/hm^2。

1.2 试验设计

2008年10月初，本试验结合大青山林区营林生产计划，在白云小试区米老排人工纯林主伐地对林木树高、胸径和冠幅等生长状况进行本底调查。同年，12月底皆伐后，在选取皆伐地中上部阳坡位置，采用PVC塑料管打桩设置边界，调查林地内每木伐桩的基径、伐桩高度，并进行标记与统一编号。

2011年3~5月，对林地内的米老排伐桩萌芽更新的生长情况进行调查，包含伐桩天然更新的萌芽条数量、高度和胸径(>1.3m)等生长指标。在皆伐林地内，米老排伐桩正常生长有萌芽条，每个伐桩萌芽条数量至少6~7株，多则十几株以上。根据米老排伐桩萌芽条的生长状况进行试验设计，设置4个处理水平，以及3个重复。处理N1：伐桩保留1株生长健壮萌芽条；处理N2：伐桩保留2株生长健壮萌芽条；处理N3：伐桩保留3株生长健壮萌芽条，对照处理(CK)：伐桩保留全部萌芽条。采用10个正常萌发植株的米老排伐桩为1个重复，即每个试验处理选取30个伐桩。同时，定期除去上述3个非对照组处理措施的伐桩新生萌芽条，确保其萌芽条保留数量不变。

1.3 试验调查与数据分析

2013年3~5月，调查伐桩保留萌芽条植株高度、胸径等生长测量指标，采用收获法对植株树干、枝条和叶片生物量进行鲜重称量，同时，将外业所取样品置于80℃烘干箱内烘干至恒重，并称取其重量，用于比较不同伐桩保留萌芽植株的生物量差别。

本文按照随机区组设计对米老排人工林伐桩萌芽条植株适应性进行调查，同时，利用方差分析和Duncan-LSD法对4个处理水平伐桩保留萌芽条高度、胸径和生物量的差异性进行检验。

2 结果与分析

2.1 伐桩萌芽条保留数量对生长量的影响

从图1可以看出，米老排伐桩萌芽条保留数量对高生长的影响有所差异，其中保留1株生长健壮萌芽条(N1)的高生长均值最大，而对照处理(CK)的高生长均值最小。伐桩萌芽条高生长大小表现为：保留1株生长健壮萌芽条(N1)>保留2株生长健壮萌芽条

(N2)>保留 3 株生长健壮萌芽条(N3)>保留全部萌芽条(CK)，试验结果表明伐桩保留 1 株生长健壮萌芽条对其高生长有明显的促进作用。经方差分析表明，4 个处理水平间伐桩保留萌芽条高生长差异显著(F=5.74，P=0.038)，其中 N1 萌芽条高生长显著高于其他 3 个处理水平，N3 与 N2、CK 差异不显著，而后两者间差异显著。

由图 1 可知，米老排伐桩萌芽条保留数量对胸径生长的影响与高生长反应情况基本一致，保留 1 株生长健壮萌芽条(N1)的胸径生长均值最大，对照处理(CK)的胸径生长均值最小。伐桩保留萌芽条的胸径生长，均值大小依次为：N1>N2>N3>CK，表明伐桩保留 1 株生长健壮萌芽条对胸径生长有明显的增大作用。方差分析显示，4 个处理水平间伐桩保留萌芽条胸径生长差异达到极显著水平(F=7.03，P=0.008)。

综上所述，不同留萌处理措施，米老排伐桩萌芽条生长差异显著(P<0.05)，随着米老排伐桩萌芽条保留数量的增加，萌芽条的高度与胸径生长呈减小趋势，保留 1 株萌芽条更具有生长优势。

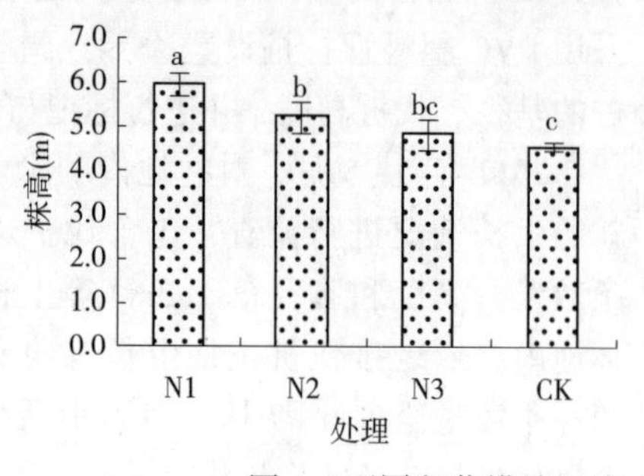

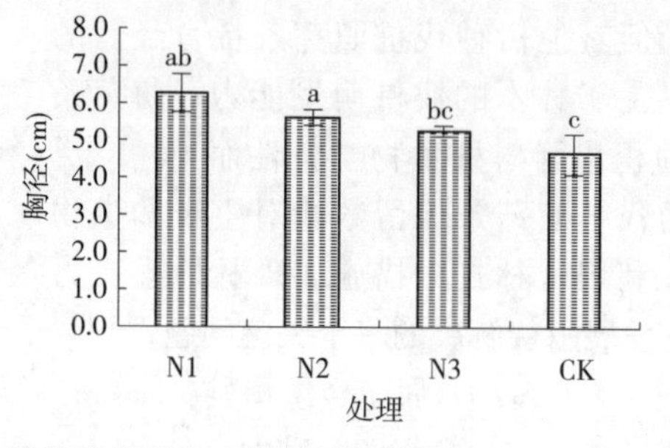

图 1 不同留萌措施对米老排伐桩萌芽条生长量的影响

2.2 伐桩萌芽条保留数量对生物量的影响

由表 1 可知，不同留萌处理措施，与对照处理相比较，伐桩保留 1 株、2 株和 3 株萌芽条总生物量分别增加 24.29%、36.44% 和 15.83%，其大小排列依次为 N2>N1>N3>CK，表明合理的留萌措施有利于米老排伐桩萌芽条生物量的积累；4 个处理水平间 N1 与 N2 主干生物量和叶生物量分配差异不显著，与 N3、CK 差异显著，而后两者间差异亦不显著；N1 与 N2 萌芽条侧枝生物量分配差异不显著，与 N3、CK 差异显著。

不同留萌处理措施，随着伐桩萌芽条保留数量的增加，萌芽条器官生物量大小均表现为主干>枝条>叶；萌芽条的总生物量呈现先增加后减小的趋势，保留 2 株萌芽条的总生物量积累最大，但保留 1 株萌芽条个体生物量积累明显高于其他处理的单株生物量。

表 1 不同留萌措施对米老排伐桩萌芽条生物量的影响

处理	叶生物量		枝条生物量		主干生物量		总生物量(kg)	
	M±SD(kg)	CV(%)	M±SD(kg)	CV(%)	M±SD(kg)	CV(%)	M±SD(kg)	CV(%)
N1	0.78±0.23a	29.49	1.63±0.25a	15.34	3.32±0.34a	10.24	5.73±0.69ab	12.04
N2	0.83±0.17a	20.48	2.01±0.34a	16.92	3.45±0.33a	9.57	6.29±0.54a	8.59
N3	0.69±0.21b	30.43	1.38±0.29b	21.01	3.27±0.41b	12.54	5.34±0.72b	13.48
CK	0.57±0.14b	24.56	1.26±0.32c	25.40	2.78±0.36b	12.95	4.61±0.67c	14.53

注：M±SD 为均值±标准差，CV 为变异系数，同列不同小字母表示差异显著(P<0.05)。

3 结论与讨论

研究表明，不同留萌处理措施下，米老排伐桩萌芽条生长差异显著(P<0.05)，伐桩萌芽条的高生长与胸径大小随着保留数量增加呈减小的趋势，保留 1 株萌芽条植株更具生长优势。这与栓皮栎[9]、白栎(*Quercus fabri*)[16]伐桩萌芽条保留数量的增加，其平均高度生长与胸径(基径)大小都呈减小趋势以及杉木(*Cunninghamia lanceolata*)[17]人工林伐桩除萌处理有利于促进杉木伐桩保留的萌芽条株高、基径生长是一致的，原因可能与伐桩萌芽更新过程中，如果不及时进行人工除萌处理，其可能持续好几代仍会萌发大量的新萌芽条而产生大量丛生植株，降低光照、水分和养分等有限资源的利用效率，致使伐桩萌芽条生长相对缓慢，甚至出现伐后林地生产力难以恢复。以往其他学者研究发现，伐桩休眠芽位于树皮内表层[15]，而大多数树种伐桩休眠芽的萌发能力很强，其能够由伐桩休眠芽萌发不定芽生长形

成大量丛生植株，且休眠芽持续萌发能力因其树种生物学特性、立地条件与采伐季节而异[18,19]，因此，需要对非对照组伐桩萌芽条进行定期除萌，其是否对保留萌芽条的生长与生物量的积累产生影响，有待进一步深入研究。

以往研究认为，10 年以内伐桩萌芽条植株尚处于旺盛生长期[20,21]。本研究发现，不同留萌处理措施，与对照组相比，更有利于处于旺盛生长期的萌芽条生物量的积累。本研究结果与栓皮栎[9]、麻栎(*Quercus acutissima*)[22]伐桩留萌处理对生物量的积累研究报道基本相似，他们认为伐桩萌芽条的高度和胸径生长与生物量的积累关系密切，经过留萌处理后，伐桩萌芽条生物量的积累都得到显著提高。究其原因，可能与林木伐桩保留少量萌芽条植株，既提高了伐桩的养分利用效率，同时，减少萌芽条对光照和生长空间的竞争而迅速生长，从而有利于伐桩萌芽条生物量的积累。

研究显示，随着米老排伐桩萌芽条保留数量的增加，萌芽条的总生物量均呈现出先增加后减小的趋势。以往其他树种研究中，栓皮栎伐桩萌芽条生物量的积累，亦出现伐桩萌芽条保留量的增加而呈现先增加后减小的规律[9]，与本研究的结论基本一致。这可能与伐桩保留较少萌芽条时，植株个体间对资源获取与生长空间的竞争不够激烈；随着萌芽条保留数量达到一定限度，大量的萌芽条分散了伐桩有限的资源，个体间对水肥激烈竞争，使得大部分萌芽条处于弱势而枯萎，不利于伐桩萌芽条生长及生物量的积累。

在米老排伐桩萌芽更新的前期研究中，庞圣江等分析了伐桩高度及基径对萌芽条的萌发数量、高度和胸径生长的影响[15]，试验区域与本研究试验地位于相同采伐林地。本研究对米老排伐桩萌芽条保留数量对其生长量与生物量的影响进行探讨，进一步充实了米老排人工林伐后萌芽更新的研究成果，同时，有助于提高米老排伐后的抚育经营水平，促进林地生产力恢复。综上所述，建议在米老排伐桩萌芽更新过程中，萌芽条生长 1~2 年进行植株的密度调控处理，如培养薪炭林为主，每个伐桩保留 1~2 株生长健壮萌芽条；若需培育木材为目的，每个伐桩保留 1 株生长健壮萌芽条为宜。

参考文献

[1]伊力塔，韩海荣．山西灵空山林区辽东栎萌芽更新规律研究[J]．林业资源管理，2007(4)：57-61.

[2]马闯，张文辉，薛瑶琴，等．邻体竞争和环境因子对栓皮栎伐桩萌苗表型特征的影响[J]．西北农林科技大学学报(自然科学版)，2011，39(10)：71-80.

[3]杨海林．杨树留桩萌蘖更新培育中小径材技术[J]．河北林业科技，2010(6)：88-88.

[4]丁伯让．栎树薪炭林的造林技术及开发利用[J]．安徽林业科技，2003(1)：15-17.

[5]BULLARD M J, Mustill S J, Mcmillan S D, et al. Yield improvements through modification of planting density and harvest frequency in short rotation coppice Salix, spp. -1. Yield response in two morphologically diverse varieties [J]. Biomass & Bioenergy, 2002, 22(1)：15-25.

[6]LOCKHART B R, CHAMBERS J L. Cherrybark oak stump sprout survival and development five years following plantation thinning in the lower Mississippi alluvial valley, USA[J]. New Forests, 2007, 33(2)：183-192.

[7]JOHANSSON T. Sprouting ability and biomass production of downy and silver birch stumps of different diameters [J]. Biomass and Bioenergy, 2008, 32(10)：944-951.

[8]李荣，张文辉，何景峰，等．辽东栎伐桩萌苗的发育规律[J]．林业科学，2012，48(3)：82-87.

[9]易青春，张文辉，唐德瑞．栓皮栎伐桩萌苗保留量对其生长和生物量积累的影响[J]．林业科学，2013，49(7)：34-39.

[10]叶镜中，孙多．森林经营学[M]．北京：中国林业出版社，1989：19-27.

[11]李景文，刘世英，王清海，等．三江平原低山丘陵区水曲柳无性更新研究[J]．植物研究，2000，20(2)：215-220.

[12]汪炳根，热带、南亚热带林木种质资源保存与评价[J]．广西林业科学，2001(30)：49-51.

[13]郭文福，蔡道雄，贾宏炎，等．米老排人工林生长规律的研究[J]．林业科学研究，2006，19(5)：585-589.

[14]白灵海，唐继新，明安刚，等．广西大青山米老排人工林经济效益分析[J]．林业科学研究，2011，24(6)：784-787.

[15]庞圣江，张培，刘福妹，等．大青山林区米老排人工林伐桩萌芽更新研究[J]．西北林学院学报，2016，31(6)：153-156.

[16]张文武，温佐吾，袁廷汉．除萌留壮和幼林密度调控对白栎次生林生长的影响[J]．山地农业生物学报，2009，28(1)：9-13.

[17]郑腾桂．不同除萌方式对杉木迹地萌芽更新生长量的影响[J]．安徽农学通报，2013(23)：68-69.

[18]叶镜中，姜志林．杉木休眠芽生物学特性的研究[J]．南京林业大学学报：自然科学版，1989(1)：50-53.

[19]孙黎黎，张文辉，何景峰，等．黄土高原丘陵沟壑区不同生境条件下柠条人工种群无性繁殖与更新研究

[J]. 西北林学院学报, 2010, 25(01): 1-6.
[20]叶镜中. 杉木萌芽更新[J]. 南京林业大学学报: 自然科学版, 2007, 31(2): 1-4.
[21]罗伟祥, 刘广全, 唐德瑞, 等. 渭北黄土高原杜仲人工林生长量生物量研究[J]. 西北林学院学报, 1994, (04): 22-26.
[22]FANG S, LIU Z, CAO Y, et al. Sprout development, biomass accumulation and fuelwood characteristics from coppiced plantations of*Quercus acutissima* [J]. Biomass & Bioenergy, 2011, 35(7): 3104-3114.

[原载: 西北林学院学报, 2017, 32(06)]

马尾松产脂量与树体因子关系研究

安　宁[1,2]　丁贵杰[1]　谌红辉[2]　农　友[2]　黄德卫[2]

([1]贵州大学造林生态研究所，贵州贵阳　550025；
[2]中国林业科学研究院热带林业实验中心，广西凭祥　532600)

摘　要　采用长期定位观测，定时、定株收获的方法，研究了马尾松胸径、树高、冠幅和年龄对产脂量的影响。研究结果表明：①产脂量随着径级的增大而增加，二者呈正相关关系，不同年龄同径级产脂量差异不显著。②产脂量有随着树高的增大而增加的规律。③产脂量与各树体因子相关性分析得出：胸径>冠幅>树高>枝下高，相关系数分别为0.749、0.686、0.545和-0.147。④产脂量随着年龄的增大而增加，且4种年龄的产脂量存在显著差异。

关键词　马尾松；产脂量；胸径；树高

Study on the Relation between Resin Yield and Tree Factors of *Pinus massoniana*

AN Ning[1,2], DING Guijie[1], CHEN Honghui[2], NONG You[2], Huang Dewei[2]

*([1]The Institute of Silviculture and Ecology in Guizhou University, Guiyang 550025, Guizhou, China
[2]The Experimental Centre of Tropical Forestry, CAF Pingxiang 532600, Guangxi, China)*

Abstract: Based on long-term positioning observation, timing and plant harvest, the influence of diameter class, tree height , crown width and age were studied on resin yield of *Pinus massoniana*. The results showed that: ①resin yield increased with increase of diameter class, resin yield of different age have no significant difference in similar diamete classr . ②resin yield increased with increase of the tree height . ③ the correlation analysis between resin yield and tree body factors was found: diameter class > crown width> tree height>helght under branch, with correlation coefficient of 0.749, 0.686, 0.545 and -0.147 respectively. ④ resin yield increased with the increase of age, and there were significant differences in the amount of resin yield in 4 ages.

Key words: *Pinus massoniana*; resin yield; diameter class; tree height

松脂是松树针叶经过光合作用生成的糖类，经过一系列复杂的生化反应，在薄壁细胞基金项目：热带马尾松高产脂及材脂兼用林培育技术研究(2015BAD09B0102-3)、八桂学者岗位项目“松树资源培育及产业化关键技术创新”、广西友谊关森林生态系统国家定位观测站基金资助中产生的代谢产物。主要由树脂酸和萜烃组成，此外还含有少量杂质和水分。松脂经过蒸馏去杂质可制成松香和松节油，二者均是重要的化学工业原料，松香广泛应用于造纸、食品、医疗和电子等领域，松节油可用于合成香料和溶剂等[1]。影响产脂量大小的因素有气象因子、营林措施、采脂技术以及遗传特性等[2-5]。

马尾松(*Pinus massoniana*)是我国南方主要的造林树种，具有分布广、面积大、速生丰产等特点，同时又是采割松脂的主要树种[6]。近年随着南方珍贵树种的发展及病虫害的频发，造成马尾松可采脂人工林面积的减少，松脂产量已不能满足林产化工业的需求，提高产脂林的经营技术水平及营建马尾松产脂原料林势在必行。以往对马尾松的研究都集中在栽培技术、生理生化以及遗传育种等方面[7-14]，而在马尾松人工林产脂方面所做的研究相对薄弱[15-18]，尤其在马尾松产脂量与测树因子之间的关系鲜有报道[19]，在生产中发现，马尾松不同单株产脂量存在显著差异，因此，有必要对产脂量的差异与树体因子之间是否存在相关性进行研究，研究结果将完善马尾松高产脂优树选择的技术指标，丰富马尾松产脂林培育技术体系。

1　材料与方法

1.1　试验地概况

试验林设在广西友谊关森林生态系统国家定位观测站区域内的伏波实验场，为人工营造的马尾松纯林，前茬杉木，主伐后明火炼山。地理位置106°43′E、22°06′N，海拔500~550m，低山地貌，年平均气温19.5℃，年降水量1500mm，属于南亚热带季风气候区。土壤为花岗岩发育成的红壤，土层厚度>1m，腐殖质厚度>10cm。

1.2　试验设计及方法

在18年林分中将密度调控为A450株/hm^2，B600株/hm^2，C750株/hm^2，D900株/hm^2，E1050株/hm^2，重复3次，共15个小区，每个小区面积600m^2。在14年、16年、18年、20年人工林选取立地条件相近的林分，密度调控为900株/hm^2，重复3次，共12个小区，每个小区面积600m^2。在密度试验林及不同年龄试验林选取14径级、16径级、18径级、20径级、22径级、24径级、26径级、28径级、30径级进行产脂量对比试验。收获时单株称重，采用"下降式"采脂法，具体方法见参考文献[18]。

1.3　数据处理

采用Excel 2007软件对数据进行统计分析并绘制图表，运用SPSS 16.0对数据进行方差分析[20]。

2　结果与分析

2.1　胸径对产脂量的影响

由表1可以看出，单株年产脂量在5种密度和4个年龄中都随着径级的增大而增加，同径级产脂量在不同密度和不同年龄中差异不显著。密度B(600株/hm^2)从14径级到30径级，产脂量依次为0.70kg、1.58kg、2.30kg、2.53kg、3.19kg、3.71kg、4.23kg、5.01kg和5.97kg，产脂量随着径级增大而增加，在其他密度中存在同样规律；14径级在5种密度年产脂量分别为0.79kg、0.70kg、0.72kg、0.78kg和0.73kg，之间差异不显著，产脂量在其他径级存在同样规律；不同年龄产脂量在同径级中差异不显著。呈现此规律是因为，在相同的采割负荷率下采脂，同径级割破的树脂道数量基本相同，因而产脂量差异不显著；而径级越大，割破的树脂道数量越多，因而产脂量越大。

表1　不同密度和不同年龄各径级产脂量

径级	密度					年龄			
	A(kg)	B(kg)	C(kg)	D(kg)	E(kg)	14年(kg)	16年(kg)	18年(kg)	20年(kg)
14	0.79±0.01	0.70±0.02	0.72±0.04	0.78±0.03	0.73±0.03	0.64±0.06	0.71±0.07	0.73±0.05	–
16	1.49±0.12	1.58±0.11	1.52±0.03	1.33±0.17	1.79±0.12	1.43±0.10	1.52±0.09	1.55±0.13	–
18	2.27±0.15	2.30±0.19	2.40±0.05	2.25±0.11	1.89±0.23	1.94±0.17	2.15±0.15	2.18±0.18	2.01±0.22
20	2.39±0.20	2.53±0.13	2.43±0.12	2.65±0.22	2.61±0.21	2.23±0.21	2.47±0.28	2.56±0.17	2.42±0.22
22	3.42±0.17	3.19±0.25	2.53±0.18	3.31±0.09	3.32±0.14	3.11±0.24	3.39±0.19	3.05±0.13	3.65±0.13
24	3.89±0.18	3.71±0.22	3.68±0.17	3.90±0.23	3.57±0.19	3.71±0.20	3.65±0.17	3.72±0.24	3.90±0.29
26	4.15±0.23	4.23±0.15	4.08±0.19	4.46±0.09	3.91±0.15	4.09±0.19	4.37±0.23	4.15±0.19	4.73±0.11
28	4.36±0.22	5.01±0.29	4.47±0.26	4.82±0.11	5.19±0.27	4.62±0.29	4.73±0.31	4.83±0.21	4.98±0.28
30	6.32±0.17	5.97±0.25	5.82±0.22	6.13±0.29	5.56±0.23	5.83±0.16	6.16±0.27	5.84±0.18	6.23±0.21

不同密度各径级产脂量进行方差分析得出(表2)：各径级产脂量存在显著差异(*Sig.* =0.000<0.05)，相关分析得出：18径级与20径级差异不显著(*Sig.* =0.821>0.05)，其他径级之间均存在显著差异；对不同年龄各径级产脂量方差分析得出(表2)：各径级产脂量存在显著差异(*Sig.* =0.000<<0.05)，18径级与20径级差异不显著(*Sig.* =0.207>0.05)，而其他径级之间均存在显著差异。

表 2 不同密度和不同年龄各径级产脂量方差分析及相关分析

方差分析			相关分析		
	密度(A-E)	年龄(14~20 年)		密度(A-E) 18~20 径级	年龄(14~20 年) 18~20 径级
F	253.58	482.31	*MD*	-0.3000	-0.3300
Sig.	0.000	0.000	*Sig.*	0.821	0.207

注：表中相关分析仅列出差异不显著的对比。

2.2 树高对产脂量的影响

在年龄相同的林分，取树高 12.0~16.9m 进行分析，12 ~ 16m 年产脂量依次为 2.49kg、2.71kg、2.63kg、3.78kg 和 3.79kg，整体规律为产脂量随着树高的增大而增加(图 1)。这是因为在树体因子中，胸径对产脂量的影响最大，在其他因素一致的情况下，树高越高，说明树木营养健康状况更好，生长越旺盛，而作为新陈代谢产物的松脂产量也越大。

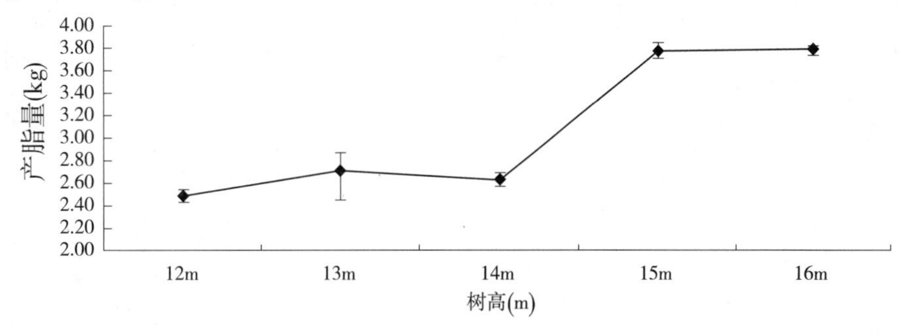

图 1 不同树高马尾松产脂量

注：12m(12.0~12.9m)，13m(13.0~13.9m)，14m(14.0~14.9m)，15m(15.0~15.9m)，16m(16.0~16.9m)。

2.3 冠幅对产脂量的影响

相关分析得出(表 3)，剔除胸径、树高和枝下高的影响，产脂量与冠幅相关系数为 0.308，*Sig.* =0.04<0.05，说明不同冠幅产脂量有差异，但未达到显著水平。冠幅大小能反映树木的营养状况，冠幅越大，说明个体竞争力更强，能获得更多的光照，进而新陈代谢产物的产脂量也越大。

表 3 产脂量与胸径、树高、冠幅和枝下高的相关分析

		产量	胸径	树高	冠幅	枝下高
产量		1.000	0.749	0.545	0.686	-0.147
	Sig.	—	0.000	0.000	0.000	0.168
	df	0	87	87	87	87
胸径		0.749	1.000	0.651	0.674	0.001
	Sig.	0.000	—	0.000	0.000	0.990
	df	87	0	87	87	87
树高		0.545	0.651	1.000	0.425	0.214
	Sig.	0.000	0.000	—	0.000	0.044
	df	87	87	0	87	87
冠幅		0.686	0.674	0.425	1.000	-0.272
	Sig.	0.000	0.000	0.000	—	0.010
	df	87	87	87	0	87
枝下高		-0.147	0.001	0.214	-0.272	1.000
	Sig.	0.168	0.990	0.044	0.010	—
	df	87	87	87	87	0

(续)

		产量	胸径	树高	冠幅	枝下高
胸径 & 树高 & 枝下高	产量	1.000			0.308	
	Sig.	—			0.04	
	df	0			84	
	冠幅	0.308			1.000	
	Sig.	0.04			—	
	df	84			0	

产脂量与胸径、树高、冠幅和枝下高的相关性由大到小为：胸径>冠幅>树高>枝下高，相关系数依次为：0.749、0.686、0.545 和-0.147。胸径在各树体因子中对产脂量的影响最大，胸径、树高和冠幅两两相关系数也都大于 0.400，说明各因子之间关系密切。而产脂量与枝下高呈负相关，相关系数为-0.147。说明枝下高越高，产脂量反而越低，这是因为在同样树高情况下，枝下高越高，冠高越小，能有效进行光合作用的树冠表面积就越小，因此产脂量也越低。

2.4 年龄对产脂量的影响

从图 2 和图 3 中可以看出，单位面积(hm^2)年产脂量和单株年产脂量都随着年龄的增大而增加，呈正相关关系。14 年每公顷年产脂量为 1173.84kg，16 年、18 年和 20 年相比 14 年分别增长了 47.03%、45.67%和 56.21%；从 14 年到 20 年，单株年产脂量依次为：1.67kg、2.87kg、2.84kg 和 3.05kg。这是因为在立地条件相一致的环境中，年龄越大，树木生长量越大，林分平均径级越大，在相同的采割负荷率下，割破的树脂道越多，因而产脂量也越大；同时年龄越大，林分平均树高越高，可接受更好的光照，光合作用越旺盛，树木的生长势越强，作为代谢产物的松脂产量也越多。

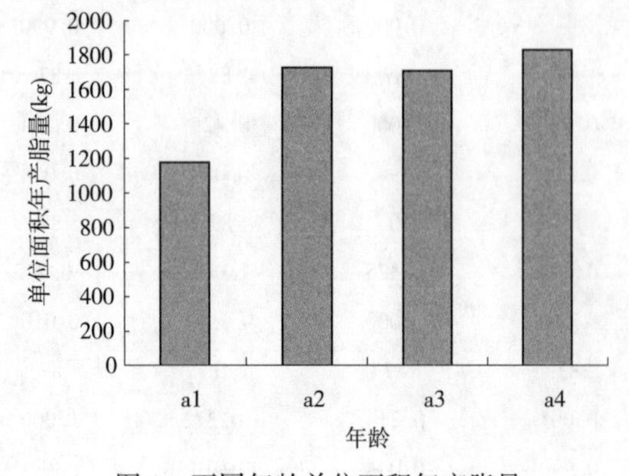

图 2 不同年龄单位面积年产脂量

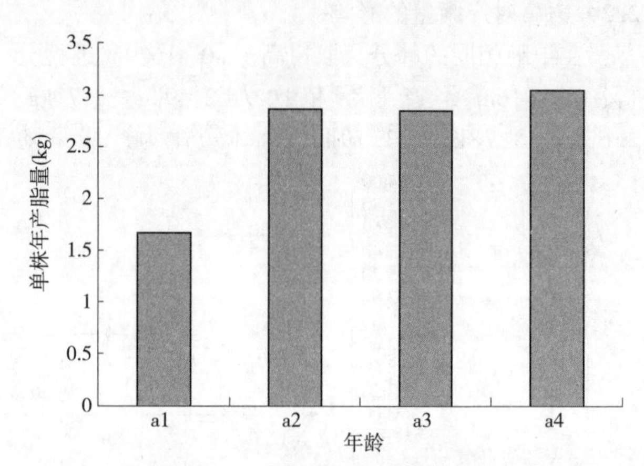

图 3 不同年龄单株年产脂量

从表 4 分析得出：不同年龄的产脂量存在显著差异(*Sig.* = 0.000<0.05)。多重比较结果可知，16 年与 18 年，16 年与 20 年产脂量差异不显著，其他年龄之间均产脂量均存在显著差异。16 年与 18 年在单位面积产脂量和单株产脂量上都无明显差异，原因在中等立地及相同的抚育措施下，二者生长量数据相近，因而产脂量也差异很小。

表 4 不同年龄多重比较检验表

林龄	14~16 年	14~18 年	14~20 年	16~18 年	16~20 年	18~20 年
MD	-183.98*	-178.63*	-219.99*	5.35	-36.01	-41.36*
Sig.	0.000	0.000	0.000	0.823	0.158	0.000

3 结论与讨论

松属群体家系间在胸径、树高、产脂力等方面都存在丰富的变异[21-23]，而产脂力与胸径、树高等生长性状间存在较高的正相关[24]。同径级的产脂量不受密度和年龄的影响，即不同密度和不同年龄的马尾松，同径级的产脂量差异很小。产脂量与径级呈正相关，随着径级的增大而增加，与前人研究结果相一致[25-28]。马尾松过小径级进行采割松脂会严重影响林木的生长，过大径级采割松脂又会造成资源的浪费。结合以往的研究确定马尾松采脂林达到

18径级就可以进行采割松脂，实现松脂效益的最大化。树高对产脂量的影响要低于胸径和冠幅[3]，对产脂量的影响是伴随着胸径和冠幅的影响，产脂量随着树高的增高而增加。试验中只考虑了树冠的大小，并未考虑树冠的均匀性对产脂量的影响，需要在以后的试验中进行验证。对4种树体因子进行相关系分析，相关性由大到小依次为：胸径>冠幅>树高>枝下高，相关系数依次为：0.749、0.686、0.545和-0.147，胸径对产脂量的影响最大，反映出在生产中产脂林合理保留密度的重要性，同时，胸径、冠幅、树高和枝下高均可作为马尾松高产脂优树选择的参考指标。

马尾松产脂量与采脂年龄呈正相关，随着年龄的增大而增加。年龄对产脂量的影响，是通过对胸径、树高、冠幅等方面的影响实现的。因为马尾松年龄越大，意味着生长量更大，树体各项指标越好，因而产脂量也越高[4]。在生产实际中，采脂年龄过小进行采割松脂对树木影响较大，往往不能获得较高的松脂效益，反而会造成林木生长的衰退，甚至死亡。综合考虑生产成本及经济效益，目前一致认为马尾松采脂林在18年开始采割松脂为宜[25]。

参考文献

[1]宋湛谦．二十一世纪世界松香松节油产业发展趋势和对策[J]．林产化工通迅，2000，34(1)：16-21.

[2]王长新．马尾松采脂量的相关因子分析[J]．河南科技大学学报(农学版)，2004，24(03)：22-25.

[3]蔡树威，龙伟，杨章旗．马尾松不同种源采脂量与树体因子关系的研究[J]．广西林业科学，2006，12(35)增刊：18-19.

[4]李爱民，全妙华，牛友芽，等．马尾松松脂产量影响因素研究进展[J]．湖南林业科技，2008，35(01)：58-59.

[5]魏永成，刘青华，周志春，等．不同产脂量马尾松无性系木质部树脂道结构差异[J]．林业科学，2016，52(07)：38-45.

[6]周政贤．中国马尾松[M]．北京：中国林业出版社，2001：14.

[7]谌红辉，丁贵杰．马尾松造林密度效应研究[J]．林业科学，2004，40(01)：92-98.

[8]丁贵杰．马尾松人工纸浆材林采伐年龄初步研究[J]．林业科学，2000，36(1)：15-20.

[9]周志春，李光荣，黄光霖，等．马尾松木材化学组分的遗传控制及其对木材育种的意义[J]．林业科学，2000，36(2)：110-115.

[10]姚瑞玲，丁贵杰．不同密度马尾松人工林凋落物及养分归还量的年变化特征[J]．南京林业大学，2006，30(05)：83-86.

[11]姜鹏，孟京辉，陆元昌，等．马尾松近自然改造初期的混交度与分布格局[J]．西北林学院学报，2014，29(05)：147-150+155.

[12]安宁，丁贵杰，谌红辉．林分密度及施肥对马尾松产脂量的影响[J]．中南林业科技大学学报，2010，30(09)：46-50.

[13]陆晓辉，丁贵杰，陆德辉．人工调控措施对马尾松凋落叶分解速率的影响[J]．西北林学院学报，2017，32(01)：25-29+53.

[14]丁波，丁贵杰，张耀荣．密度调控对马尾松人工林生态系统碳储量的影响[J]．西北林学院学报，2016，31(03)：197-203.

[15]杨章旗．马尾松不同年龄产脂量及松香组分分析[J]．林业科学，2014，50(06)：147-151.

[16]梁忠云，陈海燕，沈美英，等．广西优良品种马尾松松脂的化学组成[J]．林产化工通迅，2002，36(05)：8-10.

[17]安宁，丁贵杰．广西马尾松松脂的化学组成研究[J]．中南林业科技大学学报，2012，32(03)：59-62.

[18]祝志勇，于建云．生态因子对马尾松产脂的影响初探[J]．华东森林经，2002，16(03)：36-37.

[19]陈广财，佘济云，陆禹，等．高峰林场马尾松产脂量的影响因子分析[J]．中南林业科技大学学报，2016，36(06)：40-45.

[20]卢纹岱．SPSS for Windows 统计分析[M]．北京：电子工业出版社，2002.

[21]ROBERDS, J H, STROM B L, HAIN F P, et al. Estimates of genetic parameters for oleoresin and growth traits in juvenile loblolly pine[J]. Can J for Res, 2003, 33(12)：2469-2476.

[22] TADESSE W, NANOS N, AUNON F J, et al. Evaluation of high resin yielders of Pinus pinaster Ait. [J]. For Gen, 2001, 8(4)：271-278

[23]FRIES A, ERICSSON T, GREF R. High heritability of wood extractives in Pinus sylvestris progeny tests[J]. Can J For Res, 2000, 30(11)：1707-1713

[24]刘青华，周志春，范辉华，等．马尾松产脂力与生长性状的家系变异及优良家系早期选择[J]．林业科学研究，2013，26(6)：686-691

[25]舒文波，杨章旗．马尾松不同采脂年龄和径级产脂量变化特点研究[J]．中南林业科技大学学报，2011，31(11)：39-43.

[26]朱永安，刘海仓，刘菊花．湿地松产脂量与直径的相关关系研究[J]．湖南林业科技，2001，28(2)：41-43.

[27]翁海龙，贾红亮，陈宏伟，等．思茅松高产脂优树产脂量相关因子分析[J]．东北林业大学学报，2008，36(11)：69-70.

[28]谢善高，郑小贤，刘洪，等．马尾松产脂量密度效应的初步研究[J]．福建林业科技，2009，36(01)：31-34.

[原载：西北林学院学报，2018，33(03)]

马尾松人工林强度采伐后套种阔叶树种的生长动态

曾　冀[1,2]　雷渊才[2]　唐继新[1]　贾宏炎[1]　蔡道雄[1]

([1]中国林业科学研究院热带林业实验中心，广西凭祥　532600；
[2]中国林业科学研究院资源信息研究所，北京　100091)

摘　要　研究马尾松人工林强度采伐后套种的不同树种的生长动态规律，筛选适宜的套种树种，为马尾松人工林近自然经营提供技术支撑。2007 年 10 月于 14 年生马尾松人工林采用 4 个强度进行采伐，即保留密度分别为 225(Ⅰ)、300(Ⅱ)、375(Ⅲ)和 450(Ⅳ)株/hm^2，2008 年 2 月于林下均匀套种大叶栎、红椎、灰木莲、香梓楠、格木等 5 个乡土阔叶树种，2008 年底调查套种阔叶树的胸径、树高和冠幅等生长指标，此后至 2016 年每两年测定一次，应用方差分析和 Duncan 多重比较分析这些生长指标对不同强度采伐的动态响应。采伐强度显著影响林下套种阔叶树的生长，其中大叶栎和灰木莲的胸径、树高和冠幅以及红椎的胸径和冠幅的生长随保留密度增大而减小，而红椎、格木的树高生长受采伐强度影响不大；格木的胸径和冠幅以及香梓楠的树高和冠幅生长在套种第 7 年之前受采伐强度影响不大，此后其生长随保留密度增大而减小；香梓楠的胸径生长则一直随保留密度增大而增大。大叶栎的胸径、树高和冠幅以及灰木莲的树高和冠幅、红椎的树高生长高峰出现在套种后第 3 年；红椎、灰木莲、香梓楠的胸径生长高峰出现在第 5 年；格木的胸径、树高和冠幅以及香梓楠的树高和冠幅、红椎的冠幅生长高峰出现在套种后第 9 年。马尾松人工林强度采伐后套种阔叶树的生长动态表现为大叶栎>红椎>灰木莲>香梓楠>格木。大叶栎、红椎、灰木莲的生长随采伐强度的增大而增大，而香梓楠、格木受采伐强度的影响不显著。综合比较 5 种套种树种的生长特性，在桂西南开展马尾松中龄林近自然化改造，选用大叶栎、红椎、灰木莲进行林下套种，宜采用采伐强度Ⅰ、Ⅱ，而套种香梓楠、格木则宜采用采伐强度Ⅲ、Ⅳ。

关键词　马尾松；套种；阔叶树；生长动态

Growth Dynamics of Hardwood Species Inter-planted Under Intensively Thinned *Pinus massoniana* Plantations

ZENG Ji[1,2], LEI Yuancai[2], TANG Jixin[1], JIA HongYan[1], CAI Daoxiong[1]

([1]*Experimental Centre of Tropical Forestry*, *CAF*, *Pingxiang* 532600, *Guangxi*, *China*;
[2]*Research Institute of Forest Resources Information Techniques*, *CAF*, *Beijing* 100091, *China*)

Abstract: The Growth Dynamics of Different Tree Species Interplanted Under *Pinus massoniana* With Four Thinning Intensities Were Investigated In Order To Determine The Suitable Intercropping Species In Various Types of Stand Structure, Which Will Provide A Technical Support For The Close-To-Natural Management of *Pinus massoniana* Plantation. Fourteen-Year-Old *Pinus massoniana* Plantation Was Thinned In October 2007 And The Reserved Densities Were 225 (Ⅰ), 300 (Ⅱ), 375 (Ⅲ) And 450 (Ⅳ) Stems Per Hacter, Respectively. Five Indigenous Broadleaf Species Were Then Interplanted Meanly Under The *Pinus massoniana* Plantation In February 2008 And The Diameter At Breast Height (Dbh) Or Root Collar Diameter, Tree Height And Crown Width Were Investigated Every Two Years For Eight Consecutive Years Since 2008. The Analyses of Variance And Duncan Multiple' S Comparison Were Conducted To Examine The Growth Dynamic Response of Intercropped Tree Species To The Thinning Intensity. There Was Significant Effect of Thinning Intensity On The Growth of Understory Interplanting Broad-Leaved Tree Species. For Example, The Dbh, Tree Height And Crown Width of *Quercus griffithii* And *Manglietia glauca*, And The Dbh And Crown Width of *Castanopsis hystrix* Decreased Significantly With The Increase of Reserved Density, While The Tree Height of *Castanopsis hystrix* And *Erythrophleum fordii* Were Little

Influenced By The Thinning Intensity. The Dbh And Crown Width of *Erythrophleum fordii*, And Tree height And Crown Width of *Michelia hedyosperma* Were Also Declined With The Increase of Reserved Density Although The Thinning Intensity Had No Significant Influence On The Growth Initially. Contrarily, The Dbh of *Michelia hedyosperma* Increased With The Increasing Reserved Density. Additionally, The Dbh, Tree Height And Crown Width of *Quercus griffithii*, The Tree Height And Crown Width of *Manglietia glauca*, And Tree Height of *Castanopsis hystrix* Reached The Maximum At The Third Year After Interplanting, While The Dbh of *Castanopsis hystrix*, *Manglietia glauca* And *Michelia hedyosperma* Reached The Peak Value At The Fifth Year After Interplanting. The Dbh, Tree Height And Crown Width of *Erythrophleum fordii*, The Tree Height And Crown Width of *Michelia hedyosperma*, And The Crown Width of *Castanopsis hystrix* Peaked At Ninth Year After Interplanting. Among The Broad-Leaf Tree Species Interplanted Under The *Pinus Massoniana* Plantation After Thinning, The Best Growth Was *Quercus griffithii*, Followed By *Castanopsis hystrix*, *Manglietia glauca*, *Michelia hedyosperma* And *Erythrophleum fordii*. The Growth of *Quercus griffithii*, *Castanopsis hystrix* And *Manglietia glauca* Increased As The Thinning Intensity Increased, While The Growth of *Michelia hedyosperma* And *Erythrophleum fordii* Were Not Significantly Affected By The Thinning Intensity. The Result Suggested That *Quercus griffithii*, *Castanopsis hystrix* And *Manglietia glauca* Were Suitable Intercropping Species In The Thinning Intensity (Ⅰ, Ⅱ) And The *Michelia hedyosperma* And *Erythrophleum fordii* Were Suitable Intercropping Species In The Thinning Intensity (Ⅲ, Ⅳ) of Fifteen-Year-Old *Pinus massoniana* Plantation In Southwest Guangxi.

Key words: *Pinus massoniana*; interplantation; hardwood species; growth dynamics

马尾松（*Pinus massoniana* Lamb.）是我国南方主要造林树种，具有适应性强、速生丰产、用途广、经济价值高等特点，在林业建设中发挥着重要作用[1]。然而，由于大面积成片纯林经营，导致马尾松人工林病虫害、火灾频发，林分生产力下降，严重制约其可持续经营[2]。因此，选择适宜的混交树种进行阔叶化改造，形成针阔混交林，增加群落结构层次和生物多样性，对提高其林分稳定性、维持林地生产力具有重要意义[3-5]。

我国有关马尾松林下套种阔叶树种已有较多的研究报道，如，樊后保等[6,7]开展了25年生马尾松林下套种6个阔叶树种的试验研究，发现16年后6种马尾松-阔叶树混交林的生物量显著高于马尾松纯林，与套种阔叶树增加了林分光能利用率有关。丁敏等[8]比较了24年生的马尾松林下套种4个栲属树种16年生时的生长表现，认为马尾松林下套种的4个栲属树种的树高、胸径连年生长量高峰值出现年龄因树种而异，生长速度以格氏栲（*Castanopsis kawakamii*）最快，拉氏栲（*Castanopsis lamontii*）最慢。楚秀丽等[9]在40年生马尾松林下套种闽楠，第8年的观测结果表明，上层马尾松形成的弱光环境对闽楠生长有促进作用。王良衍等[10]应用木荷、小叶青冈和红楠对次生马尾松林进行了改造，10年后形成了较为稳定的针阔复层林结构，林下植被由阳性向阴性或中性转变。然而，对于马尾松林下套种阔叶树种缺乏较长期的动态研究，且尚未见有关马尾松系列强度采伐后套种阔叶树种的研究报道。

自2005年以来，中国林业科学研究院热带林业实验中心（以下简称“热林中心”）陆续开展了马尾松人工林近自然化改造的探索。借鉴全国马尾松林下套种阔叶树的经验，本研究于2008年初在14年生马尾松林系列强度采伐后，开展套种乡土阔叶树种试验，通过定期生长观测，揭示不同强度采伐条件下5种阔叶树种的生长动态，为完善马尾松人工林阔叶化改造与近自然经营提供理论依据。

1　试验地概况

试验地位于广西凭祥市热林中心伏波实验场（22°03″N，106°51″E），海拔为400m。该地属于南亚热带季风气候，年平均气温20.5~21.7℃，年降水量1200~1500mm，主要集中在4~9月，年日照时数1218~1620h，年蒸发量1261~1388mm，有霜期3~5d；土壤为砖红壤，由花岗岩发育而成的。

参试马尾松人工纯林营建于1993年2月，种源为宁明桐棉，采用一年生裸根苗造林，株行距为2m×2.5m，未施肥。1999年底和2003年底分别进行了透光伐（强度约20%）和抚育性采伐（强度约30%）。

2　研究方法

2.1　试验设计

2007年10月，选择立地条件和马尾松生长表现基本一致的地段，实施系列强度采伐后开展林下套种阔叶树试验。试验采用完全随机区组设计，设置80%（Ⅰ）、73%（Ⅱ）、66%（Ⅲ）和59%（Ⅳ）4个采伐强度，其保留木密度分别为225、300、375和450株/hm^2，4次重复，共16个小区，小区面积为1500m^2。2008年2月按照4m×5m株行距均匀随机套种大叶栎（*Castanopsis fissa* Rehd. et Wils.）、红椎

(*C. hystrix* Miq.)、格木(*Erythrophleum fordii* Oliv.)、灰木莲(*Manglietia glauca* Blume)和香梓楠(*Michelia hedyosperma* Law)等阔叶树种，穴规格为 30cm ×30cm × 40cm，采用1年生苗造林，由于马尾松保留密度不同，套种阔叶树的株数为 450~525 株/hm^2，5个树种在每个小区是随机排列，尽量保证每个小区内各树种数量基本一致。

2.2 生长观测

2008年12月，于每个小区内每个阔叶树种进行挂牌标记、每木检尺，应用 VERTEX 超声波测高器调查树高，分东、南、西、北4个方向测定冠幅。此后每隔2年于年底(2016年为8月)进行生长观测。

2.3 数据处理

采用单因素方差分析和 Duncan 多重比较检验系列强度采伐的马尾松林下套种阔叶树的胸径、树高和冠幅生长变化，应用 SPSS16.0 软件进行数据分析。

3 结果与分析

3.1 胸径生长动态

方差分析结果表明(表1)：在同一采伐强度处理下，大叶栎、格木、红椎、灰木莲、香梓楠5个树种间各年龄的胸径生长均差异显著($P<0.05$)。大叶栎胸径生长高峰出现在第3年，其连年生长量为 1.93cm，此后逐年减缓。格木胸径生长量则逐年增大，第9年的连年生长量为 1.51cm。红椎、灰木莲、香梓楠的胸径生长高峰出现在第5年，连年生长量分别为 1.48cm、1.65cm 和 0.97cm。马尾松林下套种的阔叶树胸径生长速度为大叶栎>灰木莲>红椎>香梓楠>格木。

4个采伐强度处理间比较，大叶栎和灰木莲在套种当年地径生长差异不显著；在套种后第3、5、7和9年，胸径生长差异显著，呈现采伐强度越大，胸径生长越快的规律。格木在套种当年处理Ⅲ的地径显著高于处理Ⅰ、Ⅱ；随着时间的推移，到第3、5、7年时4个采伐处理对格木胸径生长无显著影响；而到第9年时，采伐处理Ⅲ的格木胸径生长显著高于处理Ⅳ。红椎在套种当年处理Ⅰ、Ⅳ的地径显著高于处理Ⅱ；第3年时，处理Ⅰ的红椎胸径显著大于其余3个处理；而到第5、7、9年时，处理Ⅰ、Ⅱ、Ⅲ的红椎胸径较处理Ⅳ生长显著增大。香梓楠在套种当年和第3年，采伐强度Ⅳ的胸径(地径)生长显著高于其余3个处理；套种第5、7、9年时，采伐强度Ⅲ、Ⅳ的胸径生长显著高于处理Ⅰ、Ⅱ。

表1 马尾松林下套种阔叶树种的胸径生长表现

树种	采伐处理	套种阔叶树种的年龄(年)				
		1*	3	5	7	9
大叶栎	Ⅰ	1.06±0.16a	5.24±0.31a	10.04±0.58a	11.12±0.57ab	16.25±0.61a
	Ⅱ	0.94±0.07a	4.95±0.23ab	9.03±0.41ab	11.79±0.54a	15.27±0.58ab
	Ⅲ	1.06±0.11a	4.84±0.22ab	8.27±0.41bc	10.56±0.61ab	13.63±0.53ab
	Ⅳ	1.05±0.11a	4.52±0.32b	7.33±0.51c	10.15±0.62b	12.21±0.53b
格木	Ⅰ	0.61±0.05b	1.42±0.16a	1.97±0.09ab	2.6±0.37a	5.70±0.26ab
	Ⅱ	0.58±0.05b	1.50±0.12a	2.57±0.23a	2.93±0.46a	5.93±0.24ab
	Ⅲ	0.79±0.05a	1.68±0.12a	2.01±0.06ab	2.69±0.24a	6.33±0.38a
	Ⅳ	0.68±0.07ab	1.45±0.12a	1.54±0.24b	2.55±0.33a	4.85±0.15b
红椎	Ⅰ	0.53±0.05a	2.63±0.35a	5.64±0.37a	6.18±0.34ab	8.16±0.28ab
	Ⅱ	0.23±0.08b	2.20±0.18b	4.90±0.34b	5.86±0.81ab	7.83±0.48ab
	Ⅲ	0.38±0.04ab	1.86±0.21b	5.61±0.23a	6.94±0.48a	8.96±0.44a
	Ⅳ	0.46±0.04a	1.54±0.12c	3.92±0.42c	4.96±0.39b	6.67±0.35b
灰木莲	Ⅰ	0.90±0.06a	3.03±0.18a	6.61±0.50a	8.31±0.57a	10.51±0.46a
	Ⅱ	0.97±0.12a	2.98±0.30a	6.93±0.57a	8.10±0.42a	10.56±0.59a
	Ⅲ	0.88±0.06a	3.07±0.27a	6.68±0.46a	9.07±0.56a	10.13±0.52a
	Ⅳ	0.92±0.07a	1.95±0.25b	4.02±0.44b	5.49±0.48b	8.87±0.49b

（续）

树种	采伐处理	套种阔叶树种的年龄(年)				
		1＊	3	5	7	9
香梓楠	Ⅰ	0.82±0.14b	1.43±0.32b	3.25±0.50b	4.65±0.35ab	6.00±0.21b
	Ⅱ	0.78±0.11b	1.53±0.29b	3.40±0.10b	4.30±0.20b	5.75±0.25b
	Ⅲ	0.75±0.08b	1.45±0.22b	3.55±0.45ab	5.36±0.31a	7.47±0.27a
	Ⅳ	1.03±0.08a	1.82±0.17a	3.80±0.34a	5.22±0.40a	7.17±0.34a

注：＊此列为地径；表中字母为Duncan多重比较结果，处理间具相同字母表示差异不显著（$P \geq 0.05$），字母不同表示差异显著（$P < 0.05$）。4个采伐强度处理为：Ⅰ、采伐强度80%，马尾松保留密度225株/hm²；Ⅱ、73%，300株/hm²；Ⅲ、66%，375株/hm²；Ⅳ、59%，450株/hm²。下同。

同一年龄比较，采伐强度处理显著影响各树种的胸径生长。大叶栎在套种后第3、5、9年，采伐处理Ⅰ显著优于Ⅳ；第7年，采伐处理Ⅱ显著优于Ⅳ。格木套种后第5年，处理Ⅱ显著优于处理Ⅳ；第9年，处理Ⅲ显著优于处理Ⅳ。红椎在套种后第3年，处理Ⅰ显著优于其他3个处理；第5年，处理Ⅰ、Ⅲ显著优于处理Ⅱ、Ⅳ；第7、9年，处理Ⅲ显著优于其他3个处理。灰木莲套种后第3、5、7、9年，处理Ⅳ显著低于其他3个处理。香梓楠在套种后第3年，处理Ⅳ显著优于其他3个处理；第5年，处理Ⅳ显著优于处理Ⅰ、Ⅱ；第7年，处理Ⅲ、Ⅳ显著优于处理Ⅱ；第9年，处理Ⅲ、Ⅳ显著优于处理Ⅰ、Ⅱ。

3.2 树高生长动态

由表2可知，大叶栎、红椎、灰木莲前期树高生长较快，第3年即达生长高峰，其连年生长量分别为2.17m、1.36m和1.12m，此后逐年减缓；格木、香梓楠第7年之后树高生长较快，第9年达生长最大值，其连年生长量分别为1.50m和0.98m。5个阔叶树种的树高生长速度表现为大叶栎>红椎>灰木莲>香梓楠>格木。

表2 马尾松林下套种阔叶树的树高生长表现

树种	采伐处理	套种阔叶树种的年龄(年)				
		1	3	5	7	9
大叶栎	Ⅰ	1.25±0.07a	6.00±0.39a	9.13±0.54a	10.13±0.66a	13.82±0.54a
	Ⅱ	1.16±0.07a	5.33±0.19b	8.82±0.25a	11.15±0.29a	13.23±0.41b
	Ⅲ	1.23±0.10a	5.49±0.37b	8.23±0.45a	10.75±0.61a	12.98±0.64b
	Ⅳ	1.09±0.08a	5.27±0.28b	6.91±0.54b	9.13±0.40b	13.13±0.66b
格木	Ⅰ	0.52±0.05ab	2.02±0.22a	2.07±0.12ab	3.48±0.47ab	6.22±0.42ab
	Ⅱ	0.48±0.03b	1.88±0.15a	2.42±0.22a	3.70±0.47a	6.80±0.32a
	Ⅲ	0.62±0.03a	2.03±0.14a	2.63±0.09a	3.75±0.31a	6.93±0.35a
	Ⅳ	0.52±0.04ab	1.37±0.18b	1.51±0.26b	2.50±0.36b	5.45±0.35b
红椎	Ⅰ	0.65±0.06a	3.79±0.31a	6.55±0.52a	7.82±0.29a	8.56±0.27ab
	Ⅱ	0.43±0.02b	3.57±0.25ab	5.00±0.41b	6.98±0.30ab	8.53±0.41ab
	Ⅲ	0.60±0.06ab	2.94±0.24b	5.48±0.33ab	7.31±0.55ab	9.12±0.32a
	Ⅳ	0.56±0.03ab	2.83±0.15b	4.33±0.37c	5.84±0.36b	7.37±0.38b
灰木莲	Ⅰ	0.85±0.06ab	3.48±0.16a	6.34±0.47a	7.83±0.37a	8.57±0.30a
	Ⅱ	0.93±0.10a	3.24±0.45a	5.49±0.35b	7.04±0.42a	8.88±0.46a
	Ⅲ	0.95±0.07a	3.28±0.42a	5.94±0.49ab	7.56±0.58a	8.96±0.64a
	Ⅳ	0.78±0.06b	2.54±0.22b	3.41±0.41c	4.88±0.44b	7.43±0.50b
香梓楠	Ⅰ	0.65±0.07ab	2.00±0.13a	3.25±0.45a	4.90±0.28a	7.55±0.25a
	Ⅱ	0.50±0.07b	2.05±0.26a	3.32±0.20a	4.80±0.20a	7.20±0.30a
	Ⅲ	0.69±0.05ab	2.11±0.18a	3.35±0.25a	4.72±0.21a	6.13±0.26ab
	Ⅳ	0.80±0.05a	2.14±0.15a	2.99±0.17b	4.27±0.17b	5.68±0.33b

比较4个采伐强度处理的林分，其套种的5个阔叶树的树高生长亦存在显著差异（表2）。大叶栎在套种当年时各处理树高生长差异不显著；第3和9年时，采伐处理Ⅰ的树高显著大于其余3个处理；第5、7年时，采伐处理Ⅰ、Ⅱ、Ⅲ的树高显著高于处理Ⅳ。格木在套种当年时处理Ⅲ的树高显著大于处理Ⅱ；随着时间的推移，到第3、5、7和9年时，采伐处理Ⅰ、Ⅱ、Ⅲ的树高显著大于处理Ⅳ。红椎在套种当年时处理Ⅰ的树高显著大于处理Ⅱ；第3年时，处理Ⅰ下套种红椎的树高显著大于处理Ⅲ、Ⅳ；而第5、7、9年时，处理Ⅰ、Ⅱ、Ⅲ下套种红椎的树高较处理Ⅳ生长显著增大。灰木莲树高生长动态与大叶栎相似，在套种后9年间，采伐处理Ⅰ、Ⅱ、Ⅲ的树高显著大于处理Ⅳ。香梓楠在套种当年时，采伐强度Ⅳ的树高生长显著高于其余3个处理；第3年时，各处理间没有显著差异；第5、7、9年时，采伐强度Ⅰ、Ⅱ、Ⅲ的树高生长显著大于处理Ⅳ。

3.3 冠幅生长动态

同一采伐强度下，林下套种阔叶树种间冠幅生长的比较，其表现为大叶栎>红椎>灰木莲>香梓楠>格木(表3)。大叶栎和灰木莲的冠幅生长高峰都在第3年，其生长量分别为1.14m和0.86m；而格木、红椎、香梓楠的冠幅生长高峰出现在第9年，其生长量分别为1.31m、1.04m、0.73m。

马尾松采伐后林下套种的阔叶树，由于树种的特性差异，冠幅生长受不同采伐保留密度影响显著(表3)。大叶栎属于速生宽冠型先锋树种，对光照要求高，套种当年以及第3和9年，采伐处理Ⅰ、Ⅱ、Ⅲ的大叶栎冠幅显著大于处理Ⅳ，第5、7年，处理Ⅰ、Ⅱ、Ⅲ均大于处理Ⅳ。格木属于地带性顶极树种，前期适度遮阴环境更利于其生长；套种当年和第3年，格木冠幅在4个采伐处理下差异不显著，而第5、7年，采伐处理Ⅰ、Ⅱ格木的冠幅显著大于处

表3 马尾松林下套种阔叶树的冠幅生长表现

树种	采伐处理	套种阔叶树种的年龄(年)				
		1	3	5	7	9
大叶栎	Ⅰ	0.87±0.19a	3.36±0.40a	4.05±0.37a	5.23±0.50b	7.71±0.62a
	Ⅱ	0.70±0.11ab	2.97±0.11ab	4.53±0.18a	6.38±0.26a	7.46±0.21ab
	Ⅲ	0.86±0.17a	3.08±0.13ab	4.12±0.19a	5.05±0.23b	7.27±0.39ab
	Ⅳ	0.61±0.13b	2.76±0.13b	3.43±0.29b	4.48±0.38c	6.79±0.49b
格木	Ⅰ	0.38±0.07a	1.11±0.12a	1.56±0.08a	2.19±0.22a	5.13±0.24a
	Ⅱ	0.23±0.03b	1.12±0.06a	1.64±0.44a	2.35±0.41a	4.83±0.17ab
	Ⅲ	0.33±0.05a	1.01±0.11a	1.40±0.06ab	1.86±0.12ab	4.78±0.22ab
	Ⅳ	0.28±0.07ab	1.05±0.22a	1.25±0.15b	1.68±0.21b	3.80±0.20b
红椎	Ⅰ	0.24±0.03a	2.04±0.17a	3.27±0.22a	4.02±0.19ab	6.04±0.14ab
	Ⅱ	0.21±0.04a	1.88±0.12ab	3.46±0.23a	4.18±0.24ab	6.41±0.14a
	Ⅲ	0.24±0.03a	1.46±0.10b	3.52±0.24a	4.34±0.34a	6.67±0.26a
	Ⅳ	0.20±0.03a	1.89±0.12ab	2.63±0.28b	3.45±0.29b	5.17±0.23b
灰木莲	Ⅰ	0.18±0.02b	2.32±0.12a	3.22±0.30ab	4.49±0.23ab	5.30±0.16a
	Ⅱ	0.35±0.14a	2.15±0.19ab	3.10±0.19ab	3.90±0.27b	5.22±0.23a
	Ⅲ	0.21±0.02ab	2.04±0.15ab	3.53±0.18a	4.86±0.19a	5.55±0.20a
	Ⅳ	0.31±0.05ab	1.4±0.16b	2.60±0.26b	3.09±0.28c	4.91±0.31b
香梓楠	Ⅰ	0.18±0.04b	1.43±0.17ab	1.94±0.06a	3.27±0.18a	5.17±0.17a
	Ⅱ	0.11±0.02c	1.55±0.23ab	2.01±0.01a	2.99±0.01a	4.10±0.06b
	Ⅲ	0.27±0.05a	1.00±0.12b	2.04±0.09a	2.97±0.19a	4.72±0.16a
	Ⅳ	0.33±0.05a	1.74±0.13a	2.20±0.13a	2.96±0.14a	4.00±0.22b

理Ⅳ，第9年，采伐处理Ⅰ、Ⅱ、Ⅲ格木的冠幅显著大于处理Ⅳ。红椎是速生宽冠型顶极树种，前期生长需要足够的光照，套种当年红椎冠幅在4个采伐处理下无明显差异，随着时间的推移，其冠幅随着采伐强度的加大而增大。灰木莲属于速生宽冠形树种，套种当年时其冠幅随采伐强度的加大而减小；第3年以后，采伐处理Ⅰ、Ⅱ、Ⅲ灰木莲的冠幅显著大于处理Ⅳ。香梓楠属于慢生窄冠型树种，套种当年和第3年，采伐处理Ⅳ下香梓楠的冠幅大于其余3个处理，第5、7年时，4个处理对香梓楠的冠幅无显著影响，第9年时，采伐处理Ⅰ下香梓楠的冠幅显著大于处理Ⅳ。

4　结论与讨论

马尾松人工林通过林下套种阔叶树种实施阔叶化改造，形成针阔混交林，树种选择是其关键所在[11-13]。本研究中，从马尾松中龄林下套种的5个阔叶树种的生长表现可以看出，先锋种大叶栎生长最快，地带性顶极种格木生长最慢。大叶栎、灰木莲、红椎的胸径、树高和冠幅生长随着采伐强度的增大而增大，而格木、香梓楠的生长则与采伐强度相关性不大。

欧建德等[14,15]对马尾松林下套种南方红豆杉的研究表明，林下套种有利于红豆杉形成圆锥形树冠和通直的树干；马尾松保持合理的林分密度，形成适宜红豆杉的弱光环境，是林下套种成功的关键。周志春等[16]对马尾松次生林经40%~50%强度的择伐利用6年后，林分密度达250株/hm^2，林下套种的地带性常绿阔叶树种快速生长。

大叶栎为阳性树种，在光照条件充足的情况下生长快；其胸径、树高、冠幅的年均生长量高峰出现在套种后第1~3年。红椎和灰木莲属于中度耐阴树种，其胸径、树高、冠幅的年均生长量高峰出现在套种后第3~5年。格木、香梓楠为耐阴树种，而且生长相对缓慢[17]；格木的胸径、树高、冠幅年均生长量以及香梓楠树高、冠幅年均生长量在观测的第7~9年为最大，尚不知其是否达到生长高峰；香梓楠的胸径年均生长量则相对复杂，采伐强度处理Ⅰ和Ⅱ在套种后第3~5年达到生长高峰，而处理Ⅲ和Ⅳ在第3~5和7~9年均生长量约1.0cm/年，亦无法判断其生长高峰是否出现。这些树种间的生长动态差异与其耐阴性以及强度采伐后上层马尾松的树冠生长动态密切相关。马尾松树冠在采伐后第1~3年即呈现快速生长，林冠层逐渐得到恢复[18,19]，林下光照减弱，从而导致阳性树种大叶栎在套种后第1~3年生长最快，中度耐阴树种红椎和灰木莲在套种后第3~5年生长最快，而格木和香梓楠则整体上在7~9年生长快。这与5树种生长随采伐强度的变化规律是一致的，即随着马尾松采伐强度的增大，大叶栎、红椎和灰木莲各年龄段的生长均加快，而格木、香梓楠则减慢。

套种树种的生长差异除了树种本身的遗传品质外，林下光环境也对树木的生长具有重要影响[20]。王希华等通过研究认为，选择适生的接近演替顶级的优势树种对马尾松人工纯林进行改造，能够加速植被的演替进程，实现常绿阔叶林的快速恢复[21]。上层马尾松和下层阔叶树都需要适时间伐，让下层阔叶树得到充足的光照，下层阔叶树保留长势好的树木，使林分结构趋于合理[22,23]。本研究经过强度采伐马尾松纯林，套种阔叶树后采用近自然经营，是构造马尾松复层林的一种尝试。从胸径、树高和冠幅生长表现看，其表现为大叶栎>红椎>灰木莲>香梓楠>格木。结合其生长动态分析可知，大叶栎、红椎、灰木莲在采伐处理Ⅰ(225株/hm^2)下生长较好，香梓楠则在采伐处理Ⅳ(450株/hm^2)下生长较好，采伐强度对格木前期生长影响不显著，格木生长后期需光量逐渐增强，则在采伐处理Ⅰ下生长较好。桂西南15年生的马尾松人工林在采伐强度(Ⅰ、Ⅱ)中选择大叶栎、红椎、灰木莲套种，而在采伐强度(Ⅲ、Ⅳ)中宜选择香梓楠、格木作为套种树种。本研究的观测时间有限，套种阔叶树大多还未进入林冠层，固定样地观测将继续每两年一次，随着阔叶树进入马尾松林冠层后会产生种间竞争，种间关系和林内小环境的变化将是下一步研究的重点。

参考文献

[1] 丁贵杰，王鹏程．马尾松人工林生物量及生产力变化规律研究Ⅱ．不同林龄生物量及生产力[J]．林业科学研究，2002，15(1)：54-60.

[2] 盛炜彤．国外工业人工林培育的目标及技术途径[J]．世界林业研究，1992，5(4)：75-83.

[3] 盛炜彤．我国人工用材林发展中的生态问题及治理对策[J]．世界林业研究，1995，8(2)：51-55.

[4] WORRELL R，HAMPSON A. The influence of some forest operations on the sustainable management of forest soils--a review[J]．Forestry，1997，70(1)：61-85.

[5] PRETZSCH P H，BIBER P. Tree species mixing can increase maximum stand density[J]．Canadian Journal of Forest Research，2016，3(1)：1179-1193.

[6] 樊后保，李燕燕，苏兵强，等．马尾松-阔叶树混交异龄林生物量与生产力分配格局[J]．生态学报，2006，

26(8): 2463-2473.

[7] 樊后保, 刘文飞, 李燕燕, 等. 应用层次分析法评价闽西北山地马尾松-阔叶树混交林的综合效益[J]. 山地学报, 2009, 27(3): 257-264.

[8] 丁敏, 倪荣新, 毛轩平. 马尾松林下套种阔叶树生长状况初报[J]. 浙江农林大学学报, 2012, 29(3): 463-466.

[9] 楚秀丽, 刘青华, 范辉华, 等. 不同生境、造林模式闽楠人工林生长及林分分化[J]. 林业科学研究, 2014, 27(4): 445-453.

[10] 王良衍, 杨晓东, 曹立光. 次生马尾松、金钱松混交林的针阔异龄混交林改造成效研究[J]. 浙江林业科技, 2013, 33(2): 47-51.

[11] 韩锦春, 李宏开. 马尾松混交林混交模式的多层次综合评判[J]. 植物生态学报, 2000, 24(4): 498-501.

[12] CHEN F, YUAN Y J, YU S L, et al. Influence of climate warming and resin collection on the growth of Masson pine (*Pinus massoniana*) in a subtropical forest, southern China[J]. Trees, 2015, 29(5): 1-8.

[13] KUANG Y W, WEN D Z, ZHOU G Y, et al. Reconstruction of soil pH by dendrochemistry of Masson pine at two forested sites in the Pearl River Delta, South China [J]. Annals of Forest Science, 2008, 65(8): 804-810.

[14] 欧建德, 吴志庄. 林下套种及坡位和弱光环境对南方红豆杉人工林早期生长及林分分化的影响[J]. 东北林业大学学报, 2016, 44(10): 12-16.

[15] 欧建德, 吴志庄. 林下套种对南方红豆杉树冠形态结构及干形变化的影响[J]. 西南林业大学学报, 2016, 36(5): 106-110.

[16] 周志春, 徐高福, 金国庆, 等. 择伐经营后马尾松次生林阔叶树的生长与群落恢复[J]. 林业科学研究, 2004, 17(4): 420-426.

[17] 刘洋, 亢新刚, 郭艳荣, 等. 异龄林生长动态研究进展[J]. 西北林学院学报, 2012, 27(6): 146-151.

[18] 曾冀, 雷渊才, 贾宏炎, 等. 桂西南马尾松人工林生长对不同强度采伐的动态响应[J]. 林业科学研究, 2017, 30(2): 335-341.

[19] 潘登, 张合平, 潘高, 等. 基于 Rothermel 的南亚热带马尾松人工林潜在火行为研究[J]. 中南林业科技大学学报, 2017(6): 14-23.

[20] 舒骏, 江斌, 成向荣, 等. 马尾松复层林伴生树种生长及对土壤养分的影响[J]. 中国农学通报, 2013, 29(10): 82-86.

[21] 王希华, 宋永昌, 王良衍. 马尾松林恢复为常绿阔叶林的研究[J]. 生态学杂志, 2001, 20(1): 30-32.

[22] 陈昌雄, 陈平留, 肖才生, 等. 人工马尾松复层混交林林分结构规律的研究[J]. 林业科学, 2001, 37(s1): 205-207.

[23] 孟勇, 艾文胜, 杨明, 等. 上阔下竹复合经营模式对毛竹生长的影响[J]. 经济林研究, 2016, 34(3): 135-141.

[原载: 中南林业科技大学学报, 2018, 38(03)]

灰木莲二次开合开花动态与雌雄异熟特征

潘丽琴[1,2,3] 郝 建[1,2,4] 徐建民[2] 陈建全[1,4] 卢立华[1,4] 刘志龙[1,4]

([1]中国林业科学研究院热带林业实验中心，广西凭祥 532600；
[2]中国林业科学研究院热带林业研究所，广西广州 510520；
[3]西南林业大学，云南昆明 650204；
[4]广西友谊关森林生态系统国家定位观测研究站，广西凭祥 532600)

摘 要 为了研究灰木莲在中国引种地区的开花及雌雄异熟特征对其生殖的影响，以中国林业科学研究院热带林业实验中心树木园、白云实验场、伏波实验场的灰木莲人工林作为观测对象，调查研究灰木莲在该地区的花期物候、开花动态、花期不同阶段的花粉活力及柱头可授性。结果表明：①在引种地区广西凭祥，灰木莲的花期主要集中在3月上旬至5月下旬，个别单株1月下旬就开花，偶尔出现1年中2次花期，第2次花期从10月中旬持续到12月上旬。②在花期不同阶段，由于气温条件不同，灰木莲的开花表现不同模式的二次开合现象，开花各进程发生的时间以及二次展花间隔时间差异较大。③灰木莲在单花内表现为雌雄异熟，雌蕊先熟，雌蕊位于雄蕊上部，在时间和空间上均避免了自花传粉发生。灰木莲在单株水平上为同步集中式开花，同一植株第1轮花的花粉释放时间正好为第2轮花发生初次展花柱头活力最强的时间，2轮花的雌雄性征表达均较强，重叠可达1~6h，保证它在有限的可授期内授粉成功。④灰木莲异株异花、同株异花及自花授粉皆能正常结实，异株异花授粉结实率明显较同株异花和自花授粉高，表明其繁育习性是以异交为主，但存在自交亲和的现象；灰木莲不同花期的授粉结实率以低温的始花期最低，过渡期和盛花期较高，说明低温不利于其受精及胚胎发育。

关键词 植物学；灰木莲；开花动态；二次开合；雌雄异熟

Flowering Dynamic and Dichogamy Characteristics of *Manglietia glauca* with Two-times Opening-closure Flowering

PAN Liqin[1,2,3], HAO Jian[1,2,4], XU Jianmin[2], CHEN Jianquan[1,4], LU Lihua[1,4], LIU Zhilong[1,4]

(*[1]Experimental Center of Tropical Forestry, Chinese Academy of Forestry, Pingxiang* 532600, *Guangxi, China*;
[2]Research Institute of Tropical Forestry, Chinese Academy of Forestry, Guangzhou 510520, *Guangdong, China*;
[3]Southwest Forestry University, Kunming 650204, *Yunnan, China*;
[4]Guangxi Youyiguan Forest Ecosystem Research Station, Pingxiang 532600, *Guangxi, China*)

Abstract: In order to study the influence of flowering dynamic and dichogamy characteristics on sexual reproduction of *Manglietia glauca* in the introduced area, the *M. glauca* man-made forest of Baiyun, Fubo experimental field and forest garden in Experimental Center of Tropical Forestry of Chinese Forest Academy were selected as study area. On the base of the observation and investigation of flowering phenology and flowering dynamic of *M. glanca*, pollen and stigma viability in different flowering stages of *M. glauca* were tested. The results showed as following: ①The *M. glanca* started to bloom in late January, and reached full bloom in late March which lasted to early May in the introduced area in Pingxiang, Guangxi. The *M. glancah* bloom twice a year occasionally and the second flowering phase last from middle October to early December. ②There were great difference in the pattern of two-times opening-closure flowering, the time in each flowering process and interval time in two-times flowering opening due to different temperature. ③ *M. glanca* showed dichogamy and protogyny in its flowering process, and the pistil was located in the upper stamens and avoid the possibility of self-pollination occurred both in time and space. *M. glanca* bloomed for centralized synchronization in plant level. When the flower pollen released during the first round flowering, the pistil viability of second round flowering was strongest. The time both male and female sexuality of strongest expression overlapped up to 1-6h, and the time guaranteed pollination success in its limited receptivity period. ④After artificial xenogamy, geitonogamy and self-pollination , the fruits and seeds could be set and the seed setting rate of xenogamy was higher than others, indicating that *M. glauca* was given priority to out-crossing, but self-compatible to some degree. The seed setting rate was lowest in initial flowering period with the lowest temperature, but the seed setting rate was higher in transition period and bloom stage with higher temperature, which indicated that high temperature was more beneficial to fertilization and embryo development than low temperature.

Key words: botany; *Manglietia glanca*; flowering dynamic; two-times opening-closure flowering; dichogamy

植物的开花方式和生殖同步性是影响植物繁殖适合度的重要因子，是决定植物生殖成功的重要因素[1]。对大部分植物来说，开花通常是一次性绽放，之后通过花的闭合或花瓣的凋谢来终止开花进程，而有些植物的开花过程，开放和闭合是交替进行[2]，称之为“二次开合”。多种木兰科植物存在此种开花方式，如广玉兰(*Magnolia grandiflora*)[3]、白玉花(*M. denudata*)[4]、厚叶木莲(*Manglietia pachyphylla*)[5]、大果木莲(*M. grandis*)[6]、中缅木莲(*M. hookeri*)[6]。该种开花方式受到诸如温度、湿度和光照等气候因子的影响，其中温度条件在控制开花中起主导作用[7,8]。灰木莲(*Manglietia glanca*)原产越南及印度尼西亚，木兰科木莲属(*Manglietia*)植物，其干形通直、抗逆性强、生长快、木材纹理细致及易加工的特性[9]，是一种极具有发展潜力的速生用材及园林绿化树种。自1960年以来，灰木莲在中国西南和华南地区进行引种并规模化种植[10]。目前，有关灰木莲的研究主要集中于栽培繁育技术[11]、早期生长适应性[12]、材性[13]以及生态效应等方面。在引种地灰木莲开花后自然结实率极低[11,14]，而针对该现象的研究报道较少[14,15]。为了探究灰木莲开花及雌雄异熟特征对其生殖成功的影响，本文着重研究灰木莲的花部结构、不同花期的二次开合开花进程、花粉活力和柱头可授性，以探究灰木莲二次开合开花及雌雄异熟特征对其生殖的影响，也为深入了解木莲属的系统进化或演化过程提供资料。

1　研究地区概况和试验材料

试验点设在中国林业科学研究院热带林业实验中心，位于广西壮族自治区凭祥市，属南亚热带季风气候区，年均气温为21.5℃，≥10℃积温7500℃，年降水量1220~1400mm，降水集中期为4月中旬至8月末，年均日照约1260h。以伏波实验场、白云实验场和中心树木园栽培的灰木莲人工林为研究对象，各林分造林苗木均为1.5年生的实生苗，种源来自越南尚河，无病虫侵害，健康状况良好，基本情况见表1。观测区域花期内气温因子基本情况见表2。

表1　观测林分基本情况

人工林	造林年度	树高(m)	胸径(cm)	密度(株/hm²)	海拔(m)
伏波实验场	2002	16.69±1.13	26.13±1.46	800	240
白云实验场	1998	18.45±3.75	28.70±5.59	1200	540
中心树木园	2003	24.65±2.85	32.40±4.19	1000	630

表2　观测区域环境气温基本情况

观测年份	不同月份的气温(℃)				
	1月	2月	3月	4月	5月
2014	10~20	12~18	16~21	21~28	24~32
2015	11~19	13~20	16~22	19~28	24~32
2016	11~19	9~20	15~25	21~31	22~31

2　研究方法

2.1　林分水平物候的观测

2014—2016年连续3年对灰木莲林分进行定点、定时观察并记录其开花物候。林分水平1%~10%的个体开花时视为始花期，10%~50%的个体开花时为过渡期，50%~95%的个体达到开花高峰时为盛花期，95%植株开花结束时为终花期。

2.2　花部结构及开花动态的观测

在伏波实验场、白云实验场和中心树木园内试验点各选3株标准株，从青蕾期开始观察其开花动态，每个开花阶段分别标记10朵花，观察开花进程中的花部、柱头和花粉的形态变化特征，用佳能(Cannon)数码相机拍照，花部解剖结构在奥林帕斯(Olympus)体视显微镜成像系统下观察并拍照。

2.3 花粉活力及柱头可授性检测

在标准株上的上中下3个冠层各选取3个标准枝，每个标准枝于青蕾期选取15个花蕾。于始花期、过渡期和盛花期，分别选取3朵未散粉的花朵带回室内，人工挑出雄蕊，收集花粉，用于测定不同花期阶段花粉活力；于始花期、过渡期和盛花期，分别标记5朵青蕾，于露白(T1)、初次展花(T2)、闭合(T3)、完全展花(T4)和衰败期(T5)等5个阶段分别采集花粉用于测定单花不同开花阶段花粉活力，同时取柱头样品用于柱头可授性检测。

2.3.1 花粉活力检测

根据正交设计的花粉离体萌发试验结果，采用最佳萌发效果的培养液对各开花阶段花粉进行离体萌发培养，萌发条件为：质量浓度为6%蔗糖+0.05%硼酸$_3$+0.008%氯化钙，pH为6，温度25℃。花粉萌发6h后，把培养皿放置在倒置显微镜观察并拍照，选择5个视野统计花粉的萌发率。

2.3.2 对柱头可授性检测

采用联苯胺—过氧化氢法[16]检测，联苯胺-过氧化氢反应液为V(1%联苯胺溶液)：V(3%过氧化氢溶液)：V(水)=4：11：22的混合液。分别采各开花阶段花朵9朵，切除花被片，将雌蕊群浸入含有联苯胺—过氧化氢反应液的10ml透明样品瓶中。若柱头具有可授性，则反应液呈现蓝色，并伴有大量气泡，反之则无。根据雌蕊群柱头所冒出气泡数量和染为蓝色的柱头数量判断其可授性。将柱头可授性分为5个等级[17]：“-”为不具有可授性，气泡出现慢且少，颜色无变化；“+/-”极少部分柱头有可授性，“+”为具有较弱的可授性，表示少部分柱头有可授性，大部分没有，气泡出现慢且少，着色慢、颜色浅；“++”为具有较强的可授性，气泡出现快且多，着色快、颜色浅；“+++”为具有很强的可授性，气泡出现快且多，着色快、颜色深。

2.4 不同花期阶段灰木莲人工授粉

在标准株的上中下3个冠层各选取5个标准枝，在青蕾期每个标准枝选取30个花蕾。在始花期、过渡期和盛花期，分别标记10朵花，在初次展花阶段进行自花、同株异花和异株异花授粉；于盛花期按各单花露白阶段、初次展花阶段、(长)闭合阶段和完全展花阶段进行异株异花授粉，每个标准枝各阶段分别授粉6朵花。授粉后2个月统计结实率。单个果实明显膨大，子房数量占其果实总子房数量50%以上为成功结实。

3 结果与分析

3.1 灰木莲花部结构与群体开花物候

灰木莲为两性单花，展花直径15~25cm，花被片乳白色，厚肉质，汤勺形，长6~10cm，宽5~7cm，共6~8片；多心皮的雌蕊群在上，雄蕊群聚合于雌蕊群下方基部的花托上。雌蕊群由45~85枚单生心皮组成，心皮绿色，短纺锤形，长0.4~0.6cm，紧密组合呈锥形，每心皮内含双排胚珠8~12颗，成熟柱头向外微卷曲；雄蕊群着生于雌蕊群的下部，高度仅达到雌蕊群的下缘，单个雄蕊长0.7~1.1cm，宽0.8~1.5mm，乳白色，两对细长的花粉囊平行分布于花药的两侧。

图1 盛花期灰木莲二次开合开花动态

A. 露白；B. 初次展花；C. 长闭合中间阶段；D. 长闭合阶段；E. 短闭合；F. 完全展花

在中心树木园(海拔为240m)，灰木莲花期主要从3月上旬持续至5月上旬，少数从1月下旬开始，仅个别单株出现1年中2次花期，第2次花期从10月中旬持续到12月上旬，花量极少；第1次花期3月上中旬为始花期，3月下旬到4月中旬为盛花期，4月下旬到5月上旬为终花期。灰木莲在高海拔(600m左右)比低海拔(240m)的花期推迟15~20d。在花期内，环境温度分别为始花期的10~20℃、过渡期18~25℃和盛花期的21~30℃，不同气温下灰木莲表现出不同的开花方式。

3.2 灰木莲不同气温单花二次开合开花时序特征

灰木莲开花过程存在二次开合的开花现象(图1)。先后经历：①露白(T_1)：内部白色花被片开始松动，绿色外苞片露出花被片，形成白绿相间的花蕾。②初次展花(T_2)：花被片初次展开，外苞片和内层花被片同时展开30°~50°。③闭合阶段(T_3)：最外层花苞片沿花柄向下展开180°，白色花瓣向相反方向收拢至完全闭合，多发生于18：00~24：00，根据气温高低存在2种闭合方式，高温时闭合时间为10~12h，气温较低时闭合时间为48~72h。④完全展花(T_4)：花被片再次展开约90°，不再闭合，雄蕊群松散脱落，花粉散出，多发生于长闭合形成白蕾后第2天17：00~20：00；在盛花期，完全展花前会出现一次短闭合阶段。⑤衰败阶段(T_5)：完全展花的第2天，花被片逐渐褐化、萎蔫、凋落，至完全凋落3~7d。灰木莲在不同花期阶段的开花方式特征见表3。

表3　灰木莲不同气温单花二次开合开花时序特征

开花阶段	特征	始花期	过渡期	盛花期
白蕾期	花被片松动，绿色苞片开口，花被片露出，形成白绿相间的花蕾	第1次展花前2~3d	第1次展花前1d	第1次展花前1d
初次展花	绿色苞片和花被片同时展开30°~50°	发生于18：00~19：00	发生于18：00~19：00	发生于18：00~19：00
闭合开始	绿色苞片向下展开至与花柄方向平行，花被片向雌蕊群聚拢闭合，形成白色花蕾	发生于20：00~24：00，初次展花后4~6h	发生于18：00~21：00，初次展花后1~2h	发生于18：00~20：00，初次展花后0.5~1.0h
长闭合	花被片闭合，雄蕊群紧贴花柱	持续1~3d	持续14~20h	持续10~12h
短闭合	最外层1~2片花被片展开10°~20°，约0.5h后再次闭合	—	—	发生于16：30~17：30
完全展花	花被片展开90°后不再闭合，雄蕊群松散脱落，花粉散出	发生于17：30~18：00	发生于17：30~18：00	发生于17：30~18：00，第2次闭合后的1h
萎蔫	花被片边缘褐化向内微卷	完全展花后的1d	完全展花后的1d	完全展花后的1d
凋落	花被片凋落	完全展花后4~5d	完全展花后3~4d	完全展花后2~3d

3.3 灰木莲不同花期花粉活力和柱头可授性特征

由表4可知，灰木莲花粉活力随时间推移而下降，盛花期时花粉活力最高，平均活力为72.13%，其次为过渡期63.15%，始花期最低仅为43.28%。盛花期时，因环境温度最高(21~30℃)，散粉后花粉活力急剧下降，散粉后18h时，花粉已完全失活。而在环境温度较低的过渡期(18~25℃)及始花期(10~20℃)，散粉后花粉活力维持时间相对较长，分别达到24h、36h。

表4　灰木莲不同花期阶段的花粉活力

花期	散粉后不同时间的花粉活力(%)								
	0	6	12	18	24	30	36	42	48h
始花期	43.28	40.37	33.58	28.70	22.17	18.42	-	-	-
过渡期	63.15	36.19	28.30	20.10	-	-	-	-	-
盛花期	72.13	32.32	13.58	-	-	-	-	-	-

灰木莲在始花期的完全展花、过渡期的初次展花及盛花期的白蕾期柱头黏液分泌量最多；单花完全展花时，始花期与过渡期的柱头均有少量的黏液，而盛花期的柱头黏液已变干。经过联苯胺—过氧化氢检测，柱头黏液分泌量较多时，柱头颜色多呈乳白色和淡黄色，柱头可授性较强，柱头在纯白色和褐(黑)色时，黏液较少，可授性均较弱(表5)。柱头在过渡期时的初次开花和完全展花时、盛花期时的白蕾期及初次展花时、始花期时的完全展花时，具有较强的可授性，其余时期柱头可授性较弱或基本失活。

灰木莲雌性先于雄性表达的特征稳定，交配类型为雌先型。不同气温条件下，雌雄性成熟的间隔时间有所差异，气温越高雌性表达发生时间越早，雌雄异熟的间隔时间越短；雄性表达在初次展花时均已表达，至完全展花时，花粉活力最强。盛花期露白阶段时柱头可授性最高，雌性表达在青蕾阶段就已开始，而雄性特征还未表达，当完全展花花粉完全散开成熟时，雌性特征最弱，雌雄异熟间隔时间约为2d；过渡期初次展花时柱头可授性最强，但花粉未完全成熟，完全展花时花粉活力最高，柱头仍具可授性，雌雄异熟的间隔时间为24~36h；始花期闭合阶段时柱头可授性最强，但花粉活力较弱，完全展花时花粉活力和柱头可授性均较高，雌雄异熟的间隔时间为2~3d。

表5 灰木莲不同花期阶段的柱头可授性

花期	柱头颜色及分泌黏液量				
	T_1	T_2	T_3	T_4	T_5
始花期	纯白色，无	纯白色，较少	乳白色，少	淡黄色，多	黄褐色，无
过渡期	乳白色，多	乳白色，丰富	淡黄色，较多	黄色，少	褐色，无
盛花期	白色，丰富	黄色，多	黄褐色，少	褐色，无	黑褐色，无

花期	柱头可授性					花粉活力(%)				
	T_1	T_2	T_3	T_4	T_5	T_1	T_2	T_3	T_4	T_5
始花期	+/-	++	++	+++	+/-	3	32	39	43	15
过渡期	++	+++	++	++	-	11	42	54	65	0
盛花期	++	+++	++	+/-	-	22	53	60	62	0

注：T_1露白阶段，T_2初次展花，T_3闭合阶段，T_4完全展花，T_5衰败期；+++表示柱头具最强可授性；++柱头具较强可授性；+表示具较弱可授性；“+/-”极少部分柱头有可授性；-表示柱头不具可授性。

3.4 灰木莲花期不同阶段单株二次开合花过程可授性重叠期

根据图2，灰木莲在盛花期开花时，前一轮花完全展花的散粉期和后一轮花初次展花的可授期重叠，在这个重叠期里，前一轮完全展花，花粉自然散开，花粉活力也较强，第2轮花初次展时柱头可授性较强，但花粉未散开，为异花授粉提供极好的机会；

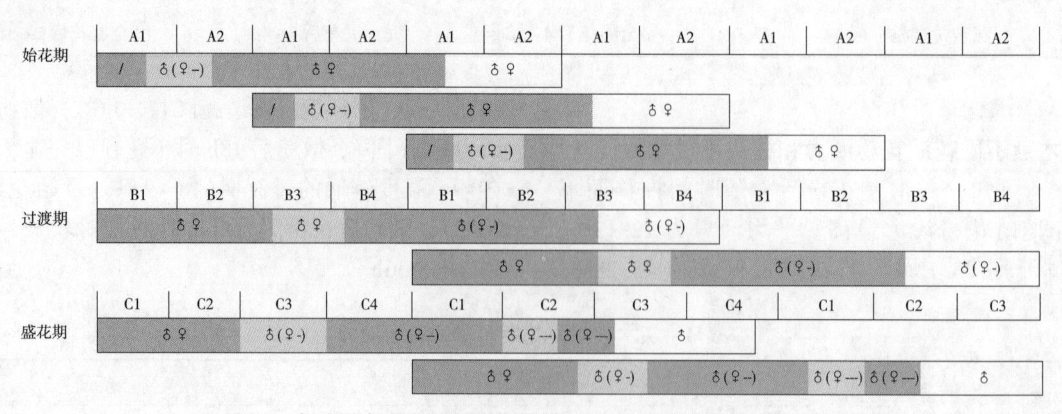

图2 灰木莲每轮开花在不同时间的柱头可授性与花粉活力

注：闭合，半展花，完全展花；♂花粉活力强，♀柱头可授性强，♀-柱头可授性一般，♀-柱头可授性弱，♀-柱头可授性较弱。始花期：A1为12：00~23：59，A2为24：00~11：59；过渡期：B1为6：00~11：59，B2为12：00~17：59，B3为18：00~23：59，B4为24：00~5：59；盛花期：C1为12：00~17：59，C2为18：00~23：59，C3为24：00~5：59，C4为6：00~11：59。

完全展花花粉散出时，同一花朵柱头可授性极低，发生自花授粉的概率极低；在过渡期开花时，异花和同花授粉均可能发生，前一轮花完全展花的散粉期和后一轮花初次展花的散粉期重叠期花粉活力和柱头可授性均较强，为异花授粉提供可能；此时同一花朵花粉活性和柱头可授性均较强，自花授粉也可发生；始花期开花时，第1轮完全展花的散粉期与第2轮初次展花的可授期没有重叠期，与第3轮初次展花时可授期重叠，可能发生异花授粉；而第1轮完全展花时同一花朵柱头可授性与花粉活力均很强，有利于自花授粉的发生。

3.5 不同花期阶段灰木莲人工授粉的成功率

图3显示：灰木莲在初次展花进行自花、同株异花和异株异花种方式均能成功授粉，但异株异花授粉成功率明显高于自花和同株异花授粉，异株异花在不同花期人工授粉平均成功率为57.67%，同株异花授粉平均成功率为45.00%，自花授粉的平均成功率为5.67%，表明灰木莲的繁育习性是以异交为主，但存在自交亲和的现象；始花期、过渡期和盛花期人工授粉平均成功率分别为26.00%、39.33%和43.00%，表明气温对授粉成功率有所影响，相对于过渡期(18~25℃)与盛花期(21~30℃)，在始花期(10~20℃)时授粉率最低，说明低温不利于灰木莲的受精及胚胎发育。

灰木莲在开花过程中的可授性是不断变化的，在初次展花前1~2d，即露白期柱头已具有可授性，完全展花后，柱头可授性急剧下降，1~6h后完全失活。在不同开花阶段通过异株异花方式进行人工授粉的授粉成功率如图4所示，闭合阶段与完全展花阶段的平均授粉成功率最高，分别为52.33%、49.67%，露白阶段的平均授粉成功率最低，为21.33%，与上述各阶段柱头可授性检测结果基本一致。

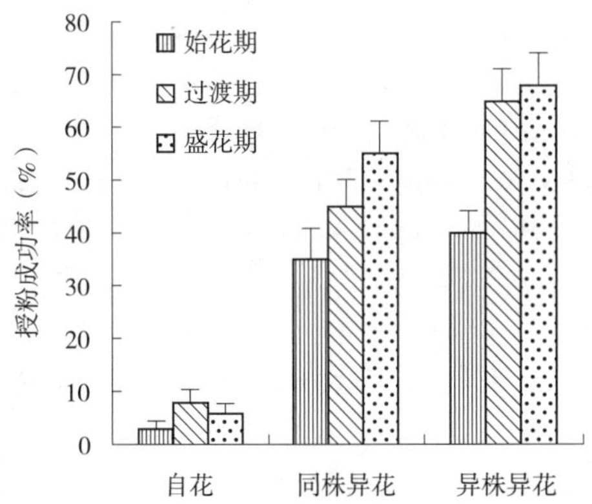

图3 不同授粉方式对灰木莲授粉成功率影响

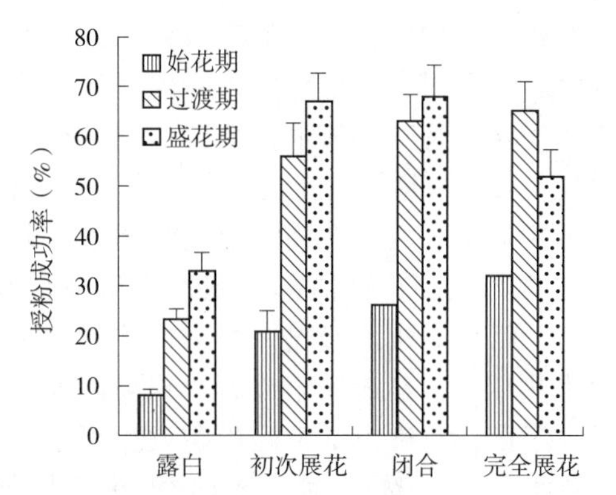

图4 不同授粉阶段对灰木莲授粉成功率的影响

4 结论与讨论

4.1 灰木莲二次开合开花的生物学意义

灰木莲的开花特征与多种木兰科树种相似[4,18]，在花期不同阶段，灰木莲的开花方式受温度影响而产生开合交替不同的开花模式，存在二次开合现象。有关开花过程中的闭合现象有几个解释：①保护花朵不被昆虫吃食[19]。②给传粉昆虫提供庇护场所，特别是对于一些直立闭合的花朵，能增加携带着花粉的甲虫类的传粉概率[20]。③对于在白天发生闭合的花朵，则是为了保护花粉不被散落或使花粉保持一定的湿度以维持更好的活力[21]。这种开花方式还可以保护花粉不易被风吹走，利于昆虫传粉[6]，利于花内形成小温室促进散落在柱头的发粉萌发[3]。

灰木莲在单花内表现为雌雄异熟，雌蕊先熟，雌蕊位于雄蕊上部，在时间和空间上均不利于自花传粉。单花水平上，花朵在初次展花时雌雄性征均已表达，但花粉仍未释放，待完全展花，花粉释放时，雌性征减弱或完全失活，避免了自花传粉发生的可能；在植株水平上，灰木莲表现为异步雌雄异熟且同步集中式开花，各轮花二次开合交替出现，前一轮花的完全展花也即雄性表达时，正好为下一轮花初次展花也即雌性表达时，此时两轮开花雌雄性征均较强，雌雄表达重叠可达1~6h，这种重叠和集中式开花可使植株开花时有较多的花粉资源进行同株异花和异株异花授粉，保证其在有限的可授期内授粉成功。这种开花方式下，传粉者的数量是否充足、环境条件是否允许传粉者在花期内进行有效的传粉活动成为灰木莲在引种地生殖成功的关键因素。

4.2 雌雄异熟与二次开合的关系

花被片展开的动力是由于花部内外温度差异导致花被片两面细胞扩张程度不均产生的[2]。木兰属植物在开花进程中存在开花生热效应，使花器官内部温度明显高于环境温度而促使开花[4]。由于木兰属存在雌雄异熟的特征，在雌蕊成熟时促使其初次展花，之后花朵闭合，雄蕊成熟时促使第二次完全展花[3]。气温较高时，灰木莲雌蕊成熟较快，短时间花内聚集较多的热量，当花内积温高于外部温度，发生第一次展花，之后花内温度迅速下降，外部温度高于花内温度，花瓣内部细胞发生可逆性收缩，花瓣闭合，闭合后雄蕊经过一段时间成熟，在花内部再次生热积温，花瓣完全展开，花瓣内面细胞拉升到最大限度，花辨内部细胞不再发生可逆性收缩而使花朵闭合。气温较低时，第1次展花发生的时间较晚，2次展花间隔期也更长。

4.3 灰木莲开花特征与生殖适合度

植物的开花时间与生殖同步性可以在多方面影响其生殖成功与否[1]，灰木莲在广西凭祥地区的花期主要集中在3月上旬至5月下旬，个别单株1月下旬出现开花，偶尔出现1年中2次花期，第2次花期从10月中旬持续到12月上旬。灰木莲不同花期经历的气温不同，表现出不同的二次开合开花特征，不同开花阶段发生和持续的时间差异较大。在温度较高的过渡期和始花期，单花雌雄表达间隔期较短，单株上各轮花雌雄表达重叠时间也相应缩短，但开花量较多，传粉者数量增加，传粉效率更高。而在温度最低的始花期，开花量和传粉者均较少，但各开花阶段持续时间较长，各轮开花的雌雄性表达重叠的时间较长，这可能是为较少传粉者提供较长传粉时间的生殖适应性策略。

参考文献

[1] O'NEIL P. Selection on flowering time: an adaptive fitness surface for nonexistent character combinations [J]. Ecology, 1999, 80(3): 806-820.

[2] van DOORN W G, van MEETEREN U. Flower opening and closure: a review [J]. J Exp Bot, 2003, 54 (389): 1801-1812.

[3] 孟希，王若涵，谢磊，等．广玉兰开花动态与雌雄异熟机制的研究[J]．北京林业大学学报，2011，33(4)：63-69.

[4] 王若涵．木兰属生殖生物学研究及系统演化表征探析[D]．北京：北京林业大学，2010.

[5] 杨晓丽．濒危植物厚叶木莲的保护生物学研究[D]．北京：中国科学院大学，2013.

[6] 付玉嫔，陈少瑜，吴涛．濒危植物大果木莲与中缅木莲的花部特征及繁育系统比较[J]．东北林业大学学报，2010，38(4)：6-10.

[7] BYNUM M R, SMITH W K. Floral movements in response to thunderstorms improve reproductive effort in the alpine species Gentianaalgida (Gentianaceae) [J]. Am J Bot, 2001, 88(6): 1088-1095.

[8] 舒素芳，毛俊萱，蔡敏．白玉兰始花期与气象因子的关系分析[J]．浙江农业学报，2013，25(2)：248-251.

[9] 刘玉壶．中国木兰[M]．北京：科学技术出版社，2003：142-143.

[10] 王克建，蔡子良．热带树种栽培技术[M]．南宁：广西科学技术出版社，2008：71-72.

[11] 乔梦吉．广西优良珍贵树种灰木莲的组织培养[J]．广西农业科学，2013，44(6)：989-993.

[12] 卢立华，何日明，农瑞红，等．坡位对灰木莲生长的影响[J]．林业科学研究，2012，25(6)：789-794.

[13] 陈松武．人工林灰木莲木材材性及其薄木贴面工艺研究[D]．南宁：广西大学，2011.

[14] 招礼军，韦善华，朱栗琼，等．灰木莲的开花特性及繁育系统的研究[J]．西部林业科学，2015(2)：24-28.

[15] 招礼军，龙永宁，朱栗琼，等．灰木莲花粉的萌发率和生活力[J]．广西林业科学，2014，43(4)：405-408.

[16] 刘林德，张萍，张丽，等．锦带花的花粉活力、柱头可授性及传粉者的观察[J]．西北植物学报，2004，24(8)：1431-1434.

[17] 张瑞，李晖，彭方仁，等．薄壳山核桃开花特征与可授性研究[J]．南京林业大学学报(自然科学版)，2014，38(3)：50-54.

[18] PAN Y Z, LIANG H X, GONG X. Studies on reproductive biology and endangerment mechanism of the endangered plant *Manglietia aromatica*[J]. Acta Bot Sin, 2003, 45(3): 311-316.

[19] STIRTON C H. Nocturnal petal movements in the Asteraceae. [J]. BOTHALIA, 1983, 14(3/4): 1003-1006.

[20] GOLDBLATT P, BERNHARDT P, MANNING J C. Pollination of petaloid geophytes by monkey beetles (Scarabaeidae: Rutelinae: Hopliini) in southern Africa [J]. Ann Missouri Bot Garden, 1998, 85(2): 215-230.

[21] DAFNI A. Autumnal and winter pollination adaptations under Mediterranean conditions [J]. Bocconea, 1996, 5(1): 171-181.

［原载：浙江农林大学学报，2018，35(01)］

灰木莲花期物候观测及生殖构件分布

潘丽琴[1,2,3] 郝 建[1,2] 徐建民[2] 郭文福[1] 陈建全[1] 杨桂芳[1]

([1]中国林业科学研究院热带林业实验中心，广西凭祥 532600；
[2]中国林业科学研究院热带林业研究所，广东广州 510520；
[3]西南林业大学，云南昆明 650204)

摘 要 为了解灰木莲(*Manglietia glauca* Blume)引种到广西凭祥地区花期物候及生殖构件分布状况，为其在引种地败育机制的研究奠定基础。选择中国林业科学研究院热带林业实验中心树木园、白云实验场、伏波实验场的灰木莲人工林为观测对象，参考 Dafni 方法，从林分到单株水平调查研究灰木莲在广西凭祥地区的花期物候、开花特征及生殖构件分布，跟踪观察单花开花动态进程，统计花部组件大小、数量及位置等特征。结果表明：灰木莲花为子房上位的两性花，雄蕊群着生于雌蕊群下部，雄蕊短小，其高度仅达到雌蕊群下缘；灰木莲花芽到开花过程可分为混合芽阶段、花芽阶段、花蕾露白阶段、白蕾阶段、展花阶段、花瓣脱落阶段 6 个阶段，花被片展开存在二次开合现象；在适生区，灰木莲开花特征稳定，开花强度中等，不同海拔高度灰木莲林分的花期物候存在差异，林分的生殖构件在林内、缘分布差异显著，且生殖构件的败育率极高。灰木莲引种到广西凭祥地区花期物候稳定，开花同步性较高，但是花果转化率极低。

关键词 灰木莲；开花特征；花期物候；生殖构件

The Flowering Phenology and Distribution of Reproductive Modules of *Manglietia glauca* Blume

PAN Liqin[2,3], HAO Jian[1,2], XU Jianmin[2], GUO Wenfu[1], CHEN Jianquan[1]
YANG Guifang[1]

([1]*Experimental Center of Tropical Forestry, Chinese Academy of Forestry, Pingxiang* 532600, *Guangxi, China*;
[2]*Research Institute of Tropical Forestry, Chinese Academy of Forestry, Guangzhou*, 510520, *Guangdong, China*;
[3]*Southwest Forestry University, Kunming* 650204, *Yunnan, China*)

Abstract: The flowering phenology and distribution of reproductive modules of *Manglietia glauca* Blume was studied to provide a basis for the research of its abortion mechanism in introduction area. The stand of *M. glauca* Blume in Arboretum, Baiyun experimental site and Fubo experimental site of the Experimental Center of Tropical Forestry in Pingxiang were chosen as observation object. Reference Dafni's method, the flowering phenology, flowering characteristics and reproductive modules distribution of *M. glauca* Blume were investigated from stand to single plant level, while flowering dynamic was observed, and the size, number and location of floral components were calculated. [Results] The results showed that the flower was bisexual flower. Androecium was borne in the bottom of the gynoecium, its height only reached the lower edge of the gynoecium. Flowering process could be divided into six stages, such as mixed bud, flower bud, white alabastrum appearing, white alabastrum, blooming and falling. Two-times opening/closure flowering was observed during blossoming process. In suitable region, flowering characteristics of *M. glauca* Blume was relatively stable, with moderate intensity blossoming. The effect of altitude on flowering phenology was obvious. The distribution of reproductive modules was also significantly different between forest edge and interior, fruit setting rate was low. According to the result of observation, the flowering phase of *M. glauca* Blume was tranquil, flowering synchronization index was higher, but conversion rate of blossom into bear fruit was extremely low in Pingxiang, Guangxi.

Key words: *Manglietia glauca*; flowering characteristics; flowering phenology; reproductive modules

被子植物花朵的结构特征是其有性生殖与环境相适应的表现[1]，因此，植物花朵的形态结构特征、开花进程及花期物候等与其生态进化之间紧密联系在一起[2]，观测花期物候是研究植物生殖生态学的重要内容。有关植物开花动态的研究已有许多报道，其中，在生态异质生境中及种群内的变化动态是研究热点之一[3~5]。高等植物的花、花序、果实和种子及着生这些器官的生殖枝被称作生殖构件[6]，而每种植物生殖构件的分布格局和数量变化具有一定的规律性[7]，主要由植物自身遗传及植物所处环境因子决定[8]。

灰木莲(*Manglietia glauca* Blume)为木兰科木莲属常绿阔叶大乔木，其干形通直、树形优美、花大洁白有芳香，是一种较好的速生用材及园林绿化树种。灰木莲原产越南及印度尼西亚，1960年，我国从越南引种在中国林科院夏石树木园试种[9]，生长表现良好[10]，具有较好的发展前景。在广东、云南、福建等地相继成功开展了灰木莲的引种栽培[11]。在原产地越南，灰木莲可以正常结实，引种到中国后开花结实率极低[12]，严重制约其大规模发展。因此，本文从灰木莲在广西凭祥地区的开花进程、花期物候特征及生殖构件分布等方面开展研究，探索其生殖生态特征，以期为灰木莲在引种地败育机制研究积累资料。

1 试验地概况

试验地设置于广西凭祥市(21°57′47″~22°19′27″N；106°39′50″~106°59′30″E)中国林业科学研究院热带林业实验中心树木园、白云实验场和伏波实验场，属南亚热带季风气候区，年均气温21.5℃，≥10℃积温7500℃，年降水量1220~1400mm，降水集中期为4月中旬至8月末，年均日照约1260h。

选择中国林业科学研究院热带林业实验中心树木园(ZX)、白云实验场(BY)、伏波实验场(FB)灰木莲人工林作为观测对象。造林所用苗木均是用越南老街省宝安县收集种子培育的1年生苗。各林分无病虫侵害，健康状况良好，基本情况见表1。

表1 观测林分基本情况

林分	造林时间(年)	树高(m)	胸径(cm)	保留密度(tree/hm²)	海拔(m)
ZX	2002	16.69±1.13	26.13±1.46	800	240
BY	1998	18.45±3.75	28.70±5.59	1200	540
FB	2003	24.65±2.85	32.40±4.19	1000	630

2 研究方法

2.1 开花物候观测

参照Dafni[13]方法观测灰木莲的开花物候。2014年1~5月、10~12月，2015年1~5月、10~12月，2016年1~5月，从林分和单株水平，对中国林业科学研究院热带林业实验中心树木园、白云实验场、伏波实验场3个生境下的灰木莲人工林的开花物候进行定期观测。

(1)单株水平开花前，在中国林业科学研究院热带林业实验中心树木园、白云实验场、伏波实验场分别选择5棵、30棵、30棵单株进行标记，观察记录始花日期(单株开花数达5%的开花日期为始花期)及当日花数，开花盛花期(单株开花数≥50%)及当日花数、持续时间，终花期(95%的花已开放)及当日花数、平均开花振幅、相对开花强度和开花同步性(同步指数)。

(2)林分水平5%的个体开花为种群始花期，50%的个体达到开花高峰时为种群开花高峰期，95%的植株开花结束时为种群花期结束。

用“花数/株·d”来表示平均开花振幅[14,15]，描述单位时间开花数。

单株相对开花强度指单株开花高峰日开放的花朵数量与该种群中植株在其开花高峰日产生的单株最大花数之比[16]。

根据文献[16]用同步指数(Si)检测开花同步性高低。

$$S_i = \frac{1}{n-1}\left(\frac{1}{f_i}\right)\sum_{j=1}^{n} e_{j\neq i}$$

式中：n为样地中个体总数；f_i为个体i开花的总时间(d)；e_j为个体i和j花期重叠时间(d)；S_i的变异范围为0~1，“0”表示种群内个体花期无重叠，“1”则表示完全重叠。

2.2 花部形态与开花动态

分别于2014年及2015年灰木莲盛花期，在中国林业科学研究院热带林业实验中心(以下简称热林中心)树木园固定选择5株灰木莲定时观察开花动态。记录花蕾期花蕾发育的形态变化、花被片张开到花被片全部脱落的时间及特征、雌蕊群和雄蕊群的形态特征及动态变化等。取新鲜花样，带回室内置于Olympus SZX7体视镜下观察、拍照、测量与统计花部组件大小、数量及位置等指标。

2.3 生殖构件调查

在3个海拔高度的灰木莲人工林中，分别选择

10棵标准木，林内、林缘各5棵。每棵间距20m以上，分上、中、下三个冠层选标准枝，在花期记录生殖枝、花、花芽数量及花、花芽在生殖枝上的位置。

2.4 数据处理与分析

采用Excel2013进行数据整理作图，用DPS V14.50进行单因素方差分析，采用LSD法进行多重比较。

3 结果与分析

3.1 灰木莲花部形态特征

灰木莲花属两性单花，生于枝顶。花蕾期，花蕾外部有2层苞片，厚纸质，多呈绿色，在展开前期逐渐变成褐色。花开后，直径15~25cm，具香味，无花盘或蜜腺，花被片乳白或绿白色，厚肉质，汤勺形，长6~10cm，宽5~7cm，共6片；雌蕊群由45~55枚单生心皮组成，心皮绿色短纺锤形，长0.4~0.6cm，呈锥形紧密组合在一起，柱头微向外卷曲；雄蕊长0.7~1.1cm，宽0.08~0.15cm，乳白色，花药内向开裂。雄蕊群着生于雌蕊群的下部，雄蕊短小，其高度仅达雌蕊群的下缘(图1)。

图1 灰木莲花部形态

A. 展开的花朵；B. 花部结构

3.2 灰木莲开花动态

灰木莲花芽到开花，过程可分为6个阶段(图2)：①混合芽阶段。该阶段芽以混合芽形式存在，花被片外具有2层内层苞片，之外还存在4~5层外苞片，每层外苞片内包有一对新叶，此阶段可持续45~50d。②花芽阶段。待混合芽外层苞片脱落完毕，花被片外仅剩一层内苞片，此时的芽称为花芽，此阶段持续25~30d。③花蕾露白阶段。花芽个体发育已足够大，内苞片停止生长，花被片生长舒展将内苞片顶开，花蕾顶部露出一点白色的花被片，此阶段可持续1~2d。④白蕾阶段。外苞片脱落，外层绿色花被片完全展开，露出6片白色花被片并抱合组成的花蕾，称为白蕾，此时花蕾内柱头向外卷曲，带有透明液体，至花被片完全展开成花瓣为止，此阶段需1~3d，在此阶段花被片展开经历第一次开放与长闭合。⑤展花阶段。在形成白蕾第二天，花瓣出现松动，其中1~2片花瓣微开(持续3~5h)，此时雄蕊群开始松动，柱头黏液量达到最多；至17:00微开的花瓣再次闭合(短闭合)，之后1~2h花瓣完全展开，同时花粉从花药中散出，雄蕊群完全脱落，至花瓣脱落为止；自花瓣完全展开柱头黏液颜色逐渐加深，变成深褐色胶体，该阶段持续4~6h；⑥花瓣脱落阶段。花瓣颜色由白色逐渐变成褐色脱落。如遇阴雨天气，温度降低，开花进程将推迟1~2d。

图2 灰木莲开花动态

A. 混合芽阶段；B. 花芽阶段；C. 花蕾露白阶段；D. 白蕾阶段；E. 展花阶段；F. 花瓣脱落阶段

3.3　灰木莲开花物候

表2表明：热林中心树木园(ZX)(海拔240m)的灰木莲花期主要从3月上旬持续至5月上旬，少数从1月下旬开始，偶尔出现一年中2次花期，第2次花期从10月中旬持续到12月上旬；第1次花期3月上中旬为始花期，3月下旬到4月中旬为盛花期，4月下旬到5月上旬为终花期。白云实验场白云山(BY)(海拔540m)的灰木莲林分的开花物候始花期比热林中心树木园相应推迟15~20d，盛花期推迟8~10d，终花期推迟6~7d；伏波实验场(FB)(海拔630m)的灰木莲林分的开花物候始花期比热林中心树木园相应推迟18~25d，盛花期推迟12~15d，终花期推迟8~10d。灰木莲开花相对集中，同步指数较高。

表2　灰木莲林分水平开花物候

林分	始花期		花期持续时间(d)	盛花期		平均开花振幅(flower/株·d)	终花期		平均开花同步性指数
	日期	平均花数(朵)		日期	平均花数(朵)		日期	平均花数(朵)	
ZX	03-10	11	53	04-05	102	21.65	05-05	5	0.87±0.03a
BY	03-25	8	46	04-12	63	14.31	05-10	3	0.68±0.07b
FB	03-28	7	51	04-25	80	16.87	05-15	2	0.71±0.04c

注：表中字母表示各处理间的LSD多重比较结果。

由图3可知：热林中心树木园(ZX)、白云实验场(BY)、伏波实验场(FB)灰木莲林分的开花进程基本相似，均呈上升至高峰期后下降趋势。与热林中心树木园林分相比，白云实验场、伏波实验场灰木莲林分达到开花高峰期比较缓慢。4月是广西凭祥灰木莲开花高峰期。

灰木莲在广西凭祥地区个体相对开花强度(图4)表明，在该地区有2个主要分布频度范围，分别在30%～50%、70%～80%，开花强度中等。

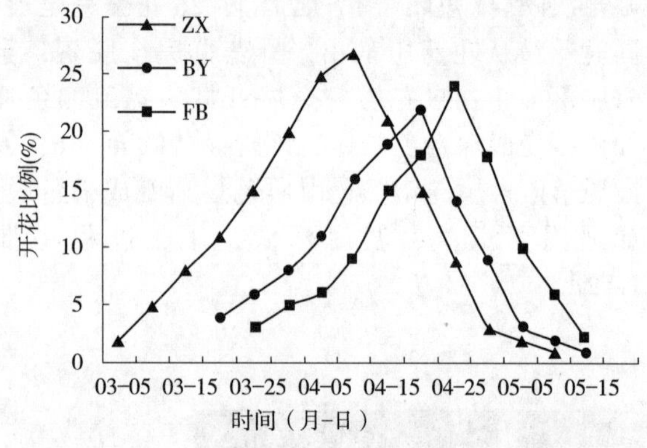

图3　不同灰木莲林分的开花物候曲线

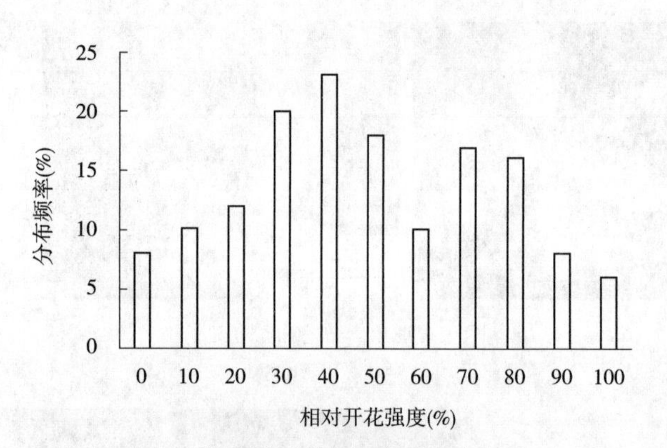

图4　灰木莲个体相对开花强度分布

3.4　生殖构件

热林中心树木园、白云实验场、伏波实验场不同海拔灰木莲林分的花枝/枝、花/花枝、果和果/花统计如表3所示。表3表明：热林中心树木园(ZX)(海拔240m)与伏波实验场(FB)(海拔630m)灰木莲林分的花枝/枝差异不显著，白云实验场(BY)(海拔540m)与热林中心树木园和伏波实验场灰木莲林分的花枝/枝差异显著；3个林分的花/花枝间差异不显著。从林分密度分析，随林分密度的升高，灰木莲林分的花枝/枝和花/花枝呈下降趋势。灰木莲生殖构件的败育率极高，调查观察发现，只有伏波实验场灰木莲有零星结实，其他林分均未发现结实。

表3　不同灰木莲林分生殖构件分布

林分	花枝/枝	花/花枝	果	果/花
ZX	0.81±0.23a	6±2.13	0	0
BY	0.65±0.26b	5±1.56	0	0
FB	0.67±0.13a	6±2.51	3±1.33	0.01±0.008

由图5可知：在林缘和林内，灰木莲树冠内花枝数量差异显著($P<0.05$)，3个立地条件下，林缘植株生殖个体上的花枝均比林内的高，且林缘和林内花枝上的单花数差异显著($P<0.05$)，林缘花枝单花数量明显比林内的多。

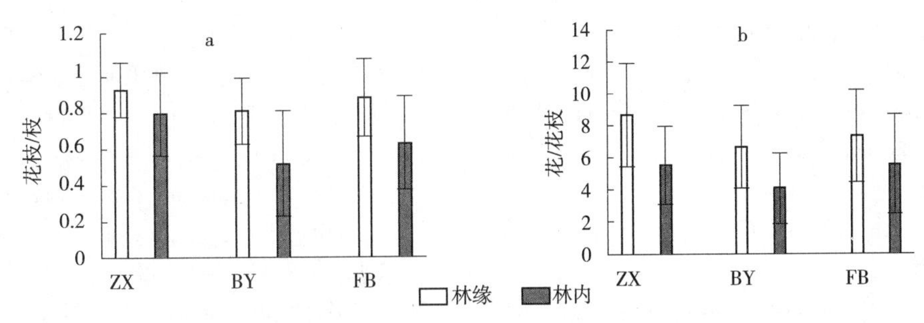

图 5 不同生境中灰木莲生殖构件的变化

4 讨论

花部结构特点、展花方式及时间是影响植物有性生殖成功与否的重要因素[17]。传粉是种子植物受精的必经阶段，传粉系统由花粉、柱头、传粉媒介三部分组成[18]。灰木莲为子房上位的两性花，雄蕊群着生于雌蕊群的下部，雄蕊短小，其高度仅达雌蕊群的下缘，花粉无法自然散落到雌蕊柱头上，即使到达柱头，其花粉数量不足，使胚珠难以受精，不利于自然结实。灰木莲花部结构特点决定了虫媒传粉是其有效传粉途径。

植物的开花物候不仅与其遗传特性及其类群的系统发生有关[19]，还与环境密切联系[20,21]。灰木莲引种到广西凭祥，海拔越高，花期开始越晚，表明环境因子对灰木莲花期物候影响较大。灰木莲单花开放时间相对固定，表明其单花开花节律受遗传因子的影响较大。灰木莲开花强度中等，但开花相对集中，同步指数较高，这与招礼军等[12]对广西南宁地区引种灰木莲的研究结果一致。单花寿命与开花期间的环境条件有关，在始花期，气温较低，且阴雨天气较多，单花寿命较长；在盛花期，气温较高，阴雨天气减少，单花寿命缩短。

根据不同海拔灰木莲林分的生殖构件分布结果分析表明：在适生范围内，海拔高度不是影响灰木莲生殖构件分布的关键因子。从林分密度分析，随着林分密度的升高，灰木莲林分的花枝/枝和花/花枝呈下降趋势，且林缘单株生殖构件数量显著高于林内植株，这表明林分密度影响单株的生长空间，导致树体接受光照不同[22-24]。植物开花数量超过最终成熟果实的数量是普遍现象[25]，但灰木莲引种到广西凭祥地区生殖构件的败育率极高，花果转化率近乎为零。影响果实败育的因子有内部因素[26]，也有外部因素[27]。因此，为探明灰木莲在引种地种实败育机制，将进一步开展灰木莲雌雄配子体发育、传粉媒介等方面的研究。

5 结论

(1)灰木莲在引种地区广西凭祥热林中心，中心树木园(ZX)(海拔 240m)的花期从 3 月上旬持续至 5 月上旬，偶尔出现一年 2 次花期，第 2 次花期从 10 月中旬持续到 12 月上旬；第 1 次花期的始花期为 3 月上中旬，3 月下旬到 4 月中旬为盛花期，4 月下旬到 5 月上旬为终花期。白云实验场白云山(BY)(海拔 540m)始花期的开花物候比热林中心树木园的相应推迟 15~20d，盛花期推迟 8~10d，终花期推迟 6~7d；伏波实验场(FB)(海拔 630m)的开花物候始花期比热林中心树木园相应推迟 18~25d，盛花期推迟 12~15d，终花期推迟 8~10d。

(2)灰木莲单花开花时间较稳定，均为 17：00~18：00。灰木莲开花强度中等，分别出现在 30%~50%、70%~80%，但开花相对集中，同步指数较高。单花寿命与开花期间的环境条件有关，在始花期，气温较低，且阴雨天气较多，单花寿命一般 2~3d；在盛花期，气温较高，阴雨天气减少，单花寿命缩短为 1~2d。

(3)灰木莲的生殖构件分布主要受林分密度的影响，海拔高度不是影响灰木莲生殖构件分布的关键因子。随着林分密度的升高，灰木莲林分的花枝/枝和花/花枝呈下降趋势，林缘单株生殖构件数量显着高于林内植株，这表明林分密度影响单株的生长空间，导致树体接受光照不同。

参考文献

[1]刘林德，祝宁，申家恒．刺五加，短梗五加的开花动态及繁育系统的比较研究[J]．生态学报，2002，22(7)：1041-1048.

[2]DAFNI A. Pollination ecology[M]. IRL Press Ltd，1992：1-57.

[3]陈波，达良俊，宋永昌．常绿阔叶树种栲树开花物候动态及花的空间配置[J]．植物生态学报，2003，27(2)：249-255.

[4]肖宜安，何平，李晓红．濒危植物长柄双花木开花物候

与生殖特性[J]. 生态学报, 2004, 24(1): 14-21.

[5]BOSCH J, RETANA J, CERDÁ X. Flowering phenology, floral traits and pollinator composition in a herbaceous Mediterranean plant community[J]. Oecologia, 1997, 109(4): 583-591.

[6]方炎明, 樊汝汶. 植物生殖生态学[M]. 济南: 山东大学出版社, 1996: 70.

[7]PRIMACK R B, LLOYD D G. Andromonoecy in the New Zealand Montane Shrub Manuka, *Leptospermum scoparium* (Myrtaceae)[J]. American Journal of Botany, 1980, 67(3): 361-368.

[8]HENDRY A P, DAY T. Population structure attributable to reproductive time: isolation by time and adaptation by time[J]. Molecular ecology, 2005, 14(4): 901-916.

[9]王克建, 蔡子良. 热带树种栽培技术[M]. 广西: 广西科学技术出版社, 2008 : 71-72.

[10]卢立华, 何日明, 农瑞红, 等. 坡位对灰木莲生长的影响[J]. 林业科学研究, 2012, 25(6): 789-794.

[11]刘玉壶. 中国木兰[M]. 北京: 科学技术出版社, 2003: 142-143.

[12]招礼军, 韦善华, 朱栗琼, 等. 灰木莲的开花特性及繁育系统的研究[J]. 西部林业科学, 2015(2): 24-28.

[13]DAFNI A. Pollination ecology: a practical approach[M]. Oxford: Oxford University Press, 1992.

[14]MCINTOSH M E. Flowering phenology and reproductive output in two sister species of Ferocactus (Cactaceae)[J]. Plant Ecology, 2002, 159(1): 1-13.

[15]HERRERA J. Flowering and fruiting phenology in the coastal shrublands of Doñana, south Spain[J]. Vegetatio, 1986, 68(2): 91-98.

[16]陈香, 胡雪华, 陆耀东, 等. 中国特有植物血水草开花物候与生殖特性[J]. 生态学杂志, 2011, 30(9): 1915-1920.

[17]何爽, 张爱勤, 夏荣, 等. 新疆不同生态区域苜蓿花粉败育情况及影响因素的细胞学研究[J]. 草业学报, 2011(4): 153-158.

[18]马玉心. 兴安鹿蹄草的传粉生态学研究[D]. 哈尔滨: 东北林业大学, 2007.

[19]OLLERTON J, DIAZ A. Evidence for stabilising selection acting on flowering time in*Arum maculatum* (Araceae): the influence of phylogeny on adaptation[J]. Oecologia, 1999, 119(3): 340-348.

[20]RATHCKE B, LACEY E P. Phenological patterns of terrestrial plants[J]. Annual Review of Ecology and Systematics, 1985: 179-214.

[21]李新蓉, 谭敦炎, 郭江. 迁地保护条件下两种沙冬青的开花物候比较研究[J]. 生物多样性, 2006, 14(3): 241-249.

[22]郭连金, 李梅, 林盛. 香果树种群开花物候、生殖构件特征及其影响因子分析[J]. 林业科学研究, 2015, 28(06): 788-796.

[23]赵志刚, 程伟, 郭俊杰, 等. 西南桦花序生长和开花物候特征[J]. 林业科学研究, 2011, 24(3): 385-389.

[24]CHAVES Ó M, AVALOS G. Is the inverse leafing phenology of the dry forest understory shrub*Jacquinia nervosa* (Theophrastaceae) a strategy to escape herbivory? [J]. Revista de biología tropical, 2006, 54(3): 951-963.

[25]STEPHENSON A G. Flower and fruit abortion: proximate causes and ultimate functions[J]. Annual review of ecology and systematics, 1981, 12: 253-279.

[26]PÍAS B, GUITIÁN P. Breeding system and pollen limitation in the masting tree*Sorbus aucuparia* L. (Rosaceae) in the NW Iberian Peninsula[J]. Acta Oecologica, 2006, 29(1): 97-103.

[27]ARNOLD A E, MEJÍA L C, KYLLO D, et al. Fungal endophytes limit pathogen damage in a tropical tree[J]. Proceedings of the National Academy of Sciences, 2003, 100(26): 15649-15654.

[原载: 林业科学研究, 2018, 31(02)]

密度对米老排萌生幼龄林生长及直径分布的影响

唐继新[1,2] 贾宏炎[1] 曾 冀[1] 蔡道雄[1] 韦叶桥[1] 农良书[1] 雷渊才[2]

([1]中国林业科学研究院热带林业实验中心，广西凭祥 532600；
[2]中国林业科学研究院资源信息研究所，北京 100091)

摘 要 伐桩萌条保留密度是萌生林培育经营的关键技术。本研究旨在探索密度对米老排萌生林生长及直径分布的影响，进而确定米老排萌生林的合理经营密度。以米老排人工林皆伐迹地伐桩萌条为试验对象，采用单因素随机区组试验设计，设计了3种留萌处理(T_{CA}：保留1株/桩，T_{CB}：保留2株/桩，T_{CC}：保留3株/桩)，3次重复，共9块固定样地；基于固定样地5年的连续观测数据，对3种不同密度米老排萌生幼林的生长及直径分布进行研究。结果表明：①除萌后1~3.5年，密度对萌生林平均的胸径、树高和单株材积的生长影响不显著，对萌生林优势木树高生长量影响不显著，对萌生林优势木单株材积的总生长量影响显著($P<0.05$)，对萌生林优势木胸径总生长量与单株材积连年生长量影响极显著($P<0.01$)。②萌生林平均胸径、平均树高、优势木平均胸径和优势木平均树高的连年生长量峰值均出现在第2年；林分平均胸径的平均生长量曲线与连年生长量曲线的相交时间为3~4年，而林分平均树高的平均生长量曲线与连年生长量曲线的相交时间则在2~5年间。③第2年后，萌生林优势高和平均树高的连年生长量呈明显递减趋势。④萌生林直径分布遵从单峰正偏山状分布，处于竞争的自然稀疏后期，使用Weibull分布函数对萌生林的直径分布拟合效果良好。[结论]前5年为萌生林径向生长和高生长的旺盛期，合理保留密度及幼林抚育对萌生林的径向生长极为重要。每伐桩保留1~2株萌条的除萌措施，对提高萌生林优势木比例作用明显。

关键词 密度；米老排；天然更新；萌生更新；除萌；幼林生长；直径分布

Effect of Density on Growth and Diameter Distribution of Young Coppice Stands of *Mytilaria laosensis* with Different Densities

TANG Jixin[1,2], JIA Hongyan[1], ZENGJi[1], CAI Daoxiong[1], WEI Yeqiao[1], NONG Liangshu[1], LEI Yuancai[2]

([1]*Experimental Center of Tropical Forestry*, *Chinese Academy of Forestry*, *Pingxiang* 532600, *Guangxi*, *China*;
[2]*Research Institute of Forest Resources Information Techniques*, *CAF*, *Beijing* 100091, *China*)

Abstract: Reserved density of the stump sprouts is the key technology in coppice cultivation. It is very important to explore the effect of density on growth and diameter distribution of coppice stands of *Mytilaria laosensis*, as in order to get reasonable density for cultivating technology with the stands. Using single factor randomized block designs, the experiment with the stump sprouts of *Mytilaria laosensis* in the clear-cut land of *Mytilaria laosensis* plantation, in three different levels of T_{CA} (reserved 1 sprout each stump), T_{CB}(reserved 2 sprouts each stump), T_{CC}(reserved 3 sprouts each stump) and repeated three times, was conducted with 5 years continuous observation data in 9 fixed plots to study the growth and diameter distribution of the coppice stands affected by density. ①During 1~3.5 years after sprouting thinning, total growth of the average diameter at breast height (DBH), average height, and average individual volume of coppices in the stands were not significantly affected by density. The growth of tree height for dominant trees of the sprout stands was not significantly affected by density. The total growth of individual volume for dominant trees of sprout stands was significantly affected by the density ($P<0.05$), the annual increment of individual volume and the DBH growth of dominant trees were very significantly affected by density($P<0.01$). ②All the peak values of annual increment of average DBH of stands, average tree height of stands, average DBH of top tree, and average tree height of the stands occurred in the second year, the intersected times of the average growth and annual growth of the stand DBH were during third year to fourth year, but the intersected times to the average tree height of the stands were during second year to fifth year. For the dominant trees, both intersected time of average growth and annual increment growth of DBH and tree height were very similar to the growth of stands. ③After the second year, both annual increment growth of tree height with the dominant tree and the average tree obviously decreased. (4) All diameter distributions of coppices in the stands were part mountain shape distribution, which indicated that they were located in the late competition period. It had excellent fitting effects using the Weibull function to fit diameter distribution of coppices in the stands. It is fast growth period for the radial and height growth of the stands before 5 years old, reasonable density and tending are extremely vital to the stands. Thinning measures within two sprouts in each stump can drastically increase the proportion of superior trees in coppices in the stands.

Key words: density; *Mytilaria laosensis*; natural regeneration; sprouting regeneration; sprout thinning; growth of young stands; diameter distribution

受"里约赫尔辛基进程"森林认证及森林近自然经营的影响，"利用天然更新实现林分皆伐迹地的下一代更新"已引起了广泛的国际关注[1]。相对于人工更新，作为天然更新重要构成的萌生更新及萌生更新林，依自然力可快速实现森林的更新及恢复，更新及恢复的过程无需育苗、炼山、整地、挖坑和造林，具有更新时间短、水土流失少、森林地被物受破坏轻、轮伐期短、经营投资少、见效快等优点[2-5]。因此，对具萌生更新潜力的速生优良树种，开展萌生更新特性及萌生林的培育利用研究，既可实现其人工林采伐迹地环保及高效更新，也能缩短其迹地更新林的轮伐期和提升经营效益，还能快速发展其短周期工业原料林以缓解"天然林保护工程实施后林产工业用材供需不平衡的矛盾"。米老排(*Mytilaria laosensis*)又名壳菜果、三角枫、山桐油，为金缕梅科壳菜果属常绿阔叶乔木，天然分布于我国广东、广西和云南等地，是我国南亚热带区域适生范围广的优良用材树种，具有速生、干形通直、出材率高、材质优良、改良土壤、萌生力强等特性[6-10]。当前，针对米老排伐桩萌生更新特性已有少量研究[11,12]，但有关米老排萌生林的经营技术研究还尚未见报道，而伐桩萌条的保留密度是萌生林培育利用的关键技术。因此，基于短周期工业原料林的经营目标，本研究旨在探索"密度对米老排萌生幼龄林生长及直径分布影响的规律"，进而为米老排萌生林的优化经营提供技术参考。

1 研究区概况

研究区位于广西崇左凭祥市的中国林业科学研究院热带林业实验中心(下文简称"热林中心")哨平实验场39林班2经营班(106°53′13″~106°53′33″E，22°05′12″~22°05′22″N，面积7.4hm²)，属南亚热带季风气候区，干湿季明显，太阳总辐射强，全年日照时数1500~1600h，年均气温21.0℃，≥10℃积温7000~7500℃，最热月平均气温27.5℃，最冷月平均气温13.0℃，极端最低气温-1.0℃，年均降水量1380mm，年蒸发量1370mm。海拔230m，坡度20°~35°，属低山丘陵；土壤为花岗岩发育而成的赤红壤，土层较厚[13]；有机质在土壤A层和B层的质量分数分别为3.71%、1.75%，全氮0.171%，全磷0.030%，pH4.1~4.6。本研究米老排萌生林迹地前茬林为米老排人工林，造林时间1984年春，造林苗木为裸根实生苗，造林密度2500株/hm²，皆伐时间为2011年底(伐前，当年林分种子雨已完全散落)；林分皆伐前经过1次透光伐及1次抚育间伐(株数间伐强度25%~35%)，伐前林分平均胸径、树高、立木株数密度和郁闭度分别为20.2cm、22.2m、825株/hm²和0.9，林分枯落物层厚度5~10cm。林下灌木主要有山苍子(*Litsea cubeba*)、香港大沙叶(*Pavetta hongkongensis*)和三桠苦(*Evodia lepta*)等，草本主要有粽叶芦(*Thysanolaena maxima*)、五节芒(*Miscanthus floridulus*)、山姜(*Alpinia japonica*)等。在皆伐作业后1.5年，迹地天然更新状况良好(调查样格2m×2m，样格合计调查面积1200m²)。米老排种子天然更新幼树树高≥1.3m的平均更新密度为11633.5株/hm²，平均更新频度为90.7%；米老排伐桩平均高为10cm，伐桩萌发率为100%。迹地各处理本底数据详见表1内2013年7月调查数据；迹地伐桩萌生更新与种子更新主要为次年更新。

2 研究方法

2.1 试验设计

2013年春，在研究区米老排皆伐迹地，选择海拔、坡向、坡位和坡度相近的地块为一区组，对米老排伐桩萌条，采用单因素随机区组的试验设计，按基部健壮、无损伤与长势优良的条件，设计3种留萌处理(T_{CA}：保留1株/桩，T_{CB}：保留

2 株/桩，T_{CC}：保留 3 株/桩)，3 个区组(每区组 3 种处理，即 3 个小区)，共 9 个矩形小区(每小区面积为 625m^2)，各小区的分布如图 1 所示。此外，鉴于迹地米老排的伐桩密度低、伐桩分布非完全均匀及种子更新状况良好，对样地内(纵向或横向)间距≥4m 的相邻伐桩，在其中间择优保留 1 株米老排种子更新苗。

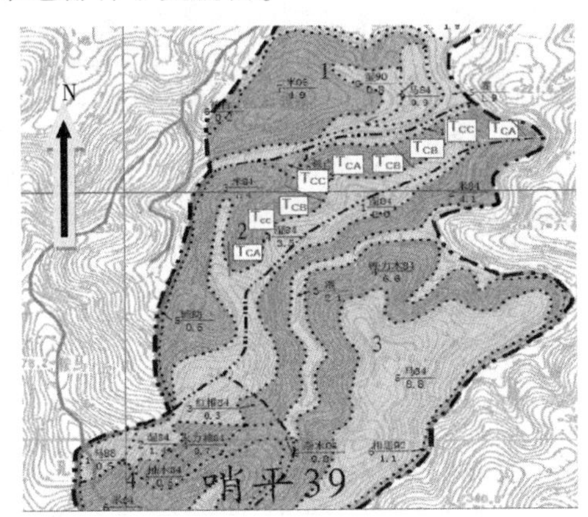

图 1 不同处理米老排萌生幼林样地分布

注：T_{CA}：保留 1 株/桩，T_{CB}：保留 2 株/桩，T_{CC}：保留 3 株/桩；米：米老排，湿：湿地松，马：马尾松；铁力木，红椎，火力楠，相思，柚木；灌：灌木，杂木，辅助。

2.2 样地调查方法与数据处理

在迹地伐桩萌条及种子更新苗定株作业后，对样地保留木编号，调查胸径和树高等。迹地除萌及定株作业和本底调查的完成时间分别为 2013 年 6 月和 2013 年 7 月。2014—2016 年，连续 3 年对样地保留木进行生长观测。基于短周期工业原料林的经营目标，本文着重针对迹地米老排萌生林的生长进行分析，仅在统计迹地林分蓄积量时将种子更新木的生长量纳入分析。萌生林的生长状况如表 1 及表 2 所示，其胸径(d)、树高(h)、单株材积(v)均为样地算术平均值，单株材积计算公式[14]：$v=0.683297\times10^{-4}d^{1.926256}h^{0.8840614}$，萌生林蓄积量为林分内所有萌生木单株材积的累计值，林分蓄积量为林分所有林木单株材积的累计值。为分析萌生林优势木的活力，每个处理在各样地取 2 株萌生优势木。选取的优势木为长势优良、干形通直、主干无分叉、无病虫害的优良林木个体。米老排萌生幼林第 1 年、第 2 年和第 3 年的生长量数据根据表 1 或表 2 的实测数据，用内插法按比例计算求出。

2.3 数据统计分析和林分径阶分布模型拟合

数据统计分析采用数据处理系统软件(DPS14.5)完成，多重比较采用 LSD 法。林分径阶分布拟合软件采用 Eviews8.0 实现，拟合方程为三参数 Weibull 分布密度函数。拟合方程如下：

$$F(\chi)=\begin{cases}0 & x\leqslant a\\ \frac{c}{b}(\frac{x-a}{b})c-1\exp[-(\frac{x-a}{b})^{c}] & x>a,\ b>0,\ c>0\end{cases}$$

式中：a 为位置参数(表示林分最小直径)；b 为尺度参数(表示林分直径分布范围)；c 为形状指数；χ 为阶中值(本研究径阶距为 2cm)。当 $a=0$ 时，三参数 Weibull 分布密度函数降为二参数 Weibull 分布密度函数；当 $c<1.0$ 时，为倒 J 型分布；当 c 值为 1~3.6 时，为正偏山状分布；当 $c=3.6$ 时，分布近似正态分布；当 $c\to\infty$ 时，为单点分布[15]。

3 结果与分析

3.1 密度对米老排萌生幼林平均生长的影响

3.1.1 密度对萌生幼林平均胸径生长的影响

由表 1 分析可知：在前 5 年(即除萌作业后 3.5 年内)，不同密度处理萌生林平均胸径的总生长量和连年生长量均有一定差异，但差异均不显著($P>0.05$)。由图 2 可知：从第 3 年起，T_{CA}处理胸径的连年生长量和平均生长量始终高于其他处理；各处理胸径的连年生长量峰值出现在第 2 年，此后各处理连年生长量开始明显下降；所有处理萌生林胸径的平均生长量与连年生长均在前 4 年相交，基本呈现伐桩萌条保留密度越低，平均生长量与连年生长量相交呈越晚的趋势；除 T_{CC}处理外，在 2~5 年间各处理萌生林平均胸径的连年生长量大于 1cm，表明此阶段萌生林仍处于速生期。

3.1.2 密度对萌生幼林平均树高生长的影响

由表 1 可知：除了定株后的第 1 年，T_{CA}与 T_{CC}处理树高连年生长量差异极显著($P<0.01$)，观测期内各处理间的总生长量和连年生长量差异均不显著($P>0.05$)，表明密度对萌生林树高生长的影响不明显。由图 3 可知：所有处理的树高连年生长量的峰值均出现在第 2 年；所有处理树高的连年生长量与平均生长量均在前 5 年相交；在前 5 年各处理林分平均树高连年生长量均高于 1.1m，表明此阶段萌生林高生长处于速生期。

3.1.3 密度对萌生幼林平均单株材积生长的影响

由表 1 可知：除萌后 2.5~3.5 年间，T_{CA}处理平均的单株材积总生长量和连年生长量开始高于其他处理，密度对不同处理萌生林平均单株材积生长有一定影响，但各处理间差异均不显著($P>0.05$)；除萌 2.5 年后，T_{CA}与 T_{CB}处理的连年生长开始加速，而T_{CC}处理的连年生长则相对平缓(图4)；在除萌后第3.5年，萌生林平均单株材积连年生长量与林分密度成负相关，即林分密度越大其连年生长量越低(图4)。

表1　不同处理米老排萌生更新幼龄林平均生长表现及方差分析

林龄(年)	除萌与定株后时间(年)	调查时间(年-月)	处理	林分密度(株/hm²)	萌条密度(株/hm²)	萌条平均胸径(cm)		萌条平均树高(m)		萌条平均单株材积(×10⁻²/m³)		萌生林蓄积(m³/hm²)		林分蓄积(m³/hm²)	
						总生长量	连年生长量	总生长量	连年生长量	总生长量	连年生长量	总生长量	连年生长量	总生长量	连年生长量
1.5	0	2013-07	T_{CA}	2493	825	1.82±0.49aA	1.22±0.33aA	2.63±0.50aA	1.75±0.34aA	0.06±0.04aA	0.04±0.03aA	0.47±0.31aA	0.31±0.21aA	0.75±0.13cB	0.50±0.09cB
			T_{CB}	2858	1650	1.76±0.49aA	1.17±0.33aA	2.56±0.66aA	1.71±0.44aA	0.05±0.03aA	0.03±0.02aA	0.88±0.50aA	0.59±0.34aA	1.07±0.10bB	0.71±0.07bB
			T_{CC}	3 427	2 485	2.07±0.52aA	1.38±0.35aA	2.86±0.32aA	1.91±0.21aA	0.08±0.04aA	0.05±0.03aA	1.91±1.00aA	1.27±0.67aA	2.18±0.17aA	1.45±0.11aA
2.5	1	2014-06	T_{CA}	2483	815	5.12±1.07aA	3.30±0.58aA	4.48±0.62aA	1.85±0.12bB	0.63±0.34aA	0.58±0.31aA	4.84±1.05bA	4.37±1.04bA	10.44±1.20bA	9.70±1.30aA
			T_{CB}	2 850	1 642	4.64±0.85aA	2.87±0.40aA	4.68±1.04aA	2.11±0.39bAB	0.54±0.26aA	0.49±0.23aA	10.02±4.92abA	9.14±4.42abA	12.88±5.00abA	11.81±4.98aA
			T_{CC}	3 425	2 483	5.16±0.63aA	3.08±0.27aA	5.65±0.32aA	2.79±0.04aA	0.76±0.21aA	0.68±0.17aA	18.70±3.06aA	16.80±2.06aA	21.45±4.34aA	19.27±4.23aA
4	2.5	2015-11	T_{CA}	2 408	794	7.84±1.49aA	1.81±0.28aA	7.53±1.26aA	2.03±0.43aA	2.27±1.20aA	1.09±0.85aA	15.15±3.06bA	6.87±1.43aA	35.43±6.99aA	16.66±3.87aA
			T_{CB}	2 633	1 475	6.60±1.54aA	1.31±0.52aA	7.23±1.45aA	1.70±0.38aA	1.61±0.85aA	0.71±0.43aA	24.61±12.07abA	9.72±4.79aA	33.77±11.86aA	13.93±4.68aA
			T_{CC}	3 258	2 325	7.09±0.99aA	1.29±0.24aA	8.22±0.80aA	1.71±0.32aA	1.97±0.66aA	0.80±0.30aA	44.72±11.15aA	17.34±5.40aA	53.91±15.11aA	21.64±7.19aA
5	3.5	2016-11	T_{CA}	2400	786	9.37±1.57aA	1.53±0.13aA	9.27±1.31aA	1.75±0.09aA	3.80±1.75aA	1.53±0.56aA	24.31±6.05bA	9.16±3.16bA	62.55±9.49aA	27.12±2.89aA
			T_{CB}	2575	1450	7.87±1.81aA	1.27±0.38aA	8.94±1.50aA	1.70±0.39aA	2.70±1.46aA	1.09±0.65aA	41.07±16.69abA	16.46±5.55abA	58.24±14.30aA	24.47±5.10aA
			T_{CC}	3 217	2 292	8.03±1.19aA	0.94±0.25aA	9.79±0.88aA	1.57±0.08aA	2.91±1.02aA	0.95±0.37aA	65.92±15.40aA	21.20±4.89aA	80.36±20.55aA	26.46±5.95aA

注：表内数据为林分样地统计均值±标准差；同列不同小写字母表示同一年相同因子的不同处理间差异显著($P<0.05$)；同列不同大写字母表示同一年相同因子的不同处理间差异极显著($P<0.01$)。下同。

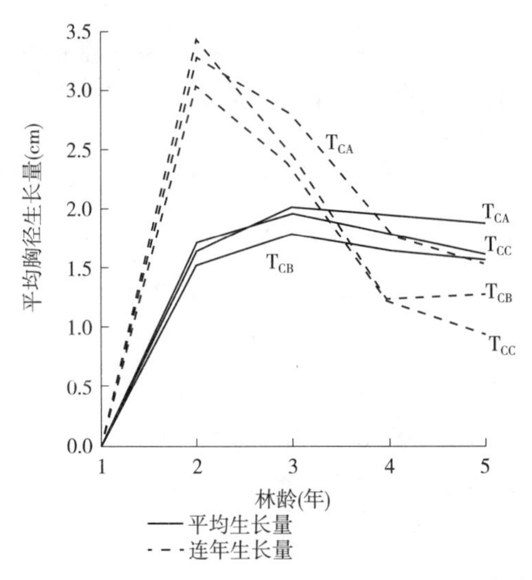

图 2 不同密度萌生幼林平均胸径生长过程曲线

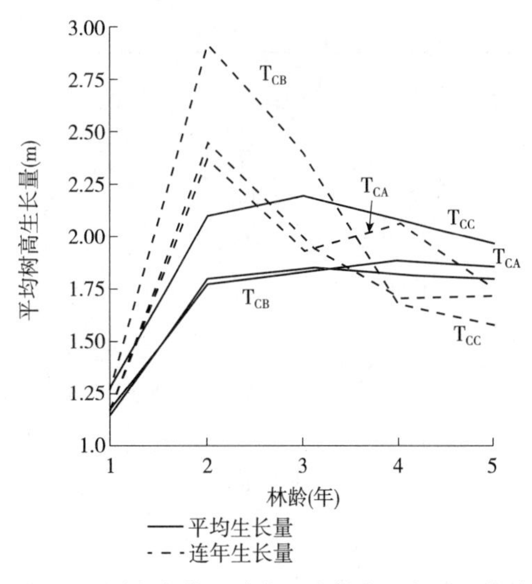

图 3 不同密度萌生幼林平均树高生长过程曲线

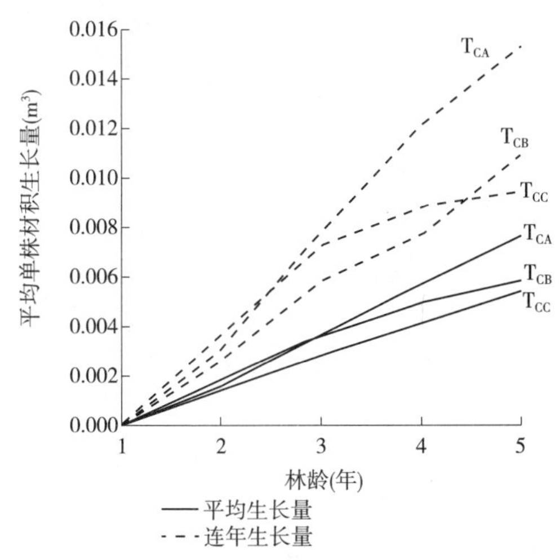

图 4 不同密度萌生幼林平均单株材积生长过程曲线

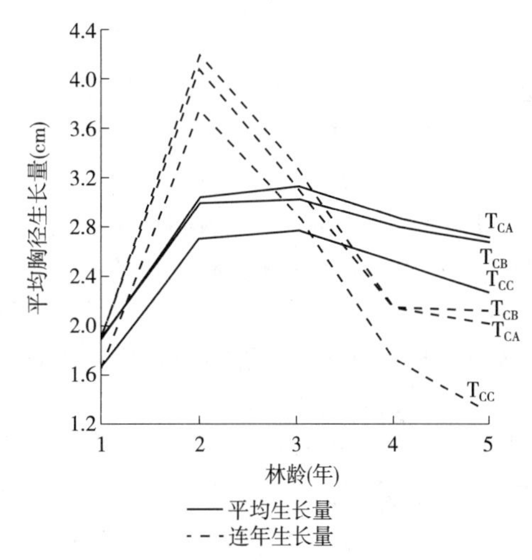

图 5 不同密度萌生幼林优势木胸径生长过程曲线

3.2 密度对萌生更新幼林优势木的生长影响

3.2.1 密度对萌生幼林优势木胸径生长的影响

由表 2 可知：除萌后 1～2.5 年，密度对优势木胸径的生长影响不显著，但除萌后第 3.5 年，密度对优势木胸径总生长量的影响极显著($P<0.01$)，T_{CA}与T_{CB}处理的总生长量高于T_{CC}处理为极显著($P<0.01$)。由图 5 可知：所有处理优势木胸径生长量(连年生长量和平均生长量)的峰值均出现在第 2 年，此后其生长速度开始明显趋缓；所有处理萌生林优势木胸径的平均生长量曲线与连年生长量曲线均在前 4 年相交，此相交时间早于林分平均胸径的相交时间；在前 5 年，萌生林优势木胸径处于速生期，各处理连年生长量大于 1.2cm，平均生长量高于 1.6cm。

表 2 不同处理米老排萌生更新幼龄林优势木生长表现及方差分析

林龄(年)	定株后时间(年)	调查时间(年-月)	处理	萌条密度(株/hm²)	平均胸径(cm)		平均树高(m)		平均单株材($\times10^{-2}/m^3$)	
					总生长量	连年生长量	总生长量	连年生长量	总生长量	连年生长量
1.5	0	2013-06	T_{CA}	825	4.23±0.49aA	2.82±0.33aA	4.08±0.38aA	2.72±0.25aA	0.39±0.12aA	0.26±0.08aA
			T_{CB}	1 650	4.25±0.93aA	2.83±0.62aA	4.52±0.88aA	3.01±0.59aA	0.44±0.21aA	0.29±0.14aA
			T_{CC}	2 485	3.72±0.63aA	2.48±0.42aA	3.98±1.08aA	2.66±0.72aA	0.31±0.17aA	0.21±0.11aA

（续）

林龄（年）	定株后时间（年）	调查时间（年-月）	处理	萌条密度（株/hm²）	平均胸径(cm)		平均树高(m)		平均单株材(×10⁻²/m³)	
					总生长量	连年生长量	总生长量	连年生长量	总生长量	连年生长量
2.5	1.0	2014-06	T_{CA}	815	8.10±0.31aA	3.87±0.19aA	6.35±0.15aA	2.27±0.31aA	1.97±0.18aA aA	1.58±0.06aA
			T_{CB}	1 642	7.82±0.20aA	3.57±0.81aA	6.78±1.13aA	2.27±0.84aA	1.95±0.36aA	1.51±0.38aA
			T_{CC}	2 483	7.23±0.65aA	3.52±0.68aA	6.48±1.36aA	2.50±1.28aA	1.65±0.57aA	1.34±0.49aA
4.0	2.5	2015-12	T_{CA}	794	11.48±0.76aA	2.26±0.30aA	9.78±1.20aA	2.29±0.76aA	5.72±1.34aA	2.50±1.17aA
			T_{CB}	1 475	11.22±0.47aA	2.27±0.25aA	9.80±1.16aA	2.01±0.45aA	5.44±0.99aA	2.33±0.43aA
			T_{CC}	2 325	10.00±0.49aA	1.84±0.17aA	9.88±0.52aA	2.27±0.57aA	4.39±0.60aA	1.82±0.06aA
5.0	3.5	2016-12	T_{CA}	786	13.48±0.75aA	2.00±0.17abA	11.55±1.10aA	1.77±0.20aA	8.99±1.69aA	3.27±0.46aAB
			T_{CB}	1 450	13.32±0.41aA	2.10±0.31aA	12.05±0.44aA	2.25±0.83aA	9.04±0.54aA	3.60±0.57aA
			T_{CC}	2 292	11.88±1.27bB	1.27±0.54bA	11.72±0.76aA	1.83±0.33aA	6.40±0.56bA	2.01±0.57bB

3.2.2　密度对萌生幼林优势木树高生长的影响

与萌生林平均树高的生长类似，在第2年后，优势木树高平均生长量呈明显递减趋势，但各处理树高连年生长量仍大于1.70m，表明其树高生长仍处于旺盛期(图6，表2)，萌生林优势高生长的影响差异不显著($P>0.05$)。

3.2.3　密度对萌生幼林优势木平均单株材积生长的影响

由表2可知：在除萌后1~2.5年，各处理优势木单株材积的总生长量虽有一定差异，但差异不显著($P>0.05$)；但在除萌后第3.5年，T_{CA}与T_{CB}处理的平均单株材积总生长量显著($P<0.05$)高于T_{CC}；在前5年，T_{CA}与T_{CB}处理单株材积的平均生长量与连年生长量差异不显著($P>0.05$)。与萌生林平均单株材积生长类似，在除萌2.5年后，T_{CA}与T_{CB}处理的连年生长开始加速，而T_{CC}处理的连年生长则相对平缓(图7)。

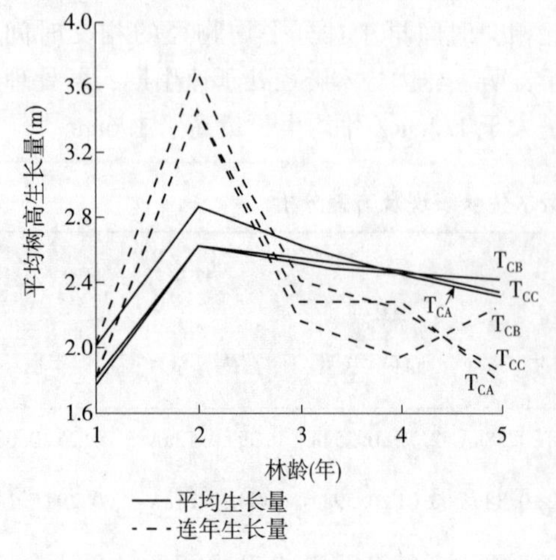

图6　不同密度萌生幼林优势木树高生长过程曲线

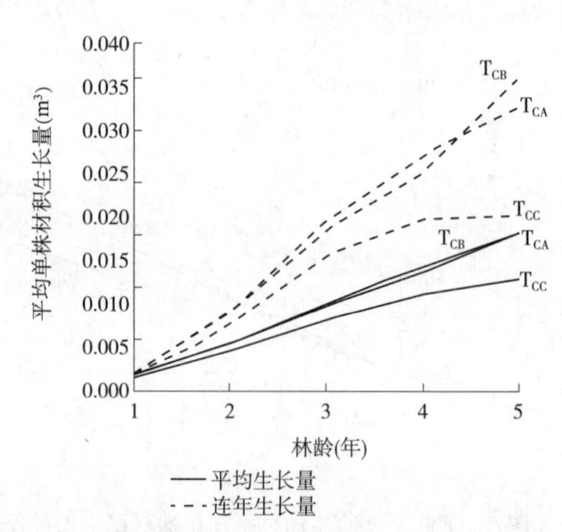

图7　不同密度萌生幼林优势木单株材积生长过程曲线

3.3　密度对蓄积量生长的影响

3.3.1　密度对萌生幼林蓄积的影响

由表1分析可知：受伐桩萌条保留密度的影响，T_{CA}处理各年萌生林的蓄积量均低于其他处理，但T_{CA}处理与T_{CB}处理各年萌生林的蓄积量差异均不显著($P>0.05$)；与T_{CC}处理相比，除了定株作业当年，T_{CA}处理与T_{CC}处理的差异不显著($P>0.05$)，在定株作业后1-3.5年，T_{CA}处理的萌生林蓄积量均显著低于T_{CC}处理($P<0.05$)。

3.3.2　密度对林分蓄积生长量的影响

由表1及图8综合可知：在林分蓄积量的总生长量方面，除作业后的第1年T_{CA}与T_{CC}处理林分蓄积量(包含种子更新木材积)的总生长量差异显著($P<0.05$)外，定株作业第2.5年后，虽然各处理蓄积量的总生长量有一定差异，但各处理的差异已经不显

著($P>0.05$)；在蓄积量的连年生长量方面，除定株作业第1年内T_{CA}处理的蓄积连年生长量低于T_{CB}与T_{CC}处理，在定株作业后第2.5年时T_{CA}处理的蓄积连年生长量开始高于T_{CB}处理，而T_{CC}处理连年生长量的增速则放缓(图8)；至定株作业后第3.5年T_{CA}处理的连年生长量开始高于T_{CC}处理。

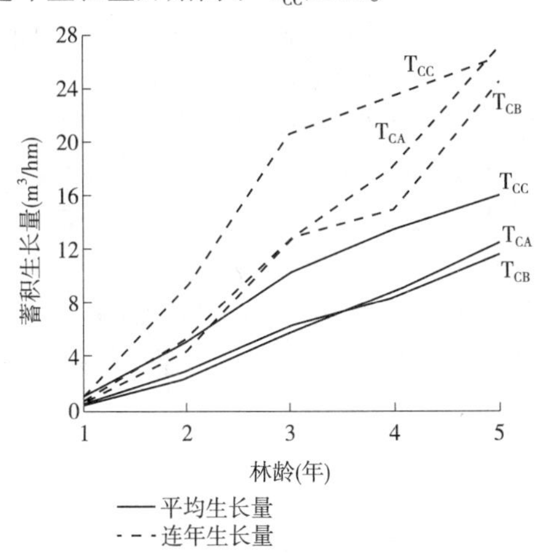

图8 不同处理林分蓄积量生长过程曲线

3.4 密度对萌生幼林直径分布结构的影响及模型拟合

3.4.1 密度对萌生幼林直径分布结构的影响

由图9可知：在除萌后1~3.5年，不同处理萌生林径阶分布范围受密度影响不明显，但同年不同处理优势木的比例不同。T_{CA}处理优势木的比例明显高于T_{CB}与T_{CC}处理。如：在除萌的第1年、第2.5年和第3.5年，萌生林对应优势木胸径的最低临界值分别为7.23cm、10.00cm、11.88cm(表2)，在同一年龄T_{CA}处理优势木的百分比明显高于其他处理；且在除萌后第2.5年与第3年，T_{CA}与其余处理的径阶分配曲线，其在相同年龄高于优势木径阶临界值部分的相交面积(即T_{CA}处理优势木比例与其他处理优势木比例的差值)明显比除萌第1年对应的相交面积大，表明除萌措施提高优势木比例的作用效果明显。在除萌后第3.5年，优势木所占的分配比例与密度呈负相关，即$T_{CA}>T_{CB}>T_{CC}$。除萌后3.5年内，各处理的径阶分布峰值的比例逐年降低；除萌后第3.5年，T_{CA}处理径阶分布峰值对应的径阶值开始高于T_{CB}与T_{CC}处理的对应值。

3.4.2 萌生幼林直径分布结构的模型拟合

径阶分布结构是林分生长稳定性及其株间竞争的主要指标，了解和掌握林分的径阶分布结构，能为林分经营管理提供理论依据[16]。大量的研究表明，Weibull分布函数在林分的直径分布拟合中有较大的适应性和灵活性[15]，本研究直接运用Weibull分布函数对5年生米老排萌生林径阶分布结构进行拟合。由表3分析可知：除T_{CA}处理的拟合优度稍低于0.7(原因可能为T_{CA}处理建模的样本数相对偏少)，其他处理的拟合优度均大于0.7，表明模型的拟合效果良好；三种处理林分拟合模型的参数C值均在1~3.6年间，林分径阶分布均为单峰正偏山状分布，表明其林分处于竞争期的自然稀疏后期[15]。

表3 5年生米老排萌生林径阶分布回归模型拟合

处理	样本数	参数			R^2
		A	B	C	
T_{CA}	95	0	10.92606±0.680041	3.197556±0.502770	0.689237
T_{CB}	174	0	8.469957±0.382323	2.011993±0.139311	0.881419
T_{CC}	275	0	8.725357±0.298281	3.287058±0.291325	0.916055

注：表内参数均值±标准差为模型参数估计值在0.05显著水平下的标准误($P<0.05$)。

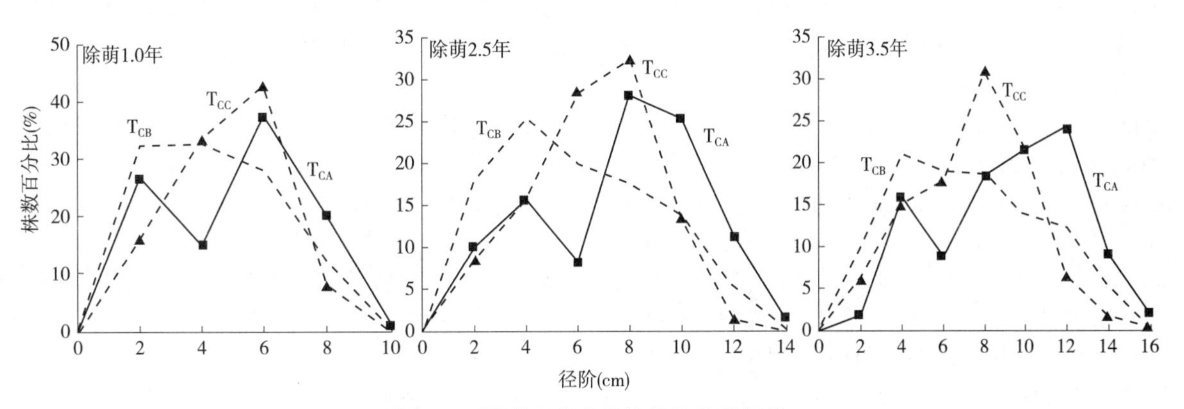

图9 不同处理萌生幼林径阶分配规律

4 结论与讨论

4.1 讨论

（1）本研究5年生米老排萌生林的生长状况（平均胸径7.87~9.37cm、平均树高8.94~9.79m）与热林中心同龄（林龄包含苗龄）不同密度米老排实生人工林（平均胸径6.27~8.64cm、树高8.71~9.00m）[17]相比，虽然前者土壤肥力（有机质含量）明显低于后者（有机质含量A层5.72%~6.80%），但前者的生长量却明显优于后者，这与前人的研究结果即在树种相同及立地条件相近的前提下，萌生林前期的生长速度比实生林快[2,18]一致。

（2）米老排萌生林平均胸径和平均树高的生长规律即林分胸径或树高的平均生长量与连年生长量相交的时间均在前5年，这与前人对米老排实生人工林的生长研究结果[17]一致。

（3）传统经营米老排人工林主伐时林分密度通常为450~600株/hm²，皆伐作业后，迹地伐桩的密度低，若仅靠迹地伐桩的萌生更新，其迹地萌生更新苗的分布频度还远未达到采伐迹地天然更新的良级标准[19]（健壮苗有效更新株数>3000株/hm²，更新频度>60%）。因此，在当年米老排林分种子雨散落完成后，及时采伐林分及清理迹地灌草和剩余物，充分促进和利用其采伐迹地的萌生更新与种子天然更新，这是确保皆伐迹地成功天然更新的关键。

（4）不同处理5年生米老排萌生林林分的蓄积量的高低排序为$T_{CC}>T_{CA}>T_{CB}$；林分平均胸径的大小排序为$T_{CA}>T_{CC}>T_{CB}$；林分优势木比例含量最高的处理为T_{CA}处理。由此可知：若米老排萌生林的经营目标是短周期的纤维原料林（对林木的径级无要求，全杆材可利用），采取TCC处理经营为最佳（林分可收获的木材纤维产量最高）；而若林分经营的目标是短周期的胶合板原料林（大径级的规格材越多越好），则选择T_{CA}处理经营更适宜（林分的平均胸径最大，且可生产大径级规格材的比例最高）。

（5）优势木为林分中最有活力的林木，是代表林分立地质量和生产力的重要指标，也是森林全周期多功能作业法设计的关键核心参数。因此，研究分析米老排萌生林各处理优势木的胸径、树高、单株材积的生长是必要的。

（6）米老排（实生）人工林的数量成熟年龄大于15年[8,20]，由于本研究试验林分的观测时间较短，且目前研究区还没有10年以上的萌生林，因此在采伐收获前，不同处理米老排萌生林的成熟龄（经济成熟、数量成熟或工艺成熟）在什么时间节点？在现有或给定的木材市场中，哪种米老排萌生更新林的经济效益更高？与米老排实生更新林（种子天然更新林或人工植苗更新林）相比，萌生林相对速生的优势能延续到什么时间节点等？这些问题仍需今后进一步研究。

4.2 结论

（1）除萌后3~5年，密度对萌生林的平均生长及其优势木树高和生长量影响不显著，对萌生林优势木单株材积的总生长量影响显著（$P<0.05$），对萌生林优势木胸径总生长量及单株材积连年生长量的影响极显著。

（2）萌生林平均胸径与优势木胸径的连年生长量峰值均出现在第2年，两者的平均生长量与连年生长量相交的时间均在前4年。前5年为萌生林径向生长的旺盛期，伐桩萌条的合理保留密度及幼林抚育对萌生林的径向生长极为重要。

（3）萌生林平均树高与优势木树高的连年生长量峰值均出现在第2年，前者树高平均连年生长量与平均生长量的相交时间在前5年，后者的相交时间在前3年；密度对萌生林树高生长量的影响不显著。

（4）萌生林直径分布为近似单峰正偏山状分布，处于竞争期的自然稀疏后期，使用Weibull分布函数拟合效果良好；每伐桩保留1~2株萌条的除萌经营措施，对提升萌生林优势木比例的作用明显。

参考文献

[1]STOKES V, KERR G. Long-term growth and yield effects of respacing natural regeneration of Sitka spruce in Britain [J]. European Journal of Forest Research, 2013, 132: 351-362.

[2]黄旺志，郑芳，李党法，等．豫南杉木萌芽林生长规律的研究[J]．信阳师范学院学报（自然科学版），2006，19(2)：191-194.

[3]田晓萍．杉木萌芽更新的研究[D]．福州：福建农林大学，2008，04.

[4]陈沐，曹敏，林露湘．木本植物萌生更新研究进展[J]．生态学杂志，2007，26(7)：1114-1117.

[5]李荣，张文辉，何景峰，等．辽东栎伐桩萌苗的发育规律[J]．林业科学，2012，48(3)：82-87.

[6]白灵海，唐继新，明安刚，等．广西大青山米老排人工林经济效益分析[J]．林业科学研究，2011，24(6)：784-787.

[7]袁洁．8个米老排天然群体的遗传多样性研究[D]．北京：中国林业科学研究院，2014，06.

[8]郭文福，蔡道雄，贾宏炎，等．米老排人工林生长规律

的研究[J]. 林业科学研究, 2006, 19(5): 585-589.
[9]王克建, 蔡子良. 热带树种栽培技术[M]. 南宁: 广西科学技术出版社, 2008. 8.
[10]郭文福. 米老排人工林生长与立地的关系[J]. 林业科学研究, 2009, 22(6): 835-839.
[11]庞圣江, 张培, 刘福妹, 等. 大青山林区米老排人工林伐桩萌芽更新研究[J]. 西北林学院学报, 2016, 31(6): 153-156.
[12]张显强. 米老排人工林萌芽更新研究[D]. 南宁: 广西大学, 2016, 11.
[13]汪炳根, 卢立华. 广西大青山实验基地森林立地评价与适地适树研究[J]. 林业科学研究, 1998, 11(1): 78-85.
[14]陈永富, 郭文福, 黄镜光. 米老排立木材积表及地位指数的编制[J]. 林业科学研究, 1991, 4(增刊): 116-119.
[15]孟宪宇. 测树学[M]. 3版. 北京: 中国林业出版社, 2006.
[16]楚秀丽, 刘青华, 范辉华, 等. 不同生境、造林模式闽楠人工林生长及林分分化[J]. 林业科学研究, 2014, 27(4): 445-453.
[17]李炎香, 谭天泳, 黄镜光, 等. 米老排造林密度试验初报[J]. 林业科学研究, 1988, 1(2): 206-212.
[18]孙长忠, 王开运, 任兴俄, 等. 渭北黄土高原刺槐萌生林生长状况的调查研究[J]. 西北林学院学报, 1993, 8(2): 36-40.
[19]于汝元, 黄晓鹤, 哈达, 等. 东北林区森林更新调查及评定中的问题[J]. 北京林业大学学报, 1990, 12(增刊3): 46-50.
[20]吴淑玲. 米老排在漳州长泰生长规律初步研究[J]. 防护林科技, 2016, (8): 22-25.

[原载: 北京林业大学学报, 2018, 40(05)]

南亚热带米老排人工林皆伐迹地天然更新研究

唐继新[1,2]　贾宏炎[1]　曾　冀[1]　李忠国[1]　庞圣江[1]　郝　建[1]　赵　总[1]
([1]中国林业科学研究院热带林业实验中心，广西凭祥　532600
[2]中国林业科学研究院资源信息研究所，北京　100091)

摘　要　基于南亚热带米老排人工林落种后皆伐迹地的天然更新调查资料，对其皆伐迹地天然更新幼树的树种、起源、数量、空间格局、分布频度和树高结构进行研究，结果表明：①皆伐迹地上绝大多数的天然更新树种为米老排树种。②在0~10m空间尺度内，米老排种子更新幼树的分布为聚集分布($P<0.01$)。③米老排种子更新幼树的数量和频度，可达到森林天然更新的良级标准，亦符合采伐迹地人工更新成林的验收标准。④种子天然更新幼树树高的分布近似正偏山状的weibull密度函数。⑤在米老排人工林皆伐作业后1.5年左右，对采伐迹地天然更新幼树进行间苗定株较为适宜。⑥对米老排人工林适时适地的采用小面积皆伐、非炼山方式清理迹地剩余物、保护和利用迹地天然更新幼树的方式，可有效实现采伐迹地的森林更新。
关键词　米老排人工林；皆伐迹地；天然更新；空间分布格局；树高结构

Natural Regeneration on Clearcut Land of *Mytilaria laosensis* Plantation in South Subtropical Area of China

TANG Jixin[1,2], JIA Hongyan[1], ZENG ji[1], LI Zhongguo[1], PANG Shengjiang[1], HAO Jian[1], ZHAO Zong[1]

([1]*Experimental Center of Tropical Forestry*, *CAF*, *Pingxiang* 532600, *Guangxi*, *China*;
[2]*Research Institute of Forest Resources Information Techniques*, *CAF*, *Beijing* 100091, *China*)

Abstract: Based on surveyed data of natural regeneration on clearcut land of *Mytilaria laosensis* plantation, while which seeds were matured and fell, in the south subtropical of china. Tree species, origin, quantity, spatial distribution pattern, distribution frequency and height structure of the natural regeneration on the clearcut land, which were analyzed, and results showed that: ① on the clearcut land, the most majority of tree species of natural regeneration is *Mytilaria laosensis* tree species. ②bellow the scale of 10m, natural regeneration seedling was aggregation distribution($P<0.01$). ③the quantity and frequency of seedling, which not only can reach the standard of a good level of the forest natural regeneration, but also meet the acceptance criteria of artificial regeneration on the clearcut land. ④the distribution density function of tree height class of natural regeneration with sapling seed, which is approximate to Weibull density function with partial mountain shape. ⑤ it is a good time to thinning the natural regeneration seedlings, in about 1.5 years after clear cutting of the plantation. ⑥ it can made effective natural regeneration on the small area clear-cutting land of *Mytilaria laosensis* plantation, while seed maturity and falling, by using clear-cutting, no burning and manually disposing of forest harvesting slash on clearcut overland, protecting and utilizing natural regeneration of sapling on the clearcut land.
Key words: *Mytilaria laosensis* plantation; clearcut land; natural regeneration; spatial distribution pattern; tree height structure

依靠自然力而实现森林再生的天然更新林，具有生物多样性高、育林成本低、更新过程地表水土流失少、森林群落结构稳定等优点[1,3]；然而，森林天然更新的顺利完成并非易事，尤其是种子更新的障碍更为突出，从种子的产生、扩散、萌发，到幼苗的定居，及幼树的建成等阶段，每一阶段都面临着自然因素和人为活动的干扰，且任何环节的干扰都能致使其更新发生障碍[4]。因此，天然更新一直是森林经营、森林生态系统研究的热点内容[5,6]，然而以往对森林天然更新的研究，主要集中在天然林方面，对人工林天然更新的研究相对较少[7,8]，还未能满足我国人工林可持续经营的需要。因此，对促进和利用人工林的天然更新进行研究，具有重要的理论和实践意义。

米老排(*Mytilaria laosensis*)又名壳菜果、三角枫、山桐油，为金缕梅科壳菜果属常绿阔叶乔木，天然分布于我国广东、广西和云南以及越南和老挝等地，是我国南亚热带区域适生范围广的用材树种，具有速生、干形通直圆满、材质优良、改良土壤、萌生力强、落种于林缘及空旷地宜萌发等特性[9,14]。广西和广东是米老排人工林的主要经营种植区，经营面积已达 2500hm^2[10]，目前不少林分已进入可主伐利用的成熟期，基于米老排树种的生物学特性，如何实现米老排树种伐后迹地的高效更新？对米老排人工林的可持续经营具有重要的理论及现实意义。自上世纪 80 年代起，国内不少研究机构及学者对米老排树种进行了系统研究，已基本摸清了其生物形态、生长、材性、制浆性能等，并解决了其采种育苗、造林选地、初植密度和林分经营等技术问题[11]，然而有关“米老排人工林采伐迹地的天然更新研究”，目前还尚未见报道。本研究的实施旨在探索米老排人工林采伐迹地天然更新幼树的树种、起源、数量、分布和树高结构规律，为促进、保护和利用米老排人工林迹地的天然更新的经营提供理论依据。

1 研究区概况及研究方法

1.1 研究区概况

研究区位于广西凭祥市的中国林业科学研究院热带林业实验中心(下文简称热林中心)哨平实验场 39 林班(22°5′10″N，106°53′14″E)，属南亚热带季风气候区，干湿季明显，太阳总辐射强，全年日照时数 1500～1600h，年均气温 21.0℃，≥10℃积温 7000～7500℃，最热月平均气温 27.5℃，最冷月平均气温 13.0℃，极端最低气温 -1.0℃，年均降水量 1380mm，年蒸发量 1370mm。海拔 230m，属低山丘陵；土壤为花岗岩发育而成的赤红壤，土层较厚。更新迹地前茬林为米老排人工纯林，造林时间 1984 年春，造林苗木为裸根实生苗，造林密度 2500 株/hm^2；前茬林皆伐前经过 1 次卫生伐及 1 次抚育间伐(株数间伐强度 20%～30%)，皆伐时间为 2011 年底(林分皆伐后，人工清理采伐剩余物，未炼山)，伐前林分的平均胸径 20.2cm、平均树高 18.7m 和郁闭度为 0.9；皆伐前，林分枯落物层厚度 5～10cm，林下灌木主要有山苍子(*Litsea cubeba*)、香港大沙叶(*Pavetta hongkongensis*)和三椏苦(*Evodia lepta*)等，草本主要有粽叶芦(*Thysanolaena maxima*)、五节芒(*Miscanthus floridulus*)、山姜(*Alpinia japonica*)等。

1.2 样地调查方法

2013 年春，经过实地踏查，在皆伐天然更新迹地上，选择具有代表性的地块，布设 2 块矩形标准地：标准地Ⅰ(面积：30m×20m；坡向：正西；坡度：25°；坡位：全坡)，标准地Ⅱ(面积：30m×20m；坡向：西北；坡度：25°；坡位：全坡)。标准地调查采用相邻网格法，即先将标准样地划分为 6 个 10m×10m 的样方，再将 10m×10m 的样方划分 4 个为 5m×5m 的小样方，并以 5m×5m 小样方为调查单元，调查样方内所有树高≥0.3m 天然更新的幼树，幼树调查指标：树种、起源、数量、树高、胸径(或地径)、坐标。

1.3 空间格局及更新频度分析方法

空间格局分析。本研究空间格局的分析全部采用空间点格局分析法，Ripley′s 函数[15]的定义为：

$$\hat{K}(r) = A\sum_{i}^{n}\sum_{j}^{n} w_{ij}I_r(i, j)/n^2$$

$$i, j = 1, 2, \cdots, i \neq j, d_{ij} \leq d \qquad (1)$$

式中：n 为样方内植物个体数；r 为距离尺度；d_{ij}是植物个体 i 和 j 之间的距离；A 为样地面积；w_{ij}是以 i 为圆心；以 j 为半径的圆周长落在样地内的长度与该周长的比例的倒数；n 为样地中个体总数。$I_r(i, j)$ 当 $d_{ij} \leq r$ 时其取值为 1；当 $d_{ij} > r$ 时取值为 0[15-17]。

$$\hat{L}(r) = \sqrt{\hat{K(r)}/\pi} - r \qquad (2)$$

当 $\hat{L}(r)<0$ 时，为均匀分布；当 $\hat{L}(r)=0$ 时，为随机分布；当 $\hat{L}(r)>0$ 时，为聚集分布；$\hat{L}(r)$ 的置信区间采用 Monte-Carlo 方法求得，当研究置信水平取 99%，要对每个样方的个体数模拟 100 次；当种群表现为聚集分布时，把偏离随机置信区间的最大值作为聚集强度指标，聚集尺度为对应聚集强度的尺度，而聚集规模为以聚集强度为半径的圆[16,17]。

更新频度分析。结合热林中心营林实践，米老排的造林密度一般为 2m×2m，为评价皆伐迹地米老排种子更新幼树相对传统人工造林的空间分布质量，基于标准地幼树的数量、生长和坐标等调查信息，在计算机上重新以 2m×2m 的小样方将标准样地进行分割，统计标准样地小样方(2m×2m)的更新频度，更新频度=有更新幼树的样方数量/标准地样方总数。

2 结果与分析

2.1 种子天然更新研究

2.1.1 更新分布研究

基于皆伐迹地种子更新幼树(高≥0.3m)调查数据，及种群空间格局的计算式(公式1和2)，可绘制标准样地种子更新幼树的分布图(图1)和空间点格局分析结果图(图2)。由图1和图2分析可知，米老排皆伐迹地种子更新幼树的密度较高，在0~10m的研究尺度内，其更新种群显著偏离随机分布，并呈聚集分布($P<0.01$)。其中，标准样地Ⅰ、Ⅱ的聚集强度分别为0.31和0.20，对应的聚集尺度r分别为5m和2m。

2.1.2 更新数量、频度和生长分析

为摸清采伐迹地种子更新幼树的数量、频度、生长等状况，及评价采伐迹地的种子更新幼树能否成林。由表1分析可知：标准地内树高(h)≥0.3m种子更新幼树平均的更新密度、更新频度、更新高度分别为19941.5株/hm^2、96.0%和1.60m，表明米老排皆伐迹地种子更新幼树的密度、分布和生长状况均较好，不仅高于国家森林天然更新数量的良级标准(10000株/hm^2)[18]，亦优于“李复春针对采伐迹地天然更新提出的数量和频度的良级标准”(有效更新株数>3000株/hm^2，频度>60%)[18]。此外，参照热林中心采伐迹地人工更新成林验收的主要标准(苗木保存率>80%，幼林平均高大于1.5m。)，及热林中心米老排不同密度育苗试验研究(表2)，标准地内树高>1.3m更新幼树平均的更新密度、更新频度、更新高度和更新胸径分别为11633.5株/hm^2、90.7%、2.08m和1.18cm，表明迹地种子更新幼树的质量，不仅达到了采伐迹地人工更新成林的主要验收标准，亦优于苗圃其不同密度育苗处理同龄苗木的生长质量(苗圃苗木树高和地径的试验均值分别为1.58~1.80m、1.22~1.52cm)。

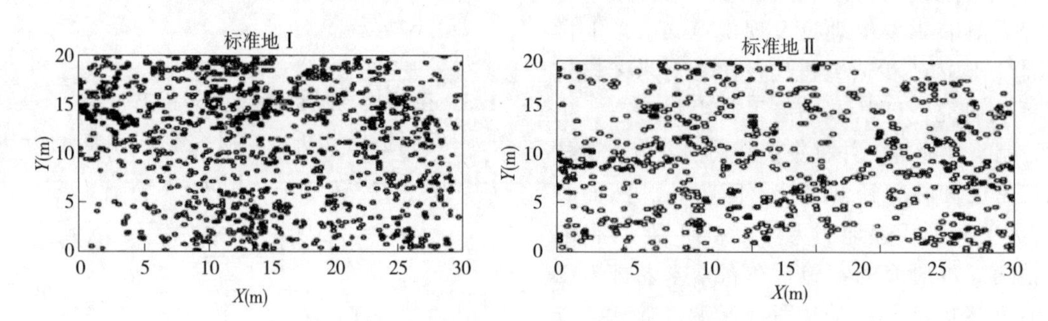

图1 米老排皆伐迹地种子更新幼树分布

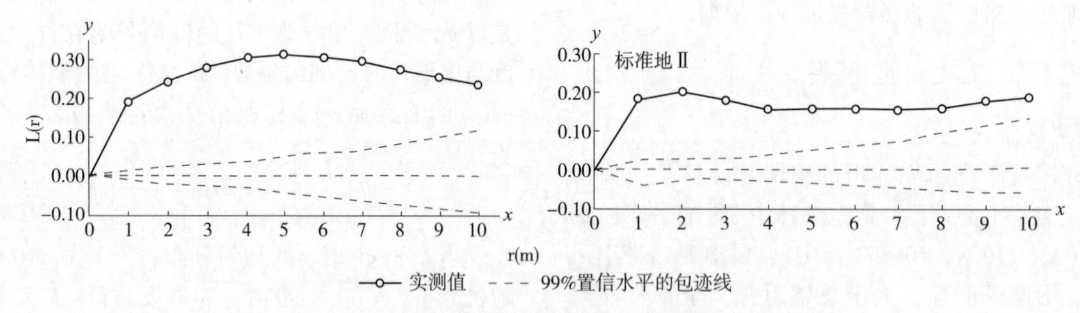

图2 米老排皆伐迹地种子更新幼树空间点格局分析结果

表1 米老排人工林皆伐迹地种子更新状况

标准地	树高≥0.3m			树高>1.3m			
	更新密度(株/hm^2)	更新频度(%)	更新平均高(m)	更新密度(株/hm)	更新频度(%)	更新平均高(m)	更新平均胸径(cm)
Ⅰ	22633	98.0	1.47	11950	88.3	2.00	1.16
Ⅱ	17250	94.0	1.72	11317	93.0	2.16	1.20
平均值	19941.5	96.0	1.60	11633.5	90.7	2.08	1.18

注：表中数据为米老排人工林皆伐后1.5年迹地种子更新状况，其更新林平均郁闭度为0.6；样地主要天然更新树种为米老排，偶见树种有红椎(*Castanpsis hystrix*)和枫香(*Liquidambar formosana*)。

树种有红椎(*Castanpsis hystrix*)和枫香(*Liquidambar formosana*)。

表 2　不同密度育苗试验的米老排苗木生长状况

处理	密度(株/m^2)	苗龄(年)	苗木平均高(m)	苗木平均地径 (cm)
1	25	1.5	1.72	1.35
2	20	1.5	1.61	1.30
3	15	1.5	1.80	1.52
4	10	1.5	1.58	1.22

注：表内数据来源于中国林业科学研究院广西大青山实验局(即“热林中心”的前身)和热带林业科学研究所的“米老排丰产林组装技术的研究专项研究报告集——米老排壮苗培育技术的研究”。

2.1.3　更新群落树高结构分析

由图 3 分析可知：标准地内种子更新幼树树高≥0.3m 的树高分布近似正偏山状的 weibull 密度函数[19]，标准地Ⅰ和标准地Ⅱ树高分布函数的形状参数 c 值分别为 2.263 和 2.231，尺度参数 b 值分别为 1.66 和 1.94；标准样地种子更新幼树树高分布频率的规律：在一定范围内，树高的分布频率先随树高值增加而增加，随后再逐渐减少，其中标准地Ⅰ与标准地Ⅱ树高分布频率增减的理论期望临界值分别为 1.66m、1.94m(即对应标准样地树高分布函数的尺度参数值 b)。标准样地树高分布频率的数据表明：在米老排人工林皆伐作业后 1.5 年时，其采伐迹地的种子更新幼树树高分布最高频率的区间为 1.66～1.94m，此时采伐迹地大部分的种子更新幼树树高已基本高出采伐迹地的杂灌草面，结合表 1 标准样地种子更新幼树树高>1.3m 的更新数量、更新频度、平均树高和平均胸径数值进行综合分析，故在采伐作业后 1.5 年左右对皆伐迹地的种子更新幼树进行间苗定株，是较为适合的。

2.2　萌生更新分析

由表 3 分析可知：在林分皆伐作业后 1.5 年时，迹地米老排伐桩的萌发及萌条的生长状况极好，伐桩萌条的萌发率、平均高、平均胸径的均值分别为 100%、2.94m 和 2.10cm，与表 1 相比，伐桩萌条的胸径和树高生长显著优于同期种子更新幼树的生长；有关研究表明，当伐桩萌条高在 1.5～2.0m 时定株更有利[20]，若对采伐迹地实行短周期工业原料林的经营目标，为降低伐桩除萌的生产成本，及避免伐桩萌条过多不利于萌条的生长，故米老排伐桩的定株工作最好为 1～1.5 年。

表 3　标准地米老排伐桩萌生更新状况

标准地	伐桩数	伐桩萌发率(%)	萌条平均高(m)	萌条平均胸径(cm)
Ⅰ	88	100	2.91	2.16
Ⅱ	91	100	2.96	2.05
平均值	89.5	100	2.94	2.10

注：因米老排伐桩萌条较多(数量少则十几株，多则几十株)，若对伐桩所有萌条进行调查，工作量过大，参照萌芽林经营实践(大多伐桩保留 1～3 株萌条)，故本研究对样地伐桩萌条高度排列前 3 的进行统计。

2.3　天然更新林经济效益分析

相对传统人工造林，米老排皆伐迹地天然更新林的生长及分布状况良好，不需再对采伐迹地人工造林，可间接节省炼山、清山、挖坑、整地、定植、造林基肥和苗木培育等环节的造林成本。有研究表明米老排人工林传统的造林模式每公顷需 5435 元[9]，且随着未来劳力的紧缺呈逐年高涨的趋势。因此，促进和利用米老排人工林迹地的天然更新，营林成本节省明显，经济效益提升显著。

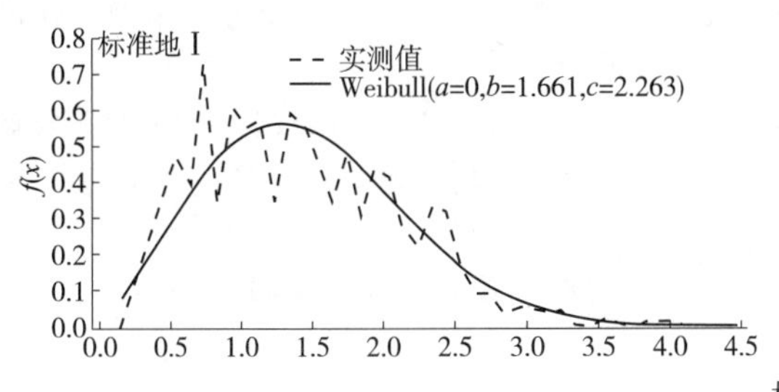

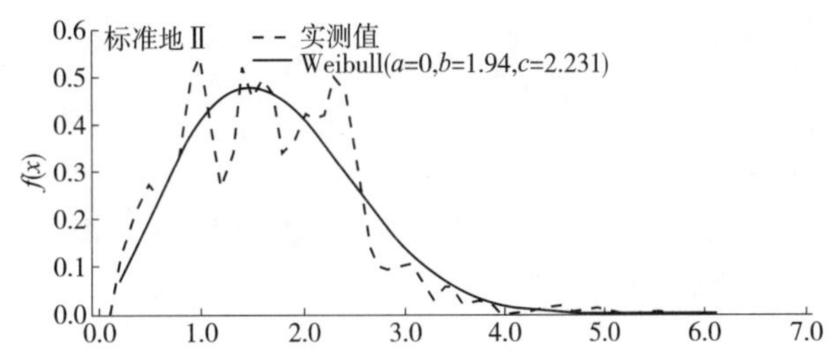

树高(m)

图 3　米老排皆伐迹地种子更新幼树树高概率分布

3　讨论与结论

(1)在皆伐作业后1.5年时，米老排皆伐迹地的主要天然更新树种为米老排树种，在0~10m的尺度范围内，种子更新幼树的空间分布为聚集分布，且种子天然更新幼树的树高分布近似正偏山状的weibull密度函数。

(2)在皆伐作业后1.5年时，皆伐迹地米老排种子的更新状况良好(样地树高≥0.3m更新幼树平均的更新密度和更新频度分别为19941.5株/hm²、96.0%)，符合到采伐迹地森林天然更新的良级标准，亦可达到采伐迹地人工更新成林的验收标准；米老排伐桩萌条(树高和胸径)的生长优于种子更新幼树的生长状况。若对采伐迹地实行大径材的培育目标，此时除萌及对种子更新幼树进行定株，是较为适宜的。因伐桩萌条的生长快于种子更新幼树的生长，过迟除萌伐桩萌条会对种子更新幼树的生长不利，而过早除萌则迹地杂草灌木会较多，亦不利于迹地种子的萌发及生长。

(3)相对传统人工更新的造林方式，充分利用米老排树种种子的成熟期、落种期、天然更新特性(落种在林缘及空旷地易萌发、伐桩易萌生)的生物学特性，对适宜皆伐的米老排人工商品林适时采用小面积皆伐、非炼山清理采伐迹地剩余物、保护和利用采伐迹地的天然更新幼树，可有效实现采伐迹地森林的更新。

(4)对米老排人工林皆伐迹地的天然更新幼林，实行大径材的经营模式还是短周期工业原料林的经营模式，每种经营模式的全周期经营参数(不同起源更新林适宜的密度及其生长过程、抚育间伐时间及强度、营林收益等)如何？此外，米老排皆伐迹地种子更新的机理(即影响米老排种子天然更新的关键因子)是什么？目前还尚不清楚，因此，本研究具有一定的局限性，这些问题还有待后续的进一步研究。

参考文献

[1] 王娜，郝清玉. 森林天然更新影响因子研究进展[J]. 广东农业科学，2012(6)：68-70.

[2] 田晓萍. 杉木萌芽更新的研究[D]. 福州：福建农林大学，2008.

[3] 李喜霞，刘明国，李海春. 朝阳地区人工采伐带内油松天然更新研究[J]. 沈阳农业大学学报，2003，34(1)：35-39.

[4] 朱教君，刘足根，王贺新. 辽东山区长白落叶松人工林天然更新障碍分析[J]. 应用生态学报，2008，19(4)：695-703.

[5] 韩有志，王政权. 森林更新与空间异质性[J]. 应用生态学报，2002，13(5)：615-619.

[6] 张群，范少辉，沈海龙，等. 次生林林木空间结构等对红松幼树生长的影响[J]. 林业科学研究，2004，17(4)：405-412.

[7] 连相汝，鲁法典，刘成杰，等. 我国人工林天然更新研究进展[J]. 世界林业研究，2013，26(6)：52-58.

[8] 罗梅，郑小贤. 金沟岭林场落叶松人工林天然更新动态研究[J]. 中南林业科技大学学报，2016，36(9)：7-11.

[9] 白灵海，唐继新，明安刚，等. 广西大青山米老排人工林经济效益分析[J]. 林业科学研究，2011，24(6)：784-787.

[10] 袁洁. 米老排的选育与多样性研究[D]. 北京：中国林业科学研究院，2014.

[11] 郭文福，蔡道雄，贾宏炎，等. 米老排人工林生长规律的研究[J]. 林业科学研究，2006，19(5)：585-589.

[12] 王克建，蔡子良. 热带树种栽培技术[M]. 南宁：广西科学技术出版社，2008.8.

[13] 熊炀. 赣南米老排天然更新调查[J]. 现代园艺，2015(2)：18.

[14] 王娜，郝清玉. 森林天然更新影响因子研究进展[J]. 广东农业科学，2012(6)：68-70.

[15] 张金屯. 植物种群空间分布的点格局分析[J]. 植物生态学报，1998，22(4)：344-349.

[16] 常静，潘存德，师瑞锋. 梭梭-白梭梭群落优势种种群分布格局及其种间关系分析[J]. 新疆农业大学学报，2006，29(2)：26-29.

[17] 李明辉，何风华，潘存德. 天山云杉天然林不同林层的空间格局和空间关联性[J]. 生态学报，2011，31(3)：0620-0628.

[18] 于汝元，黄晓鹤，哈达，等. 东北林区森林更新调查及评定中的问题[J]. 北京林业大学学报，1990，12(增刊3)：46-50.

[19] 孟宪宇. 测树学[M].3版. 北京：中国林业出版社，2006.

[20] 音振兴，李文栋，余洋. 尾巨桉萌芽林定株高度试验研究[J]. 桉树科技，2012，29(2)：21-23.

[原载：中南林业科技大学学报，2018，38(03)]

米老排人工林皆伐迹地种子天然更新林密度调控效应

唐继新[1,2] 雷渊才[2] 曾 冀[1] 李忠国[1] 李武志[1] 农良书[1] 赵 总[1]

([1]中国林业科学研究院热带林业实验中心，广西凭祥 532600；
[2]中国林业科学研究院资源信息研究所，北京 100091)

摘 要 为探索米老排皆伐迹地种子天然更新幼林生长及密度效应，基于单因素随机区组试验设计，研究了米老排皆伐迹地不同密度种子天然更新幼林生长状况，结果表明：①在间苗作业后1~3.5年，未间苗抚育林分(ck)与间苗抚育林分在平均胸径、平均树高和平均单株材积生长量(总生长量与连年生长量)上差异显著，但各处理林分优势木的胸径、树高和单株材积的总生长量差异不显著，间苗处理对提高林分平均生长水平和减小林分径阶分化程度的作用明显。②间苗抚育林分平均生长量曲线与连年生长量曲线的相交时间与密度呈负相关；未间苗抚育林分平均胸径连年生长量的峰值出现第2年，间苗抚育林分出现在第2~4年。前5年为种子更新林径向生长的旺盛期，间苗抚育保留合理密度对林分的径向生长极为重要。③未间苗抚育林分平均树高的连年生长量与平均生长量在第2~3年相交，而间苗处理平均树高的连年生长量与平均生长量在第5年仍未相交；林分优势木前5年的高生长处于旺盛期。④未间苗抚育和间苗抚育的林分径阶分布曲线类型不同，前者呈倒J型分布，后者近似正态分布，间苗处理改变林分径阶分布的类型、峰值和分化程度作用明显。⑤在间苗抚育后第3.5年，密度较低林分的平均生长量优于密度较高处理林分。⑥充分利用米老排树种种子成熟期、落种期和天然更新特性，通过科学的采伐期、采伐方式、迹地剩余物清理方式和迹地更新林的间苗抚育措施，可有效实现采伐迹地天然更新和促进其更新林的生长。

关键词 森林培育学；米老排；皆伐迹地；种子；天然更新；幼林生长；密度效应；间苗抚育

Tree Density and Growth of *Mytilaria laosensis* Stands with Natural Seed Regeneration on Clearcut Land

TANG Jixin[1,2], LEI Yuancai[2], ZENG Ji[1], LI Zhongguo[1], LI Wuzhi[1], NONG Liangshu[1], ZHAO Zong[1]

([1]*Experimental Center of Tropical Forestry*, *Chinese Academy of Forestry*, *Pingxiang* 532600, *Guangxi*, *China*;
[2]*Research Institute of Forest Resources Information Techniques*, *CAF*, *Beijing* 100091, *China*)

Abstract: To explore the effect of density on growth of young *Mytilaria laosensis* stands with natural seed regeneration on clearcut land, a single factor randomized block design experiment with natural spacing (ck treatment, three replications) and spaced seedling(T_A, T_B and T_C treatments, three replications) stands was established, and variance analysis of the stands were analysed with Least Significant Difference methods in Data Processing System(14.5), to test diameter at breast height (DBH), tree height, and single volume. Results showed that: ① between ck and spaced seedling stands, during 3.5 years after spacing, stand density significantly decreased ($P<0.01$) the average growth of total growth and annual increment of DBH, tree height, and single volume in the ck stands types; whereas, the total growth of DBH, tree height, and single volume of dominant trees with both stands, were not significantly affected ($P>0.05$). Spaced treatments (finished in 1.5 years old of seedling, remain density in hectare: T_A 1650, T_B 2800, and T_C 2500) significantly improved ($P<0.01$) the average growth of the stands and reduced differentiation of diameter grades in the stands. ② The higher stand density (CK), the earlier intersection time of the average growth curve and the annual growth curve for average DBH or tree height of the stand was appealed. Also, the peak value of annual growth for average DBH of ck stands occurred in the second year, and the peak value of spaced stands was during the second and 4th year. Before the 5th year, it was fast growth period for the stands with natural seed, so spacing was very important with radial growth of the stands. ③ The time of intersection for curves of average growth and annual growth of tree height in the ck stands was between the second and third year, but spaced stands did not intersect before the 5th year. Growth for tree height of dominant trees in all stands before the 5th year was fast (the annual increment of top height in each treatment was more than 1.49 m in each year). ④ Different types of stands had different diameter distribution curves; the ck stand had an inverted J distribution curve, but the spaced stands had an approximately normal distribution. ⑤ In the spaced stands, after 3.5 years, the average growth for DBH, tree height, and volume of less dense forest stands (T_A treatment stands) was better than those with higher density treatments (T_C or T_B stands). ⑥ Thus, understanding of the mature period for seeds, the seed falling-stage, and natural regeneration characteristics of *Mytilaria laosensis* could enhance the scientific cutting time, cutting methods, clean ways for removing residues, shrubs, and herbaceous plants, and methods for spacing seedlings to enable effective natural regeneration on cutover land thereby promoting forest growth.

Key words: silviculture; *Mytilaria laosensis*; clearcut land; seed; natural regeneration; growth of young stands; effect of density; spaced planting

基于森林近自然经营的理论和技术，利用各种自然力，改善森林的结构和功能，充分保护、促进和利用林下的天然更新，减少森林经营的人力、物力和财力投入，是森林近自然化经营的重要原则，也是森林可持续经营的有效途径[1,2]。受“里约赫尔辛基进程”森林认证及森林近自然经营要求的影响，“利用天然更新实现林分皆伐迹地的下一代更新”已引起了广泛的国际关注[3]，相对于传统人工造林更新，利用天然更新实现采伐迹地的更新，其更新及恢复的过程无需育苗、炼山、整地、挖坑和造林，具有物种和结构多样性高、更新成本低、更新过程地表水土流失少等优点[3,10]。虽然天然更新具有如此多的优点，但国内以往对森林天然更新的研究主要集中在天然林方面，有关人工林天然更新的研究报道较少[11]，还未能满足中国人工林健康可持续经营的需要。故对皆伐作业具有可有效天然更新的潜力树种，运用近自然森林经营的理论和技术，研究制定科学的经营措施，促进和利用其采伐迹地的天然更新，对丰富我国人工林健康可持续经营的理论和实践具有重要意义。米老排(*Mytilaria laosensis*)又名壳菜果、三角枫、山桐油，为金缕梅科(Hamamelidaceae)壳菜果属(*Mytilaria*)常绿阔叶乔木，天然分布于中国广东、广西和云南以及越南和老挝等地，是中国南亚热带区域适生范围广的用材树种，具有速生、干形通直圆满、材质优良、改良土壤、萌生力强、落种于林缘及空旷地宜萌发等特性[12,15]。基于米老排树种的天然更新特性，本研究旨在探索米老排人工林皆伐迹地种子天然更新幼林的生长及其密度效应规律，为米老排皆伐作业法天然更新的间苗抚育提供技术支撑。

1 研究区概况

研究区位于广西凭祥市的中国林业科学研究院热带林业实验中心(以下简称热林中心)哨平实验场39林班2经营班(22°05′12″~22°05′22″N，106°53′13″~106°53′33″E)，属南亚热带季风区，干湿季明显；太阳总辐射强，全年日照时数1500~1600h，年均气温21.0℃，≥10℃积温7000~7500℃，最热月平均气温27.5℃，最冷月平均气温13.0℃，极端最低气温-1.0℃；年均降水量1380mm，年蒸发量1370mm。海拔230m，属低山丘陵；土壤为花岗岩发育而成的赤红壤，土层较厚[16]；有机质在土壤A层、B层的质量分数含量分别为3.71%、1.75%，全氮质量分数0.171%，全磷质量分数0.030%，pH4.1~4.6。更新迹地前茬林为米老排人工纯林，造林时间1984年春，造林苗木为裸根实生苗，造林密度2500株/hm²；皆伐前经过1次透光伐及1次抚育间伐(株数间伐强度20%~30%)，皆伐时间为2011年底(皆伐作业前，当年的米老排林分种子雨已基本完全散落；林分皆伐与集材后，人工清理采伐剩余物，未炼山)。伐前林分平均胸径20.2cm，平均树高22.2m，立木株数密度1492株/hm²，郁闭度0.9；林分枯落物层厚度5~10cm，林下灌木主要有山苍子(*Litsea cubeba*)、香港大沙叶(*Pavetta hongkongensis*)和三桠苦(*Evodia lepta*)等，草本主要有粽叶芦(*Thysanolaena maxima*)、五节芒(*Miscanthus floridulus*)、山姜(*Alpinia japonica*)等。皆伐后，迹地种子天然更新苗木主要为次年更新，更新的树种主要为米老排。

2 研究方法

2.1 实验设计及样地调查方法

2013年春，经对米老排皆伐迹地天然更新实地的踏查，选择代表性地块，按单因素随机区组设计，设计间苗抚育试验，以不间苗作为对照组(CK)，处理组间苗密度分别为1650株/hm²(T_A)，2800株·hm²(T_B)，/2500株/hm²(T_C)，3重复/组，共12个面积为 400~600m^2/个矩形样地。

迹地苗木间苗抚育定株方法：将固定样地均匀划分为 5m×5m 的样格，以空间分布均匀、无虫害和长势优良的苗木的条件选择保留木。

本底数据调查方法：①间苗抚育林分，对保留木进行编号，并测量胸径和树高；②未间苗抚育林分，随机抽取 4~7 个(400m^2抽 4 个，600m^2抽 7 个)样格为固定观测样格，对固定观测样格内所有高度大于 0.3m 的林木进行编号，并测量胸径和树高。

迹地种子更新林间苗抚育和本底调查的完成时间分别为 2013 年 6 月和 2013 年 7 月。2014~2016 年，连续 3 年对固定样地林木进行定株连续生长观测，记录各年林分的生长及密度状况(表 1)。胸径、树高、单株材积均为样地林分统计的平均值，单株材积[17]计算公式：

$$v = 0.683297 \times 10^{-4} d^{1.926256} h^{0.8840614}$$

式中：d 表示林分平均胸径；h 表示林分平均树高。

2.2 数据分析软件及方法

数据统计分析及作图分别采用 DPS(V14.5)和 EVIEWS 8.0 软件，方差分析后的多重比较采用 LSD 法。

3 结果与分析

3.1 间苗对林分平均生长的影响

3.1.1 间苗对林分平均胸径生长的影响

表 1 为不同间苗密度下，米老排种子天然更新林平均生长及方差分析结果，由表 1 可知：从第 4 年(间苗抚育后第 2.5 年)起，各处理林分平均胸径的总生长量与林分密度呈负相关。对平均胸径生长作图(图 1)可知，T_A 处理平均胸径连年生长量的峰值在第 4 年；而其余处理的峰值均在第 2~3 年。除 T_A 处理外，所有处理林分的平均生长量曲线与连年生长量曲线均在前 5 年相交，呈现林分密度越高，平均生长量曲线与连年生长量曲线相交越早(表明在相同的立地条件下，林分经营密度越高其胸径的连年生长量下降越快)。间苗抚育后 1.0~3.5 年，对照组平均胸径的生长量(连年生长量和总生长量)明显低于间苗处理对应的生长量，差异极显著($P<0.01$)；而 T_A 与 T_C 处理间胸径总生长量的差异仅在间苗后的第 1 年出现，其差异随间苗时间的推移而逐渐消除，这可能与间苗抚育经营引起的误差有关。抚育间苗后的 3.5 年，随着间苗时间的推移，T_A 处理与其他处理在胸径连年生长量的差异越来越显著，如在抚育间苗后第 2.5 年，仅有 T_A 与 T_C 处理的差异显著，而到间苗后第 3.5 年，T_A 处理与任何处理的连年生长差异均达极显著水平($P<0.01$)。迹地林分抚育经营密度对胸径的生长影响极为显著，此结果与前人研究结果一致[18]。

3.1.2 间苗对林分平均树高生长的影响

基于表 1 和图 2 的分析可知与对照相比，间苗抚育处理树高的连年生长量与平均生长量均未相交(图 1)。对照组连年生长量的峰值在第 2 年，T_B 处理首次连年生长量的峰值在第 3 年(此后，连年生长量先下降，随后再上升)；而 T_A 和 T_C 处理的连年生长量在第 5 年仍未达峰值。除间苗后的第 1 年，T_A 与 T_C 处理树高生长总量差异显著外，观测期内间苗抚育处理间的生长总量差异均不显著($P>0.05$)。与未间苗林分相比，间苗 2.5 年后，林分平均树高的总生长量和连年生长量的差异均达极显著水平($P<0.01$)。上述现象的产生，可能与对照林分密度过高，林木空间生长竞争异常激烈有关；对间苗处理组而言，仅 T_A 与 T_B 处理在间苗抚育后第 3.5 年其树高连年生长量差异显著($P<0.05$)。

3.1.3 间苗对林分平均单株材积生长的影响

由表 1 和图 1(图 1，其第 1 年、第 2 年和第 3 年的生长量数据，根据表 1 数据，用插入法计算得出)可知，抚育间苗后 1~3.5 年，密度对不同处理林分平均单株材积总生长量和连年生长量的影响极为显著($P<0.01$)，间苗处理的平均单株材积生长量显著高于对照。在间苗后第 2.5 年，间苗处理间平均单株材积总生长量的差异不显著($P>0.05$)；间苗后第 3.5 年，T_A处理与 T_B处理的总生长量差异显著($P<0.05$)，T_A与其余间苗处理间平均材积连年生长量的差异均极显著($P<0.01$)。

表 1 不同密度米老排种子天然更新林平均生长及方差分析

林龄(年)	间苗时间(年)	处理	密度(株/hm^2)	胸径(cm)		树高(m)		单株材积($\times 10^{-2} m^3$/株)	
				总生长量	连年增量	总生长量	连年增量	总生长量	连年增量
1.5	0	T_A	1650	1.18±0.10 aA	0.79±0.06 aA	2.01±0.05 aA	1.34±0.04 aA	0.02±0.0 0 aA	0.01±0.00 aA
		T_B	2800	1.23±0.17 aA	0.82±0.11 aA	2.13±0.15 aA	1.42±0.10 aA	0.02±0.01 aA	0.01±0 .00aA
		T_C	2500	1.19±0.12 aA	0.79±0.08 aA	2.14±0.05 aA	1.43±0.04 aA	0.02±0.00 aA	0.01±0 .00aA
		CK	27871	1.10±0.16 aA	0.73±0.11 aA	1.97±0.15 aA	1.31±0.10 aA±	0.02±0.01 aA	0.01±0 .00aA

(续)

林龄（年）	间苗时间（年）	处理	密度（株/hm²）	胸径(cm)		树高(m)		单株材积($\times10^{-2}m^3$/株)	
				总生长量	连年增量	总生长量	连年增量	总生长量	连年增量
2.5	1.0	T_A	1650	3.20±0.15 bA	2.02±0.24 aA	3.55±0.09 bA	1.53±0.08 aA	0.20±0.03 bB	0.18±0.03 bB
		T_B	2800	3.40±0.21 abA	2.17±0.38 aA	3.88±0.20 abA	1.75±0.24 aA	0.24±0.04 bAB	0.22±0.05 bAB
		T_C	2500	3.85±0.20 aA	2.66±0.32 aA	4.08±0.23 aA	1.94±0.23 aA	0.32±0.05 aA	0.30±0.05 aA
		CK	27100	1.94±29 cB	0.84±0.40 bB	3.49±0.31 bA	1.53±0.43 aA±	0. 08±0.02 cC	0.06±0.03 cC±
4.0	2.5	T_A	1650	6.19±0.47 aA	1.99±0.40 aA	6.85±0.53 aA	2.20±0.41 aA	1.27±0.28 aA	0.72±0.20 aA
		T_B	2800	5.81±0.11 aA	1.61±0.09 abA	7.00±0.52 aA	2.08±0.33 aA	1.13±0.11 aA	0.59±0.06 aA
		T_C	2500	6.10±0.33 aA	1.50±0.10 bA	7.13±0.56 aA	2.03±0.23 aA	1.27±0.21 aA	0.63±0.11 aA
		CK	25067	2.38±0.34 bB	0.29±0.07 cB	4.45±0.29 bB	0.64±0.16 bB	0.14±0. 05 aB	0.04±0.02 bB±
5.0	3.5	T_A	1650	8.04±0.44 aA	1.85±0.08 aA	9.43±0.40 aA	2.58±0.19 aA	2.77±0.40 aA	1.50±0.14 aA
		T_B	2800	7.15±0.05 bA	1.34±0.09 bB	9.12±0.39 aA	2.12±0.15 bA	2.13±0.10 bA	1.00±0.05 bB
		T_C	2500	7.47±0.29 abA	1.37±0.06 bB	9.33±0.72 aA	2.20±0.18 abA	2.38±0.35 abA	1.11±0.12 bB
		CK	23893	2.66±0.34 cB	0.28±0.06 cC	5.13±0.29 bB	0.68±0.19 cB	0.19±0.0 5 cB	0.05±0.02 cC

注：数据为均值±标准差；同列不同小写字母表示同一年相同因子的不同处理间差异显著($P<0.05$)；同列不同大写字母表示同一年相同因子的不同处理间差异极显著($P<0.01$)；CK 即对照，数据为样地对应树高大于 1.3m 的统计数据。

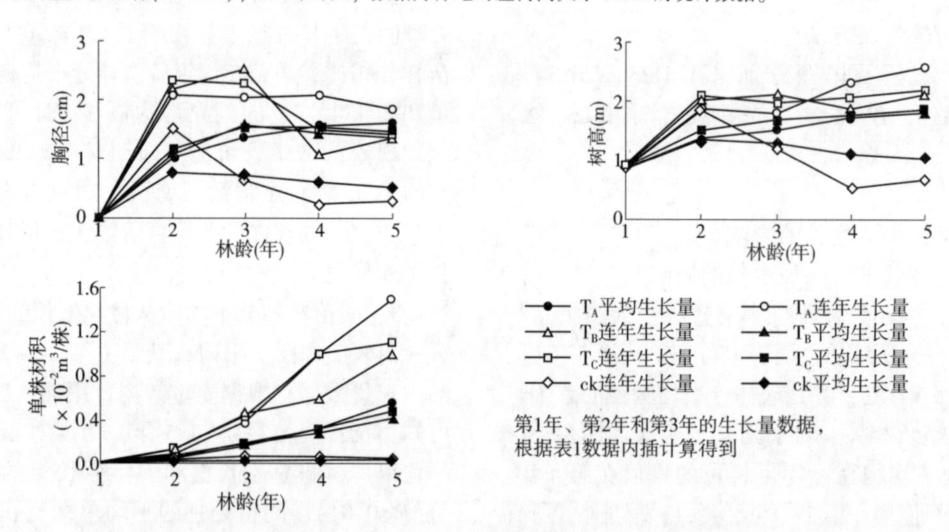

图 1 不同密度林分幼龄林平均胸径、平均树高、平均单株材积生长过程曲线

3.2 间苗对林分优势木生长的影响

3.2.1 间苗对林分优势木径向生长的影响

由表 2 及图 2 可知：各处理优势木胸径连年生长量及平均生长量的峰值出现在第 2~4 年，此后生长速度趋缓。间苗抚育后的 1.5~3.5 年，间苗处理优势木胸径连年生长量高于对照；在间苗后 3.5 年内，间苗密度对优势木胸径的生长有一定影响，但各密度处理间胸径总生长量的差异不显著。所有处理林分优势木胸径的平均生长量与连年生长量均在前 5 年相交，基本表现为林分密度越高，平均生长量与连年生长量相交越早，与林分平均胸径的生长规律类似。

3.2.2 间苗对林分优势木高生长的影响

在前 5 年，各处理优势木树高连年生长量呈一定波动性，但各年树高年连年生长量均大于 1.4m，表明其高生长仍处于旺盛期；所有处理优势木树高的平均生长量与连年生长均在前 5 年多次相交(图 2)，间苗密度对林分优势高生长的影响不显著。

3.2.3 间苗对林分优势木单株材积生长的影响

与林分优势木的胸径、树高生长类似，在抚育间苗后 1~3.5 年，各处理优势木单株材积的总生长量虽有一定差异，但差异不显著；在间苗 1 年后，间苗处理单株材积的连年生长量开始大于对照(图 2)。

3.2.4　间苗对林分优势木高径比的影响

由表 2 可知：在间苗后第 1 年，间苗与未间苗处理优势木的高径比差异不显著；至间苗后第 2.5 年时，仅有 T_A处理与对照的高径比差异显著((*P*<0.05)；而至间苗后的第 3.5 年时，对照与任何间苗处理的高径比差异都显著(*P*<0.05)。此表明，间苗抚育措施对降低林分优势木的高径比作用明显。

表 2　不同密度米老排种子天然更新幼林优势木生长及方差分析

林龄(年)	间苗时间(年)	处理	密度(株/hm²)	胸径(cm)	树高(m)	单株材积(×10⁻²m³/株)	高径比
1.5	0	T_A	1650	2.94±0.59 aA	3.24±0.15 aA	0.16±0.06 aA	113.54±23.76 aA
		T_B	2800	3.09±0.16 aA	3.44±0.46 aA	0.18±0.03 aA	111.16±12.88 aA
		T_C	2500	3.09±0.25 aA	3.37±0.20 aA	0.16±0.02 aA	109.43±3.80 aA
		CK	27871	3.14±0.83 aA	3.61±0.48 aA	0.21±0.12 aA	118.64±21.99 aA
2.5	1.0	T_A	1650	4.91±0.52 aA	4.83±0.31 aA	0.56±0.15 aA	98.77±5.53 aA
		T_B	2800	6.00±0.60 aA	5.52±0.95 aA	1.00±0.33 aA	91.67±7.90 aA
		T_C	2500	5.97±0.35 aA	5.12±0.50 aA	0.91±0.17 aA	85.79±4.66 aA
		CK	27100	5.78±1.18 aA	5.88±0.58 aA	1.01±0.48 aA	103.22±11.17 aA
4.0	2.5	T_A	1650	8.73±1.07 aA	8.06±0.73 aA	2.87±0.94 aA	92.50±3.68 bA
		T_B	2800	9.21±0.62 aA	9.24±0.46 aA	3.54±0.57 aA	100.45±4.01 abA
		T_C	2500	8.87±0.65 aA	8.67±0.38 aA	3.11±0.55 aA	97.93±3.08 abA
		CK	25067	8.03±1.04 aA	8.13±0.71 aA	2.47±0.82 aA	101.6±4.21 aA
5.0	3.5	T_A	1650	10.84±0.72 aA	10.64±0.63 aA	5.49±1.00 aA	98.19±0.68 bA
		T_B	2800	11.10±0.64 aA	11.01±0.17 aA	5.89±0.63 aA	99.46±6.33 bA
		T_C	2500	10.78±0.75 aA	10.88±1.09 aA	5.55±1.22 aA	100.84±4.84 bA
		CK	23893	9.30±1.25 aA	10.77±1.07 aA	4.19±1.45 aA	116.31±8.51 aA

注：各处理取 3 株/样地优势木作为统计对象，对照取 2 株/样地优势木作为统计对象。

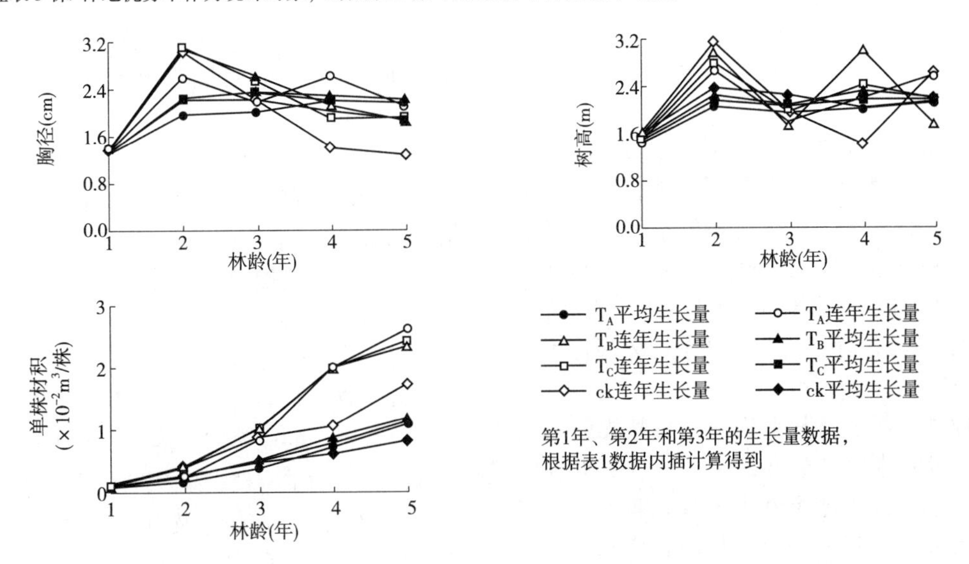

图 2　不同密度优势木平均胸径、平均树高、平均单株材积生长过程曲线

3.3　间苗对林分径阶结构的影响

由表 3 分析可知：未间苗处理林分的径阶分布呈典型的倒 J 型分布，而间苗处理林分的径阶分布为近似正态分布。在同一年龄不同密度处理下径阶分布的范围不同，T_B较其他处理径阶的分布范围更广，T_A的径阶分布相对其他处理更集中。从间苗后第 2.5 年起，对照组径阶分布的极大值开始小于间苗处理，

间苗抚育后第3.5年，对照组径阶分布的极大和极小值均小于间苗处理。间苗抚育后1~3.5年，对照组林木60%以上的径阶分布在1~2cm，表明未间苗处理林分林木的进阶极缓慢。在间苗抚育后第1年，间苗处理间径阶分布曲线峰值的对应径阶值无差异，但在间苗抚育后第3.5年，T_A处理径阶分布曲线峰值对应的径阶值开始高于T_B与T_C处理的对应值。同一年龄，各处理胸径的大小分化程度随密度增大而提高，小密度林分的林木较均匀，且"大径木"比例相对较高。

表3 米老排种子天然更新林幼林径阶分配受密度的影响

径阶(cm)	间苗1年后径阶分布百分比(%)				间苗2.5年后径阶分布百分比(%)				间苗3.5年后径阶分布百分比(%)			
	T_A	T_B	T_C	CK	T_A	T_B	T_C	CK	T_A	T_B	T_C	CK
1	6.52	6.51	0.34	40.74	—	0.78	—	34.99	—	—	—	30.27
2	21.74	13.02	10.44	32.53	0.87	1.82	1.37	28.92	—	0.53	0.69	30.27
3	25.22	32.55	23.91	18.40	6.52	6.25	3.42	14.68	0.89	2.92	1.72	13.97
4	32.61	31.51	34.68	6.24	7.83	11.98	6.85	10.60	2.67	6.37	1.38	8.73
5	13.04	14.58	25.25	1.75	11.74	16.93	20.55	6.07	5.78	8.49	6.90	7.57
6	0.87	1.30	4.38	0.22	23.04	26.30	24.66	2.98	8.89	12.20	17.24	4.66
7	—	0.26	1.01	0.11	26.52	22.14	25.68	1.32	12.89	21.75	19.66	2.21
8	—	0.00	—	—	18.70	11.20	12.67	0.33	21.78	23.61	22.07	1.63
9	—	0.26	—	—	3.04	2.08	3.77	0.11	22.67	14.85	18.28	0.58
10	—	—	—	—	1.74	0.26	1.03	—	20.44	7.16	7.93	0.00
11	—	—	—	—	—	0.26	—	—	2.67	1.59	2.41	0.12
12	—	—	—	—	—	—	—	—	1.33	0.27	1.38	—
13	—	—	—	—	—	—	—	—	—	0.00	0.34	—
14	—	—	—	—	—	—	—	—	—	0.27	—	—

注："—"表对应处理对应径阶的株数百分比为0。

4 结论与讨论

4.1 结论

与对照组相比，间苗处理对林分优势木胸径、树高和单株材积的总生长量影响不显著，对林分平均胸径、平均树高和平均单株材积的总生长量与连年生长量的影响显著；间苗处理可明显提高林分平均整体生长水平，减小林分径阶分化程度，提高"大径木"百分比，减缓林分平均胸径和平均树高连年生长量的衰减速度。

米老排种子天然更新林平均胸径与其优势木胸径的连年生长量峰值均出现第2~4年，所有处理优势木胸径平均生长量与连年生长相交时间均在前5年，此阶段为种子更新林径向生长的旺盛期，间苗抚育保留合理密度对林分的径向生长极为重要。

未间苗处理林分平均树高的连年生长量与平均生长量曲线在第2~3年相交，间苗处理平均树高的连年生长量与平均生长量曲线在第5年仍未相交；林分优势木前5年的高生长处于旺盛期。

间苗处理林分优势木树高平均生长量与连年生长量曲线的相交时间早于其林分平均树高对应曲线的相交时间。

未间苗处理林分径阶分布呈倒J型分布，间苗处理林分径阶分布近似正态分布，间苗处理相对"大径木"比例较高；表明间苗处理改变林分径阶分布的类型、峰值和分化程度作用明显。

4.2 讨论

基于森林近自然经营的理论和技术，利用米老排人工林种子的成熟期、落种期和天然更新特性(在林内及林缘的空旷裸露地易于天然更新)，择机对米老排人工林进行采伐，及对其采伐更新迹地实施科学的间苗抚育，可实现采伐迹地种子的天然更新和种子更新林的高效生长，节省传统人工造林环节的挖坑、整地、植苗等人力及物力的投入，避免或减轻传统造林方式对森林环境(植被和土壤等)的破坏。由于本研究试验林分间苗抚育时间较短，而在采伐收获前，哪种间苗抚育措施林分更好，种子天然更新林的生长与传统人工更新林的生长孰优孰劣，这仍需后续的进一步研究。

参考文献

[1]曾伟生．近自然森林经营是提高我国森林质量的可行途径[J]．林业资源管理．2009，(2)：6-11.
[2] 陆元昌．近自然森林经营的理论与实践[M]．北京：科学出版社，2006.
[3] BARBOUR R J, BAILEY R E, COOK J E. Evaluation of relative density, diameter growth and stem form in a red spruce (*Picea rubens*) stand 15 years after precommercial thinning [J]. Can J For Res, 1992, 22(2): 229-238.
[4] POTHIER D. Twenty - year results of precommercial thinning in a balsam fir stand [J]. For Ecol Manag, 2002, 168(1-3): 177-186.
[5] RUHA T, VARMOLA M. Precommercial thinning in naturally regenerated scots pine stands in northern Finland [J]. Silva Fennina, 1997, 31(4): 401-415.
[6] ULVCRONA K A, CLAESSON S, SAHLÉN K, et al. The effects of timing of pre - commercial thinning and stand density on stem form and branch characteristics of Pinus sylvestris [J]. Forestry, 2007, 80(3): 323-325.
[7] STOKES V, KERR G. Long-term growth and yield effects of respacing natural regeneration of Sitka spruce in Britain [J]. Eur J For Res, 2013, 132(2): 351-362.
[8] 王娜，郝清玉．森林天然更新影响因子研究进展[J]．广东农业科学，2012(6)：68-70.
[9] 田晓萍．杉木萌芽更新的研究[D]．福州：福建农林大学，2008.
[10] 李喜霞，刘明国，李海春．朝阳地区人工采伐带内油松天然更新研究[J]．沈阳农业大学学报，2003，34(1)：35-39.
[11] 连相汝，鲁法典，刘成杰，等．我国人工林天然更新研究进展[J]．世界林业研究，2013，26(6)：52-58.
[12] 白灵海，唐继新，明安刚，等．广西大青山米老排人工林经济效益分析[J]．林业科学研究，2011，24(6)：784-787.
[13] 袁洁．8个米老排天然群体的遗传多样性研究[D]．北京：中国林业科学研究院，2014.
[14] 郭文福，蔡道雄，贾宏炎，等．米老排人工林生长规律的研究[J]．林业科学研究，2006，19(5)：585-589.
[15] 王克建，蔡子良．热带树种栽培技术[M]．南宁：广西科学技术出版社，2008. 8.
[16] 汪炳根，卢立华．广西大青山实验基地森林立地评价与适地适树研究[J]．林业科学研究，1998，11(1)：78-85.
[17] 陈永富，郭文福，黄镜光．米老排立木材积表及地位指数的编制[J]．林业科学研究，1991，4(增刊)：116-119.
[18] 李炎香，谭天泳，黄镜光，等．米老排造林密度试验初报[J]．林业科学研究，1988，1(2)：206-212

[原载：浙江农林大学学报，2018，35(03)]

红椎天然更新及其影响因子研究

赵 总[1,2] 贾宏炎[2] 蔡道雄[2] 庞圣江[2] 安 宁[2] 刘 勇[1]

([1]北京林业大学林学院，北京 100083；

[2]中国林业科学研究院热带林业实验中心，广西凭祥 532600)

摘 要 开展基于近自然化森林经营技术的红椎人工林天然更新研究，为解决长期困扰红椎人工林的生长质量差、林分稳定性低及经营成本过高等问题提供依据。利用样地调查法对广西凭祥热林中心林区3种林分(红椎人工林、针阔人工混交林及马尾松人工林)中的红椎天然更新进行了调查，并采取室内试验方法对红椎种子萌发进行测定，通过多元回归统计及方差分析方法对数据进行分析。红椎在所有林分中更新频度及密度均超过其他树种，在针叶树人工林下红椎更新最差，更新密度为625株/hm²，而在针阔混交林下更新密度相对较低3673株/hm²，在红椎人工林下更新最好21231株/hm²；林分中凋落物层厚度与草本盖度是影响红椎更新的重要环境因子，红椎更新密度与两者呈明显负相关；当与母树的距离$S \leqslant 5m$时红椎出现更新不良，周围仅出现7.5%的更新幼苗。经红椎种子萌发试验测定，马尾松未分解凋落叶水浸液质量比为1∶10时，对红椎种子发芽产生明显抑制作用。试验结果说明草本盖度、凋落物层厚度和凋落叶中化感物质的抑制作用及母树的缺失可能是导致红椎在针叶树人工林下更新不良的主要原因。

关键词 红椎；人工林；天然更新；抑制作用；化感物质

The Regeneration and Influence Factors of *Castanopsis hystrix*

ZHAO Zong[1,2], JIA Hongyan[2], CAI Daoxiong[2], PANG Shengjiang[2], AN Ning[2], LIU Yong[1]

(*[1]Schoolof Forestry, Beijing Forestry University, Beijing* 100083, *China*;

[2]The Experimental Center of Tropical Forestry, Chinese Academy of Forestry Sciences, Pingxiang 532600, *Guangxi*, *China*)

Abstract: Carrying out the research for regeneration of *Castanopsis hystrix* plantation is an important approach to solve the problems of bad growth, low stand stability, high management cost of *Castanopsis hystrix* plantation. Sample-plot survey was conducted for *Castanopsis hystrix*' s regeneration of three type plantations (*Castanopsis hystrix* plantationn、Coniferous broadleaved artificial mixed forest and *Pinus massoniana* plantation) in the experimental forest, Experimental Center of Tropical Forestry, Pingxiang, Guangxi province of Southern China, while the indoor testing was conducted for seed germination of *Castanopsis hystrix* and the research data were analysed by multiple regression and variance analysis. The result showed that the reproduction frequency and dencity of *Castanopsis hystrix* were higher than other tree species, and it had the lowest regeneration density under the coniferous plantation 62511 plants/ha, relatively low regeneration density under the conifer and broadleaf mixed plantation 3673 plants/ha, but the highest under its own forest 21231 plants/ha; The litter thickness and herb coverage were the important environmental factors influenced the regeneration of *Castanopsis hystrix*, they negatively correlated with regeneration density. The regeneration was not good When the distance from the seed tree was short than 5m ($S \leqslant 5m$), the proportion of regenerated *Castanopsis hystrix* plants was only 7.5%. Seed germination of *Castanopsis hystrix* was significantly inhibited by soaking solution of leaf litter with the mass and water ration of 1∶10. Thus the herb coverage, litter thickness, inhibiting effect of allelochemical in the leaf litter and dificiency of mother plant were the main reasons that may lead to the bad regeneration of *Castanopsis hystrix* under coniferous plantation.

Key words: *Castanopsis hystrix*; plantation; natural regeneration; inhibiting effect; allelochemical

天然更新是森林资源再生产的一个自然的生物生态学过程[1]，其中包括种子的生产、传播及幼苗的生成与定居等关键环节[2]。不同森林群落类型更新的所有环节，受某一树种特性、种间关系、立地条件及外界干扰等因素的影响，其过程极为复杂[3,4]。了解森林更新与环境因素之间的关系，对科学进行森林经营管理及实现人工林经营过程的近自然化具有重要意义。

红椎(*Castanopsis hystrix*)是我国热带亚热带地区生长速度较快的珍贵用材树种，自然分布主要集中在福建、广西、广东、湖南、江西等省(自治区)。因其材性材质兼优，过去一直以来人们对其过分的采伐利用，致使该树种资源面临毁灭性破坏的危险。因此，对该树种资源开展培育技术方面的研究，就显得十分迫切和必要。近年来国内外学者针对不同森林群落类型更新组成、数量、机理以及影响因子等[3-4,5-9]开展了广泛研究。如桂西南喀斯特山地蚬木(*Excentrodendron hsienmu*)天然更新受物种的生物学特性，种内、种间竞争，生境异质性及人为干扰等因素的综合影响[10]。辽东栎(*Quercus liaotungensis*)更新所需的最佳郁闭度为0.6~0.8[11]。适宜的林内环境有利于树种的萌发、生长，坡向、坡位、坡度等立地因子的变化间接影响光、热、水等条件，进而影响到更新的数量和质量[12-14]。有研究认为，红椎在自身林下更新良好，其更新与环境因素影响有关。邓硕坤等[5]研究发现，坡位是影响红椎更新的一个主要环境因子。然而，我们调查发现，红椎在针叶树人工林下更新不良，其原因尚未见报道，说明关于红椎天然更新与环境因素之间的关系还缺少深入研究。本文通过对不同林分下红椎天然更新状况进行调查分析，并对其在针叶林下更新不良的影响因素进行探讨，旨在为红椎天然更新及其近自然化森林培育与经营提供科学依据。

1 研究区概况

试验地位于中国林业科学研究院热带林业实验中心哨平实验场林区，22°03′N，106°53′E。属南亚热带季风气候，海拔350~380m，年均气温21.5℃，年均降水量为1200~1400mm，≥10℃年积温为7500℃。地貌类型属低山丘陵，地带性土壤为砖红壤。森林顶级群落为马尾松林。从1980—1984年相继营造了一定面积的红椎人工林，这些人工林与已有的马尾松人工林呈镶嵌分布格局。林下植被乔木树种有红椎、火力楠(*Michelia macclurei*)、印度栲(*Castanopsis indica*)、毛桐(*Mallotus barbatus*)、破布木(*Cordia dichotoma*)、漆树(*Toxicodendron succedaneum*)、琴叶榕(*Ficus pandurata*)等；灌木树种有杜茎山(*Maesa japonica*)、九节(*Psychotria rubra*)、大青(*Clerodendrum cyrtophyllum*)、酸藤子(*Embelia sessiliflora*)等；草本种类有弓果黍(*Cyrtococcum patens*)、山菅兰(*Dianella ensifolia*)、金毛狗(*Cibotium barometz*)、玉叶金花(*Mussaenda pubescens*)等。

2 材料与方法

2.1 红椎天然更新调查

调查选择了3个林分类型，天然更新调查样地状况(表1)，调查时间为2017年4~10月，采用样地调查法，在所选的3个林分中分别在山体的上、中、下部设置观测样地，并在每个样地内随机布设5个5m×5m的小样方，完成样地设置后，记录样地的林分生长状况(如种源、林龄、郁闭度、灌草种类和盖度及结实状况等)及其所处的立地环境条件(如海拔、坡度、坡位、坡向、凋落物层及腐殖质层厚等)。然后在每个小样方内开展天然更新调查，记录小样方内红椎更新幼苗株数和其他乔木种类及株数。对每个样地内更新的红椎幼苗随机选取10株，实测其与最近红椎母树之间的空间距离。

更新频度=(更新幼苗出现样方÷调查样方数)×100%

表1 不同林分天然更新调查样地状况

林分类型	样地数(块)	样地面积(m^2)	林龄(年)	主要组成树种
红椎人工林	20	20×20	33~37	红椎、印度栲、火力楠等
针阔人工混交林	30	20×20	30~42	马尾松、红椎、印度栲、毛桐、破布木等
马尾松人工林	20	20×20	32~47	马尾松、红椎、漆树、琴叶榕等

2.2 红椎和马尾松凋落叶水浸液影响红椎种子发芽试验

凋落叶提取液的配制：于2017年3月在热带林业实验中心哨平实验林场采集红椎和马尾松凋落叶，将其分为半分解(叶子部分已腐坏)和未分解(叶形及新鲜度保持完好)2种。在实验室内将采集的凋落叶用蒸馏水浸泡72h，然后用滤纸进行过滤，配成质量比为W(凋落叶干质量)：W(蒸馏水)=1：10原液，保存在5℃的冷柜中备用。

红椎种子萌发试验：该试验是在发芽培养箱中进行，其中发芽床由滤纸、玻璃培养皿及脱脂棉条组成。将半分解和未分解2种凋落叶按以下配比分别制成凋落叶(W)：蒸馏水(W)=1：10、1：20、1：40、1：70的水浸液，另设对照(纯蒸馏水)共5个处理，每个处理4次重复，每个重复50粒种子。试验过程发芽温度始终保持在常温25℃。试验前分别将红椎和马尾松凋落叶的各种配比水浸液浸泡红椎种子3d，试验过程各个处理的发芽床仅补充蒸馏水，发芽试验观测结束后，开始计算种子发芽率。

2.3 数据处理方法

采用SPSS 20.0和Excel 2007统计分析软件对相关研究数据进行分析，其中红椎更新密度与环境因子相关性采取多元回归模块进行分析，采用方差分析模块对不同质量比红椎和马尾松凋落叶水浸液对红椎种子萌发影响之间的差异进行检验，如差异显著，则采取Duncan法多重比较。

3 结果与分析

3.1 不同林分下红椎的更新效果

通过调查3种人工林分的天然更新状况，结果表明，红椎人工林、针阔人工混交林、马尾松人工林中，其林下更新总株数分别为22 707、6 536、1 947株/hm^2(表2)。3种林分中出现更新的乔木树种有红椎、印度栲、火力楠、马尾松、毛桐、破布木、琴叶榕、漆树等。在各个林分中红椎的更新频度和密度都是最高。在红椎人工林下，红椎更新最好，幼苗株数为21231株/hm^2，占更新总株数的93.5%；针阔人工混交林中红椎更新相对较差，幼苗株数为3673株/hm^2，约占更新总株数的56.2%；而马尾松人工林下红椎更新最差，仅为625株/hm^2，占该林分更新总株数的32.1%。不同林分中红椎更新幼苗数量差别较大，更新最好的红椎人工林中其幼苗数量是马尾松人工林中的34.0倍，这表明红椎在自身林分内更新表现非常好，但在马尾松林下却更新不良，这可能是两方面的原因：第一，林分中缺少红椎母树，导致散落林分内的种子数量少；第二，由于红椎种子较大，其扩散能力远不及小种子强，且易被动物取食，从而导致从相邻林分落入马尾松林分内的红椎种子数量极有限，而品质好的种子数量则更少。因此，不利于马尾松林下红椎的更新。其他乔木树种在马尾松林下更新也较差。

表2 3种林分类型林下树种天然更新状况调查结果

主要更新树种	红椎人工林		针阔人工混交林		马尾松人工林	
	更新频度(%)	更新株数(株/hm^2)	更新频度(%)	更新株数(株/hm^2)	更新频度(%)	更新株数(株/hm^2)
红椎	100	21231	42	3673	16	625
火力楠	8	689	13	973	0	0
马尾松	0	0	9	181	5	261
印度栲	11	237	6	402	3	160
毛桐	5	103	10	470	7	358
破布木	6	141	15	337	4	229
漆树	9	259	12	391	6	117
琴叶榕	3	47	8	109	10	197
合计	142	22707	115	6536	51	1947

3.2 红椎天然更新与环境因子的关系

采用多元回归统计分析法分析了红椎幼苗更新密度与环境影响因子间的相关性(表4)。在所调查的环境因子中，仅有草本盖度、郁闭度和凋落物层厚与红椎更新的相关性显著($P<0.01$)，其他因子与红椎更新的相关关系不明显。红椎的更新密度与林分郁闭度呈正相关，与凋落物层厚及草本盖度呈负相关。此次所调查的林分中，马尾松人工林下的凋落物层厚度为3~6cm，针阔人工混交林下凋落物层厚为3cm左右，且两者的草本盖度比较接近，前者为82.3%，后者为85.4%(表3)，但马尾松人工林下红椎更新幼苗数量却仅有针阔人工混交林下更新的17.0%(表2)。这说明除了草本盖度是明显影响红椎更新的因子外，凋落物层厚度也是制约和影响红椎更新的重要因子。由于马尾松人工林凋落物层厚度明显大于针阔人工混交林及红椎人工林，其对种子的阻隔作用大于后两者，使得种子难以接触到土壤表层，因此马尾松林下红椎更新幼苗数量远低于针阔混交林和红椎林。

表 3　不同林分类型的环境因子调查状况

环境因子	红椎人工林	针阔人工混交林	马尾松人工林
郁闭度	0.65	0.73	0.40
草本盖度(%)	15.8	85.4	82.3
灌木盖度(%)	31.0	50.6	42.5
腐殖质层厚度(cm)	5~10	3~6	2~5
凋落物层厚度(cm)	1~2	3	3~6

表 4　红椎更新密度与环境因子相关性分析

立地环境因子	回归系数	偏相关系数	复相关系数
坡度	230.0	0.1027	0.5026 *
郁闭度	15331.6	0.3421 *	回归方程常数项为:
草本盖度	-214.2	-0.4609 *	B_0=-16274.5
灌木盖度	-103.6	-0.1068	
腐殖质层厚度	2720.1	0.1326	
凋落物层厚度	4615.3	-0.3120 *	
上坡	10214.0		
中坡	13608.2		
下坡	18156.4		

注：偏相关检验 $R_{0.05}(70)=0.1617$，$R_{0.01}(70)=0.2315$；复相关检验 $R_{0.05}(70,\ 5)=0.2838$，$R_{0.01}(70,\ 5)=0.3428$。* 表示相关性显著。

3.3　红椎更新幼苗的分布规律

对 3 个林分类型中选取的 700 株红椎更新幼苗与最近母树之间的距离进行了调查(表 5)，平均结果显示：在距离母树 $S<5$m 的范围内，周围仅出现 7.5%的红椎更新幼苗；当与母树的距离在 5m≤S<10m 的范围时，更新幼苗出现的频率为 26.1%；在距离母树 10m≤S<15m、15m≤S<20m、20m≤S<25m、25m≤S<30m 的部位，红椎幼苗更新的频率分别为 68.3%、57.6%、43.7%、25.4%。3 种不同林分其林下红椎更新幼苗与最近母树的距离分布均表现出相同的变化趋势。由此看出，在距离 $S>15$m 时幼苗更新频率开始出现逐渐下降趋势。与母树距离 5m 的范围内，更新频率仅有 7.5%，这表明红椎在自身林冠下幼苗的更新频率水平较低，但在距母树 $S>30$m 处仍有 38.9%的更新幼苗出现。此处出现一定数量的红椎更新苗，与其种子库格局发生改变不无关系。红椎种子体积大，仅靠风力传播或地表径流等外力作用是难以顺利到达距离母树 30m 外的地块上，很大程度上是由于鼠类动物对种子的取食搬运所致。动物不仅是种子的捕食者，而且也是种子的传播者，会使种子库格局发生改变，从而影响或改变更新幼苗的分布动态[15]。

表 5　不同林分类型红椎幼苗更新的频率变化

与最近母树的距离(S)	红椎幼苗的更新频率(平均值)(%)			总平均值(%)
	红椎人工林	针阔人工混交林	马尾松人工林	
S<5m	10.1	8.0	4.4	7.5
5m≤S<10m	50.6	18.7	9.0	26.1
10m≤S<15m	100.0	74.8	30.1	68.3
15m≤S<20m	90.5	56.4	25.9	57.6
20m≤S<25m	69.0	47.8	14.3	43.7

（续）

与最近母树的距离(S)	红椎幼苗的更新频率(平均值)(%)			总平均值(%)
	红椎人工林	针阔人工混交林	马尾松人工林	
25m≤S<30m	43.5	22.3	10.4	25.4
S>30m	61.0	35.2	20.5	38.9

3.4 红椎和马尾松凋落叶水浸液对红椎种子萌发的影响

发芽试验结果表明(表6)，不同质量比的红椎和马尾松未分解及半分解凋落叶浸泡液对红椎种子萌发率的抑制作用显著，其中，马尾松未分解凋落叶水浸液X(凋落叶W)：Y(蒸馏水W)＝1：10的抑制作用最强，发芽率明显低于其他处理，比对照低17.6%；1：20和1：40两者的种子发芽率差异不显著(P>0.05，表6)，但明显低于1：70和对照。马尾松半分解凋落叶各质量比1：10、1：20、1：40之间种子发芽率差异显著，三者明显低于1：70和对照，而1：70的水浸液种子发芽率与对照无显著性差异。

红椎未分解及半分解凋落叶水浸液对自身种子发芽率影响的分析表明，未分解凋落叶3种质比X：Y＝1：10、1：20及1：70水浸液的种子发芽率无明显差异(P>0.05)，但均显著低于对照(P<0.05)。另外，红椎半分解凋落叶不同质量比水浸液对自身种子萌发的抑制作用也达到显著差异，而1：20及1：40这两个水平无显著差异(P>0.05)，1：70的水浸液对种子萌发的抑制作用最大，4个处理均显著低于对照水平。

表6 红椎与马尾松凋落叶水浸液对红椎种子萌发的影响

不同质量比水平	红椎种子平均发芽率(%)			
	马尾松未分解凋落叶浸泡液	马尾松半分解凋落叶浸泡液	红椎未分解凋落叶浸泡液	红椎半分解凋落叶浸泡液
X：Y＝1：10	22.1d	24.5d	30.8c	32.5c
X：Y＝1：20	26.4c	28.2c	31.6bc	34.7b
X：Y＝1：40	27.3c	31.9b	33.7b	36.3b
X：Y＝1：70	33.6b	38.1a	29.5c	30.4d
对照 Contrast	39.7a	39.7a	39.7a	39.7a

注：X表示凋落叶(M)，Y表示蒸馏水(M)；多重比较结果以小写字母表示，不同小写字母表示水平间差异显著(P<0.05)。

4 讨论

对研究区3种林分中红椎的天然更新状况进行调查后发现，红椎更新幼苗数量及频度均远远超出其他乔木树种，这表明红椎是一个具有较强天然更新能力的珍贵树种。邓硕坤等[5]对该地区红椎人工林天然更新影响因素调查认为，红椎在自身林下具有较好的更新优势，这与本研究结果相近。红椎自身较强的天然更新能力可能与其种子扩散密度大，土壤种子库中种子活力较高有关。尽管红椎在该区更新优势明显，但不同林分下红椎更新密度存在较大的差异。本次所调查的3种林分中，红椎人工林下天然更新普遍较好，针阔混交林和马尾松人工林下更新效果相对较差，这与前期一些相关的研究结果相似[16,17]。这可能是受林分郁闭度、林下草本盖度及凋落物层厚度等的影响。环境因子中的郁闭度与凋落物层和灌草盖度之间是相互影响的关系。森林天然更新受林分结构影响的主要因子是郁闭度，其对林内光、热、水和气等条件的改变可直接影响到天然更新过程。红椎人工林、针阔人工混交林分郁闭度明显大于马尾松人工林，因此，前两者林分的郁闭度更有利于红椎幼苗更新。有学者认为，郁闭的林分中，由于林下植被稀疏，种间竞争压力减小，因此利于幼苗更新[18]。黄忠良等[19]认为，林下灌草植被对更新树种更新苗的生成和定居及幼树的生长产生竞争压力的原因是由于其与更新树种竞争有限的林下光照、温度、水分和养分等资源所造成。因此，灌草盖度在一定程度上影响了森林的天然更新。也有研究认为，凋落物层过厚可导致某些树种更新幼苗数量明显减少[6,20-21]。森林凋落物具有保湿、保温、遮阳和机械阻隔及养分供给等作用，对林木种子的萌发、幼苗的定居和生长起双重影响效应[22-24]。

由于受到凋落物的阻隔影响，更新林木种子无法与土壤表面接触，因而缺乏种子发芽所需的水分和养分，极大的降低了种子的萌发率，使幼苗丧失更多成功定居的机会，从而影响天然更新质量[25]。由于马尾松针叶难以分解，易形成较厚的凋落物层，才导致其林下红椎更新不良。本研究表明，红椎更新数量与郁闭度呈明显的正相关关系，而与凋落物层厚度和草本盖度呈极显著的负相关，这与针阔混交林、马尾松林下较高的盖度(前者为85.4%，后者为82.3%)和红椎林下较低的草本盖度(15.8%)以及三者之间的郁闭度(红椎林0.65、针阔混交林0.73及马尾松林0.40)的研究结果相符。

倘若红椎在林下的天然更新主要受草本盖度及凋落物的影响，那么马尾松林下和针阔混交林下因草本盖度较接近，其林下红椎的更新数量也应相近，但实际红椎在马尾松林下的更新数量仅为针阔混交林下的17.0%，由此可见，红椎在马尾松林下更新不良，除受郁闭度、草本盖度及凋落物影响外，可能还与其他因素的影响有关，如林内腐殖质层及种子传播距离等。樊后保等[26]认为，林下厚腐殖质层及土层有利于森林天然更新。这与本研究中林内腐殖质层厚度(红椎人工林5~10cm>针阔人工混交林3~6cm>马尾松人工林2~5cm)，其林下红椎更新最好是红椎自身林分，其次是针阔人工混交林，马尾松人工林次之的研究结果是吻合的。有研究者对水曲柳种子散落格局进行过研究，有72.5%~96.4%的种子散落在距母树10m以内的范围[27]。然而，本研究的结果表明，距红椎母树10m以内的范围，有80%~95%的种子散落其中，但仅有50.6%的更新苗分布，说明尽管在红椎母树周围有大量的种子散落，但在自身林冠下却更新不良。此外，在距离红椎母树$S>30$m处，仍出现38.9%的更新幼苗，这可能也与种子传播距离的负密度制约有关。由于林冠下散落的红椎种子是密集重叠分布，很大一部分种子接触不到土壤，同时受林冠庇荫的影响，因而出现更新幼苗稀少的现象。在远离红椎母树的地块上又有一定数量的更新苗出现，这可能与散落种子的密度及格局有关。在距母树30m处出现的红椎种子，基本上是在外力搬运作用下到达的，此处的种子密度相比母树林冠下及周围的要小得多，因此种子接触土壤的机会更大，而且在相对开阔的林隙或林窗下，温度和光照条件更好，更有利于种子萌发。因此，此处红椎更新幼苗数量比母树林冠下的多是完全有可能的。

红椎和马尾松凋落叶水浸液对红椎种子萌发率影响的试验表明，马尾松凋落叶水浸液对红椎种子的萌发均产生显著的抑制作用。这说明马尾松凋落物中可能含有某些化感物质，这些物质可产生一些如乙醇、有机酸(乙酸和酪酸类)、醛及醚等之类的植物毒素[28]，这些植物毒素一般会对种子萌发、幼苗生长产生强烈抑制作用[29,30]。前期的一些研究也证实，影响森林天然更新的化学因子中，对树木种子发芽和幼苗幼树生长起障碍的是植物的化感作用[31]。如辽东栎和油松凋落物浸泡液，达到一定浓度时可抑制种子的萌发，而浓度较低时，则抑制作用不明显[32]。林木的凋落物可释放一些含有化学毒性成分的化感物质而导致其他树种在自身林下更新不良[33,34]。虽然通过试验方法能检测出林木凋落物中含有对更新产生影响的毒性物质，但在森林复杂的环境中，天然更新对凋落物所产生的毒素的影响效应，有可能受到凋落物其他效应的干扰影响。因此，红椎在马尾松林下更新受凋落物中化感物质影响程度的问题尚待进一步深入研究探讨。

5　结论

(1)红椎是本地区森林中极具天然更新优势的一个珍贵阔叶树种，更新频度及密度均超过其他树种，在自身林下更新最好(21231株/hm^2)，而在针阔混交林下更新密度相对较低(3673株/hm^2)，在针叶树人工林下红椎更新最差，更新密度为625株/hm^2。

(2)红椎更新密度与立地环境因子之间的相关性结果表明，红椎更新与郁闭度呈正相关，与凋落物层厚及草本盖度呈负相关。

(3)红椎更新幼苗的分布规律为，距母树$S<5$m的范围内，林冠下及周围仅有7.5%的更新幼苗分布；与母树的距离为$5\text{m}\leq S<10\text{m}$时，更新幼苗出现的频率为26.1%；当距离母树在$10\text{m}\leq S<15\text{m}$、$15\text{m}\leq S<20\text{m}$、$20\text{m}\leq S<25\text{m}$、$25\text{m}\leq S<30\text{m}$的地块上，幼苗出现的频率分别为68.3%、57.6%、43.7%、25.4%。不同林分内红椎更新幼苗与最近母树的距离分布均表现出相同的变化趋势。但在距母树$S>30$m的地块上仍有38.9%的更新幼苗出现。

(4)马尾松未分解及半分解凋落叶4种质量比1∶10、1∶20、1∶40、1∶70水浸液对红椎种子萌发率抑制作用显著。其中未分解凋落叶水浸液质比1∶10的抑制作用最强。红椎未分解和半分解凋落叶不同质比水浸液对自身种子萌发的抑制作用亦显著。

参考文献

[1]王贺新，李根柱，于冬梅，等．枯枝落叶层对森林天然更新的障碍[J]．生态学杂志，2008，27(1)：83-88.

[2]柏广新，张彦东．水曲柳天然更新及其影响因子[J]．东北林业大学学报，2013，41(1)：7-13.

[3]WILD J，KOPECKY M，SVOBODA M，et al. Spatial patterns with memory：tree regeneration after stand-replacing disturbance in *Picea abies* mountain forests[J]．Journal of Vegetation Science，2014，25(6)：1327-1340.

[4] GAMA J R V，BOTELHO S A，BENTES-GAMA M M. Floristic composition and natural regeneration of a secondary low floodplain forest in the Amazonian estuary[J]．Revista Arvore，2002，26(5)：559-566.

[5]邓硕坤，廖树寿，黄柏华，等．广西凭祥红椎人工林天然更新影响因素初探[J]．广西林业科学，2013，42(1)：48-51.

[6]刘明国，殷有，孔繁轼，等. 辽西半干旱地区油松人工林天然更新的影响因子研究[J]．沈阳农业大学学报，2014，45(4)：418-423.

[7]张树梓，李梅，张树彬，等．塞罕坝华北落叶松人工林天然更新影响因子分析[J]．生态学报，2015，35(16)：1-12.

[8] KRAMER K，BRANG P，BACHOFEN H，et al. Site factors are more important than salvage logging for tree regeneration after wind disturbance in Central European forests[J]．Forest Ecology and Management，2014，331：116-128.

[9]闫淑君，洪伟，林勇明，等．闽江口琅岐岛风景区朴树种群天然更新特征[J]．林业科学，2013，49(4)：147-151.

[10]欧芷阳，苏志尧，彭玉华，等．桂西南喀斯特山地蚬木幼龄植株的天然更新[J]．应用生态学报，2013，24(9)：2440-2446.

[11]张巧明．辽东栎林天然更新特征的研究[D]．杨凌：西北农林科技大学，2007.

[12]谢帆，王素珍．井冈山区常绿阔叶林更新动态的研究[J]．应用生态学报，1991，2(1)：1-7.

[13]徐振邦，代力民，陈吉泉，等．长白山红松阔叶混交林森林天然更新条件的研究[J]．生态学报，2001，21(9)：1413-1420.

[14]BURGHARD von Lupke. Silvi cultural methods of oak regeneration with special respect to shade tolerant mixed species[J]．For Ecol Manage，1998，106(1)：19-26.

[15]MURALI K S，KAVITHA A，HARISH R P. Spatial patterns of tree and shrub species diversity in Savandurga State Forest，Karnataka[J]．Current Science，2003，84：808-813.

[16]刘炜洋，陈国富，张彦东．不同林分内水曲柳天然更新及影响因子研究[J]．华东森林经理，2010，24(4)：19-23.

[17]邓硕坤．桂西南红椎和马尾松人工林中红椎天然更新的研究[D]．南宁：广西大学，2013，14-20.

[18]HOLMES T H. Woodland canopy structure and the light response of *Juvenile Quercus Lobata*[J]．American Journal of Botany，1995，82(11)：1432-1442.

[19]黄忠良．影响季风常绿阔叶林幼苗定居的主要因素[J]．热带亚热带植物学报，2001，9(2)：123-128.

[20]王娜，郝清玉．森林天然更新影响因子研究进展[J]．广东农业科学，2012，(6)：67-70.

[21]刁淑清，韩国君，张树乐，等．水曲柳种子更新调查研究[J]．吉林林业科技，2010，24(4)：19-23.

[22]潘开文，何静，吴宁．森林凋落物对林地微生境的影响[J]．应用生态学报，2004，15(1)：153-158.

[23]FALCELLI J M，PICKETT S T A. Plant litter：its dynamics and effects on plant community structure[J]．Ecology，1991，57(1)：1-3.

[24]GOLLEY F B. Structure and function of an old field broom sedge community[J]．Ecological Monographs，1965，35(1)：113-137.

[25]吴承祯，洪伟，姜志林，等．我国森林凋落物研究进展[J]．江西农业大学学报，2000，22(3)：405-410.

[26]樊后保，臧润国，李德志．蒙古栎种群天然更新的研究[J]．生态学杂志，1996，15(3)：15-20.

[27]韩有志，王政权．天然次生林中水曲柳种子的扩散格局[J]．植物生态学报，2002，26(1)：51-57.

[28]van der VALK AG. The impact of litter and annual plants on recruitment from seed bank of a lacustrine wetland [J]．Aquatic Botany，1986，24(1)：13-26.

[29]LODHI M A K. Role of allelopathy as expressed by dominating trees in a low land forest in controlling productivity and pattern of herbacecus growth[J]．American Journal of Botany，1976，63(1)：1-8.

[30]汪思龙，廖利平．杉木火力楠混交林养分归还与生产力[J]．应用生态学报，1997，8(4)：347-352.

[31]贾黎明，翟明普，尹伟伦，等．油松白桦混交林中生化他感作用的生物测定[J]．北京林业大学学报，1996，18(4)：1-8.

[32]贾黎明，翟明普，尹伟伦．油松和辽东栎混交林植化相克研究[J]．林业科学，1995，31(6)：491-498.

[33]李登武，王冬梅，姚文旭．油松的自毒作用及其生态学意义[J]．林业科学，2010，46(11)：175-178.

[34]曾淼．柠条锦鸡儿凋落物提取液对自身种子萌发和幼苗生长的影响[J]．河南农业科学，2015，44(6)：90-95.

[原载：北京林业大学学报，2018，40(11)]

广西大青山柚木人工林生长过程研究

贾宏炎[1,2]

([1]中国林业科学研究院热带林业实验中心，广西凭祥 532600；
[2]广西友谊关森林生态系统国家定位观测研究站，广西凭祥 532600)

摘 要 研究柚木人工林生长过程及其与气象因子的相关性，为柚木抚育经营提供理论依据。以优良、中等和差3种生长类型的30余年生柚木人工林为对象，基于样地调查，选取优势木、平均木、被压木进行树干解析，对比分析其生长过程，应用灰色关联分析法揭示气象因子对柚木生长的影响。3种生长类型林分柚木胸径、树高和材积生长过程基本一致，各分级木的生长过程亦相类似，其胸径平均和连年生长量随年龄呈现先增加后逐渐降低的趋势，树高生长整体上呈下降趋势，材积生长则呈递增趋势。然而无论优良林分的林木还是各类型林分的优势木，其胸径、材积平均和连年生长量大，速生期持续时间长，生长衰减慢，而其树高生长量的优势相对不明显；30余年生时柚木尚未达数量成熟龄。各类型林分间柚木生长与气象因子关系的差异仅体现在胸径，优良林分胸径连年生长量主要受极端低温影响，而中等和差林分则与年均降水量相关性最大；各分级木间柚木生长与气象因子的关系无明显差异；影响树高和材积连年生长量的最主要气象因子分别为年均降水量和年均气温。无论生长类型间还是分级木间比较，柚木人工林生长过程整体趋势基本一致，其差异主要体现在生长量大小和快速生长期长短。约30年生柚木人工林仍未达到数量成熟，后期抚育经营对于其优质大径材高效培育仍不可忽视。

关键词 柚木；生长过程；生长类型；分级木；气象因子；灰色关联分析

Growth Process of *Tectona grandis* Plantations in Daqing Mountain, Guangxi

JIA Hongyan[1,2]

([1]*Experimental Center of Tropical Forestry*, *CAF*, *Pingxiang* 532600, *Guangxi*, *China*;
[2]*Guangxi Youyiguan Forest Ecosystem Research Station*, *Pingxiang* 532600, *Guangxi*, *China*)

Abstract: Growth process of *Tectona grandis* plantations was investigated and its relationship with meteorological factors was analyzed so as to provide evidences for plantation management of this species. Based on investigation of sampling plots, dominant, mean and suppressed *T. grandis* trees of about 30 years old were sampled from high, medium and low growth types of its plantations, stem analysis was then conducted for each sampled tree, and growth process of these trees were compared between growth types. Grey correlation analysis was further applied to explore effects of meteorological factors on annual tree growth of this species. Growth process of stem diameter at breast height (DBH), tree height and individual volume of *T. grandis* did not differ significantly among three growth types and three classes of trees. The mean and current annual increments of DBH increased with tree age initially and then decreased, while decreased for those of height and increased for individual tree volume. Dominant trees or trees in high growth type of plantations had high mean and current annual growth, long rapid growth duration and slow growth declination of DBH and volume, while their heights growth difference were not relatively higher. 30-year-old trees had not reached quantitative maturity. The difference of relationships between tree growth and meteorological factors among three growth types was only found on DBH. The main factor influencing current annual increment of DBH was extreme low temperature at high growth type, while mean annual precipitation at medium and low growth types. There was no significant difference of relationships between tree growth and meteorological factors among three classes of trees. The main factors influencing current annual increments of tree height and individual volume were mean annual precipitation and mean annual air temperature, respectively. Growth processes of *T. grandis* plantations are basically consistent in three growth types of plantations as well as for three grades of trees, their differences mainly demonstrate in growth performance and length of rapid growth. Due to the fact that *T. grandis* plantation do not reach quantitative maturity at the age of about 30 years yet, tending and management are still important at the late stage for high-quality large-sized timber production of this species.

Key words: *Tectona grandis*; growth process; growth types; tree grading; meteorological factors; grey correlation analysis

林木生长过程是林木于不同年龄生长状态的直接表现，研究林木生长过程对于林分合理抚育经营有着重要指导意义。林木生长过程数据通常由定期连续观测或树干解析获取，树干解析对于未连续观测林分的生长过程重建尤为重要[1-3]。林木生长过程不仅受林分密度和抚育措施[4-6]、立地条件[7,8]等显著影响，亦与林分生长类型[9,10]、林木分级地位[11-13]等紧密相关。而生长类型通常通过立地指数进行划分[9,10]，两者具有共通性。许景伟等[9]对沙岸黑松(*Pinus thunbergii* Parlatore)海防林及邱治军等[10]对徐淮平原杨树(*Populus* spp.)农田防护林的研究表明：优良林分生长表现好，数量成熟龄明显推后。李晓庆等[14]和张东北等[7]比较不同立地条件下杉木[*Cunninghamia lanceolata*(Lamb.) Hook.]的生长过程得出，优良立地杉木胸径、树高、材积的总生长量、平均生长量和最大连年生长量均较高，而且速生期持续时间较长。对于林分中不同分级木（优势木、平均木、被压木）生长过程研究，国内外多数研究侧重于优势木或平均木生长过程及其模拟分析[12,15-17]，而开展分级木间的比较研究则不多见，如玉宝等[18]对大兴安岭兴安落叶松[*Larix gmelinii*(Ruprecht) Kuzeneva]天然林各分级木的生长过程及相互之间的转化率进行了研究。

气象条件是影响林木分布和生长的重要因素，温度、水分等气象因子对林分的组成、分布和生长具有重要的影响。以往研究表明，不同地区、生境、海拔、生长阶段的林木生长及林分生产力对气象因子的响应各不相同[19-21]，研究一个树种不同生长类型林分的优势木或平均木以及相同生长类型林分各分级木的生长过程与气象因子关系，将为其经营决策提供理论指导。

柚木(*Tectona grandis* L. f)是世界名贵的用材树种之一，其心材呈黄褐色或暗褐色，颜色优美，材质坚韧耐腐，结构致密，具有不翘不裂，易于加工、用途广泛等特点，是国际市场上最受欢迎的阔叶材之一[22]。众所周知，我国引种柚木已有180多年的历史，柚木对立地和气候条件要求较高，如何在有限的适宜范围内对其进行高效培育是柚木研究的重点。近年来，Perez[2]通过树干解析构建柚木胸径、树高和材积生长方程并与现有的生长方程进行对比，得出树干解析是一种有效的构建柚木生长数据库的工具，杜健等[23,24]开展柚木人工林立地质量评价。但国内对于柚木的生长过程研究未见报道，尤其是不同分级木的生长及气象因子的影响。因此，本研究以30余年生柚木人工林为对象，开展不同生长类型林分的平均木以及同一生长类型林分各分级木生长过程的分析及其与气象因子关系的研究，为柚木合理栽培及抚育经营提供理论依据。

1 材料与方法

1.1 研究地概况

所调查的柚木人工林位于广西壮族自治区凭祥市中国林业科学研究院热带林业实验中心(21°57′~22°19′N，106°39′~106°59′E)，属北热带季风气候，年均气温约22℃，年均降水量1550mm，干湿季明显，降水主要集中于5~9月，约占全年总降水量的3/4。土壤类型以砖红壤和红壤为主。

表1 广西大青山14个柚木样地的基本信息

样地	海拔(m)	坡(°)	坡位	坡向	造林时间	造林密度(株/hm²)
BY1	232	30	下	东南	1981	2500
BY2	230	35	下	东南	1981	2500
BY3	350	40	下	西北	1981	2500
BY4	349	40	下	西北	1981	2500
QS1	199	0	下	西北	1982	2500
QS2	185	5	下	北	1982	2500
QS3	153	16	下	北	1982	2500

（续）

样地	海拔(m)	坡(°)	坡位	坡向	造林时间	造林密度(株/hm²)
QS4	164	18	下	北	1982	2500
QS5	146	25	下	北	1982	2500
QSCL1	150	20	下	北	1982	1667
QSCL2	150	20	下	北	1982	1667
QSCL3	150	20	下	北	1982	1667
QSCL4	150	20	下	北	1982	1667
XS1	199	11	下	东	1982	2500

1.2 研究方法

2013 年底至 2014 年初，于上述柚木林分内设置 14 块 20m×30m 的样地，其中青山实验场平而河边 5 块，茶陋站 4 块，白云实验场 4 块，夏石树木园 1 块。除青山茶陋站造林密度为 1667 株/hm²外，其余所有调查林分的造林密度为 2500 株/hm²，经多次间伐后所有样地现存株数约为 360 株/hm²，样地具体信息如表 1 所示。

测定每个样地内所有柚木的胸径、树高、枝下高以及四个方向的冠幅，其生长统计数据如表 2 所示。根据调查结果于每个样地内各选出优势木或亚优势木、平均木和被压木各 1~2 株。由于青山茶陋站沟边柚木属于长期保存的采种母树林，每个样地仅选择一株平均木进行树干解析，共选择 14 株优势木或亚优势木、19 株平均木、12 株被压木作为解析木。所选优势木或亚优势木的胸径为 24.3~32.9cm，树高 20.9~27.4m；平均木(不包括青山茶陋站沟边解析木)胸径为 21.0~26.8cm，树高 19.2~23.4m；被压木胸径为 11.7~24.6cm，树高 9.3~20.7m；青山茶陋站沟边 4 株平均木的胸径为 29.0~35.7cm，树高 20.7~27.4m。

表 2　广西大青山 14 个样地的柚木生长表现

样地	平均胸径(cm)	平均树高(m)	平均枝下高(m)	平均冠幅(m)	优势木平均高(m)	生长类型
BY1	27.05 (081)	18.90 (0.50)	7.25 (0.56)	5.27 (0.30)	21.56 (0.31)	差
BY2	25.91 (1.26)	16.93 (0.84)	6.61 (0.69)	5.46 (0.50)	20.70 (0.86)	差
BY3	28.48 (1.01)	18.77 (0.53)	9.33 (0.66)	4.93 (0.21)	22.26 (0.27)	中等
BY4	27.57 (0.51)	18.42 (0.30)	8.24 (0.34)	5.13 (0.15)	22.69 (0.27)	中等
QS1	23.71 (0.91)	16.94 (0.58)	9.45 (0.53)	6.78 (0.24)	19.90 (0.25)	差
QS2	26.07 (0.86)	19.40 (0.68)	10.79 (0.62)	7.80 (0.68)	22.98 (0.61)	中等
QS3	25.46 (0.90)	19.97 (0.62)	9.49 (0.40)	7.12 (0.56)	22.76 (0.30)	中等
QS4	25.17 (0.59)	22.23 (0.39)	12.75 (0.58)	5.19 (0.24)	25.00 (0.53)	中等
QS5	23.46 (0.78)	17.71 (0.53)	9.22 (0.46)	6.40 (0.34)	20.12 (0.38)	差
QSCL1	39.60 (1.35)	22.52 (0.56)	6.51 (0.79)	10.10 (0.52)	25.70 (0.13)	优良
QSCL2	41.05 (2.06)	23.14 (0.80)	9.06 (0.71)	10.15 (0.55)	25.98 (0.48)	优良
QSCL3	34.04 (1.27)	24.88 (0.76)	14.00 (1.51)	8.62 (0.45)	28.26 (0.55)	优良
QSCL4	45.66 (1.52)	25.28 (1.04)	10.25 (0.93)	10.11 (0.50)	31.04 (0.59)	优良
XS1	22.13 (1.26)	15.78 (0.64)	8.13 (0.51)	5.01 (0.37)	20.15 (0.40)	差

解析木伐倒后按常规方法进行树干解析，分别于根茎处以上 0.3m、1.3m、2m 高度处截取圆盘，此后按 2m 区分段截取圆盘，圆盘厚度约 5cm，每个圆盘表明高度位置及北向。将圆盘带回实验室内晾干，然后对每个圆盘进行打磨和判读，用电子游标卡尺测量每个圆盘四个方向的带皮半径以及每个年轮四

个方向至髓心的距离（0.01mm）。

1.3　数据处理

鉴于调查林分年龄基本一致，利用林分优势木平均高比较其生产力。根据样地调查数据于每个样地内选取5株优势木或亚优势木，计算其树高平均值，按照优势木平均高通过系统聚类分析将柚木林分划分为优良（$H>25$m）、中等（22m$<H\leqslant$25m）、差（$H\leqslant$22m）3种生长类型（表2）。基于树干解析数据，应用统计之林（Forstat2.0）软件解析木各年龄时胸径、树高和材积的连年、平均生长量等指标[25]，进而分别按照生长类型、分级木进行对比分析。比较不同类型林分间柚木生长过程差异时，仅对抽取的平均木进行分析；由于青山茶陋站沟边柚木林分仅选择平均木开展树干解析，且林分生长状况明显优于其他林分，因此未将其纳入分级木生长过程比较分析。

运用灰色系统理论的关联分析法[26]分析柚木生长过程与气象因子年际变化间的关系。具体而言，分别按照生长类型、分级木计算每个年龄时胸径、树高和材积连年生长量与对应年份各气象因子的关联度，找出气象因子对各生长类型林分以及各分级木胸径、树高和材积生长的影响顺序。气象因子数据来源于凭祥市局，主要包括1981-2013年间每年的年平均气温、极端高温、极端低温以及年平均降水量等。

2　结果与分析

2.1　柚木胸径生长过程

胸径平均生长量在造林后几年内迅速增加，优良林分柚木胸径平均生长量于12年左右时达到最大，超过1.2cm/年；此后开始缓慢下降，约20年内胸径平均生长量仍超过1cm/年。而对于中等和差林分而言，其胸径平均生长量最大值出现在造林后6年左右，此后亦开始逐年下降，其下降速率明显高于优良林分，其胸径平均生长量最终维持在约0.7cm/年（图1A）。

对于胸径连年生长量而言，不同年份间均存在较大差异，特别是优良林分，其连年生长量变化最为剧烈。从造林后第5年开始，优良林分胸径连年生长量一直明显高于中等和差林分，约10年时差异达最大值（达1cm/年）。此后随着连年生长量的降低差异逐渐减小，造林后第20年，各生长类型林分的柚木胸径连年生长量基本上趋于一致，约0.6cm/年（图1B）。

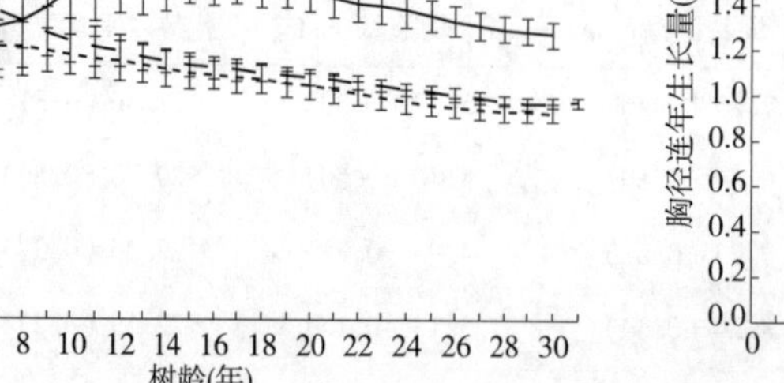

图1　3种生长类型林分的柚木胸径生长过程

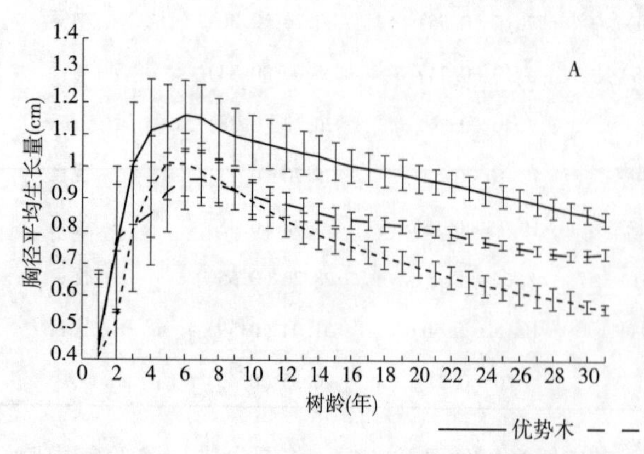
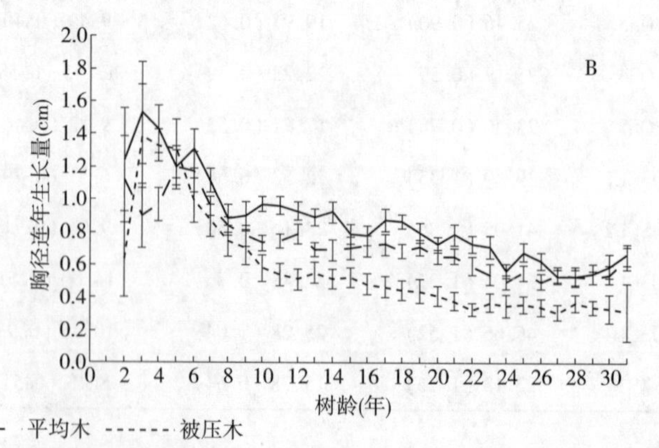

图2　3种级别柚木的胸径生长过程

各分级木间比较，造林后优势木的胸径平均和连年生长量一直高于平均木和被压木，但是3种分级木的最大胸径平均生长量均出现于造林后第5年，连年生长量亦均于造林后3~5年达到最大。此后，被压木的平均和连年生长量急剧下降，而优势木和平均木下降缓慢，造林后第 10 年左右，被压木与平均木、优势木的胸径平均和连年生长量差异随树龄增加逐渐增大，而平均木和优势木间平均生长量的差异值约为 0.2cm/年，仅最后几年有减小趋势(图 2A)。30 余年内，优势木胸径连年生长量亦均略高于平均木，27 年后，优势木和平均木胸径连年生长量均约 0.8cm/年(图 2B)。

2.2 柚木树高生长过程

从柚木树高生长过程可知(图 3A、B)，树高平均生长量和连年生长量总体呈现下降趋势，且 3 种生长类型林分间树高生长差异较小。造林前 7 年是柚木高生长的高峰期，各类型林分的树高连年生长量均在 1m/年以上。其中优良林分的最高可达 1.8m/年，而且此后第 8~17 年内，其树高连年生长量仍维持在 1m/年左右；对于中等和差林分，造林 5 年后其树高连年生长量开始迅速下降，此阶段亦是优良林分与中等、差林分树高平均生长量表现出差异的时期。差林分树高最低连年生长量出现于造林 18 年后，约为 0.1m/年，而中等和优良林分出现于造林后 23~24 年时，此后均又随年龄增加逐渐回升，于 28 年时达到一个生长小高峰。约 30 年生时，优良、中等和差林分的柚木树高平均生长量分别约为 1.0m/年、0.8m/年和 0.6m/年。

各分级木树高连年和平均生长量亦随树龄增加均呈下降趋势，前 7 年优势木和平均木连年生长量超过 1m/年，被压木的树高平均和连年生长量均于造林后 4 年时开始明显低于平均木和优势木，而平均木和优势木的树高连年和平均生长量差异极小(图 4A、B)。柚木优势木和平均木树高连年生长量最低值亦出现于造林后 24 年，而被压木则在造林后 20 年。各分级木均于造林后 28 年时出现一次树高生长高峰。

2.3 柚木材积生长过程

柚木材积平均生长量和连年生长量随年龄增加总体均呈上升趋势，优良林分材积平均生长量和连年生长量分别于造林后约 10 年和 7 年时即明显大于中等和差林分，且随树龄增加差异均逐渐增大。造林 10 年后，优良林分材积连年生长量均大于 $0.02m^3$/年，最高可超过 $0.04m^3$/年，而中等和差林分的材积连年生长量一直维持在 $0.010 \sim 0.015m^3$/年，仅 30 年后才超过 $0.02m^3$/年。此外，优良林分材积连年生长量后期变化幅度要明显高于中等和差林分，而中等林分材积平均和连年生长量与差林分差异很小，均仅稍高于差林分。

通过对 3 种生长类型林分柚木连年生长量与平均生长量作图分析可知，当前材积连年生长量和平均生长量曲线均尚未出现交叉(图 6)，且两者差异随树龄增加呈增大趋势，说明当前 3 种类型林分均远未达数量成熟。

3 种分级木材积平均和连年生长量亦随树龄增加而逐渐增大(图 7)，但是被压木在造林 10 年后，材积连年和平均生长量随树龄增大增量甚微，其材积连年生长量一直约为 $0.005m^3$/年，平均生长量亦介于 $0.004 \sim 0.006m^3$/年之间，远低于后期优势木和平均木平均生长量($0.017m^3$/年和 $0.011m^3$/年)和连年生长量($0.03m^3$/年和 $0.025m^3$/年)，此时亦是被压木与平均木材积生长出现分化的时间，而优势木与平均木、被压木生长量在造林后第 4 年即表现出差异。此外，不同分级木间材积生长差异亦随树龄增大逐渐增大。

各分级木材积连年生长量与平均生长量在 30 余年时亦仍未出现交叉(图 8)，即当前林龄的 3 种分级木均尚未达数量成熟。但从变化趋势来看，被压木达到数量成熟龄的时间可能要早于平均木和优势木。

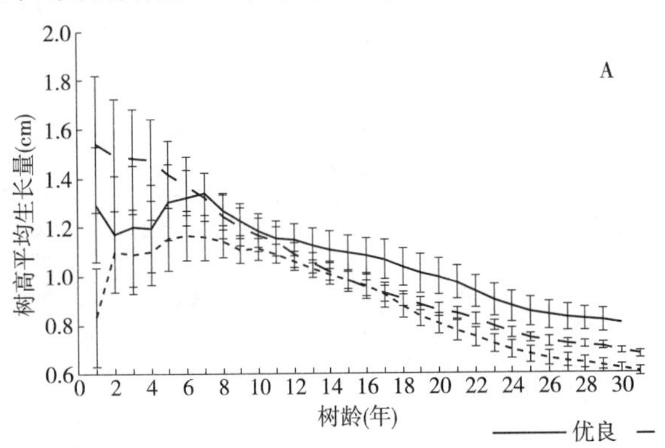

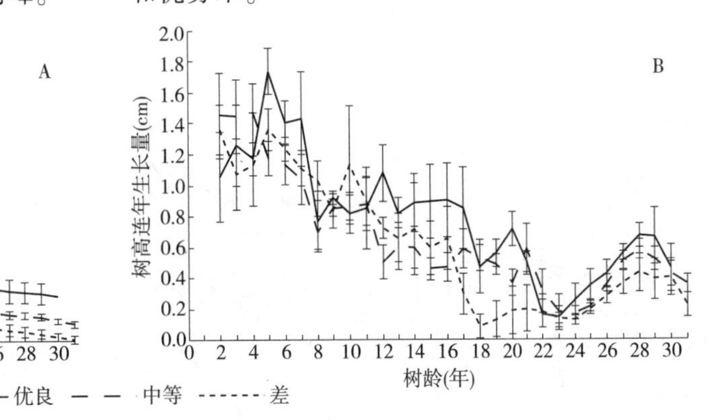

图 3 3 种生长类型林分的柚木树高生长过程

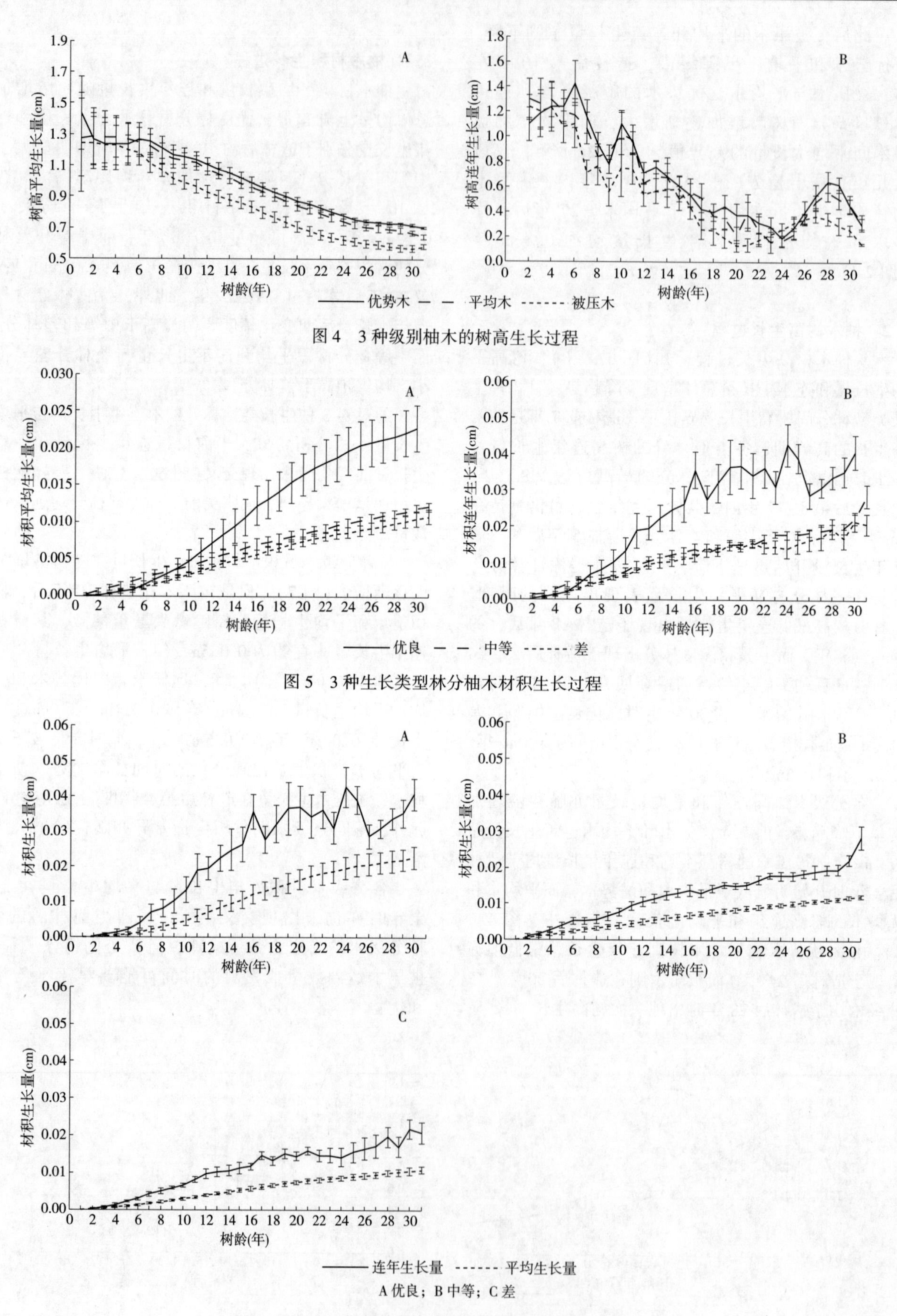

图 4　3 种级别柚木的树高生长过程

图 5　3 种生长类型林分柚木材积生长过程

图 6　3 种生长类型林分柚木材积连年和平均生长量

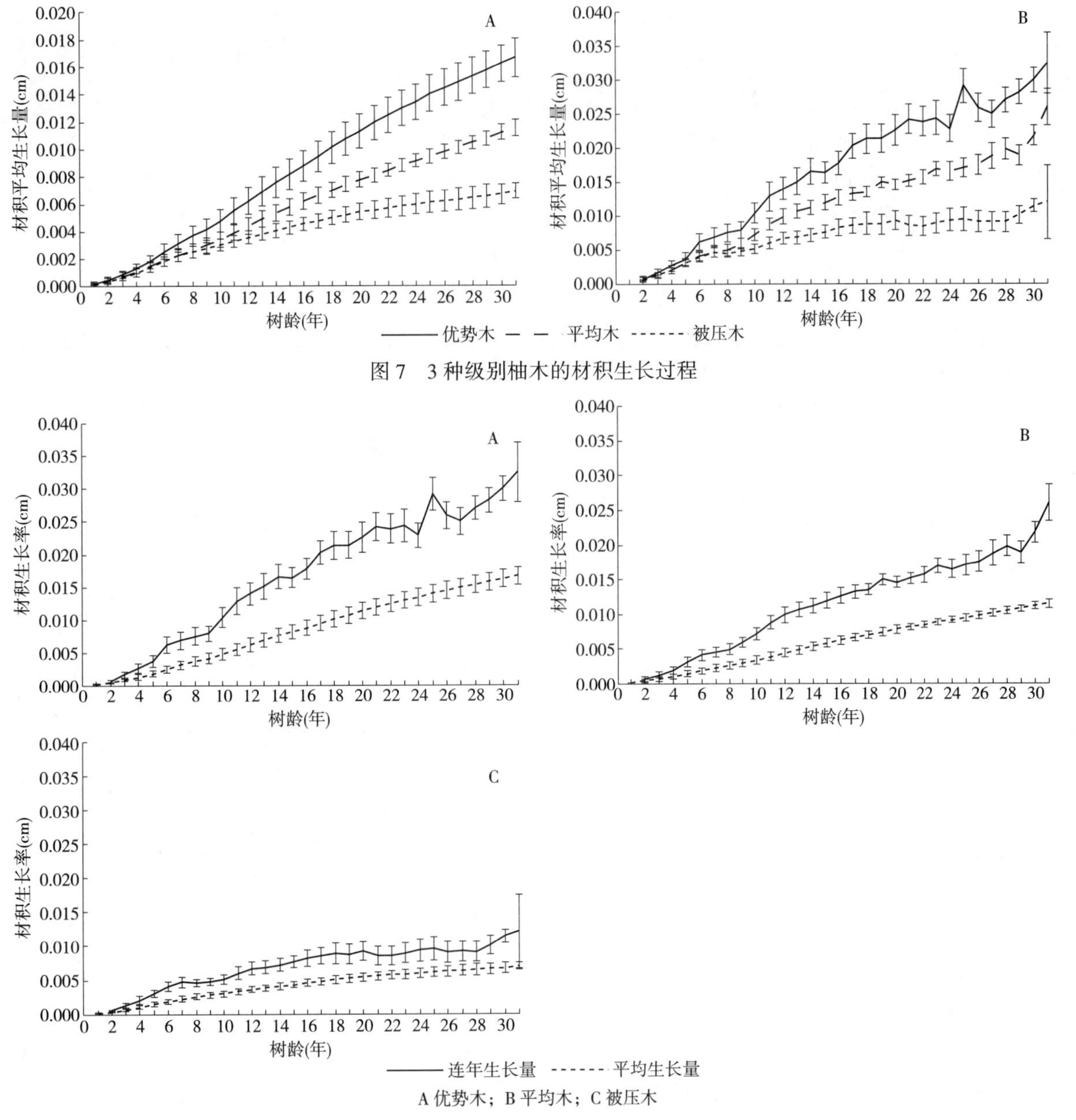

图 7　3 种级别柚木的材积生长过程

图 8　3 种级别柚木材积连年生长量和平均生长量

2.4　柚木生长与气象因子的关系

由灰色关联度分析结果可知：影响柚木优良林分胸径连年生长量最主要的气象因子为极端低温，其次为年均降水量；而对于中等和差林分而言，年均降水量为最主要的气象因子(表 3)。气象因子影响柚木树高和材积生长的强弱顺序在林分生长类型间无明显差异，各类型林分影响树高生长最主要气象因子均为年均降水量，其次为极端低温；对于材积生长而言，其重要性排序则为：年均气温 > 极端高温 > 极端低温 > 年均降水量(表 3)。

表 3　不同生长类型林分柚木生长与气象因子的关联度

生长类型	连年生长量	气象因子			
		极端高温	极端低温	年均温	年均降水量
优良	胸径(cm)	0.61960	0.6815	0.6282	0.6550
	树高(m)	0.6553	0.6737	0.6724	0.6952
	材积(m^3)	0.6922	0.6209	0.7328	0.6053

（续）

生长类型	连年生长量	气象因子			
		极端高温	极端低温	年均温	年均降水量
中等	胸径(cm)	0.7282	0.7400	0.7532	0.7940
	树高(m)	0.6515	0.6856	0.6441	0.7189
	材积(m^3)	0.7311	0.6599	0.7431	0.6086
差	胸径(cm)	0.6143	0.63740	0.6371	0.7081
	树高(m)	0.6072	0.6168	0.5846	0.7071
	材积(m^3)	0.7210	0.6624	0.7571	0.6242

注：气象因子数据由当地气象部门提供。下同。

影响优势木、平均木和被压木胸径、树高生长的最主要气象因子均为年均降雨量，而影响柚木材积连年生长量的气象因子重要性排序为：年均气温>极端高温>极端低温>年均降雨量（表4）。

表4 3种级别柚木生长与气象因子的关联度

林木级别	连年生长量	气象因子			
		极端高温	极端低温	年均温	年均降雨量
优势木	胸径(cm)	0.7267	0.7213	0.7162	0.7908
	树高(m)	0.6135	0.6215	0.5860	0.7098
	材积(m^3)	0.7338	0.6549	0.7440	0.6108
平均木	胸径(cm)	0.6586	0.7013	0.6994	0.7421
	树高(m)	0.6377	0.6498	0.6025	0.7256
	材积(m^3)	0.7241	0.6569	0.7414	0.6079
被压木	胸径(cm)	0.6596	0.6871	0.6654	0.7422
	树高(m)	0.6278	0.64250	0.6055	0.7181
	材积(m^3)	0.7382	0.6595	0.7468	0.6171

3 讨论

考虑到优良林分仅选取平均木进行树干解析，各类型(优良、中等、差)林分间的比较分析仅针对平均木，而且分级木(优势木、平均木、被压木)间的差异分析仅针对中等和差立地，这些是本研究的局限所在。尽管如此，仍然得到一些较为有益的结果。总体来看，不同类型林分间以及分级木间柚木胸径、树高、材积生长量差异较大，但是其生长过程均无明显差异。然而，无论优良林分的平均木，还是中等和差林分的优势木，其平均和连年生长量大、高峰期长、生长衰减慢，与以往对于杉木[7,14]、刺槐(*Robinia pseudoacacia* Linn.)[8]等人工林的研究结果相似。究其原因，主要与林分内单株对养分、水分和生长空间等的竞争有关。优良林分相对于中等和差林分具有相对充足的养分、水分供给以支撑其持续生长；对于优势木而言，由于其位于林分上层，根系亦更为发达，使其对生长空间以及养分和水分的竞争更有优势。此亦可从本研究中生长过程与气象因子的相关性得到验证：在中等和差林分，年平均降雨量是柚木胸径生长的主要影响因子；而在优良林分，由于水肥条件相对较好，极端低温是胸径生长的主要影响因子。此外，在本研究中，优良林分的造林密度相对较低，亦是其胸径前期快速生长持续时间较长的原因所在。

各生长类型林分间以及各分级木间树高平均和连年生长量差异小于其胸径和材积生长量，但不同类型林分间树高生长差异大于分级木间差异。从生长后期(24年后)可以看出，最后一次间伐(2009—2011)后树高连年生长量出现快速上升，而胸径和材积连年生长量变化则存在滞后性，说明柚木高生长对间伐引起的林分环境因素变化的响应更为敏感，刘彬彬等[27]对64年生小叶青冈[*Cyclobalanopsis myrsinaefolia*(Blume) Oersted]的研究亦发现，相较于胸径和材积生长，其高生长对环境因子变化的响应更加灵敏。

与胸径和树高生长过程不同，柚木材积生长量随年龄增大均呈上升趋势，这主要与柚木生长特性有关。本研究所涉及的所有气象因子中，年均气温是影响柚木材积连年生长量的最主要因素，由此可以推测，年积温对于材积生长最为重要。结合胸径、树高、材积生长与气象因子的相关性可知，柚木适宜种植于气温高、降水量大的环境中，此结论与柚木的生存习性相吻合[22]。30余年生时，无论优势木、平均木、被压木，还是优良、中等和差林分，柚木材积连年生长量和平均生长量曲线均未出现相交，且两者差异仍有继续增大的趋势，说明此时柚木尚远未达到其数量成熟龄，不宜过早皆伐利用，后期抚育经营仍具有很高的价值。

由中等和差林分柚木生长过程可以看出，造林6~8年后，其胸径和树高生长量即出现明显下降，材积增速变缓，说明此时林分竞争效应开始显现，需要进行第一次间伐，通过伐除被压木，延长其快速生长期。对于优良林分，造林约16年后，其胸径生长才出现下降趋势，材积连年生长量增速放缓，需要进行首次间伐。优良林分较中等和差林分生长高峰期出现时间晚、持续时间长的原因除与地力有关外，造林密度与林分保存率亦是其潜在原因。

4 结论

柚木胸径、树高和材积生长过程，无论生长类型(优良、中等、差)间的差异，还是分级木(优势木、平均木、被压木)间的差异，主要在于生长量大小以及快速生长期长短，其整体变化趋势基本上一致。柚木宜选择水热条件较好的立地造林，优良林分柚木首次间伐时间宜于造林16年后。柚木30余年生时尚未达数量成熟龄，不宜主伐，需适时加强抚育经营以促进其生长，从而实现优质大径材的高效培育。

参考文献

[1] GARCIA O. Comparing and combining stem analysis and permanent sample plot data in site index models [J]. Forest Sciece, 2005, 51: 277-283.

[2] PEREZ D. Growth and volume equations developed from stem analysis for *Tectona grandis* in Costa Rica [J]. Journal of Tropical Forest Science, 2008, 20(1): 66-75.

[3] MACHADO S D A, SILVA L C R D, FIGURA M A, et al. Comparison of methods for estimating heights from complete stem analysis data for *Pinus taeda* [J]. Ciência Florestal, 2010, 20(1): 45-55.

[4] 童书振, 盛炜彤, 张建国. 杉木林分密度效应研究 [J]. 林业科学研究, 2002, 15(1): 66-75.

[5] ZHANG J, OLIVER W W, RITCHIE M W. Effect of stand densities on stand dynamics in white fir (*Abies concolor*) forests in northeast California, USA [J]. Forest Ecology and Management, 2007, 244: 50-59.

[6] 王春胜, 赵志刚, 曾冀, 等. 广西凭祥西南桦中幼林林木生长过程与造林密度的关系 [J]. 林业科学研究, 2013, 26(2): 257-262.

[7] 张东北, 周启忠, 赵忠北, 等. 立地条件对杉木生长过程的影响 [J]. 福建林学院学报, 1996, 16(1): 86-88.

[8] 彭鸿, MOSANDI R. 立地和人为干扰对渭北黄土高原刺槐人工林个体生长过程的影响 [J]. 山东农业大学学报(自然科学版), 2003, 34(1): 44-49.

[9] 许景伟, 李琪, 王卫东, 等. 沙岸黑松海防林防护成熟期及更新年龄的研究 [J]. 林业科学, 2003, 39(2): 91-97.

[10] 邱治军, 胡海波. 徐淮平原杨树农田防护林的成熟龄与更新龄 [J]. 南京林业大学学报(自然科学版), 2005, 29(1): 46-50.

[11] 洪玲霞. 初植密度、间伐对杉木林分优势高生长过程的影响 [J]. 林业科学研究, 1997, 10(4): 448-452.

[12] 陈英, 杨华, 李伟, 等. 北京地区侧柏人工林标准木生长过程研究 [J]. 西北林学院学报, 2012, 27(5): 153-157.

[13] 赵西平, 余建基, 张超男, 等. 黑龙江省帽儿山林场天然白桦林优势生长过程研究 [J]. 西北林学院学报, 2014, 29(2): 191-195.

[14] 李晓庆, 周俊宏, 郑勇平. 不同立地指数级杉木生长过程的数学模拟 [J]. 浙江林学院学报, 1991, 8(3): 281-287.

[15] MILIOS E. The influence of stand development process on the height and volume growth of dominant *Fagus sylvatica* L. sl trees in the central Rhodope Mountains of north-eastern Greece [J]. Forestry, 2004, 77: 17-26.

[16] SANCHEZ-GONZALEZ M, TOME M, MONTERO G. Modelling height and diameter growth of dominant cork oak trees in Spain [J]. Annals of Forest Science. 2005, 62, 633-643.

[17] 林文树, 穆丹, 王丽平, 等. 针阔混交林不同演替阶段表层土壤理化性质与优势林木生长的相关性 [J]. 林业科学, 2016, 52(5): 17-25.

[18] 玉宝, 王立明. 兴安落叶松天然林分级木生长特性分析 [J]. 林业科学研究, 2007, 20(4): 452-457.

[19] 侯爱敏, 周国逸, 彭少麟. 鼎湖山马尾松径向生长动态与气候因子的关系 [J]. 应用生态学报, 2003, 14(4): 637-639.

[20] 程瑞梅, 封晓辉, 肖文发, 等. 北亚热带马尾松净生产力对气候变化的响应 [J]. 生态学报, 2011, 31

(8)：2086-2095.

[21] 张丽云，邓湘雯，雷相东，等．不同生长阶段马尾松生产力与气候因子的关系［J］．生态学杂志，2013，32(5)：1104-1110.

[22] 梁坤南，周再知，马华明，等．我国珍贵树种柚木人工林发展现状、对策与展望［J］．福建林业科技，2011，38(4)：173-178.

[23] 杜健，梁坤南，周树平，等．不同地区柚木人工林生长及土壤理化性质的研究［J］．林业科学研究，2016a，29(6)：854-860.

[24] 杜健，梁坤南，周再知，等．云南西双版纳柚木人工林立地类型划分及评价［J］．林业科学，2016b，52(9)：1-10.

[25] 唐守正，郎奎建，李海奎．统计和生物数学模型计算：ForStat 教程［M］．北京：科学出版社，2009.

[26] 刘贤谦，师光禄，张厉燕．应用灰色关联度分析关键因子的研究［J］．林业科学，1996，32(5)：447-453.

[27] 刘彬彬，楼炉焕，刘广宁，等．浙江省小叶青冈生长过程的研究［J］．浙江农林大学学报，2013，30(4)：517-522.

［原载：林业科学研究，2019，32(01)］

密度调控对米老排中龄人工林生长的影响

唐继新[1,2] 贾宏炎[1] 王 科[3] 曾 冀[1] 郑 路[1] 王亚南[1] 杨保国[1]
([1]中国林业科学研究院热带林业实验中心，广西凭祥 532600；
[2]中国林业科学研究院资源信息研究所，北京 100091；
[3]广西林业勘测设计院，广西南宁 530001)

摘 要 密度是影响林分生产力的关键因素之一，分析密度调控对米老排中龄林生长的影响，进而为其间伐密度调控提供决策依据。以南亚热带中等立地两种不同密度调控的米老排中龄林为对象，按优势木、中等木、被压木的条件选取了28株标准木(每林分各14株)，基于2m区分段的中央断面积树干解析法和双侧t检验的统计分析法，对不同调控密度下米老排林分的平均木、优势木和林分蓄积等生长过程进行对比分析。米老排径向生长的缓慢期在第1~2年，速生期在第3~10年，衰减期在第14年后。树高的早期速生特性明显，连年生长量呈多峰状，速生期主要在第2~6年。平均木与优势木材积生长的缓慢期均在前6年，从第8年起均进入速生期；密度对平均木材积连年生长与林分数量成熟时间的影响显著，哨平试验林(在第12年经过生长伐1次，伐后林分最终密度为1200株/hm^2)数量成熟在第24年，而青山试验林(分别在第12年、第17年、第25年经过3次生长伐后，林分最终密度为520株/hm^2)直至第34年仍未达到数量成熟。中弱度间伐(株数间伐强度<30%)对中龄林蓄积总生长量的影响不显著，对林分蓄积连年生长量短期有一定影响；强度间伐(株数间伐强度>30%)对中龄林蓄积总生长量与连年生长量的影响显著($P<0.05$)。在第14年后，米老排树种的实验形数趋于稳定。林分密度调控在520~1200株/hm^{2}的范围内，密度调控措施对米老排平均木的胸径和材积的生长影响显著($P<0.05$)，对林分树高和平均实验形数的影响不显著，对优势木的胸径与材积的短期生长影响显著($P<0.05$)，对其长期生长的影响不显著，对减小林分径阶分化及提高大径木比例的作用明显。

关键词 米老排；密度调控；中龄林；人工林；平均木；优势木；生长；双侧*t*检验

Effect of Density Regulation on Growth of *Mytilaria laosensis* Plantation with Middle Age

TANG Jixin[1,2], JIA Hongyan[1], WANG Ke[3], ZENG Ji[1], ZHENG Lu[1], WANG Yanan[1], YANG Baoguo[1]

([1]*Experimental Center of Tropical Forestry, Chinese Academy of Forestry, Pingxiang* 532600, *Guangxi, China*;
[2] *Research Institute of Forest Resources Information Techniques, CAF, Beijing* 100091, *China*;
[3] *Guangxi Forestry Inventory and Planning Institute, Nanning* 530001, *Guangxi, China*)

Abstract: Stand density regulation can promote the growth of trees, improve forest quality as well as forest stand structure, and play the key role in the technology of forest multi-function. Thus, reasonable stand density regulation can affect the a-chievement of a forest culture´s goals and influence the full extent of the forest for greater bene-fits. To obtain a reasonable thinning measure for mid-maturation *Mytilaria laosensis* plantations, it is important to ascertain the effects of density regula-tion on plantation growth. Based on 28 analytic trees (14 analytic trees in each stand), including dominant trees, medium trees, and pressed trees, the growth of dominant trees, mean trees, and stand volume in two mid-maturation Mytilaria laosensis plantations (Qingshan stand and Shaoping stand) with different density regulation in the south subtropical area of China were compared. Comparisons were made using the tree stem ana-lytic method of the middle section in 2m and Student´s t-test was conducted with data processing system software (DPS14.5). The Qingshan stand had an afforestation time in the spring of 1982, a planted density of 2500 plants per ectare, and experienced one lighting cutting (in the 7th year) and three accretion cuttings (in the 12th, 17th, and 27th year; the stem thinning intensity was between 21%～42%) after afforestation; after which the forest stand density was 520 plants per hectare. For the Shaoping stand, planting time was in the spring of 1984, afforestation density was 2500 plants per hectare, and the stand experienced once lighting cutting (in the 7th year) and one accretion cutting (in the 12th year; the stem thinning intensity was 27%) ; after thinning the stand density was 1200 plants per hectare. ①The radial slow-growing period occurred during the first 1～2 years, and the fast-growing stage occurred at 3～10 years of age (annual increment of diameter at breast height was during 0.72～2.45cm) , with attenuation starting at 14 years of age. ②Tree height exhibited distinct fast growing characteristics in the early stages of growth, spanning from the 2nd to the 6th year (annual growth of tree height was between 1.30～1.75m) , and the annual increment of tree height took on a multimodality. ③The stock volume slow growth period of the medium trees and the dominant trees was in the first 6 years, giving way to a fast-growth period in the 8th year. Density regulation had a significant impact on the annual volume increment of the medium trees, as well as the stand quantitative maturity ages. The Shaoping stand reached quantity maturity in its 24th year, but for the Qingshan stand, it was not reached until the 34th year. ④The intermediate and weak thinning operations, with the stem thinning intensity less than 30%, had no significant influence on the total growth of stand volume of the middle-aged stands, but could influence the annual increment of stand volume to some extent over a short time period. The high-intensity thinning practices, with the stem thinning intensity greater than 30%, had a significant influence on the total growth and the annual increment of stand volume. ⑤After 14 years of age, the experimental form factor of the species tended to be stable (mean value between 0.41～0.42), and the step form level was Ⅲ-Ⅳ. When the stem density fell into the range from 520 to 1 200 stems per hectare, the growth of tree height and the experimental form factor were not significantly affected by density regulation. The DBH growth and the stock volume growth of the mean trees were obviously affected by density regulation ($P<0.05$). The short-term DBH growth and stock volume growth of the dominant trees were also significantly affected by density regulation ($P<0.05$) , but the influence was not significant in the long run. Density regulation can reduce the stand diameter order differentiation and can increase the percentage of large diameter stems. Dominant trees are in the forest's upper layer, being the most dynamic in the forest, the effect of stand density on tree height growth was very small. Thus, the tree height growth process can be used as key process parameters for the full cycle multi-function forest silviculture system design under different site conditions. Based on business objectives, timber market expectations, and other information, stand density regulation can control the stand maturity period, and decrease the risk of forest management.

Key words: *Mytilaria laosensis*; density regulation; middle age stand; plantation; mean tree; dominant trees; growth; student's t test

人工林是森林资源的重要构成部分，在木材生产、环境改善、景观建设和减缓气候变化等方面发挥着重要作用[1]。20世纪90年代以来，虽然中国人工林建设取得了举世瞩目的成就，但也面临着结构不合理、质量不高、生态功能低下等问题[2]。改善人工林的结构(树种结构、年龄结构、空间结构等)，精准提升人工林的质量，增强人工林的多种生态服务功能(固碳释氧、涵养水源、防风固沙和减缓气候变化等)，是中国人工林多功能可持续经营亟须解决的关键问题。林分密度调控是促进林木生长，提高林木质量，改善林分结构，发挥森林多效益的关键技术[3-5]，合理的林分密度调控(即造林密度、间伐时间、间伐强度、间伐次数等[6])，关系着森林培育目标的实现，也影响森林多效益的发挥[5,7]。为改善人工林的结构，提升人工林的质量及发挥人工林的多种效益，国内众多学者对人工林的密度调控进行了深入的研究，并取得了相应的成果[7]，然而以往对人工林密度调控的研究主要集中在松、杉、桉、杨类树种人工林较多[8-14]，对乡土速生优良阔叶树种人工林的密度调控研究较少，尤其对乡土速生优良阔叶树种中龄人工林的密度调控研究更少，已不能满足我国社会对乡土速生优良阔叶树种人工林建设与发展的需要。米老排(*Mytilaria laosensis*)为金缕梅科壳菜果属常绿阔叶乔木，天然分布于我国广东、广西和云南等地，是我国南亚热带区域适生范围极广的优良用材树种，具有速生、干形通直、出材率高、材质优良、改良土壤等特性[15-19]。有关米老排人工幼龄林的密度调控(不同的造林密度和幼龄林抚育间伐)研究[20,21]已有报道，但有关该树种中龄林的间伐密度调控研究，还鲜见报道，而中龄林密度调控是当前米老排人工林营林实践亟待解决的关键技术问题之一。因此，本研究以中国林业科学研究院热带林业实验中心(下文简称热林中心)中等立地的两种不同间伐密度调控米老排中龄林(两林分立地质量相近)为对象，通过分析间伐密度调控措施对米老排中龄林生长的影响，为米老排中龄林的密度调控决策提供参考。

1　材料与方法

1.1　研究区概况

研究地位于广西友谊关森林生态系统国家定位观测研究站热林中心站点(21°57′50″~22°19′29″N，106°40′20″~106°59′14″E)，气候属于南亚热带季风区，年均气温20.5~21.5℃，≥10℃积温6000~7500℃，月平均气温≥22.0℃的有6个月，最热月平均气温27.5℃，极端最高气温39.8℃；最冷月平均气温12.5~13.5℃，极端最低气温-1.5℃。年降水量1400mm，年蒸发量1260mm；全年的日照时间1200~1300h。

1.2　试验林概况

青山试验林(下文简称QS)位于热林中心青山实验场16林班，海拔540m，坡向为西北向，坡位为中上坡，面积14.6hm²；土壤为酸性火山岩发育的红壤，土层较厚，立地类型属于Ⅱ类，土壤有机质在A层的质量分数为3.68%~4.32%。造林时间为1982年春，造林苗木为1年生裸根实生苗，造林密度2500株/hm²，造林后抚育1~3年，经历了1次透光伐(第7年)和3次生长伐(第12年、第17年、第27年，株数间伐强度21%~42%)。2015年9月，林分平均郁闭度0.9，平均密度(520±32.44)株/hm²，平均胸径(28.38±0.86)cm，平均树高(26.60±0.47)m。林下植被主要有水东哥(*Saurauia tristyla*)、杜茎山(*Maesa japonica*)、等。解析木采伐时间为2015年9月。

哨平试验林(下文简称SP)位于热林中心哨平实验场39林班，海拔为300m，面积3.7hm²；坡向为正东，坡位为中下坡；土壤为由花岗岩发育而成的赤红壤，土层较厚，立地类型属于Ⅱ类，土壤有机质在A层和B层的质量分数分别为4.27%~4.44%和0.96%~1.29%[22]。造林时间为1984年春，造林苗木为1年生裸根实生苗，造林密度2500株/hm²，造林后抚育1~3年；造林后为抚育间伐经历了1次透光伐(第7年)和1次生长伐(第12年，株数间伐强度27%)。2016年11月，林分平均郁闭度0.95，平均密度(1200±197.16)株/hm²，平均胸径(23.39±2.13)cm，平均树高(27.08±1.71)m；林分枯落物层厚度5~10cm，林下植被稀少，主要有山苍子(*Litsea cubeba*)、三桠苦(*Evodia lepta*)、粽叶芦(*Thysanolaena maxima*)等。解析木采伐时间为2016年11月。

1.3　研究方法

数据采集。在上述米老排试验林中，各布设有代表性临时样地3个，共布设临时样地6个(每个样地面积为600~667m²，试验林概况中各地点试验林的郁闭度、密度和生长概况均为3个调查样地的均值)，基于林分乔木的每木检尺信息，按林木生长分级法划分的优势木、中等木、被压木条件各选取一定数量标准木，采用中央断面积法，以2m为区分段对所选标准木进行树干解析[23]。各地点试验林3个调查样地的径阶分布及解析木状况详见表1。

表1　不同林分林木径级分布及解析木情况

试验林	密度(株/hm²)	林木分级	径阶(cm)	径阶株数	径阶株数百分比(%)	解析木株数	解析木平均胸径(cm)	解析木平均高(m)
青山QS	520	被压木	≤16	2	1.96	—	—	—
			18	2	1.96	—	—	—
			20	4	3.92	—	—	—
			22	6	5.88	—	—	—
			24	9	8.82	1	23.00	21.20
		中等木	26	20	19.61	1	25.10	22.60
			28	21	20.59	4	28.33	27.48
			30	13	12.75	2	29.55	23.40
		优势木	32	11	10.78	3	32.05	29.48
			34	7	6.86	2	33.20	25.45
			36	2	1.96	1	35.60	31.80
			38	4	3.92	—	—	—
			40	1	0.98	—	—	—
		合计		102	100	14	—	—

（续）

试验林	密度（株/hm²）	林木分级	径阶（cm）	径阶株数	径阶株数百分比（%）	解析木株数	解析木平均胸径（cm）	解析木平均高（m）
哨平 SP	1200	被压木	≤10	3	1.40	—	—	—
			12	5	1.87	—	—	—
			14	7	3.74	1	14.60	22.00
			16	15	7.01	1	16.20	20.00
		中等木	18	17	7.01	1	18.20	23.30
			20	32	14.49	1	20.00	25.40
			22	27	12.62	2	22.00	27.35
			24	37	17.76	2	24.15	29.10
			26	33	16.36	1	27.70	28.80
			28	17	7.01	1	28.10	29.30
		优势木	30	7	3.74	1	29.70	27.10
			32	6	3.27	1	32.10	29.10
			34	6	2.80	2	33.40	28.10
			36	0	0.93	—	—	—
			38	2	1.40	—	—	—
		合计		214	100	14	—	—

注：表内不同地点林分的径阶分布为3个调查样地所有林木的合计数；“—”表未伐取解析木。

数据分析。林分平均木生长量为样地平均木的均值(QS平均木样本4株，SP平均木样本4株)。林分优势木生长量取样地优势木的均值，林分蓄积量为样地所有分级林木材积的累加值。

实验形数计算公式为[23]：

$$f_{\partial}=\frac{V}{g_{1.3}\times(h+3)}$$

式中：f_{∂}、V、$g_{1.3}$、h 分别为实验形数、树干材积、胸高断面积和树高。

不同林分同等级林木(平均木或优势木)的生长差异分析采用双侧t检验法，数据统计分析基于数据处理系统软件(DPS14.5)完成。

2　结果与分析

2.1　密度对米老排中龄林分平均木生长的影响

2.1.1　密度对平均木胸径生长的影响

不同密度平均木胸径连年生长与平均生长过程见图1A。由图1A分析可知：试验林平均木径向生长的缓慢期在第1~2年，速生期在第3~10年，衰减期在第14年后；连年生长量的峰值在第4~6年；连年生长量与平均生长量在第8~10年相交。分析生长表现及双侧t检测结果(表2)可知：在第4~18年间，两试验林平均木绝大多数年份的平均生长量大于1cm；QS试验林与SP试验林胸径平均生长量的峰值分别为1.3cm和1.5cm；在20~33年间的多数年份，不同间伐试验林平均木的连年生长量差异显著($P<0.05$)；前18年，不同间伐试验林的平均木胸径总生长量差异不显著；而至第20年时，两试验林平均木胸径总生长量的差异开始显著($P<0.05$)，且在第22年后，其生长差异均达极显著水平($P<0.01$)。

2.1.2　密度对平均木树高生长的影响

不同密度平均木树高连年生长与平均生长过程见图1B。由图1B可知：两试验林平均木树高生长速生期主要在第2~6年(该阶段连年生长量>1.2m，明显高于其他阶段连年生长量)；连年生长量呈多峰状，连年生长量与平均生长量首次相交在第6~8年间，从第24年后生长趋缓。由表2可知：在前33年，不同间伐试验林平均木树高总生长量差异不显著($P>0.05$)，连年生长量基本差异不显著(在观测期内，仅有第33年的差异显著)。

2.1.3　密度对平均木单株材积生长的影响

不同密度平均木单株材积连年生长与平均生长过程见图1C。由图1C可知：平均木材积的生长，前6年为缓慢期，从第8年起进入速生期(SP试验林速生期在第8~22年，QS试验林第8年后则一直处于速生阶段)；SP试验林平均木材积连年生长量峰值在第14年，连年生长量与平均生长量在第24年相交，试验林材积生长已数量成熟；QS试验林平均木材积连年生量峰值在第30年，直到前34年其连年生长量与平均生长量仍未相交。由表2可知：不同试验林平均木材积的连年生长量，在前18年基本差异不显著($P>0.05$)，而在第20~33年内绝大多数林龄段的生长量差异则显著(SP试验林呈明显递减趋势，QS试验林的增长则相对平稳)；第12年后，QS试验林材积总生长量开始高于SP试验林，但直到第26年两林分的平均木材积总生长量差异才显著($P<0.05$)。

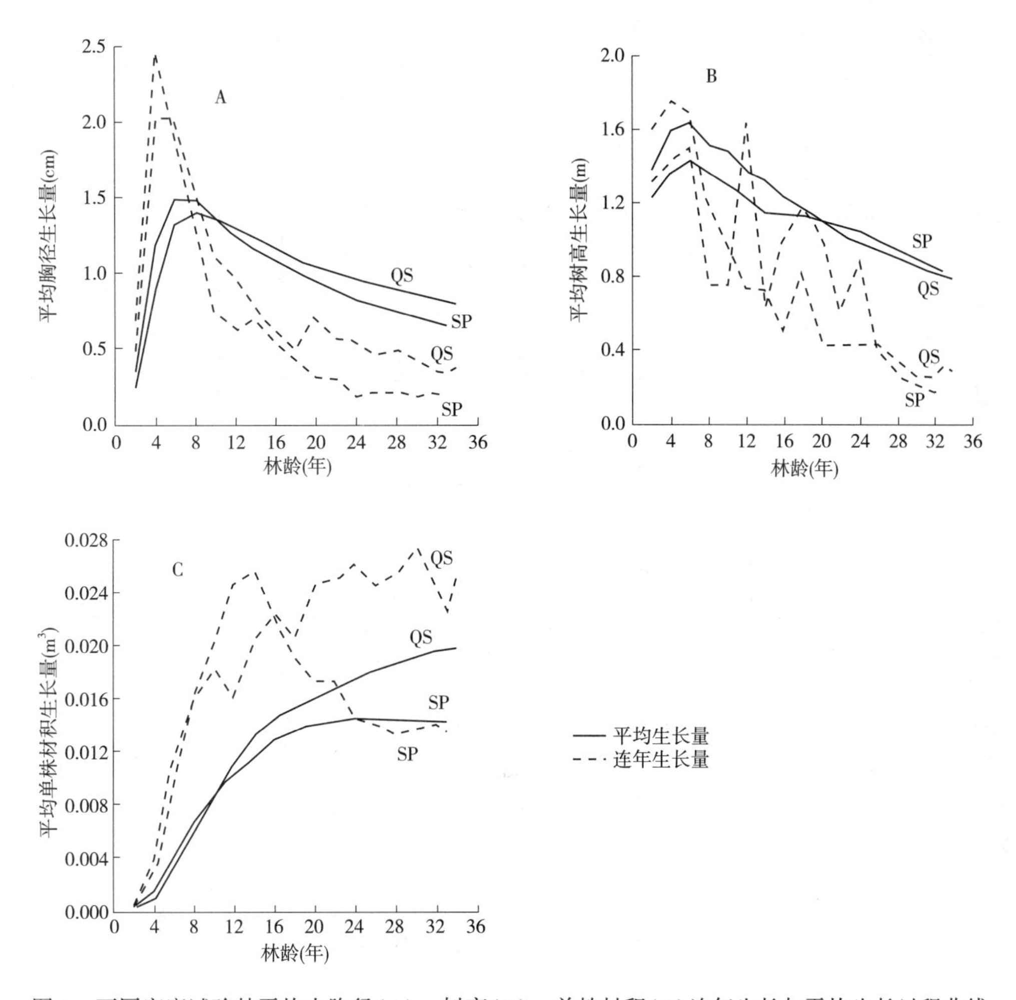

图1 不同密度试验林平均木胸径(A)、树高(B)、单株材积(C)连年生长与平均生长过程曲线

2.2 密度对米老排中龄人工林分优势木生长的影响

2.2.1 密度对优势木胸径生长的影响

不同密度试验林优势木胸径连年生长与平均生长过程见图2A。由图2A可知：优势木径向生长的缓慢期在第1~2年，速生期在第3~10年，衰减期在第14年后；连年生长量在第4年达峰值，连年生长量与平均生长量在第8~10年间相交。由表2可知：在速生期内，优势木胸径的连年生长量基本大于1.3cm；在第4~14年间，其胸径的连年生长量基本大于1cm；在衰减期，其胸径仍保持较高的连年生长量(0.5~1.0cm)；从第18年起，虽然QS试验林的胸径连年生长量均高于SP试验林，但其胸径连年生长量的差异达显著水平仅在第22年出现($P<0.01$)，且其胸径总生长量在绝大多数林龄段内的差异不显著(其仅在第20年时差异显著，$P<0.05$)；在研究期内，间伐密度调控对试验林优势木胸径总生长量的影响不显著。

2.2.2 密度对优势木树高生长的影响

不同密度试验林优势木树高连年生长与平均生长过程见图2B。由图2B可知：优势木树高的速生期主要在第2~6年，该阶段连年生长量基本大于1.50m，连年生长量的最高峰值在第2年，在第18年后树高生长趋缓(连年生长量基本≤0.50m)；树高平均生长量在第2~4年间达峰值(≥1.70m)。在研究期内，不同密度试验林优势木树高生长过程有一定的差异，但差异不显著(表2)。

表 2　不同密度米老排人工林生长表现及双侧 t 检验分析

林龄(年)	试验林	密度(株/hm²)	平均木胸径(cm)		平均木树高(m)		平均木单株材积($\times10^{-2}/m^3$)		优势木胸径(cm)		优势木树高(m)		优势木单株材积($\times10^{-2}/m^3$)		林分蓄积(m^3/hm^2)	
			总生长量	连年生长量	总生长量	连年生长量	总生长量	连年生长量	总生长量	连年生长量	总生长量	连年生长量	总生长量	连年生长量	总生长量	连年生长量
2	QS	2250	0.48±0.32	0.48±0.32	2.75±1.10	1.60±0.84	0.01±0.01	0.01±0.01	0.65±0.37	0.65±0.37	3.33±0.55	2.03±0.55	0.01±0.01	0.01±0.02	0.23±0.00**	0.23±0.00**
	SP	2250	0.68±0.53	0.68±0.53	2.48±0.94	1.31±0.71	0.03±0.03	0.03±0.03	0.75±0.13	0.75±0.13	4.10±1.00	2.80±1.00	0.01±0.00	0.01±0.01	0.45±0.00**	0.45±0.00**
4	QS	2250	3.65±0.95	2.02±0.35	6.35±0.96	1.75±0.50	0.33±0.14	0.26±0.10	4.58±1.26	2.25±0.40	6.60±1.00	1.50±0.45	0.54±0.30	0.41±0.19	5.48±1.11**	3.83±1.03**
	SP	2250	4.78±1.64	2.45±0.40	5.43±1.36	1.43±0.69	0.55±0.33	0.39±0.20	4.68±1.09	2.15±0.44	7.10±0.58	1.25±0.50	0.66±0.43	0.47±0.30	11.33±1.50**	8.17±1.13**
6	QS	2250	7.93±1.25	2.03±0.54	9.78±1.36	1.68±0.65	1.83±0.59	0.93±0.33	8.70±1.19	1.90±0.43	10.00±1.14	1.60±0.55	2.51±0.81	1.16±0.33	30.60±3.46	15.75±1.47
	SP	2250	8.85±1.03	1.87±0.49	8.60±1.41	1.50±0.58	2.44±1.09	1.18±0.60	8.95±2.00	2.10±0.73	11.10±1.00	2.50±1.00	3.16±1.67	1.61±0.85	45.44±12.21	19.18±9.25
8	QS	1650	11.10±1.67	1.50±0.50	12.10±1.91	1.17±0.57	4.73±1.33	1.64±0.57	11.58±1.31	1.40±0.23	12.74±1.20	1.34±0.61	5.83±1.48	1.92±0.45	60.94±4.95	22.28±1.51
	SP	1650	11.80±1.46	1.35±0.31	10.85±2.02	0.75±0.29	5.26±2.34	1.57±0.60	13.35±2.78	2.00±0.98	13.93±1.42	1.15±0.60	8.76±3.43	2.99±1.32	85.80±16.34	26.23±5.73
10	QS	1650	13.58±1.73	1.10±0.18	14.83±1.95	0.98±0.69	8.61±2.26	2.01±0.48	14.38±1.24	1.33±0.25	15.54±1.29	1.28±0.67	11.24±2.24	2.89±0.44*	115.78±9.80	29.15±2.52
	SP	1650	13.60±1.49	0.72±0.36	13.00±2.40	0.75±0.29	8.92±3.26	1.82±0.65	16.28±3.48	2.00±0.42	16.03±1.41	1.18±0.57	16.00±5.43	4.02±0.90*	141.85±28.06	28.60±6.26
12	QS	1300	15.53±1.48	0.95±0.17	16.48±2.53	0.73±0.21	13.28±2.70	2.48±0.27*	16.68±1.11	1.00±0.09	18.06±1.55	1.26±0.69	17.86±3.10	3.40±0.58	149.02±12.59	−2.12±0.57
	SP	1200	14.85±1.85	0.60±0.24	14.73±2.53	1.63±0.75	12.29±4.51	1.59±0.69*	18.73±3.60	1.10±0.14	17.85±1.71	0.92±0.15	24.52±6.61	4.15±0.84	144.63±13.36	−28.08±21.24
14	QS	1300	17.15±1.33	0.77±0.10	18.33±2.33	0.73±0.32	18.41±3.25	2.56±0.39	18.55±1.12	0.92±0.16	20.88±1.38	1.28±0.70	25.10±4.14*	3.75±0.57	210.86±18.79	31.37±3.46*
	SP	1200	16.13±1.85	0.70±0.14	15.98±2.69	0.63±0.25	16.10±5.50	2.03±0.63	20.85±3.35	1.00±0.26	20.68±1.25	1.33±0.83	34.10±7.71*	4.85±0.96	191.44±18.18	23.96±2.38*
16	QS	1300	18.48±1.00	0.60±0.18	19.73±2.45	0.50±0.08	23.24±3.13	2.25±0.21	20.13±1.10	0.71±0.12	22.32±1.34	0.48±0.13	32.48±4.78*	3.60±0.62	271.09±25.81	29.16±3.70
	SP	1200	17.23±1.52	0.53±0.17	17.83±2.41	1.00±0.71	20.47±6.10	2.23±0.36	22.53±2.65	0.88±0.46	22.18±1.18	0.73±0.34	42.93±8.24*	4.39±1.29	239.19±22.12	23.52±1.96
18	QS	900	19.43±1.04	0.48±0.18	21.08±2.07	0.83±0.78	27.30±3.45	2.05±0.15	21.57±1.21	0.74±0.15	23.26±1.15	0.48±0.13	39.68±5.82*	3.60±0.79	224.46±23.33*	18.42±2.62
	SP	1200	18.08±1.30	0.43±0.19	20.53±3.08	1.18±0.57	24.52±6.45	1.90±0.13	23.70±2.07	0.55±0.29	23.30±0.96	0.55±0.30	50.84±8.45*	4.00±1.42	285.60±24.31*	22.70±1.67

（续）

林龄（年）	试验林	密度（株/hm²）	平均木胸径(cm)		平均木树高(m)		平均木单株材积（×10⁻²/m³）		优势木胸径(cm)		优势木树高(m)		优势木单株材积（×10⁻²/m³）		林分蓄积(m³/hm²)	
			总生长量	连年生长量	总生长量	连年生长量	总生长量	连年生长量	总生长量	连年生长量	总生长量	连年生长量	总生长量	连年生长量	总生长量	连年生长量
20	QS	900	20.78±0.96*	0.70±0.18*	21.90±2.17	0.42±0.21	32.13±4.07	2.45±0.33*	22.88±1.06*	0.61±0.13	24.12±1.32	0.46±0.18	46.64±6.16*	3.40±0.48	264.87±27.38*	20.76±1.75
	SP	1200	18.73±1.04*	0.30±0.14*	22.28±3.13	1.03±0.69	27.92±6.03	1.73±0.27*	24.75±1.51*	0.52±0.26	24.15±0.93	0.47±0.17	58.25±9.17*	3.41±0.69	328.20±26.21*	20.95±0.69
22	QS	900	21.95±0.62**	0.57±0.22	22.75±2.30	0.42±0.21	36.98±3.98	2.49±0.69	24.23±1.03	0.70±0.06**	25.06±1.58	0.48±0.27	53.33±6.28	3.42±0.32	308.34±30.87	21.75±1.92
	SP	1200	19.30±0.91**	0.30±0.08	23.75±2.10	0.60±0.29	31.50±5.88	1.73±0.17	25.60±1.25	0.42±0.15**	25.10±1.00	0.47±0.17	64.11±9.85*	2.87±0.64	367.12±20.43	18.68±0.75
24	QS	900	23.03±0.34**	0.53±0.13**	23.53±2.33	0.43±0.21	42.14±4.16	2.62±0.47**	24.52±0.83	0.59±0.15	25.70±1.68	0.53±0.24	61.15±6.08	3.95±0.62	351.48±35.22	21.90±2.01*
	SP	1200	19.70±0.85**	0.17±0.05**	24.93±1.85	0.88±0.77	34.45±5.83	1.44±0.29**	26.48±1.16	0.43±0.17	25.88±0.69	0.40±0.16	69.96±10.94	2.97±0.89	401.92±29.11	17.08±0.94*
26	QS	900	23.93±0.28**	0.43±0.19	24.50±2.61	0.42±0.21	47.12±4.61*	2.45±0.73*	26.73±0.88	0.56±0.16	26.73±1.71	0.50±0.11**	68.46±6.00	3.61±0.46	396.36±39.98	21.99±1.97*
	SP	1200	20.18±0.83**	0.20±0.00	25.85±1.48	0.37±0.24	37.28±6.06*	1.40±0.25*	27.43±1.20	0.45±0.17	26.53±0.59	0.28±0.10**	76.51±12.76	3.22±0.95	437.47±31.43	17.89±1.13*
28	QS	520	24.73±0.26**	0.48±0.10**	25.23±2.70	0.35±0.13	52.07±2.56*	2.52±0.49**	27.87±1.01	0.59±0.21	27.65±1.86	0.42±0.21	75.92±6.00	3.86±0.38	255.41±22.57**	12.93±1.04*
	SP	1200	20.58±0.83**	0.20±0.08**	26.53±1.52	0.25±0.10	40.00±3.27*	1.32±0.31**	28.35±1.19	0.45±0.19	27.03±0.71	0.25±0.06	83.08±14.57	3.22±0.99	471.47±34.28**	16.60±1.52*
30	QS	520	25.55±0.29**	0.42±0.05**	25.63±2.70	0.25±0.10	57.35±5.96*	2.74±0.48**	28.98±1.13	0.56±0.10	28.33±1.92	0.30±0.15	84.25±6.99	4.42±0.73	282.42±24.82**	13.95±1.11*
	SP	1200	20.95±0.83**	0.17±0.05**	26.93±1.52	0.20±0.08	42.75±7.24*	1.36±0.38**	29.23±1.33	0.48±0.22	27.43±0.71	0.23±0.10	89.55±16.79	3.31±1.15	505.85±37.22**	17.34±1.58*
32	QS	520	26.25±0.39**	0.35±0.13	26.18±2.73	0.25±0.06	62.24±3.27*	2.40±0.33*	30.07±1.37	0.49±0.15	28.97±1.95	0.27±0.14	92.60±8.32	4.01±0.70	309.53±27.00**	12.96±1.11*
	SP	1200	21.40±0.94**	0.20±0.08	27.30±1.47	0.17±0.05	45.44±8.17*	1.39±0.59*	30.13±1.58	0.45±0.21	27.88±0.85	0.20±0.08	96.14±18.64	3.21±1.19	540.21±40.56**	17.22±1.58*
33	QS	520	26.58±0.38**	0.33±0.05*	26.50±2.73	0.32±0.05*	64.48±6.85*	2.24±0.34*	30.57±1.51	0.50±0.15	29.38±2.01	0.41±0.21	96.56±9.08	3.96±0.84	321.84±27.95**	12.31±1.00**
	SP	1200	21.58±0.93**	0.18±0.10*	27.48±1.44	0.18±0.10*	46.78±8.59*	1.34±0.47*	30.58±1.77	0.45±0.19	28.08±0.87	0.20±0.08	99.56±19.69	3.42±1.09	558.61±42.30**	18.40±1.76**
34	QS	520	26.95±0.37	0.37±0.13	26.78±2.73	0.28±0.10	67.03±7.06	2.55±0.45	31.12±1.68	0.55±0.18	29.77±2.06	0.39±0.19	100.54±9.85	3.98±0.83	334.54±28.93	12.70±1.06

注：表内数据为统计均值±标准差；*与**分别表示在相同林龄下，不同林分相同因子的生长量差异显著($P<0.05$)和差异极显著($P<0.01$)；胸径值为去皮胸径；蓄积连年生长量为负数，为当年(或上年)间伐所致。

2.2.3　密度对优势木单株材积生长的影响

不同密度试验林优势木单株材积连年生长与平均生长过程见图2C。由图2C可知：与平均木生长类似，优势木单株材积生长的缓慢期在前6年，从第8年起，两试验林优势木材积生长均进入速生期(连年生长量基本>0.020m³)，直到第33年仍保持高速生长；SP试验林材积连年生长量的峰值在第14年，QS试验林材积连年生长量的峰值在第32年；QS试验林连年生长量曲线与平均生长量曲线仍未见未相交，SP试验林连年生长量与平均生长量在先是在第22年后相交，随后连年生长量又开始高于平均生长量。由表2可知：在前33年内，不同试验林优势木材积连年生长量基本差异不显著，仅在第10年差异显著；在不同林龄段，不同试验林优势木材积总生长量的表现不同：先是差异不显著，然后是差异显著，最后是差异又不显著。

2.3　密度对米老排中龄人工林分蓄积生长量的影响

由表2可知：在林分蓄积的总生长量方面，两试验林在第2~4年、第18~20年、第28~33年的时段内差异显著，在其余时段内差异不显著。在林分蓄积量的连年生长量方面，从第18年起，SP试验林蓄积的连年生长量表现为先是递减(第18~22年间)，然后再稳定在某一特定区间波动(第24~33年)；而QS试验林蓄积的连年生长量，在第2次与第3次间伐期间(第18~26年)，表现为稳定的增长，且在第24~26年的连年生长量显著高于SP试验林，但到第3次间伐作业(强度间伐)后，受林分密度影响，其连年生长量则显著低于SP试验林蓄积的连年生长量，且未能恢复到第3次间伐前的连年生长量水平。

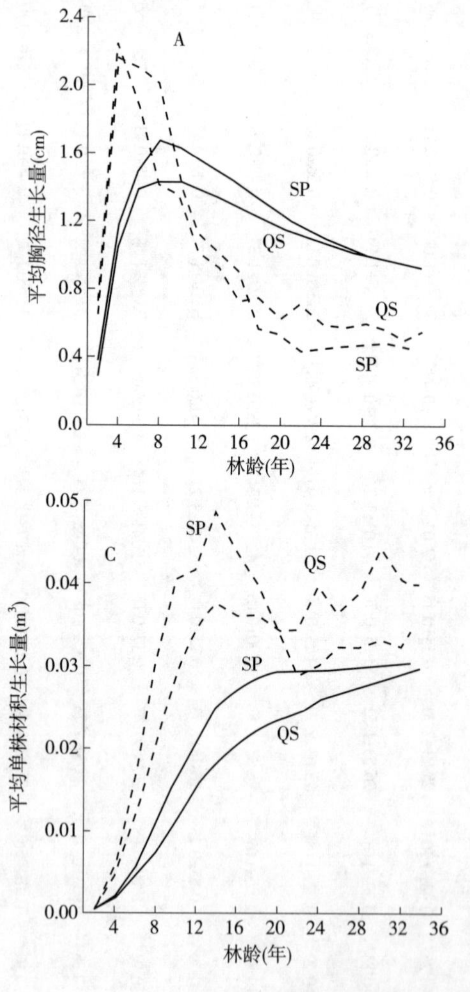

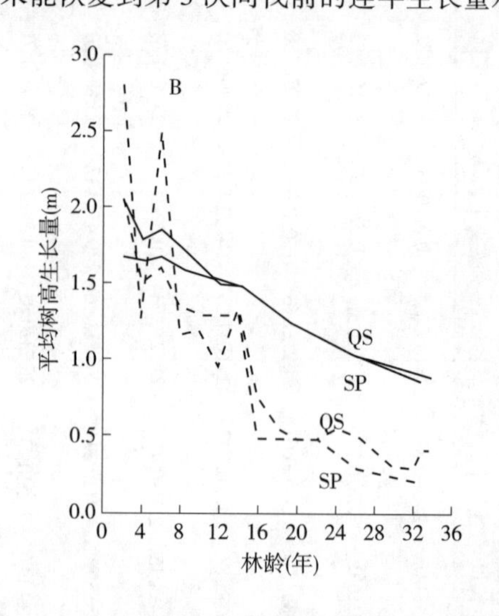

图2　不同密度试验林优势木胸径(A)、树高(B)、单株材积(C)连年生长与平均生长过程曲线

2.4　密度对米老排中龄人工林平均实验形数的影响

由表3可知：米老排平均的实验形数受林龄的影响较大，在前6年，两试验林米老排树种的平均实验形数基本先随树龄的增加而递减，随后再大致随林龄的增加而递增(第8~14年)，并在第14年后趋于

表 3 不同密度林分米老排树种平均实验形数的动态变化及双侧 t 检验分析

林龄(年)	试验林	密度(株/hm²)	平均实验形数	林龄(年)	试验林	密度(株/hm²)	平均实验形数
2	QS	2250	0.64±0.40	20	QS	900	0.42±0.03
	SP	2250	0.69±0.73		SP	1200	0.42±0.05
4	QS	2250	0.34±0.04	22	QS	900	0.41±0.03
	SP	2250	0.78±1.19		SP	1200	0.42±0.05
6	QS	2 250	0.31±0.05	24	QS	900	0.41±0.03
	SP	2 250	0.35±0.06		SP	1200	0.41±0.04
8	QS	1650	0.34±0.04	26	QS	900	0.41±0.03
	SP	1650	0.37±0.05		SP	1200	0.41±0.04
10	QS	1650	0.36±0.03**	28	QS	520	0.41±0.03
	SP	1650	0.40±0.03**		SP	1200	0.41±0.04
12	QS	1300	0.38±0.03	30	QS	520	0.41±0.03
	SP	1200	0.40±0.05		SP	1200	0.41±0.04
14	QS	1300	0.40±0.04	32	QS	520	0.41±0.03
	SP	1200	0.40±0.06		SP	1200	0.41±0.04
16	QS	1300	0.41±0.04	33	QS	520	0.41±0.03
	SP	1200	0.41±0.05		SP	1200	0.41±0.04
18	QS	900	0.42±0.04	34	QS	520	0.41±0.03
	SP	1200	0.42±0.05		SP	/	/

注：表内数据为(所有解析木的)统计均值±标准差；* 与 ** 分别表示在相同林龄下，不同林分的实验形数的差异显著($P<0.05$)和差异极显著($P<0.01$)。

稳定状态(均值为 0.41～0.42，干形级属Ⅲ－Ⅳ级[23])；而在相同的林龄条件下(在同一年)，无论两试验林的密度是否相同，两林分绝大多数相同林龄时的平均实验形数差异均不显著(仅有第 10 年时差异极显著，$P<0.01$)，这表明林分密度对米老排树种平均实验形数的影响不显著。

2.5 密度对米老排中龄人工林分径阶分布的影响

由表 1 可知：SP 试验林径阶分布的范围相对较广，径阶分布的极大值和极小值均小于 QS 试验林，林分以中小径材居多(≤24cm 林木的比例为 65.9%)；QS 试验林径阶分布相对较集中，最大径阶株数百分比所在的径阶值大于 SP 试验林的对应值，林分的大径木(≥26cm 的林木)对应的数量与比例(79 株，77.45%)均高于 SP 试验林的对应值(73 株，34.11%)。

3 讨论

(1)在南亚热带的中等立地，米老排胸径生长的缓慢期在第 1～2 年，速生期在第 3～10 年，衰减期在第 14 年后。间伐密度调控对平均木胸径总生长量与连年生长量影响显著，对优势木胸径长期的生长影响不显著，对提升林分大径材比例的作用明显。米老排树高早期速生特性明显，连年生长量呈多峰状，速生期主要在第 2～6 年；林分最终密度在 520～1200 株/hm²范围内，密度调控对平均木与优势木的树高生长影响不显著。

(2)平均木与优势木材积生长的缓慢期均在第 6 年前，速生期均在第 8 年后；密度对平均木材积连年生长与林分数量成熟时间影响明显，对优势木材积的短期生长有一定影响，但对其长期的生长影响不显著。中弱度间伐(株数间伐强度<30%)对米老排中龄林蓄积总生长量的影响不显著，对其林分蓄积连年生长量短期有一定影响；强度间伐(株数间伐强度 > 30%)对米老排中龄林蓄积的总生长量与连年生长量的影响显著。在第 14 年后米老排树种的实验形数趋于稳定，密度对其平均实验形数的影响不明显；间伐经营措施对减小林分径阶分化及提高大径材比例的作用明显。

(3)该研究米老排平均木，其胸径与树高的最大平均生长量出现的时间(即其平均生长量与连年生长量相交的时间)与李炎香对米老排幼林生长的研究结果(在相同密度下，其平均胸径对应的相交时间在第4~5年，平均树高对应的相交时间在第3~4年)[20]相比，本研究平均木胸径与树高，其对应的相交时间均晚于后者3~4年。上述结果差异的原因，既与立地的差异有关(后者有机质在A层含量5.72%~6.80%，在B层3.83%，土壤肥力相对较高)，也与不同研究方法的系统误差有关(本研究的方法为树干解析法，李炎香研究的方法为固定样地立木定株观测法)。在测树中，一般围尺测径值总是大于轮尺测径值(围尺测径值为数学期望值，轮尺测径值是与测量方向有关的随机变数)[24]，解析木圆盘的测量工具为直尺(与轮尺的性质相同)，解析木圆盘的测径值为去皮直径，此外解析木圆盘制作完成后存在木材干缩(立木采伐后，由生材到气干状态，会产生一定的收缩，米老排木材弦向与径向的干缩率分别为508%、2.08%[25])，故解析木圆盘的测径值总是相对小于立木的测径值，从而导致树干解析法与固定样地立木定株观测法的结果存在一定的系统误差。

(4)该研究林分数量成熟的研究结果：不同密度林分的数量成熟期不同，其中SP试验林(密度1200株/hm^2)的数量成熟龄在第24年，而QS试验林(第18~26年的林分密度为900株/hm^2，第26~34年的林分密度为520株/hm^2)直至第34年也仍未达数量成熟。此与吴淑玲[26]对福建漳州米老排人工林研究结果(林分密度750株/hm^2，林分第23年未达数量成熟)类似，与郭文福等[17]的研究结果(林分密度在1250~3000株/hm^2的数量成熟龄在第15~17年，立地相对较好与密度相对较低林分数量成熟可在第22~25年)相近，和林能庆等[27]对福建龙岩市米老排人工林的研究结果(密度1280株/hm^2，林分在第26年时未达数量成熟)不同。此表明，区域、密度和立地条件均对相同树种林分的数量成熟有较大影响，因此基于不同的立地水平、经营目标和木材市场预期等，制定合理的林分密度经营措施，调控林分的数量成熟期，减少森林经营的风险，非常重要。

优势木处于林分的上层，为林分中最有活力的林木，其树高生长受林分密度的影响小，因此将其生长过程作为不同立地森林全周期多功能作业法设计的关键过程参数，是科学的。

参考文献

[1]刘世荣，杨予静，王晖．中国人工林经营发展战略与对策：从追求木材产量的单一目标经营转向提升生态系统服务质量和效益的多目标经营[J]．生态学报，2018 38(1)：1-10.

[2]陈幸良，巨茜，林昆仑．中国人工林发展现状、问题与对策[J]．世界林业研究，2014，27(6)：54-59.

[3] 段劼，马履一，贾黎明，等．北京地区侧柏人工林密度效应[J]．生态学报，2010，30(12)：3206-3214.

[4] 胡凌，商侃侃，张庆费，等．密度调控对香樟人工林林木生长及空间分布的影响[J]．西北林学院学报，2014，29(2)：20-25.

[5]于世川，张文辉，尤健健，等．抚育间伐对黄龙山辽东栎林木形质的影响[J]．林业科学，2017，53(11)：104-113.

[6]张阳峰．造林密度对米老排凋落物量及养分归还特征的影响[D]．北京：中国林业科学研究院，2017．

[7]谌红辉，方升佐，贵杰，等．马尾松间伐的密度效应[J]．林业科学，2010，46(5)：84-90.

[8] 段爱国，张建国，童书振，等．杉木人工林林分直径结构动态变化及其密度效应的研究[J]．林业科学研究，2004，17(2)：178-182.

[9]童书振，盛炜彤，张建国．杉木林分密度效应研究[J]．林业科学研究，2002，15(1)：66-75.

[10]刘青华，周志春，张开明，等．造林密度对不同马尾松种源生长和木材基本密度的影响[J]．林业科学，2010，46(9)：58-63.

[11] 张连金，惠刚盈，孙长忠．马尾松人工林首次间伐年龄的研究[J]．中南林业科技大学学报，2011，31(6)：22-27.

[12] 李国雷，刘勇，吕瑞恒，等．华北落叶松人工林密度调控对林下植被发育的作用过程[J]．北京林业大学学报，2009，31(1)：19-31.

[13] 田新辉，孙荣喜，李军，等．107杨人工林密度对林木生长的影响[J]．林业科学，2011，47(3)：184-188.

[14] 冯秋红，吴晓龙，徐峥静茹，等．密度调控对川西山地云杉人工林地被物及土壤水文特征的影响[J]．南京林业大学学报(自然科学版)，2018，42(1)：98-104.

[15] 白灵海，唐继新，明安刚，等．广西大青山米老排人工林经济效益分析[J]．林业科学研究，2011，24(6)：784-787.

[16] 袁洁．8个米老排天然群体的遗传多样性研究[D]．北京：中国林业科学研究院，2014，06.

[17] 郭文福，蔡道雄，贾宏炎，等．米老排人工林生长规律的研究[J]．林业科学研究，2006，19(5)：585-589.

[18]王克建．热带树种栽培技术[M]．南宁：广西科学技术出版社，2008.

[19]郭文福．米老排人工林生长与立地的关系[J]．林业

科学研究, 2009, 22(6): 835-839.
[20]李炎香, 谭天泳, 黄镜光, 等. 米老排造林密度试验初报[J]. 林业科学研究, 1988, 1(2): 206-212.
[21]郭文福, 黄镜光. 米老排抚育间伐研究[J]. 林业科学研究, 1991, 4(增刊): 76-81.
[22]汪炳根, 卢立华. 同一立地营造不同树种林木生长与土壤理化性质变化的研究[J]. 林业科学研究, 1995, 8(3): 334-339.
[23]孟宪宇. 测树学[M].3版. 北京: 中国林业出版社, 2006.
[24]唐守正. 围尺测径和轮尺测径的理论比较[J]. 林业勘测设计, 1977(3): 23-26.
[25]梁善庆, 罗建举. 人工林米老排木材的物理力学性质[J]. 中南林业科技大学学报, 2007, 5(10): 97-116.
[26]吴淑玲. 米老排在漳州长泰生长规律初步研究[J]. 防护林科技, 2016(8): 22-25.
[27]林能庆, 洪永辉, 李蔚倩, 等. 米老排人工林生长规律及生长模型拟合研究[J]. 林业勘察设计, 2017(3): 22-28.

[原载: 南京林业大学学报(自然科学版), 2019, 43(01)]

柚木冠幅与树高、胸径的回归分析

郝 建[1] 贾宏炎[1] 杨保国[1] 黄旭光[1] 黄桂华[2] 莫世宇[1] 苏贤业[1] 蔡道雄[1]
([1]中国林业科学研究院热带林业实验中心，广西凭祥 532600；
[2]中国林业科学研究院热带林业研究所，广东广州 510520)

摘 要 为了考察柚木生长因子与冠幅的关系，准确反映各生长因子与冠幅之间的关系，建立冠幅预测模型，为柚木人工林目标树经营提供理论依据。以广西凭祥、云南德宏、云南景洪、海南乐东4个地区不同林龄阶段的柚木人工林中优势木为研究对象，以胸径、树高、冠长、林龄4个因子作为变量与冠幅进行回归分析，筛选关键因子建立柚木冠幅生长的预测模型。研究结果表明，胸径($R=0.5342$，$P=0.0001$)、树高($R=0.1798$，$P=0.0026$)是影响柚木冠幅的关键因子；胸径、树高与冠幅的一元回归方程：冠幅与胸径$y=15.7893x+1.84766$($F=516.4180$，$P=0.0001$)，冠幅与树高$y=0.3717x-0.60189$($F=174.2954$，$P=0.0001$)。并应用胸径、树高2个关键因子与冠幅建立回归模型：$y=13.5658x_1+0.1064x_2+0.35866$($F=279.5048$，$P=0.0001$)，计算结果与实际测量结果差异性较小($F=0.0140$，$P=0.9072$)。可以根据目标树的培育目标胸径、树高因子，利用该模型来预测该目标树的冠幅，从而确定单位面积内保留目标树的数量。

关键词 柚木；冠幅；回归分析；预测模型

Regression Analysis of Teak Crown Growth with Tree Height and DBH

HAO Jian[1], JIA Hongyan[1], YANG Baoguo[1], HUANG Xuguang[1],
HUANG Guihua[2], MO Shiyu[1], SU Xianye[1], CAI Daoxiong[1]

([1]*Experimental Center of Tropical Forestry, Chinese Academy of Forestry, Pingxiang 532600, Guangxi, China*;
[2]*Research Institute of Tropical Forestry, Chinese Academy of Forestry, Guangzhou 510520, Guangdong, China*)

Abstract: In order to investigate the relationship between teak growth factors and crown width, the relationship between growth factors and crown width was accurately reflected, and the crown growth prediction model was established to provide a theoretical basis for teak plantation target tree management. Taking the dominant trees in the teak plantations at different forest age in Pingxiang Guangxi, Dehong Yunnan, Jinghong Yunnan and Ledong Hainan as the research objects, the four factors of DBH, tree height, crown length and forest age were used as variables and regression analysis was performed with the crown width, to selecte the key factors and establish a predictive model for teak crown growth. The results showed that the DBH ($R=0.5342$, $P=0.0001$) and tree height ($R=0.1798$, $P=0.0026$) were the key factors affecting the crown width of teak. The one-dimensional regression equations for DBH, tree height with crown width were established: crown width with the DBH $y=15.7893x+1.84766$ ($F=516.4180$, $P=0.0001$), crown width with tree height $y=0.3717x-0.60189$ ($F=174.2954$, $P=0.0001$). The regression model was established by applying two key factors of breast diameter and tree height with crown width: $y=13.5658x1+0.1064x2+0.35866$ ($F=279.5048$, $P=0.0001$), and the difference between the calculation result and the actual measurement result was small ($F=0.0140$, $P=0.9072$). The model could be used to predict the crown width of the target tree according to the breast diameter and tree height of target tree, thereby determining the number of target trees retained per unit area.

Key words: *Tectona grandis*; crown width; regression analysis; prediction model

柚木(*Tectona grandis* L. f.)是世界名贵用材树种，具有材质优良、纹理美观、易于加工等特点[1,2]。我国引进柚木栽培历史悠久[2]，在云南、海南、广东、广西、贵州和福建都有较大规模的种植。

因柚木对立地条件要求的特殊性，传统单一林分水平经营，既不符合柚木的生物学和生态学特性，也不适应当前社会经济发展的需求，不利于柚木人工林的可持续经营[2]。随着经济的高速发展，社会对木材的需求结构发生了巨大变化，大径级木材已成为一种短缺资源[3,4]。而以培育大径级木材为主要目的的目标树经营技术是解决这一问题的有效途径。

传统的林分水平的森林经营没有明确把单株生长量及木材质量纳入到经营体系中，未涉及如何提升木材价值[5,6]。目标树经营是以目标树直径生长为标准，根据直径生长评估目标树是否需要调整竞争强度、释放生长空间[7]。冠幅是反映树木生长的重要指标，也是反映森林生长收获情况的重要变量[8]。树木冠幅与胸径之间存在着密切的关系[9]。根据冠幅生长可以预测林分内单株的生长量[10]。而冠幅与胸径生长均受树木生长空间的限制[11]。根据培育单株目标胸径如何科学调整单株目标树的生长空间，是目标树经营的重要内容。利用树冠直径与胸径模型可以确定目标树间的最佳距离，从而科学有效地预测单位面积目标树的最大数量[11]。由于胸径小、大值处的线性偏离，冠幅和胸径之间的真正关系是“S”形，而“S”形曲线的中间3/4部分属于线性关系，因此，对于实际应用线性模型是合适的[12,13]。有的学者引入立地质量[14,15]、林龄[16]、树高[17]、树冠比例[18]、冠长[19]等因子与冠幅进行拟合分析，比较后认为胸径是预测冠幅最相关因子[15,20,21]。因此，为了考察柚木生长因子与冠幅的关系，准确反映不同生长指标与冠幅生长之间的关系，以广西凭祥、云南德宏、云南景洪、海南乐东不同林龄阶段的柚木人工林为研究对象，用回归分析方法考察柚木胸径、树高对冠幅的影响，以期为建立更好的冠幅预测模型，为柚木人工林目标树经营提供理论依据。

1 研究区概况

广西凭祥研究地点位于广西凭祥市中国林业科学研究院热带林业实验中心青山实验场。该地区处于南亚热带季风气候区域内的西南部，与北热带北缘毗邻，属湿润半湿润气候。境内日照充足，雨量充沛，干湿季节明显，光、水、热资源丰富。年均气温为20.5~21.7℃，极端高温为40.3℃，极端低温为-1.5℃；≥10℃活动积温为6000~7600℃。年均降水量为1200~1500mm，相对湿度为80%~84%。主要地貌类型以低山丘陵为主，坡度以25°~30°为多。地带性土壤为砖红壤，土层厚度大于1m。

云南德宏研究点位于畹町林场，属南亚热带山地湿润季风气候。年平均气温21.0℃，最高气温36.0℃，最低气温1.2℃。日照充足，年均日照时数为2343.4小时。年平均降水量1400~1700mm，年均相对湿度为79%。观测区海拔800~1000m，最低海拔778m，城区海拔830m。

云南景洪研究地点位于中国科学院西双版纳热带植物园，属北回归线以南的热带北缘区，受印度洋季风气候控制。年气温18~22℃，最冷月均温8.8~15.6℃，≥10℃的活动积温5062~8000℃，海拔800m以下地区的活动积温皆在7500℃以上；年降水量1100~1900mm，雨季5~10月，气候温暖、湿润，长夏无冬，夏热多雨，秋春相连且为期较短。地带性土壤以砖红壤和赤红壤为主。

海南乐东研究地点位于中国林科院热带林业研究所尖峰岭试验站，属热带季风气候，年平均温度24.5℃，极端最高气温38.1℃，极端最低气温5.0℃；年降水量1500mm，11月至翌年5月为月降水量<30mm的干旱季节。土壤为花岗岩发育的褐色砖红壤，pH5.6~6.0。

表1 各地区不同林龄林分统计信息

地区	树龄（年）	初植密度（株/hm²）	保留密度（株/hm²）	坡向	坡度	树高（m）	胸径（cm）	冠长（m）	冠幅（m）	冠幅/胸径（cm）	样地数量
广西凭祥	12	1667	1100	南	15	14.14±2.46	17.66±2.02	9.83±1.49	3.42±0.79	19.18±3.34	6
	17	1667	1100	西南	23	18.01±1.28	28.24±3.28	11.15±1.69	6.55±0.93	23.20±1.72	4
	22	1667	1050	东南	20	22.10±3.22	31.44±3.77	13.56±2.51	7.70±0.83	24.70±2.82	4
	25	1667	1050	西南	26	20.28±2.88	31.26±7.78	10.67±2.68	6.84±1.57	22.11±2.69	6
	31	1667	750	西南	25	22.69±2.34	32.69±6.22	10.88±2.71	7.20±1.72	22.04±3.64	4
	35	1667	600	南	15	24.18±2.30	37.17±3.69	12.58±1.53	8.89±1.53	24.11±4.69	6
	45	1667	450	西南	3	23.98±1.84	53.98±7.05	14.13±1.67	10.61±2.09	19.67±3.08	3

（续）

地区	树龄（年）	初植密度（株/hm²）	保留密度（株/hm²）	坡向	坡度	树高（m）	胸径（cm）	冠长（m）	冠幅（m）	冠幅/胸径（cm）	样地数量
云南德宏	32	1667	600	西南	18	22. 14±2. 80	36. 53±5. 86	12. 77±3. 2 7	6. 78±1. 71	18. 66±3. 84	6
	44	1667	450	南	25	23. 99±2. 38	44. 99±9. 26	11. 47±2. 93	8. 91±2. 54	19. 66±2. 95	6
	51	1667	450	西南	28	24. 00±2. 00	41. 69±6. 12	13. 10±3. 21	7. 70±2. 44	18. 57±5. 92	6
云南景洪	40	1667	450	南	3	23. 38±2. 18	40. 99±5. 37	12. 81±2. 21	10. 31±1. 85	19. 03±1. 78	3
	47	1667	450	南	2	24. 67±2. 01	52. 23±8. 30	12. 61±2. 19	8. 85±0. 55	17. 42±2. 31	3
海南乐东	40	1667	450	南	15	24. 18±1. 34	40. 51±2. 74	13. 10±2. 23	7. 01±0. 56	17. 39±2. 01	3

2 研究方法

目标树经营主要以林分中的优势木为对象。因此，分别在广西凭祥中国林业科学研究院热带林业实验中心、云南德宏畹町林场、云南景洪、海南乐东4个地区(地理位置如图1所示)，根据不同林龄阶段的柚木人工林的生长状况分别设置3~6个20m×30m调查样地。测定样地中优势木的胸径、树高、冠下高及冠幅。

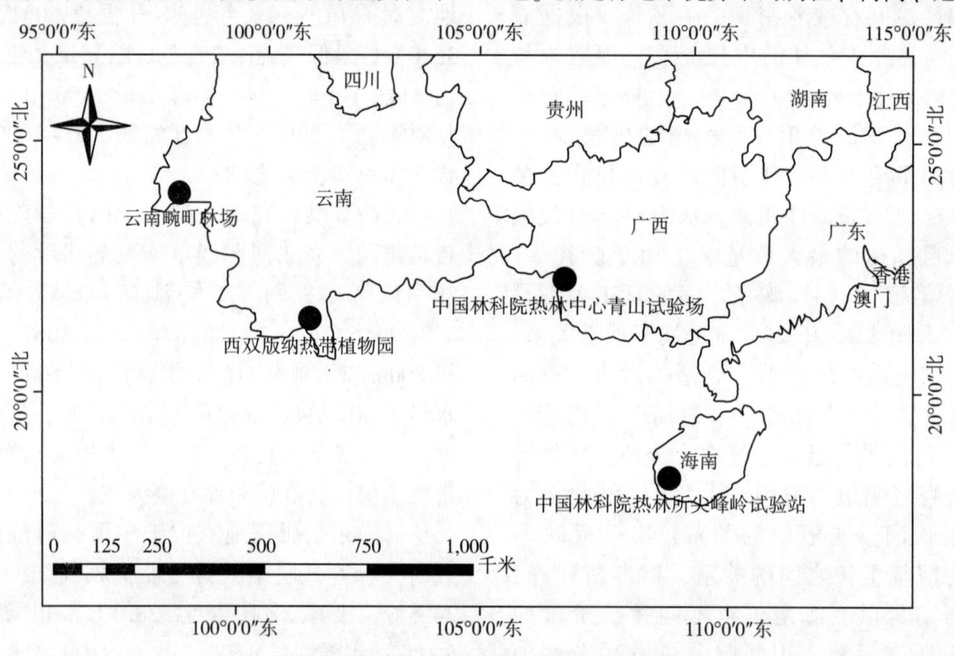

图1　柚木人工林观测样地的地理位置

2.1　指标测定

胸径：利用胸径尺测定胸径(0. 1cm)。

树高、冠下高：利用Haglof VERTEX. IV超声波测高器测量树高(0. 1m)及冠下高(0. 1m)；冠长(m)=树高-冠下高。

冠幅：用8根标杆准确标出树冠的边缘，然后站在树干处利用超声波测高器、简易罗盘仪，分别测定树干到每个标杆的距离(0. 1m)、角度(1°)。根据距离和角度计算出树冠面积(八边形)，再利用圆形面积公式计算出树冠直径。

2.2　数据分析

采用Microsoft Office Excel 2013对收集数据进行统计分析，DPS 14. 50数据处理软件进行回归分析。

3　结果与分析

3.1　林分类型基本信息

各地区不同林龄柚木林分中优势木的胸径、树高、冠长和冠幅等信息见表1。由表可知，林龄达到20年时，柚木树高生长趋于稳定，各地区柚木树高最大值在22. 56~26. 48m。不同林龄段柚木的胸径增幅不同，呈现出快(小于20年)—缓慢(20~30年)—快(30~50年)的变化趋势。冠长与林分密度相关，经过不同林龄阶段的间伐等经营措施，冠长分布在8. 00~16. 50m，占树高1/2左右。冠幅生长变化趋势与树高类似，林龄小于20年冠幅增长速度较快，之后趋缓。冠幅与胸径比值呈现出先增后降的趋势，先从12年林龄的19增加至35年的24左右，后降低至50年的18左右。

3.2 影响冠幅的关键因子确定与分析

以胸径、树高、冠长、林龄4个因子作为变量与冠幅进行回归分析。各变量的回归系数、偏相关系数及检验结果如表2所示。

表2 各因子与冠幅的回归参数

项目	变量			
	胸径	树高	冠长	林龄
回归系数	0.1270	0.0920	0.0286	0.0117
标准回归系数	0.6480	0.1535	0.0363	0.0509
偏相关系数	0.5342	0.1798	0.0472	0.0538
T 值	10.4995	3.0357	0.7853	0.8958
P 值	0.0001	0.0026	0.433	0.3711

偏相关系数最大的变量是胸径，为0.5342，其次是树高的0.1798，最小的是冠长与林龄，分别为0.0472和0.0538。t 检验结果显示胸径、树高同冠幅的关系达到极显著水平($P<0.01$)，且胸径、树高与冠幅均呈正相关关系；而冠长、林龄同冠幅关系不显著($P>0.05$)。因此选择胸径、树高这2个因子作为预测柚木冠幅的关键因子。

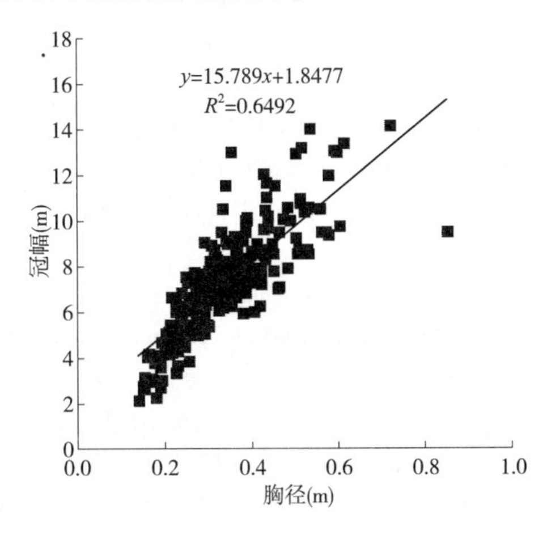

图2 冠幅与胸径的关系

冠幅与胸径的回归方程为：$y=15.7893x+1.84766$(如图2所示)，F 检验的结果显示相关性达到了极显著水平($F=516.4180$，$P=0.0001$)。冠幅与树高的回归方程为：$y=0.3717x-0.60189$(如图3所示)，F 检验的结果显示相关性达到了极显著水平($F=174.2954$，$P=0.0001$)。

3.3 建立回归模型

应用胸径和树高2个关键因子建立与冠幅的回归

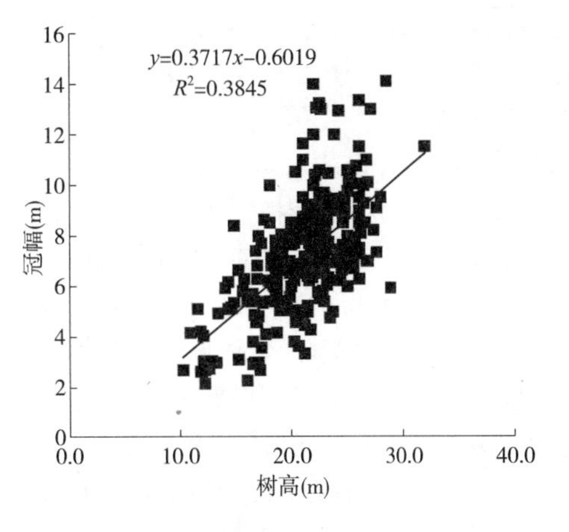

图3 冠幅与树高的关系

模型为：$y=13.5658x_1+0.1064x_2+0.35866$($x_1$—胸径，$x_2$—树高)。经 F 检验的结果显示相关性达到了极显著水平($F=279.5048$，$P=0.0001$)，模型决定系数 $R^2=0.669272$，利用该模型对测量数据进行计算，拟合效果如图4所示。将计算结果与实际测量结果进行均值检验，显示两者之间差异性较小($F=0.0140$，$P=0.9072$)。因此，在柚木人工林目标树经营中，可以根据目标树培养目标胸径、树高，利用该模型来预测该目标树的冠幅。从而确定单位面积内保留目标树的数量。

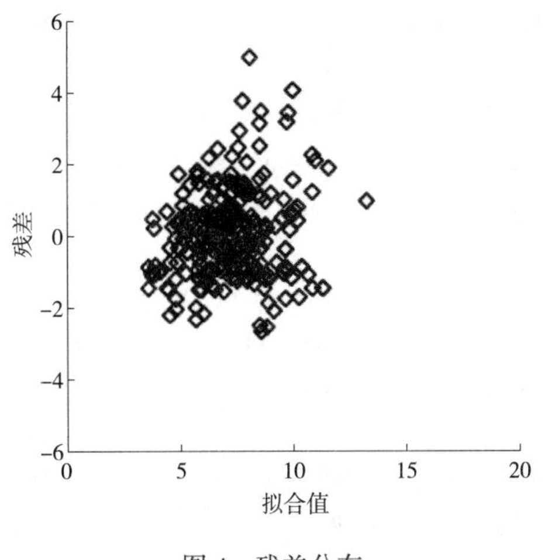

图4 残差分布

4 结论与讨论

根据各地区不同林龄柚木林分中优势木的胸径、树高、冠长和冠幅等信息分析，林龄达到20年时，柚木树高生长趋于稳定，各地区柚木树高最大值在22.56~26.48m；不同林龄段柚木的胸径增幅不同，呈现出“快(<20年)—缓慢(20~30年)—快(30~50年)”的变化趋势；不同林龄阶段的柚木冠长占树高1/2左右，冠幅生长趋势与树高类似，林龄小于20年冠幅增长速度较快，之后

趋缓。冠幅与胸径比值呈现出先增后降的趋势，从12年林龄的19增加至35年的24左右，后降低50年的18左右。

林木胸径显著影响冠幅生长[8,9,22]。许多研究利用胸径、树高等因子来模拟冠幅生长[15,21,23]。本研究通过回归分析，从胸径、树高、冠长、林龄4个因子中筛选出胸径($R=0.5342$，$P=0.0001$)、树高($R=0.1798$，$P=0.0026$)这2个影响柚木冠幅的关键因子。分别建立与冠幅的一元回归方程：冠幅与胸径 $y=15.7893x+1.84766$($F=516.4180$，$P=0.0001$)，冠幅与树高 $y=0.3717x-0.60189$($F=174.2954$，$P=0.0001$)。而有的学者认为，由于冠幅与林分密度相关，只利用胸径、树高等生长因子来模拟冠幅生长显然不够，应该引入相关生长竞争因子进行模拟[24]。笔者认为，从林分水平模拟冠幅生长，应该引入生长竞争因子；而模拟林分中优势木的冠幅生长可以不引入，因为优势木高度一般处于林分上层，受密度影响相对较小，且目标树经营过程中，影响目标树生长的干扰树均被伐除，林分密度引起生长竞争的影响被降到最小。

应用胸径和树高2个关键因子建立与冠幅的回归模型为：$y=13.5658x_1+0.1064x_2+0.35866$($F=279.5048$，$P=0.0001$，$R^2=0.669272$,)，将计算结果与实际测量结果进行方差分析，显示两者之间差异性较小($F=0.0140$，$P=0.9072$)。在柚木人工林目标树经营中，可以根据目标树的培育目标胸径、树高因子，利用该模型来预测该目标树的冠幅，从而确定单位面积内保留目标树的数量。

参考文献

[1] 马华明，梁坤南，周再知．我国柚木的研究与发展[J]．林业科学研究，2003，16(6)：768-773.

[2] 梁坤南，周再知，马华明，等．我国珍贵树种柚木人工林发展现状、对策与展望[J]．福建林业科技，2011，38(4)：173-178.

[3] 江泽慧．中国森林资源与可持续发展[M]．北京：科学出版社，2007.

[4] 张金文．巨尾桉大径材间伐试验研究[J]．林业科学研究，2008，21(4)：464 - 468.

[5] HOULLIER F，LEBAN J M，COLIN F. Linking growth modelling to timber quality assessment for Norway spruce [J]. Forest Ecology and Management，1995，74(1-3)：91-102.

[6] SPIECKER H. Silvicultural management in maintaining biodiversity and resistance of forests in Europe—temperate zone[J]. Journal of Environmental Management，2003，67(1)：55-65.

[7] ABETZ P，KLADTKE J. The Target Tree Management System：Die Z-Baum-Kontrollmethode[J]. Forstwissenschaftliches Centralblatt，2010，121(2)：73-82.

[8] 雷相东，张则路，陈晓光．长白落叶松等几个树种冠幅预测模型的研究[J]．北京林业大学学报，2006，28(6)：75-79.

[9] JOHNSON P S，SHIFLEY S R，ROGERS R. The ecology and silviculture of oaks[M]. New York：CABI，2009.

[10] NUTTO L，SPATHELF P，SELING I. Management of individual tree diameter growth and implications for pruning for Brazilian*Eucalyptus grandis* Hill ex Maiden[J]. Floresta，2006，36(3)：351-368.

[11] FOLI E G，ALDER D，MILLER H G，et al. Modelling growing space requirements for some tropical forest tree species[J]. Forest Ecology and Management，2003，173：79-88.

[12] DAWKINS H C. Crown diameters：their relation to bole diameter in tropical forest trees[J]. The Commonwealth Forestry Review，1963：318-333.

[13] SCHULER J，CUENI J，SPIECKER H，et al. Modelling growing space of four valuable broadleaved tree species in southern China[J]. Forest Science and Practice，2013，15(3)：167-178.

[14] BULLOCK S H. Developmental patterns of tree dimensions in a neotropical deciduous forest[J]. Biotropica，2000，32(1)：42-52.

[15] NORTON D A，COCHRANE C H，Reay S D. Crown-stem dimension relationships in two New Zealand native forests[J]. New Zealand Journal of Botany，2005，43(3)：673-678.

[16] Schuler J K. Astentwicklung und Astreinigung in Abhängigkeit vom Dickenwachstum bei Buche (Fagus sylvatica L.) und Eiche (Quercus petraea (Matt.) Liebl.; Quercus robur L.)[D]. Dissertation. -Freiburg (Breisgau). Online：http：//www. freidok. uni-freiburg. de/volltexte/8396/. Accessed 15 Jan 2013. Google Scholar，2011.

[17] 卢妮妮，王新杰，张鹏，等．不同林龄杉木胸径树高与冠幅的通径分析[J]．东北林业大学学报，2015，43(4)：12-16.

[18] 符利勇，孙华，张会儒，等．不同郁闭度下胸高直径对杉木冠幅特征因子的影响[J]．生态学报，2013，33(8)：2434-2443.

[19] 张鹏，王新杰，高志雄，等．将乐地区马尾松最优冠幅模型研究[J]．西北林学院学报，2015，30(4)：94-98.

[20] GILL S J，BIGING G S，MURPHY E C. Modeling conifer tree crown radius and estimating canopy cover

[J]. Forest Ecology and Management, 2000, 126(3): 405-416.

[21] HUMMEL S. Height, diameter and crown dimensions of Cordia alliodora associated with tree density[J]. Forest Ecology and Management, 2000, 127: 31-40.

[22] 石小龙, 杜彦昌, 王鹏, 等. 小陇山油松人工林林冠指标相关性研究[J]. 西北林学院学报, 2018, 33(3): 67-73.

[23] GROTE R. Estimation of crown radii and crown projection area from stem size and tree position[J]. Annals of Forest Science, 2003, 60(5): 393-402.

[24] 覃阳平, 张怀清, 陈永富, 等. 基于简单竞争指数的杉木人工林树冠形状模拟[J]. 林业科学研究, 2014, 27(3): 363-366.

[原载: 西北林学院学报, 2019, 34(03)]

坡位梯级微变化对红椎林木生长的影响

郭文福[1,2] 郝 建[1,2] 韦菊玲[1,2]

([1]中国林业科学研究院热带林业实验中心，广西凭祥 532600；
[2]广西友谊关森林生态系统国家定位观测研究站，广西凭祥 532600)

摘 要 在广西凭祥中国林科院热带林业实验中心伏波实验场14年生红椎人工林中设置顺山行状样地，进行小尺度坡位梯级微变化对林木生长和土壤性质变化影响的研究，为珍贵树种大径材培育的造林立地选择提供精准技术。研究结果表明：①林木平均树高和胸径随坡位梯级上升而降低，不同坡位梯级对林木树高和胸径的影响呈显著差异。②土壤大量营养元素、pH、有机质等指标与坡位高度呈负相关关系，即坡位越高各指标数值越小，而pH、全P和全K含量的不同坡位梯级间差异达到显著水平。③春季土壤旱情较为轻微情况下，不同坡位梯级间的含水率差异不大。建议根据不同树种对水肥立地条件要求的差异，当坡面长度在60m以上应分上下坡位安排不同树种，超过100m时应分3个坡位段安排不同树种。

关键词 坡位梯级；红椎；生长；线状样地；坡位效应

The Influence of the Step-change of Slope Position on the Growth of *Castanopsis hystrix*

GUO Wenfu[1,2], HAO Jian[1,2], WEI Juling[1,2]

([1]*Experimental Center of Tropical Forestry*, *Chinese Academy of Forestry*, *Pingxiang* 532600, *Guangxi*, *China*;
[2]*Guangxi Youyiguan Forest Ecosystem Research Station*, *Pingxiang* 532600, *Guangxi*, *China*)

Abstract: Linear sampling plots was setup with a 14-years-old plantation of *Castanopsis hystrix in* Fubo Experimental Station of Experimental Center of Tropical Forestry, Chinese Academy of Forestry in Guangxi, China. The effects of small scale slope level and micro change on tree growth and soil properties were studied to provide precise techniques for the selection of afforestation site for the cultivation of large diameter timber of valuable tree species. The result showed that: ① the tree height and DBH decreased with the increase of slope position, and the influence of different slope positions on tree height and DBH was significantly different. ②Soil nutrient elements, pH and soil organic matter were negatively correlated with slope height. That is, the higher the slope position, the smaller the value of each index, and the difference of pH, total P and K content between different slope levels reached a significant level. ③Under the mild soil drought in spring, the difference of water content among different slope levels was not significant. It is suggested that according to the difference of site conditions of water and fertilizer for different tree species, different tree species should be arranged in the upper and lower slope positions when the slope length is more than 60m, and different tree species should be arranged in three slope positions when the slope length is more than 100m.

Key words: step-change of slope position; *Castanopsis hystrix*; growth; linear plot; slope position effect of growth

红椎(*Castanopsis hystrix* A. DC.)又名红锥，树体通直，树高可达30m，胸径100cm以上，是我国南亚热带常绿阔叶林主要建群种，也是该区域重要乡土珍贵用材和多用途树种[1-4]，自20世纪60年代以来，广西、广东和福建等地已营造红椎人工林5万hm^2以上。近年来红椎被列为国家储备林培育大径材的重要树种，在华南各地得到大力发展[5]。该树种生态适应性强，造林成活率较高，但林木生长对土壤肥力及水分反应十分敏感，其中坡位因子对林木生长产生显著影响[4,6-14]。优越的立地条件是高效培育大径材的基础，目前主要乡土珍贵树种的造林适生立地研究仅在适宜气候区、母岩母质和土壤种类和较大尺度的地形因子等方面有一些研究[2,4,7,10]，而基于坡位更小尺度的微小变化对土壤性质及林木生长研究尚未见报道。本文通过对不同坡位梯级小尺度立地变化对红椎林木生长与土壤性质影响的研究，探讨微型立地因子对林木生长的影响，为红椎等珍贵树种大径材培育造林立地精细选择提供技术依据。

1 试验区概况

试验区为红椎主要自然分布区及人工林种植主产区，位于广西凭祥市热带林业实验中心伏波实验林区，22°02′01″N，106°50′19″E。海拔高600m，属北热带季风气候区，湿润半湿润气候；光热条件极好，降水充沛，但夏湿冬干，10月至翌年3月为干季，4~9月为湿季；年平均气温19.9℃，年降水量1400mm，年蒸发量1200mm，相对湿度80%~84%，土壤为花岗岩发育的红壤，土层厚度大于1m。试验地为采取顺山行状混交方式营造的14年红椎×马尾松混交林，造林初植密度1667株/hm²，红椎和马尾松的混交比例是1∶2，红椎平均树高14.5m，平均胸径16.5cm，坡度约25°，坡长70m，即垂直高差共29.6m。林下主要灌木主要有盐肤木(*Rhus chinensis*)、大沙叶(*Aporosa chinensis*)等，金毛狗脊(*Cibotium barometz*)和铁线蕨(*Adiantum capillus-veneris*)，草本有飞机草(*Eupatorium oderatum*)、铁芒萁(*Dicranopteris linearis*)、五节芒(*Miscanthus floridulus*)等。

2 研究方法

本研究分别于2015年12月和2016年2月分别在上述14年红椎林中进行不同坡位梯级林木生长调查和土壤理化性质调查。

2.1 不同坡位梯级林木生长调查

采取线状样地调查法，在同一坡面相同林分内共设3个样地(即3个重复)，各样地均为走向与等高线垂直行状红椎林木。每样地行间的水平距离相隔约50m。调查时，每个样地按从下至上(表1的坡位梯级1至23)顺序观测不同坡位梯级上每株林木的树高和胸径(表1)。样地内红椎林木上下株间水平距离为3m，每行调查林木均为23株。3个重复共观测林木69株。使用Vertex VL5测高仪测树高，精度为0.1m；胸径围尺量胸高直径，精度0.1cm。

表1 不同梯级林木胸径和树高生长情况

坡位梯级	胸径(cm)				树种(m)			
	重复1	重复2	重复3	平均	重复1	重复2	重复3	平均
1	23.2	18.5	15.0	18.9	19.1	15.3	14.6	17.0
2	16.8	12.8	13.8	14.5	19.0	12.7	13.2	14.8
3	20.5	15.0	13.2	16.2	17.9	14.5	13.3	15.5
4	14.8	14.5	19.0	16.1	17.8	14.0	14.5	15.6
5	19.7	11.2	10.0	13.6	18.5	12.3	11.4	14.0
6	14.8	17.5	15.0	15.8	13.6	13.0	11.8	13.5
7	14.7	14.5	14.1	14.4	14.0	13.7	13.2	13.8
8	18.6	10.3	17.9	15.6	13.5	11.0	12.0	13.0
9	16.7	18.6	15.0	16.8	12.3	11.6	12.4	13.3
10	16.0	8.7	16.4	13.7	14.8	10.1	12.6	12.8
11	14.0	18.3	11.4	14.6	15.2	12.7	10.1	13.1
12	15.0	11.8	14.0	13.6	14.2	10.6	11.7	12.5
13	10.0	18.0	14.1	14.0	12.2	11.6	13.0	12.7
14	14.0	16.4	20.1	16.8	13.8	12.8	13.5	14.2
15	18.2	14.2	12.9	15.1	14.9	11.5	11.8	13.3
16	17.1	14.0	17.3	16.1	14.7	13.5	12.2	14.1

（续）

坡位梯级	胸径(cm)				树种(m)			
	重复1	重复2	重复3	平均	重复1	重复2	重复3	平均
17	12.1	14.5	16.2	14.3	11.9	12.5	11.7	12.6
18	14.0	12.3	14.3	13.5	12.7	11.2	10.2	11.9
19	16.2	13.8	17.5	15.8	13.2	10.6	12.8	13.1
20	15.3	14.4	13.6	14.4	13.2	12.0	11.9	12.9
21	13.0	14.3	18.1	15.1	10.9	11.9	11.9	12.5
22	12.1	11.8	16.6	13.5	10.4	11.0	11.9	11.7
23	11.1	13.3	14.0	12.8	11.0	12.9	11.1	12.0

2.2 不同坡位梯级土壤调查

土壤调查点与林木调查样地相同，均为线性样地。采样点沿红椎调查样行从下坡至上坡按每隔5m一个采样点，共采集了13个样点，即13个坡位梯级样点的土样。采样点须离开树根部水平方向1m左右，受人工干扰较小、代表性较好的原坡土，采样深度为0~40cm，向下方每隔10cm分别采1个土样50g，然后将4个小采点样品混合后取200g带回实验室做化验。按土壤常规分析方法化验样品的pH、有机质、全N、全P、全K及速效N和速效P。土壤含水率取样点同上，在表土层5~20cm范围内混合取土样约100g置于密封铝盒，并拿回实验室采用烘干法测定土壤含水率。

2.3 数据处理

调查数据的方差分析、相关及回归分析等统计处理及作图，使用DPS(VER15.10)数据处理系统软件进行[15]。

3 研究结果

3.1 林木胸径生长与坡位梯级关系

对3个线状样地，23个不同坡位梯级每株林木胸径数据进行方差分析，结果表明，不同坡位梯级对林木胸径的影响呈显著差异(表2)。

表2 不同坡位梯级林木胸径方差分析表

变异来源	平方和	自由度	均方	F值	P值
区组间	51.5177	2	25.7588	1.97	0.1501
处理间	620.2165	22	24.8087	1.897	0.027*
误差	653.7823	50	13.0756		
总变异	1325.517	74			

注：*表示显著。

上述数据进一步进行一元线性回归分析，发现林木胸径与坡位梯级变化呈线性关系，即 $D=17.0527-0.0754X$，相关系数为0.6036，在置信水平 $a=0.05$ 情况下，达极显著水平($P=0.007$)，详见图1。

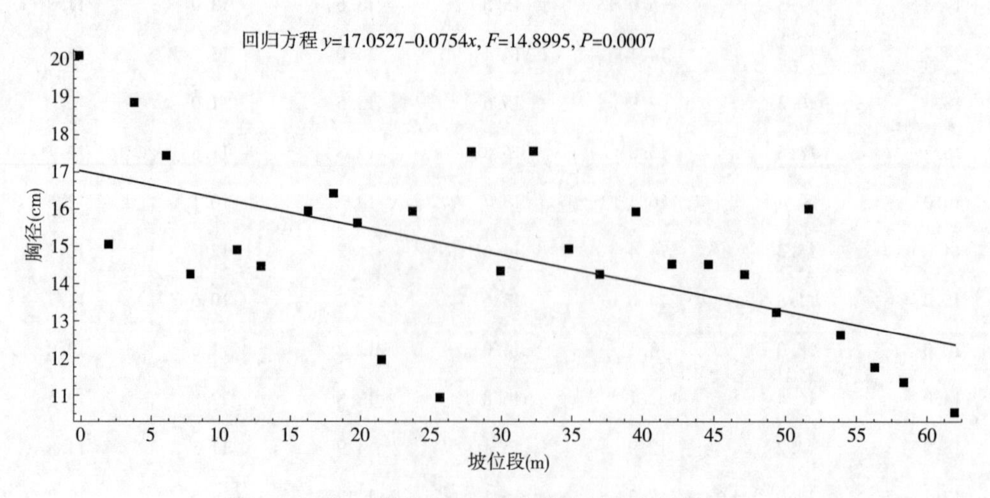

图1 红椎14年林分不同坡位梯级林木胸径值分布

3.2 林木树高生长与坡位梯级的关系

对处于不同坡位梯级的林木树高进行方差分析，结果显示不同坡位梯级林木间树高差异达到极显著水平（$P<0.001$）（表3）。

表3 不同坡位梯级林木树高方差分析表

变异来源	平方和	自由度	均方	F值	P值
区组间	27.8229	2	13.9115	6.009	0.0045 *
处理间	214.3225	22	8.2432	3.561	0 * *
误差	120.3773	50	2.3149		
总变异	362.5227	74			

注：* 表示显著；* * 表示极显著。

数据经回归分析，不同坡位梯级林木树高（y）与坡位梯级相对垂直高差（数值越大，坡位越高）的一元线性回归分析结果，得到树高线性方程 $y=15.0755-0.0795x$（图2），相关系数为0.8817，显著水平达极显著（$P=0.0000$），决定系数=0.777，方程拟合程度较高。

3.3 不同坡位梯级的土壤性质

3.3.1 不同坡位梯级土壤主要化学指标的变化

线形样地按每相隔5m坡长的13个采样点按从低至高坡位梯级，各项土壤样品化验数据如表4。

13个坡位梯级的各项化学分析指标见表3。经一元回归分析，各指标与坡位呈负相关关系（见表5），即坡位梯级的垂直高差越高，各指标的数值越小，而pH、全P、全K的坡位效应较为明显，经相关分析达到显著水平以上（$P<0.05$），其中pH和全K与坡位梯级相关关系达到极显著水平。全K坡位梯级的一元相关回归方程的决定值达0.7419。pH的坡位效应曲线见图3。

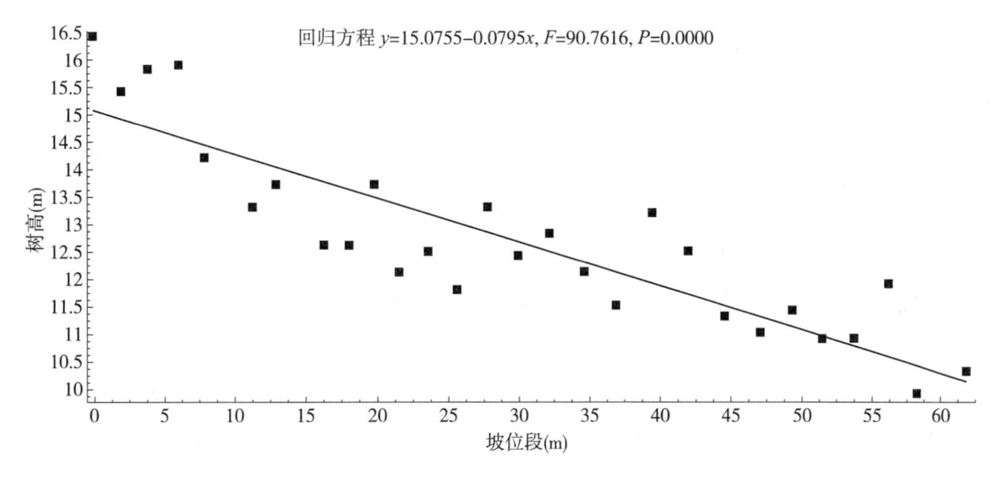

图2 红椎14年林分不同坡位梯级林木树高生长变化

表4 不同坡位梯级土壤化学性质分析结果

序号	样品编号	pH	有机质(g/kg)	全氮(g/kg)	全钾(g/kg)	全磷(g/kg)	碱解氮(mg/kg)	速效磷(mg/kg)
1	A1	4.25	38.933	1.398	0.922	0.256	99.010	1.137
2	A6	4.24	27.377	1.330	0.756	0.259	84.158	1.151
3	A11	4.28	31.378	1.352	0.759	0.283	84.866	1.049
4	A16	4.29	28.807	1.150	0.609	0.233	77.793	1.020
5	A21	4.3	26.822	1.172	0.584	0.278	77.086	0.991
6	A26	4.26	38.510	1.533	0.615	0.311	86.280	1.311
7	A31	4.21	42.008	1.330	0.682	0.285	102.546	1.587
8	A36	4.23	33.917	1.352	0.615	0.278	99.010	1.296
9	A41	4.23	33.121	1.352	0.628	0.233	96.181	1.238
10	A46	4.22	23.770	0.992	0.635	0.218	64.356	0.962
11	A51	4.21	30.539	1.150	0.456	0.226	70.721	1.296
12	A56	4.21	32.009	1.127	0.370	0.192	84.866	1.180
13	A61	4.21	32.422	1.375	0.430	0.216	78.501	1.616

表5　土壤主要化学分析指标与坡位一元回归分析结果

分析因子	相关系数	显著水平	决定系数	剩余标准差	回归方程
pH	-0.6998	0.0077**	0.4897	0.0240	$y=4.2819-0.0012x$
有机质	-0.1087	0.7237	0.0118	5.3895	$y=33.2922-0.0290x$
全N	-0.3280	0.2739	0.1076	0.1451	$y=1.3646-0.0025x$
全磷	-0.6001	0.0301*	0.3601	0.0288	$y=0.2885-0.0011x$
全K	-0.8613	0.0002**	0.7419	0.0781	$y=0.8478-0.0065x$

注：* 显著水平，** 极显著水平。

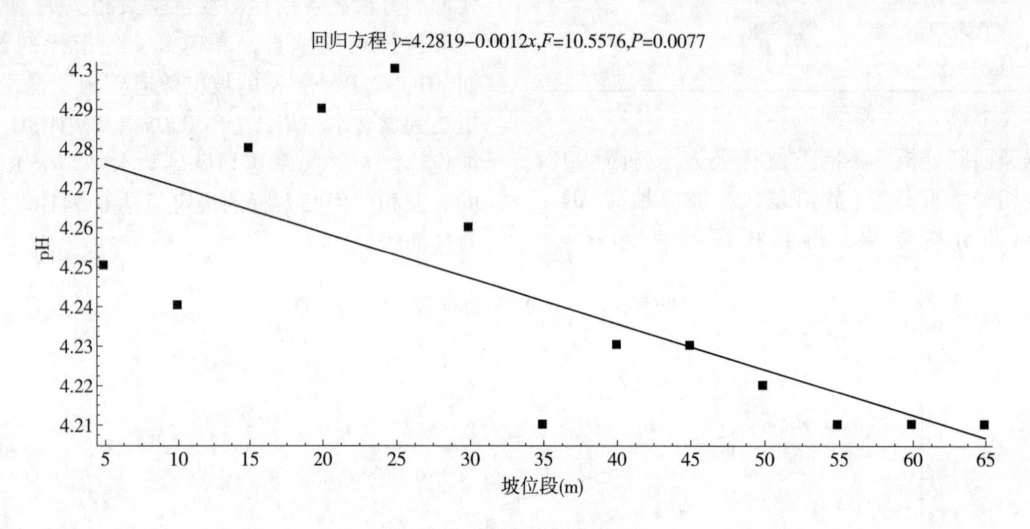

图3　pH的坡位效应曲线

3.3.2　土壤表土含水率不同坡位梯级变化

不同坡位梯级(由下坡向上坡顺序)土壤样品的含水率观测结果见表6。经相关性分析，表土含水率与坡位段高低相关性较低(相关系数 $R=0.1397$，显著水平 $P=0.6491$，决定系数=0.0195)。此次测定于2016年2月16日的单次测定，观测日为最近一次降雨(24mm)后连续12个无雨日进行的，因土壤未出现旱情，故坡位梯级变化的表土含水率差异不大。

表6　不同坡位梯级表土含水率测定结果

坡位梯级	1	2	3	4	5	6	7	8	9	10	11	12	13
含水率(%)	24.4	26.6	28.2	24.3	25.9	25.3	19.9	29.2	26.0	25.2	26.8	26.6	26.6

4　结论与讨论

研究进一步表明，坡位微小变化对红椎树高和径级生长影响显著，说明该树种是喜肥、喜湿的造林树种，对造林立地的水肥条件反应非常敏感。土壤有机质、酸碱度及主要养分也存在显著的梯级变化，并且有土壤大量营养元素、pH、有机质等指标与坡位呈负相关关系，即坡位越高，其各指标的数值越小，而pH、全P、全K的坡位效应较为明显。全K坡位一元相关回归方程的决定值达0.7419。因为下坡以及沟谷地带是山地土壤营养元素和水分富集的微地形，微生物活动也较为活跃，土层质地疏松肥沃，因而适合红椎这样喜肥喜水的树种生长。根据坡位梯级变化与胸径树种生长影响呈线性关系的规律，将坡面较长的立地分段规划设定不同造林目标及施业方法。建议坡面长度超过60m的至少应分2个坡位段，即上坡和下坡，分别经营类似红椎等喜水肥树种并培育大径材用材，上坡则经营类似马尾松等较耐瘠薄的树种，培育中小径材；超过100m的坡面应分3段布局不同造林树种或造林模式。

参考文献

[1] 廖培来，朱积余．广西名优经济树种[M]．北京：中国林业出版社，2006.

[2] 朱积余．红椎速生丰产造林技术的研究[J]．广西林业

科学，1991，20(4) ：175 -180.

[3] 刘恩，王晖，刘世荣．南亚热带不同林龄红椎人工林碳贮量与碳固定特征[J]．应用生态学报，2012，23(2)：335-340.

[4] 丘小军，朱积余，蒋燚等．红椎的天然分布与适生条件研究[J]．广西农业生物科学，2006，25(2)：175-179.

[5] 郭文福，蔡道雄．红椎大径材近自然培育适用性分析[J]．湖南林业科技，2015，42(1)：79-82.

[6] 黄承标，杨茂好，牙美华等．桂西北红椎次生林生长及土壤的理化性质[J]．南京林业大学学报(自然科学版)，2004，28(2)：54-56.

[7] 卢立华，汪炳根，何日明．立地与栽培模式对红椎生长的影响[J]．林业科学研究 1999，12 (5) ：519-523.

[8] 郑威，谭一波，唐 洁，等．不同坡位下红椎人工林的土壤呼吸特征[J]．中南林业科技大学学报，2015，35(6)：11-14.

[9] 黄全能．红椎天然林生长规律与生物量的调查研究[J]．福建林业科技，1998，25 (2) ：20-23.

[10]卢立华，冯益明，农友等．基于林班尺度的森林立地类型划分与质量评价[J]．林业资源管理，2018(2)：48-57.

[11]黄全能，陈东华，代全林，等．红椎天然林土壤理化性质及水源涵养功能的研究[J]．福建林业科技，2001，28 (2) ：17-19.

[12]钱永平，李宝银，吴承祯等．生态环境因子对阔叶林质量的影响——坡位与林分蓄积关系的研究[J]．林业资源管理，2009(1)：80-83.

[13]王明怀，陈建新．红椎等 8 个阔叶树种抗旱生理指标比较及光合作用特征[J]．广东林业科技，2005，21(2)：1-5.

[14]林俊平．红椎不同模式造林效果分析 J]．福建林业科技，2002，9(3) ：59 -61.

[15]唐启义．DPS 数据处理系统：第一卷基础统计及实验设计[M]．4 版．北京：科学出版社，2017.

[原载：林业资源管理，2019(01)]

不同林分格木人工林目标树生长性状差异

郝　建　曾　冀　郭文福

（中国林业科学研究院热带林业实验中心，广西凭祥　532600）

摘　要　为研究格木人工林中个体间的差异，给格木个体选优提供理论基础，选取中国林业科学研究院热带林业实验中心白云试验场及哨平试验场的1982年定植的格木人工林中目标树为研究对象，测量胸径、树高枝下高以及林分中个体的分类等级，分析各林分中个体之间在各个测定指标中的离散系数和4个林分之间的生长差异。结果表明：4个格木林分中优势个体的分类等级及枝下高的离散系数较大，最高达到0.220、0.360，树高和胸径的差异相对较小，分别达到0.151、0.127；哨平试验场的格木的树高及枝下高生长分别极显著和显著大于其他3个林分，胸径生长与白云1相差1.71cm，与其他2个林分无显著差异，分类等级无显著差异，说明哨平试验场的格木人工林在4个林分中生长最好。因此，加大格木林个体选优工作以及无性繁殖研究力度对解决格木个体生长差异大和资源稀少的现状具有重要意义。

关键词　格木；生长；目标树；离散系数

［原载：中国农学通报，2010，26(19)］

灰木莲生物学特性及引种栽培

曾　冀　卢立华　贾宏炎

（中国林业科学研究院热带林业实验中心，广西凭祥　532600）

摘　要　本文重点介绍了灰木莲的生物学生态学特性、苗木培育、栽培技术以及在不同引种地的生长情况。

关键词　灰木莲；生物学特性；引种；栽培

［原载：林业实用技术，2010(10)］

柚木寒害调查与防护措施

李运兴[1] 梁坤南[2] 马华明[2] 周再知[2] 刘 云[1] 莫慧华[1]

([1]中国林业科学研究院热带林业实验中心，广西凭祥 532600；
[2]中国林业科学研究院热带林业研究所，广东广州 510000)

摘 要 通过对柚木幼苗、1 年生施肥试验林与混交林、2~40 年生柚木林按 3 个寒害级进行了调查。结果表明：持续 4.5℃以下的低温，会出现柚木苗期和幼龄林期寒害；裸地培育的移植苗轻于营养袋苗，高径比小的粗壮苗木不易受害；1~2 年生幼林易受寒害，3 年生以上的林分寒害轻微；不同施肥处理寒害差异显著，施 Ca 可提高柚木个体的抗寒能力，而施 N 则会加重树木的枯干程度；1 年生混交林处理间柚木寒害差异不显著，各树种御寒能力不同，马尾松、观光木、火力楠、格木、红椎 5 个树种大于铁力木、柚木、铁刀木。

关键词 柚木寒害；苗木；幼林；施肥处理；混交林

[原载：林业实用技术，2010(03)]

立地条件对川滇桤木生长的影响

赵　总[1]　张兆国[2]　谢德兵[3]　张玉娟[3]　李祖智[3]
([1]中国林业科学研究院热带林业实验中心，广西凭祥　532600；
[2]西南林业大学资源学院，云南昆明　650224；[3]重庆市黔江区林业局，重庆　409000)

摘　要　根据立地条件不同设置样地，对样地内生长的5年生川滇桤木幼林树高和胸径生长量进行测量，选择标准木进行树干解析并对数据进行比较分析。结果表明：不同的立地条件对川滇桤木的树高、胸径、材积生长量的生长有显著影响，其中水湿条件在促进川滇桤木生长方面作用明显。

关键词　川滇桤木；立地；生长量；标准木

[原载：江苏农业科学，2011，39(06)]

马尾松近熟林的间伐效应

白灵海　谌红辉　温恒辉　安　宁　蒙明君

（中国林业科学研究院热带林业实验中心，广西凭祥　532600）

摘　要　利用16~21年生马尾松近熟林间伐密度试验林的观测资料，探讨了不同间伐强度对林分生长与效益的影响。结果表明：不同间伐保存密度（450、600、750、900、1036株/hm^2）对林分生长、林分结构及材种规格均有显著影响。其中林分胸径、单株材积、冠幅随密度增大而减小，密度对树高生长无显著影响；不同间伐密度的林分蓄积量与出材量随着时间推移差异变小；间伐有利于提高大径级林木株数比率从而提高材种规格与效益。根据21年生马尾松人工林生长规律与效益评价，16~21年生近熟林间伐密度保存在600~900株/hm^2效益较好。

关键词　马尾松；近熟林；间伐密度；密度效应；经济评价

［原载：林业科技开发，2012，26(06)］

马尾松锯材干燥中试研究

白灵海[1]　刁海林[1,2]　唐贤明[2]　唐继新[1]　赖玉海[2]　罗建举[2]
([1] 中国林业科学研究院热带林业实验中心，广西凭祥　532600；[2] 广西大学林学院，广西南宁　530000)

摘　要　进行马尾松锯材干燥中级试验，验证中试干燥基准的正确性，为制定合理的适于实际生产的马尾松干燥工艺提供依据。应用生产中广泛使用的强制气流循环普通干燥窑，在给定的中试干燥基准下对马尾松锯材实施干燥中试试验，并按照国家标准(GB/T6491—1999)进行干燥质量评定。试验结果表明：锯材的平均终含水率为3.55%；厚度含水率偏差为2.01%；应力指标4.5%；平均顺弯度为0.39%，横弯度为0.18%；翘弯度为1.36%；扭曲度为0.46%；纵裂度为0.77%；截面收缩率为0.89%；无内裂。干燥至含水率12%时所用时间为106h(4.4d)。马尾松锯材总体干燥质量良好，达到木制品生产对马尾松干燥质量的要求。

关键词　马尾松；锯材；干燥基准；中试

［原载：林业实用技术，2012(12)］

3个珍贵阔叶树种大树移栽效果评价

陈 琳 农瑞红 贾宏炎

（中国林业科学研究院热带林业实验中心，广西凭祥 532600）

摘 要 本文以格木、柚木和降香黄檀等3个珍贵阔叶树种中幼林的间伐木为材料开展大树移栽，依据移栽后的成活率、生长表现、观赏性以及健康状况评价其移栽效果，结果表明：3个树种的大树移栽成活率均在98%以上，移栽3年后所有林木均生长良好，但格木和降香黄檀的冠幅较柚木舒展，其园林绿化效果明显优于柚木。

关键词 珍贵阔叶树种；大树移栽；效果评价

[原载：现代农业科技，2012(02)]

珍贵树种山地造林新技术——丛植混交林营造法

郭文福　蔡道雄　贾宏炎　曾　冀　郝　建

（中国林业科学研究院热带林业实验中心，广西凭祥　532600）

摘　要　针对珍贵树种用材林经营周期长、良种使用率低、培育及经营技术欠缺，较多树种尚无法进行大面积山地造林的现状，本文在借鉴一般用材树种丛植造林法技术和经验的基础上，创新应用目标树经营法等森林定向培育技术，研制了适合珍贵树种大面积山地造林的丛植混交林法，包括"针阔同丛混交法""阔丛针单混交法"和"阔叶异丛混交法"共3种丛植混交造林法。本文主要介绍此三种丛植混交法的树种配置技术及应用特点。

关键词　珍贵树种；造林；目标树；丛植法；混交林；丛内间伐法

［原载：林业实用技术，2012(01)］

人工整枝技术在用材林培育中的应用研究进展

郝　建　李武志　陈厚荣　赵　樟　郭文福

（中国林业科学研究院热带林业实验中心，广西凭祥　532600）

摘　要　人工整枝技术对提高林木的木材材质，增加树干的圆满度，促进林木生长具有重要作用。世界上林业发达国家早已对人工整枝技术进行了系统研究。我国对人工整枝技术重视度不够。目前我国还是以传统的普通木材生产为主，所以我国的木材品质还有较大的提升空间。本文论述了人工整枝技术的发展历程、在用材林培育中的意义、开展人工整枝所需的技术措施及相关研究进展，分析了人工整枝对树木生长的影响。为人工整枝技术研究与应用的开展提供借鉴。

关键词　人工整枝；用材林；自然整枝

［原载：林业科技开发，2012，26(05)］

热林中心边境生物防火林带的营造与管理

张显强　农良书　劳庆祥　赵　樟

（中国林业科学研究院热带林业实验中心，广西凭祥　532600）

摘　要　生物防火林带与传统的放火线相比，不仅能发挥良好的防火功能，还能发挥涵养水源和水土保持的作用。当森林火灾的发生时，生物防火林带能起到阻隔地表火蔓延，拦截飞火，减弱火势，切断火源，为人工扑救赢得宝贵时间。热林中心林区与越南接壤53.6km，因此营建高质量、高标准的生物防火林带对边境一带林区安全显得尤为重要。

关键词　生物防火林带；营造；管理措施

［原载：农业与技术，2012，32(07)］

林相结构调整对马尾松毛虫发生的影响

张显强　陶　怡

（中国林业科学研究院热带林业实验中心，广西凭祥　532600）

摘　要　针对林相结构调整对马尾松毛虫发生的影响，中国林业科学研究院热带林业实验中心对10~25年生马尾松开展了长期的调查研究。通过针对性的营林措施，主要通过培育针阔混交林、调整树种结构与提高林分质量为主的林相结构调整措施，以应对林区马尾松毛虫的发生。从结果可知，林相结构调整在一定程度上破坏了松毛虫的生活环境，起到保护天敌、控制松毛虫发生发展的作用。

关键词　马尾松毛虫防治；营林措施；林相调整；森林经营

［原载：吉林农业，2012(05)］

广西凭祥红椎人工林天然更新影响因素初探

邓硕坤[1,2] 廖树寿[2] 黄柏华[2] 蔡道雄[2] 温远光[1]

([1]广西大学林学院，广西南宁 530005；[2]中国林业科学研究院热带林业实验中心，广西凭祥 532600)

摘 要 设置典型样地，对广西凭祥28~32年生红椎（*Castanopsis hystrix*）人工林林下红椎天然更新及其影响因素进行调查和分析。结果表明：林下幼苗、幼树、小树的平均密度分别为20097.1、920.0、2308.6株/hm^2。小树的平均胸径和树高分别为3.10cm和4.49m，幼树平均胸径和树高分别为1.52 cm和2.77m，其天然更新良好，说明开展近自然经营是可行的。林下红椎的更新状况与上层林木的生长表现密切相关，上层红椎乔木表现越好，其幼苗越多，但其小树、幼树越少，而且生长表现越差，可能与光环境以及养分竞争差异有关；坡位是影响红椎更新的一个主要环境因子。

关键词 红椎；人工林；天然更新；坡位；相关分析

［原载：广西林业科学，2013，42(01)］

尾巨桉2种留萌密度幼林生长比较

陶 怡[1] 陈厚荣[1,2] 明安刚[1] 温远光[2] 全昭孔[1] 周国福[2,3]

([1]中国林业科学研究院热带林业实验中心，广西凭祥 532600；[2]广西大学林学院，广西南宁 530004；[3]广西国有东门林场，广西扶绥 532108)

摘 要 为阐明留萌密度对尾巨桉萌芽林生长量的影响，分4个生长时期(0.3年生、1.4年生、2.3年生和3.3年生)对广西东门林场尾巨桉2种留萌密度(留萌1株和留萌2株)的萌芽林的生长量进行了调查研究。结果表明：2种留萌密度下桉树萌芽林的胸径，树高及林分蓄积量均有不同程度的差异。留萌1株的萌芽林平均树高和平均胸径均显著高于留萌2株的萌芽林，而且这种差异会随林龄的增加而增加；而留萌2株的林分蓄积量表现出高于留萌1株的萌芽林的态势，但差异随着林龄的增加而减小。

关键词 留萌株数；尾巨桉；生长量；林龄

[原载：广西林业科学，2013，42(04)]

丘陵采伐索道集材技术研究

白灵海[1] 赵章荣[2] 蔡道雄[1] 傅万四[2] 唐继新[1]
([1]中国林业科学研究院热带林业实验中心，广西凭祥 532600；[2]国家林业局北京林业机械研究所，北京 100020)

摘 要 丘陵地区集材是木材采伐中占用劳动力最多的一个环节，选择一种合适的集运方式是实现木材采伐高效作业的关键。通过对常用的索道集材系统进行分析，研究不同木材集运作业功能指标，并将地面集材方式与索道集材方式进行对比。结果表明：索道集材为地势复杂山区集材作业的一个优选方式。

关键词 索道；集材；山区

［原载：林业机械与木工设备，2014，42(10)］

陡坡山地索道集材作业效率的研究

曾　冀[1]　蔡道雄[1]　刁海林[2]　唐继新[1]　白灵海[1]　黄建友[1]　张显强[1]
([1]中国林业科学研究院热带林业实验中心，广西凭祥　532600；
[2]广西大学林学院，广西南宁　530004)

摘　要　20世纪50年代，我国从国外引进的林业索道技术，并在南方和西南林区得到广泛应用，索道集材能降低人力成本，提高效率，降低作业人员的危险。研究表明：马尾松人工林索道集材作业效率最高已达6.79m^3/h，与传统人力集材相比，索道集材是人力集材效率的4.3倍，成本只有人力集材的3/8；索道集材的时间效率最高为89.89%，其效率还有很大的提升空间，延误的时间80.7%是可以避免的。本研究旨在引进国外先进的采运技术及设备，推动我国的林业集材设备发展，实现我国林业集材机械化，推进林业现代化进程。
关键词　马尾松人工林；索道集材系统；作业效率

[原载：木材加工机械，2014，25(05)]

顶果木抗旱生理特性研究

郝　建[1]　全昭孔[1]　农　志[1]　刘志龙[1]　马小峰[2]　郭文福[1]
([1]中国林业科学研究院热带林业实验中心，广西凭祥　532600；[2]凭祥出入境检验检疫局，广西凭祥　532600)

摘　要　为了评价顶果木(*Acrocarpus fraxinifolius*)的耐旱性，采取模拟自然极度干旱条件方法，利用盆栽方式，研究顶果木在干旱胁迫期间的生理特性。结果表明：在水分胁迫期间，顶果木叶片的叶绿素 a、b 含量出现先升高后降低的趋势，而类胡萝卜素含量出现增加趋势，叶绿素 a 的含量明显高于叶绿素 b 和类胡萝卜素的含量；顶果木叶片的可溶性糖、可溶性蛋白质、游离脯氨酸含量出现先增加后降低的趋势；受水分胁迫的影响，顶果木叶片质膜透性越来越大。干旱胁迫开始在停止供水后的第 11～16 天的某一天，干旱胁迫的临界土壤含水量在 20%左右；第 21 天顶果木各种生理机能开始丧失，已经达到顶果木的萎蔫系数(12.99%)。
关键词　顶果木；干旱胁迫；耐旱性；生理特性

[原载：中国农学通报，2014，30(01)]

除草、松土及抹芽整枝措施对柚木生长的影响

李运兴[1] 梁坤南[2]
([1]中国林业科学研究院热带林业实验中心，广西凭祥 532600；
[2]中国林业科学研究院热带林业研究所，广东广州 510000)

摘 要 为确定除草、松土和抹芽整枝等造林抚育措施对柚木生长的影响，对柚木进行该项试验。结果表明：不同除草松土方式、不同抹芽整枝处理对树高、胸径生长差异显著。

关键词 柚木；除草；松土；抹芽；整枝；生长

［原载：林业实用技术，2014(12)］

马尾松高产脂优树选择及高产脂林培育

安　宁[1,2]　丁贵杰[1]　谌红辉[2]　农　志[2]　黄柏华[2]

([1]贵州大学造林生态所，贵州贵阳　550025；[2]中国林业科学研究院热带林业实验中心，广西凭祥　532600)

摘　要　马尾松分布广泛，可脂材兼用，松脂制成的松香和松节油是重要的化工原料，由于近年珍贵阔叶树的大面积种植以及松毛虫危害，马尾松林面积逐年下降，可采脂树逐年减少，产脂量已不能满足林产化工业的需求，营建马尾松高产脂人工林是发展的必然趋势。从良种选用、造林地选择、整地、苗木培育及栽植、密度控制、林地施肥、抚育管理、混交林模式和病虫害防治等方面系统地概述了马尾松优树选择的标准与方法以及高产脂马尾松林培育，指出了存在问题，并对今后的研究进行了展望。

关键词　马尾松；高产脂；优树选择；培育

[原载：贵州农业科学，2015，43(02)]

红椎大径材近自然培育适用性分析

郭文福　蔡道雄

（中国林业科学研究院热带林业实验中心，广西凭祥　532600）

摘　要　根据近自然林经营理念及原则，结合红椎生物-生态学特性及木材利用特性，系统分析红椎采取近自然林经营技术培育大径材的适用性。结论认为，红椎为南亚热带地带性常绿阔叶林顶极群落的主要建群种，可以采取近自然林目标树单株培育技术，高效培育大径材多功能人工林。

关键词　红椎；大径材；目标树；近自然林；适用性

［原载：湖南林业科技，2015，42(01)］

除杂灌和松土抚育措施对柚木及其林下植被的影响

李运兴[1] 明安刚[1] 王亚南[1] 陈海生[2] 覃福龙[2] 陈茂日[2]
([1]中国林业科学研究院热带林业实验中心，广西凭祥 532600；
[2]广西农垦国有龙北总场，广西龙州 532400)

摘　要　为了解除杂灌、松土抚育措施对柚木、林下植被的影响及柚木与林下植被的相互作用，笔者于造林后第2年在柚木幼林中布设了除杂灌(全砍与全铲)和松土(松土与不松土)抚育试验。结果表明：不同除杂灌、松土处理措施对柚木生长和林下植被生长影响显著。①全砍与全铲除杂灌处理相比，柚木树高、胸径分别减少39.2%和39.5%；林下灌木高度、盖度和干重分别增加32.6%、76.2%和184.0%，草本则分别增加36.8%、5.8%和39.3%；灌木和草本种类分别减少16.7%和50.0%。②松土与不松土措施相比，柚木树高、胸径分别提高13.5%和11.0%；对林下灌木及草本的生长也有显著影响，但较除杂灌措施程度小。③柚木树高和胸径生长与林下草本盖度存在显著负相关关系。

关键词　柚木；除杂灌；松土；林下植被；草本盖度

[原载：中国农学通报，2015，31(34)]

红椎目标树选择技术规程的研制

郭文福　蔡道雄　贾宏炎　郝　建　刘志龙

（中国林业科学研究院热带林业实验中心，广西凭祥　532600）

摘　要　介绍了采取近自然林经营理论及其技术方法研制旨在高效培育大径级珍贵用材的红椎目标树选择技术规程，包括规程制定目的意义、研制原则和规程的主要技术参数及内容，简要评价了这一行业标准预期的经济效益，提出了贯彻该标准的措施和建议。

关键词　红椎；目标树；近自然林经营；规程

［原载：湖南林业科技，2016，43(01)］

我国南方人工林近自然化改造模式和效益分析
——基于中国林业科学研究院热带林业实验中心实践研究

刘志龙[1,4] 蔡道雄[1] 贾宏炎[1] 曾祥谓[2] 陆元昌[3]
([1]中国林业科学研究院热带林业实验中心，广西凭祥 532600；[2]中国林学会，北京 10009；
[3]中国林业科学研究院资源信息研究所，北京 100091；
[4]广西友谊关森林生态系统国家定位观测研究站，广西凭祥 532600)

摘 要 针对我国南方长期大面积连片经营人工纯林存在的问题，本文介绍了“近自然林业”的起源、引进、概念和基本原则，总结中国林业科学研究院热带林业实验中心多年来人工林近自然化改造的主要模式、改造过程和初步效益，分析了推广应用前景，以期为我国南方人工林近自然化改造提供实例借鉴和科学依据。
关键词 近自然化改造；改造模式；效益分析

[原载：林业科技通讯，2016(10)]

谈广西区直国有林场林地被侵占的治理对策

许基煌

（中国林业科学研究院热带林业实验中心，广西凭祥 532600）

摘 要 广西是中国最重要的商品林基地和林业生态大省之一，林业占有重要地位，区直国有林场在生态文明建设中具有重要地位。由于诸多原因，广西国有林场林地被大量非法侵占，且有愈演愈烈之势。收复被占有国有林地，尚有许多问题亟待解决。要从加强顶层设计，理顺国有林场与驻场森林公安的关系，发挥森林公安森林资源管护的主导作用，组织严厉打击非法占用林地等涉林违法犯罪专项行动等方面入手，综合治理。

关键词 广西；区直国有林场；林地；非法侵占；对策

［原载：国家林业局管理干部学院学报，2017，16(02)］

珍贵树种格木研究进展

杨保国[1,2] 刘士玲[1,2] 郝 建[1,2] 庞圣江[1,2] 张 培[1,2]
([1] 中国林业科学研究院热带林业实验中心，广西凭祥 532600)
([2] 广西友谊关森林生态系统国家定位观测研究站，广西凭祥 532600)

摘 要 格木(*Erythrophleum fordii*)是我国南亚热带地区的乡土阔叶树种，具有重要的经济价值、观赏价值、生态效益和药用价值。文中从生物学和生态学特性、栽培技术、种质资源及生态效益等方面阐述了国内外格木的研究现状，分析了当前我国格木研究中存在的问题，提出了未来优先研究和发展的方向。
关键词 格木；种质资源；生长；生态效益

[原载：广西林业科学，2017，46(02)]

山白兰的抗旱性评价

李武志　陈建全　梁福江　蒙明君　谢安德　潘启龙　郝　建

（中国林业科学研究院热带林业实验中心，广西凭祥　532600）

摘　要　为了考察山白兰苗木的抗旱性，设置不同浓度聚乙二醇(PEG6000)处理，模拟干旱胁迫试验，测定苗木生长及生理生化指标，采用隶属函数值法综合评价山白兰的抗旱性。结果表明：①山白兰的生长增量随着PEG6000胁迫浓度的增大而减少，胁迫浓度越大对苗木生长量的抑制作用越大。②随着胁迫浓度的增加，山白兰叶片的SOD、POD活性均呈先上升后下降的变化趋势，均高于CK。在浓度10%处SOD、POD活性达到最大值。③山白兰的细胞膜透性均随着PEG6000胁迫浓度的增加而增加，当PEG6000浓度为5%时，与对照差异不显著($P>0.05$)，当浓度高于5%时，膜透性急剧增加，高浓度胁迫对山白兰膜透性的伤害程度较高。④山白兰的MDA含量变化趋势与细胞膜透性的变化趋势相一致，山白兰的MDA含量与PEG6000胁迫浓度呈极显著的负相关。⑤当胁迫浓度低于10%时，脯氨酸含量增幅较小，高于10%时，增幅变大。脯氨酸含量与细胞膜透性、MDA含量呈显著或极显著正相关。通过模糊隶属函数法综合评价，山白兰的隶属综合值为0.472，属于中抗旱型。

关键词　山白兰；干旱胁迫；生长；生理特性

［原载：广西林业科学，2018，47(02)］

对热林中心森林经营的再认识

孙文胜

（中国林业科学研究院热带林业实验中心，广西凭祥 532600）

摘 要 中国林业科学研究院热带林业实验中心(以下简称“热林中心”)成立以来森林经营取得一些成绩，但在树种结构、森林经营管理水平等方面还存在一些问题。今后热林中心森林经营的思路是：以保护发展森林资源、改善生态环境和推进生态文明和美丽中国建设为宗旨，以森林可持续经营理论和森林经营规划为依据，以培育健康、稳定、高效的森林生态系统和提供更多更好的优质林产品为目标，建立以珍贵树种大径材培育为主导的多功能近自然森林经营技术体系，通过严格保护、积极发展、科学经营、持续利用森林资源，不断提升森林资源的数量和质量，稳步增强森林生态系统的整体功能，充分发挥森林资源的多种效益，实现林业可持续发展。

关键词 热林中心；森林经营；森林质量

[原载：国家林业局管理干部学院学报，2018，17(04)]

广西马尾松毛虫危害现状与防治对策

陶　怡　莫慧华　明财道　明安刚

（中国林业科学研究院热带林业实验中心，广西凭祥　532600）

摘　要　马尾松是广西主要造林树种之一，马尾松人工林的发展给广西人民带来了巨大的经济效益和生态效益，但马尾松人工林极易受到马尾松毛虫(*Dendrolimus punctatus* Walker)的危害。马尾松毛虫是广西地区历史性森林害虫之一，更是马尾松人工林主要虫害来源之一，严重影响广西马尾松人工林的发展。基于此，本文笔者对广西马尾松毛虫危害现状与防止对策进行探讨，旨在为广西地区松树的安全生产提供参考。

关键词　马尾松毛虫；松树；危害现状；防治对策；广西

［原载：农业与技术，2018，38(02)］

南亚热带人工林近自然高效可持续经营模式

韦菊玲 郭文福 雷丽群 刘福妹 邓硕坤 庞圣江

(中国林业科学研究院热带林业实验中心，广西凭祥 532600)

摘 要 介绍近自然森林经营的基本概念及其以乡土树种营建当地森林、针阔混交经营、复层异龄经营、目标树经营等近自然经营的基本原则。基于地处广西的中国林业科学研究院热带林业实验中心人工林经营中存在的主要问题，如以马尾松和杉木为主的针叶纯林面积比重过大，树种单一，林分质量不高，林地生产力较低等问题，自2002的起，采用降香黄檀—桉复层混交林、西南桦—红椎异龄混交林、马尾松(或杉木)针叶林阔叶化等近自然林经营模式，通过对林分高强度间伐、促进优良树种天然更新，以及在林冠下种植土沉香、红椎、紫檀、擎天树等较耐阴的长寿先锋树种，使群落树种组成、年龄结构趋于合理，形成了异龄复层混交林，使森林天然更新机制得到恢复，促进了人工林近自然经营高效可持续发展。

关键词 近自然经营；可持续经营模式；异龄复层混交林；南亚热带人工林

[原载：林业调查规划，2018，43(03)]

热带珍稀树种铁力木资源可持续经营对策

陈建全[1] 黄 婷[2] 马小峰[3] 李武志[1] 农瑞红[1]

([1]中国林业科学研究院热带林业实验中心，广西凭祥 532600；
[2]广西大学行健文理学院，广西南宁 530004
[3]广西钦州农业学校，广西钦州 535000)

摘 要 铁力木为我国的一级珍贵保护树种，主要分布在热带地区，具有非常优越的发展潜力。对铁力木资源进行有效分析与研究，提出了可持续经营的对策措施。

关键词 铁力木；可持续发展；对策研究

［原载：中国林业经济，2019(01)］

大红酸枝——交趾黄檀木材材性研究进展

刘福妹[1,2] 韦菊玲[1] 庞圣江[1] 劳庆祥[1] 谌红辉[1] 洪 舟[3]

([1]中国林业科学研究院热带林业实验中心，广西凭祥 532600；[2]广西友谊关森林生态系统国家定位观测研究站，广西凭祥 532600；[3]中国林业科学研究院热带林业研究所，广东广州 510000)

摘 要 交趾黄檀是顶级红木用材树种之一，又称为大红酸枝、老红木。本文从木材解剖特性、化学组成、木材价值等方面论述了交趾黄檀木材材性特点和价值，并归纳总结了区分辨别交趾黄檀木材的方法，为进一步开展交趾黄檀木材材性相关研究工作提供参考和依据。

关键词 交趾黄檀；木材材性；木材成分；辨别方法

［原载：林业科技通讯，2019(01)］

红椎人工林不同季节造林试验研究初报

韦菊玲[1,2]　刘福妹[1]　雷丽群[2]　郭文福[1]　郝　建[1]　吴方成[1]
([1]中国林业科学研究院热带林业实验中心，广西凭祥　532600；
[2]广西友谊关森林生态系统国家定位观测研究站，广西凭祥　532600)

摘　要　通过春季、夏季、秋季不同季节红椎(*Castanopsis hystrix* A. DC.)造林对比实验，对红椎造林成活率、幼树生长(高径)、树木形质生长和健康状况进行调查评价。试验结果表明：红椎在春季造林成活率最高，夏季造林次之。迹地全光造林夏季红椎平均胸径和平均高等调查因子比春季高。秋季造林生长状况最差；与米老排林冠下更新补植相比，迹地全光种植红椎生长较快较好。

关键词　红椎；不同季节；造林；试验

[原载：林业科技通讯，2019(03)]

第三部分

热带南亚热带主要人工林生态系统服务功能研究

Liparis guangxiensis sp. nov. (Malaxideae: Orchidaceae) from China

FENG Changlin[1], JIN Xiaohua[2]

([1]*Experimental Centre of Tropical Forestry, Chinese Academy of Forestry, Pingxiang, CN 532600, Guangxi, China.*
[2]*Herbarium (PE), Inst. of Botany, Chinese Academy of Sciences, Nanxinchun 20, CN 100093, Beijing, China.*)

A new orchid species, *Liparis guangxiensis* C. L. Feng & X. H. Jin, is described and illustrated from Guangxi, China. It is a widespread epiphyte in the limestone region at elevations of 200~1400m in tropical evergreen forest. *Liparis guangxiensis* is similar to *L. rhombea* sharing a similar habit, a bifid lip and a column without wings, but differs by having a compressed peduncle, lateral sepals forming in 'V' below the lip and an oblong lip with a callus at the base.

Liparis L. C. Richard is a medium-sized cosmopolitan orchid genus of about 250~350 species, mainly distributed in the tropics with some species extending into the temperate regions and quite a few found in the alpine zone[1-9]. Morphologically, *Liparis* differs from its relatives by its resupinate flowers and the curved column. Our own observations indicate that some characters, such as the folding pattern of the lip in the bud and the types of inflorescence, can also be used to distinguish these genera. Recently, however, many 'aberrant species' or infrageneric sections have been separated from *Liparis* s. l. as new genera based on morphological characters[10-12]. Here, *Liparis* is treated in a wide sence since a sound phylogenetic analysis of the whole complex is still wanting and the segregate genera are based on incomplete sampling.

There are about 54 species of *Liparis* s. l. in China, of which about 20 are endemic[4,8,13,14]. During our taxonomic revision of *Liparis* and fieldwork in southern China, a new species was discovered as described below.

***Liparis guangxiensis* C. L. Feng et X. H. Jin sp. nov. Figure** 1)

Liparidi rhombeae J. J. Sm. affinis, sed pedunculo compresso et labio oblongo basi calloso diversa.

Type: China, Guangxi Zhuangzhu Autonomous Province, Leye County, on limestone rocks in forest, 500~1100m a. s. l., 8 Nov 2009, Feng C. L. 091108 (holotype: PE!).

Epiphytic, 13~20cm tall. Pseudo-bulbs clustered on the rhizome, ovate to compressed-ovate, more or less adhering to the rhizome, 1. 7~3. 5cm tall, 0. 8~1. 5cm in diameter, with one leaf on top. Leaf oblanceolate or linear-lanceolate, articulate, deep green, 11~20cm long, 1. 5~2. 2cm wide, acute, base narrowing into a 7~10mm long petiole which is V-shaped in cross-section. Inflorescence 9. 0~17. 5cm long, 14~40 flowered; peduncle compressed, without in-fertile bracts; fertile bracts narrowly lanceolate, 2~4mm long; ovary and pedicel 3~5mm long; flowers yellowish green, 5~7mm across; dorsal sepal lanceolate, obtuse, veins indistinct, 3~4mm long, 0. 4mm wide; lateral sepals reflexed and forming a 'V' below the lip, lanceolate, acute, veins indistinct, 3~4mm long, 0. 4mm wide; petals linear, veins indistinct, 3. 2~4. 0mm long, 0. 25~0. 30mm wide; lip oblong, recurved at the base, the centre of the base with a yellow, thickened oblong spot with a transverse callus at the base of the thickened spot, apex bi-lobed and obscurely dentate, 3. 5~4. 2mm long, 1. 5~2. 0mm wide at apex; column 2. 5~3. 5mm long, without wings; anther cap acute. Capsule obovate, 4~5mm long, 3. 0~3. 2mm in diameter.

The flowering season is in Oct-Nov and the fruiting season in May-Jun.

Additional specimens examined (*paratypes*)

China, Yunnan Province, Hekou County, on limestone rocks in forest, 28 Nov 1992, Tsi Z. H. 92-101 (PE!); Guangxi Zhuangzhu Autonomous Region, Jingxi, 20 Sep 1935, Ko S. P. 55778 (KUN!).

Similar species

Liparis guangxiensis *resembles* L. rhombea *J. J. Smith by the ovate-compressed pseudo-bulb with one leaf, the bifid lip and the almost naked column without wings. However,* Liparis guangxiensis *is readily distinguished from the latter by having a distinctly compressed peduncle, an oblong lip with a callus at the base, and lateral sepals forming a 'V' below the lip.*

Acknowledgements – This study was funded by a public project of the Ministry of Science and Technology of China (2005DIB6J144) and a grant from the National Natural Science Foundation of China (30600037).

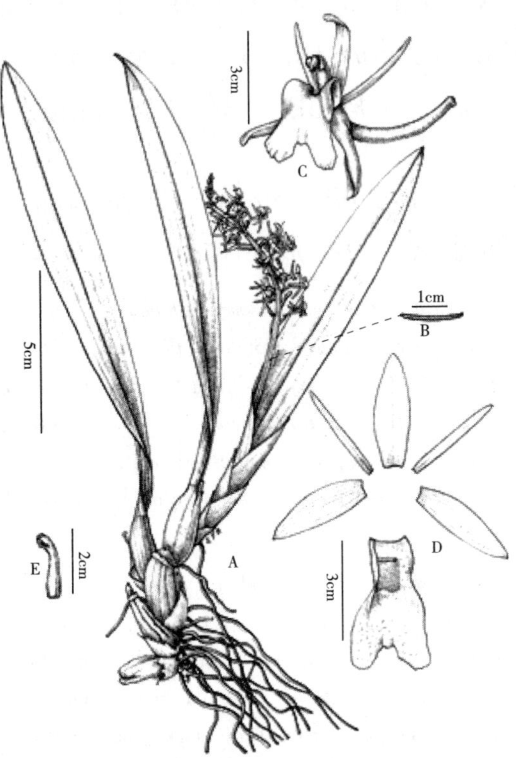

Figure 1 Liparis guangxiensis C. L. Feng & X. H. Jin sp. nov.

A. habit of plant; B. transverse section of peduncle; C. front view of flower; D. spals, petals and lip; E. column. Drawn by Ying-Bao Sun from the holotype

REFERENCES

[1] BOSE T K, BHATTACHARJEE S K, *Liparis* L. C. Rich. – In: Bose, T. K. and B hattacharjee, S. K. (eds), Orchids of India [M]. (2nd ed.). Naya Prokash, 1999: 303–318.

[2] RIDLEY H N. A monograph of the genus *Liparis* [J]. J. Linn. Soc. Bot. 1886, 22: 244–297.

[3] SEIDENFADEN G. Orchid genera in Thailand IV. *Liparis* L. C. Rich[J]. Dansk Bot. Ark, 1976, 31: 5–105.

[4] SEIDENFADEN, G. The orchids of Indochina[J]. Opera Bot, 1992, 114: 128–145.

[5] CHEN S C. *Liparis* L. C. Rich. –In: Chen, S. C. (ed.), Flora Reipublicae popularis sinicae 18 [M]. Science Press, 1999: 54–106.

[6] COMBER J B. Orchids of Sumatra. Natural History Publications (Borneo), Kota Kinabalu, in assoc. with R. Bot. Gard. Kew, Natl Parks Board, Singapore Bot. Gard. 2001.

[7] PEARCE N R, CRIBB P J. The orchids of Bhutan. R. Bot. Gard. Edinburgh, R. Gov. of Bhutan, 2002.

[8] PRIDGEON A M, et al. 2005. Genera Orchidacearum. Vol. 4. Epidendroideae, part one. -Publisher ?

[9] CHEN S C, et al. *Liparis* L. C. Rich. –In: Chen, S. C. et al. (eds), Flora of China 25[J]. Science Press, Miss. Bot. Gard. Press, 2009: 211–228.

[10] MARGONSKA H B, SZLACHETKO D L. *Alatiliparis* (Orchidaceae: Malaxidinae), a new orchid genus with two species from Sumatra[J]. Ann. Bot. Fenn, 2001, 38: 77–81.

[11] MARGONSKA H B SZLACHETKO D L. *Disticholiparis*, a new genus of subtribe Malaxidinae[J]. Orchidee (Hamburg), 2004, 55: 175–179.

[12] LIU Z J, et al. *Ypsilorchis* and Ypsilorchidinae, a new genus and a new subtribe of Orchidaceae[J]. J. Syst. Evol, 2008, 46: 622–627.

[13] ORMEROD, P. Orchidaceous additions to the flora of China and Vietnam [J]. – Taiwania, 2007, 52: 307–314.

[14] YANG P H, et al. *Liparisangustioblonga* sp. nov. (Malaxideae: Orchidaceae) from Shannxi, China[J]. Nord. J. Bot, 2009, 27: 348–350.

[原载: Nordic Journal of Botany, 2010, 28(6)]

The Impact of Near Natural Forest Management on the Carbon Stock and Sequestration Potential of *Pinus massoniana* (Lamb.) and *Cunninghamia lanceolata* (Lamb.) Hook. Plantations

MING Angang[1], YANG Yujing[2], LIU Shirong[3], NONG You[1], LI Hua[1], TAO Yi[1], SUN Dongjing[1], LEI Liqun[1], ZENG Ji[1], AN Ning[1]

([1]*Experimental Center of Tropical Forestry, Chinese Academy of Forestry, Guangxi Youyiguan Forest Ecosystem Research Station, Pingxiang* 532600, *Guangxi, China*; [2]*Hubei Key Laboratory of Regional Development and Environmental Response; Faculty of Resources and Environmental Sciences, Hubei University, Wuhan* 430062, *Hubei, China*; [3]*Key Laboratory of Forest Ecology and Environment, State Forestry Administration; Institute of Forest Ecology, Environment and Protection, Chinese Academy of Forestry, Beijing* 100091, *china*)

China Received: 13 June 2019; Accepted: date; Published: date

Abstract: Quantifying the impact of forest management on carbon (C) stock is important for evaluating and enhancing the ability of plantations to mitigate climate change. Near natural forest management (NNFM), through species enrichment planting in single species plantations, structural adjustment, and understory protection, is widely used in plantation management. However, its long-term effect on the forest ecosystem C stock remains unclear. We therefore selected two typical coniferous plantations in southwest China, *Pinus massoniana* (Lamb.) and *Cunninghamia lanceolate* (Lamb.) Hook., to explore the effects of long-term NNFM on ecosystem C storage. The C content and stock of different components in the pure plantations of *P. massoniana* (PCK) and *C. lanceolata* (CCK), and their corresponding near natural managed forests (PCN and CCN, respectively) were investigated during a 8 years of NNFM beginning in 2008. In 2016, there was no change in the vegetation C content, while soil C content in the 0~20cm and 20~40cm layers significantly increased, compared to the pure forests. In the *P. massoniana* and *C. lanceolata* plantations, NNFM increased the ecosystem C stock by 31.8% and 24.3%, respectively. Overall, the total C stock of soil and arborous layer accounted for 98.2%~99.4% of the whole ecosystem C stock. The increase in the biomass of the retained and underplanted trees led to a greater increase in the arborous C stock in the near natural forests than in the controls. The NNFM exhibited an increasingly positive correlation with the ecosystem C stock over time. Long-term NNFM enhances ecosystem C sequestration by increasing tree growth rate at individual and stand scales, and likely changing the litter decomposition rate resulting from shifts in species composition and stand density. These results indicated that NNFM plays a positive role in achieving multi-objective silviculture and climate change mitigation.

Key words: near natural forest management; *Pinus massoniana*; *Cunninghamia lanceolata*; plantation; carbon allocation; climate change

1 INTRODUCION

Recently, climate change has become a major issue creating global concern [1]. It has been widely recognized that rational plantation management can mitigate climate change by enhancing its carbon(C) sequestration capacity [2]. Understanding the impact of forest management on the C stock in different components of the forest ecosystem is critical for evaluating and enhancing the C sequestration potential of plantations.

China boasts the largestplantation area in the world, 63% of which is located in southern subtropical regions[3]. However, over 70% of the subtropical plantations consist of pure stands of coniferous species, dominated by *Pinus massoniana* (Lamb.), *Cunninghamia lanceolata* (Lamb.) Hook, and short-rotation exotic species like *Eucalyptus* spp. The biodiversity, forest biomass and productivity associated with pure stands of *P. massoniana* and *C. lanceolata* are lower than with mixed forests, thus C - sequestration capacity and microbial community diversity are both limited in pure stands[4-6]. Some pure coniferous plantations even potentially cause soil acidification[7]. Furthermore, problems such as auto-toxicity, nutrient deficiency, and understory competition have been observed in areas reforested with several rotations of the same species[8,9].

These problems arise from the traditional plantation management. For example, the traditional management of *P. massoniana* plantations in China is clear cutting with a rotation of 29 years, followed by prescribed burning. In contrast, species enrichment planting in single species plantations to form coniferous broad-leaved mixed forest, is now becoming a promising silvicultural approach[10]. Near natural forest management (NNFM), focused on multi-functional management and multi-quality products, is widely practiced[11]. Stand density of the original forest is firstly reduced by thinning, and fast-growing tree species are then underplanted, and the pure even-aged coniferous forest is gradually transformed to uneven-aged coniferous broad-leaved mixed forest[12]. The NNFM abandons clear cutting and prescribed burning, and increases forest productivity, soil fertility, and biodiversity[12]. Previous studies, however, have failed to address NNFM impacts on C sequestration in the whole ecosystem.

Since thevegetation and soil pools are the two largest components of C stock in forest ecosystems, they essentially determine the total ecosystem C stock. Species structure, composition, and forest age are the main influencing factors on the forest C stock[13]. Our previous study showed that intensive, intermediate and mild thinning increased the C stock of the arborous layer by 11.47%, 11.78% and 14.49% in a *P. massoniana* plantation, respectively[14]. However, the effects of thinning on forest soil C stocks are controversial[15,16]. In a Norway spruce stand, thinning from 3190 to densities of 2070, 1100 and 820 trees per ha did not affect the organic layer and mineral soil C stock[17]. However, the C stocks in the surface soil of red pine stands in Minnesota decreased in thinning regimes with 10%, 25% and 35% basal area removal, but not in stands where 50% of the basal area was removed[18].

Many studies have demonstrated the effects of species enrichment planting on C stock. The planting cannot only directly affect above-ground productivity, but also influence soil C stock by affecting the quality and decomposition of litter. Compared with pure coniferous plantations, native broad-leaved mixtures increase plant diversity, vegetation and soil C stocks[19,20]. The soil C concentrations and stocks were affected in pure stands of Norway spruce and mixed species stands[21]. In our study site, soil C stock in the 0~20cm layer in a mixed *P. massoniana* and *Castanopsis hystrix* plantation was 14.3% higher than that in the *P. massoniana* pure plantation[20]. Examples of underplanting with nitrogen-fixing species in planted forests to enhance the productivity of soil and vegetation are also widely documented[22]. However, no significant difference was found among the soil C stocks of the pure and mixed forests of *Erythrophleum fordii* and *P. massoniana*[23].

Because NNFMinvolves mixed forest establishment and thinning, it inevitably shifts the vegetation community composition and structure, thus altering the production and composition of litter as the main source of soil C. Therefore, NNFM is very likely to affect C processes and stocks in forest ecosystems. A 3-year NNFM through *C. hystrix* and *Michelia hedyosperma* significantly reduced the soil C contents in the 0~20, 20~40 and 40~60cm layers in a *P. massoniana* plantation, while slightly increased those in a *C. lanceolata* plantation[24]. Conversely, the effects of NNFM on soil C stock in pure *Fagus sylvatica* and *Picea abies* plantations varied with soil nutrient content[25]. Our previous research showed that soil CO_2 emissions in *P. massoniana* and *C. lanceolata* plantations were increased by NNFM[26]. Therefore, there are uncertainties on how soil C stock responds to NNFM. Although NNFM is one promising option to improve extensive pure coniferous plantation, its long-term effect on the ecosystem C stock and its allocation, and the underlying mechanisms remain unclear.

We therefore selected two typical coniferous plantations in subtropical China, *P. massoniana* and *C. lanceolata*, to explore the effects of long-term NNFM (i. e., thinning and species enrichment planting) on ecosystem C stock in subtropical *P. massoniana* and *C. lanceolata* plantations, and the underlying mechanisms. The C content and stock in different above- and below-ground ecosystem components were investigated. We hypothesized that: 1) the changes in tree species composition and stand density that are induced by NNFM increase the C stocks of the vegetation and soil, and 2) the ecosystem C stock is enhanced due to the increased C stock in both vegetation and soil layers. This study could provide an empirical and theoretical basis for multi-objective silviculture and ecosystem C management in subtropical China.

2 MATERIALS ABD METHODS

2.1 Study Site

This study was conducted at the Guangxi Youyiguan Forest Ecosystem Research Station, the Experimental Center of Tropical Forestry, Chinese Academy of Forestry (22°10′ N, 106°50′ E, Pingxiang, Guangxi, China). It is one of the forest ecology research stations under the jurisdiction of the State Forestry and Grassland Administration. The site has a subtropical monsoon climate, with a semi-humid climate and obvious dry and wet seasons. The annual sunshine duration is 1200~1600h. Precipitation is abundant, with an annual average of 1200 ~ 1500mm, mainly from April to September. The annual evaporation is 1200 ~ 1400mm, the relative humidity is 80% ~ 84%, and the average annual temperature is 20.5~21.7℃. The main types of landforms are low hills and hills. The soil is mainly composed of laterite and red soil based on the Chinese soil classification; this is classified as ferralsols in the World Reference Base for Soil Resources. Soil depth is generally greater than 80cm. Subtropical evergreen broad-leaved forests comprise the local vegetation.

There are nearly 20000hm^2 of various plantation types in the Experimental Center of Tropical Forestry. *P. massoniana* and *C. lanceolata* are the main coniferous tree species. Native broad-leaved tree species include *Quercus griffithii* (Hook. f. and Thomson ex Miq.), *Erythrophleum fordii* Oliver, *Castanopsis hystrix* Miq., *Mytilaria laosensis* Lecomte., *Betula alnoides* Buch.-Ham. ex D. Don, and *Dalbergia lanceolata* Zipp. ex Span. Among these species, *E. fordii* and *D. lanceolata* are nitrogen-fixing trees, and *Q. griffithii* is a fast-growing broad-leaved tree species with strong natural regeneration abilities. The near natural management of pure plantations of *P. massoniana* and *C. lanceolata* with *E. fordii* and *Q. griffithii* has been widely applied, as it not only meets the need for short-period timbers and valuable large-diameter logs, but also realises the natural regeneration of native broad-leaved species and achieves the goal of near natural management.

2.2 Experimental Design

A single-factor and two-level stochastic block design was used. There were four blocks representing four replicates. Four forest types were set up in each block: near natural *P. massoniana* plantation (PCN), unimproved *P. massoniana* pure plantation (PCK), near natural *C. lanceolata* plantation (CCN), and unimproved *C. lanceolata* pure plantation (CCK). There were thus a total of sixteen 0.5hm^2 experimental plots.

The pure plantations of *P. massoniana* and *C. lanceolata* were established in 1993 with an initial planting density of 2500 trees/hm^2 after the clear-cutting of *C. lanceolata*. The coniferous plantations were improved by planting *Q. griffithii* and *E. fordii* in 2008. The detailed management processes for the plantations are described in Table 1 and in our previous work[26]. Presently, the improved plantations are uneven-aged mixed stands with multilayer structures.

In 2016, eight years after the NNFM, we made field survey and took plant and soil samples to determine the C stock of the four forest ecosystems. The average diameter at breast height (DBH) and average tree height of *Q. griffithii* were 14.7cm and 15.4m, respectively, and the average DBH and average tree height of *E. fordii* were 5.2cm and 6.3m, respectively.

Table 1 Basic information and management history of the four plantations

Year	Management	Plantation type			
		PCK*	PCN*	CCK*	CCN*
1993	Afforestation	2500 trees/hm^2	2500 trees/hm^2	2500 trees/hm^2	2500 trees/hm^2
1993—1995	Tending for new plantations	6 times	6 times	6 times	6 times
2000	Released thinning	1600 trees/hm^2	1600 trees/hm^2	1600 trees/hm^2	1600 trees/hm^2
2004	Increment felling	1200 trees/hm^2	1200 trees/hm^2	1200 trees/hm^2	1200 trees/hm^2

(续)

Year	Management	Plantation type			
		PCK*	PCN*	CCK*	CCN*
2007	Intensity thinning	No 1200 trees/hm²	Yes 600 trees/hm²	No 1200 trees/hm²	Yes 600 trees/hm²
2008	Complementary planting	No	Planting *Q. griffithii* and *E. fordii* with 300trees/hm² respectively	No	Planting *Q. griffithii* and *E. fordii* with 300trees/hm² respectively
2009	Tending	No	2 times	No	2 times
2016	Average DBH*	22.2 ± 1.3cm for *P. massoniana*	32.2 ± 1.6cm for *P. massoniana*	17.1 ± 2.1cm for *C. lanceolata*	22.3 ± 0.8cm for *C. lanceolata*
2016	Average height	16.7 ± 0.5m for *P. massoniana*	17.3 ± 0.7m for *P. massoniana*	17.1 ± 0.4m for *C. lanceolata*	17.2 ± 0.4m for *C. lanceolata*

Notes: * PCK, PCN, CCK, and CCN represent the pure and near natural managed *P. massoniana* plantation, and the pure and near natural managed *C. lanceolata* plantation, respectively. * DBH represents diameter at breast height.

2.3 Sampling, Measurement and Statistical Analysis

2.3.1 Determination of Tree Biomass

In each year from 2007 to 2016, the C content and stock of each component in the forest ecosystem were measured. One 30m×30m subplot was established in each of the 16 plots. All trees in the subplots were inventoried. The biomass of *P. massoniana*, *C. lanceolata* and *E. fordii* were calculated using existing biomass equations in the research area[14,27,28]. The biomass of *Q. griffithii*was calculated using a newly developed equation (Table 2). The DBH distribution diagram was drawn after all trees were tallied, and fresh weight of each organ (i.e., stem, bark, branch, leaf and root) was measured by selecting 9 sample trees in each 2cm interval in the DBH range.

After weighingall the fresh samples, approximately 200g subsamples were taken from each organ and dried to constant weight at 65℃ to calculate dry mass as follows:

where W_0, W_1, and W_2 is the fresh weight of the sample, the dry mass of the subsample, and the dry mass of the sample, respectively.

Table 2 Biomass allometric equations of *Q. griffithii*

Organ	Regression equation	Number of sampled trees	R^2	*F* value	*P* value
Stem	$W=0.027(D^2H)-0.125$*	9	0.981	379.405	<0.001
Branch	$W=0.013(D^2H)-0.354$*	9	0.911	72.487	<0.001
Leaf	$W=0.004(D^2H)+0.169$*	9	0.979	332.336	<0.001
Root	$W=0.009(D^2H)-0.357$*	9	0.863	44.145	<0.001
Whole tree	$W=0.054(D^2H)-0.666$*	9	0.969	225.052	<0.001

Notes: * W, D and H represent dry mass, DBH and plant height, respectively.

2.3.2 Measurement of Understory Vegetation Biomass and Litter Quantity

Above-and below-ground fresh weight of shrubs and herbs was determined using destructive sampling techniques (i.e., total harvesting, including roots). The sampling was conducted in five randomly selected 2m×2m subplots within each plot. To measure the un-decomposed and semi-decomposed biomass of the litter, the branches, leaves, flowers, and fruits of all plants were sampled from five 1m × 1m subplots in each plot. Approximately 200g of each sample was dried to constant weight at 65℃ to calculate dry mass using equation (1).

2.3.3 Soil Sampling

Five soil core samples were collected from each plot at depths of 0~20cm, 20~40cm, 40~60cm, 60~80cm, and 80 ~ 100cm, and then combined according to soil depth. After carefully removing the fine roots, stones and organic materials, each sample was then air dried to determine the C content. Soil bulk density was measured using the cutting ring method[29].

2.3.4 Determination of C Content and Stock

The C content and stock were measured for all the plant and soil samples. The C contents were analyzed using the potassium dichromate oxidation method, with 0.8mol/L $K_2Cr_2O_7-H_2SO_4$ solution[29]. The vegetation and soil C stock was calculated as follows:

$$Sp = Wp \times Cp$$

$$Ss = \sum_{i=1}^{n} Ti \times Bi \times Ci$$

$$Se = Sp + Ss$$

where Sp, Ss and Se is the vegetation, soil and ecosystem C stock, respectively. Wp is the plant dry mass per hectare. Cp is the plant C content. Ti, Bi, and Ci is the thickness, bulk density, and C content of the i-th soil layer, respectively, and n is the number of soil layer.

Thevegetation C stock includes the arborous and ground layers. The arborous layer includes the main storey (i.e., *P. massoniana* and *C. lanceolata* trees that were retained after thinning) and underwood layer (i.e., the underplanted *Q. griffithii* and *E. fordii* and natural regenerated seedlings), while the ground layer includes shrubs, herbs, and litter (including branches, leaves, flowers, and fruit of all the plants in the plot).

2.3.5 Statistical Analysis

One-way ANOVA followed by a Duncan test (95% confidence level) was performed to analyze the effects of NNFM on the C content and stock in the forest ecosystem. The heterogeneity of variance was tested, and the original data were normalized by log-transformation or standardization prior to analysis when necessary. The ANOVA model was expressed as:

$$V_{ijkl} = \mu + B_i + S_j \times T_k + \varepsilon_{ijkl}$$

where V_{ijk} represents the lth variation (i.e., the C content and stock of different ecosystem components) under ith block (B), jth plant species (S, *P. massoniana* and *C. lanceolata*) and kth treatment (T, control and NNFM), μ is the mean of each corresponding variation, and ε_{ijkl} is the unobserved error component.

Multiple stepwise linear regression analysis was used to determine the contributions of C stock of each ecosystem component (i.e., main storey, underwood, shrub, herb, litter layer, and soil layers of 0~20cm, 20~40cm, 40~60cm, 60~80cm, and 80~100cm) to the variations in ecosystem C stock. All the analyses were performed using R (version 3.5.3).

3 RESULTS

3.1 C stock of Each Component in the Forest Ecosystem

After 8 years of NNFM, no significant difference was detected in the C content of the organs of *P. massoniana* and *C. lanceolata* between the near natural and unimproved forests ($P > 0.05$, Table 3).

Compared with the control, NNFM significantly increased the C content of the aboveground of the shrub layer in the *C. lanceolata* plantations by 17.5% (Table 4). However, the NNFM signific-antly reduced the C content of the un-decomposed components of the litter layer in the two plantations and the semi-decomposed litter in the *P. massoniana* plantation.

Table 3 C content of different organs of *P. massoniana* and *C. lanceolata* (mean ± SE, n = 4, g/kg)

Organ	PCK*	PCN*	CCK*	CCN*
Stem	476.6 ± 16.0a	481.2 ± 30.2a	486.4 ± 19.4a	488.2 ± 14.3a
Bark	475.3 ± 13.8a	488.1 ± 6.9a	459.3 ± 12.7b	464.1 ± 12.4b
Branch	465.4 ± 18.2a	470.2 ± 11.9a	460.5 ± 18.5a	456.0 ± 11.6a
Leaf	491.7 ± 13.1b	479.6 ± 10.7b	513.3 ± 15.7a	513.0 ± 17.9a
Root	425.6 ± 14.3b	426.3 ± 12.8b	442.8 ± 10.2a	448.2 ± 14.2a

Notes: * PCK, PCN, CCK, and CCN represent the pure and near natural managed *P. massoniana* plantation, and the pure and near natural managed *C. lanceolata* plantation, respectively. Values with different letters indicate significant plantation effects at $P < 0.05$. Data collected in 2016 were shown.

Table 4 C content of the different components in the underground layer of the four plantation ecosystems (mean ± SE, n = 4, g/kg)

Layer	Component	PCK*	PCN*	CCK*	CCN*
Shrub layer	Above-ground	435. 3 ± 43. 5a	414. 2 ± 19. 2a	365. 3 ± 34. 2b	429. 4 ± 24. 6a
	Below-ground	442. 0 ± 29. 7a	426. 8 ± 34. 1a	407. 0 ± 26. 7a	438. 3 ± 36. 1a
Herb layer	Above-ground	420. 7 ± 21. 9a	400. 7 ± 11. 5a	419. 3 ± 19. 4a	400. 3 ± 13. 7a
	Below-ground	343. 1 ± 31. 8a	345. 0 ± 31. 1a	342. 5 ± 23. 7a	341. 0 ± 12. 3a
Litter layer	Un-decomposed	496. 4 ± 16. 6a	435. 4 ± 37. 8b	491. 6 ± 14. 1a	428. 6 ± 33. 2b
	Semi-decomposed	434. 6 ± 26. 1a	413. 9 ± 38. 2b	416. 4 ± 21. 8b	394. 9 ± 18. 7b

Notes: * PCK, PCN, CCK, and CCN represent the pure and near natural managed *P. massoniana* plantation, and the pure and near natural managed *C. lanceolata* plantation, respectively. Values with different letters indicate significant plantation effects at P< 0. 05. Data collected in 2016 were shown.

The soil C content declined significantly with soil depth. Though not significantly affecting the C stock of deep soil, NNFM significantly increased soil C content at 0~20cm and 20~40cm in the *P. massoniana* and *C. lanceolata* plantations (Figure 1).

3. 2 Ecosystem C Stock and Its Allocation

By 2016, in the *P. massoniana* plantation, NNFM had significantly increased the C stocks of the arborous layer and its components, herb layer, 0 ~ 20cm, 20~40cm, 40~60cm, and 0~100cm soil layers, and the ecosystem C stock (Table 5). In contrast, in the *C. lanceolata* plantation, NNFM significantly increased the C stocks of the underwood and arborous layer, 0~20cm, 20~40cm, and 0~100cm soil layer, and the ecosystem C stock, but reduced that of the shrub and herb layers.

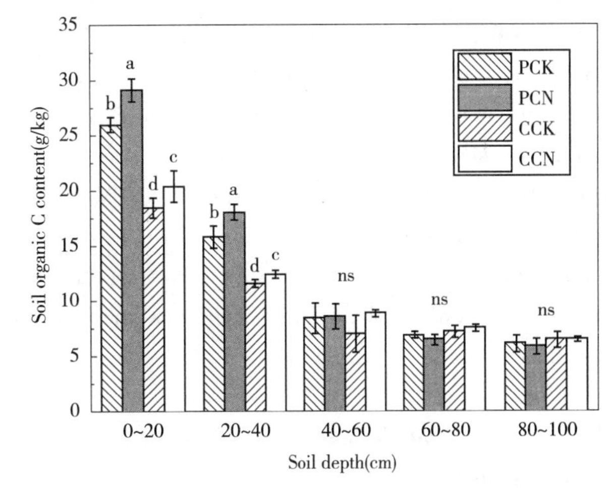

Figure 1. Soil C content at different depth in the four plantations (mean±SE, n = 4). PCK, PCN, CCK, and CCN represent the pure and near natural managed *P. massoniana* plantation, and the pure and near natural managed *C. lanceolata* plantation, respectively. Values with different letters indicate significant plantation effects at P< 0. 05. Data collected in 2016 were shown. ns, P> 0. 05.

Table 5 C stock of the different components in the four plantation ecosystems (mean ± SE, n = 4, t/hm^2)

Layer	Component	PCK*	PCN*	CCK*	CCN*
Arborous layer	Main storey	120. 44 ± 9. 71b	151. 62 ± 11. 4a	48. 15 ± 12. 03c	39. 33 ± 4. 37c
	Underwood	0. 23 ± 0. 01d	26. 06 ± 1. 41b	0. 32 ± 0. 02c	29. 68 ± 1. 60a
	Sum	120. 67 ± 10. 91b	177. 68 ± 12. 35a	48. 47 ± 13. 17d	69. 01 ± 6. 12c
Ground layer	Shrub	0. 16 ± 0. 03ab	0. 14 ± 0. 02ab	0. 22 ± 0. 04a	0. 10 ± 0. 05b
	Herb	0. 15 ± 0. 02 b	0. 08 ± 0. 02c	0. 24 ± 0. 03a	0. 10 ± 0. 05bc
	Litter	2. 02 ± 0. 14a	1. 81 ± 0. 13ab	1. 75 ± 0. 09b	1. 78 ± 0. 22b
	Sum	2. 33 ± 0. 23a	2. 04 ± 0. 11a	2. 21 ± 0. 21a	1. 98 ± 0. 17a
Soil layer	0~20cm	55. 40 ± 3. 36bc	66. 80 ± 4. 07a	51. 09 ± 3. 05c	63. 18 ± 3. 72 ab
	20~40cm	36. 80 ± 2. 75b	45. 86 ± 3. 33a	35. 94 ± 2. 49b	46. 80 ± 3. 04a
	40~60cm	22. 25 ± 1. 70b	26. 28 ± 2. 06a	20. 98 ± 1. 54ab	24. 57 ± 1. 88a
	60~80cm	19. 97 ± 1. 56a	24. 03 ± 1. 89a	22. 11 ± 1. 41a	20. 50 ± 1. 72a

(续)

Layer	Component	PCK*	PCN*	CCK*	CCN*
	80~100cm	15.50 ± 1.29a	17.06 ± 1.56a	15.33 ± 1.17a	17.81 ± 1.42a
	Sum	149.92 ± 5.52b	180.03 ± 6.69a	145.45 ± 5.00b	172.86 ± 6.10a
Ecosystem	Total	272.93 ± 13.63c	359.75 ± 15.74a	196.14 ± 14.94d	243.84 ± 0.12b

Notes: * PCK, PCN, CCK, and CCN represent the pure and near natural managed *P. massoniana* plantation, and the pure and near natural managed *C. lanceolata* plantation, respectively. Values with different letters indicate significant plantation effects at $P< 0.05$. Data collected in 2016 were shown.

From 2008 to 2016, the C stocks of the four forest ecosystems all continuously increased (Figure 2). The annual rate of increase in the near natural *P. massoniana* and *C. lanceolata* plantation (22.64 and 14.17t/hm^2 · a, respectively) was significantly higher than that of the controls (8.54 and 4.62t/hm^2 · a, respectively). The total C stock of each near natural forest began to overtake that of the unimproved forests from 2011. The NNFM exhibited an increasingly positive impact on the ecosystem C stock over time. In 2016, after 8 years of NNFM, the C stock of the transformed *P. massoniana* and *C. lanceolata* forests was 359.75 t/hm^2 and 243.84 t/hm^2 respectively, which was 31.8% and 24.3% higher than their corresponding controls.

3.3 Relationship between Ecosystem C Stock and Its Components

Overall, theecosystem C stock was significantly and positively affected by the C stocks of the main storey and underwood layer, and the 0~20cm soil layer (R^2 = 0.994, Table 6). Only the C stock in PCK had a significant positive correlation with that of underwood layer (R^2 = 0.965), while the C stock in PCN was positively correlated with that of main storey and 0~20cm soil layer (R^2 = 0.998). The C stock in CCK and CCN was positively correlated with the C stock of the main storey (R^2 = 0.911) and 0~20cm soil layer (R^2 = 0.963), respectively.

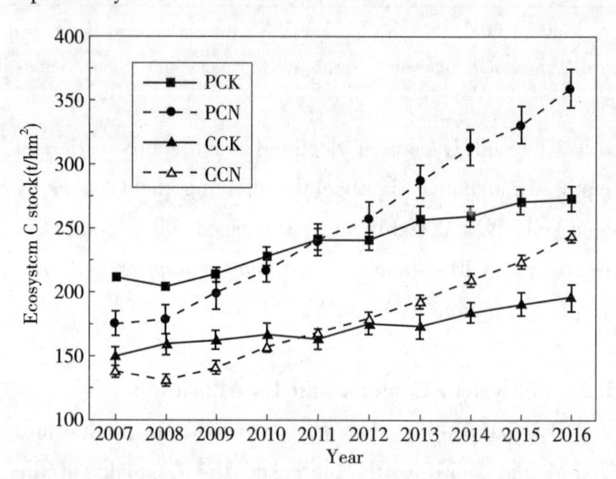

Figure 2 Dynamics of ecosystem C stock in the four plantations (mean ± SE, n = 4). PCK, PCN, CCK, and CCN represent the pure and near natural managed *P. massoniana* plantation, and the pure and near natural managed *C. lanceolata* plantation, respectively. The arborous and soil layer stored 12.2%~49.4% and 50.0%~86.6% of the whole C stock in the ecosystem respectively, with a sum of 98.2%~99.4%. Meanwhile, the C stock of the arborous layer accounted for 89.1%~98.9% of the vegetation layer (Figure 3). From 2008 to 2011, the arborous layer in each near natural plantation stored less C than the control. However, 2015 and 2016 saw an increase of C stock in the arborous and vegetation layers in the near natural forests compared to the controls (Figure 3a, c).

Table 6 Models of regressions between ecosystem C stock and its components in the four plantations

Plantation	Equation	R^2	F value	P value
PCK*	$Y=302.754x2+205.341$*	0.965	250.677	0.000
PCN*	$Y=1.402x1+1.106x3+72.259$*	0.998	2617.328	0.000
CCK*	$Y=1.588x1+114.941$*	0.911	92.199	0.000
CCN*	$Y=3.468x3+22.321$*	0.963	233.080	0.000
Total	$Y=1.006x1+1.354x2+1.623x3+64.72$*	0.994	2224.522	0.000

Notes: * PCK, PCN, CCK, and CCN represent the pure and near natural managed *P. massoniana* plantation, and the pure and near natural managed *C. lanceolata* plantation, respectively. * x1, x2, x3, and Y represent the C stock of main storey layer, underwood layer, 0~20cm soil layer, and ecosystem, respectively.

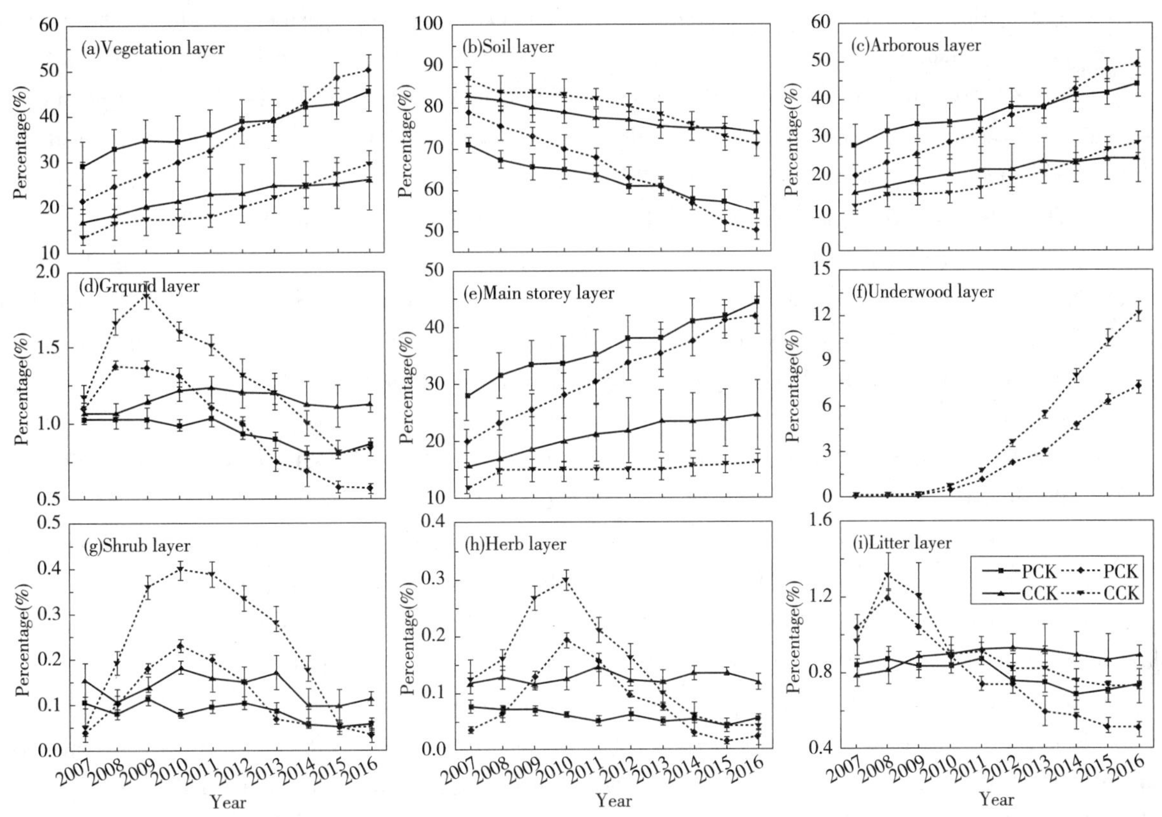

Figure 3 Dynamics of C stock percentage of (a) vegetation layer; (b) soil layer; (c) arborous layer; (d) ground layer; (e) main storey layer; (f) underwood layer; (g) shrub layer; (h) herb layer; and (i) litter layer in the four plantations (mean ± SE, *n* = 4). PCK, PCN, CCK, and CCN represent the pure and near natural managed *P. massoniana* plantation, and the pure and near natural managed *C. lanceolata* plantation, respectively. Main storey layer was dominated by *P. massoniana* and *C. lanceolata*, whereas underwood layer was dominated by *E. fordii* and *Q. griffithii* underplanted in 2008 and the natural regenerated seedlings.

4 DISCUSSION

4.1 Effects of NNFM on Vegetation C Stock

Theforest ecosystem C stock included the arborous and ground layer C stocks, with the former accounting for 95.64%~98.87% (Figure 3). Therefore, the C sequestration capacity of vegetation largely depends on the arborous layer, which is similar to previous studies [30]. Any changes in the growth and C content of plants may alter the vegetation C stock in a forest ecosystem. Because NNFM did not affect the C content of the *P. massoniana* and *C. lanceolata* plant components (Table 3), the differences in the vegetation C stock between the near natural and controlled plantations came from the positive effects of NNFM on the biomass of dominant tree species in the arborous layer at a stand-scale. However, the stand-scale increase was highly related to the increase in the growth rate of the retained and underplanted trees under NNFM.

In the near natural managed forest, the intensethinning reduced the original stand density, and greatly improved the growth of the retained trees through release of growing space. A previous study also showed improved stem growth by thinning in a spruce forest [31]. In our study, after thinning, the increase in the annual C stock of the retained trees was 13.8 and 2.63t/hm^2 a in the *P. massoniana* and *C. lanceolata* plantations, respectively; whereas it was only 7.0 and 2.52t/hm^2 a in their corresponding controls (data not shown). This is consistent with a previous study showing that thinning increased the C stock of the arborous layer [14]. In addition, during NNFM, underplanted species usually have a high growth rate and consequently cause an increase in the rate of C stock accumulation [12]. This might also cause the rapid increase in the arborous C stock in the near natural forests. In our study, native fast-growing species *E. fordii* and *Q. griffithii* were planted during NNFM. The increase in the annual C stock which they caused was 3.3 and 3.7t/hm^2 a in the *P. massoniana* and *C. lanceolata* plantations, respectively (data not shown). Because the increase in the rate of underwood C stock accumulation was only 0.02 and 0.04 t/hm^2 a in the two control forests, the vegetation C stock that was nearly the sum of the original and underplanted tree C stock was increased by NNFM (Table 5). These results suggest that tree species allocation and vegetation structure optimization are important for enhancing the vegetation C stock. Increasing biomass through facilitating plant growth and planting trees with high C density is an effective means to achieve this aim.

4.2 Effects of NNFM on Soil C Stock

Soilhas the largest C pool in the forest ecosystem, accounting for 50.0%~86.6% in our present study (Figure 3). The soil C stock is affected by the soil C content, bulk density and soil thickness. In our study, NNFM did not affect soil bulk density (data not shown), but increased the soil C content in the 0~20cm and 20~40cm layers in the *P. massoniana* and *C. lanceolata* plantation, respectively (Figure 1). Consequently, the soil C stocks at 0~20cm and 20~40cm in the near natural forests were significantly greater than that in the unimproved stands (Table 5). This indicated that the NNFM of *P. massoniana* and *C. lanceolata* plantations could enhance the C sequestration potential of the top soil. However, the soil C contents at 0~20cm, 20~40cm and 40~60 cm in a *P. massoniana* plantation were reduced after a 3-year NNFM by *C. hystrix* and *M. hedyosperma*, whereas they increased slightly in a *C. lanceolata* plantation [24]. These differing results suggest uncertainties with respect to how NNFM affects soil C content. It is very likely due to the differences in management approach, time period, and vegetation composition.

Numerous studies have confirmed that changing vegetation structure and litter composition can alter soil C content[2,15]. In our study, neither of the C content of the component in the vegetation layer (i. e., main storey, underwood, shrub, herb, and litter) was positively correlated with the soil C content. Therefore, other factors affecting soil C content, including the trait of litter and root [32], structure and activity of soil microbial community[33], and other soil physical and chemical properties[34], may lead to the increase in soil C content induced by NNFM. One study has indicated that NNFM can accelerate the decomposition rate of plant litter, and therefore altering the accumulation of soil C [35]. The modification of tree species structure can change the composition and quality of roots and litter, and alter the soil microbial community, which accelerates litter decomposi-tion and increases the soil C content [36]. Broad-leaved species underplanting also improved the litter quality and its decomposition rate in *P. massoniana* plantations[4]. Further studies are therefore required on exploring the drivers of the higher C content in topsoil under NNFM, and the relations between the dynamics of the vegetation community structure and soil C.

4.3 Long-Term Effects of NNFM on Ecosystem C Stock

The arborous and soil layer had the largest C stocks inthe forest [4], contributing over 98% to the ecosystem C stock (Figure 3). As a result, they controlled the ecosystem C stock and its dynamics in the four plantations. Multiple regression analysis also indicated that the ecosystem C stock was influenced by the C stocks of the main storey, underwood, and 0~20cm soil layer (Table 6). Due to the decline in soil C content with soil depth (Figure 1), and little change in the soil bulk density, the soil C stock was concentrated in the topsoil. In 2016, the 8-year NNFM had increased the C stock in the main storey, underwood and 0~20cm soil layer in the *P. massoniana* plantation, and the C stock in the main storey and 0~20cm soil layer in the *C. lanceolata* plantation. Thus, NNFM significantly increased the total C stock of each of the two coniferous plantations (Table 5). However, there was little deadwood in our forests, and we did not measure its C stock. Including the C stock of the deadwood would slightly increase the total ecosystem C stock.

Forest age is another key factor affecting the C stock and its allocation in plantations [30,37]. In China, either of the forest biomass or soil C storage is increased exponentially over the stand age [38]. Similarly, the NNFM exhibited long-term dynamic impacts on forest ecosystem C stock and its allocation (Figures 2 and 3). During the study period, the proportion of vegetation C stock to that of the ecosystem showed an increasing trend, while the proportion of soil C stock was downward (Figure 3). This is because the vegetation C stock is less stable than that of soil, and increases with plant growth. At the initial stage of NNFM, the vegetation C stock was lower than the control stands due to the thinning treatment. With the extension of time, the C stock loss from thinning was replaced by the rapid growth of retained and underplanted trees. This led to a significantly greater rate of increase in arborous and vegetation layer C stocks in the near natural forests than in their controls. Finally, it resulted in an increasingly positive correlation between NNFM and C stock in the ecosystem over time (Figure 2). Meanwhile, combined with our previous findings [26], it can be inferred that increasing the stability of soil C will further boost this positive correlation. These results indicate that NNFM is a promising way to enhance long-term C sequestration in forest ecosystems.

5 Conclusions

The 8-year period of near natural forest management increased the C stock and sequestration potential of the *P. massoniana* and *C. lanceolata* plantations. This can be attributed to the enhanced C stock in the arborous and 0~20cm soil layer. The improvement in species diversity and stand density increased the individual and stand-scale growth rate, thereby increasing the vegetation C stock. The litter decomposition likely changed to increase the topsoil C stock. Our study indicates that NNFM plays a positive role in enhancing the forest C sink function. Increasing soil C stability and plant biomass through facilitating tree growth is the main way to increase the total C stock in near natural managed plantations.

Author Contributions: A. M. collected data and drafted the manuscript. Y. Y. revised the manuscript and participated in analyzing the experiment data. S. L. conceived and designed the work. Y. N., H. L., Y. T., D. S., L. L., J. Z., and N. A. participated in collecting the experiment data. All the authors contributed to carrying out additional analyses and finalizing this paper.

Funding: This study was supported by the fundamental research funds of CAF (CAFYBB2019MA003), 13th Five-Year National Key Technology R&D Program (2017YFD0600304), and Guangxi forestry science and technology projects (Document of Guangxi forestry department [2016] No. 37).

Acknowledgments: We gratefully acknowledge the help of Hui Wang from Institute of Forest Ecology, Environment and Protection, Chinese Academy of Forestry.

Conflicts of Interest: The authors declare no conflict of interest.

REFERENCES

[1] IPCC. Climate change 2014: Synthesis report. Contribution of working groups I, II, and III to the fifth assessment report of the intergovernmental panel on climate change[J]. Geneva, Switzerland, 2014.

[2] NOORMETS A, EPRON D, DOMEC J C, et al. Effects of forest management on productivity and carbon sequestration: A review and hypothesis[J]. Forest Ecol. Manag., 2015, 355: 124-140.

[3] LIU S, WU S, WANG H. Managing planted forests for multiple uses under a changing environment in China[J]. NZ J. Forestry Sci., 2014, 44: S3.

[4] HE Y, QIN L, LI Z, et al. Carbon storage capacity of monoculture and mixed-species plantations in subtropical China[J]. Forest Ecol. Manag., 2013, 295: 193-198.

[5] WANG H, LIU S, MO J, et al. Soil organic carbon stock and chemical composition in four plantations of indigenous tree species in subtropical China[J]. Ecol. Res., 2010, 25: 1071-1079.

[6] LIU L, DUAN Z, XU M, et al. Effect of monospecific and mixed *Cunninghamia lanceolata* plantations on microbial community and two functional genes involved in nitrogen cycling[J]. Plant Soil, 2010, 327: 413-428.

[7] JACOBSON S. Addition of stabilized wood ashes to Swedish coniferous stands on mineral soils - effects on stem growth and needle nutrient concentrations[J]. Silva Fenn., 2003, 37: 437-450.

[8] TIAN D, XIANG W, CHEN X, et al. A long-term evaluation of biomass production in first and second rotations of Chinese fir plantations at the same site[J]. Forestry, 2011, 84: 411-418.

[9] ZHANG W, WANG S. Effects of NH_4^+ and NO_3 on litter and soil organic carbon decomposition in a Chinese fir plantation forest in South China[J]. Soil Biol. Biochem., 2012, 47: 116-122.

[9] WANG H, LIU S, WANG J, et al. Mixed-species plantation with *Pinus massoniana* and *Castanopsis hystrix* accelerates C loss in recalcitrant coniferous litter but slows C loss in labile broadleaf litter in southern China[J]. Forest Ecol. Manag., 2018, 422: 207-213.

[10] EMBORG J, CHRISTENSEN M, HEILMANNCLAUSEN J. The structural dynamics of Suserup Skov, a near-natural temperate deciduous forest in Denmark[J]. Forest Ecol. Manag., 2000, 126: 173-189.

[11] BRUNET J, FRITZ Ö, RICHNAU G. Biodiversity in European beech forests-a review with recommendations for sustainable forest management[J]. Ecological Bulletins, 2010: 77-94.

[12] GALIK C S, JACKSON R B. Risks to forest carbon offset projects in a changing climate. Forest Ecol[J]. Manag. 2009, 257: 2209-2216.

[13] MING A, ZHANG Z, CHEN H, et al. Effects of thinning on the biomass and carbon storage in *Pinus massoniana* plantation[J]. Scientia Silvae Sinicae, 2013, 49: 1-6.

[14] CLARKE N, GUNDERSEN P, JÖNSSON-BELYAZID U, et al. Influence of different tree-harvesting intensities on forest soil carbon stocks in boreal and northern temperate forest ecosystems[J]. Forest Ecol. Manag., 2015, 351: 9-19.

[15] ZHANG X, GUAN D, LI W, et al. The effects of forest thinning on soil carbon stocks and dynamics: A meta-analysis[J]. Forest Ecol. Manag., 2018, 429: 36-43.

[16] NILSEN P, STRAND L T. Thinning intensity effects on carbon and nitrogen stores and fluxes in a Norway spruce [*Picea abies* (L.) Karst.] stand after 33 years[J]. Forest Ecol. Manag., 2008, 256: 201-208.

[17] JURGENSEN M, TARPEY R, PICKENS J, et al. Long-term effect of silvicultural thinnings on soil carbon and nitrogen pools[J]. Soil Sci. Soc. Am. J. , 2012, 76: 1418.

[18] MING A, JIA H, ZHAO J, et al. Above- and below-ground carbon stocks in an indigenous tree (*Mytilaria laosensis*) plantation chronosequence in subtropical China [J]. Plos One, 2014, 9: e109730.

[19] WANG H, LIU S, WANG J, et al. Effects of tree species mixture on soil organic carbon stocks and greenhouse gas fluxes in subtropical plantations in China[J]. Forest Ecol. Manag., 2013, 300: 4-13.

[20] BERGER T W, NEUBAUER C, GLATZEL G. Factors controlling soil carbon and nitrogen stores in pure stands of Norway spruce (*Picea abies*) and mixed species stands in Austria[J]. Forest Ecol. Manag., 2002, 159: 3-14.

[21] MAO R, ZENG D, AI G, et al. Soil microbiological and chemical effects of a nitrogen-fixing shrub in poplar plantations in semi-arid region of Northeast China[J]. Eur. J. Soil Biol. 2010, 46: 325-329.

[22] DA L, ZUOMIN S, WEIXIA W, et al. Carbon and nitrogen storage in monoculture and mixed young plantation stands of *Erythrophleum fordii* and *Pinus massoniana* in subtropical China [J]. 缺期刊信息, 2015, 35: 6051-6059.

[23] HE Y, LIANG X, QIN L, et al. Community characteristics and soil properties of coniferous plantation forest monocultures in the early stages after close-to-nature transformation management in southern subtropical China[J]. Acta Ecologica Sinica, 2013, 33: 2484-2495.

[24] BERGER T W, INSELSBACHER E, MUTSCH F, et al. Nutrient cycling and soil leaching in eighteen pure and mixed stands of beech (*Fagus sylvatica*) and spruce (*Picea abies*) [J]. Forest Ecol. Manag., 2009, 258: 2578-2592.

[25] MING A, YANG Y, LIU S, et al. Effects of near natural forest management on soil greenhouse gas flux in *Pinus massoniana* (Lamb.) and *Cunninghamia lanceolata* (Lamb.) Hook. plantations[J]. Forests, 2018, 9: 229.

[26] KANG B, LIU S, CAI D, et al. Characteristics of biomass, carbon accumulation and its spatial distribution in *Cunninghamia lanceolata* forest ecosystem in low subtropical area [J]. Scientia Silvae Sinicae, 2009, 45: 147-153.

[27] MING A, LIU S, NONG Y, et al. Comparison of carbon storage in juvenile monoculture and mixed plantation stands of three common broadleaved tree species in subtropical China[J]. Acta Ecologica Sinica , 2015, 35: 180-188.

[28] PANSU M, GAUTHEYROU J. Handbook of Soil Analysis. Mineralogical, Organic and Inorganic Methods [M]. Springer: Berlin, Heidelberg, New York, 2006.

[29] PEICHL M, ARAIN M A. Above-and belowground ecosystem biomass and carbon pools in an age-sequence of temperate pine plantation forests[J]. Agr. Forest Meteorol., 2006, 140: 51-63.

[30] NICOLL B, CONNOLLY T, GARDINER B. Changes in spruce growth and biomass allocation following thinning and guying treatments[J]. Forests, 2019, 10: 253.

[31] ANGST G, MESSINGER J, GREINER M, et al. Soil organic carbon stocks in topsoil and subsoil controlled by parent material, carbon input in the rhizosphere, and microbial-derived compounds[J]. Soil Biol. Biochem., 2018, 122: 19-30.

[32] TRIVEDI P, ANDERSON I C, SINGH B K. Microbial modulators of soil carbon storage: integrating genomic and metabolic knowledge for global prediction[J]. Trends Microbiol, 2013, 21, 641-651.

[33] WIESMEIER M, URBANSKI L, HOBLEY E, et al. Soil organic carbon storage as a key function of soils - A review of drivers and indicators at various scales[J]. Geoderma, 2019, 333: 149-162.

[33] JANDL R, LINDNER M, VESTERDAL L, et al. How strongly can forest management influence soil carbon sequestration? [J]. Geoderma , 2007, 137: 253-268.

[34] HUANG X, LIU S, WANG H, et al. Changes of soil microbial biomass carbon and community composition through mixing nitrogen-fixing species with *Eucalyptus urophylla* in subtropical China[J]. Soil Biol. Biochem.,

2014, 73: 42-48.

[35] CAO J, WANG X, TIAN Y, et al. Pattern of carbon allocation across three different stages of stand development of a Chinese pine (*Pinus tabulaeformis*) forest[J]. Ecol. Res., 2012, 27: 883-892.

[36] TANG X, ZHAO X, BAI Y, et al. Carbon pools in China's terrestrial ecosystems: New estimates based on an intensive field survey[J]. PNAS, 2018, 115: 4021-4026.

[原载：Forests 2019, 10, 626]

Above-and Below-Ground Carbon Stocks in an Indigenous Tree (*Mytilaria laosensis*) Plantation Chronosequence in Subtropical China

MING Angang[1,2,3], JIA Hongyan[1,3], ZHAO Jinlong[4], Tao Yi[1], LI Yuanfa[5]

([1] *Experimental Center of Tropical Forestry, Chinese Academy of Forestry, Pingxiang* 532600, *Guangxi, China*;
[2] *Institute of Forest Ecology, Environment and Protection, Chinese Academy of Forestry, Beijing* 100093, *China*;
[3] *Guangxi Youyiguan Forest Ecosystem Research Station, Pingxiang* 532600, *Guangxi, China*;
[4] *College of Forestry, Beijing Forestry University, Beijing* 10083, *China*;
[5] *College of Forestry, Guangxi University, Nanning* 530004, *Guangxi, China*)

Abstract: More than 60% of the total area of tree plantations in China is in subtropical, and over 70% of subtropical plantations consist of pure stands of coniferous species. Because of the poor ecosystem services provided by pure coniferous plantations and the ecological instability of these stands, a movement is under way to promote indigenous broadleaf plantation cultivation as a promising alternative. However, little is known about the carbon (C) stocks in indigenous broadleaf plantations and their dependence on stand age. Thus, we studied above- and below-ground biomass and C stocks in a chronosequence of Mytilaria laosensis plantations in subtropical China; stands were 7, 10, 18, 23, 29 and 33 years old. Our assessments included tree, shrub, herb and litter layers. We used plot-level inventories and destructive tree sampling to determine vegetation C stocks. We also measured soil C stocks by analyses of soil profiles to 100cm depth. C stocks in the tree layer dominated the above-ground ecosystem C pool across the chronosequence. C stocks increased with age from 7 to 29 years and plateaued thereafter due to a reduction in tree growth rates. Minor C stocks were found in the shrub and herb layers of the all six plantations and their temporal fluctuations were relatively small. C stocks in the litter and soil layers increased with stand age. Total above-ground ecosystem C also increased with stand age. Most increases in C stocks in below-ground and total ecosystems were attributable to increases in soil C content and tree biomass. Therefore, considerations of C sequestration potential in indigenous broadleaf plantations must take stand age into account.

INTODUCTION

Biomass and carbon (C) stocks in forest ecosystems play important roles in the global C cycle[1, 2, 3, 4]. Trees and soils are components of forest ecosystems that provide the largest potential for C storage[3, 5, 6, 7, 8, 9]. Increasing global C sequestration through enlargement of the proportion of forested land on the planet has been suggested as an effective measure for mitigating elevated concentrations of atmospheric carbon dioxide[10, 11, 12]. As the area of natural stands has decreased in recent decades, tree plantations have become increasingly important components of the planet's forest resources. Commercial plantations are now a central issue in sustainable forest management across the globe. Well-designed, multi-purpose plantations can reduce pressure on natural forests, restore some ecological services provided by natural forests and mitigate climate change through direct C sequestration[13].

China's large plantation programme is assuming an increasingly significant role in C sequestration from the atmosphere. The total land area under tree plantations has reached $6.2\times10^7 hm^2$ and now accounts for 31.8% of the total forested landscape in the country[14]. The largest proportion (63%) of the total plantation area in China is located in subtropical regions, which provide hot and humid conditions appropriate for tree growth[15]. Most of these subtropical plantations consist of stands containing either a single coniferous species or an exotic tree (e.g. *Pinus massoniana*, *Cunninghamia lanceolata*, *Eucalyptus*)[15]. The creation of monospecific stands of trees that are not native to a landscape carries a high risk of consequential ecological damage, such as decrease in ecosystem stability and outbreak of diseases and insect pests[16, 17, 18, 19, 20]. As a result, alternative plantations of indigenous broadleaf species are spreading in this region of China and in neighbouring countries[21, 22, 23, 24, 25, 26, 27].

Mytilaria laosensis, an indigenous broadleaf tree species, has potential in the afforestation of subtropical China. It grows rapidly and is strongly adaptable; the trunk is straight and the wood has desirable properties for the economic production of high - value timber. The species occurs naturally in western Guangdong, south-western Guangxi and south-eastern Yunnan. It is also indigenous to Vietnam and Laos. *M. laosensis* is expected to become a major afforestation species in subtropical China and beyond[28, 29, 30]. Its growth patterns, biomass production and wood properties, and the physical and chemical properties of the soils in which it grows have been reported in earlier literature[28, 31, 32, 33]. However, information on biomass and C stocks in *M. laosensis* stands is still lacking. According to previous investigations, C stock size in plantations (especially biomass C) is related not only to tree species, site conditions and soil properties[34, 35, 36], but also to stand age . According to the previous studies, the C stock of *Castanopsis hystrix* plantations and *Erythrophleum fordii* plantation in subtropical China were increased with the increase in stand age[38,39]. And in the other region, stand age can also remarkably affect the C stocks of plantations' ecosystem[10, 37]. However, Effects of diverse factors, including stand age, on C sequestration by *M. laosensis* plantation are poorly documented[33].

Here, we provide first measurements of the C stock across an age sequence of six *M. laosensis* plantation stands. Specifically, the objectives of this study were (1) to document the changes of the sizes and proportional contributions of plantation C pools as stands aged in the early decades following plantation establishment, and (2) to provide baseline information for forest biomass and C estimations focussing on indigenous broadleaf plantations in subtropical regions.

MATERIALS AND METHODS

Study site and plot establishment

Ethics Statement

This research was conducted in Experimental Center of Tropical Forestry, Chinese Academy of Forestry (ECTF for short), This study was also supported by this center. We confirmed that the location is not privately owned and the sampling of soils and plants was approved by ECTF. We also confirmed that the field studies did not involve endangered or protected species.

Study site

The study site is located in the Experimental Centre of Tropical Forestry at the Chinese Academy of Forestry location in Pingxiang City, Guangxi Zhuang Autonomous Region, China (22°02′~22°19′N, 106°43′~106°52′E). The region is within a semi-humid southern subtropical monsoon climate zone that has defined dry and wet seasons. The dry season extends from October to the following March, and the wet season from April to September. The annual mean precipitation at the site is 1200~1500mm, annual average evaporation is 1261~1388mm and the relative humidity is 80%~84%. The annual mean temperature is 21℃, with a mean monthly minimum of 12.1℃ and a mean monthly maximum of 26.3℃. The landscape is largely comprised of low mountains and hills at elevations of 350~650m. The soils at the study site are categorised as red soils by the Chinese soil classification procedure; this category is equivalent to Oxisol in the USDA Soil Taxonomy. Most developed from granite and had a sandy texture[40, 41, 42].

Plot establishment

In 2013, six adjacent plantations with different stand ages were selected based on similarities in topography, soil texture, management methodology, environmental conditions and previous vegetation composition (dominated by *C. lanceolata*). We identified a chronosequence of *M. laosensis* stands that were 7, 10, 18, 23, 29 and 33 years of age. All six stands were located within 15km of one another. They were established in 2006, 2003, 1995, 1990, 1984 and 1980, after the clear-cutting of previous *C. lanceolata* vegetation. Stand characteristics are summarised in Table 1.

During the summer of 2013, four sampling plots (each 30m×20m) were established at random locations in each of the six stands. In each of these plots, we measured the diameter at breast height (DBH, diameter at breast height) of individual trees using a diameter tape, and measured heights using a Hag-löf-VERTEX IV clinometer. Five subplots containing shrubs (each 2m×2m) and five subplots of herbaceous vegetation and litter (each 1m×1m) were established at random locations within each sampling plot (20 subplots in total in the stands of the same age). Plant species, numbers, heights and coverages were recorded; litter was collected from litter subplots (each 1m×1m). Environmental factors including altitude, slope, aspect and slope position were also recorded.

Table 1 Site properties and vegetation characteristics of the six plantation stands studied (values are means ± SE; $n = 4$)

	7 yr old	10 yr old	18 yr old	23 yr old	29 yr old	33 yr old
Altitude (m)	360~520	450~530	450~540	550~650	450~550	350~500
Slope aspect	North-western	Northern	North-western	Northern	Northern	North-western
Slope gradient (°)	36.1 ± 2.4	37.7 ± 3.1	34.2 ± 1.8	32.2 ± 2.2	32.8 ± 2.6	28.9 ± 1.7
DBH (cm)	13.6± 0.7	16.2± 1.2	20.2± 1.2	23.4± 1.7	26.0± 2.3	27.3± 2.7
Tree height (m)	13.6± 1.4	14.9± 1.1	16.1± 1.7	17.8± 2.7	21.9± 3.1	20.5± 1.9
Stem density (trees/hm^2)	1500± 21	1224± 18	911± 12	721± 11	675± 8	671± 10
Main understorey species	*M · laosensis*, *Cunninghamia lanceolata*	*M. laosensis*, *Thysanolaena maxima*	*M. laosensi*, *T. maxima*	*M. laosensis*, *C. lanceolata*	*M. laosensis*, *C. lanceolata*	*M. laosensis*, *T. maxima*

MEASUREMENTS

Tree biomass

On the basis of the DBH and height measurements in the sampling plots, we selected and harvested six sample trees from different diameter classes in each of the six stands for biomass measurements (36 trees in total). The above-ground portions of the trees were divided into 2cm sections for measurement. We measured the fresh weights of stems, bark, branches and leaves. The below-ground portions of the sample trees were dug out and examined using the open cut method. We measured the fresh weights of the stump roots, thick roots (diameter >2.0cm), medium-thick roots (diameter 0.5~2.0cm) and small roots (diameter<0.5cm). Organ samples were collected (200g of each organ) and oven-dried at a temperature of 65℃ to constant weight to calculate the moisture contents and dry weights. We built regression models for the different organs to estimate tree biomass (using data from the 18 sample trees).

Understorey vegetation and litter biomass

We used a destructive harvesting method to measure the biomass in the above-ground and below-ground portions of the shrub and herbaceous layers[43]. The fresh masses of these two portions were obtained directly by weighing. After oven-drying to constant weight at 65℃, we weighed subsamples and calculated the respective dry weights from determinations of moisture contents. The components of understorey vegetation were then separated and measured. We weighed litter material that had not decomposed or was semi-decomposed at the same time. The litter samples were oven dried at 65℃ and weighed.

C content

Samples of the above-ground and below-ground components of the sample trees (*M. laosensis*), shrubs, herbaceous plants and litter were dried, ground and sieved in the laboratory. These samples were then bottled for later chemical analysis. In total, 288 soil-sampling points were selected within the 24 sampling plots in each of the six stands. Soil pits were dug to a depth of 100cm and samples were collected randomly from four depth horizons: 0~10cm, 10~30cm, 30~50cm and 50~100cm. Soil samples from the same depth horizon in the same stand were mixed in equal proportions and the mixtures were air-dried at room temperature (25℃). The samples were then ground and passed through a 2mm-mesh sieve to remove coarse living roots and gravel; they were then ground in a mill in preparation for sieving through a 0.25mm mesh before chemical analysis. A soil-sample cutting ring ($100cm^3$) was used to collect samples of undisturbed soil from different horizons. These samples were taken to the laboratory for measurements of soil bulk density using the cutting ring method. We measured the C content of the tree component samples, understorey vegetation, litter by vario Macro Elemental Analyzer (Elementar Analysensysteme GmbH, Germany), but the soil organic carbon was established by the oil-bath $K_2Cr_2O_7$ titration method.

C storage

The C stocks (C in biomass per unit area of land surface) in the vegetation and litter biomass were determined by multiplying C content by biomass (dry mass per unit area of land surface). The C stocks per unit area of land in each of the soil horizons were calculated by multiplying soil bulk density at a chosen soil depth by the C content at that depth. Total soil C stocks were computed by summing the stocks in each soil horizon.

Statistical analysis

We used one-way ANOVA to test for differences in the C content and C stock among plantations of different ages. The dependent parameters were normally distributed and homoscedastic. All analyses were performed using the following software: Microsoft Excel 2007 and SPSS (ver. 13.0; SPSS, Chicago, IL) for Windows. Statistical significance was detected at $P < 0.05$.

RESULTS

C content in plantation stands

C content in the vegetation and litter layer

The C contents of the component organs differed significantly among the six stands ($P < 0.05$) and fell into the following rank order: leaf > stem > coarse root > medial root > bark > small root > branch > stump root > fruit. C content was not significantly different between medium roots and bark ($P < 0.05$). The C contents of above-ground components of the shrub and herb layers in all six stands were higher than those of their below-ground components. In the litter layer, the C content of the undecomposed portion was higher than that of the semi-decomposed portion (Table 2).

The average C content did not differ significantly within plantation components among stands of different ages ($P > 0.05$). No obvious pattern relationships were detected between C contents and increasing stand age.

Table 2 Carbon contents in the vegetation components and litter layers of six differently aged plantation stands (values are means ± SE; n = 4g/kg)

Layer	Components	7 yr old	10 yr old	18 yr old	23 yr old	29 yr old	33 yr old	Mean
Tree layer	Stem	527.8 ± 14.1^{Eb}	536.9 ± 12.7^{Cb}	545.6 ± 16.4^{Bab}	520.7 ± 29.3^{Cb}	556.7 ± 15.2^{Ba}	569.1 ± 34.7^{ABa}	542.8 ± 20.4^{Cb}
	Bark	524.4 ± 11.6^{Fa}	534.8 ± 31.2^{Ca}	554.1 ± 24.8^{ABa}	506.8 ± 13.7^{Da}	543.2 ± 19.4^{Ca}	517.9 ± 18.8^{Da}	530.2 ± 22.3^{Da}
	Branches	534.1 ± 27.6^{Da}	512.5 ± 21.8^{Ea}	532.2 ± 31.5^{Ca}	498.8 ± 33.2^{Eb}	516.9 ± 19.7^{Ea}	542.3 ± 41.2^{Ba}	522.8 ± 28.6^{Ea}
	Leaves	573.6 ± 44.7^{Aa}	525.9 ± 31.2^{Ca}	547.1 ± 28.8^{Ba}	560.2 ± 42.9^{Aa}	568.3 ± 44.4^{Aa}	586.1 ± 36.1^{Aa}	560.2 ± 38.4^{Aa}
	Fruit	512.4 ± 17.2^{Ha}	498.8 ± 21.8^{Fa}	505.2 ± 15.4^{Ea}	489.7 ± 17.7^{Fa}	504.3 ± 21.9^{Fa}	488.4 ± 20.4^{Ea}	499.8 ± 18.8^{Ha}
	Stump roots	514.3 ± 22.2^{Ga}	533.7 ± 27.4^{Ca}	498.2 ± 23.7^{Fb}	509.6 ± 14.9^{Dab}	532.1 ± 20.8^{Da}	508.1 ± 24.6^{DEa}	516.0 ± 23.8^{Fa}
	Coarse roots	543.4 ± 31.8^{Ca}	554.1 ± 37.1^{Aa}	527.6 ± 30.4^{Ca}	538.7 ± 28.6^{Bab}	520.8 ± 27.3^{Eb}	558.4 ± 40.2^{Ba}	540.5 ± 33.6^{Cab}
	Medium roots	517.9 ± 30.2^{Fa}	541.7 ± 37.7^{Ba}	532.8 ± 24.9^{Ca}	557.4 ± 40.2^{Aa}	531.0 ± 23.4^{Da}	520.2 ± 28.7^{Ca}	533.5 ± 32.6^{Da}
	Small roots	544.1 ± 23.8^{Ca}	537.0 ± 31.9^{BCa}	515.6 ± 21.3^{Da}	520.4 ± 19.4^{Ca}	536.3 ± 33.1^{CDa}	525.4 ± 29.2^{Ca}	529.8 ± 25.2^{Da}
	Average of roots	529.9 ± 26.8^{Ea}	541.6 ± 34.2^{Ba}	518.6 ± 21.7^{Da}	531.5 ± 23.7^{BCa}	530.1 ± 27.4^{Da}	533.7 ± 28.9^{Ca}	530.9 ± 28.3^{Da}
	Average of trees	538.0 ± 25.5^{Da}	530.3 ± 30.4^{Ca}	539.5 ± 27.1^{BCa}	523.6 ± 34.2^{Ca}	543.0 ± 27.3^{Ca}	549.9 ± 26.5^{Ba}	537.4 ± 29.7^{CDa}
Shrub layer	Above ground	544.2 ± 32.2^{Ca}	516.7 ± 21.7^{Da}	545.2 ± 31.8^{Ba}	534.8 ± 25.5^{Ba}	541.7 ± 27.4^{Ca}	522.2 ± 18.9^{Ca}	534.1 ± 25.8^{Da}
	Below ground	522.3 ± 17.3^{Fa}	522.4 ± 19.1^{CDa}	497.5 ± 20.4^{Fa}	508.4 ± 24.3^{Da}	516.8 ± 19.4^{Ea}	519.2 ± 9.7^{CDa}	514.4 ± 18.3^{FGa}
	Average	533.3 ± 28.3^{Da}	519.6 ± 19.6^{Da}	521.4 ± 25.3^{Ca}	521.6 ± 14.8^{Ca}	529.3 ± 25.4^{Da}	520.7 ± 13.7^{Ca}	524.3 ± 22.9^{DEa}
Herb layer	Above ground	513.3 ± 31.7^{Ha}	532.8 ± 28.8^{Ca}	527.9 ± 35.4^{Ca}	508.4 ± 31.0^{Da}	529.1 ± 19.8^{Da}	511.7 ± 27.6^{Da}	520.5 ± 30.7^{Ea}
	Below ground	518.4 ± 27.2^{Fa}	509.3 ± 19.7^{Ea}	497.6 ± 16.9^{Fa}	509.8 ± 21.3^{Da}	524.2 ± 24.7^{DEa}	510.5 ± 18.8^{Da}	511.6 ± 20.6^{Ga}
	Average	515.9 ± 29.7^{Ga}	521.05 ± 25.4^{Da}	512.8 ± 27.3^{Ea}	509.1 ± 26.8^{Da}	526.7 ± 22.3^{Da}	511.1 ± 25.2^{Da}	516.1 ± 26.4^{Fa}
Litter layer	Undecomposed	557.8 ± 32.2^{Ba}	578.4 ± 40.4^{Aa}	562.1 ± 34.2^{Aa}	550.9 ± 27.9^{ABa}	564.2 ± 40.1^{Aa}	578.5 ± 31.2^{Aa}	565.3 ± 35.7^{A}
	Semi-decomposed	554.2 ± 29.9^{Ba}	543.8 ± 31.0^{Ba}	542.2 ± 27.6^{Ba}	537.1 ± 30.4^{Ba}	529.7 ± 27.7^{Da}	547.0 ± 29.4^{Ba}	542.3 ± 28.3^{C}

(续)

Layer	Components	7 yr old	10 yr old	18 yr old	23 yr old	29 yr old	33 yr old	Mean
	Average	556.0 ± 31.3^{Ba}	561.1 ± 36.4^{Aa}	552.2 ± 30.9^{Ba}	544.0 ± 28.7^{Ba}	547.0 ± 35.1^{Ca}	562.8 ± 30.1^{Ba}	553.8 ± 31.4^{B}

Notes: Different Capital letters in the same list indicate significant pairwise differences within stand ages between components, and different lower-cases indicate significant pairwise differences within components between stand ages (multiple comparisons test; $P < 0.05$).

C content in the soil layer

The C content changed markedly with increasing soil depth in all six stands ($P < 0.01$; Table 3). The value in the topsoil (0~10cm depth) was 46.1% higher than the average at 100cm depth.

The C contents of the two shallowest soil layers (0~10cm and 10~30cm), the horizon at 50~100cm and values obtained by summing across all soil layers increased significantly with increasing stand age ($P < 0.05$). The C content in the 3050cm soil layer, however, did not differ significantly among different ages ($P > 0.05$).

Table 3 Carbon contents by soil depth in six differently aged Mytilaria laosensis plantation stands (values are means ± SE, $n = 12$; g/kg).

Soil depth (cm)	7 yr old	10 yr old	18 yr old	23 yr old	29 yr old	33 yr old	Mean
0~10	28.1± 4.2^{A}	27.4 ± 2.8^{A}	29.6 ± 2.9^{A}	31.7 ± 2.4^{AB}	39.7 ± 4.3^{C}	47.1 ± 5.7^{D}	33.9 ± 3.4
10~30	23.2 ± 2.7^{A}	24.7 ± 2.1^{A}	26.3 ± 1.7^{B}	26.6 ± 2.3^{B}	27.7 ± 1.8^{B}	31.0 ± 2.5^{C}	26.6 ± 2.4
30~50	18.9 ± 2.4^{A}	19.4 ± 2.7^{A}	18.6 ± 1.9^{A}	19.0 ± 2.6^{A}	19.4 ± 3.1^{A}	20.5 ± 3.4^{A}	19.3 ± 3.0
50~100	10.9 ± 1.5^{A}	11.3 ± 1.7^{A}	13.2 ± 2.3^{B}	12.8 ± 2.0^{B}	14.4 ± 2.4^{C}	15.4 ± 2.7^{C}	13.0 ± 2.3
Mean	20.3 ± 2.5	20.7 ± 2.3	21.9 ± 2.7	22.5 ± 2.4	25.3 ± 3.2	28.5 ± 3.7	23.2 ± 2.7

Notes: Different capital letters indicate significant pairwise differences within soil depths between stand ages (multiple comparisons test; $P < 0.05$).

C stocks in plantation stands

Biomass and C stocks in tree layers

The allometric relationship between the biomass of the tree organs (W) and DBH (D) andheight (H) were best-fitted with equations in the form $W = a(D^2H)^b$. The F-tests showed that all regressions were highly significant ($P < 0.01$). Biomass calculations based on these allometric equations were used in the estimations of C stocks detailed in Table 4.

Table 4 Individual biomass regressions models for Mytilaria laosensis trees ($n = 18$ for all models); W, biomass; D, DBH; H, height

Organ	Allometric equation	R^2	F-value	P
Stem	$W_s = 0.1740(D^2H)^{0.7661}$	0.9196	104.3812	<0.0001
Bark	$W_{ba} = 0.0220(D^2H)^{0.7081}$	0.7191	58.6375	<0.0001
Branches	$W_{br} = 0.0002(D^2H)^{1.2696}$	0.6291	11.2124	0.0065
Leaves	$W_l = 0.00003(D)^{1.2634}$	0.8091	41.0306	0.0001
Roots	$W_r = 0.0094(D^2H)^{0.9538}$	0.7247	14.7872	0.0027
Total tree	$W_t = 0.1536(D^2H)^{0.8268}$	0.9049	64.8073	<0.0001

Figuer 1a depicts C stocks in stands of different ages and their allocation among component organs. The total C stocks in the trees were 77.3, 91.3, 104.1, 114.6, 153.0 and 156.2 t/hm^2 in 7-, 10-, 18-, 23-, 29- and 33-year-old stands, respectively. Thus, stocks rapidly increased with age from 7 to 29 years, but plateaued thereafter. C stocks in stems made up 74.0%, 72.1%, 69.7%, 67.7%, 65.3% and 65.2% of the total tree content in 7-, 10-, 18-, 23-, 29- and 33-year-old stands, respectively. Furthermore, trends in the C stocks of stems, roots, bark, branches and leaves tracked those of total tree C stocks across all stand ages.

The allocation of C stocks to stems and bark decreased with stand age from 7 to 29 years; stem and bark allocations were similar in 29- and 33-year-old stands. In contrast, C allocations to branches, leaves and roots increased with stand age from 7 to 29 years, but were similar in 29- and 33-year-old stands.

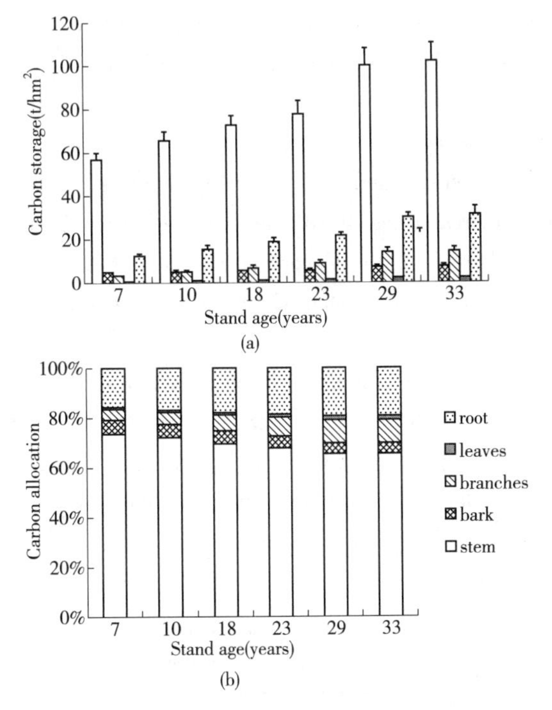

Figure1 Carbon stocks and their allocation to tree components in six differently aged *Mytilaria laosensis* stands. Values in (a) are means ± SE, $n=4$.

Notes: Different lowercase letters above the bars indicate significant differences between means (multiple comparisons test; $P<0.05$).

C stocks in the shrub, herb and litter layers

The C stock levels of each layer of the six plantation stands are shown in Figure 2. Small proportions of biomass and C were measured in the shrub, herb and litter layers. The summed C contents of the shrub, herb and litter layers made up 2.9%, 3.0%, 3.6%, 4.0%, 3.2% and 2.6% of the total C in 7-, 10-, 18-, 23-, 29- and 33-year-old stands, respectively. The above-ground C stocks in the shrub layers of the six stands were higher than below-ground shrub stocks. However, above-ground biomass and C stocks in the herb layer were lower than those below ground. The undecomposed biomass and C stocks in the litter layer were higher than those in the semi-decomposed portion; the highest average C stock in the undecomposed litter was 5.1-fold higher than that in the semi-decomposed portion.

Forest ground vegetation C stocks were correlated with stand age across the entire plantation chronosequence. The shrub C stocks increased with stand age from 7 to 23 years ($P<0.05$), but decreased with stand age from 23 to 33 years ($P<0.05$). Herb C stocks increased with stand age from 7 to 29 years, but decreased thereafter. C stocks in semi-decomposed portions of the litter layer were not related to stand age, but those in undecomposed portions and in the combined litter layer increased remarkably with increasing stand age ($P<0.05$). We therefore predict that the C stock in the litter layer will increase continually as the stands become older.

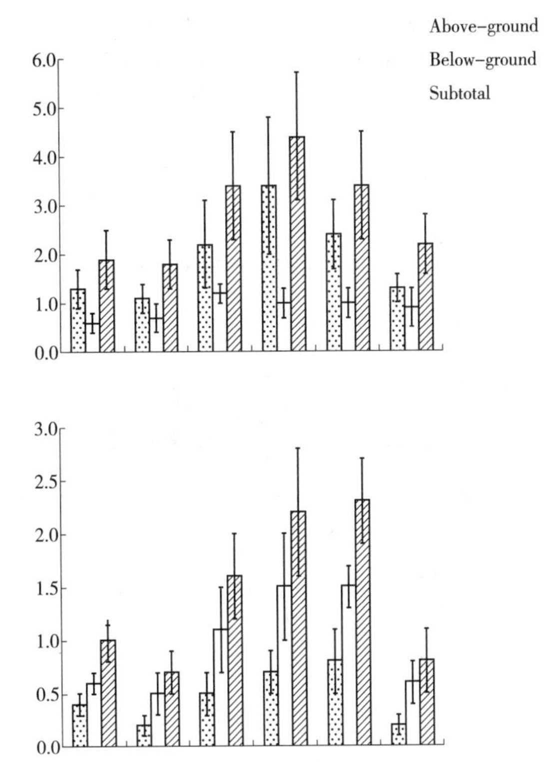

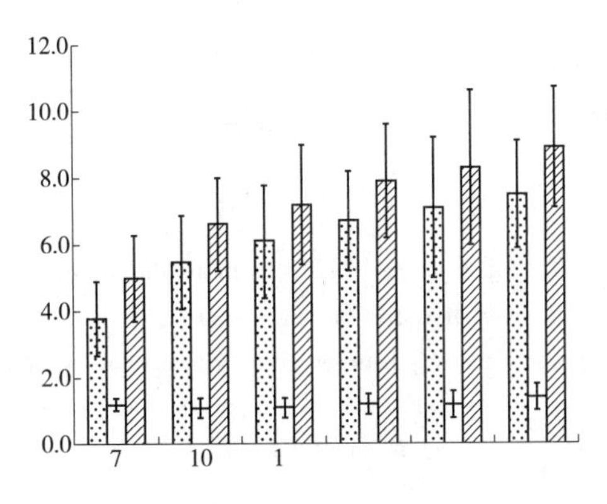

C stock in the soil layer

C stocks decreased with increasing soil depth even though the soil bulk density increased with depth. Figure 3 depicts trends in the soil layer C stock across the *M. laosensis* stand age sequence. The top soil (0 ~ 10cm) and deeper soil (50 ~ 100cm) stocks followed an increasing trend with stand age ($P<0.05$), especially in the older stands (the difference between 29- and 33-year-old stands was especially significant; $P<0.01$). Although soil C stocks in the 10 ~ 30cm and 30 ~ 50cm horizons increased with age, the relationship was not significant (Figure 3a).

Summed C stocks from 0 to 100 cm soil depth were 198.2, 199.6, 222.8, 227.9, 278.7 and 292.0 t/hm^2 in the 7-, 10-, 18-, 23-, 29- and 33-year-old stands, respectively. This linear trend was significant (Figure 3b).

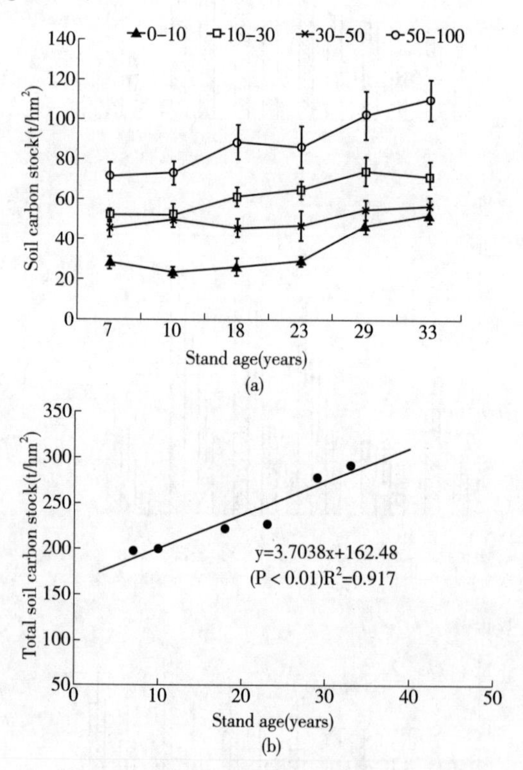

Figure 3 Soil layer carbon stocks in six differently aged Mytilaria laosensis stands

(a) and the linear relationship between soil carbon stocks and stand age (b) Values are means ± SE n = 12. Soil depth ranges in the key to (a) are in cm units.

C stock in the plantation ecosystem

Table 5 summarises individual ecosystem C stocks measured within each of the six stands. The rank order of C stock proportions across the six stands was as follows: soil layer (62.6% ~ 70.0%) > tree layer (27.3% ~ 33.9%) > litter layer (1.8% ~ 2.2%) > shrub layer (0.5% ~ 1.3%) > herb layer (0.2% ~ 0.6%). Averaging across stands, 65.3% of the total C was in the soil and 31.5% in the trees (Figure 4b).

Figure 4a shows changes in C stocks in above-ground, below-ground and total ecosystem components with increasing stand age. Above-ground ecosystem C stocks increased as stand age increased from 7 to 29 years; thereafter, the C content changed little. Below-ground and total ecosystem C stocks increased across the entire age range.

The above-ground to below-ground ecosystem C stock ratios were 0.338, 0.388, 0.391, 0.417, 0.431 and 0.419 in the 7-, 10-, 18-, 23-, 29- and 33-year-old stands, respectively; the ratios increased gradually with age due to the accumulation of above-ground C in tree biomass.

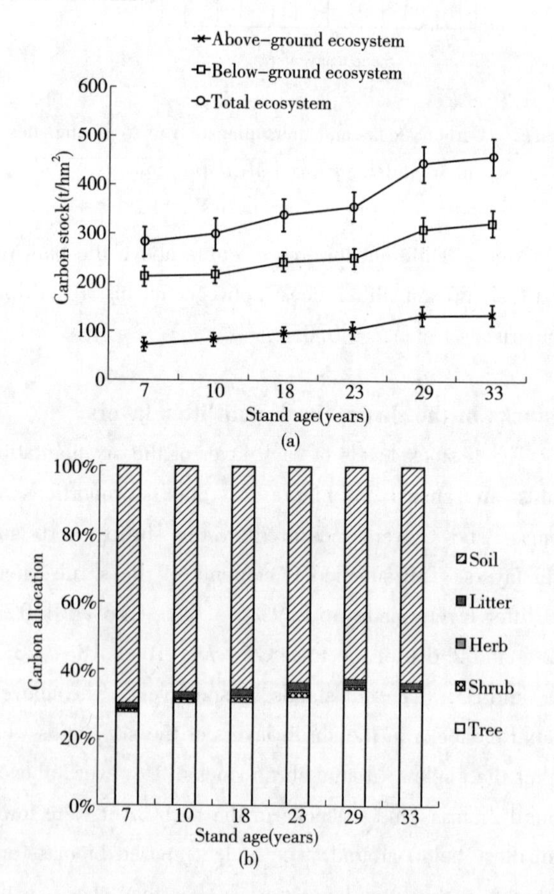

Figure 4 Carbon stocks in ecosystem components (a) and their proportional allocation (b) in six differently aged Mytilaria laosensis stands. Values in (a) are means ± SE, $n=4$

DISCUSSION

C content

C content in forest ecosystems varies by forest type. Tree species and site conditions are related to the C content[35, 43]. Across six *M. laosensis* stands with different ages, we detected a significant difference in mean C contents among different tree organs. No significant effect of stand age was detected within individual tree organs in the *M. laosensis* plantations, which corroborates the findings of studies on other species[38, 45].

The C contents of the various vegetation layers in the same stand fell into the following rank order: trees > shrubs > herbaceous plants (Table 2). Trees are probably high ranking because they can synthesise and accumulate more organic matter than can other types of vegetation[46].

The C content of litter varies with many factors, such as tree species, litter productivity, decomposition rate and microenvironment[35,44]. Litter reportedly decomposes at a considerably faster rate in broadleaf forests than in coniferous stands; thus, the standing stock of C is lower in broadleaf forests [35]. We measured a mean litter C content of 553. 8 g/kg in *M. laosensis* stands. This value is considerably higher than that in *P. massoniana* stands (505. 9 g/kg) in subtropical China[43], probably because the leaves of *M. laosensis*, which were the main component of the litter layer, are very leathery and refractory[47].

The mean C content of the soil layers in each stand decreased as the depth increased. The C content in the topsoil was higher than that in the deeper soil because the organic C produced from the decomposition of litter and root systems near the ground surface had entered the topsoil first, as demonstrated in other studies[10, 48, 49]. The C contents of the top two soil layers (0～10cm and 10～30cm), the deepest layer (50～100cm) and across all soil layers significantly increased with increasing stand age, probably due to the increasing litter productivity in older stands. This finding helps to explain the increase in below-ground C stocks as stands age.

C stocks in plantation stands

Estimating the C stock pools stored in age-sequenced plantations may contribute to forest management for C sequestration. C stocks in trees depend on stand density, biomass and relative C contents in the tissues. We found that C contents were positively correlated with C stocks (Table 5). The tree C stocks increased rapidly with age from 7 to 29 years, but plateaued thereafter, perhaps due to declining tree growth rates. Most previous studies have reported increases throughout the growth phases of trees[10, 38]. The laws of growth may vary among tree species, For example, *Euclyptus urophylla×E. grandis* grew faster in its early years, but turned slowly later, On the contrary, Castanopsis hystrix grew faster in the later stage than that in the early years[35, 38,39, 45], that is to say, different tree species show different growth charactersAs to *Mytilaria laosensis* , itneeds 25～30 years to reach the largest yield stage[28]. This variation may account for the difference between *Mytilaria laosensis* plantations and most other stands.

Only small portions of biomass and C were sequestered in the shrub, herb and litter layers, which accounted for only 2. 6%～4. 0% of the total. We detected a correlation between forest ground vegetation C stock and stand age across the entire chronosequence. C stocks in ground vegetation increased in the early stages of tree stand development, but decreased in older stands in concert with changes in tree canopy cover and stand density. Nevertheless, no obvious common patterns have been detected in studies on ground vegetation C storage. In the litter layer, the undecomposed and the total litter C stock decreased markedly with increasing stand age, which probably accounts for increases in soil C content with increasing stand age.

Soil was the largest C pool in the six stands that we studied. Soil C stock depends on stand age, the physical and chemical properties of the soil, forest type, litter productivity and litter decomposition rate[6, 35, 43, 50]. We found a significant linear relationship between total soil C stocks across the 0～100cm depth range and stand age. Hence, the proportion of C stock below ground will probably increase over protracted periods in the future life of the plantation ecosystems we studied.

Proportions of plantation ecosystem C stocks in the six stands fell into the following rank order: soil layer > tree layer > litter layer > shrub layer > herb layer. Soil and tree biomass harboured the largest C pools in the ecosystems, accounting for 96. 8% of the total. These findings are congruent with previous studies[51, 52, 53, 54].

Above-ground ecosystem C stocks increased during

the early stages of stand development and plateaued after 29 years due to a deceleration in tree growth. However, below-ground and total ecosystem C stocks increased with the stand age during the whole chronosequence we studied.

Table 5 Carbon stocks and their allocation in six differently aged Mytilaria laosensis stands (values are means ± SE, $n=4$; t/hm^2)

Layers	Components	7 yr old	10 yr old	18 yr old	23 yr old	29 yr old	33 yr old
tree layer	stem	56.8± 3.4^{F}	65.5± 4.2^{E}	72.3± 4.7^{D}	77.4± 6.3^{C}	99.8± 8.5^{B}	101.6± 9.1^{A}
	bark	4.5± 0.5^{D}	5.0± 1.0^{C}	5.4± 0.6^{B}	5.6± 0.8^{B}	7.1± 0.4^{A}	7.2± 1.3^{A}
	branches	3.2± 0.4^{E}	4.7± 0.7^{D}	6.7± 1.3^{C}	8.7± 1.4^{B}	13.9± 1.7^{AB}	14.4± 1.7^{A}
	leaves	0.5± 0.2^{E}	0.7± 0.1^{D}	1.0± 0.1^{C}	1.3± 0.1^{B}	2.1± 0.2^{AB}	2.2± 0.3^{A}
	total above-ground tree	65.0± 2.9^{E}	75.9± 4.5^{D}	85.4± 4.9^{C}	93.0± 5.8^{B}	122.9± 9.2^{AB}	125.4 ± 10.4^{A}
	root	12.3± 0.9^{E}	15.4± 1.7^{D}	18.7± 2.1^{C}	21.6± 1.4^{B}	30.1± 2.3^{AB}	30.8± 4.2^{A}
	subtotal	77.3± 4.4^{E}	91.3± 4.7^{D}	104.1± 5.1^{C}	114.6± 6.2^{B}	153.0± 8.4^{AB}	156.2± 10.3^{A}
shrub layer	above-ground	1.3± 0.4^{C}	1.1± 0.3^{D}	2.2± 0.9^{B}	3.4± 1.4^{A}	2.4± 0.7^{B}	1.3± 0.3^{C}
	below-ground	0.6± 0.2^{D}	0.7± 0.3^{C}	1.2± 0.2^{A}	1.0± 0.3^{AB}	1.0± 0.3^{AB}	0.9± 0.4^{B}
	subtotal	1.9± 0.6^{D}	1.8± 0.5^{D}	3.4± 1.1^{B}	4.4± 1.3^{A}	3.4± 1.1^{B}	2.2± 0.6^{C}
herb layer	above-ground	0.4± 0.1^{B}	0.2± 0.1^{C}	0.5± 0.4^{B}	0.7± 0.2^{A}	0.8± 0.3^{A}	0.2± 0.1^{C}
	below-ground	0.6± 0.1^{C}	0.5± 0.2^{C}	1.1± 0.4^{B}	1.5± 0.5^{A}	1.5± 0.2^{A}	0.6± 0.2^{C}
	subtotal	1.0± 0.2^{C}	0.7± 0.2^{D}	1.6± 0.4^{B}	2.2± 0.6^{A}	2.3± 0.4^{A}	0.8± 0.3^{D}
litter layer	undecomposed	3.8± 0.1.1^{D}	5.5± 1.4^{C}	6.1± 1.7^{B}	6.7± 1.5^{B}	7.1± 2.1^{A}	7.5± 1.6^{A}
	semi-decomposed	1.2± 0.2^{B}	1.1± 0.3^{B}	1.1± 0.3^{B}	1.2± 0.3^{B}	1.2± 0.4^{B}	1.4± 0.4^{A}
	subtotal	5.0± 1.3^{E}	6.6± 1.4^{D}	7.2± 1.8^{C}	7.9± 1.7^{B}	8.3± 2.3^{AB}	8.9± 1.8^{A}
soil layer	0~10	28.5± 3.2^{E}	23.9± 2.4^{F}	26.3± 4.2^{D}	29.1± 2.5^{C}	46.9± 3.4^{B}	52.6± 4.1^{A}
	10~30	52.2± 3.4^{D}	52.1± 5.8^{D}	61.7± 4.8^{C}	65.1± 4.6^{B}	74.3± 7.4^{A}	72.5± 6.7^{A}
	30~50	46.1± 5.1^{C}	50.3± 4.4^{B}	45.9± 4.7^{C}	47.4± 7.4^{BC}	55.2± 4.6^{A}	57.0± 4.1^{A}
	50~100	71.4± 7.4^{E}	73.3± 5.7^{E}	88.8± 8.6^{D}	86.3± 10.4^{C}	102.3± 12.3^{B}	109.9± 10.3^{A}
	subtotal	198.2± 12.4^{E}	199.6± 10.3^{E}	222.7± 11.4^{D}	227.9± 13.2^{C}	278.7± 17.3^{B}	292.0± 16.4^{A}
	above-ground ecosystem	71.7± 14.2^{E}	83.8± 12.4^{D}	95.3± 14.2^{C}	105.0± 17.6^{B}	134.4± 18.8^{A}	135.8± 20.4^{A}
	below-ground ecosystem	211.7± 20.4^{F}	216.2± 17.8^{E}	243.7± 19.4^{D}	252.0± 21.3^{C}	311.3± 25.6^{B}	324.3± 27.3^{A}
	total ecosystem	283.4± 30.4^{F}	300.0± 34.1^{E}	339.0± 33.7^{D}	357.0± 29.8^{C}	445.7± 37.1^{B}	460.1± 36.4^{A}

Notes: Different Capital letters in the same row indicate significant pairwise differences within components between stand ages; (multiple comparisons test; $P < 0.05$).

Conclusions

Stand age is a major determinant of C stocks in plantations. Both the C stocks and their distributions among plantation ecosystem components were affected by stand age. We found no significant differences in C contents in above-ground components among stand ages, but the soil C content increased with increasing stand age. Tree C was the largest above-ground ecosystem fraction, which contributed 25.7% to the total ecosystem C stocks in all six stands. The soil fraction was the largest C pool across plantations. C stocks across the 0 ~ 100cm soil depth range increased across the entire chronosequence. They were significantly linearly related to tree growth. The increase in above-ground tree biomass with increasing stand age significantly affected the above-ground ecosystem C stock size. The increases in below-ground ecosystem C stocks through a 7-year to 33-year stand chronosequence were mainly attributable to increases in soil organic C. Thus, one must take into account the successional development in forest ecosystem C pools when estimating C sink potentials over the complete life cycle of plantation stands.

Acknowledgements

The authors would like to thank Hui Wang, Dongjing Sun, Ji Zeng, You Nong, Haolong Yu, Jixin Tang, Henghui Wen, Yi Tao, Riming He, Hai Chen and Dewei Huang for their assistance in field sampling and data collection. We also thank Zhaoying Li, Lili Li and Bin He for their help in laboratory chemical analyses. We also acknowledge the helpful comments and suggestions of manuscript reviewers.

Author Contributions

Conceived and designed the experiments: Angang Ming, Yuanfa Li , Yuanfa Li.

Performed the experiments: Hongyan Jia, Angang Ming.

Analyzed the data: Yi Tao , Yuanfa Li.

Contributed reagents/materials/analysis tools: Yi Tao , Jilong Zhao, Hongyan Jia.

Wrote the manuscript: Angang Ming, Hongyan Jia, Yuanfa Li.

Funding

Our research was financially Supported by the Fundamental Research Funds for the Central Non - profit Research Institution of CAF(No. CAFYBB2014QA033), Nature Science Foundation of Guangxi (No. 2014jj BA30073), The Ministry of Science and Technology (2012BAD22B01), National Science Foundation of China (No. 31400542) and the Director Foundation Project of the Experimental Centre of Tropical Forestry, Chinese Academy of Forestry (No. RL2011-02).

REFERENCES

[1] CHOI S D, LEE K, CHANG Y S. Large rate of uptake of atmospheric carbon dioxide by planted forest biomass in Korea [J]. Global Biogeochemistry Cycle, 2002, 16: 1089.

[2] GOODALE C L, APPS M J, BIRDSEY R A, et al. Forest carbon sinks in the Northern Hemisphere[J]. Ecological Applications, 2002, 12: 891-899.

[3] HOUGHTON R A. Aboveground forest biomass and the global carbon balance[J]. Global Change Biology, 2005, 11: 945-958.

[4] LITTON C M, RAICH, J W, RYAN M G . Carbon allocation in forest ecosystems [J]. Global Change Biology, 2007, 13: 2089-2109.

[5] BROWN S. Measuring carbon in forests: current status and future challenges[J]. Environmental Science and Pollution Research, 2002, 116: 363-372.

[6] GOWER S T. Patterns and mechanisms of the forest carbon cycle[J]. Annual Review of Environment and Resource, 2003, 28: 169-204.

[7] HOUGHTON R A. Balancing the global carbon budget[J]. Annual Review of Earth and Planetary Sciences, 2007, 35: 313-347.

[8] KURZ W A, BEUKEMA S J, APPS M J . Estimation of root biomass and dynamics for the carbon budget model of the Canadian forest sector[J]. Canadian Journal of Forest Research, 1996, 26: 1973-1979.

[9] VOGT K. Carbon budgets of temperate forest ecosystems [J]. Tree Physiology, 1991, 9: 69-86.

[10] PEICHL M, ARAIN M A. Above- and belowground ecosystem biomass and carbon pools in an age-sequence of temperate pine plantation forests [J]. Agricultural and Forest Meteorology, 2006, 140: 51-63.

[11] PEICHL M, ARAIN M A. Allometry and partitioning of above- and belowground tree biomass in an age-sequence of white pine forests [J]. Forest Ecology and Management, 2007, 253: 68-80.

[12] TAYLOR A R, WANG J R, CHEN HYH. Carbon storage in a chronosequence of red spruce (*Picea rubens*) forests in central Nova Scotia[J]. Canada. Canadian Journal of Forest Research, 2007, 37: 2260-2269.

[13] PAQUETTE A, MESSIER C. The role of plantations in managing the world′s forests in the Anthropocene [J]. Frontiers in the Ecology and Environment, 2010, 8(1): 27-36.

[14] Department of Forest Resources Management, SFA. The 7th National forest inventory and status of forest resources [J]. Forest Ecology and Management, 2010, 1: 3-10.

[15] SFA (State Forestry Administration) . China's Forestry 1999 - 2005 [M]. Beijing: China Forestry Publishing House, 2007.

[16] HE Y, LI Z, CHEN J, et al . Sustainable management of planted forests in China: comprehensive evaluation, development recommendation and action framework [J]. Chinese Forest Science and Technology, 2008, 7: 1-15.

[17] JAGGER P . The role of trees for sustainable management of less favored lands: the case of Eucalyptus in Ethiopia

[J]. Forest Policy and Economics, 2003, 5: 83-95.

[18] PENG S L, WANG D X, ZHAO H, et al. Discussion the status quality of plantation and near nature forestry management in China[J]. Journal of Northwest Forestry University, 2008, 23: 184-188.

[19] POORE M E, FRIES C. The ecological effects of Eucalyptus[J]. FAO Forestry, 1985, 59: 1-97.

[20] WANG H, LIU S R, MO J M, et al. Soil organic carbon stock and chemical composition in four plantations of indigenous tree species in subtropical China[J]. Ecological Research, 2010a, 25: 1071-1079.

[21] BORKEN W, BEESE F. Soil carbon dioxide efflux in pure and mixed stands of oak and beech following removal of organic horizons[J]. Canadian Journal of Forest Research, 2005, 35: 2756-2764.

[22] BORKEN W, BEESe F. Methane and nitrous oxide fluxes of soils in pure and mixed stands of European beech and Norway spruce[J]. European Journal of Soil Science, 2006, 57: 617-625.

[23] KRAENZEL M, CASTILLO A, MOORE T, et al. Carbon storage of harvest-age teak (Tectona grandis) plantations, Panama [J]. Forest Ecology and Management, 2002, 5863: 1-13.

[24] LACLAU P. Biomass and carbon sequestration of ponderosa pine plantations and native cypress forests in northwest Patagonia[J]. Forest Ecology and Management, 2003, 180: 317-333.

[25] VESTERDAL L, SCHMIDT I K, CALLESEN I, et al. Carbon and nitrogen in forest floor and mineral soil under six common European tree species. Forest Ecology and Management, 2008, 255: 35-48.

[26] WANG H, LIU S R, MO J M, et al. Soil-atmosphere exchange of greenhouse gases in subtropical plantations of indigenous tree species. Plant and Soil, 2010b, 335: 213-227.

[27] YUAN S F, REN H, LIU N, et al. Can thinning of overstorey trees and planting of native tree saplings increase the establishment of native trees in exotic Acacia plantations in South China? [J]. Journal of Tropical Forest Science, 2013, 25: 79-95.

[28] GUO W F, CAI D X, JIA H Y, et al. Growth laws of Mytilaria laosensis plantation [J]. Forest Research, 2006, 19: 585-589.

[29] LI Y X, TAN T Y, HUANG J G, et al. A preliminary report on the densities of *Mytilaria laosensis* plantation [J]. Forest Research, 1988, 1: 206-212.

[30] LIANG, R L. Current situation of Guangxi indigenous broadleaf species resource and their development countermeasures[J]. Guangxi Forest Science, 2007, 36: 5-9.

[31] LIN D X, HAN J F, XIAO Z Q, et al. Mytilaria laosensis improved on the soil physical & chemical properties [J]. Journal of Fujian College of Forestry, 2000, 20: 62-65.

[32] LIN J G, ZHANG X Z, WENG X. Effects of site conditions on growth and wood quality of Mytilaria laosensis plantations[J]. Journal of Plant Resources and Environment, 2004, 13: 50-54.

[33] MING A G, JIA H Y, TAO Y, et al. Biomass and its allocation in 28-year-old Mytilaria laosensis plantation in southwest Guangxi [J]. Chinese Journal of Ecology, 2012, 31: 1050-1056.

[34] JONARD M, ANDRE F, JONARD F, et al. Soil carbon dioxide efflux in pure and mixed stands of oak and beech [J]. Annual of Forest Science, 2007, 64: 141-150.

[35] KANG B, LIU S, ZHANG G, et al. Carbon accumulation and distribution in Pinus massoniana and Cunninghamia lanceolata mixed forest ecosystem in Daqingshan, Guangxi of China [J]. Acta Ecological Sinica, 2006, 26: 1321-1329.

[36] KASEL S, BENNETT T L. Land-use history, forest conversion, and soil organic carbon in pine plantations and native forests of south eastern Australia [J]. Geoderma, 2007, 137: 401-413.

[37] LI X, YI M J, SON Y, et al. Biomass and Carbon Storage in an Age-Sequence of Korean Pine (*Pinus koraiensis*) Plantation Forests in Central Korea[J]. Journal of Plant Biology, 2011, 54 (1): 33-42.

[38] LIU E, WANG H, LIU S R. Characteristics of carbon storage and sequestration in different age beech (*Castanopsis hystrix*) plantations in south subtropical area of China[J]. Chinese Journal of Applied Ecology, 2012, 23: 335-340.

[39] MING A G, JIA H Y, TIAN Z W, et al. Characteristics of carbon storage and its allocation in Erythrophleum fordii plantations with different ages[J]. Chinese Journal of Applied Ecology, 2014, 25(4): 940-946.

[40] Soil Survey Staff of Usda. Keys to Soil Taxonomy[R]. United States Department of Agriculture (USDA), Natural Resources Conservation Service, Washington, DC, USA. 2006.

[41] State Soil Survey Service Of China. China Soil[M]. Beijing: China Agricultural Press, 1998.

[42] WANG, W X, SHI Z M, LUO D, et al. Carbon and nitrogen storage under different plantations in subtropical south China [J]. Acta Ecological Sinica, 2013, 33: 925-933.

[43] HE Y J, QIN L, LI Z Y, et al. Carbon storage capacity

of monoculture and mixed - species plantations in subtropical China[J]. Forest Ecology and Management, 2013, 295: 193-198.

[44] ZHOU Y, YU Z, ZHAO S. Carbon storage and budget of major Chinese forest types [J]. Acta Phytoecology. Sinicea, 2000, 24, 518-522.

[45] LIANG H W, WEN Y G, WEN L H, et al. Effects of continuous cropping on the carbon storage of Eucalyptus urophylla × E. grandis short rotations plantations [J]. Acta Ecological Sinica, 2009, 29: 4242-4250.

[46] CLEVELAND C, TOWNSEND A, TAYLOR P, et al. Relationships among net primary productivity, nutrients and climate in tropical rain forests: a pan - tropical analysis[J]. Ecology Letters, 2011, 14: 939 -947.

[47] LU L H, CAI D X, JIA H Y, et al. Annual variations of nutrient concentration of the foliage litters from seven stands in the southern subtropical area[J]. Scientia Silvae Sinicae, 2009, 45: 1-6.

[48] TIAN D, YIN G, FANG X, et al. Carbon density, storage and spatial distribution under different 'Grain for Green' patterns in Huitong, Hunan province[J]. Acta Ecological Sinica, 2010, 30: 6297-6308.

[49] ZHANG H, GUAN D S, SONG M W. Biomass and carbon storage of Eucalyptus and Acacia plantations in the Pearl River Delta, South China[J]. Forest Ecology and Management, 2012, 277: 90-97.

[50] JANDL R, LINDNER M, VESTERDAL L, et al. How strongly can forest management influence soil carbon sequestration? [J]. Geoderma, 2007, 137: 253-268.

[51] GRIGAL D F, OHMANN L F. Carbon storage in upland forests of the Lake States[J]. Soil Science Society of America Journal, 1992, 56: 935-943.

[52] WANG H, LIU S R, WANG J X, et al. Effects of tree species mixture on soil organic carbon stocks and greenhouse gas fluxes in subtropical plantations in China[J]. Forest Ecology and Management, 2013, 300: 43-52.

[53] WANG FM, XU X, ZOU B, et al. Biomass Accumulation and Carbon Sequestration in Four Different Aged Casuarina equisetifolia Coastal Shelterbelt Plantations in South China[J]. Plos One, 2013, 8(10): e77449.

[54] Zhao J L, Kang F F, Wang L X, et al. Patterns of Biomass and Carbon Distribution across a Chronosequence of Chinese Pine (*Pinus tabulaeformis*) Forests [J]. Plos One, 2014, 9(4): e94966.

[原载: Plos One , 2014, 9(10)]

Reponses of *Castanopsis hystrix* Seedlings to Macronrient Imbalances: Geowth, Photosynthetic Pigments and Foliar Nutrient Interactions

CHEN Lin[1], JIA HongYan[1], BERNARD Dell[2], GUO Wenfu[1], CAI Daoxiong[1], ZENG Jie[3]

([1]*Experimental Center of Tropical Forestry, CAF, Pingxiang 532600, Guangxi, China;*
[2]*Institute for Sustainable Ecosystems, Murdoch University, Perth, Australia;*
[3]*Research Institute of Tropical Forestry, CAF, Guangzhou 510520, Guangdong, China*)

Abstract: Imbalanced fertilization of nursery stock can lead to nutrient disorders which affect plant quality and productivity. *Castanopsis hystrix* Miq. is a plantation hardwood species that is being widely established in south China. Nutrient disorders are common in planting stock and difficult to manage, as the causes are unknown. Therefore, macronutrient deletion fertilizer treatments were applied to *C. hystrix* seedlings and foliar symptoms, growth performance, leaf area, photosynthetic pigments and foliar nutrient interactions were determined. The appearance of foliar symptoms including chlorosis, marginal scorching, necrotic spotting and leaf malformation will be useful for the initial diagnosis of fertilizer imbalances in this species in forest nurseries. Vector analysis revealed that foliar nutrient interactions tended to be enhanced under nutrient imbalance, and these findings may provide a preliminary guideline for confirming macronutrient disorders of this species in nursery production.

Keywords: *Castanopsis hystrix*; foliar symptom; growth performance; pigment composition; foliar nutrient interaction

1 INTRODUCTION

Macronutrients are important constituents of many molecules, such as amino acids, nucleic acids, protein, and chlorophyll; therefore they play an important role in energy homeostasis, signaling, osmotic balance, enzyme activation and protein regulation[1-3]. Imbalanced fertilization due to poor nursery management or financial constraints can result in distinctive foliar symptoms and nutrient disorders in plants (Figure 1), which not only decreases the plant growth rate and quality, but also causes water and soil pollution[4,5] and may enhance pests and disease impacts. Thus, detecting deficiencies or excesses is important for plant health and survival if timely corrective action is to be undertaken. Many studies have reported that plants adapt to nutrient imbalances through changing the morphology and physiological functions[6-7]. There are large differences exhibited among species and even within the same species, due to different ages, stock-type patterns, sample parts or the form of nutrients applied[8-11]. A visible foliar symptom approach is useful to quickly and economically examine the limiting nutrition of plant growth[12]. However, it must coordinate with chemical analyses of plant tissue, soil analysis, as well as different types of mathematical analysis, in order to improve the accuracy of nutrient diagnosis, since the foliar symptoms are often confused by soil type, water, cultivation, soil pH and nutrient interactions[13]. Vector analysis is often used to improve detection of nutrient interactions in plants by simultaneously assessing growth and nutrients in a graph[14,15].

Castanopsis hystrix Miq. is a broadleaved evergreen species distributed in subtropical forests, ranging from the eastern Himalayas of Nepal, Bhutan, and northeastern India, across Indochina, southern China, and Taiwan[16]. It prefers moist, acid and neutral soils, and young trees grow better under semi-shade. *Castanopsis hystrix* is also an excellent timber species because of its hard and flexible wood and beautiful texture. Hence, this species has been recommended for reforestation and commercial production in south China. However, knowledge about inorganic nutrient requirements of this species is scarce and problems exist in nursery production that appears to be related to fertilizer use. The aim of this present work is to determine morphological and physiological responses of *Castanopsis hystrix* seedlings to macronutrient deficiencies (–N, –P,

–K, –Ca, –Mg and –S), and nutrient surplus (+N and +P) in terms of visual foliar symptoms, growth performance, leaf area, photosynthetic pigments and foliar nutrient interactions. The results of this study will be developed as a preliminary guide for diagnosing nutritional disorders of this species in forest nurseries.

Figure 1 Foliar symptoms in a panoramic view (a) and close-up (b) of *Castanopsis hystrix* seedlings in the nursery due to imbalanced fertilization

2 MATERIALS AND METHODS

A sand pot culture of *Castanopsis hystrix* seedlings was carried out in the greenhouse of the Experimental Centre of Tropical Forestry located at Pingxiang City, Guangxi, China on August 24th, 2011. The mean daytime temperature in the greenhouse varied from 22℃ to 25℃ and relative humility from 60% to 80%. Quarz sand of 1 ~ 2mm diameter was used as the growing media. It was rinsed with running water, soaked in 1% HCl solution for 24 hours, then washed with distilled water and finally wind dried before use. Healthy *C. hystrix* seedlings of 17 cm height were selected and transplanted to 17.5cm × 11cm×12cm plastic pots filled with approximately 1.7kg squarz sand. The irrigation regime followed the method of Chen *et al.* [17] and was maintained consistently for all seedlings during the experimental period. To avoid the leaching of water and fertilizer, two plastic bags were placed in the pots. Every one or two weeks, a 0.1% solution of beta – cypermethrin was sprayed to the leaves to prevent disease.

The plants were arranged in a randomized complete block design with nine treatments: complete nutrient mix (Control), minus nitrogen (–N), plus extra nitrogen (+N), minus phosphorus (–P), plus extra phosphorus (+P), minus potassium (–K), minus calcium (–Ca), minus magnesium (–Mg) and minus sulfur (–S) nutrient solutions (Table 1). The experiment was replicated four times and each treatment replication (experimental unit) consisted of 20 seedlings for a total of 720 seedlings. At two weeks after transplanting (WAT), seedlings were supplied with a 50ml of treatment nutrient solutions once a week, which continued for 10 weeks when the foliar symptoms were fully exhibited.

Table 1 Composition of the nutrient solutions used in the experiment

No.	Chemical composition (ml/L)	Control	–N	+N	–P	+P	–K	–Ca	–Mg	–S
1	1mol/L KNO_3	5	–	5	6	5	–	5	6	6
2	1mol/L $Ca(NO_3)_2 4H_2O$	5	–	45	4	5	5	–	4	4
3	1mol/L $MgSO_4 7H_2O$	2	2	2	2	2	2	2	–	–
4	1mol/L KH_2PO_4	1	–	1	–	–	–	1	1	1
5	0.5mol/L K_2SO_4	–	5	–	–	–	–	–	3	–
6	0.5mol/L $Ca(H_2PO_4)_2 \cdot H_2O$	–	10	–	–	20	10	–	–	–
7	0.01mol/L $CaSO_4 2H_2O$	–	20	–	–	–	–	–	–	–

（续）

No.	Chemical composition (ml/L)	Control	-N	+N	-P	+P	-K	-Ca	-Mg	-S
8	1mol/L $Mg(NO_3)_2 6H_2O$	–	–	–	–	–	–	–	–	2
9	0.05mol/L Fe-EDTA	2	2	2	2	2	2	2	2	2
10	Arnon micronutrient mixture	1	1	1	1	1	1	1	1	1

Abstracted from Shen and Mao[18].

Foliar symptoms were recorded and photographed regularly. The root collar diameter and height of all seedlings were investigated at the end of 2, 5, 8, 11 and 14 WAT. Seven seedlings were randomly selected to investigate the initial biomass (mean 0.77g/seedling) before fertilization. At the end of the experiment, five seedlings were randomly selected from each experimental unit and washed gently with tap water to remove the sand. The plants were partitioned into leaves, stems and roots. The leaf samples were scanned using a Canon Image Class MF 4150 scanner and analyzed by Adobe Photoshop CS5 to obtain leaf photographs and pixels, respectively. Leaf area could be calculated according to the significantly positive relationships between the leaf area and pixel. Plant parts were oven dried at 105℃ for 15 minutes, and then at 70℃ for 48 hours to determine the dry mass. Leaves of each plant were combined by treatment replication and ground for subsequent chemical analysis. Foliar total N was determined by the diffusion method[16], and total P, K, Ca, Mg and S by inductively coupled plasma optical emission spectrometry[19]. Another three plants in each experimental unit were randomly selected and the third or fourth leaf from the apex of each plant was collected. The contents of chlorophyll a (Chla), chlorophyll b (Chlb) and carotenoid (Car) in fully expanded leaves were estimated spectrophotometrically in 80% acetone extract[20]. The total content of chlorophyll (chlorophyll a and b), and ratios of chlorophyll a and b (Chla/Chlb) and carotenoid and chlorophyll (Car/Chl) were also calculated.

A one-way analysis of variance (ANOVA) was conducted to study the effects of the six macronutrients on growth, foliar nutrient status, leaf area and pigment composition by SPSS 16.0 (2003). Significant means were ranked by Duncan's multiple range tests at 0.05 level. Vector analysis was used to interpret possible synergistic and antagonistic interactions between ions by simultaneously comparing leaf dry mass, leaf nutrient concentration and leaf nutrient content between complete and imbalanced treatments (Figure 2[14]).

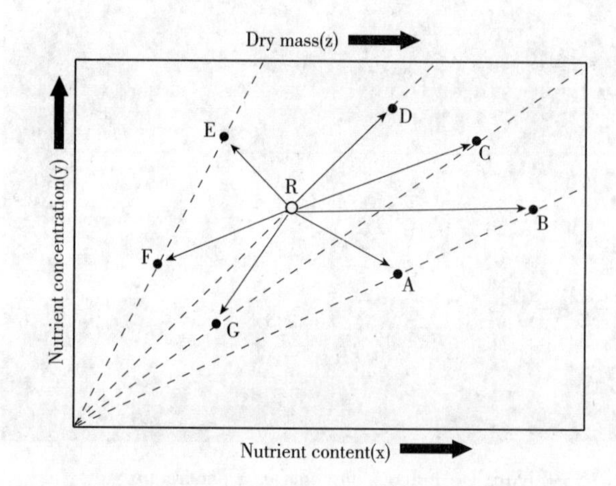

Figure 2 Vector interpretation of directional changes in nutrient content (x), nutrient concentration (y) and dry mass (z) of plants.

Vector shift	Relative change in			Interpretation	Possible diagnosis
	z	y	x		
A	+	-	+	Dilution	Growth dilution
B	+	0	+	Sufficiency	Steady-state
C	+	+	+	Deficiency	Limiting
D	0	+	+	Luxury consumption	Accumulation
E	-	+ +	±	Excess	Toxic accumulation
F	-	-	-	Excess	Antagonistic
G	0, +	-	-	Depletion	Retranslocation

Reference point (*R*) represents status of seedlings normalized to 100. Diagnosis is based on vector shifts (A to G) which characterize an increase (+), decrease (-) or no change (0) in dry mass and nutrient status relative to the reference status as described in the box below[21].

3 RESULTS

3.1 Growth

The foliar symptoms of *Castanopsis hystrix* seedlings began showing differences between treatments at WAT 6. Both N and K deficiencies induced leaf yellowing, marginal scorching, necrotic spotting and leaf curling, while P deficiency resulted in some light-green and yellow-green mottles appearing on the lower leaves. Similar foliar symptoms were observed in the +N and +P seedlings, such as distinctive discoloration between the veins, dark-brown blotches, marginal scorching or entire leaves twisting upward. Fading from light-green to yellow occurred both in the upper leaves of -Ca and -S seedlings, while chlorosis between the green midrib and main veins of leaves was the typical symptom of -Mg seedlings.

Except for -S plants, nutrient imbalance resulted in a significant decline in the root collar diameter when compared to the control ($P<0.05$, Table 2). In addition, the heights of these plants, except for -Ca, -Mg and -S treatments, were significantly lower than that of the control. As for the biomass, +N and +P treatments were significantly lower than the control, while the other treatments were not different from the control. The -N and -K treatments had only a significantly higher root/shoot ratio than the control.

Table 2 Effect of treatments on root collar diameter, height, biomass and root/shoot ratio of *Castanopsis hystrix* seedlings

Treatment	Root collar diameter (mm)	Height (cm)	Biomass (g/seedling)	Root/shoot ratio
Control	2.84 (0.05) a	19.9 (0.25) a	1.29 (0.06) ab	0.5564(0.04) bc
-N	2.49 (0.03) d	18.1 (0.18) ef	1.14 (0.04) abcd	0.76 (0.07) a
+N	2.58 (0.04) cd	18.8 (0.26) cd	0.93 (0.12) d	0.46 (0.02) c
-P	2.64 (0.04) c	19.0 (0.20) bc	1.36 (0.11) a	0.48 (0.02) c
+P	2.50 (0.04) d	17.6 (0.16) f	0.95 (0.06) cd	0.66 (0.02) ab
-K	2.61 (0.04) cd	18.3 (0.17) de	1.04 (0.05) bcd	0.72 (0.06) a
-Ca	2.64 (0.04) c	19.6 (0.27) ab	1.23 (0.11) abc	0.52 (0.03) c
-Mg	2.67 (0.04) bc	19.6 (0.24) ab	1.26 (0.14) ab	0.55 (0.04) bc
-S	2.78 (0.05) ab	19.7 (0.22) ab	1.23 (0.06) abc	0.56 (0.02) bc

Means in the same column followed by the same letter are not significantly different at the 5% level by Duncan's Multiple Range Test at 14 weeks after transplanting. The figures in parentheses are standard errors.

3.2 Photosynthetic Pigments

The -N and +P treatments induced an increase in the Car/Chl ratio, but a decrease in other pigment composition (Table 3). The +N and -K treatments had no significant effect on the Chlb content, Chla/Chlb or Car/Chl ratios, while the total chlorophyll and carotenoid contents were significantly lower than the control due to the reduced Chla content. However, the -P, -Ca, -Mg and -S treatments had no significant influence on the pigment compositions. This was similar to the leaf area of -P, -Ca, -Mg and -S treatments which were same as the control, while the remaining treatments resulted in a significant decline in leaf area ($P<0.05$, Figure 3).

Table 3 Effect of treatments on photosynthetic pigments of *Castanopsis hystrix* seedlings

Treatment	Chla (mg/g)	Chlb (mg/g)	Chla/Chlb	Chl (mg/g)	Car (mg/g)	Car/Chl
Control	1.610 (0.26) a	0.521 (0.10) a	3.138 (0.09) a	2.131 (0.36) a	0.565 (0.08) a	0.268 (0.01) b
-N	0.486 (0.05) e	0.197 (0.03) d	2.486 (0.14) c	0.684 (0.08) e	0.205 (0.02) e	0.300 (0.01) a

（续）

Treatment	Chla (mg/g)	Chlb (mg/g)	Chla/Chlb	Chl (mg/g)	Car (mg/g)	Car/Chl
+N	1.137 (0.14) bc	0.377 (0.05) abc	3.033 (0.17) ab	1.514 (0.19) bc	0.409 (0.05) bcd	0.271 (0.00) b
-P	1.516 (0.06) ab	0.478 (0.02) a	3.176 (0.07) a	1.994 (0.07) ab	0.548 (0.02) ab	0.275 (0.00) b
+P	0.686 (0.14) de	0.251 (0.04) cd	2.671 (0.21) bc	0.937 (0.18) de	0.276 (0.04) de	0.303 (0.02) a
-K	0.944 (0.16) cd	0.313 (0.04) bcd	2.989 (0.12) ab	1.257 (0.20) cd	0.352 (0.05) cd	0.284 (0.01) ab
-Ca	1.264 (0.13) abc	0.422 (0.06) ab	3.036 (0.11) ab	1.686 (0.18) abc	0.449 (0.05) abc	0.266 (0.00) b
-Mg	1.450 (0.12) ab	0.449 (0.04) ab	3.239 (0.06) a	1.900 (0.17) ab	0.512 (0.05) ab	0.269 (0.00) b
-S	1.422 (0.05) ab	0.481 (0.02) a	2.958 (0.04) ab	1.903 (0.07) ab	0.520 (0.02) ab	0.273 (0.00) b

Chla, Chlb, Chl and Car refer to contents of chlorophyll a, b, total chlorophyll and carotenoid, respectively; Chla/Chlb and Car/Chl are ratios of chlorophyll a and b contents and carotenoid and chlorophyll contents. Means in the same column followed by the same letter are not significantly different at the 5% level by Duncan's Multiple Range Test at 14 weeks after transplanting. The figures in parentheses are standard errors.

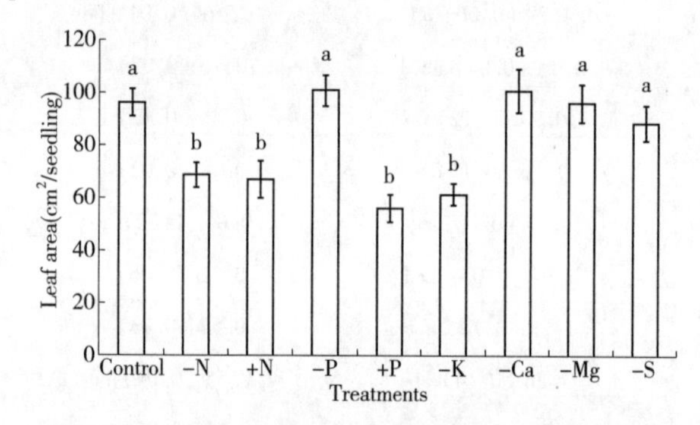

Figure 3 Effects of treatment on leaf area of *Castanopsis hystrix* seedlings

Paired treatments with the same letter are not significantly different at the 5% level by Duncan's Multiple Range Test at 14 weeks after transplanting. Error bars represent standard errors.

3.3 Foliar Nutrient Interactions

The -N treatment induced N limitation (Figure 4a) because N showed one of the major relative response of all nutrients, registering a positive, positive and positive shift with respect to leaf dry mass, leaf nutrient concentration and leaf nutrient content, respectively. And P excess (Figure 4a) rather than N limitation occurred in the treatment because P was the most responsive element and P excess occurred associated with increased leaf P concentration (311%) and leaf P content (199%) but diminished leaf dry mass (27%) in -N seedlings compared with those of the control seedlings. Similarly, K limitation and P excess were apparent in the -K monogram (Figure 4e) because they were the major vectors on the one hand, on the other hand their orientations were positive in three shifts or in a positive, negative and negative shift exhibited by three parameters (z, y, and x, respectively). The P excesses in -N and -K seedlings could be explained by high P supply in -N and -K nutrient solutions in this study (Table 1).

The leaf N, P, K, Ca, Mg and S contents of the control were 8.09, 1.32, 7.76, 6.52, 1.04 and 0.93mg respectively. The leaf N, P, K, Ca, Mg and S concentrations of the control were 13.456, 2.177, 12.885, 10.751, 1.712 and 1.547g/kg respectively. Status of the reference treatment (open symbols) at 14 weeks after transplanting was normalized to 100 to facilitate relative comparisons on a common base. The two vectors in each monogram represent the shift and identity of the most and second most responsive nutrients. Diagnosis mainly relied on the orientation of these two vectors and associated shifts (differences) in plant nutrient concentration, nutrient content and dry mass[21,22].

The +N application not only resulted in N toxicity but also Ca toxicity in the +N monogram (Figure 4b), and Ca toxicity in +N seedlings could be explained by high Ca supply in +N nutrient solutions in this study (Table 1). The +P treatment induced P toxicity (Figure 4d) associated with reduced leaf dry mass (37%) but increased leaf P concentration (168%) and leaf P uptake (70%). Furthermore, an antagonistic interaction of K (Figure 4d) occurred associated with a decline in leaf K concentration (12%), leaf K content (45%) and leaf dry mass

(37%), respectively.

Luxury consumptions of P, Ca, Mg and S (Figure 4c, 4f, 4g and 4h) were found in the corresponding monograms, since leaf dry mass had no significant differences between these treatments and the control (results not presented), but an increases in respective leaf nutrient concentration and leaf nutrient content occurred as the control compared to each of these nutrient deficiency treatments, suggesting that *C. hystrix* may be less sensitive to these nutrient elements and their amounts in seedlings at the beginning were enough for seedling growth in the study period. Additionally, antagonistic interactions of Mg, P and K (Figure 4c; Figure 4f; Figure 4g) and asymmetric interaction of Ca (Figure 4h) were as the second responsive nutrient diagnoses induced by the deficiencies of P, Ca, Mg and S, respectively. As demonstrated above, the foliar nutrient interactions tended to be enhanced under nutrient imbalances.

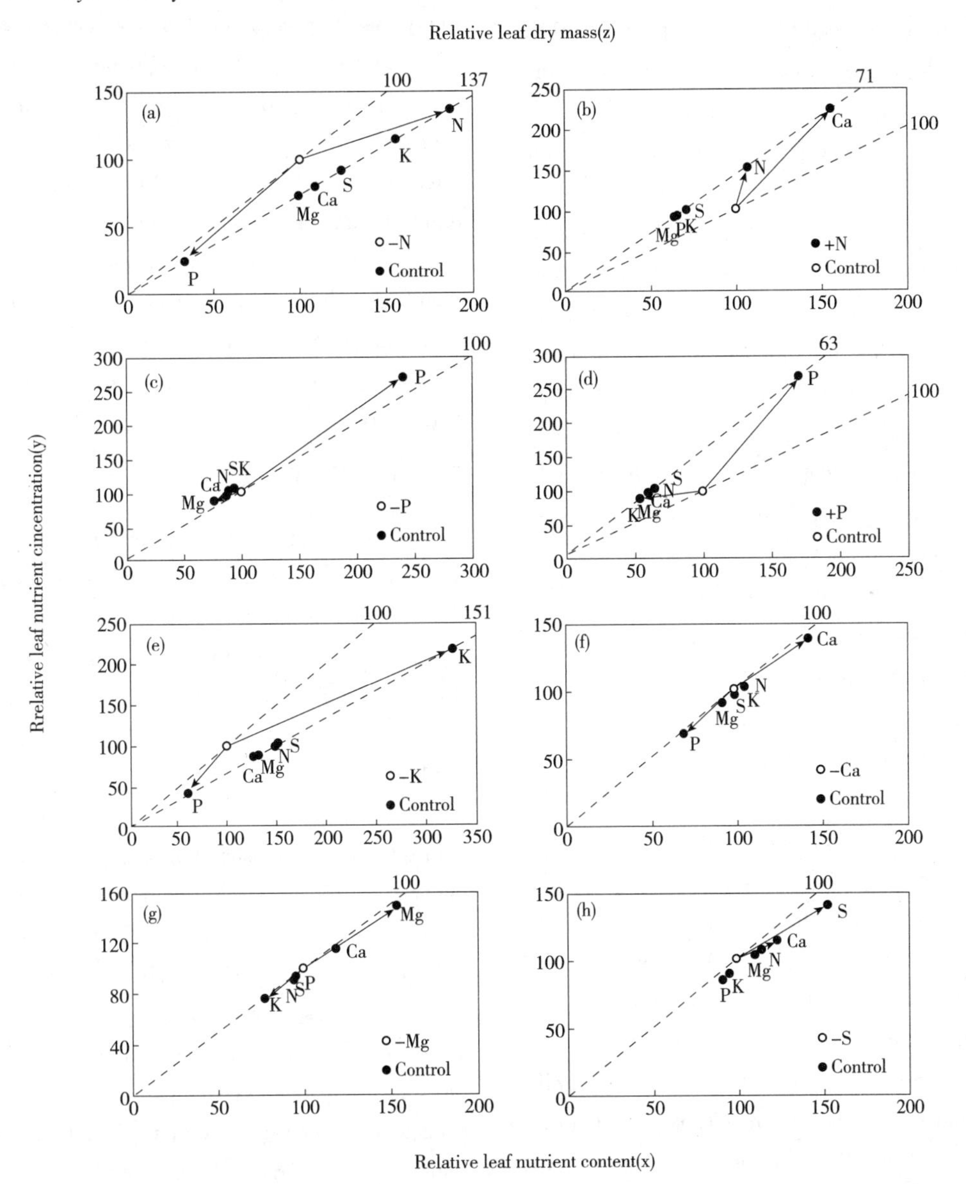

Figure 4 Vector monograms of relative changes in the leaf nutrient content (x), leaf nutrient concentration (y) and leaf dry mass (z) between *Castanopsis hystrix* seedlings supplied with different nutrient solutions.

4 DISCUSSION

Induction of N and K deficiencies resulted in a significant decline in seedling height and root collar diameter, and leaf area (Table 2; Figure 3), as found previously in other plants, such as *Tectona grandis*[23],

Pistacia lentiscus[24] and *Vitellaria paradoxa*[25]. There was no influence of N and K deficiencies on biomass (Table 2), due perhaps to the relatively slow growth rate of this species. On the contrary, the root/shoot ratio was significantly higher than those of the control (Table 2), probably because the biomass was preferentially allocated to the below-ground root in order to absorb more water and nutrients when seedling encountered nutrient deficiency[26]. The –N and –K treatments caused a significant reduction in pigment composition but an increase in the Car/Chl ratio (Table 3), thus obvious foliar symptoms were observed at the end of the experiment.

Deficiency of P reduced seedling height and root collar diameter (Table 2) in line with results of Gopikumar and Varghese[23]. However there were no effect of P deficiency on the biomass, root/shoot ratio, leaf area and pigment composition because *C. hystrix* may be less sensitive to P and initial amounts of P in seedlings were enough for their growth in the study period. Other studies have found changes in root architecture development and root exudation if P is deficient[27-29], whether the same changes occurred in this species needs to be confirmed in a further study.

Optimum nutrient quantity is very important for plant growth[30]. In this study, N and P excesses depressed seedling height, root collar diameter, biomass and leaf area (Table 2; Figure 3). This may be related to the N and P toxicities diagnosed by vector analysis (Figure 4b, 4d), while they had no effect on the root/shoot ratio (Table 2). Moreover, the Chla, total Chlorophyll and carotenoid contents of +N and +P seedlings were significantly lower than those of control, but the Car/Chl ratios were not affected or were increased (Table 3), suggesting that N and P excesses had a negative impact on growth as well as pigment composition in the study period[8,31].

The height of the –Ca and –Mg plants were significantly lower than that of the control (Table 2), however there were no evidence that biomass, root/shoot ratio, leaf area and pigment compositions were depressed (Table 2; Table 3; Figure 3). This differs from the report for *Picea abies*[9] and *Khaya ivorensis*[19] in which Ca and Mg deficiencies decreased seedling height, root collar diameter or chlorophyll content, probably because of the relatively less sensitive responses to lower Ca and Mg amounts in *Castanopsis hystrix* seedlings during the study period, which was confirmed by vector analysis (Figure 4f, 4g).

Sulfur deficiency had little impact on the various growth indexes, leaf area and pigment composition (Table 2; Table 3; Figure 3). This contrasts with some studies reporting that S deficiency resulted in growth retardation and reduction of chlorophyll content[13,32], which may be explained by less sensitive response to lower S amounts in the seedlings during the study period (Figure 4h), in the same way as the –P, –Ca and –Mg treatments. It is likely that only mild S deficiency occurred in the current trial and a longer trial is needed before deficiency can be well expressed. Although our study determined the responses of *Castanopsis hystrix* seedlings to macronutrient imbalances in morphology and foliar nutrient interactions, plant tolerance mechanisms under nutrient imbalance conditions also involve biochemistry and metabolism, therefore multi-level approaches should be used to understand the exact relationship between nutrient supply and metabolic adjustment in further studies as suggested by other workers[33-35].

5 CONCLUSIONS

In conclusion, *Castanopsis hystrix* seedlings are quite sensitive to changes in fertilization composition and thus the management of fertilizer type, composition and rate needs to be carefully managed in forest nurseries producing seedlings for outplanting. Visual foliar symptoms were obvious under –N, +N, +P and –K treatments, and seedling height, root collar diameter, leaf area, Chla, total chlorophyll and carotenoid contents were all decreased in comparison with those of the control. However, no significant specific differences were detected in most of the growth indexes, leaf area and photosynthetic pigments under –P, –Ca, –Mg and –S treatments. The nutrient diagnoses by vector analysis matched visual foliar symptoms, growth performance and nutritional responses induced by treatments, suggesting that vector analysis is a good tool to analyse foliar nutrient interactions and thus improve the accuracy of nutrient diagnosis. There is a need for a complete set of foliar symptoms for all essential nutrients for this species, as well as guidelines for nutrient status of leaves of known age for deficient, healthy and toxic conditions.

6 ACKNOWLEDGMENTS

We thank to Cai Lan Meng for assistance in the nursery and Le Su Yang and Bin Yu for nutrient determinations. This study was financially supported by the Fund of Director of Experimental Center of Tropical Forestry, Chinese Academy of Forestry (Project No.: RL2011-5).

REFERENCES

[1] ASHLEY M K, GRANT M, GRABOV A. Plant responses to potassium deficiencies: a role for potassium transport proteins[J]. Journal of Experimental Botany, 2006, 57: 425-436.

[2] MAATHUIS F J. Physiological functions of mineral macronutrients[J]. Current Opinion in Plant Biology, 2009, 12: 250-258.

[3] WARREN C R. How does P affect photosynthesis and metabolite profiles of *Eucalyptus globulus*? [J]. Tree Physiology, 2011, 31: 727-739.

[4] BARRETT D R, J E D. Fox. *Santalum album*: kernel composition, morphological and nutrient characteristics of pre-parasitic seedlings under various nutrient regimes[J]. Annals of Botany, 1997, 79: 59-66.

[5] GLOSER V, P SEDLAEK, GLOSER J. Consequences of nitrogen deficiency induced by low external N concentration and by patchy N supply in *Picea abies* and *Thuja occid-entalis*[J]. Trees, 2009, 23: 1-9.

[6] VERBRUGGEN N, HERMANS C. Physiological and molecular responses to magnesium nutritional imbalance in plants[J]. Plant and Soil, 2013, 368: 87-99.

[7] AFROUSHEH M, ARDALAN M, HOKMABADI H, et al. Nutrient deficiency disorders in *Pistacia vera* seedling rootstock in relation to eco-physiological, biochemical characteristics and uptake pattern of nutrients[J]. Scientia Horticulturae, 2010, 124: 141-148.

[8] SHEDIEY E, DELL B, GROVE T. Diagnosis of nitrogen deficiency and toxicity of *Eucalyptus globulus* seedlings by foliar analysis[J]. Plant and Soil, 1995, 177: 183-189.

[9] MEHNE-JAKOBS B. Magnesium deficiency treatment causes reductions in photosynthesis of well-nourished Norway spruce[J]. Trees, 1996, 10: 293-300.

[10] FOLK R S, GROSSNICKLE S C. Stock-type patterns of phosphorus uptake, retranslocation, net photosynthesis and morphological development in interior spruce seedlings [J]. New Forests, 2000, 19: 27-49.

[11] KOPRIVA S, HARTMANN T, MASSARO G, et al. Regulation of sulfate assimilation by nitrogen and sulfur nutrition in poplar trees[J]. Trees, 2004, 18: 320-326.

[12] LANDIS T D, HAASE D L, DUMROESE R K. Plant nutrient testing and analysis in forest and conservation nurseries [C]. RK, Dumroese, LE Riley, and TD Landis, (tech. coords). National proceedings: Forest and Conservation Nursery Associations - 2004. Proc. RMRS-P-35. USDA Forest Service, Rocky Mountain Forest and Range Experimental Station. Fort Collins, Colo, 2005, 76-83.

[13] HESSE H, NIKIFOROVA V, GAKIERE B, et al. Molecular analysis and control of cysteine biosynthesis: integration of nitrogen and sulphur metabolism[J]. Journal of Experimental Botany, 2004, 55: 1283-1292.

[14] HAASE D L, ROSE R. Vector analysis and its use for interpreting plant nutrient shifts in response to silvicultural treatments[J]. Forest Science, 1995, 41: 54-66.

[15] LTEIF A, WHALEN J K, BRADLEY R L, et al. Diagnostic tools to evaluate the foliar nutrition and growth of *Hybrid poplars*[J]. Canadian Journal of Forest Research, 2008, 38: 2138-2147.

[16] HUANG Y F, ZHUANG X Y. Seedling silvicultural techniques of species in southern China[M]. Beijing: Chinese Forestry Publishing House, 2007.

[17] CHEN L, ZENG J, XU D P, et al. Macronutrient deficiency symptoms in *Betula alnoides* seedlings[J]. Journal of Tropical Forest Science, 2010, 22: 403-413.

[18] SHEN J B, MAO D R (eds.). Research methods of plant nutrition, third edition[M]. Beijing: China Agricultural University Press. 2011, 17-18.

[19] JEYANNY V, RASIP A G Ab, RASIDAH K W, et al. Effects of macronutrient deficiencies on the growth and vigour of *Khaya ivorensis* seedlings[J]. Journal of Tropical Forest Science, 2009, 21: 73-80.

[20] LICHTENTHALER H K, WELLBURN A R. Determination of total carotenoids and chlorophyll a and b of leaf extracts in different solvents[J]. Biochemical Society Transactions, 1983, 603: 591- 592.

[21] SALIFU K F, TIMMER V R. Optimizing nitrogen loading of *Picea mariana* seedlings during nursery culture[J]. Canadian Journal of Forest Research, 2003, 33: 1287-1294.

[22] SALIFU K F, JACOBS D F. Characterizing fertility targets and multi-element interactions in nursery culture of *Quercus rubra* seedlings[J]. Annals of Forest Science, 2006, 63: 231-237.

[23] GOPIKUMAR K, V VARGHESE. Sand culture studies of teak (*Tectona grandis*) in relation to nutritional deficiency symptoms, growth and vigour[J]. Journal of Tropical Forest Science, 2004, 16: 46-61.

[24] TRUBAT R, CORTINA J, VILAGROSA A. Plant morphology and root hydraulics are altered by nutrient defi-

ciency in *Pistacia lentiscus* (L.)[J]. Trees, 2006, 20: 334-339.

[25] UGESE F D, BAIYERI K P, MBAH B N. Expressions of macronutrient deficiency in seedlings of the shea butter tree (*Vitellaria paradoxa* C. F. Gaertn.)[J]. Journal of Agricultural Technology, 2012, 8: 1051-1058.

[26] ERICSSON T. Growth and root: root ratio of seedlings in relation to nutrient availability[J]. Plant and Soil, 1995, 168-169: 205-214.

[27] ROUACHED H, ARPAT A B, POIRIER Y. Regulation of phosphate starvation responses in plants: signaling players and cross-talk[J]. Molecular plant, 2010, 3: 288-299.

[28] NIU Y F, CHAI R S, JING L, et al. Responses of root architecture development to low phosphorus availability: a review[J]. Annals of Botany, 2013, 112: 391-408.

[29] CARVALHAIS L C, DENNIS P G, FEDSOEYENKO D, et al. Root exudation of sugars, amino acids, and organic acids by maize as affected by nitrogen, phosphorus, potassium, and iron deficiency[J]. Journal of Plant Nutrition and Soil Science, 2011, 174: 3-11.

[30] WHITE P J, BROWN P H. Plant nutrition for sustainable development and global health[J]. Annals of Botany, 2010, 105: 1073-1080.

[31] BURNETT S E, ZHANG D, STACK L B, et al. Effects of phosphorus on morphology and foliar nutrient concentrations of hydroponically grown *Scaevola aemula* R. Br. 'Whirlwind Blue'[J]. HortScience, 2008, 43: 902-905.

[32] BROWDER J F, NIEMIERA A X, HARRIS J R, et al. Wright. Growth response of container-grown pin oak and japanese maple seedlings to sulfur fertilization [J]. HortScience, 2005, 40: 1524-1528.

[33] SCHACHTMAN D P, SHIN R. Nutrient sensing and signaling: NPKS[J]. Annual Review of Plant Biology, 2007, 58: 47-69.

[34] WILLIAMS L, SALT D E. The plant ionome coming into focus[J]. Current Opinion in Plant Biology, 2009, 12: 247-249.

[35] AMYMANN A, BLATT M R. Regulation of macronutrient transport[J]. New Phytologist, 2009, 181: 35-52.

[原载: Journal of Plant Nutrition, 2016]

Effects of Near Natural Forest Management on Soil Greenhouse Gas Flux in *Pinus massoniana* and *Cunninghamia lanceolata* Plantations

MING Angang[1,3], YANG Yujing[2,3], LIU Shirong[2,3], WANG Hui[2], LI Yuanfa[4], LI Hua[1,3], NONG You[1,3], CAI Daoxiong[1,3], JIA Hongyan[1,3], TAO Yi[1,3], SUN Dongjing[1,3]

([1]*Experimental Center of Tropical Forestry*, *Chinese Academy of Forestry*, *Pingxiang* 532600, *Guangxi China*;
[2]*Key Laboratory of Forest Ecology and Environment*, *State Forestry Administration*;
Institute of Forest Ecology, *Environment and Protection*, *Chinese Academy of Forestry*, *Beijing* 100091, *China*;
[3]*Guangxi Youyiguan Forest Ecosystem Research Station*, *Pingxiang* 532600, *Guangxi China.*
[4]*College of Forestry*, *Guangxi University*, *Nanning*, 530004, *Guangxi China*)

Abstract: Greenhouse gases are the main cause of global warming, and forest soil plays an important role in greenhouse gas flux. Near natural forest management is one of the most promising options for improving the function of forests as carbon sinks. However, its effects on greenhouse gas emission are not yet clear. It is therefore necessary to characterise the effects of near natural forest management on greenhouse gas emission and soil carbon management in plantation ecosystems. We analysed the influence of near natural management on the fluxes of three major greenhouse gases (carbon dioxide [CO_2], methane [CH_4], and nitrous oxide [N_2O]) in *Pinus massoniana* and *Cunninghamia lanceolata* plantations. The average emission rates of CO_2 and N_2O in the near natural plantations were higher than those in the corresponding unimproved pure plantations of *P. massoniana* and *C. lanceolata*, and the average absorption rate of CH_4 in the pure plantations was lower than that in the near natural plantations. The differences in the CO_2 emission rates between plantations could be explained by differences in the C: N ratio of fine root. The differences in the N_2O emission rates could be attributed to differences in soil available N content and the C: N ratio of leaf litter, while the differences in CH_4 uptake rate could be explained by differences in the C: N ratio of leaf litter only. Near natural forest management negatively affected the soil greenhouse gas emissions in *P. massoniana* and *C. lanceolata* plantations. The potential impact of greenhouse gas flux should be considered when selecting tree species for enrichment planting.

Keywords: near natural forest management; *Pinus massoniana* plantation; *Cunninghamia lanceolata* plantation; soil greenhouse gas flux

1 INTRODUCTION

Increased emissions of greenhouse gases, dominated by carbon dioxide (CO_2), methane (CH_4), and nitrous oxide (N_2O), are the main cause of global climate change[1]. Most greenhouse gases in the atmosphere are produced and absorbed by soil[2]. Forest soils have the largest carbon pool in terrestrial ecosystems owing to soil respiration processes, mainly root respiration, microbial respiration, and soil animal respiration[3]. N_2O is released from soil to the atmosphere through microbe-regulated nitrification and denitrification [4], while forest soil usually serves as the absorption sink for atmosphere CH_4[5]. About 6% of global CH_4 is absorbed through soil processes by methanogenic bacteria[6-7]. The global warming potential of CH_4 and N_2O is 25 and 298 times, respectively, larger than that of CO_2, although they are much less abundant than CO_2 in the atmosphere[8]. Therefore, a comprehensive understanding of the rates of greenhouse gas emission and absorption and their key influencing factors in forest soils is critical to assessing the contribution of forest ecosystems to global climate change[9-10].

Near natural forest management, one of the most promising options for plantation sivilculture, has received widespread attention in recent years[11]. In the principle of near natural forest management, pure plantations are transformed into near natural forests through a series of management strategies, according to the structure and succession of natural forests. The strategies include species introduction, structural adjustment, natural regeneration promotion, and understory protection. Thus, the management of coniferous plantations has significant impacts on the structure, tree species composition, and regeneration of the forests[12,13]. Tree species are considered to alter the soil environment (including soil temperature and moisture), soil physical and chemical properties, and soil biological processes by influencing the composition and quality of the stand root system, canopy, litter, and fine roots[14-15]. As a result, soil greenhouse gas flux is greatly impacted by the composition of tree species. For example, the soil CH_4 flux of *Populus tremula*, *Picea asperata*, and pine forests in Europe differs significantly[16]. Menyailo and Hungate[17] observed higher CH_4 consumption in aspen, birch and spruce forest soils compared to Scots and Arolla pine forest soils in Siberia. However, average CH_4 uptake rates in the mixed and the pure beech plantations were about twice as large as that in the pure spruce plantation[18]. Soil CO_2 efflux was accelerated after conversion from secondary oak forest to pine plantation in southeastern China[19]. Mature pine plantation soil emits 1.5 and 2.5 times of CO_2 than the mature beech and the Douglas fir[20]. Studies have also shown significant differences in soil respiration rates among 16 tree species in the tropics, with an emission flux of 2.8 to 6.8μmol/m^2 · s[21].

Although forest soil-atmosphere greenhouse gas exchange in temperate and tropical regions has been studied in depth[5,22-24], little is known about it in the southern subtropical forests. There is a growing need locally and abroad to reduce greenhouse gas emissions from forests through plantation management. However, few studies have examined plantation management strategies for manipulating soil greenhouse gas flux. Near natural management of coniferous plantation involves the transformation of even-aged pure stands of coniferous species into uneven-aged mixed broad-leaved forests, but it is not well known on how this strategy affects the emission and absorption of greenhouse gases. Therefore, subtropical near natural *Pinus massoniana* plantation (P[CN]) and unimproved pure stand of *P. massoniana* (P[CK]), as well as near natural *Cunninghamia lanceolata* plantation (C[CN]) and unimproved pure *C. lanceolata* stand (C[CK]) were selected in southern China. The objective of this study is to examine the effects of near natural forest management on soil-atmosphere greenhouse gas exchange and the main factors influencing these processes. The present study provides a theoretical basis for the multi-objective and sustainable management of plantations in southern subtropical regions.

2 MATERIALS AND MERHODS

2.1 Study site description

The study site is located in the Experimental Center of Tropical Forestry, Chinese Academy of Forestry (Pingxiang, Guangxi, China). It is one of the forest ecology study stations under the jurisdiction of the State Forestry Administration (22°10′N, 106°50′E). The site is within the southwestern region in the subtropical monsoon climate, with a semi-humid climate andobvious dry and wet seasons. The annual duration of sunshine is 1200 to 1600h. The precipitation is abundant, with an annual average precipitation of 1200 to 1500mm, mainly from April to September each year. The annual evaporation is 1200~1400mm, the relative humidity is 80% ~ 84%, and the average annual temperature is 20.5~21.7℃. The main types of landforms are low hills and hills. The soil is mainly composed of laterite and red soil based on Chinese soil classification, which are classified as ferralsols in the World Reference Base for Soil Resources. The soil thickness is generally higher than 80cm. Subtropical evergreen broad-leaved forests comprise the local vegetation.

There are nearly 20000hm^2 of various plantation types in the Experimental Center of Tropical Forestry. *P. massoniana* and *C. lanceolata* are the main coniferous tree species. Native broad-leaved tree species include *Quercus griffithii*, *Erythrophleum fordii*, *Castanopsis hystrix*, *Mytilaria laosensis*, *Betula alnosensis*, and *Dalbergia lanceolata*. The main alien tree species are eucalyptus and *Tectona grandis*. Among these species, *E. fordii* and *D. lanceolata* are nitrogen-fixing trees, and *Q. griffithii* is a fast-growing broad-leaved tree species with great natural regeneration ability. The near natural management of pure plantations of *P. massoniana* and *C. lanceolata* with *E. fordii* and *Q. griffithii* is widely applied in the Center, for it not only meets the need for short-period timbers and precious large-diameter timbers but also realises the natural regeneration of native broad-leaved species and achieves the goal of near natural management.

2.2 Experimental design

A single-factor and two-level stochastic block design was used for the present experiment. There were four blocks representing four replicates. Four forest types were set up in each block: near natural *P. massoniana* plantation (P[CN]), unimproved *P. massoniana* plantation (P[CK]), near natural *C. lanceolata* plantation (C[CN]), and unimproved pure *C. lanceolata* plantation (C[CK]). There were thus a total of 16 experimental plots, and the area of each experimental plot was 0.5hm^2.

The pure plantations of *P. massoniana* and *C. lanceolata* were established in 1993 after the clear-cutting of *C. lanceolata*, with an initial planting density of 2500trees/ hm^2. Felling and afforestation were repeated a total of six times within the first 3 years after initial afforestation. The release felling was carried out in the seventh year, and the first-increment felling was carried out in the 11th year, retaining a density of 1200trees/ hm^2. In 2007, near natural management was carried out, and the main management strategies included reducing the intensity of intermediate felling of pure stands of *P. massoniana* and *C. lanceolata* forests while simultaneously preserving natural regeneration (the retention density was 600trees/hm^2). In early 2008, *Q. griffithii* and *E. fordii* were replanted after the intermediate felling of *P. massoniana* and *C. lanceolata*, and the density of native replanted tree species was 600 trees/hm^2(the average density of *Q. griffithii* and *E. fordii* was 300 trees/hm^2, respectively). Uneven-aged mixed broad-leaved forests with a total density of 1200trees/ hm^2 was formed. During the whole processes, pure plantations of *P. massoniana* and *C. lanceolata* were maintained as controls, whose total density was keeping as 1200trees/hm^{2}At present, the improved plantations have become uneven-aged mixed stands with multilayer structures. A survey carried out in 2016 showed that the average diameter at breast height (*DBH*) and average tree height of *Q. griffithii* were 14.7cm and 15.4m, respectively, and the average *DBH* and average tree height of *E. fordii* were 5.2cm and 6.3m, respectively. The management processes for the four forests are shown in Table 1.

Table 1 Basic information andmanagement history of the four plantations

Year	Management	Plantation type			
		P(CK)	P(CN)	C(CK)	C(CN)
1993	Afforestation	2500trees hm^2	2500trees/hm^2	2500tree/hm^2	2500trees/hm^2
1993-1995	Tending for new plantations	6times	6times	6times	6times
2000	Released thinning	1600 trees/hm^2	1600trees/hm^2	1600trees/hm^2	1600trees/hm^2
2004	Increment felling	1200trees/hm^2	1200trees/hm^2	1200trees/hm^2	1200trees/hm^2
2007	Intensity thinning	No 1200trees/hm^2	Yes 600trees/hm^2	No 1200 t trees/hm^2	Yes 600 t trees/hm^2
2008	Complementary planting	No	Planting*Q. griffithii* and *E. fordii* with 300 t trees/hm^2 respectively	No	Planting*Q. griffithii* and *E. fordii* with 300 trees/hm^2 respectively
2009	Tending	No	2 times	No	2 times
2016	Average *DBH*	22.2±1.3 cm for *P. massoniana*	32.2±1.6 cm for *P. massoniana*	17.1±2.1 cm for *C. lanceolata*	22.3±0.8 cm for *C. lanceolata*
2016	Average height	16.7±0.5m for *P. massoniana*	17.3±0.7m for *P. massoniana*	17.1±0.4m for *C. lanceolata*	17.2±0.4m for *C. lanceolata*

2.3 Measurement and statistical analysis

2.3.1 Soil CO_2, N_2O, and CH_4 measurement

The sampling and analysis of three main greenhouse gases (N_2O, CH_4, and CO_2) in soils were performed using the static chamber method and gas chromatography[25]. Three static boxes were randomly set in each plot of P(CN), P(CK), C(CN), and C(CK). The static box was 25cm in diameter and 30cm in height. A gas extraction valve and a small fan (8cm in diameter) were installed at the top of the box to facilitate uniform gas mixing during sampling. The bottom of the box was buried in the ground at a depth of 5cm 2 or 3 months before initial sampling[21]. From October 2014 to September 2015, sampling in all four plantations (a total of 16 plots) was completed from 9: 00 a. m. to 11: 00 a. m. on one day at the end of each month, and measured values were used to calculate the average daily gas exchange flux[2]. At each sampling period, 100ml gas samples were taken from static boxes with a medical syringe and timed with a stopwatch. The gas was sampled at 0, 15 and 30min intervals. Three gas samples at each chamber were collected. The sample was injected into a polyethylene polythene sampling bag, cryopreserved, and sent back to the laboratory for measurement. We analysed gas samples for their N_2O, CH_4, and CO_2 concentrations using a gas chromatograph (Agilent 4890D, Agilent, Santa Clara, CA, USA). The flux of N_2O, CH_4, and CO_2 was calculated using the following formula:

$$F = \rho \times \frac{V}{A} \times \frac{P}{P_0} \times \frac{T_0}{T} \times \frac{dC_1}{dt} \tag{1}$$

where F is the mass change of the gas in the observation box per area and per unit time, ρ is the density of the measured gas in the standard state, V is the gas volume in the box, A is the area covered by the box, P is the atmospheric pressure at the sampling point, T is the absolute temperature at the time of sampling, $dC1/dt$ is the linear slope of gas concentration over time during the sampling, and P_0 and T_0 are the atmospheric pressure and absolute temperature in the standard state, respectively.

2. 3. 2 Micro-environmental data measurement

Temperature and atmospheric pressure were measured with a thermometer and a barometer at the same time as sampling. The temperature of the soil at a depth of 5cm was measured with a portable digital thermometer. Soil moisture (volumetric water content) at a depth of 5cm was measured with an MPKit hygrometer and converted into water filled pore space (*WFPS*) using the following formula:

$$WFPS(\%) = \frac{Vol}{1\frac{bd}{2.65}} \tag{2}$$

where *bd* is bulk density, *vol* is volumetric water content, and 2. 65 is the density of quartz.

2. 3. 3 Soil and litterfall sampling and measurements

After the fresh and semi-decomposed litter residueat the upper surface of soil was stripped from the woodland near each static box in the four plantations, twelve soil samples at a depth of 0 to 10cm were randomly collected using a stainless steel soil auger with an inner diameter of 8. 7cm. These samples were placed in mixed sample bags for preservation. The soil samples were then taken back to the laboratory to remove coarse roots, rubble, and other impurities using a 2mm aperture screen and air dried for physical and chemical analysis.

Six 1m×1m leaf litterfall collectors made of nylon gauze (1mm aperture) were set up randomly in the woodland near each static box in the four plantations. Leaf litterfall was collected once a month, and the leaves, branches, skin, and fruits were picked and sorted by tree species and organ and dried at 65℃ to a constant weight. A total of 12 collections of litterfall samples were prepared over the course of a year.

2. 3. 4 Fine root sampling and measurements

Fine root biomass was determined by the continuous soil drilling method. Fine roots (diameter <2mm) were sampled in the 0~10cm soil layer using a stainless steel soil auger with a diameter of 8. 7cm for sorting and collection. Twelve soil drillings collected for fine roots biomass determination were carefully sorted out at random at the end of each bimonthly period in a sample plot of the four different plantations. In each plantation, the fine root samples were collected for six times throughout each year during the experiment. The fine root samples were weighed after drying at 65℃ to a constant weight. The average fine root biomass of the six sampling periods was used as the average fine root biomass[26].

2. 3. 5 Biogeochemical properties analysis of plant and soil samples

Soil bulk density was measured using the volumetric ring during field sampling. Soil pH value was measured using glass electrodes after leaching the soil with 1 mol/L KCl solution. Organic C contents of the soil, litterfall, and fine root samples were determined by the potassium dichromate external heating method, and total N was determined by the Kjeldahl method. Soil ammonium and nitrate N contents were determined by spectrophotometry. Soil available N was analyzed through quantification of alkali-hydrolysable N in a Conway diffusion unit with Devarda' s alloy in the outer chamber and boric acid-indicator solution in the inner chamber[27]. Soil total P was measured by inductively-coupled plasma optical-emission spectrometry (ICP-OES). Soil microbial biomass C and N were determined by the fumigation-extraction method[28].

2.3.6 Statistical analysis

A one-way analysis of variance (ANOVA) was employed to determine the differences among the annual mean fluxes of soil greenhouse gases, as well as the biogeochemical properties of soil and plant samples in different plantations. Regression models were used to analyse the correlation between soil greenhouse gas flux and soil temperature and soil moisture in the four plantations. Multiple linear regression analyses were used to determine the main factors influencing differences in soil greenhouse gas flux among the four plantations. All of the data in the study followed a normal distribution and satisfied the test of homogeneity of variance. We performed statistical analyses using Windows SPSS 19.0. Statistical significance was determined at a threshold of $P<0.05$.

3 RESULTS

3.1 Soil temperature and moisture

Soil temperature and *WFPS* in the four plantations varied seasonally. The soil was cooler and drier during November 2014 and February 2015, whereas the soil was warmer and more humid from March 2015 to August 2015 (Figure 1). The sampling period in December 2014 was unusual in that it was a short wet period within the cool-dry season. January 2015 and July 2015 could be classified as within the cool-dry season and warm-humid season, respectively.

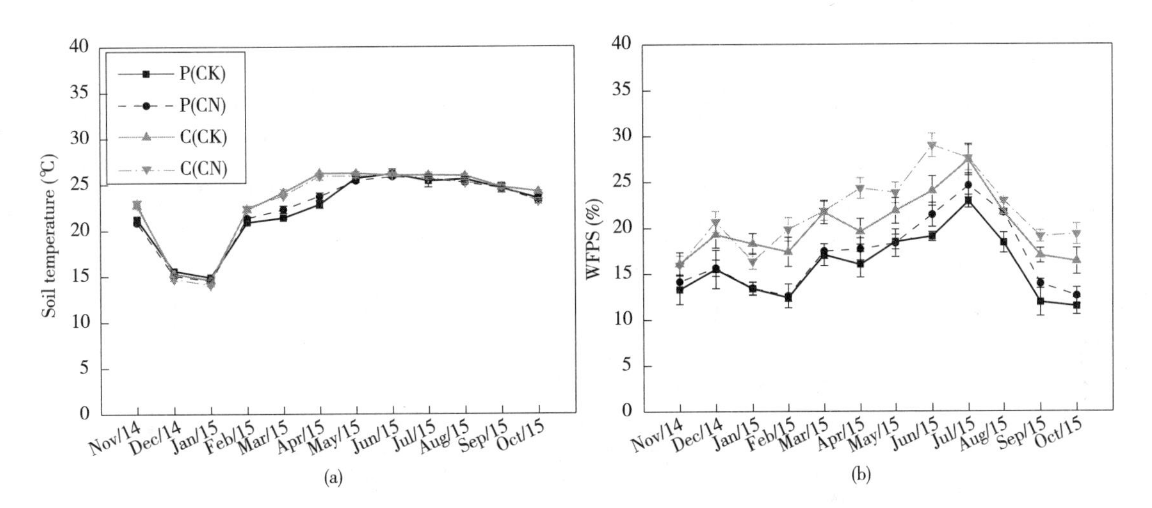

Figure 1 Seasonal patterns of soil temperature (a) and soil water filled pore space (WFPS) (b) in the four plantations

3.2 Seasonal variation in soil greenhouse gas flux

The soil CO_2 and N_2O emission rates in the four plantations showed significant seasonal variations. The CO_2 emission rate was highest in July, when it was hot and humid, but lowest in January, during the dry season. All plantations had similar seasonal patterns for N_2O emission and CH_4 uptake (Figure 2).

Soil CO_2 emission rates were positively correlated with soil temperature and soil moisture (Figure 3a and 3b), but the correlation between CO_2 flux and soil moisture was significant in P(CN) only (Table 2). N_2O emission was significantly and positively correlated with soil temperature in both P(CK) and C(CK) (Figure 3c). However, no significant correlation was found between soil N_2O flux and soil moisture (Figure 3d and Table 2).

CH_4 flux had a significant correlation with soil temperature in P(CK) only (Figure 3e). In the near natural and pure *C. lanceolata* plantations, soil CH_4 uptake rates decreased with seasonal increases in soil moisture (Figure 3f and Table 2).

When combining soil temperature and moisture in a regression model, significant relations were detected for the CO_2 flux in C(CK) and C(CN), N_2O flux in P(CK), and CH_4 flux in each pure forest (Table 2).

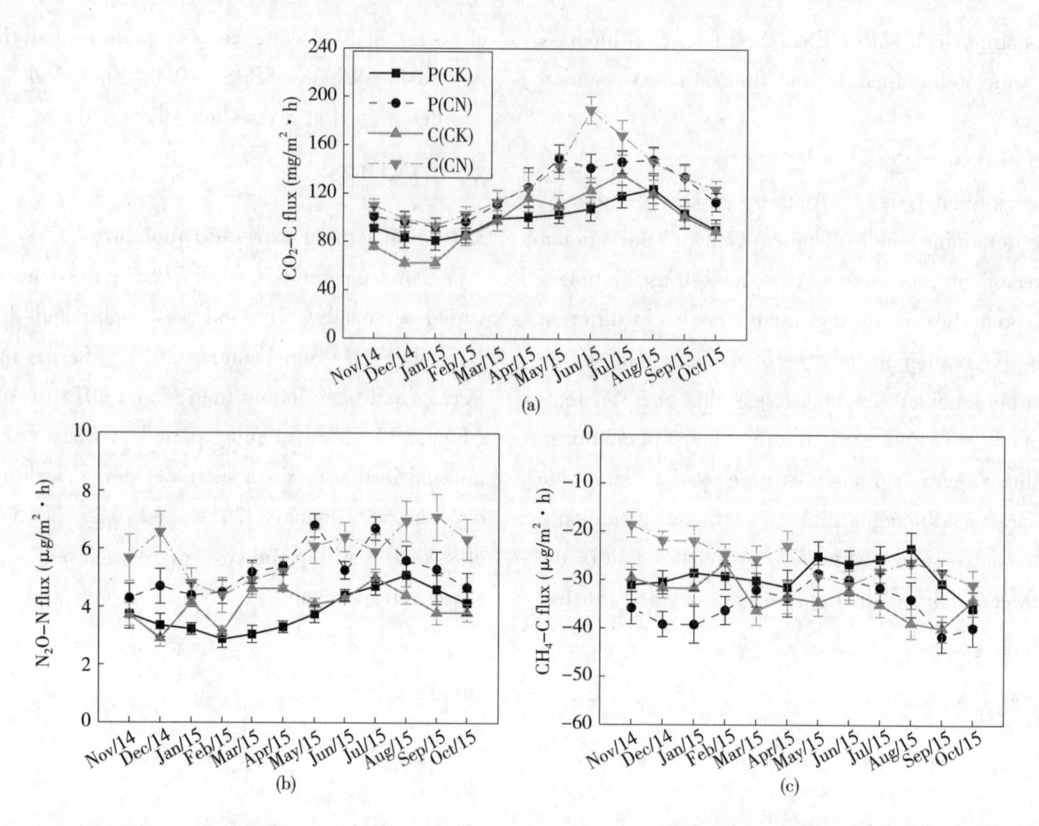

Figure 2 Seasonal patterns of soil CO_2, N_2O, and CH_4 flux in the four plantations

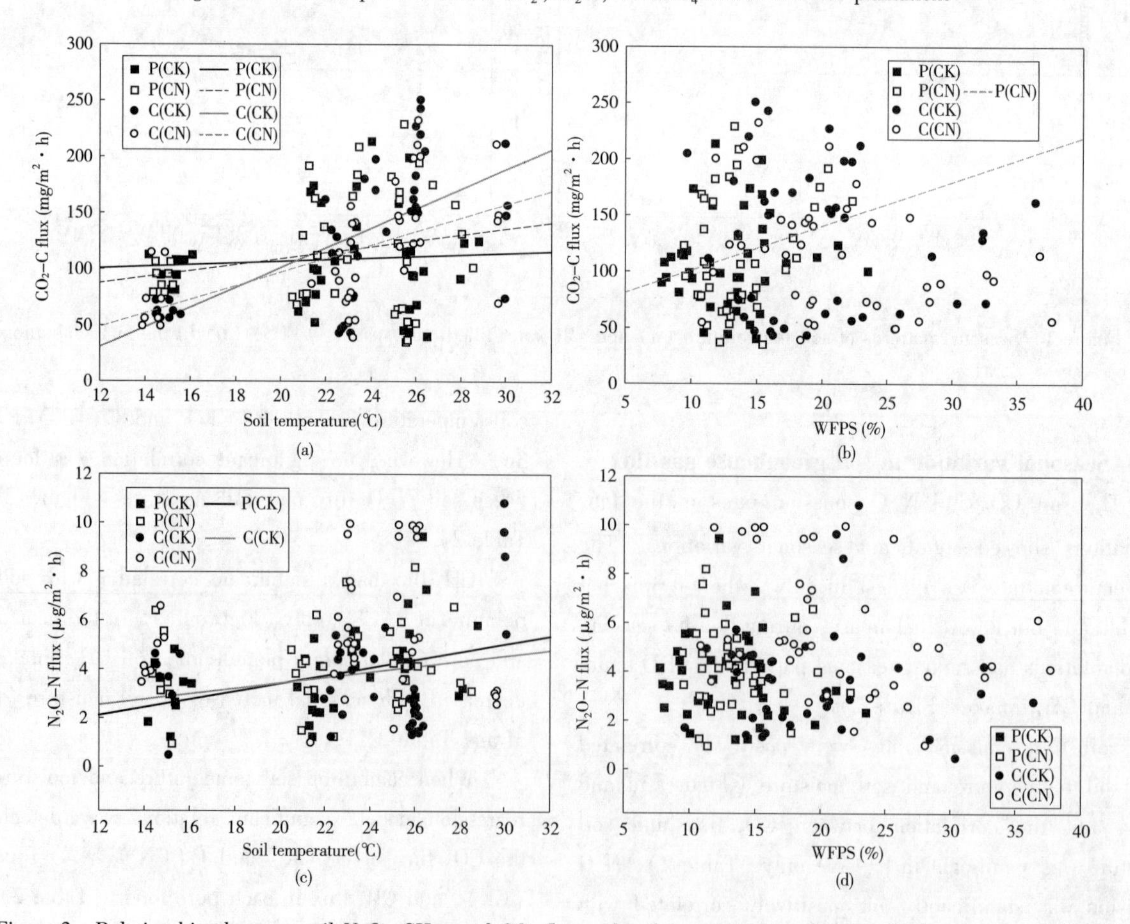

Figure 3 Relationships between soil N_2O, CH_4, and CO_2 flux and soil temperature and water filled pore space (WFPS) in the four plantations. Significant correlations were shown in solid and dashed lines (P<0.05)

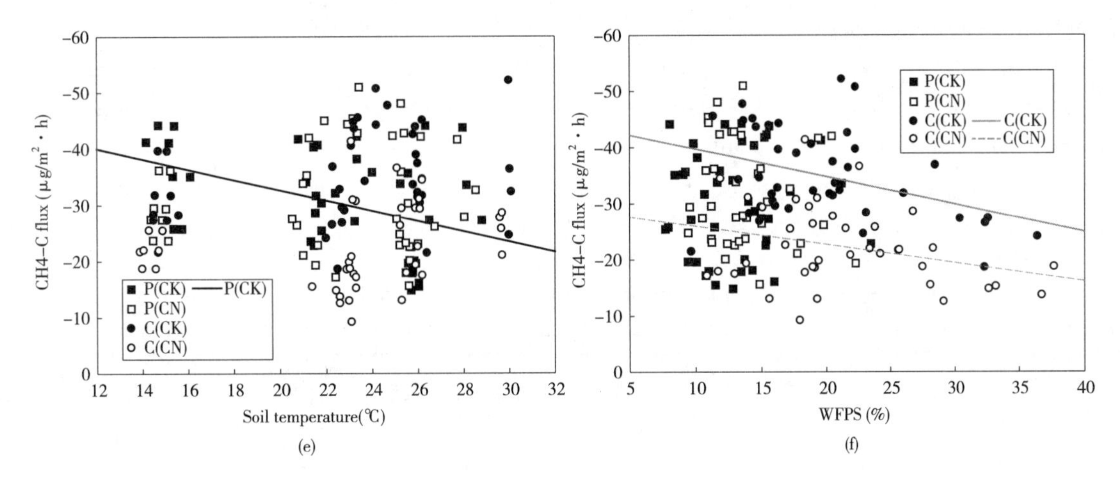

Figure 3 Relationships between soil N_2O, CH_4, and CO_2 flux and soil temperature and water filled pore space (WFPS) in the four plantations. Significant correlations were shown in solid and dashed lines ($P<0.05$)(续)

Table 2 Models, coefficients of determination (R^2) and p-values of regressions between soil greenhouse gas flux and soil temperature (T) and WFPS (W) in the four plantations. The rows of "T+W" represent the models considering both T and W, while others are those using T and W separately

Plantation type	P(CK)	P(CN)	C(CK)	C(CN)
		CO_2-C flux (mg/m² h)		
T(℃)	$CO_2=0.71T+92.31$	$CO_2=2.67T+55.83$	$CO_2=9.14T+80.44$	$CO_2=6.05T+24.15$
	$R^2=0.11$, $P<0.05$	$R^2=0.15$, $P<0.05$	$R^2=0.37$, $P<0.001$	$R^2=0.30$, $P<0.001$
W(%)	$R^2=0.01$, $P=0.47$	$CO_2=3.92W+61.71$	$R^2=0.06$, $P=0.61$	$R^2=0.04$, $P=0.28$
		$R^2=0.13$, $P<0.05$		
T(℃)+W(%)	$R^2=0.01$, $P=0.80$	$R^2=0.09$, $P=0.17$	$CO_2=9.19T+0.98W-103.51$	$CO_2=5.34T-1.28W+19.03$
			$R^2=0.41$, $P<0.001$	$R^2=0.33$, $P<0.01$
		N_2O-N flux (μg/m² h)		
T(℃)	$N_2O=0.16T+0.17$	$R^2=0.02$, $P=0.43$	$N_2O=0.11T+1.19$	$R^2=0.00$, $P=0.77$
	$R^2=0.16$, $P<0.01$		$R^2=0.16$, $P<0.05$	
W(%)	$R^2=0.03$, $P=0.297$	$R^2=0.01$, $P=0.54$	$R^2=0.01$, $P=0.90$	$R^2=0.00$, $P=0.77$
T(℃)+W(%)	$N_2O=0.19T-0.13W+1.32$	$R^2=0.03$, $P=0.54$	$R^2=0.06$, $P=0.30$	$R^2=0.15$, $P=0.05$
	$R^2=0.22$, $P<0.01$			
		CH_4-C flux (μg/m² h)		
T(℃)	$CH_4=0.92T-51.07$	$R^2=0.01$, $P=0.13$	$R^2=0.05$, $P=0.15$	$R^2=0.04$, $P=0.25$
	$R^2=0.17$, $P<0.01$			
W(%)	$R^2=0.00$, $P=0.998$	$R^2=0.01$, $P=0.454$	$CH_4=0.49W-44.75$	$CH_4=0.24W-29.21$
			$R^2=0.15$, $P<0.01$	$R^2=0.10$, $P<0.05$
T(℃)+W(%)	$CH_4=0.96T-0.23W-48.95$	$R^2=0.03$, $P=0.56$	$CH_4=-0.23T+0.44W-38.50$	$R^2=0.09$, $P=0.16$
	$R^2=0.18$, $P<0.05$		$R^2=0.16$, $P<0.05$	

3.3 The effects of plantation type on soil greenhouse gas flux

Near natural management had significant effects on the annual average emission rate of soil CO_2 and N_2O, and the uptake rate of soil CH_4 (Table 3). The soil CO_2 emission rate in the near natural *P. massoniana* plantation was 17.7% higher than that in the control forest, and the soil CO_2 emission rate of the near natural *C. lanceolata plantation was* 14.5% higher than control. This indicates that the soil CO_2 emission rates for *P. massoniana* and *C. lanceolata* plantations were accelerated by near natural management. Compared with the control forests, the near natural management enhanced the annual average soil N_2O emission rate by 19.4% and 47.4% in the *P. massoniana* and *C. lanceolata* plantation, respectively. Therefore, the soil N_2O emission rates for *P. massoniana* and *C. lanceolata* plantations increased as a result of near natural management.

Table3 Annual average flux of soil greenhouse gas in the four plantations. Data are shown as means ± standard errors (n=4). Values designated by the different letters within each variable are significant at P<0.05

Plantation type	P(CK)	P(CN)	C(CK)	C(CN)
CO_2-C flux (mg/m^2 h)	103.3±9.7cd	121.6±4.8ab	112.4±8.9bc	128.7±5.0a
N_2O-N flux (μg/m^2 h)	3.6±0.1cd	4.3±0.5b	3.8±0.2bc	5.6±1.1a
CH_4-C flux (μg/m^2 h)	−34.7±1.7c	−27.2±1.6b	−34.9±2.8c	−22.4±1.8a

The average soil CH_4 flux was negative for all the four plantations, which indicates that all the forest soils were functioning as CH_4 sinks. The annual average soil CH_4 uptake rate for the near natural plantations was 21.6% and 55.8% lower than the corresponding controls, as for *P. massoniana* and *C. lanceolata*, respectively (Table 3). Therefore, near natural management reduces the soil CH_4 uptake rate of *P. massoniana* and *C. lanceolata* plantations.

3.4 Main influencing factors on soil greenhouse gas flux

Compared with control, the near naturalmanagement of each plantation increased the fine root biomass, soil temperature, pH, and the contents of soil organic C, available N, NH_{4+}-N, NO_{3-}-N, microbial biomass C, and microbial biomass N, while reduced the C: N of leaf litter and fine root, as well as soil total P and C: N (P<0.05, Table 4).

Table 4 The biogeochemical properties in the four plantations. Data are shown as means ± standard errors (n=4). Values designated by the different letters within each variable are significant at P< 0.05

Properties	P(CK)	P(CN)	C(CK)	C(CN)
Litterfall quantity (t/hm^2 r)	10.23±0.94a	10.84 ± 0.49a	9.02 ± 0.19b	9.54±0.34b
Fine root biomass(t/hm^2)	0.81±0.07b	1.36±0.22a	0.64±0.26b	1.33±0.28a
C: Nof leaf litter	48.07±4.82c	37.49±4.77d	68.13±8.12a	52.70±6.92b
C: Nof fine root	57.53±10.7a	39.70±5.70c	55.38±3.30a	45.70±4.40b
Soilporosity (%)	56.80±2.83a	56.04±2.58a	49.05 ±4.99b	45.17 ±4.86b
Soil temperature (℃)	22.15±0.12d	22.47±0.17c	22.73±0.04b	23.04±0.03a
SoilWFPS (%)	13.06±0.56b	13.67±0.49b	19.91±1.00a	21.28±1.06a
Soil pH	4.18±0.04d	4.31±0.08c	4.67±0.07b	4.91±0.20a
Soil organicC(g/kg)	25.99±1.32b	29.15±2.42a	17.24±1.85d	21.61±2.58c
Soil total N(g/kg)	2.58±0.04	3.28±0.12	2.29±0.15	3.32±0.13
Soil available N (mg/kg)	94.37±3.94b	103.32±5.62a	77.0±9.07c	96.25±7.27ab
Soil totalP (g/kg)	0.28±0.01a	0.25±0.02b	0.24±0.03b	0.21±0.01c
Soil C: N	17.06±0.50a	15.34±0.72c	16.42±0.14b	15.16±0.46c
SoilNH_{4+}-N content (mg/kg)	20.30±2.07b	26.67±3.35a	18.44±2.17b	24.56±4.02a
SoilNO_{3-}-N content (mg/kg)	21.97±1.83b	25.00 ± 2.21a	18.36±2.28b	24.65 ±4.19a

(续)

Properties	P(CK)	P(CN)	C(CK)	C(CN)
Soil microbial biomassC (mg/kg)	301.12±24.54b	388.12±11.76a	234.44±29.49c	312.50±32.51b
Soil microbial biomass N (mg/kg)	39.07±6.59bc	53.30±8.11a	36.40±6.45c	46.51±4.21ab

To explain the observed variations in annual average soil greenhouse gas flux among the plantations, the first "stepwise" multiple linear regression model was performed by using all the tested biogeochemical properties in the plantations. The model performed on CO_2 emission indicated that soil temperature and C: N ratio of fine root explained 77.4% of the variation in the soil CO_2 emission rate among the plantations (R^2 = 0.774, <0.001; Table 5). Other independent variables, such as the C: N ratio of leaf litter, soil organic C, soil pH, and soil nitrogen content, were excluded in the model owing to their non-significance or evidence of multicollinearity. The C: N ratio of fine root was negatively correlated with the annual average soil CO_2 emission rate, whereas soil temperature was positively correlated with the annual average soil CO_2 emission rate (Table 5). This indicates that the annual average CO_2 uptake rate increases with increasing soil temperature and decreasing C: N ratio of fine root.

Table 5 Results of multiple linear regression analysis of biogeochemical parameters and annual average soil greenhouse gas flux in the four plantations

Parameters	Models
CO_2-C flux (mg/m^2 · h) (Y_1) C: N ratio of fine root (X_1) Soil temperature (℃) (X_2)	$Y_1=-0.707X_1+16.2X_2-217.0$, $R^2=0.774$, $P<0.001$
N_2O-N flux (μg /m^2 · h) (Y_2) C: N ratioof leaf litter (X_3) Soil available N (mg /kg) (X_4)	$Y_2=-0.044X_3+0.16X_4+5.886$, $R^2=0.693$, $P<0.001$
CH_4-C flux (μg/m^2 · h) (Y_3) C: N ratioof leaf litter (X_5)	$Y_3=0.343X_5$~6.026, $R^2=0.624$, $P<0.001$

Another multiple linear regression model examined the variation in the average soil N_2O flux among the four plantations showed that the C: N ratio of leaf litter and soil available N explained 69.3% of the variation in the annual average soil N_2O emission rate ($R^2=0.693$, $P<0.001$; Table 5). The annual average soil N_2O emission rate was negatively correlated with the C: N ratio of leaf litter but positively correlated withsoil available N content. This indicates that the annual average N_2O emission rate increases with decreasing C: N ratio in leaf litter and increasing soil available N content.

A final multiple linear regression model showed that the C: N ratio of leaf litter was the only variable that explained a significant proportion (62.4%) of the variation in the annual average soil CH_4 uptake rate among the plantations (R^2 = 0.624, $P<0.001$; Table 5). The annual average soil CH_4 flux was positively correlated with the C: N ratio of leaf litter.

4 CONCLUSIONS

Near natural management increased the average soil CO_2 and N_2O emission rates in *P. massoniana* and *C. lanceolata* plantations and reduced average soil CH_4 absorption rates. The differences in the CO_2 emission rate among plantations can be attributed mainly to the C: N ratio of fine root, whereas the differences in the N_2O emission rate can be attributed to soil available N content and the C: N ratio of leaf litter. The variation in the CH_4 uptake rate can be attributed only to the C: N ratio of leaf litter. The results of the present study show that near natural management of *P. massoniana* and *C. lanceolata* plantations may increase the emission of greenhouse gases in subtropical China. Therefore, plantation enrichment strategies should take into account potential impacts on greenhouse gas flux. Other research is needed to evaluate the effects of near natural forest management on global climate change.

Acknowledgments: We are grateful to Ji Zeng, Zhang Zhao, Zhaoying Li, Lili Li, Wenlian Zhou, Yuan He, Zhongguo Li, Hai Chen and Yanling Huang of Chinese Academy of Forestry's Experimental Center of Tropical Forestry for their assistance in field sampling and data collection. We also gratefully acknowledge the support from the Agriculture College, Guangxi University. This study was funded by the fundamental research funds of CAF (CAFYBB2014QA033), Guangxi forestry science and technology projects (Document of Guangxi forestry department [2016] No. 37), Guangxi Natural Science Foundation (2014GXNSFBA118100), International cooperation projects (2015DFA31440) and China's National Natural Science Foundation (31470627).

Author Contributions: Angang Ming analyzed data and drafted the manuscript. Yujing Yang revised the manuscript and participated in collecting the experiment data. Shirong Liu conceived and designed the work. Hui Wang was involved in planning of study and designing of the work. Yuanfa Li, Daoxiong Cai, and Hongyan Jia contributed to technical advice and refine the ideas of this paper. The remaining authors contributed to carrying out additional analyses and finalizing this paper.

Conflicts of Interest: The authors declare no conflict of interest.

REFERENCES

[1] IPCC. Climate change 2014: Synthesis report. Contribution of working groups I, II, and III to the fifth assessment report of the intergovernmental panel on climate change[J]. In. Geneva, Switzerland, 2014.

[2] TANG X, LIU S, ZHOU G, ZHANG D C. Soil-atmospheric exchange of CO_2, CH_4, and N_2O in three subtropical forest ecosystems in southern China[J]. Glob. Chang. Biol., 2006, 12: 546-560.

[3] LAL, R. Forest soils and carbon sequestration[J]. For. Ecol. Manag., 2005, 220: 242-258.

[4] BUTTERBACH-BAHL K, GASCHE R, BREUER L, et al. Fluxes of NO and N_2O from temperate forest soils: impact of forest type, N deposition and of liming on the NO and N_2O emissions[J]. Nutr. Cycl. Agroecosys., 1997, 48, 79-90.

[5] PITZ S, MEGONIGAL J P. Temperate forest methane sink diminished by tree emissions[J]. New Phytol., 2017, 214: 1432-1439.

[6] MER J L, ROGER P. Production, oxidation, emission and consumption of methane by soils: A review[J]. Eur. J. Soil Biol, 2001, 37, 25-50.

[7] BODELIER P L E, LAANBROEK H J. Nitrogen as a regulatory factor of methane oxidation in soils and sediments [J]. FEMS Microbiol. Ecol., 2004, 47: 265-277.

[8] SOLOMON S, QIN D, MANNING M, et al. Changes in atmospheric constituents and in radiative forcing. In: Climate change 2007: The physical science basis. Contribution of working group I to the fourth assessment report of the intergovernmental panel on climate change[M]. Cambridge: Cambridge University Press, 2007.

[9] BUTTERBACH-BAHL K, GASCHE R, WILLIBALD G, et al. Exchange of N-gases at the Höglwald Forest - A summary[J]. Plant Soil, 2002, 240, 117-123.

[10] GUO L B, GIFFORD R M. Soil carbon stocks and land use change: a meta analysis[J]. Glob. Chang. Biol., 2002, 8: 345-360.

[11] WU C, WEI X, MO Q, et al. Effects of stand origin and near-natural restoration on the stock and structural composition of fallen trees in mid-subtropical forests[J]. Forests, 2015, 6: 4439-4450.

[12] EMBORG J, CHRISTENSEN M, HEILMANNCLAUSEN J. The structural dynamics of Suserup Skov, a near-natural temperate deciduous forest in Denmark[J]. For. Ecol. Manag., 2000, 126: 173-189.

[13] WANG G, LIU F. The influence of gap creation on the regeneration of *Pinus tabuliformis* planted forest and its role in the near-natural cultivation strategy for planted forest management[J]. For. Ecol. Manag., 2011, 262, 413-423.

[14] JONARD M, ANDRÉ F, JONARD F, et al. Soil carbon dioxide efflux in pure and mixed stands of oak and beech [J]. Ann. For. Sci., 2007, 64: 141-150.

[15] ULLAH S, FRASIER R, KING L, et al. Potential fluxes of N_2O and CH_4 from soils of three forest types in Eastern Canada [J]. Soil Biol. Biochem., 2008, 40: 986-994.

[16] AMANDA M, DAN P, ANGELA B H. Methane and nitrous oxide emissions from mature forest stands in the boreal forest, Saskatchewan, Canada[J]. For. Ecol. Manag., 2009, 258: 1073-1083.

[17] MENYAILO O V, HUNGATE B A. Interactive effects of tree species and soil moisture on methane consumption [J]. Soil Biol. Biochem., 2003, 35: 625-628.

[18] BORKEN W, BEESE F. Methane and nitrous oxide fluxes of soils in pure and mixed stands of European beech and Norway spruce[J]. Eur. J. Soil Sci., 2006, 57: 617-625.

[19] SHI Z, LI Y, WANG S, et al. Accelerated soil CO_2 efflux after conversion from secondary oak forest to pine plantation in southeastern China[J]. Ecol. Res. , 2009, 24: 1257-1265.

[20] BARRENA I, MENÉNDEZ S, DUÑABEITIA M, et al. Greenhouse gas fluxes (CO_2, N_2O and CH_4) from forest soils in the Basque Country: Comparison of different tree species and growth stages[J]. For. Ecol. Manag., 2013, 310: 600-611.

[21] BRÉCHET L, PONTON S, ROY J, et al. Do tree species characteristics influence soil respiration in tropical forests? A test based on 16 tree species planted in monospecific plots[J]. Plant Soil , 2009, 319: 235-246.

[22] LEITNER S, SAE-TUN O, KRANZINGER L, et al. Contribution of litter layer to soil greenhouse gas emissions in a temperate beech forest[J]. Plant Soil , 2016, 403: 455-469.

[23] YAMULKI S, MORISON J I L. Annual greenhouse gas fluxes from a temperate deciduous oak forest floor[J]. Forestry , 2017, 90: 1-12.

[24] DANIEL L, WHENDEEL S, MATTEO D. Temporal dynamics in soil oxygen and greenhouse gases in two humid tropical forests[J]. Ecosystems , 2011, 14: 171-182.

[25] WANG H, LIU S, WANG J, et al. Effects of tree species mixture on soil organic carbon stocks and greenhouse gas fluxes in subtropical plantations in China[J]. For. Ecol. Manag., 2013, 300: 4-13.

[26] JANSSENS I A, SAMPSON D A, CURIELYUSTE J, et al. The carbon cost of fine root turnover in a Scots pine forest[J]. For. Ecol. Manag. , 2002, 168: 231-240.

[27] SHEN J, LI R, ZHANG F, et al. Crop yields, soil fertility and phosphorus fractions in response to long-term fertilization under the rice monoculture system on a calcareous soil[J]. Field Crop Res., 2004, 86: 225-238.

[28] HUANG X, LIU S, WANG H, et al. Changes of soil microbial biomass carbon and community composition through mixing nitrogen-fixing species with *Eucalyptus urophylla* in subtropical China[J]. Soil Biol. Biochem. , 2014, 73: 42-48.

[29] JANSSENS I A, LANKREIJER H, MATTEUCCI G, et al. Productivity overshadows temperature in determining soil and ecosystem respiration across European forests[J]. Glob. Chang. Biol. , 2001, 7: 269-278.

[30] EPRON D, BOSC A, BONAL D, et al. Spatial variation of soil respiration across a topographic gradient in a tropical rain forest in French Guiana[J]. J. Trop. Ecol. , 2006, 22: 565-574.

[31] LIVESLEY S J, KIESE R, MIEHLE P, et al. Soil-atmosphere exchange of greenhouse gases in a *Eucalyptus marginata* woodland, a clover-grass pasture, and *Pinus radiata* and *Eucalyptus globulus* plantations[J]. Glob. Chang. Biol. , 2009, 15: 425-440.

[32] XU X, HIRATA E. Decomposition patterns of leaf litter of seven common canopy species in a subtropical forest: N and P dynamics[J]. Plant Soil , 2005, 273: 279-289.

[33] ZHANG Y, GUO S, LIU Q, et al. Responses of soil respiration to land use conversions in degraded ecosystem of the semi-arid Loess Plateau[J]. Ecol. Eng. , 2015, 74: 196-205.

[34] HU X, LIU L, ZHU B, et al. Asynchronous responses of soil carbon dioxide, nitrous oxide emissions and net nitrogen mineralization to enhanced fine root input[J]. Soil Biol. Biochem. , 2016, 92: 67-78.

[35] ROSENKRANZ P, BRÜGGEMANN N. Soil N and C trace gas fluxes and microbial soil N turnover in a sessile oak (*Quercus petraea* (Matt.) Liebl.) forest in Hungary[J]. Plant Soil , 2006, 286: 301-322.

[36] GUNDERSEN P, CHRISTIANSEN J R, ALBERTI G, et al. The response of methane and nitrous oxide fluxes to forest change in Europe[J]. Biogeosciences , 2012, 9: 3999-4012.

[37] PILEGAARD K, SKIBA U, AMBUS P, et al. Factors controlling regional differences in forest soil emission of nitrogen oxides (NO and N_2O) [J]. Biogeosciences , 2006, 3: 651-661.

[38] MORISHITA T, SAKATA T, TAKAHASHI M, et al. Methane uptake and nitrous oxide emission in Japanese forest soils and their relationship to soil and vegetation types[J]. Soil Sci. Plant Nutr., 2007, 53: 678-691.

[39] WESLIEN P, KLEMEDTSSON A Å K, BÖRJESSON A G, et al. Strong pH influence on N_2O and CH_4 fluxes from forested organic soils[J]. Eur. J. Soil Sci. , 2009, 60: 311-320.

[40] ROWLINGS D W, GRACE P R, KIESE R, et al. Environmental factors controlling temporal and spatial variability in the soil - atmosphere exchange of CO_2, CH_4 and N_2O from an Australian subtropical rainforest [J]. Glob. Chang. Biol., 2012, 18: 726-738.

[41] GÜTLEIN A, GERSCHLAUER F, KIKOTI I, et al. Impacts of climate and land use on N_2O and CH4 fluxes from tropical ecosystems in the Mt. Kilimanjaro region, Tanzania [J]. Glob. Chang. Biol. , 2018, 24: 1239-1255.

[42] WEN Y, CORRE M D, SCHRELL W, et al. Gross N_2O emission and gross N_2O uptake in soils under temperate spruce and beech forests [J]. Soil Biol. Biochem. , 2017, 112: 228-236.

[43] WERNER C, KIESE R, BUTTERBACH-BAHL K. Soil-

atmosphere exchange of N_2O, CH_4, and CO_2 and controlling environmental factors for tropical rain forest sites in western Kenya [J]. J. Geophys. Res., 2007, 112: D3308.

[44] VERCHOT L V, DAVIDSON E A, CATTÂNIO J H, et al. Land-use change and biogeochemical controls of methane fluxes in soils of eastern Amazonia[J]. Ecosystems, 2000, 3: 41-56.

[45] BÁRCENA T G D, IMPERIO L, GUNDERSEN P, et al. Conversion of cropland to forest increases soil CH_4 oxidation and abundance of CH_4 oxidizing bacteria with stand age[J]. Appl. Soil Ecol., 2014, 79: 49-58.

[原载：Forests 2018, 9, 229]

封山对石灰岩山地植被恢复的影响

卢立华[1] 贾宏炎[1] 冯昌林[1] 何日明[1] 黄文龙[2] 冼少达[2] 史作民[3]
([1]中国林业科学研究院热带林业实验中心，广西凭祥 532600；
[2]广西田阳县林业局，广西田阳 533600；
[3]中国林业科学研究院森林生态环境与保护研究所，北京 100091)

摘　要　在广西田阳县开展了封山育林对石灰岩山地植被恢复的影响研究。结果表明：封山3年后石灰岩山地植被恢复明显，2个试验地的物种多样性、重要值及植被覆盖率都发生了不同程度的改变。其中：A试验地的物种数增加了13种，增量达59.09%；重要值降多升少，植物在群落中的地位没有发生根本的改变；植被覆盖率提高了40%。B试验地的物种数增加4种，增量仅为19.05%；在重要值中，木本及藤本植物降多升少，它们在群落中的地位没有发生明显改变，而草本植物则升多降少，部分植物的地位已发生了根本改变；植被覆盖率达到了100%。

关键词　封山；石灰岩山地；植被恢复

Effects of Hillside Closing on Vegetation Restoration in Limestone Mountains

LU Lihua[1], JIA Hongyan[1], FENG Changlin[1], HE Riming[1], HUANG Wenlong[2], XIAN Shaoda[2], SHI Zuomin[3]

([1]*Tropical forestry experimental center of Chinese academy of forestry*, *Pingxiang* 532600, *Guangxi*, *China*;
[2]*Forestry bureau of Tianyang county*, *Guangxi*, *Tianyang* 533600, *Guangxi*, *China*;
[3]*Institute of forest ecology*, *environment and Protection*, *Chinese academy of forestry*,
key lab on forest ecology and environment of state forestry administration, *Beijing* 100091, *China*)

Abstract: Effects of hillside closing on vegetation restoration in different sites in lime stone mountains were studied in Tianyang county, Guangxi Autonomous Region. The results showed that the effects of hillside closing on vegetation restoration were obvious in different sites after 3 years. There were 13 new plant species occurring in experimental site A and the increase ratio was 50.09%. The importance value of majority species decreased and minority species increased though their significance had not changed obviously in community. Total vegetation coverage increased 40% obviously. There were only 4 new plant species occurring in experimental B which w as better site and the increase ratio was only 19.05% . Although importance value of most woody and liana species decreased, their significance had not changed obviously in community. But importance value of most herb species increased and significance of part of them had changed radically in community. The first most dominant species *Eupatorium odoratum* Linn. changed to the second most dominant while *Microstegium vagans* (N ees) A. Camus changed to be the first most important from the second most important species. The vegetation coverage as 100%.

Key words: hillside closing; limestone mountains; vegetation restoration

岩溶地貌(亦称喀斯特地貌)是世界上分布较广泛的一种地貌类型。我国的岩溶地貌分布广泛，岩溶面积占国土面积35.93%[1]，主要分布在我国南方的贵州、云南、广西、湖南、四川、湖北、广东及重庆等地，以及北方的山西、山东、河南、河北一带。据估测，我国岩溶地区中严重石漠化的面积达4163万 km^2，短期内具有潜在石漠化趋势的土地8176万 km^2，尤其以贵州高原为中心，形成云南、贵州、广西大面积连片的西南岩溶区，面积约28万 km^2，是我国连片面积最大、石漠化最严重的区域[2]。广西是岩溶（以石灰岩溶为主）面积分布较广和石漠化最严重的省区之一，达8195万 km^2，占全区总面积37.8%[3]。由于岩溶区群众长期采取刀耕火种，广种薄收的耕作方式，导致了石漠化的加剧，致使严重石漠化地区已丧失了人类生存的基本条件，因此，治理石漠化已十分必要和迫切。

目前，我国在石漠化治理方面已取得了不少的治理经验和成熟技术[4-13]。在石漠化治理中，最核心的问题是如何尽可能快地恢复石漠化地区的森林植被，通过采取“封、造、管、改”综合措施是恢复森林植被、修复森林生态系统的最佳选择[14]。封山育林是恢复和建设植被最省钱、省工的办法[15]。本文通过在气候条件较一致的石灰岩山地中选择土壤条件相对较好的地段与土壤条件相对较差的地段同时开展封山对植被恢复影响的对比试验，探讨在不同立地条件下封山育林对石灰岩山地植被恢复的影响，以期为我国石灰岩地区的植被恢复及生态重建提供依据及技术经验。

1　研究区概况

研究地点位于广西田阳县那坡镇永常村，地理位置为23°29′8″~24°7′6″N，106°22′7″~107°8′22″E，属南亚热带季风气候，年平均气温21.5~22.1℃，≥10℃年积温7600~7835℃，年降水量1100.8~1374.2mm，年蒸发量1572.6~1930.2mm，年均日照时数1911.9h；无霜期352d，土壤为石灰岩发育的棕色石灰土。

选择土壤条件差异较大，在实施封山育林试验前其植被种类及植被覆盖度也不一致的不同地段的东南坡分别设置试验地A和B，开展封山对植被恢复影响研究。试验地相隔500m左右，故试验地A、B所处的周边环境条件，尤其是植被条件对它们在封山育林植被恢复过程中的影响较为一致。从表1可见，A试验地土壤厚度及土壤养分含量等都比B试验地差。

表1　A、B试验地土壤情况

试验地	土层情况	pH	有机质(g/kg)	全量(g/kg)			速效(mg/kg)		
				N	P_2O_5	K_2O	N	P	K
A	土层浅薄，3个土壤剖面的平均厚度22.8cm，土石相间明显，土被不连续	6.98	37.22	2.74	1.05	6.68	132.5	0.3	26.4
B	土层较深，3个土壤剖面的平均厚度33.6cm，土被相对连续	7.15	60.64	3.42	1.78	10.54	246.2	1.9	43.2

2　研究方法

2005年3月，在试验地A，B中分别设置调查样地，每个样地面积为20m×30m，3次重复；然后在每个样地中心及四角各设置1个5m×5m的植被调查样方，在调查样地和样方的4个角埋小水泥桩作为标记，分别调查记录样方内植物的种类、每种植物的株数及盖度。然后采取了一致的封山育林措施，包括在试验区内停止一切人为的干预活动，严格禁止人、畜进入试验区。2008年7月采用同样的调查方法分别对A、B试验地进行封山后的植被调查。对试验地内乔灌草藤的相对密度、相对盖度、相对频度以及重要值进行计算，公式分别为：

相对密度=（某种植物的个体数/全部植物种类个体的总数）×100%

相对盖度=（某种植物的盖度/全部植物种类盖度的总和）×100%

相对频度=（某种植物的频度/全部植物种类频度的总和）×100%

重要值=相对密度+相对频度+相对盖度

3　结果与分析

3.1　封山对植被恢复的影响

封山前后A、B试验地的植物种类变化详见表2、表3。从表2可见，A在封山前有植物22种，其中：木本及藤本植物14种，草本植物8种；经封山3年多后，植物的种类增加到35种，其中：木本及藤本植物21种，草本植物14种，封山后植物种类比封山前多13种，增加了59.09%；在所增加的植物种类中：木本及藤本植物增加了7种，草本植物增加了6种，与封山前相比分别增加了50.00%和75.00%。

表 2 A 试验地封山前后植物种类及植被总盖度的变化

生长型	植物种名		学名
	封育前	封育后	
木本及藤本植物	番石榴	番石榴	*Psidium guajava* L.
	盐芙木	盐芙木	*Rhus chinensis* Mill.
	排钱树	排钱树	*Desmodium pulchellum* (L.) Benth.
	红背山麻杆	红背山麻杆	*Alchornea treioides* Muell. Arg.
	潺槁树	潺槁树	*Litsea glutinosa*(Lour.) C. B. Rob.
	老虎刺	老虎刺	*Pterolobium punctatum* Hemsl.
	粗叶悬钩子	粗叶悬钩子	*Runus alceaefolius* Poir.
	假木豆	假木豆	*Desmodium triangulare* (Retz.)Merr.
	灰毛桨果楝	灰毛桨果楝	*Cipadessa cinerascens* Hand. -Mazz
	毛桐	毛桐	*Mallotus barbatus*(Wall.) Muel. l -Arg.
	藤构	藤构	*Broussonetia kaempferi* Sieb.
	黑面神	黑面神	*Breynia fruticosa*(Linn.) Hook. f.
	白饭树	白饭树	*Fluggea virosa* Baill.
	五色梅	五色梅	*Lantana camara*Linn
		黄皮	*Clausena lansium* (Lour.)Skeels
		木鳖子	*Momordica cochinchinensis* (Lour.) Spreng.
		朴树	*Celtis sinensis*
		鸡屎藤	*Paederia scandens* (Lour.)Merr.
		地桃花	*Urena lobata* Linn.
		铁线莲	*Clematis florida* Thunb.
		酸藤子	*Embelia laeta*(Linn.) Mez
草本植物	黄茅	黄茅	*Heteropogon contortus* (Linn.) P. Beauv.
	飞机草	飞机草	*Eupatorium odoratum* Linn.
	鬼针草	鬼针草	*Bidens bipinnata* L.
	鞭叶铁线蕨	鞭叶铁线蕨	*Adiantum caudatum* Linn.
	五节芒	五节芒	*Miscanthus floridulus* (Lab.) arb.
	肾蕨	肾蕨	*Nephnolepis Cordifoolia*Presl
	毛蕨	毛蕨	*Cyclosorus interruptus*(Illd.) H. Ito.
	芒	芒	*Miscanthus sinensis*Anderss.
		地宝兰	*Geodorum densiflorum* (Lam.) Schltr.
		荩草	*Arthraxon hispidus* (Trin.) Makino
		苍耳	*Xanthium sibiricum* Patrin.
		一点红	*Emilia sonchifolia* (L.) DC.
		金丝草	*Pogonatherum crinitum*(Thunb.) Kunth
		千里光	*Senecio scandens* Buch. -Ham
总盖度(%)	50.0	90.0	

表3　B 试验地封山前后植物种类及植被总盖度的变化

生长型	植物种名		学名
	封育前	封育后	
木本及藤本植物	扁担杆	扁担杆	*Greia biloba* G. Don
	盐肤木	盐芙木	*Rhus chinensis* Mill.
	灰毛浆果楝	灰毛浆果楝	*Cipadessa cinerascens* Hand. -Mazz
	排钱树	排钱树	*Desmodium pulchellum*(L.) Benth.
	红背山麻杆	红背山麻杆	*Alchornea treioides* (Benth.) Muell. Arg.
	山芝麻	山芝麻	*Helicteres angustifolia* Linn.
	番石榴	番石榴	*Psidium guajava* L.
	了哥王	了哥王	*ikstroemia indica* (L.) C. A. Mey
	老虎刺	老虎刺	*Pterolobium punctatum* Hemsl.
	藤构	藤构	*Broussonetia kaempferi* Sieb.
	粗叶悬钩子	粗叶悬钩子	*Runus alceaefolius* Poir.
	粗糠柴	粗糠柴	*Mallotus philippmensis* (Lam.) Muell. -Arg
	潺槁树	潺槁树	*Litsea glutinosa* (Lour.) C. B. Rob.
	酸藤子	酸藤子	*Embelia laeta* (Linn.) Mez
	五色梅	五色梅	*Lantana camara* Linn
		木鳖子	*Momordica cochinchinensis*(Lour.) Spreng.
		白花丹	*Plumbago zeylanica* Linn.
		假木豆	*Desmodium triangulare* (Retz.)Merr.
草本植物	蔓生莠竹	蔓生莠竹	*Microstegium vagans*(Nees) A. Camus
	珍珠矛	珍珠矛	*Scleria levis* Retz.
	五节芒	五节芒	*Miscanthus floridulus* (Lab.) arb.
	飞机草	飞机草	*Eupatorium odoratum*Linn.
	马连安	马连安	*Agrimonia pilosa* Ledeb
	石山棕	石山棕	*Guihaia argyrata* (S. K. Lee) S. K
		千里光	*Senecio scandens* Buch. -Ham
总盖度(%)	70.0	100.0	

从表3可见，B 试验地封山前的植物种类有21种，其中：木本及藤本植物15种，草本植物6种；封山后，植物的种类增加到25种，其中：木本及藤本植物18种，草本植物7种。封山后植物的种类有所增加，但增幅不大，仅比封山前多了4种，增量仅为19.05%；在增加的4种植物中，木本植物增加3种，增加了33.33%，草本植物增加1种，增量仅16.67%。

封山前A、B试验地的植物物种丰富程度较为接近，分别为22种和21种。对木本及藤本植物：A试验地为14种，B试验地为15种，其共有的木本及藤本植物达到了10种，种类一致率分别为71.43%和66.67%；对草本植物：A试验地有8种，B试验地有6种，但共有的草本植物却不多，仅为2种，一致率仅分别为25.00%和33.33%。

经封山后，A、B试验地的物种多样性程度都有所增加，尤其以A试验地增加较多，达13种，而B试验地仅增加了4种，A试验地的增加量是B试验地的3.25倍，这主要由于A试验地的植被覆盖率较低，裸地较多。封山后，通过风、鸟等方式带入的种子，掉落后大多能直接与土壤接触，这对维持或延长种子的发芽能力较为有利，当种子发芽条件得到满足后，即能及时地发芽、扎根于土壤中。此外，土壤种子库中的种子在获得萌芽条件发芽出土后，所经受的物种竞争压力也较少，容易生长，故其物种的增加量较多。而B试验地刚好相反，封山前植被的覆盖度已较高，加之土壤条件较好，封山后没有了人畜等的干扰和破坏，植物的生长要比土壤条件较差的A试验地快，植被覆盖率的提高迅速并很快地实现全覆盖，使其从封山时的裸地较少逐渐发展至没有裸地，这样外界传进去的种子容易被植物的枝叶等架在空中，尤其是大粒种，被架空的可能性更大，这样在风吹日晒下容易丧失发芽能力。此外，植被盖度过高后，一个新物种要从种子库中萌芽、生长并掘起，须要克服众多物种的竞争。众所共知，在所有植物竞争相同的资源和空间时，植物间越靠近，对资源和空间的竞争就越强烈。这就是说，每个个体所面临的竞争强度受到其有效面积的影响。种的分布越聚集，种间竞争就越弱，而种内竞争越强烈[16]，导致了A试验地的物种增加数量明显多于B试验地。

从A、B试验地植被的状况还可发现，无论是封山前还是封山后，木本及藤本植物的种都多于草本植物，这与桂西南石漠化山地土壤种子库以草本为主[17]的研究结论没能得到正相关的结果。这一方面由于石漠化山地生境严酷，种子落地后很少能及时得到维持种子活力的有利条件，木本和藤本植物的种子个体一般较大，且其油脂或淀粉含量较高，不利于种子生活力的保存，故寿命较短；而草本植物种子个体一般较小，油脂或淀粉含量均较低，较有利于种子生活力的保存而寿命较长[18]，导致了其土壤种子库以草本为主。可见，在土壤种子库中木本及藤本植物的种子密度比草本的低并不一定是因其种子来源少所造成，而很可能是因种子的特点导致种子寿命较短所致。至于A、B试验地中的木本及藤本植物的种类多于草本植物，主要因木本及藤本植物粒大种子富含的营养物质多，种子发芽快，发芽力强，萌发出来的幼苗粗壮，抵御恶劣环境的能力强。而粒小的种子虽然种子量大，萌发出来的幼苗较多，但相对于粒大的种子会较弱小，有研究证明，在拥挤的种群中，往往是最小的个体受到密度制约死亡的威胁[19]，可见，在植物群落的生存竞争中，木本及藤本植物应具有比草本更强的竞争力。

3.2 封山对物种重要值的影响

A、B试验地封山前后植物特征值见表4、表5。从封山前后各物种特征值的变化情况不难发现，随着封山后植物种类的增加及各种植物生长的变化，导致了封山前已存在的植物，经封山育林后其特征值也发生了变化。从表4可见，在A试验地中，封山前已存在的植物，经封山后它们的重要值以下降为多，上升为少，从整体看重要值的增减幅度还都不是太大，说明群落中物种数及植物的生长虽已发生了变化，但是，变化还不够大，还不能明显改变植物在群落中的地位，这表明了植物群落中物种之间的竞争关系还没有形成或还不激烈，各物种仍基本处于自由发展阶段。

从表5可见，B试验地封山前后各物种的重要值变化为：木本及藤本以下降多，上升少，且重要值的升降没能明显改变植物在群落中的重要地位，但有2种植物重要值的变化值得重视，一种是扁担杆，其重要值下降了15.61，降幅达24.72%，另一种是山芝麻，其重要值的增加较为突出，封山前仅7.63，封山后增加到了27.69，增加了20.06，比封山前增加了2.63倍。而在草本植物中，则升多降少，尤其是蔓生莠竹，其重要值从90.12提高到了160.62，增加了70.50，增幅为78.22%，并由原来在群落中的次重要地位上升为最重要地位，而飞机草刚好相反，其重要值从159.15下降到了73.98，下降了88.65，降幅为55.70%，并由原来在群落中的最重要地位下降至次重要地位。这预示着B试验地中物种之间的竞争关系已经显现，尤其是草本植被，其竞争关系已经较激烈，出现了强阳性物种由优势种向消退种的方向发展，而阴生性物种已由弱变强，并向优势种地位演替的格局。其中最典型的是飞机草和蔓生莠竹，随着上层木本植被覆盖率的不断提高，强阳性草本植物飞机草的生长明显受到了抑制，在调查样地中随处可见到其枯死的个体，它的覆盖度也出现了明显的下降，由封山前的30%，下降至封山后的10%，而耐阴性较强的蔓生莠竹刚好相反，它的生长十分繁茂，覆盖率已从封山前的10%，上升至70%，已经成为了草本植物中的绝对优势种群。从B试验地中2种典型生境指示植物优势地位的更替表明，经封山后它的小气候环境已发生了改变，已由原来较为干热的环境转变成相对阴湿的环境。而在A试验地中，由于植被还没有实现全覆盖，各个物种都能够较为正常的生长，物种之间仍处于共存共荣的阶段，还没有形成明显的竞争关系。可见，在土壤条件好、封山前植被覆盖度高的区域，封山后植物的竞争关系比在土壤条件差、封山前植被覆盖率低的区域形成得更早对环境改善的能力也更强。

3.3 封山育林对植被覆盖率的影响

从表2、表3还可以看出，封山育林能够明显提高石灰岩山地植被的覆盖率，其中A试验地的植被覆盖率从封山前50%提高到封山后的90%，提高了40%；而B试验地则从封山育林前的70%提高到了100%，已经实现了全覆盖。如果仅从植被覆盖率增加的绝对数量上看，A试验地的植被覆盖率增加量高于B试验地，但是，从封山前后各物种覆盖度及重要值的变化情况不难发现，B试验地的植被生长要快于A试验地。说明土壤条件优越，封山前植被覆盖度高的区域，封山后植物的生长会更快，植被恢复也更迅速，更容易实现全覆盖。

表4 A试验地封山前后植物特征值的变化

植物种名		相对频度		相对密度		相对盖度		重要值	
		封山前	封山后	封山前	封山后	封山前	封山后	封山前	封山后
木本及藤本植物	番石榴	15.29	11.02	51.61	34.05	55.48	44.37	122.38	89.44
	盐肤木	7.06	4.72	2.15	3.49	1.28	2.26	10.49	10.47
	排钱树	14.12	10.24	25.09	20.27	10.67	8.29	49.88	38.80
	红背山麻杆	9.41	6.30	2.87	11.13	0.85	2.51	13.13	19.94
	潺槁树	5.88	3.94	1.43	1.16	0.43	0.96	7.74	6.06
	老虎刺	2.35	1.57	0.72	0.66	0.85	1.35	3.92	3.58
	粗叶悬钩子	7.06	5.51	1.43	1.66	17.07	20.25	25.56	27.42
	假木豆	5.88	7.09	2.87	13.95	0.21	0.77	8.96	21.81
	灰毛浆果楝	7.06	6.30	3.58	3.82	1.28	1.97	11.92	12.09
	毛桐	4.71	3.15	1.08	1.00	0.09	0.23	5.88	4.38
	藤构	7.06	4.72	2.51	1.66	0.85	0.98	10.42	7.36
	黑面神	3.53	2.36	0.36	0.17	0.04	0.02	3.93	2.55
	白饭树	4.71	3.15	0.72	0.33	0.21	0.29	5.64	3.77
	五色梅	5.88	3.94	3.58	1.99	10.67	5.83	20.13	11.76
	木鳖子		1.57		0.33		0.04		1.94
	朴树		2.36		0.17		0.10		2.63
	地桃花		3.94		0.17		0.02		4.13
	铁线莲		3.15		0.17		0.39		3.71
	酸藤子		7.09		2.66		8.68		18.43
	黄皮		0.79		0.17		0.02		0.98
	鸡屎藤		7.09		1.00		0.68		8.77
草本植物	黄茅	31.71	17.07	72.46	69.44	23.94	15.46	128.11	101.97
	飞机草	12.19	6.10	0.97	0.69	3.99	2.15	17.15	8.94
	鬼针草	4.88	2.44	1.45	1.16	7.98	4.30	14.31	7.90
	鞭叶铁线蕨	9.76	4.88	1.45	1.16	1.60	0.95	12.81	6.99
	五节芒	17.07	13.41	9.66	11.57	53.19	64.43	79.92	89.41
	肾蕨	12.19	7.32	5.80	4.63	1.33	0.86	19.32	12.81
	毛蕨	4.88	2.44	5.31	3.47	2.66	1.72	12.85	7.63
	芒	7.32	6.10	2.90	1.85	5.32	2.58	15.54	10.53
	地宝兰		4.88		0.23		0.09		5.20
	荩草		10.98		3.47		5.15		19.60
	苍耳		2.44		0.23		0.09		2.76

（续）

植物种名		相对频度		相对密度		相对盖度		重要值	
		封山前	封山后	封山前	封山后	封山前	封山后	封山前	封山后
草本植物	一点红		3.66		0.46		0.09		4.21
	金丝草		9.76		1.39		1.72		12.87
	千里光		8.54		0.23		0.43		9.20

表5 B试验点封山前后植物特征值的变化

植物种名		相对频度		相对密度		相对盖度		重要值	
		封山前	封山后	封山前	封山后	封山前	封山后	封山前	封山后
木本及藤本植物	扁担杆	18.18	12.75	17.91	13.63	27.06	21.16	63.15	47.54
	盐芙木	10.61	6.86	8.96	7.15	15.82	11.56	35.39	25.57
	灰毛浆果楝	12.12	10.78	7.46	7.79	10.41	10.86	29.99	29.43
	排钱树	9.09	5.88	10.45	6.49	2.50	2.83	22.04	15.20
	红背山麻杆	4.55	8.82	5.97	5.85	0.21	2.66	10.73	17.33
	山芝麻	3.03	4.90	4.48	20.13	0.12	2.66	7.63	27.69
	番石榴	15.15	10.78	23.88	18.18	41.63	40.89	80.66	69.85
	了哥王	1.52	0.99	1.49	0.65	0.04	0.02	3.05	1.66
	老虎刺	1.52	1.96	1.49	0.65	0.04	0.02	3.05	2.63
	藤构	3.03	2.94	2.99	1.30	0.08	0.04	6.10	4.28
	粗叶悬钩子	3.03	4.90	2.99	2.60	0.42	0.53	6.44	8.03
	粗糠柴	3.03	3.92	2.99	1.30	0.42	0.35	6.44	5.57
	潺槁树	4.55	7.85	4.48	5.19	0.42	1.08	9.45	14.12
	酸藤子	3.03	1.96	1.49	0.65	0.42	1.78	4.94	4.39
	五色梅	7.56	6.86	2.99	1.95	0.42	1.78	10.97	10.59
	木鳖子		0.99		0.65		1.08		2.72
	白花丹		1.96		1.95		0.35		4.26
	野蚂蝗		4.90		3.89		0.35		9.14
草本植物	蔓生莠竹	40.00	38.79	24.21	64.59	25.91	57.24	90.12	160.62
	珍珠矛	8.57	7.75	4.84	6.46	0.43	1.27	13.84	15.48
	五节芒	11.43	10.19	7.26	10.76	8.64	16.53	27.33	37.48
	飞机草	31.43	33.62	62.95	17.21	64.77	23.15	159.15	73.98
	石山棕	2.86	2.59	0.24	0.11	0.04	0.77	3.14	3.47
	马连安	5.71	5.18	0.48	0.22	0.22	0.51	6.41	5.91
	千里光		1.88		0.65		0.53		3.06

4 结论

（1）石灰岩山地生境恶劣，适于其生长繁衍的植物种类少，但研究表明其土壤种子库相当丰富[20,21]。此外，即使是石漠化严重的区域也有一定种类、数量的植物残根留存，具备了植被自然恢复的基础，可以通过采取封山育林措施实现其生态重建。

（2）土壤条件的优劣对封山后植被的恢复影响明显，土壤条件好，物种的恢复与生长较快，植被覆盖率提高也较迅速，实现全覆盖的时间缩短，植物竞争关系的形成也较快；反之亦然。

(3)封山前植被覆盖率的高低对封山后物种的恢复与侵入影响明显。封山前植被覆盖率较低的区域，封山后其物种的恢复或侵入会更有利，物种增加的数量会较多。而封山前植被覆盖率较高的区域，封山后其物种的恢复或侵入会较难，物种增加的数量也会较少。

(4)封山育林对物种的重要值会产生影响。土壤条件较差、封山前植被覆盖率较低的，封山3年多后其原有植物重要值以降多升少，且升降幅度还不能改变物种在群落中的重要地位；土壤条件较好、封山前植被覆盖率较高的，封山后其原有木本及藤本植物的重要值以降多升少，升降幅度同样不能明显改变物种在群落中的地位，但对草本植物则升多降少，并有部分物种改变其在群落中的地位。

参考文献

[1]马遵平，谢泽氡．南方岩溶区植被自然演替恢复研究综述[J]．四川林勘设计，2006(1)：1-6.
[2]贺庆棠，陆佩玲．中国岩溶山地石漠化问题与对策研究[J]．北京林业大学学报，2006，28(1)：117-120.
[3]韦茂繁．广西石漠化及其对策[J]．广西大学学报(哲学社会科学版)，2002，24(2)：42-47.
[4]李品荣，常恩福，陈强，等．滇东南岩溶地区石质山封山育林效果初探[J]．云南林业科技，2001，97(4)：13-17.
[5]李阳兵，王世杰，容丽．西南岩溶山地石漠化及生态恢复研究展望[J]．生态学杂志，2004，23(6)：84-88.
[6]袁春，周常萍，童立强，等．贵州土地石漠化的形成原因及其治理对策[J]．现代地质，2003，17(2)：181-185.
[7]王世杰，李阳兵，李瑞玲．喀斯特石漠化的形成背景演化与治理[J]．第四纪研究，2003，23(6)：657-666.
[8]唐秀玲，何新华，彭宏祥．广西石山土地石漠化的成因及防治对策[J]．资源开发与市场，2003，19(3)：154-156.
[9]苏维词．贵州喀斯特生态系统的脆弱性及其对策[J]．中国水土保持科学，2004，2(3)：64-69.
[10]祝小科，朱守谦．喀斯特石质山地封山育林效果分析[J]．林业科技，2001，26(6)：1-4.
[11]王方芳，傅松玲．石灰岩山地容器育苗及造林技术[J]．安徽农学通报，2005，11(16)：110-111.
[12]唐兰芳．石灰岩地区生态公益林营造技术[J]．广东林业科技，2005，21(1)：56-59.
[13]敖惠修，何道泉．粤北石灰岩山地的造林树种及造林技术[J]．广东林业科技，1994(1)：16-19.
[14]付兆雯．石漠化治理的有效途径[J]．中国林业，2008(5)：55.
[15]王永安．封山育林的生态经济作用[J]．世界林业研究，2000，13(3)：19-24.
[16]何艺玲，傅懋毅．人工林林下植被的研究现状[J]．林业科学研究，2002，15(6)：727-733.
[17]吕仕洪，陆树华，欧祖兰，等．桂西南石漠化山地土壤种子库的基本特征及植被恢复对策[J]．植物资源与环境学报，2007，16(1)：6-11.
[18]黄忠良，孔国辉，魏平，等．南亚热带森林不同演替阶段土壤种子库的初步研究[J]．热带亚热带植物学报，1996，4(4)：42-49.
[19]李博，陈家宽，AR沃金森．植物竞争研究进展[J]．植物学通报，1998，15(4)：18-29.
[20]沈有信，江洁，陈胜国，等．滇东南岩溶山地退化植被土壤种子库的储量与组成[J]．植物生态学报，2004，28(1)：101-106.
[21]龙翠玲．贵州茂兰喀斯特森林土壤种子库研究[G]//朱守谦．喀斯特森林生态研究(Ⅲ)．贵阳：贵州科技出版社，2003：265-275.

[原载：林业资源管理，2010(03)]

尾巨桉纯林土壤浸提液对4种作物生理的影响

郝 建[1,2] 陈厚荣[2] 王凌晖[1] 秦武明[1] 曾 冀[2] 张明慧[3]

([1]广西大学林学院，广西南宁 530005；

[2]中国林业科学研究院热带林业实验中心，广西凭祥 532600；

[3]广西国营派阳山林场，广西宁明 532500)

摘 要 用尾巨桉(*Eucalyptus urophylla* × *E. grandis*)纯林土壤水浸提液处理菜心(*Brassica parachinensis*)、早熟白菜(*Brassica pekinensis*)、水稻(*Oryza sativa*)和萝卜(*Raphanus sativus*)幼苗，研究浸提液对这几种作物的化感作用。结果表明：浸提液对菜心、早熟白菜、水稻和萝卜叶绿素质量分数、脯氨酸质量分数、可溶性糖质量分数均有显著影响。随着浸提液浓度的升高，受体植物叶片的叶绿素质量分数均降低；脯氨酸质量分数呈现先升高后降低的趋势；菜心、早熟白菜、萝卜可溶性糖质量分数也是呈下降趋势，而水稻的可溶性糖质量分数却先升高后降低。根据所测定指标综合判断，水稻比其他几种作物对尾巨桉纯林土壤水浸提液中化感物质的抗性要强。这说明尾巨桉纯林土壤中存在某些化感物质，因受体植物种类不同其化感效应有所差别。

关键词 尾巨桉；化感作用；生理特性；土壤

Physiological Responses to Four Crops in Aqueous Extracts From Soil of Pure *Eucalyptus urophylla*×*E. grandis* Forests

HAO Jian[1,2], CHEN Hourong[2], WANG Linghui[1], QIN Wuming[1], ZENG Ji[2], ZHANG Minghui[3]

([1]*College of Forestry*, *Guangxi University*, *Nanning* 530004, *Guangxi*, *China*;

[2]*Experimental Center of Tropical Forestry*, *Pingxiang* 532600, *Guangxi*, *China*;

[3]*Guangxi Paiyangshan Forest Centre*; *Ningming* 532500, *Guangxi*, *China*)

Abstract: The allelopathy of *Eucalyptus urophylla*×*E. grandis* on several crops was investigated by treating aqueous extracts from soil of pure *Eucalyptus urophylla*×*E. grandis* forests to the growth medium of *Brassica parachinensis*, *Brassica pekinensis*, *Oryza sativa*, *Raphanus sativus*. The results showed that the water extracts had extremely significant effects on the chlorophyll content, proline content and soluble sugar content of *Brassica parachinensis*, *Brassica pekinensis*, *Oryza sativa* and *Raphanus sativus*. With the increase of the aqueous extracts concentration, receptor plant leaves lowered their chlorophyll content. The content of proline rose at first, then decreased. The contents of soluble sugar from *Brassica parachinensis*, *Br assica pekinensis*, *Oryza sativa* and *Raphanus sativus* also declined. The content of soluble sugar of rice rose at first, but then decreased. According to measuring indexes and comprehensive judgment, the allelochemicals resistance of rice was stronger than that of other crops in aqueous extracts from soil of pure *Eucalyptus urophylla*×*E. grandis* forests. From the results, it was speculated that there were some allelochemicals in the soil of pure *Eucalyptus urophylla*×*E. grandis* forests, the allelopathic effects on varied plant species were different.

Key words: *Eucalyptus urophylla*×*E. grandis*; allelopathy; physiological characteristics; soil

化感作用是指一种植物(包括微生物)产生并释放于环境的生化物质，对另一种植物产生直接或间接的相生或相克的作用[1,2]。关于植物化感作用国内外学者做了大量研究，油松(*Pinus tabulaeformis*)和辽东栎(*Quercus liaotungensis*)混交林中化感作用的作用机理研究[3]、马尾松(*Pinus massoniana*)根化感物质的生物活性评价与物质鉴定[4]、三裂叶蟛蜞菊(*Wedelia trilobata*)化感作用研究[5]、木麻黄(*Casuarina equisetifolia*)水浸液对其幼苗生长的影响[6]、落叶松(*Larix gmelinii*)水提物对胡桃楸(*Juglans mandshurica*)化感作用的生物测定[7]等。目前桉树人工林的发展，在学术界和社会上存在许多争论，化感作用作为焦点之一受到广泛关注[8,9]。有关桉树化感作用方面已开展了一些研究，尾巨桉(*Eucalyptus urophylla*×*E. grandis*)叶片水浸提液化感作用的生物评价[10]、尾叶桉(*Eucalyptus urophylla*)抑制银合欢(*Leucaena leucocephala*)幼苗生长[11]、巨尾桉枝叶的水浸提物影响水稻(*Oryza sativa*)和菜苔(*Brassica parachinensis*)种子萌芽[12]、巨尾桉影响小麦(*Triticum aestivum*)种子发芽及幼苗生长[13]、艮叶山桉(*Eucalyptus pulverlenta*)抑制独行菜(*Lepidium apetalum*)萌发及幼苗的光合作用[14]。以往学者对桉树化感作用的研究大多停留在对受体植物种子萌发和幼苗生长的影响方面，而对受体植物的生理作用影响研究较少。故本文以尾巨桉为研究对象作用，研究尾巨桉纯林土壤水浸提液对广西主要4种作物：菜心(*Brassica parachinensis*)、早熟白菜(*Brassica pekinensis*)、水稻和萝卜(*Raphanus sativus*)幼苗生理指标的影响，对该树种化感作用作出探讨。

1 材料与方法

1.1 试验材料

供体材料：尾巨桉纯林土壤，采于广西林业科技示范园5年生尾巨桉纯林。

受体材料：菜心、早熟白菜、水稻和萝卜种子均购于南宁市农业科技市场。

1.2 试验方法

1.2.1 尾巨桉纯林土壤水浸提液的制备

所采土样风干后研磨过20目筛子，称取1.28g材料与10ml去离子水混合[15]，振荡浸提24h，用双层滤纸过滤，制得质量浓度为128.0g/L的土壤水浸提液，然后分别稀释成0.5g/L、2.0g/L、8.0g/L、32.0g/L、128.0g/L等5种质量浓度[16]，最后用0.1mol/L的HCl和0.1mol/L的NaOH溶液将浸提液的pH调到6.5~7.0[17,18]，置4℃冰箱中备用。

1.2.2 尾巨桉纯林土壤水浸提液对几种农作物生理的影响

将受体植物种子用10g/L高锰酸钾消毒10min，用30℃温水浸种1h。然后分别点播在基质中(珍珠岩：砂子=3：1)。放在25℃，30μmol/m²·s 12h光照条件下培养。萌发后，每周统一添加各浸提液。在受体植物长出第一对真叶之前，每3d统一定量喷施1/2Hoagland营养液，之后每2d统一定量喷施Hoagland营养液，培养期间根据基质湿度每2d定量喷洒适量蒸馏水。处理20d后取样测定叶绿素质量分数，脯氨酸，可溶性糖质量分数[19]。

1.2.3 数据分析

采用SAS软件对试验结果进行方差分析及LSD多重比较。

2 结果与分析

2.1 尾巨桉纯林土壤水浸提液对几种农作物叶绿素质量分数的影响

叶片中叶绿素质量分数是反映作物光合能力的重要指标。从表1数据可以看出，各质量浓度土壤水浸提液处理的农作物叶片叶绿素质量分数均低于对照，且随着浸提液质量浓度的升高，叶绿素质量分数逐渐降低。经方差分析，各质量浓度浸提液对白菜($F=30.14$，$P<0.0001$)、菜心($F=30.05$，$P<0.0001$)、水稻($F=6.66$，$P=0.0041$)、萝卜($F=24.95$，$P<0.0001$)叶片叶绿素质量分数的影响均极显著。除0.5g/L的浸提液处理的白菜、水稻与对照差异显著，其余质量浓度处理的各农作物与其对照差异均极显著。

表1 尾巨桉纯林土壤水浸提液对农作物叶绿素质量分数的影响

处理浓度(g/L)	农作物(mg/g)			
	白菜	菜心	水稻	萝卜
0(对照)	0.9573Aa	1.6922Aa	0.7535Aa	1.2480Aa
0.5	0.8805ABb	1.4966Bb	0.7366ABa	1.1480Bb
2.0	0.8068BCc	1.5065Bb	0.6314BCb	1.0823BCb
8.0	0.7446CDd	1.3905Bc	0.6344BCb	0.9878CDc
32.0	0.7239Dd	1.2065Cd	0.6117Cb	0.9750Dc

（续）

处理浓度(g/L)	农作物(mg/g)			
	白菜	菜心	水稻	萝卜
128.0	0.6322Ee	1.1420Cd	0.5623Cb	0.8975Dd
F值	30.14	30.05	6.66	24.95
P	<0.0001	<0.0001	0.0041	<0.0001

说明：同行数据后不同小写字母表示差异达到0.05显著水平，不同大写字母表示差异达到0.01极显著水平。

2.2　尾巨桉纯林土壤水浸提液对几种农作物叶片脯氨酸质量分数的影响

从表2可知，各质量浓度尾巨桉纯林土壤水浸提液对4种农作物叶片脯氨酸质量分数均有促进作用，且促进作用随着浸提液质量浓度的升高而增强，当质量浓度达到32.0g/L时，水稻、萝卜和白菜3种作物叶片脯氨酸质量分数出现下降趋势，当浓度达到8.0g/L时，菜心叶片脯氨酸质量分数呈下降趋势。

各质量浓度尾巨桉纯林土壤水浸提液对白菜($F=59.54$，$P<0.0001$)、菜心($F=23.59$，$P<0.0001$)、水稻($F=7.29$，$P=0.0029$)、萝卜($F=7.07$，$P=0.0032$)叶片脯氨酸质量分数的影响极显著。各质量浓度土壤水浸提液处理的白菜叶片脯氨酸质量分数与对照差异均极显著；除0.5g/L和128.0g/L土壤水浸提液处理的菜心叶片脯氨酸质量分数与对照差异不显著，其余质量浓度处理与对照差异均极显著；除0.5g/L和2.0g/L土壤水浸提液处理的水稻叶片脯氨酸质量分数与对照差异显著，其余质量浓度处理与对照差异均极显著；各质量浓度尾巨桉纯林土壤水浸提液极显著，0.5g/L和2.0g/L的尾巨桉纯林土壤水浸提液对萝卜叶片脯氨酸质量分数影响不显著，128.0g/L的土壤水浸提液影响显著，其余质量浓度影响均极显著。

表2　浸提液对农作物脯氨酸质量分数的影响

处理浓度(g/L)	农作物(mg/g)			
	白菜	菜心	水稻	萝卜
0(对照)	0.00346Dd	0.00538Cd	0.00807Db	0.00871Cb
0.5	0.00573Cc	0.00582Ccd	0.00893CDb	0.00881Cb
2.0	0.00584Cc	0.00727Bb	0.00926BCDb	0.00913BCb
8.0	0.00712Bb	0.00821Aa	0.01182ABCa	0.01060ABa
32.0	0.00855Aa	0.00721Bb	0.01273Aa	0.01156Aa
128.0	0.00626Cc	0.00601Cc	0.01206ABa	0.00905BCg
F值	59.54	23.59	7.29	7.07
P	<0.000 1	<0.000 1	0.002 9	0.003 2

注：同行数据后不同小写字母表示差异达到0.05显著水平，不同大写字母表示差异达到0.01极显著水平。

2.3　尾巨桉纯林土壤水浸提液对几种农作物可溶性糖质量分数的影响

从表3可知，水稻叶片可溶性糖质量分数随着土壤水浸提液质量浓度的升高逐渐增加，当质量浓度达到128.0g/L时，可溶性糖质量分数下降；白菜、菜心和萝卜叶片可溶性糖质量分数随着土壤水浸提液质量浓度的升高逐渐降低。

各质量浓度土壤水浸提液对白菜($F=282.24$，$P<0.0001$)、菜心($F=55.87$，$P<0.0001$)、水稻($F=16.25$，$P=0.0001$、萝卜($F=803.53$，$P<0.0001$)叶片可溶性糖质量分数影响均极显著。各质量浓度尾巨桉纯林土壤水浸提液处理白菜、菜心与萝卜叶片可溶性糖质量分数与各对照差异均极显著；除128.0g/L土壤水浸提液处理的水稻叶片可溶性糖质量分数与对照差异极显著，其余质量浓度处理与对照差异均不显著。

表 3　浸提液对农作物可溶性糖质量分数的影响

处理浓度(g/L)	农作物(mg/g)			
	白菜	菜心	水稻	萝卜
0(对照)	6.8485Aa	4.2556Aa	13.7557Aa	6.3299Aa
0.5	2.8384Bb	2.4761Bb	13.5511Aa	3.4096Bb
2.0	2.1969Cc	1.9384Cc	14.2481Aa	2.4096Cd
8.0	1.9495CDc	1.8532Ccd	14.3136Aa	2.5755Cc
32.0	2.0510Cc	1.7399Ccd	14.905Aa	2.3978Cd
128.0	1.5479Dd	1.5510Cd	9.8376Bb	2.0978De
F 值	282.24	55.87	16.25	803.53
P	<0.0001	<0.0001	0.0001	<0.0001

注：同行数据后不同小写字母表示差异达到 0.05 显著水平，不同大写字母表示差异达到 0.01 极显著水平。

3　讨论

叶绿素质量分数是影响植物光合作用的重要因子。尾巨桉纯林土壤水浸提液水对水稻、萝卜和白菜叶绿素质量分数均有显著的抑制作用，且随着浸提液质量浓度的升高，抑制作用增强，这和前人的研究结果相似。化感物质对植物体光合作用的影响主要表现为叶绿素质量分数和光合速率的降低[20]，如银胶菊(*Parthenium hysterophorus*)倍半萜烯类化感物质处理胜红蓟(*Ageratum conyzoides*)幼苗后，叶绿素质量分数显著降低[21]，Yang 等研究表明阿魏酸和香豆酸处理水稻后，水稻幼苗叶绿素合成过程中的镁—螯合酶受到抑制[22]。

通常情况下，脯氨酸被作为一个反映植物体抗性指标使用。从研究结果看，几种受体植物叶片的脯氨酸质量分数均有促进作用，且促进作用随着浓度的升高而增强，具有一定规律性，与曹成有等对瑞香狼毒(*Stellera*)(*chamaejasma*)根提取液对被处理植物的脯氨酸质量分数提高[23]相似。当浸提液质量浓度达到 32.0g/L 时，水稻、萝卜和白菜等 3 种作物叶片脯氨酸质量分数出现下降趋势，当质量浓度达到 8.0g/L 时，菜心叶片脯氨酸质量分数呈下降趋势。这可能反映出，在浸提液低质量浓度条件下几种受体植物体内产生脯氨酸抵抗化感物质的伤害，而当化感物质质量浓度达到一定高度后，受体植物体内脯氨酸的合成机制可能受到了破坏，不能继续合成脯氨酸抗逆，所以分别出现质量分数下降的趋势。从结果也可以反映出，菜心的抗逆性可能弱于其他几种作物。

可溶性糖既是渗透调节剂，也是合成其他有机溶质的碳架和能量的来源。白菜、菜心和萝卜叶片可溶性糖质量分数随着土壤水浸提液质量浓度的升高，逐渐降低，这与曹成有等用瑞香狼毒根提取液处理植物，被处理植物的可溶性糖质量分数却降低[23]结果相似；但水稻叶片可溶性糖质量分数随着浸提液质量浓度的升高，逐渐增加，当浸提液质量浓度达到 128.0g/L 时，叶片可溶性糖质量分数下降。此结果可能表明水稻的抗化感物质的能力较强，可溶性糖在水稻受到化感物质胁迫时起到一定的调节作用，当化感物质浓度达到一定高度后，合成机制同其他机能一样受到破坏，从而导致质量分数降低；其他几种作物对化感物质的抗性较差，可溶性糖合成受阻，从而质量分数下降。

由此可见，尾巨桉纯林土壤水浸提液对不同作物的化感作用强度相异。从所测定的指标综合判断，水稻比其他几种作物对尾巨桉纯林土壤水浸提液中化感物质的抗性要强。

参考文献

[1]RICE E L. Allelopathy[M]. New Yoke: Academic Press, 1984: 422.
[2]林思祖，杜玲，曹光球．化感作用在林业中的研究进展及应用前景[J]．福建林学院学报，2002，22(2)：184-88.
[3]贾黎明，翟明普，冯长红．化感作用物对油松幼苗生长及光合作用的影响[J]．北京林业大学学报，2003，25(4)：6-10.
[4]曹光球，林思祖，王爱萍，等．马尾松根化感物质的生物活性评价与物质鉴定[J]．应用与环境生物学报，2005，11(6)：686-689.
[5]聂呈荣，曾任森，黎华寿，等．三裂叶蟛蜞菊对菜心化感作用的生理机理[J]．华南农业大学学报(自然科

学版)，2003，24(4)：106-107.

[6]林武星，洪伟．木麻黄水浸液对其幼苗生长的影响[J]. 江西农业大学学报，2005，27(1)：46-51.

[7]杨立学．落叶松水提物对胡桃楸化感作用的生物测定[J]. 东北林业大学学报，2006，34(2)：15-17.

[8]沈国舫，翟明普．混交林研究——全国混交林与树种关系学术讨论会论文集[M]. 北京：中国林业出版社，1997.

[9]白嘉雨，甘四明．桉树人工林的社会、经济和生态问题[J]. 世界林业研究，1996，9(2)：63-68.

[10]秦武明，郝建，王凌晖，等．尾巨桉叶片水浸提液化感作用的生物评价[J]. 福建林学院学报，2008，28(3)：257-261.

[11]曾任森，李蓬为．窿缘桉和尾叶桉的化感作用研究[J]. 华南农业大学学报，1997，18(1)：6-10.

[12]赵绍文，王凌晖，蒋欢军，等．巨尾桉枝叶水浸提液对3种作物种子萌发的影响[J]. 广西科学院学报，2000，16(1)：14-17.

[13]廖建良，宋冠华，曾令达．巨尾桉叶片水提液对小麦幼苗生长的影响[J]. 惠州大学学报，2000，20(4)：50-52.

[14] BOLTE M L, BOWERS J. CROW W D, et al. Germination inhibitor from *Eucalyptus pulverlenta*[J]. Agric Biol Chem, 1984, 48(2)：373-376.

[15]孔垂华．植物化感作用研究中应注意的问题[J]. 应用生态学报，1998，9(3)：332-336.

[16] WANG D L, ZHU X R. Research on allelopathy of Ambrosia artemisiifolia[J]. Acta Ecol Sin, 1996, 20(1)：11-19.

[17]曹光球，林思祖，杜玲，等．阿魏酸与肉桂酸对杉木化感作用的生物评价[J]. 中国生态农业学报，2003，11(2)：8-10.

[18]陈龙池，廖利平，汪思龙，等．香草醛和对羟基苯甲酸对杉木幼苗生理特性的影响[J]. 应用生态学报，2002，13(10)：1291-1294.

[19]中国科学院上海植物生理研究所．现代植物生理学实验指南[M]. 北京：科学出版社，2004.

[20] HEJL A M, EINHELLIG P A, RASMUSSEN J A. Effects of juglone on growth, photosynthesis, and respiration[J]. J Chem Ecol, 1993, 19(3)：559-568.

[21] SINGH H P, BATISH D R, KOHLI R K, et al. Effect of parthenin-asesquiterpene lactose from *Parthenium hysterophorus*-on early growth and physiology of *Ageratum conyzoides*[J]. J Chem Ecol, 2002, 28(11)：2169-2179.

[22] YANG C M, CHANG I F, LIN S J, CHOU C H, Effects of three allelopathic phenolics on chlorophyll accumulation of rice(*Oryzas ativa*) seedlings：Ⅱ. Stimulation of consumption-orientation [J]. Bot Bull Acad Sin, 2004, 45：119-125.

[23]曹成有，富瑶，王文星，等．瑞香狼毒根提取液对植物种子萌发的抑制作用[J]. 东北大学学报(自然科学版)，2007，28(5)：729-732.

[原载：浙江农林大学学报，2011，28(05)]

尾巨桉纯林土壤化感效应的生物评价

郝　建[1,2]　王凌晖[2]　秦武明[2]

([1] 中国林业科学研究院热带林业实验中心，广西凭祥　532600；

[2]广西大学林学院，广西南宁　530005)

摘　要　用不同浓度的尾巨桉(*Eucalyptus urophylla*×*E. grandis*)纯林土壤水浸提液处理菜心(*Brassica parachinensis* L. H. Bariley)、白菜(*B. pekinensis* L.)、水稻(*Oryza sativa* L.)、萝卜(*Raphanus sativus* L.)的种子和幼苗，综合评价尾巨桉纯林土壤的化感作用。结果表明，尾巨桉纯林土壤对所选作物有化感作用，浸提液浓度不同，其化感效应不同，受体植物不同，化感效应差别也较大，浓度越低化感作用越小，而随着溶液浓度的增大化感抑制作用逐渐增强。

关键词　尾巨桉；土壤；化感作用；水浸提液

Bioassay of Allelopathic Effects of Pure Forest Soil of *Eucalyptus urophylla*×*E. grandis*

HAO Jian[1,2], WANG LingHui[2], QIN WuMing[2]

([1]*Experimental Center of Tropical Forestry*, *Pingxiang* 532600 *Guangxi*, *China*;

[2]*College of Forestry*, *Guangxi University*, *Nanning* 530004 *Guangxi*, *China*)

Abstract: Treating the seeds and seedlings of *Brassica parachinensis* L. H. Bariley、*B. pekinensis* L.、*Oryza sativa* L.、*Raphanus sativus* L. with soil water extraction of *Eucalyptus urophylla* × *E. grandis* pure forest, synthetically evaluated the allelopathy of *Eucalyptus urophylla* × *E. grandis* pure forest's soil. The results showed that the soil of *Eucalyptus urophylla*×*E. grandis* pure forest had the allelopathy to the selected crops, different concentration showed different effects, performance of different treated crops were different. specifically expressed that low concentration took less allelopathy, and with the increase of concentration of extraction the allelopathy increased.

Key words: *Eucalyptus urophylla*×*E. grandis*; soil; allelopathy; aqueous extracts

化感作用是指一种植物(包括微生物)产生并释放于环境的生化物质，对另一种植物产生直接或间接的相生或相克的作用[1,2]。一个重要表现是植物通过分泌和释放有毒化学物质对其他植物的生理、生长起抑制或促进作用。关于林木化感作用国外的研究起步较早，国内学者也做了大量研究，不少研究表明植物化感作用是影响林分生产力及生物多样性的重要因素之一，而化感作用方式及作用程度也随着供、受体植物种类的不同而异，因此不同受体植物被利用对许多林木的化感作用进行生物评价，如火炬树根、叶浸提液对紫穗槐、侧柏、火炬树种子萌发效应[3]，木麻黄水浸液对其幼苗生长的影响[4]，核桃鲜叶挥发油对小麦、绿豆、黄瓜、萝卜种子化感作用研究[5]，3个品系橡胶根、茎、叶水浸液对花生、西瓜、玉米3种受体作物种子及其幼苗生长的影响[6]，不同季节草地早熟禾的化感作用研究[7]，天山云杉针叶提取物对种子萌发和幼苗生长的自毒作用[8]，落叶松水提物对胡桃楸化感作用的生物测定[9]等。

随着桉树人工林的不断发展，学术界和社会上引发了许多争论，争议的焦点是桉树人工林生态问题[10,11]，主要包括桉树是否过度消耗养分和水分、桉树化感作用是否会减少生物多样性。对桉树化感作用方面已开展了一些研究，研究对象主要有：尾叶桉[12]、窿缘桉[13]、巨尾桉[14,15]、巨桉[16]、良叶山桉[17]、邓恩桉[18]、刚果12号桉[19]。可见，对不同品系桉树化感作用的研究已初成规模，而作为南方速生桉品种之一的尾巨桉(*Eucalyptus urophylla*×*E. grandis*)化感作用研究未见报道。因此本研究以选取生物测定常用的受体植物同时又是广西主要常见的几种作物作为受体材料：菜心(*Brassica parachinensis* L. H. Bariley)、白菜(*B. pekinensis* L.)、水稻(*Oryza sativa* L.)、萝卜(*Raphanus sativus* L.)，分析尾巨桉纯林土壤水浸提液对受体植物种子的萌发、苗期生理、生长等指标影响，对尾巨桉纯林土壤的化感作用作出生物评价。

1 材料与方法

1.1 试验材料

供体材料：采用"S"型方法布设采样，在广西林业科技示范园取5年生尾巨桉纯林的林下土壤。

受体材料：水稻、萝卜、白菜和菜心种子，均购于南宁市农业科技市场。

1.2 试验方法

在自然界中，化感物质主要通过雨水和雾滴等的淋溶进入土壤而发生化感作用[20]，因此本试验采用最接近自然状态的常温水浸提法。将所采尾巨桉纯林土壤风干后研磨，过20目筛子，分别称取1.28kg材料与10L蒸馏水混合(直接制得浓度12.8%的原液)，浸提3d。用双层滤纸过滤，制得浓度为12.8%(质量浓度)的水浸提液，然后分别稀释成0.05%、0.2%、0.8%、3.2%、12.8%共5种浓度，用0.1N的HCl和0.1N的NaOH溶液调节浸提液的pH在6.5~7.0[21,22]，于4℃冰箱保存备用。

将受体植物种子用45℃的热水浸泡10min，用0.5%高锰酸钾消毒30min，分别用各不同浓度的浸提液浸泡60min。在直径10cm的培养皿中垫2张滤纸，选取浸泡过的饱满受试植物种子50粒，整齐排列于培养皿中的滤纸上，每个处理设3个重复，分别加入各供试液5ml，对照(CK)加蒸馏水5ml，置于25℃、80%相对湿度、光照强度1500lx、光照12h的条件下培养。试验过程中，及时补充相应的水浸提液，保持滤纸湿润。每天观察、记录其萌发情况，计算发芽率。

将受体植物种子用1%高锰酸钾消毒10min，用30℃温水浸种1h后，分别点播在基质中(珍珠岩：沙子=3：1)。放在25℃，1800~2000lx，12h光照条件下培养。受体植物长出第一对真叶前，每3d喷施1/2Hoagland营养液，之后每2d喷施Hoagland营养液。每周补充浸提液，蒸馏水作对照，根据基质湿度每2d定量喷洒适量蒸馏水。

处理40d后测定各作物叶片的叶绿素a+b含量、脯氨酸含量、可溶性糖含量、可溶性蛋白质含量。

实验结束后分别取整个植株在80℃的烘箱内烘干至恒重，以g为单位计算单株生物量烘干重。

1.3 数据分析

采用DPS V6.55软件对试验结果进行统计分析。

2 结果分析

2.1 尾巨桉纯林土壤水浸提液对农作物种子发芽率的影响

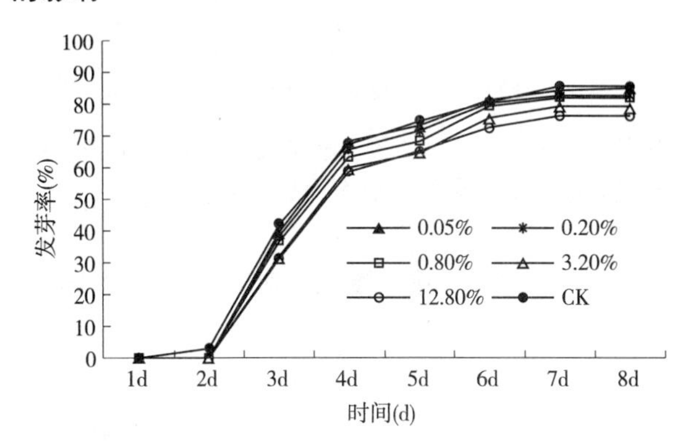

图1 纯林土壤水浸提液对水稻发芽率影响

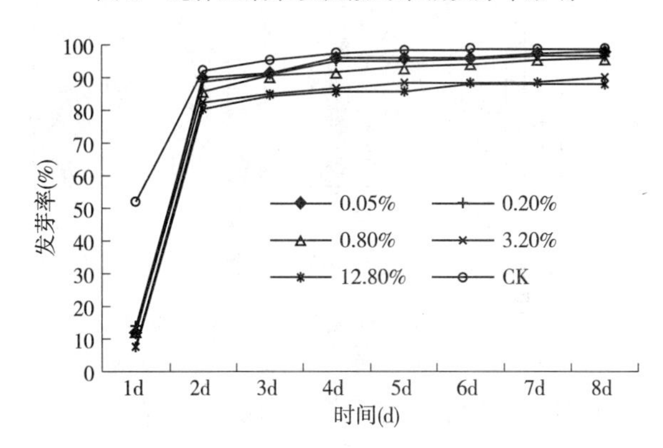

图2 纯林土壤水浸提液对菜心发芽率影响

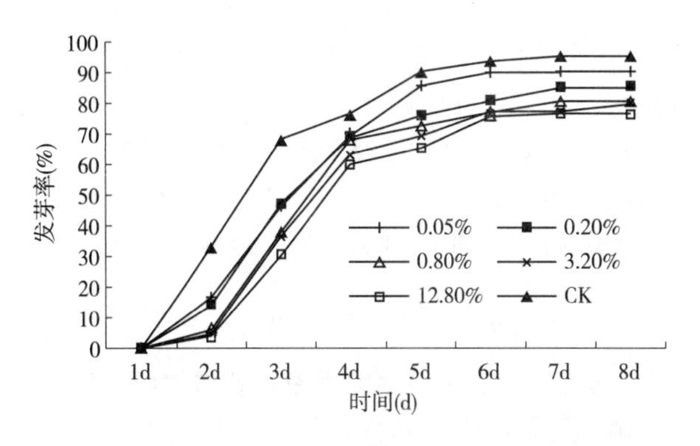

图3 纯林土壤水浸提液对白菜发芽率影响

从上图可知，尾巨桉纯林土壤水浸提液处理的水稻、菜心、白菜、萝卜种子的萌发进程来看，在同一个时段内，随着浸提液浓度的升高，农作物种子的发芽率降低。方差分析表明，各浓度尾巨桉纯林土壤水浸提液对几种作物种子发芽率均有极显著地

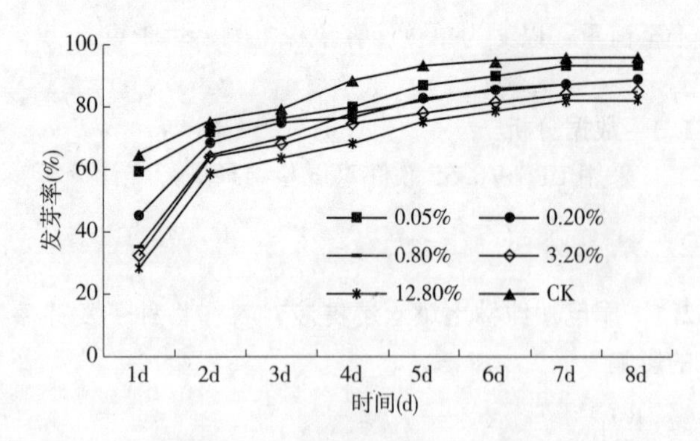

图4　纯林土壤水浸提液对萝卜发芽率影响

影响($P<0.01$)。低浓度处理的水稻发芽率与对照相差较小，浓度越高抑制作用越明显。

2.2　尾巨桉纯林土壤水浸提液对农作物蛋白质含量的影响

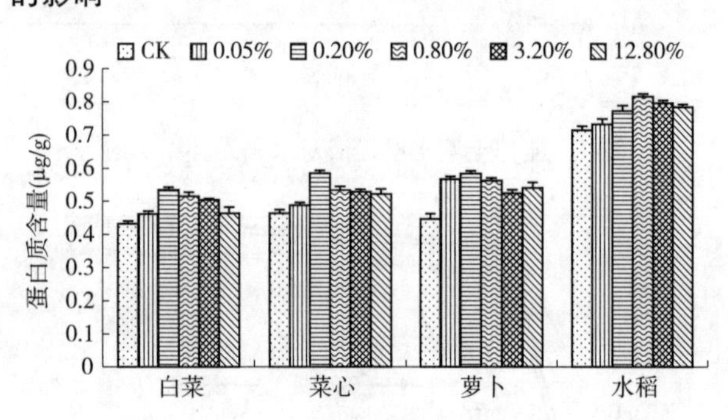

图5　纯林土壤水浸提液对几种作物蛋白质含量影响

由图5可知，尾巨桉纯林土壤水浸提液处理的白菜、菜心、萝卜和水稻的蛋白质含量均随着浸提液浓度的增大而增加，当浸提液的浓度达到一定高度，白菜、菜心、萝卜和水稻的蛋白质含量相应的开始下降。方差分析结果表明，尾巨桉纯林土壤水浸提液处理及对照的白菜、菜心、萝卜和水稻的蛋白质含量之间分别存在极显著差异(白菜：$F=25.18$，$P=0.0001$；菜心：$F=30.05$，$P=0.0001$；萝卜：$F=34.19$，$P=0.0001$；水稻：$F=20.30$，$P=0.0001$)。各作物对照处理的蛋白质含量均极显著低于浸提液处理。

2.3　尾巨桉纯林土壤水浸提液对农作物脯氨酸含量的影响

由图6可知，尾巨桉纯林土壤水浸提液处理的白菜、萝卜、菜心、水稻的脯氨酸含量均随浸提液浓度的增大呈现出先增加后下降的趋势。方差分析结果表明，尾巨桉纯林土壤水浸提液处理的白菜、萝卜、菜心、水稻的脯氨酸含量及对照之间存在极显著差异(白菜：$F=312.17$，$P=0.0001$；萝卜：$F=13.52$，$P=0.0001$；菜心：$F=116.86$，$P=0.0001$；水稻：$F=32.03$，$P=0.0001$)。

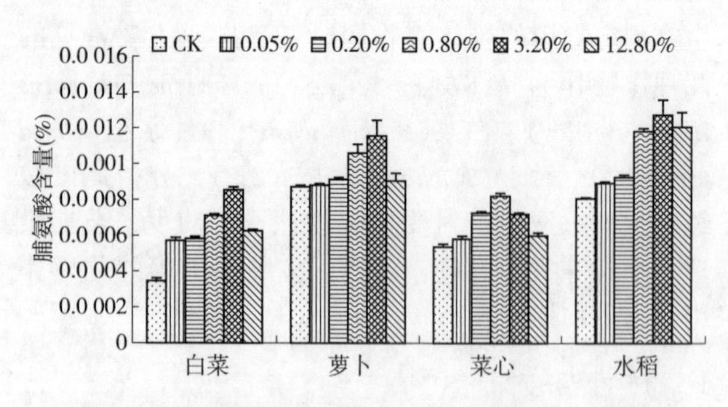

图6　纯林土壤水浸提液对几种作物脯氨酸含量影响

2.4　尾巨桉纯林土壤水浸提液对农作物可溶性糖含量的影响

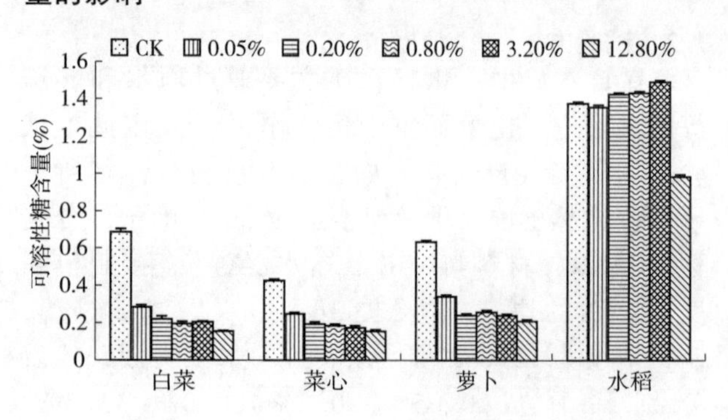

图7　纯林土壤水浸提液对几种作物可溶性糖含量影响

尾巨桉纯林土壤水浸提液处理的白菜、菜心、萝卜、水稻的可溶性糖含量及对照之间存在极显著差异(白菜：$F=519.23$，$P=0.0001$；菜心：$F=231.01$，$P=0.0001$；萝卜：$F=643.01$，$P=0.0001$；水稻：$F=711.80$，$P=0.0001$)。由图7可知，随着尾巨桉纯林土壤水浸提液浓度的增加，浸提液处理的白菜、萝卜、菜心的可溶性糖含量总体呈下降趋势；水稻的可溶性糖含量均随浸提液浓度的增大而增加，当浸提液浓度为3.2%时，水稻的可溶性糖含量达到几个浓度处理中的最大值1.49%，之后开始降低。

2.5　尾巨桉纯林土壤水浸提液对农作物叶绿素含量的影响

由图8可知，尾巨桉纯林土壤水浸提液处理的白菜、菜心、萝卜、水稻的叶绿素含量随着浸提液处理浓度的增大而呈下降趋势，高浓度尾巨桉纯林土壤水浸提液处理的白菜、菜心、萝卜、水稻的叶绿素含量分别最低。方差分析结果表明，尾巨桉纯林土壤水浸提液处理的白菜、菜心、萝卜、水稻的叶绿素量及对照之间存在极显著的差异(白菜：$F=$

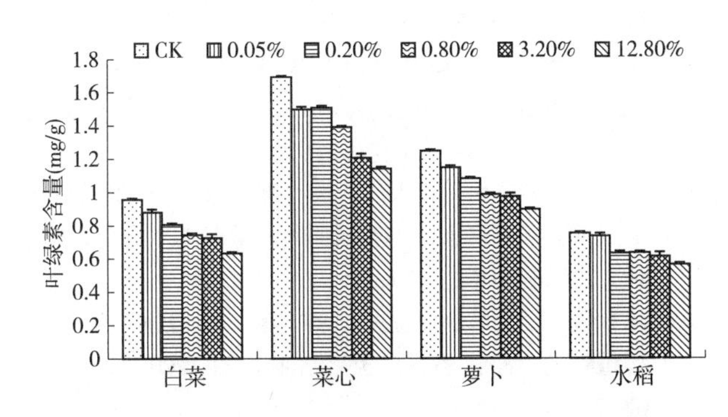

图8　纯林土壤水浸提对几种农作物叶绿素含量影响

116.95，P=0.0001；菜心：F=386.02，P=0.0001；萝卜：F=236.60，P=0.0001；水稻：F=51.63，P=0.0001)。白菜、菜心、萝卜和水稻的对照叶绿素含量均极显著高于不同浓度浸提液处理的叶绿素含量。

2.6　尾巨桉纯林土壤水浸提液对农作物生物量(干重)的影响

从表1所示，尾巨桉纯林土壤水浸提液处理的白菜、菜心、萝卜、水稻的生物量均小于对照处理，存在极显著的抑制作用(P<0.01)，浓度越高抑制作用越强。

表1　浸提液对几种苗木生物量影响的方差分析

受体植物	处理						方差分析	
	CK	0.05%	0.20%	0.80%	3.20%	12.80%	F值	P值
白菜	0.89Aa	0.87Aab	0.83ABb	0.77BCc	0.72CDcd	0.67Dd	23.80	0.0001
菜心	1.02Aa	0.97ABb	0.94BCb	0.88CDc	0.83DEd	0.78Ee	33.16	0.0001
萝卜	1.14Aa	1.1ABab	1.07Bb	1.00Cc	0.96CDc	0.9Dd	36.25	0.0001
水稻	0.81Aa	0.8Aa	0.77ABa	0.70BCb	0.67Cbc	0.64Cc	16.99	0.0001

3　结论与讨论

试验结果表明，尾巨桉纯林土壤水浸提液对菜心、白菜、萝卜、水稻的种子萌发及苗期生长具有不同程度的抑制。叶绿素含量是影响植物光合作用的重要因子，尾巨桉纯林土壤水浸提液对受体植物叶绿素含量具有抑制作用。前人研究表明化感物质对植物体光合作用的影响主要表现为叶绿素含量和光合速率的降低，如银胶菊倍半萜烯类化感物质处理胜红蓟幼苗后，叶绿素含量显著降低[23]；阿魏酸和香豆酸抑制水稻幼苗叶绿素合成过程中的镁螯合酶活性[24]。

在光合作用的主要因子受到抑制的条件下，植物体的其他生理机能肯定要相应的出现变化或异常现象，所以受体植物体内作为光合作用衍变的产物，既是渗透调节剂，也是合成其他有机溶质的碳架和能量的来源的可溶性糖含量相应的减少，而水稻的可溶性糖含量却表现出特殊的现象：随着浸提液浓度增加而增加，这是否与水稻相对于其他受体植物特殊的解剖结构对应激环境适应能力较强的缘故，还有待进一步研究其中的奥妙。曹成有等研究表明，瑞香狼毒根提取液会降低被处理植物体内的可溶性糖含量[25]。

植物体新陈代谢的重要部分——蛋白质代谢同样受到影响。大多研究认为几乎所有的酚酸类物质都降低了磷向DNA和RNA的整合，除香豆酸和香草酸外的其他酚酸类物质还抑制甲硫氨酸向蛋白质整合。尾巨桉纯林土壤水浸提液处理的受体植物叶片的蛋白质含量均随着浸提液浓度的升高而升高，当浓度达到一定高度，含量开始下降，导致的原因可能是植物体内核酸代谢受到影响。

目前，植物体内脯氨酸含量被作为植物抗逆性指标利用，曹成有等的研究表明，瑞香狼毒根提取液促使被处理植物的脯氨酸升高[25]。本研究中发现，受体植物的脯氨酸含量在处理液的作用下均呈现出随着浓度升高先生后降的趋势。已有研究表明桉树含有多种被认为是化感物质的化合物，包括醇、酸、酯和芳香族等化合物，酚类和萜类是高等植物主要的化感物质，酚类物质能够影响植物幼苗的多种生理活动，前人在这方面做了大量研究[16,18,26,27]。从试验的结果综合分析，尾巨桉纯林土壤可能具有化感作用的物质基础。

本试验仅在实验室条件下采用生物检测方法测定尾巨桉纯林土壤水浸提液对几种受体作物种子萌发和幼苗生长的影响，尚未考虑自然条件下，不同环境因子、林龄及无性系对尾巨桉他感作用的影响。因此，在下一步研究中，将结合光照、温度、空气及径流等因子对不同品系速生桉的化感作用进行综合评价，筛选出生长好、化感作用弱的桉树无性系。

参考文献

[1] RICE E L. Allelopathy[M]. New Yoke: Academic Press, 1984: 422.

[2] 林思祖, 杜玲, 曹光球. 化感作用在林业中的研究进展及应用前景[J]. 福建林学院学报, 2002, 22(2): 184-88.

[3] 吴长虹, 翟明普. 火炬树化感作用的初步研究[J]. 西北林学院学报, 2008, 23(6): 162-165.

[4] 林武星, 洪伟, 叶功富. 木麻黄水浸液对其幼苗生长的影响[J]. 江西农业大学学报, 2005, 27(1): 46-51.

[5] 张凤云, 翟梅枝, 贾彩霞, 等. 核桃鲜叶挥发油化感作用初步研究[J]. 西北林学院学报, 2005, 20(2): 144-146.

[6] 杨华庚, 陈惠娟. 橡胶树他感作用初步研究[J]. 华南热带农业大学学报, 2005, 11(4): 1-4.

[7] 翟梅枝, 张凤云, 田治国, 等. 不同季节草地早熟禾的化感作用研究[J]. 西北林学院学报, 2006, 21(6): 154-157.

[8] 黄闽敏, 潘存德, 罗侠, 等. 天山云杉针叶提取物对种子萌发和幼苗生长的自毒作用[J]. 新疆农业大学学报, 2005, 28(3): 30-34.

[9] 杨立学. 落叶松水提物对胡桃楸化感作用的生物测定[J]. 东北林业大学学报, 2006, 34(2): 15-17.

[10] 侯元兆. 科学地认识我国南方发展桉树速生丰产林问题[J]. 世界林业研究, 2006(3): 54-61.

[11] 刘小香, 谢龙莲, 陈秋波, 等. 桉树化感作用研究进展[J]. 热带农业科学, 200(2): 71-76.

[12] 黄卓烈, 林韶湘, 谭绍满, 等. 尾叶桉等植物叶提取液对几种植物插条生根和种子萌发的影响[J]. 林业科学研究, 1997, 10(5): 546-550.

[13] 曾任森, 李蓬为. 窿缘桉和尾叶桉的化感作用研究[J]. 华南农业大学学报, 1997, 18(1): 6-10.

[14] 赵绍文, 王凌晖, 蒋欢军, 等. 巨尾桉枝叶水浸提液对3种作物种子萌发的影响[J]. 广西科学院学报, 2000, 16(1): 14-17.

[15] 廖建良, 宋冠华, 曾令达. 巨尾桉叶片水提液对小麦幼苗生长的影响[J]. 惠州大学学报: 自然科学版, 2000, 20(4): 50-52.

[16] 王晗光, 张健, 杨婉身, 等. 巨桉根系和根系土壤化感物质的研究[J]. 四川师范大学学报(自然科学版), 2006, 29(3): 368-371.

[17] BOLTE M L, BOWERS J, CROW W D, et al. Germination inhibitor from *Eucalyptus pulverlenta*[J]. Agric Biol Chem, 1984, 48(2): 373-376.

[18] 田玉红, 刘雄民, 周永红, 等. 邓恩桉叶挥发性成分的提取及分析[J]. 南京林业大学学报(自然科学版), 2006, 30(2): 55-58.

[19] 陈秋波, 王真辉, 林位夫, 等. 刚果12号桉对4种豆科植物的化感作用[J]. 热带作物学报, 2003, 24(3): 67-72.

[20] JUAN JIMENEZ - OSORNIO F M V Z. Allelopathy Acticity of *Chenopodium ambrosioides* L [J]. Biochemical Systematics and Ecology, 1996, 24(3): 195-205.

[21] 曹光球, 林思祖, 杜玲, 等. 阿魏酸与肉桂酸对杉木化感作用的生物评价[J]. 中国生态农业学报, 2003, 11 (2) : 8 - 10.

[22] 陈龙池, 廖利平, 汪思龙, 等. 香草醛和对羟基苯甲酸对杉木幼苗生理特性的影响[J]. 应用生态学报, 2002, 13 (10) : 1291 - 1294.

[23] SINGH H P, BATISH D R, KOHLI R K, et al. Variance analyses of parthenin-asesquiterpene lactose from Parthenium hysterophorus - On Early Growth and Physiology of Ageratum conyzoides[J]. Journal of Chemical Ecology, 2002, 28(11): 2169-2179.

[24] YANG C M, CHANG I F, LIN S J, et al. Variance analysess of three allelopathic phenolics on chlorophyll accumulation of rice(*Oryzas ativa*) seedlings: Ⅱ. Stimulation of consumption-orientation. Bot. Bull[J]. Acad. Sin, 2004, 45: 119-125.

[25] 曹成有, 富瑶, 王文星, 等. 瑞香狼毒根提取液对植物种子萌发的抑制作用[J]. 东北大学学报(自然科学版), 2007, 28(5): 729-732.

[26] 田玉红, 刘雄民, 陶明有. 巨尾桉叶挥发性成分的提取及成分分析[J]. 广西科学院学报, 2006, 22(5): 466-468.

[27] 王晗光, 张健, 杨婉身, 等. 气相色谱-质谱法分析巨桉叶的挥发性化感成分[J]. 四川农业大学学报, 2006, 24(1): 51-54.

[原载: 西北林学院学报, 2011, 26(04)]

气象因子对马尾松松脂产量影响的研究

刘 云

（中国林业科学研究院热带林业实验中心，广西凭祥 532600）

摘 要 通过对凭祥市马尾松林分连续4年产脂量的调查，研究气候因子的变化对其影响程度，并分析其相关性。结果表明：马尾松产脂量在年变化中最高值出现在气温高、降雨量强度中等的月份，或该月份的前后，但短期间气候变化的影响程度不突出。相对于年变化，年间变化对产脂量的影响较突出。产脂量与年温度变化呈显著正相关，与年降雨量呈不显著的负相关，年降雨量通过与年温度的显著负相关来改变温度条件，从而间接影响产脂量。

关键词 温度；降雨量；产脂量；相关性

The Effect of Meteorological Factors on the Oleoresin Output of *Pinus massonian*

LIU Yun

（*Experimental Center of Tropical Forestry*，*Chinese Academy of Forestry*，*Pingxiang* 532600，*Guangxi*，*China*）

Abstract：By the investigation of oleoresin output for four years running in masson pine forest of Pinxiang city，the effect of meteorological factors and the correlation were analyzed. The results were as follows：during a year，the maximum value of oleoresin output was present in the month which has high temperature and moderate rainfall，or before and after that. But the effect of climate in short-term fluctuation was non-significant. By contrast，Interannual fluctuation has prominent influence. Oleoresin output showed significantly positive correlation with temperature while non-significantly negative correlation with rainfall. The oleoresin output was changed by the negative interaction between temperature and rainfall.

Key words：temperature；rainfall；oleoresin output；correlation

马尾松是我国脂材兼用的乡土树种，其产脂量约占全国松脂产量的90%，在松脂主产区，马尾松被作为采脂"经济林"来经营[1]。松脂产业是广西具有资源优势的产业，其各项经济指标均居全国首位。广西松脂产业在2003年生产总值达20亿~21亿元，行业产值28亿~30亿元，同时为广西地方各级财政提供约2亿元的税收。目前广西已采脂的马尾松蓄积量为2537.27万m^3，占可采总蓄积30.14%[2]。

松脂是松树通过光合作用形成的糖类，再经过复杂的生物化学反应和一系列中间产物，进一步形成了萜烯和树脂酸。因此可以把松脂视为松树的光合作用产物[3-5]，其产脂量受到树体因子与生态因子等影响较大[6]。本试验通过研究气候因子中月均温度、年均温度、积温、月均降水量、年均降水量和总降水量等因子对马尾松产脂量的影响，为提高和改善马尾松采脂经营措施、科学预测马尾松产脂量与其经济效益提供重要依据。

1 试验地概况

中国林业科学研究院热带林业实验中心地处广西西南部，地跨凭祥、龙州、宁明两县一市，紧靠中越边境，边境线蜿蜒曲折，总长150km。地理坐标为21°57′47″~22°19′27″N，106°39′50″~106°59′30″E。属南亚热带季风型半湿润—湿润气候，太阳辐射强烈，热量充足，雨量充沛，年均气温20.5~21.7℃，年均降水量1200~1500mm，年均相对湿度80%~84%。干湿季节较为明显。该中心地处低山丘陵区。最高海拔1045.9m，一般山峰500~800m。丘陵区所占的比重为72.5%，低山、中山合占27.5%，林地坡度较大，坡度大于26°的约占总面积的50%。

该中心有丰富的马尾松资源，现有松类（小班平均胸径≥5cm）面积为8190.0hm^2，株数1148.02万株。符合采脂条件（小班平均胸径≥20cm）的松类面积为1957.6hm^2，株数187.06万株，分别占松类面积和株数的23.90%和16.29%。按每公顷年产脂1.2t计算，每年可采脂2400t。

2　研究方法

试验地设置在该中心青山实验场27林班，地貌类型为低山，海拔375m，坡向南向，中坡，坡度15°，枯枝落叶层厚度为4cm，腐蚀质层2cm，土层厚度>80cm，石砾含量<20%，土壤种类为红壤。马尾松林造林时间为1984年，设置样地数2个，密度分别为416株/hm^2和366株/hm^2，样地面积为20m×30m=600m^2。

采脂方法为下降式浅割薄修采脂，割面负荷率40%，每年的割面长度为18~20cm，侧沟夹角80°~90°。上述试验从2004年开始布置，2005年4月开始调查记载松脂产量，往后每年5~11月进行采脂并称重记录采脂重量。2005-2008年的气象资料来源于《广西统计年鉴》。结合该地区温度、降水量、产脂量的变化趋势以及它们之间的相关性，分析气候因子对马尾松产脂量的影响。

3　结果与分析

3.1　马尾松松脂产量的连续4年月变化分析

图1为2005—2008年采脂时间5~11月的松脂月产量。由图可见4年中的产脂量每年各采脂月中呈波动变化，大部分年份采脂中期产脂量最高，首月或尾月较低。将表1的气象数据与图1松脂月产量变化结合分析可见，2005年月温度最高为5月和7月，最低11月，降水量最高为6月，最低为10月，产脂量最高峰出现在6月；2006年最高温度出现在6、7月，最低温度出现在11月，降水量最高为7月，最低为10月，产脂高峰出现在8月；2007年最高温度出现在6、7月，最高降水量在6、8月，最低温度和最小降水量出现在11月，但产脂量却表现出5~8月产量较低，9~11月产量较高，并且最高产量出现在11月；2008年最高温度为7、8月，最强降水为8、9月，最低温度出现在11月，最低降水出现在10月，产脂量最高出现在6月，但8、9、10、11月的产脂量差异较小。

可见产脂量较少的月份多为温度较低、降雨较少的月份，产脂较多的月份多出现在高温、降水多的月份或其前后的月份，但从总体情况看来，马尾松月产脂量变化趋势与月平均温度、月降水量的变化情况趋势不明显，可见短期气候的变化对产脂量的影响较小，且影响效果不明显。

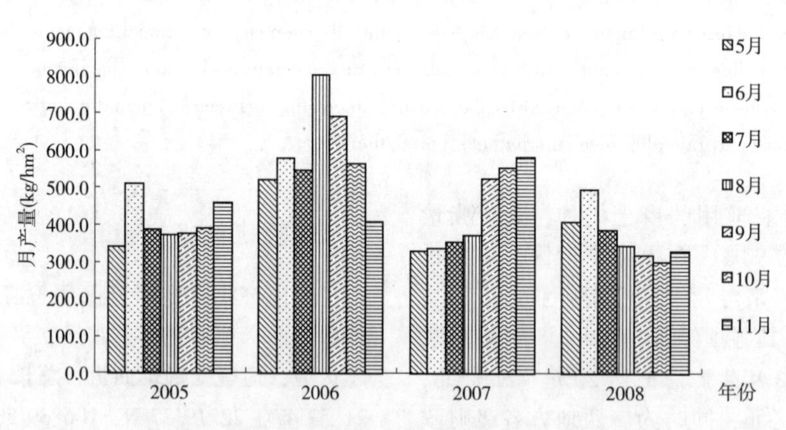

图1　2005-2008年松脂月产量

表1　凭祥市2005-2008年采脂月份气象变化

年份 \ 月份	温度(℃)				降雨量(mm)			
	2005	2006	2007	2008	2005	2006	2007	2008
5月	29.4	26.8	26.7	27.3	130.5	185.8	114.1	59.7
6月	28.5	28.0	29.6	27.7	356.7	155.2	238.4	211.0
7月	29.2	28.7	29.5	28.6	95.9	345.1	103.8	166.2
8月	28.6	27.9	28.6	28.3	86.4	130.8	234.8	278.2
9月	27.7	27.2	26.5	27.9	149.3	38.3	94.4	382.7

（续）

年份＼月份	温度(℃)				降雨量(mm)			
	2005	2006	2007	2008	2005	2006	2007	2008
10月	24.9	27.4	24.8	25.5	0.8	9.6	31.4	47.9
11月	20.8	22.2	18.7	19.0	116.6	54.7	23.4	158.3
年均	22.7	23.4	23.2	22.4	1123.5	1055.1	992.5	1502.0
总量	8302.3	8539.2	8486.1	8194.5	34067.5	32210.9	30270.7	45619.6

3.2 气象因子对马尾松年产脂量的影响

3.2.1 年度温度对马尾松产脂量的影响分析

从图2、图3可见，随着年均温度、年度积温升高，年产脂量增大。2006年平均温度和年度积温分别达到4年最高23.4℃和8539.2℃，而松脂年产量也在2006年达到4218.8kg/hm^2；2008年平均温度和年度积温分别为4年中最低22.4℃和8194.5℃，该年的松脂年产量也为4年最低2545.4kg/hm^2。随着气温升高树木自身生命活动加强，新陈代谢活动旺盛，松脂产量随之提高；而当气温降低时，细胞酶活性下降，新陈代谢强度降低，松脂产量也下降[6]，再而，松树体液流动速度是随着气温的升高而加快，马尾松泌脂能力也随着气温的升高而增强[7]。

3.2.2 年度降水量对马尾松产脂量的影响分析

图4、图5分别为年均降水量、年总降水量和年产脂量变化的趋势。2008年均降水量和年总降水量为4年中最大值，分别为1502.0mm和45619.6mm，年产脂量为4年最小值2545.4kg/hm^2；年均降水量和年总降水量最低值出现在2007年，而产脂量最高值出现在2006年。降水量大小顺序为2008年>2005年>2006年>2007年，而产脂量高低顺序为2006年>2007年>2005年>2008年，可见降水量不仅通过干扰马尾松产脂生理过程而影响松脂产量，而且通过降水量与其他环境因子，包括气温、光照强度、相对湿度和养分淋溶等的协同或者拮抗作用间接的影响产值效果。一定程度的降雨强度有利于马尾松的光合作用，加强了其生命活动和新陈代谢，从而提升了产脂能力，充足的水分吸收也利于树脂的流动和溢出；降雨较少除了影响树木产脂生理外，也会使割面及侧沟因过于干燥而影响采脂作业[8]；降雨强度过大，地面养分的流失量增大，土壤空隙含水量增大，同时也减缓了土壤中的根呼吸，使树木养分吸收减少，同时强降雨降低了环境气温，从而降低了马尾松产脂量。

3.2.3 气象因子与马尾松年产脂量的相关性分析

利用分析软件对年均温度、年积温、年均降水量、年总降水量和产脂量进行相关性分析，并由相关系数$\rho=0$临界表可知$F_{0.05}(7-2)=0.7545$，$F_{0.01}(7-2)=0.8745$[9]，分析结果如表2所示，年产脂量与年均温度、年积温呈显著正相关，与年均降水量和年总降水量呈负相关，但相关性未到达显著水平，年均温度、年积温与年均降水量和年总降水呈显著负相关。

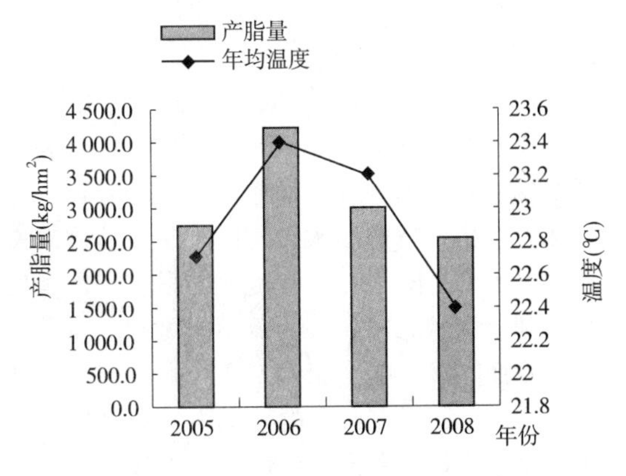

图2　年均温度对松脂产量的影响

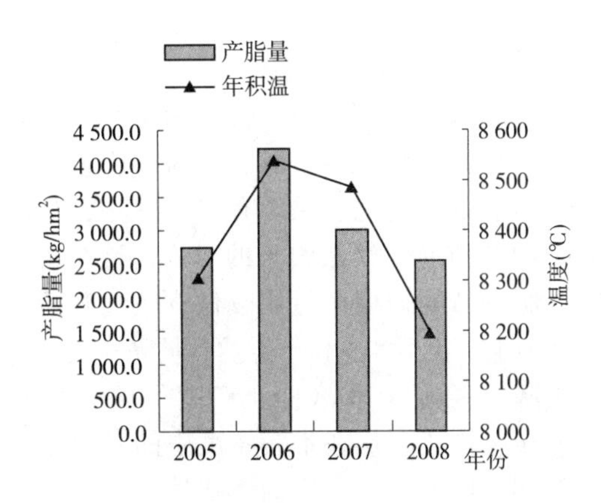

图3　年积温对松脂产量的影响

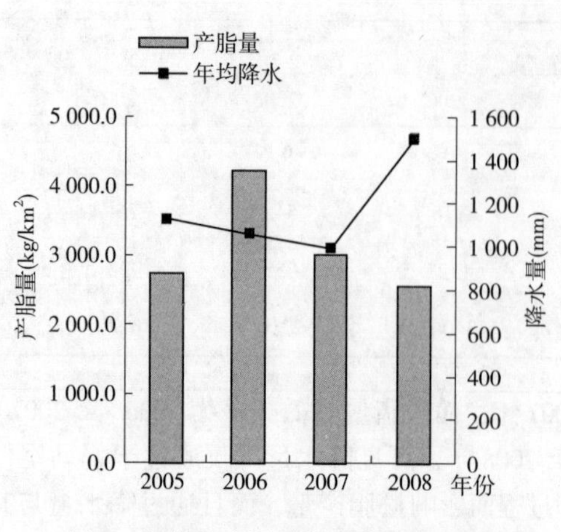

图 4 年均降水量对松脂产量的影响

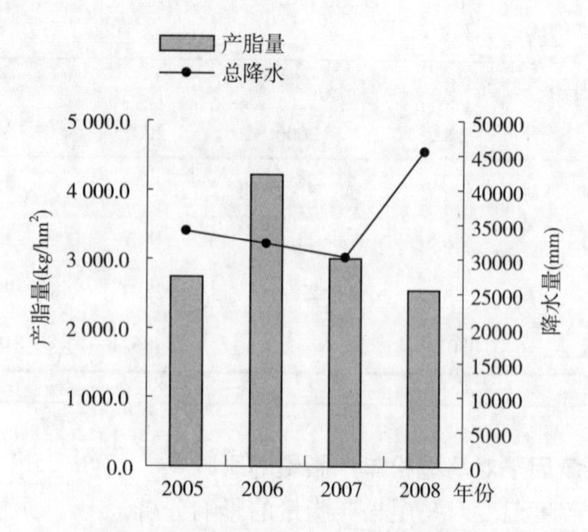

图 5 年总降水量对松脂产量的影响

表 2 各气象因子与产脂量相关性分析

	年均温度	年积温	年均降水量	年总降水量	年产脂量
年均温度	1.000				
年积温	0.999**	1.000			
年均降水量	-0.851*	-0.864*	1.000		
年总降水量	-0.845*	-0.858*	1.000**	1.000	
年产脂量	0.838*	0.816*	-0.561	-0.554	1.000

注：* 表示相关性显著，** 表示相关性极显著。

可见年产脂量随年温度升高而增加，气温对其的影响是显著而直接的；而随着年降水量的增多，年产脂量下降，但影响并不显著，降水量对其的影响是间接而复杂的；再由温度和降水量的显著负相关来看，降水量对产脂量的拮抗作用主要通过改变气温而间接影响产脂效果。所以松脂年产量的预测工作中，高产脂量最有可能出现在气温较高、降水适中的年份，低产脂量可能出现在低温干旱或者低温强降水的年份。

4 结论与讨论

(1)通过对马尾松连续 4 年产脂的月变化趋势及各年份每月气象数据的综合分析，认为产脂量的变动与气温、降水量的变化有关。有研究报道，马尾松在一年中的产值受到气象因子的影响，呈抛物线波动，在产脂季节的初期产量比较低，随着时间的推移产量逐渐上升，至 8 月份产量达到最高，之后产量又逐渐下降，直到 10 月底停止产脂[6,10,11]。但在凭祥地区，马尾松产脂量在不同年度中的月变化并不呈抛物线形，但也具有一定的规律性，即最高多为气温高、降水量强度中等的月份，但也有可能出现在该月份的前后；而产脂量较少的月份一般为气温较低，降水较小的采脂前期或后期。可见气象因子对马尾松产脂有着重要的影响，而且这种影响的效果不止停留在当月，而是随着因子的动态变化而延续更长的一段时间。正确的掌握马尾松产脂年变化规律，可以更科学的指导实际生产和规划更合理的采割松脂的起止时间。

(2)在连续 4 年的马尾松松脂年产量与各年份平均温度、总积温、平均降水量和总降水量的变化的相关性分析中，松脂产量和年份平均温度、总积温变化趋势相一致，随着年温度的升高，产脂量增大，其相关系数达到显著正相关水平，可见温度对产脂量有着重要的影响，且全年气温变化的影响效果比月气温变化的影响效果突出；而松脂产量和年份平均降水量、总降水量变化趋势不一致，随着年降水量的变化，产脂量高低起伏，其相关系数呈负相关但未达到显著水平，降水对松脂产量的影响是复杂的。并由温度和降水量之间显著的负相关可知，降水量通过改变温度而间接影响产脂量。又因水分充足可提高植物生命活动，所以在一定的范围内，降水量增大应可促进植物的生理过程，其在促进作用和抑制作用之间的临界阈值，需要更多的实验数据支持和更深入的调查研究。如能明确气象因子对马尾松产脂的影响以及因子之间的相互关系，将对马尾松产脂量及经济效益的预测工作起重要的促进作用。

参考文献

[1] 谢善高，郑小贤，刘洪．马尾松产脂量密度效应的初步研究[J]．福建林业科技，2009，36(1)：31-34.

[2] 杨章旗．广西松脂原料林发展现状、存在问题与对策研究[J]．广西林业科学，2007，36(3)：143-146.

[3] 王长新．马尾松采脂量的相关因子分析[J]．河南科技大学学报(农学版)，2004，24(3)：22-25.

[4] 李爱民，全妙华，牛友芽，等．马尾松松脂产量影响因素研究进展．湖南林业科技，2008，35(1)：58-59，55.

[5] 郭宇渭，赵文书，王福斌，等．思茅松林分生物环境因子与产脂量关系研究．云南林业科技，1993(4)：11-16.

[6] 安宁，桂杰．主要气象因子对马尾松松脂产量的影响初报[J]．山地农业生物学报，2010，29(2)：177-180.

[7] 谢善高．广西马尾松采脂林综合经营技术及其产业化模式研究[D]．北京：北京林业大学，2009.

[8] 祝志勇，于建云．生态因子对马尾松产脂的影响初探[J]．华东森林经理，2002，16(3)：36-37.

[9] 洪伟，吴承祯．试验设计与分析——原理·操作·案例[M]．北京：中国林业出版社，2004：309.

[10] 王以珊，曾令海，罗敏，等．马尾松天然林采脂试验分析[J]．广东林业科技，2002，18(2)：1-4.

[11] 王之颖，赵仁富．气象因子对湿地松松脂产量的影响[J]．林业科技开发，2001，15(6)：18-19.

[原载：中南林业科技大学学报，2011，31(06)]

桂西南米老排人工林单株生物量回归模型

明安刚[1] 唐继新[1] 于浩龙[1] 史作民[2] 卢立华[1] 贾宏炎[1] 蔡道雄[1]
([1] 中国林业科学研究院热带林业实验中心，广西凭祥 532600；
[2]中国林业科学研究院森林生态环境与保护研究所国家林业局森林生态环境重点实验室，北京 100091)

摘 要 通过对桂西南大青山林区28年生米老排(*Mytilaria laosensis*)人工林林分进行每木检尺和生物量的测定，建立了米老排各器官生物量与胸径、树高和胸径平方乘树高(D^2H)的相关关系；分别选用幂函数等5种模型，用回归分析方法对米老排人工林单株生物量模型进行了拟合。结果表明：树叶和树根生物量分别与胸径和树高的相关关系最显著，而树干、树枝、树皮和全株的生物量都与D^2H的相关关系最为显著。胸径、树高和D^2H与各器官生物量拟合的模型中，全株、树干和树皮的拟合效果最好，树叶和树根的拟合效果中等，而树枝的拟合效果较差；除树皮外，各器官均以幂指数模型的拟合效果最好。

关键词 生物量；回归模型；米老排；桂西南

Individual Biomass Regression Model of *Mytilaria laosensis* in Southwest of Guangxi

MING Angang[1], TANG Jixin[1], YU Haolong[1], SHI Zuomin[2], LU Lihua[1], JIA Hongyan[1], CAI Daoxiong[1]

([1]*Experimental Center of Tropical Forestry*, *CAF*, *Pingxiang* 532600, *Guangxi*, *China*;
[2]*Institute of Forest Ecology*, *Environment and Protection*, *Key lab on Forest Ecology and Environmental Sciences of State Forestry Administration*, *CAF*, *Beijing* 100091, *China*)

Abstract: Based on the inventory data of a 28-years-old *Mytilaria laosensis* plantation and the biomass measurements of the mean sample trees, the correlation between diameter at breast height (DBH), tree height (H), D^2H and different parts of an individual tree were analyzed. Five biomass models, including power function model of individual tree of Mytilaria laosensis in southwest of Guangxi province were established by regression analysis. The results showed that the relationships between stem, branch, bark, total biomass and D^2H were significant while the relationships between leaf biomass and DBH and between root biomass and H were significant. In the models in which the independent variable is the DBH (D), tree height (H) and D^2H respectively, total tree, trunk, and bark biomass recession models were fitted best, leaf and root biomass recession models were fitted moderately, and branch biomass recession model were fitted worst. Power function model was the best regression model for fitting biomass of the most parts of an individual tree except bark biomass.

Key words: biomass; regression models; *Mytilaria laosensis*; southwest of Guangxi

生物量是人工林生态系统的功能指标，是评估人工林固碳能力和固碳效益的基础[1]，因此，森林生物量和生产力的研究与森林碳汇功能研究紧密结合起来，已成为全球环境问题新的研究热点[2,3]。米老排(*Mytilaria laosensis*)又名壳菜果、三角枫，为金缕梅科米老排属的常绿大乔木，是我国南亚热带地区优良乡土速生阔叶树种，具生长快、干形好、材质优良、用途广泛等优良特性，有着重要的经济效益、社会效益和生态效益。米老排天然分布在我国广东西部、广西西南部、云南南部及越南和老挝等地[4-8]。桂西南是米老排人工林在南亚热带地区重要的种源地，利用林木每木检尺材料和一定数量样木的生物量实测数据为基础，建立米老排单株生物量模型，对估计桂西南地区米老排人工林生物量，科学评价米老排人工林固碳功能，具有重要的科学意义和应用价值。本文通过曲线估计，建立了桂西南米老排人工林单株各器官与树高(H)、胸径(D)及 D^2H 的关系式，以期为今后对桂西南米老排生物量和固碳功能的研究提供借鉴。

1 研究区概况

试验地位于广西凭祥市中国林业科学研究院热带林业实验中心伏波实验场，地理位置为106°39′50″~106°59′30″E，21°57′47″~21°19′27″N，海拔600m。属南亚热带季风气候，年平均气温19.5~21.0℃，年降水量1400mm左右，≥10℃活动积温6000~7600℃，土壤为花岗岩发育成的山地红壤，土层厚度>100cm。试验地于1981进行块状整地，1982年春季造林，初植密度为1500株/hm²，造林当年秋季抚育除草1次，以后每年4~6月和9月各除草松土1次，连续3年，直至郁闭成林。成林后开始间伐，间伐后的郁闭度一般保持在0.6左右，目前林分郁闭度为0.9。

2009年调查当年，林分的保留密度为625株/hm²，平均胸径26.3cm，平均树高23.3m。林下植被总盖度为10%，主要有米老排、杜茎山(*Maesa japonica*)、酸藤子(*Embelia laeta*)、金毛狗(*Cibotium barometz*)、五节芒(*Miscanthus floridulu*)、铁芒萁(*Dicranopteris linearis*)等。

2 研究方法

2.1 林木调查及生物量的测定

2009年11月在林分中选取坡度均匀且人为干扰较小的林地作为试验样地。在上、中、下3个坡位分别设置1个面积为400m²的方形样地，对每个样地进行每木检尺，调查胸径、树高等因子，绘制胸径分布图。根据胸径分布情况，以2cm为一个径阶，在样地的外围按不同径阶共选取标准木13株，将所有的标准木伐倒，按2m一个区分段分别称重并取样，树枝、树叶分上中下3层分别称重并取样；采用"全挖法"挖掘样木根系，称重并取样；鲜样在85℃烘干至恒温，根据含水率将各器官鲜重换算成干重。

2.2 生物量模型的选择

关于立木生物量方程，由Kittredge[9]引进的相对生长方程目前被大多数人接受，且认为是一种较为理想的基本模型，其表达式为：$Y=aX^b$，简称为CAR(Constant Allometric Ratio)，但国内外对此也有些不同的看法[10-13]。为了验证CAR模型对米老排生物量的适用性，选择更适合米老排生物量预测的相关模型，本文在分析米老排各器官与胸径、树高及 D^2H 的相关关系显著度的基础上，分别用 $Y=aX^b$，$Y=ae^{bx}$，$Y=ax+b$，$Y=a\ln(x)+b$ 和 $Y=a_1x^{b1}+a_2x^{b2}+a_3x^{b3}\cdots+c$ 等5种模型对各器官生物量方程进行拟合，比较并选出拟合效果最好的模型作为预测米老排各器官及全株的生物量模型。

3 结果与分析

3.1 米老排人工林单株树高 胸径及 D^2H 与各器官生物量的相关性

根据米老排标准木的调查资料，对米老排单株胸径、树高和 D^2H 与各器官生物量进行相关性分析。从表1可以得出，桂西南米老排单株胸径与树高、D^2H、树干生物量、树皮生物量、树叶生物量、树根生物量和全株生物量呈极显著相关，与树枝生物量呈显著相关；树高与胸径、D^2H、树干生物量、树皮生物量、树根生物量和全株生物量呈极显著相关，与树枝和树叶生物量呈显著相关；而 D^2H 与米老排各器官生物量和全株生物量均呈极显著相关。从树高、胸径、D^2H 与各变量之间相关矩阵中可以看出，桂西南米老排各器官生物量与树高、胸径和 D^2H 的密切程度为：D^2H>胸径>树高，且胸径和树高都与全株生物量和树干生物量的相关关系最为密切，相关系数 r 值均在0.8735以上。

表1 树高、胸径、D^2H 与各变量之间的相关矩阵①

项目	D	H	D^2H	W_s	W_{ba}	W_{br}	W_l	W_r	W_t
D	1.0000								
H	0.7936**	1.0000							
D^2H	0.9820**	0.8439**	1.0000						
W_s	0.9005**	0.8735**	0.9511**	1.0000					
W_{ba}	0.8759**	0.7408**	0.9176**	0.8853**	1.0000				
W_{br}	0.7058*	0.6209*	0.7105**	0.8138**	0.5939	1.0000			
W_l	0.8932**	0.6269*	0.8880**	0.8131**	0.8941**	0.6562*	1.0000		
W_r	0.7145**	0.8529**	0.7573**	0.8361**	0.7745**	0.7293**	0.6790*	1.0000	

（续）

项目	D	H	D^2H	W_s	W_{ba}	W_{br}	W_l	W_r	W_t
W_t	0.8812**	0.8749**	0.9246**	0.9857**	0.8759**	0.8568**	0.8149**	0.9064**	1.0000

注：*表示在 $P<0.05$ 的水平上显著，**表示在 $P<0.01$ 的水平上显著；W_s 表示树干生物量(kg)，W_{ba} 表示树皮生物量(kg)，W_{br} 表示树枝生物量(kg)，W_l 表示树叶生物量(kg)，W_r 表示树根生物量(kg)，W_t 表示全株生物量(kg)；H 表示树高(m)，D 表示胸径(cm)，D^2H 表示胸径平方与树高的乘积。

3.2 米老排单株各器官生物量的回归模型

3.2.1 树干生物量模型

通过对米老排人工林胸径、树高及 D^2H 3 个自变量与树干生物量(W_s)进行回归分析，筛选出 3 个与树干生物量间相关关系最紧密的回归方程。由表 2 可以看出，胸径、树高和 D^2H 与树干生物量间均呈幂函数关系，且都有较好的拟合效果(R^2 值分别为 0.8788、0.8076、0.9196，并且都满足相伴概率值 $P<0.001$)。相比之下，3 个自变量中，以 D^2H 与树干生物量的拟合效果最好。因此，利用方程 $W_s=0.1740(D^2H)^{0.7661}$ 估算米老排人工林树干生物量具有更高的准确度。

表 2　树干生物量回归模型

自变量	回归方程	判定系数 R^2	F 值	相伴概率值
D	$W_s=0.6769D^{1.8761}$	0.8788	47.1569	0.0000
H	$W_s=0.0144H^{3.0648}$	0.8076	35.4091	0.0001
D^2H	$W_s=0.1740(D^2H)^{0.7661}$	0.9196	104.3812	0.0000

3.2.2 树枝生物量模型

对 3 个自变量与树枝生物量(W_{br})进行回归分析，筛选出 3 个与树枝生物量间相关显著度最高的方程。从表 3 可以得出，3 个自变量与树枝生物量间均呈幂函数关系，但拟合效果一般(R^2 值分别为 0.6251、0.4778、0.6291，虽然相关性都达到了显著水平，但相伴概率值都大于 0.001)。相比之下，3 个自变量中，树枝生物量与树高的拟合效果较差($R^2=0.4778<0.5$，相伴概率值大于 0.01)，与胸径和 D^2H 拟合效果相对较好，判定系数相近，但与 D^2H 的拟合效果稍优于与胸径的拟合。实践中可以利用方程 $W_{br}=0.0002(D^2H)^{1.2696}$ 来估算米老排树枝生物量。

表 3　树枝生物量回归模型

自变量	回归方程	判定系数 R^2	F 值	相伴概率值
D	$W_{br}=0.0015D^{3.1701}$	0.6251	10.9218	0.0070
H	$W_{br}=1.0\times10^{-5}H^{4.7233}$	0.4778	6.9011	0.0235
D^2H	$W_{br}=0.0002(D^2H)^{1.2696}$	0.6291	11.2124	0.0065

3.2.3 树皮生物量模型

对树皮生物量(W_{ba})回归分析结果显示，米老排树皮生物量与胸径、树高和 D^2H 之间都达到了显著相关。由表 4 可知，树皮生物量与胸径和 D^2H 呈二次曲线关系，与树高呈幂函数关系。对 3 个回归方程进行比较，与树高的拟合效果较差，与胸径的拟合效果较好，而与 D^2H 的拟合效果最佳，F 值最大，判定系数最接近 1，相伴概率值最小，因而实践中可以使用方程 $W_{ba}=3\times10^{-8}(D^2H)^2+0.0001(D^2H)+9.7883$ 来估算米老排树皮生物量。

表 4　树皮生物量回归模型

自变量	回归方程	判定系数 R^2	F 值	相伴概率值
D	$W_{ba}=0.1051D^2-3.4574D+40.525$	0.8463	36.2662	0.0001
H	$W_{ba}=1.1476H^{0.1138}$	0.5827	13.3748	0.0038
D^2H	$W_{ba}=3\times10^{-8}(D^2H)^2+0.0001(D^2H)+9.7883$	0.8668	58.6376	0.0000

3.2.4 树叶生物量模型

从表 5 可看出，米老排树叶生物量(W_l)与胸径、树高和 D^2H 之间都达到了显著相关，且与各自变量均呈幂函数关系。对 3 个回归方程进行比较，树叶生物量与树高的相关显著度较低，相伴概率值 0.0219>0.01，相关关系没达到极显著水平，方程的拟合效果不佳($R^2=0.5050$)；而与胸径的拟合效果最好($R^2=0.8491$)，与 D^2H 也有较好的拟合效果($R^2=0.8091$)，但不如与胸径的拟合度高，实践中可以使用方程 $W_l=0.0002D^{3.2304}$ 来估算米老排树叶生物量。

表 5　树叶生物量回归模型

自变量	回归方程	判定系数 R^2	F 值	相伴概率值
D	$W_l=0.0002D^{3.2304}$	0.8429	43.3948	0.0000
H	$W_l=6.0\times10^{-6}H^{4.2610}$	0.5050	7.1210	0.0219
D^2H	$W_l=3.0\times10^{-5}(D^2H)^{1.2634}$	0.8091	41.0306	0.0001

3.2.5　树根生物量模型

对米老排树根生物量(W_r)进行回归分析并筛选出与胸径、树高和 D^2H 3 个自变量拟合度较高的生物量方程。从表 6 可以看出，米老排树根生物量与各自变量均呈幂函数关系。对 3 个回归方程进行比较得出，树根生物量与树高 H 的相关关系最为显著，相伴概率值 0.0002<0.001，方程的拟合效果最佳($R^2=8150$)；而与胸径和 D^2H 的相关关系远不如与树高密切，相伴概率值分别为 0.0061 和 0.0027，均大于 0.001。与胸径和 D^2H 的拟合效果也不如树高($R_D{}^2=0.6422$，$R_{(D2H)}{}^2=0.7247$)，因此，建议使用方程 $W_r=8.0\times10^{-5}H^{4.3179}$ 来估算米老排树根生物量。

表 6　树根生物量回归模型

自变量	回归方程	判定系数 R^2	F 值	相伴概率值
D	$W_r=0.0668D^{2.2492}$	0.6422	11.4698	0.0061
H	$W_r=8.0\times10^{-5}H^{4.3179}$	0.8150	29.3630	0.0002
D^2H	$W_r=0.0094(D^2H)^{0.9538}$	0.7247	14.7872	0.0027

3.3　全株生物量模型

全株生物量(W_t)包括乔木地上和地下两个部分，即各器官生物量的总和，是人工林群落生物量的主要组成部分，在人工林生态系统植被生物量中占有最大的比例。全株生物量方程建立的效果直接影响到人工林生物量和生产力乃至碳储量测算的准确度，甚至影响到人工林固碳功能评估的可信度，因此，全株生物量模型的有效模拟对人工林生物量和碳储量的估测至关重要。

本文对米老排人工林胸径、树高及 D^2H 与全株生物量进行了回归分析，并筛选出 3 个与全株生物量间相关显著的方程，由表 7 可以看出，胸径、树高和 D^2H 与全株生物量间均呈幂函数关系，且都有较好的拟合效果(R^2值分别为 0.8573、0.8194、0.9049；相伴概率值均小于 0.001)，因此，这 3 个回归方程都可以较好地用来预测米老排全株生物量。但相比之下，3 个自变量中，以 D^2H 与全株生物量的回归方程中的 F 值最大、判定系数最接近 1、相伴概率值最小，在 3 个回归方程中，拟合效果最好。因而，为了能让生物量的估测尽可能接近实际值，实践中使用方程 $W_s=0.1536(D^2H)^{0.8268}$ 来估算米老排全株生物量。

表 7　全株生物量回归模型

自变量	回归方程	判定系数 R^2	F 值	相伴概率值
D	$W_t=0.6841D^{2.0158}$	0.8573	38.2042	0.0001
H	$W_t=0.0089H^{3.3584}$	0.8194	35.8884	0.0001
D^2H	$W_t=0.1536(D^2H)^{0.8268}$	0.9049	64.8073	0.0000

4　结论与讨论

(1)对胸径、树高和 D^2H 3 个自变量与各器官生物量相关性分析结果中可知，桂西南米老排各器官生物量与树高、胸径和 D^2H 的密切程度为：D^2H>胸径>树高，且胸径和树高都与全株生物量和树干生物量的相关关系最为密切，相关系数 r 值都在 0.8735 以上。这与李斌等人在贺兰山对天然油松林的研究结果基本一致[14]。由此可知，对人工林而言，自变量与单株生物量之间的相关关系与天然林相比，没有明显差异。

(2)分别用胸径、树高和D^2H 3个自变量与各器官生物量拟合的方程中，除树皮与自变量之间呈二次曲线关系外，其他各器官和全株生物量与各自变量之间均呈很好的幂函数关系，这一结论有力地支持了Kittredge提出的CAR模型在单株生物量的适用性[8]。在对各器官的生物量方程中，树枝的生物量拟合方程判定系数小，拟合效果不佳，可能与采样过程中造成的误差有关。

(3)在桂西南地区，实践中建议使用的米老排各器官的生物量回归方程分别是：

树干 $W_s=0.1740(D^2H)^{0.7661}$

树枝 $W_{br}=0.0002(D^2H)^{1.2696}$

树皮 $W_{ba}=3\times10^{-8}(D^2H)^2+0.0001(D^2H)+9.7883$

树叶 $W_l=0.0002D^{3.2304}$

树根 $W_r=8.0\times10^{-5}H^{4.3179}$

全株 $W_t=0.1536(D^2H)^{0.8268}$。

所建议使用的方程均具有统计学意义，可用来预测各器官的生物量。适用范围为胸径15.9~30.9cm、树高16.3~28.4m的米老排纯林，适用区域为桂西南地区及周边的南亚热带地区。在本研究中，树干生物量和全株生物量与胸径、树高和D^2H均具有极显著的相关关系和较好的拟合效果，实践中可以结合实际需要，任意选择所给出的生物量拟合方程来预算全株生物量和树干生物量。

(4)本文只针对大青山28年米老排成熟林进行了研究，对于桂西南其他地区及其他年龄阶段的林分，尤其是米老排幼林单株生物量的回归模型有待进一步研究。

参考文献

[1] 李娜，黄从德．川西亚高山针叶林生物量遥感估算模型研究[J]．林业资源管理，2008(3)：100-104.

[2] 方精云．中国森林生产力及其对全球气候变化的影响[J]．植物生态学报，2000，24(5)：513-517.

[3] 郭永清，郎南军，杨旭，等．膏桐人工林单木生物量回归模型研究[J]．浙江林业科技，2009，29(6)：35-37.

[4] 明安刚，于浩龙，陈厚荣，等．广西大青山米老排人工林土壤物理性质[J]．林业科技开发，2010，24(5)：71-74.

[5] 郭文福，蔡道雄，贾宏炎，等．米老排人工林生长规律的研究[J]．林业科学研究，2006，19(5)：585-589.

[6] 林金国，张兴正，翁闲．立地条件对米老排人工林生长和材质的影响[J]．植物资源与环境学报，2004，13(3)：50-54.

[7] 景跃波，杨德军，马赛宇，等．热带速生树种米老排的育苗与造林[J]．林业实用技术，2008(1)：21-23.

[8] 卢立华，贾宏炎，何日明，等．南亚热带6种人工林凋落物的初步研究[J]．林业科学研究，2008，21(3)：346-352.

[9] KITTREDGE J. Estimation of the amount of foliage of trees and stand [J]. J. For，1944，42：905-912.

[10] RUARK G A. Comparison of constant and variable allomtric ratios for estimating Populus tremuloides biomass [J]. Forest Science，1987，33：294-300.

[11] 胥辉．两种生物量模型的比较[J]．西南林学院学报，2003，23(2)：36-39.

[12] GERON C D. Comparison of constant and variable allometric ratios for predicting foliar biomass of various tree genera[J]. Can. J. For. Res，1988，18：1298-1304.

[13] 刘志刚．华北落叶松人工林生物量及生产力的研究[J]．北京林业大学学报，1992，14(增)：114-123.

[14] 刘斌，刘建军，任军辉，等．贺兰山天然油松林单株生物量回归模型的研究[J]．西北林学报，2010，25(6)：69-74.

[原载：林业资源管理，2011(06)]

广西雅长林区野生兰科植物资源现状与保护策略

冯昌林[1] 邓振海[2] 蔡道雄[1] 吴天贵[2] 贾宏炎[1] 白灵海[1] 赵祖壮[2] 苏 勇[3]

([1]中国林业科学研究院热带林业实验中心，广西凭祥 532600；
[2]广西雅长兰科植物国家级自然保护区管理局，广西乐业 533209；
[3]广西南宁树木园，广西南宁 530031）

摘 要 广西雅长林区野生兰科植物资源丰富，约有52属156种，其中雅长保护区几乎每座山头都有兰科植物的分布，但其种类与居群基株数量等分布不均匀，且不连续，呈现破碎化现象，其中面积较大的局部密集分布区有16处。文中根据野生兰科植物的资源现状及生物生态学特性，针对目前兰科植物保育存在的问题，分析人畜干扰及自然条件变化等不利因素，从兰科植物维持机制角度出发，提出雅长野生兰科植物的保护应以就地保护为主、迁地保育为辅的保育措施与策略：保护区应加快建设的步伐，构建野生兰科植物种质资源基因库，同时对密集分布区在人畜干扰大的区域，优先拉设铁丝网围护，并派专人巡护，重点保护，确保野生兰科植物及基因库安全。

关键词 野生兰科植物；资源现状；保护策略；雅长林区；广西

Current Status and Conservation Strategies of Wild Orchid Resources in Guangxi Yachang Forests

FENG ChangLin[1], DENG ZhenHai[2], CAI DaoXiong[1], WU TianGui[2], JIA HongYan[1], BAI LinHai[1], ZHAO ZuZhuang[2], SU Yong[3]

([1] Chinese Academy of Forestry Experimental Centre of Tropical Forest, Pingxiang 532600, Guangxi, China;
[2] Yachang Wild Orchids National Nature Reserve Administration, Leye 533209, Guangxi, China;
[3] Guangxi Nanning Arboretum, Nanning 530031, Guangxi, China)

Abstract: The Yachang forest region is rich in wild orchid diversity, with 156 orchid species belonging to 52 genera distributed in the hills throughout the Yachang Orchid Nature Reserve. The orchids´ distribution in the Yachang region is, however, uneven and naturally or unnaturally fragmented. In particular, there are sixteen distribution hotspots with large concentrations of orchid species and populations within the region. The continuing existence of these rich wild orchid resources are threatened by disturbance from human activities, livestock grazing, and other fast-changing environmental conditions within the Yachang region. We recommend that the conservation strategies of the wild orchids in Yachang should include primarily of in-situ conservation measures, supplemented by ex-situ conservation. Yachang Orchid Nature Reserve should complete and improve its protection facilities as soon as possible, including the completion of a wild orchid germplasm bank, construction of fences around orchid-rich areas that aresubject to heavy herbivory from livestock, and intensification of foot-patrols in these areas.

Key words: wild orchids; resources status; conservation strategy; Yachang forests; Guangxi

兰科(Orchidaceae)植物种类繁多，全世界有700属20000多种，是具有重要经济价值和开发前景的多用途植物，许多种类为著名观赏和名贵药用植物。主要分布于热带和亚热带地区，少数分布于温带地区，我国有171属1247种[1]。兰科植物生长环境多样，其生活型有地生、半地生、附生及腐生等4种类型。

近年来，全世界兰花市场日益繁荣，观赏及药用兰科植物的年贸易额达千亿美元，市场上销售的兰花有不少为野生兰，全球野生兰科植物资源受到严重破坏，在中国，经济价值较高的兰属(*Cymbidium*)、兜兰属(*Paphiopedilum*)、石斛属(*Dendrobium*)等遭破坏最为严重，像霍山石斛(*Dendrobium huoshanense*)、铁皮石斛(*Dendrobium officinale*)已被滥采到"极度濒危"的程度。虽然全球所有签订《濒危野生动植物种国际贸易公约》(CITES)保护文件的国家，可减轻或禁止野生兰的国际贸易，但许多野生兰科植物并未被列入 CITES 附录，致使野生兰科植物在全世界受干扰和破坏仍为所有植物之首，濒临灭绝的种类最多。进入 21 世纪以来，我国十分重视野生兰科植物资源的保护，采取多种多样的保护措施，并取得了较大的成就。2005 年 4 月，广西有关部门把雅长林区野生兰科植物分布比较集中的区域规划和建设成为全国第一处以兰科植物为主要保护对象的"广西雅长兰科植物自治区级自然保护区"；2006 年 1 月，国家科技部立项建设全国第一个国家级野生兰科植物种质基因库——雅长野生兰科植物种质基因库；2008 年，雅长林区所在的乐业县被中国野生植物保护协会授予"中国兰花之乡"荣誉称号；2009 年，雅长兰科植物自然保护区被国务院批准晋升为国家级自然保护区——广西雅长兰科植物国家级自然保护区(以下简称雅长保护区)；2010 年广西把兰科植物所有野生种类列入《广西壮族自治区重点保护植物名录(第一批)》。这些政策、法规和措施的实施，使雅长野生兰科植物的保护有法可依，对其野生资源的保护和培育具有深远的影响[2,3]。

我们在开展雅长林区野生兰科植物资源调查研究的基础上，根据野生兰科植物的资源现状及生物生态学特性，针对兰科植物保育目前存在的问题，分析人畜干扰及自然条件变化等不利因素，从兰科植物维持机制角度，提出雅长林区区域野生兰科植物的保护策略。

1 雅长林区地理位置

雅长林区位于广西桂西北地区，处于乐业县、田林县及贵州省望谟县、册亨县交界，地理坐标介于24°37′~25°00′N 和 106°08′~106°27′E，具有面积较大的喀斯特地貌。

2 雅长林区野生兰科植物资源现状

2.1 野生兰科植物种类繁多，资源丰富

广西是中国野生兰科植物种类最多的省区之一，约有 109 属 397 种[2]，具有十分重要的兰科植物资源优势。雅长林区兰科植物种类繁多，资源丰富，约有 52 属 156 种(含 4 变种)[2-9]，分别占广西属数、种数[2]的 47.7%和 39.29%，居广西野生兰科植物主要分布区的首位。

2.2 兰科植物资源分布不均匀，呈块状或零星分布

雅长林区野生兰科植物分布主要以雅长保护区为分布中心，区域大尺度上资源分布不均匀，呈零星或块状分布。零星分布面积大，密集块状分布区面积相对较少。在雅长保护区内几乎每座山头都有兰科植物分布，总体呈现零星分布格局。零星分布区域种类少，居群基株数量少，而块状分布居群种类多，基株数量巨大(表 1)。此外，林区面积较大的局部密集分布区还有店子上、黄狼洞、拉雅大峡谷及金矿石壁等 16 处，面积达 915hm²。

2.3 兰科植物居群数量大，资源量丰富

雅长林区野生兰科植物局部密集分布区多，居群数量大，基株个体数量巨大，资源量丰富。林区中梳帽卷瓣兰(*Bulbophyllum andersonii*)、广东石豆兰(*B. kwangtungense*)、栗鳞贝母兰(*Coelogyneflaccida*)、流苏贝母兰(*C. fimbriata*)、莎叶兰(*Cymbidium cyperifolium*)、足茎毛兰(*Eria coromai*)、白棉毛兰(*E. lasiopetala*)、棱唇毛兰(*Erhomboidalis*)、平卧曲唇兰(*Panisea cavaliere*)、带叶兜兰(*Paphiopedilum hirsutissimun*)、长瓣兜兰(*P. dianthus*)、云南石仙桃(*Pholidota yunnanenss*)、台湾香荚兰(*Vanilla somai*)等呈片块状密集分布居群基株数量巨大。如保护区风岩洞密集分布区域，面积约 10hm²，分布的兰科植物多达 35 种，分株达 50 多万珠(表 1)。其中莎叶兰、流苏贝母兰带叶兜兰广东石豆兰、台湾香荚兰等居群堪称世界最大野生居群。大居群不仅是兰科植物重要的基因库，而且填补了中国野生兰科植物大居群缺乏的缺憾，具有极高的科研价值。

表 1 雅长林区风岩洞兰科植物资源

基株数(丛)	种名	总株数(株)	分布情况
≥1001	*Cymbidium cyperifolium*、*Paphiopedilum hirsutissimum*	22180	密集或零星不均匀分布
501~1000	*Coelogyne fimbriata*、*Pholidota yunnanensis*	412854	

（续）

基株数(丛)	种名	总株数(株)	分布情况
101～500	*Bulbophyllum andersonii*、*B. odoratissimum*、*Cymbidium lancifolium*、*Liparis esquirolii*、*L. bootanensis*、*L. chapaensis*、*Malaxis biaurita*、*Oberonia myosurus*、*Panisea cavalerei*	47093	密集或零星不均匀分布
51～100	*Eria rhombodalis*、*Malaxis acuminata*、*Paphiopedilum dianthum*	455	
11～50	*Cheirostylis chinensis*、*Bulbophyllum tianguii*、*Eria corneri*、*E. coromaria*、*Kingidium braceanum*、*Luisia teres*、*Malaxis purpurea*、*Phaius tankervilliae*	17753	
1～10	*Cleisostoma paniculatum*、*Cymbidium macrorhizon*、*Dendrobium fimbriatum*、*Eria spicata*、*Eulophia zollingeri*、*Flickingeria albopurea*、*Habenaria ciliolaris*、*H. dentata*、*Liparis nervosa*、*Paphiopedilum micranthum*、*Spianthes sinensis*	114	
合计	35 种约 7840 个基株(丛)	500449	

2.4 林区兰科植物物种的分布以低密度种为主

根据雅长野生兰科植物分布情况及遇见率高低，可将物种丰富度划分为：高(+++)、中(++)、低(+)、罕见(-)4 个等级(表 2)。丰富度“高”的有 21 种(占总种数的 13.46%)，丰富度“中”的有 25 种(占 16.03%)，丰富度“低”的有 66 种(占 42.31%)，“罕见”的有 44 种(占 28.21%)。可见，雅长林区野生兰科植物其丰富度等级以“低”及“罕见”种占多数，即林区兰科植物物种的分布以低密度种为主。

表 2　雅长林区野生兰科植物资源状况

等级序号	种　名	种数与资源量级别
1	*Bulbophyllum andersonii*、*B. longibrachiatum*、*B. kwangtungense*、*Cleisostoma williamsonii*、*Cheirostylis chinensis*、*Coelogyne fimbriata*、*Cymbidium cyperifolium*、*C. goeringii*、*C. lancifolium*、*Dendrobium chrysanthum*、*D. loddigesii*、*Eria coronaria*、*E. lasiopetala*、*E. rhombodalis*、*Malaxis biaurita*、*Liparis esquirolii*、*Luisia teres*、*Panisea cavalerei*、*Paphiopedilum dianthum*、*P. hirsutissimum*、*Pholidota yunnanensis* 等	约 21 种，丰富度高(+++)
2	*Bletilla ochracea*、*B. striata*、*Bulbophyllum ambrosia*、*B. odoratissimum*、*Calanthe argentro-striata*、*C. davidii*、*C. triplicata*、*Cleisostoma paniculatum*、*Coelogyne flaccida*、*Cymbidium qiubeiensis*、*Dendrobium aduncum*、*D. aphyllum*、*D. fimbriatum*、*Eria corneri*、*Eulophia zollingeri*、*Habenaria ciliolaris*、*H. dentata*、*Malaxis acuminata*、*M. latifolia*、*M. purpurea*、*Pholidota missionariorum*、*Vanda concolor*、*Vanilla somai* 等	约 25 种，丰富度中(++)
3	*Acanthephippium sylhetense*、*Bletilla formosana*、*Calanthe alismaefolia*、*C. hancockii*、*C. reflexa*、*C. sylvatica*、*Cheirostylis yunnanensis*、*Cleisostoma menghaiense*、*C. nangongense*、*Cymbidium aloifolium*、*C. bicolor* subsp. *obtusum*、*C. ensifolium*、*C. faberi*、*C. floribundum*、*C. geoeringii* var. *longibracteatum*、*C. goeringii* var. *serratum*、*C. kanran*、*Dendrobium aurantiacum* var. *denneanum*、*D. devonianum*、*D. hancockii*、*D. henryi*、*D. hercoglossum*、*D. limdleyi*、*D. lohohense*、*D. williamsonii*、*Eria obvia*、*E. spicata*、*Eulophia flava*、*Flickingeria albopurea*、*F. calocephala*、*Geodorum densiflorum*、*G. recurvum*、*Goodyera brachystegia*、*Habenaria davidii*、*H. fordii*、*H. petelotii*、*Herminium lanceum*、*Kingidium braceanum*、*Liparis bootanensis*、*L. cordifolia*、*L. disitans*、*L. japonica*、*L. nervosa*、*L. viridiflora*、*Nervilia fordii*、*N. plicatao*、*Oberonia ensiformis*、*O. myosurus*、*Pholidota cantonensis*、*P. leveilleana*、*Pleione yunnsnensis*、*Thelasis pygmaca*、*Vandopsis gigantea*、*Zeuxine goodyeroides* 等	约 66 种，丰富度低(+)
4	*Aerides rosea*、*Anoectochilus elwesii*、*A. mouleinensis*、*A. roxburghii*、*Aphyllorchis montana*、*Apostasia odorata*、*Arundina graminifolia*、*Bulbophyllum tianguii*、*Cephalanthera nanlingensis*、*Cremastra appendiculate*、*Cymbidium macrorhizon*、*C. nanulum*、*C. sinense*、*C. tracyanum*、*Cypripedium henryi*、*Dendrobium officinale*、*Epipactis helleborine*、*Eria clausa*、*Flickingeria angustifolia*、*Galeola lindleyana*、*Gastrodia eleta*、*Geodorum eulophioides*、*Goodyera henryi*、*Lecanorchis multiflora*、*Liparis guangxiensis*、*Oreorchis patens*、*Pachystoma pubescens*、*Paphiopedilum micranthum*、*Peristylus affinis*、*P. bulleyi*、*P. flagellifer*、*P. mannii*、*Phaius flavus*、*P. tankervilliae*、*Pogonia japonica*、*Robiquetia succisa*、*Spathoglottis pubescens*、*Spianthes sinensis*、*Tainia angustifolia*、*T. macrantha* 等	约 44 种，丰富度罕见(-)

2.5 地理成分复杂，热带性质明显

雅长林区分布的野生兰科植物有52属156种，其地理成分复杂多样，热带性质明显。根据植物属的分布区类型划分[10]，可分为10个类型3个变型[6]（表3）。起源于新、旧世界的热带、亚热带甚至温带分布的成分都有，热带成分有39属，占总属数的75.00%，其中热带亚洲（印度—马来西亚）分布的属占32.69%；温带主要为北温带分布、旧世界温带分布、东亚和北美间断分布及东亚分布类型约有10属占总属数的19.23%；中国—喜马拉雅分布1属占总属数的1.92%；世界广泛分布类型的属数有2属，占总属数的3.85%。热带性与温带性所含的属比值（R/T）为3.9（R/T为5.0[6]），雅长林区兰科植物亚热带与广泛分布类型均较少，缺热带亚洲和热带美洲间断分布、温带亚洲分布、地中海区、西亚至中亚分布类型。可见雅长分布的兰科植物以热带性为主，且分布的兰科植物与地中海地区和中亚地区的联系十分微弱。

表3 雅长林区兰科植物地理成分统计

分布区类型代码	分布区类型	属数	占总属数（%）	主要属
1	世界广布	2	3.85	*Liparis*、*Malaxis*
2	泛热带分布	4	7.69	*Bulbophyllum*、*Calanthe*、*Eulophia*
4	旧世界热带分布	4	7.69	*Nervilia*、*Oberonia*、*Phaius*、*Zeuxine*
4-1	热带亚洲、非洲、大洋洲间断分布	1	1.92	*Galeola*
5	热带亚洲至热带大洋洲分布	10	19.23	*Cleisostoma*、*Cymbidium*、*Eria*、*Gastrodia*、*Geodorum*、*Peristylus*、*Pholidota*
6	热带亚洲至热带非洲分布	2	3.85	*Cheirostylis*、*Spathoglottis*
7	热带亚洲（印度—马来西亚）分布	17	32.69	*Coelogyne*、*Dendrobium*、*Flickingeria*、*Goodyera*、*Luisia*、*Paphiopedilum*、*Thelasis*、*Vanda*、*Vandopsis*
7-2	热带印度至华南分布	1	1.92	*Pleione*
8	北温带分布	5	9.62	*Cephalanthera*、*Cypripedium*、*Epipactis*、*Habenaria*、*Spianthes*
9	东亚和北美间断分布	1	1.92	*Pogonia*
10	旧世界温带分布	1	1.92	*Herminium*
14	东亚分布	3	5.77	*Bletilla*、*Cremastra*、*Oreorchis*
14-1	中国—喜马拉雅分布	1	1.92	*Panisea*
	合计	52	100	

2.6 分布广泛，垂直分布以中上坡为主

雅长林区野生兰科植物分布广泛，分布不均匀。垂直分布从山脚到山顶都有分布，其分布种类、多度因海拔的不同而有较大差异（表4、图1），并受土壤类型的影响[11]。兰科植物种类及居群内基株较多的为中上坡，山脚及山顶较少。但兰科植物多数种类分布于海拔700~1200m，与文献记载相一致[6,7]，有些种类适应性广，从低海拔至高海拔均有分布（表4、图1）。

表4 雅长林区不同海拔高度主要分布兰科植物

海拔高度（m）	主要分布兰科植物种类		垂直分布跨度大的种类
	石灰岩石质土壤及附生类	红黄壤等	
400~800	*Aerides rosea*、*Calanthe davidii*、*Cleisostoma nangongense*、*Cymbidium bicolor* subsp. *obtusum*、*C. floribundum*、*C. tracyanum*、*Dendrobium limdleyi*、*Eria corneri*、*Liparis bootanensis*、*Vanda concolor*、*Vandopsis gigantea* 等	*Malaxis acuminata*、*M. latifolia*、*Geodorum densiflorum*、*G. eulophioides*、*G. recurvum* 等	

（续）

海拔高度（m）	主要分布兰科植物种类		垂直分布跨度大的种类
	石灰岩石质土壤及附生类	红黄壤等	
800～1200	*Bulbophyllum*、*Dendrobium*、*Coelogyne*、*Eria*、*Eulophia zollingeri*、*Flickingeria*、*Kingidium braceanum*、*Oberonia*、*Panisea cavalerei*、*Paphiopedilum*、*Pholidota*、*Vanilla somai* 等	*Aphyllorchis montana*、*Cymbidium sinense*、*C. tortisepalum* var. *longibracteatum*、*Eulophia*、*Habenaria*、*Lecanorchis multiflora*、*Malaxis*、*Nervilia*、*Oreorchis*、*Tainia* 等	*Anoectochilus*、*Bletilla*、*Calanthe reflexa*、*Cymbidium ensifolium*、*C. goeringii*、*C. goeringii* var. *serratum*、*Dendrobium loddigesii*、*Goodyera schlechtendaliana* 等
1200～1500	*Bulbophyllum*、*Dendrobium Gastrodia eleta* 等	*Cymbidium kanran*、*Galeola lindleyana*、*Peristylus* 等	
1500～1750		*Cymbidium kanran*、*Herminium*、*Peristylus*、*Pleione hookeriana*、*Zeuxine goodyeroides* 等	

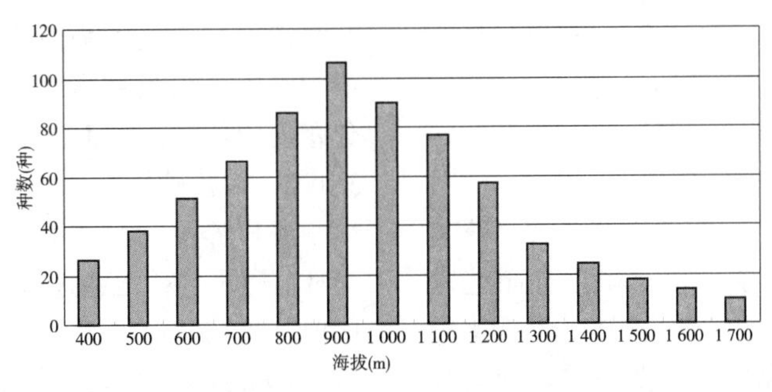

图 1　雅长林区野生兰科植物垂直分布与海拔关系

2.7　特有种类多

雅长林区地处云南、贵州、广西三省（自治区）交界处，地形地貌复杂多样，在典型的喀斯特石山区，兰科植物丰富多样，且特有性较强，其中中国特有 31 种[1,12-17]，广西新记录属杓兰属有 *Cypripedium* 1 属，广西新记录种 20 种，雅长特有种 3 种[1,2,12,13]（表 5）。

2.7.1　雅长林区内中国特有种

雅长林区野生兰科植物中国特有种共有 31 种（表 5）。

表 5　雅长林区野生兰科植物的中国特有种的分布区类型

特有种类型	分布亚型	种名	种数	占中国特有种的比例（%）
中国特有	滇黔桂特有	*Bulbophyllum longibrachiatum*、*Cymbidium qiubeiensis*、*Liparis esquirolii*、*Panisea cavalerei*、*Paphiopedilum dianthum*、*Pholidota leveilleana*、*P. missionariorum*	7	22.58
	华南至华西南特有	*Calanthe hancockii*、*Cephalanthera nanlingensis*、*Cleisostoma nangongense*、*Cymbidium nanulum*、*C. tortisepalum* var. *longibracteatum*、*Cypripedium henryi*、*Dendrobium hancockii*、*Flickingeria calocephala*、*Habenaria davidii*、*Peristylus bulleyi*、*Vanda concolor*	11	35.48
	华南至华东特有	*Bulbophyllum kwangtungense*、*Calanthe argentro-striata*、*Cymbidium floribundum*、*Dendrobium lohohense*、*D. officinale*、*Eria rhombodalis*、*Pholidota cantonensis*	7	22.58
	中国广布	*Bletilla ochracea*、*Calanthe davidii*、*Habenaria ciliolaris*	3	9.68
	雅长特有	*Bulbophyllum tianguii*、*Liparis guangxiensis*、*Geodorum eulophioides*	3	9.68
	合计		31	100

2.7.2 雅长林区内广西兰科植物新记录种

雅长林区内发现广西新记录种有：坛花兰(*Acanthephippium sylhetense*)、天贵卷瓣兰(*Bulbophyllum tianguii*)、南岭头蕊兰(*Cephalanthera nanlingensis*)、珍珠矮(*Cymbidium nanulum*)、西藏虎头兰(*C. tracyanum*)、绿花杓兰(*Cypripedium henryi*)、小花火烧兰(*Epipactis helleborine*)、滇金石斛(*Flickingeria albopurea*)、红头金石斛(*F. calocephala*)、贵州地宝兰(*Geodorum eulophioides*)、多花地宝兰(*G. recurvum*)、莲座叶斑叶兰(*Goodyera brachystegia*)、多花盂兰(*Lecanorchis multiflora*)、贵州羊耳蒜(*Liparis esquirolii*)、广西羊耳蒜(*L. guangxiensis*)、山兰(*Oreorchis patens*)、条叶阔蕊兰(*Peristylus bulleyi*)、鞭须阔蕊兰(*P. flagellifer*)、纤茎阔蕊兰(*P. mannii*)、台湾香荚兰等20种。

2.7.3 雅长特有种

雅长林区特有种有：天贵卷瓣兰、广西羊耳蒜、贵州地宝兰3种(表5)，其中贵州地宝兰花大且花瓣呈玫瑰红色，具有很高的观赏价值，在花卉园艺产业中的价值不可估量。贵州地宝兰自1921年被德国植物分类学家Schlechter根据在贵州罗甸采集到的唯一的一份标本命名及描述后，后人就没有在野外发现过该种，以至于我国植物学分类家在编写中国植物志时，只能参考德国植物分类学家的描述来记录该种。根据原记载，贵州地宝兰在罗甸海拔500m左右的河谷地带有少量分布，至今整整90年，贵州境内再也没有发现其踪迹，可能雅长是贵州地宝兰目前唯一的活体分布区，具有重要的科学价值。

2.8 生态类型丰富多样，生活型较全，以地生与附生类为主

雅长林区野生兰科植物生态类型丰富多样。生活型分为地生74种、附生70种、半附生6种及腐生6种[1,4-20]，以地生和附生为主。地生类主要生长在褐红土、红壤、黄壤、黑色石灰土和棕色石灰土壤上。林区土壤理化性质为pH4.28~8.05，土壤疏松，总孔隙度变幅为24.19%~68.98%，平均值为50.79%，富含腐殖质，有机质含量为17.64~506.44g/kg，平均值为100.70g/kg[11]。附生类多附着在岩石、石壁或阔叶树上生长，多具贮藏水分的茎或假鳞茎，耐旱性极强，附生在石灰岩上的种类多为喜钙种类；多数种类具较强的耐阴性，可充分利用林内弱漫射光；但绶草(*Spianthes sinensis*)、地宝兰(*Geodorum densiflorum*)、多花地宝兰、贵州地宝兰、小花火烧兰等少数种类可耐强直射光，可见其生态类型丰富。

3 雅长林区野生兰科植物受威胁的主要因素

3.1 人畜干扰破坏因素

人畜直接干扰破坏可直接影响到兰科植物种群(基株、植株)数量的扩大及兰科植物的生存和维持。

3.1.1 人为滥挖乱采野生兰科植物，是兰科植物生存和维持面临的最大威胁

兰科植物中许多种类具有较高的观赏或药用价值，有些种类两者兼备，易受人为滥挖乱采。野生铁皮石斛与天麻(*Gastrodia eleta*)最为严重，野外已很少见到，资源濒临枯竭。2010年起野生兰科植物已有法律保护，应加强对市场收购和贸易的监管，对盗卖野生兰科植物者绳之以法，同时林区群众的文化程度偏低，普法还有一段很长的路要走。

3.1.2 家畜破坏

农户养殖的马、牛、羊如无人看管，满山放养，对林区野生兰科植物危害严重。尤其是混生在草丛中的白芨属(*Bletilla*)、美冠兰属(*Eulophia*)、地宝兰属(*Geodorum*)、玉风花属(*Habenaria*)、叉唇角盘兰(*Herminium lanceum*)、阔叶沼兰(*Malaxis latifolia*)、阔蕊兰属(*Peristylus*)、苞舌兰(*Spathoglottis pubescens*)、绶草及兰属等地生兰类的兰科植物，其叶子常被咬光。

3.2 兰科植物栖息地丧失的因素

森林生态系统被破坏、土地被侵占及改变土地用途等可造成兰科植物栖息地丧失。群众经营耕作农作物、林业部门木材生产的采伐及更新、抚育等经营活动，直接或间接导致兰科植物短暂或长期性的栖息地丧失。干扰破坏后的兰科植物更新极为困难，甚至无法更新。

3.3 传粉昆虫的减少及其栖息地的破坏

传粉昆虫的人为捕捉作菜食用及其栖息地的破坏，是传粉昆虫减少的直接原因。如人为捕吃蜂幼虫和蜂蜜，造成蜂类数量的急剧减少，传粉者的减少造成兰科植物结实率极低，甚至不能结实，进而造成兰科植物居群数量增长慢，自然更新少。

3.4 极端恶劣气候条件影响

随着全球气候变暖的变化，雅长气候条件更加复杂，也给野生兰科植物造成了一定的负面影响，如火灾、旱灾、冻害等自然灾害，对野生兰科植物影响较大[6,23]。

4 雅长野生兰科植物资源与保护策略

根据雅长林区野生兰科植物的资源现状及生物生态学特点，针对人畜干扰及气候变化等不利的威胁因素，我们从兰科植物维持机制角度，提出了雅长兰科植物的保护措施和保育策略[2,3,6,23]。

(1)加快雅长兰科植物保护区的建设，营建种质基因保存库，增强保护基础设施建设。加快雅长兰科植物国家级自然保护区的建设，使雅长林区野生兰科植物密集分布区域得重点保护，加强保护力度。雅长兰科植物国家级自然保护区是我国第一处以野生兰科植物作为保护对象的兰科植物保护区，是国家生态环境保护和建设的一部分，也是我国野生珍稀濒危动植物保护建设工程的一部分，雅长保护区管理局建设是雅长野生兰科植物资源得到有效保护的保证。雅长兰科植物保护区的建设应围绕国家生态环境保护和建设的总规划，加快雅长兰科植物保护规划工程的建设步伐，完善保护区边界林权的稳定，促进保护区管理局开展保护和管理工作，营建兰科植物种质基因库，拯救和迁地保育栖息地被破坏或丧失的野生兰科植物，对局部兰科植物密集分布的区域，特别是人畜干扰大的区域加强保护设施的建设，如拉设铁丝网围护，并有护林员巡逻管护等，注意加强护林防火工作，确保野生兰科植物的长期维持。

(2)全民参与、普遍保护。雅长林区野生兰科植物的保护以就地保护为主，迁地保育为辅。就地全面普遍保护，以雅长保护区及雅长林场为重点保护区域，确保野生兰科植物及其栖息地不受人畜的干扰和破坏。

(3)加强林区保护管理工作的组织领导。加强雅长林区兰科植物保护管理工作的组织领导，是雅长林区兰科植物保护的当务之急。林区面积大，居民复杂，须加强雅长兰科植物国家级自然保护区管理局、乐业县政府、国有雅长林场、乡镇政府及村委会等多部门的保护管理工作的组织领导，加强各单位部门间对兰科植物保护工作的协作。

(4)加强宣传，确保保护区林地不被侵占，努力提高社会对野生动植物的保护意识。加强野生兰科植物等珍稀动植物的保护知识普及与保护区保护意义的宣传教育，增强全民兰科植物保护的责任感和使命感，确保保护区林地不被侵占。结合国家珍稀濒危植物保护法规法律，制定雅长林区野生兰科植物保护条例，努力提高全社会对野生兰科植物等动植物的保护意识。

(5)加强野生兰科植物的科学研究，加快培养兰科植物保育高级专业人才。科学技术是第一生产力，在保护区总体基础建设过程中，必须加强野生兰科植物的科学研究，以促进保护区兰科植物种群的维持和扩大。

①加强野生兰科植物生物生态学特性的研究。②加强野生兰科植物种群维持机制研究，优先研究小种群维持与扩张技术。保护区目前已知罕见种类达39种之多，应优先制定和开展贵州地宝兰、天贵卷瓣兰、广西羊耳蒜、多花盂兰、无叶兰(*Aphyllorchis montana*)、大根兰(*Cymbidium macrorhizon*)、坛花兰、尖囊兰(*Kingidium braceanum*)、毛萼山珊瑚(*Galeola lindleyana*)等小种群维持技术的研究。③加快野生兰科植物动态监测研究，建立种群动态长期监测的研究体系。④加快苗木繁育、仿生栽培及可持续开发利用的研究。加大兰科植物苗木繁育设施的投入，加快苗木的繁育，苗木培育应以种子无菌播种育苗为主，其他繁殖方法为辅[21]，并配套仿生栽培技术，加快可持续利用的研究和技术推广，可产生显著的社会、生态和经济效益。⑤通过科学研究、业务培训，加快兰科植物保育高级专业人才的培养。

(6)加强兰科植物传粉昆虫的保护。对兰科植物授粉昆虫的保护，是兰科植物种群扩大的关键，加强森林生态系统与传粉昆虫栖息地的保护，将有利于传粉昆虫的维持。应加快发展种植多种蜜源植物，特别是秋冬季开花的经济植物，如油菜(*Brassica capestris*)等，其既可保证昆虫食源，又能增加经济收入，对本区域保持较高的生物多样性及兰科植物种群扩张都具有十分重要的意义。

(7)开展和制定对严重自然灾害的应急措施和方案，确保野生兰科植物的维持。全球气候变暖有可能使雅长本来恶劣的气候条件复杂化[3,22]，极端气候可能导致自然灾害趋向严重化，特别是火灾、旱灾、冻害、暴雨及洪涝等异常气候。根据不同的自然灾害，科学制定灾后促进野生兰科植物恢复生长的应急措施，对兰科植物的维持具有十分重要意义。

致谢：中国科学院植物研究所郎楷永、金效华、罗毅波及广西植物研究所刘演等研究员帮助鉴定部分标本，中国林科院热带林业实验中心明安刚、孙文胜，广西雅长林场周千淞、梁才及雅长兰科植物国家级自然保护区管理局韦新莲、蓝玉甜、罗玉婷、黄岚等许多职工参加了本文外业调查工作，广西林业厅冯碧燕处长、刘杰恩副站长、广西大学陈宝善副校长、广西林业勘测设计院谭伟福副院长及美国Liu Hong 教授(Florida International University, Fairchild Tropical Botanical Garden)对本文的研究工作给予大力帮助，在此表示衷心感谢!

参考文献

[1]郎楷永，陈心启，罗毅波，等．中国植物志：第 17 卷[M]．北京：科学出版社，1999：1-551.

[2] 覃海宁，刘演．广西植物名录[M]．北京：科学出版社，2010：1-625.

[3]LIU H, FENG C L, LUO Y B, et al. Potential challenges of climate change to orchid conservation in a wild orchid hotspot in Southwestern China[J]. Bot. Rev., 2010(76): 174-192.

[4] FENG Changlin, JIN Xiaohua. *Liparis guangxiensis* sp. nov. (Malaxideae: Orchidaceae) from China [J]. Nord J. Bot., 2010, 28: 1-2: 697-698.

[5]HU Aiqun, TIAN Huaizhen, XIANG Fuwu. *Cephalanthera nanlingensis* (Orchidacere), a New Species from Guangdong, China[J]. Novon, 2009, 19: 56-58.

[6]和太平，彭定人，邓荣艳，等．广西雅长林区兰科植物区系分析[J]．广西农业生物科学，2007，26(3)：215-220.

[7]和太平，彭定人，黎德丘，等. 广西雅长林区兰科植物多样性研究[J]. 广西植物，2007，27(4)：590-595.

[8]郎楷永，罗敦．中国兰科石豆兰属一新种[J]．武汉植物学研究，2007，25(6)：558-560.

[9]赖家业，林少芳，何荣，等．广西雅长兰科植物自然保护区石斛属植物资源保护与利用[J]．安徽农业科学，2008，36(5)：1824-1829.

[10]吴征镒．中国种子植物属的分布区类型[J]．云南植物研究，1991(增刊Ⅳ)：1-139.

[11]黄承标，冯昌林，李保平，等. 广西雅长兰科植物分布区土壤理化性质[J]．东北林业大学学报，2010，38(1)：56-59.

[12]陈心启，吉占和，郎楷永，等．中国植物志：第 18 卷[M]．北京：科学出版社，1999：1-463.

[13]吉占和，陈心启，罗毅波，等．中国植物志：第 19 卷[M]．北京：科学出版社，1999：1-485.

[14]陈心启，吉占和，罗毅波．中国野生兰科植物彩色图鉴[M]．北京：科学出版社，1999.

[15]陈心启，吉占和．中国兰花全书[M]．北京：中国林业出版社，1998.

[16]金效华，吉占和，覃海宁，等．广西兰科植物增补[J]．广西植物，2002，22(5)：388-389.

[17]中国科学院植物研究所．中国高等植物图鉴：第 5 册[M]．北京：科学出版社，1983：602-772.

[18]陈灵芝．中国的生物多样性现状及其保护对策[M]．北京：科学出版社，1993：1-243.

[19]中国生物多样性保护行动计划总报告编写组．中国生物多样性保护行动计划[M]．北京：中国环境科学出版社，1994.

[20]方精云，沈泽昊，唐志尧，等．“中国山地植物物种多样性调查计划”及若干技术规范[J]．生物多样性，2004，12(1)：5-9.

[21]谌红辉，冯昌林，吴天贵，等．铁皮石斛组培快繁技术研究[J]．林业实用技术，2008，11：9-10.

[22]黄承标. 广西森林气候与森林水文特征[M]．北京：现代教育出版社，2010：1-173.

[23]朱华．中国植物区系研究文献中存在的几个问题[J]．云南植物研究，2007，29(5)：489-491.

[原载：林业资源管理，2011(06)]

桂西南 28 年生米老排人工林生物量及其分配特征

明安刚[1] 贾宏炎[1] 陶 怡[1] 卢立华[1] 苏建苗[1] 史作民[2]

([1]中国林业科学研究院热带林业实验中心，广西凭祥 532600

[2]中国林业科学研究院森林生态环境与保护研究所，北京 100091)

摘 要 应用相对生长法对桂西南 28 年生米老排人工林生物量及其分配特征进行了研究。结果表明：28 年生米老排人工林生物量为 281.47t/hm²，生态系统生物量分配格局为乔木层(97.89%)>凋落物层(1.87%)>灌木层(0.16%)>草本层(0.08%)；其中，乔木层生物量为 275.54 t/hm²，其生物量在各器官的分配规律为：树干(63.01%)>树根(21.01%)>树枝(9.64%)>树皮(4.38%)>树叶(1.72%)>果实(0.25%)；乔木生物量的径级分布接近正态分布，生物量主要集中在径级为 25~29cm 的林木，占乔木层生物量总量的 48.15%。28 年生米老排人工林林分年均净生产力为 15.61t/hm²，各组分净生产力大小顺序为乔木层(81.50%)>凋落物层(16.82%)>灌木层(0.98%)>草本层(0.70%)；乔木层年均净生产力为 12.72t/hm²，各器官净生产力大小顺序为树干(48.76%)>树叶(18.64%)>树根(16.26%)>树枝(7.46%)>果实(5.50%)>树皮(3.39%)。

关键词 桂西南；米老排人工林；生物量；分配

Biomass and Its Allocation in a 28-year-old *Mytilaria laosensis* Plantation in Southwest of Guangxi

MING Angang[1], JIA Hongyan[1], TAO Yi[1], LU Lihua[1], SU Jianmiao[1], SHI Zuomin[2]

([1]*Experimental Center of Tropical Forestry*, *CAF*, *Pingxiang* 532600, *Guangxi*, *China*;

[2]*Institute of Forest Ecology*, *Environment and Protection*, *CAF*, *Beijing* 100091, *China*)

Abstract: By the methods of plot sampling and allometric dimension, this paper studied the biomass and its allocation in a 28-year-old *Mytilaria laosensis* plantation in southwest Guangxi. The biomass in the plantation was 281.47 t/hm², and the biomass allocation was in the order of tree layer (97.89%) > litter layer (1.87%) > shrub layer (0.16%) > herb layer (0.08%). The biomass in the tree layer was 275.54 t/hm², and the biomass allocation was in the sequence of stem (63.01%) > root (21.01%) > branch (9.64%) > bark (4.38%) > leaf (1.72%) > fruit (0.25%). The biomass of the trees with different diameter at breast height (*DBH*) was approximately in normal distribution, and that of the trees with 25~29cm *DBH* accounted for 48.15% of the total. The mean annual net productivity of the plantation was 15.61t/hm²/a, and the net productivity of different components of the plantation was tree layer (81.50%) > litter layer (16.82%) > shrub layer (0.98%) > herb layer (0.70%). The mean annual net productivity of the tree layer was 12.72 t/hm²/a, and the relative proportion of different tree organs was stem (48.76%) > leaf (18.64%) > root (16.26%) > branch (7.46%) > fruit (5.50%) > bark (3.39%).

Key words: Southwest of Guangxi; *Mytilaria laosensis* plantation; biomass; distribution

生物量是人工林生态系统的功能指标，是研究人工林生态系统物质循环和能量流动的基础，是评估人工林固碳能力和固碳效益的主要内容[1]。从 20 世纪 60 年代开始，森林生物量的研究就受到林学界和生态学界的高度重视。到 20 世纪末，生物量的研究进展迅速，尤其是近十年来，人工林生物量的研究与人工林生态系统碳汇功能研究紧密结合起来，成为全球环境问题新的研究热点，受到国内外广泛关注[2-5]。

随着人们对人工林生物量研究的不断深入，涉及的树种越来越广泛，国外近几年报道的树种有辐射松(*Pinus radiata*)[6]、肯宁南洋杉(*Araucaria cun-*

ninghamii)[7]、蓝桉(*Eucalyptus globulus*)[8]、夏栎(*Quercus robur*)[9]、美洲栗(*Castanea dentata*)[10]、狄氏黄胆木(*Nauclea diderrichii*)[11]、棉白杨(*Populus Deltoides*)[12]等。

与国外相比，国内有关人工林生物量的研究起步较晚，70年代后期开始对一些主要用材树种如杉木、马尾松、尾叶桉、杨树等树种的生物量和生产力进行了研究[12]，90年代以来，人工林生态系统的生物量研究日益增多，并逐步涉及一些经济树种如油茶、毛竹、锥栗等人工林的生物量研究[13-15]。最近10年来报道的主要树种除传统的松、杉、桉以外，还有小黑杨[16]、樟树[17]、楠木[18]、福建柏[19,20]等。

米老排(*Mytilaria laosensis*)为金缕梅科米老排属的常绿大乔木，是分布在南亚热带地区的主要乡土珍贵阔叶树种之一，天然分布于广东西部、广西西南部和云南南部，越南和老挝也有天然分布。米老排生长较快，干形通直，材质优良，用途广泛，是优质的造纸、家具和建筑原料，具有重要的经济效益和生态环境效益[20]。广西是米老排主要的人工种植区，近几年发展迅速，将来有可能发展成为广西主要造林树种之一。目前，对米老排的研究，主要集中在生长规律及木材材性等方面[20,21]，而对米老排人工林生物量及生产力较少。本文对桂西南28年生米老排人工林生物量进行了研究，为今后米老排人工林的合理经营及固碳效益的研究提供理论参考。

1 材料与方法

1.1 试验地概况

试验地位于广西凭祥市中国林业科学研究院热带林业实验中心伏波实验场，106°51′30″~106°52′36″E，22°02′17″~22°02′23″N，海拔510~580m，属南亚热带季风气候，年平均气温19.5~21.0℃，年降水量1400mm左右，土壤为花岗岩发育成的山地红壤，土层厚度在100cm以上。

试验地于1981进行块状整地，1982年春季造林，初植密度为1500株/hm²，造林当年秋季抚育除草1次，以后每年4~6月和9月各除草松土1次，连续3年，直至郁闭成林。成林后1994年进行第一次间伐，2003年进行第二次间伐，间伐后的郁闭度一般保持在0.6左右；调查当年，林分的保留密度为621株/hm²。林下植被盖度总盖度为10%，主要物种有九节(*Psychotria rubra*)、杜茎山(*Maesa japonica*)、酸藤子(*Embelia laeta*)、金毛狗(*Cibotium barometz*)、五节芒(*Miscanthus floridulu*)、铁芒萁(*Dicranopteris linearis*)等。

1.2 研究方法

1.2.1 样地设置

选取坡度均匀且人为干扰较小的林地作试验样区，按坡位的上、中、下各设置1个样区，共3个样区，(样区编号分别为1，2，3)各样区设3个面积为400m²的方形样地(每个样区的调查面积为1200m²)。

表1 28年生米老排人工林各样区乔木层群落结构参数

样区	坡位	坡度(°)	胸径(cm)		树高(m)		乔木密度(t/hm²)
			范围	平均	范围	平均	
Ⅰ	上	33.3	15.9~31.4	24.7±3.6	17.3~27.3	21.8±3.2	600
Ⅱ	中	35.3	17.1~33.5	25.6+3.1	17.7~28.3	22.7±2.7	622
Ⅲ	下	37.6	16.9~30.7	26.3±2.8	15.5~27.8	23.3±2.9	625

将每个固定样地分成4个10m×10m的小样方，并以小样方为单位对每个样地进行每木检尺，调查胸径、树高等因子，得到28年生米老排人工林3个样区乔木层群落组成结构(表1)

1.2.2 林木调查及生物量的测定

根据固定样地每木检尺结果，按径级选取标准木13株进行生物量测定。样木伐倒后，地上部分按不同器官分级测定树干、树皮、树枝、树叶、果实的鲜重，其中，树枝再细分为粗枝(>5.0cm)、中枝(1.0~5.0cm)、细枝(<1.0cm)，分别测定其鲜重；地下部分用“全挖法”分别测定主根、大侧根(>2.0cm)、中侧根(0.5~2.0cm)、细侧根(<0.5cm)的鲜重。同时，按器官及枝条和根系径级的不同采集伐倒木的分析样品400g左右，带回实验室在80℃烘干至恒重，计算含水率并将各器官的鲜重换算成干重。

根据13株标准木的各器官生物量的实测数据，分别用$Y=ax^b$、$Y=ae^{bx}$、$Y=ax+b$、$Y=a\ln x+b$和$Y=a_1x^{b1}+a_2x^{b2}+a_3x^{b3}\cdots+c$等5种模型建立米老排各器官生物量与胸径、树高及$D^2H$之间的回归方程，比较并选出拟合效果最好的模型作为预测米老排各器官及全株的生物量方程(表2)。

表2 米老排人工林单株生物量回归方程

器官	回归方程	R^2	F	*Sig.*
树干	$W_s=0.1740(D^2H)^{0.7661}$	0.9196	104.3812	0.0000
树皮	$W_{ba}=3\times10^{-8}(D^2H)^2+0.0001(D^2H)+9.7883$	0.8668	58.6376	0.0000
树枝	$W_{br}=0.0002(D^2H)^{1.2696}$	0.6291	11.2124	0.0065
树叶	$W_l=0.0002D^{3.2304}$	0.8429	43.3948	0.0000
果实	$W_f=3\times10^{-7}(D^2H)^{1.5626}$	0.6659	9.4598	0.0106
树根	$W_r=0.0094(D^2H)^{0.9538}$	0.7247	14.7872	0.0027
全株	$W_t=0.1536(D^2H)^{0.8268}$	0.9049	64.8073	0.0000

1.2.3 林下植被生物量及凋落物现存量的测定

在每个20m×20m样方中按对角线选取2个5m×5m的小样方，记录每个样方内植物的种类，灌木和草本植物均分地上和地下部分，采用“全挖法”实测生物量，同种植物的相同器官取混合样品，将林下植被各样品带回在80℃烘干至恒重，计算含水率并将鲜重换算成干重；同时在每个样方内的4个10m×10m的小样方内各随机选取1个1×1m的小样方，按未分解、半分解2个层次，测定现存凋落物的干质量。

2 结果与分析

2.1 米老排人工林林分生物量及其分配

从表3可见，28年生米老排人工林林分总生物量为281.47t/hm²，各组分生物量的大小顺序为：乔木层>凋落物层>灌木层>草本层。其中，以乔木层生物量最大，为275.54t/hm²，占据了林分生物量总量的97.89%，其次是凋落物层，生物量为5.25t/hm²，占1.87%；以林下植被生物量最小，灌木层与草本层在林分生物量总量中所占比例的总和仅为0.24%。

表3 米老排人工林林分生物量及其分配

组分	乔木层	灌木层	草本层	凋落物层	合计
生物量(t/hm²)	275.54	0.46	0.22	5.25	281.47
百分比(%)	97.89	0.16	0.08	1.87	100.00

2.2 米老排人工林乔木生物量及其分配

2.2.1 米老排人工林乔木生物量及其在各器官的分配

由表4可知，28年生米老排人工林乔木生物量为275.54t/hm²，在树体各部分中，树干的生物量最高，占63.01%，其次为树根，而果实生物量最小，不到乔木总生物量的1%。在乔木层中各器官生物量大小顺序为：树干>树根>树枝>树皮>树叶>果实，乔木生物量集中在树干，这有利于米老排人工林光合作用产物在树干中的积累，提高林分出材率。

乔木地上部分生物量为217.66t/hm²，地下部分57.88t/hm²，分别占乔木总生物量的78.99%和21.01%，地下部分生物量为地上部分生物量的0.27倍。表明米老排根系发达，具有较强的吸收和储存能力。

表4 米老排人工林乔木生物量

项目	树干	树皮	树叶	果实	树枝	地上部分	树根	乔木合计	地下/地上
生物量(t/hm²)	173.61	12.06	4.74	0.70	26.55	217.66	57.88	275.54	0.27
所占百分比(%)	63.01	4.38	1.72	0.25	9.64	78.99	21.01	100	

2.2.2 米老排人工林不同径级乔木生物量的分配

用于试验的米老排人工林虽为同龄人工林，但由于各林木自身的遗传特性和所处局部环境条件的不同，使胸径生长产生了一定差异，这种差异造成了各径级的密度和生物量的不同(表5)，由表5可以看出：林分的总密度为621株/hm²，以径级为25~27cm的立木密度最高，达174株/hm²，占乔木总密度的27.78%，其次是径级为27~29cm的乔木，密度占总数的20.37%；各径级的密度分配整体呈现出由中间向两端逐渐减小的规律，近似正态分布(图1)。

在生物量的分配中，不同径级乔木生物量的分配也呈近似正态分布，径级为25~27cm的立木生物量最大，为79.43t/hm²，占林分乔木总生物量的28.83%，然后往两端逐渐降低；生物量主要集中在径级为21~31cm的立木中，密度和生物量分别占据了乔木总密度及乔木总生物量的87.04%和89.52%，而小于17cm及大于31cm的立木在乔木层中较少，二者的加和尚不足乔木密度和生物量总量的10%。由此可见，米老排人工林林木生长均匀，被压木和优势木的分布均较小，由平均木构成乔木了主林层(图1)。

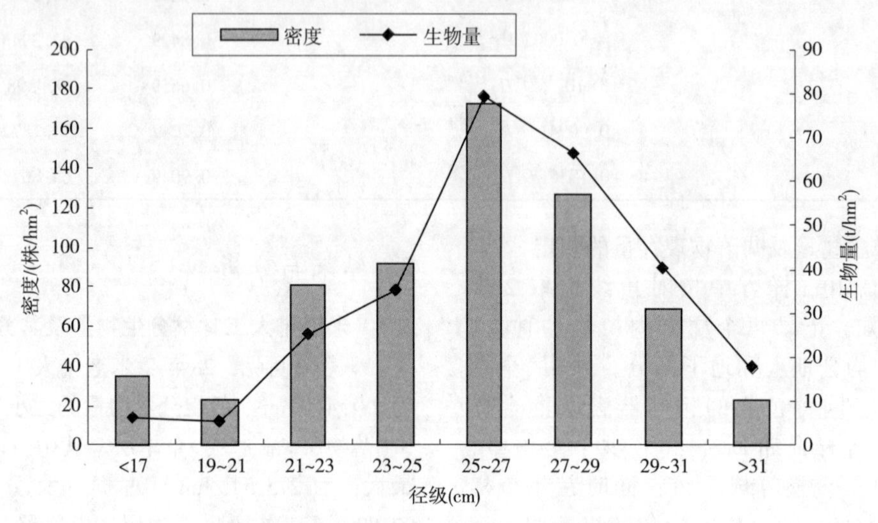

图1 28年生米老排人工林生物量的径级分配

从表5还可以看出：在乔木各器官生物量的分配中，随着胸径的增加，单株平均生物量、树枝、树叶、果实和根系的生物量所占比例均随之增加，树干所占比例小幅降低，而树皮所占比例基本持平，保持在4.5%左右。表明大径级米老排立木枝叶果实生长更加旺盛，利用更多的枝叶争取光照，以满足生长的需要。另外大径级的立木根系所占比例有所提高，这可能是由于米老排的生长对水肥条件要求较高，随着林木的增长，不断加速根系的生长，以更好地获取和利用水肥环境促进树体生长的需要，而树干所占比例随胸径的增加而减小，反映出树干积累的生物量所占比例随胸径的增加而减弱，但减小的幅度不大。多重比较结果显示，除 *DBH*<19cm 与 *DBH*>29cm 的林木之间树干生物量所占比例存在显著关系外(P<0. 05)，其他各径级之间并无显著性差异(*P*>0. 05)。

表5 米老排人工林乔木层不同径级各器官生物量及其分配

径级(cm)	单株平均生物量(t)	密度(tree/hm²)	所占百分比(%)	树干		树皮		树叶		树枝		果实		树根		合计	
				生物量(t/hm²)	百分比(%)	生物量(t/hm²)	百分比(%)	生物量(t/hm²)	百分比(%)	生物量(t/hm²)	百分比(%)	生物量(t/hm²)	百分比(%)	生物量(t/hm²)	百分比(%)	生物量(t/hm²)	百分比(%)
<17	0.18	34.48	5.56	4.20	68.40	0.38	6.19	0.06	0.98	0.36	5.86	0.01	0.16	1.13	18.40	6.14	2.23
19~21	0.24	22.99	3.70	3.66	67.23	0.28	5.13	0.07	1.37	0.37	6.83	0.01	0.14	1.05	19.30	5.44	1.97
21~23	0.31	80.46	12.96	16.62	65.74	1.13	4.48	0.38	1.50	2.01	7.96	0.05	0.18	5.09	20.15	25.28	9.17
23~25	0.38	91.95	14.81	22.77	64.60	1.49	4.23	0.53	1.50	3.09	8.78	0.07	0.21	7.29	20.68	35.25	12.79
25~27	0.46	172.41	27.78	50.25	63.26	3.33	4.20	1.28	1.61	7.62	9.59	0.20	0.25	16.75	21.09	79.43	28.83
27~29	0.53	126.44	20.37	41.27	62.17	2.82	4.25	1.22	1.83	6.74	10.16	0.18	0.27	14.15	21.31	66.38	24.09
29~31	0.58	68.97	11.11	24.62	61.08	1.78	4.43	0.82	2.03	4.31	10.69	0.12	0.30	8.65	21.46	40.30	14.63
>31	0.75	22.99	3.70	10.22	59.02	0.83	4.82	0.38	2.21	2.04	11.79	0.06	0.36	3.78	21.81	17.31	6.28
合计	620.69	100	173.61		12.06		4.74		26.55		0.70		57.88		275.54	100.00	

2.2.3 米老排人工林乔木枝条和根系生物量的分配

米老排人工林根系生物量为57.88t/hm²，占乔木生物量总量的21.01%。根系生物量分配大小顺序为：主根>大侧根>中侧根>细侧根。主根在根系中生物量最大，3个样区平均主根生物量高达41.29t/hm²，占根系生物量总量的72.74%，其次是大侧根，占18.28%，中侧根和细侧根的生物量很小，仅占根系生物量总量的7.99%和0.99%（表6）。可见，米老排发达的根系中，90%以上用来支持和稳定树体，而用于水分和养分吸收的根系不足10%。

28年生米老排人工林枝条生物量总量为26.1t/hm²，占林分乔木生物量总量的9.64%，以中枝生物量最大，占枝条生物量的48.69%；其次是粗枝，占31.38%，而以细枝的生物量最小，仅占枝条生物量的19.93%。细枝多为当年生嫩枝，是树叶的主要着生器官。由此可知，米老排80%以上的老枝用以形成和支持树冠，而只有20%的嫩枝用于叶的生长和萌发。

2.3 米老排人工林林下植被生物量及凋落物现存量的分配

2.3.1 林下植被生物量及其分配

28年生米老排人工林林下植被生物量较小，仅为0.684t/hm²，其中灌木生物量为0.461t/hm²，草本植物生物量为0.223t/hm²，分别占林下植被生物量总量的67.40%和32.60%；地上部分生物量为0.501t/hm²，占林下植被生物量的73.09%，地下生物量为0.183t/hm²，占26.91%（见表7）。28年米老排人工林虽然林分密度不高，仅为621株/hm²，但林冠生长茂盛，枝叶发达，林分郁闭度高，林下光照不足，灌草植被植被难以获得足够的光照来维持生长，因而，植被稀少，覆盖度低，林下植被生物量也很小。

表6 米老排人工林乔木枝条和根系生物量的分配

项目		根系					枝条			
		主根	大侧根（>2.0cm）	中侧根（0.5~2.0cm）	细侧根（<0.5cm）	合计	粗枝（>5.0cm）	中枝（1.0~5.0cm）	细枝（<1.0cm）	合计
生物量（t/hm²）	Plot1	36.77	9.24	4.04	0.50	50.55	7.09	11.00	4.50	22.58
	Plot2	42.37	10.65	4.65	0.57	58.24	8.48	13.16	5.39	27.04
	plot 3	44.72	11.24	4.91	0.61	61.48	8.92	13.84	5.67	28.43
	Mean	41.29	10.38	4.53	0.56	56.76	8.16	12.67	5.18	26.01
所占百分比（%）		72.74	18.28	7.99	0.99	100.00	31.38	48.69	19.93	100.00

表7 米老排人工林林下植被生物量及分配

组分	地上部分		地下部分		合计	
	生物量（t/hm²）	百分比（%）	生物量（t/hm²）	百分比（%）	生物量（t/hm²）	百分比（%）
灌木层	0.34	73.54	0.12	26.46	0.46	67.40
草本层	0.16	72.65	0.06	27.35	0.22	32.60
合计	0.50	73.09	0.18	26.91	0.68	100.00

2.3.2 凋落物现存量及其分配

由表8可知：米老排林下凋落物层较厚，现存量较大，达5.25t/hm²，其中，未分解凋落物现存量为3.14t/hm²，半分解的2.11t/hm²，分别占凋落物现存量总量的59.81%和40.19%。

表8 米老排人工林凋落物现存量及分配

组分	生物量（t/hm²）	百分比（%）
未分解	3.14	59.81
半分解	2.11	40.19
合 计	5.25	100.00

3 结论与讨论

本研究结果显示28年生米老排人工林生物量总量为281.47t/hm²，在各组分的分配大小顺序为乔木层（97.89%）>凋落物层（1.87%）>灌木层（0.16%）>草本层（0.08%），其中，以乔木层生物量最大，占据林分生物量的97.89%，而福建明溪县15年生米老排的生物量总量为195.97t/hm²，乔木生物量所占比例为99.39%[22]。由此可以看出：米老排人工林生物量受林龄的影响极为显著，林分生物量随着林龄的增长而增长。前人对不同林龄的杉木[23,24]、马尾松[25]、小黑杨[16]、福建柏[19]等人工林生物量的研究中，也得出类似的结论。

然而，本研究与吴庆锥对米老排生物量的研究相比，虽然林龄有所差异，但各林龄的乔木生物量在林分生物量总量所占比例均最大，且在97%以上，由此可知，米老排人工林林下群落结构单一，植被盖度极小，这主要受米老排人工林林冠发达，郁闭度高的影响。但由于米老排林分凋落物量较大，年总凋落量达7.096t/hm^2，[26]现存量达5.25t/hm^2，因而，凋落物生物量占据着米老排林分生物量中除乔木以外的大部分，是林下植被生物量的7.79倍。较大的凋落物量对米老排人工林土壤的改良，地力维持及腐殖质的形成极为有利。但也有可能会影响林下种子的萌发，造成林下植物多样性的降低。

米老排乔木生物量为275.54t/hm^2，其中，以树干的生物量最高，占63.01%。在乔木层中各器官生物量分配大小顺序为：树干（63.01%）>树根（21.01%）>树枝（9.64%）>树皮（4.38%）>树叶（1.72%）>果实（0.25%）。树干和树根在乔木生物量中占据84.02%，表明树干和根系是米老排人工林光合作用产物最主要的存储器官，乔木生物量在树干上的集中对米老排大径材培育提供有利条件，发达的根系有利于树体对水分和营养的吸收，为保持树体的快速生长提供了物质和能量补给。米老排根系和枝条分别占乔木生物量总量的21.01%和9.64%，根系的分配大小顺序为：主根>大侧根（>2.0cm）>中侧根（0.5~2.0cm）>细侧根（<0.5cm），而枝条为：中枝（1.0~5.0cm）>粗枝（>5.0cm）>细枝（<0.5cm），中枝在枝条中分配越多，树冠越零散，获取光照的叶面积就越大，这样有利于米老排形成较大的林冠，以争取充分的光照条件。

乔木生物量在各径级的分配呈现近似的正态分布分布规律，生物量主要集中在径级为21~31cm的立木中，占据了乔木生物量总量的89.52%，且以径级为25~29cm的乔木生物量最高，几乎占据林分乔木生物量的一半（48.15%）。说明米老排人工林各立木虽受个体遗传特性及生长环境的影响而出现了差异，但绝大多数生长表现比较均匀，林分仍以平均木形成主林层。这以结果与江泽慧等人对华北沙地小黑杨的研究结论基本一致[16]。

米老排林下植被及凋落物在林分生物量的分配均较少，总体不到3%，但林下植被对米老排人工林林分水分涵养和土壤的改良有着重要作用，也是林分生物量和碳储量的重要组成部分。

本文仅对28年生米老排人工林生物量做了研究，至于不同年龄阶段米老排人工林生物量的变化情况以及米老排人工林生物量的垂直分布尚无明确结论，有待进一步研究。

参考文献

[1] 李娜，黄从德．川西亚高山针叶林生物量遥感估算模型研究[J]．林业资源管理，2008，3：100~104

[2] 方精云．中国森林生产力及其对全球气候变化的影响[J]．植物生态学报，2000，24(5)：513-517

[3] LUXMOORE R J, THARP M L., POST W M. Simulated biomass and soil carbon of loblolly pine and cottonwood plantations across a thermal gradient in southeastern United States[J]. Forest Ecology and Management, 2008, 254: 291-299.

[4] PEICHL M, ARAIN M A. Above-and below ground ecosystem biomass and carbon pools in an age-sequence of temperate pine plantation forests [J]. Agricultural and Forest Meteorology, 2006, 140: 51-63.

[5] DOSSA E L, FERNANDES ECM, REID WS, et al. Above- and belowground biomass, nutrient and carbon stocks contrasting an open-grown and a shaded coffee plantation [J]. Agroforest Syst, 2008, 72: 103-115.

[6] BI HQ, LING YS, JOHN, et al. Additive prediction of aboveground biomass for*Pinus radiate* (D. Don) plantations[J]. Forest Ecology and Management, 2010, 259: 2301-2314.

[7] CHEN C R, XU Z H, BLUMFIELD T J, et al. Soil microbial biomass during the early establishment of hoop pine plantation: seasonal variation and impacts of site preparation[J]. Forest Ecology and Management, 2003, 186: 213-225.

[8] KIDANU S, MAMO T, STROOSNIJDER L. Biomass production of Eucalyptus boundary plantations and their effect on crop productivity on Ethiopian highland Vertisols[J]. Agroforestry Forum, 2005, 63: 281-290.

[9] MAGNUS Lof AB. Root spatial distribution and biomass partitioning in *Quercus robur* L. seedlings: the effects of mounding site preparation in oak plantations[J]. Eur J Forest Res, 2010, 129: 603-612.

[10] JACOBS D F, SELIG M F, SEVEREID L R. Above-ground carbon biomass of plantation-grown American chestnut(*Castanea dentata*) in absence of blight[J]. Forest Ecology and Management , 2009, 258: 288-294.

[11] ONYEKWELU J C. Growth, biomass yield and biomass functions for plantation-grow Nauclea diderrichii (de wild) in the humid tropical rainforest zone of south-western Nigeria [J]. Bioresource Technology, 2007, 98:

2679-2687.

[12] LEE K H, JOSE S. Soil respiration, fine root production, and microbial biomass in cottonwood and loblolly pine plantation along a nitrogen fertilization gradient [J]. Forest Ecology and Management, 2003, 185: 263-273.

[13] 陈辉，何方．锥栗人工林生物量和生产力研究[J]. Ⅰ. 生物量研究．中南林学院学报，2000，20(4)：6-10.

[14] 何方，王义强，谭晓风，等．油茶生物量与养分循环的研究[J]. 林业科学，1996，32(5)：403-410.

[15] 陈辉，洪伟，兰斌，等．冰北毛竹生物和生产力的研究[J]. 林业科学，1998，34(专刊)：60-64.

[16] 江泽慧，范少辉，冯慧想，等．华北小黑杨人工林生物量及其分配规律[J]. 林业科学，2007，43(11)：15-20.

[17] 姚迎九，康文星，田大伦．18年生樟树人工林生物量的结构与分布[J]. 中南林学院学报，2003，23(1)：1-5.

[18] 彭龙福．35年生楠木人工林生物量及生产力的研究[J]. 福建林学院学报，2003，23(2)：128-131.

[19] 杨宗武，谭芳林，肖祥希，等．福建柏人工林生物量的研究[J]. 林业科学，2000，36(专刊1)：120-124.

[20] 将宗垲．建柏人工林生物量的研究[J]. 东北林业大学学报，2000，28(6)：16-19.

[21] 郭文福，蔡道雄，贾宏炎，等．米老排人工林生长规律的研究[J]. 林业科学研究，2006，19(5)：585-589.

[22] 林金国，张兴正，翁闲．立地条件对米老排人工林生长和材质的影响[J]. 植物资源与环境学报，2004，13(3)：50-54.

[23] 吴庆锥．米老排人工林生物量研究[J]. 福建林业科技，2005，32(3)：125-129.

[24] 吴文德，范少辉，王国礼，等．不同立地条件下不同林龄杉木人工林生物量研究．Ⅱ. 林分生产力分析[J]. 林业科学研究，1996，9(专刊)：86-91

[25] 贵杰，王鹏程．马尾松人工林生物量与生产力变化规律研究Ⅱ. 不同龄林生物量及生产力[J]. 林业科学研究，2002，15(1)：54-60.

[26] 卢立华，贾宏炎，何日明等．南亚热带6种人工林凋落物的初步研究[J]. 林业科学研究，2008，21(3)：346-352.

[原载：生态学杂志，2012，31(05)]

米老排人工林碳素积累特征及其分配格局

明安刚[1] 贾宏炎[1] 陶 怡[1] 陆 毅[1] 于浩龙[1] 卢立华[1] 蔡道雄[1] 史作民[2]
([1] 中国林业科学研究院热带林业实验中心，广西凭祥 532600；
[2]中国林业科学研究院森林生态环境与保护研究所，北京 100091)

摘 要 在生物量调查的基础上，对桂西南地区28年生米老排人工林生态系统的碳素积累特征及分配格局进行了研究。结果表明：米老排各器官碳含量在516.0~560.2g/kg，大小排序为：树叶(560.2g/kg)>树干(542.8g/kg)>树根(530.9g/kg)>树皮(530.8g/kg)>树枝(522.8g/kg)；土壤碳含量以表土层最高，且随土层深度的增加而降低；米老排人工林乔木层碳贮量为147.90t/hm²，其中，树干占乔木层碳贮量的63.72%；米老排人工林生态系统碳贮量为285.36t/hm²，各组分的分配顺序为乔木层>土壤层>凋落物层>灌木层>草本层；其中，植被层碳贮量为土壤层(0~100cm)的1.1倍。

关键词 米老排人工林；碳含量；碳贮量；分配格局

Characteristics of Carbon Accumulation and Allocation Pattern in *Mytilaria laosensis* Plantation

MING Angang[1], JIA Hongyan[1], TAO Yi[1], LU Yi, LU Lihua[1], YU Haolong[1], CAI Daoxiong[1], SHI Zuomin[2]

([1]*Experimental Center of Tropical Forestry, Chinese Academy of Forestry, Pingxiang* 532600, *Guangxi, China*;
[2]*Institute of Forest Ecology, Environment and Protection, Chinese Academy of Forestry, Beijing* 100091, *China*)

Abstract: Based on the biomass survey, this paper studied the characteristics of carbon accumulation and allocation pattern of a 28 -year-old*Mytilaria laosensis* plantation ecosystem in southwest Guangxi of South China. The carbon content in different organs of *M. laosensis*Was in the range of 522.8-560.2 g/kg, with the order of leaf (560.2 g/kg) > stem (542.8 g/kg) >root (530.9 g/kg) > bark (530.8 g/kg) > branch (522.8 g/kg). The soil carbon content was the highest in surfacelayer, and decreased with increasing depth. The carbon storage in tree layer was 147.90 t /hm², of which, stem accounted for 63.72% of the total. The total carbon storage of the *M. laosensis* plantation ecosystem was 285.36 t / hm², and the allocation of the carbon storage in different components was tree layer > soil layer > litter layer > shrub lay-er > herb layer, with the carbon storage in vegetable layer being 1.1 times of that in soil layer(0~100 cm). The annual net carbon storage of the tree layer was 6.87 t /hm²/ a, which was dominated by the above ground part.

Key words: *Mytilaria laosensis* plantation; carbon content; carbon storage; allocation pattern.

CO_2是温室气体中最主要的成分之一，自从工业革命以来，由于化石燃料的大量燃烧和土地利用方式的改变，大气中的CO_2的浓度逐年升高[1]。近年来，有学者认为，大气中的CO_2浓度以每年1.9ml/m³的线性速度增加[2]，全球气候变暖对人类的影响越来越显著，已成为国际社会最关注的全球性环境问题，对CO_2的排放、吸收和固定已经成为全球变化研究的热点和前沿[3-5]。森林作为陆地生态系统的主体，维持着全球86%的植被碳库[6]和73%的土壤碳库[7]，在调节全球气候变暖，减缓大气中CO_2等温室气体浓度的上升以及调节全球气候等方面有着不可替代的作用。方精云和陈安平[8]、王效科和冯宗炜[9,10]、刘国华等[11]和周玉荣等[12]分别利用我国森林资源清查结果，结合我国森林生态系统生物量与生产力研究的基础，推算了我国近50年来森林碳库及其动态。近十几年来，随着绿化工程及退耕还林工程的实施，人工林面积在森林总面积占的比重快速增加。人工林在吸收和固定CO_2及减缓全球气候变暖等方面作用日益显著，并引起国内外众多学者的重视。有研究者对不同的树种人工林碳含量、碳贮量及其空间分布格局进行了深入研究[13-20]。这些成果为森林碳汇功能的研究做出了积极贡献。

米老排(*Mytilaria laosensis*)为金缕梅科米老排属的常绿大乔木，是分布在南亚热带地区的主要乡土珍贵阔叶树种之一。近几年，米老排人工林发展迅速，将很有可能发展成为南亚热带主要造林树种之一。目前，有关米老排的研究，主要集中在生长规律、木材材性、土壤物理性质、开发利用及生物量方面[21-26]，本文在前人的研究基础之上，对位于南亚热带地区的广西大青山28年生米老排人工林生态系统碳素积累及其空间分布情况进行了研究，为在区域或国家尺度上估算森林生态系统碳库及碳平衡的估算提供基础数据，及在生态服务功能的科学评价提供科学依据。

1 研究地区与研究方法

1.1 试验地概况

试验地位于广西凭祥市中国林业科学研究院热带林业实验中心伏波实验场(106°51′30″~106°52′36″E，22°02′17″~22°02′23″N)，海拔510~580m，属南亚热带季风气候，年平均气温19.5~21.0℃，年降水量1400mm左右，土壤为花岗岩发育成的山地红壤，土层厚度在100cm以上。

试验地于1981进行块状整地，1982年春季造林，初植密度为1500株/hm^2，造林当年秋季抚育除草1次，以后每年4~6月和9月各除草松土1次，连续3年，直至郁闭成林。成林后1994年进行第1次间伐，2003年进行第2次间伐，间伐后的郁闭度一般保持在0.6左右；2010年8月对林分进行调查，调查当年林分的保留密度为621株/hm^2。林下植被总盖度为10%，主要物种有九节(*Psychotria rubra*)、杜茎山(*Maesa japonica*)、酸藤子(*Embelia laeta*)、金毛狗(*Cibotium barometz*)、五节芒(*Miscanthus floridulu*)和铁芒萁(*Dicranopteris linearis*)等。

1.2 研究方法

1.2.1 样地设置

选取坡度均匀且人为干扰较小的林地作试验样区，按坡位的上、中、下设置3个样区(样区编号分别为1、2、3)，每个样区再沿等高线平行设3个面积为400m^2的方形样地(3个样区共9个样方，每个样方的面积为20m×20m)。对每个样地进行每木检尺，调查胸径、树高等因子，得到28年生米老排人工林3个样区乔木层群落组成结构(表1)。

表1 28年生米老排人工林各样区乔木层群落结构参数

样区	坡位	坡度(°)	胸径(cm)		树高(m)		乔木密度(t/hm^2)
			范围	平均	范围	平均	
Ⅰ	上	33.3	15.9~31.4	24.7±3.6	17.3~27.3	21.8±3.2	600
Ⅱ	中	35.3	17.1~33.5	25.6+3.1	17.7~28.3	22.7±2.7	622
Ⅲ	下	37.6	16.9~30.7	26.3±2.8	15.5~27.8	23.3±2.9	625

1.2.2 林木调查及生物量的测定

根据固定样地每木检尺结果，按径级(2cm)选取标准木13株进行生物量测定。样木伐倒后，地上部分按不同器官分级测定树干、树皮、树枝、树叶、果实的鲜重，其中，树枝再细分为粗枝(>5.0cm)、中枝(1.0~5.0cm)、细枝(<1.0cm)，分别测定其鲜重；地下部分用“全挖法”分别测定主根、粗根(>2.0cm)、中根(0.5~2.0cm)、细根(<0.5cm)的鲜重。同时，按器官及枝条和根系径级的不同采集伐倒木的分析样品400g左右，带回实验室在80℃烘干至恒重，计算含水率并将各器官的鲜重换算成干重。

根据13株标准木的各器官生物量的实测数据，以明安刚等[25]建立的28年生米老排生物量方程为基础，新增自变量胸径(D)与各器官生物量之间的关系，对原有的生物量方程进行修正，建立米老排各器官生物量与胸径(D)及胸径平方与树高的乘积(D^2H)之间的相对生长回归方程，用以估测米老排各器官及全株的生物量(表2)。

表2 米老排人工林单株生物量回归方程

器官	自变量	相对生长方程	判定系数 R^2	F值	P值
树干	D	$W_s=0.6769D^{1.8761}$	0.8788	47.1569	0.0000
	D^2H	$W_s=0.1740(D^2H)^{0.7661}$	0.9196	104.3812	0.0000

（续）

器官	自变量	相对生长方程	判定系数 R^2	F 值	P 值
树皮	D	$W_{ba}=0.0712D^{1.7602}$	0.7080	36.2662	0.0001
	D^2H	$W_{ba}=0.0220(D^2H)^{0.7081}$	0.7191	58.6375	0.0000
树枝	D	$W_{br}=0.0015D^{3.1701}$	0.6215	10.9218	0.0070
	D^2H	$W_{br}=0.0002(D^2H)^{1.2696}$	0.6291	11.2124	0.0065
树叶	D	$W_l=0.0002D^{3.2304}$	0.8429	43.3948	0.0000
	D^2H	$W_l=0.00003(D)^{1.2634}$	0.8091	41.0306	0.0001
树根	D	$W_r=0.0668(D^2H)^{3.2304}$	0.8429	43.3948	0.0000
	D^2H	$W_r=0.0094(D^2H)^{0.9538}$	0.7247	14.7872	0.0027
全株	D	$W_t=0.6841D^{2.0158}$	0.8573	38.2042	0.0001
	D^2H	$W_t=0.1536(D^2H)^{0.8268}$	0.9049	64.8073	0.0000

1.2.3 林下植被生物量和凋落物现存量的测定

在每个20m×20m样方中按对角线选取2个5m×5m的小样方，记录每个样方内植物的种类，灌木和草本植物均分地上和地下部分，采用"收获法"实测生物量，同种植物的相同器官取混合样品，将林下植被各样品带回在80℃烘干至恒重，计算含水率并将鲜重换算成干重；同时在每个20m×20m样方内按对角线位置选取4个1m×1m的小样方，按未分解、半分解2个层次，测定现存凋落物的现存量。

1.2.4 样品的采集与分析

在各样区随机挖取3个土壤剖面，按照0~10cm、10~30cm、30~50cm和50~100cm将土壤分为4个层次，先用环刀进行取样，用于测定各层次土壤容重，再在各层取混合分析样品200g左右，用于含碳量的测定。

植物分析样品于70℃烘干，土壤样品置于室内风干，然后磨碎。所有样品均采用重铬酸钾-水合加热法测定有机碳含量。

1.2.5 碳贮量的计算方法

碳贮量根据单位面积生物量乘以有机碳含量而求得，土壤碳贮量则是土壤有机碳含量、土壤容重与土壤厚度三者的乘积而得。实测米老排人工林各层土壤容重值可得，0~10cm、10~30cm、30~50cm和50~100cm土层土壤容重的平均值依次为1.18、1.34、1.42和1.42g/cm^3。

2 结果与分析

2.1 米老排人工林生态系统各组分碳含量

2.1.1 乔木层碳含量

米老排乔木各器官碳含量各异(表3)，树叶含量最高，为560.2g/kg，其次为树干，为542.8g/kg，树根、树皮中等，果实最低；各器官碳含量高低顺序为树叶(560.2g/kg)>树干(542.8g/kg)>树根(530.9g/kg)>树皮(530.8g/kg)>树枝(522.8g/kg)，根系碳含量的高低顺序为粗根(540.5g/kg)>中根(533.5g/kg)>细根(529.8g/kg)>主根(516.0g/kg)；乔木各器官碳含量的平均值为537.4g/kg，变化幅度为516.0~560.2g/kg，样本的变异系数为0.4%~4.6%，各器官碳含量差异不显著($P>0.05$)。

表3 米老排不同器官的碳含量

器官	碳含量(g/kg)	标准差	变异系数(%)
树干	542.8	10.6	2.0
树皮	530.2	9.5	1.8
树枝	522.8	2.6	0.4
树叶	560.2	7.0	1.3
果实	499.8	1.9	0.4
主根	516	5.4	1.0
粗根	540.5	24.6	4.6

（续）

器官	碳含量(g/kg)	标准差	变异系数(%)
中根	533.5	16.8	3.1
细根	529.8	7.6	1.4
树根平均	530.9	8.5	1.6
器官平均	537.4	7.6	1.4

2.1.2 林下地被物的碳含量

林下地被物包括活地被物和死地被物，本研究主要测定了林下灌木层、草本层及凋落物层的碳含量(表4)。从表4可知，米老排林下地被物个层次平均碳含量以凋落物层最高，为553.83g/kg，灌木层居次，为524.28g/kg，草本层最低，为516.08g/kg；不管是灌木和草本，其地上部分碳含量均高于地下部分，未分解的凋落物碳含量高于半分解的碳含量。灌木层、草本层及凋落物层碳含量各组分间差异不显著($P>0.05$)。

2.1.3 土壤中碳含量

从表5可知，土壤碳含量以表土层(0~10cm)碳含量最高，平均值高达23.19g/kg；各土层碳含量随着土层深度的增加而降低，且差异显著($P<0.05$)，变异系数达57.47%~60.80%；同一土层在各样区碳含量差异不显著($P>0.05$)，变异系数为3.18%~12.99%。

表4 林下植被及凋落物碳含量

层次	组分	碳含量(g/kg)	标准差	变异系数(%)
灌木层	地上部分	534.13	7.15	1.34
	地下部分	514.43	6.75	1.31
	平均	524.28	12.45	2.38
草本层	地上部分	520.53	6.22	1.19
	地下部分	511.63	4.95	0.97
	平均	516.08	7.00	1.36
凋落物层	未分解	565.32	5.44	0.96
	半分解	542.34	3.65	0.67
	平均	553.83	12.87	2.32

表5 土壤中碳含量

土壤层次(cm)	碳含量(g/kg)				标准差	变异系数(%)
	样区1	样区2	样区3	平均		
0~10	20.68	23.54	25.36	23.19	2.36	10.17
10~30	13.07	12.94	12.31	12.77	0.41	3.18
30~50	7.50	8.26	9.45	8.40	0.98	11.70
50~100	5.68	7.35	6.89	6.64	0.86	12.99
平均	11.73	13.02	13.50	12.75	1.15	9.51
标准差	6.74	7.43	8.21	7.42	0.84	4.37
变异系数(%)	57.47	57.03	60.80	58.21	73.02	45.98

2.2 米老排人工林生态系统碳贮量及其分配格局

2.2.1 米老排乔木层不同器官碳贮量及其分配格局

林木各器官单位面积生物量与相应碳含量的乘积为其单位面积碳贮量，因而各器官碳贮量受生物量的正向影响(表6)，各器官中，树干的生物量最高，占林分生物量总量的61.68%，相应的树干的碳贮量在各器官中也是最高的，达94.24t/hm^2，占林分碳贮量总量的 33.02%；枝、叶及果的碳贮量甚小，分别占林分碳贮量总量的 4.86%、0.93%和 0.12%，总和不到 10%。乔木各器官碳贮量分配顺序为：树干>树根>树枝>树皮>树叶>果实。

由表 6 还可以看出，根系在乔木碳贮量中也占有较大比例，生物量和碳贮量分别为 57.88t/hm² 和 30.38t/hm²，所占比例分别为 20.56%和 10.65%；在根系中，以主根碳贮量最大，达 14.80t/hm²，占据根系碳贮量的 72.05%，其次为粗根，细根最小，根系碳贮量分配顺序为：主根>粗根>中根>细根。

2.2.2 米老排人工林生态系统各组分生物量和碳贮量及分配格局

从表 6 可以看出，28 年生米老排人工林生态系统生物量为 281.47t/hm²，其中乔木层生物量最大，为 275.54t/hm²，占生态系统生物量的 97.89%；其他组分依次是凋落物层、灌木层和草本层，分别占 1.87%、0.16%和 0.08%；在整个生态系统中，地上部分和地下部分生物量分别为 223.41t/hm² 和 58.06t/hm²，分别占林分生物量总量的 79.37%和 20.63%。

表 6 米老排人工林生态系统各组分碳贮量及其分配格局

层次	组分	生物量		碳贮量	
		(t/hm²)	(%)	(t/hm²)	(%)
乔木层	树干	173.61	61.68	94.24	33.02
	树皮	12.06	4.28	6.39	2.24
	树枝	26.55	9.43	13.88	4.86
	树叶	4.74	1.68	2.66	0.93
	果实	0.70	0.25	0.35	0.12
	主根	42.09	14.95	21.88	7.67
	粗根	10.58	3.76	5.72	2.00
	中根	4.63	1.64	2.47	0.87
	细根	0.58	0.21	0.31	0.11
	根系合计	57.88	20.56	30.38	10.65
	小计	275.54	97.89	147.90	51.83
灌木层	地上部分	0.34	0.12	0.18	0.06
	地下部分	0.12	0.04	0.06	0.02
	小计	0.46	0.16	0.24	0.08
草本层	地上部分	0.16	0.06	0.08	0.03
	地下部分	0.06	0.02	0.03	0.01
	小计	0.22	0.08	0.11	0.04
凋落物层	未分解	3.14	1.12	1.77	0.62
	半分解	2.11	0.75	1.14	0.40
	小计	5.25	1.87	2.92	1.02
土壤层	0~10cm			26.67	9.35
	10~30cm			36.52	12.80
	30~50cm			23.86	8.36
	50~100cm			47.14	16.52
	小计			134.19	47.02
地上合计		223.41	79.37	79.37	120.70
地下合计		58.06	20.63	20.63	164.66
总计		281.47	100.00	100.00	285.36

28年生米老排人工林生态系统碳贮量为285.36t/hm²，其中，乔木层碳贮量最大，为147.90t/hm²，占生态系统碳贮量总量的51.83%；其次是土壤层（0～100cm），为134.19t/hm²，占47.02%；凋落物层、灌木层和草本层碳贮量均较小，分别是2.92t/hm²、0.24t/hm²和0.11t/hm²，仅占生态系统碳贮量总量的1.02%、0.09%和0.04%；由此可见，米老排人工林生态系统碳素主要积累于乔木层和土壤层，二者占据碳贮量总量的98.85%，分配在林下植被及枯枝落叶的碳素极少，仅为1.15%。

在米老排人工林生态系统中，地上部分和地下部分碳贮量的分配分别为120.70t/hm²和164.66t/hm²，分别占生态系统碳贮量总量的42.30%和57.70%。

米老排人工林土壤碳贮量(0～100cm)为134.19t/hm²，占人工林生态系统碳贮量总量47.02%，略低于植被碳贮量。同时可以看出，土壤碳贮量随土层深度的增加而降低，其主要原因是土壤碳含量随土层深度的增加而降低。

3 讨论

28年生米老排各器官碳含量有所差异，排列顺序为树叶>树干>树根>树皮>树枝，与27年生观光木及28年秃杉等树种各器官碳含量的排列顺序不完全一致[27,28]，由此可以看出，尽管树龄相似，但各器官碳含量的排列次序会因树种的不同而存在差异。28年生米老排各器官碳含量的平均值为537.4g/kg，相比而言，高于广西27年生观光木(471.1g/kg)、8年生桉树(475.2g/kg)及13年生杉木(497.0g/kg)及湖南会同23年生杉木为463.2g/kg[27-31]平均碳含量。广西16年生湿地松及14年生马尾松均表现出平均碳含量大小的不同及各器官碳含量不同的情况[14,32]。说明各器官碳含量的大小与树种的关系极为密切，不同树种之间碳含量有明显差异，而米老排在是广西区碳含量较高的树种，是发展碳汇人工林较好的选择。

林下地被物各组分碳含量也有所差异，其中，碳含量最高的是凋落物，达553.8g/kg，不仅高于灌木层和草本层，还高于乔木各器官碳含量的平均水平(537.4g/kg)，主要原因是在米老排的凋落物中，枯叶是凋落物层的主要成分，而在米老排乔木各器官中，叶的碳含量最高。广西米老排林分凋落物碳素平均含量较本地杉木林凋落物碳素平均含量(418.7g/kg)高出32.27%[30]，主要因为米老排凋落物叶高度革质化，分解速率较低，碳素积累较多；灌木层和草本层均显示出地上部分平均碳含量高于地下部分的规律；米老排土壤碳含量较低，平均为12.75g/kg，低于广西13年生杉木林(16.6g/kg)及桉树林(14.57g/kg)平均水平[30,31]。土壤各层次碳含量呈现随土层深度的增加而降低的规律，同一土层在各试验样区无显著差异。

28年生米老排人工林生态系统碳库主要包括植被层、凋落物层和土壤层，生态系统碳贮量总量为285.36t/hm²，高于我国森林生态系统平均碳贮量(258.82t/hm²)；其中，植被层碳贮量为148.26t/hm²，是我国森林植被平均碳贮量(57.07t/hm²)[11]的2.6倍，是我国亚热带针叶林平均碳贮量(63.7t/hm²)的2.3倍[33]。米老排人工林植被层高碳贮量主要源于乔木层的高生物量及乔木层较高的碳含量。植被中乔木碳贮量为147.90t/hm²，其中，树干占乔木层碳贮量的63.72%，其次是树根，也占有较大比例，为20.54%，乔木枝、叶、皮共占乔木碳总储量的15.51%，可见，枝叶皮根等非木材器官在乔木碳总储量中占有较大比例(36.27%)，因此，在对米老排进行采伐时，采伐剩余物的处理对碳贮量的影响会较大。

在本研究中，米老排人工林土壤碳贮量(0～100cm)为134.19t/hm²，略高于广西西南部杉木林地土壤碳贮量(139.27t/hm²)[29]却低于我国森林土壤平均碳贮量(193.55t/hm²)[12]；桂西南地区米老排人工林土壤碳贮量与其植被碳贮量的比值为1∶1.1，远低于全国平均水平(1∶3.4)[12]，原究其原因，一是南亚热带较好的水热条件促使人工林乔木的快速生长，加快了乔木生物量的累积；二是土壤呼吸速率过快及乔木对土壤养分的大量吸收，导致土壤碳的流失。采伐后保留采伐剩余物，营林管理中减少水土流失，可有效保持土壤碳的累积。

参考文献

[1] TANS P P. How can global warming be traced to CO_2[J]. Scientific American, 2006, 295: 124.

[2] ARTUSO F, CHAMARD P, PIACENTINO S, et al. Influence of transport and trends in atmospheric CO_2 at Lampedusa[J]. Atmospheric Environment, 2009, 43: 3044-3055.

[3] CALDEIRA K, DUFFY P B. The role of the southern ocean in up take and storage of anthropogentic carbon dioxide[J]. Science, 2000, 287: 620-622.

[4] FANG J Y, CHEN A P, PENG C H, et al. Changes in forest biomass carbon storage in China between 1949 and 1998[J]. Science, 2001, 292: 2320-2322.

[5] NORBY R J, LUO Y. Evaluating ecosystem responses to

rising atmospheric CO_2 and global warming in a multi-factor world[J]. New Phytologist, 2006, 162: 281-293.
[6] WOODWELL G M, WHITTAKER R H, REINERS W A, et al. The biota and the world carbon budget[J]. Science, 1978, 199: 141-1467.
[7] POST W M, EMANUEL W R, ZINKE P J, et al. Soil pools and world life zone[J]. Nature, 1982, 298: 156-159.
[8] 方精云，陈安平．中国森林植被碳库的动态变化及其意义[J]．植物学报，2001，43 (9)：967-973.
[9] 王效科，冯宗炜．中国森林生态系统中植物固定大气碳的潜力[J]．生态学杂志，2000，19(4)：72-74.
[10] 王效科，冯宗炜．中国森林生态系统的植物碳贮量和碳密度研究[J]．应用生态学报，2001，12(1)：13-16.
[11] 刘国华，傅伯杰，方精云．中国森林碳动态及其对全球碳平衡的贡献[J]．生态学报，2000，20(5)：733-740.
[12] 周玉荣，于振良，赵士洞．我国主要森林生态系统碳贮量和碳平衡[J]．植物生态学报，2000，24(5)：518-522.
[13] 方晰，田大伦，项文化．速生阶段杉木人工林 C 密度、贮量和分布[J]．林业科学，2002，38(3)：14-19.
[14] 方晰，田大伦，项文化，等．不同密度湿地松人工林中碳的积累与分配[J]．浙江林学院学报，2003，20(4)：374-379.
[15] 何宗明，李丽红，王义祥，等．33 年生福建柏人工林碳库与碳吸存[J]．山地学报，2003，21 (3)：298-303.
[16] LACLAU P. Biomass and carbon sequestration of ponderosa pine plantations and native cypress forests in northwest Patagonia [J]. Forest Ecology and Management, 2003, 180: 317-333.
[17] SPECHT A, WEST PW. Estimation of biomass and sequestered carbon on farm forest plantations in northern New South Wales, Australia[J]. Biomass and Bioenergy, 2003, 25: 363-379.
[18] 雷丕锋，项文化，田大伦，等．樟树人工林生态系统碳素贮量与分布研究[J]．生态学杂志，2004，23(4)：25-30.
[19] 李轩然，刘琪，陈永瑞，等．千烟洲人工林主要树种地上生物量的估算[J]．应用生态学报，2006，17(8)：1382-1388.
[20] 马明东，江洪，刘跃建．楠木人工林生态系统生物量、碳含量、碳贮量及其分布[J]．林业科学，2008，44(3)：34-39.
[21] 林金国，张兴正，翁闲．立地条件对米老排人工林生长和材质的影响[J]．植物资源与环境学报，2004，13(3)：50-54.
[22] 郭文福，蔡道雄，贾宏炎，等．米老排人工林生长规律的研究[J]．林业科学研究，2006，19(5)：585-589.
[23] 景跃波，杨德军，马赛宇，等．热带速生树种米老排的育苗与造林[J]．林业实用技术，2008，(1)：21-23.
[24] 明安刚，于浩龙，陈厚荣，等．广西大青山米老排人工林土壤物理性质[J]．林业科技开发，2010，24(5)：71-74.
[25] 明安刚，贾宏炎，陶怡，等．桂西南 28 年生米老排人工林生物量及其分配特征[J]．生态学杂志，2011，31(5)：1050-1056.
[26] 明安刚，唐继新，于浩龙，等．桂西南米老排人工林单株生物量回归模型[J]．林业资源管理，2011，(6)：83-87.
[27] 何斌，黄寿先，招礼军，等．秃杉人工林生态系统碳素积累的动态特征[J]．林业科学，2009，45(9)：151-157.
[28] 黄松殿，吴庆标，廖克波，等．观光木人工林生态系统碳贮量及其分布格局[J]．生态学杂志，2011，30(11)：2400-2404.
[29] 陈楚莹，廖利平，汪思龙．杉木人工林生态系统碳素分配与贮量的研究[J]．应用生态学报，2000，11(增刊 1)：175-178.
[30] 康冰，刘世荣，蔡道雄，等．南亚热带杉木生态系统生物量和碳素积累及其空间分布特征[J]．林业科学，2009，45(8)：147-153.
[31] 梁宏温，温远光，温琳华，等．连栽对尾巨桉短周期人工林碳贮量的影响[J]．生态学报，2009，29(8)：4242-4250.
[32] 方晰，田大伦，胥灿辉．马尾松人工林生产与碳素动态[J]．中南林学院学报，2003，23(2)：11-15.
[33] 王绍强，周成虎，罗承文，等．中国陆地自然植被碳量空间分布探讨[J]．地理科学进展，1999，18(3)：238-244.

[原载：生态学杂志，2012，31(11)]

广西林区木霉菌多样性调查

于浩龙 赵 樟 刘 云 卢立华

（中国林业科学研究院热带林业实验中心，广西凭祥 532600）

摘 要 通过对广西林区6种林分类型的林地木霉菌多样性进行调查，采用形态学方法进行分类鉴定，共分离鉴定出木霉菌11个种。研究表明，混交林中木霉菌的分离率最高，松树林木霉菌分离率最低；海拔高度对木霉菌的丰富度影响不大。在所分离的木霉菌中，哈茨木霉（*Trichoderma harzianum* Rifai）出现频率高，是广西林区木霉菌资源的优势种类。

关键词 林分类型；木霉菌；物种多样性

Investigation of *Trichoderma* Diversity in Guangxi Regions

YU Haolong, ZHAO Zhang, LIU Yun, LU Lihua

(*Experimental Center of Tropical Forestry*, *CAF*, *Pingxiang* 532600, *Guangxi*, *China*)

Abstract: *Trichoderma* diversity in 6 types of forest stands were investigated in Guangxi, and 11 species of *Trichoderma* were identified according to the morphological method. The result showed that *Trichoderma* isolation rate in mixed forest was the highest, and the lowest in the pine stand. There was not obvious effect on *Trichoderma* abundance on different altitudes. *T. harzianum* Rifai was dominant species whose isolation rate was higher than other species in Guangxi forest regions.

Key words: stand types; *Trichoderma*; species diversity

木霉（*Trichoderma* Pers.）在自然界的分布十分广泛，是土壤微生物的重要组成群落之一[1]，有些木霉可产生活性很强的纤维素酶、几丁质酶、木聚酶等水解酶类，被广泛地应用于生产微生物源的酶类、糖蛋白以及用于植物病害的生物防治、食品、卫生、饲料、制浆、纺织、造纸、生物基因工程和环保等领域[2]。此外，木霉属个别种也可引起蘑菇的病害，导致蘑菇生产污染，造成减产[3,4]。有些木霉菌还能通过与植物形成共生关系，来促进植物生长发育，诱导并激发植物自身免疫功能，有些木霉菌甚至还能引起人的病害[5-7]。由于木霉菌对人类的生产和生活如此重要，作为基础研究的木霉菌分类学研究一直受到人们的重视，许多学者对此做了深入的研究和总结。我国对木霉属真菌的研究主要侧重在木霉菌的应用开发方面，在分类研究方面相对较少。1986年起福建三明真菌研究所木霉研究组收集木霉菌株20余个，并对它们的形态及培养性状进行了系统研究[8]。1998年，云南农业大学王家和等从云南省大围山自然保护区土壤样品中分离到69株木霉，按Rifai-Bissett对木霉属种群分类系统，鉴定出6个集合种[9]。1999年，陈建爱等利用可溶性蛋白质聚丙烯酰胺凝胶电泳对26个木霉菌株进行特征性图谱分析，结果表明木霉可溶性蛋白质电泳图谱种内基本一致，种间差异显著，但仍有个别菌株不能很好地归类[10]。早期戴芳澜记载我国木霉属真菌5个种，分别是白色木霉［*T. album*（*T. polysporum*）］、粉绿木霉［*T. glaucum*（*T. aureoviride*）］、木素木霉［*T. lignorum*（*T. viride* Pers.）］和亚木霉（康氏木霉）；文成敬等对中国西南地区的木霉属进行了分类研究，鉴定并描述了9个集合种；章初龙等简要报道了我国木霉菌16个种，并提出了两个可能为新种的菌株；赵智慧等简要报道了东北保护地土壤中木霉菌11个种，但没有种的具体描述[11-15]。广西地处热带亚热带，地域辽阔，水热条件优越，地形地貌复杂，植被类型多样，林区枯枝落叶层深厚、湿度大，十分适宜木霉等真菌生长发育，菌类资源极为丰富[16]。作为一种重要的自然资源，目前广西对木霉菌资源的调查和分类研究比较少。为此，本研究以广西不同林分类型的林区木霉菌资源调查为切入点，旨在了解该区域林分类型对木霉菌的种类及分布特征的影响，为探索广西林区木霉菌资源的多样性及开发利用提供参考依据。

1 材料与方法

1.1 土样采集

在广西(河池市的罗城和金城江、桂林市的永福和兴安、柳州市三江和融水、百色市的田林和乐业、南宁市和凭祥市)林区选取不同林分类型的林地，包括杉木林、马尾松林、竹林、桉树林、阔叶林(主要树种为米老排)和混交林(主要为常绿阔叶树种混交)6种林分类型。每种林分分不同海拔，不同的土层深度取样(先去掉0~5cm表层土，按机械分层分为5~10cm、10~20cm、20~50cm三个层次取样)。将采集的土样置于4℃冰箱中保存备用。

1.2 木霉菌分离

木霉菌的分离采用稀释涂抹平板法[14]，将采集土样过2mm筛后制成一系列土壤悬浮液，用PDA培养基(加入链霉素)进行培养分离。

1.3 木霉菌株鉴定

采用直接从PDA培养基上挑取孢梗丛制片法鉴定。经初步筛选，将待鉴定菌株置于PDA培养基上纯化，并放入25℃恒温箱中培养。待菌落产生孢子梗丛束而又未变为深绿前，挑取分生孢子梗丛束制片，观察分生孢子梗形态；大约培养10~14d后，挑取分生孢子制片，观察孢子大小和形态。按Gams *et al.* (1998)等分类系统和Kullnig-Gradinger *et al.* (2002)等最新研究成果进行分类鉴定[17,18]。

1.4 不同林分木霉菌资源调查数据分析

实验数据处理工作通过Excel软件和SPSS18.0统计软件完成。采用One-way ANOVA进行方差检验，LSD检验进行多重比较分析。分离率=(分离出木霉菌平板数/涂抹平板数)×100%。

2 结果与分析

从广西林区6种林分中共分离鉴定出木霉菌11个种，分别是深绿木霉、绒状木霉、长枝木霉、棘孢木霉、橘绿木霉、康氏木霉、哈茨木霉、黄绿木霉、绿色木霉、螺旋木霉、钩状木霉。

2.1 木霉菌11个种的主要形态特征对比

长枝木霉(*Trichoderma longibrachiatum* Rifai)孢子较大，3.8~5.1(4.3)μm×2.3~2.8(2.6)μm，近椭圆形，新分离的孢子往往要大些。*T. longibrachiatum* 的分生孢子梗分枝比较稳定和明确。它的分枝要比这一组的 *T. citrinoviride* Bissett 和 *T. pseudokoningii* Rifai 简单，分枝更稀少，但比木霉组的 *T. atroviride* P. Karsten 复杂。*T. longibrachiatum* Rifai 的瓶梗在分生孢子梗上大多单生。

与其他种相比，橘绿木霉(*Trichoderma citrinoviride* Bissett)分生孢子较小，通常不大于3.5μm×2.4μm，瓶梗短小，基部缢缩。橘绿木霉的分生孢子梗结构与拟康氏木霉相似，和长枝木霉相比，橘绿木霉的分枝要复杂些。

绿色木霉(*Trichoderma viride* Pers.)的典型特征是分生孢子壁具有很明显的疣状突起，分生孢子近球形，4.0~4.8μm×3.5~4.2μm；不规则分枝的分生孢子梗；钩状至弯曲的、常常单生的瓶梗。而与 *T. viride* Pers. 同组的 *T. asperellum* Samuels 的分生孢子壁具有细微的棘状突起，近球形至倒卵形，分枝一致，瓶梗直。绿色木霉主要分布在温带地区，其适宜温度较低20~25℃，30℃以上不生长。

棘孢木霉(*Trichoderma asperellum* Samuels)最明显的特征是分生孢子壁具有细微的棘状突起，大小3.4~6.3μm×3.0~5.2μm，近球形至倒卵形，分枝一致，瓶梗直，而同组与其相似的绿色木霉 *T. viride* Pers. 的分生孢子壁有明显的疣状突起，近球形，有不规则分枝的分生孢子梗及钩状至弯曲的、常常单生的瓶梗。绿色木霉分布在温带，其适温较低，20~25℃，30℃以上不生长。棘孢木霉的最适温度为30℃。

深绿木霉(*Trichoderma atroviride* P. Karsten)分生孢子梗简单，孢子较大，4.2~6.6μm×2.7~3.9μm，颜色深绿。深绿木霉的分生孢子梗分枝方式、瓶梗形状大小与同组的 *T. viride* Pers.、*T. asperellum* Samuels 相似，但 *T. atroviride* Rifai 分生孢子光滑，而 *T. viride* Pers.、*T. asperellum* Samuels 的分生孢子有纹饰。但与 *T. harzianum* Rifai 的分生孢子相似，两者的区别在于分生孢子梗，并且 *T. harzianum* Rifai 的生长适温较高。深绿木霉的有性型生长在腐朽的硬木上，而无性型则常见于土壤。已知的有性型只从美国、中美洲和欧洲分离到，无性型则是世界性分布。

钩状木霉(*Trichoderma hamatum* Bon.)培养物具有明显的疱状突起，由于其孢子梗主轴具有不育的延伸丝，次级分枝粗短瓶梗粗短而拥有典型特征，容易识别；分生孢子绿色，光滑，椭圆形，3.1~4.3μm×2.2~3.3μm。该种木霉具有较强的重寄生作用。*T. hamatum* Bon. 与同具不育延伸丝的 *T. pubescens* 相比，后者分生孢子堆无色或白色，而 *T. hamatum* Bon. 为绿色，与刺状木霉（*T. erinaceum*）相比，后者在35℃时生长大于30mm，而 *T. hamatum* Bon. 在35℃ 时不生长。

康氏木霉（*Trichoderma koningii* Oud.）常见于土壤和腐烂的有机物，与 *T. harzianum* Rifai、*T. atroviride* P. Karsten、*T. viride* Pers.、*T. asperellum* Samuels 等近似种比较，*T. koningii* 的显著特征是形成长方形至窄椭圆形的分生孢子，大小 3.5～5.4(4.6)μm×2.4～3.3(2.6)μm。在分枝繁多、分枝系统似金字塔的3个木霉种中，*T. koningii* Oud. 以它较大而椭圆、甚至长方形的分生孢子容易与 *T. harzianum* Rifai、*T. aureoviride* Rifai 区别。

黄绿木霉（*Trichoderma aureoviride* Rifai）与 *T. harzianum* Rifai 和 *T. koningii* Oud. 一样具有复杂的分枝系统，但黄绿木霉的分生孢子倒卵形，且菌落反面变成污黄色或黄褐色。分生孢子壁光滑，倒卵形，基部平截，淡黄绿色，2.4～3.4μm×1.9～2.4(2.3)μm。*T. harzianum* Rifai 和 *T. koningii* Oud. 由于不产生色素，菌落反面不改颜色，这两个种的分生孢子形状也与 *T. aureoviride* Rifai 不同。

哈茨木霉（*Trichoderma harzianum* Rifai）是木霉属内最常见的一个“种”，广泛分布于土壤中[19]。哈茨木霉（*T. harzianum* Rifai）对许多病原真菌具有较强的拮抗作用，对其用于植物病理生物防治的可能性已进行了大量的研究，它在生物防治中的地位不容忽视。与绿色木霉、棘孢木霉相比，哈茨木霉的分生孢子壁光滑，2.8～3.2μm×2.5～2.8μm，而绿色木霉、棘孢木霉的分生孢子壁粗糙。与深绿木霉（*T. atroviride* P. Karsten）相比，*T. harzianum* Rifai 生长适温较宽，生长速度快，在35℃能生长良好。而 *T. atroviride* P. Karsten 在35℃是生长受抑制，且生长速度较慢。

螺旋木霉（*Trichoderma spirale* Bissett）在培养皿边缘形成紧密的疱状突起，并在体视显微镜下可见明显的螺旋状不育延伸丝是其区别于其他木霉的最大特征。

绒状木霉（*Trichoderma velutinum* Bissett）与哈茨木霉（*T. harzianum* Rifai）相似，都能产生致密的疱状突起，两者的区别在于孢子的大小，绒状木霉孢子2.1～2.7μm×1.4～2.4μm，而哈茨木霉的分生孢子较大。

2.2 不同林分类型木霉菌的分布情况

从表1中看出不同林分类型的木霉的出现频率有以下特点：哈茨木霉出现频率最高，在各种林分都存在，说明哈茨木霉是优势种；出现频率最低的是螺旋木霉和绒状木霉，出现频率都是16.7%。在杉木林、米老排、马尾松林三种林分里存在的木霉种类相差不大，竹林林分中培养出的木霉种类最少，仅为4种；出现种类最多的是混交林。从表中可以看出，大部分的木霉可以普遍的培养出来，适应环境力比较强；而有一部分种类的木霉很少出现，其原因有待于进一步的研究。6种林分中，混交林有机质丰富，土壤中有机质、纤维素等含量高，其木霉出现频率均高于其他林分。

表1　不同林分中木霉种类的分布

木霉种类	林分						出现频率(%)
	杉木林	米老排	马尾松林	竹林	混交林	桉树林	
长枝木霉	+	+	+	+	+		83.3
橘绿木霉		+	+		+	+	66.7
绿色木霉	+	+			+		50
棘孢木霉		+	+		+	+	66.7
深绿木霉	+	+	+		+		66.7
钩状木霉			+	+	+		50
康氏木霉	+	+		+	+		66.7
黄绿木霉	+	+	+		+	+	83.3
哈茨木霉	+	+	+	+	+	+	100
螺旋木霉						+	16.7
绒状木霉						+	16.7
合计(种)	6	8	7	4	9	6	

2.3 不同林分类型木霉菌的分离情况

表2 不同林分类型木霉菌方差分析

	平方和	自由度	均方	*F*	*Sig.*
组间	3602.222	5	720.444	4.936	.002
组内	4378.333	30	145.944		
总计	7980.556	35			

从图1中看出不同林分类型的木霉的分离率有以下特点：方差分析6种林分木霉菌分离率差异极显著($P<0.01$)，混交林的木霉分离率最高，达到41.3%；其次为米老排和杉木林；桉树林的木霉分离率最低，只有14.0%。

LSD多重比较表明，杉木林与混交林之间存在显著性差异；米老排与马尾松林、竹林之间存在显著性差异，与桉树林之间存在极显著性差异；马尾松林与米老排之间存在显著性差异，与混交林之间存在极显著性差异；竹林与米老排之间存在显著性差异，与混交林之间存在极显著性差异；混交林与杉木林之间存在显著性差异，与马尾松林、桉树林、竹林之间存在极显著差异；桉树林与米老排、混交林之间存在极显著性差异。

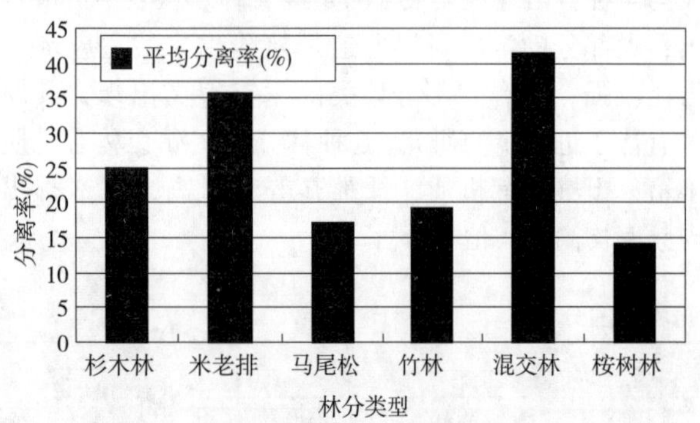

图1 不同林分类型下木霉菌平均分离率

2.4 海拔高度对木霉菌种群分布的影响

表3 各个海拔区间的木霉菌分离率方差分析

	平方和	自由度	均方	*F*	*Sig.*
组间	1225.889	5	245.178	.940	.469
组内	7823.333	30	260.778		
总计	9049.222	35			

从图2可以看出，广西林区海拔100~2100m范围内，各种不同海拔之间的林地木霉菌的分离率有以下特点：木霉菌平均分离率随着海拔的升高几乎没有变化，方差分析海拔对木霉菌的影响差异不显著($P>0.05$)。总的来说林地木霉菌分离率在各个海拔都比较高。说明海拔高度对木霉的丰富度影响不大。

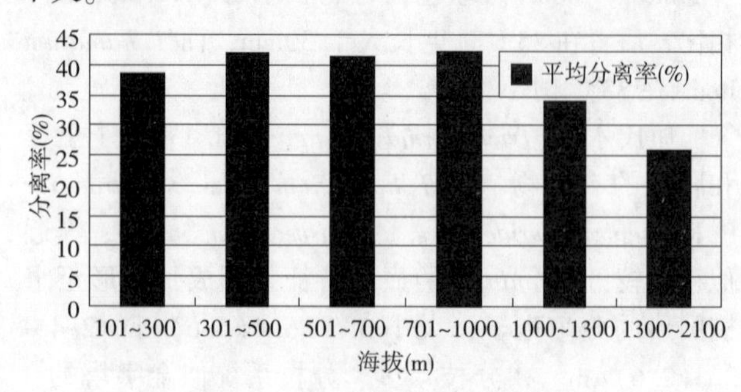

图2 不同海拔木霉菌分离率

3 结论与讨论

从对广西林区6种林分类型林地木霉菌资源调查结果可知，广西林区木霉菌资源极其丰富，在各种林分中都能够分离到。在所鉴定出的11种木霉中，哈茨木霉是出现频率及分离率最高的种，有研究表明，哈茨木霉是木霉属内最常见的一个“种”，广泛分布于土壤中[19]。而螺旋木霉和绒状木霉的出现频率为最低，原因可能是这两种木霉对环境条件要求比较苛刻，但具体原因有待于进一步研究探索。6种林分中，混交林下枯枝落叶多，有机质丰富，水肥分充足，为木霉菌的生存提供了有利的环境条件，所以混交林木霉种类及分离率都是最高，说明了混交林对保护木霉菌的生物多样性有很好的作用。分离率最低的为马尾松林和桉树林。在松树林下，林下植被少，枯枝落叶多为松针及松枝，含有松油等难于分解的物质，并且其分解物中可能分泌了某种物质抑制了木霉菌的生长，从而造成了木霉丰富度低。桉树林一直是争议比较大的一个树种[20]，其原因仍有待于深入的研究。从海拔对木霉的影响来看，木霉对海拔高度要求不是很严格，在所研究的海拔高度范围内，木霉基本没有太大的差别。

从形态学上对木霉菌进行分类鉴定，是鉴定木霉菌株的一种有效方法，一些常见菌株完全可以通过形态学加以识别。但有的菌株存在很大的变异性，有的菌株在人工培养基中很难被培养，所以单存靠形态学方法很难全部鉴定，且准确度不是很高，这就需要借助于分子系统学的手段和方法对木霉进行更加全面的分析，从而筛选出更多有利的木霉菌株[21-24]。

本研究仅仅从木霉菌资源的物种多样性去探索了广西林区木霉菌的存在情况，从6种林分共分离出11种木霉，与国内已报道的种相同[8-15]。木霉菌的更深入研究应结合于其生物学及其生防机制等方面的应用，从而为该真菌的开发利用提供更可靠的依据。

参考文献

[1]DOMSCH K H, GAMS W, ANDERSOM T H. Compendium of soil fungi [M]. London: Acadimic press, 1980: 794-809.

[2]KUBICEK C P, EVELEIGH D E, ESTERBAUER H, et al. *Trichoderma reesei* Cellulases: Biochemistry genetics physiology and application[C]//Senior D J. Production and application of xylanases from *Trichoderma harzianum* Vienna: Royal Society of Chemistry, 1990: 47-60.

[3]SEABY D A. Infection of mushroom compost by*Trichoderma* species[J]. Mushroom Journal, 1987, 179: 355-361.

[4]SEABY D A. Differentation of *Trichoderma* taxa associated with mushroom production [J]. Plant Pathology, 1996, 45: 905-912.

[5]YEDIDIA I, BENHAMOU N, KAPULNIK Y, et al. Induction and accumulation of PR proteins activity during early stages of root colonization by the mycoporasite *Trichoderma harzianum* strain T-203[J]. Plant Physiology and Biochemistry, 2000, 38: 863-873

[6]YEDIDIA I, SHORESH M, KERERN Z, et al. Concomitant induction of systemic resistance to *pseudomonas syringae pv. lachrymans* in cucumber by *Trichoderma asperellum*(T-203) accumulation of Phytoalexins[J]. Applied and Environmental Microbiology, 2003, 69 (12): 7343-7353

[7] KUHLS K, LIECKFELDT E, BORNER T, et al. Molecular reidentification of human pathogenic *Trichoderma* isolates as *Trichoderma longibrachiatum* and *Trichoderma citrinoviride* Med. Mycol., 1999, 37: 25-33.

[8]REHNER S A, SAMUELS G[J]. Canad. J. Bot, 1995, 73(Suuple. 1): 816-823.

[9]唐嘉义,王家和,等.云南大围山自然保护区木霉菌多样性与 RAPD 分析[J]. 微生物学杂志, 1998, 10(增刊): 48-53.

[10]陈建爱,等.木霉可溶性蛋白电泳分析[J]. 中国生物防治, 1999, 25(2): 77-80.

[11]戴芳澜.中国真菌总汇[M]. 北京:科学出版社, 1979.

[12]章初龙,徐同.我国河北、浙江、云南及西藏木霉种记述[J]. 菌物学报, 2005, 24(2): 184-192.

[13]章初龙,徐同.木霉属分类研究进展[J]. 云南农业大学学报, 2000, 15(3): 269-273.

[14]文成敬,陶家风,陈文瑞.中国西南地区木霉分类研究[J]. 真菌学报, 1993, 12(2): 118-130.

[15]ZHAO Zhihui, SUN Xiaodong, YANG Ruixiu, et al. Diversity of *Trichoderma* in greenhouse soil[J]. Journal of Zhejiang University (Agric. & Life Sci.), 2004, 30(4): 467.

[16]于浩龙,韦继光,苏小兰.广西北部林区木霉菌资源初探[J]. 广西农业科学, 2010, 41(7): 703-706.

[17]GAMS W, BISSETT J. Morphology and identification of *Trichoderma* [A], Kubicek CP, Harman G E, *Trichoderma* and *Gliocladium*, volume 1 Basic biology, taxonomy and genetics [C]. London: Taylor & Francis Led., 1998, 3-34

[18]KULLNIG-GRADINGER C M, SZAKACS G, KUBICEK C P. *Phylogeny* and evolution of the genus *Trichoderma*: a multi gene approach Mycol. Res., 2002, 106 (7): 757-767.

[19]RIFAI MA, WEBSTER J, 1969. A revision of the genus *Trichoderm. Mycol Pap*, 116: 1-56.

[20]温远光,刘世荣,陈放.连栽对桉树人工林下物种多样性的影响[J]. 应用生态学报, 2005, 16(9): 1667-1671.

[21]MACH R L, ZEILINGER S, KRISTUFEK D, et al. Ca^{2+}-calmodulin antagonists interfere with xylanase formation and secretion in *Trichoderma reesei*[J]. Biochimica et Biophysica Acta, 1998, 1403: 281-289.

[22]MACH R L, ZEILINGER S. Regulation of gene expression in industrial fungi: *Trichoderma*[J]. Applied Microbiological Biotechnology, 2003, 60: 515-522.

[23]SEIBOTH B, HAKOLA S, MACH R L, et al. Role of four major cellulases in triggering of cellulose gene expression by cellulose in *Trichoderma reesei*[J]. Journal of Bacteriology, 1997, 179(17): 5318-5320.

[24]MUKHERJEE P K. TgaA and TgaB, in the antagonism of plant pathogens by *Trichoderma virens*[J]. Applied and Environmental Microbiology, 2004, 70(1): 542-549.

[原载:西北林学院学报, 2012, 27(02)]

我国森林地表凋落物现存量及养分特征

郑　路　卢立华

（中国林业科学研究院热带林业实验中心，广西凭祥　532600）

摘　要　归纳总结了我国森林地表凋落物现存量、养分含量及贮量的变化规律以及凋落物现存量的研究方法。结果表明：我国对地表凋落物的研究主要集中在亚热带，基本上涵盖了亚热带的各种林型，而热带和温带研究相对较少。从热带到亚热带和温带，随着纬度的增高，凋落物现存量增加，其平均值分别为 4.62t/hm²，28.44t/hm² 和 68.90t/hm²；L 层所占比例逐渐增大，F 和 H 层比例逐渐减小。此外，海拔、林型、林龄、群落演替及采伐强度等均影响凋落物现存量。地表凋落物的元素含量以有机 C 最高，其次为 N 或 Ca，再其次是 K 和 Mg，P 含量最低。养分贮量具有明显的纬度地带性，即从热带、亚热带到温带，养分贮量逐渐增大，其范围大致是热带在 68.7～147.9kg/hm²，平均为 117.6kg/hm²；亚热带在 15.1～1504.4kg/hm²，平均为 199.5kg/hm²；温带在 367.2～1499kg/hm²，平均为 767.7kg/hm²。

关键词　地表凋落物；凋落物现存量；养分贮量；研究方法

Standing Crop and Nutrient Characteristics of Forest Floor Litter in China

ZHENG Lu，LU Lihua

(*Experimental Centre of Tropical Forestry*，*Chinese Academy of Forestry*，*Pingxiang* 532600，*Guangxi*，*China*)

Abstract：The variation of litter standing crop，nutrient contents and the methods of examining storage of forest floor litter in China were summarized. Researches of forest littere were mainly concentrated on sub-tropical forests，covering various forest types，in these regions，little was found on tropical and temperate forests. Litter standing crop increased with the increase of the latitude，and the average values in tropical，subtropical and temperate forest were 4.62t/hm² 28.44t/hm² and 68.90t/hm²，respectively；and the proportion of L layer increased，F and H layer gradually reduced from tropical to subtropical and temperate forests. In addition，litter standing crop was also influenced by elevation. forest types，forest age，vegetation succession and cutting intensity etc. The order of nutrient contents were the following，or ganic carbon>nitrogen or calcium>calcium or magnesium>phosphorus. Nutrient storage had obvious features in latitudinal zone，increasing gradually from tropical forests to temperate. forests. The scopes of nutrient storage were from 68.7 to 147.9kg/hm²，with an average value of 117.16kg/hm² in tropical forests，from 15.1 to 1 504.4kg/with an average value of 199.5kg/hm² in subtropical forests，from 367.2 to1499.4kg/hm² with an average value of 767.7kg/hm² in temperate forests .

Key words：floor litter；litter standing crop；nutrient storage；research method

森林凋落物是森林植物在其生长发育过程中新陈代谢的产物，是生态系统主要养分库[1]，对维持森林生态系统的物质循环、能量流动及信息传递功能都具有重要的作用[2]。凋落物的积累与分解是森林生态系统中最为关键的生态过程之一[3]，是控制森林结构和生态系统功能的一个复杂而重要的因素[4]。目前有关凋落物的研究，主要集中在凋落物产量和凋落物分解两个方面，并已有多篇综述性文章发表[5-7]。但凋落物研究的另一个重要方面——地表凋落物现存量(或称凋落物积累量)研究相对较少，多散见于各类文献之中，未见综述性的报导文章。为推动我国森林凋落物的研究水平，可持续利用森林资源，有必要对我国森林地表凋落物现存量方面已取得的研究成果进行总结。

1 凋落物现存量研究方法

1.1 取样方法

多以样地或样线法进行地表凋落物小样方取样。取样面积及重复次数没有统一标准，取样面积从0.05m^2(0.02m×0.25m)到1m^2(1m×1m)[8,9]，其中以0.1m^2(0.40m×0.25m)和0.25m^2(0.5m×0.5m)采用较多[10-14]。肖瑜[15](1989)对此开展过专门研究，认为设置样方面积以0.5~1.0m^2为佳。取样重复5到15个不等，以设10个小样方居多。关于取样方式，主要采用不考虑取样位置的林分内随机设置样点[15-19]，也有机械布点法[20,21]或S形路线的方法[22]。多数只取一次样，取样时间或在生长季初始期(5月)[10]，或在秋冬季[23]。杨玉盛等[24]在对格氏栲天然林与人工林枯枝落叶层碳库及养分库研究时，分1月(冬)、4月(春)、7月(夏)、10月(秋)四次进行凋落物现存量取样，其结果为：凋落物现存量随季节波动，格氏栲林春季凋落物现存量最大，杉木人工林凋落物现存量在夏季达最高值。

此外，有学者提出由于森林生态系统异质性现象的广泛存在，导致上述不考虑取样位置的方法所获得的凋落物数据可能存在空间自相关，即出现样本不独立的现象，从而引起平均值未必能够如实反映凋落物的现存量。因此，孙志虎等[25]采用地统计学的变异函数分析方法定量研究了18~40年生长白落叶松人工林凋落物的空间异质性特征，在此基础上利用地统计学的克里格内插法结合定积分，对落叶松人工林凋落物的现存量进行了估测。并认为采用地统计学的克里格空间插值，结合多元回归和定积分的方法，可以实现落叶松人工林凋落物现存量的估计。

1.2 地表凋落物的分层

大多数研究将地表凋落物分为三层，即未分解层(L层)、半分解层(F层)和完全分解层(或称腐殖质层，H层)[14,23]，或分为两层，即未分解层和半分解层[24]。分层的标准为：L层，由新鲜枯落物组成，保持原有形态，颜色变化不明显，质地坚硬，外表无分解的痕迹；F层，叶无完整外观轮廓，多数枯落物已经粉碎，叶肉被分解成碎屑；H层，已不能辨识原形。分层后一般对未分解层和半分解层的凋落物按枯叶、枯枝、花(果)和其他等进行分组，由此计算不同组分的凋落物现存量(或称凋落物贮量)。

2 地表凋落物现存量

2.1 凋落物现存量及影响因素

表1列出我国主要森林类型凋落物现存量。可以看出，凋落物现存量的研究主要集中在亚热带，基本上涵盖了亚热带的各种林型，热带和温带研究相对较少。

从热带、亚热带到温带，随着纬度的增大，凋落物现存量增加，其平均值分别为4.62t/hm^2，28.44t/hm^2。在同一纬度，随海拔高度的增加，凋落物现存量增加。如从低海拔的常绿阔叶林到高海拔的暗针叶林，凋落物现存量从11.24t/hm^2到113.94t/hm^2。不同森林类型比较，在亚热带，阔叶林凋落物现存量高于针叶林；在温带，表现出针叶林凋落物现存量大于阔叶林。多数研究结果表明针阔混交林凋落物现存量普遍高于针叶林。天然林凋落物现存量高于人工林。

凋落物现存量与林龄基本呈正相关关系，即随着林龄增大，凋落物现存量增多，当达到一定林龄后，林地内植物种类增多，尤其是阔叶树种类和数量的增多，加快了凋落物分解的速率，此时凋落物现存量开始下降[26]。

此外，随着群落演替的正向进行，凋落物现存量逐渐增大。例如有学者研究亚热带箭竹—冷云杉暗针叶林，发现随着亚高山箭竹群落向箭竹—冷云杉林顶级群落恢复演替，林下枯枝落叶层贮量逐渐增大[23]。

不同强度采伐也影响着凋落物现存量，有学者研究采伐作业10年后林地上凋落物的现存量，其排列顺序为：未采伐>弱度>中度>强度>极强度>皆伐，即随着采伐强度的增大，林地凋落物现存量呈递减趋势[27]。

2.2 凋落物各亚层分布格局及组成

从表1还可以看出，地表凋落物各亚层分布格局受地带性因子、林龄和森林类型等影响表现不一致，总体表现出由亚热带到温带，随纬度升高，L层所占比例逐渐增大，F和H层比例逐渐减小；随林龄增大，针叶林及温带森林表现出L层所占比例逐渐减小，F和H层比例逐渐增大；不同森林类型比较，针叶林未分解层所占比例普遍高于阔叶林。

表 1 我国主要森林类型凋落物现存量

气候带	森林类型	林龄（年）	郁闭度	凋落物现存量（t/hm^2）	组分比(%)			各亚层百分比(%)			资料来源
					枯叶	枯枝	其他	L	F	H	
热带	假柿木姜子—印度栲次生林	23	0.9	4.68	29.07	36.33	37.6				
	白背桐—假柿木姜子次生林	35	0.8	5.17	29.91	29.03	41.06				[28]
	绒毛番龙眼—千果榄仁原始林	100	0.9	4.02	28.16	28.9	42.94				
亚热带	云杉人工林	40	0.8	36.4				30.16	69.84		
	桦木次生林	40	0.6	124.32				12.68	87.32		[29]
	云杉—岷江冷杉原始林	100	0.6	145.06				29.15	70.85		
	箭竹—云杉、冷杉天	160	0.8	46.34				19.1	35.88	45.02	
	箭竹—桦木—云冷杉天然林	90	0.5	25.77				23.54	30.03	46.38	[23]
	箭竹—桦木次生林	60	0.1	6.5				38.15	61.85		
	青冈栎天然林		0.8	11.24				46.98	53.02		
	五裂槭—多毛椴天然林		0.8	15.54				24.13	75.87		
	铁杉—红桦天然林		0.9	57.5				10.61	8.75	80.64	[30]
	岷江冷杉天然林		0.7	113.94				13.38	32.09	54.53	
	格氏栲天然林	150	0.9	8.99	64.96	31.59	3.45	36.26	63.74		[24]
	格氏栲人工林	40	0.9	7.56	61.38	37.83	0.79	20.9	79.1		
	栓皮栎次生林	中龄		10.7				44.4	55.6		[8]
	青冈天然林			14.14				29.77	33.45	36.78	[21]
	锥栗—荷木—厚壳	400	0.9	8.97				43.64	28.41	27.95	[31]
	滇青冈—元江栲天然林		0.6	24.61				23.41	30.35	46.24	[32]
	云南松天然林	30~60	0.6	21.59				25.2	38.77	36.03	
	云南松天然林	40~60	0.6	28.95				24.34	41.55	34.1	
	马尾松人工林	8	5.1	91	8	1		50.3	21.9	27.8	
	马尾松人工林	14		11.48	81.8	8.5	9.7	51.3	19.5	29.2	[33]
	马尾松人工林	23		15.54	82.3	12.7	5	36.6	32.1	31.3	
	马尾松人工林	38		4.99	82.5	11.5	6	34.8	27.6	37.6	
	马尾松人工林	41		5.02	53.03	23.09	23.88	86.25	13.75		[9]
	杉木人工林	27		0.8	3.53	48	40.8	11.2			[18]
	杉木人工林	40		0.8	4.81	38.05	42.62	19.33	43.87	56.13	[24]
	马尾松—甜槠—木荷次生林	60	0.9	6.15	73.6	13.6	12.8				[27]
	马尾松—青栲—拉氏栲人工林	28		11.14	60.14	25.91	13.95	59.47	40.53		[9]
温带	急尖长苞冷杉原始林	200	0.7	5.86				36.89	36.71	26.4	[12]
	落叶松人工林	28		25.61	69.17	27.87	2.96	69.85	30.15		[34]
	油松—蒙古栎人工林	30		11.13	90.82	7.13	2.05	61.24	38.76		[35]
	油松人工林	30		8.26	83.11	11.67	5.22	58.95	41.05		
	油松人工林	28	0.7	17.95	91.6	5.7	2.7	37.88	62.12		[36]
	山杨次生林	中龄	0.7	8.34	79.5	18.1	2.4				[42]
	新疆落叶松—天山云杉天然林	100	0.6	135.7				3.83	8.7	87.47	[11]
	天山云杉天然林	80	0.6	124.1				3.47	17.24	79.29	
	天山云杉天然林	140	0.7	222.6				4.27	24.21	71.52	
	椴树—红松天然林			12.7					28.35	71.65	[37]

地表凋落物组成，因受物种特性和研究侧重点的影响，不同学者细分程度不同。为便于统计，本文将其分为三类，即枯叶、枯枝和其他(落皮、落花、落果等均归为其他)。由表1可见，地表凋落物的组成，热带以其他组分所占比例较高，其次为枯叶和枯枝；亚热带和温带均以枯叶为主，所占比例为38.1%~91.6%，其次为枯枝，占5.7%~42.6%，其他组分比例较低，为0.8%~42.9%。

3 地表凋落物养分含量及贮量

3.1 养分含量

地表凋落物养分元素的含量以有机C最高，达323.3~540.0g/kg；其次为N元素或Ca元素，因森林类型不同、区域不同而有差异，总体趋势是南方森林地表凋落物N元素含量高于Ca元素含量，分别为5.3~19.9g/kg和2.2~15.9g/kg，北方森林地表凋落物Ca元素含量高于N元素含量，分别为5.4~21.9g/kg和6.7~10.7g/kg；再其次是K元素和mg元素，分别为1.3~9.2g/kg和0.3~2.1g/kg，北方森林略高于南方森林；P元素含量最低，为0.2~1.8g/kg[11,16,21,23,28,31,32,33]。

地表凋落物不同分解亚层，元素含量变化呈现不同规律。多数研究表明从L层、F层到H层，随着地表凋落物分解程度的增加，N元素含量升高，在个别淋溶强烈地区会出现相反的趋势[22]。有机C含量变化随地表凋落物分解程度的增加而下降。C/N值变化与有机C含量变化趋势一致。

P、K、Ca、mg等元素在地表凋落物各亚层自上而下的含量变化特征不一致，有些林分随着分解进行，营养元素含量升高，即各营养元素在L层到H层逐渐积累[23]；有些林分营养元素含量随着凋落物的分解而下降[12,31]；另有一些林分营养元素含量在地表凋落物各亚层互有增减，如刘文耀等[32]在研究了滇中常绿阔叶林及云南松林死地被物养分动态后认为由L亚层到H亚层，P、K含量略有增加，而Ca和mg含量则出现下降的趋势；于明坚等[21]的研究结果表明Ca在L、F亚层的含量明显高亚层，而mg、P的含量从上到下呈梯度增加的趋势，各亚层K含量变化不大。总体趋势是南方森林易于淋洗，地表凋落物各亚层随分解程度的增加，养分元素含量逐渐降低，北方森林易于积累，养分元素含量逐渐升高；针叶林易于积累，阔叶林趋于释放。

地表凋落物不同组分间元素含量具有一定差异，总体上表现出C含量在枯枝部分较高，枯叶中的N、P、K、Ca和mg含量高于其他部分，即地表凋落物养分主要集中在落叶之中。

3.2 养分贮量

由地表凋落物形成的枯枝落叶层是一个庞大的养分贮存库，这些营养元素，是森林有机质和矿质元素的重要来源，随着凋落物的分解作用，将贮存的养分逐渐归还给林地。因此它们在保持地力，提高森林生产力中起着特别重要的作用。枯枝落叶层养分的贮量取决于本身的养分含量及凋落物现存量。

表2可见，地表凋落物养分贮量同样具有明显的纬度地带性，即从热带、亚热带到温带，养分贮量逐渐增大，其范围大致是热带在68.7~147.9kg/hm^2，平均为117.kg/hm^2；亚热带在15.1~1504.4kg/hm^2，因不同森林类型，变动幅度很大，平均为199.5kg/hm^2；温带在367.2~1499.4kg/hm^2，平均为767.7kg/hm^2。地表凋落物不同元素的贮量，以有机C贮量最高，达743.16~4023.19kg/hm^2。其次为N或Ca贮量，阔叶林地表凋落物中N贮量一般大于Ca贮量，针叶林因树种不同而有差异，如杉木林地表凋落物N贮量大于Ca贮量，侧柏林地表凋落物N贮量小于Ca贮量。再其次是K和mg元素，其贮量分别是5.4~285.7kg/hm^{2}和2.5~125.8kg/hm^2。P元素的贮量最低，为1.9~54.6kg/hm^2。

表2 地表凋落物养分贮量

气候带	森林类型	林龄(年)	郁闭度	养分储量(kg/hm^2)						合计	资料来源
				C	N	P	K	CA	Mg		
热带	假柿木姜子—印度栲次生林	23	0.9	1743.6	74	2.8	8.8	31.2			[28]
	白背桐—假柿木姜子次生林	35	0.8	2012.7	85.8	3.3	6.1	52.7			
	绒毛番龙眼—千果榄仁原始林	100	0.9	1759.4	56.9	2.7	7.1	58.5			
	粗果相思—雷州1号桉林				48.13	2.2	5.14	44.09	29.63	129.19	[38]
	雷州1号桉林				24.64	1.94	4.1	28.79	9.26	68.73	

（续）

气候带	森林类型	林龄（年）	郁闭度	养分储量（kg/hm^2）						合计	资料来源
				C	N	P	K	CA	Mg		
亚热带	青冈天然林				218.57	4.03	25.86	93.43	15.62	357.51	[21]
	锥栗—荷木—厚壳桂天然林		0.9		102.08	4.83	48.94	17.38	10.43	183.66	[31]
	箭竹—云杉、冷杉天然林	160	0.8		553.14	54.63	164.75	606.12	125.78	1504.42	[23]
	箭竹—桦木—云冷杉天然林	90	0.5		269.45	23.61	96.31	367.04	79.08	835.49	
	箭竹—桦木次生林	60	0.1		68.69	7.73	27.64	21.66	11.45	137.17	
	格氏栲天然林	150	0.9	4023.2	66.41	3.8	33.36	26.79	8.06	138.42	[24]
	格氏栲人工林	40	0.9	3293	53.78	3.42	24.7	20.93	10.73	113.56	
	杉木人工林	40	0.8	2276.6	28.44	2.37	11.25	24.97	5.36	72.39	
	杉木—红荷人工林	6.5			16.85	0.67	2.75	8.78	3.07	32.12	[39]
	杉木—木荷人工林	6.5			21.64	0.85	3.31	12.01	3.41	41.22	
	杉木—火力楠人工林	6.5			26.13	0.91	3.89	14.14	2.08	47.15	
	杉木人工林	6.5			11.22	0.51	1.03	5.99	1.37	20.12	
	杉木人工林	6.5			8.2	0.31	0.62	4.9	1.02	15.05	
	马尾松人工林	8			45.31	1.93	9.54	37.62	4.16	98.56	
	马尾松人工林	14			106.35	4.08	5.43	31.06	11.01	157.93	
	马尾松人工林	23			123.82	3.88	24.99	56.95	4.73	214.37	
	马尾松人工林	38			39.82	1.26	9.52	28.53	2.48	81.61	
	马尾松—拉氏栲人工林	41(16)			50.59	11.1	17.34	85.95	21.13	186.11	[9]
	马尾松—青栲人工林	41(16)			93.99	23.56	30.31	180.86	38.28	367	
	马尾松—闽粤栲人工林	41(14)			82.51	17.08	33.03	132.14	26.77	291.53	
	马尾松—格氏栲人工林	41(16)			85.95	17.23	31.86	145.11	36.85	317	
	马尾松—苦槠人工林	41(17)			60.3	8.63	11.01	75.41	18.87	174.22	
	马尾松人工林				33.74	7.1	9.77	65.3	14.14	130.05	
温带	油松—侧柏人工林	30	0.8		106.87	19.28	7.14	220.39	13.48	367.16	[43]
	油松人工林	30	0.8		121.64	26.67	21.33	190.62	22.33	382.59	
	油松天然林	29	0.7		267.3	22.8	125.5	188.6	66	670.2	[16]
	落叶松天然林	31	0.7		360.1	31.7	173.3	236.4	60.5	862	
	云杉天然林	38	0.7		613.4	51.8	285.7	437.8	110.7	1499.4	
	侧柏天然林	28	0.7		180.5	11.1	78	253.3	91	613.9	
	落叶松人工林	21			320.92	27.37	10.67	473.76	34.52	867.24	[34]
	落叶松人工林	28			309.41	25.13	16.19	463.6	64.89	879.22	

从不同森林类型来看，地表凋落物中营养元素的总贮量进行比较，常绿阔叶林明显地高于针叶林，N、P、K、Ca、Mg 等元素的贮量均以阔叶林为高。马尾松与不同树种构成的混交林，其任一养分元素贮量均大于马尾松纯林，杉木阔叶混交林同样呈现出上述趋势。油松与侧柏混交，混交林没有显示出优越性，其贮量小于油松纯林。4 年生桉树与相思树混交林营养元素贮量为桉树纯林的 1.9 倍。天然林地表凋落物养分贮量一般大于人工林。

4　问题与展望

我国对地表凋落物现存量的研究，主要集中在亚热带，比较细致深入，涉及多种不同的森林类型，不但有针叶林、阔叶林和针阔混交林，还有天然林和人工林及不同海拔梯度的变化，对微量元素及有机化学组成也有涉及[40,41]。相比较而言，温带地表凋落物研究相对较少，也不够系统深入，研究多局限于针叶林，而阔叶林及针阔混交林研究较少。因此应加强温带地表凋落物的研究，以推动此区域森林生态系统研究的水平。

多数研究只对地表凋落物取一次样，但受种生物学特性和年内降水量、气温、风力等气候因子的综合影响，地表凋落物现存量会随季节波动[24,37]。因此笔者建议一年最少取2次样，即在凋落物现存量最大的时候1次，凋落高峰期之前1次，一般北方在5月和9月。

凋落物现存量高，一方面可提高林地水源涵养能力及养分潜在供应能力等，但也存在负面作用，即养分循环利用率低，火灾隐患突出，尤其是未分解层的大量积累。因此合理的凋落物现存量和各亚层分布格局及通过调控措施来加强或减缓凋落物分解研究应是将来对人工林管理的一个重要方向。

参考文献

[1]SCOTTNA，BINKLEY D. Foliage litter quality and annual net N mineralization：Comparison across North American forest. sites[J]. Ecologia，1997，111(2)：151-159.

[2]刘强，彭少麟．植物凋落物生态学[M]．北京：科学出版社，2010.

[3]TURNER J，SINGER M J. Nutrient distribution and cycling in sub-apline coniferous forest ecosystem[J]. Journal of applied Ecology，1976，13(1)：295-301.

[4]BERG B，MCMLAUGHERTY C. Plant litter：Decomposition，humus formation，carbon sequestration[M]. New York：Springer Verlag，2003.

[5]王风友．森林凋落量研究综述[J]．生态学进展，1989，6(2)：82-89.

[6]林波，刘庆，吴彦，等．森林凋落物研究进展[J]．生态学杂志，2004，23(1)：60-64.

[7]郭剑芬，杨玉盛，陈光水，等．森林凋落物分解研究进展[J]．林业科学，2006，42(4)：93-100.

[8]张洪江，程金花，史玉虎，等．三峡库区3种林下枯落物储量及其持水特性[J]．水土保持学报，2003，17(3)：55-58，123.

[9]樊后保，苏素霞，卢小兰，等．林下套种阔叶树的马尾松林凋落物生态学研究Ⅲ．凋落物现存量及其养分含量[J]．福建林学院学报，2003，23(3)：193-197.

[10]马志贵，王金锡．大熊猫栖息环境的森林凋落物动态研究[J]．植物生态学与地植物学学报，1993，17(2)：155-163.

[11]李叙勇，孙继坤．天山森林凋落物和枯枝落叶层的研究[J]．土壤学报，1997，34(4)：406-417.

[12]钟国辉，辛学兵．西藏色季拉山暗针叶林凋落物层化学性质研究[J]．应用生态学报，2004，15(1)：167-169.

[13]张庆费，徐绒娣．浙江天童常绿阔叶林演替过程的凋落物现存量[J]．生态学杂志，1999，188(2)：17-21.

[14]吴毅，刘文耀，沈有信，等．滇石林地质公园喀斯特山地天然林和人工林凋落物与死地被物的动态特征[J]．山地学报，2007，25(3)：317-325.

[15]肖瑜．样方面积和数量对测定次生林凋落物精度的影响[J]．生态学报，1989，9(1)：59-65.

[16]金小麒．华北地区针叶林下凋落物层化学性质的研究[J]．生态学杂志，1991，10(6)：24-29.

[17]闫俊华，周国逸，唐旭利，等．鼎湖山3种演替群落凋落物及其水分特征对比研究[J]．应用生态学报，2001，12(4)：509-512.

[18]何宗明，陈光水，刘剑斌，等．杉木林凋落物产量、分解率与储量的关系[J]．应用与环境生物学报，2003，9(4)：352-356.

[19]俞益武，吴家森．木荷林凋落物的归还动态及分解特性[J]．水土保持学报，2004，18(2)：63-65.

[20]刘传照，李景文，潘桂兰，等．小兴安岭阔叶红松林凋落物产量及动态的研究[J]．生态学杂志，1993，12(6)：29-33.

[21]于明坚，陈启常，李铭红，等．青冈常绿阔叶林死地被层和土壤性质特征的研究[J]．林业科学，1996，32(2)：103-110.

[22]魏晶，吴钢，邓红兵．长白山高山冻原生态系统凋落物养分归还功能[J]．生态学报，2004，24(10)：2211-2216.

[23]齐泽民，王开运，宋光煜，等．川西亚高山箭竹群落枯枝落叶层生物化学特性[J]．生态学报，2004，24(6)：1230-1236.

[24]杨玉盛，郭剑芬，林鹏，等．格氏栲天然林与人工林枯枝落叶层碳库及养分库[J]．生态学报，2004，24(2)：359-367.

[25]孙志虎，牟长城，张彦东．用地统计学方法估算长白落叶松人工林凋落物现存量[J]．北京林业大学学报，2008，30(4)：59-64.

[26]逯军峰，王辉，曹靖，等．不同林龄油松人工林枯枝落叶层持水性及养分含量[J]．浙江林学院学报，2007，24(3)：319-325.

[27]周新年．天然林择伐10年后凋落物现存量及养分含量[J]．林业科学，2008，44(10)：25-28.

[28]余广彬，杨效东．不同演替阶段热带森林地表凋落物

和土壤节肢动物群落特征[J]. 生物多样性，2007，15(2)：188-198.
[29]林波，刘庆，吴彦，等．亚高山针叶林人工恢复过程中凋落物动态分析[J]. 应用生态学报，2004，15(9)：1491-1496.
[30]张万儒，许本彤．山地森林土壤枯枝落叶层结构和功能的研究[J]. 土壤学报，1990，27(2)：121-131.
[31]张德强，余清发．鼎湖山季风常绿阔叶林凋落物层化学性质的研究[J]. 生态学报，1998，18(1)：96-100.
[32]刘文耀，荆桂芬．滇中常绿阔叶林及云南松林调落和死地被物中的养分动态[J]. 植物学报，1990，32(8)：637-646.
[33]田大伦，宁晓波．不同龄组马尾松林凋落物量及养分归还量研究[J]. 中南林学院学报，1995，15(2)：163-169.
[34]陈立新，陈祥伟．落叶松人工林凋落物与土壤肥力变化的研究[J]. 应用生态学报，1998，9(6)：581-586.
[35]崔建国，镡娟．辽西油松蒙古栎林下凋落物现存量及持水能力的研究[J]. 水土保持研究，2008，15(2)：154-155，158.
[36]吴钦孝，刘向东．陕北黄土丘陵区油松林枯枝落叶层蓄积量及其动态变化[J]. 林业科学，1993，29(1)：63-66.
[37]詹鸿振，刘吉春，任淑文．阔叶林红松内枯枝落叶层的生态作用[J]. 植物研究，1989. 9(1)：95-100.
[38]陈楚莹，汪思龙．人工混交林生态学[M]. 北京：科学出版社，2004：114-119.
[39]谭绍满，符国荣．杉木与阔叶树混交试验初报[J]. 植物生态学，1995，19(2)：83-191.
[40]郭剑芬，杨玉盛，陈光水，等．格氏栲天然林与人工林枯枝落叶层和粗木质残体有机化学组成研究[J]. 亚热带资源与环境学报，2008，3(3)：40-45.
[41]何帆，王得祥，雷瑞德，等．秦岭林区主要树种叶片凋落物性质的研究[J]. 西北林学院学报，2008，23(4)：30-33.
[42]吴钦孝，刘向东，苏宁虎，等．山杨次生林枯枝落叶蓄积量及其水文作用[J]. 水土保持学报，1999，26(1)：71-76.
[43]姚延梼．京西山区油松侧柏人工混交林生物量及营养元素循环的研究[J]. 北京林业大学学报，1989，1(12)：38-45.

[原载：西北林学院学报，2012，27(01)]

抚育间伐对马尾松人工林生物量与碳贮量的影响

明安刚[1,2] 张治军[3] 谌红辉[2] 张显强[2] 陶 怡[2] 苏 勇[4]
([1]中国林业科学研究院森林生态环境与保护研究所，北京 100091；[2]中国林业科学研究院热带林业实验中心，广西凭祥 532600；[3]国家林业局昆明勘察设计院，云南昆明 650216；[4]南宁树木园，广西南宁 530031)

摘 要 对广西大青山林区不同间伐强度下25年生马尾松人工林碳贮量进行研究。结果表明：间伐措施有利于提高马尾松乔木层生物量和碳贮量，不利于林下地被物和凋落物生物量和碳的累积，重度、中度和轻度间伐下马尾松林乔木层碳贮量分别高出对照11.47%、11.78%和14.49%，林下植被层分别低于对照20.82%、19.80%和0.20%，凋落物层分别低于对照15.81%、2.87%和27.31%；间伐对土壤层碳贮量和生态系统碳贮量总量无显著影响，但有降低土壤碳贮量和增加生态系统碳贮量总量的趋势，土壤碳贮量分别低于对照4.15%、1.83%和5.53%，生态系统碳贮量总量分别高出对照2.62%、4.19%和3.58%，但差异均未达到显著水平；间伐措施对各器官碳贮量在乔木层的分配无显著影响，各处理均表现为树干(53.38%~60.12%)>树根(15.70%~16.74%)>树皮(12.88%~14.38%)>树枝(7.53%~9.12%)>树叶(2.27%~2.88%)。
关键词 抚育间伐；马尾松人工林；生物量；碳贮量；分配特征

Effects of Thinning on the Biomass and Carbon Storage in *Pinus massoniana* Plantation

MING Angang[1,2], ZHANG Zhijun[1], CHEN Honghui[2], ZHANG Xianqiang[2], TAO Yi[2], SU Yong[3]
([1]*Institute of Forest Ecology, Environment and Protection, CAF, Beijing* 100091, *China*; [2]*Experimental Center of Tropical Forestry, CAF, Pingxiang* 532600, *Guangxi, China*; [3]*China Forest Exploration & Design Institute on Kunming, Kunming* 650216, *Yunnan, China*; [4]*Nanning Arboretum Nanning* 530031, *Guangxi, China*)

Abstract: The biomass and carbon storage of 25-year-old *Pinus massoniana* plantation under different thinning density was studied in Daqingshan mountain in Guangxi. The results showed that the biomass and carbon storage is increased in tree layer while decreased in the understory layer and litter layer because of thinning. Carbon storage of tree layer under Intensive thinning, intermediate thinning and mild thinning weremore than that of control 11.47%, 11.78% and 14.49%, respectively ; The understory layer less than that of control 20.82%, 19.80% and 0.20% respectively, and the litter layer less than that of control 15.81%, 2.87% and 27.31% respectively; The thinning tend to decrease carbon storage in the soil layer while increase the total carbon storage in the ecosystem, but not significant. Soil carbon storage were less than that of control4.15%, 1.83% and 5.53% respectively, total carbon storage in the ecosystem were less that that of control 2.62%, 4.19% and 3.58% respectively. Thinning have no significant effect on the distribution of carbon storage in the tree layer, different thinning density have the same distribution order that stem(53.38%~60.12%)>root(15.70%~16.74%)>bark(12.88%~14.38%)>branch(7.53%~9.12%)>leaf(2.27%~2.88%).
Key words: thinning; *Pinus massoniana* plantation; biomass; carbon storage; distribution characteristics

抚育间伐作为森林培育的重要经营措施，对森林生态系统生态服务功能和演变过程有着重要影响，在改善林木生长环境，促进林木生长，提高林分生产力及林下植物多样性，改善土壤的理化性质，优化林分结构及提高人工林生态稳定性等方面发挥着重要作用[1-11]。目前，国内外对抚育间伐的研究很多：杜纪山等(1996)和李春明等(2003)研究了抚育间伐对林分生产力的影响[12,13]；Jessica 等(2007)、

Donald 等(1999)、任立忠等(2000)、徐扬等(2008)和李春义等(2007)研究了抚育间伐对人工林植物多样性的影响[14-19]；Edith 等(2004)和徐有明等(2002)研究了抚育间伐对林木材质的影响[20,21]；熊有强等(1995)研究了不同间伐强度对杉木(*Cunninghamia lanceolata*)林林下植被发育及生物量的影响[22]。这些研究集中在抚育间伐对林木生长、木材材性和林下植物多样性的影响上，而抚育间伐对人工林碳贮量的影响研究鲜有报道。

马尾松(*Pinus massoniana*)是我国南方主要工业用材树种之一，广泛分布在16个省(自治区、直辖市)。间伐保留密度是否合理，不仅关系到林木的生长，影响经济效益，还关系到马尾松林下植物多样性，土壤养分及林分的固碳能力，进而影响马尾松人工林生态服务功能和生态效益。因此，密度调控技术一直是人工林科学管理和可持续经营中的难点和热点。本研究在广西大青山林区开展不同间伐强度下马尾松人工林生物量和碳贮量的研究，旨在探明抚育间伐对马尾松人工林生物量与碳贮量的影响，为提高马尾松人工林固碳能力和马尾松人工林的科学经营和可持续发展提供参考。

1　研究区概况

研究区位于中国林业科学研究院热带林业实验中心伏波实验场(106°52′32″E，22°01′58″N)，属广西大青山林区，海拔545m左右，坡度25°，年均气温19.9℃，年降水量1400mm。该地区与北热带北缘毗邻，属于南亚热带季风气候。土壤为花岗岩发育成的红壤，土壤厚度大于100cm。

2　研究方法

2.1　试验设计

1983年春季，对杉木采伐迹地进行明火炼山，穴状整地后，用1年生马尾松裸根苗造林，株行距为2m×1m，造林当年开始抚育，连续抚育3年，每年抚育2次。1991年春季，采用随机区组设计，开展间伐试验，设置重度、中度、轻度3个间伐强度和1个对照，共4个处理，重复4次，小区面积为1hm²，间伐后，林分的保留密度分别为1170株/hm²、2000株/hm²、2833株/hm²和3700株/hm²。2008年9月，对4个处理分别选择3个重复，共12个小区，每个小区设4块20m×30m的调查样方，共计48个样方进行林分生物量及碳贮量的调查。

2.2　马尾松乔木生物量的测定

在对样地进行每木检尺的基础上，绘制乔木胸径分布图，根据径级分布图，以2cm为一个径级，选取标准木8株，将标准木伐倒，测定干、皮、枝、叶的鲜质量，地下部分采用“全挖法”分主根和侧根2级分别测定鲜质量。各器官取分析样品200g左右，带回实验室在80℃烘干至恒重，把各器官鲜质量换算成干质量。然后用相对生长法估算乔木层生物量。根据8株标准木的胸径、树高和各器官生物量的实测数据，分别建立样木各器官的生物量(W：kg)与胸径(D：cm)和树高(H：m)的相对生长方程，用以估测林分各器官的生物量(表1)。

表1　各器官生物量相对生长方程

器官	回归方程	样本量	判定系数 R^2	F	显著性
树干	$W=0.0081(D^2H)^{1.0703}$	8	0.9806	84.0027	0.0001
树皮	$W=0.0067(D^2H)^{0.9217}$	8	0.9762	30.7255	0.0015
树枝	$W=0.00003(D^2H)^{1.4785}$	8	0.9842	7.7542	0.0318
树叶	$W=0.000004(D^2H)^{1.5686}$	8	0.9714	11.7246	0.0141
根系	$W=0.0005(D^2H)^{1.2420}$	8	0.9927	34.6277	0.0011
全株	$W=0.0085(D^2H)^{1.1234}$	8	0.9903	58.2924	0.0003

2.3　林下植被生物量及凋落物现存量的测定

按“梅花五点法”在每个固定样地布设5个1m×1m的小样方，记录每个小样方内林下植物的种名，分地上部分和地下部分，采用“收获法”分别测定其鲜质量，同种植物的相同器官取混合样品，凋落物全部测定生物量，取混合样品烘干至恒质量后，计算出各组分的干质量。

2.4　土壤样品的采集

在每个样地内选取具有典型性的地段设置3个土壤剖面，分出腐殖质层后，按照0~10cm，10~30cm和30~50cm进行机械分层，先用便携式土壤水分仪(Hydrosense CS620，CD620)测定土壤体积含水率；

再对各土层用环刀取样，每层3个重复，测定土壤密度；同层土壤取混合样。将土壤样品带回实验室风干，磨碎，过筛，装瓶用以测定土壤含碳量。

2.5 分析样品中碳含量及碳贮量的测定

植物及土壤碳含量均采用重铬酸钾-水合加热法测定，数据统计采用 Excel 和 SPSS13.0 软件包进行。植被碳贮量用单位面积生物量与碳含量的乘积估算；土壤碳贮量用土层厚度、土壤密度和各土层碳含量的乘积估算。

3 结果与分析

3.1 不同间伐强度下马尾松林乔木层生物量和碳贮量及其分配特征

对不同间伐强度下马尾松人工林乔木层生物量及碳贮量的测定结果显示(表2)，重度间伐、中度间伐、轻度间伐及对照区的马尾松林乔木层生物量分别为 246.08t/hm^2、246.45t/hm^2、252.47t/hm^2 和 220.38t/hm^2，乔木层碳贮量分别为 120.92t/hm^2、121.26t/hm^2、124.20t/hm^2 和 108.48t/hm^2。重度间伐、中度间伐和轻度间伐下马尾松林乔木生物量分别高出对照 11.62%、11.83% 和 14.56%，碳贮量分别高出对照区 11.47%、11.78% 和 14.49%，而在重度、中度和轻度间伐处理之间，以轻度间伐的乔木层生物量和碳贮量最大。从表2还可以看出，除重度间伐的树皮生物量和碳贮量略小于对照外，不同间伐强度马尾松林各器官生物量和碳贮量均高于对照。可见，间伐措施有利于马尾松林乔木层生物量和碳贮量的提高，但对各器官生物量和碳贮量在乔木层的分配无显著影响。

从乔木层各器官生物量和碳贮量的分配来看(表2)，4种间伐强度的马尾松林乔木层各器官生物量和碳贮量在乔木层的分配顺序基本相同，均以树干生物量和碳贮量的分配最大，碳贮量所占比例依次为 58.38%、59.71%、59.57% 和 60.12%；其次是根系，分别为 16.74%、15.98%、16.06% 和 15.70%，再次是树皮和树枝，树叶所占比例最小，各处理均小于3%。

3.2 不同间伐强度下马尾松林林下植被及凋落物层生物量与碳贮量

对不同间伐强度下马尾松人工林林下植被生物量及凋落物现存量的测定结果显示(表3)：重度间伐、中度间伐林下植被层生物量和碳贮量均低于对照区，而轻度间伐与对照区接近，这与营林中间伐活动对林下植被带来的干扰有关，与对照相比，其他处理尚处在不同程度的恢复期。

表2 不同间伐强度下乔木层碳贮量及分配特征

项目	碳含量(%)	重度间伐			中度间伐			轻度间伐			对照		
		生物量(t/hm^2)	碳贮量(t/hm^2)	比例(%)	生物量(t/hm^2)	碳贮量(t/hm^2)	比例(%)	生物量(t/hm^2)	碳贮量(t/hm^2)	比例(%)	生物量(t/hm^2)	碳贮量(t/hm^2)	比例(%)
树干	49.38±0.09	142.95	70.59	58.38	146.62	72.40	59.71	149.82	73.98	59.57	132.08	65.22	60.12
树枝	47.96±0.22	23.00	11.03	9.12	20.02	9.60	7.92	20.85	10.00	8.05	17.04	8.17	7.53
树叶	49.41±0.15	7.04	3.48	2.88	5.93	2.93	2.42	6.19	3.06	2.46	4.98	2.46	2.27
树皮	51.98±0.68	29.97	15.58	12.88	32.63	16.96	13.99	33.11	17.21	13.86	30.01	15.60	14.38
主根	47.10±0.13	33.67	15.86	13.12	32.23	15.18	12.52	33.21	15.64	12.59	28.34	13.35	12.31
侧根	46.39±0.11	9.44	4.38	3.62	9.03	4.19	3.46	9.29	4.31	3.47	7.93	3.68	3.39
合计		246.08	120.92	100.0	246.45	121.26	100.0	252.47	124.20	100.0	220.38	108.48	100.0

表3 不同间伐强度下林下植被及凋落物层生物量与碳贮量

组分		碳含量(%)	重度间伐		中度间伐		轻度间伐		对照	
			生物量(t/hm^2)	碳贮量(t/hm^2)	生物量(t/hm^2)	碳贮量(t/hm^2)	生物量(t/hm^2)	碳贮量(t/hm^2)	生物量(t/hm^2)	碳贮量(t/hm^2)
林下植被	地上部分	44.96±0.27	4.85±0.37	2.18±0.18	4.85±0.42	2.18±0.21	6.65±0.54	2.99±0.19	6.52±0.66	2.93±0.16
	地下部分	42.99±0.88	3.95±0.28	1.70±0.14	4.09±0.39	1.76±0.17	4.42±0.39	1.90±0.13	4.58±0.56	1.97±0.19
	小计		8.80±0.76	3.88±0.35	8.94±0.61	3.93±0.29	11.07±0.67	4.89±0.28	11.10±0.94	4.90±0.43
凋落物		49.03±0.29	8.36±0.42	4.10±0.20	9.65±0.59	4.73±0.28	7.22±0.65	3.54±0.30	9.93±0.97	4.87±0.47

而凋落物现存量与碳贮量的测定结果恰好与林下植被层相反，重度间伐和中度间伐下凋落物现存量与对照区接近，但轻度间伐下的凋落物现存量与碳贮量显著低于对照区($P<0.05$)(表3)。除轻度间伐外，其他处理间没有显著差异($P<0.05$)，表明间伐处理对凋落物现存量和碳贮量并无显著影响，轻度间伐导致凋落物现存量降低的原因还有待进一步研究。

3.3 不同间伐强度下马尾松人工林土壤碳贮量

土壤碳贮量测定结果表明(表4)，不同间伐强度间马尾松林地土壤碳贮量差异不显著($P>0.05$)。相比之下，以对照的土壤碳贮量最高，为(111.49±29.12) t/hm^2，中度间伐与对照无明显差异，为(109.45±18.86)t/hm^2，重度间伐和轻度间伐比较接近，分别为(106.86±20.67) t/hm^{2}和(105.33±26.19) t/hm^2。方差分析结果表明，不同间伐强度马尾松林地的相同土层碳贮量之间差异也不显著($P>0.05$)，可见间伐措施对马尾松林下土壤碳贮量并无显著影响，但有降低土壤碳贮量的趋势，重度、中度和轻度间伐下马尾松林土壤碳贮量分别低于对照4.15%、1.83%和5.53%。

表4 不同间伐强度下土壤碳贮量 (t/hm^2)

土层	重度间伐	中度间伐	轻度间伐	对照
腐质层	16.88± 4.46	16.15±6.00	17.28±7.86	18.23±5.11
0~10cm	38.65±4.19	37.76±5.41	36.65±5.60	37.85±9.86
10~30cm	29.85±6.31	32.01± 3.79	31.78±5.96	32.78±8.36
30~50cm	21.48±3.44	23.54±2.52	19.62±2.05	22.63±5.29
合计	106.86± 20.67	109.45±18.86	105.33±26.19	111.49±29.12

3.4 不同间伐强度下马尾松人工林生态系统总碳贮量

由表5可以看出，4种间伐强度的马尾松林生态系统碳贮量分别为235.76t/hm^2、239.37t/hm^2、237.97t/hm^{2}和229.74t/hm^2，方差分析结果显示各间伐强度下马尾松林生态系统碳贮量总量与对照之间没有显著差异($P>0.05$)，但重度、中度和轻度间伐下马尾松林生态系统碳贮量总量分别高出对照2.62%、4.19%和3.58%，表现出间伐处理有利于提高马尾松林生态系统碳贮量总量的态势。

重度、中度和轻度间伐的马尾松林生态系统均以乔木层碳贮量最大，分别占总贮量的51.29%、50.66%和52.20%；其次是土壤层，分别占45.33%、45.72%和44.26%；林下植被层和凋落物层碳贮量最小，林下植被层分别占生态系统碳贮量的1.65%、1.64%和2.05%，凋落物层分别占1.74%、1.98%和1.49%；对照以土壤碳贮量最大，占48.53%，其次是乔木层，占47.22%，林下植被层与凋落物层最小并接近，分别占2.13%和2.12%。表明间伐处理对马尾松林各组分碳贮量在生态系统碳贮量总量中的分配产生明显的影响。

不同间伐强度林下各组分碳贮量均有所差异，3种间伐强度下马尾松乔木层碳贮量均显著高于对照，重度、中度和轻度间伐分别高出对照11.47%、11.78%和14.50；土壤层碳贮量均低于对照，重度、中度和轻度间伐下林分土壤碳贮量分别低于对照4.15%、1.83%和5.53%，

但差异不显著；而林下植被层和凋落物层碳贮量低于对照，重度、中度和轻度间伐的马尾松林林下植被层碳贮量分别低于对照20.82%、19.80%和0.20%，凋落物层分别低于对照15.81%、2.87%和27.31%。除轻度间伐林下植被碳贮量与对照差异不显著外，其他差异均达到显著水平($P<0.05$)。由此可见，间伐处理有利于提高马尾松林乔木层碳贮量，而不利于林下地被物和土壤层碳的累积。

表5 不同间伐强度下马尾松人工林生态系统各组分碳贮量

组分	重度间伐		中度间伐		轻度间伐		对照	
	碳贮量(t/hm^2)	比例(%)	碳贮量(t/hm^2)	比例(%)	碳贮量(t/hm^2)	比例(%)	碳贮量(t/hm^2)	比例(%)
乔木层	120.92±14.38	51.29	121.26±7.21	50.66	124.21±4.98	52.20	108.48±10.48	47.22

（续）

组分	重度间伐		中度间伐		轻度间伐		对照	
	碳贮量（t/hm²）	比例（%）	碳贮量（t/hm²）	比例（%）	碳贮量（t/hm²）	比例（%）	碳贮量（t/hm²）	比例（%）
林下植被层	3. 88±0. 35	1. 65	3. 93±0. 29	1. 64	4. 89±0. 28	2. 05	4. 90±0. 43	2. 13
凋落物层	4. 10± 0. 20	1. 74	4. 73±0. 28	1. 98	3. 54±0. 30	1. 49	4. 87±0. 47	2. 12
土壤层	106. 86± 6. 13	45. 33	109. 45± 5. 91	45. 72	105. 33± 7. 15	44. 26	111. 49± 9. 54	48. 53
合计	235. 76±23. 71	100. 00	239. 37±26. 13	100. 00	237. 97±24. 12	100. 00	229. 74±19. 56	100. 00

4 结论与讨论

重度间伐、中度间伐、轻度间伐及对照条件下马尾松 25 年生时总碳贮量依次为 235. 762t/hm²、239. 37t/hm²、237. 97t/hm² 和 229. 74t/hm²，间伐处理有提高马尾松人工林生态系统碳贮量总量的趋势，但差异未达到显著水平。但需要指出的是：在对各处理碳贮量总量的测定和计量过程中，抚育间伐过程中被移除的碳贮量未计入碳贮量总量，因而，各间伐处理碳贮量实际值应高于本研究的测量值，但对照林碳贮量总量实际值与测量值一致。但若考虑抚育间伐中碳贮量的移除量，间伐措施对马尾松人工林碳贮量增加效应可能会更为显著。

间伐对马尾松人工林生态系统各组分碳贮量均有不同程度的影响。研究表明，间伐有利于提高乔木层碳贮量，重度间伐、中度间伐和轻度间伐马尾松林乔木层碳贮量分别高出对照区 11. 47%、11. 78% 和 14. 49%，乔木层碳贮量的升高，主要是间伐措施有利于马尾松林木的生长，提高了林分的生长量和生物量，与对照相比，重度间伐、中度间伐和轻度间伐马尾松林乔木层生物量分别高出对照 11. 62%、11. 83%和 14. 56%，各间伐处理间，以轻度间伐最有利于乔木层生物量和碳贮量的提高。间伐处理有利于提高林分生产力，这一结论在马尾松、桉树（*Eucalyptusgrandis×E. urophylla*）、红椎（*Castanopsis hystrix*）等多个树种上均有不同程度的体现，采用适当的间伐处理或密度调控措施，以提高林木的生长量，进而提高林分乔木层生物量和碳贮量，是人工林增汇技术的一种有效途径[23-26]。

研究结果显示，不同间伐强度对马尾松林各器官碳贮量在乔木层的分配没有显著影响，重度、中度和轻度间伐处理和对照处理乔木层均以树干碳贮量所占比例最大，分别为 53. 38%、59. 71%、59. 57%和 60. 12%；其次是根系，再次是树皮，然后是树枝，树叶所占比例最小。相比之下，重度间伐下林分乔木层碳贮量中，树干所占比例较小，主要是重度间伐下的林木光照充足，侧枝发达，增加了碳在枝叶上的累积，从而降低了主干在乔木生物量中的分配。

不同的间伐处理对马尾松林下土壤碳贮量并无显著影响，相比之下，各间伐处理下的土壤碳贮量低于对照，表明间伐措施有降低土壤碳贮量的可能，因而间伐虽然提高了林木生长量，但对林分生态系统碳贮量并无显著影响。间伐过程中，大量地移除木材及采伐剩余物，降低了枯落物在林地的蓄积，从而影响到土壤碳的输入，很可能是间伐降低土壤碳贮量的主要原因，但目前并不明确，需做进一步研究。

各处理马尾松林下植被碳贮量和凋落物碳贮量分别为 3. 88～4. 90t/hm² 和 4. 10～4. 87t/hm²，在生态系统碳贮量中所占的比例很小，仅为 1. 65%～2. 13% 和 1. 74%～2. 12%，但林下植被和凋落物对土壤碳贮量的积累和碳素在生态系统中的循环发挥着重要作用。研究结果表明，间伐对林下植被碳贮量和凋落物碳贮量的影响较为复杂，无明显规律，重度间伐、中度间伐林下植被层生物量和碳贮量均低于对照区，而轻度间伐与对照区接近，这可能与营林中间伐活动对林下植被带来的干扰有关。而间伐对凋落物碳贮量的影响除轻度间伐外，其他与对照无显著差异。轻度间伐导致凋落物现存量的降低，原因还有待进一步研究。

参考文献

[1]潘辉，张金文，林顺德，等．不同间伐强度对巨尾桉林分生产力的影响研究[J]．林业科学，2003，39(增刊1)：106-111.

[2]段劼，马履一，贾黎明，等．抚育间伐对侧柏人工林及林下植被生长的影响[J]．生态学报，2010，30(6)：1431-1441.

[3]谌红辉，丁贵杰．马尾松间伐的密度效应[J]．林业科学，2010，46(5)：84-91.

[4] KAMMESHEIDT L. Effect of selective logging on tree

species diversity in a seasonally wet tropical forest in Venezuela[J]. Forestarchiv, 1996, 67(1): 14-24.

[5] NIESE J N, Strong T F. Economic and tree diversity tradeoffs in managed northern hardwoods[J]. Canadian Journal of Forest Research, 1992, 22(11): 1807-1813.

[6] 马履一，王希群，甘敬，等．北京市森林经营的基本原则刍议[J]．北京林业大学学报，2006，28(4)：159-163.

[7] 毛志宏，朱教君，留足根，等．间伐对落叶松人工林内草本植物多样性及其组成的影响[J]．生态学志，2006，25(10)：1201-1207.

[8] PAUL A M. Managing for forest health[J]. Journal of Forestry, 2002, 100(7): 22-27.

[9] ZARMORCH S J, BOCHLOLD W A, STOLTE K W. Using crown condition variables as indicators of frorest health[J]. Canadian Journal of Forest Reseach, 2004, 34(5): 1057-1070.

[10] 雷向东，陆元昌，张会儒，等．抚育间伐对落叶松云冷杉混交林的影响[J]．林业科学，2005，41(4)：78-85.

[11] 曹云，杨劼，宋炳煜，等．人工抚育措施对油松生长及结构特征的影响[J]．应用生态学报，2005，16(3)：397-402.

[12] 杜纪山，唐守正．抚育间伐对林分生长的效应及模型研究[J]．北京林业大学学报，1996，18(1)：79-83.

[13] 李春明，杜纪山，张会儒．抚育间伐对森林生长的影响及其模型研究[J]．林业科学研究，2003，16(5)：636-641.

[14] JESSICA K A, DEBORAH L M, GEORGE W T. Changes in understory vegetation and soil characteristics following silvicultural activities in a southeastern mixed pine forest[J]. Journal of the Torrey Botanical Society, 2007, 134(4): 489-504.

[15] DONALD A, LIGUORI, DENISE A, et al. Plant diversity in managed forests: understory responses to thinning and fertilization [J]. Ecological Applications, 1999 (9): 864-879.

[16] 任立忠，罗菊春，李新彬．抚育采伐对山杨次生林植物多样性影响的研究[J]．北京林业大学学报，2000，22(4)：14-17.

[17] 徐扬，刘勇，李国雷，等．间伐强度对油松中龄林人工林林下植物多样性的影响[J]．南京林业大学学报(自然科学版)，2008，32(3)：135-138.

[18] 李春义，马履一，王希群，等．抚育间伐对北京山区侧柏人工林林下植物多样性的短期影响[J]．北京林业大学学报，2007，29(3)：60-66.

[19] 李春义，马履一，徐昕．抚育间伐对森林生物多样性研究进展[J]．世界林业研究，2006，19(6)：27-23.

[20] EDITH G, JEAN C H, GERARD N. The influence of site quality, silviculture and region on wood density mixed model in *Quercus petraea* Liebl[J]. Forest Ecology and Management, 2004, 189: 111-121.

[21] 徐有明，林汉，魏柏松，等．间伐强度对湿地松人工林木材质量的影响效应[J]．东北林业大学学报，2002，30(2)：38-42.

[22] 熊有强，盛伟彤，曾满生．不同间伐强度杉木林下植被发育及生物量研究[J]．林业科学研究，1995，8(4)：408-412.

[23] 谌红辉，方升佐，丁贵杰，等．马尾松间伐的密度效应[J]．林业科学，2010，46(5)：84-91.

[24] 李江才．杉木红椎混交林间伐试验研究[J]．江西农业大学学报，2003，25(10)：119-123.

[25] 张金文．巨尾按大径材间伐试验研究[J]．林业科学研究，2008，21(4)：464-468.

[26] 张水松，陈长发，吴克选，等．杉木林间伐强度试验年生长效应的研究[J]．林业科学，2005，41(5)：56-65.

[原载：林业科学，2013，49(10)]

南亚热带不同树种人工林碳素密度

郑　路[1,2]　蔡道雄[1,2]　卢立华[1,2]　明安刚[1,2]　李朝英[1]　李丽利[1]

([1]中国林业科学研究院热带林业实验中心，广西凭祥　532600；
[2]广西友谊关森林生态系统定位观测研究站，广西凭祥　532600)

摘　要　采用干烧法对南亚热带6个树种人工林不同器官及生物组分进行碳素密度测定。结果表明：树木不同器官之间碳素密度不同，所测6个树种中以树叶和树干最高，其次为树枝，树根和干皮最低；树种之间以马尾松碳素密度最高，米老排、红椎等阔叶树种较低。林下灌木不同器官间碳素密度差异显著，其规律与乔木类似，以叶和枝碳素密度最高，根最低；林分间随着林分密度增加，林下灌木碳素密度呈下降趋势。林下草本碳素密度的分布规律是地上部分大于地下部分；不同林分间，碳素密度变化规律与灌木相似。由未分解、半分解到完全分解，地表凋落物碳素密度快速下降。林分内不同生物组分之间，碳素密度从高到低依次为乔木>灌木>草本>地表凋落物。

关键词　南亚热带；人工林；碳素密度；生物组分

Carbon Density of Different Species Plantationin the Sub-tropical Area of China

ZHENG Lu[1,2], CAI Daoxiong[1,2], LU Lihua[1,2],
MING Angang[1,2], LI Zhaoying[1], LI Lili[1]

([1]*Experimental Center of Tropical Forestry*, *Chinese Academy of Forestry*, *Pingxiang* 532600, *Guangxi*, *China*;
[2]*Guangxi Youyiguan Forest Ecosystem Research Station*, *Pingxiang* 532600, *Guangxi*, *China*)

Abstract: Carbon density in trees, shrubs, herbs, and litters of six species plantation in the sub-tropical area of China was measured using the combustion method. Our results showed that the carbon density varied in different parts of trees, with the highest density in leaves and trunk, followed by branches, and the lowest density in roots and bark. The highest density was observed in species *Pinus massoniana*, and the lower density was detected in species *Mytilaria laosensis* and *Castanopsis hystrix*. The carbon density in different parts of understory shrubs was also measured and significant differences was found among the parts: high carbon density was found in leaves and branches, and with the lowest density in roots. Our results also suggested that the carbon density of understory shrubs decreased with increasing stand density. A higher carbon density was found in the aboveground portion of understory herbs than that of the belowground. The carbon density in the floor litter was rapidly declined from the litter layer, fermentation layer to humus layer. Our results also indicated that in plantation forests, the carbon density was sequenced by trees>shrubs>herbs>floor litters.

Key words: South sub-tropical area; plantation; carbon density; stand components.

森林生态系统是地球上除海洋之外最大的碳库[1]，同时，在调节陆地生态系统与大气碳库之间的碳交换中也起着巨大的“生物泵”作用[2,3]。由于它的巨大碳库及碳交换的活跃性，森林生态系统在维护全球碳平衡中起着举足轻重的作用[4]。为了正确评估森林在全球碳平衡中的作用，了解森林生态系统碳循环过程，森林的碳动态研究正日益成为人们关注的重点[5,6]。

目前，对森林碳贮量及其与大气 CO_2的交换通量的估计，主要有样地清查法、涡度相关法及应用遥感技术的模型模拟法3种[7]，其中以样地清查法应用最为普遍，即通过设立典型样地，直接或间接测定森林中的植被、枯落物等的生物量再乘以生物量中碳元素的密度推算而得[8,9]。因此，森林群落的生物量及其组成成分的碳素密度是研究森林碳贮量与碳通量的两个关键因子，对他们的准确测定或估计是估算森林群落或森林生态系统以及区域和国家森林碳贮量及通量的基础。当前，我国对不同区域及不同森林群落类型生物量与生产力的研究已有大量报道[10,11]，而对森林群落不同组分碳素密度的研究报道并不多见[12]，难以满足我国丰富的森林群落类型精确估算碳动态的要求。本研究以我国南亚热带6种树种人工林为研究对象，实测了乔木、灌木、草本和地表凋落物不同器官及组分的碳素密度，分析了乔、灌、草及地表凋落物不同组分碳素密度特征，为降低森林碳汇估算中的不确定性，为不同尺度估算森林碳储量提供合理的参数和数据分析资料。

1 研究地区与研究方法

1.1 研究区概况

研究区位于中国林业科学研究院热带林业实验中心夏石那造大山（21°57′47″N，106°59′30″E），海拔350m，年均温21.5℃，≥10℃积温7500℃，年降水量1220~1380mm，年蒸发量1370~1390mm，干湿季节明显，雨季(4~9月)降水占年总降水量的85%左右，旱季(10月至翌年3月)降水仅占年降水量的15%左右，土壤为花岗岩母质发育的赤红壤。于1984年2月，在同一坡面营造了米老排林、云南石梓林、火力楠林、红椎林(各约1hm^2)，在对坡营造铁力木林(6.6hm^2)，山脊营造马尾松林(8.8hm^2)。定植时米老排、石梓、马尾松的苗龄为1年，其他为2年，造林密度为1.67m×1.67m。栽植前3年人工铲草抚育，郁闭成林后自然生长。林分概况见表1。

1.2 研究方法

2011年10月，选择米老排、云南石梓、火力楠、红椎、铁力木和马尾松6种人工纯林，每树种随机选择平均标准木3株，分别树叶、枝、主干、干皮和根取样(树叶和枝于树冠外围南面中下部取样，干皮选择树干南面距地面1.2m处取样，主干用生长锥取树干距地面1.2m处样，挖取树木南面侧根取样)。每树种人工林下灌木、草本和地表凋落物分别设置样方取样，各3个重复。灌木按叶、枝、根分器官取样。草本分别地上部分(茎叶)、地下部分(根)取样。地表凋落物按未分解、半分解和完全分解分层取样。所有样品带回实验室65℃烘干后粉碎备用。

植物样品碳素密度测定主要有湿烧法和干烧法两种，其中干烧法误差不超过0.3%，分析精度远高于湿烧法[13]。本研究所有样品碳素密度测定均采用干烧法，具体使用日本岛津总有机碳分析仪TOC-L/SSM5000进行样品分析。

1.3 数据处理

采用单因素方差分析(one-way ANOVA)检测器官间及生物组分间的差异性。使用LSD法进行多重比较。

数据分析及相关计算用Microsoft Excel 2007和SPSS 19.0软件完成。

表1 林分基本情况表

森林类型	平均树高(m)	平均胸径(cm)	密度(株/hm^2)	郁闭度	林下优势灌草种类
米老排林	18.7	20.2	1492	0.9	米老排、大沙叶、海金沙、扇叶铁线蕨、鞭叶铁线蕨
红椎林	19.1	25.4	525	0.8	红椎、三叉苦、扇叶铁线蕨、半边旗
火力楠林	19.4	18.1	1167	0.8	红椎、酸藤子、火力楠、九节、鞭叶铁线蕨、山姜、粽叶芦、半边旗
云南石梓林	15.7	18.8	675	0.6	玉叶金花、杜茎山、山芝麻、铁芒萁、蔓生莠竹、五节芒
铁力木林	14.0	16.4	842	0.9	铁力木、三叉苦、琴叶榕、铁芒萁、扇叶铁线蕨、蕨
马尾松林	16.6	19.1	1142	0.7	大沙叶、红椎、桃金娘、铁芒萁、五节芒、弓果黍

注：表中植物拉丁学名：大沙叶 *Pavetta hongkongensis*；三叉苦 *Evodia lepta*；山芝麻 *Helictercs angustifolia*；杜茎山 *Maesa japonica*；琴叶榕 Ficus lyrata；桃金娘 Rhodomyrtus tomentosa；九节 *Psychotriarubra*；酸藤子 *Embelia laeta*；海金沙 *Lygodium japonicum*；玉叶金花 *Mussaenda pubescens*；半边旗 *Pteris semipinnata*；山姜 *Alpina japonica*；粽叶芦 *Thysanolaena maxima*；五节芒 *Miscanthus floridulu*；弓果黍 *Cyrtococcum patens*；蔓生莠竹 *Microstegium vagans*；扇叶铁线蕨 *Adiantum flabellulatum*；鞭叶铁线蕨 *Adiantum caudatum*；蕨 *Pteridium aquilinum*；铁芒萁 *Dicranopteris dichotoma*。

2 结果与分析

2.1 各树种的器官碳素密度特征

树木叶、枝、干、皮和根等器官碳素密度测定结果(表2)表明，不同器官其碳素密度不同，并且差异显著，如火力楠主干与干皮的碳素密度相差为73.2g/kg，铁力木树叶与干皮碳素密度相差为72.1g/kg。虽然不同树种间，其器官碳素密度排列顺序略有不同，但总体呈现规律是以树叶和主干的碳素密度较高，六个树种平均为494.3g/kg和491.4g/kg，其次是树枝，平均为470.2g/kg，树根和干皮的碳素密度较低，平均为460.9g/kg和458.1g/kg。

不同树种间碳素密度亦有差异，以针叶树种马尾松最高，平均为495.0g/kg，其他5种阔叶树之间差异较小，在461.5~478.6g/kg(表2)。

2.2 灌木碳素密度

由表3看出，灌木不同器官间碳素密度差异显著，其规律与乔木类似，以叶和枝碳素密度最高，根最低，其相差在12.6~76.0g/kg。不同林分间林下灌木碳素密度也有显著差异，其中以云南石梓林下灌木碳素密度最高，平均为478.7g/kg，其次是马尾松林，平均为466.2g/kg，米老排林最低，平均为433.0g/kg。

表2 不同树种器官碳素密度 (g/kg)

树种	树叶	树枝	主干	干皮	树根
米老排	500.5±4.64a	477.5±3.29bc	484.20±2.88b	434.3±2.99d	473.9±8.77c
红椎	473.8±4.76b	476.4±4.22b	495.1±5.36a	469.8±7.20b	452.1±9.30c
火力楠	490.2±6.32a	464.2±8.15b	501.2±5.75a	428.1±4.95c	452.6±10.91b
云南石梓	457.0±3.95b	448.0±4.56c	484.3±4.08a	457.8±5.39b	460.3±2.35b
铁力木	524.5±1.87a	472.0±2.51c	487.7±2.53b	452.4±3.03d	456.3±0.15d
马尾松	519.8±0.10a	483.0±5.30d	495.9±4.62c	506.4±3.40b	470.0±5.82e

注：数据为平均值±标准差。同行不同字母表示器官间差异显著($P<0.05$)。下表同。

表3 不同林分类型灌木碳素密度 (g/kg)

林分类型	叶	枝	灌根	平均
米老排林	462.2±11.58a	450.4±8.05a	386.3±27.30b	433.0±3.40c
红椎林	470.3±11.32a	478.4±6.79a	441.8±4.79b	463.5±2.02b
火力楠林	455.1±18.93a	458.2±16.10a	423.2±11.23b	445.5±11.17c
云南石梓林	487.8±8.04a	480.5±2.91a	467.8±20.08a	478.7±11.67a
铁力木林	479.0±20.76a	459.0±17.43a	445.6±15.15a	461.2±4.19b
马尾松林	477.8±13.12a	473.0±14.78ab	447.8±9.63b	466.2±11.76b

2.3 草本碳素密度

图1表明，林下草本植物地上部分和地下部分碳素密度不同，甚至差异显著，不同林分均表现为地上部分碳素密度高于地下部分，其相差在14.9~77.3g/kg。不同林分间比较，林下草本植物碳素密度也有差异，最高的为马尾松林，草本植物地上地下平均碳素密度为449.2g/kg，最低的为铁力木林，地上地下平均碳素密度为354.5g/kg。

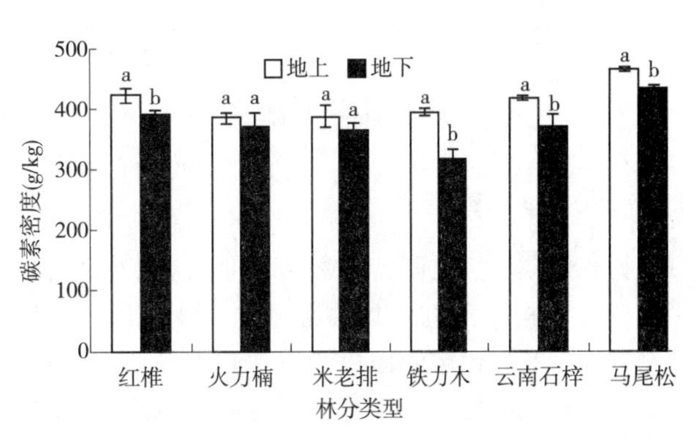

图1 林下草本碳素密度

2.4 地表凋落物碳素密度

图2看出，地表凋落物不同的分解层，碳素密度变化很大，且差异显著，从未分解层、半分解层到完全分解层，随着分解的进行，碳素密度大幅度下降，由未分解层的447.2~496.2g/kg到完全分解层的68.5~279.2g/kg。不同林分类型间地表凋落物碳素密度亦有差异，其中以马尾松林最高，3层平均为408.7g/kg，云南石梓林最低，3层平均为307.1g/kg。

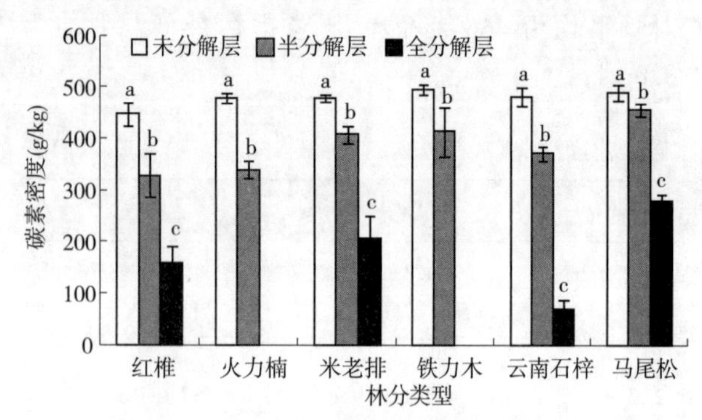

图2 地表凋落物碳素密度

2.5 林分不同组分碳素密度比较

图3表明，林分内乔木、灌木、草本和地表凋落物4种不同组成成分，碳素密度不同，其变化规律是乔木>灌木>草本>地表凋落物，且相互之间均差异显著。其中乔木碳素密度最高，平均为475.6g/kg，地表凋落物最低，平均为346.3g/kg。

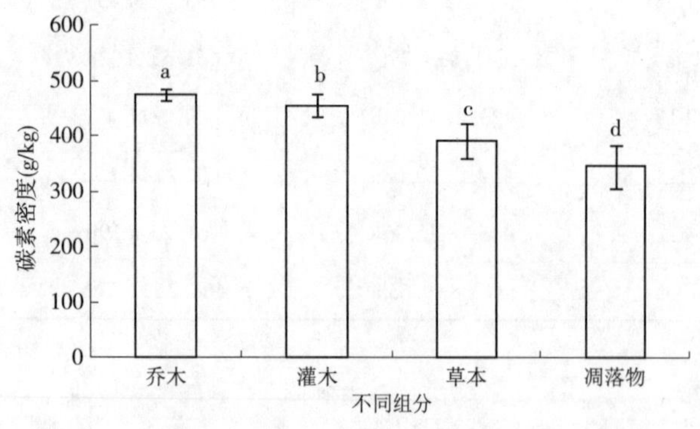

图3 林分不同组分碳素密度

3 讨论与结论

树木不同器官之间碳素密度不同，所测6个树种中以树叶和树干最高，其次为树枝，树根和干皮最低。这与生长在温带的杨树[14]、亚热带的杉木(*Cunninghamia lanceolata*)、湿地松(*Pinus elliotti*)以及热带的橡胶(*Hevea brasiliensis*)等树种具有相似的器官碳素分配规律[15-17]。可能因为树叶是树木的主要同化器官，通过光合作用吸收CO_2，并转化为稳定的碳水化合物[18]，树干中含有较高的纤维素和木质素[19]，因此这两个器官碳素密度较高；而根系通过呼吸作用为植物生长发育提供能量，同时通过消耗地上光合作用产生的碳素营养来带动整个植株的代谢[20,21]，故碳素密度相对较低。

树种之间碳素密度比较，以马尾松等针叶树种最高，米老排、红椎等阔叶树种较低。这与我国北方及南方其他地区树种之间碳素分布规律一致[22,23]，即针叶树种碳素密度高于阔叶树种，马尾松相对其他阔叶树种高出幅度在3.3%~6.8%。国际上常用平均含碳率转换系数为0.45或0.5，本研究6个树种平均碳素密度为450.0~500.0g/kg，如以0.45作为转换系数，无论对针叶树或阔叶树结果都明显偏小，如以0.5作为转换系数，除马尾松比较准确外，其他阔叶树都将明显偏大。因此更准确的估算应该是分树种采用不同的含碳率转换系数[24-26]。

不同林分间，林下灌木碳素密度也有差异，其中以云南石梓林最高，米老排林最低。此外亦可发现云南石梓林下灌木叶、枝碳素密度显著高于根，而米老排林下灌木叶、枝和根等器官间碳素密度差异不显著。由表1可知，云南石梓林自然稀疏强烈，郁闭度较低(0.6)，林下光照较强，自然生长以喜光灌木为主；而米老排林密度大，郁闭度较高(0.9)，林下光照弱，仅稀疏生长耐阴灌木。由此说明，乔木层树种不同，林下生境不同，适合生长的灌木植物种类也不同，使得林下灌木碳素密度产生差异，并且随着林分密度的增加，林下灌木碳素密度降低，灌木地上地下器官间碳素密度差异逐步减小。

林下草本植物碳素密度在不同林分间，因生境不同，生长草本种类不同，使得碳素密度有差异，其规律与灌木类似，即随着林分密度增加，林下草本碳素密度降低，其地上部分碳素密度大于地下部分。

地表凋落物，由未分解层、半分解层到完全分解层，碳素密度快速下降，在不同层次间，碳素密度变幅在6.9%~81.5%。因此，建议在计算地表凋落物碳贮量时，最好能分层取样，分层计算，以提高估算精度，尤其是地表凋落物现存量较大的林分[27]。另外，在我国南方，由于温度高，湿度大，凋落物分解较北方[28]，尤其是一些阔叶树种[29]，所以在火力楠和铁力木林下，未能取到完全分解层的凋落物样。

林分不同生物组分之间，以乔木碳素密度最高，其次为灌木，之后是草本，以地表凋落物最低。这一规律与大多数研究结果相同[29,30]，但也有研究报道地表凋落物碳素密度高于灌木和草本[30]，这很可能是取样时仅取了未分解层，而忽略了半分解层和完全分解层，使得研究结果产生偏差。另外，6 个树种林分不同组分碳素密度变异系数比较发现，其规律为乔木(2.49%)<灌木(4.11%)<草本(8.35%)<地表凋落物(11.13%)，说明乔木树种之间碳素密度变化幅度相对较小；灌木和草本由于受林下环境条件的影响，林分间碳素密度变化相对较大；地表凋落物除树种不同使碳素密度有差异外，不同树种凋落物分解速率的快慢差异使地表凋落物未分解层、半分解层和完全分解层各层所占比例不同[31]，造成了林分间地表凋落物碳素密度变幅较大。

参考文献

[1] ANDEREGGA WRL, BERRY JA, SMITH DD, et al. The roles of hydraulic and carbon stress in a widespread climate-induced forest die-off [J]. Proceeding of the National Academy of Sciences of the United States of America, 2012, 109: 233-237.

[2]方精云，陈安平．中国森林植被碳库的动态变化及其意义[J]．植物学报，2001，43(9)：967-973.

[3] WOOD T E, CAVALERI M A, REED S C. Tropical forest carbon balance in a warmer world: a critical review spanning microbial to ecosystem scale processes[J]. Biological Reviews, 2012, 87: 912-927.

[4] GOETZ S J, BOND-LAMBERTY B, LAW B E, et al. Observations and assessment of forest carbon dynamics following disturbance in North America[J]. Journal of Geophysical Research, 2012, 117: 1-17.

[5] SWIFT K. Forest carbon and management options in an uncertain climate[J]. Journal of Ecosystems and Management, 2012, 13: 1-7.

[6]王娟，陈云明，曹扬，等．子午岭辽东栎林不同组分碳含量与碳储量[J]．生态学杂志，2012，31(12)：3058-3063.

[7]曹吉鑫，田赟，王小平，等．森林碳汇的估算方法及其发展趋势[J]．生态环境学报，2009，18(5)：2001-2005.

[8]方精云，郭兆迪，朴世龙，等．1981—2000 年中国陆地植被碳汇的估算[J]．中国科学 D 辑：地球科学，2007，37(6)：804-812.

[9]陈青青，徐伟强，李胜功，等．中国南方 4 种林型乔木层地上生物量及其碳汇潜力[J]．科学通报，2012，57(13)：1119-1125.

[10]杨明，汪思龙，张伟东，等．杉木人工林生物量与养分积累动态．应用生态学报，2010，21(7)：1674-1680.

[11]张远东，刘彦春，刘世荣，等．基于年轮分析的不同恢复途径下森林乔木层生物量和蓄积量的动态变化[J]．植物生态学报，2012，36(2)：117-125.

[12]田大伦，方晰，项文化．湖南会同杉木人工林生态系统碳素密度[J]．生态学报，2004，24(11)：2382-2386.

[13]马钦彦，陈遐林，王娟，等．华北主要森林类型建群种的含碳率分析[J]．北京林业大学学报，2002，24(5)：96-100.

[14]李春平，吴斌，张宇清，等．山东郓城农田防护林杨树器官含碳率分析[J]．北京林业大学学报，2010，32(2)：74-78.

[15]肖复明，范少辉，汪思龙，等．毛竹(*Phyllostachy pubescens*)、杉木(*Cunninghamia lanceolata*)人工林生态系统碳贮量及其分配特征[J]．生态学报，2007，27(7)：2794-2801.

[16]涂洁，刘琪璟．亚热带红壤丘陵区湿地松人工林生态系统碳素贮量与分布研究[J]．江西农业大学学报，2007，29(1)：47-54.

[17]王春燕，陈秋波，彭懿，等．老龄橡胶树不同器官含碳率分析[J]．热带作物学报，2011，32(4)：587-591

[18]潘瑞炽．植物生理学[M]．北京：高等教育出版社，2008.

[19]万劲．杨树无性系及重阳木木质能源性状的研究[D]．南京：南京林业大学，2008.

[20]蒋高明．植物生理生态学[M]．北京：高等教育出版社，2004.

[21]李志霞，秦嗣军，吕德国，等．植物根系呼吸代谢及影响根系呼吸的环境因子研究进展[J]．植物生理学报，2011，47(10)：957-966.

[22]胡青，汪思龙，陈龙池，等．湖北省主要森林类型生态系统生物量与碳密度比较[J]．生态学杂志，2012，31(7)：1626-1632.

[23]郑路，卢立华．我国森林地表凋落物现存量及养分特征[J]．西北林学院学报，2012，27(1)：63-69.

[24] HOUGHTON R A, SKOLE D L, NOBRE C A, et al. Annual fluxes of carbon from deforestation and regrowth in the Brazilian Amazon[J]. Nature, 2000, 403: 301-304.

[25]周玉荣，于振良，赵士洞．我国主要森林生态系统碳贮量和碳平衡[J]．植物生态学报，2002，24(5)：518-522.

[26]张骏，袁位高，葛滢，等．浙江省生态公益林碳储量和固碳现状及潜力[J]．生态学报，2010，30(14)：3839-3848.

[27]郭剑芬，杨玉盛，陈光水，等．森林凋落物分解研究进展[J]．林业科学，2006，42(04)：93-100.

[28]张浩，庄雪影．华南 4 种乡土阔叶树种枯落叶分解能

力[J]. 生态学报, 2008, 28(5): 2395-2403.
[29]张会儒, 赵有贤, 王学力, 等. 应用线性联立方程组方法建立相容性生物量模型研究[J]. 林业资源管理, 1999, (6): 63-67.
[30]郎飞, 叶功富, 黄义雄, 等. 武夷山甜槠天然林含碳率与碳贮量研究[J]. 亚热带资源与环境学报, 2011, 7(4): 71-77.
[31]路翔, 项文化, 任辉, 等. 中亚热带四种森林凋落物及碳氮贮量比较[J]. 生态学杂志, 2012, 31(9): 2234-2240.

[原载: 生态学杂志, 2013, 32(10)]

灰木莲生长对土壤养分和气候因子的响应

卢立华　何日明　农瑞红　李忠国

（中国林业科学研究院热带林业实验中心，广西凭祥　532600）

摘　要　对原产地越南以及我国多个引种地的灰木莲的树高和胸径生长特征进行研究，分析了灰木莲生长对土壤养分和气候因子的响应。结果表明：不同种植地灰木莲的树高和胸径年均生长量差异显著。灰木莲胸径、树高的年均生长量与土壤全N、全P、速效N、速效P含量呈显著正相关，而与土壤有机质、全K、速效K含量相关性不显著，表明土壤N、P含量是影响灰木莲生长的主要养分因子。不同海拔灰木莲树高年均生长量差异显著，胸径则无显著差异。在海拔150~550m，树高年生长量随海拔升高而增大，在海拔550m处达到最大，随后降低，表明海拔550m为灰木莲引种的最适海拔。灰木莲树高、胸径年均生长量与年均温、≥10℃积温呈显著负相关关，与年降水量呈显著正相关，表明年均温、≥10℃积温和降水量是影响其生长的主要气候因子。

关键词　灰木莲；树高；胸径；土壤养分；气候因子

Responses of *Manglietia glauca* Growth to Soil Nutrients and Climatic Factors

LU Lihua, HE Riming, NONG Ruihong, LI Zhongguo

(*Experimental Center of Tropical Forestry, Chinese Academy of Forestry, Pingxiang* 532600, *Guangxi, China*)

Abstract: Tree height and diameter of breast height(DBH) as growth characteristics of *Manglietia glauca* introduced from Vietnam were measured at many sites in south China and responses of *M. glauca* growth to soil nutrients and climatic factors were analyzed in this study. Annual average increments of tree height and DBH among different planted sites had significant differences. Annual average increments of tree height and DBH had significant positive correlation with soil total N and P, available N and P, but no significant correlation with soil organic matter, total K, available K, indicating that soil N and P contents could be the main affecting factors for the growth of *M. glauca*. Annual average increment of tree height had significant difference, but annual average increment of DBH had no significant difference at different altitudes. Annual average increment of tree height increased with the altitude from 150 to 550m, the maximum was at the altitude of 550m, and then it decreased. It indicated that the most appropriate altitude for *M. glauca* introduction is 550m. Annual average increments of tree height and DBH had significant negative correlation with annual average temperature and ≥10℃ accumulated temperature, and significant positive correlation with annual average precipitation, suggesting that annual mean temperature, ≥10℃ accumulated temperature and annual average precipitation could be the main climatic factors influencing the growth of *M. glauca*.

Key words: *Manglietia glauca*; tree height; diameter of breast height(DBH); soil nutrient; climatic factor

灰木莲(*Manglietia glauca*)属木兰科常绿阔叶大乔木，原产于越南、印度尼西亚爪哇等地，为热带速生阔叶用材树种，它树形优美、四季常绿、干形通直、花大洁白有芳香，观赏价值较高，故也是优良园林绿化树种。1960年从越南引种到广西南宁等地[1]，2000年前中国仅有零星引种，研究也少见报道。但进入21世纪后，它在中国南部的发展快速，广西、广东、福建、云南等地已初具规模。研究也较为活跃，已开展种子储藏[2]、育苗[3]、引种[4-6]、抑菌、杀菌[7]及滞尘能力[8]、生长规律[9]、木材特性[10-11]、大苗移植[12]及坡位对生长影响[13]等研究。但在土壤养分及气候因子对林木生长影响方面，其他树种已有较多研究[14-18]，针对灰木莲的仍未见报道，而它对指导灰木莲发展及适地适树布局非常重要。因此，开展了灰木莲生长对土壤养分及气候因子响应研究，以期为灰木莲在我国的科学经营与发展提供依据。

1　研究区概况

研究区包括灰木莲原产地越南老街省宝安县新阳乡和尚河乡(简称：新阳、尚河)及中国主要引种地中国林业科学研究院热带林业实验中心白云实验场和哨平实验场(白云、哨平)、广西国营高峰林场(高峰)、广东肇庆北岭山林场(北岭)、广东四会市大南山(大南山)、广东南海西樵镇西岸林场(西岸)。各研究区土壤母岩均为花岗岩，且位于同一气候带，属于亚热带季风气候区，各气候因子值较相近(表1)。

表1　灰木莲研究地区主要气候因子

地点	地理位置	年均温(℃)	≥10℃积温(℃)	极端低温(℃)	年降水量(mm)
尚河、新阳	22°05′~22°30′N; 104°11′~104°38′E	22.5	7000~8000	1	1500~2000
白云、哨平	21°57′~22°19′N; 106°39′~106°59′E	21.6	7000~7600	-1	1200~1500
高峰	22°13′~23°32′N; 107°45′~108°51′E	21.6	6800~7500	-2.1	1200~1750
西岸	22°48′~23°19′N; 112°49′~113°15′E	22.2	6800~7300	-1.9	1500~2000
北岭	23°03~23°19′N; 112°23′~112°41′E	21.9	6700~7500	-1	1600~1700
大南山	23°17′~23°23′N; 112°37′~112°48′E	22.3	6800~7800	0	1500~1600

1.2　研究方法

1.2.1　试验设计

采用典型样地调查方法，选择8~10年生灰木莲林分为研究对象。试验林种源来自越南尚河，苗木为1.5年生营养苗，造林密度2m×2m，1~3年生幼林每年抚育两次，其他经营管理措施基本一致。2010年9~10月，在上述地点选择海拔300~350m，坡向均为东南坡，坡面长约100m的中下坡铺设样地。

在中国林业科学研究院热带林业实验中心伏波实验场，选择坡面、坡向、坡位一致，但海拔不同的中下坡铺设样地。表2中气象数据为热林中心在不同海拔设立的气象站点提供，为1983—1992年连续10年观测数据的平均值。每个样地内设置20m×30m样方3个，在样方内进行每木检尺，测定树高、胸径。然后计算平均树高及胸径，将平均生长量除以林龄得到年均生长量。

1.2.2　土壤样品分析

在样地内按对角线随机挖3个土壤剖面，按2个层次(0~25cm、25~60cm)分别取土样，然后按层次等量混合均匀，采用四分法取适量土壤进行分析样。采用重铬酸钾外加热法测定有机质含量；凯氏法测定全N含量；扩散法测定碱解N含量；$HClO_4$~H_2SO_4法测定全P含量；钼锑抗比色法测定速效P含量；火焰光度计法测定全K和速效K含量；酸度计法测定pH(表2和表3)。

表2　不同海拔地的气候因子及土壤养分含量

海拔(m)	年均温(℃)	≥10℃积温(℃)	年降水量(mm)	土层厚度(cm)	有机质(g/kg)	氮(g/kg)	磷(g/kg)	钾(g/kg)	速效氮(mg/kg)	速效磷(mg/kg)	速效钾(mg/kg)
150	22.1	7600	1220	0~25	20.23±2.33	1.12±0.20	0.18±0.02	8.92±0.03	91.61±12.13	0.86±0.12	71.77±5.38
				25~60	10.25±1.24	0.74±0.12	0.19±0.01	8.92±0.08	38.29±8.45	0.35±0.09	45.07±2.89

（续）

海拔（m）	年均温（℃）	≥10℃积温（℃）	年降水量（mm）	土层厚度（cm）	有机质（g/kg）	氮（g/kg）	磷（g/kg）	钾（g/kg）	速效氮（mg/kg）	速效磷（mg/kg）	速效钾（mg/kg）
250	21.5	7200	1260	0~25	18.82±2.45	1.11±0.21	0.23±0.04	8.67±0.6	118.65±13.12	1.16±0.11	32.48±2.59
				25~60	10.32±1.76	0.76±0.20	0.21±0.03	8.62±0.07	103.96±12.03	1.26±0.17	32.26±3.37
350	21.0	6900	1300	0~25	21.28±1.98	1.19±0.33	0.23±0.09	9.74±0.06	121.16±10.38	1.60±0.98	85.91±4.65
				25~60	17.64±3.86	1.02±0.24	0.21±0.08	9.02±0.04	79.10±10.10	0.55±0.32	68.69±4.03
450	20.4	6700	1350	0~25	21.99±5.83	1.21±0.21	0.24±0.08	9.07±0.07	121.73±11.21	1.16±0.43	65.25±3.69
				25~60	11.17±2.01	1.00±0.18	0.23±0.09	9.49±0.08	77.99±9.67	0.31±0.06	51.47±3.95
550	19.8	6400	1370	0~25	22.35±2.98	1.26±0.16	0.25±0.06	10.84±0.09	131.43±14.99	1.83±0.07	87.06±4.67
				25~60	9.49±1.32	0.99±0.10	0.20±0.04	8.72±0.07	77.60±8.98	0.24±0.07	41.14±2.93
650	19.3	6100	1400	0~25	25.10±4.25	1.39±0.14	0.30±0.07	11.53±0.08	158.40±24.46	1.95±0.09	88.21±6.98
				25~60	16.38±3.56	1.27±0.18	0.24±0.05	8.78±0.08	91.93±10.75	0.52±0.07	69.84±5.39

表 3 灰木莲不同种植区土壤养分含量

地点	土层厚度（cm）	pH	有机质（g/kg）	氮（g/kg）	磷（g/kg）	钾（g/kg）	速效氮（mg/kg）	速效磷（mg/kg）	速效钾（mg/kg）
北岭	0~25	4.0±0.19	29.6±2.1	1.23±0.17	0.20±0.04	19.2±1.97	122.8±20.23	0.6±0.12	82.11±9.66
	25~60	4.1±0.20	9.1±1.5	0.52±0.06	0.19±0.03	18.4±1.34	46.07±12.11	0.3±0.07	45.59±9.82
大南山	0~25	4.1±0.16	34.4±1.7	1.06±0.10	0.26±0.06	21.2±2.00	105.9±16.98	0.5±0.11	49.37±8.87
	25~60	4.0±0.16	12.1±2.9	0.61±0.07	0.22±0.05	22.9±1.86	46.07±11.43	0.3±0.08	34.26±7.84
高峰	0~25	4.2±0.16	28.8±3.6	0.94±0.08	0.23±0.03	15.6±1.19	118.2±24.35	1.0±0.68	30.48±6.82
	25~60	4.2±0.14	18.3±2.8	0.77±0.06	0.22±0.05	14.8±2.04	105.9±14.69	1.2±0.31	34.26±7.39
西岸	0~25	4.2±0.17	20.2±3.9	1.11±0.91	0.18±0.04	8.92±1.21	90.61±11.36	0.6±0.05	70.77±12.43
	25~60	4.1±0.18	10.2±1.9	0.73±0.09	0.20±0.06	8.97±0.97	38.39±11.02	0.3±0.06	43.07±6.92
尚河	0~25	5.0±0.20	39.2±3.8	2.16±1.23	0.44±0.07	12.9±1.32	183.8±26.99	2.7±0.28	104.2±13.19
	25~60	4.6±0.14	12.4±2.0	1.21±0.17	0.33±0.07	18.1±1.22	76.37±22.20	0.3±0.09	60.66±11.11
新阳	0~25	4.6±0.21	23.5±2.1	1.26±1.03	0.27±0.08	7.36±1.02	118.3±15.63	1.7±0.47	84.77±10.41
	25~60	4.6±0.24	11.9±2.3	0.99±0.07	0.22±0.09	8.14±1.13	71.66±14.26	0.5±0.10	67.16±16.65
白云	0~25	4.1±0.12	31.2±2.8	2.01±1.16	0.37±0.08	19.7±1.35	154.1±34.91	1.6±0.27	85.91±9.04
	25~60	4.2±0.13	17.5±3.4	1.02±0.91	0.34±0.08	22.0±2.76	79.20±13.88	0.4±0.08	68.69±8.79
哨平	0~25	4.4±0.18	26.10±3.6	0.95±0.07	0.19±0.04	4.06±1.45	103.2±19.45	1.1±0.09	97.39±9.98
	25~60	4.2±0.20	20.57±2.9	0.73±0.71	0.22±0.06	4.05±1.64	63.64±17.83	0.8±0.18	64.10±10.22

1.3 数据处理

采用 PASW 18.0 软件对数据进行统计分析，采用 Tukey 法进行多重比较和差异性检验（$\alpha=0.05$）。图表中数据为平均值±标准差。

2 结果与分析

2.1 不同种植地区灰木莲胸径和树高的年均生长量

由图 1 可以看出，不同种植地区灰木莲胸径年均生长量达到显著差异，其中，尚河的灰木莲胸径年均生长量最高，达到 1.88cm，显著高于其他地区；其次是白云、新阳、北岭，分别为尚河的 94.2%、79.8%、72.3%；而高峰、大南山、哨平与北岭差异不显著；西岸胸径年均生长量最低，仅 1.25cm，为尚河的 66.5%。

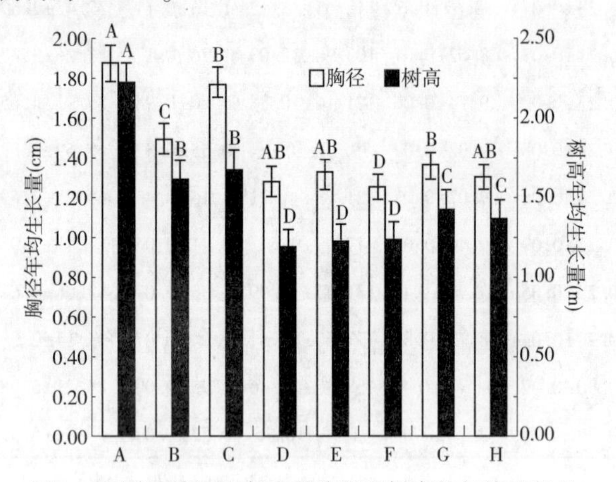

图 1　不同种植地区灰木莲胸径和树高的年均生长量

不同大写字母表示不同种植区差异显著($P<0.05$)

A：尚河；B：新阳；C：白云；D：哨平；E：高峰；F：西岸；G：北岭；H：大南山

不同种植地区灰木莲树高年均生长量达到显著差异，其中，尚河的灰木莲树高年均生长量最高，达到 2.23m，显著高于其他地区；其次是白云、新阳，分别为尚河的 75.3%、72.2%；再次是北岭和大南山，分别为尚河的 64.1%、61.0%；西岸、高峰和哨平的树高年均生长量最低，分别为尚河的 55.6%、55.2%、52.9%。

2.2　不同海拔灰木莲胸径和树高的年均生长量

由图 2 可以看出，不同海拔灰木莲胸径年均生长量未达到显著差异，表明海拔为 150~650m 时，灰木莲胸径生长受海拔影响不大。海拔为 150~550m 时，灰木莲胸径年均生长量随海拔升高而增加，依次为 1.81cm、1.82cm、1.82cm、1.83cm、1.84cm，而超过 550m 后，胸径年均生长量下降，海拔为 650m 时达到最低，为 1.77cm。

不同海拔灰木莲树高年均生长量达到显著差异，表明在海拔为 150~650m 时，灰木莲树高生长受海拔影响显著。海拔为 150~550m 时，灰木莲树高年均生长量随海拔升高而增加，依次为 1.46m、1.47m、1.55m、1.56m、1.74m，超过 550m 后，树高年均生长量下降，海拔为 650m 时达到最低，为 1.42m。

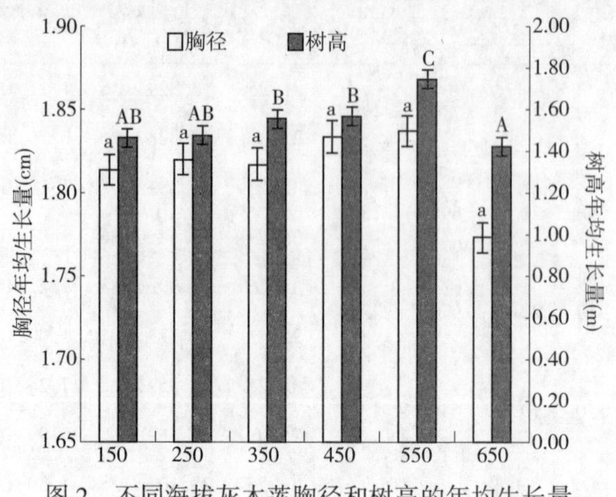

图 2　不同海拔灰木莲胸径和树高的年均生长量

不同大写字母表示不同海拔差异显著($P<0.05$)

2.3　不同种植地灰木莲胸径和树高生长与生态因子的关系

由表 4 可以看出，在不同种植地，灰木莲胸径和树高年均生长量与土壤全 N、全 P、速效 N、速效 P 呈显著正相关，与有机质、全 K 和速效 K 的相关性不显著，与主要气候因子的相关性也不显著。表明导致不同种植地灰木莲生长差异的主要因素为土壤养分因子，尤其是土壤的 N、P 含量。这与灰木莲的喜肥特点相吻合，也与陈琳等[14]对灰木莲苗期叶面施肥的研究结果一致。

2.4　不同海拔灰木莲胸径和树高生长与生态因子的关系

由表 5 可以看出，不同海拔灰木莲的胸径和树高年均生长量与土壤全 N、海拔、年均温≥10℃积温、年降水量均呈显著正相关，而与土壤有机质、速效 N、全 P、速效 P、全 K、速效 K 的相关性不显著，这表明因海拔不同导致的气候因子差异是影响灰木莲生长的主要因素，除全 N 外的土壤养分对灰木莲的生长影响不显著，其原因可能是这些土壤养分已能满足灰木莲的生长需求。

表 4　不同种植地灰木莲胸径和树高年均生长量与土壤养分和气候因子的相关系数

	有机质 (g/kg)	氮 (g/kg)	磷 (g/kg)	钾 (g/kg)	速效氮 (mg/kg)	速效磷 (mg/kg)	速效钾 (mg/kg)	年均温 (℃)	≥10℃积温 (℃)	年降水量 (mm)
Ⅰ	0.609	0.971**	0.960**	0.190	0.956*	0.877**	0.531	0.219	0.617	-0.043
Ⅱ	0.660	0.908**	0.918**	0.138	0.921**	0.871**	0.563	0.516	0.498	0.276

注：Ⅰ为胸径年均生长量；Ⅱ为树高年均生长量．* $P<0.05$；** $P<0.01$。下同。

表 5 不同海拔灰木莲胸径和树高年均生长量与土壤养分和气候因子的相关系数

	海拔 (m)	有机质 (g/kg)	氮 (g/kg)	磷 (g/kg)	钾 (g/kg)		速效磷 (mg/kg)	速效钾 (mg/kg)	年均温 (℃)	≥10℃积温 (℃)	年降水量 (mm)
Ⅰ	0.971**	0.714	0.884*	0.876	-0.382	0.867	0.744	0.303	-0.978**	-0.960**	0.954*
Ⅱ	0.914*	0.675	0.938*	0.701	-0.584	0.735	0.857	0.610	-0.917*	-0.896*	0.868

3 讨论

在植物生长所必需的大量元素中，氮和磷是许多森林生态系统生产力的最主要限制性养分因子[15]。氮是叶绿素的主要成分，它可促进叶绿素的合成，并在一定条件下提高植物的产量和品质[16]。吴楚等[17]研究表明，在氮胁迫下水曲柳(*Fraxinus mandshurica*)幼苗的净同化速率会下降，从而导致总生物量下降。磷是植物生长发育的必需元素，直接参与光合作用的同化和光合磷酸化，缺磷严重影响光合产物从叶片中输出，并在碳水化合物的代谢过程中控制碳水化合物的代谢[18]。孙华[19]研究表明，缺磷可降低菜豆光合能力、蒸腾速率和气孔导度，而施磷能使叶面积和生物量显著增加。冷华妮等[20]研究发现，缺磷时，枫香(*Liquidambar formosana*)细胞分裂和伸长迟缓，叶片生长缓慢，叶片呈现暗绿色。可见，氮、磷在植物生长中发挥着重要作用。土壤作为植物生长的物质基础，其质量的差异会导致植物呈现不同的生长状态[21]。灰木莲为速生阔叶树种，生长快，喜肥[1]，对土壤养分敏感[13]。本研究表明，在气候条件相近的环境下，土壤氮、磷含量是影响灰木莲树高和胸径生长的主要养分因子。这与赵雪梅等[22]对三倍体毛白杨(*Populus tomentosa*)无性系人工林的研究结果一致，与周志凯等[21]对杂交马褂木(*Liriodendron chinense*×*L. tulipifera*)的研究结果类似。

植物生长受气候因子的影响，而树木生长对气候因子的响应规律十分复杂[23-26]。Ettl 和 Peterson[27]研究发现，幼年高山冷杉(*Abies lasiocarpa*)受气候变化的影响较小；Carrer 和 Urbinati[28]对落叶松(*Larix decidua*)和五针松(*Pinus cembra*)的研究表明，气候和生长的关系受树木年龄的影响。本研究中，灰木莲的生长受年均温和≥10℃积温及降水量等气候因子的影响显著。植物生长虽需要水和热，但并非多多益善。温度过高或过低对植物的光合作用均不利，低温可引起植物体内活性氧代谢失调，导致生物膜结构和叶绿体结构的破坏，从而使得光合速率和光合作用下降[29-32]。高温可使叶片气孔导度降低，限制了通过气孔进入叶片组织的CO_2，从而影响叶片净光合速率[33]。同样，植物长期受干旱胁迫也会造成伤害，导致光合作用减弱，净光合速率降低[34,35]，甚至出现萎蔫、死亡。

温度和降水是影响树木生长较敏感的气候因子[36]，海拔变化，水热条件随之改变。郑征等[37]研究表明，海拔每升高 100m 气温下降 0.52℃，年降水量显著增加；Grubb 和 Whitmore[38]研究证实，海拔上升降水量增加。而温度和降水常常是相互制衡、共同作用的[39]，高温和低降水联合作用引起的干旱，会增强水分胁迫，使树木光合作用降低，影响树木生长；而低温和高降水量的联合作用对树木光合作用不利，同样会影响树木生长。本研究区具有高温多雨的气候特点，但雨量分布不均，林木生长常受季节性干旱的胁迫。隋月等[40]研究表明，在全球变暖背景下，近 50 年来华南地区夏旱、秋旱和冬旱发生的可能性增加，说明华南地区的季节性干旱不是偶然的。因此，在干旱时节，生长于较低海拔的林木受到较高温度、较少降雨的共同影响，干旱胁迫严重，对林木生长的影响较大，故生长量较低；随着海拔的上升，温度逐渐下降，降水量逐渐增加，林木受干旱胁迫的影响随之下降，对生长的影响减小，达到最适海拔时，温、湿度处于最佳状态，对生长也最有利，故生长量最高；但超过最适海拔后，则受低温和高降水量的影响，气温过低对热带树种灰木莲的光合作用不利，影响其生长。气温和降水量在不同海拔范围内对林木生长影响的差异，导致了在海拔 150~550m 范围灰木莲生长量随海拔上升而增加，而超过 550m 海拔后，则随海拔上升而下降。这与不同海拔苦楝生长研究的结论类似[41]。

参考文献

[1] WANG K J, CAI Z L. Cultivation technology of tropical tree species[M]. Nanning: Guangxi Science and Technology Press, 2008.

[2] LU L H, MENG C L, HE R M, et al. Effect of different storage treatments on storage period and germination rate of *Manglietia glauca* Seeds[J]. Seed, 2011, 30(10): 82-85.

[3]CAI D X, LI M, GUO W F, et al. Experiment on Media Selection for Container Seedling Cultivation of *Mangleitia glauca*[J]. Journal of Zhejiang Forestry Science and Technology, 2006, 26(5): 36-38.

[4]ZHANG Y H, DAI R K, OU S K. The Early Growth Performance of 6 Broadleaf Tree Specie[J]. Journal of Guangdong Forestry Science and Technology, 2004, 20(1): 43-46.

[5]ZENG J, LU L H, JIA H Y. Biological Characteristics and Introduction Cultivation of *Manglietia glanca*[J]. Practical Forestry Technology, 2010, 10: 20-21.

[6]YANG Y H, LIU M Y, CHANG S Y. Experiment on Introduction and Cultivation of *Manglietia glanca*[J]. Journal of Southwest Forestry College, 2007, 27(3): 29-32.

[7]HU X C, WU X Y, WEN H X, et al. Effects on Bacteria Control of Urban Ornamental Trees and Application in Pearl River Delta Region[J]. Journal of Chinese Urban Forestry, 2005, 6(3): 46-49.

[8]HU X C, YIN A H, WU X Y, et al. The Selecting of Superior Dust Detention Tree Species in Modern Urban Forestry of Pearl River Delta[J]. Guangdong Landscape Architecture, 2007, 3: 44-46.

[9]WEI S H, QIN J, ZHU X L, et al. Growth Regularity of *Manglietia glauca* Plantation Growing in Nanning[J]. Journal of Northwest Forestry University, 2011, 26(5): 174-178.

[10]LI J Z, LI X B, TANG T, et al. Drying characteristics of *Manglietia glauca*[J]. China Wood Industry, 2011, 25(3): 44-46.

[11]WEI S H, TANG T, FU Y L, et al. Bark percentage, Heartwood percentage and density for *Manglietia glauca*[J]. Journal of Northwest Forestry University, 2011, 26(3): 152-155.

[12]JIA H Y, NONG R H. The big tree transplanting technology of *Manglietia glanca*[J]. Guangxi Forestry Science, 2006, 35(1): 34-35.

[13]LU L H, HE R M, NONG R H, et al. Effect of slope position on the growth of *Manglietia glance*[J]. Forest Research, 2012, 25(6): 789-794.

[14]ZHAO Z J, TAN L Y, KANG D W, et al. Responses of Picea likiangensis radial growth to climate change in the Small Zhongdian area of Yunnan Province, Southwest China[J]. Chinese Journal of Applied Ecology, 2012, 23(3): 603-609.

[15]ZHANG L Y, DENG X W, LEI X D, et al. Pinus massoniana productivity at different age stages in relation to climatic factors[J]. Chinese Journal of Ecology, 2013, 32(5): 1104-1110.

[16]KANG W X, TIAN D L, ZHAO Z H, et al. Effect of hydrological process on the productivity of *Cunninghamia lanceolata*(Lamb.) Hook. Plantation[J]. Science of Soil and Water Conservation, 2008, 6(4): 71-76.

[17]HU Z D, HAO Y E. On relationship between soil nutrient content and growth of *Alstonia scholaris*[J]. Journal of Southwest Forestry College, 2007, 27(4): 7-12.

[18]LIU F D. The effect of soil nutrient around broad-leaved trees on *Phyllostachys* growth in the mixed forest[J]. Chinese Agricultural Science Bulletin, 2010, 26(9): 127-131.

[19]CHEN L, LU L H, MENG C L, et al. Effect of foliar fertilization on growth of *Manglietia glance* seedlings[J]. Seed, 2013, 32(6): 79-81.

[20]ZHANG L J, LIANG Z S. Plant physiology[J]. Beijing: Science press, 2007.

[21]XU J C, LIN Z M, LUO W, et al. The advance of inorganic effect on photosynthesis[J]. Anhui Agricultural Science Bulletin, 2007, 13(7): 23-25.

[22]SUN H. Influence of soil qualities on the photosynthesis, physiological and ecological functions of vegetable[J]. Chinese Journal of Eco—Agriculture, 2005, 13(1): 116-118.

[23]ZHOU Z K, REN X Q, PAN G Q. Effects of soil nutrition contents on growth of *Liriodendron chinense* Sarg. × *L. tulipifera* L. [J]. Journal of Central South University of Forestry and Technology, 2010, 30(12): 42-46.

[24]CAIN M L, SUBLER S, EVANS J P, et al. Sampling spatial & temporal variation in soil nitrogen availability[J]. Oecologia, 1999, 118: 397-404.

[25]FARLEY R A, FITTER A H. a. Temporal and spatial variation in soil resources in a deciduous woodland[J]. Journal of Ecology, 1999, 87: 688-696.

[26]GROSS K L, PREGITZER K S, BURTON A J. Spatial variation in nitrogen availablity in three successional plant communities[J]. Journal of Ecology, 1995, 83: 357-368.

[27]JACKSON R B, CALDWELL M M. The scale of nutrient heterogeneity around individual plants & its quantification with geostatistics[J]. Ecology, 1993, 74: 612-614.

[28]SCHLESINGER W H, RAIKES J A, HARTLEY A E, et al. On the spatial pattern of soil nutrients in desert ecosystems[J]. Ecology, 1996, 77: 364-374.

[29]WANG Q C, CHENG Y H. Response of fine roots to soil nutrient spatial heterogeneity[J]. Chinese Journal of Applied Ecology, 2004, 15(6): 1063-1068.

[30]ZHAO X M, WANG H Y, SUN X Y, et al. Annual change of soil nutrients in triploid *Populus tomentosa* Plantation and Its Correlation with Tree Growth[J]. Forest Research, 2008, 21(3): 419-423.

[31]DAI Y H, LIU X Y, MENG Q W, et al. Effect of low temperature on lipid metabolism of thylakoid membrane [J]. Chinese Bulletin of Botany, 2004, 21(4): 506-511.

[32]DUAN W, LI X G, MENG Q W, et al. Photoinhibiton mechanisms of plant under low temperature[J]. Acta Botanica Boreali-Occidentalia Sinica, 2003, 23(6): 1017-1023.

[33]HE J, LIU H X, WANG Y R, et al. Low temperature and Photosynthesis of plants[J]. Plant Physiology Communications, 1986(2): 1-6.

[34]江福英, 李延, 翁伯琦. Review on physiology of chilling stress and chilling resistance of plants[J]. Fujian Journal of Agricultural Sciences, 2002, 17(3): 190-195.

[35]YANG W P, WU R Y, DAI Y. Effect of high temperature on photosynthesis of two poplar varieties[J]. Journal of Zhejiang Forestry Science and Technology, 2009, 29(1): 31-35.

[36] CHARTZOULAKIS K, NOITSAKIS B, THERIOS I. Photosynthesis, plant growth and carbon allocation in Kiwi cv. Hayward, as influenced by water deficits[J]. Acta Horticulturae, 1993, 335, 227-234.

[37]SHAO X M, FAN J M. Past climate on west Sichuan plateau as reconstructed from ring-widths of Dragon spruce [J]. Quaternary Sciences, 1999(1): 81-89.

[38]ZHENG Z, LI Y R, ZHANG S B, et al. Influence of the altitudinal increase on water and humidity conditions, Xishuangbanna[J]. Journal of Mountain Science, 2007, 25(1): 33-38.

[39]GRUBB P J, WHITMORE T C. Acomparison of montane and low landrain forest in Ecuador. II. The climate and its effects on the distribution and physiognomy of the forest [J]. Journal of Ecology, 1966, 54: 303-333.

[40] GRAUMLICH L J. Subalpine tree growth climate and increasing carbon dioxide an assessment of recent growth trends[J]. Ecology, 1991, 72: 1-11.

[41]LIN S S, LI X L, XIAO F D, et al. The analysis of growth effect of Meliaazedarach at different altitudes in Shounigcounty[J]. Journal of Fujian Forestry Science and Technology, 2011, 38(4): 41-45.

[原载: 应用生态学报, 2014, 25(04)]

红椎经营模式对林木生长及乔木层碳储量的影响

卢立华　贾宏炎　农　友　黄德卫　明安刚　郑　路

（中国林业科学研究院热带林业实验中心，广西凭祥　532600）

摘　要　对11年生红椎纯林、红椎×西南桦及红椎×马尾松的生长、碳素密度、生物量、碳储量及其分配进行研究。结果表明：不同经营模式红椎的胸径、树高生长量及乔木层生物量、碳储量都达极显差异($P<0.01$)。红椎胸径、树高生长量及乔木层生物量、碳储量均以红椎×西南桦最高，分别达17.87cm、16.27m、127.52t/hm²、57.84t/hm²；17.87cm、16.27m、127.52t/hm²、57.84t/hm²；其次红椎纯林，分别为13.93cm、12.78m、108.67t/hm²、49.13t/hm²；红椎×马尾松最低，分别为11.57cm、12.03m、70.42t/hm²、32.22t/hm²。红椎、西南桦、马尾松的器官碳素密度范围分别为417.20~465.37g/kg、405.93~509.90g/kg、470.70~524.67g/kg；相同树种不同器官及不同树种相同器官的碳素密度都达到极显著差异($P<0.01$)。经营模式对乔木层生物量、碳储量在器官的分配有明显影响，模式间乔木层生物量、碳储量在器官的分配量达到极显著差异($P<0.01$)。它们在各器官的分配比例，红椎纯林从大到小为：干(63.39%、65.25%)、根(20.46%、18.87%)、枝(7.76%、7.76%)、皮(7.22%、7.02%)、叶(1.17%、1.10%)；红椎×马尾松从大到小为：干(63.43%、64.49%)、根(20.73%、19.31%)、皮(7.38%、7.60%)、枝(6.58%、6.58%)、叶(1.88%、2.02%)。红椎×西南桦从大到小为：干(67.53%、68.97%)、根(11.90%、11.01%)、枝(10.10%、9.99%)、皮(8.19%、7.64%)、叶(2.28%、2.39%)。

关键词　经营模式；生长量；碳储量；生物量；碳素密度

Effects of Stand Management Patterns of *Castanopsis hystrix* on Tree Growth and Carbon Storage Capacity

LU Lihua, JIA Hongyan, NONG You, HUANG Dewei, MING Angang, ZHENG Lu

(*The Experimental Center of Tropical Forestry, Chinese Academy of Forestry, Pingxiang 532600, Guangxi, China*)

Abstract: We studied the growth increment, carbon content, biomass, carbon storage and their space distribution pattern of pure *Castanopsis hystrix*, mixture of *C. hystrix* and *Betula alnoides*, and mixture of *C. hystrix* and *Pinus massoniana* of 11-year-old. The *C. hystrix* growth of DBH, tree height, total biomass and total carbon storage of the tree layer were significant different in three different stand management patterns ($P < 0.01$), in which the mixed stand of *C. hystrix* and *B. alnoides* had the highest increment of DBH and tree height total biomass and total carbon storage increments among them 17.87cm, 16.27m, 127.52t/hm² and 57.84t/hm², followed by the pure *C. hystrix* 13.93cm, 12.78m, 108.67t/hm² and 49.13t/hm², and the lowest were mixed stand of *C. hystrix* and *P. massoniana* only 11.57cm, 12.03m, 70.42t/hm² and 32.22t/hm². The carbon contents of *C. hystrix*, *B. alnoides* and *P. massoniana* ranged from 417.20 to 465.37g/kg, 405.93 to 509.90g/kg, and 470.70 to 524.67g/kg. The carbon content were significant differences between different organs within the same tree species or different tree species in the same organs($P<0.01$). The biomass and the carbon storage in different organs of the tree layer of the management patterns were significantly different($P<0.01$). The distribution proportion of biomass and carbon storage in different organs of the tree layer were not the same, three management patterns, were as follows: stem (63.39%, 65.25%) > root (20.46%, 18.87%) > branch (7.76%, 7.76%) > bark (7.22%, 7.02%) > leaf (1.17%, 1.10%) in the pure *C. hystrix* stand, stem(63.43%, 64.49%) > root(20.73%, 19.31%) > bark (7.38%, 7.60%) > branch (6.58%, 6.58%) > leaf (1.88%, 2.02%) in the mixed stand of *C. hystrix* and

P. massoniana, stem (67.53%, 68.97%) > root (11.90%, 11.01%) > branch (10.10%, 9.99%) > bark (8.19%, 7.64%) >leaf(2.28%, 2.39%) in the mixed stand of *C. hystrix* and *B. alnoides*.

Key words: management pattern; growth increment; carbon storage; biomass; carbon content

红椎(*Castanopsis hystrix*)为壳斗科栲属(*Castanopsis*)常绿乔木，具有生长快、材质优、适应性强、用途广等优良特性，为中国华南地区重要乡土珍贵阔叶用材树种和高效多用途树种[1]；它的木材价值及林分综合效益都较高[2]，适于培育大径材[3,4]。其枝叶浓密，较耐荫蔽，混生性能良好，宜营造混交林。已有研究表明，将它与杉木(*Cunninghamia lanceolata*)、马尾松(*Pinus massoniana*)等针叶树种混交效果良好[5-8]，能实现长短结合、以短养长及可持续经营的目标。故其人工林发展迅速，研究也较活跃，已开展育苗[9]、优树选择[10,11]、良种区域化试验[12]、繁育与造林技术[13,14]、木材材性[15]、生物量及其分配格局[16]、混交效果与碳储量等研究[17-19]。有学者对南亚热带地区的杉木、秃杉(*Taiwania flousiana*)、马占相思(*Acacia mangium*)、米老排(*Mytilaria laosensis*)等的生态系统碳储量进行了研究[20-30]。而红椎经营模式对红椎生长，乔木层生物量、碳储量及其分配的综合研究未见报道。笔者对11年生红椎纯林、红椎×西南桦及红椎×马尾松3种经营模式红椎的生长量及乔木层生物量、碳储量与分配格局以及不同树种器官碳素密度进行研究，为合理配置人工林经营模式及对人工林进行科学的经营管理，提高人工林的生态、经济效益提供依据。

1 研究区概况

试验地位于广西凭祥市东南部的中国林业科学研究院热带林业实验中心伏波实验场，地理位置为21°57′47″~22°19′27″N，106°39′50″~106°59′30″E。该地属南亚热带季风型半湿润—湿润气候，干湿季节明显(10月份至翌年3月份为干季，4~9月份为湿季)，太阳年总辐射439.614kJ/cm²，年日照时间1320h，年均温20.5℃，极端低温-1℃，≥10℃积温6600℃，年降水量1400mm，相对湿度83%。地貌为低山，海拔550m，坡度为25°，坡位：中下坡，坡向：东南坡。土壤为花岗岩发育的山地红壤，土层厚>5.0m，土壤养分见表1。

表1 试验地土壤养分质量分数

土层厚度(cm)	pH	有机质(g/kg)	氮(g/kg)	磷(g/kg)	钾(g/kg)	速效氮(μg)	速效磷(μg)	速效钾(μg)
0<h≤20	4.4	47.5	1.39	0.58	2.60	154.46	1.08	24.7
20<h≤60	4.5	30.2	0.96	0.46	1.86	110.32	0.86	18.3

2 材料与方法

2.1 试验设置及林分概况

在同一面坡上铺设红椎纯林、红椎×马尾松、红椎×西南桦(*Betula alnoides*)试验林，3次重复，共9个小区，随机排列，小区面积约0.3hm²，试验区总面积3hm²。为便于叙述，用模式中各树种第一个拉丁字母为代号，依次为C、CM、CB，造林密度2m×3m，穴状整地，穴规格：60cm(长)×60cm(宽)×35cm(深)，每穴施基肥250g[m(桐麸)：m(磷肥)=1：1混合沤制]。CM、CB的数量混交比例为1：2，株间混交。2002年3月用营养苗造林，造林成活率90%，造林当年7~8月份进行第1次抚育，第2年和第3年，每年的3~4月份和7~8月份各抚育1次，2010年卫生清理一次，清除枯枝、死木和严重被压木。林分现状：林分郁闭度、平均植株数量、平均胸径和树高林分C分别为0.9、1500株/hm²、13.93cm、12.78m，林分CM分别为0.85、1430株/hm²(马尾松枯死木5%左右)、9.86cm、10.16m；林分CB分别为0.85、1480株/hm²、16.31cm、16.12m。因林分郁闭度较高，故林下植被稀少，仅偶见凤尾蕨(*Pteris multifida*)、铁线蕨(*Adiantum capillusveneris*)等较耐阴植物。

2.2 研究方法

2013年5月，在各小区中设置1个20m×30m调查样地，测定幼林树高、胸径，计算平均树高和胸径。并在3种树种中各选3株平均木(即3次重复)，然后伐倒并挖根，分别取其树干、树皮、树枝、树叶、树根样，在80℃恒温下烘干至恒质量，计算含水率，然后，把器官鲜质量换算成干质量。

2.2.1 生物量

西南桦、马尾松、红椎单木器官生物量分别用何友均等[18]、明安刚等[30]、覃林等[31]生物量估测模型计算，以上建模数据均来自与本研究同地点的伏波实验场，所选建模样木胸径范围 7.2~28.0cm，涵盖了本研究红椎、西南桦、马尾松的径级范围。根据生物量模型计算器官生物量、单株平均生物量及乔木层各器官生物量、乔木层总生物量。

2.2.2 碳储量

树木器官碳素密度用有机碳测定仪(日本岛津，型号：TOC-L-ssm5000A)测定。器官碳素密度乘以对应树种、器官的生物量，得到各器官的碳储量，各器官的碳储量相加得到乔木层总碳储量。

2.3 数据分析

采用 Excel 2007 及 SPSS 17.0 进行数据处理，单因素方差分析法进行显著性比较。

3 结果与分析

3.1 经营模式对红椎生长的影响

从表 2 可见，模式间红椎的胸径、树高生长量达到极显著差异($P<0.01$)。其中以 CB 红椎胸径、树高生长量最高，分别达 17.87cm、16.27m，与 C、CM 比较都达到极显著差异；其次是 C，椎胸径、树高生长量分别为 13.93cm、12.78m，达 CB 的 77.95%、78.55%；最差为 CM，椎胸径、树高生长量分别为 11.57cm、12.03m，达 CB 的 64.75%、73.94%。这表明，红椎与西南桦混交能促进红椎生长，而红椎与马尾松混交却对红椎生长不利。这是因为，红椎、西南桦和马尾松虽都为阳性、深根树种，但红椎较耐阴，尤其在幼林期，适度遮阴对其生长有利。另外，它们根系的差异较大，根系生物量占其总生物量的比例依次为 17.16%、17.72%[31]、9.3%[18]，而总生物量差异不大，故红椎和马尾松的根系比西南桦发达。而林木根系分布与发育状况关系到能否充分利用土壤中的营养物质和混交林的成功与否[32]。红椎与西南桦混交林，因西南桦根系欠发达，它们根系相互穿插交织的概率小，对地下营养空间的争夺不剧烈，故能充分利用土壤中的营养物质。此外，在地上空间利用上，在林分郁闭初期(4~8 年生)，西南桦生长快于红椎，为红椎提供了遮阴条件，红椎处于下层，使西南桦获得了充足光照，故它们此时的种间关系表现为互利关系，能促进彼此的生长。但随着林龄的增加(8 年生后)，红椎喜光程度逐渐增强，混交林也处于高度郁闭状态，红椎开始与西南桦竞争空间资源，此时的种间关系变成了竞争关系。而西南桦对光照比红椎敏感，在高度郁闭的林分中，西南桦因光照的不足而生长明显受抑，为红椎获得生长空间提供条件，使红椎生长逐渐加快。11 年生红椎的树高已超过西南桦 0.3m，如果不及时进行人工干预(间伐)，将因种群内剧烈营养空间的竞争而致西南桦出现自然稀疏，并最终被淘汰。红椎则在整个过程处于有利地位，故其生长良好，明显优于纯林。而红椎与马尾松混交林，因它们根系都较发达，林分郁闭后，根系处于同一营养空间的概率大，对地下营养物质的争夺较为激烈。同时，上层空间的竞争也同样激烈，因红椎的生长快于强阳树种马尾松，红椎的遮挡严重影响马尾松生长，马尾松为改变被动必须与红椎进行空间资源的激烈竞争，但其树种特点决定了它在竞争中的必然劣势。对 11 年生林分的调查，马尾松已呈现严重被压状态，林木个体参差不齐，且有 5%左右的枯死率，这与火力楠和马尾松混交的结果较一致[33]。由于林分从地上到地下都存在激烈竞争，故对彼此的生长不利，致使红椎的生长不如其纯林。

表 2 不同经营模式林木生长量

经营模式	树种	胸径(cm)	树高(m)
红椎纯林(C)	红 椎	13.93Cc	12.78Bb
红椎×马尾松混交林(CM)	红 椎	11.57Bb	12.03Bb
	马尾松	8.15	8.28
红椎×西南桦混交林(CB)	红 椎	17.65Aa	16.27Aa
	西南桦	14.97	15.97

注：同列不中同大、小写字母分别表示处理间生长量差异极显著($P<0.01$)或差异显著($P<0.05$)。

3.2 树种对器官碳素密度的影响

从表 3 知，马尾松、红椎、西南桦的器官碳素密度变动范围依次为：470.70~524.67g/kg、417.20~465.37g/kg、405.93~509.90g/kg。方差分析结果显示，同树种不同器官或同器官不同树种的碳素密度都达到极显著差异($P<0.01$)。同树种不同器官碳素密度，红椎以树干最高，与树枝、树叶、树皮、树根比较均达极显著差异；西南桦和马尾松则以树叶最高，与树干、树枝、树皮、树根比较达极显著差异。不同树种相同器官的碳素密度都以马尾松为最高，与红椎、西南桦对应器官比较都达到极显著差异；而红椎与西南桦比较，仅树叶、树皮达到极显著差异，树根达到显著差异，其他器官的差异不显著。同树种不同器官碳素密度从大到小排序，马尾松：叶、皮、枝、干、根；红椎：干、枝、叶、皮、根；西南桦：叶、干、根、枝、皮。树叶、树干这两个器

官的碳素密度较高，可能与树叶为树木的主要同化器官，通过光合作用吸收 CO_2，并转化为稳定的碳水化合物[34]，树干中含有较高的纤维素和木质素[35]有关。马尾松碳素密度高于红椎、西南桦，这与针叶树种碳素密度高于阔叶树种的结论一致[36,37]。

3.3 经营模式对乔木层生物量、碳储量及分配特征的影响

从表4可知，模式间乔木层的生物量、碳储量及它们在各器官的分配量存在极显著差异($P<0.01$)。总生物量、总碳储量及它们在干、皮、枝、叶的分配量均以 CB 最高，分别达 127.52t/hm²、86.11t/hm²、10.44t/hm²、12.88t/hm²、2.91t/hm² 和 57.84t/hm²、39.89t/hm²、4.42t/hm²、5.78t/hm²、1.38t/hm²，与 C、CM 比较都达到了极显著差异。由于 BC 能明显提高林木生长量及林分的产量，故其总生物量、总碳储量及它们在干、皮、枝、叶中的分配量也会随之提高。可见，科学配置人工林经营模式，能提高人工林的产量、生物量及其碳汇能力。而根则以 C 为最高，分别达 22.23t/hm²、9.27t/hm²，与 CM、CB 比较都达到了显著差异，这与红椎为根系发达树种有关。不同模式乔木层生物量、碳储量在各器官分配量排序，C 和 CB 从大到小都为干、根、枝、皮、叶；而 CM 为干、根、皮、枝、叶；这主要由树种特性所决定。

表 3 不同树种器官碳素密度 (g/kg)

树种	碳素密度				
	树干	树枝	树叶	树皮	树根
红椎	(465.37±5.80)Cd；Aa	(451.17±5.42)Bb；Aa	(447.30±3.87)Bc；Aa	(438.3±0.81)Be；Bb	(417.20±6.07)Af；Aa
西南桦	(459.83±5.75)Cc；Aa	(439.60±5.27)Bb；Aa	(509.90±4.40)Dd；Bb	(405.93±1.99)Aa；Aa	(442.90±12.74)BCb；Ab
马尾松	(490.20±8.55)ABb；Bb	(490.47±12.59)Ab；Bb	(524.67±6.88)Cd；Bc	(506.97±3.65)BCc；Cc	(470.70±5.61)Aa；Bc

注：表中数据为平均值±标准误差；前一组同列大、小写字母分别表示处理间碳素密度差异极显著($P<0.01$)或差异显著($P<0.05$)；后一组同行大、小字母分别表示同树种不同器官碳素密度差异极显著($P<0.01$)或差异显著($P<0.05$)。

表 4 不同经营模式乔木层生物量、碳储量及其分配

器官	红椎纯林(C)				红椎×马尾松(CM)				红椎×西南桦(CB)			
	生物量		碳储量		生物量		碳储量		生物量		碳储量	
	生物量(t/hm²)	占器官比例(%)	碳储量(t/hm²)	占器官比例(%)	生物量(t/hm²)	占器官比例(%)	碳储量(t/hm²)	占器官比例(%)	生物量(t/hm²)	占器官比例(%)	碳储量(t/hm²)	占器官比例(%)
树干	68.87Aa	63.39	32.06Aa	65.25	44.67Ab	63.43	20.78Ab	64.49	86.11Cc	67.53	39.89Cc	68.97
树皮	7.83Aa	7.22	3.45Aa	7.02	5.20Ab	7.38	2.45Ab	7.60	10.44Cc	8.19	4.42Cc	7.64
树枝	8.43Aa	7.76	3.81Aa	7.76	4.63Ab	6.58	2.12Ab	6.58	12.88Cc	10.10	5.78Cc	9.99
树叶	1.26Aa	1.17	0.54Aa	1.10	1.32Aa	1.88	0.65Aa	2.02	2.91Bb	2.28	1.38Bb	2.39
树根	22.23Aa	20.46	9.27Aa	18.87	14.60Ab	20.73	6.22Ab	19.31	15.18Ab	11.90	6.37Ab	11.01
合计	108.67Aa	100.00	49.13Aa	100.00	70.42Ab	100.00	32.22Ab	100.00	127.52Bc	100.00	57.84Bc	100.00

注：同行不同大、小写字母分别表示相同项目比较差异极显著($P<0.01$)或差异显著($P<0.05$)。

4 结论与讨论

配置科学的经营是提高人工林生产力、生物量和碳储量的有效途径之一，但模式配置不当，则会产生反效果。研究显示，红椎与西南桦混交林，因其空间分布格局合理，能充分利用营养空间，故能提高林分的产量和质量，尤其对红椎生长的促进效果明显，胸径、树高生长量与红椎纯林比分别提高了28.28%、27.31%，乔木层生物量、碳储量比红椎纯林分别增加了17.35%、17.73%。这表明红椎与西南桦混交是科学的。而红椎与马尾松混交林，因其空间分布格局不合理，导致了树种之间对营养空间的激烈竞争，致使红椎和马尾松的生长都不佳，尤其是马尾松，因严重被压已出现了自然稀疏，红椎的胸径、树高生长量也比红椎纯林分别下降了20.40%、6.23%，乔木层生物量、碳储量比红椎纯林分别下低了54.32%、52.48%，表明红椎与马尾松混交不是很科学。判断混交模式是否科学，关键为

种间关系是否协调[38]。只有种间关系协调的混交林才有可能取得理想效果。但混交林的种间关系并不是一成不变的，它会随林分的发育阶段而变化，因此，在混交林的经营过程中，必须根据不同阶段林分的状况对林分进行适当调控，以确保种间关系的协调、林分的稳定及健康。

器官碳素密度与树种有关。研究表明，不同树种器官的碳素密度范围，马尾松为470.70～524.67g/kg，红椎为417.20～465.37g/kg、西南桦为405.93～509.90g/kg。针叶树种马尾松器官的碳素密度都显著或极显著高于阔叶树种红椎、西南桦对应器官。国际上常用的含碳率平均转换系数为0.45或0.5[39-41]，本研究3个树种的平均碳素密度介于405～525g/kg，无论以0.45或0.5作为转换系数，结果都会与实际有较大的偏差。因此，采用各树种对应器官的含碳率作为转换系数，才能得到准确的林分碳储量。经营模式不同，乔木层生物量、碳储量在干、皮、枝、叶、根的分配比例有异，红椎纯林生物量、碳储量依次为：63.39%、7.22%、7.76%、1.17%、20.46%和65.25%、7.02%、7.76%、1.10%、18.87%；红椎与马尾松混交林生物量、碳储量依次为：63.43%、7.38%、6.58%、1.88%、20.73%和64.49%、7.60%、6.58%、2.02%、19.31%；红椎与西南桦混交林生物量、碳储量依次为：67.53%、8.19%、10.10%、2.28%、11.90%和68.97%、7.64%、9.99%、2.39%、11.01%。其中以红椎与西南桦混交林树干生物量分配比例最高，分别比红椎纯林、红椎与马尾松混交林高了4.14%、4.10%，表明红椎与西南桦混交不仅能促进彼此的生长，而且能提高林分的出材率。

参考文献

[1]朱积余，蒋燚，梁瑞龙，等．广西红椎种源/家系造林试验研究初报[J]．西部林业科学，2005，34(4)：5-9.

[2]蔡道雄，贾宏炎，卢立华，等．论我国南亚热带珍优乡土阔叶树种大径材人工林的培育[J]．林业科学研究，2007，20(2)：165-169.

[3]郭文福，蔡道雄，贾宏炎，等．马尾松与红椎等3种阔叶树种营造混交林的生长效果[J]．林业科学研究，2010，23(6)：839-844.

[4]周诚．珍贵用材树种红椎的生物学特性与研究综述[J]．江西林业科技，2007(5)：29-31.

[5]蒋家淡．红椎杉木混交造林效果研究[J]．福建林学院学报，2002，22(4)：329-333.

[6]林俊平．红椎不同模式造林效果分析[J]．福建林业科技，2002，29(3)：59-61.

[7]卢立华，汪炳根，何日明．立地与栽培模式对红椎生长的影响[J]．林业科学研究，1999，12(5)：519-523.

[8]王宏志．中国南方混交林研究[M]．北京：中国林业出版社，1993.

[9]蒋燚，朱积余，张泽尧，等．红椎种源多点育苗试验研究[J]．广西林业科学，2005，34(4)：196-199.

[10]朱积余，蒋燚，潘文．广西红椎优树选择标准研究[J]．广西林业科学，2002，31(3)：109-113.

[11]刘光金，贾宏炎，卢立华，等．不同林龄红椎人工林优树选择技术[J]．东北林业大学学报，2014，42(5)：9-12.

[12]朱积余，蒋燚，唐玉贵．红椎良种区域化试验示范[J]．广西林业科学，2008，37(3)：115-118.

[13]潘坚．红椎的繁育与栽培[J]．林业实用技术，2003(2)：29-30.

[14]朱积余．红椎速生丰产栽培的试验研究[J]．林业科技通讯，1993(2)：8-10.

[15]吕建雄，林志运，骆秀琴，等．红椎和西南桦人工林木材干缩特性的研究[J]．北京林业大学学报，2005，27(1)：6-9.

[16]牛长海，梁宏温，温远光，等．26年生红椎人工林的生物量及其分配格局[J]．中国科技纵横，2010(20)：137-138，147.

[17]赵金龙，梁宏温，温远光，等．马尾松与红椎混交异龄林生物量分配格局[J]．中南林业科技大学学报，2011，31(2)：60-64，71.

[18]何友均，覃林，李智勇，等．西南桦纯林与西南桦×红椎混交林碳贮量比较[J]．生态学报，2012，32(23)：7586-7594.

[19]HE Y J, QIN L, LI Z Y, et al. Carbon storage capacity of monoculture and mixed-species plantations in subtropical China[J]. Forest Ecology and Management, 2013, 295: 193-198.

[20]王效科，冯宗炜，欧阳志云．中国森林生态系统的植物碳储量和碳密度研究[J]．应用生态学报，2001，12(1)：13-16.

[21]康冰，刘世荣，蔡道雄，等．南亚热带杉木生态系统生物量和碳素积累及其空间分布特征[J]．林业科学，2009，45(8)：147-153.

[22]何斌，黄寿先，招礼军，等．秃杉人工林生态系统碳素积累的动态特征[J]．林业科学，2009，45(9)：151-157.

[23]何斌，刘运华，余浩光，等．南宁马占相思人工林生态系统碳素密度与贮量[J]．林业科学，2009，45(2)：6-11.

[24]何斌，余春和，王安武，等．厚荚相思人工林碳素贮量及其空间分布[J]．南京林业大学学报(自然科学

版)，2009，33(3)：46-50.
[25]苏勇，吴庆标，施福军，等．擎天树人工林生态系统碳贮量及分布格局[J]．安徽农业科学，2011，39(9)：5271-5273.
[26]莫德祥，廖克波，吴庆标，等．山白兰人工林生态系统碳储量及空间分布特征[J]．安徽农业科学，2011，39(23)：14072-14075.
[27]WANG H，LIU S R，MO J M，et al. Soil organic carbon stock and chemical composition in four plantations of indigenous tree species in subtropical China[J]. Ecological Research，2010，25：1071-1079.
[28]ZHANG H，GUAN D S，SONG M W. Biomass and carbon storage of Eucalyptus and Acacia plantations in the Pearl River Delta，South China[J]. Forest Ecology and Management，2012，277：90-97.
[29]明安刚，贾宏炎，陶怡，等．米老排人工林碳素积累特征及其分配格局[J]．生态学杂志．2012，31(11)：2730-2735.
[30]明安刚，张治军，谌红辉，等．抚育间伐对马尾松人工林生物量与碳贮量的影响[J]．林业科学，2013，49(10)：1-6.
[31]覃林，何友均，李智勇，等．南亚热带红椎马尾松纯林及其混交林生物量和生产力分配格局[J]．林业科学，2011，47(12)：17-21.
[32]张传峰，陈双礼，张玉荣，等．杉木与光皮桦桤木混交林的研究[M]//沈国舫，翟明普．全国混交林与种间关系学术研讨会论文集．北京：中国林业出版社，1997：192-197.
[33]林星华．闽南沿海山地火力楠马尾松混交林种间关系变化规律[J]．江西农业大学学报，2001，23(3)：340-344.
[34]潘瑞炽．植物生理学[M]．北京：高等教育出版社，2008.
[35]万劲．杨树无性系及重阳木木质能源性状的研究[D]．南京：南京林业大学，2008.
[36]马钦彦，陈遐林，王摇娟，等．华北主要森林类型建群种的含碳率分析[J]．北京林业大学学报，2002，24(5)：96-100.
[37]胡青，汪思龙，陈龙池，等．湖北省主要森林类型生态系统生物量与碳密度比较[J]．生态学杂志，2012，31(7)：1626-1632.
[38]林思祖，黄世国．论中国南方近自然混交林营造[J]．世界林业研究，2001，14(2)：73-78.
[39]HOUGHTON R A，SKOLE D L，Nobre C A，et al. Annual fluxes of carbon from deforestation and regrowth in the Brazilian Amazon[J]. Nature，2000，403：301-304.
[40]周玉荣，于振良，赵士洞．我国主要森林生态系统碳贮量和碳平衡[J]．植物生态学报，2000，24(5)：518-522.
[41]张骏，袁位高，葛滢，等．浙江省生态公益林碳储量和固碳现状及潜力[J]．生态学报，2010，30(14)：3839-3848.

[原载：东北林业大学学报，2014，42(12)]

不同林龄格木人工林碳储量及其分配特征

明安刚[1,2,3]　贾宏炎[1,3]　田祖为[1,3]　陶　怡[1,3]　卢立华[1,3]　蔡道雄[1,3]　史作民[2]　王卫霞[2]

([1]中国林业科学研究院热带林业实验中心，广西凭祥　532600；[2]中国林业科学研究院森林生态环境与保护研究所，北京　100091；[3]广西友谊关森林生态定位观测研究站，广西凭祥　532600)

摘　要　在生物量调查的基础上，对广西7年、29年和32年格木人工林生态系统碳储量及其分配特征进行了研究。结果表明：格木各器官碳含量在509.0~572.4g/kg，大小顺序为：树干>树枝>树根>树皮>树叶；不同林龄格木人工林的灌木层、草本层和凋落物层碳含量无显著差异；土壤层(0~100cm)碳含量随土层深度的增加而降低，随林龄的增加而增大。7年、29年和32年格木人工林乔木层碳储量分别为21.8t/hm²、100.0t/hm²和121.6t/hm²，各器官碳储量大小顺序与碳含量一致；生态系统碳储量分别为132.6t/hm²、220.2t/hm²和242.6t/hm²，乔木层和土壤层为主要碳库，占生态系统碳储量的97%以上。乔木层碳储量分配随着林龄的增加而增大，土壤碳储量分配则减小，而林龄对灌木层、草本层和凋落物层碳储量分配的影响无明显规律。

关键词　格木人工林；林龄；碳含量；碳储量；分配

Characteristics of Carbon Storage and Its Allocation in *Erythrophleum fordii* Plantationswith Different Ages

MING Angang[1,2,3], JIA Hongyan[1,3], TIAN Zuwei[1,3], TAO Yi[1,3], LU Lihua[1,3], CAI Daoxiong[1,3], SHI Zuomin[2], WANG Weixia[2]

(*[1]Experimental Center of Tropical Forestry, Chinese Academy of Forestry, Pingxiang* 532600, *Guangxi, China*; *[2]Institute of Forest Ecology, Environment and Protection, Chinese Academy of Forestry, Beijing* 100091, *China*; *[3]Guangxi Youyiguan Forest Ecosystem Research Station, Pingxiang* 532600, *Guangxi, China*)

Abstract: Carbon storage and its allocation of 7-, 29-and 32-year-old *Erythrophleum fordii* plantations ecosystem in Guangxi were studied on the basis of biomass survey. The results showed that carbon content in different organs of *E. fordii* was 509.0~572.4g/kg, and ranks as stem>branch>root>bark>leaf. Carbon content showed no significant difference among the shrub, herb and litter layers in *E. fordii* plantations with different ages. Carbon content in soil layer (0~100cm) decreased with increasing soil depth, but increased with increasing stand ages. The carbon storage in arbor layer was 21.8, 100.0 and 121.6t/hm² in 7-, 29-and 32-year-old stands, respectively, and had the same allocation in different organs with carbon content. Carbon storage in ecosystem was 132.6, 220.2 and 242.6t/hm² in 7-, 29-and 32-year-old stands, respectively. Arbor laycr and soil layer were the main carbon pools, which accounted for more than 97% of carbon storage in the ecosystem. Carbon storage allocation increased in arbor layer but decreased in soil layer with increasing stand ages. The influence of stand ages on carbon storage allocation in shrub, herb and litter layers was not obvious regular.

Key words: *Erythrophleum fordiis* plantation; stand age; carbon content; carbon storage; allocation.

自工业革命以来，由于化石燃料的大量燃烧和土地利用方式的改变，大气中CO_2浓度逐年升高[1]，CO_2的排放、吸收和固定是全球气候变化研究的重要内容[2-4]。森林作为陆地生态系统的主体，维持着全球86%的植被碳库和73%的土壤碳库，在调节全球气候、减缓大气中CO_2等温室气体浓度上升等方面具有不可替代的作用[5,6]。方精云和陈安平[7]、王效科和冯宗炜[8]、刘国华等[9]和周玉荣等[10]分别利用我国森林资源清查结果，结合森林生态系统生物量与生产力的研究，估算了近50年来我国森林碳库及其动态，为评价北半球中高纬度地区碳库和我国森林碳汇功能奠定了基础。当前，造林和再造林作为一种新增碳汇的主要途径，已受到学术界的高度重视[11]。人工林在吸收和固定 CO_2 及减缓全球气候变暖等方面发挥着重要作用，并日益引起人们的广泛关注。最近十年，诸多学者对不同树种人工林的碳含量、碳储量及其空间分布格局进行了深入研究[12-22]，为森林碳汇功能的研究做出了积极贡献。

格木(*Erythrophleum fordii*)为苏木科格木属常绿乔木，生长速度较慢，木材硬而亮，纹理致密，是我国南亚热带珍贵用材树种。格木主要分布于我国广东、广西、浙江、福建和台湾等地，是我国著名的硬木之一，与蚬木(*Excetrodendron hsiemvu*)、金丝李(*Garcinia paucinervis*)并称为“广西三大硬木”。近几年，广西格木人工林发展迅速，逐渐成为主要造林树种之一。目前，有关格木的研究主要集中在繁殖特性、种子发育和生物量方面[23-26]。本文对广西大青山林区7年、29年和32年格木人工林生态系统碳储量及其分配特征进行了研究，为区域尺度上估算森林生态系统碳库及碳平衡提供基础数据和科学参考。

1 研究地区与研究方法

1.1 研究区概况

研究区位于广西凭祥市中国林业科学研究院热带林业实验中心(106°39′50″~106°59′30″E，21°57′47″~22°19′27″N)，属南亚热带季风气候，年均温19.5~21.0℃，太阳年总辐射439.61kJ/cm²，年日照时数1218~1620h，≥10℃积温6000~7600℃，年降水量1400mm，土壤为花岗岩发育成的山地红壤，土层厚度在100cm以上。

选取7年、29年和32年格木人工林为研究对象，3种人工林均是在杉木(*Cunninghamia lanceolata*)人工林皆伐炼山后，经块状整地营建的人工纯林。各林分初植密度均为1750株/hm²，7年格木人工林由于尚未开始间伐，林冠郁闭，林下植被较少，仅有少量的五节芒(*Miscanthus floridulus*)和铁芒萁(*Dicranopteris dichotoma*)，盖度为5%；29年和32年格木人工林因间伐2次(间伐时间均为造林后第7年和第15年)，林分郁闭度较7年生幼林郁闭度小，林下草本、灌木较多，以杜茎山(*Maesa japonica*)、酸藤子(*Embelia laeta*)、玉叶金花(*Mussaenda pubuscens*)、五节芒等为优势种，盖度为55%。

2012年8月，在7年、29年和32年格木人工林中，选取坡面均匀，人为干扰相对较少的区域，按坡位分别随机设置5个20m×20m样地，共计15个样地。对每个样方内的树木进行每木检尺，调查胸径、树高等指标。林分基本情况见表1。

表1 不同林龄格木人工林地概况

林龄(年)	坡度(°)	密度(ind/hm²)	郁闭度	胸径(cm)	树高(m)
7	26.8	1750±48	0.9	7.1±1.8	6.6±1.2
29	27.3	476±23	0.7	21.5+4.2	16.8±2.7
32	29.1	491±26	0.8	22.7±4.8	17.7±2.9

1.2 研究方法

1.2.1 林木调查及生物量的测定

根据样方每木检尺的结果，按径级(2cm)选取标准木18株(7年林分7株，29年林分5株，32年林分6株)进行乔木生物量测定。样木伐倒后，地上部分测定树干、树皮、树枝、树叶等器官的鲜质量；地下部分用“全挖法”测定(18株标准木的根系主要分布在0~85cm土层)，将林木的根系由上至下全部挖出，把根系表面的土壤清理干净，再按照经级大小将根系分为主根、大侧根(>2.0cm)、中侧根(0.5~2.0cm)和细侧根(<0.5cm)5个组分，分别测定鲜质量。同时，按不同器官和根系组分采集伐倒木的分析样品各4份，每份样品400g，带回实验室在65℃下烘干至恒量，称干质量。计算含水率后，将各器官的鲜质量换算成干质量。

根据18株标准木生物量的实测数据，建立格木各器官生物量(用 W 表示器官重量)与胸径(D)及胸径平方与树高的乘积(D^2H)之间的相对生长方程，用以估测格木各器官的生物量。由于格木幼林分叉较多，密度较大，林分郁闭度高，树高难以准确测量；因此，在生物量方程的建立中，用7年林分中选取的7株标准木建立7年格木人工林各器官生物量与胸径的一元生物量方程，而29年和32年林分生物量方程可用这2个林分中选取的11株标准木建立各器官生物量与“D^2H”之间通用的二元生物量方程(表2)。

表2 格木人工林生物量方程

器官	7(年)			29~32(年)		
	回归方程	R^2	F	回归方程	R^2	F
树干	$W=0.1957D^{2.0341}$	0.99**	529.63	$W=0.0315(D^2H)^{0.9737}$	0.99**	1944.62

（续）

器官	7(年)			29~32(年)		
	回归方程	R^2	F	回归方程	R^2	F
树皮	$W=0.0431D^{1.9442}$	0.99**	778.56	$W=0.0110(D^2H)^{0.8580}$	0.99**	1933.42
树枝	$W=0.0020D^{3.8719}$	0.92**	34.94	$W=0.0055(D^2H)^{1.0628}$	0.92**	118.79
树叶	$W=0.0126D^{2.8032}$	0.95**	57.01	$W=0.0662(D^2H)^{0.6064}$	0.87**	144.10
树根	$W=0.0340D^{2.3429}$	0.94**	22.81	$W=0.0072(D^2H)^{1.0243}$	0.99**	318.25
全株	$W=0.1938D^{2.4503}$	0.99**	118.19	$W=0.0740(D^2H)^{0.9549}$	0.99**	943.48

注：** $P<0.01$。

1.2.2 林下植被生物量和凋落物现存量的测定

在每个20m×20m样方中，按对角线选取2个5m×5m小样方，记录小样方内灌木和草本植物的种类，并采用"收获法"测地上和地下部分生物量。将同种植物相同器官混合，取样品4份带回实验室在65℃下烘干至恒量，称干质量。计算含水率后，将鲜质量换算成干质量。在每个20m×20m样方中，按对角线选取4个1m×1m小样方，按未分解、半分解，测定凋落物现存量。

1.2.3 土壤样品的采集

在各样方随机挖取3个土壤剖面，按照0~10cm、10~30cm、30~50cm和50~100cm将土壤分为4个土层，用环刀取样，测定各土层土壤容重。各土层土样混合，取200g用于含碳量的测定。

1.2.4 碳含量测定和碳储量计算

将植物样品于65℃下烘干，土壤样品置于室内风干，磨碎。植物和土壤样品均采用重铬酸钾-水合加热法测定有机碳含量。植物碳储量=有机碳含量×单位面积生物量，土壤碳储量=土壤有机碳含量×土壤容重×土壤厚度。

1.3 数据处理

采用SPSS 13.0软件对数据进行统计分析，方差分析和差异显著性检验（$\alpha=0.05$）。采用Excel和PS软件作图。图表数据为平均数±标准差。

2 结果与分析

2.1 格木人工林生态系统各组分碳含量

2.1.1 乔木层碳含量

不同林龄格木各器官碳含量在509.0~572.4g/kg，且随着林龄的增加呈增大趋势；不同林龄间各器官碳含量差异不显著。不同器官间碳含量明显不同，以树干最高，树叶最低，大小顺序为：树干>树枝>树根>树皮>树叶，除树皮和树枝之间碳含量差异不显著外，其他各器官之间差异均显著（表3）。

表3 不同林龄格木各器官碳含量

器官	林龄(年)		
	7	29	32
树干	561.7±10.2Aa	566.1±13.2Aa	572.4±16.5Aa
树皮	534.8±8.5Da	541.9±10.6Ca	537.7±7.7Da
树枝	549.8±7.4Ba	552.9±8.2Ba	562.3±10.4Ba
树叶	512.4±9.1Ea	509.0±10.5Da	520.6±13.2Ea
树根	537.7±5.2Ca	544.9±10.2Ca	543.2±9.8Ca

注：不同大写字母表示器官间差异显著，不同小写字母表示林龄间差异显著（$P<0.05$）。

2.1.2 林下地被物碳含量

林下地被物包括活地被物和死地被物，本研究主要测定了林下灌木层、草本层及凋落物层的碳含量。从表4可知，格木人工林下地被物各层次平均碳含量以凋落物层最高，灌木层其次，草本层最低；灌木层和草本层地上部分碳含量均高于地下部分，未分解的凋落物碳含量高于半分解的碳含量。林龄对灌木层、草本层及凋落物层的碳含量均无显著影响。

表4 不同林龄格木人工林下植被及凋落物碳含量

（g/kg，$n=4$）

层次	组分	林龄(年)		
		7	29	32
灌木层	地上部分	481.4±10.6Aa	478.4±8.4Aa	477.6±11.9Aa
	地下部分	444.3±6.8Da	437.2±6.9Da	431.8±8.7Da
草本层	地上部分	479.1±12.3Ba	484.4±10.5Ba	485.7±13.8Ba
	地下部分	402.4±10.2Ea	393.3±7.9Ea	393.5±9.6Ea
凋落物	未分解	516.4±13.5Aa	527.3±8.3Aa	487.8±15.0Ba
	半分解	464.8±9.3Ca	445.3±5.8Da	457.2±11.4Ca

注：不同大写字母表示组分间差异显著，不同小写字母表示林龄间差异显著（$P<0.05$）。

2.1.3 土壤层碳含量

从表5可知，土壤碳含量以表土层(0~10cm)最高，平均高达(23.4±2.0)g/kg；各林龄土壤碳含量随着土层深度的增加而降低，且差异显著；同一土层不同林龄间碳含量差异显著，碳含量随着林龄的增加而增大。

表5 不同林龄格木人工林各土层土壤碳含量

(g/kg，$n=5$)

土层(cm)	林龄(年)		
	7	29	32
0~10	21.4±1.7Ac	23.5±2.3Ab	25.3±2.7Aa
10~30	12.7±0.8Ba	14.4±0.7Bb	16.2±1.1Ba
30~50	8.2±0.5Cb	7.5±0.4Cb	9.6±0.7Ca
50~100	5.8±0.3Db	5.9±0.3Db	6.4±0.4Da

注：不同大写字母表示土层间差异显著，不同小写字母表示林龄间差异显著($P<0.05$)。

2.2 格木人工林生态系统各组分碳储量及其分配

2.2.1 乔木层碳储量及其分配

不同林龄格木人工林乔木层各器官碳储量及其分配特征与各器官生物量的分配情况有密切关系。从图1可知，不同林龄格木人工林乔木层各器官生物量随林龄增加而增大；树干、树枝和树根生物量分配随林龄增加而增大，树皮和树叶生物量分配随林龄增加而减小。

乔木层各器官碳储量随林龄的变化趋势与生物量变化特征较为一致，随着林龄的增加，各器官碳储量显著增大。7年、29年和32年格木人工林乔木层碳储量分别为21.8t/hm²、100.0t/hm²和121.6t/hm²。各器官碳储量在乔木层的分配以树干最高，7年、29年和32年格木人工林树干碳储量分别占乔木层碳储量的47.5%、52.2%和52.2%；树皮和树叶碳储量所占比例最小，平均<10%。乔木层各器官碳储量分配大小顺序为：树干>树枝>树根>树皮>树叶。林龄对乔木层各器官碳储量分配有显著影响，随着林龄的增加，碳储量在树干、树枝和树根的分配增加，而在树皮和树叶的分配下降。

2.2.2 林下地被物碳储量及其分配

从表6可知，7年、29年和32年格木人工林下灌木层、草本层和凋落物层等林下地被物碳储量均较小，分别为3.0t/hm²、5.9t/hm²和4.5t/hm²。29年和32年林分灌木层碳储量高于7年林分，而29年林分草本层碳储量高于7年和32年林分，其原因可能是林分郁闭度差异所致。格木人工林凋落物碳储量随林龄增加而增大。

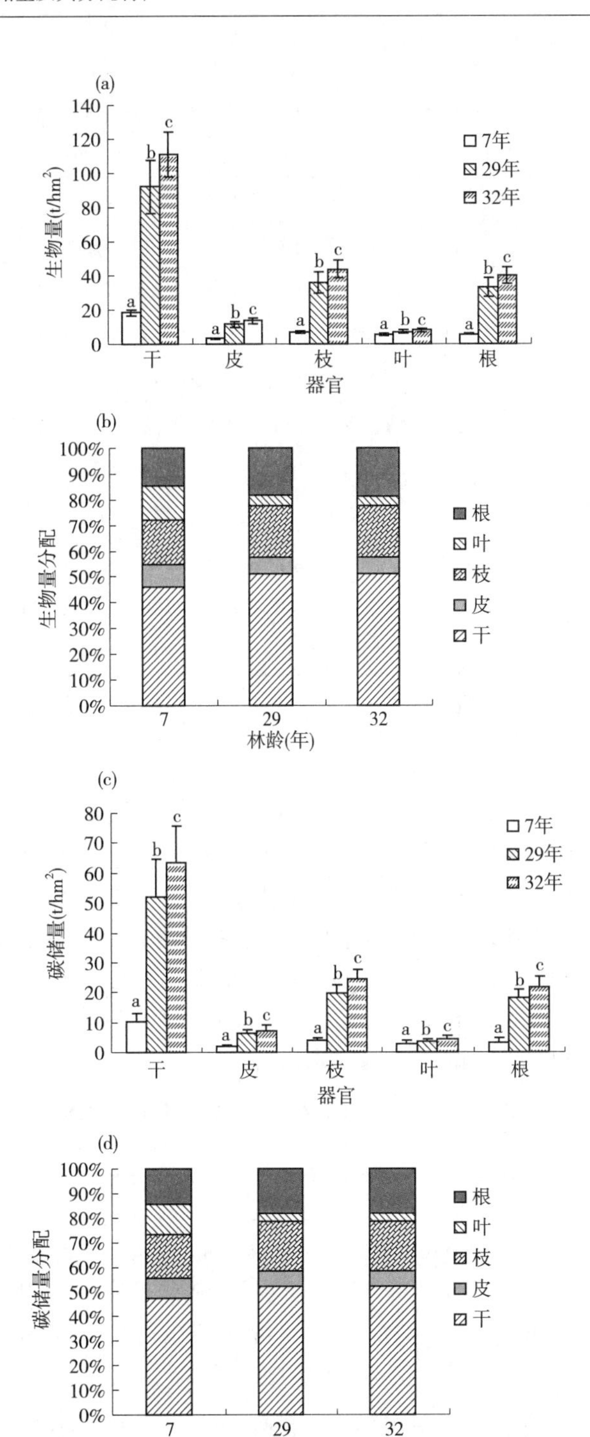

图1 不同林龄格木人工林乔木层各器官碳储量及其分配

不同小写字母表示林龄间差异显著($P<0.05$)

3个层次中，碳储量大小顺序为：凋落物层>灌木层>草本层。在灌木层和草本层中，地上部分碳储量均大于地下部分，但草本层植物地上与地下部分碳储量差异不显著；凋落物层中，未分解的凋落物碳储量显著高于半分解的，约为半分解凋落物碳储量的3倍。

表6 不同林龄格木人工林林下植被和凋落物碳储量

(t/hm^2, $n=10$)

层次	组分	林龄(年)		
		7	29	32
灌木层	地上	0.60±0.19Bb	1.55±0.36Aa	1.28±0.34Aa
	地下	0.29±0.13Cb	0.65±0.22Ca	0.53±0.13Ba
	合计	0.89±0.31	2.20±0.53	1.81±0.40
草本层	地上	0.47±0.16Bb	1.34±0.27Ba	0.51±0.24Bb
	地下	0.41±0.14Bb	0.75±0.17Ca	0.39±0.07b
	合计	0.88±0.26	2.09±0.42	0.90±0.31
凋落物层	未分解	0.94±0.22Ab	1.24±0.32Ba	1.39±0.33Aa
	半分解	0.27±0.07Cb	0.35±0.14Db	0.45±0.20Ba
	合计	1.21±0.24	1.59±0.39	1.84±0.48

注：不同大写字母表示组分间差异显著，不同小写字母表示林龄间差异显著($P<0.05$)。

2.2.3 土壤碳储量及其分配

不同林龄格木人工林各土层土壤碳储量随林龄增加而增大，尤其是表土层(0~10cm 和 10~30cm)，29年和32年林分土壤碳储量显著高于7年林分。在3个不同林龄的林分中，土壤碳储量均随土层深度的增加而显著降低，随土层深度的变化趋势与土壤碳含量的变化规律一致。土壤碳含量与碳储量随土层深度的变化情况不受林龄的影响。

表7 不同林龄格木人工林各土层土壤碳储量

(t/hm^2, $n=15$)

土层(cm)	林龄(年)		
	7	29	32
0~10	18.73±1.82a	20.54±2.06b	20.66±2.31b
10~30	27.75±2.67a	33.00±3.68b	33.11±3.08b
30~50	20.07±2.19a	20.10±1.88a	21.35±2.42a
50~100	41.30±5.77a	40.74±7.17a	41.39±4.88a
合计 Sum	107.85±7.57	114.38±8.92	116.51±10.34

注：不同小写字母表示林龄间差异显著($P<0.05$)。

2.3 格木人工林生态系统碳储量及其分配

从图2可知，7年、29年和32年格木人工林生态系统碳储量分别为132.6t/hm²、220.2t/hm²和242.6t/hm²，以乔木层和土壤层为主要碳库，二者占格木人工林生态系统碳储量的97%以上，而灌木层、草本层和凋落物层的碳储量所占比例<3%。

林龄对格木人工林生态系统碳储量有显著影响，乔木层、凋落物层和土壤层碳储量均随林龄的增加而增大，灌木层和草本层碳储量则无明显变化规律。林龄对格木人工林生态系统各组分碳储量分配存在不同的影响，其中，乔木层碳储量分配随林龄的增加而增大，而土壤层碳储量分配随林龄的增加而减小，而灌木层、草本层和凋落物层碳储量的分配随林龄的增加无明显规律。

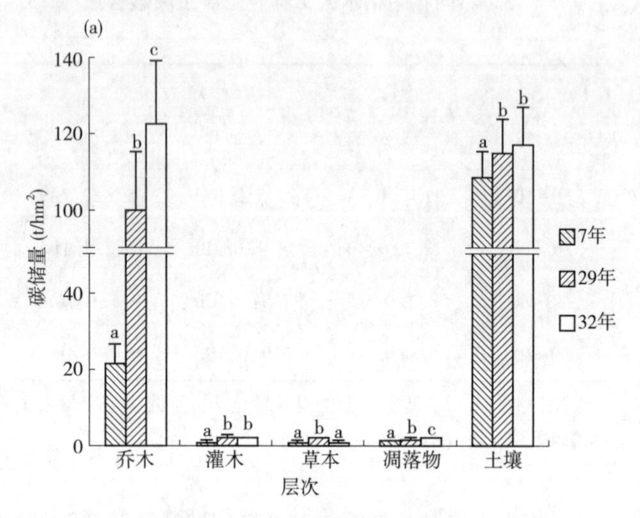

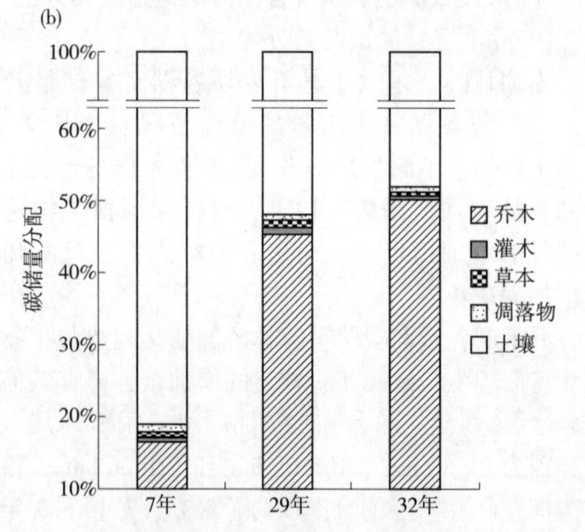

图2 不同林龄格木人工林生态系统碳储量及其分配

3 讨论

广西大青山林区7年、29年和32年格木人工林中，格木各器官碳含量不同，大小顺序为树干>树枝>树根>树皮>树叶，与同一地区的马尾松(*Pinus massoniana*)[27]、杉木(*Cunninghamia lanceolata*)[27]、米老排[28](*Mytilaria laosensis*)和红椎(*Castnopsis hystrix*)[20,26]等树种器官碳含量的排序不完全一致，可见，即使在同一地区，树木各器官碳含量大小因树种不同而存在差异。7年、29年和32年格木人工林平均碳含量为543.1g/kg高于广西观光木(*Tsoongiodendron odorum*)(471.1g/kg)、桉树(*Eucalyptus urophylla × E. grandis)(475.2g/kg)和杉木(497.0g/kg)[29-33]，也高于国际通用的树木平均碳含量(0.5g/kg)及热带32个树种的平均碳含量(444.0~494.5g/kg)[34]。这表明各器官碳含量大小与树种的关系极为密切，不同树种的碳含量有明显差异，而格木是广西碳含量较高的树种，是发展碳汇林较好的树种之一。

格木人工林下地被物各层次平均碳含量大小顺序为：凋落物层>灌木层>草本层；凋落物层碳含量高于灌木层，其原因可能是凋落物层的主要组分是枯落枝，而树枝在格木人工林乔木各器官中碳含量较高，仅次于树干。未分解的凋落物碳含量较半分解的凋落物高，这可能由凋落物在分解过程中碳释放的数量和速率的差异导致。灌木层和草本层植被碳含量均表现为地上部分大于地下部分，其原因主要是因为植物体木质化程度不同。一般而言，木质化程度越高，植物体碳累积量越高。

7年、29年和32年格木人工林的土壤层(0~100cm)平均碳含量为13.1g/kg，高于同地区米老排人工林[28]，低于13年杉木林(16.6g/kg)和桉树林(14.57g/kg)[31,32]，表明不同树种对土壤碳累积过程的影响不同。然而，本研究中，格木人工林的土壤碳含量随林龄增加而增大，因此延长格木人工林的轮伐期，可能有利于土壤碳累积。但这一结论尚需更多的研究，因为我国南方土壤空间异质性较大，不同年龄阶段格木林土壤碳含量的差异并不能完全排除空间异质性的影响。

格木人工林生态系统碳库主要包括植被层、凋落物层和土壤层，7年、29年和32年3个林龄的格木人工林生态系统平均碳储量为198.5t/hm^2。其中，植被层碳储量为84.1t/hm^2，是我国森林植被平均碳储量(57.07t/hm^2)的1.47倍，是我国亚热带针叶林平均碳储量(63.7t/hm^2)的1.32倍[35]。格木人工林植被层高碳储量主要源于乔木层的高生物量及乔木层较高的碳含量。然而，格木人工林生态系统碳储量低于我国森林生态系统平均碳储量(258.82t/hm^2)[10]，这与格木人工林土壤层相对较低的平均碳含量有关。本研究中，格木人工林土壤碳储量(0~100cm)为112.9t/hm^2，仅为我国森林土壤平均碳储量(193.55t/hm^2)的58.3%，其原因可能是在计算土壤碳储量时，不同研究中土层深度不同，也可能是由于树种的不同引起了林分地上和地下部分凋落物的输入组分的不同，进而影响了土壤碳的固持，具体原因需进一步研究。

参考文献

[1] TANS P P. How can global warming be traced to CO_2[J]. Scientific America, 2006, 295: 124.

[2] FANG J Y, CHEN A P, PENG C H, et al. Changes in forest biomass carbon storage in China between 1949 and 1998[J]. Science, 2001, 292: 2320-2322.

[3] CALDEIRA K, DUFFY P B. The role of the southern ocean in up take and storage of anthropogentic carbon dioxide [J]. Science, 2000, 287: 620-622.

[4] NORBY R J, LUO Y. Evaluating ecosystem responses to rising atmospheric CO_2 and global warming in a multi-factor world[J]. New Phytologist, 2006, 162: 281-293.

[5] WOODWELL G M, WHITTAKER R H, REINERS W A, et al. The biota and the world carbon budget[J]. Science, 1978, 199: 141-1467.

[6] POST W M, EMANUEL W R, ZINKE P J, et al. Soil carbon pools and world life zone[J]. Nature, 1982, 298: 156-159.

[7] FANG J Y, CHEN A P. Dynamic forest biomass carbon pools in China and their significance[J]. Acta Botanica Sinica, 2001, 43(9): 967-973(in Chinese).

[8] WANG X K, FENG Z W. The potential to sequester atmospheric carbon through forest ecosystems in China [J]. Chinese Journal of Ecology, 2000, 19(4): 72-74(in Chinese).

[9] LIU G H, FU B J, FANG J Y. Carbon dynamics of Chinese forests and its contribution to global carbon balance [J]. Acta Ecologica Sinica, 2000, 20(5): 733-740(in Chinese).

[10] ZHOU Y R, YU Z L, ZHAO S D. Carbon storage and budget of major Chinese forest types[J]. Acta Phytoecologica Sinica, 2000, 24(5): 518-522(in Chinese).

[11] ONIGKEIT J, SONNTAG M, ALCAMO J. Carbon plantation in the IMAGE model-moder description and scenarios. Center for Environmental Systems Research, University of Kassek, Germany, 2000, 35-113.

[12] LACLAU P. Biomass and carbon sequestration of ponderosa pine plantations and native cypress forests in northwest Patagonia[J]. Forest Ecology and Management, 2003, 180: 317-333.

[13] SPECHT A, WEST P W. Estimation of biomass and sequestered carbon on farm forest plantations in northern New South Wales, Australia[J]. Biomass and Bioenergy, 2003, 25: 363-379.

[14] FANG X, TIAN D L, XIANG W H. Density, storage and distribution of carbon in Chinese Fir Plantation at fast growing stage[J]. Scientia Silvae Sinicae, 2002, 38(3): 14-19(in Chinese).

[15]LEI P F, XIANG W H, TIAN D L, et al. Carbon storage and distribution in Cinnamomum camphor plantation [J]. Chinese Journal of Ecology, 2004, 23(4): 25-30 (in Chinese).

[16]MA M D, JIANG H, LIU Y J. Biomass, carbon content, carbon storage and their vertical distribution of phoebe bourmei artificial stand [J]. Scientia Silvae Sinicae, 2008, 44(3): 34-39(in Chinese).

[17]WEI P, LI X W, FAN C, et al. Fine root biomass and carbon storage in surface soil of Cinnamomum camphora plantation in rainy area of west China[J]. Chinese Journal of Applied Ecology, 2013, 24(10): 2755-2762(in Chinese).

[18]LI X R, LIU Q J, CHEN Y R, et al. Aboveground biomass of three conifers in Qi anyanzhou plantation[J]. Chinese Journal of Applied Ecology, 2006, 17(8): 1382-1388(in Chinese).

[19]LIU E, LIY S R. The research of carbon storage and distribution feature of the *Mytilaria laosensis* plantation in south subtropical area[J]. Acta Ecologica Sinica, 2012, 32(16): 5103-5109(in Chinese).

[20]HE Y J, QIN L, LI Z Y, et al. Carbon storage capacity of a *Betula alnoides* stand and a mixed *Betula alnoides* × *Castanopsis hystrix*stand in Southern Subtropical China: a comparison study[J]. Acta Ecologica Sinica, 2012, 32(23): 7586-7594(in Chinese).

[21]HE Y J, QIN L, LI Z Y, LIANG X Y, et al. Carbon storage capacity of monoculture and mixed-species plantations in subtropical China[J]. Forest Ecology and Management, 2013, 295: 193-198.

[22]WANG X L, WANG A, SHI H H, et al. Carbon storage of *Pinus thunbergii* and *Robinia pseudoacacia* plantations on Nanchangshan Island, Changdao County of Shandong Province, China [J]. Chinese Journal of Applied Ecology, 2013, 24(5): 1263-1268. (in Chinese).

[23]CHEN R Z, FU J R. Physiological studies on the seed democracy and germination of Erythrophloeum fordii [J]. Scientia Silvae Sinicae, 20(1): 35-41(in Chinese).

[24]HUANG Z L, GUO G Z, ZHANG Z P. A study about endangered mechanism of *Erythrophleum fordii* [J]. Acta Ecologica Sinica, 1997, 17(6): 671-676(in Chinese).

[25]YI W M, ZHANG Z P, DING M M, et al. Biomass and efficiency of radiation utilization in *Erythrophloeum fordii* community[J]. Acta Ecologica Sinica, 2000, 20(2): 397-403(in Chinese).

[26]WANG W X, SHI Z M, LUO D, et al. Carbon and nitrogen storage under different plantations in subtropical south China[J]. Acta Ecologica Sinica, 2013, 33(3): 0925-0933(in Chinese).

[27]KANG B, LIU S R, ZHANG G J, Chang J G, Wen Y G, Ma J M, Hao W F. Carbon accumulation and distribution in Pinus massoniana and *Cunninghamia lanceolata* mixed forest ecosystem in Daqingshan, Guangxi of China [J]. Acta Ecologica Sinica, 2006, 26(5): 1321-1329 (in Chinese).

[28]MING A G, JIA H Y, TAO Y, et al. Characteristics of carbon accumulation and allocation pattern in *Mytilaria laosensis* plantation [J]. Chinese Journal of Ecology, 2012, 31(11): 2730-2735(in Chinese).

[29]CHEN C Y, LIAO L P, WANG S L. Study of carbon stocks and allocation in *Cunninghamia lanceolata* plantations [J]. Chinese Journal of Applied Ecology, 2000, 11(sup1): 175-178(in Chinese).

[30]HE B, HUANG S X, ZHAO L J, et al. Dynamic characteristics of carbon accumulation in *Taiwania flousiana* plantation ecosystem[J]. Scientia Silvae Sinicae, 2009, 45(9): 151-157(in Chinese).

[31]KANG B, LIU S R, CAI D X, et al. Characteristics of biomass, carbon accumulation and its spatial distribution in *Cunninghamia lanceolata* forest ecosystem in low subtropical Area[J]. Scientia Silvae Sinicae, 2009, 45(8): 147-153(in Chinese).

[32]LIANG H W, WEN Y G, WEN L H, et al. Effects of continuous cropping on the carbon storage of *Eucalyptus urophylla*×*E. grandis* short-rotations plantations[J]. Acta Ecologica Sinica, 2009, 29(8): 4242-4250(in Chinese).

[33]HUANG S D, WU Q B, LIAO K B, et al. Carbon storage and its allocation in an artificial *Tsoongiodendron odorum* ecosystem in southern subtropical region of China [J]. Chinese Journal of Ecology, 2011, 30(11): 2400-2404(in Chinese).

[34]ELIAS M, POTVIN C. Assessing inter-and intra-specific variation in trunk carbon concentration for 32 neotropical tree species[J]. Forest Ecology and management, 2006, 222: 279-295.

[35]WANG S Q, ZHOU C H, LUO C W. Studying carbon storage spatial distribution of terrestrial natural vegetation in China[J]. Progress in Geography, 1999, 18(3): 238-244(in Chinese).

[原载：应用生态学报，2014，25(04)]

南亚热带不同树种人工林生态系统碳库特征

郑　路[1,2]　蔡道雄[1,2]　卢立华[1,2]　明安刚[1,2]　李朝英[1]

([1]中国林业科学研究院热带林业实验中心，广西凭祥　532600；
[2]广西友谊关森林生态系统国家定位观测研究站，广西凭祥　532600)

摘　要　研究比较了我国南亚热带5个树种人工林生态系统的碳储量及分配格局，结果表明：在相似的生境条件下，林龄和经营管理措施相同，不同树种人工林生态系统碳储量表现出较大差异，其中以火力楠林具有最大的储碳能力，其碳储量为359.43t/hm^2，其次是米老排林，为319.80t/hm^2，红椎林、马尾松林和铁力木林碳储量差异不大，分别为225.87t/hm^2、222.43t/hm^{2}和207.81t/hm^2。乔木层与土壤层是森林生态系统碳储量的主体，占生态系统碳储量总量的95%以上。不同树种林分乔木层碳储量以米老排林最高，为188.09t/hm^2，其次是火力楠林，为176.44t/hm^2，再其次是红椎林，为102.56t/hm^2，马尾松林和铁力木林最低，分别为84.59t/hm^{2}和84.01t/hm^2。树种不同，树木器官碳储量分配各不相同，但是均以树干最高，其次为根或枝，再次为干皮，叶最低。林下灌木层和草本层碳储量分别为0.036~1.163t/hm^{2}和0.027~0.913t/hm^2，地表凋落物层碳储量在9.54~2.37t/hm^2。不同树种人工林100cm厚土壤有机碳储量以火力楠林最大，为179.59t/hm^2，而米老排林、红椎林、铁力木林和马尾松林之间差异不大，均在120t/hm^{2}左右。研究结果表明，南亚热带人工林具有较高的储碳能力，可发展成为高碳汇人工林基地。

关键词　南亚热带；人工林；米老排；火力楠；碳储量

Carbon Pool of Different Species Plantation Ecosystems in Lower Subtropical Area of China

ZHENG Lu[1,2], CAI Daoxiong[1,2], LU Lihua[1,2], MING Angang[1,2], LI Zhaoying[1]

([1]*Experimental Center of Tropical Forestry*, *Chinese Academy of Forestry*, *Pingxiang* 532600, *Guangxi*, *China*;
[2]*Guangxi Youyiguan Forest Ecosystem Research Station*, *Pingxiang* 532600, *Guangxi*, *China*)

Abstract: We made a comparative study of the carbon storage and its distribution of five species plantations located at the Shaopin experimental farm of the Experimental Centre of Tropical Forestry in Pingxiang, Guangxi, China. The purpose was to evaluate the potential carbon sinks of large sized trees in plantations, and to provide more perspective about commercial forest plantations that are being adaptively managed for timber production objectives in conjunction with carbon storage objectives. The results showed that the carbon storage of different species plantation systems was differences that under similar habitat, the same stand age (27-year old) and same management history. Among them, the *Michelia macelurei* stand has the greatest capacity to store carbon, its carbon storage was 359.43t/hm^2. The next was the *Mytilaria laosensis* stand, for 319.80t/hm^2. Carbon storages of the *Castanopsis hystrix* stand, the *Pinus massoniana* stand and the *Mesua ferrea* stand was 225.87t/hm^2, 222.43t/hm^2 and 207.81t/hm^2, and was insignificant. The majority of carbon storage was found in the tree layer and soil layer, accounting for over 95% of the total carbon storage of ecosystems. Litter floor contributed 0.87% ~ 4.29% carbon storage of ecosystems. Understory shrubs and herbs were in small contribution to ecosystem carbon storage, up to only 0.5%.

Carbon storage of tree layer showed larger differences at different species forests, which the *Mytilaria laosensis* stand was the highest, for 188.09t/hm^2, followed by the *Michelia macelurei* stand, for 176.44t/hm^2, and then followed by the *Castanopsis hystrix* stand, for 102.56t / hm^2, the *Pinus massoniana* stand and the *Mesua ferrea* stand were the lowest, for 84.59t/hm^2 and 84.01t/hm^2. Different species of trees had the different organ carbon allocation, but they are all the same law, that the trunk's was the highest, followed by the root's or the branch's, and then followed by the bark's, the leave's was the lowest. Carbon storage of shrub layer and herb layer was 0.036~1.163t/hm^2 and 0.027~0.913t/hm^2, that of litter floor was between 9.54t/hm^2 and 2.37t/hm^2.

In the case of the soil, the carbon was stored mainly in the 0 to 50 cm layer, which accounted for 61.33% to 64.69% of the carbon stored in 100 cm depth of the soil. Carbon storage declined as the depth of soil increased. The carbon storage in the 0 to 100 cm soil layer of the *Michelia macelurei* stand was the greatest, 179.59t/hm^2, higher than that of the other stands, and that of the *Mytilaria laosensis* stand, *Castanopsis hystrix* stand, *Mesua ferrea* stand and *Pinus massoniana* stand was all most the same, about 120t/hm^2.

These findings suggest that the plantations in lower subtropical China have high biomass productivity and higher carbon storage capacity, particularly the *Michelia macelurei* stand and the *Mytilaria laosensis* stand, because of abundant light and heat resources. Therefore, when we become the lower subtropical China into efficient artificial timber base, while also develop it into a carbon sink plantations to give full play to China's forestry response to climate change in the current and future.

Key words: lower subtropical China; plantation forestry; *Mytilaria laosensis*; *Michelia macelurei*; carbon storage

森林生态系统是地球上除海洋之外最大的碳库[1,2]，目前森林包含的碳储量约占陆地生物圈地上碳储量的80%和地下碳储量的40%[3,4]。森林作为陆地生态系统的主体，不但能贮存大量的碳，在调节陆地生态系统与大气碳库之间的碳交换中也起着巨大的“生物泵”作用[5,6]，平均每7年陆地植被就可消耗掉大气中全部的CO_2，其中70%的交换发生在森林生态系[7,8]。由于它的巨大碳库及碳交换的活跃性，森林生态系统在维护全球碳平衡中起着举足轻重的作用[9]，森林与气候变化的关系也逐渐成为人类关注的焦点[10-12]。

随着全球社会经济的快速发展以及改善生态环境的需要，众多国家都制定和实施了长期造林计划，世界人工林面积迅速扩大，日益成为全球森林的重要组成部分[13,14]。我国大面积的人工林至今已有几十年的历史，南方人工林已成为该区域森林的重要组成部分，有的地方甚至成为森林的主体[15,16]。人工林在调节全球碳平衡、减缓大气中CO_2等温室气体浓度上升以及维护全球气候等方面具有不可替代的作用[17,18]。本研究以中国林业科学研究院热带林业实验中心的人工林为研究对象，通过对马尾松、米老排、红椎等5个树种人工林生态系统碳储量的研究，了解和掌握南亚热带人工林生态系统的储碳能力和碳库特征，期待能够为我国林分尺度的森林生态系统碳汇功能研究提供部分基础研究数据，为在全球气候变化的条件下中国区域生态环境建设、制定区域森林生态系统碳汇管理对策以及为中国政府参与世界“碳汇贸易”谈判提供依据和参考。

1 研究地区与研究方法

1.1 研究区概况

研究区位于中国林业科学研究院热带林业实验中心夏石那造大山(21°57′47″N，106°59′30″E)，海拔350m，年均温21.5℃，≥10℃积温7500℃，年降水量1220~1380mm，年蒸发量1370~1390mm，干湿季节明显，雨季(4~9月)降水占年总降水量的85%左右，旱季(10月至翌年3月)降水仅占年降水量的15%左右，土壤为花岗岩母质发育的赤红壤。于1984年2月，在同一坡面营造了米老排林、火力楠林、红椎林，在对坡营造铁力木林，在山脊营造马尾松林。定植时米老排、马尾松的苗龄为1年，其他为2年，造林株行距为1.67m×1.67m。栽植后，前三年人工铲草抚育，郁闭成林后自然生长。林分概况见表1。

表1 林分基本情况表

森林类型	平均树高(m)	平均胸径(cm)	林分密度(trees/hm^2)	郁闭度	林下优势灌草种类
米老排林	18.7	20.2	1492	0.9	米老排、大沙叶、海金沙、扇叶铁线蕨、鞭叶铁线蕨
红椎林	19.1	25.4	525	0.8	红椎、三叉苦、扇叶铁线蕨、半边旗
火力楠林	19.4	18.1	1167	0.8	红椎、酸藤子、火力楠、九节、鞭叶铁线蕨、山姜、粽叶芦

（续）

森林类型	平均树高（m）	平均胸径（cm）	林分密度（trees/hm²）	郁闭度	林下优势灌草种类
铁力木林	14.0	16.4	842	0.9	铁力木、三叉苦、琴叶榕、铁芒萁、扇叶铁线蕨、蕨
马尾松林	16.6	19.1	1142	0.7	大沙叶、红椎、桃金娘、铁芒萁、五节芒、弓果黍

注：表中植物拉丁学名：大沙叶 *Pavetta hongkongensis*；三叉苦 *Evodia lepta*；山芝麻 *Helictercs angustifolia*；杜茎山 *Maesa japonica*；琴叶榕 *Ficus lyrata*；桃金娘 *Rhodomyrtus tomentosa*；九节 *Psychotriarubra*；酸藤子 *Embelia laeta*；海金沙 *Lygodium japonicum*；玉叶金花 *Mussaenda pubescens*；半边旗 *Pteris semipinnata*；山姜 *Alpina japonica*；粽叶芦 *Thysanolaena maxima*；五节芒 *Miscanthus floridulu*；弓果黍 *Cyrtococcum patens*；蔓生莠竹 *Microstegium vagans*；扇叶铁线蕨 *Adiantum flabellulatum*；鞭叶铁线蕨 *Adiantum caudatum*；蕨 *Pteridium aquilinum*；铁芒萁 *Dicranopteris dichotoma*。

1.2 研究方法

1.2.1 样地设置及植物样取样

2011 年 10 月，选择米老排、红椎、火力楠、铁力木和马尾松 5 种人工纯林，每树种设 20m×20m 乔木调查样方 3 块，以胸径 5cm 为起测径级，每木检尺测树高和胸径(树高用瑞典生产的 Vertex Ⅳ 树木超声波测高测距仪测量)。每样方选一株标准木，分别树叶、枝、主干、干皮和根取样，树叶和枝于树冠外围南面中下部取样，干皮选择树干南面距地面 1.2m 处取样，主干用生长锥取树干距地面 1.2m 处样，挖取树木南面侧根取样。样品带回实验室 65℃ 烘干并粉碎。在每个乔木样方对角位置设两个 5m×5m 灌木样方，按叶、枝、根分器官收获样方内所有灌木，称鲜重，然后按各器官分别取样，置于 65℃ 烘箱中烘至恒重，求算灌木层生物量并粉碎备用。在每个乔木样方沿对角线等距离设 3 个 1m×1m 草本样方，分别地上、地下部分收获样方内所有草本，称鲜重，65℃ 烘干计算草本层生物量并粉碎备用。在草本样方内按未分解、半分解和完全分解分三层全部收获地表凋落物样，称鲜重，65℃ 烘干计算凋落物现存量，同时样品粉碎备用。

1.2.2 土壤调查及取样

每树种林地内随机挖 3 个 100cm 深土壤剖面，每剖面按 0～10cm、10～20cm、20～30cm、30～50cm、50～100cm 五层机械分层取样，环刀法测不同深度土壤容重。

1.2.3 乔木生物量计算方法

结合样地调查和建立的单木相对生长方程来计算不同树种生物量，其中米老排、红椎、火力楠和马尾松单木相对生长方程采用已有文献资料和硕博士毕业论文[19-23]。铁力木采用尚未正式发表的自建单木分器官相对生长方程：

$$W_s = 0.0465(D^2H)^{0.9192} \quad R^2 = 0.9792$$

$$W_{ba} = 0.0322(D^2H)^{0.6949} \quad R^2 = 0.9144$$

$$W_{br} = 0.0042D^{3.4565} \quad R^2 = 0.9708$$

$$W_l = 0.0013D^{3.2245} \quad R^2 = 0.9378$$

$$W_r = 0.0344D^{2.4009} \quad R^2 = 0.9485$$

式中：W_s、W_{ba}、W_{br}、W_l 和 W_r 分别为树干、干皮、枝、叶和根的生物量。

1.2.4 有机碳含量测定

所有烘干粉碎后样品采用日本岛津总有机碳分析仪 TOC-L/SSM5000A 进行样品有机碳含量测定。

1.3 数据处理

数据整理、计算与作图均采用 Microsoft Excel 2007 软件进行。

植物体碳储量=植物体生物量×植物体含碳率；

土壤有机碳储量=土壤有机碳含量×土壤容重×土层厚度/10。

2 结果与分析

2.1 不同树种人工林乔木层碳储量特征

表 2 可见，在相似的生境条件下，林龄和经营管理措施相同，不同树种林分乔木层碳储量表现出较大差异，其中以米老排林最高，为 188.09t/hm²，其次是火力楠林，为 176.44t/hm²，再次为红椎林，为 102.56t/hm²，马尾松林和铁力木林最低，分别为 84.59t/hm² 和 84.01t/hm²。最高的米老排林乔木层碳储量是最低铁力木林的 2.24 倍。树种不同，树木器官碳储量分配各不相同，但是均以树干最高，所占比例在 43.36%～66.17%，其次为根(占比 12.55%～24.50%)或枝(占比 7.84%～33.40%)，再其次为干皮，占比 4.33%～13.54%，叶所占比例最低，仅占乔木层的 1.54%～5.91%。

表 2 不同树种人工林乔木层碳的分配 (t/hm^2)

森林类型	树干	干皮	枝	叶	根	合计
米老排林	124.46±2.59 (66.17)	9.15±1.24 (4.87)	14.75±2.45 (7.84)	2.90±0.45 (1.54)	36.83±2.40 (19.58)	188.09±4.87 (100)
红椎林	62.47±1.91 (60.91)	6.26±0.24 (6.11)	17.42±1.85 (16.99)	3.53±0.55 (3.44)	12.87±0.98 (12.55)	102.56±2.83 (100)
火力楠林	92.88±6.57 (52.64)	10.21±0.72 (5.79)	21.31±0.99 (12.08)	8.81±0.24 (4.99)	43.23±2.16 (24.50)	176.44±10.40 (100)
铁力木林	36.43±2.00 (43.36)	3.64±0.22 (4.33)	28.06±3.52 (33.40)	4.97±0.55 (5.91)	10.92±0.63 (13.00)	84.01±5.65 (100)
马尾松林	50.42±3.14 (59.60)	11.45±0.82 (13.54)	6.93±1.03 (8.19)	2.23±0.39 (2.64)	13.56±1.21 (16.03)	84.59±5.68 (100)

注：数值为平均值±标准差；括号内数据为分配比；下同。

2.2 不同树种人工林林下灌草碳分配

虽然林下灌木层和草本层碳储量相比乔木层均较低，灌木层平均为 0.429t/hm^2，草本层平均为 0.236t/hm^2，但不同林分间仍表现出极大差异(表 3)。灌木层碳储量以红椎林最高，为 1.163t/hm^2，其次为马尾松林，为 0.606t/hm^2，最低为铁力木林，仅 0.036t/hm^2，最高和最低相差达 32.3 倍。草本层碳储量以马尾松林最高，为 0.913t/hm^2，其次为红椎林，为 0.172t/hm^2，火力楠、米老排和铁力木林最低，均在 0.030t/hm^{2}左右，最高的马尾松林是最低铁力木林的 33.81 倍。灌木层器官碳储量主要分配在根中，平均占 50.23%，其次为枝，平均占比为 39.29%，叶较低，仅占 10.48%；草本层碳储量除马尾松林外，均为地下部分高于地上部分，地下占比平均为 64.35%，地上占 35.65%。

表 3 不同树种人工林灌木层和草本层碳分配 (t/hm^2)

森林类型	灌木层				草本层		
	枝	叶	根	合计	地上	地下	合计
米老排林	0.034±0.034 (38.23)	0.005±0.004 (5.99)	0.050±0.050 (55.78)	0.090±0.087bc (100)	0.011±0.009 (35.66)	0.020±0.016 (64.34)	0.031±0.025 (100)
红椎林	0.505±0.409 (43.44)	0.176±0.135 (15.17)	0.481±0.418 (41.39)	1.163±0.854a (100)	0.071±0.058 (41.41)	0.101±0.115 (58.59)	0.172±0.169 (100)
火力楠林	0.103±0.073 (41.38)	0.015±0.018 (6.22)	0.130±0.111 (52.40)	0.249±0.193bc (100)	0.013±0.010 (34.26)	0.026±0.020 (65.74)	0.039±0.028 (100)
铁力木林	0.010±0.008 (28.25)	0.003±0.001 (7.21)	0.023±0.029 (64.54)	0.036±0.028c (100)	0.008±0.006 (31.28)	0.019±0.009 (68.72)	0.027±0.011 (100)
马尾松林	0.274±0.156 (45.13)	0.108±0.048 (17.81)	0.225±0.228 (37.06)	0.606±0.414b (100)	0.507±0.357 (55.58)	0.406±0.212 (44.42)	0.913±0.525 (100)

2.3 不同树种人工林地表凋落物碳储量

不同林分间比较，地表凋落物碳储量以马尾松林最高，为 9.54t/hm^2，明显高于其他 4 个林分，其次为红椎林和米老排林，分别为 4.77t/hm^{2}和 4.31t/hm^2，火力楠林和铁力木林最低，分别为 3.12t/hm^{2}和 2.37t/hm^2，最高的马尾松林是最低铁力木林的 4.03 倍。以米老排、红椎等为代表的阔叶树林分，碳素主要储存在未分解层和半分解层，完全分解层碳储量较低(火力楠林和铁力木林由于凋落物分解较快，未见完全分解层)；马尾松林相反，碳素主要储存在完全分解层，其次为未分解层，半分解层储量较低。

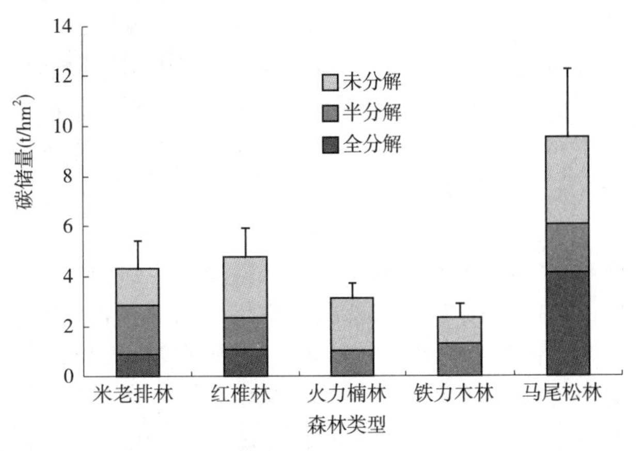

图 1　5 种人工林地表凋落物碳储量

2.4　不同树种人工林土壤碳含量和碳储量

由表 4 可见，对于不同树种人工林，土壤有机碳含量差异较大，但在整个采集剖面上，均表现出随着土壤深度增加，有机碳含量逐步减少，其中以表土层(0~10cm)有机碳含量最高，底层(50~100cm)有机碳含量最低，表明土壤有机碳含量具有明显的垂直递减特征。从林地土壤碳储量空间分布来看，土壤有机碳主要储存在上层，其中，0~50cm 层的有机碳储量占 1m 厚土层的 61.33%~64.69%。不同树种人工林 1m 厚土壤有机碳储量以火力楠林最大，明显高于其他林分，而米老排林、红椎林、铁力木林和马尾松林之间相差不大，均在 120t/hm² 左右。

表 4　不同树种人工林土壤有机碳含量及储量　(g/kg，t/hm²)

森林类型	0~10cm		10~20cm		20~30cm		30~50cm		50~100cm		碳储量合计
	碳含量	碳储量	碳含量	碳储量	碳含量	碳储量	碳含量	碳储量	碳含量	碳储量	
米老排林	23.38 ±5.32	28.29 ±6.44	14.79 ±2.51	19.15 ±3.25	10.08 ±1.57	12.56 ±1.95	7.94 ±0.62	20.65 ±1.61	7.22 ±0.56	46.63 ±3.65	127.28 ±2.49
红椎林	21.03 ±1.73	23.18 ±1.91	16.14 ±3.39	18.51 ±3.89	11.50 ±3.51	14.45 ±4.41	7.26 ±0.64	18.48 ±1.62	6.58 ±0.82	42.59 ±5.29	117.21 ±16.46
火力楠林	29.23 ±6.06	30.89 ±6.41	20.14 ±4.04	25.50 ±5.12	15.81 ±2.99	20.30 ±3.84	12.05 ±3.42	33.45 ±9.50	10.51 ±0.74	69.45 ±4.92	179.59 ±24.41
铁力木林	22.77 ±2.68	26.52 ±3.21	13.28 ±1.00	16.72 ±1.26	9.39 ±1.61	12.78 ±2.19	7.52 ±1.16	20.55 ±3.18	6.83 ±0.72	44.79 ±4.69	121.36 ±12.02
马尾松林	26.28 ±5.35	28.00 ±5.71	14.52 ±5.11	19.3 ±6.81	9.89 ±2.74	13.87 ±3.84	7.64 ±0.41	20.82 ±1.12	6.53 ±0.58	44.76 ±3.97	126.78 ±12.55

2.5　不同树种人工林生态系统碳储量

图 2 可见，综合乔木、林下植被、地表凋落物和土壤，不同树种人工林生态系统碳储量有较大差异，其中以火力楠林具有最大的储碳能力，其碳储量为 359.43t/hm²，其次为米老排林，为 319.80t/hm²，其他三个树种人工林碳储量差异不大，在 207.81~225.87t/hm²。最高的火力楠林碳储量比最低的铁力木林高 72.96%，相差为 151.62t/hm²。

从分配格局(图 2)来看，乔木层与土壤层是森林生态系统碳储量的主体，占生态系统碳储量总量的 95%以上。其中，米老排林生态系统碳储量以乔木层贡献率最大，为 58.81%，土壤层贡献率为 39.80%；火力楠林生态系统碳储量乔木层和土壤层贡献率基本相当，分别为 49.09%和 49.96%；其他树种生态系统碳储量均以土壤层贡献率最大，在 51.89%~58.40%，乔木层贡献率在 38.03%~45.40%。地表凋落物对生态系统碳储量的贡献率在 0.87%~4.29%，林下灌木与草本对生态系统碳储量贡献微小，最多仅占生态系统碳储量的 0.5%。

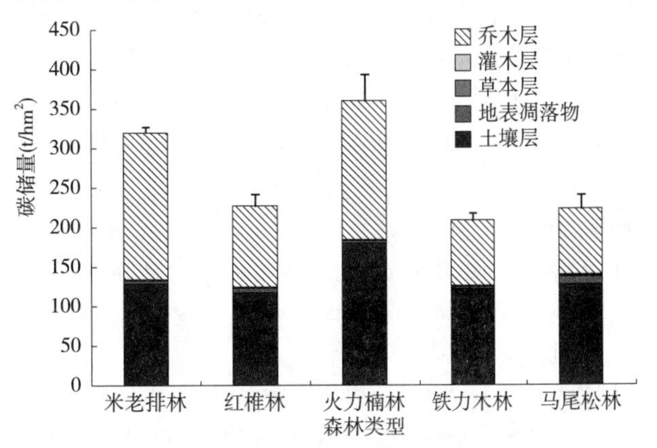

图 2　5 种人工林生态系统碳储量

3 结论与讨论

生境条件相似，林龄相同，经营管理措施一致，不同树种林分乔木层碳储量有较大差异，最高的米老排林是最低铁力木林的2.24倍，相差达104.08 t/hm^2。这主要与树木形态特征和生态习性密切相关，如米老排为当地乡土树种，对本地环境适应良好，人工栽种后成活率和保存率较高，加之冠幅较小，属中性偏阳树种，耐荫蔽[24]，生长后期自疏强度低，故保留有较大的林分密度；此外，该树种树体高大，主干通直，极为速生，单株生物量也较大，所以有高的林分生物量，相应的表现出高的碳储量。铁力木材质坚硬，斧锯难入，是我国热带地区著名的硬材、珍贵材树种，但生长缓慢[25]，由表1可以看出，27年生时，林分平均树高仅14.0m，平均胸径16.4cm，远低于其他树种，所以生物量不高，相应的碳储量较低。因此，在营造碳汇林时需充分了解不同树种的生长特性和生态习性，选择合适树种，充分发挥人工林固碳效能。

树种不同，树木器官碳储量分配各不相同，这是因为树木器官碳储量分配与树木形态关系密切，如米老排树干通直圆满，枝条较细，冠幅较小，故树干碳素占比最高，为66.17%，叶占比最低，仅为1.54%。马尾松树皮较厚，有铜皮松(薄皮)和铁皮松(厚皮)之分[26]，所以干皮占比较其他树高，达13.54%。铁力木树冠圆锥形，冠高比较大，树冠下部侧枝发达，故枝占比较其他树高，达33.40%。火力楠主根不明显，但侧根发达[27]，所以根占比较高，为24.50%。尽管如此，树木器官碳素分配仍表现出以下规律，即树干最高，其次为根或枝，再次为干皮，叶最低。

本研究表明，尽管林下灌木层、草本层和地表凋落物层的碳储量在不同林分间表现出极大的差异，如灌木层最高的红椎林是最低铁力木林的32.3倍，草本层最高和最低相差达33.81倍，地表凋落物层最高和最低也相差达4.03倍，但5种人工林生态系统碳储量在207.81~359.43 t/hm^2，并没有相应的表现出几倍的差异。这是由于灌木层、草本层和地表凋落物层在整个森林生态系统中所占份额极少，三层相加占比不到5%，故对森林生态系统碳储量影响很小。由此看出，南亚热带人工林生态系统碳储量主要由乔木层和土壤层占主导，两者相加占生态系统碳储量总量的95%以上。

火力楠林和米老排林是两个高碳汇树种人工林，经过27年的生长，其生态系统碳储量达到359.43 t/hm^2 和319.80 t/hm^2，远高于我国平均森林碳储量[28]，亦高于同纬度地带[29]和热带天然林平均碳储量[30]。分析比较可以看出，火力楠林的高碳储量主要来自乔木层和土壤层，尤其是土壤层，其土壤层的碳储量明显高于其他四个林分。这可能归功于火力楠发达的根系，表2可见，火力楠根系碳储量为43.23 t/hm^2，占乔木层碳储量的24.5%，无论是绝对值还是相对值在5个树种中均最高。根系死亡后残留在土壤中，腐烂分解成为土壤中的碳，因此越是发达的根系，对土壤碳的贡献越大[31,32]。米老排林的高碳储量主要来自乔木层，乔木层碳储量不但在5个林分中最高，而且在米老排人工林生态系统碳储量中贡献最大，占生态系统碳储量的58.81%，同时亦可发现米老排根系碳储量占乔木层的19.58%，虽低于火力楠，但高于其他三个树种。由此我们可以认为，满足高碳汇树种人工林的条件，一方面是树体高大，生长迅速，能快速积累地上有机物质，另一方面要有发达的根系，通过根系的新陈代谢，持续不断地为土壤提供碳素。

由于南亚热带具有丰富的光热资源，气候暖热、夏长冬短，雨量充沛，光照丰富，且雨热同季[33]，使得南亚热带人工林具有较高的生产力和生物量，以及较高的储碳能力[34]，尤其是本研究中的火力楠和米老排人工林，无论是植被碳储量还是土壤碳储量均高于我国北温带、温带和中亚热带人工林[35-37]。因此，将我国南亚热带建设成为高效人工用材林基地的同时，亦可发展成为高碳汇人工林基地，以充分发挥我国林业在当前和未来应对气候变化的国家战略中的独特作用。

参考文献

[1] PAN Y D, BIRDSEY R A, FANG J Y, et al. A large and persistent carbon sink in the world's forests[J]. Science, 2011, 333: 988-993.

[2] ANDEREGG W R L, BERRY J A, SMITH D D, et al. The roles of hydraulic and carbon stress in a widespread climate-induced forest die-off[J]. Proceeding of the National Academy of Sciences of the United States of America, 2012, 109(1): 233-237.

[3] DIXON R K, BROWN S, HOUGHTON R A, et al. Carbon pools and flux of global forest ecosystem[J]. Science, 1994, 263(4): 185-190.

[4] 刘世荣，王晖，栾军伟. 中国森林土壤碳储量与土壤碳过程研究进展[J]. 生态学报，2011，31(19)：5437-5448.

[5] 方精云，陈安平. 中国森林植被碳库的动态变化及其

意义[J]. 植物学报，2001，43(9)：967-973.

[6]WOOD T E，CAVALERI M A，REED S C. Tropical forest carbon balance in a warmer world：a critical review spanning microbial to ecosystem scale processes[J]. Biological Reviews，2012，87(14)：912-927.

[7]SCHROEDER P. Carbon storage potential of short rotation tropical tree plantations [J]. Forest Ecology and Management，1992，50，31-41.

[8]李新宇，唐海萍 陆地植被的固碳功能与适用于碳贸易的生物固碳方式[J]. 植物生态学报，2006，30(2)：200-209.

[9]GOETZ S J，BOND-LAMBERTY B，LAW B E，et al. Observations and assessment of forest carbon dynamics following disturbance in North America[J]. Journal of Geophysical Research：Biogeosciences，2012，117(G2)：1-17.

[10]KATHLEEN E S，WILIAM J P，ERIC A D，et al. Long-term changes in forest carbon under temperature and nitrogen amendments in a temperate northern hardwood forest [J]. Global Change Biology，2013，19(8)：2389-2400.

[11]SWIFT K. Forest carbon and management options in an uncertain climate [J]. Journal of Ecosystems and Management，2012. 13(1)：1-7.

[12]陈青青，徐伟强，李胜功，等. 中国南方 4 种林型乔木层地上生物量及其碳汇潜力[J]. 科学通报，2012，57(13)：1119-1125.

[13]FAO. Key findings of global forest resources assessment 2010[R]. Rome：FAO，2010.

[14]张明，李智勇，何友均. 人工林与绿色经济[J]. 世界林业研究，2013，26(1)：7-11.

[15]马泽清，王辉民，王绍强，等. 雨雪冰冻灾害对中亚热带人工林的影响——以江西省千烟洲为例[J]. 植物生态学报，2010，34(2)：204-212.

[16]刘庆，尹华军，程新颖，等. 中国人工林生态系统的可持续更新问题与对策[J]. 世界林业研究，2010，23(1)：71-75.

[17]HUANG Z Q，HE Z M，WAN X H，et al. Harvest residue management effects on tree growth and ecosystem carbon in a Chinese fir plantation in subtropical China[J]. Plant and Soil，2013，364(1-2)：303-314.

[18]LIA X D，SONB Y M，LEEB K H，et al. Biomass and carbon storage in an age-sequence of Japanese red pine (*Pinus densiflora*) forests in central Korea[J]. Forest Science and Technology，2013，9(1)：39-44.

[19]明安刚，贾宏炎，陶怡，等. 桂西南 28 年生米老排人工林生物量及其分配特征[J]. 生态学杂志，2012，31(5)：1050-1056.

[20]覃林，何友均，李智勇，等. 南亚热带红椎马尾松纯林及其混交林生物量和生产力分配格局[J]. 林业科学，2011，47(12)：17-21.

[21]赵凯. 福建柏火力楠人工纯林及其混交林碳储量的研究[D]. 福州：福建农林大学，2010.

[22]齐之尧，马家禧，李顺明. 火力楠人工林生物量、生产力的研究[J]. 生态学杂志，1985，4(2)：17-30.

[23]张治军. 广西造林再造林固碳成本效益研究[D]. 北京：中国林业科学研究院，2009.

[24]黄正暾，王顺峰，姜仪民，等. 米老排的研究进展及其开发利用前景[J]. 广西农业科学，2009，40(9)：1220-1223.

[25]朱积余，廖培来. 广西名优经济树种[M]. 北京：中国林业出版社，2006，84-86.

[26]秦国峰，周志春. 中国马尾松优良种质资源[M]. 北京：中国林业出版社，2012，25.

[27]梁有祥，韦中绵，玉桂成，等. 桂东南地区火力楠人工林生物量研究[J]. 林业技开发，2010，24(5)：45-49.

[28]周玉荣，于振良，赵士洞. 我国主要森林生态系统碳贮量和碳平衡[J]. 植物生态学报，2000，24(5)：518-522.

[29]方运霆，莫江明，彭少麟，等. 森林演替在南亚热带森林生态系统碳吸存中的作用[J]. 生态学报，2003，23(9)：1685-1694.

[30]吕晓涛. 西双版纳热带季雨林碳储量的研究[D]. 昆明：中国科学院西双版纳热带植物园，2006：42-43.

[31]王晖. 南亚热带四种人工林土壤碳固持及其主要相关过程研究[D]. 北京：中国林业科学研究院，2010，38-43.

[32]万晓华，黄志群，何宗明，等. 阔叶和杉木人工林对土壤碳氮库的影响比较[J]. 应用生态学报，2013，24(2)：345-350.

[33]况雪源，苏志，涂方旭. 广西气候区划[J]. 广西科学，2007，14(3)：278-283.

[34]YOU J H，QIN L，LI Z Y，et al. Carbon storage capacity of monoculture and mixed-species plantations in subtropical China[J]. Forest Ecology and Management，2013(295)：193-198.

[35]马炜，孙玉军，郭孝玉，等. 不同林龄长白落叶松人工林碳储量[J]. 生态学报，2010，30(17)：4659-4667.

[36]杨玉姣，陈云明，曹扬. 黄土丘陵区油松人工林生态系统碳密度及其分配[J]. 生态学报，2014，34(3)：360-368.

[37]邸月宝，王辉民，马泽清，等. 亚热带森林生态系统不同重建方式下碳储量及其分配格局[J]. 科学通报，2012，57(17)：1553-1561.

[原载：中南林业科技大学学报，2014，34(12)]

南亚热带不同树种人工林生物量及其分配格局

郑 路[1,2] 蔡道雄[1,2] 卢立华[1,2] 明安刚[1,2] 于浩龙[1,2] 李忠国[1]

([1]中国林业科学研究院热带林业实验中心，广西凭祥 532600；
[2]广西友谊关森林生态系统国家定位观测研究站，广西凭祥 532600)

摘 要 通过收获法和建立的单木相对生长方程研究了南亚热带5种树种人工林乔、灌、草不同组分的生物量及其分配。结果表明：在立地条件相似、林龄和经营管理措施相同的情况下，不同树种人工林生物量有较大差异，表现为米老排林(404.95t/hm^2)>火力楠林(376.61t/hm^2)>马尾松林(239.94t/hm^2)>红椎林(231.01t/hm^2)>铁力木林(181.06t/hm^2)。林分生物量空间分布格局以乔木层为主，占总生物量的87.71%~97.86%，其次为地表凋落物层，占1.96%~10.90%，灌木层和草本层最低，仅占0.02%~1.09%。林分乔木层各器官的生物量分配格局总体呈现出树干生物量所占比例最大，根或枝所占比例次之，再其次是干皮，叶生物量最低，但具体到某一树种又略有差异，体现出共性中又有个性。林下灌木层、草本层和地表凋落物层生物量在不同林分间均差异很大，但表现出相似的规律，即以马尾松林最高，红椎林其次，米老排林、火力楠林和铁力木林较低。

关键词 南亚热带；人工林；生物量；空间分布

Biomass Allocation of Different Species Plantations in the Sub-tropical Area of China

ZHENG Lu[1,2], CAI Daoxiong[1,2], LU Lihua[1,2], MING Angang[1,2], YU Haolong[1,2], LI Zhongguo[1]

(*[1]Experiment Center of Tropical Forestry, Chinese Academy of Forestry, Pingxiang 532600, Guangxi, China;*
[2]Guangxi Youyiguan Forest Ecosystem Research Station, Pingxiang 532600, Guangxi, China)

Abstract: Allocation pattern of biomass in five plantations was studied by harvesting method and established allometric equations in south subtropics of China. The results showed that: the biomass of different species plantations were quite different in the same circumstances of similar site conditions, age and management measures, showing that *Mytilaria laosensis* stand(404.95t/hm^2) > *Michelia macclurei* stand(376.61t/hm^2) > *Pinus massoniana* stand (239.94t/hm^2) > *Castanopsis hystrix* stand (231.01t/hm^2) > *Mesua ferrea* stand(181.06t/hm^2). Biomass spatial pattern dominated by tree layer, which accounted for 87.71%~97.86%, followed by the litter layer, accounting for 1.96%~10.90%, shrub and herb layers minimum, accounting for only 0.02%~1.09%. Organs biomass pattern of tree layer showed that proportion of stem biomass was the largest, followed by the root′s or branche's, and then followed by the bark′s, the lowest was the leaf′s overall. But to a particular species, there was different slightly, reflecting personality among commonality. Biomass of shrub layer, herb layer, and floor litter among different stands were very different, but showed a similar law, that *Pinus massoniana* stand was the highest, *Castanopsis hystrix* secondly, *Mytilaria laosensis* stand, *Michelia macclurei* stand and *Mesua ferrea* stand were lower.

Key words: south sub-tropical area; plantation; biomass; spatial distribution

森林生物量是量度森林结构和功能变化的重要指标[1-3]，并为生态系统的碳汇和碳素循环研究提供关键数据，在碳循环、全球气候变化研究中起到重要作用[4-6]。我国森林生物量的测定开始于20世纪70年代末80年代初[7,8]，之后，各地对我国主要森林类型的生物量都有测定[9-11]，这些研究大大推进了我国森林生物量及相关的生态系统生态学和全球变化研究的开展[12,13]。我国是世界上人工林保存面积最大的国家，南方人工林已成为该区域森林的重要组成部分，有的地方甚至成为森林的主体[14,15]，重视对人工林生态系统生物量和生产力研究意义重大。

位于广西凭祥市的中国林业科学研究院热带林业实验中心的人工林群落，在南亚热带区域人工林中具有广泛的代表性。近年来有学者对该地区的杉木、米老排等人工林生物量开展了研究[16,17]，也有学者对纯林和混交林生物量及生产力进行了研究[18]，但是缺少多树种人工林生物量的比较研究，对于不同树种人工林生物量有何差异尚不清楚。本文选取中国林业科学研究院热带林业实验中心在相近立地条件下同一年栽种，并且经营管理措施相同的5个树种人工林开展生物量及其分配格局的比较研究，以揭示我国南亚热带人工林物质生产规律，为进一步研究南亚热带人工林能量转化、物质循环及准确评估南亚热带人工林碳汇潜力提供基本数据。

1 研究区概况与研究方法

1.1 研究区概况

研究区位于中国林业科学研究院热带林业实验中心夏石那造大山(21°57′47″N，106°59′30″E)，海拔350m，年均温21.5℃，≥10℃积温7500℃，年降水量1220~1380mm，年蒸发量1370~1390mm，干湿季节明显，雨季(4~9月)降水占年总降水量的85%左右，旱季(10月至翌年3月)降水仅占年降水量的15%左右，土壤为花岗岩母质发育的赤红壤。于1984年2月，在同一坡面营造了米老排林、火力楠林、红椎林，在对坡营造铁力木林，在山脊营造马尾松林。定植时米老排、马尾松的苗龄为1年，其他为2年，造林株行距为1.67m×1.67m。栽植前3年人工铲草抚育，郁闭成林后自然生长。米老排和铁力木林下植物稀少且分布不均，灌木主要有大沙叶(*Pavetta hongkongensis* Brem.)、海金沙[*Lygodium japonicum* (Thunb.) Sw.]、三叉苦[*Evodia lepta* (Spreng.) Merr.]和琴叶榕(*Ficus pandurata* Hance)等，草本主要有扇叶铁线蕨(*Adiantum flabellulatum* L. Sp.)、鞭叶铁线蕨(*Adiantum caudatum* Linn.)等；红椎林的林下天然更新红椎幼苗较多，平均株高1.2m，灌木主要有三叉苦等，草本主要有扇叶铁线蕨和半边旗(*Pteris semipinnata* L. Sp.)等；火力楠林的林下灌木主要有酸藤子[*Embelia laeta* (Linn.) Mez]和九节[*Psychotria rubra* (Lour.) Poir.]等，草本主要有鞭叶铁线蕨、山姜(*Alpina japonica* Miq.)和粽叶芦[*Thysanolaena maxima* (Roxb.) Kuntze]等；马尾松林因较稀疏，故林下植被较多，灌木主要有大沙叶、桃金娘[*Rhodomyrtus tomentosa*(Ait.) Hassk.]等，草本主要有铁芒萁[*Dicranopteris linearis* (Burm.) Underw.]、五节芒[*Miscanthus floridulus* (Lab.) Warb. ex Schum et Laut.]和弓果黍[*Cyrtococcum patens*(Linn.) A. Camus]等。林分概况见表1。

表1 林分基本情况表

森林类型	坡向	坡度(°)	平均树高(m)	平均胸径(cm)	林分密度(株/hm²)	郁闭度
米老排林	东南	35	18.7±4.12	20.2±5.44	1492	0.9
红椎林	东	30	19.1±3.85	25.4±7.51	525	0.8
火力楠林	东	30	19.4±6.08	18.1±6.56	1167	0.8
铁力木林	西	30	14.0±1.24	16.4±2.70	842	0.9
马尾松林	北	15	16.6±2.45	19.1±4.27	1142	0.7

1.2 研究方法

样地设置及取样：2011年10月，选择米老排、红椎、火力楠、铁力木和马尾松5种人工纯林，每树种按上、中、下三个坡位设20m×20m乔木调查样方三块，以胸径5cm为起测径级，每木检尺测树高和胸径(树高用瑞典生产的VertexⅣ树木超声波测高、测距仪测量)。在每个乔木样方对角位置设两个5m×5m灌木样方，按叶、枝、根分器官收获样方内所有灌木，实验室称鲜重，65℃烘干计算灌木层生物量。在每个乔木样方沿对角线等距离设三个1m×1m草本样方，分别地上、地下部分收获样方内所有草本，实验室称鲜重，65℃烘干计算草本层生物量。在草本样方内按未分解、半分解和完全分解三层分别收获地表凋落物样，实验室称鲜重，65℃烘干计算凋落物现存量。

乔木生物量计算方法：结合样地调查和建立的单木相对生长方程来计算不同树种生物量，其中米老排、红椎、火力楠和马尾松单木相对生长方程采用已发表的文献资料和硕博士毕业论文[17-21]。根据铁力木林乔木样方每木检尺结果，选取径阶标准木10株伐倒后用分层切割法将样树分成干、干皮、枝、叶和根5个组分，收获法测定铁力木各组分生物量，以此建立铁力木单木相对生长方程：

$Ws = 0.0465(D^2H)0.9192 \quad R^2 = 0.9792$

$Wba = 0.0322(D^2H)0.6949 \quad R^2 = 0.9144$

$Wbr = 0.0042\ D\ 3.4565 \quad R^2 = 0.9780$

$Wl = 0.0013\ D\ 3.2245 \quad R^2 = 0.9378$

$Wr = 0.0344\ D\ 2.4009 \quad R^2 = 0.9485$

式中：Ws、Wba、Wbr、Wl 和 Wr 分别为树干、干皮、枝、叶和根的生物量。

2 结果分析

2.1 乔木层生物量

由表2可以看出，林龄相同，在相似的立地条件下，不同树种人工林乔木层生物量有很大差异，以米老排林和火力楠林较高，红椎林其次，铁力木林和马尾松林较低，其中，米老排林比红椎林、铁力木林和马尾松林分别高出85.14%、123.54%和128.50%。林分乔木层各器官的生物量分配格局总体呈现出树干生物量所占比例最大，在40.16%~65.49%，根或枝所占比例次之(7.87%~26.21%)，再其次是干皮(4.18%~12.80%)，叶生物量最低(1.48%~5.70%)。但具体到某一树种又略有差异，如多数树种根生物量高于枝，但红椎和铁力木的枝生物量高于根；马尾松的干皮生物量较多，高于枝；铁力木的叶生物量高于干皮。体现出共性中又有个性。地下与地上生物量的比值在0.16~0.35。

表2 不同树种人工林乔木层生物量 (t/hm^2)

林分类型	树干	干皮	枝	叶	根	合计	根冠比
米老排	257.03 (65.49%)	21.07 (5.37%)	30.88 (7.87%)	5.80 (1.48%)	77.71 (19.80%)	392.50 (100%)	0.25
火力楠	185.30 (49.50%	23.85 (6.37%)	45.91 (12.66%)	17.97 (5.26%)	95.50 (26.21%)	368.54 (100%)	0.35
红椎	126.17 (59.93%)	13.34 (6.37%)	36.57 (16.42%)	7.45 (3.29%)	28.48 (13.98%)	212.00 (100%)	0.16
铁力木	74.69 (40.16%)	8.05 (4.18%)	59.45 (36.38%)	9.47 (5.70%)	23.93 (13.57%)	175.58 (100%)	0.16
马尾松	101.67 (58.74%)	22.61 (12.80%)	14.34 (8.75%)	4.30 (2.65%)	28.85 (17.06%)	171.77 (100%)	0.20

2.2 林下植被生物量

不同林分间林下灌木层生物量差异很大(表3)，其中，以红椎林和马尾松林较高，均在$1t/hm^2$以上，火力楠林和米老排林较低，铁力木林最低，不到$0.1t/hm^2$，最高的红椎林灌木层生物量是最低的铁力木林的32倍。灌木层不同器官生物量分配格局，除马尾松林外，均以根所占比例最大，在43.22%~65.41%，其次为枝(27.79%~41.90%)，叶所占比例最低(5.35%~14.88%)；马尾松林灌木层不同器官生物量空间分配为枝(44.28%)>根(38.42%)>叶(17.30%)。灌木层地下地上生物量之比以马尾松林最低，铁力木林最高。

不同林分间林下草本层生物量同样表现出巨大差异，以马尾松林最高，超过$2t/hm^2$，红椎林和火力楠林较低，米老排林和铁力木林最低，均不到$0.1t/hm^2$，最高的马尾松林草本层生物量是最低的铁力木林的25倍。除马尾松林外，草本层地下生物量均高于地上生物量。草本层生物量根冠比与灌木层表现出同样的规律，以马尾松林最低，铁力木林最高。

表 3 不同树种人工林灌木层和草本层生物量 (t/hm^2)

林分类型	灌木层					草本层			
	枝	叶	根	合计	根冠比	地上	地下	合计	根冠比
米老排	0.077 (35.04%)	0.012 (5.35%)	0.130 (59.61%)	0.219 (100%)	1.48	0.029 (34.52%)	0.055 (65.48%)	0.084 (100%)	1.92
火力楠	0.224 (39.64%)	0.034 (6.00%)	0.308 (54.36%)	0.566 (100%)	1.19	0.035 (33.65%)	0.069 (66.35%)	0.104 (100%)	2.00
红椎	1.056 (41.90%)	0.375 (14.88%)	1.090 (43.22%)	2.521 (100%)	0.76	0.169 (39.49%)	0.259 (60.51%)	0.428 (100%)	1.53
铁力木	0.022 (27.79%)	0.005 (6.80%)	0.052 (65.41%)	0.080 (100%)	1.89	0.022 (27.50%)	0.059 (72.50%)	0.080 (100%)	2.73
马尾松	0.579 (44.28%)	0.226 (17.30%)	0.502 (38.42%)	1.307 (100%)	0.62	1.091 (53.82%)	0.936 (46.18%)	2.027 (100%)	0.86

2.3 地表凋落物现存量

地表凋落物现存量以马尾松林最高(图 1)，达 26.16t/hm^2，其次为红椎林和米老排林，分别为 16.06t/hm^2和 12.15t/hm^2，以火力楠林和铁力木林最低，仅为 7.40t/hm^2和 5.32t/hm^2。马尾松林和红椎林以完全分解凋落物所占比例最大，分别为 56.85% 和 42.20%，其次为未分解凋落物(27.12% 和 33.42%)，半分解凋落物所占比例最低(16.03%和 24.39%)。米老排林为半分解凋落物现存量(39.92%)>完全分解(35.21%)>未分解(24.87%)。火力楠林和铁力木林因凋落物分解较快，完全分解层已和土壤融为一体，故只有未分解和半分解凋落物，其中火力楠林未分解凋落物现存量大于半分解，而铁力木林正相反。

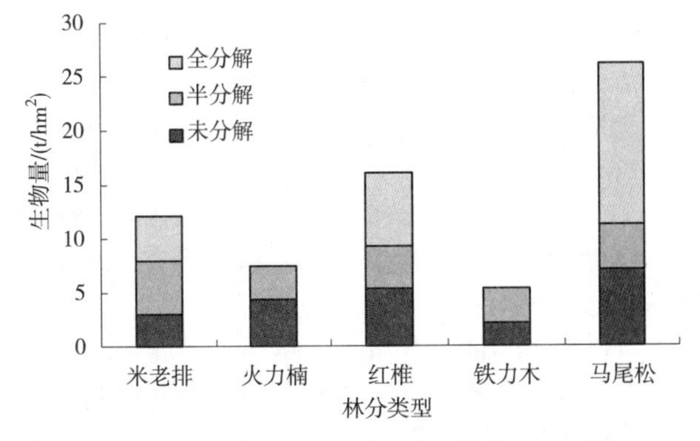

图 1 不同树种人工林地表凋落物现存量

2.4 林分生物量及其分配格局

林分生物量由乔木层、灌木层、草本层和地表凋落物层组成，从图 2 可以看出，在立地条件相似，林龄和经营管理措施相同的情况下，不同树种人工林生物量有较大差异，表现为米老排林(404.95t/hm^2)>火力楠林(376.61t/hm^2)>马尾松林(239.94t/hm^2)>红椎林(231.01t/hm^2)>铁力木林(181.06t/hm^2)，其中最高的米老排林生物量比最低的铁力木林生物量高出 123.66%。5 种人工林生物量的空间分布格局基本一致，以乔木层为主，占总生物量的 87.71% ~ 97.86%，其次为地表凋落物层，占 1.96% ~ 10.90%，灌木层和草本层最低，仅占 0.02%~1.09%。

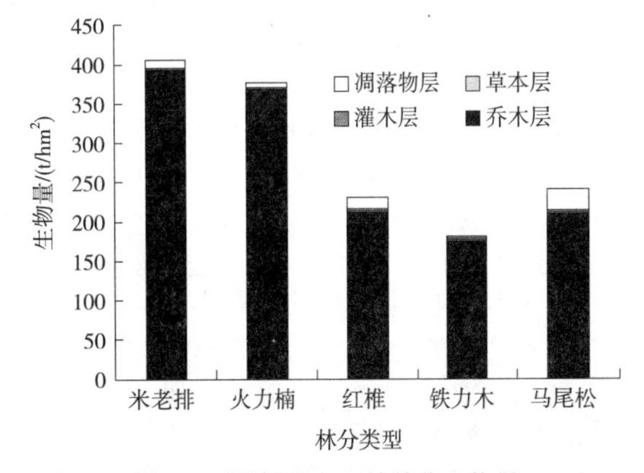

图 2 不同树种人工林林分生物量

3 讨论

已有研究表明人工林生物量与其地理位置、气候、林龄和立地条件等因素密切相关[22-25]。本研究结果进一步表明，尽管林龄相同，立地条件相近，经营管理措施相同，但由于树种不同，仍使得人工林生物量表现出极大差异，其中最高的米老排林生物量是最低铁力木林的 2.24 倍。产生这种差异的主要原因是由于乔木层是林分生物量的主体，一般占总生物量的85%以上，林分生物量主要由乔木层主导。而不同乔木树种形态特征、生态习性和生长特性各异，使得树木生长有快有慢，林分自疏有强有弱，如表1中红椎林的平均树高和平均胸径比铁力木林高36.43%和54.88%，米老排林的保留密度是铁力木林的1.77倍。因此造成乔木层生物量差异较大，从而使林分生物量产生较大差异，且随着生长时间的延长，这种差异愈加明显。

植物生物量分配受生物因子(物种、植株大小、年龄等)和环境因子(光照、水分和养分等)影响[26,27]。植物生物量分配格局的差异是植物生理和生态因素共同作用的结果[28]。本研究显示，乔木生物量分配格局主要由树种形态特征所决定，如米老排为常绿高大乔木，树干通直圆满，故主干生物量所占比例较多(65.49%)，铁力木为常绿大乔木，树冠圆锥形，冠大荫浓，故枝和叶生物量所占比例相对其他树种高(36.38%和5.70%)；根冠比在0.16~0.35，表明乔木生物量主要集中在地上部分。灌木和草本生物量分配主要受环境因子的影响，特别是林下光环境的影响。由表1和表3可以看出，灌木和草本的地下地上生物量比值随着林下光照强度的减弱而增大，如马尾松林(郁闭度0.7)灌木层的根冠比为0.62，铁力木林(郁闭度0.9)灌木层的根冠比为1.89。由乔木、灌木到草本，地下地上生物量比值快速增大，对于林下草本，生物量更多的分配在地下部分。

林分生物量分配格局比较发现，乔木层占林分生物量较大时，其林下灌木层和草本层的生物量则较小，即乔木层和林下其他层间在生物量上有一种互补效应。这是因为乔木层生物量比例的提高，反映了该森林的郁闭度大，林下光照弱，不利于下木层植被的生长发育。即下木层生物量受乔木层控制。对于人工林，乔木层生物量所占比例常大于天然林[29]，林下灌木和草本生物量所占比例偏低，不利于人工林的健康稳定和可持续发展。如何合理进行人工林结构调控，使乔、灌、草生物量达到合适比例，是人工林经营需要进一步研究的问题。

参考文献

[1]冯宗炜，王效科，吴刚．中国森林生态系统的生物量和生产力[M]．北京：科学出版社，1999.

[2] ANDERSON N，JONES J G，PAGE-DUMROESE D，et al. A comparison of producer gas，biochar，and activated carbon from two distributed scale thermochemical conversion systems used to process forest biomass [J]. Energies，2013，6(1)：164-183.

[3] QIU S，BELL R W，HOBBS R J，et al. Estimating nutrient budgets for prescribed thinning in a regrowth eucalyptus forest in south-west Australia [J]. Forestry，2012，85(1)：51-61.

[4] AHERNE J，POSCH M，FORSIUS M，et al. Impacts of forest biomass removal on soil nutrient status under climate change：a catchment-based modelling study for Finland [J]. Biogeochemistry，2012，107(1)：471-488.

[5] COOMES D A，HOLDAWAY R J，KOBE R K，et al. A general integrative framework for modelling woody biomass production and carbon sequestration rates in forests [J]. Journal of Ecology，2012，100(1)：42-64.

[6] ZHANG C，JU W，CHEN J M，et al. China's forest biomass carbon sink based on seven inventories from 1973 to 2008[J]. Climatic Change，2013，118(3-4)：933-948.

[7]陈炳浩，陈楚莹．沙地红皮云杉森林群落生物量和生产力的初步研究[J]．林业科学，1980，16(4)：269-278.

[8]冯宗炜，陈楚莹，张家武，等．湖南会同地区马尾松林生物量的测定[J]．林业科学，1982，18(2)：127-134.

[9]李意德，曾庆波，吴仲民，等．尖峰岭热带山地雨林生物量的初步研究[J]．植物生态学与地植物学学报，1992，16(4)：293-300.

[10]任海，彭少麟，向言词．鹤山马占相思人工林的生物量和净初级生产力[J]．植物生态学报，2000，24(1)：18-21.

[11]王洪岩，王文杰，邱岭，等．兴安落叶松林生物量，地表枯落物量及土壤有机碳储量随林分生长的变化差异[J]．生态学报，2012，32(3)：833-843.

[12]赵敏．中国主要森林生态系统碳储量和碳收支评估[D]．北京：中国科学院植物研究所，2004.

[13]方精云，郭兆迪，朴世龙，等．1981—2000年中国陆地植被碳汇的估算[J]．中国科学：D辑，2007，37(6)：804-812.

[14]刘庆，尹华军，程新颖，等．中国人工林生态系统的可持续更新问题与对策[J]．世界林业研究，2010，23(1)：71-75.

[15]马泽清，王辉民，王绍强，等．雨雪冰冻灾害对中亚热带人工林的影响——以江西省千烟洲为例[J]．植物生态学报，2010，34(2)：204-212.

[16]康冰，刘世荣，蔡道雄，等．南亚热带杉木生态系统生物量和碳素积累及其空间分布特征[J]．林业科学，2009，45(8)：147-153.

[17]明安刚，贾宏炎，陶怡，等．桂西南28年生米老排人工林生物量及其分配特征[J]．生态学杂志，2012，

31(5)：1050-1056.

[18]覃林，何友均，李智勇，等．南亚热带红椎马尾松纯林及其混交林生物量和生产力分配格局[J]. 林业科学，2011，47(12)：17-21.

[19]赵凯．福建柏火力楠人工纯林及其混交林碳储量的研究[D]. 福州：福建农林大学，2010.

[20]齐之尧，马家禧，李顺明．火力楠人工林生物量、生产力的研究[J]. 生态学杂志，1985，4(2)：17-30.

[21]张治军．广西造林再造林固碳成本效益研究[D]. 北京：中国林业科学研究院，2009.

[22]唐罗忠，刘志龙，虞木奎，等．两种立地条件下麻栎人工林地上部分养分的积累和分配[J]. 植物生态学报，2010，34(6)：661-670.

[23]程瑞梅，封晓辉，肖文发，等．北亚热带马尾松净生产力对气候变化的响应[J]. 生态学报，2011，31(8)：2086-2095.

[24]宋曰钦，翟明普，贾黎明．不同树龄三倍体毛白杨生物量分布规律[J]. 东北林业大学学报，2010，38(1)：1-3.

[25]刘延惠，王彦辉，于澎涛，等．六盘山主要植被类型的生物量及其分配[J]. 林业科学研究，2011，24(4)：443-452.

[26]ENQUIST B J, NIKLAS K J. Global allocation rules for patterns of biomass partitioning in seed plants [J]. Science, 2002, 295(5559): 1517-1520.

[27]马维玲，石培礼，李文华，等．青藏高原高寒草甸植株性状和生物量分配的海拔梯度变异[J]. 中国科学：生命科学，2010，40(6)：533-543.

[28]MCCARTHY M C, ENQUIST B J. Consistency between an allometric approach and optimal partitioning theory in global patterns of plant biomass allocation[J]. Functional Ecology, 2007, 21(4): 713-720.

[29]方江平．西藏南伊沟林芝云杉林生物量与生产力研究[J]. 林业科学研究，2012，25(5)：582-589.

[原载：林业科学研究，2014，27(04)]

南亚热带3种阔叶树种人工幼龄纯林及其混交林碳贮量比

明安刚[1,2,3] 刘世荣[2] 农 友[1,3] 蔡道雄[1,3] 贾宏炎[1,3] 黄德卫[1,3] 王群能[1,3] 农 志[1,3]

([1]中国林业科学研究院热带林业实验中心，广西凭祥 532600；[2]中国林业科学研究院森林生态环境与保护研究所，北京 100091；[3]广西友谊关森林生态系统国家定位观测研究站，广西凭祥 532600)

摘 要 如何通过优化造林模式来提高人工林生态系统碳贮量已受到广泛关注。本文以南亚热带8年生格木纯林(PE)、红椎纯林(PC)、米老排纯林(PM)及格木×红椎×米老排混交林(MECM)生态系统为研究对象，对其碳贮量及其分配特征进行了比较研究。结果表明：格木、红椎和米老排不同器官平均碳含量分别为512.4~561.7g/kg，474.2~553.4g/kg和512.8~556.3g/kg。相同树种不同器官之间碳含量差异显著($P<0.05$)。各器官碳含量的平均值大小顺序为格木(539.3g/kg)>米老排(532.7g/kg)>红椎(515.3g/kg)。不同林分间，灌木层、草本层和凋落物层碳含量均以米老排纯林最高，混交林(MECM)居次，红椎纯林和格木纯林最低；不同林分之间的土壤碳含量差异显著($P<0.05$)，0~10cm，10~30cm，30~50cm和50~100cm土壤碳含量均以米老排纯林最高，红椎纯林居次，格木纯林和混交林(MECM)土壤碳含量最低。生态系统碳贮量大小顺序为米老排(308.0t/hm^2)>混交林(182.8t/hm^2)>红椎纯林(180.2t/hm^2)>格木纯林(135.2t/hm^2)，相同组分不同林分间以及相同林分的不同组分间均存在显著差异($P<0.05$)，但混交林与红椎纯林间碳贮量总量无显著差异($P>0.05$)。造林模式对人工林碳贮量及其分配有显著影响，营建混交林有利于红椎和格木地上碳的累积，不利于土壤碳的固定，而营建纯林既有利于米老排生物量碳的吸收，也有利于土壤碳的固定。因而，对碳汇林造林模式的选择，应根据树种固碳特性而定。

关键词 格木；红椎；米老排；人工幼林；造林模式；碳贮量

Comparison of Carbon Storage in Juvenile Monoculture and Mixture Plantation Stands of Three Common Broadleaved Tree Species in Subtropical China

MING Angang[1,2,3], LIU Shirong[2], NONG You[1,3], CAI Daoxiong[1,3], JIA Hongyan[1,3], HUANG Dewei[1,3], WANG Qunneng[1,3], NONG Zhi[1,3]

([1]*Experiment Center of Tropical Forestry*, *Chinese Academy of Forestry*, *Pingxiang* 532600, *Guangxi*, *China*; [2]*Institute of Forest Ecology*, *Environment and Protection*, *Chinese Academy of Forestry*, *Beijing* 100091, *P. R. China*; [3]*Guangxi Youyiguan Forest Ecosystem Research Station*, *Pingxiang* 532600, *Guangxi*, *China*)

Abstract: Much attention was paid to improve the carbon storage of the plantation ecosystem by optimizing afforestation pattern. Carbon storage and its allocation of 8-year-old *Erythrophleum fordii* stand(PE) *Castanopsis hystrix* stand(PC), *Mytilaria laosensis* stand(PM) and a mixed of *E. fordii* × *C. hystrix* × *M. laosensis* stand(MECM) in subtropical China were studied in this paper. The results showed thatthe average carbon content of different organs in *E. fordii*, *C. hystrix* and *M. laosensis* were respectively 509.0~572.4g/kg, 474.2~553.4g/kg and 512.8~556.3g/kg. Significant differences were found among different organs in the same tree species. In the same organs, the average carbon content in different stands ranks as *E. fordii*(539.3g/kg)> *M. laosensis*(532.7g/kg)> *C. hystrix*(515.3g/kg). In shrub, herb and litter layer, PM has the highest carbon content in the four stands, that of MECM was the second highest, while, PC and PE have the lowest value. Soil content in the 0-10cm, 10-30cm, 30-50cm and 50-100cm showed significant difference among the four stands, PM has the highest carbon content in the four soil layer, PC was the second, MECM and PE were the lowest. The carbon storage in the ecosystem were ranked as PM(308.0t/hm^2)> MECM(182.8t/hm^2)> PC(180.2t/hm^2)> PE(135.2t/hm^2), Significant differences were found among different components in the same stand, and also among different stands in the same component($P<0.05$), but no significant differences was found in total ecosystem carbon storage between MECM and PE($P>0.05$). Afforestation mode can remarkably affect the carbon storage and allocation for plantations. Mixed stand will benefit for the increase of aboveground biomass carbon for *C. hystrix* and *E. fordii*, but not good for the accumination of soil carbon; for *M. laosensis*, both the sequestration of biomass carbon and soil carbon can benefit from the monoculture mode. Therefore, we should select the afforestation mode on the basis of the carbon accumination characteristics of tree species for carbon sequestration forest.

Key words: *Erythrophleum fordii*; *Castanopsis hystrix*; *Mytilaria laosensis*; young plantations; afforestation mode; carbon storage

工业革命以后，由于化石燃料的大量使用和土地利用方式的变化，大气中CO_2浓度连年升高[1]，CO_2的吸收、固定和排放过程是全球气候变化研究的重要内容[2-4]。当前，世界上近$4\times10^9hm^2$森林中储存了860Pg碳，且每年可以从大气吸收2.4Pg碳，折合成CO_2为8.8Pg[5-8]。因而，作为陆地生态系统的主体，森林在储存CO_2、调节全球气候、减缓全球气候变暖方面的作用不可替代[9,10]。当前，造林和再造林作为一种新增碳汇的主要途径，已受到学术界的高度重视[11-13]。人工林在吸收和固定CO_2及减缓全球气候变暖等方面发挥着重要作用，并日益引起人们的广泛关注。为了更好地利用科学经营的方式减缓全球气候变化，需要对不同经营模式的人工林固碳能力和潜力有深入的认识和科学的评估[14,15]。

最近十年，诸多学者对不同树种人工林的碳含量、碳贮量及其空间分布格局进行了深入研究[16-25]，为森林碳汇功能的研究做出了积极贡献。也有学者对不同造林模式的人工林生物量和碳贮量进行了研究，发现造林模式对人工林碳贮量有重要影响，纯林与混交林地上、地下碳贮量都有明显差异，但并未得出一致的结论，He和Wang等人的研究认为红椎×马尾松混交林土壤碳贮量高于马尾松纯林的，也高于红椎纯林[13,24]，而何友均等人的研究发现西南桦纯林生态系统碳贮量高于西南桦×红椎混交林[23]。由此可见，造林模式对人工林碳贮量的究竟会产生怎样的影响，仍有相当大的不确定性，有进一步研究的必要。

格木(*Erythrophleum fordii*)、红椎(*Castanopsis hystrix*)和米老排(*Mytilaria laosensis*)是我国南亚热带地区乡土阔叶树种，也是适合在该地区培养大径材的珍优造林树种[23]。其中，格木是我国著名的硬木之一，与蚬木(*Excetrodendron hsiemvu*)、金丝李(*Garcinia paucinervis*)并称为“广西三大硬木”；红椎和米老排是分布在南亚热带地区优良速生的用材树种，米老排人工林还具有较强的水源涵养功能[26]。近几年，格木、红椎和米老排3个树种人工林发展迅速，逐渐成为南亚热带地区主要的乡土阔叶造林树种。目前，已有学者对不同龄林的格木、米老排和红椎人工林碳贮量的进行了研究，发现林龄对格木、米老排和红椎人工林碳贮量有显著影响，人工林碳贮量随林龄的增加而增加[22-24,27]。但种植模式对3种人工林固碳功能和固碳潜力的影响如何，学术界尚缺乏这方面的了解和认识，急需对不同种植模式的人工林碳贮量进行研究。本文对南亚热带中国林业科学研究院热带林业实验中心林区8年生格木、红椎和米老排幼龄人工纯林和混交林生态系统碳贮量及其分配特征进行了研究，旨在进一步阐明造林模式对人工林固碳能力与潜力影响，为区域尺度上科学评估森林生态系统碳库及碳平衡提供基础数据和理论依据。

1 研究地区与研究方法

1.1 研究区概况

研究区位于广西凭祥市中国林业科学研究院热带林业实验中心(106°39′50″~106°59′30″E，21°57′47″~22°19′27″N)，属南亚热带季风气候区的西南部，与北热带北缘毗邻，为湿润半湿润气候。干湿季节明显(10月至翌年3月份为干季，4~9月份为湿季)，太阳年总辐射439.61kJ/cm^2，年日照时数1218~1620h，年均温20.5~21.7℃，极端高温40.3℃，极端低温-1.5℃，≥10℃积温6000~7600℃，年降水量1400mm，相对湿度80%~84%；地貌类型以低山丘陵为主，海拔400~650m，地带性土壤为花岗岩发育成的山地红壤，土层厚度在100cm以上。

青山实验场于2005年4月营造了格木纯林(PE)、米老排纯林(PM)、红椎纯林(PC)和格木×红椎×米老排混交林(MECM，以下简称混交林)为研究对象，4种林分均是在杉木人工林皆伐炼山后，经块状整地营建的人工幼龄林。造林当年和翌年各抚育1次，直至郁闭，目前未曾间伐。各林分初植密度均为2000株/hm^2，混交林为行间混交，混交比例为格木：红椎：米老排=1：1：1，4种人工林中，红椎自然稀疏较多，林下灌草植被丰富，灌草层盖度为75%；其次为混交林，盖度为15%，格木米老排纯林灌草盖度均小于15%。林下植被以杜茎山(*Maesa japonica*)、酸藤子(*Embelia laeta*)、玉叶金花(*Mussaenda* pubuscens)、五节芒(*Miscanthus floridulu*)等。

2013年9月，在4种个林分中，选取坡面均匀，人为干扰相对较少的区域，按坡位分别随机设置4个20m×20m样地，共计16个样地。对每个样方内的树木进行每木检尺，调查胸径、树高等指标。林分基本情况见表1。

表1 不同造林模式人工林的基本特征

林分类型	坡度(°)	密度(trees/hm²)	郁闭度	胸径(cm)	树高(m)
格木纯林	26.8±2.3	1725±41	0.9	7.1±1.8	6.6±1.2
红椎纯林	31.4±2.7	1548±27	0.7	7.7+2.2	8.4±2.7
米老排纯林	27.1±1.8	1693±36	0.9	13.6±4.8	13.6±2.9
格木×红椎×米老排混交林	24.4±2.1	1772±47	0.9	9.7±3.6	9.5±3.0

1.2 研究方法

1.2.1 林木调查及生物量的测定

根据每木检尺的结果，作出各林分胸径分布图，依据径级分布情况，在格木、红椎和米老排纯林中按大、中、小3个径级分别选取平均木3株(每个树种各有9株平均木，共选平均木27株)，将平均木伐倒并挖完全部根系后，按器官分别测定树干、树皮、树枝、树叶和树根的鲜重。同时，按不同器官采集伐倒木的分析样品各4份，每份样品400g，带回实验室在65℃下烘干至恒量，称干质量。计算含水率后，将各器官的鲜质量换算成干质量，各林分乔木各器官生物量按3个径级平均木各器官干重与样地株数的乘积的加和进行计算。

1.2.2 林下植被生物量和凋落物现存量的测定

在每个20m×20m样方中的左上角和右下角选取2个5m×5m小样方，记录小样方内灌木和草本植物的种类，并采用"收获法"测地上和地下部分生物量(每个林分3个20m×20m样方，共6个5m×5m小样方用于植被生物量的测定)。将同种植物相同器官混合取样300g左右，每个植物样品取4份带回实验室在65℃下烘干至恒量，称干质量。计算含水率后，将鲜质量换算成干质量。在每个20m×20m样方中的4个角落各选取1个1m×1m小样方，按未分解、半分解，测定凋落物现存量，并按不同组分取样各200g左右，用于含水率和碳含量的测定。

1.2.3 土壤样品的采集

在各样方随机挖取3个土壤剖面，按照0~10cm、10~30cm、30~50cm和50~100cm将土壤分为4个土层(林地土壤厚度>100cm)，用环刀取样，测定各土层土壤容重。各土层土样混合，取200g用于含碳量的测定。

1.2.4 碳含量测定和碳贮量计算

将植物样品于65℃下烘干，土壤样品置于室内风干，磨碎。植物和土壤样品均采用重铬酸钾-水合加热法测定有机碳含量。

植物碳贮量=有机碳含量×单位面积生物量

土壤碳贮量=土壤有机碳含量×土壤容重×土壤厚度。

1.3 数据处理

采用SPSS 13.0软件对数据进行统计分析，方差分析和差异显著性检验($\alpha=0.05$)。采用Excel和PS软件作图。图表数据为平均数±标准误。

2 结果与分析

2.1 不同人工林生态系统各组分碳含量

2.1.1 乔木层碳含量

方差分析表明：相同树种不同器官之间和相同器官不同树种之间均存在显著差异($P<0.05$)。不同树种各器官碳含量高低顺序有所不同，格木不同器官碳含量的排列顺序为：树干>树枝>树根>树皮>树叶，红椎为：树皮>树叶>树干>树枝>树根，米老排为：树叶>树干>树根>树皮>树枝。从整体上看，碳含量的均值为格木(539.3g/kg)>米老排(532.7g/kg)>红椎(515.3g/kg)。

表2 不同树种各器官碳含量 (g/kg)

器官	格木	红椎	米老排
树干	561.7±10.2Aa	526.7±14.0Bc	540.2±11.6Bb
树皮	534.8±8.5Db	553.4±10.0Aa	522.2±9.5Db
树枝	549.8±7.4Bb	487.1±11.2Cc	512.8±12.6Eb
树叶	512.4±9.1Ec	534.8±14.5Bb	556.3±16.0Aa
树根	537.7±5.2Ca	474.6±16.8Dc	531.9±10.1Cb

注：不同大写字母表示相同树种不同器官间差异显著，不同小写字母表示相同器官不同树种间差异显著。

2.1.2　林下地被物碳含量

格木人工林下地被物各层次平均碳含量以凋落物层最高，灌木层居次，草本层最低；灌木层和草本层地上部分碳含量均高于地下部分，未分解的凋落物碳含量高于半分解的碳含量。不同林分间，灌木层、草本层和凋落物层碳含量均以米老排纯林最高，混交林居次，红椎和格木纯林最低，且4个林分间地被层含碳量差异多数达到显著水平($P<0.05$)(见表3)。

表3　不同林分林下植被和凋落物碳含量　(g/kg)

层次	组分	格木纯林	红椎纯林	米老排纯林	混交林
灌木层	地上部分	494.4±10.3b	488.7±9.0c	512.5±17.7a	501.3±11.4 b
	地下部分	445.8±7.4d	454.9±10.1c	486.3±9.7a	466.6±8.7b
草本层	地上部分	476.3±11.2a	437.5±6.9c	477.1±10.1a	468.7±7.9b
	地下部分	411.5±12.1b	407.8±8.8b	424.7±7.9a	427.8±14.2a
凋落物	未分解	517.2±9.4a	504.1±11.8b	523.8±6.4a	520.4±4.7a
	半分解	471.4±20.3b	452.4±14.7c	490.1±17.2a	473.5±11.8b

注：同一行不同字母表示不同林分间碳含量差异显著($P<0.05$)。

2.1.3　土壤层碳含量

4种林分的土壤碳含量均以表土层(0~10cm)最高，在20.8~28.1g/kg。随着土层深度的增加，土壤碳含量显著降低($P<0.05$)；而相同深度的土层中，不同树种之间的土壤碳含量差异显著($P<0.05$)，0~10cm、10~30cm、30~50cm和50~100cm土壤碳含量均以米老排纯林最高，红椎纯林居次，格木纯林和混交林土壤碳含量较低，米老排纯林和红椎纯林0~100cm土壤平均含碳量分别是混交林的1.8倍和1.3倍，而格木纯林和混交林平均土壤碳含量之间无显著差异($P>0.05$)。

表4　不同林分土壤碳含量　(g/kg)

土层深度	格木纯林	红椎纯林	米老排纯林	混交林
0~10cm	21.4±1.4c	25.1±1.8b	28.1±2.7a	20.8±1.7c
10~30cm	12.7±0.7c	16.8±1.2b	23.2±2.0a	10.4±0.8d
30~50cm	8.2±0.7c	10.2±0.9b	18.9±1.7a	7.8±0.8c
50~100cm	5.8±0.4d	8.7±0.7b	10.9±0.9a	7.1±0.7c
土壤平均	12.0± 0.8c	15.2± 1.3b	20.3±1.9a	11.5± 1.0c

注：同一行不同字母表示不同林分间碳含量差异显著($P<0.05$)。

2.2　格木人工林生态系统各组分碳贮量及其分配

2.2.1　乔木层碳贮量及其分配

乔木层碳贮量以米老排纯林最大，为87.2t/hm^2，混交林居次，红椎纯林碳贮量最小，格木纯林碳贮量较红椎纯林大，但远比混交林低(表5)。混交林碳贮量是米老排纯林的45.2%，却分别是格木和红椎纯林的1.8和2.1倍。可见，对米老排而言，相比与其他阔叶树种混交，营建纯林更有利于乔木生物量的生长和碳素的累积。

不同林分各器官碳贮量在乔木层的分配均以树干最高，树干碳贮量在4种林分中的分配顺序为：米老排纯林>混交林>红椎纯林>格木纯林；树根和树皮与树干的分配大小顺序相同。树枝和树叶在乔木层碳贮量的分配顺序有所不同，格木纯林树枝和树叶碳贮量在乔木层有较高的分配，分别占乔木层碳贮量的17.4%和12.6%。总体上看，不同林分间各器官在乔木层碳贮量的分配顺序不同，格木纯林为树干>树根>树枝>树叶>树皮；红椎和米老排纯林为树干>树根>树皮>树枝>树叶；混交林为树干>树根>树皮=树枝>树叶，混交林中树皮和树枝在乔木中的分配差异不显著($P>0.05$)。

表5　乔木层各器官碳贮量及其分配(t/hm^2)

器官	格木纯林	红椎纯林	米老排纯林	格木×红椎×米老排混交林
树干	10.2(47.3)	10.8(54.5)	60.1(68.9)	25.8(63.1)
树皮	1.8(8.3)	1.5(7.7)	5.0(5.7)	2.9(7.0)
树枝	3.7(17.4)	1.3(6.6)	4.6(5.3)	2.9(7.0)
树叶	2.7(12.6)	1.1(5.6)	2.7(3.1)	2.3(5.5)
树根	3.1(14.4)	5.1(25.6)	14.8(17.0)	7.1(17.4)
合计	21.5(100.0)	19.8(100.0)	87.2(100.0)	41.0(100.0)

注：括号内数据为各器官碳贮量占乔木碳贮量总量的百分比。

2.2.2　林下地被物碳贮量及其分配

从表6可知，人工林下地被物碳贮量均较小，为2.439~11.077t/hm^2。相同组分不同林分间以及相同林分的不同组分间均存在显著差异($P<0.05$)，灌草

层碳贮量大小顺序为红椎纯林>混交林>格木纯林>米老排纯林；凋落物层为米老排>混交林>格木纯林>红椎纯林；地被层碳贮量总量大小顺序为米老排纯林>混交林>红椎纯林>格木纯林。

不同林分各组分碳贮量分配顺序也有所差异，格木纯林和米老排纯林大小顺序为凋落物层>灌木层>草本层，红椎纯林为草本层>凋落物层>灌木层，而混交林为凋落物层>草本层>灌木层。在灌木层中，地上部分碳贮量均大于地下部分，但草本层植物地上与地下部分碳贮量差异不显著；凋落物层中，未分解的凋落物碳贮量显著高于半分解的，约为半分解凋落物碳贮量的2倍。

2.2.3　土壤碳贮量及其分配

各林分相同土层厚度的土壤平均碳贮量随土层深度增加而降低，变化趋势与土壤碳含量随土层深度的变化一致，土壤碳贮量主要集中在0~30cm的表土层，格木纯林、红椎纯林、米老排纯林和混交林0~30cm土碳贮量分别占0~100cm土壤碳贮量的44.7%、43.7%、42.0%和41.3%（表7）。

从表7可以看出，不同林分间土壤碳贮量差异显著，0~100cm土壤碳贮量以米老排纯林最高，红椎纯林居次，格木纯林最低。各土层碳贮量大小顺序为米老排纯林>红椎纯林>混交林>格木纯林，除红椎纯林与混交林0~10cm和30~50cm土层，格木纯林与混交林的10~30cm和30~50cm土层差异不显著外，其他各土层林分间土壤碳贮量差异均达显著水平($P<0.05$)。

表6　林下地被物碳贮量　　(t/hm^2)

层次	组分	格木纯林	红椎纯林	米老排纯林	混交林
灌木层	地上	0.220±0.073c	0.698±0.159a	0.076±0.012d	0.354±0.037b
	地下	0.094±0.028c	0.295±0.067a	0.019±0.005d	0.164±0.014b
	小计	0.314±0.086c	0.993±0.189a	0.096±0.013d	0.518±0.045b
草本层	地上	0.021±0.008c	2.380±0.681a	0.002±0.001d	1.257±0.242b
	地下	0.030±0.008c	1.513±0.236a	0.001±0.000d	1.307±0.317b
	小计	0.051±0.012c	3.894±0.788a	0.003±0.001d	2.564±0.510b
凋落物层	未分解	1.393±0.237c	0.691±0.087d	6.945±1.144a	2.419±0.314b
	半分解	0.682±0.170c	0.333±0.042d	4.033±1.275a	1.263±0.208b
	小计	2.075±0.284c	1.025±0.109d	10.978±2.087a	3.682±0.481b
地被层总计		2.439±0.441d	5.911±1.004c	11.077±2.090a	6.764±0.913b

注：同行中不同小写字母表示林分间差异显著($P<0.05$)。

表7　不同林分各土层土壤碳贮量　　(t/hm^2)

土层深度	格木纯林	红椎纯林	米老排纯林	格木×红椎×米老排混交林
0~10cm	19.6± 1.8c	26.9± 2.8b	31.9± 3.4a	27.2± 3.1b
10~30cm	30.1± 3.5c	40.6± 4.1b	56.3± 3.9a	28.7± 4.7c
30~50cm	20.1± 4.1c	26.4± 3.7b	50.0± 5.7a	23.8± 2.5bc
50~100cm	41.3± 5.5d	60.5± 6.3b	71.6± 7.1a	55.5± 4.7c
合计	111.1± 12.7d	154.4± 14.5b	209.8± 18.4a	135.2± 13.3c

注：同行中不同小写字母表示林分间差异显著($P<0.05$)。

2.3　人工林生态系统碳贮量及其分配

格木纯林、红椎纯林、米老排纯林和混交林生态系统碳贮量分别是135.2t/hm^2、180.2t/hm^2、308.0t/hm^2和182.8t/hm^2，混交林碳贮量显著高于格木纯林，为格木纯林碳贮量的1.6倍，却显著低于米老排纯林，仅为米老排纯林碳贮量的0.6倍，而混交林与红椎纯林碳贮量总量无显著差异($P>0.05$)。相比于红椎纯林，混交林有较高的乔木碳贮量，但混交林土壤碳贮量和灌草层碳贮量均低于红椎纯林。可见，米老排纯林在4种林分中具有最高的碳贮量，不仅表现在生态系统碳贮量总量上，同时具有最高的植被碳贮量，土壤碳贮量和地被层碳贮量。

乔木层和土壤层为各林分主要碳库，二者占各人工林生态系统碳贮量的96.3%~98.2%，而灌木层、草本层和凋落物层的碳贮量所占比例<4%。不同林分相同组分碳贮量在生态系统分配有所不同，乔木层碳贮量在生态系统碳分配的大小顺序为米老排(28.3%)>混交林(22.4%)>格木纯林(16.0%)>红椎纯林(11.0%)，土壤碳贮量的分配顺序为红椎纯林(85.7%)>格木纯林(82.2%)>混交林(73.9%)>米老排纯林(68.1%)。

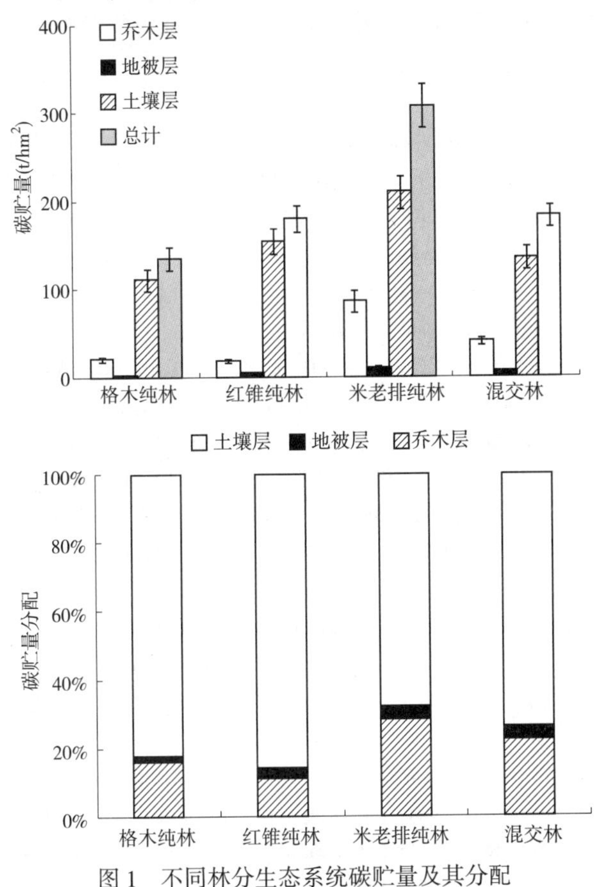

图1 不同林分生态系统碳贮量及其分配

3 结论与讨论

8年生格木、红椎和米老排各器官碳含量的平均值分别为539.3g/kg、532.7g/kg和515.3g/kg，高于广西巨尾桉(*Eucalyptus urophylla*×*E. grandis*)(475.2g/kg)和杉木(497.0g/kg)[28-31]，也高于国际通用的树木平均碳含量(0.5g/kg)及热带32个树种的平均碳含量(444.0~494.5g/kg)[32]。相比之下，3个树种碳含量平均值的大小顺序为格木>米老排>红椎，可见，即使在同一地区，树木各器官碳含量大小因树种不同而存在差异。另外，相同树种的不同器官碳含量差异显著($P<0.05$)，不同树种之间，各器官碳含量的排列顺序也有所不同。可见树种是影响乔木器官含碳量的重要因素之一，这可能与树种本身的生理特性相关。

不同林分间，灌木层、草本层和凋落物层碳含量的大小顺序为米老排纯林>混交林>红椎纯林>格木纯林，这与不同林分间灌草植被及凋落物的组成密切相关，米老排纯林中灌草植被较少，林下植被主要以米老排幼苗为主，凋落物以米老排落叶为主分，而米老排叶的碳含量极高(556.3g/kg)[28]，自然以米老排幼苗为主的林下植被和以米老排凋落叶为主的凋落物层碳含量较其他林分高。

不同林分间的土壤碳含量差异显著($P<0.05$)，以米老排纯林最高，红椎纯林居次，格木纯林和混交林土壤碳含量较低。米老排纯林和红椎纯林土壤碳含量较高的主要原因是由于具有较高的土壤碳输入。米老排纯林是通过较多的凋落物增加土壤碳输入，而红椎纯林是因为自然稀疏强度较大，较多的光照条件和较大的生长空间使得林下植被得以快速生长，丰富的林下灌草植被也增加了其土壤的碳输入。混交林和格木纯林因郁闭度较大，林下植被生长受限，且没有足够的凋落物覆盖，因而缺乏土壤碳源，造成土壤碳含量低于米老排和红椎纯林。

本研究表明，不同林分生态系统碳贮量有所不同，米老排纯林在4种林分中具有最高的碳贮量，高于混交林碳贮量总量的68.5%，一方面是米老排纯林中米老排生长较为迅速，生物量较高，而乔木层在整个植被层碳储量中占的比例较高，这样就增加了植被层碳的累积量；另一方面米老排纯林丰富的凋落物不仅增加了地被层碳贮量，而且由于增加了土壤凋落物的输入，进而增加了土壤碳贮量的累积；而混交林由于林分树种组成的改变，引起乔木层生物量生长的变化，3个主要乔木树种的平均生物量较米老排纯林小，从而导致混交林乔木层生物量碳低于米老排纯林。这一结论与诸多前人研究所得出的混交林的碳贮量大于纯林的结论并不一致[13,23,24]，营建纯林与混交林，其林分碳贮量大小，同所选择的造林树种有关，并非混交林固碳能力始终高于纯林。

乔木层和土壤层为各林分主要碳库，二者占格木人工林生态系统碳贮量的96.3%~98.2%，而灌木层、草本层和凋落物层的碳贮量所占比例<4%。不同林分相同组分碳贮量在生态系统分配有所不同，乔木层碳贮量在生态系统碳分配的大小顺序为米老排纯林(28.3%)>混交林(22.4%)>格木纯林(16.0%)>红椎纯林(11.0%)，这与明安刚对米老排和格木近熟林乔木层(28年生)碳贮量在人工林生态系统中的分配结果明显不同(28年生米老排和29年生格木人工林乔木层碳贮量占分别占生态系统碳贮量总量的51.8%和52.4%)[27,28]，与27年生红椎人工林乔木层碳贮量所占比例(32.8%)也有所不同[33]。这表明人工林碳贮量各组分在生态系统碳贮量总量中的分配与林龄密切相关，随着林龄的增加，乔木层碳贮量在生态系统中的分配也随之增加。

四种人工林中，格木纯林碳贮量最小，为135.2t/hm^2，仅相当于米老排纯林碳贮量(308.0t/hm^2)和混交林碳贮量(182.8t/hm^2)的43.9%和74.0%，主要原因有二，一是相对于红椎和米老排，格木生长较慢，生物量生长较慢，因而乔木层碳贮量较小；二是由于格木人工林较低的土壤碳含量，直接减少了土壤碳的累积。

总体上看，乔木层碳贮量大小顺序为米老排纯林>混交林>红椎纯林>格木纯林，而土壤层碳贮量大小顺序为米老排纯林>红椎纯林>混交林>格木纯林，且差异显著($P<0.05$)。由此可见，造林模式对人工林碳贮量及其分配有显著影响。如果选用红椎和格木这两个树种营建碳汇林，选用混交模式更有利于地上碳的累积，但不利于土壤碳的固定；而对米老排而言，营建纯林不仅有利于生物量碳的吸收，也有利于土壤碳的累积。因而，对碳汇林造林模式的选择，应根据树种生物学及固碳特性而定。

参考文献

[1]TANS P P. How can global warming be traced to CO_2[J]. Scientific America，2006，295：124.

[2]FANG J Y，CHEN A P，PENG C H，et al. Changes in forest biomass carbon storage in China between 1949 and 1998[J]. Science，2001，292：2320-2322.

[3]CALDEIRA K，DUFFY P B. The role of the southern ocean in up take and storage of anthropogentic carbon dioxide[J]. Science，2000，287：620-622.

[4] NORBY R J，LUO Y. Evaluating ecosystem responses to rising atmospheric CO_2 and global warming in a multi-factor world[J]. New Phytologist，2006，162：281-293.

[5] BROWN S. Present and potential roles of forests in the global climate change debate[J]. Unasylva，1996，47：3-10.

[6]LUN F，LI W H，LIU Y. Complete forest carbon cycle and budget in China 1999-2008[J]. Forest Ecology and Management，2012，264：81-89.

[7] HOOVER C M，LEAK W B，KEEL B G. Benchmark carbon stocks from old-growth forests in northern New England，USA[J]. Forest Ecology and Management，2012，266：108-114.

[8]PAN Y D，BIRDSEY R A，FANG J Y，et al. A large and persistent carbon sink in the world's forests[J]. Science，2011，333：988-993.

[9]WOODWELL G M，WHITTAKER R H，REINERS W A，et al. The biota and the world carbon budget[J]. Science，1978，199：141-1467.

[10]POST W M，EMANUEL W R，ZINKE P J，et al. Soil carbon pools and world life zone[J]. Nature，1982，298：156-159.

[11]张小全，李怒云，武曙红．中国实施清洁发展机制造林和再造林项目的可行性和潜力[J]．林业科学，2005，41(5)：139-143.

[12]武曙红，张小全，李俊清．CDM造林或再造林项目的基线问题[J]．林业科学，2006，41(4)：112-116.

[13]WANG H，LIU S R，WANG J X，et al. Effects of tree species mixture on soil organic carbon stocks and greenhouse gas fluxes in subtropical plantations in China[J]. Forest Ecology and Management，2013，300：43-52.

[14]WANG S，CHEN J M，JU W M，et al. Carbon sinks and sources in China's forests during 1901—2001[J]. Journal of Environmental Management，2007，85：524-537.

[15]SALIMON CI，PUTZ F E，et al. Estimating state-wide biomass carbon stocks for a REDD plan In Acre，Brazil[J]. Forest Ecology and Management，2011，262(3)：555-560.

[16]LACLAU P. Biomass and carbon sequestration of ponderosa pine plantations and native cypress forests in northwest Patagonia[J]. Forest Ecology and Management. 2003，180：317-333.

[17]SPECHT A，WEST P W. Estimation of biomass and sequestered carbon on farm forest plantations in northern New South Wales，Australia[J]. Biomass and Bioenergy，2003，25：363-379.

[18]方晰，田大伦，项文化．速生阶段杉木人工林碳素密度、贮量和分布[J]．林业科学，2002，38(3)：14-19.

[19]马明东，江洪，刘跃建．楠木人工林生态系统生物量、碳含量、碳贮量及其分布[J]．林业科学，2008，44(3)：34-39.

[20]方晰，田大伦，项文化，等．不同密度湿地松人工林中碳的积累与分配[J]．浙江林学院学报 2003，20(4)：374-379.

[21]张国庆，黄从德，郭恒，等．不同密度马尾松人工林生态系统碳储量空间分布格局[J]．浙江林业科技，2007，27(6)：10-14.

[22]刘恩，刘世荣南亚热带米老排人工林碳贮量及其分配特征[J]．生态学报，2012，32(16)：5103-5109.

[23]何友均，覃林，李智勇，等．西南桦纯林与西南桦×红椎混交林碳贮量比较[J]．生态学报，2012，32(23)：7586-7594.

[24]HE Y J，QIN L，LI Z Y，et al. Carbon storage capacity of monoculture and mixed-species plantations in subtropical China[J]. Forest Ecology and Management，2013，295：193-198.

[25]王晓丽，王媛，石洪华，等．山东省长岛县南长山岛黑松和刺槐人工林的碳储量[J]．应用生态学报，2013，24(5)：1263-1268.

[26]郭文福，蔡道雄，贾宏炎，等．米老排人工林生长规律的研究[J]．林业科学研究，2006，19(5)：585-589.

[27]明安刚，贾宏炎，陶怡，等．米老排人工林碳素累积特征及其分配格局[J]．生态学杂志，2012，31(11)：2730-2735.

[28]明安刚，贾宏炎，田祖为，等．不同林龄格木人工林碳储量及其分配特征[J]．应用生态学报，2014，25(4)：940-946.

[29]何斌，黄寿先，招礼军，等．秃杉人工林生态系统碳素积累的动态特征[J]．林业科学，2009，45(9)：151-157.

[30]康冰，刘世荣，蔡道雄，等．南亚热带杉木生态系统生物量和碳素积累及其空间分布特征[J]．林业科学，2009，45(8)：147-153.

[31]梁宏温，温远光，温琳华，等．连栽对尾巨桉短周期人工林碳贮量的影响[J]．生态学报，2009，29(8)：4242-4250.

[32]Elias M，Potvin C. Assessing inter-and intra-specific variation in trunk carbon concentration for 32 neotropical tree species. Forest Ecology and management，2006，222：279-295.

[33]刘恩，王晖，刘世荣．南亚热带不同林龄红椎人工林碳贮量与碳固定特征[J]．应用生态学报，2012，23(2)：335-340.

[原载：生态学报，2015，35(01)]

铁力木人工林生物量与碳储量及其分配特征

明安刚[1,2,3] 郑 路[1,3] 麻 静[1] 陶 怡[1] 劳庆祥[1] 卢立华[1,3]

([1]中国林业科学研究院热带林业实验中心，广西凭祥 532600；
[2]中国林业科学研究院森林生态环境与保护研究所，北京 100091；
[3]广西友谊关森林生态系统国家定位观测研究站，广西凭祥 532600)

摘 要 本文在样方调查和实测生物量的基础上，采用相对生长法对28年生铁力木人工林碳储量及其分配特征进行了研究。结果表明：铁力木各器官碳含量在452.4~524.5g/kg，大小排序为：树叶>树干>树枝>树根>树皮；土壤碳含量以表土层最高，且随土层深度的增加而降低；铁力木人工林乔木层生物量和碳储量分别为165.8t/hm^{2}和79.3t/hm^2，分配顺序均为树干>树枝>树根>树叶>树皮；铁力木人工林生态系统生物量与碳储量分别为173.5t/hm^{2}和203.1t/hm^2，生物量的分配主要集中在乔木层(95.6%)，碳储量的分配顺序为土壤层(59.3%)>乔木层(39.0%)>地被层(1.7%)；林下植被碳含量为地上部分>地下部分，而生物量和碳储量的分配均为地上部分<地下部分。

关键词 铁力木人工林；碳含量；生物量；碳储量；分配特征

Biomass, Carbon Stock and Allocation Characteristics in *Mesua ferrea* Plantation

MING Angang[1,2,3], ZHENG Lu[1,3], MA Jing[1], TAO Yi[1], LAO Qingxiang[1], LU Lihua[1,3]

([1]*Experimental Center of Tropical Forestry*, *Chinese Academy of Forestry*, *Pingxiang* 532600, *Guangxi*, *China*;
[2]*Institute of Forest Ecology*, *Environment and Protection*, *Chinese Academy of Forestry*, *Beijing* 100091, *P. R. China*;
[3]*Guangxi Youyiguan Forest Ecosystem Research Station*, *Pingxiang* 532600, *Guangxi*, *China*)

Abstract: Based on the quadrat survey and actual biomass measurement in a 28-year-old Mesua ferrea plantation in Experimental Center of Tropical Forestry located at Pingxiang City of Guangxi Zhuang Autonomous Region, the carbon stock and its allocation characteristics were studied by the allometric method. The results showed that carbon contents in different organs of M. ferrea ranged between 452.4 and 524.5g/kg, following the order of leaf>stem>branch>root>bark. Soil carbon content in the surface layer was the highest, and it was reduced with the increase of soil depth. The biomass and carbon stock of tree layer in M. ferrea plantation were 165.8t/hm^2 and 79.3t/hm^2, and the allocation of biomass ranked as stem>branch>root>leaf>bark, the same as the allocation of the carbon stock in the tree layer. The biomass of M. ferrea plantation ecosystem was 173.5t/hm^2 and carbon stock 203.1t/hm^2, the biomass in the plantation was mainly distributed in the tree layer(95.6%), and the allocation of carbon stock in different components was soil layer(59.3%)> tree layer(39.0%)> ground layer(1.7%). The carbon content aboveground was greater than that underground in shrub layer and herb layer; however, the biomass and carbon stock aboveground were less than those underground.

Key words: *Mesua ferrea* plantation; carbon content; biomass; carbon stock; allocation characteristics

工业化革命以后，由于人类活动(主要是化石燃料的燃烧和土地利用变化)，大气中以CO_2为主要成分的温室气体(主要有CO_2、CH_4和N_2O)的浓度逐年升高[1]。研究表明，近年来大气中的CO_2浓度正以1.9ml/(m^3·年)的线性速度增加[2]，全球气候变化给人类的生存和发展带来了显著影响，受到生态学界和林学界的普遍关注[3-5]。森林是陆地生态系统的主体，维持着全球86%的植被碳库[6]和73%的土壤碳库[7]，在缓解温室气体浓度上升及气候变暖方面发挥着重要作用。方精云等[8]、王效科等[9-10]、刘国华等[11]和周玉荣等[12]利用全国森林资源清查数据，研究了我国森林碳库及其动态变化特征，为我国森林碳汇的研究奠定了基础。最近20年，随着人工林的快速发展，人工林碳库成为在全球植被碳库的重要组成部分，在减少温室气体排放，增加 CO_2 的吸收与固定，进而缓解全球气候变暖等方面的作用更加显著。有关常见的造林树种人工林的生物量和碳储量，前人已有较多的研究[13-23]，并为森林碳汇功能的研究做出了积极贡献。

铁力木(*Mesua ferrea*)是藤黄科(Guttiferae)铁力木属常绿乔木。干形通直，材质坚硬，是亚洲热带著名的硬木之一。铁力木为阳性树种，但幼树喜阴，原产于亚洲热带地区，广泛分布于印度，在越南、缅甸和孟加拉等国的山地也有分布。我国引种栽培已有500年的历史，铁力木主产于云南南部边缘地区海拔450~1300m一带的河谷坡地或丘陵斜坡上[24]，广西藤县、容县、凭祥和广东信宜有也少量分布。随着社会的进步和发展，铁力木等珍贵用材供求矛盾日益加剧，研究与发展这一极具发展前景的珍贵乡土树种，已成为我国南亚热带地区珍贵树种发展和林业产业结构调整的迫切需要。目前，有关铁力木的研究，主要集中在生长规律、育苗技术和经营对策上[25-27]，本文在前人研究基础上，对位于南亚热带地区的中国林业科学研究院热带林业实验中心林区的28年生铁力木人工林生物量和碳储量及其分配特征进行了研究，旨在为区域尺度上人工林碳汇潜力的科学评估提供数据支撑。

1 试验区概况与研究方法

1.1 试验区概况

试验区设在中国林业科学研究院热带林业实验中心哨平实验场39林班(21°57′47″N，106°59′30″E)，位于广西凭祥市境内，海拔350m，属南亚热带季风气候，年平均气温19.5~21.5℃，≥10℃积温7500℃，年降水量1220~1380mm，年蒸发量1370~1390mm，干湿季节明显，土壤为花岗岩发育成的山地红壤，土层厚度在100cm以上。

1984年2月，在杉木采伐迹地上营造了铁力木人工林。定植时铁力木苗龄为2年生，造林株行距为1.67m×1.67m。栽植前3年人工铲草抚育，郁闭成林后自然生长。2012年11月对林分进行生物量调查和采样。调查当年林下植物稀少且分布不均，灌木主要有大沙叶(*Pavetta hongkongensis*)、三叉苦(*Evodia lepta*)、海金沙(*Lygodium japonicum*)和琴叶榕(*Ficus lyrata*)等，草本主要有扇叶铁线蕨(*Adiantum flabellulatum*)、鞭叶铁线蕨(*Adiantum caudatum*)等。

1.2 方法

1.2.1 样地设置

在林分坡度较为缓和的核心区域，按照坡位从上到下设置3个试验样区，分别为样区Ⅰ、样区Ⅱ和样区Ⅲ，在每个样区布设3个30m×20m的样方(每个样方投影面积为600m²，同一样区的3个样方分布在不同的坡向，3个样区各有3个样方，共9个样方)。每个样地胸径>5cm的乔木进行每木检尺，实测每个个体的胸径、树高等因子。3个样区乔木层群落组成结构如表1所示。

1.2.2 乔木层生物量的测定

结合样地每木检尺数据，建立林木胸径分布图，以2cm为一个径级共选取11株标准木进行生物量测定。样木伐倒后，按照不同器官对干、皮、枝、叶和根的鲜质量进行称重，同时，按器官分别取样400g左右，在65℃烘干后推算各器官干质量。

利用11株样木各器官生物量的实测数据(表2)，对各器官生物量与胸径(D)和树高(H)及胸径平方与树高的乘积(D^2H)等因子进行相关分析，发现铁力木各器官生物量均与胸径 D 的相关性最为密切(相关系数在0.8854~0.9528)，而与树高的相关性较低，相关系数仅在0.1049~0.6166。因而，本研究选择胸径(D)作为拟合铁力木各器官生物量方程的自变量。分别用 $W=aD^b$、$W=ae^{bD}$、$W=aD+b$ 和 $W=aln D+b$(式中，W 为各器官生物量；D为样木胸径；a，b，c 为方程参数)等4种模型建立铁力木各器官生物量与胸径(D)之间的回归方程，用判定系数(R^2)、F 值和显著性 P 值等评价模型的拟合优度，比较并筛选出拟合效果最好的模型作为预测铁力木各器官生物量方程，结果表明该树种的各器官生物量均以幂函数($W=aD^b$)模型拟合效果最佳，其判定系数(R^2)在0.8516~0.9708，F 检验相关性均达到极显著差异($P<0.01$)。因而，据此利用“直接拟合幂函数”的方法建立了28年生铁力木各器官生物量(W)与胸径(D)之间的生物量模型，用以估算铁力木各器官的生物量(表3)。

表1 28年生铁力木人工林各样区乔木检尺情况

样区	坡位	坡度(°)	胸径(cm)		树高(m)		乔木密度(tree/hm²)
			范围	平均	范围	平均	
Ⅰ	上	33.7	11.7~23.1	15.5±2.7	11.6~16.5	14.3±1.1	808

（续）

样区	坡位	坡度(°)	胸径(cm)		树高(m)		乔木密度(tree/hm²)
			范围	平均	范围	平均	
Ⅱ	中	35.3	11.5~20.4	16.2+1.3	10.2~15.6	13.7±1.3	800
Ⅲ	下	34.6	12.8~22.8	17.1±2.4	9.3~15.9	13.9±1.3	820

表 2　28 年生铁力木 11 株不同径级标准木基本参数及各器官生物量原始数据

标准木编号	胸径(cm)	地径(cm)	树高(m)	枝下高(m)	树干(kg)	树皮(kg)	树枝(kg)	树叶(kg)	树根(kg)	全株(kg)
1	6.2	8.7	10.5	4.3	10.3	1.7	1.9	0.4	2.8	17.1
2	8.4	11.8	12.4	5.7	23.7	3.6	6.6	1.2	5.7	40.9
3	10.2	14.1	14.2	8.3	45.4	6.4	13.4	3.4	9.6	78.2
4	12.1	16.4	14.6	6.1	60.7	8.2	28.1	3.6	13.6	114.2
5	14.2	17.8	14.6	5.3	80.1	9.0	46.0	7.6	15.5	158.2
6	15.5	19.6	15.8	8.6	87.1	10.5	47.8	5.5	23.6	174.6
7	16.3	19.1	11.6	4.7	65.2	7.4	58.6	16.8	43.3	191.4
8	17.3	21.4	14.5	6.4	92.7	10.6	113.6	15.4	32.3	264.7
9	18.2	23.1	17.1	7.7	122.1	13.6	107.2	10.5	27.9	281.2
10	19.2	23.8	13.4	5.8	119.5	12.9	108.3	16.8	37.4	294.9
11	21.3	25.2	14.8	6.8	155.8	10.7	108.2	24.5	63.8	363.0

表 3　铁力木各器官生物量回归方程

器官	相对生长方程	N	R^2	F	P
树干	$W_s=0.315\ 6D^{2.030}$	11	0.941 3	78.831 4	0.000 0
树皮	$W_{ba}=0.145\ 9D^{1.5106}$	11	0.851 6	21.616 5	0.000 8
树枝	$W_{br}=0.004\ 2D^{3..4565}$	11	0.970 8	45.176 7	0.000 1
树叶	$W_l=0.001\ 3D^{3.2245}$	11	0.937 8	28.505 4	0.000 8
树根	$W_r=0.034\ 4D^{2.4009}$	11	0.948 5	29.036 9	0.000 7

注：W_s、W_{ba}、W_{br}、W_l和 W_r分别为树干、树皮、树枝、树叶和树根的生物量。W_s、W_{ba}、W_{br}、W_l and W_r。

1.2.3　林下植被生物量和凋落物现存量的测定

在每 30m×20m 的样方中选取 5 个 2m×2m 的子样方，调查子样方内灌草植被的种名、数量和盖度等群落因子，然后，将灌木和草本植物按照地上部分和地下部分分别实测鲜质量，并取不同组分的混合样品 400g 在 65℃烘干后，换算各组分的干质量；同时在每个 30m×20m 的样方内随机布设 5 个 1m×1m 的子样方，用于凋落物现存量的测定，凋落物现存量分成未分解和半分解 2 个组分分别称重，并按不同组分各取样 400g 恒干后用于干质量的换算。

1.2.4　样品的采集与分析

在每个 30m×20m 的样方内随机挖取 3 个土壤剖面（每个样区 3 个样方，共 27 个土壤剖面），按照 0~10cm、10 ~ 20cm、20 ~ 30cm、30 ~ 50cm、和 50 ~ 100cm 将土壤层分成 5 个土层，在各土层取混合样品 200g 左右，经风干，过筛后测定土壤有机碳含量。同时，各层取环刀样品 3 个，用于土壤密度的测定。

1.2.5　含碳量的测定

所有植物样品和土壤样品有机碳含量均用 TOC-L/SSM 5000 总有机碳分析仪进行测定。

1.2.6　碳储量的计算方法

植被碳储量=单位面积生物量×碳含量

土壤碳储量=土壤有机碳含量×土壤密度×土层厚度。

1.2.7　数据分析

采用单因素方差分析（one-wayANOVA）检验器官

间及各组分间的差异显著性；使用 LSD 法进行多重比较；数据分析及生物量方程的拟合用 Microsoft Excel 2007 和 SPSS 19.0 软件完成。

2 结果与分析

2.1 铁力木人工林生态系统各组分生物量与碳储量及其分配

2.1.1 乔木层生物量与碳储量

由表 4 可知，铁力木人工林乔木层个体平均碳含量为 478.6g/kg，在 452.4～524.5g/kg，不同器官碳含量有所差异，以树叶含量最高，树干和树枝居次，树根和树皮最低。

乔木层生物量为 165.8t/hm^2，以树干和树枝最高，二者占乔木层生物量总量的 76.3%，各器官生物量的分配顺序为树干>树枝>树根>树叶>树皮。乔木层碳储量总量为 79.3t/hm^2，分配顺序与生物量的分配顺序一致，为树干>树枝>树根>树叶>树皮，且各器官生物量与碳储量的差异均达到显著水平(P<0.05)。

2.1.2 林下地被物的生物量与碳储量

测定结果显示(表 5)，地被物各层次碳含量平均值为未分解的凋落物>灌木层>草本层；且灌木和草本层均为地上部分>地下部分，凋落物为未分解层>半分解层。灌木层、草本层及凋落物层碳含量在各组分间存在显著差异(P<0.05)。

地被层生物量和碳储量总量分别为 7.670t/hm^2 和 3.302t/hm^2，以凋落物为主，分别占地被层生物量总量的 97.9% 和 98.1%，且半分解凋落物生物量和碳储量均显著大于未分解的，分别是未分解的 3.5 倍和 2.9 倍(P<0.05)；灌木层和草本层生物量和碳储量均较小，且地上部分生物量和碳储量显著小于地下部分(P<0.05)。

表 4 乔木各器官生物量与碳储量

器官	碳含量(g/kg)	生物量(t/hm^2)	百分比(%)	碳储量(t/hm^2)	百分比(%)
树干	487.7±2.5b	73.7±9.2a	44.5	35.9±3.4a	45.3
树皮	452.4±3.0d	8.0±0.7e	4.8	3.6±0.4e	4.5
树枝	472.0±2.5c	52.8±11.1b	31.8	24.9±5.3b	31.4
树叶	524.5±1.8a	8.6±1.8d	5.2	4.5±1.0d	5.7
树根	456.3±0.1d	22.7±3.3c	13.7	10.4±1.7c	13.1
合计		165.8±18.7	100.0	79.3±11.4	100.0

注：同列之间不同字母表示在 P<0.05 水平上差异显著。下同。

表 5 林下地被物生物量与碳储量

层次	组分	碳含量(%)	生物量(t/hm^2)	百分比(%)	碳储量(t/hm^2)	百分比(%)
灌木层	地上部分	469.0±18.8b	0.025±0.007d	0.33	0.012±0.003d	0.36
	地下部分	445.6±15.1c	0.052±0.017c	0.68	0.023±0.009c	0.70
	小计		0.077±0.022	1.00	0.035±0.011	1.06
草本层	地上部分	393.1±5.4e	0.022±0.007d	0.29	0.009±0.003d	0.27
	地下部分	315.8± 15.3f	0.059±0.012c	0.77	0.019±0.007c	0.58
	小计		0.081±0.020	1.06	0.027±0.008	0.82
凋落物层	未分解	496.2±9.9a	1.661± 0.217b	21.66	0.824±0.143b	24.95
	半分解	412.9±48.3d	5.851± 0.848a	76.28	2.416±0.427a	73.17
	小计		7.512± 1.159	97.94	3.240±0.474	98.12
合计			7.670± 1.166	100.00	3.302±0.481	100.00

2.1.3 土壤层碳含量与碳储量

从表 6 可知，土壤碳含量以 0～10cm 最高，为 22.8g/kg；各土层碳含量随着土层深度的增加而降低，且差异显著(P<0.05)；土壤密度随土层深度的增加呈现出增加的态势，且 0～30cm 土层差异显著(P<0.05)，0～100cm 土壤碳储量为 120.5t/hm^2，土壤 10cm 厚度的平均碳储量随土层深度的增加显著降低(P<0.05)。

表 6　土壤层碳含量与碳储量

土层深度(cm)	碳含量(g/kg)	土壤密度(g/cm)	土层厚度(cm)	碳储量(t/hm^2)
0~10	22. 8±2. 7a	1. 154c	10	26. 3±3. 1
10~20	13. 3±1. 0b	1. 250b	10	16. 7±1. 1
20~30	9. 4±1. 6c	1. 351a	10	12. 7±1. 1
30~50	7. 5±1. 2d	1. 360a	20	20. 4±1. 7
50~100	6. 8±0. 7e	1. 306a	50	44. 4±3. 2
合计 Total				120. 5±5. 7

2. 2　铁力木人工林生态系统碳储量及其分配格局

从表 7 可以看出，28 年生铁力木人工林生态系统生物量为 173. 5t/hm^2，其中乔木层生物量远大于地被层，为 165. 8t/hm^2，占生态系统生物量总量的 95. 6%，地被层的灌木，草本及凋落物总和仅占 4. 4%；地被层生物量分配顺序为凋落物层>灌木层>草本层；在铁力木人工林生态系统中，地上部分和地下部分生物量分别为 150. 7t/hm^{2}和 22. 8t/hm^2，分别占林分人工林生物量总量的 86. 8%和 13. 1%。

表 7　铁力木人工林生态系统各组分生物量与碳储量

层次	组分	生物量(t/hm^2)	百分比(%)	碳储量(t/hm^2)	百分比(%)
乔木层	地上部分	143. 1±19. 4	82. 5	68. 9±9. 7	33. 9
	地下部分	22. 7±3. 3	13. 0	10. 4±1. 5	5. 1
	小计	165. 8±21. 7	95. 6	79. 3±11. 0	39. 0
地被层	地上部分	7. 559±1. 168	4. 4	3. 260±0. 476	1. 6
	地下部分	0. 111±0. 023	0. 1	0. 042±0. 011	0. 1
	小计	7. 670±1. 166	4. 4	3. 302±0. 481	1. 7
土壤层				120. 5±5. 7	59. 3
生态系统	地上部分	150. 7±19. 7	86. 9	72. 2±8. 2	35. 5
	地下部分	22. 8±3. 3	13. 1	130. 9±7. 2	64. 5
	小计	173. 5±20. 8	100. 0	203. 1±14. 9	100. 0

注：表中百分比为各地被层组分占地被物总生物量和碳储量的百分比。

28 年生铁力木人工林生态系统碳储量为 203. 1t/hm^2，其中，土壤层(0~100cm)碳储量最大，为 120. 5t/hm^2，占生态系统碳储量总量的 59. 3%；其次是乔木层，为 79. 3t/hm^2，占 39. 0%；地被层碳储量最小，仅占 1. 7%；由此可见，铁力木人工林碳储量主要分布在乔木层和土壤层，二者共占据林分碳储量总量的 98. 4%，仅有极小部分的碳素分布在林下植被和枯枝落叶层。

在林分生态系统中，地上部分碳储量为 72. 2t/hm^2，占林分储量总量的 35. 5%，地下部分碳储量为 130. 9t/hm^2，占总量的 64. 5%。

3　讨论与结论

28 年生铁力木各器官碳含量的平均值为 478. 6g/kg，相比而言，同广西 27 年生观光木(471. 1g/kg)，8 年生桉树(475. 2g/kg)，16 年生湿地松及 14 年生马尾松均平均碳含量[28-31]。说明树种对乔木碳含量的影响显著，不同树种碳含量即使在同一地区也存在明显差异。28 年生铁力木人工林乔木层不同器官碳之间含量差异显著，碳含量由高到低依次为树叶>树干>树枝>树根>树皮，与 27 年生观光木(树干>树枝>树根>树叶>树皮)[30]，28 年生米老排(树叶>树干>树根>树皮>树枝)[32]及 28 年秃杉(树皮>树枝>树干>树根>树叶)等树种各器官碳含量的排列顺序不完全一致[30,32,33]，由此可以看出，即便林龄相差不大，但碳含量在器官间的大小顺序也会因树种的不同而不同。

广西铁力木人工林林分凋落物碳含量(454. 6g/kg)高于本地杉木人工林(418. 7g/kg)[34]，低于本地米老排人工林(553. 8g/kg)，这可能与树种叶片的角质化和革质化程度有关，一般凋落物分解速率会随其革质化或者角质化程度的不同产生差异，革质化或角质化程度越高就越有利于碳素的累积。灌木层和草本层地上部分平均碳含量均高于地下部分，这一结论与多数树种研究的结论一致[22,35]。铁力木土壤碳含量较低(9. 5g/kg)，低于同一地区杉木人工林(16. 6g/kg)及桉树人工林的(14. 57g/kg)平均值[29,34]，可能是铁力木人工林土壤碳输入的限制，也可能与人工林种植之前土壤碳含量大小不同有关。28 年生铁力木人工林生态系统碳储量总量为 203. 1t/hm^2，较我国森林生态系统平均碳储量(258. 82t/hm^2)低；其中，植被层碳储量为 79. 4t/hm^2，是我国森林植被平均碳储量(57. 07t/hm^2)[12]的 1. 4 倍，是我国亚热带针叶林平均碳储量(63. 7t/hm^2)的 1. 2 倍[37]。铁力木人工林较高的植被层碳储量主要因为其乔木层具有的较高的生物量和碳含量。乔木碳库是铁力木人工林地上碳库的主要部分，其中，树干占乔木层生物量和碳储量的 44. 5%和 45. 3%，其次是树枝，也占有较大比例，为 31. 8%；生物量及碳储量碳在树根的分配较低，仅为 13. 7%和 13. 1%；根冠比为 0. 16，显著低于该地区 28 年米老排(0. 26)[23]和 29 年生格木的根冠比(0. 22)[36]。这主要由铁力木的固有生长特性所致，铁力木树体呈塔形，其发达的枝叶系统形成了较大的树冠，但其根系并不发达，且在土壤中的分布较浅，发达的枝叶和不发达的根系共同造成了铁力木较低的根冠比。铁力木人工林发达的枝叶系统导致林分较大的郁闭度(0.95 左右)，也是导致林下植被稀少，灌草层生物量和碳储量分配较低的重要原因。

铁力木人工林 0~100cm 土壤碳储量为 120.5t/hm²，相对于本地区杉木林地土壤碳储量(139.27t/hm²)[34]和我国森林土壤平均碳储量(193.55t/hm²)[12]来说有所偏低；广西铁力木人工林土壤与植被碳储量的比值为 1∶1.5，远低于全国平均水平(1∶3.4)[12]，可能是铁力木人工林土壤呼吸通量过大，或者林分土壤受到了碳输入的限制，从而影响了土壤碳的累积，但具体原因有待进一步研究。

参考文献

[1]TANS P P. How can global warming be traced to CO_2[J]. Scientific American, 2006, 295: 124.

[2]ARTUSO F, CHAMARD P, PIACENTINO S, et al. Influence of transport and trends in atmospheric CO_2 at Lampedusa[J]. Atmospheric Environment, 2009, 43: 3044-3055.

[3]CALDEIRA K, DUFFY P B. The role of the southern ocean in up take and storage of anthropogenic carbon dioxide[J]. Science, 2000, 287: 620-622.

[4]FANG J Y, CHEN A P, PENG C H, et al. Changes in forest biomass carbon storage in China between 1949 and 1998[J]. Science, 2001, 292: 2320-2322.

[5] NORBY R J, LUO Y. Evaluating ecosystem responses to rising atmospheric CO_2 and global warming in a multi-factor world[J]. New Phytologist, 2006, 162: 281-293.

[6]WOODWELL G M, WHITTAKER R H, REINERS W A, et al. The biota and the world carbon budget[J]. Science, 1978, 199: 141-1467.

[7]POST W M, EMANUEL W R, ZINKE P J, et al. Soil pools and world life zone[J]. 1982, Nature, 298: 156-159.

[8]方精云，陈安平．中国森林植被碳库的动态变化及其意义[J]. 植物学报, 2001, 43(9): 967-973.

[9]王效科，冯宗炜．中国森林生态系统中植物固定大气碳的潜力[J]. 生态学杂志, 2000, 19(4): 72-74.

[10]王效科，冯宗炜．中国森林生态系统的植物碳储量和碳密度研究[J]. 应用生态学报, 2001, 12(1): 13-16.

[11]刘国华，傅伯杰，方精云．中国森林碳动态及其对全球碳平衡的贡献[J]. 生态学报, 2000, 20(5): 733-740.

[12]周玉荣，于振良，赵士洞．我国主要森林生态系统碳储量和碳平衡[J]. 植物生态学报, 2000, 24(5): 518-522.

[13]方晰，田大伦，项文化．速生阶段杉木人工林 C 密度、贮量和分布[J]. 林业科学, 2002, 38(3): 14-19.

[14]何宗明，李丽红，王义祥，等.33 年生福建柏人工林碳库与碳吸存[J]. 山地学报, 2003, 21(3): 298-303.

[15]LACLAU P. Biomass and carbon sequestration of ponderosa pine plantations and native cypress forests in northwest Patagonia[J]. Forest Ecology and Management, 2003, 180: 317-333.

[16]SPECHT A, WEST P W, Estimation of biomass and sequestered carbon on farm forest plantations in northern New South Wales, Australia[J]. Biomass and Bioenergy, 2003, 25: 363-379.

[17]雷丕锋，项文化，田大伦，等．樟树人工林生态系统碳素贮量与分布研究[J]. 生态学杂志, 2004, 23(4): 25-30.

[18]李轩然，刘琪，陈永瑞，等．千烟洲人工林主要树种地上生物量的估算[J]. 应用生态学报, 2006, 17(8): 1382-1388.

[19]马明东，江洪，刘跃建．楠木人工林生态系统生物量、碳含量、碳储量及其分布[J]. 林业科学, 2008, 44(3): 34-39.

[20]刘恩，刘世荣．南亚热带米老排人工林碳储量及其分配特征[J]. 生态学报, 2012, 32(16): 5103-5109.

[21]明安刚，张治军，张显强，等．抚育间伐对马尾松人工林生物量与碳储量的影响[J]. 林业科学, 2013, 49(10): 1-6.

[22]郑路，蔡道雄，卢立华，等，南亚热带不同树种人工林碳素密度[J]. 生态学杂志, 2013, 32(10): 2654-2658.

[23]明安刚，贾宏炎，陶怡，等．桂西南 28 年生米老排人工林生物量及其分配特征[J]. 生态学杂志, 2012, 31(5): 1050-1056.

[24]中国科学院昆明植物研究所．西双版纳植物名录[M]. 昆明：云南人民出版社, 1983.

[25]杨德军，邱琼，文进，等．热带珍稀多用途树种铁力木育苗与造林[J]. 林业实用技术, 2008(6): 38-39.

[26]王卫斌，史鸿飞，张劲峰．热带珍稀树种铁力木资源可持续经营对策研究[J]. 林业资源管理, 2002(6): 35-38.

[27]刘勐．铁力木育苗技术[J]. 广西林业科学, 2008, 37(3): 159-160.

[28]陈楚莹，廖利平，汪思龙．杉木人工林生态系统碳素分配与贮量的研究[J]. 应用生态学报, 2000, 11(增刊1): 175-178.

[29]梁宏温，温远光，温琳华，等．连栽对尾巨桉短周期

人工林碳储量的影响[J]. 生态学报，2009，29(8)：4242-4250.

[30]黄松殿，吴庆标，廖克波，等. 观光木人工林生态系统碳储量及其分布格局[J]. 生态学杂志，2011，30(11)：2400-2404.

[31]方晰，田大伦，胥灿辉. 马尾松人工林生产与碳素动态[J]. 中南林学院学报，2003，23(2)：11-15.

[32]明安刚，贾宏炎，陶怡，等. 米老排人工林碳素积累特征及其分配格局[J]. 生态学杂志，2012，31(11)：2730-2735.

[33]何斌，黄寿先，招礼军，等. 秃杉人工林生态系统碳素积累的动态特征[J]. 林业科学，2009，45(9)：151-157.

[34]康冰，刘世荣，蔡道雄，等. 南亚热带杉木生态系统生物量和碳素积累及其空间分布特征[J]. 林业科学，2009，45(8)：147-153.

[35]王卫霞，史作民，罗达，等. 我国南亚热带几种人工林生态系统碳氮储量[J]. 生态学报，2013，33(3)：0925-0933.

[36]明安刚，贾宏炎，田祖为，等. 不同林龄格木人工林碳储量及其分配特征[J]. 应用生态学报，2014，25(4)：940-946.

[37]王绍强，周成虎，罗承文，等. 中国陆地自然植被碳量空间分布探讨[J]. 地理科学进展，1999，，18(3)：238-244.

[38]覃林，何友均，李智勇，等. 南亚热带红椎马尾松纯林及其混交林生物量和生产力分配格局[J]. 林业科学，2011，47(12)：17-21.

[原载：北京林业大学学报，2015，37(02)]

广西大青山次生林的群落特征及主要乔木种群的空间分布格局

农　友[1,2]　郑　路[1,2]　贾宏炎[1,2]　卢立华[1,2]　黄德卫[1,2]　黄柏华[1,2]　雷丽群[1,2]

([1]中国林业科学研究院热带林业实验中心，广西凭祥　532600；
[2]广西友谊关森林生态系统国家定位观测研究站，广西凭祥　532600)

摘　要　本文依托广西友谊关森林生态系统国家定位观测研究站设置的 $1hm^2$ 长期监测样地，分析了森林群落的物种组成、区系特征及乔木种群的径级结构，并用点格局分析方法的 O-ring 统计对主要乔木植物种群的空间分布格局及其空间关联性进行了研究，旨在深入探讨该区域次生林的物种多样性特点及其维持机制，分析影响优势树种空间分布格局的可能因素。大青山次生林共调查到植物 109 种，其中乔木 58 种、灌木 29 种、草本 22 种，樟科为样地内物种最丰富的科，区系组成以泛热带成分为主；乔木树种的径级结构接近倒“J”形，主要种群的个体集中分布于中小径级范围(1~5cm)，林分结构合理，更新良好。从物种多度、胸高断面积和重要值来看，大叶栎(*Quercus griffithii*)和锈毛梭子果(*Eberhardtia aurata*)是群落中的共优种，其个体数占总个体数的 30.8%，鸭公树(*Neolitsea chuii*)、广东琼楠(*Beilschmiedia fordii*)和尖连蕊茶(*Camellia cuspidata*)为群落的主要伴生树种。用完全随机模型不排除生境异质性的条件下，主要种群多数呈聚集分布；用异质性随机模型排除生境异质性的条件下，主要种群的聚集程度显著下降，仅在小尺度上呈聚集分布；主要树种均在一定尺度上表现为两两间正相关，且在 0~50m 的大部分尺度上显示出相互独立的特点，没有表现出负相关。

关键词　空间分布格局；空间相关性；物种组成；次生林；南亚热带

Community Characteristics and Spatial Distribution of Common Secondary Forest Tree Species in the Daqing Mountains, Southwest Guangxi, China

NONG You[1,2], ZHENG Lu[1,2], JIA Hongyan[1,2], LU Lihua[1,2], HUANG Dewei[1,2], HUANG Bohua[1,2], LEI Liqun[1,2]

([1]*Experimental Center of Tropical Forestry, Chinese Academy of Forestry, Pingxiang* 532600, *Guangxi, China*
[2]*Guangxi Youyiguan Forest Ecosystem National Research Station, Pingxiang* 532600, *Guangxi, China*)

Abstract: In order to determine species characteristics and the factors affecting distribution patterns of dominant tree species of secondary forests in the Daqing Mountains, we analyzed community composition, DBH class structure, species distribution and spatial correlations using a point pattern analysis of O-ring statistical method. There were a total of 109 species including 58 tree species, 29 shrub species and 22 herb species. The floristic composition was pan-tropical with Quercus griffithii and Eberhardtia aurata as the co-dominant species. These two species accounted for 30.8% of the total number of individuals. Neolitsea chuii, Beilschmiedia fordii and Camellia cuspidate were the secondary tree species of the community. The DBH class structure of the populations had a reverse J-shaped pattern with a greater number of small diameter(1~5cm) individuals. Species spatial distribution with environmental heterogeneity was aggregated but decreased significantly with increasing spatial scales. Spatial distribution and heterogeneity were independent at most spatial scales among the main species populations.

Key words: spatial distribution; spatial correlation; species composition; secondary forest; subtropical

群落组成与空间格局研究为了解物种共存机制提供了重要信息[1,2]，有助于认识种群特征、种群间以及种群与环境之间相互作用的基本规律[3,4]。植物种群空间分布格局是种群个体在群落中的空间分布状况，它取决于自身和群落环境[5]，反映了种群物种、个体大小、分布等在空间上的相互关系，是种群的重要属性之一，在一定程度上影响着种群的发展[6,7]。物种生长、繁殖、死亡、资源利用及对干扰的反应等均受到种群空间格局的显著影响[8]。物种间不同的空间关系导致了群落结构的不同，决定了物种间的竞争及空间分布格局，直接影响到不同层次树种的空间分布类型及种子散布能力，与群落的更新机制密切相关[9]，能在一定程度上解释群落结构的发展历史和环境变化过程，进而预测植被演替的趋势[10,11]。植物种群的空间分布格局往往与研究的空间尺度密切相关[12,13]，同时又受到生境异质性和扩散限制的影响，在较小的尺度上可能是由种内竞争、种间竞争、种子扩散限制等因素决定，而在较大的尺度上则可能取决于种群分布区的环境异质性[12,14]。物种组成、群落结构、空间分布格局和生物多样性保护等研究往往依托于固定样地建设所提供的平台[15-17]。

大青山位于广西西南部，其次生林以亚热带常绿阔叶树种为主，康冰等[18]对其种群演替动态进行了研究，但有关其主要物种组成及其空间分布格局至今未见相关报道。南亚热带区域保存较好的次生常绿阔叶林已不多，研究其物种组成和空间分布格局并揭示其演变规律，对该区域植物多样性保护及南亚热带人工针叶纯林的近自然化改造等均具有重要意义。基于此，本文依托广西大青山次生林 1hm^2 固定监测样地，分析其物种组成、区系特征、乔木种群的径级结构、主要乔木树种的空间分布格局及其空间关联性，旨在深入探讨该区域物种多样性特点及其维持机制，分析影响优势树种空间分布格局的可能因素。同时，友谊关森林生态系统国家定位观测研究站是一个新建的以南亚热带为核心研究对象的生态站，背景资料奇缺，尤其缺乏植物群落结构与物种组成及多样性等方面的相关资料。填补这一空白，为今后开展相关研究提供必要的参考也是本研究的一个目的。

1 研究方法

1.1 研究区概况

研究区域为友谊关森林生态系统国家定位观测研究站在大青山建立的 1hm^2 固定监测样地（22°18′15.17″N，106°41′50.50″E），该区域处南亚热带南缘，与北热带毗邻，属于南亚热带季风气候区。年平均气温 21.5℃，最冷月（1 月）平均气温 13.5℃，最热月（7 月）平均气温 27.6℃，≥10℃的年积温 7500℃，年降雨量 1400mm[18]。大青山主峰海拔 1045m，地带性土壤为中酸性火山岩和花岗岩发育而成的砖红壤（含紫色土），土层平均厚度 0.5～1.0m[19]。大青山主峰地形复杂，地势陡峭，具有气温高、雨量充沛、湿度大、人为干扰较少的特点。

1.2 样地构建及野外调查

2013 年，经过实地踏查，选择大青山次生林保存较好的山顶东北坡，参照巴拿马 Barro ColoradoIsland（BCI）50hm^2 热带雨林样地的技术规范[20]，结合当地地形地貌，设置东西宽 100m、南北长 100m，面积为 1hm^2 的固定监测样地（海拔 960～1040m）进行长期观测。用全站仪将样地划分成 25 个 20m×20m 的样方，再把每个样方划分成 4 个 10m×10m 和 16 个 5m×5m 的小样方，每个 20m×20m 样方的 4 个角用带有样方编号的水泥桩作永久标记。

以 5m×5m 的小样方为基本调查单元，对每个 DBH≥1.0cm 的木本植物挂铝牌标记，记录种名、胸径、树高、坐标等。每个 20m×20m 的样方中，选择东北角及西南角的两个 5m×5m 小样方作为灌草层的补充调查样方，记录灌木、DBH<1cm 的乔木幼苗和草本植物的种名、个体数、高度、盖度等数据。全部野外工作于 2013 年 12 月至 2014 年 1 月进行。

1.3 方法

1.3.1 物种多样性的测定

物种丰富度指数（S）= 样方内出现的物种数目

$$\text{Shannon-Wiener 指数 } H' = -\sum_{i=1}^{S} P_i \ln P_i$$

$$\text{Simpson 指数 } D = 1 - \sum_{i=1}^{S} P_i^2$$

$$\text{Pielou 均匀度指数 } J_{sw} = H'/\ln S$$

式中：$P_i = N_i/N$，即某个物种的相对多度；N_i 为种 i 的株数；N 为种 i 所在样方所有物种的总株数[21]。

1.3.2 乔木重要值的测定

重要值（IV）= 相对多度（%）+相对频度（%）+相对胸高断面积（%）

重要值范围 0～300%。

1.3.3 地理区系及径级的划分

物种的地理区系参考吴征镒[22]的分类标准。参考王磊等[23]和郭垚鑫等[24]的方法，并结合实际调查，将乔木径级划分为 10 级：Ⅰ级 1.0～2.5cm；Ⅱ级 2.5～5.0cm；Ⅲ级 5.0～10.0cm；Ⅳ级 10.0～

15.0cm；Ⅴ级 15.0～20.0cm；Ⅵ级 20.0～25.0cm；Ⅶ级 25.0～30.0cm；Ⅷ级 30.0～40.0cm；Ⅸ级 40.0～50.0cm；Ⅹ级≥50.0cm。

1.3.4 点格局分析

本文采用 Wiegand-Molloney's *O-ring* 统计方法分析一定尺度范围内优势种的空间分布格局。在 Ripley'sK 函数和 Mark 相关函数的基础上，用半径为 r，宽度为 w 的圆环替代 Ripley'sK 函数计算中所使用的半径为 r 的圆，并根据圆环内的点平均数目，量化出一定距离等级上物种的空间分布。其原理和计算过程详见文献[13,25]。

O-ring 统计方法要求慎重选择零模型（null model）。本文选择完全空间随机模型（complete spatial randomness，CSR）[26]作为零假设来检验包含生境异质性的优势种群分布格局；采用异质性 Possion 过程（heterogeneous passion process，HP）模型对排除生境异质性效应的优势种群分布格局进行检验。同时，利用双变量统计分析比较样地内不同种群两两之间的空间关联性。因小尺度的聚集可能是植物之间相互作用引起的，而大于 10m 的尺度上，如果树种呈聚集分布，则可能是生境异质性效应在起作用[27]。因此，本文采用异质性 Possion 过程模型消除点密度不均后再进行结果检验，用移动窗口方法（moving window）使空间点的随机分布限定在半径<10m 的范围内。

本文的 *O-ring* 统计分析通过 Programita（2014 版）软件完成。研究尺度 0～50m，设定栅格大小为 1m×1m，圆环宽度为 3m。根据相应的零假设模型，为了提高植物种群空间分布的分析精度，应用 MonteCarlo 循环 99 次，产生置信度为 99%的包迹线以检验点格局分析的显著性。根据 MonteCarlo 模拟结果，对于空间分布，若 $O(r)$ 值位于上包迹线之上，则为聚集分布；若位于上下包迹线之间，则为随机分布；若位于下包迹线之下，则为均匀分布。对于种间相关性，若 $O_{ii}(r)$ 值位于上包迹线之上，则二者空间上正关联；若位于上下包迹线之间，表明二者相互独立；若位于下包迹线之下，则二者空间上负关联。

2 结果

2.1 群落特征及其物种多样性

共调查到植物 109 种，其中，乔木 3026 株，隶属于 29 科 45 属 58 种；灌木 390 株，隶属于 22 科 29 属 29 种；草本 4782 株，隶属于 15 科 22 属 22 种。樟科为物种最丰富的科。样地内木本植物共包括 9 个植物区系成分，其组成以泛热带成分为主，同时混有东亚（热带、亚热带）及热带南美间断及旧世界热带等成分。物种丰富度、Shannon-Wiener 指数、Simpson 指数和 J_{sw} 均匀度指数等均表现出乔木层>灌木层>草本层的规律（表 1）。乔木树种的径级分布呈近似倒“J”形，即随着径级的增加，株数逐渐减少（图 1）。主要种群的个体集中分布于中小径级范围（1～5cm）（图 2），林分内小径木和中径木较多，大径木较少，Ⅰ、Ⅱ径级的个体数量占总个体数的 58.67%。按树种重要值（IV）排序，样地中 IV≥10%的树种有 12 种（表 2），这些物种的多度与胸高断面积分别占样地总多度与总胸高断面积的 76.31%与 77.66%。

表 1 大青山次生林群落层次及其多样性

层次	物种丰富度指数 S	Shannon-Wiener 指数 H'	Simpson 指数 D	均匀度指数 J_{sw}
乔木层	58	2.95	0.91	0.73
灌木层	29	2.25	0.82	0.70
草本层	22	2.02	0.78	0.65

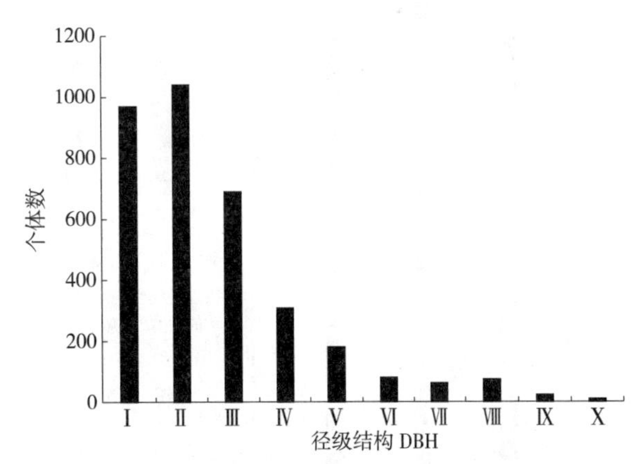

图 1 大青山天然次生林乔木径级结构

2.2 主要乔木种群的空间分布格局

样地中种群的分布点图可直观反映不同种群的分布情况。大叶栎（*Quercus griffithii*）、锈毛梭子果（*Eberhardtia aurata*）和尖连蕊茶（*Camellia cuspidata*）的个体数较多，聚集程度也较明显。8 个主要树种中，罗浮锥（*Castanopsis faberi*）和鸭公树（*Neolitsea chuii*）在样地西北部的聚集程度最为明显，其他树种则在样地东北部的聚集程度较高（图 3）。

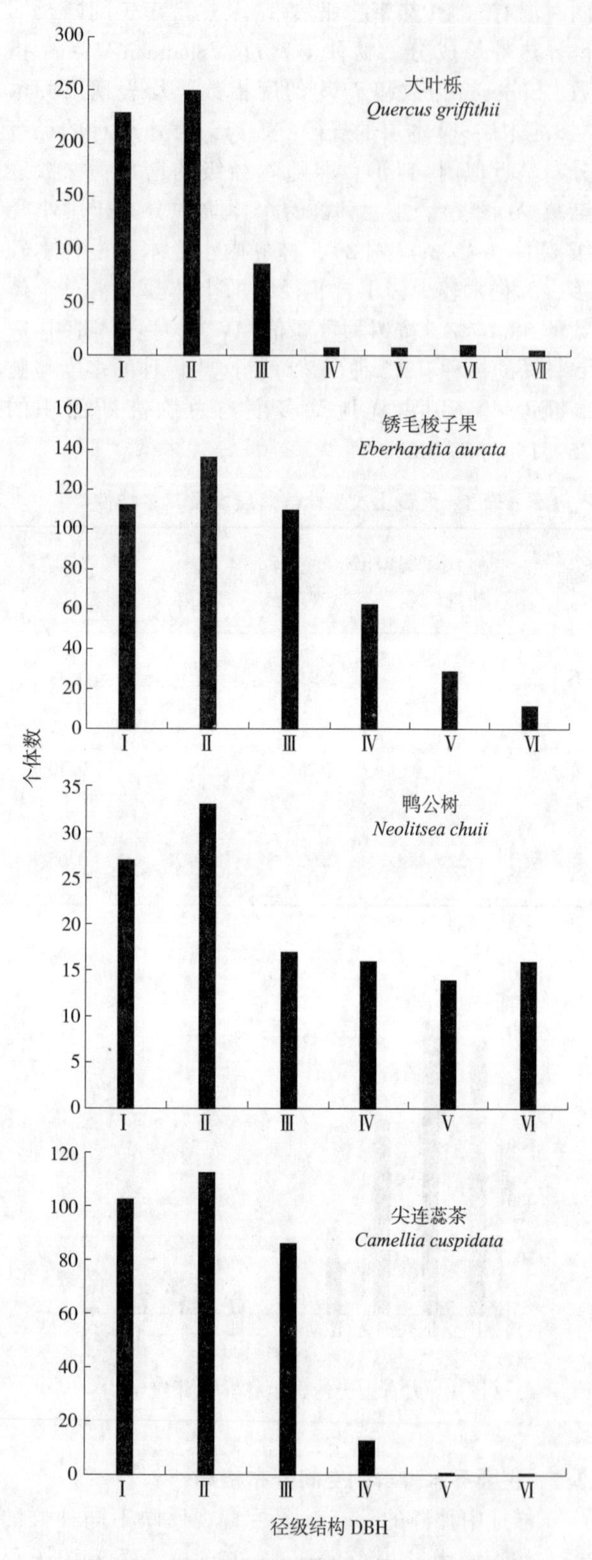

图2 大青山次生林主要乔木种群径级结构

2.2.1 生境异质性对空间分布格局的影响

结果表明，样地中树种分布受生境异质性的影响很大。8个主要树种中除木姜子(*Litsea pungens*)在10~15m、4~50m的尺度上呈随机分布外，其余树种均在0~50m的尺度上呈聚集分布(表3)。

表2 大青山次生林主要树种重要值

种名	个体数	胸高断面积（cm²）	重要值（%）
大叶栎 *Quercus griffithii*	594	29800.92	33.59
锈毛梭子果 *Eberhardtia aurata*	459	26591.51	28.83
鹿角椎 *Castanopsis lamontii*	107	44791.11	21.98
尖连蕊茶 *Camellia cuspidata*	318	10046.76	17.69
木姜子 *Litsea pungens*	75	22724.85	13.58
鸭公树 *Neolitsea chuii*	123	16414.89	13.50
广东琼楠 *Beilschmiedia fordii*	121	13834.18	13.34
罗浮锥 *Castanopsis faberi*	39	27981.65	13.16
环鳞烟斗柯 *Lithocarpus corneus*	127	15114.14	13.02
桂南木莲 *Manglietia chingii*	75	20219.74	12.77
柠檬金花茶 *Camellia limonia*	142	6196.79	11.00
海南山龙眼 *Helicia hainanensis*	129	5881.87	10.28

2.2.2 排除生境异质性的空间分布格局

排除生境异质性的情况下，该样地内8个主要树种的分布均呈现一致的规律，即在小尺度下为聚集分布，在大尺度下为随机分布，并在0~50m的大部分尺度上主要呈随机分布(图4)。其中，鸭公树在1m时，其集聚程度最大，$O(r)$值为2.28；木姜子和锈毛梭子果在2m时种群聚集程度最大，$O(r)$值分别为2.34和1.40；广东琼楠(*Beilschmiedia fordii*)和罗浮锥在3m时种群聚集程度最大，$O(r)$值分别为2.78和5.60；鹿角椎(*Castanopsis lamontii*)在5m时种群集聚程度最大，$O(r)$值为4.02；大叶栎和尖连蕊茶在7m时其集聚程度最大，$O(r)$值分别为2.28和2.77。

2.3 主要种群的空间关联性

在1hm²固定监测样地中，选取重要值≥15.0的所有物种大叶栎、锈毛梭子果、鹿角椎、尖连蕊茶进行空间关联性分析。

*O-ring*统计分析表明，主要树种两两间的空间关联性各异，但均在一定尺度上表现为正相关，且在0~50m的大部分尺度上呈相互独立的特点，没有表现出负相关(图5)。其中，大叶栎—尖连蕊茶最大正关联强度为2.29；大叶栎—鹿角椎最大正关联强度为4.53；大叶栎—锈毛梭子果最大正关联强度为2.40；鹿角椎—尖连蕊茶最大正关联强度为3.19；鹿角椎—锈毛梭子果最大正关联强度为3.52；锈毛梭子果—尖连蕊茶最大正关联强度为2.80。

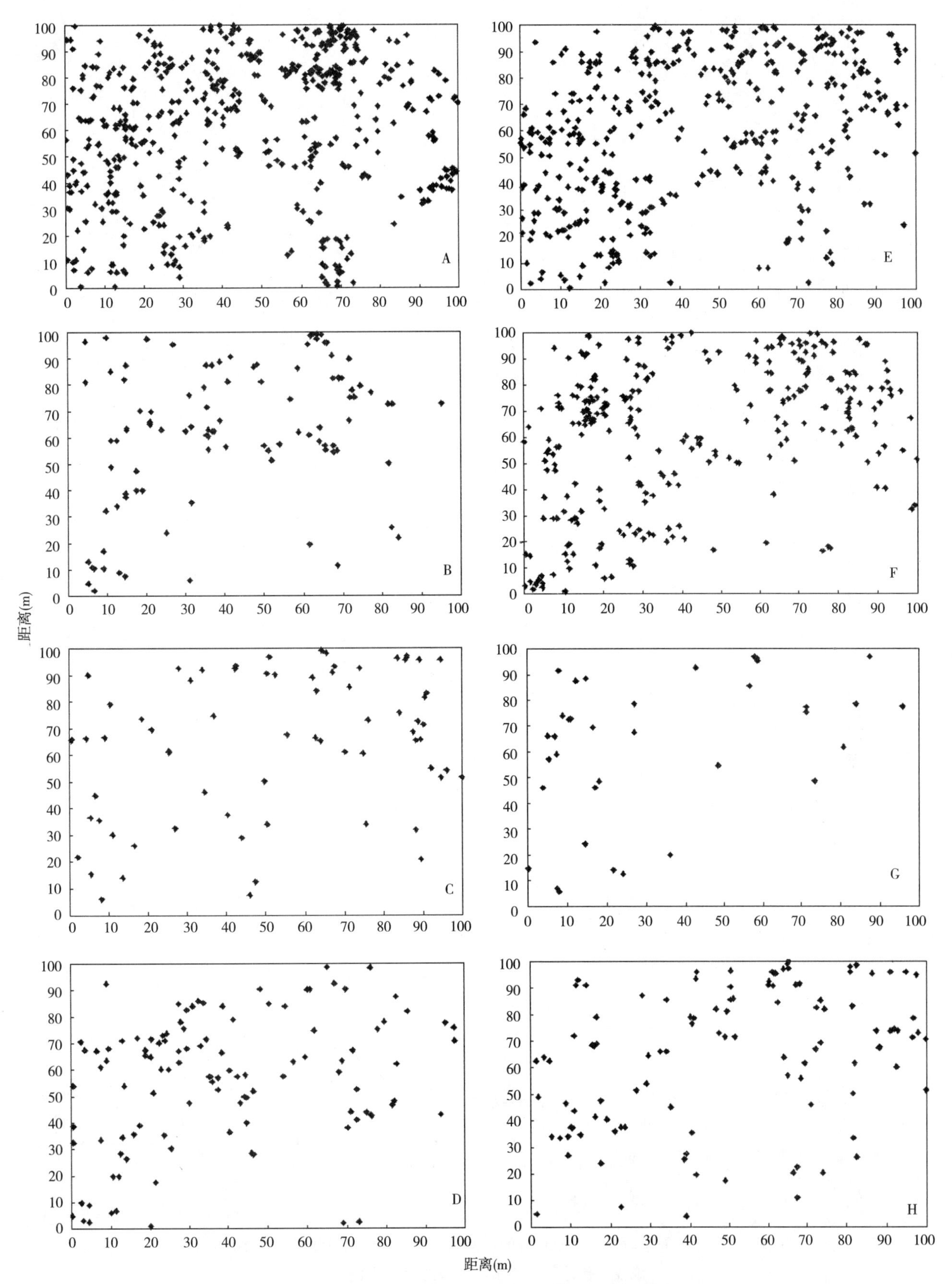

图 3 大青山次生林主要物种在 $1hm^2$ 样地中的分布

A. 大叶栎；B. 鹿角椎；C. 木姜子；D. 鸭公树；E. 锈毛梭子果；F. 尖连蕊茶；G. 罗浮锥；H. 广东琼楠

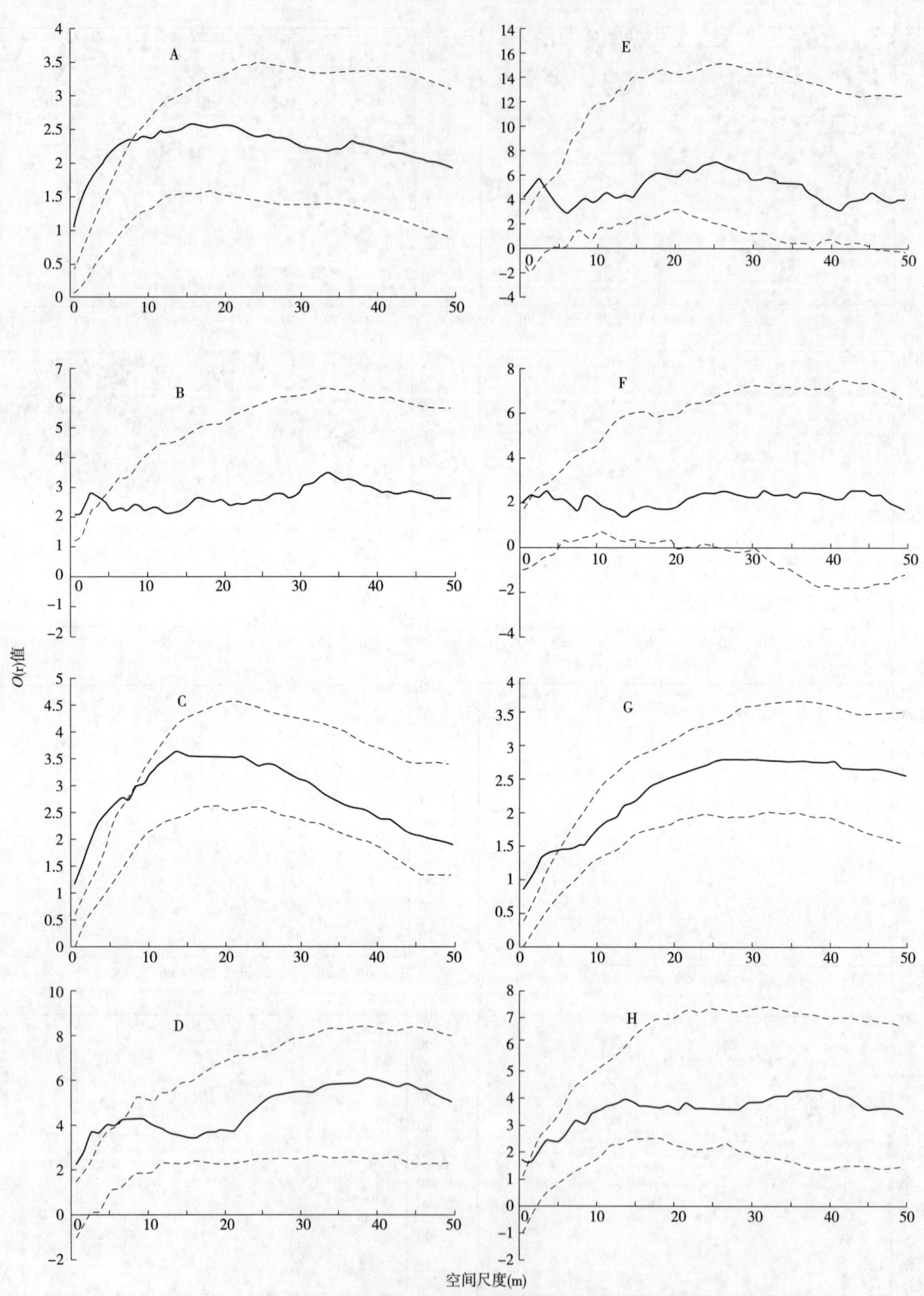

图 4　排除生境异质性后主要种群的点格局分析。实线为分析数据的 $O(r)$ 值，虚线为包迹线，表示所模拟的 99%置信区间

A. 大叶栎；B. 广东琼楠；C. 尖连蕊茶；D. 鹿角椎；E. 罗浮锥；F. 木姜子；G. 锈毛梭子果；H. 鸭公树

表 3 生境异质性对主要树种空间分布格局的影响

种名	尺度(m)				
	0~10	11~20	21~30	31~40	41~50
大叶栎 *Quercus griffithii*	a	a	a	a	a
广东琼楠 *Beilschmiedia fordii*	a	a	a	a	a
尖连蕊茶 *Camellia cuspidata*	a	a	a	a	a
鹿角椎 *Castanopsis lamontii*	a	a	a	a	a
罗浮锥 *Castanopsis faberi*	a	a	a	a	a
木姜子 *Litsea pungens*	a(+)	a(0)	a	a	a(+)
锈毛梭子果 *Eberhardtia aurata*	a	a	a	a	a
鸭公树 *Neolitsea chuii*	a	a	a	a	a

注：a，聚集分布；a(+)，研究尺度上的聚集分布大于随机分布；a(0)，研究尺度上的聚集分布等于随机分布。

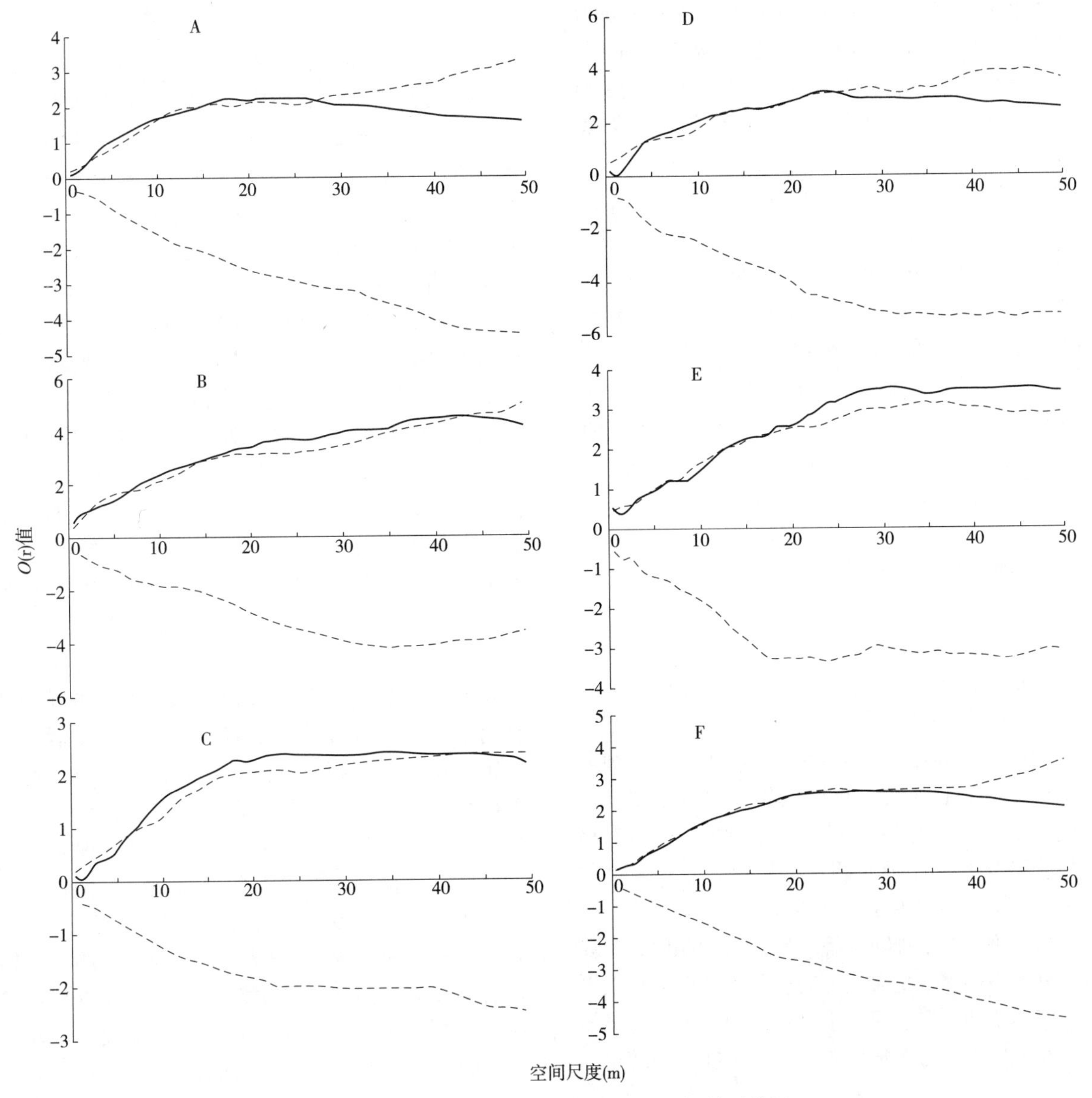

图 5 大青山天然次生林主要种群的种间关系分析

实线为分析数据的 $O_{12}(r)$ 值，虚线为包迹线，表示所模拟的 99%置信区间

A. 大叶栎—尖连蕊茶；B. 大叶栎—鹿角椎；C. 大叶栎—锈毛梭子果；

D. 鹿角椎—尖连蕊茶；E. 鹿角椎—锈毛梭子果；F. 锈毛梭子果—尖连蕊茶

3 讨论

3.1 物种组成与乔木径级结构

大青山天然次生林样地树种组成丰富，乔木树种较多，灌木、草本种类较少，区系组成以泛热带成分为主，符合亚热带物种组成的一般规律[28-30]。乔木层中，从物种多度、胸高断面积和重要值看，大叶栎和锈毛梭子果为群落的优势种；鸭公树、广东琼楠和尖连蕊茶等为主要伴生种，主要分布于林分的中下层，这些伴生树种虽然重要值较大，但胸径较小、个体数较少。

径级结构是衡量植物群落稳定性和生长发育状况的重要指标。样地内 DBH≥1cm 的树木的总胸高断面积为 31.31m^2/hm^2，乔木最大胸径为 71.9cm(罗浮锥)。乔木群落的径级结构接近倒"J"形，主要种群的个体集中分布于中小径级范围，林分结构合理，更新状况良好，说明群落处于稳定增长的状态。从径级结构可以看出，DBH≤10cm 的个体数量居多，占 78.95%。样地中 DBH>10cm 的个体比例为 21.05%，高于 CTFS(Centerfor Tropical Forest Science)中热带雨林区的大样地，如 BCI、Pasoh、Sinharaja(均<10%)[31-33]，高于南亚热带季风气候林区的大样地，如鼎湖山(Dinghushan, DHS)(16.42%)[34]，而低于 CTFS 中热带季雨林区的大样地，如 HKK(Huai Kha Khaeng, Tailand)(27.10%)、Mudumalai(58.97%)[15]，体现了其南亚热带季风气候的特征。

3.2 主要种群的空间分布格局

种群分布格局是物种与环境长期相互作用的结果，生物因子和非生物因子均可影响种群的分布格局[24,35]。研究表明，生境异质性[36,37]和限制性扩散[38]是导致物种聚集分布的重要因素[39]。在较小尺度上，物种的分布格局主要由种内种间竞争、种子的扩散机制决定；而较大尺度上的空间分布格局则更多受到生境异质性的影响[9,24]。物种的空间分布格局一般受到亲代种子散布习性的影响[7]，小尺度的聚集分布可能与树种的密度和传播特性有关[10,40]。

本研究中，在不排除生境异质性的影响时，主要种群的分布格局多数呈聚集分布；用异质性随机模型排除生境异质性的影响后，主要种群的聚集程度显著下降，在小尺度下为聚集分布，在大尺度下为随机分布，并以随机分布为主。说明南亚热带次生林物种的空间分布受到环境异质性影响。大青山次生林主要树种如大叶栎、锈毛梭子果、鹿角椎、罗浮锥等的果实均较大，掉落的种子因重力作用，多在林缘或母树周围处聚集分布，这也是导致这些种群呈现聚集分布的一个重要因素。同时，壳斗科植物种子是啮齿类动物的取食对象，动物取食、搬运、贮藏种子等的过程可能是导致种群在大尺度下呈随机分布的因素之一。大青山次生林的空间分布与环境异质性、生物因子(如种子散布方式或生物学特性、动物的取食干扰)等密切相关，但应如何定量分析不同组分、不同因子的贡献大小，是一个值得探讨的问题。

3.3 主要种群的空间关联性

种间关联是群落形成、演化的基础和重要的数量、结构指标，也是种间关系的一种表现形式和群落分类的依据[41]。优势种间的关联关系反映了其空间依赖性[42]。伴生树种与群落的优势树种保持互利共生的关系，对群落的稳定和物种多样性的维持具有重要的作用[43]。

大青山次生林主要树种均在一定尺度上表现为两两间正相关，但在大多尺度上呈现相互独立的特点。居于林木上层的优势种与下层的伴生树种(如大叶栎、鹿角椎、锈毛梭子果与尖连蕊茶之间)在个别中小尺度上为正关联，说明它们之间有一定的依赖性；但在大尺度上相互独立，说明它们之间没有明显的种间竞争，其生态位重叠较少。上层优势种如大叶栎、鹿角椎、锈毛梭子果两两之间在大多尺度上为正关联，体现了它们利用资源的相似性和较大的生态位重叠。

一般而言，群落处于演替初期时，物种间的关联程度往往较低，甚至会出现较大程度的负关联，种间竞争较为激烈[44]；随着群落演替的发展，物种间正关联程度将会逐步增大；当演替到顶极阶段时，群落结构及种类组成将逐渐趋于稳定，种间关系表现为明显正关联[23,24]。本次研究的所有种对大多在不同尺度上呈现相互独立的特点，说明大青山次生林植物群落结构及种类组成还不是很稳定，应该是一个演替前期的先锋群落。随着演替的继续进行，今后树种之间的竞争将更加激烈、物种更替现象更频繁，在经营管理过程中，可采取择伐等手段，缓和种间矛盾，人为促进天然更新。

参考文献

[1] LOREAU M, NAEEM S, INCHAUSTI P, et al. Biodiversity and ecosystem functioning: current knowledge and future challenges[J]. Science, 2001, 294: 804-808.

[2] TILMAN D, REICH PB, KNOPS JMH. Biodiversity and ecosystem stability in a decade-long grassland experiment [J]. Nature, 2006, 441: 629-632.

[3] HE F, DUNCAN RP. Density-dependent effects on tree

survival in an old-growth Douglas fir forest[J]. Journal of Ecology, 2000, 88: 676-688.

[4]JOHN R, DALLING J W, HARMS K E, et al. Soil nutrients influence spatial distributions of tropical tree species [M]. Proceedings of the National Academy of Sciences, USA, 2007, 104: 864-869.

[5]ZHANG J T, MENG D P. Spatial pattern analysis of individuals in different age-classes of Larix gmelinii in Luyashan Mountain, China[J]. Acta ecological Sinica, 2004, 24: 35-40.

[6]STERNER R W, RIBIC C A, SCHATZ G E. Testing for life historical changes in spatial patterns of four tropical tree species[J]. Journal of Ecology, 1986, 74: 621-633.

[7]CAO G X, ZHONG Z C, LIU Y, et al. The study of distribution pattern of Camellia rosthorniana population in Jinyun Mountain. Journal of Biology, 2003, 20: 10-12.

[8]HE F, LEGENDRE P, LAFRANKIE J V. Distribution patterns of tree species in a Malaysian tropical rain forest [J]. Journal of Vegetation Science, 1997, 8: 105-114.

[9]HAO ZQ, ZHANG J, SONG B, et al. Vertical structure and spatial associations of dominant tree species in an old growth temperate forest [J] . Forest Ecology and Management, 2007, 252: 1-11.

[10]HUBBELL S P. Tree dispersion, abundance and diversity in a tropical dry forest [J] . Science, 1979, 203: 1299-1309.

[11]HU Y B, HUI G Y, QI J Z, et al. Analysis of the spatial structure of nature Korean pine broadleaved forest. Forest Research, 2003, 16: 523-530.

[12]HARMS K E, CONDIT R, HUBBELL S P, et al. Habitat associations of trees and shrubs in a 50-ha neotropical forest plot[J]. Journal of Ecology, 2001, 89: 947-959.

[13]WIEGAND T, MOLONEY K A. Rings circles and null-models for point pattern analysis in ecology[J]. Oikos, 2004, 104: 209-229.

[14]LIN Y C, CHANG L W, YANG K C, et al. Point patterns of tree distribution determined by habitat heterogeneity and dispersal limitation [J] . Oecologia, 2011, 165: 175-184.

[15]CONDIT R, ASHTON P S, BAKER P, et al. Spatial patterns in the distribution of tropical tree species [J]. Science, 2000, 288: 1414-1418.

[16]MA K P. Large scale permanent plots: important platform for long term research on biodiversity in forest ecosystem [J]. Journal of Plant Ecology, 2008, 32: 237-237.

[17]LEGENDRE P, MI X, REN H, et al. Partitioning beta diversity in a subtropical broad-leaved forest of China [J]. Ecology, 2009, 90: 663-674.

[18]KANG B, LIU S R, WEN Y G, et al. Population dynamics during succession of secondary natural forest in Daqingshan, Guangxi, China [J] . Journal of Plant Ecology, 2006, 30: 931-940.

[19]HUANG C B, LU L H, WEN Y G, et al. Vertical distribution of main meteorological elements in Daqingshan forest zone of Guangxi[J]. Guizhou Agricultural Scinences, 2011, 39, 90-95.

[20]CONDIT R. Tropical Forest Census Plots: Method and Results from Barro Colorado Island, Panama and a Comparison with Other Plot[M]. Springer Verlag, Berlin, 1998.

[21]MAGURRAN A E. Ecological Diversity and Its Measurement [J] . Princeton university press, Princeton, 1988.

[22]WU Z Y. The areal-types of Chinese genera of seed plants [J]. Acta Botanica Yunnanica, 1991, 4(Suppl.), 1-139.

[23]WANG L, SUN Q W, HAO C Y, et al. Point pattern analysis of different age-class Taxus chinensis var. mairei individuals in mountainous area of southern Anhui Province[J]. Chinese Journal of Applied Ecology, 2010, 21, 272-278.

[24]GUO Y X, KANG B, LI G, et al. Species composition and point pattern analysis of standing trees in secondary Betula albosinensis forest in Xiaolongshan of west Qinling Mountains [J] . Chinese Journal of Applied Ecology, 2011, 22, 2799-2806.

[25]WIEGAND T, MOLONEY K A. A Handbook of Spatial Point Pattern Analysis in Ecology. Chapman and Hall/CRC press, Boca Raton, FL, 2014.

[26]GETZIN S, WIEGAND T, WIEGAND K, HE F. Heterogeneity influences spatial patterns and demographics in forest stands [J] . Journal of Ecology, 2008, 96, 807-820.

[27]STOYAN D, PENTTINEN A. Recent applications of point process methods in forestry statistics [J] . Statistical Science, 2000, 15, 61-78.

[28]XIE Z S, GU Y K, CHEN B G, et al. Species diversity of the natural forest communities in Nanling National Nature Reserve, Guang Dong[J]. Journal of South China Agricultural University, 1998, 19, 61-66.

[29]LIAO C Z, HONG W, WU C Z, et al. Study on the spatial of species diversity for the subtropical evergreen broadleaf forest in Fujian Province[J]. Guihaia, 2003, 23, 517-522.

[30]MENG F H. Study on Species Diversity and Conservation the Flora of Yuanbaoshan Nature Reserve[D]. PhD dissertation, Guangxi Nomal University, Guilin, 2006.

[31]KNIGHT D H. A phytosociological analysis of species-rich

tropical forest on Barro Colorado Island, Panama [J]. Ecological Monographs, 1975, 45, 259-284.

[32] MANOKARAN N, LaFrankie Jr JV. Stand structure of Pasoh Forest Reserve, a lowland rain forest in Peninsular Malaysia[J]. Journal of Tropical Forest Science, 1990, 3, 14-24.

[33] GUNATILLEKE S. Ecology of Sinharaja Rain Forest and the Forest Dynamics Plot in Sri Lanka's Natural World Heritage Site[J]. WHT Publications, Sri Lanka, 2004.

[34] YE W H, CAO H L, HUANG Z L, et al. Community structure of a 20 hm2 lower subtropical evergreen broad-leaved forest plot in Dinghushan, China[J]. Journal of Plant Ecology(Chinese Version), 2008, 32, 274-286.

[35] ZHANG Y T, LI J Z, CHANG S L, et al. Spatial distribution pattern of Picea schrenkiana var. tianshanica population and its relationships with topographic factors in middle part of Tianshan Mountain[J]. Chinese Journal of Applied Ecology, 2011, 22, 2799-2806.

[36] HARMS K E, WRIGHT J S, CALDERÓN O, et al. Pervasive density-dependent recruitment enhances seedling diversity in a tropical forest [J]. Nature, 2000, 404, 493-495.

[37] QUEENBOROUGH S A, BURSLEM DFRP, GARWOOD N C, et al. Habitat niche partitioning by 16 species of Myristicaceae in Amazonian Ecuador[J]. Plant Ecology, 2007, 192, 193-207.

[38] GRUBB P. Maintenance of species-richness in plant communities: importance of regeneration niche[J]. Biological Reviews, 1977, 52, 107-145.

[39] ZHANG J, SONG B, LI B H, et al. Spatial patterns and associations of six congeneric species in an old-growth temperate forest[J]. Acta Oecologica, 2010, 36, 29-38.

[40] MURRELL D J. On the emergent spatial structure of size-structured populations: when does self-thinning lead to a reduction in clustering [J]. Journal of Ecology, 2009, 97, 256-266.

[41] YUE Y J, YU X X, LI G T, et al. Spatial structure of Quercus mongolica forest in Beijing Songshan Mountain Nature Reserve[J]. Chinese Journal of Applied Ecology, 2009, 20, 1811-1816.

[42] TANG M P, ZHOU G M, SHI Y J, et al. Spatial patterns in evergreen broadleaved forest in Tianmu Mountain, China [J]. Journal of Plant Ecology, 2006, 30, 743-752.

[43] AN H J. Study on the Spatial Structure of the Broad-leaved Korean Pine Forest[D]. Beijing Forestry University, Beijing, 2003.

[44] GUO Z L, MA Y D, ZHENG J P, et al. Biodiversity of tree species, their populations' spatial distribution pattern and interspecific association in mixed deciduous broadleaved forest in Changbai Mountain [J]. Chinese Journal of Applied Ecology, 2004, 15, 2013-2018.

[原载：生物多样性，2015，23(0)]

西南桦和尾巨桉凋落叶分解及其与土壤性质的相关性

郝　建[1,2]　莫慧华[1,2]　黄弼昌[3]　周燕萍[3]　蔡道雄[1,2]

([1]中国林业科学研究院热带林业实验中心，广西凭祥　532600；[2]广西友谊关森林生态系统国家定位观测研究站，广西凭祥　532699；[3]天峨县林朵林场，广西天峨　547300)

摘　要　探讨南亚热带西南桦和尾巨桉人工纯林的凋落叶分解动态及其与土壤化学性质之间的相关关系。采用原位分解袋法研究凋落叶的分解过程。表明：西南桦、尾巨桉人工林凋落叶分解系数分别为0.96/年和0.88/年。在为期12个月的分解实验中，2种凋落叶有机C含量在整个分解过程中呈逐渐下降的趋势；全K含量和C/N在分解前期迅速下降，之后趋于平缓；全N含量和全P含量在整个分解过程中呈逐渐上升趋势；2种凋落叶N/P则呈先升高后下降的趋势。无论是分解前期还是分解后期，凋落叶质量损失与N含量均呈显著正相关(前期 $R=0.877$；后期 $R=0.855$)，与C/N均呈显著负相关(前期 $R=-0.735$；后期 $R=-0.697$)。与尾巨桉林地土壤性质相比，西南桦凋落叶分解提高了林地0~10cm、10~20cm土壤有机C、全N、全P、全K、N/P($P<0.05$)，而对20~30cm土壤有机C、全K、pH、C/N、N/P未产生显著影响($P>0.05$)。相关分析表明：凋落叶初始有机C含量与土壤有机C、全N、全P、全K、N/P显著相关；凋落叶初始全N含量与土壤全N、pH、C/N显著相关。凋落叶分解速率主要受凋落叶自身养分含量的影响，即使在造林的初级阶段，不同的凋落叶养分含量对土壤养分状况的影响也不同。

关键词　凋落叶；养分含量；分解速率；化学性质；南亚热带

Relationships Between Leaf Litter Decomposition and Soil Properties in *Betula alnoides* and *Eucalyptus urophylla*×*E. grandis*

HAO Jian[1,2], MO Huihua[1,2], HUANG Bichang[3], ZHOU Yanping[3], CAI Daoxiong[1,2]

([1]*Experimental Center of Tropical Forestry*, *Chinese Academy of Forestry*, *Pingxiang* 532600, *Guangxi*, *China*; [2]*Guangxi Youyiguan Forest Ecosystem Research Station*, *Pingxiang* 532699, *Guangxi*, *China*; [3]*Linduo Forest Farm of Tiane County*, *Tiane* 547300, *Guangxi*, *China*)

Abstract: The decomposition dynamics of leaf litter and its relationship with soil chemical properties in two young plantation stands(monocultures of *Betula alnoides* and *Eucalyptus urophylla* × *E. grandis*) were studied in subtropical China. The decomposition processes of leaf litter were measured using mesh nylon bag method. The decomposition coefficients of leaf litter of *B. alnoides* and *E. urophylla* × *E. grandis* were 0.96/year and 0.88/year, respectively. During the 12-month decomposition, the organic carbon contents declined gradually in the two leaf litter. Total K and C/N were rapidly decrease at the early stage, and then tended towards stability thereafter. Total N and total P in the two leaf litter increased gradually, whereas the N/P increased firstly and then decreased throughout the entire decomposition. In both early and late phase of decomposition, dry mass loss of the leaf litter was correlated positively with N contents($R=0.877$ and 0.855, respectively), and a negative relation was observed with C/N ratio($R=-0.735$ and -0.697, respectively). Compared with the soil properties in *E. urophylla*×*E. grandis*, the *B. alnoides* had significantly greater contents of soil organic carbon(SOC), total N, total P, total K and N/P ratio in 0~10 and 10~20 cm soil depth. However, litter decomposition had no significant effects on SOC, total K, pH, C/N ratio and N/P ratio in 20~30 cm soil depth. The correlation analysis revealed that the organic C of leaf litter was significantly related to SOC, total N, total P, total K and N/P ratio, whereas total N of leaf litter was significantly related to soil total N, pH and C/N ratio. An implication of these findings is that the decomposition rates of leaf litter were mainly affected by nutrient content. Even in the initial growth phase of these plantations, different nutrient content of leaf litter display different effects on soil nutrient status.

Key words：leaf litter；decomposition rate；nutrient content；chemical properties；subtropical China

森林凋落物通过分解参与森林生态系统物质循环和能量转换，逐步把养分输入给土壤[1]，影响土壤的理化性质、养分及生物活性。凋落物分解速率的高低在一定程度上影响了土壤的养分状况[2,3]，加快其分解，可促进养分循环，改善土壤肥力。

在我国南亚热带地区，人工造林、再造林已成为森林培育和经营的重要方式。然而，随着大规模、持续单一人工针叶林如马尾松（*Pinus massoniana* Lamb.）和杉木［*Cunninghamia lanceolata*（Lamb.）Hook.］或桉树（*Eucalyptus* sp.）等外来树种短周期工业用材林的发展，造成了诸如生物多样性减少、土壤退化、生态系统稳定性降低等问题[4]。为促进人工林的多目标经营，提高人工林的生态功能和经济价值，许多乡土珍贵阔叶树种如西南桦（*Betula alnoides* Buch. -Ham. ex D. Don）、格木（*Erythrophleum fordii* Oliv.）、红椎（*Castanopsis hystrix* Miq.）等，逐渐被用于亚热带人工林营建的生产实践中[5]。近年来，有关不同林分对土壤养分影响方面的研究在国内外已有报道，然而，对乡土珍贵树种和外来树种用于人工林营建后凋落物养分状况与土壤养分关系的研究仍相对缺乏。为此，本文以南亚热带具有相同经营历史与立地条件的乡土珍贵树种西南桦和外来树种尾巨桉（*Eucalyptus urophylla* × *E. grandis*）人工林为对象，研究比较不同人工林凋落物养分状况和土壤养分含量及凋落物养分含量对土壤性质的影响，旨在更深入认识该地区不同人工林生态系统的生态功能，以期更有效地对人工林进行经营管理。

1　研究地点与方法

1.1　研究区概况

研究地点位于广西天鹅县林朵林场，海拔600～900m，属亚热带季风气候。年均最高气温37.9℃，最低气温2.9℃，年平均气温20.9℃，年均积温7 475.2℃，平均日照时数1232.2h，年平均降水量1253.6mm，年平均无霜期336d。林场造林地土壤为砂页岩发育而成的黄壤、黄红壤和红壤，大部分林地土层厚度约100cm，表土层厚度10～30cm，土壤质地多为壤土或轻壤土，结构疏松。植被类型为南亚热带季雨林植被带，乔木主要有杉木、桉树、马尾松、八角（*Illicium verum* Hook. f.）等；灌木有鸭脚木（*Schefflera minutistellata* Merr. ex Li）、野牡丹（*Melastoma candidum* D. Don）、山黄麻［*Trema tomentosa*（Roxb.）Hara］等；草本有铁芒萁［*Dicranopteris linearis*（Burm.）Underw.］、五节芒［*Miscanthus floridulus*（Lab.）Warb. ex Schum. et Laut.］、黄毛草［*Pogontherum paniceum*（Lam.）Hack.］、龙须草［*Eulaliosis binate*（Retz.）C. E. Hubb.］等[6]。

选择2007年以该区域主要的造林、再造林树种营建的西南桦纯林和尾巨桉纯林人工林生态系统为研究对象，其中，西南桦是广西重要的优良乡土树种，具有速生、适应性强、材性上等和经济效益好等特点；桉树是世界著名的速生树种，也是世界重要的硬质阔叶树之一，具有适应性强，生长快、产量高、周期短、材油兼备、用途广泛等优良性状，是增加农民收入、壮大林产业的一条有效途径。造林前2种林分造林地均为杉木采伐迹地，立地条件基本一致，造林后的森林经营管理方式相同。林分基本情况见表1。

1.2　研究方法

1.2.1　试验设计

2012年5月，分别在西南桦和尾巨桉2种人工林内，各设置大小为20m×20m的固定样地4个，每个固定样地间至少间隔12m以上。2012年6月，从西南桦和尾巨桉各自林下收集新近自然凋落、上层未分解的凋落叶样品（为避免破坏样地，凋落叶样品均收集自固定样地以外），带回实验室后放置于地板上风干至恒质量[7]。分别称取10g风干重量的2种凋落叶样品装入尼龙网质分解袋（分解袋规格为孔径1mm，尺寸25cm×25cm），然后将装好袋的2种凋落叶样品于同一天放回到初始样地中[8]。按样地坡度大小将凋落叶袋倾斜放置，并用铁针固定。每个固定样地分5个点放置凋落叶袋，每个点放置5袋，每个树种总计放置100袋。

1.2.2　凋落叶样品采集与测定

2012年6月至2013年6月间，每隔3个月从西南桦和尾巨桉各自样地中随机取回20袋凋落叶分解袋（4个固定样地×5袋/每次每个样地），用镊子小心地清除侵入到分解袋内的土壤颗粒、植物根和菌体等，带回实验室置入70℃烘箱中烘干至恒质量，称量干质量并计算干质量残留率和分解速率[8]。随后将样品磨粉过0.25mm细筛，用于凋落物养分含量的测定。凋落叶有机C含量采用重铬酸钾氧化—外加热法；全N采用凯氏定氮法；P、K含量用等离子发射光谱法测定[9]。

表 1 研究样地林分基本情况

林分类型	林龄(年)	树高(m)	胸径(cm)	密度(株/hm²)	海拔(m)	坡向	坡度(°)
西南桦	5	12.69±1.13	13.13±1.46	1335	700	半阳	25
尾巨桉	5	16.45±3.75	14.70±4.19	1245	650	半阳	25

1.2.3 土壤样品采集与理化性质分析

凋落叶分解试验结束后(2013 年 6 月)，在每个固定样地内(采样点位于凋落叶分解袋之外)用土钻(直径 8cm)采集深度 0~10cm、10~20cm、20~30cm 的 3 个层次土壤各 5 钻，除去动植物残体和石块等杂质，并将同一固定样地内同层的 5 钻土壤充分混合为 1 个土样后装入塑料袋中带回实验室。采集的土样在实验室内自然风干后一分为二，其中一份过 2mm 土壤筛后用于土壤 pH 测定；另外一份过 0.25mm 细筛，用于其他土壤理化性质的分析。土样总有机 C、全 N、全 P 和全 K 含量测定方法同凋落物测定方法，土壤 pH(水土比 2.5：1)采用玻璃电极测定。

1.3 数据处理与分析

采用常用的 Olson[10] 单指数模型计算西南桦和尾巨桉凋落叶分解质量损失系数：

$$X_t/X_0 = e^{-kt}$$

式中：X_t表示分解时间 t 时刻的凋落叶残体质量；X_0表示凋落叶初始质量(kg)；e 是自然对数的底；k 表示凋落物的分解系数；t 是分解时间(月)。

采用单因素方差分析(one-wayANOVA)和多重比较(LSD)检验不同林分之间凋落叶养分含量、分解速率、分解系数 k 和土壤化学性质之间的差异。凋落叶养分含量与土壤化学性质之间的关系采用 Spearman 相关分析法进行分析。所有方差分析均在统计分析软件 SPSS 16.0 中进行，显著性水平设为 $\alpha = 0.05$。采用 Sigmaplot 10.0 作图。

2 结果与分析

2.1 凋落叶分解残存率的动态变化与分解系数

从图 1 可看出：西南桦和尾巨桉凋落叶的分解可分为 2 个阶段，在分解的前 9 个月，凋落叶残存率下降较快，即为凋落叶的快速失重期；在分解的后 3 个月，凋落叶残存率下降缓慢，即为凋落叶的慢速失重期。经过 12 个月的分解，西南桦和尾巨桉残存率分别为 38.2% 和 41.5%。在整个分解期间，西南桦凋落叶的分解速率始终高于尾巨桉凋落叶的($P <$ 0.05)，西南桦凋落叶分解系数 k 为 0.96/年，明显高于尾巨桉凋落叶的 0.88/年(图 2)。

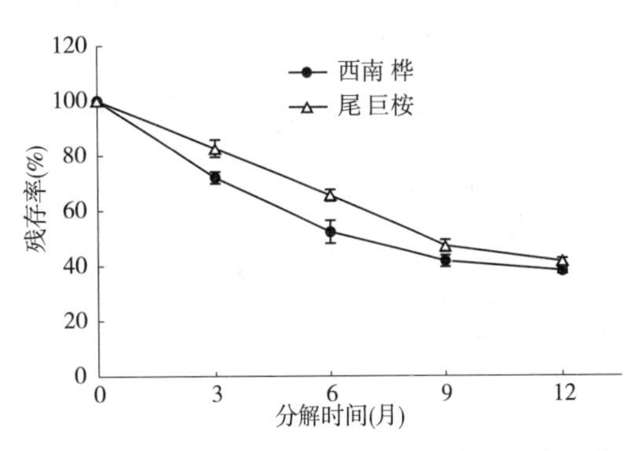

图 1 西南桦和尾巨桉凋落叶残存率的动态变化

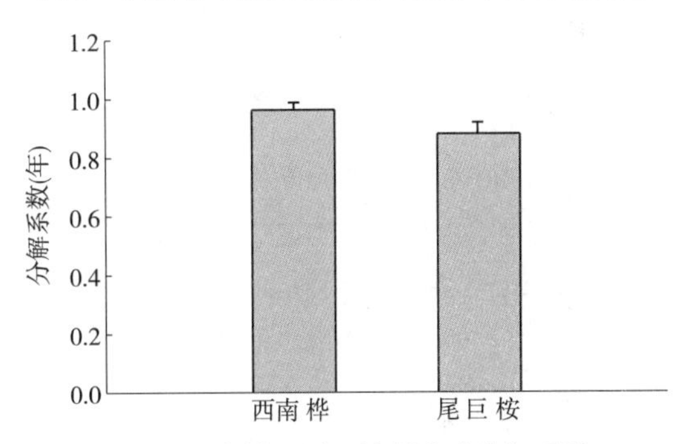

图 2 西南桦和尾巨桉凋落叶分解系数

2.2 凋落叶在分解过程中的养分动态变化

在为期 12 个月的分解试验中，西南桦和尾巨桉 2 种凋落叶养分含量呈不同的变化趋势，其中，2 种凋落叶有机 C 含量在整个分解过程中呈逐渐下降的趋势，分解末期西南桦和尾巨桉凋落叶有机 C 含量较初始 C 含量分别下降 37.7% 和 39.5%；全 K 含量和 C/N 比在分解前期(前 6 个月)迅速下降，之后趋于平缓；全 N 含量和全 P 含量在分解前期呈逐渐上升趋势，之后略有下降，分解末期西南桦和尾巨桉凋落叶全 N 含量较初始 N 含量分别升高 20.4% 和 30.1%，全 P 含量较初始 P 含量分别升高 11.4% 和 34.1%；2 种凋落叶 N/P 比则呈先升高后下降的趋势。

整个分解期间，不同树种凋落叶养分含量的动态变化也存在差异，其中，西南桦凋落叶有机 C 含量始终高于尾巨桉；2 种凋落叶初始全 N 含量差异不大，但随着分解时间的延长，西南桦凋落叶全 N 含量高于尾巨桉，到分解末期西南桦全 N 含量反而低于尾巨桉；在分解前 3 个月，西南桦凋落叶全 P 含量、全K含量和C/N比高于尾巨桉，而在分解的后6个月，西南桦凋落叶全P含量、全K含量和C/N比低于尾巨桉。

2.3 土壤化学性质

表2表明：除土壤C/N外，土壤各化学性质指标均随土层深度增加而降低；林分对土壤化学性质也产生显著影响，从0~10cm和10~20cm土层看，西南桦土壤有机C、全N、全P、全K、N/P比均显著高于尾巨桉(P<0.05)；2种林分土壤pH和C/N仅在10~20cm土层差异显著(P<0.05)；从20~30cm土层看，西南桦林地土壤全N和全P含量显著高于尾巨桉(P<0.05)，而林分对该层土壤有机C、全K、pH、C/N、N/P未产生显著影响(P>0.05)。

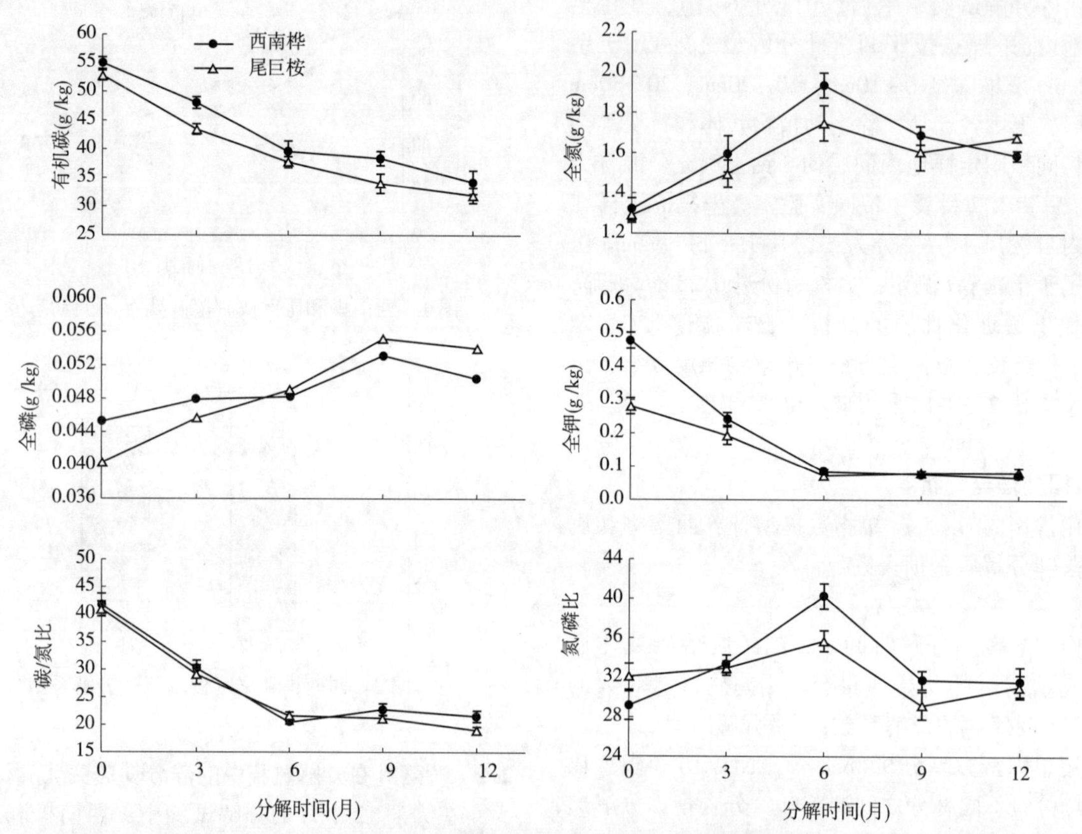

图3 西南桦和尾巨桉凋落叶分解过程中养分的动态变化

表2 西南桦和尾巨桉林地0~30cm土壤化学性质

土壤化学性质	西南桦			尾巨桉		
	0~10cm	10~20cm	20~30cm	0~10cm	10~20cm	20~30cm
有机C	34.15±1.39aA	24.98±1.55bA	18.12±1.59cA	31.33±1.03aB	21.88±1.73bB	16.90±0.55cA
全N	2.14±0.05aA	1.27±0.07bA	0.86±0.08cA	2.05±0.07aB	1.08±0.05bB	0.76±0.04cB
全P	0.49±0.02aA	0.40±0.01bA	0.34±0.02cA	0.40±0.01aB	0.30±0.03bB	0.25±0.03cB
全K	16.70±0.57aA	14.46±0.84bA	11.68±0.45cA	15.06±0.32aB	13.03±0.41bB	10.93±0.80cA
pH	4.56±0.10aA	4.51±0.10aA	4.45±0.11aA	4.39±0.18aA	4.37±0.14aB	4.36±0.04aA
C/N	15.05±0.94cA	23.29±0.32aA	21.00±0.76bA	15.28±0.76bA	20.20±0.77aB	22.36±1.92aA
N/P	4.38±0.17aB	2.71±0.17bB	2.58±0.43bA	5.18±0.03aA	3.64±0.48bA	3.07±0.45bA

注：同一林分同一指标的不同小写字母表示土层间差异显著(P<0.05)；同一土层同一指标的不同大写字母表示不同林分间差异显著(P<0.05)。表中数据均为四次重复的平均值±标准误。

2.4　凋落叶养分性状对凋落物分解的影响

西南桦、尾巨桉凋落物养分性状不同具有不同的分解速率，而凋落叶不同的养分特征与分解过程中质量损失的相关关系也不同。表3表明：分解前期，凋落叶质量损失与N含量和N/P显著正相关(R分别为0.877和0.812)，与C/N显著负相关($R=-0.735$)；而分解后期，凋落叶质量损失与N含量显著正相关($R=0.855$)，与C/N显著负相关($R=-0.697$)。

2.5　凋落叶养分含量对土壤化学性质的影响

2种人工林的初始凋落叶养分含量与0~10cm土壤化学性质间的相关分析结果(表4)表明：凋落叶初始有机C含量与土壤有机C、全N、全P、全K、N/P显著或极显著相关；凋落叶初始全N含量与土壤全N、pH值显著或极显著相关；凋落叶初始全P含量和N/P与土壤全P、全K、N/P显著或极显著相关；凋落叶初始全K含量则与土壤全N、全P、全K、pH、N/P显著或极显著相关；而凋落叶初始C/N比与土壤各化学性质(除全N外)间均无显著相关性。

3　结论与讨论

分解系数k的生态学意义即为凋落物分解速率的快慢，k值越大，分解越快。本研究中，西南桦和尾巨桉2种凋落叶分解系数k分别为0.96/年和0.88/年，西南桦凋落叶分解比尾巨桉快。宋新章等[11]研究我国鼎湖山小叶青冈栎[*Cyclobalanopsis gracilis* (Rehd. et Wils.) Cheng et T. Hong]和毛竹[*Phyllostachys heterocycla* (Carr.) Mitford cv. *Pubescens*]2种凋落叶分解特征发现，其k值分别为0.89/年和1.05/年，与本研究结果的k值差异不大。一般认为，在大尺度的气候带下，气候因素如年均气温(MAT)、年均降水(MAP)、实际蒸散(AET)等对凋落物的分解起主要的控制作用。如唐仕姗等[12]研究发现，我国森林生态系统凋落叶分解系数k值为0.13~1.80/年，与郭忠玲等[13]研究发现的0.10~2.17/年差异不大，而全球陆地生态系统的分解系数k值变化较大，为0.006~4.993/年[14]。造成分解速率产生差异的主要原因在于全球陆地森林生态系统森林类型多样、地理地貌特征丰富、自然气候条件复杂、环境因子空间异质性大和复杂性高，因而，凋落物分解速率的空间异质性也大。然而，笔者的研究对象为南亚热带同一地区2种类型的林分，这2种森林生态系统仅仅是全球陆地生态系统的一部分，因而其k值也在此范围内。此外，我国森林凋落叶分解速率随气候带的不同呈规律性变化，即分解速率从大到小依次为热带>亚热带>温带[12]。如刘颖等[15]在研究我国温带4种森林类型凋落物分解动态时得出，红松云冷杉林、阔叶红松林、岳桦云冷杉林、岳桦林的凋落物分解系数为0.25~0.47/年，明显低于本研究的k值。这也进一步证实了亚热带森林凋落叶分解速率明显高于温带的结论。

表3　凋落叶质量损失与养分含量的相关关系

凋落叶分解	C	N	P	K	C/N	N/P
分解前期(前6个月)	−0.512(0.112)	0.877(0.022)	−0.190(0.718)	0.426(0.184)	−0.735(0.044)	0.812(0.037)
分解后期(后6个月)	0.353(0.493)	0.855(0.030)	−0.268(0.607)	−0.482(0.136)	−0.697(0.049)	−0.320(0.537)

注：表中数值为Spearman's相关系数，括号中数值为显著性水平。

表4　初始凋落叶养分含量与0~10cm土壤化学性质间的相关系数

凋落叶初始养分含量	土壤性质						
	有机C	全N	全P	全K	pH	C/N	N/P
有机C	0.578*	0.586*	0.898**	0.528*	0.469	−0.236	−0.910**
全N	−0.028	0.880**	0.483	0.213	0.572*	−0.387	−0.290
全P	0.455	0.465	0.714**	0.923**	0.341	0.112	−0.745**
全K	0.220	0.509*	0.886**	0.926**	0.559*	−0.140	−0.946**
C/N	0.153	−0.633*	0.195	0.191	−0.183	0.410	−0.381
N/P	−0.485	−0.048	−0.490*	−0.825**	−0.028	−0.427	0.611*

注：* $P<0.05$；** $P<0.01$。

2种森林凋落叶在整个分解过程中，有机C浓度始终呈逐渐下降的趋势，这可能是由于该地区年均温始终处于相对较高的水平，即使是冬天，地温也能达到10℃左右[16]。在较高的温度下，与枯落物分解有关的动物和微生物活性、酶活性等始终维持在较高的水平，因而，凋落叶的有机C始终处于分解释放状态。2种凋落叶N和P的养分动态均呈先富集后释放的现象，但富集阶段持续时间的长短因养分元素的不同而不同。凋落叶N浓度在分解的前6个月内迅速积累，后6个月则迅速释放，但分解末期时凋落叶N浓度仍高于初始N浓度。凋落叶P浓度在分解的前9个月迅速积累，后3个月逐渐释放，分解末期时凋落叶P浓度也高于初始P浓度，这与游巍斌等[17]的研究结果有差异。游巍斌等研究发现，凋落叶N浓度在分解后期出现富集现象，P则处于波动的富集状态，并认为凋落物P分解与气候因子密切相关，尤其是温度和湿度。本研究中，凋落叶K浓度在分解的前6个月迅速释放，后6个月趋于平稳，可能是因为分解的前6个月(即2012年6月至2012年11月)主要为该区的雨季，较高的温湿度等环境条件使凋落叶分解速率加快(图1)，因而K的释放也较快；而分解的后6个月(即2012年12月至2013年5月)主要为该区的干季，因而，凋落叶K的释放也较慢。相关研究表明，凋落物K浓度随着分解时间的延长呈单调递减趋势，呈淋溶—释放模式[18]。

影响凋落物分解的因素众多，除环境影响因子外，凋落物初始养分特征和其养分归还速度也具重要影响[19]。本研究中，西南桦和尾巨桉2种凋落叶初始养分特征存在显著差异，其中，2种凋落叶的N含量和C/N的差异是影响凋落叶分解最主要的控制因子(表3)。已有研究表明，在凋落叶分解过程中，N素与微生物生长繁殖关系密切，环境中N量越高，微生物的繁殖越快，活性越强[20]。研究表明，凋落物中N、P、K含量越多，其养分分解归还越快[21]，反之越慢[22]。Xu等[23]研究发现，在凋落物分解初期(前3~4个月)，凋落物的干质量损失与N含量显著正相关，与木质素/N和C/N显著负相关；在凋落物分解后期(1~2年)，凋落物的干质量损失与N含量显著正相关，与木质素含量、木质素/N和C/N显著负相关。本研究结果也表明，在凋落物分解前期(前6个月)，凋落物分解与N含量和N/P显著正相关，与C/N显著负相关；而凋落物分解后期(6~12个月)，其分解速率与N含量也显著正相关，与C/N仍显著负相关(表4)。因本研究未探讨凋落叶中难分解物质如木质素和纤维素等含量对凋落叶分解的影响，故有必要开展更长久的凋落叶分解实验来观察凋落叶的养分释放动态。

西南桦和尾巨桉2种人工林凋落叶对土壤化学性质的影响显著(表2)，且二者之间存在一定相关关系。凋落物是森林生态系统土壤养分的重要来源[24-26]，其养分含量和分解过程对森林土壤肥力有重要影响[27]。凋落叶有机C含量与土壤有机C含量间呈显著正相关，原因在于西南桦凋落叶中有机C含量较高，分解较快，释放到土壤中的有机C也越多。Ohrui等[28]发现，凋落物的C/N与土壤N矿化呈负相关。这与笔者的研究结果一致，凋落叶C/N与土壤N含量显著负相关(表4)。N矿化速率和土壤N的输入受凋落叶中N含量的影响[29]，本研究结果也显示，土壤全N含量与凋落叶中N含量显著正相关。此外，Moore等[30]研究表明，凋落物养分的释放模式显著影响土壤表层的N、P等养分含量，凋落物中N、P等含量越高，土壤养分越易于富集[31]。本研究也表明，凋落叶中的P含量与土壤P含量极显著正相关。

综上所述，在我国南亚热带地区，西南桦凋落叶的分解速率显著大于尾巨桉。在为期12个月的分解试验中，2种凋落叶中各元素含量呈不同的动态变化，有机C含量在整个分解过程中呈逐渐下降的趋势；全K含量和C/N在分解前期迅速下降，之后趋于平缓；全N含量和全P含量在分解前期呈逐渐上升趋势，之后略有下降；2种凋落叶N/P则呈先升高后下降的趋势。无论是分解前期还是分解后期，凋落叶分解与凋落叶中的N含量呈显著正相关，与C/N呈显著负相关。通过凋落叶分解过程中养分的释放，显著影响了林地的土壤养分水平，西南桦林地土壤有机C、全N、全K含量等显著高于尾巨桉。本研究从凋落叶的分解过程及其与土壤养分的相关关系探讨了不同树种对林地土壤养分状况的影响，更深入地认识了该地区不同树种人工林生态系统的生态功能，为未来更有效地进行人工林经营管理提供了科学参考。

参考文献

[1] BERG B. Litter decomposition and organic matter turnover in northern forest soils [J]. Forest Ecology and Management, 2000, 133(1): 13-22.

[2] WANG G G, KLINKA K. White spruce foliar nutrient concentrations in relation to tree growth and soil nutrient amounts [J]. Forest Ecology and Management, 1997, 98(1): 89-99.

[3] 葛晓改，肖文发，曾立雄，等．不同林龄马尾松凋落物基质质量与土壤养分的关系[J]. 生态学报，2012,

32(3)：852-862.

[4]WANG H，LIU S，WANG J，et al. Effects of tree species mixture on soil organic carbon stocks and greenhouse gas fluxes in subtropical plantations in China[J]. Forest Ecology and Management，2013，300：4-13.

[5]梁瑞龙．广西乡土阔叶树种资源现状及其发展对策[J]. 广西林业科学，2007，36(1)：5-9.

[6]周燕萍，庞正轰，黄弼昌，等．林朵林场西南桦造林试验[J]. 广西科学，2012，19(2)：192-195.

[7]MO J，BROWN S，XUE J，et al. Response of litter decomposition to simulated N deposition in disturbed，rehabilitated and mature forests in subtropical China[J]. Plant and Soil，2006，282(1-2)：135-151.

[8]OSTERTAG R，MARÍN-SPIOTTA E，SILVER W L，et al. Litterfall and decomposition in relation to soil carbon pools along a secondary forest chronosequence in Puerto Rico[J]. Ecosystems，2008，11(5)：701-714.

[9]鲍士旦．土壤农化分析[M]. 南京：中国农业出版社，2000.

[10]OLSON J S. Energy storage and the balance of producers and decomposers in ecological systems[J]. Ecology，1963，44(2)：322-331.

[11]宋新章，江洪，马元丹，等．中国东部气候带凋落物分解特征——气候和基质质量的综合影响[J]. 生态学报，2009，29(10)：5219-5226.

[12]唐仕姗，杨万勤，殷睿，等．中国森林生态系统凋落叶分解速率的分布特征及其控制因子[J]. 植物生态学报，2014，38(6)：529-539.

[13]郭忠玲，郑金萍，马元丹，等．长白山各植被带主要树种凋落物分解速率及模型模拟的试验研究[J]. 生态学报，2006，26(4)：1037-1046.

[14]ZHANG D，HUI D，Luo Y，et al. Rates of litter decomposition in terrestrial ecosystems：global patterns and controlling factors[J]. Journal of Plant Ecology，2008，1(2)：85-93.

[15]刘颖，武耀祥，韩士杰，等．长白山四种森林类型凋落物分解动态[J]. 生态学杂志，2009，28(3)：400-404.

[16]罗达，史作民，唐敬超，等．南亚热带乡土树种人工纯林及混交林土壤微生物群落结构[J]. 应用生态学报，2014，25(9)：2543-2550.

[17]游巍斌，刘勇生，何东进，等．武夷山风景名胜区不同天然林凋落物分解特征[J]. 四川农业大学学报，2010，28(2)：141-147.

[18]刘颖，韩士杰，林鹿．长白山 4 种森林凋落物分解过程中养分动态变化[J]. 东北林业大学学报，2009，37(8)：28-30.

[19]MEIER C L，BOWMAN W D. Links between plant litter chemistry，species diversity，and below-ground ecosystem function[J]. Proceedings of the National Academy of Sciences，2008，105(50)：19780-19785.

[20]KRASHEVSKA V，SANDMANN D，MARAUN M，et al. Consequences of exclusion of precipitation on microorganisms and microbial consumers in montane tropical rainforests[J]. Oecologia，2012，170(4)：1067-1076.

[21]MENDONÇA E S，STOTT D E. Characteristics and decomposition rates of pruning residues from a shaded coffee system in Southeastern Brazil[J]. Agroforestry Systems，2003，57(2)：117-125.

[22]JENSEN L S，SALO T，PALMASON F，et al. Influence of biochemical quality on C and N mineralisation from a broad variety of plant materials in soil[J]. Plant and Soil，2005，273(1-2)：307-326.

[23]XU X，HIRATA E. Decomposition patterns of leaf litter of seven common canopy species in a subtropical forest：N and P dynamics[J]. Plant and Soil，2005，273(1-2)：279-289.

[24]AERTS R，de CALUWE H. Nutritional and plant-mediated controls on leaf litter decomposition of Carex species[J]. Ecology，1997，78(1)：244-260.

[25]潘开文，何静，吴宁．森林凋落物对林地微生境的影响[J]. 应用生态学报，2004，15(1)：153-158.

[26]MAISTO G，DE MARCO A，MEOLA A，et al. Nutrient dynamics in litter mixtures of four Mediterranean maquis species decomposing in situ[J]. Soil Biology and Biochemistry，2011，43(3)：520-530.

[27]解宪丽，孙波，周慧珍，等．中国土壤有机碳密度和储量的估算与空间分布分析[J]. 土壤学报，2004，41(1)：35-43.

[28]OHRUI K，MITCHELL M J，BISCHOFF J M. Effect of landscape position on N mineralization and nitrification in a forested watershed in the Adirondack Mountains of New York[J]. Canadian Journal of Forest Research，1999，29(4)：497-508.

[29]陈印平，赵丽华，吴越华，等．森林凋落物与土壤质量的互作效应研究[J]. 世界科技研究与发展，2006，27(4)：88-94.

[30]MOORE T R，TROFYMOW J A，PRESCOTT C E，et al. Patterns of carbon，nitrogen and phosphorus dynamics in decomposing foliar litter in Canadian forests[J]. Ecosystems，2006，9(1)：46-62.

[31]MOORE T R，TROFYMOW J A，PRESCOTT C E，et al. Nature and nurture in the dynamics of C，N and P during litter decomposition in Canadian forests[J]. Plant and Soil，2011，339(1-2)：163-175.

[原载：林业科学研究，2016，29(02)]

南亚热带红椎、杉木纯林与混交林碳贮量比较

明安刚[1,2,3] 刘世荣[2] 莫慧华[1,3] 蔡道雄[1,3] 农 友[1,3] 曾 冀[1,3] 李 华[1,3] 陶 怡[1,3]

([1]中国林业科学研究院热带林业实验中心，广西凭祥 532600；[2]中国林业科学研究院森林生态环境与保护研究所，北京 100091；[3]广西友谊关森林生态国家定位观测研究站，广西凭祥 532600)

摘 要 造林再造林作为新增碳汇的一种有效途径，受到国际社会的广泛关注。如何通过改变林分树种组成，优化造林模式提高人工林生态系统碳贮量已成为国内外学者关注的重点。本文通过样方调查和生物量实测相结合的方法，对南亚热带26年生红椎纯林(PCH)、杉木纯林(PCL)及红椎×杉木混交林(MCC)生态系统各组分碳含量、碳贮量及其分配特征进行了比较研究。结果表明：杉木、红椎各器官平均碳含量分别为492.1~545.7g/kg和486.7~524.1g/kg。相同树种不同器官以及不同树种的相同器官间碳含量差异显著($P<0.05$)。红椎各器官碳含量的平均值(521.3g/kg)高于杉木(504.7g/kg)。不同林分间地被物碳含量大小顺序为PCH>MCC>PCL；不同树种之间的土壤碳含量差异显著($P<0.05$)，0~100cm土壤平均碳含量为PCL>MCC>PCH。生态系统碳贮量大小顺序为PCL(169.49t/hm^2)>MCC(141.18t/hm^2)>PCL(129.20t/hm^2)，相同组分不同林分以及相同林分的不同组分碳贮量均存在显著差异($P<0.05$)。造林模式对人工林碳贮量及其分配规律有显著影响，营建混交林有利于红椎生物量和土壤碳的累积，而营建纯林有利于杉木人工林生物量碳的吸收，也有利于土壤碳的固定。因而，混交林的固碳功能未必高于纯林，在选择碳汇林的造林模式时，应以充分考虑不同树种的固碳特性。

关键词 南亚热带；红椎；杉木；纯林；混交林；碳贮量

Comparison of Carbon Storage in Pure Stand and Mixed Stand for *Castanopsis hystrix* and *Cunninghamia lanceolata* in Subtropical China

MING Angang[1,2,3], LIU Shirong[2], MO Huihua[1,3], CAI Daoxiong[1,3], NONG You[1,3], ZENG Ji[1,3], LI Hua[1,3], TAO Yi[1,3]

(*[1]Experimental Center of Tropical Forestry, Chinese Academy of Forestry, Pingxiang* 532600, *Guangxi, China*; *[2]Institute of Forest Ecology, Environment and Protection, Chinese Academy of Forestry, Beijing* 100091, *China*; *[3]Guangxi Youyiguan Forest Ecosystem Research Station, Pingxiang* 532600, *Guangxi, China*)

Abstract: Afforestation and re-afforestation were paid more attentions as one of the new methods for carbon sequestration, it was very important to improve the carbon sequestration by optimizing the mode of afforestation and adjusting the forest tree species composition. Carbon content and carbon storage of different components including tree layer, shrub layer, herb layer, litter layer and soil layer and their allocations of 26-year-old *Castanopsis hystrix* stand(PCH), *Cuninghamia lanceolata* stand(PCL) and a mixed *C. hystrix* × *C. lanceolata* stand(MCC) in subtropical China were studied with the method of Quadrat survey combinated with actually measure the biomass. The results showed that the average carbon content of different organs in *C. laoceolata* and *C. hystrix* were 492.1 ~ 545.7g/kg and 486.7 ~ 524.1g/kg. Significant differences were found among different organs in the same tree species. The average carbon content of *C. hystrix* (539.3g/kg) was higher than *C. laoceolata*. The order of carbon content of ground cover was PCH>MCC>PCL. Soil content in the 0-100cm showed significant difference among the three stands, which was ranked as PCL>MCC>PCH. The carbon storage in the total ecosystem was ranked as PCL(169.49t/hm^2) > MCC(141.18t/hm^2) > PCH(129.20t/hm^2). Significant differences were found among different components in the same stand, and also among different stands in the same component($P<0.05$). Afforestation mode can remarkably affected the carbon storage and it's allocation for plantations, mixed stand benefited due to the increased biomass carbon and soil carbon for *C. hystrix*; both can benefits for the biomass and soil carbon for *C. laoceolata* by the monoculture mode. Therefore, we should select the afforestation mode on the basis of the carbon acumination characteristics of tree species in terms of carbon sequestration forest because mixed stand do not always fix more carbon compare to the pure forest.

Key words: subtropical China; *C. hystrix* ; *C. laoceolata*; pure forest; mixed forest; carbon storage

自工业革命以来，由于化石燃料的大量燃烧和土地利用方式的改变，大气中 CO_2浓度不断升高[1]，CO_2的排放、吸收和固定是全球气候变化研究的热点[2-4]。目前，近 $4\times10^9hm^2$森林中储存了 860Pg 碳，而且每年可以从大气吸收 2.4Pg 碳，折合 8.8PgCO_2[5-8]。因而，作为陆地生态系统的主体，森林在储存 CO_2，调节全球气候、减缓全球气候变化方面具有不可替代的作用[9,10]。当前，造林和再造林作为一种新增碳汇的主要途径，已受到学术界的高度重视[11-13]。人工林在吸收和固定 CO_2及减缓全球气候变暖等方面发挥着重要作用，并引起人们的广泛关注。为了更好地利用科学经营的方式减缓全球气候变化，需要对不同造林模式的人工林固碳能力与潜力有深入的认识和科学的评估[14,15]。

最近十年，诸多学者对不同树种、不同林龄及不同密度人工林的碳含量、碳贮量及其空间分布格局进行了深入研究[16-26]，发现人工林碳贮量随着林龄的增加而增加[18,22,26]，林分密度对林分碳贮量的影响的研究得出的结论有所不同，方晰等人得出湿地松人工林碳贮量随林龄的增加而增加[20]，而张国庆等人对马尾松人工林碳贮量的研究得出了相反的结论[21]，认为马尾松人工林碳贮量随林龄的增加而减少。这些研究为森林碳汇功能的研究做出了积极贡献。近些年也有学者对不同造林模式的人工林生物量和碳贮量进行了研究，发现造林模式对人工林碳贮量有显著影响，纯林与混交林地上、地下碳贮量都有明显差异，但不同学者的研究得出的结论并不一致[13,23]，He 和 Wang 等人的研究认为红椎×马尾松混交林土壤碳贮量高于马尾松纯林的，也高于红椎纯林[13,24]，而何友均等人的研究发现西南桦纯林生态系统碳贮量高于西南桦×红椎混交林[23]，由此可见，造林模式对人工林碳贮量的究竟会产生怎样的影响，仍有相当大的不确定性，有进一步研究的必要。

红椎（*Castanopsis hystrix*）和杉木（*Cunninghamia lanceolata*）都是我国南亚热带地区主要造林树种，也是适合在该地区培养大径材的用材树种[23]。其中，红椎是分布在南亚热带地区珍贵乡土阔叶树种，是替代大面积针叶人工林较为理想的高价值乡土阔叶树种之一[27]。目前，已有学者对不同林龄红椎人工林以及红椎与马尾松混交林碳贮量进行了研究[24,28]，但对红椎纯林及其与杉木混交林碳贮量的比较研究了解甚少。本文对南亚热带中国林业科学研究院热林业实验中心林区 26 年生红椎、杉木纯林及其二者混交林生态系统碳贮量及其分配特征进行了比较研究，旨在进一步阐明造林模式对人工林固碳能力与潜力影响，为区域尺度上科学评估人工林生态系统碳库及碳平衡提供基础数据和理论依据，为碳汇林的营建和人工林可持续经营提供科学的理论指导。

1 研究地区与研究方法

1.1 研究区概况

研究区位于广西凭祥市中国林业科学研究院热带林业实验中心（106°39′50″~106°59′30″ E，21°57′47″~22°19′27″N），属南亚热带季风气候区域内的西南部，与北热带北缘毗邻。干湿季节明显（10 月至翌年 3 月份为干季，4~9 月份为湿季），太阳年总辐射 439.61kJ/cm^2，年日照时数 1218~1620h，年均温 19.5~21.0℃，极端高温 40.3℃，极端低温-1.5℃，≥10℃积温 6000~7600℃，年降水量 1400mm，相对湿度 80%~84%；地貌类型以低山丘陵为主，海拔 400~650m，地带性土壤为花岗岩发育成的山地红壤，土层厚度在 100cm 以上。

哨平实验场于 1987 年 4 月营造了红椎纯林（PCH）、杉木纯林（PCL）和红椎×杉木混交林（MCC，以下简称混交林）为研究对象，调查当年，3 种林分的林龄均为 26 年，且都是在马尾松（*Pinus massoniana*）人工林皆伐炼山后，经块状整地营建的人工林。造林当年和翌年各进行常规抚育 2 次，直至郁闭，目前未曾间伐。各林分初植密度均为 2000 株/hm^2，混交林为行间混交，混交比例为红椎：杉木=1：1。调查当年，红椎纯林枝叶茂盛，冠幅较大，林分郁闭度高，但林分自然稀疏较多，保留密度较小。但因林冠郁闭较大，林下植被稀少，灌草层盖度为 5%；其次为混交林，盖度为 25%，杉木纯林随保留密度最大，但郁闭度最小，灌草植被盖度最大，达 55%。林下植被灌木主要有红椎、九节（*Psychotria rubra*）、大沙叶（*Pavetta hongkongensis* ）、酸藤子（*Embelia laeta*）、玉叶金花（*Mussaenda pubuscens*）、草本植物以扇叶铁线蕨（*Adiantum flabellulatum*）和半边旗(*Pteris semipinnata*)为主要优势种。

2013年9月，在3种林分中，选取坡面均匀，人为干扰相对较少的区域，按坡位分别随机设置4个20m×20m样地，共计12个样地。对每个样方内的树木进行每木检尺，调查胸径、树高等指标。林分基本情况见表1。

表1 3种人工林地概况

林分类型		土壤类型	坡度(°)	海拔(m)	小地形	坡向	郁闭度	密度(trees/hm^2)	胸径(cm)	树高(m)
红椎纯林	22°2′58″N, 106°53′40″E	山地红壤	31.4±2.7	274±24	中坡山地	南坡	0.95	1716.7±194.2	11.6+0.6	10.9±0.8
杉木纯林	22°3′41″N, 106°52′08″E	山地红壤	34.1±1.8	248±21	中坡山地	南坡	0.85	2341.7±450.2	14.3±1.2	17.2±1.5
混交林 红椎	22°3′14″N, 106°53′58″E	山地红壤	32.7±2.1	241±18	中坡山地	南坡	0.90	937.5±197.5	12.1±1.4	12.6±1.8
混交林 杉木								1045.8±220.8	13.0±2.7	14.3±1.8

1.2 研究方法

1.2.1 林木调查及生物量的测定

根据样方每木检尺的结果，用He等人在该地区建立的红椎生物量方程估算红椎各器官生物量[24]，用康冰等人在该地区建立的杉木生物量方程估算样方内杉木各器官生物量[29]。

1.2.2 林下植被生物量和凋落物现存量的测定

在每个20m×20m样方中，按梅花形布点设置5个2m×2m小样方和5个1m×1m小样方，记录2m×2m小样方内灌木和草本植物的种类，并采用“样方收获法”分别测定灌木层和草本层地上和地下部分生物量。在1m×1m小样方内按未分解、半分解组分分别测定凋落物鲜重。同时，取各组分样品带回实验室在65℃烘干至恒重，计算干重。

1.2.3 植物和土壤样品的采集

在乔木每木检尺和林下地被物生物量调查的同时，分别采集红椎和杉木不同器官(干、皮、枝、叶和根)灌木层、草本植层和凋落物层样品4份，经烘干、粉碎、过筛后以备碳含量的测定。

在每个20m×20m样方中，按梅花形布点挖取5个土壤剖面，按照0~10cm、10~30cm、30~50cm和50~100cm将土壤分为4个土层分别采集土壤样品400g左右，各剖面的同层土样取混合样。将样品带回实验室自然风干后碾碎过筛，用于土壤含碳量的测定。同时，用铝盒和100cm^{3}的环刀取样，以用于含水率和土壤容重的测定。

1.2.4 碳含量测定和碳贮量计算

植物和土壤样品均采用重铬酸钾—水合加热法测定有机碳含量。植物碳贮量=有机碳含量×单位面积生物量，土壤碳贮量=土壤有机碳含量×土壤容重×土壤厚度。

1.3 数据处理

采用SPSS 13.0软件对数据进行统计分析，方差分析和差异显著性检验($\alpha=0.05$)。采用Excel 2007和PS软件作图。

2 结果与分析

2.1 不同人工林生态系统各组分碳含量

2.1.1 乔木层碳含量

26年生红椎和杉木各器官碳含量分别在492.1~545.7g/kg和486.7~524.1g/kg之间(表2)。方差分析表明：相同树种不同器官之间碳含量有所不同，除红椎的干和叶以及杉木的皮和叶之间差异不显著外，其他各器官间均存在显著差异($P<0.05$)。不同树种各器官碳含量高低顺序也不尽相同，红椎不同器官碳含量的排列顺序为：树皮>树干>树叶>树枝>树根；杉木为：树皮>树叶>树干>树枝>树根。不同树种的相同器官碳含量也存在显著差异(树叶除外)($P<0.05$)，从整体上看，26年生红椎各器官碳含量的平均值高于杉木。

表2 不同树种各器官碳含量 (g/kg，Mean+SD)

树种	干	皮	枝	叶	根	平均
红椎	531.4±14.2Ba	545.7±21.4Aa	512.8±20.4Ca	524.7±15.6Ba	492.1±16.7Da	521.3±17.7a
杉木	502.4±23.4Bb	524.1±13.6Ab	491.6±17.4Cb	518.5±25.3Aa	486.7±19.7Db	504.7±19.3b

注：同行中不同大写字母表示相同树种不同器官间差异显著，同列中不同小写字母表示相同器官不同树种间差异显著($P<0.05$)。

2.1.2 林下地被物碳含量

对不同林分地被层的不同组分碳含量的测定结果显示，不同林分相同组分碳含量有所差异，表现为红椎纯林>混交林>杉木纯林，方差分析结果表明：不同林分间灌木层，草本层地上部分和凋落物的未分解部分碳含量差异达显著水平($P<0.05$)；3 种人工林林下地被物各层次平均碳含量以凋落物层最高，灌木层居次，草本层最低；灌木层和草本层地上部分碳含量均高于地下部分，未分解的凋落物碳含量高于半分解的碳含量。

2.1.3 土壤层碳含量

从表 3 可知，3 种林分的土壤碳含量均以表土层(0~10cm)最高，为 9.18~12.19g/kg。随着土层深度的增加，土壤碳含量显著降低($P<0.05$)；林分间土壤碳含量在不同土层深度大小顺序有所不同，0~10cm 和 10~30cm 土层中，杉木纯林和混交林碳含量均显著高于红椎纯林，而 30~50cm 和 50~100cm 土层中，碳含量大小顺序为杉木纯林>红椎纯林>混交林，但仅有杉木纯林和混交林间碳含量存在显著差异($P<0.05$)。

而相同深度的土层中，不同林分之间的土壤碳含量不同，0~100cm 土壤平均碳含量均以杉木纯林最高，混交林居次，红椎纯林土壤碳含量最低，红椎纯林与杉木纯林间土壤平均碳含量差异显著($P<0.05$)，杉木纯林 0~100cm 土壤平均含碳量比红椎纯林高出 31.3%，红椎×杉木混交林土壤平均碳含量与杉木纯林和红椎纯林均无显著差异($P>0.05$)。

2.2 不同人工林生态系统各组分碳贮量及其分配

2.2.1 乔木层碳贮量及其分配

乔木层生物量和碳贮量均以杉木纯林最大，混交林居次，红椎纯林最小(表 4)。从表 4 可以看出，红椎×杉木混交林的生物量和碳贮量分别是杉木纯林的 87.68%和 89.84%，却是红椎纯林的 1.09 和 1.08 倍。可见，对红椎而言，与杉木营建混交林更有利于乔木生物量的生长和碳素的累积；而对杉木而言，营建纯林更有利于林分生物量生长和碳贮量的增加。

从图 1 可以看出，不同林分各器官生物量和碳贮量在乔木层的分配均以树干最高(59.9%~63.1% 和 60.1%~64.3%)，树干生物量和碳贮量在 3 种林分中的分配顺序为：红椎纯林(63.1，64.3)>混交林(61.8，62.5)>杉木纯林(59.9，60.1)；总体上看，不同林分各器官在乔木层碳贮量的分配顺序不同，杉木纯林为：树干>树根>树枝>树皮>树叶；红椎纯林和混交林为：树干>树根>树皮>树枝>树叶，红椎和杉木营造混交林减少了枝条在乔木层中的碳分配，增加了树皮的碳分配，但差异不显著($P>0.05$)。

表 3 林下植被、凋落物及土壤碳含量 (g/kg)

层次	组分	红椎纯林	杉木纯林	混交林
灌木层	地上部分	519.24±13.66a	497.74±15.12c	506.21±9.44b
	地下部分	488.27±12.17a	458.19±10.37c	468.84±8.18b
	平均	503.76±10.43a	477.97±22.20c	487.53±19.81b
草本层	地上部分	458.28±7.74a	437.88±11.42b	441.72±8.43b
	地下部分	425.17±8.13a	398.78±9.38b	392.75±7.58b
	平均	441.73±10.32a	418.33±5.79b	417.24±10.47b
凋落物	未分解	532.18±17.98a	501.39±3.74c	517.13±10.43b
	半分解	474.23±21.35a	389.43±12.28c	437.64±6.78b
	平均	503.21±17.43a	445.41±19.77c	477.39±12.63b
土壤层	0~10cm	9.18±2.54b	11.06±3.17a	12.19±3.22a
	10~30cm	4.79±1.47b	8.44±2.01a	7.66±2.18a
	30~50cm	4.86±1.44ab	5.80±2.24a	4.34±1.19bc
	50~100cm	3.91±1.31ab	4.55±1.16a	3.31±0.74bc
	土壤平均	5.68±2.37bc	7.46±2.90a	6.87±4.00ab

注：PCH：红椎纯林，PCL：杉木纯林，MCC：红椎×杉木混交林，下同；同一行不同字母表示不同林分间碳含量差异显著($P<0.05$)。

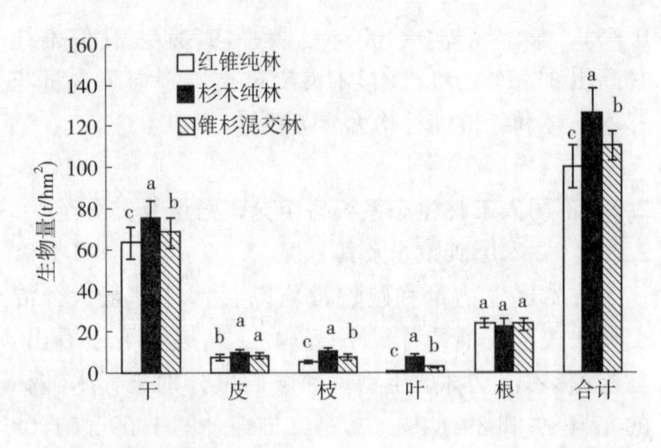

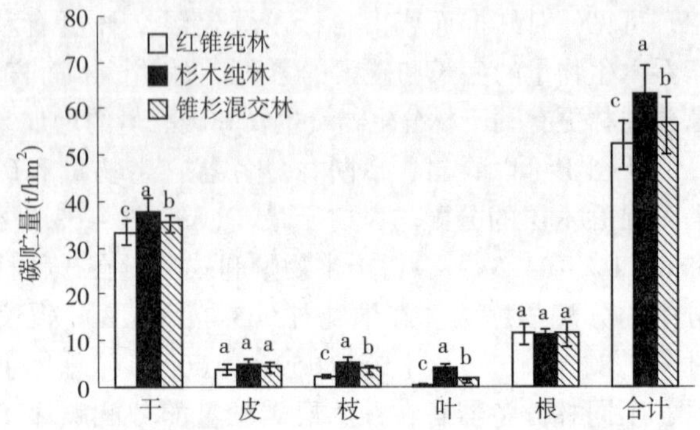

图1 不同林分乔木层生物量和碳贮量分配

不同小写字母表示不同林分相同组分生物量，碳贮量差异显著

2.2.2 林下地被物碳贮量及其分配

林下地被物碳贮量包括灌木层、草本层和凋落物层三个层次，从表4可知，3种林分地被物碳贮量均较小，为1.48~8.56t/hm²。不同林分间相同组分碳贮量存在显著差异(P<0.05)，灌木层、草本层和凋落物层碳贮量大小顺序均为杉木纯林>混交林>红椎纯林。

相同林分不同组分碳贮量分配顺序也有所差异，3种林分不同层次碳贮量分配顺序均为灌木层>凋落物层>草本层。在灌木层和草本层中，地上部分碳贮量均大于地下部分；凋落物层中，未分解的凋落物碳贮量显著高于半分解的，约为半分解凋落物碳贮量的2.5~3.1倍。

2.2.3 土壤碳贮量及其分配

各林分相同土层厚度的土壤平均碳贮量随土层深度增加而降低，变化趋势与土壤碳含量随土层深度的变化一致，土壤碳贮量主要集中在0~30cm的表土层，红椎纯林、杉木纯林和混交林0~30cm土碳贮量分别占0~100cm土壤碳贮量的36.6%、43.9%和49.1%(见表4)。

从表4可以看出，不同林分间土壤碳贮量差异显著，0~100cm土壤碳贮量以杉木纯林最高，混交林林居次，红椎纯林最低。多重比较结果显示各林分见土壤碳贮量差异均达显著水平(P<0.05)。

2.3 人工林生态系统碳贮量及其分配

红椎纯林、杉木纯林和混交林生态系统碳贮量总量分别是129.20t/hm²、169.49t/hm²和141.18t/hm²，混交林碳贮量显著高于红椎纯林，低于杉木纯林，且差异显著(P<0.05)。混交林碳贮量总量高于红椎纯林9.3%，却低于杉木纯林20.1%。红椎纯林、杉木纯林和混交林生态系统植被碳贮量总量分别为53.83t/hm²，71.62t/hm²和61.34t/hm²，杉木纯林具有最高植被碳贮量，显著高于红椎纯林，也显著高于混交林(表4)。

表4 3种人工林生态系统生物量、碳贮量及其分布 (t/hm²)

层次	组分	红椎纯林		杉木纯林		混交林	
		生物量	碳贮量	生物量	碳贮量	生物量	碳贮量
乔木层	地上部分	76.43	40.61	103.01	51.97	86.91	45.20
	地下部分	23.84	11.74	22.79	11.09	23.36	11.45
	小计	100.3	52.35	125.8	63.06	110.3	56.65
灌木层	地上部分	1.24	0.64	8.44	4.20	5.02	2.54
	地下部分	0.18	0.09	2.31	1.06	0.73	0.34
	小计	1.42	0.73	10.75	5.26	5.75	2.88
草本层	地上部分	0.11	0.05	1.44	0.63	0.27	0.12
	地下部分	0.08	0.03	0.96	0.38	0.19	0.08
	小计	0.19	0.08	2.40	1.01	0.46	0.19
凋落物	未分解	0.92	0.49	3.45	1.73	2.23	1.15
	半分解	0.37	0.18	1.42	0.55	1.04	0.46
	小计	1.29	0.67	4.87	2.28	3.28	1.61

（续）

层次	组分	红椎纯林		杉木纯林		混交林	
		生物量	碳贮量	生物量	碳贮量	生物量	碳贮量
土壤层	0~10cm		12.50		16.26		14.66
	10~30cm		15.09		26.74		24.56
	30~50cm		15.53		18.45		14.27
	50~100cm		32.25		36.42		26.36
	小计		75.37		97.88		79.84
	总计	103.17	129.20	143.81	169.49	119.75	141.18

由表4可以看出，乔木层和土壤层为各林分主要碳库，二者占各人工林生态系统碳贮量的94.9%~98.8%以上，而灌木层、草本层和凋落物层的碳贮量总和仅占1.2%~5.1%。各层次碳贮量在生态系统中的分配顺序均为土壤层(56.5%~58.3%)>乔木层(37.2%~40.5%)>凋落物层(1.1%~5.0%)>灌木层(0.6%~3.1%)>草本层(0.1%~0.6%)。不同林分相同组分碳贮量在生态系统分配有所差异，乔木层碳贮量在生态系统碳分配的大小顺序为红椎纯林(40.5%)>混交林(40.1%)>杉木纯林(37.2%)，土壤碳贮量的分配顺序为红椎纯林(58.3%)>杉木纯林(57.7%)>混交林(56.5%)。

3　结论与讨论

26年生红椎和杉木各器官碳含量的平均值分别为521.3g/kg和504.7g/kg，高于广西26年生楠木(493.1g/kg)以及28年生秃杉(491.9g/kg)[30,31]，也高于国际通用的树木平均碳含量(0.5g/kg)及热带32个树种的平均碳含量(444.0~494.5g/kg)[32]。可见，即使在同一地区，树木各器官碳含量大小因树种不同而存在差异。另外，相同树种的不同器官碳含量差异显著($P<0.05$)，不同树种之间，各器官碳含量的排列顺序也不尽相同。可见树种是影响乔木器官含碳量的重要因素之一，这可能与树种本身的生理特性相关。

相比之下，红椎碳含量平均值显著高于杉木，本研究发现，杉木纯林乔木层生物量高出红椎纯林25.4%，但碳贮量仅高出红椎纯林20.5%，这是因为杉木虽然具有较高的生物量，但红椎因具有较高的碳含量，从而减小了与杉木纯林之间碳贮量的差异。

不同林分间，灌木层、草本层和凋落物层碳含量的大小顺序为红椎纯林>混交林>杉木纯林，这与不同林分间灌草植被及凋落物的组成密切相关，红椎纯林中灌草植被较少，林下植被主要以天然更新的红椎幼苗为主，凋落物以红椎落叶为主要成分，而红椎叶的碳含量较高(524.7g/kg)，因而红椎纯林林下植被和凋落物的碳含量高于杉木纯林和混交林。

不同林分间的土壤碳含量大小顺序为杉木纯林>混交林>红椎纯林。这与3种林分间地被物碳含量大小顺序恰好相反，但与地被层生物量大小顺序一致。杉木纯林的凋落叶虽然分解较慢，不易将碳素分解进入土壤，但其林下较多的灌草植被及地下凋落物均可较快地增加土壤碳输入。混交林和和纯林因郁闭度较大，林下植被生长受限，且没有足够的凋落物覆盖，因而缺乏土壤碳源，造成土壤碳含量低于杉木纯林。这一结果表明决定土壤有机碳含量的关键因素很可能是林下植被生物量和地下凋落物(主要是死根)的大小，而非地被层碳含量的高低。

本研究表明，不同林分生态系统碳贮量差异显著($P<0.05$)，杉木纯林在3种林分中具有最高的碳贮量，高于混交林碳贮量总量的31.2%。原因有三：一是杉木在前20年生长较快，杉木纯林乔木层生物量的快速生长，迅速增加了植被层生物量碳的累积；二是杉木纯林丰富的灌草植被不仅增加了地被层碳贮量，而且由于凋落物输入的增加，进而增加了土壤碳的累积；三是较轻的自然稀疏强度为乔木层维持了较高的保留密度，也是乔木层碳贮量高于混交林和红椎纯林的重要原因。然而，混交林由于林分树种组成的改变，引起乔木层生物量生长的变化，2个主要乔木树种的平均生物量较杉木纯林小，从而导致混交林乔木层生物量碳低于杉木纯林。这一结论与诸多前人研究所得出的混交林的碳贮量大于纯林的结论并不一致[14,23,24]，影响乔木层生物量生长和碳贮量累积的因素较多，不仅与造林模式有关，还取决于纯林树种的选择和混交林树种的配置方式。

3种人工林中，以红椎纯林碳贮量最小，为135.2t/hm^2，仅相当于杉木纯林和混交林碳贮量的76.2%和91.5%，主要原因有二，一是相对于杉木纯林和混交林，红椎土壤有机碳含量最低，而土壤有机碳含量较小的差异就会引起土壤碳贮量较大的不同，红椎纯林较低的土壤有机碳含量直接导致了红椎纯林较低的土壤碳贮量。其二，红椎纯林自然稀疏严重，保留乔木密度较小，再加上林冠郁闭度高，林下植被和凋落物少，共同导致了其生物量碳的累积。

本研究的结果表明，就红椎和杉木两个树种而言，营造纯林有利于杉木人工林生态系统碳的累积，而不利于红椎生物量的累积，而营造红椎和杉木的混交林有利于红椎生物量碳和土壤碳的累积，而不利于杉木对碳素的吸收和固定。从表1也可以看出，混交林中，红椎胸径及树高生长量均高于红椎纯林中红椎的胸径和树高的生长量，杉木恰好相反，由此可见，造林模式对人工林生物量碳和土壤碳的累积均有影响。就造林后的前26年而言，营林中若选取红椎营建碳汇林，选择混交模式更有利于林分碳的累积；而对杉木而言，营建纯林对碳的吸存更有利。至于26年以后，杉木生长量是否会慢慢进入衰退期，而红椎的生长量在一定时期后会超越杉木，从而导致后期各林分碳贮量大小顺序开始发生变化，需要进一步观测和研究。但从目前研究的结果来看，至少有一点需要引起人们的关注，无论林分碳贮量发生怎样的变化，营林中对碳汇林造林模式的选择，不仅需要考虑树种固碳特性，还应选择科学的树种搭配。

参考文献

[1]POPE J. How can global warming be traced to CO_2[J]. Scientific America, 2006, 295(6): 124-124.

[2]FANG J Y, CHEN A P, PENG C H, et al. Changes in forest biomass carbon storage in China between 1949 and 1998[J]. Science, 2001, 292(5525): 2320-2322.

[3]CALDEIRA K, DUFFY P B. The role of the southern ocean in up take and storage of anthropogenic carbon dioxide. Science[J], 2000, 287(5453): 620-622.

[4]NORBY R J, LUO Y Q. Evaluating ecosystem responses to rising atmospheric CO_2 and global warming in a multi-factor world[J]. New Phytologist, 2006, 162(2): 281-293.

[5]BROWN S. Present and potential roles of forests in the global climate change debate[J]. Unasylva, 1996, 47(185): 3-10.

[6]LUN F, LI W H, LIU Y. Complete forest carbon cycle and budget in China, 1999-2008[J]. Forest Ecology and Management, 2012, 264: 81-89.

[7]HOOVER C M, LEAK W B, KEEL B G. Benchmark carbon stocks from old-growth forests in northern New England, USA[J]. Forest Ecology and Management, 2012, 266: 108-114.

[8]PAN Y D, BIRDSEY R A, FANG J Y, et al. A large and persistent carbon sink in the world's forests[J]. Science, 2011, 333(6045): 988-993.

[9]WOODWELL G M, WHITTAKER R H, REINERS W A, et al. The biota and the world carbon budget[J]. Science, 1978, 199(4325): 141-146.

[10]POST W M, EMANUEL W R, ZINKE P J, et al. Soil carbon pools and world life zones[J]. Nature, 1982, 298(5870): 156-159.

[11]张小全，李怒云，武曙红．中国实施清洁发展机制造林和再造林项目的可行性和潜力[J]．林业科学，2005，41(5)：139-143.

[12]WU S H, ZHANG X Q, LI J Q. Baseline issues for forest-based carbon sink project on Clean Development Mechanism(CDM)[J]. Scientia Silvae Sinicae, 2006, 42(4): 112-116.

[13]WANG H, LIU S R, WANG J X, et al. Effects of tree species mixture on soil organic carbon stocks and greenhouse gas fluxes in subtropical plantations in China[J]. Forest Ecology and Management, 2013, 300: 4-13.

[14]WANG S, CHEN J M, JU W M, et al. Carbon sinks and sources in China's forests during 1901-2001[J]. Journal of Environmental Management, 2007, 85(3): 524-537.

[15]SALIMON C I, PUTZ F E, MENEZES-FILHO L, et al. Estimating state-wide biomass carbon stocks for a REDD plan In Acre, Brazil[J]. Forest Ecology and Management, 2011, 262(3): 555-560.

[16]LACLAU P. Biomass and carbon sequestration of ponderosa pine plantations and native cypress forests in northwest Patagonia[J]. Forest Ecology and Management, 2003, 180(1-3): 317-333.

[17]SPECHT A, WEST P W. Estimation of biomass and sequestered carbon on farm forest plantations in northern New South Wales, Australia[J]. Biomass and Bioenergy, 2003, 25(4): 363-379.

[18]方晰，田大伦，项文化．速生阶段杉木人工林碳素密度、贮量和分布[J]．林业科学，2002，38(3)：14-19.

[19]马明东，江洪，刘跃建．楠木人工林生态系统生物量、碳含量、碳贮量及其分布[J]．林业科学，2008，44(3)：34-39.

[20]张国庆，黄从德，郭恒，等．不同密度马尾松人工林生态系统碳储量空间分布格局[J]．浙江林业科技，2007，27(6)：10-14.

[21]方晰，田大伦，项文化，等．不同密度湿地松人工林

中碳的积累与分配[J]. 浙江林学院学报，2003，20(4)：374-379.

[22]刘恩，刘世荣. 南亚热带米老排人工林碳贮量及其分配特征[J]. 生态学报，2012，32(16)：5103-5109.

[23]何友均，覃林，李智勇，等. 西南桦纯林与西南桦×红椎混交林碳贮量比较[J]. 生态学报，2012，32(23)：7586-7594.

[24]HE Y J，QIN L，LI Z Y，et al. Carbon storage capacity of monoculture and mixed-species plantations in subtropical China[J]. Forest Ecology and Management，2013，295：193-198.

[25]明安刚，贾宏炎，陶怡，等. 米老排人工林碳素积累特征及其分配格局[J]. 生态学杂志，2012，31(11)：2730-2735.

[26]明安刚，贾宏炎，田祖为，等. 不同林龄格木人工林碳储量及其分配特征[J]. 应用生态学报，2014，25(4)：940-946.

[27]朱积余，蒋燊，潘文. 广西红椎优树选择标准研究[J]. 广西林业科学，2002，31(3)：109-113.

[28]刘恩，王晖，刘世荣. 南亚热带不同林龄红椎人工林碳贮量与碳固定特征[J]. 应用生态学报，2012，23(2)：335-340.

[29]康冰，刘世荣，蔡道雄，等. 南亚热带杉木生态系统生物量和碳素积累及其空间分布特征[J]. 林业科学，2009，45(8)：147-153.

[30]何斌，黄寿先，招礼军，等. 秃杉人工林生态系统碳素积累的动态特征[J]. 林业科学，2009，45(9)：151-157.

[31]LIANG H W，WEN Y G，WEN L H，et al. Effects of continuous cropping on the carbon storage of*Eucalyptus urophylla*×*E. grandis* short-rotations plantations[J]. Acta Ecologica Sinica，2009，29(8)：4242-4250.

[32]ELIAS M，POTVIN C. Assessing inter-and intra-specific variation in trunk carbon concentration for 32 neotropical tree species[J]. Canadian Journal of Forest Research，2003，33(6)：1039-1045.

[原载：生态学报，2016，36(01)]

广西大青山南亚热带森林植物群落的种间联结性

农　友[1,2]　郑　路[1,2]　贾宏炎[1,2]　卢立华[1,2]　明安刚[1,2]

([1]中国林业科学研究院热带林业实验中心，广西凭祥　532600；
[2]广西友谊关森林生态系统国家定位观测研究站，广西凭祥　532600)

摘　要　广西大青山南亚热带天然次生林植物群落种间联结性的研究尚未见报道，本文用方差比率法(VR)和基于2×2联列表，通过X^2统计量检验、共同出现百分率(PC)、联结系数(AC)，对其主要乔木层17个树种、灌木层8个树种、草本层12个树种的种间联结性进行了研究。方差分析表明，乔木层的总体种间联结性表现为显著正关联，灌木层的总体种间联结性表现为不显著负关联，草本层的总体种间联结性表现为不显著正关联。不同检验结果表明，该区域植物群落主要乔木树种间具有正联结性的种对较多，大多数乔木树种的种间联结性较紧密；主要灌木树种的种间联结性较松散，草本层主要种间表现为弱联结性。物种生态习性、群落演替阶段等因素可能是造成这一特性的主要原因。

关键词　大青山；优势种；种间联结性；天然次生林；南亚热带

Interspecific Association Between South Subtropical Forest Plant Community Species in Daqingshan in Guangxi

NONG You[1,2]，ZHENG Lu[1,2]，JIA Hongyan[1,2]，LU Lihua[1,2]，MING Angang[1,2]

([1]*Experimental Center of Tropical Forestry*，*Chinese Academy of Forestry*，*Pingxiang* 532600，*Guangxi*，*China*
[2]*Guangxi Youyiguan Forest Ecosystem Research Station*，*Pingxiang* 532600，*Guangxi*，*China*)

Abstract：The natural secondary forest in southwest of Guangxi is rich in typical forest communities；however，little is known about the interspecific associations of the dominate species. Hence，our objective was to investigate how the dominant species in the communities interact with each other. We sampled 10000m^2 in the natural secondary forest，selected 17 tree species，8 shrub species，and 12 herb species according to importance values and studied interspecific associations by using variance ratio (VR) analysis and X^2 - tests，common occurrence percentage (PC)，and coupling coefficient (AC). There was a significant positive correlation of overall association between trees，an insignificant negative correlation between shrubs，and an insignificant positive correlation between herbs. X^2-tests showed that there was a positive association for 81 pairs and a negative association for 19 pairs and 5 pairs did not have a relationship between trees. There was a positive association for 14 pairs and a negative association for 12 pairs and 2 pairs did not have a relationship between shrubs. A positive association was found for 32 pairs and a negative association for 24 pairs and 5 pairs did not show a relationship between herbs. The main tree species are connected more closely whereas main shrub species association is loose and main herb species association is week. The main causes for different associations were assumed to be ecological habits，community succession stage among others.

Key words：Daqingshan；dominate species；interspecific association；natural secondary forest；south subtropical

种间联结是指不同种类在空间分布上的相互关联性，作为两个物种出现的相似性尺度，是以物种的存在与否为依据，是一种定性的数据[1]，是不同物种在不同生境中相互影响、相互作用形成的有机联系的反映[2]，通常是由群落生境的差异影响了物种的分布引起的，它反映了物种相互排斥或相互吸引的性质[3]。了解种群间的联结性，探讨植物对环境的适应能力[4]，有助于了解森林群落数量结构特征和种群对环境资源利用的相似性，对正确认识群落中各个物种之间的相互作用以及群落的组成、结构、功能和分类有重要的指导意义[1,5]，并能为植物的经营管理、自然植被恢复和生物多样性保护提供理论依据[1,6,7]。

广西大青山地处广西西南部，对该地区的研究主要集中在人工林[8-11]，天然次生林的研究较少[12]，种间联结性的研究尚未见报道。开展大青山天然次生林植物群落种间联结性的研究，可以看出大青山天然次生林植物群落的演替现状及趋势，了解演替过程中群落种对间的相互作用，可为进一步研究南亚热带天然次生林植物群落演替规律、维持和保护其物种多样性奠定基础，也可为该地区人工林的树种选择与配置提供数据支撑，进而对人工林的近自然经营提供一定的理论参考。

1 研究区自然条件

研究区域地处南亚热带南缘，与北热带毗邻，属于南亚热带季风气候[13]。位于21°57′47″~22°19′27″N，106°39′50″~106°59′30″E。年平均气温21.5℃，最冷月(1月)平均气温为13.5℃，最热月(7月)平均气温为27.6℃，≥10℃年积温7500℃，年降水量1400mm。大青山主峰海拔为1045m，河谷高度在130~150m。地带性土壤为中酸性火山岩和花岗岩发育而成的砖红壤(含紫色土)，土层平均厚度为0.5~1.0m[14]。大青山主峰地形复杂，气温高，雨量充沛，湿度大，地形陡峭，人为干扰较少。本次调查的群落位于大青山海拔900~1040m的山地，组成植物以常绿阔叶树种为主。

2 研究方法

2.1 样地构建及野外调查

2013年，经过实地踏查，选择大青山天然次生林保存较好的大青山山顶东北坡，海拔960~1040m，参照巴拿马Barro Colorado Island(BCI)50hm²热带雨林样地的技术规范及中国森林生物多样性监测网络统一的调查方法，结合当地地形地貌，将样地设置为1.0hm²的固定样地，进行长期观测。用全站仪将1.0hm²(东西宽100m、南北长100m)样地划分出25个20m×20m的样方，再把每个20m×20m的样方划分成4个10m×10m和16个5m×5m的小样方，每个20m×20m样方的4个角用带有样方编号的水泥桩作永久标记。

野外调查以5m×5m小样方为基本调查单元，对每个DBH≥1.0cm的木本植物挂铝牌标记，记录种名、胸径、树高、坐标等。每个20m×20m的样方中，选择东北角及西南角共2个5m×5m小样方作为灌草层的调查样方，记录DBH<1.0cm灌木和草本植物的种名、个体数、高度、盖度等数据。调查植物共136种，其中乔木83种、灌木31种、草本22种。全部野外工作于2013年12月至2014年1月完成。

2.2 重要值的测定

乔木重要值(IV)=相对多度(%)+相对频度(%)+相对胸高断面积(%)

灌草重要值(IV)=相对多度(%)+相对频度(%)+相对盖度(%)

重要值范围0~300%。

2.3 总体联结性的测定

用方差比率法(VR)确定总体的联结性[15]，用W检验关联的显著性[16]。计算公式：

$$\delta_T^2 = \sum_{i=1}^{S} P_i(1-P_i),\ P_i = n_i/N$$

$$S_T^2 = 1/N \sum_{j=1}^{N} (T_j - t)^2$$

$$VR = S_T^2/\delta_T^2$$

$$W = N \cdot VR$$

式中：S为物种总数；N为样方总数；T_j为样方j内出现的物种总数；n_i为物种i出现的样方总数；t为样方中种的平均数，即$t=(T_1+T_2+\cdots\cdots+T_n)/N$。公式在假设条件下，$VR$的期望值是1，当$VR=1$时，符合所有种间无关联的假设，当$VR>1$时，种间为净的正关联，当$VR<1$时，种间为净的负关联。

2.4 种间联结性测定

基于建立的2×2联列表，以X^2统计量为基础，结合共同出现百分率PC及联结系数AC确定种间联结性[1,17]。2×2联列表样式如下表1。

表1 2×2联列表样式

		物种B		
		出现的样方数	不出现的样方数	
物种A	出现的样方数	a	b	a+b
	不出现的样方数	c	d	c+d
		a+c	b+d	a+b+c+d

2.4.1 X^2统计量的计算及其检验

首先将物种在样方中出现与否的原始数矩阵转化为$S \times N$(S为种数，N为样方数)形式的二元数据矩阵，即0，1矩阵(0为不出现，1为出现)。然后分别对种对构建2×2联列表，并统计a、b、c、d的值，其中 a 为两个种 A 和 B 均出现样方数，b 为种 B 出现种 A 不出现的样方数，c 为种 A 出现种 B 不出现的样方数，d 为两个种 A 和 B 都不出现的样方数[5]。假设种 A 和种 B 相互独立，没有关联，可以通过 X^2 值来检验这一假设。由于取样为非连续性取样，用 Yates 连续校正公式计算[1]，即：

$$X^2=[(|(ad-bc)|-N/2)^2\times N]/[(a+b)(b+c)(c+d)(a+c)]\text{（自由度}p=1\text{）}$$

式中：N 为样方总数，当 $ad>bc$ 时，为正联结；当 ad<bc 时，为负联结。若 $P>0.05$，即当 $X^2<3.841$ 时，种对相互独立的假设成立，它们独立分布，即为中性联结；若 $P<0.01$，即当 $X^2>6.635$ 时，种对相互独立的假设不成立，种间联结为极显著；若 $0.01<P<0.05$，即当 $3.841<X^2<6.635$ 时，种间联结为显著。

2.4.2　共同出现百分率 *PC*

PC 值用来测定种间的正联结程度，公式为：

$$PC=a/(a+b+c)$$

PC 值的范围介于 0 到 1 之间，其值越接近于 1，表明种对间的正关联程度越大，无关联时为 0[5,18]。

2.4.3　联结系数 *AC*

联结系数 *AC* 用来进一步检验由 X^2 所测定出的结果及说明种间联结程度。

若 $ad\geq bc$，

则 $AC=(ad-bc)/[(a+b)(b+d)]$

若 $bc>ad$ 且 $d\geq a$，

则 $AC=(ad-bc)/[(a+b)(a+c)]$

若 $bc>ad$ 且 $d<a$，

则 $AC=(ad-bc)/[(b+d)(d+c)]$

AC 值的范围介于-1 到 1 之间，其值越接近于 1，表明物种的正联结性越强；其值越接近于-1，表明物种的负联结性越强；其值为 0 时，种对间表现出相互独立的特点[5,18]。

2.5　数据处理

所有数据均在 Excel 2003（Microsoft Corporation）和 SPSS 18.0 等软件平台下进行处理分析和制图。

3　结果与分析

3.1　种群的重要值

由于植物种类较多，根据每个物种的重要值排序，选取重要值较大的种群进行了种间联结性的计算和分析。参与分析的物种包括乔木 17 种、灌木 8 种、草本 12 种。选取物种的重要值如表 2 所示。

表 2　主要种群的重要值

编号	乔木层植物名称	重要值 *IV*(%)	编号	灌木层植物名称	重要值 *IV*(%)	编号	草本层植物名称	重要值 *IV*(%)
1	大叶栎 *Quercus griffithii*	30.40	1	大叶栎 *Quercus griffithii*	81.12	1	阔片短肠蕨 *Allantodia matthewii*	112.75
2	锈毛梭子果 *Eberhardtia aurata*	26.03	2	菝葜 *Smilax china*	38.42	2	楼梯草 *Elatostema involucratum*	29.70
3	鹿角椎 *Castanopsis lamontii*	20.63	3	海南山龙眼 *Heliciahainanensis*	31.98	3	高秆珍珠茅 *Scleria terrestris*	27.08
4	尖连蕊茶 *Camellia cuspidata*	15.71	4	锈毛梭子果 *Eberhardtia aurata*	31.34	4	扁柄沿阶草 *Ophiopogon compressus*	23.11
5	木姜子 *Litsea pungens*	12.49	5	尖连蕊茶 *Camellia cuspidata*	25.53	5	苦竹 *Pleioblastus amarus*	20.87
6	罗浮椎 *Castanopsis faberi*	12.36	6	柠檬金花茶 *Camellia limonia*	14.47	6	广西省藤 *Calamus guangxiensis*	16.35
7	鸭公树 *Neolitsea chuii*	12.19	7	鸭公树 *Neolitsea chuii*	13.07	7	华山姜 *Alpinia chinensis*	14.01

(续)

编号	乔木层植物名称	重要值 IV(%)	编号	灌木层植物名称	重要值 IV(%)	编号	草本层植物名称	重要值 IV(%)
8	广东琼楠 *Beilschmiedia fordii*	11.91	8	桃叶珊瑚 *Aucuba chinensis*	10.51	8	镰羽贯众 *Cyrtomium balansae*	9.81
9	环鳞烟斗柯 *Lithocarpus corneus*	11.74				9	灰绿耳蕨 *Polystichum eximium*	9.69
10	桂南木莲 *Manglietia chingii*	11.69				10	蜘蛛抱蛋 *Aspidistra elatior*	7.74
11	柠檬金花茶 *Camellia limonia*	9.63				11	短穗鱼尾葵 *Caryota mitis*	7.68
12	海南山龙眼 *Helicia hainanensis*	9.00				12	狭翅巢蕨 *Neottopteris antrophyoides*	5.86
13	腺叶桂樱 *Laurocerasus phaeosticta*	8.70						
14	白背桐 *Mallotus apelta*	7.09						
15	桃叶珊瑚 *Aucuba chinensis*	6.73						
16	革叶算盘子 *Glochidion daltonii*	6.28						
17	青藤公 *Ficus langkokensis*	5.81						

3.2 群落的总体联结性

根据调查数据，计算种群间的总体联结性，结果表明(表3)：大青山天然次生林植物群落乔木、灌木、草本3个层次的优势种种间联结性存在明显差异。

乔木层的 $VR=6.451>1$，表明总体上种间存在一定的正联结，其显著性统计量 W 为161.277，以自由度查询相应的 X^2 值[19]，即 $X^{2\,0.05(25)}=14.61$，$X^{2\,0.95(25)}=37.65$，W 值落入 $X^{2\,0.05(25)}$ 与 $X^{2\,0.95(25)}$ 区间之外，且大于 $X^{2\,0.95(25)}$ 值，表明乔木层群落总体上的正联结程度达到显著水平；灌木层的 $VR=0.839<1$，表明总体上种间存在一定的负联结，其显著性统计量 W 为20.969，W 值落入 $X^{2\,0.05(25)}$ 与 $X^{2\,0.95(25)}$ 之间，表明灌木层群落总体上的负联结程度不显著；草木层的 $VR=1.151>1$，表明总体上种间存在一定的正联结，其显著性统计量 W 为28.93，W 值落入 $X^{2\,0.05(25)}$ 与 $X^{2\,0.95(25)}$ 之间，表明草本层群落总体上的正联结程度不显著。

表3 种群间的总体联结性

层次	方差比率	检验统计量	X^2 临界值($X^{2\,0.95(N)}$，$X^{2\,0.05(N)}$) $X^{2-tests}$	测度结果
乔木层	6.451	161.25	(14.61，37.65)	显著正关联

（续）

层次	方差比率	检验统计量	X^2 临界值($X^{2\,0.95(N)}$，$X^{2\,0.05(N)}$) $X^{2\text{-tests}}$	测度结果
灌木层	0.839	21.00	(14.61，37.65)	不显著负关联
草本层	1.151	28.93	(14.61，37.65)	不显著正关联

3.3　主要种群的联结性

3.3.1　乔木层优势种群的联结性

大青山天然次生林乔木层优势种群间的共同出现百分率(PC)半矩阵图、联结系数AC半矩阵图、X^2检验数据半矩阵如下图1和表4所示。由于锈毛梭子果和广东琼楠均出现在所有的调查样方中，与之配对的共31个种对无法计算出X^2值及AC值。

X^2结果显示(表4)，有20个种对的种间联结达到显著或极显著水平($3.841 < X^2 < 6.635$ 或 $X^2 > 6.635$)，这些种对的物种大部分来自群落的中上层，他们在群落所处的位置相似，存在一定的互利关系。PC值分析结果(图1A)显示，所有种对间均呈现正的联结性，其中广东琼楠和锈毛梭子果、白背桐和鸭公树、尖连蕊茶和腺叶桂樱这3个种对间存在极强的正联结性(PC值接近1)。AC值分析结果显示(图1B)，正联结种对数为81个，负联结种对数为19，5个种对表现出完全相互独立的特点($AC=0$)。其中，有6个种对出现强的正联结，白背桐和鸭公树、桃叶珊瑚和鸭公树的AC值均为1，11个种对为强的负联结($AC=-1$)。

X^2检验结合共同出现百分率(PC)、联结系数AC的分析结果表明，检验中正联结种对数较多，在105个种对中，达到显著和极显著正联结的种对20个；不显著联结的种对数为85个，其中正联结61个，占总对数的71.77%，负联结19个，占22.35%，无联结5个，占5.88%，这与总体种间联结性检验的方差比率VR值的结果一致，乔木层优势种群正联结性明显。

表4　乔木层优势种群X^2检验数据半矩阵

1																
—	2															
1.76	—	3														
2.68	—	4.44	4													
9.00	—	0.46	0.00	5												
0.45	—	2.34	1.04	1.04	6											
0.76	—	0.16	4.64	0.85	1.01	7										
—	—	—	—	—	—	—	8									
0.29	—	0.00	0.07	0.07	0.20	0.65	—	9								
0.07	—	0.88	6.25	0.00	1.04	0.38	—	0.07	10							
0.76	—	3.40	0.38	0.38	1.01	0.46	—	0.65	0.38	11						
6.51	—	3.40	4.64	0.38	0.06	1.47	—	0.65	0.38	0.46	12					
2.68	—	4.44	14.1	1.56	4.17	4.64	—	1.19	1.56	0.38	4.64	13				
2.68	—	0.88	6.25	0.00	1.04	13.6	—	0.07	1.56	0.85	4.64	6.25	14			
2.43	—	2.34	4.17	0.00	2.78	5.11	—	2.43	1.04	0.06	1.01	4.17	9.38	15		
0.02	—	0.11	0.45	0.20	1.19	2.53	—	0.02	0.45	0.05	1.33	0.45	0.45	1.19	16	
1.14	—	0.11	0.20	3.17	1.19	1.33	—	1.14	0.20	8.77	1.33	0.45	1.69	0.03	0.00	17

注：1. 大叶栎；2. 锈毛梭子果；3. 鹿角椎；4. 尖连蕊茶；5. 木姜子；6. 罗浮椎；7. 鸭公树；8. 广东琼楠；9. 环鳞烟斗柯；10. 桂南木莲；11. 柠檬金花茶；12. 海南山龙眼；13. 腺叶桂樱；14. 白背桐；15. 桃叶珊瑚；16. 革叶算盘子；17. 青藤公。

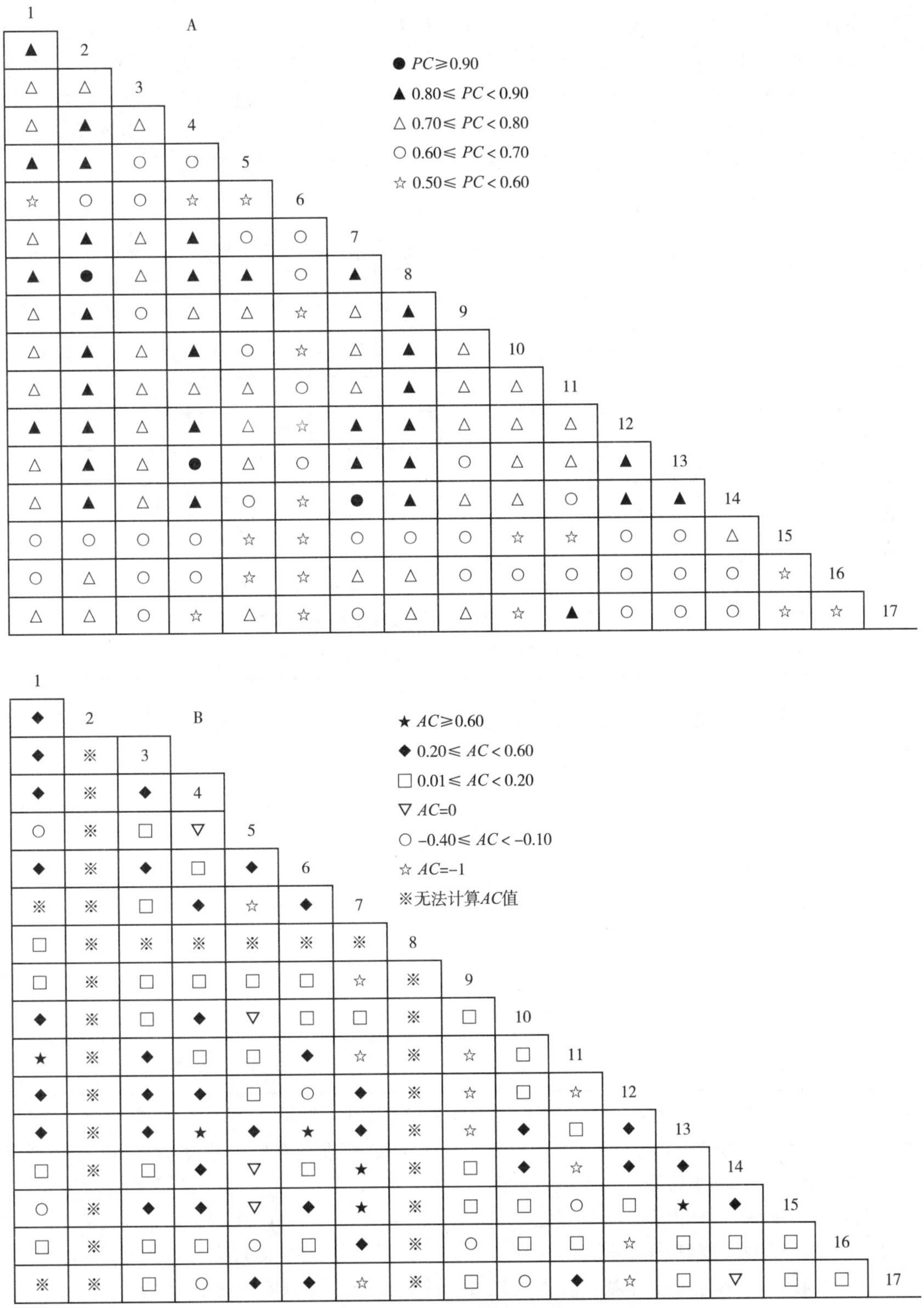

A

● $PC \geq 0.90$
▲ $0.80 \leq PC < 0.90$
△ $0.70 \leq PC < 0.80$
○ $0.60 \leq PC < 0.70$
☆ $0.50 \leq PC < 0.60$

	1	2	3	4	5	6	7	8	9	10	11	12	13	14	15	16
2	▲															
3	△	△														
4	△	▲	△													
5	▲	▲	○	○												
6	☆	○	○	☆	☆											
7	△	▲	△	▲	○	○										
8	▲	●	△	▲	▲	○	▲									
9	△	▲	○	△	△	☆	△	▲								
10	△	▲	△	▲	○	☆	△	▲	△							
11	△	▲	△	△	△	○	△	▲	△	△						
12	▲	▲	△	▲	△	☆	▲	▲	△	△	△					
13	△	▲	△	●	△	○	▲	▲	○	△	△	▲				
14	△	▲	△	▲	○	☆	●	▲	△	△	○	▲	▲			
15	○	○	○	○	☆	☆	○	○	○	☆	☆	○	○	△		
16	○	△	○	○	☆	☆	△	△	○	○	○	○	○	○	☆	
17	△	△	○	☆	△	☆	○	△	△	☆	▲	○	○	○	☆	☆

B

★ $AC \geq 0.60$
◆ $0.20 \leq AC < 0.60$
□ $0.01 \leq AC < 0.20$
▽ $AC=0$
○ $-0.40 \leq AC < -0.10$
☆ $AC=-1$
※无法计算 AC 值

	1	2	3	4	5	6	7	8	9	10	11	12	13	14	15	16
2	◆															
3	◆	※														
4	◆	※	◆													
5	○	※	□	▽												
6	◆	※	◆	□	◆											
7	※	※	□	◆	☆	◆										
8	□	※	※	※	※	※	※									
9	□	※	□	□	□	□	☆	※								
10	◆	※	□	◆	▽	□	□	※	□							
11	★	※	◆	□	□	◆	☆	※	☆	□						
12	◆	※	◆	◆	□	○	◆	※	☆	□	☆					
13	◆	※	◆	★	◆	★	◆	※	☆	◆	□	◆				
14	□	※	□	◆	▽	□	★	※	□	◆	☆	◆	◆			
15	○	※	◆	◆	▽	◆	★	※	□	□	○	□	★	◆		
16	□	※	□	□	○	□	◆	※	○	□	□	☆	□	□	□	
17	※	※	□	○	◆	◆	☆	※	□	○	◆	☆	□	▽	□	□

图 1　乔木层优势种群种间联结性半矩阵图

注：1. 大叶栎；2. 锈毛梭子果；3. 鹿角椎；4. 尖连蕊茶；5. 木姜子；6. 罗浮椎；7. 鸭公树；8. 广东琼楠；9. 环鳞烟斗柯；10. 桂南木莲；11. 柠檬金花茶；12. 海南山龙眼；13. 腺叶桂樱；14. 白背桐；15. 桃叶珊瑚；16. 革叶算盘子；17. 青藤公。

3.3.2　灌木层优势种群种对的联结性

大青山天然次生林优势灌木种群间的共同出现百分率(PC)半矩阵图、联结系数 AC 半矩阵图及 X^2 统计数阵如图 2 和表 5 所示。

X^2 结果显示(表5)，灌木层28个种对中，只有菝葜和海南山龙眼1个种对的种间联结达到显著水平($3.841<X^2<6.635$)；PC值分析结果(图2A)显示，所有种对间均呈现正的联结性，其中，大叶栎与海南山龙眼、锈毛梭子果、尖连蕊茶及海南山龙眼与锈毛梭子果、尖连蕊茶组成的5个种对有较强的联结性($PC\geq0.50$)，其他种对为弱联结性($PC<0.50$)；AC值分析结果显示(图2B)，14个种对为正关联，12个种对为负关联，大叶栎与海南山龙眼、尖连蕊茶和鸭公树2个种对无关联($AC=0$)。不同检验结果表明，灌木种间的联结性较弱。

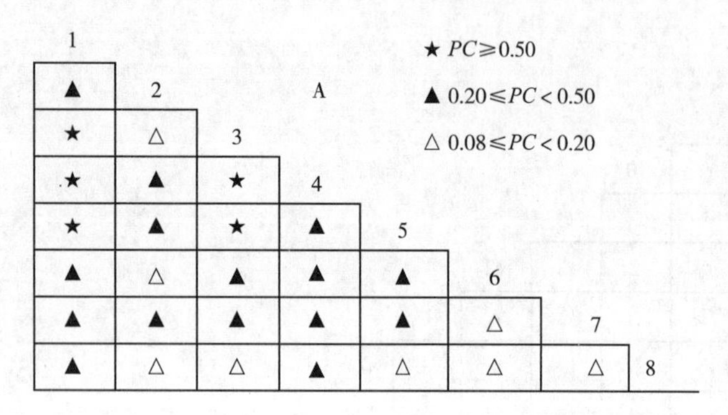

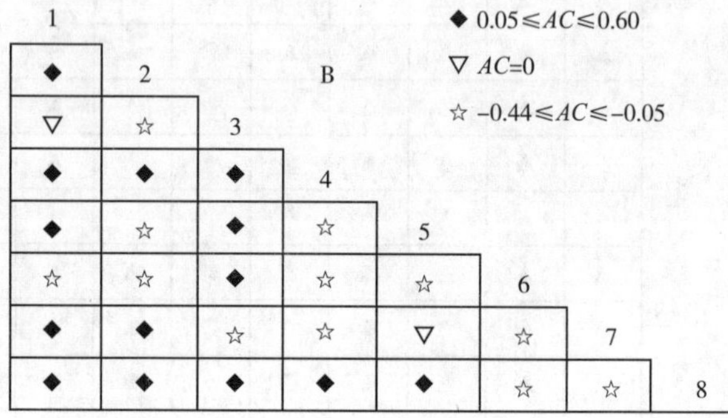

图2 灌木层优势种群种间联结性半矩阵图

注：A为PC值；B为AC值；1. 大叶栎；2. 菝葜；3. 海南山龙眼；4. 锈毛梭子果；5. 尖连蕊茶；6. 柠檬金花茶；7. 鸭公树；8. 桃叶珊瑚。

表5 灌木层优势种群 X^2 检验数据半矩阵

1							
3.52	2						
0.00	4.17	3					
3.17	1.99	0.03	4				
1.04	0.12	0.69	0.53	5			
1.56	1.16	0.26	1.89	1.42	6		
1.04	0.12	0.69	0.03	0.00	0.26	7	
1.19	0.41	0.45	1.85	0.45	0.25	0.45	8

注：1. 大叶栎；2. 菝葜；3. 海南山龙眼；4. 锈毛梭子果；5. 尖连蕊茶；6. 柠檬金花茶；7. 鸭公树；8. 桃叶珊瑚。

3.3.3 草本层优势种群种对的关联性

由表6和图3可知，扁柄沿阶草和蜘蛛抱蛋之间存在显著的正联结；灰绿耳蕨和短穗鱼尾葵之间存在极显著的正联结；广西省藤和华山姜之间存在显著的负联结性；高秆珍珠茅与灰绿耳蕨、短穗鱼尾葵、苦竹和华山姜、狭翅巢蕨之间的PC值为0和AC值为-1，说明它们之间无关联；苦竹和短穗鱼尾葵之间完全独立($AC=0$)。在众多种对中，仅有阔片短肠蕨和扁柄沿阶草、楼梯草和阔片短肠蕨、灰绿耳蕨和短穗鱼尾葵、楼梯草和扁柄沿阶草4个种对的共同出现百分率PC在较高的水平($PC\geq0.50$)，说明他们的正联结性较强，除此之外，25的种对联结系数AC在0上下波动($-0.1<AC<0.1$)，联结性较为松散。

草本层66个种对中，达到显著和极显著联结的种对5个，其中极显著正联结1个，显著正联结1个，显著负联结3个；不显著联结的种对数为61个，其中正联结32个，占总对数的52.46%，负联结24个，占39.34%，无联结5个，占8.20%。分析结果显示，草本层呈现出弱的正联结性，与总体种间联结性检验的方差比率VR值的结果一致。

A

▲ 0.50≤ PC≤0.75
■ 0.20≤ PC<0.50
★ 0.06≤ PC<0.20
□ PC=0

	1	2	3	4	5	6	7	8	9	10	11
2	▲										
3	■	■									
4	▲	▲	■								
5	■	■	★	■							
6	■	★	■	■	■						
7	■	■	■	■	□	★					
8	■	■	■	■	★	■	■				
9	■	★	□	■	■	★	■	★			
10	■	■	★	■	★	★	■	■	★		
11	■	★	□	■	★	★	■	★	▲	★	
12	■	★	■	■	□	★	■	■	★	★	★

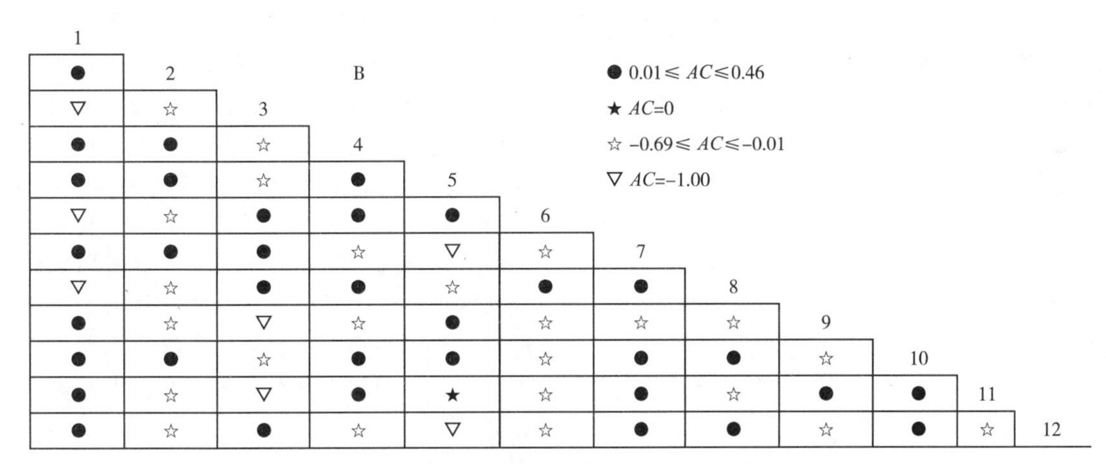

图 3　草本层优势种群种间联结性半矩阵图

注：A 为 PC 值；B 为 AC 值；1. 阔片短肠蕨；2. 楼梯草；3. 高秆珍珠茅；4. 扁柄沿阶草；5. 苦竹；6. 广西省藤；7. 华山姜；8. 镰羽贯众；9. 灰绿耳蕨；10. 蜘蛛抱蛋；11. 短穗鱼尾葵；12. 狭翅巢蕨。

表 6　草本层优势种群 X^2 检验数据矩半矩阵

1											
0. 96	2										
1. 85	0. 07	3									
2. 68	1. 47	1. 89	4								
0. 26	0. 36	0. 69	0. 20	5							
1. 56	2. 16	0. 12	0. 53	1. 04	6						
0. 82	0. 34	0. 76	0. 68	4. 91	3. 90	7					
1. 85	0. 07	2. 33	0. 23	0. 69	1. 42	0. 00	8				
0. 41	0. 10	5. 47	0. 00	3. 17	0. 53	0. 01	1. 99	9			
0. 49	0. 99	0. 62	4. 58	0. 18	3. 71	0. 17	0. 01	0. 05	10		
0. 26	0. 16	3. 52	0. 20	0. 00	1. 04	0. 65	0. 69	8. 38	0. 18	11	
0. 33	3. 11	0. 67	0. 11	1. 97	1. 79	1. 65	0. 67	0. 50	0. 01	0. 05	12

注：1. 阔片短肠蕨；2. 楼梯草；3. 高秆珍珠茅；4. 扁柄沿阶草；5. 苦竹；6. 广西省藤；7. 华山姜；8. 镰羽贯众；9. 灰绿耳蕨；10. 蜘蛛抱蛋；11. 短穗鱼尾葵；12. 狭翅巢蕨。

有研究表明，应用方差比率法对优势种群间的总体联结性进行检验，再以 X^2 检验为基础，结合 *PC*、*AC* 等值进行分析，能达到较好的效果[20-22]。这些方法依赖于种存在与否的二元数据，其测定结果受研究的尺度(样方大小)、取样方法(数目多少)影响较大[5]。本文先用方差比率法对优势种进行总体联结性分析，再运用 X^2 统计数阵、共同出现百分率(*PC*)、联结系数 *AC* 对种对间的联结性进行测定，发现，*AC* 值和 *PC* 值虽能反映种间联结性的相对强弱，但当两个种都存在于所有的样方中的情况下(锈毛梭子果、广东琼楠)，X^2 统计量及联结系数 *AC* 无法计算。张金屯[1]的研究表明，在这种情况下，我们就要使用数量数据，比如多度、盖度等，一般通过计算种间相关系数来衡量两种间的相关程度。种间联结性与物种的生物学特性、生态适应性等因素有关，其测定值只是说明竞争的结果或现状，不能揭示其原因及过程。

4 讨论与结论

4.1 优势种群的总体联结

总体上优势种群具有正联结反映了群落具有较强的稳定性[21,23,24]。本研究对大青山天然次生林优势种群间的总体联结性分析结果表明，乔木层群落总体上的正联结程度达到显著水平；灌木层群落总体上的负联结程度不显著；草本层群落总体上的正联结程度不显著。说明，乔木优势种群对所处的综合生长环境有着相似的反应，生态位在一定程度上出现重叠，乔木群落具有一定的稳定性，并且这种稳定性将随着演替的进行得到逐步加强；但林下的灌草层种间联结关系松散，暗示灌草层还处于不稳定的演替阶段，物种有一定的独立性，种类和数量均处于波动的状态。

这种联结的松散性可能与大青山天然次生林目前的发展阶段及物种本身的生态学特性有关，调查发现，大青山天然次生林林下灌木和草本的种类、数量、分布范围均不及乔木树种，其物种间相遇的概率较低。今后可加强对种间联结性内在机制及物种与环境之间的耦合关系等方面的研究，进一步从物种生态适应性及生态位、群落结构与动态方面深入了解群落的总体联结性。

4.2 优势种群种对间的联结性

优势种群间的联结性分析，可看出种群间的相互作用及群落组成的动态。植物之间存在直接或间接的相互影响[16]，正联结表明种对间存在至少对一方有利的作用，负联结表明种对间存在至少不利于一方的相互作用机制[7]。两个物种的正联结程度越高说明它们所需的生长环境条件越相似；反之则表明对生长环境条件的需求有所不同[5]。在同一环境条件下，正的联结，可能在某种程度上指示相互作用的存在对一方或双方种是有利的，例如互惠共生或资源划分方面的互补；负的联结，可能表明不利于一方或双方的相互作用，例如种间竞争、干扰[25]。竞争只是生物进化过程中出现的阶段性现象，生物进化的发展方向终将是生物与生物之间以及生物与环境之间的协同[26]。许多研究认为，一个种对相关性大，是因为它们具有相似的生物学特性和生态适应性，对环境的适应能力、对资源利用能力、对群落所起的功能作用等均有一致性，因此，它们的关系应该是稳定的[2,27]。植物群落内物种间的种间关系与该群落所处的演替阶段密切相关，在群落的演替初期，物种间尚未形成一定的种间关系；演替中期，物种间主要表现为竞争关系；群落演替到达中生阶段时，物种间的关系较为复杂，群落内部处于同一层次的物种由于对生长环境需求相似而表现出种间联结[16,28]。种间联结程度与生态位重叠值之间是密切相关的，种间正联结程度越高，生态位重叠程度也越高[29]。

在生长环境及竞争一致的条件下，具有相同或相似生态习性的物种往往共同出现在同一群落中，其种间联结性一般表现为显著的正相关，这在前人的研究中得到了验证[4,21,22]，本研究结果进一步证实了这一规律。例如，乔木层具有显著正联结性的种对由尖连蕊茶、腺叶桂樱、白背桐、鸭公树、海南山龙眼等乔木中下层植物组成，这些种一般生于林内潮湿环境内，它们对生境和资源要求有一定的相似性，具有一定的耐阴性，它们之间的显著正联结暗示它们具有较大的生态位重叠；木姜子与白背桐、桂南木莲、桃叶珊瑚、尖连蕊茶、白背桐与青藤公 5 个种对相互独立，它们的生态位重叠度也较小，可能是由于它们对环境具有不同的生态适应性或相互分离的生态位所致，这与其生态习性有关，如木姜子喜光，而尖连蕊茶、桃叶珊瑚性喜温暖湿润的气候环境，耐阴性强。

相对于乔木层，灌草层的联结性则表现出松散及较弱的特性。由于研究区域的生态环境变化并不大，灌木层中具有较大重要值的物种多为上层乔木的优势种的幼苗，个体数量并不大，且他们在研究区域的分布具有一定的随机性，具有较强联结性的种都是具有较大个体数及频度的乔木层幼苗，故其联结性也较大。例如，大叶栎与海南山龙眼、锈毛梭子果、尖连蕊茶及海南山龙眼与锈毛梭子果、尖连蕊茶组成的5个种对有较强的联结性。草本层一般居于林下，一般都呈现聚集分布，容易受到林分内小环境的影响，如土壤、水分、郁闭度等的变化都会引起其个体数、种类及分布的变化，因此，草本层的种间联结性呈现较弱联结性可能是物种对生境的不同偏好或相异的生物学特性所致。例如，高杆珍珠茅与灰绿耳蕨、短穗鱼尾葵，苦竹和华山姜、狭翅巢蕨、短穗鱼尾葵5个种对表现出相互独立的特点。所以，灌草层之间总体的联结性较松散是符合实际的。

通过上述分析表明，大青山天然次生林植物群落乔木树种间具有较强的正联结性，大多数乔木树种的种间联结性较紧密；主要灌木树种的种间联结性较松散，草本层主要种间表现为弱联结性。物种生态习性、群落演替阶段等因素可能是造成这一特性的主要原因。大青山天然次生林乔木层的建群种大叶栎、锈毛梭子果、鹿角椎及其伴生种尖连蕊茶、鸭公树之间存在较强的联结性，它们的个体数及频度都较大，群落中尚未存在具更强的竞争潜力的物种，群落处于较稳定的发展阶段，预示着这些种群将在一定时期内占据并主导大青山天然次生林整个群落的演替方向。未来应加强对其自然环境的管理和保护，如果有必要，可适度人工干预，促进天然次生林的演替。大叶栎在大青山天然次生林中占据优势，野外调查时发现，大叶栎母株下的幼苗个体数量较多，自然更新良好，暗示该种群具有较好的自我更新能力，在群落中具有一定的稳定性且处于动态的发展阶段之中，正在不断更新。因此，大叶栎在该地区人工林树种选择上是一个值得考虑的物种。

参考文献

[1]张金屯．数量生态学[M]．北京：科学出版社，2004.

[2]GREIG-SMITH P. Quantitative Plant Ecology[M]. California：University of California Press，1983.

[3]周纪纶．植物种群生态学[M]．北京：高等教育出版社，1992.

[4]邓福英，臧润国．海南岛热带山地雨林天然次生林的功能群划分[J]．生态学报，2007，27(08)：3240-3249.

[5]王伯荪，彭少麟．南亚热带常绿阔叶林种间联结测定技术研究——Ⅰ．种间联结测试的探讨与修正[J]．植物生态学报，1985(04)：274-285.

[6]宋永昌．植被生态学[M]．上海：华东师范大学出版社，2001.

[7]许涵，黄久香，唐光大，等．南昆山观光木所在群落优势树种的种间联结性[J]．华南农业大学学报，2008，29(01)：57-62.

[8]杨继镐，汪炳根．广西大青山森林植物分布立地适宜性与土壤关系的研究[J]．林业科学，1990，26(03)：209-218.

[9]康冰，刘世荣，蔡道雄，等．2009. 马尾松人工林林分密度对林下植被及土壤性质的影响[J]．应用生态学报，2009，20(10)：2323-2331.

[10]康冰，刘世荣，蔡道雄，等．2010 南亚热带不同植被恢复模式下土壤理化性质[J]．应用生态学报，2010，21(10)：2479-2486.

[11]明安刚，张治军，谌红辉，等．抚育间伐对马尾松人工林生物量与碳贮量的影响[J]．林业科学，2013，49(10)：1-6.

[12]康冰，刘世荣，温远光，等．广西大青山南亚热带次生林演替过程的种群动态[J]．植物生态学报，2006，30(06)：931-940.

[13]黄承标，卢立华，温远光，等．大青山林区不同海拔高度主要气象要素的变化[J]．贵州农业科学，2011，39(01)：90-95.

[14]康冰．广西大青山退化森林植被特征及生态恢复研究[D]．西安：西北农林科技大学，2007.

[15]SCHLUTER D. A variance test for detecting species associations，with some example applications[J]．Ecology，1984，65(3)：998-1005.

[16]周先叶，王伯荪，李鸣光，等．广东黑石顶自然保护区森林次生演替过程中群落的种间联结性分析[J]．植物生态学报，2000，24(03)：332-339.

[17]DICE L R . Measures of the amount of ecologic association between species[J]. Ecology，1945，26(3)：297-302.

[18]郭志华，卓正大，陈洁，等．庐山常绿阔叶、落叶阔叶混交林乔木种群种间联结性研究[J]．植物生态学报，1997，21(05)：424-432.

[19]杜荣骞．生物统计学[M]．北京：高等教育出版社，2014.

[20]刘金福，洪伟，樊后保，等．天然格氏栲林乔木层种群种间关联性研究[J]．林业科学，2001，37(04)：117-123.

[21]邓贤兰，刘玉成，吴杨．井冈山自然保护区栲属群落优势种群的种间联结关系研究[J]．植物生态学报，2003，27(04)：531-536.

[22]史作民，刘世荣，程瑞梅，等．宝天曼落叶阔叶林种间联结性研究[J]．林业科学，2001，37(02)：29-35.

[23]杨一川，庄平．峨眉山峨眉栲，华木荷群落研究[J].

植物生态学报，1994，18(02)：105-120.
[24]杜道林，刘玉成，李睿．缙云山亚热带栲树林优势种群间联结性研究[J]．植物生态学报，1995，19(02)：149-157.
[25]考克斯．普通生态学实验手册[M]．北京：科学出版社，1979.
[26]王德利，高莹．竞争进化与协同进化[J]．生态学杂志，2005，24(10)：1182-1186.
[27]张金屯，焦蓉．关帝山神尾沟森林群落木本植物种间联结性与相关性研究[J]．植物研究，2003，23(04)：458-463.
[28]PETERS H A. Neighbor-regulated mortality：the influence of positive and negative density dependence on tree populations in species-rich tropical forests[J]. Ecology Letters，2003，6(8)：757-765.
[29]李帅锋，刘万德，苏建荣，等．季风常绿阔叶林不同恢复阶段乔木优势种群生态位和种间联结[J]．生态学杂志，2011，30(03)：508-515.

[原载：广西植物，2016，36(07)]

广西石灰岩地区单枝竹属资源的分布和变异

潘启龙[1] 刘光金[1] 蔡道雄[1] 卢立华[1] 杨桂芳[1] 郭起荣[2]

([1]中国林业科学研究院热带林业实验中心，广西凭祥 532600；[2]国际竹藤中心，
国家林业局竹藤科学与技术重点开放实验室，北京 100102)

摘 要 单枝竹属为石漠化山地生态维护功能竹种，深入了解其种质资源的分布和变异特征是对其进行保育及开发的基础。本研究对广西单枝竹属资源及其分布开展调查，并进行了秆型、秆高、叶长、叶宽形态特征分析。结果表明，广西单枝竹属竹种资源丰富，在25个天然居群共收集单枝竹、芸香竹和箭秆竹等2个种、1个变种的106份种质；单枝竹属竹种在广西集中分布在21°57′~25°57′N，111°06′~106°29′E，海拔100~800m的岩溶地貌区，呈现箭秆竹、单枝竹、芸香竹自北向南天然分布规律，单枝竹、芸香竹分布存在重合现象。秆型存在4种类型：①直立且秆梢挺直型。②直立但秆梢钓丝状下垂型。③攀缘型。④匍匐型。秆高、叶长、叶宽形态差异显著，单枝竹变异系数由高到低依次为叶宽、秆高、叶长；芸香竹和箭秆竹依次均为秆高、叶宽、叶长。单枝竹属资源及其表型变异丰富，具有种质挖掘的潜力。

关键词 单枝竹属；资源分布；秆型；形态特征

Economic Bamboo Species in Guangxi Province, China-Genus *Bonia*: Resources Distribution and Morphological Charateristics

PAN Qilong[1], LIU Guangjin[1], CAI Daoxiong[1], LU Lihua[1],
YANG Guifang[1], GUO Qirong[2]

(*[1]The Experimental Centre of Tropical Forestry, CAF, Pingxiang* 532600, *Guangxi, China*;
[2]International Centre for Bamboo and Rattan, Beijing 100102, *China*)

Abstract: Bamboo species of genus *Bonia* have important ecological maintenance functions for rocky desertification mountain, a thorough understanding of its germplasm resources distribution and variation characteristics are on the basis of its conservation and development. We studied the resources and distribution of *Bonia* in Guangxi, and 4 different traits were analyzed to do morphological charateristics variation research on culm type、culm hight、leaf length、leaf width. The results showed that bamboo species of *Bonia* were diffusely distributed in Guangxi province, with the distribution centre locates in the lower and medium mountain areas at the altitude of 100~800m above sea level, ranging in 21°57′~25°57′N, 111°06′~106°29′E. 106 germplasm resources including 2 species (*Bonia saxatilis*、*B. amplexicaulis*) and 1 varietas (*B. saxatilis* var. *solida*) were collected in 25 geographical population. There are 4 different culm types: ①straight type. ②straight, but top prolapsed. ③climbing type. ④Significant difference were showed within 4 different traits. The variable coefficient of *B. saxatilis* showed leaf width > culm height > leaf length. The variable coefficient of *B. amplexicaulis* and *B. saxatilis* var. *solida* showed culm height>leaf width>leaf length. Due to the phenotypic variation and its rich resources, *Bonia* have the germplasm mining potential.

Key words: *Bonia*; resources distribution; culm type; phenotypic variation

世界已知竹种超过1500种，我国超过600种，中国竹林面积601万hm^2[1-3]。单枝竹属(*Bonia* Balansa)因单分枝、合轴丛生、叶基近截形等形态和特殊的地理分布特征而独立1属。1988年，国内贾良智等原发表为1个新属 *Monocladus*，在1996年版《中国植物志》中文版第9卷第1分册中，可见其属名为 *Monocladus*

Chia *et al*.；后经夏念和等的研究[4]，*Monocladus* 为 Balansa 在 1890 年发表的 *Bonia* 的异名，遂在 2006 年出版的《中国植物志》英文及修订版《Flora of China》第 22 卷中，更正单枝竹属名为 *Bonia* Balansa。

单枝竹属现已知有 5 种 1 变种：单枝竹(*B. saxatilis*)、芸香竹(*B. amplexicaulis*)、响子竹(*B. levigata*)、小花单枝竹(*B. parvifloscula*)、北部湾单枝竹(*B. tonkinensis*)，变种箣秆竹(*B. saxatilis* var. *solida*)[5,6]，主要分布于广西、广东、海南和越南等喀斯特地貌山区[5]，新的研究发现在贵州以及滇桂边界地区也有分布[6,7]，仅北部湾单枝竹分布越南。

单枝竹属是在石灰岩山地具有重要生态服务功能竹种，但长期以来，关于用材竹种、笋用竹种的研究较多，但关于单枝竹属的研究报道较少。新近见单枝竹叶片气孔特征，染色体观测，核糖体 ITS 序列，DNA 条形码和芸香竹分株育苗技术[8-13]等方面的研究报道。

中国科学院华南植物园曾开展过单枝竹属物种保育工作，但目前文献仅见 1992 年福建华安竹类植物园异地保存单枝竹属 2 个种：单枝竹和芸香竹[14]。本研究中在对广西石灰岩山地单枝竹属竹子种质资源的调查、采集和保育的基础上，对秆型特征等进行了整理，以期望对我国西南、华南和越南北部的石灰岩山体喀斯特山区的石漠化治理，林下植被恢复有参考意义。

1 材料与方法

1.1 种质收集

广西地处南疆(104°28′~112°04′E，20°54′~26°23′N)，气候温湿，年均气温 16.0~23.0℃，年均降水量 1070mm 以上，属中、南亚热带季风气候区。境内喀斯特地貌密集，石灰岩山地散状、片状分布，是单枝竹属植物集中分布区。2012 年 3 月至 2013 年 12 月，对广西 25 个县石灰岩山地天然次生林中的单枝竹属竹种进行了实地踏查和种质收集，为保证种质的遗传变异多样性，充分考虑田园、山脉、江河、水库自然地理隔离等来确定采样点，每个采样点收集 1 份种质，共计收集 106 份。

1.2 性状选择和测定

单枝竹属竹种为丛生竹种，秆高、胸径测定为丛内秆高、胸径的平均值。随机采取 5~10 片完整叶片测量叶长、叶宽。因长期生境等的变异，秆型在分布区以 4 种类型进行统计：①直立且秆梢挺直型。②直立但秆梢钓丝状下垂型。③攀缘型，秆长而柔软，不能全部直立，在稠密林间攀附树枝。④匍匐型，秆纤细、柔软，完全不能直立，仅能匍匐或悬垂岩边，或攀附灌丛上。

1.3 统计分析方法

利用软件 SPSS 13.0 进行单枝竹属竹种 25 个居群 106 份样丛秆高、胸径、叶长、叶宽性状差异性分析与相关性分析。

2 结果与分析

2.1 分布区域

经研究发现，广西单枝竹属资源丰富，分布范围较广，集中分布在广西北部、西北部、西部、西南部、中部的石灰岩山区，共发现芸香竹、单枝竹、箣秆竹 2 个种 1 个变种，呈现箣秆竹、单枝竹、芸香竹自北向南天然分布的规律，其地理分布具体见表 1。其中，箣秆竹仅在桂林 8 个县发现，中心分布区为 24°34′~25°57′N，110°31′~111°03′E，海拔 179~587m 的岩溶地貌区；芸香竹分布在 5 个县，中心分布区 21°57′~23°41′N，106°47′~108°12′，海拔 215~663m，呈连续分布。单枝竹分布在 14 个县，分布范围最广，中心分布区 22°45′~25°10′N，106°29′~110°53′E，海拔 100~587m 的岩溶地貌区，呈现桂北、桂西北、桂西、桂西南环状连续分布。

表 1 芸香竹、单枝竹、箣秆竹地理分布

竹种	采样地点	样本数	纬度 N	经度 E	海拔(m)	秆型
	凭祥	7	21°57′	106°47′	290	①④
	马山	4	23°41′	108°12′	228	②④
	龙州	6	22°11′	106°52′	179	①
芸香竹	大新	2	22°45′	107°12′	587	②④
	天等	4	23°01′	107°10′	427	②
	马山	1	23°41′	108°12′	228	②
	田东	1	23°34′	106°58′	318	②

（续）

竹种	采样地点	样本数	纬度 N	经度 E	海拔(m)	秆型
芸香竹	平果	3	23°23′	107°34′	183	①
	大新	2	22°45′	107°12′	587	①
	东兰	8	24°20′	107°23′	195	②
	田阳	1	23°34′	106°58′	318	②
单枝竹	巴马	8	24°13′	107°26′	574	②
	靖西	3	23°01′	106°29′	663	①
	柳州	1	24°13′	109°26′	100	①
	鹿寨	5	24°47′	109°37′	188	②
	宜州	3	24°32′	108°30′	162	①④
	河池	9	24°42′	108°09′	205	①③④
	阳朔	1	25°10′	110°53′	215	②
	大化	2	23°55′	107°43′	233	②
箭秆竹	兴安	1	25°35′	110°37′	200	①
	全州	7	25°57′	111°03′	174	①
	灌阳	5	25°38′	111°06′	339	①
	贡城	3	25°10′	110°53′	215	①
	平乐	5	24°37′	110°55′	272	①
	荔浦	4	24°34′	110°31′	146	①
	阳朔	5	25°10′	110°53′	215	①
	临桂	3	25°38′	110°03′	366	①

2.2 秆型类型

单枝竹属竹种秆型通直，以①型和②型为主，少数为2种或3种秆型俱有。箭秆竹梢头挺立，无弯曲，秆型直立，属①型；芸香竹秆在龙州表现为①型，在天等表现为②型，在凭祥①型和④型混生一处，在马山、大新②型和④型均有；单枝竹在4个县表现为①型，在8个县表现为②型，在宜州①型和④型混生，在河池①型、③型和④型混生。

2.3 丛态特性

对芸香竹、单枝竹、箭秆竹秆高、叶长、叶宽3个表型性状统计分析，结果见表2。芸香竹秆高最大值为8.5m，为最小值(1.0m)的8.5倍；叶长最大值41.6cm，为最小值(14.7cm)的2.8倍；叶宽最大值7.2cm，为最小值(1.9cm)的3.8倍。单枝竹秆高最大值3.6m，为最小值(0.4m)的9倍；叶长最大值44.1cm，为最小值7.4cm的5.96倍；叶宽最大值8cm，为最小值(2cm)的4倍。箭秆竹秆高最大值3.6m，为最小值（0.6m）的6倍；叶长最大值45.7cm，为最小值(3cm)的13.2倍；叶宽最大值8.1cm，为最小值(2.5cm)的3.24倍。芸香竹秆高变异系数为57.099%，叶宽变异系数27.621%，叶长变异系数16.826%，秆高变异系数>叶宽变异系数>叶长变异系数；箭秆竹秆高变异系数为31.912%，叶宽变异系数为22.161%，叶长变异系数为18.018%，表现为秆高变异系数>叶宽变异系数>叶长变异系数；而单枝竹叶宽变异系数为44.528%，秆高变异系数为29.160%，叶长变异系数为28.789%，表现为叶宽变异系数>秆高变异系数>叶长变异系数。

表2　芸香竹、单枝竹、箭秆竹表型性状统计分析

竹种	秆高(m)				叶长(cm)				叶宽(cm)			
	最小值	最大值	平均值	标准差	最小值	最大值	平均值	标准差	最小值	最大值	平均值	标准差
芸香竹	1.000	8.500	2.411	1.377	14.700	41.600	28.473	4.791	1.900	7.200	4.268	1.179
单枝竹	0.400	3.600	2.092	0.610	7.400	44.100	25.817	7.432	2.000	8.000	3.128	1.393
箭秆竹	0.600	3.600	2.203	0.703	3.000	45.700	28.154	5.073	2.500	8.100	5.160	1.144

对单枝竹属竹种秆高、叶长、叶宽进行统计分析(表3)，结果表明各表型性状在不同纬度和海拔的天然次生林分间均存在显著的差异。芸香竹凭祥居群形态高大，最大值为8.5m，龙州、天等居群次之，大新居群最小，仅为1.1m；天等居群叶长最长，最大值为41.6cm，龙州、凭祥居群次之，大新居群最小，最小值为14.1cm；龙州居群叶宽最大，最大值为7.2cm，凭祥、马山居群次之，大新居群最小，仅为1cm。单枝竹秆高最大为东兰居群，最大值为3.6m，田阳、巴马、河池、平果、鹿寨居群次之，达3.0m以上，大新居群最小仅为0.8m；叶长最大为宜州居群，最长44.1cm，田阳、东兰、鹿寨居群次之，达40.0cm以上，最小为大新居群，仅为15.3cm；叶宽最大为东兰居群，最大值为8cm，鹿寨、巴马、阳朔、田阳居群次之，达6.0cm以上，大新居群最小，仅为2.3cm。箭秆竹秆高、叶长、叶宽性状差异小，阳朔居群秆高最大，最大值为3.6m，灌阳居群最小，仅为2.4m，其余均达3.0m以上，临桂居群叶长最大，最大值为45.7cm，平乐居群最小，为34.7cm，贡城居群叶宽最大，最大为8.1cm，临桂居群最小，仅为6.2cm。

表3 芸香竹、单枝竹、箭秆竹的表型变异

竹种	采样地点	秆高(m)				叶长(cm)				叶宽(cm)			
		最小值	最大值	平均值	标准差	最小值	最大值	平均值	标准差	最小值	最大值	平均值	标准差
芸香竹	凭祥	1.100	8.500	2.700	0.300	14.700	38.200	26.233	0.445	1.900	7.100	3.570	0.097
	马山	1.600	2.600	1.967	0.133	22.100	31.800	28.200	0.736	3.900	6.400	5.133	0.194
	龙州	1.100	3.600	2.400	0.096	18.600	38.400	30.100	0.552	2.600	7.200	4.678	0.129
	大新	1.000	1.700	1.167	0.037	14.100	21.600	15.533	0.789	1.000	4.500	3.867	0.166
	天等	1.400	3.100	2.300	0.072	14.700	41.600	28.571	0.336	1.200	4.200	2.470	0.083
单枝竹	马山	1.600	2.500	2.300	0.169	19.700	35.200	27.533	1.137	3.300	5.400	4.175	0.165
	田东	1.500	2.600	2.100	0.149	21.000	28.000	23.300	1.105	3.300	4.800	4.167	0.185
	平果	0.400	3.500	2.050	0.157	22.800	29.600	26.600	0.845	2.700	3.700	3.300	0.116
	大新	0.400	0.800	0.600	0.058	10.200	15.300	13.500	0.895	1.200	2.300	1.700	0.186
	东兰	0.600	3.600	2.157	0.072	24.600	43.000	31.200	1.221	1.900	8.000	4.050	0.380
	田阳	1.500	3.400	2.167	0.181	24.600	40.700	30.300	1.513	2.100	8.000	3.150	0.641
	巴马	1.400	3.400	2.273	0.048	10.300	39.000	26.200	0.948	0.800	6.400	2.570	0.151
	靖西	0.400	3.400	2.350	0.181	15.500	27.000	18.833	1.141	0.800	3.000	1.700	0.179
	柳州	0.600	1.900	1.375	0.111	10.200	20.500	15.367	0.954	1.200	2.300	1.567	0.113
	鹿寨	0.600	3.100	1.975	0.115	23.800	43.600	31.850	0.798	2.600	6.100	4.450	0.184
	宜州	1.400	2.900	2.175	0.074	12.200	44.100	30.500	1.563	0.500	5.500	3.267	0.215
	河池	0.400	3.400	2.300	0.083	7.400	32.700	23.100	0.438	0.200	3.900	2.468	0.052
	阳朔	1.600	2.600	2.300	0.105	18.400	36.100	25.900	1.671	3.000	7.300	5.100	0.441
	大化	1.600	2.600	2.300	0.095	16.800	21.600	19.300	0.504	1.500	2.400	1.980	0.085
箭秆竹	兴安	0.600	3.100	1.925	0.100	19.800	38.600	29.133	0.500	3.400	7.100	4.800	0.110
	全州	0.600	3.100	2.033	0.074	3.000	39.700	28.500	0.680	3.600	7.800	5.433	0.131
	灌阳	0.600	2.400	1.357	0.102	1.100	36.600	25.700	2.368	3.800	8.000	6.000	0.232
	贡城	2.300	3.600	3.075	0.078	17.900	40.600	27.633	0.668	3.100	8.100	5.360	0.136
	平乐	1.100	3.600	2.500	0.155	20.400	34.700	27.200	0.655	3.800	7.600	5.200	0.176
	荔浦	1.400	3.100	2.189	0.061	17.900	39.300	28.733	0.659	2.500	7.900	4.800	0.177
	阳朔	1.600	3.600	2.580	0.081	14.600	36.100	25.300	0.619	2.700	7.300	4.233	0.163
	临桂	0.900	3.400	2.067	0.110	20.100	45.700	30.150	0.854	2.700	6.200	4.760	0.133

3 结论与讨论

3.1 资源分布

广西单枝竹属竹种资源丰富，分布范围较广，集中分布在21°57′~25°57′N，111°06′~106°29′E，海拔100~800m的岩溶地貌区，呈现箣秆竹、单枝竹、芸香竹自北向南天然分布规律，单枝竹、芸香竹部分分布区存在交错现象。根据刘金荣[15]等人对广西岩溶地貌类型演化的研究，竹种集中分布在桂东北—桂中—桂西南峰林平原和桂西—桂西南峰丛洼地类型区。箣秆竹中心分布区为24°34′~25°57′N，110°31′~111°03′E，海拔179~587m的桂东北山间盆地谷地岩溶地貌区（峰丛洼地及峰林平原区）；芸香竹中心分布区为21°57′~23°41′N，106°47′~108°12′E，海拔215~663m的桂西南低山中山地岩溶地貌区（峰丛洼地区）；单枝竹中心分布区为22°45′~25°10′N，106°29′~110°53′E，海拔100~587m的桂西北高原斜坡岩溶地貌区、桂中盆地岩溶地貌区（峰丛洼地及丘陵谷地区）、桂西、中低山地岩溶地貌区（峰丛洼地区）、桂西南低、中山地岩溶地貌区（峰丛洼地区）。徐明锋等[16]经研究发现，全磷、全钾及土壤有机质是天然林植物种类组成与分布的限制因子，与南方砖红壤缺磷和有机质的状况相吻合。箣秆竹、单枝竹、芸香竹的分布范围可一定程度上反映出该区域喀斯特地貌的土壤养分状况。喀斯特山区峰峦起伏，形成天然屏障，以及分布区内河流、耕地等自然隔离，推测单枝竹属植物在中南半岛喀斯特地区应有资源分布。

3.2 秆型变异

单枝竹和芸香竹在其分布区内存在多种秆型混生现象，与谷志佳[17]发现的巨龙竹在分布区内由南向北秆型呈现弯曲型、混生型、通直型变化相似，存在多种秆型，因单枝竹为小型竹种，秆型还存在攀援型和匍匐型，其生长受分布区气候与立地环境条件的影响，秆型表现为较强的可塑性。杨佳俊等[18]在对水竹天然林的研究中也发现，在不同光照环境下，水竹形态特征表现出明显的适应性变化，不同坡向间差异显著，证明立地条件是形态可塑性的主导因子之一。而庄若楠等[19]经研究发现，毛竹秆长、总节数、去梢后长度等秆型特征都是较稳定的指标，受立地土壤养分水平的影响较小，秆形的分化可能与海拔、坡向、坡度等立地条件更为密切。分布在石灰岩山地天然次生林下不同地点的单枝竹、芸香竹秆型分化明显。李鹏等进行了巨龙竹种不同变异类型的RAPD和通直型不同地理种源遗传分化的ISSR分析，研究发现秆型可能受遗传基因调控[20-21]。单枝竹属秆型变化的遗传因子值得进一步研究。

3.3 叶型变异

单枝竹属种内、种间叶型差异显著。单枝竹叶鞘具长纤毛，而其他种叶鞘光滑无纤毛；芸香竹叶下表皮起白霜，具软毛，响子竹叶基圆形，不抱茎，叶下表皮浅绿色，光滑无毛；叶鞘和叶舌均长有纤毛的为箣竿竹，叶鞘和叶舌无纤毛的为北部湾单枝竹；而小花单枝竹叶鞘、叶耳不显著。叶表型性状受遗传效应和气候、海拔、经纬度、土壤、水分等环境效应的影响，是揭示天然居群遗传变异及其格局的有效途径[22,23]。广西单枝竹属种间差异显著，箣竿竹叶长、叶宽最大，为宽大披针形；而芸香竹叶长，为短、宽披针形；单枝竹叶长居中，叶宽最窄，呈狭长披针形。种内不同居群间受气候、纬度和海拔等环境因素影响，叶表型变异显著。芸香竹等居群叶长最长，龙州、凭祥居群次之，大新居群最小；龙州居群叶宽最大，凭祥、马山居群次之，大新居群最小。单枝竹叶长最大为宜州居群，田阳、东兰、鹿寨居群次之，最小为大新居群；叶宽最大为东兰居群，鹿寨、巴马、阳朔、田阳居群次之，大新居群最小。箣竿竹竿高、叶长、叶宽性状差异小，临桂居群叶长最大，平乐居群最小，恭城居群叶宽最大，临桂居群最小。从单枝竹属秆型、叶型可见，其天然居群遗传多样性丰富，有待试验证实。

3.4 前景展望

单枝竹属竹种现已知广泛分布于广西、广东、贵州、云南等省、区的喀斯特地貌山区[4,6]，是喀斯特地貌山区长势优良竹种。其根系发达，紧附石岩山地土壤，少数能扎入岩石缝隙，是喀斯特山区困难立地防护和治理的优良竹种，生态服务功能价值显著，该属是否为竹类植物在石灰岩地区分化出来的特有属具有科学研究意义，该属植物对喀斯特地区生态恢复具有实际意义。该属竹种经济价值较明显，对于改善山区生产生活有积极作用。其竹叶富含蛋白质，是牲畜重要饲料；其竹秆纤细、柔韧，常用以编制、造纸。箣秆竹因竹叶清香民间常用于传统蒸煮粽子等。

喀斯特地区单枝竹属资源及其表型变异丰富，具有种质挖掘的潜力。综合表型、生理和现代分子生物学技术，借鉴经济林包括竹类已有的种质资源遗传分析成果[24-38]等，可以选育和引种生态、经济和产业功能更加优良的种质（种源、家系、无性系、品种）。

参考文献

[1]郭起荣，张莹，冉洪，等．竹子基因调查分析报告[J]．世界竹藤通讯，2015，13(2)：11-14.

[2]郭起荣，廉超，冯云，等．中国竹种名称的互通[J]．竹子研究汇刊，2015，34(1)：15-20.

[3]国家林业局．中国森林资源报告(2009-2013)[M]．北京：中国林业出版社，2014.

[4]XIA N H. A study of the genus Bonia(Gramineae：Bambusoideae)[J]. Kew Bull，1996，51(3)：565-569.

[5]Flora of China Editorial Committee. Flora of China(vol 22)[M]. Beijing：Science Press and Missouri Botanical Garden Press，2006.

[6]李德铢，郭振华．云南竹亚科一些属种的增订[J]．云南植物研究，2000，22(1)：43-46.

[7]史军义，易同培，王海涛．贵州竹子新报道[J]．植物研究，2007，28(6)：645-650.

[8]董蕾，曹洪麟，叶万辉，等．5种喀斯特生境植物叶片解剖结构特征[J]．应用与环境生物学报，2011，17(5)：747-749.

[9]李秀兰，林汝顺，冯学琳，等．中国部分丛生竹类染色体数目报道[J]．植物分类学报，2001，39(5)：433-442.

[10]CHRIS MA S，LI D Z，XIA N H. New combinations for Chinese bamboos(Poaceae，Bambuseae)[J]. Novon，2005，15(4)：599-601

[11]HUI X S，SU P G，MING Y J. The evolution and utility of ribosomal ITS sequences in*Bambusinae* and related species：divergence，pseudogenes，and implications for phylogeny[J]. Journal of genetics，2012，91(2)：1-11.

[12]张玉霄，许宇星，马朋飞，等．竹亚科叶绿体DNA条形码筛选[J]．植物分类与资源学报，2013，35(6)：743-750.

[13]刘光金，李洪果，卢立华，等．石灰岩山地芸香竹分株育苗基质和时间的选择[J]．经济林研究，2014，32(4)：144-146.

[14]邹跃国．福建华安竹类植物园种质资源异地保存与分析[J]．世界竹藤通讯，2006，4(4)：23-26.

[15]刘金荣，黄国彬，黄学灵，等．广西区域热带岩溶地貌不同类型的演化浅议[J]．中国岩溶，2001，20(4)：247-252.

[16]徐明锋，胡砚秋，李文斌，等．土壤养分对亚热带天然林物种分布的影响[J]．中南林业科技大学学报，2014，34(9)：91-97.

[17]谷志佳，杨汉奇，孙茂盛，等．巨龙竹资源分布特点及其开花结实现象[J]．林业科学研究，2012，25(1)：1-5.

[18]杨佳俊，董文渊，唐海龙，等．坡向对水竹天然林形态可塑性的影响[J]．西南林业大学学报，2015，35(3)：20-24.

[19]庄若楠，金爱武．施肥对毛竹秆型特征的影响[J]．中南林业科技大学学报，2013，33(1)：80-84.

[20]李鹏，杜凡，普晓兰，等．巨龙竹种下不同变异类型得RAPD分析[J]．云南植物研究，2004，26(3)：290-296.

[21]杨汉奇，阮桢媛，田波，等．通直型巨龙竹资源不同地理种源遗传分化的ISSR分析[J]．浙江林学院学报，2010，27(1)：81-86.

[22]曾杰，郑海水，甘四明，等．广西西南桦天然居群的表型变异[J]．林业科学，2005，3(2)：59-65.

[23]李文英，顾万春．蒙古栎天然群体表型多样性研究[J]．林业科学，2005，41(1)：49-56.

[24]骆文华，代文娟，刘建，等．广西火桐自然种群和迁地保护种群的遗传多样性比较[J]．中南林业科技大学学报，2015，35(2)：66-71.

[25]左慧．野生平榛种质资源的形态特征变异分析[J]．中南林业科技大学学报，2015，35(7)：31-36.

[26]郭起荣，任立宁，牟少华，等．毛竹种质分子鉴别SRAP，AFLP，ISSR联合分析[J]．江西农业大学学报，2010，32(5)：0982-0986.

[27]HSIAO J Y，RIESEBERG L H. Population Genetic Structure of Yushania niitakayamensis(Bambusoideae，Poacese)in Taiwan[J]. Molecular Ecology，1994，3(3)：201-208.

[28]廖国华．毛竹黄、白笋个体的RAPD指纹图谱分析[J]．福建农业学报，2003，18(1)：46-49.

[29]ISAGI Y，SHIMADA K，KUSHIMA H，et al. Clonal structure and flowering traits of a bamboo[*Phyllostachys pubescens*(Mazel)Ohwi]stand grown from a simultaneous flowering as revealed by AFLP analysis[J]. Molecular Ecology，2004，3(7)：2017-2021.

[30]周建梅．开花毛竹的基因流及F_1代的种质发掘[D]．北京：中国林业科学研究院，2014.

[31]师丽华，杨光耀，林新春，等．毛竹种下等级的RAPD研究[J]．南京林业大学学报，2002，5(3)：65-68.

[32]任艳军．刚竹属(*Phyllostachys*)部分观赏竹种间亲缘关系的RAPD标记研究[D]．雅安：四川农业大学，2004.

[33]阮晓赛．毛竹种源及栽培变种遗传变异的AFLP和ISSR分析[D]．杭州：浙江林学院，2008.

[34]郭小勤，李犇，阮晓赛，等．利用 ACGM 分子标记研究 10 个毛竹不同栽培变种的遗传多样性[J]．林业科学，2009，45(4)：28-32.
[35]刘贯水．毛竹 SSR 位点引物开发及部分竹种系统学分析[D]．北京：中国林业科学研究院，2010.
[36]杨茹，孙志娟，项艳．刚竹属 14 个品种遗传多样性的 ISSR 分析[J]．竹子研究汇刊，2010，29(4)：11-14.
[37]ZHAO H，YANG L，PENG Z，et al. Developing genome-wide microsatellite markers of bamboo and their applications on molecular marker assisted taxonomy for accessions in the genus *Phyllostachys*[J]. Scientific reports，2015，5：1-10，doi：10.1038/srep08018
[38]冉洪，张莹，胡陶，等．经济树种全基因组测序成果要报[J]．经济林研究，2015，33(2)：149-157.

[原载：经济林研究，2016，34(01)]

氮、磷、钾对灰木莲幼苗生长和光合作用的影响

陈 琳 卢立华 蒙彩兰

（中国林业科学研究院热带林业实验中心，广西凭祥 532600）

摘 要 采用 $L_{18}(3^7)$ 正交设计，开展灰木莲幼苗 N、P、K 配方施肥研究，依据幼苗生长表现和光合作用筛选其适宜 N、P、K 施肥配方。结果表明：处理 8（N0.668g/株、$P_2O_5$1g/株和 K_2O0.668g/株）和 18（N0.668g/株、$P_2O_5$2g/株和 K_2O0.334g/株）的生长和光合表现最优，其苗高、地径、总生物量、叶片数、净光合速率、气孔导度和蒸腾速率分别比对照高 314%、88%、598%、282%、83%、56%和 37%，而根冠比和胞间 CO_2 浓度则比对照低 70%和 14%。N 和 P 显著影响灰木莲幼苗的生长和光合作用，而 K 对灰木莲幼苗的叶片数和光合作用具有显著影响。影响灰木莲苗高和根冠比的因素次序为 N>P>K，影响其他生长指标和光合参数的因素次序为 P>N>K。

关键词 灰木莲；配方施肥；生长表现；光合作用

Combined Effects of Nitrogen, Phosphorus and Potassium on Growth and Photosynthesis of *Magnoliaceae glanca* Seedlings

CHEN Lin, LU Lihua, MENG Cailan

(*Experimental Center of Tropical Forestry*, *Chinese Academy of Forestry*, *Pingxiang* 532600, *Guangxi*, *China*)

Abstract: A fertilization trail by $L_{18}(3^7)$ orthogonal design was conducted to determine the optimal nitrogen, phosphorus and potassium requirements of *Magnoliaceae glance* seedlings in terms of the growth and photosynthesis. The best growth and photosynthesis rates were found in treatments No. 8 (N 0.668g/seedling, P_2O_5 1g/seedling and K_2O 0.668g/seedling) and No. 18 (N 0.668g/seedling, P_2O_5 2g/seedling and K_2O 0.334g/seedling), in which the seedling height, root collar diameter, total bi/omass, number of the leaves, net photosynthetic rate, stomatal conductance and transpiration rate were 314%, 88%, 598%, 282%, 83%, 56% and 37% higher than those of the control, respectively. However, the root and shoot ratio and intercellular CO_2 concentration were 70% and 14% lower than those of the control, separately. Nitrogen and phosphorus had profound effects on all growth and photosynthesis indexes of *M. glance* seedlings, while potassium only had a significant effect on the number of leaves and photosynthesis. The rank order of nutrient elements affecting the height and root and shoot ratio was N>P>K. For other growth and photosynthesis indexes, the rank order was P>N>K.

Key words: *Magnoliaceae glanca*; fomula fertilization; growth performance; photosynthesis

平衡施肥是苗木培育的关键技术之一[1]，适宜施肥配方不仅有利于苗木生长和生理效应，而且能够促进幼林初期生长[2-5]。赵燕等[6]采用 $L_9(3^4)$ 正交设计，研究了 N、P、K 配施对毛白杨（*Populus tomentosa*）杂种无性系苗木光合生理的影响，发现施肥能显著提高毛白杨幼苗的光合效率，最佳施肥处理是 N7mg/株、P_2O_5 2.25mg/株，K_2O0.5g/株。胡厚臻等[7]采用正交设计，研究了 N、P、K 对刨花润楠（*Machilus pauhoi*）幼苗生长和光合生理的影响，发现施肥量为 N240mg/株、P36mg/株、K162mg/株对刨花润楠的生长和生理指标的促进作用最显著。蔡雅桥[8]以 2 年生钩栗（*Castanopsis tibetana*）苗为材料，采用 N、P、K 三因素四水平正交试验设计，研究了配方施肥对钩栗苗木生长和生理特性的影响，发现当 N、P、K 施肥量分别为 6g/株、6g/株和 3g/株时，钩栗苗木的生长表现最优。由于不同树种、苗龄、或者同一树种不同无性系对 N、P、K 的响应不一致，因此有必要针对特定树种、苗龄或无性系开展其施

肥配方研究。

灰木莲(*Magnoliaceae glanca*)原产越南和印度尼西亚，我国广东、海南和广西于20世纪60年代初开始引种。因其树形美观，被广泛用于庭园观赏，此外其木材纹理细致、易加工，亦被用于建筑、家具制作和胶合板生产等[9]。目前，国内开展了灰木莲苗木的养分分布格局以及叶面施肥研究[10,11]，然而对灰木莲幼苗N、P、K施肥配方研究鲜见报道。正交试验设计具有减少试验次数，节约人力物力，在因素水平偏选的情况下可逐步优化，最终达到优化苗木施肥方案之目的[12]。因此，本研究采用$L_{18}(3^7)$正交设计，开展灰木莲幼苗N、P、K施肥配方研究，揭示灰木莲幼苗生长和光合作用对N、P、K的响应，确定适宜N、P、K施用量，为灰木莲幼苗养分管理及壮苗培育提供科学依据。

1　材料与方法

1.1　试验材料

以灰木莲实生苗为供试材料，在中国林业科学研究院热带林业实验中心温室内开展灰木莲苗木N、P、K施肥研究。育苗用种为通过人工授粉获得的种子，育苗基质为黄心土、沤制树皮、森林表土的混合基质(7∶1.5∶1.5，体积比)，育苗容器为17.5cm×11cm×12cm(上口径×下口径×高)塑料盆，为了防止水肥流失，盆内套有双层白色塑料袋。2013年2月28日，选择高约4cm且生长健康苗木移栽至事先装好基质的塑料盆中，初始浇水量为田间持水量的90%，此后控制基质含水量在最大田间持水量的60%~90%，最大田间持水量的确定参见Timmer & Armstrong的方法[13]。

1.2　试验设计

采用N、P、K3因素3水平$L_{18}(3^7)$正交设计，18个处理，以不施肥作为对照(表1)，重复3次，每个小区9盆，每盆1株，共计486株。N为尿素(N≥46.4%)，P为钙镁磷肥(P_2O_5≥18%)，K为氯化钾(K_2O≥60%)。P作为基肥一次性施入基质中，移苗后约1个月，开始追施N和K肥，每10d施1次，共18次。第1~9次(2013年3月28至6月16日)，每次施入尿素和氯化钾分别为0、32.7、65.4mg/株和0、25.3、50.6mg/株；第10~13次(6月26日至7月26日)，每次施入尿素和氯化钾分别为0、65.4、130.8mg/株和0、50.6、101.2mg/株；第14~18次(8月5日至9月14日)，每次施入尿素和氯化钾分别为0、32.7、65.4mg/株和0、25.3、50.6mg/株。

表1　$L_{18}(3^7)$正交设计

处理	水平组合	N/(g/株)	P_2O_5/(g/株)	K_2O/(g/株)
1(CK)	$N_1P_1K_1$	0	0	0
2	$N_1P_2K_2$	0	1	0.334
3	$N_1P_3K_3$	0	2	0.668
4	$N_2P_1K_1$	0.334	0	0
5	$N_2P_2K_2$	0.334	1	0.334
6	$N_2P_3K_3$	0.334	2	0.668
7	$N_3P_1K_2$	0.668	0	0.334
8	$N_3P_2K_3$	0.668	1	0.668
9	$N_3P_3K_1$	0.668	2	0
10	$N_1P_1K_3$	0	0	0.668
11	$N_1P_2K_1$	0	1	0
12	$N_1P_3K_2$	0	2	0.334
13	$N_2P_1K_2$	0.334	0	0.334
14	$N_2P_2K_3$	0.334	1	0.668
15	$N_2P_3K_1$	0.334	2	0
16	$N_3P_1K_3$	0.668	0	0.668
17	$N_3P_2K_1$	0.668	1	0
18	$N_3P_3K_2$	0.668	2	0.334

1.3　生长调查和光合测定

试验开始前，随机选取18株平均苗，测定苗木的初始生物量为0.48g/株。2013年10月3日，即施肥结束约3周后，调查苗高、地径和叶片数。另外每个小区随机选取3株健康平均木，于105℃杀青15min，70℃烘干48h至恒重后，称地下和地上部分生物量。

每个小区随机选择3株健康平均木，每株选取1片健康完满叶，利用Li-6400光合仪测定叶片的净光合速率、气孔导度、胞间CO_2浓度和蒸腾速率，测定期间平均气温33℃，人工光强为1200μmol/m^2·s，CO_2浓度为380μmol/mol。

1.4　统计分析

应用SPSS 16.0软件的One-Way ANOVA和Gene-aral Linear Model程序对灰木莲幼苗的生长指标和光合参数进行单因素和多因素方差分析，揭示不同施肥处理对幼苗生长和光合作用的影响以及N、P和K主效应，并对差异显著的指标进行Duncan多重比较。由于N、P、K 3因素依次安排在$L_{18}(3^7)$正交设计表的前3列，因此无法对元素之间交互效应进行统计分析。

2 结果与分析

2.1 氮、磷、钾对灰木莲幼苗生长的影响

从不同施肥处理灰木莲幼苗的生长表现(表2)来看，处理8的生长表现最优，其次是处理18和17。处理8的苗高、地径、地下部分生物量、地上部分生物量、总生物量和叶片数最大，分别比对照增加了315%、84%、244%、935%、592%和300%($P<0.05$)，而根冠比比对照减少了67%($P<0.05$)；处理18和17的苗高、地径、地下部分生物量、地上部分生物量、总生物量和根冠比与处理8均无显著差异($P>0.05$)，但叶片数分别比处理8减少了9%和18%($P<0.05$)。

N和P显著影响灰木莲幼苗的高、地径、地下部分生物量、地上部分生物量、总生物量和叶片数($P<0.01$)，而K仅显著影响幼苗的叶片数($P<0.01$)。对各因素不同水平的均值(K_1、K_2和K_3)进行多重比较，得到灰木莲幼苗生长的最优水平组合为$N_3P_2K_3$和$N_3P_2K_1$(表3)，即处理8和处理17，与灰木莲幼苗生长实际最佳处理组合基本一致。虽然处理18($N_3P_3K_2$)的各项生长指标亦相对较优，但是该处理并未出现在最优水平组合中，这是因为最优水平组合仅考虑了N、P、K主效应，未考虑到因素之间的交互效应。

比较各因素极差R值，影响灰木莲幼苗地径、地下部分生物量、地上部分生物量、总生物量和叶片数的因素次序为P>N>K，而影响苗高和根冠比的因素次序为N>P>K(表3)。

表2 不同施肥处理下灰木莲幼苗的生长表现

处理	苗高(cm)	地径(mm)	地下部分生物量(g/株)	地上部分生物量(g/株)	总生物量(g/株)	根冠比	叶片数
1	22.48(±0.65)g	6.39(±0.11)g	2.72(±0.12)cd	2.77(±0.48)e	5.49(±0.02)hi	1.04(±0.51)a	11(±0.29)de
2	33.33(±1.10)f	7.67(±0.20)de	3.81(±0.43)c	5.21(±1.02)de	9.01(±0.01)gh	0.75(±1.44)ab	13(±0.57)de
3	32.62(±1.33)f	7.67(±0.29)de	2.71(±0.41)cd	4.06(±0.12)de	6.77(±0.03)ghi	0.68(±0.30)abc	12(±0.64)de
4	44.67(±2.21)e	7.19(±0.15)ef	3.36(±0.20)cd	6.82(±0.37)d	10.18(±0.01)g	0.49(±0.53)bcd	12(±0.87)de
5	84.42(±2.47)b	10.76(±0.17)c	8.97(±0.71)ab	22.97(±1.08)c	31.94(±0.03)cde	0.39(±0.76)def	31(±1.53)c
6	78.87(±2.53)b	10.97(±0.27)c	7.71(±0.27)ab	20.56(±1.66)c	28.27(±0.01)ef	0.38(±1.93)def	30(±1.79)c
7	43.57(±1.75)e	6.61(±0.18)fg	1.79(±0.24)d	2.80(±0.31)e	4.58(±0.03)i	0.67(±0.08)abc	5(±0.24)f
8	93.30(±3.00)a	11.74(±0.22)ab	9.35(±1.52)a	28.66(±2.15)a	38.01(±0.06)ab	0.34(±1.51)f	44(±2.61)a
9	65.45(±3.36)c	10.74(±0.31)c	7.16(±0.58)b	20.54(±1.41)c	27.70(±0.03)f	0.35(±1.09)ef	31(±2.55)c
10	23.96(±1.02)g	6.47(±0.15)g	2.08(±0.04)cd	2.85(±0.33)e	4.92(±0.01)i	0.75(±0.30)abc	11(±0.45)de
11	31.04(±0.98)f	7.45(±0.21)e	3.72(±0.38)c	6.02(±0.84)de	9.74(±0.01)g	0.62(±1.20)abc	14(±0.56)d
12	37.35(±1.38)f	8.12(±0.15)d	3.62(±0.08)cd	5.74(±0.56)de	9.36(±0.02)gh	0.65(±0.48)abc	15(±0.41)d
13	51.89(±1.43)d	7.18(±0.16)ef	3.00(±0.22)cd	5.76(±0.76)de	8.76(±0.01)gh	0.53(±0.98)abc	9(±0.63)ef
14	81.91(±3.57)b	11.20(±0.19)bc	9.37(±0.34)a	20.99(±1.71)c	30.36(±0.02)def	0.45(±1.81)cde	29(±1.85)c
15	82.91(±2.71)b	10.79(±0.24)c	8.89(±1.34)ab	23.27(±0.03)bc	32.15(±0.04)cd	0.38(±1.36)def	27(±2.24)c
16	45.40(±1.86)e	6.67(±0.20)fg	1.75(±0.30)d	3.04(±0.49)de	4.78(±0.01)i	0.58(±0.74)abc	5(±0.42)f
17	91.52(±2.26)a	11.20(±0.18)bc	7.90(±0.28)ab	26.68(±1.06)ab	34.58(±0.00)bc	0.30(±1.34)f	36(±2.64)b
18	92.75(±3.56)a	12.25(±0.21)a	8.70(±0.29)ab	29.99(±2.73)a	38.68(±0.03)a	0.29(±2.53)f	40(±2.03)b

注：数值为平均值(±标准误)，同一列中含相同小写字母表示差异不显著($P>0.05$)，不含相同小写字母表示差异显著($P<0.05$)，下同。

表3 不同施肥处理灰木莲幼苗生长的极差分析和多重比较

因素	指标	苗高(cm)	地径(mm)	地下部分生物量(g/株)	地上部分生物量(g/株)	总生物量(g/株)	根冠比	叶片数
N	K_1	30(±1.12)B	7.32(±0.10)B	3.11(±0.36)B	4.44(±1.19)B	7.55(±1.46)B	0.75(±0.03)A	13(±0.76)C
	K_2	71(±1.14)A	9.66(±0.11)A	6.88(±0.36)A	16.73(±1.19)A	23.61(±1.46)A	0.44(±0.03)B	23(±0.78)B
	K_3	72(±1.15)A	9.86(±0.11)A	6.11(±0.36)A	18.62(±1.19)A	24.72(±1.46)A	0.42(±0.03)B	27(±0.81)A
	R	42	2.54	3.77	14.18	17.17	0.33	14

（续）

因素	指标	苗高(cm)	地径(mm)	地下部分生物量(g/株)	地上部分生物量(g/株)	总生物量(g/株)	根冠比	叶片数
P	K_1	39(±1.10)C	6.75(±1.10)B	2.45(±0.36)B	4.01(±1.19)B	6.45(±1.46)B	0.68(±0.03)A	9(±0.77)B
	K_2	69(±1.17)A	9.97(±1.11)A	7.19(±0.36)A	18.42(±1.19)A	25.61(±1.46)A	0.48(±0.03)B	27(±0.80)A
	K_3	65(±1.14)B	10.12(±1.11)A	6.47(±0.36)A	17.36(±1.19)A	23.82(±1.46)A	0.46(±0.03)B	26(±0.78)A
	R	30	3.37	4.74	14.41	19.16	0.22	18
K	K_1	57(±1.13)	8.97(±0.10)	5.63(±0.36)	14.35(±1.19)	19.97(±1.46)	0.53(±0.03)	22(±0.78)AB
	K_2	57(±1.12)	8.76(±0.10)	4.98(±0.36)	12.08(±1.19)	17.06(±1.46)	0.55(±0.03)	19(±0.77)B
	K_3	59(±1.15)	9.11(±0.11)	5.49(±0.36)	13.36(±1.19)	18.85(±1.46)	0.53(±0.03)	22(±0.80)A
	R	2	0.35	0.65	2.27	2.91	0.02	3

注：数值为平均值(±标准误)，同一列中含相同大写字母表示差异不显著($P>0.05$)，不含相同大写字母表示差异极显著($P<0.01$)，K_1、K_2、K_3代表各水平均值，R 表示水平均值的极差，下同。

2.2 氮、磷、钾对灰木莲幼苗光合作用的影响

从不同施肥处理灰木莲幼苗的光合作用来看(表4)，处理8的光合表现最优，其次是处理14和处理18。处理8的净光合速率、气孔导度和蒸腾速率分别比对照增加了100%、63%和40%($P<0.05$)，而胞间CO_2浓度比对照减少了14%($P<0.05$)；处理14的净光合速率比处理8减少了15%($P<0.05$)，而气孔导度和蒸腾速率分别比处理8增加了8%和11%($P<0.05$)，胞间CO_2浓度与处理8无显著差异($P>0.05$)；处理18的气孔导度、胞间CO_2浓度和蒸腾速率与处理8均无显著差异($P>0.05$)，但净光合速率比处理8减少了17%($P<0.05$)。

表4 不同施肥处理下灰木莲幼苗的光合作用

处理	净光合速率(μmol/m²·s)	气孔导度(μmol/m²·s)	胞间CO_2浓度(μmol/m²·s)	蒸腾速率(μmol/m²·s)
1	4.38(±0.12)gh	0.08(±0.00)gh	273.53(±2.23)cdef	1.58(±0.03)gh
2	4.53(±0.10)g	0.08(±0.00)gh	263.16(±2.76)fg	1.63(±0.05)fg
3	3.93(±0.09)h	0.07(±0.00)hij	256.41(±3.52)g	1.39(±0.05)hi
4	2.85(±0.11)i	0.08(±0.00)hi	290.13(±4.11)b	1.54(±0.07)gh
5	6.27(±0.23)cd	0.12(±0.01)cd	263.61(±2.28)fg	1.91(±0.10)de
6	6.21(±0.16)d	0.11(±0.01)cde	256.71(±3.14)g	1.90(±0.07)de
7	1.99(±0.13)j	0.04(±0.00)k	280.78(±5.07)c	0.89(±0.04)k
8	8.78(±0.15)a	0.13(±0.01)ab	234.59(±3.93)ij	2.21(±0.09)b
9	7.29(±0.24)b	0.10(±0.00)def	230.17(±2.85)j	1.83(±0.08)def
10	2.82(±0.08)i	0.06(±0.00)j	278.58(±2.70)cd	1.14(±0.06)j
11	5.02(±0.19)f	0.09(±0.00)fg	265.53(±2.60)efg	1.75(±0.06)efg
12	3.28(±0.09)i	0.07(±0.00)ij	275.24(±2.80)cde	1.21(±0.04)ij
13	2.82(±0.13)i	0.05(±0.00)j	269.67(±2.36)def	1.17(±0.06)j
14	7.46(±0.10)b	0.14(±0.00)a	266.74(±2.64)efg	2.46(±0.06)a
15	6.71(±0.10)c	0.10(±0.00)ef	245.56(±1.68)h	1.99(±0.06)cd
16	2.23(±0.20)j	0.07(±0.01)hij	301.54(±4.33)a	1.33(±0.08)ij
17	5.71(±0.29)e	0.10(±0.01)def	243.77(±4.82)hi	1.88(±0.12)de
18	7.29(±0.29)b	0.12(±0.01)bc	238.31(±4.15)hij	2.13(±0.12)bc

N、P和K显著影响灰木莲幼苗的净光合速率、气孔导度、胞间CO_2浓度和蒸腾速率($P<0.01$)。对各因素不同水平的均值(K_1、K_2和K_3)进行多重比较，得到灰木莲幼苗光合作用的最优水平组合为$N_2P_2K_1$和$N_2P_2K_3$(表5)。$N_2P_2K_3$即处理14，其苗木实际光合表现亦较优，而由于正交设计为不完全设计，$N_2P_2K_1$并未出现在已设置的处理中，因此该组合的光合表现有待今后验证。处理8($N_3P_2K_3$)和18($N_3P_3K_2$)的实际光合表现较优，但是它们并未出现在最优水平组合中，这是因为最优水平组合未考虑到因素之间的交互效应。

比较各因素极差*R*值，影响灰木莲幼苗净光合速率、气孔导度、胞间CO_2浓度和蒸腾速率的因素从大到小依次为P>N>K(表5)。

表5　不同施肥处理灰木莲幼苗光合作用的极差分析和多重比较

因素	光合指标	净光合速率(μmol/m²·s)	气孔导度(mol/m²·s)	胞间CO_2浓度(μmol/mol)	蒸腾速率(mmol/m²·s)
N	K_1	3.99(±0.08)B	0.07(±0.00)B	268.74(±1.44)A	1.45(±0.03)C
	K_2	5.39(±0.08)A	0.10(±0.00)A	265.40(±1.44)A	1.83(±0.03)A
	K_3	5.55(±0.08)A	0.10(±0.00)A	254.86(±1.44)B	1.71(±0.03)B
	R	1.56	0.03	13.88	0.38
P	K_1	2.85(±0.08)C	0.06(±0.00)C	282.37(±1.44)A	1.27(±0.03)C
	K_2	6.29(±0.08)A	0.11(±0.00)A	256.23(±1.44)B	1.97(±0.03)A
	K_3	5.79(±0.08)B	0.10(±0.00)B	250.40(±1.44)C	1.74(±0.03)B
	R	3.44	0.05	31.97	0.70
K	K_1	5.33(±0.08)A	0.09(±0.00)A	258.11(±1.44)B	1.76(±0.03)A
	K_2	4.36(±0.08)B	0.08(±0.00)B	265.13(±1.44)A	1.49(±0.03)B
	K_3	5.24(±0.08)A	0.10(±0.00)A	265.76(±1.44)A	1.74(±0.03)A
	R	0.97	0.02	7.65	0.27

3　结论与讨论

N和P显著影响灰木莲的苗高、地径、地下部分、地上部分和总生物量、叶片数，而K仅对灰木莲幼苗的叶片数具有显著影响，可能因为基质中含有的钾已经满足了灰木莲幼苗的对钾的需求，因此在灰木莲苗木的养分管理中，应该根据基质中养分的实际情况，合理搭配N、P、K比例。N、P、K配施处理对毛白杨(*Populus tomentosa*)杂种无性系各项生长指标影响显著，以N影响最大，其次是P，K的影响最小[6]。N、P、K对金叶榆(*Ulmus pumila*)地径的影响显著，其中P的影响最大，其次为N和K[2]。本研究中，苗高和根冠比方面，影响最大的因素是N，其次是P和K；对于其他生长指标而言，影响最大的因素是P，其次是N和K，可见不同树种苗木或同一树种不同指标对N、P、K的响应存在差异。

N、P、K不仅显著影响灰木莲幼苗的生长，而且影响苗木的净光合速率，可能随着N、P和K含量的增加，气孔导度、叶肉细胞光合活性、叶绿素和可溶性糖含量也随之增加，从而促进了苗木的光合作用[6,7]。银杏(*Ginkgo biloba*)苗木的净光合速率、气孔导度、水分利用效率、羧化效率等参数亦随着N、P、K施用量的增加不断增加，但超过一定阈值后反而下降[14]。N、P、K对灰木莲光合作用的影响效应，以P的影响最大，其次是N和K，这与欧洲鹅耳枥(*Carpinus betulus*)盆栽幼苗光合特性的影响因素次序相同[15]，但是与银杏苗木光合速率的影响因素次序不同[14]，可能与树种的光合养分利用特性有关。

综合生长和光合表现，灰木莲幼苗的最佳施肥处理为8($N_3P_2K_3$，N0.668g/株、P_2O_5 1g/株，K_2O 0.668g/株)和18($N_3P_3K_2$，N0.668g/株、P_2O_5 2g/株，K_2O 0.334g/株)，其苗高、地径、总生物量、叶片数、净光合速率、气孔导度、蒸腾速率比对照平均高314%、88%、598%、282%、83%、56%和37%，而根冠比和胞间CO_2浓度则比对照分别低70%和14%。指数施肥比恒量施肥和3段式施肥更加提高北美鹅掌楸(*Liriodendron tulipifera*)和日本落叶松(*Larix leptolepis*)苗木养分利用率[16]。等量施肥、指数施肥和反指数施肥均有效促进欧洲云杉(*Picea abies*)无性系生长、生物量分配、叶绿素含量、光合特性和根系特征值，其中指数施肥效果最佳[17]，因此今后进一步探讨不同施肥方式对灰木莲苗木生长和生理特性影响以提高其养分利用率、降低成本、减少环境污染。

参考文献

[1]左海军，马履一，王梓，等．苗木施肥技术及其发展趋势[J]．世界林业研究，2010，23(3)：39-43.

[2]邓华平，王正超，耿赓．不同氮，磷，钾量对金叶榆容器苗生长效果的比较[J]．中南林业科技大学学报，2009，29(5)：62-66.

[3]肖良俊，宁德鲁，李勇杰，等．配方施肥对八角幼林生长结实的影响[J]．西北林学院学报，2012，27(3)：87-90.

[4]蔡伟建，窦霄，高捍东，等．氮磷钾配比施肥对杂交鹅掌楸幼林初期生长的影响[J]．南京林业大学学报(自然科学版)，2011，35(4)：27-33.

[5]常冀原，张斌，许琪，等．氮、磷、钾对江南油杉形态及生理变化的影响[J]．中南林业科技大学学报，2015，35(5)：46-50.

[6]赵燕，王辉，李吉跃．氮、磷、钾对毛白杨幼苗光合生理的影响[J]．西北林学院学报，2015，30(5)：34-38.

[7]胡厚臻，侯文娟，潘启龙，等．配方施肥对刨花润楠幼苗生长和光合生理的影响[J]．西北林学院学报，2015，30(6)：39-45.

[8]蔡雅桥，许德琼，陈松，等．配方施肥对钩栗生长和生理特性的影响[J]．中南林业科技大学学报，2016，36(3)：33-37，95.

[9]蔡道雄，黎明，郭文福，等．灰木莲容器苗培育基质筛选试验[J]．浙江林业科技，2007，26(5)：36-38.

[10]田雪琴，谭家得，陆耀东，等．6种阔叶树幼苗养分分布格局[J]．广东林业科技，2015，43(11)：32-36.

[11]陈琳，卢立华，蒙彩兰，等．叶面施肥对灰木莲幼苗生长的影响[J]．种子，2013，32(6)：79-81.

[12]张里千．交互作用和多因素最优化—澄清一个历史性误会[J]．数理统计与管理，1991，3：28-35.

[13]TIMMER V R，ARMSTRONG G. Growth and nutrition of containerized *Pinus resinosa* seedlings at varying moisture regimes[J]. New Forests，1989，3(2)：171-180.

[14]吴家胜，张往祥，曹福亮．氮磷钾对银杏苗生长和生理特性的影响[J]．南京林业大学学报(自然科学版)，2003，27(1)：63-66.

[15]钱燕萍，祝遵凌．配方施肥对欧洲鹅耳枥幼苗光合生理的影响[J]．东北林业大学学报，2015，43(11)：32-36.

[16]PARK B B，CHO M S，LEE S W，et al. Minimizing nutrient leaching and improving nutrient use efficiency of *Liriodendron tulipifera* and *Larix leptolepis* in a container nursery system[J]. New Forests，2012，43(1)：57-68.

[17]王燕，晏紫依，苏艳，等．不同施肥方法对欧洲云杉生长生理和根系形态的影响[J]．西北林学院学报，2015，30(6)：15-21.

[原载：西北林学院学报，2017，32(02)]

不同林龄马尾松人工林土壤碳氮磷生态化学计量特征

雷丽群[1,2] 卢立华[1,2] 农 友[1,2] 明安刚[1,2] 刘士玲[1,2] 何 远[1,2]

([1]中国林业科学研究院热带林业实验研究中心，广西凭祥 532600；
[2]广西友谊关森林生态系统国家定位观测研究站，广西凭祥 532600)

摘 要 研究了马尾松从幼龄林至成熟林生长序列中的土壤有机C、全N、全P含量及其生态化学计量特征，以丰富该区域马尾松生态系统生态化学计量学领域的基础研究。以广西凭祥4个林龄(6年、16年、23年、35年)马尾松人工林为研究对象，每个林龄选取3块林分，每个林分内设置一个400㎡的调查样地，按照0~20、20~40、40~60cm三层土层取样，测定其有机C含量、全N含量、全P含量，并计算它们之间的比值。结果表明：4个林龄马尾松人工林0~60cm土壤有机C含量为4.91~10.60g/kg、全N含量为0.58~0.91g/kg、全P含量为0.17~0.23g/kg，土壤有机C含量、全N含量均随林龄的递增先降低后增加，随土层加深持续降低；土壤全P含量在林龄和土层间均无显著性变化；林龄对土壤C∶N、N∶P有极显著的影响($P=0.001$，$P=0.000$)，土层对土壤C∶P、N∶P有显著性影响($P=0.000$，$P=0.014$)。土壤有机C、全N含量从成熟林阶段开始回升；N在不同林龄间和不同土层间的变化是土壤N∶P变化的主要原因；土壤C∶N、C∶P主要受有机C的影响。

关键词 马尾松人工林；林龄；碳氮磷；生态化学计量

Stoichiometry Characterization of Soil C, N and P of *Pinus massoniana* Plantations at Different Aged Stages

LEI Liqun[1,2], LU Lihua[1,2], NONG You[1,2], MING Angang[1,2],
LIU Shiling[1,2], HE Yuan[1,2]

([1]*Experimental Center of Tropical Forestry, Chinese Academy of Forestry, Pingxiang* 532600, *Guangxi, China*;
[2]*Guangxi Youyiguan Forest Ecosystem Research Station, Pingxiang* 532600, *Guangxi, China*)

Abstract: To investigate concentrations and stoichiometric ratio of soil organic C, total soil N and P in the Pinus massoniana plantations from young to mature stages in order to enrich the basic research of Pinus massoniana ecosystem in stoichiometric field. Four Pinus massoniana plantations at different ages (6-, 16-, 23-and 35-year-old) in Pingxiang Guangxi were selected as the research object, three plots in size of $400m^2$ were set up for each of four plantations. Soil samples were collected from 0~20, 20~40, 40~60cm depths respectively for measuring organic C, and total N and total P, and the ratio among them was estimated. The concentration of four plantations were 4.91~10.60g/kg for soil organic C, 0.58~0.91g/kg for total soil N and 0.17~0.27g/kg for total soil P, respectively. The concentrations of soil organic C and total N increased at first and then decreased as stand age increased, while decreased with the increasing soil depths. No significant difference was found in the soil total P between different stand ages and soil depths. Stand ages have statistically significant effect on soil C∶N and N∶P ($P=0.001$, $P=0.000$), soil layers have significant effect on soil C∶P and N∶P ($P=0.000$, $P=0.014$). Soil organic C, total N content began to pick up from the mature stage. The change of N between different stand ages and soil depths was the key factors to change soil N∶P. Soil C∶N, C∶P was mainly affected by organic C.

Key words: *Pinus massoniana* plantation; stand age; CNP; stoichiometry

近10年，生态化学计量学为研究分子、细胞、有机体、种群、生态系统等不同尺度的生物能力平衡和多重化学元素平衡，以及元素平衡对生态交互作用的影响等提供了崭新视角[1]。它着重于强调有机体的主要组成元素(特别是C、N、P)的比值关系，认为元素的比值对有机体的关键特征及其对资源种类和数量的需求均有决定性作用[2]。土壤作为植物养分的主要来源，对植物的生长发育以及生态系统服务功能有着重要的调控作用[3]。探讨森林生态系统土壤的生态化学计量特征，了解森林生态系统养分元素循环过程、养分限制性关系及其对全球气候变化的响应与反馈，对提升森林生态系统服务功能和森林可持续经营均有着重要的意义[4-7]。

土壤CNP比是有机质或其他成分中的C素与N素、P素总质量的比值，是土壤有机质组成和质量程度的一个重要指标，反映了土壤内部CNP的循环特征[8]。曹娟等[9]研究了湖南会同3个林龄(7年、17年、25年)杉木人工林的土壤C、N、P特征，发现随着造林时间增加，土壤有机C、全N、全P含量逐渐增加，土壤的C∶N和C∶P主要受土壤有机C影响。崔宁浩等[10]对5年、14年、39年三个林龄的马尾松人工林的研究亦表明，土壤C、N、P含量随林龄增加而增加，马尾松人工林受N和P的共同限制，但林龄对N、P养分限制的影响并不显著。不同学者的研究结论各异，曾凡鹏等[11]对辽东山区的落叶松群落的研究则表明，土壤C、N、P含量随着林龄的增加而降低，地力呈逐渐衰退的趋势，C∶N和C∶P随林龄变化显著。可见，土壤C、N、P含量及生态化学计量比随林龄变化的趋势，以及养分元素间的限制性关系仍有相当大的不确定性，需要进一步的研究探讨。

马尾松(*Pinus massoniana* Lamb.)是我国南方主要的造林用材树种，在林业生产及森林生态系统中占有及其重要的地位，具有耐干旱、适应性强、速生、优质等特点[12]。本文以位于凭祥市的中国林业科学研究院热带林业实验中心林区的4个林龄(6年、16年、23年、35年)马尾松人工林为研究对象，分析了马尾松从幼龄林至成熟林生长序列中的土壤有机C、全N、全P含量及其生态化学计量比格局，探讨养分元素随林龄的变化趋势及它们之间的制约关系，以丰富该区域马尾松生态系统生态化学计量学领域的基础研究。

1 研究地区与研究方法

1.1 研究区概况

研究区位于广西凭祥市中国林业科学研究院热带林业实验中心林区(106°50′E，22°10′N)，属南亚热带季风型气候区。年平均温度20.8~22.9℃，年降水量1400mm左右，全年降水量呈正态分布，7~8月份达至峰值，高达500~600mm，雨热同季[13]。土壤为花岗岩发育而成的山地红壤，土层厚度达100cm以上，土壤pH均值为4.5左右，属酸性土。

选取成土母岩、地形地貌、土壤类型、海拔、坡度等立地条件较为一致的马尾松人工林(6年、16年、23年、35年)为研究对象，每个林龄选取3块林分，每个林分内设置面积400m²的调查样地。4个林龄初植密度均为2940株/hm²左右，造林后前3年进行幼林抚育，抚育措施均一致，第10年左右进行第一次间伐，第20年左右进行第二次间伐，第25年左右进行第三次间伐。根据广西用材林林组划分标准[14]，将所选的4个林龄马尾松人工林划分为：幼龄林(6年)、中龄林(16年)、近熟林(23年)、成熟林(35年)。4个林龄马尾松样地的基本情况如下(表1)。林下灌木主要有玉叶金花(*Mussaenda pubescens* Ait. f.)、酸藤子[*Embelia laeta*(L.)Mez]、越南悬钩子(*Rubus cochinchinensis* Tratt.)、红皮水锦树[*Wendlandia tinctoria*(Roxb.)DC. subsp. *intermedia* (How) W. C. Chen]、大沙叶(*Pavetta arenosa* Lour.)等，草本主要有铁芒萁[*Dicranopteris linearis*(Burm.)Underw.]、半边旗(*Pteris semipinnata* L.)、金毛狗[*Cibotium barometz*(L.)J. Sm.]、扇叶铁线蕨(*Adiantum flabellulatum* L.)等。

表1 样地基本情况

样地号	林龄(年)	经度	纬度	海拔(m)	坡度(°)	平均胸径(cm)	平均树高(m)	密度(株/hm²)
1	6	106°54′40.32″	22°02′16.19″	291	15	8.03	5.17	2797
2	6	106°54′46.54″	22°03′33.53″	304	15	9.15	5.23	2940
3	6	106°52′33.77″	22°03′45.64″	497	26	7.32	5.93	2955
4	16	106°54′36.63″	22°01′27.51″	262	30	14.09	11.46	1186

（续）

样地号	林龄（年）	经度	纬度	海拔（m）	坡度（°）	平均胸径（cm）	平均树高（m）	密度（株/hm²）
5	16	106°54′26.24″	22°01′34.78″	282	25	14.57	10.76	1356
6	16	106°54′47.60″	22°01′35.10″	250	20	15.86	11.31	1186
7	23	106°55′07.48″	22°03′39.45″	275	23	20.69	15.61	545
8	23	106°54′59.46″	22°03′44.53″	241	30	22.14	16.26	600
9	23	106°54′46.54″	22°03′33.53″	252	30	22.14	14.87	576
10	35	106°53′55.82″	22°02′50.14″	297	30	25.05	19.78	455
11	35	106°53′49.66″	22°02′41.16″	306	25	26.01	19.96	450
12	35	106°54′04.73″	22°02′48.28″	270	25	24.02	20.05	448

1.2 研究方法

1.2.1 土壤采集

2013年7~9月在每个调查样地内选择代表性地段，挖取一个1m左右深的土壤坡面，按0~20、20~40、40~60cm土层混合取样，取土500g，带回实验室。自然风干后，磨细过0.15mm孔径筛，用于测定土壤有机碳、全氮、全磷含量。

1.2.2 样品测定

土壤有机碳采用重铬酸钾—水合加热法测定；土壤全氮采用凯氏定氮法测定；土壤全磷采用钼锑抗比色法测定。

1.3 数据处理

在Excel 2013中进行数据统计及绘图，在SPSS16.0中进行单因素方差分析(One-way ANOVA)和多重比较分析(LSD)，采用Pearson进行相关性分析。

2 结果与分析

2.1 不同林龄马尾松人工林土壤C、N、P含量变化

由表2可知，4个林龄0~60cm土壤有机C含量均值分别为10.60g/kg、5.76g/kg、4.39g/kg、4.91g/kg，表现为随林龄先降低后增加，6年林分显著高于16年林分，16年林分显著高于23年和35年林分，23年、35年林分间无显著差异。4个林龄土壤有机C含量在0~20、20~40、40~60cm土层均值分别为8.88g/kg、5.87g/kg、4.50g/kg，表现为随土层增加显著降低，40~60cm土层显著低于20~40cm土层，20~40cm土层显著低于0~20cm土层。同一土层不同林龄，0~20cm土层土壤有机C表现为6年林分显著高于其他3个林分，16年林分次之(7.61g/kg)，23年林分最小(5.73g/kg)；20~40、40~60cm土层土壤有机C均表现为23年、35年林分显著低于6年和16年林分，6年林分显著高于16年林分。同一林龄不同土层，6年、16年、23年林分土壤有机C含量均表现为0~20cm土层显著高于20~40cm土层，20~40cm土层显著高于40~60cm土层；35年林分0~20cm土层有机C含量显著高于20~40、40~60cm土层。

由表2可知，4个林龄0~60cm土壤N含量均值分别为0.91g/kg、0.69g/kg、0.58g/kg、0.79g/kg，表现为随林龄先降低后增加，6年、35年林分显著高于16年林分和23年林分，16年林分显著高于23年林分。4个林龄土壤N含量在0~20、20~40、40~60cm土层均值分别为0.93g/kg、0.71g/kg、0.59g/kg，表现为随土层增加显著降低。同一土层不同林龄，0~20cm土层土壤N含量表现为16年、35年林分显著高于23年林分，显著低于6年林分；20~40cm土层土壤N含量表现为16年、23年林分显著低于6年林分和35年林分，6年林分显著高于35年林分；40~60cm土层土壤N含量表现为16年、23年林分显著低于6年林分和35年林分，其余无显著差异。同一林龄不同土层，6年、16年、35年林分土壤N含量均表现为0~20cm土层显著高于20~40cm土层，20~40cm土层显著高于40~60cm土层；23年林分土壤N含量表现为0~20、20~40cm土层显著高于40~60cm土层。

由表2可知，4个林龄0~60cm土壤P含量均值分别为0.21g/kg、0.17g/kg、0.17g/kg、0.17g/kg，表现为6年略高于其他3个林龄林分，但无显著差异。4个林龄土壤P含量在0~20、20~40、40~60cm土层均值分别为0.19g/kg、0.18g/kg、0.18g/kg，土壤P含量随土层变化不显著。同一土层不同林龄或同一林龄不同土层间土壤全P含量均无显著性差异。

表 2　不同林龄马尾松人工林土壤 C、N、P 含量变化特征　(g/kg)

元素	林龄(年)	土层			
		0~20cm	20~40cm	40~60cm	平均
TOC	6	15.56±1.20Aa	9.56±0.50Ab	6.69±0.53Ac	10.60±1.01A
	16	7.61±0.42Ba	5.54±0.38Bb	4.14±0.29Bc	5.76±0.60B
	23	5.73±0.24Ca	4.10±0.35Cb	3.33±0.18Cc	4.39±0.50C
	35	6.61±0.28Da	4.29±0.21Cb	3.84±0.57Cb	4.91±0.56C
	Mean	8.88±0.59a	5.87±0.57b	4.50±0.61c	
TN	6	1.23±0.07Aa	0.83±0.04Ab	0.68±0.14Ac	0.91±0.15A
	16	0.89±0.10Ba	0.65±0.10Cb	0.53±0.07Bc	0.69±0.07B
	23	0.69±0.14Ca	0.59±0.03Ca	0.47±0.03Bb	0.58±0.05C
	35	0.91±0.05Ba	0.78±0.01Bb	0.68±0.03Ac	0.79±0.04A
	Mean	0.93±0.09a	0.71±0.06b	0.59±0.06c	
TP	6	0.23±0.05Aa	0.21±0.04Aa	0.18±0.02Aa	0.21±0.04A
	16	0.17±0.01Aa	0.17±0.02Aa	0.18±0.01Aa	0.17±0.01A
	23	0.17±0.03Aa	0.17±0.03Aa	0.17±0.03Aa	0.17±0.01A
	35	0.17±0.02Aa	0.18±0.02Aa	0.17±0.02Aa	0.17±0.01A
	Mean	0.19±0.02a	0.18±0.02a	0.18±0.01a	

注：同元素同行不同小写字母表示该元素在不同土层中差异显著($P<0.05$)，同元素同列不同大写字母表示该元素在不同林龄间差异显著($P<0.05$)。

2.2　不同林龄马尾松人工林土壤 C、N、P 的化学计量特征

由图 1 可知，4 个林龄马尾松人工林 0~60cm 土层土壤 C∶N 为 5.54~12.67，随林龄增加逐渐降低。单因素方差分析表明，林龄对土壤 C∶N 有极显著的影响($P=0.001$)，土层对土壤 C∶N 无显著性影响($P=0.562$)。多重比较表明，0~20cm 土层土壤 C∶N 表现为 6 年林分显著高于其他 3 个林分，20~40cm 土层土壤 C∶N 表现为 23 年、35 年林分显著低于 6 年和 16 年林分，6 年林分显著高于 16 年林分，40~60cm 土层土壤 C∶N 表现为 6 年林最大，16 年林分次之，35 年林分最小，23 年林分与 16 年、35 年均无显著差异。就同一林龄不同土层而言，6 年、16 年各土层间土壤 C∶N 无显著差异，23 年、35 年林分均表现为 0~20cm 土层显著高于 20~40、40~60cm 土层。

由图 1 可知，4 个林龄马尾松人工林 0~60cm 土壤土层 C∶P 为 19.75~49.91，随林龄增加逐渐降低，至成熟林略有提高。单因素方差分析表明，林龄对土壤 C∶P 无显著性影响($P=0.132$)，土层对土壤 C∶P 有极显著性影响($P=0.000$)。多重比较表明，0~20cm 土层土壤 C∶P 表现为 6 年林分显著高于 23 年、35 年林分，与 16 年林分无显著差异，20~40cm 土层土壤 C∶P 表现为 6 年、16 年林显著高于 23 年和 35 年林分，40~60cm 土层土壤 C∶P 表现为 6 年、16 年林显著高于 23 年，与 35 年林分无显著差异。就同一林龄不同土层而言，6 年、16 年林分土壤 C∶P 均表现为 0~20cm 土层显著高于 20~40cm 土层，20~40cm 显著高于 40~60cm 土层，23 年、35 年林分土壤 C∶P 均表现为 0~20cm 土层显著高于 20~40、40~60cm 土层。

由图 1 可知，4 个林龄马尾松人工林 0~60cm 土层土壤 N∶P 为 2.85~5.61，随林龄增加呈"N"字形变化。单因素方差分析表明，林龄、土层对土壤 N∶P 均有显著性影响($P=0.014$，$P=0.000$)。多重比较表明，0~20cm 土层土壤 N∶P 表现为 16 年、35 年林分显著高于 6 年、23 年林分，20~40cm 土层土壤 N∶P 表现为 16 年、23 年林分显著高于 6 年林分，显著低于 35 年林分，40~60cm 土层土壤 N∶P 表现为 35 年显著高于其他 3 个林分。

2.3　不同林龄马尾松人工林土壤 C、N、P 含量及化学计量比的相关性

由表 3 相关性分析结果表明，土壤 C∶N 与有机 C 含量有极显著的相关性，与全 N 含量相关性不显著，土壤 C∶P 与有机 C 含量有极显著相关性，与全 P 含量相关性不显著，说明研究区马尾松人工林土壤 C∶N、土壤 C∶P 主要受有机 C 的影响。土壤 N∶P 与全 N 含量呈现极显著的相关性，与有全 P 含量相关性不显著。

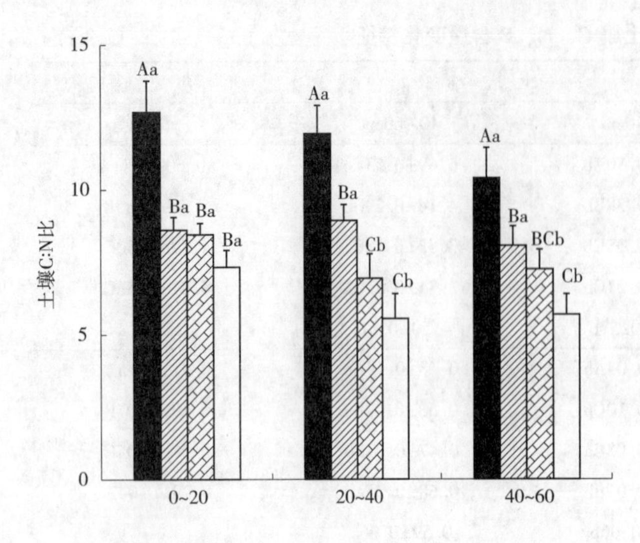

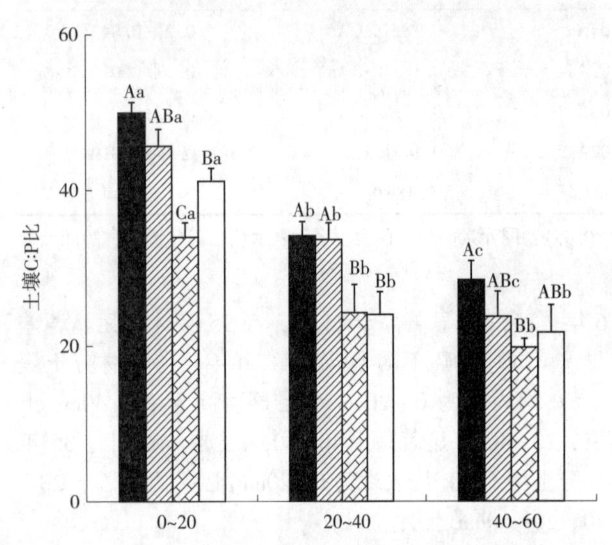

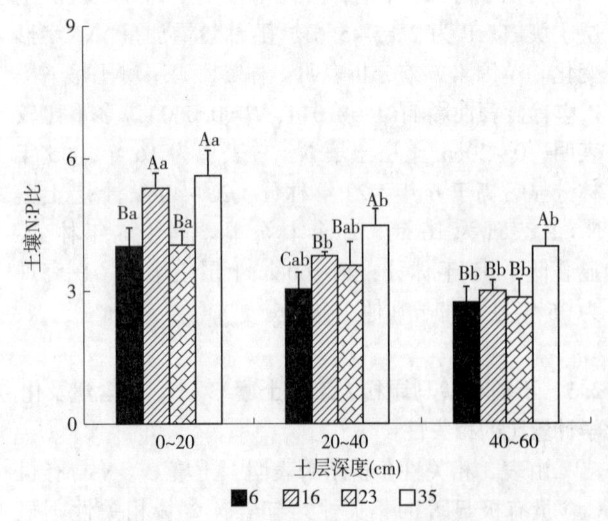

图1　4个林龄马尾松人工林各土层土壤C：N、C：P、N：P的变化特征

注：同一图组不同大写字母表示该土层元素计量比在不同林龄间差异显著($P<0.05$)，同一图例不同小写字母表示该林龄不同土层间元素计量比差异显著($P<0.05$)。

表3　马尾松人工林土壤C、N、P含量及化学计量比的相关性

项目	C	N	P	C：N	C：P	N：P
C	1	0.799**	0.826**	0.686**	0.734**	0.071
N		1	0.740**	0.142	0.520**	0.453**
P			1	0.454**	0.266	−0.246
C：N				1	0.661**	−0.343*
C：P					1	0.440**
N：P						1

3　讨论

本研究结果表明，马尾松人工林土壤有机C、全N含量随林龄增加逐渐降低，至成熟林有所回升，但仍低于幼龄林或中龄林，这与一些学者的研究结论不同。崔宁浩等[10]对马尾松人工林(5年、14年、39年)的研究以及吴明等[15]对杉木人工林(10年、20年、30年)的研究均表明，土壤有机C和全N含量随林龄的增大而增加。杨会侠等[16]对7年、17年、31年、51年，4个林龄的马尾松人工林发育过程中的养分动态格局进行了研究，认为幼龄林阶段林分主要在于构建树冠，净生产力较低，加之林分密度大，凋落物丰富，土壤养分高归还、高吸收、周转时间最短，养分消耗相对较小，土壤中仍然保留着较高的养分含量；中林龄至近成熟林阶段，林分净生产力大幅度提高，生物量大量积累，而进一步的间伐使得凋落物量减少，土壤养分处于高吸收或低归还，林地养分极度被消耗，17生林分是耗费地力最大的年龄阶段；成熟林阶段，由于林木生长速度下降，凋落物归还有限，对土壤养分处于低吸收、低归还状态，土壤养分消耗相对较小，有利于林地土壤养分的积累。本研究支持此论点，虽然马尾松人工林土壤养分在成熟林阶段有所回升，但由于中龄林—近熟林的茂盛生长期对土壤养分的大量消耗，马尾松人工林地力仍有不断衰退的趋势，其中16年生中龄林土壤肥力消耗最大。从土层变化上来看，马尾松人工林土壤有机C、全N含量随土层增加逐渐降低，这符合以往的研究结论[17-20]。土壤C、N主要来源于凋落物的归还，首先在表层富集，之后经淋溶向下迁移扩散，因此，在土层上表现为“倒金字塔”的变化特征。4个林龄土壤全P含量在年龄和土层间均无显著性差异，数值为0.17~0.28g/kg，普遍低于其他研究区，且非常稳定。主要原因有两点：首先土壤P素的主要获得方式是岩石风化，而岩石风化需要漫长的时间才能完成，且在0~60cm土层中的变化非常有限[21,22]，其次研究区处于低纬度的南亚热带，土壤全P含量本身偏低[23,24]。从研究区土壤P素特征来看，该地区植被生长过程可能受到P的限制。

土壤C∶N∶P是土壤有机质组成和质量程度的一个重要指标，可用于判定土壤C、N、P的矿化作用和固持作用[8]。研究区4个林龄马尾松人工林土壤C∶N∶P分别为37∶3∶1、34∶4∶1、26∶3∶1、29∶5∶1，低于我国土壤C∶N∶P均值60∶5∶1。C∶N是衡量土壤C、N营养平衡状况的指标，较低的C∶N比表明土壤有机质具有更快的矿化速率[25]。Bengtsson等[26]指出，当土壤C∶N比值较高时，微生物需要输入N素来满足自身的生长需要，当土壤C∶N比值较低时，超过微生物生长所需要的N素就会被释放到土壤中。研究区4个林龄马尾松人工林土壤C∶N平均值为8.48，低于我国土壤C∶N平均值11.9，表现为6年幼龄林(11.66)>16年中龄林(8.53)>23年近熟林(7.54)>35年成熟林(6.18)，不同林龄土壤C∶N时间差异显著，可能原因是由于随着树木年龄的累积，原来土壤中丰富的C、N等结构性养分被大量消耗，造成土壤C∶N持续性降低。

土壤C∶P是衡量微生物矿化土壤有机物质释放P或从环境中吸收固持P素潜力的一个指标，较低的C∶P比是土壤P有效性高的一个指标[8,9,27]。贾宇等[28]指出，当土壤C∶P>200时，微生物C素大幅度增加，竞争土壤中的速效磷，P素发生净固持作用，当土壤C∶P<200时，会出现土壤微生物的C素短暂性增加，P素发生净矿化作用。研究区4个林龄马尾松人工林土壤C∶P均值为31.66，小于200更低于我国土壤C∶P平均值105，表现为6年幼龄林(37.47)>16年中龄林(34.31)>35年成熟林(28.93)>23年近熟林(25.97)，林龄对土壤C∶P无显著性影响，说明随着林龄的增加，土壤中可利用性P素仍然非常有限。

土壤N∶P可用作N养分限制、饱和的诊断指标，指示植物生长过程中土壤营养成分的供应情况[8,9]。研究区4个林龄马尾松人工林土壤N∶P均值为3.87，低于全国土壤N∶P平均值5.20，表现为35年成熟林(4.69)>16年中龄林(4.04)>23年近熟林(3.49)>6年幼龄林(3.27)。林龄、土层对土壤N∶P均有显著性影响，但4个林龄马尾松人工林土壤P素含量偏低且非常稳定，相关性结果亦表明土壤N∶P与全N含量呈现极显著的相关性，与全P含量相关性不显著，说明N在不同林龄间和不同土层间的变化是土壤N∶P变化的主要原因。N的变化跟N的来源密切相关，主要有以下3个途径：①凋落物归还。植物有机组织中的氮通过凋落物归还到土壤中，受凋落量和分解速率的共同影响。卢立华等[29]在同纬度地带对人工林凋落物的研究表明，马尾松人工林(18年，900株/hm^2)年凋落物量高达5580.07kg/hm^2，而杨会侠等[30]发现马尾松人工林幼龄林至成熟林凋落物N素年归还量分别为40.03kg/hm^2、34.60kg/hm^2、29.43kg/hm^2、34.17kg/hm^2，表现为从7年生林分降低至31年生林分，到51年生时又有所增加，与本文全N含量在林龄间的变化规律一致。②生物固氮。即依靠自生和共生的固氮菌将气态氮固定为含氮有机化合物，再通过微生物及共生植物，直接或间接进入土壤。③大气降水。大气中的气态氮经降水淋溶，部分被森林的林冠层截留，最后随径流进入土壤中，成为土壤氮的经常性来源之一。不同地域、不同林型导致土壤最后实际接收到的氮含量有所差异。周光益等[31]对酸雨区的21年和40年马尾松冠层淋溶规律的研究发现，两者穿透雨硝态氮年沉积量分别为33.90kg/hm^2、32.37kg/hm^2，周国逸等[32]在鼎湖山的研究甚至表明，降水氮沉降量已远超出植物生长所需要的氮含量。张捷等[33]在宜宾的研究则指出，18年马尾松穿透雨硝态氮、铵态氮7个月的输入量分别仅为1.159kg/hm^2、0.729kg/hm^2，32a仅为1.228kg/hm^2、0.741kg/hm^2。本研究区处于非酸雨区，通过大气降水获得的氮输入应该非常有限，更接近于后者的研究结论。综上所述，可以认为本研究区马尾松生态系统中N素的来源主要依靠凋落物的归还和营林中的人为干扰(即通过引入林下固氮类豆科植物或者人为施加氮肥)。研究表明，植物体内的N∶P能明确植物群落生长过程中受到哪种元素的限制作用[21]，本研究对植物的N、P含量尚未涉足，今后应该开展土壤与植物N、P化学计量特征的相关研究。

4 结论

(1)土壤有机C、全N含量从成熟林阶段开始回升。

(2)N在不同林龄间和不同土层间的变化是土壤N∶P变化的主要原因。

(3)有机C是影响土壤C∶N、C∶P的重要因素。

参考文献

[1]程滨，赵永军，张文广，等．生态化学计量学研究进展[J]．生态学报，2010，30(6)：1628-1637.

[2]曾德慧，陈广生．生态化学计量学：复杂生命系统奥秘的探索[J]．植物生态学报，2005，29(6)：1007-1019.

[3]喻林华，方晰，项文化，等．2016．亚热带4种林分类型枯落物层和土壤层的碳氮磷化学计量特征[J]．林业科学，52(10)：10-21.

[4]赵亚芳，徐福利，王渭玲，等．华北落叶松根茎叶碳氮磷含量及其化学计量学特征的季节变化[J]．植物学

报，2014，49(5)：560-568.

[5]YANG Y H，FANG J Y，GUO D L，et al. Vertical patterns of soil carbon，nitrogen and carbon：nitrogen stoichiometry in Tibetan grasslands[J]. Cancer epidemiology，biomarkers & prevention：a publication of the American Association for Cancer Research，cosponsored by the American Society of Preventive Oncology，2010，9(6)：631-3.

[6]JING A N. Effect of Forest and Farm on Vertical Patterns of Soil Carbon，Nitrogen and Other Parameters in Northeast China[J]. Bulletin of Botanical Research，2012，32(3)：331-338.

[7]牛瑞龙，高星，徐福利，等．秦岭中幼林龄华北落叶松针叶与土壤的碳氮磷生态化学计量特征[J]. 生态学报，2016，36(22).

[8]王绍强，于贵瑞．生态系统碳氮磷元素的生态化学计量学特征[J]. 生态学报，2008，28(8)：3937-3947.

[9]曹娟，闫文德，项文化，等．湖南会同3个林龄杉木人工林土壤碳、氮、磷化学计量特征[J]. 林业科学，2015，51(7)：1-8.

[10]崔宁洁，刘小兵，张丹桔，等．不同林龄马尾松(*Pinus massoniana*)人工林碳氮磷分配格局及化学计量特征[J]. 生态环境学报，2014，23(2)：188-195.

[11]曾凡鹏，迟光宇，陈欣，等．辽东山区不同林龄落叶松人工林土壤-根系C：N：P生态化学计量特征[J]. 生态学杂志，2016，35(8)：1819-1825.

[12]周政贤．中国马尾松[M]. 北京：中国林业出版社，2001：53-60.

[13]廖静姝，唐昌秀，黄喆敏．凭祥市近45年气候变化特征分析[J]. 农家之友(理论版)，2010，12：33-35+41.

[14]韩畅，宋敏，杜虎，等．广西不同林龄杉木、马尾松人工林根系生物量及碳储量特征[J]. 生态学报，2017，07：1-8.

[15]吴明，邵学新，周纯亮，等．中亚热带典型人工林土壤质量演变及其环境意义[J]. 生态学杂志，2009，28(9)：1813-1817.

[16]杨会侠，汪思龙，范冰，等．马尾松人工林发育过程中的养分动态[J]. 应用生态学报，2010，21(8)：1907-1914.

[17]王维奇，徐玲琳，曾从盛，等．河口湿地植物活体-枯落物-土壤的碳氮磷生态化学计量特征[J]. 生态学报，2011，23：134-139.

[18]MAISTO G，DE MARCO A，MEOLA A，et al. Nutrient dynamics in litter mixtures of four Mediterranean maquis species decomposing in situ[J]. Soil Biology and Biochemistry，2011，43(3)：520-530.

[19]苗娟，周传艳，李世杰，等．不同林龄云南松林土壤有机碳和全氮积累特征[J]. 应用生态学报，2014，03：625-631.

[20]秦娟，唐心红，杨雪梅．马尾松不同林型对土壤理化性质的影响[J]. 生态环境学报，2013，04：598-604.

[21]胡耀升，么旭阳，刘艳红．长白山森林不同演替阶段植物与土壤氮磷的化学计量特征[J]. 应用生态学报，2014，03：632-638.

[22]刘兴诏，周国逸，张德强，等．南亚热带森林不同演替阶段植物与土壤中N、P的化学计量特征[J]. 植物生态学报，2010，34(1)：64-71.

[23]ZHANG C，TIAN H Q，LIU J Y，et al. Pools and distributions of soil phosphorus in China[J]. Global Biogeochemical Cycles，2005，19(1)：347-354.

[24]KELLOGG L E，BRIDGHAM S D. Phosphorus retention and movement across an ombrotrophic-minerotrophic peatland gradient[J]. Biogeochemistry，2003，63(3)：299-315.

[25]MAJDA H，OHRVIK J. Interactive effects of soil warming and fertilization on root production，mortality in Norway spruce stand in Northern Sweden[J]. Global Change Biology，2004，10(2)：182-188.

[26]BENGTSSON G，BENGTSON P，MANSSON K F. Gross nitrogen mineralization-，immobilization-，and nitrification rates as a function of soil C/N ratio and microbial activity[J]. Soil Biology and Biochemistry，2003，35(1)：143-154.

[27]刘万德，苏建荣，李帅锋，等．云南普洱季风常绿阔叶林演替系列植物和土壤C、N、P化学计量特征[J]. 生态学报，2010，23：6581-6590.

[28]贾宇，徐炳成，李凤民，等．半干旱黄土丘陵区苜蓿人工草地土壤磷素有效性及对生产力的响应[J]. 生态学报，2007，27(1)：42-47.

[29]卢立华，贾宏炎，何日明，等．南亚热带6种人工林凋落物的初步研究[J]. 林业科学研究，2008，21(3)：346-352.

[30]杨会侠，汪思龙，范冰，等．不同林龄马尾松人工林年凋落量与养分归还动态[J]. 生态学杂志，2010，29(12)：2334-2340.

[31]周光益，徐义刚，吴仲民，等．广州市酸雨对不同森林冠层淋溶规律的研究[J]. 林业科学研究，2000，13(6)：598-607.

[32]周国逸，闫俊华．鼎湖山区域大气降水特征和物质元素输入对森林生态系统存在和发育的影响[J]. 生态学报，2001，21(12)：2002-2012.

[33]张捷，刘洋，张健，等．马尾松人工林林冠层对氮、磷、硫的截留效应[J]. 水土保持学报，2014，28(4)：37-43.

[原载：林业科学研究，2017，30(06)]

土壤颗粒悬液搅拌对土壤质地分析的影响

李朝英[1] 郑 路[1,2]

([1]中国林业科学研究院热带林业实验中心，广西凭祥 532600，
[2]广西友谊关森林生态系统国家定位观测研究站，广西凭祥 532600)

摘 要 为了探究土壤颗粒悬液搅拌对土壤质地分析的影响。采用比重法测定土壤机械组成，在搅拌不同行程、搅拌不同次数、土壤颗粒悬液放置不同时间后搅拌、不同测定时点搅拌的多种条件下所测结果进行比较分析。结果表明，搅拌行程、搅拌次数、土壤颗粒悬液放置后搅拌对黏粒含量测定无影响，对粉砂粒和砂粒含量测定均有影响。搅棒从液面下行至沉降筒底搅拌，有利于颗粒充分运动及均匀分布，黏粒、粉砂粒、砂粒含量测定稳定，精密度良好；土壤颗粒以氢氧化钠为分散剂清洗后搅拌30次，黏粒、粉砂粒、砂粒含量测定准确，精密度良好；悬液静置时间大于1h，搅拌30次所测粉砂粒和砂粒含量的精密度呈下降趋势，搅拌45次所测粉砂粒、砂粒、黏粒含量稳定，精密度良好，与洗后即搅拌30次的测定结果一致，两者的相关性良好。在第二测定时点增加搅拌1min对测定结果无影响。

关键词 土壤颗粒；悬液；搅拌；质地分析

Effects of Stirring Soil Particle Suspension on Texture Analysis of Soil

LI Zhaoying[1], ZHENG Lu[1,2]

([1]*Experimental Center of Tropical Forestry*, *Chinese Academy of Forestry*, *Pingxiang* 532600, *Guangxi*, *China*;
[2]*Guangxi Youyiguan Forest Ecosystem The National Research Station*, *Pingxiang* 532600, *Guangxi*, *China*)

Abstract: In order to investigate the effect of soil particle suspension on soil texture analysis, the comparison and analysis were carried out under different conditions, such as stirring stroke, stirring number, stirring when soil particle suspension was placed at different time and stirring at different determination time, in the determination of soil mechanical composition by gravimetric method. The results show that the stirring stroke, stirring number and the stirring when soil particle suspension was placed at different time after washing had no effect on clay particle content, had an impact on the silt and sand content . Stirring rod stirred from the liquid level down to the bottom of the precipitation tube is conducive to the full movement of particles and uniform distribution, Determination of the clay, silt and sand content was stable, precision was good. Measurement of clay, silt and sand content were accurate and precision was good when soil particles using sodium hydroxide as dispersing agent were cleaned and stirred 30 times. At the condition of the suspension setting time was larger than 1h, measurement of silt and sand content showed anunstable trend, and poor precision when it was stirred 30 times, measurement of silt sand and clay content was stable, precision was good when it was stirred 45 times, and the results were consistent with that stirring 30 times after washing. the correlation between the two results were good. Added stirring 1min at the second test points, there were no impact to the results.

Key words: soil particle; suspension; stir; texture analysis

土壤机械组成分析即测定土壤的不同粒径颗粒含量，用于评价土壤基本性质，揭示其形成环境，对于水土保持、环境演变等领域定量化推论提供基础数据。同时为指导耕作，施肥、土壤修复与改良，推动土地生产力可持续发展有着重要的现实意义[1-3]。土壤机械组成测定的传统方法有比重法和吸管法[4-6]，主要检测步骤均包含颗粒分散与清洗、悬液搅拌及测定几部分[7,8]。现期有研究讨论分析了土壤颗粒分散效果对测定结果的影响，提出振荡分散、超声分散等多种方法以替代国标所述煮沸分散颗粒

法，求得检测简便，准确高效[9,10]。对于悬液搅拌，国标与林业行标指出搅拌次数(30次/60s)，对搅拌行程、搅拌前土壤颗粒悬液静置时间等无过多描述，现有文献中未见有关搅拌对测定结果影响的报道，所述搅拌时机却有两种做法，一种是搅拌1min后，即完成三个时点测定；另一种是搅拌1min，完成第一测定时点测定，再搅拌1min后，完成第二、三测定时点测定。两种做法对测定结果的影响无从得知[11-13]。上述情况难以为准确测定土壤机械组成提供可行的参考指导。为此，本实验以林业行标所述比重法为基准，探讨搅拌行程、悬液静置时间、搅拌次数以及搅拌时机对土壤机械组成测定结果的影响，以期对土壤机械组成测定方法的搅拌条件进行优化完善与补充，减少人为误差，提高准确性，为实验室准确测定土壤机械组成提供可行的参考与指导。

1 材料与方法

1.1 样品采集区环境

本次实验土壤样品于2015年8月采集于广西友谊关森林生态系统国家定位观测研究站于大青山固定监测样地(22°18′15.17″N，106°41′50.50″E)。该区处于南亚热带南缘，与北热带毗邻，属于南亚热带季风气候区。年平均气温21.5℃，最冷月(1月)平均气温13.5℃，最热月(7月)平均气温27.6℃，大于10℃的年积温7500℃，年降水量1400mm，属于南亚热带季风气候区。地带性土壤为中酸性火山岩和花岗岩发育而成的砖红壤(含紫色土)，土层平均厚度0.5~1.0m。其地形地貌复杂，气候高温高湿，雨量充沛。

1.2 样品处理

在固定监测样地的3个5m×5m样地中分别在0~20cm，20~40cm两个剖面层各取1个土样，共计6个样。风干后研磨过10目筛后装入有标识的聚丙烯塑料袋中密封保存于阴凉处。

1.3 实验仪器及试剂

甲种密度计(刻度0~60g/L)、1L沉降筒、搅棒、60目筛、40目筛、18目筛。

20g/L的氢氧化钠、无水乙醇。

1.4 实验方法

6个土壤样品的机械组成分析采用比重法，在机械组成分析中粒级划分采用国际制。因国标所述煮沸法分散土壤颗粒较繁琐，效率低。本实验采用振荡法分散土壤颗粒。

1.4.1 6个样品各称取50.00g入250ml塑料瓶中，加入20g/L氢氧化钠溶液50ml和200ml蒸馏水振荡30min，将土壤悬浊液倒至1L沉降筒上的60目筛中，用蒸馏水少量多次冲洗筛上颗粒至洁净无泥。筛上洗净的颗粒放入105℃烘箱中，烘干后倒入组合筛(从上至下依次为18目、40目、60目筛)筛分处理，分别将18目筛上粒径>1mm的颗粒，40目筛上粒径0.5~1mm的颗粒，60目筛上粒径0.25~0.5mm的颗粒称量记录，土壤颗粒清洗液定容至1L后即用搅拌棒从液面行至筒底上下搅拌30次/60s，在静置30s、4.5min、8h时分别放入比重计读数。以上实验各重复3次。

1.4.2 6个样品各称取50.00g入250ml塑料瓶中，按1.4.1所述加入分散剂、振荡、过筛烘干称重及清洗液定容，土壤颗粒清洗液由搅棒从液面到距筒底10cm处之间搅拌(后简称搅拌不到筒底)，按1.4.1所述测定。以上实验各重复3次。

1.4.3 6个样品各称取50.00g入250ml塑料瓶中，共称3组，按1.4.1所述加入分散剂、振荡、过筛烘干称重及清洗液定容，第1组土壤颗粒清洗液静置1h，第2组土壤颗粒清洗液静置2h，分别按1.4.1所述搅拌测定。第3组土壤颗粒清洗液静置2h，1.4.1所述搅拌测定，但搅拌45个来回/60s。以上实验各重复3次。

2 结果与分析

2.1 搅拌行程对测定结果的影响

由图1、图2、图3可见，两种搅拌行程所测黏粒、粉砂粒及砂粒含量变化趋于一致，但搅拌到筒底的黏粒含量、粉砂粒含量大于搅拌不到筒底的，砂粒含量小于搅拌不到筒底的。两者黏粒含量差异偏小，粉砂粒和砂粒含量差异偏大。

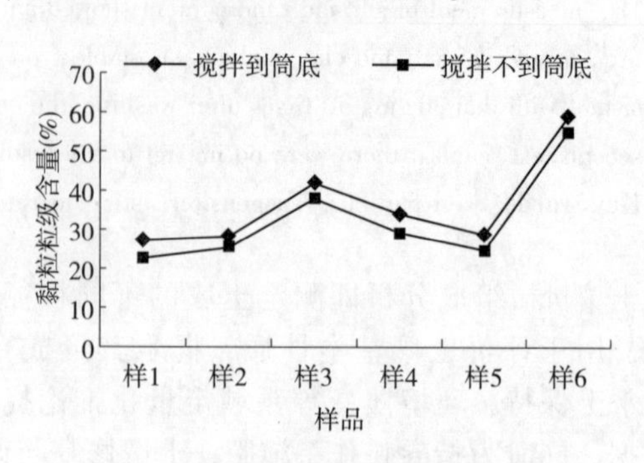

图1 搅拌不同行程测定的黏粒含量

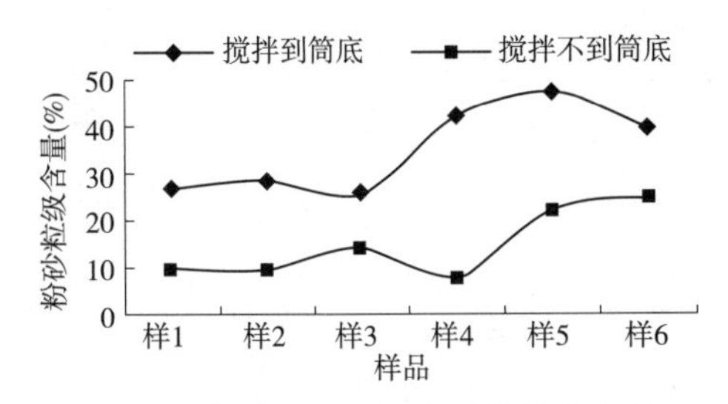

图 2 搅拌不同行程测定的粉砂粒含量

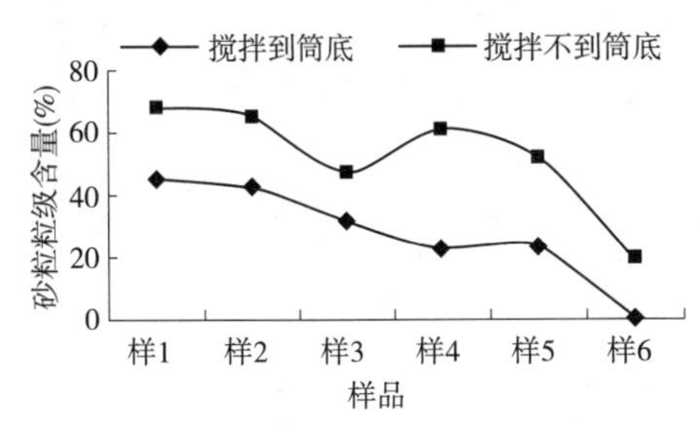

图 3 搅拌不同行程测定的砂粒含量

由表 1 可见，搅拌到筒底所测 6 个样品的黏粒、粉砂粒、砂粒含量 CV 均小于 6%。搅拌不到筒底所测黏粒含量 CV 为 2.12%~7.22%，粉砂粒含量与砂粒含量 CV 大于 15%。两种搅拌行程的检测结果进行质地判定分析，样品 1、3、6 的质地相同，样品 2、4、5 的质地判定不相同，占 6 个样品量的 50%。

搅拌到筒底使悬液不同深度的大小颗粒充分运动，满足 Stoke's 定律的假设条件，所测结果精密度及准确性良好，且具代表性，是适宜的搅拌行程。搅拌不到筒底所测结果精密度差，尤其是粉砂粒和砂粒含量的精密度明显降低。这是因为悬液中的粉砂粒及砂粒未能充分运动，测定结果不稳定且缺乏代表性。故选择搅拌到筒底的行程为宜。

2.2 悬液静置时间及搅拌次数对测定结果的影响

由图 4、图 5、图 6 可见，悬液静置 1h 搅拌 30 次所测黏粒、粉砂粒与砂粒与洗后即搅的趋于一致。悬液静置 2h 搅拌 30 次所测黏粒与洗后即搅的趋于一致，所测粉砂粒和砂粒含量与洗后即搅的不一致，其中粉砂粒有偏低趋势，砂粒有偏高趋势。静置 2h 后搅拌 45 次所测粉砂粒和砂粒含量与洗后即搅的趋于一致土壤颗粒洗后即搅拌，颗粒充分分散，在悬液中均匀分布，达到最佳测定状态，所测结果准确稳定。上述可见，随着悬液静置时间延长，粉砂粒与砂粒难以达到最佳测定状态，影响测定的准确性。这说明悬液状态随着静置时间延长而有变化，影响搅拌效果，即颗粒的均匀分布。悬液静置 1h 搅拌 30 次可使不同颗粒均匀分布，达到最佳测定状态，所测结果保持准确；悬液静置 2h 搅拌 30 次使黏粒均匀分布，未使粉砂粒和砂粒均匀分布，测定结果的准确性下降。悬液静置 2h 搅拌 45 次可使不同颗粒在悬液中均匀分布，达到最佳测定状态，保证检测准确。这表明增加搅拌次数可克服悬液状态变化对测定的干扰而保证测定准确。

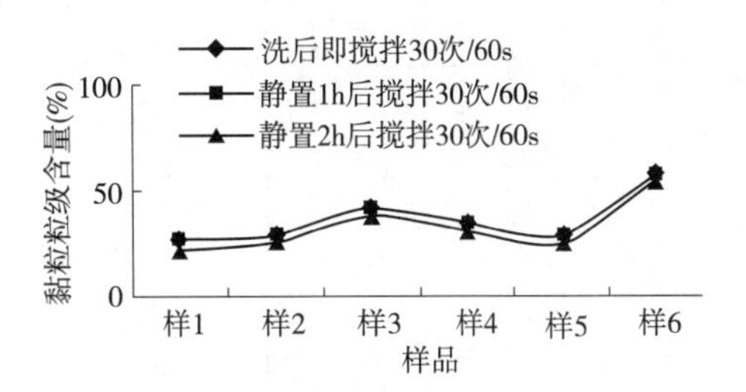

图 4 悬液静置前后的黏粒含量

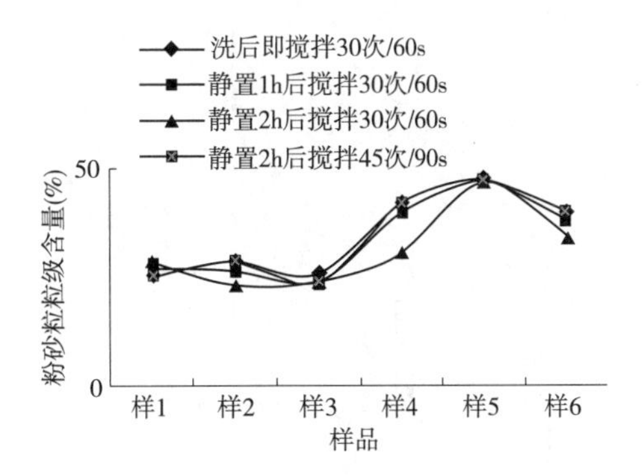

图 5 悬液静置前后的粉砂粒含量

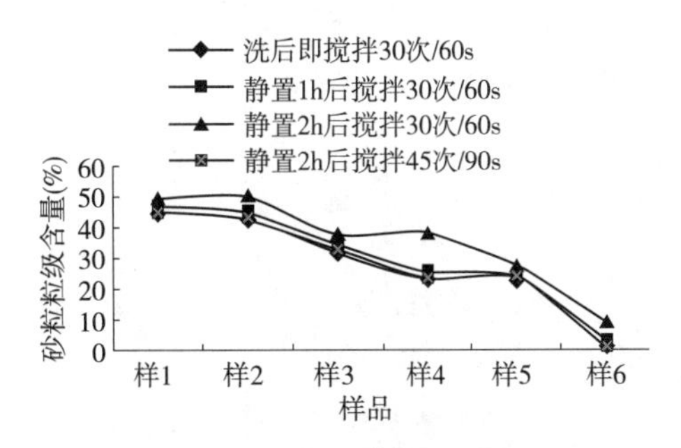

图 6 悬液静置前后的砂粒含量

由表 2 可见，静置 2h 后搅拌 45 次所测黏粒、粉砂粒及砂粒含量 CV 与土壤颗粒洗后即搅拌 30 次的 CV 均小于 6%，两者测定结果精密度良好。悬液静置 1h 搅拌 30 次所测黏粒含量 CV 小于 5%，精密度与洗后即搅的相当，砂粒与粉砂粒含量的 CV 小于 7%。可见，悬液静置 1h 后搅拌 30 次所测黏粒含量的精密度良好，砂粒和粉砂粒精密度有下降趋势，但还能满足一般的检测分析需要。静置 2h 后搅拌 45 次、洗后即搅拌 30 次与悬液静置 1h 搅拌 30 次测定结果进行质地判断分析，结果一致。

表 1　搅拌不同行程的测定结果

样品	不同粒径颗粒含量(%)								CV(%)					
	搅棒到筒底				搅棒到距筒底 10cm 处				搅棒到筒底			搅棒到距筒底 10cm 处		
	<0. 002mm 黏粒粒级含量(%)	0. 05~2mm 粉砂粒级含量(%)	0. 25~2mm 砂粒含量(%)	质地判定	<0. 002mm 黏粒粒级含量(%)	0. 05~2mm 粉砂粒级含量(%)	0. 25~2mm 砂粒含量(%)	质地判定	<0. 002mm 黏粒粒级含量(%)	0. 05~2mm 粉砂粒级含量(%)	0. 25~2mm 砂粒含量(%)	<0. 002mm 黏粒粒级含量(%)	0. 05~2mm 粉砂粒级含量(%)	0. 25~2mm 砂粒含量(%)
样 1	27. 15+ 1. 09	26. 95+ 1. 35	45. 18+ 1. 98	砂质黏壤	22. 57+ 1. 63	9. 95+ 2. 07	67. 96+ 1. 11	砂质黏壤土	4. 03	4. 99	4. 39	7. 22	20. 83	1. 74
样 2	28. 15+ 0. 12	28. 53+ 1. 50	42. 59+ 2. 07	黏质壤土	25. 28+ 0. 96	9. 94+ 3. 59	66. 04+ 3. 42	砂质黏壤土	0. 41	5. 27	4. 87	3. 80	36. 08	5. 6
样 3	41. 83+ 1. 14	25. 84+ 1. 36	31. 6+ 1. 48	黏土	37. 83+ 1. 14	14. 16+ 3. 53	47. 28+ 4. 32	黏土	2. 73	5. 26	4. 67	2. 73	24. 95	9. 97
样 4	33. 95+ 0. 45	42. 23+ 1. 77	23. 1+ 1. 25	黏质壤土	29. 23+ 0. 62	7. 88+ 5. 49	62. 17+ 5. 37	砂质黏壤土	1. 31	4. 18	5. 43	2. 12	69. 63	9. 39
样 5	28. 77+ 1. 21	47. 25+ 1. 40	23. 25+ 1. 20	黏质壤土	24. 77+ 1. 32	22. 38+ 5. 48	52. 12+ 5. 22	砂质黏壤土	4. 16	2. 96	5. 15	5. 33	24. 5	10. 84
样 6	59. 1+ 0. 12	39. 68+ 0. 68	0. 49+ 0. 01	黏土	55. 24+ 3. 11	25. 08+ 2. 34	18. 96+ 2. 39	黏土	0. 2	1. 72	1. 26	5. 63	9. 32	16. 0

表 2 悬液静置前后搅拌及搅拌不同次数的测定结果

样品	不同粒径颗粒含量(%)								CV(%)					
	静置 1h 后搅拌 30 次/60s				静置 2h 后搅拌 45 次/90s				静置 1h 后搅拌 30 次/60s			静置 2h 后搅拌 45 次/90s		
	<0.002mm 黏粒粒级含量(%)	0.05~2mm 粉砂粒级含量(%)	0.25~2mm 砂粒含量(%)	质地判断	<0.002mm 黏粒粒级含量(%)	0.05~2mm 粉砂粒级含量(%)	0.25~2mm 砂粒含量(%)	质地判断	<0.002mm 黏粒粒级含量(%)	0.05~2mm 粉砂粒级含量(%)	0.25~2mm 砂粒含量(%)	<0.002mm 黏粒粒级含量(%)	0.05~2mm 粉砂粒级含量(%)	0.25~2mm 砂粒含量(%)
样 1	26.11+1.20	27.22+1.87	46.66+2.28	砂质黏壤土	27.39+1.20	25.28+1.10	46.59+0.24	砂质黏壤土	4.59	6.87	4.89	4.37	4.34	0.51
样 2	29.07+1.07	26.35+1.79	44.57+2.73	黏质壤土	28.63+1.33	28.73+1.71	41.89+2.39	黏质壤土	3.68	6.79	6.13	4.65	5.94	5.71
样 3	41.35+1.20	24.19+1.67	34.45+2.24	黏土	42.59	23.76+1.20	32.91+1.19	黏土	2.9	6.90	6.5	4.87	5.04	3.63
样 4	34.43+0.12	40.03+2.25	25.53+1.67	黏质壤土	34.09+0.21	42.07+2.17	23.10+1.00	黏质壤土	0.35	5.62	6.54	0.61	5.15	4.32
样 5	28.97+1.20	46.96+1.04	24.06+1.03	黏质壤土	28.63+1.33	47.39+0.24	23.24+1.20	黏质壤土	4.14	2.22	4.28	4.65	0.50	5.15
样 6	58.92+0.11	38.13+2.45	2.94+0.19	黏土	58.59+0.52	39.81+0.39	0.86+0.05	黏土	0.19	6.42	6.46	0.88	0.97	5.88

综上说明，搅拌效果与悬液静置时间长短及搅拌次数有关。土壤样品逐个清洗即搅的做法对测定结果的影响最小。但在同时检测多个样品时，土壤样品逐个清洗即搅的做法不便于检测，操作繁琐，易出错。多个样品清洗后，一并搅拌测定的做法更简捷方便。结合检测实际情况，本实验提出悬液静置时间小于 1h 为宜，静置时间长于 1h 后搅拌次数 45 次，这不仅检测准确且精密度良好，且更适用于实际检测分析。

2.3　不同测定时点前搅拌对测定结果的影响

由图 7、图 8、图 9 可见，搅拌 1min，在三个测定时点所测黏粒含量、粉砂含量、砂粒含量与在前两个测定时点前均搅拌 1min 所测结果一致。第二测定时点前增加一次搅拌，对第二、三个时点的测定结果无影响。

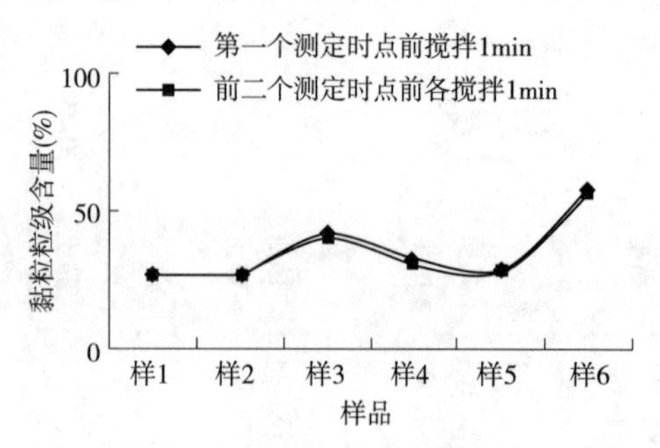

图 7　第一个测定时点和前两个测定时点前搅拌测定黏粒含量

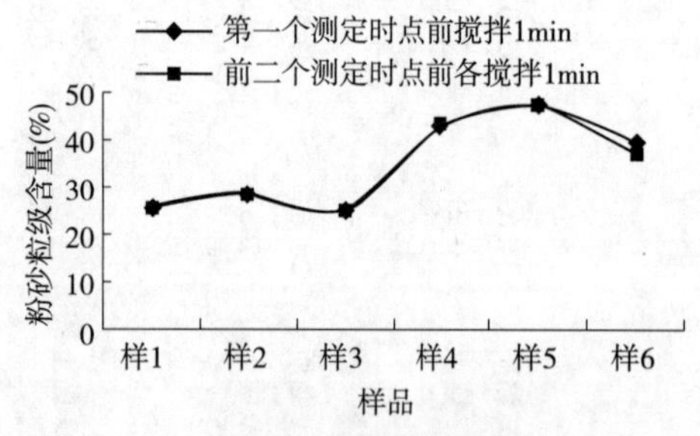

图 8　第一个测定时点和前两个测定时点前搅拌测定粉粒含量

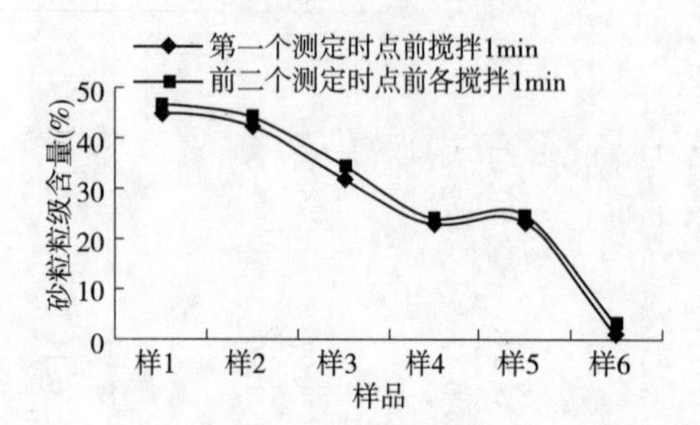

图 9　第一个测定时点和前两个测定时点前搅拌测定砂粒含量

由表 3 可见，第一个测定时点前搅拌 1min 与前两个测定时点前各搅拌 1min 所测黏粒、粉砂粒及砂粒含量一致，测定结果的 CV 均小于 6%，两者精密度良好。这说明第一个测定时点前搅拌 1min 已使颗粒充分运动后，第二个测定时点搅拌与否对测定结果无影响，这种做法费时费力，易产生人为差错。故搅拌 1min 后，在三个测定时点进行测定即可。

3　结论

根据 Stoke's 定律的原理，土壤悬液搅拌是促使不同颗粒均匀分布的重要步骤，是影响不同颗粒含量准确测定的关键因素之一。本实验提出搅拌适宜行程应与悬液深度相当，悬液静置时间小于 1h，当悬液静置时间大于 1h 时搅拌 45 次。第一个时点前搅拌 1min，分别在三个测定时点测定为宜[17]。本实验所述方法测定准确可靠，避免重复操作，有利于减少误差，适用于多个样品的同时检测。

由于氢氧化钠分散剂对土壤颗粒分散效果的持久性有一定时限，悬液静置过程中，土壤颗粒受悬液电解质、pH 等影响不断聚合絮凝，悬液状态有所变化，国标或行标所述 30 次搅拌不能使聚合絮凝的颗粒分散至单粒体状，颗粒未均匀分布，测定结果准确性下降[14-16]。45 次搅拌更易于破坏颗粒的聚合絮凝状态，分散颗粒，达到最佳测定状态。因此本实验首次提出悬液静置时限对测定结果的影响，并提出增加搅拌次数的做法。

砂粒(粒径 2～0.05mm)与粉砂粒(粒径 0.05～0.022mm)粒径大，在 1～5min 之间快速沉降；黏粒(粒径<0.002mm)粒径小，沉降时间长达 5～6h[18,19]。因此，粒径大、沉降快的粉砂粒与砂粒在搅拌行程及搅拌次数不足时不易均匀分布，从而更易影响测定结果。悬液中颗粒未达到均匀分布的最佳测定状态，所测粒径<0.05mm 的砂粒含量及黏粒偏小，粒径 0.25～0.05mm 砂粒含量由上述测定结果计算所得值偏高，故砂粒含量偏高，黏粒及粉砂粒含量偏低。

现有文献主要讨论分析温度、沉降筒规格、分散方法、比重计读数等对土壤机械组成测定的影响[20,21]，本实验所述悬液搅拌行程和搅拌时机的阐述，对现有土壤机械组成检测方法的理解进行明确，所述悬液静置时限及搅拌次数是对现有土壤机械组成检测方法的检测条件进行了补充及优化，从而解决实际问题，减少误差，提高检测准确性。本实验提出的方法对指导实验室准确检测批量样品的土壤机械组成有着重要的现实意义。

表 3 第一个测定时点前搅拌 1min 与前两个测定时点前搅拌 1min 的测定结果的比较

样品	第一个测定时点前搅拌 1min				前两个测定时点前各搅拌 1min				第一个测定时点前搅拌 1min			前两个测定时点前各搅拌 1min		
	<0. 002mm 黏粒粒级含量(%)	0. 05~0. 002mm 粉砂粒级含量(%)	0. 25~2mm 砂粒含量(%)	质地判定	<0. 002mm 黏粒粒级含量(%)	0. 05~0. 002mm 粉砂粒级含量(%)	0. 25~2mm 砂粒含量(%)	判定	<0. 002mm 黏粒粒级含量(%)	0. 05~0. 002mm 粉砂粒级含量(%)	0. 25~2mm 砂粒含量(%)	<0. 002mm 黏粒粒级含量(%)	0. 05~0. 002mm 粉砂粒级含量(%)	0. 25~2mm 砂粒含量(%)
样 1	27. 15+ 1. 09	26. 95+ 1. 35	45. 18+ 1. 98	砂质黏壤	26. 69+ 0. 86	25. 78+ 1. 23	46. 80+ 2. 29	砂质黏壤	4. 03	4. 99	4. 39	3. 23	4. 58	4. 89
样 2	28. 15+ 0. 12	28. 53+ 1. 50	42. 59+ 2. 07	黏质壤土	27. 19+ 0. 70	28. 21+ 1. 23	43. 87+ 1. 79	黏质壤土	0. 41	5. 27	4. 87	2. 56	4. 37	4. 08
样 3	41. 83+ 1. 14	25. 84+ 1. 36	31. 6+ 1. 48	黏土	40. 28+ 1. 72	24. 99+ 0. 75	34. 00+ 1. 40	黏土	2. 73	5. 26	4. 67	4. 28	3. 02	4. 11
样 4	33. 95+ 0. 45	42. 23+ 1. 77	23. 1+ 1. 25	黏质壤土	30. 85+ 0. 92	43. 19+ 1. 71	25. 23+ 1. 43	黏质壤土	1. 31	4. 18	5. 43	2. 97	3. 95	5. 66
样 5	28. 77+ 1. 21	47. 25+ 1. 40	23. 25+ 1. 20	黏质壤土	28. 09+ 0. 87	46. 52+ 2. 54	24. 66+ 1. 06	黏质壤土	4. 16	2. 96	5. 15	3. 09	5. 46	4. 31
样 6	59. 1+ 0. 12	39. 68+ 0. 68	0. 49+ 0. 01	黏土	56. 23+ 1. 43	37. 59+ 1. 81	5. 45+ 0. 14	黏土	0. 2	1. 72	1. 26	2. 54	4. 81	2. 57

参考文献

[1]依丽燕．土壤物理研究法[M]．北京大学出版社，2009(8)：19-43.

[2]武天云，李凤民，钱佩源．利用离心法进行土壤颗粒分级[J]．应用生态学，2004，15(3)：477-481.

[3]刘广通，海春义，李占宏．应用吸管法进行沙土机械组成分析的实验研究[J]．水土保持研究，2007，14(2)：121-123.

[4]王成燕，张丽君．激光法传统方法比较测定土壤粒度的研究[J]．环境与发展，2010，22(6)：57-62.

[5]马艳霞，冯秀丽，叶银灿，等．比重计法和吸管法粒度分析比较[J]．海洋科学，2002，26(6)：63-67.

[6]曹玉鹏，邓永降，洪振舜．激光法与比重计法粒度分布相关性试验研究[J]．东南大学学报，2012，42(4)：755-760.

[7]GB7845-87，中华人民共和国国家标准 森林土壤颗粒组成(机械组成)测定方法[S].

[8]LY/T 1225-1999，中华人民共和国林业部行业标准 森林土壤颗粒组成(机械组成)测定方法[S].

[9]章明奎．不同预处理方法对高铁土壤颗粒分析结果的影响[J]．浙江农业大学学报，1996(1)：94-97.

[10]刘雪琴，黄元仿．应用激光粒度仪分析土壤机械组成的实验研究[J]．土壤通报，2005，36(4)：579-582.

[11]NY/T1121.3-2006 中华人民共和国农业部行业标准 土壤机械组成测定方法[S].

[12]杨乐苏，于彬．比重计法测定酸性土壤机械组成的方法改进[J]．河南林业科技，2005，25(3)：43-44，52.

[13]陈丽琼．比重计法测定土壤颗粒组成的研究[J]．环境科学导刊，2010，29(4)：97-99.

[14]胡琼英，兰叶青，薛家骅．土壤胶体稳定性影响因素[J]．土壤，1996(6)：290-294.

[15]覃宏华，丁武泉，张智，等．不同环境条件对土壤细颗粒絮凝沉降的影响研究[J]．人民长江，2013，(9)：72-74.

[16]施鲁莎，邓东升，别学清，等．国外比重计法颗粒分析试验相关性研究[J]．三峡大学学报(自然科学版)，2015，37(4)：39-43.

[17]王敏，洪健．对密度计沉降法颗粒分析试验的一点改进[J]．山西建筑，2009，35(16)：173-174.

[18]张建根，刘学芹，宋胜虎．浅析颗粒分析中比重计读数校正[J]．矿产勘查，2006，9(7)：31-32.

[19]李兴林，李福春．用沉降法和激光法测定土壤粒度对比研究[J]．土壤，2011，43(1)：130-134.

[20]曹晓娟．关于颗粒分析试验的误差分析和对策研究[J]．铁道工程学报，2014(2)：38-40.

[21]刘学芹．颗粒分析试验成果影响因素分析[J]．城市道桥与防洪，2013(7)：304-305.

[原载：水土保持通报，2017，37(06)]

广西大青山西南桦人工林藤本植物区系分析

刘士玲[1,2] 杨保国[1] 贾宏炎[1] 庞圣江 张 培[1] 黄柏华[1]
([1]中国林业科学研究院热带林业实验中心，广西凭祥 532600；[2]广西友谊关森林生态系统国家定位观测研究站，广西凭祥 532600)

摘 要 分析广西大青山西南桦(*Betula alnoides*)人工林藤本植物区系特点，为开展人工林藤本植物研究提供参考依据。通过野外样地调查和查阅相关文献资料，分析广西大青山西南桦人工林野生藤本植物区系的物种组成、生活型、攀缘方式和地理分布类型。大青山西南桦林下共有野生藤本20科27属30种，其中双子叶藤本植物具有明显优势，占该群落藤本植物总种数的93.33%，蕨类藤本和裸子藤本较贫乏，均仅有1科1属1种；在科属构成中，单种科和单种属比较集中；木质藤本是大青山西南桦群落的优势藤本，占总种数的70.00%；藤本生活型以高位芽植物为主，占总种数的60.00%，其余各生活型所占比例均较小；攀缘类藤本以缠绕类种类最多，占总种数的60.00%，其次是卷曲类，搭靠类和吸固类所占比例较小；分布区类型以热带地理成分为主。西南桦人工林藤本植物物种种类较多，生活型和地理分布类型较复杂，有较大的开发利用价值。

关键词 西南桦；野生藤本植物；物种组成；生活型；地理分布

Floristic Analysis for Vines in *Betula alnoides* Plantation in Daqingshan, Guangxi

LIU Shiling[1,2], YANG Baoguo[1], JIA Hongyan[1], PANG Shengjiang[1], ZHANG Pei[1], HUANG Bohua[1]
([1]*Experimental Center of Tropical Forestry, Chinese Academy of Forestry, Pingxiang* 532600, *Guangxi, China*; [2]*Guangxi Youyiguan Forest Ecosystem National Research Station, Pingxiang* 532600, *Guangxi, China*)

Abstract: Flora characteristics of vines in *Betula alnoides* plantation in Daqingshan were studied in order to provide reference for the research of vines in plantation forests. Species composition, life forms, climbing styles, gepgraphic distribution of wild vine resources in Daqingshan were studied using the methods of field investigation and literature survey. There were 30 species of wild vines, belonging to 27genera in 20 families, most of which were dicotyledonous plants, accounting for 93. 33% of the total. Only one families, one genera and one species of pteridophyte were found in the region, so did gymnosperm. The wild vines were composed mainly of singular families with singular genera. Woody vines were dominant vines in *B. alnoides* plantation in Daqingshan, making up 70. 00% of the total. In terms of life-form, phanerophyte vines outnumber others, accounting for 60. 00% of total types, and the proportions of other forms were small. . In terms of climbing styles, twining vines were the main type, accounting for 60. 00% of the total, followed bycurling vines, while the minority styles include hookingand and adhering vines. In terms of geographic distributin types, the vines were mainly tropical plants. In *B. alnoides* plantations, there are a variety of wild vine species with compolicated life forms and distribution types. There are great development and utilization value.

Key words: *Betula alnoides*; wild vine; species composition; life form; gepgraphic distribution

0　引言

【研究意义】藤本植物是热带、亚热带森林植被物种多样性和生物量的重要贡献者和森林结构中重要的外貌特征体现者[1,2]，也是群落动态的重要影响因素之一[3]。西南桦(*Betula alnoides*)是我国热带、南亚热带地区造林面积最大的乡土阔叶树种，具有较高的经济价值和生态价值[4,5]。根据2014年二类调查数据统计，大青山林区内现有西南桦人工林约300hm²，在其经营过程中，藤本植物经常作为有害植物来清除，缺乏对其重要的经济、药用、观赏和生态等价值的深入认识。因此，分析广西大青山西南桦人工林藤本植物的特点，对全面认识西南桦人工林藤本植物群落的植被性质和特征及进一步保护和利用藤本植物资源具有一定的现实意义。【前人研究进展】许多学者对藤本植物尤其是天然林群落中藤本植物进行了较深入地研究，但主要集中在生态适应性、生理生态学特性、植物多样性和开发利用价值等方面。在生态适应性方面，蔡永立和宋永昌[6]研究发现，天童常绿阔叶林藤本植物的叶片结构在不同种类及同种藤本不同植物体之间均具有一定的差异；梁松洁等[7]研究发现，北方地区藤本类忍冬叶片表皮形态和解剖结构与生态适应性之间有很强的相关性。在生理生态学特性方面，夏江宝等[8]研究发现，不同土壤水肥条件下紫藤叶片的*Pn*、*Tr*及WUE对土壤湿度和光照强度的变化具有明显的阈值响应；周军等[9]研究发现，细胞膜透性和丙二醛含量增加速度与藤本植物抗旱能力成反比，而脯氨酸含量增加速度与藤本植物抗旱能力成正比；周杰良等[10]研究发现，蔓长春花对Cd污染的耐性较强，具备了镉超富集植物的基本特征。在植物多样性方面，Yan和Qi[11]研究发现，湖南壶瓶山有藤本植物330种，隶属于44科111属；Hu和Li[12]研究发现，欧亚大陆和北美有藤本植物6659种，隶属于101科809属；肖之强等[13]研究发现，铜壁关自然保护区有藤本植物676种，隶属于65科209属。在开发利用价值方面，刘晓铃等[14]和王业社等[15]研究发现，藤本植物具有水土保持、改良土壤和美化环境等生态功能，具有重大开发利用价值。【本研究切入点】目前，国内对西南桦人工林内藤本植物区系特点的研究鲜见报道。【拟解决关键问题】调查分析广西大青山西南桦人工林群落藤本植物的物种多样性、攀缘习性及地理分布区类型等，全面认识其群落的植被性质和特征，为开展人工林藤本植物研究提供参考依据。

1　材料与方法

1.1　研究区域概况

研究区域为广西凭祥市中国林业科学研究院热带林业实验中心科学试验示范林区(106°41′~106°52′E，22°01′~24°16′N)，海拔190~640m；属典型南亚热带季风气候，年均降水量1200~1600mm，相对湿度80%~84%，降水主要集中在4~10月；年平均气温21.5℃。土壤主要为花岗岩发育的红壤和赤红壤，风化壳厚度6m以上，土壤呈酸性(pH4.3~5.1)。该区西南桦人工林林分类型为西南桦纯林或混交林，混交模式主要是西南桦×马尾松(*Pinus massoniana*)、西南桦×杉木(*Cunninghamia lanceolata*)、西南桦×红椎(*Castanopsis hystrix*)等。

1.2　研究方法

对广西大青山试验示范林区内西南桦人工林的藤本植物进行实地调查，结合不同的群落类型，采用典型选样，样地面积为20m×30m，共47个样地，分别详细记录调查地点、群落类型及藤本植物种类、生活习性、攀缘方式、生境、坡度、坡位、坡向、海拔及人为干扰情况等。根据调查结果并参考相关文献资料[6,16,17]建立广西大青山西南桦人工林野生藤本植物资源数据库，对其物种组成、生活型、攀缘方式和地理分布类型等进行分析。

2　结果与分析

2.1　广西大青山西南桦人工林藤本植物物种组成

由表1和表2可知，广西大青山西南桦人工林共有野生藤本植物30种，隶属于20科27属。其中，被子植物有18科25属28种，均为双子叶植物，占藤本植物总种数的93.33%；蕨类藤本和裸子藤本相对较少，分别仅有1科1属1种，各占藤本植物总种数的3.33%。该群落藤本植物中，全为木质藤本的科有13科，全为草质藤本的科仅有4科，其中防己科(Menispermaceae)包含木质和草质藤本的比例均等，茜草科(Rubiaceae)和菊科(Compositae)中木质藤本为草质藤本种数的2倍。说明大青山西南桦人工林中双子叶藤本植物具有明显的优势，且以木质藤本植物为构成该区西南桦人工林藤本植物的主要类群。

表 1 西南桦人工林藤本植物的科、属和种组成

科名	属数	种数	
		木质藤本	草质藤本
百合科 Liliaceae	1	2	—
蝶形花科 Fabaceae	1	1	—
豆科 Leguminosae	2	2	—
防己科 Menispermaceae	2	1	1
海金沙科 Lygodiaceae	1	—	1
夹竹桃科 Apocynaceae	1	1	—
桔梗科 Campanulaceae	1	—	1
菊科 Compositae	3	2	1
莲叶桐科 Hernandiaceae	1	1	—
马钱科 Loganiaceae	1	1	—
买麻藤科 GnetaceaeLindl	1	1	—
毛茛科 Ranunculaceae	1	—	2
猕猴桃科 Actinidiaceae	1	1	—
木通科 Lardizabalaceae	1	1	—
葡萄科 Vitaceae	1	—	1
茜草科 Rubiaceae	2	2	1
蔷薇科 Rosaceae	1	1	—
五桠果科 Dilleniaceae	1	1	—
旋花科 Convolvulaceae	2	2	—
紫金牛科 Myrsinaceae	2	2	—

表 2 西南桦人工林藤本植物的分类及数量特征

类群	蕨类植物		裸子植物		被子植物	
	数量	百分比(%)	数量	百分比(%)	数量	百分比(%)
科	1	5.00	1	5.00	18	90.00
属	1	3.70	1	3.70	25	92.60
种	1	3.33	1	3.33	28	93.33
草质藤本	1	3.33	0	0	7	23.33
木质藤本	0	0	1	3.33	21	70.00

以科所含的种数进行统计，西南桦人工林藤本植物资源中含 3 种的科有 2 科，即茜草科(Rubiaceae)和菊科(Compositae)，分别占该区西南桦群落藤本植物总科数的 10.00%和总种数的 20.00%；含 2 种的科有 6 科，即百合科(Liliaceae)、豆科(Leguminosae)、防己科(Menispermaceae)、毛茛科(Ranunculaceae)、旋花科(Convolvulaceae)和紫金牛科(Myrsinaceae)，分别占该群落藤本植物总科数的 30.00%和总种数的 40.00%；单种科有 12 科，所占比例达该群落藤本植物总科数的 60.00%和总种数的 40.00%。说明广西大青山西南桦人工林野生藤本植物主要属于单种科，寡种科(2 科、3 科)较少，无大种科。

以属所含的种数进行统计，西南桦人工林藤本植物资源中含2种的属有3属，即菝葜属(*Smilax*)、铁线莲属(*Clematis*)和玉叶金花属(*Mussaenda*)，分别占该区该群落藤本植物总属数的11.11%和总种数的20.00%；单种属有24属，如猕猴桃属(*Actinidia*)、酸藤子属(*Embelia*)和悬钩子属(*Rubus*)等，占该区该群落藤本植物总属数的88.89%和总种数的80.00%。说明广西大青山西南桦人工林野生藤本植物单种属具有明显优势，含2种属的优势次之，在一定程度上丰富了该群落的植物多样性。

2.2 广西大青山西南桦人工林藤本植物的生活型

植物的生活型是植物对生境条件长期适应而在外貌上反映出来的植物类型。根据蔡永立和宋永昌[3]修订的藤本植物生活型系统，对广西大青山西南桦人工林野生藤本植物生活型进行划分和分析，结果见表3。

表3 野生藤本植物生活型统计

类型	各生活型数量					
	高位芽	地上芽	地面芽	地下芽	一年生	合计
缠绕类	13	3	2	—	—	18
卷曲类	2	1	—	4	—	7
叶柄卷曲类	—	—	—	2	—	2
卷须类	2	1	—	2	—	5
搭靠类	2	—	1	—	1	4
钩搭类	1	—	—	—	—	1
枝搭类	—	—	1	—	1	2
刺搭类	1	—	—	—	—	1
吸固类	1	—	—	—	—	1
不定根类	1	—	—	—	—	1
合计	18	4	3	4	1	30

该区西南桦人工林群落高位芽藤本具有明显的优势，有18种，占总种数的60.00%，具有明显优势，其次是地上芽和地下芽藤本，其种数和所占比例均分别为4种和13.33%，地面芽藤本有3种，占总种数的10.00%，一年生藤本仅1种，仅占3.33%的比例。从藤本植物的质地特征可知(表2)，该群落藤本植物木质和草质藤本分别有22种、8种，所占比例分别为73.33%和26.67%，说明木质藤本是大青山西南桦人工林的优势藤本。

由表3可知，广西大青山西南桦群落藤本植物攀缘类型有缠绕、卷曲、搭靠和吸固类4种类型，其中，缠绕类种类最多，有18种，占总种数的60.00%。在缠绕类藤本植物中，高位芽植物所占比例最高，有13种，占高位芽藤本植物的72.22%；地上芽植物有3种，占地上芽藤本植物的75.00%；地面芽植物有2种，无地下芽和1年生植物。卷曲类藤本植物有7种，占总种数的23.33%，且以地下芽植物为主(4种)，其次是高位芽(2种)和地上芽(1种)植物，其中，卷须类有5种，占藤本植物总种数的16.67%，如乌蔹莓(*Cayratia japonica*)、菝葜(*Smilax china*)和抱茎菝葜(*Smilax ocreata*)等；叶柄卷曲类有2种，占总种数的6.67%，均为地下芽植物，如铁线莲属(*Clematis*)。搭靠类藤本植物有4种，占总种数的13.33%，包括钩搭类(1种)、枝搭类(2种)和刺搭类(1种)3种类型。吸固类藤本植物所占比例最小，仅络石(*Trachelospermum jasminoides*)1种，为不定根类。

2.3 藤本植物地理分布

在植物分类学上，属的形态特征和分布区较稳定，能较好地反映植物系统发育过程中的进化分化情况和地区性特征，故适宜以属为单位研究植物的分布区类型[18]。根据陈景艳等[16]的蕨类植物分布区类型和吴征镒[17]的中国种子植物分布区类型的划分标准，将广西大青山西南桦人工林野生藤本植物的分布型统计于表4。

表4 藤本植物属的分布区类型

分布区类型及亚型	属数	占属的百分率(%)
1 世界分布	2	7.40
2 泛热带分布	11	40.74
4 旧世界热带分布	5	18.52
5 热带亚洲至热带大洋洲分布	2	7.40
7 热带亚洲(印度-马来西亚)分布	2	7.40
7-1 爪哇(或苏门答腊)、喜马拉雅间断或星散分布到华南、西南	1	3.70
8 北温带分布	1	3.70
9 东亚和北美洲间断分布	1	3.70
14 东亚分布	1	3.70
15 中国特有种分布	1	3.70

从属的分布区类型看，热带成分占优势，有21属，占总属数的77.78%，说明广西大青山西南桦人工林野生藤本植物区系热带属性较为突出。在热带分布类型中，泛热带分布所占比例最大，有11属，如菝葜属、钩藤属(*Uncaria Schreber*)、崖豆藤属(*Millettia Wight*)、买麻藤属(*Gnetum*)和青藤属(*Illigera*)等；旧世界热带分布所占比例次之，包含玉叶金花属、乌蔹莓属(*Cayratia*)、艾纳香属(*Blumea*)、信筒子(酸筒子)属(*Embelia Burm*)和千金藤属(*Stephania*)5属；其余几种热带成分较少。温带成分有3属，占总属数的11.11%，为该区第二大分布类型，其中东亚成分有络石属(*Trachelospermum*)和猕猴桃属(*Actinidia*)2属，北温带分布型有蒲公英属(*Taraxacum*)。此外，世界分布型有2属，即铁线莲属和悬钩子属(*Rubus*)；中国特有分布型仅有大血藤属(*Sargentodoxa*)，是中国植物区系中最有代表的特有成分之一，大多数为第三纪古老孑遗植物[19]。

3 讨论

广西大青山西南桦人工林野生藤本植物有20科27属30种，其中被子植物有18科25属28种，且均为双子叶植物，蕨类藤本和裸子藤本相对较贫乏，与王业社等[15]对湖南城步野生藤本植物资源、魏宗贤等[18]对庐山野生藤本植物资源的研究结果相似。本研究中，藤本植物主要集中于单种科、属，寡种科、属较少，无大种科、属，与广西大青山降水量大(1200~1600mm)，海拔跨度较大(190~640m)的自然地理特征密切相关，且西南桦人工林长期受人为干扰，一定程度的干扰丰富了藤本植物的多样性。单种属占有较大比率，说明藤本植物分化和变异较大，且单种属多为古特有属[17]，反映了广西大青山藤本植物区系成分的古老性特点。

西南桦人工林野生藤本植物以热带成分占优势，温带成分也占有一定的比例，可能与广西大青山处于亚热带南缘，具有明显的热带亲缘性，但受一定的温带成分影响有关。西南桦群落藤本植物攀缘类型有缠绕、卷曲、搭靠和吸固类4种类型，与蔡永立和宋永昌[3]的研究结果相似。本研究中，缠绕类藤本植物占总种数的60.00%，明显多于其他生活型藤本植物，与颜立红等[19]的研究结果一致。植物界的攀缘机制起源于热带，故保存了较多最原始的缠绕类攀缘方式，但亚热带森林是热带森林的衍生物，兼具了较进化的卷曲类、搭靠类等攀缘机制类型[11]。

陈亚军等[20]研究表明，木质藤本在树木生长和更新、物种多样性维持和森林生态系统动态变化过程中发挥着重要作用，王业社等[15]研究认为，藤本植物也具有经济、药用、食用和观赏等价值。广西大青山林区人工林藤本植物资源丰富，且以木质藤本为优势藤本，但对该林区藤本植物利用价值的研究尚属空白，今后在加强对藤本植物基础研究的同时，应加强对藤本植物资源合理保护和开发利用的研究。

4 结论

广西大青山西南桦林下野生藤本植物种类较多，生活型和地理分布类型较复杂，有较大的开发利用价值。

参考文献

[1]PUTZ F E. The natural history of lianas on Barro Colordo Island, Panama[J]. Ecology, 1984, 65: 1713-1724.

[2]SCHNITZER S A, BRONGERS F. The ecology of lianas and their role in forests[J]. Trend in Ecology & Evolution, 2002, 17(5): 223-230.

[3]蔡永立，宋永昌．藤本植物生活型系统的修订及中国亚热带东部藤本植物的生活型分析[J]．生态学报，2000，20(5)：808-814.

[4]何友均，覃林，李智勇，等．西南桦纯林与西南桦×红椎混交林碳贮量比较[J]．生态学报，2012，32(23)：7586-7594.

[5]孟梦，陈宏伟，刘永刚，等．西双版纳西南桦、山桂花人工林水源涵养效能研究[J]．云南林业科技，2002，3：46-49.

[6]蔡永立，宋永昌．浙江天童常绿阔叶林藤本植物的适应生态学Ⅰ叶片解剖特征的比较[J]．植物生态学报，2001，25(1)：90-98.

[7]梁松洁，张金政，张启翔，等．北方地区藤本类忍冬叶表皮结构及其生态适应性比较研究[J]．植物研究，2004，24(4)：434-438.

[8]夏江宝，张光灿，刘刚，等．不同土壤水分条件下紫藤叶片生理参数的光响应[J]．应用生态学报，2007，18(1)：30-34.

[9]周军，武金翠，杜宝明，等．4种藤本植物的抗旱性比较[J]．江苏农业学报，2016，32(3)：674-679.

[10]周杰良，葛大兵，李树战，等．藤本植物中具镉超积累特征植物的筛选[J]．林业科学研究，2016，29(4)：515-520.

[11]YAN L H, QI J C. Vine diversity of Huping Mountain in Hunan Province[J]. Scientia Silvae Sincae, 2007, 43(6): 20-26.

[12]HU L, LI M G. Diversity and distribution of climbing plants in Eurasia and North Africa. In: Biodiversity of Lianas (ed. Parthasarathy N)[M]. Springer, Switzerland, 2015: 57-79.

[13]肖之强，马晨晨，代俊，等．铜壁关自然保护区藤本植物多样性研究[J]．热带亚热带植物学报，2016，24(4)：437-443.

[14]刘晓铃，谢树莲，陈丽．山西历山自然保护区藤本植物资源研究[J]．山西大学学报(自然科学版)，2007，30(4)：544-549.

[15]王业社，陈立军，杨贤均，等．湖南城步野生藤本植物资源及开发利用研究[J]．草业学报，2015，24(8)：11-23.

[16]陈景艳，邓伦秀，邹胜北．贵州蕨类植物属的分布区类型及区系特征[J]．贵州林业科技，2013，41(4)：19-23.

[17]吴征镒．中国种子植物属分布区类型[J]．云南植物研究(增刊Ⅳ)，1991：1-139.

[18]魏宗贤，宋满珍，牛艳丽，等．庐山地区野生藤本植物区系与生活型[J]．浙江农林大学学报，2013，30(4)：505-510.

[19]颜立红，祁经承，彭春良．湖南湖北藤本植物物种多样性和生态特征[J]．林业科学，2006，42(11)：17-22.

[20]陈亚军，陈军文，蔡志全．木质藤本及其在热带森林中的生态学功能[J]．植物学通报，2007，24(2)：240-249.

[原载：南方农业学报，2017，48(05)]

我国森林生态系统枯落物现存量研究进展

刘士玲[1,2] 郑金萍[1] 范春楠[1] 杨保国[2] 郭忠玲[1]
([1]北华大学，吉林 132000；[2]中国林业科学研究院热带林业实验中心，广西凭祥 532600)

摘 要 在参考大量相关文献的基础上，对我国森林生态系统枯落物现存量研究进展情况进行了归纳和总结。资料表明，我国枯落物现存量的研究主要集中在亚热带和温带，热带相对较少。枯落物现存量的研究主要采用收获法，但样方布设、面积、数量及取样时间等尚没有统一的规定。在一个大的区域尺度上，气候是影响其变化的主导因子，不同气候带森林枯落物现存量的顺序为温带>亚热带>热带，森林枯落物现存量还受到林分起源、群落组成和结构、群落发育阶段、土壤、地形条件及干扰的影响。此外，在全球变化的情景下，对我国森林枯落物现存量研究中存在的问题及今后研究的方向进行了探讨。

关键词 森林生态系统；枯落物；现存量；影响因素；中国

Research Progress in Litter Accumulation of Forest Ecosystem in China

LIU Shiling[1,2], ZHENG Jinping[1], FAN Chunnan[1], YANG Baoguo[2], GUO Zhongling[1]
([1] *Beihua University, Jilin 132000, Jilin, China*; [2] *Experimental Center of Tropical Forestry, Chinese Academy of Forestry, Pingxiang 532600, Guangxi, China*)

Abstract: In this paper, the authors summarized and concluded the research progress on litter accumulation of forest ecosystem in China based on literatures review. The results showed that the current researches of litter accumulation are mainly focused on subtropical and temperate forests, with litter attention put to researches on tropical forests. Harvest is the main research method adopted now, but there is no unified standard of the quadrant layout, area, quantity as well as for sampling time. In a large scale, climate is the leading influencing factor affecting litter accumulation in forest ecosystems, and litter accumulation in forest of different climatic zones could be ordered as follows: temperate forest > subtropical forest > tropical forest. Forest litter accumulation is also influenced by origin, community development, community composition and structure, soil, terrain condition and interference. This paper also discussed the problems found in the research on forest litter accumulation in China under global climate change and explored the directions of future litter accumulation research.

Key words: forest ecosystems; litter; accumulation; influencing factor; China

森林枯落物是指由植物地上部分产生并归还到地表的所有有机物质的总称。有关森林枯落物的研究已从枯落物本身特征(如枯落物量、分解及影响因子)逐渐深入到其生态功能的研究等方面。对森林枯落物现存量(即单位面积林地上所积累的森林枯落物量)变化特征的研究，是了解森林生态系统养分循环过程的基础，有利于通过对地上枯落物的管理调控其生态系统功能和服务功能。随着全球碳循环问题受到广泛关注，枯落物现存量及其分解过程成为解译森林生态系统碳循环过程不确定性的重要途径。从现有文献来看，我国森林枯落物现存量的研究方法比较混乱，相关内容的总结相当匮乏。因此，非常有必要对我国枯落物现存量的研究进展情况进行梳理。

1 文献状况

本文收集了我国1989年以来公开发表的涉及森林枯落物内容的文献247篇，研究区域涉及我国热带、亚热带和温带，各气候区域涉及文献数及森林地表枯落物现存量统计结果见表1。

表1 枯落物现存量文献状况

气候带	枯落物现存量(t/hm²)	文献数(篇)
热带	1.93~15.90	14
亚热带	0.27~246.94	125
温带	0.35~246.00	108

从表1可以看出，我国森林枯落物现存量的研究主要集中在亚热带和温带，热带相对较少。经统计，森林枯落物现存量的变化范围为0.27~246.94t/hm^2，并有随纬度增大，现存量增加的趋势，其中温带森林枯落物现存量平均为17.62t/hm^2，亚热带枯落物现存量为14.73t/hm^2，而热带森林枯落物现存量仅为5.94t/hm^{2}左右。郑路和卢立华[1]也曾对我国温带、亚热带和热带的森林地表枯落物现存量进行过综述，结果分别为68.90t/hm^2、28.44t/hm^{2}和4.62t/hm^2，亦表现出随纬度升高而增加的趋势；但相应纬度带枯落物现存量的数值与本文统计结果存在一定差异，这可能是由于与他们仅归纳了不同气候带主要的森林类型所致，也可能与本文参考的文献时间跨度差异有关。

2 研究方法

目前，国内对于枯落物现存量的研究主要采用样方收获法，少数采用其他方法，如孙志虎等[2]使用了地统计学的克里格内插法结合定积分的方法。作为最常用的收获法，在具体实施过程中，所确定的取样时间及分层、样方空间布设方法、样方大小及其数量等都有较大的不同。

2.1 取样时间及分层

森林枯落物现存量的取样时间以及是否分层取样，是影响枯落物现存量精确估测的重要因素。多数学者对地表枯落物的研究只进行一次取样，取样时间在生长季初始期、凋落高峰期或森林生长结束期等。因枯落物现存量呈现随季节波动的现象，也有学者对枯落物进行了多次取样，以平均值或模型统计的结果作为现存量的指标[3,4]。从对森林枯落物现存量估测的精确度来看，多次取样的方式更具说服力。

如果仅测定枯落物的量，直接收集全部的枯落物即可。但基于研究目的不同，研究者会对样方内的枯落物采用分层取样，采用这类方法的文献占50.10%。其中，分2层取样文献最多，占该类文献的38.10%，且9成多是采用未分解层和半分解层进行取样；4层取样所占比例最少，仅占该类文献数的0.60%。

采用分层取样的方法尽管比较费时费力，但通过统计不同分层的厚度及其现存量，或进一步对未分解层和半分解层中枯落物的组分进行分组，统计不同组分的现存量，可以为进一步分析枯落物来源及其与地上植被的关系、枯落物的分解和养分循环速率提供更多的数据参考依据。

2.2 样方布设、样方大小及数量

森林枯落物现存量的样方布设、取样面积及数量尚没有统一的标准，这就为区域尺度上枯落物现存量的研究带来不确定性。取样样方在空间上的布设有随机设置样点法、机械布点法、S形路线法等。具体采用哪种方法没有优劣之分，主要依研究区域特征、枯落物在空间异质性等而定。

根据文献统计，确定的取样面积有18种之多，基本为正方形，大小介于0.01~4m^2。其中取样面积为1m×1m最多，占文献的43.50%；其次为0.50m×0.50m，占22.60%；其他还有0.1m^{2}和0.04m^{2}等。取样面积大小往往要考虑取样数量、研究林分发育年龄、群落组成和结构复杂程度等因素，取样数量多、幼龄、相对单纯的林分面积可小些，当然也要考虑到工作量。

取样样方数量为2~64个，其中多数采用的是3或5个样方。取样样方数量的确定需要满足统计和精度要求，同时也要综合考虑到取样样方面积大小、取样林分面积的大小、立地条件等因素。例如，何斌等[5]根据样地的立地条件差别在一次取样过程中确定不同样地的样方取样数量。肖瑜[6]对同一样地不同测量精度下各样方面积所需最少样方数的研究表明，随着精度提高和样方面积增大，所需样方数量减少。但目前有关取样样方面积、样方数量与样地面积之间的关系还缺乏研究。

3 影响枯落物现存量的因素

枯落物现存量主要取决于枯落量、分解速率及积累年限，主要受气候影响。此外，还受林分起源、群落发育阶段、群落组成和结构、地形条件、土壤及干扰等的影响。

3.1 气候因素

邹碧等[7]研究认为，在区域尺度上，气候特征是决定地表枯落物现存量的关键因素。其中，温度和湿度的作用最为突出，对枯落物现存量的影响主要体现在枯落量和分解速率2个方面，水热条件优越则枯落物现存量少[8]。

温度对枯落现存量的影响表现在纬度上，其作用于植物的生长和演替，改变森林类型，影响枯落量和分解速率，进而影响枯落物现存量，这点在本文“文献状况”部分已有表述。研究文献表明，不同气候带森林枯落量及其分解速率大小均为热带>亚热带>温带[9]。由此可见，热带较低的枯落物现存量源

于其较高的枯落量和较快的分解速率，而温带的高枯落物现存量则主要源于其低分解速率，亚热带枯落量和分解速率均中等，所以枯落物现存量处于两者之间。湿度对地表枯落物的累积作用主要体现为降水对其分解的影响。降水可以通过制约枯落物化学成分淋溶的物理过程直接影响枯落物的分解速率，也可以通过影响微生物和土壤动物的活动等间接影响其分解。

3.2 林分起源

一般天然林的枯落物输入量要显著大于人工林，天然林人为干扰程度相对较低，植物多样性高，森林树种组成和群落结构保存较好，林地地表有较多的枯落物输入，枯落物年归还量大、分解快，具有良好自培肥地力的能力[10]。人工林树种单一，间伐、清林等人为干扰较为严重，对森林枯枝落叶的组成和数量产生较大的影响。杨玉盛等[4]对不同林分起源格氏栲林枯落物现存量的研究结果也证明了这点。由此可见，若天然林转化为人工林，就会显著降低地表枯落物现存量，势必会对林地水土保持、涵养水源的功能产生长远影响。因此应提倡保护天然林，营造混交林，对现有人工纯林通过栽植伴生树种等途径进行近自然化改造。

3.3 群落组成和结构

枯落物现存量与森林群落组成、生长状况都有直接的关系。在气候条件相同的情况下，不同林龄、森林类型之间枯落物现存量存在显著差异。即使同一群落类型，由于调查季节的不同，枯落物现存量也可能存在明显差异。

国内学者对不同森林群落的枯落物现存量进行了大量的研究。多数研究表明，针叶林和针阔混交林枯落物现存量均大于阔叶林，但也有研究表明阔叶林枯落物现存量大于针叶林[11-14]。郑路和卢立华[1]分析认为，这种变化与气候带有一定关系，在亚热带，阔叶林枯落物现存量高于针叶林，而在温带则表现出针叶林枯落物现存量大于阔叶林。对于针阔混交林与针叶纯林枯落物现存量的大小研究，也尚未得出较为一致的结论。

林分密度和郁闭度的差异会引起林分内水热条件和生物量的变化，影响枯落物的累积量。对于同一森林类型而言，在一定范围内，林分密度增加，郁闭度增大，枯落物现存量增大。如在同一气候带条件下测定的川滇高山栎林枯落物现存量，薛建辉等[13]的测定结果(25.2 t/hm^2)要高于张远东等[15]的(12.12t/hm^2)，原因可能与前者有较高的林分密度和郁闭度(0.8)有关，而后者林分密度和郁闭度(0.5~0.7)较低。张峰等[16]研究的油松林枯落物现存量与魏强等[12]的差异也可能是相似的原因。然而，当林分密度超出一定限度，林木之间的竞争就会加剧，随之生长发育也会受限，使得枯落物现存量减少[17]。

3.4 群落发育阶段

目前，关于林龄与枯落物现存量的关系，较为一致的结论是枯落物现存量与林龄关系紧密，表现为随林龄增大，枯落物现存量先增大后趋于平缓或减小[5,18]。然而，由于立地条件和林分特征不同，林分衰老的过程存在较大差异。因此，尽管林龄影响枯落物的现存量的观点大家都已经形成共识，但要说明林龄与枯落物现存量的关系，就必须严格地控制其他因素。

森林枯落物现存量与森林生态系统的演替进程密切相关。演替主要是通过不同演替阶段林分类型及林下小气候的变化对枯落物现存量产生影响。从现有文献来看，目前研究不同演替阶段森林枯落物现存量，主要运用次生演替的空间序列代替时间序列的研究方法[19]。例如，以马尾松为演替初期的森林群落向常绿阔叶林群落的演替[11,19,20]，阔叶林群落间的演替及阔叶林群落向针阔混交林群落的演替[8,21]等。从研究结果看，有随森林类型向顶级群落常绿阔叶林演替，林下枯落物现存量逐渐增大的现象[19,20]；也有随群落演替枯落物现存量表现出先增大后减小或逐渐减小的趋势，在顶级群落阶段最小的结论[8,11,21]。显然，要想精确估测演替与枯落物现存量之间的定量关系，长期定位观测是今后必不可少的环节。

3.5 土壤因素

土壤类型的差异导致土壤理化性质和土壤生物类群的不同。有研究表明，枯落物现存量与土壤总空隙度、全氮、全磷、全钾、水分、pH 呈正相关，与土壤容重和 C 贮量呈负相关，尤其受土壤温度影响[22,23]。土壤微生物与枯落物现存量的相关研究甚为缺乏，现有研究只是发现土壤微生物量与枯落物现存量呈极显著的负相关关系[8]。由于土壤动物不仅能粉碎枯落物，增大枯落物比表面积，且其排泄的粪便能有效地改善土壤理化性质，使枯落物更容易分解，是导致枯落物现存量分解的主要因素。枯落物现存量亦与土壤动物种类和数量密切相关，潘开文等[24]研究发现枯落物的现存量，尤其是呈未分解状态的数量，与土壤动物的生物量、总个体密度都表现为显著的正相关。枯落物现存量与大型土壤动物群落生物量亦呈显著正相关[25]。杨赵和杨效东[26]研究发现，枯落物现存量与土壤节肢动物群落和主要类群密度呈显著的相关关系。

3.6　地型条件

3.6.1　海拔

海拔的变化常伴随着温度、降水、光照、土壤等许多因子的改变，实质上海拔与枯落物现存量的关系极为复杂，它是环境、生物综合作用的结果。低海拔处虽然枯落量大，但分解速率较快，加之人为活动较频繁，枯落物现存量较少；中山地带最适合植物生长，且人类活动减少，枯落物现存量较高；达到一定海拔，气候变冷，降水减少，植被减少，枯落量减少，虽然枯落物分解很慢，但枯落物现存量并不大。刘颖等[27]对长白山4种森林类型的枯落物现存量的研究表明，同一纬度枯落物现存量随海拔梯度增加呈现先升高后降低趋势。张万儒等[28]对海拔为1200~3300m的4块垂直植被带试验林地的研究，刘蕾等[29]对神农架海拔梯度上4种典型森林以及董金相[30]对戴云山黄山松林枯落物现存量的研究也得到了类似结果。

3.6.2　其他地型因素

坡位主要通过影响土壤水分和养分的再分配影响植被生长，进而对枯落物现存量产生影响。刘刚等[31]认为不同坡位的枯落物现存量大小顺序依次为：山脊>上坡>中坡>下坡>平地；李科[32]认为，麻竹林地坡中、坡下的枯落物现存量都显著高于坡上部；刘中奇等[33]认为，沟坡的枯落物现存量大于梁坡的。

坡度不同，导致土壤水分和坡面上的植被存在差异，且降水形成的径流对枯落物产生直接的推冲作用。潘复静等[34]和刘中奇等[33]都发现，坡度对枯落物现存量有显著影响。肖瑜[6]对坡度与取样面积的关系的研究也间接表明，坡度对枯落物现存量有影响。因此，在实际工作中，应根据林地的坡度状况及取样面积，适当地增减取样数目，从而提高估测林地枯落物现存量的精度。

不同坡向使光照、水分和土壤理化性质等存在较大差异，生境异质性引起植物的生长状况与植被组成的变化，进而导致枯落物现存量的不同[34]。董金相[30]在研究戴云山黄山松枯落物现存量时发现，北坡的枯落物现存量明显高于南坡；刘中奇等[33]对阴坡的枯落物现存量大于阳坡的研究结论也证明了这一观点。

3.7　干扰因素

森林干扰可能导致生物多样性、林地生产力及林内生境条件发生变化，继而导致枯落物现存量发生明显改变。就目前研究结果来说，干扰因强度的不同，对植物群落的影响有正反2个方面，在受到强度干扰时，植物物种丰富度较低，地面枯落物累积减少，而长期缺乏干扰的群落枯落物积累会增加。

经营方式和强度对枯落物现存量都有较大的影响。例如，竹林在不进行采伐和采笋的粗放经营状态下枯落物现存量较高[35]，经过低产改造和封山育林的林分枯落物现存量也会明显增高[31]，而经过抚育和管护的林分枯落物现存量则有降低的趋势，特别是随着采伐强度的增大，林地枯落物现存量递减尤为明显[36,37]。异常的冰雪灾害也可以通过增加非正常枯落物或造成的林冠折损引起林地光照增强，改变地表枯落物的现存量[38,39]；林火作用使森林枯落物现存量明显减少[40]。

4　问题与展望

枯落物与气候、土壤、植物等紧密相关，全球变化导致的这些因子的变化将影响枯落物与环境、生物的作用关系，并进一步影响其生态作用。鉴于全球变化的背景和当前的研究进展，关于枯落物现存量的研究应从以下几个方面进一步改进和完善。

4.1　调查方法进一步科学化

首先，目前国内对于森林枯落物现存量的估测仍主要采用经典的收获法，但在实际操作过程中调查时间、取样点的数量、面积和空间布设还缺乏科学的量化技术体系，甚至过多地考虑成本而忽视了精度要求，由此导致同一群落不同研究者测定结果的不确定性大，不能真实反映实际状况，结论缺乏可比性。其次，森林枯落物现存量存在一定的季节动态，由于取样时间的不同也会导致调查结果产生偏差；尽管也有研究者通过定期的动态测定研究枯落物现存量，但在如何统计和界定年平均现存量方面方法不多。因此，今后枯落物研究如何确定相对统一的研究方法、且建立不同空间和时间尺度下枯落物调查规范势在必行，以提高研究结果的科学性和参考价值。

4.2 重视多因子交互作用

我国对森林地表枯落物现存量的研究多集中在其组分、量和持水能力等方面，而在其影响机制方面的研究还很薄弱。森林枯落物现存量的变化实质浓缩了森林生态系统的各生态过程，包括树种组成和结构、生物多样性的复杂程度、林分发育阶段、环境状况的改变、干扰因素的种类和强度等。在自然状态下，各种因子同时作用于枯落物现存量，现存量的大小是多因素共同作用的结果，如单独研究每个因子的作用往往会忽略因子间的互作效应，因此多个因子对枯落物现存量的共同影响亟须采用更先进和统一的方法开展更长期、深入的研究。只有准确了解多因子对森林枯落物现存量影响的综合作用机制，才能为固碳森林的林分结构改造及森林优化经营等提供重要科学依据。

4.3 加强枯落物在全球变化下的响应研究

目前，虽有不少关于全球变化对枯落物影响的研究，但由于时间尺度和其他干扰因素的影响，研究得往往不够深入，如气候变暖是如何引起生态系统水热条件改变，使枯落量和枯落物分解速率发生相应的变化[41]，进而引起枯落物现存量的变化？诸如此类问题都有待于进一步加强研究。因此，在今后的研究中应充分利用区域或全球范围的生态系统定位站网络，并与近红外光谱技术、遥感技术、地理信息系统等相结合，逐步实现对我国枯落物现存量的动态实时监测，获得可比性强的数据进行综合分析，以形成我国枯落物现存量的总体数量分布格局，为全面、深刻理解并准确预估我国森林生态系统对全球变化的响应提供基础支撑。

参考文献

[1]郑路，卢立华．我国森林地表凋落物现存量及养分特征[J]．西北林学院学报，2012，27(1)：63-69.

[2]孙志虎，牟长城，张彦东．地统计学方法在长白落叶松人工林凋落物现存量估测中的应用[J]．生物数学学报，2007，22(4)：703-710.

[3]郑金萍，郭忠玲，徐程扬，等．长白山主要次生林的枯落物现存量组成及持水特性[J]．林业科学研究，2011，24(6)：736-742.

[4]杨玉盛，郭剑芬，林鹏，等．格氏栲天然林与人工林枯枝落叶层碳库及养分库[J]．生态学报，2004，24(4)：359-367.

[5]何斌，黄承标，韦家国，等．不同林龄秃杉人工林凋落物储量及其持水特性[J]．东北林业大学学报，2009，37(3)：44-46.

[6]肖瑜．样方面积和数目对测定次生林凋落物精度的影响[J]．生态学报，1989，9(1)：59-65.

[7]邹碧，李志安，丁永祯，等．南亚热带4种人工林凋落物动态特征[J]．生态学报，2006，26(3)：715-721.

[8]黄宗胜，符裕红，喻理飞．喀斯特森林植被自然恢复中凋落物现存量及其碳库特征演化[J]．林业科学研究，2013，26(1)：8-14.

[9]刘强，彭少麟．植物凋落物生态学[M]．北京：科学出版社，2010，26-28.

[10]YANG Y S，LING P，GUO J F，et al. Litter production，nutrient return and leaf-litter decomposition in natural and monoculture plantation forests of Castanopsis kawakamii in subtropical China[J]．Acta Ecologica Sinica，2003，23(7)：1278-1289.

[11]闫俊华，周国逸，唐旭利，等．鼎湖山3种演替群落凋落物及其水分特征对比研究[J]．应用生态学报，2001，12(4)：509-512.

[12]魏强，凌雷，张广忠，等．甘肃兴隆山主要森林类型凋落物累积量及持水特性[J]．应用生态学报，2011，22(10)：2589-2598.

[13]薛建辉，郝奇林，何常清，等．岷江上游两种亚高山林分枯落物层水文特征研究[J]．生态学报，2009，23(3)：168-172.

[14]殷沙，赵芳，欧阳巡志，等．马尾松木荷不同比例混交林枯落物和土壤持水性能比较分析[J]．江西农业大学学报，2015，37(3)：454-460.

[15]张远东，刘世荣，马姜明．川西高山和亚高山灌丛的地被物及土壤持水性能[J]．生态学报，2006，26(9)：2775-2782.

[16]张峰，彭祚登，安永兴，等．北京西山主要造林树种林下枯落物的持水特性[J]．林业科学，2010，46(10)：6-14.

[17]赵磊，王兵，蔡体久，等．江西大岗山不同密度杉木林枯落物持水与土壤贮水能力研究[J]．水土保持学报，2013，27(1)：203-208.

[18]彭云，丁贵杰．不同林龄马尾松林枯落物储量及其持水性能[J]．南京林业大学学报(自然科学版)，2008，32(4)：43-46.

[19]张庆费，徐绒娣．浙江天童常绿阔叶林演替过程的凋落物现存量[J]．生态学杂志，1999，18(2)：17-21.

[20]雷云飞，张卓文，苏开君，等．流溪河森林各演替阶段凋落物层的水文特性[J]．中南林业科技大学学报，2007，27(6)：38-43.

[21]余广彬，杨效东．不同演替阶段热带森林地表凋落物

和土壤节肢动物群落特征[J]. 生物多样性，2007，15(2)：188-198.
[22]路翔. 中亚热带四种森林凋落物及碳氮贮量比较[D]. 湖南：中南林业科技大学，2012.
[23]逮军锋. 不同林龄油松人工林凋落物及其对土壤理化性质的影响研究[D]. 兰州：甘肃农业大学，2007.
[24]潘开文，何静，吴宁. 森林凋落物对林地微生境的影响[J]. 应用生态学报，2004，15(1)：153-158.
[25]肖以华，佟富春，杨昌腾，等. 冰雪灾害后的粤北森林大型土壤动物功能类群[J]. 林业科学，2010，46(7)：99-105.
[26]杨赵，杨效东. 哀牢山不同类型亚热带森林地表凋落物及土壤节肢动物群落特征[J]. 应用生态学报，2011，22(11)：3011-3020.
[27]刘颖，韩士杰，林鹿. 长白山4种森林凋落物分解过程中养分动态变化[J]. 东北林业大学学报，2009，37(8)：28-30.
[28]张万儒，许本彤，杨承栋，等. 山地森林土壤枯枝落叶层结构和功能的研究[J]. 土壤学报，1990，27(2)：121-130.
[29]刘蕾，申国珍，陈芳清，等. 神农架海拔梯度上4种典型森林凋落物现存量及其养分循环动态[J]. 生态学报，2012，32(7)：2142-2149.
[30]董金相. 戴云山黄山松林碳储量及其影响因子研究[D]. 福建：福建农林大学，2012.
[31]刘刚，朱剑云，叶永昌，等. 东莞主要森林群落凋落物碳储量及其空间分布[J]. 山地学报，2010，28(1)：69-75.
[32]李科，曾小毕，李智彪，等. 成都市退耕还竹地不同坡位凋落物及土壤的持水特性研究[J]. 四川林业科技，2011，32(4)：96-98.
[33]刘中奇，朱清科，邝高明，等. 半干旱黄土丘陵沟壑区封禁流域植被枯落物分布规律研究[J]. 草业科学，2010，27(4)：20-24.
[34]潘复静，张伟，王克林，等. 典型喀斯特峰丛洼地植被群落凋落物 C：N：P 生态化学计量特征[J]. 生态学报，2011，31(2)：335-343.
[35]赵雨虹，范少辉，夏晨. 亚热带4种常绿阔叶林林分枯落物储量及持水功能研究[J]. 南京林业大学学报(自然科学版)，2012，39(6)：93-98.
[36]周新年，巫志龙，郑丽凤，等. 天然林择伐10年后凋落物现存量及其养分含量[J]. 林业科学，2008，44(10)：25-28.
[37]田国恒. 不同间伐抚育强度对华北落叶松人工林林下凋落物的影响研究[J]. 山东林业科技，2014(3)：70-72.
[38]骆土寿，张国平，吴仲民，等. 雨雪冰冻灾害对广东杨东山十二度水保护区常绿与落叶混交林凋落物的影响[J]. 林业科学，2008，44(11)：177-183.
[39]徐涵湄，阮宏华. 雪灾对武夷山毛竹林凋落物分解和养分释放的影响[J]. 南京林业大学学报(自然科学版)，2010，34(3)：131-135.
[40]黄铄淇，胡慧蓉，韩钊龙，等. 林火对昆明人工林凋落物和表层土壤碳氮的影响[J]. 四川农业大学学报，2014，32(1)：18-22.
[41]CHEN H，MARK E H，TIAN H Q. Effects of global change on litter decomposition in terrestrial ecosystems. Acta Ecological Sinica，2001，21(9)：1549-1563.

[原载：世界林业研究，2017，30(01)]

近自然化改造对桂南马尾松和杉木人工林结构特征的影响

刘志龙[1,2] 明安刚[1] 贾宏炎[1] 蔡道雄[1] 马 跃[1] 王亚南[1] 孙冬婧[1]

([1]中国林业科学研究院热带林业实验中心，广西凭祥 532600；
[2]广西友谊关森林生态系统国家定位观测研究站，广西凭祥 532600)

摘 要 通过近自然化改造改变林分组成和结构，进而影响林木生长和林分稳定性，研究近自然化改造对林分结构变化的影响，对指导人工林质量提升和珍贵树种大径材培育具有重要意义。以广西凭祥市1993年造林并于2008年开始近自然化改造的马尾松和杉木人工林为研究对象，运用混交度、大小比数和角尺度参数分析近自然化改造后林分结构特征的变化。结果表明：①对照林分的直径结构遵从正态分布，近自然化改造林分表现出向倒“J”型过渡的特征，对照林分的树高结构呈“单峰”型，而近自然化改造林分则呈“双峰”型。②近自然化改造明显提高了林分混交度，马尾松和杉木对照林分的平均混交度由0.00和0.16分别提高到自然化改造林分的0.82和0.89。③近自然化改造对林分角尺度影响较小，杉木林从均匀分布状态向团状分布转变，但马尾松林一直处于随机分布状态。④近自然化改造增加了优势和亚优势的个体比例，对照林分总体处于中庸生长状态，马尾松和杉木近自然化改造林分的平均大小比数分别为0.40和0.46。从林分非空间结构角度看，近自然化改造林分逐步摆脱了人工纯林零度或弱度混交、水平分布均匀和中庸状态的结构特征。研究区补植树种和天然更新树种已成功地在林分中更新和生长，促进了林分混交度、树种多样性和林分空间结构的优化，林分向异龄复层混交林方向发展。

关键词 近自然化改造；林分结构；角尺度；混交度；大小比数

Effects of Close-to-nature Transformation on Structure Characteristics of *Pinus massoniana* and *Cunninghamia lanceolata* Plantations

LIU Zhilong[1,2], MING Angang[1], JIA Hongyan[1], CAI Daoxiong[1], MA Yue[1], WANG Yanan[1], SUN Dongjing[1]

([1]*Experimental Center of Tropical Forestry*, *CAF*, *Pingxiang* 532600, *Guangxi*, *China*;
[2]*Guangxi Youyiguan Forest Ecosystem Research Station*, *Pingxiang* 532600, *Guangxi*, *China*)

Abstract: The close-to-nature transformation changes the forest composition and stand structure and thus affects the tree growth and stand stability. Therefore, a study on the effects of close-to-nature transformation on stand structure has important significance in improving the quality of degrading plantations and producing large diameter timber. This study focused on *Pinus massoniana* and *Cunninghamia lanceolata* plantations, examining the plantation and close-to-nature transformation years, 1993 and 2008 respectively. Three parameters (mingling, uniform angle index and neighbourhood comparison) were considered to compare the changes in stand spatial structure before and after close-to-nature transformation. The results showed that: ①The diameter of control stands followed normal distribution, while the close-to-nature transformation stands showed the reversal "J" type. The tree height of control stands was a single peak type, whereas that of close-to-nature transformation plantations showed a double peak. ②The close-to-nature transformation improved stand mingling significantly, the average mingling increased from 0.00 and 0.16 to 0.82 and 0.89 in the *P. massoniana* and *C. lanceolata* control stands and their close-to-nature stands respectively. ③Close-to-nature transformation has little effects on the uniform angle index, it changed from uniform distribution to reunion distribution in *C. lanceolata* stands, however it has been in a random distribution state in *P. massoniana* stands. ④The number of dominant and sub-dominant individuals were increased after close-to-nature transformation. The average neighborhood comparison of *P. massoniana* and *C. lanceolata* close-to-nature stands as 0.40 and 0.46 respectively, whereas there were generally medium individuals in the control stands.

Key words: close-to-nature transformation; stand structure; uniform angel index; mingling; neighborhood comparison

近年来，我国已成为世界上人工林保存面积最大的国家[1]。由于历史条件和经济发展水平的限制，我国南方长期大面积连片经营马尾松(*Pinus massoniana*)和杉木(*Cunninghamia lanceolata*)人工纯林，出现质量下降、病虫害严重、地力衰退、生物多样性下降和生态服务功能降低等问题[2-6]。我国南方水热条件优越，发展珍贵树种具有得天独厚的自然条件，且随着国家对生态环境保护需求，森林多功能可持续经营已成为森林经营的主要发展方向[7,8]。基于这一现实，如何将珍贵树种引入现有大面积马尾松或杉木人工纯林，开展林分改造和质量提升是当前迫切需要解决的问题。德国等一些国家长期的实践证明，近自然森林经营是实现森林的服务功能和木材生产功能之间利益平衡的森林经营方法[9-11]。我国一些学者对此也开展了研究和探索，相关研究结果已逐渐被接受和应用[12-15]。

森林经营的目的不仅是提供木材，更是维护整个生态系统的健康和稳定，培育健康稳定的森林[16-19]，其核心是强调创建或维护最佳林分结构[20]。惠刚盈等系统地提出了结构化森林经营理论和基于相邻木关系的森林空间结构量化分析方法[21,22]，其他学者进行了一些研究和实践[23-28]，但这些研究多集中在分析北方天然林、天然次生林或人工林，而对南方主要典型人工林空间结构研究少见报道。本研究以桂南马尾松和杉木人工林为研究对象，运用混交度、大小比数和角尺度参数对林分近自然化改造后林分结构变化进行比较分析，旨在为该地区培育健康稳定的森林和制定合理人工林结构优化措施提供科学依据。

1 材料与方法

1.1 研究地区概况

研究区位处我国亚热带南缘的广西壮族自治区凭祥市(106°41′~106°59′E，21°57′~22°16′N)，属南亚热带季风气候。该区地带性植被为季节性雨林，年均气温为20.5~21.7℃，极端高温40.3℃，极端低温-1.5℃，≥10℃活动积温6000~7600℃，年均降水量1200~1500mm，年蒸发量1261~1388mm，相对湿度80%~84%，地带性土壤为赤红壤。

选取的代表性林分位于凭祥市中国林业科学研究院热带林业实验中心伏波实验场(106°39′50″~106°59′30″E，21°57′47″~22°19′27″N)，1993年营建马尾松和杉木人工纯林，造林密度为3000株/hm²，造林后连续3年进行新造林抚育，第7年透光伐，第11年和第14年进行2次抚育间伐，保留密度为375~450株/hm²。2008年开始近自然化改造，通过抚育间伐并林下补植1年生乡土阔叶树种，补植密度为750株/hm²，马尾松林下补植红椎(*Castanopsis hystrix*)和香梓楠(*Michelia hedyosperma*)，杉木林下补植大叶栎(*Quercus griffithii*)和格木(*Erythrophloeum fordii*)，补植后连续3年抚育管理。林分基本情况详见表1。

1.2 试验设计及标准地调查

试验采用完全随机区组设计，2个处理为马尾松和杉木林下补植珍贵树种，对照为马尾松和杉木人工纯林(传统经营，不补植珍贵树种)，4次重复，共设16个标准地。

2015年，在林分中设立了半径为11.29m的圆形标准地(面积400m²)，对样地内胸径大于5cm的林木按从样地中心向外沿顺时针方向的次序对树木进行编号和调查。采用DQL-12Z型正像森林罗盘仪测定方位角(以样地中心点为参照到对象林木顺时针方向的角度)，采用激光测距仪测定距离(树干胸径处与样地中心连线的水平距离)。调查内容包括三类：①林分基本情况：海拔、坡向、郁闭度、健康度、经营历史和干扰和森林退化程度等。②单株木测树因子：胸径、树高、枝下高、生活力、层次、起源、损伤情况、干形质量等。③幼树和灌木：幼树调查要记录5m×5m样方内每株幼树(胸径小于5cm或树高大于30cm的林木)的树种名称和高度，所有灌木种类、株数、平均高和盖度。

表1 林分基本状况

林分类型	主林层			亚林层			灌木层	
	树高(m)	胸径(cm)	密度(株/hm²)	树高(m)	胸径(cm)	密度(株/hm²)	高度(cm)	盖度(%)
马尾松人工纯林	17.2±1.20	20.4±3.93	1150±23				1.2±1.01	3.6±5.46
杉木人工纯林	16.7±2.56	19.2±3.68	850±12	7.2±1.72	7.7±2.47	300±5	1.1±0.37	4.0±8.93
马尾松近自然化改造林	16.4±3.82	25.7±6.98	625±16	7.9±2.49	8.3±1.58	350±11	1.1±1.47	2.6±3.33
杉木近自然化改造林	17.2±1.82	22.6±3.70	450±11	12.5±4.82	12.0±5.77	775±16	0.4±0.02	2.3±1.89

1.3 空间结构参数计算及分析

以样地调查数据为基础，通过数据转化，各树木位置坐标分别为：$X=L\times\sin(\alpha\times\pi/180)$，$Y=L\times\cos(\alpha\times\pi/180)$；$L$ 为林木距圆心的距离。利用空间结构分析软件 Winkelmass 1.0 进行数据处理，设置 2m 缓冲区，以参照树及其周围 4 株相邻木组成的结构单元为基础，计算样地内全部单木的混交度、大小比数和角尺度。

(1)混交度。树种混交度用以描述树种的空间隔离程度，表明了任意一个树的最近相邻木为非同种的概率，即参照树的 n 株最近相邻木中与参照树不属同种的个体所占的比例[29]，其计算公式为：

$$M_i=\frac{1}{n}\sum_{j=1}^{n}V_{ij}$$

式中：M_i 为第 i 株参照树的混交度；n 为最近相邻木株数；V_{ij} 为一个离散性的变量。Mi 可能取值为 0.00、0.25、0.50、0.75、和 1.00，其意义分别为零度、弱度、中度、强度、极强度混交。

(2)大小比数。大小比数用以量化参照树与其相邻木的大小相对关系，以大于参照树的相邻木数占所考察的全部相邻木的比例表示，即大于参照树的相邻木占 n 株最近相邻木的株数比例[30]，其计算公式为：

$$U_i=\frac{1}{n}\sum_{j=1}^{n}K_{ij}$$

式中：U_i 为参照树 i 的大小比数；n 为最近相邻木的株数；K_{ij} 是一个离散性的变量。U_i 可能取值 0.00、0.25、0.50、0.75、和 1.00，其意义分别为优势、亚优势、中庸、劣势、极劣势地位。

(3)角尺度。角尺度用以林木个体在水平地面上的分布形式，用相邻木围绕参照树的均匀性来判定林木分布格局。任意 2 个邻接最近相邻木的夹角有 2 个，小角为 α，大角为 β，标准角是衡量参照树周围 4 株最近相邻木分布均匀性的标准，理论推导出的标准角 α_0 应为 72°。角尺度的定义为小角 α 小于标准角 α_0 的个数占所考察的 n 个夹角的比例[31]，其计算公式为：

$$W_i=\frac{1}{n}\sum_{j=1}^{n}Z_{ij}$$

式中：W_i 为第 i 株参照树的角尺度；Z_{ij} 为一个离散性的变量。W_i 可能取值 0.00、0.25、0.50、0.75、和 1.00，其意义分别为绝对均匀、均匀、随机、不均匀、团状分布。

2 结果与分析

2.1 林分非空间结构变化

2.1.1 林分直径结构

由图 1 可以看出，对照林分以中径级 14~26cm 为主且分布比较集中，总体上表现出同龄人工纯林的直径分布特征，遵从正态分布(图 1A、1B)。近自然化改造林分中小径级林木比例大幅增加，6cm 径级的树种主要为罗浮柿(*Diospyros morrisiana*)、毛桐(*Mallotus barbatus*)等天然更新树种，表现出一定的异龄林直径分布特征和向倒“J”型特征发展的趋势。

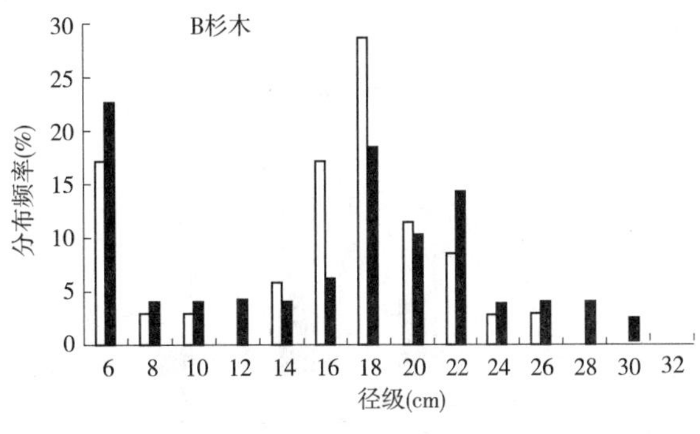

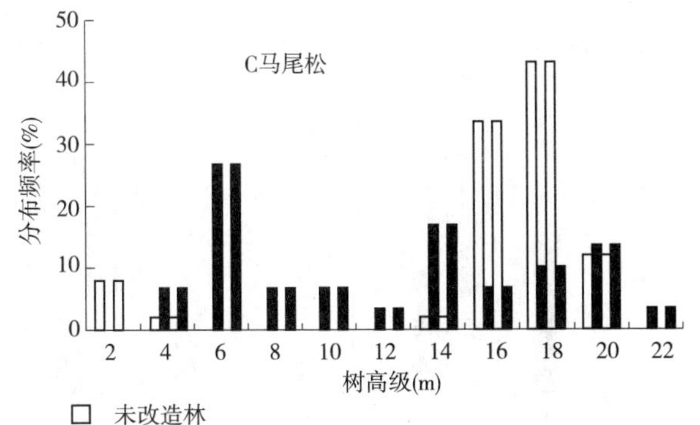

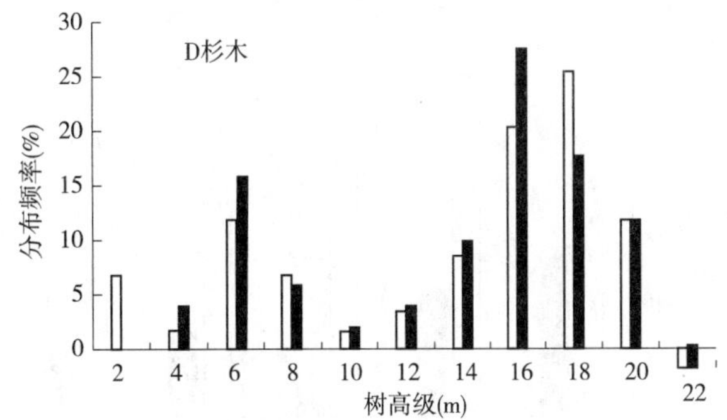

图 1 不同林分径级结构及树高结构变化

2.1.2 林分树高结构

由林分树高结构变化可以看出，对照林分的树高结构呈“单峰”型，主要分布在16~20m，杉木在6~8m(主要为林下天然更新乔木和灌木)也有少量分布(图1C、1D)。近自然化改造林分的树高呈“双峰”型，原有的马尾松和杉木为主林层，分布在16~18m，补植树种和天然更新树种为次林层，主要分布在4~12m，表现出逐步进入主林层的趋势，初步形成了林分的垂直结构。

2.2 林分空间结构变化

2.2.1 林分混交度

马尾松近自然化改造林分中，补植树种(红椎和香梓楠)和黄毛榕(*Ficus esquiroliana*)等原生树种逐步进入主林层，弱度混交、中度混交、强度混交和极强度混交四个等级上的个体数比例都有不同程度的增加，其中，极强度混交占到了总数的25%，林分平均混交度从改造前的0.00增加到改造后的0.82。杉木对照林分中罗浮柿、麻楝(*Chukrasia tabularis*)等天然更新树种占一定比例，混交度以零度混交为主，弱度混交、中度混交、强度混交也占有一定比例，近自然化改造林分以强度混交和极强度混交为主，分别占39.29%和50%，林分平均混交度也由改造前的0.16上升为改造后的0.89(图2A、2B)。

对各树种进行树种混交程度分析，结果见表2。在对照林分中，马尾松和杉木平均混交度分别为0.00和0.07，在近自然化改造林分中的混交度分别为0.58和0.83，属于中度混交和强度混交，说明近自然化改造后，占绝对优势的马尾松和杉木出现单种聚集现象大幅降低。黄毛榕、灰毛浆果楝(*Cipadessa cinerascens*)、麻楝、三叉苦(*Evodia lepta*)等天然更新的树种的零度混交比例为零，均为极强度混交，说明其株数较少且散生在马尾松和杉木周围，相邻木均为不同树种。

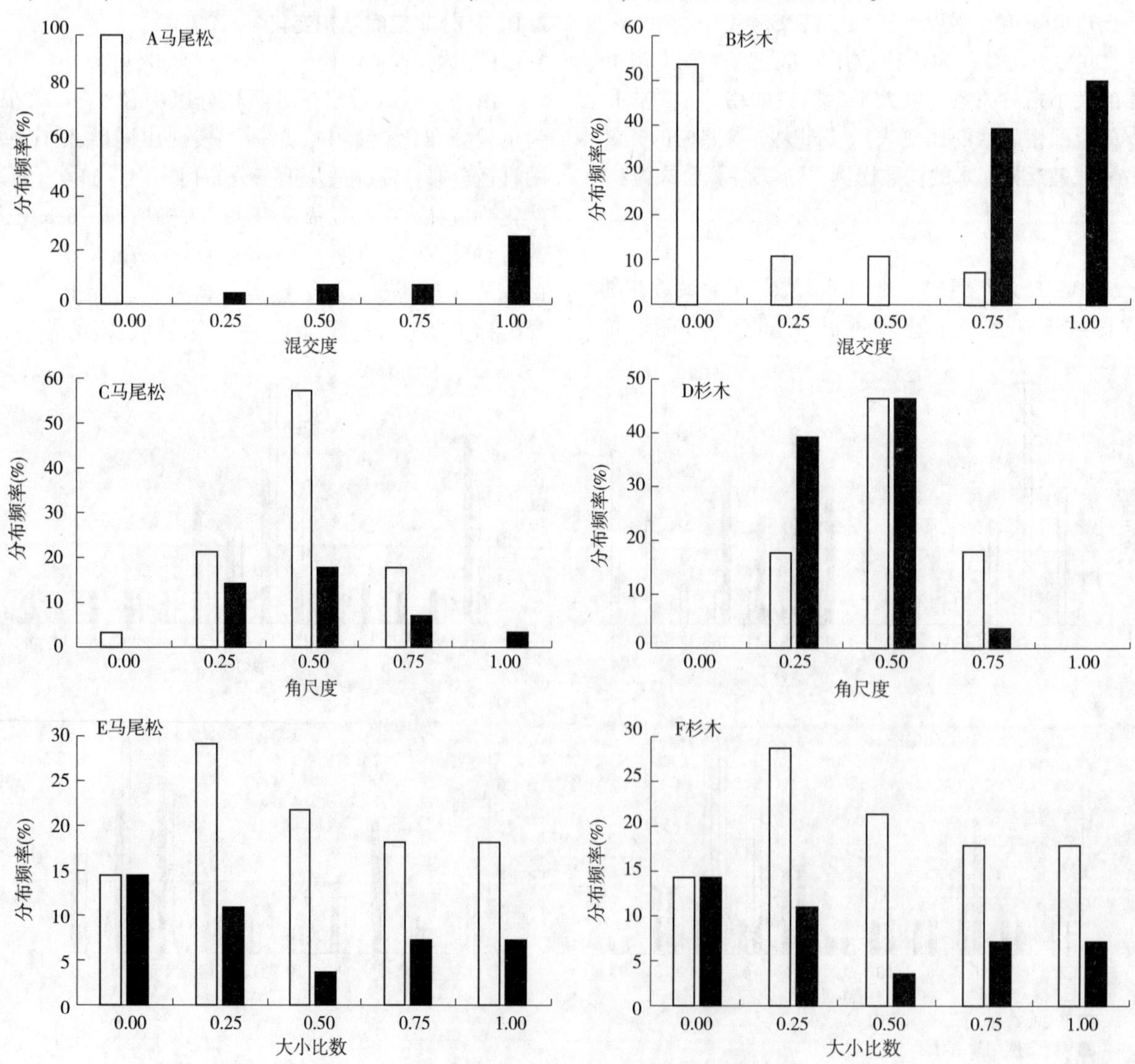

图2 不同林分混交度、角尺度及大小比数变化

表 2 各树种混交度及大小比数的分布

林分类型	树种	混交度					均值	大小比数					均值
		0.00	0.25	0.50	0.75	1.00		0.00	0.25	0.50	0.75	1.00	
马尾松人工纯林	马尾松	0.00	0.00	0.00	0.00	0.00	0.00	0.19	0.27	0.12	0.15	0.27	0.57
杉木人工纯林	杉木	0.77	0.12	0.12	0.00	0.00	0.07	0.23	0.35	0.04	0.19	0.19	0.52
	罗浮柿	0.00	0.00	0.00	1.00	0.00	0.75	0.00	0.00	0.00	0.50	0.50	
马尾松近自然化改造林	马尾松	0.00	0.22	0.44	0.11	0.22	0.58	0.44	0.33	0.00	0.11	0.11	0.40
	红椎	0.00	0.00	0.00	0.33	0.67	0.92	0.00	0.00	0.00	0.00	1.00	
	香梓楠	0.00	0.00	0.00	0.00	1.00	1.00	0.00	0.67	0.33	0.00	0.00	
	黄毛榕	0.00	0.00	0.00	0.00	1.00	1.00	0.00	0.00	0.00	0.00	1.00	
	大叶栎	0.00	0.00	0.00	0.00	1.00	1.00	0.00	0.00	0.00	0.67	0.33	
杉木近自然化改造林	杉木	0.00	0.00	0.00	0.70	0.30	0.83	0.40	0.27	0.05	0.27	0.00	0.46
	大叶栎	0.00	0.00	0.00	0.73	0.27	0.82	0.00	0.00	0.00	0.40	0.60	
	格木	0.00	0.00	0.00	0.00	1.00	1.00	0.00	0.00	0.00	0.50	0.50	
	灰毛浆果楝	0.00	0.00	0.00	0.00	1.00	1.00	0.00	0.00	0.00	1.00	0.00	
	麻楝	0.00	0.00	0.00	0.00	1.00	1.00	0.00	0.00	0.00	0.50	0.50	
	三叉苦	0.00	0.00	0.00	0.00	1.00	1.00	0.70	0.30	0.00	0.00	0.00	

2.2.2 林分角尺度

由于近自然化改造过程中，伐除干扰树主要参照目标树的情况来确定，没有考虑其角度关系，林分各级角尺度上的变化不大(图 2C、2D)。马尾松林分中随机的个体数比例都最高，但由对照林分的 57.14%下降为近自然化改造林分的 41.67%，团状分布增加到 8.33%，两种林分平均角尺度为 0.49 和 0.50，取值范围属于 0.475~0.517，说明林木总体上处于随机分布的状态[31]。杉木对照林分平均角尺度为 0.40<0.475，林分表现出均匀分布状态，符合人工林的特点。近自然化改造林分平均角尺度为 0.55>0.517，林分表现出团状分布。

3.2.3 林分大小比数

不同林分大小比数变化见图 2E、2F，可以看出，马尾松对照林分的平均大小比为 0.57，各种大小比数频率分布比较均匀，整体处于中庸状态。近自然化改造林分中处于优势和亚优势的个体比例由 14.29%上升 33.33%，中庸个体比例由 21.43%下降 8.33%，处于受压生长状态的劣势和极劣势的个体比例变化不大，说明近自然化改造促进了保留木的生长，增加了优质大径材的比例。杉木对照林分的大小比数平均值为 0.52，中庸个体比例仅占 4.35%，分中处于优势生长状态(优势和亚优势)和受压生长状态(劣势和极劣势)的个体比例均为 47.83%，说明林分稳定性比较差。改造后，优势木和中庸个体比例明显上升至 28.00%和 20.00%，而其他类型林木个体比例均表现下降的趋势，林分平均大小比数也由改造前的 0.52 降至改造后的 0.46，说明林分逐步趋于稳定的状态。

对各树种进行大小比数进行分析，结果见表 2。在对照林分总体表现为中庸状态，马尾松和杉木的优势木的比例分别为 19%和 23%，而近自然化改造林分的比例分别为 44%和 40%，说明近自然化改造大幅提升马尾松和杉木优势木的比例。补植树种的大小比数表现不同状态，同为马尾松林下补植树种，香梓楠的亚优势木比例达 67%，而红椎均处于极劣势状态。杉木林下补植的大叶栎和格木处于劣势和极劣势状态，天然更新树种也处于相同状态，说明需要择伐部分马尾松和杉木，进一步调整林分结构，为林下补植树种和天然树种提供优良的生长环境。

3 讨论

3.1 近自然化改造对林分非空间结构的影响

对照林分的树种组成比较简单，乔木层只有马尾松或杉木，由于郁闭度比较高，天然更新树种很难快速生长并进入主林层，林下灌木和草本更新缓慢。近自然化改造后，改善了林分结构，补植的乡土珍贵树种和天然更新树种有了更多的生长空间，占领了林下主要空间并逐步进入主林层，优势度均呈上升趋势，林下灌木和草本的数量和盖度有所降低。于洪光等[32]研究红松针阔混交林结构化森林经营的结果发现，珍贵树种及主要伴生树种的优势度均呈上升状态，这与此次研究的结果相似。

对照林分的直径分布表现出同龄人工纯林特征，遵从正态分布。近自然化改造林分中小径级的补植树种和天然更新树种比例大幅增加，两种林分中6~12cm径级的比例分别为51.75%和34.69%，表现出向异龄林直径分布倒"J"型的特征过渡的状态。对照林分的树高级结构呈"单峰"型，马尾松或杉木占据主林层(16~20m)，在杉木林中还有一些林下天然更新乔木和灌木，主要分布在2~8m。近自然化改造林分的树高级呈"双峰"型，主要是由于补植树种和天然更新树种生长，形成亚林层，主要分布在4~12m，表现出逐步进入主林层的趋势。

3.2 近自然化改造对林分空间结构的影响

近自然化改造有利于提高林分混交度，林分平均混交度由对照林分的零度混交提升到强度混交，分析认为，近自然化改造过程中，改善了林分空间和光环境，补植树种和天然更新树种逐步进入主林层，不仅丰富了树种的组成，而且提升了林分的混交程度[33,34]。就单个树种而言，近自然化改造林分中马尾松和杉木的混交度平均值分别为0.58和0.83，属于中度混交和强度混交，占绝对优势的马尾松和杉木出现单种聚集现象大幅降低，此外，林隙补植树种和天然更新的树种均为极强度混交。

近自然化改造对马尾松林分角尺度影响不大，林木总体上处于随机分布的状态。杉木对照林分的平均角尺度为0.40<0.475，林木总体上处于均匀分布状态，符合典型人工纯林的特点。改造林分的平均角尺度为0.55>0.517，林分表现出团状分布。顶级群落的水平分布格局应为随机分布，林分团状分布格局需要向随机分布调整[23]。因此，对杉木林应结合下周期的干扰树伐除，运用角尺度分析结果来调整林木的分布格局，尽可能保留取值为0.5的立木，建议优先采伐挤在一个角或同一侧，且与保留木同种的相邻木，使林分分布格局向随机分布转变，优化林分空间结构。

马尾松近自然化改造林分中处于优势状态的个体比例增加，中庸状态比例下降，受压状态比例变化不大，说明近自然化改造促进了保留木的生长，增加了优质大径材的比例。相对马尾松林，杉木林大小比数的变化存在一定差异，主要表现在改造后优势木和中庸个体比例明显上升，而其他类型林木个体比例均表现下降的趋势，说明林分逐步趋于稳定的状态。这主要是近自然化改造过程中，伐除了干扰树及部分中小径级林木，同时，补植树种和天然更新树种的增加，提高其马尾松或杉木的优势等级，导致处于优势状态的个体比例显著增加，这与万丽等[26]的研究结果一致。林下补植和天然更新树种基本处于劣势和极劣势状态，需要择伐部分马尾松和杉木，进一步调整林分结构，为其提供优良的生长环境。

从林分非空间结构角度看，近自然化改造林分提高了林分树种组成，径级结构向异龄林直径分布倒"J"型的特征过渡。从林分空间结构角度看，近自然化改造林分逐步摆脱了人工纯林零度或弱度混交、水平分布均匀和中庸状态的结构特征。补植树种和天然更新树种已成功地在林分中更新和生长，促进了林分混交度、树种多样性的提高和林分空间结构的优化，林分向异龄复层混交林方向发展。近期森林经营中，伐除部分马尾松和杉木，为林下补植和天然更新树种提供优良的生长环境。在以后的经营中，应定期对林分的分布格局、树种隔离程度和林木的大小分化程度等特征进行动态的综合评价，通过合理的经营规划和措施，从而实现林分空间结构的优化，诱导林分结构向健康稳定的方向发展，最终实现森林可持续经营。

参考文献

[1]陈幸良，巨茜，林昆仑．中国人工林发展现状、问题与对策[J]．世界林业研究，2014，27(6)：54-59.

[2]杨承栋，孙启武，焦如珍，等．大青山一二代马尾松土壤性质变化与地力衰退关系的研究[J]．土壤学报，2003，40(2)：267-273.

[3]田大伦，沈燕，康文星，等．连栽第1和第2代杉木人工林养分循环的比较[J]．生态学报，2011，31(17)：5025-5032.

[4]丁应祥，田野，戚玲．连栽杉木人工林生产力的模拟与预测[J]．南京林业大学学报，2000，24(3)：21-25.

[5]罗云建，张小全．杉木(*Cunninghamia lanceolata*)连栽地力退化和杉阔混交林的土壤改良作用[J]．生态学报，2007，27(2)：715-724.

[6]杨玉盛，陈光水，黄宝龙．杉木多世代连栽的土壤水分和养分变化[J]．南京林业大学学报，2000，24(2)：25-28.

[7]侯元兆，陈统爱．我国结合次生林经营发展珍贵用材树种的战略利益[J]．世界林业研究，2008，21(2)：49-52.

[8]曾祥谓，樊宝敏，张怀清，等．我国多功能森林经营的理论探索与对策研究[J]．林业资源管理，2013，

(2)：10-16.

[9]MOSTAFA Moradi，MOHAMMAD R，MARVIE Mohadjer，et al. Over-mature beech trees(Fagus orientalis Lipsky) and close-to-nature forestry in northern Iran[J]. Journal of Forestry Research，2012，23(2)：289-294.

[10]LARSEN J B，NIELSEN A B. Nature-based forest management：Where are we going? Elaborating forest development types in and with practice[J]. Forest Ecology and Management，2007，238：107-117.

[11]KEVIN L O. What is close-to-nature silviculture in a changing world[J]. Forestry，2016，89：1-6.

[12]陆元昌．近自然森林经营的理论与实践[M]. 北京：科学出版社，2006.

[13]林思祖，黄世国．论中国南方近自然混交林营造[J]. 世界林业研究，2001，14(2)：73-78.

[14]何友均，梁星云，覃林，等．南亚热带人工针叶纯林近自然改造早期对群落特征和土壤性质的影响[J]. 生态学报，2013，33(8)：2484-2495.

[15]罗应华，孙冬婧，林建勇，等．马尾松人工林近自然化改造对植物自然更新及物种多样性的影响[J]. 2013，33(19)：6154-6162.

[16]赵中华．基于林分状态特征的森林自然度评价研究[D]. 北京：中国林业科学研究院，2009.

[17]刘于鹤，林进．新形势下森林经营工作的思考[J]. 林业经济，2013，(1)：3-9.

[18]王宏，陆元昌．木材储备基地及其多功能森林经营建设对策[J]. 世界林业研究，2013，26(1)：12-17.

[19]赵良平．森林生态系统健康理论的形成与实践[J]. 南京林业大学学报(自然科学版)，2007，31(3)：1-7.

[20]惠刚盈，胡艳波，赵中华．再论"结构化森林经营"[J]. 世界林业研究，2009，22(1)：14-19.

[21]惠刚盈，赵中华，胡艳波．结构化森林经营技术指南[M]. 北京：中国林业出版社，2010.

[22]惠刚盈，KLAUS von Gadow，胡艳波，等．结构化森林经营[M]. 北京：中国林业出版社，2007.

[23]赵中华，惠刚盈，胡艳波，等．结构化森林经营方法在阔叶红松林中的应用[J]. 林业科学研究，2013，26(4)：467-472.

[24]赵中华，袁士云，惠刚盈，等．经营措施对林分空间结构特征的影响[J]. 西北农林科技大学学报(自然科学版)，2008，36(7)：135-142.

[25]巫志龙，周成军，周新年，等．杉阔混交人工林林分空间结构分析[J]. 林业科学研究，2013，26(5)：609-615.

[26]万丽，刘昀东，丁国栋，等．密度调控对油松人工林空间结构的影响[J]. 四川农业大学学报，2013，31(1)：27-31.

[27]张连金，胡艳波，赵中华，等．北京九龙山侧柏人工林空间结构多样性[J]. 生态学杂志，2015，34(1)：60-69.

[28]胡艳波．基于结构优化森林经营的天然异龄林空间结构优化模型研究[D]. 北京：中国林业科学研究院，2010.

[29]惠刚盈，胡艳波．混交林树种空间隔离程度表达方式的研究[J]. 林业科学研究，2001，14(1)：23-27.

[30]惠刚盈，胡艳波．角尺度在林分空间结构调整中的应用[J]. 林业资源管理，2006，4(2)：31-35.

[31]惠刚盈，KLAUSVON Gadow，MATTHIAS Albert. 一个新的林分空间结构参数—大小比数[J]. 林业科学研究，1999，12(1)：1-6.

[32]于洪光，高海涛，张义涛．结构化森林经营技术要点与应用[J]. 吉林林业科技，2014，43(6)：16-19.

[33]姜鹏，孟京辉，陆元昌，等．马尾松近自然改造初期的混交度与分布格局[J]. 西北林学院学报，2014，30(5)：147-150.

[34]郝云庆，王金锡，王启和，等．柳杉人工林近自然改造过程中林分空间的结构变化[J]. 四川农业大学学报，2008，26(1)：48-52.

[原载：南京林业大学学报(自然科学版)，2017，41(04)]

近自然化改造对马尾松和杉木人工林生物量及其分配的影响

明安刚[1,2,3] 刘世荣[2,3] 李 华[1,3] 曾 冀[1,3] 孙冬婧[1,3]
雷丽群[1,3] 蒙明君[1,3] 陶 怡[1,3] 明财道[1,3]

([1]中国林业科学研究院热带林业实验中心，广西凭祥 532600；[2]中国林业科学研究院森林生态环境与保护研究所，北京 100091；[3]广西友谊关森林生态系统国家定位观测研究站，广西凭祥 532600)

摘 要 近自然化改造作为森林新增碳汇的最有希望的选择之一，将如何通过改变林分结构影响林分生物量和生产力进而影响林分固碳能力和潜力目前尚不清楚，因此，了解近自然化改造对人工林生物量及其分配的影响，对人工林生态系统碳管理具有重要意义。以马尾松近自然化改造林[P(CN)]、马尾松未改造纯林[P(CK)]、杉木近自然改造林[C(CN)]和杉木未改造纯林[C(CK)]4种人工林为研究对象，采用样方调查和生物量实测的方法，分析4种林分生物量差异，旨在揭示近自然化改造对马尾松和杉木人工林生物量及其分配的影响。结果表明：马尾松杉木人工林近自然化改造通过调整林分结构显著提升马尾松和杉木人工林生物量和生产力，8年后马尾松和杉木林分生物量分别增加46.71%和37.24%。乔木层生物量在林分生物量总量中占主导地位(95.48%~98.82%)，并对林分生态系统总生物量变化起决定性作用。林分生物量和生产力的增加主要因为近自然化改造改变了林分群落结构，进而提高了乔木层生产力。研究结果表明，合理的经营措施不仅可以改善林分结构，提升林分生产力，并可为增强植被固碳能力创造有利条件。

关键词 近自然化改造；马尾松和杉木人工林；生物量；生产力；分配

Effects of Close-to-nature Transformation on Biomass and Its Allocation in *Pinus massoniana* and *Cunninghamia lanceolata* Plantations

MING Angang[1,2,3], LIU Shirong[2,3], LI Hua[1,3], ZENG Ji[1,3], SUN Dongjing[1,3], LEI Liqun[1,3], MENG Mingjun[1,3], TAO Yi[1,3], MING Caidao[1,3]

(*[1]Experimental Center of Tropical Forestry, CAF, Pingxiang* 532600, *Guangxi, China*; *[2]Institute of Forest Ecology, Environment and Protection, Chinese Academy of Forestry, Beijing* 100091, *China*
[3]Guangxi Youyiguan Forest EcosystemResearch Station, Pingxiang 532600, *Guangxi, China*)

Abstract: Close-to-nature transformation is considered one of the most promising options for creating new forest carbon sinks, but the mechanism by which it influences the biomass by changing the forest structure and thereby influencing the ability and potential of the forest for carbon sequestration remains unclear. Therefore, there is an urgent need to understand these key effects of close-to-nature transformation on the biomass for carbon management in plantation ecosystems. Based on a close-to-nature forest of *Pinus massoniana* plantation [P(CN)] and an unimproved pure stand of *P. massoniana* [P(CK)], and a close-to-nature stand of *Cunninghamia lanceolata* [C(CN)] and an unimproved pure stand of *C. lanceolata* [C(CK)] as the research objects, the biomass and allocation difference of the four forest types were studied using the method of quadrat sampling combined with biomass measurement, aiming to reveal the influence of close-to-nature transformation on forest biomass and its allocation patterns in *P. massoniana* and *C. lanceolata* plantation. The results indicated that the biomass and productivity of *P. massoniana* and *C. lanceolata* plantations can be significantly increased by close-to-nature transformation, and the biomass of *P. massoniana* and *C. lanceolata* forest stands can be increased by 46.71% and 37.24%, respectively, after 8 years. The biomass of the arborous layer dominates the total biomass (95.48%~98.82%), which plays a vital role in the overall change in forest stand ecosystem biomass. The increase in biomass and productivity in the forest is mainly due to the change in the forest stand community structure, which increases the productivity of the arborous layer. Taken together, the results indicate that reasonable management measures can not only improve stand structure and productivity, but also create favorable conditions that enhance vegetation carbon fixation capacity and potential.

Key words: close-to-nature transformation; *Pinus massoniana* and *Cunninghamia lanceolata* plantations; biomass; productivity; allocation

生物量和生产力是反映人工林群落结构组成的重要指标，较高的林分生物量和生产力是人工林生态系统健康和活力的重要体现。同时，生物量和生产力也是研究人工林生态系统碳循环的基础，是评估人工林固碳潜力的主要内容和反映人工林生态系统的服务功能的重要指标之一[1,2]。因而，准确测算人工林生物量与生产力，对人工林的健康经营和生态系统碳管理具有重要意义。

当前，人工林生物量受到国内外广泛关注[3-5]，国外近些年报道的树种有辐射松(*Pinus radiata*)、美洲栗(*Castanea americana*)、肯宁南洋杉(*Araucaria cunninghamii*)、夏栎(*Quercus robur*)、蓝桉(*Eucalyptus globulus* subsp. *globulus*)、狄氏黄胆木(*Nauclea diderrichii*)和绵白杨(*Populus tomentosa*)等[6-12]；国内对人工林生物量和生产力早期研究的树种主要是马尾松(*Pinus massoniana*)、杉木(*Cunninghamia lanceolata*)、桉树(*Eucalyptus* sp.)等一些用材树种[13-16]，最近10年来报道的主要树种既包括较多的松、杉、桉等传统造林树种[17,18]，还有樟树(*Cinnamomum camphora*)[19]、小黑杨(*Populus nigra*)[20]、楠木(*Phoebe zhennan*)[21]、红椎(*Castanopsis hystrix*)[22]、米老排(*Mytilaria laosensis*)[23]、格木(*Erythrophleum fordii*)[24]等乡土阔叶树种。

我国南亚热带地区，人工林主要以马尾松、杉木和桉树等短周期树种为主导[25]，其中，针叶纯林占72%。人工针叶纯林的树种结构单一，径级结构和垂直结构简单，再加上不合理的经营方式，不仅造成生物多样性锐减，而且降低人工林生物量和生产力，进而影响人工林的固碳能力和潜力[26-28]。因而，通过疏伐补植的措施将针叶纯林改造成针阔异龄混交的近自然林，正逐渐成为替代大面积针叶人工纯林最有希望的选择途径之一[29,30]。针叶人工林近自然化改造在提升地力，提高林分生产力、碳储量和物种多样性等方面发挥重要作用[30-32]。马尾松、杉木人工林经过近自然经营措施(疏伐、补植)，如何通过改变林分结构和群落组成改变林分的生物量和生产力，进而影响林分植被碳储量，有待进一步深入研究。

本研究以四种不同经营方式的人工林：马尾松未改造纯林[P(CK)]、马尾松近自然化改造林[P(CN)]、杉木未改造纯林[C(CK)]和杉木近自然化改造林[C(CN)]为研究对象，通过对不同林分人工林生物量和生产力及其动态变化规律进行研究，弄清近自然经营对马尾松和杉木人工林生物量和生产力的影响及动态，旨在为南亚热带人工林生态系统碳管理和可持续经营提供科学依据。

1 材料与方法

1.1 研究区概况

研究地点位于广西壮族自治区凭祥市中国林业科学研究院热带林业实验中心(以下简称中国林科院热林中心)(22°10′N，106°50′E)，是国家林业局管辖的森林生态定位研究站之一。该地区属于南亚热带季风气候区域内的西南部，属湿润半湿润气候，干湿季分明。境内光照充足，全年日照时数1200~1600h；降水充沛，年平均降水1200~1500mm，主要发生在每年4~9月；年蒸发量1200~1400mm；相对湿度80%~84%；年平均气温20.5~21.7℃。主要地貌类型以低山丘陵为主；土壤以砖红壤和红壤为主，其次为紫色土；成土母岩主要有泥岩夹砂岩、砾状灰岩、花岗岩和石灰岩等，土层厚度在80cm以上，南亚热带常绿阔叶林是站区地带性植被。

中国林科院热林中心分布不同类型的人工林近2万hm^2，针叶树种主要有马尾松和杉木，乡土阔叶树种主要有大叶栎(*Quercus griffithii*)、格木、红椎、米老排、西南桦(*Betula alnosensis*)和降香黄檀(*Dalbergia lanceolata*)，外来树种主要有桉树和柚木(*Tectona grandis*)，其中，格木和降香黄檀为固氮树种。而大叶栎为速生阔叶树种，具有较好的天然更新能力。用格木和大叶栎改造马尾松和杉木纯林，既适合短周期用材和大径级珍贵用材的需要，又可以实现乡土阔叶树种的自然更新，达到近自然经营的目的。

1.2 试验林概况

试验林均为1993年在杉木采伐迹地上营造的马尾松和杉木纯林，初植密度为2500株/hm^2，造林后连续铲草抚育3年共6次，第7年透光伐抚育，第11年第一次抚育间伐，保留密度1200株/hm^2。2007年开始实施近自然化改造，主要改造措施是在保护天然更新的同时，对马尾松纯林进行疏伐(保留密度为600株/hm^2），2008年初，在疏伐后的马尾松和杉木林下1∶1均匀补植大叶栎和格木，补植乡土树种密度为600株/hm^2（格木和大叶栎密度均为300株/hm^2），形成总密度为1200株/hm^2的针阔异龄混交林。同时保留与总密度一致（1200株/hm^2）的未实施近自然改造的马尾松和杉木纯林为对照。4种林分均设置4个重复。目前，被改造的林分已经郁闭，已演替成具有明显复层结构的针阔异龄混交林。2016年调查结果显示，马尾松、杉木及补植的大叶栎和格木全部存活，大叶栎平均胸径和平均树高分别为13.7cm和14.6m，格木平均胸径和平均树高分别为5.2cm和6.3m。4种林分基本情况与经营历史见表1所示。

1.3 试验设计

本试验采用单因素两水平的随机区组设计，共设4个区组，即为4个真重复，每个区组各设置4种林分类型，即马尾松近自然化改造林、马尾松未改造纯林、杉木近自然改造林和杉木未改造纯林。4个林分类型，4个重复共16个试验小区，每个试验小区面积为0.5hm^2，小区之间的间隔在100m以内。

2 研究方法

2.1 乔木生物量的测定

4种林分涉及的4个主要树种的生物量均采用径级平均木法测定，杉木、马尾松和格木选择热林中心适合本研究的已有的生物量方程进行估算[17,24,33]，并新建大叶栎生物量模型对大叶栎生物量及其各器官进行估算。根据样方每木检尺的调查结果，作出胸径分布图，再对每个树种按径级（2cm）选取标准木9株进行乔木生物量测定。样木伐倒后，地上部分按不同器官测定树干、树皮、树枝、树叶的鲜重；地下部分用"全挖法"测定根系的鲜重，收集全部直径>2mm的根系。同时，按器官采集植物样品200g左右，带回实验室在65℃烘箱中烘干至恒重，计算含水率并将各器官的鲜重换算成干重。

根据9株标准木生物量的实测数据，建立大叶栎各器官生物量（W）与胸径（D）及胸径平方与树高的乘积（D^2H）之间的相对生长方程，用以估测大叶栎各器官的生物量（表2）。

表1 4种林分基本情况与经营历史

年份（年）	项目	林分类型			
		马尾松对照林	马尾松改造林	杉木对照林	杉木改造林
1993	造林	初植密度2500株/hm^2，2年生容器苗	初植密度2500株/hm^2，2年生容器苗	初植密度2500株/hm^2，2年生容器苗	初植密度2500株/hm^2，2年生容器苗
1993—1995	新造林抚育	全铲抚育6次	全铲抚育6次	全铲抚育6次	全铲抚育6次
2000	透光伐	透光伐	透光伐	透光伐	透光伐
2004	生长伐	生长伐，保留1200株/hm^2	生长伐，保留1200株/hm^2	生长伐，保留1200株/hm^2	生长伐，保留1200株/hm^2
2007	强度间伐	不间伐，保留1200株/hm^2	强度间伐，保留600株/hm^2	不间伐，保留1200株/hm^2	强度间伐，保留600株/hm^2
2008	林下补植	不补植	均匀补植大叶栎、格木各300株/hm^2	不补植	均匀补植大叶栎、格木各300株/hm^2
2009	改造林抚育	不抚育	抚育2次	不抚育	抚育2次

表2 热林中心伏波实验场1林班大叶栎单株生物量相对生长方程

器官	回归方程	样本量	判定系数R^2	F值	显著性
干皮	$W=0.027(D^2H)-0.125$	9	0.981	379.405	0.000
树枝	$W=0.013(D^2H)-0.354$	9	0.911	72.487	0.000
树叶	$W=0.004(D^2H)+0.169$	9	0.979	332.336	0.000
根系	$W=0.009(D^2H)-0.357$	9	0.863	44.145	0.000
全株	$W=0.054(D^2H)-0.666$	9	0.969	225.052	0.000

注：W为各器官生物量；D为胸径；H为树高。

2.2 林下植被生物量及凋落物现存量的测定

按“梅花五点法”在每个固定样地布设5个2m×2m个固定的小样方，记录每个小样方内的植物种名，分地上部分和地下部分，采用“收获法”分别测定其鲜重，同种植物的相同器官取混合样品，凋落物全部测定生物量，取混合样品烘干至恒重后，计算出各组分的干重。

2.3 凋落物产量的测定

在每个样地随机布设6个1m×1m凋落物收集器，每个月底收集一次凋落物，1年为1个周期，共收集12次。每月收集的凋落物样品先按器官(落叶、落枝、落皮、落果和杂物)分别测量鲜重，所选落叶再按树种区分并测定鲜重。然后按树种和器官取样烘干后，计算凋落物干重和全年凋落物产量。

2.4 统计分析

用单因素方差分析(one Way-ANOVA)来检验不同林分、不同组分之间生物量的差异性。用一元回归分析(Simple regression)模拟大叶栎各器官生物量回归方程。用SPSS 19.0(SPSS, Inc, Chicago, IL)完成统计分析。作图利用OringinPro 9.0和SigmaPlot10.0软件完成。

3 结果与分析

3.1 不同林分乔木生长量

表3反映了2016年，即近自然化改造8年后，4种林分保留的松杉和补植树种的平均树高和平均胸径情况。由表3可知，改造后，马尾松和杉木平均胸径显著高于对照林，改造后的马尾松和杉木平均胸径分别高于对照林的45.0%和30.4%，可见，近自然化改造显著提高了马尾松和杉木的胸径生长量，但同样是改造8年后，马尾松平均胸径高出杉木平均胸径的44.4%，而四种林分马尾松和杉木的平均树高均无显著差异(P<0.05)，此外，由表3我们还可以看出，无论是大叶栎，还是格木，在马尾松和杉木改造林中，平均胸径和平均树高均无显著差异(P<0.05)，表明，补植树种的生长量并未受到林冠层树种不同的影响。

表3 2016年4种林分平均胸径和平均树高比较

树种	项目	林分类型			
		马尾松对照林	马尾松改造林	杉木对照林	杉木改造林
马尾松	平均胸径(cm)	22.2±1.3b	32.2±1.6a		
	平均树高(m)	16.7±0.5a	17.3±0.7a		
杉木	平均胸径(cm)			17.1±2.1b	22.3±0.8a
	平均树高(m)			17.1±0.4a	17.2±0.4a
大叶栎	平均胸径(cm)		13.4±1.7a		14.2±1.2a
	平均树高(m)		14.8±1.4a		14.4±1.5a
格木	平均胸径(cm)		5.1±1.1a		5.3±0.8a
	平均树高(m)		6.1±0.4a		6.5±0.6a

注：数据为平均值±标准误差；同一个指标变量的不同字母代表差异显著(P<0.05)。

3.2 不同林分生物量及其分配特征

3.2.1 乔木层生物量及其分配特征

图1显示了近自然化改造后的第8年(2016年)不同林分乔木层及其各器官生物量情况，马尾松对照林、马尾松改造林、杉木对照林、杉木改造林乔木层生物量总量分别为257.1、380.1、97.4、135.5t/hm^2，无论是马尾松、还是杉木，改造林乔木层生物量总量均高于其对照林，近自然化改造8年后，马尾松和杉木人工林乔木层生物量分别提高47.8%和39.1%。除树皮外，改造林P(CN)和C(CN)的树干、树枝、树叶和树根生物量均显著高于其对照林分P(CK)和C(CK)。无论改造与否，马尾松林乔木生物量P(CK)和P(CN)均显著高于C(CK)和C(CN)。

由图2可以看出，树干是四种林分生物量的主体部分，在乔木生物量的分配比例最大，占乔木生物量总量的55.4%~59.9%，其次是根系，再次是枝条和树皮，叶片的生物量分配最小，不同林分树枝和树皮的分配比例有所差异，但树枝和树皮生物量分配比例的差异均不显著(*P*>0.05)。4种林分乔木层各器官生物量的分配也有所差异，改造林P(CN)和C(CN)枝条和叶片生物量的分配显著高于对照林P(CK)和C(CK)，但改造林P(CN)和C(CN)树干和树皮的生物量分配比例低于对照林P(CK)和C(CK)，表明近自然化改造可以显著促进叶片和枝条的发育，而减少树干和树皮在林木生物量中的分配。而在不同的树种之间，无论改造与否，马尾松林P(CN)和P(CK)叶片的分配低于杉木林C(CN)和C(CK)，而树皮的分配比例高于杉木林。表明树种不同，不同器官生物量在林木中的分配比例也存在差异。

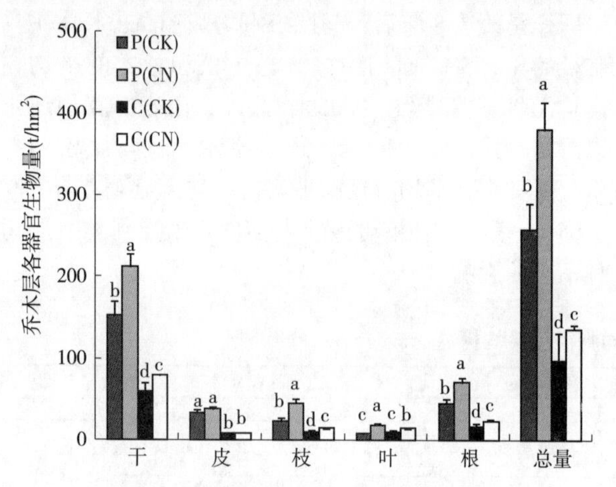

图1　不同林分乔木层各器官生物量

注：不同小写字母表示不同林分相同组分生物量差异显著(*P*<0.05)；P(CK)：马尾松对照林；P(CN)：马尾松改造林；C(CK)：杉木对照林；C(CN)：杉木改造林。下同。

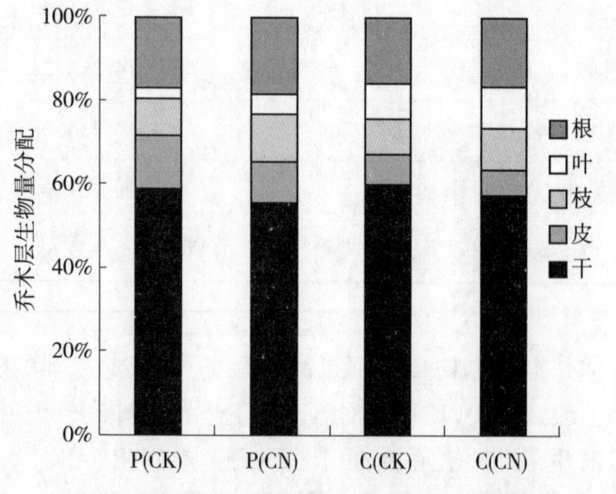

图2　不同林分乔木层各器官生物量分配

3.2.2　地被层生物量及其分配特征

在改造后的第8年，P(CK)、P(CN)、C(CK)、C(CN)4种林分灌木层生物量总量在0.287~0.329t/hm²之间，除C(CK)和C(CN)之间存在显著差异外，其他林分间灌木层生物量总量均无显著差异(*P*>0.05)，灌木层地下部分虽然存在显著差异，但并无明显规律，地上部分生物量在4种林分间均无显著差异(图3)。表明近自然化改造后第8年，杉木林灌木层生物量有所下降，而马尾松林并无显著变化。

草本层生物量总量为0.130~0.645t/hm²。改造之后，马尾松和杉木人工林草本层及各组分生物量均有所降低，且马尾松林差异达到显著水平(*P*<0.05)。在马尾松和杉木林之间，草本层生物量及其各组分生物量也存在显著差异，杉木林C(CK)和C(CN)草本层生物量显著高于马尾松林P(CK)和P(CN)(*P*<0.05)(图3)。表明近自然化改造处理和树种均对草本层生物量有显著影响。

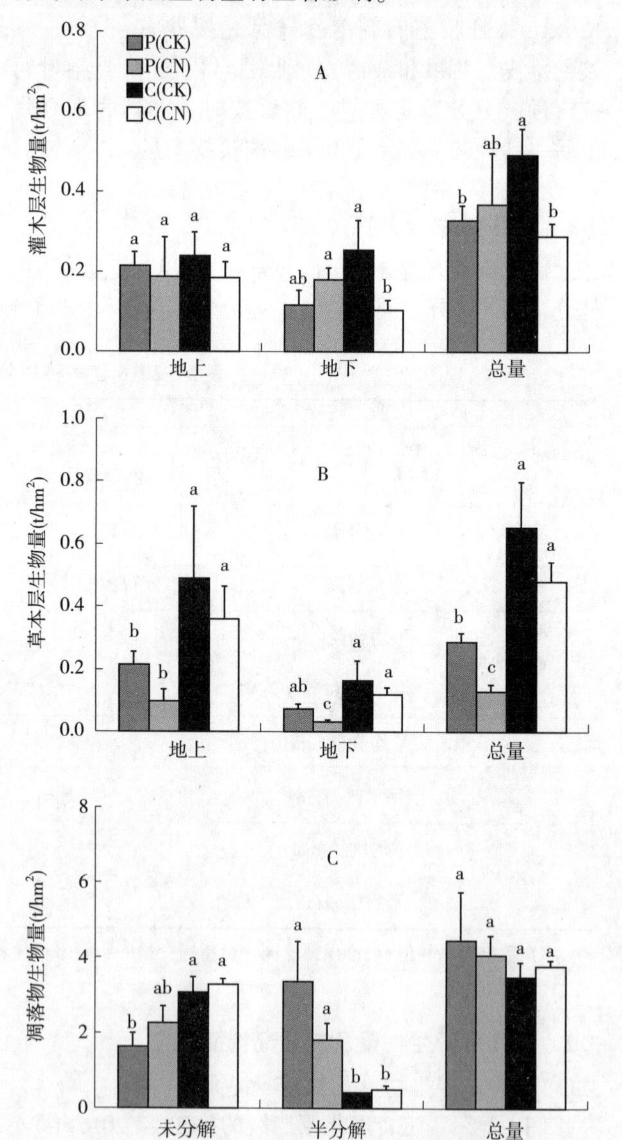

图3　不同林分地被层各器官生物量

近自然化改造实施8年后，4种林分凋落物现存量总量为3.370~4.465t/hm²，不同林分间凋落物现存量总量及各组分现存量与对照林P(CK)和C(CK)之间均无显著差异(*P*>0.05)，但未分解凋落物和半分解凋落物现存量在马尾松林 P(CK)、P(CN)和杉木林 C(CK)、C(CN)之间存在显著差异。马尾松林半分解组分现存量显著高于杉木林，而未分解组分现存量显著低于杉木林(图 3)。表明近自然化改造对马尾松和杉木人工林凋落物现存量总量均无显著影响。

3.2.3 林分生物量及其分配

表 4 显示了 2016 年 4 种人工林生态系统和各组分生物量及其分配情况，由表 4 可以看出，近自然化改造实施 8 年后，改造林分 P(CN)和 C(CN)生态系统生物量总量比对照林 P(CK)和 C(CK)分别高出 46.71%和 37.24%。其中，改造林乔木层生物量比对照林分别高出 47.84%和 39.08%。而灌木层、草本层生物量和凋落物现存量在改造 8 年后变化情况不同，或增或减，或无显著变化。表明，近自然化改造可显著提升乔木层生物量和林分生物量总量。此外，无论改造与否，马尾松林 P(CK)和 P(CN)林分生物量总量也显著高于杉木林 C(CK)和 C(CN)。

4 种林分各组分在生态系统生物量总量中的分配比例，均以乔木层最大，凋落物层居次，灌木层和草本层最小，仅乔木层生物量就占据林分生物量总量的 95.48%~98.82%，而凋落层现存量、灌木层和草本层三者生物量总和仅占林分生物量总量的 1.18%~4.52%。由此可以看出，4 种林分生态系统生物量基本由乔木层主导，乔木层生物量的变化对各林分生态系统生物量总的变化起着决定性的作用。

表 4 不同林分各组分生物量及其分配

层次	马尾松对照林	马尾松改造林	杉木对照林	杉木改造林
乔木层	257.13±32.67(98.06)	380.14±33.30(98.82)	97.45±33.63(95.48)	135.53±5.31(96.76)
灌木层	0.33±0.04(0.13)	0.37±0.10(0.10)	0.49±0.07(0.40)	0.29±0.03(0.21)
草本层	0.29 ±0.03(0.11)	0.13 ±0.02(0.03)	0.65 ±0.15(0.64)	0.48±0.06(0.33)
凋落物层	4.47 ±1.28(1.70)	4.05 ±0.53(1.05)	3.47±0.42(3.40)	3.78±0.15(2.70)
生态系统	262.21±24.70(100.00)	384.69±23.62(100.00)	102.06±14.76(100.00)	140.07±3.45(100.00)

注：数据为平均值±标准误差；括号内数据为各层次生物量占林分生物量总量的百分比。

此外，相同组分生物量的分配在不同林分间也有所差异，各组分生物量分配的差异不仅出现在改造林和对照林之间，同时也出现在马尾松和杉木两个树种之间，近自然化改造增加了乔木层生物量的分配，而降低了地被层(包括灌木层、草本层和凋落物层)生物量的分配，马尾松林乔木层生物量的分配高于杉木林，而地被层生物量的分配低于杉木林。

3.3 不同林分乔木层年净生产力

表 5 显示了 4 种林分 2007—2016 年间乔木层的年净生产力，乔木生产力采用 2007—2016 年共 9 年的年净生物量增量的平均值与年凋落量的加和来计算。由表 5 可以看出，4 种林分乔木层生产力存在显著差异，改造林 P(CN)和 C(CN)高于对照林年净生产力的 56.3%和 26.8%，仅乔木层年平均生物量净增量就分别高于对照林的 92.1%和 64.9%。由此可见，经过近自然化改造，林分乔木生产力大幅提升，且马尾松人工林，近自然化改造对生产力的促进效益更加明显，改造后，乔木生产力提升到改造前的 1.9 倍。

表 5 不同林分年净生产力(2007—2016 年)

组分	马尾松对照林	马尾松改造林	杉木对照林	杉木改造林
乔木层	15.2±1.4b	29.2±2.6a	5.7±2.3d	9.4±0.1c
凋落量	10.2±2.2a	10.5±1.8a	9.2±0.7a	9.5±1.1a
合计	25.4±3.1b	39.7±3.9a	14.9±2.4d	18.9±1.0c

注：数据为平均值±标准误差；同一个指标变量的不同字母代表差异显著($P<0.05$)。

由表 5 还可以看出，马尾松和杉木人工林近自然化改造后，林分生产力的提高主要归因于乔木生产力的大幅提升。因为各林分年凋落量并无显著差异，而且林分其他组分，诸如灌木层和草本层，因其生物量在林分所占的比例极小，对林分生产力并无显著影响。因而，乔木生物量的快速增长才是林分生产力提升的直接原因。

4 结论与讨论

森林生物量和生产力的影响因素较多，气候、土壤、水热条件、森林类型都会对森林生产力产生影响[5,34,35]，而对人工林来说树种、林龄、造林模式及经营管理活动都会影响到林分的生物量和生产力[18,36-39]，人工林近自然化改造，通过强度间伐和林下补植两种关键措施调整林分的群落结构，必然对生物量产生影响。本研究结果显示近自然化改造8年后，马尾松和杉木人工林林分生物量增加了46.71%和37.24%，年净生产力提高56.3%和26.8%。由此看来，近自然化改造对林分生长量的促进效果极为明显，乔木生长量的增加直接增加了林分生物量和生产力。生物量增加的速度加快的主要原因是补植珍贵乡土阔叶树种后，林分结构发生改变，林分的密度结构、树种结构、层次结构发生改变，一方面促使主林层林木生长量和生物量的快速增长，另一方面促进补植树种的快速增长，二者共同促进乔木层和林分生物量的快速增加。此外，本研究还发现，无论改造与否，马尾松人工林乔木生物量都远高于杉木林，主要原因是南亚热带地区是杉木分布的边缘产区，20年后，林木生长速度已经过了高峰期，甚至生活力开始下降，而马尾松是该地区的乡土树种，作为马尾松的主产区，马尾松在南亚热带地区具有最适宜的生长环境，23年生的马尾松林正值生长高峰期，因而无论改造与否，马尾松林的乔木生物量远高于杉木林。

从林分生物量的分配上看，林分生物量总量的95.48%~98.82%由乔木层贡献，表明乔木层生物量和生产力的提高对林分生物量增加起着主导作用。而地被层生物量虽然在林分间有显著差异，但由于在林分生物量中所占的比重极小，对林分生物量总量变化的贡献可忽略不计，这与先前的研究结论较为一致[2,3,20]。这一结果同时反映了人工林林分结构单一，林下植被稀少，乔木层的高度郁闭，限制了林下植被的生长和发育，近自然化改造尽管改善了林分乔木层的结构组成，大幅提升了乔木层的生物量，但对林下植被层生物量的贡献极其有限。

本研究中，4种林分乔木层生物量的分配顺序为树干>树根>枝条>树皮>叶片，这一结果虽然与国内外多数树种人工林的结果相似[2,20,24,40]，然而，不同经营模式的林分间，相同树种的相同器官在林分中的分配比例有显著差异，本研究结果发现近自然化改造可以显著促进叶片和枝条的发育，而减少树干和树皮在林木生物量中的分配。这表明，近自然化改造以后，尤其是强度间伐和林下补植的干扰，林分结构得以改善，林木个体获得较好的生长条件和生长空间而快速生长。为增加生态系统光合生产力以满足林木个体生长的需求，林木个体通过增加枝条和叶片的数量(增加叶面积)和比重以达到增加光合生产力的目的。

如上所述，在人工林生物量和生产力的影响因子中，林龄对人工林生物量的影响是非常关键的[4,41,42]。林龄不仅影响林分生物量的总量，而且影响到林分生物量在各组分的分配[4]。林分生物量随林龄增长动态变化是反映森林生态系统活力的重要指标，尤其是生态系统长期生物量动态特征的研究，对森林健康程度和人工林可持续经营的研究意义重大[5,34]。本研究反映的是2016年对林分各组分生物量调查的结果，即近自然化改造8年后的林分生物量，因而得出的生物量结果是阶段性的，无法代表近自然化改造对林分生物量的影响过程。然而近自然经营对林分生物量的影响是长期的，动态的，尤其是林下灌木层和草本层生物量是往往受林下植被多样性的影响，而光照、林分郁闭度是影响林下植物多样性的重要影响因子[43]。而且，近自然化改造初期，由于间伐和补植作业对林地的扰动，可能导致林下植被生物量的减少，而后期由于间伐打开了空间，林下植被生物量又开始增加，直至补植树种进入次林层，再次郁闭，林下生物量可能又会下降，因而，林下植被生物量可能呈现一定的波动性。因此，准确认识近自然化改造对林下植被生物量的影响，需要对林分作连年观测，了解其动态变化规律。国内对人工林生物量和生产力的研究大多局限在不同人工林短期生物量和生产力的比较上，或者利用空间代替时间的方法研究生物量和生产力的动态特征，缺乏对人工林生物量生产力的定位观测研究[23,24,41,42,44-46]。因而，准确而充分的认识近自然经营对人工林生物量及其分配的影响规律，需要开展长期定位观测研究，以便通过生物量动态变化规律深入了解近自然化改造对生物量及生产力的影响。

参考文献

[1] LUXMOORE R J, THARP M L, POST W M. Simulated biomass and soil carbon of loblolly pine and cottonwood plantations across a thermal gradient in southeastern United States[J]. Forest Ecology and Management, 2008, 254(2): 291-299.

[2] RIZVI R H, DHYANI S K, YADAV R S, et al. Biomass production and carbon stock of poplar agroforestry systems in Yamunanagar and Saharanpur districts of northwestern India[J]. Current Science, 2011, 100(5): 736-742.

[3] DOSSA E L, FERNANDES E C M, REID W S, et al. Above-and belowground biomass, nutrient and carbon stocks contrasting an open-grown and a shaded coffee plantation[J]. Agroforestry Systems, 2008, 72(2): 103-115.

[4]PEICHL M, ARAIN M A. Above-and belowground ecosystem biomass and carbon pools in an age-sequence of temperate pine plantation forests[J]. Agricultural and Forest Meteorology, 2006, 140(1/4): 51-63.

[5]方精云. 中国森林生产力及其对全球气候变化的响应[J]. 植物生态学报, 2000, 24(5): 513-517.

[6]BI H Q, LONG Y S, TURNER J, et al. Additive prediction of aboveground biomass for Pinus radiata (D. Don) plantations[J]. Forest Ecology and Management, 2010, 259(12): 2301-2314.

[7]BOLTE A, LÖF M. Root spatial distribution and biomass partitioning in Quercus robur L. seedlings: the effects of mounding site preparation in oak plantations[J]. European Journal of Forest Research, 2010, 129(4): 603-612.

[8]CHEN C R, XU Z H, BLUMFIELD T J, et al. Soil microbial biomass during the early establishment of hoop pine plantation: seasonal variation and impacts of site preparation[J]. Forest Ecology and Management, 2003, 186(1/3): 213-225.

[9]JACOBS D F, SELIG M F, SEVEREID L R. Aboveground carbon biomass of plantation-grown American chestnut(*Castanea dentata*) in absence of blight[J]. Forest Ecology and Management, 2009, 258(3): 288-294.

[10]KIDANU S, MAMO T, STROOSNIJDER L. Biomass production of Eucalyptus boundary plantations and their effect on crop productivity on Ethiopian highland vertisols[J]. Agroforestry Systems, 2005, 63(3): 281-290.

[11]LEE K H, JOSE S. Soil respiration, fine root production, and microbial biomass in cottonwood and loblolly pine plantations along a nitrogen fertilization gradient[J]. Forest Ecology and Management, 2003, 185(3): 263-273.

[12]ONYEKWELU J C. Growth, biomass yield and biomass functions for plantation-grown Nauclea diderrichii(de wild) in the humid tropical rainforest zone of south-western Nigeria[J]. Bioresource Technology, 2007, 98(14): 2679-2687.

[13]陈辉, 洪伟, 兰斌, 等. 闽北毛竹生物量与生产力的研究[J]. 林业科学, 1998, 34(S1): 60-64.

[14]刘茜. 不同龄组马尾松人工林生物量及生产力的研究[J]. 中南林学院学报, 1996, 16(4): 47-51.

[15]温远光, 梁宏温, 招礼军, 等. 尾叶桉人工林生物量和生产力的研究[J]. 热带亚热带植物学报, 2000, 8(2): 123-127.

[16]温远光, 梁宏温, 蒋海平. 广西杉木人工林生物量及分配规律的研究[J]. 广西农业大学学报, 1995, 14(1): 55-64.

[17]明安刚, 张治军, 谌红辉, 等. 抚育间伐对马尾松人工林生物量与碳贮量的影响[J]. 林业科学, 2013, 49(10): 1-6.

[18]杜虎, 曾馥平, 王克林, 等. 中国南方3种主要人工林生物量和生产力的动态变化[J]. 生态学报, 2014, 34(10): 2712-2724.

[19]姚迎九, 康文星, 田大伦. 18年生樟树人工林生物量的结构与分布[J]. 中南林学院学报, 2003, 23(1): 1-5.

[20]江泽慧, 范少辉, 冯慧想, 等. 华北沙地小黑杨人工林生物量及其分配规律[J]. 林业科学, 2007, 43(11): 15-20.

[21]彭龙福. 35年生楠木人工林生物量及生产力的研究[J]. 福建林学院学报, 2003, 23(2): 128-131.

[22]覃林, 何友均, 李智勇, 等. 南亚热带红椎马尾松纯林及其混交林生物量和生产力分配格局[J]. 林业科学, 2011, 47(12): 17-21.

[23]MING A G, JIA H Y, ZHAO J L, et al. Above-and below-ground carbon stocks in an indigenous tree(*Mytilaria laosensis*) plantation chronosequence in subtropical China[J]. PLoS One, 2014, 9(10): e109730.

[24]明安刚, 贾宏炎, 田祖为, 等. 不同林龄格木人工林碳储量及其分配特征[J]. 应用生态学报, 2014, 25(4): 940-946.

[25]蔡道雄, 郭文福, 贾宏炎, 等. 南亚热带优良珍贵阔叶人工林的经营模式[J]. 林业资源管理, 2007, (2): 11-14, 65-65.

[26]WANG H, LIU S R, MO J M, et al. Soil organic carbon stock and chemical composition in four plantations of indigenous tree species in subtropical China[J]. Ecological Research, 2010, 25(6): 1071-1079.

[27]康冰, 刘世荣, 蔡道雄, 等. 马尾松人工林林分密度对林下植被及土壤性质的影响[J]. 应用生态学报, 2009, 20(10): 2323-2331.

[28]VESTERDAL L, SCHMIDT I K, CALLESEN I, et al. Carbon and nitrogen in forest floor and mineral soil under six common European tree species[J]. Forest Ecology and Management, 2008, 255(1): 35-48.

[29]WANG H, LIU S R, WANG J X, et al. Effects of tree species mixture on soil organic carbon stocks and greenhouse gas fluxes in subtropical plantations in China[J]. Forest Ecology and Management, 2013, 300: 4-13.

[30]CARNEVALE N J, MONTAGNINI F. Facilitating regeneration of secondary forests with the use of mixed and pure plantations of indigenous tree species[J]. Forest Ecology and Management, 2002, 163(1/3): 217-227.

[31]SCHÜTZ J P. Development of close to nature forestry and the role of ProSilva Europe[J]. Zbornik Gozdarstva in Lesarstva, 2011, (94): 39-42.

[32]陆元昌, 张守攻, 雷相东, 等. 人工林近自然化改造的理论基础和实施技术[J]. 世界林业研究, 2009, 22(1): 20-27.

[33]康冰, 刘世荣, 蔡道雄, 等. 南亚热带杉木生态系统

生物量和碳素积累及其空间分布特征[J]. 林业科学，2009，45(8)：147-153.

[34]巨文珍，农胜奇. 森林生物量研究进展[J]. 西南林业大学学报，2011，31(2)：78-83，89-89.

[35]林金国，张兴正，翁闲. 立地条件对米老排人工林生长和材质的影响[J]. 植物资源与环境学报，2004，13(3)：50-54.

[36]蔡秀珠. 人为干扰强度对马尾松人工林生物量及其分配的影响[J]. 中南林学院学报，2006，26(2)：51-54.

[37]崔秋芳，赵佳宝，陈家林，等. 不同林龄阶段的松栎混交人工林碳储量研究[J]. 生态环境学报，2015，24(12)：1944-1949.

[38]方升佐，李光友，李同顺，等. 经营措施对青檀人工林生物量及檀皮产量的影响[J]. 植物资源与环境学报，2001，10(1)：21-24.

[39]郑路，蔡道雄，卢立华，等. 南亚热带不同树种人工林生物量及其分配格局[J]. 林业科学研究，2014，27(4)：454-458.

[40]赵金龙，梁宏温，温远光，等. 马尾松与红椎混交异龄林生物量分配格局[J]. 中南林业科技大学学报，2011，31(2)：60-64，71-71.

[41]CAO J X，WANG X P，TIAN Y，et al. Pattern of carbon allocation across three different stages of stand development of a Chinese pine(*Pinus tabulaeformis*) forest [J]. Ecological Research，2012，27(5)：883-892.

[42]ZHAO J L，KANG F F，WANG L X，et al. Patterns of biomass and carbon distribution across a chronosequence of Chinese pine(*Pinus tabulaeformis*) forests[J]. PLoS One，2014，9(4)：e94966.

[43]尤业明，徐佳玉，蔡道雄，等. 广西凭祥不同年龄红椎林林下植物物种多样性及其环境解释[J]. 生态学报，2016，36(1)：164-172.

[44]丁贵杰，王鹏程. 马尾松人工林生物量及生产力变化规律研究Ⅱ. 不同林龄生物量及生产力[J]. 林业科学研究，2002，15(1)：54-60.

[45]段爱国，张建国，何彩云，等. 杉木人工林生物量变化规律的研究[J]. 林业科学研究，2005，18(2)：125-132.

[46]田大伦，项文化，闫文德. 马尾松与湿地松人工林生物量动态及养分循环特征[J]. 生态学报，2004，24(10)：2207-2210.

[原载：生态学报，2017，37(23)]

桂西南岩溶区不同恢复模式群落生物量及林下植物多样性

农 友[1,2] 卢立华[1,2] 贾宏炎[1,2] 雷丽群[1,2] 明安刚[1,2] 李 华[1,2] 王亚南[1,2] 韦菊玲[1,2]
([1]中国林业科学研究院热带林业实验中心，广西凭祥 532600)
([2]广西友谊关森林生态系统国家定位观测研究站，广西凭祥 532600)

摘 要 研究桂西南岩溶区不同恢复模式群落的生物量及其林下植物多样性的特点，为该区域的生物多样性保护、生态功能恢复效果评价积累基础数据。以桂西南岩溶区4种不同恢复模式为研究对象，采用样方法对林下植物多样性进行研究；采用收获法研究灌木层与草本层的地上、地下生物量以及凋落物层现存量；采用异速生长模型来估算乔木层的地上生物量，并参考IPCC根茎比来量化乔木的地下生物量。结果表明：共调查记录林下植物85种，隶属于46科81属；其中灌木植物25科46属50种；草本植物21科35属35种；不同恢复模式群落生物量的变化趋势为自然恢复林(166.65t/hm^2)>任豆林(48.61t/hm^2)>吊丝竹林(36.52t/hm^2)>灌草坡(0.95t/hm^2)。不同恢复模式灌木层物种丰富度差异不显著，最高的为灌草坡(16种)，其次为任豆林(15种)，最低为自然恢复林(12种)；草本层物种丰富度最高的为任豆林(12种)，其次为灌草坡(10种)，最低为自然恢复林(4种)。自然恢复林乔木不同组分生物量与任豆林、吊丝竹林之间差异显著(P<0.05)；灌草坡的灌木、草本地上生物量与吊丝竹林、任豆林、自然恢复林之间差异显著(P<0.05)。吊丝竹林、任豆林的灌草生物量表现为草本层>灌木层，而灌草坡与自然恢复林的灌草生物量则表现为灌木层>草本层。结合本次研究，对西南岩溶区4种不同恢复模式提出了参考管理措施。

关键词 生物量；植物多样性；岩溶区；恢复模式

The Community Biomass and Understory Plant Diversity of Different Vegetation Restoration of Karst Region in Southwest of Guangxi

NONG You[1,2], LU Lihua[1,2], JIA Hongyan[1,2], LEI Liqun[1,2], MING Angang[1,2], LI Hua[1,2], Wang Yanan[1,2], WEI Juling[1,2]
([1]*Experimental Center of Tropical Forestry, Chinese Academy of Forestry, Pingxiang* 532600, *Guangxi, China*)
([2]*Guangxi Youyiguan Forest Ecosystem Research Station, Pingxiang* 532600, *Guangxi, China*)

Abstract: The study investigated the community biomass and understory plant species diversity in four different recovery models in karst region of Southwest Guangxi in order to accumulate the basic information for biodiversity conservation and restoration of ecological function in the region. Four different recovery models were selected as the research object, and the understory plant species diversity were studied by samples, the ground and underground biomass of shrub layer and herb layer was studied by the harvest method; the arbor layer biomass were studied by using allometic models and the ratio of root and stem recommended by IPCC. 85 species of 81 genera, 47 families were recorded, among of which, 50 species of 46 genera, 25 families were shrubs and 35 species of 35 genera, 22 families were herds. The biomass of different vegetation restoration models showed as, the highest was the natural restoration forests(166.65 t/hm^2) and then the *Zenia insignis* forest(48.61 t/hm^2), then was the *Dendrocalamus minor* forest(36.52 t/hm^2), and the lowest was the shrub grassland (0.95 t/hm^2). The shrub species richness had no significant difference between different models, the highest richness was the shrub grassland(16 species), and then the *Zenia insignis* forest(15 species), the lowest was the natural restoration forests(12 species); the highest in herb was the *Zenia insignis* forest(12 species), and then the shrub grassland(10 species), the lowest was the natural restoration forests(4 species). In the tree layer, biomass of different components of the natural restoration forest was significant difference with the *Zenia insignis* and the *Dendrocalamus minor* forest(P<0.05);

The aboveground biomass in the shrub grassland was significant difference with the others($P<0.05$); The litter biomass was not significant($P<0.05$); The biomass of herb layer was higher than the shrub layer in the *Zenia insignis* and *Dendrocalamus minor* forest; but in the natural restoration forest and shrub grassland it showed the opposite; The precaution management for four different restoration models in the karst area of Southwest of Guangxi was put forward based on this result.

Key words: biomass; plant diversity; karst area; vegetation restoration model

在当前可持续发展和生态环境改善日益重视的社会背景与发展趋势下，作为生态系统最重要功能的生物量和生物多样性研究已成为生态学研究中的热点[1-5]。生物量是生态系统生产力基础和功能的主要表现形式[6]。物种多样性与群落的功能过程密切相关，是生物多样性最为基础和关键的层次[7]。研究表明，林下植被作为森林生态系统的一个重要组成部分，在促进养分循环和维护森林立地质量方面起着重要作用[8-11]。对于退化生态系统恢复成功的标准，物种多样性、群落生物量是其中非常重要的评价指标[12,13]。因此，开展不同区域、不同树种、不同模式森林生态系统群落生物量及物种多样性的研究仍具有必要性和紧迫性。

西南岩溶区的位置非常特殊，它处于长江、珠江、澜沧江等水系的中上游，其脆弱的生态环境不仅威胁下游地区生态系统的安全，还严重制约着西南地区经济社会的可持续发展。岩溶植被的恢复与重建成为我国西南岩溶地区退化生态系统恢复与重建研究的难点与重点[14]。目前，对西南岩溶区开展的研究主要集中在石漠化治理[15-16]、水土流失[17,18]、土壤[17,19]等方面，但对于岩溶生态系统植物多样性及生物量研究报道还不多见，已有工作主要针对不同年龄系列或不同演替阶段的群落[20-22]，对不同土地利用方式或恢复模式的生物多样性、生物量的变化缺乏探讨。岩溶区生态重建或生态恢复的关键是森林植被的重建与恢复，岩溶区植被恢复的主要驱动力是植物的多样性[23]，而植物多样性的恢复是退化生态系统恢复与重建效果的重要考核指标，不同植被恢复模式产生的生态效益不同，因此，对该区不同恢复模式下植物群落生物量及物种多样性的研究显得尤为重要。

本文以桂西南岩溶区4种不同恢复模式为研究对象，对其群落生物量及林下植被多样性进行了研究。拟揭示该区不同恢复模式群落生物量及林下植物多样性的特点，有助于科学评价植被恢复效果，为该区域的生物多样性保护、生态功能恢复效果评价提供理论依据和基础数据。

1 研究区概况及研究方法

1.1 研究区概况

研究区位于广西西南部的天等县，属南亚热带季风气候，地理坐标为107°11′E，23°09′N。全县面积2159.25km²，以低山丘陵为主，山地面积1696.42km²，占总面积77.98%。其中土山占总面积的22.60%；石灰岩山地占总面积的41.50%；硅质灰岩山地占总面积的10.31%；半土半石山占总面积的3.57%。全县地势西南高东北低，最高海拔1073.7m，最低海拔263m。春末至初秋多受偏南气流影响，气温高，湿度大，降水量多。冬季受北方寒潮影响，气温偏低，湿度小，水量少。年平均太阳总辐射量为100.6Kcal/cm²，年均气温为20.5℃，年均降水量1459.1mm。土壤主要由第四纪红土、砂页岩、河流冲积、洪积、棕色石灰土、紫色岩、硅质岩等7种母质发育而成。其中砂页岩母质最多，占59.7%，次为硅质岩母质、棕色石灰土母质，各占18.8%和17.9%[24]。

1.2 研究方法

1.2.1 样地设置及调查

2015年5月，在林分踏查的基础上，选取海拔、坡度、坡向等环境背景条件基本一致的四种模式，在每一种模式中各设置3个面积20m×30m样方。在样方四个角及中心点设置面积5m×5m的灌木、草本样方5个；在样方内随机设置2m×2m的灌草生物量收获样方3个，1m×1m的凋落物样方3个。共完成乔木样方12个，灌木样方60个，草本样方60个。

调查内容包括样地内乔木的种类、高度、胸径；灌木、草本的种类、株树、高度、盖度。采用样方收获法，测定灌木层与草本层的地上和地下生物量，以及凋落物层现存量。对各组分进行取样，所取样品均带回实验室在65℃烘干至恒重，计算含水率并将鲜重换算成干重。烘干样品经粉碎、过筛后装瓶，用于有机碳含量的测定。

表1 样地概况

样地号	恢复模式	坡度(°)	坡向	海拔(m)
R1	吊丝竹林	21±2	东南	340±4
R2	灌草坡	18±3	东南	415±10
R3	任豆林	20±5	东南	440±7
R4	自然恢复林	16±2	东南	515±2

注：表中数据为平均值±标准差。

1.2.2 物种多样性

按灌木层、草本层描述各个样地的生物多样性特点。物种多样性测定指标采用物种丰富度指标 S、Shannon-wiener 指数、Simpson 指数、J_{sw} 均匀度指数，其具体计算方法见文献[25,26]。重要值(IV)采用宋永昌计算方法[27]。

1.2.3 生物量测定

目前，对生物量的研究多采用直接收获法，考虑到当地不准采伐，因此，本研究利用汪珍川等建立的广西主要树种(组)异速生长模型来估算乔木层的地上生物量[28]，并参考 IPCC 根茎比来量化乔木的地下生物量[29]。

1.2.4 数据处理

样地的生物量和多样性指数数据用 Excel2010 统计，基于 SPSS18.0，采用单因素 ANOVA 进行方差分析。

2 结果与分析

2.1 不同恢复模式的林下植物多样性

植物多样性是植物群落组成和结构的重要指标，它可以反映群落的组织化水平，是影响群落生物量和生产力的重要因子。根据物种多样性指数计算公式对乔木层、灌木层及草本层进行计算，得到各样地的植物多样性指数。不同恢复模式植物多样性各指数如表 2 所示。

表 2 不同演替阶段次生林植物多样性

类型	样地号	物种丰富度 S	Shannon-wiener 指数	Simpson 指数	J_{sw} 均匀度指数
灌木层	R1	14±4	2.3733±0.1897ab	0.8920±0.1433ab	0.8993±0.0098bcd
	R2	15±3	2.1363± 0.1843ab	0.8431±0.0783ab	0.7889±0.0320bcd
	R3	16±1	2.2814±0.0602a	0.8780± 0.0064a	0.8228±0.0052bcd
	R4	12±3	1.3792±0.1991b	0.6524±0.0639b	0.5550± 0.0148a
草木层	R1	7±2ab	0.9241±0.1865	0.4165±0.0622	0.4749±0.0222
	R2	12±2a	1.4571±0.1919	0.6489±0.0267	0.5864±0.0047
	R3	10±1ab	1.5406±0.1886	0.7335±0.0647	0.6691±0.0393
	R4	4±2b	0.9817±0.1616	0.5467±0.1684	0.7082±0.0309

注：表中数据为平均值±标准差。R1 为吊丝竹林；R2 为任豆林；R3 为灌草坡；R4 为自然恢复林；同一行相同字母表示差异不显著，不同字母表示差异显著($P<0.05$)。

本研究共调查林下植物 85 种，隶属于 46 科 81 属；其中灌木植物 25 科 46 属 50 种；草本植物 21 科 35 属 35 种。样地灌木层植物主要由番石榴(*Psidium guajava* Linn.)、红背山麻杆[*Alchornea trewioides* (Benth.) Muell. Arg.]、潺槁木姜子[*Litsea glutinosa*(Lour.) C. B. Rob.]、灰毛浆果楝[*Cipadessa cinerascens* (Pellegr.) Hand.-Mazz.]、构棘[*Cudrania cochinchinensis*(Lour.) Kudo et Masam.]等组成；草本层草本植物主要由水蔗草(*Apluda mutica* Linn.)、五节芒[*Miscanthus floridulus* (Lab.) Warb. ex Schum. et Laut.]、肾蕨[*Nephrolepis auriculata* (L.) Trimen]、艾草(*Artemisia argyi* Levl. et Van.)等组成。

2.1.1 灌木层植物多样性比较

不同恢复模式，灌木层物种丰富度差异不显著，最高的为灌草坡($S=16$)，其次为任豆林($S=15$)，最低为自然恢复林($S=12$)；Shannon-wiener 指数、Simpson 指数、J_{sw} 均匀度指数均为吊丝竹林>灌草坡>任豆林>自然恢复林。灌草坡与自然恢复林的 Shannon-wiener 指数、Simpson 指数差异显著($P<0.05$)；自然恢复林与吊丝竹林、灌草坡、任豆林之间的 J_{sw} 均匀度指数差异显著($P<0.05$)。

2.1.2 草本层植物多样性比较

不同恢复模式，草本层物种丰富度最高的为任豆林($S=12$)，其次为灌草坡($S=10$)，最低为自然恢复林($S=4$)，其中，任豆林与自然恢复林之间差异显著；Shannon-wiener 指数、Simpson 指数均为灌草坡>任豆林>自然恢复林>吊丝竹林；J_{sw} 均匀度指数自然恢复林>灌草坡>任豆林>吊丝竹林。不同恢复模式的 Shannon-wiener 指数、Simpson 指数、J_{sw} 均匀度指数差异不显著($P<0.05$)。

2.2 不同恢复模式林下植物的物种重要值

表 3 列出了不同恢复模式林下植物重要值较大的物种，可以看出，不同恢复模式林下主要植物种类组成不同，但大部分都是一些耐旱性、石生性、喜钙性的植物，如红背山麻杆、灰毛浆果楝、茶条木(*Delavaya toxocarpa* Franch.)等，这与岩溶区岩石裸露率大，土层浅薄而干燥有关。

2.3 不同恢复模式的群落生物量

生物量是一个有机体或群落在一定时间内积累的干物质量，是表征其结构及功能的重要参数。不同恢复模式群落生物量如表4所示。不同植被恢复模式群落总生物量的变化趋势为自然恢复林(166.65t/hm^2)>任豆林(48.61t/hm^2)>吊丝竹林(36.53t/hm^2)>灌草坡(0.95t/hm^2)。乔木层中，自然恢复林不同组分生物量与任豆林、吊丝竹林之间差异显著($P<0.05$)；灌草坡的灌木、草本地上生物量与吊丝竹林、任豆林、自然恢复林之间差异显著($P<0.05$)；凋落物之间的差异不显著($P<0.05$)。吊丝竹林、任豆林表现为草本层>灌木层；灌草坡与自然恢复林表现为，灌木层>草本层。

表3 不同演替阶段次生林植物物种重要值

样地号	灌木		草本	
	种名	重要值	种名	重要值
R1	番石榴	48.70±2.85	水蔗草	178.90±9.82
	灰毛浆果楝	44.58±1.21	类芦	36.60±2.70
	潺槁木姜子	38.07±2.38	假杜鹃	28.64±1.63
	红背山麻杆	38.07±1.57	飞机草	23.89±2.33
	小果叶下珠	25.40±3.11	蜈蚣凤尾蕨	11.94±1.18
R2	红背山麻杆	67.46±3.12	弓果黍	122.06±2.37
	苎麻	41.54±2.30	肾蕨	52.84±3.25
	香椿	41.27±4.24	艾草	44.76±2.10
	灰毛浆果楝	40.69±2.28	小窃衣	17.65±1.54
	簕仔树	21.36±1.81	假臭草	16.27±1.28
R3	红背山麻杆	50.08±4.47	五节芒	148.78±5.73
	番石榴	46.34±2.90	水蔗草	61.16±6.29
	潺槁木姜子	39.70±3.75	白茅	38.91±2.63
	茶条木	31.60±2.29	金丝草	14.84±1.40
	扁担杆	29.54±2.66	斑茅	8.95±1.32
R4	灰毛浆果楝	106.61±7.27	肾蕨	175.79±5.86
	红背山麻杆	84.62±3.67	弓果黍	50.93±3.16
	金樱子	23.85±1.94	艾草	36.72±1.93
	雀梅藤	18.46±2.08	五节芒	36.57±1.86
	黑面神	18.33±1.86		

注：表中数据为平均值±标准差。R1为吊丝竹林；R2为任豆林；R3为灌草坡；R4为自然恢复林。

表4 不同植被恢复模式群落生物量

层次	组分	恢复模式			
		R1	R2	R3	R4
乔木层	树干(t/hm^2)	17.42±1.05a	31.19±1.19a	—	112.34±1.70b
	树枝(t/hm^2)	2.84±0.64a	9.10±1.44a	—	24.65±1.04b
	树叶(t/hm^2)	7.70±1.83a	0.18±0.03b	—	4.40±0.21c
	地上(t/hm^2)	27.96±2.73a	40.46±1.46a	—	141.40±1.93b
	地下(t/hm^2)	7.85±1.10a	7.31±1.60a	—	23.58±1.55b
	林分(t/hm^2)	35.81±3.79a	47.77±4.02a	—	164.97±6.47b

（续）

层次	组分	恢复模式			
		R1	R2	R3	R4
灌木层	地上(t/hm²)	0.10±0.01bcd	0.12±0.08bcd	0.35±0.04a	1.01±0.02bcd
	地下(t/hm²)	0.09±0.01	0.14±0.02	0.33±0.02	0.10±0.01
	小计(t/hm²)	0.19±0.01	0.25±0.03	0.68±0.02	1.11±0.04
草本层	地上(t/hm²)	0.17±0.05bcd	0.23±0.07bcd	0.07±0.02a	0.22±0.04bcd
	地下(t/hm²)	0.16±0.02	0.25±0.03	0.08±0.01	0.20±0.02
	小计(t/hm²)	0.33±0.05	0.49±0.02	0.15±0.03	0.41±0.05
凋落物层	凋落物(t/hm²)	0.21±0.01	0.10±0.03	0.13±0.01	0.16±0.03
	总生物量(t/hm²)	36.53±3.91	48.61±4.10	0.95±0.04	166.65±6.60

注：表中数据为平均值±标准差。R1 为吊丝竹林；R2 为任豆林；R3 为灌草坡；R4 为自然恢复林；同一行相同字母表示差异不显著，不同字母表示差异显著($P<0.05$)。

3 讨论

本次研究共记录到林下植物 85 种，隶属于 47 科 81 属。与国内一些非岩溶区相关研究对比，可以发现桂西南岩溶区物种丰富度不算高。例如，浙江古田山 24hm² 样地[30]、广东鼎湖山[31] 和云南西双版纳[32] 20hm² 样地分别有 159、210 和 468 种。这与桂西南岩溶区生境的特殊性、结构的多样性与复杂性、小生境的高度异质性等密切相关[33,34]。

不同恢复模式，林下灌木和草本植物的组成有所不同。灌木层物种丰富度最高的为灌草坡，其次为任豆林，最低为自然恢复林；草本层物种丰富度最高的为任豆林，其次为灌草坡，最低为自然恢复林。灌木层林下物种多样性高于草本层，这与群落结构越复杂，物种多样性指数越高的结论一致[35]。

不同恢复模式，不同组分生物量有所差异，群落总生物量、乔木生物量、灌木生物量最大的均为自然恢复林；草本层生物量最大的为任豆林；凋落物层生物量最大的为吊丝竹林；吊丝竹林、任豆林的灌草生物量表现为草本层>灌木层，而灌草坡与自然恢复林的灌草生物量则表现为灌木层>草本层。因此，在进行石漠化治理时应重视灌草的作用，注重对灌草的保护和利用[36]。人工恢复模式(吊丝竹、任豆林)群落生物量比灌草坡群落生物量高 453.14%~592.55%，说明，人工干预在退化生态系统的恢复中对系统功能的尽快恢复起着重要的作用。相关分析结果表明，林下灌木层、草本层物种丰富度与生物量无显著相关性($R^2=0.02$)，对于它们两者之间的相互关系有待进一步研究。

灌草坡群落是植被遭严重破坏后恢复的较初始阶段，以耐干旱贫瘠的灌草植物为主，偶有自然生长的乔木。此模式以红背山麻杆、番石榴、潺槁木姜子、茶条木、扁担杆(*Grewia biloba* G. Don)为优势种，比较耐旱的五节芒、水蔗草、白茅[*Imperata cylindrica*(L.) Beauv.]、金丝草[*Pogonatherum crinitum*(Thunb.) Kunth]、斑茅(*Saccharum arundinaceum* Retz.)为常见草本。对于此群落，首先应采取禁牧保护措施，减少人为干扰，避免进一步的退化，其次可以人为引种一些抗旱性较强的乔木，使其尽快得到恢复，促进植物群落的迅速形成。

任豆林和吊丝竹林都是经过人工干预后产生的群落，草本层多样性大于灌木层。任豆林群落这一恢复模式乔木层密度相对比较固定，上层乔木对林下造成一定的郁闭，林下小环境相对湿润，从而为大量草本提供了生存条件，有大量肾蕨着生于石缝中，弓果黍[*Cyrtococcum patens*(L.) A. Camus]、艾草、小窃衣[*Torilis japonica*(Houtt.) DC.]等则生于土坑中；林窗下，红背山麻杆、苎麻[*Boehmeria nivea*(L.) Gaudich.]、灰毛浆果楝、簕仔树(*Mimosa sepiaria* Benth.)等喜光或耐阴灌木占据一定比例。吊丝竹恢复模式对林地造成一定的郁闭，但林内环境较为干燥，存在大量的未分解竹叶，土壤表层生长着大量竹根，此模式下林下植物都是耐干旱贫瘠的物种，如灌木层的番石榴、灰毛浆果楝、潺槁木姜子、红背山麻杆、小果叶下珠(*Phyllanthus reticulatus* Poir.)；草本层的水蔗草、类芦[*Neyraudia reynaudiana*(Kunth) Keng ex Hitchc.]、假杜鹃(*Barleria cristata* L.)、飞机草(*Eupatorium odoratum* L.)、蜈蚣凤尾蕨(*Pteris vittata* L.)等。对于这两个群落，要适当进行间伐，保持合理的密度，有利于林下植被的恢复和林木自身的生长。

自然恢复林这一恢复模式是经过灌草坡演替而来，生长着少量的小乔木，但仍然保留着灌草坡的特性。群落中灌木种的丰富度较低，群落优势种主要集中于少数几个优势物种，因而群落的灌木多样性指数与均匀度指数均较低。林下植被主要以一些耐干旱贫瘠的灌草植物为主，如灌木层的红背山麻杆、雀梅藤[*Sageretia thea*(Osbeck)Johnst.]、金樱子(*Rosa laevigata* Michx.)等，草本层的五节芒、肾蕨等。群落多样性指标仍较低，稳定性较差，要达到相对合理的群落结构还需要经历较长的演替时间和过程。对于此群落，应该保持现状，以封山保育为主，顺其自然发展。

4　结论

(1)桂西南岩溶区物种不算丰富，本次研究共调查林下植物 85 种，隶属于 47 科 81 属。

(2)不同恢复模式群落生物量的变化趋势为自然恢复林(166.65t/hm^2)>任豆林(48.61t/hm^2)>吊丝竹林(36.52t/hm^2)>灌草坡(0.95t/hm^2)。

(3)适度的人工干预在退化生态系统的恢复中，通过生物多样性和植被的恢复，促进对系统实体功能的恢复起着重要的作用。

(4)结合本次研究，对西南岩溶区 4 种不同恢复模式提出了参考管理措施。

参考文献

[1] BENGTSSON J. Which species? What kind of diversity? Which ecosystem function? Some problems in studies of relations between biodiversity and ecosystem function[J]. Applied Soil Ecology, 1998, 10(3): 191-199.

[2] HEDLUND K, SANTA Regina I, Van der PUTTEN W H, et al. Plant species diversity, plant biomass and responses of the soil community on abandoned land across Europe: idiosyncrasy or above-belowground time lags[J]. Oikos, 2003, 103(1): 45-58.

[3] CARDINALE B J, WRIGHT J P, CADOTTE M W, et al. Impacts of plant diversity on biomass production increase through time because of species complementarity[J]. Proceedings of the national academy of sciences, 2007, 104(46): 18123-18128.

[4] CARDINALE B J, VENAIL P, GROSS K, et al. Further re-analyses looking for effects of phylogenetic diversity on community biomass and stability[J]. Functional Ecology, 2015, 29(12): 1607-1610.

[5] YUAN F, WU J, LI A, et al. Spatial patterns of soil nutrients, plant diversity, and aboveground biomass in the Inner Mongolia grassland: Before and after a biodiversity removal experiment[J]. Landscape Ecology, 2015, 30(9): 1737-1750.

[6] 杨利民，周广胜，李建东．松嫩平原草地群落物种多样性与生产力关系的研究[J]．植物生态学报，2002，26(5)：589-593.

[7] 王伯荪，王昌伟，彭少麟．生物多样性刍议[J]．中山大学学报(自然科学版)2005，44(6)：68-70.

[8] 方海波，田大伦．杉木人工林间伐后林下植被生物量的研究[J]．中南林业科技大学学报，1998(1)：5-9.

[9] KUME A, SATOMURA T, TSUBOI N, et al. Effects of understory vegetation on the ecophysiological characteristics of an overstory pine, Pinus densiflora[J]. Forest Ecology & Management, 2003, 176(1-3): 195-203.

[10] CHASTAIN R A, CURRIE W S, TOWNSEND P A. Carbon sequestration and nutrient cycling implications of the evergreen understory layer in Appalachian forests[J]. Forest Ecology & Management, 2006, 231(1-3): 63-77.

[11] GILLIAM F S. The Ecological Significance of the herbaceous layer in temperate forest ecosystems[J]. Bioscience, 2007, 57(57): 845-858.

[12] 任海，彭少麟．恢复生态学导论[M]．北京：科学出版社，2001，19.

[13] NAGARAJA B C, SOMASHEKAR R K, RAJ M B. Tree species diversity and composition in logged and unlogged rainforest of Kudremukh National Park, South India[J]. Journal of Environmental Biology, 2005, 26(4): 627-634.

[14] 贺庆棠，陆佩玲．中国岩溶山地 石漠化问题与对策研究[J]．北京林业大学学报，2006，28(1)：117-120.

[15] 苏维词．中国西南岩溶山区石漠化治理的优化模式及对策[J]．水土保持学报，2002，16(5)：24-27.

[16] 中国科学院学部．关于推进西南岩溶地区石漠化综合治理的若干建议[J]．地球科学进展，2003，18(4)：489-492.

[17] 曹建华，蒋忠诚，杨德生，等．我国西南岩溶区土壤侵蚀强度分级标准研究[J]．中国水土保持科学，2008，6(6)：1-7.

[18] 周蕊，方荣杰．西南岩溶区的水土保持措施体系构建[J]．中国水土保持，2012(3)：7-9.

[19] 王巨，谢世友，戴国富．西南岩溶区土壤生态系统退化研究[J]．中国农学通报，2011，27(32)：181-185.

[20] 温远光，雷丽群，朱宏光，等．广西马山岩溶植被年龄序列的群落特征[J]．生态学报，2013，(18)：5723-5730.

[21] 朱宏光，蓝嘉川，刘虹等．广西马山岩溶次生林群落生物量和碳储量[J]．生态学报，2015，(08)：

2616-2621.
[22]雷丽群．广西马山岩溶植被不同演替阶段的群落结构与环境因子的关系[D]．南宁：广西大学，2014.
[23]李明辉，彭少麟，申卫军，等．景观生态学与退化生态系统恢复[J]．生态学报，2003，23(8)：1622-1628.
[24]张万富．天等县志[M]．南宁：广西人民出版社1991.
[25]刘灿然，马克平．生物群落多样性的测度方法[J]．生态学报，1997(6)：601-610.
[26]马克平，刘玉明．生物群落多样性的测度方法：Ⅰα多样性的测度方法(下)[J]．生物多样性，1994(3)：162-168.
[27]宋永昌．2001. 植被生态学[M]．上海：华东师范大学出版社．
[28]汪珍川，杜虎，宋同清，等．广西主要树种(组)异速生长模型及森林生物量特征[J]．生态学报 2015，35(13)：4462-4472.
[29]AMSTEL AV. IPCC 2006 Guidelines for National Greenhouse Gas Inventories[M]．2006.
[30]祝燕，赵谷风，张俪文，等．古田山中亚热带常绿阔叶林动态监测样地——群落组成与结构[J]．植物生态学报，2008，32(2)：262-273.
[31]叶万辉，曹洪麟，黄忠良，等．鼎湖山南亚热带常绿阔叶林 20 公顷样地群落特征研究[J]．植物生态学报，2008，32(2)：274-286.
[32]兰国玉，胡跃华，曹敏，等．西双版纳热带森林动态监测样地——树种组成与空间分布格局[J]．植物生态学报，2008，32(2)：287-298.
[33]王德炉，朱守谦，黄宝龙．贵州喀斯特区石漠化过程中植被特征的变化[J]．南京林业大学学报(自然科学版)，2003，27(3)：26-30.
[34]俞筱押，李玉辉，马遵平．云南石林喀斯特小生境木本植物多样性特征[J]．山地学报，2007，25(4)：438-447.
[35]刘方炎，李昆，张春华，等．金沙江干热河谷植被恢复初期的群落特征[J]．南京林业大学学报(自然科学版)，2007，31(6)：129-132.
[36]司彬，姚小华，任华东，等．滇东喀斯特植被恢复演替过程中物种多样性研究[J]．西南大学学报(自然科学版)，2009，31(1)：132-139.

[原载：林业科学研究，2017，30(02)]

南亚热带红椎和马尾松人工林生长对穿透雨减少的响应

陈 琳[1,2,3] 刘世荣[4] 温远光[1] 曾 冀[2] 李 华[2] 杨予静[4]

([1]广西大学，广西南宁 530004；[2]中国林业科学研究院热带林业实验中心，广西凭祥 532600；[3]广西友谊关森林生态系统国家定位观测研究站，广西凭祥 532600；[4]中国林业科学研究院森林生态环境与保护研究所，北京 100091)

摘 要 为了预测和评估全球气候变化背景下森林生长和生产力对降水格局变化和季节性干旱的响应，以南亚热带红椎和马尾松人工林为对象，设置穿透雨减少50%和不减雨(对照)处理，开展连续3年(2015—2017年)的模拟试验，研究降雨减少对人工林胸径生长、凋落物量和叶面积指数的影响。结果表明：与对照相比，穿透雨减少导致红椎2017年胸径增长量显著降低31.8%，而对马尾松无影响；红椎叶面积指数平均降低8.8%，马尾松叶面积指数降低7.2%或者不变；红椎林2015年枝凋落量和2017年凋落物总量分别增加29.6%和35.8%，马尾松林2015年其他树种(除了马尾松以外)叶凋落量则显著减少50.7%，而其他凋落物组分无显著变化。短期穿透雨减少对人工林产生了胁迫作用，这种作用存在年际变异和树种差异。

关键词 穿透雨减少；红椎；马尾松；胸径生长；凋落物量；叶面积指数

Growth Responses of *Castanopsis hystrix* and *Pinus massoniana* Plantations to Throughfall Reduction in Subtropical China

CHEN Lin[1,2,3], LIU Shirong[4], WEN Yuanguang[1], ZENG Ji[2], LI Hua[2], YANG Yujing[4]

(*[1]Guangxi University, Nanning* 530004, *Guangxi China*; *[2]Experimental Center of Tropical Forestry, Chinese Academy of Forestry, Pingxiang* 532600, *Guangxi, China*; *[3]Youyiguan Forest Ecosystem Research Station, Pingxiang* 532600, *Guangxi, China*; *[4]Institute of Forest Ecology, Environment and Protection, Chinese Academy of Forestry, Beijing* 100091, *China*)

Abstract: To better predict and evaluate responses of tree growth and forest productivity to change of precipitation pattern and seasonal drought under global climate change scenarios, throughfall reduction experiments including 50% of throughfall and natural rainfall(control) treatments were conducted in *Castanopsis hystrix* and *Pinus massoniana* plantations of warm subtropical region over a three-year period(2015-2017). Diameter at breast height(DBH), litterfall amount and leaf area index of both plantations were investigated, respectively. The results showed that throughfall reduction resulted in a 31.8% decrease of annual increment of DBH of *C. hystrix* in 2017 and had no significant impact on that of *P. massoniana*. Leaf area index under throughfall reduction decreased by 8.8% in *C. hystrix* plantation and decreased by 7.2% or remained unchanged in *P. massoniana* plantation. Branch litterfall in 2015 and total litterfall in 2017 of *C. hystrix* increased by 29.6% and 35.8% by throughfall reduction, but leaf litterfall of other tree species(except for *P. massoniana*) in *P. massoniana* plantation declined by 50.7% in 2015, with no significant differences for other litterfall components. In conclusion, throughfall reduction had consequences of drought stress in both *C. hystrix* and *P. massoniana* plantations, with inter-annual variation and inter-specific differences.

Key words: throughfall reduction; *Castanopsis hystrix*; *Pinus massonana*; diameter growth; litterfall production; leaf area index

气候变化导致全球降水格局发生改变，干旱等极端降水频率和强度不断增加[1]，全球大面积森林出现生长和抗性下降，甚至死亡等现象，从而影响了森林生态系统生产力和生物多样性，反过来加剧气候变化[2-4]。目前，森林生态系统对降水变化的响应已成为各国关注的焦点，而模拟降水变化的野外控制试验是了解森林生态系统结构与功能对降水变化响应的有效途径。Hoover 等[5]系统总结了全球干旱模拟试验，发现研究地点主要集中在欧洲和北美，其中森林生态系统研究仅占 17.5%，而且大部分研究时间为 1~4 年，截雨比例以 50%为主。针对不同地区的降雨特点合理设计截雨比例，加强开展典型生态系统长期干旱模拟试验，特别是森林生态系统，对于提高预测未来降水变化对生态系统影响的准确性具有重要意义。

树木对干旱的适应性研究包括解剖结构、形态、生长、生物量分配、光合生理等方面[6-10]。平均生长速率因其容易测定、成本低，常作为树木生长和死亡的预测指标，一般树木通过减缓生长速率来适应干旱[11]，然而其影响程度受纬度、干旱类型、林分类型、林分密度、树种和树木年龄等因素的影响[4,12-15]。叶面积指数是树木生长又一个敏感性指标，特别是在森林群落尺度上反映森林群落的生产力[16]，干旱通过改变叶片寿命和凋落物量，从而影响叶面积指数[17,18]。凋落物是森林生态系统的重要组成部分，凋落物分解为植物生长、土壤生物活动提供物质和能量。干旱可能导致森林凋落物量不变、增加或者减少，这与树种、凋落物组分、降水强度和时间以及其他气候因子密切相关[19-22]。

根据亚热带地区常绿阔叶林的长期监测结果，增温和干旱显著影响亚热带森林树木生长、死亡和更新，林分的树种组成逐渐由以少量大树为主向以大量小树为主转变，树木平均胸径和森林生物量显著降低，表明亚热带森林正面临气候变化威胁，抵御能力下降[23,24]。因此，本研究选择我国南亚热带地区 2 个代表性的人工林类型——红椎(*Castanopsis hystrix*)和马尾松(*Pinus massoniana*)人工林为对象，设置穿透雨减少 50%和不减雨(对照)，连续测定胸径生长、不同组分凋落物量(叶、其他树种叶、枝、皮、果和碎屑)和叶面积指数，分析红椎和马尾松胸径生长、凋落物量和叶面积指数对短期穿透雨减少的响应，为预测未来降水变化对该区域人工林生态系统生产力的影响提供科学依据。

1 研究地区与研究方法

1.1 研究区概况

本研究于中国林业科学研究院热带林业实验中心伏波实验场(22°10′N，106°50′E)的红椎和马尾松人工林林冠下进行，海拔 550m。属于南亚热带半湿润—湿润气候，干湿季节明显，年平均气温 21.7℃，极端最高气温 40.0℃，极端最低气温—1.2℃，年降水量 1361mm，78%降水量发生在 5~10 月(1965—2015 年)。土壤是花岗岩风化形成的红壤，质地为砂质壤土。

红椎和马尾松均为 1983 年种植在杉木采伐迹地上新造林地，2015 年 1 月 2 个林分情况见表 1。红椎林下有天然更新的红椎萌芽，灌木有黄毛五月茶(*Antidesma fordii*)、九节(*Psychotria rubra*)、酸藤子(*Embelia laeta*)、单叶省藤(*Calamus simplicifolius*)、毒根斑鸠菊(*Vernonia andersonii*)等，草本植物有扇叶铁线蕨(*Adiantum flabellulatum*)、淡竹叶(*Lophatherum gracile*)、铁芒萁(*Dicranopteris linearis*)和金毛狗(*Cibotium barometz*)等。马尾松林下乔木主要有少量的大叶栎(*Quercus griffithii*)、杉木(*Cunninghamia lanceolata*)和火力楠(*Michelia macclurei*)，灌木有玉叶金花(*Mussaenda pubescens*)、柏拉木(*Blastus cavaleriei*)、越南悬钩子(*Rubus cochinchinensis*)和菝葜(*Smilax china*)等，草本植物有乌毛蕨(*Blechnum orientale*)、弓果黍(*Cyrtococcum patens*)和小花露籽草(*Ottochloa nodosa*)等。

表 1 红椎和马尾松人工林的林分密度、树高、胸径和灌草盖度

项目	红椎		马尾松	
	对照	处理	对照	处理
林分密度(trees/hm^2)	325.0±25.0a	342.0±67.0a	242.0±38.0a	292.0±29.0a
平均树高(m)	19.1±0.7a	20.8±0.5a	20.0±0.4a	17.7±0.3b
平均胸径(cm)	26.3±0.7a	27.3±0.7a	31.8±0.9a	30.8±1.1a
胸高断面积(m^2/hm^2)	17.7±1.6a	20.6±3.1a	20.5±1.2a	23.3±1.2a
灌木盖度(%)	8.2±1.5a	12.9±7.6a	21.0±10.0a	17.8±1.2a
草本盖度(%)	1.1±0.4b	2.4±0.2a	16.1±7.9a	15.9±3.4a

注：不同字母表示处理间差异显著($P<0.05$)。

1.2 试验设计

分别于红椎和马尾松人工林林冠下地形、地貌等生境条件相对均一的条件下，设置穿透雨减少50%和不减雨(对照)两个处理，每处理3个重复，2个树种共12块样地，样地大小为20m×20m，样地间距大于20m。处理时间为2012年9月至2017年12月，本研究时间为2015年1月至2017年12月。样地如图1所示，每个减雨样地由198块透明薄膜(长3m×宽0.3m)组成，薄膜安装在距地面1~1.5m(上坡)和3~3.5m(下坡)的不锈钢架上，截获约50%穿透雨通过样地内的导水槽(长4m×上口宽0.4m×下口宽0.2m×深0.4m)和样地周围的排水沟(深70cm×宽40cm)排出样地。在减雨样地顺坡上、左、右方向埋入1m深PVC膜，并修建挡水墙(宽0.1m，高出地面0.3m，入土深度1m)以阻止样地外的地表径流和壤中流进入样地内。对照样地设相同的不锈钢架和挡水墙，上面覆盖2~3mm孔径的白色尼龙网，每月清理薄膜、纱网和导水槽上的凋落物，原位归还于样地内。

图1 红椎和马尾松人工林样地

CK：对照；T：处理；A：红椎；B：马尾松。下同。

1.3 气温、降水量和5cm土壤温湿度测定

利用广西友谊关森林生态系统国家定位观测研究站获取气温和降水量。采用HoBo H21土壤水分监测仪(Onset Computer Corp.，美国)自动记录5cm深土壤湿度和温度，1h读取1次。每块样地3个土壤水分监测仪，随机分布在样地内。

1.4 胸径和叶面积指数测定

2015年1月，对样地内全部乔木进行编号，安装胸径生长环，分别于2015年1月、2016年1月、2016年12月、2017年12月调查胸径。

每个样地随机选取5个固定点，利用佳能EOS60D相机(Cannon Inc.，日本)，在距离地面3m处拍摄乔木冠层图像，图像采集时间为2015年4月至2016年3月和2017年1~12月，阴天拍摄，每月1~2次，采用Hemiview冠层分析软件(Delta-TDevices Ltd.，英国)计算叶面积指数。

1.5 凋落物量测定

2015年1月至2017年12月，每月收集1次凋落物。每个样地设置5个1m×1m收集筐，距地面1m，随机分布。将收集的凋落物区分叶、其他树种叶、枝、皮、果、碎屑等，于70℃烘箱烘干48h至恒量，称干质量。由于2016年马尾松凋落物数据不完整，故不作分析。

1.6 数据处理

采用Excel2003和SPSS16.0软件对数据进行统计分析。采用独立样本t检验进行单因素方差分析($\alpha=0.05$)。利用Excel2003软件作图。图表中数据为平均值±标准误。

2 结果与分析

2.1 气温、降水量和5cm土壤温湿度

由图2可以看出，2015—2017年，年均气温和年均降水量相对稳定，3年平均值分别为20℃和1015mm。气温的年内月变化呈单峰曲线，其中5~10月气温最高(24℃)，1月、2月和12月气温最低(13℃)，而降水量的年内月波动较大，其中5~10月降水量占总降水量的64.3%。同一树种不同处理之间5cm土壤温湿度的月动态相似，并且5cm土壤温度与气温的月动态一致，而5cm土壤湿度的月波动平缓，无明显峰值。

由表2可以看出，2015年，穿透雨减少导致红椎5cm土壤湿度显著降低21.4%，而对红椎5cm土壤温度、马尾松5cm土壤温湿度无影响。2016和2017年，红椎减雨处理的5cm土壤湿度比对照分别减少了16.9%和22.3%，而其5cm土壤温度比对照分别增加1.2%和0.6%。对于马尾松，减雨处理的5cm土壤湿度比对照分别减少8.1%和8.3%，而其5cm土壤温度比对照分别增加4.2%和0.8%。

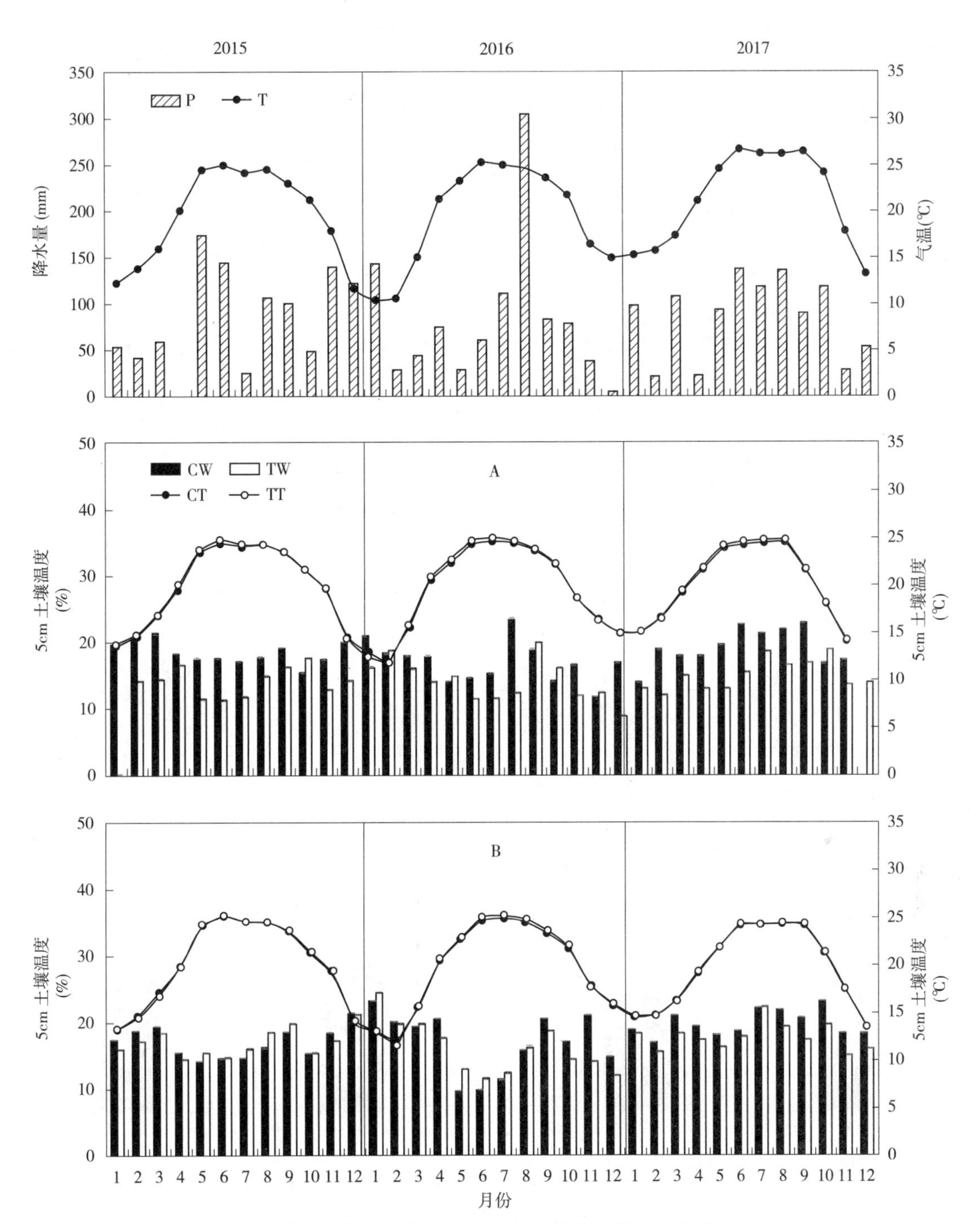

图 2 研究期间气温、降水量和 5cm 土壤温度和含水量的月动态

P：降水量；T：气温；CW：对照土壤含水量；TW：减雨土壤含水量；CT：对照土壤温度；TT：减雨土壤温度。

表 2 红椎和马尾松人工林对照和处理样地 5cm 土壤温度和含水量

年份（年）	项目	红椎		马尾松	
		对照	处理	对照	处理
2015	SW(%)	18.4±0.0a	14.4±0.0b	17.0±0.0a	17.0±0.0a
	ST(℃)	20.0±0.0a	20.1±0.0a	20.3±0.0a	20.4±0.0a
2016	SW(%)	17.8±0.0a	14.8±0.0b	17.6±0.0a	16.2±0.0b
	ST(℃)	21.2±0.0b	21.5±0.0a	19.3±0.0b	20.2±0.0a

（续）

年份（年）	项目	红椎		马尾松	
		对照	处理	对照	处理
2017	SW(%)	17.2±0.0a	13.3±0.0b	18.8±0.0a	17.2±0.0b
	ST(℃)	17.5±0.0b	17.6±0.0a	18.0±0.0b	18.1±0.0a

注：SW 为土壤含水量；ST 为土壤温度。

2.2 树木胸径生长和叶面积指数

由表 3 可以看出，同一年份同一树种不同处理之间胸径差异不显著。对于平均连年胸径增长量，同一年份同一树种表现出减雨处理小于对照的趋势，其中，2017 年减雨处理红椎人工林平均连年胸径增长量比对照显著降低 31.8%。

由图 3 可以看出，2015 年 4 月至 2016 年 3 月，穿透雨减少导致红椎和马尾松的叶面积指数均显著下降，均降低了 7.2%；2017 年 1~12 月，红椎减雨处理的叶面积指数比对照显著降低了 10.3%，而马尾松减雨处理的叶面积指数与对照之间无显著差异。

表 3 红椎和马尾松人工林胸径和连年胸径增长量

年份	项目	红椎		马尾松	
		对照	处理	对照	处理
2015	平均胸径(cm)	26.9±0.8a	27.8±0.8a	32.6±1.0a	31.3±1.1a
	平均连年胸径增长量(cm)	0.6±0.1a	0.5±0.1a	0.8±0.1a	0.7±0.1a
2016	平均胸径(cm)	27.5±0.8a	28.3±0.8a	33.3±1.0a	31.9±1.1a
	平均连年胸径增长量(cm)	0.6±0.1a	0.5±0.1a	0.7±0.1a	0.6±0.1a
2017	平均胸径(cm)	27.9±0.8a	28.7±0.8a	33.9±1.0a	32.8±1.2a
	平均连年胸径增长量(cm)	0.5±0.1a	0.3±0.0b	0.9±0.1a	0.7±0.1a

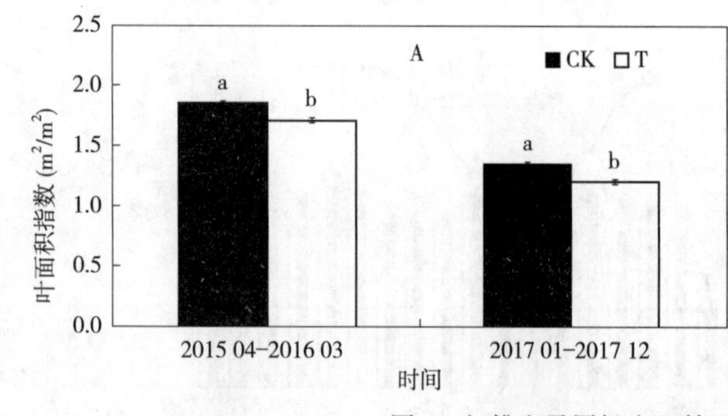

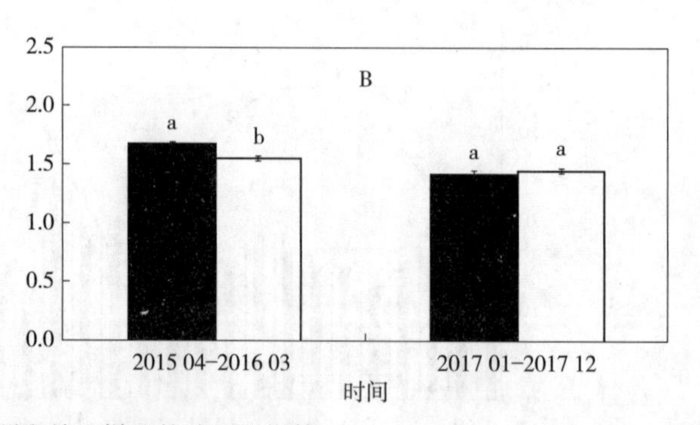

图 3 红椎和马尾松人工林对照和处理样地的叶面积指数

注：不同字母表示处理间差异显著($P<0.05$)。

2.3 凋落物量及其组分

由图 4 可以看出，同一树种不同处理之间凋落物量及其组分的月动态基本一致，凋落物总量的年内月变化呈单峰或者双峰曲线，最小值出现的时间均为 1 或 2 月，而最大值出现的时间因年份和树种而异，如红椎最大值出现在 1、3、4 或 12 月，而马尾松最大值出现在 5、6 或 12 月。叶凋落物量的月变化呈单峰曲线，并与凋落物总量的月动态相似。红椎果凋落量的最大值分别出现在 12 月或 1 月，而其他凋落物组分的年内月动态波动平缓，无明显凋落峰值。

由表 4 可以看出，凋落物各组分从大到小的次序因树种而异，其中红椎依次为：叶>碎屑>枝≥果>皮≥其他树种叶；马尾松依次为：叶>其他树种叶≥碎屑>枝≥皮>果。穿透雨减少导致红椎 2015 年枝凋落量和 2017 年凋落物总量比对照分别显著增加 29.6% 和 35.8%，而对红椎凋落物其他组分影响不显著。穿透雨减少导致马尾松 2015 年其他树种叶凋落量比对照显著减少 50.7%，而对马尾松凋落物其他组分影响不显著。

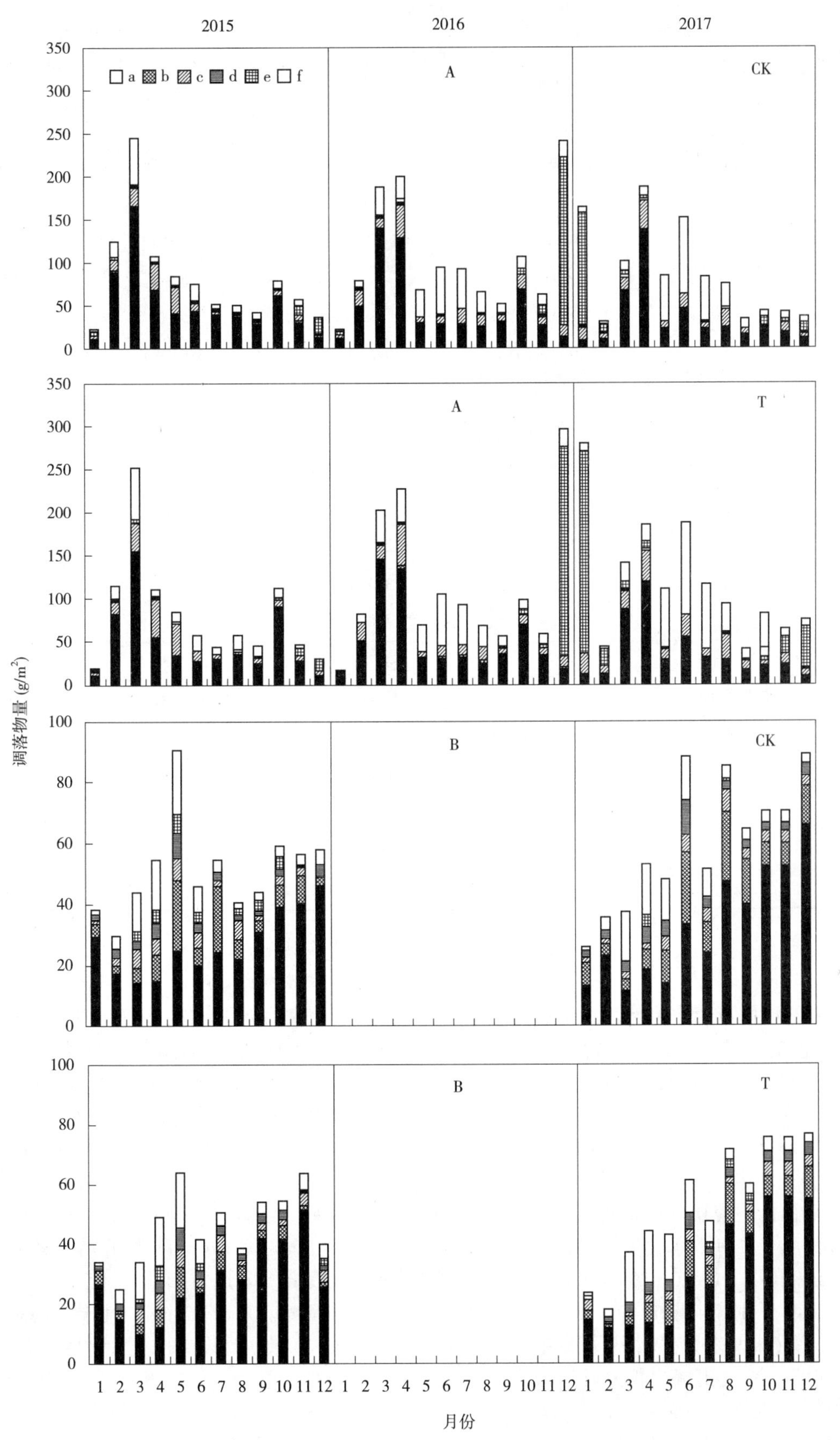

图4　红椎和马尾松人工林凋落物量的月动态变化

注：a. 叶；b. 其他树种叶；c. 枝；d. 皮；e. 果；f. 碎屑。

表4 红椎和马尾松人工林年凋落物量 (g/m²)

年份(年)	组分	红椎		马尾松	
		对照	处理	对照	处理
2015	叶	617±27a	564±24a	321±27a	329±29a
	其他树种叶	11±10a	13±8a	103±16a	50±9b
	枝	142±5b	184±7a	44±10a	39±6a
	皮	12±3a	12±3a	37±5a	34±5a
	果	40±4a	29±8a	25±4a	11±4a
	碎屑	152±11a	168±4a	85±4a	84±9a
	总量	961±22a	948±26a	576±38a	519±38a
2016	叶	571±28a	606±49a	—	—
	其他树种叶	9±9a	21±15a	—	—
	枝	168±5a	174±9a	—	—
	皮	20±9a	9±5a	—	—
	果	216±61a	255±81a	—	—
	碎屑	285±44a	309±27a	—	—
	总量	1270±66a	1373±110a	—	—
2017	叶	404±60a	427±61a	392±31a	375±50a
	其他树种叶	6±6a	12±10a	133±37a	87±23a
	枝	152±20a	204±9a	44±4a	37±8a
	皮	15±8a	14±8a	49±6a	38±5a
	果	170±39a	340±93a	5±3a	6±3a
	碎屑	283±30a	380±44a	95±4a	92±9a
	总量	984±88b	1336±82a	691±18a	611±45a

3 讨论

3.1 穿透雨减少对胸径生长和叶面积指数的影响

降水变化通过改变土壤湿度和温度、大气压亏缺和植物气孔导度等，间接影响树木生长[25]。本研究中，减雨后第5年(2017年)，红椎的平均连年胸径增长量显著降低31.8%，而红椎的胸径、马尾松的平均胸径和平均连年胸径增长量均不变，这表明红椎生长对减雨的响应比马尾松更敏感。其原因可能是，一方面与土壤有效水含量有关，如减少50%穿透雨导致红椎0~5、0~20、20~40、40~60和60~80cm土层湿度分别下降21.4%、8.2%、13.0%、22.0%和19.2%(未发表数据)，从而影响了红椎的胸径生长。但是，穿透雨减少仅使马尾松5cm土壤湿度显著下降约8.2%，因而未达到影响马尾松胸径生长的程度；另一方面与树种对水分的竞争与利用以及适应干旱的能力有关。相对于红椎，马尾松更加耐旱，因而在短期穿透雨减少50%的情况下，马尾松能够保持胸径生长量不变。这只是短期试验结果，随着截雨时间的延长，红椎胸径生长受影响程度是否一直大于马尾松，还有待进一步研究。Martin-Stpaul等[17]研究发现，减少28%穿透雨处理7年对64年生、68年生和82年生冬青栎(*Quercus ilex*)的胸径无显著影响。Samuelson等[18]研究表明，减少30%穿透雨2年对7年生火炬松(*Pinus taeda*)的胸径和胸径增长量生长无影响。然而，Mackay等[26]模拟极端干旱发现，生长季减少90%穿透雨导致加拿大70年生白松(*Pinus strobus*)当年胸径增长量下降17%，可能因为极端干旱严重影响树木吸收地下水的能力，其影响程度远大于长期试验性干旱的影响[25,27]。另外，树木年龄也是影响树木对干旱响应的重要因素。Nepstad等[28]发现，亚马逊热带雨林穿透雨去除处理8个月时间，导致当年小树(胸径<10cm，树高>15m)胸径增长量减少20%，而对大树(胸径大于10cm，树高大于15m)胸径增长量无影响，这是因为大树的根系比小树发达，能够利用更多的地下水。

随着干旱时间的延长，树木可利用水不断减少，树木利用深层土壤水分的能力也不断减弱[29,30]，此时大树生长受抑制，死亡率增加[31]。

植物通过叶片与大气进行水和能量交换，叶面积指数是植物在干旱胁迫下立地水平上光合能力的反映。叶面积指数的大小由凋落叶和新生叶两部分决定。Martin-Stpaul 等[17]研究结果表明，减少 28%穿透雨处理第 7 年，地中海冬青栎叶面积指数下降 18%，主要由凋落叶产量下降 20%引起。Samuelson 等[18]研究发现，减少 30%穿透雨第 2 年，美国火炬松叶面积指数最大值下降 13%，主要由减雨第 1 年新叶伸展降低 8%所致。本研究中，减少 50%穿透雨第 3 年和第 5 年(2015 和 2017 年)，红椎的叶面积指数分别降低 7.2%和 10.3%，马尾松的叶面积指数降低了 7.2%或者不变，其中红椎叶面积指数下降可能由新叶减少引起的，因为其叶凋落量不变，而马尾松可能由凋落叶减少或者不变所致，因为 2015 年马尾松其他叶凋落量比对照减少了 50.7%，而 2017 年马尾松叶凋落量不变。与本研究结果一致，Nepstad 等[28]研究发现，穿透雨去除 5 个月和 8 个月导致地中海热带雨林叶面积指数降低 0.6~0.8m²/m²。Limousin 等[32]发现，减少 29%降雨输入 4 年后，导致法国冬青栎叶面积指数降低 0.33m²/m²，尽管第一年无影响。

3.2 穿透雨减少对凋落物量及其组分的影响

凋落物是森林地上生物量的重要组成部分，包括生长(叶、枝)和生殖(花、果)器官，降水通过改变凋落物量和组成来影响树木生长和更新，从而影响森林生态系统服务功能。Moser 等[20]在印度尼西亚的热带雨林内开展了为期 2 年的野外降水控制试验，发现减少 60%穿透雨 8 个月和减少 80%穿透雨 17 个月未对叶凋落量和其他凋落物量产生影响，表明短期试验性干旱不会影响森林凋落物量。Liu 等[12]在地中海开展了长达 15 年的野外截雨试验，发现减少 30%穿透雨导致冬青栎、高山红景天(*Phillyrea latifolia*)和垂花树莓(*Arbutus unedo*)的叶和枝凋落物量显著增加，而高山红景天受影响程度小于其他树种，可能与其根和树干等水力传导结构有关。然而，Rowland 等[22]研究表明，减少 50%穿透雨 15 年导致热带森林在减雨前 4 年花和果实、叶片和凋落物总量分别下降 54%、13%和 12%；减雨超过 10 年，叶片和凋落物总量仍然减少，而花和果实凋落量反而增加，表明干旱胁迫初期树木优先投资叶片以最大限度地获得光合产物，而后期树木通过更多地将碳水化合物分配到生殖器官以适应长期干旱[9]。本研究中，减少 50%穿透雨导致红椎 2015 年枝凋落量和 2017 年凋落物总量分别增加了 29.6%和 35.8%，其中红椎 2017 年凋落物总量增加主要由当年枝、果和碎屑凋落量增加引起，分别比对照增加 33.9%、100.0%和 34.2%，可能由于减雨引起木质部栓塞，从而导致枝、果和碎屑凋落物量增加。不同的是，减少 50%穿透雨仅导致马尾松 2015 年其他树种叶凋落量显著减少，原因可能是减雨条件下其他树种通过提高叶片凋落量以减少叶面积和叶片蒸腾，维持其在干旱胁迫下的水分传导，而减少 50%穿透雨对马尾松凋落物其他组分无影响，这与马尾松胸径生长对减雨的响应结果一致。由于森林凋落物量及其组分的年内、年际动态除了与林分自身的生物学特性有关，还受温度、降雨等气候因子的影响，因而加强森林凋落量及其组分的长期动态监测以消除年度波动而非降水处理带来的影响，从而提高森林凋落物量预测的准确性。

综上所述，减雨处理对红椎胸径生长的影响比马尾松大；减雨处理导致红椎的叶面积指数显著降低，马尾松叶面积指数显著降低或不变；减雨条件下红椎枝凋落量和总凋落物趋于增加，马尾松叶凋落量趋于减少或不变，表明短期穿透雨减少对红椎和马尾松人工林均产生了胁迫作用，但是存在年际变异和树种差异。

参考文献

[1]STOCKER T F, QIN D, PLATTNER G K, et al. Climate Change 2013: The physical science basis. Contribution of working group Ⅰ to the fifth assessment report of the intergovernmental panel on climate change[M]. New York, US: Cambridege Press, 2013.

[2]VANONI M, BUGMANN H, NÖTZLI M, et al. Quantifying the effects of drought on abrupt growth decreases of major tree species in Switzerland[J]. Ecology and Evolution, 2016, 6: 3555-3570.

[3]CHEN L, HUANG J G, ALAM S A, et al. Drought causes reduced growth of trembling aspen in western Canada[J]. Global Change Biology, 2017, 23: 2887-2902.

[4]SCHWALM C R, ANDEREGG W R L, MICHALAK A M, et al. Global patterns of drought recovery[J]. Nature, 2017, 548: 202-205.

[5]HOOVER D L, WILCOX K R, YOUNG K E. Experimental droughts with rainout shelters: a methodological review[J]. Ecosphere, 2018, 9: e02088.

[6]RASCHER U, BOBICH E G, LIN G H, et al. Functional

diversity of photosynthesis during drought in a model tropical rainforest-the contributions of leaf area, photosynthetic electron transport and stomatal conductance to reduction in net ecosystem carbon exchange [J]. Plant, Cell and Environment, 2004, 27: 1239-1256.

[7] STAHL U, KATTGE J, REU B, et al. Whole-plant trait spectra of North American woody plant species reflect fundamental ecological strategies [J]. Ecosphere, 2013, 4: 1-28.

[8] CHENG R M, LIU Z S, FENG X-H, et al. Advances in research on the effect of climatic change on xylem growth of trees [J]. Cientia Silvae Sinicae, 2015, 51 (6): 147-154.

[9] EZIZ A, YAN Z, TIAN D, et al. Drought effect on plant biomass allocation: A meta-analysis [J]. Ecology and Evolution, 2017, 7: 11002-11010.

[10] LIU C, SUN P S, LIU S R, et al. Leaf photosynthetic pigment seasonal dynamic of *Quercus aliena* var. *acuteserrata* and its spectral reflectance response under throughfall elimination [J]. Chinese Journal of Applied Ecology, 2017, 28(4): 1077-1086.

[11] SUAREZ M L, GHERMANDI L, KITZBERGER T. Factors predisposing episodic drought-induced tree mortality in *Nothofagus*-site, climatic sensitivity and growth trends [J]. Journal of Ecology, 2004, 92: 954-966.

[12] LIU D, OGAYA R, BARBETA A, et al. Contrasting impacts of continuous moderate drought and episodic severe droughts on the aboveground-biomass increment and litterfall of three coexisting Mediterranean woody species [J]. Global Change Biology, 2015, 21: 4196-4209.

[13] GÖRANSSON H, BAMBRICK M T, GODBOLD D L. Overyielding of temperate deciduous tree mixtures is maintained under throughfall reduction [J]. Plant and Soil, 2016, 408: 285-298.

[14] GLEASON K E, BRADFORD J B, BOTTERO A, et al. Competition amplifies drought stress in forests across broad climatic and compositional gradients [J]. Ecosphere, 2017, 8: e01849.

[15] GREENWOOD S, RUIZ-BENITO P, MARTÍNEZ-VILALTA J, et al. Tree mortality across biomes is promoted by drought intensity, lower wood density and higher specific leaf area [J]. Ecology Letters, 2017, 20: 539-553.

[16] GLATTHORN J, PICHLER V, HAUCK M, et al. Effects of forest management on stand leaf area: Comparing beech production and primeval forests in Slovakia [J]. Forest Ecology and Management, 2017, 389: 76-85.

[17] MARTIN-STPAUL N K, LIMOUSIN J M, VOGT-SCHILB H, et al. The temporal response to drought in a Mediterranean evergreen tree: comparing a regional precipitation gradient and a throughfall exclusion experiment [J]. Global Change Biology, 2013, 19: 2413-2426.

[18] SAMUELSON L J, PELL C J, STOKES T A, et al. Two-year throughfall and fertilization effects on leaf physiology and growth of loblolly pine in the Georgia piedmont [J]. Forest Ecology and Management, 2014, 330: 29-37.

[19] CASCO S L, GALASSI M E, MARI E K A, et al. Linking hydrologic regime, rainfall and leaf litter fall in a riverine forest within the Ramsar Site Humedales Chaco (Argentina) [J]. Ecohydrology, 2016, 9: 773-781.

[20] MOSER G, SCHULDT B, HERTEL D, et al. Replicated throughfall exclusion experiment in an Indonesian perhumid rainforest: wood production, litter fall and fine root growth under simulated drought [J]. Global Change Biology, 2014, 20: 1481-1497.

[21] GU Y, WANG F, CHEN P S, et al. Effects of long-term nitrogen addition and precipitation decreasing on the litterfall production of broadleaved Korean pine forest in Changbai Mountains of northeastern China [J]. Journal of Beijing Forestry University, 2017, 39(4): 29-37 (in Chinese).

[22] ROWLAND L, da COSTA A C L, OLIVEIRA A A R, et al. Shock and stabilisation following long-term drought in tropical forest from 15 years of litterfall dynamics [J]. Journal of Ecology, 2018. doi. org/10. 1111/1365 - 2745. 12931.

[23] ZHOU G, PENG C, LI Y, et al. A climate change-induced threat to the ecological resilience of a subtropical monsoon evergreen broad-leaved forest in Southern China [J]. Global Change Biology, 2013, 19: 1197-1210.

[24] WANG W, WANG J, LIU X, et al. Decadal drought deaccelerated the increasing trend of annual net primary production in tropical or subtropical forests in southern China [J]. Scientific Reports, 2016, 6: 28640.

[25] WU X, LIU H, LI X, et al. Differentiating drought legacy effects on vegetation growth over the temperate Northern Hemisphere [J]. Global Chang Biology, 2018, 24: 504-516.

[26] MACKAY S L, ARAIN M A, KHOMIK M, et al. The impact of induced drought on transpiration and growth in a temperate pine plantation forest [J]. Hydrological Processes, 2012, 26: 1779-1791.

[27] BARBETA A, MEJÍA-CHANG M, OGAYA R, et al. The combined effects of a long-term experimental drought and an extreme drought on the use of plant-water sources in a Mediterranean forest [J]. Global Change Biology, 2015, 21: 1213-1225.

[28] NEPSTAD D C, MOUTINHO P, DIAS-FILHO M B, et al. The effects of partial throughfall exclusion on canopy processes, aboveground production, and biogeochemistry of an Amazon forest[J]. Journal of Geophysical Research: Atmospheres, 2002, 107: 8085.

[29] NEPSTAD D C, TOHVER I M, RAY D, et al. Mortality of large trees and lianas following experimental drought in an Amazon forest[J]. Ecology, 2007, 88: 2259-2269.

[30] MARKEWITZ D, DEVINE S, DAVIDSON E A, et al. Soil moisture depletion under simulated drought in the Amazon: impacts on deep root uptake[J]. New Phytologist, 2010, 187: 592-607.

[31] COSTA A C L D, GALBRAITH D, ALMEIDA S, et al. Effect of 7 yr of experimental drought on vegetation dynamics and biomass storage of an eastern Amazonian rainforest[J]. New Phytologist, 2010, 187: 579-591.

[32] LIMOUSIN J M, RAMBAL S, OURCIVAL J M, et al. Long-term transpiration change with rainfall decline in a Mediterranean *Quercus ilex* forest[J]. Global Change Biology, 2009, 15: 2163-2175.

[原载：应用生态学报，2018，29(07)]

马尾松人工林采伐剩余物生物量及养分贮量

李 华[1,2] 郑 路[1,2] 李朝英[1,2] 卢立华[1,2] 明安刚[1,2] 农 友[1,2] 孙冬婧[1,2]

([1]中国林业科学研究院热带林业实验中心，广西凭祥 532600；
[2]广西友谊关森林生态系统国家定位观测研究站，广西凭祥 532600)

摘 要 通过对马尾松人工林采伐剩余物各组分生物量和养分贮量分配特征的研究，为其地力维护研究提供基础数据，为人工林的科学经营和生态管理提供理论依据。在广西南部马尾松人工林皆伐林地采用样方收获法获取采伐剩余物各组分生物量，测定其养分含量并计算养分贮量。结果表明：马尾松人工纯林皆伐后林地采伐剩余物生物量为39.1t/hm^2，碳、氮、磷、钾、钙和镁贮量分别为18303、101.2、8.3、73.4、96.0和24.7kg/hm^2。不同组分间比较，生物量和碳贮量均是小枝最高(分别占总量的25.3%和23.3%)，其次是大枝(21.7%和21.1%)和主根(17.1%和18.5%)，粗根(11.5%和12.4%)和叶(9.2%和9.1%)也较高；氮、磷、钾、钙和镁贮量排在前三位的组分均是叶、小枝和大枝(三者之和分别占各养分总量的70.5%、76.5%、72.2%、76.2%和72.6%)，其次为主根和粗根；而中根、小根和细根无论是生物量还是各养分贮量均很低。马尾松人工林采伐剩余物的生物量和养分储量庞大，尤其是残留在地表的枝和叶，因此保留采伐剩余物的林地更新方式对于维护其林地生产力具有重要意义。

关键词 南亚热带；采伐迹地；生物量方程；养分含量

Biomass and Nutrient Storage of Logging Residues of *Pinus massoniana* Plantation

LI Hua[1,2], ZHENG Lu[1,2], LI Zhaoying[1,2], LU Lihua[1,2],
MING Angang[1,2], NONG You[1,2], SUN Dongjing[1,2]

(*[1]Experimental center of tropical forestry, Chinese academy of forestry, Pingxiang 532600 Guangxi, China;*
[2]Guangxi Youyiguan Forest Ecosystem Research Station, Pingxiang 532600 Guangxi, China)

Abstract: To understand soil productivity maintenance and provide reliable scientific basis for plantations management, biomass and nutrient allocation in different components of logging residues were examined in the clearcutting land of a 31-year-old *Pinus massoniana* plantation. We measured biomass and nutrient concentration of logging residue's components, and then calculated nutrient storage. The results showed that the total biomass of logging residues was 39.1 t/hm^2, and the storages of carbon, nitrogen, phosphorus, potassium, calcium and magnesium were 18303, 101.2, 8.3, 73.4, 96.0 and 24.7 kg/hm^2, respectively. Among different components, the biomass and carbon storage were in the order of sprig (accounting for 25.3% and 23.3% of total biomass and carbon storage, respectively) > branch (21.7% and 21.1%) and taproots (17.1% and 18.5) > coarse root (11.5% and 12.4%) and leaf (9.2% and 9.1%). Nitrogen, phosphorus, potassium, calcium and magnesium were mainly stored in sprig, branch and leaf (the sum of those components were 70.5%, 76.5%, 72.2%, 76.2% and 72.6%, respectively), followed by taproot and coarse root. The middle-sized root, small root and fine root had the lowest biomass and nutrient storage. The biomass and nutrient storage of logging residues were enormous, especially in branche and leaf. Therefore, retaining logging residues during forest regeneration is important for forest productivity maintenance.

Key words: subtropical China; clear cutting land; biomass equation; nutrient concentration.

我国人工林发展迅速，面积居世界首位，在生态功能维护和经济发展中发挥着重大的作用，并随着时间的推移，其地位和作用还将不断提升[1]。但由于缺乏科学的理论指导和实用的可持续经营技术，过分追求经济利益，致使部分人工林出现较为明显的地力衰退，导致长期生产力降低，阻碍人工林的可持续经营和生态服务功能的发挥。

采伐剩余物是森林采伐后，在采伐林地上残留下来的不适于加工为经济材、薪炭材和小规格材的所有废弃物[2]。我国南方人工林经营中仍然沿用传统的皆伐后"炼山"(人为放火烧掉采伐剩余物和地被物)的更新造林方式。已有研究表明，与炼山相比，保留采伐剩余物能显著减少林地水、土、肥的流失，且在皆伐后的前两年效果最好[3]。分析认为，一方面树木采伐后，其枝、叶、皮等采伐剩余物堆积在土层表面，在地表形成一个保护层，可减弱雨滴对土壤颗粒的分散力，同时增加地表糙率，促进降水入渗，并降低流量的挟沙能力[3-6]；另一方面深入土壤的伐根对土壤颗粒的机械缠绕和网结作用，可防止土体内团聚体雨水分散，下移填塞土壤大空隙，而且当根系腐烂后，根系生长所形成的大空隙便留在土体内，亦可增加土壤的入渗性能[7]。此外，保留采伐剩余物还可为土壤生物提供食物，而生物活动(如白蚁、蚂蚁、蚯蚓、蜚蠊、蜘蛛等)也能在土壤内形成大空隙，增加土壤入渗[8,9]。保留采伐剩余物不仅能增加土壤入渗，减少林地水土肥的流失，还能显著增加土壤表层(0~10cm)有机碳含量[10]，提高土壤碱解氮和有效钾含量[11]；增加土壤可溶性有机碳、可溶性有机氮含量，提高土壤碳矿化速率[12]，降低土壤容重，提高 pH，有利于维护土壤肥力和促进林木生长[13]。但以上报道仅研究了保留采伐剩余物产生的作用和功能，而有关采伐剩余物的生物量和养分含量等的研究较少。本研究以马尾松人工林皆伐迹地残留的采伐剩余物为研究对象，开展了马尾松采伐剩余物生物量、养分含量和贮量的研究，为采伐剩余物的分解及地力维护研究提供基础数据，为人工林的科学经营和生态管理提供理论依据。

1 研究地区与研究方法

1.1 研究区概况

研究地点位于广西友谊关森林生态系统国家定位观测研究站热林中心站点，地理位置为 106°48′12″E，22°5′0″N，属南亚热带季风气候。其干湿季节明显，年均温度 21.0℃，≥10℃积温 7600℃，年均降水量 1400mm，多集中在 4~9 月。研究区为低山丘陵地貌，坡度 20°~30°，海拔 260m，土壤为赤红壤，土层厚度大于 100cm。

调查样地面积 15.8hm^2，经营历史：于 1982 年 2 月全面整地后，挖穴定植一年生马尾松纯林(当地种源)，初植密度3000株/hm^2，分别在 1989、1993、1997 和 2002 年进行透光伐和 3 次生长伐，保留密度为 450 株/hm^2，从 2003 年开始采割松脂，连续割脂至 2013 年，并在 2013 年夏秋季皆伐。皆伐前马尾松林郁闭度为 0.7，平均树高 15.2m，平均胸径 25.4cm，林分密度为 442 株/hm^2。皆伐后伐根保留高度为 4.2~8.8cm，平均为 5.1cm。

1.2 研究方法

样地设置及采样方法：在马尾松人工林皆伐迹地，沿等高线 S 形布点取样，样方大小为 5m×5m，10 个重复。收集样方内除伐根外的所有采伐剩余物，分别叶、小枝(直径在 25mm 以下)、大枝(直径在 25mm 以上)和树皮称湿重，取样后装入塑封袋内带回实验室称量鲜重，在 90℃下杀青 30min，再在 70℃下烘干至恒重，称其干重，计算含水率，由此推算单位面积上残留在地表的采伐剩余物生物量。另外随机设置 5 个 10m×10m 样方，调查样方内所有根桩直径、高度和数量。样方内所有伐根，采用分层挖掘法，根据林木的冠幅和现存密度以及前期调查，将根系挖掘的水平范围确定为以树干基部为中心，半径为 1.5m，垂直深度 1m。每挖掘 0.2m 的深度，分别将伐根按根桩(伐根地表以上的部分)、主根和各级侧根称重并取样，侧根按照细根(根径小于 2mm)、小根(根径 2~5mm)、中根(根径 5~20mm)、大根(根径 20~50mm)和粗根(根径大于 50mm)进行分级。样品带回实验室称量鲜重，90℃下杀青 30min，再在 70℃下烘干至恒重，称干重，计算含水率，由此推算单位面积伐根生物量。并在调查伐桩直径(平均值为 29.4cm，最大值为 38.0cm，最小值为 20.0cm)和伐根数量(21 株)的基础上，以伐根各组分生物量(W：kg)为因变量，以伐桩直径(D：cm)为自变量，采用数学模型对马尾松常规经营林分皆伐后单株伐根各组分生物量与伐桩直径进行拟合，发现幂函数模型相关系数最大，F 检验值也最大。故以幂函数模型拟合，结果见表 1。

养分测定方法：所有样品粉碎后，采用重铬酸钾—硫酸氧化法测定有机碳含量；采用浓硫酸双氧水消煮法消煮样品后，凯氏定氮法测定氮含量，钒钼黄比色法测定磷含量，火焰光度计法测定钾含量。

植物样品经干灰化后用稀盐酸煮沸，用原子吸收分光光度计法测定钙和镁含量[14]。

1.3 数据处理

采用 Microsoft Excel 2007 进行数据处理和图表绘制，采用 SPSS 19.0 统计软件包进行单因素方差分析(One-Way ANOVA)和回归分析，不同组分间多重比较采用 LSD 法。

2 结果与分析

2.1 马尾松采伐剩余物生物量

2.1.1 采伐剩余物各组分生物量

从图 1 可以看出，南亚热带 31 年生马尾松纯林皆伐后，采伐剩余物总量为 39.07t/hm²，其中残留在地表的生物量(叶、枝、皮)有 23.63t/hm²，占总生物量的 60.5%，是伐根生物量 1.53 倍。采伐剩余物各组分生物量由大到小的排列顺序为：小枝>大枝>主根>粗根、叶>根桩、皮、大根>中根、小根、细根。其中小枝生物量高达 9.88t/hm²、占总生物量 25.3%，大枝为 8.48t/hm²、21.7%，主根 6.70t/hm²、17.1%，粗根 4.49t/hm²、11.5%，叶 3.59t/hm²、9.2%，根桩 2.42t/hm²、6.1%，皮 1.67t/hm²、4.3%，大根 1.40t/hm²、3.6%。而中根、小根和细根生物量之和仅为 0.42t/hm²，仅占总生物量的 1.1%，尤其是小根和细根，其生物量每公顷仅几十千克，两者之和仅占总生物量的 0.1%。

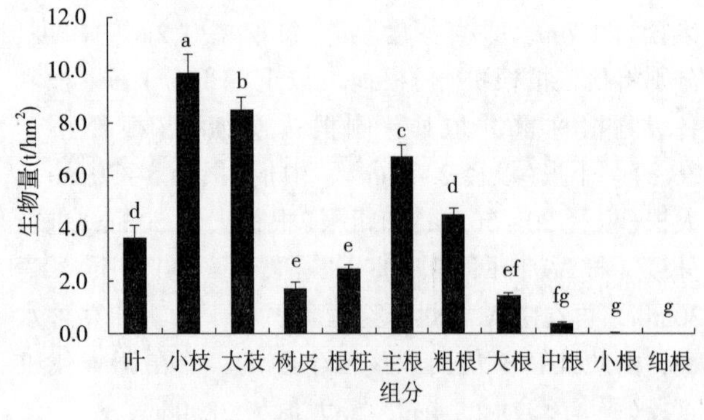

图 1 采伐剩余物不同组分生物量

注：不同字母表示采伐剩余物不同组分间生物量存在显著差异($P<0.05$)。

2.1.2 伐根各组分生物量与伐桩直径的相关关系

以伐根各组分生物量(W)为因变量，以伐桩直径(D)为自变量，采用数学模型拟合，幂函数模型拟合的相关系数和 F 值均最大，其拟合结果见表 1。从表 1 可知，当地常规经营的马尾松纯林皆伐后，单株伐根各组分生物量与伐桩直径密切相关。其中根桩、主根、粗根和整个伐根生物量与伐桩直径所拟合的 4 个回归模型的相关系数介于 0.76~0.89，大根和中根拟合模型的相关系数分别为 0.59 和 0.58，均高于自由度为 19 的极显著相关系数的临界值(0.55)，其回归模型的 F 统计量值介于 9.53~69.72，均高于 $F_{0.01(1,19)}$ 的临界值(8.18)；小根和细根拟合模型的相关系数分别为 0.52 和 0.54，其显著性仅达到显著相关($r_{0.05}=0.43$)，F 值也高于 $F_{0.05}$ 的临界值(4.38)。说明除小根和细根外，其他伐根各组分及伐根总生物量的拟合精度均较高，其生物量均可用幂函数拟合。

表 1 单株伐根各组分生物量与伐桩直径的相互关系

伐根组分	回归模型	相关系数	自由度	F 值	显著性
根桩	$W=0.0013\times D^{2.5039}$	0.77	19	28.20	0.00
主根	$W=0.0079\times D^{2.2646}$	0.76	19	25.37	0.00
粗根	$W=8E-05\times D^{3.4999}$	0.89	19	69.72	0.00
大根	$W=0.018\times D^{1.564}$	0.59	19	10.04	0.01
中根	$W=0.0025\times D^{1.7496}$	0.58	19	9.53	0.01
小根	$W=0.0007\times D^{1.3963}$	0.52	19	7.03	0.02
细根	$W=0.0001\times D^{1.534}$	0.54	19	7.67	0.01
伐根	$W=0.0104\times D^{2.4357}$	0.85	19	47.69	0.00

2.2 马尾松采伐剩余物的养分含量

马尾松采伐剩余物不同组分间养分含量差异显著(表 2)，各组分碳含量介于 414~506g/kg，除细根外，伐根的各组分碳含量均显著高于叶、大枝等地上部分，而小枝、皮和细根的碳含量最低。各组分氮含量介于 1.28~11.53g/kg，不同组分间差异达近 10 倍。其中叶的氮含量远高于其他组分，其次是细根，小根和皮亦有较高的氮含量，其余各组分的氮含量则无显著差异，在 1.50g/kg 左右。磷含量依然是叶最高，细根次之，小枝和小根含量也较高，大枝、皮、中根、粗根和主根的磷含量较低，均在 0.15g/kg 左右，而大根和根桩的磷含量最低。各组分钾含量介于 0.57~4.97g/kg，其中叶和细根钾含量最高，其次是小根，中根、粗根和枝也较高，皮和大根次之，主根和根桩最低。钙含量则以皮最高，其次是小枝、细根和叶，根桩、主根、粗根和大根的钙含量最低，在 0.95g/kg 左右。镁含量依然以叶最高，细根次之，小根、中根和小枝镁含量也较高，根桩、主根和大根最低。由表 2 综合比较可看出，在马尾松采伐剩余物中，除钙含量外，叶中其他矿质养分含量均显著高于枝、皮及伐根等组分，与叶相反，根桩的氮、磷、钾、钙和镁含量均最低，主根、粗根和大根与根桩相似，也具有低的矿质元素含量和高的碳含量，但细根具有较高的矿质元素含量和低的碳含量。

2.3 马尾松采伐剩余物的养分贮量

马尾松人工林皆伐后，留在林地的采伐剩余物贮存着大量的碳和矿质元素。由表3可知，采伐剩余物中碳的总贮量为18303kg/hm^2，其中残留在地表的枝、叶和皮中的碳贮量为10507kg/hm^2，占总量的57.4%，是伐根的1.3倍；氮素总贮量为101.2kg/hm^2，残留在地表的贮量为77.3kg/hm^2，占总量的76.4%，是伐根的3.2倍；磷素总贮量为8.3kg/hm^2，残留在地表的贮量为6.6kg/hm^2，占总量的79.6%，是伐根的3.9倍；钾素总贮量为73.4kg/hm^2，残留在地表的贮量为55.1kg/hm^2，占总量的75.1%，是伐根的3.0倍；钙素总贮量为96.0kg/hm^2，残留在地表的贮量为80.8kg/hm^2，占总量的84.2%，是伐根的5.3倍；镁素总贮量为24.7kg/hm^2，残留在地表的贮量为18.9kg/hm^2，占总量的76.3%，是伐根的3.2倍。

表2 马尾松采伐剩余物各组分养分含量

组分	碳(g/kg)	氮(g/kg)	磷(g/kg)	钾(g/kg)	钙(g/kg)	镁(g/kg)
叶	468.42±11.34bc	11.53±1.47a	0.69±0.09a	4.97±0.15a	3.03±0.30c	1.48±0.09a
皮	420.40±22.51d	3.61±0.85c	0.16±0.03def	1.35±0.43d	4.72±1.64a	0.54±0.09de
小枝	430.40±21.78d	1.54±0.25d	0.23±0.05c	1.92±0.48c	4.10±1.24ab	0.72±0.17cd
大枝	453.34±26.09c	1.78±0.53d	0.18±0.07cde	1.86±0.11c	2.40±0.53cd	0.63±0.23cde
根桩	505.35±7.72a	1.28±0.43d	0.08±0.00h	0.57±0.06e	0.80±0.20e	0.26±0.08f
主根	506.79±8.66a	1.44±0.52d	0.11±0.03fgh	0.71±0.19e	1.04±0.14e	0.31±0.11f
粗根	503.55±6.10a	1.76±0.45d	0.12±0.02efgh	2.10±0.63c	0.98±0.33e	0.53±0.12e
大根	504.46±9.69a	1.65±0.39d	0.10±0.02gh	1.19±0.39d	0.75±0.09e	0.36±0.03f
中根	503.20±10.51a	2.43±0.08d	0.15±0.01efg	2.25±0.14c	1.69±0.40de	0.72±0.07cd
小根	485.07±7.88ab	3.71±0.48c	0.22±0.01cd	3.49±0.18b	2.55±0.26cd	0.81±0.12c
细根	414.35±11.63d	5.70±1.17b	0.35±0.03b	4.57±0.21a	3.19±0.52bc	1.27±0.07b

注：数值为平均值±标准差。同列不同字母表示同一养分含量在采伐剩余物不同组分间差异显著($P<0.05$)。

表3 马尾松采伐剩余物养分贮量

组分	碳(kg/hm^2)	氮(kg/hm^2)	磷(kg/hm^2)	钾(kg/hm^2)	钙(kg/hm^2)	镁(kg/hm^2)
叶	1675±449d	40.80±10.67a	2.46±0.81a	17.78±4.95a	10.96±3.32c	5.27±1.41b
皮	704±247ef	6.01±2.52cde	0.26±0.07cd	2.13±0.61c	7.68±3.39cd	0.93±0.40cde
小枝	4270±761a	15.42±4.48b	2.37±0.77a	19.43±7.20a	41.78±16.70a	7.27±2.53a
大枝	3859±604ab	15.10±4.88b	1.51±0.68b	15.80±2.06a	20.36±5.39b	5.42±2.43b
根桩	1224±187de	3.05±0.92de	0.20±0.02cd	1.09±0.60c	1.90±0.42d	0.62±0.19cde
主根	3392±480b	9.33±2.08c	0.74±0.33c	4.88±1.99c	7.05±1.72cd	2.00±0.51cd
粗根	2262±250c	7.97±2.52cd	0.55±0.13cd	9.64±3.78b	4.43±1.71cd	2.42±0.73c
大根	708±140ef	2.39±1.06de	0.14±0.05cd	1.64±0.50c	1.05±0.25d	0.50±0.08de
中根	190±32fg	0.92±0.17e	0.06±0.01d	0.86±0.20c	0.64±0.22d	0.27±0.07de
小根	16±4g	0.12±0.04e	0.01±0.00d	0.12±0.03c	0.08±0.03d	0.03±0.00e
细根	4±0g	0.06±0.01e	0.00±0.00d	0.04±0.01c	0.03±0.00d	0.01±0.00e
地上部合计	10507±1082A	77.3±11.8A	6.6±1.6A	55.1±11.1A	80.8±25.8A	18.9±5.4A
伐根合计	7796±946B	23.8±3.0B	1.7±0.5B	18.3±6.0B	15.2±3.5B	5.9±0.2B
总计	18303±1717	101.2±13.1	8.3±1.9	73.4±14.3	96.0±27.0	24.7±5.3

注：数值为平均值±标准差。同列不同小写字母表示同一养分贮量在采伐剩余物不同组分间差异显著($P<0.05$)；同列不同大写字母表示同一养分在采伐剩余物的地上部和伐根间差异显著($P<0.05$)。

对表3分析发现，碳贮量各组分由高到低的排列顺序为：小枝>大枝、主根>粗根>叶、根桩>大根、皮>中根、小根和细根，其中小枝占总量的23.3%，大枝占21.1%，主根占18.5%，粗根占12.4%，叶和根桩分别占9.1%和6.7%，大根和皮分别占3.9%和3.8%，而中根、小根和细根这三者之和仅占1.1%。从氮、磷、钾、钙、镁5种矿质元素贮存总量来看，采伐剩余物的各组分排列顺序为：小枝、叶>大枝>粗根、主根、皮>根桩、大根>中根、小根、细根，其中小枝占28.4%，叶占25.5%，大枝占19.2%，粗根、主根和皮分别占8.2%、7.9%和5.6%，根桩和大根分别占2.3%和1.9%，而中根、小根和细根三者之和仅占1.1%。

进一步对各矿质元素分析发现，氮、磷、钾、钙和镁5种元素贮量排在前三位的组分均为叶、小枝和大枝。叶、小枝和大枝分别占氮素总贮量的40.3%、15.2%和14.9%，三者之和占总氮量的70.5%；分别占磷素总贮量的29.7%、28.6%和18.2%，之和占总量的76.5%；分别占钾素总贮量的24.2%、26.5%和21.5%，之和占72.2%；分别占钙素总贮量的11.4%、43.8%和21.2%，之和占76.2%，分别占镁素总贮量的21.3%、29.4%和21.9%，之和占72.6%。而对伐根来说，各矿质养分主要贮存在主根和粗根中，两者之和分别占氮、磷、钾、钙和镁总量的17.1%、15.5%、19.8%、12.0%和17.9%。可见，采伐剩余物的各矿质养分主要贮存在残留于地表的枝和叶中。

3 讨论

我国南方山地山高坡陡，土层浅薄，降水量集中且强度大，极易引起水土流失，如果再加上经营措施不当，将会造成严重的水土流失[15]。与炼山相比，保留采伐剩余物能显著减少林地水土流失。炼山后第一年水土流失量分别是保留采伐剩余物的2.3和13倍[16]，11和88倍[17]；马庆祥等[3]在福建长期定位观测研究也表明，在皆伐炼山方式下更新的林地前两年水、土流失量尤为严重，可分别高达5179.0m^3/hm^2和35.1t/hm^2，而保留采伐剩余物的林地水、土流失量分别是631.2m^3/hm^2和0.6t/hm^2，降低了8.2和58.5倍。林地保留采伐剩余物的作用和功能与凋落物的相似，均能减少水土流失、促进降水入渗，为土壤生物提供碳源和养分。然而两者来源却不同，采伐剩余物是上一代林分收获后残留下来的不具经济价值的废弃物，而凋落物是树木生长过程中由地上部分产生并归还到地面的新陈代谢的产物，因此采伐剩余物和凋落物的组分不同，各组分生物量及占比、养分含量等差异也较大。本实验对31年生马尾松人工林采伐剩余物的研究显示，叶、枝和皮等残留在土壤表面的采伐剩余物生物量为23.63t/hm^2，是同区域同树种年凋落物量的2.1~6.5倍[18-20]，是凋落物现存量的1.8~4.7倍[18]。其中树枝的生物量最高，小枝和大枝生物量分别是9.88和8.48t/hm^2，分别占采伐剩余物地上部生物量的41.8%和35.9%，可见，残留在地表的采伐剩余物有77.7%都是树枝，而凋落物中树枝仅占1.1%~3.0%；树叶在凋落物中的比例高达61.6%~84.5%[18,19]，但残留在地面的采伐剩余物中只有15.2%。与杨会侠等[19]对不同年龄马尾松凋落物的研究结果相比，采伐剩余物中叶的氮磷钾和镁含量明显高于凋落物的，分别高出44.8%、56.8%、77.2%和60.9%，而枝中的氮磷含量则较凋落物的降低83.6%和66.7%。采伐剩余物中残留在地表的叶、枝和皮中的氮磷钾钙和镁贮量分别是77.34、6.61、55.15、80.77和18.88kg/hm^2；与年凋落物中矿质元素归还量相比，氮磷钾钙镁分别提高0.9~1.6倍、1.2~2.7倍、3.5~10.5倍、0.8~2.0倍和2.5~6.5倍[19]；与现存地表凋落物中各矿质养分贮量相比，氮磷钾钙镁分别提高0.6~1.7、1.6~5.2、2.2~8.6、1.4~2.8和1.7~7.6倍[18]。可见，人工林皆伐后，地表残留着大量的采伐剩余物，储存着大量的矿质养分。

除地上部的枝叶皮等，伐根也是采伐剩余物的重要组成部分，包括地上部分的根桩和深入地下的根系。伐根不仅能影响森林更新、野生动物的栖息、真菌的寄生，还能影响土壤的稳定性和养分循环等[21-25]，是森林生态系统的重要组成部分。目前对伐根的研究主要集中在伐根的分解、养分的释放，以及与真菌、土壤动物等方面，而有关伐根组分及生物量的研究很少。我们在马尾松皆伐后，采用挖掘法研究了伐根各组分的生物量及养分贮量，结果表明：伐根生物量为15.44t/hm^2，占采伐剩余物总生物量的39.5%，其中主根生物量最大，远高于其他根系组分，每公顷高达6.70t，占伐根生物量的43.4%，其次依次是粗根、根桩和大根，生物量分别为4.49、2.42和1.40t/hm^2，分别占伐根生物量的29.1%、15.7%和9.1%，中根、小根和细根生物量很低，三者之和仅0.42t/hm^2，占伐根生物量的2.7%。伐根中碳贮量为7796kg/hm^2，占采伐剩余物总碳量的42.6%；伐根中氮、磷、钾、钙和镁贮量分别是23.8、1.7、18.3、15.2和5.9kg/hm^2，均显著低于残留在地表的采伐剩余物，其分别占采伐剩余物中氮、磷、钾、钙和镁贮量的23.6%、20.4%、24.9%、15.8%和23.7%。

尽管伐根的生物量和养分贮量显著低于残留在地表的采伐剩余物，但因伐根深入土壤，无论炼山与否均能保存下来，而且伐根分解是一个缓慢的过程，可以持续向土壤生物及伐根周围的土壤提供碳、氮等养分，同时随着伐根的分解，还可在土壤中形成大的空隙，促进降水入渗。目前有关伐根生物量的研究很少，而且在根系生物量研究中，因其测定费时费力，多采用数学模型进行拟合[26-28]。因此，我们在马尾松皆伐林地布设样方，实测样方内所有伐桩直径和伐根各组分生物量，在此基础上采用数学模型对伐根各组分生物量与伐桩直径进行拟合。拟合结果表明，除小根和细根生物量与伐桩直径相关系数较低，呈显著相关外，其他伐根各组分生物量与伐桩直径均呈极显著相关。说明在测定伐桩直径的情况下，采用本文拟合方程，可估算主根、根桩、粗根、大根、中根以及整个伐根的生物量。另外，与土芯法(土钻法)相比，文中细根生物量的实测值和拟合方程估算值均偏低，可能的原因有：我们在测定伐根各组分生物量时采用挖掘法，在此过程中尽管我们仔细挑拣，但因细根(根径小于2mm)细小，可能存在遗漏；此外，我们研究的对象是生长31年、割完脂、要采伐的马尾松，树的生长势弱，可能导致了细根生物量减少。

项文化等[29]对不同年龄马尾松人工林的研究发现，叶、枝和皮生物量之和占总生物量的20.5%~37.2%，伐根占总生物量的11.0%~14.3%；叶、枝和皮中氮磷钾钙镁贮量分别占相应总贮量的54.3%~80.6%、62.8%~89.6%、52.3%~74.7%、50.4%~65.0%和27.7%~67.7%，伐根中氮磷钾钙镁贮量分别占相应总贮量的8.4%~24.2%、8.7%~16.6%、12.6%~25.3%、12.7%~15.6%和9.8%~29.1%。李淑仪等[30]研究结果也显示，桉树人工林采伐时，树木吸收累积养分的45%贮存在采伐剩余物中。我们对广西南部31年生马尾松人工林的研究也表明，皆伐后林地上残留着大量的采伐剩余物，其生物量为39.07t/hm^2，氮、磷、钾、钙、镁矿质元素贮量分别是101.2、8.3、73.4、96.0和24.7kg/hm^2。可见，林木采伐后，林地残留的采伐剩余物生物量庞大，而且通过采伐剩余物归还林地的矿质养分量也较大，而另一方面，在进入衰老之前，因林木养分年吸收量远高于凋落物中的年归还量，土壤养分长期处在严重的消耗状态，土壤肥力不断下降[31]。因此，从养分平衡的角度来看，合理利用采伐剩余物对人工林地力维持至关重要。

杨承栋[32]认为土壤有机质量和质的下降是制约我国人工林林木生长的关键因子，同时指出采伐剩余物和林地枯落物是土壤有机质的重要来源。而皆伐迹地人工更新林地的前3年因幼树未能郁闭成林，也不能产生较多的凋落物，在此阶段采伐剩余物是林地土壤有机质和养分元素的主要来源，因此在林业生产中应合理利用采伐剩余物，以达到长期维护地力、保持生产力的作用。

参考文献

[1]国家林业局森林资源管理司．中国森林资源第七次清查结果及其分析[J]．林业经济，2010，(2)：66-72.

[2]高金铎．谈采伐剩余物[J]．内蒙古林业，1982，(10)：35.

[3]马庆祥，俞新妥，何智英，等．不同林地清理方式对杉木幼林生态系统水土流失的影响[J]．自然资源学报，1996，11(1)：33-40.

[4]FACELLI J M，PICKETT S TA. Plant litter：its dynamics and effects on plant community structure[J]．Botanical Review，1991，57：1-32.

[5]高光耀，傅伯杰，吕一河，等．干旱半干旱区坡面覆被格局的水土流失效应研究进展[J]．生态学报，2013，33(1)：12-22.

[6]李强，周道玮，陈笑莹．地上枯落物的累积、分解及其在陆地生态系统中的作用[J]．生态学报，2014，34(14)：3807-3819.

[7]王力，邵明安，王全九．林地土壤水分运动研究评述[J]．林业科学，2005，41(2)：147-153.

[8]SARR M，AGBOGBA C，RUSSELL-SMITH A，et al. Effects of soil faunal activity and woody shrubs on water infiltration rates in a semi-arid fallow of Senegal[J]．Applied Soil Ecology，2001，16：283-290.

[9]HARDEN C P，SCRUGGS P D. Infiltration on mountain slopes：A comparison of three environments[J]．Geomorphology，2003，55：5-24.

[10]吴君君，杨智杰，刘小飞，等．火烧和保留采伐剩余物对土壤有机碳矿化的影响[J]．土壤学报，2015，52(1)：203-211.

[11]薛立，向文静，何跃君，等．不同林地清理方式对杉木林土壤肥力的影响[J]．应用生态学报，2005，16(8)：1417-142.

[12]吴波波，郭剑芬，吴君君，等．采伐剩余物对林地表层土壤生化特性和酶活性的影响[J]．生态学报，2014，34(7)：1645-1653.

[13]何宗明，范少辉，卢镜铭．立地管理措施对2代6年生杉木林生长的影响[J]．林业科学，2006，42(11)：47-51.

[14]鲍士旦．土壤农化分析[M]．北京：中国农业出版社，2000.

[15]赵其国．我国南方当前水土流失与生态安全中值得重视的问题[J]．水土保持通报，2006，26(2)：1-8.

[16]林思祖，俞新妥，吴擢溪，等．炼山对水土流失影响的初步研究[J]．热带林业科技，1987，4：51-54.

[17]俞新妥，杨玉盛，何智英，等．炼山对杉木人工林生态系统的影响 I 炼山初期林地水土流失的初步研究[J]．福建林学院学报，1989，9(3)：238-255.

[18]田大伦，宁晓波．不同龄组马尾松林凋落物量及养分归还量研究[J]．中南林学院学报，1995，15(2)：163-169.

[19]杨会侠，汪思龙，范冰，等．不同林龄马尾松人工林年凋落物量与养分归还动态[J]．生态学杂志，2010，29(12)：2334-2340.

[20]徐旺明，闫文德，李洁冰，等．亚热带4种森林凋落物量及其动态特征[J]．生态学报，2013，33(23)：7570-7575.

[21]ANDERS L，AKE L. Cut high stumps of spruce，birch，aspen and oak as breeding substrates for saproxylic beetles[J]. Forest Ecology and Management，2004，203：1-20.

[22]法蕾，李文宁．试述粗木质残体及伐根分解的研究进展[J]．地质技术经济管理，2004，26(6)：54-57.

[23]法蕾，赵秀海．阔叶红松林山杨伐根分解过程中真菌动态研究[J]．北京林业大学学报，2006，28(supp.2)：90-94.

[24]孟春，庞凤艳．伐根研究进展[J]．森林工程，2005，21(4)：11-13.

[25]田子珩，张春雨，赵秀海．红松伐根分解过程中土壤动物动态[J]．生态学杂志，2007，26(2)：286-290.

[26]张艳杰，温佐吾．不同造林密度马尾松人工林的根系生物量[J]．林业科学，2011，47(3)：75-81.

[27]明安刚，张治军，谌红辉，等．抚育间伐对马尾松人工林生物量与碳贮量的影响[J]．林业科学，2013，49(10)：1-6.

[28]贾呈鑫卓，李帅锋，苏建荣，等．思茅松人工林根系特征与生物量分配[J]．生态学杂志，2017，36(1)：21-28.

[29]项文化，田大伦．不同年龄阶段马尾松人工林养分循环的研究[J]．植物生态学报，2002，26(1)：89-95.

[30]李淑仪，钟继洪，莫晓勇，等．桉树土壤与营养研究(M)．广州：广东科技出版社，2007，58-59.

[31]陈楚莹，张家武，周崇莲，等．改善杉木人工林的林地质量和提高生产力的研究[J]．应用生态学报，1990，1(2)：97-106.

[32]杨承栋．我国人工林土壤有机质的量和质下降是制约林木生长的关键因子[J]．林业科学，2016，52(12)：1-12.

[原载：生态学杂志，2018，37(03)]

广西大青山土壤水热特征及其对气象要素的响应

刘士玲[1,2] 杨保国[1,2] 卢立华[1,2] 张 培[1,2] 雷丽群[1,2] 何 远[1,2] 郑 路[1,2]
([1]中国林业科学研究院热带林业实验中心，广西凭祥 532600；
[2]广西友谊关森林生态系统国家定位观测研究站，广西凭祥 532600)

摘 要 利用广西友谊关森林生态系统国家定位观测研究站自动气象站的观测数据，分析了广西大青山地区土壤水热的变化特征及其与气象要素的统计关系，结果表明：①土壤温度的年内变化特征与气温一致，呈"单峰型"，最低值出现在1月，最高值在7月，但存在滞后期，滞后时间随土层深度增加而增加。②土壤水分年内变化复杂，受各月气温、降水的影响而呈现出一定的波动性，1~4月份土壤含水率较高，4月下旬土壤含水率迅速降低，至7月底处于全年较低值，变化幅度随土层深度加深而减小。③影响土壤温度的主要气象要素为大气温度，土壤水分主要受降水和饱和水汽压影响。月尺度上这些气象要素对土壤水热的影响更加显著。④土壤温度和土壤水分为显著的负相关关系。研究结论为土壤水热对气候变化的响应提供参考资料，也为本区域森林经营管理提供科学依据。

关键词 土壤水分；土壤温度；气象要素；大青山

Soil Moisture and Temperature Characteristics and the Responses to Meteorological Factors in Daqingshan, Guangxi

LIU Shiling[1,2], YANG Baoguo[1,2], LU Li-hua[1,2], ZHANG Pei[1,2],
LEI Li-qun[1,2], HE Yuan[1,2], ZHENH Lu[1,2]
([1]*Experimental Center of Tropical Forestry, CAF, Pingxiang, 532600, Guangxi, China*;
[2]*Guangxi Youyiguan Forest Ecosystem Research Station, Pingxiang, 532600, Guangxi, China*)

Abstract: By using Automatic meteorological station of Guangxi Youyiguan Forest Ecosystem National Research Station, we analyzed the relationships between soil temperature, soil moisture dynamics and other meteorological factors in Daqingshan. The results were shown as follows: ①the changing characteristics of the soil temperature are largely consistent with the air temperature, and the seasonal dynamics of them showed one peak, the lowest value appeared in January, the highest value in July, while a lag phase between them was observed which increased with soil depth. ②Soil moisture dynamics is much more complex compared with the temperature dynamics, and strongly affected by the precipitation and tempreture. The soil moisture content was high from January to April, and it decreased rapidly in late April, reaching a year-long low value at the end of July. Soil moisture dynamics decreases with soil depth. ③On different time scales, Soil temperature is highly related with the air temperature, while the soil moisture is also highly related with precipitation and Saturated vapor pressure differences(VPD). Better statistical relationships between soil temperature, soil moisture dynamics and meteorological factors was observed at month-scale rather than a daily time scale. ④Soil temperature and moisture shows an obvious negative correlation relationship. The research conclusion could provide reference material for the response of soil water and heat to climate change and provide the scientifc basis for forest management in this region.

Key words: soil moisture; soil temperature; meteorological factors; Daqingshan

地表土壤的水热动态过程和变化机制是陆面过程的重要研究内容[1,2]，其动态影响着不同生态系统土壤的生物和非生物的生态学过程，包括植物的生产量、分解、矿化、水热能量平衡[3]，但它们又受气候、地形、地上植被的盖度和数量、土壤自身的物理特性等因素控制[4]。森林土壤是森林生态系统发挥水源涵养功能的主要载体，水热在土壤中的分布特征和动态变化对植被生态系统具有重要意义。

关于土壤水热国内外开展了诸多研究，如唐振兴等[2]、牛赟等[5,6]研究了祁连山土壤水热的垂直分层变化特征、空间变化特征及其影响因素；Koster等[7]研究了土壤水分与降水在大尺度上的耦合关系；Western等[8]研究了Australia Tarrawarra流域土壤水分的时空变异结构及尺度效应。以上研究虽然对土壤水热变化过程有了一定认识，但对于南亚热带林区的土壤水热变化仍缺乏点尺度的长期观测资料，尤其缺乏土壤水热与气象要素的同步观测，该区域点尺度上土壤水热变化及其与气候变化内在关联的研究也不够。本文在广西大青山林区利用友谊关森林生态系统国家定位观测研究站建立的自动气象观测站，对土壤水热、气象要素进行定点连续观测。通过对数据的分析，旨在了解南亚热带大青山林区土壤水热的变化特征及影响土壤水热的主要气象要素，以期为全球变化情景下，土壤水热对气候变化的响应提供参考，也为本区域林业可持续发展提供科学依据和基础数据。

1　研究区概况

广西友谊关森林生态系统国家定位观测研究站(22°03′49″N，106°50′37″E)位于广西大青山林区，属南亚热带半湿润—湿润季风气候，有明显的干湿季。年均气温21.5℃，≥10℃积温6000~7600℃，年均降水量1200~1500mm，年蒸发量1261~1388mm，相对湿度80%~84%。地貌类型以低山丘陵为主，海拔430~680m。地带性土壤为由花岗岩发育而成的赤红壤，土层厚度100~150cm，腐殖质层厚度5~10cm。

2　研究方法

2.1　观测设备

根据国家气象局编制的《自动气象站建设技术要求》和《地面气象观测规范》，在林外建立自动气象观测站，主要观测大气温湿度、风速、风向、降水、辐射等常规气象要素和地面下5、15、30和50cm的土壤要素。该气象站采用的是CR1000数据采集器(Campbell，美国)。传感器类型：温度传感器(109，Campbell，美国)、风速风向传感器(034B，MetOne，美国)、翻斗式雨量筒(TE525MM，Campbell，美国)、辐射传感器(LI200X，Li-Cor，美国)、土壤水分传感器(CS616，Vaisala & Campbell，美国)、土壤温度传感器(HMP45C，Campbell，美国)等。各要素数据采集步长均为5min。

2.2　数据处理与分析

本文所用的是2016年观测的土壤要素和气象要素数据。其中土壤要素为4个深度的土壤体积含水率(即Sdw5、Sdw15、Sdw30、Sdw50)和土壤温度(即ST5、ST15、ST30、ST50)。气象要素为大气温度(Ta)、相对湿度(RH)、大气降水(P)、总辐射(Eg)和饱和水汽压(VPD)。

饱和水汽压的计算公式如下：

$$VPD = 0.611 exp \frac{17.27T}{T+237.3}(1-RH)$$

式中：T(℃)为空气温度；RH(%)为相对湿度。

对数据进行修订校正后将各要素的5min步长观测数据加以整合，进行数据的统计分析。本文所用软件为：SPSS 19.0、Sigmaplot 13.0。

3　结果与分析

3.1　气温与降水变化

图1为自动气象站记录的步长为5min的降水和气温数据经整理后的月变化趋势图。降水主要集中于1月、6~9月，其中8月份降水量312.6mm为全年月降水量的最大值，而10月份降水量较少，12月份无降水。从气温曲线可以看出，在5~9月期间气温处于全年的极大值阶段。

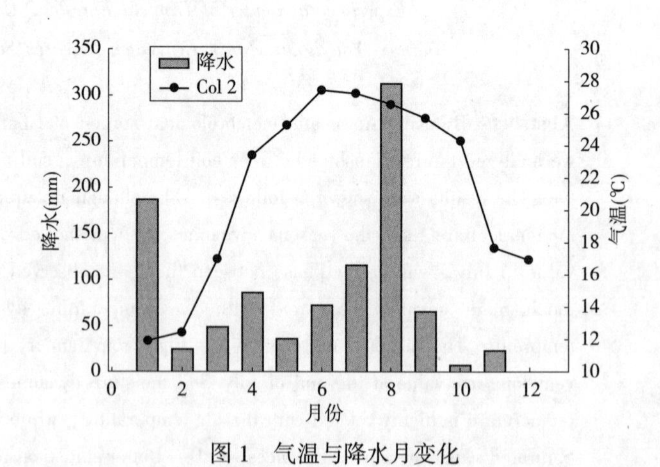

图1　气温与降水月变化

3.2　土壤水热变化特征

图2和图3分别给出了土壤温度和水分的变化趋势，从图中可知浅层土壤水热曲线在5~9月处于极值阶段，这与气温曲线(图1)的变化特征相似，由于降水主要集中于这些月份，且受气温的影响，所以土壤水热变化在此期间波动较显著。

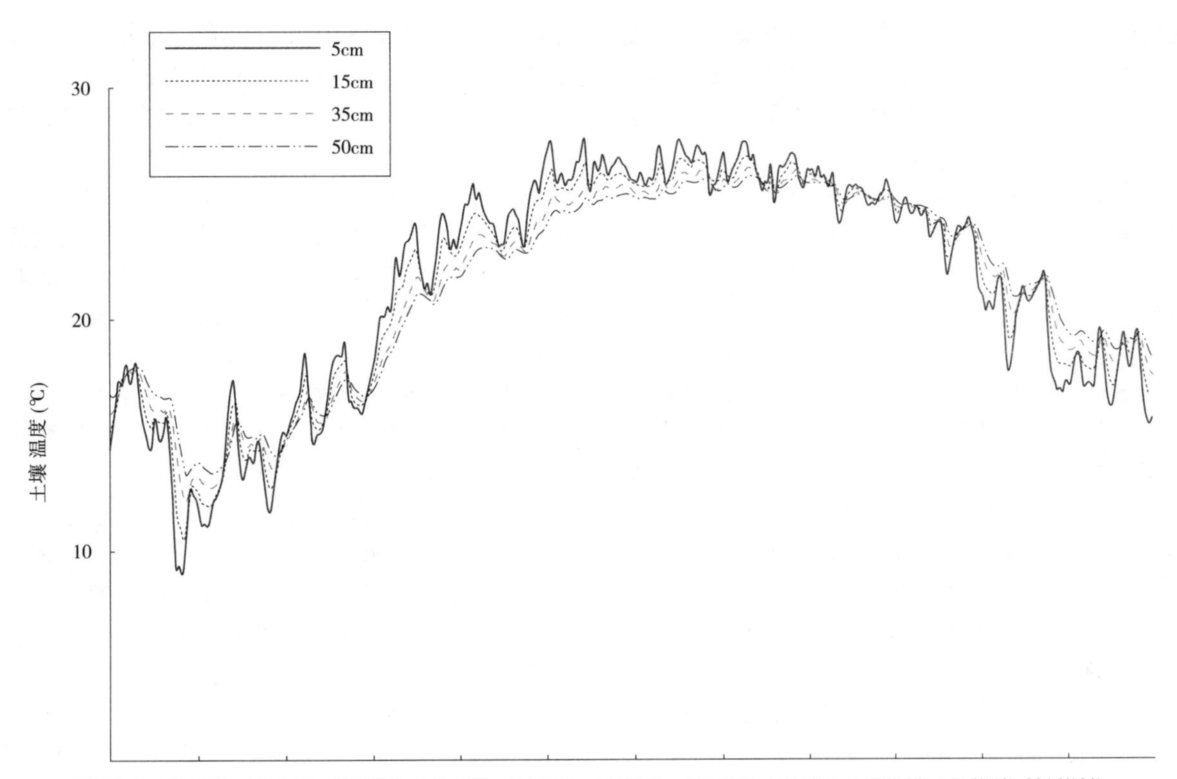

图 2 土壤温度日变化特征

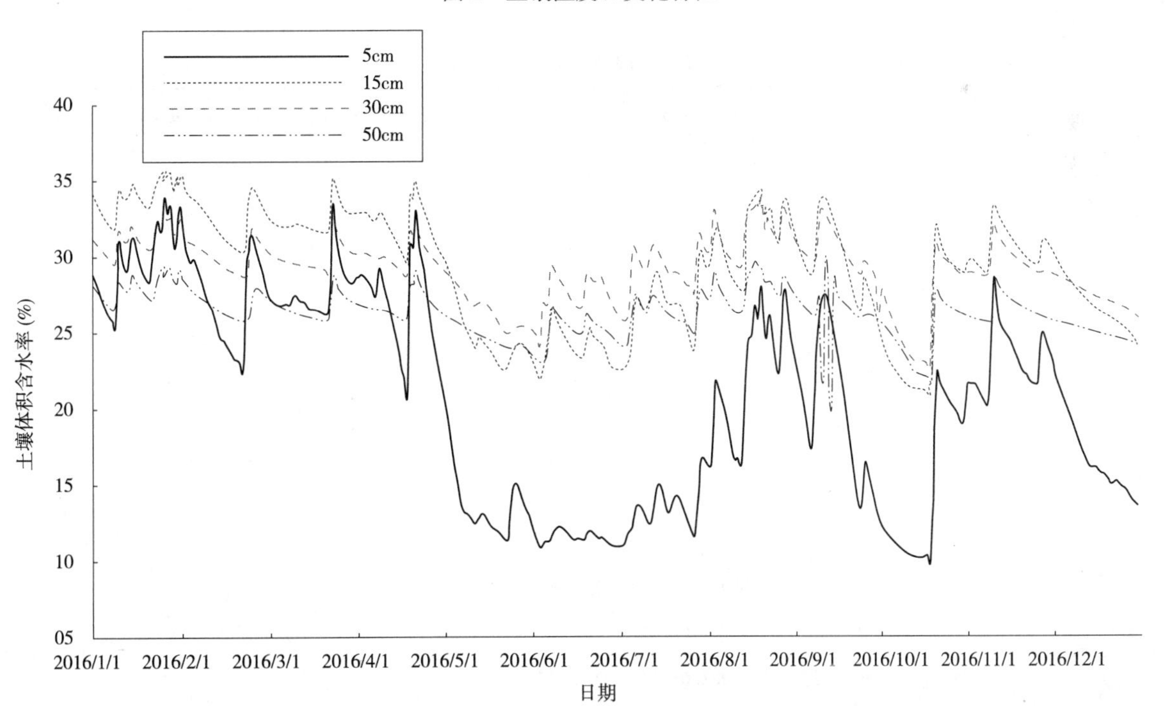

图 3 土壤含水率日变化特征

土壤温度的年内变化特征与气温一致，呈“单峰型”，土壤温度最低值出现在 1 月，最高值出现在 7 月。整体上各层土壤温度变化较为一致，土壤温度年内变化幅度随深度增加而减小，且土壤温度的变化周期与大气温度相比存在一定的滞后期，滞后时间随深度增加而增加，这主要是因为土壤对太阳辐射有显著削弱作用。

土壤温度的垂直变化特征为：3～9 月土壤温度随深度增加而降低，1 月、2 月、10～12 月随深度增加而升高。造成这种差异的原因在于，土壤温度主要受气温的影响，3~9 月气温较高，土壤表面吸收太阳辐射而增温，并通过热传导向深处传递热量，愈是处在下层的土壤，获得的热量愈少，故土壤温度随深度增加而降低；1 月、2 月、10~12 月气温较低，地表接收的太阳辐射小于地表长波辐射，地表温度下降，当温度下降到比深层的温度低时，热量将由深层向地表传输，但由于从下层得到的能量不足以抵消上层的降温，因此土壤温度随深度增加而升高。

从土壤体积含水率变化特征曲线(图 3)可以看出，整体上各层土壤水分变化较为一致，土壤水分年内变化幅度随深度加深而减小。1~4 月份土壤水分较大，4 月下旬开始土壤水分迅速减小，至 7 月底虽然降水使土壤水分有一定的波动，但土壤水分整体偏低，5cm 层土壤水分表现更为显著，15cm、30cm、50cm 土壤水分维持在相对稳定的水平，但均处于全年较低值，这是因为该时期降水量不高，但大气温度高，蒸发量大，表层土壤水分受蒸发量影响较大，深层土壤受影响程度较小。此外，生长季植物根系的水分吸收也是影响土壤水分的重要因素。8~9 月上旬降水量较大且降水次数较多，降水的补给使得土壤水分迅速提高，但由于气温较高、蒸发量较大，降水过后土壤水分又迅速减小，从而表现出该时期土壤水分较高，但又具迅速升降的变化趋势。9 月中旬至 10 月份中旬土壤水分又出现低谷，这是因为该月份无降水，加之气温较高、蒸发量较大，土壤水分持续走低。11~12 月除降水对土壤水分阶段性增加外，土壤水分整体上呈逐渐下降的趋势。

从土壤温度和土壤水分的变化关系图中，我们很难清晰地看出土壤温度与水分之间的关系，为了明确各土层土壤温度与土壤水分之间的关系，对各土层的土壤温度和水分数据进行了统计分析(表 1)。

从土壤水热的相互关系来看，各层土壤温度与土壤水分之间存在显著负相关关系，相关性随土壤深度增加表现为先增大后减小再增大的趋势，30cm 土壤温度与土壤水分相关性最低。各土层土壤温度都具有极显著的正相关关系，随土壤深度增加，浅层土壤温度与深层土壤温度的相关系数减小，这是因为土壤温度主要受气温影响，随深度增加土壤温度的分子热传导作用逐渐减弱。各土层土壤水分亦都具有极显著的正相关关系，从影响程度大小来看，由于地表水进入土壤后的下渗作用，随土壤深度增加，浅层土壤水分与深层土壤水分的相关系数亦减小。

表 1 土壤水热相关性分析

土壤要素	5cm 土壤温度	15m 土壤温度	30cm 土壤温度	50cm 土壤温度	5cm 土壤水分	15cm 土壤水分	30cm 土壤水分	50cm 土壤水分
5cm 土壤温度	1							
15m 土壤温度	0.998**	1						
30cm 土壤温度	0.987**	0.994**	1					
50cm 土壤温度	0.973**	0.984**	0.997**	1				
5cm 土壤水分	-0.249**	-0.249**	-0.239**	-0.226**	1			
15cm 土壤水分	-0.247**	-0.253**	-0.256**	-0.253**	0.934**	1		
30cm 土壤水分	-0.185**	-0.195**	-0.212**	-0.221**	0.811**	0.903**	1	
50cm 土壤水分	-0.347**	-0.360**	-0.384**	-0.397**	0.725**	0.837**	0.952**	1

注：* 在 $P<0.05$ 水平上显著相关；** 在 $P<0.01$ 水平上显著相关。

3.3 土壤水热对气象要素的响应

通过多元逐步回归对土壤水分、温度和气象要素气温、降水、辐射、饱和水汽压差、风速进行分析，筛选出影响土壤水分和温度的主要气象因子。以气象站日观测数据为基础，给出了不同深度土壤温度、含水率与主要气象控制因子的关系图(图4、图5)。

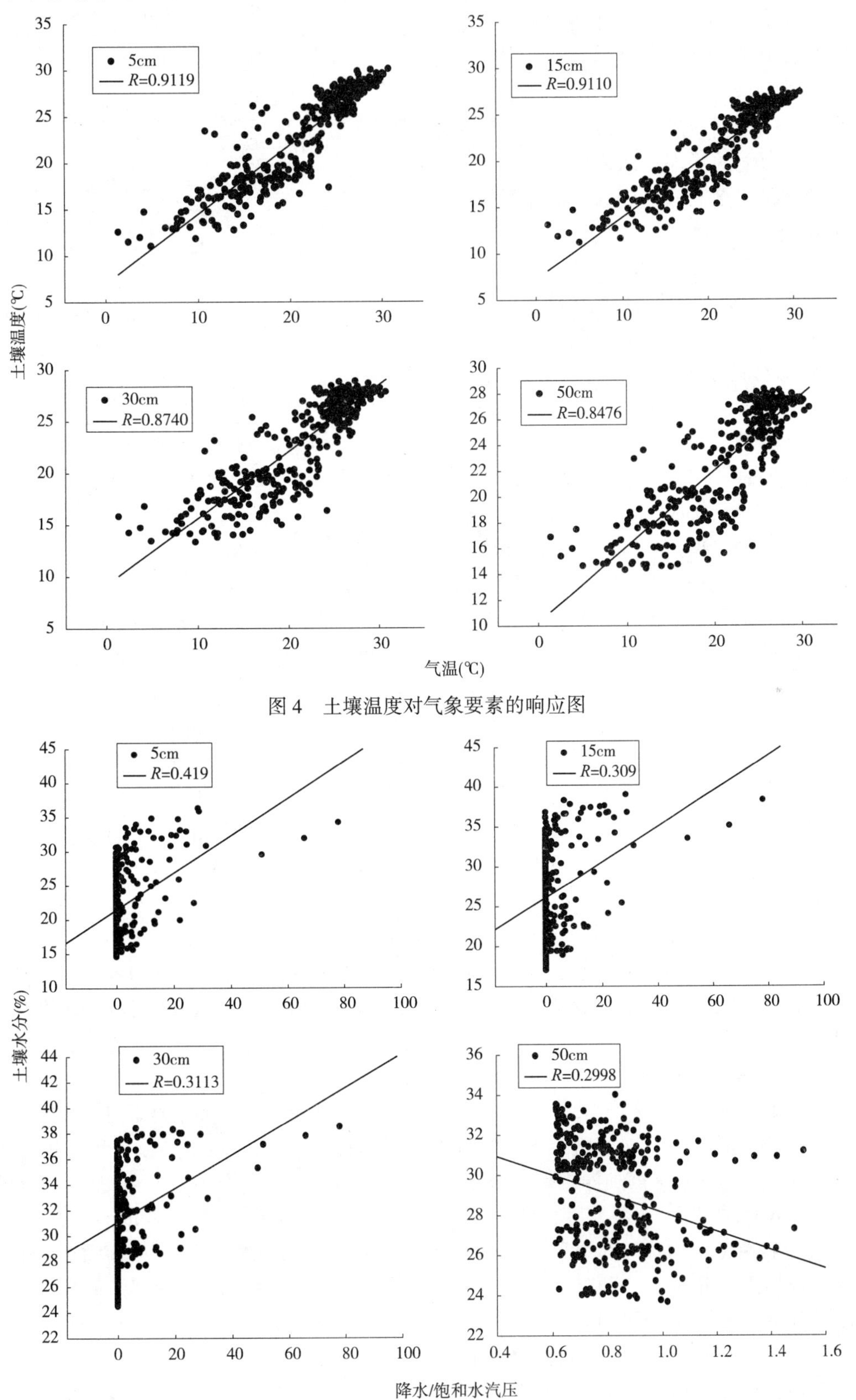

图4 土壤温度对气象要素的响应图

图5 土壤水分对气象要素的响应图

由土壤温度对气象要素的响应图4可知，土壤温度的主要气象控制因子为气温，为正相关关系。不同深度土壤温度对气温的响应强度存在一定差异，随土壤深度增加，二者之间相关系数变小，土壤温度变化对气温的响应强度减弱，5cm土壤温度与气温的相关性最好为0.9119，50cm土壤温度与气温的相关性最差为0.8476。这主要是因为，太阳辐射通过加热浅层土壤并以热传导和热对流的方式传递给深层土壤，随着土层深度的增加，无论热传导，还是热对流所携带的能量逐渐衰减，造成土壤温度随深度增加而逐渐降低，所以浅层土壤受气温变化的影响较大，深层土壤受气温变化的影响较小。

图5给出了对土壤水分有较大影响的降水、饱和水汽压差与土壤水分响应图，从图中可以看出5cm、15cm、30cm土壤水分的主要气象控制因子为降水，二者表现为正相关关系，其中5cm土壤水分与降水的相关性最好为0.4189，随土壤深度增加二者关系明显减弱，15cm、30cm土壤水分与降水相关性仅为0.3085、0.3113；而影响50cm土壤水分的主要气象因子为饱和水汽压，表现为负相关，相关系数为0.2998。土壤水分的主要来源是降水，浅层土壤水分主要受降水的影响，深层土壤水分由于被上层土壤的截留以及植物根系与枯落物对土壤结构和理化性质的改善能力降低，其土壤密度增大，容重增大，土壤孔隙度减小，水分在土壤中的流动能力减弱，因此，随土壤深度增加，土壤水分对降水的响应逐渐减弱。土壤水分与降水的相关性比预期偏低，可能是因为本文统计的是点尺度上的短时间条件下气象要素对土壤水热的影响，在日变化水平上降水只是作为土壤水分来源的输入参量，土壤水分对降水的响应存在滞后性，且受到地表径流、土壤特征、植被及其他气象要素的影响，这可能弱化了降雨对表层土壤水分的作用，随着土壤深度加深，这种弱化作用表现得更加明显。

为进一步分析土壤水热的主要气象控制因子，本文给出了月尺度上的土壤水热对气象要素的响应统计结果(表2和表3)。月尺度结果表明土壤温度的气象控制因子为气温，随土壤深度加深判定系数 R^2 逐渐减小，这与日尺度的结果一致，但月尺度上的判定系数 R^2 均优于相应土层的日尺度结果；土壤水分月尺度的气象控制因子除50cm为饱和水汽压差，5cm、15cm和30cm的气象控制因子均为降水，这亦与日尺度的结果一致，且月尺度的判定系数 R^2 也优于日尺度结果。由此可见，月尺度上的主要气象要素对土壤水热的影响更加显著。

表2 土壤温度与主要气象因子的统计关系

时间	指标	5cm	15cm	30cm	50cm
日平均	气象因子	T	T	T	T
	判定系数	0.8316	0.8299	0.7639	0.7184
月平均	气象因子	T	T	T	T
	判定系数	0.9652	0.9515	0.9113	0.8719

注：结果均通过显著性水平0.01检验，表中T表示气温(℃)。

表3 土壤水分与主要气象因子的统计关系

时间	指标	5cm	15cm	30cm	50cm
日平均	气象因子	P	P	P	VPD
	判定系数	0.1755	0.0952	0.0969	0.0899
月平均	气象因子	P	P	P	VPD
	判定系数	0.5074	0.4475	0.3392	0.2656

注：结果均通过显著性水平0.01检验，表中T，P，VPD分别表示气温(℃)和降水(mm)饱和水汽压差(hPa)。

4 结论与讨论

4.1 结论

本文利用广西友谊关生态定位站建立的自动气象站对该区各气象要素和土壤温度、水分进行了长期连续定位观测，分析土壤温度、水分的变化特征，建立土壤温度、水分相关关系，阐明影响土壤温度、水分的主要气象要素，主要得出以下结论：

(1)土壤温度的年内变化特征与气温一致，呈“单峰型”。整体上各层土壤温度变化较为一致，土壤温度年内变化幅度随深度增加而减小，且土壤温度的变化周期与大气温度相比存在一定的滞后期，滞后时间随深度增加而增加。

(2)土壤水分年内变化复杂，受各月气温、降水的影响而呈现出一定的波动性。整体上各层土壤水分变化较为一致，土壤水分年内变化幅度随深度加深而减小。

(3)土壤水热与气象要素在月尺度上的统计关系好于日尺度。两尺度上控制土壤温度的主要气象因子为气温；5cm、15cm、30cm土壤水分的主要气象控制因子为降水，二者表现为正相关关系，影响50cm土壤水分的主要气象因子为饱和水汽压。

(4)土壤温度与水分呈显著的线性负相关。

4.2 讨论

(1)唐振兴等[2]、车宗玺等[9]研究表明土壤温度年内变化与气温一致，但存在滞后期，滞后时间随深度增加而增加，这与本文研究结果一致。土壤温度对气象要素的响应关系，本文研究结果表明，在不同时间尺度上，影响土壤温度的主要气象要素均为大气温度；这与唐振兴等[3]的研究结果略有差异，其主要原因是本文的结果主要是基于气象观测站的所观测的土层50cm以上的土壤数据，而未对深层土壤温度进行观测，深层土壤温度变化在物理机制上考虑主要受上下层土壤热量平衡影响，今后应加强对深层土壤温度对气象要素的响应研究。

(2)已有研究[10-11]表明降水是影响土壤含水量变化的主要因素。本文土壤水分对气象要素的响应研究表明，不同时间尺度上，5cm、15cm、30cm土壤水分的主要控制气象因素为降水，50cm土壤水分控制要素为饱和大气压，且土壤水分的变化与降水呈正相关，这与唐振兴等[2]的研究结果存在一定的差异。牛赟等[12]认为影响土壤水分垂直变化的直接因素是降水和土壤植被的土壤水分消耗，引起这些变化的间接因素是土壤空间属性的变化及其植被类型对土壤水分的影响。在今后研究中，要注重土壤水分空间变化机理及各因素的综合影响程度等研究。

(3)在已有的文献中土壤温度和土壤水分具有显著负相关关系[9,13,14]或负相关不显著[15]，王红梅等[16]认为土壤温度和土壤水分的相关性表现为小采样粒度(0.5m×0.5m，1m ×1m)的显著负相关性，而在较大采样粒度(2m×2m，5m×5m)表现为显著的正相关关系，因此，土壤水热特征的关系研究应确定合理的采样粒度和尺度。本文利用自动气象观测站所观测的土壤温度和水分数据进行统计分析，土壤温度和土壤水分表现为极显著的负相关关系。随着生态监测科技的发展，利用自动气象观测站的土壤含水量和温度监测仪对土壤水分和温度进行实时动态自动监测，有效地避免了以往研究者所采用不同的采样粒度或尺度所产生的结果的差异性。

(4)我国亚热带和热带湿润区年降水量很大，土壤水分的变化影响土壤的导热率和热容量，进而影响土壤导温率，使得土壤温度与气温、降水之间的响应关系非常复杂[17]，土壤水分对降水和温度的响应亦非常复杂[18]，杨永辉等[19]、张一平等[20]利用实验和模型结合探讨温度与降水变化对土壤水分或温度的影响的方法值得借鉴。在全球气候变暖情境下，森林土壤温湿度的特征、变化规律是否发生改变?在将来的科研工作中，我们要更加关注土壤水热在空间上的年际和季节变化及对气象要素的响应特征和响应程度，为探索全球气候变暖与土壤水热在空间上的响应关系，为森林经营提供参考资料。此外，影响土壤呼吸的主要环境因子是温度和土壤水分，全球温度上升将促进土壤呼吸，加速土壤中储存的碳向大气中排放，进而加剧全球变化。因此伴随增温导致的土壤水热变化对土壤呼吸的影响也是下一步研究中需要关注的问题。

参考文献

[1]SUN G, NOORMETS A, GAVAZZI M J, et al. Energy and water balance of two contrasting loblolly pine plantations on the lower coastal plain of North Carolina, USA[J]. Forest Ecology and Management, 2010, 259(7): 1299-1310.

[2]唐振兴，何志斌，刘鹄．祁连山中段林草交错带土壤水热特征及其对气象要素的响应[J]. 生态学报，2012，32(4)：1056-1065.

[3]BUBIERJ L, CRILL P M, MOORE T R. Seasonal patternsand controlon net ecosystem CO_2 exchange in a boreal peatland complex[J]. Global Biogeochemical Cycles, 1999, 12(4): 703-714.

[4]STROYAN H, DE-POLLIH, BOHMS, et al. Spatial heterogeneity of soil respiration and related properties at the plant scale[J]. Plant Soil, 2000, 222(1-2): 203-214.

[5]牛赟，刘贤德，吕一河，等．祁连山大野口流域土壤水热垂直分层变化特征分析[J]. 干旱区资源与环境，2016，30(6)：141-146.

[6]牛赟，刘贤德，苗毓鑫，等．祁连山大野口流域土壤水热空间变化特征研究[J]. 冰川冻土，2015，37(5)：1354-1360.

[7]KOSTER R D, SUAREZ M J, HIGGINS R W, et al. Observational evidence that soil moisture variations affect precipitation. GeophysicalResearch Letters, 2003, 30(5): 1241.

[8]WESTERN A W, BILSCHL G, GRAYSON R B, et al. Geostatistical characterization of soil moisture patterns in the tarrawarra catchment. Journal of Hydrology, 1998, 205: 20-37.

[9]车宗玺，李进军，汪有奎，等．祁连山西段草地土壤温度、水分变化特征[J]. 生态学报，2018，38(1)：1-7.

[10]巩合德，张一平，刘玉洪，等．哀牢山常绿阔叶林土壤水分动态变化[J]. 东北林业大学学报，2008，36(1)：53-54.

[11]祁栋灵，孙瑞，谢贵水，等．海南西部低割龄橡胶林土壤水分季节变化特征及其对气象因子响应研究初报[J]. 生态科学，2017，36(6)：44-48.

[12]牛赟，刘贤德，吕一河，等．祁连山大野口流域土壤水分空间变化特征研究[J]. 中南林业科技大学学报，2016，36(10)：94-100.

[13]BOND L B, BROWN K M, CAROL G. Spatial Dynamics of soil moisture and temperature in a black spruce boreal chronose quence[J]. Canadian Journal of Forest Research, 2006, 369(11): 2794-2802.

[14]REDDING T E, HOPE G D, FORTIN M J. Spatial pattern of soil temperature and across subalpine forest-clear cut edges in the southern interior of British Columbia[J]. Canadian Journal of Soil Science, 2003, 83(1): 121-130.

[15]BOND L B, WANG C, GOWER S T. Spatial-temporal measurement and modeling of boreal forest soil temperature [J]. Agricultural and Forest Meteorology, 2005, 131 (1-2): 27-40.

[16]王红梅, 王堃, 米佳, 等. 北方农牧交错带沽源农田-草地界面土壤水热空间特征[J]. 生态学报, 2009, 29(12): 6589-6599.

[17]张慧智, 史学正, 于东升, 等. 中国土壤温度的季节性变化及其区域分异研究[J]. 土壤学报, 2009, 46(2): 227-234.

[18]CAMARGO, J L C, KAPOS V. Complex edge effects on soil moisture and microclimate in central Amazonian forest [J]. Journal of Tropical Ecology, 1995, 11 (2): 205-221.

[19]杨永辉, 渡边正孝, 王智平, 等. 气候变化对太行山土壤水分及植被的影响[J]. 地理学报, 2004, 59 (1): 56-63.

[20]张一平, 武传胜, 梁乃申, 等. 哀牢山亚热带常绿阔叶林森林土壤温湿特征及其对温度升高的响应[J]. 生态学报, 2015, 35(22): 7418-7425.

[原载: 中南林业科技大学学报, 2018, 38(11)]

南亚热带不同演替阶段次生林植物多样性及乔木生物量

农 友[1,2] 卢立华[1,2] 游建华[3] 雷丽群[1,2] 王亚南[1,2] 李 华[1,2] 杨桂芳[1,2]

([1]中国林业科学研究院热带林业实验中心，广西凭祥 532600;[2]广西友谊关森林生态系统国家定位观测研究站，广西凭祥 532600;[3]广西崇左市凤凰山林场，广西崇左 532200)

摘 要 以南亚热带5个年龄系列次生林为研究对象，通过时空互换的方法对其植物多样性及乔木生物量进行了研究，对比分析不同演替阶段次生林之间的植物多样性及乔木生物量差异，为次生林的管理提供数据参考。结果表明：①共调查植物69种，隶属于39科55属；随着演替的进行，物种多样性呈波动的不稳定状态，乔、灌、草各层物种组成及多样性也发生相应的变化；乔木层植物的物种丰富度S、Shannon-wiener指数、Simpson指数、J_{sw}均匀度指数均呈现先降低后升高再降低的规律；林下灌、草层植物的物种丰富度S呈现前期平稳，中期升高，后期下降的趋势；林下灌、草层植物的Shannon-wiener指数、Simpson指数、J_{sw}均匀度指数均呈现先降低后升高再降低的规律。②通过异速生长方程计算，乔木生物量随演替的进行呈现先升高后降低再趋于稳定的状态，演替进行到30年时，乔木生物量最大(296.91t/hm^2)。③方差分析结果表明，不同演替阶段不同层次的物种丰富度S差异不显著($P>0.05$)，但Shannon-wiener指数、Simpson指数、J_{sw}均匀度指数差异显著($P<0.05$)；乔木层不同组分生物量差异不显著($P>0.05$)。

关键词 南亚热带；次生林；植物多样性；生物量

The Plant Diversity and Biomass of Trees in Different Successional Stages of Secondary Forest of South Subtropical

NONG You[1,2], LU Lihua[1,2], YOU Jianhua[3], LEI Liqun[1,2], WANG Yanan[1,2], LI Hua[1,2], YANG Guifang[1,2]

([1]*Experimental Center of Tropical Forestry*, *Chinese Academy of Forestry*, *Pingxiang* 532600, *Guangxi*, *China*; [2]*Guangxi Youyiguan Forest Ecosystem Research Station*, *Pingxiang* 532600, *Guangxi*, *China*; [3]*Guangxi Fenghuangshan Forest farm*, *Chongzuo* 532200, *Guangxi*, *China*)

Abstract: In order to compared the plant diversity and biomass of trees between different successional stages of secondary forest of south subtropical, we used the spatiotemporal interchange method to study and analysis them with five different ages of secondary forest. The results showed that: ①there were total 69 species of plants which belonged to 39 families and 55 genera. The species diversity was unstable with the change of succession, and the species composition of different layers also changed accordingly, but the changed was very slow. Among them, the plant species richness, Shannon-Wiener index, Simpson index, J_{sw} evenness index had showed decreased first and then increased and then decreased of tree layer. The species richness of shrub and grass kept a stable level during the early stage, and then increased in the middle stage, and then decreased. The Shannon-Wiener index, Simpson index, J_{sw} evenness index had showed decreased first and then increased and then decreased of shrub and grass layer. ②The tree biomass was first increased and then decreased and then kept a stable level through the calculation of growth equation, and when the succession was carried out in 30 years, the tree biomass was the largest(296.91t/hm^2). ③Analysis of variance showed that the species richness had no significant difference between different successional stages, but the Shannon-Wiener index, Simpson index, J_{sw} evenness index were significantly difference($P<0.05$). Different components had no significant difference among different successional stages of secondary forest($P>0.05$).

Key words: subtropical; secondary forest; plant diversity; biomass

森林生物量是生态学研究的重要内容之一，也是研究整个生态系统结构、功能、能量流动和物质循环的重要参数[1,2]。物种多样性与群落功能密切相关，植物多样性的恢复程度是衡量植被和生态系统恢复效果最重要的指标之一[3]。森林植被作为森林生态系统的一个重要组成部分，在促进养分循环和维护森林立地生产力方面起着关键作用[4-6]。在全球变化的大背景下，环境问题日益为人们所重视，作为生态系统最主要功能的生物量和生物多样性已成为生态学研究中的热点[7-11]。

不同气候区各种典型地带性植被的生物量和物种多样性是群落对环境长期适应的结果[12]，与之相关的研究报道也很多[9,13]，但大多集中在人工林和天然林，次生林较为少见[14-16]。演替最早由欧洲的Spurr[17]于1952年提出，国内的学者对其开展了大量的研究，主要集中在草地[18-20]、天然林[21,22]，而对不同演替阶段次生林开展的研究主要集中在土壤[23,24]、微生物[25]、群落稳定性[26]、生物量[27,28]等方面。南亚热带次生林在全球森林生态系统中有着特殊作用和功能，因此，研究南亚热带次生林自然恢复过程中林下植物多样性和乔木生物量及其变化规律，掌握和了解不同演替阶段次生林的生物量及生物多样性的动态变化过程，对于了解该地区次生林自然恢复过程中物种多样性及乔木生物量的变化以及合理改造和有效利用次生林、促进其正向演替具有重要意义。

本文以南亚热带5个年龄系列次生林为研究对象，对其林下植物多样性及乔木生物量进行了研究，初步揭示该区次生林自然恢复过程中物种多样性及乔木生物量的变化，有助于科学评价植被恢复效果，为该区域的生物多样性保护和生态功能恢复实践提供基础数据。

1　研究地区及研究方法

1.1　研究区概况

研究区位于广西西南部的凤凰山林场，属南亚热带季风气候，107°38′~107°47′E，22°47′~22°57′N。属低山丘陵地貌，海拔多在300~600m，最高海拔(西大明山)1071.2m，最低海拔150m，坡度为15°~40°。年平均气温22°C，最高温度36°C，最低温度1°C，年平均降水量1500mm，雨季多集中6~9月，雨水期分布较为平衡。气候温和，雨量充沛，热量丰富，无霜期长。植物种类繁多，乔木树种主要有马尾松(*Pinus massoniana*)、杉木(*Cunninghamia lanceolata*)、樟树(*Cinnamomum camphora*)、枫香(*Liquidambar formosana*)、荷木(*Schima superba*)、南酸枣(*Choerospondias axillaris*)、黄杞(*Engelhardia roxburghiana*)等；灌木植物主要有桃金娘(*Rhodomyrtus tomentosa*)、木姜子(*Litsea pungens*)、番石榴(*Psidium guajava*)、野牡丹(*Melastoma malabathricum*)、算盘子(*Glochidion puberum*)等；草本植物主要有铁芒萁(*Dicranopteris linearis*)、五节芒(*Miscanthus floridulus*)、蔓生莠竹(*Microstegium fasciculatum*)、乌毛蕨(*Blechnum orientale*)等。

1.2　研究方法

1.2.1　样地设置及调查

在林分踏查的基础上，根据林场现有资料，选取海拔、坡度、坡向等环境条件基本一致的次生林，按林分(林龄分别为10、30、40、55、65年)分别设置3个20m×30m样方。在乔木样方四个角及中心点设置面积为5m×5m的灌草样方5个。共完成乔木样方15个，灌草样方75个。调查内容包括样方内乔木种类、胸径、树高；灌木和草本种类、株数、高度、盖度。调查中，乔木的起测胸径为≥2cm，胸径<2cm的乔木种类归为灌木层。样地信息如表1所示。

表1　样地概况

样地号	林龄(年)	坡度(°)	坡向	密度(株/hm^2)	平均胸径(cm)	平均树高(m)
S1	10	26	东南	733	12.4±6.20	9.5±3.19
S2	30	22	东南	1388	10.5±4.20	9.6±2.66
S3	40	25	东南	927	14.4±8.12	11.1±4.31
S4	55	28	东南	594	17.1±11.00	12.0±5.32
S5	65	21	东南	622	17.4±8.12	12.9±4.06

注：表中数据为平均值±标准差。

1.2.2　物种多样性测定

按乔木层、灌木层、草本层描述各个样地的生物多样性特点。物种多样性指标采用物种丰富度指数、Shannon-wiener指数、Simpson指数、J_{sw}均匀度指数。多样性指数的具体计算方法见文献[29,30]。重要值(*IV*)采用宋永昌计算方法[31]，重要值范围0~300%。

1.2.3　生物量测定

生物量异速生长方程能够克服收获法费时、费力和破坏性大的缺点，建立易测因子(胸径、树高等)和林木不同组分(树叶、树枝、树干和树根等)生物量之间的数量关系[32]，是森林生态系统生物量和净生产力估测的一种普遍且可靠的方法[33]。目前，对生物量的研究多采用直接收获法，考虑到砍伐会对次生林造成较大干扰，因此，本研究利用汪珍川等[34]建立的广西主要树种(组)异速生长模型来估算乔木层的地上生物量。

1.2.4 数据处理

调查数据用 Excel 2010 统计，基于 SPSS18.0，采用单因素 ANOVA 分析不同演替系列次生林林下植物多样性、乔木生物量差异。

2 结果与分析

2.1 不同演替阶段次生林植物多样性

本研究共调查植物 69 种，隶属于 40 科 55 属，其中，乔木层植物 54 种，隶属于 29 科 43 属；灌木层植物 9 种，隶属于 8 科 9 属；草本层植物 10 种，隶属于 8 科 9 属。样地乔木层植物主要由白楸(*Mallotus paniculatus*)、山鸡椒(*Litsea cubeba*)、中平树(*Macaranga denticulate*)、南酸枣(*Choerospondias axillaris*)、西南木荷(*Schima wallichii*)、山黄麻(*Trema tomentosa*)等组成；灌木层植物主要由柏拉木(*Blastus cochinchinensis*)、猴耳环(*Pithecellobium clypearia*)、杜茎山(*Maesa japonica*)、九节(*Psychotria rubra*)、罗伞树(*Ardisia quinquegona*)等组成；草本层草本植物主要由乌毛蕨(*Blechnum orientale*)、阔片短肠蕨(*Allantodia matthewii*)、砂仁(*Amomum villosum*)、铁芒萁(*Dicranopteris linearis*)、粽叶芦(*Thysanolaena maxima*)等组成。不同演替阶段次生林植物多样性如表 2 所示。

表 2 不同演替阶段次生林植物多样性

类型	样地号	物种丰富度	Shannon-wiener 指数	Simpson 指数	J_{sw} 均匀度指数
乔木层	S1	26	2.4610abc	0.8720ac	0.7553a
	S2	21	1.6276bc	0.6685bc	0.5346b
	S3	24	2.4718ac	0.8871ac	0.7778a
	S4	20	2.5664ac	0.9003ac	0.8567a
	S5	21	2.4624abc	0.8801ac	0.8088a
灌木层	S1	4	1.3322b	0.7200bc	0.9610b
	S2	4	0.9515bc	0.5263b	0.6864ac
	S3	4	1.3322b	0.7200bc	0.9610b
	S4	5	1.4942a	0.7500a	0.9284b
	S5	3	0.5851bc	0.3158bc	0.5325ac
草本层	S1	3	1.0889b	0.6600b	0.9912b
	S2	3	0.8487b	0.4938b	0.7725bc
	S3	3	1.0901b	0.6612b	0.9922b
	S4	6	1.6581a	0.7880a	0.9254b
	S5	0	0bc	0bc	0a

注：S1 林龄为 10 年；S2 为 30 年；S3 为 40 年；S4 为 55 年；S5 为 65 年；相同类型的同一列，相同字母表示差异不显著，不同字母表示差异显著(P<0.05)。

2.1.1 乔木层植物多样性比较

不同演替阶段次生林，乔木层物种丰富度呈现先降低后升高再降低再升高的波动状态，随着演替的进行，物种丰富度总体呈现降低的趋势；Shannon-wiener 指数、Simpson 指数、J_{sw} 均匀度指数均呈现先降低后升高再降低的规律，最高均为 S4。方差分析结果显示，不同演替阶段次生林之间的乔木物种丰富度差异不显著；S2 和 S3、S4 之间的 Shannon-wiener 指数差异显著；S2 和 S1、S3、S4 之间的 Simpson 指数差异显著；S2 和 S1、S3、S4、S5 之间的均匀度指数差异显著(P<0.05)。

2.1.2 灌木层植物多样性比较

不同演替阶段次生林，灌木层物种丰富度呈现先降低后升高再降低的状态，但变化十分缓慢，演替前期没有变化，中期升高，后期降低；Shannon-wiener 指数、Simpson 指数、J_{sw} 均匀度指数均呈现先降低后升高再降低的规律，最高均为 S4。方差分析结果显示，不同演替阶段次生林之间的灌木物种丰富度差异不显著；S4 和 S1、S2、S3、S5 之间的 Shannon-wiener 指数、Simpson 指数差异显著；S2 和 S1、S3、S4 之间的 Simpson 指数差异显著；S2、S5 和 S1、S3、S4 之间的均匀度指数差异显著(P<0.05)。

2.1.3 草本层植物多样性比较

不同演替阶段次生林，草本层物种丰富度呈现早期保持不变，中期升高，后期降低的趋势，演替进行到S5时，草本呈现消亡的状态；Shannon-wiener指数、Simpson指数、J_{sw}均匀度指数均呈现先降低后升高再降低的规律，最高均为S4。方差分析结果显示，不同演替阶段次生林之间的物种丰富度差异不显著；S2、S5分别其他阶段之间的Shannon-wiener指数、Simpson指数、均匀度指数差异显著($P<0.05$)。

2.2 不同演替阶段次生林主要物种重要值

不同演替阶段次生林主要植物物种的重要值如表3所示。次生林群落的主要植物种类组成随着其演替进程发生明显变化。在恢复前期的30年内，乔木层主要由山鸡椒、山黄麻等阳性树种组成，恢复到40年后，乔木层主要由西南木荷、鸭脚木、南酸枣等较耐阴的树种组成；灌木层的变化亦是如此，恢复初期由酸藤子(*Embelia laeta*)、三椏苦(*Evodia lepta*)组成，到后期则由耐阴性较好的柏拉木、九节、罗伞树等占据主要优势；草本层的变化十分明显，恢复前期一直由一些耐阴性较好的蕨类组成，如乌毛蕨、薄叶卷柏(*Selaginella delicatula*)等，到后期，当灌木层的柏拉木占据优势，林下郁闭度大大增加，草本层植物完全消失。

表3 不同演替阶段次生林植物物种重要值

样地号	乔木		灌木		草本	
	种名	重要值(%)	种名	重要值(%)	种名	重要值(%)
S1	山鸡椒	57.33	酸藤子	90.00	砂仁	126.67
	白楸	39.01	猴耳环	70.00	阔片短肠蕨	103.33
	山黄麻	29.14	三椏苦	70.00	山菅兰	70.00
	千年桐	25.98	白楸	70.00		
	中平树	24.98				
S2	中平树	93.46	柏拉木	135.22	薄叶卷柏	157.14
	白楸	86.17	杜茎山	86.61	半边旗	84.13
	山鸡椒	22.44	罗伞树	42.03	乌毛蕨	58.73
	润楠	11.96	青藤公	36.15		
	橄榄	9.08				
S3	西南木荷	50.05	润楠	97.26	乌毛蕨	132.20
	鸭脚木	43.15	猴耳环	93.39	山菅兰	86.36
	白楸	28.52	鸭脚木	61.13	粽叶芦	81.44
	山乌桕	28.47	罗伞树	48.23		
	罗浮柿	26.49				
S4	大叶栎	38.68	九节	53.62	铁芒萁	95.79
	厚鳞柯	33.10	罗伞树	41.12	乌毛蕨	52.97
	西南木荷	30.47	鸭脚木	39.40	薄叶卷柏	50.03
	鸭脚木	30.42	罗浮柿	39.40	山菅兰	39.16
	润楠	25.18			淡竹叶	35.30
S5	黄毛榕	52.10	柏拉木	207.87		
	南酸枣	44.71	九节	49.25		
	橄榄	33.11	罗伞树	42.89		
	鸭脚木	21.21				
	白楸	20.70				

2.3 不同演替阶段次生林乔木生物量

不同演替阶段次生林乔木生物量如表4所示。乔木层的生物量随着演替的进行，呈现先升高后降低再升高再降低的趋势。其中，当次生林演替进行到30年时，其乔木生物量最大，为296.91t/hm^2。不同演替阶段，树叶生物量表现为，S2>S3>S5>S4>S1；树枝、树干、树根生物量均表现为，S2>S4>S3>S5>S1。方差分析结果显示，不同演替阶段不同组分生物量差异不显著；乔木总的生物量差异不显著(P>0.05)。

表4 不同演替阶段次生林乔木生物量

样地号	树叶 (t/hm^2)	树枝 (t/hm^2)	树干 (t/hm^2)	树根 (t/hm^2)	总和 (t/hm^2)
S1	2.58a	12.11a	40.55a	9.72a	64.96a
S2	4.35a	35.48a	232.52a	24.56a	296.91a
S3	4.20a	23.52a	91.16a	18.22a	137.11a
S4	3.63a	24.40a	105.10a	18.32a	151.45a
S5	3.87a	23.23a	91.00a	17.86a	135.96a

注：同一列相同字母表示差异不显著(P<0.05)。

3 结论与讨论

本次研究共调查植物69种，隶属于39科55属，其中，乔木层植物54种，隶属于29科43属；灌木层植物9种，隶属于8科9属；草本层植物10种，隶属于8科9属。不同演替阶段次生林植物物种多样性表现为乔木层>灌木层>草本层，物种多样性随着演替的进行呈先降低再升高再降低的波动状态，乔、灌、草各层物种组成也发生相应的变化。植物群落的稳定性与植物多样性密切相关[35-37]，次生林乔木种类较丰富，灌草层的物种相对单一，说明，次生林乔木层植物在维持群落的稳定性方面相对灌草层植物发挥更大作用。

余作岳[38]研究表明，随着林龄的增加，地带性物种会陆续侵入群落，其物种数目和个体数量也会持续增加，森林的林下植被表现为先发展后消退的过程。本研究与其相符：在次生林演替初期，由于群落的优势种主要集中于少数几个先锋物种，如山鸡椒、山黄麻、白楸等，因而群落的多样性指数与均匀度指数均较低，当演替进行到55年，群落内乔木层、灌木层优势种的优势度降低，因此，多样性指数与均匀度相对增加。群落中不同层次植物多样性的变化是不同步的，植物种类的更替时间也不一样，演替过程中，多样性变化幅度较大的为草本层，其次为灌木层，这与灌草层植物容易受到群落微环境影响的生理特性有关。本研究发现，不同演替阶段次生林的草本层、灌木层与乔木层植物多样性达到最大的时间均为55年，当演替进行到65年时，林下耐阴灌木如柏拉木、罗伞树等占据优势，林下郁闭度大大增加，草本层呈现消亡的状态。

生物量可以直接反映森林结构和功能的变化，对生物量的研究有助于更能了解林分的结构、物质的累积等规律[39]。不同演替阶段不同组分生物量之间差异不显著(P>0.05)，当次生林恢复到30年时，其乔木生物量最大，为296.91t/hm^2，当次生林恢复到65年时，乔木生物量降低为135.96t/hm^2。调查发现，当演替进行到30年时，乔木的密度最大(1388株/hm^2)，随着演替的进行，乔木密度开始急剧下降，演替进行到40年后，其乔木密度大幅度下降(594株/hm^2)，所以乔木层生物量呈现先升高后降低的趋势，说明不同演替阶段次生林乔木生物量与活立木的密度密切相关，这与李高飞等[40]对中国不同气候带各类型森林的生物量和净第一性生产力的研究结论一致。

结合上述研究结果，为了林下植被的恢复和林木自身的生长，当次生林恢复到30年时，建议进行适当间伐，保持合理的密度。演替本身是一个长期动态的过程，而本文仅对次生林自然恢复的前65年进行研究，且采用的是空间代替时间的方法，想要真正要弄清楚次生林的演替过程，更加深入了解次生林演替过程中群落结构、物种多样性和生物量的变化规律，需要通过对同一林分进行长期的数据积累与分析。对于具体如何调节次生林的林分郁闭度，使其物种多样性指数更高、生物量更大，森林生态效能更充分地发挥，应是未来的研究方向。

参考文献

[1] SALDARRIAGA J G. Evaluation of regression models for above-ground biomass determination in Amazon rainforest [J]. Journal of Tropical Ecology, 1994, 10: 207-218.
[2] GARKOTI S C. Estimates of biomass and primary productivity in a high-altitude maple forest of the west central Himalayas [J]. Ecological Research, 2008, 23: 41-49.
[3] 温远光，陈放，刘世荣，等. 广西桉树人工林物种多样性与生物量关系[J]. 林业科学，2008，44(4)：14-19.
[4] KUME A, SATOMURA T, TSUBOI N, et al. Effects of understory vegetation on the ecophysiological characteristics of an overstory pine, Pinus densiflora[J]. Forest Ecology

& Management，2003，176：195-203.

[5]CHASTAIN R A，CURRIE W S，TOWNSEND P A. Carbon sequestration and nutrient cycling implications of the evergreen understory layer in Appalachian forests [J]. Forest Ecology & Management，2006，231：63-77.

[6]GILLIAM F S. The Ecological Significance of the Herbaceous Layer in Temperate Forest Ecosystems [J]. Bioscience，2007，57：845-858.

[7]BENGTSSON J. Which species? What kind of diversity? Which ecosystem function? Some problems in studies of relations between biodiversity and ecosystem function [J]. Applied Soil Ecology，1998，10：191-199.

[8]HEDLUND K，SANTA REGINA I，VANDER PYTTEN W H，et al. Plant species diversity，plant biomass and responses of the soil community on abandoned land across Europe：idiosyncracy or above-belowground time lags[J]. Oikos，2003，103：45-58.

[9]CARDINAIL B J，WRIGHT J P，CADOTTE M W，et al. Impacts of plant diversity on Biomass production increase through time because of species complementarity[J]. Proceedings of the national academy of sciences，2007，104：18123-18128.

[10]CARDINAIL B J，VENAIL P，GROSS K，et al. Further re-analyses looking for effects of phylogenetic diversity on community biomass and stability[J]. Functional Ecology，2015，29：1607-1610.

[11]YUAN F，WU J，LI A，et al. Spatial patterns of soil nutrients，plant diversity，and aboveground biomass in the Inner Mongolia grassland：Before and after a biodiversity removal experiment[J]. Landscape Ecology，2015，30：1737-1750.

[12]王斌，杨校生.4种典型地带性植被生物量与物种多样性比较[J]. 福建林学院学报，2009，29：345-350.

[13]FOX J W. The long-term relationship between plant diversity and total plant biomass depends on the mechanism maintaining diversity[J]. Oikos，2003，102：630-640.

[14]贾亚运，周丽丽，吴鹏飞，等. 不同发育阶段杉木人工林林下植被的多样性[J]. 森林与环境学报，2016，36(01)：36-41.

[15]BECKNELL J M，POWERS J S. Stand age and soils as drivers of plant functional traits and aboveground biomass in secondary tropical dry forest[J]. Canadian Journal of Forest Research，2014，44：604-613.

[16]LASKY J R，URIARTE M，BOUKILI V K，et al. The relationship between tree biodiversity and biomass dynamics changes with tropical forest succession[J]. Ecology letters，2014，17：1158-1167.

[17]SPURR S H. Origin of the concept of forest succession[J]. Ecology，1952，33：426-427.

[18]辛晓平，单保庆. 恢复演替中草地斑块动态及尺度转换分析[J]. 生态学报，2000，20(4)：587-593.

[19]王鑫厅，王炜，梁存柱. 典型草原退化群落不同恢复演替阶段羊草种群空间格局的比较[J]. 植物生态学报，2009，33(1)：63-70.

[20]李立新，陈英智. 退化演替对高山草地植被和土壤理化特性影响[J]. 水土保持通报，2015，35(4)：40-42.

[21]张庆费，由文辉，宋永昌. 浙江天童植物群落演替对土壤化学性质的影响[J]. 应用生态学报，1999，10(1)：19-22.

[22]张德强，叶万辉，余清发，等. 鼎湖山演替系列中代表性森林凋落物研究. 生态学报[J]，2000，20(6)：938-944.

[23]梁国华，吴建平，熊鑫，等. 南亚热带不同演替阶段森林土壤呼吸对模拟酸雨的响应[J]. 生态学杂志，2016，35(1)：125-134.

[24]张雪，韩士杰，王树起，等. 长白山白桦林不同演替阶段土壤有机碳组分的变化[J]. 生态学杂志，2016，35(2)：282-289.

[25]彭艳，李心清，程建中，等. 贵阳次生林不同演替阶段土壤微生物生物量与反硝化酶活性的研究[J]. 地球化学，2009，38(6)：600-606.

[26]张梦弢，张青，亢新刚，等. 长白山云冷杉林不同演替阶段群落稳定性[J]. 应用生态学报，2015，26(6)：1609-1616.

[27]徐伟强，周璋，赵厚本，等. 南亚热带3种常绿阔叶次生林的生物量结构和固碳现状[J]. 生态环境学报，2015，24(12)：1938-1943.

[28]朱宏光，蓝嘉川，刘虹，等. 广西马山岩溶次生林群落生物量和碳储量[J]. 生态学报，2015，35(8)：2616-2621.

[29]马克平，刘玉明. 生物群落多样性的测度方法：Ⅰα多样性的测度方法(下)[J]. 生物多样性，1994，3(3)：162-168.

[30]刘灿然，马克平. 生物群落多样性的测度方法 V. 生物群落物种数目的估计方法[J]. 生态学报，1997，17(6)：601-610.

[31]宋永昌. 植被生态学[M]. 上海：华东师范大学出版社，2001.

[32]刘琪璟. 嵌套式回归建立树木生物量模型[J]. 植物生态学报，2009，33(2)：331-337.

[33]汪金松，张春雨，范秀华，等. 臭冷杉生物量分配格局及异速生长模型[J]. 生态学报，2011，31(14)：3918-3927.

[34]汪珍川，杜虎，宋同清，等. 广西主要树种(组)异速生长模型及森林生物量特征[J]. 生态学报，2015，35(13)：4462-4472.

[35]TILMAN D，DOWNING J A. Biodiversity and Stability in

Grasslands[J]. Nature, 1994, 367: 363-365.

[36] LEHMAN C L, TILMAN D. Biodiversity, Stability, and Productivity in Competitive Communities [J]. American Naturalist, 2000, 156: 534-552.

[37] PFISTERER A B, JOSHI J, SCHMID B, et al. Rapid decay of diversity-productivity relationships after invasion of experimental plant communities [J]. Basic & Applied Ecology, 2004, 5: 5-14.

[38] 余作岳. 热带亚热带退化生态系统植被恢复生态学研究[M]. 广东: 广东科技出版社, 1996.

[39] 黄石嘉, 李铁华, 文仕知, 等. 青冈栎的生长规律及生物量分布格局研究[J]. 中南林业科技大学学报, 2017, 37(3): 57-62.

[40] 李高飞, 任海. 中国不同气候带各类型森林的生物量和净第一性生产力[J]. 热带地理, 2004, 24(4): 306-310.

[原载: 中南林业科技大学学报, 2018, 38(11)]

桂西北喀斯特山区4种森林表土土壤有机碳含量及其养分分布特征

庞圣江　杨保国　刘士玲　刘福妹　张　培　黄柏华

（中国林业科学研究院热带林业实验中心，广西凭祥　532600）

摘　要　研究桂西北喀斯特山区4种森林植被类型表层(0~20cm)土壤有机碳含量和土壤养分特征，与国内喀斯特土壤有机碳的研究成果进行比较分析，为该区域天然林管护、植被恢复和生态重建提供参考。结果表明：4种森林植被类型表土SOC含量表现为：常绿阔叶林(151.31g/kg)>常绿落叶阔叶林(145.33g/kg)>针阔叶混交林(90.61g/kg)>灌草丛(86.92g/kg)；其中，常绿阔叶林、常绿落叶阔叶林与针阔叶混交林表层SOC含量差异不显著($P>0.05$)，而与灌草丛之间差异显著($P<0.05$)。土壤SOC含量与速效N、速效K和全N呈显著或极显著正相关，表明土壤氮素积累与土壤有机碳的输入关系密切。表土SOC含量明显高于该研究区域内耕地(19.36g/kg)、退耕还草(15.31g/kg)和退耕还林地(18.89g/kg)以及受干扰林地(46.14g/kg)等不同土地利用类型，也高于贵州开阳和贵州关岭等地区森林植被类型，与云南曲靖不同森林群落土壤SOC含量也存在一定的差异。

关键词　土壤有机碳；土壤养分；植被类型；森林；喀斯特山区

The Distribution of Organic Carbon and Soil Nutrients Under Four Forest Types in Karst Mountain Areas of Northwest Guangxi, China

PANG Shengjiang, YANG Baoguo, LIU Shiling, LIU Fumei, ZHANG Pei, HUANG Bohua

(*Experimental Center of Tropical Forestry*, *CAF*, *Pingxiang* 536000, *Guangxi*, *China*)

Abstract: This paper studied the variation characteristics of surface layer soil organic carbon(SOC) content and soil nutrient, comparative analysis of soil organic carbon under different vegetation types in Karst mountain areas, provided reference for the protection, vegetation restoration and ecological construction to the natural forest. The results showed that the average SOC content of the topsoil are characterized by evergreen broad-leaved forest (151.31g/kg) > evergreen deciduous broad-leaved forest (145.33g/kg) > coniferous and broadleaved mixed forest (90.61g/kg) > shrub-grassland (86.92g/kg), But statistically there was no significant difference between the four forest types, were significantly correlated with SOC of Shrub-grassland. The content of soil organic carbon and total N, total P, available N, available K showed significant or extremely significant positive correlation, interpretation there was closely related to soil nitrogen accumulation and soil SOC input. In this study, the content of SOC under different forest types was significantly higher than the cultivated land(19.36g/kg), replacing cropland with pasture(15.31g/kg), conversion of cropland to forest(18.89g/kg) and the disturbed forest(46.14g/kg) in Karst mountains area, there were existed some differences among the content of SOC with other places forest types.

Key words: soil organic carbon; soil nutrient; vegetation types; forest; karst mountain areas

土壤有机碳(Soil organic carbon)是土壤碳库的重要组成部分，在调节森林土壤理化性质、提供植物养分、改善土壤结构和生态性状等方面产生深刻影响[1]。土壤有机碳也是土壤微生物主要营养元素和能量来源，它能够促进土壤生态系统中C、N和P等养分化合物的转化和循环，影响土壤性质以及养分的供给能力[2,3]。由于森林土壤碳库约占全球土壤碳库73%[4]，其在全球碳循环过程中起着不可代替的作用；森林植被与土壤有机碳关系密切，但不同植被类型土壤有机碳间却存在较大的差异，主要受气候、植被类型、林分状况、土壤类型和研究方法等方面的影响[5]，对不同森林类型土壤 SOC 含量与养分特征加以研究显得尤为必要，为森林土壤质量评价、土壤有机碳管理以及应对全球气候变化均具有重要意义。

我国是喀斯特分布面积最大的国家，主要集中于广西、贵州和云南等地，属于典型的生态脆弱区[6]。自然条件下，喀斯特山区成土速率慢、土层较为浅薄、土壤持水能力差[7]；因此，该地区森林植被一旦遭受破坏，土壤流失严重，容易出现石漠化，进而加剧喀斯特地区生态环境的恶化。截至目前，国内学者围绕喀斯特山区森林植被和土壤开展了大量研究，比如，在不同土地利用方式对土壤水分、容重以及有机碳空间异质性[8-10]等方面进行研究探索，这些研究成果有效地遏制喀斯特山区植被迅速退化、石漠化，并对当地生态环境的改善起到了积极作用。不过，以往研究多侧重于植被生长或者森林植被受干扰对土壤物理性质及土壤有机碳稳定性的研究[7,11,12]，而二者的关联纽带却没有引起足够重视；在全球气候变化背景下，通过野外实测获得更多土壤基础数据，也有利于精确地评价喀斯特山区森林植被类型土壤的碳汇功能。

基于上述喀斯特山区森林群落土壤有机碳的研究现状，本文作者拟开展桂西北喀斯特山区 4 种主要森林类型（灌草丛、针阔叶混交林、常绿落叶阔叶林、常绿阔叶林）表层土壤有机碳及养分特征研究，比较分析其表土 SOC 含量和肥力变化，并与国内喀斯特山区不同森林植被类型表土 SOC 含量的研究成果进行比较分析，为该区域天然林管护、生态系统植被恢复与生态重建提供依据。

1 材料与方法

1.1 研究区概况

研究区位于广西西北部乐业县雅长林区，地理坐标为 106°16′~106°22′E，24°42′~24°56′N。该区域位于云贵高原东南麓，属于我国西南典型的喀斯特地区。研究区域属南亚热带季风气候区，无霜期较长，年平均温度 18℃左右，极端高温 35℃，极端低温-5℃，年均降水量 1100~1500mm，但季节分布不均匀，主要集中在 4~9 月。土壤类型为石灰岩发育的棕色石灰土，土壤厚度 10~30cm，土层浅薄、土地植被不连片，但土壤有机质含量较高。研究区域的植被类型属于南亚热带原生性喀斯特森林，本研究选取的森林植被类型包括灌草丛，针阔叶混交林，常绿-落叶阔叶林，常绿阔叶林。4 种森林植被群落概况，详见表 1。

表 1 不同植被类型的样地基本概况

森林类型	优势种群	海拔(m)	坡度(°)	坡向	坡位	郁闭度	土壤类型
灌草丛（Ⅰ）	1 鼠刺、草决明、艾蒿、胜红蓟	870	20	WN	中上	0.2	棕色石灰土
	2 石岩枫、余甘子、扁穗莎草、珍珠茅	820	27	WN	中	0.3	棕色石灰土
	3 盐肤木、鼠刺、山蚂蝗、五节芒	1258	26	E	上	0.3	棕色石灰土
	4 化香、野牡丹、五节芒、白茅	1074	20	W	中上	0.2	棕色石灰土
	5 盐肤木、化香、雀梅、珍珠茅	930	30	E	中上	0.3	棕色石灰土
针阔混交林（Ⅱ）	1 细叶云南松、青冈栎、水锦树、麻栎	988	40	S	中上	0.7	棕色石灰土
	2 细叶云南松、栓皮栎、水锦树、槲栎	732	20	EN	中	0.9	棕色石灰土
	3 细叶云南松、栓皮栎、青冈栎、山合欢	795	25	W	中	0.8	棕色石灰土
	4 铁坚油杉、栓皮栎、青冈栎、化香	862	26	EN	中	0.8	棕色石灰土
	5 铁坚油杉、青冈栎、鹅耳枥、盐肤木	915	33	ES	上	0.9	棕色石灰土
	6 铁坚油杉、青冈栎、鹅耳枥、化香	730	30	WN	中	0.9	棕色石灰土
常绿-落叶阔叶林(Ⅲ)	1 青冈栎、酸枣、密榴木、柳叶润楠	951	42	WN	中	0.9	棕色石灰土
	2 鹅耳枥、朴树、密花树、化香	1107	40	WN	中	0.9	棕色石灰土
	3 青冈栎、光皮桦、朴树、千年桐	1307	35	ES	中	0.9	棕色石灰土
	4 蒙自桤木、光皮桦、枫香、云南樱桃	1315	30	WN	中上	0.9	棕色石灰土
	5 蒙自桤木、朴树、光皮桦、毛红椿	1391	40	ES	上	0.8	棕色石灰土
	6 润楠、朴树、枫香、云南樱桃	1550	30	E	上	0.9	棕色石灰土

（续）

森林类型	优势种群	海拔(m)	坡度(°)	坡向	坡位	郁闭度	土壤类型
常绿阔叶林（Ⅳ）	1 朴树、润楠、阴香、大果山香圆	1048	25	WN	中	0.9	棕色石灰土
	2 青冈栎、铁榄、乌冈栎、密花树	1080	35	WN	中	0.9	棕色石灰土
	3 青冈栎、朴树、密花树、化香	778	34	ES	中	0.8	棕色石灰土
	4 青冈栎、铁榄、密花树、化香	1105	31	WS	中上	0.8	棕色石灰土
	5 青冈栎、西南八角枫、潺槁木、川桂	1204	30	WN	上	0.9	棕色石灰土
	6 青冈栎、朴树、润楠、铁榄	1123	37	WN	中上	0.8	棕色石灰土

注：鼠刺(*Itea chinensis*)、草决明(*Celosia argentea*)、艾蒿(*Artemisia lavandulifolia*)、胜红蓟(*Ageratum conyzoides*)、石岩枫(*Mallotus repandus*)、余甘子(*Phyllanthus emblica*)、扁穗莎草(*Cyperus compressus*)、珍珠茅(*Scleria levis*)、盐肤木(*Rhus chinensis*)、山蚂蝗(*Desmodium oxyphyllum*)、五节芒(*Miscanthus floridulus*)、化香(*Platycarya strobilacea*)、野牡丹(*Melastoma malabathricum*)、白茅(*Imperata cylindrica*)、雀梅(*Sageretia thea*)、细叶云南松(*Pinus yunnanensis* var. *tenuifolia*)、青冈栎(*Cyclobalanopsis glaucoides*)、水锦树(*Wendlandia tinctoria* subsp. *intermedia*)、麻栎(*Quercus acutissima*)、栓皮栎(*Quercus variabilis*)、槲栎(*Quercus aliena*)、山合欢(*Albizia odoratissima*)、铁坚油杉(*Keteleeria davidiana*)、鹅耳枥(*Carpinus turczaninowii*)、酸枣(*Choerospondias axillaris*)、密榴木(*Miliusa balansae*)、柳叶润楠(*Machilus salicina*)、朴树(*Celtis sinensis*)、密花树(*Rapanea neriifolia*)、光皮桦(*Betula luminifera*)、千年桐(*Vernicia montana*)、蒙自桤木(*Alnus nepalensis*)、枫香(*Liquidambar formosana*)、云南樱桃(*Cerasus yunnanensis*)、毛红椿(*Toona ciliata* Roem. var. *pubescens*)、润楠(*Machilus nanmu*)、阴香(*Cinnamomum burmannii*)、大果山香圆(*Turpinia pomifera*)、铁榄(*Sinosideroxylon pedunculatum*)、乌冈栎(*Quercus phillyraeoides*)、西南八角枫(*Alangium faberi*)、潺槁木(*Litsea glutinosa*)、川桂(*Cinnamomum wilsonii*)。

1.2　土壤样品采集

2012 年 9~11 月，在研究区域踏查的基础上，根据植被代表性和典型性原则，本研究选取 4 种不同森林类型，每种森林类型分别设置面积 20m×30m 标准样地，其中，灌草丛 5 个，针阔混交林 6 个，常绿—落叶阔叶林 6 个，常绿阔叶林 6 个，共计 23 个。用网格划分法将样地划分为 6 个 10m×10m 乔木样方，调查胸径(*DBH*)≥1cm 的乔木，记录乔木种类、株高、胸径(对株高<1.3m 的幼树仅测树高)、冠幅和生长势等。每个标准样地内分别于样地四角及中心适当位置设置采样点 5 个，每个采样点按三角形取 3 个土样混合均匀，并取土壤环刀；土壤采集时除去表层土上的凋落物和植物，取样深度为 0~20cm，将所取样品风干，研磨并标号记录装入密封塑料袋，组成待测样品。灌草丛 25 个，针阔混交林 30 个，常绿—落叶阔叶林 30 个，常绿阔叶林 30 个，共采集 115 份土壤样品。

1.3　样品测试与分析

将土样剔除石粒、动植物残体和根系等杂物，风干后研磨并过筛，装自封袋待用。采用环刀法测定土壤密度和烘干法(105℃，12h)测定土壤含水量《LY/1215-1999》；土壤 pH 用酸度计测定《LY/1239-1999》；有机碳含量采用 $K_2Cr_2O_7$ 氧化-外加热法测定《LY/1237-1999》，速效 N 用碱解扩散法测定《LY/1229-1999》，速效 P 采用 HCI-H_2SO_4 浸提-钼锑抗比色法测定《LY/1233-1999》，速效 K 采用 CH_3COONH_4 浸提-火焰光度计法测定《LY/1236-1999》；全 N 含量采用浓 H_2SO_4-$HClO_4$ 消化法-自动凯氏定氮仪(KDY-9830，KETUO)测定《LY/1228-1999》，全 P 含量采用 NaOH 熔融-钼锑抗比色-紫外分光光度法测定《LY/1232-1999》，全 K 采用 NaOH 熔融-火焰光度计法测定《LY/1234-1999》[13]。每个样品重复测定 3 次取均值。

采用 Excel 2007 对土壤数据进行初步整理，利用 SPSS16.0 对土壤数据进行统计分析，单因素方差分析(one-way ANOVA)以及差异性显著性检验[11]，采用 Pearson 相关性进行相关关系分析。

2　结果与分析

2.1　不同植被类型表层土壤有机碳及养分特征

4 种森林类型表土(0~20cm)SOC 含量的测定结果如表 2 所示。从表 2 可以看出，表土 SOC 含量大小依次为：常绿阔叶林(151.31g/kg)>常绿落叶阔叶林(145.33g/kg)>针阔叶混交林(90.61g/kg)>灌草丛(86.92g/kg)。其中，常绿阔叶林、常绿落叶阔叶林与针阔叶混交林表层土壤有机碳含量差异不显著($P>0.05$)，而与灌草丛之间差异显著($P<0.05$)；由于植物群落结构与物种组成差异，在植被—土壤生态系统相互作用下，立地因子产生较大差异，从而对相应的森林植被类型表层土壤有机碳状况造成影响。

由表 2 可知，4 种森林类型表层土壤养分中速效

N、速效P、速效K和全N质量分数与土壤SOC含量的变化趋势一致，大小顺序依次为：常绿阔叶林>常绿-落叶阔叶林>针阔叶混交林>灌草丛。其中，常绿-落叶阔叶林和针阔叶混交林土壤速效N含量差异不显著($P>0.05$)，与其他2种森林类型差异显著($P<0.05$)。常绿阔叶林和常绿—落叶阔叶林速效P含量差异不显著($P>0.05$)，与其他2种森林类型差异显著($P<0.05$)，而常绿-落叶阔叶林与针阔叶混交林差异亦不显著($P>0.05$)。灌草丛与其他3种森林类型的土壤速效K和全N差异显著($P<0.05$)，而后者之间差异不显著($P>0.05$)。土壤全P和全K的质量分数变化趋势均表现为：常绿阔叶林>针阔叶混交林>常绿-落叶阔叶林>灌草丛；其中，灌草丛与其他3种类型的土壤全K含量差异显著($P<0.05$)，而后者之间差异不显著($P>0.05$)，4种森林类型的表层土壤全P含量差异不显著($P>0.05$)。

表2　不同植被类型土壤有机碳含量及养分特征

植被类型	样地代码	SOC (g/kg)	速效N (mg/kg)	速效P (mg/kg)	速效K (mg/kg)	全N (g/kg)	全P (g/kg)	全K (g/kg)
	1	66.74±13.46	167.9±13.99	4.9±0.28	29.7±0.51	2.97±0.39	1.74±0.08	8.38±0.65
	2	103.49±27.87	215.9±17.92	2.8±0.13	27.7±1.18	2.14±0.20	1.96±0.12	9.16±0.72
Ⅰ	3	107.99±24.28	195.3±11.04	3.6±0.29	39.6±1.53	3.26±0.93	1.62±0.14	9.47±1.08
	4	112.78±25.89	108.9±31.67	4.1±0.37	48.4±0.96	3.67±0.08	2.07±0.09	8.94±0.82
	5	43.58±10.06	184.1±18.16	3.3±0.17	45.1±1.85	3.15±0.07	1.72±0.26a	10.64±0.63
	均值	86.92±30.30a	174.42±40.58a	3.74±0.80a	38.10±9.17a	3.04±0.56a	1.82±0.19a	9.28±0.96a
	1	73.91±29.58	202.6±18.10	4.3±0.11	85.4±0.77	4.11±0.53	1.87±0.17	8.03±0.33
	2	72.44±27.12	291.4±49.66	7.5±0.56	80.6±5.30	5.46±0.23	1.73±0.32	11.09±0.27
	3	86.56±20.45	275.2±45.29	4.3±0.24	87.5±2.96	5.01±0.16	2.01±0.60	9.34±1.17
Ⅱ	4	104.92±10.88	372.4±21.43	5.3±0.33	90.1±2.12	4.42±0.46	2.12±0.08	13.12±0.88
	5	52.67±43.96	218.8±11.55	6.7±0.46	105.3±1.07	3.99±0.24	2.68±0.10	12.82±1.11
	6	153.16±54.38	347.6±12.04	5.2±0.08	95.4±2.85	4.39±0.13	2.46±0.18	12.48±0.93
	均值	90.61±35.16ab	284.53±67.78ab	5.55±1.30ab	90.72±8.67b	4.56±0.56ab	2.15±0.36a	11.15±2.07b
	1	110.99±31.79	231.9±25.27	8.4±0.64	124.4±1.21	4.15±0.18	2.04±0.09	10.34±0.98
	2	97.30±59.05	317.6±18.48	6.7±0.40	90.6±4.56	4.04±0.15	1.94±0.05	13.41±1.52
	3	186.97±54.77	554.2±74.33	10.2±0.42	82.2±3.49	5.49±0.27	1.61±0.12	9.87±0.76
Ⅲ	4	142.85±6.32	418.2±35.72	9.3±0.21b	111.2±3.37	5.37±0.43	2.23±0.27	9.62±1.04
	5	156.87±11.52	391.7±16.95	6.6±0.17	107.2±1.60	4.43±0.67	1.77±0.16	12.93±0.51
	6	177.02±32.81	347.8±14.92	7.8±0.66	94.8±2.69	5.38±0.20	2.48±0.35	10.24±0.69
	均值	145.33±35.68bc	376.90±108.40bc	8.17±1.43bc	101.73±15.41bc	4.81±0.67b	2.01±0.31a	11.07±1.66b
	1	188.42±26.68	445.2±39.64	10.7±0.12	130.9±11.04	4.55±0.47	2.32±0.41	14.57±0.64
	2	140.04±18.72	741.9±52.87	9.1±0.28	157.4±7.95	5.76±1.79	2.14±0.11	11.14±1.25
	3	87.42±75.88	646.9±25.11	9.3±0.03	84.7±6.63	5.29±0.22	1.93±0.04	13.35±1.19
Ⅳ	4	195.71±37.39	549.7±45.39	14.8±0.19	95.4±0.97	4.32±0.68	1.73±0.29	9.78±0.72
	5	128.47±21.84	343.7±21.98	12.9±0.63	82.6±1.54	4.84±0.41	1.85±0.40	8.67±0.17
	6	167.80±16.61	627.9±64.34	11.5±0.34	97.5±1.30	4.73±0.08	2.23±0.13	10.49±1.46
	均值	151.31±40.87bc	559.22±145.13c	11.38±2.19cd	108.08±29.74bc	4.92±0.53b	2.03±0.23a	11.33±2.23b

注：表中数据为平均值±标准差，同列中不同植被类型之间，用不同字母表示存在显著差异($P<0.05$，ANOVA，$df=5$)。

2.2 不同植被类型土壤有机碳与土壤养分相关关系

森林植被类型因植物种类组成、植被盖度状况、群落结构等差异，导致土壤理化性质发生变化，进而影响土壤养分含量及其变异性。从表3可以看出，桂西北喀斯特山区4森林植被类型SOC含量与速效N、全N存在极显著正相关($P<0.01$)；灌草丛SOC含量与速效P、全P呈正相关关系($P<0.05$)；4种森林植被类型SOC含量与速效K呈显著正相关($P<0.05$)，而针阔叶混交林和常绿—落叶阔叶林SOC含量与全K呈负相关关系($P<0.05$)。

不同森林类型表层土壤SOC含量与速效N、速效K和全N呈显著或极显著正相关，表明土壤氮素积累与土壤SOC的输入关系密切。以往研究认为，土壤养分的变化可通过影响微生物活动、凋落物分解和有机碳组分以及有机碳的矿化速率来影响土壤SOC含量，特别是土壤氮素的富集可有效地促进森林凋落物的形成和林下植被生物量的积累，从而有利于SOC积累[14]。

表3　土壤有机碳含量与土壤养分的相关分析

植被类型	速效N	速效P	速效K	全N	全P	全K
灌草丛	0.714**	0.612*	0.907*	0.836**	0.582*	-0.604
针阔叶混交林	0.856**	0.222	0.811**	0.868**	0.802	-0.449*
常绿-落叶阔叶林	0.742**	0.831	0.850**	0.940**	0.595	-0.680*
常绿阔叶林	0.929**	0.754	0.912**	0.745**	0.421	-0.822

注：** 表示极显著相关($P<0.01$)，* 表示显著相关($P<0.05$)。

3　结论与讨论

土壤有机碳主要来源于枯落物、根系分泌物和细根分解产生的碎屑[15]，而枯落物与根系分泌物及其被微生物分解形成的有机碳首先进入表层土壤，使得林地土壤SOC含量具有明显的“表聚作用”[6]。研究表明，桂西北喀斯特山区4种植被类型的表土具有丰富的土壤有机碳，但不同植被类型土壤SOC含量存在较大差异，土壤SOC含量大小依次为：常绿阔叶林(151.31g/kg)>常绿落叶阔叶林(145.33g/kg)>针阔叶混交林(90.61g/kg)>灌草丛(86.92g/kg)；其中，常绿阔叶林、常绿落叶阔叶林与针阔叶混交林表层土壤有机碳含量差异不显著($P>0.05$)，而与灌草丛之间差异显著($P<0.05$)；这与原生阔叶林土壤SOC含量显著高于针阔叶混交林与灌草丛[16-18]，与本研究结果基本一致。调查亦发现，灌草丛的植被建群种以灌木和草本为主，土壤SOC含量主要来源是少量地表枯落物和残根，不利于有机碳形成和积累[19]；针叶混交林，树种增多，地表枯落物包含有阔叶树种和较难被微生物分解利用硬质针叶[20]，细根生物量增加[21]，使得土壤SOC含量的积累逐渐增多；而阔叶林(常绿—落叶阔叶林和常绿阔叶林)，地表枯落物以枯枝阔叶为主，且细根分布密集，质地松软，易于微生物分解，有利于表土土壤有机碳的长期积累，并维持稳定的土壤碳循环，因此，明显高于其他2种植被类型土壤SOC含量，表明该地区植被类型越是复杂，其表土土壤SOC含量越高。

研究结果显示，常绿阔叶林和常绿—落叶阔叶林这2种植被类型，土壤SOC含量增幅不明显，且差异不显著。究其原因，其一，常绿—落叶阔叶林处于生长旺盛的阶段，林内活跃的生命活动使得枯落物量、养分含量较大(表2)，由于微生物对枯落物、细根的分解速率较快，有利于土壤SOC含量在较短时间内得到大量积累；其二，常绿阔叶林是经过漫长的演替而来，林分生长趋缓，由于林分的自疏作用，也会使枯落物量以及养分归还量有所减少，养分循环速率降低，土壤有机碳积累和周转趋于稳定状态，尽管如此，长期的原始积累仍使得常绿阔叶林表土存在着丰富的有机质和养分，这些因子的综合作用，导致常绿阔叶林与常绿-落叶阔叶林土壤SOC含量差异不显著。说明土壤SOC含量能够反映植被群落的分布格局和未来的演替趋势，并影响土壤潜在生产力[22]，因此，在特定的气候条件下，森林随着植被生长，土壤碳素形态和碳库之间会逐渐达到动态的稳定状态，可视为生态系统功能发挥的重要标志。

本研究中，从土壤有效肥力(速效N、速效P和速效K)含量大小的状况来看，常绿阔叶林>常绿—落叶阔叶林>针阔叶混交林>灌草丛，这与于扬等[23]对于喀斯特峰丛洼地不同生态系统土壤肥力变化特征的研究结论相似，他们研究认为，随着森林植被的恢复，枯枝落叶等残留物的增加，在各种微生物分解作用下，分解的有机质以及含氮化合物进入土壤，土壤有效肥力得到明显的提高。相关性分析结果显示，4种植被类型的土壤剖面中，土壤SOC含量与速效N、全N、速效K均呈极显著或显著正相关，其中与全N的相关系数最大，与速效N的相关性次之。这4种植被类型土壤SOC含量和全N、速效N表现出显著相关性，表明土壤氮素积累与土壤有机质的输入关系密切。但也有研究发现，土壤质地、土壤持水特性、温度等因素对土壤有机碳分布也有着非常重要的影响[24,25]。

本研究中，4种森林类型的土壤SOC含量与其他喀斯特山区林地土壤SOC含量相比(表4)，明显高于袁海伟等[8]对桂西北喀斯特峰丛洼地耕地(19.36g/kg)、退耕还草(15.31g/kg)和退耕还林(18.89g/kg)以及受干扰林地(46.14g/kg)等不同土地利用类型；同时，也高于贵州开阳[15]和贵州关岭[6]等森林植被类型土壤SOC含量；与云南曲靖不同森林植被群落土壤SOC含量相比[19]，也存在一定差异。说明桂西北喀斯特山区原始天然林植被与土壤养分之间达到了良好的平衡状态，建议应加强荒山荒地和残次林抚育管理，改善森林环境，增加植物物种多样性，才能保障植物、土壤养分和微生物之间良好的协调关系，促进该区域植被迅速恢复和生态重建。

由于不同植被类型土壤SOC含量的影响因素较为复杂，而且土壤SOC含量和分布，受成土母质、土壤理化性质、立地因子和植被类型等综合影响，这些因子必然会导致土壤SOC积累和周转方面产生差异；同时，研究学者所采用的研究方法、研究尺度、林分状况、土壤类型以及区域性环境因子(温度、降水量)等因素亦存在较大差异[26]。因此，不同地区的植被类型、立地条件及其利用方式的改变等，对森林土壤有机碳的影响及其机制仍有待深入研究。

参考文献

[1]徐江兵，何园球，李成亮，等．不同施肥处理红壤生物活性有机碳变化及与有机碳组分的关系[J]．土壤，2007，39(4)：627-632.

[2]HU S，COLEMAN D C，CARROLL CR，et al. Labile soil carbon pools in subtropical forest and agricultural ecosystems as influenced by management practices and vegetation types [J]. Agriculture，Ecosystems& Environment，1997，65(1)：69-78.

[3]熊红福，王世杰，容丽，等．普定喀斯特地区不同演替阶段植物群落凋落物动态[J]．生态学杂志，2013，32(4)：802-806.

[4]HOUGHTON R A. Land-use change and the carbon cycle [J]. Global Change Biology，1995，1(4)：275-287.

[5]王绍强，周成虎，李克让，等．中国土壤有机碳库及空间分布特征分析[J]．地理学报，2000，5：533-544.

[6]龙健，廖洪凯，李娟，等．基于冗余分析的典型喀斯特山区土壤-石漠化关系研究[J]．环境科学，2012，33(6)：2131-2138.

[7]魏亚伟，苏以荣，陈香碧，等．人为干扰对喀斯特土壤团聚体及其有机碳稳定性的影响[J]．应用生态学报，2011，22(4)：971-978.

[8]袁海伟，苏以荣，郑华，等．喀斯特峰丛洼地不同土地利用类型土壤有机碳和氮素分布特征[J]．生态学杂志，2007，26(10)：1579-1584.

[9]张伟，陈洪松，王克林，等．桂西北喀斯特洼地土壤有机碳和速效磷的空间变异[J]．生态学报，2007，27(12)：5168-5175.

[10]严俊华，周传燕，文安邦，等．贵州喀斯特石漠化过程中的土壤有机碳与容重关系[J]．热带南亚热带植物学报，2011，19(3)：273-278.

[11]廖洪凯，龙健．喀斯特山区不同植被类型土壤有机碳的变化[J]．应用生态学报，2011，22(9)：2253-2258.

[12]吴鹏，陈骏，崔迎春，等．茂兰喀斯特植被主要演替群落土壤有机碳研究[J]．中南林业科技大学学报，2012，32(12)：181-186.

[13]国家林业局．森林土壤分析方法[S]．北京：中国标准出版社，1999.

[14]杨金艳，王传宽．东北东部森林生态系统土壤碳贮量和碳通量[J]．生态学报，2005，25(11)：2875-2882.

[15]丁访军，潘忠松，周凤娇，等．黔中喀斯特地区3种林型土壤有机碳含量及垂直分布特征[J]．水土保持学报，2012，26(1)：161-169.

[16]苏永中，赵哈林．土壤有机碳储量、影响因素及其环境效应的研究进展[J]．中国沙漠，2002，22(3)：220-228.

[17]杨曾奖，曾杰，徐大平，等．森林枯枝落叶分解及其影响因素[J]．生态环境学报，2007，16(2)：649-654.

[18]刘聪，项文化，田大伦，等．中亚热带森林植物多样性增加导致细根生物量"超产"[J]．植物生态学报，2011，35(5)：539-550.

[19]曹向文，赵洋毅，熊好琴，等．滇东喀斯特石漠化地区不同植被模式土壤酶活性与有机碳[J]．东北林业大学学报，2015，43(11)：79-97.

[20]张剑，汪思龙，王清奎，等．不同森林植被下土壤活性有机碳含量及其季节变化[J]．中国生态农业学报，2009，17(1)：41-47.

[21]梁启鹏，余新晓，庞卓，等．不同林分土壤有机碳密度研究[J]．生态环境学报，2010，19(4)：889-893.
[22]刘聪，项文化，田大伦，等．中亚热带森林植物多样性增加导致细根生物量“超产”[J]．植物生态学报，2011，35(5)：539-550.
[23]于扬，杜虎，宋同清，等．喀斯特峰丛洼地不同生态系统的土壤肥力变化特征[J]．生态学报，2013，33(23)：7455-7466.
[24]丁正亮，王雷，刘西军，等．安徽霍山毛竹林生产力及其土壤养分的特点[J]．经济林研究，2011，29(1)：72-76.
[25]廖洪凯，龙健，李娟，等．西南地区喀斯特干热河谷地带不同植被类型下小生境土壤碳氮分布特征[J]．土壤，2012，44(3)：421-428.
[26]韦宏民，何斌，梁运，等．不同板栗-农间作模式对土壤理化性质的影响[J]．经济林研究，2014(3)：150-153.

[原载：中南林业科技大学学报，2018，38(04)]

广西大青山西南桦人工林草本优势种群生态位研究

庞圣江[1,2] 张 培[1] 杨保国[1] 刘士玲[1] 邓硕坤[1] 贾宏炎[1]
([1]中国林业科学研究院热带林业实验中心，广西凭祥 532600；
[2]广西友谊关森林生态系统国家定位观测研究站，广西凭祥 532600)

摘 要 根据野外159个样方的调查数据，采用Shannon-wiener、Levins生态位宽度和Levins生态位重叠指数，对广西大青山林区西南桦人工林草本优势种群的生态位特征进行了研究．结果表明：①5种西南桦林分中五节芒、铁线蕨的生态位宽度较大，Shannon-wiener与Levins指数分别为2.4094～3.1924、2.2284～3.1982和0.1219～0.7485、0.2879～0.4672，说明它们在西南桦林分内分布广、资源利用具有较大优势，对环境的适应能力较强。②草本优势种生态相似性相对较低，生态位重叠范围主要在0～0.0500，约占优势草本物种总数量98%以上，表明草本优势种群之间竞争不激烈，群落处于稳定状态。③从生态响应速率来看，西南桦×马尾松混交林内铁芒萁的发展性最强，在未来的演替过程中，可能形成优势种群占据草本层的格局。

关键词 西南桦；草本植物；优势种群；生态位；人工林

Niche Characteristics of Dominant Herbage Populations within *Betula alnoides* Plantations in Daqingshan Mountain Areas of Guangxi

PANG Sheng-jiang[1,2], ZHANG Pei[1], YANG Baoguo[1], LIU Shiling[1], DENG Shuokun[1], JIA Hongyan[1]
([1]*Experimental Center of Tropical Forestry*, *CAF*, *Pingxiang* 536000, *Guangxi*, *China*;
[2]*Guangxi Youyiguan Forest Ecosystem Research Station*, *Pingxiang* 532600, *Guangxi*, *China*)

Abstract: Based on the data of 159 plots obtained from investigation, the niche breadth and overlapping of dominant herbage populations within *Betula alnoides* plantations were studied by using Shannon-wiener、Levin niche breadth index and Levins niche overlapping index. The results showed that①The niche breadths values of *Miscanthus floridulus* and *Adiantum flabellulatum* were relatively larger, Shannon-wiener indices were 2.4094～3.1924、2.2284～3.1982, Levins indices were 0.1219～0.7485、0.2879～0.4672, respectively, which showed that they are all widely distributed within the community, having strong ability to adapt to the environment. ②It was niche proportional similarity value relatively low in all dominant herbage species, niche proportional similarity value ranges were 0～0.0500, accounting for more than 98%, indicated that low degrees of niche overlap found in this study suggested weak interspecies competition among dominant species and therefore a potentially stable plant community within *Betula alnoides* plantation. ③From the eco-response rate, aggression of *Dicranopteris linearis* was strongest, and few species would monopolize they will develop as the dominant species of herb layer in *Betula alnoides*×*Pinus massoniana* mixed plantation for future development.

Key words: *Betula alnoides*; herbage plants; dominant populations; niche; plantation

生态位(Niche)是植物群落生物多样性、种间竞争与共存的研究基础，属于种群生态学研究的核心领域[1,2]。究其本质而言，生态位是物种在特定生态环境条件下，物种与环境二者之间相互作用的特征表现[3]。它既能有效地反映物种在生态系统的功能和地位，揭示植物种群的种间关系、群落结构和及其与环境之间的耦合关系，亦有助于认识种群利用自然资源的地位和作用，对群落的稳定性和生物多样性恢复都具有重要意义[4]。

目前，国内外学者已从植物种群生态位理论[5]、生态位测度[6]和具体应用[7]等方面做了大量研究，在种间关系、群落结构和生物多样性及其形成机制等方面得到广泛应用；如，刘帅华等[8]利用Levins与Pianka公式分析云杉(*Picea asperata*)纯林、白桦(*Betula platyphylla*)纯林以及云杉-白桦混交林3种人工林的草本层优势种发现，表针阔混交林的草本层优势种生态位所代表的群落结构更为合理，混交林的配置模式更适合在当地推广；王伟伟等[9]对青海地区不同树种配置的人工林草本群落生态位研究也表明，针阔叶混交林更有利于草本植物群落的生长和发育；潘高等[10]研究认为，处于植被演替过程中的针叶林和针阔混交林下草本植物生态位重叠程度较高，对资源的利用存在较大的空间异质性，但群落结构仍处于不稳定的动态变化。然而，有关人工林下植物优势种群之间的相互关系怎样，各草本优势种群的地位和功能又是如何，是否会对人工林群落结构产生影响，这些问题是否对人工林生态系统的稳定性及生物多样性产生影响，应引起学者们共同关注。

西南桦(*Betula alnoides*)为桦木科(Betulaceae)桦木属(*Betula*)的落叶大乔木；其具有速生、干形通直、木材纹理美观、适应性广和耐贫瘠干旱等优良特性，天然分布于我国广西、云南和西藏等地。西南桦人工林是近年来南亚热带地区发展较快、种植面积广、生态和经济效益较好的人工林生态系统之一[11]。草本植物是乡土珍贵树种西南桦林下植被的重要组成部分，在促进人工林养分循环和维护森林立地生产力方面起着不可代替的作用。因此，本研究运用生态位研究理论，对广西大青山林区西南桦人工林草本植物优势种群的生态位宽度、生态位重叠程度和生态响应进行定量研究，将有助于揭示草本优势种在群落中的生态适应性、功能和地位，为解释西南桦人工林群落结构、物种竞争与共存等问题提供理论依据，亦为西南桦人工林可持续经营中，林下植被如何配置和调节，有利于生物多样性的恢复以及建立稳定的生态系统提供参考。

1　研究方法

1.1　研究地概况

大青山林区位于广西凭祥市热带林业实验中心(21°57~22°19′N，106°40′~106°59′E)。该林区属于南亚热带季风气候，全年日照时数1200~1600h。年均气温20.5~21.7℃，极端高温40.3℃，极端低温-1.5℃；年均降水量为1200~1500mm，季节分布差异较大，主要集中在4~9月，相对湿度80%~84%。该林区原生植被有季雨林和常绿阔叶林，西南桦主要种植于海拔190~670m，土壤类型主要为山地红壤，土层深厚、肥沃，土层厚度均在1m以上。

1.2　样地设置与数据采集

2015年5月以来，在广西大青山林区营林资料和野外全面勘查的基础上，选择该林区具有代表性的西南桦林地，设置面积为20m×30m的标准样地，同时，在每个样地内沿对角线的两个角和中心位置，分别设置3个5m×5m的样方，用于调查林下的植物群落。其中，西南桦纯林调查9个样地27个草本样方，西南桦×红椎(*Castanopsis hystrix*)混交林17个样地51个草本样方，西南桦×灰木莲(*Manglietia glauca*)混交林6个样地18个草本样方，西南桦×杉木(*Cunninghamia lanceolata*)混交林15个样地45个草本样方和西南桦×马尾松(*Pinus massoniana*)混交林6个样地18个草本样方，共53块标准样地159个草本样方。调查记录样方的草本植物群落中物种数量、株高和盖度等特征，参考植物种重要值选取每种西南桦人工林内前10个草本优势种进行生态位分析，重要值公式=(相对密度+相对频度+相对盖度)/3。

1.3　数据分析

1.3.1　生态位宽度

生态位宽度采用Shannon-wiener指数和Levins指数计算公式对数据处理分析：

$$\text{Shannon-wiener 指数：} B_{(SW)i} = -\sum_{j=1}^{r}(P_{ij}\ln P_{ij})$$

$$\text{Levins 指数：} B_{(L)i} = \frac{1}{r\sum_{j=1}^{r} P_{ij}^{\ 2}}$$

式中：$B_{(SW)i}$为物种i的Shannon-wiener生态位宽度指数；$B_{(L)i}$为物种i的Levins生态位宽指数；P_{ij}为物种i在第j资源位占它利用全部资源位的百分率比例；r为资源位总数(样方数)。$P_{ij}=n_{ij}/N_i$，而$N_i=\sum n_{ij}$，n_{ij}为物种i在第j样方的数量特征值(如密度、盖度和重要值等)，本研究为物种i在第j个样方内的密度。

1.3.1　生态位重叠

生态位重叠测度公式为：

$$L_{ih} = B_{(L)i}\sum_{j=1}^{r} P_{ij}P_{hj}$$

$$L_{hi} = B_{(L)h}\sum_{j=1}^{r} P_{ij}P_{hj}$$

式中：L_{ih}为物种i与物种h的生态位重叠指数；L_{hi}为物种h与物种i的生态位重叠指数；$B_{(L)i}$为物种i的Levins生态位宽指数；$B_{(L)h}$为物种h的Levins生态位宽度指数。

1.3.3 生态响应

生态位响应的测定公式：

$$R=B_{(L)}/\triangle L_{ih}(i=h)$$

式中：R 为生态效应速率；$B_{(L)}$ 为 Levins 生态位宽指数[12,13]。

2 结果与分析

2.1 生态位宽度

西南桦林下草本植物优势种 Shannon-wiener 和 Levins 生态位宽度指数计算结果，见表 1。从表 1 中可以看出，不同树种配置模式下，草本优势种间具有的生态位宽度各不相同；在同一种植模式下，草本植物优势种间具有的生态位宽度亦是有差异的。在西南桦纯林中，五节芒、弓果黍和铁线蕨等生态位宽度指数较大，对资源的利用能力较强，是草本层优势种；在西南桦×红椎混交林中，铁线蕨的生态位宽度值最大，半边旗和乌毛蕨的生态位宽度值次之，而大多数草本种群的生态位宽度值较小，生态幅度较窄；西南桦×灰木莲混交林中，五节芒和蔓生莠竹的生态位宽度值较大，说明这 2 个草本植物对资源的利用较为充分，在对资源的竞争中处于优势地位；西南桦×杉木混交林中，生态宽度值较大的草本植物分别为五节芒、铁线蕨和半边旗，表明这几个物种是该种植模式优势种；在西南桦×马尾松混交林中，五节芒、团叶鳞始蕨、铁线蕨和蔓生莠竹等为草本植物，对资源的竞争中较为强势。

由表 1 可知，相同的草本植物，可出现在不同种植模式的西南桦人工林中，如弓果黍、五节芒、铁芒萁、半边旗和荩草等 5 种草本植物在西南桦林分内均有分布，说明这些草本植物种对当地环境的适应性较强。而在不同的西南桦人工林内，同一植物的生态位宽度指数变化较大，弓果黍在西南桦纯林的生态位宽度值最大，而在西南桦×红椎混交林和西南桦×杉木混交林的生态位宽度较小；五节芒在纯林和西南桦×马尾松混交林的生态位宽度较大，而在西南桦×红椎混交林的生态位宽度较小；铁芒萁在西南桦×红椎混交林的生态位宽度较小，在其他 4 个西南桦林内生态位宽度较为居中；而荩草在 5 种西南桦林内的生态位宽度均处于较低地位。

综上所述，5 种西南桦林内草本优势种 Shannon-wiener 和 Levins 生态位宽度指数有所差异，但总体的大小变化趋势较为一致性。此外，五节芒、铁线蕨等生态位宽度值较大的物种，说明它们在西南桦林分内分布广、资源利用具有较大优势，对环境的适应能力较强。

表 1 西南桦人工林草本优势种生态位宽度

林分类型	物种	生态位宽度	
		Shannon-wiener	Levins
西南桦纯林	弓果黍 *Cyrtococcum patens*	2.7276	0.4703
	蔓生莠竹 *Microstegium fasciculatum*	1.4456	0.1468
	五节芒 *Miscanthus floridulus*	2.8845	0.5125
	露籽草 *Ottochloa nodosa* var.	1.2799	0.1235
	铁芒萁 *Dicranopteris linearis*	2.0045	0.2315
	粽叶芦 *Thysanolaena latifolia*	2.1219	0.2525
	淡竹叶 *Lophatherum gracile*	2.0634	0.2960
	铁线蕨 *Adiantum flabellulatum*	2.4843	0.3766
	半边旗 *Pteris semipinnata*	2.2741	0.3029
	荩草 *Arthraxon hispidus*	1.3378	0.1249
西南桦×红椎混交林	露籽草 *Ottochloa nodosa* var.	0.0479	0.0199
	弓果黍 *Cyrtococcum patens*	1.0695	0.0490
	淡竹叶 *Lophatherum gracile*	2.2983	0.0980
	五节芒 *Miscanthus floridulus*	2.3307	0.1219
	半边旗 *Pteris semipinnata*	2.6362	0.1768
	铁线蕨 *Adiantum flabellulatum*	2.9280	0.2879
	荩草 *Arthraxon hispidus*	1.8318	0.0946
	铁芒萁 *Dicranopteris linearis*	1.4871	0.0635
	乌毛蕨 *Adiantum flabellulatum*	2.6342	0.2161
	华南毛蕨 *Cyclosorus parasiticus*	1.7359	0.0857
西南桦×灰木莲混交林	蔓生莠竹 *Microstegium fasciculatum*	2.1549	0.4634
	弓果黍 *Cyrtococcum patens*	1.3669	0.1717
	铁芒萁 *Dicranopteris linearis*	1.3492	0.1500
	荩草 *Arthraxon hispidus*	1.4198	0.1729
	淡竹叶 *Lophatherum gracile*	1.0485	0.1527
	五节芒 *Miscanthus floridulus*	2.4094	0.5197
	华南毛蕨 *Cyclosorus parasiticus*	0.301	0.0640
	乌毛蕨 *Adiantum flabellulatum*	1.2701	0.1631
	江南卷柏 *Selaginella moellendorffii*	0.9679	0.1351
	半边旗 *Pteris semipinnata*	1.4186	0.1988
西南桦×杉木混交林	蔓生秀竹 *Microstegium fasciculatum*	2.1368	0.1240
	半边旗 *Pteris semipinnata*	2.8312	0.2662
	铁芒萁 *Dicranopteris linearis*	2.4966	0.2388
	五节芒 *Miscanthus floridulus*	3.1924	0.4438
	棕叶芦 *Thysanolaena latifolia*	1.8537	0.1265
	弓果黍 *Cyrtococcum patens*	0.4011	0.1966
	铁线蕨 *Adiantum flabellulatum*	3.1982	0.4409

（续）

林分类型	物种	生态位宽度	
		Shannon-wiener	Levins
西南桦×杉木混交林	荩草 *Arthraxon hispidus*	2.0455	0.1452
	团叶鳞始蕨 *Lindsaea orbiculata*	2.7598	0.2908
	乌毛蕨 *Adiantum flabellulatum*	2.2067	0.1306
西南桦×马尾松混交林	蔓生莠竹 *Microstegium fasciculatum*	2.2496	0.4394
	铁芒萁 *Dicranopteris linearis*	2.0538	0.3593
	弓果黍 *Cyrtococcum patens*	1.7308	0.2441
	五节芒 *Miscanthus floridulus*	2.7500	0.7485
	海金沙 *Lygodium japonicum*	0.8670	0.1042
	荩草 *Arthraxon hispidus*	1.2028	0.1499
	半边旗 *Pteris semipinnata*	2.1113	0.3271
	团叶鳞始蕨 *Lindsaea orbiculata*	2.4511	0.5926
	铁线蕨 *Adiantum flabellulatum*	2.2284	0.4671
	淡竹叶 *Lophatherum gracile*	1.8963	0.2921

2.2　生态位重叠

草本植物优势种的生态位重叠值（表2至表6）表明，生态位宽度指数较高的物种，与其他种群存在较高的生态重叠，如五节芒的分布较为广泛，利用资源的能力较强，进而与其他草本优势种分布空间有着较大的生态位重叠。此外，一些种植模式中，生态位宽度较窄的草本优势种亦可能出现较高的重叠度，如露籽草在纯林和西南桦×红椎混交林的生态位宽度较窄（表1），与其他草本优势种有着较大的生态重叠值；西南桦纯林内，露籽草-弓果黍的生态位重叠值为0.0307，露籽草—铁线蕨的生态重叠值为0.0434；而在后者林分内，露籽草—半边旗的生态重叠度达到0.0469。

西南桦人工林内草本优势种生态位重叠程度尚处于较低水平，西南桦纯林 L_{ih} 重叠值变化为0~0.0307，L_{hi} 值变化为0~0.0434；西南桦×红椎混交林 L_{ih} 值变化为0~0.0084，L_{hi} 值变化为0~0.0469；西南桦×灰木莲混交林 L_{ih} 值变化为0~0.0538，L_{hi} 值变化为0~0.0792；西南桦×杉木混交林 L_{ih} 值变化为0~0.0237，L_{hi} 值变化为0~0.0126；西南桦×马尾松混交林 L_{ih} 值变化为0~0.0645，L_{hi} 值变化为0~0.0594；总体而言，草本优势种生态相似性相对较低，生态位重叠高度集中在0~0.0500，占草本植物总对数的98%以上。

表2　西南桦纯林草本优势种生态位重叠值

物种	弓果黍	蔓生莠竹	五节芒	露籽草	铁芒萁	粽叶芦	淡竹叶	铁线蕨	半边旗	荩草
弓果黍		0.0065	0.0190	0.0307	0.0137	0.0201	0.0092	0.0224	0.0242	0.0240
蔓生莠竹	0.0020		0.0026	0.0067	0.0020	0.0001	—	0.0025	0.0005	0.0014
五节芒	0.0207	0.0090		0.0113	0.0156	0.0109	0.0272	0.0115	0.0241	0.0093
露籽草	0.0081	0.0056	0.0027		0.0007	0.0012	0.0028	0.0142	0.0046	0.0048
铁芒萁	0.0068	0.0032	0.0071	0.0012		0.0081	0.0158	0.0129	0.0082	—
粽叶芦	0.0108	0.0002	0.0054	0.0025	0.0088		0.0074	0.0077	0.0036	0.0163
淡竹叶	0.0058	—	0.0157	0.0067	0.0202	0.0086		0.0108	0.0202	0.0108
铁线蕨	0.0179	0.0064	0.0084	0.0434	0.0210	0.0115	0.0137		0.0209	0.0096
半边旗	0.0156	0.0011	0.0142	0.0114	0.0107	0.0043	0.0207	0.0168		0.0122
荩草	0.0064	0.0012	0.0023	0.0048	—	0.0081	0.0045	0.0032	0.0050	

表3　西南桦×红椎混交林主要草本优势种生态位重叠值

物种	露籽草	弓果黍	淡竹叶	五节芒	半边旗	铁线蕨	荩草	铁芒萁	乌毛蕨	华南毛蕨
露籽草		—	—	0.0014	0.0053	—	—	0.0010	0.0001	0.0015
弓果黍	—		—	—	—	—	0.0015	0.0006	0.0021	0.0008
淡竹叶	—	—		0.0004	0.0003	0.0033	0.0063	0.0003	0.0021	0.0028
五节芒	0.0084	—	0.0005		0.0038	0.0005	0.0002	0.0053	0.0019	0.0007
半边旗	0.0469	—	0.0006	0.0055		0.0020	0.0031	0.0047	0.0053	0.0054

（续）

物种	露籽草	弓果黍	淡竹叶	五节芒	半边旗	铁线蕨	荩草	铁芒萁	乌毛蕨	华南毛蕨
铁线蕨	—	—	0.0096	0.0012	0.0033		0.0053	0.0084	0.0057	0.0037
荩草	—	0.0055	0.0061	0.0001	0.0016	0.0017		—	0.0028	0.0002
铁芒萁	0.0031	0.0033	0.0002	0.0028	0.0017	0.0019	—		0.0015	0.0002
乌毛蕨	0.0016	0.0091	0.0046	0.0034	0.0065	0.0043	0.0063	0.0050		0.0043
华南毛蕨	0.0064	0.0014	0.0024	0.0005	0.0026	0.0011	0.0002	0.0003	0.0017	

表 4　西南桦×灰木莲混交林草本优势种生态位重叠值

物种	蔓生秀竹	弓果黍	铁芒萁	荩草	淡竹叶	五节芒	华南毛蕨	乌毛蕨	江南卷柏	半边旗
蔓生莠竹		0.0045	0.0001	0.0290	—	0.0538	—	0.0041	—	0.0087
弓果黍	0.0017		0.0043	0.0100	0.0122	0.0708	0.0275	0.0203	0.0193	—
铁芒萁	0.0000	0.0038		0.0017	0.0056	0.0099	0.0050	0.0048	0.0047	0.0026
荩草	0.0108	0.0100	0.0019		0.0091	0.0441	0.0326	0.0216	0.0133	0.0024
淡竹叶	—	0.0109	0.0057	0.0081		0.0431	0.0316	0.0375	0.0489	0.0188
五节芒	0.0074	0.0263	0.0042	0.0163	0.0180		0.0110	0.0144	0.0183	0.0075
华南毛蕨	—	0.0103	0.0021	0.0121	0.0132	0.0110		0.0311	0.0230	0.0006
乌毛蕨	0.0014	0.0193	0.0052	0.0204	0.0400	0.0368	0.0792		0.0461	0.0105
江南卷柏	0.0000	0.0152	0.0042	0.0104	0.0432	0.0389	0.0486	0.0382		0.0049
半边旗	0.0037	0.0000	0.0035	0.0027	0.0244	0.0235	0.0018	0.0128	0.0071	

表 5　西南桦×杉木混交林草本优势种生态位重叠值

物种	蔓生秀竹	半边旗	铁芒萁	五节芒	棕叶芦	弓果黍	铁线蕨	荩草	团叶鳞始蕨	乌毛蕨
蔓生秀竹		0.0005	0.0017	0.0012	0.0013	0.0002	0.0009	0.0013	0.0008	0.0036
半边旗	0.0011		0.0034	0.0068	0.0074	0.0112	0.0067	0.0059	0.0033	0.0057
铁芒萁	0.0032	0.0030		0.0052	—	0.0102	0.0064	0.0039	0.0077	0.0018
五节芒	0.0043	0.0114	0.0098		0.0237	0.0102	0.0127	0.0065	0.0108	0.0060
棕叶芦	0.0014	0.0035	—	0.0067		0.0001	0.0033	0.0009	0.0003	0.0001
弓果黍	0.0003	0.0083	0.0084	0.0045	0.0001		0.0048	0.0081	0.0056	0.0029
铁线蕨	0.0033	0.0111	0.0118	0.0126	0.0115	0.0108		0.0066	0.0127	0.0078
荩草	0.0016	0.0032	0.0024	0.0021	0.0010	0.0060	0.0022		0.0052	0.0015
团叶鳞始蕨	0.0019	0.0036	0.0094	0.0071	0.0006	0.0083	0.0084	0.0105		0.0143
乌毛蕨	0.0038	0.0028	0.0010	0.0018	0.0001	0.0020	0.0023	0.0013	0.0064	

表 6　西南桦×马尾松混交林草本优势种生态位重叠值

物种	蔓生秀竹	铁芒萁	弓果黍	五节芒	海金沙	荩草	半边旗	团叶鳞始蕨	铁线蕨	淡竹叶
蔓生莠竹		0.0089	0.0028	0.0099	0.0008	0.0063	0.0076	0.0144	0.0049	0.0065
铁芒萁	0.0073		0.0077	0.0285	0.0645	0.0029	0.0053	0.0134	0.0108	0.0035
弓果黍	0.0015	0.0053		0.0036	0.0066	0.0095	0.0242	0.0082	0.0167	0.0139
五节芒	0.0168	0.0594	0.0015		—	0.0312	0.0348	0.0440	0.0366	0.0178
海金沙	0.0002	0.0187	0.0041	—		0.0007	0.0013	0.0036	0.0007	0.0034
荩草	0.0022	0.0012	0.0291	0.0063	0.0010		0.0030	0.0061	0.0031	0.0020

（续）

物种	蔓生莠竹	铁芒萁	弓果黍	五节芒	海金沙	荩草	半边旗	团叶鳞始蕨	铁线蕨	淡竹叶
半边旗	0.0057	0.0049	0.0324	0.0152	0.0042	0.0066		0.0233	0.0252	0.0131
团叶鳞始蕨	0.0195	0.0221	0.0200	0.0348	0.0207	0.0240	0.0422		0.0297	0.0223
铁线蕨	0.0053	0.0140	0.0320	0.0228	0.0033	0.0095	0.0359	0.0234		0.0165
淡竹叶	0.0043	0.0029	0.0166	0.0069	0.0097	0.0039	0.0117	0.0110	0.0103	

2.3　生态位响应

草本优势种表现出的发展型或衰退型，是对西南桦林内生态环境因子的综合响应，其生态响应速率（R值）详见表7。

从表7可以看出，西南桦纯林中，仅有3个草本植物种群为发展型，分别为弓果黍（6.2103）、五节芒（8.2237）和铁线蕨（16.8952）；蔓生莠竹、露籽草、铁芒萁等其他7个属于衰退型种群，随着对资源利用竞争加剧，淡竹叶（−117.6499）和半边旗（−69.1493）生态空间衰退的趋势更为明显。西南桦×红椎混交林内，发展性最强的是五节芒（20.2158），其次是铁线蕨（11.3851）和乌毛蕨（9.8705），半边旗（3.3799）的发展性最弱；其他6种属于衰退性种群，生态空间均呈不断缩小的趋势。西南桦×灰木莲混交林中，铁芒萁（22.2363）为发展性最强，但草本群落对资源利用的竞争强度尚处于较低水平。西南桦×杉木混交林中，半边旗（30.3765）的发展性最强，而蔓生莠竹、铁芒萁、粽叶芦等6个衰退型种群，在资源的竞争中逐渐处于劣势地位。在西南桦×马尾松混交林中，蔓生莠竹（−853.9157）的衰退最为明显，在资源竞争中处于被动地位，其生存的生态空间有严重退缩的趋势；铁芒萁（53.6588）的发展性最强，在未来的演替过程中，可能形成优势种群占据草本层的格局。

表7　西南桦人工林草本优势种发展或衰退状况

林分类型	物种	ΔLih	R	物种	ΔLih	R
西南桦纯林	弓果黍	0.0757	6.2103	粽叶芦	−0.0103	−24.4426
	蔓生莠竹	−0.0153	−9.5760	淡竹叶	−0.0025	−117.6499
	五节芒	0.0623	8.2237	铁线蕨	0.0223	16.8952
	露籽草	−0.0739	−1.6706	半边旗	−0.0044	−69.1493
	铁芒萁	−0.0296	−7.8326	荩草	−0.0529	−2.3610
西南桦×红椎混交林	露籽草	−0.0571	−0.3490	铁线蕨	0.0253	11.3851
	弓果黍	−0.0144	−3.4010	荩草	−0.0088	−10.7824
	淡竹叶	−0.0085	−11.4647	铁芒萁	−0.0137	−4.6308
	五节芒	0.0060	20.2158	乌毛蕨	0.0219	9.8705
	半边旗	0.0523	3.3799	华南毛蕨	−0.0030	−28.5943
西南桦×灰木莲混交林	蔓生莠竹	0.0751	6.1714	五节芒	−0.2084	−0.3059
	弓果黍	0.0659	2.6052	华南毛蕨	−0.1338	−0.4784
	铁芒萁	0.0067	22.2363	乌毛蕨	0.0742	2.1980
	荩草	0.0353	4.8936	江南卷柏	0.0228	5.9344
	淡竹叶	0.0219	6.9778	半边旗	0.0237	8.3816
西南桦×杉木混交林	蔓生秀竹	−0.0093	−28.6385	弓果黍	−0.0158	−7.8686
	半边旗	0.0041	30.3765	铁线蕨	0.0405	3.0584
	铁芒萁	−0.0063	−19.8250	荩草	−0.0199	−6.2278
	五节芒	0.0472	2.6263	团叶鳞始蕨	0.0112	11.0282
	粽叶芦	−0.0295	−4.1983	乌毛蕨	−0.0223	−5.5544

（续）

林分类型	物种	△*Lih*	*R*	物种	△*Lih*	*R*
西南桦×马尾松混交林	蔓生莠竹	-0.0005	-853.9157	荩草	-0.0629	-2.3813
	铁芒萁	0.0067	53.6588	半边旗	-0.0356	-9.1781
	弓果黍	-0.0567	-4.3036	团叶鳞始蕨	0.0878	6.7509
	五节芒	0.1358	5.5118	铁线蕨	0.0329	14.2153
	海金沙	-0.0775	-1.3442	淡竹叶	-0.0050	-58.8645

3 结论与讨论

研究发现，不同树种配置模式下，西南桦人工林内草本优势种间具有的生态位宽度各不相同；在同一种植模式下，草本植物优势种间具有的生态位宽度亦是有差异的；这与王伟伟等[9]对云杉林、白桦林、云杉—白桦混交林草本优势种群生态位研究结果相似，他们研究认为草本优势种群中广生态幅物种的差异，可能与物种自身的生物生态学特性及立地条件差异有关。草本优势种 Shannon-wiener 和 Levins 生态位宽度指数有所差异，但总体的大小变化趋势较为一致性，即二者的大小变化是相同的，草本优势种生态位宽度值较大，说明对资源的利用较为充分，在对资源的竞争中处于优势地位。西南桦林分内五节芒、铁线蕨的生态位宽度指数值较大，说明它们自然分布广、资源利用具有较大优势，对环境的适应能力较强；在未来的演替过程中，其对有限资源的竞争能力有可能得到进一步增强，使其他竞争能力相对较弱的物种，难以适应严苛生存环境，成为少数优势种占据草本层的格局；王伟伟等[14]对青海云杉人工林草本层生态位及生态响应研究也得出相类似观点。本研究调查亦发现，五节芒、铁线蕨等生态位宽度指数较高的草本优势种，由于其结实量大、萌蘖能力强、根系密布以及分布广泛等生物生态学特性[15-17]，在西南桦人工林下生长密集、盖度大，往往形成连片的单优势种群落，表现出较强的生态适应性。

草本优势种生态位重叠分析结果表明，五节芒等生态位宽度指数较高的物种，与其他草本优势种群存在较高的生态重叠；但露籽草等生态位宽度较窄的草本优势种，与其他草本优势种群亦可能出现较高的重叠度，这与陈艳瑞等[18]对人工固沙林演替过程中优势种群生态位变化特征以及徐德静等[19]对桫椤（*Alsophila spinulosa*）群落优势种群生态位研究结论基本一致。究其原因，主要存在以下可能：一是优势种间对资源利用的相似性而出现相互促进的关系，二是优势种间共享资源亦存在相互竞争的关系，三是生态位宽度指数值较大的优势种，由于自身生物生态学特性的存在一定的差异，对资源的需求则表现出明显的不同，导致优势种间的生态位重叠程度不一定很高。

若在资源有限时，生态位重叠既反映物种间的竞争程度，同时亦表征二者之间的生态相似性；在资源极为丰富时，生态位重叠值并不能有效反映物种间的竞争关系，而只能说明二者的生态相似性或者占据着相似的生存空间[20]。本研究发现，西南桦人工林内草本优势种的生态相似性相对较低，生态位重叠指数值绝大部分小于 0.0500，表明草本层植物群落尚处于比较稳定的状态，这与吴友贵等[21]对百山祖常绿阔叶林优势种群的生态位研究结论相似，他们研究认为生态位重叠值较小时，优势种间对资源的竞争不激烈，群落的稳定性就越大；但生态位重叠程度较高并不一定能够说明存在激烈的竞争关系。

从草本优势种群的△*Lih* 来看：在 5 种西南桦人工林内，生态位宽度指数较大，具有较广的资源利用谱的草本优势种群大多数属于发展阶段，如西南桦纯林中弓果黍和五节芒，西南桦×红椎混交林中五节芒、铁线蕨和乌毛蕨，西南桦×灰木莲混交林和西南桦×杉木混交林中半边旗以及西南桦×马尾松混交林中铁芒萁等；生态位宽度指数较小，对生存环境适应性较差的草本优势种大多数属于衰退型种群，如西南桦纯林、西南桦×红椎混交林和西南桦×马尾松混交林中淡竹叶，西南桦×灰木莲混交林五节芒和华南毛蕨，西南桦×杉木混交林中蔓生莠竹和铁芒萁等。从草本优势种的生态响应速率来看：铁线蕨（西南桦纯林）、五节芒（西南桦×红椎混交林）、铁芒萁（西南桦×灰木莲混交林和西南桦×马尾松混交林）和半边旗（西南桦×杉木混交林）的发展性是最强的，明显超过其他草本优势种群，可以预见，在未来的演替过程中，这些草本优势种将会逐渐代替其他衰退型的草本优势种群，成为大青山林区 5 种不同西南桦人工林下草本层的优势种群，而诸如淡竹叶（西南桦纯林和西南桦×马尾松混交林）、蔓生莠竹（西南桦×

马尾松混交林）等生态位较窄的物种，将会逐渐消亡。从西南桦人工林生态系统的生物多样性恢复的角度来考虑，如何使生态位宽度较窄的草本植物保存下来而不被取代，最有效的办法就是提高其生态位宽度，具体包括采取合理的人为干扰措施，改变林内环境状态，增加生态位宽度较窄草本植物的可利用资源；其次，减少生态位宽度较大的物种数量，达到降低其对资源的竞争能力。

随着乡土树种西南桦种植业的快速发展，草本层物种对西南桦人工林生态系统的影响是不可忽视的，在未来的演替过程中，草本植物种类和数量的此消彼长，对大青山林区西南桦人工林的生物多样性有着重要的作用。通过对草本优势种群生态位和生态响应的研究，了解不同植物的生态幅度及其对资源利用能力的差异，在营林实践的过程中，结合不同草本植物的发展趋势，采取合理人为抚育措施能够更好地促进西南桦人工林的高效、健康的发展。

参考文献

[1] WHITTAKER R H, LEVIN S A, ROOT R B. Niche, habitat and ecotope[J]. American Naturalist, 1973, 107: 321-338.

[2]杨利民，周广胜，王国宏．草地群落物种多样性维持机制的研究Ⅱ物种实现生态位[J]．植物生态学报，2001，25(5)：634-638.

[3] LEIBOLD M A. The niche concept revisited mechanistic model and community contex [J]. Ecology, 1995, 76 (5): 1371-1382.

[4] MUSTSHINDA C M, O'HARA R B. Integrating the niche and neutral perspectives on community structure and dynamics[J]. Oecologia, 2011, 166: 241-251.

[5] WESTMAN W E. Measuring realized niche spaces climatic response of chaparral and coastal sage scrub[J]. Ecology, 1991, 72: 1678-1684.

[6] SYLVAIN D, DANIEL C, CLEMENTINE G C. Niche separation in community analysis: a new method [J]. Ecology, 2000, 81: 2914-2927.

[7] THUILLER W, GASSO'N, PINA J, et al. Ecological niche and species traits: key drivers of regional plant invader assemblages[J]. Biol Invasions, 2012, 14: 1963-1980

[8]刘帅华，贺康宁，董梅，等．青海大通不同人工林林下草本群落生态位特征[J]．四川农业大学学报，2012，30(2)：167-173.

[9]王伟伟，杨海龙，贺康宁，等．青海高寒区不同人工林配置下草本群落生态位研究[J]．水土保持研究，2012. 19(3)：156-165.

[10]潘高，张合平，潘登．南方红壤丘陵区3种森林群落内主要草本植物种群生态位特征[J]．草业科学，2015，32(12)：2094-2106.

[11]曾杰，郭文福，赵志刚，等．我国西南桦研究的回顾与展望[J]．林业科学研究，2006，19(3)：379-384.

[12]张金屯．数量生态学[M]．北京：科学出版社，2004.

[13]钟宇，张健，刘泉波，等．巨桉人工林草本层主要种群的生态位分析[J]．草业科学，2010，19(4)：16-21.

[14]王伟伟，杨海龙，贺康宁，等．祁连山青海云杉人工林草本层生态位及生态响应研究[J]．草地学报，2012，20(4)：626-630.

[15]吴征镒，俞德浚，等．中国植物志[M]．北京：科学出版社，1990.

[16]叶福钧，黄显标．我国南方芒箕草山的科学改良 I. 铁芒萁根状茎、叶芽生长特点[J]．草业科学，1993. 10(3)：54-57.

[17]陈慧娟，宁祖林，张卓文．五节芒生物学特性及能量生产动态变化[J]．草业科学，2012，21(6)：252-257.

[18]陈艳瑞，尹林克．人工防风固沙林演替中群落组成和优势种群生态位变化特征[J]．植物生态学报，2008，32(5)：1126-1133.

[19]徐德静，王鹏鹏，何跃军，等．黔北丹霞地貌桫椤群落优势种群生态位研究[J]．植物研究，2014，34(5)：612-618.

[20]陈旭．长白山哈泥泥炭地七种苔藓植物种间联结和生态位研究[D]．长春：东北师范大学，2008.

[21]吴友贵，叶珍林，周荣飞，等．百山祖常绿阔叶林优势种群的生态位[J]．广西植物，2016，36(2)：186-192.

［原载：中南林业科技大学学报，2018，38(06)］

广西大青山西南桦人工林林下植物优势种群分布格局

庞圣江[1,2] 张 培[1] 杨保国[1] 刘士玲[1] 农 友[1] 贾宏炎

([1]中国林业科学研究院热带林业实验中心，广西凭祥 532600；
[2]广西友谊关森林生态系统国家定位观测研究站，广西凭祥 532600)

摘 要 以广西大青山林区西南桦纯林、西南桦×灰木莲混交林和西南桦×马尾松混交林作为研究对象，调查林下植物种类、数量和盖度等植被因子，应用重要值和扩散系数(DI)、负二项式分布(K)和聚块性指数(PAI)等测定指标，探讨3种造林模式对西南桦人工林林下植物优势种群分布格局的影响。结果表明：①西南桦纯林植物种类最多，西南桦×马尾松混交林次之，西南桦×灰木莲混交林中物种数量最少。②林下共有幼树优势种漆树，在西南桦×马尾松混交林呈聚集分布，但聚集程度不够明显，在纯林和西南桦×灰木莲混交林为均匀分布；灌木九节和藤本植物玉叶金花，在西南桦×灰木莲混交林和西南桦×马尾松混交林呈强烈聚集分布，二者在纯林聚集程度不明显；弓果黍和五节芒均在3种西南桦林内均呈强烈聚集分布。③3种造林模式对西南桦林下植物优势种及其分布格局的影响存在差异；我们建议适当立地选择、减少西南桦纯林比例，宜选取马尾松作为伴生树种混交造林，采取合理的营林措施，改善林分环境，提高林分异质性，从而促进西南桦人工林健康、高效的可持续经营。

关键词 西南桦；林下植物；优势种群；分布格局；人工林

Effects of Different Stand Types on Spatial Distribution Characteristics of Understory Plants in *Betula alnoides* Plantations at Mountain Daqingshan, Guangxi

PANG Shengjiang, ZHANG Pei, YANG Baoguo, LIU Shiling, NONG You, JIA Hongyan

([1]*Experimental Center of Tropical Forestry*, *CAF*, *Pingxiang* 532600, *Guangxi*, *China*;
[2]*Guangxi Youyiguan Forest Ecosystem Research Station*, *Pingxiang* 532600, *Guangxi*, *china*)

Abstract: The pure plantation of *Betula alnoides*, mixed plantations between *Betula alnoides* and *Manglietia glauca*or *Pinus massoniana*in Daqingshan Mountains area were selected to study the effect of afforestation modes on the distribution pattern of dominant population in the understory of . *Betula alnoides* plantation by investigating the species, number and coverage of understory plants, means of important values and diffusivity coefficient (DI), negative binomial distribution (K) and clustering index (PAI). The results show that: ①the pure plantation of *Betula alnoides* had the largest number of plant species, followed by mixed plantation between *Betula alnoides* and *Pinus massoniana*, and mixed plantation of *Betula alnoides* and *Manglietia glauca*. ② The dominant species *Toxicodendron delavayi* was distributed aggregately in the understory of *Betula alnoides*×*Pinus massoniana* mixed plantation. However, it was evenly distributed in the other plantations. *Psychotria asiatica* and liane *Mussaenda pubescens*were clumped significantly both in *Betulaalnoides*×*Manglietia glauca* and *Betula alnoides*×*Pinus massoniana* mixed plantations, while the aggregation intensity was weaker in *Betula alnoides* pure plantation. Moreover, *Cyrtococcum patens* and *Miscanthus floridulus* showed clumped pattern in all plantations. ③There were significant different in the understory dominant species and their distribution among three kinds of afforestation modes. Therefore, some measurements were taken, such as selecting proper site, reducing the proportion of *Betula alnoides* pure forest, selecting the *Pinus massoniana* mixed, to improve the forest environment and heterogeneity, and thus promote the healthy and efficient development of *Betula alnoides* plantation.

Key words: *Betula alnoides*; understory plants; dominant species; distribution pattern; plantation

种群分布格局是植物群落的重要结构特征[1]， 能够直接反映植物个体在水平空间的分布状况[2]，

是植物自身生物学特性与自然环境长期相互作用的结果[3]。由于植物对环境长期的选择与适应，群落中优势种或者伴生种分布格局，随着环境条件的改变而变化。因此，对植物群落分布格局的研究，不仅可以了解植物的生物生态学特性，揭示种群的空间分布结构以及种群与群落的动态变化趋势[4,5]，亦有助于种内与种间关系的确定、种群与环境互作机制以及理解森林的协同进化过程[6]，因此，种群空间分布格局一直是生态领域的研究热点之一[7-9]。

林下植物作为人工林群落的重要组成，不仅促进凋落物的分解，并且具有增强土壤养分归还的能力[10]。由于植被层次现象明显，林下植物群落中幼树、灌木、藤本和草本植物有着各自的优势种群；它们能够迅速地调整个体的形态结构，提高环境的适应能力，为人工林生态系统的长期稳定[11,12]、资源利用的最大化[13-16]等方面起着不可替代的作用。关于林下植物种群分布格局的研究报道，多以优势种群作为对象，开展不同立地、干扰措施对种群空间分布格局的影响及其生态学意义。如，Beatty (1984)研究北美糖槭(*Acer saccharinum*)-水青冈(*Fagus grandifolia*)阔叶林更新树种分布格局发现，加拿大铁杉(*Tsuga canadensis*)幼树种群对微地形的长期适应与选择产生特定微生境，导致其呈斑片状的聚集分布[17]；Gálhidy *et al*. (2006)研究发现，欧洲山毛榉(*Fagus sylvatica*)林隙环境差异，改变了林内光照、土壤水分，致使草本植物优势种呈不同空间分布格局[18]；王成等(2004)研究认为，带状间伐有利于促进赤松(*Pinus densiflora*)人工林天然更新和幼苗生长，且赤松幼树更新表现为聚集分布[19]；樊艳荣等(2013)研究不同林下植被干扰对毛竹(*Phyllostachys edulis*)林下植物优势种群分布格局的影响，结果表明，多年未除草、劈山除草和除草剂除草对毛竹(*Phyllostachys edulis*)林植物优势种及其空间分布格局的影响是有差异的；如杉木(*Cunninghamia lanceolata*)前在两种措施中呈聚集分布，而在后者呈随机分布，地菍(*Melastoma dodecandrum*)和芒萁(*Dicranopteris dichotoma*)在3种毛竹林中均呈强烈聚集分布[3]。目前，关于不同造林模式下植物优势种空间分布格局的研究鲜见报道，由于人工配植树种发育程度的差异，为林下植物群落提供了空间异质性较大的生存环境。同时，林下植物对人工林生态系统稳定性和长期林地肥力的维持作用独特，对不同造林模式的人工林下植物优势种空间分布格局进行研究，掌握林下植物如何充分利用人工配置树种在协同进化过程空缺出的生态资源，从而使整个人工林生态系统的结构与功能趋于稳定。

西南桦(*Betula alnoides*)为桦木科(Betulaceae)桦木属(*Betula*)乔木树种，因其具有速生、适应性广、材质优良以及经济价值高等优良特性，已成为我国热带、南亚热带地区林业生产的重要乡土珍贵阔叶树种，现有人工林面积已逾15万 hm^2[20]。随着西南桦种植业的规模化发展，西南桦人工林对当地林区生态环境的恢复与保护以及木材生产已占有非常重要地位。因此，本研究选取广西大青山林区西南桦纯林、西南桦×灰木莲(*Manglietia glauca*)混交林和西南桦×马尾松(*Pinus massoniana*)混交林为研究对象，调查林下植物种类、数量和盖度等，揭示不同造林模式的西南桦人工林林下植物优势种分布格局的影响规律，为西南桦人工林的可持续经营提供参考。

1 研究方法

1.1 研究地概况

大青山林区位于广西凭祥市中国林业科学研究院热带林业实验中心(简称热林中心)，地理坐标为21°58′~22°19′N，106°40′~106°59′E。该研究区属于南亚热带季风气候区，年均温21℃，年均降水量约1400mm，主要集中于4~9月。自20世纪70年代末至今，热林中心在国内最先开展西南桦引种、驯化、良种选育和栽培等研究，不同造林模式中，包括有西南桦纯林、西南桦×灰木莲混交林和西南桦×马尾松混交林等；林下植物种类丰富，幼树以西南木荷(*Schima wallichii*)、漆树(*Toxicodendron delavayi*)、中平树(*Macaranga denticulata*)、山杜英(*Elaeocarpus sylvestris*)和枫香(*Liquidambar formosana*)等为主；常见灌木有银柴(*Aporosa dioica*)、琴叶榕(*Ficus pandurata*)、九节(*Psychotria asiatica*)、粗叶悬钩子(*Rubus alceifolius*)和三叉苦(*Melicope pteleifolia*)等；藤本植物有玉叶金花(*Mussaenda pubescens*)、毒根斑鸠菊(*Vernonia cumingiana*)、厚叶玉叶金花(*Mussaenda erosa*)和酸藤子(*Embelia laeta*)等；草本植物以弓果黍(*Cyrtococcum patens*)、五节芒(*Miscanthus floridulus*)、蔓生莠竹(*Microstegium fasciculatum*)和铁芒萁(*Dicranopteris linearis*)等为主。土壤类型为山地红壤。

1.2 样地设置与数据采集

2015年5月以来，根据热林中心营林资料以及在野外勘查现有西南桦林的基础上，采用典型样地调查法，即用森林罗盘仪设置面积为 20m×30m 标准调查样地 20 个，在每个样地内沿对角线的两个角和中心位置，分别设置 3 个 5m×5m 的样方，用于林下植被调查，包含幼树、灌木和藤本以及草本植物的名称、数量和盖度等。其中，西南桦纯林 8 个样地 24 个草本样方，西南桦×灰木莲混交林 6 个样地 18 个草本样方，西南桦×马尾松混交林 6 个样地 18 个草本样方。同时，调查记录主林层林木树高、胸径、郁闭度、海拔、坡位和坡度等，详见表 1。

表 1 研究地基本概况

造林模式	样地（个）	林龄（年）	密度（株/hm^2）	$\bar{D}$(m)	$\bar{H}$(m)	海拔（m）	坡位	坡向	坡度（°）	郁闭度
西南桦纯林	8	20	522	19.89	15.36	250~440	中坡	西北坡/北坡	18~31	0.7
西南桦×灰木莲混交林	6	16	566	16.84	13.69	480~670	中上坡	北坡/东北坡	16~28	0.7
西南桦×马尾松混交林	6	14	640	13.91	11.83	190~480	中上坡	北坡/西北坡	19~26	0.8

1.3 数据分析

本研究重要值和种群分布格局采用的计算公式，具体如下：

重要值 IV=相对密度+相对频度+相对优势度(相对基盖度)/3 (1)

$$扩散系数\ DI=S^2/\bar{x} \quad (2)$$

式中：S^2 为种群多度的方差，DI>1 多度平均值。若 DI >1，种群为聚集分布；DI = 1，种群为随机分布；DI <1，种群为均匀分布。为了检验种群分布是否偏离 Poisson 分布，需要进行 t 检验。

$$负二项式分布\ K=\frac{\bar{x}^2}{s^2-\bar{x}} \quad (3)$$

式中：K 为聚集强度，K 值越大，聚集程度越低；K 值越小，聚集程度越高。当 K>0 时，种群聚集分布；当 K<0 时，为均匀分布；当 $K\to\infty$（一般在 8 以上）时，为随机分布。

$$聚块性指标\ PAI=m/\bar{x} \quad (4)$$

式中：m 表示生物个体在一个样方中的平均相邻个体数，反映了样方内生物个体的拥挤程度，数值越大聚集强度越大，表示一个个体受其他个体的拥挤效应越大[21]。

应用 Excel 2007 对原始数据进行初步整理，用 SPSS 16.0 软件对西南桦纯林、西南桦×灰木莲混交林和西南桦×马尾松混交林下植物种类、数量、盖度和重要值以及分布格局等待测指标进行统计分析。

2 结果与分析

2.1 林下植物优势种及其重要值

从表 2 可以看出，大青山林区 3 种西南桦林下植物种类较为丰富，调查记录到 58 科 107 属 126 种，限于篇幅，仅列出林下植物优势种。其中，西南桦纯林植物种类最多(93 种)，西南桦×马尾松混交林次之(78 种)，西南桦×灰木莲混交林中物种数量最少(52 种)。纯林幼树优势种为西南木荷、漆树，西南桦×灰木莲混交林中为山杜英和漆树，西南桦×马尾松混交林中为漆树、枫香和中平树；纯林灌木优势种为银柴和九节，西南桦×灰木莲混交林中为九节、粗叶悬钩子和三叉苦，西南桦×马尾松混交林中为银柴、琴叶榕和九节；纯林藤本植物优势种为玉叶金花和毒根斑鸠菊，西南桦×灰木莲混交林中为玉叶金花、毒根斑鸠菊和厚叶玉叶金花，西南桦×马尾松混交林中为玉叶金花、毒根斑鸠菊和络石；纯林草本优势种为弓果黍和五节芒，西南桦×灰木莲混交林中为弓果黍、五节芒和蔓生莠竹，西南桦×马尾松混交林中为弓果黍、五节芒、蔓生莠竹和铁芒萁。

上述 3 种造林模式的西南桦人工林中，林下植物优势种的重要值分别占西南桦纯林、西南桦×灰木莲混交林和西南桦×马尾松混交林的 56%、67%、74.5%，表明林下植物优势种在西南桦人工林群落结构与发展过程中的重要作用和地位。3 种西南桦人工林林下共有幼树优势种为漆树，灌木优势种为九节，藤本植物为玉叶金花和毒根斑鸠菊，草本植物为弓果黍和五节芒。

表 2 西南桦人工林下植物主要物种组成

层次	物种	科	属	重要值		
				PF	BF	CBF
幼树	西南木荷 *Schima wallichii*	山茶科 Theaceae	木荷属 *Schima*	64.44	—	—
	漆树 *Toxicodendron succedaneum*	漆树科 Anacardiaceae	漆树属 *Toxicodendron*	49.94	45.96	86.64
	中平树 *Macaranga denticulata*	大戟科 Euphorbiaceae	血桐属 *Macaranga*	25.67	11.77	49.73
	乌桕 *Triadica cochinchinensis*	大戟科 Euphorbiaceae	乌桕属 *Triadica*	24.80	—	—
	毛桐 *Mallotus barbatus*	大戟科 Euphorbiaceae	野桐属 *Mallotus*	19.22	—	21.11
	苹果榕 *Ficus oligodon*	桑科 Moraceae	榕属 *Ficus*	19.18	23.93	—
	鸭脚木 *Schefflera minutistellata*	五加科 Araliaceae	南鸭掌柴属 *Schefflera*	13.63	—	—
	潺槁木姜子 *Litsea glutinosa*	樟科 Lauraceae	木姜子属 *Litsea*	11.15	—	—
	山杜英 *Elaeocarpus sylvestris*	杜英科 Elaeocarpaceae	杜英属 *Elaeocarpus*	—	40.74	—
	方叶五月茶 *Antidesma ghaesembilla*	大戟科 Euphorbiaceae	五月茶属 *Antidesma*	—	21.95	—
	黄毛五月茶 *Antidesma fordii*	大戟科 Euphorbiaceae	五月茶属 *Antidesma*	—	24.74	—
	假柿木姜子 *Litsea monopetala*	樟科 Lauraceae	木姜子属 *Litsea*	—	22.33	—
	枫香 *Liquidambar formosana*	金缕梅科 Hamamelidaceae	枫香树属 *Liquidambar*	—	—	53.24
灌木层	银柴 *Aporosa dioica*	大戟科 Euphorbiaceae	银柴属 *Aporosa*	54.95	—	42.06
	琴叶榕 *Ficus pandurata*	桑科 Moraceae	榕属 *Ficus*	27.17	—	39.44
	九节 *Psychotria asiatica*	茜草科 Rubiaceae	九节属 *Psychotria*	34.42	52.23	33.67
	越南悬钩子 *Rubus cochinchinensis*	蔷薇科 Rosaceae	悬钩子属 *Rubus*	28.72	—	—
	毛果算盘子 *Glochidion eriocarpum*	大戟科 Euphorbiaceae	算盘子属 *Glochidion*	19.94	—	17.29
	粗叶榕 *Ficus hirta*	桑科 Moraceae	榕属 *Ficus*	16.89	10.90	21.80
	粗叶悬钩子 *Rubus alceifolius*	蔷薇科 Rosaceae	悬钩子属 *Rubus*	—	41.50	—
	三叉苦 *Melicope pteleifolia*	芸香科 Rutaceae	密茱萸属 *Melicope*	—	36.12	16.49
	大青 *Clerodendrum cyrtophyllum*	马鞭草科 Verbenaceae	大青属 *Clerodendrum*	—	22.97	21.77
	杜茎山 *Maesa japonica*	紫金牛科 Ardisiaceae	杜茎山属 *Maesa*	—	15.38	21.26
	楤木 *Aralia chinensis*	五加科 Araliaceae	楤木属 *Aralia*	—	11.67	—

（续）

层次	物种	科	属	重要值		
				PF	BF	CBF
藤本植物	玉叶金花 *Mussaenda pubescens*	茜草科 Rubiaceae	玉叶金花属 *Mussaenda*	56.05	74.1	62.51
	毒根斑鸠菊 *Vernonia cumingiana*	菊科 Compositae	斑鸠菊属 *Vernonia*	51.23	53.57	48.53
	细圆藤 *Pericampylus glaucus*	防己科 Menispermaceae	细圆藤属 *Pericampylus*	32.97	17.09	16.96
	柱果铁线莲 *Clematis uncinata*	毛茛科 Ranunculaceae	铁线莲属 *Clematis*	29.14	—	29.41
	络石 *Trachelospermum jasminoides*	夹竹桃科 Apocynaceae	络石属 *Trachelospermum*	25.23	26.03	46.06
	马莲鞍 *Streptocaulon juventas*	萝藦科 Asclepiadaceae	马莲鞍属 *Streptocaulon*	11.65	—	—
	厚叶玉叶金花 *Mussaenda erosa*	茜草科 Rubiaceae	玉叶金花属 *Mussaenda*	—	44.67	—
	厚果崖豆藤 *Millettia pachycarpa*	豆科 Fabaceae	崖豆藤属 *Millettia*	—	21.91	—
	菝葜 *Smilax china*	菝葜科 Smilacaceae	菝葜属 *Smilax*	—	16.17	18.11
	乌蔹莓 *Cayratia japonica*	葡萄科 Vitaceae	乌蔹莓属 *Cayratia*	—	12.21	—
	酸藤子 *Embelia laeta*	紫金牛科 Myrsinaceae	酸藤子属 *Embelia*	—	—	23.69
草本层	弓果黍 *Cyrtococcum patens*	禾本科 Poaceae	弓果黍属 *Cyrtococcum*	94.60	72.80	71.81
	五节芒 *Miscanthus floridulus*	禾本科 Poaceae	芒属 *Miscanthus*	49.36	57.47	63.38
	粽叶芦 *Thysanolaena latifolia*	禾本科 Poaceae	粽叶芦属 *Thysanolaena*	19.95	—	—
	扇叶铁线蕨 *Adiantum flabellulatum*	铁线蕨科 Adiantaceae	铁线蕨属 *Adiantum*	9.54	—	9.72
	飞机草 *Chromolaena odorata*	菊科 Compositae	飞机草属 *Chromolaena*	6.22	—	6.03
	淡竹叶 *Lophatherum gracile*	禾本科 Poaceae	淡竹叶属 *Lophatherum*	5.72	—	7.47
	半边旗 *Pteris semipinnata*	凤尾蕨科 Pteridaceae	凤尾蕨属 *Pteris*	5.32	18.22	11.02
	海金沙 *Lygodium japonicum*	海金沙科 Lygodiaceae	海金沙属 *Lygodium*	5.31	—	15.55
	蔓生莠竹 *Microstegium fasciculatum*	禾本科 Poaceae	莠竹属 *Microstegium*	—	163.33	138.6
	铁芒萁 *Dicranopteris linearis*	里白科 Gleicheniaceae	芒萁属 *Dicranopteris*	—	24.60	73.43
	荩草 *Arthraxon hispidus*	禾本科 Poaceae	荩草属 *Arthraxon*	—	12.82	—
	乌毛蕨 *Adiantum flabellulatum*	乌毛蕨科 Blechnaceae	乌毛蕨属 *Adiantum*	—	8.61	8.82
	阔片短肠蕨 *Allantodia matthewii*	蹄盖蕨科 Athyriaceae	短肠蕨属 *Allantodia*	—	6.38	—
	浆果薹草 *Carex baccans*	莎草科 Cyperaceae	薹草属 *Carex*	—	5.6	—
	团叶鳞始蕨 *Lindsaea orbiculata*	鳞始蕨科 Lindsaeaceae	鳞始蕨属 *Lindsaea*	—	—	10.04

注：PF 为西南桦纯林，BF 西南桦×灰木莲混交林，CBF 为西南桦×马尾松混交林，“ - ”不存在数据，下同。

2.2　林下植物优势种分布格局

由表3可知，3种造林模式的西南桦人工林中，林下共有幼树优势种漆树，在西南桦×马尾松混交林的扩散系数（DI）>1，负二项参数（K=5.64）>0，聚块性指标（PAI）>1，$t<t_{(17)}$呈聚集分布，但聚集程度不够明显；在纯林和西南桦×灰木莲混交林中为均匀分布。共有灌木优势种九节和藤本植物玉叶金花，在西南桦×灰木莲混交林和西南桦×马尾松混交林中K均小于2，远远小于8，呈强烈聚集分布；二者在纯林中则表现为强度较弱的聚集分布。共有草本优势种弓果黍和五节芒均呈强烈聚集分布，K值均小于2.5。在纯林中聚集强度五节芒高于弓果黍，在西南桦×灰木莲混交林和西南桦×马尾松混交林则是弓果黍高于五节芒。说明五节芒在纯林中对空间的占有程度较高，种群处于扩张阶段；而弓果黍在后二者亦处于侵占和定居阶段。

其他非共有林下植物优势种中，纯林幼树西南木荷为均匀分布，灌木银柴为聚集分布。在西南桦×灰木莲混交林，植物优势种山杜英、粗叶悬钩子、三叉苦和叶玉叶金花均为聚集分布，但聚集程度较低；蔓生莠竹则呈强烈聚集分布。在西南桦×马尾松混交林，植物优势种铁芒萁和蔓生莠竹为强烈聚集分布，枫香和琴叶榕为聚集分布，中平树、银柴和络石均为强度较弱的聚集分布。

3种造林模式的西南桦人工林中，林下植物优势种聚集分布占所有优势种，比例分别为62.5%，81.8%、92.3%；随机或趋向随机分布的植物优势种，比例分别为12.5%、9.1%、7.7%；均匀分布的植物优势种，比例分别为25%、9.1%、0%，即西南桦×马尾松混交林无均匀分布的植物优势种。

表3　西南桦人工林下植物优势种分布格局

层次	物种	扩散系数(DI)			t			负二项分布(K)			聚块性指数(PAI)			分布类型		
		PF	BF	CBF	PF	BF	CBF	PF	BF	CBF	PF	BF	CBF	PF	BF	CBF
乔木树种	西南木荷	0.22	—	—	-1.12	—	—	-2.38	—	—	0.58	—	—	U	—	—
	漆树	0.28	0.15	1.34	-1.03	-1.22	0.49	-1.93	-1.47	5.64	0.48	0.32	1.18	U	U	CL
	中平树	—	—	1.63	—	—	1.24	—	—	3.85	—	—	1.26	—	—	CL
	山杜英	—	1.80	—	—	1.15	—	—	6.25	—	—	1.16	—	—	CL	—
	枫香	—	—	4.72	—	—	5.34	—	—	1.01	—	—	1.99	—	—	CL
灌木层	银柴	2.88	—	1.41	2.69	—	0.58	1.22	—	8.54	1.82	—	1.12	CL	—	R
	琴叶榕	—	—	8.38	—	—	10.60	—	—	2.19	—	—	1.46	—	—	CL
	九节	1.89	4.69	3.04	9.48	5.29	2.93	4.27	1.98	1.96	1.23	1.50	1.51	CL	CL	CL
	粗叶悬钩子	—	2.97	—	—	2.84	—	—	2.48	—	—	1.40	—	—	CL	—
	三叉苦	—	3.72	—	—	3.90	—	—	1.15	—	—	1.87	1.03	—	CL	—
藤本植物	玉叶金花	1.26	11.33	4.42	0.37	14.83	4.91	13.84	0.67	1.36	1.07	2.49	1.74	R	CL	CL
	毒根斑鸠菊	7.82	1.93	2.31	9.79	1.34	1.89	0.50	3.36	2.47	2.99	1.19	1.40	CL	CL	CL
	络石	—	—	2.40	—	—	9.20	—	—	3.96	—	—	1.25	—	—	CL
	厚叶玉叶金花	—	1.74	—	—	1.06	—	—	4.65	—	—	1.22	—	—	CL	—
草本层	弓果黍	219.49	186.62	72.87	313.71	266.52	103.19	2.13	1.07	1.23	1.47	1.94	1.81	CL	CL	CL
	五节芒	51.31	6.28	16.47	72.24	7.58	22.22	1.13	2.40	2.20	1.88	1.42	1.46	CL	CL	CL
	蔓生莠竹	—	152.87	1082.48	—	218.06	1552.80	—	12.73	0.60	—	1.08	2.65	—	R	CL
	铁芒萁	—	—	85.33	—	—	121.08	—	—	1.18	—	—	1.85	—	—	CL

注：分布格局类型，CL聚集分布；R随机分布；U均匀分布。t（0.05，17）= 2.110，t（0.01，17）= 2.898。

4 结论与讨论

本研究结果显示，3 种造林模式的西南桦人工林中，西南桦纯林植物种类最多，西南桦×马尾松混交林次之，西南桦×灰木莲混交林中物种数量最少，这与以往学者对西南桦纯林[22]、西南桦×马尾松混交林[23]以及西南桦×山桂花混交林[24]林下植物种类数量的研究结果是一致的。原因可能与西南桦为落叶乔木树种，营造纯林时，冠幅较小且不连续，冬春落叶期间，林内光照条件较好，为植物生长与繁殖提供了良好的生存空间，林下植物种类最为丰富；西南桦×马尾松混交造林，与西南桦×速生阔叶树种(灰木莲、山桂花等)比较，林冠层盖度相对较小，形成大量的林隙环境，为土壤种子库和地上植被的天然更新以及更多物种的入侵和定居创造了良好的条件，致使林下植物种类相对较多，物种多样性恢复的效果较好。由于发育时间较短，3 种西南桦人工林群落结构层次尚未明显，特别是主林层均为造林树种，但林下幼树、灌木、藤本和草本植物组成极为丰富。在近自然经营的理念下，采取合理的干扰措施，促进林下幼树、灌木生长发育，将使主林层结构进一步分化，有助于形成西南桦人工林复层群落结构。

在本研究中，林下共有幼树优势种漆树，在西南桦×马尾松混交林呈聚集分布，但聚集程度不够明显，在纯林和西南桦×灰木莲混交林为均匀分布；灌木九节和藤本植物玉叶金花，在西南桦×灰木莲混交林和西南桦×马尾松混交林呈强烈聚集分布，二者在纯林聚集程度不明显；弓果黍和五节芒均在 3 种西南桦林内均呈强烈聚集分布。说明 3 种造林模式的西南桦人工林中，林下植物优势种有随机、均匀和聚集分布 3 种类型；其中，林下植物优势种聚集分布占所有优势种的比例达 60%以上，与上官铁梁等(1988)对山西绵山植被优势种群分布格局的研究[25]以及庞圣江等(2015)对细叶云南松种群空间分布格局的研究结论基本一致[26]，他们研究认为，绝大多数植物种群的分布格局类型为聚集分布。究其原因，可能与植物种群个体的生物和生态学特性密切相关，由于植物种群在自然环境分布的偶然性、地形差异和空间距离等因素，致使植物种群侵入某一领地的定居过程是非同步的点状或者非均匀的斑块状，造成植物种群的空间分布异质性。也有学者研究认为，植物种群的空间分布格局，一方面取决于植物个体的生物学特性和种间竞争关系等一系列生理生态过程，另一方面与植物群落环境，如林窗大小、土壤养分和水分资源分布等密切相关[27,28]。有关西南桦林下植物优势种分布格局的影响，尚待从植物生物学特性、种间竞争、土壤肥力以及林分环境异质性等方面开展更深层次的研究。

以往研究发现，植物种群分布格局聚集强度的变化，在某种程度上可用于表征种群的动态变化，同时，与植物群落结构的变化和演替发展相关联[29]。3 种造林模式的西南桦人工林中，林下植物共有优势种漆树、九节、玉叶金花、毒根斑鸠菊、弓果黍和五节芒，在不同群落的空间分布格局以及聚集强度方面的差异；造成上述原因，与植物群落的自身特性、干扰特征、资源和生境状况的差异等有关[30-32]。由于西南桦与不同伴生树种造林时，改变了西南桦林内植物种类组成、光照、土壤肥力和水分等生境状况，致使种内以及种间竞争关系发生变化[22]。西南桦纯林呈均匀分布的林下植物优势种占所有优势种的比例最高，表明其生境相对均匀，一些植物优势种已经完成对部分空间的侵占，能够使种群长期保持稳定的状态。而西南桦×灰木莲混交林与西南桦×马尾松混交林中，强烈聚集的植物优势种占所有优势种的比例较高，说明大多数植物优势种群处于侵入与定居或者扩张阶段。聚集强度差异的结果表明，针对不同西南桦造林模式，林下植物优势种为了获取充足的资源，已经进化出不同的生存策略或者适应机制。有学者研究指出，植物优势种群处于侵入与定居或者得到进一步扩散的阶段时，具有极高聚集程度；当种群定居成功并不断地繁衍增长，侵占大部分空间以后，聚集程度就会相应地下降[29]。

综上所述，3 种造林模式对西南桦林下植物优势种及其分布格局的影响存在差异；我们建议适当立地选择、减少西南桦纯林比例，宜选取马尾松作为伴生树种混交造林，采取合理的营林措施，改善林分环境，提高林分异质性，从而促进西南桦人工林健康、高效的可持续经营。

参考文献

[1]康永祥，康博文，刘建军，等．陕北黄土高原文冠果群落结构及物种多样性[J]．生态学报，2010，30(16)：4328-4339.

[2]樊艳荣，陈双林，林华，等．不同林下植被干扰措施对毛竹林下植物种群分布格局的影响[J]．生物多样性，2013，21(6)：709-714.

[3]彭少麟．南亚热带森林群落动态学[M]．科学出版社，1996：84-89.

[4]BROWN J R，STUTH J W. How herbivory affects grazing

tolerant and sensitive grasses in a central Texas grassland: integrating plant response across hierarchical levels [J]. Oikos, 1993, 67(2): 291-297.
[5]刘振国，李镇清．不同放牧强度下冷蒿种群小尺度空间格局[J]．生态学报，2004，24(2)：227-234.
[6]封磊，洪伟，吴承祯，等．杉木-拟赤杨人工混交林种内、种间竞争强度研究[J]．热带亚热带植物学报，2004，12(1)：46-50.
[7]玉宝，乌吉斯古楞，王百田，等．兴安落叶松天然林不同林分结构林木水平分布格局特征研究[J]．林业科学研究，2010，23(1)：83-88.
[8]柴勇，朱华，孟广涛，等．云南哀牢山古茶树群落优势树种的种群结构与分布格局[J]．林业科学研究，2011，24(3)：277-284.
[9]赵中华，惠刚盈，胡艳波，等.2种类型阔叶红松林优势种群空间分布格局及其关联性[J]．林业科学研究，2011，24(5)：554-562.
[10]郭琦，王新杰．不同混交模式杉木人工林林下植被生物量与土壤物理性质研究[J]．中南林业科技大学学报，2014，(05)：70-74.
[11]姚茂和，盛炜彤，熊有强．杉木林下植被对立地的指示意义[J]．林业科学，1992，28(3)：208-212.
[12]袁正科，田育新，李锡泉，等．缓坡梯土幼林林下植被覆盖与水土流失[J]．中南林业科技大学学报，2002，22(2)：21-24.
[13]HUFFMAN D W, MOORE M M. Responses of Fendler ceanothus to overstory thinning, prescribed fire, and drought in an Arizona ponderosa pine forest [J]. Forest Ecology and Management, 2004, 198(1-3): 105-115.
[14]TAYLOR A H, JANG S W, ZHAO L J, et al. Regeneration patterns and tree species coexistence in old-growth Abies-Picea forests in southwestern China [J]. Forest Ecology and Management, 2006, 223(1-3): 303-317.
[15]褚建民，卢琦，崔向慧，等．人工林林下植被多样性研究进展[J]．世界林业研究，2007，20(3)：9-12.
[16]尹伟伦．全球森林与环境关系研究进展[J]．森林与环境学报，2015，35(1)：1-7.
[17]BEATTY S W. Influence of microtopography and canopy species on spatial patterns of forest understory plants [J]. Ecology, 1984, 65(5): 1406-1419.
[18]GÁLHIDY L, MIHÓk B, HAGYÓ A, et al. Effects of gap size and associated changes in light and soil moisture on the understorey vegetation of a Hungarian beech forest [J]. Plant Ecology, 2006, 183(1): 133-145.
[19]王成，刘继生，张美淑，等．带状间伐促进赤松人工林天然更新效果的研究——天然更新的综合效果与分布格局[J]．延边大学农学学报，2004，26(2)：77-80.
[20]曾杰，郭文福，赵志刚，等．我国西南桦研究的回顾与展望[J]．林业科学研究，2006，19(3)：379-384.
[21]张金屯．植物种群空间分布的点格局分析[J]．植物生态学报，1998，22(4)：344-349.
[22]李品荣，曾觉民，陈强，等．西南桦人工纯林与混交林群落学特征比较[J]．东北林业大学学报，2007，35(5)：14-16.
[23]杨保国，刘士玲，贾宏炎，等．广西大青山西南桦人工林群落结构特征[J]．广西林业科学，2016，45(4)：414-418.
[24]李莲芳，刘永刚，孟梦，等．热带山桂花、西南桦人工幼林植物区系成分分析[J]．广西植物，2007，27(3)：414-419.
[25]上官铁梁，张峰．山西绵山植被优势种群的分布格局与种间联结的研究[J]．植物科学学报，1988，6(4)：357-364.
[26]庞圣江，张培，贾宏炎，等．细叶云南松群落物种多样性与种群空间分布格局[J]．南方农业学报，2015，46(4)：645-651.
[27]刘聪，朱教君，吴祥云，等．辽东山区次生林不同大小林窗土壤养分特征[J]．东北林业大学学报，2011，39(1)：79-81.
[28]张金屯．植物种群空间分布的点格局分析[J]．植物生态学报，1998，22(4)：344-349.
[29]张继义，赵哈林．科尔沁沙地草地植被恢复演替进程中群落优势种群空间分布格局研究[J]．生态学杂志，2004，23(2)：1-6.
[30]罗菊春，王庆锁．干扰对天然红松林植物多样性的影响[J]．林业科学，1997，33(6)：498-503.
[31]马云波，牛聪傑，许中旗，等．不同造林模式对铁尾矿地土壤性质的影响[J]．水土保持学报，2015(3)：242-248.
[32]邓娇娇，周永斌，杨立新，等．落叶松和水曲柳带状混交对土壤微生物群落功能多样性的影响[J]．生态学杂志，2016，35(10)：2684-2691.

[原载：中南林业科技大学学报，2018，38(07)]

广西大青山西南桦人工林林下植物多样性与稳定性

庞圣江[1] 张 培[1] 杨保国[1] 唐 诚[2] 曾 杰[2] 贾宏炎[1]
([1]中国林业科学研究院热带林业实验中心，广西凭祥 532600；
[2]中国林业科学研究院热带林业研究所，广东广州 510520)

摘 要 以广西大青山西南桦人工林为研究对象，根据林分类型设置53块20m×30m典型样地，在群落学的调查基础上，分析西南桦人工林植物多样性和群落稳定性，揭示了人工植被的动态变化特征。结果表明：①大青山林区西南桦人工林，记录到林下植物种类76科139属164种；植物优势种重要值所占比例较大(西南桦纯林49.77%，西南桦×红椎混交林58.40%，西南桦×灰木莲混交林50.04%，西南桦×杉木混交林58.17%和西南桦×马尾松混交林70.18%)，其在群落结构和演替发展中占据优势地位。②灌木层植物种类最多，Shannon-Wiener指数和Simpson指数均表现为灌木层>乔木树种>藤本植物>草本层。③利用改进的M. Godron稳定性测定方法测得，人工林下植物群落均处一定的稳定状态，说明植物群落的物种多样性越大，群落结构越复杂，有利于维持群落的稳定。

关键词 西南桦；林下植被；群落；多样性；稳定性；人工林

Species Diversity and Community Stability of *Betula alnoides* Plantations in Daqingshan Mountain Areas of Guangxi

PANG Shengjiang[1], ZHANG Pei[1], YANG Baoguo[1], TANG Cheng[2], ZENG jie[2], JIA Hongyan[1]
(*[1]Experimental Center of Tropical Forestry, CAF, Pingxiang*, 536000, *Guangxi, China*;
[2]Research Institute of Tropical Forestry, CAF, Guangzhou, 510520, *Guangdong, China*)

Abstract: The 53 plots of 20m×30m were set up according to stand types of *Betula alnoides* plantations in Daqingshan forest region, there were selected to study the species diversity and community stability of the plantations, In order to explore the dynamic characteristics and change mechanism of artificial vegetation. The results showed that①A total of 164 species belonging to 139 genera and 78 families were recorded across forest types, there were a large proportion in the importance values of dominant species (the pure plantations 49.77%, *Betula alnoides*×*Castanopsis hystrix* mixed plantations 58.40%, *Betula alnoides*×*Manglietia glauca* mixed plantations 50.04%, *Betula alnoides*×*Cunninghamia lanceolata* mixed plantations 50.04%, *Betula alnoides*×*Pinus massoniana* mixed plantations 70.18%), its occupied the favorable position in structure of community and succession. ② Shrub plants was the most abundant species in *Betula alnoides* plantations, Shannon-Wiener index and Simpson index showed shrub layer >arbor species > vine species> herb layer. ③Using the modified method of *M. Gordon* to measure the stability of 4 communities, the plant communities was stable state in the artificial forest, with more species diversity of plant communities, forest structure is more complex and stability is better.

Key words: *Betula alnoides*; understory plant; community; diversity; stability; plantation

物种多样性和稳定性是植物群落结构特征的两个基本属性[1]；物种多样性是植物群落结构与功能复杂性的一种度量[2]，稳定性则是表现群落结构与功能的一种综合特征[3]。随着全球生态环境问题日益突出，森林生态系统稳定性问题，即维持天然林的稳定性和建立稳定的人工林植被，显然，已成为林业生态领域研究的热点问题之一[4-6]。我国是人工林种植面积最大的国家，人工林面积已超 5.3×10^7 hm^2[7]，人工林的大规模发展，在一定程度上缓解和满足了人类的木材需要，对生态环境的保护与恢复发挥了巨大作用。但是，由于长期以来缺乏科学的理论指导和有效的可持续经营技术等管理措施，这些人工林普遍存在地力衰退、病虫害加剧、生物多样性和经济效益下降一系列生态问题[8-11]。因此，开展人工林植物多样性与稳定性研究显得尤为重要，对揭示人工林植物群落的动态特征和变化规律、造林树种的选取以及建立稳定的人工林生态系统具有重要意义。

西南桦(*Betula alnoides*)为桦木科(Betulaceae)桦木属(*Betula*)乔木树种，天然分布于我国广西、云南、贵州和西藏等地。又因速生、适应性广、材性优良和经济价值高等特性，现已成为我国热带、南亚热带地区人工种植的重要乡土珍贵树种之一，人工林面积已超过15万 hm^2[12]。随着西南桦种植业的大规模发展，经营模式的集约化、盲目造林等，在一定程度上影响了林下植被生长发育，导致了一些地方人工林生态系统的稳定性较差[12-14]。鉴于林下植被在人工林生态系统中的重要作用，以及西南桦在我国南方乡土珍贵树种木材生产中的重要地位。本研究以广西大青山林区5种造林模式的西南桦人工林为研究对象，对西南桦纯林、西南桦×红椎混交林、西南桦×灰木莲混交林、西南桦杉木混交林和西南桦×马尾松混交林植被群落物种多样性和群落稳定性进行研究，揭示珍贵树种西南桦人工林下植被的动态特征及其稳定性的影响，为西南桦种植业的健康、高效发展提供科学依据

1　研究地概况

广西大青山林区位于广西凭祥市中国林业科学研究院热带林业实验中心，坐标为21°58′~22°19′N、106°40′~106°59′E。研究区属南亚热带季风气候区，年均气温20~22℃，年均降水量为1200~1600mm，季节分布差异大，集中于4~10月。根据中国林业科学研究院热带林业实验中心营林资料和野外勘查现有西南桦林分的基础上，该林区的西南桦人工林按不同种植模式，主要包括西南桦纯林和西南桦×红椎(*Castanopsis hystrix*)、西南桦×灰木莲(*Manglietia glauca*)、西南桦×杉木(*Cunninghamia lanceolata*)、西南桦×马尾松(*Pinus massoniana*)混交林等5种林分类型，海拔分布190~670m，土壤类型以山地红壤为主，土层厚度在1m以上。

2　研究方法

2.1　样地设置与调查

采用标准地调查法，设置面积为20m×30m的调查样地53个；其中，西南桦纯林9个，西南桦×红椎混交林17个，西南桦×灰木莲混交林6个，西南桦×杉木15个，西南桦×马尾松6个。调查内容包括：记录乔木树种名称、株数、胸径和树高等。同时，在每个样地内沿对角线的两个角和中心位置，分别设置3个5m×5m的样方，用于林下植被调查，包含幼树树种名称、株数、胸径、树高和盖度；灌木植物名称、株数、树高和盖度，藤本植物种类、数量、株高和盖度以及草本植物的种类、数量、株高和盖度。记录调查样地的林分类型、林龄、主林层郁闭度、海拔、坡位和坡度等。同时，对样地周边植被进行全面踏察记录。样地基本概况，详见表1。

表1　不同植被模式基本概况

编号	群落类型	林龄(年)	林分密度(株/hm^2)	$\bar{D}$(cm)	$\bar{H}$(m)	郁闭度	海拔(m)	坡位	坡向	坡度(°)
Ⅰ	西南桦纯林	16~20	522	19.89	15.36	0.7	250~440	L/M	SW/SE	18~31
Ⅱ	西南桦×红椎混交林	8~14	668	15.16	12.95	0.8	460~640	L/M	S/SE/SW	24~42
Ⅲ	西南桦×灰木莲混交林	16	566	16.84	13.69	0.7	480~670	M	N/NS/NE	16~28
Ⅳ	西南桦×杉木混交林	12~17	865	12.10	10.42	0.9	390~550	L/M	S/SE/SW	14~39
Ⅴ	西南桦×马尾松混交林	14	640	10.91	7.83	0.8	190~480	M	N/NW	9~26

注：$\bar{D}$平均胸径，$\bar{H}$平均树高。

2.2　数据处理与分析

林下植物群落种类组成及优势度采用其重要值统计，具体公式如下：

$$重要值=(相对密度+相对频度+相对盖度)/3 \quad (1)$$

林下植物物种多样性的分析选取丰富度指数(R)、Shannon-Wiener指数(H)和Simpson指数(D)计算，其表达式如下：

$$R=S(样方内物种总数) \quad (2)$$

$$\text{Shannon-Wiener 指数：} H=-\sum_{i=1}^{S}P_i\ln P_i \quad (3)$$

$$\text{Simpson 指数：} D=1-\sum_{i=1}^{S}P_i^{\ 2} \quad (4)$$

式中：S为物种总数；Pi为物种i的个体数所占

总个体数的比例[15]。

林下植物群落的稳定性应用郑元润改进的M. Godron稳定性测定方法[16]，并用Excel和SPSS 16.0对数据处理及分析，并进行平滑曲线的模拟，具体公式如下：

平滑曲线模拟公式：$y=ax^2+bx+c$ (6)

直线方程：$y=100-c$ (7)

将(6)代入(7)求得 x 解为：

$$x=\frac{-(b+1)\pm\sqrt{(b+1)^2-4a(c-100)}}{2a} \quad (8)$$

上述方程求解一个介于0~100之间，另一个求解大于100，根据研究情况，交点 x 轴的坐标应采用在0~100区间。

3 结果与分析

3.1 林下植物物种组成及重要值

本研究对西南桦人工林植物物种组成进行了较为详细的调查，在所有的样方中，共发现记录林下植物种类76科139属164种。限于篇幅，仅列出各层次植物优势种及其重要值(表2)。从表2可以看出，西南桦纯林乔木优势种为西南木荷、漆树，西南桦×红椎混交林中为火力楠和肉桂，西南桦×灰木莲混交林中为山杜英和中平树，西南桦×杉木混交林中为油茶和漆树，西南桦×马尾松混交林中为漆树、中平树和枫香。灌木优势种在西南桦纯林中为银柴和九节，西南桦×红椎混交林中为杜茎山和肉桂，西南桦×灰木莲混交林和西南桦×杉木混交林中均为九节，西南桦×马尾松混交林中为银柴、九节和琴叶榕。藤本植物优势种在西南桦纯林和西南桦×红灰木莲混交林中为玉叶金花，西南桦×红椎混交林中为玉叶金花、酸藤子和菝葜，西南桦×杉木混交林中为玉叶金花、毒根斑鸠菊和厚叶玉叶金花，西南桦×马尾松混交林中为玉叶金花和毒根斑鸠菊。草本植物优势种在西南桦纯林中为弓果黍和五节芒，西南桦×红椎混交林中为弓果黍、五节芒、淡竹叶和半边旗，西南桦×灰木莲混交林、西南桦×杉木混交林和西南桦×马尾松混交林中为均为蔓生莠竹。

上述在5个西南桦林分中植物优势种重要值所占比例较大(西南桦纯林49.77%，西南桦×红椎混交林58.40%，西南桦×灰木莲混交林50.04%，Ⅳ西南桦×杉木混交林58.17%和西南桦×马尾松混交林70.18%)，其在群落结构和演替发展中占据优势地位。

表2 西南桦人工林植物物种组成

层次	物种	重要值				
		Ⅰ	Ⅱ	Ⅲ	Ⅳ	Ⅴ
乔木树种	西南木荷 *Schima wallichii*	62.98	17.48		—	—
	漆树 *Toxicodendron delavayi*	41.01	12.53	31.77	44.41	86.64
	乌桕 *Triadica cochinchinensis*	35.95	4.50	—	11.56	7.94
	中平树 *Macaranga denticulata*	30.53	—	45.96	9.52	49.73
	鸭脚木 *Schefflera minutistellata*	21.09	10.85	6.92	31.10	—
	火力楠 *Michelia foveolata*	—	59.71	—	—	—
	肉桂 *Cinnamomum cassia*	—	45.69	—	—	—
	山杜英 *Elaeocarpus sylvestris*	4.50	—	40.74	—	34.01
	油茶 *Camellia oleifera*	—	—	15.97	55.64	—
	枫香 *Liquidambar formosana*	—	—	—	—	53.24
灌木层	银柴 *Aporosa dioica*	52.80	3.44	2.20	17.06	42.06
	九节 *Psychotria asiatica*	31.93	25.62	52.23	51.52	33.67
	琴叶榕 *Ficus pandurata*	28.40	—	—	—	39.44
	杜茎山 *Maesa japonica*	14.71	80.31	15.38	12.26	21.26
	三叉苦 *Melicope pteleifolia*	0.93	28.75	30.12	21.90	16.49
	大青 *Clerodendrum cyrtophyllum*	6.70	12.32	22.97	25.69	21.77

（续）

层次	物种	重要值				
		Ⅰ	Ⅱ	Ⅲ	Ⅳ	Ⅴ
藤本植物	玉叶金花 *Mussaenda pubescens*	50.69	49.60	74.10	46.61	62.51
	毒根斑鸠菊 *Vernonia cumingiana*	36.39	21.79	33.57	57.74	48.53
	厚叶玉叶金花 *Mussaenda erosa*	10.30	16.82	34.67	51.10	2.30
	酸藤子 *Embelia laeta*	2.70	37.19	10.77	10.65	23.69
	菝葜 *Smilax china*	—	36.55	16.17	10.57	18.11
草本层	弓果黍 *Cyrtococcum patens*	87.41	30.30	17.80	38.38	11.81
	五节芒 *Miscanthus floridulus*	52.14	34.95	17.47	24.56	23.38
	蔓生莠竹 *Microstegium fasciculatum*	32.32	5.21	163.33	85.97	138.60
	粽叶芦 *Thysanolaena latifolia*	19.74	10.60	2.22	17.82	2.88
	铁芒萁 *Dicranopteris linearis*	17.56	13.32	24.60	18.05	33.43
	淡竹叶 *Lophatherum gracile*	9.49	30.64	5.50	10.46	7.47
	半边旗 *Pteris semipinnata*	9.29	29.60	6.22	22.98	11.02

3.2 林下植物多样性分析

林下植物物种丰富度（S）、Shannon-Wiener 指数（H）和 Simpson 指数（D）的计算结果，详见表3。由表3可知，西南桦纯林中物种数量为93种，西南桦×红椎混交林中物种数量为105种，西南桦×灰木莲混交林中物种数量为72种，西南桦×杉木混交林中物种数量为94种，西南桦×马尾松混交林中物种数量为78种；表明大青山林区植物种类比较多，群落结构较为复杂，但植物物种丰富度存在一定的差异，可能与调查样地数量的选取有关。

5种西南桦林分中，丰富度（S）、Shannon-Wiener 指数（H）和 Simpson 指数（D）均以灌木层最大；其余各植物群落层次除西南桦×红椎混交林物种丰富度外，另外4种人工林都表现出比较一致的顺序，由此说明，灌木层植物种类最多，乔木树种次之，藤本植物和草本层的多样性最低。各植物群落层次 Shannon-Wiener 指数和 Simpson 指数均表现为灌木层>乔木树种>藤本植物>草本层。

表3 不同林分类型物种多样性

林分类型	层次	丰富度（S）	Shannon-Wiener 指数（H）	Simpson 指数（D）
Ⅰ	乔木树种	17	2.41	0.89
	灌木层	36	2.77	0.91
	藤本植物	19	2.47	0.89
	草本层	21	1.75	0.72
Ⅱ	乔木树种	23	2.57	0.89
	灌木层	33	2.68	0.89
	藤本植物	17	2.27	0.88
	草本层	31	2.33	0.85
Ⅲ	乔木树种	12	2.30	0.88
	灌木层	24	2.50	0.89
	藤本植物	15	2.26	0.86
	草本层	21	0.80	0.31

（续）

林分类型	层次	丰富度(S)	Shannon-Wiener 指数(H)	Simpson 指数(D)
Ⅳ	乔木树种	15	2.34	0.88
	灌木层	37	2.77	0.90
	藤本植物	19	2.20	0.86
	草本层	23	1.95	0.75
Ⅴ	乔木树种	12	2.24	0.82
	灌木层	30	2.57	0.88
	藤本植物	15	2.12	0.84
	草本层	21	1.19	0.46

3.3 林下植物群落的稳定性

由表4可知，5个西南桦人工林中植物群落的交点坐标值与理论稳定点(20，80)较为近，即现阶段5种西南桦人工林中植物群落均处于一定的稳定状态。5种西南桦人工林植物群落之间的植物种类各有所差异，相较而言，西南桦纯林的林下植物群落的稳定性最高，而西南桦×杉木混交林与西南桦×马尾松混交林的稳定性略高于西南桦×红椎混交林与西南桦×灰木莲混交林。说明植物群落的物种多样性越大，群落结构越复杂，有利于维持群落的稳定。显然，植物群落维持稳定的机制与造林配置树种的生物学特性、繁殖更新和种群间的竞争等密切相关。

表4 不同群落稳定性分析结果

林分类型	曲线类型	相关系数 R^2	交点坐标	P 值	结果
Ⅰ	$y=-0.0133x^2+2.1335x+13.131$	0.9849	20.11/79.89	$P<0.01$	稳定
Ⅱ	$y=-0.0128x^2+2.225x+18.597$	0.9698	21.04/78.96	$P<0.01$	稳定
Ⅲ	$y=-0.109x^2+1.9355x+10.924$	0.9896	26.09/73.91	$P<0.01$	稳定
Ⅳ	$y=-0.0138x^2+2.1598x+15.22$	0.9764	18.39/81.61	$P<0.01$	稳定
Ⅴ	$y=-0.0138x^2+2.2141x+9.7103$	0.9858	19.60/80.40	$P<0.01$	稳定

4 结论与讨论

调查发现，广西大青山5种西南桦人工林下植物种类，共76科139属164种。其中，西南桦纯林的植物种类为96种，西南桦×红椎混交林为104种，西南桦×灰木莲混交林为75种，西南桦×杉木混交林为94种，西南桦×马尾松混交林为78种，说明利用乡土珍贵阔叶树种西南桦人工造林，林下植物种类丰富，人工干扰后的植被恢复效果较好。这与李莲芳等(2007)对云南西双版纳西南桦人工幼林植物群落的研究结果基本一致[17]，他们调查发现，热带地区的西南桦纯林内植物种类高达113种、西南桦×山桂花混交林为106种，明显高于该地区的山地雨林75种和季风常绿阔叶林58种；王卫斌等(2009)对13年生西南桦人工林植物物种多样性的研究也得出相类似结论[18]，即其灌木层和草本植物等林下植物多样性均超过山地雨林，林下植物物种多样性得到良好的恢复。这可能与西南桦为落叶树种，树冠较小且不连续，冬春落叶过程中，林内光照条件充足，为林下植物的生长发育提供了良好的条件；同时，在西南桦人工造林初期，所形成的大面积林隙空地，土壤种子库中的种子直接参与林下植被的更新，以及为外来物种的侵入与定居创造了机会，致使林下植物的丰富度均较高。

在所调查西南桦林内，由于植物生理特性及其生态习性的差异，乔灌树种大多数以阳性植物占据优势地位，如西南木荷、银柴、中平树和山杜英、毒根斑鸠菊等；而草本植物以弓果黍、粽叶芦、蔓生莠竹和半边旗等禾本科和蕨类植物为主，说明在不同造林模式下，西南桦林内植物群落对光照和水分条件的依赖性较强。本研究的结果与李品荣等(2007)对西南桦人工纯林与混交林群落学特征研究结论基本一致[19]，他们研究发现，西南桦林分内植物群落的光照生态谱以阳生种类为主，水分生态谱则以中生植物种类占据优势地位。喻阳华等(2015)对赤水河上游杉木、马尾松林下植物多样性研究也得出相类似结论[20]，其研究结果表明，光照是影响林下植物种类、数量和多样性的重要生态因素。植物多样性指数中，Shannon-Wiener 指数、Simpson 指数和 Pielou 指数基本表现为，乔木树种、灌木和藤本植物大于草本植物，说明现阶段的林下植物群落处于侵占或者扩散阶段，草本植物对资源的获取，竞争压力较大；表明林下植物多样性与西南桦林分的生长发育关系密切，及时的营林抚育措施，是提高西南桦人工林下植物物种多样性和维持人工林生态系统稳定性的有效途径。

本研究中，利用郑元润改进的 M. Godron 稳定性测定方法，测得 5 个西南桦人工林植物种群交点坐标值与理论稳定点(20，80)较为接近。有学者研究指出，植物种群维持稳定性机制的主要途径是繁殖更新，植物优势种能否顺利地繁殖更新是群落维持稳定的基本前提[21]，由此表明，广西大青山西南桦人工林下植物优势种已经顺利完成新空间的侵占和定居过程，植物优势种群处于稳定的繁衍阶段。这与张鹏等(2016)对北京山区落叶松林、栓皮栎林和油松林等 3 种典型人工林的植物群落结构及稳定性的研究基本一致[22]，他们研究亦认为，植物物种多样性越高，群落结构越复杂，其稳定性就越强。也有学者认为，长苞铁杉天然林中，林隙的存在植物物种多样性的增加，反而降低了群落的稳定性[23]，与本研究结论不一致，这可能与所选取不同的研究对象有关，致使林隙外比林隙内植物群落稳定性较高，长苞铁杉群落处于一个亚稳定的状态；而西南桦林分中植物种群的能够顺利完成繁殖更新，物种多样性的增加及其生态功能的恢复，使得人工林群落趋于比较稳定的状态。在今后西南桦人工林群落的稳定性研究中，可考虑引入诸如主林层树种，土壤肥力、病虫害等更多的指标进行分析。

参考文献

[1]MCNAUGHTO S J. Diversity and stability of ecological communities：A comment on the role of empiricism in ecology[J]. The American Naturalist，1977，111：515-525.

[2]高贤明，马克平，陈灵芝．暖温带若干落叶阔叶林群落物种多样性及其与群落动态的关系[J]．植物生态学报，2001，25(3)：283-290.

[3]冶民生，吴斌，关文彬，等．岷江上游植物群落稳定性研究[J]．水土保持研究，2009，16(1)：259-263.

[4]马风云．生态系统稳定性若干问题研究评述[J]．中国沙漠，2002，22 (4)：401-407.

[5]高润梅，石晓东，郭跃东．山西文峪河上游河岸林群落稳定性评价[J]．Journal of Plant Ecology(植物生态学报(英文版)，2012，36(6)：491-503.

[6]杨瑞红，赵成义，王新军，等．固沙植物群落稳定性研究[J]．中南林业科技大学学报自然科学版，2015，(11)：128-135.

[7]彭舜磊，王得祥，赵辉，等．我国人工林现状与近自然经营途径探讨[J]．西北林学院学报，2008，23(2)：184-188.

[8]张鼎华，叶章发，李宝福．杉木、马尾松轮作对林地土壤肥力和林木生长的影响[J]．林业科学，2001，37(5)：10-15.

[9]张昌顺，李昆．人工林地力的衰退与维护研究综述[J]．世界林业研究，2005，18(1)：17-21.

[10]庞圣江，唐诚，张培，等．广西大青山西南桦人工林拟木蠹蛾为害的影响因子[J]．东北林业大学学报，2016，44(11)：85-88.

[11]彭姣，赵运林，徐正刚，等．常德油茶次生林林下植被物种多样性调查[J]．中南林业科技大学学报自然科学版，2016，(12)：68-73.

[12]曾杰，郭文福，赵志刚，等．我国西南桦研究的回顾与展望[J]．林业科学研究，2006，19(3)：379-384.

[13]蒋云东，周凤林，周云，等．西南桦人工林土壤养分含量变化规律研究[J]．云南林业科技，1999(2)：27-31.

[14]李江，陈宏伟，冯弦．云南热区几种阔叶人工林 C 储量的研究[J]．广西植物，2003，23(4)：294-298.

[15]方精云，王襄平，沈泽昊，等．植物群落清查的主要内容、方法和技术规范[J]．生物多样性，2009，17(6)：533-548.

[16]郑元润．森林群落稳定性研究方法初探[J]．林业科学，2000，36(5)：28-32.

[17]李莲芳，刘永刚，孟梦，等．热带山桂花、西南桦人工幼林植物区系成分分析[J]．广西植物，2007，27(03)：414-419.

[18]王卫斌，杨德军，曹建新．西南桦人工林植物多样性与相似性比较[J]．南京林业大学学报自然科学版，2009，33(5)：32-36.

[19]李品荣，曾觉民，陈强，等．西南桦人工纯林与混交林群落学特征比较[J]．东北林业大学学报，2007，35(5)：14-16.

[20]喻阳华，李飒，严令，等．赤水河上游次生林林窗特征及森林更新[J]．森林与环境学报，2015，35(3)：265-271.
[21]张继义，赵哈林．植被(植物群落)稳定性研究评述[J]．生态学杂志，2003，22(4)：42-48.
[22]张鹏，王新杰，王勇，等．北京山3种典型人工林群落结构及稳定性[J]．东北林业大学学报，2016，44(1)：1-5.
[23]钱莲文，吴承祯，洪伟，等．长苞铁杉林林隙内外群落多样性及稳定性[J]．福建林学院学报，2005，25(3)：202-205.

[原载：中南林业科技大学学报，2018，38(02)]

基于林班尺度的森林立地类型划分与质量评价

卢立华[1,2]　冯益明[3]　农　友[1,2]　李　华[1,2]　农良书[1,2]　孙冬婧[1,2]　黄德卫[1,2]　明安刚[1,2]

（[1] 中国林业科学研究院热带林业实验中心，广西凭祥　532600；
[2]广西友谊关森林生态系统国家定位观测研究站，广西凭祥　532600；
[3]中国林业科学研究院荒漠化研究所，北京　100091）

摘　要　在广西凭祥中国林业科学研究院热带林业实验中心哨平实验场，以林班为单元，按 7hm²/块密度布设样地调查获取立地因子与林木生长信息，以主成分分析确定立地类型划分主导因子，依据六级分类系统及利用 ArcGIS 分析叠加技术得到立地类型及图表，综合评价立地质量。结果表明：①研究区划分 2 个立地类型小区，3 个立地类型组，88 个立地类型，其中：立地质量Ⅰ级为零；Ⅱ，Ⅲ，Ⅳ，Ⅴ级立地类型数、面积及占总面积百分数依次为 4 个，48. 87hm²，1. 44%；11 个，106. 42hm²，3. 14%；52 个，2799. 21hm²，82. 65%；19 个，432. 36hm²，12. 77%。②立地质量中等及以上立地类型 15 个，面积 155. 29hm²，占 4. 58%；而较差或差达 73 个，3231. 57hm²，占 95. 42%。

关键词　林班尺度；立地类型划分；立地质量评价；地位指数；肥力指标

The Classification and Quality Evaluation on Forest Site Types Using Compartment As the Research Scale

LU Lihua[1,2], FENG Yiming[3], Nong You[1,2], LI Hua[1,2], NONG Liangshu[1,2], SUN Dongjing[1,2], HUANG Dewei[1,2], MING Angang[1,2]

([1]*Experimental Center of Tropical Forestry, Chinese Academy of Forestry, pingxiang 532600, Guangxi, China;*
[2]*Guangxi Youyiguan Forest Ecosystem Research Station, Pingxiang 532600, Guangxi China;*
[3]*Institute of Desertification Studies, Chinese Academy of Forestry, Beijing 100091, China*)

Abstract: Sampling plots with a density of 7hm²/plot were distributed and set up based on the results of National Forest Resource Inventory in the Shaoping Experimental Station of Experimental Center of Tropical Forestry, Chinese Academy of Forestry in Guangxi China. After previous site selection on the forest map, on site investigations and plot setting were conducted for acquisition site properties. the site types classification guiding factors found out through the Principal Components Analysis method. According to the "Six-level classification system" from "Forest site classification in China", the data from field study were overlaid in the ArcGIS software to obtain forest site types and related figures, and the site quality was evaluated through comprehensive evaluation method. Results show: There were 88 site types which were included in 2 blocks and 3 groups in the study area, being pertained to 5 levels of site quality as following: level I: none; level II: 4 site types with total area of 48. 87hm², accounting for 1. 44%; level III: 11 site types with total area of 106. 42 hm², accounting for 3. 14%; level IV: 52 site types with total area of 2799. 21hm², accounting for 82. 65%; and level V: 19 site types with total area of 432. 36 hm², accounting for12. 77%. There were 15 site types with the site quality level of middle or above, accounting for a total area of 155. 29 hm²or 4. 58% in the whole study area, while the site type with the site quality level of low or lower was 73 with the area of 3231. 57 hm² accounting for 95. 42%.

Key words: compartment scale; site type classification; site quality evaluation; site index; fertility index

立地分类是将生态学上相近的立地进行组合称之，立地质量评价是对立地的宜林性或潜在生产力进行判断、预测或量化[1,2]。立地分类、立地质量评价都是林业重要的基础性工作，对了解、掌握林地的立地条件、造林规划设计、适地适树布局树种、森林更新、地力维持和经营管理等提供了支持。因此，国内外研究甚众[3-8]，已形成了较成熟的技术方法体系。如：在立地类型划分方面已形成了综合多因子法、遥感结合GIS技术分类等[9,10]；对立地质量评价有林分因子直接评价法、环境因子间接评价法以及两者综合评价法等[11-13]。RS和GIS等技术在立地研究中的广泛应用[14,15]，使立地信息获取、类型划分与成图等更容易、快捷与准确，并在立地类型划分及质量评价的信息获取方面实现了多元化。如除传统野外调查、分析获取[16]外，还可利用森林资源二类调查数据结合补充样地调查、基于森林资源一类清查数据建模、利用森林资源二类调查数据结合RS技术、采用BP人工神经网络模型、应用"3S"技术等获取[17-21]。这些技术的应用为森林立地研究奠定了坚实基础。

虽然森林立地类型划分与质量评价研究已取得令人瞩目的成绩，但也还存在一些问题。其中在经营单位级的立地研究中多为大尺度的宏观研究，致其研究成果直接应用于林业生产经营时，常感觉精准度偏低，达不到精准利用林地与科学经营的要求。而以林地区划单位为尺度的研究未见报道。为此，本文开展了以林地区划单位"林班"为尺度的森林立地分类与质量评价研究，以期为林地资源的精准、合理、高效、可持续利用及人工林的科学经营与林分质量的精准提升提供依据和技术支撑。

1 研究区概况

研究区位于广西凭祥市中国林业科学研究院热带林业实验中心哨平实验场(21°57′47″~22°19′27″N，106°39′50″~106°59′30″E)。年均温20.5~21.7℃，≥10℃积温6000~7600℃，年均降水量1200~1500mm，年蒸发量1260~1390mm，相对湿度80%~84%；属南亚热带半湿润—湿润气候区；林地面积3386.86hm^2，海拔140~565m；母岩(母质)有紫色砂岩、花岗岩、洪积母质；土类有紫色土、砖红壤性红壤和红壤，其中紫色土占88.54%；原生植被为季雨林和常绿阔叶林；地带性乔木有人面子(*Dracontomelon duperreanum*)、乌榄(*Canarium pimela*)、风吹楠(*Horsfieldia glabra*)等，灌木主要有盐肤木(*Rhus chinensis*)、桃金娘(*Rhodomyrtus tomentosa*)、大沙叶(*Aporosa chinensis*)等，草本有飞机草(*Eupatorium oderatum*)、铁芒萁(*Dicranopteris linearis*)、五节芒(*Miscanthus floridulus*)等；人工林以马尾松(*Pinus massoniana*)为主(约占80%)，其余树种有巨尾桉(*Eucalyptus grandis*×*Eucalyptus urophylla*)、红椎(*Castanopsis hystrix*)等(约占10%)。

2 研究方法

2.1 样地确定与布设

在林班中，按7hm^2/块布设样地。林班样地数=该林班面积(hm^2)÷7(四舍五入取整)。为提高立地类型划分及质量评价的精准度，对地形变化较复杂的局部地段，样地密度进行了适当加密，全实验场样地数达500块。依各林班样地数，在1∶10000林相图上兼顾树种、坡位布点，调查时选代表性位置布设20m×20m样地。

2.2 样地调查与土壤分析

调查母岩、土类、海拔、坡位、坡度、坡向、植被、腐殖质层厚度、土层厚度、树高和胸径等，挖土壤标准剖面进行调查，并按土壤发生层次取样分析容重，有机质，pH，N、P、K全量与速效养分，土壤孔隙度，持水量，机械组成等。分析方法按国标执行。

2.3 立地主导因子选择

根据野外调查信息与分析数据及参考文献[16-22]，本研究选取地貌、海拔、母质、土壤类型、坡向、坡位、坡度、腐殖质层厚度、土壤厚度、枯枝落叶层厚度、土壤质地、非毛管孔隙度、毛管孔隙度和总孔隙度共14个因子通过主成分分析筛选主导因子。

2.4 技术应用与评价

(1)立地类型图勾绘。利用1∶10000林相图，在实地依据立地主导因子对立地类型图进行勾绘。

(2)立地类型划分。按六级分类系统[23]划分立地类型，利用ArcGIS分析叠加技术获得立地类型并成图。

(3)立地质量评价。采用立地生产力与土壤质量相结合的综合方法评价立地质量。立地生产力采用地位指数法，按100株/hm^{2}选取优势木[24]。以马尾松地位指数评价立地生产力；土壤质量采用数值化综合法评价[25,26]。

3 结果分析

3.1 立地类型

3.1.1 立地主导因子筛选

对初选立地因子数量化后利用SPSS 17.0软件进行计算，获得特征值和方差贡献率及主成分累计贡献率(表1)。

表1　协方差特征根和方差贡献率

变量	特征值	百分率(%)	累计百分率(%)
1	3.373	24.090	24.090
2	1.511	10.793	34.884
3	1.391	9.935	44.819
4	1.205	8.609	53.428
5	1.114	7.957	61.385
6	1.038	7.415	68.800
7	0.941	6.718	75.518
8	0.936	6.688	82.207
9	0.840	6.003	88.209
10	0.689	4.922	93.132
11	0.635	4.534	97.666
12	0.212	1.513	99.179
13	0.115	0.821	100.000
14	0.000	0.000	100.000

从表1可知，前8个主成分的累积贡献率超过80%。因此，选它们为主导因子已基本包含全部立地因子信息。经对8个主成分的得分分析，得到14个因子的载荷值(表2)。从每个主成分中不重复的选取一个载荷绝对值最高因子作为主导因子，结果为土类、母岩(母质)、土层厚、坡度、坡位、腐殖质层厚、地貌、海拔。

依据研究区情况及调查数据对所选定的立地主导因子进行分级(表3)。

表2　主成分载荷值

主成分	地貌	海拔(m)	母岩(母质)	土类	坡向	坡位	坡度(°)	腐殖质层厚度(cm)	土层厚(cm)	枯落层厚(cm)	土壤黏粒含量(%)	非毛管孔隙度(%)	毛管孔隙度(%)	总孔隙度(%)
X(1)	0.72	0.68	−0.63	−0.79	0.48	−0.13	0.03	−0.10	0.03	0.16	0.10	0.50	0.50	0.75
X(2)	0.26	0.41	0.61	0.38	−0.20	0.05	−0.07	−0.31	0.14	0.19	0.11	−0.46	0.54	0.17
X(3)	0.12	0.09	−0.11	−0.16	0.07	0.01	−0.11	0.23	0.66	0.63	−0.50	−0.25	−0.15	−0.30
X(4)	−0.39	−0.26	−0.11	0.05	−0.34	−0.35	0.44	0.29	0.04	0.19	−0.27	0.08	0.42	0.40
X(5)	0.46	−0.27	0.13	0.14	0.09	0.51	−0.33	−0.24	0.39	0.05	−0.08	−0.24	0.06	0.37
X(6)	−0.06	0.01	−0.16	−0.01	−0.04	0.41	0.71	−0.35	0.04	0.33	0.28	−0.01	−0.12	−0.10
X(7)	0.38	0.13	0.33	0.27	0.29	−0.05	0.17	0.58	0.10	0.19	0.18	0.27	−0.14	0.08
X(8)	0.18	0.52	0.12	0.03	−0.14	0.42	0.22	0.31	−0.22	−0.25	−0.25	0.08	−0.02	0.04

表3　立地类型划分因子与分级标准

分类单位	立地类型小区	立地类型组	立地类型				
主导因子	地貌海拔(m)	母岩母质	坡度(°)	土层厚(cm)	腐殖质层厚(cm)	坡位	土类
分级标准	山丘>350 丘陵≤350	花岗岩 紫色砂岩 洪积母质	缓≤10 斜10~30 陡≥30	厚≥80 中50~80 薄≤50	厚≥20 中10~20 薄≤10	上坡 中坡 下坡 全坡	砖红壤性红壤 红壤 紫色土

3.1.2 立地类型划分

依据六级分类系统，本研究区属华南亚热带热立地区域，粤桂沿海丘陵台地立地区，西部石灰岩丘陵台地立地亚区。立地类型小区依据地貌、海拔的差异分为丘陵和山丘两小区；立地类型组以母岩(母质)为依据，分为花岗岩、紫色砂岩、洪积母质3个组。

将1：10000林相上所勾绘的立地类型图搬到1：10000电子化矢量图上，把立地类型矢量图与立地主导因子和分级标准在Arc GIS中分析叠加得立地类型名称、面积及相关图表。结果表明，研究区共划分为88个立地类型(表4)。

表4 哨平实验场的立地类型及立地质量

立地类型名称	代号	生产力等级	土壤质量等级	立地质量等级	面积(hm²)	占总面积(%)
山丘花岗岩斜坡厚土层厚腐殖质下坡红壤	T37					
丘陵花岗岩斜坡厚土层厚腐殖质下坡砖红壤性红壤	T43	Ⅰ	Ⅱ	Ⅱ	48.87	1.44
丘陵花岗岩缓坡厚土层厚腐殖质下坡砖红壤性红壤	T65					
丘陵花岗岩斜坡厚土层厚腐殖质下坡红壤	T71					
山丘花岗岩斜坡厚土层厚腐殖质上坡红壤	T35					
山丘花岗岩斜坡厚土层中腐殖质下坡红壤	T36					
丘陵花岗岩斜坡厚土层中腐殖质下坡砖红壤性红壤	T40					
山丘花岗岩斜坡厚土层中腐殖质下坡砖红壤性红壤	T55					
丘陵花岗岩斜坡厚土层厚腐殖质上坡砖红壤性红壤	T56					
丘陵花岗岩缓坡厚土层中腐殖质下坡砖红壤性红壤	T63	Ⅱ	Ⅲ	Ⅲ	106.42	3.14
丘陵花岗岩斜坡厚土层厚腐殖质全坡砖红壤性红壤	T66					
丘陵花岗岩缓坡厚土层厚腐殖质全坡砖红壤性红壤	T69					
丘陵花岗岩斜坡厚土层厚腐殖质上坡红壤	T70					
山丘花岗岩斜坡厚土层厚腐殖质上坡砖红壤性红壤	T74					
丘陵花岗岩斜坡厚土层中腐殖质下坡红壤	T75					
丘陵紫色砂岩斜坡厚土层厚腐殖质下坡紫色土	T1					
丘陵紫色砂岩斜坡厚土层中腐殖质下坡紫色土	T2					
丘陵紫色砂岩缓坡厚土层厚腐殖质下坡紫色土	T10					
山丘紫色砂岩斜坡厚土层厚腐殖质下坡紫色土	T21					
丘陵洪积母质缓坡厚土层中腐殖质下坡砖红壤性红壤	T25					
山丘花岗岩斜坡厚土层中腐殖质上坡红壤	T38					
丘陵花岗岩斜坡厚土层中腐殖质上坡砖红壤性红壤	T46	Ⅲ	Ⅳ	Ⅳ	950.15	28.05
山丘花岗岩斜坡厚土层中腐殖质上坡砖红壤性红壤	T60					
山丘花岗岩斜坡中土层中腐殖质下坡红壤	T64					
山丘花岗岩斜坡中土层中腐殖质上坡红壤	T72					
山丘紫色砂岩缓坡厚土层厚腐殖质下坡紫色土	T81					
丘陵花岗岩缓坡厚土层中腐殖质上坡砖红壤性红壤	T87					

（续）

立地类型名称	代号	生产力等级	土壤质量等级	立地质量等级	面积（hm^2）	占总面积（%）
丘陵紫色砂岩斜坡厚土层厚腐殖质上坡紫色土	T3					
丘陵紫色砂岩斜坡厚土层中腐殖质上坡紫色土	T4					
丘陵紫色砂岩斜坡中土层中腐殖质下坡紫色土	T5					
丘陵紫色砂岩缓坡厚土层中腐殖质全坡紫色土	T6					
丘陵紫色砂岩缓坡厚土层厚腐殖质全坡紫色土	T7					
丘陵紫色砂岩缓坡厚土层中腐殖质下坡紫色土	T9					
丘陵紫色砂岩斜坡中土层厚腐殖质下坡紫色土	T11					
丘陵紫色砂岩缓坡厚土层中腐殖质上坡紫色土	T12					
丘陵紫色砂岩缓坡中土层中腐殖质下坡紫色土	T13					
丘陵紫色砂岩斜坡厚土层厚腐殖质全坡紫色土	T14					
丘陵紫色砂岩缓坡厚土层厚腐殖质上坡紫色土	T15					
丘陵紫色砂岩斜坡厚土层中腐殖质全坡紫色土	T16					
丘陵洪积母质缓坡厚土层中腐殖质全坡砖红壤性红壤	T18					
丘陵紫色砂岩斜坡厚土层厚腐殖质中坡紫色土	T20					
丘陵洪积母质缓坡厚土层厚腐殖质全坡砖红壤性红壤	T24					
山丘紫色砂岩斜坡厚土层中腐殖质下坡紫色土	T26					
丘陵紫色砂岩斜坡厚土层薄腐殖质下坡紫色土	T27					
丘陵紫色砂岩斜坡厚土层中腐殖质中坡紫色土	T29					
山丘紫色砂岩斜坡厚土层厚腐殖质上坡紫色土	T30	Ⅳ	Ⅳ	Ⅳ	1849.06	54.60
山丘紫色砂岩斜坡厚土层中腐殖质上坡紫色土	T33					
丘陵紫色砂岩缓坡中土层厚腐殖质下坡紫色土	T34					
丘陵紫色砂岩斜坡厚土层薄腐殖质上坡紫色土	T39					
丘陵紫色砂岩缓坡厚土层薄腐殖质下坡紫色土	T44					
丘陵紫色砂岩斜坡中土层厚腐殖质中坡紫色土	T45					
丘陵洪积母质斜坡厚土层中腐殖质下坡砖红壤性红壤	T47					
山丘洪积母质斜坡厚土层中腐殖质上坡砖红壤性红壤	T49					
丘陵洪积母质斜坡厚土层厚腐殖质全坡砖红壤性红壤	T53					
丘陵紫色砂岩缓坡厚土层薄腐殖质全坡紫色土	T54					
山丘紫色砂岩斜坡厚土层中腐殖质中坡紫色土	T57					
丘陵紫色砂岩缓坡中土层薄腐殖质下坡紫色土	T58					
丘陵紫色砂岩缓坡厚土层薄腐殖质上坡紫色土	T59					
山丘紫色砂岩缓坡厚土层厚腐殖质上坡紫色土	T61					
丘陵洪积母质陡坡厚土层中腐殖质下坡砖红壤性红壤	T62					
丘陵洪积母质斜坡厚土层中腐殖质全坡砖红壤性红壤	T67					
山丘紫色砂岩斜坡厚土层薄腐殖质下坡紫色土	T68					
丘陵紫色砂岩斜坡厚土层薄腐殖质全坡紫色土	T73					
山丘紫色砂岩斜坡厚土层薄腐殖质上坡紫色土	T76					

（续）

立地类型名称	代号	生产力等级	土壤质量等级	立地质量等级	面积（hm^2）	占总面积（%）
山丘紫色砂岩缓坡厚土层中腐殖质上坡紫色土	T78					
山丘洪积母质斜坡厚土层中腐殖质中坡砖红壤性红壤	T79					
山丘紫色砂岩缓坡厚土层中腐殖质下坡紫色土	T84	Ⅳ	Ⅳ	Ⅳ	1849.06	54.60
丘陵洪积母质斜坡厚土层中腐殖质中坡砖红壤性红壤	T85					
丘陵洪积母质陡坡厚土层中腐殖质中坡砖红壤性红壤	T88					
丘陵紫色砂岩斜坡中土层中腐殖质上坡紫色土	T8					
丘陵紫色砂岩斜坡中土层厚腐殖质上坡紫色土	T17					
丘陵紫色砂岩缓坡中土层中腐殖质全坡紫色土	T19					
丘陵紫色砂岩缓坡中土层中腐殖质上坡紫色土	T22					
丘陵紫色砂岩斜坡中土层中腐殖质全坡紫色土	T23					
丘陵洪积母质缓坡厚土层薄腐殖质全坡砖红壤性红壤	T28					
丘陵紫色砂岩斜坡中土层中腐殖质中坡紫色土	T31					
丘陵紫色砂岩缓坡中土层厚腐殖质全坡紫色土	T32					
丘陵紫色砂岩缓坡薄土层中腐殖质全坡紫色土	T41					
丘陵紫色砂岩斜坡中土层厚腐殖质全坡紫色土	T42	Ⅴ	Ⅴ	Ⅴ	432.36	12.77
丘陵紫色砂岩斜坡薄土层中腐殖质全坡紫色土	T48					
丘陵紫色砂岩斜坡薄土层中腐殖质上坡紫色土	T50					
丘陵紫色砂岩斜坡薄土层中腐殖质下坡紫色土	T51					
丘陵紫色砂岩缓坡中土层厚腐殖质上坡紫色土	T52					
山丘紫色砂岩斜坡中土层厚腐殖质上坡紫色土	T77					
丘陵洪积母质缓坡中土层中腐殖质全坡砖红壤性红壤	T80					
丘陵紫色砂岩缓坡薄土层中腐殖质下坡紫色土	T82					
丘陵紫色砂岩缓坡薄土层中腐殖质上坡紫色土	T83					
丘陵紫色砂岩斜坡薄土层薄腐殖质下坡紫色土	T86					
				合计	3386.86	100

3.2 立地生产力评价

因马尾松占总面积的 80%，且遍及所有立地类型，故用其地位指数对立地类型的立地生产力进行评价。以 5~40 年生马尾松优势木平均高建立地位指数模型(图 1)。

应用传统导向曲线法求得马尾松导向曲线方程为：

$$H = 30.66 \times [1 - \exp(-2.46 \times A)]^{0.7}$$

式中：H 为优势木平均高；A 为林龄。

由导向曲线方程求得马尾松标准年龄(20 年)地位指数为：10，12，14，16，18，20，22，24，26。将它们分成 5 等评价立地生产力(表 5)。

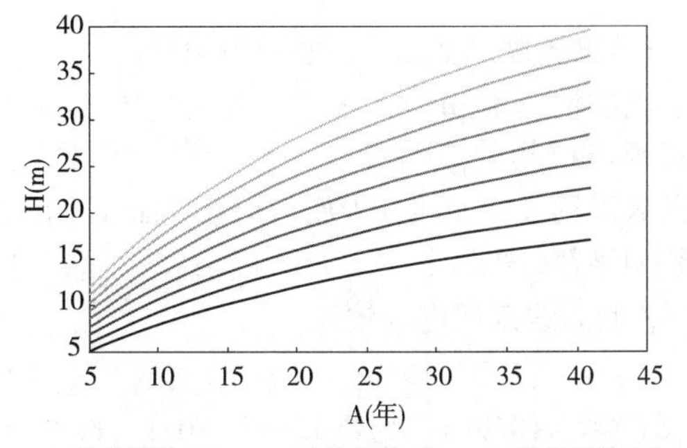

图 1 马尾松地位指数曲线

注：横坐标 A 为林龄；纵坐标 H 为优势木平均高。

表5 立地生产力评价结果

项目	地位指数				
	24~26	20~22	18	14~16	10~12
生产力等级	Ⅰ(高)	Ⅱ(较高)	Ⅲ(中)	Ⅳ(较低)	Ⅴ(低)
立地类型(个)	4	11	12	42	19
面积(hm^2)	48.77	106.35	950.01	1849.23	432.50
占总面积(%)	1.44	3.14	28.05	54.60	12.77

由表5可知：立地生产力Ⅰ,Ⅱ,Ⅲ,Ⅳ,Ⅴ级的立地类型数(个)、面积(hm^2)及占总面积百分比(%)分别为4、48.77、1.44；11、106.35、3.14；12、950.01、28.05；42、1849.23、54.60；19、432.5、12.77。立地生产力较高及以上、中等、较低及以下的立地类型数、面积、占总面积比例分别为：15个、155.12hm^2、4.58%；12个、950.01hm^2、28.05%；61个、2281.28hm^2、67.37%。表明研究区以低生产力林地占多数。这与紫色土占88.54%有关，因紫色土属幼年土，有机质贫乏，且土层多浅薄，黏重，致林分生产力较低。

3.3 土壤质量评价

遵循重要性、易获取性、差异性、稳定性等原则，结合土壤分析数据及参考文献资料[26-28]，以A层土壤厚度，容重，粘粒含量，pH，有机质，N、P、K全量及速效养分为评价指标，经统计分析所得指标值如表6所示。

表6 A层土壤指标值

项 目	厚度(cm)	容重(g/cm^3)	黏粒含量(%)	pH	有机质(g/kg)	全量养分(g/kg)			速效养分(mg/kg)		
						N	P	K	N	P	K
平均值	21.13	1.44	31.95	4.80	17.46	0.93	0.19	8.38	79.86	1.50	47.70
最小值	7.00	1.00	18.10	3.81	6.65	0.42	0.05	0.25	27.34	0.05	6.22
最大值	46.00	1.70	51.50	6.36	43.57	1.84	0.43	25.34	151.02	7.02	179.90
标准差	6.44	0.13	7.45	0.42	7.58	0.25	0.06	5.32	25.20	1.28	29.97
变异系数	0.30	0.09	0.23	0.09	0.43	0.27	0.30	0.63	0.32	0.85	0.63

3.3.1 隶属度函数曲线转折点取值确定

因酸性土壤pH，有机质，N、P、K全量及速效养分等与植物效应曲线呈S型[25,27]，因此，其隶属度函数采用S型曲线，为便于计算将曲线型函数转为折线型函数(图2)。

S型曲线隶属度函数为：

$$U(X)=\begin{cases}1.0 & x\geqslant x2\\ 0.9(x-x1)/(x2-x1)+0.1 & x1\leqslant x<x2\\ 0.1 & x<x1\end{cases}$$

折线图见图2。

据表6和参考中国第二次土壤普查养分分级标准及相关研究[25,28]，确定各指标转折点取值(表7)。

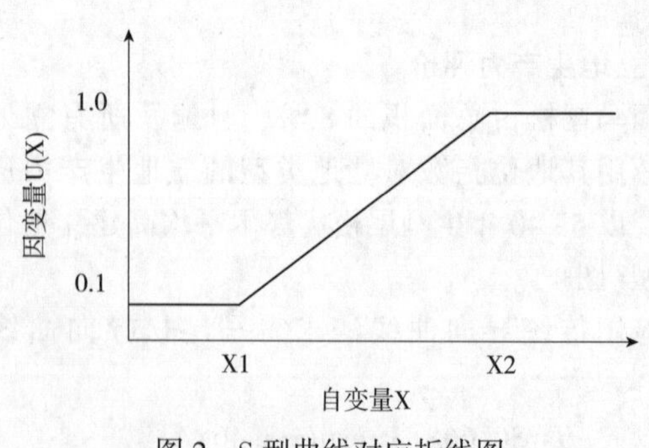

图2 S型曲线对应折线图

土壤容重和黏粒含量与植物生长呈抛物线型，隶属度函数亦采用抛物线型[25,29]，并将抛物线图转化为折线(图3)。

表 7 S 型隶属度函数曲线转折点取值

项目	土层厚度(cm)	pH	有机质(g/kg)	全量养分(g/kg)			速效养分(mg/kg)		
				N	P	K	N	P	K
X1	10	4.5	10	0.75	0.1	9.0	60	5	50
X2	25	6.0	30	1.5	1.5	20	120	20	150

隶属度函数为：

$$U(X)=\begin{cases}0.9(x-x4)/(x3-x4)+0.1 & x3<x\leqslant x4\\ 1.0 & x2\leqslant x\leqslant x3\\ 0.9(x-x1)/(x2-x1)+0.1 & x1\leqslant x<x2\\ 0.1 & x<x1 \text{ 或 } x>x4\end{cases}$$

折线图为：

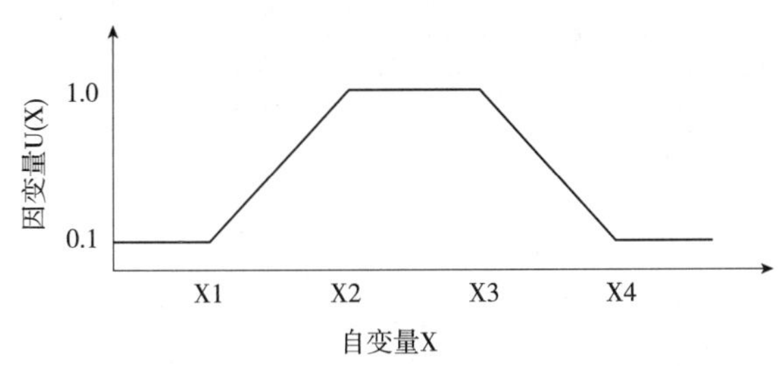

图 3 抛物线型曲线对应折线图

从表 6 及参考相关研究[28]，确定转折点取值如表 8 所示。

表 8 抛物线型隶属度函数曲线转折点取值

项目	土壤容重(g/cm^3)	黏粒含量(%)
X1	1.10	10
X2	1.25	20
X3	1.30	40
X4	1.50	65

3.3.2 隶属度值计算

利用转折点数值及隶属度函数，求各评价因子的隶属度值(*Ni*)，*Ni* 为 0.1~1.0 的数值。将它们的平均值绘制综合肥力指数 11 边形雷达图(图 4)。图中的轴上点离原点越近，属性状态越差[30]，对土壤肥力的作用分值越小，反之亦然；各点围成的 11 边形面积代表参评因子所组成评价对象的整体状态，面积越大，土壤质量越好，反之越差。

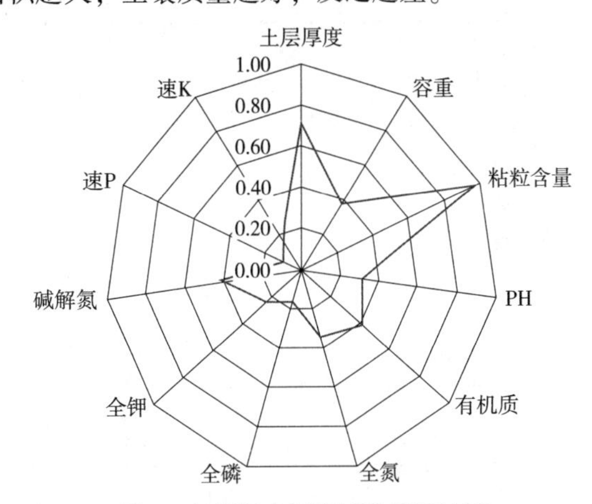

图 4 各指标隶属度函数值雷达图

由图 4 可知：①速效磷离原点最近，对土壤肥力的贡献最小，最缺乏；全磷次之，再次速效钾和全钾，数值均小于 0.20，都是土壤肥力的主要限制因子。②全氮、碱解氮、有机质、pH 和土壤容重的隶属度值在 0.4 左右，处于中下状态；土层厚和粘粒含量都大于 0.7，属较好。③各指标值围成的 11 边形面积不及它们最佳状态面积的 1/3，表明土壤整体质量较差。

3.3.3 权重值计算

以主成分分析求各评价因子权重值 *Wi*，结果如表 9 所示。

表 9 评价因子方差及权重

指标名称	土层厚度	土壤容重	黏粒含量	pH	有机质	全氮	全磷	全钾	碱解氮	速效磷	速效钾
公因子方差	0.987	0.683	0.661	0.726	0.803	0.647	0.609	0.837	0.722	0.561	0.701
权重值	0.124	0.086	0.083	0.091	0.101	0.082	0.077	0.105	0.091	0.071	0.088

3.3.4 土壤肥力综合指标值计算与质量评价

依据加乘法则，求土壤肥力综合指标值(*IFI*)，它为≤1.0 的值。*IFI* 越大表示土壤肥力越好，反之亦然。计算公式为：

$$IFI = Wi\times Ni$$

式中：*Ni*，*Wi* 分别表示第 *i* 种养分隶属度值和权重值。

经对 500 个样地的土壤肥力综合指标进行统计分析，所得结果如表 10 所示。

表 10　土壤肥力综合评价指标值

平均值	最小值	最大值	标准差	变异系数
0.28	0.15	0.70	0.07	0.26

将肥力指标值(0~1.0)均分为5个等级评价土壤质量(表11)。表11显示：①土壤质量Ⅰ级为零，Ⅱ，Ⅲ，Ⅳ，Ⅴ级立地类型数(个)、面积(hm^2)、占总面积(%)分别为4、48.77、1.44；11、106.35、3.14；54、2799.24、82.65；19、432.50、12.77。②土壤质量较差及差占95.42%，中等及以上的仅占4.58%。表明研究区土壤质量整体较差，与前面的直观分析结果一致。以上结果主要与研究区地处高温多雨的南亚热带，有机质分解快，土壤养分因冲刷淋溶流失严重，致土壤P，K养分贫乏，有机质、N较低，容重较高，土壤结构不良等有关。

3.4　立地质量评价

将立地生产力和土壤质量评价结果进行综合评价，两者等级不一时，以低者为结果。

表 11　土壤质量评价标准与结果

评价指标等级	Ⅰ	Ⅱ	Ⅲ	Ⅳ	Ⅴ
	0.81~1.0	0.61~0.8	0.41~0.6	0.21~0.4	0.0~0.2
土壤质量	优	良	中等	较差	差
立地类型数	0	4	11	54	19
面积(hm^2)	0	48.77	106.35	2799.24	432.50
占总面积(%)	0	1.44	3.14	82.65	12.77

表 12　立地质量评价结果

生产力	Ⅰ(高)	Ⅰ(高)	Ⅱ(较高)	Ⅲ-Ⅳ(中-较低)	Ⅴ(低)
土壤质量	Ⅰ(优)	Ⅱ(良)	Ⅲ(中等)	Ⅳ(较差)	Ⅴ(差)
立地质量	Ⅰ(优)	Ⅱ(良)	Ⅲ(中等)	Ⅳ(较差)	Ⅴ(差)
立地类型(个)	0	4	11	52	19
面积(hm^2)	0	48.77	106.35	2799.24	432.50
占总面积(%)	0	1.44	3.14	82.65	12.77

由表12可知：立地质量Ⅰ级为零，Ⅱ，Ⅲ，Ⅳ，Ⅴ级立地类型数(个)、面积(hm^2)、占总面积(%)分别为4、48.77、1.44；11、106.35、3.14；52、2799.24、82.65；19、432.50、12.77。其中：中等及以上立地类型有15个，面积155.29hm^2，占总面积4.58%；而较差和差73个，3231.57hm^2，占95.42%。出现了有高生产力林分而无优质土壤的情况，这主要与评价立地生产力树种马尾松具有强阳、低营养特点[31]，以及耐瘠薄，光、水、热充足，土壤稍差(良好)也能高产有关。这表明，采用单一评价方法可能会出现误判，综合法则更可靠。

4　结论

经主成分分析确定了土类、母岩(母质)、土层厚、坡度、坡位、腐殖质层厚、地貌、海拔8个因子为立地类型划分主导因子，并依据调查数据及相关文献确定它们的分级标准，将勾绘的立地类型图及立地分类主导因子和分级标准在Arc GIS进行分析叠加，结果显示，研究区共划分为88个立地类型。

立地质量综合评价结果表明，立地质量Ⅰ级为零，Ⅱ，Ⅲ，Ⅳ，Ⅴ级的立地类型数(个)、面积(hm^2)、占总面积(%)分别为4、48.77、1.44；11、106.35、3.14；52、2799.24、82.65；19、432.50、12.77。立地质量中等及以上的立地类型仅15个，面积155.29hm^2，占总面积的4.58%；而较差和差73个，3231.57hm^2，占95.42%。土壤整体立地质量，尤其缺P和K，有机质和N也偏低，容重则偏高，土壤理化性质不良，生产力低的立地占绝对多数。

在林地科学利用上：①立地质量Ⅱ，海拔350m以下的林地应发展质优价高的热带珍贵树种，如降香黄檀(*Dalbergia odorifera*)、交趾黄檀(*Dalbergia cochinchinensis*)、格木(*Erythrophleum fordii*)等，沟谷地以发展柚木(*Tectona grandis*)为佳；而海拔350m以上林地，可种植红椎、火力楠(*Michelia macclurei*)、香梓楠(*Michelia hedyosperma*)等。②Ⅲ级林地，可种植红椎、火力楠、云南石梓(*Gmelina arborea*)、观光木(*Michelia odora*)等。③Ⅳ—Ⅴ级林地，以种植马尾松、铁坚杉(*Keteleeria davidiana*)等耐瘠薄树种为宜。④非水源地及非公益林，可适当发展桉树，或选择速生固N树种，如马占相思(*Acacia mangium*)、银合欢(*Leucaena leucocephala*)等作为先锋树种，待土壤改良后再种植格木、红椎等珍贵树种。

参考文献

[1]沈国舫．森林培育学[M]．北京：中国林业出版社，2001.

[2]SKOVSGARRD J P，VANCLAY J K. Forest site productivity：a review of the evolution of dendrometric concepts for even-aged stands[J]. forestry，2008，81(1)：13-31.

[3]李艳洁，周来，靳爱仙，等．东京城林业局落叶松人工林立地质量评价[J]．林业资源管理，2017(2)：53-57.

[4]徐罗，亢新刚，郭韦韦，等．天然云冷杉针阔混交林立地质量评价[J]．北京林业大学学报，2016，38(5)：11-22.

[5] CHARLES J E，JOHN H T. Site quality evaluation of loblolly pine on the South Carolina Lower Coastal Plain，USA[J]. Journal of Forestry Research，2008，19(3) ：187-192.

[6]CURT T，BOUCHARD M，AGRECH G. Predicting site index of Douglas-fir plantations from ecological variables in the Massif Central area of France[J]. Forest Ecology and Management，2001，149(1/2/3)：61-74.

[7]范金顺，高兆蔚，蔡元晃，等．福建省森林立地分类与质量评价[J]．林业勘察设计，2012(1)：1-5.

[8]杜健，梁坤南，周再知，等．云南西双版纳柚木人工林立地类型划分及评价[J]．林业科学，2016，52(9)：1-10.

[9]张复礼．宿松县森林立地类型的划分和评价[J]．安徽林业科技，2004(2)：10 -11.

[10]余其芬，唐德瑞，董有福．基于遥感与地理信息系统的森林立地分类研究[J]．西北林学院学报，2003，18(2)：87-90.

[11]孟宪宇．测树学[M]．3版．北京：中国林业出版社，2006：98-112.

[12]顾云春，李永武，杨承栋．森林立地分类与评价的立地要素原理与方法[M]．北京：科学出版社，1993：74，129-132.

[13]杨文姬，王秀茹．国内立地质量评价研究浅析[J]．水土保持研究，2004，11(3) ：289-292.

[14]丰绪霞，刘兆刚，张海玉，等．基于RS和GIS帽儿山林场森林立地分类及质量评价[J]．东北林业大学学报，2010，38(8)：27-30.

[15]王永昌，张金池．基于遥感技术的云台山立地分类及质量评价[J]．南京林业大学学报：自然科学版，2007，31(1)：85-89.

[16]汪炳根，卢立华．广西大青山实验基地森林立地评价与适地适树研究[J]．林业科学研究，1998，1(4)：78-85.

[17]李志先，邓绍林．林朵林场杉木立地分类及质量评价研究[J]．广西农业大学学报，1995，14(3)：230-234.

[18]黄国胜，马炜，王雪军，等．基于一类清查数据的福建省立地质量评价技术[J]．东北林业大学学报，2014，36(3)：1-8.

[19]何瑞珍，孟庆法，刘志术，等．基于RS和GIS技术的森林立地类型分类研究——以河南省商城县国营黄柏山林场为例[J]．河南科学，2010，28(7)：799-803.

[20]巩垠熙，高原，仇琪，等．基于遥感影像的神经网络立地质量评价研究[J]．中南林业科技大学学报，2013，33(10)：42-47.

[21]张晓丽，游先祥．应用“3S”技术进行北京市森林立地分类和立地质量评价的研究[J]．遥感学报，1998，2(4)：292-295.

[22]卢立华，何日明，农瑞红，等．灰木莲生长对土壤养分和气候因子的响应[J]．应用生态学报，2014，25(4)：961-966.

[23]詹昭宁，王国祥，李芬兰，等．中国森林立地分类[M]．北京：中国林业出版社，1995.

[24] MEMG Shawnx ，HUANG Shongming，YANG Yuqing，et al．Evalution of population-averaged and subject-specific approaches for modeling the dominant or co-dominant height of lodgepole pine trees [J]. Canadian Journal of Forest Research，2009，39(6)：1148-1158.

[25]骆伯胜，钟继洪，陈俊坚．土壤肥力数值化综合评价研究[J]．土壤，2004，36(1)：104-106.

[26]阚文杰，吴启堂．一个定量综合评价土壤肥力的方法初探[J]．土壤通报，1994，25(6)：245-247.

[27]吴玉红，田霄鸿，同延安．等．基于主成分分析的土壤肥力综合指数评价[J]．生态学杂志，2010，29

(1)：173-180.

[28]刘世平，陈后庆，聂新涛，等．稻麦两熟制不同耕作方式与秸秆还田土壤肥力的综合评价[J]．农业工程学报，2008，24(5)：51-56.

[29]王建国，杨林章．模糊数学在土壤质量评价中的应用研究[J]．土壤学报，2001，38(2)：176 -183.

[30]高慧璇．应用多元统计分析[M]．北京：北京大学出版社，2005.

[31]秦国峰．马尾松地理种源[M]．杭州：浙江大学出版社，2003.

[原载：林业资源管理，2018(02)]

马尾松林木养分含量特征及其与产脂量的关系

安 宁[1,2] 贾宏炎[2] 谌红辉[2] 李朝英[2] 丁贵杰[1]
([1]贵州大学林学院/贵州省森林资源与环境研究中心，贵州贵阳 550025；
[2]中国林业科学研究院热带林业实验中心，广西凭祥 532600)

摘 要 分析不同产脂类型马尾松的养分含量差异，探讨养分含量与产脂量的关系，为马尾松产脂林的科学施肥提供理论依据。以广西凭祥26年生马尾松人工林为研究对象，分析不同产脂类型(低产脂类型和高产脂类型)马尾松针叶、枝、皮的N、P、K、Ca、Mg含量及其变化情况，运用方差分析及相关性分析，揭示不同产脂类型马尾松的养分含量与产脂量的相关性。结果表明：①N、K、Ca含量在马尾松不同器官间差异显著，而P和Mg含量差异不显著；针叶中N、P、K、Mg含量最高，皮中Ca含量最高。②在1年针叶和2年针叶中，各养分含量表现为N>K>Ca>Mg>P，1年针叶和2年针叶N、K、Ca、Mg含量在不同产脂类型马尾松间差异显著；除N含量是低产脂类型高于高产脂类型外，P、K、Ca、Mg含量均是低产脂类型低于高产脂类型。③在不同产脂类型马尾松间，枝和皮的养分含量均表现为Ca>N>K>Mg>P，Ca含量在不同产脂类型间差异显著，N、P、K、Mg含量差异不显著，5种养分含量均表现为低产脂类型低于高产脂类型。④马尾松产脂量与1年针叶和2年针叶养分含量关系密切，其中与K、Ca、Mg含量呈极显著正相关，与N含量呈极显著负相关。不同产脂类型马尾松的养分特点不同，养分含量对产脂量存在影响。与产脂量相关性较高的是1年针叶和2年针叶K、Ca、Mg、N含量。

关键词 马尾松；林木养分；产脂量；针叶林

Nutrient Characteristics Forest of *Pinus massoniana* and the Relationship with Resin Yields

AN Ning[1,2], JIA Hongyan[2], CHEN Honghui[2], LI Zhaoying[2], DING Guijie[1]
([1]*College of Forestry/Institute for forest resources & environment of Guizhou*, *Guizhou University*, *Guiyang* 550025, *Guizhou*, *China*；[2]*Experimental Centre of Tropical Forestry*, *CAF*, *Pingxiang* 532600, *Guangxi*, *China*)

Abstract: Through analyzing the nutrient characteristics of different resin individual of *Pinus massoniana*, expored the relationship between nutrient characteristics and the resin yields for improving forest of *pinus massoniana* resin yields provides a theoretical basis for scientific fertilization. 26 years' plantations of *Pinus massoniana* in Pingxiang Guangxi were chosen as research objects, the changes and contents of N, P, K, Ca and Mg in foliar, branch, bark in different resin yields were analyzed. Using variance analysis and correlation analysis, explore the correlations between nutrient characteristics and the resin yields. There were significant differences in the contents of N, K and Ca in different organs, while P and Mg were not significantly different. The highest contents of N, P, K and Mg were in the needles, and the highest Ca was in the bark. Nutrient contents in needles were N>K>Ca>Mg>P. The concentrations of N, K, Ca and Mg in needles of 1 year and 2 years were significantly different among different resin yields. The contents of N in low resin yields was higher than that in high resin yields, whine the contents of K, Ca, and Mg in low resin yields were lower than those in high resin yields. The features of nutrients were the same in the branch and bark, which were Ca>N>K>Mg>P. The content of Ca was significant difference in different resin yields, while N, P, K and Mg didn't show significantly different. The characteristics of 5 nutrient contents were lower in low resin yield than those of high resin yield. The resin yields were closely related to the nutrient contents of needles of 1 year and 2 years, which was significantly positively correlated with K, Ca and Mg, and was significantly negatively correlated with N. The nutrient characteristics of different resin individual of *Pinus massoniana* were different, and the nutrient characteristics had an effect on the resin yields. Through analyzing the correlations between nutrient characteristics of different resin individual organs of *Pinus massoniana*. The high positive correlations with resin yields are contens of K and Ca in 1 year needle, conten of Ca in 2 years needle, contens of N in 1 year needle and ontens of N in 2 years needle. These provide a theoretical basis for the early selection of the high resin yields trees, as well as the scientific fertilization of the resin yields forest of *Pinus massoniana*.

Key words: *Pinus massoniana*; resin yield; needle; nutrient conten

松脂经过蒸馏和去除杂质可制成松香和松节油，两者均为重要的化工原料。松香广泛应用于胶黏剂、油墨、涂料、造纸施胶剂、合成橡胶、表面活性剂、肥皂、食品、医药和电子等工业领域，松节油可用于合成香料、精细化学品、溶剂和清洁剂等，使用松香和松节油的行业有300多个[1]。我国是世界上第一大松脂生产大国，年产量和出口量均居世界首位。我国可以供采脂的树种有22个[2]，而马尾松(*Pinus massoniana*)一个树种的产脂量约占全国产量的70%。

国家“八五”以来，针对马尾松以及其产脂性状从优树选择、种子园建立、保留密度、合理施肥等方面进行了大量的研究工作[3-7]。近年来，随着南方珍贵阔叶树种的大规模种植和病虫害的频发，造成马尾松可采脂林面积逐渐减少，而市场对松脂原料资源需求日益增长，营建高质量马尾松采脂原料林势在必行。对于马尾松养分含量的相关研究多集中在营养元素的生物循环以及养分循环[8,9]。树木叶片的养分含量能够反映出树木对养分吸收、储存、转移和利用情况[10-12]，树木叶片的养分含量也会对植物的生长和光合作用产生影响[13,14]，马尾松针叶养分含量研究大多针对针叶凋落前后含量以及再吸收特征上[15,16]。生产中发现，马尾松单株产脂量存在显著的个体差异，这种差异是否体现在马尾松针叶、枝和皮的养分含量上并未见报道。本文通过分析马尾松不同产脂类型的养分含量特点及其与产脂量的关系，掌握高产脂类型的养分含量特点，旨在为完善马尾松高产脂优树的选择方法，以及提高马尾松产脂林经营管理水平提供科学依据。

1　材料与方法

1.1　试验地概况及样品的采集与处理

试验林设在广西友谊关森林生态系统国家定位观测站区域内的青山实验场，为1991年营造的马尾松人工林，广西宁明桐棉种源，郁闭度0.7，平均胸径27.8cm，平均树高20.7m，现有林分密度450株/hm^2。前茬为杉木人工林，主伐后明火炼山。地理位置106°43′E，22°06′N，海拔500~550m，低山地貌，年平均气温19.5℃，年降水量1500mm，属于南亚热带季风气候区。土壤为花岗岩发育成的红壤，土层厚度>1m，腐殖质厚度>10cm，林下植被为芒萁(*Dicranopteris dichotoma*)、五节芒(*Miscanthus floridulus*)等。

2017年5月，在林分中选取相同径级、相同生长势、但产脂量差别较大的两组(高产脂型和低产脂型)林木进行研究，每组各选30株，概况见表1。在树冠中上部向阳面采集三级枝上完整无病虫害的1年生和2年生新鲜针叶，均采集2针一束。在树冠中上部向阳面相同的三级枝上采集树枝，在树冠中上部向阳面相同一级枝的二级枝上采集树皮。采集的材料用密封袋分别标记。样品采集后，置于90℃烘箱中杀青30min，然后在65℃下继续烘干48h，之后取样进行粉碎，过0.5mm尼龙网筛后，装入密封袋中，标记并保存至干燥器中以备化学分析。

1.2　针叶、枝和皮养分元素测定

试样采用浓H_2SO_4-$HClO_4$法消煮后测定N、P、K、Ca、Mg含量，N含量采用靛酚蓝比色法测定，P含量采用钼锑抗比色法测定，K含量采用火焰光度法测定，Ca、Mg含量采用EDTA络合滴定法测定[18,19]。

表1　不同产脂类型样木概况

类型	平均胸径(cm)	平均树高(m)	平均冠幅(m)	平均枝下高(m)	平均产脂量(kg)
低产脂型	27.7	20.5	3.7	12.6	2.45
高产脂型	27.8	20.8	4.0	13.0	4.28

1.3 数据处理

采用 Excel 2007 软件对数据进行统计分析并绘制图表，运用 SPSS 16.0 软件对数据进行方差分析及相关性分析[20]。

2 结果与分析

2.1 针叶、枝和皮中养分的分配特征

从不同器官养分含量(图 1)可以看出，1 年、2 年生针叶各养分含量非常相近，含量规律较一致，均表现为 N>K>Ca>Mg>P。枝养分含量特征为 Ca>N>K>Mg>P，皮养分含量特征为 Ca>N>K>Mg>P。N、P、K、Ca 和 Mg 含量在 1 年和 2 年针叶中均差异不显著。N 含量在不同器官间差异显著，在针叶中含量相对较高且变化较小，1 年、2 年生针叶 N 含量分别为 13.64g/kg 和 13.83g/kg，而枝和皮 N 含量相对较低，分别为 4.68g/kg 和 4.43g/kg。P 在 5 种养分中含量最低，且不同器官间含量差异不显著，1 年、2 年生针叶 P 含量分别为 0.75g/kg 和 0.73g/kg，枝和皮含量分别为 0.29g/kg 和 0.28g/kg；K 含量在不同器官间差异不显著，2 年生针叶 K 含量略高于 1 年针叶，含量仅相差 0.09g/kg，枝和皮的 K 含量相近，枝的 K 含量为 3.35g/kg，皮的 K 含量为 3.48g/kg；Ca 含量在不同器官间差异显著，含量特征表现为皮>枝>2 年生针叶>1 年生针叶，含量依次为 12.53g/kg、8.52g/kg、5.24g/kg 和 4.97g/kg；Mg 含量在不同器官间差异不显著，相比针叶的含量，皮含量略高，而枝含量略低。

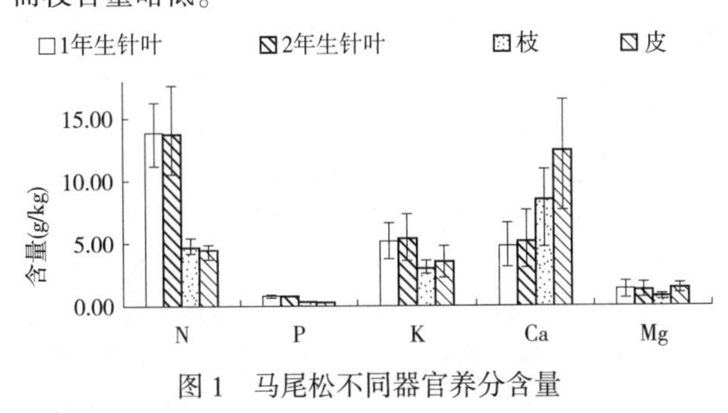

图 1 马尾松不同器官养分含量

2.2 不同产脂量类型针叶养分含量特征

对比不同产脂量个体针叶养分含量(图 2，图 3)，发现 N 含量在不同产脂类型 1 年、2 年针叶间均差异显著($P<0.05$)，均表现为低产脂类型高于高产脂类型，N 含量在不同产脂类型 1 年针叶中相差 2.97g/kg，在 2 年针叶间相差 2.78g/kg；P 含量在不同产脂类型 1 年、2 年针叶间均差异不显著，高产脂类型 1 年针叶含量仅比低产脂类型高 0.01g/kg，在不同产脂类型 2 年针叶中含量相同；K 含量在不同产脂类型 1 年、2 年针叶间均差异显著($P<0.05$)，均表现为低产脂类型低于高产脂类型，在 1 年针叶中含量相差 2.53g/kg，在 2 年针叶中含量相差 2.25g/kg；Ca 含量在不同产脂类型 1 年、2 年针叶间差异显著($P<0.05$)，均表现为低产脂类型低于高产脂类型，在 1 年和 2 年针叶中分别相差 2.46g/kg 和 2.38g/kg。Mg 含量在不同产脂类型 1 年、2 年针叶间均差异显著($P<0.05$)，均表现为低产脂类型低于高产脂类型，在 1 年针叶中相差 0.33g/kg，在 2 年针叶相差 0.42g/kg。

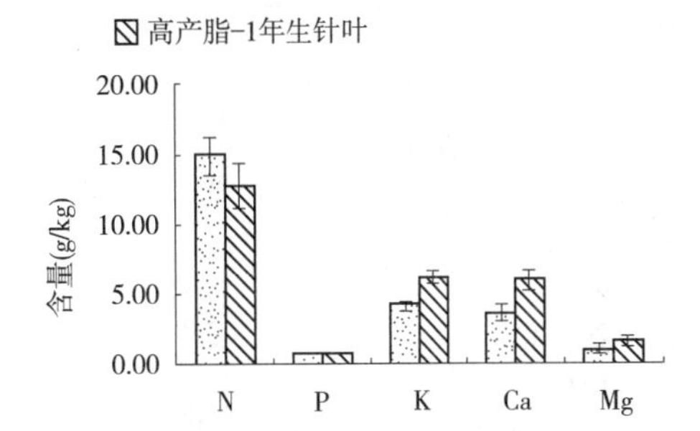

图 2 不同产脂类型 1 年生针叶养分含量

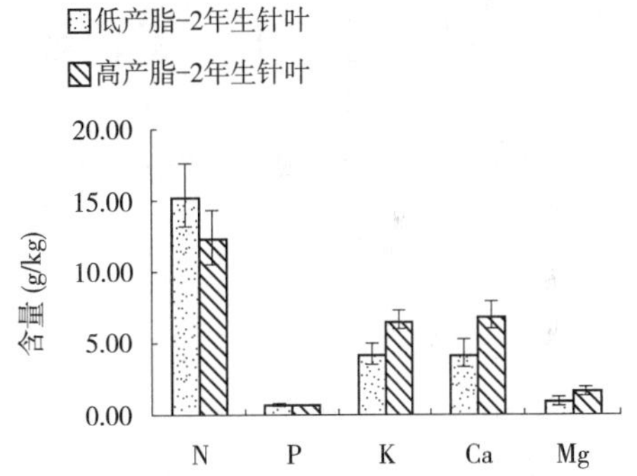

图 3 不同产脂类型 2 年生针叶养分含量

2.3 不同产脂量类型的枝养分含量特征

从图 4 中可以看出，5 种养分在枝中的含量规律表现为 Ca>N>K>Mg>P。N、P、K 和 Mg 含量在不同产脂类型枝中均差异不显著($P>0.05$)，均表现为低产脂类型低于高产脂类型，含量分别相差 0.17g/kg、0.01g/kg、0.34g/kg 和 0.08g/kg。Ca 含量在不同产脂类型枝中差异显著($P<0.05$)，表现为低产脂个体低于高产脂个体，低产脂个体平均含量 7.01g/kg，高产脂个体平均含量 9.83g/kg，相差 2.82g/kg。

2.4 不同产脂量类型的皮养分含量特征

从图 5 中可以看出，5 种养分在皮中的含量规律表现为 Ca>N>K>Mg>P。N、P、K 和 Mg 含量在不同产脂类型皮中均差异不显著($P>0.05$)，均表现为低产脂类型低于高产脂类型，含量分别相差 0.09g/kg、

0.01g/kg、0.29g/kg 和 0.07g/kg。Ca 含量在不同产脂类型皮中差异显著($P<0.05$)，表现为低产脂个体低于高产脂个体，含量相差 3.56g/kg。由上可以看出，养分在不同产脂类型枝和皮的含量特点相同。

表 2　产脂量与不同器官养分含量的相关性

	1 年生针叶 N	2 年生针叶 N	枝 N	皮 N	1 年生针叶 P	2 年生针叶 P	枝 P
产脂量	-0.817**	-0.752**	0.327	-0.581**	-0.071	-0.080	-0.073

	皮 P	1 年生针叶 K	2 年生针叶 K	枝 K	皮 K	1 年生针叶 Ca	2 年生针叶 Ca
产脂量	-0.093	0.835**	0.773	0.392	0.498*	0.824**	0.821**

	枝 Ca	皮 Ca	1 年生针叶 Mg	2 年生针叶 Mg	枝 Mg	皮 Mg
产脂量	0.632*	0.618*	0.820**	0.764**	0.225	0.116

备注：** 表示 1%水平差异显著，* 表示 5%水平差异显著。

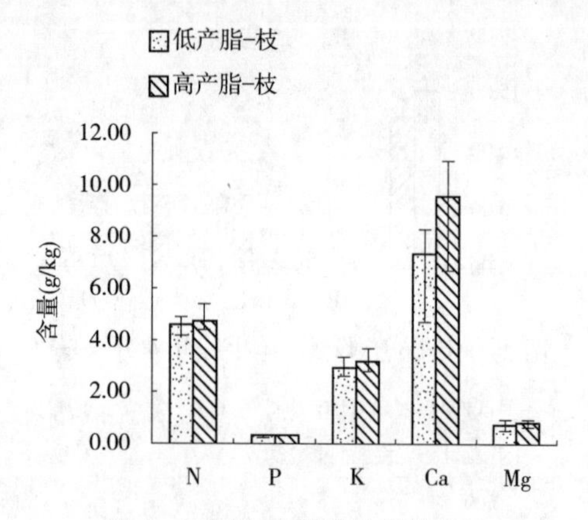

图 4　不同产脂类型枝养分含量

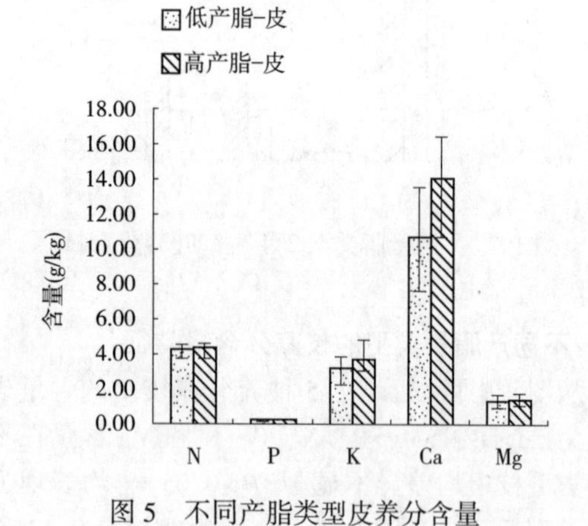

图 5　不同产脂类型皮养分含量

2.5　产脂量与针叶、枝和皮养分含量的相关性

表 2 中可以看出，马尾松产脂量在不同器官间，与 1 年针叶和 2 年针叶相关性较高，相关系数绝对值均大于 0.752。在不同养分间，马尾松产脂量与 N、K、Ca 和 Mg 含量相关性较高，相关系数绝对值均大于 0.764。与产脂量相关性达到极显著水平($P<0.01$)的有 1 年针叶的 N、K、Ca 和 Mg 含量，2 年针叶的 Ca 和 Mg 含量，以及皮的 N 含量。与产脂量正相关高的是 1 年针叶 K 含量和 Ca 含量、2 年针叶 Ca 含量，1 年针叶 Mg 含量，相关系数分别为 0.835、0.824、0.821 和 0.820。与产脂量负相关高的是 1 年针叶和 2 年生针叶的 N 含量产脂量与 1 年、2 年生针叶 K、Ca 和 Mg 含量，负相关最高的是 1 年针叶和 2 年针叶 N 含量，相关系数分别为-0.835 和-0.752。综上可以看出，马尾松植株产脂量与 1 年针叶和 2 年针叶的养分含量关系极为密切。

3　讨论

马尾松等松属家系间在胸径、树高、产脂力等方面都存在丰富的变异[21-23]，而产脂力与胸径、树高等生长性状间存在较高的正相关关系，所以在其高产脂优树选择时，多将胸径、树高等生长性状作为参考指标，往往忽视了马尾松有机物含量特征、树脂道结构特征以及养分特征。针叶通过光合作用产生的碳水化合物经过一系列生物化学反应，形成代谢产物松脂，因此叶片养分含量可以直接反映树木的健康状况，以及对养分吸收、储存和利用的能力[24,25]。

本研究中 1 年针叶 N 含量 13.64g/kg，2 年针叶 N 含量 13.83g/kg，与陈剑锋[26]报道的 13.90g/kg 接近，低于项文化[27]报道的 14.80g/kg，而高于崔宁杰[28]报道的 10.00g/kg，说明在不同地域马尾松针叶 N 含量存在差异，造成这种原因可能是立地的差异，土壤中 N 含量不同[29]，另外也可能与种源及树龄等因素有关。生产中发现，在相同林分中，某些同经级的单株产脂量存在显著的差异，本研究即选取了差异显著的林木作为研究材料。结果表明不同产脂类型 1 年和 2 年针叶 N 含量均差异显著，表现为低产脂类型的含量高于高产脂类型，相关分析也表明产脂量与 N 含量呈显著负相关关系。K、Ca 和 Mg 含量均与产脂量呈正相关关系，而 N 含量与 P、K、Ca 和 Mg 含量均为负相关，说明针叶的 N 含量，可能会影响 P、K、Ca 和 Mg 元素在松脂合成过程中的利用，进而影响了松脂产量。P 是核酸、核苷酸、辅酶的重要组成，在 ATP 的反应中起关键作用，同时在糖类、蛋白质、和脂肪等代谢中也起着重要的作用。南方土壤中有效 P 含量普遍较低，本研究发现 P 是各器官 5 种养分中含量最少的，且差异也较小，不同产脂类型 1 年针叶和 2 年针叶含量为 0.73~0.75g/kg，枝和皮中含量更低，仅为 0.27~0.29 g/kg。K、Ca 和 Mg 含量在不同产脂类型 1 年和 2 年针叶均差异显著，均表现为低产脂类型低于高产脂类型，可能的原因是 K 可以调节细胞内适宜的渗透压和体液的酸碱平衡，是光合作用多种酶的活化剂，能提高酶的活性，并参与细胞内糖和蛋白质的代谢，能够促进碳水化合物代谢并加速同化产物向贮藏器官中运输。Ca 是细胞壁的重要组成，能够促进细胞分裂，增强根系的吸收；Mg 也是酶的激活剂，同时也是构成叶绿素的重要组成。K、Ca 和 Mg 的作用与植物光合作用能力息息相关，在健康的个体中，这 3 种养分含量能反映出光合能力和根吸收能力，进而影响马尾松产脂量。枝、皮中 Ca 含量最高，P 含量最低，形成这样结果可能是因为 N、P、K、Mg 在植物体内可以参与养分循环并多次利用[30,31]，而 Ca 不参与养分循环且只能被利用一次[32]，所以在枝和皮中含量最高。

N、K、Ca 和 Mg 含量在不同产脂个体 1 年、2 年针叶均存在显著差异，高产脂类型 1 年、2 年针叶的 K、Ca 和 Mg 含量均高于低产脂类型，而 N 含量则低于低产脂类型，P 含量两者相近。Ca 含量在不同产脂类型枝和皮差异显著，均表现为低产脂类型低于高产脂类型，N、P、K 和 Mg 含量在不同产值类型枝和皮中均差异不显著。

4 结论

对 60 株不同产脂类型马尾松林木的养分特点分析表明，不同产脂类型养分特点不同，且养分含量对产脂量存在影响。产脂量与 1 年针叶和 2 年针叶中的养分含量关系最为密切，其中与 1 年针叶和 2 年针叶的 K、Ca 和 Mg 含量均达到极显著正相关，而与 N 为极显著的负相关。相同养分在 1 年针叶和 2 年针叶间相关性都较高，而在枝和皮间相关性较弱。高产脂类型针叶 K、Ca 和 Mg 含量相对较高，而 N 含量则低于低产脂类型。这为下一步进行马尾松高产脂优树的早期选择，以及提高马尾松产脂林产量进行科学施肥提供了理论基础。

参考文献

[1]刘玉春. 1995-2000 年中国松香的生产、消费和发展趋势[J]. 林产化工通迅，2001，35(5)：31-33.

[2]杨章旗. 广西松脂原料林发展现状，存在问题与对策研究[J]. 广西林业科学，2007，36(3)：143-146.

[3]丁贵杰. 马尾松人工纸浆材林采伐年龄初步研究[J]. 林业科学，2000，36(1)：15-20.

[4]李瑞霞，闵建刚，彭婷婷，等. 间伐对马尾松人工林植被物种多样性的影响[J]. 西北农林科技大学学报(自然科学版)，2013，41(3)：61-68.

[5]LAI H L，WANG Z R，JIANG Rui-rong. Branching and growth of plantings in fifth year of a seedling seed orchard of Masson pine (*Pinus massoniana* Lamb.) [J]. Journal of Forestry Research，2002，13 (01)：28-32.

[6]安宁，丁贵杰，谌红辉. 林分密度及施肥对马尾松产脂量的影响[J]. 中南林业科技大学学报，2010，30 (09)：46-50.

[7]舒文波，杨章旗，兰富. 施肥对马尾松中龄林生长量和产脂的影响[J]. 福建林学院学报，2009，29(02)：97-102.

[8]马婷瑶，翟凯燕，金雪梅，等. 间伐对马尾松人工林土壤活性氮组分的影响[J]. 西北农林科技大学学报(自然科学版)，2017，45(12)：44-53.

[9]杨会侠，汪思龙，范冰，等. 不同林龄马尾松人工林年凋落量与养分归还动态[J]. 生态学杂志，2010，29 (12)：2334-2340.

[10]理永霞，茶正早，罗微，等. 3 种桉树幼苗叶片养分变化及其转移特性[J]. 林业科学，2009，45(1)：152-157.

[11]薛立，徐燕，吴敏，等. 4 种阔叶树种叶中氮和磷的季节动态及其转移[J]. 生态学报，2005，25(3)：520-526.

[12]曾德慧，陈广生，陈伏生，等. 不同林龄樟子松叶片养分含量及其再吸收效率[J]. 林业科学，2005，41 (5)：21-27.

[13]贾瑞丰，尹光天，杨锦昌，等. 不同氮素水平对红厚壳幼苗生长及光合特性的影响[J]. 林业科学研究，2012，25(1)：23-29.

[14]ABROL Y P，CHATTERJEE S R，KUMAR P A，et al. Improvement in nitrogen use efficiency：physiological and molecular approaches [J]. Curr Sci，1999，76：1357-1364.

[15]李荣华，汪思龙，王清奎. 不同林龄马尾松针叶凋落前后养分含量及回收特征[J]. 应用生态学报，2008，19(7)：1443-1447.

[16]宁秋蕊，李守中，姜良超，等. 亚热带红壤侵蚀区马尾松针叶养分含量及再吸收特征[J]. 生态学报，2016，36(12)：3510-3517.

[17]刘青华，周志春，范辉华，等. 马尾松产脂力与生长性状的家系变异及优良家系早期选择[J]. 林业科学研

究，2013，26(6)：686-691.

[18]国家林业局．LY-T 1271-1999 森林植物与森林枯枝落叶层全氮、磷、钾、钠、钙、镁的测定[M]．北京：中国标准出版社，1999.

[19]国家林业局．LY-T 1270-1999 森林植物与森林枯枝落叶层全硅、铁、铝、钙、镁、钾、钠、磷、硫、锰、铜、锌的测定[M]．北京：中国标准出版社，1999.

[20]卢纹岱．SPSS for Windows 统计分析[M]．北京：电子工业出版社，2002：1-592.

[21]FRIES A，ERICSSON T，GREF R. High heritability of wood extractives in Pinus sylvestris progeny tests[J]．Can J For Res，2000，30(11)：1707-1713.

[22]ROBERDS，J H，STROM B L，HAIN F P，et al. Estimates of genetic parameters for oleoresin and growth traits in juvenile loblolly pine[J]．Can J for Res，2003，33(12)：2469-2476.

[23]TADESSE W，NANOS N，AUNON F J，et al. Evaluation of high resin yielders of *Pinus pinaster* Ait.[J]．For Gen，2001，8(4)：271-278.

[24]HUANG J J，WANG X H，YAN E R. Leaf nutrient concentration，nutrient resorption and litter decomposition in an evergreen broad-leaved forest in eastern China[J]．Forest Ecology and Management，2007，239(1/3)：150-158.

[25]KOBE R K，LEPCZYK C A，IYER M. Resorption efficiency decreases with increasing green leaf nutrients in a global data set[J]．Ecology，2005，86(10)：2780-2792.

[26]陈剑锋，侯恩庆，张玲玲，等．福建省马尾松和杉木针叶中 7 种营养元素含量特征[J]．热带亚热带植物学报，2016，24(6)：595-602.

[27]项文化，田大伦．不同年龄阶段马尾松人工林养分循环的研究[J]．植物生态学报，2002，26(1)：89-95.

[28]崔宁洁，刘小兵，张丹桔，等．不同林龄马尾松人工林碳氮磷分配格局及化学计量特征[J]．生态环境学报，2014，23(2)：188-195.

[29]赵琼，刘兴宇，胡亚林，等．氮添加对兴安落叶松养分分配和再吸收效率的影响[J]．林业科学，2010，46(5)：14-19.

[30]NORRIS M D，REICH P B. Modest enhancement of nitrogen conservation via retranslocation in response to gradients in N supply and leaf N status[J]．Plant and Soil，2009，316(1/2)：193-204.

[31]YUAN Z Y，CHEN H Y H. Negative effects of fertilization on plant nutrient resorption[J]．Ecology，2015，96(2)：373-380.

[32]郭峰，周运超．不同密度马尾松林针叶养分含量及其转移特征[J]．南京林业大学学报(自然科学版)，2010，34(4)：93-96.

[原载：西北农林科技大学学报(自然科学版)，2019，47(07)]

马尾松人工林小气候调节效应

刘士玲[1,2] 杨保国[1,2] 郑 路[1,2] 雷丽群[1,2] 庞圣江[1,2] 张 培[1,2] 李 华[1,2] 韦菊玲[1,2]
([1]中国林业科学研究院热带林业实验中心，广西凭祥 532600；
[2]广西友谊关森林生态系统国家定位观测研究站，广西凭祥 532600)

摘 要 通过对比分析马尾松林外内小气候特征，为其生态效应研究提供依据。利用马尾松人工林内外两个相同的地面气象站气象资料，分析比较两站点的气温、湿度等小气候因子的日变化和月变化特征。结果表明：①林内外气温、土壤温度、空气湿度的差值均为白天大于夜晚，马尾松林白天具有更好的降温和增湿作用。②马尾松林内年均气温和土壤温度分别比林外空旷地低1.1℃、2.8℃，年均空气相对湿度比空旷地高4.7%，林内高湿日数多于林外空旷地，而低湿日数少于空旷地。③林内外月均气温、空气湿度差值表现为：夏季>春季>秋季>冬季。④林内外土壤温度季节变化规律一致：夏季>秋季>春季>冬季，与日变化相似，土壤温度越向下振幅越小。
关键词 马尾松；马尾松人工林；小气候特征；小气候调节效应

Microclimate Regulation Effect of *Pinus massoniana* Plantation

LIU Shiling[1,2], YANG Baoguo[1,2], ZHENG Lu[1,2], LEI Liqun[1,2], PANG Shengjiang
ZHANG Pei[1,2], LI Hua[1,2], WEI Juling[1,2]
(*[1]Experimental Center of Tropical Forestry, CAF, Pingxiang, 532600, Guangxi, China;*
[2]Guangxi Youyiguan Forest Ecosystem Research Station, Pingxiang, 532600, Guangxi, China)

Abstract: The microclimate characteristics of *Pinus massoniana* forest were analyzed, which provided evidence for the ecological benefit of *P. massoniana* forest. Based on the meteorological data of two identical ground meteorological stations established inside & outside the *P. massoniana* plantation , the daily variation and monthly variation characteristics of forest climate such as temperature and humidity of the two sites were analyzed and compared. The results show that: ①The differences in air temperature, relative humidity and soil temperature during the daytime between the two sites were much higher than that at night time, the *P. massoniana* forest had better cooling and humidifying effect in daytime. ②The average annual temperature and soil temperature of *P. massoniana* forest were 1.1℃ and 2.8℃ lower than those of open land outside the forest respectively, the average annual relative humidity of air was 4.7% higher than that of open land, and the number of days of high humidity in forests was more than that of open land outside forest, while the number of days of low humidity was less than that of open land. ③ The difference order of monthly mean temperature and air humidity inside and outside forests was summer > spring > autumn > winter. ④ The seasonal variations of soil temperature inside and outside the forest was the same: summer > autumn > spring > winter, this was similar to the diurnal variation, with soil depth increasing the lower the soil temperature was.
Key words: *Pinus massoniana*; *Pinus massoniana* plantation; microclimate characteristics; microclimate regulation effect

森林小气候具有协调生物与环境关系的作用，是评价森林系统综合效益的重要特征之一[1]。无论是开展森林生物的生理生态特征、养分循环和水文功能等方面的研究，还是开展其他森林生态系统结构、功能等方面的研究，都不可缺少对森林小气候的观测[2]。传统的小气候研究注重对气象资料的简单描述，对小气候的时间和空间变异把握不足[3]。以往国内对森林小气候观测多为人工观测，局限于特定时间段的小气候特征分析，而缺少连续高频率的观测数据作为研究支撑，难以对森林的气候状况变化进行全面分析和把握。

马尾松(*Pinus massoniana*)是我国主要的用材与造林树种之一，分布区域广且适应性强。近年来，国内外关于马尾松的研究主要集中在材性[4]、种子园[5]、外生菌根真菌[6]、毛虫发生规律与防治[7]、地力维护[8]等方面，鲜有利用定点连续高频率的观测数据研究马尾松林小气候特征。本文基于广西友谊关森林生态系统国家定位观测研究站的自动气象站对气象因子的高精度观测，以不同的尺度，对比马尾松林与空旷地气象因子差异，既可丰富关于马尾松生理生态方面的研究内容，也可揭示广西大青山地区不同下垫面水热的变化规律，同时也为本区域森林经营提供科学依据和基础数据。

1 研究区概况

研究区域位于广西友谊关森林生态系统国家定位观测研究站实验区范围内(106°41′~106°59′E，21°57′~22°16′N)，属南亚热带半湿润-湿润季风气候，有明显的干湿季。年均气温 21.5~21.7℃，≥10℃积温 6000~7600℃，年降水量 1200~1500mm，年蒸发量 1261~1388mm，相对湿度 80%~84%。地貌类型主要是低山丘陵，海拔 430~680m。土壤为由花岗岩发育而成的赤红壤，土层厚度 100~150cm。

林内自动气象站安装在马尾松林，林龄为 29 年，树高约 19.1m；灌木层以大沙叶(*Pavetta arenosa*)、古钩藤(*Cryptolepis buchananii*)为主；草本层主要有铁芒萁(*Dicranopteris dichotoma*)、半边旗(*Pteris semipinnata*)、蔓生莠竹(*Microstegium fasciculatum*)、弓果黍(*Cyrtococcum patens*)等。林外气象站放置在热带林业实验中心伏波实验场红星站(距马尾松林<1.5km)，根据国家气象局编制的《自动气象站建设技术要求》和《地面气象观测规范》建立。

2 研究方法

2.1 观测设备

利用自动气象站进行定点对比观测，主要观测蒸发、降水、空气温湿度、风速、风向、辐射等常规气象要素和 5、15、30、50cm 的土壤要素。该气象站采用的是 CR1000 数据采集器(Campbell，美国)，各要素数据采集步长均为 5min，传感器类型见表 1。通过分析和对比林内外气象因子的日变化及月变化特征来探讨其时空变化规律。

表 1 传感器的主要特征

传感器	型号	范围	精度	产地
空气温度(℃)	HMP45C	−39.2~60	±0.2	美国
空气湿度(%)	HMP45C	0.8~100	±2	美国
土壤水分(%)	CS616	—	±2.5	美国
土壤温度(℃)	109	−50~70	±0.2	美国

2.2 数据处理

采用气象划分法划分春(阳历 3~5 月)、夏(阳历 6~8 月)、秋(阳历 9~11 月)、冬季(阳历 12 月至翌年 2 月)，且冬、春、夏、秋季的代表月份分别为 1、4、7、10 月。该区地处南亚热带，干湿季水热条件差异较大，因此选取干季 1 月份和湿季 7 月份分析林内外小气候日变化的差异。

统计 14 时相对湿度≥80%和≤30%的出现日数，用于空气干、湿小气候特征的研究。以≥80%作为高湿指标，≤30%作为低湿指标[9]。

选取 2016 年全年数据进行分析。采用 Microsoft Excel 2007 对修订校正后的各要素的 5min 步长观测数据加以整合，采用 SigmaPlot 11.0 完成图形分析。

3 结果与分析

3.1 小气候因子日变化规律

3.1.1 空气温度日变化

林内外气温日变化在 1 月(图 1A)和 7 月(图 1B)均呈现同步的倒 U 型曲线，7：00 气温开始升高，15：00 达到日气温最大值，随后逐步降低。林内日气温低于林外，1、7 月份林内外气温日振幅分别为 2.25℃、3.00℃和 4.59℃、6.04℃。从干、湿季林内外气温差值变化趋势图(图 1C)可以看出，相同时刻林内外差值表现为白天温差大于夜晚，湿季林内外气温差值大于相同时刻干季的差值，但波动幅度与峰值出现时间不尽相同。1 月份林内外气温差值从 6：00(0.55℃)的最低值开始上升，13：00(1.30℃)~16：00(1.35℃)差值相似且为日差值最大的时段，随后16：00迅速减小至20：00(0.74℃)后林外气温差值曲线又趋于平稳，日平均温差为0.86℃。7月份差值曲线从7：00(0.69℃)的最低值开始上升，12：00达到最高温差2.17 ℃，随后16：00开始逐渐减小至21：00(0.74℃)，日平均温差为1.34℃。

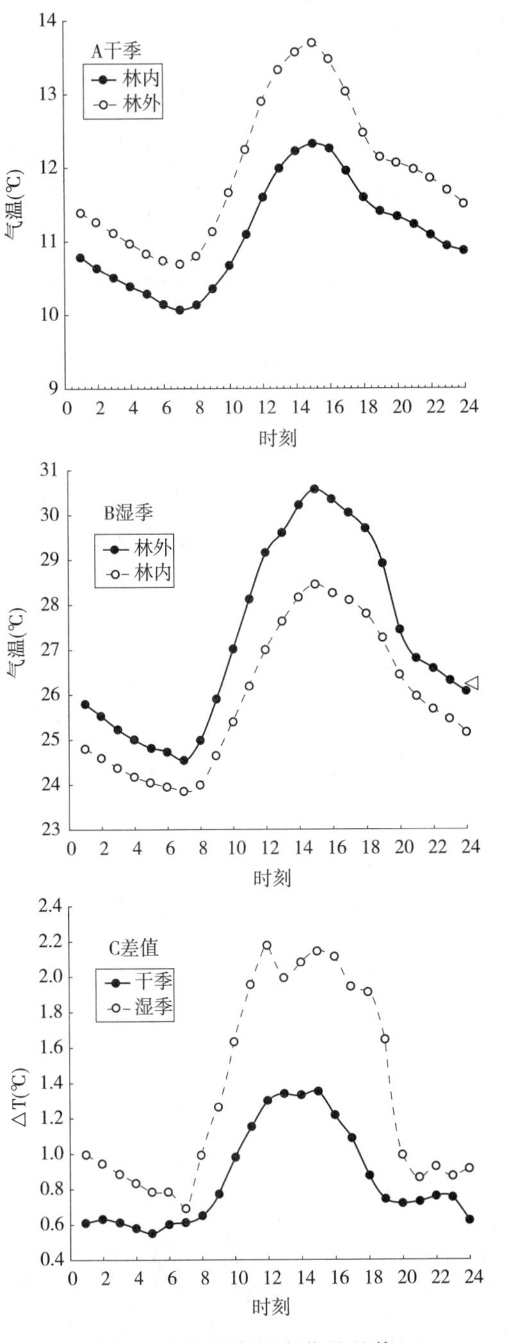

图1 空气温度日变化及差值

3.1.2 空气相对湿度日变化

空气相对湿度日变化在1月(图2A)和7月(图2B)均呈现同步的U型曲线，二者差值曲线同气温差值曲线相似，为倒U型曲线，峰值均出现在清晨时分，最低值在15：00~16：00，干湿季空旷地和马尾松林内相对湿度日振幅分别为8.54%、5.82%和29.00%、17.79%。干湿季空气相对湿度差值曲线变化趋势相似，湿度差白天大于夜晚，马尾松林内相对湿度均大于同时刻空旷地，湿季空气相对湿度差值曲线振幅大于干季，湿度差的较大值出现在10：00~19：00。干湿季日相对湿度差值最大值分别为3.96%(13：00)、16.56(15：00)，最小值分别为1.17%(08：00)、5.36%(07：00)，平均差值分别为2.32%、10.42%。

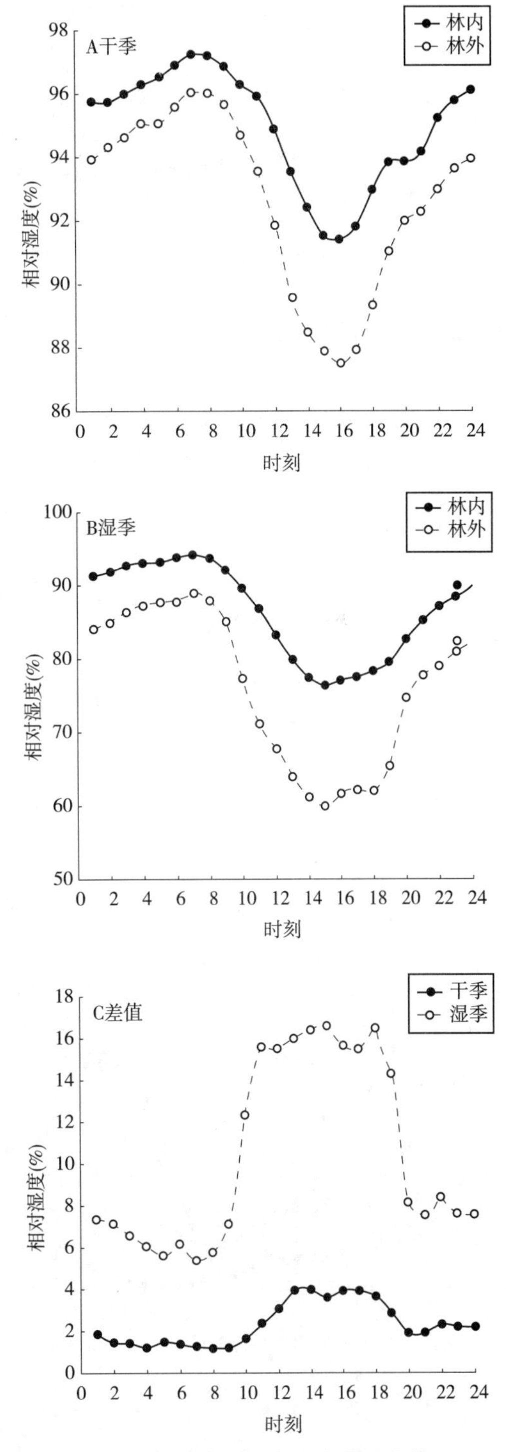

图2 空气相对湿度日变化及差值

3.1.3　土壤温度日变化

干、湿季马尾松林内外土壤温度的时态变化趋势基本一致，但林内各层土壤温度稍低于林外空旷地。从季节上来看，湿季土壤温度的变化幅度较大。林内外土壤温度曲线最值出现的时间随土壤深度的增加而逐渐向后推移，即表现为迟滞效应(图3)，且随土壤深度的增加土壤温度的变化幅度减小。5cm 土壤温度日振幅最大，干季林内外日振幅分别为 0.50℃、0.58℃，湿季林内外日振幅分别为 0.86℃、1.83℃，50cm 土壤温度日振幅近似为0。土壤温度的垂直变化表现为1月份随土壤深度的增加温度增加，7月份则相反。土壤温度差值曲线变化幅度表现为湿季>干季，干湿季土壤温度差值变化幅度均随土壤深度的增加而减小。

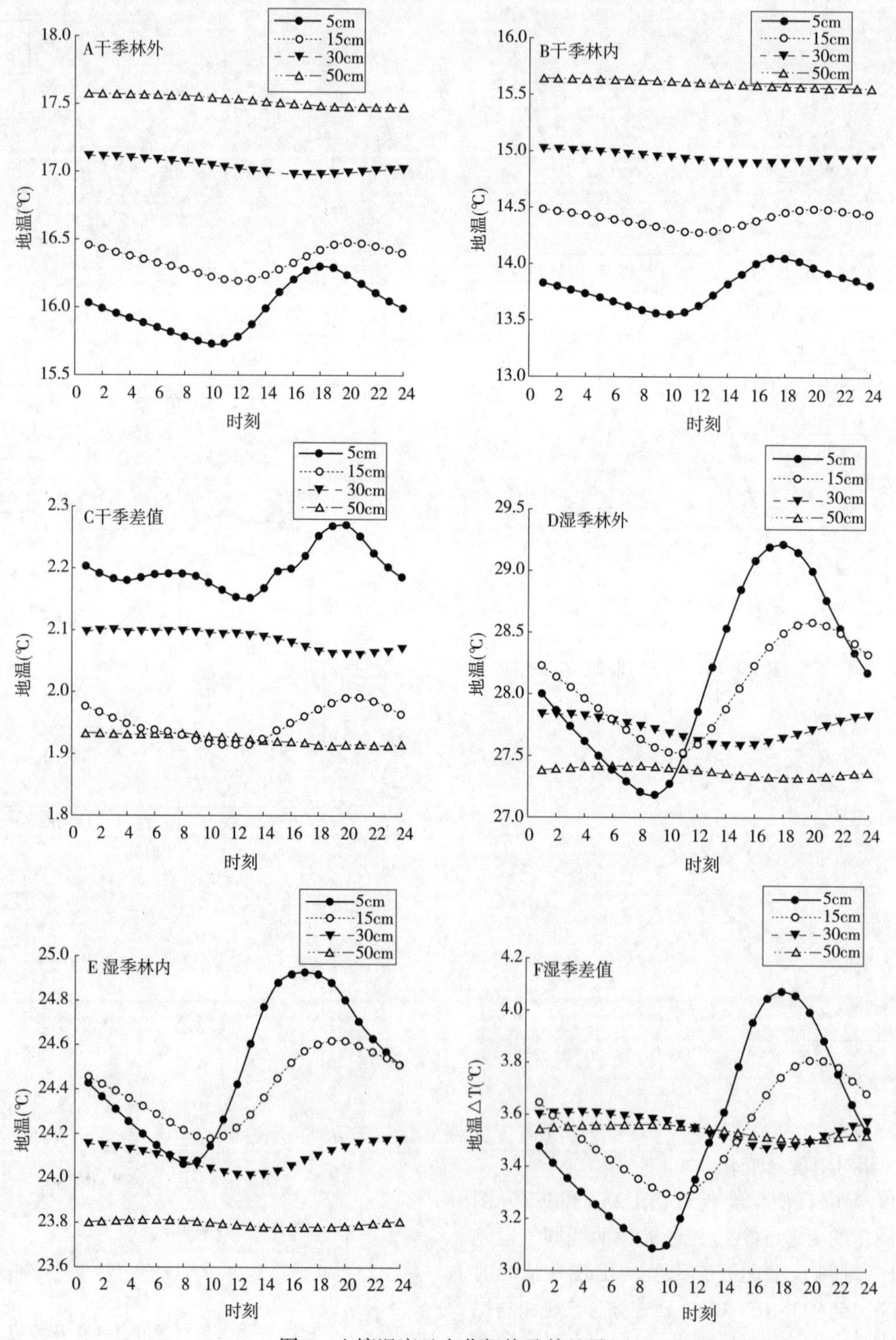

图3　土壤温度日变化规律及其差异

3.2 小气候因子月变化规律

3.2.1 空气温度月变化特征

2016年马尾松林林内年平均气温为20.3℃，空旷地年平均气温为21.4℃，月气温差为0.8~1.3℃，二者差异不显著($P>0.05$)(表1)。从季节上来看，马尾松林与空旷地的月均温差值表现出夏季>春季>秋季>冬季的规律，其中3~4月平均气温的变化幅度最大。从林内外空气温度月极值差异比较可以看出，林外空气温度月极端高温和极端低温均高于林内。林内外极端温度差异表现出一定的季节变化规律，即极端高温季节变化趋势为：夏季>春季>秋季>冬季，林内外极端高温差异显著($P<0.05$)；极端低温季节变化趋势为：秋季>春季>夏季>冬季，其差异不显著($P>0.05$)。马尾松林各月气温月均日较差均小于林外，且二者差异极显著($P<0.01$)，对比不同月份林分内外月均日较差的差值可以看出，其不同季节变化幅度大小为：夏季(2.03℃)>秋季(1.67℃)>春季(1.23℃)>冬季(1.13℃)，夏季高温时马尾松林内温度相对林外较低，降温幅度较大，可有效地防止叶片灼伤；冬季低温时林内气温降低幅度变小，仅为夏季的55.6%，又起防寒作用，保障了马尾松林内稳定的气温变化，为林木生长提供了良好的条件。

表1 马尾松林内外月气温比较

月份	月均气温(℃)		差值	月极端高温(℃)		月极端低温(℃)		月均日较差(℃)	
	林内	林外		林内	林外	林内	林外	林内	林外
1	11.1	11.9	0.9	22.7	25.1	-0.9	-0.4	4.2	4.9
2	11.7	12.5	0.8	30.1	31.2	3.9	4.9	5.9	6.6
3	15.9	17.0	1.0	24.6	27.0	6.5	7.7	4.1	5.0
4	22.2	23.4	1.2	30.8	33.2	14.1	15.2	5.5	6.6
5	24.1	25.3	1.2	31.6	33.7	17.1	17.7	5.9	7.6
6	26.1	27.5	1.3	34.5	37.0	20.6	21.0	6.7	8.6
7	25.9	27.2	1.3	32.9	35.5	20.8	22.1	6.1	7.9
8	25.4	26.6	1.2	32.8	36.1	20.7	21.6	5.8	8.2
9	24.7	25.7	1.0	30.2	33.0	19.7	20.8	5.1	7.2
10	23.0	24.0	1.0	29.3	32.0	14.2	15.3	5.6	8.0
11	17.5	18.3	0.8	27.3	28.8	8.3	9.6	4.7	5.6
12	16.1	16.9	0.8	26.5	28.5	7.8	8.2	6.8	8.8

3.2.2 空气相对湿度月变化特征

由表2可知，该区林内外相对湿度较大，均大于70%。马尾松林、空旷地年平均相对湿度分别为85.1%、80.4%，林内、外空气相对湿度差异性不显著($P>0.05$)。通过统计14时相对湿度≥80%和≤30%的出现日数发现，高湿日数林内多于空旷地，低湿日数林内少于空旷地。林内外不同季节空气相对湿度差值排序为：夏季>春季>秋季>冬季，这与空气温度差值季节变化规律相同，主要是因为生长季林木快速生长，蒸腾作用较强和林冠较大，对气流交换的阻挡效果较好，此外，春夏季雨水较多，使得夏春两季森林提高空气湿度的效果更为明显。

相对湿度林内较高，这是因为林木可以通过发达的根系从深层土壤吸取水分供林木蒸腾消耗，且林冠较大，对气流交换的阻挡效果较好，林地蒸发、植物蒸腾的水蒸气不易向外扩散，导致林内的空气湿度较大。

表2 林内外月相对湿度和高湿低湿日数的比较

月份	月平均相对湿(%)		差值	高湿日数(d)		低湿日数(d)	
	林内	林外		林内	林外	林内	林外
1	92.6	95.8	-3.2	23	26	0	0
2	72.0	74.7	-2.7	8	11	2	2

（续）

月份	月平均相对湿(%)		差值	高湿日数(d)		低湿日数(d)	
	林内	林外		林内	林外	林内	林外
3	86.2	90.5	-4.3	18	22	0	0
4	79.2	89.3	-10.1	10	18	1	0
5	80.2	85.0	-4.8	7	12	0	0
6	80.2	84.9	-4.7	6	8	0	0
7	76.0	87.4	-11.3	5	15	0	0
8	84.6	89.0	-4.4	11	15	0	0
9	82.4	83.6	-1.2	5	11	0	0
10	75.0	79.5	-4.5	4	8	0	0
11	84.0	87.4	-3.5	8	17	0	0
12	70.9	74.0	-3.1	2	5	0	0

3.2.3 土壤温度月变化特征

由图4可知，马尾松林与空旷地土壤温度季节变化规律一致：夏季>秋季>春季>冬季。与日变化相似，随土层加深土壤温度变化幅度减小。林外空旷地各层土壤温度月振幅从14.1℃下降到12.1℃，马尾松林内则从12.0℃下降到10.3℃，由此可见，马尾松林具有较好的稳定土温的能力。林内外月均土壤温差为1.7~2.1℃，略高于月均气温差。林内外同层土壤温差年变化表现出先增大，6月份达到最大值后，又趋于减小，在12月份达到最小值，即干湿季差异更为显著（$P<0.01$）。

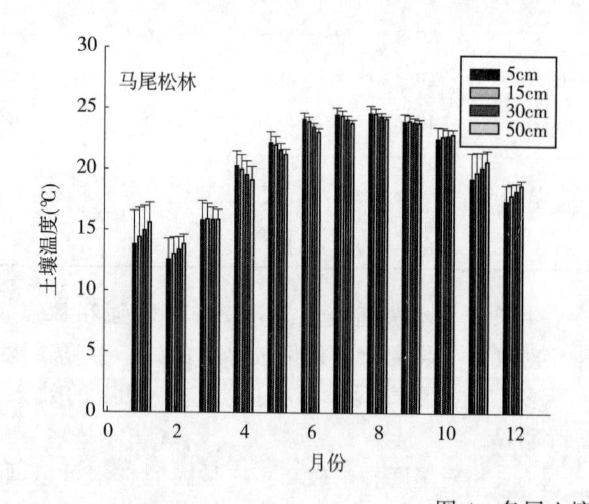

图4 各层土壤温度月变化

4 讨论与结论

4.1 结论

(1)林内外气温日变化在1月和7月均呈现同步的倒U型曲线，日气温林外高于林内。比较温度日较差可知，7月份林内外气温差值大于相同时刻1月份的差值，白天温差大于夜晚。从气温月变化来看，月均气温林外>林内，二者气温均呈现夏季>春季>秋季>冬季的规律。

(2)1月和7月林内外空气湿度日变化为同步的U型曲线，二者差值变化为倒U型曲线，日振幅马尾松林大于林外。林内相对湿度均大于同时刻空旷地，林内年均空气相对湿度比空旷地高4.7%，湿度差均是白天大于夜晚。月均空气湿度林外<林内，从林内外空气湿度的季节变化规律来看，为夏季>春季>秋季>冬季。

(3)干湿季林外与马尾松林内土壤温度的时态变化趋势基本一致，且林内外土壤温度曲线均表现出一定的迟滞效应，这种效应随土壤深度增加而逐渐减弱。土壤温度日变化和月变化均表现为林外各层高于林内，林内外同层土壤温差月变化表现出先增大后减小的趋势；从季节变化规律来看，马尾松林内外土壤温度均表现为夏季>秋季>春季>冬季。

4.2 讨论

由于森林的存在使得林内与林外相比，热量交换在时间和空间上发生了显著的改变，从而使热量在林内重新分配，使林内温度发生了变化。本研究马尾松林年均温、年均湿度分别为20.3℃、85.1%，这与南亚热带鼎湖山处于中期演替阶段的沟谷雨林小气候（年均温20.4℃、年均湿度87%）[10]极为相近，说明本研究马尾松林在改善森林气候功能方面已达到南亚热带鼎湖山演替中期阶段森林群落的水平，具有较好的调温保湿效应。沈运扩[11]研究表明，白蜡和柽柳人工林分别在春季和秋季具有保温作用，在夏季均表现为降温作用；蒋丹丹[12]通过对石灰岩山地两种模式人工林小气候比较研究，发现四季气温值均为林内<林外；这与本文马尾松林各月气温日较差均小于空旷地和四季气温均为林内<林外的研究结果一致，马尾松林夏季的降温作用更为显著，而冬季又能够一定程度上稳定林内温度，起到防寒作用。但这与周和锋等[13]杭州湾生态绿地小气候冬季林内温度高于林外的研究结果有所不同，这可能因为植被类型和研究区域地理位置的差异所造成的。

林内外空气湿度日变化在呈现同步的U型曲线，但不同生长季节的湿度最值和变化幅度有差异，这与闫文德和田大伦[14]研究的樟树人工林空气湿度的结果一致。闫俊华等[15]研究发现鼎湖山常绿阔叶林内外，空气相对湿度在春夏两季较大，秋冬两季较小，这与本文及徐云蕾等[16]林内外月均空气湿度差值排序夏季>春季>秋季>冬季的研究结果一致。马尾松林内外干季和湿季相对湿度差分别为10.42%、2.32%，相对于其他气候带而言，亚热带地区湿季的高温高降水将导致森林与空旷地的相对湿度差异变小，因此在水热资源相对缺乏的干季能表现出更好的增湿效应。从全年空气湿度来看，马尾松林高湿日数较多、低湿日数较少，表明马尾松林具有较好的保湿作用。

此外，由于小气候观测研究工作是一项长期的、系统的研究工作，今后应集中多年观测数据进行系统的研究，同时开展混交林与纯林小气候的对比研究、纯林及近自然化改造林分小气候的对比研究，更加全面观测与分析马尾松林小气候特征，从而为研究生态系统能量平衡、探讨生物与环境的关系打下良好的基础。

参考文献

[1]薛雪，杨静，郑云峰，等．南京城市杂交马褂木林小气候特征研究[J]．水土保持研究，2016，23(04)：225，226-232.

[2]林永标，申卫军，彭少麟，等．南亚热带鹤山三种人工林小气候效应对比[J]．生态学报，2003，23(08)：1660-1661.

[3] CARLSON D W，GROOT A. Microclimate of clear-cut，forestinterior and small opening in trembling aspen forest [J]. Agricultural and Forest Meteorology，1997，87：313-329.

[4]程曦依，李芸，全鹏，等．马尾松锯材常压过热蒸汽干燥脱脂特性研究[J]．中南林业科技大学学报，2017，37(06)：108-113.

[5]张国洲，谢维斌．我国马尾松种子园研究进展[J]．安徽农业大学学报，2012，39(01)：84-87.

[6]张珍明，张家春，何云松，等．马尾松外生菌根真菌研究进展[J]．耕作与栽培，2016，(02)：66-68，72.

[7]胡树恒，王军旺，李珍．马尾松毛虫发生规律与防治研究进展[J]．农业与技术，2013，33(03)：109-110.

[8]李明军，杜明凤，聂朝俊．马尾松人工林地力维护研究进展[J]．世界林业研究，2014，27(05)：31-36.

[9]王树廷，王伯民．气象资料的整理和统计方法[M]．北京：气象出版社，1982.

[10]欧阳旭，李跃林，张倩媚．鼎湖山针阔叶混交林小气候调节效应[J]．生态学杂志，2014，33(03)：575- 582.

[11]沈运扩．滨海盐碱地白蜡与柽柳人工林小气候效应研究[D]．山东：山东农业大学，2014.

[12]蒋丹丹，万福绪，黄文庆．石灰岩山地两种模式人工林小气候比较[J]．水土保持研究，2015，22(01)：110-114

[13]周和锋，邵学新，房聪玲，等．杭州湾滨海生态绿地小气候效应研究[J]．华东森林经理，2012，26(01)：34-38.

[14]闫文德，田大伦．樟树人工林小气候特征研究[J]．西北林学院学报，2006，21(2)：30-34.

[15]闫俊华，周国逸，韦琴．鼎湖山季风常绿阔叶林小气候特征分析[J]．武汉植物学研究，2000，18(05)：397-404.

[16]徐云蕾，曾馥平，宋同清，等．喀斯特峰丛洼地次生林小气候特征[J]．农业现代化研究，2012，33(02)：242-244.

［原载：中南林业科技大学学报，2019，39(02)］

马尾松与红椎纯林及混交林生态系统碳储量研究

卢立华[1,2] 郭文福[1,2] 蔡道雄[1,2] 贾宏炎[1,2] 陈 琳[1,2] 农 友[1,2]
雷丽群[1,2] 李 华[1,2] 刘士玲[1,2] 杨保国[1,2]

([1]中国林业科学研究院热带林业实验中心，广西凭祥 532600；
[2]广西友谊关森林生态系统国家定位观测研究站，广西凭祥 532600；)

摘 要 为了解树种与造林模式对人工林生态系统碳储量的影响，在南亚热带相同立地上，采用样地调查方法，对33年生马尾松纯林(PCL)、红椎纯林(CCL)、马尾松×红椎同龄混交林(PCM)生态系统碳储量进行了研究。结果表明：PCM、PCL、CCL人工林生态系统碳储量顺次为：235.38t/hm^2、196.40t/hm^2、144.59t/hm^2，处理间差异显著($P<0.05$)，混交林生态系统碳储量显著高于纯林，马尾松纯林显著高于红椎纯林。PCM、PCL、CCL人工林生态系统碳储量在不同层次的分配比例，乔木层分别为：53.70%、54.05%、33.02%；地被物层为：1.47%、2.06%、1.37%；土壤层为：44.83%、43.89%、65.61%。生态系统碳储量在不同层次分配比例排序，PCM、PCL都为乔木层>土壤层>地被物层；CCL为土壤层>乔木层>地被物层，乔木层和土壤层碳储量占生态系统碳储量的97.94%~98.63%，地被物层仅占1.37%~2.06%。

关键词 马尾松；红椎；纯林；混交林；碳储量比较

Study on Carbon Storage of Monoculture and Mixed Plantation of *Pinus massoniana* and *Castanopsis hystrix*

LU Lihua[1,2], GUO Wenfu[1,2], CAI Daoxiong[1,2], JIA Hongyan[1,2], CHEN Lin[1,2], NONG You[1,2], LEI Liqun[1,2], LI Hua[1,2], LIU Shiling[1,2], YANG Baoguo[1,2]

([1]*Experimental Center of Tropical Forestry, Chinese Academy of Forestry, Pingxiang 532600, Guangxi, China*;
[2]*Youyiguan Research Station for Localization Observation of Forest Ecosystem, Pingxiang 532600, Guangxi, China*)

Abstract: In order to understand the effects of tree species and planting patterns on carbon storage in plantation ecosystem, the carbon storage in 33-year-old pure *Pinus massoniana* forest (PCL), pure *Castanopsis hystrix* forest (CCL) and their even-age mixed forest (PCM) were studied by plot investigation. The results showed that there were significant differences in the carbon storage of PCM, PCL and CCL plantation ecosystems ($P<0.05$). The carbon storage contents were 235.93 t/hm^2, 194.87 t/hm^2 and 146.12t/hm^2 respectively. The carbon storage in mixed plantation ecosystem was significantly higher than that in pure plantation, while in pure plantation, PCL was significantly higher than that of CCL. The distribution ratio of carbon storage in different tree species and models was different in different layers, in the arbor layer the ratio was 53.81%、54.05%、33.02%; in the ground cover was 1.46%、2.06%、1.37%; and in the soil layer was 44.73%、43.89%、65.61%. The distribution ratio of PCM, PCL was the highest in the arbor layer, followed by the soil layer and then the ground cover layer. While the soil layer was the highest, followed by the arbor layer and then the ground cover layer of CCL The proportion of carbon storage in the arbor layer and soil layer is 97.93%~98.63% of the total carbon storage in ecosystem, while that in ground cover layer is only 1.37%~2.06%.

Key words: *Pinus massoniana*; *Castanopsis hystrix*; monoculture plantation; mixed plantation; carbon storage

全球气候变暖早被确证[1]，它的变暖对植物种群、生态系统乃至整个生物圈都将产生重大影响[2-4]。在导致气候变暖的诸多因素中，温室气体排放占90%以上[5]，而CO_2为最重要的温室气体，因此，对CO_2的吸收、固定与减排成为了当今全球气候变化研究热点之一[6-8]。

森林作为陆地上最大生态系统，在固存 CO_2、调节气候、减缓气候变化等方面具有不可替代的作用[9,10]，对维护全球大气碳平衡亦十分关键[11]，而造林和再造林是固碳增汇最重要的途径，备受学术界重视[12-14]。但我国宜林荒山地资源相当限，依靠扩大造林面积增汇空间不大，而从我国第八次全国森林资源清查公布的结果(2009-2013 年)：中国人工林面积 6933 万 hm^2，为全球人工林面积最大国家；人工林平均蓄积量 $52.76m^3/hm^2$，仅为世界平均蓄积量 $131m^3/hm^2$的 40.27%。表明，我国多数人工林的生产力仍十分低下，同时也表明，通过选择适宜树种或模式对现有低产人工林进行提产增效改造空间巨大，故在固碳增汇策略上，既要重视造林和再造林，更应重视低产林改造。

在造林和再造林中选择适宜树种与模式是提高人工林固碳增汇的重要决策[15-17]。对现有人工林，除经营措施对固碳、增汇影响明显外[18]，通过对低产林的适度疏伐，然后在林下补植 1 个或多个适生耐阴树种，将原来人工纯林改造成异龄多树种混交林或近自然林，能有效提高林分的固碳、增汇量[19,20]，因此，人工林固碳、增汇研究较活跃。霍常富等研究了杉木(*Cunninghamia lanceolata*)人工林生态系统蓄积和碳数量成熟龄关系，表明杉木林分蓄积成熟龄为 18~27 年，生态系统碳成熟龄为 16~35 年[21]；刘灵等开展了樟子松 *Pinus sylvestris* 不同家系固碳研究，证明不同家系樟子松地上部碳储量差异明显[22]；刘之洲等在喀斯特地区研究了三种针叶人工林碳储量，表明，湿地松(*Pinus elliottii*)人工林生态系统碳储量最高，马尾松(*Pinus massoniana*)人工林次之，马尾松天然林最低[23]；陈雷等研究了银杏(*Ginkgo biloba*)不同复合经营系统碳储量，结果表明，银杏复合系统碳储量都高于农地[24]；何友均等对西南桦(*Betula alnoides*)纯林和西南桦×红椎(*Castanopsis hystrix*)混交林及明安刚等对红椎、杉木纯林及红椎×杉木混交林的研究都表明，纯林生态系统碳储量高于混交林[25,26]；You 等对红椎、马尾松纯林及红椎×马尾松混交林和卢立华等对红椎纯林、红椎×马尾松和红椎×西南桦混交林的研究，结果都为混交林生态系统碳储量高于纯林[27,28]。表明，纯林与混交林碳储量比较研究因树种、林龄、密度等的不同结果有异，尚需进行深入研究。

马尾松和红椎都为南亚热带乡土树种，You 等已对 26 年生马尾松、红椎纯林及它们同龄混交林进行了研究[27]，但 33 年生相同处理人工林生态系统碳储量的情况如何，仍未能回答。本文对 33 年生马尾松、红椎纯林及它们同龄混交林生态系统碳储量进行研究，旨在了解和掌握林龄对人工纯林及混交林生态系统碳储量的影响，为科学经营人工用材林及高效碳汇林等提供依据，为碳计量及碳贸易等提供更丰富数据与支撑。

1 研究区概况

研究区位于广西西南部的凭祥市中国林业科学研究院热带林业实验中心伏波实验场(106°39′50″~106°59′30″E，21°57′47″~22°19′27″N)，属南亚热带季风气候区，干湿季节明显(10 月至翌年 3 月为干季，4~9 月为湿季)，年均温 19.5~21.0℃，极端高温 40.3℃，极端低温-1.5℃，≥10℃积温 6000~7600 ℃，年降水量 1200~1400 mm，相对湿度 80%~84%。土壤为中酸性火山岩发育的砖红壤性红壤。主要地带性植被，灌木以桃金娘(*Rhodomyrtus tomentosa*)、大沙叶(*Pavetta hongkongensis*)、酸藤子(*Embelia laeta*)等为优势种，草本以五节芒(*Miscanthus floridulus*)、蔓生莠竹(*Microstegium vagans*)、铁芒萁(*Dicranopteris dichotoma*)等为优势种。

2 研究方法

2.1 试验设置

设置红椎纯林(代号 CCL 下同)、马尾松纯林(PCL)、马尾松×红椎(马尾松：红椎=1：1)同龄株间混交林(PCM)3 种处理。随机区组排列，3 次重复，小区面积约 $0.7hm^2$。3 种处理造林密度都为 2500 株/hm^2，于 1983 年 3 月在马尾松采伐迹地上同时用马尾松或红椎 Ⅰ 级裸根苗造林，造林后 3 种林分的经营管理措施一致。

2.2 林木调查及生物量测定

2016 年 4 月在 33 年生的 3 种林分各小区中，选代表性位置铺设 1 块 20m×30m 样地，共 9 块样地。然后，对样地内胸径大于 5cm 树木每木检尺，测定树木的胸径和树高，试验林现状见表 1。

表 1 林分概况

林分类型	坡度(°)	海拔(m)	坡位	郁闭度	林龄(年)	密度(株/hm^2)	胸径(cm)	树高(m)	生物量(t/hm^2)
PCL	33	550	中	0.85	33	433	28.90	18.33	201.39

（续）

林分类型		坡度（°）	海拔（m）	坡位	郁闭度	林龄（年）	密度（株/hm²）	胸径（cm）	树高（m）	生物量（t/hm²）
CCL		30	550	中	0.95	33	430	20.77	15.30	97.93
PCM	马尾松	30	550	中	0.90	33	223	24.93	17.83	250.35
	红椎						225	30.70	20.93	

将样地调查取得每木胸径、树高数据，利用He等[29]、明安刚等[30]在相同研究区建立的红椎、马尾松生物量方程估算马尾松、红椎乔木层各器官单位面积生物量。采用明安刚等测定获得的红椎[26]、马尾松[30]乔木各器官碳密度计算红椎、马尾松各器官单位面积碳储量。

2.3 林下植被生物量与凋落物现存量及碳密度测定

在各样地中间及四角各设置1个2m×2m样方，在每个样方中各设置1个m×1m小样方，每个样地共5个样方，5个小样方。对各样方分别调查与记录样方内灌木、草本种类后，采用收获法分别测定各样方灌木、草本层的地上、地下部生物量。在每一小样方中按未分解、半分解分别测定凋落物重量。同时，分别取灌木层、草本层、凋落层分析样各约200g测定它们含水量和碳密度。

2.4 土壤样品采集与测定

在样地中，选代表性位置各挖1个土壤剖面，按0~20cm、20~40cm、40~60cm 3个层次分别采集土壤分析样各约400g测定土壤碳密度。同时，在各土层中用铝盒取土约20g和用100cm³环刀取一环刀样，分别测定各土层测定含水量和土壤容重。

2.5 碳密度测定与碳储量计算

植物和土壤有机碳密度测定采用重铬酸钾-水合加热法。乔木层碳储量计算公式：

$$C_t=[(D_{st}\times M_{st})+(D_{ba}\times M_{ba})+(D_{br}\times M_{br})+(D_{le}\times M_{le})+(D_{ro}\times M_{ro})]\div 10^3$$

式中：C_t为乔木层碳储量t/hm²；D_{st}、D_{ba}、D_{br}、D_{le}、D_{ro}分别为干、皮、枝、叶、根碳密度g/kg；M_{st}、M_{ba}、M_{br}、M_{le}、M_{ro}分别为干、皮、枝、叶、根生物量t/hm²。

灌木层碳储量计算公式：

$$C_s=[(D_{sa}\times M_{sa})+(D_{su}\times M_{su})]\div 10^3$$

式中：C_s为灌木层碳储量t/hm²；D_{sa}、D_{su}、M_{sa}、M_{su}分别为灌木地上、地下部碳密度g/kg、地上、地下部生物量t/hm²。

草本层碳储量计算公式：

$$C_h=[(D_{ha}\times M_{ha})+(D_{hu}\times M_{hu})]\div 10^3$$

式中：C_h为草本层碳储量t/hm²；D_{ha}、D_{hu}、M_{ha}、M_{hu}分别为草本地上、地下部碳密度g/kg、地上、地下部生物量t/hm²。

凋落层碳储量计算公式：

$$C_l=[(D_{ln}\times M_{ln})+(D_{lh}\times M_{lh})]\div 10^3$$

式中：C_l为凋落层碳储量t/hm²；D_{ln}、D_{lh}、M_{ln}、M_{lh}分别为未分解、半分解凋落物碳密度g/kg、未分解、半分解凋落物现存量t/hm²。

土壤碳储量计算公式：

$$C=[(S_a\times V_a)+(S_m\times V_m)+(S_u\times V_u)]\times 2$$

式中：C为土壤碳储量t/hm²；S_a、S_m、S_u、V_a、V_m、V_u分别为上、中、下土层的有机碳密度g/kg、上、中、下土层容重t/m³。

2.6 数据处理

采用Excel 2007进行数据处理和作图，用SPSS 17.0分析软件进行方差分析和显著性检验。

2 结果与分析

2.1 不同人工林生态系统各组分碳储量及分配

2.1.1 乔木层碳储量及分配

从图1可见，乔木层碳储量以PCM最高，达126.40t/hm²；PCL次之，106.16t/hm²，CCL最低，仅47.74t/hm²，处理间差异显著（$P<0.05$）。PCM的乔木层碳储量与PCL、CCL的比分别高了19.07%、164.77%，PCL与CCL比高了122.37%，多重比较，PCM、PCL、CCL乔木层碳储量处理之间相互比较都差异显著。表明，提高林分的乔木层碳储量，以马尾松×红椎混交林（PCM）优于马尾松、红椎纯林，在纯林中则以马尾松纯（PCL）林优于红椎纯林（CCL）。图1显示，乔木层碳储量在干、皮、枝、叶、根的分配量：PCM分别为73.89t/hm²、10.39t/hm²、20.76t/hm²、4.90t/hm²、16.46t/hm²；CCL分别为29.99t/hm²、3.35t/hm²、6.11t/hm²、1.12t/hm²、7.17t/hm²，PCL分别为59.18t/hm²、11.93t/hm²、12.04t/hm²、4.04t/hm²、18.97t/hm²。同处理不同器官或同器官不同处理碳储量都差异显著（$P<0.05$）。

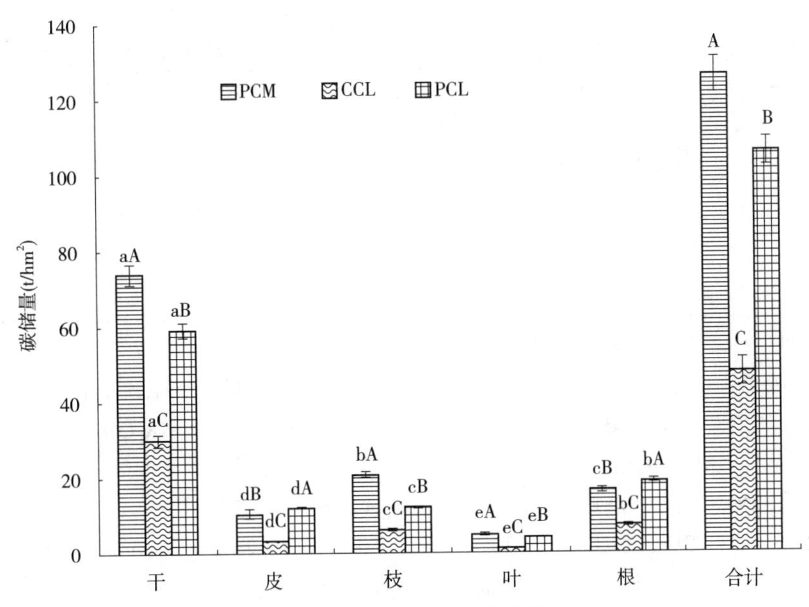

图 1 不同处理乔木器官碳储量

注：PCM、CCL、PCL 分别为马尾松×红椎混交林、红椎纯林、马尾松纯林。

同处理不同小写字母表示器官碳储量差异显著；同器官不同大写字母表示处理间碳储量差异显著($P<0.05$)。

表 2 见发现，3 种林分乔木层碳储量在干、皮、枝、叶、根器官的分配比例均以树干最高，达 55.75%~62.83%，树叶最低，仅 2.33%~3.86%，皮、枝、根分别为 7.01%~11.23%、11.34%~16.36%、12.97%~17.78%。在干、皮、枝、叶、根器官中，碳储量分配比例最高的处理顺次为 CCL、PCL、PCM、PCM、PCL。同器官不同处理碳储量分配比例比较，除叶中 PCM 与 PCL 比较差异不显著外，其他器官处理之间比较都差异显著。乔木层碳储量在各器官分配比例排序因林分不同有异，PCM 为树干>树枝>树根>树皮>树叶，CCL、PCL 都为树干>树根>树枝>树皮>树叶。表明，马尾松与红椎同龄混交提高了枝碳的分配比例，而降低了根碳的分配比例。

表 2 3 种处理乔木器官碳分配 (%)

处理	干	皮	枝	叶	根
PCM	58.23±0.79a B	8.58±0.86 d B	16.36±0.06 b A	3.86±0.08 e A	12.97±0.14 c C
CCL	62.83±0.36a A	7.01±0.08 d C	12.78±0.74c B	2.33±0.17 e B	15.04±0.48 b B
PCL	55.75±0.14a C	11.23±0.09 cA	11.34±0.12 c C	3.80±0.05 d A	17.78±0.06 b A

注：同行不同小写字母及同列不同大写字母，表示差异显著($P<0.05$)，下同。

2.1.2 林下地被物碳储量及分配

本文林下地被物包括了灌木层、草本层和凋落物层 3 个层次。从表 3 见，3 种林分地被物碳储量为 1.98~4.04t/hm²，远低于乔木层碳储量。地被物碳储量以 PCL(4.04t/hm²) > PCM(3.45t/hm²) > CCL(1.98t/hm²)。方差分析表明，处理间差异显著($P<0.05$)，其中：PCL、PCM 与 CCL 比较都差异显著。地被物碳储量在不同层次分配量差异显著($P<0.05$)，且都以凋落物层>灌木层>草本层，凋落物层碳储量与灌木层、草本层比较都差异显著。灌、草层碳储量在组分中分配量都以地上部高于地下部，3 种处理地被物碳储量在不同植物层次分配量排序，灌木层、草本层和凋落物层都为 PCL> PCM> CCL。

表 3 不同处理林下植被及凋落物的碳储量 (t/hm²)

层次	组分	CCL	PCL	PCM
灌木层	地上部	0.23±0.02b	0.41±0.05a	0.24±0.03
	地下部	0.12±0.04b	0.26±0.02a	0.16±0.03b
	小计	0.35±0.03bB	0.67±0.05aB	0.40±0.01bB

（续）

层次	组分	CCL	PCL	PCM
草本层	地上部	0.19±0.04b	0.32±0.04a	0.26±0.06ab
	地下部	0.11±0.04b	0.19±0.03a	0.13±0.04ab
	小计	0.30±0.07cC	0.51±0.02aC	0.39±0.03bC
凋落物层		1.33±0.06cA	2.86±0.09aA	2.66±0.06bA
地被物合计		1.98±0.16b	4.04±0.09a	3.45±0.04a

注：同行不同小写字母表示差异显著（$P<0.05$），下同。

2.1.3　土壤碳储量及分配

表4可见，CCL、PCL、PCM土壤碳储量顺次为94.87t/hm^2、86.20t/hm^2、105.53t/hm^2，处理间差异不显著（$P>0.05$）。但PCM的土壤碳储量比CCL、PCL分别高了11.24%、22.42%，CCL比PCL高了10.06%。各处理土层碳储量都随土层深度增加而降低，其中：0~20cm与20~40cm和40~60cm比较都差异显著（$P<0.05$），20~40cm与40~60cm比较，只有PCL处理差异显著。同土层不同处理土壤碳储量比较差异都不显著（$P>0.05$），3种处理土层碳储量及土壤碳储量均以PCM>CCL>PCL，以混交林高于纯林，红椎纯林高于马尾松纯林。

土壤碳储量在0~20cm、20~40cm、40~60cm 3个土层分配量，CCL分别为52.14t/hm^2、26.16t/hm^2、16.57t/hm^2，分别占土壤碳储量的54.96%、27.57%、17.47%；PCL为48.64t/hm^2、23.59t/hm^2、13.91t/hm^2，占56.43%、27.37%、16.20%；PCM为59.69t/hm^2、27.78t/hm^2、18.06t/hm^2，占56.56%、26.33%、17.11%。其中：0~20cm土层碳储量占土壤碳储量的54.96%~56.56%；20~40cm占26.33%~27.57%；40~60cm占16.20%~17.47%，不同处理土壤碳储量在相同土层分配比例相当接近。

2.2　人工林生态系统碳储量及其分配

2.2.1　人工林生态系统碳储量

表5可见，3种人工林生态系统碳储量以PCM最高，达235.38t/hm^2，其次PCL，196.40t/hm^2，最低CCL，仅144.59t/hm^2，其中：PCM比PCL、CCL分别高了19.85%、62.58%，PCL比CCL高了35.83%。处理间差异显著（$P<0.05$）。表明，适宜树种与模式能有效提高人工林生态系统碳储量。

表4　不同处理土壤碳储量

（t/hm^2）

土壤层次	CCL	PCL	PCM
0~20cm	52.14±4.54aA	48.64±5.67aA	59.69±8.46aA
20~40cm	26.16±8.74aB	23.59±4.13aB	27.78±5.95aB
40~60cm	16.57±2.61aB	13.97±4.32aC	18.06±2.71aB
合计	94.87±14.70a	86.20±4.13a	105.53±14.50a

表5　不同处理生态系统碳储量

层次	红椎	马尾松	马尾松×红椎
乔木层	47.74(33.02)±4.38c	106.16(54.05)±4.30b	126.40(53.70)±6.98a
地被物层	1.98(1.37)±0.16b	4.04(2.06)±0.09a	3.45(1.47)±0.04a
土壤层	94.87(65.61)±14.70a	86.20(43.89)±5.46a	105.53(44.83)±14.50a
生态系统	144.59(100.0)±10.14c	196.40(100.00)±2.14b	235.38(100.00)±21.37a

注：表中括号前数据为碳储量，单位：t/hm^2，括号内数据为其碳储量占生态系统碳储量的百分数(%)。

2.2.2　人工林生态系统碳储量在各层次分配比例

表5显示，3种人工林生态系统碳储量在不同植物层次分配比例各异，CCL、PCL、PCM在各层次的分配比例分别为：乔木层33.02%、54.05%、53.70%；土壤层65.61%、43.89%、44.83%；地被

物层1.37%、2.06%、1.47%。乔木层和土壤层碳储量占生态系统碳储量的97.94%~98.63%。碳储量在不同层次分配比例变幅：乔木层33.02%~54.05%，极差21.03%；土壤层43.89%~65.61%，极差21.72%；地被物层1.37%~2.06%，极差0.69%。以土壤层变幅最大，乔木层次之，地被物层最小。

3 讨论

森林植被碳储量受生物量、碳含量及树种生物生态学特性等众多因素的影响[29]。本研究的马尾松×红椎同龄混交林，因在改善林分内光照条件、调节林内温度、湿度及提高光能利用率、林下灌草多样性及凋落物量、改善土壤理化性质、恢复地力等多个方面优于纯林，致混交林比纯林更有利于林木生长[31]，故其乔木层碳储量最高，达126.40t/hm^2，与马尾松(106.16t/hm^2)、红椎(47.74t/hm^2)纯林比较都差异显著。表明，科学的混交经营模式能提高人工林乔木层碳储量。此外，树种不同，乔木层碳储量差异明显，本研究的马尾松纯林乔木层碳储量显著高于红椎纯林，主要与它们树种特点差异较大有关。红椎适生于水肥条件优越的立地，属喜肥、喜湿速生树种[33]，而马尾松的适应性较广，属耐干旱、瘠薄低营养速生树种[34]，它们在相同立地上种植，因树种适宜性的差异，生长量差异较大。从表1见，红椎胸径、树高、生物量分别为：18.77cm、14.30m、97.93 t/hm^2；马尾松分别为28.90cm、18.33m、201.39t/hm^2，马尾松胸径、树高、生物量与红椎比分别高了53.97%、28.18%、105.65%，马尾松乔木层碳储量也比高红椎了122.37%。与You等26年生相同处理林分的结论一致[27]，不同的是，26年生马尾松×红椎同龄混交林中的马尾松和红椎，它们的胸径、树高生长量受混交的影响大于33年生，3种林分乔木层的碳储量也都以33年生显著高于26年生，随林龄增加乔木层碳储量明显提高，与徐慧芳等[32]研究结论一致。

灌木层、草本层碳储量与林分郁闭度关系密切。红椎为阔叶树种，叶片较大，枝叶繁茂，其林分郁闭度最高(表1)，林下灌木和草本则较稀少(盖度10%)，灌、草的碳储量也最低；马尾松为针树种，叶片细小，林分郁闭度最低，其林下灌木、草本最多(盖度40%)，灌、草的碳储量也最高；马尾松×红椎混交林，它们林分的结构为双林层，郁闭度界于马尾松、红椎纯林之间，林下灌木、草本量居中（盖度20%），碳储量亦居中。

凋落物层碳储量主要受林分凋落物量与凋落物分解速度影响。研究证明，林分凋落物量以红椎纯林高于马尾松纯林[35]，混交林高于纯林[36]，故3种林分凋落物量以PCM>CCL>PCL。样地调查获得的凋落物现存量数据为PCL(6.31t/hm^2)>PCM(5.87t/hm^2)>CCL(2.94t/hm^2)，林分凋落物量排序与凋落物现存量排序不同，表明各林分凋落物的分解速度不同。而凋落物分解速度主要受凋落物C/N影响，研究证明，由针叶马尾松林向针阔叶混交林再向常绿阔叶林演替凋落物C/N比、木质素含量逐步降低，凋落物分解常数逐渐增大[37]，据此，本研究3种林分凋落物C/N排序为：PCL>PCM>CCL，这与凋落物现存量排序(PCL>PCM>CCL)相吻合。可见，林分凋落物分解速度以CCL最快，PCM次之，PCL最慢。

森林土壤有机碳储量受凋落物、细根、动物残体等多种因素影响。凋落物分解后以腐殖质形式首先进入表土层，植物细根和动物残体等也主要集中于表土层，因此，表土层有机碳储量是决定土壤碳储量最重要层次。研究证明，土壤碳主要集中于0~30cm土层，并随深度增加而降低[29]。本研究显示，PCL、CCL、PCM处理0~20cm土层的有机碳储量分别占其土壤碳储量(0~60cm)的56.43%、54.96%、56.56%，表层的碳储量都超过了土壤碳储量的一半，土层碳储量都随土层深度增加而降低，与何友均等[29]研究结论一致。

树种与模式对人工林生态系统碳储量影响明显。人工林生态系统碳储量主要来自乔木层和土壤层，乔木层碳储量高低主要取决于林分生长优劣；森林土壤碳储量主要来自森林凋落物的分解、对大气中含碳气体的吸收、大气含碳物质的沉降及岩石的风化，其中凋落物是森林土壤中碳的主要来源[38]。3种林分的生长以PCM优于PCL和CCL，PCL优于CCL，它们乔木层碳储量排序为PCM(126.40t/hm^2)>PCL(106.16t/hm^2)>CCL(47.74t/hm^2)；林分凋落物量以PCM>CCL>PCL，土壤碳储量亦为PCM(105.53t/hm^2)>CCL(94.87t/hm^2)>PCL(86.20t/hm^2)，PCM乔木层和土壤层碳储量都最高，故其生态系统碳储量也最高，而PCL、CCL乔木层和土壤层碳储量则互有高低，但乔木层和土壤层碳储量之和以PCL高于CCL，故生态系统碳储量以PCL高于CCL。

4 结论

综上研究，树种与模式对人工林生态系统碳储量的影响明显，其中：以马尾松×红椎同龄混交林生态系统碳储量显著高于马尾松、红椎纯林，马尾松纯林生态系统碳储量显著高于红椎纯林。因此，在营建高产人工林及高效碳汇林时，应优先选用科学的混交模式造林；如需营造纯林则宜选用适应性广、速生的树种。

参考文献

[1] Intergovernmental panel on climate change: The physical science basis-summary for policymakers of the working group 1 report [M]. Cambridge: Cambridge University Press, 2007.

[2] PIELKE R J, WIGLEY T, GREEN C. Dangerous assumptions[J]. Nature, 2008, 452, 531-532.

[3] SOLOMON S, PLATTNER G K, KNUTTI R, et al. Irreversible climate change due to carbon dioxide emissions [J]. Proceedings of the National Academy of Sciences of UAS, 2009, 106: 1704-1709.

[4] 周广胜，邢雪荣，王辉民．植被在全球气候变化中的作用[J]．植物学通报，1995，12：190-194.

[5] 张志强，曲建升，曾静静．温室气体排放评价指标及其定量分析[J]．地理学报，2008，63(7)：693-702.

[6] FANG J Y, CHEN A P, PENG C H, et al. Changes in forest biomass carbon storage in China between 1949 and 1998[J]. Science, 2001, 292(5525): 2320-2322.

[7] CALDEIRA K, DUFFY P B. The role of the southern ocean in uptake and storage of anthropogenic carbon dioxide[J]. Science, 2000, 287(5453): 620-622.

[8] NORBY R J, LUO Y Q. Evaluating ecosystem responses to rising atmospheric co_2 and global warming in a multi-factor world[J]. New Phytologist, 2006, 162(2): 281-293.

[9] WOODWELL G M, WHITTAKER R H, REINERS W A, et al. The biota and the world carbon budget[J]. Science, 1978, 199(4325): 141-146.

[10] POST W M, EMANUEL W R, ZINKE P J, et al. Soil carbon pools and world life zones [J]. Nature, 1982, 298(5870): 156-159.

[11] HOUGHTON R A. Aboveground forest biomass and the global carbon balance[J]. Global Change Biology, 2005, 17(2): 279-296.

[12] 张小全，李怒云，武曙红．森林、林业活动与温室气体的减排增汇 [J]．林业科学，2005，41(5)：139-143.

[13] WU S H, ZHANG X Q, LI J Q. Baseline issues for forest-based carbon sink project on Clean Development Mechanism (CDM) [J]. Scientia Silvae Sinicae, 2006, 42(4): 112-116.

[14] WANG H, LIU S R, WANG J X, et al. Effects of tree species mixture on soil organic carbon stocks and greenhouse gas fluxes in subtropical plantations in China[J]. Forest Ecology and Management, 2013, 300: 4-13.

[15] VALLET P, MEREDIEU C, SEYNAVE I. Species substitution for carbon storage: Sessile oak versus Corsican pine in France as a case study[J]. Forest Ecology and Management, 2009, 257: 1314-1323.

[16] SCHULP C, NABUURS G, VERBUG P, et al. Effect of tree species on carbon stocks in forest floor and mineral soil and implications for soil carbon inventories[J]. Forest Ecology and Management, 2008, 256: 482-490.

[17] ZHENG , OUYANG Z, XU W. Variation of carbon storage by different reforestation types in the hilly red soil region of southern China[J]. Forest Ecology and Management, 2008, 255: 1113-1121.

[18] CHRISTOPHER S G, ROBERT B J. Risks to forest carbon offset projects in a changing climate[J]. Forest Ecology and Management, 2009, 257(11): 2209-2216.

[19] 樊后保，李燕燕，苏兵强，等．马尾松。阔叶树混交异龄林生物量与生产力分配格局[J]．生态学报，2006(08)：2463-2473.

[20] 明安刚，刘世荣，李华，等．近自然化改造对马尾松和杉木人工林生物量及其分配的影响[J]．生态学报，2017，37(23)：7883-7842.

[21] 霍常富，王朋，陈龙池，等．杉木人工林蓄积量和生态系统碳数量成熟龄的关系[J]．中南林业科技大学学报，2018(09)：94-99.

[22] 刘灵，张含，国张明，等．樟子松含碳量家系变异与高碳汇家系选择[J]．中南林业科技大学学报，2017，37(02)：44-49.

[23] 刘之洲，宁晨，闫文德，等．喀斯特地区三种针叶林林分生物量及碳储量研究[J]．中南林业科技大学学报，2017，37(10)：105-111.

[24] 陈雷，孙冰，曹福亮，等．不同银杏复合经营系统对碳储量的影响[J]．中南林业科技大学学报，2017，37(10)：112-117.

[25] 何友均，覃林，李智勇，等．西南桦纯林与西南桦×红椎混交林碳贮量比较[J]．生态学报，2012，32(23)：7586-7594.

[26] 明安刚，刘世荣，莫慧华，等．南亚热带红椎、杉木纯林与混交林碳贮量比较[J]．生态学报，2016，36(1)：244-251.

[27] YOU Y M, HUANG X M, ZHU H G, et al. Positive interactions between *Pinus massoniana* and *Castanopsis hystrix* species in the uneven-aged mixed plantations can produce more ecosystem carbon in subtropical China [J]. Forest Ecology and Management, 2018, 410: 193-200.

[28] 卢立华，贾宏炎，农友，等．红椎经营模式对林木生

长及乔木层碳储量的影响[J]. 东北林业大学学报，2014，42(12)：63-66.

[29]HE Y J，QIN L，Li Z Y，et al. Carbon storage capacity of monoculture and mixed-species plantations in subtropical China[J]. Forest Ecology and Management，2013，295：193-198.

[30]明安刚，张治军，谌红辉，等. 抚育间伐对马尾松人工林生物量与碳贮量的影响[J]. 林业科学，2013，49(10)：1-6.

[31]郭琦，王新杰. 不同混交模式杉木人工林林下植被生物量与土壤物理性质研究[J]. 中南林业科技大学学报，2014，34(5)：70-74.

[32]徐慧芳，宋同清，黄国勤，等. 广西不同林龄马尾松碳储量及分配格局. 农业现代化研究，2016，37(1)：195-203.

[33]卢立华，汪炳根，何日明. 立地与栽培模式对红椎生长的影响[J]. 林业科学研究，1999，12（5）：519-523.

[34]秦国峰. 马尾松地理种源[M]. 杭州：浙江大学出版社，2003.

[35]卢立华，贾宏炎，何日明，等. 南亚热带6种人工林凋落物的初步研究[J]. 林业科学研究，2008，21(3)：346-352.

[36]樊后保，李燕燕，孙新，等. 马尾松纯林及其与阔叶树混交林的凋落量与养分通量[J]. 应用与环境生物学报，2005，11(5)：521-527.

[37]HUANG Y H，LI Y L，XIOA Y，et al. Controls of litter quality on the carbon sink in soils through partitioning the products of decomposing litter in a forest succession series in south China[J]. Forest Ecology and Management，2011，261(7)：1170-1177.

[38]郑顺安，常庆瑞. 黄土高原不同类型人工林对土壤肥力的影响[J]. 西北农林科技大学学报，2006，34(2)：119-123.

[原载：中南林业科技大学学报，2019，39(07)]

广西大青山米老排人工林土壤物理性质

明安刚[1]　于浩龙[1]　陈厚荣[1]　卢立华[1]　蔡道雄[1]　温恒辉[2]

([1] 中国林业科学研究院热带林业实验中心，广西凭祥　532600；[2] 广西大学林学院，广西南宁　530004)

摘　要　用环刀法分3个不同坡位和4个土层深度对广西大青山米老排人工林土壤体积质量、含水率、持水量、孔隙度等物理性质进行了研究，结果表明：大青山米老排人工林土壤体积质量、土壤水分及土壤空隙的差异主要体现在表土层上，表层以下各土层之间差异不显著；不同的坡位对米老排表土层的土壤物理性质影响显著，但对表层以下其他土层无显著影响。

关键词　米老排人工林；土壤物理性质；环刀法

［原载：林业科技开发，2010，24(05)］

不同郁闭度林茶复合模式对茶树光合日变化的影响

马 跃[1] 刘志龙[1] 虞木奎[2] 徐庆玲[3] 傅 强[3] 赵 樟[1]

([1]中国林业科学研究院热带林业实验中心，广西凭祥 532600；

[2]中国林业科学研究院亚热带林业研究所，浙江富阳 311400；

[3]广西壮族自治区林业勘测设计院，广西南宁 530011)

摘 要 茶叶的产量和品质与生长环境关系非常密切，为探明林分不同郁闭度对茶树光合日变化的影响，本研究以不同郁闭度的栾树(*Koelreuteria paniculata*)-茶(*Camellia sinensis*)复合系统中茶树为材料，纯茶园为对照(CK)，利用便携式LI-6400光合作用测定仪研究茶树分别在0.3(T_1)、0.6(T_2)和0.9(T_3)郁闭度条件下的光合作用参数日变化。结果表明：随郁闭度增加，茶树光合有效辐射(*PAR*)和叶片温度(T_l)日均值和差值呈递减趋势，日变化呈单峰曲线；茶树净光合速率(*Pn*)日均值随郁闭度增加呈递减趋势，均表现双峰型曲线；CK和T_1的蒸腾速率(*Tr*)和气孔导度(*Gs*)日变化午间受到明显的抑制且呈双峰曲线，T_2和T_3呈单峰型曲线。

关键词 林茶复合系统；光合；郁闭度；日变化

[原载：中国农学通报，2011，27(16)]

测定植物全氮的 H_2SO_4-H_2O_2 消煮法改进

李朝英[1,2] 郑 路[1,2] 卢立华[1,2] 李丽利[1]
([1]中国林业科学研究院热带林业实验中心，广西凭祥 532600；
[2]广西友谊关森林生态系统国家定位观测研究站，广西凭祥 532600)

摘 要 为了快速准确测定批量植物全氮，提高检测效率，对测定植物全氮的 H_2SO_4-H_2O_2 消煮法提出改进。通过反复实验，摸索出在消煮过程只加 1 次 H_2O_2 的改进方法。结果表明：改进方法测定氮含量的变异系数（*CV*）小于5%，回收率在95%~105%。与原 H_2SO_4-H_2O_2 消煮法相比，节约用时50~80min/批样，减少 H_2O_2 消耗0.38~0.68 ml/样，检测结果无显著性差异。改进方法的精密度及准确性符合检测分析要求，同时缩短了植物全氮检测用时，H_2O_2 耗用量相对较少，效率明显提升，适宜批量植物全氮的检测。

关键词 植物全氮；H_2SO_4-H_2O_2 消煮法；改进

[原载：中国农学通报，2014，30(06)]

广西乡土用材树种米老排资源分布

刘光金　张　培　贾宏炎　蔡道雄　农　志　劳庆祥

（中国林业科学研究院热带林业实验中心，广西凭祥　532600）

摘　要　保护和利用优异米老排种质资源。开展米老排种质资源分布调查，收集82株优势木，并对其种实特性进行了分析。米老排分布区域狭窄，零星分布于龙州、靖西、德保、那坡、东兴、上思、容县、凭祥8个县，集中分布在21°32′~23°11′N，105°51′~110°34′E，海拔179~1202m的低中山区。采种优势木胸径与树高之间呈现极显著的正相关性，相关系数为0.26，S型曲线拟合度最高，相关系数为0.749。米老排结实量大，果实为蒴果，10月中旬至11月上旬是种子成熟期和收集期。该研究为进 一步筛选和鉴定优良种源提供基础资料。

关键词　米老排；资源分布；种实特性

［原载：安徽农业科学，2015，43(11)］

桂西北不同森林类型土壤生态化学计量特征

庞圣江 张 培 贾宏炎 杨保国 邓硕坤 冯昌林 王庆灵

（中国林业科学研究院热带林业实验中心，广西凭祥 532600）

摘 要 以广西雅长林区3种森林类型(针阔混交林、常绿—叶阔叶混交林、常绿阔叶林)的土壤作为研究对象，对土壤C、N、P、K等养分含量及其比例的影响状况等进行研究，探讨土壤养分变化和化学计量特征。研究结果表明：随着森林的演替，3种森林类型植被下土壤C、N、P和K含量中，N素(全N和有效N)含量不断积累；有机C、P和K含量则表现出一致性的先增加后减少趋势，即演替初期低—中期高—后期低的变化规律；C/N呈逐渐降低趋势，均值大小依次为：针阔混交林>常绿—落叶阔叶混交林>常绿阔叶林；而C/P、C/K、N/P、N/K和P/K则均呈升高趋势，均值大小依次为：常绿阔叶林>常绿—落叶阔叶混交林>针阔混交林。相关性分析表明，土壤有机C与N/P和N/K都呈极显著的正相关关系；全N与C/P呈极显著的正相关关系，与C/K具有显著正相关关系；全P与C/K、N/K呈极显著的正相关关系。

关键词 森林生态系统；土壤养分；C/N；化学计量特征

[原载：中国农学通报，2015，31(01)]

米老排人工纯林枯落物厚度对种子萌发的影响

于浩龙[1] 邓硕坤[1] 苏小兰[2] 赵 樟[1] 卢立华[1] 苏俊萌[2] 欧 军[2]
([1] 中国林业科学研究院热带林业实验中心，广西凭祥 532600；
[2]广西壮族自治区国有派阳山林场，广西崇左 532200)

摘 要 在温室大棚内，采用29年生米老排人工纯林里采摘的成熟种子，以经过碳化处理的树皮为基质，播种后分别以枯落物厚度0、2、4、6、8、10、12、20cm覆盖；以先在基质表面覆盖6cm厚的枯落物，然后在枯落物上播种为对照；模拟研究枯落物厚度对米老排种子萌发的影响。结果表明：对照和枯落物0cm处理的种子均没有发芽，其他处理的种子均有发芽。其中，以米老排枯落物层厚度为6cm时，米老排种子的发芽率最高，随着枯落物厚度的进一步增加，米老排种子萌发率都逐渐降低；同时，在枯落物层厚度6cm、8cm和10cm处理时萌芽时间最长，其他处理萌芽率均很快稳定。

关键词 米老排；人工纯林；自然更新；枯落物厚度；种子萌发

[原载：福建林业科技，2015，42(04)]

森林生态系统水化学效应研究综述

雷丽群[1,2] 韦菊玲[1] 农 友[1,2] 杨桂芳[1] 卢立华[1,2] 何 远[1,2]
([1] 中国林业科学研究院热带林业实验研究中心，广西凭祥 532600
[2]广西友谊关森林生态系统国家定位观测研究站，广西凭祥 532600)

摘 要 综述了国内有关森林生态系统水化学效应的主要研究成果，概述了pH和SO_4^{2-}、NO_3^-、K^+、Na^+、Ca^{2+}、Mg^{2+}六种离子在森林林冠层、枯落物层和土壤层三个层次的变化特征及响应机制。指出在今后的工作中应继续深化有关作用机理的研究，特别是需要在系统层面上进行多因子的探讨，加强如何应对环境酸化问题以及酸化问题恢复措施的相关研究，重视森林生态系统水化学效应研究成果的运用。

关键词 水化学效应；森林生态系统；盐基离子；综述

［原载：林业调查规划，2016，41(06)］

基于 DNS 比色法的土壤蔗糖酶检测条件选择

李朝英[1,2]　郑　路[1,2]
([1]中国林业科学研究院热带林业实验中心；[2]广西友谊关森林生态系统国家定位观测研究站，广西凭祥　532600)

摘　要　为了准确测定土壤蔗糖酶含量，分析了不同波长、显色剂加入量、甲苯加入量、加热时间、显色稳定时间等条件对3，5-二硝基水杨酸(3，5-dinitrosalicylic acid，DNS)比色法测定土壤蔗糖酶含量的影响。结果表明：当波长为510nm，显色剂用量为1.5 ml，甲苯加入量0.1ml，水浴时间5min，待测液存放时间小于4h，显色稳定时间为0.5~2h时，标准曲线在0.25~2mg范围内线性良好($R^2=0.9991$)，土壤蔗糖酶含量的变异系数(CV)小于9%，回收率在96%~109%内，检测结果的精密度和准确性较高。上述检测方法安全、经济、高效，适用于实验室批量土壤的蔗糖酶含量测定。

关键词　DNS 比色法；土壤；蔗糖酶

[原载：中国农学通报，2016，32(27)]

崇左市石灰岩地区适生树种及造林方法

卢 寰

（中国林业科学研究院热带林业实验中心，广西凭祥 532600）

摘 要 崇左市山地多，平地少，石灰岩地层占大部分。由于人为因素和自然因素的破坏，石质荒漠化土地面积越来越扩大，阻碍了崇左市社会经济的发展，必须要加快推进“荒漠化、石漠化、水土流失治理”措施。介绍了适合桂西南石灰岩地区生长且人工栽种容易成活的一些树种及其造林方法，提出加强石灰岩地区森林管护的一些措施。

关键词 造林树种；造林方法；森林管护措施；广西崇左；石灰岩地区

［原载：现代农业科技，2016(11)］

广西大青山西南桦人工林群落结构特征

杨保国 刘士玲 贾宏炎 庞圣江 张 培 黄旭光 韦菊玲 杨桂芳

（中国林业科学研究院热带林业实验中心，广西凭祥 532600）

摘 要 通过对西南桦纯林、西南桦+马尾松、西南桦+杉木和西南桦+红椎混交林4种群落类型的群落特征进行分析。结果表明：西南桦人工林群落林下物种数量丰富，除西南桦+马尾松混交林(55)外，西南桦纯林(106)、西南桦+杉木(102)和西南桦+红椎(119)混交林已超过同气候带天然林群落内林下物种数量；林下物种组成种类相似，但其分布具有一定的差异性和不均匀性。从林分密度和径级结构来看，三种混交林中西南桦对林分密度的贡献度小于其对林分胸高断面积的贡献度；且西南桦径级分布相对混交树种更偏向对大径级分布，这说明西南桦比混交树种更具有竞争优势，有利于西南桦大径材的培育。

关键词 西南桦；群落结构；物种组成；林分密度；径级结构

[原载：广西林业科学，2016，45(04)]

总有机碳分析仪测定植物总有机碳含量的方法

李朝英[1,2] 郑 路[1,2]

([1]中国林业科学研究院热带林业实验中心，广西凭祥 532600；
[2]广西友谊关森林生态系统国家定位观测研究站，广西凭祥 532600)

摘 要 为了摸索TOC(总有机碳)分析仪测定植物TOC含量的方法、保证植物碳素研究获得准确可靠数据提供参考，对不同进样量和不同类型植物样品TC(总碳)、IC(无机碳)含量的测定结果进行分析探讨。结果表明，不同进样量测定的TC含量存在不同差异，进样量在45.55mg，TC含量的精密度与准确性较高。植物的IC含量为0，TC=TOC。TC含量在1.8929mg，TC的回归方程为$Y_{TC}=133.7X+77.431$($r^2=0.999$)，回收率为101.24%；IC的回归方程为$Y_{IC}=108.5X+56.307$($r^2=0.9995$)，回收率为99.24%。以上述方法测定植物TOC的方法准确高效，适用于不同类型、不同含量的植物样品的TOC测定。

关键词 总有机碳(TOC)分析仪；植物；TOC；碳素研究；进样量；回收率

[原载：江苏农业科学，2017，45(09)]

岩溶山地造林关键技术

卢立华[1,2] 农 友[1,2] 李 华[1,2]
([1] 中国林业科学研究院热带林业实验中心，广西凭祥 532600；
[2]广西友谊关森林生态系统国家定位观测研究站，广西凭祥 532600)

摘 要 对岩溶山地造林中的树种选择、育苗、密度控制、整地、种植与经营管理等关键技术进行了系统、全面的总结，以期为岩溶区的生态恢复与重建提供参考与技术支撑。

关键词 岩溶山地；造林；树种选择；关键技术

[原载：林业科技通讯，2017(01)]

硫酸钡比浊改进法在测定森林地表径流中硫酸根含量的应用

李朝英[1] 郑 路[1,2]
([1]中国林业科学研究院热带林业实验中心，广西凭祥 532600；
[2]广西友谊关森林生态系统国家定位观测研究站，广西凭祥 532600)

摘 要 为了准确高效地测定地表径流中硫酸根含量，该试验探讨阿拉伯胶稳定剂用量、加入酸量、氯化钡状态等对硫酸钡比浊体系的影响。结果表明：加入0.25%阿拉伯胶1.5ml，加入0.5%硝酸溶液0.5ml，加入氯化钡颗粒0.2g，待测液摇匀静置15~60min，阿拉伯胶稳定剂存放30d，线性范围在0~8mg/L，R^2为0.999，线性良好。实际检测样品的变异系数<12%，回收率在90%~110%，检测结果的精密度与准确性较高。与国标方法及其他改进方法相比，该试验方法操作简单，检测高效准确，适用于批量水样的硫酸根含量的测定。
关键词 比浊改进法；森林地表径流；硫酸根含量

[原载：环境保护科学，2018，44(05)]

土壤颗粒粒径及进样量对 TOC 含量测定精度的影响

李朝英[1] 郑 路[1,2]

([1]中国林业科学研究院热带林业实验中心，广西凭祥 532600；

[2]广西友谊关森林生态系统国家定位观测研究站，广西凭祥 532600)

摘 要 为了准确高效地测定土壤有机碳(TOC)含量，研究了土壤颗粒粒径及进样量对 TOC 分析仪测定土壤 TOC 含量的影响，比较分析了 TOC 分析仪法与重铬酸钾外加热法-沸水浴测定 TOC 含量的差异。结果表明：过 60 目筛的颗粒进样量在 175~275mg，所测 TOC 含量与过 100 目筛的颗粒无显著性差异，TOC 含量的变异系数(CV)<3%，加标回收率在 98%~102%，检测结果的精密度及准确性良好。酸性土壤的无机碳(IC)为 0，总碳(TC)= TOC；TOC 分析仪与重铬酸钾外加热法-沸水浴的测定结果无显著性差异，两者相关性良好。TOC 分析仪对有机碳含量高、中、低样品的检测结果准确稳定，重铬酸钾外加热法-沸水浴测定 TOC 含量>15mg/g 样品的准确性与 TOC 分析仪法相当，测定 TOC 含量 5~15mg/g 样品的准确性有所下降。

关键词 土壤；粒径；进样量；TOC 含量；测定精度

[原载：上海农业学报，2018，34(05)]

搅拌仪在土壤机械组成测定中的应用

李朝英[1,2] 郑 路[1,2]

([1]中国林业科学研究院热带林业实验中心，广西凭祥 532600；
[2]广西友谊关森林生态系统国家定位观测研究站，广西凭祥 532600)

摘 要 为了提高土壤机械组成测定精度，降低劳动强度，本文使用比重法对搅拌仪处理的土壤进行了土壤机械组成测定，同时与人工搅拌的测定结果进行比较。结果表明：搅拌仪摇臂长度根据最佳搅拌行程确定，搅拌行程一致且搅拌充分；摇臂转速调整为1周/2s，上下搅拌30次/min，搅拌速度均匀一致，符合国标要求；2种搅拌方式所测土壤中黏粒和粉砂粒含量无显著差异，砂粒含量有显著差异。搅拌仪操作方便，一次可同时搅拌多个样品，工作强度降低，检测准确高效，可用于批量土壤样品的机械组成检测分析。

关键词 土壤机械组成；比重法；搅拌仪；人工搅拌

[原载：现代农业科技，2018(20)]

珍稀濒危树种格木更新障碍研究进展

韦菊玲　刘福妹　雷丽群　邓硕坤　农　友　刘士玲

（中国林业科学研究院热带林业实验中心，广西凭祥　532600）

摘　要　从自然原因和人为原因两方面详细阐述格木出现天然更新障碍的原因。格木生境要求苛刻，如温暖、光照强和水分好等；格木更新依赖种子，其种子不透水性和抑制物质导致难以萌发，且幼苗易遭病菌和虫蛀，这些均为格木天然更新能力弱的原因。人们过度砍伐也成为格木种群更新障碍的原因之一。针对格木濒危现状和原因，提出加强对格木原地保护，做好迁地保护工作，开展培育技术方面的研究等5方面格木种群资源科学保护对策。提出今后格木重点研究方向为格木天然更新障碍机制研究和格木大径材近自然高效培育技术研究。

关键词　格木；珍稀濒危树种；天然更新；更新障碍；种质资源保护；近自然森林经营

［原载：林业调查规划，2018，43(04)］

盐酸浸提-AAS 法测定植物钾钙镁含量的检测条件优化与改进

李朝英[1,2]　郑　路[1,2]

([1] 中国林业科学研究院热带林业实验中心，广西凭祥　532600；[2] 广西友谊关森林生态系统国家定位观测研究站，广西凭祥　532600)

摘　要　为了优化盐酸浸提-AAS 法的检测条件，在植物颗粒不同粒径、振荡时间、浸提温度、浸提方式及不同盐酸浓度浸提条件下，以盐酸浸提法测定钾、钙、镁含量。结果表明：盐酸浸提法适宜采用过 60 目筛的植物颗粒，以室温盐酸浸提，适宜振荡时间为 90 min。以热盐酸浸提，适宜采用 55℃ 热盐酸振荡 60 min，盐酸适宜浓度为 1mol/L，盐酸浸提样品的适宜方式为振荡。试验提出的盐酸浸提-AAS 方法准确高效，经济安全，操作方便，适用于批量样品的检测分析。

关键词　盐酸浸提-AAS；植物；钾；钙；镁

［原载：中国农学通报，2019，35(09)］

范氏法与王玉万法植物纤维素测定方法探讨

李朝英[1,2] 郑 路[1,2] 莫世宇[1]
([1]中国林业科学研究院热带林业实验中心，广西凭祥 532600；
[2]广西友谊关森林生态系统国家定位观测研究站，广西凭祥 532600)

摘 要 为了以范氏法准确测定植物纤维素，探讨消煮时间和硫酸浓度对测定结果的影响。结果表明：适宜消煮时间为1.5~2h，硫酸适宜浓度为72%~75%，有利于充分水解多糖及纤维素，准确测定纤维素。范氏法与王玉万法以重量法所测纤维素无显著性差异，两种方法以重量法所测纤维素与蒽酮法均有显著性差异。两种方法以重量法所测纤维素的精密度高于蒽酮法。重量法操作简单，不易产生误差。蒽酮法操作繁琐，检测干扰多，易引入误差。范氏法与王玉万法的样品处理程序不同，但两者所测结果趋于一致，有良好的相关性，均适用于植物纤维素的日常检测分析。

关键词 植物；纤维素；范氏法；王玉万法

[原载：浙江农业科学，2019，60(03)]

用范氏法测定多类植物中酸性洗涤木质素的影响因素研究

李朝英[1] 郑 路[1,2] 杨文娟[1] 莫世宇[1]

([1]中国林业科学研究院热带林业实验中心，广西凭祥 532600；

[2]广西友谊关森林生态系统国家定位观测研究站，广西凭祥 532600)

摘 要 探讨范氏法测定植物中酸性洗涤木质素(ADL)的影响因素，对不同消煮时间、不同硫酸浓度、不同浸泡时间的测定结果进行比较分析。结果表明：消煮适宜时间为30min，硫酸适宜浓度为72%~75%，浸泡适宜时间为3~4h，多糖及纤维素水解完全，ADL测定准确。人工检测结果变异系数(CV)在2.23%~6.66%，仪器法的CV在0.15%~1.10%，人工检测与仪器法所测结果无显著性差异，人工检测易出现人为偏差，精密度低于仪器法。范氏法与王玉万法所测结果无显著性差异，两种方法适用于植物ADL的测定。

关键词 范氏法；植物；酸性洗涤木质素；影响因素

[原载：江西农业学报，2019，31(03)]

植物种群天然更新障碍机制研究进展

韦菊玲[1] 刘福妹[1] 雷丽群[2] 陈世清[3] 李洪果[1]
([1] 中国林业科学研究院热带林业实验中心，广西凭祥 532600
[2]广西友谊关森林生态系统国家定位观测研究站，广西凭祥 532600)；
[3]华南农业大学林学与风景园林学院，广东广州 510642)

摘 要 在研究了国内外大量文献基础上，从种子生产障碍、种子传播障碍和幼苗幼树建成障碍三个阶段对天然更新障碍机制进行了综述。对种子生产来说，具有生活力的种子缺乏是影响森林天然更新障碍的重要因素。种子雨和土壤种子库动态变化在一定程度上造成种子传播限制；分别从种子的输入(即种子雨)；种子的输出(种子库的消耗)；种子的留存等方面阐述了种子库的变化对更新的影响。对建成障碍来说，光、温度和水分等环境因子对于幼苗幼树建成具有重要的影响；枯落叶层可通过生物因子(动物侵害和微生物致病)、化学因子(化感作用)和物理因子(物理阻断)等对植物更新造成负面影响；不同的植被类型对植物天然更新的影响不同，植被类型(如竹林)是不适宜植物更新的环境。最后提出了今后该领域的研究展望：在植物天然更新障碍时，采取适当营林措施促进更新尤为重要；造成传播障碍中涉及动物行为的过程极其复杂，深化种子传播障碍机制结合动物行为的定量研究，进而建立植物天然更新障碍因子及其影响机制模型应是未来的一个研究方向。

关键词 天然更新；更新障碍；机制；生产障碍；传播障碍；建成障碍

[原载：林业科技通讯，2019(06)]

附 录

热带林业实验中心依托单位学术论文情况一览表(2010-2019年)

序号	作者	题名	刊名	年	第一完成单位
1	Fang Shengzuo, Liu Zhilong, Cao Yida, Liu Dong, Yu Mukui, Tang Luozhong	Sprout development, biomass accumulation and fuelwood characteristics from coppiced plantations of *Quercus acutissima*	Biomass& Bioenergy	2011	Nanjing Forestry University
2	Liu Zhilong, Fang Shengzuo, Liu Dong, Yu Mukui, Tang Luozhong	Influence of thinning time and density on sprout development, biomass production and energy stocks of sawtooth oak stumps	Forest Ecology and Management	2011	Nanjing Forestry University
3	Liu Hong, Feng Changlin, Chen Baoshan, Wang Zhongsheng, Xie Xiaoqing, Deng Zhenghai, Wei Xinlian, Liu Shiyong, Zhang Zibin, Luo Yibo	Overcoming extreme weather challenges: Successful but variable assisted colonization of wild orchids in southwestern China	Biological Conservation	2012	Florida International University
4	Zhang Xiongqing, Lei Yuancai, Cai Daoxiong, Liu Fengqiang	Predicting tree recruitment with negative binomial mixture models	Forest Ecology and Management	2012	Research Institute of Forest Resource Information Techniques, CAF
5	Wang Hui, Liu Shirong, Wang Jingxin, Shi Zuomin, Lu Lihua, Guo Wenfu, Jia Hongyan, Cai Daoxiong	Dynamics and speciation of organic carbon during decomposition of leaf litter and fine roots in four subtropical plantations of China	Forest Ecology and Management	2013	Institute of Forest Ecology Environment and Protection, CAF
6	Wang Hui, Liu Shirong, Wang Jingxin, Shi Zuomin, Lu Lihua, Zeng Ji, Ming Angang, Tang Jixin, Yu Haolong	Effects of tree species mixture on soil organic carbon stocks and greenhouse gas fluxes in subtropical plantations in China	Forest Ecology and Management	2013	Institute of Forest Ecology Environment and Protection, CAF
7	Huang Xueman, Liu Shirong, Wang Hui, Hu Zongda, Li Zhongguo, You Yeming	Changes of soil microbial biomass carbon and community composition through mixing nitrogen-fixing species with *Eucalyptus urophylla* in subtropical China	Soil Biology & Biochemistry	2014	Institute of Forest Ecology Environment and Protection, CAF

（续）

序号	作者	题名	刊名	年	第一完成单位
8	Meng Jinghui, Lu Yuanchang, Zeng Ji	Transformation of a degraded *Pinus massoniana* plantation into a mixed-species irregular forest: Impacts on stand structure and growth in southern China	Forests	2014	Beijing Forestry University
9	Hackenberg Jan, Marc Wassenberg, Heinrich Spiecker, Sun Dongjing	Non destructive method for biomass prediction combining TLS derived tree volume and wood density	Forests	2015	University of Freiburg
10	Jiang Jun, Lu Yuanchang, Pang Lifeng, Liu Xianzhao, Cai Daoxiong, Xing Haitao	Short-term effects of the management intensities on structure dynamic in monoculture forests of southern subtropical China	Tropical Conservation Science	2015	Beijing Forestry University
11	Wang Chunsheng, Zhao Zhigang, Sebastian Hein , Zeng Ji, Johanna Schuler , Guo Junjie, Guo Wenfu, Zeng Jie	Effect of planting density on knot attributes and branch occlusion of *Betula alnoides* under natural pruning in Southern China	Forests	2015	Research Institute of Tropical Forestry, CAF
12	Wang Hui, Liu Shirong, Scott X. Chang, Wang Jingxin, Shi Zuomin, Huang Xueman, Wen Yuan, Lu Lihua, Cai Daoxiong	Soil microbial community composition rather than litter quality is linked with soil organic carbon chemical composition in plantations in subtropical China	Journal of Soils and Sediments	2015	Institute of Forest Ecology Environment and Protection, CAF
13	Marc Wassenberg, Haw-Shi Chiu, Guo Wenfu, Heinrich Spiecker	Analysis of wood density profiles of tree stems: incorporating vertical variations to optimize wood sampling strategies for density and biomass estimations	Tree-structure and Function	2015	Albert-Ludwigs University
14	Wang Hui, Liu Shirong, Wang Jingxin, Shi Zuomin, Xu Jia, Hong Pizheng, Ming Angang, Yu Haolong, Chen Lin, Lu Lihua, Cai Daoxiong	Differential effects of conifer and broadleaf litter inputs on soil organic carbon chemical composition through altered soil microbial community composition	Scientific Reports	2016	Institute of Forest Ecology Environment and Protection, CAF
15	Wen Yuan, Jamie L. Schuler, Liu Shirong, Mou Pu, Wang Hui, Yu Haolong, Wang Jingxin	Soil carbon dynamics in a *Pinus massoniana* plantation following clear-cutting and slash removal	Journal of Plant Ecology	2016	Beijing Normal University
16	Guo Yili, Wang Bin, Li Dongxing, Azim U. Mallik, Xiang Wusheng, Ding Tao, Wen Shujun, Lu Shuhua, Huang Fuzhao, He Yunlin, Li Xiankun	Effects of topography and spatial processes on structuring tree species composition in a diverse heterogeneous tropical karst seasonal rainforest	Flora	2017	Guangxi Institute of Botany, Chinese Academy of Sciences

（续）

序号	作者	题名	刊名	年	第一完成单位
17	Guo Yili, Wang Bin, Azim U. Mallik, Huang Fuzhao, Xiang Wusheng, Ding Tao, Wen Shujun, Lu Shuhua, Li Dongxing, He Yunlin, and Li Xiankun	Topographic species-habitat associations of tree species in a heterogeneous tropical karst seasonal rain forest, China	Journal of Plant Ecology	2017	Guangxi Institute of Botany, Chinese Academy of Sciences
18	Tang, Cheng, Wang Chunsheng, Pang Shengjiang, Zhao Zhigang, Guo Junjie, Lei, YuanCai, Zeng Jie	Stem taper equations for *Betula Alnoides* in South China	Journal of Tropical Forest Science	2017	Research Institute of Tropical Forestry, CAF
19	Wang Chunsheng, Zeng Ji, Sebastian Hein, Zhao Zhigang, Guo Junjie, Zeng Jie	Crown and branch attributes of mid-aged *Betula alnoides* plantations in response to planting density	Scandinavian Journal of Forest Research	2017	Research Institute of Tropical Forestry, CAF
20	Wang Hui, Liu Shirong, Wang Jingxin, Li Dejun, Shi Zuomin, Liu Yanchun, Xu Jia, Hong Pizheng, Yu Haolong, Zhao Zhang, Ming Angang, Lu Lihua, Cai Daoxiong	Contrasting responses of heterotrophic and root-dependent respiration to soil warming in a subtropical plantation	Agricultural and Forest Meteorology	2017	Institute of Forest Ecology Environment and Protection, CAF
21	Wen Yuan, Marife D. Corre, Christine Rachow, Chen Lin, Edzo Veldkamp	Nitrous oxide emissions from stems of alder, beech and spruce in a temperate forest	Plant and Soil	2017	University of Goettingen
22	Wen Ding, Ning Liping, Xing Huizi, Wei Zhiting, Zhou Qian, Nong Ruihong, Chen Jianquan	Essential oil extracted from leaf of *Phoebe bournei* (Hemsl.) yang: chemical constituents, antitumor, antibacterial, hypoglycemic activities	Natural Product Research	2018	Sichuan Agricultural University
23	Guo Yili, Xiang Wusheng, Wang Bin, Li Dongxing, Mallik Azim U, Chen Han Y H, Huang Fuzhao, Ding Tao, Wen Shujun, Lu Shuhua, Li Xiankun	Partitioning beta diversity in a tropical karst seasonal rainforest in Southern China	Scientific Reports	2018	Guangxi Institute of Botany, Chinese Academy of Sciences
24	Wang Hui, Liu Shirong, Wang Jingxin, You Yeming, Yang Yujing, Shi Zuomin, Huang Xueman, Zheng Lu, Li Zhaoying, Ming Angang, Lu Lihua, Cai Daoxiong,	Mixed-species plantation with*Pinus massoniana* and *Castanopsis hystrix* accelerates C loss in recalcitrant coniferous litter but slows C loss in labile broadleaf litter in southern China	Forest Ecology and Management	2018	Institute of Forest Ecology Environment and Protection, CAF

（续）

序号	作者	题名	刊名	年	第一完成单位
25	You Yeming, Huang Xueman, Zhu Hongguang, Liu Shirong, Liang Hongwen, Wen Yuanguang, Wang Hui, Cai Daoxiong, Ye Duo,	Positive interactions between*Pinus massoniana* and *Castanopsis hystrix* species in the uneven-aged mixed plantations can produce more ecosystem carbon in subtropical China	Forest Ecology and Management	2018	Guangxi University
26	Zhang Xiao, Liu Shirong, Huang Yongtao, Fu Shenglei, Wang Jingxin, Ming Angang, Li Xiangzhen, Yao Minjie, Li Huan	Tree species mixture inhibits soil organic carbon mineralization accompanied by decreased r-selected bacteria	Plant and Soil	2018	Henan University
27	Zhang Xiao, Huang Yongtao, Liu Shirong, Fu Shenglei, Ming Angang, Li Xiangzhen, Yao Minjie, Li Huan, Tian Chao	Mixture of tree species enhances stability of the soil bacterial community through phylogenetic diversity	European Journal of Soil Science	2018	Henan University
28	Chen Lin, Wen Yuanguang, Zeng Ji, Wang Hui, Wang Jingxin, Bernard Dell, Liu Shirong	Differential responses of net N mineralization and nitrification to throughfall reduction in a*Castanopsis hystrix* plantation in southern China	Forest Ecosystems	2019	Guangxi University
29	Wang Hui, Liu Shirong, Andreas Schindlbacher, Wang Jingxin, Yang Yujing, Song Zhanchao, You Yeming, Shi Zuomin, Li Zhaoying, Chen Lin, Ming Angang, Lu Lihua, Cai Daoxiong	Experimental warming reduced topsoil carbon content and increased soil bacterial diversity in a subtr opical planted forest	Soil Biology and Biochemistry	2019	Institute of Forest Ecology Environment and Protection, CAF
30	Yang Yujing, Liu Shirong, Wang Hui, Chen Lin, Lihua Lu. Lu, Cai Daoxiong	Reduction in throughfall reduces soil aggregate stability in two subtropical plantations	European Journal of Soil Science	2019	Institute of Forest Ecology Environment and Protection, CAF
31	Liu Fumei, Hong Zhou, Yang Zengjiang, Zhang Ningnan, Liu Xiaojin, Xu Daping	De novo transcriptome analysis of *Dalbergia odorifera* T. Chen (Fabaceae) and transferability of SSR markers developed from the transcriptome	Forests	2019	Northeast Forestry University
32	Liu Fumei, Hong Zhou, Xu Daping, Jia Hongyan, Zhang Ningnan, Liu Xiaojin, Yang Zengjiang, Lu Mengzhu	Genetic diversity of the endangered *Dalbergia odorifera* aevealed by SSR markers	Forests	2019	Research Institute of Tropical Forestry, CAF

（续）

序号	作者	题名	刊名	年	第一完成单位
33	Liu Fumei，Zhang Ningnan，Liu Xiaojin，Yang Zengjiang，Jia Hongyan，Xu Daping	Genetic diversity and population structure analysis of *Dalbergia odorifera* germplasm and development of a core collection using microsatellite markers	Genes	2019	Research Institute of Tropical Forestry，CAF
34	Zhang Pu，He Youjun，Feng Yiming，Rafael De La Torre，Jia Hongyan，Tang Jixin，Frederick Cubbage	An analysis of potential investment returns of planted forests in South China	New Forests	2019	Institute of Desertifcation Studies，CAF
35	黄承标，冯昌林，李保平，邓振海，罗玉婷，蓝玉甜	广西雅长兰科植物分布区土壤理化性质	东北林业大学学报	2010	广西大学
36	康冰，刘世荣，蔡道雄，卢立华，何日明，高妍夏，迪玮峙	南亚热带不同植被恢复模式下土壤理化性质	应用生态学报	2010	西北农林科技大学
37	李荣生，尹光天，林彦，卢立华，曾冀，姜艳，杨锦昌，邹文涛	蛇皮果育苗基质的选择	林业实用技术	2010	中国林业科学研究院热带林业研究所
38	林开勤，赵志刚，郭俊杰，曾杰，曾冀，郭文福	西南桦嫩枝扦插繁殖试验	种子	2010	中国林业科学研究院热带林业研究所
39	覃先林，张子辉，李增元，田祖为，熊有强，杨德付，张瑞	林火预警及实现方法	北京林业大学学报	2010	中国林业科学研究院资源信息研究所
40	吴雪琼，覃先林，李程，田祖为，熊有强，杨德付，张瑞	我国林火监测体系现状分析	内蒙古林业调查设计	2010	中国林业科学研究院资源信息研究所
41	杨锦昌，尹光天，冯昌林，李荣生，邹文涛	单叶省藤人工林茎长分布模型的研制	浙江林业科技	2010	中国林业科学研究院热带林业研究所
42	袁东，姜忠斌，蔡道雄，白灵海，贾宏炎，卢立华，许基煌，刁海林	棕榈藤采割机械研制与采割新技术	木材加工机械	2010	国家林业局北京林业机械研究所
43	郑海水，曾杰，黎明，翁启杰，卢立华	不同种源西南桦在热带山区生长表现	广西林业科学	2010	中国林业科学研究院热带林业研究所
44	车腾腾，冯益明，蔡道雄，张万幸，张显强，吴春争	热带林业实验中心人工林区景观格局变化分析	浙江农林大学学报	2011	中国林业科学研究院荒漠化研究所
45	刁海林，蔡道雄，姜金英，唐继新，高伟	红椎干燥特性研究	安徽农业科学	2011	广西大学
46	刁海林，蔡道雄，罗建举，莫玉婵，唐继新，高伟	米老排干燥特性研究	湖北农业科学	2011	广西大学
47	刁海林，袁东，蔡道雄，袁全平，贾宏炎，白灵海，卢立华，许基煌	便携式棕榈藤采收机设计	林业实用技术	2011	广西大学
48	段永照，郑路	衰弱园林树木的保护与复壮技术	北方园艺	2011	新疆农业职业技术学院

（续）

序号	作者	题名	刊名	年	第一完成单位
49	何佩云，谌红辉	连栽马尾松林根际与非根际土壤养分及酶活性研究	浙江林业科技	2011	贵州师范大学
50	何佩云，丁贵杰，谌红辉	连栽马尾松人工林土壤肥力比较研究	林业科学研究	2011	贵州大学
51	黄承标，卢立华，温远光，蔡道雄，吕广阳，黄竞	大青山林区不同海拔高度主要气象要素的变化	贵州农业科学	2011	广西大学
52	黄承标，韦国富，冯昌林，覃文更，谭卫宁，莫桂蝶	桂西北喀斯特典型森林植被的小气候特征	广东农业科学	2011	广西大学
53	苏文会，彭颖，范少辉，卢立华，张文元	车筒竹不同育苗方法比较研究	竹子研究汇刊	2011	国际竹藤网络中心
54	杨玉海，郑路，段永照	干旱区人工防护林带不同林分凋落叶分解及养分释放	应用生态学报	2011	荒漠与绿洲生态国家重点实验室
55	袁东，姜忠斌，蔡道雄	棕榈藤采收方法与新型采藤机械的研制	世界竹藤通讯	2011	国家林业局北京林业机械研究所
56	赵金龙，梁宏温，温远光，卢立华，明安刚，苏建苗	马尾松与红椎混交异龄林生物量分配格局	中南林业科技大学学报	2011	广西大学
57	钟秋平，蔡子良，王森，万细瑞，朱小毛，赵学民，龚洪恩	油茶容器育苗基质配方的研究	中南林业科技大学学报	2011	中国林业科学研究院亚热带林业实验中心
58	朱宏光，李燕群，苏建苗，王家妍，农友，刘正富，赵金龙，林建勇，温远光	不同水分条件对蚬木幼苗生长的影响	安徽农业科学	2011	广西大学
59	朱宏光，刘正富，苏建苗，农友，王家妍，李燕群，赵金龙，林建勇，温远光	广西北热带岩溶区蚬木幼苗的光合与蒸腾特性	安徽农业科学	2011	广西大学
60	朱宏光，农友，苏建苗，王家妍，李燕群，刘正富，赵金龙，林建勇，温远光	广西热带岩溶区苹婆幼树光合蒸腾特性研究	安徽农业科学	2011	广西大学
61	陈新美，雷渊才，张雄清，贾宏炎	样本量对 MaxEnt 模型预测物种分布精度和稳定性的影响	林业科学	2012	中国林业科学研究院资源信息研究所
62	刁海林，白灵海，罗建举，余天华，唐继新，高伟	马尾松干燥特性研究	湖北农业科学	2012	广西大学
63	刁海林，蔡道雄，罗建举，程仁春，唐继新	红椎 3 向弹性常数研究	中国农学通报	2012	广西大学
64	刁海林，蔡道雄，袁东，白灵海，卢立华	手动机械采藤法及其工艺研究	林业实用技术	2012	广西大学
65	刁海林，蔡道雄，袁东，贾宏炎，白灵海，卢立华，许基煌	车辆采藤法及其工艺研究	林业实用技术	2012	广西大学

（续）

序号	作者	题名	刊名	年	第一完成单位
66	刁海林，姜金英，蔡道雄，唐继新，罗建举	米老排三向弹性常数研究	中国农学通报	2012	广西大学
67	刁海林，唐继新，唐贤明，郑华庭，白灵海，罗建举	米老排锯材干燥中试研究	西北林学院学报	2012	广西大学
68	刁海林，袁东，蔡道雄，姜忠斌	行星轮式棕榈藤采收机设计	木材加工机械	2012	广西大学
69	范少辉，彭颖，苏文会，卢立华，张文元	箣竹不同方法育苗效果研究	西北林学院学报	2012	国际竹藤网络中心
70	葛乐，成向荣，段溪，虞木奎，刘志龙	施肥对麻栎人工林碳密度及休眠期土壤呼吸的影响	生态学杂志	2012	中国林业科学研究院亚热带林业研究所
71	何佩云，丁贵杰，谌红辉	马尾松连栽对根际与非根际土壤微量元素及微生物的影响	辽宁师范大学学报(自然科学版)	2012	贵州师范大学
72	何佩云，丁贵杰，谌红辉	第1代和第2代马尾松林土壤微生物及生化作用比较	浙江农林大学学报	2012	贵州师范大学
73	何佩云，丁贵杰，谌红辉	一二代不同林龄马尾松林分材积生长过程及林分结构比较	江苏农业科学	2012	贵州师范大学
74	何佩云，丁贵杰，谌红辉	第1、2代马尾松人工林林下植被的多样性比较	中南林业科技大学学报	2012	贵州师范大学
75	何佩云，丁贵杰，谌红辉	1、2代马尾松人工林林分生长特性比较	辽宁林业科技	2012	贵州师范大学
76	黄承标，覃文更，冯昌林，韦国富，王肇钦，莫桂蝶	木论保护区喀斯特森林凋落物的产量及其养分归还量	贵州农业科学	2012	广西大学
77	卢翠香，项东云，陈健波，徐峰，任世奇，刘媛，唐继新	人工林马尾松管胞形态特征及其变异	浙江农林大学学报	2012	广西林业科学研究院
78	吴国欣，王凌晖，梁惠萍，李远发，郝建	氮磷钾配比施肥对降香黄檀苗木生长及生理的影响	浙江农林大学学报	2012	广西林业勘测设计院
79	吴英，张万幸，张丽琼，伍静	基于DEM的地形与植被分布关联分析	东北林业大学学报	2012	广西大学
80	刁海林，白灵海，唐贤明，莫理，唐继新，罗建举	红椎锯材中试干燥工艺	浙江农林大学学报	2013	广西大学
81	刁海林，罗建举，蔡道雄	红椎木材蠕变特性研究	西北林学院学报	2013	广西大学
82	范春楠，庞圣江，郑金萍，李兵，郭忠玲	长白山林区14种幼树生物量估测模型	北京林业大学学报	2013	东北林业大学
83	何佩云，丁贵杰，谌红辉	1、2代不同林龄马尾松人工林土壤微量元素及酶活性	东北林业大学学报	2013	贵州师范大学

（续）

序号	作者	题名	刊名	年	第一完成单位
84	刘福妹，穆怀志，刘子嘉，李志新，姜静，刘桂丰	用秋水仙素处理不同家系白桦种子诱导四倍体的研究	北京林业大学学报	2013	东北林业大学
85	龙涛，蓝嘉川，陈厚荣，王家妍，谢益君，陆晓明，雷丽群，朱宏光，温远光，蔡道雄	采伐和炼山对马尾松林土壤微生物多样性的影响	南方农业学报	2013	广西大学
86	王斌，梅秀英，汪阳东，胡延杰，杨锦昌，刘志龙，刘玉鹏	林业科研人员评价指标体系构建及权重分析	科研管理	2013	中国林业科学研究院亚热带林业研究所
87	王春胜，赵志刚，曾冀，郭俊杰，沙二，郭文福，曾杰，郑海水，	广西凭祥西南桦中幼林林木生长过程与造林密度的关系	林业科学研究	2013	中国林业科学研究院亚热带林业研究所
88	王卫霞，史作民，罗达，刘世荣，卢立华，	南亚热带 3 种人工林土壤微生物生物量和微生物群落结构特征	应用生态学报	2013	中国林业科学研究院森林生态环境与保护研究所
89	王卫霞，史作民，罗达，刘世荣，卢立华，明安刚，于浩龙	我国南亚热带几种人工林生态系统碳氮储量	生态学报	2013	中国林业科学研究院森林生态环境与保护研究所
90	赵志刚，王敏，曾冀，莫洪荣，郭俊杰，曾杰	珍稀树种格木蛀梢害虫的种类鉴定与发生规律初报	环境昆虫学报	2013	中国林业科学研究院亚热带林业研究所
91	刁海林，蔡道雄，贾宏炎，白灵海，唐继新	自行式缆索起重集材机集材作业工艺研究	中南林业科技大学学报	2014	广西大学
92	黄弼昌，谌红辉，吴龙墩，孙浩忠，潘启龙，李洪果，马跃	光皮桦无性系单株材积生长量测定与评价	林业实用技术	2014	天峨县林朵林场
93	李婷婷，陆元昌，庞丽峰，张显强，王霞，刘宪钊，姜俊	杉木人工林近自然经营的初步效果	林业科学	2014	中国林业科学研究院资源信息研究所
94	李婷婷，陆元昌，张显强，王霞，庞丽峰，刘宪钊，姜俊	经营的马尾松森林类型发育演替阶段量化指标研究	北京林业大学学报	2014	中国林业科学研究院资源信息研究所
95	刘世荣，温远光，蔡道雄，朱宏光，黄雪蔓，李晓琼	气候变化对森林的影响与多尺度适应性管理研究进展	广西科学	2014	中国林业科学研究院森林生态环境与保护研究所
96	罗达，史作民，唐敬超，刘世荣，卢立华	南亚热带乡土树种人工纯林及混交林土壤微生物群落结构	应用生态学报	2014	中国林业科学研究院森林生态环境与保护研究所
97	温远光，左花，朱宏光，刘世荣，梁宏温，黄雪蔓，李晓琼，罗应华	连栽对桉树人工林植被盖度、物种多样性及功能群的影响	广西科学	2014	广西大学
98	吴忠锋，杨锦昌，成铁龙，尹光天，许涵，刘志龙	海南油楠的重要生物学特性及产油特征	林业科学	2014	中国林业科学研究院热带林业研究所

（续）

序号	作者	题名	刊名	年	第一完成单位
99	朱宏光，陆晓明，温远光，蔡道雄，刘世荣，李晓琼，黄雪蔓，周晓果，李婉舒	马尾松人工林近自然化改造对林分生长的影响	广西科学	2014	广西大学
100	朱积余，申文辉，蒋燚，卢立华，谭一波，刘秀	红椎家系遗传变异与优良家系选择	热带亚热带植物学报	2014	广西林业科学研究院
101	毕会涛，付梦瑶，杨红震，凡琳洁，黄旭光，高贤明	老君山自然保护区乔木层生物量动态研究	河南农业大学学报	2015	河南农业大学
102	程静，欧阳旭，黄德卫，刘世忠，张德强，李跃林	鼎湖山针阔叶混交林4种优势树种树干液流特征	生态学报	2015	中国科学院华南植物园
103	刁海林，蔡道雄，唐继新，李开成，罗建举	水浸时效对红椎锯材性能的影响	中南林业科技大学学报	2015	广西大学
104	刁海林，罗建举，蔡道雄	水浸时效对红椎木材蠕变特性的影响	西北林学院学报	2015	广西大学
105	胡厚臻，侯文娟，潘启龙，周袁慧子，王艺锦，王凌晖，滕维超	配方施肥对刨花润楠幼苗生长和光合生理的影响	西北林学院学报	2015	广西大学
106	罗达，史作民，王卫霞，刘世荣，卢立华，明安刚，于浩龙	南亚热带格木、马尾松幼龄人工纯林及其混交林生态系统碳氮储量	生态学报	2015	中国林业科学研究院森林生态环境与保护研究所
107	莫柳园，覃德文，邓硕坤	广西立体生态林业经营及经营模式研究——以广西国有钦廉林场为例	经济研究导刊	2015	广西国有钦廉林场
108	庞丽峰，贾宏炎，陆元昌，牛常海，符利勇	分段削度方程2种估计方法比较	林业科学	2015	中国林业科学研究院资源信息研究所
109	王楠，李芳东，马跃，叶金山	泡桐优良无性系“TF33”干材表型性状杂种优势研究	林业科学研究	2015	国家林业局泡桐研究开发中心
110	韦录，谌红辉，农淑霞，蒙彩兰，赵光新，游彬，农良书	光皮桦无性系扦插育苗技术研究	林业科技通讯	2015	广西崇左市凤凰山林场
111	温远光，杨柳，朱宏光，蔡道雄，尤业明，贾宏炎，周晓果，严理	除草剂对桉树×降香黄檀混交林下植物功能群的影响	广西科学	2015	广西大学
112	徐基平，郑路，李艳红，邱文成	克拉玛依城市防护绿地稳定性现状评价与分析	安徽农业科学	2015	克拉玛依区园林局
113	尤业明，黄雪蔓，朱宏光，蓝嘉川，蔡道雄，温远光	间伐强度对杉木林下植物物种多样性和结构组成的影响	广西科学	2015	广西大学
114	张苏芳，张真，孔祥波，王鸿斌，蔡道雄，陶怡	马尾松毛虫常灾区、偶灾区和无灾区松针挥发物特征	林业科学	2015	中国林业科学研究院森林生态环境与保护研究所

（续）

序号	作者	题名	刊名	年	第一完成单位
115	陈超凡，吴水荣，覃林，范垚城，谭玲，郭文福	南亚热带典型乡土阔叶人工林土壤微生物碳源利用及功能多样性	生态学杂志	2016	广西大学
116	洪丕征，刘世荣，王晖，于浩龙	南亚热带红椎和格木人工幼龄林土壤微生物群落结构特征	生态学报	2016	中国林业科学研究院森林生态环境与保护研究所
117	洪丕征，刘世荣，王晖，于浩龙	正常和环割条件下不同形态氮素添加对红椎幼苗光合特性的影响	生态学报	2016	中国林业科学研究院森林生态环境与保护研究所
118	洪丕征，刘世荣，于浩龙，郝建	模拟氮沉降对红椎人工幼龄林土壤微生物生物量和微生物群落结构的影响	山东大学学报(理学版)	2016	中国林业科学研究院森林生态环境与保护研究所
119	黄雪蔓，尤业明，蓝嘉川，刘世荣，贾宏炎，朱宏光，温远光	不同间伐强度对杉木人工林碳储量及其分配的影响	生态学报	2016	广西大学
120	李嫦宪，黄德卫	林木种子活力研究意义及进展	绿色科技	2016	南宁青秀山风景名胜旅游开发有限责任公司
121	李良，张胜军，赵总	川滇桤木优树选择方法研究	内蒙古林业调查设计	2016	新疆质检院
122	庞丽峰，贾宏炎，陆元昌，符利勇	热带地区几个珍贵树种干形曲线研究	南京林业大学学报(自然科学版)	2016	中国林业科学研究院资源信息研究所
123	孙浩永，孙浩忠，劳庆祥，潘启龙	GPS在设计种子园配置图中的应用	林业科技通讯	2016	广西天峨县林业局
124	谭长强，彭玉华，杨桂芳，申文辉，何琴飞，欧芷阳	极小种群广西青梅群落凋落物及土壤理化特征	广西林业科学	2016	广西林业科学研究院
125	谢鹏芳，谢敏，裴志涛，黄旭光，黄家荣，周俊朝	林分直径分布的Weibull参数估计系统研制	河南农业大学学报	2016	河南农业大学
126	邢海涛，陆元昌，刘宪钊，王晓明，贾宏炎，曾冀	基于近自然改造的马尾松林分竞争强度研究	北京林业大学学报	2016	中国林业科学研究院资源信息研究所
127	徐振国，黄大勇，郭起荣，李忠国	麻竹地上器官生物量分配及其模型优化	中南林业科技大学学报	2016	广西林业科学研究院
128	尤业明，徐佳玉，蔡道雄，刘世荣，朱宏光，温远光	广西凭祥不同年龄红椎林林下植物物种多样性及其环境解释	生态学报	2016	广西大学
129	周再知，梁坤南，马华明，黄桂华，李运兴	酸性土壤上不同钙肥及用量对柚木苗生长的影响	西部林业科学	2016	中国林业科学研究院热带林业研究所
130	李勇，劳庆祥，孙浩忠，潘启龙，李洪果	广西光皮桦无性系在湖南安化的引种初报	林业科技通讯	2017	安化县林业局

（续）

序号	作者	题名	刊名	年	第一完成单位
131	李远发，明安刚，叶绍明，吕成群，惠刚盈	红松次生混交林中针叶树与阔叶树的结构差异	西北植物学报	2017	广西大学
132	刘丹，刘士玲，郭忠玲，崔晓阳，范春楠	水冬瓜赤杨种群生命表与时间序列分析	北京林业大学学报	2017	北华大学
133	覃林，马雪珍，吴水荣，谭玲，范垚城，郭文福	南亚热带典型乡土阔叶人工林与桉树人工林土壤微生物量氮及可溶性氮特征	应用与环境生物学报	2017	广西大学
134	唐诚，王春胜，庞圣江，黄日逸，曾杰	西南桦人工林树皮厚度模型模拟	林业科学	2017	中国林业科学研究院热带林业研究所
135	王晓明，陆元昌，邢海涛，贾宏炎，刘宪钊，谢阳生，曾冀	马尾松-大叶栎混交林皆伐作业法的设计	林业资源管理	2017	中国林业科学研究院资源信息研究所
136	徐振国，黄大勇，郭起荣，李忠国	麻竹开花前后营养元素分配格局及动态变化规律	广西林业科学	2017	广西林业科学研究院
137	徐振国，黄大勇，郭起荣，冉洪，李忠国	麻竹开花前后主要营养元素变化研究	世界竹藤通讯	2017	广西林业科学研究院
138	周晓果，温远光，朱宏光，王磊，李晓琼	大明山常绿阔叶林冠层垂直结构与林下植物更新	应用生态学报	2017	广西大学
139	艾娟娟，厚凌宇，邵国栋，李智超，卢立华，李朝英，孙启武	林业废弃物基质配方特性及其对柚木生长的影响	浙江农林大学学报	2018	中国林业科学研究院林业研究所
140	艾娟娟，厚凌宇，邵国栋，李智超，蒙彩兰，卢立华，孙启武	不同林业废弃物配方基质的理化性质及其对西桦幼苗生长效应的综合评价	植物资源与环境学报	2018	中国林业科学研究院林业研究所
141	陈臣，徐道春，李武志，李文彬，姚立红，张保卫	基于土壤三组分重构模型的导热系数研究	森林工程	2018	北京林业大学
142	郝中明，吴水荣，覃林，谭玲，郭文福	广西南部地区不同林龄马尾松及其混交林理化性质研究	广东农业科学	2018	广西大学
143	何佩云，谌红辉	连栽马尾松林单株不同器官含水率及生物量的比较	江苏农业科学	2018	贵州师范大学
144	洪舟，刘福妹，张宁南，杨曾奖，徐大平	降香黄檀生长性状家系间变异与优良家系初选	南京林业大学学报(自然科学版)	2018	中国林业科学研究院热带林业研究所
145	胡小燕，段爱国，张建国，杜海伦，张雄清，郭文福，郭光智，	南亚热带杉木人工成熟林密度对土壤养分效应研究	林业科学研究	2018	中国林业科学研究院林业研究所
146	黄婷，李钊，陈建全，罗银花，罗应华，	柳州市行道树资源调查与结构特征分析	广西植物	2018	广西大学行健文理学院
147	陆艳仙，周晓果，陶彦良，赵明威，朱宏光，温远光，蔡道雄，贾宏炎，李晓琼	炼山和套种绿肥对桉树人工林下植物和土壤 C：N：P 化学计量特征的影响	广西科学	2018	广西大学

（续）

序号	作者	题名	刊名	年	第一完成单位
148	吕万英，黄丽荣	关于如何加强事业单位工资管理的研究	中国乡镇企业会计	2018	广西国有高峰林场
149	彭雯，谭玲，明安刚，何友均，覃林	南亚热带典型人工纯林土壤剖面细菌群落组成差异分析	土壤通报	2018	广西大学
150	唐诚，王春胜，庞圣江，贾宏炎，曾杰	广西大青山西南桦人工林立地类型划分及评价	西北林学院学报	2018	中国林业科学研究院热带林业研究所
151	唐诚，王春胜，庞圣江，王欢，曾杰	广西大青山西南桦人工林土壤养分特征及其与立地生产力的关系	林业科学研究	2018	中国林业科学研究院热带林业研究所
152	陶彦良，周晓果，温远光，朱宏光，陆艳仙，杜氏清闲，赵明威，蔡道雄，贾宏炎，李武志，李晓琼	不同林地清理方式下生物炭和氮添加对桉树红椎混交林土壤养分的影响	广西科学	2018	广西大学
153	温远光，严宇航，陶彦良，蔡道雄，周晓果，朱宏光，贾宏炎，明安刚，赵明威，陆艳仙，严理，李武志	不同林地清理和培肥措施对桉树人工林植物多样性的影响	广西科学	2018	广西大学
154	杨予静，刘世荣，陈琳，王晖	马尾松人工林土壤和团聚体有机碳矿化对穿透雨减少的响应	应用生态学报	2018	中国林业科学研究院森林生态环境与保护研究所
155	杨予静，刘世荣，陈琳，王晖，卢立华	模拟降雨减少对马尾松人工林凋落物量及其化学性质的短期影响	生态学报	2018	中国林业科学研究院森林生态环境与保护研究所
156	尤业明，吴溪玭，明安刚，刘通，陈永康，朱宏光，温远光，廖树寿，黄雪蔓	南亚热带典型针叶林改造成乡土阔叶林后林下植物功能群的变化及其环境解释	生态学杂志	2018	广西大学
157	常英英，梁立雄，高亚南，吴晓娟，鲁俊倩，王庆灵，丁昌俊，张伟溪，苏晓华，张冰玉	转多基因1年生库安托杨对溃疡病的抗性分析	林业科学研究	2019	中国林业科学研究院林业研究所
158	洪舟，刘福妹，杨曾奖，曾杰，张宁南，林国雄，徐大平	5个泰国种源大果紫檀的早期生长及材性分析	南京林业大学学报（自然科学版）	2019	中国林业科学研究院热带林业研究所
159	李运筹，周晓果，温远光，朱宏光，覃志伟，蔡道雄，贾宏炎，李晓琼，尤业明	南亚热带马尾松×红椎混交林红椎天然更新及环境解释	广西科学	2019	广西大学
160	闵惠琳，宋维峰	全福庄小流域不同土地利用类型地表水氮浓度动态特征	水土保持研究	2019	西南林业大学

（续）

序号	作者	题名	刊名	年	第一完成单位
161	申文辉，谭长强，劳庆祥，郝海坤，曹艳云，郑威，黄志玲	红椎优树二代家系抗逆性苗期选择研究	西南林业大学学报(自然科学)	2019	广西林业科学研究院
162	覃志伟，周晓果，温远光，朱宏光，李海燕，阮友维，蔡道雄，贾宏炎，黄雪蔓，尤业明	去除和添加凋落物对马尾松×红椎混交林土壤呼吸的影响	广西科学	2019	广西大学
163	温远光，李海燕，周晓果，朱宏光，李运筹，蔡道雄，贾宏炎，黄雪蔓，尤业明	马尾松与红椎异龄混交林对土壤微生物群落结构和功能的影响	广西科学	2019	广西大学